Birkhäuser

Modeling and Simulation in Science, Engineering and Technology

Angelo Morro · Claudio Giorgi

Mathematical Modelling of Continuum Physics

Birkhäuser

Angelo Morro
DIBRIS
University of Genova
Genova, Italy

Claudio Giorgi
DICATAM
University of Brescia
Brescia, Italy

ISSN 2164-3679 ISSN 2164-3725 (electronic)
Modeling and Simulation in Science, Engineering and Technology
ISBN 978-3-031-20813-3 ISBN 978-3-031-20814-0 (eBook)
https://doi.org/10.1007/978-3-031-20814-0

Mathematics Subject Classification: 35Q30, 35Q74, 35Q79, 70B10, 74-XX, 76-XX, 78A25, 80A05, 80A17, 80A22

This book is published under the imprint Birkhäuser, www.birkhauser-science.com by the registered company Springer Nature Switzerland AG
The registered company address is: Gewerbestrasse 11, 6330 Cham, Switzerland

Preface

This book provides a unified treatment of continuum physics. A systematic approach to the balance equations is elaborated for wide-ranging classes of materials (classical continua, micropolar continua, mixtures, electromagnetic continua) following the lines of the encyclopedic handbook articles of Truesdell and Toupin and Truesdell and Noll. As is standard in Rational Thermodynamics, the constitutive properties are required to obey the objectivity principle and to be consistent with the second law of thermodynamics. Yet here a rather new approach is developed by viewing the entropy production as a constitutive function per se as is the case for the entropy and the entropy flux. While this does not determine any new result for simple materials, it proves conceptually and practically advantageous in the modelling of nonlinear phenomena such as those occurring in hysteretic continua, e.g. in plasticity, electromagnetism, and physics of shape-memory alloys.

The book is suitable for engineers, physicists, and mathematicians. The derivations of the sought results are fairly detailed through careful proofs. Though a wide variety of subjects are examined, the contents are developed so as to get a self-contained and consistent presentation of the various topics.

Part I reviews the kinematics of continuous bodies and illustrates the general setting of balance laws. Kinematics treats essential preliminaries to continuum physics such as reference and current configurations, transport relations, singular surfaces, objectivity and objective time derivatives. Next, a chapter on balance equations develops the balance laws of mass, linear momentum, angular momentum, energy, entropy (Clausius–Duhem inequality), and the balance laws in electromagnetism.

Part II is first devoted to the general requirements of constitutive models. In this sense, emphasis is given to the application of objectivity (the constitutive equations must be invariant under changes of frame) and consistency with the second law of thermodynamics (the constitutive equations must satisfy the restrictions placed by the entropy inequality). Next, a review is given of common models of simple materials, namely materials described by the first-order gradient of deformation, velocity, and temperature. In this framework, detailed descriptions are given of (thermoelastic, elastic, and dissipative) solids and (elastic, thermoelastic, viscous, and Newtonian) fluids.

A wide variety of constitutive models is investigated in Part III, consisting of separate chapters of non-simple materials. The chapter on rate-type models reviews the rheological devices and next shows some schemes within the Eulerian and the Lagrangian description with emphasis on objective time derivatives. The chapter on materials with memory begins with a general setting of (fading) memory and next shows memory models for thermoelasticity, heat conduction, viscoelasticity, electromagnetic solids, and modelling via fractional derivatives. Next, the modelling of aging (thermoelastic, rate-type, and thermo-viscoelastic) materials is exhibited in connection with the thermodynamic restrictions. Also, materials of higher-order grade are examined in the form of fluids and solids or models via interstitial working. The possible consistency of the hyperstress with the standard balance laws is shown to hold. A chapter is devoted to mixtures; in addition to balance equations for the constituents and the whole mixture, some aspects are investigated such as models of diffusion, Soret and Dufour effects, immiscible mixtures, and models for the whole mixture. Micropolar media are modelled as materials with a physical internal structure. Each point of the continuum is viewed as a body with a finite number of degrees of freedom. The balance laws are then reviewed so that the internal degrees of freedom show up in the balance of orientational momentum and energy. The model, so established, is then applied to the description of liquid crystals and nanofluids. Porous materials are described as mixtures (solid–liquid or solid–void). Also, porous materials with double porosity are established. The chapter on electromagnetism of continuous media describes a number of phenomena. The interaction of the electromagnetic field with deformation is examined within electroelasticity, magnetoelasticity, and plasma theory. This in turn allows us to examine magneto-, electro-, and mechanical-optical effects. Memory effects are modelled in quite a general setting. Nonlinearity effects are especially framed within micromagnetics and ferrofluids. Chiral media and ferrites are investigated in detail also to show the optical activity effects on linearly- and circularly-polarized waves. Finally, superconductivity and superfluidity are developed in a common framework of mixtures of reacting fluids; normal electrons and superconducting electrons in one case, normal fluid and superfluid in the other.

Hysteretic effects and phase transitions are developed in Part IV. Hysteresis is modelled in ferroelectrics, ferromagnetism, and plasticity. In all of these contexts, the modelling is performed by having recourse to the entropy production as a (non-negative) constitutive function. It follows that the free energy governs the anhysteretic behaviour, while the entropy production characterizes the hysteretic properties. Phase transitions are described in different ways. A transition may occur at a sharp interface between two different phases; the jump conditions across the interface govern the transition. Instead, a transition may occur in a diffuse region where the pertinent fields change continuously; in essence, the transition region is occupied by a mixture of constituents in different phases. A scheme for phase transitions and hysteretic effects in shape-memory alloys is also outlined.

Notation

We use mostly direct notation and the recourse to index notation is made only when, otherwise, the pertinent expression might become ambiguous. The symbols of continuum mechanics are used, e.g. in the books of Truesdell and Noll [426] and Gurtin et al [214]. Yet difficulties arise whenever any well-established symbol has a different meaning in continuum mechanics and in electromagnetism. For instance, **E** is the Green-St Venant strain tensor in continuum mechanics and the electric field in electromagnetism while **D** is the stretching tensor in continuum mechanics and the electric displacement in electromagnetism. To avoid the introduction of newly defined symbols, we have maintained the standard symbols whenever the pertinent chapter makes it clear which meaning has to be assigned to the symbol. There are places where the stretching tensor and the electric displacement occur simultaneously; in such cases, we use $\mathfrak{D}$ for the stretching tensor. Another difficulty arises in connection with the vector $\boldsymbol{\omega}$; it is currently used for angular velocity and vorticity. To avoid any ambiguities, we have maintained the symbol $\boldsymbol{\omega}$ for the angular velocity and have used $\boldsymbol{\varpi}$ for the vorticity (i.e. $\nabla \times \mathbf{v}$).

Throughout, lightface letters indicate scalars, while boldface letters indicate vectors or tensors. In the mechanical context, lowercase (uppercase) boldface letters indicate vectors (tensors). In electromagnetism, following the literature, the (electric and magnetic) vectors are denoted by uppercase letters, e.g. **E**, **B**. Moreover, Lin is the set of all tensors, Lin^+ is the set of all tensors with positive determinant, Sym is the set of all symmetric tensors, Skw is the set of all skew tensors, Psym is the set of all symmetric, positive definite tensors, Orth is the set of all orthogonal tensors, Orth^+ is the set of all rotations (all orthogonal tensors with positive determinant). For the benefit of the reader an Appendix "Notes on vectors and tensors" reviews the essential contents of algebra and analysis of vector and tensor functions.

Genova, Italy — Angelo Morro
Brescia, Italy — Claudio Giorgi

Contents

Part I
Basic Principles and Balance Equations

Chapter 1
Kinematics

This chapter deals with the kinematics of deformable bodies. Both deformation and motion of a body are developed by using the reference configuration; the position vector in the reference configuration is the operative label of the points of the body.

The basic relations so determined for deformation and motion are essential to the next chapters. Attention is addressed to the topics of objectivity and objective time derivatives, thus establishing a general framework that proves remarkable in the description of material properties in terms of time derivatives. This framework shows the connection between various known objective time derivatives (Jaumann, Green–Naghdi, Cotter–Rivlin, Oldroyd, Truesdell). Transport relations are obtained for convecting (or non-convecting) sets, thus establishing basic properties for the derivation of (local) balance equations and jump conditions for discontinuous fields. Hence, the kinematical and the geometric relations are derived for singular surfaces which provide a general setting for the investigation of discontinuity waves. Moreover, the transport theorems for surface integrals are established thus leading, in particular, to the convected time derivative.

1.1 Frames of Reference and Configurations

A frame of reference or observer $\mathscr{F}$ is an arbitrary set of rigidly fixed axes relative to which the position of points is determined. For simplicity, the chosen fixed axes are taken to be orthogonal and hence the corresponding unit vectors are an orthonormal basis; denote by $\{\mathbf{e}_1, \mathbf{e}_2, \mathbf{e}_3\}$ the chosen orthonormal basis. The time at which an event takes place can be specified relative to a particular event taken as a reference. An event is then a pair $\{P, t\}$ consisting of a point P, in the three-dimensional Euclidean space $\mathscr{E}$, and a time t. A chosen position O, origin, is taken as reference and hence we can identify each position P with the position vector $\mathbf{x}$ relative to the origin,

A. Morro and C. Giorgi, *Mathematical Modelling of Continuum Physics*,
Modeling and Simulation in Science, Engineering and Technology,
https://doi.org/10.1007/978-3-031-20814-0_1

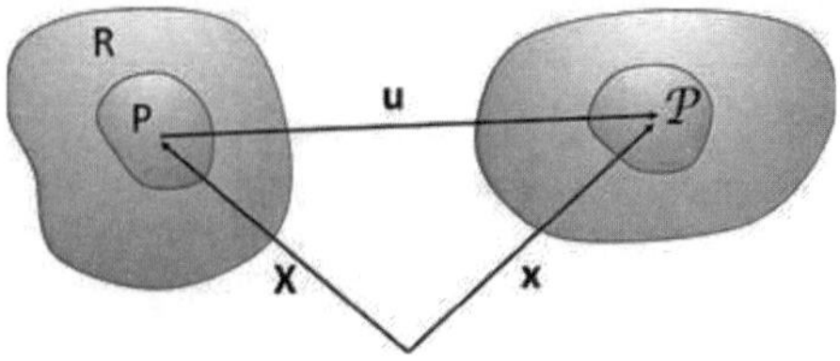

Fig. 1.1 A motion χ maps the point, labelled by the position vector $\mathbf{X} \in \mathrm{R}$, to the position vector $\mathbf{x} = \chi(\mathbf{X}, t)$

$\mathbf{x} = P - O$. Hence, a frame of reference $\mathscr{F}$ may be viewed as the Cartesian product of the set $\{\mathscr{E}, O, \{\mathbf{e}_1, \mathbf{e}_2, \mathbf{e}_3\}\}$ and the real time axis $\mathbb{R}$.

A change of frame is a 1–1 mapping of space time onto itself such that distances, time intervals, and temporal order are preserved. An event $\{\mathbf{x}, t\}$ and its image $\{\mathbf{x}^*, t^*\}$ under a change of frame are related by rigid transformations and a time shift. Chosen possibly different origins O, O^*, for the two observers, $\mathscr{F}, \mathscr{F}^*$, we can express a change of frame by

$$\mathbf{x}^* = \mathbf{c}(t) + \mathbf{Q}(t)\mathbf{x}, \qquad t^* = t - a,$$

where $a \in \mathbb{R}$, $\mathbf{c}(t) = O - O^*$, and $\mathbf{Q}(t)$ is a rotation tensor, $\det \mathbf{Q}(t) = 1$.

Bodies occupy regions of the Euclidean space $\mathscr{E}$. To describe the evolution of a body, we need to label the points.[1] This is accomplished by labelling the points of the body by the position in a reference configuration (or placement) $\mathrm{R} \subset \mathscr{E}$; sometimes the body is identified with the reference configuration [216]. Upon choosing an origin, the positions in R are associated with a position vector; to make it apparent that we are dealing with the reference configuration the vector is denoted by a uppercase boldface letter, most often $\mathbf{X}$. By definition, the reference configuration is independent of time. We denote by ∇ the gradient operator in $\mathscr{E}$ and by ∇_R the gradient operator in R (Fig. 1.1).

The motion of the body is described by saying the position of any point in $\mathscr{E}$ as a function of time. Granted a choice of origin also for the position in the current configuration, the current position vector, say $\mathbf{x}$, is then a function of the point (at) $\mathbf{X}$ and time t. Hence, a *motion* of the body is a (smooth) function χ that assigns to each position (vector) in R and time t a position

$$\mathbf{x} = \chi(\mathbf{X}, t)$$

in space; we say that $\mathbf{X}$ is a material vector while $\mathbf{x}$ is a spatial vector. Symbolically, $\chi(\mathrm{R}, t) \subset \mathscr{E}$ denotes the current configuration (at time t). The reference configuration R might be the configuration at a chosen initial time t_0 but need not be so. If $\mathbf{X}$ is fixed then $\chi(\mathbf{X}, t)$ describes the motion of the pertinent point in time. Hence

$$\dot{\chi}(\mathbf{X}, t) := \partial_t \chi(\mathbf{X}, t), \qquad \ddot{\chi}(\mathbf{X}, t) := \partial_t^2 \chi(\mathbf{X}, t),$$

[1] The term points is used, instead of particles, to make it clear that no internal structure is ascribed.

are the velocity and acceleration of the point $\mathbf{X}$ at time t. If, instead, t is fixed then χ is a mapping from R to the current configuration, at time t.

For brevity, throughout $\mathbf{X} \in \mathrm{R}$ means that $\mathbf{X}$ is the position vector of a point in R. Likewise, $\mathbf{x} \in \mathscr{D} \subset \mathscr{E}$ means that $\mathbf{x}$ is the position vector of a point in $\mathscr{D}$.

1.2 Deformation

To begin with, we restrict attention to a fixed time t and hence $\chi(\mathbf{X}, t)$, as a function of $\mathbf{X}$, describes the *deformation* at time t. Hence, the dependence on the parameter t is understood and we let

$$\mathbf{x} = \chi(\mathbf{X}).$$

It is assumed that $\chi(\mathbf{X})$ is a 1–1 mapping so that two material points, $\mathbf{X} \neq \mathbf{Y}$, cannot occupy the same position in the current configuration, $\mathbf{x} \neq \mathbf{y}$. Assume also that χ is differentiable. Hence, there is a tensor $\mathbf{F}(\mathbf{X})$ such that

$$\chi(\mathbf{Y}) - \chi(\mathbf{X}) = \mathbf{F}\,(\mathbf{Y} - \mathbf{X}) + o(|\mathbf{Y} - \mathbf{X}|),$$

$\mathbf{F}$ being a tensor function of the position $\mathbf{X}$ and possibly of t; $\mathbf{F}$ is said to be the *deformation gradient.* To within $o(|\mathbf{Y} - \mathbf{X}|)$, $\mathbf{F}$ maps material vectors, $\mathbf{Y} - \mathbf{X}$, to spatial vectors, $\chi(\mathbf{Y}) - \chi(\mathbf{X})$. If χ depends linearly on $\mathbf{X}$ then $\mathbf{F}$ is independent of $\mathbf{X}$,

$$\chi(\mathbf{Y}) - \chi(\mathbf{X}) = \mathbf{F}\,(\mathbf{Y} - \mathbf{X}),$$

and the deformation is said to be *homogeneous.* Hence, in general, a deformation approaches the corresponding homogeneous deformation, with $\mathbf{F} = \mathbf{F}(\mathbf{X})$; the closer $\mathbf{Y}$ is to $\mathbf{X}$ the more χ approaches a homogeneous deformation.

In components,[2] there is a matrix $F \in \mathbb{R}^{3\times3}$ such that, for any two points $\mathbf{X}, \mathbf{Y} \in \mathrm{R}$,

$$\chi_h(\mathbf{Y}) - \chi_h(\mathbf{X}) = F_{hK}(\mathbf{X})(Y_K - X_K) + o(|\mathbf{Y} - \mathbf{X}|), \qquad h = 1, 2, 3.$$

The entries $\{F_{hK}\}$ are the components of the deformation gradient

$$\mathbf{F} = \nabla_R\, \chi, \qquad F_{hK} = \partial_{X_K} \chi_h.$$

To determine the effects of deformation on lengths, areas, and volumes, it is convenient to establish some properties of the deformation gradient. First we observe that the invertibility of χ implies

$$J := \det \mathbf{F} \neq 0.$$

[2] As in well-known textbooks [152, 428, 429], to make the notation more explicit, throughout in suffix notation, we use capital indices for quantities related to the reference configuration.

Indeed since the trivial deformation

$$\mathbf{x} = \mathbf{X}$$

gives $J = 1$ we may regard the deformation as a continuous process and hence we assume

$$J > 0.$$

Also, we let

$$j = \det \mathbf{F}^{-1}$$

and hence $J\, j = 1$.

As any invertible tensor, the deformation gradient $\mathbf{F}$ satisfies the *polar decomposition*

$$\mathbf{F} = \mathbf{R}\mathbf{U} = \mathbf{V}\mathbf{R},$$

where $\mathbf{R} \in$ Orth while $\mathbf{U}, \mathbf{V} \in$ Psym.

To show this property, we first observe that $\mathbf{F}^T\mathbf{F} \in$ Psym in that, for any vector $\mathbf{w}$,

$$\mathbf{w} \cdot \mathbf{F}^T\mathbf{F}\mathbf{w} = (\mathbf{F}\mathbf{w}) \cdot (\mathbf{F}\mathbf{w}) \geq 0$$

and, since $\mathbf{F}$ is invertible,

$$\mathbf{F}\mathbf{w} = \mathbf{0} \quad \Longleftrightarrow \quad \mathbf{w} = \mathbf{0}.$$

We can then define[3] $\mathbf{U} = \sqrt{\mathbf{F}^T\mathbf{F}}$. Since $\mathbf{F}^T\mathbf{F} \in$ Psym then the eigenvalues, say $\{\mu_i\}$, are positive; let $\{\mathbf{N}_i\}$ be the eigenvectors and then $\mathbf{F}^T\mathbf{F} = \sum_{i=1}^3 \mu_i \mathbf{N}_i \otimes \mathbf{N}_i$. Hence $\mathbf{U} = \sqrt{\mathbf{F}^T\mathbf{F}}$ is defined as $\mathbf{U} = \sum_{i=1}^3 \lambda_i \mathbf{N}_i \otimes \mathbf{N}_i$, $\lambda_i = \sqrt{\mu_i}$. Hence also $\mathbf{U}$ is invertible and $\mathbf{U}^{-1} = \sum_{i=1}^3 (1/\lambda_i)\mathbf{N}_i \otimes \mathbf{N}_i$. We can then write $\mathbf{F} = \mathbf{R}\mathbf{U}$ and take it as the definition of $\mathbf{R}$; we find that

$$\mathbf{R} = \mathbf{F}\mathbf{U}^{-1}, \qquad \mathbf{R}^T\mathbf{R} = \mathbf{U}^{-1}\mathbf{F}^T\mathbf{F}\mathbf{U}^{-1} = \mathbf{1}$$

and, if $J > 0$ then $\det \mathbf{R} = 1$. Accordingly, $\mathbf{R}$ is a rotation if $J > 0$ and is orthogonal if merely $J \neq 0$. Given $\mathbf{U} : \mathbf{U}^2 = \mathbf{F}^T\mathbf{F}$ the tensor $\mathbf{R}$ is unique in that by

$$\mathbf{R}\mathbf{U} = \hat{\mathbf{R}}\mathbf{U}$$

it follows $\mathbf{R} = \hat{\mathbf{R}}$.

Likewise, since $\mathbf{F}\mathbf{F}^T \in$ Psym we can define $\mathbf{V} \in$ Psym such that

$$\mathbf{F}\mathbf{F}^T = \mathbf{V}^2.$$

---

[3] The uniqueness of $\mathbf{U}$ is proved in [216], p. 32.

Let $\mathbf{F} = \mathbf{V}\tilde{\mathbf{R}}$. It follows $\det \tilde{\mathbf{R}} = 1$ if $J > 0$. Now by

$$\mathbf{F} = \mathbf{V}\tilde{\mathbf{R}} = \tilde{\mathbf{R}}\tilde{\mathbf{R}}^T\mathbf{V}\tilde{\mathbf{R}} = \tilde{\mathbf{R}}(\tilde{\mathbf{R}}^T\mathbf{V}\tilde{\mathbf{R}})$$

and the uniqueness of the decomposition **RU**, it follows

$$\tilde{\mathbf{R}} = \mathbf{R}, \qquad \mathbf{V} = \mathbf{R}\mathbf{U}\mathbf{R}^T.$$

We can then write the following statement.

Theorem 1.1 (Polar decomposition) *If* $\mathbf{F}$ *is an invertible tensor with* $\det \mathbf{F} > 0$ *then there are unique, symmetric, positive-definite tensors* $\mathbf{U}$ *and* $\mathbf{V}$ *and a rotation* $\mathbf{R}$ *such that* $\mathbf{F} = \mathbf{R}\mathbf{U} = \mathbf{V}\mathbf{R}$.

We say that $\mathbf{F} = \mathbf{R}\mathbf{U}$ and $\mathbf{F} = \mathbf{V}\mathbf{R}$ are the right and left polar decompositions of $\mathbf{F}$.

A positive-definite symmetric tensor represents a state of pure stretches along three mutually orthogonal axes. Therefore, the polar decomposition means that any (homogeneous) deformation may be viewed as the result of a pure stretch $\mathbf{U}$ and a rotation $\mathbf{R}$ or the same rotation $\mathbf{R}$ followed by the stretch $\mathbf{V}$. Accordingly, $\mathbf{U}$ is called the *right stretch tensor* and $\mathbf{V}$ the *left stretch tensor.* In calculations, it proves more convenient to use the *right and left Cauchy–Green tensors*

$$\mathbf{C} = \mathbf{U}^2 = \mathbf{F}^T\mathbf{F}, \qquad \mathbf{B} = \mathbf{V}^2 = \mathbf{F}\mathbf{F}^T.$$

When $\mathbf{F}$ is a rotation, $\mathbf{F} = \mathbf{R}$, $\mathbf{F}^T\mathbf{F} = \mathbf{1}$. Also $\mathbf{F} = \mathbf{1}$ if the deformation is the identity transformation $\mathbf{x} = \mathbf{X}$. It is then useful to use the *Green-St. Venant strain tensor*

$$\mathbf{E} = \tfrac{1}{2}(\mathbf{C} - \mathbf{1}).$$

The vector

$$\mathbf{u} = \mathbf{x} - \mathbf{X} = \chi(\mathbf{X}) - \mathbf{X}$$

is called the *displacement* of the point (at $\mathbf{X}$). The *displacement gradient* is the tensor[4]

$$\mathbf{H} = \nabla_R\mathbf{u} = \mathbf{F} - \mathbf{1}.$$

Hence

$$\mathbf{C} = (\mathbf{H} + \mathbf{1})^T(\mathbf{H} + \mathbf{1}) = \mathbf{1} + \mathbf{H}^T + \mathbf{H} + \mathbf{H}^T\mathbf{H}$$

and

$$\mathbf{E} = \tfrac{1}{2}(\mathbf{H} + \mathbf{H}^T) + \tfrac{1}{2}\mathbf{H}^T\mathbf{H}.$$

---

[4] To follow the standard notation in continuum mechanics, the symbols **B**, **H**, **E**, and **D** are used. When **B**, **H**, **E**, and **D** are used within electromagnetism they denote the magnetic induction, the magnetic field, the electric field, and the electric displacement.

It is useful sometimes to consider the linear part $\mathcal{E}$ of $\mathbf{E}$, i.e.

$$\mathcal{E} := \frac{1}{2}(\mathbf{H}^T + \mathbf{H});$$

we refer to $\mathcal{E}$ as the *strain tensor.* Since

$$\mathbf{E} = \mathcal{E} + \mathbf{H}^T\mathbf{H}$$

then for small deformations, that is when $\|\mathbf{H}\| \ll 1$, we can use the approximation $\mathbf{E} \simeq \mathcal{E}$.

Further

$$\mathcal{E} = \text{sym}\mathbf{H} = \text{sym}[\nabla\mathbf{u}(\mathbf{1} + \mathbf{H})].$$

The *infinitesimal strain tensor* is defined by

$$\varepsilon := \text{sym}\nabla\mathbf{u}.$$

Hence, for small deformations $\mathbf{E} \simeq \mathcal{E} \simeq \varepsilon$.

It is worth observing that $\mathbf{C}$ and $\mathbf{B}$ admit the same principal invariants. By direct calculations, it follows that

$$\text{tr}\,\mathbf{C} = \text{tr}\,\mathbf{F}^T\mathbf{F} = \mathbf{F}\cdot\mathbf{F}, \qquad \text{tr}\,\mathbf{B} = \text{tr}\,\mathbf{F}\mathbf{F}^T = \mathbf{F}\cdot\mathbf{F},$$

$$\text{tr}\,\mathbf{C}^2 = \text{tr}\,(\mathbf{F}^T\mathbf{F}\mathbf{F}^T\mathbf{F}) = (\mathbf{F}^T\mathbf{F})\cdot(\mathbf{F}^T\mathbf{F}), \quad \text{tr}\,\mathbf{B}^2 = \text{tr}\,(\mathbf{F}\mathbf{F}^T\mathbf{F}\mathbf{F}^T) = (\mathbf{F}^T\mathbf{F})\cdot(\mathbf{F}^T\mathbf{F}),$$

$$\det\mathbf{C} = \det\mathbf{F}^T\mathbf{F} = (\det\mathbf{F})^2, \qquad \det\mathbf{B} = \det\mathbf{F}\mathbf{F}^T = (\det\mathbf{F})^2.$$

Hence, we find the common values of the principal invariants

$$I_1 = \text{tr}\,\mathbf{C} = \text{tr}\,\mathbf{B},$$

$$I_2 = \frac{1}{2}[(\text{tr}\,\mathbf{C})^2 - \text{tr}\,\mathbf{C}^2] = \frac{1}{2}[(\text{tr}\,\mathbf{B})^2 - \text{tr}\,\mathbf{B}^2],$$

$$I_3 = \det\mathbf{C} = \det\mathbf{B}.$$

By the polar decomposition $\mathbf{F} = \mathbf{R}\mathbf{U} = \mathbf{V}\mathbf{R}$, it follows that, associated with a deformation gradient $\mathbf{F}$, there are various rotation-independent tensors related to $\mathbf{U}$ and $\mathbf{V}$. The right Cauchy–Green deformation tensor (or Green's deformation tensor) is defined by $\mathbf{C} = \mathbf{U}^2 = \mathbf{F}^T\mathbf{F}$. The inverse $\mathbf{C}^{-1}$ is denoted by $\mathcal{F}$,

$$\mathcal{F} = \mathbf{C}^{-1} = \mathbf{F}^{-1}\mathbf{F}^{-T}.$$

As to $\mathbf{V}$, the left Cauchy–Green deformation tensor is denoted by $\mathbf{B}$ and defined by $\mathbf{B} = \mathbf{V}^2 = \mathbf{F}\mathbf{F}^T$. The inverse $\mathbf{B}^{-1}$ is denoted by $\mathcal{C}$,

$$\mathcal{C} = \mathbf{B}^{-1} = \mathbf{F}^{-T}\mathbf{F}^{-1}.$$

In the literature, $\mathcal{F}$ is called the Finger tensor (and denoted by $\mathbf{f}$). The inverse $\mathcal{C}$ of $\mathbf{B}$ was introduced by Cauchy and is denoted by $\mathbf{c}$; as with tensors, we prefer to use a capital letter ($\mathcal{C}$ instead of $\mathbf{c}$). In addition, $\mathcal{C}$ is called Piola tensor, and Finger tensor in the fluid dynamics literature. We avoid the use of Piola tensor for $\mathcal{C}$ because the Piola, or Piola-Kirchhoff, tensor is a well-known stress tensor of continuum mechanics.

We now examine the spectral representation of $\mathbf{F}$. The common eigenvalues of $\mathbf{C} = \mathbf{U}^2$ and $\mathbf{B} = \mathbf{V}^2$ are λ_i^2, $i = 1, 2, 3$. Hence

$$\mathbf{C} = \sum_{i=1}^{3} \lambda_i^2 \mathbf{N}_i \otimes \mathbf{N}_i, \qquad \mathbf{B} = \sum_{i=1}^{3} \lambda_i^2 \mathbf{n}_i \otimes \mathbf{n}_i$$

and

$$\mathbf{U} = \sum_{i=1}^{3} \lambda_i \mathbf{N}_i \otimes \mathbf{N}_i, \qquad \mathbf{V} = \sum_{i=1}^{3} \lambda_i \mathbf{n}_i \otimes \mathbf{n}_i = \sum_{i=1}^{3} \lambda_i \mathbf{R}\mathbf{N}_i \otimes \mathbf{R}\mathbf{N}_i.$$

Since $\mathbf{n}_i = \mathbf{R}\mathbf{N}_i$, we can write

$$\mathbf{R} = \sum_{j=1}^{3} \mathbf{n}_j \otimes \mathbf{N}_j;$$

$\mathbf{R}$ is the active rotation $\mathcal{R}$ as described in Sect. 1.4.1.

Hence, $\mathbf{F}$ is represented in the form

$$\mathbf{F} = \mathbf{R}\mathbf{U} = \left(\sum_{j=1}^{3} \mathbf{n}_j \otimes \mathbf{N}_j\right)\left(\sum_{i=1}^{3} \lambda_i \mathbf{N}_i \otimes \mathbf{N}_i\right) = \sum_{j=1}^{3} \lambda_j \mathbf{n}_j \otimes \mathbf{N}_j.$$

From the spectral representation, we can derive the matrix representation. If the deformation is a rotation, then $\mathbf{U} = \mathbf{1}$. Let $\mathbf{R}$ be given by a rotation of the angle θ around $\mathbf{n}_3 = \mathbf{N}_3$. Since $\mathbf{R} = \sum_{j=1}^{3} \mathbf{n}_j \otimes \mathbf{N}_j$, we find

$$R_{hk} = \mathbf{N}_h \cdot \mathbf{R}\mathbf{N}_k = \mathbf{N}_h \cdot \mathbf{n}_j \ \mathbf{N}_j \cdot \mathbf{N}_k = \mathbf{N}_h \cdot \mathbf{n}_k.$$

It follows that

$$R = \begin{bmatrix} \cos\theta & -\sin\theta & 0 \\ \sin\theta & \cos\theta & 0 \\ 0 & 0 & 1 \end{bmatrix}.$$

The Green-St Venant strain tensor $\mathbf{E}$ describes the deviation of the present configuration from the reference one, where $\mathbf{F} = \mathbf{1}$ and then also $\mathbf{C} = \mathbf{1}$. Hence, a better description of deformation might involve $\mathbf{E}$. Now

$$\mathbf{F}^{-T}(\mathbf{C} - \mathbf{1})\mathbf{F}^{-1} = \mathbf{F}^{-T}(\mathbf{F}^T\mathbf{F} - \mathbf{1})\mathbf{F}^{-1} = \mathbf{1} - \mathbf{F}^{-T}\mathbf{F}^{-1} = \mathbf{1} - \mathbf{B}^{-1}.$$

Hence, letting

$$\mathbf{E}_A := \tfrac{1}{2}(\mathbf{1} - \mathbf{B}^{-1})$$

we have

$$\mathbf{E}_A = \mathbf{F}^{-T}\mathbf{E}\mathbf{F}^{-1}.$$

The tensor $\mathbf{E}_A$ is called the Eulerian Almansi strain tensor.

1.2.1 Effects on Lengths, Areas, and Volumes

Distances and Lengths

Look at two position vectors $\mathbf{X}, \mathbf{Y} \in \mathrm{R}$ and the corresponding spatial position vectors $\mathbf{x} = \chi(\mathbf{X}), \mathbf{y} = \chi(\mathbf{Y})$ in the deformed region $\chi(\mathrm{R}) \subset \mathscr{E}$. It is

$$\mathbf{y} - \mathbf{x} = \mathbf{F}(\mathbf{Y} - \mathbf{X}) + o(|\mathbf{Y} - \mathbf{X}|),$$

$\mathbf{F}$ being evaluated at $\mathbf{X}$. To within $o(|\mathbf{Y} - \mathbf{X}|)$, we have

$$\mathbf{y} - \mathbf{x} = \mathbf{F}(\mathbf{Y} - \mathbf{X}) \tag{1.1}$$

and hence

$$(\mathbf{y} - \mathbf{x}) \cdot (\mathbf{y} - \mathbf{x}) = [\mathbf{F}(\mathbf{Y} - \mathbf{X})] \cdot [\mathbf{F}(\mathbf{Y} - \mathbf{X})] = (\mathbf{Y} - \mathbf{X}) \cdot \mathbf{F}^T\mathbf{F}(\mathbf{Y} - \mathbf{X}) = (\mathbf{Y} - \mathbf{X}) \cdot \mathbf{C}(\mathbf{Y} - \mathbf{X}).$$

Then the distance $l = |\mathbf{y} - \mathbf{x}|$ between the points in the deformed region is given by

$$l^2 = (\mathbf{Y} - \mathbf{X}) \cdot \mathbf{C}(\mathbf{Y} - \mathbf{X}); \tag{1.2}$$

the length l depends on the material vector $\mathbf{Y} - \mathbf{X}$ and not merely on the length $|\mathbf{Y} - \mathbf{X}|$. That is why the right Cauchy–Green tensor $\mathbf{C}$ is viewed as a metric tensor.

If we are interested also in the direction of $\mathbf{y} - \mathbf{x}$, then we consider (1.1) and observe that the unit vector $\mathbf{e} = (\mathbf{Y} - \mathbf{X})/|\mathbf{Y} - \mathbf{X}|$ is mapped to

$$\frac{\mathbf{y} - \mathbf{x}}{|\mathbf{Y} - \mathbf{X}|} = \mathbf{F}\mathbf{e}.$$

The ratio of the lengths

$$\frac{|\mathbf{y} - \mathbf{x}|}{|\mathbf{Y} - \mathbf{X}|} = |\mathbf{F}\mathbf{e}|,$$

is consistent with the action of deformation via the right Cauchy–Green tensor $\mathbf{C}$,

$$|\mathbf{y} - \mathbf{x}|^2 = |\mathbf{Y} - \mathbf{X}|^2\, \mathbf{e} \cdot \mathbf{C}\mathbf{e} = |\mathbf{Y} - \mathbf{X}|^2\, \mathbf{e} \cdot \mathbf{F}^T\mathbf{F}\mathbf{e} = |\mathbf{Y} - \mathbf{X}|^2\, |\mathbf{F}\mathbf{e}|^2.$$

The deviation of the deformation from the identity deformation is well described by $\mathbf{E}$. To get a detailed effect of $\mathbf{E}$, we let $\mathbf{y}-\mathbf{x}=\mathbf{F}(\mathbf{Y}-\mathbf{X})$ and find

$$|\mathbf{y}-\mathbf{x}|^2-|\mathbf{Y}-\mathbf{X}|^2=(\mathbf{Y}-\mathbf{X})\cdot(\mathbf{C}-\mathbf{1})(\mathbf{Y}-\mathbf{X}).$$

Since $\mathbf{Y}-\mathbf{X}=\mathbf{F}^{-T}(\mathbf{y}-\mathbf{x})$ we find

$$|\mathbf{y}-\mathbf{x}|^2-|\mathbf{Y}-\mathbf{X}|^2=\mathbf{F}^{-1}(\mathbf{y}-\mathbf{x})\cdot(\mathbf{C}-\mathbf{1})\mathbf{F}^{-1}(\mathbf{y}-\mathbf{x})=(\mathbf{y}-\mathbf{x})\cdot\mathbf{F}^{-T}(\mathbf{C}-\mathbf{1})\mathbf{F}^{-1}(\mathbf{y}-\mathbf{x}).$$

Hence

$$|\mathbf{y}-\mathbf{x}|^2-|\mathbf{Y}-\mathbf{X}|^2=(\mathbf{Y}-\mathbf{X})\cdot 2\mathbf{E}(\mathbf{Y}-\mathbf{X})=(\mathbf{y}-\mathbf{x})\cdot 2\mathbf{E}_A(\mathbf{y}-\mathbf{x}).$$

The Green-St. Venant tensor $\mathbf{E}$ yields the difference $|\mathbf{y}-\mathbf{x}|^2-|\mathbf{Y}-\mathbf{X}|^2$ in terms of the vector $\mathbf{Y}-\mathbf{X}$ while the Eulerian Almansi tensor, $\mathbf{E}_A$ or $\mathbf{e}$, yields the difference in terms of $\mathbf{y}-\mathbf{x}$.

We now examine the length of a curve. Let C be a curve in R represented parametrically as

$$\mathbf{Y}=\hat{\mathbf{Y}}(\lambda),\qquad \lambda\in[a,b].$$

By deformation, C is mapped to $\mathcal{C}$ given by

$$\mathbf{Y}(\lambda)\ \rightarrow\ \mathbf{y}(\lambda)=\chi(\mathbf{Y}(\lambda)).$$

The line integral of a continuous function f defined on $\mathcal{C}$ reads

$$\int_a^b f(\mathbf{y}(\lambda))|\mathbf{y}'(\lambda)|d\lambda=\int_a^b f(\mathbf{y}(\chi(\lambda)))|\chi'(\mathbf{Y}(\lambda))|d\lambda$$

where $'$ means derivative with respect to the parameter λ. Since $\chi'(\mathbf{Y}(\lambda))=\mathbf{F}(\mathbf{Y}(\lambda))\mathbf{Y}'(\lambda)$ then

$$\int_a^b f(\mathbf{y}(\lambda))|\mathbf{y}'(\lambda)|d\lambda=\int_a^b f(\chi(\mathbf{Y}(\lambda)))|\mathbf{F}(\mathbf{Y}(\lambda))\mathbf{Y}'(\lambda)|d\lambda.$$

Let $f=1$ and observe that

$$\mathcal{L}=\int_a^b|\mathbf{y}'(\lambda)|d\lambda$$

is the length of $\mathcal{C}$ while $\int_a^b|\mathbf{Y}'(\lambda)|d\lambda$ is the length of C. Hence

$$\mathcal{L}=\int_a^b|\mathbf{F}(\mathbf{Y}(\lambda))\mathbf{Y}'(\lambda)|d\lambda.$$

By means of the polar decomposition, we can write

$$|\mathbf{F}\,\mathbf{Y}'| = (\mathbf{R}\,\mathbf{U}\,\mathbf{Y}' \cdot \mathbf{R}\,\mathbf{U}\,\mathbf{Y}')^{1/2}(\mathbf{R}^T\,\mathbf{R}\,\mathbf{U}\,\mathbf{Y}' \cdot \mathbf{U}\,\mathbf{Y}')^{1/2} = |\mathbf{U}\,\mathbf{Y}'|.$$

As a consequence

$$\mathcal{L} = \int_a^b |\mathbf{U}(\mathbf{Y}(\lambda))\mathbf{Y}'(\lambda)|d\lambda,$$

so that the length is unchanged if $\mathbf{F}$ is (locally) a pure rotation, $\mathbf{U} = \mathbf{1}$. This shows that $\mathbf{U} \neq \mathbf{1}$ does not preserve lengths.

A deformation that preserves the distance between points is said to be rigid; χ is *rigid* if

$$|\chi(\mathbf{Y}) - \chi(\mathbf{X})| = |\mathbf{Y} - \mathbf{X}|, \qquad \forall\,\mathbf{Y}, \mathbf{X} \in \mathcal{R}. \tag{1.3}$$

By (1.3), it follows

$$[\chi(\mathbf{Y}) - \chi(\mathbf{X})] \cdot [\chi(\mathbf{Y}) - \chi(\mathbf{X})] = (\mathbf{Y} - \mathbf{X}) \cdot (\mathbf{Y} - \mathbf{X}).$$

Differentiation with respect to $\mathbf{X}$ and $\mathbf{Y}$ yields

$$\mathbf{F}^T(\mathbf{Y})\mathbf{F}(\mathbf{X}) = \mathbf{1}.$$

Letting $\mathbf{X} = \mathbf{Y}$ we find

$$\mathbf{F}^T(\mathbf{X})\mathbf{F}(\mathbf{X}) = \mathbf{1}$$

and hence $\mathbf{F}$ is a rotation, say $\mathbf{R}$, at any point of the body. Consequently

$$\mathbf{F}(\mathbf{Y})\mathbf{F}^T(\mathbf{Y})\mathbf{F}(\mathbf{X}) = \mathbf{F}(\mathbf{Y})$$

implies $\mathbf{F}(\mathbf{X}) = \mathbf{F}(\mathbf{Y})$. Hence, in a rigid deformation $\mathbf{F}$ is a rotation $\mathbf{R}$, independent of the position. Moreover, by integration of

$$\nabla_R \chi = \mathbf{R}$$

we have

$$\chi(\mathbf{Y}) = \chi(\mathbf{X}) + \mathbf{R}(\mathbf{Y} - \mathbf{X}). \tag{1.4}$$

It is of interest to examine the (approximate) description of small rigid deformations. By (1.4), since $\mathbf{R} = \mathbf{F}$ then letting $\chi(\mathbf{Y}) = \mathbf{Y} + \mathbf{u}(\mathbf{Y})$, we have

$$\mathbf{u}(\mathbf{Y}) - \mathbf{u}(\mathbf{X}) = (-\mathbf{1} + \mathbf{F})(\mathbf{Y} - \mathbf{X}).$$

By $\mathbf{F}^T\mathbf{F} = \mathbf{C} = \mathbf{U}^2 = \mathbf{1}$, it follows

$$(\mathbf{1} + \mathbf{H}^T)(\mathbf{1} + \mathbf{H}) = \mathbf{1}, \qquad \mathbf{H}^T + \mathbf{H} = \mathbf{0}$$

where $\mathbf{H}^T\mathbf{H}$ has been neglected in the last relation. In this approximation $\mathbf{F} - \mathbf{1} = \mathbf{H} \in \text{Skw}$. As a further consequence, it follows the *projection property*

$$[\mathbf{u}(\mathbf{Y}) - \mathbf{u}(\mathbf{X})] \cdot (\mathbf{Y} - \mathbf{X}) = 0, \qquad \forall \mathbf{Y}, \mathbf{X} \in \mathrm{R}.$$

Volumes

To establish the effect on volumes, we consider a sub-region $\mathrm{P} \subseteq \mathrm{R}$ and let $\chi(\mathrm{P}) \subset \mathscr{E}$ be the region produced by deformation. Let f be a function defined at the points of the body so that

$$f(\mathbf{y}) = f(\chi(\mathbf{Y})) =: F(\mathbf{Y}),$$

where $\mathbf{Y} \in \mathrm{P}$. We can view

$$\mathbf{Y} \to \mathbf{y} = \chi(\mathbf{Y}), \qquad (Y_1, Y_2, Y_3) \to y_i = \chi_i(Y_1, Y_2, Y_3), \; i = 1, 2, 3,$$

as a change of variables. Hence

$$\int_{\chi(\mathrm{P})} f(\mathbf{y}) dv = \int_{\mathrm{P}} F(\mathbf{Y}) J(\mathbf{Y}) dv_R. \tag{1.5}$$

Let $f = 1$. Then it follows

$$\text{vol}\chi(\mathrm{P}) = \int_{\chi(\mathrm{P})} 1\, dv = \int_{\mathrm{P}} J(\mathbf{Y}) dv_R.$$

Now, for any $\mathbf{X} \in \mathrm{P}$, we can write

$$\int_{\mathrm{P}} J(\mathbf{Y}) dV = \int_{\mathrm{P}} J(\mathbf{X}) dv_R + \int_{\mathrm{P}} [J(\mathbf{Y}) - J(\mathbf{X})] dv_R.$$

Granted the continuity of J, if d is the diameter of P, we have

$$\int_{\mathrm{P}} J(\mathbf{Y}) dV = [J(\mathbf{X}) + O(d)] \int_{\mathrm{P}} 1\, dV$$

and hence

$$\text{vol}\, \chi(\mathrm{P}) = [J(\mathbf{X}) + O(d)] \text{vol}\, \mathrm{P}.$$

At the leading order, as $d \to 0$,

$$\text{vol}\, \chi(\mathrm{P}) = J \, \text{vol}\, \mathrm{P}.$$

If $J = 1$ the volume is constant. Now, since $\mathbf{F} = \mathbf{1} + \mathbf{H}$ then

$$J = \det[\mathbf{1} + \mathbf{H}] = 1 + \text{tr}\, \mathbf{H} + o(|\mathbf{H}|).$$

In the approximation of small deformations $|o(\mathbf{H})| < |\mathrm{tr}\,\mathbf{H}|$ and hence the requirement $J = 1$ implies

$$\mathrm{tr}\,\mathbf{H} = \mathrm{tr}\,\mathcal{E} = 0, \qquad \nabla_R \cdot \mathbf{u} = 0. \tag{1.6}$$

Deformation of Normals and Areas

Consider a surface $\mathrm{S} \subset \mathrm{R}$ and let $\mathcal{S} = \chi(\mathrm{S})$ be the deformed surface. Let $\mathbf{X}$ be (the position vector of) a point in S and let $\mathbf{x} = \chi(\mathbf{X})$. A curve $\mathrm{C} \subset \mathrm{S}$ is represented parametrically as

$$\mathbf{Y} = \hat{\mathbf{Y}}(\lambda), \qquad \hat{\mathbf{Y}}(\lambda_0) = \mathbf{X}.$$

The vector $\hat{\mathbf{Y}}'(\lambda_0)$ is tangent to C at $\mathbf{X}$. Via the deformation function χ, the image of C is a curve $\mathcal{C}$ in $\mathcal{S}$, $\mathcal{C} = \chi(\mathrm{C})$, described by

$$\hat{\mathbf{y}}(\lambda) = \chi(\hat{\mathbf{Y}}(\lambda)), \qquad \hat{\mathbf{y}}(\lambda_0) = \mathbf{x}.$$

Hence $\hat{\mathbf{y}}'(\lambda_0)$ is tangent to $\mathcal{C}$ at $\hat{\mathbf{y}}(\lambda_0) = \mathbf{x}$. The derivative with respect to λ of the composite function $\hat{\mathbf{y}}(\lambda)$ results in

$$\hat{\mathbf{y}}'(\lambda_0) = \mathbf{F}\hat{\mathbf{Y}}'(\lambda_0);$$

this is the relation between corresponding tangent vectors in S and $\mathcal{S} = \chi(\mathrm{S})$. This means that if $\mathbf{t}_R$ and $\mathbf{t}$ are corresponding tangent vectors (not necessarily unit vectors) then

$$\mathbf{t} = \mathbf{F}\mathbf{t}_R.$$

At the chosen points ($\mathbf{X}$ and $\mathbf{x}$) in S and $\mathcal{S}$ the (not necessarily unit) normals $\mathbf{n}_R$ and $\mathbf{n}$ are subject to

$$\mathbf{n}_R \cdot \mathbf{t}_R = 0, \qquad \mathbf{n} \cdot \mathbf{t} = 0,$$

for any tangent vectors $\mathbf{t}_R$, $\mathbf{t}$; the directions of $\mathbf{n}_R$, $\mathbf{n}$ are unique. Now,

$$0 = \mathbf{n} \cdot \mathbf{t} = \mathbf{n} \cdot \mathbf{F}\mathbf{t}_R = (\mathbf{F}^T\mathbf{n}) \cdot \mathbf{t}_R.$$

Hence we can define

$$\mathbf{n}_R = \mathbf{F}^T\mathbf{n}, \qquad \mathbf{n} = \mathbf{F}^{-T}\mathbf{n}_R.$$

Quite naturally, it is technically more convenient to have unit normals $\mathbf{n}_R$, $\mathbf{n}$ and to know the relation between areas in S and $\mathcal{S}$. This is accomplished as follows.

The surface S is described parametrically by $\mathbf{Y} = \phi(u, v)$ while $\mathcal{S} = \chi(\mathrm{S})$ is described by $\mathbf{y} = \chi(\phi(u, v))$, where $(u, v) \in D \subset \mathbb{R}^2$. Hence we have

$$\mathbf{n}_R = \frac{\partial_u \phi \times \partial_v \phi}{|\partial_u \phi \times \partial_v \phi|}, \qquad \mathbf{n} = \frac{\partial_u \chi(\phi) \times \partial_v \chi(\phi)}{|\partial_u \chi(\phi) \times \partial_v \chi(\phi)|}.$$

In view of the definition of surface integral we can write

$$\int_{\mathcal{S}} g\, \mathbf{n}\, da = \int_D g(u,v)\partial_u \chi(\phi(u,v)) \times \partial_v \chi(\phi(u,v))\, du dv,$$

$$\int_{\mathrm{S}} G\, \mathbf{n}_R\, da_R = \int_D G(u,v)\partial_u \phi(u,v) \times \partial_v \phi(u,v)\, du dv,$$

To compute $\partial_u \chi(\phi(u,v)) \times \partial_v \chi(\phi(u,v))$ we observe that

$$[\partial_u \chi(\phi(u,v)) \times \partial_v \chi(\phi(u,v))]_i = \epsilon_{ihk} F_{hP} F_{kQ} \partial_u \phi_P \partial_v \phi_Q.$$

Now, by

$$\epsilon_{ihk} F_{iN} F_{hP} F_{kQ} = \epsilon_{NPQ} J$$

we obtain

$$\epsilon_{ihk} F_{hP} F_{kQ} = J F^{-1}_{Ni} \epsilon_{NPQ}.$$

Consequently it follows

$$[\partial_u \chi(\phi(u,v)) \times \partial_v \chi(\phi(u,v))]_i = J\, F^{-1}_{Ni} \epsilon_{NPQ} \partial_u \phi_P \partial_v \phi_Q,$$

whence

$$\partial_u \chi(\phi(u,v)) \times \partial_v \chi(\phi(u,v)) = J\, \mathbf{F}^{-T} [\partial_u \phi(u,v) \times \partial_v \phi(u,v)].$$

Thus

$$\int_{\mathcal{S}} g\, \mathbf{n}\, da = \int_D g(u,v)\partial_u \chi(\phi(u,v)) \times \partial_v \chi(\phi(u,v))\, du dv$$
$$= \int_D g(u,v) J \mathbf{F}^{-T} [\partial_u \chi(\phi(u,v)) \times \partial_v \chi(\phi(u,v))] du dv = \int_S g\, J\, \mathbf{F}^{-T}\, \mathbf{n}_R\, da_R.$$

This result indicates how to express the integral of $g\mathbf{n}$ on $\mathcal{S}$ in terms of an integral on S. Moreover, let $g = 1$ and denote by $\mathcal{A}$, A the areas of $\mathcal{S}$, S. Hence we observe that, at the leading order when D shrinks to a point, the desired relations for normals and areas follow in the form

$$\mathbf{n}\, \mathcal{A} = J\, \mathbf{F}^{-T} \mathbf{n}_R\, A, \qquad \mathcal{A} = J\, |\mathbf{F}^{-T} \mathbf{n}_R|\, A. \tag{1.7}$$

Two remarks are in order. Mnemonically we might say that the connection between the integrals on S and $\chi(\mathrm{S})$ is obtained by letting

$$\mathbf{n}\, da = J\, \mathbf{F}^{-T}\, \mathbf{n}_R\, da_R.$$

This relation is known as *Nanson formula*.[5] The second remark is about the relation between areas. By (1.7), we might say that

$$\jmath = J\,|\mathbf{F}^{-T}\mathbf{n}_R|$$

is the *areal Jacobian*. In [216], via a different procedure and the choice $\mathbf{e}_3 = \mathbf{n}_R$, the equivalent result $\jmath = |\mathbf{F}\mathbf{e}_1 \times \mathbf{F}\mathbf{e}_2|$ is obtained.

We have proved that, for any function g on $\chi(\mathrm{R})$,

$$\int_{\mathcal{S}} g\,\mathbf{n}\,da = \int_{\mathrm{S}} g\,J\mathbf{F}^{-T}\mathbf{n}_R\,da_R.$$

If g is replaced with a vector $\mathbf{g}$, or likewise with a tensor, then we have

$$\int_{\mathcal{S}} \mathbf{g}\cdot\mathbf{n}\,da = \int_{\mathrm{S}} J\,\mathbf{g}\cdot(\mathbf{F}^{-T}\mathbf{n}_R)da_R = \int_{\mathrm{S}} J(\mathbf{g}\mathbf{F}^{-T})\cdot\mathbf{n}_R\,da_R.$$

Hence, letting

$$\mathbf{g}_R := J\,\mathbf{g}\,\mathbf{F}^{-T} \tag{1.8}$$

we have

$$\int_{\mathcal{S}} \mathbf{g}\cdot\mathbf{n}\,da = \int_{\mathrm{S}} \mathbf{g}_R\cdot\mathbf{n}_R\,da_R.$$

As a comment, let $\mathcal{S}$, S be closed surfaces and denote by $\mathcal{P}$, P the regions bounded by $\mathcal{S}$, S. In view of the divergence theorem we can write

$$\int_{\mathcal{P}} \nabla\cdot\mathbf{g}\,dv = \int_{\mathrm{P}} \nabla_R\cdot\mathbf{g}_R\,dv_R.$$

Since $\int_{\mathcal{P}} \nabla\cdot\mathbf{g}\,dv = \int_{\mathrm{P}} J\,\nabla\cdot\mathbf{g}\,dv_R$ then the arbitrariness of P implies

$$\nabla_R\cdot\mathbf{g}_R = J\,\nabla\cdot\mathbf{g}. \tag{1.9}$$

1.2.2 Some Identities Involving the Deformation Gradient

Given an invertible matrix F, the entries of F^{-1} depend on the entries of F such that

$$\partial_{F_{mK}} F^{-1}_{Mk} = -F^{-1}_{Kk}F^{-1}_{Mm}. \tag{1.10}$$

Since $F^{-1}_{Mk}F_{kP} = \delta_{MP}$ we observe

$$0 = \partial_{F_{mK}}(F^{-1}_{Mk}F_{kP}) = (\partial_{F_{mK}}F^{-1}_{Mk})F_{kP} + F^{-1}_{Mk}\delta_{km}\delta_{KP}$$

[5] See, e.g., [429], p. 20; [147], p. 71.

whence, upon multiplying by F^{-1}_{Pl},

$$\delta_{kl}(\partial_{F_{mK}} F^{-1}_{Mk}) = -F^{-1}_{Pl} F^{-1}_{Mk} \delta_{km} \delta_{KP}.$$

The result (1.10) follows.

The assumed invertibility of the deformation gradient implies that (1.10) holds for the entries of the deformation gradient matrix.

In dealing with the connection between fields in R and χ(R) the identities

$$\partial_{X_K}(J F^{-1}_{Ki}) = 0, \qquad \partial_{x_k}(J^{-1} F_{kH}) = 0 \tag{1.11}$$

prove convenient.[6] A direct proof can be given by viewing $J F^{-1}_{Ki}$ as a function of the entries F_{mH} and observing

$$\partial_{X_K}(J F^{-1}_{Ki}) = \partial_{F_{mH}}(J F^{-1}_{Ki})\partial_{X_K} F_{mH} = [(\partial_{F_{mH}} J) F^{-1}_{Ki} + J \partial_{F_{mH}} F^{-1}_{Ki}]\partial_{X_K} F_{mH}.$$

Since, by (A.16),

$$\partial_{F_{mH}} J = J F^{-1}_{Hm},$$

in view of (1.10) we have

$$\partial_{X_K}(J F^{-1}_{Ki}) = (J F^{-1}_{Hm} F^{-1}_{Ki} - J F^{-1}_{Hi} F^{-1}_{Km})\partial_{X_K} F_{mH}.$$

Now

$$\partial_{X_K} F_{mH} = \partial_{X_K}\partial_{X_H} f_m = \partial_{X_H} F_{mK}.$$

Hence, for each m and i, in the product

$$(J F^{-1}_{Hm} F^{-1}_{Ki} - J F^{-1}_{Hi} F^{-1}_{Km})\partial_{X_K} F_{mH}$$

the factor $(J F^{-1}_{Hm} F^{-1}_{Ki} - J F^{-1}_{Hi} F^{-1}_{Km})$ is skew in HK whereas $\partial_{X_K} F_{mH}$ is symmetric; the sum over H and K vanishes, and hence $\partial_{X_K}(J F^{-1}_{Ki}) = 0$, which completes the proof. The second identity in (1.11) follows by duality, that is by interchanging **x** and **X**.

Identity (1.11) follows also from (1.9). By the relation between **g** and $\mathbf{g}_R$ we have

$$\nabla_R \cdot \mathbf{g}_R = \partial_{X_K}(J g_i F^{-T}_{iK}) = J \partial_{X_K} g_i F^{-1}_{Ki} + g_i \partial_{X_K}(J F^{-1}_{Ki}) = J \nabla \cdot \mathbf{g} + g_i \partial_{X_K}(J F^{-1}_{Ki}).$$

By (1.9) we have $\nabla_R \cdot \mathbf{g}_R = J \nabla \cdot \mathbf{g}$ and hence a nonzero g_i implies (1.11).

---

[6] These identities were first given by Piola and Jacobi [429], p. 18.

1.3 Motion

A motion is described by a function χ on R $\times \mathbb{R}$ such that $\mathbf{x} = \chi(\mathbf{X}, t)$ is the position (vector) at time t of the point at $\mathbf{X}$ in the reference configuration R. At any time t, $\chi(\mathbf{X}, t)$ describes a deformation such that R $\mapsto \mathcal{R}_t \subset \mathcal{E}$. Consistently the function is assumed to be invertible in that $\mathbf{X} = \chi^{-1}(\mathbf{x}, t)$ associates with each time t and spatial position $\mathbf{x} \in \mathcal{R}_t$ the position $\mathbf{X}$ in the reference configuration. The function χ is assumed to be differentiable, on R; the deformation gradient $\mathbf{F} = \nabla_R \chi(\mathbf{X}, t)$ is then a function on R $\times \mathbb{R}$. Moreover it is assumed that $J = \det \mathbf{F} > 0$. The function χ is assumed to be a C^2 function with respect to time t.

For any set of points in P $\subseteq$ R we let $\mathcal{P}_t = \chi_t(\mathrm{P}) = \chi(\mathrm{P}, t)$ denote the region occupied by the points of P at time t; this can be the case of sub-regions, surfaces, curves. We then say that P deforms to $\mathcal{P}_t$, at time t, and that a time-dependent spatial set $\mathcal{P}_t$ *convects with the body* if there is a set P of material points such that $\mathcal{P}_t = \chi_t(\mathrm{P})$ for all t.

Let φ be a scalar, vector, or tensor quantity. By $\varphi = \phi(\mathbf{x}, t)$ we mean a function providing φ, at time t, in a spatial domain $D \in \mathcal{E}$. If φ is defined on a set of points of the body then

$$\phi(\mathbf{x}, t) = \phi(\chi(\mathbf{X}, t)) =: \Phi(\mathbf{X}, t),$$

where $\mathbf{x} \in \mathcal{P}_t$, $\mathbf{X} \in \mathrm{P} \subseteq \mathrm{R}$; we say that ϕ is the *spatial description* of φ while Φ is the *material description.* The connection between gradients in the two descriptions is given by the chain rule so that

$$\partial_{X_K} \Phi = \partial_{x_i} \phi \, \partial_{X_K} \chi_i, \qquad \nabla_R \Phi = \nabla \phi \, \mathbf{F} = \mathbf{F}^T \nabla \phi.$$

Things are more involved about the time derivative in that the derivatives

$$\partial_t \phi(\mathbf{x}, t), \qquad \partial_t \Phi(\mathbf{X}, t)$$

are in general unequal. Indeed, by means of the chain rule

$$\partial_t \Phi(\mathbf{X}, t) = \partial_t \phi(\chi(\mathbf{X}, t), t) = \partial_t \phi(\mathbf{x}, t) + \partial_{x_i} \phi(\mathbf{x}, t) \, \partial_t \chi_i(\mathbf{X}, t).$$

Since $\partial_t \chi(\mathbf{X}, t)$ is the velocity, $\mathbf{v}$, of the point $\mathbf{X}$ then

$$\partial_t \Phi = \partial_t \phi + \mathbf{v} \cdot \nabla \phi.$$

To simplify the notation, customarily only one symbol is used, here ϕ, and we let $\dot{\phi}$ stand for $\partial_t \Phi$ so that

$$\dot{\phi} = \partial_t \phi + \mathbf{v} \cdot \nabla \phi. \tag{1.12}$$

It is common to say that $\dot{\phi}$ is the *material time derivative*[7] and $\partial_t \phi$ is the *spatial time derivative.* By definition, $\dot{\phi}$ is the derivative at $\mathbf{X}$ fixed namely the derivative observed by following the motion of the material point $\mathbf{X}$. Instead, $\partial_t \phi$ is the time derivative relative to an observer fixed in space, at the point $\mathbf{x}$. The difference between the two derivatives, $\mathbf{v} \cdot \nabla \phi$, is called the convective part and is due to the motion of the continuum.

Denote by $\mathbf{v}(\mathbf{x}, t)$ the velocity field in the spatial description. The spatial tensor field

$$\mathbf{L} = \nabla \mathbf{v}, \qquad L_{ij} = \partial_{x_j} v_i,$$

is called the *velocity gradient.* To find the connection with the velocity gradient in the material description we assume χ is a C^2 function and observe that

$$\dot{F}_{iK} = \partial_t F_{iK}(\mathbf{X}, t) = \partial_{X_K} \partial_t \chi_i(\mathbf{X}, t) = \partial_{x_j} v_i(\mathbf{x}, t) \partial_{X_K} \chi_j(\mathbf{X}, t) = L_{ij} F_{jK}.$$

Hence we have the identity

$$\dot{\mathbf{F}} = \mathbf{L}\mathbf{F}.$$

To determine the material time derivative of $J = \det \mathbf{F}$ we apply (A.17) to obtain

$$\dot{J} = J \,\mathrm{tr}\, (\dot{\mathbf{F}} \mathbf{F}^{-1}) = J \,\mathrm{tr}\, \mathbf{L}.$$

Since

$$\mathrm{tr}\, \mathbf{L} = \nabla \cdot \mathbf{v}$$

then

$$\dot{J} = J \nabla \cdot \mathbf{v}. \tag{1.13}$$

The inverse $j = 1/J$ satisfies

$$(1/J)^{\cdot} = -(1/J^2) \dot{J}$$

and hence

$$(j)^{\cdot} = -j \nabla \cdot \mathbf{v}. \tag{1.14}$$

Since $^{\cdot} = \partial_t + \mathbf{v} \cdot \nabla$ then (1.14) can be written

$$\partial_t j + \nabla \cdot (j\mathbf{v}) = 0.$$

If $\nabla \cdot \mathbf{v} = 0$ then J is constant and the volume is constant. Conversely if J is constant then (1.13) becomes

$$0 = J F^{-1}_{Ki} \dot{H}_{iK}.$$

---

[7] In the literature other adjectives are used for the material time derivative: advective, convective, hydrodynamic, Lagrangian, substantial, total.

In the approximation of small deformations $J F^{-1}_{Ki} \to \delta_{iK}$ and hence

$$0 \simeq \operatorname{tr} \dot{\mathbf{H}} = \nabla_R \cdot \dot{\mathbf{u}} = \overline{\nabla_R \cdot \mathbf{u}}^{\boldsymbol{\cdot}}.$$

If $\mathbf{u}(\mathbf{X}, t_0) = \mathbf{0}$ and hence $\nabla_R \cdot \mathbf{u} = 0$ at some time t_0 then, consistent with (1.6), it follows $\nabla_R \cdot \mathbf{u} = 0$.

1.3.1 Stretching and Spin Tensors

Let $\mathbf{v}(\mathbf{x}, t)$ be the spatial description of the velocity field $\mathbf{v}$. If, as we assume it to be, $\mathbf{v}$ is differentiable then

$$\mathbf{v}(\mathbf{y}) - \mathbf{v}(\mathbf{x}) = \mathbf{L}(\mathbf{x})(\mathbf{y} - \mathbf{x}) + o(|\mathbf{y} - \mathbf{x}|),$$

the dependence of $\mathbf{v}$ and $\mathbf{L}$ on t being understood and not written. Let $\mathbf{D}$ and $\mathbf{W}$ be the symmetric and skew parts of $\mathbf{L}$,

$$\mathbf{D} = \tfrac{1}{2}(\mathbf{L} + \mathbf{L}^T), \qquad \mathbf{W} = \tfrac{1}{2}(\mathbf{L} - \mathbf{L}^T).$$

Then $\mathbf{L} = \mathbf{D} + \mathbf{W}$ and

$$\mathbf{v}(\mathbf{y}) - \mathbf{v}(\mathbf{x}) = \mathbf{W}(\mathbf{x})(\mathbf{y} - \mathbf{x}) + \mathbf{D}(\mathbf{x})(\mathbf{y} - \mathbf{x}) + o(|\mathbf{y} - \mathbf{x}|).$$

The tensors $\mathbf{W}$ and $\mathbf{D}$ are called the *spin* and the *stretching*.

It is worth establishing a connection between $\mathbf{L}$ and the tensors $\mathbf{U}, \mathbf{V}, \mathbf{R}$ of the polar decomposition. Substitution of $\mathbf{F}$ with $\mathbf{RU}$ yields

$$\mathbf{L} = \dot{\mathbf{F}}\mathbf{F}^{-1} = (\dot{\mathbf{R}}\mathbf{U} + \mathbf{R}\dot{\mathbf{U}})\mathbf{U}^{-1}\mathbf{R}^T = \dot{\mathbf{R}}\mathbf{R}^T + \mathbf{R}\dot{\mathbf{U}}\mathbf{U}^{-1}\mathbf{R}^T.$$

Now, time differentiation of $\mathbf{R}\mathbf{R}^T = \mathbf{1}$ results in

$$\dot{\mathbf{R}}\mathbf{R}^T = -\mathbf{R}\dot{\mathbf{R}}^T = -(\dot{\mathbf{R}}\mathbf{R}^T)^T.$$

This implies that $\dot{\mathbf{R}}\mathbf{R}^T \in \mathrm{Skw}$. Yet, in general $\mathbf{W} \neq \dot{\mathbf{R}}\mathbf{R}^T$; indeed

$$\mathbf{D} = \mathbf{R}[\mathrm{sym}(\dot{\mathbf{U}}\mathbf{U}^{-1})]\mathbf{R}^T, \qquad \mathbf{W} = \dot{\mathbf{R}}\mathbf{R}^T + \mathbf{R}[\mathrm{skw}(\dot{\mathbf{U}}\mathbf{U}^{-1})]\mathbf{R}^T.$$

The stretching $\mathbf{D}$ (as well as the spin $\mathbf{W}$ and the velocity gradient $\mathbf{L}$), maps spatial vectors to spatial vectors, while the Green-St.Venant strain $\mathbf{E} = \frac{1}{2}(\mathbf{C} - \mathbf{1})$ maps material vectors to material vectors. Nevertheless, there is a direct connection between $\mathbf{D}$ and $\dot{\mathbf{E}}$. By direct substitutions, we have

$$2\mathbf{F}^T\mathbf{D}\mathbf{F} = \mathbf{F}^T(\mathbf{L}+\mathbf{L}^T)\mathbf{F} = \mathbf{F}^T(\dot{\mathbf{F}}\mathbf{F}^{-1}+\mathbf{F}^{-1}\dot{\mathbf{F}}^T)\mathbf{F} = \mathbf{F}^T\dot{\mathbf{F}}+\dot{\mathbf{F}}^T\mathbf{F} = 2\,\dot{\mathbf{E}}.$$

Hence, we get the identity

$$\mathbf{F}^T\mathbf{D}\mathbf{F} = \dot{\mathbf{E}}.$$

A motion χ is *rigid* if, at each time t, for any two points $\mathbf{X}$, $\mathbf{Y}$ of the body

$$\partial_t|\chi(\mathbf{Y},t) - \chi(\mathbf{X},t)| = 0.$$

The time derivative yields

$$(\mathbf{y}-\mathbf{x})\cdot[\mathbf{v}(\mathbf{y},t)-\mathbf{v}(\mathbf{x},t)] = 0,$$

being understood that $\mathbf{x} = \chi(\mathbf{X},t),\ \mathbf{y} = \chi(\mathbf{Y},t)$. The gradient with respect to $\mathbf{x}$ and next the gradient with respect to $\mathbf{y}$ result in

$$\mathbf{v}(\mathbf{x},t) - \mathbf{v}(\mathbf{y},t) - \mathbf{L}^T(\mathbf{x},t)(\mathbf{y}-\mathbf{x}) = \mathbf{0},$$

$$\mathbf{L}(\mathbf{y},t) + \mathbf{L}^T(\mathbf{x},t) = \mathbf{0}.$$

The continuity of $\mathbf{L}$ allows us to let $\mathbf{y} = \mathbf{x}$ and to obtain $\mathbf{L} \in \mathrm{Skw}$. Back to $\mathbf{L}(\mathbf{y},t) = \mathbf{L}(\mathbf{x},t)$ it follows that $\mathbf{L} = \mathbf{W}$ is independent of the position and hence $\mathbf{L} = \mathbf{W}(t)$. Thus

$$\mathbf{v}(\mathbf{y},t) = \mathbf{v}(\mathbf{x},t) + \mathbf{W}(t)(\mathbf{y}-\mathbf{x}).$$

We know[8] that there is a 1–1 correspondence between skew tensors and axial vectors. Let $\mathbf{w}$ be the axial vector associated to $\mathbf{W}$; we have $\mathbf{W} = \mathbf{w}\times$ and

$$w_j = -\tfrac{1}{2}\epsilon_{jpq}W_{pq} = -\tfrac{1}{4}\epsilon_{jpq}(\partial_{x_q}v_p - \partial_{x_p}v_q) = \tfrac{1}{2}(\nabla\times\mathbf{v})_j.$$

Thus, in a rigid motion

$$\mathbf{v}(\mathbf{y},t) = \mathbf{v}(\mathbf{x},t) + \tfrac{1}{2}(\nabla\times\mathbf{v})\times(\mathbf{y}-\mathbf{x}),$$

$\nabla\times\mathbf{v}$ being a function of time t.

The vector

$$\varpi = \nabla\times\mathbf{v}$$

is called the *vorticity*. Hence we can also write[9]

$$\mathbf{W} = \tfrac{1}{2}\varpi\times\ .$$

[8] See Appendix A.

[9] In the literature the vorticity $\nabla\times\mathbf{v}$ is denoted by ω. To avoid ambiguities with the angular velocity, throughout we denote the vorticity $\nabla\times\mathbf{v}$ by ϖ rather than by ω.

In a rigid motion, the vorticity is independent of the position.

A motion is *irrotational* if the spin $\mathbf{W}$, as well as the vorticity $\boldsymbol{\varpi}$, is zero at any point and time.

For any velocity field, the velocity gradient $\mathbf{L}$ and the vorticity $\boldsymbol{\varpi}$ satisfy the identity

$$\mathbf{L}\,\boldsymbol{\varpi} = \mathbf{D}\,\boldsymbol{\varpi}. \tag{1.15}$$

The proof follows by observing that

$$\mathbf{L}\,\boldsymbol{\varpi} = \mathbf{W}\,\boldsymbol{\varpi} + \mathbf{D}\,\boldsymbol{\varpi} = \tfrac{1}{2}\boldsymbol{\varpi} \times \boldsymbol{\varpi} + \mathbf{D}\,\boldsymbol{\varpi} = \mathbf{D}\,\boldsymbol{\varpi}.$$

1.4 Rotations and Angular Velocities

Consider a conventionally fixed frame of reference associated with orthonormal base vectors $(\mathbf{e}_1, \mathbf{e}_2, \mathbf{e}_3)$ and a rotating frame of reference associated with orthonormal base vectors $(\mathbf{k}_1, \mathbf{k}_2, \mathbf{k}_3)$. Let $R(t)$ be the time-dependent rotation matrix defined by

$$R_{ij}(t) = \mathbf{k}_i(t) \cdot \mathbf{e}_j. \tag{1.16}$$

Hence, the two bases are related by R in the form

$$\mathbf{k}_i = (\mathbf{k}_i \cdot \mathbf{e}_j)\mathbf{e}_j = R_{ij}\mathbf{e}_j.$$

For any vector $\mathbf{u}$, we have

$$\mathbf{u} \cdot \mathbf{k}_i = \mathbf{u} \cdot R_{ij}\mathbf{e}_j = R_{ij}\mathbf{u} \cdot \mathbf{e}_j.$$

Letting $u_i := \mathbf{u} \cdot \mathbf{k}_i$ and $U_j := \mathbf{u} \cdot \mathbf{e}_j$ we conclude that

$$u_i = R_{ij}U_j.$$

This transformation of vector components shows that the result of the rotation is geometrically that of the rotation of axes. We can then say that the definition (1.16) characterizes *passive rotations* or rotations of the axes.

Owing to the orthonormal property of the two bases, R is orthogonal

$$\delta_{ij} = \mathbf{k}_i \cdot \mathbf{k}_j = R_{ih}R_{jk}\mathbf{e}_h \cdot \mathbf{e}_k = R_{ik}R_{jk}.$$

Upon time differentiation, it follows

$$\dot{\mathbf{k}}_i = \dot{R}_{ij}\mathbf{e}_j = \dot{R}_{ij}R^T_{jh}\mathbf{k}_h = \mathcal{O}_{ih}\mathbf{k}_h, \tag{1.17}$$

where

$$\mathcal{O} = \dot{R}R^T \in \text{Skw}.$$

Let

$$\boldsymbol{\omega} := \tfrac{1}{2}\textstyle\sum_i \mathbf{k}_i \times \dot{\mathbf{k}}_i. \tag{1.18}$$

Since

$$\mathbf{k}_h \cdot \mathbf{k}_i = \delta_{ih}, \qquad \mathbf{k}_h \cdot \dot{\mathbf{k}}_i = -\dot{\mathbf{k}}_h \cdot \mathbf{k}_i$$

it follows that

$$\begin{aligned}\boldsymbol{\omega} \times \mathbf{k}_h &= -\tfrac{1}{2}\textstyle\sum_i \mathbf{k}_h \times (\mathbf{k}_i \times \dot{\mathbf{k}}_i) = -\tfrac{1}{2}\sum_i [(\mathbf{k}_h \cdot \dot{\mathbf{k}}_i)\mathbf{k}_i - (\mathbf{k}_h \cdot \mathbf{k}_i)\dot{\mathbf{k}}_i] \\ &= \tfrac{1}{2}\textstyle\sum_i [(\mathbf{k}_h \cdot \dot{\mathbf{k}}_i)\mathbf{k}_i - (\mathbf{k}_h \cdot \mathbf{k}_i)\dot{\mathbf{k}}_i] = \tfrac{1}{2}\sum_i [(\dot{\mathbf{k}}_h \cdot \mathbf{k}_i)\mathbf{k}_i + \delta_{hi}\dot{\mathbf{k}}_i) = \dot{\mathbf{k}}_h.\end{aligned}$$

Thus, we have

$$\dot{\mathbf{k}}_h = \boldsymbol{\omega} \times \mathbf{k}_h, \qquad h = 1, 2, 3. \tag{1.19}$$

Equation (1.19) is sometimes called *Poisson relation* (of kinematics) for the rate of change of the corotational unit vectors. Indeed, by (1.19), we state that $\boldsymbol{\omega}$ is the *angular velocity* of $\{\mathbf{k}_j\}$ relative to $\{\mathbf{e}_j\}$. This view is consistent with the classical relation

$$\Big(\frac{d\mathbf{u}}{dt}\Big)_{\text{space}} = \Big(\frac{d\mathbf{u}}{dt}\Big)_{\text{body}} + \boldsymbol{\omega} \times \mathbf{u},$$

where $(d\mathbf{u}/dt)_{\text{body}}$ is the derivative of $\mathbf{u}$ as seen from the rotating frame. Formally,

$$\dot{\mathbf{u}} = \dot{u}_i\mathbf{k}_i + u_i\dot{\mathbf{k}}_i = \dot{u}_i\mathbf{k}_i + u_i\boldsymbol{\omega} \times \mathbf{k}_i = \dot{u}_i\mathbf{k}_i + \boldsymbol{\omega} \times \mathbf{u},$$

thus showing that

$$\Big(\frac{d\mathbf{u}}{dt}\Big)_{\text{space}} = \dot{\mathbf{u}}, \qquad \Big(\frac{d\mathbf{u}}{dt}\Big)_{\text{body}} = \dot{u}_i\mathbf{k}_i.$$

In view of (1.17), we find the connection between the angular velocity $\boldsymbol{\omega}$ and the rotation matrix R. Upon substitution of $\dot{\mathbf{k}}_i$ from (1.17) into (1.18), it follows

$$\boldsymbol{\omega} = \tfrac{1}{2}\textstyle\sum_i \mathbf{k}_i \times \mathcal{O}_{ih}\mathbf{k}_h = \tfrac{1}{2}\epsilon_{lih}\mathcal{O}_{ih}\mathbf{k}_l,$$

whence

$$\boldsymbol{\omega} \cdot \mathbf{k}_l = \tfrac{1}{2}\epsilon_{lih}\mathcal{O}_{ih}.$$

Relative to the (fixed) basis $\{\mathbf{e}_j\}$ we have

$$\boldsymbol{\omega} = \omega_h\mathbf{k}_h = \omega_h R_{hj}\mathbf{e}_j = \tfrac{1}{2}R_{hj}\epsilon_{hip}\mathcal{O}_{ip}\mathbf{e}_j, \tag{1.20}$$

whence

$$\boldsymbol{\omega} \cdot \mathbf{e}_j = \tfrac{1}{2}R^T_{jh}\epsilon_{hip}\mathcal{O}_{ip}.$$

It is worth remarking that in both cases, $\boldsymbol{\omega}$ involves $\frac{1}{2}\epsilon_{hip}\mathcal{O}_{ip}$ which is the opposite of the axial vector (h-component) of $\mathcal{O}$.

Sometimes the angular velocity is defined via the right-hand rule, i.e. when the fingers of the right hand curl in the direction of the rotation, the thumb points in the direction of the angular velocity vector. This means that, relative to the right-handed basis $(\mathbf{e}_1, \mathbf{e}_2, \mathbf{e}_3)$, in a counterclockwise rotation in the plane $\{\mathbf{e}_1, \mathbf{e}_2\}$, the angular velocity is directed along $\mathbf{k}_3 = \mathbf{e}_3$. Let θ be the angle of rotation, then

$$R = \begin{bmatrix} \cos\theta & \sin\theta & 0 \\ -\sin\theta & \cos\theta & 0 \\ 0 & 0 & 1 \end{bmatrix}, \qquad \dot{R}R^T = \dot{\theta}\begin{bmatrix} 0 & 1 & 0 \\ -1 & 0 & 0 \\ 0 & 0 & 0 \end{bmatrix}.$$

Hence, as expected, the relation $\boldsymbol{\omega} \cdot \mathbf{k}_i = \frac{1}{2}\epsilon_{ijk}(\dot{R}R^T)_{jk}$ results in

$$\boldsymbol{\omega} = \dot{\theta}\,\mathbf{k}_3.$$

The matrix R so defined determines the new coordinates of a point once the axes are rotated; R is said to represent a passive rotation (or passive transformation).

1.4.1 Active Rotations

Sometimes in continuum physics, and usually in linear algebra, rotations are defined by transforming points, or rotating vectors, instead of axes. Hence, in the change of basis $\{\mathbf{e}_i\} \to \{\mathbf{k}_i\}$, we might define the rotation $\mathcal{R}$ in the form

$$\mathcal{R}\mathbf{e}_i = \mathbf{k}_i.$$

This definition characterizes $\mathcal{R}$ as an *active rotation.* We now establish some properties of active rotations and the connection between the matrices of passive and active rotations.

Consider the transformation $\{\mathbf{e}_i\} \to \{\mathbf{k}_i\}$; depending on the representation, we can write

$$\mathbf{k}_i = R_{ik}\mathbf{e}_k, \qquad \mathcal{R}\mathbf{e}_i = \mathbf{k}_i.$$

Hence, R is the matrix of the passive rotation. Concerning $\mathcal{R}$, we observe that the orthonormality of $\{\mathbf{k}_i\}$ implies

$$\delta_{ij} = \mathbf{k}_i \cdot \mathbf{k}_j = \mathcal{R}\mathbf{e}_i \cdot \mathcal{R}\mathbf{e}_j = \mathbf{e}_i \cdot \mathcal{R}^T\mathcal{R}\mathbf{e}_j;$$

the orthonormality of $\{\mathbf{e}_i\}$ implies that $\mathcal{R}$ is orthogonal, $\mathcal{R}^T\mathcal{R} = \mathbf{1}$. Since $\mathbf{k}_i = R_{ij}\mathbf{e}_j$ then inner multiplication by $\mathbf{e}_k$ of

$$\boldsymbol{\mathcal{R}}\mathbf{e}_i = R_{ij}\mathbf{e}_j$$

implies

$$\mathcal{R}_{ki} = \mathbf{e}_k \cdot \boldsymbol{\mathcal{R}}\mathbf{e}_i = R_{ik};$$

the matrix $\mathcal{R}$ associated with $\boldsymbol{\mathcal{R}}$, in the basis $\{\mathbf{e}_i\}$, is the transpose of R. Now observe that

$$\boldsymbol{\omega} = \tfrac{1}{2}\textstyle\sum_i \mathbf{k}_i \times \dot{\mathbf{k}}_i = \tfrac{1}{2}\sum_i \boldsymbol{\mathcal{R}}\mathbf{e}_i \times \dot{\boldsymbol{\mathcal{R}}}\mathbf{e}_i.$$

Since $(\boldsymbol{\mathcal{R}}\mathbf{e}_p)_h = \mathbf{e}_h \cdot \boldsymbol{\mathcal{R}}\mathbf{e}_p = \mathcal{R}_{hp}$ then letting $\mathcal{A}_{jh} = \dot{\mathcal{R}}_{jp}\mathcal{R}_{hp}$ we find

$$\boldsymbol{\omega} = \tfrac{1}{2}\textstyle\sum_i \boldsymbol{\mathcal{R}}\mathbf{e}_i \times \dot{\boldsymbol{\mathcal{R}}}\mathbf{e}_i = \mathbf{e}_l \tfrac{1}{2}\epsilon_{lhj}\mathcal{R}_{hp}\dot{\mathcal{R}}_{jp} = \mathbf{e}_l \tfrac{1}{2}\epsilon_{lhj}\mathcal{A}_{jh} = \mathbf{a}_{\mathcal{A}}, \tag{1.21}$$

$\mathbf{a}_{\mathcal{A}}$ being the axial vector of $\mathcal{A} = \dot{\mathcal{R}}\mathcal{R}^T$.

As a check of consistency of (1.20) and (1.21), we now show that the same result for $\boldsymbol{\omega}$ follows by computation via the passive rotation matrix R and the active rotation matrix $\mathcal{R}$, namely

$$\boldsymbol{\omega} = \tfrac{1}{2}R_{hl}\epsilon_{hip}\mathcal{O}_{ip}\mathbf{e}_l = \tfrac{1}{2}\epsilon_{lhj}\mathcal{A}_{jh}\mathbf{e}_l. \tag{1.22}$$

We start with the passive matrix R, recall that $R = \mathcal{R}^T$, and observe

$$\omega_l = \tfrac{1}{2}R_{hl}\epsilon_{hip}\mathcal{O}_{ip} = \tfrac{1}{2}R_{hl}\epsilon_{hip}\dot{R}_{ik}R_{pk} = \tfrac{1}{2}\mathcal{R}_{lh}\epsilon_{hip}\dot{\mathcal{R}}_{ki}\mathcal{R}_{kp}.$$

Since

$$\mathcal{R}_{lh}\epsilon_{hip}\mathcal{R}_{kp} = \epsilon_{ljk}\mathcal{R}_{ji}$$

then

$$\mathcal{R}_{lh}\epsilon_{hip}\dot{\mathcal{R}}_{ki}\mathcal{R}_{kp} = \epsilon_{ljk}\dot{\mathcal{R}}_{ki}\mathcal{R}_{ji} = \epsilon_{ljk}\mathcal{A}_{kj},$$

and the conclusion follows.

In addition, we observe that

$$\dot{\mathbf{k}}_i = (\dot{\boldsymbol{\mathcal{R}}}\boldsymbol{\mathcal{R}}^T)\mathbf{k}_i.$$

Borrowing from the evolution of a triad of eigenvectors [219, 225], we may refer to $\boldsymbol{\mathcal{A}} = \dot{\boldsymbol{\mathcal{R}}}\boldsymbol{\mathcal{R}}^T$ as the *twirl tensor* of the triad $\{\mathbf{k}_i\}$. Hence, in terms of the twirl tensor, we have

$$\dot{\mathbf{k}}_i = \boldsymbol{\mathcal{A}}\mathbf{k}_i = \boldsymbol{\omega} \times \mathbf{k}_i.$$

1.5 Transport Relations for Convecting Sets

We now compute the effects of motion on integrals over sets convecting with the body.

Theorem 1.2 (Reynolds' transport) *Let* P *be any sub-region of the reference configuration* R *and let* $\mathcal{P}_t = \chi_t(\mathrm{P})$. *Let* ϕ *and* ϕ_R *be the spatial and material descriptions of a field and let* $\partial_t \phi_R$, $\partial_t J$ *be continuous functions. Then*

$$\frac{d}{dt}\int_{\mathcal{P}_t} \phi\, dv = \int_{\mathcal{P}_t} (\dot{\phi} + \phi \nabla \cdot \mathbf{v}) dv. \tag{1.23}$$

Proof By the change of variables $\mathbf{x} \to \mathbf{X} = \chi^{-1}(\mathbf{x}, t)$, we have

$$\int_{\mathcal{P}_t} \phi(\mathbf{x}, t)\, dv = \int_{\mathrm{P}} \phi_R(\mathbf{X}, t) J(\mathbf{X}, t)\, dv_R.$$

The sub-region P is time independent, and hence by the continuity of $\partial_t \phi_R(\mathbf{X}, t)$ and $\partial_t J(\mathbf{X}, t)$, we deduce

$$\frac{d}{dt}\int_{\mathrm{P}} \phi_R\, J\, dv_R = \int_{\mathrm{P}} [\partial_t \phi_R\, J + \phi_R \partial_t J] dv_R.$$

In view of (1.13), we find

$$\frac{d}{dt}\int_{\mathrm{P}} \phi_R\, J\, dv_R = \int_{\mathrm{P}} (\dot{\phi} + \phi \nabla \cdot \mathbf{v}) J dv_R = \int_{\mathcal{P}_t} (\dot{\phi} + \phi \nabla \cdot \mathbf{v}) dv,$$

whence (1.23) follows. □

Since $\dot{\phi} = \partial_t \phi + \mathbf{v} \cdot \nabla \phi$, then

$$\dot{\phi} + \phi \nabla \cdot \mathbf{v} = \partial_t \phi + \mathbf{v} \cdot \nabla \phi + \phi \nabla \cdot \mathbf{v}.$$

Now, for any scalar, vector, or tensor ϕ, we have

$$\mathbf{v} \cdot \nabla \phi + \phi \nabla \cdot \mathbf{v} = \nabla \cdot (\phi \mathbf{v}) := \partial_{x_i}(\phi v_i).$$

With this meaning of $\nabla \cdot (\phi \mathbf{v})$ in mind, we can write (1.23) in the form

$$\frac{d}{dt}\int_{\mathcal{P}_t} \phi\, dv = \int_{\mathcal{P}_t} [\partial_t \phi + \nabla \cdot (\phi \mathbf{v})] dv. \tag{1.24}$$

Moreover, by means of the divergence theorem, we can write the alternative form

$$\frac{d}{dt}\int_{\mathcal{P}_t} \phi\, dv = \int_{\mathcal{P}_t} \partial_t \phi\, dv + \int_{\partial \mathcal{P}_t} \phi \mathbf{v} \cdot \mathbf{n}\, da. \tag{1.25}$$

Some comments are in order about the transport theorem. Equation (1.25) shows that the time rate of $\int_{\mathcal{P}_t} \phi\, dv$ consists of two contributions; $\int_{\mathcal{P}_t} \partial_t \phi\, dv$ is the time rate determined by the derivative of ϕ within the region $\mathcal{P}_t$ while $\int_{\partial \mathcal{P}_t} \phi \mathbf{v} \cdot \mathbf{n}\, da$ is the contribution due to the flux of ϕ across the boundary $\partial \mathcal{P}_t$. Now observe that $\text{vol}\, \mathcal{P}_t = \int_{\mathcal{P}_t} dv$. Hence, letting $\phi = 1$ in (1.25) and (1.24), we find

$$\frac{d}{dt} \text{vol}\, \mathcal{P}_t = \int_{\partial \mathcal{P}_t} \mathbf{v} \cdot \mathbf{n}\, da = \int_{\mathcal{P}_t} \nabla \cdot \mathbf{v}\, dv.$$

A motion is said to be *isochoric* if

$$\frac{d}{dt} \text{vol}\, \mathcal{P}_t = 0,$$

for all spatial regions convecting with the body. Thus, the motion is isochoric if $\nabla \cdot \mathbf{v} = 0$ or rather the flux of $\mathbf{v}$ across the boundary is zero.

We now let C be a closed material curve and let $\mathcal{C}_t = \chi(\mathrm{C}, t)$ be the corresponding spatial curve determined by motion. Let $\mathbf{X} = \mathbf{X}(\lambda)$, $\lambda \in [a, b]$, be the representation of C and hence $\mathbf{x} = \mathbf{x}(\lambda, t) = \chi(\mathbf{X}(\lambda), t)$ be the representation of $\mathcal{C}_t$. The integral

$$\Gamma(t) = \int_{\mathcal{C}_t} \mathbf{v}(\mathbf{x}, t) \cdot d\mathbf{x} := \int_a^b \mathbf{v}(\mathbf{x}(\lambda, t), t) \cdot \partial_\lambda \mathbf{x}(\lambda, t) d\lambda$$

is the *circulation* of $\mathbf{v}$ around $\mathcal{C}_t$. To compute the derivative of Γ we observe that

$$\dot{\Gamma}(t) = \int_a^b \partial_t [\mathbf{v}(\mathbf{x}(\lambda, t), t) \cdot \partial_\lambda \mathbf{x}(\lambda, t)] d\lambda,$$

where ∂_t is meant as the derivative at fixed λ. Therefore

$$\partial_t \mathbf{v}(\mathbf{x}(\lambda, t), t) = \partial_t \mathbf{v}(\mathbf{x}, t) + \partial_{x_i} \mathbf{v}(\mathbf{x}, t) v_i(\mathbf{x}, t) = \dot{\mathbf{v}}(\mathbf{x}, t)$$

where $\mathbf{x}$ stands for $\mathbf{x}(\lambda, t)$. Now

$$\partial_t \partial_\lambda \mathbf{x}(\lambda, t) = \partial_\lambda \partial_t \mathbf{x}(\lambda, t) = \partial_\lambda \mathbf{v}(\mathbf{x}(\lambda, t), t).$$

Since $\mathcal{C}_t$ is closed, then

$$\int_a^b \mathbf{v} \cdot \partial_\lambda \mathbf{v} d\lambda = \tfrac{1}{2}[\mathbf{v}^2(\mathbf{x}(b, t), t) - \mathbf{v}^2(\mathbf{x}(a, t), t)] = 0.$$

Hence, it follows the *circulation-transport relation*

$$\dot{\Gamma}(t) = \int_a^b [\dot{\mathbf{v}}(\mathbf{x}(\lambda, t), t) \cdot \partial_\lambda \mathbf{x}(\lambda, t)] d\lambda.$$

The motion is said to preserve circulation if $\dot{\Gamma} = 0$ for every closed material curve and time t. If $\dot{\mathbf{v}}$ is the gradient of a potential, $\dot{\mathbf{v}} = \nabla\varphi$, then

$$\int_a^b [\dot{\mathbf{v}}(\mathbf{x}(\lambda, t), t) \cdot \partial_\lambda \mathbf{x}(\lambda, t)] d\lambda = \int_a^b [\nabla\varphi(\mathbf{x}(\lambda, t), t) \cdot \partial_\lambda \mathbf{x}(\lambda, t)] d\lambda = \int_a^b \partial_\lambda \varphi(\lambda, t) d\lambda = 0.$$

This result traces back to Kelvin: if the acceleration is the gradient of a potential then the motion preserves circulation.

1.5.1 Identities About the Total Time Derivative of Spatial Gradients

In applications, we are faced with the total time derivative of spatial gradients, say $\dot{\overline{\nabla\phi}}$, for any C^2 function $\phi(\mathbf{x}, t)$, $\mathbf{x} \in \Omega$, $t \in \mathbb{R}$, with scalar, vector, or tensor values. The evaluation is simplified by using the identity

$$\dot{\overline{\nabla\phi}} = \nabla\dot{\phi} - \mathbf{L}^T \nabla\phi. \tag{1.26}$$

To prove (1.26), we observe that, for any C^2 function ϕ,

$$\dot{\overline{\nabla\phi}} = \partial_t \nabla\phi + (\mathbf{v} \cdot \nabla)\nabla\phi = \nabla\partial_t\phi + (\mathbf{v} \cdot \nabla)\nabla\phi.$$

Since

$$\mathbf{v} \cdot \nabla\, \nabla\phi = \nabla(\mathbf{v} \cdot \nabla\phi) - (\mathbf{L}^T \nabla)\phi$$

we have

$$\dot{\overline{\nabla\phi}} = \nabla\partial_t\phi + \nabla(\mathbf{v} \cdot \nabla\phi) - (\mathbf{L}^T \nabla)\phi$$

whence we obtain (1.26).

Equivalently, if ϕ is a scalar, then (1.26) can also be written in the form

$$\dot{\overline{\nabla\phi}} = \nabla\dot{\phi} - \nabla\phi\mathbf{L}.$$

If ϕ is a vector, $\mathbf{w}$, or a tensor, $\mathbf{A}$, we can write the identity (1.26) in the forms

$$\dot{\overline{\nabla\mathbf{w}}} = \nabla\dot{\mathbf{w}} - (\mathbf{L}^T \nabla)\mathbf{w}, \qquad \dot{\overline{\partial_{x_i} w_j}} = \partial_{x_i} \dot{w}_j - L_{ki} \partial_{x_k} w_j. \tag{1.27}$$

$$\dot{\overline{\nabla\mathbf{A}}} = \nabla\dot{\mathbf{A}} - (\mathbf{L}^T \nabla)\mathbf{A}, \qquad \dot{\overline{\partial_{x_i} A_{jk}}} = \partial_{x_i} \dot{A}_{jk} - L_{pi} \partial_{x_p} A_{jk}. \tag{1.28}$$

If higher-order gradients are involved, we can iterate the procedure by starting from (1.26). For definiteness, look at second-order gradients. Let $\phi = \nabla\psi$ and assume ψ is a C^3 function on $\Omega \times \mathbb{R}$. By

$$\overline{\nabla\nabla\psi}^{\,\cdot} = \nabla\overline{\nabla\psi}^{\,\cdot} - (\mathbf{L}^T\nabla)(\nabla\psi),$$

in indicial notation, we have

$$\begin{aligned}\overline{\partial_{x_i}\partial_{x_k}\psi}^{\,\cdot} &= \partial_{x_i}\overline{\partial_{x_k}\psi}^{\,\cdot} - L_{hi}\partial_{x_h}\partial_{x_k}\psi\\ &= \partial_{x_i}\partial_{x_k}\dot{\psi} - L_{hk}\partial_{x_i}\partial_{x_h}\psi - (\partial_{x_i}L_{hk})\partial_{x_h}\psi - L_{hi}\partial_{x_h}\partial_{x_k}\psi\\ &= \partial_{x_i}\partial_{x_k}\dot{\psi} - (\partial_{x_i}\partial_{x_k}v_h)\partial_{x_h}\psi - (L_{hk}\partial_{x_h}\partial_{x_i}\psi + L_{hi}\partial_{x_h}\partial_{x_k}\psi).\end{aligned}$$

Thus, in direct notation, we can write the identity for the second-order gradient in the form

$$\overline{\nabla\nabla\psi}^{\,\cdot} = \nabla\nabla\dot{\psi} - \nabla\psi(\nabla\nabla\mathbf{v}) - 2\,\mathrm{sym}(\mathbf{L}^T\nabla)\nabla\psi). \tag{1.29}$$

We now apply (1.27) to $\mathbf{w} = \mathbf{v}$ and find

$$\overline{\partial_{x_i}v_j}^{\,\cdot} = \partial_{x_i}\dot{v}_j - L^T_{ik}\partial_{x_k}v_j$$

whence

$$\dot{\mathbf{L}} = \boldsymbol{\Xi} - \mathbf{L}\,\mathbf{L}, \qquad \Xi_{ji} := \partial_{x_i}\dot{v}_j. \tag{1.30}$$

Iteration of the procedure allows us to obtain $\ddot{\mathbf{L}}$. By

$$\ddot{L}_{ji} = \big(\overline{\partial_{x_i}v_j}^{\,\cdot}\big)^{\cdot} = (\partial_{x_i}\dot{v}_j)^{\cdot} - (L_{jk}L_{ki})^{\cdot} = \partial_{x_i}\ddot{v}_j - L^T_{ik}\partial_{x_k}\dot{v}_j - \dot{L}_{jk}L_{ki} - L_{jk}\dot{L}_{ki}$$

and use of (1.30), we have

$$\ddot{L}_{ji} = \partial_{x_i}\ddot{v}_j - L^T_{ik}\partial_{x_k}\dot{v}_j - (\partial_{x_k}\dot{v}_j - L^T_{kp}\partial_{x_p}v_j)L_{ki} - L_{jk}(\partial_{x_i}\dot{v}_k - L^T_{ip}\partial_{x_p}v_k)$$

whence

$$\ddot{\mathbf{L}} = \boldsymbol{\Lambda} - 2\,\boldsymbol{\Xi}\,\mathbf{L} - \mathbf{L}\,\boldsymbol{\Xi} - 2\,\mathbf{L}\,\mathbf{L}\,\mathbf{L}, \qquad \Lambda_{ji} := \partial_{x_i}\ddot{v}_j. \tag{1.31}$$

It is also of interest to evaluate $\ddot{\mathbf{F}}$. Time differentiation of $\dot{\mathbf{F}} = \mathbf{L}\mathbf{F}$ and use of (1.30) yield

$$\ddot{\mathbf{F}} = \dot{\mathbf{L}}\mathbf{F} + \mathbf{L}\dot{\mathbf{F}} = (\nabla\dot{\mathbf{v}} - \mathbf{L}\mathbf{L})\mathbf{F} + \mathbf{L}\mathbf{L}\mathbf{F}.$$

Hence

$$\ddot{\mathbf{F}}\mathbf{F}^{-1} = \nabla\dot{\mathbf{v}}. \tag{1.32}$$

Alternatively, we might have observed that

$$\partial_{x_i}\dot{v}_j = \partial_{x_i}\ddot{\chi}_j = \partial_{X_K}\ddot{\chi}_j F^{-1}_{Ki} = \ddot{F}_{jK}F^{-1}_{Ki}, \qquad \nabla\dot{\mathbf{v}} = \ddot{\mathbf{F}}\mathbf{F}^{-1}.$$

1.5.2 Transport Relations for Vorticity and Velocity Gradient

The material time derivative is the derivative as seen by an observer in motion with the velocity of the continuum. In applications, we are often required to compute the total time derivative in particular of velocity, vorticity, and velocity gradient.

We first apply the relation (1.12) to the velocity field $\mathbf{v}$ and observe that

$$(\mathbf{v}\cdot\nabla)\mathbf{v} = \mathbf{L}\mathbf{v}.$$

Hence, it follows

$$\dot{\mathbf{v}} = \partial_t\mathbf{v} + \mathbf{L}\mathbf{v}.$$

Now

$$(\mathbf{L}-\mathbf{L}^T)\mathbf{v} = 2\mathbf{W}\mathbf{v}, \qquad \mathbf{L}\mathbf{v} = 2\mathbf{W}\mathbf{v} + \mathbf{L}^T\mathbf{v}.$$

Furthermore

$$\mathbf{L}^T\mathbf{v} = \nabla\tfrac{1}{2}\mathbf{v}^2, \qquad 2\mathbf{W}\mathbf{v} = \boldsymbol{\varpi}\times\mathbf{v}.$$

Hence, we obtain

$$\dot{\mathbf{v}} = \partial_t\mathbf{v} + \nabla\tfrac{1}{2}\mathbf{v}^2 + \boldsymbol{\varpi}\times\mathbf{v}. \tag{1.33}$$

The curl of (1.33) and the identity $\nabla\times\nabla\mathbf{v}^2 = \mathbf{0}$ imply that

$$\nabla\times\dot{\mathbf{v}} = \partial_t\boldsymbol{\varpi} + \nabla\times(\boldsymbol{\varpi}\times\mathbf{v}).$$

Now, in suffix notation

$$[\nabla\times(\boldsymbol{\varpi}\times\mathbf{v})]_i = \epsilon_{ijk}\epsilon_{kpq}\partial_{x_j}(\varpi_p v_q) = \partial_{x_j}(\varpi_i v_j) - \partial_{x_j}(\varpi_j v_i).$$

Moreover

$$\nabla\cdot\boldsymbol{\varpi} = \nabla\cdot\nabla\times\mathbf{v} = 0.$$

Hence, it follows

$$\nabla\times(\boldsymbol{\varpi}\times\mathbf{v}) = (\mathbf{v}\cdot\nabla)\boldsymbol{\varpi} + (\nabla\cdot\mathbf{v})\boldsymbol{\varpi} - \mathbf{L}\boldsymbol{\varpi}.$$

Consequently, we obtain a transport relation for $\boldsymbol{\varpi}$ in the form

$$\dot{\boldsymbol{\varpi}} - \mathbf{L}\boldsymbol{\varpi} + (\nabla\cdot\mathbf{v})\boldsymbol{\varpi} = \nabla\times\dot{\mathbf{v}}. \tag{1.34}$$

An analogous transport relation can be derived for the spin tensor $\mathbf{W}$. Since $2\mathbf{W} = \mathbf{L} - \mathbf{L}^T = \dot{\mathbf{F}}\mathbf{F}^{-1} - \mathbf{F}^{-T}\dot{\mathbf{F}}^T$, then

$$2\mathbf{F}^T\mathbf{W}\mathbf{F} = \mathbf{F}^T(\dot{\mathbf{F}}\mathbf{F}^{-1} - \mathbf{F}^{-T}\dot{\mathbf{F}}^T)\mathbf{F} = \mathbf{F}^T\dot{\mathbf{F}} - \dot{\mathbf{F}}^T\mathbf{F}.$$

In view of (1.32), we obtain

$$2\overline{\mathbf{F}^T\mathbf{W}\mathbf{F}}^{\boldsymbol{\cdot}} = \mathbf{F}^T\ddot{\mathbf{F}} - \ddot{\mathbf{F}}^T\mathbf{F} = \mathbf{F}^T(\ddot{\mathbf{F}}\mathbf{F}^{-1} - \mathbf{F}^{-T}\ddot{\mathbf{F}}^T)\mathbf{F} = \mathbf{F}^T[\nabla\dot{\mathbf{v}} - (\nabla\dot{\mathbf{v}})^T]\mathbf{F},$$

whence

$$\overline{\mathbf{F}^T\mathbf{W}\mathbf{F}}^{\boldsymbol{\cdot}} = \mathbf{F}^T\tfrac{1}{2}[\nabla\dot{\mathbf{v}} - (\nabla\dot{\mathbf{v}})^T]\mathbf{F}. \tag{1.35}$$

Two remarkable consequences follow from (1.35). First assume *the acceleration is the gradient of a scalar* function φ,

$$\dot{\mathbf{v}} = \nabla\varphi.$$

Hence

$$\nabla\dot{\mathbf{v}} - (\nabla\dot{\mathbf{v}})^T = \nabla\nabla\varphi - (\nabla\nabla\varphi)^T = \mathbf{0}.$$

By (1.35), it follows that $\mathbf{F}^T\mathbf{W}\mathbf{F}$ is constant in time,

$$\mathbf{F}^T\mathbf{W}\mathbf{F} = \mathbf{K}(\mathbf{X}), \qquad \mathbf{W} = \mathbf{F}^{-T}(\mathbf{X}, t)\mathbf{K}(\mathbf{X})\mathbf{F}^{-1}(\mathbf{X}, t).$$

If $\mathbf{W}(\mathbf{X}, t_0) = \mathbf{0}$ at some time t_0, then $\mathbf{K}(\mathbf{X}) = \mathbf{0}$ and $\mathbf{W}(\mathbf{X}, t) = \mathbf{0}$ at any time t. If, further, $\mathbf{W}(\mathbf{X}, t_0) = \mathbf{0}$ for any $\mathbf{X} \in \mathrm{R}$, then

$$\mathbf{W} = \mathbf{0}, \qquad \boldsymbol{\varpi} = \mathbf{0}$$

at any point $\mathbf{X}$ and time t. This allows us to conclude that[10] *if the acceleration is the gradient of a scalar field, then the motion is irrotational, provided that it is at some point in time.*

The transport relation for the spin $\mathbf{W}$ is a further consequence of (1.35). Direct evaluation of the time derivative and substitution of $\dot{\mathbf{F}} = \mathbf{L}\mathbf{F}$ gives

$$2\overline{\mathbf{F}^T\mathbf{W}\mathbf{F}}^{\boldsymbol{\cdot}} = 2\mathbf{F}^T(\dot{\mathbf{W}} + \mathbf{L}^T\mathbf{W} + \mathbf{W}\mathbf{L})\mathbf{F}.$$

Moreover

$$\mathbf{L}^T\mathbf{W} + \mathbf{W}\mathbf{L} = \mathbf{D}\mathbf{W} + \mathbf{W}\mathbf{D}.$$

Hence, (1.35) becomes

---

[10] The result is due to Lagrange and Cauchy.

$$\dot{\mathbf{W}} + \mathbf{DW} + \mathbf{WD} = \tfrac{1}{2}(\nabla\dot{\mathbf{v}} - (\nabla\dot{\mathbf{v}})^T). \tag{1.36}$$

The transport relation (1.34) for the vorticity $\boldsymbol{\varpi}$ is expected to be equivalent to the transport relation (1.36) for the spin $\mathbf{W}$. To prove the expected equivalence, we multiply (1.34) by $-\frac{1}{2}\epsilon_{pqi}$ to obtain

$$\dot{W}_{pq} + \tfrac{1}{2}\epsilon_{pqi}[L_{ij}\varpi_j - (\nabla\cdot\mathbf{v})\varpi_i] = \tfrac{1}{2}(\partial_{x_q}\dot{v}_p - \partial_{x_p}\dot{v}_q).$$

Hence, we have to show that

$$\tfrac{1}{2}\epsilon_{pqi}[L_{ij}\varpi_j - (\nabla\cdot\mathbf{v})\varpi_i] = D_{pk}W_{kq} + W_{pk}D_{kq}.$$

Since $\mathbf{W}\boldsymbol{\varpi} = \mathbf{0}$ and $W_{kq} = -\frac{1}{2}\epsilon_{kqi}\,\varpi_i$, then we need to prove that

$$A_{pq} := \tfrac{1}{2}\epsilon_{pqi}[D_{ij}\varpi_j - (\nabla\cdot\mathbf{v})\omega_i] - \tfrac{1}{2}\epsilon_{qki}D_{kp}\varpi_i + \tfrac{1}{2}\epsilon_{pki}D_{kq}\varpi_i = 0.$$

By definition, A_{pq} is skew, $A_{pq} = -A_{qp}$. Hence $A_{pq} = 0$, for every pair p, q, if and only if

$$\epsilon_{pqh}A_{pq} = 0, \qquad h = 1, 2, 3.$$

Now

$$2\epsilon_{pqh}A_{pq} = (\delta_{qq}\delta_{hi} - \delta_{qi}\delta_{hq})(D_{ij}\varpi_j - \nabla\cdot\mathbf{v}\,\varpi_i)$$
$$+(\delta_{pk}\delta_{hi} - \delta_{pi}\delta_{hk})D_{kp}\varpi_i + (\delta_{qk}\delta_{hi} - \delta_{qi}\delta_{hk})D_{kq}\varpi_i = 0$$

identically. The equivalence of (1.34) and (1.36) is then proved. □

1.6 Transport Relations for Non-Convecting Sets

There are cases where the boundary of a set (sub-region, surface) is not convecting with the body so that the velocity $\boldsymbol{\nu}$ of the boundary need not be equal to the velocity $\mathbf{v}$ of the material points. We then consider a sub-region Ω and establish the following property (Fig. 1.2).

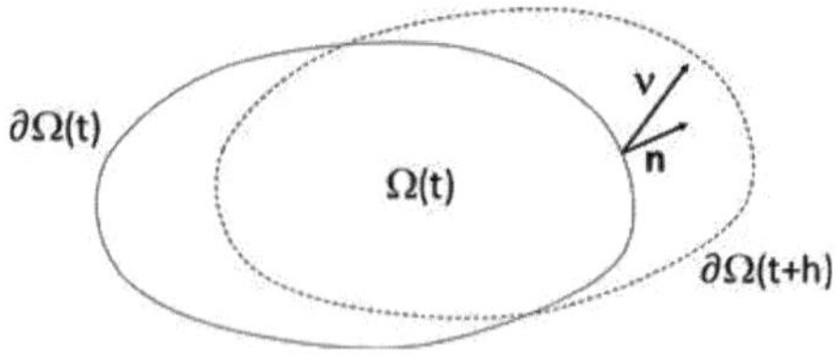

Fig. 1.2 The region $\Omega(t)$ is not convected by the body

Theorem 1.3 (generalized transport) *Let $\Omega \subset \mathscr{E}$ be a time-dependent region. If ϕ and $\partial_t \phi$ are continuous on Ω then*

$$\frac{d}{dt}\int_{\Omega} \phi\, dv = \int_{\Omega} \partial_t \phi\, dv + \int_{\partial\Omega} \phi \boldsymbol{\nu} \cdot \mathbf{n}\, da.$$

Proof We compute the derivative as the limit of the difference quotient

$$\frac{d}{dt}\int_{\Omega(t)} \phi\, dv = \lim_{h\to 0} \frac{1}{h}\Big[\int_{\Omega(t+h)} \phi(\mathbf{x}, t+h)\, dv - \int_{\Omega(t)} \phi(\mathbf{x}, t)\, dv\Big].$$

Upon some rearrangements, we have

$$\begin{aligned}\int_{\Omega(t+h)} \phi(\mathbf{x}, t+h)\, dv - \int_{\Omega(t)} \phi(\mathbf{x}, t)\, dv &= \int_{\Omega(t)} [\phi(\mathbf{x}, t+h) - \phi(\mathbf{x}, t)] dv \\ &+ \int_{\Omega(t+h)} \phi(\mathbf{x}, t+h)\, dv - \int_{\Omega(t)} \phi(\mathbf{x}, t+h)\, dv\end{aligned}$$

Now

$$\phi(\mathbf{x}, t+h) - \phi(\mathbf{x}, t) = \partial_t \phi(\mathbf{x}, \xi) h, \qquad \xi \in (t, t+h).$$

By the uniform continuity of $\partial_t \phi$, we can write

$$|\partial_t \phi(\mathbf{x}, \xi) - \partial_t \phi(\mathbf{x}, t)| \le \epsilon(h)$$

where $\epsilon(h) \to 0$ as $h \to 0$. Consequently, it follows

$$\lim_{h\to 0} \frac{1}{h}\int_{\Omega(t)} [\phi(\mathbf{x}, t+h) - \phi(\mathbf{x}, t)]\, dv = \int_{\Omega(t)} \partial_t \phi(\mathbf{x}, t)\, dv.$$

The boundary $\partial\Omega(t)$ moves with velocity $\boldsymbol{\nu}$, possibly not continuous. Also, in view of the continuity of $\phi(\mathbf{x}, \cdot)$, we have

$$\int_{\Omega(t+h)} \phi(\mathbf{x}, t+h)\, dv - \int_{\Omega(t)} \phi(\mathbf{x}, t+h)\, dv = h \int_{\partial\Omega(t)} \phi(\mathbf{x}, t) \boldsymbol{\nu} \cdot \mathbf{n}\, da + o(h)$$

where $|o(h)| \le \delta(h) h$ and $\delta(h) \to 0$ as $h \to 0$. Hence the conclusion follows. □

As a remark, this generalized transport holds, in particular, if $\boldsymbol{\nu}$ is continuous on $\partial\Omega$. In such a case, we might consider a continuous field $\boldsymbol{\nu}$ on Ω of a fictitious underlying continuum and say that Reynolds' transport holds so that also

$$\frac{d}{dt}\int_{\Omega} \phi\, dv = \int_{\Omega} [\dot{\phi} + \phi \nabla \cdot \boldsymbol{\nu}] dv$$

holds. Yet this view is hardly operative because $\boldsymbol{\nu}$ is not the real velocity of the continuum.

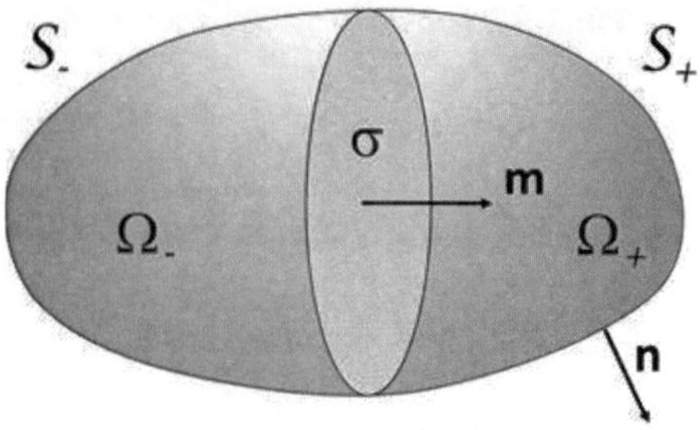

Fig. 1.3 The surface $\sigma(t)$ divides Ω into the regions Ω_+, Ω_- and forms a common boundary between them

1.6.1 Transport Theorems for Discontinuous Fields

Let $\Omega \subset \mathscr{E}$ be a region possibly dependent on the time $t \in \mathbb{R}$. A surface $\sigma(t)$ divides Ω into the regions Ω_+, Ω_- and forms a common boundary between them so that $\Omega = \Omega_- \cup \sigma \cup \Omega_+$, $\Omega_- \cap \Omega_+ = \emptyset$. Let $S_\pm = \partial\Omega_\pm \setminus \sigma$ and hence $\partial\Omega = S_+ \cup S_-$. Let $\mathbf{m}$ be the unit normal of σ, directed towards Ω_+. Thus, on σ the outward (unit) normal of Ω_- is $\mathbf{m}$ whereas the outward normal to Ω_+ is $-\mathbf{m}$. Let $\boldsymbol{\nu}$ be the velocity of (the points of) σ and $\mathbf{n}$ be the unit normal to $\partial\Omega$ (Fig. 1.3).

Let $\phi(\cdot,\cdot)$ be a function on $\Omega \times \mathbb{R}$, whose values may be scalars, vectors, or tensors of any order, such that $\phi(\cdot, t)$ is continuous within the regions Ω_+ and Ω_- and let $\phi(\mathbf{x}, t)$ have definite limits ϕ_+ and ϕ_- as $\mathbf{x}$ approaches a point on σ from paths entirely within the regions Ω_+ and Ω_-, respectively

$$\phi_\pm(\mathbf{y}, t) = \lim_{\mathbf{x}\to\mathbf{y}} \phi(\mathbf{x}, t), \qquad \mathbf{x} \in \Omega_\pm.$$

The surface $\sigma(t)$ is said to be *singular* with respect to $\phi(\cdot, t)$ at time t if

$$[\![\phi]\!](\mathbf{y}, t) := \phi_-(\mathbf{y}, t) - \phi_+(\mathbf{y}, t) \neq 0, \qquad \mathbf{y} \in \sigma(t).$$

By the divergence theorem, for a vector function $\mathbf{w} \in C^1(D)$, $D \subset \mathscr{E}$, we have

$$\int_D \nabla \cdot \mathbf{w}\, dv = \int_{\partial D} \mathbf{w} \cdot \mathbf{n}\, da,$$

$\mathbf{n}$ being the unit outward normal to ∂D. A generalization is required if the domain, Ω, contains a singular surface.

Theorem 1.4 (divergence, singular surface) *If* $\mathbf{w}$ *is a continuously differentiable vector function in* Ω_- *and* Ω_+ *and* σ *is a singular surface for* $\mathbf{w}$ *then*

$$\int_{\Omega\setminus\sigma} \nabla \cdot \mathbf{w}\, dv = \int_{\partial\Omega} \mathbf{w} \cdot \mathbf{n}\, da + \int_\sigma [\![\mathbf{w}]\!] \cdot \mathbf{m}\, da. \tag{1.37}$$

Proof Observe

$$\int_{\Omega\backslash\sigma} \nabla\cdot\mathbf{w}\,dv = \int_{\Omega_+} \nabla\cdot\mathbf{w}\,dv + \int_{\Omega_-} \nabla\cdot\mathbf{w}\,dv.$$

In $\Omega_\pm$, $\nabla\cdot\mathbf{w}$ is continuous and the divergence theorem applies to obtain

$$\int_{\Omega_+} \nabla\cdot\mathbf{w}\,dv = \int_{S_+} \mathbf{w}\cdot\mathbf{n}\,da - \int_{\sigma} \mathbf{w}_+\cdot\mathbf{m}\,da,$$

$$\int_{\Omega_-} \nabla\cdot\mathbf{w}\,dv = \int_{S_-} \mathbf{w}\cdot\mathbf{n}\,da + \int_{\sigma} \mathbf{w}_-\cdot\mathbf{m}\,da.$$

Summation provides the result. □

Theorem 1.5 (transport of discontinuous fields) *If ϕ and the velocity $\mathbf{v}$ are continuously differentiable functions in Ω_- and Ω_+ while $\boldsymbol{\nu}$ is the velocity of σ then*

$$\frac{d}{dt}\int_{\Omega\backslash\sigma} \phi\,dv = \int_{\Omega\backslash\sigma} \{\partial_t\phi + \nabla\cdot(\phi\mathbf{v})\}dv + \int_{\sigma} [\![\phi(\boldsymbol{\nu}-\mathbf{v})]\!]\cdot\mathbf{m}\,da. \tag{1.38}$$

Proof Since ϕ and $\mathbf{v}$ are C^1 functions in Ω_- and Ω_+ we apply the generalized transport theorem to obtain

$$\frac{d}{dt}\int_{\Omega_+} \phi\,dv = \int_{\Omega_+} \partial_t\phi\,dv + \int_{S_+} \phi\mathbf{v}\cdot\mathbf{n}\,da - \int_{\sigma} \phi_+\boldsymbol{\nu}\cdot\mathbf{m}\,da,$$

$$\frac{d}{dt}\int_{\Omega_-} \phi\,dv = \int_{\Omega_-} \partial_t\phi\,dv + \int_{S_-} \phi\mathbf{v}\cdot\mathbf{n}\,da + \int_{\sigma} \phi_-\boldsymbol{\nu}\cdot\mathbf{m}\,da.$$

By adding these relations, we find

$$\frac{d}{dt}\int_{\Omega\backslash\sigma} \phi\,dv = \int_{\Omega\backslash\sigma} \partial_t\phi\,dv + \int_{\partial\Omega} \phi\mathbf{v}\cdot\mathbf{n}\,da + \int_{\sigma} [\![\phi]\!]\boldsymbol{\nu}\cdot\mathbf{m}\,da.$$

As to the integral over $\partial\Omega$, we apply the divergence theorem to obtain

$$\int_{\partial\Omega} \phi\mathbf{v}\cdot\mathbf{n}\,da = \int_{\Omega\backslash\sigma} \nabla\cdot(\phi\mathbf{v})\,dv - \int_{\sigma} [\![\phi\mathbf{v}]\!]\cdot\mathbf{m}\,da.$$

Substitution and the observation that

$$[\![\phi]\!]\boldsymbol{\nu} = [\![\phi\boldsymbol{\nu}]\!]$$

lead to the result (1.38). □

The general relation (1.38) reduces to Reynolds' transport relations (1.23) or (1.24)–(1.25) if $\boldsymbol{\nu} = \mathbf{v}$, even though $[\![\phi]\!] \neq 0$, that is if the discontinuity surface σ

moves with the velocity of the continuum. Instead, if $[\![\phi]\!] = 0$ then $[\![\phi\boldsymbol{\nu}]\!] = \mathbf{0}$ and

$$\int_{\sigma} [\![\phi(\boldsymbol{\nu} - \mathbf{v})]\!] \cdot \mathbf{m}\, da = -\int_{\sigma} \phi [\![\mathbf{v}]\!] \cdot \mathbf{m}\, da.$$

The normal speed $\boldsymbol{\nu} \cdot \mathbf{m}$ is called the speed of displacement and is a measure of the speed with which the surface $\sigma(t)$ traverses in space. The quantity

$$U = (\boldsymbol{\nu} - \mathbf{v}) \cdot \mathbf{m} \tag{1.39}$$

is called the local speed of propagation; it is a measure of the normal speed of the surface $\sigma(t)$ with respect to the material points that are instantaneously situated upon it.

1.7 Kinematics of Singular Surfaces

Let P be a region occupied by the body in the reference configuration R. As with Ω in the spatial description, we let a surface $\Sigma(t)$ divide P into the regions P_+, P_- so that $P = P_- \cup \Sigma \cup P_+, P_- \cap P_+ = 0$. Following is an outline of the geometrical and kinematical properties of $\Sigma(t)$ as a singular surface[11] for a function ϕ.

The surface $\Sigma(t)$ is represented by the relation

$$\mathbf{X} = \mathbf{Y}(\Delta, \Gamma, t),$$

where Δ and Γ are a pair of surface parameters. Hence, a point on the surface $\Sigma(t)$ is identified by Δ and Γ. The velocity $\mathbf{U}$ of a point on $\Sigma(t)$ is defined by

$$\mathbf{U}_R(\Delta, \Gamma, t) = \partial_t \mathbf{Y}(\Delta, \Gamma, t).$$

Let $G(\mathbf{X}, t) = 0$ represent the surface $\Sigma(t)$. The unit normal to $\Sigma(t)$ is given by

$$\mathbf{n}_R = \frac{\nabla_R G}{|\nabla_R G|}.$$

We call the quantity U_R, defined by

$$U_R := \mathbf{U}_R \cdot \mathbf{n}_R = -\frac{\partial_t G}{|\nabla_R G|},$$

the speed of propagation of the surface $\Sigma(t)$. Hence, U_R is a measure of the speed with which the surface $\Sigma(t)$ traverses the material. The speed of propagation is an inherent property of the surface and all possible velocities $\mathbf{U}_R$ of the surface have the

[11] See, e.g. [429], p. 173–194; [89, 163, 289], Sect. 8.8.

same normal component U_R. A surface $\Sigma(t)$ that is singular with respect to some quantity and has nonzero speed of propagation

$$U_R(\mathbf{X}, t) \neq 0,$$

is said to be a *wave.*

We now examine the corresponding surface $\sigma(t)$ induced by the motion of the continuum $\mathbf{x} = \chi(\mathbf{X}, t)$, $\mathbf{X} \in \mathrm{R}$. Letting

$$G(\mathbf{X}, t) = G(\chi^{-1}(\mathbf{x}, t), t) =: g(\mathbf{x}, t)$$

we say that

$$g(\mathbf{x}, t) = 0$$

is the representation of the surface $\sigma(t)$, in the current configuration $\chi(\mathrm{R}, t)$. The two surfaces $\Sigma(t)$ and $\sigma(t)$ are the duals of each other; they are the material representation and the spatial representation of the singular surface. The surface $\sigma(t)$ has the parametric representation

$$\mathbf{x} = \mathbf{y}(\delta, \gamma, t)$$

where δ and γ are a pair of surface parameters.

The unit normal $\mathbf{n}$ to the surface $\sigma(t)$ is given by

$$\mathbf{n} = \frac{\nabla g}{|\nabla g|}.$$

The normal speed u_n of $\sigma(t)$, called the *speed of displacement* is given by

$$u_n = -\frac{\partial_t g}{|\nabla g|}.$$

We may select (the sign of) g so that $\mathbf{n}$ points in the direction of propagation. The speed of displacement u_n is a measure of the speed with which the surface $\sigma(t)$ traverses in space; it is also independent of the choice of parameterization of $\sigma(t)$.

Since

$$g(\chi(\mathbf{X}, t), t) = G(\mathbf{X}, t) \tag{1.40}$$

we have

$$\nabla g\, \mathbf{F} = \nabla_R G.$$

Consequently

$$\mathbf{F}^T \frac{\nabla g}{|\nabla g|} |\nabla g| = \nabla_R G \tag{1.41}$$

whence the absolute value gives

$$\frac{|\nabla g|}{|\nabla_R G|} = \frac{1}{|\mathbf{F}^T\mathbf{n}|}.$$

We now determine the connection between the normals $\mathbf{n}$, $\mathbf{n}_R$. By (1.41), we have

$$\mathbf{n}_R = \mathbf{F}^T\mathbf{n}\,\frac{|\nabla g|}{|\nabla_R G|} = \frac{\mathbf{F}^T\mathbf{n}}{|\mathbf{F}^T\mathbf{n}|}.$$

Likewise, we have the dual relation

$$\mathbf{n} = \mathbf{F}^{-T}\mathbf{n}_R\,\frac{|\nabla_R G|}{|\nabla g|} = \frac{\mathbf{F}^{-T}\mathbf{n}_R}{|\mathbf{F}^{-T}\mathbf{n}_R|}.$$

To find the connection between the speeds u_n, U_R, we look at (1.40) and observe

$$\partial_t G = \partial_t g + \dot{\mathbf{x}} \cdot \nabla g.$$

As a consequence

$$U_R = -\frac{\partial_t G}{|\nabla_R G|} = -\frac{\partial_t g + \dot{\mathbf{x}} \cdot \nabla g}{|\mathbf{F}^T\mathbf{n}|\,|\nabla g|} = \frac{1}{|\mathbf{F}^T\mathbf{n}|}(u_n - \dot{\mathbf{x}} \cdot \mathbf{n}),$$

whence

$$U_R = \frac{1}{|\mathbf{F}^T\mathbf{n}|}U, \qquad U = u_n - \dot{\mathbf{x}} \cdot \mathbf{n}.$$

Consistent with (1.39),[12] the quantity U is called the local speed of propagation and is the speed of the surface $\sigma(t)$ with respect to the material points.

1.7.1 Geometrical Conditions of Compatibility

Let $\phi(\cdot, t)$ be a function which is continuous in the interiors of Ω_+ and Ω_- and let $\sigma(t)$ be the common boundary between them. The theory of singular surfaces rests upon

Lemma 1.1 (Hadamard) *Let ϕ be defined and continuously differentiable on σ and let ϕ_+ and $\nabla\phi_+$ be the limit of ϕ and $\nabla\phi$ as σ is approached from Ω_+. If $\mathbf{x} = \mathbf{x}(s)$ is a smooth curve upon σ, then*

$$\frac{d\phi_+}{ds} = \nabla\phi_+ \cdot \frac{d\mathbf{x}}{ds}.$$

[12] Where $\mathbf{m}$ stands for $\mathbf{n}$ and $\nu \cdot \mathbf{m}$ for u_n.

The proof is immediate. If ϕ_+ is the function on the $+$ side of σ, then differentiation of

$$\phi_+ = \phi_+(\mathbf{x}(s))$$

yields the result. □

Applying Hadamard's lemma to each side of σ, we have

$$\frac{d\phi_+}{ds} = \nabla\phi_+ \cdot \frac{d\mathbf{x}}{ds}, \qquad \frac{d\phi_-}{ds} = \nabla\phi_- \cdot \frac{d\mathbf{x}}{ds}.$$

Subtraction gives

$$\frac{d}{ds}[\![\phi]\!] = [\![\nabla\phi]\!] \cdot \frac{d\mathbf{x}}{ds} = [\![\nabla\phi \cdot \frac{d\mathbf{x}}{ds}]\!].$$

This formula means that the tangential derivative of the jump is the jump of the tangential derivative. Since the values of ϕ in Ω_+ and Ω_- are unrelated to one another, the limiting values of the normal derivatives of ϕ on the two sides of σ need not be connected and hence $[\![d\phi/dn]\!]$ is unrestricted.

Likewise Hadamard's lemma, letting $\mathbf{x}(\delta, \gamma, t)$ be the parameterization of a curve on σ, we have

$$\partial_\gamma[\![\phi]\!] = [\![\nabla\phi]\!] \cdot \partial_\gamma\mathbf{x}.$$

In particular, if ϕ is continuous across σ or $[\![\phi]\!]$ is constant, then

$$[\![\nabla\phi]\!] \cdot \partial_\gamma\mathbf{x} = 0.$$

Thus, we have

Theorem 1.6 (Maxwell) *If* $[\![\phi]\!] = 0$ *or* $[\![\phi]\!]$ *is constant on* σ, *then*

$$[\![\nabla\phi]\!] = b\mathbf{n}, \qquad b = [\![\mathbf{n} \cdot \nabla\phi]\!] = [\![\frac{d\phi}{dn}]\!].$$

If $\phi = \mathbf{w}$, a vector, then

$$[\![\nabla\mathbf{w}]\!] = \mathbf{n} \otimes \mathbf{b}, \qquad \mathbf{b} = [\![\mathbf{n} \cdot \nabla\mathbf{w}]\!] = [\![\frac{d\mathbf{w}}{dn}]\!].$$

Hence, we have

$$[\![\nabla \times \mathbf{w}]\!] = \mathbf{n} \times \mathbf{b}, \qquad [\![\nabla \cdot \mathbf{w}]\!] = \mathbf{n} \cdot \mathbf{b}, \qquad \mathbf{b} = \mathbf{n}[\![\nabla \cdot \mathbf{w}]\!] - \mathbf{n} \times [\![\nabla \times \mathbf{w}]\!].$$

This is phrased by saying that the longitudinal and transversal jumps of the gradient of a continuous vector are the jumps of its divergence and curl, respectively (Weingarten's first theorem) ([425, 443], Sect. 11.4).

Assume now σ is a singular surface for the second-order derivatives of ϕ, that is $[\![\nabla\phi]\!] = \mathbf{0}$, whereas $[\![\nabla\nabla\phi]\!] \neq \mathbf{0}$. By Hadamard's lemma

$$\partial_\gamma [\![\nabla\phi]\!] = [\![\nabla\nabla\phi]\!]\partial_\gamma \mathbf{x}.$$

The assumption $[\![\nabla\phi]\!] = \mathbf{0}$ implies that

$$[\![\nabla\nabla\phi]\!]\,\mathbf{t} = \mathbf{0}$$

for every vector $\mathbf{t}$ tangent to σ. Hence there is a vector $\mathbf{w}$ such that

$$[\![\nabla\nabla\phi]\!] = \mathbf{w} \otimes \mathbf{n}.$$

The symmetry of $\nabla\nabla\phi$ implies that $\mathbf{w} = a\,\mathbf{n}$ and then

$$[\![\nabla\nabla\phi]\!] = a\,\mathbf{n} \otimes \mathbf{n}.$$

We can view a as the amplitude of the discontinuity. Indeed, inner multiplying by $\mathbf{n} \otimes \mathbf{n}$ gives

$$a = [\![\frac{d^2\phi}{dn^2}]\!].$$

Given a vector, $\boldsymbol{\phi}$ say, for each component ϕ_i, we have a replaced by a_i and hence

$$[\![\nabla\nabla\boldsymbol{\phi}]\!] = \mathbf{a}\,\mathbf{n} \otimes \mathbf{n}.$$

We then find that $\mathbf{a}$ is the second-order normal derivative so that

$$[\![\nabla\nabla\boldsymbol{\phi}]\!] = [\![\frac{d^2\boldsymbol{\phi}}{dn^2}]\!]\,\mathbf{n} \otimes \mathbf{n}.$$

If $\mathbf{a} \parallel \mathbf{n}$ the singularity is said to be longitudinal, if $\mathbf{a} \perp \mathbf{n}$ transversal.

The dual properties hold across $\Sigma(t)$, with $\Phi(\mathbf{X}, t) = \Phi(\chi^{-1}(\mathbf{x}, t)) = \phi(\mathbf{x}, t)$, in the reference configuration. If $[\![\Phi]\!] = 0$ or $[\![\Phi]\!]$ is constant, then

$$[\![\nabla_R \Phi]\!] = B\,\mathbf{n}_R, \qquad B = [\![\mathbf{n}_R \cdot \nabla_R \Phi]\!] = [\![\frac{d\Phi}{dn_R}]\!].$$

If $[\![\nabla_R \Phi]\!] = \mathbf{0}$, whereas $[\![\nabla_R \nabla_R \Phi]\!] \neq \mathbf{0}$, then

$$[\![\nabla_R \nabla_R \Phi]\!] = A\,\mathbf{n}_R \otimes \mathbf{n}_R.$$

1.7.2 Kinematical Conditions of Compatibility

Let $g(\mathbf{x}, t) = 0$ be the equation of the moving surface $\sigma(t)$ and let $\mathbf{x} = \mathbf{x}(\delta, \gamma, t)$ be a parameterization of the points of $\sigma(t)$. Differentiation with respect to t gives

$$\partial_t g + \mathbf{u} \cdot \nabla g = 0, \qquad \mathbf{u} = \partial_t \mathbf{x}(\delta, \gamma, t);$$

$\mathbf{u}$ is the velocity of the surface point identified by the pair of parameters δ, γ. Dividing by $|\nabla g|$ and letting

$$u_n = \mathbf{u} \cdot \mathbf{n} = \mathbf{u} \cdot \frac{\nabla g}{|\nabla g|}$$

we obtain

$$u_n = -\frac{\partial_t g}{|\nabla g|}.$$

Since the right-hand side involves the function g alone, and then is independent of the selected parameterization, the quantity u_n too is independent of the parameterization. Accordingly, all possible velocities $\mathbf{u}$ of the moving surface have the same normal component u_n, which is called the speed of displacement of the surface. The particular choice

$$\mathbf{u} = u_n \mathbf{n}$$

makes $\mathbf{u}$ normal to the surface. This velocity is then called the normal velocity of the surface.

An observer moving with the normal velocity $u_n \mathbf{n}$ will encounter points on the surface having surface coordinates, γ_1, γ_2 say, which vary in time in a proper way. Now if $\mathbf{x} = \mathbf{x}(\gamma_1, \gamma_2, t)$, and γ_1, γ_2 vary in time, then the velocity is given by

$$\mathbf{v} = \partial_t \mathbf{x} + \partial_t \gamma_i \, \partial_{\gamma_i} \mathbf{x}.$$

If the observer moves with the normal velocity, then the derivatives $\partial_t \gamma_i$ are subject to

$$\partial_t \mathbf{x} + \partial_t \gamma_i \, \partial_{\gamma_i} \mathbf{x} = u_n \mathbf{n}. \tag{1.42}$$

If $f(\gamma_1, \gamma_2, t)$ is a function defined on the moving surface we denote by $\delta f / \delta t$ the displacement derivative that is the rate of change of f as apparent to an observer moving with the normal velocity. Hence

$$\frac{\delta f}{\delta t} = \partial_t f + \partial_t \gamma_i \, \partial_{\gamma_i} f,$$

where $\partial_t \gamma_1$ and ∂_{γ_2} are such that (1.42) holds. Now let

$$f(\gamma_1, \gamma_2, t) = \hat{f}(\mathbf{x}(\gamma_1, \gamma_2, t), t).$$

Then

$$\partial_t f = \partial_t \hat{f} + \partial_t \mathbf{x} \cdot \nabla \hat{f}, \qquad \partial_{\gamma_i} f = \partial_{\gamma_i} \mathbf{x} \cdot \nabla \hat{f}.$$

Consequently

$$\frac{\delta \hat{f}}{\delta t} \equiv \frac{\delta f}{\delta t} = \partial_t \hat{f} + (\partial_t \mathbf{x} + \partial_t \gamma_i \, \partial_{\gamma_i} \mathbf{x}) \cdot \nabla \hat{f} = \partial_t \hat{f} + u_n \mathbf{n} \cdot \nabla \hat{f}.$$

Let ϕ_+ (ϕ_-) be a function defined on the whole region Ω_+ (Ω_-). The displacement derivative of ϕ_+ reads

$$\frac{\delta \phi_+}{\delta t} = \partial_t \phi_+ + u_n \mathbf{n} \cdot \nabla \phi_+$$

and the like holds for ϕ_-. Of course

$$\partial_t \phi_\pm = (\partial_t \phi)_\pm, \qquad \nabla \phi_\pm = (\nabla \phi)_\pm.$$

Subtracting the relation for ϕ_+ from that for ϕ_-, we obtain

$$\frac{\delta}{\delta t}[\![\phi]\!] = [\![\partial_t \phi]\!] + u_n [\![\mathbf{n} \cdot \nabla \phi]\!], \tag{1.43}$$

which is named kinematical condition of compatibility. In the special case where ϕ is continuous across σ Eq. (1.43) reduces to

$$[\![\partial_t \phi]\!] = -u_n [\![\mathbf{n} \cdot \nabla \phi]\!]. \tag{1.44}$$

In particular, across a stationary surface ($u_n = 0$) that is singular with respect to $\nabla \phi$ but not with respect to ϕ we have $[\![\partial_t \phi]\!] = 0$.

If $[\![\phi]\!] = 0$, then the geometrical condition of compatibility and (1.44) yield

$$[\![\nabla \phi]\!] = b\mathbf{n}, \qquad [\![\partial_t \phi]\!] = -u_n b.$$

For practical purposes, it is useful to determine a relation for the jump of the Lagrangian derivative $\dot{\phi}$ within the Eulerian description. Since $\dot{\phi} = \partial_t \phi + \mathbf{v} \cdot \nabla \phi$ then

$$[\![\dot{\phi}]\!] = [\![\partial_t \phi]\!] + [\![\mathbf{v} \cdot \nabla \phi]\!].$$

If $[\![\phi]\!] = 0$ then by (1.44) it follows

$$[\![\dot{\phi}]\!] = -[\![(u_n \mathbf{n} - \mathbf{v}) \cdot \nabla \phi]\!].$$

If, in particular, $[\![\mathbf{v}]\!] = \mathbf{0}$, then

$$[\![\dot{\phi}]\!] = -(u_n \mathbf{n} - \mathbf{v}) \cdot [\![\nabla \phi]\!] = -(u_n \mathbf{n} - \mathbf{v}) \cdot \mathbf{n} \, [\![\mathbf{n} \cdot \nabla \phi]\!]$$

whence

$$[\![\dot{\phi}]\!] = -U [\![\mathbf{n} \cdot \nabla \phi]\!], \qquad U := u_n - \mathbf{v} \cdot \mathbf{n}, \tag{1.45}$$

U being called the local speed of propagation.

Further, kinematical conditions of compatibility hold for higher-order derivatives. For example, applying (1.43) to $\partial_t \phi$ and $\nabla \phi$, we have

$$\frac{\delta}{\delta t}[\![\partial_t \phi]\!] = [\![\partial_t^2 \phi]\!] + u_n [\![\mathbf{n} \cdot \nabla \partial_t \phi]\!],$$

$$\frac{\delta}{\delta t}[\![\nabla \phi]\!] = [\![\nabla \partial_t \phi]\!] + u_n [\![\mathbf{n} \cdot \nabla \nabla \phi]\!].$$

We say that $\sigma(t)$ is a propagating singular surface of order n, with respect to ϕ, if ϕ and the derivatives up to order $n-1$ are continuous functions everywhere whereas the derivatives of order n and higher have jump discontinuities across $\sigma(t)$ but are continuous functions everywhere else. Singular surfaces of order 0 and 1 are said to be strong singularities; if ϕ is the position vector such is the case of contact discontinuities and shock waves. Singular surfaces of order 2 or higher are said to be weak singularities.

If σ is a singular surface of order 2 with respect to ϕ, then, by Maxwell's theorem, we have

$$[\![\nabla \nabla \phi]\!] = a\, \mathbf{n} \otimes \mathbf{n}.$$

By the kinematical condition of compatibility

$$[\![\partial_t \nabla \phi]\!] = -u_n [\![\mathbf{n} \cdot \nabla \nabla \phi]\!] = -u_n\, a\, \mathbf{n},$$

and

$$[\![\partial_t^2 \phi]\!] = -u_n \mathbf{n} \cdot [\![\partial_t \nabla \phi]\!] = u_n^2\, a.$$

The dual properties hold for singularities across $\Sigma(t)$ in the reference configuration. Let

$$F(\Gamma_1, \Gamma_2, t) = \hat{F}(\mathbf{X}(\Gamma_1, \Gamma_2, t), t)$$

Then

$$\partial_t F = \partial_t \hat{F} + \partial_t \mathbf{X} \cdot \nabla_R \hat{F}, \qquad \partial_{\Gamma_i} F = \partial_{\Gamma_i} \mathbf{X} \cdot \nabla_R \hat{F}.$$

If $F(\Gamma_1, \Gamma_2, t)$ is a function defined on the moving surface we denote by $\delta F/\delta t$ the displacement derivative that is the rate of change of F as apparent to an observer moving with the normal velocity U_R. Hence

$$\frac{\delta F}{\delta t} = \partial_t F + \partial_t \Gamma_i\, \partial_{\Gamma_i} F,$$

and

$$\frac{\delta \hat{F}}{\delta t} \equiv \frac{\delta F}{\delta t} = \partial_t \hat{F} + (\partial_t \mathbf{X} + \partial_t \Gamma_i\, \partial_{\Gamma_i} \mathbf{X}) \cdot \nabla_R \hat{F} = \partial_t \hat{F} + U_R \mathbf{n}_R \cdot \nabla_R \hat{F},$$

where $U_R \mathbf{n}_R = \partial_t \mathbf{X} + \partial_t \Gamma_i \, \partial_{\Gamma_i} \mathbf{X}$ by definition of the displacement derivative. Accordingly, for any function $\Phi(\mathbf{X}(\Gamma_1, \Gamma_2, t), t)$ we consider the limit functions Φ_+, Φ_- as the current point approaches $\Sigma(t)$ from R_+, R_-. Subtraction of the corresponding relation for Φ_-, Φ_+ gives

$$\frac{\delta [\![\Phi]\!]}{\delta t} = [\![\partial_t \Phi]\!] + U_R [\![\mathbf{n}_R \cdot \nabla_R \Phi]\!], \tag{1.46}$$

that is the kinematical condition of compatibility in the reference configuration. In the case where Φ is continuous across Σ, Eq. (1.46) reduces to

$$[\![\partial_t \Phi]\!] = -U_R [\![\mathbf{n}_R \cdot \nabla_R \Phi]\!]. \tag{1.47}$$

Moreover, by the geometrical condition of compatibility, we have

$$[\![\nabla_R \Phi]\!] = B \, \mathbf{n}_R, \qquad [\![\partial_t \Phi]\!] = -U_R \, B.$$

Singular surfaces $\Sigma(t)$ of order n are defined as with the corresponding surface $\sigma(t)$ in the Eulerian description. If Σ is a singular surface of order 2, with respect to Φ, then by Maxwell's theorem and the kinematical condition of compatibility (1.47), we have

$$[\![\nabla_R \nabla_R \Phi]\!] = A \, \mathbf{n}_R \otimes \mathbf{n}_R, \qquad [\![\partial_t \nabla_R \Phi]\!] = -U_R A \mathbf{n}_R, \qquad [\![\partial_t^2 \Phi]\!] = U_R^2 A.$$

If the function Φ is the motion χ, then we have the following results

$$[\![\nabla_R \nabla_R \chi]\!] = [\![\nabla_R \mathbf{F}]\!] = \mathbf{A} \, \mathbf{n}_R \otimes \mathbf{n}_R, \qquad [\![\nabla_R \dot{\chi}]\!] = [\![\dot{\mathbf{F}}]\!] = -U_R \mathbf{A} \mathbf{n}_R, \qquad [\![\ddot{\chi}]\!] = U_R^2 \mathbf{A},$$

which are referred to as Hugoniot's compatibility conditions [425], Sect. 11.4. These compatibility relations can be given a form appropriate for the Eulerian description, involving the normal $\mathbf{n}$ and the speed U of the surface $\sigma(t)$. Since

$$\mathbf{n}_R = \frac{\mathbf{F}^T \mathbf{n}}{|\mathbf{F}^T \mathbf{n}|}, \qquad U_R = \frac{1}{|\mathbf{F}^T \mathbf{n}|} U,$$

letting

$$\mathbf{a} = \frac{1}{|\mathbf{F}^T \mathbf{n}|^2} \mathbf{A}$$

we obtain

$$[\![\nabla_R \mathbf{F}]\!] = \mathbf{a} \, \mathbf{F}^T \mathbf{n} \otimes \mathbf{F}^T \mathbf{n}, \qquad [\![\dot{\mathbf{F}}]\!] = -U \, \mathbf{a} \, \mathbf{F}^T \mathbf{n}, \qquad [\![\ddot{\chi}]\!] = U^2 \, \mathbf{a}.$$

For the sake of generality, we now let Σ be a singular surface of order $p \geq 1$. For definiteness, we let the function Φ be the motion χ. Hence, order p means that the $(p-1)$-th order derivatives, of $\chi(\mathbf{X}, t)$, are continuous whereas the p-th order derivatives suffer jump discontinuities across Σ. Consequently

$$[\![\partial_t^{p-1}\chi]\!] = \mathbf{0}, \qquad [\![\underbrace{\nabla_R \cdots \nabla_R}_{p-1}\chi]\!] = \mathbf{0}.$$

By Maxwell's theorem[13]

$$[\![\underbrace{\nabla_R \cdots \nabla_R}_{p}\chi]\!] = [\![\partial_{n_R}\underbrace{\nabla_R \cdots \nabla_R}_{p-1}\chi]\!]\mathbf{n}_R = [\![\underbrace{\nabla_R \cdots \nabla_R}_{p-1}\partial_{n_R}\chi]\!]\mathbf{n}_R.$$

Iterating the procedure, we find

$$[\![\underbrace{\nabla_R \cdots \nabla_R}_{p}\chi]\!] = [\![\partial_{n_R}^p\chi]\!]\,\underbrace{\mathbf{n}_R \otimes \cdots \otimes \mathbf{n}_R}_{p}.$$

By the kinematical condition of compatibility, we obtain

$$[\![\partial_t^p\chi]\!] = -U_R[\![\partial_{n_R}\partial_t^{p-1}\chi]\!] = -U_R[\![\partial_t^{p-1}\partial_{n_R}\chi]\!].$$

Applying the kinematical condition of compatibility $p-1$ more times, we find

$$[\![\partial_t^p\chi]\!] = (-U_R)^p\,[\![\partial_{n_R}^p\chi]\!].$$

These relations are traced back to Hadamard [425]. We can then write the following *Higher-order compatibility condition.* At a singular surface of order $p \geq 1$ for the motion, the jumps of space and time derivatives are related by

$$(-U_R)^p\,[\![\,\underbrace{\nabla_R \cdots \nabla_R}_{p}\chi]\!] = [\![\partial_t^p\chi]\!]\,\underbrace{\mathbf{n}_R \otimes \cdots \otimes \mathbf{n}_R}_{p}.$$

A further relation follows from the equation of motion, in the material description, as $p \geq 2$. Apply ∂_t^{p-2} to (2.28) so that

$$\rho_R\,\partial_t^{p-1}\,\mathbf{v} = \partial_t^{p-2}\,\nabla_R \cdot \mathbf{T}_R + \rho_R\,\partial_t^{p-2}\,\mathbf{b},$$

and, subject to $[\![\partial_t^{p-2}\,\mathbf{b}]\!] = 0$, take the jumps to have

$$\rho_R\,[\![\partial_t^{p-1}\,\mathbf{v}]\!] = [\![\partial_t^{p-2}\,\nabla_R \cdot \mathbf{T}_R]\!].$$

Now

$$[\![\partial_t^{p-2}\,\nabla_R \cdot \mathbf{T}_R]\!] = [\![\nabla_R \cdot \partial_t^{p-2}\mathbf{T}_R]\!] = -\frac{1}{U_R}[\![\partial_t^{p-1}\,\mathbf{T}_R]\!]\mathbf{n}_R$$

[13] For ease in writing we use ∂_n and ∂_{n_R} in place of $d/dn = \mathbf{n} \cdot \nabla$ and $d/dn_R = \mathbf{n}_R \cdot \nabla_R$.

and

$$[\![\partial_t^{p-1}\mathbf{v}]\!] = [\![\partial_t^p\chi]\!] = (-U_R)^p\,[\![\partial_{n_R}^p\chi]\!].$$

Hence, we have

$$[\![\partial_t^{p-1}\mathbf{T}_R]\!]\mathbf{n}_R = \rho_R(-U_R)^{p+1}\,[\![\partial_{n_R}^p\chi]\!].$$

1.7.3 Jumps at Acoustic Waves

In applications, it may happen that we need the geometrical description in the current configuration, at the surface $\sigma(t)$, whereas because of the balance equations, we may use the total time derivative. To fix ideas, let σ be a singular surface of order 1 with respect to ϕ,

$$[\![\phi]\!] = 0, \qquad [\![\nabla\phi]\!] \neq \mathbf{0}.$$

By Maxwell's theorem

$$[\![\nabla\phi]\!] = b\,\mathbf{n}, \qquad b = [\![\partial_n\phi]\!], \qquad \partial_n\phi := \mathbf{n}\cdot\nabla\phi.$$

An acoustic wave (or acceleration wave) is a moving singular surface $\sigma(t)$ such that

$$[\![\mathbf{v}]\!] = \mathbf{0}, \qquad [\![\nabla\mathbf{v}]\!] \neq \mathbf{0},$$

across $\sigma(t)$. Hence Maxwell's theorem implies that

$$[\![\nabla\mathbf{v}]\!] = \mathbf{a}\otimes\mathbf{n}, \qquad \mathbf{a} := [\![\partial_n\mathbf{v}]\!].$$

Hence, it follows that

$$2[\![\mathbf{D}]\!] = \mathbf{a}\otimes\mathbf{n}+\mathbf{n}\otimes\mathbf{a}, \quad [\![\nabla\cdot\mathbf{v}]\!] = \mathbf{a}\cdot\mathbf{n}, \quad [\![\nabla\times\mathbf{v}]\!] = \mathbf{n}\times\mathbf{a}, \quad 2[\![\mathbf{W}]\!] = \mathbf{a}\otimes\mathbf{n}-\mathbf{n}\otimes\mathbf{a}.$$

By the kinematical condition of compatibility, we can write

$$[\![\partial_t\mathbf{v}]\!] = -u_n[\![\mathbf{n}\cdot\nabla\mathbf{v}]\!] = -u_n\,\mathbf{a}.$$

Now

$$[\![\dot{\mathbf{v}}]\!] = [\![\partial_t\mathbf{v}+\mathbf{v}\cdot\nabla\mathbf{v}]\!] = [\![\partial_t\mathbf{v}]\!]+\mathbf{v}\cdot[\![\nabla\mathbf{v}]\!] = (-u_n\mathbf{n}+\mathbf{v})\cdot[\![\nabla\mathbf{v}]\!] = (-u_n+\mathbf{v}\cdot\mathbf{n})\mathbf{a}.$$

Consequently, we have

$$[\![\dot{\mathbf{v}}]\!] = -U[\![\partial_n\mathbf{v}]\!]. \tag{1.48}$$

In terms of $[\![\partial_n^2\chi]\!]$, we have

$$[\![\nabla\mathbf{v}]\!] = \mathbf{n}\otimes[\![\partial_n\mathbf{v}]\!] = -U\mathbf{n}\otimes[\![\partial_n^2\chi]\!] = -U\mathbf{n}\otimes\mathbf{c}, \qquad \mathbf{c} := [\![\partial_n^2\chi]\!],$$

in suffix notation $[\![\partial_{x_i} v_j]\!] = n_i[\![\partial_n v_j]\!]$. Hence, we find

$$2[\![\mathbf{D}]\!] = -U(\mathbf{c}\otimes\mathbf{n}+\mathbf{n}\otimes\mathbf{c}), \qquad 2[\![\mathbf{W}]\!] = -U(\mathbf{c}\otimes\mathbf{n}-\mathbf{n}\otimes\mathbf{a}),$$

$$[\![\nabla\cdot\mathbf{v}]\!] = -U\mathbf{c}\cdot\mathbf{n}, \qquad [\![\nabla\times\mathbf{v}]\!] = -U\mathbf{n}\times\mathbf{c},$$

$$[\![\ddot{\chi}]\!] = -U[\![\partial_n\dot{\chi}]\!] = U^2[\![\partial_n^2\chi]\!],$$

the relations for $\nabla\cdot\mathbf{v}$ and $\mathbf{W}$ being the content of the *Weingarten-Hadamard theorem* [425].

1.8 Transport Theorems for Surface Integrals

Let $\mathcal{S}$ be a surface in the three-dimensional space $\mathscr{E}$, possibly dependent on the time $t\in\mathbb{R}$. The dependence on t may be induced by the motion of the points of a body or by a given velocity field of $\mathcal{S}$ as a purely geometrical surface.

We first compute the time derivative of a surface integral, over a time-dependent surface $\mathcal{S}$, when the time dependence is induced by a continuous velocity field of the underlying body. We then say that $\mathcal{S}$ convects with the body. Let $\mathcal{S}$ consist of points of a body undergoing a velocity field $\mathbf{v}\in C^1(\mathcal{S})\cap C(\bar{\mathcal{S}})$. The surface $\mathcal{S}$ is the image of S through the motion of the body,

$$\mathcal{S} = \chi(\mathrm{S},t), \qquad \mathcal{S}\ni\mathbf{x} = \chi(\mathbf{X},t), \quad \mathbf{X} = \chi^{-1}(\mathbf{x},t)\in\mathrm{S},$$

for any time t. In light of the Nanson formula, we can write

$$\int_{\mathcal{S}}\phi\,\mathbf{n}\,da = \int_{\mathrm{S}}\phi\,J\,\mathbf{F}^{-T}\mathbf{n}_R\,da_R,$$

where $\mathbf{n}$ is the unit normal to $\mathcal{S}$ and $\mathbf{n}_R$ is the unit normal to S. The surface S is time independent. Hence

$$\frac{d}{dt}\int_{\mathcal{S}}\phi\,\mathbf{n}\,da = \int_{\mathrm{S}}\overline{\phi\,J\,\mathbf{F}^{-T}}^{\boldsymbol{\cdot}}\,\mathbf{n}_R\,da_R.$$

Since

$$\dot{J} = J\,\nabla\cdot\mathbf{v}, \qquad \overline{\mathbf{F}^{-T}}^{\boldsymbol{\cdot}} = -\mathbf{F}^{-T}\dot{\mathbf{F}}^T\mathbf{F}^{-T}$$

and $\mathbf{F}^{-T}\dot{\mathbf{F}}^T = \mathbf{L}^T$ we find

$$\frac{d}{dt}\int_S \phi\,\mathbf{n}\,da = \int_S (\dot{\phi}\,J\,\mathbf{F}^{-T} + \phi\,J\nabla\cdot\mathbf{v}\,\mathbf{F}^{-T} - \phi\,J\,\mathbf{F}^{-T}\dot{\mathbf{F}}^T\mathbf{F}^{-T})\,\mathbf{n}_R\,da_R$$
$$= \int_S \{\dot{\phi}\mathbf{n} + \phi[(\nabla\cdot\mathbf{v})\mathbf{1} - \mathbf{L}^T]\mathbf{n}\}da.$$

Now we let the boundary $\partial\mathcal{S}$ of $\mathcal{S}$ depend on time via a velocity field $\boldsymbol{\nu}$, on $\partial\mathcal{S}$. Then the boundary is not convecting and moves with velocity $\boldsymbol{\nu} - \mathbf{v}$, possibly discontinuous, relative to the material points and we say that $\mathcal{S}$ is not convecting with the body. The area produced by the relative velocity $\boldsymbol{\nu} - \mathbf{v}$, per unit length of the boundary, is $|(\boldsymbol{\nu} - \mathbf{v}) \times \mathbf{t}|$ and $(\boldsymbol{\nu} - \mathbf{v}) \times \mathbf{t}$ is orthogonal to the surface. Hence, the additional contribution to the time derivative is given by

$$\int_{\partial\mathcal{S}} \phi\,(\boldsymbol{\nu} - \mathbf{v}) \times \mathbf{t}\,dl.$$

Consequently, for a generic motion of the boundary $\partial\mathcal{S}$, we have

$$\frac{d}{dt}\int_S \phi\,\mathbf{n}\,da = \int_S \{\dot{\phi}\mathbf{n} + \phi[(\nabla\cdot\mathbf{v})\mathbf{1} - \mathbf{L}^T]\mathbf{n}\}da + \int_{\partial\mathcal{S}} \phi\,(\boldsymbol{\nu} - \mathbf{v}) \times \mathbf{t}\,dl.$$

Of course the integral on $\partial\mathcal{S}$ vanishes if the boundary consists of material points, in which case $\boldsymbol{\nu} = \mathbf{v}$.

With a view to applications, we now let ϕ be a vector, say $\phi = \mathbf{w}$, and hence $\phi\,\mathbf{n} = \mathbf{w}\cdot\mathbf{n}$ or a tensor $\mathbf{A}$ so that $\phi\mathbf{n} = \mathbf{A}\mathbf{n}$. Since

$$\dot{\mathbf{w}}\cdot\mathbf{n} + \mathbf{w}\cdot\mathbf{n}\,\nabla\cdot\mathbf{v} - \mathbf{w}\cdot\mathbf{L}^T\mathbf{n} = [\dot{\mathbf{w}} + (\nabla\cdot\mathbf{v})\mathbf{w} - (\mathbf{w}\cdot\nabla)\mathbf{v}]\cdot\mathbf{n}$$

then

$$\frac{d}{dt}\int_S \mathbf{w}\cdot\mathbf{n}\,da = \int_S \overset{\diamond}{\mathbf{w}}\cdot\mathbf{n}\,da + \int_{\partial\mathcal{S}} \mathbf{w}\cdot(\boldsymbol{\nu} - \mathbf{v}) \times \mathbf{t}\,dl, \tag{1.49}$$

where

$$\overset{\diamond}{\mathbf{w}} := \dot{\mathbf{w}} + (\nabla\cdot\mathbf{v})\mathbf{w} - (\mathbf{w}\cdot\nabla)\mathbf{v} = \dot{\mathbf{w}} - \mathbf{L}\mathbf{w} + (\nabla\cdot\mathbf{v})\mathbf{w};$$

$\overset{\diamond}{\mathbf{w}}$ is referred to as the convected time derivative of the vector $\mathbf{w}$. If $\phi\mathbf{n} = \mathbf{A}\mathbf{n}$, then we can formally replace $\dot{\phi}\mathbf{n} + \phi[(\nabla\cdot\mathbf{v})\mathbf{1} - \mathbf{L}^T]\mathbf{n}$ with $[\dot{\mathbf{A}} - \mathbf{A}\mathbf{L}^T + (\nabla\cdot\mathbf{v})\mathbf{A}]\mathbf{n}$ to obtain

$$\frac{d}{dt}\int_S \mathbf{A}\,\mathbf{n}\,da = \int_S \overset{\diamond}{\mathbf{A}}\;\mathbf{n}\,da + \int_{\partial\mathcal{S}} \mathbf{A}\,(\boldsymbol{\nu} - \mathbf{v}) \times \mathbf{t}\,dl,$$

where

$$\overset{\diamond}{\mathbf{A}} := \dot{\mathbf{A}} - \mathbf{A}\mathbf{L}^T + (\nabla\cdot\mathbf{v})\mathbf{A}.$$

The symbol $\overset{\diamond}{\mathbf{A}}$ is referred to as the convected time derivative of the tensor $\mathbf{A}$. If S is convecting with the body, then

$$\frac{d}{dt}\int_{\mathcal{S}} \mathbf{A}\,\mathbf{n}\,da = \int_{\mathcal{S}} \overset{\diamond}{\mathbf{A}}\;\mathbf{n}\,da.$$

Vectors and tensors are subject to the same convected time derivative; the common expression of the derivative is due to the evolution of the surface.

We add a property of vectors. In view of the identity

$$\overset{\diamond}{\mathbf{w}} = \partial_t \mathbf{w} + \nabla \times (\mathbf{w} \times \mathbf{v}) + \mathbf{v}\nabla \cdot \mathbf{w}, \tag{1.50}$$

by applying Stokes' theorem, we have

$$\frac{d}{dt}\int_{\mathcal{S}} \mathbf{w}\cdot\mathbf{n}\,da = \int_{\mathcal{S}} (\partial_t \mathbf{w} + \mathbf{v}\nabla\cdot\mathbf{w})\cdot\mathbf{n}\,da + \int_{\partial\mathcal{S}} \mathbf{w}\times\mathbf{v}\cdot\mathbf{t}\,dl + \int_{\partial\mathcal{S}} \mathbf{w}\cdot(\boldsymbol{\nu}-\mathbf{v})\times\mathbf{t}\,dl,$$

whence

$$\frac{d}{dt}\int_{\mathcal{S}} \mathbf{w}\cdot\mathbf{n}\,da = \int_{\mathcal{S}} (\partial_t \mathbf{w} + \mathbf{v}\nabla\cdot\mathbf{w})\cdot\mathbf{n}\,da + \int_{\partial\mathcal{S}} \mathbf{w}\times\boldsymbol{\nu}\cdot\mathbf{t}\,dl. \tag{1.51}$$

If $\mathbf{w}$ is divergence-free, i.e. $\nabla \cdot \mathbf{w} = 0$, then

$$\frac{d}{dt}\int_{\mathcal{S}} \mathbf{w}\cdot\mathbf{n}\,da = \int_{\mathcal{S}} \partial_t \mathbf{w}\cdot\mathbf{n}\,da + \int_{\partial\mathcal{S}} \mathbf{w}\times\boldsymbol{\nu}\cdot\mathbf{t}\,dl.$$

1.8.1 Transport Theorems for Discontinuous Fields

We now allow for the occurrence of a line of discontinuity, say γ, which divides $\mathcal{S}$ in two surfaces $\mathcal{S}_+, \mathcal{S}_-$, and take the unit tangent $\mathbf{t}_\gamma$ to γ such that $\mathbf{t}_\gamma \times \mathbf{n}$ is along the direction from $\mathcal{S}_-$ to $\mathcal{S}_+$. The surface $\mathcal{S}$ consists of points of the body, moving with the velocity field $\mathbf{v}$. The line γ, instead, is moving with a velocity $\boldsymbol{\nu}$ possibly different from the velocity $\mathbf{v}$ of the points of $\mathcal{S}$. The vector field $\mathbf{w}$ suffer a jump discontinuity across γ. We let

$$[\![\mathbf{w}]\!] = \mathbf{w}_- - \mathbf{w}_+,$$

$\mathbf{w}_-, \mathbf{w}_+$ at a point $\mathbf{x} \in \gamma$ being the limit values of $\mathbf{w}(\mathbf{y}, t)$ as $\mathbf{y}$ approaches $\mathbf{x}$ within $\mathcal{S}_-, \mathcal{S}_+$, respectively (Fig. 1.4).

We have

$$\frac{d}{dt}\int_{\mathcal{S}_+} \mathbf{w}\cdot\mathbf{n}\,da = \int_{\mathcal{S}_+} (\partial_t \mathbf{w} + \mathbf{v}\,\nabla\cdot\mathbf{w})\cdot\mathbf{n}\,da + \int_{\partial\mathcal{S}_+\setminus\gamma} \mathbf{w}\cdot\mathbf{v}\times\mathbf{t}\,dl - \int_{\gamma} \mathbf{w}_+\cdot\boldsymbol{\nu}\times\mathbf{t}_\gamma\,dl,$$

$$\frac{d}{dt}\int_{\mathcal{S}_-} \mathbf{w}\cdot\mathbf{n}\,da = \int_{\mathcal{S}_-} (\partial_t \mathbf{w} + \mathbf{v}\,\nabla\cdot\mathbf{w})\cdot\mathbf{n}\,da + \int_{\partial\mathcal{S}_-\setminus\gamma} \mathbf{w}\cdot\mathbf{v}\times\mathbf{t}\,dl + \int_{\gamma} \mathbf{w}_-\cdot\boldsymbol{\nu}\times\mathbf{t}_\gamma\,dl.$$

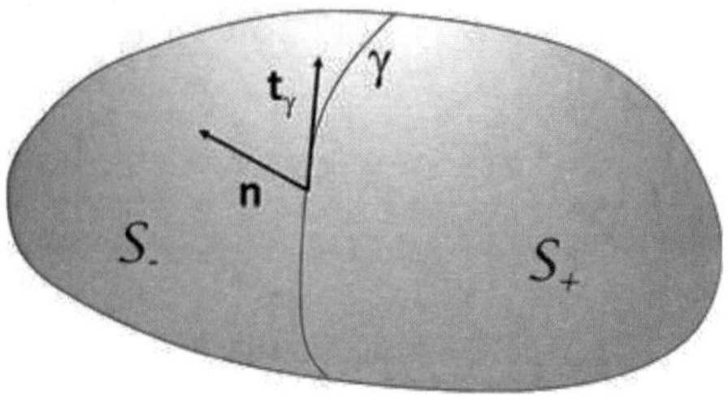

Fig. 1.4 The line γ divides the surface $\mathcal{S}$ in two surfaces $\mathcal{S}_-$, $\mathcal{S}_+$ and is the common boundary between them

Summation gives

$$\frac{d}{dt}\int_{\mathcal{S}\setminus\gamma} \mathbf{w}\cdot\mathbf{n}\,da = \int_{\mathcal{S}\setminus\gamma} (\partial_t\mathbf{w} + \mathbf{v}\,\nabla\cdot\mathbf{w})\cdot\mathbf{n}\,da + \int_{\partial\mathcal{S}} \mathbf{w}\times\mathbf{v}\cdot\mathbf{t}\,dl + \int_\gamma [\![\mathbf{w}\times\boldsymbol{\nu}]\!]\cdot\mathbf{t}_\gamma\,dl$$

where the identity $\mathbf{w}\cdot\mathbf{v}\times\mathbf{t} = \mathbf{w}\times\mathbf{v}\cdot\mathbf{t}$ and the analogue with $\boldsymbol{\nu}$ have been used. Likewise, we apply Stokes' theorem to $\mathcal{S}_+$ and $\mathcal{S}_-$ to obtain

$$\int_{\mathcal{S}_+} \nabla\times\mathbf{u}\cdot\mathbf{n}\,da = \int_{\partial\mathcal{S}_+\setminus\gamma} \mathbf{u}_+\cdot\mathbf{t}\,dl - \int_\gamma \mathbf{u}_+\cdot\mathbf{t}_\gamma\,dl,$$

$$\int_{\mathcal{S}_-} \nabla\times\mathbf{u}\cdot\mathbf{n}\,da = \int_{\partial\mathcal{S}_-\setminus\gamma} \mathbf{u}_-\cdot\mathbf{t}\,dl + \int_\gamma \mathbf{u}_-\cdot\mathbf{t}_\gamma\,dl,$$

for any vector field $\mathbf{u}$ of class C^1 in S_+ and S_-. The summation results in

$$\int_{\mathcal{S}\setminus\gamma} \nabla\times\mathbf{u}\cdot\mathbf{n}\,da = \int_{\partial\mathcal{S}} \mathbf{u}\cdot\mathbf{t}\,dl + \int_\gamma [\![\mathbf{u}]\!]\cdot\mathbf{t}_\gamma\,dl, \tag{1.52}$$

that is the *generalized Stokes theorem* for a surface $\mathcal{S}$ including a discontinuity line γ.

Identify $\mathbf{u}$ with $\mathbf{w}\times\mathbf{v}$ and observe

$$\int_{\partial\mathcal{S}} \mathbf{w}\times\mathbf{v}\cdot\mathbf{t}\,dl = \int_{\mathcal{S}\setminus\gamma} \nabla\times(\mathbf{w}\times\mathbf{v})\cdot\mathbf{n}\,da - \int_\gamma [\![\mathbf{w}\times\mathbf{v}]\!]\cdot\mathbf{t}_\gamma\,dl.$$

Replacing the integral over $\partial\mathcal{S}$, we obtain

$$\frac{d}{dt}\int_{\mathcal{S}\setminus\gamma} \mathbf{w}\cdot\mathbf{n}\,da = \int_{\mathcal{S}\setminus\gamma} (\partial_t\mathbf{w} + \mathbf{v}\,\nabla\cdot\mathbf{w})\cdot\mathbf{n}\,da + \int_{\mathcal{S}\setminus\gamma} \nabla\times(\mathbf{w}\times\mathbf{v})\cdot\mathbf{n}\,da + \int_\gamma [\![\mathbf{w}\times(\boldsymbol{\nu}-\mathbf{v})]\!]\cdot\mathbf{t}_\gamma\,dl$$

whence

$$\frac{d}{dt}\int_{\mathcal{S}\setminus\gamma} \mathbf{w}\cdot\mathbf{n}\,da = \int_{\mathcal{S}\setminus\gamma} \overset{\diamond}{\mathbf{w}}\cdot\mathbf{n}\,da + \int_\gamma [\![\mathbf{w}\times(\boldsymbol{\nu}-\mathbf{v})]\!]\cdot\mathbf{t}_\gamma\,dl. \tag{1.53}$$

Equation (1.53) gives the time derivative of the flux of $\mathbf{w}$ across a time-dependent surface $\mathcal{S}$.

1.9 Objectivity

Consider two frames of reference $\mathscr{F}$, $\mathscr{F}^*$, possibly with different origins O, O^*. An event $\{\mathbf{x}, t\}$ and its image $\{\mathbf{x}^*, t^*\}$ under a change of frame are related by rigid transformations and a time shift so that

$$\mathbf{x}^* = \mathbf{c}(t) + \mathbf{Q}(t)\mathbf{x}, \qquad t^* = t - a, \tag{1.54}$$

where $a \in \mathbb{R}$, $\mathbf{c}(t) = O - O^*$, and $\mathbf{Q}(t)$ is a rotation tensor, $\det \mathbf{Q}(t) = 1$. In connection with classical kinematics, $\mathscr{F}$ might be the relative observer and $\mathscr{F}^*$ the absolute observer.

Under a change of frame, *objective* (or frame-indifferent) scalars remain unchanged. Objective vectors change according to the transformation

$$\mathbf{v}^* = \mathbf{Q}\mathbf{v}, \tag{1.55}$$

the dependence of $\mathbf{Q}$ on time being understood. Let $\mathbf{S}$ be a second-order objective tensor, that is a linear transformation of objective vectors. Let $\mathbf{w} = \mathbf{S}\mathbf{v}$ and let $\mathbf{v}^* = \mathbf{Q}\mathbf{v}$, $\mathbf{w}^* = \mathbf{Q}\mathbf{w}$ the transforms under the change of frame whence $\mathbf{v} = \mathbf{Q}^T\mathbf{v}^*$. Then

$$\mathbf{w}^* = \mathbf{Q}\mathbf{S}\mathbf{v} = \mathbf{Q}\mathbf{S}\mathbf{Q}^T\mathbf{v}^*,$$

whence

$$\mathbf{S}^* = \mathbf{Q}\mathbf{S}\mathbf{Q}^T. \tag{1.56}$$

Objective scalars are invariant whereas objective vectors and objective tensors transform according to (1.55) and (1.56) under a change of frame.

Consider a motion of a body $\mathcal{B}$ and assume that, relative to a frame of reference $\mathscr{F}$, the motion is given by

$$\mathbf{x} = \chi(\mathbf{X}, t).$$

In another frame of reference $\mathscr{F}^*$, the same motion is described by

$$\mathbf{x}^* = \chi^*(\mathbf{X}, t^*).$$

If the change from $\mathscr{F}$ to $\mathscr{F}^*$ is described by (1.54), then we have

$$\mathbf{x}^* = \chi^*(\mathbf{X}, t^*) = \mathbf{c}(t) + \mathbf{Q}(t)\chi(\mathbf{X}, t), \qquad t^* = t - a.$$

Understanding the time dependence, we can write

$$\mathbf{x}^* = \mathbf{c} + \mathbf{Q}\mathbf{x}.$$

Time differentiation, at fixed $\mathbf{X}$, yields

$$\overline{\mathbf{x}^*}^{\cdot} = \mathbf{Q}\dot{\mathbf{x}} + \dot{\mathbf{c}} + \boldsymbol{\Omega}(\mathbf{x}^* - \mathbf{c}),$$

$$\overline{\mathbf{x}^*}^{\cdot\cdot} = \mathbf{Q}\ddot{\mathbf{x}} + \ddot{\mathbf{c}} + (\dot{\boldsymbol{\Omega}} - \boldsymbol{\Omega}^2)(\mathbf{x}^* - \mathbf{c}) + 2\boldsymbol{\Omega}(\overline{\mathbf{x}^*}^{\cdot} - \dot{\mathbf{c}}),$$

where

$$\boldsymbol{\Omega} = \dot{\mathbf{Q}}\mathbf{Q}^T \tag{1.57}$$

is a skew tensor. Only for constant $\mathbf{c}$ and $\mathbf{Q}$ are the velocity $\dot{\mathbf{x}}$ and the acceleration $\ddot{\mathbf{x}}$ objective, i.e. $\overline{\mathbf{x}^*}^{\cdot} = \mathbf{Q}\dot{\mathbf{x}}$, $\overline{\mathbf{x}^*}^{\cdot\cdot} = \mathbf{Q}\ddot{\mathbf{x}}$. Otherwise they are not and hence velocity and acceleration are regarded as frame-dependent vectors. In addition, these formulae provide the classical composition law for velocities and accelerations.

By the skew character of $\boldsymbol{\Omega}$, we can consider the axial vector $\mathbf{a}_\Omega$ such that

$$\boldsymbol{\Omega}(\mathbf{x}^* - \mathbf{c}) = \mathbf{a}_\Omega \times (\mathbf{x}^* - \mathbf{c})$$

and the like for $(\dot{\boldsymbol{\Omega}} - \boldsymbol{\Omega}^2)(\mathbf{x}^* - \mathbf{c}) + 2\boldsymbol{\Omega}(\overline{\mathbf{x}^*}^{\cdot} - \dot{\mathbf{c}})$. Now, to identify the kinematical meaning of $\mathbf{a}_\Omega$ we let $\dot{\mathbf{x}} = \mathbf{0}$ and $\mathbf{c} = \mathbf{0}$ so that

$$\overline{\mathbf{x}^*}^{\cdot} = \mathbf{a}_\Omega \times \mathbf{x}^*.$$

Hence, since $\mathbf{x}$ is co-rotating ($\dot{\mathbf{x}} = \mathbf{0}$), then $\mathbf{x}^*$ plays the role of any of the co-rotating vectors of (1.19) so that

$$\dot{\mathbf{k}}_i = \omega \times \mathbf{k}_i.$$

This allows us to identify $\mathbf{a}_\Omega$ with the angular velocity of $\mathscr{F}$ with respect to $\mathscr{F}^*$.

We then let

$$\overline{\mathbf{x}^*}^{\cdot} = \mathbf{Q}\dot{\mathbf{x}} + \dot{\mathbf{c}} + \omega \times (\mathbf{x}^* - \mathbf{c}),$$

$$\overline{\mathbf{x}^*}^{\cdot\cdot} = \mathbf{Q}\ddot{\mathbf{x}} + \ddot{\mathbf{c}} + \dot{\omega}(\mathbf{x}^* - \mathbf{c}) - \omega \times [\omega \times (\mathbf{x}^* - \mathbf{c})] + 2\omega \times (\overline{\mathbf{x}^*}^{\cdot} - \dot{\mathbf{c}}),$$

and regard $\dot{\mathbf{c}} + \omega \times (\mathbf{x}^* - \mathbf{c})$ as the drag velocity, $\ddot{\mathbf{c}} + \dot{\omega} \times (\mathbf{x}^* - \mathbf{c}) - \omega \times [\omega \times (\mathbf{x}^* - \mathbf{c})]$ as the drag acceleration, and $2\omega \times (\overline{\mathbf{x}^*}^{\cdot} - \dot{\mathbf{c}})$ as the Coriolis acceleration.

To complete the kinematic properties under changes of frame, we now investigate the transformation of the angular velocity. Let $\mathbf{Q}$ be the orthogonal tensor associated with the change of reference $\mathscr{F} \to \mathscr{F}^*$; the vectors $\{\mathbf{k}_i\}$ in $\mathscr{F}$ appear as $\{\mathbf{Q}\mathbf{k}_i\}$ in $\mathscr{F}^*$. Let

$$\mathbf{k}_i \to \mathbf{Q}\mathbf{k}_i, \qquad \mathbf{Q}\mathbf{Q}^T = \mathbf{1},$$

and denote by ω^* the angular velocity of $\{\mathbf{Q}\mathbf{k}_i\}$ relative to $\mathscr{F}^*$. Hence, by definition (1.18)

$$\omega^* = \tfrac{1}{2}\textstyle\sum_i \mathbf{Q}\mathbf{k}_i \times \overline{\mathbf{Q}\mathbf{k}_i}^{\cdot} = \tfrac{1}{2}\sum_i \mathbf{Q}\mathbf{k}_i \times \mathbf{Q}\dot{\mathbf{k}}_i + \tfrac{1}{2}\sum_i \mathbf{Q}\mathbf{k}_i \times \dot{\mathbf{Q}}\mathbf{k}_i.$$

Observe that, for any pair of vectors $\mathbf{u}$, $\mathbf{v}$, the i-component relative to $\{\mathbf{e}_j\}$ of $\mathbf{Q}\mathbf{u} \times \mathbf{Q}\mathbf{v}$ is given by

$$[\mathbf{Qu} \times \mathbf{Qv}]_i = \epsilon_{ipq} Q_{ph} u_h Q_{qk} v_k.$$

Since

$$\epsilon_{ipq} Q_{ph} Q_{qk} = \epsilon_{jhk} Q_{ij}.$$

then

$$[\mathbf{Qu} \times \mathbf{Qv}]_i = Q_{ij} \epsilon_{jhk} u_h v_k = [\mathbf{Q}(\mathbf{u} \times \mathbf{v})]_i.$$

Consequently

$$\tfrac{1}{2} \textstyle\sum_i \mathbf{Qk}_i \times \mathbf{Q}\dot{\mathbf{k}}_i = \mathbf{Q} \tfrac{1}{2} \textstyle\sum_i \mathbf{k}_i \times \dot{\mathbf{k}}_i = \mathbf{Q}\boldsymbol{\omega}.$$

Now observe that

$$\tfrac{1}{2} \textstyle\sum_i \mathbf{Qk}_i \times \dot{\mathbf{Q}}\mathbf{k}_i = \tfrac{1}{2} \textstyle\sum_i \mathbf{Q} R_{ip} \mathbf{e}_p \times \dot{\mathbf{Q}} R_{iq} \mathbf{e}_q = \tfrac{1}{2} \textstyle\sum_p \mathbf{Qe}_p \times \dot{\mathbf{Q}}\mathbf{e}_p$$

and

$$(\mathbf{Qe}_p)_h = Q_{hp}.$$

Hence, we have

$$\tfrac{1}{2} \textstyle\sum_i \mathbf{Qk}_i \times \dot{\mathbf{Q}}\mathbf{k}_i = \mathbf{e}_l \tfrac{1}{2} \epsilon_{lhj} Q_{hp} \dot{Q}_{jp} = \mathbf{e}_l \tfrac{1}{2} \epsilon_{lhj} \Omega_{jh} = \mathbf{a}_\Omega,$$

$\mathbf{a}_\Omega$ being the axial vector of $\boldsymbol{\Omega} = \dot{\mathbf{Q}}\mathbf{Q}^T$. Thus, under the change of frame $\mathscr{F} \to \mathscr{F}^*$ associated with the orthogonal tensor $\mathbf{Q}$ the angular velocity obeys the transformation law

$$\boldsymbol{\omega}^* = \mathbf{Q}\boldsymbol{\omega} + \mathbf{a}_\Omega. \tag{1.58}$$

As with the velocity, by (1.58) we can say that the absolute angular velocity $\boldsymbol{\omega}^*$ is the sum of ($\mathbf{Q}$ applied to) the relative angular velocity, $\boldsymbol{\omega}$, and the drag angular velocity, $\mathbf{a}_\Omega$.

1.9.1 Transformation Rules for Kinematic Fields

For formal simplicity, we let $t^* = t$ and hence the motion of $\mathbf{X}$ in $\mathscr{F}^*$ is related to that in $\mathscr{F}$ by

$$\chi^*(\mathbf{X}, t) = \mathbf{c}(t) + \mathbf{Q}(t)\chi(\mathbf{X}, t). \tag{1.59}$$

The gradient with respect to $\mathbf{X}$ yields the transformation law

$$\mathbf{F}^*(\mathbf{X}, t) = \mathbf{Q}(t)\mathbf{F}(\mathbf{X}, t)$$

for the deformation gradient. Time differentiating these relations, we find

$$\overline{\chi^*}^{\boldsymbol{\cdot}}(\mathbf{X}, t) = \dot{\mathbf{c}}(t) + \mathbf{Q}(t)\dot{\chi}(\mathbf{X}, t) + \dot{\mathbf{Q}}(t)\chi(\mathbf{X}, t), \tag{1.60}$$

$$\overline{\mathbf{F}^*}^{\boldsymbol{\cdot}}(\mathbf{X}, t) = \mathbf{Q}(t)\dot{\mathbf{F}}(\mathbf{X}, t) + \dot{\mathbf{Q}}(t)\mathbf{F}(\mathbf{X}, t).$$

Hence, $\mathbf{v} = \dot{\chi}$ and $\dot{\mathbf{F}}$ are not objective, whereas $\mathbf{F}$ transforms as a (objective) vector, not as a tensor.

Time differentiation of fields $\phi(\mathbf{x}, t)$ in the Eulerian, or spatial, description is more involved because $\mathbf{x}$ is no longer invariant while $\mathbf{X}$ is. Consider the velocity field with $\mathbf{v}(\mathbf{x}, t)$, $\mathbf{v}^*(\mathbf{x}^*, t)$, in the two frames $\mathscr{F}$, $\mathscr{F}^*$, with $\mathbf{x}, \mathbf{x}^*$ corresponding to the same material point $\mathbf{X}$, namely $\mathbf{x} = \chi(\mathbf{X}, t)$, $\mathbf{x}^* = \chi^*(\mathbf{X}, t)$. Hence, by (1.59), we have

$$\partial_{\mathbf{x}}\mathbf{x}^* = \mathbf{Q}$$

and, by (1.60), we can write

$$\mathbf{v}^*(\mathbf{x}^*, t) = \mathbf{Q}(t)\mathbf{v}(\mathbf{x}, t) + \dot{\mathbf{c}}(t) + \dot{\mathbf{Q}}(t)\mathbf{x}. \tag{1.61}$$

We compute the velocity gradient in $\mathscr{F}^*$, i.e. $\mathbf{L}^* = \partial_{\mathbf{x}^*}\mathbf{v}^*(\mathbf{x}^*, t)$. By the chain rule, we have

$$\partial_{\mathbf{x}}\mathbf{v}^*(\mathbf{x}^*, t) = \partial_{\mathbf{x}^*}\mathbf{v}^*(\mathbf{x}^*, t)\partial_{\mathbf{x}}\mathbf{x}^* = \mathbf{Q}\partial_{\mathbf{x}}\mathbf{v}(\mathbf{x}, t) + \dot{\mathbf{Q}},$$

whence

$$\mathbf{L}^*\mathbf{Q} = [(\mathbf{Q}\mathbf{L}\mathbf{Q}^T) + \mathbf{\Omega}]\mathbf{Q}.$$

It follows that the velocity gradient transforms according to

$$\mathbf{L}^*(\mathbf{x}^*, t) = \mathbf{Q}(t)\mathbf{L}(\mathbf{x}, t)\mathbf{Q}^T(t) + \mathbf{\Omega}(t). \tag{1.62}$$

Replace $\mathbf{L}$ with $\mathbf{D} + \mathbf{W}$, $\mathbf{D} = \mathrm{sym}\mathbf{L}$, $\mathbf{W} = \mathrm{skw}\mathbf{L}$ to obtain

$$\mathbf{Q}\mathbf{L}\mathbf{Q}^T = \mathbf{Q}\mathbf{D}\mathbf{Q}^T + \mathbf{Q}\mathbf{W}\mathbf{Q}^T,$$

and

$$(\mathbf{Q}\mathbf{D}\mathbf{Q}^T)^T = \mathbf{Q}\mathbf{D}\mathbf{Q}^T, \qquad (\mathbf{Q}\mathbf{W}\mathbf{Q}^T)^T = -\mathbf{Q}\mathbf{W}\mathbf{Q}^T.$$

Hence, the symmetric and the skew parts of (1.62), along with the obvious definitions $\mathbf{D}^* = \mathrm{sym}\mathbf{L}^*$, $\mathbf{W}^* = \mathrm{skw}\mathbf{L}^*$, result in

$$\mathbf{D}^*(\mathbf{x}^*, t) = \mathbf{Q}(t)\mathbf{D}(\mathbf{x}, t)\mathbf{Q}^T(t),$$

$$\mathbf{W}^*(\mathbf{x}^*, t) = \mathbf{Q}(t)\mathbf{W}(\mathbf{x}, t)\mathbf{Q}^T(t) + \mathbf{\Omega}(t).$$

Thus, $\mathbf{D}$ is objective, whereas $\mathbf{L}$ and $\mathbf{W}$ are not.

By the transformation law $\mathbf{F}^* = \mathbf{QF}$, we derive easily the transformation laws for the stretch fields $\mathbf{U}$ and $\mathbf{C}$ and the strain field $\mathbf{E} = (\mathbf{C} - \mathbf{1})/2$. Since $\mathbf{C} = \mathbf{U}^2 = \mathbf{F}^T\mathbf{F}$ we have

$$\mathbf{C}^* = (\mathbf{QF})^T\mathbf{QF} = \mathbf{F}^T\mathbf{F} = \mathbf{C}.$$

Consequently, $\mathbf{C}$ an hence also $\mathbf{U}$ and $\mathbf{E}$ are invariant. Instead the left Cauchy–Green tensor $\mathbf{B} = \mathbf{FF}^T$ satisfies

$$\mathbf{B}^* = (\mathbf{QF})(\mathbf{QF})^T = \mathbf{QFF}^T\mathbf{Q}^T = \mathbf{QBQ}^T$$

and hence $\mathbf{B}$ is a (objective) tensor.

In view of the polar decomposition $\mathbf{F} = \mathbf{RU} = \mathbf{VR}$ and of the transformation law $\mathbf{F}^* = \mathbf{QF}$, we have

$$\mathbf{R}^*\mathbf{U} = \mathbf{QRU}, \qquad \mathbf{V}^*\mathbf{R}^* = \mathbf{QVR}$$

whence

$$\mathbf{R}^* = \mathbf{QR}, \qquad \mathbf{V}^* = \mathbf{QVQ}^T.$$

To sum up, we say that

$$\mathbf{U}^* = \mathbf{U}, \quad \mathbf{C}^* = \mathbf{C}, \quad \mathbf{E}^* = \mathbf{E},$$

$$\mathbf{V}^* = \mathbf{QVQ}^T, \quad \mathbf{B}^* = \mathbf{QBQ}^T, \quad \mathbf{D}^* = \mathbf{QDQ}^T,$$

$$\mathbf{F}^* = \mathbf{QF}, \quad \mathbf{R}^* = \mathbf{QR}, \quad \mathbf{L}^* = \mathbf{QLQ}^T + \boldsymbol{\Omega}, \quad \mathbf{W}^* = \mathbf{QWQ}^T + \boldsymbol{\Omega}, \tag{1.63}$$

Hence, $\mathbf{U}$, $\mathbf{C}$, and $\mathbf{E}$ are invariant, $\mathbf{V}$, $\mathbf{B}$, and $\mathbf{D}$ are objective (tensors), $\mathbf{F}$ and $\mathbf{R}$ are objective as vectors whereas $\mathbf{L}$ and $\mathbf{W}$ are not objective.

Velocity Differences

In view of (1.61), we have

$$\mathbf{v}^*(\mathbf{x}^*, t) = \mathbf{Q}(t)\mathbf{v}(\mathbf{x}, t) + \dot{\mathbf{c}}(t) + \boldsymbol{\Omega}(t)(\mathbf{x}^* - \mathbf{c}).$$

If $\mathbf{x} = \chi(\mathbf{X}, t)$ and $\mathbf{y} = \chi(\mathbf{Y}, t)$, we obtain

$$\mathbf{v}^*(\mathbf{X}, t) = \mathbf{Q}(t)\mathbf{v}(\mathbf{X}, t) + \dot{\mathbf{c}}(t) + \boldsymbol{\Omega}(t)[\mathbf{x}^*(\mathbf{X}, t) - \mathbf{c}],$$

$$\mathbf{v}^*(\mathbf{Y}, t) = \mathbf{Q}(t)\mathbf{v}(\mathbf{Y}, t) + \dot{\mathbf{c}}(t) + \boldsymbol{\Omega}(t)[\mathbf{x}^*(\mathbf{Y}, t) - \mathbf{c}(t)].$$

Subtracting the two equations, we find

$$\mathbf{v}^*(\mathbf{X}, t) - \mathbf{v}^*(\mathbf{Y}, t) = \mathbf{Q}(t)[\mathbf{v}(\mathbf{X}, t) - \mathbf{v}(\mathbf{Y}, t)] + \boldsymbol{\Omega}(t)[\chi^*(\mathbf{X}, t) - \chi^*(\mathbf{Y}, t)].$$

Hence, velocity differences are not objective unless $\mathbf{X} = \mathbf{Y}$ in which case this relation degenerates. Yet, in mixtures, different particles (of different constituents) may occupy the same position (see Chap. 11). In that case, we may have

$$\mathbf{x}_\alpha = \chi_\alpha(\mathbf{X}, t), \qquad \mathbf{x}_\beta = \chi_\beta(\mathbf{Y}, t), \quad \text{subject to } \mathbf{x}_\alpha(t) = \mathbf{x}_\beta(t).$$

Since $\mathbf{x}_\alpha = \mathbf{x}_\beta \implies \mathbf{x}^*_\alpha = \mathbf{x}^*_\beta$ then

$$\mathbf{v}^*_\alpha(\mathbf{X}, t) - \mathbf{v}^*_\beta(\mathbf{Y}, t) = \mathbf{Q}(t)[\mathbf{v}_\alpha(\mathbf{X}, t) - \mathbf{v}_\beta(\mathbf{Y}, t)].$$

Velocity differences of particles in the same position are objective.

1.9.2 Time Derivatives Relative to Generic Frames of Reference

For any function

$$\phi(\mathbf{x}, t) = \phi(\chi(\mathbf{X}, t), t) =: \Phi(\mathbf{X}, t)$$

the derivatives

$$\partial_t \phi \quad \text{and} \quad \dot{\phi} = \partial_t \Phi$$

denote the Eulerian (or local or spatial) time derivative and the Lagrangian (or material) time derivative. Hence $\partial_t \phi$ represents the derivative of ϕ at a given point in space whereas $\dot{\phi}$ represents the derivative of ϕ holding the material point fixed; $\dot{\phi}$ is the derivative relative to the observer moving with the velocity of the material point. We now examine further time derivatives of interest in continuum physics.

As in Sect. 1.4, consider a conventionally fixed frame of reference associated with orthonormal base vectors $(\mathbf{e}_1, \mathbf{e}_2, \mathbf{e}_3)$ and a rotating frame of reference associated with orthonormal base vectors $(\mathbf{k}_1, \mathbf{k}_2, \mathbf{k}_3)$. Consistent with (1.54), we let $\mathbf{k}_i = \mathbf{Q}\mathbf{e}_i$. Hence, it follows $\dot{\mathbf{k}}_i = \dot{\mathbf{Q}}\mathbf{e}_i = \dot{\mathbf{Q}}\mathbf{Q}^T\mathbf{k}_i$, whence

$$\dot{\mathbf{k}}_i = \boldsymbol{\omega} \times \mathbf{k}_i,$$

where $\boldsymbol{\omega}$ is defined by (1.18) and equals the axial vector $\mathbf{a}_\Omega$ of $\boldsymbol{\Omega} = \dot{\mathbf{Q}}\mathbf{Q}^T$. Hence for any vector function $\mathbf{u}(\mathbf{X}, t)$ we have

$$\dot{\mathbf{u}} = \overline{u_i \mathbf{k}_i}^{\boldsymbol{\cdot}} = \dot{u}_i \mathbf{k}_i + u_i \dot{\mathbf{k}}_i = \dot{u}_i \mathbf{k}_i + u_i \boldsymbol{\omega} \times \mathbf{k}_i = \dot{u}_i \mathbf{k}_i + \boldsymbol{\omega} \times \mathbf{u}$$

Consequently, $\dot{u}_i \mathbf{k}_i$, namely the derivative of $\mathbf{u}$ with respect to the frame at rest with $\{\mathbf{k}_j\}$, is given by

$$\dot{u}_i \mathbf{k}_i = \dot{\mathbf{u}} - \boldsymbol{\omega} \times \mathbf{u} = \dot{\mathbf{u}} - \boldsymbol{\Omega}\mathbf{u}.$$

Likewise, we determine the derivative of a tensor field. Let

$$A_{ij} := \mathbf{k}_i \cdot \mathbf{A}\mathbf{k}_j$$

be the ij-component of the tensor $\mathbf{A}$ relative to the basis $\{\mathbf{k}_i\}$. A direct differentiation, subject to $\dot{\mathbf{k}}_i = \boldsymbol{\Omega}\mathbf{k}_i$, implies that

$$\dot{A}_{ij} = \mathbf{k}_i \cdot \dot{\mathbf{A}}\mathbf{k}_j + \boldsymbol{\Omega}\mathbf{k}_i \cdot \mathbf{A}\mathbf{k}_j + \mathbf{k}_i \cdot \mathbf{A}\boldsymbol{\Omega}\mathbf{k}_j.$$

Now,

$$\boldsymbol{\Omega}\mathbf{k}_i \cdot \mathbf{A}\mathbf{k}_j = \mathbf{k}_i\boldsymbol{\Omega}^T \cdot \mathbf{A}\mathbf{k}_j = \mathbf{k}_i \cdot \boldsymbol{\Omega}^T\mathbf{A}\mathbf{k}_j = (\boldsymbol{\Omega}^T\mathbf{A})_{ij}$$

and

$$\mathbf{k}_i \cdot \mathbf{A}\boldsymbol{\Omega}\mathbf{k}_j = \mathbf{k}_i \cdot (\mathbf{A}\boldsymbol{\Omega})\mathbf{k}_j = (\mathbf{A}\boldsymbol{\Omega})_{ij}.$$

Hence

$$\dot{A}_{ij}\mathbf{k}_i \otimes \mathbf{k}_j = \dot{\mathbf{A}} - \boldsymbol{\Omega}\mathbf{A} - \mathbf{A}\boldsymbol{\Omega}^T$$

is the rate of change of $\mathbf{A}$ with respect to the frame which is spinning along with the basis $\{\mathbf{k}_i\}$.

A particular derivative relative to the set of axes spinning with a suitable tensor $\boldsymbol{\Omega} = \dot{\mathbf{Q}}\mathbf{Q}^T$ arises by choosing $\mathbf{Q} = \mathbf{R}$, $\mathbf{R}$ being the rotation tensor given by the polar decomposition. Letting

$$\boldsymbol{\Omega} = \mathbf{Z} := \dot{\mathbf{R}}\mathbf{R}^T$$

we denote the derivative by a superposed $\odot$ and write

$$\overset{\odot}{\mathbf{u}} = \dot{\mathbf{u}} - \mathbf{Z}\mathbf{u}, \qquad \overset{\odot}{\mathbf{A}} = \dot{\mathbf{A}} - \mathbf{Z}\,\mathbf{A} - \mathbf{A}\mathbf{Z}^T.$$

Owing to [207], this derivative is named after Green and Naghdi.

In a conceptually analogous way, we now look for the time derivative relative to non-orthonormal bases. Let $\mathbf{X}$ be referred to coordinates $\{X^i\}$ and let

$$\mathbf{g}_i = \partial_{X^i}\boldsymbol{\chi}(\mathbf{X}, t).$$

Hence

$$\dot{\mathbf{g}}_i = \partial_t\,\partial_{X^i}\boldsymbol{\chi} = \partial_{X^i}\,\partial_t\boldsymbol{\chi} = \partial_{X^i}\mathbf{v} = \partial_{x_h}\mathbf{v}\partial_{X^i}\chi_h = (\partial_{X^i}\chi_h\partial_{x_h})\mathbf{v} = (\partial_{X^i}\boldsymbol{\chi} \cdot \nabla)\mathbf{v} = \mathbf{L}\,\mathbf{g}_i.$$

To compute $\dot{\mathbf{g}}^i$, we observe that by

$$\mathbf{g}^k \cdot \mathbf{g}_i = \delta_{ik}$$

it follows

$$\dot{\mathbf{g}}^k \cdot \mathbf{g}_i = -\mathbf{g}^k \cdot \dot{\mathbf{g}}_i.$$

Multiplication by $\mathbf{g}^i$ and summation over i results in

$$\dot{\mathbf{g}}^k = -(\mathbf{g}^k \cdot \dot{\mathbf{g}}_i)\mathbf{g}^i = -(\mathbf{g}^k \cdot \mathbf{L})\mathbf{g}_i)\mathbf{g}^i = -(\mathbf{g}_i \cdot \mathbf{L}^T\mathbf{g}^k)\mathbf{g}^i = -\mathbf{L}^T\mathbf{g}^k.$$

Accordingly, covariant and contravariant vectors satisfy

$$\dot{\mathbf{g}}_i = \mathbf{L}\mathbf{g}_i, \qquad \dot{\mathbf{g}}^i = -\mathbf{L}^T\mathbf{g}^i.$$

As an application, we now look for the time derivative of covariant and contravariant components of a tensor. By

$$A_{ij} = \mathbf{g}_i \cdot \mathbf{A}\mathbf{g}_j, \qquad \dot{\mathbf{g}}_i = \mathbf{L}\mathbf{g}_i$$

we have

$$\dot{A}_{ij} = \dot{\mathbf{g}}_i \cdot \mathbf{A}\mathbf{g}_j + \mathbf{g}_i \cdot \dot{\mathbf{A}}\mathbf{g}_j + \mathbf{g}_i \cdot \mathbf{A}\dot{\mathbf{g}}_j = \mathbf{g}_i \cdot (\mathbf{L}^T\mathbf{A} + \dot{\mathbf{A}} + \mathbf{A}\mathbf{L})\mathbf{g}_j.$$

We define the *covariant* or *lower convected* derivative of $\mathbf{A}$ as

$$\overset{\triangle}{\mathbf{A}} := \mathbf{g}_i\, \dot{A}_{ij}\, \mathbf{g}_j$$

and hence we have

$$\overset{\triangle}{\mathbf{A}} = \dot{\mathbf{A}} + \mathbf{L}^T\mathbf{A} + \mathbf{A}\mathbf{L}.$$

The covariant derivative $\overset{\triangle}{\mathbf{A}}$ is also referred to as *Cotter–Rivlin rate* [108].

Likewise we consider the contravariant components

$$A^{ij} = \mathbf{g}^i \cdot \mathbf{A}\mathbf{g}^j, \qquad \dot{\mathbf{g}}^i = -\mathbf{L}^T\mathbf{g}^i.$$

Hence, we have

$$\dot{A}^{ij} = \dot{\mathbf{g}}^i \cdot \mathbf{A}\mathbf{g}^j + \mathbf{g}^i \cdot \dot{\mathbf{A}}\mathbf{g}^j + \mathbf{g}^i \cdot \mathbf{A}\dot{\mathbf{g}}^j = \mathbf{g}^i \cdot (-\mathbf{L}\mathbf{A} + \dot{\mathbf{A}} - \mathbf{A}\mathbf{L}^T)\mathbf{g}^j.$$

Upon defining the *contravariant* or *upper convected* derivative[14] of $\mathbf{A}$ as

$$\overset{\triangledown}{\mathbf{A}} := \dot{A}^{ij}\mathbf{g}_i \otimes \mathbf{g}_j$$

we have

$$\overset{\triangledown}{\mathbf{A}} = \dot{\mathbf{A}} - \mathbf{L}\mathbf{A} - \mathbf{A}\mathbf{L}^T.$$

---

[14] Or Lie derivative.

It is of interest that the mean of the two convected derivatives is just the *Jaumann–Zaremba derivative* $\overset{\circ}{\mathbf{A}}$ ([428], p. 36),

$$\tfrac{1}{2}(\overset{\vartriangle}{\mathbf{A}} + \overset{\triangledown}{\mathbf{A}}) = \dot{\mathbf{A}} + \mathbf{W}^T\mathbf{A} + \mathbf{A}\mathbf{W} =: \overset{\circ}{\mathbf{A}} \,.$$

Moreover, it follows that

$$\overset{\vartriangle}{\mathbf{A}} - \overset{\triangledown}{\mathbf{A}} = 2\,\mathbf{D}\mathbf{A} - 2\mathbf{A}\,\mathbf{D}.$$

Within the models of fluids, the Oldroyd-B model is taken to describe the flow of viscoelastic fluids and involves the upper convected derivative. The model traces back to Oldroyd [348] and that is why the convected derivatives are named after Oldroyd.

Further, a variant of the Oldroyd derivative is that of the (upper convected) Oldroyd derivative of a vector density and of a tensor density. They are denoted and defined as

$$\overset{\square}{\mathbf{u}} = \dot{\mathbf{u}} - \mathbf{L}\mathbf{u} + (\nabla\cdot\mathbf{v})\mathbf{u}, \qquad \overset{\square}{\mathbf{A}} = \dot{\mathbf{A}} - \mathbf{L}\mathbf{A} - \mathbf{A}\mathbf{L}^T + (\nabla\cdot\mathbf{v})\mathbf{A}.$$

This derivative is referred to as the *Truesdell rate* of vectors and tensors [424].

In a more direct way, we can obtain time derivatives by considering the following problem. Let $(\mathbf{e}_1, \mathbf{e}_2, \mathbf{e}_3)$ be a conventionally fixed system of orthonormal base vectors and denote by $(\mathbf{k}_1, \mathbf{k}_2, \mathbf{k}_3)$ a system of base vectors. The two bases are taken to be related by[15]

$$\mathbf{k}_i(t) = \mathbf{K}(t)\mathbf{e}_i, \;\; i = 1, 2, 3,$$

the dependence of $\mathbf{K}$ on the point being understood and not written. Two cases are of interest, namely

$$\mathbf{K} = \mathbf{F}, \qquad \mathbf{K} = J^{-1}\mathbf{F}.$$

If $\mathbf{K} = \mathbf{F}$ then the vectors $\{\mathbf{k}_i\}$ evolve as the space vectors in a deformation and they are said to convect as a tangent. If $\mathbf{K} = J^{-1}\mathbf{F}$ then $\det\mathbf{K} = 1$ and the vectors $\{\mathbf{k}_i\}$ are a system of unitary base vectors. In both cases, the basis $\{\mathbf{k}_i\}$ need not be orthonormal. Since $\dot{\mathbf{F}} = \mathbf{L}\mathbf{F}$ and $\dot{J} = J\nabla\cdot\mathbf{v}$, if $\mathbf{k}_i = \mathbf{F}\mathbf{e}_i$ then

$$\dot{\mathbf{k}}_i = \dot{\mathbf{F}}\mathbf{e}_i = \mathbf{L}\mathbf{F}\mathbf{e}_i = \mathbf{L}\mathbf{k}_i$$

while, if $\mathbf{K} = J^{-1}\mathbf{F}$,

$$\dot{\mathbf{k}}_i = (J^{-1}\dot{\mathbf{F}} - \dot{J}J^{-2}\dot{\mathbf{F}})\mathbf{e}_i = (\mathbf{L}J^{-1}\mathbf{F} - \nabla\cdot\mathbf{v}J^{-1}\mathbf{F})\mathbf{e}_i = (\mathbf{L} - (\nabla\cdot\mathbf{v})\mathbf{1})\mathbf{k}_i.$$

Let

$$\mathbf{M} = \mathbf{L} \qquad \text{or} \qquad \mathbf{M} = \mathbf{L} - (\nabla\cdot\mathbf{v})\mathbf{1}$$

---

[15] Here we apply again the active transformation in that the base-transformation tensor $\mathbf{K}$ maps the initial vector into the corresponding final vector.

as appropriate. Hence

$$\dot{\mathbf{k}}_i = \mathbf{M}\,\mathbf{k}_i.$$

Let $\{\mathbf{k}^i\}$ be the corresponding dual basis so that

$$\mathbf{k}_i \cdot \mathbf{k}^j = \delta_{ij}.$$

Hence

$$0 = \overline{\mathbf{k}_i \cdot \mathbf{k}^j}^{\bullet} = \mathbf{k}_i \cdot \dot{\mathbf{k}}^j + \dot{\mathbf{k}}_i \cdot \mathbf{k}^j = \mathbf{k}_i \cdot \dot{\mathbf{k}}^j + \mathbf{M}\,\mathbf{k}_i \cdot \mathbf{k}^j = \mathbf{k}_i \cdot (\dot{\mathbf{k}}^j + \mathbf{M}^T\,\mathbf{k}^j).$$

Since $\{\mathbf{k}_i\}$ is a basis, then it follows

$$\dot{\mathbf{k}}^j = -\mathbf{M}^T\,\mathbf{k}^j.$$

Consider the covariant components $\{u_i\}$ of a vector $\mathbf{u}$, so that $\mathbf{u} = u_i\mathbf{k}^i$. Represent the vector $\mathbf{u}$ in the dual basis, $\mathbf{u} = u_i\mathbf{k}^i$. Hence

$$\dot{\mathbf{u}} = \dot{u}_i\mathbf{k}^i + u_i\dot{\mathbf{k}}^i = \dot{u}_i\mathbf{k}^i - u_i\mathbf{M}^T\mathbf{k}^i = \dot{u}_i\mathbf{k}^i - \mathbf{M}^T\mathbf{u}.$$

Now $\dot{u}_i\mathbf{k}^i$ is the derivative of $\mathbf{u}$ relative to the dual basis $\{\mathbf{k}^i\}$ and we have

$$\dot{u}_i\mathbf{k}^i = \dot{\mathbf{u}} + \mathbf{M}^T\mathbf{u}.$$

Likewise, we compute the derivative of a tensor field,

$$\mathbf{A} = A_{ij}\mathbf{k}^i \otimes \mathbf{k}^j, \qquad A_{ij} = \mathbf{k}_i \cdot \mathbf{A}\mathbf{k}_j.$$

Time differentiation of A_{ij} gives

$$\begin{aligned}\dot{A}_{ij} = \dot{\mathbf{k}}_i \cdot \mathbf{A}\mathbf{k}_j + \mathbf{k}_i \cdot \dot{\mathbf{A}}\mathbf{k}_j + \mathbf{k}_i \cdot \mathbf{A}\dot{\mathbf{k}}_j &= \mathbf{M}\mathbf{k}_i \cdot \mathbf{A}\mathbf{k}_j + \mathbf{k}_i \cdot \mathbf{M}\mathbf{A}\mathbf{k}_j + \mathbf{k}_i \cdot \mathbf{A}\mathbf{M}\mathbf{k}_j \\ &= \mathbf{k}_i \cdot (\dot{\mathbf{A}} + \mathbf{M}^T\mathbf{A} + \mathbf{A}\mathbf{M})\mathbf{k}_j\end{aligned}$$

whence

$$\dot{A}_{ij}\mathbf{k}^i \otimes \mathbf{k}^j = \dot{\mathbf{A}} + \mathbf{M}^T\mathbf{A} + \mathbf{A}\mathbf{M}.$$

If instead we consider the contravariant components,

$$u^i = \mathbf{u} \cdot \mathbf{k}^i, \qquad A^{ij} = \mathbf{k}^i \cdot \mathbf{A}\mathbf{k}^j$$

then we have

$$\dot{u}^i = \dot{\mathbf{u}} \cdot \mathbf{k}^i + \mathbf{u} \cdot \dot{\mathbf{k}}^i = \dot{\mathbf{u}} \cdot \mathbf{k}^i - \mathbf{u} \cdot \mathbf{M}^T\mathbf{k}^i = (\dot{\mathbf{u}} - \mathbf{M}\mathbf{u}) \cdot \mathbf{k}^i,$$

$$\dot{A}^{ij} = \dot{\mathbf{k}}^i \cdot \mathbf{A}\mathbf{k}^j + \mathbf{k}^i \cdot \dot{\mathbf{A}}\mathbf{k}^j + \mathbf{k}^i \cdot \mathbf{A}\dot{\mathbf{k}}^j = -\mathbf{M}^T\mathbf{k}^i \cdot \mathbf{A}\mathbf{k}^j + \mathbf{k}^i \cdot \dot{\mathbf{A}}\mathbf{k}^j - \mathbf{k}^i \cdot \mathbf{A}\mathbf{M}^T\mathbf{k}^j$$
$$= \mathbf{k}^i \cdot (\dot{\mathbf{A}} - \mathbf{M}\mathbf{A} - \mathbf{A}\mathbf{M}^T)\mathbf{k}^j,$$

whence

$$\dot{w}^i\mathbf{k}_i = \dot{\mathbf{w}} - \mathbf{M}\mathbf{w}, \qquad \dot{A}^{ij}\mathbf{k}_i \otimes \mathbf{k}_j = \dot{\mathbf{A}} - \mathbf{M}\mathbf{A} - \mathbf{A}\mathbf{M}^T.$$

Letting $\mathbf{M}$ be replaced by $\mathbf{L}$ or $\mathbf{L} - (\nabla \cdot \mathbf{v})\mathbf{1}$ we obtain the rate of vectors and tensors relative to the bases $\{\mathbf{k}_i\}$ and $\{\mathbf{k}^i\}$,

$$\dot{u}_i\mathbf{k}^i = \dot{\mathbf{u}} + \mathbf{L}^T\mathbf{u}, \qquad \dot{A}_{ij}\mathbf{k}^i \otimes \mathbf{k}^j = \dot{\mathbf{A}} + \mathbf{L}^T\mathbf{A} + \mathbf{A}\mathbf{L},$$

$$\dot{u}^i\mathbf{k}_i = \dot{\mathbf{u}} - \mathbf{L}\mathbf{u}, \qquad \dot{A}^{ij}\mathbf{k}_i \otimes \mathbf{k}_j = \dot{\mathbf{A}} - \mathbf{L}\mathbf{A} - \mathbf{A}\mathbf{L}^T$$

or

$$\dot{u}_i\mathbf{k}^i = \dot{\mathbf{u}} + \mathbf{L}^T\mathbf{u} - (\nabla \cdot \mathbf{v})\mathbf{u}, \qquad \dot{A}^{ij}\mathbf{k}_i \otimes \mathbf{k}_j = \dot{\mathbf{A}} + \mathbf{L}^T\mathbf{A} + \mathbf{A}\mathbf{L} - 2(\nabla \cdot \mathbf{v})\mathbf{A},$$

$$\dot{u}^i\mathbf{k}_i = \dot{\mathbf{u}} - \mathbf{L}\mathbf{u} + (\nabla \cdot \mathbf{v})\mathbf{u}, \qquad \dot{A}^{ij}\mathbf{k}_i \otimes \mathbf{k}_j = \dot{\mathbf{A}} - \mathbf{L}\mathbf{A} - \mathbf{A}\mathbf{L}^T + 2(\nabla \cdot \mathbf{v})\mathbf{A}.$$

To sum up, we now give a list of rates of vectors ($\mathbf{u}$) and tensors ($\mathbf{A}$).

	Rate of vectors	Rate of tensors
Green–Naghdi rate	$\overset{\odot}{\mathbf{u}} = \dot{\mathbf{u}} - \mathbf{Z}\mathbf{u}$	$\overset{\odot}{\mathbf{A}} = \dot{\mathbf{A}} - \mathbf{Z}\mathbf{A} - \mathbf{A}\mathbf{Z}^T$
Cotter–Rivlin rate	$\overset{\triangle}{\mathbf{u}} = \dot{\mathbf{u}} + \mathbf{L}^T\mathbf{u}$	$\overset{\triangle}{\mathbf{A}} = \dot{\mathbf{A}} + \mathbf{L}^T\mathbf{A} + \mathbf{A}\mathbf{L}$
Oldroyd rate	$\overset{\triangledown}{\mathbf{u}} = \dot{\mathbf{u}} - \mathbf{L}\mathbf{u}$	$\overset{\triangledown}{\mathbf{A}} = \dot{\mathbf{A}} - \mathbf{L}\mathbf{A} - \mathbf{A}\mathbf{L}^T$
Truesdell rate	$\overset{\square}{\mathbf{u}} = \dot{\mathbf{u}} - \mathbf{L}\mathbf{u} + (\nabla \cdot \mathbf{v})\mathbf{u}$	$\overset{\square}{\mathbf{A}} = \dot{\mathbf{A}} - \mathbf{L}\mathbf{A} - \mathbf{A}\mathbf{L}^T + (\nabla \cdot \mathbf{v})\mathbf{A}$
$\mathbf{F}$-rate	$\overset{\triangleright}{\mathbf{u}} = \dot{\mathbf{u}} + \mathbf{L}^T\mathbf{u} - (\nabla \cdot \mathbf{v})\mathbf{u}$	$\overset{\triangleright}{\mathbf{A}} = \dot{\mathbf{A}} + \mathbf{L}^T\mathbf{A} + \mathbf{A}\mathbf{L} - 2(\nabla \cdot \mathbf{v})\mathbf{A}$
$J^{-1}\mathbf{F}$-rate	$\overset{\triangleleft}{\mathbf{u}} = \dot{\mathbf{u}} - \mathbf{L}\mathbf{u} + (\nabla \cdot \mathbf{v})\mathbf{u}$	$\overset{\triangleleft}{\mathbf{A}} = \dot{\mathbf{A}} - \mathbf{L}\mathbf{A} - \mathbf{A}\mathbf{L}^T + 2(\nabla \cdot \mathbf{v})\mathbf{A}$

In the Green–Naghdi rate, $\mathbf{Z}$ is the spin tensor associated with the rotation $\mathbf{R}$ of the polar decomposition $\mathbf{F} = \mathbf{R}\mathbf{U}$. The Cotter–Rivlin rate $\overset{\triangle}{\mathbf{u}}$, $\overset{\triangle}{\mathbf{A}}$ is also referred to as covariant rate and is the rate induced by the evolution of the covariant components. Likewise, the Oldroyd rate $\overset{\triangledown}{\mathbf{u}}$, $\overset{\triangledown}{\mathbf{A}}$ is also referred to as contravariant rate. The Truesdell rate $\overset{\square}{\mathbf{A}}$ is motivated by a property of the second Piola stress which is shown in the next chapter.

The convected rate $\overset{\diamond}{\mathbf{u}}$, $\overset{\diamond}{\mathbf{A}}$ arises in connection with the surface integral of vectors and tensors,

$$\overset{\diamond}{\mathbf{u}} = \dot{\mathbf{u}} - \mathbf{u}\mathbf{L}^T + (\nabla \cdot \mathbf{v})\mathbf{u}, \qquad \overset{\diamond}{\mathbf{A}} = \dot{\mathbf{A}} - \mathbf{A}\mathbf{L}^T + (\nabla \cdot \mathbf{v})\mathbf{A}.$$

The convected time derivative $\overset{\diamond}{\mathbf{u}}$ coincides with the Truesdell rate $\overset{\square}{\mathbf{u}}$, in the case of vectors.

In *rigid motions,* $\mathbf{D} = \mathbf{0}$ and then $\mathbf{L} = \mathbf{W}$. Moreover, $\mathbf{U} = \mathbf{1}$ and then

$$\mathbf{Z} = \dot{\mathbf{R}}\mathbf{R}^T = \dot{\mathbf{F}}\mathbf{F}^T = \mathbf{L}\mathbf{F}\mathbf{F}^T = \mathbf{L} = \mathbf{W}.$$

Consequently, all of the derivatives in the table coincide with the Jaumann–Zaremba derivative,[16]

	Rate of vectors	Rate of tensors
Jaumann–Zaremba rate	$\overset{\circ}{\mathbf{u}} = \dot{\mathbf{u}} - \mathbf{W}\mathbf{u}$	$\overset{\circ}{\mathbf{A}} = \dot{\mathbf{A}} - \mathbf{W}\mathbf{A} - \mathbf{A}\mathbf{W}^T$

This means that differences between the derivatives arise in non-rigid motions of the continuum.

Both the Jaumann–Zaremba rate and the Green–Naghdi rate are the rate, of vectors and tensors, relative to an observer rotating with the angular velocity associated with the spin tensors $\mathbf{W}$ and $\mathbf{Z}$, respectively. To avoid ambiguities, though, here[17] the corotational derivative is identified with the Jaumann–Zaremba rate.

In the next section, we define and investigate the form of objective time derivatives in connection with the requirements of (rate-type) constitutive equations. We find that the derivatives so established are particular cases of a wide class of objective time derivatives.

1.10 Objective Time Derivatives

Let $\mathbf{u}$ be any vector field. The values $\mathbf{u}^*$ in $\mathscr{F}^*$ and $\mathbf{u}$ in $\mathscr{F}$, at the same point $\mathbf{X} \in R$ and time $t \in \mathbb{R}$, are related by

$$\mathbf{u}^* = \mathbf{Q}\mathbf{u}.$$

We are interested in the material time derivative in which we follow the evolution of a chosen material point. The natural view is to fix the point via the position vector $\mathbf{X} \in R$, and hence we look at the vectors $\mathbf{u}$ within the Lagrangian description $\mathbf{u} = \tilde{\mathbf{u}}(\mathbf{X}, t)$. Hence $\mathbf{u}^*(\mathbf{X}, t) = \mathbf{Q}\tilde{\mathbf{u}}(\mathbf{X}, t)$ and

$$\dot{\overline{\mathbf{u}^*}} = \dot{\overline{\mathbf{Q}\mathbf{u}}} = \dot{\mathbf{Q}}\mathbf{u} + \mathbf{Q}\dot{\mathbf{u}}$$

[16] Throughout the Jaumann–Zaremba derivative (or rate) is also denoted as Jaumann derivative or corotational derivative.

[17] As, e.g., in [216].

whence

$$\overline{\mathbf{u}^*}^{\boldsymbol{\cdot}} = \dot{\mathbf{Q}}\mathbf{u} + \mathbf{Q}\dot{\mathbf{u}}.$$

Any other time derivative might be considered, provided it is understood that the derivative of $\mathbf{u}^*$ and that of $\mathbf{u}$ are considered at the same point of the body.

Throughout this section

$$\boldsymbol{\Omega} := \dot{\mathbf{Q}}\mathbf{Q}^T,$$

$\mathbf{Q}$ being related to the change of frame. For later convenience we observe that

$$\dot{\mathbf{Q}} = \dot{\mathbf{Q}}\mathbf{Q}^T\mathbf{Q} = \boldsymbol{\Omega}\mathbf{Q}$$

and hence

$$\dot{\mathbf{Q}}\mathbf{u} = \boldsymbol{\Omega}\mathbf{Q}\mathbf{u} = \boldsymbol{\Omega}\mathbf{u}^*$$

so that

$$\overline{\mathbf{u}^*}^{\boldsymbol{\cdot}} = \boldsymbol{\Omega}\mathbf{u}^* + \mathbf{Q}\dot{\mathbf{u}}.$$

1.10.1 General Form of Objective Derivatives

A derivation [26, 224] ∂ of a vector algebra $\mathscr{V}$, over $\mathbb{R}$, is a rule $\partial : \mathscr{V} \to \mathscr{V}$ such that, for every $\mathbf{u}, \mathbf{w} \in \mathscr{V}$ and $\alpha, \beta \in \mathbb{R}$,

$$\partial(\alpha\mathbf{u} + \beta\mathbf{w}) = \alpha\partial\mathbf{u} + \beta\partial\mathbf{w},$$

$$\partial(\mathbf{u} \otimes \mathbf{w}) = (\partial\mathbf{u}) \otimes \mathbf{w} + \mathbf{u} \otimes (\partial\mathbf{w}).$$

These conditions are referred to as the linearity and the Leibniz rule.

Owing to linearity, a derivation ∂ such that $\partial\mathbf{u}$ has the meaning of a time derivative is considered in the form

$$\partial\mathbf{u} = \dot{\mathbf{u}} - \boldsymbol{\Lambda}\mathbf{u} \tag{1.64}$$

where $\boldsymbol{\Lambda} : \mathscr{V} \to \mathscr{V}$ is a function on R $\times$ $\mathbb{R}$. Accordingly

$$\partial(\mathbf{u} \otimes \mathbf{w}) = (\dot{\mathbf{u}} - \boldsymbol{\Lambda}\mathbf{u}) \otimes \mathbf{w} + \mathbf{u} \otimes (\dot{\mathbf{w}} - \boldsymbol{\Lambda}\mathbf{w}).$$

The function $\boldsymbol{\Lambda}$ may depend on properties of the continuum and hence may depend on the chosen reference. Accordingly a possible frame dependence is allowed so that

$$\partial(\mathbf{u}^*) = \mathbf{u}^* - \boldsymbol{\Lambda}^*\mathbf{u}^*, \qquad \partial\mathbf{u} = \mathbf{u} - \boldsymbol{\Lambda}\mathbf{u}.$$

A derivative $\partial\mathbf{u}$ of a vector $\mathbf{u}$ is objective if, under the Euclidean transformation (1.54), it is

$$\mathbf{Q}\partial\mathbf{u} = (\partial\mathbf{u})^* = \partial(\mathbf{u}^*) = \partial(\mathbf{Q}\mathbf{u}), \quad \mathbf{u} \in \mathscr{V}.$$

The following statement gives a characteristic condition for the set of operators $\mathbf{\Lambda}$ to make (1.64) an objective derivative.

Proposition 1.1 *The derivative (1.64) of a vector* $\mathbf{u}$ *is objective if and only if*

$$\mathbf{\Lambda}^* = \mathbf{Q}\mathbf{\Lambda}\mathbf{Q}^T + \mathbf{\Omega}. \tag{1.65}$$

Proof If $\partial\mathbf{u} = \dot{\mathbf{u}} - \mathbf{\Lambda}\mathbf{u}$ is objective then necessarily

$$\overset{\cdot}{\overline{\mathbf{Q}\mathbf{u}}} - \mathbf{\Lambda}^*\mathbf{Q}\mathbf{u} = \partial\mathbf{u}^* = (\partial\mathbf{u})^* = \mathbf{Q}(\dot{\mathbf{u}} - \mathbf{\Lambda}\mathbf{u}).$$

Hence, we have

$$\dot{\mathbf{Q}}\mathbf{u} - \mathbf{\Lambda}^*\mathbf{Q}\mathbf{u} = -\mathbf{Q}\mathbf{\Lambda}\mathbf{u}.$$

The arbitrariness of $\mathbf{u}$ implies that

$$\mathbf{\Lambda}^*\mathbf{Q} = \mathbf{Q}\mathbf{\Lambda} + \dot{\mathbf{Q}}.$$

Right multiplication by $\mathbf{Q}^T$ leads to the conclusion (1.65).

Conversely, if $\mathbf{\Lambda}^* = \mathbf{Q}\mathbf{\Lambda}\mathbf{Q}^T + \mathbf{\Omega}$ then

$$\overset{\cdot}{\overline{\mathbf{u}^*}} - \mathbf{\Lambda}^*\mathbf{u}^* = \overset{\cdot}{\overline{\mathbf{Q}\mathbf{u}}} - \mathbf{Q}\mathbf{\Lambda}\mathbf{u} - \mathbf{\Omega}\mathbf{u} = \mathbf{Q}(\dot{\mathbf{u}} - \mathbf{\Lambda}\mathbf{u}),$$

that is $\partial\mathbf{u}^* = (\partial\mathbf{u})^*$. □

To determine the transformation law of the derivative of a tensor we now consider the derivative of a dyadic product $\mathbf{u} \otimes \mathbf{w}$.

Proposition 1.2 *The derivative* $\partial(\mathbf{u} \otimes \mathbf{w})$ *is objective if and only if (1.65) holds.*

Proof The objectivity requirement

$$[\partial(\mathbf{u} \otimes \mathbf{w})]^* = \partial(\mathbf{u}^* \otimes \mathbf{w}^*)$$

results in

$$\mathbf{Q}(\dot{\mathbf{u}} - \mathbf{\Lambda}\mathbf{u}) \otimes \mathbf{Q}\mathbf{w} + \mathbf{Q}\mathbf{u} \otimes \mathbf{Q}(\dot{\mathbf{w}} - \mathbf{\Lambda}\mathbf{w}) = (\overset{\cdot}{\overline{\mathbf{Q}\mathbf{u}}} - \mathbf{\Lambda}^*\mathbf{Q}\mathbf{u}) \otimes \mathbf{Q}\mathbf{w} + \mathbf{Q}\mathbf{u} \otimes (\overset{\cdot}{\overline{\mathbf{Q}\mathbf{w}}} - \mathbf{\Lambda}^*\mathbf{Q}\mathbf{w}).$$

Upon some rearrangements, by the arbitrariness of $\mathbf{u}$ and $\mathbf{w}$ we find

$$\dot{\mathbf{Q}}\mathbf{u} - \mathbf{\Lambda}^*\mathbf{Q}\mathbf{u} + \mathbf{Q}\mathbf{\Lambda}\mathbf{u} = \mathbf{0},$$

whence the transformation law (1.65) follows. Conversely, if (1.65) holds, then the objectivity requirement is satisfied. □

The derivative of a dyadic product yields

$$\partial(\mathbf{u}\otimes\mathbf{w}) = (\dot{\mathbf{u}} - \boldsymbol{\Lambda}\mathbf{u})\otimes\mathbf{w} + \mathbf{u}\otimes(\dot{\mathbf{w}} - \boldsymbol{\Lambda}\mathbf{w}) = \overset{\cdot}{\overline{\mathbf{u}\otimes\mathbf{w}}} - \boldsymbol{\Lambda}(\mathbf{u}\otimes\mathbf{w}) - (\mathbf{u}\otimes\mathbf{w})\boldsymbol{\Lambda}^T.$$

Hence, we define the derivative of a tensor $\mathbf{K}$ in the form

$$\partial\mathbf{K} := \dot{\mathbf{K}} - \boldsymbol{\Lambda}\mathbf{K} - \mathbf{K}\boldsymbol{\Lambda}^T.$$

With this derivation ∂ we may proceed and determine the corresponding properties.

Proposition 1.3 *If $\boldsymbol{\Lambda}$ satisfies (1.65), then the derivative $\partial\mathbf{K}$ is objective.*

Proof By direct substitutions, we find

$$\begin{aligned}\partial(\mathbf{K}^*) &= \overset{\cdot}{\overline{\mathbf{K}^*}} - \boldsymbol{\Lambda}^*\mathbf{K}^* - \mathbf{K}^*\boldsymbol{\Lambda}^{*T} \\ &= \overset{\cdot}{\overline{\mathbf{Q}\mathbf{K}\mathbf{Q}^T}} - (\mathbf{Q}\boldsymbol{\Lambda}\mathbf{Q}^T + \boldsymbol{\Omega})\mathbf{Q}\mathbf{K}\mathbf{Q}^T - \mathbf{Q}\mathbf{K}\mathbf{Q}^T(\mathbf{Q}\boldsymbol{\Lambda}\mathbf{Q}^T + \boldsymbol{\Omega})^T \\ &= \mathbf{Q}(\dot{\mathbf{K}} - \boldsymbol{\Lambda}\mathbf{K} - \mathbf{K}\boldsymbol{\Lambda}^T)\mathbf{Q}^T = (\partial\mathbf{K})^*\end{aligned}$$

and the conclusion follows. □

Based on the transformation property (1.65), we observe that the operator $\boldsymbol{\Lambda}$ is non-unique as is shown by the following

Proposition 1.4 *If $\tilde{\boldsymbol{\Lambda}}$ satisfies (1.65) and $\mathbf{B}$ is any tensor, namely*

$$\tilde{\boldsymbol{\Lambda}}^* = \mathbf{Q}\tilde{\boldsymbol{\Lambda}}\mathbf{Q}^T + \boldsymbol{\Omega}, \qquad \mathbf{B}^* = \mathbf{Q}\mathbf{B}\mathbf{Q}^T,$$

then $\boldsymbol{\Lambda} = \tilde{\boldsymbol{\Lambda}} + \mathbf{B}$ satisfies (1.65).

Proof A direct substitution leads to

$$\boldsymbol{\Lambda}^* = \tilde{\boldsymbol{\Lambda}}^* + \mathbf{B}^* = \mathbf{Q}\tilde{\boldsymbol{\Lambda}}\mathbf{Q}^T + \boldsymbol{\Omega} + \mathbf{Q}\mathbf{B}\mathbf{Q}^T = \mathbf{Q}(\tilde{\boldsymbol{\Lambda}} + \mathbf{B})\mathbf{Q}^T + \boldsymbol{\Omega} = \mathbf{Q}\boldsymbol{\Lambda}\mathbf{Q}^T + \boldsymbol{\Omega},$$

which proves the assertion. □

As a consequence, given an operator satisfying (1.65), we obtain infinitely many operators satisfying (1.65) by merely adding any tensor $\mathbf{B}$.

The derivative of vectors and tensors need not be the same in that $\partial\mathbf{u}$ and $\partial\mathbf{K}$ may involve different operators $\boldsymbol{\Lambda}$. What is more, the derivative of a tensor $\partial\mathbf{K}$ may involve two different operators $\boldsymbol{\Lambda}$ at the same time. To see this, let $\boldsymbol{\Lambda}_1$ and $\boldsymbol{\Lambda}_2$ satisfy (1.65) and consider the derivative

$$\partial\mathbf{K} = \dot{\mathbf{K}} - \boldsymbol{\Lambda}_1\mathbf{K} - \mathbf{K}\boldsymbol{\Lambda}_2^T. \tag{1.66}$$

Now

$$(\partial\mathbf{K}^*) = \overline{\mathbf{QKQ}^T}^{\cdot} - (\mathbf{Q}\boldsymbol{\Lambda}_1\mathbf{Q}^T + \boldsymbol{\Omega})\mathbf{QKQ}^T - \mathbf{QKQ}^T(\mathbf{Q}\boldsymbol{\Lambda}_2\mathbf{Q}^T + \boldsymbol{\Omega})^T$$
$$= \mathbf{Q}(\dot{\mathbf{K}} - \boldsymbol{\Lambda}_1\mathbf{K} - \mathbf{K}\boldsymbol{\Lambda}_2^T)\mathbf{Q}^T = (\partial\mathbf{K})^*$$

and hence the derivative (1.66) is objective.

It is worth remarking that if $\boldsymbol{\Lambda}_1$ and $\boldsymbol{\Lambda}_2$ satisfy (1.65), then $\boldsymbol{\Lambda} = \boldsymbol{\Lambda}_1 + \boldsymbol{\Lambda}_2$ does not satisfy (1.65). This is so because

$$(\boldsymbol{\Lambda}_1 + \boldsymbol{\Lambda}_2)^* = \boldsymbol{\Lambda}_1^* + \boldsymbol{\Lambda}_2^* = \mathbf{Q}\boldsymbol{\Lambda}_1\mathbf{Q}^T + \boldsymbol{\Omega} + \mathbf{Q}\boldsymbol{\Lambda}_1\mathbf{Q}^T + \boldsymbol{\Omega} = \mathbf{Q}(\boldsymbol{\Lambda}_1 + \boldsymbol{\Lambda}_2)\mathbf{Q}^T + 2\boldsymbol{\Omega}.$$

Objective time derivatives are now established by showing that continuum physics provides infinitely many operators $\boldsymbol{\Lambda}$ satisfying (1.65). By

$$\mathbf{L}^* = \mathbf{QLQ}^T + \boldsymbol{\Omega},$$

whence

$$\mathbf{W}^* = \mathbf{QWQ}^T + \boldsymbol{\Omega}, \qquad \mathbf{D}^* = \mathbf{QDQ}^T,$$

it follows that the operators

$$\boldsymbol{\Lambda} = \mathbf{W} + \lambda\mathbf{D} + \nu(\nabla\cdot\mathbf{v})\mathbf{1} \tag{1.67}$$

constitute a two-parameters family of operators satisfying (1.65). On the other hand, letting

$$\mathbf{Z} = \dot{\mathbf{R}}\mathbf{R}^T \in \mathrm{Skw}, \qquad \mathbf{R} = \mathbf{FU}^{-1},$$

we have

$$\mathbf{Z}^* = \overline{\mathbf{QR}}^{\cdot}(\mathbf{QR})^T = \dot{\mathbf{Q}}\mathbf{Q}^T + \mathbf{Q}\dot{\mathbf{R}}\mathbf{R}^T\mathbf{Q}^T = \mathbf{QZQ}^T + \boldsymbol{\Omega}.$$

Hence $\mathbf{Z}$ satisfies (1.65). As a consequence, also

$$\boldsymbol{\Lambda} = \mathbf{Z} + \lambda\mathbf{D} + \nu(\nabla\cdot\mathbf{v})\mathbf{1} \tag{1.68}$$

is a two-parameters family of operators satisfying (1.65).

In the Jaumann–Zaremba derivative $\boldsymbol{\Lambda} = \mathbf{W}$; in the Cotter–Rivlin derivative $\boldsymbol{\Lambda} = \mathbf{W} - \mathbf{D}$, $(\lambda = -1, \nu = 0)$; in the Oldroyd derivative $\boldsymbol{\Lambda} = \mathbf{W} + \mathbf{D}$, $(\lambda = 1, \nu = 0)$; in the Truesdell derivative $\boldsymbol{\Lambda} = \mathbf{W} + \mathbf{D} - (\nabla\cdot\mathbf{v})\mathbf{1}$, $(\lambda = 1, \nu = -1)$ for vectors and $\boldsymbol{\Lambda} = \mathbf{W} + \mathbf{D} - \frac{1}{2}(\nabla\cdot\mathbf{v})\mathbf{1}$, $(\lambda = 1, \nu = -\frac{1}{2})$ for tensors; in the $\mathbf{F}$-rate $\boldsymbol{\Lambda} = \mathbf{W} - \mathbf{D} + (\nabla\cdot\mathbf{v})\mathbf{1}$, $(\lambda = -1, \nu = 1)$; in the $J^{-1}\mathbf{F}$-rate $\boldsymbol{\Lambda} = \mathbf{W} + \mathbf{D} - (\nabla\cdot\mathbf{v})\mathbf{1}$, $(\lambda = 1, \nu = -1)$. Instead, in the Green–Naghdi derivative, $\boldsymbol{\Lambda} = \mathbf{Z}$.

There are models of material behaviour where the derivative is a superposition of objective derivatives in the form

$$\overset{\diamond}{\mathbf{A}} = \sum_k \nu_k \, \partial_k \mathbf{A}, \qquad \sum_k \nu_k = 1.$$

Consider the objective derivatives

$$\partial_k \mathbf{A} := \dot{\mathbf{A}} - \mathbf{W}\mathbf{A} + \mathbf{A}\mathbf{W} + \xi_k(\mathbf{D}\mathbf{A} + \mathbf{A}\mathbf{D}), \tag{1.69}$$

$\{\xi_k\}$ being real numbers; this is the case for the derivatives examined in the previous section.

Proposition 1.5 *A superposition of the objective derivatives (1.69) is objective.*

Proof The superposition results in

$$\overset{\diamond}{\mathbf{A}} = (\sum_k \nu_k)\dot{\mathbf{A}} - (\sum_k \nu_k)\mathbf{W}\mathbf{A} + (\sum_k \nu_k)\mathbf{A}\mathbf{W} + (\sum_k \nu_k \xi_k)(\mathbf{D}\mathbf{A} + \mathbf{A}\mathbf{D}),$$

whence

$$\overset{\diamond}{\mathbf{A}} = \dot{\mathbf{A}} - \mathbf{W}\mathbf{A} + \mathbf{A}\mathbf{W} + (\sum_k \nu_k \xi_k)(\mathbf{D}\mathbf{A} + \mathbf{A}\mathbf{D}).$$

Since $\dot{\mathbf{A}} - \mathbf{W}\mathbf{A} + \mathbf{A}\mathbf{W}$ and $\mathbf{D}\mathbf{A} + \mathbf{A}\mathbf{D}$ are objectives, then the conclusion follows. □

Objective Derivatives and Trace Conservation

It is of interest to examine the operator (1.67) (and (1.68)) in connection with the trace of tensors, say $\mathbf{A}$, under the derivative

$$\partial \mathbf{A} = \dot{\mathbf{A}} - \mathbf{\Lambda}\mathbf{A} - \mathbf{A}\mathbf{\Lambda}^T.$$

In view of (1.67), we have the generic (objective) derivative

$$\overset{\boxdot}{\mathbf{A}} := \dot{\mathbf{A}} - \mathbf{W}\mathbf{A} - \mathbf{A}\mathbf{W}^T - \lambda(\mathbf{D}\mathbf{A} + \mathbf{A}\mathbf{D}) - \nu(\nabla \cdot \mathbf{v})\mathbf{A}.$$

For any tensor $\mathbf{A}$

$$\mathrm{tr}\,(\mathbf{A}\mathbf{W} - \mathbf{W}\mathbf{A}) = A_{ij}W_{ji} - W_{ij}A_{ji} = 0.$$

Instead,

$$\mathrm{tr}\,(\mathbf{D}\mathbf{A} + \mathbf{A}\mathbf{D}) = D_{ij}A_{ji} + A_{ij}D_{ji} = 2\mathrm{tr}\,\mathbf{D}\mathbf{A},$$

and $\mathrm{tr}\,\mathbf{D}\mathbf{A} = 0$ if $\mathbf{A} \in \mathrm{Skw}$. Hence it follows that

$$\mathrm{tr}\ \overset{\boxdot}{\mathbf{A}} = \overline{\mathrm{tr}\,\mathbf{A}}^{\,\cdot} - 2\lambda\mathrm{tr}\,(\mathbf{D}\mathbf{A}) - \nu(\nabla \cdot \mathbf{v})\mathrm{tr}\,\mathbf{A}.$$

Consequently

$$\operatorname{tr}\mathbf{A} = 0 \quad \Longrightarrow \quad \operatorname{tr}\ \overset{\square}{\mathbf{A}} = 0,$$

if $\lambda = 0$ or $\mathbf{A} \in \mathrm{Skw}$. The trace-free condition is preserved by the derivative

$$\overset{\square}{\mathbf{A}} = \dot{\mathbf{A}} - \mathbf{WA} - \mathbf{AW}^T - \nu(\nabla \cdot \mathbf{v})\mathbf{A},$$

for any scalar function ν. Hence, this happens for the corotational derivative modified by the addition of the term $\nu(\nabla \cdot \mathbf{v})\mathbf{A}$ for any function ν and any velocity field $\mathbf{v}$.

Derivatives of the Deformation Gradient

Particular results follow about the derivatives of the deformation gradient, as a consequence of the relation $\dot{\mathbf{F}} = \mathbf{LF}$ and of the transformation law $\mathbf{F}^* = \mathbf{QF}$.

If we apply the Oldroyd rate, it follows

$$\overset{\triangledown}{\mathbf{F}} = \dot{\mathbf{F}} - \mathbf{LF} = \mathbf{0}.$$

The deformation gradient is constant relative to the Oldroyd rate. Consequently, the left Cauchy–Green tensor $\mathbf{B}$ too is constant relative to the Oldroyd rate in that

$$\overset{\triangledown}{\mathbf{B}} = \dot{\mathbf{B}} - \mathbf{LB} - \mathbf{BL}^T = \dot{\mathbf{F}}\mathbf{F}^T + \mathbf{F}\dot{\mathbf{F}}^T - \mathbf{LB} - \mathbf{BL}^T = \mathbf{LFF}^T + \mathbf{FF}^T\mathbf{L}^T - \mathbf{LFF}^T - \mathbf{FF}^T\mathbf{L}^T = \mathbf{0}.$$

By applying the Jaumann–Zaremba rate, it follows

$$\overset{\circ}{\mathbf{F}} = \dot{\mathbf{F}} - \mathbf{WF} = \dot{\mathbf{F}} - \tfrac{1}{2}(\mathbf{L} - \mathbf{L}^T)\mathbf{F} = \mathbf{D}\,\mathbf{F}.$$

In rigid motions $\mathbf{F} = \mathbf{R}, \mathbf{D} = \mathbf{0}$, and hence $\overset{\circ}{\mathbf{F}} = \mathbf{0}$. Thus, in rigid motions, $\mathbf{F}$ is constant relative to the corotational derivative. Moreover, the Finger tensor $\mathbf{B}^{-1}$ is constant relative to Cotter–Rivlin rate in that

$$(\mathbf{B}^{-1})^{\cdot} = -\mathbf{L}^T\mathbf{B}^{-1} - \mathbf{B}^{-1}\mathbf{L}$$

and hence $\overset{\vartriangle}{\overline{\mathbf{B}^{-1}}} = \mathbf{0}$.

By applying the Truesdell rate, it follows

$$\overset{\square}{\mathbf{F}} = \dot{\mathbf{F}} - \mathbf{LF} + (\nabla \cdot \mathbf{v})\mathbf{F} = (\nabla \cdot \mathbf{v})\mathbf{F}.$$

The Truesdell derivative of $\mathbf{F}$, $\overset{\square}{\mathbf{F}}$, is zero for isochoric deformations.

1.10.2 Rivlin–Eriksen Tensors

Rivlin–Ericksen tensors [380] are sometimes applied in the modelling of materials with memory properties via suitable dependences on higher-order derivatives. We start with the observation that the right Cauchy–Green tensor $\mathbf{C} = \mathbf{F}^T\mathbf{F}$ is invariant under a Euclidean transformation (1.54). Hence, a time derivative of any order of $\mathbf{C}$ is invariant too (and hence objective). The first-order derivative results in

$$\dot{\mathbf{C}} = \mathbf{F}^T\, 2\,\mathbf{D}\,\mathbf{F}.$$

Let $\mathbf{A}_1 = 2\,\mathbf{D}$ and hence

$$\dot{\mathbf{C}} = \mathbf{F}^T\, \mathbf{A}_1\,\mathbf{F}.$$

A further differentiation yields

$$\ddot{\mathbf{C}} = \mathbf{F}^T(\dot{\mathbf{A}}_1 + \mathbf{L}^T\mathbf{A}_1 + \mathbf{A}_1\mathbf{L})\mathbf{F},$$

whence

$$\ddot{\mathbf{C}} = \mathbf{F}^T\, \mathbf{A}_2\,\mathbf{F}, \qquad \mathbf{A}_2 := \dot{\mathbf{A}}_1 + \mathbf{L}^T\mathbf{A}_1 + \mathbf{A}_1\mathbf{L}.$$

Denote the n-th order time derivative as

$$\overset{(n)}{\mathbf{C}} = \mathbf{F}^T\, \mathbf{A}_n\,\mathbf{F}.$$

A further differentiation yields

$$\overset{(n+1)}{\mathbf{C}} = \mathbf{F}^T\, \mathbf{A}_{n+1}\,\mathbf{F}, \qquad \mathbf{A}_{n+1} := \dot{\mathbf{A}}_n + \mathbf{L}^T\mathbf{A}_n + \mathbf{A}_n\mathbf{L}.$$

We have thus proved by induction that

$$\overset{(n)}{\mathbf{C}} = \mathbf{F}^T\, \mathbf{A}_n\,\mathbf{F}, \qquad \mathbf{A}_n := \dot{\mathbf{A}}_{n-1} + \mathbf{L}^T\mathbf{A}_{n-1} + \mathbf{A}_{n-1}\mathbf{L},$$

and $\mathbf{A}_1 = 2\,\mathbf{D}$. The sequence of tensors $\{\mathbf{A}_n\}$ so determined constitutes the set of Rivlin–Ericksen tensors.

The tensors $\{\mathbf{A}_n\}$ are symmetric. This statement is proved by observing that

$$\mathbf{A}_1 = 2\,\mathbf{D} \in \mathrm{Sym}, \qquad \mathbf{A}_n \in \mathrm{Sym} \implies \mathbf{A}_{n+1} \in \mathrm{Sym}.$$

This property in turn follows by direct check,

$$\mathbf{A}_{n+1}^T = \dot{\mathbf{A}}_n + (\mathbf{L}^T\mathbf{A}_n)^T + (\mathbf{A}_n\mathbf{L})^T = \dot{\mathbf{A}}_n + \mathbf{L}^T\mathbf{A}_n + \mathbf{A}_n\mathbf{L} = \mathbf{A}_{n+1}.$$

It is apparent that Rivlin–Ericksen tensors might be defined by

$$\mathbf{A}_1 = 2\,\mathbf{D}, \qquad \mathbf{A}_{n+1} = \overset{\triangle}{\mathbf{A}}_n, \qquad n = 1, 2, ...,$$

where the superposed $\triangle$ denotes the Cotter–Rivlin derivative. Also in this way we conclude that $\{\overset{\triangle}{\mathbf{A}}_n\}$ is a sequence of objective tensors.

With this view we might define other sequences of Rivlin–Ericksen-like tensors by merely assuming another time derivative,

$$\mathbf{A}_1 = 2\,\mathbf{D}, \qquad \mathbf{A}_{n+1} = \mathcal{D}_t \mathbf{A}_n,$$

$\mathcal{D}_t$ denoting an objective time derivative. This would be the case with the sequences

$$\{\overset{\triangledown}{\mathbf{A}}_n\}, \qquad \{\overset{\circ}{\mathbf{A}}_n\},$$

with $\mathbf{A}_1 = 2\,\mathbf{D}$ for both sequences, where the Oldroyd and the Jaumann–Zaremba rates are involved.

1.10.3 Spins and Angular Velocities Under Euclidean Transformations

We now examine the properties of spins, angular velocities, and axial vectors under Euclidean transformations such that $\mathbf{x} \rightarrow \mathbf{x}^*$ in the the form

$$\mathbf{x}^* = \mathbf{c}(t) + \mathbf{Q}(t)\mathbf{x},$$

where $\mathbf{Q}$ is a rotation.

As it happens with micropolar media, Sect. 11.1, at each point of the body an angular velocity $\boldsymbol{\omega}$ is defined as that of the particle at the point. The balance of energy then involves a power proportional to the difference $\boldsymbol{\omega} - \mathbf{w}$, where $\mathbf{w}$ is the axial vector of the spin $\mathbf{W}$. It is then natural here to investigate the possible objectivity of $\boldsymbol{\omega}$, $\mathbf{w}$, and $\boldsymbol{\omega} - \mathbf{w}$.

By (1.63), the spin $\mathbf{W}$ obeys the transformation law

$$\mathbf{W}^* = \mathbf{Q}\mathbf{W}\mathbf{Q}^T + \boldsymbol{\Omega}.$$

We then compute the axial vector $\mathbf{w}$ of $\mathbf{W}$ to have

$$\mathbf{w}^* = \mathbf{e}_l \tfrac{1}{2} \epsilon_{lhj} Q_{jp} W_{pk} Q^T_{kh} + \mathbf{a}_{\boldsymbol{\Omega}}.$$

Since

$$\epsilon_{lhj} Q_{jp} Q_{hk} = \epsilon_{ikp} Q_{li}$$

then

$$\mathbf{e}_l \tfrac{1}{2}\epsilon_{lhj} Q_{jp} W_{pk} Q_{kh}^T = \mathbf{e}_l Q_{li} w_i = \mathbf{Q}\mathbf{w}.$$

Hence, we obtain

$$\mathbf{w}^* = \mathbf{Q}\mathbf{w} + \mathbf{a}_{\Omega}. \tag{1.70}$$

Now, the angular velocity obeys the transformation law (1.58), $\boldsymbol{\omega}^* = \mathbf{Q}\boldsymbol{\omega} + \mathbf{a}_{\Omega}$. Thus, both the angular velocity $\boldsymbol{\omega}$ and the axial vector $\mathbf{w}$ are not objective. Yet the difference $\boldsymbol{\omega} - \mathbf{w}$ satisfies

$$\boldsymbol{\omega}^* - \mathbf{w}^* = \mathbf{Q}(\boldsymbol{\omega} - \mathbf{w}) \tag{1.71}$$

and then it is objective.

A further question is in order about the objectivity of the velocity gradients. First look at the material gradient $\nabla_R \mathbf{w}$. Since $\nabla_R \boldsymbol{\Omega} = \mathbf{0}$ then (1.70) implies that

$$(\partial_{X_K} w_i)^* = Q_{ir}\, \partial_{X_K} w_r.$$

Thus, $\nabla_R \mathbf{w}$ is an objective vector. As to the spatial gradient we observe

$$(\partial_{x_k} w_i)^* = \partial_{x_k^*} w_i^* = \partial_{x_j} w_i^* \partial_{x_k^*} x_j = Q_{jk}^{-1} \partial_{x_j} w_i^*.$$

Hence, we have

$$(\partial_{x_k} w_i)^* = Q_{kj} \partial_{x_j} (Q_{ir} w_r + \epsilon_{ijk} \Omega_{kj}) = Q_{kj} Q_{ir} \partial_{x_j} w_r.$$

We then conclude that $\nabla \mathbf{w}$ is an objective tensor.

Chapter 2
Balance Equations

Borrowing from the mechanics of systems of particles, the balance equations for mass, linear momentum, and angular momentum are established. Next, the balance of energy is obtained after having realized that energy in general requires a more general framework including quantities of non-mechanical character. This is done first by considering fields in the spatial (Eulerian) description for regions that convect with the body. As is customary, the starting point is the pertinent axiom about an arbitrary region (global law), and hence the corresponding local balance equation is derived. The key step of the mathematical development is Reynolds' transport theorem, mainly for regions convecting with the body.

The balance of entropy is made formal by assuming that the entropy rate, deprived of the divergence of a pertinent flux and of an entropy supply, is non-negative, this quantity being, by definition, the entropy production.

Next, after a brief introduction to electromagnetic fields, balance equations are established for electromagnetism in deformable bodies. The approach stems from the physical insights about particle behaviour. Hence, balances are obtained for linear and angular momentum, energy, and entropy. Moreover, jump relations are derived for balance properties across singular surfaces.

2.1 Balance of Mass

The (assumed) conservation of mass, irrespective of the motion and the size of a body, is expressed by the following

Axiom. The mass of any convecting subregion of a body is constant in time. To represent mathematically the content of the axiom, we let $\rho(\mathbf{x}, t)$ be the (positive-valued) mass density in the spatial description so that the mass $\mathcal{M}(\mathcal{P}_t)$, within the region $\mathcal{P}_t = \chi_t(\mathrm{P})$, is given by

$$\mathcal{M}(\mathcal{P}_t) = \int_{\mathcal{P}_t} \rho \, dv.$$

A. Morro and C. Giorgi, *Mathematical Modelling of Continuum Physics*,
Modeling and Simulation in Science, Engineering and Technology,
https://doi.org/10.1007/978-3-031-20814-0_2

The axiom then reads: if $\mathcal{P}_t = \chi_t(\mathrm{P})$, then

$$\int_{\mathcal{P}_t} \rho \, dv \quad \text{is constant in time.}$$

Denote by $\rho_R(\mathbf{X})$ the mass density in the reference configuration so that

$$\mathcal{M}(\mathrm{P}) = \int_{\mathrm{P}} \rho_R \, dv_R.$$

Hence, if $\mathcal{P}_t = \chi_t(\mathrm{P})$, the constancy of mass means

$$\int_{\mathcal{P}_t} \rho \, dv = \int_{\mathrm{P}} \rho_R \, dv_R. \tag{2.1}$$

Otherwise, the constancy of mass is expressed by

$$\frac{d}{dt} \int_{\mathcal{P}_t} \rho \, dv = 0. \tag{2.2}$$

By a change of variables, $\mathcal{P}_t \ni \mathbf{x} \to \mathbf{X} \in \mathrm{P}$, we have

$$\int_{\mathcal{P}_t} \rho \, dv = \int_{\mathrm{P}} \rho \, J \, dv_R.$$

Hence, (2.1) results in

$$\int_{\mathrm{P}} (\rho_R - \rho \, J) dv_R = 0.$$

The arbitrariness of P and the continuity of the integrand imply that

$$\rho_R - \rho \, J = 0, \qquad \forall \mathbf{X} \in \mathrm{P}, \; t \in \mathbb{R}.$$

Consequently, the product $\rho \, J$ is constant in time and

$$\rho = \frac{\rho_R}{J}. \tag{2.3}$$

The mass density ρ in the current configuration $\mathcal{P}_t$ is $1/J$ times the mass density ρ_R in the reference configuration. Since ρ_R depends only on $\mathbf{X}$ then (2.3) shows that ρ depends on time as J^{-1}.

If instead we start with (2.2), we assume ρJ is a C^1 function on P and, by means of Reynolds' transport theorem, we have

$$0 = \frac{d}{dt} \int_{\mathcal{P}_t} \rho \, dv = \int_{\mathcal{P}_t} (\dot{\rho} + \rho \nabla \cdot \mathbf{v}) dv.$$

The arbitrariness of P implies that

$$\dot{\rho} + \rho \nabla \cdot \mathbf{v} = 0, \qquad \partial_t \rho + \nabla \cdot (\rho \mathbf{v}) = 0. \tag{2.4}$$

The differential equations (2.4) are referred to as the *continuity equations* for the mass.

By (2.4), the specific volume (volume per unit mass)

$$\upsilon := \frac{1}{\rho}$$

satisfies

$$\dot{\upsilon} = \upsilon \nabla \cdot \mathbf{v}.$$

If $\nabla \cdot \mathbf{v} = 0$ both ρ and υ are constant along the path of any point (isochoric motions). In incompressible bodies, only isochoric motions are possible and the velocity field is subject to the constraint $\nabla \cdot \mathbf{v} = 0$.

The result (2.4) about the mass density allows us to restate Reynolds' transport theorem in a more convenient form. Let $\phi = \rho\varphi$ so that ϕ is a density, per unit volume, whereas φ is the corresponding density, per unit mass. Upon direct substitution and use of (2.4), it follows

$$\dot{\phi} + \phi \nabla \cdot \mathbf{v} = (\dot{\rho} + \rho \nabla \cdot \mathbf{v})\varphi + \rho\dot{\varphi} = \rho\dot{\varphi}.$$

Hence, by (1.23), we obtain Reynolds' transport theorem in the form

$$\frac{d}{dt} \int_{\mathcal{P}_t} \rho \varphi \, dv = \int_{\mathcal{P}_t} \rho \dot{\varphi} \, dv. \tag{2.5}$$

Equation (2.5) holds for any C^1 density φ irrespective of the scalar, vector, or tensor character of φ.

As a comment, we observe that, owing to (2.4), there is a direct connection between the Oldroyd rate and the Truesdell rate, for vectors and tensors. Since

$$\dot{\overline{\mathbf{u}/\rho}} = \frac{1}{\rho}[\dot{\mathbf{u}} + \mathbf{u} \nabla \cdot \mathbf{v}], \qquad \dot{\overline{\mathbf{A}/\rho}} = \frac{1}{\rho}[\dot{\mathbf{A}} + \mathbf{A} \nabla \cdot \mathbf{v}]$$

then we have

$$\overset{\square}{\mathbf{u}} = \rho \, \overset{\triangledown}{\overline{\mathbf{u}/\rho}}, \qquad \overset{\square}{\mathbf{A}} = \rho \, \overset{\triangledown}{\overline{\mathbf{A}/\rho}}\,.$$

Instead

$$\overset{\square}{\mathbf{u}} \neq \rho \, \overset{\triangleleft}{\overline{\mathbf{u}/\rho}}, \qquad \overset{\square}{\mathbf{A}} \neq \rho \, \overset{\triangleleft}{\overline{\mathbf{A}/\rho}}\,.$$

In isochoric motions, $\nabla \cdot \mathbf{v} = 0$, the $\mathbf{F}$-rate coincides with the Cotter–Rivlin rate, while the $J^{-1}\mathbf{F}$-rate coincides with the Truesdell rate and the Oldroyd rate.

Referential densities

For any density field φ, per unit mass, $\rho\varphi$ is the corresponding field per unit volume and

$$\varphi_R := \rho_R \varphi = J \rho \varphi$$

is the density field per unit volume in the reference configuration. We then call φ_R the *referential density* field.

2.2 Balance of Linear Momentum

Throughout the dynamic equations of mechanics, i.e. evolution equations of linear momentum and angular momentum are assumed to hold in *inertial frames.* For a system of particles, dynamics is based on the assumption that the linear momentum $\mathfrak{P}$ changes in time according to

$$\dot{\mathfrak{P}} = \mathfrak{F}$$

where $\mathfrak{F}$ is the total force acting on the system. For a continuous body, we let

$$\mathfrak{P}(\mathcal{P}_t) := \int_{\mathcal{P}_t} \rho \mathbf{v} \, dv.$$

The force $\mathfrak{F}(\mathcal{P}_t)$ acting on $\mathcal{P}_t$ consists of two terms:
body forces exerted on the interior points of $\mathcal{P}_t$,
contact forces through the boundary $\partial\mathcal{P}_t$ of $\mathcal{P}_t$.

Body forces are modelled by assuming that there is a force density $\mathbf{b}$ (per unit mass) so that the whole body force on $\mathcal{P}_t$ is

$$\int_{\mathcal{P}_t} \rho \mathbf{b} \, dv,$$

where the dependence of ρ and $\mathbf{b}$ on $(\mathbf{x}, t)$ is understood

To model the contact forces, we consider a surface $\mathcal{S}$ in the current configuration, passing through $\mathbf{x}$, at time t, with unit normal $\mathbf{n}$ (Fig. 2.1). Denote by positive side of $\mathcal{S}$ that on the portion of space into which $\mathbf{n}$ points. By *Cauchy's hypothesis,* there is a *surface-traction field* $\mathbf{t}$ such that $\mathbf{t}(\mathbf{x}, \mathbf{n}, t)$ is the force exerted on the positive side of $\mathcal{S}$, at $\mathbf{x}$, per unit area, at time t. Hence we let

$$\int_{\mathcal{S}} \mathbf{t}(\mathbf{x}, \mathbf{n}, t) da$$

be the force on $\mathcal{S}$ at time t. The whole force on $\mathcal{P}_t$ is then given by

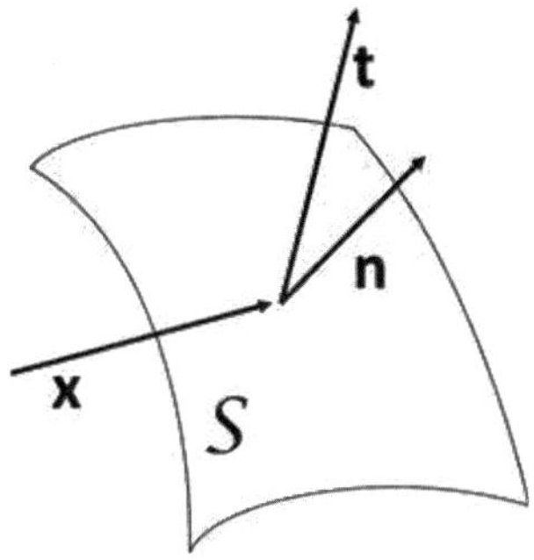

Fig. 2.1 The vector **t** is the force per unit area of $\mathcal{S}$ at the point **x**

$$\mathfrak{F}(\mathcal{P}_t) = \int_{\mathcal{P}_t} \rho \mathbf{b}\, dv + \int_{\partial \mathcal{P}_t} \mathbf{t}\, da.$$

Incidentally, two tangent surfaces at $\mathbf{x}$ at time t, are subject to the same traction $\mathbf{t}(\mathbf{x}, \mathbf{n}, t)$.

With this assumption on $\mathfrak{P}$ and $\mathfrak{F}$ in a continuous body, we state
Axiom. The motion of a continuous body is governed by the equation

$$\dot{\mathfrak{P}}(\mathcal{P}_t) = \mathfrak{F}(\mathcal{P}_t), \tag{2.6}$$

for any convecting region $\mathcal{P}_t$. By the transport theorem (2.5), we can write (2.6) in the form

$$\int_{\mathcal{P}_t} \rho(\dot{\mathbf{v}} - \mathbf{b}) dv = \int_{\partial \mathcal{P}_t} \mathbf{t}\, da. \tag{2.7}$$

As with any equality between volume and surface integrals, this condition cannot hold unless the integrand $\mathbf{t}$ is appropriate. To show that a restriction on $\mathbf{t}$ is required, we argue as follows. Let $\rho(\dot{\mathbf{v}} - \mathbf{b})$ be bounded, say $|\rho(\dot{\mathbf{v}} - \mathbf{b})| < \kappa$. We then find that

$$\Big| \int_{\mathcal{P}_t} \rho(\dot{\mathbf{v}} - \mathbf{b}) dv \Big| \le \int_{\mathcal{P}_t} |\rho(\dot{\mathbf{v}} - \mathbf{b})| dv < \kappa \operatorname{vol} \mathcal{P}_t.$$

Hence

$$\Big| \int_{\partial \mathcal{P}_t} \mathbf{t}\, da \Big| < \kappa \operatorname{vol} \mathcal{P}_t$$

and this implies the requirement

$$\frac{1}{\operatorname{area} \partial \mathcal{P}_t} \int_{\partial \mathcal{P}_t} \mathbf{t}\, da \to 0 \quad \text{as} \quad \frac{\operatorname{vol} \mathcal{P}_t}{\operatorname{area} \partial \mathcal{P}_t} \to 0. \tag{2.8}$$

For formal simplicity, we omit writing the dependence on time t. The explicit requirement on $\mathbf{t}$ is determined by means of the following

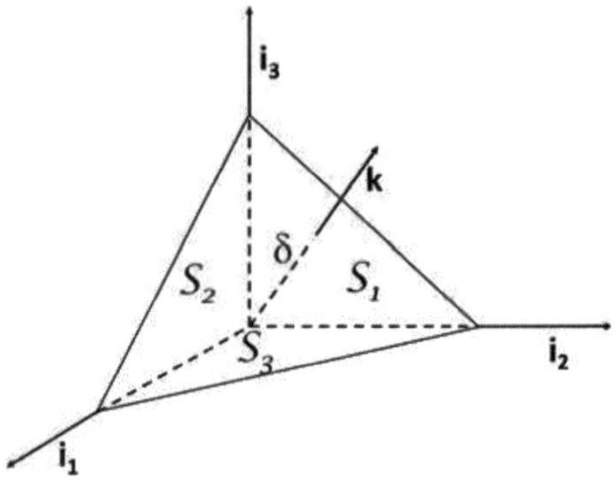

Fig. 2.2 The tetrahedron $\mathscr{T}$ employed in the proof of Cauchy's theorem

Theorem 2.1 (Cauchy) *If* $\mathbf{t}(\mathbf{x}, \mathbf{n})$ *is a continuous function and* $\rho(\dot{\mathbf{v}} - \mathbf{b})$ *is bounded then the condition (2.8) implies the existence of a tensor field* $\mathbf{T}$ *such that*

$$\mathbf{t}(\mathbf{x}, \mathbf{n}) = \mathbf{T}(\mathbf{x})\,\mathbf{n}. \tag{2.9}$$

Proof Given any $\mathbf{x} \in \mathcal{R}_t \setminus \partial\mathcal{R}_t$ and an orthonormal basis $(\mathbf{i}_1, \mathbf{i}_2, \mathbf{i}_3)$, we choose a tetrahedron $\mathscr{T}$ such that $\partial\mathscr{T} = \mathcal{S}_1 \cup \mathcal{S}_2 \cup \mathcal{S}_3 \cup \mathcal{S}_\delta$ where $\mathcal{S}_1, \mathcal{S}_2, \mathcal{S}_3$ are the faces in the coordinate planes through $\mathbf{x}$ while $\mathcal{S}_\delta$ is the oblique face, opposite to the vertex $\mathbf{x}$. The distance δ of $\mathcal{S}_\delta$ from $\mathbf{x}$ is so small that $\mathscr{T} \subset \mathcal{R}_t \setminus \partial\mathcal{R}_t$. The unit outward normals to $\mathcal{S}_1, \mathcal{S}_2, \mathcal{S}_3, \mathcal{S}_\delta$ are $-\mathbf{i}_1, -\mathbf{i}_2, -\mathbf{i}_3, \mathbf{k}$. It is $\mathbf{k} \cdot \mathbf{i}_j > 0$, $j = 1, 2, 3$ (Fig. 2.2).

Let A_1, A_2, A_3, A_δ denote the areas of $\mathcal{S}_1, \mathcal{S}_2, \mathcal{S}_3, \mathcal{S}_\delta$. Since $A_j/A_\delta < 1$, $j = 1, 2, 3$, then

$$\text{area}\,\partial\mathscr{T} = A_\delta\big(1 + \textstyle\sum_j \frac{A_j}{A_\delta}\big) = \lambda A_\delta, \quad 1 < \lambda < 4,$$

and hence the requirement (2.8) yields

$$\frac{1}{A_\delta} \int_{\partial\mathscr{T}} \mathbf{t}\, da \to 0 \quad \text{as} \quad \delta \to 0. \tag{2.10}$$

Let $\mathbf{y} \in \partial\mathscr{T}$. Upon substitution of

$$\mathbf{t}(\mathbf{y}, \mathbf{n}) = \mathbf{t}(\mathbf{x}, \mathbf{n}) + [\mathbf{t}(\mathbf{y}, \mathbf{n}) - \mathbf{t}(\mathbf{x}, \mathbf{n})], \qquad \mathbf{n} = -\mathbf{i}_1, -\mathbf{i}_2, -\mathbf{i}_3, \mathbf{k},$$

it follows that

$$\int_{\partial\mathscr{T}} \mathbf{t}(\mathbf{y}, \mathbf{n}) da = \textstyle\sum_j \mathbf{t}(\mathbf{x}, -\mathbf{i}_j) A_j + \mathbf{t}(\mathbf{x}, \mathbf{k}) A_\delta + \hat{\mathbf{t}},$$

where

$$\hat{\mathbf{t}} = \textstyle\sum_j \int_{\mathcal{S}_j} [\mathbf{t}(\mathbf{y}, -\mathbf{i}_j) - \mathbf{t}(\mathbf{x}, -\mathbf{i}_j)] da + \int_{\mathcal{S}_\delta} [\mathbf{t}(\mathbf{y}, \mathbf{k}) - \mathbf{t}(\mathbf{x}, \mathbf{k})] da.$$

We now estimate $\hat{\mathbf{t}}$. Let d be the maximum distance of a point of $\partial\mathscr{T}$ from $\mathbf{x}$. Then, by the continuity of $\mathbf{t}(\cdot, \mathbf{n})$, for any $\epsilon > 0$ there is $\tilde{d}$ such that $d < \tilde{d}$ implies $|\mathbf{t}(\mathbf{y}, -\mathbf{i}_j) - \mathbf{t}(\mathbf{x}, -\mathbf{i}_j)| < \epsilon$ and $|\mathbf{t}(\mathbf{y}, \mathbf{k}) - \mathbf{t}(\mathbf{x}, \mathbf{k})| < \epsilon$. Hence $d < \tilde{d}$ implies that

$$|\hat{\mathbf{t}}| < \epsilon(\textstyle\sum_j A_j + A_\delta) = \epsilon\,\lambda\,A_\delta.$$

Since $\delta \to 0$ implies $d \to 0$ then a suitably small δ makes ϵ as small as we please. Consequently,

$$\frac{1}{A_\delta}\hat{\mathbf{t}} = O(\delta),$$

whence

$$\frac{1}{A_\delta}\int_{\partial\mathscr{T}} \mathbf{t}\,da = \textstyle\sum_j \dfrac{A_j}{A_\delta}\mathbf{t}(\mathbf{x}, -\mathbf{i}_j) + \mathbf{t}(\mathbf{x}, \mathbf{k}) + O(\delta).$$

The requirement (2.10) implies that

$$\textstyle\sum_j \dfrac{A_j}{A_\delta}\mathbf{t}(\mathbf{x}, -\mathbf{i}_j) + \mathbf{t}(\mathbf{x}, \mathbf{k}) = \mathbf{0}. \tag{2.11}$$

By the continuity of $\mathbf{t}(\mathbf{x}, \mathbf{n})$, we can evaluate the limit of (2.11) as $\mathbf{k} \to \mathbf{i}_1$. It follows

$$\mathbf{t}(\mathbf{x}, -\mathbf{i}_1) + \mathbf{t}(\mathbf{x}, \mathbf{i}_1) = \mathbf{0}.$$

The arbitrariness of $\mathbf{i}_1$ implies that

$$\mathbf{t}(\mathbf{x}, -\mathbf{n}) = -\mathbf{t}(\mathbf{x}, \mathbf{n}) \tag{2.12}$$

for any unit normal $\mathbf{n}$.

Now observe that

$$\frac{A_j}{A_\delta} = \mathbf{k} \cdot \mathbf{i}_j.$$

To prove this relation, we consider, e.g. S_δ and S_3 (Fig. 2.3). The ratio A_3/A_δ equals the ratio of the heights, h_3/h_δ relative to the common segment of the two triangles.

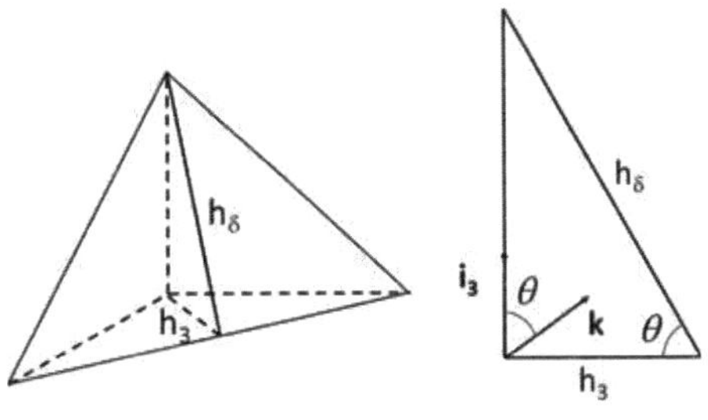

Fig. 2.3 A section of the tetrahedron $\mathscr{T}$ to prove that $A_3/A = \mathbf{k} \cdot \mathbf{i}_3$

If θ is the angle between the two heights, then $h_3/h_\delta = \cos\theta = \mathbf{i}_3 \cdot \mathbf{k}$. The analogue holds for A_j/A_δ, $j = 1, 2$. Hence, by use of (2.12), Eq. (2.11) results in

$$\mathbf{t}(\mathbf{x}, \mathbf{k}) = \sum_j \mathbf{t}(\mathbf{x}, \mathbf{i}_j)\, \mathbf{k} \cdot \mathbf{i}_j. \tag{2.13}$$

Equation (2.13) is derived upon the restriction $\mathbf{k} \cdot \mathbf{i}_j > 0$, $j = 1, 2, 3$. We show that (2.13) holds without any restriction on $\mathbf{k}$. Let $\{\mathbf{f}_j\}$ be a new basis such that

$$\mathbf{i}_j = (\operatorname{sgn} \mathbf{k} \cdot \mathbf{f}_j)\mathbf{f}_j.$$

By definition, $\mathbf{k} \cdot \mathbf{i}_j = (\operatorname{sgn} \mathbf{k} \cdot \mathbf{f}_j)\mathbf{k} \cdot \mathbf{f}_j > 0$, as required, while no restriction is placed on $\mathbf{k} \cdot \mathbf{f}_j$. By (2.13) it follows

$$\begin{aligned}\mathbf{t}(\mathbf{x}, \mathbf{k}) &= \sum_j \mathbf{t}(\mathbf{x}, \mathbf{i}_j)\, \mathbf{k} \cdot \mathbf{i}_j = \sum_j \mathbf{t}(\mathbf{x}, \operatorname{sgn}(\mathbf{k} \cdot \mathbf{f}_j)\mathbf{f}_j) \operatorname{sgn}(\mathbf{k} \cdot \mathbf{f}_j)\mathbf{k} \cdot \mathbf{f}_j \\ &= \sum_j \operatorname{sgn}(\mathbf{k} \cdot \mathbf{f}_j)\mathbf{t}(\mathbf{x}, \mathbf{f}_j) \operatorname{sgn}(\mathbf{k} \cdot \mathbf{f}_j)\mathbf{k} \cdot \mathbf{f}_j = \sum_j \mathbf{t}(\mathbf{x}, \mathbf{f}_j)\, \mathbf{k} \cdot \mathbf{f}_j.\end{aligned}$$

Hence Eq. (2.13) holds for any basis. In particular it holds relative to the basis $\{\mathbf{e}_i\}$ chosen for the representation of vectors and tensors so that

$$\mathbf{t}(\mathbf{x}, \mathbf{n}) = \sum_i \mathbf{t}(\mathbf{x}, \mathbf{e}_i)\, \mathbf{n} \cdot \mathbf{e}_i,$$

for any unit vector $\mathbf{n}$. This means that

$$\mathbf{t}(\mathbf{x}, \mathbf{n}) = [\sum_i \mathbf{t}(\mathbf{x}, \mathbf{e}_i) \otimes \mathbf{e}_i]\mathbf{n}$$

and the conclusion follows with

$$\mathbf{T}(\mathbf{x}) = \sum_i \mathbf{t}(\mathbf{x}, \mathbf{e}_i) \otimes \mathbf{e}_i.$$

□

The intermediate result (2.12) has a remarkable mechanical meaning. Since $\mathbf{t}(\mathbf{n})$ and $\mathbf{t}(-\mathbf{n})$ are the tractions on opposite sides (at a point) of a surface, then (2.12) means that $\mathbf{t}(\mathbf{n})$ and $\mathbf{t}(-\mathbf{n})$ are opposite to each other. This property may be regarded as *Newton's law of action and reaction.* This result is consistent with the fact that letting $\mathbf{k}$ approach any of the vectors $\{\mathbf{i}_j\}$ makes the tetrahedron to approach a membrane with vanishing thickness and hence the volume integral approaches zero. This in turn means that (2.12) expresses the equilibrium of surface forces.

Conceptually, we started with the observation that equalities between surface and volume integrals cannot hold for generic integrands. By Cauchy's theorem, the traction $\mathbf{t}(\mathbf{n})$ is required to depend on $\mathbf{n}$ in the form (2.9). A natural question now arises as to whether Eq. (2.9) makes in fact the whole contact force be equal to a volume integral (sufficient condition) for any boundary surface $\partial\mathcal{P}_t$. Let $\mathbf{a}$ be any constant vector and consider

$$\mathbf{a} \cdot \int_{\partial \mathcal{P}_t} \mathbf{T} \mathbf{n} \, da = \int_{\partial \mathcal{P}_t} (\mathbf{T}^T \mathbf{a}) \cdot \mathbf{n} \, da.$$

By the divergence theorem

$$\int_{\partial \mathcal{P}_t} (\mathbf{T}^T \mathbf{a}) \cdot \mathbf{n} \, da = \int_{\mathcal{P}_t} \nabla \cdot (\mathbf{T}^T \mathbf{a}) \, dv = \int_{\mathcal{P}_t} \partial_{x_j} T_{hj} a_h \, dv = \mathbf{a} \cdot \int_{\mathcal{P}_t} \nabla \cdot \mathbf{T} \, dv$$

provided we define

$$\nabla \cdot \mathbf{T} := \partial_{x_j} T_{hj} \, \mathbf{e}_h.$$

Hence, the balance of linear momentum (2.7) can be written in the form

$$\int_{\mathcal{P}_t} \rho(\dot{\mathbf{v}} - \mathbf{b}) dv = \int_{\mathcal{P}_t} \nabla \cdot \mathbf{T} \, dv$$

or

$$\int_{\mathcal{P}_t} [\rho(\dot{\mathbf{v}} - \mathbf{b}) - \nabla \cdot \mathbf{T}] dv = 0;$$

thanks to the existence of **T** now the balance of linear momentum is expressed by two volume integrals. The arbitrariness of the convecting region $\mathcal{P}_t$ and the assumed continuity of the integrand imply the local balance equation

$$\rho \, \dot{\mathbf{v}} = \nabla \cdot \mathbf{T} + \rho \, \mathbf{b}, \tag{2.14}$$

that is the local *equation of motion*. In suffix notation Eq. (2.14) reads

$$\rho \, \dot{v}_i = \partial_{x_j} T_{ij} + \rho \, b_i.$$

The tensor **T** is called the *stress tensor*. Since **Tn** is the force per unit area, on a positive side of a surface, then **T n** can be decomposed into the sum of two terms. One is the *normal traction*

$$(\mathbf{n} \cdot \mathbf{T}\mathbf{n})\mathbf{n} = (\mathbf{n} \otimes \mathbf{n})\mathbf{T}\mathbf{n}.$$

The other one is the *shear traction*

$$(\mathbf{1} - \mathbf{n} \otimes \mathbf{n})\mathbf{T}\mathbf{n}.$$

Consequently, $\mathbf{T}\mathbf{e}_j$ is the traction on the surface orthogonal to $\mathbf{e}_j$ and $\mathbf{e}_i \cdot \mathbf{T}\mathbf{e}_j$ is the i-th component. Hence

$$T_{ij} := \mathbf{e}_i \cdot \mathbf{T}\mathbf{e}_j$$

is the i-th component of the traction on the positive side of the surface orthogonal to $\mathbf{e}_j$.

By Eq. (1.33), we can write the equation of motion in the form

$$\rho(\partial_t \mathbf{v} + \nabla \tfrac{1}{2}\mathbf{v}^2 + \boldsymbol{\varpi} \times \mathbf{v}) = \nabla \cdot \mathbf{T} + \rho \mathbf{b},$$

where $\boldsymbol{\varpi} = \nabla \times \mathbf{v}$ is the vorticity.

2.3 Balance of Angular Momentum

Let $\bar{O}$ be any point in space and let $\mathbf{r}$ be the position vector of a point of the body with respect to $\bar{O}$. We say that $\bar{O}$ is a *base point* (or reference point). Let $\mathbf{o}$ be the position vector of $\bar{O}$ so that $\mathbf{x} = \mathbf{o} + \mathbf{r}$. We let $\bar{O}$ be fixed in space and hence $\dot{\mathbf{o}} = \mathbf{0}$.

For a system of particles, the angular momentum (or moment of momentum) $\mathfrak{L}$ changes in time by the action of the moment of forces, or torque, $\mathfrak{T}$ according to

$$\dot{\mathfrak{L}}_{\bar{O}} = \mathfrak{T}_{\bar{O}}.$$

Based on this equation, we state the following
Axiom. The motion of a continuous body is governed by the equation

$$\dot{\mathfrak{L}}_{\bar{O}}(\mathcal{P}_t) = \mathfrak{T}_{\bar{O}}(\mathcal{P}_t), \tag{2.15}$$

for any convecting region $\mathcal{P}_t$.

For a continuous body, we let

$$\mathfrak{L}_{\bar{O}}(\mathcal{P}_t) := \int_{\mathcal{P}_t} \mathbf{r} \times \rho \mathbf{v}\, dv.$$

The moment (or torque) $\mathfrak{T}_{\bar{O}}(\mathcal{P}_t)$ acting on $\mathcal{P}_t$ consists of the terms due to the body force $\mathbf{b}$ and the traction $\mathbf{T}\,\mathbf{n}$,

$$\mathfrak{T}_{\bar{O}}(\mathcal{P}_t) = \int_{\partial\mathcal{P}_t} \mathbf{r} \times \mathbf{T}\,\mathbf{n}\, da + \int_{\mathcal{P}_t} \mathbf{r} \times \rho \mathbf{b}\, dv.$$

Equation (2.15) can then be written in the form

$$\frac{d}{dt}\int_{\mathcal{P}_t} \mathbf{r} \times \rho \mathbf{v}\, dv = \int_{\partial\mathcal{P}_t} \mathbf{r} \times \mathbf{T}\,\mathbf{n}\, da + \int_{\mathcal{P}_t} \mathbf{r} \times \rho \mathbf{b}\, dv. \tag{2.16}$$

First observe that

$$\overline{\mathbf{r} \times \mathbf{v}}^{\boldsymbol{\cdot}} = \dot{\mathbf{r}} \times \mathbf{v} + \mathbf{r} \times \dot{\mathbf{v}} = \mathbf{v} \times \mathbf{v} + \mathbf{r} \times \dot{\mathbf{v}} = \mathbf{r} \times \dot{\mathbf{v}}.$$

Thus, by Reynolds' transport theorem

$$\frac{d}{dt}\int_{\mathcal{P}_t} \mathbf{r} \times \rho \mathbf{v}\, dv = \int_{\mathcal{P}_t} \mathbf{r} \times \rho \dot{\mathbf{v}}\, dv.$$

The surface integral can be rearranged by observing that

$$\mathbf{r} \times \mathbf{T}\,\mathbf{n} = \epsilon_{ijk} r_j T_{kh} n_h \mathbf{e}_i$$

and hence

$$\int_{\partial \mathcal{P}_t} \mathbf{r} \times \mathbf{T}\,\mathbf{n}\, da = \epsilon_{ijk} \mathbf{e}_i \int_{\partial \mathcal{P}_t} r_j T_{kh} n_h \, da.$$

By the divergence theorem, it follows

$$\int_{\partial \mathcal{P}_t} r_j T_{kh} n_h \, da = \int_{\mathcal{P}_t} \partial_{x_h}(r_j T_{kh}) dv = \int_{\mathcal{P}_t} (\delta_{hj} T_{kh} + r_j \partial_{x_h} T_{kh}) dv.$$

Consequently, we have

$$\epsilon_{ijk} \mathbf{e}_i \int_{\partial \mathcal{P}_t} r_j T_{kh} n_h \, da = \int_{\mathcal{P}_t} (\boldsymbol{\Upsilon} + \mathbf{r} \times \nabla \cdot \mathbf{T}) dv, \qquad \boldsymbol{\Upsilon} := \epsilon_{ijk} T_{kj} \mathbf{e}_i,$$

and Eq. (2.16) yields

$$\int_{\mathcal{P}_t} [\mathbf{r} \times (\rho \dot{\mathbf{v}} - \nabla \cdot \mathbf{T} - \rho \mathbf{b}) - \boldsymbol{\Upsilon}] dv = \mathbf{0}.$$

In view of the equation of motion (2.14), it follows

$$\int_{\mathcal{P}_t} \boldsymbol{\Upsilon}\, dv = \mathbf{0}.$$

The arbitrariness of $\mathcal{P}_t$ and the assumed continuity of $\boldsymbol{\Upsilon}$ imply[1]

$$\boldsymbol{\Upsilon} = \mathbf{0}. \tag{2.17}$$

Conversely, if (2.14) and (2.17) hold, then (2.7) and (2.16) hold too. Equation (2.17) means that

$$\epsilon_{ijk} T_{kj} = 0, \qquad i = 1, 2, 3.$$

Hence $T_{12} = T_{21}, T_{13} = T_{31}, T_{23} = T_{32}$ or

$$\mathbf{T} = \mathbf{T}^T.$$

To sum up, we can write

Proposition 2.1 *For smooth fields* $\rho, \mathbf{v}, \mathbf{b}, \mathbf{T}$, *the global balance Eqs. (2.7) and (2.16) imply that the local balance Eqs. (2.14) and (2.17) hold.*

---

[1] Observe $\boldsymbol{\Upsilon}$ is twice the axial vector of skw$\mathbf{T}$.

For completeness, we now determine the balance of angular momentum relative to a moving base point. Let B be the moving base point and let $\mathbf{r}_B$ be the position vector of a point relative to B. Let $\mathbf{d}$ be the position vector of B, $\mathbf{d} = B - \bar{O}$, so that $\mathbf{r} = \mathbf{d} + \mathbf{r}_B$ is the position vector relative to $\bar{O}$, supposed to be fixed. First observe

$$\mathfrak{T}_{\bar{O}} = \int_{\mathcal{P}_t} \mathbf{r} \times \rho\mathbf{b}\, dv + \int_{\partial\mathcal{P}_t} \mathbf{r} \times \mathbf{T}\mathbf{n}\, da = \int_{\mathcal{P}_t} \mathbf{r}_B \times \rho\mathbf{b}\, dv + \int_{\partial\mathcal{P}_t} \mathbf{r}_B \times \mathbf{T}\mathbf{n}\, da + \mathbf{d} \times \mathfrak{F}(\mathcal{P}_t),$$

namely

$$\mathfrak{T}_{\bar{O}} = \mathfrak{T}_B + \mathbf{d} \times \mathfrak{F}(\mathcal{P}_t).$$

Now, by

$$\mathfrak{L}_{\bar{O}}(\mathcal{P}_t) = \int_{\mathcal{P}_t} \mathbf{r} \times \rho\mathbf{v}\, dv = \int_{\mathcal{P}_t} \mathbf{r}_B \times \rho\mathbf{v}\, dv + \mathbf{d} \times \int_{\mathcal{P}_t} \rho\mathbf{v}\, dv$$

we obtain

$$\frac{d}{dt} \int_{\mathcal{P}_t} \mathbf{r}_B \times \rho\mathbf{v}\, dv = \frac{d}{dt}\mathfrak{L}_{\bar{O}}(\mathcal{P}_t) - \mathbf{v}_B \times \int_{\mathcal{P}_t} \rho\mathbf{v}\, dv - \mathbf{d} \times \int_{\mathcal{P}_t} \rho\dot{\mathbf{v}}\, dv.$$

Consequently

$$\frac{d}{dt}\mathfrak{L}_B(\mathcal{P}_t) = \mathfrak{T}_{\bar{O}} - \mathbf{d} \times \mathfrak{F}(\mathcal{P}_t) - \mathbf{v}_B \times \mathfrak{P},$$

whence

$$\frac{d}{dt}\mathfrak{L}_B(\mathcal{P}_t) = \mathfrak{T}_B - \mathbf{v}_B \times \mathfrak{P},$$

as it happens in the classical mechanics of systems of particles.

As is the case in electromagnetism, there are continua subject to couple fields. This means that there is a couple density field,[2] per unit mass, $\mathbf{l}$ such that

$$\int_{\mathcal{P}_t} \rho\mathbf{l}\, dv$$

is the couple acting on $\mathcal{P}_t$. In this circumstance, the balance of angular momentum involves the new term $\rho\mathbf{l}$ so that

$$\frac{d}{dt} \int_{\mathcal{P}_t} \mathbf{r} \times \rho\mathbf{v}\, dv = \int_{\partial\mathcal{P}_t} \mathbf{r} \times \mathbf{T}\mathbf{n}\, da + \int_{\mathcal{P}_t} \rho[\mathbf{l} + \mathbf{r} \times \mathbf{b}] dv.$$

By repeating the previous procedure, we obtain

$$\int_{\mathcal{P}_t} (\rho\mathbf{l} + \boldsymbol{\Upsilon}) dv = \mathbf{0},$$

[2] Or couple per unit volume $\mathbf{t} = \rho\mathbf{l}$.

whence

$$\rho\,\mathbf{l} + \boldsymbol{\Upsilon} = \mathbf{0}.$$

Hence, $\mathbf{T}$ is (no longer symmetric and is) required to satisfy the condition

$$\epsilon_{ijk} T_{kj} + \rho l_i = 0, \qquad i = 1, 2, 3.$$

As an example, if $\rho\,\mathbf{l} = \mathbf{M} \times \mathbf{B}$ then

$$\epsilon_{ijk}(T_{kj} + M_j B_k) = 0.$$

Consequently

$$\operatorname{skw}\mathbf{T} = \operatorname{skw}\mathbf{M} \otimes \mathbf{B}$$

or

$$\mathbf{T} + \mathbf{B} \otimes \mathbf{M} \in \mathrm{Sym}.$$

2.4 Balance of Energy

Consider a system of N particles subject to the equations of motion

$$m_i \dot{\mathbf{v}}_i = \mathbf{f}_i, \qquad i = 1, 2, ..., N,$$

$\mathbf{f}_i$ denoting the force acting on the i-th particle. Inner multiply by $\mathbf{v}_i$,

$$m_i \mathbf{v}_i \cdot \dot{\mathbf{v}}_i = \mathbf{v}_i \cdot \mathbf{f}_i,$$

and sum over i to obtain

$$\frac{d}{dt}\tfrac{1}{2}\textstyle\sum_i m_i \mathbf{v}_i^2 = \sum_i \mathbf{v}_i \cdot \mathbf{f}_i;$$

the derivative of the kinetic energy is equal to the power of the forces applied to the system. This might suggest that an analogous balance law occurs in continuum mechanics provided appropriate definitions are considered for the kinetic energy and the power.

The kinetic energy $K(\mathcal{P}_t)$ of the convecting subregion $\mathcal{P}_t$ is defined by

$$K(\mathcal{P}_t) = \int_{\mathcal{P}_t} \tfrac{1}{2}\rho \mathbf{v}^2\, dv.$$

The power $\Pi(\mathcal{P}_t)$ on $\mathcal{P}_t$ of the body force and traction fields is defined as

$$\Pi(\mathcal{P}_t) = \int_{\mathcal{P}_t} \rho\mathbf{b} \cdot \mathbf{v}\, dv + \int_{\partial\mathcal{P}_t} \mathbf{v} \cdot \mathbf{T}\mathbf{n}\, da.$$

We might expect that any subregion obeys the balance equation

$$\frac{d}{dt}K(\mathcal{P}_t) = \Pi(\mathcal{P}_t). \tag{2.18}$$

Now, by Reynolds' transport theorem

$$\frac{d}{dt}K(\mathcal{P}_t) = \int_{\mathcal{P}_t} \rho\,\mathbf{v}\cdot\dot{\mathbf{v}}\,dv.$$

We employ the identity

$$\mathbf{v}\cdot(\nabla\cdot\mathbf{T}) = v_i\partial_{x_j}T_{ij} = \partial_{x_j}(v_iT_{ij}) - T_{ij}\partial_{x_j}v_i = \nabla\cdot(\mathbf{vT}) - \mathbf{T}\cdot\mathbf{L}$$

and observe that since $\mathbf{T}$ is symmetric then

$$\mathbf{T}\cdot\mathbf{L} = \mathbf{T}\cdot\mathbf{D}.$$

Consequently, upon substitution of $\rho\dot{\mathbf{v}} = \nabla\cdot\mathbf{T} + \rho\,\mathbf{b}$, it follows

$$\frac{d}{dt}K(\mathcal{P}_t) = \int_{\mathcal{P}_t} \rho\,\mathbf{b}\cdot\mathbf{v}\,dv + \int_{\partial\mathcal{P}_t} \mathbf{v}\cdot\mathbf{Tn}da - \int_{\mathcal{P}_t} \mathbf{T}\cdot\mathbf{D}\,dv.$$

Hence, we deduce

$$\frac{d}{dt}K(\mathcal{P}_t) = \Pi(\mathcal{P}_t) - \int_{\mathcal{P}_t} \mathbf{T}\cdot\mathbf{D}\,dv.$$

This relation is the content of the *kinetic energy theorem* (or work-energy theorem or *living forces* method according to Euler). This in turn implies that Eq. (2.18) holds only if $\mathbf{T}\cdot\mathbf{D} = 0$.

Alternatively, if we start from (2.18)

$$\frac{d}{dt}\int_{\mathcal{P}_t} \tfrac{1}{2}\rho\,\mathbf{v}^2\,dv = \int_{\mathcal{P}_t} \rho\,\mathbf{b}\cdot\mathbf{v}\,dv + \int_{\partial\mathcal{P}_t} \mathbf{v}\cdot\mathbf{Tn}\,da$$

and employ the divergence theorem so that

$$\int_{\partial\mathcal{P}_t} \mathbf{v}\cdot\mathbf{Tn}da = \int_{\mathcal{P}_t}[\mathbf{v}\cdot(\nabla\cdot\mathbf{T}) + \mathbf{T}\cdot\mathbf{D}]dv$$

then we obtain

$$\int_{\mathcal{P}_t} [\mathbf{v}\cdot(\rho\dot{\mathbf{v}} - \rho\mathbf{b} - \nabla\cdot\mathbf{T}) + \mathbf{T}\cdot\mathbf{D}]dv = 0.$$

Owing to the equation of motion (2.14), it follows

$$\int_{\mathcal{P}_t} \mathbf{T}\cdot\mathbf{D}\,dv = 0, \qquad \mathbf{T}\cdot\mathbf{D} = 0.$$

Both views show that a seemingly redundant term $\mathbf{T}\cdot\mathbf{D}$ occurs in the balance of energy. Incidentally, for an inviscid incompressible fluid $\mathbf{T}\cdot\mathbf{D} = 0$ but this is merely a particular case of continuous body.

Conceptually, a simple and natural way to overcome this difficulty is to regard (2.18) as an overly poor scheme in that a correct balance of energy should involve also fields of non-mechanical character. In this sense, we proceed by keeping the mathematical structure of the time derivative of a volume integral equal to the sum of a volume integral and a surface integral

$$\frac{d}{dt}\int_{\mathcal{P}_t} \rho u\, dv = \int_{\mathcal{P}_t} \rho\beta\, dv + \int_{\partial\mathcal{P}_t} s\, da.$$

We let

$$u = \tfrac{1}{2}\mathbf{v}^2 + \varepsilon, \qquad \beta = \mathbf{b}\cdot\mathbf{v} + r, \qquad s = \mathbf{v}\cdot\mathbf{Tn} + h$$

and so ε is the energy density (per unit mass), r is the power supply, and h is the power per unit area, all of them in addition to the corresponding mechanical term $(\frac{1}{2}\mathbf{v}^2, \mathbf{b}\cdot\mathbf{v}, \mathbf{v}\cdot\mathbf{Tn})$.

Based on these arguments, we assume the existence of fields ε, r, h and state the following

Axiom. The time evolution of energy of a continuous body is governed by the equation

$$\frac{d}{dt}\int_{\mathcal{P}_t} \rho(\tfrac{1}{2}\mathbf{v}^2 + \varepsilon)dv = \int_{\mathcal{P}_t} \rho(\mathbf{b}\cdot\mathbf{v} + r)dv + \int_{\partial\mathcal{P}_t} (\mathbf{v}\cdot\mathbf{Tn} + h)da, \tag{2.19}$$

for any convecting region $\mathcal{P}_t$. This axiom can be viewed as the *first principle of thermodynamics;* see, e.g. [111]. As we show shortly, the first law of thermodynamics is a consequence of the axiom (2.19).

We first observe that (2.19) involves both volume and surface integrals. On the basis of Cauchy's theorem, we expect that the function h be subject to an appropriate requirement. Yet it is convenient to simplify the form of (2.19) by means of Reynolds' theorem and the equation of motion. From

$$\frac{d}{dt}\int_{\mathcal{P}_t} \rho(\tfrac{1}{2}\mathbf{v}^2 + \varepsilon)dv - \int_{\mathcal{P}_t} \rho(\mathbf{b}\cdot\mathbf{v} + r)dv - \int_{\partial\mathcal{P}_t} \mathbf{v}\cdot\mathbf{Tn}\, da = \int_{\partial\mathcal{P}_t} h\, da$$

it follows

$$\int_{\mathcal{P}_t} (\rho\dot{\varepsilon} - \rho r - \mathbf{T}\cdot\mathbf{D})dv = \int_{\partial\mathcal{P}_t} h\, da. \tag{2.20}$$

If $\rho(\dot{\varepsilon} - r) - \mathbf{T}\cdot\mathbf{D}$ is bounded and $h(\mathbf{x}, \mathbf{n}, t)$ is continuous, then by arguing as in Sect. 2.2, we can state the necessary condition

$$\frac{1}{\text{area}\,\partial\mathcal{P}_t}\int_{\partial\mathcal{P}_t} h\,da \to 0 \quad \text{as} \quad \frac{\text{vol}\mathcal{P}_t}{\text{area}\,\partial\mathcal{P}_t} \to 0.$$

We can then develop the analogue of Cauchy's theorem, step by step, to obtain

$$h(\mathbf{x}, \mathbf{n}) = [\textstyle\sum_i h(\mathbf{x}, \mathbf{e}_i)\mathbf{e}_i] \cdot \mathbf{n}.$$

This proves that there exists a vector function, say

$$\mathbf{q}(\mathbf{x}) = -\textstyle\sum_i h(\mathbf{x}, \mathbf{e}_i)\mathbf{e}_i,$$

such that $h = -\mathbf{q} \cdot \mathbf{n}$.

Substitution of h in (2.20) and use of the divergence theorem yield

$$\int_{\mathcal{P}_t} (\rho\dot{\varepsilon} - \rho r - \mathbf{T} \cdot \mathbf{D} + \nabla \cdot \mathbf{q}) dv = 0,$$

whence

$$\rho\dot{\varepsilon} = \rho r + \mathbf{T} \cdot \mathbf{D} - \nabla \cdot \mathbf{q}. \tag{2.21}$$

As we will see in the next chapters, the balance of energy comprises further terms (powers) if e.g. electromagnetic fields occur or the body has a non-simple character (internal structure).

2.4.1 First Law of Thermodynamics

After the mathematical derivation of (2.21), it is natural to inquiry about the physical relevance. For any subregion $\mathcal{P}_t$, integration of (2.21) and use of Reynolds' theorem yield

$$\frac{d}{dt}\int_{\mathcal{P}_t} \rho\varepsilon\, dv = \int_{\mathcal{P}_t} (\rho r + \mathbf{T} \cdot \mathbf{D}) dv - \int_{\partial\mathcal{P}_t} \mathbf{q} \cdot \mathbf{n}\, da. \tag{2.22}$$

Again we have a balance law where the time derivative of a set function is given by a surface integral and a volume integral. Now let

$$U(\mathcal{P}_t) = \int_{\mathcal{P}_t} \rho\varepsilon\, dv, \qquad \mathcal{W}(\mathcal{P}_t) = \int_{\mathcal{P}_t} \mathbf{T} \cdot \mathbf{D}\, dv, \qquad \mathcal{Q}(\mathcal{P}_t) = \int_{\mathcal{P}_t} \rho r\, dv - \int_{\partial\mathcal{P}_t} \mathbf{q} \cdot \mathbf{n}\, da.$$

The same equation, in the form

$$\frac{d}{dt} U(\mathcal{P}_t) = \mathcal{W}(\mathcal{P}_t) + \mathcal{Q}(\mathcal{P}_t), \tag{2.23}$$

allows us to view $\mathcal{W}$ as a power of mechanical character and $\mathcal{Q}$ as a power of non-mechanical character. Hence, $\mathbf{T} \cdot \mathbf{D}$ is the power density of mechanical character that contributes to the rate of energy $dU(\mathcal{P}_t)/dt$. In the purely-mechanical balance (2.18), the power density $\mathbf{T} \cdot \mathbf{D}$ could not contribute to any balance law.

The power $\mathcal{Q}(\mathcal{P}_t)$ is the *heat flow* ; the model involves a heat flux vector $\mathbf{q}$ and a scalar heat supply ρr. The occurrence of the $-$ sign in front of the surface integral arises merely by the definition of $\mathbf{q}$. By (2.23), with $r = 0$ and $\mathbf{D} = \mathbf{0}$ (rigid motions), we have

$$\frac{d}{dt} U(\mathcal{P}_t) = - \int_{\partial \mathcal{P}_t} \mathbf{q} \cdot \mathbf{n} \, da.$$

If $\mathbf{q} \cdot \mathbf{n} < 0$ then $dU(\mathcal{P}_t)/dt > 0$ and vice versa. Since $\mathbf{n}$ is the (unit) outward normal to $\partial \mathcal{P}_t$ then $\mathbf{q} \cdot \mathbf{n} < 0$ when $\mathbf{q}$ points into $\mathcal{P}_t$. This allows $\mathbf{q}$ to be viewed as a flux of energy (heat flux); if $\mathbf{q}$ points into $\mathcal{P}_t$ then $dU(\mathcal{P}_t)/dt > 0$ and the energy $U(\mathcal{P}_t)$ increases.

As to the mechanical power $\mathcal{W}$, it is worth considering the simplest case when

$$\mathbf{T} = -p \, \mathbf{1}, \qquad p > 0,$$

which happens in inviscid fluids. Then, in view of (2.4),

$$\mathbf{T} \cdot \mathbf{D} = -p \, \nabla \cdot \mathbf{v} = p \, \frac{\dot{\rho}}{\rho}.$$

If, further, ρ is uniform (independent of the position) then $\rho = M/V$, M being the mass and V being the volume of $\mathcal{P}_t$. Since M is constant, then

$$\frac{\dot{\rho}}{\rho} = -\frac{\dot{V}}{V}, \qquad \mathbf{T} \cdot \mathbf{D} = -p\dot{V}.$$

If p is uniform too, then

$$\int_{\mathcal{P}_t} \mathbf{T} \cdot \mathbf{D} \, dv = V \, \mathbf{T} \cdot \mathbf{D} = -p\dot{V}.$$

Hence, (2.23) becomes

$$\frac{d}{dt} U(\mathcal{P}_t) = -p\dot{V} + \mathcal{Q}(\mathcal{P}_t).$$

In the well-known terminology of classical physics, Eq. (2.23) is the *first law of thermodynamics* . In the continuum physics framework, Eqs. (2.23) and (2.21) are the global and local forms of the first law.

In the physical literature, the formal statement of the first law is a bit more involved. For instance, in [449], Chap. 4, for "a process involving only infinitesimal changes" the first law is written in the form

$$dU = đQ - đW,$$

(or $dE = \delta Q - \delta W$ in [316], Sect. 2.5), the use of $đ$ instead of d being due to the fact that the infinitesimal amount is an inexact differential [449] or because Q and W are not properties [316]. If "the infinitesimal process is quasi-static", then $đW = p\,dV$, which is consistent with $-p\dot{V}$ as the mechanical power if p is uniform. In both cases, the $-$ sign in front of $đW$ is due to the view that $đW$ is energy transferred out of the system.

External and Internal Powers

The literature shows balances of energy in terms of external and internal powers. To illustrate the concept, we go back to the balance of kinetic energy and observe that, in view of the equation of motion

$$\frac{d}{dt} K(\mathcal{P}_t) = \int_{\mathcal{P}_t} \mathbf{v} \cdot \mathbf{T}\mathbf{n}\, da - \int_{\mathcal{P}_t} \mathbf{T} \cdot \mathbf{D}\, dv + \int_{\mathcal{P}_t} \rho \mathbf{b} \cdot \mathbf{v}\, dv. \tag{2.24}$$

The field $\mathbf{T} \cdot \mathbf{D}$ is said to be the power expended within $\mathcal{P}_t$ and then the integral is referred to as the *internal power.* What we denoted by $\Pi(\mathcal{P}_t)$ instead is regarded as external power. Hence, upon the form[3]

$$\underbrace{\int_{\partial\mathcal{P}_t} \mathbf{v} \cdot \mathbf{T}\mathbf{n}\, da + \int_{\mathcal{P}_t} \rho \mathbf{b} \cdot \mathbf{v}\, dv}_{\text{external power}} = \underbrace{\int_{\mathcal{P}_t} \mathbf{T} \cdot \mathbf{D}\, dv}_{\text{internal power}} + \underbrace{\frac{d}{dt} \int_{\mathcal{P}_t} \tfrac{1}{2} \rho \mathbf{v}^2\, dv}_{\text{kinetic—energy rate}}$$

of the balance of (kinetic) energy, the view is stated that the external power expended on $\mathcal{P}_t$ is balanced by the sum of the internal power expended within $\mathcal{P}_t$ and the time derivative of the kinetic energy of $\mathcal{P}_t$.

2.4.2 Balance of Virtual Power

The literature exhibits a remarkable attention to the use of principles of virtual power in various contexts.[4] Here we show a motivation of the principle based on the balance of kinetic energy.

The balance of kinetic energy (2.24) can be written in the form

$$\int_{\partial\mathcal{P}_t} \mathbf{v} \cdot \mathbf{T}\mathbf{n}\, da + \int_{\mathcal{P}_t} \rho \mathbf{v} \cdot (\mathbf{b} - \dot{\mathbf{v}}) dv = \int_{\mathcal{P}_t} \mathbf{T} \cdot \mathbf{D}\, dv.$$

[3] See, e.g., [216], p. 143.

[4] See, e.g. the book [189] and the review paper [302]; see also [367].

The left-hand side is viewed as the external power, while the right-hand side as the internal power. Now drop the suffix t to $\mathcal{P}_t$ in that the assumption of convecting region is here inessential. Moreover we write the (internal) power $\mathbf{T} \cdot \mathbf{D}$ as $\mathbf{T} \cdot \mathbf{L}$, which corresponds to ignoring the symmetry condition $\mathbf{T} \in \mathrm{Sym}$. Further, we ignore Cauchy's relation $\mathbf{t}(\mathbf{n}) = \mathbf{T}\mathbf{n}$. Hence, we consider the power balance

$$\int_{\partial\mathcal{P}} \mathbf{v} \cdot \mathbf{t}(\mathbf{n})\, da + \int_{\mathcal{P}} \rho\mathbf{v} \cdot (\mathbf{b} - \dot{\mathbf{v}}) dv = \int_{\mathcal{P}} \mathbf{T} \cdot \mathbf{L}\, dv.$$

Now, instead of the effective fields $\mathbf{v}$ and $\mathbf{L}$, we insert the virtual velocity $\tilde{\mathbf{v}}$, regarded as an *arbitrary* vector field; $\tilde{\mathbf{v}}$ need not coincide with the velocity field, though it may, and need not have the dimension of a velocity. Hence, let

$$\mathcal{W}(\mathcal{P}, \tilde{\mathbf{v}}) := \int_{\partial\mathcal{P}} \tilde{\mathbf{v}} \cdot \mathbf{t}(\mathbf{n})\, da + \int_{\mathcal{P}} \rho\tilde{\mathbf{v}} \cdot (\mathbf{b} - \dot{\mathbf{v}}) dv, \qquad \mathcal{I}(\mathcal{P}, \tilde{\mathbf{v}}) := \int_{\mathcal{P}} \mathbf{T} \cdot \nabla\tilde{\mathbf{v}}\, dv.$$

be the virtual expenditures of external and internal power.

Principle of virtual power. *For any time t and region $\mathcal{P}$ of the body the virtual power balance*

$$\mathcal{W}(\mathcal{P}, \tilde{\mathbf{v}}) = \mathcal{I}(\mathcal{P}, \tilde{\mathbf{v}}),$$

holds for any arbitrary function $\tilde{\mathbf{v}}(\mathbf{x})$, $\mathbf{x} \in \mathcal{P}$.

We now show the consequences of the principle. Observe that

$$\int_{\mathcal{P}} \mathbf{T} \cdot \nabla\tilde{\mathbf{v}}\, dv = \int_{\partial\mathcal{P}} \tilde{\mathbf{v}} \cdot \mathbf{T}\mathbf{n}\, da - \int_{\mathcal{P}} (\nabla \cdot \mathbf{T}) \cdot \tilde{\mathbf{v}}\, dv.$$

Hence $\mathcal{W}(\mathcal{P}, \tilde{\mathbf{v}}) = \mathcal{I}(\mathcal{P}, \tilde{\mathbf{v}})$ results in

$$\int_{\mathcal{P}} (\nabla \cdot \mathbf{T} + \rho\mathbf{b} - \rho\dot{\mathbf{v}}) \cdot \tilde{\mathbf{v}}\, dv + \int_{\partial\mathcal{P}} (\mathbf{t}(\mathbf{n}) - \mathbf{T}\mathbf{n}) \cdot \tilde{\mathbf{v}}\, da = 0.$$

By the arbitrariness of $\tilde{\mathbf{v}}$, we first let $\tilde{\mathbf{v}} = \mathbf{0}$ on the boundary $\partial\mathcal{P}$. Consequently

$$\int_{\mathcal{P}} (\nabla \cdot \mathbf{T} + \rho\mathbf{b} - \rho\dot{\mathbf{v}}) \cdot \tilde{\mathbf{v}}\, dv = \mathbf{0}.$$

Now, for any $\mathbf{x} \in \mathcal{P} \setminus \partial\mathcal{P}$, we let $\tilde{\mathbf{v}} = \alpha(\mathbf{y})[\nabla \cdot \mathbf{T} + \rho\mathbf{b} - \rho\dot{\mathbf{v}}]$, α being positive in an open neighbourhood $\mathcal{N}$ of $\mathbf{x}$, $\mathcal{N} \subset \mathcal{P}$, and equal to zero outside. It follows $\nabla \cdot \mathbf{T} + \rho\mathbf{b} - \rho\dot{\mathbf{v}} = \mathbf{0}$ as $\mathbf{y} \in \mathcal{N}$. The arbitrariness of $\mathbf{x}$ implies that

$$\nabla \cdot \mathbf{T} + \rho\mathbf{b} - \rho\dot{\mathbf{v}} = \mathbf{0}$$

in $\mathcal{P}$. Thus

$$\int_{\partial\mathcal{P}} (\mathbf{t}(\mathbf{n}) - \mathbf{T}\mathbf{n}) \cdot \tilde{\mathbf{v}}\, da = 0.$$

The arbitrariness of $\tilde{\mathbf{v}}$ implies that $\mathbf{t}(\mathbf{n}) = \mathbf{T}\mathbf{n}$ on $\partial\mathcal{P}$. So, the principle implies the equation of motion and Cauchy's relation

$$\nabla \cdot \mathbf{T} + \rho\mathbf{b} - \rho\dot{\mathbf{v}} = \mathbf{0}, \qquad \mathbf{t}(\mathbf{n}) = \mathbf{T}\mathbf{n}.$$

Let $\tilde{\mathbf{x}}$ be any motion, $\dot{\tilde{\mathbf{x}}} = \tilde{\mathbf{v}}$. By a Euclidean transformation, $\tilde{\mathbf{x}}^*$ and $\tilde{\mathbf{v}}^*$ become

$$\tilde{\mathbf{x}}^* = \mathbf{c} + \mathbf{Q}\tilde{\mathbf{x}}, \qquad \tilde{\mathbf{v}}^* = \dot{\mathbf{c}} + \dot{\mathbf{Q}}\tilde{\mathbf{x}} + \mathbf{Q}\tilde{\mathbf{v}}.$$

The corresponding velocity gradients are related by

$$\tilde{\mathbf{L}}^* = \mathbf{Q}\tilde{\mathbf{L}}\mathbf{Q}^T + \boldsymbol{\Omega},$$

where $\boldsymbol{\Omega} = \dot{\mathbf{Q}}\mathbf{Q}^T \in \mathrm{Skw}$. Since

$$\det \partial_{\mathbf{x}^*}\mathbf{x} = \det \mathbf{Q}^T = 1,$$

the invariance of $\mathcal{I}(\mathcal{P}, \tilde{\mathbf{v}})$ under Euclidean transformations results in

$$\mathbf{T} \cdot \tilde{\mathbf{L}} = \mathbf{T}^* \cdot (\mathbf{Q}\tilde{\mathbf{L}}\mathbf{Q}^T + \boldsymbol{\Omega}) = \mathbf{Q}^T\mathbf{T}^*\mathbf{Q} \cdot \tilde{\mathbf{L}} + \mathbf{T}^* \cdot \boldsymbol{\Omega}.$$

First let $\mathbf{Q}$ be constant so that $\boldsymbol{\Omega} = \mathbf{0}$. Hence, the invariance of $\mathcal{I}(\mathcal{P}, \tilde{\mathbf{v}})$ results in

$$(\mathbf{T} - \mathbf{Q}^T\mathbf{T}^*\mathbf{Q}) \cdot \tilde{\mathbf{L}} = 0.$$

The arbitrariness of $\tilde{\mathbf{L}}$ implies that

$$\mathbf{T}^* = \mathbf{Q}\mathbf{T}\mathbf{Q}^T$$

for all rotations $\mathbf{Q}$; consequently $\mathbf{T}$ is a (objective) tensor. The remaining condition

$$\mathbf{T}^* \cdot \boldsymbol{\Omega} = 0$$

for any skew tensor $\boldsymbol{\Omega}$ implies that $\mathbf{T}^*$, and hence $\mathbf{T}$, is symmetric.

To sum up, the principle of virtual power, that is $\mathcal{W}(\mathcal{P}, \tilde{\mathbf{v}}) = \mathcal{I}(\mathcal{P}, \tilde{\mathbf{v}})$ for any subregion $\mathcal{P}$ and any field $\tilde{\mathbf{v}}$, implies
Cauchy's relation $\mathbf{t}(\mathbf{n}) = \mathbf{T}\mathbf{n}$, the (local) equation of motion $\rho\dot{\mathbf{v}} = \nabla \cdot \mathbf{T} + \rho\mathbf{b}$, the symmetry of $\mathbf{T}$, and the (objective) tensor character of the stress, $\mathbf{T}^* = \mathbf{Q}\mathbf{T}\mathbf{Q}^T$, for any rotation $\mathbf{Q}$. The fact that the principle of virtual power yields eventually the boundary condition ($\mathbf{t}(\mathbf{n}) = \mathbf{T}\mathbf{n}$), the local balance equation, the symmetry and the objective character of $\mathbf{T}$ might suggest the recourse to this principle, especially in connection with more involved or non-classical settings. Yet a question arises about the consistent statement of a principle of virtual power. It seems then that, for a correct modelling of material behaviour, a thorough analysis of the balance equations is unavoidable.

2.5 Material Forms of the Balance Equations

The balance laws considered so far are expressed in the form

$$\frac{d}{dt}\int_{\mathcal{P}_t} \rho\varphi \, dv = \int_{\mathcal{P}_t} \beta \, dv + \int_{\partial\mathcal{P}_t} s \, da,$$

where φ is a density (per unit mass) while β and s represent the density fields, per unit volume and unit area, that determine the evolution of φ. As in Cauchy's theorem, we can prove that s has to be linear in the unit normal $\mathbf{n}$ to the boundary; let $\mathbf{h}$ be such that $s = \mathbf{h} \cdot \mathbf{n}$. Hence

$$\frac{d}{dt}\int_{\mathcal{P}_t} \rho\varphi \, dv = \int_{\mathcal{P}_t} \beta \, dv + \int_{\partial\mathcal{P}_t} \mathbf{h} \cdot \mathbf{n} \, da. \tag{2.25}$$

If φ is a scalar (vector), then β is a scalar (vector) and $\mathbf{h}$ is a vector (tensor). This is the generic form of a balance law in the spatial (or Eulerian) description in that it involves functions $\varphi(\mathbf{x}, t)$, where $\mathbf{x} \in \mathcal{P}_t \subseteq \mathcal{R}_t \subset \mathcal{E}$. Here we establish the dual form of balance equations in the material (or Lagrangian or referential) description that is the corresponding law relative to $\mathrm{P} = \chi^{-1}(\mathcal{P}_t) \subseteq \mathrm{R}$.

The key point is the transformation of integrals induced by the change of variable $\mathbf{x} \to \mathbf{X} = \chi^{-1}(\mathbf{x}, t)$. As shown in Sect. 1.2.1

$$\int_{\mathcal{S}_t} g \, \mathbf{n} \, da = \int_{\mathrm{S}} g \, J \, \mathbf{F}^{-T} \mathbf{n}_R \, da_R,$$

where $\mathcal{S}_t = \chi(\mathrm{S}, t)$. Hence, by the change of variable $\mathbf{x} \to \mathbf{X} = \chi^{-1}(\mathbf{x}, t)$ the generic balance equation (2.25) can be written in the form

$$\frac{d}{dt}\int_{\mathrm{P}} \rho_R \varphi \, dv_R = \int_{\mathrm{P}} J \, \beta \, dv_R + \int_{\partial\mathrm{P}} \mathbf{h} \cdot J \, \mathbf{F}^{-T} \mathbf{n}_R \, da_R,$$

where $\rho_R = J \, \rho$. It is advantageous in calculations that $\rho_R = \rho_R(\mathbf{X})$.

Observe that

$$\mathbf{h} \cdot \mathbf{F}^{-T} \mathbf{n}_R = \mathbf{h} \mathbf{F}^{-T} \cdot \mathbf{n}_R$$

and let

$$\mathbf{h}_R := J \, \mathbf{h} \mathbf{F}^{-T}. \tag{2.26}$$

Hence the divergence theorem, the arbitrariness of P and the assumed continuity of the integrands imply

$$\rho_R \dot{\varphi} = J \, \beta + \nabla_R \cdot \mathbf{h}_R.$$

Thus, there is the following 1–1 correspondence between balance equations in the spatial and the material descriptions

$$\rho\dot{\varphi} = \beta + \nabla \cdot \mathbf{h} \quad \Longleftrightarrow \quad \rho_R\dot{\varphi} = J\,\beta + \nabla_R \cdot \mathbf{h}_R, \tag{2.27}$$

$\mathbf{h}$, $\mathbf{h}_R$ being related by (2.26). The correspondence (2.27) follows also directly multiplying $\rho\dot{\varphi} = \beta + \nabla \cdot \mathbf{h}$ by J and observing that $\rho J = \rho_R$ and

$$\partial_{X_K} h_K^R = h_i\, \partial_{X_K}(J F_{Ki}^{-1}) + J F_{Ki}^{-1}(\partial_{x_p} h_i) F_{pK} = h_i\, \partial_{X_K}(J F_{Ki}^{-1}) + J \partial_{x_i} h_i.$$

By (1.11), $\partial_{X_K}(J F_{Ki}^{-1}) = 0$ and hence (2.27) follows.

The equation of motion is obtained by letting

$$\varphi = \mathbf{v}, \qquad \beta = \rho\mathbf{b}, \qquad \mathbf{h} = \mathbf{T}.$$

By (2.26) and (2.27), in the material formulation, we have

$$\rho_R\dot{\mathbf{v}} = \rho_R\mathbf{b} + \nabla_R \cdot \mathbf{T}_R, \tag{2.28}$$

where, consistently with (1.8),

$$\mathbf{T}_R := J\,\mathbf{T}\mathbf{F}^{-T}.$$

The tensor $\mathbf{T}_R$ denotes the contact force, per unit area, in the reference configuration and is referred to as the *first Piola–Kirchhoff stress tensor* ([428], Sect. 43A) or merely *Piola stress* ([216], Sect. 24.1). The suffix R emphasizes the meaning associated with the reference configuration.[5] The Piola stress $\mathbf{T}_R$ is not in general a symmetric tensor. Indeed, since

$$\mathbf{T} = J^{-1}\mathbf{T}_R\mathbf{F}^T,$$

the symmetry of the Cauchy stress $\mathbf{T}$ implies that

$$\mathbf{T}_R\mathbf{F}^T = \mathbf{F}\mathbf{T}_R^T.$$

Look now at the balance of energy in the form (2.22)

$$\frac{d}{dt}\int_{\mathcal{P}_t} \rho\,\varepsilon\, dv = \int_{\mathcal{P}_t} (\rho r + \mathbf{T}\cdot\mathbf{D}) dv - \int_{\mathcal{P}_t} \mathbf{q}\cdot\mathbf{n}\, da.$$

so that

$$\varphi = \varepsilon, \qquad \beta = \rho r + \mathbf{T}\cdot\mathbf{D}, \qquad \mathbf{h} = -\mathbf{q}.$$

Letting

$$\mathbf{q}_R = J\,\mathbf{q}\,\mathbf{F}^{-T}$$

[5] The symbols $\mathbf{S}$ and $\mathbf{P}$ are also used in the literature for the Piola stress. The Piola stress is also called nominal stress or engineering stress.

we obtain the balance of energy in the material form

$$\rho_R\dot{\varepsilon} = \rho_R r + J\,\mathbf{T}\cdot\mathbf{D} - \nabla_R\cdot\mathbf{q}_R.$$

The vector $\mathbf{q}_R$ denotes the heat flux, per unit area, in the reference configuration. The quantity $J\,\mathbf{T}\cdot\mathbf{D}$ is the mechanical power, per unit volume in the reference configuration. It is then natural to look for a representation involving the Piola stress $\mathbf{T}_R$ rather than the Cauchy stress $\mathbf{T}$. Now, by obvious identities, we can write

$$\begin{aligned} J\,\mathbf{T}\cdot\mathbf{D} &= J\tfrac{1}{2}\mathbf{T}\cdot(\mathbf{L}+\mathbf{L}^T) = J\tfrac{1}{2}\mathbf{T}\cdot(\dot{\mathbf{F}}\mathbf{F}^{-1}+\mathbf{F}^{-T}\dot{\mathbf{F}}^T)\\ &= J\tfrac{1}{2}\mathbf{T}\cdot\mathbf{F}^{-T}(\mathbf{F}^T\dot{\mathbf{F}}+\dot{\mathbf{F}}^T\mathbf{F})\mathbf{F}^{-1} = \tfrac{1}{2}(J\,\mathbf{F}^{-1}\mathbf{T}\mathbf{F}^{-T})\cdot(\mathbf{F}^T\dot{\mathbf{F}}+\dot{\mathbf{F}}^T\mathbf{F}) \end{aligned}$$

This suggests that we define the stress tensor

$$\mathbf{T}_{RR} := J\,\mathbf{F}^{-1}\mathbf{T}\mathbf{F}^{-T}.$$

Since

$$\mathbf{F}^T\dot{\mathbf{F}}+\dot{\mathbf{F}}^T\mathbf{F} = \dot{\mathbf{C}}$$

we can write

$$J\,\mathbf{T}\cdot\mathbf{D} = \tfrac{1}{2}\mathbf{T}_{RR}\cdot\dot{\mathbf{C}}.$$

Hence, the local form of the balance of energy becomes

$$\rho_R\dot{\varepsilon} = \rho_R r + \tfrac{1}{2}\mathbf{T}_{RR}\cdot\dot{\mathbf{C}} - \nabla_R\cdot\mathbf{q}_R. \tag{2.29}$$

The symmetry of $\mathbf{T}$ implies that of $\mathbf{T}_{RR}$,

$$\mathbf{T}^T_{RR} = (J\,\mathbf{F}^{-1}\mathbf{T}\mathbf{F}^{-T})^T = J\,\mathbf{F}^{-1}\mathbf{T}^T\mathbf{F}^{-T} = J\,\mathbf{F}^{-1}\mathbf{T}\mathbf{F}^{-T} = \mathbf{T}_{RR}.$$

The tensor $\mathbf{T}_{RR}$ is called *second Piola stress* [216] or *second Piola-Kirchhoff stress* [428]. The Kirchhoff stress tensor $\mathbf{T}_K$ is defined as $\mathbf{T}_K = J\mathbf{T}$. One reason for its use is that the stress power in the reference configuration is merely $\mathbf{T}_K\cdot\mathbf{D}$.

A further form of the mechanical power follows by observing that

$$\tfrac{1}{2}J\mathbf{T}\cdot(\dot{\mathbf{F}}\mathbf{F}^{-1}+\mathbf{F}^{-T}\dot{\mathbf{F}}^T) = \tfrac{1}{2}J(\mathbf{T}\mathbf{F}^{-T}+\mathbf{T}^T\mathbf{F}^{-T})\cdot\dot{\mathbf{F}} = J\mathbf{T}\mathbf{F}^{-T}\cdot\dot{\mathbf{F}}.$$

Hence

$$J\mathbf{T}\cdot\mathbf{D} = \mathbf{T}_R\cdot\dot{\mathbf{F}}$$

and $\mathbf{T}_R\cdot\dot{\mathbf{F}}$ is the sought representation of the mechanical power, which involves the first Piola stress $\mathbf{T}_R$. Therefore, we can also write the balance of energy in the form

$$\rho_R\dot{\varepsilon} = \rho_R r + \mathbf{T}_R\cdot\dot{\mathbf{F}} - \nabla_R\cdot\mathbf{q}_R. \tag{2.30}$$

It is worth remarking that if $\mathbf{T}$ is non-symmetric, then

$$\mathbf{T} \cdot \mathbf{L} = \mathbf{T} \cdot (\dot{\mathbf{F}}\mathbf{F}^{-1}) = (\mathbf{T}\mathbf{F}^{-T}) \cdot \dot{\mathbf{F}} = J^{-1}\mathbf{T}_R \cdot \dot{\mathbf{F}}.$$

Piola stresses enjoy interesting properties in connection with the objective time derivatives. Consider the Piola stress $\mathbf{T}_R = J\mathbf{T}\mathbf{F}^{-T}$ and observe that

$$\mathbf{F}^{-T}\mathbf{F}^T = \mathbf{1}, \qquad \overline{\mathbf{F}^{-T}}^{\,\cdot} = -\mathbf{F}^{-T}\dot{\mathbf{F}}^T\mathbf{F}^{-T}, \qquad \dot{\mathbf{F}}^T = \mathbf{F}^T\mathbf{L}^T.$$

Hence, we find that

$$\dot{\mathbf{T}}_R = J(\dot{\mathbf{T}} - \mathbf{T}\mathbf{L}^T + (\nabla \cdot \mathbf{v})\mathbf{T})\mathbf{F}^{-T}.$$

Likewise, since

$$\mathbf{F}^{-1}\mathbf{F} = \mathbf{1}, \qquad \overline{\mathbf{F}^{-1}}^{\,\cdot} = -\mathbf{F}^{-1}\dot{\mathbf{F}}\mathbf{F}^{-1}, \qquad \dot{\mathbf{F}} = \mathbf{L}\mathbf{F},$$

then time differentiation of the second Piola stress $\mathbf{T}_{RR} = J\mathbf{F}^{-1}\mathbf{T}\mathbf{F}^{-T}$ results in

$$\dot{\mathbf{T}}_{RR} = J\mathbf{F}^{-1}(\dot{\mathbf{T}} - \mathbf{L}\mathbf{T} - \mathbf{T}\mathbf{L}^T + (\nabla \cdot \mathbf{v})\mathbf{T})\mathbf{F}^{-T}.$$

We then observe that

$$\dot{\mathbf{T}}_{RR} = J\mathbf{F}^{-1}\overset{\square}{\mathbf{T}}\mathbf{F}^{-T};$$

the operator mapping $\mathbf{T}_{RR}$ to $\mathbf{T}$ maps $\dot{\mathbf{T}}_{RR}$ to the Truesdell derivative of the Cauchy stress, $\overset{\square}{\mathbf{T}}$. In addition, we remark that $\overset{\square}{\mathbf{T}} = \partial_t \mathbf{T} + \mathcal{L}_{\mathbf{v}} J\mathbf{T}$, $\mathcal{L}_{\mathbf{v}}$ being the Lie derivative of the contravariant components of $J\mathbf{T}$ relative to the velocity $\mathbf{v}$.

The interpretation of $\dot{\mathbf{T}}_R$ is less direct. Observe that, if $\mathbf{T}$ is symmetric

$$(\mathbf{T}\mathbf{L}^T)_{ij} = L_{jk}T_{ki}.$$

If we replace $\mathbf{T}$ with a vector $\mathbf{u}$, then

$$\mathbf{u}\mathbf{L}^T = \mathbf{L}\mathbf{u}.$$

Consequently, $\dot{\mathbf{T}} - \mathbf{T}\mathbf{L}^T + (\nabla \cdot \mathbf{v})\mathbf{T}$ is the Truesdell rate of $\mathbf{T}$ if $\mathbf{T}$ is a vector. This is not surprising in that $\mathbf{T}_R^* = \mathbf{Q}\mathbf{T}_R$ under a Euclidean transformation.

It is worth summarizing some properties of the power of the stress. In the Eulerian description, the power π, per unit volume, is

$$\pi = \mathbf{T} \cdot \mathbf{L};$$

if $\mathbf{T} = \mathbf{T}^T$, then $\mathbf{T} \cdot \mathbf{L} = \mathbf{T} \cdot \mathbf{D}$. In the material description, the power π_R, per unit volume in the reference configuration, is

$$\pi_R = J\pi = \mathbf{T}_R \cdot \dot{\mathbf{F}}.$$

If $\mathbf{T} = \mathbf{T}^T$, then

$$\pi = \mathbf{T} \cdot \mathbf{L} = \mathbf{T} \cdot \mathbf{D}, \qquad \pi_R = \mathbf{T}_R \cdot \dot{\mathbf{F}} = \tfrac{1}{2}\mathbf{T}_{RR} \cdot \dot{\mathbf{C}} = \mathbf{T}_{RR} \cdot \dot{\mathbf{E}}.$$

Sometimes π occurs in the literature as $\mathbf{T} \cdot \dot{\mathbf{E}}$ or $\mathbf{T} \cdot \dot{\varepsilon}$. Both expressions are linear approximations of π.

2.5.1 Equation of Motion in Pre-Stressed Materials

Consider a body occupying a region R in the reference configuration. Under the action of a body force $\tilde{\mathbf{b}}$ the body occupies a region $\tilde{\mathrm{R}}$. Let $\tilde{\mathbf{T}}_R$ and $\rho_0\tilde{\mathbf{b}}$ be the first Piola stress and the body force density such that the body is at equilibrium

$$\nabla_{\mathbf{X}} \cdot \tilde{\mathbf{T}}_R + \rho_0\tilde{\mathbf{b}} = \mathbf{0}.$$

Because of a different body force $\mathbf{b}$ or because of a superimposed motion, the body occupies the time-dependent region $\mathcal{R}_t$; as in Fig. 2.4, the point $\mathbf{X}$ occupies the position $\tilde{\mathbf{x}}(\mathbf{X})$ in $\tilde{\mathrm{R}}$ and $\mathbf{x}(\mathbf{X}, t)$ in $\mathcal{R}_t$. The equation of motion takes the form

$$\rho_0\dot{\mathbf{v}} = \nabla_{\mathbf{X}} \cdot \mathbf{T}_R + \rho_0\mathbf{b}.$$

At the same point $\mathbf{X}$, the difference yields

$$\rho_0\dot{\mathbf{v}} = \nabla_{\mathbf{X}} \cdot \mathcal{T}_R + \rho_0(\mathbf{b} - \tilde{\mathbf{b}}), \qquad \mathcal{T}_R := \mathbf{T}_R - \tilde{\mathbf{T}}_R; \tag{2.31}$$

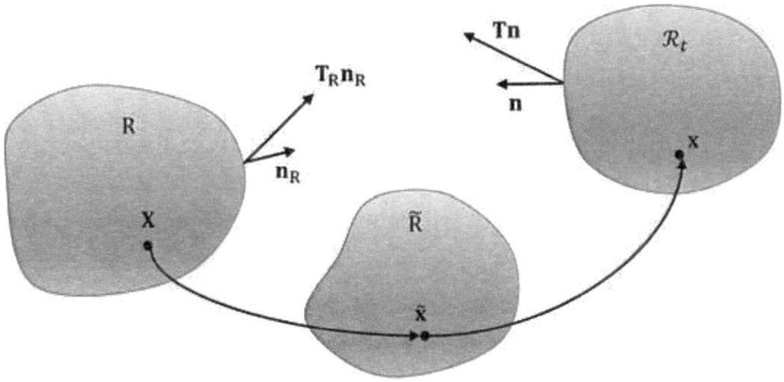

Fig. 2.4 The reference configuration R is mapped into a pre-stressed configuration $\tilde{\mathrm{R}}$ and next into the current configuration $\mathcal{R}_t$

the motion is determined by the variation $\mathcal{T}_R$ of the Piola stress and that of the body force. For simplicity, ρ_0 is used both in $\mathcal{R}_t$ and $\tilde{\mathcal{R}}$ and let $\rho_0\tilde{\mathbf{b}}$ be the force per unit volume applied in $\tilde{\mathcal{R}}$.

We now look for the Eulerian version of the equation of motion. Denote by $\tilde{\mathbf{F}}$, $\mathbf{F}$ the deformation gradients and by $\tilde{\mathbf{H}}$, $\mathbf{H}$ the displacement gradients relative to the deformations $\mathrm{R} \to \tilde{\mathrm{R}}$, $\mathrm{R} \to \mathcal{R}_t$. Moreover, let $\tilde{J} = \det \tilde{\mathbf{F}}$, $J = \det \mathbf{F}$. The corresponding Cauchy stresses $\tilde{\mathbf{T}}$, $\mathbf{T}$ in $\tilde{\mathrm{R}}$, $\mathcal{R}_t$ are given by

$$\tilde{\mathbf{T}} = \frac{1}{\tilde{J}}\tilde{\mathbf{T}}_R\tilde{\mathbf{F}}^T, \qquad \mathbf{T} = \frac{1}{J}\mathbf{T}_R\mathbf{F}^T.$$

Denote by $\mathcal{F}$ and $\mathcal{H}$ the deformation gradient and the displacement gradient associated with $\tilde{\mathrm{R}} \to \mathcal{R}_t$. Also, let $\mathcal{J} = \det \mathcal{F}$. Hence

$$\mathbf{F} = \mathcal{F}\tilde{\mathbf{F}}, \quad \mathcal{F} = \mathbf{1} + \mathcal{H}, \quad \mathcal{J} = \det(\mathbf{1} + \mathcal{H}) = 1 + \operatorname{tr}\mathcal{H} + o(|\mathcal{H}|).$$

It follows that

$$\mathbf{T} = \frac{1}{\mathcal{J}\tilde{J}}(\tilde{\mathbf{T}}_R + \mathcal{T}_R)\tilde{\mathbf{F}}^T\mathcal{F}^T = \frac{1}{\mathcal{J}}\tilde{\mathbf{T}}\mathcal{F}^T + \frac{1}{\mathcal{J}\tilde{J}}(\tilde{\mathbf{T}}_R + \mathcal{T}_R)\tilde{\mathbf{F}}^T\mathcal{F}^T + \mathcal{T} + o(|\mathcal{H}|)$$

where $\mathcal{T} = (1/\mathcal{J}\tilde{J})\mathcal{T}_R\tilde{\mathbf{F}}^T\mathcal{F}^T$. Since

$$\frac{1}{\mathcal{J}}\tilde{\mathbf{T}}\mathcal{F}^T = \tilde{\mathbf{T}} + \tilde{\mathbf{T}}\mathcal{F}^T - \tilde{\mathbf{T}}\nabla \cdot \mathcal{H} + o(|\mathcal{H}|)$$

then

$$\mathbf{T} - \tilde{\mathbf{T}} = \tilde{\mathbf{T}}\mathcal{F}^T - \tilde{\mathbf{T}}\nabla \cdot \mathcal{H} + \mathcal{T} + o(|\mathcal{H}|).$$

The variation $\mathbf{T} - \tilde{\mathbf{T}}$ of the Cauchy stress $\mathbf{T}$ consists of two terms; $\mathcal{T}$ is the Eulerian analogue of the variation $\mathcal{T}_R$ of the Piola stress while $\tilde{\mathbf{T}}\mathcal{F}^T - \tilde{\mathbf{T}}\nabla \cdot \mathcal{H}$ is the linear approximation of the variation $\mathbf{T} - \tilde{\mathbf{T}}$ induced by the pre-stress $\tilde{\mathbf{T}}_R$. While $\tilde{\mathbf{T}}\mathcal{F}^T - \tilde{\mathbf{T}}\nabla \cdot \mathcal{H}$ is due only to deformation, $\mathcal{T}$ depends on the constitutive properties because so does $\mathcal{T}_R$.

We now consider the equation of motion (2.31). First we observe that

$$J^{-1}\partial_{X_K}\mathcal{T}_R^{iK} = J^{-1}\partial_{x_j}\mathcal{T}_R^{iK})\partial_{X_K}x_j = \partial_{x_j}(J^{-1}\mathcal{T}_R^{iK}F_{jK})$$

where the identity $(1.11)_2$ has been used. Since $J^{-1}\mathcal{T}_R\mathbf{F}^T = \mathcal{T}$ and $\rho_0/J = \rho$, we can write (2.31) in the form

$$\rho\dot{\mathbf{v}} = \nabla_{\mathbf{x}} \cdot \mathcal{T} + \rho(\mathbf{b} - \tilde{\mathbf{b}}).$$

This shows that, in the Eulerian formulation, the effective stress governing the motion is $\mathcal{T}$, the transform of $\mathcal{T}_R = \mathbf{T}_R - \tilde{\mathbf{T}}_R$. If, for definiteness, we let $\mathcal{T}_R$ be linear and

$$\mathcal{T}_R^{iK} = A_{iKl\alpha}\partial_{\tilde{x}_\alpha} u_l = (A_{iKl\alpha}\mathcal{F}_{j\alpha})\partial_{x_j} u_l F_{hK}$$

then we have

$$\frac{1}{J}(\mathcal{T}_R \mathbf{F}^T)_{ij} = B_{ihlj}\partial_{x_j} u_l, \qquad B_{ihlj} = A_{iKl\alpha}\mathcal{F}_{j\alpha}F_{hK}.$$

To the linear order in $\mathbf{u}$, we find the equation of motion in the form

$$\rho\partial_t^2 u_i = B_{ihlj}\partial_{x_h}\partial_{x_j} u_l + (\partial_{x_h} B_{ihlj})\partial_{x_j} u_l + u_j \partial_{x_j} b_i.$$

2.5.2 Lie Derivative of the Cauchy Stress

The Lie derivative of the stress $\mathbf{T}$ is also defined by

$$\mathcal{L}\mathbf{T} = J^{-1}\mathbf{F}\overline{J\mathbf{F}^{-1}\mathbf{T}\mathbf{F}^{-T}}^{\boldsymbol{\cdot}}\mathbf{F}^T,$$

$J\mathbf{F}^{-1}\mathbf{T}\mathbf{F}^{-T}$ being the second Piola stress $\mathbf{T}_{RR}$. Since

$$\overline{J\mathbf{F}^{-1}\mathbf{T}\mathbf{F}^{-T}}^{\boldsymbol{\cdot}} = J\mathbf{F}^{-1}(\dot{\mathbf{T}} - \dot{\mathbf{F}}\mathbf{F}^{-1}\mathbf{T} - \mathbf{T}\mathbf{F}^{-T}\dot{\mathbf{F}}^T + (\nabla\cdot\mathbf{v})\mathbf{T})\mathbf{F}^{-T}$$

then

$$\mathcal{L}\mathbf{T} = \dot{\mathbf{T}} - \mathbf{L}\mathbf{T} - \mathbf{T}\mathbf{L}^T + (\nabla\cdot\mathbf{v})\mathbf{T};$$

accordingly, the Lie derivative of the Cauchy stress coincides with the Truesdell derivative.

We now look at deformations subject to the identity stretch, $\mathbf{U} = \mathbf{1}$. With reference to the polar decomposition, we let

$$\mathbf{F} = \mathbf{R}, \qquad \mathbf{U} = \mathbf{1}.$$

Then $J = \det \mathbf{F} = \det \mathbf{R} = 1$. The derivative is given by

$$\mathcal{L}\mathbf{T} = \mathbf{R}\,\overline{\mathbf{R}^T\mathbf{T}\mathbf{R}}^{\boldsymbol{\cdot}}\,\mathbf{R}^T.$$

Since

$$\overline{\mathbf{R}^T\mathbf{T}\mathbf{R}}^{\boldsymbol{\cdot}} = \dot{\mathbf{R}}^T\mathbf{T}\mathbf{R} + \mathbf{R}^T\dot{\mathbf{T}}\mathbf{R} + \mathbf{R}^T\mathbf{T}\dot{\mathbf{R}} = \mathbf{R}^T\mathbf{R}\,\dot{\mathbf{R}}^T\mathbf{T}\mathbf{R} + \mathbf{R}^T\dot{\mathbf{T}}\mathbf{R} + \mathbf{R}^T\mathbf{T}\dot{\mathbf{R}}\,\mathbf{R}^T\mathbf{R},$$

by means of $\mathbf{Z} = \dot{\mathbf{R}}\mathbf{R}^T = -\mathbf{Z}^T$ we find

$$\mathcal{L}\mathbf{T} = \dot{\mathbf{T}} - \mathbf{Z}\mathbf{T} - \mathbf{T}\mathbf{Z}^T,$$

that is the Green–Naghdi derivative.

2.5.3 Eshelby Stress and Linear-Angular Momentum

By the equation of motion in the (material) form (2.28), right multiplication by $\mathbf{F}$ results in

$$\rho_R \dot{\mathbf{v}} \mathbf{F} = \rho_R \mathbf{b} \mathbf{F} + (\nabla_R \cdot \mathbf{T}_R)\mathbf{F}.$$

In view of the identities

$$\rho_R \dot{\mathbf{v}} \mathbf{F} = (\rho_R \mathbf{v} \mathbf{F})^{\cdot} - \rho_R \mathbf{v} \dot{\mathbf{F}}, \qquad \rho_R \mathbf{v} \dot{\mathbf{F}} = \rho_R \nabla_R (\tfrac{1}{2} \mathbf{v}^2),$$

$$(\nabla_R \cdot \mathbf{T}_R)\mathbf{F} = \nabla_R \cdot (\mathbf{F}^T \mathbf{T}_R) - \mathbf{T}_R \nabla_R \mathbf{F},$$

we obtain[6]

$$(\rho_R \mathbf{v} \mathbf{F})^{\cdot} = \nabla_R \cdot (\mathbf{F}^T \mathbf{T}_R) + \rho_R \nabla_R (\tfrac{1}{2} \mathbf{v}^2) - \mathbf{T}_R \nabla_R \mathbf{F} + \rho_R \mathbf{b} \mathbf{F}.$$

If, as in hyperelastic solids, $\mathbf{T}_R$ is given by an energy function, say ψ, such that

$$\mathbf{T}_R = \rho_R \partial_{\mathbf{F}} \psi,$$

then

$$\mathbf{T}_R \nabla_R \mathbf{F} = \rho_R \nabla_R \psi.$$

The balance of linear momentum then can be written in the form

$$(\rho_R \mathbf{v} \mathbf{F})^{\cdot} = \nabla_R \cdot (\mathbf{F}^T \mathbf{T}_R) + \rho_R \nabla_R L + \rho_R \mathbf{b} \mathbf{F}, \tag{2.32}$$

where

$$L = \tfrac{1}{2} \mathbf{v}^2 - \psi.$$

Equation (2.32) can be viewed as an equation of motion where $\rho_R \mathbf{v} \mathbf{F}$ is the linear momentum and $\mathbf{F}^T \mathbf{T}_R$ is the stress tensor. The effective body force comprises two terms, $\rho_R \nabla_R L$ arising from the energy per unit mass L, $\rho_R \mathbf{b} \mathbf{F}$ being the body force. The stress $\mathbf{F}^T \mathbf{T}_R$ is similar to the second Piola stress (with $\mathbf{F}^T$ in place of $\mathbf{F}^{-1}$). By (2.32), we can view

$$\mathbf{T}_{\text{eff}} = \mathbf{F}^T \mathbf{T}_R + \rho_R L \, \mathbf{1}$$

as the effective stress. Correspondingly the vector $\rho_R \mathbf{v} \mathbf{F}$ is called the material linear momentum.

The definition of $\mathbf{T}_{\text{eff}}$ as it stands is consistent if ρ_R is constant. Sometimes definitions are given with differences of the signs for $\mathbf{T}_{\text{eff}}$ and, consistently, for the linear momentum.

---

[6] In component form, $(\mathbf{T}_R \nabla_R \mathbf{F})_Q = T_{hK} \partial_{X_K} F_{hQ} = T_{hK} \partial_{X_Q} F_{hK}$.

Often the effective stress is employed in static conditions in which case

$$\mathbf{T}_{\text{eff}} = \mathbf{F}^T \mathbf{T}_R - \rho_R \psi \mathbf{1}$$

and (2.32) becomes an equilibrium condition; if $\mathbf{b} = \mathbf{0}$ then (2.32) reduces to

$$\nabla_R \cdot (\mathbf{F}^T \mathbf{T}_R - \rho_R \psi) = \mathbf{0}.$$

This in turn shows why usually the stress

$$\mathbf{T}_E := \rho_R \psi \mathbf{1} - \mathbf{F}^T \mathbf{T}_R \tag{2.33}$$

is involved; owing to the decisive contributions of Eshelby to the subject [153, 154] the tensor (2.33) is usually referred to as *Eshelby (material) stress*. The original motivation and main applications of the Eshelby stress are within equilibrium problems. Concerning the motivation [154], let $\mathcal{S}$ be the boundary of a region in the current configuration. The vanishing of the couple on the material inside $\mathcal{S}$ is expressed by

$$\int_{\mathcal{S}} \mathbf{r} \times \mathbf{T}\mathbf{n}\, da = \mathbf{0},$$

$\mathbf{r}$ being the position vector relative to a fixed base point. In the reference configuration, with $\mathrm{S} = \mathbf{f}^{-1}(\mathcal{S})$, we have

$$\int_{\mathrm{S}} \mathbf{r} \times \mathbf{T}_R \mathbf{n}_R\, da_R = \mathbf{0}.$$

Since

$$(\mathbf{r} \times \mathbf{T}_R \mathbf{n}_R)_p = \epsilon_{plj} r_l T^R_{jK} n^R_K$$

it follows

$$\epsilon_{plj} \int_{\mathrm{S}} r_l T^R_{jK} n^R_K\, da_R = 0, \qquad p = 1, 2, 3,$$

whence

$$\int_{\mathrm{S}} (r_l T^R_{jK} - r_j T^R_{lK}) n^R_K\, da_R = 0, \qquad l, j = 1, 2, 3.$$

Let $\mathbf{r} = \mathbf{d} + \mathbf{x} = \mathbf{d} + \mathbf{X} + \mathbf{u}$, $\mathbf{d}$ being any constant vector. By applying the divergence theorem and observing that

$$\nabla_R \cdot \mathbf{T}_R = \mathbf{0}$$

as a condition for equilibrium, it follows from

$$\int_{\mathrm{P}} \partial_{X_K} (r_l T^R_{jK} - r_j T^R_{lK}) dv_R = 0, \qquad l, j = 1, 2, 3,$$

that

$$\mathbf{T}_R - \mathbf{T}_R^T = \mathbf{H}\mathbf{T}_R^T - \mathbf{T}_R\mathbf{H}^T$$

or, equivalently,

$$\mathbf{F}\mathbf{T}_R^T = \mathbf{T}_R\mathbf{F}^T.$$

This conclusion is not surprising. Since $\mathbf{T}_R\mathbf{F}^T = J^{-1}\mathbf{T}\mathbf{F}^{-T}\mathbf{F}^T = J^{-1}\mathbf{T}$, we have found the known result that the vanishing of the couple implies the symmetry of the Cauchy stress.

Suppose we look for a function $W(\mathbf{u}, \nabla_R\mathbf{u}, \mathbf{X})$ making

$$\mathcal{F} = \int_{\mathrm{P}} W(\mathbf{u}, \nabla_R\mathbf{u}, \mathbf{X})dv_R$$

stationary. Then we consider the vector function $\mathbf{u} + \alpha\mathbf{g}$ on P and require that

$$\partial_\alpha \int_{\mathrm{P}} W(\mathbf{u} + \alpha\mathbf{g}, \nabla_R(\mathbf{u} + \alpha\mathbf{g}), \mathbf{X})dv_R = 0$$

subject to $\mathbf{g} = \mathbf{0}$ on ∂P. Since

$$\partial_\alpha W = \partial_{\mathbf{u}} W \cdot \mathbf{g} + \partial_{\nabla_R\mathbf{u}} W \cdot \nabla_R\mathbf{g} = \partial_{\mathbf{u}} W - \nabla_R \cdot \partial_{\nabla_R\mathbf{u}} W) \cdot \mathbf{g} + \nabla_R \cdot (\partial_{\nabla_R\mathbf{u}} W\, \mathbf{g})$$

and $\mathbf{g} = \mathbf{0}$ on ∂P the continuity of the integrand allows us to say that $\mathcal{F}$ is stationary at $\alpha = 0$ if and only if

$$\partial_{\mathbf{u}} W - \nabla_R \cdot \partial_{\nabla_R\mathbf{u}} W = \mathbf{0}. \tag{2.34}$$

We now assume the displacement function $\mathbf{u}(\mathbf{X})$ satisfies the differential equation (2.34). To compute the gradient $\nabla_R W$, we observe that

$$(\nabla_R W)_L = \partial_{u_i} W H_{iL} + \partial_{H_{iK}} W \partial_{X_L} H_{iK} + \partial_{X_L} W,$$

where $\partial_{X_L} W$ denotes the partial derivative. Since $\partial_{u_i} = \partial_{X_K}\partial_{H_{iK}} W$, then it follows

$$(\nabla_R W)_L = \partial_{X_K}(H_{Li}^T \partial_{H_{iK}} W) + \partial_{X_L} W.$$

Since $\nabla_R W = \nabla_R \cdot (W\mathbf{1})$, letting[7]

$$\mathbf{T}_E = W\mathbf{1} - \mathbf{H}^T\partial_{\mathbf{H}} W \tag{2.35}$$

we find that

$$\nabla_R \cdot \mathbf{T}_E = \partial_{\mathbf{X}} W.$$

In homogeneous bodies, $\partial_{\mathbf{X}} W = \mathbf{0}$, and hence the stationary condition is

[7] The literature shows the use of both (2.33), with $\mathbf{F}$, and (2.35), with $\mathbf{H}$, for the Eshelby tensor.

$$\nabla_R \cdot \mathbf{T}_E = \mathbf{0}. \tag{2.36}$$

The formal structure of $\mathbf{T}_E$, with W viewed as a Lagrangian, justifies the name of *energy-momentum tensor* given to $\mathbf{T}_E$ by Eshelby [154].

Despite the rather qualitative introduction of $\mathbf{T}_E$, applications of $\mathbf{T}_E$ are essentially based on the property (2.36). The integral

$$\mathfrak{F} = \int_S \mathbf{T}_E \mathbf{n}_R \, da_R$$

is zero when taken over a closed surface S within which the material is homogeneous. If defects (or inhomogeneities) occur inside S, then the integral should be nonzero. Moreover, if S_1, S_2 are two closed surfaces with, e.g. S_2 inside S_1 and the region between S_2 and S_1 is homogeneous, then we conclude that

$$\int_{S_1} \mathbf{T}_E \, \mathbf{n}_R \, da_R = \int_{S_2} \mathbf{T}_E \, \mathbf{n}_R \, da_R$$

and say that the integral is path- (surface-) independent. Accordingly, the vector $\mathfrak{F}$ is said to be the *force on the defect,* which means a measure of the inhomogeneity.

2.6 Balance of Entropy

It is assumed that there exists the *absolute temperature* field $\theta(\mathbf{x}, t)$, for any $\mathbf{x} \in \mathcal{R}_t \subset \mathscr{E}$ and $t \in \mathbb{R}$. The values of θ are strictly positive.

In the thermodynamics of homogeneous processes, it is assumed that there exists a function $S(t)$ such that the Clausius–Planck inequality

$$\theta \dot{S} \geq \mathcal{Q}$$

holds where $\mathcal{Q}$ is the (positive or negative) heat transferred to the body per unit time. Consequently

$$\mathcal{Q} = 0 \quad \Rightarrow \quad \dot{S} \geq 0,$$

that is, in isolated bodies the entropy cannot decrease.

Since we are dealing with non-homogeneous processes, we let the pertinent variables be functions of the position $\mathbf{x}$ and time t. Assume that there exists a specific entropy, per unit mass, η such that the entropy $S(\mathcal{P}_t)$ of any subregion $\mathcal{P}_t$ is given by

$$S(\mathcal{P}_t) = \int_{\mathcal{P}_t} \rho \eta \, dv.$$

Hence, the Clausius–Planck inequality may suggest that

$$\frac{d}{dt}\int_{\mathcal{P}_t} \rho\eta\, dv \geq -\int_{\partial\mathcal{P}_t} \frac{\mathbf{q}}{\theta}\cdot\mathbf{n}\, da + \int_{\mathcal{P}_t} \frac{\rho r}{\theta}\, dv. \tag{2.37}$$

Inequality (2.37) is called the Clausius-Duhem inequality. Computation of the time derivative, the continuity of the functions $\rho\dot{\eta}$, $\rho r/\theta$, and $\nabla\cdot(\mathbf{q}/\theta)$, and the arbitrariness of $\mathcal{P}_t$ yield the local Clausius–Duhem inequality

$$\rho\dot{\eta} \geq -\nabla\cdot\frac{\mathbf{q}}{\theta} + \frac{\rho r}{\theta}. \tag{2.38}$$

A *thermodynamic process* is described by the set of functions occurring in the balance equations and in the Clausius–Duhem inequality. By way of example, if no other fields occur, a thermodynamic process is given by the functions ρ, $\mathbf{v}$, $\mathbf{T}$, $\mathbf{b}$, ε, $\mathbf{q}$, r, η, θ on $\mathcal{R}_t \times \mathbb{R}$. Such a set of functions is called a thermodynamic process if it is compatible with the balance equations of mass, momentum, and energy.

As to (2.37) and (2.38), we might say that $\mathbf{q}/\theta$ is the entropy flux and $\rho r/\theta$ is the entropy supply. A more general statement of the second law is adopted correctly by Müller [337] by allowing the entropy flux, $\mathbf{j}$ say, to be an unknown function which thus enters the set of the thermodynamic process. The Clausius–Duhem inequality is then generalized by letting $\mathbf{j}$ be an unknown vector function to be determined so that the inequality

$$\rho\dot{\eta} + \nabla\cdot\mathbf{j} - \frac{\rho r}{\theta} \geq 0, \tag{2.39}$$

holds. Hence, a thermodynamic process is said to be admissible if it is compatible with (2.39), (see Chap. 3).

Inequality (2.39) allows the balance of entropy to be stated in a form that we regard as the second law of thermodynamics.

Second law of thermodynamics. *For every admissible thermodynamic process, inequality* (2.39) *must hold for all times t and points* $\mathbf{x}$.

Since the compatibility with the balance equations is satisfied, by definition, by any thermodynamic process, the requirement of the second law places restrictions on the effective evolutions and hence selects those which are physically admissible.

There are cases where we can prove that $\mathbf{j}$ equals $\mathbf{q}/\theta$. Instead, in more involved models, e.g. when higher-order derivatives occur, $\mathbf{j}$ need not equal $\mathbf{q}/\theta$. It is convenient to let

$$\mathbf{k} = \mathbf{j} - \frac{\mathbf{q}}{\theta},$$

$\mathbf{k}$ being termed extra-entropy flux. Hence, (2.39) takes the form

$$\rho\dot{\eta} + \frac{1}{\theta}(\nabla\cdot\mathbf{q} - \rho r) - \frac{1}{\theta^2}\mathbf{q}\cdot\nabla\theta + \nabla\cdot\mathbf{k} \geq 0.$$

In terms of the Helmholtz free energy $\psi = \varepsilon - \theta\eta$, we obtain

$$-\rho\dot{\psi}-\rho\eta\dot{\theta}+\mathbf{T}\cdot\mathbf{D}-\frac{1}{\theta}\mathbf{q}\cdot\nabla\theta+\theta\nabla\cdot\mathbf{k}\geq 0. \tag{2.40}$$

This is the form of the second law inequality[8] we apply whenever the balance of energy (2.21) holds.

The view that $\mathbf{j}$ need not equal $\mathbf{q}/\theta$ generalizes the local form of the second law for the body under consideration. Yet we may argue that the entropy flux $\mathbf{j}$ enters the second law inequality in the divergence form. This allows us to say that the additive flux $\mathbf{k}$ is physically admissible provided the net contribution to the whole body, in the region Ω

$$\int_{\Omega}\nabla\cdot\mathbf{k}\,dv=\int_{\partial\Omega}\mathbf{k}\cdot\mathbf{n}\,da$$

vanishes. That is why it is assumed

$$\int_{\partial\Omega}\mathbf{k}\cdot\mathbf{n}\,da=0.$$

By (2.38), we have

$$\dot{\eta}\geq-\frac{1}{\rho}\nabla\cdot\frac{\mathbf{q}}{\theta}+\frac{r}{\theta}.$$

Integration, at a fixed point $\mathbf{X}$, on $[t_0,t_1]$ results in

$$\eta(t_1)-\eta(t_0)\geq\int_{t_0}^{t_1}\Big[-\frac{1}{\rho}\nabla\cdot\frac{\mathbf{q}}{\theta}+\frac{r}{\theta}\Big]dt.$$

In a cyclic process, the final values equal the initial values. In particular $\eta(t_1)=\eta(t_0)$. Therefore, *in any cyclic process at a point* $\mathbf{X}$ *the inequality*

$$\int_{t_0}^{t_1}\Big[-\frac{1}{\rho}\nabla\cdot\frac{\mathbf{q}}{\theta}+\frac{r}{\theta}\Big]dt\leq 0$$

holds. This inequality is sometimes referred to as *Clausius inequality.* The conceptual advantage of this statement is that entropy is not used as a primitive concept or is not assumed to exist from the outset [115].

The conceptual role of the second law, as a restriction on the admissible thermodynamic processes, is maintained in the following more general setting. Consider the inequality (2.39), or possibly (2.38), and define the *entropy production* σ, per unit volume and unit time, as

$$0\leq\rho\dot{\eta}+\nabla\cdot\mathbf{j}-\frac{\rho r}{\theta}=\sigma. \tag{2.41}$$

[8] Also referred to as Clausius–Duhem inequality or entropy inequality or local dissipation inequality.

The previous statement of the second law can then be expressed as follows: *For every admissible thermodynamic process, the entropy production must be non-negative for all times t and points* **x**. Owing to the definition of the entropy production, the two statements look fully equivalent.

The statement in terms of the entropy production might become more general as follows.

Generalized second law of thermodynamics. *The entropy production σ is a non-negative valued constitutive function for every admissible thermodynamic process for all times t and points* **x**.

The constitutive dependence of the entropy production may happen in two ways. First σ coincides with what follows from the definition. As a remarkable example, in connection with Newtonian fluids, see Sect. 5.6, where the dissipative stress $\mathbf{T}_{\rm dis}$ and the heat flux $\mathbf{q}$ are assumed in the form $\mathbf{T}_{\rm dis} = 2\mu\mathbf{D}_0 + \zeta(\operatorname{tr}\mathbf{D})\mathbf{1}$, $\mathbf{q} = -\kappa\nabla\theta$, with $\mu, \zeta, \kappa > 0$, it follows that

$$\theta\sigma = 2\mu\mathbf{D}_0 \cdot \mathbf{D}_0 + \zeta(\operatorname{tr}\mathbf{D})^2 - \frac{1}{\theta}\kappa\nabla\theta\cdot\nabla\theta.$$

Secondly, σ is assumed to have a constitutive dependence per se subject to being positive valued and satisfying the equality in (2.41). This view proves profitable in the modelling of hysteresis such as in ferroelectric, ferromagnetic, and plastic materials.

In [427], lecture 2, the internal dissipation δ is considered which amounts to letting $\delta = \theta\sigma/\rho$. The requirement

$$\delta \geq 0$$

is referred to as the *Planck inequality*. Hence, the assumption $\sigma \geq 0$ is equivalent to the Planck inequality.

Remark 2.1 The statement of the second law in terms of the (specific) entropy production ($\gamma = \sigma/\rho$) traces back to Coleman and Noll [93], though Planck inequality involves the entropy production via the internal dissipation δ. The definition of the entropy production σ allows us to write

$$\rho\dot{\eta} + \nabla\cdot\mathbf{j} - \frac{\rho r}{\theta} - \sigma = 0.$$

This has motivated the writing of the balance of entropy as an equality (see, e.g. [209]). It is worth emphasizing that ignoring the non-negative character of σ may lead to unphysical constitutive equations.

Remark 2.2 The thermodynamic scheme degenerates when the constitutive properties are independent of the temperature. In this case, the classical relation between entropy and Helmholtz free energy does not hold and a consistent scheme takes r

and $\mathbf{q}$ to be zero. This view is developed in Sect. 5 in connection with the elastic fluid.

We now go back to (2.40). Integration over any convecting region $\mathcal{P}_t$ gives

$$-\frac{d}{dt}\int_{\mathcal{P}_t} \rho\psi\, dv - \int_{\mathcal{P}_t} (\rho\eta\dot{\theta} + \frac{1}{\theta}\mathbf{q}\cdot\nabla\theta) dv + \int_{\mathcal{P}_t} \mathbf{T}\cdot\mathbf{D}\, dv + \int_{\mathcal{P}_t} \theta\nabla\cdot\mathbf{k} \geq 0.$$

By (2.24), we can replace the integral of $\mathbf{T}\cdot\mathbf{D}$ to obtain

$$\int_{\partial\mathcal{P}_t} \mathbf{v}\cdot\mathbf{T}\mathbf{n}\, da + \int_{\mathcal{P}_t} \rho\mathbf{b}\cdot\mathbf{v}\, dv - \frac{d}{dt}\int_{\mathcal{P}_t} \rho(\psi + \tfrac{1}{2}\mathbf{v}^2) dv - \int_{\mathcal{P}_t} (\rho\eta\dot{\theta} + \frac{1}{\theta}\mathbf{q}\cdot\nabla\theta - \theta\nabla\cdot\mathbf{k}) dv \geq 0.$$

An approximation is sometimes used in the literature by assuming that thermal influences be negligible. Apart from the contribution of the extra-entropy flux $\mathbf{k}$, very often disregarded, the last integral is assumed to vanish. Hence, it follows the *free-energy imbalance*

$$\int_{\partial\mathcal{P}_t} \mathbf{v}\cdot\mathbf{T}\mathbf{n}\, da + \int_{\mathcal{P}_t} \rho\mathbf{b}\cdot\mathbf{v}\, dv - \frac{d}{dt}\int_{\mathcal{P}_t} \rho(\psi + \tfrac{1}{2}\mathbf{v}^2) dv \geq 0,$$

the left-hand side being viewed as the dissipation; in words, the net free energy and kinetic energy $d\int_{\mathcal{P}_t} \rho(\psi + \frac{1}{2}\mathbf{v}^2) dv/dt$ is balanced by the conventional expended power $\int_{\partial\mathcal{P}_t} \mathbf{v}\cdot\mathbf{T}\mathbf{n}\, da + \int_{\mathcal{P}_t} \rho\mathbf{b}\cdot\mathbf{v}\, dv$ plus the dissipation ([216], Chap. 29).

2.6.1 Material Formulation of the Second Law

As with any balance equation, we can state the second law of thermodynamics in the material description. Let P be the inverse image of the region $\mathcal{P}_t$, $\mathrm{P} = \chi^{-1}(\mathcal{P}_t, t)$, convecting with the body. By Nanson's formula, inequality (2.37) can be given the form

$$\frac{d}{dt}\int_{\mathrm{P}} \rho_R\eta\, dv_R \geq -\int_{\partial\mathrm{P}} J\frac{\mathbf{q}}{\theta}\cdot\mathbf{F}^{-T}\mathbf{n}_R\, da_R + \int_{\mathrm{P}} \rho_R r\, dv_R,$$

where $\rho_R = \rho J$. In terms of the heat flux vector in the material description

$$\mathbf{q}_R = J\mathbf{F}^{-1}\mathbf{q},$$

we can write

$$\frac{d}{dt}\int_{\mathrm{P}} \rho_R\eta\, dv_R \geq -\int_{\partial\mathrm{P}} \frac{\mathbf{q}_R}{\theta}\cdot\mathbf{n}_R\, da_R + \int_{\mathrm{P}} \rho_R r\, dv_R.$$

The divergence theorem and the arbitrariness of $\mathcal{P}$ yield the local inequality

$$\rho_R \dot{\eta} \geq -\nabla_R \cdot \frac{\mathbf{q}_R}{\theta} + \frac{\rho_R r}{\theta}, \tag{2.42}$$

which can be viewed as the Clausius–Duhem inequality in the material description. Hence, the second law of thermodynamics may be stated by saying that *for every admissible thermodynamic process,* expressed by functions on R $\times \mathbb{R}$, *the inequality* (2.42) *must hold for all times t and point* $\mathbf{X}$. This statement of the second law amounts to assuming that $\mathbf{q}/\theta$ is the entropy flux in the spatial description and hence $\mathbf{q}_R/\theta$, with $\mathbf{q}_R = J\mathbf{F}^{-1}\mathbf{q}$, is the entropy flux in the material description. The analogue of Müller's assumption about the second law would be to let $\mathbf{j}_R$, with $\mathbf{j}_R = J\mathbf{F}^{-1}\mathbf{j}$, be an unknown vector function, of $\mathbf{X}$ and t, such that

$$\rho_R \dot{\eta} \geq -\nabla_R \cdot \mathbf{j}_R + \frac{\rho_R r}{\theta}$$

holds for every thermodynamic process.

As with the spatial description, it is convenient to let

$$\mathbf{j}_R = \frac{\mathbf{q}_R}{\theta} + \mathbf{k}_R$$

and to regard $\mathbf{k}_R$ as the extra-entropy flux. Upon substitution of $\mathbf{j}_R$, the entropy inequality can be written in the form

$$\rho_R \theta \dot{\eta} \geq \rho_R r - \nabla_R \cdot \mathbf{q}_R + \frac{1}{\theta} \mathbf{q}_R \cdot \nabla_R \theta - \theta \nabla_R \cdot \mathbf{k}_R .$$

Substitution for $\rho_R r - \nabla_R \cdot \mathbf{q}_R$ from the balance of energy (2.29) and use of the free energy $\psi = \varepsilon - \theta\eta$ yield the material form of the Clausius–Duhem inequality,

$$-\rho_R(\dot{\psi} + \eta\dot{\theta}) + \tfrac{1}{2}\mathbf{T}_{RR} \cdot \dot{\mathbf{C}} - \frac{1}{\theta}\mathbf{q}_R \cdot \nabla_R \theta + \theta \nabla_R \cdot \mathbf{k}_R \geq 0. \tag{2.43}$$

For formal convenience, we consider the referential densities

$$\varepsilon_R = \rho_R \varepsilon, \qquad \eta_R = \rho_R \eta, \qquad \psi_R = \rho_R \psi,$$

so that ε_R, η_R, and ψ_R are the internal energy, the entropy, and the free energy per unit volume in the reference configuration. Moreover, let $r_R = \rho_R r$. Since ρ_R is constant in time, the balance of energy and the Clausius–Duhem inequality are given the form

$$\dot{\varepsilon}_R = \tfrac{1}{2}\mathbf{T}_{RR} \cdot \dot{\mathbf{C}} - \nabla_R \cdot \mathbf{q}_R + r_R,$$

$$-(\dot{\psi}_R + \eta_R\dot{\theta}) + \tfrac{1}{2}\mathbf{T}_{RR} \cdot \dot{\mathbf{C}} - \frac{1}{\theta}\mathbf{q}_R \cdot \nabla_R \theta + \theta \nabla_R \cdot \mathbf{k}_R \geq 0. \tag{2.44}$$

If $\mathbf{T}$ is non-symmetric then the power $\mathbf{T} \cdot \mathbf{L}$ becomes

$$\mathbf{T} \cdot \mathbf{L} = \mathbf{T} \cdot (\dot{\mathbf{F}}\mathbf{F}^{-1}) = (\mathbf{T}\mathbf{F}^{-T}) \cdot \dot{\mathbf{F}} = J^{-1}\mathbf{T}_R \cdot \dot{\mathbf{F}}.$$

2.6.2 Third Law of Thermodynamics

The statement of the second law involves the entropy η only via the derivative $\dot{\eta}$. Hence, we might compute the entropy to within a constant though the undetermined value of the constant might be of no consequence. Yet physical arguments motivate the way of determining the value of the constant in the computation of the entropy. This determination is given by Nernst's principle or Nernst's heat theorem or third law of thermodynamics. It is not really a theorem; we take it as an axiom in the form of Planck's statement.
Third law of thermodynamics. *The entropy of a system is zero at absolute zero.* This means that the entropy of a system depends only on the temperature as absolute zero is approached.

2.6.3 Exergy

The balance of energy, or the first law of thermodynamics, is viewed as the conservation of energy; energy is conserved, it cannot be destroyed. Instead, as we see in a while, *exergy* is a quantity that is not conserved and depends on the process and the environment.

We start from the balance of energy (2.21) and let $\mathbf{T} = -p\mathbf{1} + \hat{\mathbf{T}}$. Since $\nabla \cdot \mathbf{v} = -\dot{\rho}/\rho = \dot{v}/v$ we can write the balance of energy in the form

$$\rho\dot{\varepsilon} = \rho r - \nabla \cdot \mathbf{q} - \rho\, p\, \dot{v} + \hat{\mathbf{T}} \cdot \mathbf{D}.$$

Substitution of $\rho r - \nabla \cdot \mathbf{q}$ in (2.38) results in

$$\hat{\mathbf{T}} \cdot \mathbf{D} \leq -\rho(\dot{\varepsilon} - \theta\dot{\eta} + p\, \dot{v}) - \frac{1}{\theta}\mathbf{q} \cdot \nabla\theta.$$

Neglect $\mathbf{q} \cdot \nabla\theta/\theta$ (no irreversibility). The left-hand side is the mechanical power and the inequality provides a bound for it. Upon integration over $[t_0, t] \times \Omega$, Ω being the region occupied by the body, we have

$$\int_{t_0}^{t}\int_{\Omega} \hat{\mathbf{T}} \cdot \mathbf{D}\, dv\, d\tau \leq U(t) - U(t_0) + \int_{t_0}^{t}\int_{\Omega} \rho(\theta\, \dot{\eta} - p\, \dot{v}) dv\, d\tau,$$

where $U = \int_\Omega \rho\,\varepsilon\, dv$ is the internal energy of the body. If, further, the pressure p and the temperature θ are kept constant, $p = p_0, \theta = \theta_0$,then

$$\int_{t_0}^{t}\int_\Omega \hat{\mathbf{T}} \cdot \mathbf{D}\, dv\, d\tau \le -\{U(t) - U(t_0) - \theta_0[S(t) - S(t_0)] + p_0[V(t) - V(t_0)]\},$$

where $S = \int_\Omega \rho\,\eta\, dv$ and $V = \int_\Omega 1\, dv$ are the entropy and the volume of the body. The work available is then bounded by the right-hand side. That is why the exergy E of the system is defined by [316]

$$E = U(t) - U(t_0) - \theta_0[S(t) - S(t_0)] + p_0[V(t) - V(t_0)].$$

2.6.4 *Entropy Production in Stochastic Kinetics*

In statistical mechanics, the time evolution of the probability density function, of the velocity of a particle, is governed by the Fokker–Planck equation, based on the view that particles are subject to force fields (drag forces) and random forces. The Fokker–Planck equation is only an approximation to the Boltzmann model [33]: the dynamics of a particle, subject to a (external) force field $\mathbf{F}(\mathbf{x})$ and a fluctuating force $\boldsymbol{\xi}(t)$, with amplitude a, is governed by the equation of motion

$$m\ddot{\mathbf{x}} = -\gamma\dot{\mathbf{x}} + \mathbf{F}(\mathbf{x}) + a\boldsymbol{\xi}(t),$$

where γ is the frictional coefficient. Near equilibrium, the force $\gamma\dot{\mathbf{x}}$ is assumed to be predominant on the inertial force $m\ddot{\mathbf{x}}$ and then the equation is considered in the (approximate) form

$$\gamma\dot{\mathbf{x}} = \mathbf{F}(\mathbf{x}) + a\boldsymbol{\xi}(t).$$

This equation of motion is associated with the Fokker–Planck equation

$$\partial_t p = -\nabla \cdot (\frac{\mathbf{F}}{\gamma} p) + \Delta(\frac{a^2}{2\gamma^2} p)$$

for the probability density p. Letting $D = a^2/2\gamma^2$ we can write the *Smoluchowski equation*

$$\partial_t p = -\nabla \cdot \mathbf{J}, \qquad \mathbf{J} = -D\nabla p + \frac{1}{\gamma}\mathbf{F}p,$$

where $\mathbf{J}$ plays the role of probability current density. If $\mathbf{J}$ is viewed as a diffusion flux then $-D\nabla p$ is the analogue of Fick's model while $\mathbf{F}p/\gamma$ is an additional term due to the force field. If $\mathbf{F}$ is conservative, $\mathbf{F} = -\nabla U$ then $\mathbf{J} = -(D\nabla p + p\nabla U/\gamma)$.

2.7 Second Law and Phase-Field Models

Phase transitions describe the transformation of a substance from one phase to another, a phase denoting thermodynamic system with common physical states of matter. It is then necessary to identify at least one physical quantity which distinguishes the two phases. In the Landau theory, such a quantity is called an order parameter while quite often it is called a phase field. Denote the phase field by φ and assume it is a scalar. Phase transformations are modelled in detail in Chap. 16. Here, we outline the presumably far-reaching model of phase fields.

Different approaches occur in the literature. Here we point out a procedure often referred to as Ginzburg theory. Now, by (2.40), it follows that at constant temperature, with no deformation and no heat conduction, $\rho\dot{\psi} \le 0$. Consequently

$$\frac{d}{dt}\int_\Omega \rho\psi dv \le 0.$$

The free energy is then a decreasing function of time.

Suppose that a fixed constant temperature θ is maintained in the region $\Omega \in \mathscr{E}$ and consider the free energy in the form of the *Landau–Ginzburg functional*

$$\Psi(\varphi) = \int_\Omega [F(\varphi, \theta) + \tfrac{1}{2}\gamma(\varphi, \theta)|\nabla\varphi|^2]dv,$$

γ being a positive-valued function. The gradient term accounts for the interfacial energy; in multiphase systems the phase boundaries consist of small transition layers of finite nonzero thickness across which energy changes significantly.

The functional Ψ is assumed to have a minimum at equilibrium and hence φ is required to satisfy the Euler–Lagrange equation of $\psi = F(\varphi, \theta) + \frac{1}{2}\gamma(\varphi, \theta)|\nabla\varphi|^2$,

$$\delta_\varphi \psi = 0,$$

at any point $\mathbf{x} \in \Omega$. Hence, we have

$$\partial_\varphi F + \tfrac{1}{2}\partial_\varphi \gamma\,|\nabla\varphi|^2 - \nabla\cdot(\gamma\nabla\varphi) = 0$$

at the sought equilibrium configuration.

An additional equation is needed to determine the evolution of φ and a rate-type equation seems to be the natural scheme. If φ represents a conserved quantity, like mass density of a phase, then the evolution has to satisfy an appropriate conservation law. That is why the literature distinguishes models for conserved or non-conserved phase fields.

The (opposite of the) variational derivative $\delta_\varphi\psi$ is interpreted as a generalized thermodynamic force that tends to decrease the total free energy. Hence, for non-conserving dynamics, a natural evolution equation is assumed by letting $\partial_t\varphi$ be proportional to $-\delta_\varphi\psi$ so that

$$\partial_t \varphi = k[\nabla \cdot (\gamma \nabla \varphi) - \partial_\varphi F - \tfrac{1}{2}\partial_\varphi \gamma \, |\nabla \varphi|^2]$$

k being a positive-valued function of φ and θ. This equation is referred to as Cahn–Allen equation [6]. Consequently, subject to the boundary condition $\mathbf{n} \cdot \nabla \varphi = 0$ at $\mathbf{x} \in \partial\Omega$, we find

$$\frac{d\Psi}{dt} = \int_\Omega \delta_\varphi \psi \partial_t \varphi dv = -\int_\Omega k(\delta_\varphi \psi)^2 dv \leq 0$$

and hence the free energy is proved to decrease in time.

If instead φ is a conserved field, then we let

$$\frac{d}{dt}\int_\Omega \varphi \, dv = 0.$$

Borrowing from the conservation of mass (or charge), we then assume that

$$\frac{d}{dt}\int_\mathcal{P} \varphi \, dv = -\int_{\partial\mathcal{P}} \mathbf{J} \cdot \mathbf{n} \, da$$

for any subregion $\mathcal{P} \subseteq \Omega$ and an appropriate flux $\mathbf{J}$. Hence at any point $\mathbf{x} \in \Omega$ the continuity equation

$$\partial_t \varphi = -\nabla \cdot \mathbf{J}$$

is required to hold. If

$$\mathbf{J} = -\tilde{k} \nabla \delta_\varphi \psi,$$

$\tilde{k}$ being a function of φ and θ, then the evolution of φ is governed by

$$\partial_t \varphi = \nabla \cdot (\tilde{k} \nabla \delta_\varphi \psi) = \nabla \cdot \{\tilde{k} \nabla [-\nabla(\gamma \nabla \varphi) + \partial_\varphi F + \tfrac{1}{2}\partial_\varphi \gamma |\nabla \varphi|^2]\};$$

if $\tilde{k}$ and γ are constants then this equation simplifies to the Cahn-Hilliard equation [74]

$$\partial_t \varphi = \tilde{k}(\Delta \partial_\varphi - \gamma \Delta^2 \varphi).$$

As the boundary condition, we assume that

$$\mathbf{n} \cdot \nabla \varphi = 0, \qquad \mathbf{n} \cdot \nabla \delta_\varphi \psi = 0 \qquad \forall \mathbf{x} \in \partial\Omega.$$

Hence, we obtain the conservation property

$$\frac{d}{dt}\int_\Omega \varphi \, dv = \int_\Omega \partial_t \varphi \, dv = \int_\Omega \nabla \cdot (\tilde{k} \nabla \delta_\varphi \psi) dv = \int_{\partial\Omega} \tilde{k} \mathbf{n} \cdot \nabla \delta_\varphi \psi \, da = 0$$

and the decay property,

$$\frac{d}{dt}\int_\Omega \psi \, dv = \int_\Omega \delta_\varphi \psi \partial_t \varphi \, dv = \int_\Omega \delta_\varphi \psi \nabla \cdot (\tilde{k} \nabla \delta_\varphi \psi) dv = -\int_\Omega \tilde{k} |\nabla \delta_\varphi \psi|^2 dv \leq 0.$$

2.8 Bernoulli's Law and Balance Equations for Fluids

Bernoulli's law is a property of the equation of motion of fluids. Yet often arguments are given on the basis of particular properties of energy. To emphasize the effective properties of fluid motions, we now review some forms of the balance of energy and next we establish the general form of Bernoulli's law.

By the balance of energy in the form (2.19), with $h = -\mathbf{q} \cdot \mathbf{n}$, Reynolds' transport theorem and the divergence theorem result in

$$\rho \overline{(\varepsilon + \tfrac{1}{2}\mathbf{v}^2)}^{\boldsymbol{\cdot}} = \rho \mathbf{b} \cdot \mathbf{v} + \rho r + \nabla \cdot (\mathbf{T}\mathbf{v} - \mathbf{q}).$$

Now for any function, f say, we employ the continuity equation (2.4) to obtain

$$\rho \dot{f} = \rho \partial_t f + \rho \mathbf{v} \cdot \nabla f = \partial_t(\rho f) + \nabla \cdot (\rho f \mathbf{v}) - f(\partial_t \rho + \nabla \cdot (\rho \mathbf{v})) = \partial_t(\rho f) + \nabla \cdot (\rho f \mathbf{v}).$$

Moreover let $p = -\frac{1}{3}\mathrm{tr}\,\mathbf{T}$ so that $\mathbf{T} = \mathbf{T}_0 - p\mathbf{1}$ and p is the pressure. Consequently the balance of energy can be written in the form

$$\partial_t[\rho(\varepsilon + \tfrac{1}{2}\mathbf{v}^2)] = -\nabla \cdot [\rho(\varepsilon + \tfrac{1}{2}\mathbf{v}^2)\mathbf{v}] + \rho r - \nabla \cdot \mathbf{q} + \nabla \cdot [(\mathbf{T}_0 - p\mathbf{1})\mathbf{v}] + \rho \mathbf{b} \cdot \mathbf{v} \tag{2.45}$$

As is customary, let $\mathbf{b} = \mathbf{g}$, the acceleration gravity. Let z be the vertical coordinate being taken as positive upward. Hence $\mathbf{g} = -g\nabla z = -\nabla(gz)$ and

$$\rho \mathbf{g} \cdot \mathbf{v} = -\rho \mathbf{v} \cdot \nabla(gz) = -\nabla \cdot (\rho g z \mathbf{v}) + gz\nabla \cdot (\rho \mathbf{v}).$$

By the continuity equation (2.4), we can replace $\nabla \cdot (\rho \mathbf{v})$ with $-\partial_t \rho$; hence letting $\partial_t z = 0$, we conclude

$$\rho \mathbf{g} \cdot \mathbf{v} = -\nabla \cdot (\rho g z \mathbf{v}) - \partial_t(\rho g z).$$

Upon this condition, the balance of energy can be written in the form

$$\partial_t[\rho(\varepsilon + \tfrac{1}{2}\mathbf{v}^2 + gz)] = -\nabla \cdot [\rho(\varepsilon + \tfrac{1}{2}\mathbf{v}^2 + gz)\mathbf{v}] + \rho r - \nabla \cdot \mathbf{q} + \nabla \cdot [(\mathbf{T}_0 - p\mathbf{1})\mathbf{v}]. \tag{2.46}$$

On the basis of Eq. (2.46), $\rho(\varepsilon + \frac{1}{2}\mathbf{v}^2 + gz)$ is often referred to as the *total* energy density.

Another interpretation follows from (2.45) by observing that the power $\nabla \cdot (-p\mathbf{v})$ can be viewed as arising from a flux so that we can write the equivalent equation

$$\partial_t[\rho(\varepsilon + \tfrac{1}{2}\mathbf{v}^2)] = -\nabla \cdot [\rho(\varepsilon + \frac{p}{\rho} + \tfrac{1}{2}\mathbf{v}^2)\mathbf{v}] + \rho r - \nabla \cdot \mathbf{q} + \nabla \cdot (\mathbf{T}_0\mathbf{v}) + \rho \mathbf{g} \cdot \mathbf{v}. \tag{2.47}$$

and $\varepsilon + p/\rho$ is regarded as the enthalpy density, per unit mass. That is why sometimes $\varepsilon + \frac{1}{2}\mathbf{v}^2$ is said to be the energy density of the fluid, whereas $\rho(\varepsilon + p/\rho + \frac{1}{2}\mathbf{v}^2)\mathbf{v}$ is viewed as the energy flux (see, e.g. [273], Chap. 1).

2.8.1 Bernoulli's Law

Perhaps regarding the internal energy density ε as a function of the mass density ρ, and letting ρ be constant (incompressible fluids), often the view that

$$\rho(\tfrac{1}{2}\mathbf{v}^2 + gz)$$

is the effective energy is at the basis of the (heuristic) derivations of Bernoulli's law. Yet Bernoulli's law holds for compressible inviscid fluids, where $\mathbf{T} = -p\,\mathbf{1}$, and the derivation has to be based on the equation of motion.

Upon the observation that

$$\mathbf{g} = -g\nabla z,$$

z being positive upward, we can write the equation of motion as

$$\rho\dot{\mathbf{v}} = -\nabla p - \rho\nabla(gz).$$

Divide by ρ to have

$$\partial_t\mathbf{v} + (\mathbf{v}\cdot\nabla)\mathbf{v} = -\frac{1}{\rho}\nabla p - \nabla(gz). \tag{2.48}$$

Denote by $\mathbf{e}$ the unit vector of $\mathbf{v}$, that is $\mathbf{v} = v\mathbf{e}$, $v = |\mathbf{v}|$. Now, $\mathbf{e}\cdot\nabla$ is the derivative along $\mathbf{e}$; we let

$$\mathbf{v}\cdot\nabla = v\partial_\xi, \qquad \partial_\xi = \mathbf{e}\cdot\nabla,$$

so that ξ is the arclength along the pertinent streamline,

$$\mathbf{x} = \mathbf{x}(\xi, t): \quad \partial_\xi\mathbf{x} = \mathbf{e}.$$

At any point $\mathbf{x}$, and time t, ∂_ξ is the derivative along the tangent to the streamline at the point $\mathbf{x}$, and time t. Hence, we have

$$(\mathbf{v}\cdot\nabla)\mathbf{v} = v\partial_\xi\mathbf{v}, \qquad \mathbf{e}\cdot\partial_\xi\mathbf{v} = \partial_\xi v.$$

Assume the fluid undergoes a stationary motion,

$$\partial_t\rho = 0, \qquad \partial_t\mathbf{v} = \mathbf{0}.$$

Inner multiplication of (2.48) by $\mathbf{e}$ results in

$$v\partial_\xi v = -\frac{1}{\rho}\partial_\xi p - \partial_\xi gz. \tag{2.49}$$

Now, assume the fluid is barotropic, that is p and ρ are related by an invertible function

$$p = p(\rho).$$

Hence, we define

$$h(p) := \int_{p_0}^{p} \frac{1}{\rho(\tilde{p})} d\tilde{p}.$$

We have

$$\nabla h = \frac{1}{\rho}\nabla p, \qquad \partial_\xi h = \frac{1}{\rho}\partial_\xi p.$$

Consequently, (2.49) amounts to

$$\partial_\xi(\tfrac{1}{2}v^2 + h + gz) = 0.$$

If, further, the fluid is incompressible, then

$$h(p) = \frac{p}{\rho} + \text{ constant}$$

and hence

$$\tfrac{1}{2}v^2 + \frac{p}{\rho} + gz = \text{ constant}.$$

We then state Bernoulli's law by means of the following

Theorem 2.2 (Bernoulli's law) *Along a streamline of an inviscid, barotropic fluid, undergoing a stationary motion*

$$\tfrac{1}{2}v^2 + h + gz = \text{ constant}.$$

Dividing by g, we have

$$\tfrac{1}{2}v^2/g + h/g + z = \text{ constant}.$$

Each term has the dimensions of a length; hence h/g is referred to as pressure head.

2.8.2 Variational Derivation of the Equation of Motion

We show how the equation of motion can be derived in a variational setting. For definiteness, we let the fluid be incompressible,[9] so that $\nabla \cdot \mathbf{v} = 0$ and ρ is a constant. Consider the Lagrangian density

[9] More involved variational formulations for (compressible) fluids are given in [27].

$$\mathscr{L} = \tfrac{1}{2}\rho\mathbf{v}^2 + \rho\mathbf{b}\cdot\mathbf{x}.$$

We let the motion $\chi(\mathbf{X}, t)$ be the unknown function, and hence we write the Lagrangian L in the form

$$L = \int_0^T\int_\Omega \mathscr{L}(\chi, p)(\mathbf{X}, t)\, dv_R\, dt, \qquad \mathscr{L} = \tfrac{1}{2}\rho(\partial_t\chi)^2 + \rho\mathbf{b}\cdot\chi + p(J-1),$$

where $J = \det\nabla\chi$ and $p(\mathbf{X}, t)$ is a Lagrange multiplier to account for the incompressibility constraint. The extremality condition for χ and p and appropriate conditions on the boundary $\partial\Omega$ and at $t = 0, T$ imply the Euler–Lagrange equations

$$\partial_\chi\mathscr{L} - \partial_t\partial_{\partial_t\chi}\mathscr{L} - \nabla\cdot\partial_{\mathbf{F}}\mathscr{L} = \mathbf{0}, \qquad \partial_p\mathscr{L} = 0.$$

Observe that, in suffix notation

$$\partial_{F_{iK}} J = JF^{-1}_{Ki}, \qquad \partial_{X_K}(pJF^{-1}_{Ki}) = JF^{-1}_{Ki}\partial_{X_K} p = J\partial_{x_i} p.$$

Hence, the Euler–Lagrange equations become

$$\rho\mathbf{b} - \rho\dot{\mathbf{v}} - J\nabla p = \mathbf{0}, \qquad J - 1 = 0,$$

whence

$$\rho\dot{\mathbf{v}} = -\nabla p + \rho\mathbf{b}, \qquad J = 1;$$

the multiplier p takes the meaning of pressure.

2.9 Balance in a Control Volume

Mathematically, a control volume is a fixed region in space. We then look for the formulation of balance laws in control volumes associated with balance laws in convecting regions.

Let ϕ be any density field, per unit volume. The balance law for ϕ is written by letting two contributions occur, namely a volume term β and a surface term s such that

$$\frac{d}{dt}\int_{\mathcal{P}_t} \phi\, dv = \int_{\mathcal{P}_t} \beta\, dv + \int_{\partial\mathcal{P}_t} s\, da.$$

For technical convenience, we let $\varphi = \phi/\rho$. As in Cauchy's theorem, we can argue that s has to be linear in the unit normal $\mathbf{n}$ to the boundary; let $\mathbf{h}$ be such that $s = \mathbf{h}\cdot\mathbf{n}$. Hence

$$\frac{d}{dt}\int_{\mathcal{P}_t} \rho\varphi\, dv = \int_{\mathcal{P}_t} \beta\, dv + \int_{\partial\mathcal{P}_t} \mathbf{h}\cdot\mathbf{n}\, da. \tag{2.50}$$

If φ is a scalar (vector), then β is a scalar (vector) and $\mathbf{h}$ is a vector (tensor). Upon substituting and employing the transport theorem, we find

$$\int_{\mathcal{P}_t} \rho\dot{\varphi}\, dv = \int_{\mathcal{P}_t} \beta\, dv + \int_{\partial\mathcal{P}_t} \mathbf{h}\cdot\mathbf{n}\, da.$$

Since $\rho\dot{\varphi} = \partial_t(\rho\varphi) + \nabla\cdot(\rho\varphi\mathbf{v})$, by means of the divergence theorem, we obtain

$$\int_{\mathcal{P}_t} \partial_t(\rho\varphi) dv = \int_{\mathcal{P}_t} \beta\, dv + \int_{\partial\mathcal{P}_t} (\mathbf{h} - \rho\varphi\mathbf{v})\cdot\mathbf{n}\, da.$$

Since $\mathcal{P}_t$ is an arbitrary region in $\mathscr{E}$, we let $\mathcal{P}_t = \Omega$ and keep Ω fixed in space. Hence

$$\int_{\Omega} \partial_t(\rho\varphi) dv = \frac{d}{dt}\int_{\Omega} \rho\varphi\, dv.$$

Consequently, the balance law can be written in the form

$$\frac{d}{dt}\int_{\Omega} \rho\varphi\, dv = \int_{\Omega} \beta\, dv + \int_{\partial\Omega} (\mathbf{h} - \rho\varphi\mathbf{v})\cdot\mathbf{n}\, da. \tag{2.51}$$

It is apparent from (2.50) and (2.51) that, while the balance law of $\rho\varphi$ in a convecting region involves the source terms β and $\mathbf{h}\cdot\mathbf{n}$, the corresponding balance law in a fixed region (control volume) involves also a vector (or tensor) flux $\rho\varphi\mathbf{v}$.

Let $\partial\Omega_+$ and $\partial\Omega_-$ be the subsets of $\partial\Omega$ where $\mathbf{v}\cdot\mathbf{n} < 0$ (mass enters) and $\mathbf{v}\cdot\mathbf{n} > 0$ (mass exits), respectively. Hence, we let

$$\Phi_{\text{cv}} := \int_{\Omega} \rho\varphi\, dv, \qquad \check{\Phi}_i := -\int_{\partial\Omega_+} \rho\varphi\mathbf{v}\cdot\mathbf{n}\, da, \qquad \check{\Phi}_e := \int_{\partial\Omega_-} \rho\varphi\mathbf{v}\cdot\mathbf{n}\, da.$$

Consequently

$$\frac{d\Phi_{\text{cv}}}{dt} = \int_{\Omega} \beta\, dv + \int_{\partial\Omega} \mathbf{h}\cdot\mathbf{n}\, da + \check{\Phi}_i - \check{\Phi}_e.$$

This means that the time derivative of the content Φ_{cv}, within the control volume, depends not only on the physical sources β and $s = \mathbf{h}\cdot\mathbf{n}$, but also on the flow through the boundary of $\rho\varphi$ at the rate $\check{\Phi}_i - \check{\Phi}_e$.

A customary approximation used in the technical literature amounts to assuming that φ is quite constant so that

$$\check{\Phi}_i \simeq -\varphi_i \int_{\partial\Omega_+} \rho\mathbf{v}\cdot\mathbf{n}\, da, \qquad \check{\Phi}_e \simeq \varphi_e \int_{\partial\Omega_-} \rho\mathbf{v}\cdot\mathbf{n}\, da.$$

The simplest balance law is that for mass. We let $\varphi = 1$ and $\beta = 0$, $\mathbf{h} = \mathbf{0}$. Equation (2.51) reduces to

$$\frac{dm_{\text{cv}}}{dt} = -\int_{\partial\Omega} \rho\mathbf{v}\cdot\mathbf{n}\,da, \qquad m_{\text{cv}} := \int_{\Omega} \rho\,dv.$$

This equation indicates that the rate at which mass changes within Ω is equal to the net inflow (or outflow) of mass through the boundary $\partial\Omega$. Indeed, let

$$\check{m}_i := -\int_{\partial\Omega_+} \rho\mathbf{v}\cdot\mathbf{n}\,da, \qquad \check{m}_e := \int_{\partial\Omega_-} \rho\mathbf{v}\cdot\mathbf{n}\,da;$$

$\check{m}_i$ and $\check{m}_e$ are the rates at which mass flows into, or outside, Ω. Hence[10]

$$\frac{dm_{\text{cv}}}{dt} = \check{m}_i - \check{m}_e.$$

Moreover, the generic balance equation (2.51) can be written in the (approximate) form

$$\frac{d\Phi_{\text{cv}}}{dt} = \int_{\Omega} \beta\,dv + \int_{\partial\Omega} \mathbf{h}\cdot\mathbf{n}\,da + \varphi_i\check{m}_i - \varphi_e\check{m}_e.$$

The balance of linear momentum follows via the identifications

$$\varphi = \mathbf{v}, \qquad \beta = \rho\,\mathbf{b}, \qquad \mathbf{h} = \mathbf{T}.$$

The balance equation

$$\frac{d\mathfrak{P}_{\text{cv}}}{dt} = \int_{\Omega} \rho\,\mathbf{b}\,dv + \int_{\partial\Omega} \mathbf{T}\mathbf{n}\,da - \int_{\partial\Omega} \rho\,\mathbf{v}\,\mathbf{v}\cdot\mathbf{n}\,da$$

is approximated by

$$\frac{d\mathfrak{P}_{\text{cv}}}{dt} = \int_{\Omega} \rho\,\mathbf{b}\,dv + \int_{\partial\Omega} \mathbf{T}\mathbf{n}\,da + \mathbf{v}_i\check{m}_i - \mathbf{v}_e\check{m}_e.$$

The balance of energy is expressed in various ways depending on the definition of the energy E_{cv} within the control volume. Following the balance law (2.45), we make the identifications

$$E_{\text{cv}} = \int_{\Omega} \rho(\varepsilon + \tfrac{1}{2}\mathbf{v}^2)dv, \qquad \beta = \rho(\mathbf{b}\cdot\mathbf{v} + r), \qquad s = (\mathbf{T}\,\mathbf{v} - \mathbf{q})\cdot\mathbf{n}.$$

The correct balance equation

$$\frac{dE_{\text{cv}}}{dt} = \int_{\Omega} \rho(\mathbf{b}\cdot\mathbf{v} + r)dv + \int_{\partial\Omega} (\mathbf{T}\,\mathbf{v} - \mathbf{q})\cdot\mathbf{n}\,da - \int_{\partial\Omega} \rho(\varepsilon + \tfrac{1}{2}\mathbf{v}^2)\mathbf{v}\cdot\mathbf{n}\,da$$

---

[10] The literature often indicates the two contributions by $\dot{m}_i$ and $\dot{m}_e$. We avoid using $\dot{m}_i$ and $\dot{m}_e$ which would seem the derivatives of $m_i(t)$ and $m_e(t)$; formally we might determine $m_i(t)$ by $m_i(t) - m_i(t_0) = \int_{t_0}^{t} \check{m}_i(\tau)d\tau$.

simplifies to the approximate equation

$$\frac{dE_{cv}}{dt} = \int_{\Omega} \rho(\mathbf{b} \cdot \mathbf{v} + r) dv + \int_{\partial\Omega} (\mathbf{T}\,\mathbf{v} - \mathbf{q}) \cdot \mathbf{n}\, da + \check{m}_i(\varepsilon + \tfrac{1}{2}\mathbf{v}^2)_i - \check{m}_e(\varepsilon + \tfrac{1}{2}\mathbf{v}^2)_e.$$

Otherwise, letting $\mathbf{T} = \hat{\mathbf{T}} - p\mathbf{1}$, $p > 0$ being the pressure, we can write

$$\frac{dE_{cv}}{dt} = \int_{\Omega} \rho(\mathbf{b} \cdot \mathbf{v} + r) dv + \int_{\partial\Omega} (\hat{\mathbf{T}}\,\mathbf{v} - \mathbf{q}) \cdot \mathbf{n}\, da - \int_{\partial\Omega} \rho(\varepsilon + \frac{p}{\rho} + \tfrac{1}{2}\mathbf{v}^2)\mathbf{v} \cdot \mathbf{n}\, da.$$

Hence, the approximation results in

$$\frac{dE_{cv}}{dt} = \int_{\Omega} \rho(\mathbf{b} \cdot \mathbf{v} + r) dv + \int_{\partial\Omega} (\hat{\mathbf{T}}\,\mathbf{v} - \mathbf{q}) \cdot \mathbf{n}\, da + \check{m}_i(\varepsilon + \frac{p}{\rho} + \tfrac{1}{2}\mathbf{v}^2)_i - \check{m}_e(\varepsilon + \frac{p}{\rho} + \tfrac{1}{2}\mathbf{v}^2)_e.$$

We can then say that the balance of E_{cv} involves the heat and work powers

$$\int_{\Omega} \rho r\, dv - \int_{\partial\Omega} \mathbf{q} \cdot \mathbf{n}\, da, \qquad \int_{\Omega} \rho \mathbf{b} \cdot \mathbf{v}\, dv + \int_{\partial\Omega} (\hat{\mathbf{T}}\,\mathbf{v}) \cdot \mathbf{n}\, da,$$

and the enthalpy inflow, $\check{m}_i(\varepsilon + p/\rho + \frac{1}{2}\mathbf{v}^2)_i$, and outflow, $\check{m}_e(\varepsilon + p/\rho + \frac{1}{2}\mathbf{v}^2)_e$. Really, the enthalpy (per unit mass) is $\varepsilon + p/\rho$ and hence the inflow and outflow involve the enthalpy along with the kinetic term $\frac{1}{2}\mathbf{v}^2$.

If instead we consider the balance of energy (2.46), then we let

$$U_{cv} = \int_{\Omega} \rho(\varepsilon + \tfrac{1}{2}\mathbf{v}^2 + gz) dv, \qquad \beta = \rho r, \qquad s = (\mathbf{T}\,\mathbf{v} - \mathbf{q}) \cdot \mathbf{n}$$

and state the balance law in the form

$$\frac{dU_{cv}}{dt} = \int_{\Omega} \rho r\, dv + \int_{\partial\Omega} (\mathbf{T}\,\mathbf{v} - \mathbf{q}) \cdot \mathbf{n}\, da - \int_{\partial\Omega} \rho(\varepsilon + \tfrac{1}{2}\mathbf{v}^2 + gz)\mathbf{v}\, da.$$

Hence, the approximate form reads (see, e.g. (4.9) of [316])

$$\frac{dU_{cv}}{dt} = \int_{\Omega} \rho r\, dv + \int_{\partial\Omega} (\mathbf{T}\,\mathbf{v} - \mathbf{q}) \cdot \mathbf{n}\, da + \check{m}_i(\varepsilon + \tfrac{1}{2}\mathbf{v}^2 + gz)_i - \check{m}_e(\varepsilon + \tfrac{1}{2}\mathbf{v}^2 + gz)_e.$$

2.10 Basic Principles in Electromagnetism

Throughout, we use the rationalized MKSA system of units or SI units (see, e.g., [230, 401]). The electric field $\mathbf{E}$ and the magnetic induction $\mathbf{B}$ are defined operatively by means of the force on a charge. In fact, a particle with charge $\mathfrak{q}$ and velocity $\mathbf{v}$ experiences a force $\mathbf{F} = \mathfrak{q}(\mathbf{E} + \mathbf{v} \times \mathbf{B})$. Then measuring the force $\mathbf{F}$ on a test particle with different velocities determines the fields **E** and **B** at the location of the particle. The magnetic induction **B** can also be measured through the torque on a dipole.

We can express the basic principles of electromagnetism in terms of charge (and current), electric field, and magnetic induction. These principles are essentially based on Coulomb's inverse square law of force between charges, Ampère's findings about the interaction of current elements, and Faraday's experiments on variable magnetic fields. Following are the basic conceptual steps and axioms of electromagnetism (in free space).

2.10.1 Balance Equations in Free Space

As with conservation of mass, conservation of charge can be stated by saying that if q is the charge density then

$$\frac{d}{dt}\int_{\mathcal{P}_t} q\, dv = 0,$$

$\mathcal{P}_t$ being any time-dependent region convecting with the set of charges. Hence we find the continuity equation

$$\partial_t q + \nabla \cdot (q\mathbf{v}) = 0,$$

$\mathbf{v}$ being the velocity of the charges. It is understood that q is the volume charge density of positive (ions) or negative (ions, electrons) free charges. Alternatively, we may let $\mathbf{J} = q\mathbf{v}$ denote the current density vector and say that

$$I_S = \int_S \mathbf{J} \cdot \mathbf{n}\, da$$

is the current flowing through S in the direction given by $\mathbf{n}$. While the continuity equation reads

$$\partial_t q + \nabla \cdot \mathbf{J} = 0, \tag{2.52}$$

the law of conservation of charge can be given the global form

$$\frac{d}{dt}\int_{\Omega} q\, dv = -\int_{\partial\Omega} \mathbf{J} \cdot \mathbf{n}\, da,$$

Ω being any region fixed in space.

Coulomb's and Gauss' laws. In 1785, Coulomb measured the forces that charges exert on each other. It resulted that the force between the charges varies directly as the product of their magnitudes and inversely as the square of the distance between them.[11] If $\mathbf{x}_1, \mathbf{x}_2$ are the position vectors of the charges $\mathfrak{q}_1, \mathfrak{q}_2$, then the force exerted by $\mathfrak{q}_2$ on $\mathfrak{q}_1$ is

$$\mathbf{F}_{\mathfrak{q}_2} = \frac{\mathfrak{q}_1\mathfrak{q}_2}{4\pi\epsilon_0} \frac{\mathbf{x}_1 - \mathbf{x}_2}{|\mathbf{x}_1 - \mathbf{x}_2|^3}$$

and hence the electric field produced by $\mathfrak{q}_2$ at $\mathbf{x}_1$ is

$$\mathbf{E}(\mathbf{x}_1) = \frac{\mathfrak{q}_2}{4\pi\epsilon_0} \frac{\mathbf{x}_1 - \mathbf{x}_2}{|\mathbf{x}_1 - \mathbf{x}_2|^3}.$$

This indicates that the field produced by a continuous distribution of charges in the region Ω, with charge density q, is given by

$$\mathbf{E}(\mathbf{x}) = \frac{1}{4\pi\epsilon_0} \int_\Omega q(\mathbf{r}) \frac{\mathbf{x} - \mathbf{r}}{|\mathbf{x} - \mathbf{r}|^3} dv_r$$

where v_r denotes that $\mathbf{r}$ is the variable of integration. Let $\nabla_\mathbf{x}$ and $\nabla_\mathbf{r}$ be the gradients with respect to $\mathbf{x}$ and $\mathbf{r}$. To evaluate $\nabla_\mathbf{x} \cdot \mathbf{E}(\mathbf{x})$, we observe

$$\nabla_\mathbf{x} \cdot \frac{\mathbf{x} - \mathbf{r}}{|\mathbf{x} - \mathbf{r}|^3} = -\nabla_\mathbf{r} \cdot \frac{\mathbf{x} - \mathbf{r}}{|\mathbf{x} - \mathbf{r}|^3}, \qquad \nabla_\mathbf{r} \cdot \frac{\mathbf{x} - \mathbf{r}}{|\mathbf{x} - \mathbf{r}|^3} = 0, \quad |\mathbf{x} - \mathbf{r}| > 0.$$

Hence, letting $\Omega_R \subset \Omega$ be a sphere of radius R centered at $\mathbf{x}$ we have

$$4\pi\epsilon_0 \nabla_\mathbf{x} \cdot \mathbf{E}(\mathbf{x}) = -\int_\Omega q(\mathbf{r}) \nabla_\mathbf{r} \cdot \frac{\mathbf{x} - \mathbf{r}}{|\mathbf{x} - \mathbf{r}|^3} dv_r = -\int_{\Omega_R} q(\mathbf{r}) \nabla_\mathbf{r} \cdot \frac{\mathbf{x} - \mathbf{r}}{|\mathbf{x} - \mathbf{r}|^3} dv_r.$$

Replace $q(\mathbf{r})$ by $q(\mathbf{x}) + [q(\mathbf{r}) - q(\mathbf{x})]$ and observe that, by the divergence theorem,

$$\int_{\Omega_R} q(\mathbf{x}) \nabla_\mathbf{r} \cdot \frac{\mathbf{x} - \mathbf{r}}{|\mathbf{x} - \mathbf{r}|^3} dv_r = q(\mathbf{x}) \int_{\partial\Omega_R} \frac{\mathbf{x} - \mathbf{r}}{|\mathbf{x} - \mathbf{r}|^3} \cdot \frac{\mathbf{r} - \mathbf{x}}{|\mathbf{x} - \mathbf{r}|} da = -4\pi q(\mathbf{x}).$$

Further

$$\begin{aligned}\int_{\Omega_R} [q(\mathbf{r}) - q(\mathbf{x})] \nabla_\mathbf{r} \cdot \frac{\mathbf{x} - \mathbf{r}}{|\mathbf{x} - \mathbf{r}|^3} dv_r &= \int_{\Omega_R} \nabla_\mathbf{r} \cdot [q(\mathbf{r}) - q(\mathbf{x})] \frac{\mathbf{x} - \mathbf{r}}{|\mathbf{x} - \mathbf{r}|^3} dv_r - \int_{\Omega_R} \frac{\mathbf{x} - \mathbf{r}}{|\mathbf{x} - \mathbf{r}|^3} \cdot \nabla_\mathbf{r} [q(\mathbf{r}) - q(\mathbf{x})] dv_r \\ &= \frac{1}{R^2} \int_{\partial\Omega_R} [q(\mathbf{r}) - q(\mathbf{x})] da + \int_0^R \int_0^\pi \int_0^{2\pi} \frac{\mathbf{r} - \mathbf{x}}{|\mathbf{r} - \mathbf{x}|} \cdot \nabla_\mathbf{r} q(\mathbf{r}) d\rho\, d\theta\, d\phi.\end{aligned}$$

The boundedness of $\nabla_\mathbf{r} q(\mathbf{r})$ and the continuity of $q(\mathbf{r})$ imply that this integral approaches zero as $R \to 0$. Hence, we find that

$$\nabla \cdot \mathbf{E} = \frac{1}{\epsilon_0} q.$$

---

[11] In SI units the constant of proportionality is equal to $1/4\pi\epsilon_0$, ϵ_0 being the permittivity of free space, $\epsilon_0 = 8.854 \times 10^{-12}\, \mathrm{C}^2/\mathrm{N} \cdot \mathrm{m}^2$.

Based on this local equation, if S is the boundary of the region V, by the divergence theorem, we have

$$\int_{\partial V} \mathbf{E}\cdot\mathbf{n}\,da = \frac{1}{\epsilon_0}\int_V q\,dv.$$

Biot-Savart and Ampère's laws. Denote by $\mathbf{B}$ the *magnetic induction* (also named the *magnetic flux density*). The operative definition of $\mathbf{B}$ is provided by measuring the mechanical torque $\mathfrak{T}$ exerted on a magnetic dipole. If $\mathfrak{m}$ is the magnetic moment, then $\mathfrak{T}$ is given by[12]

$$\mathfrak{T} = \mathfrak{m}\times\mathbf{B}.$$

In 1819, Oersted observed that wires carrying electric currents produced deflections of permanent magnetic dipoles. Next, Biot and Savart (1820) and Ampère (1820–1825) established the basic experimental laws relating $\mathbf{B}$ to the currents and determined also the law of force between currents. If I is the current flowing along the (closed) curve C, then the magnetic induction, at $\mathbf{x}$, is given by the Biot–Savart law expressed by the line integral

$$\mathbf{B}(\mathbf{x}) = \frac{\mu_0}{4\pi}I\int_C \frac{\mathbf{t}(\mathbf{r})\times(\mathbf{x}-\mathbf{r})}{|\mathbf{x}-\mathbf{r}|^3}ds,$$

where[13] $\mathbf{t}$ is the unit tangent vector to C. Now, since $\mathbf{t}$ is independent of $\mathbf{x}$,

$$\partial_{x_i}\epsilon_{ijk}t_j\frac{x_k-r_k}{|\mathbf{x}-\mathbf{r}|^3} = \epsilon_{ijk}t_j\Big(\frac{\delta_{ki}}{|\mathbf{x}-\mathbf{r}|^3} - 3\frac{(x_k-r_k)(x_i-r_i)}{|\mathbf{x}-\mathbf{r}|^5}\Big) = 0.$$

Hence, it follows that

$$\nabla\cdot\mathbf{B} = 0.$$

Now look at a surface S and denote by $\mathbf{J}$ the current density vector so that

$$I = \int_S \mathbf{J}\cdot\mathbf{n}\,da$$

is the current flowing through S. Ampère's (circuital) law is written in the form

$$\int_{\partial S} \mathbf{B}\cdot\mathbf{t}\,ds = \mu_0\int_S \mathbf{J}\cdot\mathbf{n}\,da.$$

The arbitrariness of S and use of Stokes' theorem yield

$$\nabla\times\mathbf{B} = \mu_0\mathbf{J}.$$

[12] Often $\mathbf{N}$ and μ are used in place of $\mathfrak{T}$ and $\mathfrak{m}$ in the physical literature.

[13] The constant μ_0 is the permeability of free space, $\mu_0 = 4\pi\times 10^{-7}\mathrm{N/A^2}$.

Faraday's law. In 1831, Faraday observed that a time-varying induction **B** produces a current in a closed loop of wire. According to Lenz's law, the induced current is in such a direction as to oppose the change of flux through the wire.

Let S be a surface and let

$$\Phi = \int_S \mathbf{B} \cdot \mathbf{n}\, da$$

be the magnetic flux. The electromotive force around the circuit is

$$\mathcal{E}_C = \int_C \boldsymbol{\mathcal{E}} \cdot \mathbf{t}\, ds,$$

where $\boldsymbol{\mathcal{E}}$ is the electric field at the point of the wire. Faraday's (and Lenz's) observations are summed up in the mathematical law

$$\int_C \boldsymbol{\mathcal{E}} \cdot \mathbf{t}\, ds = -\frac{d}{dt} \int_S \mathbf{B} \cdot \mathbf{n}\, da. \tag{2.53}$$

If the wire C is moving with velocity field $\mathbf{v}$ then, by (1.49) and (1.50), it follows

$$\frac{d}{dt} \int_S \mathbf{B} \cdot \mathbf{n}\, da = \int_S [\partial_t \mathbf{B} + \nabla \times (\mathbf{B} \times \mathbf{v}) + \mathbf{v} \nabla \cdot \mathbf{B}] \cdot \mathbf{n}\, da.$$

As a consequence, Faraday's law (2.53) and the divergence-free condition $\nabla \cdot \mathbf{B} = 0$ result in

$$\int_C \boldsymbol{\mathcal{E}} \cdot \mathbf{t}\, ds = -\int_S [\partial_t \mathbf{B} + \nabla \times (\mathbf{B} \times \mathbf{v})] \cdot \mathbf{n}\, da.$$

Hence, in view of Stokes' theorem, we have

$$\int_C [\boldsymbol{\mathcal{E}} + \mathbf{B} \times \mathbf{v}] \cdot \mathbf{t}\, ds = -\int_S \partial_t \mathbf{B} \cdot \mathbf{n}\, da.$$

Now $\boldsymbol{\mathcal{E}} + \mathbf{B} \times \mathbf{v} = \mathbf{E}$, that is the electric field in the chosen reference frame. Consequently, by a further use of Stokes' theorem, we find the equation

$$\nabla \times \mathbf{E} = -\partial_t \mathbf{B},$$

where **E** and **B** are the fields in the chosen reference frame.

As a check of consistency, if the wire C, and hence S, is fixed in space, then we have

$$\int_C \mathbf{E} \cdot \mathbf{t}\, ds = -\int_S \partial_t \mathbf{B} \cdot \mathbf{n}\, da,$$

E and **B** being the fields as seen in the reference of the wire. The arbitrariness of S and use of Stokes' theorem result again in the equation $\nabla \times \mathbf{E} = -\partial_t \mathbf{B}$.

By way of summary, the mathematical results so obtained, which are a consequence of the experiments performed in the first half of 1800s, can be written in the form of equations

$$\begin{array}{ll}
\int_{\partial\Omega} \mathbf{E}\cdot\mathbf{n}\,da = (1/\epsilon_0)\int_\Omega q\,dv & \nabla\cdot\mathbf{E} = q/\epsilon_0 \\
\int_{\partial\Omega} \mathbf{B}\cdot\mathbf{n}\,da = 0 & \nabla\cdot\mathbf{B} = 0 \\
\int_{\partial S} \mathbf{B}\cdot\mathbf{t}\,ds = \mu_0\int_S \mathbf{J}\cdot\mathbf{n}\,da & \nabla\times\mathbf{B} = \mu_0\mathbf{J} \\
\int_{\partial S} \mathbf{E}\cdot\mathbf{t}\,ds = -\frac{d}{dt}\int_S \mathbf{B}\cdot\mathbf{n}\,da & \nabla\times\mathbf{E} = -\partial_t\mathbf{B},
\end{array}$$

which can be viewed as the statements of Coulomb's law, absence of free magnetic poles, Ampère's law, and Faraday's law.

2.10.2 Balance Laws in Matter

Electromagnetism in matter involves also the polarization $\mathbf{P}$ and the magnetization $\mathbf{M}$. For compactness of presentation, balance laws and constitutive properties are given in a unified scheme.

An externally applied electric field may cause microscopic separations of the centres of positive and negative charges which thus behave as dipoles of charges. This phenomenon is referred to as the polarization of the (dielectric) material. The tendency of the dipoles to align with the electric field causes charges to appear at the surfaces of the material. Hence, Gauss' law in matter is stated by saying that, for any region Ω

$$\int_{\partial\Omega} \mathbf{E}\cdot\mathbf{n}\,da = \frac{1}{\epsilon_0}\int_\Omega q\,dv + \frac{1}{\epsilon_0}\int_{\partial\Omega} \lambda_p\,da,$$

where q is the density, per unit volume, of free charges while λ_p is the charge density, per unit area, due to polarization. By the generalized Cauchy's theorem, there exists a vector field, say $-\mathbf{P}$, such that

$$\int_{\partial\Omega} \lambda_p\,da = -\int_{\partial\Omega} \mathbf{P}\cdot\mathbf{n}\,da = -\int_\Omega \nabla\cdot\mathbf{P}\,dv.$$

The negative sign allows us to view $\mathbf{P}$ as the polarization vector, namely the dipole moment per unit volume. Letting

$$\mathbf{D} = \epsilon_0\mathbf{E} + \mathbf{P}$$

we can write Gauss' law in the form

$$\int_{\partial\Omega} \mathbf{D}\cdot\mathbf{n}\,da = \int_\Omega q\,dv.$$

The vector $\mathbf{D}$ is termed *electric displacement* or *electric flux density.* This view is the mathematical basis for the occurrence of the polarization vector $\mathbf{P}$. A physical picture of the polarization field is gained by looking at an assembly of dipoles. An electric dipole consists of a positive and a negative charge, $\mathfrak{q}$ and $-\mathfrak{q}$, and $\mathbf{d}$ is the displacement from the position of $-\mathfrak{q}$ to that of $\mathfrak{q}$. We denote by $\mathfrak{p} = \mathfrak{q}\,\mathbf{d}$ the vector which is then directed from the negative to the positive charge; we say that $\mathfrak{p} = \mathfrak{q}\,\mathbf{d}$ is the *electric dipole moment* and $\mathbf{d}$ is the displacement vector. Assume the dipoles have the same length $|\mathbf{d}| = c$. Let $f(\mathbf{x}, \mathbf{d}, t)$ be the distribution function, on the sphere Σ of radius c, of $\mathbf{d}$ at $(\mathbf{x}, t)$. Hence, we define the dipole moment density in the form

$$\mathbf{p}(\mathbf{x}, t) = \langle \mathfrak{p} \rangle = \mathfrak{q}\,\langle \mathbf{d} \rangle = \mathfrak{q} \int_{\Sigma} \mathbf{d}\, f(\mathbf{x}, \mathbf{d}, t) da,$$

that is the first moment of f. Letting $n(\mathbf{x}, t)$ be the number density of dipoles we let

$$\mathbf{P} = n\mathbf{p} = n\mathfrak{q}\langle \mathbf{d} \rangle$$

be the *polarization vector.* Outside possible surface and line discontinuities, $\mathbf{p}$ is taken to be a continuous function of $\mathbf{x}$ and t. We associate a constant number of dipoles to each atom or molecule. This in turn justifies the assumption that n is a constant times the mass density ρ. As a consequence, n satisfies the balance equation

$$\frac{\dot{n}}{n} = \frac{\dot{\rho}}{\rho} = -\nabla \cdot \mathbf{v}.$$

An atom has a magnetic moment with orbital and spin contributions. For atoms or ions containing many electrons, the electrons in closed shells pair up with their spins and orbital angular momenta opposite each other, thus producing a zero net magnetic moment. Atoms with an odd number of electrons must have at least one unpaired electron and a spin magnetic moment of at least one Bohr magneton μ_B,

$$\mu_B = \frac{e\hbar}{2m_e},$$

where e and m_e are the (positive) charge and the mass of the electron. The total magnetic moment of an atom is the vector sum of the orbital and spin magnetic moments. An unpaired outer electron can contribute both an orbital moment and a spin moment. The orbital moment is about the same order of magnitude as the Bohr magneton. The nucleus of an atom also has a magnetic moment associated with protons and neutrons. Yet, owing to their masses, the magnetic moment of a proton or neutron is small compared with the magnetic moment of an electron, by a factor of about 10^{-3}.

The *magnetization* $\mathbf{M}$ is the magnetic moment of the substance per unit volume. Let $\mathbf{B}_0$ be the magnetic induction produced by, e.g. a current-carrying conductor. If the region is filled with a magnetic substance, the total magnetic induction is

$\mathbf{B} = \mathbf{B}_0 + \mathbf{B}_m$, where $\mathbf{B}_m$ is the field produced by the magnetic substance. We let $\mathbf{B}_m = \mu_0 \mathbf{M}$ and hence

$$\mathbf{B} = \mathbf{B}_0 + \mu_0 \mathbf{M}.$$

By letting $\mathbf{H} = (\mathbf{B}/\mu_0) - \mathbf{M}$, we can write

$$\mathbf{B} = \mu_0(\mathbf{H} + \mathbf{M}).$$

The field $\mathbf{M}$ is defined as the *magnetization*[14] while $\mathbf{H}$ is termed the *magnetic intensity*[15]; $\mathbf{H}$ is then $1/\mu_0$ times the magnetic induction $\mathbf{B}_0$ in the absence of matter.

In a large class of substances, the magnetization $\mathbf{M}$ is proportional to the magnetic intensity $\mathbf{H}$,

$$\mathbf{M} = \chi \mathbf{H},$$

χ being called the magnetic susceptibility. If $\chi > 0$, the substance is said to be paramagnetic, if $\chi < 0$ the substance is said to be diamagnetic. Hence

$$\mathbf{B} = \mu_m \mathbf{H}, \qquad \mu_m = \mu_0(1 + \chi).$$

We can view $\mathbf{B} = (1 + \chi)\mathbf{B}_0$ as the magnetic induction in the material and hence $\mathbf{H} = \mathbf{B}_0/\mu_0$ as the magnetic field.

Since usually $|\chi| \ll 1$ then diamagnetism is such a weak property that its effects are not observable in everyday life. All conductors exhibit an effective diamagnetism when they experience a changing magnetic field. The Lorentz force on electrons causes them to circulate around forming eddy currents. Such currents in turn produce an induced magnetic field opposite to the applied field. In the limit case $\chi = -1$, the material is called a perfect diamagnet and all magnetic fields applied to it are expelled. Both in diamagnetic and paramagnetic substances $\mu_m \geq 0$. As a consequence, $\mathbf{B}$ is in the same direction as $\mathbf{H}$; in paramagnetic substances $\mathbf{M} \cdot \mathbf{H} > 0$, whereas in diamagnetic substances $\mathbf{M} \cdot \mathbf{H} < 0$. For ferromagnetic substances, μ_m is several thousand times μ_0 but is not a constant.

A paramagnetic substance has a positive, but small, susceptibility which is due to the presence of atoms, or ions, with permanent magnetic dipole moments. These dipoles are randomly oriented in the absence of an external magnetic field and tend to align with an applied magnetic field. This alignment is contrasted by thermal motion which tends to randomize the dipole orientations. This competition is made formal by Curie–Weiss law, whereby

$$M = C \frac{B}{\mu_0(\theta - \theta_c)},$$

[14] Or magnetic polarization or magnetic moment density.

[15] Or magnetic field (strength).

C being the Curie constant and θ_c the Curie temperature; the magnetization increases with increasing the induction B and with decreasing the temperature θ. As $\theta < \theta_c$, the behaviour of M is nonlinear (Sect. 15.3.1).

So as to establish balance equations, we observe that an atom as a whole has a small magnetic moment but the random orientation of the atoms in a larger sample produces an average zero magnetic moment. When an external field is applied there is a small torque on each atomic moment and these moments tend to become aligned with the external field. Accordingly, an atom is similar to a current loop and can thus be considered to have magnetic dipole moments. The current, however, reflects the movement of bound charges (orbital electrons, electron spin, and nuclear spin). We then let the current density be the sum of the free current density $\mathbf{J}$ and the bound current density $\mathbf{J}_b$. To model $\mathbf{J}_b$ we observe that the current loops in the material result in a circulating current around the periphery of the material. We then let

$$\mathbf{J}_b = \nabla \times \mathbf{M}$$

and modify Ampère's law as

$$\int_{\partial S} \mathbf{B} \cdot \mathbf{t}\, ds = \mu_0 \int_S \mathbf{J} \cdot \mathbf{n}\, da + \mu_0 \int_{\partial S} \mathbf{M} \cdot \mathbf{t}\, ds,$$

where $\mathbf{J}$ is the free current density. Since $\mathbf{B} = \mu_0(\mathbf{H} + \mathbf{M})$ we have

$$\int_{\partial S} \mathbf{H} \cdot \mathbf{t}\, ds = \int_S \mathbf{J} \cdot \mathbf{n}\, da.$$

The balance equations in matter are then given the form

$$\int_{\partial\Omega} \mathbf{D} \cdot \mathbf{n}\, da = \int_\Omega q\, dv \qquad \nabla \cdot \mathbf{D} = q,$$

$$\int_{\partial S} \mathbf{H} \cdot \mathbf{t}\, ds = \int_S \mathbf{J} \cdot \mathbf{n}\, da \qquad \nabla \times \mathbf{H} = \mathbf{J}.$$

It is naturally claimed that $\mathbf{E}$ and $\mathbf{B}$ are the fundamental fields while $\mathbf{D}$ and $\mathbf{H}$ are the derived fields.

Incidentally, a physical picture of $\mathbf{M}$ arises by considering an assembly of magnetic dipoles with moment

$$\mathfrak{m} = \mu_0 \mathbf{s}, \qquad |\mathbf{s}| = c$$

and then letting

$$\mathbf{m} = \mu_0 \langle \mathbf{s} \rangle, \qquad \mathbf{M} = n\mathbf{m}.$$

It was Maxwell [306] who compiled the known results obtained by Ampère, Faraday, Gauss, Coulomb, and others and made an important addition to one of them by overcoming an inconsistency. All these laws, but Faraday's law, were derived from steady-state observations. Hence, it is natural to expect that such equations requires some improvements for time-dependent fields. The faulty equation is Ampère's law.

The divergence of $\nabla \times \mathbf{H} = \mathbf{J}$ gives

$$\nabla \cdot \mathbf{J} = 0,$$

which holds only for steady-state problems. In general, the continuity condition and Gauss' law imply that

$$\nabla \cdot \mathbf{J} = -\partial_t q = -\nabla \cdot \partial_t \mathbf{D}.$$

Hence, Ampère's law is made consistent by replacing $\mathbf{J}$ with

$$\mathbf{J} + \partial_t \mathbf{D}.$$

The whole set of differential equations become

$$\nabla \cdot \mathbf{D} = q, \qquad \nabla \times \mathbf{H} = \mathbf{J} + \partial_t \mathbf{D}, \tag{2.54}$$

$$\nabla \cdot \mathbf{B} = 0, \qquad \nabla \times \mathbf{E} + \partial_t \mathbf{B} = \mathbf{0}. \tag{2.55}$$

Equations (2.54)–(2.55) are known as *Maxwell's equations.* Maxwell called the added term $\partial_t \mathbf{D}$ the *displacement current.* The form (2.54)–(2.55) of Maxwell's equations is usually referred to as Minkowski formulation.[16]

The Field H

Even within the MKS system of units, the literature shows different definitions of $\mathbf{M}$ and $\mathbf{H}$. By analogy with the definition of $\mathbf{E} + \mathbf{P}/\epsilon_0$ as the field $\mathbf{D}/\epsilon_0$ then $\mathbf{B} - \mu_0\mathbf{M}$ is defined as the vector, say $\tilde{\mathbf{H}}$, so that $\mathbf{B} = \mu_0\mathbf{M} + \tilde{\mathbf{H}}$. Instead, as the majority does in the literature, we have chosen $\mathbf{H} = \tilde{\mathbf{H}}/\mu_0$ as the new field so that $\mathbf{B} = \mu_0(\mathbf{M} + \mathbf{H})$. Others (e.g. [372]) define $\mathbf{M}$ and $\mathbf{H}$, say $\hat{\mathbf{M}}$ and $\hat{\mathbf{H}}$, so that $\mathbf{B} = \hat{\mathbf{M}} + \hat{\mathbf{H}}$. Of course, the physical units of $\mathbf{H}$, $\tilde{\mathbf{H}}$ and those of $\mathbf{M}$, $\hat{\mathbf{M}}$ are different.

2.11 The Invariant Form of Maxwell's Equations

According to the special theory of relativity:
(i) the laws of physics are the same in all systems which move with uniform velocity relative to one another,
(ii) the speed of light in free space[17] is c whatever the relative uniform motion of source and observer.

---

[16] See, e.g. [357]. An analysis of the connection between various formulations of Maxwell's equations is given in [381].

[17] The speed of light in free space, c, is a universal physical constant. Its exact value is defined as 299792458 metres per second. It is exact because, by the international agreement, a metre is defined as the length of the path travelled by light in free space during a time interval of 1/299792458 second.

We now show how these principles affect the transformation laws of coordinates relative to two different inertial frames $\mathscr{F}$, $\mathscr{F}'$.

Let $Oxyz$ and $O'x'y'z'$ be Cartesian coordinate systems in $\mathscr{F}$ and $\mathscr{F}'$. The coordinates are chosen so that $\mathscr{F}'$ is moving with uniform speed v relative to $\mathscr{F}$ along both Ox and $O'x'$. Moreover let $t = 0$ and $t' = 0$ when O and O' coincide. There is no relative motion in the y and z directions and hence these coordinates are not affected by the transformation law

$$y' = y, \qquad z' = z.$$

Since $x' = 0$ when $x = 0, t = 0$, it is reasonable to assume that

$$x' = \gamma(x - vt)$$

where γ is a dimensionless constant possibly dependent on v. It is expected that the Galilei transformation $x' = x - vt$ is recovered at the limit of small speeds, relative to c, and hence it is expected that $\gamma \to 1$ as $v/c \to 0$. By (i), the inverse law

$$x = \gamma(x' + vt')$$

has to hold.

Let a light pulse be emitted at the origin of $\mathscr{F}$ at $t = 0$. By (ii), we have $x = ct$, $x' = ct'$. Hence

$$x = ct \iff x' = ct'.$$

As a consequence, we can write

$$0 = x' - ct' = \gamma(x - vt) - ct' = \gamma(ct - vt) - ct'$$

whence

$$t' = \gamma(t - \frac{v}{c}t).$$

Also, by $x = \gamma(x' + vt')$, we have

$$ct = \gamma(ct' + vt') = \gamma(c + v)t'.$$

Upon substitution of t', it follows

$$ct = \gamma(c + v)\gamma(t - \frac{v}{c}t) = \frac{\gamma^2}{c}(c^2 - v^2)t$$

whence

$$\gamma^2 = \frac{c^2}{c^2 - v^2}.$$

Since $\gamma > 0$ we have

$$\gamma = (1 - v^2/c^2)^{-1/2}.$$

Now, that we have determined γ, we replace x' to obtain

$$x = \gamma(x' + vt') = \gamma[\gamma(x - vt) + vt']$$

whence

$$t' = \gamma(t - vx/c^2).$$

The set of relations

$$x' = \gamma(x - vt), \quad y' = y, \quad z' = z, \quad t' = \gamma(t - vx/c^2) \tag{2.56}$$

constitutes the *Lorentz* transformation. The inverse transformation, that is x, y, z, t in terms of x', y', z', t', follows by replacing v by $-v$ and by interchanging primed and unprimed coordinates,

$$x = \gamma(x' + vt'), \quad y = y', \quad z = z', \quad t = \gamma(t' + vx'/c^2).$$

By (2.56), it follows that

$$x'^2 + y'^2 + z'^2 - c^2 t'^2 = x^2 + y^2 + z^2 - c^2 t^2.$$

Hence, letting $x^1 = x, x^2 = y, x^3 = z, x^4 = ct$, we conclude that the invariant distance of events x^α is given by

$$g_{\alpha\beta} x^\alpha x^\beta,$$

where the metric tensor $\{g_{\alpha\beta}\}$ is given by

$$\{g_{\alpha\beta}\} = \mathrm{diag}[1, 1, 1, -1].$$

Consequently

$$\{x_\alpha\} = [x, y, z, -ct], \qquad \{g^{\alpha\beta}\} = \mathrm{diag}[1, 1, 1, -1].$$

If $v^2 \ll c^2$, then γ approaches unity and the Lorentz transformation approaches the *Galilean transformation*

$$x' = x - vt, \quad y' = y, \quad z' = z, \quad t' = t.$$

By (i), any physical law must be put in a form which is unaltered by a Lorentz transformation. Now, in the four-dimensional setting, we can express the Lorentz transformation as

$$x'^\alpha = \Lambda^\alpha_\beta x^\beta,$$

the matrix Λ being given by

$$\Lambda^{\alpha}_{\beta} = \begin{bmatrix} \gamma & 0\,0 & -\gamma v/c \\ 0 & 1\,0 & 0 \\ 0 & 0\,1 & 0 \\ -\gamma v/c & 0\,0 & \gamma \end{bmatrix}.$$

We then show that Maxwell's equations (2.54)–(2.55) can be written in tensor form. By (2.55), there exist a vector potential $\mathbf{A}$ and a scalar potential V such that

$$\mathbf{B} = \nabla \times \mathbf{A}, \qquad \mathbf{E} = -\partial_t \mathbf{A} - \nabla V.$$

Regard $A^{\alpha} = (\mathbf{A}, V/c)$ as a four-vector; we next check the consistency of this assumption while

$$A_1 = A^1, \quad A_2 = A^2, \quad A_3 = A^3, \quad A_4 = -V/c.$$

Define the skew-symmetric tensor

$$F_{\alpha\beta} = \partial_{x^{\alpha}} A_{\beta} - \partial_{x^{\beta}} A_{\alpha}, \quad \alpha, \beta = 1, 2, 3, 4,$$

that is

$$F_{\alpha\beta} = -F_{\beta\alpha}, \quad F_{hk} = \epsilon_{hki} B_i, \quad F_{h4} = E_h/c, \quad h, k = 1, 2, 3.$$

In matrix form it is

$$F_{\alpha\beta} = \begin{bmatrix} 0 & B_3 & -B_2 & E_1/c \\ -B_3 & 0 & B_1 & E_2/c \\ B_2 & -B_1 & 0 & E_3/c \\ -E_1/c & -E_2/c & -E_3/c & 0 \end{bmatrix}.$$

The two equations (2.55) may then be given the tensor form

$$\partial_{x^{\sigma}} F_{\alpha\beta} + \partial_{x^{\alpha}} F_{\beta\sigma} + \partial_{x^{\beta}} F_{\sigma\alpha} = 0, \tag{2.57}$$

where α, β, σ are any three of $1, 2, 3, 4$. Indeed, if $(\alpha, \beta, \sigma) = (1, 2, 3)$, then (2.57) gives $\nabla \cdot \mathbf{B} = 0$. If, instead, $(\alpha, \beta, \sigma) = (1, 2, 4), (3, 1, 4), (2, 3, 4)$, then we obtain the components of $\nabla \times \mathbf{E} + \partial_t \mathbf{B} = \mathbf{0}$.

Let $G^{\alpha\beta}$ be the skew-symmetric tensor defined by

$$G^{hk} = \epsilon_{hki} H_i, \quad G^{h4} = -c D_h,$$

that is

$$G^{\alpha\beta} = \begin{bmatrix} 0 & H_3 & -H_2 & -cD_1 \\ -H_3 & 0 & H_1 & -cD_2 \\ H_2 & -H_1 & 0 & -cD_3 \\ cD_1 & cD_2 & cD_3 & 0 \end{bmatrix},$$

and J^α the four-vector given by

$$J^\alpha = (J_1, J_2, J_3, c\rho).$$

Equation (2.55) can be written in the form

$$\partial_{x^\beta} G^{\alpha\beta} = J^\alpha. \tag{2.58}$$

If $\alpha = 4$, we obtain $\nabla \cdot \mathbf{D} = \rho$. If $\alpha = 1, 2, 3$, we obtain the components of $\nabla \times \mathbf{H} - \partial_t \mathbf{D} = \mathbf{J}$. Moreover, by the skew-symmetry of G, it follows that

$$0 = \partial_{x^\alpha} \partial_{x^\beta} G^{\alpha\beta} = \partial_{x^\alpha} J^\alpha, \tag{2.59}$$

that is the continuity equation (2.52).

The tensor form (2.57)–(2.58) of Maxwell's equations implies that they hold invariant under a Lorentz transformation. The same is true for the continuity equation (2.59). Moreover, the tensor representation of **E**, **B**, **D**, **H** allows us to determine how these fields change relative to different observers. First, since F involves only **E** and **B**, then it follows that $\mathbf{E}'$, $\mathbf{B}'$, the fields relative to $\mathscr{F}'$, are linear combinations of **E** and **B**. Analogously, the form of G implies that $\mathbf{H}'$, $\mathbf{D}'$, the fields relative to $\mathscr{F}'$, are linear combinations of **H** and **D**.

By

$$F^{\alpha\beta} = g^{\alpha\mu} g^{\beta\nu} F_{\mu\nu}$$

it follows that $-F^{4i} = F^{i4} = -F_{4i} = F_{4i}$, $i = 1, 2, 3$, and the like for G. Instead, $F^{ij} = F_{ij}, i, j = 1, 2, 3$. Let $F'^{\alpha\beta}$ and $G'^{\alpha\beta}$ be the values of $F^{\alpha\beta}$ and $G^{\alpha\beta}$ at $\mathscr{F}'$. By

$$F'^{\alpha\beta} = \Lambda^\alpha_\mu \Lambda^\beta_\nu F^{\mu\nu}$$

and the like for G we find the transformation laws

$$\mathbf{E}' = \gamma \mathbf{E} + (1 - \gamma)(\mathbf{E} \cdot \mathbf{v})\mathbf{v}/v^2 + \gamma \mathbf{v} \times \mathbf{B},$$

$$\mathbf{B}' = \gamma \mathbf{B} + (1 - \gamma)(\mathbf{B} \cdot \mathbf{v})\mathbf{v}/v^2 - \gamma \mathbf{v} \times \mathbf{E}/c^2,$$

and

$$\mathbf{D}' = \gamma \mathbf{D} + (1 - \gamma)(\mathbf{D} \cdot \mathbf{v})\mathbf{v}/v^2 + \gamma \mathbf{v} \times \mathbf{H}/c^2,$$

$$\mathbf{H}' = \gamma \mathbf{H} + (1 - \gamma)(\mathbf{H} \cdot \mathbf{v})\mathbf{v}/v^2 - \gamma \mathbf{v} \times \mathbf{D}.$$

Moreover, $J'^{\alpha} = \Lambda^{\alpha}_{\beta} J^{\beta}$ gives

$$\mathbf{J}' = \mathbf{J} - (1-\gamma)(\mathbf{J}\cdot\mathbf{v})\mathbf{v}/v^2 - \gamma q\mathbf{v}, \qquad q' = \gamma(q - \mathbf{J}\cdot\mathbf{v}/c^2).$$

It is of interest to look at the transformation properties under the approximation of small speeds, $|v| \ll c$, where $\gamma \simeq 1$. The $1/c^2$ terms are often disregarded and hence the *approximations*

$$\mathbf{J}' \simeq \mathbf{J} - q\mathbf{v}, \qquad q' \simeq q.$$

are considered in the literature. This means that q is (roughly) invariant, whereas $\mathbf{J}'$ and $\mathbf{J}$ differ by the convective term $q\mathbf{v}$. The fields $\mathbf{E}, \mathbf{B}, \mathbf{D}, \mathbf{H}$ in $\mathscr{F}, \mathscr{F}'$ are approximately related by

$$\mathbf{E}' \simeq \mathbf{E} + \mathbf{v}\times\mathbf{B}, \quad \mathbf{B}' \simeq \mathbf{B}, \quad \mathbf{D}' \simeq \mathbf{D}, \quad \mathbf{H}' \simeq \mathbf{H} - \mathbf{v}\times\mathbf{D}.$$

If $\mathscr{F}'$ is (momentarily) at rest with the charge, then $\mathbf{E}'$ is the electric field at rest with the charge. At $\mathscr{F}'$, a charge q at rest experiences a force $\mathsf{q}\mathbf{E}' = \mathsf{q}(\mathbf{E} + \mathbf{v}\times\mathbf{B})$, $\mathbf{E}, \mathbf{B}$ being the fields at $\mathscr{F}$. This shows, and confirms the experimental evidence, that a charge q in motion with velocity $\mathbf{v}$ experiences a force $\mathsf{q}(\mathbf{E} + \mathbf{v}\times\mathbf{B})$, called the Lorentz force.

Instead of the primed quantities, henceforth we use $\mathcal{E}, \mathcal{B}, \mathcal{D}, \mathcal{H}, \mathcal{J}$; the *approximations* can then be taken in the form

$$\mathcal{E} = \mathbf{E} + \mathbf{v}\times\mathbf{B}, \quad \mathcal{B} = \mathbf{B}, \quad \mathcal{D} = \mathbf{D}, \quad \mathcal{H} = \mathbf{H} - \mathbf{v}\times\mathbf{D}, \quad \mathcal{J} = \mathbf{J} - q\mathbf{v}.$$

The approximations for $\mathcal{B}$ and $\mathcal{D}$ are justified in several ways [354, 409] with purely Galilean arguments. Assume the body is rigid and let

$$\nabla'\times\mathcal{E} = -\partial_{t'}\mathcal{B}.$$

Hence

$$\partial_{t'}\mathcal{B} = (\partial_t + \mathbf{v}\cdot\nabla)\mathcal{B},$$

$\mathbf{v}$ being the velocity of $\mathscr{F}'$ with respect to $\mathscr{F}$. Since $\nabla\cdot\mathcal{B} = 0$ and $\nabla\mathbf{v} = \mathbf{0}$, then we have

$$\nabla\times(\mathbf{v}\times\mathcal{B}) = -(\mathbf{v}\cdot\nabla)\mathcal{B}.$$

Hence, since $\nabla = \nabla'$, we can write

$$\nabla\times(\mathcal{E} - \mathbf{v}\times\mathcal{B}) = -\partial_t\mathcal{B}.$$

This is Maxwell's equation with respect to $\mathscr{F}$ if $\mathcal{B} = \mathbf{B}$ and $\mathbf{E} = \mathcal{E} - \mathbf{v}\times\mathbf{B}$. A similar argument leads to $\mathcal{D} = \mathbf{D}$.

2.12 Maxwell's Equations for the Fields at the Rest Frame

We now review Maxwell's equations by means of the fields at the local rest frame of the body. Let the surface S and the contour $C = \partial S$ of S be in motion with velocity $\mathbf{v}$, possibly varying along C. By (1.51) we have

$$\frac{d}{dt}\int_S \mathbf{D}\cdot\mathbf{n}\,da = \int_S [\partial_t\mathbf{D} + \nabla\times(\mathbf{D}\times\mathbf{v}) + \mathbf{v}\nabla\cdot\mathbf{D}]\cdot\mathbf{n}\,da = -\int_{\partial S}\mathbf{v}\times\mathbf{D}\cdot\mathbf{t}\,ds + \int_S(\partial_t\mathbf{D} + q\mathbf{v})\cdot\mathbf{n}\,da$$

Moreover

$$\mathcal{H} + \mathbf{v}\times\mathbf{D} = \mathbf{H},$$

is the magnetic intensity in the chosen reference frame whereas $\mathcal{H}$ is the value at the contour $C = \partial S$. Also

$$\mathcal{J} = \mathbf{J} - q\mathbf{v}$$

is the current density relative to the observer at rest with the body while $\mathbf{J} = \mathcal{J} + q\mathbf{v}$ comprises the convective term $q\mathbf{v}$. Hence, Maxwell's equations in integral form can be written as

$$\int_{\partial\Omega}\mathbf{D}\cdot\mathbf{n}\,da = \int_\Omega q\,dv, \qquad \int_{\partial S}\mathcal{H}\cdot\mathbf{t}\,ds = \int_S \mathcal{J}\cdot\mathbf{n}\,da + \frac{d}{dt}\int_S\mathbf{D}\cdot\mathbf{n}\,da, \tag{2.60}$$

$$\int_{\partial\Omega}\mathbf{B}\cdot\mathbf{n}\,da = 0, \qquad \int_{\partial S}\mathcal{E}\cdot\mathbf{t}\,ds = -\frac{d}{dt}\int_S\mathbf{B}\cdot\mathbf{n}\,da. \tag{2.61}$$

Equations (2.60)–(2.61) imply (2.54)–(2.55). In the next section, we give a full account of the transformation properties of $\mathbf{E}, \mathbf{B}, \mathbf{D}, \mathbf{H}$ thus providing a better understanding of the connection between (2.60)–(2.61) and (2.54)–(2.55). Equations (2.54)–(2.55) are relative to a chosen reference frame, while (2.60)–(2.61) involve the fields relative to the observer at rest with the body (S and ∂S).

For any C^1 vector field $\mathbf{w}$

$$\frac{d}{dt}\int_S\mathbf{w}\cdot\mathbf{n}\,da = \int_S \overset{\diamond}{\mathbf{w}}\cdot\mathbf{n}\,da.$$

Then the arbitrariness of S and Ω in (2.60)–(2.61) implies

$$\nabla\cdot\mathbf{D} = q, \qquad \nabla\times\mathcal{H} = \mathcal{J} + \overset{\diamond}{\mathbf{D}}, \tag{2.62}$$

$$\nabla\cdot\mathbf{B} = 0, \qquad \nabla\times\mathcal{E} = -\overset{\diamond}{\mathbf{B}}. \tag{2.63}$$

As it must be, substitution of $\mathcal{H} = \mathbf{H} - \mathbf{v}\times\mathbf{D}$ and $\mathcal{E} = \mathbf{E} + \mathbf{v}\times\mathbf{B}$ and the definition $\overset{\diamond}{\mathbf{w}} = \dot{\mathbf{w}} - \mathbf{L}\mathbf{w} + (\nabla\cdot\mathbf{v})\mathbf{w}$ imply that Maxwell's equations (2.62)–(2.63) are equivalent to Maxwell's equations (2.54)–(2.55).

2.13 Material Description of Electromagnetic Fields

The material description of the electromagnetic fields is obtained by means of the integral forms of Maxwell's equations. The fields so obtained are referred to as Lagrangian fields; here they are denoted by the subscript R.

By Gauss' law relative to the region $\mathcal{P}_t = \chi(\mathrm{P}, t)$, we have

$$\frac{1}{\epsilon_0}\int_{\mathrm{P}} q\, J\, dv_R = \frac{1}{\epsilon_0}\int_{\mathcal{P}_t} q\, dv = \int_{\partial\mathcal{P}_t} \mathbf{D}\cdot\mathbf{n}\, da = \int_{\partial\mathrm{P}} \mathbf{D}\cdot J\mathbf{F}^{-T}\mathbf{n}_R\, da_R.$$

Hence, we have Gauss' law in the material description

$$\frac{1}{\epsilon_0}\int_{\mathrm{P}} q_R\, dv = \int_{\partial\mathrm{P}} \mathbf{D}_R\cdot\mathbf{n}_R\, da_R$$

if

$$q_R = J\, q, \qquad \mathbf{D}_R = J\,\mathbf{F}^{-1}\mathbf{D}.$$

For a curve $C = \partial S_t$ parameterized by $\gamma \in [a, b]$, $\mathbf{x} = \chi(\mathbf{X}(\gamma))$, we have $\partial_\gamma\chi = \mathbf{F}\partial_\gamma\mathbf{X}$ and hence

$$\int_a^b \mathbf{E}\cdot\mathbf{F}\partial_\gamma\mathbf{X}\, d\gamma = \int_a^b \mathbf{E}\cdot\partial_\gamma\chi\, d\gamma = -\frac{d}{dt}\int_{S_t}\mathbf{B}\cdot\mathbf{n}\, da = -\frac{d}{dt}\int_{\mathrm{S}}\mathbf{B}\cdot J\mathbf{F}^{-T}\mathbf{n}_R\, da_R.$$

Hence, we have

$$\mathbf{E}_R = \mathbf{E}\mathbf{F}, \qquad \mathbf{B}_R = J\mathbf{F}^{-1}\mathbf{B}.$$

Likewise, Ampere's law yields

$$\begin{aligned}\int_a^b \mathbf{H}\cdot\mathbf{F}\partial_\gamma\mathbf{X}\, d\gamma = \int_a^b \mathbf{H}\cdot\partial_\gamma\chi\, d\gamma &= \int_{S_t}\mathbf{J}\cdot\mathbf{n}\, da + \frac{d}{dt}\int_{S_t}\mathbf{D}\cdot\mathbf{n}\, da \\ &= \int_{S_t}\mathbf{J}\cdot J\mathbf{F}^{-T}\mathbf{n}_R\, da_R + \frac{d}{dt}\int_{\mathrm{S}}\mathbf{D}\cdot J\mathbf{F}^{-T}\mathbf{n}_R\, da_R.\end{aligned}$$

Hence, we find that

$$\mathbf{H}_R = \mathbf{H}\mathbf{F}, \qquad \mathbf{J}_R = J\mathbf{F}^{-1}\mathbf{J}$$

and, again, $\mathbf{D}_R = J\mathbf{F}^{-1}\mathbf{D}$. Moreover, we let $\mathbf{P}$ and $\mathbf{M}$ transform like $\mathbf{D}$ and $\mathbf{B}$. Hence, we have

$$\mathbf{E}_R = \mathbf{E}\mathbf{F},\ \mathbf{H}_R = \mathbf{H}\mathbf{F};\quad \mathbf{D}_R = J\mathbf{F}^{-1}\mathbf{D},\ \mathbf{P}_R = J\mathbf{F}^{-1}\mathbf{P},\ \mathbf{B}_R = J\mathbf{F}^{-1}\mathbf{B},\ \mathbf{M}_R = J\mathbf{F}^{-1}\mathbf{M}.$$

It then follows that

$$\mathbf{D}_R = \epsilon_0 J\mathbf{F}^{-1}\mathbf{E}_R\mathbf{F}^{-1} + \mathbf{P}_R, \qquad \mathbf{B}_R = \mu_0(J\mathbf{F}^{-1}\mathbf{H}_R\mathbf{F}^{-1} + \mathbf{M}_R).$$

The material description of Maxwell's equations is then formally invariant to the spatial description, in that

$$\nabla_R \cdot \mathbf{D}_R = q_R, \qquad \nabla_R \cdot \mathbf{B}_R = 0,$$

$$\nabla_R \times \mathbf{H}_R = \mathbf{J}_R + \partial_t \mathbf{D}_R, \qquad \nabla_R \times \mathbf{E}_R = -\partial_t \mathbf{B}_R,$$

hold along with the continuity equation

$$\partial_t q_R + \nabla_R \cdot \mathbf{J}_R = 0.$$

To complete the scheme, we have to specify the constitutive relations. The relations

$$\mathbf{D} = \epsilon_0 \mathbf{E} + \mathbf{P}, \qquad \mathbf{B} = \mu_0(\mathbf{H} + \mathbf{M})$$

provide two of these fields in terms of the other ones. Moreover, in *linear dielectrics* $\mathbf{P}$ is a (linear) function of $\mathbf{E}$. In view of a possible anisotropy, $\mathbf{P}$ and $\mathbf{E}$ are allowed to be non-collinear and hence we let

$$\mathbf{P} = \epsilon_0 \chi_e \mathbf{E}, \qquad \mathbf{D} = \epsilon_0(\mathbf{1} + \chi_e)\mathbf{E},$$

where χ_e is the positive-definite bulk *electric susceptibility*[18] tensor. Likewise, in *linear paramagnets*, $\mathbf{M}$ is assumed to be a linear function of $\mathbf{H}$ so that

$$\mathbf{M} = \chi_m \mathbf{H}, \qquad \mathbf{B} = \mu_0(\mathbf{1} + \chi_m)\mathbf{H},$$

χ_m being the positive-definite bulk *magnetic susceptibility* tensor. Equivalently, we can evaluate $\mathbf{M}$ and $\mathbf{H}$ in terms of $\mathbf{B}$ in the form

$$\mathbf{M} = \frac{1}{\mu_0}\chi_b \mathbf{B}, \qquad \mathbf{H} = \frac{1}{\mu_0}(\mathbf{1} - \chi_b)\mathbf{B}, \qquad \chi_b := \chi_m(\mathbf{1} + \chi_m)^{-1}.$$

Both χ_e and χ_m (and hence χ_b) are allowed to depend on temperature and deformation. Now we see that different definitions are allowed for $\mathbf{P}_R$ and $\mathbf{M}_R$ and here we indicate three possibilities.

First we take the view that $\mathbf{D}(\mathbf{E})$ and $\mathbf{B}(\mathbf{H})$ keep the same form in the reference configuration. Since $\mathbf{D}_R = J\mathbf{F}^{-1}\mathbf{D}$ and $\mathbf{E}_R = \mathbf{E}\mathbf{F}$ then $\mathbf{D} = \epsilon_0(\mathbf{1} + \chi_e)\mathbf{E}$ implies

$$\mathbf{D}_R = \epsilon_0(J\mathbf{F}^{-1}\mathbf{F}^{-T} + J\mathbf{F}^{-1}\chi_e\mathbf{F}^{-T})\mathbf{E}_R.$$

---

[18] Also $\epsilon_0\chi_e$ is called electric susceptibility.

The analogue holds for **B** and **H**. Accordingly, if the relation $\mathbf{P} = \epsilon_0 \chi_e \mathbf{E}$ has to hold in the reference configuration then we conclude that

$$\mathbf{P}_R = \epsilon_0 J \mathbf{F}^{-1} \chi_e \mathbf{F}^{-T} \mathbf{E}_R = \epsilon_0 J \mathbf{F}^{-1} \chi_e \mathbf{E}$$

and hence

$$\mathbf{D}_R = \epsilon_0 J \mathbf{C}^{-1} \mathbf{E}_R + \mathbf{P}_R, \qquad \mathbf{P}_R = \epsilon_0 \chi_{eR} \mathbf{E}_R, \qquad \chi_{eR} = J \mathbf{F}^{-1} \chi_e \mathbf{F}^{-T}.$$

Likewise, since $\mathbf{B}_R = J\mathbf{F}^{-1}\mathbf{B}$, $\mathbf{H}_R = \mathbf{HF}$, and $\mathbf{B} = \mu_0(\mathbf{H} + \mathbf{M})$, $\mathbf{M} = \chi_m \mathbf{H}$ then

$$\mathbf{B}_R = \mu_0 J \mathbf{F}^{-1} (\mathbf{H} + \mathbf{M}) = \mu_0 J \mathbf{C}^{-1} \mathbf{H}_R + \mu_0 J \mathbf{F}^{-1} \chi_m \mathbf{F}^{-T} \mathbf{H}_R.$$

Hence, we have

$$\mathbf{B}_R = \mu_0 (J \mathbf{C}^{-1} \mathbf{H}_R + \mathbf{M}_R), \qquad \mathbf{M}_R = \chi_{mR} \mathbf{H}_R, \qquad \chi_{mR} = J \mathbf{F}^{-1} \chi_m \mathbf{F}^{-T}.$$

Accordingly, $\mathbf{P}_R = \chi_{eR} \mathbf{E}_R$ and $\mathbf{M}_R = \chi_{mR} \mathbf{H}_R$ provide the polarization $\mathbf{P}_R$ and the magnetization $\mathbf{M}_R$ in terms of the fields $\mathbf{E}_R$, $\mathbf{H}_R$ via the susceptibilities χ_{eR}, χ_{mR}. Because of the deformation, χ_{eR} and χ_{mR} show anisotropic dependences of $\mathbf{P}_R(\mathbf{E}_R)$ and $\mathbf{M}_R(\mathbf{H}_R)$ even though χ_e and χ_m are isotropic.

Another possibility arises from the view, or the assumption, that **D**, **E**, **P** and **B**, **H**, **M** in the reference configuration are related as they are in the spatial description, so that

$$\mathbf{D}_R = \boldsymbol{\epsilon}_{0R} \mathbf{E}_R + \mathbf{P}_R, \qquad \mathbf{B}_R = \boldsymbol{\mu}_{0R} (\mathbf{H}_R + \mathbf{M}_R)$$

for appropriate tensors $\boldsymbol{\epsilon}_{0R}$, $\boldsymbol{\mu}_{0R}$. Since

$$\mathbf{D}_R = J \mathbf{F}^{-1} (\epsilon_0 \mathbf{E} + \mathbf{P}) = \epsilon_0 J \mathbf{C}^{-1} \mathbf{E}_R + J \mathbf{F}^{-1} \mathbf{P}$$

and

$$\mathbf{B}_R = J \mathbf{F}^{-1} \mu_0 (\mathbf{H} + \mathbf{M}) = \mu_0 J \mathbf{C}^{-1} (\mathbf{H}_R + \mathbf{C} \mathbf{F}^{-1} \mathbf{M})$$

then it follows that

$$\mathbf{P}_R = J \mathbf{F}^{-1} \mathbf{P}, \qquad \mathbf{M}_R = \mathbf{C} \mathbf{F}^{-1} \mathbf{M} = \mathbf{F}^T \mathbf{M} = \mathbf{MF},$$

and

$$\boldsymbol{\epsilon}_{0R} = \epsilon_0 J \mathbf{C}^{-1}, \qquad \boldsymbol{\mu}_{0R} = \mu_0 J \mathbf{C}^{-1}.$$

This scheme arises without any assumption on the possible linearity of $\mathbf{P}(\mathbf{E})$ and $\mathbf{M}(\mathbf{H})$. Incidentally, this scheme implies that **P** and **D** have the same transformation law as is the case for **M** and **H**.

A third scheme arises by regarding **E** and **B** as the primitive variables so that the constitutive functions are

$$\mathbf{P} = \mathbf{P}(\mathbf{E}, \mathbf{B}, ...), \qquad \mathbf{M} = \mathbf{M}(\mathbf{E}, \mathbf{B}, ...)$$

in the spatial description and

$$\mathbf{P}_R = \mathbf{D}_R(\mathbf{E}_R, \mathbf{B}_R, ...), \qquad \mathbf{M}_R = \mathbf{M}_R(\mathbf{E}_R, \mathbf{B}_R, ...),$$

in the material description. We then determine **P** and **M** via

$$\mathbf{P} = \mathbf{D}(\mathbf{E}, \mathbf{B}, ...) - \epsilon_0 \mathbf{E}, \qquad \mathbf{M} = \frac{1}{\mu_0}\mathbf{B} - \mathbf{H}(\mathbf{E}, \mathbf{B}, ...).$$

By the transformation laws of **D**, **E** and **B**, **H**, we obtain

$$\mathbf{P} = \frac{1}{J}\mathbf{F}\mathbf{D}_R - \epsilon_0 \mathbf{F}^{-T}\mathbf{E}_R = \frac{1}{J}\mathbf{F}(\mathbf{D}_R - \epsilon_0 J\mathbf{C}^{-1}\mathbf{E}_R),$$

$$\mathbf{M} = \frac{1}{\mu_0 J}\mathbf{F}\mathbf{B}_R - \mathbf{F}^{-T}\mathbf{H}_R = \frac{1}{J}\mathbf{F}(\frac{1}{\mu_0}\mathbf{B}_R - J\mathbf{C}^{-1}\mathbf{H}_R) = \mathbf{F}^{-T}(\frac{1}{\mu_0 J}\mathbf{C}\mathbf{B}_R - \mathbf{H}_R).$$

Accordingly

$$J\mathbf{F}^{-1}\mathbf{P} = \mathbf{D}_R - \epsilon_0 J\mathbf{C}^{-1}\mathbf{E}_R,$$

$$\mathbf{F}^T\mathbf{M} = \frac{1}{\mu_0 J}\mathbf{C}\mathbf{B}_R - \mathbf{H}_R$$

allow us to define

$$\mathbf{P}_R = J\mathbf{F}^{-1}\mathbf{P}, \qquad \mathbf{M}_R = \mathbf{F}^T\mathbf{M}$$

and hence to state the constitutive relations in the material description as

$$\mathbf{D}_R = \boldsymbol{\epsilon}_{0R}\mathbf{E}_R + \mathbf{P}_R, \qquad \boldsymbol{\epsilon}_{0R} = \epsilon_0 J\mathbf{C}^{-1},$$

$$\mathbf{H}_R = \boldsymbol{\mu}_{0R}^{-1}\mathbf{B}_R - \mathbf{M}_R, \qquad \boldsymbol{\mu}_{0R} = \mu_0 J\mathbf{C}^{-1}.$$

As a consequence, we have also the transformation laws

$$\epsilon_0 \to \boldsymbol{\epsilon}_{0R} = \epsilon_0 J\mathbf{C}^{-1}, \qquad \mu_0 \to \boldsymbol{\mu}_{0R} = \mu_0 J\mathbf{C}^{-1}.$$

Conversely, we might start with the definitions

$$\mathbf{P}_R = \mathbf{D}_R(\mathbf{E}_R, \mathbf{B}_R, ...) - \epsilon_0\mathbf{E}_R, \qquad \mathbf{M}_R = \frac{1}{\mu_0}\mathbf{B}_R - \mathbf{H}_R(\mathbf{E}_R, \mathbf{B}_R, ...).$$

Hence, we have

$$\mathbf{P}_R = J\mathbf{F}^{-1}\mathbf{D} - \epsilon_0 \mathbf{F}^T \mathbf{E} = J\mathbf{F}^{-1}(\mathbf{D} - \frac{\epsilon_0}{J}\mathbf{F}\mathbf{F}^T \mathbf{E}),$$

$$\mathbf{M}_R = \frac{J}{\mu_0}\mathbf{F}^{-1}\mathbf{B} - \mathbf{F}^T \mathbf{H} = \mathbf{F}^T (\frac{J}{\mu_0}\mathbf{F}^{-T}\mathbf{F}^{-1}\mathbf{B} - \mathbf{H}).$$

As a consequence, the relations

$$\mathbf{D} = \epsilon_0 \mathbf{E} + \mathbf{P}, \qquad \mathbf{H} = \mu_0^{-1}\mathbf{B} - \mathbf{M}$$

holds provided we let again

$$\mathbf{P}_R = J\mathbf{F}^{-1}\mathbf{P}, \qquad \mathbf{M}_R = \mathbf{F}^T \mathbf{M} \tag{2.64}$$

whereas

$$\epsilon_0 = \frac{\epsilon_0}{J}\mathbf{F}\mathbf{F}^T, \qquad \mu_0 = \frac{\mu_0}{J}\mathbf{F}\mathbf{F}^T.$$

The second and third views lead to the transformation (2.64) as the appropriate ones for **P** and **M** in passing from the spatial description to the material description. As to the permittivity and permeability tensors (of free space), it looks consistent to let

$$\epsilon_{0R} = J\mathbf{F}^{-1}\epsilon_0\mathbf{F}^{-T}, \qquad \mu_{0R} = J\mathbf{F}^{-1}\mu_0\mathbf{F}^{-T}.$$

Things are simpler in *rigid motions*. Since $\mathbf{F} = \mathbf{R}$, $\mathbf{F}\mathbf{F}^T = \mathbf{F}^T\mathbf{F} = \mathbf{1}$, and $J = 1$, then we have

$$\epsilon_{0R} = \epsilon_0 = \epsilon_0 \mathbf{1}, \qquad \mu_{0R} = \mu_0 = \mu_0 \mathbf{1}.$$

In all the three schemes, we obtain the invariance properties

$$\mathbf{E}_R = \mathbf{E}, \quad \mathbf{P}_R = \mathbf{P}, \quad \mathbf{D}_R = \mathbf{D}, \quad \mathbf{B}_R = \mathbf{B}, \quad \mathbf{M}_R = \mathbf{M}, \quad \mathbf{H}_R = \mathbf{H}$$

and hence

$$\mathbf{D}_R = \epsilon_0 \mathbf{E}_R + \mathbf{P}_R, \qquad \mathbf{B}_R = \mu_0(\mathbf{H}_R + \mathbf{M}_R),$$

$$\mathbf{D} = \epsilon_0 \mathbf{E} + \mathbf{P}, \qquad \mathbf{B} = \mu_0(\mathbf{H} + \mathbf{M}).$$

2.14 Formulations of Electromagnetism in Matter

The subject of electromagnetism in matter is quite controversial, the origin of the controversy being related to different views about the selection of the primary fields. This in turn leads to seemingly different balance equations in matter.

Seemingly, the literature shows different basic laws about force and torque produced by electromagnetic fields in matter. Most often the Lorentz force **F** on a charge q and the torque $\mathfrak{T}$ on a magnetic dipole $\mathfrak{m}$ are (e.g. [377], p. 212; [230], p. 168)

$$\mathsf{F} = \mathsf{q}(\mathbf{E} + \mathbf{v} \times \mathbf{B}), \qquad \mathfrak{T} = \mathfrak{m} \times \mathbf{B}. \tag{2.65}$$

Yet the Lorentz force is also taken in the form ([357], p. 48)

$$\mathsf{F} = \mathsf{q}(\mathbf{E} + \mathbf{v} \times \mu_0 \mathbf{H}). \tag{2.66}$$

The view underlying (2.66) is that in free space $\mathbf{B} = \mu_0 \mathbf{H}$ and in matter the reference magnetic field is maintained $\mu_0 \mathbf{H}$. Of course, due to a nonzero magnetization $\mathbf{M}$, in matter $\mathbf{B} \neq \mu_0 \mathbf{H}$. We now examine the Chu formulation and the Amperian formulation of Maxwell's equations.[19]

Throughout, Maxwell's equations are considered in the form

$$\nabla \times \mathbf{E} = -\partial_t \mathbf{B}, \qquad \nabla \times \mathbf{H} = \mathbf{J} + \partial_t \mathbf{D},$$

$$\nabla \cdot \mathbf{D} = q, \qquad \nabla \cdot \mathbf{B} = 0,$$

along with the relations $\mathbf{D} = \epsilon_0 \mathbf{E} + \mathbf{P}$, $\mathbf{B} = \mu_0(\mathbf{H} + \mathbf{M})$. This form is referred to as *Minkowski formulation.* It was the first formulation for moving media, all four field variables $\mathbf{E}, \mathbf{D}, \mathbf{H}, \mathbf{B}$ being present both in free space (where $\mathbf{D} = \epsilon_0 \mathbf{E}$, $\mathbf{B} = \mu_0 \mathbf{H}$) and in matter, although in free space two of them are redundant. In matter, six significant fields occur and hence constitutive equations are needed. We now examine two formulations that sometimes are applied in the literature.

To see how other formulations can be determined, we observe that, in free space, Maxwell's equations can be written in the form

$$\nabla \times \mathbf{E} = -\mu_0 \partial_t \mathbf{H}, \qquad \nabla \times \mathbf{H} = \epsilon_0 \partial_t \mathbf{E} + \mathbf{J},$$

$$\epsilon_0 \nabla \cdot \mathbf{E} = q, \qquad \mu_0 \nabla \cdot \mathbf{H} = 0.$$

Often, in the literature, $\mathbf{J}$ and q are written as $\mathbf{J}_f$ and q_f to make it evident that they are the current density and the charge density generated by free charges (mainly electrons).

Though these equations are generally valid, in any formulation, they allow one to say that the primitive fields are $\mathbf{E}, \mathbf{H}$ while one is led to viewing $\mathbf{E}$ and $\mathbf{B}$ as the primitive fields in the Minkowski formulation. Consistently, this might be the origin of the different forms of Lorentz force in (2.65) and (2.66).

We now examine two well-known formulations and contrast them with the Minkowski formulation.

[19] See [357], Chap. 7, and [165]. The equivalence of these formulations is investigated by [381] in the relativistic context.

Chu formulation. It is based on the view that only two field quantities are necessary in free space[20] and these quantities are **E** and **H**. In matter, the source terms q and $\mathbf{J}$ are supplemented by a current density $\mathbf{J}_p$ and charge density q_p produced by the polarization of the medium. By analogy, a magnetic current density $\mathbf{J}_m$ and a magnetic charge density q_m are allowed. Hence, the equations become

$$\nabla \times \mathbf{E}_c = -\mu_0 \partial_t \mathbf{H}_c - \mathbf{J}_m, \qquad \nabla \times \mathbf{H}_c = \epsilon_0 \partial_t \mathbf{E}_c + \mathbf{J}_c + \mathbf{J}_p,$$

$$\epsilon_0 \nabla \cdot \mathbf{E}_c = q + q_p, \qquad \mu_0 \nabla \cdot \mathbf{H}_c = q_m;$$

the subscript C labels fields pertaining to the Chu formulation. To determine $\mathbf{J}_p$, we consider a surface S and a picture where two types of motion of dipoles occur around S; roughly, a motion makes dipoles flowing perpendicularly to S, the other one makes the dipoles moving along a curve C on S. The polarization current I_p is then made formal as

$$I_p = \int_S \partial_t \mathbf{P} \cdot \mathbf{n} da + \int_C \mathbf{P} \times \mathbf{v} \cdot \mathbf{t} ds = \int_S (\partial_t \mathbf{P} + \nabla \times \mathbf{P} \times \mathbf{v}) \cdot \mathbf{n}\, da.$$

This motivates the definition

$$\mathbf{J}_p = \partial_t \mathbf{P} + \nabla \times (\mathbf{P} \times \mathbf{v}).$$

By requiring that $\nabla \cdot \mathbf{J}_p + \partial_t q_p = 0$, it follows that $q_p = -\nabla \cdot \mathbf{P}$. By analogy, $\mathbf{J}_m$ and q_m are

$$\mathbf{J}_m = \partial_t (\mu_0 \mathbf{M}) + \nabla \times (\mu_0 \mathbf{M} \times \mathbf{v}), \qquad q_m = -\nabla \cdot (\mu_0 \mathbf{M}).$$

Hence, Maxwell's equations in the Chu formulation for moving polarizable and magnetizable matter are

$$\nabla \times \mathbf{E}_c = -\mu_0 \partial_t (\mathbf{H}_c + \mathbf{M}_c) - \nabla \times (\mu_0 \mathbf{M}_c \times \mathbf{v}),$$

$$\nabla \times \mathbf{H}_c = \partial_t (\epsilon_0 \mathbf{E}_c + \mathbf{P}_c) + \nabla \times (\mathbf{P}_c \times \mathbf{v}) + \mathbf{J}_c,$$

$$\nabla \cdot (\epsilon_0 \mathbf{E}_c + \mathbf{P}_c) = q, \qquad \nabla \cdot \mu_0 (\mathbf{H}_c + \mathbf{M}_c) = 0.$$

To find the connection between Chu variables and Minkowski variables, we observe that the two versions of Maxwell's equations are equivalent if

$$\mathbf{E} = \mathbf{E}_c + \mu_0 \mathbf{M}_c \times \mathbf{v}, \quad \mathbf{H} = \mathbf{H}_c - \mathbf{P}_c \times \mathbf{v}, \quad \mathbf{B} = \mu_0 (\mathbf{H}_c + \mathbf{M}_c), \quad \mathbf{D} = \epsilon_0 \mathbf{E}_c + \mathbf{P}_c,$$

and then

[20] This is true also for the Minkowski formulation.

$$\mathbf{M} = \mathbf{M}_c + \mathbf{P}_c \times \mathbf{v}, \qquad \mathbf{P} = \mathbf{P}_c - \epsilon_0 \mu_0 \mathbf{M}_c \times \mathbf{v}.$$

Conversely, we derive the Chu fields in terms of the Minkowski fields. By direct substitutions, we have

$$\mathbf{E}_C = \mathbf{E} - \mu_0 \mathbf{M}_C \times \mathbf{v} = \mathbf{E} - \mathbf{B} \times \mathbf{v} - \mu_0 \mathbf{H}_C \times \mathbf{v} = \mathbf{E} - (\mathbf{B} - \mu_0 \mathbf{H}) \times \mathbf{v} - \mathbf{v} \times \mu_0 (\mathbf{P}_C \times \mathbf{v}).$$

Since $\mathbf{P}_c \times \mathbf{v} = \mathbf{D} \times \mathbf{v} - \epsilon_0 \mathbf{E}_c \times \mathbf{v}$ then

$$\mathbf{E}_c (1 - v^2/c^2) = \mathbf{E} - (\mathbf{B} - \mu_0 \mathbf{H}) \times \mathbf{v} - \mu_0 \mathbf{v} \times (\mathbf{D} \times \mathbf{v}) - (1/c^2)(\mathbf{v} \cdot \mathbf{E})\mathbf{v},$$

where $c^2 = 1/\epsilon_0 \mu_0$. In addition

$$\frac{1}{1 - v^2/c^2} [\mathbf{E} - \frac{1}{c^2} (\mathbf{v} \cdot \mathbf{E})\mathbf{v}] = \mathbf{E} + \frac{\mathbf{v} \times (\mathbf{E} \times \mathbf{v})}{c^2 - v^2}.$$

Likewise, we determine $\mathbf{H}_c$, $\mathbf{P}_c$, $\mathbf{M}_c$. We then find

$$\mathbf{E}_c = \mathbf{E} + \frac{\mathbf{v} \times [(\mathbf{E} - \mathbf{D}/\epsilon_0) \times \mathbf{v}]}{c^2 - v^2} + \frac{\mathbf{v} \times (\mathbf{B} - \mu_0 \mathbf{H})}{1 - v^2/c^2},$$

$$\mathbf{H}_c = \mathbf{H} + \frac{\mathbf{v} \times [\mathbf{H} - \mathbf{B}/\mu_0) \times \mathbf{v}]}{c^2 - v^2} - \frac{\mathbf{v} \times (\mathbf{D} - \epsilon_0 \mathbf{E})}{1 - v^2/c^2},$$

$$\mathbf{P}_c = \mathbf{D} - \epsilon_0 \mathbf{E} + \frac{\mathbf{v} \times [(\mathbf{D} - \epsilon_0 \mathbf{E}) \times \mathbf{v}]}{c^2 - v^2} - \frac{\epsilon_0 \mathbf{v} \times (\mathbf{B} - \mu_0 \mathbf{H})}{1 - v^2/c^2},$$

$$\mu_0 \mathbf{M}_c = \mathbf{B} - \mu_0 \mathbf{H} + \frac{\mathbf{v} \times [(\mathbf{B} - \mu_0 \mathbf{H}) \times \mathbf{v}]}{c^2 - v^2} + \frac{\mu_0 \mathbf{v} \times (\mathbf{D} - \epsilon_0 \mathbf{E})}{1 - v^2/c^2}.$$

By the transformation law of electromagnetic fields, the effective fields $\mathcal{E}$ and $\mathcal{H}$ (i.e. the fields at the reference at rest at the point under consideration) are given by

$$\mathcal{E} = \mathbf{E} + \mathbf{v} \times \mathbf{B}, \qquad \mathcal{H} = \mathbf{H} - \mathbf{v} \times \mathbf{D},$$

where **E** and **B** are the Minkowski fields in the laboratory frame. In terms of the Chu variables, the effective fields are given by

$$\mathcal{E}_c = \mathbf{E} + \mathbf{v} \times \mu_0 \mathbf{H}, \qquad \mathcal{H}_c = \mathbf{H} - \mathbf{v} \times \epsilon_0 \mathbf{E}.$$

Amperian formulation. It is based on the view that, in free space, only two fields are necessary, namely **E** and **B**. The effects of polarization and magnetization are modelled by having mind distributions of charges and currents in free space. We denote by the subscript A the fields in the Amperian formulation and then start with Maxwell's equations in the form

$$\nabla \times \mathbf{E}_A = -\partial_t \mathbf{B}_A, \qquad \epsilon_0 \nabla \cdot \mathbf{E}_A = q + q_{pm},$$

$$\nabla \times \mathbf{B}_A = \mu_0(\mathbf{J} + \mathbf{J}_{pm}) + \mu_0 \epsilon_0 \partial_t \mathbf{E}_A, \qquad \nabla \cdot \mathbf{B}_A = 0,$$

where $q_{pm}, \mathbf{J}_{pm}$ denote terms arising from polarization and magnetization. Based on arguments about the effects of current loops, the contributions to $q_{pm}, \mathbf{J}_{pm}$ are determined so that Maxwell's equations become [165]

$$\nabla \times \mathbf{E}_A = -\partial_t \mathbf{B}_A, \qquad \epsilon_0 \nabla \cdot \mathbf{E}_A = q - \nabla \cdot \mathbf{P}_A + \nabla \cdot (\mathbf{M}_A \times \mathbf{v}/c^2),$$

$$\nabla \times \mathbf{B}_A/\mu_0 = \mathbf{J} + \epsilon_0 \partial_t \mathbf{E}_A + \partial_t \mathbf{P}_A + \nabla \times (\mathbf{P}_A \times \mathbf{v}) + \nabla \times \mathbf{M}_A - \partial_t (\mathbf{M}_A \times \mathbf{v}/c^2), \qquad \nabla \cdot \mathbf{B}_A = 0.$$

Incidentally

$$\mathbf{J}_{pm} = \partial_t \mathbf{P}_A + \nabla \times (\mathbf{P}_A \times \mathbf{v}) + \nabla \times \mathbf{M}_A - \partial_t (\mathbf{M}_A \times \mathbf{v}/c^2), \qquad q_{pm} = -\nabla \cdot \mathbf{P}_A + \nabla \cdot (\mathbf{M}_A \times \mathbf{v}/c^2)$$

satisfy identically the continuity equation $\nabla \cdot \mathbf{J}_{pm} + \partial_t q_{pm} = 0$. The connection between the Chu variables, Amperian variables, and Minkowski variables are established by requiring the equivalence of the formulations. It follows that

$$\mathbf{E} = \mathbf{E}_A, \quad \mathbf{B} = \mathbf{B}_A, \quad \mathbf{H} = \mathbf{B}_A/\mu_0 - \mathbf{M}_A - \mathbf{P}_A \times \mathbf{v}, \quad \mathbf{D} = \epsilon_0 \mathbf{E}_A + \mathbf{P}_A - \mathbf{M}_A \times \mathbf{v}/c^2,$$

and then

$$\mathbf{M} = \mathbf{M}_A + \mathbf{P}_A \times \mathbf{v}, \qquad \mathbf{P} = \mathbf{P}_A - \mathbf{M}_A \times \mathbf{v}/c^2.$$

Equalling the Minkowski variables in terms of Chu variables and Amperian variables, we find

$$\mathbf{E}_A = \mathbf{E}_C + \mu_0 \mathbf{M}_C \times \mathbf{v}, \quad \mathbf{B}_A = \mu_0 \mathbf{H}_C + \mu_0 \mathbf{M}_C, \quad \mathbf{P}_A = \mathbf{P}_C, \quad \mathbf{M}_A = \mathbf{M}_C,$$

and then

$$\mathbf{D}_A = \mathbf{D}_C + \epsilon_0 \mu_0 \mathbf{M}_C \times \mathbf{v}, \qquad \mathbf{H}_A = \mathbf{H}_C.$$

Since $\mathbf{E} = \mathbf{E}_A$, $\mathbf{B} = \mathbf{B}_A$, $\mathbf{H} = \mathbf{B}_A/\mu_0 - \mathbf{M}_A - \mathbf{P}_A \times \mathbf{v}$, and $\mathbf{D} = \epsilon_0 \mathbf{E}_A + \mathbf{P}_A + \mathbf{v} \times \mathbf{M}/c^2$ then the effective fields in terms of Amperian variables are

$$\mathcal{E}_A = \mathbf{E}_A + \mathbf{v} \times \mathbf{B}_A, \qquad \mathcal{H}_A = \mathbf{H}_A - \mathbf{v} \times (\epsilon_0 \mathbf{E}_A + \mathbf{v} \times \mathbf{M}_A/c^2).$$

2.15 Poynting's Theorem

A classical relation in electromagnetism, known as Poynting's theorem, is often framed within a balance of energy. To obtain this relation, inner multiply Ampère's law by **E**,

$$\mathbf{E} \cdot \nabla \times \mathbf{H} = \mathbf{J} \cdot \mathbf{E} + \mathbf{E} \cdot \partial_t \mathbf{D}.$$

By the vector identity

$$\nabla \cdot (\mathbf{E} \times \mathbf{H}) = -\mathbf{E} \cdot \nabla \times \mathbf{H} + \mathbf{H} \cdot \nabla \times \mathbf{E}$$

and Faraday's law, $\nabla \times \mathbf{H} = -\partial_t \mathbf{B}$, we obtain

$$\mathbf{H} \cdot \partial_t \mathbf{B} + \mathbf{E} \cdot \partial_t \mathbf{D} = -\mathbf{J} \cdot \mathbf{E} - \nabla \cdot (\mathbf{E} \times \mathbf{H}). \tag{2.67}$$

Equation (2.67) is the content of Poynting's theorem; the vector $\mathbf{E} \times \mathbf{H}$ is known as the *Poynting vector.*

If there is a function $u(\mathbf{x}, t)$ such that

$$\partial_t u = \mathbf{H} \cdot \partial_t \mathbf{B} + \mathbf{E} \cdot \partial_t \mathbf{D} \tag{2.68}$$

then

$$\partial_t u = -\mathbf{J} \cdot \mathbf{E} - \nabla \cdot (\mathbf{E} \times \mathbf{H}) \tag{2.69}$$

can be interpreted as a balance of energy where u is the energy density, $\mathbf{J} \cdot \mathbf{E}$ is the power (dissipated) per unit volume and $\mathbf{E} \times \mathbf{H}$ is the energy flux vector. In integral form, we have

$$\frac{d}{dt} \int_\Omega u \, dv = -\int_\Omega \mathbf{J} \cdot \mathbf{E} \, dv - \int_{\partial\Omega} \mathbf{E} \times \mathbf{H} \cdot \mathbf{n} \, da.$$

The interpretation of the Poynting vector as an energy flux is subject to the existence of the function u. Now, if

$$\mathbf{D} = \epsilon \mathbf{E}, \qquad \mathbf{B} = \mu \mathbf{H},$$

where ϵ, μ are possibly functions of the position, then $u(\mathbf{x}, t)$ is given by

$$u = \tfrac{1}{2}(\epsilon \mathbf{E}^2 + \mu \mathbf{H}^2).$$

Equation (2.69) is often regarded as a balance law for electromagnetic energy. Yet Eq. (2.69) is not a balance law; it is an identity satisfied by all fields that are solutions of Maxwell's equations provided u satisfies (2.68).

2.16 Balance Equations in Electromagnetism

Balance equations for electromagnetic materials show different formulations in the literature. A formal apparent reason is due to the different formulations of Maxwell's equations (Minkowski, Chu, Amperian). Yet even in a selected formulation, like the

Minkowski one adopted here, subjective positions occur about the pertinent roles of fields in matter. Hence, we first motivate the present selection of fields in matter in connection with the basic balance laws.

Look at the Lorentz force on electric charges. In free space, the Lorentz force (2.65) is well established; possibly one might write $\mathsf{F} = \mathfrak{q}(\mathbf{E} + \mathbf{v} \times \mu_0 \mathbf{H})$ thanks to the equality $\mathbf{B} = \mu_0 \mathbf{H}$ in free space. Yet in matter $\mathbf{B} \neq \mu_0 \mathbf{H}$ and hence the two formulae $\mathsf{F} = \mathfrak{q}(\mathbf{E} + \mathbf{v} \times \mathbf{B})$ and $\mathsf{F} = \mathfrak{q}(\mathbf{E} + \mathbf{v} \times \mu_0 \mathbf{H})$ are inequivalent.

Following the historical, experimental, and conceptual development of Maxwell's equations, it emerges that **B** and **E** are physically the fundamental fields ([168], Sect. 36.6). Furthermore, we may take the view that, owing to the elementary structure of charged particles (protons or electrons), the particles are subject to the laws valid in the free space and hence (2.65) are maintained in matter. This view is consistent, e.g. with the model of ferromagnetism in connection with the spinning electron; while the torque on the spinning electron is taken to obey (2.65) then the evolutions of the magnetic dipole $\mathfrak{m}$ is assumed to be governed by (e.g. [235], Sect. 12.3)

$$\dot{\mathfrak{m}} = \gamma \mathfrak{m} \times \mathbf{B}.$$

The question now emerges as to the appropriate or effective magnetic field in magnetic media ($\mathbf{M} \neq \mathbf{0}$). Maxwell's equations in the form

Electrostatics	Static ferromagnetism
$\nabla \cdot \mathbf{D} = q$	$\nabla \cdot \mathbf{B} = 0$
$\nabla \times \mathbf{E} = \mathbf{0}$	$\nabla \times \mathbf{H} = \mathbf{0}$

show mathematical correspondences between the two sets of equations with the analogy between **E** and **H** ([168], p. 36.12)

We observe that while $\mathbf{E} + \mathbf{v} \times \mathbf{B}$ is the electric field in the rest frame of the body, $\mathcal{E} = \mathbf{E} + \mathbf{v} \times \mathbf{B}$, the magnetic intensity in the rest frame is $\mathcal{H} = \mathbf{H} - \mathbf{v} \times \mathbf{D}$. Hence, we might pursue the analogy between **E** and **H** by deriving the force on a magnetic charge $\mathfrak{q}^*$ in the form

$$\mathsf{F}_M = \mu_0 \mathfrak{q}^* \mathcal{H}, \qquad \mathsf{F}_P = \mathfrak{q} \mathcal{E}.$$

Yet electric monopoles exist in the form of particles that have a positive or negative electric charge. The analogy breaks down when we try to find the magnetic counterpart for the electric charge. While we can find electric monopoles in the form of charged particles, we have never observed magnetic monopoles. Magnets exist only in the form of dipoles with a north and a south end. When a bar magnet is split into two pieces, we don't get a separate north part and a south part. Rather we get two new, smaller magnets, each with a north and south end. Nevertheless, we pursue the analogy in a mathematical way so as to derive forces and torques on electric and magnetic dipoles and then to determine forces and torques associated with the polarization **P** and the magnetization **M**. The present approach is then based on this assumption.

Assumption. In matter and in free space, an electric charge $\mathfrak{q}$ experience the Lorentz force in the form $\mathsf{F} = \mathfrak{q}(\mathbf{E} + \mathbf{v} \times \mathbf{B})$, while a magnetic charge $\mathfrak{q}^*$ experience the force $\mathsf{F} = \mu_0 \mathfrak{q}^*(\mathbf{H} - \mathbf{v} \times \mathbf{D})$.

2.16.1 Forces, Torques, and Energy Supply

To determine the force, per unit volume, due to polarization, we start with the force on a single electric dipole. Following the model of the electric dipole, let $-\mathfrak{q}$ be the negative (point) charge, at $\mathbf{x}$, and $\mathfrak{q}$ the positive charge, at $\mathbf{x} + \mathbf{d}$. We denote the velocities at $\mathbf{x}$ and $\mathbf{x} + \mathbf{d}$ by $\mathbf{v}$ and $\mathbf{v} + \dot{\mathbf{d}}$. By evaluating the Lorentz force on the charges of the dipole, we have

$$\mathsf{F}^- = -\mathfrak{q}\,\mathbf{E}(\mathbf{x}) - \mathfrak{q}\,\mathbf{v} \times \mathbf{B}(\mathbf{x}),$$

$$\mathsf{F}^+ = \mathfrak{q}\,\mathbf{E}(\mathbf{x} + \mathbf{d}) + (\mathfrak{q}\,\mathbf{v} + \mathfrak{q}\,\dot{\mathbf{d}}) \times \mathbf{B}(\mathbf{x} + \mathbf{d}).$$

Evaluate F^+ up to linear terms in $\mathbf{d}$. The sum of the two forces $\mathsf{F}^+ + \mathsf{F}^-$, at $\mathbf{x}$, then becomes

$$\mathsf{F}^+ + \mathsf{F}^- = (\mathfrak{p} \cdot \nabla)\mathbf{E} + \mathbf{v} \times (\mathfrak{p} \cdot \nabla)\mathbf{B} + \dot{\mathfrak{p}} \times \mathbf{B},$$

where $\mathfrak{p} = \mathfrak{q}\,\mathbf{d}$ is the electric dipole moment.

We let $\mathbf{p}$ be the average of $\mathfrak{p}$, that is $\mathbf{p} = \langle \mathfrak{p} \rangle = \mathfrak{q}\langle \mathbf{d} \rangle$, and define the dielectric polarization as

$$\mathbf{P} = n\mathbf{p} = n\mathfrak{q}\langle \mathbf{d} \rangle.$$

Moreover, we let $\langle \dot{\mathbf{d}} \rangle = \langle \mathbf{d} \rangle^{\cdot}$ and hence

$$n\mathfrak{q}\langle \dot{\mathbf{d}} \rangle = n\dot{\mathbf{p}} = n(\mathbf{P}/n)^{\cdot} = \dot{\mathbf{P}} - \mathbf{P}\,\dot{n}/n.$$

Since, by assumption, $\dot{n}/n = \dot{\rho}/\rho$ then we can write

$$n\mathfrak{q}\langle \dot{\mathbf{d}} \rangle = \rho\dot{\pi}, \qquad \pi := \mathbf{P}/\rho.$$

The force per unit volume is denoted by $\mathbf{f}_P$; by definition, we let

$$\mathbf{f}_P := n\langle \mathsf{F}^+ + \mathsf{F}^- \rangle.$$

Hence, $\mathbf{f}_P$ can be given the form[21]

$$\mathbf{f}_P = (\mathbf{P} \cdot \nabla)\mathbf{E} + \mathbf{v} \times (\mathbf{P} \cdot \nabla)\mathbf{B} + \rho\dot{\pi} \times \mathbf{B}. \tag{2.70}$$

[21] Equation (2.70) coincides with the analogous Eq. (5.7a) of [354], except for $\mu_0\mathbf{H}$ in place of $\mathbf{B}$ due to the choice of the Chu formulation.

The use of $\mathbf{B}$, instead of $\mu_0\mathbf{H}$ as is done in [354], allows $\mathbf{f}_P$ to be written in terms of the fields $\boldsymbol{\mathcal{E}}, \boldsymbol{\mathcal{B}}$ at the rest frame. Since $\mathbf{v} \times (\mathbf{P} \cdot \nabla)\mathbf{B} = \mathbf{P} \cdot \nabla(\mathbf{v} \times \mathbf{B}) - [(\mathbf{P} \cdot \nabla)\mathbf{v}] \times \mathbf{B}$ and $\pi = \mathbf{P}/\rho$ then

$$\begin{aligned}\mathbf{f}_P &= (\mathbf{P} \cdot \nabla)\mathbf{E} + \mathbf{P} \cdot \nabla(\mathbf{v} \times \mathbf{B}) - [(\mathbf{P} \cdot \nabla)\mathbf{v}] \times \mathbf{B} + \rho\dot{\pi} \times \mathbf{B} \\ &= (\mathbf{P} \cdot \nabla)\boldsymbol{\mathcal{E}} + \rho(\dot{\pi} - \mathbf{L}\pi) \times \mathbf{B} = (\mathbf{P} \cdot \nabla)\boldsymbol{\mathcal{E}} + \rho \overset{\triangledown}{\pi} \times \boldsymbol{\mathcal{B}}.\end{aligned}$$

Also, we observe that

$$\rho \overset{\triangledown}{\pi} = \rho(\mathbf{P}/\rho)^{\cdot} - \mathbf{L}\mathbf{P} = \overset{\square}{\mathbf{P}}$$

and hence

$$\mathbf{f}_P = (\mathbf{P} \cdot \nabla)\boldsymbol{\mathcal{E}} + \overset{\square}{\mathbf{P}} \times \mathbf{B}.$$

Consequently, we can write the force density $\mathbf{f}_P$ in terms of the Oldroyd and the Truesdell rates

$$\mathbf{f}_P = (\mathbf{P} \cdot \nabla)\boldsymbol{\mathcal{E}} + \rho \overset{\triangledown}{\pi} \times \boldsymbol{\mathcal{B}} = (\mathbf{P} \cdot \nabla)\boldsymbol{\mathcal{E}} + \overset{\square}{\mathbf{P}} \times \boldsymbol{\mathcal{B}}.$$

As a comment, the vector $(\mathbf{P} \cdot \nabla)\mathbf{E}$ is often referred to as Kelvin polarization force density.[22]

The representation (2.70) of the polarization force density comprises also the term $\rho\dot{\pi} \times \mathbf{B}$ which is of interest in time-dependent fields. Similar results occur in the literature, e.g. in the form [255]

$$\mathsf{F} = (\mathbf{p} \cdot \nabla)\mathbf{E} + \dot{\mathbf{p}} \times \mathbf{B}.$$

The torque on an electric dipole is

$$\mathfrak{T}^+ = \mathbf{d} \times \mathsf{F}^+$$

and hence, to within linear terms in $\mathfrak{p}$

$$\mathfrak{T}^+ = \mathfrak{p} \times \mathbf{E} + \mathfrak{p} \times (\mathbf{v} \times \mathbf{B}).$$

The corresponding torque per unit volume,

$$\tau_P := n\langle \mathfrak{T}^+ \rangle,$$

is given by

$$\tau_P = \mathbf{P} \times (\mathbf{E} + \mathbf{v} \times \mathbf{B}) = \mathbf{P} \times \boldsymbol{\mathcal{E}}.$$

---

[22] Likewise, the vector $\mu_0(\mathbf{M} \cdot \nabla)\mathbf{H}$, we will find shortly, is referred to as Kelvin magnetization force density.

A magnetic dipole moment is thought to be generated by two possible mechanisms. One way is by a small loop of electric current, called an *Ampèrian* magnetic dipole. Another way is by a pair of magnetic monopoles of opposite magnetic charge, called a *Gilbertian* magnetic dipole,[23] though elementary magnetic monopoles seem to be so far unobserved.

To determine body forces and torques due to magnetization, we consider Gilbertian dipoles and then proceed by analogy with the electric polarization by computing the force and the torque on a magnetic charge $\mathfrak{q}^*$ by starting with

$$\mathsf{F} = \mathfrak{q}^* \mathcal{H} = \mathfrak{q}^*(\mathbf{H} - \mathbf{v} \times \mathbf{D});$$

Hence, on a magnetic dipole, we have the forces

$$\mathsf{F}^-(\mathbf{x}) = -q^*\mathbf{H}(\mathbf{x}) + \mathfrak{q}^*\mathbf{v} \times \mathbf{D}(\mathbf{x}),$$

$$\mathsf{F}^+(\mathbf{x}) = q^*\mathbf{H}(\mathbf{x}+\mathbf{d}) - (\mathfrak{q}^*\mathbf{v} + \mathfrak{q}^*\dot{\mathbf{d}}) \times \mathbf{D}(\mathbf{x}+\mathbf{d}).$$

Within linear terms in $\mathbf{d}$, $\dot{\mathbf{d}}$, it follows that

$$\mathsf{F}^+ + \mathsf{F}^- = (\mathfrak{m} \cdot \nabla)\mathbf{H} - \mathbf{v} \times (\mathfrak{m} \cdot \nabla)\mathbf{D} - \mathfrak{q}^*\dot{\mathbf{d}} \times \mathbf{D},$$

where $\mathfrak{m} = \mathfrak{q}^*\mathbf{d}$ is the magnetic dipole moment. We then define the magnetic polarization $\mathbf{M}$ such that

$$\mu_0\mathbf{M} = n\langle\mathfrak{m}\rangle = n\mathfrak{q}^*\langle\mathbf{d}\rangle.$$

Hence, letting $\langle\dot{\mathbf{d}}\rangle = \langle\mathbf{d}\rangle^{\cdot}$, we have

$$nq^*\langle\mathbf{d}\rangle^{\cdot} = n(\mu_0\mathbf{M}/n)^{\cdot} = \mu_0\dot{\mathbf{M}} - \mu_0\mathbf{M}\frac{\dot{n}}{n}.$$

Further, since $\dot{n}/n = \dot{\rho}/\rho$, then we have

$$n\mathfrak{q}^*\langle\mathbf{d}\rangle^{\cdot} = \rho\mu_0\dot{\mathfrak{m}}, \qquad \mathbf{m} := \frac{\mathbf{M}}{\rho}.$$

Define the magnetic force per unit volume, $\mathbf{f}_M$, in the form

$$\mathbf{f}_M := \langle\mathsf{F}^+ + \mathsf{F}^-\rangle.$$

Hence we obtain

$$\mathbf{f}_M = \mu_0(\mathbf{M} \cdot \nabla)\mathbf{H} - \mu_0\mathbf{v} \times (\mathbf{M} \cdot \nabla)\mathbf{D} - \rho\mu_0\dot{\mathfrak{m}} \times \mathbf{D}. \tag{2.71}$$

---

[23] Due to William Gilbert's modelling of elementary magnetism in sixteenth century.

Again, we can express the force density in terms of the fields at the rest frame. Since $\mathbf{H} - \mu_0 \mathbf{v} \times \mathbf{D} = \boldsymbol{\mathcal{H}}$, we obtain

$$\mathbf{f}_M = \mu_0 (\mathbf{M} \cdot \nabla)\boldsymbol{\mathcal{H}} - \rho\mu_0(\dot{\mathbf{m}} - \mathbf{L}\mathbf{m}) \times \mathbf{D}.$$

Hence, it follows that

$$\mathbf{f}_M = \mu_0 (\mathbf{M} \cdot \nabla)\boldsymbol{\mathcal{H}} - \rho\mu_0 \overset{\triangledown}{\mathbf{m}} \times \boldsymbol{\mathcal{D}} = \mu_0 (\mathbf{M} \cdot \nabla)\boldsymbol{\mathcal{H}} - \mu_0 \overset{\square}{\mathbf{M}} \times \boldsymbol{\mathcal{D}},$$

where $\overset{\triangledown}{\mathbf{m}}$ is the Oldroyd rate and $\overset{\square}{\mathbf{M}}$ is the Truesdell rate.

As to (2.71), we observe that if $\mathbf{v} = \mathbf{0}$, then $\dot{\mathbf{m}} = \partial_t \mathbf{m}$ and

$$\mathbf{f}_M = \rho\mu_0 (\mathbf{m} \cdot \nabla)\mathbf{H} - \rho\mu_0 \partial_t \mathbf{m} \times \mathbf{D}$$

is the expression of the force on a magnetic dipole at rest in the laboratory frame. Restrict attention to free space, where $\mathbf{H} = \mathbf{B}/\mu_0$ and $\mathbf{D} = \epsilon_0 \mathbf{E}$. Consequently, we can write $\mathbf{f}_M$ in the form

$$\mathbf{f}_M = \rho\mu_0 (\mathbf{m} \cdot \nabla)\mathbf{H} - \rho\epsilon_0\mu_0 \partial_t \mathbf{m} \times \mathbf{E}.$$

Otherwise, in light of Maxwell's equation $\nabla \times \mathbf{H} = \mathbf{J} + \partial_t \mathbf{D}$, we have

$$\mathbf{f}_M = \rho\mu_0 (\mathbf{m} \cdot \nabla)\mathbf{H} - \rho\mu_0 \partial_t (\mathbf{m} \times \mathbf{D}) + \rho\mu_0 \mathbf{m} \times (\mathbf{J} + \nabla \times \mathbf{H}).$$

If, further, $\mathbf{J} = \mathbf{0}$ and $\nabla \mathbf{m} = \mathbf{0}$, then it follows

$$\mathbf{f}_M = \rho\mu_0 \nabla(\mathbf{m} \cdot \mathbf{H}) - \rho\mu_0 \partial_t (\mathbf{m} \times \mathbf{D}).$$

In free space, this relation can be written as

$$\mathbf{f}_M = \rho \nabla(\mathbf{m} \cdot \mathbf{B}) - \rho\epsilon_0\mu_0 \partial_t (\mathbf{m} \times \mathbf{E}).$$

This result establishes a direct connection with a representation of the force often given in the literature. Since magnetic dipoles tend to align in the direction of the magnetic field, the potential energy is assumed to be $U = -\mathbf{m} \cdot \mathbf{H}$ and then $-\nabla U = \nabla(\mathbf{m} \cdot \mathbf{H})$ is viewed as the force on $\mathbf{m}$. The additive term $-\rho\epsilon_0\mu_0\partial_t(\mathbf{m} \times \mathbf{E})$ is then regarded as *hidden momentum of the magnetization*[24] $\mathbf{m}$.

If the body is also an electric conductor, then we may view the charges as subject to the Lorentz force. Let $\mathbf{v}$ be the velocity of body, at the pertinent point, and $\mathbf{v}_c$ the velocity of the charges relative to the body. Hence, we write the Lorentz force in the form

$$q[\mathbf{E} + (\mathbf{v} + \mathbf{v}_c) \times \mathbf{B}] = q(\mathbf{E} + \mathbf{v} \times \mathbf{B}) + q\mathbf{v}_c \times \mathbf{B} = q\boldsymbol{\mathcal{E}} + \boldsymbol{\mathcal{J}} \times \mathbf{B},$$

[24] Usually with respect to the magnetic moment $\mathbf{m}$; see, e.g. [255, 396, 430].

q being the charge density and $\mathcal{J}$ the current density relative to the reference at rest. The force $\mathsf{F}_\mathbf{J}$ associated with the current density field $\mathbf{J}$ in the region Ω is then

$$\mathsf{F}_\mathbf{J} = \int_\Omega \mathbf{J} \times \mathbf{B}\, dv.$$

Correspondingly, the torque $\mathfrak{T}_\mathbf{J}$ can be written in the form

$$\mathfrak{T}_\mathbf{J} = \int_\Omega \mathbf{x} \times (\mathbf{J} \times \mathbf{B}) dv.$$

Very often the literature[25] takes the force, on the magnetic dipole with moment $\mathfrak{m}$, in the form $\mathsf{F} = \nabla(\mathfrak{m} \cdot \mathbf{B})$ on the basis of the view that the potential energy is $U = -\mathfrak{m} \cdot \mathbf{B}$ and the force then would be $-\nabla U$. If $\mathfrak{m}$ is independent of the position and $\mathbf{B}$ is irrotational, then

$$\mathsf{F} = \nabla(\mathfrak{m}_i B_i) = \mathfrak{m}_i \nabla B_i = \mathfrak{m}_i \partial_{x_i} \mathbf{B} = (\mathfrak{m} \cdot \nabla)\mathbf{B}.$$

If instead we start from

$$\mathbf{f} = \mu_0 \nabla(\mathbf{M} \cdot \mathbf{H}),$$

then it is very restrictive to assume $\mathbf{M}$ as independent of the position (and $\mathbf{H}$ or $\mathbf{B}$ as irrotational).

To obtain the expression for the torque $\mathfrak{T}_\mathfrak{m}$ on a magnetic dipole, we merely observe that the magnetic moment shows the tendency of aligning with the applied magnetic field. Hence, we assume a magnetic dipole $\mathfrak{m}$ at rest in matter is subject to the torque

$$\mathfrak{T}_\mathfrak{m} = \mu_0 \mathfrak{m} \times \mathcal{H}.$$

The torque per unit volume, due to magnetization, is then

$$\tau_M = n \langle \mathfrak{T}_\mathfrak{m} \rangle = \mu_0 \mathbf{M} \times \mathcal{H}.$$

Thus, the total torque per unit volume is

$$\tau = \mathbf{P} \times \mathcal{E} + \mu_0 \mathbf{M} \times \mathcal{H}.$$

The computation of power on electric and magnetic dipoles is deeply dependent on the physical modelling. Start with electric dipoles and look at a dipole as two charges, $-\mathrm{q}$ and q, displaced by $\mathbf{d}$, moving with velocities $\mathbf{v}$ and $\mathbf{v} + \dot{\mathbf{d}}$. Hence, we write the power, on a dipole, as

[25] E.g. [230], p. 185.

$$\begin{aligned}&\mathsf{F}^- \cdot \mathbf{v} + \mathsf{F}^+ \cdot (\mathbf{v} + \dot{\mathbf{d}}) \\ &= -\mathrm{q}[\mathbf{E}(\mathbf{x}) + \mathbf{v} \times \mathbf{B}] \cdot \mathbf{v} + \mathrm{q}[\mathbf{E}(\mathbf{x} + \mathbf{d}) + (\mathbf{v} + \dot{\mathbf{d}}) \times \mathbf{B}(\mathbf{x} + \mathbf{d})] \cdot (\mathbf{v} + \dot{\mathbf{d}}) \\ &= \mathrm{q}\mathbf{E}(\mathbf{x}) \cdot \dot{\mathbf{d}} + \mathrm{q}(\mathbf{d} \cdot \nabla)\mathbf{E}(\mathbf{x}) \cdot \mathbf{v} + \mathrm{q}[(\mathbf{d} \cdot \nabla)\mathbf{E}] \cdot \dot{\mathbf{d}}.\end{aligned}$$

The last term is nonlinear in $\mathbf{d}, \dot{\mathbf{d}}$; the limit as $\mathbf{d}, \dot{\mathbf{d}} \to 0$ with $\mathrm{q}\mathbf{d}$ and $\mathrm{q}\dot{\mathbf{d}}$ fixed shows that the third term is negligible. With this view, we then obtain

$$\langle \mathsf{F}^- \cdot \mathbf{v} + \mathsf{F}^+ \cdot (\mathbf{v} + \dot{\mathbf{d}}) \rangle = \mathbf{E} \cdot \dot{\mathbf{p}} + [(\mathbf{p} \cdot \nabla)\mathbf{E}] \cdot \mathbf{v}.$$

Define the power per unit volume, w_P, by

$$w_P = n\langle \mathbf{f}^- \cdot \mathbf{v} + \mathbf{f}^+ \cdot (\mathbf{v} + \dot{\mathbf{d}}) \rangle.$$

Hence, we obtain

$$w_P = \rho \mathbf{E} \cdot \big(\mathbf{P}/\rho\big)^{\cdot} + [(\mathbf{P} \cdot \nabla)\mathbf{E}] \cdot \mathbf{v}.$$

The power on the magnetic dipoles is sometimes taken as [354]

$$w_M = \rho\mu_0 \mathbf{H} \cdot \big(\mathbf{M}/\rho\big)^{\cdot} + [\mu_0(\mathbf{M} \cdot \nabla)\mathbf{H}] \cdot \mathbf{v}$$

by appealing to *similarity*. In addition to similarity with $\rho \mathbf{E} \cdot \big(\mathbf{P}/\rho\big)^{\cdot}$, the power $\rho\mu_0\mathbf{H} \cdot \big(\mathbf{M}/\rho\big)^{\cdot}$ is also justified by the view that $-\mathfrak{m} \cdot \mathbf{H}$ is the energy of the dipole moment $\mathfrak{m}$ in the field $\mathbf{H}$ and then $\mathbf{H} \cdot \dot{\mathfrak{m}}$ is the power expended to vary the moment $\mathfrak{m}$. Further

$$\langle \dot{\mathfrak{m}} \rangle = \langle \mathfrak{m} \rangle^{\cdot} = \mu_0 \rho (\mathbf{M}/\rho)^{\cdot}$$

and then we find the power $\rho\mu_0\mathbf{H} \cdot \big(\mathbf{M}/\rho\big)^{\cdot}$.

To sum up, we represent the body forces $\mathbf{f}_P, \mathbf{f}_M$, the body couples $\boldsymbol{\tau}_P, \boldsymbol{\tau}_M$, and the powers w_P, w_M in the form

$$\mathbf{f}_P = (\mathbf{P} \cdot \nabla)\mathbf{E} + \mathbf{v} \times (\mathbf{P} \cdot \nabla)\mathbf{B} + \rho\dot{\boldsymbol{\pi}} \times \mathbf{B}, \quad \mathbf{f}_M = \mu_0(\mathbf{M} \cdot \nabla)\mathbf{H} - \mu_0\mathbf{v} \times (\mathbf{M} \cdot \nabla)\mathbf{D} - \rho\mu_0\dot{\mathbf{m}} \times \mathbf{D},$$

$$\boldsymbol{\tau}_P = \mathbf{P} \times \boldsymbol{\mathcal{E}}, \qquad \boldsymbol{\tau}_M = \mu_0 \mathbf{M} \times \boldsymbol{\mathcal{H}},$$

$$w_P = [(\mathbf{P} \cdot \nabla)\mathbf{E}] \cdot \mathbf{v} + \rho\mathbf{E} \cdot \dot{\boldsymbol{\pi}}, \qquad w_M = \mu_0[(\mathbf{M} \cdot \nabla)\mathbf{H}] \cdot \mathbf{v} + \mu_0\rho\mathbf{H} \cdot \dot{\mathbf{m}}.$$

To complete the scheme, we determine the force and power associated with the electric current. The free charges, with charge density q, are taken to flow with velocity $\mathbf{v}_c$, relative to the body. Hence, we let $\mathbf{J} = q(\mathbf{v} + \mathbf{v}_c)$ be the current density. By the Lorentz force, the free charges are then subject to a force density

$$\mathbf{f}_c = q[\mathbf{E} + (\mathbf{v} + \mathbf{v}_c) \times \mathbf{B}] = q\boldsymbol{\mathcal{E}} + \boldsymbol{\mathcal{J}} \times \mathbf{B},$$

where $\mathcal{J} = \mathbf{J} - q\mathbf{v} = q\mathbf{v}_c$. The power density w_c is obtained via the observation

$$w_c = q[\mathbf{E} + (\mathbf{v} + \mathbf{v}_c) \times \mathbf{B}] \cdot (\mathbf{v} + \mathbf{v}_c) = q\mathbf{E} \cdot (\mathbf{v} + \mathbf{v}_c) = q\mathbf{E} \cdot \mathbf{v} + q\mathbf{E} \cdot \mathbf{v}_c.$$

Since $q\mathbf{E} \cdot \mathbf{v} + q\mathbf{E} \cdot \mathbf{v}_c = q\boldsymbol{\mathcal{E}} \cdot \mathbf{v} + \mathbf{E} \cdot \mathcal{J}$, then it follows

$$w_c - \mathbf{f}_c \cdot \mathbf{v} = q\boldsymbol{\mathcal{E}} \cdot \mathbf{v} + \mathbf{E} \cdot \mathcal{J} - q\boldsymbol{\mathcal{E}} \cdot \mathbf{v} - (\mathbf{E} + \mathbf{v} \times \mathbf{B}) \cdot \mathcal{J} = \boldsymbol{\mathcal{E}} \cdot \mathcal{J}.$$

To state the balance of angular momentum, we observe that, since no rotational structure occurs, then we require that the total torque vanishes, namely

$$\boldsymbol{\tau}_{\text{mech}} + \mathbf{P} \times \boldsymbol{\mathcal{E}} + \mu_0 \mathbf{M} \times \boldsymbol{\mathcal{H}} = \mathbf{0}$$

or $\text{skw}\mathbf{T} = \text{skw}(\mathbf{P} \otimes \boldsymbol{\mathcal{E}} + \mu_0 \mathbf{M} \otimes \boldsymbol{\mathcal{H}})$.

We can then state the balance equations of linear momentum, angular momentum, and energy in the form

$$\begin{aligned} \rho\dot{\mathbf{v}} &= \nabla \cdot \mathbf{T} + \rho\mathbf{b} + (\mathbf{P} \cdot \nabla)\mathbf{E} + \mathbf{v} \times (\mathbf{P} \cdot \nabla)\mathbf{B} + \rho\dot{\boldsymbol{\pi}} \times \mathbf{B} \\ &+\mu_0(\mathbf{M} \cdot \nabla)\mathbf{H} - \mu_0\mathbf{v} \times (\mathbf{M} \cdot \nabla)\mathbf{D} - \mu_0\rho\dot{\mathbf{m}} \times \mathbf{D} + q\boldsymbol{\mathcal{E}} + \mathcal{J} \times \mathbf{B}, \end{aligned} \tag{2.72}$$

$$\text{skw}\mathbf{T} = \text{skw}(\mathbf{P} \otimes \boldsymbol{\mathcal{E}} + \mu_0 \mathbf{M} \otimes \boldsymbol{\mathcal{H}}), \tag{2.73}$$

$$\begin{aligned} \rho(\varepsilon + \tfrac{1}{2}\mathbf{v}^2)^{\cdot} &= \rho\mathbf{b} \cdot \mathbf{v} + \nabla \cdot (\mathbf{v}\mathbf{T}) - \nabla \cdot \mathbf{q} + \rho r + \rho\mathbf{E} \cdot \dot{\boldsymbol{\pi}} + \mu_0\rho\mathbf{H} \cdot \dot{\mathbf{m}} \\ &+[(\mathbf{P} \cdot \nabla)\mathbf{E}] \cdot \mathbf{v} + \mu_0[(\mathbf{M} \cdot \nabla)\mathbf{H}] \cdot \mathbf{v} + q\boldsymbol{\mathcal{E}} \cdot \mathbf{v} + \mathbf{E} \cdot \mathcal{J}, \end{aligned} \tag{2.74}$$

where $\mathbf{b}$ is the mechanical force per unit mass. In light of (2.72), we obtain[26]

$$\rho\dot{\varepsilon} = \boldsymbol{\mathcal{E}} \cdot \mathcal{J} + \rho\boldsymbol{\mathcal{E}} \cdot \dot{\boldsymbol{\pi}} + \mu_0\rho\boldsymbol{\mathcal{H}} \cdot \dot{\mathbf{m}} + \mathbf{T} \cdot \mathbf{L} - \nabla \cdot \mathbf{q} + \rho r. \tag{2.75}$$

Equation (2.75) involves the electric field $\boldsymbol{\mathcal{E}}$ and the magnetic intensity $\boldsymbol{\mathcal{H}}$ at the frame $\hat{\mathscr{F}}$ locally at rest with the body. Also, $\mathcal{J} = q\mathbf{v}_c$ is the current density relative to $\hat{\mathscr{F}}$.

By analogy with the first principle (2.19) and the first law (2.22), we may view (2.74) as a principle of conservation of energy (mechanical + electromagnetic). Once the consequences of (2.72) are evaluated, the balance law of electromagnetic energy (2.75) follows.

For undeformable media, at rest, Eq. (2.75) simplifies to

$$\rho\dot{\varepsilon} = \mathbf{E} \cdot \mathbf{J} + \dot{\mathbf{P}} \cdot \mathbf{E} + \mu_0\dot{\mathbf{M}} \cdot \mathbf{H} - \nabla \cdot \mathbf{q} + \rho r. \tag{2.76}$$

---

[26] Equation (2.75) coincides with the analogous one in [354] and is in agreement with that assumed in [357], Eq. (7.35).

Quite often, the balance of energy for undeformable media is taken as a direct consequence of Poynting's theorem whereby $\mathbf{E} \times \mathbf{H}$ is viewed as the energy flux of electromagnetic character. With this view, the balance is expressed by

$$\dot{u}_0 = -\nabla \cdot (\mathbf{q} + \mathbf{E} \times \mathbf{H}) + \rho r,$$

where u_0 is the energy per unit volume and ρ is a constant; u_0 is the energy density (per unit volume) free from the kinetic energy $\frac{1}{2}\rho \mathbf{v}^2$. The identity

$$\nabla \cdot (\mathbf{E} \times \mathbf{H}) = \mathbf{H} \cdot \nabla \times \mathbf{E} - \mathbf{E} \cdot \nabla \times \mathbf{H}$$

and Maxwell's equations yield

$$\nabla \cdot (\mathbf{E} \times \mathbf{H}) = -\mathbf{E} \cdot \mathbf{J} - \mathbf{E} \cdot \dot{\mathbf{D}} - \mathbf{H} \cdot \dot{\mathbf{B}}.$$

Hence, the balance of energy becomes

$$\dot{u}_0 = \mathbf{E} \cdot \mathbf{J} + \mathbf{E} \cdot \dot{\mathbf{D}} + \mathbf{H} \cdot \dot{\mathbf{B}} - \nabla \cdot \mathbf{q} + \rho r. \tag{2.77}$$

Now, substitution of

$$\mathbf{D} = \epsilon_0 \mathbf{E} + \mathbf{P}, \qquad \mathbf{B} = \mu_0 (\mathbf{H} + \mathbf{M})$$

and some rearrangements give (2.76) provided $\rho\varepsilon$ and u_0 are related by

$$\rho\varepsilon = u_0 - \tfrac{1}{2}[\epsilon_0 \mathbf{E}^2 + \mu_0 \mathbf{H}^2]. \tag{2.78}$$

Hence, (2.76) and (2.77) are equivalent provided ε and u_0 are connected by (2.78). This result holds irrespective of any constitutive equation for the energy density. We may then view Eq. (2.77) as the balance of the internal + electromagnetic energy and (2.76) as the balance of internal energy.

An alternative form of Poynting's theorem is based on the form (2.62)–(2.63) of Maxwell's equations. Since $\nabla \cdot (\mathcal{E} \times \mathcal{H}) = \mathcal{H} \cdot \nabla \times \mathcal{E} - \mathcal{E} \cdot \nabla \times \mathcal{H}$, by (2.62)–(2.63), we obtain

$$\nabla \cdot (\mathbf{E} \times \mathbf{H}) = -\mathcal{E} \cdot \mathcal{J} - \mathcal{H} \cdot \overset{\diamond}{\mathbf{B}} - \mathcal{E} \cdot \overset{\diamond}{\mathbf{D}}\,.$$

If the balance of energy is assumed in the form

$$\rho\dot{\varepsilon} = \rho r - \nabla \cdot (\mathbf{q} + \mathcal{E} \times \mathcal{H}) + ...,$$

the dots representing powers of different character, then we have

$$\rho\dot{\varepsilon} = \rho r - \nabla \cdot \mathbf{q} + \mathcal{E} \cdot \mathcal{J} + \mathcal{H} \cdot \overset{\diamond}{\mathbf{B}} + \mathcal{E} \cdot \overset{\diamond}{\mathbf{D}} + ...,$$

which is inequivalent to (2.76) and (2.75). This is a further indication that Poynting's theorem has not to be viewed as a balance law.

2.16.2 Angular Momentum and Magnetic Moment

By (2.73), we infer that a magnetically polarized body experiences a torque $\mu_0 \mathbf{M} \times \mathcal{H}$; a dual behaviour happens for an electrically polarized body under the action of an electric field. The body is then in (rotational) equilibrium if a mechanical torque is applied so that the resulting torque vanishes

$$\boldsymbol{\tau}_{\text{mech}} + \mu_0 \mathbf{M} \times \mathcal{H} = \mathbf{0}.$$

If the mechanical torque is produced by a deformation, then we might observe that, at equilibrium, a magnetic field produces a deformation.

The Einstein–de Haas effect [143] arises from experimental observations where a change in the applied magnetic field causes the body to rotate. In the classical experiment, a suspended (ferromagnetic) cylindrical specimen is subject to the magnetic field produced by a magnetizing coil. The magnetic field is along the axis of the cylinder as well as the magnetic moment $\mathbf{M}$. Hence, $\text{skw}(\mathbf{M} \otimes \mathcal{H}) = \mathbf{0}$ and no rotation is caused by the applied magnetic field. We then look for an explanation by means of microscopic considerations.

The (total) angular momentum $\boldsymbol{\mathfrak{L}}$ is the sum of contributions from spin, orbit, and crystal lattice motions

$$\boldsymbol{\mathfrak{L}} = \boldsymbol{\mathfrak{L}}_{\text{spin}} + \boldsymbol{\mathfrak{L}}_{\text{orbit}} + \boldsymbol{\mathfrak{L}}_{\text{lattice}}.$$

The observation shows the changes of $\boldsymbol{\mathfrak{L}}_{\text{lattice}}$. Likewise, the magnetic moment $\mathbf{M}$ consists of three terms

$$\mathbf{M} = \mathbf{M}_{\text{spin}} + \mathbf{M}_{\text{orbit}} + \mathbf{M}_{\text{lattice}}.$$

It is assumed that the lattice contribution is negligible because of the relatively large mass, and too slow rotation, of the positive ions [256].

Let $\boldsymbol{\mathfrak{L}}$ and $\mathbf{M}$ be parallel and denote by $\mathfrak{L}$, M the components in the common direction. First observe that

$$M = M_{\text{spin}} + M_{\text{orbit}} + M_{\text{lattice}} \simeq M_{\text{spin}} + M_{\text{orbit}}$$

and

$$\mathfrak{L} = \mathfrak{L}_{\text{spin}} + \mathfrak{L}_{\text{orbit}} + \mathfrak{L}_{\text{lattice}}.$$

The magnetization M is measured through a pickup coil around the specimen, the angular momentum is measured through the angular velocity produced. Hence, measures are possible of the ratio

$$\lambda = \frac{M_{\text{spin}} + M_{\text{orbit}}}{-\mathfrak{L}_{\text{lattice}}} = \frac{M_{\text{spin}} + M_{\text{orbit}}}{\mathfrak{L}_{\text{spin}} + \mathfrak{L}_{\text{orbit}}}.$$

Elementary arguments suggest that the ratio equals $e/2m$, the ratio between the charge and twice the mass of electrons. Instead, the measure of

$$\lambda = \frac{ge}{2m}$$

indicates that $g = 2$; g is called the magneto-mechanical factor (or Landé g-factor) and λ the gyromagnetic ratio.

If we let

$$M_{\text{orbit}}/M_{\text{spin}} = \epsilon, \qquad \mathfrak{L}_{\text{orbit}}/\mathfrak{L}_{\text{spin}} = 2\epsilon,$$

then we have

$$\lambda = \frac{M_{\text{spin}}}{\mathfrak{L}_{\text{spin}}} \frac{1 + M_{\text{orbit}}/M_{\text{spin}}}{1 + \mathfrak{L}_{\text{orbit}}/\mathfrak{L}_{\text{spin}}}; \qquad \frac{ge}{2m} = \frac{e}{m} \frac{1+\epsilon}{1+2\epsilon}.$$

As $\epsilon \ll 1$, we find

$$g \simeq 2(1 - \epsilon).$$

If only the spin term occurs, then $g = 2$.

2.16.3 Maxwell Stress Tensor

The Maxwell stress tensor is a second-order tensor used in classical electromagnetism to describe the electromagnetic forces in terms of a tensor in a way similar to the mechanical contact forces represented by the Cauchy stress tensor.

In free space, the (Lorentz) force on the charge $\mathfrak{q}$ is

$$\mathfrak{q}\mathcal{E} = \mathfrak{q}(\mathbf{E} + \mathbf{v} \times \mathbf{B})$$

and hence the (electromagnetic) force $\mathbf{f}_c$ per unit volume is given by

$$\mathbf{f}_c = n\mathfrak{q}(\mathbf{E} + \mathbf{v} \times \mathbf{B}) = q\mathbf{E} + \mathbf{J} \times \mathbf{B},$$

where $\mathbf{J} = q\mathbf{v}$. By Gauss' and Ampère's laws, q and $\mathbf{J}$ can be written

$$q = \epsilon_0 \nabla \cdot \mathbf{E}, \qquad \mathbf{J} = \frac{1}{\mu_0} \nabla \times \mathbf{B} - \epsilon_0 \partial_t \mathbf{E}.$$

Hence

$$\mathbf{f}_c = \epsilon_0(\nabla \cdot \mathbf{E})\mathbf{E} + \frac{1}{\mu_0}(\nabla \times \mathbf{B}) \times \mathbf{B} - \epsilon_0 \partial_t \mathbf{E} \times \mathbf{B}.$$

In light of Faraday's law, we find

$$\partial_t(\mathbf{E} \times \mathbf{B}) = \partial_t \mathbf{E} \times \mathbf{B} + \mathbf{E} \times \partial_t \mathbf{B} = \partial_t \mathbf{E} \times \mathbf{B} - \mathbf{E} \times (\nabla \times \mathbf{E}).$$

Consequently, we have

$$\mathbf{f}_c = \epsilon_0(\nabla \cdot \mathbf{E})\mathbf{E} + \frac{1}{\mu_0}(\nabla \times \mathbf{B}) \times \mathbf{B} - \epsilon_0 \partial_t(\mathbf{E} \times \mathbf{B}) - \epsilon_0 \mathbf{E} \times (\nabla \times \mathbf{E}).$$

Since $\nabla \cdot \mathbf{B} = 0$, then we can formally add a zero term $(\nabla \cdot \mathbf{B})\mathbf{B}/\mu_0$ to obtain

$$\mathbf{f}_c = \epsilon_0[(\nabla \cdot \mathbf{E})\mathbf{E} - \mathbf{E} \times (\nabla \times \mathbf{E})] + \frac{1}{\mu_0}[(\nabla \cdot \mathbf{B})\mathbf{B} - \mathbf{B} \times (\nabla \times \mathbf{B})] - \epsilon_0 \partial_t(\mathbf{E} \times \mathbf{B}).$$

For any vector $\mathbf{w}$, the identity

$$\mathbf{w} \times (\nabla \times \mathbf{w}) = \nabla \tfrac{1}{2}\mathbf{w}^2 - (\mathbf{w} \cdot \nabla)\mathbf{w}$$

allows us to obtain

$$(\nabla \cdot \mathbf{w})\mathbf{w} - \mathbf{w} \times (\nabla \times \mathbf{w}) = -\nabla \tfrac{1}{2}\mathbf{w}^2 + \nabla \cdot (\mathbf{w} \otimes \mathbf{w}),$$

where

$$\nabla \cdot (\mathbf{w} \otimes \mathbf{w})|_i = \partial_{x_j}(w_i w_j).$$

Consequently, we can write $\mathbf{f}_c$ in the form

$$\mathbf{f}_c = \epsilon_0[-\nabla \tfrac{1}{2}\mathbf{E}^2 + \nabla \cdot (\mathbf{E} \otimes \mathbf{E})] + \frac{1}{\mu_0}[-\nabla \tfrac{1}{2}\mathbf{B}^2 + \nabla \cdot (\mathbf{B} \otimes \mathbf{B})] - \epsilon_0 \partial_t(\mathbf{E} \times \mathbf{B}).$$

Letting

$$\boldsymbol{\sigma} := \epsilon_0(\mathbf{E} \otimes \mathbf{E} - \tfrac{1}{2}\mathbf{E}^2\, \mathbf{1}) + \frac{1}{\mu_0}(\mathbf{B} \otimes \mathbf{B} - \tfrac{1}{2}\mathbf{B}^2\, \mathbf{1})$$

we find that $\mathbf{f}_c$ can be written

$$\mathbf{f}_c = \nabla \cdot \boldsymbol{\sigma} - \epsilon_0 \partial_t(\mathbf{E} \times \mathbf{B}). \tag{2.79}$$

The tensor $\boldsymbol{\sigma}$ is referred to as the *Maxwell stress tensor.* The reason for the term "stress tensor" is due to the form of Eq. (2.79). If $\epsilon_0 \mathbf{E} \times \mathbf{B}$ is viewed as the momentum density of the electromagnetic field, then

$$\epsilon_0 \partial_t(\mathbf{E} \times \mathbf{B}) = -\mathbf{f}_c + \nabla \cdot \boldsymbol{\sigma},$$

might be interpreted as the balance of linear momentum. Yet it is rather odd that $-\mathbf{f}_c$ instead of $\mathbf{f}_c$ occurs in the balance of linear momentum with $\boldsymbol{\sigma}$ as the stress tensor. It is perhaps wiser to regard (2.79) as an identity. Incidentally

$$\epsilon_0 \, \mathbf{E} \times \mathbf{B} = \epsilon_0 \mu_0 \, \mathbf{E} \times \mathbf{H}, \qquad \mathbf{H} = \mathbf{B}/\mu_0,$$

is usually regarded as $\epsilon_0 \mu_0$ times the energy flux associated with the electromagnetic field (Poynting's theorem).

Maxwell Stress Tensor in Matter

In matter, we need to account also for the forces $\mathbf{f}_P$ and $\mathbf{f}_M$. Let $\mathbf{f} = \mathbf{f}_c + \mathbf{f}_P + \mathbf{f}_M$. As the simplest approximation, we let

$$\mathbf{f} = q\mathbf{E} + \mathbf{J} \times \mathbf{B} + (\mathbf{P} \cdot \nabla)\mathbf{E} + \mu_0 (\mathbf{M} \cdot \nabla)\mathbf{H}$$

be the whole electromagnetic body force (per unit volume). Since $\nabla \cdot \mathbf{D} = q$ and $\mathbf{D} = \epsilon_0 \mathbf{E} + \mathbf{P}$ we obtain

$$q\mathbf{E} + (\mathbf{P} \cdot \nabla)\mathbf{E} = \nabla \cdot (\mathbf{E} \otimes \mathbf{D}) - \epsilon_0 (\mathbf{E} \cdot \nabla)\mathbf{E}.$$

By a direct calculation, for any triplet of vector functions $\mathbf{u}$, $\mathbf{v}$, $\mathbf{w}$, it follows from

$$\nabla \times \mathbf{u} = \mathbf{w}$$

that

$$\partial_{x_k} u_i = \partial_{x_i} u_k - \epsilon_{ikp} w_p, \qquad v_k \partial_{x_k} u_i = v_k \partial_{x_i} u_k - (\mathbf{v} \times \mathbf{w})_i.$$

If we let $\mathbf{u} = \mathbf{E}$, $\mathbf{v} = \mathbf{E}$, and $\mathbf{w} = -\partial_t \mathbf{B}$, then it follows from Faraday's law $\nabla \times \mathbf{E} = -\partial_t \mathbf{B}$ that

$$(\mathbf{E} \cdot \nabla)\mathbf{E} = \tfrac{1}{2} \nabla \, \mathbf{E}^2 + \mathbf{E} \times \partial_t \mathbf{B}.$$

As a consequence

$$q\mathbf{E} + (\mathbf{P} \cdot \nabla)\mathbf{E} = \nabla \cdot [\mathbf{E} \otimes \mathbf{D} - \tfrac{1}{2} \epsilon_0 \mathbf{E}^2 \, \mathbf{1}] - \epsilon_0 \mathbf{E} \times \partial_t \mathbf{B}.$$

Now examine the magnetic force. By Ampère's law, we can write

$$\mathbf{J} \times \mathbf{B} = [\nabla \times (\mathbf{B}/\mu_0 - \mathbf{M}) - \partial_t \mathbf{D}] \times \mathbf{B}.$$

Since

$$[\nabla \times (\mathbf{B}/\mu_0 - \mathbf{M}) \times \mathbf{B}]_i = \frac{1}{\mu_0} B_k (\partial_{x_k} B_i - \partial_{x_i} B_k) + B_k (\partial_{x_i} M_k - \partial_{x_k} M_i)$$

it follows

$$[\mathbf{J} \times \mathbf{B} + \mu_0(\mathbf{M} \cdot \nabla)\mathbf{H}]_i = \{\nabla \cdot [\frac{1}{\mu_0}(\mathbf{B} \otimes \mathbf{B} - \tfrac{1}{2}\mathbf{B}^2\, \mathbf{1}) - \mathbf{M} \otimes \mathbf{B} + \mathbf{M} \cdot \mathbf{B}\, \mathbf{1}]\}_i$$
$$- M_k(\partial_{x_i} B_k - \mu_0 \partial_{x_k} H_i) - (\partial_t \mathbf{D} \times \mathbf{B})_i$$

Accordingly, we have

$$\mathbf{f} = \nabla \cdot [\mathbf{E} \otimes \mathbf{D} - \tfrac{1}{2}\epsilon_0 \mathbf{E}^2\, \mathbf{1} + \frac{1}{\mu_0}(\mathbf{B} \otimes \mathbf{B} - \tfrac{1}{2}\mathbf{B}^2\, \mathbf{1}) - \mathbf{M} \otimes \mathbf{B} + \mathbf{M} \cdot \mathbf{B}\, \mathbf{1} + \tfrac{1}{2}\mu_0 \mathbf{M}^2 \mathbf{1}] + \tilde{\mathbf{f}}, \tag{2.80}$$

where

$$\tilde{\mathbf{f}} = -\epsilon_0 \mathbf{E} \times \partial_t \mathbf{B} - \mu_0 M_k (\nabla H_k - \partial_{x_k} \mathbf{H}) - \partial_t \mathbf{D} \times \mathbf{B}.$$

For a polarizable and magnetizable solid, the electromagnetic force $\mathbf{f}_{em}$ can be written as the divergence of a (stress) tensor ([260], Sect. 56) in static conditions ($\partial_t \mathbf{D} = \mathbf{0}$, $\partial_t \mathbf{B} = \mathbf{0}$) provided $M_k(\nabla H_k - \partial_{x_k}\mathbf{H}) = \mathbf{0}$. In non-polarizable and non-magnetizable media ($\mathbf{P} = \mathbf{0}$, $\mathbf{M} = \mathbf{0}$), Eq. (2.80) reduces to (2.79).

As a formal particular case, we now look for a representation of $\mathbf{f}$ if $(\mathbf{P} \cdot \nabla)\mathbf{E}$ and $\mu_0(\mathbf{M} \cdot \nabla)\mathbf{H}$ are neglected. Consider

$$\mathbf{f} = \mathbf{f}_c = q\mathbf{E} + \mathbf{J} \times \mathbf{B}.$$

We replace

$$q = \nabla \cdot \mathbf{D}, \qquad \mathbf{J} = \nabla \times \mathbf{H} - \partial_t \mathbf{D}.$$

and use Maxwell's equations to obtain

$$\mathbf{f} = (\nabla \cdot \mathbf{D})\mathbf{E} + (\nabla \times \mathbf{H}) \times \mathbf{B} - \mathbf{D} \times (\nabla \times \mathbf{E}) - \partial_t (\mathbf{D} \times \mathbf{B}).$$

In view of the identity

$$(\nabla \times \mathbf{u}) \times \mathbf{w} = (\mathbf{w} \cdot \nabla)\mathbf{u} - w_k \nabla u_k$$

for any pair of vectors $\mathbf{u}$, $\mathbf{w}$, the body force $\mathbf{f}$ can be written

$$\mathbf{f} = (\nabla \cdot \mathbf{D})\mathbf{E} + (\mathbf{D} \cdot \nabla)\mathbf{E} - D_k \nabla E_k + (\mathbf{B} \cdot \nabla)\mathbf{H} - B_k \nabla H_k - \partial_t (\mathbf{D} \times \mathbf{B}).$$

Hence, since $\nabla \cdot \mathbf{B} = 0$, we have

$$\mathbf{f} = \nabla \cdot (\mathbf{E} \otimes \mathbf{D} + \mathbf{H} \otimes \mathbf{B}) - D_k \nabla E_k - B_k \nabla H_k - \partial_t (\mathbf{D} \times \mathbf{B}). \tag{2.81}$$

If, further, the linear relations

$$\mathbf{B} = \mu \mathbf{H}, \qquad \mathbf{D} = \epsilon \mathbf{E}$$

hold, where μ and ϵ are constants, then

$$D_k \nabla E_k + B_k \nabla H_k = \nabla \tfrac{1}{2}(\mathbf{D} \cdot \mathbf{E} + \mathbf{B} \cdot \mathbf{H}).$$

Subject to linearity, we can then write

$$\mathbf{f} = \nabla \cdot \{\mathbf{E} \otimes \mathbf{D} + \mathbf{H} \otimes \mathbf{B} - \tfrac{1}{2}(\mathbf{D} \cdot \mathbf{E} + \mathbf{B} \cdot \mathbf{H})\mathbf{1}\} - \partial_t (\mathbf{D} \times \mathbf{B}). \tag{2.82}$$

Letting

$$\boldsymbol{\sigma} = \mathbf{E} \otimes \mathbf{D} + \mathbf{H} \otimes \mathbf{B} - \tfrac{1}{2}(\mathbf{D} \cdot \mathbf{E} + \mathbf{B} \cdot \mathbf{H})\mathbf{1}$$

we can write (2.82) in the form

$$\mathbf{f} = \nabla \cdot \boldsymbol{\sigma} - \partial_t (\mathbf{D} \times \mathbf{B}).$$

Again, as expected, Eq. (2.82) reduces to (2.79) if $\mathbf{D} = \epsilon_0 \mathbf{E}$, $\mathbf{B} = \mu_0 \mathbf{H}$. The tensors $\boldsymbol{\sigma}$ shown in (2.80) and (2.82) are different forms of the Maxwell stress tensor in matter, subject to different models of $\mathbf{f}$.

It is worth remarking that the result (2.82) coincides with the result obtained within the relativistic four-dimensional setting (see, e.g. [237], Sect. 2.12); in that case the linearity is expressed by $G^{\mu\nu}$ being a constant multiple of $F^{\mu\nu}$.

The stress tensor is also found *if*

$$\mathbf{f} = q\mathbf{E} + \mathbf{J} \times \mathbf{B} + P_k \nabla E_k + \mu_0 M_k \nabla H_k.$$

Since

$$P_k \nabla E_k + M_k \nabla B_k = D_k \nabla E_k - \epsilon_0 E_k \nabla E_k + B_k \nabla H_k - \mu_0 H_k \nabla H_k$$

then adding $D_k \nabla E_k - \epsilon_0 E_k \nabla E_k + B_k \nabla H_k - \mu_0 H_k \nabla H_k = \mathbf{f}_c$ to the right-hand side of (2.81) we find

$$\mathbf{f} = \nabla \cdot \{\mathbf{E} \otimes \mathbf{D} + \mathbf{H} \otimes \mathbf{B}\} - \nabla \tfrac{1}{2}(\epsilon_0 \mathbf{E}^2 + \mu_0 \mathbf{H}^2) - \partial_t (\mathbf{D} \times \mathbf{B}).$$

Hence, the Maxwell stress tensor would be given by

$$\boldsymbol{\sigma} = \mathbf{E} \otimes \mathbf{D} + \mathbf{H} \otimes \mathbf{B} - \tfrac{1}{2}(\epsilon_0 \mathbf{E}^2 + \mu_0 \mathbf{H}^2)\mathbf{1}. \tag{2.83}$$

The stress tensor (2.83) is considered in [144] where it is claimed that

$$\mathbf{f} = \nabla \cdot \boldsymbol{\sigma} = (\mathbf{P} \cdot \nabla)\mathbf{E} + \mu_0 (\mathbf{M} \cdot \nabla)\mathbf{H},$$

that is the Kelvin expression for the force. As is known, this is true if there is no free charge ($q = 0, \mathbf{J} = \mathbf{0}$) and $\mathbf{B}, \mathbf{D}$ are time independent, which implies $\nabla \times \mathbf{E} = \mathbf{0}, \nabla \times \mathbf{H} = \mathbf{0}$. Accordingly

$$\partial_{x_j}\sigma_{ij} = -(\epsilon_0 E_k \partial_{x_i} E_k + \mu_0 H_k \partial_{x_i} H_k) + D_j \partial_{x_j} E_i + B_j \partial_{x_j} H_i$$
$$= -(\epsilon_0 E_k \partial_{x_k} E_i + \mu_0 H_k \partial_{x_k} H_i) + D_j \partial_{x_j} E_i + B_j \partial_{x_j} H_i = P_k \partial_{x_k} E_i + \mu_0 M_k \partial_{x_k} H_i.$$

2.16.4 Balance of Entropy

Consistent with the general form (2.39), we still state the balance of entropy in the form

$$\rho\dot{\eta} + \nabla\cdot\mathbf{j} - \frac{\rho r}{\theta} \geq 0,$$

and again, for formal convenience, the entropy flux $\mathbf{j}$ is split in the form $\mathbf{j} = \mathbf{q}/\theta + \mathbf{k}$. As the second law of thermodynamics, we assume that, for every admissible thermodynamic process, the balance of entropy must hold for all times t and points $\mathbf{x}$.

Since

$$\rho\theta\dot{\eta} + \nabla\cdot\mathbf{q} - \rho r - \frac{1}{\theta}\mathbf{q}\cdot\nabla\theta + \theta\nabla\cdot\mathbf{k} \geq 0,$$

by compatibility with the balance of energy, we evaluate $\nabla\cdot\mathbf{q} - \rho r$ from (2.75) to obtain

$$\rho\theta\dot{\eta} - \rho\dot{\varepsilon} + \rho\boldsymbol{\mathcal{E}}\cdot\dot{\boldsymbol{\pi}} + \rho\mu_0\boldsymbol{\mathcal{H}}\cdot\dot{\mathbf{m}} + \boldsymbol{\mathcal{E}}\cdot\boldsymbol{\mathcal{J}} + \mathbf{T}\cdot\mathbf{L} - \frac{1}{\theta}\mathbf{q}\cdot\nabla\theta + \theta\nabla\cdot\mathbf{k} \geq 0.$$

In terms of the free energy $\psi = \varepsilon - \theta\eta$, we can write the (second law) inequality as

$$-\rho(\dot{\psi} + \eta\dot{\theta}) + \rho\boldsymbol{\mathcal{E}}\cdot\dot{\boldsymbol{\pi}} + \rho\mu_0\boldsymbol{\mathcal{H}}\cdot\dot{\mathbf{m}} + \boldsymbol{\mathcal{E}}\cdot\boldsymbol{\mathcal{J}} + \mathbf{T}\cdot\mathbf{L} - \frac{1}{\theta}\mathbf{q}\cdot\nabla\theta + \theta\nabla\cdot\mathbf{k} \geq 0. \tag{2.84}$$

In rigid bodies, at rest, we can take $\boldsymbol{\mathcal{E}} = \mathbf{E}, \boldsymbol{\mathcal{H}} = \mathbf{H}, \boldsymbol{\mathcal{J}} = \mathbf{J}$ and

$$\dot{\boldsymbol{\pi}} = \frac{1}{\rho}\dot{\mathbf{P}}, \qquad \dot{\mathbf{m}} = \frac{1}{\rho}\dot{\mathbf{M}}.$$

Moreover we have $\mathbf{L} = \mathbf{0}$ and hence the second law inequality simplifies to

$$-\rho(\dot{\psi} + \eta\dot{\theta}) + \mathbf{E}\cdot\dot{\mathbf{P}} + \mu_0\mathbf{H}\cdot\dot{\mathbf{M}} + \mathbf{E}\cdot\mathbf{J} - \frac{1}{\theta}\mathbf{q}\cdot\nabla\theta + \theta\nabla\cdot\mathbf{k} \geq 0. \tag{2.85}$$

2.17 Conservation Laws Across a Singular Surface

Look at a conservation law in the general form

$$\frac{d}{dt}\int_{\mathcal{P}_t} \phi\, dv = \int_{\mathcal{P}_t} \beta\, dv + \int_{\partial\mathcal{P}_t} \mathbf{h}\,\mathbf{n}\, da, \tag{2.86}$$

where $\mathbf{h}$ is a vector if ϕ is a scalar and a (second-order) tensor if ϕ is a vector. Hence, $\mathbf{h}\,\mathbf{n}$ stands for $\mathbf{h} \cdot \mathbf{n}$ or the tensor composition according as ϕ is a scalar or a vector. The region $\mathcal{P}_t$ is allowed to be divided into two subregions $\mathcal{P}_t^+$, $\mathcal{P}_t^-$ by a surface $\sigma(t)$ which is singular with respect to ϕ. Let $\hat{\sigma}(t) = \sigma(t) \cap \mathcal{P}_t$. Hence, by (1.38) with Ω and σ replaced by $\mathcal{P}_t$ and $\hat{\sigma}$, we obtain

$$\int_{\mathcal{P}_t\setminus\hat{\sigma}} \{\partial_t\phi + \nabla\cdot(\phi\mathbf{v})\}dv + \int_{\hat{\sigma}} [\![\phi(\boldsymbol{\nu} - \mathbf{v})]\!] \cdot \mathbf{m}\, da = \int_{\mathcal{P}_t\setminus\hat{\sigma}} \beta\, dv + \int_{\partial\mathcal{P}_t} \mathbf{h}\,\mathbf{n}\, da. \tag{2.87}$$

Let $\partial_t\phi$, $\nabla\cdot(\phi\mathbf{v})$, and β be bounded in $\mathcal{P}_t \setminus \hat{\sigma}$ and $\mathbf{h}$ bounded at the lateral surface of the pillbox. By the arbitrariness of $\mathcal{P}_t$, we can take $\mathcal{P}_t$ as a pillbox across $\hat{\sigma}$, with one face in $\mathcal{P}_t^+$ and the other one in $\mathcal{P}_t^-$. At the limit of vanishing thickness of the pillbox, $\partial\mathcal{P}_t$ reduces to the $+$ and $-$ faces of $\hat{\sigma}$. Hence equation (2.87) reduces to

$$\int_{\hat{\sigma}} [\![\mathbf{h}\,\mathbf{m} + \phi(\boldsymbol{\nu} - \mathbf{v})\cdot\mathbf{m}]\!]\, da = 0.$$

The arbitrariness of $\hat{\sigma}$ and the continuity of the integrand imply that

$$[\![\mathbf{h}\,\mathbf{m} + \phi(\boldsymbol{\nu} - \mathbf{v})\cdot\mathbf{m}]\!] = 0.$$

By (1.39), we can write

$$[\![\mathbf{h}]\!]\mathbf{m} + [\![U\,\phi]\!] = 0, \tag{2.88}$$

$U = (\boldsymbol{\nu} - \mathbf{v})\cdot\mathbf{m}$ being the local speed of propagation of σ. At a shock front $[\![\mathbf{v}\cdot\mathbf{m}]\!] \neq 0$ and hence $[\![U]\!] \neq 0$. As a consequence, (2.88) holds with different values of U according as the $+$ and $-$ sides of the surface are considered. At a weak singularity, instead, we have $[\![\mathbf{v}]\!] = \mathbf{0}$ and hence $[\![U]\!] = 0$, which allows (2.88) to be written in the form

$$[\![\mathbf{h}]\!]\mathbf{m} + U[\![\phi]\!] = 0. \tag{2.89}$$

We can then state these results by

Theorem 2.3 (Kotchine) *Given the conservation law (2.86), if* $\partial_t\phi$, $\nabla\cdot(\phi\mathbf{v})$, β *and* $\mathbf{h}$ *are bounded in their domains then, at a singular surface,* ϕ *and* $\mathbf{h}$ *are required to satisfy (2.88). For weak singularities,* $[\![U]\!] = 0$ *and then (2.88) simplifies to (2.89).*

2.17.1 Imbalance Laws Across a Singular Surface

We now consider balance laws expressed by inequalities, say imbalance laws. The generic form (2.86) is then replaced with

$$\frac{d}{dt}\int_{\mathcal{P}_t}\phi\,dv \geq \int_{\mathcal{P}_t}\beta\,dv + \int_{\partial\mathcal{P}_t}\mathbf{h}\,\mathbf{n}\,da. \tag{2.90}$$

A remarkable example of (2.90) is the balance (or imbalance) of entropy where $\phi = \rho\eta$, $\beta = \rho r/\theta$, and $\mathbf{h}\,\mathbf{n} = -\mathbf{j}\cdot\mathbf{n}$ while $\mathbf{j} = \mathbf{q}/\theta + \mathbf{k}$.

By the theorem on the transport of discontinuous fields, see (1.38), we have the kinematic relation

$$\frac{d}{dt}\int_{\mathcal{P}_t\setminus\sigma}\phi\,dv = \int_{\mathcal{P}_t\setminus\sigma}\{\partial_t\phi + \nabla\cdot(\phi\mathbf{v})\}dv + \int_{\sigma}[\![\phi(\boldsymbol{\nu}-\mathbf{v})]\!]\cdot\mathbf{m}\,da.$$

In light of (2.90), we have

$$\int_{\mathcal{P}_t\setminus\sigma}\{\partial_t\phi + \nabla\cdot(\phi\mathbf{v})\}dv + \int_{\sigma}[\![\phi(\boldsymbol{\nu}-\mathbf{v})]\!]\cdot\mathbf{m}\,da \geq \int_{\mathcal{P}_t}\beta\,dv + \int_{\partial\mathcal{P}_t}\mathbf{h}\,\mathbf{n}\,da.$$

Hence, reducing $\mathcal{P}_t$ to a pillbox around σ while $\partial_t\phi$, $\nabla\cdot(\phi\mathbf{v})$, and β are bounded, we have

$$\int_{\sigma}[\![\phi(\boldsymbol{\nu}-\mathbf{v})]\!]\cdot\mathbf{m}\,da \geq -\int_{\sigma}[\![\mathbf{h}]\!]\cdot\mathbf{m}\,da + \mathbf{w}(\mathcal{P}_t),$$

where $\mathbf{w}(\mathcal{P}_t)$ approaches zero when the thickness of the pillbox approaches zero. Taking the limit, we have

$$\int_{\sigma}\{[\![\phi(\boldsymbol{\nu}-\mathbf{v}) + \mathbf{h}]\!]\cdot\mathbf{m}\,da \geq 0$$

By the arbitrariness of σ, we find the local inequality

$$[\![\phi(\boldsymbol{\nu}-\mathbf{v}) + \mathbf{h}]\!]\cdot\mathbf{m} \geq 0.$$

In the remarkable case of the second law (imbalance of entropy), it follows that

$$[\![\rho\eta(\mathbf{v}-\boldsymbol{\nu}) + \mathbf{j}\cdot\mathbf{m} \leq 0,$$

where $\mathbf{j} = \mathbf{q}/\theta + \mathbf{k}$.

2.17.2 Jump Conditions

The balance equations for mass, linear momentum, and energy correspond to

	ϕ	$\mathbf{h}$
Mass	ρ	0
Linear momentum	$\rho\mathbf{v}$	$\mathbf{T}$
Energy	$\rho(\varepsilon + \mathbf{v}^2/2)$	$-\mathbf{q} + \mathbf{T}\mathbf{v}$

Hence, we find the consequences of (2.88). The balance of mass gives

$$[\![\rho U]\!] = 0,$$

which means that ρU is constant across a discontinuity surface. By the balance of linear momentum, we have

$$[\![\mathbf{T}]\!]\mathbf{m} + \rho U[\![\mathbf{v}]\!] = \mathbf{0}.$$

At a weak singularity, where $[\![\mathbf{v}]\!]= \mathbf{0}$, it follows that

$$[\![\mathbf{T}]\!]\mathbf{m} = \mathbf{0},$$

which is usually referred to as Poisson's condition.

At a shock wave in inviscid fluids, we have

$$\rho U[\![\mathbf{v}]\!] - [\![p]\!]\mathbf{m} = \mathbf{0}.$$

By the balance of energy, it follows

$$[\![-\mathbf{q} + \mathbf{T}\mathbf{v}]\!] \cdot \mathbf{m} + \rho U[\![\varepsilon + \tfrac{1}{2}\mathbf{v}^2]\!] = 0.$$

At a weak singularity, $[\![\mathbf{v}]\!] = \mathbf{0}$, $[\![\mathbf{T}]\!]\mathbf{m} = \mathbf{0}$, and hence

$$-[\![\mathbf{q}]\!] \cdot \mathbf{m} + \rho U[\![\varepsilon]\!] = 0.$$

For definiteness, let $\mathbf{j} = \mathbf{q}/\theta$ be the entropy flux. The jump relation induced by the second law inequality is

$$\rho U[\![\eta]\!] - [\![\mathbf{q}/\theta]\!] \cdot \mathbf{m} \geq 0.$$

2.17.3 Rankine–Hugoniot Condition

There are systems of conservation laws in the form

$$\partial_t \phi + \nabla \cdot \boldsymbol{\mathcal{F}}(\phi) = f,$$

where $\phi, f : \Omega \times \mathbb{R} \to \mathbb{R}^n, \boldsymbol{\mathcal{F}} : \mathbb{R}^n \to \mathbb{R}^{n\times 3}$ in the unknown vector function ϕ. In component notation

$$\partial_t \phi_j + \partial_{x_k} \mathcal{F}_{jk}(\phi) = f_j, \qquad j = 1, ..., n.$$

They arise from the global balance laws

$$\frac{d}{dt} \int_{\mathcal{P}_t} \phi \, dv = \int_{\mathcal{P}_t} f \, dv + \int_{\partial \mathcal{P}_t} (-\boldsymbol{\mathcal{F}} + \phi \otimes \mathbf{v})\mathbf{n} \, da.$$

We then identify $-\boldsymbol{\mathcal{F}} + \phi \otimes \mathbf{v}$ with $\mathbf{h}$ of (2.86) to obtain

$$-[\![\boldsymbol{\mathcal{F}}]\!]\mathbf{m} + [\![\phi]\!]\boldsymbol{\nu} \cdot \mathbf{m} = 0.$$

Since $\boldsymbol{\nu} \cdot \mathbf{m}$ is the speed of displacement u_n of σ, we can write

$$[\![\boldsymbol{\mathcal{F}}]\!]\mathbf{m} = u_n [\![\phi]\!]. \tag{2.91}$$

The set of jump relations (2.91) is usually referred to as the *Rankine–Hugoniot jump condition.*[27]

Equation (2.91) involves the speed of displacement u_n which is the normal velocity of the singular surface relative to the chosen observer $\mathscr{F}$ and hence is dependent on $\mathscr{F}$. Equation (2.89), instead, involves the local speed of propagation U that is invariant under change of the observer. In this regard, it is worth checking the effect of a change of the observer on Eq. (2.91). Let (2.91) be the equation associated with $\mathscr{F}$ and denote by $\mathbf{x}$ the position of a point relative to the origin of $\mathscr{F}$. Let $\hat{\mathscr{F}}$ be the observer such that the position vector is $\mathbf{y} = \mathbf{x} - \mathbf{V}t$. Accordingly, $\mathbf{V}$ is the velocity of $\hat{\mathscr{F}}$ relative to $\mathscr{F}$. The space-time coordinates are related by

$$(\mathbf{x}, t) \Longleftrightarrow (\mathbf{y}, \tau), \qquad t = \tau, \quad \mathbf{x} = \mathbf{y} + \mathbf{V}\tau.$$

Hence

$$\partial_\tau \phi = \partial_t \phi + \mathbf{V} \cdot \nabla_{\mathbf{y}} \phi, \qquad \nabla_{\mathbf{y}} \cdot \boldsymbol{\mathcal{F}} = \nabla_{\mathbf{x}} \cdot \boldsymbol{\mathcal{F}}.$$

[27] See, e.g., [273], Sect. 87.

Relative to $(\mathbf{y}, \tau)$ the system of equations then becomes

$$\partial_\tau \phi + \nabla_{\mathbf{y}} \cdot (\boldsymbol{\mathcal{F}} - \phi \otimes \mathbf{V}) = 0.$$

The corresponding jump condition is

$$[\![\boldsymbol{\mathcal{F}} - \phi \otimes \mathbf{V}]\!]\mathbf{m} = \hat{u}_n [\![\phi]\!],$$

$\hat{u}_n$ being the speed of displacement relative to $\hat{\mathscr{F}}$. As a consequence, letting $V_n = \mathbf{V} \cdot \mathbf{m}$, we have

$$[\![\boldsymbol{\mathcal{F}}]\!]\mathbf{m} = (\hat{u}_n + V_n)[\![\phi]\!].$$

We then conclude that the flux function $\boldsymbol{\mathcal{F}}$ is dependent on the observer,

$$\boldsymbol{\mathcal{F}} \to \boldsymbol{\mathcal{F}} - \phi \otimes \mathbf{V},$$

and that the speed of displacement is just the speed relative to the chosen observer. This is confirmed by the result that, as expected,

$$u_n = \hat{u}_n + V_n.$$

2.17.4 Interaction Between Discontinuities of Different Order

Balance equations of the form

$$\rho \dot{\varphi} = \beta + \nabla \cdot \mathbf{h},$$

in the spatial description, have 1–1 correspondence with balance equations

$$\rho_R \dot{\varphi} = J\beta + \nabla_R \cdot \mathbf{h}_R,$$

in the material description, via the definition (2.26), $\mathbf{h}_R := J\,\mathbf{h}\mathbf{F}^{-T}$. Accordingly, in the material description, balance laws can be written in the form

$$\partial_t \phi + \nabla_R \cdot \boldsymbol{\mathcal{F}} = f \tag{2.92}$$

where $\phi, f : \Omega \times \mathbb{R} \to \mathbb{R}^n, \boldsymbol{\mathcal{F}} : \Omega \times \mathbb{R} \to \mathbb{R}^{n\times 3}$ and $\Omega \subseteq \mathrm{R}$; $\phi = \rho_R \varphi, \boldsymbol{\mathcal{F}} = -J\mathbf{h}\mathbf{F}^{-T}, f = J\beta$.

The arguments of the previous section apply again with the advantage that the body is at rest in Ω, formally $\mathbf{v} = \mathbf{0}$. Let $\mathbf{m}_R$ be the unit normal to the singular surface in Ω. Hence, the Rankine–Hugoniot condition

$$[\![\boldsymbol{\mathcal{F}}]\!]\mathbf{m}_R = U_R [\![\phi]\!]$$

holds with U_R being the same relative to the two subregions. Letting $\mathfrak{F} = \boldsymbol{\mathcal{F}}\mathbf{m}_R$, we can write $[\![\mathfrak{F}]\!] = U_R[\![\phi]\!]$ or

$$\mathfrak{F}_- - U_R\phi_- = \mathfrak{F}_+ - U_R\phi_+. \tag{2.93}$$

To fix ideas, we say that $[\![\phi]\!] \neq 0$ represents a strong discontinuity. We have in mind the case where $[\![\mathbf{v}]\!] \neq \mathbf{0}$; hence, for brevity, we say that $[\![\phi]\!] \neq 0$ characterizes a shock. Denote by a superposed dot the time derivative relative to the reference at rest with the shock and let X be the coordinate in the direction of $\mathbf{m}_R$. Since U_R is the speed of the shock, we have

$$\dot{\phi} = \partial_t\phi + U_R\partial_X\phi.$$

Time (dot) differentiation of (2.93) results in

$$\dot{\mathfrak{F}}_- - U_R\dot{\phi}_- = \dot{\mathfrak{F}}_+ - U_R\dot{\phi}_+ + \dot{U}_R[\![\phi]\!],$$

whence

$$\begin{aligned} &\partial_t\mathfrak{F}_- + U_R\partial_X\mathfrak{F}_- - U_R(\partial_t\phi_- + U_R\partial_X\phi_-) \\ &= \partial_t\mathfrak{F}_+ + U_R\partial_X\mathfrak{F}_+ - U_R(\partial_t\phi_+ + U_R\partial_X\phi_+) + \dot{U}_R[\![\phi]\!]. \end{aligned} \tag{2.94}$$

Weak discontinuities are characterized as those where ϕ is continuous, whereas $\partial_t\phi$, $\partial_X\phi$ suffer jump discontinuities. This is the case of acceleration (or acoustic) waves where $\mathbf{v}$ is continuous, while $\partial_t\mathbf{v}$, $\partial_X\mathbf{v}$ are not. To avoid ambiguities, here we denote by $[\,\cdot\,]$ the jump of weak discontinuities and hence

$$[\phi] = 0, \qquad [\partial_t\phi] \neq 0.$$

Accordingly, if U is the speed of the surface then the kinematical condition of compatibility implies

$$[\partial_t\phi] = -U\,[\partial_X\phi].$$

The occurrence of a weak discontinuity perturbs the values $\partial_t\mathfrak{F}_\pm$, $\partial_X\mathfrak{F}_\pm$ $\partial_t\phi_\pm$, $\partial_X\phi_\pm$, and $\dot{U}_R$. If, e.g. a weak discontinuity occurs in the region ahead, then the associated contributions satisfy

$$[\partial_t\mathfrak{F}_+] = -U[\partial_X\mathfrak{F}_+] = U[\partial_t\phi_+] = -U^2[\partial_X\phi_+]$$

The overall contribution of, e.g., the jump of $\partial_t\mathfrak{F}_+ + U_R\partial_X\mathfrak{F}_+ - U_R(\partial_t\phi_+ + U_R\partial_X\phi_+)$ is given by

$$[\partial_t\mathfrak{F}_+ + U_R\partial_X\mathfrak{F}_+ - U_R(\partial_t\phi_+ + U_R\partial_X\phi_+)] = -(U - U_R)^2[\partial_X\phi_+].$$

2.17.5 Eigenvector Form of Weak Discontinuities

Go back to the system (2.92) of balance equations. In suffix notation, we have

$$\partial_t \phi_i = \partial_{X_K} \mathcal{F}_{iK} + f_i.$$

Letting f be continuous, we can write

$$[\partial_t \phi_i] + [\partial_{X_K} \mathcal{F}_{iK}] = 0.$$

If $\boldsymbol{\mathcal{F}}$ is a function of ϕ, then

$$\partial_{X_K} \mathcal{F}_{iK} = \partial_{\phi_j} \mathcal{F}_{iK} \partial_{X_K} \phi_j$$

and hence

$$[\partial_{X_K} \mathcal{F}_{iK}] = \partial_{\phi_j} \mathcal{F}_{iK} [\partial_{X_K} \phi_j].$$

By the kinematical condition of compatibility and Maxwell's theorem, we obtain

$$-U[\partial_X \phi_i] = \partial_{\phi_j} \mathcal{F}_{iK} n_K [\partial_X \phi_j].$$

Hence, any admissible weak discontinuity $[\partial_X \phi]$ is an eigenvector of $\partial_\phi \boldsymbol{\mathcal{F}} \mathbf{m}_R$ with eigenvalue $-U$. Denote by $\mathbf{D}$ the eigenvectors, namely

$$\mathbf{D}_{\text{inc}}, \quad \mathbf{D}^-_{k_-}, \quad \mathbf{D}^+_{k_+}$$

are the incident eigenvector and the outgoing eigenvectors in the states behind (k_-) and ahead (k_+). Hence, Eq. (2.94) becomes

$$\textstyle\sum_{k_-} c_{k_-} (U_{k_-} - U_R)^2 \, \mathbf{D}_{k_-} + \sum_{k_+} c_{k_+} (U_{k_+} - U_R)^2 \, \mathbf{D}_{k_+} + [\dot{U}_R] [\![\phi]\!] = c_{\text{inc}} (U_{\text{inc}} - U_R)^2 \, \mathbf{D}_{\text{inc}},$$

where $[\dot{U}_R]$ denotes the jump of $\dot{U}_R$ induced by the weak discontinuities[28] Upon the interaction of the shock U_R, $[\![\phi]\!]$ with the incident weak discontinuity $c_{\text{inc}} \mathbf{D}_{\text{inc}}$ a number of emergent outgoing weak discontinuities $c^-_{k_-} \mathbf{D}_{k_-}$, $c^+_{k_+} \mathbf{D}_{k_+}$ arise. The solution to this problem exists and is unique if the eigenvectors $\mathbf{D}^-_{k_-}$, $\mathbf{D}^+_{k_+}$ and the jump $[\![\phi]\!]$ are n linearly independent vectors. As a consequence, for the existence and uniqueness, there must be $n-1$ (linearly independent) vectors $\mathbf{D}^-_{k_-}$, $\mathbf{D}^+_{k_+}$. Moreover, the $n-1$ speeds $U^{\pm}_{k_\pm}$ are required to satisfy

$$U^-_{n_-} \leq \ldots \leq U^-_1 \leq U_R \leq U^+_1 \leq \ldots \leq U^+_{n_+}.$$

These conditions are referred to as *Lax inequalities.*

---

[28] See [71], Sect. 15; applications are developed in [320, 321]. The jump of $\dot{U}_R$ at time t is denoted as $\Delta \dot{U}_R$ in [71]. If the interaction occurs at (X, t), we assume that $\phi(X, t_+) - \phi(X, t_-) = \phi(X_-, t) - \phi(X_+, t)$ [322].

2.17.6 Jump Conditions of Integrals

Let $\mathbf{w}$ be a scalar, vector or tensor function on $\mathbb{R}$. Consider the time-dependent integral

$$\mathbf{g}(t) = \int_0^\infty \mathbf{A}(s)\mathbf{w}(t-s)ds,$$

where $\mathbf{A}(s)$ is a tensor-valued differentiable function. Hence, we may regard $\mathbf{g}$ as a functional of $\mathbf{w}$, $\mathbf{g}(t) = \mathbf{G}(\mathbf{w}^t)$, $\mathbf{w}^t(s) = \mathbf{w}(t-s)$. The dependence of $\mathbf{g}$ on the position $\mathbf{x} \in \mathcal{R}_t$ or $\mathbf{x} \in \mathrm{R}$, is understood. We let $\mathbf{w}$ suffer a jump at time t and evaluate the jumps $[\![\mathbf{g}]\!](t)$, $[\![\partial_t \mathbf{g}]\!](t)$.

We replace the jump $[\![\mathbf{w}]\!]$, at time t, with a smooth bounded function $\mathbf{h}$ with compact support in $(t-\delta, t]$. We let $\mathbf{h}(\tau) \to [\![\mathbf{w}]\!]$ as $\tau \to t$ and look for the jumps $[\![\mathbf{g}]\!](t)$, $[\![\partial_t \mathbf{g}]\!](t)$ as $\delta \to 0$.

Let

$$\mathbf{w}(\tau) = \mathbf{w}_0(\tau) + \mathbf{h}(\tau).$$

Then the jump of $\mathbf{g}$ in passing from $\mathbf{w}_0$ to $\mathbf{w}$ is given by

$$\mathbf{G}(\mathbf{w}) - \mathbf{G}(\mathbf{w}_0) = \int_0^\delta \mathbf{A}(s)\mathbf{h}(\mathbf{x}, t-s)ds.$$

Furthermore

$$\partial_t[\mathbf{G}(\mathbf{w}) - \mathbf{G}(\mathbf{w}_0)] = \int_0^\delta \mathbf{A}(s)\partial_t \mathbf{h}(t-s)ds.$$

Upon the observation $\partial_t \mathbf{h}(t-s) = -\partial_s \mathbf{h}(t-s)$ and an integration by parts, we find

$$\partial_t[\mathbf{G}(\mathbf{w}) - \mathbf{G}(\mathbf{w}_0)] = -\{\left[\mathbf{A}(s)\mathbf{h}(t-s)\right]_0^\delta - \int_0^\delta \mathbf{A}'(s)\mathbf{h}(t-s)ds\}.$$

Thus, as $\delta \to 0$, we find

$$\mathbf{G}(\mathbf{w}) - \mathbf{G}(\mathbf{w}_0) = \mathbf{0}, \qquad \partial_t[\mathbf{G}(\mathbf{w}) - \mathbf{G}(\mathbf{w}_0)] = \mathbf{A}(0)[\![\mathbf{w}]\!]. \tag{2.95}$$

The argument can be applied with further integrations by parts to singularities of higher order [322].

We now apply these results to a singular surface for $\mathbf{w}$. Select a point $\mathbf{x}$ of the body in the current configuration (or in the reference configuration) and let

$$\mathbf{w}_-(\mathbf{x}, \tau) := \mathbf{w}(\mathbf{x}_-, \tau), \qquad \mathbf{w}_+(\mathbf{x}, \tau) := \mathbf{w}(\mathbf{x}_+, \tau), \qquad \tau \le t,$$

$\mathbf{x}_\mp$ being the limit point $\mathbf{x}$ as seen from behind or ahead of the discontinuity surface. Accordingly, we write

$$\mathbf{g}_-(\mathbf{x}, t) = \int_0^\infty \mathbf{A}(s)\mathbf{w}_-(\mathbf{x}, t - s)ds, \qquad \mathbf{g}_+(\mathbf{x}, t) = \int_0^\infty \mathbf{A}(s)\mathbf{w}_+(\mathbf{x}, t - s)ds.$$

Hence, in view of (2.95), we find

$$[\![\mathbf{g}]\!] = \mathbf{0}, \qquad [\![\partial_t \mathbf{g}]\!] = \mathbf{A}(0)[\![\mathbf{w}]\!]. \tag{2.96}$$

2.17.7 *Jump Conditions in the Reference Configuration*

The referential version of the conservation law (2.86) is obtained by the change of variable, $\mathbf{x} \to \mathbf{X}$, and use of Nanson formula. It follows that

$$\frac{d}{dt}\int_{\mathrm{P}} \phi_R \, dv_R = \int_{\mathrm{P}} \beta_R \, dv_R + \int_{\partial \mathrm{P}} \mathbf{h}_R \, \mathbf{n}_R \, da_R, \tag{2.97}$$

where

$$\phi_R = J\phi, \qquad \beta_R = J\,\beta, \qquad \mathbf{h}_R = J\,\mathbf{h}\,\mathbf{F}^{-T}.$$

By analogy with the spatial description, the region P is allowed to be divided in two subregions $\mathrm{P}_+, \mathrm{P}_-$ by a singular surface $\Sigma(t)$ which is singular with respect to ϕ_R. Let $\hat{\Sigma} = \Sigma(t) \cap \mathrm{P}$ and denote by $\mathbf{m}_R$ the unit normal to $\hat{\Sigma}$ directed from P_- to P_+. Also, denote by $\mathbf{U}_R$ the velocity of $\hat{\Sigma}$. Hence, we have

$$\frac{d}{dt}\int_{\mathrm{P}_-} \phi_R \, dv_R = \int_{\mathrm{P}_-} \dot{\phi}_R \, dv_R + \int_{\hat{\Sigma}} \phi_{R-} \mathbf{U}_R \cdot \mathbf{m}_R \, da_R$$

and

$$\frac{d}{dt}\int_{\mathrm{P}_+} \phi_R \, dv_R = \int_{\mathrm{P}_+} \dot{\phi}_R \, dv_R - \int_{\hat{\Sigma}} \phi_{R+} \mathbf{U}_R \cdot \mathbf{m}_R \, da_R.$$

Adding these relations, we find

$$\frac{d}{dt}\int_{\mathrm{P}} \phi_R \, dv_R = \int_{\mathrm{P}} \dot{\phi}_R \, dv_R + \int_{\hat{\Sigma}} [\![\phi_R]\!] \mathbf{U}_R \cdot \mathbf{m}_R \, da_R,$$

where $[\![\phi_R]\!] = \phi_{R-} - \phi_{R+}$. As a consequence of (2.97), we obtain

$$\int_{\mathrm{P}} (\dot{\phi}_R - b_R) dv_R + \int_{\hat{\Sigma}} [\![\phi_R]\!] \mathbf{U}_R \cdot \mathbf{m}_R \, da_R = \int_{\partial \mathrm{P}} \mathbf{h}_R \mathbf{m}_R da_R.$$

Now we take P as a pillbox across $\hat{\Sigma}$ with one face in P_+ and the other one in P_-. At the limit of vanishing thikness of the pillbox

$$\int_{\partial P} \mathbf{h}_R \mathbf{m}_R dA \rightarrow -\int_{\hat{\Sigma}} [\![\mathbf{h}_R]\!] \mathbf{m}_R da_R$$

while, if $\dot{\phi}_R - b_R$ is bounded, the volume integral approaches zero. Hence we obtain

$$\int_{\hat{\Sigma}} \{[\![\phi_R]\!] \mathbf{U}_R \cdot \mathbf{m}_R + [\![\mathbf{h}_R]\!] \mathbf{m}_R\} da_R = 0.$$

The arbitrariness of $\hat{\Sigma}$ allows the following statement, that is the dual of Kotchine's theorem.

Theorem 2.4 *Given the conservation law (2.97), if $\dot{\phi}_R$ and s_R are bounded then, at a singular surface, ϕ_R and $\mathbf{h}_R$ are required to satisfy*

$$[\![\phi_R]\!] \mathbf{U}_R \cdot \mathbf{m}_R + [\![\mathbf{h}_R]\!] \mathbf{m}_R = 0. \tag{2.98}$$

We now derive some consequences of (2.98). In the balance of mass $\phi = \rho$ and $\mathbf{h} = \mathbf{0}$. Hence, (2.98) becomes

$$[\![\rho_R]\!] \mathbf{U}_R \cdot \mathbf{m}_R = 0.$$

Across moving surfaces, $\mathbf{U}_R \cdot \mathbf{m}_R \neq 0$, the reference mass density ρ_R is continuous, $[\![\rho_R]\!] = 0$. Instead, $[\![\rho_R]\!] \neq 0$ is allowed if the surface is stationary in the reference configuration, $\mathbf{U}_R \cdot \mathbf{m}_R = 0$.

In the balance of linear momentum $\phi = \rho \mathbf{v}$ and $\mathbf{h} = \mathbf{T}$ and so

$$\phi_R = \rho_R \mathbf{v}, \qquad \mathbf{h}_R = \mathbf{T}_R.$$

Equation (2.98) becomes

$$[\![\rho_R \mathbf{v}]\!] \mathbf{U}_R \cdot \mathbf{m}_R + [\![\mathbf{T}_R]\!] \mathbf{m}_R = \mathbf{0}.$$

Since $[\![\rho_R]\!] \mathbf{U}_R \cdot \mathbf{m}_R = 0$, it follows that

$$\rho_R [\![\mathbf{v}]\!] \mathbf{U}_R \cdot \mathbf{m}_R + [\![\mathbf{T}_R]\!] \mathbf{m}_R = \mathbf{0}.$$

At a weak singularity, $[\![\mathbf{v}]\!] = \mathbf{0}$, we have

$$[\![\mathbf{T}_R]\!] \mathbf{m}_R = \mathbf{0},$$

that is the dual of Poisson's condition.

Look at the balance of energy, where $\phi = \rho(\varepsilon + \frac{1}{2}\mathbf{v}^2)$ and $\mathbf{h} = -\mathbf{q} + \mathbf{T}\mathbf{v}$. Hence $\phi_R = \rho_R(\varepsilon + \frac{1}{2}\mathbf{v}^2)$ and $\mathbf{h}_R = -\mathbf{q}_R + \mathbf{v}\mathbf{T}_R$. Substitution in (2.98) gives

$$\rho_R [\![\varepsilon + \tfrac{1}{2}\mathbf{v}^2]\!] \mathbf{U}_R \cdot \mathbf{m}_R + [\![\mathbf{v}\mathbf{T}_R - \mathbf{q}_R]\!] \cdot \mathbf{m}_R = 0.$$

At a weak discontinuity $[\![\mathbf{v}\mathbf{T}_R]\!] \cdot \mathbf{m}_R = 0$ and the jump condition becomes

$$\rho_R [\![\varepsilon]\!] \mathbf{U}_R \cdot \mathbf{m}_R - [\![\mathbf{q}_R]\!] \cdot \mathbf{m}_R = 0.$$

2.18 Balance Laws for Discontinuous Electromagnetic Fields

The balance laws of electromagnetic fields are reviewed by letting a discontinuity surface Υ occurring in a region Ω and a discontinuity line γ occurring in a surface S. Both Ω and S convect with the underlying body. We show how the starting global (integral) law results in a local (differential) equation along with a jump condition.
Conservation of charge. The charge within a region Ω convecting with the body is constant

$$\frac{d}{dt} \int_{\Omega \setminus \Upsilon} q \, dv = 0.$$

By (1.38), we have

$$0 = \int_{\Omega \setminus \Upsilon} [\partial_t q + \nabla \cdot (q \, \mathbf{v})] dv + \int_{\Upsilon} [\![q(\boldsymbol{\nu} - \mathbf{v})]\!] \cdot \mathbf{n}_\Upsilon \, da,$$

where $\boldsymbol{\nu}$ is the velocity of Υ, $\mathbf{v}$ is the velocity of material points. By the arbitrariness of Ω and Υ, it follows that

$$\partial_t q + \nabla \cdot (q\mathbf{v}) = 0, \quad \text{in } \Omega \setminus \Upsilon \qquad [\![q(\boldsymbol{\nu} - \mathbf{v})]\!] \cdot \mathbf{n}_\Upsilon = 0, \quad \text{in } \Upsilon.$$

Gauss' law. The flux of $\mathbf{D}$ through any closed surface S equals the charge inside S.
Let $S = \partial\Omega$ and let a surface Υ occur with a surface charge density q_s. The charge Q_Ω inside Ω is given by

$$Q_\Omega = \int_{\Omega \setminus \Upsilon} q \, dv + \int_{\Upsilon} q_s \, da.$$

The divergence theorem in the generalized form (1.37) gives

$$\int_{\partial\Omega} \mathbf{D} \cdot \mathbf{n} \, da = \int_{\Omega \setminus \Upsilon} \nabla \cdot \mathbf{D} dv - \int_{\Upsilon} [\![\mathbf{D}]\!] \cdot \mathbf{n}_\Upsilon da.$$

Gauss' law then means that

$$\int_{\Omega \setminus \Upsilon} \nabla \cdot \mathbf{D} \, dv - \int_{\Upsilon} [\![\mathbf{D}]\!] \cdot \mathbf{n}_\Upsilon \, da = \int_{\Omega \setminus \Upsilon} q \, dv + \int_{\Upsilon} q_s \, da.$$

Hence we have

$$\nabla \cdot \mathbf{D} = q \quad \text{in } \Omega \setminus \Upsilon, \qquad [\![\mathbf{D}]\!] \cdot \mathbf{n}_\Upsilon = -q_s \quad \text{in } \Upsilon.$$

Conservation of magnetic flux. The flux of **B** through any closed surface is zero.

The same procedure, as with Gauss' law, gives

$$\nabla \cdot \mathbf{B} = 0 \quad \text{in } \Omega \setminus \Upsilon, \qquad [\![\mathbf{B}]\!] \cdot \mathbf{n}_\Upsilon = 0 \quad \text{in } \Upsilon.$$

Faraday's law. If the surface S and the contour $C = \partial S$ are convecting with the body then the circulation of $\mathbf{E} + \mathbf{v} \times \mathbf{B}$ around C is equal to minus the time derivative of the flux of **B** through S.

Let γ be a discontinuity line, in S, moving with velocity $\boldsymbol{\nu}$. Faraday's law reads

$$\int_{\partial S} (\mathbf{E} + \mathbf{v} \times \mathbf{B}) \cdot \mathbf{t}\, dl = -\frac{d}{dt} \int_{S \setminus \gamma} \mathbf{B} \cdot \mathbf{n}\, da.$$

By Stokes' theorem in the generalized form (1.52), we have

$$\int_{\partial S} (\mathbf{E} + \mathbf{v} \times \mathbf{B}) \cdot \mathbf{t}\, dl = \int_{S \setminus \gamma} [\nabla \times \mathbf{E} + \nabla \times (\mathbf{v} \times \mathbf{B})] \cdot \mathbf{n}\, da - \int_\gamma [\![\mathbf{E} + \mathbf{v} \times \mathbf{B}]\!] \cdot \mathbf{t}_\gamma dl$$

while (1.53) and the condition $\nabla \cdot \mathbf{B} = 0$ in $S \setminus \gamma$ yield

$$\frac{d}{dt} \int_{S \setminus \gamma} \mathbf{B} \cdot \mathbf{n}\, da = \int_{S \setminus \gamma} [\partial_t \mathbf{B} + \nabla \times (\mathbf{B} \times \mathbf{v})] \cdot \mathbf{n}\, da + \int_\gamma [\![\mathbf{B} \times (\boldsymbol{\nu} - \mathbf{v})]\!] \cdot \mathbf{t}_\gamma\, dl.$$

Hence, Faraday's law results in

$$\int_{S \setminus \gamma} (\nabla \times \mathbf{E} + \partial_t \mathbf{B}) \cdot \mathbf{n}\, da - \int_\gamma [\![\mathbf{E} + \boldsymbol{\nu} \times \mathbf{B}]\!] \cdot \mathbf{t}_\gamma dl = 0,$$

whence

$$\nabla \times \mathbf{E} + \partial_t \mathbf{B} = 0 \quad \text{in } S \setminus \gamma, \qquad [\![\mathbf{E} + \boldsymbol{\nu} \times \mathbf{B}]\!] \cdot \mathbf{t}_\gamma = 0 \quad \text{in } \gamma.$$

Ampère–Maxwell law. The circulation of $\mathbf{H} - \mathbf{v} \times \mathbf{D}$ around the contour C of a convecting surface S is equal to the electric current through S plus the time derivative of the flux of **D** through S.

Let **J** be the electric current density vector. Because S is convecting, the net electric current through S is the flux of $\mathbf{J} - q\,\mathbf{v}$. Also, for generality, we allow a current density **i**, per unit length, to flow along a discontinuity line[29] $\gamma \subset S$. Hence, we write the Ampère–Maxwell law in the form

$$\int_{\partial S} (\mathbf{H} - \mathbf{v} \times \mathbf{D}) \cdot \mathbf{t}\, dl = \int_{S \setminus \gamma} (\mathbf{J} - q\,\mathbf{v}) \cdot \mathbf{n}\, da + \int_\gamma \mathbf{i} \cdot \mathbf{n}\, dl + \frac{d}{dt} \int_{S \setminus \gamma} \mathbf{D} \cdot \mathbf{n}\, da.$$

[29] We may view γ as the intersection of S with a surface Υ through which a surface current density flows.

By the generalized Stokes theorem (1.52) and the time derivative of a flux (1.53), we can write

$$\begin{aligned}\int_{\partial S} [\nabla \times (\mathbf{H} - \mathbf{v} \times \mathbf{D})] \cdot \mathbf{n}\, da - \int_{\gamma} [\![\mathbf{H} - \mathbf{v} \times \mathbf{D}]\!] \cdot \mathbf{t}_\gamma dl &= \int_{S\setminus\gamma} (\mathbf{J} - q\,\mathbf{v}) \cdot \mathbf{n}\, da + \int_{\gamma} \mathbf{i} \cdot \mathbf{n}\, dl \\ &+ \int_{S\setminus\gamma} [\partial_t \mathbf{D} + \nabla \times (\mathbf{D} \times \mathbf{v}) + \mathbf{v} \nabla \cdot \mathbf{D}] \cdot \mathbf{n}\, da + \int_{\gamma} [\![\mathbf{D} \times (\boldsymbol{\nu} - \mathbf{v})]\!] \cdot \mathbf{t}_\gamma\, dl.\end{aligned}$$

Upon use of the local Gauss law $\nabla \cdot \mathbf{D} = q$, it follows that

$$\nabla \times \mathbf{H} - \mathbf{J} - \partial_t \mathbf{D} = \mathbf{0} \quad \text{in } S \setminus \gamma, \qquad [\![\mathbf{H} + \mathbf{D} \times \boldsymbol{\nu}]\!] \cdot \mathbf{t}_\gamma + \mathbf{i} \cdot \mathbf{n} = 0 \quad \text{in } \gamma.$$

The whole set of Maxell's equations is then obtained. We now examine the content of the jump conditions across surfaces or lines.

2.18.1 Boundary Conditions

The jump conditions, derived above, at a discontinuity surface or line provide the appropriate conditions at the boundary surface or line of a body. To fix ideas, consider a surface Υ, or a line γ, that is the common boundary between two regions $\mathcal{R}_+$, $\mathcal{R}_-$. Denote by the subscripts $-$ and $+$ the limit values at Υ, or γ, of any quantity evaluated from inside $\mathcal{R}_-$ and $\mathcal{R}_+$. Let $\mathbf{n}$ be the normal to Υ, or γ, directed from $\mathcal{R}_-$ to $\mathcal{R}_+$.

Gauss' law implies that

$$(\mathbf{D}_+ - \mathbf{D}_-) \cdot \mathbf{n} = q_s.$$

The normal component of $\mathbf{D}$ changes discontinuously across the surface Υ by an amount equal to the surface charge density. If $q_s = 0$ then the normal component of $\mathbf{D}$ is continuous across the surface.

The conservation of magnetic flux implies that

$$\mathbf{B}_+ \cdot \mathbf{n} = \mathbf{B}_- \cdot \mathbf{n}.$$

This signifies that the normal component of $\mathbf{B}$ is continuous across a surface.

Faraday's law implies that, across a line γ

$$(\mathbf{E} + \boldsymbol{\nu} \times \mathbf{B})_+ \cdot \mathbf{t} = (\mathbf{E} + \boldsymbol{\nu} \times \mathbf{B})_- \cdot \mathbf{t}.$$

If γ is fixed, $\boldsymbol{\nu} = 0$, the tangential component of $\mathbf{E} \cdot \mathbf{t}$ is continuous across γ. If $\boldsymbol{\nu} = \mathbf{v}$, that is if γ is convected, then $(\mathbf{E} + \mathbf{v} \times \mathbf{B}) \cdot \mathbf{t}$ is continuous.

The Ampère–Maxwell law implies that

$$(\mathbf{H} + \mathbf{D} \times \boldsymbol{\nu})_+ \cdot \mathbf{t} - (\mathbf{H} + \mathbf{D} \times \boldsymbol{\nu})_- \cdot \mathbf{t} = \mathbf{i} \cdot \mathbf{n}.$$

If, in particular, the line γ is fixed, $\boldsymbol{\nu} = 0$, then

$$\mathbf{H}_+ \cdot \mathbf{t} - \mathbf{H}_- \cdot \mathbf{t} = \mathbf{i} \cdot \mathbf{n}.$$

We now look at Υ as the interface between two different media. By

$$(\mathbf{D}_+ - \mathbf{D}_-) \cdot \mathbf{n} = q_s, \qquad \mathbf{B}_+ \cdot \mathbf{n} = \mathbf{B}_- \cdot \mathbf{n}$$

we say that the normal component of $\mathbf{B}$ is continuous across the interface, whereas the jump of $\mathbf{D} \cdot \mathbf{n}$ equals the surface charge density q_s. The line γ is any line along the interface and hence $\mathbf{t}$ is any unit vector in the (tangent plane to the) interface. If, further, $\mathbf{J}_s = \mathbf{0}$, then the conditions

$$\mathbf{E}_+ \cdot \mathbf{t} = \mathbf{E}_- \cdot \mathbf{t}, \qquad \mathbf{H}_+ \cdot \mathbf{t} = \mathbf{H}_- \cdot \mathbf{t}$$

hold. Owing to the arbitrariness of $\mathbf{t}$, it follows that the tangential components of $\mathbf{E}$ and $\mathbf{H}$ are continuous across the interface between the two media.

The effect of the surface current $\mathbf{i}$ can be better understood by letting $\mathbf{n}_\Upsilon$ be normal to the interface and

$$\mathbf{t} = \mathbf{n} \times \mathbf{n}_\Upsilon,$$

$\mathbf{n}_\Upsilon$ being directed from $\mathcal{R}_-$ to $\mathcal{R}_+$. Hence, the jump condition for $\mathbf{H}$ becomes

$$(\mathbf{H}_+ - \mathbf{H}_-) \cdot \mathbf{t} = \mathbf{i} \cdot \mathbf{n}.$$

Hence

$$\mathbf{i} \cdot \mathbf{n} = (\mathbf{H}_+ - \mathbf{H}_-) \cdot \mathbf{n} \times \mathbf{n}_\Upsilon = \mathbf{n}_\Upsilon \times (\mathbf{H}_+ - \mathbf{H}_-) \cdot \mathbf{n}.$$

Since $\mathbf{t}$ and $\mathbf{n}$ are arbitrary, in the tangent plane to the interface Υ, it follows that

$$\mathbf{n}_\Upsilon \times (\mathbf{H}_+ - \mathbf{H}_-) = \mathbf{i}.$$

Moreover, if $\mathbf{E}$ is the continuous tangential electric field at Υ and $\mathbf{i} = \sigma_\Upsilon \mathbf{E}$, then the boundary condition becomes

$$\mathbf{n}_\Upsilon \times (\mathbf{H}_+ - \mathbf{H}_-) = \sigma_\Upsilon \mathbf{E}.$$

If the interface is fixed ($\boldsymbol{\nu} = \mathbf{0}$) and no surface charges and currents occur, then the boundary conditions become

$$[\![\mathbf{D}]\!] \cdot \mathbf{n} = 0, \qquad [\![\mathbf{B}]\!] \cdot \mathbf{n} = 0,$$

$$[\![\mathbf{E}]\!] \cdot \mathbf{t} = 0, \qquad [\![\mathbf{H}]\!] \cdot \mathbf{t} = 0,$$

$\mathbf{n}$ being orthogonal to the interface and $\mathbf{t}$ being any direction parallel to the interface.

Part II
Constitutive Models of Simple Materials

Chapter 3
Generalities on Constitutive Models

The balance equations of a body form an under-determined differential system, insufficient to yield specific results unless further relations are supplied. The balance equations for a continuum (free from internal structures) are the continuity equation, the equation of motion, and the balance of energy, namely five equations, for the fourteen unknowns (mass density, velocity, symmetric stress, energy density, heat flux) in the pertinent space-time domain. The insufficiency of the balance equations to solve a dynamic problem is conceptual and is consistent with the fact that different material properties are expected to provide different responses. Mathematically the material properties of the body are expressed by constitutive equations, or constitutive assumptions, which provide a model of the material behaviour. Constitutive equations are not a mere mathematical model. They have to be physically admissible, and this is ascertained through the compatibility with the objectivity principle and the second law of thermodynamics. This chapter investigates both the contents of the objectivity and the second law of thermodynamics and also their application to the characterization of constitutive equations. The second law plays the role of selecting physically admissible processes and hence admissible constitutive properties. As a new approach, the entropy production is taken to be an unknown function to be determined as any other constitutive function.

3.1 Constitutive Equations

A constitutive equation defines an ideal material which may represent a real material, within a reasonable approximation, in a region of validity of the pertinent variables (constitutive variables). In addition to insights from experimental data, constitutive equations are required to be admissible on physical grounds. Admissibility means consistency with objectivity and thermodynamic requirements. In addition,

A. Morro and C. Giorgi, *Mathematical Modelling of Continuum Physics*,
Modeling and Simulation in Science, Engineering and Technology,
https://doi.org/10.1007/978-3-031-20814-0_3

material invariance of a body requires that the response functions are form-invariant with respect to the group of transformations of the material frame of reference. Objectivity, or frame-indifference, means that *the constitutive functions (or functionals) must be form-invariant under arbitrary rigid motions of the spatial frame of reference.* Compatibility with thermodynamics means that *the functions entering the constitutive equations must satisfy the second law of thermodynamics,* subject to the constraints placed by the balance equations. Moreover a rule of equipresence is adopted in that at the outset all response functions (or functionals) are taken to depend on the same list of variables. Differences in the dependences are to be derived as a consequence of thermodynamics or motivated on the basis of the structure of the material.

For materials which have memory from the past it is assumed that the constitutive variables at distant past do not affect appreciably the values of the constitutive functionals at present. This natural assumption is made formal by specifying the mathematical properties of the functionals characterizing the memory of the material.

Simple materials were originally characterized by saying that the response (stress) functions at a point $\mathbf{X}$, at the time t, is determined by the motion in a neighbourhood of $\mathbf{X}$, up to the time t [426]. For generality and definiteness, we distinguish between materials with long memory and those with short memory. For long memory, simple materials are those with response functions, at $\mathbf{X}$ and at time t, that depend on the values of $\mathbf{F}, \theta, \nabla\theta$ at $\mathbf{X}$ up to time t. Instead, for short memory simple materials are those with response functions, at $\mathbf{X}$ and at time t, that depend on the values of $\mathbf{F}, \theta, \nabla\theta$ and the derivatives $\dot{\mathbf{F}}, \dot{\theta}, \nabla\dot{\theta}$ at $\mathbf{X}$ at time t. Roughly, in materials with short memory, the information of $\mathbf{F}, \theta, \nabla\theta$ up to time t is reduced to the values of $\mathbf{F}, \theta, \nabla\theta$ and of the time derivatives $\dot{\mathbf{F}}, \dot{\theta}, \nabla\dot{\theta}$, at the time t_-.

Objectivity and compatibility with thermodynamics are made precise and commented upon in the next sections.

3.2 Objectivity

All the equations that are used in continuum physics must satisfy some properties of invariance. This means that they should describe a physical phenomenon in a unique way, yielding unique results, which are not dependent upon the frame of reference. The material properties of a body should not depend on the observer, no matter how he moves. However, the material properties may be affected by the motion of the body and hence it is assumed that the material properties are independent of the frame of reference within the set of Galilean frames. This independence of the frame of reference is made precise by the following

Principle of objectivity. *Constitutive equations must be invariant under changes of frame of reference; frames of reference $\mathscr{F}$, $\mathscr{F}^*$ are related by a Euclidean transformation*

$$\mathbf{x}^* = \chi^*(\mathbf{X}, t^*) = \mathbf{c}(t) + \mathbf{Q}(t)\chi(\mathbf{X}, t), \qquad t^* = t - a, \tag{3.1}$$

where $\mathbf{Q}$ is a rotation, $\det \mathbf{Q} = 1$.

This assumption places restrictions on the possible constitutive equations. To establish objectivity requirements we need to know, at least by assumption, the properties of the involved scalars, vectors, and tensors under change of frame. We say that scalars α, vectors $\mathbf{u}$, and tensors $\mathbf{A}$ are objective if the transformation property is

$$\alpha^* = \alpha, \qquad \mathbf{u}^* = \mathbf{Q}\mathbf{u}, \qquad \mathbf{A}^* = \mathbf{Q}\mathbf{A}\mathbf{Q}^T.$$

Indeed, objective scalars are invariants.

The tensors $\mathbf{Q}$ entering the transformation law (3.1) are subject to

$$\mathbf{Q}\mathbf{Q}^T = \mathbf{Q}^T\mathbf{Q} = \mathbf{1}, \qquad \det \mathbf{Q} = 1$$

and are often called *proper rotations.* They constitute a nonabelian group denoted SO(3) (special orthogonal group). The orthogonal tensors $\mathbf{Q}$ with $\det \mathbf{Q} = \pm 1$ constitute a group denoted O(3); they comprise also reflections (improper rotations). A vector $\mathbf{u}$ such that

$$\mathbf{u}^* = (\det \mathbf{Q})\mathbf{Q}\mathbf{u}, \qquad \mathbf{Q} \in \mathrm{O}(3)$$

is called objective *axial* vector (or pseudovector). Instead, a vector $\mathbf{r}$ such that

$$\mathbf{r}^* = \mathbf{Q}\mathbf{r}, \qquad \mathbf{Q} \in \mathrm{O}(3)$$

is called objective vector. Physical arguments indicate that the magnetic vectors $\mathbf{B}, \mathbf{H}, \mathbf{M}$ are axial vectors. An immediate example of axial vector is the vector product. If $\mathbf{u}$, $\mathbf{v}$ are two objective vectors and $\mathbf{w} = \mathbf{u} \times \mathbf{v}$, $\mathbf{Q} = -\mathbf{1}$ then

$$\mathbf{w}^* = (-\mathbf{u}) \times (-\mathbf{v}) = \mathbf{w}$$

consistent with the transformation law $\mathbf{w}^* = (-1)(-\mathbf{1})\mathbf{w}$.

Restrictions placed by objectivity will be investigated in connection with the constitutive models of materials. Here, though, we establish some preliminary results. To fix ideas we look at the possible dependence of the stress tensor $\mathbf{T}$ and the heat flux $\mathbf{q}$ on deformation, kinematic variables, and other pertinent variables. We assume that $\mathbf{T}$ and $\mathbf{q}$ are objective and hence in the change of frame $\mathscr{F} \to \mathscr{F}^*$ they satisfy

$$\mathbf{T}^* = \mathbf{Q}\mathbf{T}\mathbf{Q}^T, \qquad \mathbf{q}^* = \mathbf{Q}\mathbf{q}.$$

As to the Cauchy stress we might start with the assumption that the traction $\mathbf{t}$ and the unit vector $\mathbf{n}$ transform (as objective vectors) according to

$$\mathbf{t}^* = \mathbf{Q}\,\mathbf{t}, \qquad \mathbf{n}^* = \mathbf{Q}\,\mathbf{n}.$$

Hence

$$\mathbf{Q}\mathbf{t} = \mathbf{t}^* = \mathbf{T}^*\mathbf{n}^* = \mathbf{T}^*\mathbf{Q}\mathbf{n}$$

implies $\mathbf{t} = (\mathbf{Q}^T\mathbf{T}^*\mathbf{Q})\mathbf{n}$ whence

$$\mathbf{T}^* = \mathbf{Q}\mathbf{T}\mathbf{Q}^T;$$

the objectivity of the traction $\mathbf{t}$ implies the objectivity of the Cauchy stress $\mathbf{T}$. Further,

$$\int_{\mathcal{P}_t^*} \mathbf{T}^*\mathbf{n}^*\, da = \int_{\mathcal{P}_t} (\mathbf{Q}\mathbf{T}\mathbf{Q}^T)\mathbf{Q}\mathbf{n}\, da = \mathbf{Q}\int_{\mathcal{P}_t} \mathbf{T}\mathbf{n}\, da.$$

Hence it follows the objectivity of the surface force [216],

$$\int_{\mathcal{P}_t^*} \mathbf{T}^*\mathbf{n}^*\, da = \mathbf{Q}\int_{\mathcal{P}_t} \mathbf{T}\mathbf{n}\, da.$$

Likewise $\mathbf{b}^* = \mathbf{Q}\mathbf{b}$ implies

$$\int_{\mathcal{P}_t^*} \rho\mathbf{b}^*\, dv = \mathbf{Q}\int_{\mathcal{P}_t} \rho\mathbf{b}\, dv.$$

The whole force on the body is then objective.

The objectivity of the heat flux $\mathbf{q}$, $\mathbf{q}^* = \mathbf{Q}\mathbf{q}$, implies the invariance of the heat flow through the surface,

$$\int_{\mathcal{P}_t^*} \mathbf{q}^* \cdot \mathbf{n}^*\, da = \int_{\mathcal{P}_t} \mathbf{q} \cdot \mathbf{n}\, da.$$

By the polar decomposition theorem,

$$\mathbf{F} = \mathbf{R}\mathbf{U},$$

where $\mathbf{R}$ is a rotation[1] and $\mathbf{U}$ is a positive definite, $\mathbf{U}^2 = \mathbf{F}^T\mathbf{F}$. Since $\mathbf{F}^* = \mathbf{Q}\mathbf{F}$ it follows that $\mathbf{U}$ is invariant and then

$$J = \det\mathbf{F} = \det\mathbf{R}\det\mathbf{U} = \det\mathbf{U}$$

is invariant too.

Since the Cauchy stress $\mathbf{T}$ is objective then the first and second Piola stresses obey the transformation laws

$$\mathbf{T}_R^* = \mathbf{Q}\mathbf{T}_R, \qquad \mathbf{T}_{RR}^* = \mathbf{T}_{RR}.$$

These properties follow by direct evaluation of $\mathbf{T}_R = J\mathbf{T}\mathbf{F}^{-T}$ and $\mathbf{T}_{RR} = J\mathbf{F}^{-1}\mathbf{T}\mathbf{F}^{-T}$,

$$\mathbf{T}_R^* = J(\mathbf{Q}\mathbf{T}\mathbf{Q}^T)\mathbf{Q}^{-T}\mathbf{F}^{-T} = \mathbf{Q}J\mathbf{T}\mathbf{F}^{-T} = \mathbf{Q}\mathbf{T}_R,$$

$$\mathbf{T}_{RR}^* = J(\mathbf{F}^{-1}\mathbf{Q}^{-1})(\mathbf{Q}\mathbf{T}\mathbf{Q}^T)\mathbf{Q}^{-T}\mathbf{F}^{-T} = J\mathbf{F}^{-1}\mathbf{T}\mathbf{F}^{-T} = \mathbf{T}_{RR}.$$

---

[1] $\det\mathbf{R} = 1$ if $\det\mathbf{F} > 0$.

Instead, the objectivity of the heat flux $\mathbf{q}$, $\mathbf{q}^* = \mathbf{Q}\mathbf{q}$, implies the invariance of the reference heat flux $\mathbf{q}_R$,

$$\mathbf{q}_R^* = (J\mathbf{q}\mathbf{F}^{-T})^* = J(\mathbf{Q}\mathbf{q})\mathbf{Q}\mathbf{F}^{-T} = J\mathbf{q}\mathbf{F}^{-T} = \mathbf{q}_R.$$

A constitutive tensor function $\mathcal{G}$ of $\mathbf{K}$ is objective if

$$\mathcal{G}(\mathbf{K}^*) = \mathbf{Q}\,\mathcal{G}(\mathbf{K})\,\mathbf{Q}^T.$$

Depending on the transformation law for $\mathbf{K}$ we can find the restrictions placed on $\mathcal{G}$ as a consequence of the arbitrariness of $\mathbf{Q}$, subject to $\det \mathbf{Q} = 1$. This is now exemplified via the main constitutive properties.

3.2.1 Dependence on the Velocity Gradient

Look at the constitutive equation

$$\mathbf{T} = \hat{\mathbf{T}}(\mathbf{L}).$$

Objectivity requires that

$$\mathbf{T}^* = \mathbf{Q}\,\mathbf{T}\,\mathbf{Q}^T = \mathbf{Q}\hat{\mathbf{T}}(\mathbf{L})\mathbf{Q}^T = \hat{\mathbf{T}}(\mathbf{L}^*) = \hat{\mathbf{T}}(\mathbf{Q}\mathbf{L}\mathbf{Q}^T + \boldsymbol{\Omega})$$

where $\boldsymbol{\Omega} = \dot{\mathbf{Q}}\mathbf{Q}^T$. Since this requirement has to hold for any time dependent rotation $\mathbf{Q}$ we choose the function $\mathbf{Q}$ such that, at the pertinent time t,

$$\mathbf{Q}(t) = \mathbf{1}, \qquad \dot{\mathbf{Q}}(t) = -\mathbf{W}(t).$$

Hence, since $\mathbf{L} - \mathbf{W} = \mathbf{D}$ we have

$$\hat{\mathbf{T}}(\mathbf{L}) = \hat{\mathbf{T}}(\mathbf{D}).$$

Consequently, by objectivity the Cauchy stress $\mathbf{T}$ is allowed to depend on the velocity gradient $\mathbf{L}$ only through the stretching $\mathbf{D}$, that is $\mathbf{T}$ cannot depend on the spin $\mathbf{W}$.

3.2.2 Dependence on the Deformation Gradient

We now examine the possible dependence of the stress on the deformation gradient. Consider the first Piola stress $\mathbf{T}_R = J\mathbf{T}\mathbf{F}^{-T}$ and let

$$\mathbf{T}_R = \hat{\mathbf{T}}_R(\mathbf{F}).$$

Since $\mathbf{T}_R^* = \mathbf{Q}\mathbf{T}_R$ then objectivity requires that

$$\hat{\mathbf{T}}_R(\mathbf{F}^*) = \mathbf{Q}\hat{\mathbf{T}}_R(\mathbf{F}),$$

the constitutive function $\hat{\mathbf{T}}_R$ being the same in $\mathscr{F}$ and $\mathscr{F}^*$. Hence

$$\hat{\mathbf{T}}_R(\mathbf{F}) = \mathbf{Q}^T\hat{\mathbf{T}}_R(\mathbf{F}^*) = \mathbf{Q}^T\hat{\mathbf{T}}_R(\mathbf{QRU}).$$

Letting $\mathbf{Q}^T = \mathbf{R} = \mathbf{F}\mathbf{U}^{-1}$ we have

$$\hat{\mathbf{T}}_R(\mathbf{F}) = \mathbf{F}\mathbf{U}^{-1}\hat{\mathbf{T}}_R(\mathbf{U}).$$

This means that, apart from the multiplicative (linear) term $\mathbf{F}$, the dependence of $\mathbf{T}_R$ on $\mathbf{F}$ is through $\mathbf{U}$. In addition we find that

$$\mathbf{T} = (\det \mathbf{U})^{-1}\mathbf{F}\mathbf{U}^{-1}\hat{\mathbf{T}}_R(\mathbf{U})\mathbf{F}^T.$$

Consequently, by the principle of objectivity the Cauchy stress $\mathbf{T}$ takes the form

$$\mathbf{T} = \mathbf{F}\hat{\mathbf{T}}(\mathbf{U})\mathbf{F}^T,$$

where $\hat{\mathbf{T}}(\mathbf{U})$ is a shorthand for $(\det \mathbf{U})^{-1}\mathbf{U}^{-1}\hat{\mathbf{T}}_R(\mathbf{U})$.

The conclusion holds unchanged if $\mathbf{T}$ depends on the deformation gradient $\mathbf{F}$ and on the temperature θ, θ being invariant.

Since the finite strain tensor

$$\mathbf{E} = \tfrac{1}{2}(\mathbf{F}^T\mathbf{F} - \mathbf{1})$$

is invariant then $\mathbf{T}_R$ cannot be merely a function of $\mathbf{E}$. Instead,

$$\mathbf{T}_{RR} = \hat{\mathbf{T}}_{RR}(\mathbf{E})$$

satisfies the invariance property of $\mathbf{T}_{RR}$ whereas $\mathbf{T}_R = \mathbf{F}\mathbf{T}_{RR}$ takes the form

$$\mathbf{T}_R = \mathbf{F}\,\hat{\mathbf{T}}_{RR}(\mathbf{E}).$$

3.2.3 *Thermoelastic Variables*

In thermoelasticity the independent variables may be $\mathbf{F}, \theta$, and $\nabla_R\,\theta$. Let ψ be a scalar, $\mathbf{q}$ be a vector, and $\mathbf{T}$ be a tensor, all of them being objective and hence ψ being invariant. Since $\mathbf{F}^* = \mathbf{QF}$ whereas θ and $\nabla_R\theta$ are invariant then

$$\psi(\mathbf{F}, \theta, \nabla_R\theta) = \psi^*(\mathbf{F}, \theta, \nabla_R\theta) = \psi(\mathbf{F}^*, \theta^*, \nabla_R\theta^*) = \psi(\mathbf{QF}, \theta, \nabla_R\theta).$$

In light of the polar decomposition, $\mathbf{F} = \mathbf{RU}$, it follows that

$$\psi(\mathbf{F}, \theta, \nabla_R\theta) = \psi(\mathbf{QRU}, \theta, \nabla_R\theta).$$

Choosing $\mathbf{Q} = \mathbf{R}^T$ we find

$$\psi(\mathbf{F}, \theta, \nabla_R\theta) = \psi(\mathbf{U}, \theta, \nabla_R\theta).$$

Likewise for $\mathbf{q}$ and $\mathbf{T}$ we have

$$\mathbf{Q}\,\mathbf{q}(\mathbf{F}, \theta, \nabla_R\theta) = \mathbf{q}(\mathbf{Q}\,\mathbf{F}, \theta, \nabla_R\theta), \qquad \mathbf{Q}\,\mathbf{T}(\mathbf{F}, \theta, \nabla_R\theta)\mathbf{Q}^T = \mathbf{T}(\mathbf{Q}\,\mathbf{F}, \theta, \nabla_R\theta),$$

whence

$$\mathbf{q}(\mathbf{F}, \theta, \nabla_R\theta) = \mathbf{R}\,\mathbf{q}(\mathbf{U}, \theta, \nabla_R\theta), \qquad \mathbf{T}(\mathbf{F}, \theta, \nabla_R\theta) = \mathbf{R}\,\mathbf{T}(\mathbf{U}, \theta, \nabla_R\theta)\mathbf{R}^T.$$

In the customary Fourier's law $\mathbf{q}$ is a (linear) function of the temperature gradient in the current configuration, $\nabla\theta$. Since

$$(\nabla\theta)^* = \partial_{\mathbf{x}^*}\theta = \mathbf{Q}\partial_{\mathbf{x}}\theta = \mathbf{Q}\,\nabla\theta$$

then objectivity requires that

$$\mathbf{Q}\,\mathbf{q}(..., \nabla\theta) = \mathbf{q}^*(..., \nabla\theta) = \mathbf{q}(..., \mathbf{Q}\,\nabla\theta),$$

the dots indicating possible further dependences. Now, let

$$\mathbf{q} = -\kappa\,\nabla\theta.$$

We have

$$\mathbf{q}^* = \mathbf{Q}\,\mathbf{q} = -\kappa\mathbf{Q}\,\nabla\theta = \mathbf{q}(\mathbf{Q}\,\nabla\theta),$$

and hence the isotropic Fourier's law is objective. In anisotropic materials we can write

$$\mathbf{q} = -\mathbf{K}\,\nabla\theta$$

where $\mathbf{K}$ is a second-order tensor, possibly dependent on $\mathbf{F}$. We have

$$\mathbf{q}^* = \mathbf{Q}\,\mathbf{q} = -\mathbf{Q}\,\mathbf{K}\,\nabla\theta, \qquad \mathbf{q}((\nabla\theta)^*) = -\mathbf{K}^*\,\mathbf{Q}\,\nabla\theta.$$

Objectivity, i.e. $\mathbf{q}^*(\nabla\theta) = \mathbf{q}((\nabla\theta)^*)$, requires that

$$\mathbf{Q}\,\mathbf{K}\,\nabla\theta = \mathbf{K}^*\,\mathbf{Q}\,\nabla\theta$$

for any vector $\nabla\theta$ whence

$$\mathbf{K}^* = \mathbf{Q}\,\mathbf{K}\,\mathbf{Q}^T.$$

Fourier's law is objective if and only if the conductivity tensor $\mathbf{K}$ is objective.

In a model of generalized thermoelasticity [288, 335], the dependence on $\dot\theta$ is also allowed. Since θ is invariant then $\dot\theta$ is invariant too and then the dependence on $\dot\theta$ is compatible with the principle of objectivity.

3.2.4 Thermo-Viscous Variables

In modelling thermo-viscous materials the dependence on the velocity gradient $\mathbf{L}$ and the temperature gradient $\nabla\theta$ is usually considered. The dependence on the mass density ρ and the temperature θ are also allowed. To fix ideas, let

$$\mathbf{T}(\rho, \theta, \mathbf{L}, \nabla\theta)$$

be the constitutive equation for the Cauchy stress tensor. Objectivity requires that

$$\mathbf{Q}\,\mathbf{T}(\rho, \theta, \mathbf{L}, \nabla\theta)\mathbf{Q}^T = \mathbf{T}(\rho, \theta, \mathbf{L}^*, (\nabla\theta)^*),$$

where the invariance of ρ and θ has been used. Since $\mathbf{L}^* = \mathbf{Q}\,\mathbf{L}\,\mathbf{Q}^T + \boldsymbol{\Omega}$ and $(\nabla\theta)^* = \mathbf{Q}\nabla\theta$ we have

$$\mathbf{Q}\,\mathbf{T}(\rho, \theta, \mathbf{L}, \nabla\theta)\mathbf{Q}^T = \mathbf{T}(\rho, \theta, \mathbf{Q}\,\mathbf{L}\,\mathbf{Q}^T + \boldsymbol{\Omega}, \mathbf{Q}\,\nabla\theta).$$

Choose $\mathbf{Q} = \mathbf{1}$. Hence

$$\mathbf{T}(\rho, \theta, \mathbf{L}, \nabla\theta) = \mathbf{T}(\rho, \theta, \mathbf{L} + \boldsymbol{\Omega}, \nabla\theta)$$

has to hold for every skew-symmetric tensor $\boldsymbol{\Omega} = \dot{\mathbf{Q}}\mathbf{Q}^T = \dot{\mathbf{Q}}$. Hence letting $\dot{\mathbf{Q}} = -\mathbf{W}$ we find that $\mathbf{T}$ cannot depend on the skew part of $\mathbf{L}$. Therefore we take $\mathbf{T}$ in the form

$$\mathbf{T} = \mathbf{T}(\rho, \theta, \mathbf{D}, \nabla\theta)$$

and objectivity requires that

$$\mathbf{Q}\,\mathbf{T}(\rho, \theta, \mathbf{D}, \nabla\theta)\mathbf{Q}^T = \mathbf{T}(\rho, \theta, \mathbf{Q}\,\mathbf{D}\,\mathbf{Q}^T, \mathbf{Q}\,\nabla\theta).$$

As an example, let

$$\mathbf{T} = 2\mu(\rho, \theta)\mathbf{D} + \lambda(\rho, \theta)(\mathrm{tr}\,\mathbf{D})\mathbf{1} + \nu(\rho, \theta)\nabla\theta \otimes \nabla\theta.$$

Then we have

$$\mathbf{Q}\,\mathbf{T}\,\mathbf{Q}^T = 2\mu\mathbf{Q}\,\mathbf{D}\,\mathbf{Q}^T + \lambda(\mathrm{tr}\,\mathbf{D})\mathbf{1} + \nu\,\mathbf{Q}\,\nabla\theta \otimes \mathbf{Q}\,\nabla\theta.$$

Since

$$\mathrm{tr}\,\mathbf{Q}\,\mathbf{D}\,\mathbf{Q}^T = \mathrm{tr}\,\mathbf{D}$$

we can write

$$\mathbf{Q}\,\mathbf{T}\,\mathbf{Q}^T = 2\mu\mathbf{Q}\,\mathbf{D}\,\mathbf{Q}^T + \lambda(\mathrm{tr}\,(\mathbf{Q}\,\mathbf{D}\,\mathbf{Q}^T))\mathbf{1} + \nu\,\mathbf{Q}\,\nabla\theta \otimes \mathbf{Q}\,\nabla\theta,$$

which makes it apparent that the example provides an objective constitutive equation.

Likewise we can show that $\mathbf{q}$ cannot depend on the skew part of $\mathbf{L}$ and that

$$\mathbf{q} = -\kappa(\rho,\theta)\nabla\theta + h(\rho,\theta)\mathbf{D}\,\nabla\theta$$

is an objective constitutive equation.

Since $\mathbf{D}$ and $\nabla\theta$ are objective, the scalars $\mathbf{D}\cdot\mathbf{D}$, $\nabla\theta\cdot\mathbf{D}\,\nabla\theta$, and $\nabla\theta\cdot\nabla\theta$ are invariant. By a direct check we have

$$(\nabla\theta)^*\cdot\mathbf{D}^*(\nabla\theta)^* = \mathbf{Q}\nabla\theta\cdot\mathbf{Q}\mathbf{D}\mathbf{Q}^T\mathbf{Q}\nabla\theta = \nabla\theta\mathbf{Q}^T\mathbf{Q}\mathbf{D}\mathbf{Q}^T\mathbf{Q}\nabla\theta = \nabla\theta\cdot\mathbf{D}\nabla\theta$$

and the like for the other terms. Hence the scalars μ, λ, ν, κ, h may depend also on $\mathrm{tr}\,\mathbf{D}$, $\mathbf{D}\cdot\mathbf{D}$, $\nabla\theta\cdot\mathbf{D}\nabla\theta$, $\nabla\theta\cdot\nabla\theta$ and meanwhile $\mathbf{T}$ and $\mathbf{q}$ are objective.

3.2.5 Dependence on Histories

The causality of natural processes may be interpreted as implying that the conditions in a body, at time t, are determined by the past history of the body and that no aspect of its future behaviour need be known. This might be phrased by saying that the effect at time t is determined by the history of the cause prior to time t. Given a function $\mathbf{F}$ on $\mathbb{R}$ we define

$$\mathbf{F}^t(s) = \mathbf{F}(t-s), \qquad s \in \mathbb{R}^+$$

and say that $\mathbf{F}^t$ is the history of $\mathbf{F}$ up to time t.

To fix ideas we take the stress tensor $\mathbf{T}$ as the effect produced by the motion and take that, at a point $\mathbf{X}$,

$$\mathbf{T}(t) = \boldsymbol{\mathcal{T}}(\mathbf{F}^t),$$

where $\mathbf{T}$ and $\mathbf{F}$ are considered at the same point $\mathbf{X}$. Following [428], p. 58, objectivity is said to require that

$$\mathbf{Q}(t)\mathbf{T}(t)\mathbf{Q}^T(t) = \boldsymbol{\mathcal{T}}(\mathbf{Q}^t\mathbf{F}^t),$$

where

$$(\mathbf{Q}^t\mathbf{F}^t)(s) = \mathbf{Q}(t-s)\mathbf{F}(t-s), \qquad s \in [0, \infty).$$

Based on this statement we can use the polar decomposition, $\mathbf{F} = \mathbf{RU}$, and take $\mathbf{Q}(t-s) = \mathbf{R}^T(t-s)$ to obtain

$$\mathbf{R}^T(t)\mathbf{T}(t)\mathbf{R}(t) = \mathscr{T}(\mathbf{U}^t),$$

whence

$$\mathbf{T}(t) = \mathbf{R}(t)\,\mathscr{T}(\mathbf{U}^t)\mathbf{R}^T(t).$$

Since $\mathbf{R} = \mathbf{FU}^{-1}$ then

$$\mathbf{T}(t) = \mathbf{F}(t)\mathbf{U}^{-1}(t)\,\mathscr{T}(\mathbf{U}^t)\mathbf{U}^{-1}(t)\mathbf{F}^T(t).$$

Let $\hat{\mathscr{T}} = \mathbf{U}^{-1}(t)\,\mathscr{T}(\mathbf{U}^t)\mathbf{U}^{-1}(t)$. Since $\hat{\mathscr{T}}$ is invariant under change of frame then it is apparent that $\mathbf{F}\,\hat{\mathscr{T}}\,\mathbf{F}^T$ is an objective tensor.

3.2.6 *Variables in Rate-Type Equations*

There are cases where the behaviour of the body is taken to be modelled by specifying the time derivative, of appropriate physical quantities, in terms of the independent variables. This is the content of the constitutive rate-type equations. Yet, in deformable bodies, the time derivative is not objective. It is then required by objectivity that rate-type equations involve objective derivatives (e.g. corotational, upper and lower convected). To fix ideas, assume the vector $\mathbf{q}$ and the tensor $\mathbf{T}$ are given by rate-type equations

$$\overset{\circ}{\mathbf{q}} = \hat{\mathbf{q}}(\theta, \mathbf{F}, \nabla_{\!R}\,\theta), \qquad \overset{\circ}{\mathbf{T}} = \hat{\mathbf{T}}(\theta, \mathbf{F}, \nabla_{\!R}\,\theta).$$

Objectivity requires that

$$\mathbf{Q}\hat{\mathbf{q}}(\theta, \mathbf{F}, \nabla_{\!R}\,\theta) = \hat{\mathbf{q}}(\theta, \mathbf{QF}, \nabla_{\!R}\,\theta), \qquad \mathbf{Q}\hat{\mathbf{T}}(\theta, \mathbf{F}, \nabla_{\!R}\,\theta)\mathbf{Q}^T = \hat{\mathbf{T}}(\theta, \mathbf{QF}, \nabla_{\!R}\,\theta).$$

The restrictions are then formally the same as those for finite equations. We then find that

$$\overset{\circ}{\mathbf{q}} = \mathbf{R}\,\hat{\mathbf{q}}(\theta, \mathbf{U}, \nabla_{\!R}\,\theta), \qquad \overset{\circ}{\mathbf{T}} = \mathbf{R}\,\hat{\mathbf{T}}(\theta, \mathbf{U}, \nabla_{\!R}\,\theta)\mathbf{R}^T.$$

3.2.7 Objectivity and Invariants

A scalar-valued function f is invariant under changes of frames. If f is a constitutive function which depends on tensors then f can depend only on tensor's invariants.

Given a 3×3 tensor $\mathbf{A}$ the three quantities

$$I_1 = \operatorname{tr}\mathbf{A}, \qquad I_2 = \tfrac{1}{2}[(\operatorname{tr}\mathbf{A})^2 - \operatorname{tr}\mathbf{A}^2], \qquad I_3 = \det\mathbf{A}$$

are invariant. To prove this we observe

$$\operatorname{tr}\mathbf{A}^* = A^*_{ii} = Q_{ik}Q_{ih}A_{kh} = \delta_{kh}A_{kh} = A_{kk} = \operatorname{tr}\mathbf{A}.$$

Likewise we prove the invariance of $\operatorname{tr}\mathbf{A}^2$. As to the determinant we have

$$\det\mathbf{A}^* = \det[\mathbf{Q}\mathbf{A}\mathbf{Q}^T] = \det\mathbf{A}.$$

Yet also the three scalars

$$\mathcal{I}_1 = \operatorname{tr}\mathbf{A}, \qquad \mathcal{I}_2 = \operatorname{tr}\mathbf{A}^2, \qquad \mathcal{I}_3 = \operatorname{tr}\mathbf{A}^3$$

are invariant, as it follows owing to the trace property. Indeed, by definition

$$I_1 = \mathcal{I}_1, \qquad I_2 = \tfrac{1}{2}(\mathcal{I}_1^2 - \mathcal{I}_2).$$

Now, by the Cayley-Hamilton theorem

$$\mathbf{A}^3 - I_1\mathbf{A}^2 + I_2\mathbf{A} - I_3\mathbf{1} = 0$$

we obtain

$$\operatorname{tr}\mathbf{A}^3 - I_1 \operatorname{tr}\mathbf{A}^2 + I_2 \operatorname{tr}\mathbf{A} = 3\, I_3,$$

whence

$$I_3 = \tfrac{1}{3}[\mathcal{I}_3 - \tfrac{3}{2}\mathcal{I}_1\mathcal{I}_2 + \tfrac{1}{2}\mathcal{I}_1^3].$$

From a physical point of view $\operatorname{tr}\mathbf{A}$ and $\det\mathbf{A}$ are more convenient in the description of constitutive properties. That is why $\mathcal{I}_1, \mathcal{I}_2, \mathcal{I}_3$ are less frequent in the literature.

For formal convenience the notation of I_1, I_2, I_3 (in place of I_A, II_A, III_A) is preferred whenever no ambiguity occurs about the tensor under consideration.

3.3 Objectivity and Euclidean Invariants in Electromagnetism

Whenever a constitutive model arises from an appropriate potential function (free energy, enthalpy) it is crucial to select the pertinent independent variables. For definiteness we examine a scalar-valued function

$$f(\mathbf{F}, \mathbf{E}, \mathbf{B})$$

of the deformation gradient $\mathbf{F}$, the electric field $\mathbf{E}$, and the induction $\mathbf{B}$. Upon a Euclidean transformation

$$\mathbf{x} \to \mathbf{x}^* = \mathbf{c}(t) + \mathbf{Q}(t)\mathbf{x}, \qquad \det \mathbf{Q} = 1,$$

$\mathbf{F}$, $\mathbf{E}$, and $\mathbf{B}$ are transformed into

$$\mathbf{F}^* = \mathbf{QF}, \quad \mathbf{E}^* = \mathbf{QE}, \quad \mathbf{B}^* = \mathbf{QB}. \tag{3.2}$$

Hence objectivity requires that

$$f(\mathbf{QF}, \mathbf{QE}, \mathbf{QB}) = f(\mathbf{F}, \mathbf{E}, \mathbf{B})$$

for any orthogonal tensor $\mathbf{Q}$.

To determine consequences of (3.2) we apply a classical argument based on the use of the polar decomposition $\mathbf{F} = \mathbf{RU}$. By selecting $\mathbf{Q} = \mathbf{R}^T$ it follows

$$f(\mathbf{F}, \mathbf{E}, \mathbf{B}) = f(\mathbf{R}^T\mathbf{RU}, \mathbf{R}^T\mathbf{E}, \mathbf{R}^T\mathbf{B}) = f(\mathbf{U}, \mathbf{U}^{-1}\mathbf{F}^T\mathbf{E}, \mathbf{U}^{-1}\mathbf{F}^T\mathbf{B}).$$

Since $\mathbf{U} = \mathbf{C}^{1/2}$ we can write

$$f = \hat{f}(\mathbf{C}, \mathbf{F}^T\mathbf{E}, \mathbf{F}^T\mathbf{B}).$$

Other possible restrictions might be obtained by selecting $\mathbf{Q}$ in different ways. Yet this is only a sufficient condition on the function $f(\mathbf{F}, \mathbf{E}, \mathbf{B})$ induced by the invariance under the rotation associated with the motion of the continuum.

We may observe that f is objective if it depends on Euclidean invariants. If $\mathbf{A}$ is any Euclidean invariant, i.e. $\mathbf{A}^* = \mathbf{A}$, then the restriction placed by objectivity,

$$f(\mathbf{A}) = f(\mathbf{A}^*)$$

is identically satisfied. Now, $\mathbf{C}$, $\mathbf{F}^T\mathbf{E}$, $\mathbf{F}^T\mathbf{B}$ are invariants in that

$$\mathbf{C}^* = \mathbf{F}^{*T}\mathbf{F}^* = \mathbf{F}^T\mathbf{Q}^T\mathbf{QF} = \mathbf{F}^T\mathbf{F} = \mathbf{C}, \quad \mathbf{F}^{*T}\mathbf{E}^* = \mathbf{F}^T\mathbf{Q}^T\mathbf{QE} = \mathbf{F}^T\mathbf{E},$$

and the like for $\mathbf{F}^T\mathbf{B}$. Further, $\mathbf{F}^{-1}\mathbf{E}$ and $\mathbf{F}^{-1}\mathbf{B}$ are invariants in that

$$\mathbf{F}^{*-1}\mathbf{E}^* = \mathbf{F}^{-1}\mathbf{Q}^{-1}\mathbf{Q}\mathbf{E} = \mathbf{F}^{-1}\mathbf{E}$$

and the like for $\mathbf{F}^{-1}\mathbf{B}$. Since $J = \det\mathbf{F} = \det\mathbf{C}^{1/2}$ also $f(J)\mathbf{F}^T\mathbf{E}$ and $f(J)\mathbf{F}^{-1}\mathbf{E}$ are invariants for any function $f(J)$. In addition, the scalars $\mathbf{E}^2$, $\mathbf{B}^2$, $\mathbf{E}\cdot\mathbf{B}$ are invariants, i.e.

$$\mathbf{E}^{*2} = (\mathbf{Q}\mathbf{E})\cdot(\mathbf{Q}\mathbf{E}) = \mathbf{E}\mathbf{Q}^T\mathbf{Q}\mathbf{E} = \mathbf{E}^2, \qquad \mathbf{E}^*\cdot\mathbf{B}^* = (\mathbf{Q}\mathbf{E})\cdot(\mathbf{Q}\mathbf{B}) = \mathbf{E}\mathbf{Q}^T\mathbf{Q}\mathbf{B} = \mathbf{E}\cdot\mathbf{B}.$$

All of the invariants are possible variables for a (objective) constitutive function f. Also functions of the invariants are invariants, e.g.

$$\mathbf{F}^T\mathbf{E}\cdot\mathbf{F}^T\mathbf{E}, \qquad (\mathbf{C}\mathbf{F}^T\mathbf{E})\cdot\mathbf{F}^T\mathbf{E}, \qquad (\mathbf{C}^2\mathbf{F}^T\mathbf{E})\cdot\mathbf{F}^T\mathbf{E}$$

are invariants. Henceforth it is understood that the dependence of a scalar function on the variables $\mathbf{F}$, $\mathbf{E}$, $\mathbf{B}$ and their derivatives is in fact given by a composite function of Euclidean invariants.

In practice, the selection of the appropriate variables is related to the particular model of the continuum. As we see, e.g. in Sect. 12.3, the dependence on $\mathbf{F}^T\mathbf{E}$ induces the occurrence of a torque and then implies $\text{skw}\mathbf{T} \neq \mathbf{0}$. Instead, no torque arises if the dependence on $\mathbf{E}$ is via $\mathbf{E}^2$.

In passing, we observe that the selection might be suggested by the physical meaning of the variables. This seems to be the case for the selection of $\mathbf{F}^T\mathbf{E}$ in [137] and $J\mathbf{F}^{-1}\mathbf{B}$ in [136]; they are the Lagrangian counterparts of $\mathbf{E}$ and $\mathbf{B}$, namely the fields occurring in the material description of Maxwell's equations. Of course the Lagrangian fields $\mathbf{F}^T\mathbf{E}$, $J\mathbf{F}^{-1}\mathbf{D}$, $J\mathbf{F}^{-1}\mathbf{P}$, $J\mathbf{F}^{-1}\mathbf{B}$, $\mathbf{F}^T\mathbf{M}$, $\mathbf{F}^T\mathbf{H}$ are Euclidean invariants.

3.4 Consistency with the Second Law

Granted the compatibility with the balance equations (mass, linear momentum, angular momentum, energy), the constitutive equations are required to be consistent with the second law of thermodynamics; the second law then places restrictions on the effective evolutions and hence selects those which are physically admissible. Here we first go back to the statement of the second law and next specify a view underlying the analysis of the consistency with the second law.

The requirement of the second law on the thermodynamic process might be stated as a requirement on the constitutive equations. The standard view can be given the following form. Through the balance of entropy, in Sect. 2.6, we have established the second law of thermodynamics by assuming that *for every admissible thermodynamic process, characterized by constitutive equations and balance equations, the inequality* (2.39) *must hold for all times t and points* $\mathbf{x}$. In the generalized form of the second law, the entropy production σ, as well as the entropy η and the entropy flux $\mathbf{j}$, is given by a constitutive function.

The more general view where also the entropy production is given by a constitutive function per se proves profitable in the modelling of hysteresis such as in ferroelectric, ferromagnetic, and plastic materials. It is worth mentioning that in [93, 335] the (rate of) entropy production σ is defined and required to be non-negative, as in (2.41). Yet no constitutive property is considered for σ. In [209] the entropy production is considered as a constitutive function but the requirement $\sigma \geq 0$ is ignored.

Granted the balance equations, the second law becomes a criterion of admissibility of the assumed constitutive functions. In this connection, while exploiting the requirement of the second law it is essential to envisage the freedom allowed in the selection of the admissible thermodynamic processes. In practical problems the supplies $\mathbf{b}$ and r are assigned in advance and hence the balance equations of motion and energy result in differential equations in the pertinent unknowns, e.g. $\rho, \theta, \mathbf{v}$.

Yet we take the view that conceptually the functions $\mathbf{b}(\mathbf{x}, t)$ and $r(\mathbf{x}, t)$ can be selected arbitrarily. Then for any value of the right-hand sides of

$$\mathbf{b} = \dot{\mathbf{v}} - \frac{1}{\rho}\nabla \cdot \mathbf{T} \quad \text{and} \quad r = \dot{\varepsilon} - \frac{1}{\rho}\mathbf{T} \cdot \mathbf{D} + \frac{1}{\rho\theta}\nabla \cdot \mathbf{q}, \tag{3.3}$$

the balance equations hold because $\mathbf{b}$ and r are just taken to be given by (3.3). Analogous claims hold for the space and time derivatives. Hence Eq. (3.3) provides the body force $\mathbf{b}$ and the heat supply r, and possibly their derivatives, that must be applied to support the process. This view on the conceptual role of the supplies traces back to Coleman and Noll [93].

Based on this view, for later use we show that, given any spatial point and any time, it is possible to find a process such that $\dot{\rho}$, $\ddot{\rho}$, $\nabla\dot{\rho}$, $\dot{\theta}$, $\ddot{\theta}$, $\nabla\dot{\theta}$, and $\mathbf{L}$, $\dot{\mathbf{L}}$, $\ddot{\mathbf{L}}$, $\nabla\mathbf{L}$ or $\dot{\mathbf{F}}$, $\ddot{\mathbf{F}}$ have arbitrarily prescribed values at that point and time. First we observe that, since

$$\dot{\rho} = -\rho\nabla \cdot \mathbf{v}, \quad \ddot{\rho} = -\dot{\rho}\nabla \cdot \mathbf{v} - \rho\dot{\overline{\nabla \cdot \mathbf{v}}}, \quad \nabla\dot{\rho} = -\nabla\rho\,\nabla \cdot \mathbf{v} - \rho\nabla\,\nabla \cdot \mathbf{v},$$

then $\dot{\rho}$, $\ddot{\rho}$, and $\nabla\dot{\rho}$ can take arbitrarily prescribed values if so is the case for $\nabla \cdot \mathbf{v}$, $\dot{\overline{\nabla \cdot \mathbf{v}}}$, and $\nabla\,\nabla \cdot \mathbf{v}$; of course $\dot{\rho}$ and $\nabla \cdot \mathbf{v}$ are not independent. This in turn is the case if $\mathbf{L}$, $\dot{\mathbf{L}}$, and $\nabla\mathbf{L}$ can take arbitrarily prescribed (tensor) values.

By the equation of motion we can write

$$\mathbf{v}(\mathbf{X}, t) = \int_{t_0}^{t} \dot{\mathbf{v}}(\mathbf{X}, \xi)d\xi + \mathbf{v}(t_0) = \int_{t_0}^{t} \mathbf{b}(\mathbf{X}, \xi)d\xi + \ldots$$

the dots denoting the remaining terms. Hence we have

$$\dot{F}_{iK}(\mathbf{X}, t) = \partial_{X_K} v_i(\mathbf{X}, t) = \partial_{X_K} \int_{t_0}^{t} b_i(\mathbf{X}, \xi)d\xi + \ldots$$

$$\ddot{F}_{iK} = \partial_{X_K}\dot{v}_i(\mathbf{X},t) = \partial_{X_K}b_i(\mathbf{X},t) + \ldots \quad .$$

Arbitrarily prescribed values of $\ddot{\mathbf{F}}$ and $\dot{\mathbf{F}}$ are allowed by appropriate values of $\nabla_R\mathbf{b}$ and of the integral of $\nabla_R\mathbf{b}$.

Within the Eulerian description we observe

$$\partial_{x_j}v_i = \partial_{x_j}\int_{t_0}^{t} b_i(\mathbf{X},\xi)d\xi + \cdots = \Big(\int_{t_0}^{t} \partial_{X_K}b_i(\mathbf{X},\xi)d\xi\Big)F^{-1}_{Ki} + \ldots .$$

Now by (1.30) and (1.31) we have

$$\dot{\mathbf{L}} = \nabla\dot{\mathbf{v}} + \cdots = \nabla\mathbf{b} + \ldots$$

and

$$\ddot{\mathbf{L}} = \nabla\ddot{\mathbf{v}} + \cdots = \nabla\dot{\mathbf{b}} + \ldots \quad .$$

Hence arbitrarily prescribed values of $\mathbf{L}$, $\dot{\mathbf{L}}$, $\ddot{\mathbf{L}}$ are allowed by appropriate values of

$$\int_{t_0}^{t} \nabla_R\mathbf{b}(\mathbf{X},\xi)d\xi, \quad \nabla\mathbf{b}, \quad \nabla\dot{\mathbf{b}}.$$

In connection with the values of $\dot\theta$, $\nabla\dot\theta$, $\ddot\theta$ we let the internal energy ε depend on the temperature θ so that $\partial_\theta\varepsilon \neq 0$. The balance equation of energy gives

$$\partial_\theta\varepsilon\,\dot\theta + \cdots = r.$$

Consequently

$$\dot\theta = \frac{r}{\partial_\theta\varepsilon} + \ldots, \quad \nabla\dot\theta = \nabla\frac{r}{\partial_\theta\varepsilon} + \ldots, \quad \ddot\theta = \Big(\frac{r}{\partial_\theta\varepsilon}\Big)^{\cdot} + \ldots \quad .$$

Thus arbitrarily prescribed values of $\dot\theta$, $\nabla\dot\theta$, and $\ddot\theta$ are allowed by appropriate values of r, ∇r, and $\dot r$.

The arbitrariness of $\mathbf{L}$ and $\dot{\mathbf{L}}$ implies that of $\mathbf{D} = \mathrm{sym}\mathbf{L}$, $\nabla\cdot\mathbf{v} = \mathrm{tr}\,\mathbf{L}$, $\mathbf{W} = \mathrm{skw}\mathbf{L}$, and $\dot{\mathbf{D}} = \mathrm{sym}\dot{\mathbf{L}}$, $\dot{\mathbf{W}} = \mathrm{skw}\dot{\mathbf{L}}$. Moreover, by (1.26) we have

$$\dot{\overline{\nabla\theta}} = \nabla\dot\theta - \mathbf{L}^T\nabla\theta = \nabla\frac{r}{\partial_\theta\varepsilon} + \ldots$$

and hence, for fixed values of $\mathbf{L}$ and $\nabla\theta$, the time derivative $\dot{\overline{\nabla\theta}}$ is allowed to take any arbitrarily prescribed value.

Especially in connection with the statement and the exploitation of the second law, the literature shows different approaches. The approach so described is often referred to as that of *rational thermodynamics* and that is the approach we follow throughout the book.

3.4.1 Exploitation of the Second Law

Granted the compatibility with the balance equations, the constitutive equations are required to be consistent with the second law of thermodynamics. We can then view the balance equations as constraints on the fields that are required to satisfy the second law (via the entropy inequality).

Let Λ be a set of variables and assume the entropy inequality takes the form

$$f(\Lambda)\alpha + g(\Lambda, \beta) \geq 0. \tag{3.4}$$

As an example, $\Lambda = (\rho, \theta, \mathbf{D}, \nabla\theta)$, $\alpha = \dot{\theta}$, $\beta = (\dot{\rho}, \dot{\mathbf{D}}, \dot{\overline{\nabla\theta}})$. If α can take arbitrary (positive or negative) values, independent of Λ and β, then (3.4) implies $f = 0$ and $g \geq 0$. Otherwise, if $f \neq 0$ then the arbitrariness of α allows $|f\alpha| > |g|$ and this in turn allows $f\alpha + g < 0$, which contradicts (3.4).

Likewise, if vectors are involved then

$$\mathbf{f} \cdot \boldsymbol{\alpha} + g \geq 0$$

implies $\mathbf{f} = \mathbf{0}$. If tensors are concerned in the form

$$\mathcal{F} \cdot \mathbf{A} + g \geq 0,$$

the arbitrariness of $\mathbf{A}$ implies $\mathcal{F} \cdot \mathbf{A} = 0$ and then $\mathcal{F} = \mathbf{0}$. Rather, if $\mathbf{A} \in \mathrm{Sym}$ (or Skw) then $\mathcal{F} \in \mathrm{Skw}$ (or Sym).

This is the essence of the so-called Coleman-Noll procedure ([93]; [216], ch. 39). It is apparent that the linearity is essential in that it allows both positive and negative values of $f\alpha$ if $f \neq 0$. To save writing, henceforth we mention the pertinent arbitrariness and understand the self-evident linearity.

If α, β, and Λ are not independent then we cannot conclude $f = 0$. Instead a representation formula holds as shown in Sect. 3.6.

Exploitation via Lagrange Multipliers

Within the classical approach described in the previous section, the balance of mass (i.e. the equation of continuity) is unavoidably a constraint while the balances of linear momentum and energy allow, e.g. for arbitrary values of $\dot{\mathbf{v}}$ and $\dot{\varepsilon}$. This is so because the body force $\mathbf{b}$ and the heat supply r are arbitrary and can be assigned to satisfy the balance equations for any chosen values of $\dot{\mathbf{v}}$ and $\dot{\varepsilon}$.

In a slightly different approach established by Liu [279], the entropy inequality

$$\rho\dot{\eta} + \nabla \cdot \phi = \partial_t(\rho\eta) + \nabla \cdot (\rho\eta\mathbf{v} + \mathbf{j}) \geq 0$$

is exploited for supply-free bodies, i.e. bodies where no body forces and no heat supplies occur. Hence, in view of the balance equations, which are considered as constraints, the entropy inequality can be written in the form

$$\partial_t(\rho\eta) + \nabla\cdot(\rho\eta\mathbf{v} + \mathbf{j}) - \Lambda_\rho[\partial_t\rho + \nabla\cdot(\rho\mathbf{v})] - \Lambda_\mathbf{v}\cdot[\partial_t(\rho\mathbf{v}) + \nabla\cdot(\rho\mathbf{v}\otimes\mathbf{v} - \mathbf{T})]$$
$$-\Lambda_\varepsilon[\partial_t(\rho(\varepsilon + \tfrac{1}{2}\mathbf{v}^2)) + \nabla\cdot(\rho(\varepsilon + \tfrac{1}{2}\mathbf{v}^2)\mathbf{v} - \mathbf{v}\mathbf{T} + \mathbf{q}))] \geq 0,$$

Λ_ρ, $\Lambda_\mathbf{v}$, Λ_ε being the Lagrange multipliers, viewed as additional constitutive functions of the chosen independent variables. For definiteness let η, ε, $\mathbf{T}$, $\mathbf{q}$ be functions of ρ, θ, $\mathbf{D}$, $\nabla\theta$. The entropy inequality becomes

$$[\eta + \rho\partial_\rho\eta - \Lambda_\rho - \Lambda_\varepsilon(\varepsilon + \tfrac{1}{2}\mathbf{v}^2 - \rho\partial_\rho\varepsilon)]\partial_t\rho + \rho(\partial_\theta\eta - \Lambda_\varepsilon\partial_\theta\varepsilon)\partial_t\theta - \rho(\Lambda_\mathbf{v} + \Lambda_\varepsilon\mathbf{v})\cdot\partial_t\mathbf{v} + \ldots$$

the dots denoting linear terms in $\dot{\mathbf{D}}$, $\dot{\overline{\nabla\theta}}$, $\nabla\rho$, $\nabla\mathbf{D}$, $\nabla\nabla\theta$. The arbitrariness and the linearity of $\partial_t\rho$, $\partial_t\theta$, $\partial_t\mathbf{v}$ imply that

$$\partial_\rho(\rho\eta) - \Lambda_\rho - \Lambda_\varepsilon[\tfrac{1}{2}\mathbf{v}^2 + \partial_\rho(\rho\varepsilon)] = 0, \qquad \partial_\theta\eta - \Lambda_\varepsilon\partial_\theta\varepsilon = 0, \qquad \Lambda_\mathbf{v} + \Lambda_\varepsilon\mathbf{v} = \mathbf{0},$$

whence we obtain formally the Lagrange multipliers Λ_ε, Λ_ρ, $\Lambda_\mathbf{v}$. Further restrictions are derived via the reduced inequality and appropriate assumptions on the dependence on $\mathbf{D}$, $\nabla\theta$. In particular, further assumptions are required to determine the dependence of the Lagrange multipliers on the independent variables.[2]

3.4.2 *Other Formulations of the Second Law*

Within the classical approach of rational thermodynamics (see, e.g. [93, 95, 99]), the free energy ψ, the entropy η, and the internal energy ε cannot depend on the temperature gradient. In a rational thermodynamics approach such dependence is allowed provided the extra-entropy flux $\mathbf{k}$ is taken to depend on the time derivative $\dot{\theta}$. Sometimes the second law is modified so that the expected dependence be allowed. Following are two formulations occurring in the literature that are devised to account for the dependence on gradients.

Extended Entropy Rate

According to [382], in our notation the entropy inequality is taken in the form

$$\dot{S} + \int_{\partial\mathcal{P}_t} \frac{\mathbf{q}}{\theta}\cdot\mathbf{n}\,da - \int_{\mathcal{P}_t} \frac{\rho r}{\theta}\,dv \geq 0$$

[2] An extensive application of the method of Lagrange multipliers is given in [338] within the theory of extended thermodynamics.

but the entropy rate is assumed to be given by

$$\dot{S} = \int_{\mathcal{P}_t} \frac{\rho}{\theta}(\theta\dot{\eta} + \nabla\theta \cdot \dot{\mathbf{w}})dv, \tag{3.5}$$

as though $\nabla\theta \cdot \dot{\mathbf{w}}/\theta$ would be an additional entropy supply, $\mathbf{w}$ being an appropriate vector field. Hence the (local) entropy inequality becomes

$$\frac{\rho}{\theta}(\theta\dot{\eta} + \nabla\theta \cdot \dot{\mathbf{w}}) + \frac{1}{\theta}\nabla \cdot \mathbf{q} - \frac{1}{\theta}\rho r - \frac{1}{\theta^2}\mathbf{q} \cdot \nabla\theta \geq 0.$$

Substitution of $\nabla \cdot \mathbf{q} - \rho r$ from the balance of energy, $\nabla \cdot \mathbf{q} - \rho r = \mathbf{T} \cdot \mathbf{D} - \rho\dot{\varepsilon}$, results in

$$\rho(\theta\dot{\eta} + \nabla\theta \cdot \dot{\mathbf{w}}) + \mathbf{T} \cdot \mathbf{D} - \rho\dot{\varepsilon} - \frac{1}{\theta}\mathbf{q} \cdot \nabla\theta \geq 0.$$

Let the internal energy be defined by

$$\varepsilon := \psi + \eta\theta + \mathbf{w} \cdot \nabla\theta. \tag{3.6}$$

It follows

$$-\rho(\dot{\psi} + \eta\dot{\theta}) - \rho\mathbf{w} \cdot \dot{\overline{\nabla\theta}} + \mathbf{T} \cdot \mathbf{D} - \frac{1}{\theta}\mathbf{q} \cdot \nabla\theta \geq 0.$$

Let ξ be a suitable variable describing a possible internal structure of the body and assume the constitutive functions depend on ρ, θ, $\nabla\theta$, $\mathbf{D}$, ξ. The entropy inequality takes the form

$$-\rho(\partial_\theta\psi + \eta)\dot{\theta} - \rho(\partial_{\nabla\theta}\psi + \mathbf{w}) \cdot \dot{\overline{\nabla\theta}} + \rho^2\partial_\rho\psi\nabla \cdot \mathbf{v} + \mathbf{T} \cdot \mathbf{D} - \rho\partial_{\mathbf{D}}\psi \cdot \dot{\mathbf{D}} - \rho\partial_\xi\psi\,\dot{\xi} - \frac{1}{\theta}\mathbf{q} \cdot \nabla\theta \geq 0.$$

The arbitrariness of $\dot{\mathbf{D}}$, $\dot{\theta}$, $\dot{\overline{\nabla\theta}}$ implies that

$$\partial_{\mathbf{D}}\psi = 0, \qquad \eta = -\partial_\theta\psi, \qquad \mathbf{w} = -\partial_{\nabla\theta}\psi$$

and

$$\rho^2\partial_\rho\psi\nabla \cdot \mathbf{v} + \mathbf{T} \cdot \mathbf{D} - \rho\partial_\xi\psi\,\dot{\xi} - \frac{1}{\theta}\mathbf{q} \cdot \nabla\theta \geq 0.$$

Further restrictions follow by specifying the constitutive equation for the rate $\dot{\xi}$.

Hence the dependence of the free energy ψ on $\nabla\theta$ is allowed if both the entropy rate $\dot{S}$ and the internal energy are defined via the ad-hoc insertion of $\dot{\mathbf{w}} \cdot \nabla\theta$ and $\mathbf{w} \cdot \nabla\theta$. We are unaware of a physical argument to justify (3.5) and (3.6).

Second Law via an Entropy Equation

Green and Naghdi [209, 210] postulate the balance of entropy in the form

$$\rho\dot{\eta} = \rho(\frac{r}{\theta} + \xi) - \nabla \cdot \frac{\mathbf{q}}{\theta},$$

where ξ is the "internal rate of production of entropy" (per unit mass). As is customary, we replace $\rho r - \nabla \cdot \mathbf{q}$ via the balance of energy and consider the free energy $\psi = \varepsilon - \theta\eta$ to find

$$-\rho(\dot{\psi} + \eta\dot{\theta}) + \mathbf{T} \cdot \mathbf{D} - \frac{1}{\theta}\mathbf{q} \cdot \nabla\theta - \rho\theta\xi = 0. \tag{3.7}$$

According to [210], once constitutive equations are specified for ψ, η, $\mathbf{q}$, ξ then (3.7) can be regarded as an identity and can be used to place restrictions on the form of the constitutive functions.[3]

We might observe that (3.7) can be written in the form

$$-\rho(\dot{\psi} + \eta\dot{\theta}) + \mathbf{T} \cdot \mathbf{D} - \frac{1}{\theta}\mathbf{q} \cdot \nabla\theta = \rho\theta\xi.$$

This condition is equivalent to the statement of the second law via the Clausius-Duhem inequality if ξ is defined by

$$\rho\xi := \rho\dot{\eta} - \frac{\rho r}{\theta} + \nabla \cdot \frac{\mathbf{q}}{\theta}$$

and we assume that $\xi \geq 0$ for any admissible process. Without the assumption $\xi \geq 0$ we might face, e.g. with the equality [210]

$$\mathbf{q} \cdot \nabla\theta + \rho\theta^2\xi = 0$$

and we could not conclude about the sign of $\mathbf{q} \cdot \nabla\theta$.

Generalized Gibbs Equation and Ginzburg-Landau Entropies

In classical thermodynamics of homogeneous processes, the relation

$$\theta dS = dU + pdV,$$

where S, U, and V are the entropy, the energy, and the volume of the system, is taken to hold for reversible changes in a closed system. For non-homogeneous processes the relation becomes

---

[3] Variants of this view occur if both the absolute temperature θ and an empirical temperature T are considered [210].

$$\theta\dot{\eta} = \dot{\varepsilon} + p\dot{v}.$$

The extension to mixtures is usually referred to as Gibbs equation.

Borrowing from the Gibbs equation, the second law is established by means of the time derivative $\dot{\eta}$ such that [118, 245]

$$\rho\dot{\eta} = -\nabla \cdot \mathbf{j} + \sigma,$$

where $\mathbf{j}$ is the entropy flux and σ is the non-negative entropy production. As we see in a while, though this view is formally correct, physically significant results follow depending on the choice of the independent variables. For formal simplicity we let the body be rigid; hence ρ is constant and we let η stand for $\rho\eta$.

Incidentally, in the modelling for phase transformation theories, as an extreme model we may introduce an interface separating the two phases. More realistically, the interfacial region may have a significant thickness and hence the gradients of the pertinent fields are likely to play an important role. This is merely an example that justifies the recourse to appropriate gradients within the independent variables.

First we let

$$\eta = f(\theta, \varphi, \mathbf{q}, \dot{\varphi}) - \tfrac{1}{2}\alpha|\nabla\theta|^2 - \tfrac{1}{2}\beta|\nabla\varphi|^2,$$

where α, β are constants and φ is an appropriate field (e.g. the phase field describing an interface). The gradients make the function to be of the Ginzburg-Landau form. Computation of $\dot{\eta}$ and some rearrangements yield

$$\dot{\eta} = (\partial_\theta f + \alpha\Delta\theta)\dot{\theta} + (\partial_\varphi f + \beta\Delta\phi)\dot{\varphi} + \partial_{\dot{\varphi}} f\ddot{\varphi} + \partial_{\mathbf{q}} f \cdot \dot{\mathbf{q}} - \nabla \cdot (\alpha\dot{\theta}\nabla\theta + \beta\dot{\varphi}\nabla\varphi).$$

Hence we may identify $\mathbf{j}$ as

$$\mathbf{j} = \alpha\dot{\theta}\nabla\theta + \beta\dot{\varphi}\nabla\varphi.$$

The remaining inequality allows $\sigma \geq 0$ if $\partial_{\dot{\varphi}} f = 0$ and

$$\dot{\varphi} = \lambda(\beta\Delta\varphi + \partial_\varphi f),$$

$$\dot{\theta} = \mu(\alpha\Delta\theta + \partial_\theta f),$$

λ, μ being non-negative functions of θ, φ. The term $\partial_{\mathbf{q}} f \cdot \dot{\mathbf{q}}$ might vanish if $\partial_{\mathbf{q}} f = \mathbf{0}$ or might be non-negative with a rate-type equation for $\mathbf{q}$, e.g. $\dot{\mathbf{q}} = \partial_{\mathbf{q}} f$. The partial differential equations for θ and ϕ have the form of Ginzburg-Landau equations [178]. This is physically reasonable for the phase field φ while the temperature θ and the heat flux $\mathbf{q}$ are expected to model heat conduction.

If instead we replace θ with ε as independent variable it follows

$$\dot{\eta} = (\partial_\varepsilon f - \nabla \cdot \partial_{\nabla\varepsilon} f)\dot{\varepsilon} + \nabla \cdot (\partial_{\nabla\varepsilon} f\dot{\varepsilon}) + \partial_{\mathbf{q}} f \cdot \dot{\mathbf{q}} + (\partial_\varphi f + \beta\Delta\varphi)\dot{\varphi} + \partial_{\dot{\varphi}} f\ddot{\varphi} - \nabla \cdot (\beta\dot{\varphi}\nabla\varphi).$$

In view of the balance of energy, $\dot{\varepsilon} = -\nabla \cdot \mathbf{q}$, we obtain

$$\dot{\eta} = \mathbf{q} \cdot \nabla(\partial_\varepsilon f - \nabla \cdot \partial_{\nabla\varepsilon} f) + \partial_{\mathbf{q}} f \cdot \dot{\mathbf{q}} + (\partial_\varphi f + \beta\Delta\varphi)\dot{\varphi} + \partial_{\dot{\varphi}} f \ddot{\varphi}$$
$$-\nabla \cdot (\beta\dot{\varphi}\nabla\varphi + (\partial_\varepsilon f - \nabla \cdot \partial_{\nabla\varepsilon} f)\mathbf{q} - \partial_{\nabla\varepsilon} f \dot{\varepsilon}).$$

As with the previous case we conclude that $\partial_{\dot{\varphi}} f = 0$ while

$$\mathbf{j} = \beta\dot{\varphi}\nabla\varphi + (\partial_\varepsilon f - \nabla \cdot \partial_{\nabla\varepsilon} f)\mathbf{q} - \partial_{\nabla\varepsilon} f \dot{\varepsilon}.$$

Now let

$$f(\varepsilon, \varphi, \mathbf{q}) = \hat{f}(\varepsilon, \varphi) - \tfrac{1}{2}\gamma\mathbf{q}^2, \qquad \eta = f(\varepsilon, \varphi, \mathbf{q}) - \tfrac{1}{2}\nu(\nabla\varepsilon)^2 - \tfrac{1}{2}\beta(\nabla\varphi)^2.$$

Hence we have

$$\sigma = \dot{\eta} + \nabla \cdot \mathbf{j} = \mathbf{q} \cdot (\partial_\varepsilon \hat{f} + \nu\Delta\varepsilon - \gamma\dot{\mathbf{q}}) + (\beta\Delta\varphi + \partial_\varphi \hat{f})\dot{\varphi}.$$

The independence of $\mathbf{q}$ and φ implies that both terms are non-negative whence

$$\dot{\varphi} = \lambda(\beta\Delta\varphi + \partial_\varphi \hat{f}), \qquad \mathbf{q} = \xi(\partial_\varepsilon \hat{f} + \nu\Delta\varepsilon - \gamma\dot{\mathbf{q}}),$$

where λ and ξ are non-negative functions of ε, φ. Hence we have

$$\mathbf{q} + \xi\gamma\dot{\mathbf{q}} = \xi(\partial_\varepsilon \hat{f} + \nu\Delta\varepsilon).$$

This is a Maxwell-Cattaneo type equation for the heat flux $\mathbf{q}$ (see, Sects. 6.2.1, 7.3, 6.6), with relaxation time $\xi\gamma$, embodying the non-local term $\xi\nu\Delta\varepsilon$.

3.4.3 Remarks on Onsager's Reciprocal Relations

As it emerges in various contexts, the exploitation of the entropy inequality eventually leads to the positive valuedness of an entropy production

$$\sigma = \mathbf{J} \cdot \mathbf{X},$$

where $\mathbf{X}$ is a proper set of scalars, vectors, tensors; the components of $\mathbf{X}$ are called forces, the associated components of $\mathbf{J}$ are called fluxes. Assume that, in *linear theories,* the constitutive equations have the form

$$\mathbf{J} = \mathbf{\Lambda}\mathbf{X},$$

where $\mathbf{\Lambda}$ is non-singular. Hence the thermodynamic requirement is

$$\sigma(\mathbf{X}) = \mathbf{X} \cdot \mathbf{\Lambda X} \geq 0$$

for any n-tuple $\mathbf{X}$. It is apparent that $\sigma \geq 0$ if and only if $\mathbf{\Lambda}_+ = \mathrm{sym}\mathbf{\Lambda}$ is positive semi-definite while the skew part $\mathbf{\Lambda}_- = \mathrm{skw}\mathbf{\Lambda}$ contributes nothing to σ.

Incidentally, in the Fourier model of heat conduction

$$\mathbf{J} = \mathbf{q}, \qquad \mathbf{X} = -\nabla \ln \theta.$$

If $\mathbf{w}$ is the axial vector of $\mathbf{\Lambda}_-$ we can take

$$\mathbf{\Lambda X} = \kappa \mathbf{1} + \mathbf{w} \times \mathbf{X}$$

and

$$\mathbf{J} \cdot \mathbf{X} = \kappa \mathbf{X}^2 \geq 0 \quad \Longrightarrow \quad \kappa \geq 0.$$

Of course we may well ask about the physical motivation of the vector $\mathbf{w}$ and hence of $\mathbf{\Lambda}_-$.

Things are more involved when different phenomena occur like in the case of heat conduction and diffusion. As we see in mixture theory, Chap. 9, for binary mixtures of non-reacting inviscid fluids the entropy inequality reduces to

$$\sigma = \mathbf{m}_1 \cdot (\mathbf{v}_2 - \mathbf{v}_1) + \mathbf{q} \cdot (-\nabla \ln \theta) \geq 0, \tag{3.8}$$

where $\mathbf{m}_1$ is the force on constituent 1 by way of constituent 2 and $\mathbf{q}$ stands for the inner part of the heat flux. In a linear theory we let $\mathbf{X} = (-\nabla \ln \theta, \mathbf{m}_1)$ and assume [314]

$$\mathbf{q} = -\kappa \nabla \ln \theta + \xi \mathbf{m}_1,$$

$$\mathbf{v}_2 - \mathbf{v}_1 = -\nu \nabla \ln \theta + \zeta \mathbf{m}_1.$$

Accordingly, $\xi \mathbf{m}_1$ represents the effect of diffusion on heat conduction (Dufour effect) while $\nu(-\nabla \ln \theta)$ is the effect of heat conduction on diffusion (Soret effect). The symmetry of the matrix $\mathbf{\Lambda}$ results in

$$\xi = \nu;$$

physically we say that the transport coefficients for the Soret and Dufour effects equate.

Two remarks are in order. In general the symmetry of $\mathbf{\Lambda}$ is not merely a consequence of thermodynamics; the general class of Onsager's reciprocal relations (in irreversible processes) are claimed to follow from the assumption of microscopic reversibility [350]; a critical review of the theory of Onsager's relations is given in [427], Chap. 7.

As a second remark we observe that the symmetry of $\mathbf{\Lambda}$, due to Onsager's relations, places no bound on the cross-coupling terms, e.g. $\xi \mathbf{m}_1$, $-\nu \nabla \ln \theta$. Now, by (3.8) it follows

$$\sigma = \zeta \mathbf{m}_1^2 + (\xi + \nu)\mathbf{m}_1 \cdot (-\nabla \ln \theta) + \kappa(\nabla \ln \theta)^2.$$

Let $\mathbf{m}_1 = y(-\nabla \ln \theta)$ whence

$$\sigma = [\zeta y^2 + (\xi + \nu)y + \kappa](\nabla \ln \theta)^2.$$

Hence $\sigma \geq 0$ if and only if

$$\zeta \geq 0, \qquad \kappa \geq 0, \qquad (\xi + \nu)^2 - 4\kappa\zeta \leq 0.$$

In words, the positive value of the entropy production requires that the cross-coupling terms are not too large relative to the direct terms; formally, the sum $\xi + \nu$ cannot be too large with respect to κ and ζ. Indeed, if by chance we use the Onsager relation $\xi = \nu$ then we find the bound

$$\nu^2 \leq \kappa\zeta.$$

3.5 Entropy Equation and Gibbs Equations

While the second law is expressed by an entropy inequality, $\rho\dot{\eta} \geq \rho r/\theta - \nabla \cdot \mathbf{j}$, it is worth looking at equations for the entropy function. The entropy equation results in a relation between entropy and energy.

Let w be a generic power, per unit volume, so that the balance of energy is

$$\rho\dot{\varepsilon} = w + \rho r - \nabla \cdot \mathbf{q}. \tag{3.9}$$

Look at a constitutive setting where the temperature θ and the variable $\boldsymbol{\alpha}$ are the independent variables, hence $\eta = \hat{\eta}(\theta, \boldsymbol{\alpha})$ and $\psi = \hat{\psi}(\theta, \boldsymbol{\alpha})$. As is most frequently proved, assume $\hat{\eta} = -\partial_\theta \hat{\psi}$. It follows that

$$\dot{\varepsilon} = \overline{\psi + \theta\eta}^{\,\cdot} = (\partial_\theta \hat{\psi} + \eta)\dot{\theta} + \partial_{\boldsymbol{\alpha}} \hat{\psi} \cdot \dot{\boldsymbol{\alpha}} + \theta\dot{\eta}.$$

Hence we find the entropy equation

$$\rho\theta\dot{\eta} = \rho\dot{\varepsilon} - \rho\partial_{\boldsymbol{\alpha}} \hat{\psi} \cdot \dot{\boldsymbol{\alpha}}. \tag{3.10}$$

Some examples are now examined. First, as is the case for fluids, let $\boldsymbol{\alpha} = \rho$, $w = \mathbf{T} \cdot \mathbf{D} = -p\nabla \cdot \mathbf{v} + \mathcal{T} \cdot \mathbf{D}$, with $p = \rho^2 \partial_\rho \hat{\psi}$. Since $\nabla \cdot \mathbf{v} = -\dot{\rho}/\rho$ then, in view of (3.10), we find the Gibbs equation

$$\rho\theta\dot{\eta} = \rho\dot{\varepsilon} - \frac{p}{\rho}\dot{\rho}. \tag{3.11}$$

By the entropy inequality,

$$\rho\theta\dot{\eta} \geq \rho r - \nabla\cdot\mathbf{q} + \frac{1}{\theta}\mathbf{q}\cdot\nabla\theta,$$

and use of (3.9) we have

$$\theta\sigma = \rho\theta\dot{\eta} - \rho r + \nabla\cdot\mathbf{q} - \frac{1}{\theta}\mathbf{q}\cdot\nabla\theta = \boldsymbol{T}\cdot\mathbf{D} - \frac{1}{\theta}\mathbf{q}\cdot\nabla\theta.$$

This shows that the entropy Eq. (3.11) holds even if heat conduction and viscosity effects occur.

Second, as is shown in the next chapter, for thermoelastic solids we have $\boldsymbol{\alpha} = \mathbf{F}$ and

$$\partial_{\mathbf{F}}\hat{\psi} = \frac{1}{J\rho}\mathbf{T}_R, \qquad w = \frac{1}{J}\mathbf{T}_R\cdot\dot{\mathbf{F}},$$

$\mathbf{T}_R$ being the first Piola stress. Hence

$$\dot{\varepsilon} = \overline{\psi + \theta\eta}^{\cdot} = \partial_\theta\hat{\psi}\dot{\theta} + \partial_{\mathbf{F}}\hat{\psi}\cdot\dot{\mathbf{F}} + \dot{\theta}\eta + \theta\dot{\eta} = \frac{1}{J\rho}\mathbf{T}_R\cdot\dot{\mathbf{F}} + \theta\dot{\eta}.$$

Hence we get the entropy equation in the form

$$\rho\theta\dot{\eta} = \rho\dot{\varepsilon} - \frac{1}{J}\mathbf{T}_R\cdot\dot{\mathbf{F}}. \tag{3.12}$$

Third, let $\boldsymbol{\alpha} = p$ while $w = -p\nabla\cdot\mathbf{v}$ and $\eta = -\partial_\theta\hat{\phi}(\theta, p)$, with $\phi = \psi + p/\rho$ being the Gibbs free energy. As we show later on, $\partial_p\hat{\phi}(\theta, p) = 1/\rho$. Now,

$$\dot{\varepsilon} = \overline{\phi + \theta\eta - p/\rho}^{\cdot} = \partial_\theta\hat{\phi}\dot{\theta} + \partial_p\hat{\phi}\dot{p} - \dot{\theta}\eta - \theta\dot{\eta} + \dot{p}/\rho - p\dot{\rho}/\rho^2 = -\theta\dot{\eta} + p\dot{\rho}/\rho^2.$$

Hence we obtain the entropy equation in the form (3.11).

As a comment we observe that entropy equations can take the same form in different continua, e.g. Eq. (3.11) for compressible fluids described by θ, p and inviscid fluids described by θ, ρ. Further, entropy- or Gibbs-equations hold even though dissipative effects ($\sigma > 0$) happen.

In terms of $\hat{\phi}$ we have

$$\partial_\theta(\hat{\phi}/\theta) = \frac{1}{\theta}\partial_\theta\hat{\phi} - \frac{1}{\theta^2}\phi = -\frac{1}{\theta}\eta - \frac{1}{\theta^2}(\varepsilon + \frac{p}{\rho} - \theta\eta).$$

Letting $h = \varepsilon + p/\rho$ be the enthalpy we can write

$$\partial_\theta\Big(\frac{\hat{\phi}}{\theta}\Big) = -\frac{h}{\theta^2};$$

this result is usually referred to as the Gibbs-Helmholtz relation.

3.6 Second Law and Representation of Constitutive Functions

Consider the Clausius-Duhem inequality in the generic form

$$-\rho(\dot{\psi}+\eta\dot{\theta})+w=\theta\sigma\geq 0, \tag{3.13}$$

where w includes e.g. the pertinent expression of the power and the classical term $-\mathbf{q}\cdot\nabla\theta/\theta^2$ of heat conduction. Let Ξ be a set of variables, e.g. $\Xi=(\theta,\alpha,\mathbf{u},\mathbf{K})$, α being a scalar, $\mathbf{u}$ a vector, $\mathbf{K}$ a tensor. To fix ideas, by assumption or by proof let $\psi=\psi(\Xi)$ while w,σ depend on Ξ and $\dot{\Xi}$. If e.g. $\dot{\mathbf{K}}$ is not arbitrary and occurs linearly in (3.13) then we can write

$$\mathbf{A_K}\cdot\dot{\mathbf{K}}=g(\theta,\alpha,\mathbf{u},\mathbf{K},\dot{\theta},\dot{\alpha},\dot{\mathbf{u}}), \tag{3.14}$$

where $\mathbf{A_K}$ is a second-order tensor. The following statement allows us to derive the general restriction placed on $\dot{\mathbf{K}}$, provided only that $\mathbf{A_K}\neq\mathbf{0}$.

Lemma 3.1 *(Representation) Given a second-order tensor* $\mathbf{A}$ *let* $\mathbf{N}=\mathbf{A}/|\mathbf{A}|$. *If* $\mathbf{Z}$ *is a second-order tensor such that only* $\mathbf{Z}\cdot\mathbf{N}$ *is known, say* $\mathbf{Z}\cdot\mathbf{N}=g$, *then*[4]

$$\mathbf{Z}=g\mathbf{N}+(\mathsf{I}-\mathbf{N}\otimes\mathbf{N})\mathbf{G} \tag{3.15}$$

for any second-order tensor $\mathbf{G}$.

Equation (3.15) is merely the representation formula (A.8) where $\mathbf{Z}\cdot\mathbf{N}$ is indicated as equal to g. Now apply the Representation Lemma to Eq. (3.14) with $\mathbf{Z}=\dot{\mathbf{K}}$, $\mathbf{A}=\mathbf{A_K}$. We find

$$\dot{\mathbf{K}}=(\dot{\mathbf{K}}\cdot\mathbf{A_K})\frac{\mathbf{A_K}}{|\mathbf{A}_\mathbf{K}^2|}+(\mathsf{I}-\frac{\mathbf{A_K}\otimes\mathbf{A_K}}{|\mathbf{A_K}|^2})\mathbf{G}=g\frac{\mathbf{A_K}}{|\mathbf{A_K}|^2}+(\mathsf{I}-\frac{\mathbf{A_K}\otimes\mathbf{A_K}}{|\mathbf{A_K}|^2})\mathbf{G}.$$

A suggestive geometrical interpretation follows by observing that

$$(\dot{\mathbf{K}}\cdot\mathbf{A_K})\frac{\mathbf{A_K}}{|\mathbf{A_K}|^2}=(\frac{\mathbf{A_K}}{|\mathbf{A_K}|}\otimes\frac{\mathbf{A_K}}{|\mathbf{A_K}|})\dot{\mathbf{K}}.$$

Hence thermodynamics places a restriction on the longitudinal part of $\dot{\mathbf{K}}$ along $\mathbf{A_K}$ but leaves the orthogonal part of $\dot{\mathbf{K}}$ to $\mathbf{A_K}$ unrestricted. A strictly analogous statement holds for $\dot{\mathbf{u}}$.

---

[4] The symbol I denotes the fourth-order identity tensor.

Chapter 4
Solids

The key property of solids is that they hold their shape because their molecules are tightly packed together. Atoms are essentially locked in position and, in the same way that a large solid holds its shape, the atoms inside of a solid are not allowed to move around too much. Electrons instead can move around and hence electromagnetic and thermal phenomena occur.

The interest here is restricted to solids as simple materials namely materials described by first-order gradients of deformation, velocity, and temperature. Due to these features of solids, the chapter develops models of linear and nonlinear thermoelastic solids. Correspondingly equations are examined for discontinuity (shock) waves and time harmonic waves. Nonlinear elastic solids are modelled and related nonlinear waves are considered thus arriving at the Christoffel equation. The linear theory of elasticity leads to classical parameters such as the elasticity tensor and the Lamé moduli and to the transverse and longitudinal waves. Next attention is addressed to models of hyperelastic and rubber-like materials. Further, different approaches to the description of dissipation in solids are examined. The concept of phonon is introduced and shown numerically by means of vibrating strings, namely elastic arrangements of atoms in solids.

4.1 Thermoelastic Solids

Thermoelasticity accounts for the interaction of the motion, within an elastic solid, with a nonuniform temperature field. Hence we say that a thermoelastic solid is characterized by constitutive functions of the variables

$$\mathbf{F}, \theta, \nabla\theta,$$

A. Morro and C. Giorgi, *Mathematical Modelling of Continuum Physics*,
Modeling and Simulation in Science, Engineering and Technology,
https://doi.org/10.1007/978-3-031-20814-0_4

that is the deformation gradient, the temperature, and the temperature gradient. The mass density ρ is related to the reference mass density ρ_R by

$$\rho \det \mathbf{F} = \rho_R.$$

By regarding ρ_R as a parameter it follows that ρ is merely a function of $\mathbf{F}$. In linear thermoelasticity ρ is regarded as a constant.

For mathematical convenience we describe the stress via the first Piola stress $\mathbf{T}_R$. Hence we say that an admissible thermodynamic process is the set of functions

$$\chi, \mathbf{F}, \theta, \psi, \eta, \mathbf{T}_R, \mathbf{q},$$

on $\Omega \times \mathbb{R}$, where χ is the motion. As always, it is understood that $\det \mathbf{F} > 0, \theta > 0$. For later convenience we show the following

Lemma 4.1 *At any point $\tilde{\mathbf{x}} \in \Omega$ and time $\tilde{t} \in \mathbb{R}$, we can select admissible thermodynamic processes with arbitrary values of* $\mathbf{F}, \theta, \nabla\theta, \dot{\mathbf{F}}, \dot{\theta}, \overline{\nabla\theta}^{\boldsymbol{\cdot}}, \nabla\mathbf{F}, \nabla\nabla\theta$.

To this end let $\mathcal{P} \subset \Omega$ be a neighbourhood of $\tilde{\mathbf{x}}$. Let $\mathbf{\Lambda}$ be a third-order tensor, $\mathbf{K} = \mathbf{K}^T$ and $\mathbf{A}$ be second order tensors, while $\mathbf{g}$ and $\mathbf{a}$ are vectors. Define the motion and the temperature,

$$\chi(\mathbf{X}, t) = \tilde{\mathbf{x}} + \tilde{\mathbf{F}}(\mathbf{X} - \tilde{\mathbf{X}}) + f(t)\mathbf{A}(\mathbf{X} - \tilde{\mathbf{X}}) + \mathbf{\Lambda}[(\mathbf{X} - \tilde{\mathbf{X}}) \otimes (\mathbf{X} - \tilde{\mathbf{X}})],$$

$$\theta(\mathbf{x}, t) = \tilde{\theta} + \alpha f(t) + [\mathbf{g} + f(t)\mathbf{a}] \cdot (\mathbf{x} - \tilde{\mathbf{x}}) + (\mathbf{x} - \tilde{\mathbf{x}}) \cdot \mathbf{K}(\mathbf{x} - \tilde{\mathbf{x}}),$$

where $\tilde{\theta} > 0$ and $\det \tilde{\mathbf{F}} > 0$. Moreover, $f(\tilde{t}) = 0$, $\dot{f}(\tilde{t}) = 1$, and $|f(t)|$ is small as we please for any t. We find $\tilde{\mathbf{x}} = \chi(\tilde{\mathbf{X}}, \tilde{t})$ and

$$\mathbf{F}(\mathbf{X}, t) = \tilde{\mathbf{F}} + f(t)\mathbf{A} + \mathbf{\Lambda}(\mathbf{X} - \tilde{\mathbf{X}}).$$

The arbitrary smallness of f and the diameter of P implies that $\det \mathbf{F} > 0$ in B_R. Moreover

$$\dot{\mathbf{F}}(\tilde{\mathbf{X}}, \tilde{t}) = \mathbf{A},$$

$$\nabla\mathbf{F}(\tilde{\mathbf{X}}, \tilde{t}) = \mathbf{\Lambda}.$$

Also

$$\mathbf{v}(\mathbf{X}, t) = \dot{f}(t)\mathbf{A}(\mathbf{X} - \tilde{\mathbf{X}})$$

so that $\mathbf{v} = 0$ at $\mathbf{X} = \tilde{\mathbf{X}}$ and hence at $\mathbf{x} = \tilde{\mathbf{x}}$. Now

$$\dot{\theta}(\tilde{\mathbf{x}}, \tilde{t}) = \alpha$$

while $\nabla\theta(\mathbf{x}, t) = \mathbf{g} + f(t)\mathbf{a} + \mathbf{K}(\mathbf{x} - \tilde{\mathbf{x}})$ whence

$$\nabla\theta(\tilde{\mathbf{x}},\tilde{t}) = \mathbf{g}, \qquad \overline{\nabla\theta}^{\boldsymbol{\cdot}}(\tilde{\mathbf{x}},\tilde{t}) = \mathbf{a}, \qquad \nabla\nabla\theta(\tilde{\mathbf{x}},\tilde{t}) = \mathbf{K},$$

which concludes the proof. □

We now derive the restrictions placed by the second law. Let ψ, η, $\mathbf{T}_R$, $\mathbf{q}$, and $\mathbf{k}$ be functions of $\mathbf{F}, \theta, \nabla\theta$. Moreover let ψ be differentiable. The result is stated as follows.

A necessary and sufficient condition that every admissible thermodynamic process obey the second law is that

$$\psi = \hat{\psi}(\mathbf{F},\theta), \qquad \mathbf{T}_R = \hat{\mathbf{T}}_R(\mathbf{F},\theta), \qquad \eta = \hat{\eta}(\mathbf{F},\theta), \tag{4.1}$$

$$\hat{\mathbf{T}}_R(\mathbf{F},\theta) = \rho_R\, \partial_{\mathbf{F}}\hat{\psi}(\mathbf{F},\theta), \qquad \hat{\eta}(\mathbf{F},\theta) = -\partial_\theta\hat{\psi}(\mathbf{F},\theta), \tag{4.2}$$

$$\mathbf{k} = \mathbf{0}, \qquad \hat{\mathbf{q}}(\mathbf{F},\theta,\nabla\theta)\cdot\nabla\theta \le 0. \tag{4.3}$$

To prove this assertion we observe that $J\mathbf{T}\cdot\mathbf{D} = \mathbf{T}_R\cdot\dot{\mathbf{F}}$ and hence the entropy inequality (2.40) can be written

$$-\rho\dot{\psi} - \rho\eta\dot{\theta} + \frac{1}{J}\mathbf{T}_R\cdot\dot{\mathbf{F}} - \frac{1}{\theta}\mathbf{q}\cdot\nabla\theta + \theta\nabla\cdot\mathbf{k} \ge 0.$$

By assumption $\psi = \psi(\mathbf{F},\theta,\nabla\theta)$. Hence the entropy inequality becomes

$$-\rho(\partial_\theta\psi + \eta)\dot{\theta} + (-\rho\partial_{\mathbf{F}}\psi + \mathbf{T}_R/J)\cdot\dot{\mathbf{F}} - \rho\partial_{\nabla\theta}\psi\cdot\overline{\nabla\theta}^{\boldsymbol{\cdot}} - \frac{1}{\theta}\mathbf{q}\cdot\nabla\theta + \theta[\partial_{\mathbf{F}}\mathbf{k}\cdot\nabla\mathbf{F} + \partial_\theta\mathbf{k}\cdot\mathbf{g} + \partial_{\nabla\theta}\mathbf{k}\cdot\nabla\nabla\theta] \ge 0.$$

In view of Lemma 4.1 we can select the process such that

$$-\rho(\partial_\theta\psi + \eta)\alpha + (-\rho\partial_{\mathbf{F}}\psi + \mathbf{T}_R/J)\cdot\mathbf{A} - \rho\partial_{\nabla\theta}\psi\cdot\mathbf{a} - \frac{1}{\theta}\mathbf{q}\cdot\mathbf{g} + \theta[\partial_{\mathbf{F}}\mathbf{k}\cdot\mathbf{\Lambda} + \partial_\theta\mathbf{k}\cdot\mathbf{g} + \partial_{\nabla\theta}\mathbf{k}\cdot\mathbf{K}] \ge 0,$$

the functions $\partial_\theta\psi, \eta, \ldots, \partial_{\nabla\theta}\psi$ being evaluated at $\tilde{\mathbf{F}}, \tilde{\theta}, \mathbf{g}$. The arbitrariness of $\alpha, \mathbf{A}, \mathbf{a}, \mathbf{\Lambda}, \mathbf{K}$ implies that the inequality holds only if (4.1), (4.2) hold with

$$\partial_{\mathbf{F}}\mathbf{k} = \mathbf{0}, \qquad \partial_{\nabla\theta}\mathbf{k} \in \mathrm{Skw}, \qquad -\frac{1}{\theta}\mathbf{q}\cdot\mathbf{g} + \theta\partial_\theta\mathbf{k}\cdot\mathbf{g} \ge 0.$$

Since $\mathbf{k}$ depends only on the scalar θ and the vector $\nabla\theta$ then we conclude that $\partial_{\nabla\theta}\mathbf{k} \in \mathrm{Skw}$ implies $\partial_{\nabla\theta}\mathbf{k} = \mathbf{0}$ and hence we have $\mathbf{k}(\theta) = \mathbf{0}$. Consequently the inequality (4.3) follows. Conversely, the validity of (4.1), (4.2), (4.3) implies that the entropy inequality (2.40) holds. □

As a comment, for a thermoelastic solid $\mathbf{T}_R$ and η are related to the free energy $\hat{\psi}(\mathbf{F},\theta)$ by (4.2) while $\hat{\mathbf{q}}(\mathbf{F},\theta,\nabla\theta)$ satisfies the heat conduction inequality $\hat{\mathbf{q}}(\mathbf{F},\theta,\nabla\theta)\cdot\nabla\theta \le 0$.

Some consequences of the thermoelastic model are of interest. By definition

$$\dot{\varepsilon} = \dot{\psi} + \theta\dot{\eta} + \eta\dot{\theta}$$

while, by (4.1) and (4.2),

$$\dot{\psi} = \mathbf{T}_R \cdot \dot{\mathbf{F}}/\rho_R - \eta\dot{\theta}.$$

Since $\psi = \varepsilon - \theta\eta$ it follows the Gibbs equation

$$\theta\dot{\eta} = \dot{\varepsilon} - \mathbf{T}_R \cdot \dot{\mathbf{F}}/\rho_R, \qquad \theta\dot{\eta}_R = \dot{\varepsilon}_R - \mathbf{T}_R \cdot \dot{\mathbf{F}}.$$

In view of the energy equation $\rho\dot{\varepsilon} = \mathbf{T}_R \cdot \dot{\mathbf{F}}/J - \nabla \cdot \mathbf{q} + \rho r$ we obtain

$$\rho\theta\dot{\eta} = -\nabla \cdot \mathbf{q} + \rho r \tag{4.4}$$

which implies that a thermodynamic process is adiabatic ($-\nabla \cdot \mathbf{q} + \rho r \equiv 0$) if and only if it is isentropic ($\dot{\eta} \equiv 0$).

By (4.2), it follows that if $\hat{\psi}$ is a C^2 function then $\hat{\mathbf{T}}_R(\mathbf{F}, \theta)/\rho_R$ and $\hat{\eta}(\mathbf{F}, \theta)$ satisfy the Maxwell relation

$$\partial_\theta \hat{\mathbf{T}}_R(\mathbf{F}, \theta)/\rho_R = -\partial_{\mathbf{F}}\hat{\eta}(\mathbf{F}, \theta).$$

The specific heat is defined classically as the amount of energy, per unit mass, required to raise the temperature by one degree (Kelvin). In the continuum scheme, we look at the internal energy per unit mass

$$\varepsilon = \hat{\varepsilon}(\mathbf{F}, \theta) = \hat{\psi}(\mathbf{F}, \theta) + \theta\hat{\eta}(\mathbf{F}, \theta)$$

and say that

$$c(\mathbf{F}, \theta) = \partial_\theta\hat{\varepsilon}(\mathbf{F}, \theta)$$

is the specific heat. Substitution for $\hat{\varepsilon}$ results in

$$c(\mathbf{F}, \theta) = \partial_\theta\hat{\psi}(\mathbf{F}, \theta) + \hat{\eta}(\mathbf{F}, \theta) + \theta\partial_\theta\hat{\eta}(\mathbf{F}, \theta),$$

whence, by (4.2),

$$c(\mathbf{F}, \theta) = \theta\partial_\theta\hat{\eta}(\mathbf{F}, \theta). \tag{4.5}$$

By (4.2) we have

$$\mathbf{T} = \frac{1}{J}\mathbf{T}_R\mathbf{F}^T = \rho\,\partial_{\mathbf{F}}\hat{\psi}(\mathbf{F}, \theta)\mathbf{F}^T.$$

The symmetry of $\mathbf{T}$ requires that

$$\partial_{\mathbf{F}}\hat{\psi}(\mathbf{F}, \theta)\mathbf{F}^T \in \mathrm{Sym}.$$

Now if ψ depends on $\mathbf{F}$ through $\mathbf{C} = \mathbf{F}^T\mathbf{F}$ then, in component form,

$$\partial_{F_{iK}}\psi\, F_{jK} = \partial_{C_{PQ}}\psi\, \partial_{F_{iK}}(F_{hP}F_{hQ})F_{jK} = \partial_{C_{PQ}}\psi\,(F_{iQ}F_{jP} + F_{iP}F_{jQ})$$

whence

$$\partial_{\mathbf{F}}\psi\, \mathbf{F}^T = 2\mathbf{F}\,\partial_{\mathbf{C}}\psi\, \mathbf{F}^T \in \mathrm{Sym}$$

and

$$\mathbf{T} = \rho\,\partial_{\mathbf{F}}\psi\, \mathbf{F}^T = 2\rho\,\mathbf{F}\,\partial_{\mathbf{C}}\psi\, \mathbf{F}^T \in \mathrm{Sym}.$$

The dependence of ψ on $\mathbf{C}$ guarantees the symmetry of the Cauchy stress.

In the material representation we can write

$$\mathbf{T}_R(\mathbf{F}, \theta, \nabla_R\theta), \qquad \mathbf{q}_R(\mathbf{F}, \theta, \nabla_R\theta).$$

Objectivity requires that

$$\mathbf{Q}\hat{\mathbf{T}}_R(\mathbf{F}, \theta, \nabla_R\theta) = \hat{\mathbf{T}}_R(\mathbf{QF}, \theta, \nabla_R\theta), \qquad \hat{\mathbf{q}}_R(\mathbf{F}, \theta, \nabla_R\theta) = \hat{\mathbf{q}}_R(\mathbf{QF}, \theta, \nabla_R\theta),$$

for any orthogonal tensor $\mathbf{Q}$. By the polar decomposition $\mathbf{F} = \mathbf{RU}$ and the choice $\mathbf{Q} = \mathbf{R}^T$ we obtain

$$\hat{\mathbf{T}}_R(\mathbf{F}, \theta, \nabla_R\theta) = \mathbf{FU}^{-1}\hat{\mathbf{T}}_R(\mathbf{U}, \theta, \nabla_R\theta), \qquad \hat{\mathbf{q}}_R(\mathbf{F}, \theta, \nabla_R\theta) = \hat{\mathbf{q}}_R(\mathbf{U}, \theta, \nabla_R\theta)$$

and hence $\mathbf{T}_{RR} = \mathbf{F}^{-1}\mathbf{T}_R$ is given by

$$\hat{\mathbf{T}}_{RR}(\mathbf{F}, \theta, \nabla_R\theta) = \tilde{\mathbf{T}}_{RR}(\mathbf{U}, \theta, \nabla_R\theta).$$

Thus the objectivity requires that $\mathbf{T}_{RR}$, $\mathbf{q}_R$ and the scalars ψ, η depends on $\mathbf{F}$ through $\mathbf{U}$ or $\mathbf{C}$.[1]

As a comment we now examine a consequence of the heat conduction inequality. Let

$$\mathbf{K}(\mathbf{F}, \theta) = -\partial_{\nabla\theta}\mathbf{q}(\mathbf{F}, \theta, \nabla\theta)\big|_{\nabla\theta=\mathbf{0}}$$

be the conductivity tensor. Hence the heat conduction inequality

$$\mathbf{q}(\mathbf{F}, \theta, \nabla\theta) \cdot \nabla\theta \le 0$$

and an expansion of $\mathbf{q}$ yield

$$\mathbf{q}(\mathbf{F}, \theta, \mathbf{0}) \cdot \nabla\theta - \nabla\theta \cdot \mathbf{K}\nabla\theta + o(|\nabla\theta|^2) \le 0.$$

This inequality holds only if

[1] See, e.g. [81, 405].

$$\mathbf{q}(\mathbf{C}, \theta, \mathbf{0}) = \mathbf{0}, \qquad \mathbf{K} > \mathbf{0}.$$

Hence, in light of the second law of thermodynamics, the heat flux vanishes whenever the temperature gradient vanishes and the conductivity tensor has to be positive definite.

A further comment is given about the entropy. Let η be a function of θ and the specific heat be constant. Then (4.5) can be written in the form

$$\eta'(\theta) = \frac{c}{\theta}.$$

Let θ_0 be a reference temperature. Upon integration, from θ_0 to θ, we have

$$\eta(\theta) - \eta(\theta_0) = c \ln(\theta/\theta_0).$$

Now, integrate

$$\psi'(\theta) = -\eta(\theta) = -\eta(\theta_0) - c \ln(\theta/\theta_0)$$

to obtain

$$\psi(\theta) - \psi(\theta_0) = -\eta(\theta_0)(\theta - \theta_0) - c\theta \ln(\theta/\theta_0) + c(\theta - \theta_0).$$

Hence $\varepsilon = \psi + \theta\eta$ takes the form

$$\varepsilon(\theta) = \varepsilon(\theta_0) + c(\theta - \theta_0).$$

Quite often in the literature a simpler representation is adopted by letting

$$\eta = c \ln \theta, \qquad \psi = c\, \theta(1 - \ln \theta), \qquad \varepsilon = c\, \theta.$$

4.1.1 Linear Theory

Qualitatively linear theories are based on the view that the reference configuration is natural and furthermore some quantities, such as deformations and temperature variations, are small. Moreover we assume that the reference configuration is stress free. To make this view formal here we let θ_0 be a reference temperature and look for linear equations in $\nabla\mathbf{u}$, $\nabla\dot{\mathbf{u}}$, $\theta - \theta_0$, $\dot{\theta}$, $\nabla\theta$. Now, since

$$\nabla_{\!R}\, \mathbf{u} = \nabla \mathbf{u}\, \mathbf{F}$$

then

$$\mathbf{F} = \mathbf{1} + \nabla_{\!R}\mathbf{u} = \mathbf{1} + \nabla \mathbf{u}\, \mathbf{F} = \mathbf{1} + \nabla \mathbf{u}(\mathbf{1} + \nabla \mathbf{u}\mathbf{F})$$

so that

$$\mathbf{F} = \mathbf{1} + \nabla\mathbf{u} + o(|\nabla\mathbf{u}|), \qquad \mathbf{C} = \mathbf{1} + 2\mathbf{E} = \mathbf{1} + 2\text{sym}\nabla\mathbf{u} + o(|\nabla\mathbf{u}|),$$

$$\dot{\mathbf{F}} = \nabla\dot{\mathbf{u}}(\mathbf{1} + O(|\nabla\mathbf{u}|), \qquad \dot{\mathbf{C}} = 2\dot{\mathbf{E}}, \qquad \dot{\mathbf{E}} = \text{sym}\nabla\dot{\mathbf{u}} + O(|\nabla\mathbf{u}|)O(|\nabla\dot{\mathbf{u}}|).$$

Thus we take the approximations

$$\mathbf{F} \approx \mathbf{1} + \nabla\mathbf{u}, \qquad \mathbf{E} \approx \text{sym}\nabla\mathbf{u} = \varepsilon, \qquad \mathbf{C} \approx \mathbf{1} + 2\text{sym}\nabla\mathbf{u},$$

$$\dot{\mathbf{F}} \approx \nabla\dot{\mathbf{u}}, \qquad \dot{\mathbf{E}} \simeq \text{sym}\nabla\dot{\mathbf{u}}.$$

By objectivity, ψ is a function of $\mathbf{F}$ via $\mathbf{C}$. Yet $\mathbf{T}_R$ need not be symmetric,

$$\mathbf{T}_R = \rho\partial_{\mathbf{F}}\psi = 2\,\rho\,\mathbf{F}\,\partial_{\mathbf{C}}\psi.$$

In the linear approximation we have $\mathbf{F} \simeq \mathbf{1}$, $\rho \simeq \rho_R$, $\mathbf{T}_{RR}$, $\mathbf{T}_R \simeq \mathbf{T}$, and

$$\mathbf{T} \simeq 2\,\rho_R\,\partial_{\mathbf{C}}\psi(\mathbf{C}, \theta).$$

For formal convenience we let ψ be a function of $\mathbf{E}$ and $\vartheta = \theta - \theta_0$. Hence we let

$$\mathsf{C} = \rho_R\partial_{\mathbf{E}}\partial_{\mathbf{E}}\tilde{\psi}(\mathbf{E}, \vartheta), \qquad \mathbf{M} = \rho_R\partial_\vartheta\partial_{\mathbf{E}}\tilde{\psi}(\mathbf{E}, \vartheta),$$

at $\mathbf{E} = \mathbf{0}$, $\vartheta = 0$. Hence we let

$$\mathbf{T} = \mathsf{C}\,\mathbf{E} + \vartheta\,\mathbf{M}$$

to within $o(|\nabla\mathbf{u}|, |\vartheta|)$. The fourth-order tensor C is called the elasticity tensor and is symmetric. The second-order tensor $\mathbf{M}$ is called the stress-temperature tensor and is symmetric too. The stress-temperature tensor $\mathbf{M}$ gives the stress resulting from a given temperature variation when $\mathbf{E} = \mathbf{0}$.

As to the heat flux, an expansion with respect to $\nabla\theta$ and the observation that $\mathbf{q} = \mathbf{0}$ when $\nabla\theta = \mathbf{0}$ provide

$$\mathbf{q} = -\mathbf{K}\nabla\theta$$

to within $o(|\nabla\theta|)$. Here $\mathbf{K}$ is the conductivity tensor and is evaluated at $\mathbf{E} = \mathbf{0}$, $\vartheta = 0$, $\nabla\theta = \mathbf{0}$.

By $\eta = \tilde{\eta}(\mathbf{E}, \vartheta)$ we obtain the time derivative $\dot{\eta}$ in the form

$$\begin{aligned}\dot{\eta} = \partial_{\mathbf{E}}\tilde{\eta}(\mathbf{E}, \vartheta) \cdot \dot{\mathbf{E}} + \partial_\vartheta\tilde{\eta}(\mathbf{E}, \vartheta)\dot{\vartheta} &= -\partial_\vartheta\tilde{\mathbf{T}}(\mathbf{E}, \vartheta) \cdot \dot{\mathbf{E}} + \partial_\vartheta\tilde{\eta}(\mathbf{E}, \vartheta)\dot{\vartheta} \\ &= -\mathbf{M} \cdot \dot{\mathbf{E}} + \partial_\vartheta\tilde{\eta}(\mathbf{E}, \vartheta)\dot{\vartheta}.\end{aligned}$$

Hence, since $\rho\theta\dot{\eta} = -\nabla \cdot \mathbf{q} + \rho r$ then we have

$$-\rho\theta \mathbf{M}\cdot\dot{\mathbf{E}} + \rho\theta\partial_\vartheta\tilde{\eta}(\mathbf{E},\vartheta)\,\dot{\vartheta} = -\nabla\cdot\mathbf{q} + \rho r.$$

We thus obtain the *basic equations of linearized thermoelasticity* in the form

$$\rho_R\ddot{\mathbf{u}} = \rho_R\mathbf{b} + \nabla\cdot\mathbf{T}, \qquad \mathbf{T} = \mathsf{C}\,\mathbf{E} + \vartheta\,\mathbf{M},$$

$$\rho_R\,c\,\dot{\vartheta} = -\nabla\cdot\mathbf{q} + \theta_0\,\mathbf{M}\cdot\dot{\mathbf{E}} + \rho_R r, \qquad \mathbf{q} = -\mathbf{K}\,\nabla\vartheta.$$

As a comment, if the elasticity tensor C is invertible we can write

$$\mathbf{E} = \mathsf{K}\,\mathbf{T} + \vartheta\,\mathbf{A}.$$

The tensors

$$\mathsf{K} = \mathsf{C}^{-1}, \qquad \mathbf{A} = -\mathsf{K}\,\mathbf{M}$$

are called the *compliance tensor* and the *thermal expansion tensor,* respectively. Hence $\mathbf{A}$ determines the strain resulting from a given temperature when the stress vanishes.

When the material is isotropic we can write[2]

$$\mathsf{C}\,\mathbf{E} = 2\mu\mathbf{E} + \lambda(\mathrm{tr}\,\mathbf{E})\mathbf{1}, \qquad \mathbf{M} = m\,\mathbf{1}, \qquad \mathbf{K} = k\,\mathbf{1}.$$

The scalars μ and λ are called the *Lamé moduli, m* is the *stress-temperature modulus,* and k is the *conductivity.* In such a case the governing equations become

$$\rho_R\ddot{\mathbf{u}} = \rho_R\mathbf{b} + \nabla\cdot\mathbf{T}, \qquad \mathbf{T} = 2\mu\,\mathbf{E} + \lambda(\mathrm{tr}\,\mathbf{E})\mathbf{1} + m\,\vartheta\,\mathbf{1},$$

$$\rho_R\,c\,\dot{\vartheta} = -\nabla\cdot\mathbf{q} + m\,\theta_0\,\mathrm{tr}\,\dot{\mathbf{E}} + \rho_R r, \qquad \mathbf{q} = -k\,\nabla\vartheta.$$

Temperature variations result in pressure tensors in that

$$\mathbf{T} = m\,\vartheta\,\mathbf{1}$$

when $\mathbf{E} = \mathbf{0}$. In addition if $\mu \neq 0$ and $3\lambda + 2\mu \neq 0$ then we obtain the strain as a function of stress and temperature in the form

$$\mathbf{E} = \frac{1}{2\mu}\mathbf{T} - \frac{\lambda}{2\mu(3\lambda + 2\mu)}(\mathrm{tr}\,\mathbf{T})\mathbf{1} + \alpha\,\vartheta\,\mathbf{1}, \tag{4.6}$$

where

$$\alpha = -\frac{m}{3\lambda + 2\mu}.$$

---

[2] The definitions of C and $\mathbf{M}$ embody a factor ρ_R (or ρ). Hence μ, λ, m too are factored by ρ_R. This in turn means that μ/ρ, λ/ρ, m/ρ keep finite, or constant, when $\rho \to 0$.

Since $\mathbf{T} = 2\mu\mathbf{E}_0 + \kappa(\text{tr}\,\mathbf{E})\mathbf{1} + m\vartheta\mathbf{1}$ then in stress-free conditions we have

$$0 = \kappa(\text{tr}\,\mathbf{E}) + m\vartheta.$$

Moreover, since

$$\text{tr}\,\mathbf{E} = \tfrac{1}{2}\text{tr}\,(\mathbf{H}^T\mathbf{H} + \mathbf{H}^T + \mathbf{H}) \simeq \nabla_R \cdot \mathbf{u} \simeq \nabla \cdot \mathbf{u}$$

then[3]

$$\nabla \cdot \mathbf{u} = \tfrac{1}{3}\alpha\vartheta.$$

Since $\nabla \cdot \mathbf{u}$ is the relative variation of the volume then $\alpha/3$ is called the coefficient of thermal expansion.[4] It is natural to assume that $\alpha > 0$, $m < 0$.

4.1.2 Free Energy and Invariants

A body is *isotropic* if every rotation is a symmetry transformation. For a thermoelastic solid this implies that

$$\psi_R(\theta, \mathbf{F}) = \psi_R(\theta, \mathbf{FQ})$$

for all rotations $\mathbf{Q}$. By the polar decomposition, $\mathbf{F} = \mathbf{VR}$, and the choice $\mathbf{Q} = \mathbf{R}^T$ we have

$$\psi_R(\theta, \mathbf{F}) = \psi_R(\theta, \mathbf{V}) = \hat{\psi}_R(\theta, \mathbf{B}),$$

$\mathbf{B}$ being the left Cauchy-Green tensor.

The referential free energy $\psi_R = \rho_R\psi$ and the first Piola stress $\mathbf{T}_R$ are related by

$$\mathbf{T}_R = \partial_{\mathbf{F}}\psi_R(\theta, \mathbf{F}) = \partial_{\mathbf{F}}\hat{\psi}_R(\theta, \mathbf{B})$$

and hence

$$\mathbf{T}_R = \partial_{\mathbf{B}}\hat{\psi}_R(\theta, \mathbf{B})\partial_{\mathbf{F}}\mathbf{B}.$$

Now,

$$\partial_{F_{iK}} B_{jh} = \partial_{F_{iK}}[F_{jP}F_{hP}] = \delta_{ij}F_{hK} + F_{jK}\delta_{ih}.$$

It then follows

$$\mathbf{T}_R = 2\partial_{\mathbf{B}}\,\hat{\psi}_R(\theta, \mathbf{B})\mathbf{F}.$$

---

[3] See, e.g. [272].

[4] A different definition is given in the literature. In view of (4.6), $\mathbf{A} = \alpha\mathbf{1}$ is called the thermal expansion tensor and then α, instead of $\alpha/3$, is called the coefficient of thermal expansion [216].

To obtain the Cauchy stress $\mathbf{T}$ we observe that $\mathbf{T} = J^{-1}\mathbf{T}_R\mathbf{F}^T$ and $J = \det \mathbf{F} = \sqrt{\det \mathbf{B}}$. Thus $\mathbf{T}$ is represented as a function of $\mathbf{B}$ (and θ) in the form

$$\mathbf{T} = \frac{2}{\sqrt{\det \mathbf{B}}}\,\partial_{\mathbf{B}}\hat{\psi}_R(\theta, \mathbf{B})\,\mathbf{B}. \tag{4.7}$$

A representation theorem [120] asserts that an isotropic scalar function of a symmetric tensor $\mathbf{B}$ may be expressed as a function of the principal invariants $I_1(\mathbf{B}) = \operatorname{tr}\mathbf{B}$, $I_2(\mathbf{B}) = \frac{1}{2}[(\operatorname{tr}\mathbf{B})^2 - \operatorname{tr}\mathbf{B}^2]$, $I_3(\mathbf{B}) = \det \mathbf{B}$. Let $I_{\mathbf{B}}$ be the triplet of invariants,

$$I_{\mathbf{B}} = (I_1(\mathbf{B}), I_2(\mathbf{B}), I_3(\mathbf{B})),$$

so that we can write

$$\psi_R = \psi_R(\theta, I_{\mathbf{B}}).$$

Hence the Cauchy stress $\mathbf{T}$ can be given the form

$$\mathbf{T} = \frac{2}{\sqrt{I_3}}\,\partial_{\mathbf{B}}\psi_R(\theta, I_{\mathbf{B}})\,\mathbf{B}.$$

Since

$$\partial_{\mathbf{B}} I_1(\mathbf{B}) = \mathbf{1}, \qquad \partial_{\mathbf{B}} I_2(\mathbf{B}) = I_1(\mathbf{B})\mathbf{1} - \mathbf{B}, \qquad \partial_{\mathbf{B}} I_3(\mathbf{B}) = I_3(\mathbf{B})\mathbf{B}^{-1},$$

we obtain

$$\mathbf{T} = \frac{2}{\sqrt{I_3}}[I_3\,\partial_{I_3}\psi_R(\theta, I_{\mathbf{B}})\,\mathbf{1} + (\partial_{I_1}\psi_R(\theta, I_{\mathbf{B}}) + I_1\partial_{I_2}\psi_R(\theta, I_{\mathbf{B}}))\mathbf{B} - \partial_{I_2}\psi_R(\theta, I_{\mathbf{B}})\mathbf{B}^2\,].$$

Now, by the Cayley-Hamilton equation

$$\mathbf{B}^3 - I_1\,\mathbf{B}^2 + I_2\,\mathbf{B} - I_3\,\mathbf{1} = 0,$$

we have

$$\mathbf{B}^2 = I_1\mathbf{B} - I_2\mathbf{1} + I_3\mathbf{B}^{-1}$$

and hence

$$\mathbf{T} = \beta_0\mathbf{1} + \beta_1\mathbf{B} + \beta_2\mathbf{B}^{-1},$$

where

$$\beta_0 = \frac{2}{\sqrt{I_3}}(\partial_{I_1}\psi_R(\theta, I_{\mathbf{B}}) + I_1\partial_{I_2}\psi_R(\theta, I_{\mathbf{B}})),\ \beta_1 = \frac{2}{\sqrt{I_3}}\partial_{I_1}\psi_R(\theta, I_{\mathbf{B}}),\ \beta_2 = -2\sqrt{I_3}\partial_{I_2}\psi_R(\theta, I_{\mathbf{B}}).$$

4.2 Elastic Solids

A solid is a continuum characterized by structural rigidity and resistance to changes of shape or volume. This is made in mathematical form by saying that the stress, as well as the other constitutive properties, is related to the deformation.

The thermal aspects, that is the dependence on the temperature or on the temperature gradient, and the heat conduction are modelled within thermoelasticity and hence ignored in elasticity. In elastic solids the Cauchy stress is a function of the deformation gradient, $\mathbf{F}$, $\mathbf{T} = \hat{\mathbf{T}}(\mathbf{F})$. By the principle of objectivity, $\hat{\mathbf{T}}$ is required to satisfy

$$\mathbf{Q}\hat{\mathbf{T}}(\mathbf{F})\mathbf{Q}^T = \hat{\mathbf{T}}(\mathbf{QF}),$$

for any orthogonal tensor $\mathbf{Q}$. By the polar decomposition theorem, $\mathbf{F} = \mathbf{RU}$, we can choose $\mathbf{Q} = \mathbf{R}$ to obtain

$$\hat{\mathbf{T}}(\mathbf{F}) = \mathbf{R}\hat{\mathbf{T}}(\mathbf{U})\mathbf{R}^T$$

whence

$$\mathbf{F}^{-1}\hat{\mathbf{T}}(\mathbf{F})\mathbf{F}^{-T} = \mathbf{U}^{-1}\hat{\mathbf{T}}(\mathbf{U})\mathbf{U}^{-1}.$$

Since

$$J = \det \mathbf{F} = \det \mathbf{U},$$

the second Piola stress $\mathbf{T}_{RR} = J\,\mathbf{F}^{-1}\,\mathbf{T}\,\mathbf{F}^{-T}$ is then a function of $\mathbf{U}$,

$$\mathbf{T}_{RR} = (\det \mathbf{U})\mathbf{U}^{-1}\,\hat{\mathbf{T}}(\mathbf{U})\,\mathbf{U}^{-1}.$$

Alternatively we may take $\mathbf{T}_{RR}$ as a function of the Cauchy-Green tensor $\mathbf{C} = \mathbf{U}^2$.

The reference free energy $\psi_R = \rho_R\psi$ is taken to depend on the deformation gradient and hence, by the principle of objectivity, turns out to be a function of $\mathbf{C}$; let $\psi_R = \tilde{\psi}_R(\mathbf{C})$. The thermodynamic restrictions are more directly established in the material description. The entropy inequality reduces to

$$-\dot{\psi}_R + \tfrac{1}{2}\mathbf{T}_{RR} \cdot \dot{\mathbf{C}} \geq 0.$$

If ψ_R is a function of $\mathbf{C}$, $\psi_R = \tilde{\psi}_R(\mathbf{C})$, then the entropy inequality can be written in the form

$$(-\partial_{\mathbf{C}}\psi_R + \tfrac{1}{2}\mathbf{T}_{RR}) \cdot \dot{\mathbf{C}} \geq 0.$$

By regarding $\mathbf{b}(\mathbf{X}, t)$ as an arbitrarily assignable body force we can take $\mathbf{F}$ and $\dot{\mathbf{F}}$ as arbitrarily prescribed, at any point $\mathbf{X}$ and time t. Hence there are motions such that $\mathbf{C} = \mathbf{F}^T\mathbf{F}$ and $\dot{\mathbf{C}} = \dot{\mathbf{F}}^T\mathbf{F} + \mathbf{F}^T\dot{\mathbf{F}}$ have arbitrarily prescribed symmetric values at any $\mathbf{X}$ and t. Consequently the entropy inequality holds for all motions if and only if

$$\mathbf{T}_{RR} = 2\,\partial_{\mathbf{C}}\tilde{\psi}_R(\mathbf{C}). \tag{4.8}$$

Since $\mathbf{T} = J^{-1}\mathbf{F}\,\mathbf{T}_{RR}\mathbf{F}^T$ then the Cauchy stress takes the form

$$\mathbf{T} = \frac{2}{J}\,\mathbf{F}\,\partial_{\mathbf{C}}\tilde{\psi}(\mathbf{C})\mathbf{F}^T.$$

Materials satisfying (4.8) are termed *hyperelastic.* If the material is hyperelastic then the dependence of $\mathbf{T}_{RR}$ on $\mathbf{C}$ has to satisfy

$$\partial_{\mathbf{C}}\tilde{\mathbf{T}}_{RR}(\mathbf{C}) = 2\partial_{\mathbf{C}}\partial_{\mathbf{C}}\tilde{\psi}_R \in \mathrm{Sym}(\mathrm{Lin}, \mathrm{Lin}).$$

If, instead, $\partial_{\mathbf{C}}\tilde{\mathbf{T}}_{RR}(\mathbf{C}) \notin \mathrm{Sym}(\mathrm{Lin}, \mathrm{Lin})$ then the stress relation (4.8) does not hold and the entropy inequality simply says that

$$\dot{\psi}_R \leq \tfrac{1}{2}\mathbf{T}_{RR} \cdot \dot{\mathbf{C}}$$

and the balance of energy yields

$$\dot{\varepsilon}_R = \dot{\psi}_R + (\theta\eta_R)^{\cdot} = \tfrac{1}{2}\mathbf{T}_{RR} \cdot \dot{\mathbf{C}} + r_R.$$

In such a case we say that $\tilde{\mathbf{T}}_{RR}(\mathbf{C})$ models an elastic solid, not a hyperelastic solid.

A reference configuration is said to be *natural* if $\tilde{\psi}(\mathbf{C})$ has a local minimum at $\mathbf{C} = \mathbf{1}$. This means that

$$\tilde{\psi}(\mathbf{C}) \geq \tilde{\psi}(\mathbf{1}), \qquad |\mathbf{C} - \mathbf{1}| < \delta,$$

for a suitable $\delta > 0$. This implies that

$$\partial_{\mathbf{C}}\tilde{\psi}(\mathbf{C}) = \mathbf{0} \quad \text{at} \quad \mathbf{C} = \mathbf{1}.$$

Hence $\mathbf{T}$, $\mathbf{T}_R$, and $\mathbf{T}_{RR}$ vanish at $\mathbf{C} = \mathbf{1}$ or $\mathbf{F} = \mathbf{1}$. In words, a natural reference configuration is stress free. Consequently,

$$\partial^2_{\mathbf{C}}\tilde{\psi}(\mathbf{1}) \quad \text{is positive semidefinite}$$

that is

$$\mathbf{A} \cdot \partial^2_{\mathbf{C}}\tilde{\psi}(\mathbf{1})\mathbf{A} \geq 0 \quad \forall \mathbf{A} \in \mathrm{Sym}.$$

Elastic materials are characterized by constitutive equations where the stress is determined by the strain. Hyperelastic materials are a particular case of elastic materials where the stress is derivable from an energy potential (of the strain). In general *hypoelastic materials* are characterized by a rate-equation ([428], p. 99)

$$\overset{\circ}{\mathbf{T}} = \mathfrak{h}(\mathbf{T}, \mathbf{D}).$$

Here we restrict attention to hypoelastic solids, in the Lagrangian description, as given by *rate-independent* constitutive equations in the form [310]

$$\mathfrak{T}(\mathbf{T}_{RR}, \mathbf{C})\,\dot{\mathbf{T}}_{RR} = \mathfrak{C}(\mathbf{T}_{RR}, \mathbf{C})\,\dot{\mathbf{C}}.$$

If the tensor $\mathfrak{T}(\mathbf{T}_{RR}, \mathbf{C})$ is invertible then we can write

$$\dot{\mathbf{T}}_{RR} = \mathfrak{H}(\mathbf{T}_{RR}, \mathbf{C})\,\dot{\mathbf{C}};$$

this case is investigated in Sect. 13.6.2.

4.2.1 Material Symmetries

We now characterize the symmetries of a solid within the reference configuration. Each symmetry results in the invariance, of the second Piola stress and the free energy, to specific symmetry transformations namely rotations about specific axes and reflections about specific planes. Hence a material symmetry consists of a class, usually called *symmetry group,* of rigid transformations, of the reference shape, which are mechanically undetectable. A symmetry transformation can be represented by an orthogonal second-order tensor $\mathbf{Q} \in \mathrm{Orth}$; it is called *rotation* when $\det \mathbf{Q} = 1$ and *reflection* when $\det \mathbf{Q} = -1$.

A set $\mathcal{G}$ of rotations forms a group if it is closed under multiplication and inversion,

$$\mathbf{Q}_1, \mathbf{Q}_2 \in \mathcal{G} \implies \mathbf{Q}_1\mathbf{Q}_2 \in \mathcal{G}, \qquad \mathbf{Q} \in \mathcal{G} \implies \mathbf{Q}^T \in \mathcal{G}.$$

The set of all rotations forms the proper orthogonal group Orth^+ and $\mathcal{G}$ is a subgroup of Orth^+.

Let $\mathcal{A}$ be a set of tensors, Lin^+, Sym, or Psym. The transform of $\mathbf{A} \in \mathcal{A}$ under the rotation $\mathbf{Q}$ is $\mathbf{QAQ}^T$. With a view to the free energy and the second Piola stress, we say that a scalar-valued function ϕ on $\mathcal{A}$ is invariant under $\mathcal{G}$ if

$$\phi(\mathbf{QAQ}^T) = \phi(\mathbf{A}) \quad \forall \mathbf{A} \in \mathcal{A}$$

and that a tensor-valued function $\mathbf{\Phi}$ on $\mathcal{A}$ is invariant under $\mathcal{G}$ if

$$\mathbf{\Phi}(\mathbf{QAQ}^T) = \mathbf{Q}\mathbf{\Phi}(\mathbf{A})\mathbf{Q}^T \quad \forall \mathbf{A} \in \mathcal{A}.$$

Let $\mathbf{X}$ be a point in the reference configuration R. If $\mathbf{F}$ is a (constant) deformation gradient then

$$\mathbf{f}_{\mathbf{F}}(\mathbf{Y}) = \mathbf{X} + \mathbf{F}(\mathbf{Y} - \mathbf{X})$$

is a homogeneous deformation. If the body is subject to the rotation $\mathbf{Q}$ around $\mathbf{X}$,

$$\mathbf{f_Q(Y) = X + Q(Y - X)}$$

and then to the homogeneous deformation $\mathbf{f_F}$ then the composite deformation is given by

$$\mathbf{Y - X} \ \stackrel{\mathbf{Q}}{\to} \ \mathbf{Q(Y - X)} \ \stackrel{\mathbf{F}}{\to} \ \mathbf{FQ(Y - X)}.$$

The corresponding deformation gradient $\mathbf{F}_2$ is

$$\mathbf{F}_2 = \mathbf{FQ}$$

and

$$\mathbf{C}_2 = \mathbf{F}_2^T\mathbf{F}_2 = \mathbf{Q}^T\mathbf{C}\mathbf{Q}, \qquad \mathbf{C} = \mathbf{F}^T\mathbf{F}.$$

If

$$\tilde{\psi}_R(\mathbf{C}) = \tilde{\psi}_R(\mathbf{Q}^T\mathbf{CQ})$$

for every $\mathbf{C} \in$ Psym then $\mathbf{Q}$ is referred to as a *symmetry transformation.* The set $\mathcal{G}$ of all symmetry transformations is referred to as the *symmetry group* for the body. That $\mathcal{G}$ is indeed a group is shown as follows. If $\mathbf{Q}_1$ and $\mathbf{Q}_2$ belong to $\mathcal{G}$ then by applying first $\mathbf{Q}_1$ and then $\mathbf{Q}_2$ we have

$$\hat{\psi}_R(\mathbf{F}) = \hat{\psi}_R(\mathbf{FQ}_1), \qquad \hat{\psi}_R(\mathbf{FQ}_1) = \hat{\psi}_R(\mathbf{F}\,\mathbf{Q}_1\mathbf{Q}_2)$$

whence

$$\hat{\psi}_R(\mathbf{F}) = \hat{\psi}_R(\mathbf{F}\,\mathbf{Q}_1\mathbf{Q}_2);$$

if $\mathbf{Q}_1, \mathbf{Q}_2 \in \mathcal{G}$ then $\mathbf{Q}_1\mathbf{Q}_2 \in \mathcal{G}$. Now let $\mathbf{Q} \in \mathcal{G}$ and look at any $\mathbf{F}$. Hence

$$\hat{\psi}_R(\mathbf{F}) = \hat{\psi}_R(\mathbf{F}\,\mathbf{Q}).$$

Letting $\mathbf{F}_1 = \mathbf{F}\,\mathbf{Q}$ we can write $\mathbf{F} = \mathbf{F}_1\mathbf{Q}^T$ and hence we have

$$\hat{\psi}_R(\mathbf{F}_1) = \hat{\psi}_R(\mathbf{F}_1\mathbf{Q}^T),$$

whereby if $\mathbf{Q} \in \mathcal{G}$ then $\mathbf{Q}^T \in \mathcal{G}$. This concludes the proof that $\mathcal{G}$ is a group.

By the stress relation $\mathbf{T}_{RR} = 2\partial_{\mathbf{C}}\tilde{\psi}_R$ it follows that the invariance of $\tilde{\psi}_R$ implies that of $\mathbf{T}_{RR}$. To show this property let $\bar{\mathbf{C}} = \mathbf{Q}^T\mathbf{CQ}$ and observe that, owing to the invariance of ψ_R,

$$\dot{\psi}_R(t) = \partial_{C_{HI}}\tilde{\psi}_R(\mathbf{C})\,\dot{C}_{HI} = \partial_{\bar{C}_{AB}}\tilde{\psi}_R(\bar{\mathbf{C}})\,\partial_{C_{HI}}\bar{C}_{AB}\,\dot{C}_{HI}.$$

Since

$$\partial_{C_{HI}}\tilde{\psi}_R(\mathbf{C}) = [\tilde{\mathbf{T}}_{RR}(\mathbf{C})]_{HI}, \qquad \partial_{\bar{C}_{AB}}\tilde{\psi}_R(\bar{\mathbf{C}}) = [\tilde{\mathbf{T}}_{RR}(\bar{\mathbf{C}})]_{AB}, \qquad \partial_{C_{HI}}\bar{C}_{AB} = Q_{HA}\,Q_{IB},$$

we can write

$$[\tilde{\mathbf{T}}_{RR}(\bar{\mathbf{C}}) - \mathbf{Q}^T \tilde{\mathbf{T}}_{RR}(\mathbf{C})\mathbf{Q}] \cdot \dot{\mathbf{C}} = 0.$$

The arbitrariness of $\dot{\mathbf{C}}$ allows us to obtain the sought transformation law for the second Piola stress,

$$\tilde{\mathbf{T}}_{RR}(\mathbf{Q}^T \mathbf{C}\,\mathbf{Q}) = \mathbf{Q}^T \tilde{\mathbf{T}}_{RR}(\mathbf{C})\mathbf{Q}.$$

If every rotation is a symmetry transformation then we say that the body is *isotropic*. Accordingly, the body is said to be isotropic if

$$\mathcal{G} = \mathrm{Orth}^+.$$

If, instead, $\mathcal{G}$ is a proper subgroup of $\mathcal{G}$ then the body is said to be *anisotropic*. In particular, if $\mathcal{G} = \mathrm{Orth}^+(\mathbf{a})$, the collection of all rotations about a fixed axis $\mathbf{a}$, the body is said to be *transversely isotropic*, in that any plane that is normal to the axis $\mathbf{a}$ is a plane of isotropy.

We now show some properties of ψ_R and $\mathbf{T}_{RR}$ for isotropic bodies. We start from the invariance property

$$\tilde{\psi}_R(\mathbf{C}) = \tilde{\psi}_R(\mathbf{Q}^T \mathbf{C}\,\mathbf{Q}),$$

for all rotations $\mathbf{Q}$. Really, the property holds by replacing $\mathbf{Q}$ by $-\mathbf{Q}$ and hence is true for all orthogonal tensors. This means that $\tilde{\psi}_R$ is an isotropic function, namely it is invariant under $\mathcal{G} = \mathrm{Orth}$.

For isotropic bodies the free energy can be represented equivalently in terms of the right and left Cauchy-Green tensors

$$\mathbf{C} = \mathbf{F}^T \mathbf{F}, \qquad \mathbf{B} = \mathbf{F}\mathbf{F}^T.$$

By the polar decomposition, $\mathbf{F} = \mathbf{R}\,\mathbf{U} = \mathbf{V}\,\mathbf{R}$, it follows that

$$\mathbf{B} = \mathbf{R}\,\mathbf{C}\,\mathbf{R}^T.$$

Consequently, letting $\mathbf{Q} = \mathbf{R}^T$ we obtain

$$\tilde{\psi}_R(\mathbf{C}) = \tilde{\psi}_R(\mathbf{B}).$$

By

$$\dot{\psi}_R = \tfrac{1}{2}\mathbf{T}_{RR} \cdot \dot{\mathbf{C}} = \mathbf{T}_R \cdot \dot{\mathbf{F}}$$

and taking $\psi_R = \tilde{\psi}_R(\mathbf{B})$ we have

$$\dot{\psi}_R = \mathbf{T}_R \cdot \dot{\mathbf{F}}, \qquad \partial_{\mathbf{B}}\tilde{\psi}_R(\mathbf{B}) \cdot \dot{\mathbf{B}} = \mathbf{T}_R \cdot \dot{\mathbf{F}}.$$

Since

$$\dot{\mathbf{B}} = \dot{\mathbf{F}}\mathbf{F}^T + \mathbf{F}\dot{\mathbf{F}}^T = 2\text{sym}(\dot{\mathbf{F}}\mathbf{F}^T)$$

and $\partial_{\mathbf{B}}\tilde{\psi}_R(\mathbf{B})$ is symmetric then

$$\partial_{\mathbf{B}}\tilde{\psi}_R(\mathbf{B}) \cdot \dot{\mathbf{B}} = 2\partial_{\mathbf{B}}\tilde{\psi}_R(\mathbf{B}) \cdot (\dot{\mathbf{F}}\mathbf{F}^T) = 2[\partial_{\mathbf{B}}\tilde{\psi}_R(\mathbf{B})\mathbf{F}] \cdot \dot{\mathbf{F}}.$$

Thus, the entropy inequality yields

$$0 \geq [2\,\partial_{\mathbf{C}}\tilde{\psi}(\mathbf{C}) - \mathbf{T}_{RR}] \cdot \dot{\mathbf{C}} = [2\,\partial_{\mathbf{B}}\tilde{\psi}_R(\mathbf{B})\mathbf{F} - \mathbf{T}_R] \cdot \dot{\mathbf{F}}.$$

The arbitrariness of $\dot{\mathbf{F}}$ implies that $\mathbf{T}_R$ is given by

$$\hat{\mathbf{T}}_R(\mathbf{F}) = 2\,\partial_{\mathbf{B}}\tilde{\psi}_R(\mathbf{B})\mathbf{F}.$$

Since $\mathbf{T} = J^{-1}\mathbf{T}_R\mathbf{F}^T$ and $J = \det\mathbf{F} = (\det\mathbf{B})^{1/2}$ then it follows that the Cauchy stress is a function of $\mathbf{B}$,

$$\tilde{\mathbf{T}}(\mathbf{B}) = \frac{2}{(\det\mathbf{B})^{1/2}}\partial_{\mathbf{B}}\tilde{\psi}_R(\mathbf{B})\mathbf{B}.$$

If $\mathbf{P} = \mathbf{M}\,\mathbf{N}$, where $\mathbf{M}, \mathbf{N}, \mathbf{P} \in \text{Sym}$, then

$$\mathbf{NP} = \mathbf{NMN} = \mathbf{N}^T\mathbf{M}^T\mathbf{N}^T = (\mathbf{NMN})^T = (\mathbf{NP})^T$$

and, hence, $\mathbf{P}$ commutes with $\mathbf{N}$. The identifications

$$\mathbf{P} = \mathbf{T}, \quad \mathbf{M} = \frac{2}{(\det\mathbf{B})^{1/2}}\partial_{\mathbf{B}}\tilde{\psi}_R(\mathbf{B}), \quad \mathbf{N} = \mathbf{B}$$

allow us to say that $\mathbf{T}$ commutes with $\mathbf{B}$,

$$\mathbf{BT} = \mathbf{TB}.$$

Let $\phi : \text{Sym} \to \mathbb{R}$. If ϕ is invariant under $\mathcal{G}$ then $\partial_{\mathbf{A}}\phi(\mathbf{A})$ is also invariant under $\mathcal{G}$. To prove this property, observe that

$$\phi(\mathbf{A}) = \phi(\mathbf{Q}^T\mathbf{A}\mathbf{Q}) \qquad \forall\mathbf{Q} \in \mathcal{G},$$

$$\partial_{\mathbf{A}}\phi(\mathbf{A}) = \partial_{\mathbf{A}}\phi(\mathbf{Q}^T\mathbf{A}\mathbf{Q}) = \partial_{\bar{\mathbf{A}}}\phi(\bar{\mathbf{A}})\,\partial_{\mathbf{A}}\bar{\mathbf{A}},$$

where $\bar{\mathbf{A}} = \mathbf{Q}^T\mathbf{A}\mathbf{Q}$. Now,

$$\partial_{A_{pq}}(\mathbf{Q}^T\mathbf{A}\mathbf{Q})_{ij} = Q_{pi}\,Q_{qj},$$

whence

$$\partial_{\bar{\mathbf{A}}}\phi(\bar{\mathbf{A}})\,\partial_{\mathbf{A}}\bar{\mathbf{A}} = \mathbf{Q}\,\partial_{\bar{\mathbf{A}}}\phi(\bar{\mathbf{A}})\,\mathbf{Q}^T.$$

Upon substitution we obtain the sought relation

$$\partial_{\bar{\mathbf{A}}}\phi(\bar{\mathbf{A}}) = \mathbf{Q}^T\,\partial_{\mathbf{A}}\phi(\mathbf{A})\mathbf{Q}.$$

As an immediate consequence, we let $\mathbf{A} = \mathbf{B}$, $\phi = \tilde{\psi}_R$, to find

$$\begin{aligned}\tilde{\mathbf{T}}(\mathbf{Q}^T\mathbf{B}\mathbf{Q}) &= \frac{2}{(\det\mathbf{B})^{1/2}}\partial_{\bar{\mathbf{B}}}\tilde{\psi}_R(\bar{\mathbf{B}})\mathbf{Q}^T\mathbf{B}\mathbf{Q}\\ &= \frac{2}{(\det\mathbf{B})^{1/2}}\mathbf{Q}^T\partial_{\mathbf{B}}\tilde{\psi}_R(\mathbf{B})\mathbf{Q}\mathbf{Q}^T\mathbf{B}\mathbf{Q} = \mathbf{Q}^T\tilde{\mathbf{T}}(\mathbf{B})\mathbf{Q},\end{aligned}$$

which means that $\tilde{\mathbf{T}}(\mathbf{B})$ is isotropic.

The isotropy of $\tilde{\mathbf{T}}(\mathbf{B})$ implies that $\mathbf{T}$ commutes with $\mathbf{C}$,

$$\mathbf{C}\,\mathbf{T} = \mathbf{T}\,\mathbf{C}.$$

For this, since $\mathbf{B} = \mathbf{R}\,\mathbf{C}\,\mathbf{R}^T$ we let $\mathbf{Q} = \mathbf{R}^T$ and find

$$\mathbf{T}\,\mathbf{B} = \tilde{\mathbf{T}}(\mathbf{B})\mathbf{B} = \tilde{\mathbf{T}}(\mathbf{R}\,\mathbf{C}\,\mathbf{R}^T)\mathbf{R}\,\mathbf{C}\,\mathbf{R}^T = \mathbf{R}\tilde{\mathbf{T}}(\mathbf{C})\mathbf{R}^T\mathbf{R}\mathbf{C}\mathbf{R}^T = \mathbf{R}\tilde{\mathbf{T}}(\mathbf{C})\mathbf{C}\mathbf{R}^T.$$

Likewise we find that

$$\mathbf{B}\,\mathbf{T} = \mathbf{R}\,\mathbf{C}\,\tilde{\mathbf{T}}(\mathbf{C})\mathbf{R}^T.$$

Since $\mathbf{B}\,\mathbf{T} = \mathbf{T}\,\mathbf{B}$ then

$$\mathbf{R}\,\mathbf{C}\,\tilde{\mathbf{T}}(\mathbf{C})\mathbf{R}^T = \mathbf{R}\tilde{\mathbf{T}}(\mathbf{C})\mathbf{C}\mathbf{R}^T,$$

whence the sought result, $\mathbf{C}\,\mathbf{T} = \mathbf{T}\,\mathbf{C}$, follows.

4.2.2 *Elastic Waves*

Nonlinear Waves

We now investigate wave solutions, in elastic materials, in the form of *singular surfaces of order 2* (acceleration waves or weak waves). Hence $\mathbf{v} = \dot{\chi}$ is continuous across the singular surface whereas $\dot{\mathbf{v}}, \mathbf{L}$ and higher-order derivatives suffer jump discontinuities across the singular surface.

It is convenient to describe the waves in the reference configuration. Denote by Σ the singular surface. By means of the first Piola stress $\mathbf{T}_R$ we can write the equation of motion in the form

$$\rho_R\dot{\mathbf{v}} = \nabla_R\cdot\mathbf{T}_R + \rho_R\mathbf{b},$$

while the constitutive equation is

$$\mathbf{T}_R = \mathbf{T}_R(\mathbf{F}).$$

We let **b** be continuous across the singular surface and hence we have the jump condition

$$\rho_R[\![\dot{\mathbf{v}}]\!] = [\![\nabla_R \cdot \mathbf{T}_R]\!].$$

We assume

$$[\![\mathbf{T}_R]\!] = \mathbf{0},$$

so that Poisson's condition is satisfied.

We now apply repeatedly Maxwell's theorem and the kinematical condition of compatibility. The continuity of $\mathbf{T}_R$ implies that there exists a second-order tensor **S** such that

$$[\![\nabla_R \mathbf{T}_R]\!] = \mathbf{S} \otimes \mathbf{n}_R, \qquad [\![\dot{\mathbf{T}}_R]\!] = -U_R \mathbf{S}, \qquad \mathbf{S} := [\![\partial_{n_R} \mathbf{T}_R]\!].$$

Since $[\![\dot{\chi}]\!] = \mathbf{0}$ then

$$[\![\ddot{\chi}]\!] = -U_R[\![\partial_{n_R}\dot{\chi}]\!] = U_R^2\,\mathbf{a}, \qquad \mathbf{a} := [\![\partial_{n_R}\chi]\!].$$

Moreover

$$[\![\dot{\mathbf{F}}]\!] = [\![\nabla_R \mathbf{v}]\!] = [\![\partial_{n_R}\mathbf{v}]\!] \otimes \mathbf{n}_R = -U_R \mathbf{a} \otimes \mathbf{n}_R.$$

Since

$$U_R[\![\nabla_R \cdot \mathbf{T}_R]\!] = -[\![\dot{\mathbf{T}}_R]\!]\mathbf{n}_R$$

then we obtain

$$[\![\dot{\mathbf{T}}_R]\!]\mathbf{n}_R = -\rho_R\, U_R^3\, \mathbf{a}.$$

The jump $[\![\dot{\mathbf{T}}_R]\!]$ is now related to **a** via the constitutive property $\mathbf{T}_R = \mathbf{T}_R(\mathbf{F})$. Denote by A the fourth-order tensor defined by

$$\mathsf{A}_{iHjK} = \partial_{F_{jK}} T^R_{iH}.$$

Since **F** is continuous, across Σ, then A is continuous too. Hence we have

$$[\![\dot{\mathbf{T}}_R]\!] = \mathsf{A}[\![\dot{\mathbf{F}}]\!].$$

Upon substitution of $[\![\dot{\mathbf{F}}]\!]$ we find

$$[\![\dot{\mathbf{T}}_R]\!]\,\mathbf{n}_R = -U_R\,\mathbf{A}\,\mathbf{a},$$

where **A** depends on $\mathbf{n}_R$ such that

$$\mathbf{A}\mathbf{a} = \mathsf{A}[\mathbf{a} \otimes \mathbf{n}_R]\mathbf{n}_R.$$

Equalling the two expressions of $[\![\dot{\mathbf{T}}_R]\!]\mathbf{n}_R$ we find the *propagation condition*

$$\mathbf{A}\,\mathbf{a} = \rho_R\, U_R^2\, \mathbf{a}.$$

The second-order tensor $\mathbf{A}$ is called the *reference acoustic tensor.* Hence the vector $\mathbf{a}$ is an eigenvector of $\mathbf{A}$ and $\rho_R U_R^2$ is the corresponding eigenvalue. This eigenvalue problem is parameterized by the direction $\mathbf{n}_R$ of the surface Σ. The directions $\mathbf{n}_R$ of the eigenvectors are also called the *acoustic axes.* The notation $\mathbf{A}(\mathbf{n})$ emphasizes that $\mathbf{A}$ depends on $\mathbf{n}$.

Likewise, taking the current configuration as reference we find that the strictly analogous propagation condition holds with ρ, u_n, $\mathbf{n}$ instead of ρ_R, U_R, $\mathbf{n}_R$.

The results so obtained are stated as follows.

Theorem 4.1 (Fresnel-Hadamard) *The vector jump* $\mathbf{a}$ *of an acceleration wave in the direction* $\mathbf{n}$ *has to be an eigenvector of the acoustic tensor* $\mathbf{A}(\mathbf{n})$. *The corresponding propagation speed* U *is given by*

$$\rho_R U_R^2 = \mathbf{w} \cdot \mathbf{A}(\mathbf{n}_R)\mathbf{w}, \qquad \mathbf{w} := \mathbf{a}/|\mathbf{a}|.$$

As any eigenvalue, $\rho_R\, U_R^2$ *is the solution to the characteristic equation*

$$(\rho_R\, U_R^2)^3 - I_1\, (\rho_R\, U_R^2)^2 + I_2\, \rho_R\, U_R^2 - I_3 = 0,$$

I_1, I_2, I_3 *being the principal invariants of* $\mathbf{A}(\mathbf{n}_R)$.

Shock waves are characterized by the singularity of the velocity field. We let $[\![\chi]\!] = \mathbf{0}$. Hence the geometrical and the kinematical conditions of compatibility imply

$$[\![\mathbf{F}]\!] = [\![\partial_{n_R}\chi]\!] \otimes \mathbf{n}_R, \qquad [\![\mathbf{v}]\!] = -U_R[\![\partial_{n_R}\chi]\!].$$

The jump condition

$$\rho_R U_R[\![\mathbf{v}]\!] + [\![\mathbf{T}_R]\!]\mathbf{n}_R = \mathbf{0}$$

takes the form

$$\rho_R U_R^2[\![\partial_{n_R}\chi]\!] - [\![\mathbf{T}_R(\nabla_R\chi)]\!]\mathbf{n}_R = \mathbf{0},$$

that is an equation in the unknown derivatives $\nabla_R\chi = \mathbf{F}$. To fix ideas let the state ahead be in the reference state, that is $\mathbf{F} = \mathbf{1}$ ahead of the wave. Hence $\mathbf{T}_R(\mathbf{1})$ is the Piola stress ahead. For the state behind we can write

$$\mathbf{T}_R = \mathbf{T}_R(\mathbf{1}) + (\partial_{F_{jk}}\mathbf{T}_R)(F_{jK} - \delta_{jK}) + o(\mathbf{F} - \mathbf{1}),$$

the derivatives $\partial_{F_{jk}}\mathbf{T}_R$ being evaluated at $\mathbf{F} = \mathbf{1}$. By neglecting the $o(\mathbf{F} - \mathbf{1})$ term we follow the linearized approximation. Hence we have

$$[\![\mathbf{T}_R]\!] = \mathsf{A}[\![\mathbf{F}]\!]$$

and

$$[\![\mathbf{T}_R]\!]\mathbf{n}_R = \mathbf{A}[\![\partial_{n_R}\chi]\!].$$

Consequently, in the linearized approximation, shock waves are governed by the propagation condition

$$(\mathbf{A}(\mathbf{n}_R) - \rho_R U_R^2 \mathbf{1})[\![\partial_{n_R}\chi]\!] = 0. \tag{4.9}$$

As with the weak waves, in the linearized approximation of $\mathbf{T}_R(\mathbf{F})$, the shock wave polarization $[\![\partial_{n_R}\chi]\!]$ is an eigenvector of $\mathbf{A}$ and $\rho_R U_R^2$ is the corresponding eigenvalue.

The propagation condition (4.9) is also called Christoffel equation while $\mathbf{A}$ is called the Green-Christoffel tensor.

Plane Progressive Waves

The body is taken to be at rest in the reference configuration so that

$$\nabla_R \cdot \mathbf{T}_R + \rho_R \mathbf{b} = \mathbf{0}.$$

A deformation is now superimposed; the variation of $\mathbf{T}_R$ is considered in the linearized approximation,

$$\mathbf{T}_R = \mathsf{A}(\mathbf{F} - \mathbf{1})$$

while ρ_R is regarded as constant. The equation of motion can then be written in terms of the displacement function $\mathbf{u}(\mathbf{X}, t)$ in the form

$$\rho_R\, \partial_t^2 \mathbf{u} = \mathsf{A}\nabla_R \mathbf{u}, \qquad \rho_R\, \partial_t^2 u_i = \mathsf{A}_{iHjK} \partial_{X_H} \partial_{X_K} u_j.$$

A plane wave propagating with phase speed c in the direction of the unit propagation vector $\mathbf{p}$ is represented by the function

$$\mathbf{u}(\mathbf{X}, t) = f(\mathbf{X} \cdot \mathbf{p} - ct)\, \mathbf{d},$$

where f is a scalar-valued function and $\mathbf{d}$ is the unit vector defining the direction of motion (or polarization). Upon substitution in the equation of motion and letting $\phi = \mathbf{X} \cdot \mathbf{p} - ct$ we find

$$[\rho_R c^2 \mathbf{1} - \mathbf{A}(\mathbf{p})]\mathbf{d}\, f''(\phi) = 0,$$

where

$$\mathbf{A}(\mathbf{p}) = \mathsf{A}\, \mathbf{p} \otimes \mathbf{p}, \qquad A_{ij} = \mathsf{A}_{iHjK} p_H p_K.$$

Accordingly, plane waves are admissible, for any function f, in the direction of propagation $\mathbf{p}$, if the polarization $\mathbf{d}$ is an eigenvector of $\mathbf{A}$ with eigenvalue $\rho_R c^2$,

$$\mathbf{A}(\mathbf{p})\,\mathbf{d} = \rho_R\, c^2\,\mathbf{d}.$$

This in turn means that waves are allowed, with direction $\mathbf{p}$ and polarization $\mathbf{d}$, if

$$0 < \mathbf{d} \cdot \mathbf{A}(\mathbf{p})\,\mathbf{d} = (\mathbf{d} \otimes \mathbf{p}) \cdot \mathsf{A}(\mathbf{d} \otimes \mathbf{p}).$$

Strong ellipticity, that is

$$(\mathbf{e} \otimes \mathbf{f}) \cdot \mathsf{A}(\mathbf{e} \otimes \mathbf{f}) > 0$$

for all vectors $\mathbf{e}$ and $\mathbf{f}$, guarantees propagation with any pair of vectors $\mathbf{p}$ and $\mathbf{d}$.

4.2.3 Linear Theory

The linear theory of elasticity is derived as an approximation of the exact theory. We let the reference configuration be natural, that is stress free, and restrict attention to linear terms in $\mathbf{F} - \mathbf{1}$, which is the content of 'small deformations' approximation.

Since

$$\mathbf{F} = \mathbf{1} + \mathbf{H}, \qquad H_{ij} = \partial_{X_j} u_i,$$

$\mathbf{u}$ being the displacement, we look for linear representations with respect to the displacement gradient $\mathbf{H}$. The fourth-order tensor

$$\mathsf{C} = 2\,\partial_{\mathbf{C}} \tilde{\mathbf{T}}_{RR}(\mathbf{C}), \quad \text{at } \mathbf{C} = \mathbf{1},$$

is referred to as the *elasticity tensor.* Since both $\mathbf{T}_{RR}$ and $\mathbf{C}$ are symmetric it follows that, in the suffix notation,

$$\mathsf{C}_{ijhk} = \mathsf{C}_{jihk} = \mathsf{C}_{ijkh}.$$

These relations are referred to as minor symmetries and hold for any elastic solid. If, further, the solid is hyperelastic then

$$\mathsf{C} = 4\,\partial^2_{\mathbf{C}} \tilde{\psi}_R(\mathbf{C}), \qquad \mathsf{C}_{ijhk} = 4\,\partial_{C_{ij}} \partial_{C_{hk}} \tilde{\psi}_R(\mathbf{C}), \quad \text{at } \mathbf{C} = \mathbf{1},$$

and hence C satisfies also the major symmetry conditions

$$\mathsf{C}_{ijhk} = \mathsf{C}_{hkij}.$$

It is usually assumed that C is positive-definite, which means that

$$\mathbf{A}\cdot\mathsf{C}\,\mathbf{A}>0 \qquad \forall\mathbf{A}\in \mathrm{Sym}\setminus\{\mathbf{0}\}.$$

This amounts to assuming that the free energy has a minimum at the reference state, $\mathbf{C}=\mathbf{1}$.

We now show that C is invariant under the symmetry group $\mathcal{G}$ of the material. Look at a time dependence $\mathbf{C}(t)$ such that

$$\mathbf{C}(0)=\mathbf{1}, \qquad \dot{\mathbf{C}}(0)=\mathbf{A}$$

and consider

$$\mathbf{T}_{RR}(t)=\tilde{\mathbf{T}}_{RR}(\mathbf{C}(t)).$$

Hence we have

$$\dot{\mathbf{T}}_{RR}=\partial_{\mathbf{C}}\tilde{\mathbf{T}}_{RR}(\mathbf{C})\,\dot{\mathbf{C}}, \qquad \dot{\mathbf{T}}_{RR}(0)=\tfrac{1}{2}\mathsf{C}\,\mathbf{A}.$$

Now apply the invariance property of $\mathbf{T}_{RR}$,

$$\mathbf{Q}^T\mathbf{T}_{RR}\mathbf{Q}=\tilde{\mathbf{T}}(\mathbf{Q}^T\,\mathbf{C}\,\mathbf{Q}).$$

Time differentiation and evaluation at $t=0$ gives

$$\mathbf{Q}^T\dot{\mathbf{T}}_{RR}(0)\mathbf{Q}=\tfrac{1}{2}\mathsf{C}\,\mathbf{Q}^T\,\dot{\mathbf{C}}(0)\mathbf{Q}$$

whence

$$\mathbf{Q}^T[\mathsf{C}\,\mathbf{A}]\mathbf{Q}=\mathsf{C}[\mathbf{Q}^T\,\mathbf{A}\,\mathbf{Q}],$$

that is the transformation law of C for any symmetry transformation $\mathbf{Q}$ and every symmetric tensor $\mathbf{A}$.

For formal convenience we let the reference configuration be stress free and hence let the second Piola stress $\mathbf{T}_{RR}(\mathbf{C})$ satisfy

$$\mathbf{T}_{RR}(\mathbf{1})=\mathbf{0}, \qquad (\partial_{\mathbf{C}}\mathbf{T}_{RR})(\mathbf{1})=\tfrac{1}{2}\mathsf{C}$$

so that

$$\mathbf{T}_{RR}(\mathbf{C})=\tfrac{1}{2}\mathsf{C}(\mathbf{C}-\mathbf{1})+o(|\mathbf{C}-\mathbf{1}|).$$

In terms of the Green-St. Venant strain tensor $\mathbf{E}=\frac{1}{2}(\mathbf{C}-\mathbf{1})$ we can write

$$\mathbf{T}_{RR}=\mathsf{C}\,\mathbf{E}+o(|\mathbf{E}|).$$

To determine linear approximations it is convenient to refer to the displacement gradient $\mathbf{H}$. Now

$$\mathbf{F}=\mathbf{1}+\mathbf{H}, \qquad J=\det\mathbf{F}=1+O(|\mathbf{H}|), \quad J^{-1}=1+O(|\mathbf{H}|)$$

and hence

$$\mathbf{T} = J^{-1}\mathbf{F}\mathbf{T}_{RR}\mathbf{F}^T = \mathbf{T}_{RR} + O(|\mathbf{H}|), \qquad \mathbf{T}_R = \mathbf{F}\mathbf{T}_{RR} = \mathbf{T}_{RR} + O(|\mathbf{H}|).$$

To within $O(|\mathbf{H}|)$ terms, the Cauchy and Piola stresses coincide. Moreover,

$$\rho = J^{-1}\rho_R = \rho_R + O(|\mathbf{H}|).$$

The expansion of $\psi_R(\mathbf{C})$ about $\mathbf{C} = \mathbf{1}$ can be written in the form

$$\psi_R(\mathbf{C}) = \psi_R(\mathbf{1}) + (\partial_{\mathbf{C}}\psi_R)(\mathbf{1})(\mathbf{C}-\mathbf{1}) + \tfrac{1}{2}(\mathbf{C}-\mathbf{1}) \cdot (\partial^2_{\mathbf{C}}\psi_R)(\mathbf{1})(\mathbf{C}-\mathbf{1}) + o(|\mathbf{H}|^2).$$

The assumptions that ψ_R and $\mathbf{T}_{RR}$ vanish at $\mathbf{C} = \mathbf{1}$ and the definition of C,

$$\tfrac{1}{2}\mathsf{C} = (\partial_{\mathbf{C}}\mathbf{T}_{RR})(\mathbf{1}) = 2(\partial^2_{\mathbf{C}}\psi_R)(\mathbf{1}),$$

allow ψ_R to be given the form

$$\psi_R(\mathbf{C}) = \tfrac{1}{8}(\mathbf{C}-\mathbf{1}) \cdot \mathsf{C}(\mathbf{C}-\mathbf{1}) + o(|\mathbf{H}|^2) = \tfrac{1}{2}\mathbf{E} \cdot \mathsf{C}\mathbf{E} + o(|\mathbf{H}|^2).$$

The *linear theory* approximation of elasticity follows by restricting attention to the leading terms in $\mathbf{H}$. We let

$$\mathbf{T} = \mathsf{C}\mathbf{E}, \quad T_{ij} = \mathsf{C}_{ijhk}E_{hk}; \qquad \psi_R = \tfrac{1}{2}\mathbf{E} \cdot \mathsf{C}\mathbf{E}, \quad \psi_R = \tfrac{1}{2}\mathsf{C}_{ijhk}E_{ij}E_{hk}$$

and hence

$$\mathbf{T} = \partial_{\mathbf{E}}\psi_R(\mathbf{E}).$$

The elasticity tensor C is assumed to be positive definite. Since

$$\partial_{x_j}T_{ij} = \partial_{X_k}T_{ij}F^{-1}_{kj} = \partial_{X_k}T_{ij}(\delta_{kj} + O(|\mathbf{H}|)) = \partial_{X_j}T_{ij} + O(|\mathbf{H}|),$$

in the linear approximation $\nabla \cdot \mathbf{T} \approx \nabla_R \cdot \mathbf{T}$. Hence we can write the equation of motion in the form

$$\rho_R\ddot{\mathbf{u}} = \nabla_R \cdot \mathbf{T} + \rho_R\mathbf{b}, \qquad \rho_R\ddot{u}_i = \partial_{X_j}T_{ij} + \rho_R b_i.$$

If the body is homogeneous then C is constant and the equation of motion simplifies to

$$\rho_R\ddot{\mathbf{u}} = \nabla_R \cdot (\mathsf{C}\mathbf{H}) + \rho_R\mathbf{b}, \qquad \rho_R\ddot{u}_i = \mathsf{C}_{ijhk}\,\partial_{X_j}\partial_{X_k}u_h + \rho_R b_i.$$

Linear Theory for Isotropic Materials

If the body is isotropic then $\mathbf{C}$ is an isotropic fourth-order tensor (see Appendix A.2) and we have

$$\mathbf{CE} = 2\,\mu\,\mathbf{E} + \lambda(\operatorname{tr}\mathbf{E})\mathbf{1},$$

where μ and λ are scalar constitutive parameters called the *Lamé moduli.* To account for the assumed positive definiteness of $\mathbf{C}$ we let $\mathbf{E}_0$ be the deviatoric part of $\mathbf{E}$ and let

$$\mathbf{E} = \mathbf{E}_0 + \tfrac{1}{3}(\operatorname{tr}\mathbf{E})\mathbf{1}.$$

Hence the requirement, for $\mathbf{E} \neq \mathbf{0}$, becomes

$$0 < \mathbf{E}\cdot\mathbf{CE} = 2\mu|\mathbf{E}|^2 + \lambda(\operatorname{tr}\mathbf{E})^2 = 2\mu|\mathbf{E}_0|^2 + \kappa(\operatorname{tr}\mathbf{E})^2,$$

where

$$\kappa = \tfrac{1}{3}(2\mu + 3\lambda).$$

If $\mathbf{E} = \mathbf{1}$, so that $\operatorname{tr}\mathbf{E} = 3$ and $\mathbf{E}_0 = \mathbf{0}$, it follows $\kappa > 0$. If, instead, $\mathbf{u} = u_1(X_2, t)\mathbf{e}_1$ then

$$\mathbf{E} = \tfrac{1}{2}\partial_{X_2}u_1(\mathbf{e}_1\otimes\mathbf{e}_2 + \mathbf{e}_2\otimes\mathbf{e}_1), \qquad \operatorname{tr}\mathbf{E} = 0, \qquad |\mathbf{E}_0|^2 = \tfrac{1}{2}|\partial_{X_2}u_1|^2.$$

Consequently we find $\mu > 0$. Thus $\mathbf{C}$ is positive definite if

$$\mu > 0, \qquad 2\mu + 3\lambda > 0.$$

With these restrictions, the stress-strain law for an isotropic body is

$$\mathbf{T} = 2\mu\mathbf{E} + \lambda(\operatorname{tr}\mathbf{E})\mathbf{1}$$

and the corresponding energy is

$$\psi_R = \mu|\mathbf{E}|^2 + \tfrac{1}{2}\lambda(\operatorname{tr}\mathbf{E})^2.$$

The stress-strain law may be inverted. Since

$$\operatorname{tr}\mathbf{T} = 2\mu\operatorname{tr}\mathbf{E} + 3\lambda\operatorname{tr}\mathbf{E}$$

then upon substitution we find

$$\mathbf{E} = \frac{1}{2\mu}\Big(\mathbf{T} - \frac{\lambda}{2\mu + 3\lambda}(\operatorname{tr}\mathbf{T})\mathbf{1}\Big).$$

In the linear approximation $\nabla\cdot\mathbf{T} \approx \nabla_R\cdot\mathbf{T}$. Now $\nabla_R\cdot\mathbf{T} = 2\mu\nabla_R\cdot\mathbf{E} + \lambda\nabla_R(\operatorname{tr}\mathbf{E})$ and

$$(2\nabla_R \cdot \mathbf{E})_i = \partial_{X_j}(\partial_{X_j} u_i + \partial_{X_i} u_j) = \Delta_R u_i + \partial_{X_i} \nabla_R \cdot \mathbf{u}.$$

Hence the equation of motion reads

$$\rho_R \ddot{\mathbf{u}} = \mu \Delta_R \mathbf{u} + (\lambda + \mu)\nabla_R(\nabla_R \cdot \mathbf{u}) + \rho_R \mathbf{b}.$$

Starting from the equation of motion

$$\rho_R \ddot{\mathbf{u}} = \nabla_R \cdot \mathbf{T} + \rho_R \mathbf{b}, \qquad \mathbf{T} = \mathsf{C}\mathbf{E},$$

and inner multiplying by $\dot{\mathbf{u}}$ we obtain

$$\rho_R(\tfrac{1}{2}\dot{\mathbf{u}}^2)^{\cdot} = \nabla_R \cdot (\mathbf{T}\dot{\mathbf{u}}) - \mathbf{T} \cdot \nabla_R \dot{\mathbf{u}} + \rho_R \mathbf{b} \cdot \dot{\mathbf{u}}.$$

Since

$$\mathbf{T} \cdot \nabla_R \dot{\mathbf{u}} = \mathbf{T} \cdot \dot{\mathbf{E}} = \tfrac{1}{2}(\mathbf{E} \cdot \mathsf{C}\mathbf{E})^{\cdot},$$

then we can write the global balance of energy in the form

$$\frac{d}{dt} \int_{\mathrm{P}} \tfrac{1}{2}(\rho_R \dot{\mathbf{u}}^2 + \mathbf{E} \cdot \mathsf{C}\mathbf{E}) dv = \int_{\partial \mathrm{P}} \mathbf{T}\mathbf{n} \cdot \dot{\mathbf{u}}\, da + \int_{\mathrm{P}} \rho_R \mathbf{b} \cdot \dot{\mathbf{u}}\, dv,$$

for any region $\mathrm{P} \subset \mathrm{R}$.

It is worth looking at the stress-strain relation in connection with three statical solutions. First we consider the *pure shear*

$$\mathbf{u}(\mathbf{X}) = \gamma X_2 \mathbf{e}_1.$$

Hence the matrices $[\mathbf{E}]$ and $[\mathbf{T}]$ take the forms

$$[\mathbf{E}] = \begin{bmatrix} 0 & \gamma & 0 \\ \gamma & 0 & 0 \\ 0 & 0 & 0 \end{bmatrix}, \qquad [\mathbf{T}] = \begin{bmatrix} 0 & \mu\gamma & 0 \\ \mu\gamma & 0 & 0 \\ 0 & 0 & 0 \end{bmatrix}.$$

This shows the connection between stress and strain through the *shear modulus* μ.

The second solution is that of *uniform compressions or expansions*,

$$\mathbf{u}(\mathbf{X}) = u\, \mathbf{X}.$$

Hence we have

$$\mathbf{E} = u\mathbf{1}, \qquad \mathbf{T} = 3(\lambda + 2\mu/3)u\, \mathbf{1}.$$

This shows the role of the *bulk modulus* $\lambda + 2\mu/3$.

The third solution is related to a *pure tension*, the stress arising from a force density σ in one direction, say $\mathbf{e}_1$,

$$[\mathbf{T}] = \begin{bmatrix} \sigma & 0 & 0 \\ 0 & 0 & 0 \\ 0 & 0 & 0 \end{bmatrix}.$$

Hence, by the strain-stress relation we find that

$$[\mathbf{E}] = \begin{bmatrix} \varepsilon & 0 & 0 \\ 0 & \ell & 0 \\ 0 & 0 & \ell \end{bmatrix},$$

where[5]

$$\varepsilon = \frac{\sigma}{E}, \qquad \ell = -\nu\varepsilon, \qquad E = \frac{\mu(2\mu + 3\lambda)}{\mu + \lambda}, \qquad \nu = \frac{\lambda}{2(\mu + \lambda)}.$$

The displacement field $\mathbf{u}(\mathbf{X})$ is then given by

$$\mathbf{u}(\mathbf{X}) = \varepsilon X_1 \mathbf{e}_1 + \ell X_2 \mathbf{e}_2 + \ell X_3 \mathbf{e}_3.$$

The material parameter E is the ratio of the tensile stress σ to the strain ε produced in the same direction; it is known as *Young's modulus.* The material parameter ν is the ratio of the transverse strain to the longitudinal strain; it is called *Poisson's ratio.* It is expected that an elastic solid should increase its length when pulled, decrease its volume when acted upon by a pressure, and respond to a positive shear strain by a positive shear stress. This is the case if

$$E > 0, \qquad 2\mu + 3\lambda > 0, \qquad \mu > 0.$$

We also might expect that $\nu > 0$ so that the body undergoes a lateral contraction when it is subject to a longitudinal stress. Yet,

$$\mu + \lambda = \tfrac{1}{3}(2\mu + 3\lambda + \mu) > 0, \qquad \frac{d\nu}{d\lambda} = \frac{\mu}{2(\mu + \lambda)^2} > 0.$$

Hence ν takes values from the minimum, at $\lambda = -2\mu/3$, and the maximum, as $\lambda \to \infty$. It then follows that

$$\nu \in (-1, \tfrac{1}{2}).$$

Thus there can be materials, called auxetics, that have a negative Poisson's ratio. When stretched, they become thicker perpendicular to the applied force. This may occur due to their particular internal structure and the way this deforms when the sample is uniaxially loaded.

Also with a view to the principal directions, consider a solid subject to the normal stresses T_1, T_2, T_3 on the faces S_1, S_2, S_3. The strains $\{E_i\}$ generated by T_1 are

---

[5] The symbol ε is used here to denote the strain.

$$E_1 = \frac{1}{E}T_1, \qquad E_{2,3} = -\frac{\nu}{E}T_1.$$

Strictly analogous relations hold for the strains generated by T_2 and T_3. Due to linearity, the joint application of T_1, T_2, T_3 results in the strains

$$E_1 = \frac{1}{E}T_1 - \frac{\nu}{E}(T_2 + T_3), \quad E_2 = \frac{1}{E}T_2 - \frac{\nu}{E}(T_1 + T_2), \quad E_3 = \frac{1}{E}T_3 - \frac{\nu}{E}(T_1 + T_2). \tag{4.10}$$

Linear Waves

For formal simplicity here we let $\mathbf{X} = \mathbf{x}$. We show how the equation of motion of linear *isotropic* elasticity,

$$\rho\ddot{\mathbf{u}} = \mu\Delta\mathbf{u} + (\lambda + \mu)\nabla(\nabla \cdot \mathbf{u})$$

allows plane wave solutions with constant phase speed. A plane wave propagating with phase speed c in the direction defined by the unit propagation vector $\mathbf{p}$ is represented by the function

$$\mathbf{u}(\mathbf{x}, t) = f(\mathbf{x} \cdot \mathbf{p} - ct)\mathbf{d},$$

where f is a scalar-valued function and $\mathbf{d}$ is the unit vector defining the direction of motion. Let $\phi = \mathbf{x} \cdot \mathbf{p} - ct$. It follows that

$$\nabla\mathbf{u} = \mathbf{d} \otimes \mathbf{p} f(\phi), \qquad \nabla \cdot \mathbf{u} = (\mathbf{p} \cdot \mathbf{d}) f'(\phi), \qquad \nabla(\nabla \cdot \mathbf{u}) = (\mathbf{p} \cdot \mathbf{d}) f''(\phi)\mathbf{p},$$

$$\Delta\mathbf{u} = f''(\phi)\mathbf{d}, \qquad \ddot{\mathbf{u}} = c^2 f''(\phi)\mathbf{d}.$$

Substitution in the equation of motion yields

$$[\mu\mathbf{d} + (\lambda + \mu)(\mathbf{p} \cdot \mathbf{d})\mathbf{p} - \rho c^2\mathbf{d}] f''(\phi) = 0,$$

which obtains for any function f if and only if

$$(\mu - \rho c^2)\mathbf{d} + (\lambda + \mu)(\mathbf{p} \cdot \mathbf{d})\mathbf{p} = \mathbf{0}.$$

A vector equation

$$\alpha\mathbf{d} + \beta\mathbf{p} = \mathbf{0}$$

holds if $\beta = 0$ and hence $\alpha = 0$ or $\beta \neq 0$ and hence $\mathbf{d} \parallel \mathbf{p}$. Look at the two solutions, subject to $\lambda + \mu \neq 0$. In the first case

$$\mathbf{p} \cdot \mathbf{d} = 0, \qquad c = c_T := (\mu/\rho)^{1/2}.$$

The motion is normal to the direction of propagation and the solution is called a *transverse wave*. Since $\mathbf{p} \cdot \mathbf{d} = 0$ then $\nabla \cdot \mathbf{u} = 0$ and that is why the solution is called an equivoluminal wave; other terms are also rotational wave, shear wave, and S-wave (secondary, shear). The displacement $\mathbf{u}$ can have any direction $\mathbf{d}$ in a plane orthogonal to the direction of propagation $\mathbf{p}$. In the second case

$$\mathbf{d} = \pm\mathbf{p}, \qquad c = c_L := [(\lambda + 2\mu)/\rho]^{1/2}.$$

The displacement $\mathbf{u}$ is then parallel to the direction of propagation and the wave is called a *longitudinal wave*. The irrotationality condition

$$\nabla \times \mathbf{u} = (\mathbf{p} \times \mathbf{d}) f'(\phi) = \mathbf{0}$$

motivates the alternative term *irrotational wave*. This solution is also called a dilatational wave, a pressure wave, or a P-wave (primary, pressure).

Since $2\mu + 3\lambda > 0$ then

$$\frac{\lambda + 2\mu}{\mu} = \tfrac{1}{3}\Big[\frac{3\lambda + 2\mu}{\mu} + 4\Big] > \tfrac{4}{3}.$$

Hence it follows

$$\frac{c_L}{c_T} = \Big[\frac{\lambda + 2\mu}{\mu}\Big]^{1/2} > 1,$$

and that is the motivation for the terms primary and secondary associated with longitudinal and transverse waves. This is true for any pair of parameters λ, μ, subject to $2\mu + 3\lambda > 0$, and then also for auxetic materials; $\nu = \lambda/2(\mu + \lambda) \in (-1, \frac{1}{2})$.

Sinusoidal or time-harmonic waves are the particular case where $f(\phi) = \sin\phi$ or $f(\phi) = \exp(i\phi)$).

Anisotropic Elastic Solids

In anisotropic solids the stiffness tensor C is more involved. As with any fourth-order tensor, the matrix associated with C has $3^4 = 81$ entries. Yet the number of independent with entries is significantly smaller. By the symmetry of the stress $\mathbf{T}$ and of the strain ε the constitutive relations

$$\mathbf{T} = \mathsf{C}\varepsilon, \qquad \mathsf{C} = 4\,\partial^2_{\mathbf{C}}\tilde{\psi}_R(\mathbf{C}),$$

imply the validity of major and minor symmetries. Hence there are at most 21 different entries. This is apparent by observing that, because of the minor symmetries, the significant entries $[\mathsf{C}_{ijkl}]$ of C are represented by a 6×6 matrix $[\mathsf{C}_{\alpha\beta}]$

$$[\mathsf{C}_{ijkl}] \iff [\mathsf{C}_{\alpha\beta}]$$

via the correspondence

$$ij = 11; 22; 33; 23, 32; 13, 31; 12, 21 \iff \alpha = 1, 2, ..., 6$$

between pairs of indices, with values from 1 to 3, and a single index with values from 1 to 6. Now, because of the major symmetries, the 6×6 matrix

$$[\mathsf{C}_{\alpha\beta}] = \begin{bmatrix} \mathsf{C}_{11} & \mathsf{C}_{12} & \mathsf{C}_{13} & \mathsf{C}_{14} & \mathsf{C}_{15} & \mathsf{C}_{16} \\ \mathsf{C}_{21} & \mathsf{C}_{22} & \mathsf{C}_{23} & \mathsf{C}_{24} & \mathsf{C}_{25} & \mathsf{C}_{26} \\ \cdots & \cdots & \cdots & \cdots & \cdots & \cdots \\ \mathsf{C}_{61} & \mathsf{C}_{62} & \mathsf{C}_{63} & \mathsf{C}_{64} & \mathsf{C}_{65} & \mathsf{C}_{66} \end{bmatrix}$$

is symmetric and hence only $6 + (36 - 6)/2 = 21$ entries are independent. Transversely isotropic solids have a single axis of symmetry and hence have 5 independent entries

$$[\mathsf{C}_{\alpha\beta}] = \begin{bmatrix} \mathsf{C}_{11} & \mathsf{C}_{11} - 2\,\mathsf{C}_{66} & \mathsf{C}_{13} & 0 & 0 & 0 \\ \mathsf{C}_{11} - 2\,\mathsf{C}_{66} & \mathsf{C}_{11} & \mathsf{C}_{13} & 0 & 0 & 0 \\ \mathsf{C}_{13} & \mathsf{C}_{13} & \mathsf{C}_{33} & 0 & 0 & 0 \\ 0 & 0 & 0 & \mathsf{C}_{44} & 0 & 0 \\ 0 & 0 & 0 & 0 & \mathsf{C}_{44} & 0 \\ 0 & 0 & 0 & 0 & 0 & \mathsf{C}_{66} \end{bmatrix}.$$

Orthotropic solids have 9 independent entries. The matrix takes the form

$$[\mathsf{C}_{\alpha\beta}] = \begin{bmatrix} \mathsf{C}_{11} & \mathsf{C}_{12} & \mathsf{C}_{13} & 0 & 0 & 0 \\ \mathsf{C}_{12} & \mathsf{C}_{22} & \mathsf{C}_{23} & 0 & 0 & 0 \\ \mathsf{C}_{13} & \mathsf{C}_{13} & \mathsf{C}_{33} & 0 & 0 & 0 \\ 0 & 0 & 0 & \mathsf{C}_{44} & 0 & 0 \\ 0 & 0 & 0 & 0 & \mathsf{C}_{55} & 0 \\ 0 & 0 & 0 & 0 & 0 & \mathsf{C}_{66} \end{bmatrix}.$$

Following Voigt notation, we let

$$\tilde{T}_1 = T_{11}, \ \tilde{T}_2 = T_{22}, \ \tilde{T}_3 = T_{33}, \ \tilde{T}_4 = T_{23}, \ \tilde{T}_5 = T_{13}, \ \tilde{T}_6 = T_{12},$$

and

$$\tilde{\varepsilon}_1 = \varepsilon_{11}, \ \tilde{\varepsilon}_2 = \varepsilon_{22}, \ \tilde{\varepsilon}_3 = \varepsilon_{33}, \ \tilde{\varepsilon}_4 = 2\varepsilon_{23}, \ \tilde{\varepsilon}_5 = 2\varepsilon_{13}, \ \tilde{\varepsilon}_6 = 2\varepsilon_{12}.$$

Hence the stress-strain relation becomes

$$\tilde{T}_\alpha = \mathsf{C}_{\alpha\beta}\,\tilde{\varepsilon}_\beta.$$

Moreover

$$T_{ij}\varepsilon_{ij} = \tilde{T}_\alpha \tilde{\varepsilon}_\alpha$$

and then $T_{ij}\varepsilon_{ij}$ takes the form of a inner product. Instead,

$$\varepsilon_{ij}\varepsilon_{ij} = \tilde{\varepsilon}_1^2 + \tilde{\varepsilon}_2^2 + \tilde{\varepsilon}_3^2 + \tfrac{1}{2}(\tilde{\varepsilon}_4^2 + \tilde{\varepsilon}_5^2 + \tilde{\varepsilon}_6^2) \neq \tilde{\varepsilon}_\alpha \tilde{\varepsilon}_\alpha,$$

thus showing that the inner product structure is not preserved.

In the Mandel (or Mandel-Voigt) notation the stress and strain components are handled in the same way by considering the sextuples

$$\tilde{T}^M = (T_{11}, T_{22}, T_{33}, \sqrt{2}\,T_{23}, \sqrt{2}\,T_{13}, \sqrt{2}\,T_{12}),$$

$$\tilde{\varepsilon}^M = (\varepsilon_{11}, \varepsilon_{22}, \varepsilon_{33}, \sqrt{2}\,\varepsilon_{23}, \sqrt{2}\,\varepsilon_{13}, \sqrt{2}\,\varepsilon_{12}).$$

Hence the inner product structure is preserved,

$$T_{ij}T_{ij} = \tilde{T}^M_\alpha \tilde{T}^M_\alpha, \qquad \varepsilon_{ij}\varepsilon_{ij} = \tilde{\varepsilon}^M_\alpha \tilde{\varepsilon}^M_\alpha, \qquad T_{ij}\varepsilon_{ij} = \tilde{T}^M_\alpha \tilde{\varepsilon}^M_\alpha,$$

and the corresponding relation

$$\tilde{T}^M_\alpha = \tilde{\mathsf{C}}^M_{\alpha\beta}\tilde{\varepsilon}^M_\beta$$

holds with

$$\tilde{\mathsf{C}}^M = \begin{bmatrix} \mathsf{C}_{1111} & \mathsf{C}_{1122} & \mathsf{C}_{1133} & \sqrt{2}\,\mathsf{C}_{1123} & \sqrt{2}\,\mathsf{C}_{1113} & \sqrt{2}\,\mathsf{C}_{1112} \\ \mathsf{C}_{2211} & \mathsf{C}_{2222} & \mathsf{C}_{2233} & \sqrt{2}\,\mathsf{C}_{2223} & \sqrt{2}\,\mathsf{C}_{2213} & \sqrt{2}\,\mathsf{C}_{2212} \\ \mathsf{C}_{3311} & \mathsf{C}_{3322} & \mathsf{C}_{3333} & \sqrt{2}\,\mathsf{C}_{3323} & \sqrt{2}\,\mathsf{C}_{3313} & \sqrt{2}\,\mathsf{C}_{3312} \\ \sqrt{2}\,\mathsf{C}_{2311} & \sqrt{2}\,\mathsf{C}_{2322} & \sqrt{2}\,\mathsf{C}_{2333} & 2\,\mathsf{C}_{2323} & 2\,\mathsf{C}_{2313} & 2\,\mathsf{C}_{2312} \\ \sqrt{2}\,\mathsf{C}_{1311} & \sqrt{2}\,\mathsf{C}_{1322} & \sqrt{2}\,\mathsf{C}_{1333} & 2\,\mathsf{C}_{1323} & 2\,\mathsf{C}_{1313} & 2\,\mathsf{C}_{1312} \\ \sqrt{2}\,\mathsf{C}_{1211} & \sqrt{2}\,\mathsf{C}_{1222} & \sqrt{2}\,\mathsf{C}_{1233} & 2\,\mathsf{C}_{1223} & 2\,\mathsf{C}_{1213} & 2\,\mathsf{C}_{1212} \end{bmatrix}.$$

From the equation of motion,

$$\rho\ddot{\mathbf{u}} = \nabla \cdot \mathbf{T} + \rho\mathbf{b},$$

in the linear approximation we have

$$\rho\partial_t^2\mathbf{u} = \nabla \cdot (\mathsf{C}\,\varepsilon) + \rho\mathbf{b},$$

ρ being a constant. Consequently,

$$(\mathbf{1}\,\partial_t^2 - \mathbf{A}(\nabla))\mathbf{u} = \mathbf{b},$$

where $\mathbf{A}(\nabla)$ is the operator given by

$$\mathbf{A}(\nabla)|_{ik} = \frac{1}{\rho}\partial_{x_j} C_{ijhk} \partial_{x_k},$$

and is referred to as the *acoustic differential operator.*

Look for plane waves, of the form

$$\mathbf{u}(\mathbf{x}, t) = \mathbf{p}\exp[i(\mathbf{k}\cdot\mathbf{x} - \omega t)],$$

where $\mathbf{p}$ is the possible direction of $\mathbf{u}$ (polarization) and $\mathbf{k}$ is the direction of propagation. These waves are solutions to the equation of motion with zero forcing ($\mathbf{b} = \mathbf{0}$) if and only if

$$(\mathbf{1}\,\omega^2 - \mathbf{A}(\mathbf{k}))\mathbf{p} = 0,$$

so that ω^2 and $\mathbf{p}$ constitute an eigenvalue/eigenvector pair of the acoustic algebraic operator

$$\mathbf{A}(\mathbf{k})|_{ih} = \frac{1}{\rho}k_j\, \mathsf{C}_{ijhl}\, k_l.$$

Let $\mathbf{n} = \mathbf{k}/|\mathbf{k}|$. The propagation condition, which is also known as the *Christoffel equation*, may be written as

$$\mathbf{A}(\mathbf{n})\mathbf{p} = c^2\mathbf{p},$$

where $c = \omega/|\mathbf{k}|$ is the phase velocity in the direction of $\mathbf{k}$.

4.2.4 Elastostatics and Compatibility of Strains

Linear elastostatic solids are investigated at equilibrium. The *linear* approximation is considered and the constitutive equation for the Cauchy stress is taken in the form

$$\mathbf{T} = \mathsf{C}\varepsilon, \qquad \varepsilon = \mathrm{sym}\nabla\mathbf{u}.$$

We look for time independent functions $\mathbf{u}(\mathbf{x})$. Consistently we assume $\rho = \rho_0$, $\rho\mathbf{b} = \hat{\mathbf{b}}$ and hence the body force is constant. The equation of motion simplifies to the equilibrium equation

$$\nabla\cdot\mathbf{T} + \hat{\mathbf{b}} = \mathbf{0};$$

in components

$$\partial_{x_j}\mathsf{C}_{ijhk}\partial_{x_k}u_h + \hat{b}_i = 0,$$

whence[6]

$$\mathsf{C}_{ijhk}\, u_{h,jk} + \hat{b}_i = 0.$$

Attention is now restricted to *isotropic* solids. Hence

$$\mathbf{T} = 2\mu\varepsilon + \lambda(\operatorname{tr}\varepsilon)\mathbf{1}, \qquad \varepsilon = \frac{1}{2\mu}(\mathbf{T} - \frac{\lambda}{2\mu + 3\lambda}(\operatorname{tr}\mathbf{T})\mathbf{1}).$$

where μ and λ are the Lamé moduli (or parameters). In components we have

$$T_{ij} = \mu(u_{i,j} + u_{j,i}) + \lambda u_{k,k}\delta_{ij}$$

and the equilibrium equation reads

$$(\mu + \lambda)u_{j,ij} + \mu u_{i,jj} = -\hat{b}_i. \tag{4.11}$$

Since $\nabla \cdot \hat{\mathbf{b}} = 0$, the divergence of the equilibrium equation results in

$$(2\mu + \lambda)u_{i,ijj} = 0.$$

Consequently $\Delta\nabla \cdot \mathbf{u} = 0$ and hence

$$\nabla\Delta\nabla \cdot \mathbf{u} = \mathbf{0}. \tag{4.12}$$

The Laplacian of (4.11) implies

$$(\mu + \lambda)u_{j,ijkk} + \mu u_{i,jjkk} = 0, \qquad (\mu + \lambda)\nabla\Delta\nabla \cdot \mathbf{u} + \mu\Delta\Delta\mathbf{u} = \mathbf{0}.$$

In view of (4.12) we find

$$\Delta\Delta\mathbf{u} = \mathbf{0};$$

the displacement $\mathbf{u}$ satisfies the *biharmonic equation*.

Compatibility Conditions

Given a (infinitesimal) strain tensor field $\varepsilon(\mathbf{x})$ we look for the conditions guaranteeing that ε is associated with a displacement field $\mathbf{u}(\mathbf{x})$. The solution of this problem traces back to Saint-Venant [398] and Beltrami [35].

Given a displacement field $\mathbf{u}$ we can write

$$\nabla\mathbf{u} = \varepsilon + \boldsymbol{\xi}, \qquad u_{i,j} = \tfrac{1}{2}(u_{i,j} + u_{j,i}) + \tfrac{1}{2}(u_{i,j} - u_{j,i}).$$

---

[6] For formal convenience here we use subscripts to denote partial derivatives, e.g. , $j = \partial_{x_j}$.

Now,

$$\xi_{ij,k} = \tfrac{1}{2}(u_{i,jk} - u_{j,ik}) = \tfrac{1}{2}(u_{i,jk} + u_{k,ij} - u_{j,ik} - u_{k,ij}) = \varepsilon_{ik,j} - \varepsilon_{jk,i}.$$

Requiring that $\xi_{ij,kl} = \xi_{ij,lk}$ we find

$$0 = \xi_{ij,kl} - \xi_{ij,lk} = \varepsilon_{ik,jl} - \varepsilon_{jk,il} + \varepsilon_{il,jk} - \varepsilon_{jl,ik}.$$

Hence we have the compatibility equations

$$\Xi_{ikjl} := \varepsilon_{ik,jl} - \varepsilon_{jk,il} - \varepsilon_{il,jk} + \varepsilon_{jl,ik} = 0.$$

Due to the symmetry conditions $\Xi_{ikjl} = \Xi_{jlik} = \Xi_{kijl} = \Xi_{iklj}$ there are only six distinct components of $\boldsymbol{\Xi} = \mathbf{0}$.

4.2.5 Vibrating Strings, Chains, and Phonons

Consider a string as a one-dimensional continuum with mass density (per unit length) ρ parameterized by $x \in [0, L]$. The string is assumed to be a plane curve and hence we let $(x, y) = (x, u(x))$ be the current point. A vanishing displacement in the x-direction implies that the x component of the traction, t_x, satisfies

$$t_x(x_2) - t_x(x_1) = 0, \quad \forall x_1, x_2 \in [0, L].$$

Let $\tau = |\mathbf{t}|$ and θ be the angle of $\mathbf{t}$ with the horizontal. Then $\tan\theta = \partial_x u$ and

$$t_x = \tau\cos\theta = \tau/\sqrt{1 + (\partial_x u)^2}.$$

We assume $|\partial_x u| \ll 1$ and then conclude that $\tau =$ constant. The balance of linear momentum in the u-direction implies

$$\partial_t \int_{x_1}^{x_2} \rho\partial_t u \, dx = t_y(x_2) - t_y(x_1) = \tau[\partial_x u(x_2) - \partial_x u(x_1)].$$

Differentiation with respect to x_2 yields the wave equation

$$\rho\partial_t^2 u = \tau\partial_x^2 u.$$

It follows that perturbations of u propagate with speed $v = \sqrt{\tau/\rho}$.

Phonons

A phonon is a collective excitation in a periodic elastic arrangement of atoms in solids. Starting from simple discrete models we show how continuum properties of motion are established quite similar to the string oscillations.

As a poor, simple model a crystal is associated with a space lattice of points (atoms or molecules). The forces between each pair of points may be characterized by a potential energy that depends on the distance. For simplicity atoms and forces are modelled by masses and springs. Let c be the elastic constant of the springs and m the mass of the atoms of a chain. Assuming interaction between nearest neighbours then the displacement $u_n(t)$ of the nth atom is governed by the differential equation

$$m\ddot{u}_n = c(u_{n+1} + u_{n-1}) - 2cu_n. \tag{4.13}$$

To fix ideas we assume the chain consists of N atoms connected with $N+1$ springs; the fixed extremes of the chain are placed at $x = 0, x = l$. At equilibrium, $x_n = na, a = l/(N+1)$.

Look for solutions in the form

$$u_n(t) = u \sin(n\beta) \sin(\omega t + \alpha); \tag{4.14}$$

the vanishing of u_n at $n = 0$ and $n = N+1$ implies that $(N+1)\beta = p\pi,\ p \in \mathbb{N}$. Substitution in (4.13) yields

$$-m\omega^2 \sin(n\beta) = c[\sin(n+1)\beta + \sin(n-1)\beta - 2\sin n\beta].$$

Since $\sin(n \pm 1)\beta = \sin n\beta \cos\beta \pm \cos n\beta \sin n\beta$ it follows that (4.14) is a solution if

$$\omega^2 = 4\frac{c}{m}\sin^2\beta/2.$$

Since $\beta = p\pi/(N+1)$ then there are solutions parameterized by p,

$$u_n^{(p)}(t) = u^{(p)} \sin(np\pi/(N+1)) \sin(\omega_p t + \alpha_p), \qquad \omega_p = 2\sqrt{c/m}\sin(p\pi/2(N+1)).$$

Only the solutions labelled by $p = 1, 2, ..., N$ are linearly independent; the solutions labelled by $p = N+2, N+3, ...$ coincide with those labelled by $p = 1, 2, ...$ Hence the general solution for u_n can be written in the form

$$u_n(t) = \sum_{p=1}^{N} u^{(p)} \sin(nk_p a) \sin(\omega_p t + \alpha_p).$$

where $k_p = p\pi/l$ and $a = l/(N+1)$. In light of the identity $\sin\theta \sin\phi = \frac{1}{2}\cos(\theta - \phi) - \frac{1}{2}\cos(\theta + \phi)$ we can write

$$u_n(t) = \tfrac{1}{2}\sum_{p=1}^{N} u^{(p)}[\cos(nk_p a - \omega_p t - \alpha_p) - \cos(nk_p a + \omega_p t + \alpha_p)].$$

Thus $u_n(t)$ is the superposition of right-propagating waves, $\cos(nk_pa - \omega_pt - \alpha_p)$, and left propagating waves, $\cos(nk_pa + \omega_pt + \alpha_p)$. Further, ω_p is the angular frequency while k_p is the wavenumber. The ratio ω_p/k_p is the phase speed v_p,

$$v_p = 2\sqrt{c/m}\,\frac{\sin(k_pa/2)}{k_p}.$$

The limit of a chain that reduces to a string is performed by letting $m = \rho a$, $c = \tau/a$, while $(N+1)a = l$. We obtain

$$v_p = \frac{2}{a}\sqrt{\frac{\tau}{\rho}}\,\frac{\sin(k_pa/2)}{k_p}.$$

The continuum limit, namely $a \to 0$, yields $v_p = \sqrt{\tau/\rho}$, as with the string oscillations.

For a number of applications the behaviour of diatomic crystals (e.g. NaCl, KCl, KI, ZnS) is well described by a linear chain of atoms of two types. Consider a chain where a mass m is placed at $x = a, 3a, ..., (2N-1)a$ and a mass M is placed at $x = 2a, 4a, ..., 2Na$. Further, the chain is fixed at the extremes $x = 0$, $x = (2N+1)a$. The $2N$ equations of motion are given the form of the system

$$\begin{aligned} &M\ddot{u}_{2n} + c(-u_{2n+1} - u_{2n-1} + 2u_{2n}) = 0, \quad n = 1, ..., N, \\ &m\ddot{u}_{2n-1} + c(-u_{2n} - u_{2n-2} + 2u_{2n-1}) = 0, \quad n = 1, ..., N. \end{aligned}$$

A solution is sought in the form

$$u_{2n}(t) = P\sin(2n\beta)\sin(\omega t + \alpha), \quad u_{2n-1}(t) = D\sin((2n-1)\beta)\sin(\omega t + \alpha),$$

with $n = 1, 2, ..., N$. Upon substitution and the observation that

$$\sin((2n \pm 1)\beta) = \sin(2n\beta)\cos\beta \pm \cos(2n\beta)\sin\beta,$$

$$\sin((2n - 1 \pm 1)\beta) = \sin((2n-1)\beta)\cos\beta \pm \cos((2n-1)\beta)\sin\beta,$$

we obtain the algebraic system

$$(-\omega^2 M + 2c)P - 2c\cos\beta\, D = 0, \qquad -2c\cos\beta\, P + (-\omega^2 m + 2c)D = 0$$

in the unknowns P, D. Non-trivial solutions hold if the corresponding determinant vanishes, namely

$$mM\omega^4 - 2c(m+M)\omega^2 + 4c^2\sin^2\beta = 0.$$

Since $M > m$ it follows that two distinct real positive values of ω^2 hold,

$$\omega^2 = c(\frac{1}{m} + \frac{1}{M}) \pm \frac{c}{mM}\{(M-m)^2 + 4mM\cos^2\beta\}^{1/2}.$$

The vanishing of u at $x = 0$ is apparent; we require the vanishing at $(2N+1)a$ to obtain

$$0 = u_{2N+1} = D\sin[(2N+1)\beta]\sin(\omega t + \alpha).$$

It follows that $(2N+1)\beta = p\pi$ whence

$$\beta = \beta_p = p\frac{\pi}{2N+1}.$$

Let

$$k_p = \frac{\beta_p}{a} = p\frac{\pi}{(2N+1)a}.$$

Only the solutions with $p = 1, 2, ..., N$ are distinct and hence there are $2N$ frequencies,

$$(\omega_p^{\pm})^2 = c(\frac{1}{m} + \frac{1}{M}) \pm \frac{c}{mM}\{(M-m)^2 + 4mM\cos^2 k_p a\}^{1/2}.$$

As p increases, from $p = 1$ to $p = N$, the frequency ω_p^+ decreases from ω_1^+ to ω_N^+,

$$\omega_1^+ = \{c(\frac{1}{m} + \frac{1}{M}) + \frac{c}{mM}[(M-m)^2 + 4mM\cos^2 k_N a]^{1/2}\}^{1/2} < [2c(\frac{1}{m} + \frac{1}{M})]^{1/2},$$

$$\omega_N^+ = \{c(\frac{1}{m} + \frac{1}{M}) + \frac{c}{mM}[(M-m)^2 + 4mM\cos^2 k_N a]^{1/2}\}^{1/2} > (\frac{2c}{m})^{1/2}.$$

Conversely, ω_p^- increases from ω_1^- to ω_N^-,

$$\omega_1^- = \{c(\frac{1}{m} + \frac{1}{M}) - \frac{c}{mM}[(M-m)^2 + 4mM\cos^2 k_N a]^{1/2}\}^{1/2} > 0,$$

$$\omega_N^- = \{c(\frac{1}{m} + \frac{1}{M}) - \frac{c}{mM}[(M-m)^2 + 4mM\cos^2 k_N a]^{1/2}\}^{1/2} < (\frac{2c}{M})^{1/2}.$$

Since

$$(\frac{2c}{M})^{1/2} < (\frac{2c}{m})^{1/2}$$

then there is a forbidden band of frequencies, $\omega \notin [(2c/M)^{1/2}, (2c/m)^{1/2}]$. The values ω_p^+ constitute the so called optical branch, the values ω_p^- the acoustic branch (Fig. 4.1).

According to quantum mechanics, microscopic vibrations in solid media are quantized. The modes of vibration are viewed as phonons and to each phonon with angular frequency ω is associated an energy equals to Planck's constant $\hbar$ times ω.

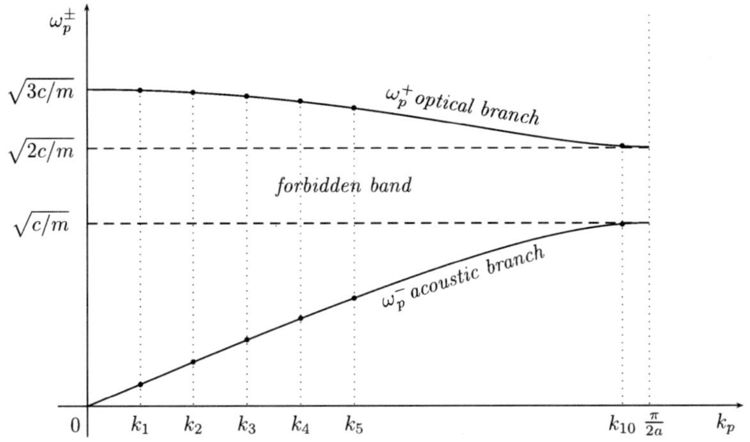

Fig. 4.1 The values ω_p^+ and ω_p^-, $p = 1, 2, ..., N$, constitute the so called optical and acoustic branches, respectively: here $N = 10$, $M = 2m$, $a = 0.5$

In the acoustic branch phonons have long wavelengths and relatively low frequencies (gigahertz). The frequencies of optical phonons are higher (terahertz) and hence with higher phonon energies.

4.3 Internal Constraints

In continuum mechanics an internal kinematical constraint is a restriction on the possible strains of a body. Usually, constraints are modelled on the basis of the notion of a bilateral, perfect and frictionless positional constraint in mass-point mechanics (see, e.g. [218, 368]).

Following [428], p. 30, and [92, 214] we say that an internal constraint is expressed in the form

$$\gamma_R(\mathbf{F}) = 0, \tag{4.15}$$

γ_R being called the constraint function relative to the reference configuration. Objectivity requires that

$$\gamma_R(\mathbf{QF}) = 0, \tag{4.16}$$

for all rotations $\mathbf{Q}$. Both $\mathbf{F}$ and $\mathbf{Q}$ are allowed to depend on time t. Let $\mathscr{C}_R$ be the set of all admissible deformation gradients (constraint set). The restriction (4.16) is expressed by

$$\mathbf{Q}\mathscr{C}_R = \mathscr{C}_R, \qquad \det \mathbf{Q} = 1.$$

As we see in a moment, the stress tensor of a constrained elastic material is determined by an admissible deformation gradient **F** only to within a stress **N** that does no work in any motion satisfying the constraint.

Time differentiation of the constraint (4.15) implies that

$$0 = \partial_{\mathbf{F}}\gamma_R \cdot \dot{\mathbf{F}} = [\mathbf{F}(\partial_{\mathbf{F}}\gamma_R)^T] \cdot \mathbf{L}^T. \tag{4.17}$$

Instead, time differentiation of (4.16) results in

$$0 = \partial_{\mathbf{F}}\gamma_R \cdot \dot{\overline{\mathbf{QF}}} = \partial_{\mathbf{F}}\gamma_R \cdot \dot{\mathbf{Q}}\mathbf{F} + \partial_{\mathbf{F}}\gamma_R \cdot \mathbf{Q}\dot{\mathbf{F}}.$$

Since $\dot{\mathbf{Q}}\mathbf{Q}^T \in \text{Skw}$, letting $\mathbf{Q}(t) = \mathbf{1}$, at time t we have

$$0 = \partial_{\mathbf{F}}\gamma_R \cdot \dot{\mathbf{Q}}\mathbf{F} + \partial_{\mathbf{F}}\gamma_R \cdot \mathbf{Q}\dot{\mathbf{F}} = [\mathbf{F}(\partial_{\mathbf{F}}\gamma_R)^T] \cdot \mathbf{W}^T + [\mathbf{F}(\partial_{\mathbf{F}}\gamma_R)^T] \cdot \mathbf{L}^T, \tag{4.18}$$

where $\mathbf{W} = \dot{\mathbf{Q}} \in \text{Skw}$. Subtracting (4.17) from (4.18) we find

$$[\mathbf{F}(\partial_{\mathbf{F}}\gamma_R)^T] \cdot \mathbf{W} = 0.$$

The arbitrariness of $\mathbf{W} \in \text{Skw}$ implies that

$$\mathbf{F}(\partial_{\mathbf{F}}\gamma_R)^T \in \text{Sym}.$$

Hence (4.17) and (4.18) require that

$$[\mathbf{F}(\partial_{\mathbf{F}}\gamma_R)^T] \cdot \mathbf{D} = 0.$$

Consequently we can say that

$$\mathbf{N} = q\mathbf{F}(\partial_{\mathbf{F}}\gamma_R)^T,$$

where q is an arbitrary scalar coefficient, is a (Cauchy) stress tensor that does no work in any motion satisfying the constraint (4.15). Therefore the effective stress tensor of a constrained (elastic) material is determined by the admissible deformation gradient **F** only to within the reactive stress **N**. The set $\mathcal{N}_R$ of all reactions **N** is called the reaction space at **F**.

The reactive stress is easily established in some special cases. Consider the constraint of incompressibility defined by

$$\gamma_R(\mathbf{F}) = \det \mathbf{F} - \mu = 0, \qquad \mu > 0.$$

Since

$$\partial_{\mathbf{F}}\gamma_R = (\det \mathbf{F})\, \mathbf{F}^{-1}$$

then

$$\mathbf{N} = -p\,\mathbf{1}, \qquad p = -q \det \mathbf{F}.$$

The scalar p may be viewed as the reactive pressure. The reaction space $\mathscr{N}_R(\mathbf{F})$ is the one-dimensional subspace in the direction of the identity $\mathbf{1}$ and hence $\mathscr{N}_R(\mathbf{F}) \in \mathrm{Sym}$.

Another example is given by the inextensibility constraint, in a direction $\mathbf{e}$; it is expressed by

$$\gamma_R(\mathbf{F}) = \mathbf{Fe} \cdot \mathbf{Fe} - \lambda = 0, \qquad \lambda > 0.$$

Since

$$\partial_{\mathbf{F}} \gamma_R = 2(\mathbf{Fe}) \otimes \mathbf{e}$$

then

$$\mathbf{N} = q(\mathbf{Fe}) \otimes (\mathbf{Fe}).$$

In general, because of the reaction $\mathbf{N}$, the constitutive relation for a constrained elastic solid may be written as

$$\mathbf{T} = \mathbf{N} + \mathbf{G}_R(\mathbf{F}),$$

where $\mathbf{G}_R$ has to be defined only for $\mathbf{F} \in \mathscr{C}_R$. Hence $\mathbf{G}_R(\mathbf{F})$ is the response function of the constrained (elastic) material relative to the reference configuration.

Another approach is now considered subject to the linear approximation of elasticity. We take the view that a uniform internal constraint is a prescription on the set of possible strains [362, 366]. For definiteness, assume the possible strains are those orthogonal to a tensor $\mathbf{V} \in \mathrm{Sym}$ in that $\mathbf{V} \cdot \mathbf{E} = 0$, $\mathbf{E}$ being the Green-St. Venant strain tensor. Then the associated constraint space $\mathscr{C}_R$ is given by

$$\mathscr{C}_R = \{\mathbf{E} \in \mathrm{Sym} : \mathbf{V} \cdot \mathbf{E} = 0\}.$$

Hence we can write

$$\mathrm{Sym} = \mathscr{C}_R \oplus \mathscr{C}_R^{\perp},$$

where $\mathscr{C}_R^{\perp}$ is the subspace spanned by $\mathbf{V}$. This in turn allows the splitting of the stress tensor into a *reactive* and an *active* part as

$$\mathbf{T} = \mathbf{N} + \mathbf{T}^{(a)}.$$

In linear elasticity $\mathbf{T}^{(a)}$ is given by a linear transformation of $\mathscr{C}_R$ into itself,

$$\mathsf{C} : \mathscr{C}_R \to \mathscr{C}_R, \qquad \mathbf{T}^{(a)} = \mathsf{C}\mathbf{E}$$

while $\mathbf{N} \in \mathscr{C}_R^{\perp}$ and hence

$$\mathbf{N} = q\,\mathbf{V}, \qquad q \in \mathbb{R}.$$

Given a smooth strain function $\mathbf{E}(t)$, $t \in [0, d]$, on $\mathscr{C}_R$ then $\dot{\mathbf{E}}(t) \in \mathscr{C}_R$ for any $t \in (0, d)$. As a consequence, reactive stresses expend no power in any admissible motion,

$$\mathbf{N} \cdot \dot{\mathbf{E}} = 0.$$

A constraint space $\mathscr{C}_R$ is proper if $\dim \mathscr{C}_R < 6$. The rigidity constraint corresponds to $\mathscr{C}_R = \{\mathbf{0}\}$ and hence $\mathscr{C}_R^\perp = \mathrm{Sym}$. Moreover, $\mathbf{T}^{(a)} = \mathbf{0}$ and hence $\mathbf{T} = \mathbf{N}$ is arbitrary in that it is not constitutively determined.

A simple example of proper internal constraint is the preservation of volume. In this case $\mathbf{V} = \mathbf{1}$ and then

$$\mathscr{C}_R^\perp = \mathrm{Sph}, \qquad \mathscr{C}_R = \mathrm{Dev} = \{\mathbf{E} \in \mathrm{Sym} : \mathbf{1} \cdot \mathbf{E} = \mathrm{tr}\,\mathbf{E} = 0\}.$$

In the linear approximation $\mathrm{tr}\,\mathbf{E} \simeq \mathrm{tr}\,\boldsymbol{\varepsilon} = \nabla \cdot \mathbf{u}$ and hence the $\mathbf{V} = \mathbf{1}$ means incompressibility. Therefore

$$\mathbf{N} = -p\,\mathbf{1},$$

where $-p$ is the undetermined scalar q. It is natural to view p as the *reactive pressure.*

Again consider the constraint of inextensibility in a direction $\mathbf{e}$, that is

$$\mathbf{Fe} \cdot \mathbf{Fe} = 1.$$

Since

$$\mathbf{Fe} \cdot \mathbf{Fe} = \mathbf{e} \cdot \mathbf{F}^T\mathbf{Fe} = \mathbf{e} \cdot \mathbf{Ce} = 1 + 2\mathbf{E} \cdot \mathbf{e} \otimes \mathbf{e}$$

then the constraint becomes

$$\mathbf{E} \cdot \mathbf{e} \otimes \mathbf{e} = 0$$

and hence $\mathbf{V} = \mathbf{e} \otimes \mathbf{e}$. Consequently,

$$\mathscr{C}_R^\perp = \{\mathbf{E} \in \mathrm{Sym} : \mathbf{E} = \lambda \mathbf{e} \otimes \mathbf{e}, \lambda \in \mathbb{R}\}, \qquad \mathscr{C}_R = \{\mathbf{E} \in \mathrm{Sym} : \mathbf{e} \cdot \mathbf{Ee} = 0\}.$$

It follows that

$$\mathbf{N} = q\,\mathbf{e} \otimes \mathbf{e}, \qquad q \in \mathbb{R}.$$

We may view $q\,\mathbf{e} \otimes \mathbf{e}$ as the reactive tension along $\mathbf{e}$.

We observe that internal constraints affect material symmetries. Indeed, the occurrence of a constraint reduces the domain of the elasticity tensor C to the constraint space $\mathscr{C}_R$ and then a rotation $\mathbf{Q} \in \mathrm{Orth}^+$ is a symmetric transformation of the material provided that $\mathbf{QE} \in \mathscr{C}_R$ for all $\mathbf{E} \in \mathscr{C}_R$. Hence, we may associate to each constraint space $\mathscr{C}_R$ a constraint group $\mathcal{G}(\mathscr{C}_R)$ consisting of all rotations which leave $\mathscr{C}_R$ invariant,

$$\mathcal{G}(\mathscr{C}_R) = \{\mathbf{Q} \in \mathrm{Orth}^+ : \mathbf{Q}\mathscr{C}_R = \mathscr{C}_R\}.$$

This group specifies the maximal material symmetry compatible with the given constraint, in the sense that the symmetry group $\mathcal{G}$ of a linearly elastic material is compatible with a constraint $\mathscr{C}_R$ if

$$\mathcal{G} \subset \mathcal{G}(\mathscr{C}_R).$$

For instance, if

$$\mathscr{C}_R = \{\mathbf{E} \in \mathrm{Sym} : \mathbf{e} \cdot \mathbf{E}\mathbf{e} = 0\}$$

then $\mathcal{G}(\mathscr{C}_R)$ contains $\mathrm{Orth}^+(\mathbf{e})$ but not Orth^+ itself. Consequently, the constraint of inextensibility is compatible with the response symmetry of transversely isotropic materials (having the same preferred direction) but is not compatible with isotropic materials.

4.4 Hyperelastic Solids and Rubber-Like Materials

By (4.7), in isotropic thermoelastic solids the Cauchy stress is given by the (referential) free energy ψ_R in the form

$$\mathbf{T} = \frac{2}{\sqrt{\det \mathbf{B}}} \partial_{\mathbf{B}} \psi_R(\theta, \mathbf{B})\, \mathbf{B}. \tag{4.19}$$

Hyperelastic solids[7] are often assumed to be characterized by (4.19) with the restrictions that ψ_R is independent of θ and the material is incompressible.

If the material is incompressible then $\det \mathbf{B} = 1$ and, moreover, there is an undetermined scalar parameter p so that

$$\mathbf{T} = 2\, \partial_{\mathbf{B}} \psi_R(\theta, \mathbf{B})\, \mathbf{B} - p\mathbf{1}.$$

Also to establish known models applied in the literature we let the free energy be given by

$$\psi_R(\theta, \mathbf{B}) = \psi_0(\theta) + W(\theta, \mathbf{B}),$$

where $W(\theta, \mathbf{B}) = 0$ when $\mathbf{B} = \mathbf{1}$.

Let $\lambda_1, \lambda_2, \lambda_3$ be the eigenvalues of the stretch tensors $\mathbf{U}$ and $\mathbf{V}$ and hence the square roots of the eigenvalues of $\mathbf{B}$ and $\mathbf{C}$. The principal invariants $\{I_i\}$ of $\mathbf{B}$ (and $\mathbf{C}$) are then related to the eigenvalues $\{\lambda_i\}$ by

$$I_1 = \lambda_1^2 + \lambda_2^2 + \lambda_3^2, \qquad I_2 = \lambda_1^2\lambda_2^2 + \lambda_2^2\lambda_3^2 + \lambda_3^2\lambda_1^2, \qquad I_3 = \lambda_1^2\,\lambda_2^2\,\lambda_3^2.$$

[7] In hyperelasticity the stress tensor is assumed to derive from a strain energy function; really the thermodynamic consistency of thermoelastic solids implies that the stress derives from the free energy function.

Moreover

$$J = \lambda_1 \lambda_2 \lambda_3.$$

Denote by

$$\bar{\mathbf{B}} = (\det \mathbf{B})^{-1/3} \mathbf{B}$$

the unimodular tensor of $\mathbf{B}$. The corresponding invariants are

$$\bar{I}_1 = J^{-2/3} I_1, \qquad \bar{I}_2 = J^{-4/3} I_2,$$

while $\bar{I}_3 = \det \bar{\mathbf{B}} = 1$. In the undeformed state, $\mathbf{F} = \mathbf{1}$, it follows

$$J = 1, \qquad \bar{I}_1 = I_1 = 3, \qquad \bar{I}_2 = I_2 = 3.$$

The generalized *Rivlin model*, or polynomial hyperelastic model, involves the free energy in terms of the invariants in the form

$$W = \sum_{i,j=0}^{N} C_{hk} (\bar{I}_1 - 3)^h (\bar{I}_2 - 3)^k + \sum_{m=0}^{M} D_m (J - 1)^{2m}, \tag{4.20}$$

where $\{C_{hk}\}, \{D_m\}$ are material parameters, possibly dependent on the temperature θ. Frequently used models in applications are particular cases of (4.20).

Neo-Hookean model

For general materials the relationship between applied stress and strain is initially linear (Hookean) but then the corresponding curve changes to nonlinear. An incompressible Neo-Hookean material is the simplest nonlinear model. The strain energy W is defined to be

$$W = C_{10}(\bar{I}_1 - 3),$$

where C_{10} is a material constant. Incompressibility implies that

$$J = \lambda_1 \lambda_2 \lambda_3 = 1.$$

Treloar model

A correction to the Neo-Hookean material is given by the Treloar model [423] where

$$W = C_{10}(\bar{I}_1 - 3) + D(J - 1)^2.$$

The occurrence of $J - 1$ means that the material is not strictly incompressible the smallness of $J - 1$ justifies the dependence of $\lambda_1, \lambda_2, \lambda_3$ via the constraint $\lambda_1 \lambda_2 \lambda_3 = 1$.

Mooney-Rivlin model

The standard model of Mooney-Rivlin material [315, 379] follows from (4.20) by letting $J = 1$ (incompressibility assumption) and $i, j \leq 1$; the model then is represented by

$$W = C_{10}(\bar{I}_1 - 3) + C_{01}(\bar{I}_2 - 3).$$

Signorini model

The Signorini model describes hyperelasticity by means of the right Cauchy-Green tensor $\mathbf{C}$ in the form

$$W = W^v(J) + W^i(\bar{I}_1, \bar{I}_2),$$

where $\bar{I}_1, \bar{I}_2$ are the invariants of the unimodular tensor $\bar{\mathbf{C}} = J^{-2/3}\mathbf{C}$. In particular, $W^v(J)$ can be taken as in the Treloar model, so that $W^v(J) = D(J-1)^2$.

Ogden model

Differently from the previous relations, Ogden model [347] is characterized by a strain energy W that depends on the principal stretch ratios $\{\lambda_i\}$ in the form

$$W = \sum_{i=1}^{N} \frac{\mu_i}{\alpha_i}(\lambda_1^{\alpha_i} + \lambda_2^{\alpha_i} + \lambda_3^{\alpha_i} - 3),$$

where μ_i, α_i are empirically determined material constants. It is mainly applied in connection with rubber and polymer and, due to the material constants, shows usually a better flexibility.

4.4.1 Compressible Mooney-Rivlin Solid

Let the referential free energy function be given by

$$\psi_R = \psi_0(\theta) + W(\theta, \bar{I}_1, \bar{I}_2, J)$$

By (4.19), to determine the Cauchy stress we need to evaluate $\partial_{\mathbf{B}} W$. We first observe that W is a composite function of $\mathbf{B}$ and

$$\partial_{\mathbf{B}} W = \partial_{\bar{I}_1} W \partial_{\mathbf{B}} \bar{I}_1 + \partial_{\bar{I}_2} W \partial_{\mathbf{B}} \bar{I}_2 + \partial_J W \partial_{\mathbf{B}} J.$$

Now,

$$\partial_{\mathbf{B}} \bar{I}_1 = \partial_{\mathbf{B}}[(\det \mathbf{B})^{-1/3} \mathrm{tr}\, \mathbf{B}] = -\tfrac{1}{3}\bar{I}_1 \mathbf{B}^{-1} + (\det \mathbf{B})^{-1/3}\mathbf{1},$$

$$\partial_{\mathbf{B}} \bar{I}_2 = \partial_{\mathbf{B}}[(\det \mathbf{B})^{-2/3} \tfrac{1}{2}[(\operatorname{tr} \mathbf{B})^2 - \operatorname{tr} \mathbf{B}^2] = -\tfrac{2}{3} \bar{I}_2 \mathbf{B}^{-1} + J^{-4/3}(\operatorname{tr} \mathbf{B}\, \mathbf{1} - \mathbf{B}),$$

$$\partial_{\mathbf{B}} J = \partial_{\mathbf{B}}(\det \mathbf{B})^{1/2} = \tfrac{1}{2}(\det \mathbf{B})^{1/2} \mathbf{B}^{-1} = \tfrac{1}{2} J \mathbf{B}^{-1}.$$

Hence we find that

$$\mathbf{T} = \frac{2}{\sqrt{\det \mathbf{B}}} \partial_{\mathbf{B}} \psi_R(\mathbf{B})\, \mathbf{B} = \frac{2}{J}\big[J^{-2/3}(\partial_{\bar{I}_1} W + \bar{I}_1 \partial_{\bar{I}_2} W)\mathbf{B} - J^{-4/3} \partial_{\bar{I}_2} W \mathbf{B}^2\big] + \big[\partial_J W - \frac{2}{3J}(\bar{I}_1 \partial_{\bar{I}_1} W + 2 \bar{I}_2 \partial_{\bar{I}_2} W)\big]\mathbf{1}. \tag{4.21}$$

By a direct check it follows that

$$\operatorname{tr}\{J^{-2/3}(\partial_{\bar{I}_1} W + \bar{I}_1 \partial_{\bar{I}_2} W)\mathbf{B} - J^{-4/3} \partial_{\bar{I}_2} W \mathbf{B}^2] - \tfrac{1}{3}(\bar{I}_1 \partial_{\bar{I}_1} W + 2 \bar{I}_2 \partial_{\bar{I}_2} W)\mathbf{1}\} = 0$$

for any function W. Consequently,

$$\operatorname{tr} \mathbf{T} = 3\, \partial_J W.$$

If we let p be defined by

$$p := -\tfrac{1}{3} \operatorname{tr} \mathbf{T}$$

then we may view p as the pressure and find that

$$p = -\partial_J W.$$

Hence it follows from (4.21) that $\mathbf{T}$ can be written in the form

$$\mathbf{T} = -p\,\mathbf{1} + \frac{2}{J}[J^{-2/3}(\partial_{\bar{I}_1} W + \bar{I}_1 \partial_{\bar{I}_2} W)\mathbf{B} - J^{-4/3} \partial_{\bar{I}_2} W \mathbf{B}^2] - \frac{2}{3J}(\bar{I}_1 \partial_{\bar{I}_1} W + 2 \bar{I}_2 \partial_{\bar{I}_2} W)\mathbf{1}. \tag{4.22}$$

or

$$\mathbf{T} = -p\mathbf{1} + \mathbf{T}_0, \qquad \mathbf{T}_0 = \frac{2}{J}\big[J^{-2/3}(\partial_{\bar{I}_1} W + \bar{I}_1 \partial_{\bar{I}_2} W)\mathbf{B} - J^{-4/3} \partial_{\bar{I}_2} W \mathbf{B}^2\big]_0.$$

We now specialize the stress relation (4.22) to the *compressible Mooney-Rivlin model*; the corresponding free energy follows from (4.20) by letting $N = 1$, $M = 1$, namely

$$W = C_1(\bar{I}_1 - 3) + C_2(\bar{I}_2 - 3) + D_1(J - 1)^2. \tag{4.23}$$

Hence

$$p = -2D_1(J - 1)$$

and (4.22) becomes

$$\mathbf{T} = -p\,\mathbf{1} + 2\,J^{-1}\big[J^{-2/3}(C_1 + \bar{I}_1 C_2)\mathbf{B} - J^{-4/3} C_2 \mathbf{B}^2\big] - \tfrac{2}{3} J^{-1}(C_1 \bar{I}_1 + 2C_2 \bar{I}_2)\mathbf{1}.$$

For an *incompressible Mooney-Rivlin material* $J = 1$ and hence $p = 0$, $\bar{I}_1 = I_1$, $\bar{I}_2 = I_2$. As a consequence it follows that $\mathbf{T} = -p\mathbf{1} + \mathbf{T}^{(a)}$, p being the reactive pressure (see Sect. 5.3) and

$$\mathbf{T}^{(a)} = 2(C_1 + I_1 C_2)\mathbf{B} - 2\,C_2\mathbf{B}^2 - \tfrac{2}{3}(C_1 I_1 + 2C_2 I_2)\mathbf{1}, \tag{4.24}$$

which satisfies the normalization condition $\mathrm{tr}\,\mathbf{T}^{(a)} = 0$. Since $J = 1$, the Cayley-Hamilton theorem implies

$$\mathbf{B}^{-1} = \mathbf{B}^2 - I_1\mathbf{B} + I_2\mathbf{1}.$$

Hence, upon replacing $\mathbf{B}^2 = \mathbf{B}^{-1} + I_1\mathbf{B} + I_2\mathbf{1}$ we find

$$\mathbf{T}^{(a)} = -p^*\mathbf{1} + 2C_1\mathbf{B} - 2C_2\mathbf{B}^{-1} \tag{4.25}$$

where

$$p^* := \tfrac{2}{3}(C_1 I_1 - C_2 I_2)$$

is not the pressure but merely the isotropic stress determined by the free energy. Since $\mathrm{tr}\,\mathbf{T}^{(a)} = 0$ then

$$\mathbf{T}^{(a)} = \mathbf{T}_0^{(a)} = 2C_1\mathbf{B}_0 - 2C_2\mathbf{B}_0^{-1}.$$

As a comment, we observe that the second Piola stress $\mathbf{T}_{RR}^{(a)}$, associated with (4.24), is given by

$$\mathbf{T}_{RR}^{(a)} = J\mathbf{F}^{-1}\mathbf{T}^{(a)}\mathbf{F}^{-T} = 2(C_1 + I_1 C_2)\mathbf{1} - 2C_2\mathbf{C} - \tfrac{2}{3}(C_1 I_1 + 2C_2 I_2)\mathbf{C}^{-1}.$$

We recall that the invariants of $\mathbf{C}$ are equal to those of $\mathbf{B}$. Hence, since $J = 1$, the Cayley-Hamilton theorem implies

$$\mathbf{C}^{-1} = \mathbf{C}^2 - I_1\mathbf{C} + I_2\mathbf{1}.$$

Upon substitution of $\mathbf{C}^{-1}$ we find

$$\mathbf{T}_{RR}^{(a)} = \tau_1\mathbf{1} + \tau_2\mathbf{C} + \tau_3\mathbf{C}^2,$$

where

$$\tau_1 = 2(C_1 + I_1 C_2) - \tfrac{2}{3} I_2 (C_1 I_1 + 2C_2 I_2), \quad \tau_2 = \tfrac{2}{3} I_1 (C_1 I_1 + 2C_2 I_2) - 2C_2, \quad \tau_3 = -\tfrac{2}{3}(C_1 I_1 + 2C_2 I_2).$$

Differently from (4.24), $\operatorname{tr} \mathbf{T}^{(a)}_{RR} \neq 0$. To show this observe that

$$\operatorname{tr} \mathbf{C}^{-1} = \operatorname{tr} \mathbf{C}^2 - I_1^2 + 3I_2.$$

Now,

$$I_2 = \tfrac{1}{2}[(\operatorname{tr} \mathbf{C})^2 - \operatorname{tr} \mathbf{C}^2], \qquad \operatorname{tr} \mathbf{C}^2 = (\operatorname{tr} \mathbf{C})^2 - 2I_2 = I_1^2 - 2I_2,$$

and hence

$$\operatorname{tr} \mathbf{C}^{-1} = I_2.$$

Consequently,

$$\operatorname{tr} \mathbf{T}^{(a)}_{RR} = 6(C_1 + I_1 C_1) - 2I_1 C_2 - \tfrac{2}{3}(C_1 I_1 + 2C_2 I_2) I_2 \neq 0.$$

It is of interest to express the Cauchy stress $\mathbf{T}$ in terms of the principal stretches $\lambda_1^2, \lambda_2^2, \lambda_3^2$. By (4.19) we have

$$T_{ii} = B_{ip}\, \partial_{B_{pi}} W = B_{ii}\, \partial_{B_{ii}} W = \lambda_i^2\, \partial_{\lambda_i^2} W, \qquad i \text{ not summed.}$$

Observe that, in general,

$$\partial_{\lambda_i} W = \partial_{\lambda_i^2} W\, 2\lambda_i$$

and hence

$$T_{ii} = \lambda_i^2\, \partial_{\lambda_i^2} W = \lambda_i\, \partial_{\lambda_i} W.$$

Thus

$$T_{11} - T_{33} = \lambda_1 \partial_{\lambda_1} W - \lambda_3 \partial_{\lambda_3} W, \qquad T_{22} - T_{33} = \lambda_2 \partial_{\lambda_2} W - \lambda_3 \partial_{\lambda_3} W.$$

For an incompressible Mooney-Rivlin material W is given by (4.23) and hence

$$W = C_1(\lambda_1^2 + \lambda_2^2 + \lambda_3^2 - 3) + C_2(\lambda_1^2\lambda_2^2 + \lambda_2^2\lambda_3^2 + \lambda_3^2\lambda_1^2 - 3),$$

subject to $\lambda_1\lambda_2\lambda_3 = 1$. Therefore we find that

$$\lambda_1\partial_{\lambda_1} W = 2C_1\lambda_1^2 + 2C_2(\frac{1}{\lambda_3^2} + \frac{1}{\lambda_2^2}), \quad \lambda_2\partial_{\lambda_2} W = 2C_1\lambda_2^2 + 2C_2(\frac{1}{\lambda_3^2} + \frac{1}{\lambda_1^2}), \quad \lambda_3\partial_{\lambda_3} W = 2C_1\lambda_3^2 + 2C_2(\frac{1}{\lambda_2^2} + \frac{1}{\lambda_1^2}).$$

Hence the stress differences can be written as

$$T_{11} - T_{33} = 2C_1(\lambda_1^2 - \lambda_3^2) - 2C_2(\frac{1}{\lambda_1^2} - \frac{1}{\lambda_3^2}), \quad T_{22} - T_{33} = 2C_1(\lambda_2^2 - \lambda_3^2) - 2C_2(\frac{1}{\lambda_2^2} - \frac{1}{\lambda_3^2}). \tag{4.26}$$

Uniaxial Tension

Let the (Mooney-Rivlin) material be subject to a uniaxial elongation (in the 1-direction, first eigenvector of **U**) so that $\lambda_1 = \lambda$, $\lambda_2 = \lambda_3 = 1/\sqrt{\lambda}$. Hence we have

$$T_{11} - T_{33} = (2C_1 + 2C_2/\lambda)(\lambda^2 - 1/\lambda) - 2C_2(1/\lambda^2 - \lambda), \qquad T_{22} - T_{33} = 0.$$

If, further, the body is free in the 2- and 3-direction then it follows

$$T_{11} = (2C_1 + 2C_2/\lambda)(\lambda^2 - 1/\lambda) - 2C_2(1/\lambda^2 - \lambda), \qquad T_{22} = T_{33} = 0.$$

Experimental results are naturally derived for the first Piola stress, that is the stress relative to the reference area. Now, from $\mathbf{T}_R = J\mathbf{T}\mathbf{F}^{-T}$, since $J = 1$ then $T^R_{11} = T_{11}/\lambda$ whence

$$T^R_{11} = 2C_1\lambda + 2C_2 - 2C_1/\lambda^2 - 2C_2/\lambda^3.$$

Upon differentiation with respect to λ we find

$$\partial_\lambda T^R_{11} = 2C_1 + 4C_1/\lambda^3 + 6C_2/\lambda^4,$$

$$\partial^2_\lambda T^R_{11} = -12C_1/\lambda^4 - 24C_2/\lambda^5.$$

Positive values of C_1 and C_2 make $T^R_{11}(\lambda)$ to be an increasing, concave function.

Equibiaxial Tension

Apply the equibiaxial tension in the directions 1 and 2 so that the principal stretches are $\lambda_1 = \lambda_2 = \lambda$. Owing to incompressibility, $\lambda_3 = 1/\lambda^2$. Substitution in (4.26) yields

$$T_{11} - T_{33} = T_{22} - T_{33} = 2C_1(\lambda^2 - 1/\lambda^4) - 2C_2(1/\lambda^2 - \lambda^4).$$

Shear Deformation

For definiteness consider a shear deformation, with a non-zero displacement in the $\mathbf{e}_1$-direction, in the form

$$\mathbf{F} = \mathbf{1} + \gamma \mathbf{e}_1 \otimes \mathbf{e}_2, \qquad F = \begin{bmatrix} 1 & \gamma & 0 \\ 0 & 1 & 0 \\ 0 & 0 & 1 \end{bmatrix}.$$

The eigenvalues r of $\mathbf{B} = \mathbf{F}\mathbf{F}^T$ are given by

$$0 = \det \begin{bmatrix} 1+\gamma^2-r & \gamma & 0 \\ \gamma & 1-r & 0 \\ 0 & 0 & 1-r \end{bmatrix} = r^2 - (2+\gamma^2)r + 1.$$

Hence we find two solutions,

$$r_1 = 1 + \tfrac{1}{2}\gamma^2 + \tfrac{1}{2}\sqrt{\gamma^4 + 4\gamma^2}, \qquad r_2 = 1 + \tfrac{1}{2}\gamma^2 - \tfrac{1}{2}\sqrt{\gamma^4 + 4\gamma^2},$$

and they are the inverse of each other. Let

$$\lambda^2 = r_1, \qquad \lambda_1 = \lambda, \;\; \lambda_2 = 1/\lambda, \;\; \lambda_3 = 1.$$

Hence

$$I_1 = \lambda^2 + 1/\lambda^2 + 1 = I_2.$$

Since

$$\mathbf{B} = \mathbf{F}\mathbf{F}^T = \begin{bmatrix} 1+\gamma^2 & \gamma & 0 \\ \gamma & 1 & 0 \\ 0 & 0 & 1 \end{bmatrix}, \qquad \mathbf{B}^{-1} = \begin{bmatrix} 1 & -\gamma & 0 \\ -\gamma & 1+\gamma^2 & 0 \\ 0 & 0 & 1 \end{bmatrix}.$$

and $\mathbf{T} = -p^* + 2C_1\mathbf{B} - 2C_2\mathbf{B}^{-1}$ then

$$\mathbf{T} = \begin{bmatrix} -p^* + 2(C_1 - C_2) + 2C_1\gamma^2 & 2(C_1 + C_2)\gamma & 0 \\ 2(C_1 + C_2)\gamma & -p^* + 2(C_1 - C_2) - 2C_2\gamma^2 & 0 \\ 0 & 0 & -p^* + 2(C_1 - C_2) \end{bmatrix},$$

where

$$p^* := \tfrac{2}{3}(C_1 I_1 - C_2 I_2) = \tfrac{2}{3}(C_1 - C_2)(3 + \gamma^2).$$

4.5 Modelling of Dissipative Solids

In modelling dissipative effects in solids analogy with the fluid models seem to be a common reference. Landau and Lifshitz[272], p. 34, model viscous dissipation by adding to the elastic stress tensor a viscosity stress tensor $\mathbf{T}^{\text{vis}}$ given by

$$\mathbf{T}^{\text{vis}} = 2\mu\dot{\varepsilon}_0 + \zeta \operatorname{tr} \dot{\varepsilon}\mathbf{1},$$

μ and ζ being the shear and bulk viscosity coefficients. In a sense this dependence is natural in that it is the strict analogue of linear elasticity, where $\mathbf{T}$ is a linear function of ε. Yet, $\dot{\varepsilon}$ is not objective and hence dissipation has to be modelled in a different way. An answer to this problem is given in [132] by considering a Piola stress (the first one) instead of the Cauchy stress.

Heat conduction is modelled ([272], p. 33) by a Fourier-like law,

$$\mathbf{q} = -\mathbf{K}\nabla\theta, \tag{4.27}$$

the solid character being represented by the (possibly) anisotropic tensor $\mathbf{K}$. For isotropic solids the standard Fourier law $\mathbf{q} = -\kappa\nabla\theta$ is assumed. Yet, to account for deformation an argument can be developed as follows.

Let for simplicity $r = 0$. The equation of balance of energy (4.4) is taken in the approximate form ($J \approx 1$)

$$\theta\dot{\eta}_R = -\nabla \cdot \mathbf{q}.$$

Applying the trace to the stress-strain relation (4.6) we obtain

$$\nabla \cdot \mathbf{u} = -kp + 3\alpha(\theta - \theta_0),$$

where θ_0 is the reference temperature, α is the thermal expansion, $p = -\mathrm{tr}\,\mathbf{T}/3$, and $k = 3/(3\lambda + 2\mu)$. Hence p is linear with respect to $\nabla \cdot \mathbf{u}$ and $\theta - \theta_0$. This suggests that we let ([272], p. 6)

$$\psi_R = \psi_0(\theta) - k\alpha(\theta - \theta_0)\nabla \cdot \mathbf{u} + \mu\boldsymbol{\varepsilon}_0 \cdot \boldsymbol{\varepsilon}_0 + \tfrac{1}{2}k(\nabla \cdot \mathbf{u})^2.$$

Since $\eta_R = -\partial_\theta\psi_R$ then

$$\eta_R = \eta_0(\theta) + k\alpha\nabla \cdot \mathbf{u}.$$

Consequently we have

$$\theta(\eta_0'\dot{\theta} + k\alpha\dot{\overline{\nabla \cdot \mathbf{u}}}) = \nabla \cdot (\kappa\nabla\theta).$$

As a rough approximation we now let the solid be in a stress-free condition, $p = 0$. Then Eq. (4.6) results in

$$\nabla \cdot \mathbf{u} = 3\alpha(\theta - \theta_0)$$

and by the balance of energy it follows

$$\theta(\eta_0' + 3k\alpha^2)\dot{\theta} = \kappa\Delta\theta.$$

Hence deformation would affect the heat equation by merely modifying the value of the coefficient of $\dot{\theta}$.

We now examine how a thermodynamically-consistent scheme can be developed for dissipation in solids.

4.5.1 A Model for Viscosity and Heat Conduction

If we look at the dependence of the stress on the deformation gradient, as is done in elasticity, by the principle of objectivity it follows that the second Piola stress $\mathbf{T}_{RR}$ can depend on $\mathbf{F}$ via the Cauchy-Green tensor $\mathbf{C}$. Now $\mathbf{C}$ is invariant under Euclidean transformations and so is the time derivative $\dot{\mathbf{C}}$. Hence both $\dot{\mathbf{T}}_{RR}$ and $\dot{\mathbf{C}}$ are objective. Indeed they both are invariant. The same properties hold for the Green-St. Venant strain $\mathbf{E} = \frac{1}{2}(\mathbf{C} - \mathbf{1})$.

A modelling of solids via a function $\mathbf{T}_{RR}(\mathbf{C}, \dot{\mathbf{C}})$, or $\mathbf{T}_{RR}(\mathbf{E}, \dot{\mathbf{E}})$, is naturally framed within a Lagrangian description. As to a Fourier-like law, we observe that

$$\mathbf{q}_R = J\mathbf{q}\mathbf{F}^{-T}, \qquad \nabla_R\theta = \nabla\theta\,\mathbf{F}$$

or

$$\mathbf{q} = J^{-1}\mathbf{F}\mathbf{q}_R, \qquad \nabla\theta = \mathbf{F}^{-T}\nabla_R\theta.$$

Hence (4.27) is replaced with

$$\mathbf{q}_R = -J\mathbf{F}^{-1}\mathbf{K}\mathbf{F}^{-T}\nabla_R\theta.$$

We can then view $J\mathbf{F}^{-1}\mathbf{K}\mathbf{F}^{-T}$ as the conductivity tensor $\mathbf{K}_R$, in the Lagrangian description, associated with the conductivity tensor $\mathbf{K}$ in the Eulerian description,

$$\mathbf{K} = J^{-1}\mathbf{F}\mathbf{K}_R\mathbf{F}^T. \tag{4.28}$$

We follow the material description and then write the entropy inequality in the form

$$-(\dot{\psi}_R + \eta_R\dot{\theta}) + \mathbf{T}_{RR}\cdot\dot{\mathbf{E}} - \frac{1}{\theta}\mathbf{q}_R\cdot\nabla_R\theta \geq 0.$$

Let $\psi_R, \eta_R, \mathbf{T}_{RR}, \mathbf{q}_R$ be functions of

$$\mathbf{E}, \theta, \dot{\mathbf{E}}, \nabla_R\theta.$$

Upon computation of $\dot{\psi}_R$ and substitution we have

$$-(\partial_\theta\psi_R + \eta_R)\dot{\theta} + (\mathbf{T}_{RR} - \partial_{\mathbf{E}}\psi_R)\cdot\dot{\mathbf{E}} - \partial_{\dot{\mathbf{E}}}\psi_R\cdot\ddot{\mathbf{E}} - \partial_{\nabla_R\theta}\psi_R\cdot\nabla_R\dot{\theta} - \frac{1}{\theta}\mathbf{q}_R\cdot\nabla_R\theta \geq 0.$$

The arbitrariness of $\ddot{\mathbf{E}}, \nabla_R\dot{\theta}, \dot{\theta}$ implies that

$$\partial_{\dot{\mathbf{E}}}\psi_R = \mathbf{0}, \qquad \partial_{\nabla_R\theta}\psi_R = \mathbf{0}, \qquad \eta_R = -\partial_\theta\psi_R.$$

Let

$$\mathbf{T}^{\mathrm{el}}_{RR}(\mathbf{E},\theta) := \mathbf{T}_{RR}(\mathbf{E},\theta,\mathbf{0},\mathbf{0}), \qquad \mathbf{T}^{\mathrm{dis}}_{RR} := \mathbf{T}_{RR}(\mathbf{E},\theta,\dot{\mathbf{E}},\nabla_R\theta) - \mathbf{T}^{\mathrm{el}}_{RR}(\mathbf{E},\theta)$$

and assume $\mathbf{T}_{RR}$ is continuous so that

$$\mathbf{T}^{\mathrm{dis}}_{RR}(\mathbf{E},\theta,\dot{\mathbf{E}},\nabla_R\theta) \to \mathbf{0} \quad \text{as} \quad \dot{\mathbf{E}},\nabla_R\theta \to \mathbf{0}.$$

Hence it follows that $\mathbf{T}^{\mathrm{el}}_{RR} = \partial_{\mathbf{E}}\psi_R$ and

$$\mathbf{T}^{\mathrm{dis}}_{RR}\cdot\dot{\mathbf{E}} - \frac{1}{\theta}\mathbf{q}_R\cdot\nabla_R\theta \geq 0.$$

If

$$\mathbf{T}^{\mathrm{dis}}_{RR} = \mathsf{A}_R(\mathbf{E},\theta)\dot{\mathbf{E}}, \qquad \mathbf{q}_R = -\mathbf{K}_R(\mathbf{E},\theta)\nabla_R\theta$$

then the reduced inequality holds if and only if

$$\mathsf{A}_R \geq \mathbf{0}, \qquad \mathbf{K}_R \geq \mathbf{0}.$$

The positive semi-definiteness of $\mathbf{K}_R = J\mathbf{F}^{-1}\mathbf{K}\mathbf{F}^{-T}$ then implies

$$\mathbf{K} = J^{-1}\mathbf{F}\mathbf{K}_R\mathbf{F}^T \geq \mathbf{0}.$$

This is so because, for any vector $\mathbf{w}$,

$$J\mathbf{w}\cdot\mathbf{K}\mathbf{w} = \mathbf{w}\cdot\mathbf{F}\mathbf{K}_R\mathbf{F}^T\mathbf{w} = (\mathbf{F}^T\mathbf{w})\cdot\mathbf{K}_R(\mathbf{F}^T\mathbf{w})$$

and the conclusion follows. Moreover, since $J = \det\mathbf{F} > 0$ then $\mathbf{K}_R$ is positive definite if and only if $\mathbf{K}$ is positive definite.

The definition (4.28) of $\mathbf{K}$, in terms of $\mathbf{K}_R$ and the deformation gradient $\mathbf{F}$, shows that even though $\mathbf{K}_R$ is independent of $\mathbf{E}$ the spatial tensor $\mathbf{K}$ is affected by deformation.

In the modelling of solids a properly selected reference configuration R is central in the description of constitutive properties. For instance, the reference configuration might be stress free and isotropic while the current configuration might have properties induced by the deformation. This suggests that the material properties be formulated in the reference configuration and next be extended to the current configuration. Consequently we describe the viscous behaviour by letting

$$\mathbf{T}^{\mathrm{dis}}_{RR} = \mathsf{A}_R(\mathbf{E},\theta)\dot{\mathbf{E}}, \qquad \mathsf{A}_R \in \mathrm{Lin}(\mathrm{Sym},\mathrm{Sym}),$$

and look for the corresponding relation

$$\mathbf{T} = \mathsf{A}\,\mathbf{D}$$

in the current configuration. The connection between $\mathbf{T}_{RR}$ and $\mathbf{T}$ is different from that between $\dot{\mathbf{E}}$ and $\mathbf{D}$. Since

$$\mathbf{T}^{\text{dis}}_{RR} = J\mathbf{F}^{-1}\mathbf{T}^{\text{dis}}\mathbf{F}^{-T}, \qquad \dot{\mathbf{E}} = \mathbf{F}^T\mathbf{D}\mathbf{F},$$

in indicial notation we have

$$J F^{-1}_{Ki} T^{\text{dis}}_{ij} F^{-1}_{Hj} = \mathsf{A}^R_{KHPQ} F^T_{Pr} D_{rs} F_{sQ},$$

whence

$$T^{\text{dis}}_{pq} = J^{-1} F_{pK} F_{qH} \mathsf{A}^R_{KHPQ} F_{rP} F_{sQ} D_{rs}.$$

In compact notation we can write

$$\mathbf{T}^{\text{dis}} = J^{-1}(\mathbf{F} \boxtimes \mathbf{F})\mathsf{A}_R(\mathbf{F}^T \boxtimes \mathbf{F}^T)\mathbf{D}.$$

Hence

$$\mathsf{A} = J^{-1}(\mathbf{F} \boxtimes \mathbf{F})\mathsf{A}_R(\mathbf{F}^T \boxtimes \mathbf{F}^T) \in \text{Lin(Sym, Sym)}.$$

Go back to the components and let

$$\mathsf{A}_{pqrs} = J^{-1} F_{pK} F_{qH} \mathsf{A}^R_{KHMN} F_{rM} F_{sN}.$$

If the deformation gradient is merely a rotation, that is $\mathbf{U} = \mathbf{1}$, $\mathbf{F} = \mathbf{Q}$, then it follows

$$\mathsf{A}_{pqrs} = Q_{pK} Q_{qH} \mathsf{A}^R_{KHMN} Q_{rM} Q_{sN}.$$

This relation is also referred to as the *basis change formula for the elasticity tensor* [65].

Now that we know the relations

$$\mathbf{K}_R \leftrightarrow \mathbf{K}, \qquad \mathsf{A}_R \leftrightarrow \mathsf{A}$$

it is worth checking the properties under Euclidean transformations. By definition, $\mathbf{K}_R$ and A_R are assumed to be invariant. Hence

$$\mathbf{K}^* = (J^{-1}\mathbf{F}\mathbf{K}_R\mathbf{F}^T)^* = J^{-1}\mathbf{F}^*\mathbf{K}_R(\mathbf{F}^*)^T = J^{-1}\mathbf{Q}\mathbf{F}\mathbf{K}_R\mathbf{F}^T\mathbf{Q}^T = \mathbf{Q}\mathbf{K}\mathbf{Q}^T.$$

Accordingly, $\mathbf{K}$ transforms as a (objective) second-order tensor. To determine the property of A we observe

$$\begin{aligned}\mathsf{A}^*_{pqrs} = J^{-1} F^*_{pK} F^*_{qH} \mathsf{A}^R_{KHMN} F^*_{rM} F^*_{sN} &= J^{-1} Q_{pi} F_{iK} Q_{qj} F_{jH} \mathsf{A}^R_{KHMN} Q_{rh} F_{hM} Q_{sk} F_{kN} \\ &= Q_{pi} Q_{qj} Q_{rh} Q_{sk} \mathsf{A}_{ijhk}\end{aligned}$$

and hence **A** transforms as a (objective) fourth-order tensor.

Assume the conductivity tensor and the viscosity tensor are isotropic[8] in the reference configuration

$$\mathbf{K}_R = \kappa \mathbf{1}, \qquad \mathbf{A}_R = 2\mu \mathbf{1} \boxtimes \mathbf{1} + \lambda \mathbf{1} \otimes \mathbf{1},$$

κ, μ, λ being possibly functions of the temperature θ. Hence

$$\mathbf{q}_R = -\kappa \nabla_R \theta, \qquad \mathbf{T}^{\text{dis}}_{RR} = 2\mu \dot{\mathbf{E}} + \lambda (\text{tr}\, \dot{\mathbf{E}}) \mathbf{1}.$$

In the current configuration the heat flux vector $\mathbf{q}$ and the dissipative Cauchy stress $\mathbf{T}^{\text{dis}}$ take the form

$$\mathbf{q} = -\kappa J^{-1} \mathbf{B} \nabla \theta, \tag{4.29}$$

$$\mathbf{T}^{\text{dis}} = 2\mu J^{-1} \mathbf{B}\,\mathbf{D}\,\mathbf{B} + \lambda J^{-1} (\mathbf{B} \cdot \mathbf{D}) \mathbf{B}. \tag{4.30}$$

Therefore, if heat conduction and viscosity are isotropic in the reference configuration then, in the current configuration, $\mathbf{q}$ and $\mathbf{T}^{\text{dis}}$ are influenced by deformation via the left Cauchy-Green tensor $\mathbf{B} = \mathbf{F}\mathbf{F}^T$. As is known, this happens as a consequence of the transformation of gradients ($\nabla_R \theta = \mathbf{F}^T \nabla \theta$, $\dot{\mathbf{E}} = \mathbf{F}^T \mathbf{D} \mathbf{F}$) and the relations $\mathbf{q} \leftrightarrow \mathbf{q}_R$, $\mathbf{T} \leftrightarrow \mathbf{T}_{RR}$. It is of interest that while $\lambda (\text{tr}\, \dot{\mathbf{E}}) \mathbf{1}$ is isotropic in R the corresponding term in $\mathcal{R}$ is a multiple of $\mathbf{B}$. Likewise the shear term $2\mu \dot{\mathbf{E}}$ is transformed into $2\mu J^{-1} \mathbf{BDB}$, not merely a term proportional to $\mathbf{D}$.

The constitutive Eqs. (4.29) and (4.30) are consistent with thermodynamics in that this is apparent in the Lagrangian description (reference configuration). Yet it is of interest to examine the thermodynamic consistency of $\mathbf{q}$ and $\mathbf{T}^{\text{dis}}$ in the Eulerian description, that is as functions of $\nabla \theta$ and $\mathbf{D}$ rather than of $\nabla_R \theta$ and $\dot{\mathbf{E}}$. Now, the function $\nabla \theta \cdot \mathbf{B} \nabla \theta$ of $\nabla \theta$ is positive definite because so is the tensor $\mathbf{B}$. Moreover, $\nabla \theta \cdot \mathbf{B} \nabla \theta = |\mathbf{F}^T \nabla \theta|^2$. Instead, $(\mathbf{BDB}) \cdot \mathbf{D} = (\mathbf{BD}) \cdot (\mathbf{DB})$ and we cannot conclude about the positive definiteness with respect to $\mathbf{D}$. However,

$$\begin{aligned}(\mathbf{BDB}) \cdot \mathbf{D} = B_{ps} D_{sr} B_{qr} D_{pq} &= F_{pK} F_{sK} D_{sr} F_{qH} F_{rH} D_{pq} \\ &= F^T_{Kp} D_{pq} F_{qH} F^T_{Ks} D_{sr} F_{rH} = |\mathbf{F}^T \mathbf{D} \mathbf{F}|^2.\end{aligned}$$

In addition, $(\mathbf{B} \cdot \mathbf{D}) \mathbf{B} \cdot \mathbf{D} = (\mathbf{B} \cdot \mathbf{D})^2$ and

$$\mathbf{B} \cdot \mathbf{D} = F_{pK} F_{qK} D_{pq} = F^T_{Kp} D_{pq} F_{qK} = \text{tr}\, (\mathbf{F}^T \mathbf{D} \mathbf{F}).$$

Let $\mathbf{G} = \mathbf{F}^T \mathbf{D} \mathbf{F}$. Then

$$[2\mu J^{-1} \mathbf{B}\,\mathbf{D}\,\mathbf{B} + \lambda J^{-1} (\mathbf{B} \cdot \mathbf{D}) \mathbf{B}] \cdot \mathbf{D} = 2\mu J^{-1} \mathbf{G} \cdot \mathbf{G} + \lambda J^{-1} (\text{tr}\, \mathbf{G})^2.$$

---

[8] See Appendix about isotropic tensors. We consider the representation (A.15) in that **A** is applied to a symmetric tensor.

Moreover, letting

$$\mathbf{G} = \mathbf{G}_0 + \tfrac{1}{3}(\mathrm{tr}\,\mathbf{G})\mathbf{1}$$

we find

$$[2\mu J^{-1}\mathbf{B}\,\mathbf{D}\,\mathbf{B} + \lambda J^{-1}(\mathbf{B}\cdot\mathbf{D})\mathbf{B}]\cdot\mathbf{D} = J^{-1}[2\mu\mathbf{G}_0\cdot\mathbf{G}_0 + (\tfrac{2}{3}\mu + \lambda)(\mathrm{tr}\,\mathbf{G})^2].$$

As a consequence

$$\mathbf{T}^{\mathrm{dis}}\cdot\mathbf{D} - \frac{1}{\theta}\mathbf{q}\cdot\nabla\theta = J^{-1}[2\mu\,\mathbf{G}_0\cdot\mathbf{G}_0 + (\tfrac{2}{3}\mu + \lambda)(\mathrm{tr}\,\mathbf{G})^2] - \frac{\kappa}{\theta J}|\mathbf{F}^T\nabla\theta|^2.$$

Hence the reduced entropy inequality holds if and only if the standard inequalities

$$\mu \geq 0, \qquad \tfrac{2}{3}\mu + \lambda \geq 0, \qquad \kappa \geq 0$$

hold.

4.6 Modelling via Dissipation Potentials

The modelling of dissipation is often established by expressing the constitutive properties via rate-type equations (see Chap. 6). The literature shows also constitutive schemes based on the notion of dissipation potential.[9] The interest in dissipation potentials seems to be the compact description of the dissipative "driving forces" via a (potential) function namely a dissipation potential.

Following is an outline of some approaches.

4.6.1 Convex Dissipation Potential

Let $\mathbf{T} = -p\mathbf{1} + \mathbf{T}^{\mathrm{dis}}$ and consider the entropy inequality in the form

$$-\rho(\dot{\psi} + \eta\dot{\theta}) - p\nabla\cdot\mathbf{v} + \mathbf{T}^{\mathrm{dis}}\cdot\mathbf{D} - \frac{1}{\theta}\mathbf{q}\cdot\nabla\theta \geq 0.$$

Moreover let the free energy $\psi = \varepsilon - \theta\eta$ be a function of ρ, θ so that the entropy inequality becomes

$$\rho(\frac{p}{\rho^2} - \partial_\rho\psi)\dot{\rho} - \rho(\eta + \partial_\theta\psi)\dot{\theta} + \mathbf{T}^{\mathrm{dis}}\cdot\mathbf{D} - \frac{1}{\theta}\mathbf{q}\cdot\nabla\theta \geq 0.$$

[9] See, e.g. [287] and references therein.

We let

$$p = \rho^2 \partial_\rho \psi, \qquad \eta = -\partial_\theta \psi$$

and hence the inequality reduces to

$$\mathcal{D} := \mathbf{T}^{\text{dis}} \cdot \mathbf{D} - \frac{1}{\theta} \mathbf{q} \cdot \nabla \theta \geq 0.$$

The dissipative stress $\mathbf{T}^{\text{dis}}$ is split in the form

$$\mathbf{T}^{\text{dis}} = p^{\text{dis}} \mathbf{1} + \mathbf{T}_0^{\text{dis}}, \qquad \text{tr}\, \mathbf{T}_0^{\text{dis}} = 0, \quad p^{\text{dis}} = \tfrac{1}{3} \text{tr}\, \mathbf{T}\text{dis}.$$

Hence

$$\mathcal{D} = p^{\text{dis}} \nabla \cdot \mathbf{v} + \mathbf{T}_0^{\text{dis}} \cdot \mathbf{D}_0 - \frac{1}{\theta} \mathbf{q} \cdot \nabla \theta.$$

Let $\mathcal{D}$ be given by a function

$$\mathcal{D} = \Upsilon(\nabla \cdot \mathbf{v}, \mathbf{D}_0, \nabla \theta),$$

and let

$$p^{\text{dis}} := \partial_{\nabla \cdot \mathbf{v}} \Upsilon, \qquad \mathbf{T}^{\text{dis}} := \partial_{\mathbf{D}_0} \Upsilon, \qquad \frac{1}{\theta} \mathbf{q} := -\partial_{\nabla \theta} \Upsilon.$$

The function $\Upsilon(\nabla \cdot \mathbf{v}, \mathbf{D}_0, \nabla \theta)$ is said to be a *potential of dissipation* if it is convex, vanishing at the origin, and non-negative.

An example in the literature [317] involves the thermo-viscous fluid. Look at the classical constitutive equations

$$\mathbf{T}^{\text{dis}} = \left(\lambda + \tfrac{2}{3}\mu\right) \nabla \cdot \mathbf{v}\, \mathbf{1} + 2\mu \mathbf{D}_0, \qquad \mathbf{q} = -\kappa \nabla \theta.$$

Hence it follows that

$$\Upsilon(\nabla \cdot \mathbf{v}, \mathbf{D}_0, \nabla \theta) = \tfrac{1}{2}\left(\lambda + \tfrac{2}{3}\mu\right)(\nabla \cdot \mathbf{v})^2 + \mu \mathbf{D}_0 \cdot \mathbf{D}_0 + \tfrac{1}{2}\, \theta\, \kappa |\nabla \theta|^2$$

is a potential of dissipation. This example with $\mathbf{T}^{\text{dis}}$ and $\mathbf{q}$ coincides in fact with the Navier-Stokes-Fourier model, Sects. 5.5 and 6.2.4.

As to solids, the constitutive Eqs. (4.29) and (4.30) satisfy

$$\mathbf{q} = -\theta \partial_{\nabla \theta} \Upsilon, \qquad \mathbf{T}^{\text{dis}} = \partial_{\mathbf{D}} \Upsilon$$

where

$$\Upsilon(\nabla \theta, \mathbf{D}) = \frac{\kappa}{2\theta J} \nabla \theta \cdot \mathbf{B} \nabla \theta + \frac{\mu}{J} \mathbf{BDBD} + \frac{\lambda}{2J} (\mathbf{B} \cdot \mathbf{D})^2.$$

Since

$$\Upsilon(\nabla\theta, \mathbf{D}) = \frac{\kappa}{2\theta J}|\mathbf{F}^T\nabla\theta|^2 + \frac{\mu}{J}|(\mathbf{F}^T\mathbf{D}\mathbf{F})_0|^2 + \frac{1}{J}(\tfrac{2}{3}\mu + \lambda)[\mathrm{tr}\,(\mathbf{F}^T\mathbf{D}\mathbf{F})]^2$$

it follows that Υ is also positive definite (as far as $\kappa > 0$, $\mu > 0$, $2\mu + 3\lambda > 0$) and vanishes at $\nabla\theta = \mathbf{0}, \mathbf{D} = \mathbf{0}$. Moreover the quadratic dependence proves convexity.

4.6.2 *Model with a Scalar Internal Variable*

A scalar internal variable is introduced to describe dissipative effects. Two views are adopted in the literature. In a case [176] a variable β is considered to describe at the macroscopic level the effects of microfractures and microcavities which result in a decrease of the material stiffness. The basic idea is that the internal variable enters the balance of energy via an appropriate power of the internal forces [175]. In the other case [301] the internal variable, say α, appears only in the free energy function; both α and $\nabla_R\alpha$ contribute to the dissipated power but do not contribute to the external power. We outline the two schemes.

By analogy with [175, 176], the balance of mass and linear momentum are left unchanged while the balance of energy is assumed in the form

$$\rho\dot{\varepsilon} = \mathbf{T}\cdot\mathbf{D} + \mathcal{H}\dot{\beta} + \mathbf{H}\cdot\nabla\dot{\beta} - \nabla\cdot\mathbf{q} + \rho r,$$

$\mathcal{H}$ being viewed as the scalar internal force (of damage) and $\mathbf{H}$ as the vector internal force. Hence substitution of $-\nabla\cdot\mathbf{q} + \rho r$ in the entropy inequality results in the modified Clausius-Duhem inequality

$$-\rho(\dot{\psi} + \eta\dot{\eta}) + \mathbf{T}\cdot\mathbf{D} + \mathcal{H}\dot{\beta} + \mathbf{H}\cdot\nabla\dot{\beta} - \frac{1}{\theta}\mathbf{q}\cdot\nabla\theta \geq 0.$$

Let

$$\mathbf{T} = \mathbf{T}^{\mathrm{en}} + \mathbf{T}^{\mathrm{dis}}, \qquad \mathcal{H} = \mathcal{H}^{\mathrm{en}} + \mathcal{H}^{\mathrm{dis}}, \qquad \mathbf{H} = \mathbf{H}^{\mathrm{en}} + \mathbf{H}^{\mathrm{dis}},$$

the purpose being to distinguish the energetic non-dissipative parts from the dissipative ones. As in the previous model we let

$$\mathbf{T}^{\mathrm{dis}} = p^{\mathrm{dis}}\mathbf{1} + \mathbf{T}_0^{\mathrm{dis}},$$

where $\mathrm{tr}\,\mathbf{T}_0^{\mathrm{dis}} = 0$. Let ψ be given by a differentiable function $\psi(\rho, \theta, \beta, \nabla\beta)$. The entropy inequality becomes

$$(\mathbf{T}^{\mathrm{en}} - \rho^2\partial_\rho\psi\mathbf{1})\cdot\mathbf{D} - \rho(\eta + \partial_\theta\psi)\dot{\theta} + (\mathcal{H}^{\mathrm{en}} - \rho\partial_\beta\psi)\dot{\beta} + (\mathbf{H}^{\mathrm{en}} - \rho\partial_{\nabla\beta}\psi)\nabla\dot{\beta}$$
$$+\rho\nabla\beta\otimes\partial_{\nabla\beta}\psi\cdot\mathbf{L} + p^{\mathrm{dis}}\nabla\cdot\mathbf{v} + \mathbf{T}_0^{\mathrm{dis}}\cdot\mathbf{D}_0 + \mathcal{H}^{\mathrm{dis}}\,\dot{\beta} + \mathbf{H}^{\mathrm{dis}}\cdot\nabla\dot{\beta} - \frac{1}{\theta}\mathbf{q}\cdot\nabla\theta \geq 0.$$

By assuming that $p^{\rm dis}$, $\mathbf{T}_0^{\rm dis}$, $\mathcal{H}^{\rm dis}$, $\mathbf{H}^{\rm dis}$, $\mathbf{q}$ approach zero as $\mathbf{D}$, $\dot\beta$, $\nabla\dot\beta$, $\nabla\theta$ approach zero it follows that

$$\mathbf{T}^{\rm en} = \rho^2\partial_\rho\psi\mathbf{1} + \rho\nabla\beta\otimes\partial_{\nabla\beta}\psi, \quad \eta = -\partial_\theta\psi, \quad \mathcal{H}^{\rm en} = \rho\partial_\beta\psi, \quad \mathbf{H}^{\rm en} = \rho\partial_{\nabla\beta}\psi,$$

and

$$\mathscr{D} := p^{\rm dis}\nabla\cdot\mathbf{v} + \mathbf{T}_0^{\rm dis}\cdot\mathbf{D}_0 + \mathcal{H}^{\rm dis}\,\dot\beta + \mathbf{H}^{\rm dis}\cdot\nabla\dot\beta - \frac{1}{\theta}\mathbf{q}\cdot\nabla\theta \geq 0,$$

while $\nabla\beta\otimes\partial_{\nabla\beta}\psi \in {\rm Sym}$. A dissipation potential $\Upsilon(\nabla\cdot\mathbf{v}, \mathbf{D}_0, \dot\beta, \nabla\dot\beta, \nabla\theta)$ holds provided

$$p^{\rm dis} = \partial_{\nabla\cdot\mathbf{v}}\Upsilon, \quad \mathbf{T}_0^{\rm dis} = \partial_{\mathbf{D}_0}\Upsilon, \quad \mathcal{H}^{\rm dis} = \partial_{\dot\beta}\Upsilon, \quad \mathbf{H}^{\rm dis} = \partial_{\nabla\dot\beta}\Upsilon, \quad \mathbf{q} = -\theta\partial_{\nabla\theta}\Upsilon$$

while

$$\partial_{\nabla\cdot\mathbf{v}}\Upsilon\nabla\cdot\mathbf{v} + \partial_{\mathbf{D}_0}\Upsilon\cdot\mathbf{D}_0 + \partial_{\dot\beta}\Upsilon\,\dot\beta + \partial_{\nabla\dot\beta}\Upsilon\cdot\nabla\dot\beta + \partial_{\nabla\theta}\Upsilon\cdot\nabla\theta \geq 0.$$

The other model with a scalar internal variable leads to the canonical balance laws; it stems from the view that the internal variable describes a purely dissipative behaviour and that it is controlled by the standard fields of stress and heat flux [301]. As a starting point, the balance equations are considered in the material description; the balance of linear momentum and energy are

$$\rho_R\dot{\mathbf{v}} = \nabla_R\cdot\mathbf{T}_R + \rho_R\mathbf{b}, \qquad \rho_R(\dot\varepsilon + \mathbf{v}\cdot\dot{\mathbf{v}}) = \nabla_R\cdot(\mathbf{v}\mathbf{T}_R - \mathbf{q}_R) + \rho_R r.$$

The second law inequality is then considered in the form

$$-\rho_R(\dot\psi + \eta\dot\theta) + \mathbf{T}_R\cdot\dot{\mathbf{F}} - \frac{1}{\theta}\mathbf{q}_R\cdot\nabla_R\theta \geq 0.$$

Let α be an internal variable and let ψ be given by

$$\psi = \psi(\mathbf{F}, \theta, \alpha, \nabla_R\alpha)$$

and the like for $\mathbf{T}_R$ and η. Hence it follows

$$\mathbf{T}_R = \partial_{\mathbf{F}}\psi, \qquad \eta = -\rho_R\partial_\theta\psi.$$

Upon the definitions

$$A := -\rho_R\partial_\alpha\psi, \qquad \mathbf{B} := -\rho_R\partial_{\nabla_R\alpha}\psi$$

we can write the reduced inequality in the form

$$\mathscr{D} := A\dot{\alpha} + \mathbf{B} \cdot \nabla_{\!R}\dot{\alpha} - \frac{1}{\theta}\mathbf{q}_R \cdot \nabla_{\!R}\theta \geq 0. \tag{4.31}$$

Hence, the free energy function ψ and the heat flux $\mathbf{q}_R$ are required to satisfy the reduced inequality (4.31). This in turn shows that the internal variable α, though occurring only in the free energy ψ, enters the modelling of dissipative effects via the reduced inequality.

In addition, the occurrence of the internal variable eventually leads to the canonical momentum balance. In view of the identities

$$\rho_R\dot{\mathbf{v}}\mathbf{F} = (\rho_R\mathbf{v}\mathbf{F})^{\cdot} - \rho_R\mathbf{v}\dot{\mathbf{F}}, \qquad \rho_R\mathbf{v}\dot{\mathbf{F}} = \nabla_{\!R}(\tfrac{1}{2}\rho_R\mathbf{v}^2),$$

$$(\nabla_{\!R} \cdot \mathbf{T}_R)\mathbf{F} = \nabla_{\!R} \cdot (\mathbf{F}^T\mathbf{T}_R) - \partial_{\mathbf{F}}\psi\, \nabla_{\!R}\mathbf{F},$$

upon right multiplication by $\mathbf{F}$ of the equation of motion (balance of linear momentum) it follows

$$(\rho_R\mathbf{v}\mathbf{F})^{\cdot} = \nabla_{\!R} \cdot (\mathbf{F}^T\mathbf{T}_R + \tfrac{1}{2}\rho_R\mathbf{v}^2\mathbf{1}) - \partial_{\mathbf{F}}\psi\, \nabla_{\!R}\mathbf{F} + \rho_R\mathbf{b}\mathbf{F}.$$

Now,

$$\begin{aligned}\nabla_{\!R}\psi(\mathbf{F}, \theta, \alpha, \nabla_{\!R}\alpha) &= \partial_{\mathbf{F}}\psi\nabla_{\!R}\mathbf{F} + \partial_\theta\psi\nabla_{\!R}\theta + \partial_\alpha\psi\nabla_{\!R}\alpha + \partial_{\nabla_{\!R}\alpha}\psi\nabla_{\!R}\nabla_{\!R}\alpha \\ &= \rho_R^{-1}[\mathbf{T}_R\nabla_{\!R}\mathbf{F} - \rho_R\eta\nabla_{\!R}\theta - A\,\nabla_{\!R}\alpha - \mathbf{B}\nabla_{\!R}\nabla_{\!R}\alpha].\end{aligned}$$

Hence, substitution of $\partial_{\mathbf{F}}\psi\, \nabla_{\!R}\mathbf{F} = \mathbf{T}_R\, \nabla_{\!R}\mathbf{F}$ allows the balance of linear momentum to be written

$$(\rho_R\mathbf{v}\mathbf{F})^{\cdot} = \nabla_{\!R} \cdot (\mathbf{F}^T\mathbf{T}_R + L\,\mathbf{1}) + \rho_R\mathbf{b}\mathbf{F} + \mathbf{f}^{\text{th}} + \mathbf{f}^{\text{intv}}, \tag{4.32}$$

where

$$L = \rho_R(\tfrac{1}{2}\mathbf{v}^2 - \psi), \qquad \mathbf{f}^{\text{th}} = -\rho_R\eta\nabla_{\!R}\theta, \qquad \mathbf{f}^{\text{intv}} = -A\,\nabla_{\!R}\alpha - \mathbf{B}\nabla_{\!R}\nabla_{\!R}\alpha.$$

The vector $\rho_R\mathbf{v}\mathbf{F}$ is the material linear momentum and the tensor $\mathbf{F}^T\mathbf{T}_R + L\,\mathbf{1}$ is the *Eshelby material stress.* The material force $\mathbf{f}^{\text{intv}}$ occurring in the balance Eq. (4.32) is due to the dependence of the free energy on the internal variable α.

Remark 4.1 Whether or not the internal variable α is observable and hence subject to a proper action in the bulk and at the boundary of any region in the body, mathematically an appropriate equation is required so that the evolution of α is determined. While physically an additional equation may be introduced by ascribing to α an angular momentum and a kinetic energy [301], a general setting for internal variables may be given within the framework of rate-type equations (see Chap. 6).

4.6.3 Model with a Vector Internal Variable

The starting point is the use of a generalized stress and a generalized strain. For definiteness, though, we let the material be non-conducting and describe the stress by the first Piola stress $\mathbf{T}_R$ and the deformation by $\mathbf{F}$. Hence we write the entropy inequality in the form

$$-\dot{\psi}_R - \eta_R\dot{\theta} + \mathbf{T}_R \cdot \dot{\mathbf{F}} \geq 0.$$

The free energy is assumed to depend on θ, $\mathbf{F}$, and a tensor-valued internal variable $\boldsymbol{\alpha}$. Upon time differentiation of $\psi_R(\theta, \mathbf{F}, \boldsymbol{\alpha})$ it follows

$$-(\partial_\theta\psi_R + \eta_R)\dot{\theta} + (\mathbf{T}_R - \partial_{\mathbf{F}}\psi_R) \cdot \dot{\mathbf{F}} - \partial_{\boldsymbol{\alpha}}\psi_R \cdot \dot{\boldsymbol{\alpha}} \geq 0.$$

If η_R and $\mathbf{T}_R$ depend on $\theta, \mathbf{F}, \boldsymbol{\alpha}$, as ψ_R does, then the inequality implies

$$\eta_R = -\partial_\theta\psi_R, \qquad \mathbf{T}_R = \partial_{\mathbf{F}}\psi_R, \qquad \partial_{\boldsymbol{\alpha}}\psi_R \cdot \dot{\boldsymbol{\alpha}} \leq 0.$$

Instead, the assumption is made that

$$\mathbf{T}_R = \mathbf{T}_R^{\text{en}} + \mathbf{T}_R^{\text{dis}}, \qquad \mathbf{T}_R^{\text{en}} = \partial_{\mathbf{F}}\psi_R,$$

and hence

$$\mathbf{T}_R^{\text{dis}} \cdot \dot{\mathbf{F}} - \partial_{\boldsymbol{\alpha}}\psi_R \cdot \dot{\boldsymbol{\alpha}} \geq 0.$$

This position is thermodynamically consistent if we let $\mathbf{T}_R^{\text{dis}} = \mathbf{T}_R^{\text{dis}}(\theta, \mathbf{F}, \boldsymbol{\alpha}, \dot{\mathbf{F}})$. In addition, the tensors

$$\mathbf{X}^{\text{en}} := \partial_{\boldsymbol{\alpha}}\psi_R, \qquad \mathbf{X}^{\text{dis}} := -\mathbf{X}^{\text{en}}$$

are defined; $\mathbf{X}^{\text{en}}$ is viewed as the energetic driving force while $\mathbf{X}^{\text{dis}}$ is the dissipative driving force [46]. The formal advantage of considering $\mathbf{X}^{\text{dis}}$ is that the reduced inequality can be expressed in the form

$$\mathbf{T}_R^{\text{dis}} \cdot \dot{\mathbf{F}} + \mathbf{X}^{\text{dis}} \cdot \dot{\boldsymbol{\alpha}} \geq 0, \tag{4.33}$$

and hence $\mathbf{T}_R^{\text{dis}} \leftrightarrow \dot{\mathbf{F}}$ and $\mathbf{X}^{\text{dis}} \leftrightarrow \dot{\boldsymbol{\alpha}}$ are viewed as dissipation conjugate pairings. Yet, by definition, $\mathbf{X}^{\text{en}}$, and $\mathbf{X}^{\text{dis}}$ too, is a function of θ, $\mathbf{F}$, $\boldsymbol{\alpha}$ and hence independent of $\dot{\boldsymbol{\alpha}}$. Henceforth the dependence on temperature is understood and not written.

A *dissipation potential* π is considered, via a function

$$\pi = \pi(\dot{\mathbf{F}}, \dot{\boldsymbol{\alpha}}),$$

such that

$$\mathbf{T}_R^{\text{dis}} = \partial_{\dot{\mathbf{F}}}\pi, \qquad \mathbf{X}^{\text{dis}} = \partial_{\dot{\boldsymbol{\alpha}}}\pi.$$

Let

$$\partial_{\dot{\mathbf{F}}}\mathbf{T}_R^{\text{dis}}, \qquad \partial_{\dot{\boldsymbol{\alpha}}}\mathbf{X}_R^{\text{dis}}$$

be invertible so that we can determine the functions

$$\dot{\mathbf{F}}(\mathbf{T}_R^{\text{dis}}, \mathbf{F}, \boldsymbol{\alpha}), \qquad \dot{\boldsymbol{\alpha}}(\mathbf{X}^{\text{dis}}, \mathbf{F}, \boldsymbol{\alpha}).$$

Next the dual dissipation potential π^*, the convex conjugate to π, is defined by

$$\pi^*(\mathbf{T}_R^{\text{dis}}, \mathbf{X}^{\text{dis}}) = \mathbf{T}_R^{\text{dis}} \cdot \dot{\mathbf{F}} + \mathbf{X}^{\text{dis}} \cdot \dot{\boldsymbol{\alpha}} - \pi(\dot{\mathbf{F}}, \dot{\boldsymbol{\alpha}}).$$

It follows that

$$\partial_{\mathbf{T}_R^{\text{dis}}}\pi^* = \dot{\mathbf{F}} + \mathbf{T}_R^{\text{dis}} \cdot \partial_{\mathbf{T}_R^{\text{dis}}}\dot{\mathbf{F}} - \mathbf{T}_R^{\text{dis}} \cdot \partial_{\mathbf{T}_R^{\text{dis}}}\dot{\mathbf{F}} = \dot{\mathbf{F}}, \quad \partial_{\mathbf{X}^{\text{dis}}}\pi^* = \dot{\boldsymbol{\alpha}} + \mathbf{X}^{\text{dis}} \cdot \partial_{\mathbf{X}^{\text{dis}}}\dot{\boldsymbol{\alpha}} - \mathbf{X}^{\text{dis}} \cdot \partial_{\mathbf{X}^{\text{dis}}}\dot{\boldsymbol{\alpha}} = \dot{\boldsymbol{\alpha}}.$$

Therefore the equations

$$\dot{\mathbf{F}} = \partial_{\mathbf{T}_R^{\text{dis}}}\pi^*, \qquad \dot{\boldsymbol{\alpha}} = \partial_{\mathbf{X}^{\text{dis}}}\pi^*,$$

determine the rate of $\mathbf{F}$ and $\boldsymbol{\alpha}$.

The dissipation potential π is assumed to be a gauge [287] in that

$$\pi(\dot{\mathbf{F}}, \dot{\boldsymbol{\alpha}}) \geq 0, \ \forall \dot{\mathbf{F}}, \dot{\boldsymbol{\alpha}}, \qquad \pi(\mathbf{0}, \mathbf{0}) = 0,$$

and π is convex and positively homogeneous. Now,

$$\mathbf{T}_R^{\text{dis}} \cdot \dot{\mathbf{F}} + \mathbf{X}^{\text{dis}} \cdot \dot{\boldsymbol{\alpha}} = \partial_{\dot{\mathbf{F}}}\pi \cdot \dot{\mathbf{F}} + \partial_{\dot{\boldsymbol{\alpha}}}\pi \cdot \dot{\boldsymbol{\alpha}} = \partial_\lambda \pi(\lambda\dot{\mathbf{F}}, \lambda\dot{\boldsymbol{\alpha}}).$$

By the positive homogeneity it follows

$$\mathbf{T}_R^{\text{dis}} \cdot \dot{\mathbf{F}} + \mathbf{X}^{\text{dis}} \cdot \dot{\boldsymbol{\alpha}} = |\lambda|^k \pi(\dot{\mathbf{F}}, \dot{\boldsymbol{\alpha}}), \quad k > 0.$$

The assumption $\pi \geq 0$ implies the inequality (4.33).

Yet, thermodynamic consistency is more severe. By definition, $\mathbf{X}^{\text{dis}}$ is independent of $\dot{\boldsymbol{\alpha}}$. Then upon the obvious integration of $\mathbf{X}^{\text{dis}} = \partial_{\dot{\boldsymbol{\alpha}}}\pi$ we find

$$\pi = \mathbf{X}^{\text{dis}}(\mathbf{F}, \boldsymbol{\alpha}) \cdot \dot{\boldsymbol{\alpha}} + f(\mathbf{F}, \dot{\mathbf{F}}, \boldsymbol{\alpha}),$$

f being an arbitrary function. This shows that π is linear with respect to $\dot{\boldsymbol{\alpha}}$. Hence the assumption that π be a gauge is contradicted by the condition

$$\partial_{\dot{\boldsymbol{\alpha}}}\pi = -\partial_{\boldsymbol{\alpha}}\psi_R.$$

Chapter 5
Fluids

Physically a fluid is defined as a substance which deforms continuously under the application of a shear stress. The term fluid generally includes both the liquid and gas phases. Some substances can be both fluid and solid. Viscoelastic fluids appear to behave similar to a solid when a sudden force is applied. Also substances with a very high viscosity such as pitch appear to behave like a solid. Here fluids are characterized via the constitutive properties. It is a common property of fluids that they exhibit elasticity in their response to compression. Fluids do not possess natural reference configurations and that is why it is more convenient to develop the pertinent equations in the spatial form.

Fluids show lack of resistance to permanent deformation, resisting only relative rates of deformation in a dissipative, frictional manner. These properties are typically a function of their inability to support a shear stress in static equilibrium.

Some models are examined in detail: elastic fluids (along with water wave theories), thermoelastic fluids, the ideal gas and some real gases, heat-conducting viscous fluids, Newtonian fluids, Stokesian fluids, generalized Newtonian fluids, viscoplastic and viscoelastic fluids, models of turbulence.

5.1 Elastic Fluids

An *elastic fluid* is a body governed by the constitutive equations

$$\varepsilon = \hat{\varepsilon}(\rho), \qquad \mathbf{T} = \hat{\mathbf{T}}(\rho).$$

Since ρ is invariant the objectivity implies that

$$\hat{\mathbf{T}}(\rho) = \mathbf{Q}\hat{\mathbf{T}}(\rho)\mathbf{Q}^T$$

A. Morro and C. Giorgi, *Mathematical Modelling of Continuum Physics*,
Modeling and Simulation in Science, Engineering and Technology,
https://doi.org/10.1007/978-3-031-20814-0_5

for all rotations $\mathbf{Q}$ and any value of ρ. Consequently $\mathbf{T}$ must be a scalar times the unit tensor $\mathbf{1}$. Since elasticity occurs only in compression, we take the scalar as $-p$, p being the pressure, and write

$$\mathbf{T} = -p\mathbf{1}, \qquad p = \hat{p}(\rho).$$

The elastic fluid is also named *piezotropic*.

The energy supply and the heat flux are taken to be zero and hence the local balance equations are

$$\dot{\rho} = -\rho\nabla\cdot\mathbf{v},$$

$$\rho\dot{\mathbf{v}} = -\nabla p + \rho\mathbf{b},$$

and

$$-\rho\dot{\varepsilon} = -p\nabla\cdot\mathbf{v}.$$

Then we can write

$$(\rho\frac{d\hat{\varepsilon}(\rho)}{d\rho} - \frac{p}{\rho})\dot{\rho} = 0. \tag{5.1}$$

By the local balance of mass, for any value of ρ we can have any value of $\dot{\rho}$ depending on the value of $\nabla\cdot\mathbf{v}$. Notice the identity

$$\nabla\cdot\dot{\mathbf{v}} = \partial_t\nabla\cdot\mathbf{v} + (\mathbf{v}\cdot\nabla)\nabla\cdot\mathbf{v} + \mathbf{L}\cdot\mathbf{L}^T.$$

Hence the equation of motion can be written in the form

$$\partial_t\nabla\cdot\mathbf{v} + (\mathbf{v}\cdot\nabla)\nabla\cdot\mathbf{v} + \mathbf{L}\cdot\mathbf{L}^T + \nabla\cdot(\frac{1}{\rho}\frac{d\hat{p}(\rho)}{d\rho}\nabla\rho) = \nabla\cdot\mathbf{b}.$$

We regard the function $\mathbf{b}(\mathbf{x}, t)$ as arbitrarily assignable. This allows us to say that for any value of ρ, we can have arbitrary values of $\nabla\cdot\mathbf{v}$, and hence of $\dot{\rho}$, while $\nabla\cdot\mathbf{b}$ takes the value required to satisfy the equation of motion. Therefore, given any spatial point and any time, it is possible to find a process such that ρ and $\dot{\rho}$ have arbitrarily prescribed values at that point and time. An equation of the form

$$f(\rho)\,\dot{\rho} = 0$$

holds for any real value of $\dot{\rho}$ and any positive value of ρ if and only if $f(\rho) = 0$. Thus the energy equation (5.1) implies the *pressure relation*

$$\hat{p}(\rho) = \rho^2\frac{d\hat{\varepsilon}(\rho)}{d\rho}.$$

The internal energy determines the pressure through the pressure relation.

To derive a property of the motion we observe

$$\frac{1}{\rho}\nabla p = \nabla\frac{p}{\rho} - p\nabla\frac{1}{\rho} = \nabla\frac{p}{\rho} + \frac{d\hat{\varepsilon}(\rho)}{d\rho}\nabla\rho = \nabla(\varepsilon + \frac{p}{\rho}).$$

If $\mathbf{b}$ is conservative, say $\mathbf{b} = -\nabla U$ and $U = gz$ for the gravity force, then the equation of motion reads

$$\dot{\mathbf{v}} = -\nabla(\varepsilon + \frac{p}{\rho} + U).$$

It follows that

$$\nabla \times \dot{\mathbf{v}} = \mathbf{0}.$$

This implies that in the motion of an elastic (or piezotropic) fluid under a conservative body force *the acceleration is the gradient of a potential.* Thus the motion preserves circulation.

Look at a steady flow, where $\partial_t \mathbf{v} = \mathbf{0}$ and $\partial_t \rho = 0$. We have

$$\dot{\mathbf{v}} = (\mathbf{v} \cdot \nabla)\mathbf{v} = \tfrac{1}{2}\nabla \mathbf{v}^2 + (\nabla \times \mathbf{v}) \times \mathbf{v}.$$

The equation of motion yields

$$\tfrac{1}{2}\nabla \mathbf{v}^2 + (\nabla \times \mathbf{v}) \times \mathbf{v} = -\nabla(\varepsilon + \frac{p}{\rho} + U).$$

Let

$$\varphi := \varepsilon + \frac{p}{\rho} + \tfrac{1}{2}\mathbf{v}^2 + U.$$

It follows that if $\nabla \times \mathbf{v} = \mathbf{0}$ then

$$\nabla\varphi = \mathbf{0}.$$

Now, irrespective of the value of $\nabla \times \mathbf{v}$,

$$\mathbf{v} \cdot \nabla\varphi = -\mathbf{v} \cdot (\nabla \times \mathbf{v}) \times \mathbf{v} = 0$$

This means that φ *is constant on streamlines.* If further the flow is irrotational then φ *is constant* on the fluid domain. This result is in fact Bernoulli's law.

In Sect. 2.8 Bernoulli's law is derived from the equation of motion and the assumption that the fluid is barotropic. Here, for elastic fluids, the equation of motion leads to Bernoulli's law in the form $\mathbf{v} \cdot \nabla\varphi = 0$. Since Bernoulli's theorem involves $\nabla\varphi$ then the two results are equivalent if

$$\varepsilon + \frac{p}{\rho} = \int_{p_0}^{p} \frac{1}{\rho(\xi)} d\xi + \text{constant}.$$

Now look at the two functions as dependent on p and observe

$$\frac{d}{dp}\big[\hat{\varepsilon}(\rho(p)) + \frac{p}{\rho(p)}\big] = \frac{d\hat{\varepsilon}}{d\rho}\frac{d\rho}{dp} + \frac{1}{\rho} - \frac{p}{\rho^2}\frac{d\rho}{dp} = \frac{1}{\rho}, \qquad \frac{d}{dp}\int_{p_0}^{p}\frac{1}{\rho(\xi)}d\xi = \frac{1}{\rho};$$

the equivalence is then proved.

5.1.1 Water Wave Theories

The simple model of elastic fluid allows us to set up significant systems of water wave equations.

Consider a homogeneous, incompressible fluid, with constant mass density ρ. By the continuity equation, the velocity $\mathbf{v}$ is solenoidal,

$$\nabla \cdot \mathbf{v} = 0.$$

Moreover, the fluid is acted upon by pressure and gravity and hence the force per unit volume is $-\nabla(p + \rho g z)$, z being the upward vertical coordinate. Then circulation is preserved and we let

$$\nabla \times \mathbf{v} = \mathbf{0}.$$

Since ρ is constant, the equation of motion can be written

$$\partial_t \mathbf{v} + \nabla \tfrac{1}{2}\mathbf{v}^2 + (\nabla \times \mathbf{v}) \times \mathbf{v} = -\nabla(\frac{p}{\rho} + gz).$$

Since $\nabla \times \mathbf{v} = \mathbf{0}$ then there is a potential field, $\mathbf{v} = \nabla\psi$, and hence

$$\nabla(\partial_t\psi + \tfrac{1}{2}\mathbf{v}^2 + \frac{p}{\rho} + gz) = \mathbf{0}.$$

Thus $\partial_t\psi + \frac{1}{2}\mathbf{v}^2 + p/\rho + gz$ is a function of time. Consistent with the literature, we let this function be zero and also neglect $\mathbf{v}^2/2$. Consequently,

$$\partial_t\psi + \frac{p}{\rho} + gz = 0.$$

Let the fluid move over an uneven bottom specified by

$$\mathbf{x} = x\mathbf{e}_1 + y\mathbf{e}_2 - h(x, y)\mathbf{e}_3,$$

where $\mathbf{e}_3$ is the upward unit vector while $x\mathbf{e}_1 + y\mathbf{e}_2$ belongs to a horizontal region D. The free surface of the fluid is described by

$$\mathbf{x} = x\mathbf{e}_1 + y\mathbf{e}_2 + \eta(x, y, t)\mathbf{e}_3,$$

η representing the free surface elevation. While $h(x, y)$ is an a-priori given function, $\eta(x, y, t)$ is unknown.

At the free surface of the fluid, the pressure is just the atmospheric pressure. Hence taking the atmospheric pressure as a reference we can take the boundary condition in the form

$$g\eta + \partial_t\psi(\eta) = 0.$$

Since $\partial_t\eta$ is the vertical velocity then

$$\partial_t\eta = \partial_z\psi(\eta).$$

Hence it follows

$$g\partial_z\psi + \partial_t^2\psi = 0 \quad \text{at} \quad z = \eta. \tag{5.2}$$

Moreover the condition $\nabla \cdot \mathbf{v} = 0$, along with $\mathbf{v} = \nabla\psi$, results in the differential equation

$$\Delta\psi = 0.$$

Deep-water waves

We let the depth be so large that we can take $z \in (-\infty, \eta(x, y)]$. We then look for a wave solution with the velocity, in the x-direction, given by the angular frequency ω and the wave number k,

$$\psi = \cos(kx - \omega t)f(z);$$

to fix ideas let $\omega, k > 0$. We determine the function f by applying $\Delta\psi = 0$ to obtain

$$f'' - k^2 f = 0.$$

Both $\exp(kz)$ and $\exp(-kz)$ are solutions; only $\exp(kz)$ is acceptable in that is bounded as $z \to -\infty$. Hence we have

$$\psi = A\exp(kz)\cos(kx - \omega t).$$

The boundary condition at the free surface $z = \eta$ implies

$$(gk - \omega^2)A\exp(k\eta)\cos(kx - \omega t) = 0.$$

As a consequence we find the propagation condition

$$\omega^2 = gk.$$

The speed of propagation (phase speed) is $U = \omega/k = g/\omega$.

If, instead, the depth h is finite then both $\exp(kz)$ and $\exp(-kz)$ are acceptable and the solution is

$$\psi = [A\exp(kz) + B\exp(-kz)]\cos(kx - \omega t).$$

We require that the vertical velocity $\partial_z\psi$ vanish at the bottom, $z = -h$. It follows

$$A\exp(-kh) - B\exp(kh) = 0.$$

We then find that

$$\psi = \gamma\cosh(k(z+h))\cos(kx - \omega t),$$

where $\gamma = A\exp(-kh)$. The condition at the free surface (5.2) yields

$$gk\sinh(k(\eta + h)) = \omega^2\cosh(k(\eta + h)).$$

Since $\eta \ll h$ we conclude that

$$\omega^2 = gk\tanh(kh).$$

The phase speed is given by

$$U = \frac{\omega}{k} = \sqrt{g/k}\sqrt{\tanh(kh)}.$$

Look at long waves namely $\lambda = 2\pi/k \gg h$ or $kh \ll 1$. Since $\tanh(kh) = kh + o(kh)$ then as $kh \ll 1$ we can take the approximation

$$U \simeq \sqrt{gh}.$$

At the limit of long waves the phase speed is independent of the frequency.

Shallow-water waves

We now follow [23, 208] and describe the fluid motion via a picture of vertical columns,

$$\mathbf{x} = \mathbf{r} + (\psi + X\varphi)\mathbf{e}_3,$$

where $\mathbf{r} = x\mathbf{e}_1 + y\mathbf{e}_2$, $\psi = (\eta - h)/2$, and $\varphi = \eta + h$. Hence φ is the height of the column, or length of the segment, while ψ is the vertical coordinate of the mean point (or centre of mass) of the column (segment). Observe that $X \in [-1/2, 1/2]$; $X = -1/2$ corresponds to the bed while $X = 1/2$ corresponds to the free surface. The velocity $\mathbf{V}$ of a point is written in the form

$$\mathbf{V} = \dot{\mathbf{r}} + (\dot{\psi} + X\dot{\xi})\mathbf{e}_3.$$

Hence $\dot{\psi}$ is the vertical velocity of the centre of mass while $X\dot{\varphi}$ is the vertical velocity, at X, relative to the centre of mass.

Let $\tilde{\nabla}$ denote the bi-dimensional gradient, $\tilde{\nabla} = \partial_x \mathbf{e}_1 + \partial_y \mathbf{e}_2$. The equations of motion are derived by considering the balance of energy for a fluid column. Let $\mathcal{P}$ denote a (cylindrical) column and let C be the closed curve given by the intersection of $\partial\mathcal{P}$ with an horizontal plane. To compute the power of pressure on $\partial\mathcal{P}$ we observe that, on the lateral surface $\partial_l\mathcal{P}$,

$$\int_{\partial_l\mathcal{P}} p\mathbf{n}\cdot\mathbf{V}\,da = \int_C \mathbf{v}\cdot\mathbf{n}\int_{-h}^{\eta} p\,dz\,ds = \int_C \Pi\mathbf{v}\cdot\mathbf{n}\,ds,$$

where s is the arclength along C and

$$\Pi = \int_{-h}^{\eta} p\,dz.$$

Moreover,

$$\mathbf{V}\cdot\mathbf{n} = [1+(\tilde{\nabla}\eta)^2]^{-1/2}\,[-\mathbf{v}\cdot\tilde{\nabla}\eta + \dot{\psi} + \tfrac{1}{2}\dot{\varphi}],$$

at the free surface ($X = 1/2$) while

$$\mathbf{V}\cdot\mathbf{n} = [1+(\tilde{\nabla}h)^2]^{-1/2}\,[\mathbf{v}\cdot\tilde{\nabla}h - \dot{\psi} + \tfrac{1}{2}\dot{\varphi}],$$

at the bottom. Hence

$$\int_{\partial\mathcal{P}} -p\mathbf{V}\cdot\mathbf{n}\,da = -\int_C \Pi\mathbf{v}\cdot\mathbf{n}\,ds + \int_D [p_a(\mathbf{v}\cdot\tilde{\nabla}\eta - \dot{\psi} - \tfrac{1}{2}\dot{\varphi}) + P(\mathbf{v}\cdot\tilde{\nabla}h + \dot{\psi} - \tfrac{1}{2}\dot{\varphi})]dxdy,$$

where D is the region enclosed by C while p_a is the atmospheric pressure and P is the pressure at the bottom.

Consistent with the balance of energy (2.46) we consider the balance equation

$$\frac{d}{dt}\int_{\mathcal{P}} \rho(\tfrac{1}{2}\mathbf{V}^2 + gz)dv = \int_{\partial\mathcal{P}} -p\mathbf{V}\cdot\mathbf{n}\,da.$$

Observe

$$\int_{\mathcal{P}} f\,dv = \int_D dxdy \int_{-h(x,y)}^{\eta(x,y,t)} f\,dz, \qquad \int_{-h(x,y)}^{\eta(x,y,t)} f(z)dz = \varphi\int_{-1/2}^{1/2} f(\psi + X\varphi)dX.$$

Hence

$$\int_{\mathcal{P}} \rho(\tfrac{1}{2}\mathbf{V}^2 + gz)dv = \int_D \tfrac{1}{2}\rho\varphi(\mathbf{v}^2 + \tfrac{1}{12}\dot{\varphi}^2 + 2g\psi)dxdy$$

and the balance of energy results in

$$\frac{d}{dt}\int_D \tfrac{1}{2}\rho\varphi(\mathbf{v}^2+\tfrac{1}{12}\dot{\varphi}^2+2g\psi)dxdy=-\int_C \Pi\mathbf{v}\cdot\mathbf{n}\,ds$$
$$+\int_D[p_a(\mathbf{v}\cdot\tilde{\nabla}\eta-\dot{\psi}-\tfrac{1}{2}\dot{\varphi})+P(\mathbf{v}\cdot\tilde{\nabla}h+\dot{\psi}-\tfrac{1}{2}\dot{\varphi})]dx\,dy, \tag{5.3}$$

We now require that the balance of energy (5.3) be invariant under a superimposed rigid motion so that the change $\mathbf{V}\to\mathbf{V}+\mathbf{U}$ leaves (5.3) unchanged. First let $\mathbf{v}\to\mathbf{v}+\mathbf{u}$, $\mathbf{u}$ being a horizontal vector.

The invariance with respect to $\mathbf{u}$ implies that

$$\frac{d}{dt}\int_D \tfrac{1}{2}\rho\varphi(2\mathbf{v}\cdot\mathbf{u}+\mathbf{u}^2)dxdy=-\mathbf{u}\cdot\int_C \Pi\mathbf{n}ds+\mathbf{u}\cdot\int_D[p_a\tilde{\nabla}\eta+P\tilde{\nabla}h]dx\,dy.$$

As to the integral on C we observe that

$$\mathbf{u}\cdot\int_C \Pi\mathbf{n}\,ds=\int_D \tilde{\nabla}\cdot(\mathbf{u}\Pi)\,dx\,dy=\mathbf{u}\cdot\int_D \tilde{\nabla}\Pi\,dx\,dy.$$

Hence, owing to the arbitrariness of $\mathbf{u}$ we deduce that

$$\frac{d}{dt}\int_D \tfrac{1}{2}\rho\varphi dx\,dy=0,$$

$$\frac{d}{dt}\int_D \rho\varphi\mathbf{v}dx\,dy=-\int_D \tilde{\nabla}\Pi\,dx\,dy+\int_D[p_a\tilde{\nabla}\eta+P\tilde{\nabla}h]dx\,dy.$$

Now let $\dot{\psi}\to\dot{\psi}+\alpha$. The invariance with respect to α and the arbitrariness of α require that

$$\frac{d}{dt}\int_D \rho\,\varphi\,g\,t\,dx\,dy=\int_D(P-p_a)dx\,dy.$$

By the Reynolds transport theorem, in the bi-dimensional domain D, we obtain the corresponding local equations in the form

$$\dot{\varphi}+\varphi\tilde{\nabla}\cdot\mathbf{v}=0, \tag{5.4}$$

$$\rho\varphi\dot{\mathbf{v}}=-\tilde{\nabla}\Pi+p_a\tilde{\nabla}\eta+P\tilde{\nabla}h, \tag{5.5}$$

$$0=P-p_a-\rho g\varphi,$$

where the superposed dot denotes the total time derivative with respect to the horizontal motion, $\dot{\varphi}=\partial_t\varphi+\mathbf{v}\cdot\tilde{\nabla}\varphi$.

The shallow-water theory is based on the assumption that the vertical component of the motion has a negligible effect on the pressure[1] p in that the depth of the water is

[1] See [400], Sect. 2.2.

sufficiently small compared with some other significant lengths (long-wave theory). This assumption amounts to assuming that the pressure p is approximately equal to the hydrostatic pressure, i.e.

$$p(z) = \rho g(\eta - z) + p_a.$$

Consequently,

$$P = \rho g(\eta + h) + p_a, \qquad \Pi = [\tfrac{1}{2}\rho g(\eta + h) + p_a](\eta + h).$$

Equations (5.4) and (5.5) then yield

$$\partial_t \eta + \tilde{\nabla} \cdot [(\eta + h)\mathbf{v}] = 0, \qquad \partial_t \mathbf{v} + (\mathbf{v} \cdot \tilde{\nabla})\mathbf{v} = -g\tilde{\nabla}\eta. \tag{5.6}$$

Equations (5.6) characterize the nonlinear shallow-water theory for the unknowns functions $\eta(x, y, t)$, $\mathbf{v}(x, y, t)$. The corresponding linearized version is given by

$$\partial_t \eta + \tilde{\nabla} \cdot (h\mathbf{v}) = 0, \qquad \partial_t \mathbf{v} = -g\tilde{\nabla}\eta.$$

Hence it follows the hyperbolic equation for the free surface elevation η,

$$\partial_t^2 \eta - g\tilde{\nabla} \cdot (h\tilde{\nabla}\eta) = 0.$$

Korteweg-de Vries equation

The Korteweg-de Vries equation (KdV) occurs in various branches of physics and is of interest mathematically in that admits particular solutions, termed solitons [284]. Here we show how the KdV can be derived from, or related to, the shallow-water theory [25].

If φ satisfies the differential equation

$$\partial_t \varphi + c_0 \partial_x \varphi = 0$$

then any function $\varphi(x - c_0 t)$ is a solution. If, instead, $c = c(x) \simeq c_0$ then we can view $\varphi(x - c_0 t)$ as merely an approximation of a solution to

$$\partial_t \varphi + \partial_x [c(x)\varphi] = 0.$$

It is convenient to apply the change of variable

$$x \to \chi = x - c_0 t$$

and argue with the coordinate χ describing the space dependence relative to the the frame moving with the velocity c_0 of the fundamental wave solution. With the purpose of accounting for long waves, slowly varying in time, the new variables

$$\zeta = \epsilon^{1/2}\chi, \qquad \tau = \epsilon^{3/2}t$$

are considered. The powers $\epsilon^{1/2}$ and $\epsilon^{3/2}$ of the parameter ϵ characterize the different weight of the derivatives under consideration; so $\partial_\chi\varphi = O(\epsilon^{-1/2})$ and $\partial_t\varphi = O(\epsilon^{-3/2})$.

Back to the nonlinear system (5.6) we restrict attention to the one-dimensional setting with variables x, t. By (5.4) and (5.5) in the one-dimensional setting we have

$$\dot{\varphi} + \varphi\partial_x v = 0, \qquad \varphi\dot{v} = -\partial_x\mathscr{P},$$

where $-\partial_x\mathscr{P}$ replaces $1/\rho$ times the right-hand side of (5.5) and v is the x-component of the velocity. For the present purpose it is assumed that $\mathscr{P} = \varphi^2\ddot{\varphi}/3 + g\varphi^2/2$. Since $\dot{\varphi} = \partial_t\varphi + v\partial_v\varphi$, and the like for v, then

$$\varphi\partial_t\varphi = \partial_t(\varphi v) - v\partial_t\varphi = \partial_t(\varphi v) + v\partial_x(\varphi v).$$

Hence we find the system of equations

$$\partial_t\varphi + \partial_x(\varphi v) = 0, \qquad \partial_t(\varphi v) + \partial_x(\varphi v^2 + \mathscr{P}) = 0.$$

Alternatively we can apply

$$\partial_t\varphi + \partial_x(v\varphi) = 0, \qquad \partial_t v + v\partial_x v + g\partial_x\varphi = 0. \tag{5.7}$$

Since

$$\partial_t = \epsilon^{3/2}\partial_\tau - c_0\epsilon^{1/2}\partial_\zeta, \qquad \partial_x = \partial_\chi = \epsilon^{1/2}\partial_\zeta,$$

then (5.7) can be written

$$\epsilon\partial_\tau\varphi + (v - c_0)\partial_\zeta\varphi + \varphi\partial_\zeta v = 0, \qquad \epsilon\partial_\tau v + (v - c_0)\partial_\zeta\varphi + \varphi^{-1}\partial_\zeta\mathscr{P} = 0.$$

Equations free of the parameter ϵ are obtained by replacing φ, v, and $\mathscr{P}$ with their formal expansions in ϵ,

$$\varphi = h_0 + \epsilon\varphi' + \epsilon^2\varphi'' + \ldots, \qquad v = \epsilon v' + \epsilon^2 v'' + \ldots, \qquad \mathscr{P} = \mathscr{P}_0 + \epsilon\mathscr{P}' + \epsilon^2\mathscr{P}'' + \ldots,$$

where

$$\mathscr{P}_0 = \tfrac{1}{2}c_0^2 h_0, \qquad \mathscr{P}' = c_0^2\varphi', \qquad \mathscr{P}'' = c_0^2\varphi'' + \tfrac{1}{2}g(\varphi')^2 + \tfrac{1}{3}c_0^2 h_0^2\partial_\zeta^2\varphi'.$$

At the lowest order of approximation we find

$$-c_0\partial_\zeta\varphi' + h_0\partial_\zeta v' = 0.$$

The obvious integration gives

$$v' = \frac{c_0}{h_0}(\varphi' + \beta).$$

Mathematically β is an arbitrary function of time. As we see in a moment, a strict connection with the Korteweg-de Vries equation follows by letting $\beta = 0$. At the order ϵ^2 we find from (5.7) that

$$\partial_\tau \varphi' + \tfrac{3}{2}\frac{c_0}{h_0}\varphi' \partial_\varsigma \varphi' + \tfrac{1}{6} c_0 h_0^2 \partial_\varsigma^3 \varphi' = 0.$$

The identification $\eta = \epsilon\varphi'$ and using the physical variables x, t result in the Korteweg-de Vries equation

$$\partial_t \eta + c_0 \partial_x \eta + \tfrac{3}{2}\frac{c_0}{h_0}\eta\, \partial_x \eta + \tfrac{1}{6} c_0 h_0^2 \partial_x^3 \eta = 0.$$

5.2 Thermoelastic Fluids

A *thermoelastic fluid* is a body governed by the constitutive equations

$$\varepsilon = \hat{\varepsilon}(\rho, \theta), \qquad \mathbf{T} = \hat{\mathbf{T}}(\rho, \theta).$$

Since ρ and θ are invariant then objectivity implies that

$$\hat{\mathbf{T}}(\rho, \theta) = \mathbf{Q}\hat{\mathbf{T}}(\rho, \theta)\mathbf{Q}^T$$

for all rotations $\mathbf{Q}$ and any value of ρ, θ. Consequently $\mathbf{T}$ must be a scalar times the unit tensor $\mathbf{1}$. Since elasticity occurs only in compression, we take the scalar as $-p$, p being the *thermodynamic pressure* and write

$$\mathbf{T} = -p\mathbf{1}, \qquad p = \hat{p}(\rho, \theta),$$

where the last expression represents the *equation of state* of the fluid. The dependence of $\hat{p}$ on ρ is usually assumed to be invertible, namely

$$\rho = \hat{\rho}(p, \theta),$$

because of the experimental evidence that the isothermal compressibility

$$\kappa = \frac{1}{\rho}\,\partial_p \hat{\rho}$$

is positive for the most fluids.[2] The local balance equations are

$$\dot{\rho} = -\rho\nabla\cdot\mathbf{v}, \quad \rho\dot{\mathbf{v}} = -\nabla p + \rho\mathbf{b}, \quad \rho\dot{\varepsilon} = -p\nabla\cdot\mathbf{v} - \nabla\cdot\mathbf{q} + \rho r.$$

Thermodynamic restrictions

By definition, ψ, η, p are functions of ρ, θ, e.g. $\psi = \hat{\psi}(\rho, \theta)$; the heat flux $\mathbf{q}$ and the extra-entropy flux $\mathbf{k}$ are allowed to depend also on $\nabla\theta$. To determine the thermodynamic restrictions we substitute $\dot{\psi} = \partial_\rho\hat{\psi}\,\dot{\rho} + \partial_\theta\hat{\psi}\,\dot{\theta}$ in the Clausius–Duhem inequality to obtain

$$-\rho(\partial_\theta\hat{\psi} + \hat{\eta})\dot{\theta} + (\rho^2\partial_\rho\hat{\psi} - \hat{p})\nabla\cdot\mathbf{v} - \frac{1}{\theta}\hat{\mathbf{q}}\cdot\nabla\theta + \theta\nabla\cdot\mathbf{k} \geq 0.$$

As always we regard the functions $\mathbf{b}(\mathbf{x}, t)$ and $r(\mathbf{x}, t)$ as arbitrarily assignable. This allows us to say that, at a point $\mathbf{x}$ and time t, the functions $\dot{\theta}$ and $\nabla\cdot\mathbf{v}$ can take arbitrary scalar values.[3] Hence the inequality holds only if

$$\hat{\eta}(\rho, \theta) = -\partial_\theta\hat{\psi}(\rho, \theta),$$

$$\hat{p}(\rho, \theta) = \rho^2\partial_\rho\hat{\psi}(\rho, \theta). \tag{5.8}$$

The former result is named *entropy relation* and the latter *pressure relation*. In particular,

$$\hat{p}(\rho, \theta) = \rho^2[\partial_\rho\hat{\varepsilon}(\rho, \theta) + \theta\partial_\rho\partial_\theta\hat{\psi}(\rho, \theta)].$$

The Clausius–Duhem inequality then simplifies to

$$-\frac{1}{\theta}\hat{\mathbf{q}}\cdot\nabla\theta + \theta\nabla\cdot\hat{\mathbf{k}} \geq 0,$$

where $\hat{\mathbf{q}}$ and $\hat{\mathbf{k}}$ are allowed to depend also on $\nabla\theta$. Since $\mathbf{k}$ is a possible function of $\rho, \theta, \nabla\theta$ then

$$\nabla\cdot\hat{\mathbf{k}} = \partial_\rho\hat{\mathbf{k}}\cdot\nabla\rho + \partial_\theta\hat{\mathbf{k}}\cdot\nabla\theta + \partial_{\nabla\theta}\hat{\mathbf{k}}\cdot\nabla\nabla\theta.$$

The arbitrariness of $\nabla\rho$, $\nabla\nabla\theta$ implies that $\partial_\rho\hat{\mathbf{k}} = \mathbf{0}$, $\partial_{\nabla\theta}\hat{\mathbf{k}} \in \mathrm{Skw}$ so that $\mathbf{k} = \mathbf{k}_0(\theta) + \mathbf{\Omega}(\theta)\nabla\theta$, $\mathbf{\Omega} \in \mathrm{Skw}$. Since no physically significant vector $\mathbf{k}_0$ and skew tensor $\mathbf{\Omega}$ occur then we let $\mathbf{k} = \mathbf{0}$. Hence we are left with the heat conduction inequality

$$\hat{\mathbf{q}}\cdot\nabla\theta \leq 0,$$

[2] Materials with the negative bulk modulus (inverse compressibility) are quite rare: see, for instance, [264].

[3] See Sect. 3.4.

a requirement for the constitutive function $\hat{\mathbf{q}}(\rho, \theta, \nabla\theta)$.

5.2.1 Gibbs Relations and Entropy Equation

As a consequence of the pressure relation (5.8) we can establish the Gibbs relations

$$\dot{\psi} = \frac{p}{\rho^2}\dot{\rho} - \eta\dot{\theta}, \qquad \dot{\varepsilon} = \frac{p}{\rho^2}\dot{\rho} + \theta\dot{\eta} \tag{5.9}$$

and the Maxwell relation

$$\partial_\theta p = -\rho^2 \partial_\rho \eta.$$

Moreover, by

$$\partial_\rho \varepsilon = \partial_\rho \psi + \theta \partial_\rho \eta = \frac{p}{\rho^2} - \frac{\theta}{\rho^2}\partial_\theta p$$

we obtain the identity

$$p = \theta \partial_\theta p + \rho^2 \partial_\rho \varepsilon$$

for the pressure function $p(\rho, \theta)$. The partial derivative

$$c_v = \partial_\theta \varepsilon(\rho, \theta)$$

is called the specific heat at constant volume (constant density).

By the Gibbs relation $(5.9)_2$ and the balance of energy we have

$$\theta\dot{\eta} + \frac{p}{\rho^2}\dot{\rho} = \frac{p}{\rho^2}\dot{\rho} + r - \frac{1}{\rho}\nabla \cdot \mathbf{q},$$

whence we have the *entropy equation*

$$\theta\dot{\eta} = r - \frac{1}{\rho}\nabla \cdot \mathbf{q}. \tag{5.10}$$

It is of interest to contrast the entropy equation (5.10) with the entropy inequality

$$\rho\dot{\eta} \geq \frac{\rho r}{\theta} - \nabla \cdot \frac{\mathbf{q}}{\theta} = \frac{\rho r}{\theta} - \frac{1}{\theta}\nabla \cdot \mathbf{q} + \frac{1}{\theta^2}\mathbf{q} \cdot \nabla\theta; \tag{5.11}$$

by (5.11) the time rate of η is greater than the rate given by (5.10) minus $|\mathbf{q} \cdot \nabla\theta|/\rho\theta^2$.

By the Gibbs relation $(5.9)_2$ and the constitutive relation for p we have

$$\theta\dot{\eta} = \partial_\rho \varepsilon\, \dot{\rho} + \partial_\theta \varepsilon\, \dot{\theta} - \frac{p}{\rho^2}\dot{\rho} = c_v\, \dot{\theta} - \frac{\theta}{\rho^2}\partial_\theta p\, \dot{\rho}$$

and hence, in light of the entropy equation (5.10),

$$\rho c_v \dot{\theta} = \frac{\theta}{\rho} \partial_\theta p \, \dot{\rho} + \rho r - \nabla \cdot \mathbf{q}.$$

For uniform fields we can multiply by the volume V to obtain a relation for the whole body. If M is the mass then

$$\rho V = M, \qquad \frac{\dot{\rho}}{\rho^2} = -\frac{\dot{V}}{M}.$$

Hence letting $S = M\,\eta$ and $C_V = M\,c_v$ we have[4]

$$\theta \dot{S} = C_V \dot{\theta} + \theta\, \partial_\theta p\, \dot{V}.$$

5.2.2 *Conjugate Variables and Maxwell's Relations*

Let $f(x, y)$ be a C^2 function on $A \subseteq \mathbb{R}^2$. Assume that f is convex in x for all y and define

$$z = \partial_x f,$$

so that z is the variable conjugate[5] to x. We can view z as a function of x, parameterized by y. The convexity of f implies that

$$\partial_x z = \partial_x^2 f > 0.$$

Hence we can define the inverse function $x = \hat{x}(z, y)$. Let

$$g(z, y) := f(\hat{x}(z, y), y) - z\,\hat{x}(z, y).$$

We find that

$$\partial_z g = \partial_x f\, \partial_z \hat{x} - x - z \partial_z \hat{x} = -x.$$

Moreover

$$\partial_y g = \partial_x f\, \partial_y \hat{x} + \partial_y f - z\, \partial_y \hat{x} = \partial_y f.$$

In summary, subject to $\partial_x^2 f > 0$, or $\partial_x^2 f < 0$,

---

[4] Apart from the occurrence of derivatives instead of differentials, this is the classical first TdS equation of thermodynamics (see, e.g. [449], Sect. (13.3).

[5] For historical reasons due to the use in analytical mechanics, the conjugate variable is usually denoted by p. Here the conjugate variable is not denoted by p to avoid possible ambiguities with the pressure.

$$z = \partial_x f, \quad g(z, y) = f(\hat{x}(z, y), y) - z\,\hat{x}(z, y) \quad \Longrightarrow \quad x = -\partial_z g, \quad \partial_y g = \partial_y f. \tag{5.12}$$

The conjugate property $z = \partial_x f$ is related to the Legendre transform $f^*(z)$ of $f(x)$ defined by

$$f^*(z) = \sup_{x \in \mathbb{R}} (z\,x - f(x)).$$

The supremum occurs at $x^*(z)$ such that the tangent to $y = f(x)$ is parallel to the straight line $y = z\,x$.

We now generalize the conjugate variable to the case where $x \in \mathbb{R}^n$, $y \in \mathbb{R}^m$ and $f(x, y)$ is a C^2 function on $A \subseteq \mathbb{R}^n \times \mathbb{R}^m$. Moreover let

$$H = \left[\partial_{x_j} \partial_{x_k} f\right], \qquad \det H \neq 0.$$

Let

$$z_k = \partial_{x_k} f.$$

The entries of the Jacobian,

$$\partial_{x_j} z_k = \partial_{x_j} \partial_{x_k} f,$$

coincide with those of H. Hence $\det H \neq 0$ allows us to write the inverse function, $x = \hat{x}(z, y)$, and to define

$$g(z, y) := f(\hat{x}(z, y), y) - z_j \hat{x}_j(z, y).$$

Hence we have

$$\partial_{z_k} g = \partial_{x_j} f\, \partial_{z_k} \hat{x}_j - x_k - z_j \partial_{z_k} \hat{x}_j = -x_k,$$

$$\partial_{y_i} g = \partial_{x_j} f\, \partial_{y_i} \hat{x}_j + \partial_{y_i} f - z_j \partial_{y_i} \hat{x}_j = \partial_{y_i} f.$$

We now go back to the entropy and pressure relations,

$$\eta(\rho, \theta) = -\partial_\theta \psi(\rho, \theta), \qquad p(\rho, \theta) = \rho^2 \partial_\rho \psi(\rho, \theta).$$

For formal convenience, and for ease of comparison with the physical literature, we use the *specific volume* $\upsilon = 1/\rho$ in place of the mass density ρ. Since

$$\partial_\upsilon = -\frac{1}{\upsilon^2} \partial_\rho = -\rho^2 \partial_\rho$$

we can use the Helmholtz free energy in the form $\psi(\upsilon, \theta)$ and write the entropy and pressure relations as

$$\eta(\upsilon, \theta) = -\partial_\theta \psi(\upsilon, \theta), \qquad p(\upsilon, \theta) = -\partial_\upsilon \psi(\upsilon, \theta).$$

To use η and υ as the independent variables we assume that $\partial_\theta \eta(\upsilon, \theta) = -\partial_\theta^2 \psi(\upsilon, \theta) \neq 0$. Hence we let $\theta = \theta(\upsilon, \eta)$, $z = \partial_\theta \psi$, and consider the conjugate function

$$\psi(\upsilon, \theta(\upsilon, \eta)) - \partial_\theta \psi \, \theta(\upsilon, \eta) = \psi(\upsilon, \theta(\upsilon, \eta)) + \eta \, \theta(\upsilon, \theta(\upsilon, \eta)) =: \varepsilon(\upsilon, \eta).$$

In view of (5.12), the identifications $x = \theta$, $z = -\eta$, $f = \psi$, $g = \varepsilon$ we have

$$\theta = \partial_\eta \varepsilon(\upsilon, \eta), \qquad p = -\partial_\upsilon \varepsilon(\upsilon, \eta).$$

To use p and η as the independent variables we assume that $\partial_\upsilon p(\upsilon, \eta) = -\partial_\upsilon^2 \psi(\upsilon, \eta) \neq 0$. Hence we let $\upsilon = \upsilon(p, \eta)$, $z = \partial_\upsilon \varepsilon$, and consider the conjugate function

$$\varepsilon(\upsilon(p, \eta), \eta) - \partial_\upsilon \varepsilon \, \upsilon(p, \eta) = \varepsilon(\upsilon(p, \eta), \eta) + p \, \upsilon(p, \eta) =: h(p, \eta).$$

The identifications $x = \upsilon$, $z = -p$, $f = \varepsilon$, $g = h$ provide

$$\upsilon = \partial_p h, \qquad \theta = \partial_\eta h.$$

Finally, to use p and θ as the independent variables we assume that $\partial_\eta \theta = \partial_\eta^2 h(p, \eta) \neq 0$. Hence we let $\eta = \eta(p, \theta)$, $z = \partial_\eta h$ and find that the conjugate function is

$$h(p, \eta(p, \theta)) - \partial_\eta h \, \eta(p, \theta) = h(p, \eta(p, \theta)) - \theta \, \eta(p, \theta) =: \mu(p, \theta).$$

Hence we obtain

$$\eta = -\partial_\theta \mu(p, \theta), \qquad \upsilon = \partial_p \mu(p, \theta).$$

The four functions so determined are the specific (per unit mass) Helmholtz free energy ψ, internal energy $\varepsilon = \psi + \theta\eta$, enthalpy $h = \varepsilon + p\upsilon$, and Gibbs free energy $\mu = \varepsilon + p\upsilon - \theta\eta = \psi + p\upsilon$. The function μ is also called the *chemical potential.* Associated with any of these four functions is a pair of natural independent variables so that the following set of relations holds.

$$\begin{cases} \psi(\upsilon, \theta) : \dot{\psi} = -\eta \, \dot{\theta} - p \, \dot{\upsilon}, & \partial_\theta p(\upsilon, \theta) = \partial_\upsilon \eta(\upsilon, \theta), \\ \varepsilon(\upsilon, \eta) : \dot{\varepsilon} = \theta \, \dot{\eta} - p \, \dot{\upsilon}, & \partial_\upsilon \theta(\upsilon, \eta) = -\partial_\eta p(\upsilon, \eta), \\ h(p, \eta) : \dot{h} = \upsilon \, \dot{p} + \theta \, \dot{\eta}, & \partial_\eta \upsilon(p, \eta) = \partial_p \theta(p, \eta), \\ \mu(p, \theta) : \dot{\mu} = \upsilon \, \dot{p} - \eta \, \dot{\theta}, & \partial_p \eta(p, \theta) = -\partial_\theta \upsilon(p, \theta). \end{cases} \tag{5.13}$$

The four equations on the right are referred to as Maxwell's relations. They follow from the assumption that $\psi, \varepsilon, h, \mu$ are C^2 functions of the pertinent pair of variables. Hence, for instance, from

$$\eta = -\partial_\theta \psi(\upsilon, \theta), \qquad p = -\partial_\upsilon \psi(\upsilon, \theta),$$

it follows that

$$\partial_v \eta(v, \theta) = -\partial_v \partial_\theta \psi(v, \theta) = -\partial_\theta \partial_v \psi(v, \theta) = \partial_\theta p(v, \theta).$$

In classical thermodynamics the systems are often regarded as homogeneous namely the fields v, θ, p, η are assumed to be independent of the position. In such a case multiplication of (5.13) by the mass M of the body gives

$$\dot{A} = -S\,\dot{\theta} - p\,\dot{V}, \qquad \dot{H} = V\,\dot{p} + \theta\,\dot{S}, \tag{5.14}$$

$$\dot{U} = \theta\,\dot{S} - p\,\dot{V}, \qquad \dot{G} = V\,\dot{p} - S\,\dot{\theta}, \tag{5.15}$$

where $A = M\,\psi,\ U = M\,\varepsilon,\ H = M\,h,\ G = M\,\mu,\ S = M\,\eta,\ V = M\,v$.

In the next section, further interesting consequences are shown to follow from the set of equations (5.13).

5.2.3 Specific Heats

By the balance of energy we have

$$\dot{\varepsilon} + p\,\dot{v} = r - v\nabla \cdot \mathbf{q}.$$

The right-hand side is the overall rate at which energy is supplied, per unit mass, by external sources, r, and by heat conduction, $-v\nabla \cdot \mathbf{q}$. By the second line in (5.13) we have

$$\dot{\varepsilon} + p\,\dot{v} = \theta\dot{\eta}.$$

Hence we define the specific heats, c_v and c_p, at constant volume and at constant pressure, as

$$c_v = \theta\,\partial_\theta \eta(v, \theta), \qquad c_p = \theta\,\partial_\theta \eta(p, \theta).$$

As a consequence we find that

$$c_v = \partial_\theta \varepsilon(v, \theta),$$

and

$$c_p = \partial_\theta \varepsilon(p, \theta) + p\,\partial_\theta v(p, \theta).$$

By the third line in (5.13) it follows that

$$c_p = \partial_\theta\,h(p, \theta).$$

This motivates why sometimes the specific heats are defined in the form[6]

[6] See, e.g. [316], Sect. 3.3.5.

$$c_v = \partial_\theta \varepsilon(v, \theta), \qquad c_p = \partial_\theta h(p, \theta).$$

If $\eta = \eta(v, \theta)$ then

$$\dot{\eta} = \partial_v \eta \, \dot{v} + \partial_\theta \eta \, \dot{\theta}.$$

From the first line in (5.13) we have $\partial_v \eta = \partial_\theta p$ and hence, upon multiplication by θ, we can write

$$\theta \, \dot{\eta} = c_v \, \dot{\theta} + \theta \, \partial_\theta p \, \dot{v},$$

which, if expressed by the differentials $d\eta, d\theta, dv$ is just the local form of the so-called first $\theta \, dS$ equation of classical thermodynamics ([449], Sect. 13.3).

An analogous relation holds if η is taken to be given by

$$\eta = \eta(\theta, p).$$

It follows at once that

$$\theta \dot{\eta} = \theta \, \partial_\theta \eta \, \dot{\theta} + \theta \, \partial_p \eta \, \dot{p}.$$

By the fourth line in (5.13) we can replace $\partial_p \eta$ with $-\partial_\theta v$. Hence

$$\theta \dot{\eta} = \theta \, \partial_\theta \eta \, \dot{\theta} - \theta \, \partial_\theta v \, \dot{p}$$

whence

$$\theta \dot{\eta} = c_p \, \dot{\theta} - \theta \, \partial_\theta v \, \dot{p},$$

that is the second $\theta \, dS$ equation of classical thermodynamics ([449], Sect. 13.4).

Since $c_p = \theta \partial_\theta \eta(\theta, p)$ it follows that the entropy can be determined in terms of $c_p(\theta)$ namely

$$\eta(\theta, p) = \int_{\theta_*}^{\theta} \frac{1}{\vartheta} c_p(\vartheta) d\vartheta,$$

(θ_*, p) being a reference state.

5.3 Ideal Gas

An ideal gas is a thermoelastic, non-conducting fluid characterized by the pressure-density relation

$$p = \frac{R\theta}{M} \rho,$$

where M is the molecular weight and R is the gas constant ($R = 8.314 \, 10^7$ erg/degrees); ideal gases may differ only by the molecular weight. Since $p = \rho^2 \partial_\rho \psi(\theta, \rho)$ we have

$$\partial_\rho \psi = \frac{R\theta}{M}\frac{1}{\rho}, \qquad \psi(\theta, \rho) = \frac{R\theta}{M}\ln\frac{\rho}{\rho_0} + \tilde{\psi}(\theta), \qquad \eta(\theta, v) = -\frac{R}{M}\ln\frac{\rho}{\rho_0} - \partial_\theta\tilde{\psi}(\theta),$$

where ρ_0 is a reference value of ρ.

To determine the function $\tilde{\psi}$ we need a further condition. It is a well-known assumption in physics that the internal energy ε of an ideal gas is a function of the temperature only and it is a standard approximation that $c_v = \partial_\theta\tilde{\varepsilon}(\theta)$ is constant. Indeed, assume $\varepsilon(\theta) = c_v\theta + \varepsilon_0$. Consequently

$$\psi - \theta\partial_\theta\psi = c_v\theta + \varepsilon_0 \tag{5.16}$$

yields

$$\tilde{\psi} - \theta\tilde{\psi}' = c_v + \varepsilon_0.$$

By means of a integrating factor μ we have

$$\mu\tilde{\psi}' - \frac{\mu}{\theta}\tilde{\psi} = -c_v\mu - \varepsilon_0\frac{\mu}{\theta}$$

It follows that $\mu\theta$ is constant, say $\mu\theta = k$. Hence we have

$$\frac{k\tilde{\psi}}{\theta} = -c_v k\ln\frac{\theta}{\theta_0} + \frac{\varepsilon_0 k}{\theta} + k\nu,$$

ν being a constant. Hence

$$\psi(\theta, v) = K\theta\ln\frac{\rho}{\rho_0} - c_v\theta\ln\frac{\theta}{\theta_0} + \nu\theta + \varepsilon_0, \qquad K = \frac{R}{M}, \tag{5.17}$$

while the requirement (5.16) is satisfied identically for any value of ν.

By (5.16) we obtain the entropy η in the form

$$\eta = K\ln\frac{\rho}{\rho_0} - c_v\ln\frac{\theta}{\theta_0} + \nu - c_v.$$

For practical convenience we may take ν such that $\eta = 0$ at the reference state (θ_0, ρ_0). This is obtained by letting $\nu = c_v$. Consequently we have

$$\psi(\theta, \rho) = K\theta\ln\frac{\rho}{\rho_0} + c_v\theta(1 - \ln\frac{\theta}{\theta_0}) + \varepsilon_0, \qquad \eta(\theta, \rho) = K\ln\frac{\rho}{\rho_0} - c_v\ln\frac{\theta}{\theta_0}.$$

Since $p = K\theta\rho$ and $p_0 = K\theta_0\rho_0$ then

$$\ln\frac{p}{p_0} = \ln\frac{\theta}{\theta_0} + \ln\frac{\rho}{\rho_0}.$$

In terms of θ and p we then have

$$\hat{\eta}(\theta, p) = (c_v + K)\ln\frac{\theta}{\theta_0} - K\ln\frac{p}{p_0}. \tag{5.18}$$

5.3.1 Specific Heats and Entropy Functions

The quantities

$$c_v = \theta\partial_\theta\eta(\theta, v), \qquad c_p = \theta\partial_\theta\hat{\eta}(\theta, p)$$

are the specific heats (per unit mass) at constant volume or at constant pressure. By means of the function $v(\theta, p)$ then c_p and c_v are related by

$$c_p = \partial_\theta\hat{\eta}(\theta, p) = \partial_\theta\eta(\theta, v(\theta, p)) + \partial_v\eta\,\partial_\theta v(\theta, p) = c_v + \theta\partial_v\eta\,\partial_\theta v(\theta, p);$$

since $\eta(\theta, v) = -\partial_\theta\psi(\theta, v)$ then

$$c_v = \partial_\theta\varepsilon(\theta, v).$$

Observe that

$$\partial_v\eta(\theta, v) = -\partial_v\partial_\theta\psi(\theta, v) = -\partial_\theta\partial_v\psi(\theta, v) = \partial_\theta p(\theta, v).$$

Hence

$$c_p - c_v = \theta\partial_\theta p\,\partial_\theta v.$$

In ideal gases

$$\partial_\theta p = \frac{R}{Mv}, \qquad \partial_\theta v = \frac{R}{Mp}, \qquad \partial_\theta p\partial_\theta v = \frac{R}{M\theta},$$

and then

$$c_p - c_v = \frac{R}{M}.$$

Alternatively we consider Eq. (5.18) and find

$$c_p = \theta\partial_\theta\hat{\eta}(\theta, p) = c_v + K.$$

5.4 Models of Real Gases

There are a number of models providing a better description of real gases. Despite the relative simplicity, the Van der Waals model holds fairly well with the liquid and vapor phases.

5.4.1 *Van der Waals Model*

Let V be the molar volume. The Van der Waals equation is

$$(p + \frac{a}{V^2})(V - b) = R\theta; \tag{5.19}$$

if $a, b = 0$ then Eq. (5.19) reduces to the equation of an ideal gas. By (5.19) we can derive the function $p(V, \theta)$ as

$$p = \frac{R\theta}{V - b} - \frac{a}{V^2}.$$

Consider the isotherms, namely the curves at constant θ. For the ideal gas isotherms are hyperbola in the $V - p$ plane. Instead Eq. (5.19) shows a monotonic behaviour only at high temperatures. Look for horizontal inflection points of (5.19). We have

$$0 = \partial_V p = -\frac{R\theta}{(V - b)^2} + \frac{2a}{V^3}, \qquad 0 = \partial_V^2 p = \frac{2R\theta}{(V - b)^3} - \frac{6a}{V^4}.$$

Solving these equations, along with (5.19), in the unknowns p, V, θ we find the unique solution

$$V = 3b, \qquad \theta = \frac{8a}{27\, b\, R}, \qquad p = \frac{a}{27\, b^2}.$$

Denote these values by V_c, θ_c, p_c. Observe that

$$p_c V_c = \frac{a}{27\, b^2} 3b = \frac{3}{8} R\theta_c.$$

Replacing $a = 3 p_c V_c^2$ and $b = V_c/3$ we can write (5.19) in the form

$$(\mathcal{P} + \frac{3}{\mathcal{V}^2})(\mathcal{V} - \frac{1}{3}) = \frac{8}{3}\vartheta,$$

where $\mathcal{P} = p/p_c$, $\mathcal{V} = V/V_c$, $\vartheta = \theta/\theta_c$. It is remarkable that this equation contains only numerical constants and then it is the same for all substances (modelled by the Van der Waals equation).

Incidentally, the ideal gas equation does not allow any inflection point; the ratio $R\theta/pV$ is constantly equal to 1.

Back to (5.19) we now look for thermodynamic potentials. Since $V = Mv$ we let $\alpha = a/M^2$ and $\beta = b/M$ so that (5.19) can be written in the form

$$p = \frac{K\theta}{v-\beta} - \frac{\alpha}{v^2}.$$

Since $p = -\partial_v \psi(\theta, v)$ then we find the Helmholtz free energy ψ in the form

$$\psi = -K\theta \ln(v-\beta) - \frac{\alpha}{v} + f(\theta).$$

Consequently the entropy $\eta = -\partial_\theta \psi(\theta, v)$ and the internal energy $\varepsilon = \psi + \theta\eta$ read

$$\eta = K \ln(v-\beta) - f'(\theta), \qquad \varepsilon = -\frac{\alpha}{v} + f(\theta) - \theta f'(\theta). \tag{5.20}$$

To determine f we consider the specific heat, at constant volume, $c_v = \partial_\theta \varepsilon(\theta, v)$. By (5.20) it follows that c_v may depend only on temperature. Hence we have

$$f(\theta) - \theta f'(\theta) = C_v(\theta) + w,$$

where w is a constant and C_v is the integral of c_v; if c_v is constant then $C_v = c_v\theta$. For formal convenience we let

$$f' - \frac{1}{\theta} f = -\frac{1}{\theta} C_v(\theta) + \frac{w}{\theta}.$$

The integration is performed in the classical way: we multiply by $\mu(\theta)$ and let $-\mu/\theta = \mu'$ to obtain $\mu\theta =$ constant, say c_0, and then

$$(\frac{c_0}{\theta} f)' = -\frac{c_0}{\theta^2} C_v(\theta) + \frac{w c_0}{\theta^2},$$

whence

$$f(\theta) = -F(\theta) + \tilde{c}\theta - w, \qquad F(\theta) = \theta \int_{\theta_0}^{\theta} \frac{1}{\xi^2} C_v(\xi) d\xi.$$

It follows that

$$\eta(\theta, v) = K \ln(v-\beta) + F'(\theta) + \hat{c}.$$

If c_v is constant then $F(\theta) = \theta c_v \ln\theta$, $F' = c_v \ln\theta + c_v$ and then

$$\eta(\theta, v) = K \ln(v-\beta) + c_v \ln\theta + \bar{c}.$$

In adiabatic, or isentropic, transformations η is constant and then we have

$$K \ln(v - \beta) + c_v \ln \theta = \text{constant},$$

whence $(v-\beta)^K \theta^{c_v} = \text{constant}$ and then

$$[\theta^{c_v}(v-\beta)^K]^{1/c_v} = \theta(v-\beta)^{K/c_v} = \text{constant}.$$

For an ideal gas we would have $\theta v^{K/c_v} = \text{constant}$.

5.4.2 *Peng-Robinson Model*

The Peng-Robinson model proves useful in modelling some liquids in addition to real gases. It is characterized by the constitutive equation

$$p = \frac{K\theta}{v-\beta} - \frac{\alpha(\theta)}{v(v+\gamma)+\gamma(v-\gamma)};$$

in the standard Peng-Robinson model $\gamma = M\beta$. In light of the identity

$$\frac{1}{v(v+\gamma)+\gamma(v-\gamma)} = \frac{1}{2\sqrt{2}\gamma}[\frac{1}{v-(\sqrt{2}-1)\gamma} - \frac{1}{v+(\sqrt{2}+1)\gamma}]$$

we rewrite the constitutive equation as

$$p = \frac{K\theta}{v-\beta} - \frac{\alpha(\theta)}{2\sqrt{2}\gamma}[\frac{1}{v-(\sqrt{2}-1)\gamma} - \frac{1}{v+(\sqrt{2}+1)\gamma}].$$

Since $p = -\partial_v\psi$ we then obtain ψ in the form

$$\psi = K\theta \ln(v-\beta) - \frac{\alpha(\theta)}{2\sqrt{2}\gamma} \ln \frac{v-(\sqrt{2}-1)\gamma}{v+(\sqrt{2}+1)\gamma} + f(\theta).$$

Consequently

$$\eta = -\partial_\theta\psi = -K \ln(v-\beta) + \frac{\alpha'(\theta)}{2\sqrt{2}\gamma} \ln \frac{v-(\sqrt{2}-1)\gamma}{v+(\sqrt{2}+1)\gamma} - f'(\theta),$$

$$\varepsilon = \frac{\theta\alpha' - \alpha}{2\sqrt{2}\gamma} \ln \frac{v-(\sqrt{2}-1)\gamma}{v+(\sqrt{2}+1)\gamma} + f(\theta) - \theta f'(\theta).$$

The specific heat $c_v = \partial_v\varepsilon(\theta, v)$ is a further property to be examined to find the unknown parameters β, γ and the functions α, f. We find that

$$c_v = \frac{\theta \alpha''(\theta)}{2\sqrt{2}\gamma} \ln \frac{v - (\sqrt{2}-1)\gamma}{v + (\sqrt{2}+1)\gamma} - \theta f''(\theta).$$

Experimental data on c_v may give estimates of the parameter γ and the functions α, f while $\beta = \gamma/M$.

Further models of real gases

There are a number of models (or equations of state) [449] that, also because of a greater number of parameters, can realize a better fit of the relation among p, v, θ. Usually the molar volume V is used instead of the specific volume v. The Dieterici model is expressed by

$$p = \frac{R\theta}{V-b} \exp(-\frac{a}{VR\theta});$$

at relatively high temperatures the approximation $\exp(-a/VR\theta) \simeq 1 - a/VR\theta$ simplifies the formal usage.

The Clausius model involves three parameters a, b, c in the form

$$(p + \frac{a}{\theta(V+c)^2})(V-b) = R\theta.$$

The Berthelot model is similar to the Van der Waals model with the change $a/V^2 \to a/\theta V^2$, namely

$$p = \frac{R\theta}{V-b} - \frac{a}{\theta V^2}.$$

The Virial model is a source of many equations; it is based on a formal development in inverse powers of V applied to the model of the ideal gas,

$$pV = R\theta\big[1 + \frac{b(\theta)}{V} + \frac{c(\theta)}{V^2} + \frac{d(\theta)}{V^3} + ...\big].$$

5.5 Heat-Conducting, Viscous Fluids

Fluids generally exhibit internal friction that opposes to the relative motion of fluid particles. Also they can support heat conduction. The relative motion is commonly described by the velocity gradient $\mathbf{L}$ and heat conduction is described by the temperature gradient $\nabla\theta$. We then describe the constitutive properties of a heat-conducting, viscous fluid by letting $\psi, \eta, \mathbf{T}, \mathbf{q}$ be functions of $\rho, \theta, \mathbf{L}, \nabla\theta$, that is

$$\mathbf{T} = \hat{\mathbf{T}}(\rho, \theta, \mathbf{L}, \nabla\theta)$$

and the like for ψ, η, $\mathbf{q}$. We first examine the consequences of objectivity. The invariance of ρ, θ, ψ, η and the objectivity of $\mathbf{T}$, $\mathbf{q}$, and $\nabla\theta$ imply that

$$\psi = \hat{\psi}(\rho, \theta, \mathbf{Q}\,\mathbf{L}\,\mathbf{Q}^T + \dot{\mathbf{Q}}\,\mathbf{Q}^T, \mathbf{Q}\,\nabla\theta),$$

$$\mathbf{Q}\,\mathbf{T}\,\mathbf{Q}^T = \hat{\mathbf{T}}(\rho, \theta, \mathbf{Q}\,\mathbf{L}\,\mathbf{Q}^T + \dot{\mathbf{Q}}\,\mathbf{Q}^T, \mathbf{Q}\,\nabla\theta),$$

$$\mathbf{Q}\,\mathbf{q} = \hat{\mathbf{q}}(\rho, \theta, \mathbf{Q}\,\mathbf{L}\,\mathbf{Q}^T + \dot{\mathbf{Q}}\,\mathbf{Q}^T, \mathbf{Q}\,\nabla\theta),$$

for any rotation $\mathbf{Q}$. For any arbitrary skew tensor $\boldsymbol{\Omega}$ we can take $\mathbf{Q}$ such that, at time t,

$$\mathbf{Q}(t) = \mathbf{1}, \qquad \dot{\mathbf{Q}}(t) = \boldsymbol{\Omega};$$

this is the case if we consider the solution to $\dot{\mathbf{Q}}(t+\tau) = \boldsymbol{\Omega}\,\mathbf{Q}(t+\tau), \quad \mathbf{Q}(t) = \mathbf{1}$. Hence the objectivity requires that

$$\psi = \hat{\psi}(\rho, \theta, \mathbf{L} + \boldsymbol{\Omega}, \nabla\theta), \quad \mathbf{T} = \hat{\mathbf{T}}(\rho, \theta, \mathbf{L} + \boldsymbol{\Omega}, \nabla\theta), \quad \mathbf{q} = \hat{\mathbf{q}}(\rho, \theta, \mathbf{L} + \boldsymbol{\Omega}, \nabla\theta),$$

for all ρ, θ, $\mathbf{L}$, $\nabla\theta$, and $\boldsymbol{\Omega} \in \mathrm{Skw}$. The arbitrariness of $\boldsymbol{\Omega}$ allows us to select $\boldsymbol{\Omega} = -\mathbf{W}$ to obtain

$$\psi = \hat{\psi}(\rho, \theta, \mathbf{D}, \nabla\theta), \quad \mathbf{T} = \hat{\mathbf{T}}(\rho, \theta, \mathbf{D}, \nabla\theta), \quad \mathbf{q} = \hat{\mathbf{q}}(\rho, \theta, \mathbf{D}, \nabla\theta).$$

Hence the constitutive relations for a heat-conducting, viscous fluid cannot include the spin as an independent variable.

As to the dependence on the stretching $\mathbf{D}$ and the temperature gradient $\nabla\theta$, the functions $\hat{\psi}$, $\hat{\mathbf{T}}$, $\hat{\mathbf{q}}$ are required to satisfy

$$\begin{aligned} \psi &= \hat{\psi}(\rho, \theta, \mathbf{Q}\,\mathbf{D}\,\mathbf{Q}^T, \mathbf{Q}\,\nabla\theta), \\ \mathbf{Q}\,\mathbf{T}\,\mathbf{Q}^T &= \hat{\mathbf{T}}(\rho, \theta, \mathbf{Q}\,\mathbf{D}\,\mathbf{Q}^T, \mathbf{Q}\,\nabla\theta), \\ \mathbf{Q}\,\mathbf{q} &= \hat{\mathbf{q}}(\rho, \theta, \mathbf{Q}\,\mathbf{D}\,\mathbf{Q}^T, \mathbf{Q}\,\nabla\theta), \end{aligned} \tag{5.21}$$

for any rotation $\mathbf{Q}$.

Pressure

To understand the properties of $\hat{\mathbf{T}}(\rho, \theta, \mathbf{D}, \nabla\theta)$ we observe that $\hat{\mathbf{T}}(\rho, \theta, \mathbf{0}, \mathbf{0})$ represents the stress in the fluid at the (locally) uniform temperature and in the absence of flow. The objectivity condition

$$\mathbf{Q}\,\hat{\mathbf{T}}(\rho, \theta, \mathbf{0}, \mathbf{0})\mathbf{Q}^T = \hat{\mathbf{T}}(\rho, \theta, \mathbf{0}, \mathbf{0})$$

for any rotation $\mathbf{Q}$ implies that $\hat{\mathbf{T}}(\rho, \theta, \mathbf{0}, \mathbf{0})$ is an isotropic tensor and, because elasticity occurs only in compression, we let

$$\hat{\mathbf{T}}(\rho, \theta, \mathbf{0}, \mathbf{0}) =: -\hat{p}(\rho, \theta)\mathbf{1}, \tag{5.22}$$

so recovering the constitutive equation of thermoelastic fluids. We may then view $\hat{p}(\rho, \theta)$ as the *thermodynamic pressure;* as we see in a moment via the pressure relation (5.23), $\hat{p}(\rho, \theta)$ is determined by the constitutive function of a thermodynamic potential. Since $\hat{p}(\rho, \theta)$ is independent of motion it is also named equilibrium pressure.

The pressure in a fluid is sometimes defined as

$$\bar{p} = -\tfrac{1}{3}\mathrm{tr}\,\mathbf{T}.$$

This definition would regard $\bar{p}$ as a function of $\mathbf{D}$ and $\nabla\theta$, in addition to ρ and θ. This pressure is then referred to as the *total pressure* or the *mean pressure* (in that it represents the mean normal stress). Although the expressions of $\hat{p}$ and $\bar{p}$ overlap for special classes of fluids, we hereafter proceed by denoting with p the thermodynamic pressure and let

$$\hat{\mathbf{T}}_{\mathrm{dis}}(\rho, \theta, \mathbf{D}, \nabla\theta) := \hat{\mathbf{T}}(\rho, \theta, \mathbf{D}, \nabla\theta) + \hat{p}(\rho, \theta)\mathbf{1}.$$

We can view $\hat{\mathbf{T}}_{\mathrm{dis}}(\rho, \theta, \mathbf{D}, \nabla\theta)$ as the *dissipative* part of the stress in that it is the stress to within the thermodynamic pressure tensor (which can be obtained from a potential and then is *conservative*). In particular,

$$\bar{p}(\rho, \theta, \mathbf{D}, \mathbf{g}) = \hat{p}(\rho, \theta) - \tfrac{1}{3}\mathrm{tr}\,\mathbf{T}_{\mathrm{dis}}(\rho, \theta, \mathbf{D}, \nabla\theta)$$

and, by definition,

$$\hat{\mathbf{T}}_{\mathrm{dis}}(\rho, \theta, \mathbf{D}, \nabla\theta) \to \mathbf{0}, \quad \bar{p}(\rho, \theta, \mathbf{D}, \nabla\theta) \to \hat{p}(\rho, \theta) \quad \text{when} \quad \mathbf{D}, \nabla\theta \to \mathbf{0}.$$

Thermodynamic restrictions

We now determine the thermodynamic restrictions on the constitutive functions $\hat{\psi}, \hat{\eta}, \hat{\mathbf{T}}_{\mathrm{dis}}, \hat{p}$, and $\hat{\mathbf{q}}$ of $\rho, \theta, \mathbf{D}, \nabla\theta$. We compute

$$\dot{\psi} = \partial_\rho\psi\,\dot{\rho} + \partial_\theta\psi\,\dot{\theta} + \partial_{\mathbf{D}}\psi \cdot \dot{\mathbf{D}} + \partial_{\nabla\theta}\psi \cdot \dot{\overline{\nabla\theta}},$$

where it is understood that ψ stands for $\hat{\psi}(\rho, \theta, \mathbf{D}, \nabla\theta)$. Upon substitution in the Clausius–Duhem inequality we obtain

$$-\rho(\partial_\theta\psi + \eta)\dot{\theta} - \rho\partial_{\mathbf{D}}\psi \cdot \dot{\mathbf{D}} - \rho\partial_{\nabla\theta}\psi \cdot \dot{\overline{\nabla\theta}} - \rho\partial_\rho\psi\,\dot{\rho} - p\nabla \cdot \mathbf{v} + \mathbf{T}_{\mathrm{dis}} \cdot \mathbf{D} - \frac{1}{\theta}\mathbf{q} \cdot \nabla\theta + \theta\nabla \cdot \mathbf{k} \geq 0.$$

We regard the functions $\mathbf{b}(\mathbf{x}, t)$ and $r(\mathbf{x}, t)$ as arbitrarily assignable. This allows us to say that, at a point $\mathbf{x}$ and time t, the functions $\dot{\theta}$, $\nabla \cdot \mathbf{v}$, $\dot{\mathbf{D}}$, and $\dot{\overline{\nabla\theta}}$ can take arbitrary scalar, symmetric tensor, and vector values.[7]

The arbitrariness of $\dot{\mathbf{D}}$, $\dot{\overline{\nabla\theta}}$, $\dot{\theta}$ implies that the inequality holds only if

$$\partial_{\mathbf{D}}\psi = 0, \qquad \partial_{\nabla\theta}\psi = 0, \qquad \eta(\rho, \theta) = -\partial_\theta\psi(\rho, \theta).$$

The inequality then simplifies to

$$(-p + \rho^2\partial_\rho\psi)\nabla \cdot \mathbf{v} + \mathbf{T}_{\text{dis}} \cdot \mathbf{D} - \frac{1}{\theta}\mathbf{q} \cdot \nabla\theta + \theta\nabla \cdot \mathbf{k} \geq 0.$$

Since $\mathbf{k}$ is a possible function of ρ, θ, $\mathbf{D}$, $\nabla\theta$ then

$$\nabla \cdot \mathbf{k} = \partial_\rho\mathbf{k} \cdot \nabla\rho + \partial_\theta\mathbf{k} \cdot \nabla\theta + \partial_{\mathbf{D}}\mathbf{k} \cdot \nabla\mathbf{D} + \partial_{\nabla\theta}\mathbf{k} \cdot \nabla\nabla\theta.$$

The arbitrariness of $\nabla\rho$, $\nabla\mathbf{D}$, $\nabla\nabla\theta$ implies that $\partial_\rho\mathbf{k} = \mathbf{0}$, $\partial_{\mathbf{D}}\mathbf{k} = \mathbf{0}$, $\partial_{\nabla\theta}\mathbf{k} \in \text{Skw}$. Hence $\mathbf{k} = \mathbf{k}_0(\theta) + \mathbf{\Omega}(\theta)\nabla\theta$, $\mathbf{\Omega} \in \text{Skw}$. For isotropic materials this implies that $\mathbf{k} = \mathbf{0}$. Since, by definition, p is independent of $\nabla\theta$ then we can restrict attention to $\nabla\theta = \mathbf{0}$ to obtain

$$(-p + \rho^2\partial_\rho\psi)\nabla \cdot \mathbf{v} + \mathbf{T}_{\text{dis}} \cdot \mathbf{D} \geq 0.$$

Since $\nabla \cdot \mathbf{v} = \text{tr}\,\mathbf{D}$, both terms involve the stretching $\mathbf{D}$. To find the consequences of the arbitrariness of $\mathbf{D}$ we replace $\mathbf{D}$ with $\lambda\mathbf{D}$, and hence $\nabla \cdot \mathbf{v}$ with $\lambda\nabla \cdot \mathbf{v}$ so that

$$\lambda(-p(\rho, \theta) + \rho^2\partial_\rho\psi(\rho, \theta))\nabla \cdot \mathbf{v} + \lambda\mathbf{T}_{\text{dis}}(\rho, \theta, \lambda\mathbf{D}, \mathbf{0}) \cdot \mathbf{D} \geq 0.$$

The analysis at small stretching tensors is made formal by looking at small values of λ. We can write the inequality as

$$\lambda(-p(\rho, \theta) + \rho^2\partial_\rho\psi(\rho, \theta))\nabla \cdot \mathbf{v} + o(\lambda) \geq 0.$$

Consequently, for suitably small values of λ we have

$$\text{sgn}\,[\lambda(-p(\rho, \theta) + \rho^2\partial_\rho\psi(\rho, \theta))\nabla \cdot \mathbf{v} + o(\lambda)] = \text{sgn}\,\lambda[-p(\rho, \theta) + \rho^2\partial_\rho\psi(\rho, \theta)]\nabla \cdot \mathbf{v}.$$

The arbitrariness of λ implies that the inequality holds only if

$$p(\rho, \theta) = \rho^2\partial_\rho\psi(\rho, \theta). \tag{5.23}$$

Hence we are left with the reduced entropy inequality

---

[7] See Sect. 3.4.

$$\mathbf{T}_{\rm dis} \cdot \mathbf{D} - \frac{1}{\theta}\mathbf{q} \cdot \nabla\theta \geq 0.$$

A comment is in order about the thermodynamic restriction $\partial_{\nabla\theta}\psi = \mathbf{0}$. This is so because the entropy inequality contains a single term,

$$\partial_{\nabla\theta}\psi \cdot \overset{\bullet}{\overline{\nabla\theta}},$$

in $\overset{\bullet}{\overline{\nabla\theta}} = \nabla\dot{\theta} - \mathbf{L}\nabla\theta$. By the balance of energy we have

$$\partial_\theta\varepsilon\,\dot{\theta} + \partial_{\nabla\theta}\varepsilon \cdot \overset{\bullet}{\overline{\nabla\theta}} = ... + r,$$

the dots indicating terms related to the constitutive model under consideration. If $\partial_{\nabla\theta}\varepsilon \neq \mathbf{0}$ then we can assign $\overset{\bullet}{\overline{\nabla\theta}}$ an arbitrary value and the balance of energy holds thanks to the appropriate value of r. If, instead, $\partial_{\nabla\theta}\varepsilon = \mathbf{0}$ then by

$$\partial_\theta\varepsilon\,\nabla\dot{\theta} = ... + \nabla r,$$

it follows that $\overset{\bullet}{\overline{\nabla\theta}}$ can be assigned an arbitrary value and the balance of energy holds thanks to the appropriate value of ∇r. In both cases the arbitrariness of $\overset{\bullet}{\overline{\nabla\theta}}$ implies the requirement $\partial_{\nabla\theta}\psi = \mathbf{0}$. If, instead, $\mathbf{k}$ depends on $\dot{\theta}$ then

$$-\rho\dot{\psi} + \theta\nabla \cdot \mathbf{k} = (-\rho\partial_{\nabla\theta}\psi + \theta\partial_{\dot{\theta}}\mathbf{k}) \cdot \nabla\dot{\theta} + ...$$

whence

$$-\rho\partial_{\nabla\theta}\psi + \theta\partial_{\dot{\theta}}\mathbf{k} = \mathbf{0}.$$

The free energy ψ can depend on $\nabla\theta$ provided only that the extra-entropy flux $\mathbf{k}$ depends on $\dot{\theta}$.

Among the constitutive functions $\mathbf{T}_{\rm dis}$ and $\mathbf{q}$ compatible with the (reduced) entropy inequality we consider the linear functions

$$\mathbf{T}_{\rm dis} = 2\mu(\rho, \theta)\mathbf{D} + \lambda(\rho, \theta)\mathrm{tr}\,\mathbf{D}\mathbf{1},$$

$$\mathbf{q} = -\kappa(\rho, \theta)\nabla\theta,$$

in $\mathbf{D}$ and $\nabla\theta$, which then are also objective. To establish the possible compatibility, look at the traceless (or deviatoric) part $\mathbf{D}_0$ of $\mathbf{D}$,

$$\mathbf{D}_0 = \mathbf{D} - \tfrac{1}{3}(\mathrm{tr}\,\mathbf{D})\mathbf{1},$$

so that

$$\mathbf{T}_{\rm dis} = 2\mu\mathbf{D}_0 + \zeta(\mathrm{tr}\,\mathbf{D})\mathbf{1}, \qquad \zeta := \lambda + \tfrac{2}{3}\mu.$$

Therefore we have

$$0 \le \mathbf{T}_{\rm dis} \cdot \mathbf{D} - \frac{1}{\theta}\mathbf{q} \cdot \nabla\theta = 2\mu \mathbf{D}_0 \cdot \mathbf{D}_0 + \zeta (\mathrm{tr}\,\mathbf{D})^2 - \frac{1}{\theta}\kappa \nabla\theta \cdot \nabla\theta.$$

The independence of $\mathbf{D}_0$, $\mathrm{tr}\,\mathbf{D}$, and $\nabla\theta$ implies that the inequality holds if and only if

$$\mu \ge 0, \quad \zeta \ge 0, \quad \kappa \ge 0.$$

The restrictions so determined are then necessary and sufficient for the validity of the second law subject to the assumption on $p(\rho, \theta)$ and $\mathbf{T}_{\rm dis}$, $\mathbf{q}$.

Some comments are in order about the model so derived. The coefficients μ and ζ are called coefficients of viscosity, notably μ is the shear viscosity and ζ is the second (or bulk or dilatational) viscosity. The independence of μ and ζ of the stretching $\mathbf{D}$ allows the model to be referred to as the *Newtonian fluid.* For rarefied gases it is customary to assume $\zeta = 0$ (referred to as Stokes relation[8]). For liquids, the assumption $\zeta > \mu$ is often made which means that $\lambda > \mu/3$.

The governing differential equations for the Newtonian fluid are

$$\dot\rho + \rho \nabla \cdot \mathbf{v} = 0,$$

$$\rho \dot{\mathbf{v}} = -\nabla p + \nabla(\zeta \mathrm{tr}\,\mathbf{D}) + 2\nabla \cdot (\mu \mathbf{D}_0) + \rho \mathbf{b},$$

$$\rho \dot\varepsilon = 2\mu \mathbf{D}_0 \cdot \mathbf{D}_0 + \zeta (\mathrm{tr}\,\mathbf{D})^2 - \nabla \cdot (\kappa \nabla\theta) + \rho r.$$

In general ρ and θ, and hence μ and ζ, are not constant throughout the fluid. If, however, the viscosities may be regarded as constant then the equation of motion becomes

$$\rho \dot{\mathbf{v}} = -\nabla p + \mu \Delta \mathbf{v} + (\zeta + \tfrac{1}{3}\mu)\nabla(\nabla \cdot \mathbf{v}) + \rho \mathbf{b},$$

while $\zeta + \mu/3 = \lambda + \mu$. This form of the equation of motion is called the *Navier–Stokes equation.*

A classical vorticity transport equation follows within the approximation that the temperature is constant throughout the fluid and the viscosities are constant. In view of the pressure relation $p = \rho^2 \partial_\rho \psi$ and of the assumption $\nabla\theta = \mathbf{0}$ we obtain

$$\nabla(\psi + \frac{p}{\rho}) = \partial_\rho \psi \,\nabla\rho - \frac{p}{\rho^2}\nabla\rho + \frac{1}{\rho}\nabla p = \frac{1}{\rho}\nabla p.$$

Hence, letting $\mathbf{b}$ be conservative, $\mathbf{b} = \nabla\varphi$, we can write the Navier–Stokes relation in the form

$$\dot{\mathbf{v}} = -\nabla(\psi + \frac{p}{\rho} - \varphi - \frac{\lambda + \mu}{\rho}\nabla \cdot \mathbf{v}) + \frac{\mu}{\rho}\Delta \mathbf{v}.$$

---

[8] See, e.g. [393], Sect. 62.

Hence $\nabla \times \dot{\mathbf{v}}$ is given by

$$\nabla \times \dot{\mathbf{v}} = \mu \nabla \times \frac{\Delta \mathbf{v}}{\rho} = \mu(\frac{\Delta \nabla \times \mathbf{v}}{\rho} - \frac{1}{\rho^2} \nabla \rho \times \Delta \mathbf{v}).$$

By (1.34)

$$\nabla \times \dot{\mathbf{v}} = \dot{\varpi} - \mathbf{L}\varpi + (\nabla \cdot \mathbf{v})\varpi) = \overset{\diamond}{\varpi},$$

$\overset{\diamond}{\varpi}$ being the convected rate, or the Truesdell rate, of the vorticity $\varpi = \nabla \times \mathbf{v}$. Hence it follows that, in isothermal conditions, ϖ satisfies the transport equation

$$\overset{\diamond}{\varpi} = \mu(\frac{\Delta \varpi}{\rho} - \frac{1}{\rho^2} \nabla \rho \times \Delta \mathbf{v}).$$

Viscosity affects the rate of change of the kinetic energy within a convecting spatial region $\mathcal{P}_t$ as well as within a fixed spatial region Ω. Inner product of the equation of motion with $\mathbf{v}$ gives

$$\rho \mathbf{v} \cdot \dot{\mathbf{v}} = \mathbf{v} \cdot \nabla \cdot \mathbf{T} + \rho \mathbf{b} \cdot \mathbf{v}.$$

Since

$$\rho \mathbf{v} \cdot \dot{\mathbf{v}} = \rho(\tfrac{1}{2}\mathbf{v}^2)^{\cdot} = \partial_t(\tfrac{1}{2}\rho \mathbf{v}^2) + \nabla \cdot (\tfrac{1}{2}\rho \mathbf{v}^2\, \mathbf{v})$$

we can write

$$\partial_t(\tfrac{1}{2}\rho \mathbf{v}^2) = \nabla \cdot (\mathbf{T}\mathbf{v} - \tfrac{1}{2}\rho \mathbf{v}^2\, \mathbf{v}) + \rho\, \mathbf{b} \cdot \mathbf{v} - \mathbf{T} \cdot \mathbf{D}.$$

Integration over a fixed region Ω and use of the divergence theorem imply

$$\partial_t \int_\Omega \tfrac{1}{2}\rho \mathbf{v}^2\, dv = \int_{\partial\Omega} (\mathbf{T}\mathbf{v} - \tfrac{1}{2}\rho \mathbf{v}^2\, \mathbf{v}) \cdot \mathbf{n}\, da + \int_\Omega (\rho\, \mathbf{b} \cdot \mathbf{v} - \mathbf{T} \cdot \mathbf{D}) dv.$$

If $\mathbf{v} = \mathbf{0}$ at the boundary $\partial\Omega$ then the rate of change is given by two terms, the power of the body force and that of the stress. Now,

$$-\mathbf{T} \cdot \mathbf{D} = p \nabla \cdot \mathbf{v} - 2\mu \mathbf{D}_0 \cdot \mathbf{D}_0 - \zeta(\nabla \cdot \mathbf{v})^2.$$

The power $p\nabla \cdot \mathbf{v}$ is positive during expansion ($\nabla \cdot \mathbf{v} > 0$) and negative during compression ($\nabla \cdot \mathbf{v} < 0$). The power of the dissipative stress, instead, is negative in any motion as a consequence of the thermodynamic requirements $\mu, \zeta \geq 0$. Consequently, the global kinetic energy decreases as a consequence of viscosity whereas, by effect of pressure, it increases in any expansion and decreases in any compression.

If, instead, we look at a convecting region $\mathcal{P}_t$, by

$$\rho(\tfrac{1}{2}\mathbf{v}^2)^{\cdot} = \mathbf{v} \cdot (\nabla \cdot \mathbf{T}) + \rho\, \mathbf{b} \cdot \mathbf{v}$$

integration over $\mathcal{P}_t$ allows us to write

$$\partial_t \int_{\mathcal{P}_t} \tfrac{1}{2}\rho \mathbf{v}^2 \, dv = \int_{\partial \mathcal{P}_t} (\mathbf{T}\mathbf{v}) \cdot \mathbf{n} \, da + \int_{\mathcal{P}_t} (\rho \mathbf{b} \cdot \mathbf{v} - \mathbf{T} \cdot \mathbf{D}) dv.$$

Again we find that viscosity results in a decay of the kinetic energy via the power $\mathbf{T} \cdot \mathbf{D}$.

5.5.1 *Stokesian Fluids*

To investigate the possible dependence of $\mathbf{T}_{\text{dis}}$ on the stretching $\mathbf{D}$, we neglect the dependence of $\mathbf{T}_{\text{dis}}$ on $\nabla\theta$ and exploit the isotropy condition (5.21) and (5.22), namely

$$\mathbf{Q}\,\mathbf{T}_{\text{dis}}\,\mathbf{Q}^T = \hat{\mathbf{T}}_{\text{dis}}(\rho, \theta, \mathbf{Q}\,\mathbf{D}\,\mathbf{Q}^T), \qquad \hat{\mathbf{T}}_{\text{dis}}(\rho, \theta, \mathbf{0}) = \mathbf{0}.$$

By adding the requirement that $\hat{\mathbf{T}}_{\text{dis}}$ is a smooth function of $\mathbf{D}$, this expression amounts to Stokes' postulates of fluidity. By means of the invariants[9] Stokes' postulates lead to the representation formula[10]

$$\mathbf{T}_{\text{dis}} = \alpha \mathbf{1} + \beta \mathbf{D} + \gamma \mathbf{D}^2$$

where

$$\alpha = \hat{\alpha}(\rho, \theta, \mathcal{I}_1, \mathcal{I}_2, \mathcal{I}_3), \quad \beta = \hat{\beta}(\rho, \theta, \mathcal{I}_1, \mathcal{I}_2, \mathcal{I}_3), \quad \gamma = \hat{\gamma}(\rho, \theta, \mathcal{I}_1, \mathcal{I}_2, \mathcal{I}_3),$$

and $\hat{\alpha}(\rho, \theta, 0, 0, 0) = 0$. A proof of this result is given also in [184] and is based on the Cayley–Hamilton theorem. Fluids whose constitutive stress function is given in the form

$$\mathbf{T} = (-p + \alpha)\mathbf{1} + \beta \mathbf{D} + \gamma \mathbf{D}^2, \tag{5.24}$$

where $p = \hat{p}(\rho, \theta)$, are usually referred to as *Stokesian fluids*. Hereafter, for ease in writing, the quantity $p - \alpha$ (which is often but improperly called pressure) will be denoted by $\bar{p}$ and called *nominal pressure*. Since $\bar{p} = -\frac{1}{3}\text{tr}\,\mathbf{T}$ then

$$\bar{p} = p - \tfrac{1}{3}\text{tr}\,\mathbf{T}_{\text{dis}}$$

whence we get the difference between the thermodynamic (or equilibrium) pressure p and the mean pressure $\bar{p}$,

$$p - \bar{p} = \alpha + \tfrac{1}{3}\beta \mathcal{I}_1 + \tfrac{1}{3}\gamma \mathcal{I}_2. \tag{5.25}$$

---

[9] We observe that $\mathcal{I}_1, \mathcal{I}_2, \mathcal{I}_3$, relative to the stretching $\mathbf{D}$, are denoted by $I_{\mathbf{D}}, II_{\mathbf{D}}, III_{\mathbf{D}}$ in [393].

[10] See, e.g. [393], p. 232, for a simple proof.

For Stokesian fluids a sufficient condition for compatibility with the (reduced) entropy inequality is given by

$$\mathbf{T}_{\text{dis}} \cdot \mathbf{D} \geq 0, \qquad \mathbf{q} \cdot \nabla\theta \leq 0,$$

which implies that the constitutive isotropic functions α, β and γ satisfy the inequality

$$\alpha\mathcal{I}_1 + \beta\mathcal{I}_2 + \gamma\mathcal{I}_3 \geq 0$$

for any admissible motion.

In the linear case,

$$\alpha = \lambda(\rho, \theta)\mathcal{I}_1, \qquad \beta = 2\mu(\rho, \theta), \qquad \gamma = 0$$

and the fluid is referred to as *Newtonian*. In *polynomial fluids* the components of the Cauchy stress are polynomials of degree $n \geq 2$ in the deformation components, namely α, β and γ are polynomials of degree n, $n - 1$ and $n - 2$, respectively, in the components of $\mathbf{D}$ and $\alpha = 0$ if $\mathbf{D} = \mathbf{0}$. Since $\mathcal{I}_1$, $\mathcal{I}_2$ and $\mathcal{I}_3$ are homogeneous of degree 1, 2 and 3 in the components of $\mathbf{D}$, respectively, then α, β and γ turn out to be polynomials of the scalar invariants of $\mathbf{D}$.

For instance, if $n = 2$ the Cauchy stress is taken in the form

$$\mathbf{T} = (-p + \lambda\mathcal{I}_1 + \lambda_1\mathcal{I}_1^2 + \lambda_2\mathcal{I}_2)\mathbf{1} + 2(\mu + \mu_1\mathcal{I}_1)\mathbf{D} + 4\nu\mathbf{D}^2, \tag{5.26}$$

where $p, \lambda, \lambda_1, \lambda_2, \mu, \mu_1$ and ν are functions of ρ, θ only. The entropy inequality requires that

$$\lambda\mathcal{I}_1^2 + \lambda_1\mathcal{I}_1^3 + \lambda_2\mathcal{I}_1\mathcal{I}_2 + 2\mu\mathcal{I}_2 + 2\mu_1\mathcal{I}_1\mathcal{I}_2 + 4\nu\mathcal{I}_3 \geq 0.$$

The total pressure $\bar{p}$ in the fluid is then expressed as a quadratic function of the scalar invariants of $\mathbf{D}$,

$$\bar{p} = p - (\lambda + \tfrac{2}{3}\mu)\mathcal{I}_1 - (\lambda_1 + \tfrac{2}{3}\mu_1)\mathcal{I}_1^2 - (\lambda_2 + \tfrac{4}{3}\nu)\mathcal{I}_2$$

If the fluid is incompressible, the constitutive relation of $\mathbf{T}$ simplifies to

$$\mathbf{T} = (-\bar{p} - \tfrac{4}{3}\nu\mathcal{I}_2(\mathbf{D}_0))\mathbf{1} + 2\mu\mathbf{D}_0 + 4\nu\mathbf{D}_0^2,$$

where $\mathcal{I}_2(\mathbf{D}_0) = \text{tr}\,\mathbf{D}_0^2$. Since ρ is constant, $\bar{p}$ is viewed as an independent variable while μ and ν are functions of θ only. Hence, the nominal pressure reduces to the total pressure $\bar{p}$ provided $\nu = 0$ and this is the case compatible with thermodynamics, as discussed in *Case II* of Sect. 5.5.3.

Higher degree stress-stretching relations can be established in a similar way.

5.5.2 Boundary Conditions

According to a classical assumption, which traces back to Stokes, a viscous fluid must adhere to a solid boundary Σ, namely

$$\mathbf{v}(\mathbf{x}, t) = \mathbf{u}(\mathbf{x}, t) \qquad \text{on } \Sigma$$

where $\mathbf{u}$ is the velocity of the boundary. This is called *no-slip condition* and is experimentally verified when moderate pressures and low surface stresses are involved, but do not apply in all cases. In rarefied viscous gases, for instance, the rarefaction effects are modelled through the partial slip at the boundary rigid wall using Maxwell's velocity slip.[11] In general, a phenomenologically reasonable slip condition is

$$v_n - u_n = 0, \qquad \mathbf{v}_t - \mathbf{u}_t = \mathfrak{s}\,\mathbf{t}_t \qquad \text{on } \Sigma$$

where $\mathbf{n}$ is the boundary outward normal vector at $\mathbf{x}$ at Σ, and

$$v_n = \mathbf{v} \cdot \mathbf{n}, \qquad t_n = \mathbf{t} \cdot \mathbf{n}, \qquad \mathbf{v}_t = \mathbf{v} - v_n \mathbf{n}, \qquad \mathbf{t}_t = \mathbf{t} - t_n \mathbf{n}.$$

The positive coefficient $\mathfrak{s}$ is called *skin friction*. Assuming that it is not a constant but vanishes at low tangential stresses and high pressures, this boundary condition reduces to no-slip at the opposite extremes. Following [393], Sect. 64, we let

$$\mathfrak{s}(t_n, \mathbf{t}_t) = \begin{cases} 0 & \text{if } \ |\mathbf{t}_t| \le \xi_0 |t_n|, \\ \mathfrak{s}_0(1 - \xi_0 |t_n / \mathbf{t}_t|) & \text{otherwise.} \end{cases} \tag{5.27}$$

where the friction coefficient ξ_0 is a dimensionless parameter which describes an empirical property of the contacting materials, as in the modelling of dynamic dry friction.[12]

To fix ideas, we now apply the no-slip assumption as the boundary condition. Let Ω be a fixed (regular) domain, $\partial\Omega$ be its fixed boundary and $\mathbf{n}$ be the outward normal at $\mathbf{x} \in \partial\Omega$. The no-slip condition reads

$$\mathbf{v} = \mathbf{0} \qquad \text{on } \partial\Omega. \tag{5.28}$$

As a consequence of Stokes' theorem, the vorticity vector is tangential to the boundary,

$$\varpi \cdot \mathbf{n} = 0 \qquad \text{on } \partial\Omega,$$

and in addition

---

[11] See, e.g. [253], Chap. 2

[12] A unified slip boundary condition, set up in [411], is applicable to a wide range of fluid flows, from rarefied gases to non-Newtonian fluids.

$$(\mathbf{n} \times \nabla) \times \mathbf{v} = \mathbf{D}\mathbf{n} - (\nabla \cdot \mathbf{v})\mathbf{n} - \tfrac{1}{2}\boldsymbol{\varpi} \times \mathbf{n} = \mathbf{0} \qquad \text{on } \partial\Omega.$$

whence it follows

$$\mathbf{D}\mathbf{n} = (\nabla \cdot \mathbf{v})\mathbf{n} + \tfrac{1}{2}\boldsymbol{\varpi} \times \mathbf{n}, \qquad \mathbf{D}^2\mathbf{n} = (\nabla \cdot \mathbf{v})^2\mathbf{n} + \tfrac{1}{2}(\nabla \cdot \mathbf{v})\boldsymbol{\varpi} \times \mathbf{n} + \tfrac{1}{2}\mathbf{D}(\boldsymbol{\varpi} \times \mathbf{n})$$

and then

$$\mathbf{D}\mathbf{n} \cdot \mathbf{n} = (\nabla \cdot \mathbf{v})^2, \qquad \mathbf{D}^2\mathbf{n} \cdot \mathbf{n} = (\nabla \cdot \mathbf{v})^2 + \tfrac{1}{2}(\boldsymbol{\varpi} \times \mathbf{n}) \cdot \mathbf{D}\mathbf{n} = (\nabla \cdot \mathbf{v})^2 + \tfrac{1}{4}(\boldsymbol{\varpi} \times \mathbf{n})^2$$

By virtue of these relations we obtain the expression of the stress vector at the fixed boundary for Stokesian fluids

$$\mathbf{t} = \mathbf{T}\mathbf{n} = (-p + \alpha + \beta\mathcal{I}_1 + \gamma\mathcal{I}_1^2)\mathbf{n} + \tfrac{1}{2}(\beta + \tfrac{4}{3}\gamma\mathcal{I}_1)\boldsymbol{\varpi} \times \mathbf{n} + \tfrac{1}{2}\gamma\mathbf{D}_0(\boldsymbol{\varpi} \times \mathbf{n})$$

and then

$$t_n = -p + \alpha + \beta\mathcal{I}_1 + \gamma\mathcal{I}_1^2 + \tfrac{1}{4}\gamma\,\varpi^2. \tag{5.29}$$

This shows that, in general, both tangential and normal stresses at the boundary depend on the local vorticity. When $\gamma = 0$, the normal stress is due to pressure and expansion, only, whereas the total stress is unaffected by local vorticity only if $\beta = \gamma = 0$. Finally, comparing t_n with the mean pressure on the boundary, we obtain

$$t_n = -\bar{p} + \tfrac{2}{3}\beta\mathcal{I}_1 + \gamma(\mathcal{I}_1^2 - \tfrac{1}{3}\mathcal{I}_2 + \tfrac{1}{4}\varpi^2). \tag{5.30}$$

5.5.3 *Incompressible Fluids*

Incompressibility means that only isochoric motions are possible, namely

$$\nabla \cdot \mathbf{v} = 0,$$

which implies that the mass density ρ is constant along streamlines. Incompressibility can be viewed as an internal linear constraint by prescribing a subspace $\mathcal{M}$ of Sym (the space of all symmetric tensors) to be interpreted[13] as the collection of all possible stretching tensors **D**. In particular, since $\nabla \cdot \mathbf{v} = \mathbf{1} \cdot \mathbf{D}$, we have

$$\mathcal{M} = \{\mathbf{D} \in \text{Sym} : \mathbf{1} \cdot \mathbf{D} = 0\}.$$

Therefore, $\mathcal{M}$ is the subspace of all traceless (deviatoric) tensors, whereas $\mathcal{M}^\perp$ is the subspace of all isotropic (spherical) tensors. According to this approach, the current stress **T** in an incompressible material is determined by the deformation process

[13] See, e.g. [42], Sect. 5.1, p. 167.

only up to an additive part, $\mathbf{T}^r$, (called *reactive*) that does no work in any admissible (isochoric) motion, namely

$$\mathbf{T}^r \cdot \mathbf{D}_0 = 0, \qquad \mathbf{D}_0 \in \mathcal{M}.$$

As a consequence, the reactive stress is spherical, $\mathbf{T}^r \in \mathcal{M}^\perp$, and there exists a time-dependent scalar field ϕ such that

$$\mathbf{T}^r = \phi \mathbf{1}.$$

Unlike the compressible case, ϕ is an arbitrary multiplier and cannot be determined by constitutive functions.

We now investigate the relationship between ϕ and the pressure. To establish the uniqueness of the decomposition

$$\mathbf{T} = \mathbf{T}^r + \mathbf{T}^a$$

we can normalize the active stress $\mathbf{T}^a = \hat{\mathbf{T}}^a(\rho, \theta, \mathbf{D}, \nabla\theta)$ by the condition ([42], Eq. (5.5))

$$\mathbf{T}^a \cdot \mathbf{T}^r = 0$$

which implies $\mathbf{T}^a \in \mathcal{M}$ for any admissible (isochoric) deformation process. Accordingly, we have

$$\mathbf{T}^r = \tfrac{1}{3}(\operatorname{tr}\mathbf{T})\,\mathbf{1} = -\bar{p}\,\mathbf{1},$$

and hence we identify $-\phi$ with the (reactive) total pressure $\bar{p}$, and then $\mathbf{T}_{\text{dis}} \neq \mathbf{T}^a$.

In view of (5.24), we finally rewrite the stress decomposition for Stokesian incompressible fluids into the form[14]

$$\mathbf{T}(\rho, \theta, \mathbf{D}_0) = -\bar{p}\,\mathbf{1} + \beta_0 \mathbf{D}_0 + \gamma_0 (\mathbf{D}_0^2 - \tfrac{1}{3}\mathcal{I}_2\,\mathbf{1}),$$

where $\mathcal{I}_2$ is relative to $\mathbf{D}_0$ and $\bar{p}$ is a parameter (Lagrangian multiplier) independent of the thermodynamic process, $\mathbf{D}_0$ is the traceless (or deviatoric) part of $\mathbf{D}$ and

$$\beta_0 = \hat{\beta}_0(\rho, \theta, \mathcal{I}_2, \mathcal{I}_3), \qquad \gamma_0 = \hat{\gamma}_0(\rho, \theta, \mathcal{I}_2, \mathcal{I}_3).$$

If we let

$$p_0(\bar{p}, \rho, \theta, \mathcal{I}_2, \mathcal{I}_3) = \bar{p} + \tfrac{1}{3}\gamma_0 \mathcal{I}_2$$

we obtain the well-known constitutive relation of *Reiner–Rivlin fluids*,

$$\mathbf{T} = -p_0 \mathbf{1} + \beta_0 \mathbf{D}_0 + \gamma_0 \mathbf{D}_0^2. \tag{5.31}$$

---

[14] $\operatorname{tr}\mathbf{D}_0^2 = \mathcal{I}_2 > 0$ for any $\mathbf{D}_0 \neq \mathbf{0}$.

Aside from the different nature of the pressure, we can state the constitutive and dynamic properties of incompressible fluids as a particular case of compressible ones. Since ρ is constant, the free energy and the entropy are given by

$$\psi = \psi(\theta), \qquad \eta(\theta) = -\partial_\theta \psi(\theta).$$

Because of the different stress decomposition, the reduced entropy inequality holds in a modified form,

$$\mathbf{T}^a \cdot \mathbf{D}_0 - \frac{1}{\theta}\mathbf{q} \cdot \nabla\theta \geq 0.$$

For Reiner–Rivlin fluids a sufficient condition is given by

$$\mathbf{T}^a \cdot \mathbf{D}_0 \geq 0, \qquad \mathbf{q} \cdot \nabla\theta \leq 0,$$

which implies

$$\beta_0 \mathcal{I}_2 + \gamma_0 \mathcal{I}_3 \geq 0. \tag{5.32}$$

Now, $\mathcal{I}_3 = 3 \det \mathbf{D}_0$. Moreover, $\mathcal{I}_2 = |\mathbf{D}_0|^2 \geq 0$ and vanishes only if $\mathbf{D}_0 = \mathbf{0}$. In particular, we examine two simple cases.

Case I. Letting $\gamma_0 = 0$ and $\beta_0 = 2\mu(\theta)$, the linear functions

$$\mathbf{T}^a = 2\mu(\theta)\mathbf{D}_0, \qquad \mathbf{q} = -\kappa(\theta)\nabla\theta,$$

are compatible with the reduced entropy inequality provided $\mu \geq 0$, $\kappa \geq 0$. This case is referred to as *incompressible Newtonian fluid* (see Sect. 5.6.5).

Case II. Let $\gamma_0 = 4\nu(\theta)$ and $\beta_0 = 2\mu(\theta)$. Then, using a special deformation process it can be proved that this model is compatible with the reduced entropy inequality provided $\nu = 0$. Indeed, by considering a shear deformation such that

$$\mathbf{D}_0 = \frac{\dot{\gamma}}{2}\begin{bmatrix} 0 & 1 & 0 \\ 1 & 0 & 0 \\ 0 & 0 & 0 \end{bmatrix}$$

we have $\mathcal{I}_2 = \frac{1}{2}\dot{\gamma}^2$ and $\mathcal{I}_3 = 0$. From (5.32) it follows $\mu\,\dot{\gamma}^2 \geq 0$ for any $\dot{\gamma}$, then $\mu \geq 0$. Now, we consider the following stretching tensors,

$$\mathbf{D}_0 = \dot{\gamma}\begin{bmatrix} 2 & 0 & 0 \\ 0 & -1 & 0 \\ 0 & 0 & -1 \end{bmatrix}, \qquad \mathbf{D}_0 = \dot{\gamma}\begin{bmatrix} -2 & 0 & 0 \\ 0 & 1 & 0 \\ 0 & 0 & 1 \end{bmatrix}.$$

For both tensors, $\mathcal{I}_2 = 6\dot{\gamma}^2$, but $\mathcal{I}_3 = 6\dot{\gamma}^3$ for the former and $\mathcal{I}_3 = -6\dot{\gamma}^3$ for the latter. Hence, from (5.32) it follows that μ and ν must satisfy for any $\dot{\gamma}$ both of the following inequalities

$$\mu + 2\nu\dot{\gamma} \geq 0, \qquad \mu - 2\nu\dot{\gamma} \geq 0$$

As a consequence, we obtain

$$\nu = 0, \qquad \mu \geq 0,$$

and we conclude that the incompressible Newtonian fluid is the only Reiner–Rivlin model with β_0 and γ_0 independent of $\mathbf{D}$ which is compatible with the entropy inequality.

5.5.4 *Oberbeck–Boussinesq Approximation*

The Oberbeck–Boussinesq approximation [60, 346] takes the fluid as incompressible in that the velocity field is assumed to be solenoidal,

$$\nabla \cdot \mathbf{v} = 0,$$

but allows the mass density to be a (linear) function of the temperature. This is somewhat contradictory because incompressibility would imply that θ and hence the mass density be constant along streamlines.

The approximation is justified as follows. Assume the fluid is at equilibrium at the reference state with mass density $\rho_0 = \rho_R$ and temperature θ_0. Hence

$$\nabla p(\rho_0) + \rho_0 \mathbf{g} = \mathbf{0}.$$

Letting $p = p_0 + \mathscr{P}$, with $p_0 = p(\rho_0)$, we can write the equation of motion in the form

$$\rho\dot{\mathbf{v}} = -\nabla\mathscr{P} + \mu\Delta\mathbf{v} + (\rho - \rho_0)\mathbf{g}.$$

By linearization,

$$\rho\dot{\mathbf{v}} \simeq \rho_0\dot{\mathbf{v}}, \qquad \rho - \rho_0 = -\alpha\rho_0(\theta - \theta_0)\mathbf{g}$$

we can write the equation of motion in the form

$$\dot{\mathbf{v}} = -\frac{1}{\rho_0}\nabla\mathscr{P} + \nu\Delta\mathbf{v} - \alpha(\theta - \theta_0)\mathbf{g},$$

where $\nu = \mu/\rho_0$ is the kinematic viscosity. The scheme is completed by letting $\mathbf{q} = -\kappa\nabla\theta$ so that the balance of energy becomes

$$\dot{\theta} = \kappa\Delta\theta + h,$$

where $h = (\rho r + \mathbf{T} \cdot \mathbf{D})/\rho c$, c being the specific heat (per unit volume).

Of course it is no more than an approximation to let the mass density be a function of temperature and hence disregarding the constancy of ρ along streamlines, required by the incompressibility assumption, $\nabla \cdot \mathbf{v} = 0$.

5.6 Newtonian Fluids

Among the constitutive functions $\mathbf{T}_{\text{dis}}$ and $\mathbf{q}$ compatible with the (reduced) entropy inequality we consider the linear functions

$$\mathbf{T}_{\text{dis}} = 2\mu(\rho, \theta)\mathbf{D} + \lambda(\rho, \theta)(\text{tr}\,\mathbf{D})\mathbf{1}, \qquad \mathbf{q} = -\kappa(\rho, \theta)\nabla\theta,$$

in $\mathbf{D}$ and $\nabla\theta$, which then are also objective. They are recovered as a special case from Stokesian fluids by letting $\gamma = 0$, $\beta = 2\mu$ and $\alpha = \lambda\mathcal{I}_1$. To establish the thermodynamic compatibility we express $\mathbf{T}_{\text{dis}}$ in the form

$$\mathbf{T}_{\text{dis}} = 2\mu\mathbf{D}_0 + \zeta(\text{tr}\,\mathbf{D})\mathbf{1}, \qquad \zeta := \lambda + \tfrac{2}{3}\mu,$$

where $\mathbf{D}_0$ is the traceless part of $\mathbf{D}$. As a consequence we have

$$0 \le \mathbf{T}_{\text{dis}} \cdot \mathbf{D} - \frac{1}{\theta}\mathbf{q} \cdot \nabla\theta = 2\mu\mathbf{D}_0 \cdot \mathbf{D}_0 + \zeta(\text{tr}\,\mathbf{D})^2 - \frac{1}{\theta}\kappa\nabla\theta \cdot \nabla\theta.$$

The independence of $\mathbf{D}_0$, $\text{tr}\,\mathbf{D}$, and $\nabla\theta$ implies that the inequality holds if and only if

$$\mu \ge 0, \quad \zeta \ge 0, \quad \kappa \ge 0.$$

The restrictions so determined are then necessary and sufficient for the validity of the second law subject to the assumption on $p(\rho, \theta)$ and $\mathbf{T}_{\text{dis}}$, $\mathbf{q}$.

Some comments are in order about the model so derived. The coefficients μ and ζ are called coefficients of viscosity, notably μ is the *shear viscosity* and ζ is the *second* (or *bulk* or *dilatational*) *viscosity*. The independence of μ and ζ of the stretching $\mathbf{D}$ allows the model to be referred to as the *Newtonian fluid*. In addition, from (5.25) the difference between pressure and mean pressure in a compressible Newtonian fluid is proportional to ζ, namely $p - \bar{p} = \zeta\mathcal{I}_1$, and then vanishes for rarefied gases (where $\zeta = 0$).

The governing differential equations for the Newtonian fluid are

$$\dot{\rho} + \rho\nabla \cdot \mathbf{v} = 0,$$

$$\rho\dot{\mathbf{v}} = -\nabla p + \nabla(\zeta\text{tr}\,\mathbf{D}) + 2\nabla \cdot (\mu\mathbf{D}_0) + \rho\mathbf{b},$$

$$\rho\dot{\varepsilon} = 2\mu\mathbf{D}_0 \cdot \mathbf{D}_0 + \zeta(\text{tr}\,\mathbf{D})^2 - \nabla \cdot (\kappa\nabla\theta) + \rho r.$$

In general ρ and θ, and hence μ and ζ, are not constant throughout the fluid. If, however, the viscosities may be regarded as constant then the equation of motion can be written as

$$\rho\dot{\mathbf{v}} = -\nabla p + \mu\Delta\mathbf{v} + (\zeta + \tfrac{1}{3}\mu)\nabla(\nabla\cdot\mathbf{v}) + \rho\mathbf{b},$$

while $\zeta + \mu/3 = \lambda + \mu$. This form of the equation of motion is called the *Navier–Stokes equation.*

Assuming a no-slip boundary condition, by virtue of (5.29) the normal and tangential stresses at a fixed rigid wall takes the form

$$t_n = -p + (\lambda + 2\mu)\mathcal{I}_1, \qquad \mathbf{t}_t = \mu\,\boldsymbol{\varpi}\times\mathbf{n}.$$

According to (5.27) and remembering that $\boldsymbol{\varpi}$ is tangential (normal to $\mathbf{n}$), for a given friction coefficient ξ_0 and a given pressure $p(\rho,\theta)$ the no-slip condition is valid provided that the velocity field $\mathbf{v}$ on the boundary satisfies

$$\mu\,|\nabla\times\mathbf{v}| \le \xi_0\,|p - (\lambda + 2\mu)\nabla\cdot\mathbf{v}|.$$

5.6.1 Models with Thermal Expansion and Pressure-Dependent Viscosities

The specific volume is assumed to depend only on temperature. Hence we assume (see [374])

$$J = f(\theta)$$

whence

$$\nabla\cdot\mathbf{v} = \alpha(\theta)\dot{\theta}, \qquad \alpha(\theta) = \frac{1}{f(\theta)}f'(\theta).$$

Thus the fluid sustains only isochoric motions in isothermal conditions. A thermodynamically consistent model is now established.

Since $\dot{\rho} = -\rho\nabla\cdot\mathbf{v}$ then

$$\frac{\dot{v}}{v} = -\frac{\dot{\rho}}{\rho} = -\alpha(\theta)\dot{\theta}.$$

Hence

$$\alpha(\theta) = \frac{1}{v(\theta)}v'(\theta)$$

is the coefficient of thermal expansion.

Take the stress $\mathbf{T}$ in the form

$$\mathbf{T} = -p\mathbf{1} + \mathcal{T}, \qquad \mathcal{T}\to\mathbf{0} \;\text{ as }\; \mathbf{D}\to\mathbf{0},$$

and hence $p = -(\operatorname{tr}\mathbf{T})/3$ as $\mathbf{D} = \mathbf{0}$. Let $\theta, p, \mathbf{D}, \nabla\theta$ be the set of independent variables while $\psi, \eta, \mathcal{T}, \mathbf{q}$ are given by constitutive functions under consideration. Observe that $p\nabla\cdot\mathbf{v} = \alpha p\dot{\theta}$. Computing $\dot{\psi}$ and substituting we obtain the Clausius–Duhem inequality in the form

$$-\rho\partial_p\psi\dot{p} - (\rho\partial_\theta\psi + \rho\eta + \alpha p)\dot{\theta} - \rho\partial_{\mathbf{D}}\psi\cdot\dot{\mathbf{D}} - \rho\partial_{\nabla\theta}\psi\cdot\dot{\overline{\nabla\theta}} + \mathcal{T}\cdot\mathbf{D} - \frac{1}{\theta}\mathbf{q}\cdot\nabla\theta \geq 0.$$

The arbitrariness of $\dot{p}$, $\dot{\mathbf{D}}$, and $\dot{\overline{\nabla\theta}}$ implies that

$$\partial_p\psi = 0, \qquad \partial_{\mathbf{D}}\psi = \mathbf{0}, \qquad \partial_{\nabla\theta}\psi = \mathbf{0}.$$

Hence $\psi = \psi(\theta, p)$ and the arbitrariness of $\dot{\theta}$ implies that

$$\eta = -\partial_\theta\psi - \frac{\alpha}{\rho}p.$$

The internal energy ε then takes the form

$$\varepsilon = \psi(\theta, p) - \theta\partial_\theta\psi(\theta, p) - \frac{\theta\alpha(\theta)p}{\rho(\theta)}.$$

Hence we are left with the reduced dissipation inequality

$$\mathcal{T}\cdot\mathbf{D} - \frac{1}{\theta}\mathbf{q}\cdot\nabla\theta \geq 0. \tag{5.33}$$

As with $\mathcal{T}$ and $\mathbf{q}$ we can improve the standard Navier–Stokes–Fourier and assume

$$\mathcal{T} = 2\mu(\theta, p)\mathbf{D} + \lambda(\theta, p)(\operatorname{tr}\mathbf{D})\mathbf{1}, \qquad \mathbf{q} = -\kappa(\theta, p)\nabla\theta.$$

Inequality (5.33) holds with the classical conditions

$$\mu \geq 0, \qquad 2\mu + 3\lambda \geq 0, \qquad \kappa \geq 0$$

for any dependence on θ and p.

Qualitative differences occur in the equation of motion and the balance of energy as a consequence of the dependences on θ, p. The equation of motion reads

$$\rho\dot{\mathbf{v}} = -\nabla p + \mu\Delta\mathbf{v} + (\mu + \lambda)\nabla(\nabla\cdot\mathbf{v}) + 2\mathbf{D}\nabla\mu + (\nabla\cdot\mathbf{v})\nabla\lambda + \rho\mathbf{b}.$$

Now observe that

$$\partial_\theta\varepsilon = -\big[\theta\partial_\theta^2\psi + \big(\frac{\theta\alpha}{\rho}\big)'p\big]$$

can be viewed as the specific heat at constant pressure; denote it by c_p. Hence the balance of energy can be expressed in the form

$$\rho c_p \dot{\theta} - \alpha\theta\dot{p} = \kappa\Delta\theta + \nabla\kappa\cdot\nabla\theta + 2\mu\mathbf{D}_0\cdot\mathbf{D}_0 + (\frac{2\mu}{3} + \lambda)(\nabla\cdot\mathbf{v})^2 + \rho r.$$

As a particular case, Newtonian incompressible fluids, with pressure-dependent viscosity, follow by letting $\nabla\cdot\mathbf{v} = 0$ and $\alpha = 0$. We now examine compressible fluids with pressure-dependent viscosities.

5.6.2 *Fluids with Pressure-Dependent Viscosities*

The fluid is described with the independent variables θ, p, $\mathbf{D}$, $\nabla\theta$. Still we let

$$\mathbf{T} = -p\mathbf{1} + \mathcal{T}, \qquad \mathcal{T} \to \mathbf{0} \text{ as } \mathbf{D} \to \mathbf{0}.$$

Consider the Clausius–Duhem inequality

$$-\rho(\dot{\psi} + \eta\dot{\theta}) - p\nabla\cdot\mathbf{v} + \mathcal{T}\cdot\mathbf{D} - \frac{1}{\theta}\mathbf{q}\cdot\nabla\theta \geq 0$$

and observe that

$$p\nabla\cdot\mathbf{v} = p\frac{\dot{v}}{v}, \qquad -\rho\dot{\psi} - p\frac{\dot{v}}{v} = -\rho[\dot{\psi} + \dot{\overline{pv}} - v\dot{p}] = -\rho\dot{\phi} + \dot{p},$$

where $\phi = \psi + pv$ is the Gibbs free energy. Hence we write the inequality in the form

$$-\rho(\dot{\phi} + \eta\dot{\theta}) + \dot{p} + \mathcal{T}\cdot\mathbf{D} - \frac{1}{\theta}\mathbf{q}\cdot\nabla\theta \geq 0.$$

Letting ϕ, η, ρ, $\mathcal{T}$, $\mathbf{q}$ be given by functions of θ, p, $\mathbf{D}$, $\nabla\theta$ and computing $\dot{\phi}$ we find

$$-\rho(\partial_\theta\phi + \eta)\dot{\theta} - (\rho\partial_p\phi - 1)\dot{p} - \rho\partial_{\mathbf{D}}\psi\cdot\dot{\mathbf{D}} - \rho\partial_{\nabla\theta}\phi\cdot\dot{\overline{\nabla\theta}} + \mathcal{T}\cdot\mathbf{D} - \frac{1}{\theta}\mathbf{q}\cdot\nabla\theta \geq 0.$$

The arbitrariness of $\dot{\mathbf{D}}$, $\dot{\overline{\nabla\theta}}$, $\dot{\theta}$, $\dot{p}$ implies

$$\partial_{\mathbf{D}}\phi = \mathbf{0}, \quad \partial_{\nabla\theta}\phi = \mathbf{0}, \quad \eta = -\partial_\theta\phi, \quad v = \partial_p\phi.$$

Letting

$$\mathcal{T} = 2\mu(\theta, p)\mathbf{D} + \lambda(\theta, p)(\mathrm{tr}\,\mathbf{D})\mathbf{1}, \qquad \mathbf{q} = -\kappa(\theta, p)\nabla\theta$$

we find that the reduced dissipation inequality $\boldsymbol{T} \cdot \mathbf{D} - (1/\theta)\mathbf{q} \cdot \nabla\theta \geq 0$ holds with the standard restrictions

$$\mu(\theta, p) \geq 0, \qquad 2\mu(\theta, p) + 3\lambda(\theta, p) \geq 0, \qquad \kappa(\theta, p) \geq 0.$$

If the fluid is incompressible then it follows

$$\eta = -\partial_\theta \psi, \quad \partial_p \psi = 0, \quad \mu(\theta, p) \geq 0, \quad \kappa(\theta, p) \geq 0.$$

Yet, instead of incompressibility the model of thermal expansion looks more realistic.

Some models of temperature- and pressure-dependent viscosities

Roughly, viscosity increases with decreasing temperature. Some examples are applied in the literature. *Poiseuille* (1840) proposed a formula for the dependence of μ on temperature by analogy with the mass-density dependence,

$$\mu(\theta) = \frac{\mu_0}{1 + \mu_1\theta + \mu_2\theta^2}.$$

This equation is also applied by replacing the absolute temperature θ with the Celsius temperature and redefining the values of the parameters μ_0, μ_1, μ_2.

Based on the Arrhenius law for the chemical reaction rates,

$$k = k_0 \exp(-E/R\theta),$$

where R is the gas constant, *De Guzman - Andrade* law [12] has been considered in the form

$$\mu = \mu_0 \exp(a/R\theta),$$

a being a positive parameter.

Viscosity increases with increasing pressure because the amount of free volume in the internal structure decreases due to compression. However, compared to the temperature influence, most viscous liquids are influenced very little by the applied pressure. For most liquids, a considerable change in pressure from 0.1 to 30 MPa causes about the same change in viscosity as a temperature change of about 1 K. Even for the enormous pressure difference of 0.1 to 200 MPa the viscosity increase for most low-molecular liquids amounts to a factor 3 to 7 only. The reason is that viscous liquids (other than gases) are almost non-compressible at low or medium pressures.

A simple model of the dependence of (shear) viscosity on temperature, pressure and mass density is given by *Andrade's* formula

$$\mu(\theta, p, \rho) = A\rho^{1/2} \exp[s(p + r^2\rho)/\theta]$$

where A, r, and s are constants. Letting $\mu_\infty(\rho) = A\rho^{1/2}$ and $a(p, \rho) = s(p + r^2\rho)$ this formula looks like a generalization of the De Guzman-Andrade law. If the fluid is regarded as incompressible then ρ is merely a parameter. In compressible fluids, where a constitutive relation gives p as a function of ρ and θ, once, e.g. p is replaced with the pertinent function $p(\rho, \theta)$ then μ becomes a function of ρ and θ.

5.6.3 Vorticity and Enstrophy Transport Equation

A classical vorticity transport equation follows for a compressible fluid within the approximation that the temperature is constant throughout the fluid and the viscosities are constant. In view of the pressure relation $p = \rho^2\partial_\rho\psi$ and the assumption $\nabla\theta = \mathbf{0}$ we obtain

$$\nabla(\psi + \frac{p}{\rho}) = \partial_\rho\psi\,\nabla\rho - \frac{p}{\rho^2}\nabla\rho + \frac{1}{\rho}\nabla p = \frac{1}{\rho}\nabla p.$$

Hence, letting $\mathbf{b}$ be conservative, $\mathbf{b} = \nabla\varphi$, we can write the Navier–Stokes relation in the form

$$\dot{\mathbf{v}} = -\nabla(\psi + \frac{p}{\rho} - \varphi - \frac{\lambda + \mu}{\rho}\nabla\cdot\mathbf{v}) + \frac{\mu}{\rho}\Delta\mathbf{v}.$$

Consequently, $\nabla\times\dot{\mathbf{v}}$ is given by

$$\nabla\times\dot{\mathbf{v}} = \mu\nabla\times\frac{\Delta\mathbf{v}}{\rho} = \mu(\frac{\Delta\nabla\times\mathbf{v}}{\rho} - \frac{1}{\rho^2}\nabla\rho\times\Delta\mathbf{v}).$$

Observe

$$\nabla\times\dot{\mathbf{v}} = \partial_t\nabla\times\mathbf{v} + \nabla\times[(\mathbf{v}\cdot\nabla)\mathbf{v}].$$

By means of the vorticity vector $\boldsymbol{\varpi} = \nabla\times\mathbf{v}$ and the identities

$$(\mathbf{v}\cdot\nabla)\mathbf{v} = 2\mathbf{W}\mathbf{v} + \nabla\tfrac{1}{2}\mathbf{v}^2, \quad 2\mathbf{W}\mathbf{v} = \boldsymbol{\varpi}\times\mathbf{v},$$

it follows

$$\nabla\times[\boldsymbol{\varpi}\times\mathbf{v}] = \nabla\times[(\mathbf{v}\cdot\nabla)\mathbf{v}].$$

By direct calculation we find

$$\nabla\times[\boldsymbol{\varpi}\times\mathbf{v}] = (\nabla\cdot\mathbf{v})\boldsymbol{\varpi} + (\mathbf{v}\cdot\nabla)\boldsymbol{\varpi} - \mathbf{L}\boldsymbol{\varpi}.$$

Thus

$$\nabla\times[(\mathbf{v}\cdot\nabla)\mathbf{v}] = (\nabla\cdot\mathbf{v})\boldsymbol{\varpi} + (\mathbf{v}\cdot\nabla)\boldsymbol{\varpi} - \mathbf{L}\,\boldsymbol{\varpi}$$

and then

$$\nabla \times \dot{\mathbf{v}} = \dot{\varpi} - \mathbf{L}\,\varpi + (\nabla \cdot \mathbf{v})\varpi = \overset{\square}{\varpi}, \tag{5.34}$$

$\overset{\square}{\varpi}$ being the Truesdell rate of ϖ. Hence in isothermal conditions the vorticity ϖ satisfies the transport equation

$$\overset{\square}{\varpi} = \mu\Big(\frac{\Delta\varpi}{\rho} - \frac{1}{\rho^2}\nabla\rho \times \Delta\mathbf{v}\Big).$$

Restrict attention to incompressible fluids. Let $\nu = \mu/\rho$. Since

$$\dot{\mathbf{v}} = -\nabla p + \mathbf{g} + \nu\Delta\mathbf{v}, \qquad \nabla \times \dot{\mathbf{v}} = \nu\Delta\varpi$$

then the vorticity transport equation (5.34) reduces to

$$\dot{\varpi} = \mathbf{L}\varpi + \nu\Delta\varpi.$$

Enstrophy is viewed as the kinetic energy associated with vorticity. Let e be the enstrophy per unit mass,

$$e = \tfrac{1}{2}|\varpi|^2.$$

Hence

$$\dot{e} = \varpi \cdot \mathbf{L}\varpi + \nu\varpi \cdot \Delta\varpi.$$

Consider the identities

$$\mathbf{u} \cdot \Delta\mathbf{u} = \partial_{x_k}(\mathbf{u} \cdot \partial_{x_k}\mathbf{u}) - (\partial_{x_k}\mathbf{u}) \cdot (\partial_{x_k}\mathbf{u}), \qquad \Delta\tfrac{1}{2}\mathbf{u}^2 = \partial_{x_k}(\mathbf{u} \cdot \partial_{x_k}\mathbf{u}),$$

and apply them to $\mathbf{u} = \varpi$. Moreover observe that $\varpi \cdot \mathbf{L}\varpi = \varpi \cdot \mathbf{D}\varpi$. Hence we find

$$\dot{e} = \varpi \cdot \mathbf{D}\varpi - \nu|\nabla\varpi|^2 + \nu\Delta e,$$

usually referred to as the enstrophy-transport equation.

5.6.4 *Viscosity and Energy Decay*

Viscosity affects the rate of change of the kinetic energy within a convecting spatial region $\mathcal{P}_t$ as well as within a fixed spatial region Ω. Inner product of the equation of motion with $\mathbf{v}$ gives

$$\rho\mathbf{v} \cdot \dot{\mathbf{v}} = \mathbf{v} \cdot \nabla \cdot \mathbf{T} + \rho\mathbf{b} \cdot \mathbf{v}.$$

Since

$$\rho\mathbf{v} \cdot \dot{\mathbf{v}} = \rho(\tfrac{1}{2}\mathbf{v}^2)^{\cdot} = \partial_t(\tfrac{1}{2}\rho\mathbf{v}^2) + \nabla \cdot (\tfrac{1}{2}\rho\mathbf{v}^2\,\mathbf{v})$$

we can write

$$\partial_t(\tfrac{1}{2}\rho\mathbf{v}^2) = \nabla\cdot(\mathbf{T}\mathbf{v} - \tfrac{1}{2}\rho\mathbf{v}^2\,\mathbf{v}) + \rho\,\mathbf{b}\cdot\mathbf{v} - \mathbf{T}\cdot\mathbf{D}.$$

Integration over a fixed region Ω and use of the divergence theorem give

$$\partial_t\int_\Omega \tfrac{1}{2}\rho\mathbf{v}^2\,dv = \int_{\partial\Omega}(\mathbf{T}\mathbf{v} - \tfrac{1}{2}\rho\mathbf{v}^2\,\mathbf{v})\cdot\mathbf{n}\,da + \int_\Omega(\rho\,\mathbf{b}\cdot\mathbf{v} - \mathbf{T}\cdot\mathbf{D})dv.$$

If $\mathbf{v} = \mathbf{0}$ at the boundary $\partial\Omega$ then the rate of change is given by two terms, the power of the body force and that of the stress. Now,

$$-\mathbf{T}\cdot\mathbf{D} = p\nabla\cdot\mathbf{v} - 2\mu\mathbf{D}_0\cdot\mathbf{D}_0 - \zeta(\nabla\cdot\mathbf{v})^2.$$

The power $p\nabla\cdot\mathbf{v}$ is positive during expansion ($\nabla\cdot\mathbf{v} > 0$) and negative during compression ($\nabla\cdot\mathbf{v} < 0$). The power of the dissipative stress, instead, is negative in any motion as a consequence of the thermodynamic requirements $\mu, \zeta \geq 0$. Hence the global kinetic energy decreases as a consequence of viscosity whereas, by effect of pressure, it increases in any expansion and decreases in any compression.

If, instead, we look at a convecting region $\mathcal{P}_t$, by

$$\rho(\tfrac{1}{2}\mathbf{v}^2)^{\cdot} = \mathbf{v}\cdot\nabla\cdot\mathbf{T} + \rho\,\mathbf{b}\cdot\mathbf{v}$$

integration over $\mathcal{P}_t$ allows us to write

$$\frac{d}{dt}\int_{\mathcal{P}_t}\tfrac{1}{2}\rho\mathbf{v}^2\,dv = \int_{\partial\mathcal{P}_t}(\mathbf{T}\mathbf{v})\cdot\mathbf{n}\,da + \int_{\mathcal{P}_t}(\rho\,\mathbf{b}\cdot\mathbf{v} - \mathbf{T}\cdot\mathbf{D})dv.$$

Again we find that viscosity results in a decay of the kinetic energy.

5.6.5 Incompressible Newtonian Fluids

According to Sect. 5.5.3, the stress constitutive function for an *incompressible Newtonian fluid* is given by

$$\mathbf{T} = -\bar{p}\,\mathbf{1} + 2\mu(\theta)\mathbf{D}_0,$$

and then nominal and total pressures are the same. The kinetic energy balance holds for any convecting region $\mathcal{P}_t$ as with compressible fluids except for the vanishing of the power $\zeta(\nabla\cdot\mathbf{v})^2$ and we have

$$\frac{d}{dt}\int_{\mathcal{P}_t}\tfrac{1}{2}\rho\mathbf{v}^2\,dv = \int_{\partial\mathcal{P}_t}(\mathbf{T}\mathbf{v})\cdot\mathbf{n}\,da + \int_{\mathcal{P}_t}(\rho\,\mathbf{b}\cdot\mathbf{v} - \mathbf{T}\cdot\mathbf{D})dv, \qquad \mathbf{T}\cdot\mathbf{D} = 2\mu\mathbf{D}_0\cdot\mathbf{D}_0.$$

Let μ be constant. Since

$$2\nabla \cdot \mathbf{D}_0 = \Delta \mathbf{v},$$

the *Navier–Stokes equation* of motion takes the form

$$\rho \dot{\mathbf{v}} = -\nabla \bar{p} + \mu \, \Delta \mathbf{v} + \rho \, \mathbf{b}.$$

The velocity $\mathbf{v}$ is also subject to

$$\nabla \cdot \mathbf{v} = 0.$$

These are the incompressible Navier–Stokes equations in the unknowns $\mathbf{v}$, $\bar{p}$.

Let the body force be conservative, $\mathbf{b} = \nabla \Pi$, so that

$$\frac{1}{\rho} \nabla \bar{p} - \mathbf{b} = \nabla \Big(\frac{\bar{p}}{\rho} - \Pi \Big).$$

The equation of motion can be written as

$$\dot{\mathbf{v}} = -\nabla \big(\frac{\bar{p}}{\rho} - \Pi \big) + \nu \Delta \mathbf{v},$$

where $\nu = \mu/\rho$. The quantity ν is called *kinematic viscosity* and it is natural to regard ν as a diffusivity for the velocity $\mathbf{v}$ in that the equation for $\mathbf{v}$ has formally the structure of the diffusion equation for $\mathbf{v}$. This feature holds also for the vorticity $\varpi = \nabla \times \mathbf{v}$. In view of (5.34) and the incompressibility condition $\nabla \cdot \mathbf{v} = 0$ we obtain

$$\dot{\varpi} - \mathbf{L} \varpi = \nu \Delta \varpi$$

or

$$\dot{\varpi} - \mathbf{D}_0 \, \varpi = \nu \Delta \varpi$$

since $\mathbf{L} = \mathbf{W} + \mathbf{D}_0$ and $\mathbf{W} \varpi = \mathbf{0}$.

Under no-slip boundary conditions, by virtue of (5.30) the normal stress of an incompressible Newtonian fluid at a fixed rigid wall equals the opposite of the mean pressure,

$$t_n = -\bar{p}.$$

Then, by virtue of (5.27), for a given friction coefficient ξ_0 and a given pressure $\bar{p}$ the no-slip condition $\mathbf{v} = \mathbf{0}$ is valid provided that the velocity field $\mathbf{v}$ on the boundary satisfies

$$|\nabla \times \mathbf{v}| \le \frac{\xi_0}{\mu} |\bar{p}|,$$

that is to say, provided that the boundary vorticity ω is low if compared to the pressure.

5.7 Generalized Newtonian Fluids

A Newtonian fluid is a fluid in which the (Cauchy) viscous stress, arising from its flow, is linearly proportional to the stretching $\mathbf{D}$. Owing to linearity, Newtonian fluids are the simplest mathematical models of fluids that account for viscosity. While water and air can be regarded as Newtonian fluids, more involved, non-Newtonian models are in order for other fluids.

Generalized Newtonian fluid models are widely adopted in the literature and account for a variety of behaviours. In the shear-thickening fluid or dilatant fluid, the viscosity appears to increase when the shear-rate increases. Corn starch dissolved in water is a common example[15]: when stirred slowly it looks milky, when stirred vigorously it feels like a very viscous liquid. In the shear-thinning fluid the viscosity appears to decrease when the shear-rate increases. A familiar example in this sense is wall paint. Another example of a shear-thinning fluid is blood. This application is highly favoured within the body, as it allows the viscosity of blood to decrease with increased shear strain rate.

A constitutive relation between stress and stretching for generalized Newtonian fluids is obtained from (5.24) by letting[16]

$$\gamma = 0, \qquad \alpha = \tilde{\alpha}(\rho, \theta, \mathcal{I}_1, \mathcal{I}_2), \qquad \beta = \tilde{\beta}(\rho, \theta, \mathcal{I}_1, \mathcal{I}_2).$$

Neglecting the dependence of p, $\tilde{\alpha}$ and $\tilde{\beta}$ on ρ and θ, such fluids are characterized by the relation

$$\mathbf{T} = [-p + \tilde{\alpha}(\mathcal{I}_1, \mathcal{I}_2)]\mathbf{1} + \tilde{\beta}(\mathcal{I}_1, \mathcal{I}_2)\mathbf{D}, \tag{5.35}$$

and are then referred to as *generalized Newtonian fluids*.

The relationship between stress and stretching for *incompressible generalized Newtonian fluids* takes the form

$$\mathbf{T} = -\bar{p}\,\mathbf{1} + \tilde{\beta}_0(\bar{p}, \theta, \mathcal{I}_2)\mathbf{D}_0, \tag{5.36}$$

where the viscosity function $\tilde{\beta}_0$ is expected to be non-negative valued and possibly depends on the mean pressure (the independent variable that replaces ρ). By properly choosing the expression of this constitutive function special rheological models for incompressible fluids have been proposed (see, e.g. [18, 318]).

Power-law model. Relative to equation (5.36), β_0 is taken in the form

$$\beta_0(\theta, \mathcal{I}_2) = 2\mu(\theta)\, \mathcal{I}_2^{\frac{\sigma-1}{2}}, \tag{5.37}$$

where $\mu \geq 0$, as required by the entropy inequality. The real parameter $\sigma > 0$ is the power-law index of the model. The constitutive relation of a Newtonian fluid is

[15] This non-Newtonian fluid is called oobleck.

[16] This form may be reasonable for real Stokesian fluids; see e.g. [18].

recovered when $\sigma = 1$. When $\sigma > 1$ the model describes a *shear-thickening* (also termed *dilatant*) fluid in which viscosity increases with the rate of shear strain, whereas the constitutive equation is *shear-thinning* when $0 < \sigma < 1$.

It is a drawback of this model that it does not match experimental results merely because the viscosity function β_0 is unbounded and vanishes as $\mathbf{D}_0 \to \mathbf{0}$. To avoid this drawback some models have been established subject to the limit conditions

$$\lim_{\mathcal{I}_2 \to 0} \beta_0 = 2\mu_0 , \qquad \lim_{\mathcal{I}_2 \to \infty} \beta_0 = 2\mu_\infty .$$

Equations

$$\frac{\beta_0}{2} = \mu_\infty + (\mu_0 - \mu_\infty)\frac{1}{1 + [\kappa^2 \mathcal{I}_2]^{(1-\sigma)/2}},$$

$$\frac{\beta_0}{2} = \mu_\infty + (\mu_0 - \mu_\infty)\,[1 + \kappa^a \mathcal{I}_2^{a/2}]^{(\sigma-1)/a}, \qquad a > 0,$$

$$\frac{\beta_0}{2} = \mu_\infty + (\mu_0 - \mu_\infty)\frac{\sinh^{-1}(\kappa\sqrt{\mathcal{I}_2})}{\kappa\sqrt{\mathcal{I}_2}},$$

are referred to as **Cross model, Carreau–Yasuda model, Powell-Eyring model.** In all of these models μ_0, μ_∞ and κ are constitutive functions of θ. In addition, κ is a time and hence $\kappa^2 \mathcal{I}_2$ is a pure number. Indeed, κ is viewed as a relaxation time; it characterizes a threshold relative to $\mathbf{D}_0$. At low values of the stretching, $\mathcal{I}_2 \ll 1/\kappa^2$, Cross fluids behave as Newtonian fluids with $\beta_0 = 2\mu_0$, at large values of the stretching, $\mathcal{I}_2 \gg 1/\kappa^2$, the behaviour is that of a power-law fluid.

It is worth considering a successful class of nonlinear models that reproduce different viscous responses of the fluid through some properly defined *shear-rate critical value*. This quantity is somewhat related to the notion of yield stress that is peculiar to viscoplastic fluids (see Sect. 5.8.1).

Casson model. The viscosity function β_0 is taken in the form

$$\beta_0 = \begin{cases} 2\mu_\infty + \tau_y/\sqrt{\mathcal{I}_2} & \text{if } \mathcal{I}_2 > \mathcal{I}_2^c, \\ 2\mu_0 & \text{otherwise}, \end{cases}$$

where τ_y is the yield stress of the fluid and $\mathcal{I}_2^c$ is the critical value of $\mathcal{I}_2$. They are related by

$$\mathcal{I}_2^c = \frac{\tau_y^2}{(\mu_0 - \mu_\infty)^2},$$

where μ_0 is the (Newtonian) viscosity at low stretching; the parameter μ_∞ is called the *plastic dynamic viscosity*. If $\tau_y = 0$ and $\mu_0 = \mu_\infty$ the model reduces to the Newtonian fluid.

A Casson fluid exhibits a yield stress $\tau_y > 0$. If a shear stress, smaller than the yield stress, is applied then the fluid behaves like a Newtonian fluid, whereas if a shear stress is greater than the yield stress the fluid moves like a power-law fluid with index $n = 0$. These properties make the Casson fluid a realistic model for the blood circulation in narrow arteries and chocolate flow.[17]

Herschel–Bulkley model. It is a model of viscosity where both the yield stress and the power-law dependence occur. The function β_0 is given by

$$\beta_0 = \begin{cases} 2\mu_\infty \mathcal{I}_2^{\frac{\sigma-1}{2}} + \tau_y/\sqrt{\mathcal{I}_2} & \text{if } \mathcal{I}_2 > \mathcal{I}_2^c, \\ 2\mu_\infty {\mathcal{I}_2^c}^{\frac{\sigma-1}{2}} + \tau_y/\sqrt{\mathcal{I}_2^c} & \text{if } \mathcal{I}_2 < \mathcal{I}_2^c, \end{cases}$$

where τ_y is the yield stress of the fluid and $\mathcal{I}_2^c$ is the shear-rate critical value, as in the Casson model. For large values of the yield stress the fluid will significantly flow in response to a large applied force. If $\sigma < 1$ the fluid is shear-thinning, whereas if $\sigma > 1$ the fluid is shear-thickening. If $\sigma = 1$ the model reduces to the Casson model.

5.8 Viscoplastic and Viscoelastic Fluids

The vast majority of non-Newtonian fluid models are concerned with simple models (like power-law, Carreau–Yasuda and Casson) which are special versions of the generalized Newtonian model. Nevertheless, a lot of composite and biological materials (e.g. polymeric solutions, melts, muds, condensed milk, glues, printing ink, emulsions, soaps, chocolate, paints, shampoos, toothpaste and tomato paste) show properties of fluid flows that don't match with Stokes' assumptions. In particular, Stokesian fluid models are unable to predict either memory or other elastic/plastic effects exhibited by these substances. The governing equations expected for non-Stokesian fluids are then much more complicated if compared to those of Newtonian fluids. Due to complexity, no single constitutive equation exhibiting all properties of such fluids is available. So, several models of non-Stokesian fluids are exhibited in the literature through the differential, integral, and rate-type categories. Roughly, we may divide them into two main classes: *viscoplastic* and *viscoelastic fluids*.

5.8.1 Viscoplastic Fluids

The viscoplastic behaviour involves a threshold value of the stress, termed *yield stress*. When the applied stress remains below this value, the fluid doesn't flow, but it is allowed to move as a rigid solid. When the applied stress increases beyond a finite threshold, termed yield stress, then the fluid exhibits a viscous (possibly nonlinear) stress–strain relationship so that it begins to flow.

[17] Remarkable applications of this model can be found in [181].

The viscoplastic behaviour is usually not observed in pure materials, but may occur in concentrated suspensions of particles. First, Bingham [43] observed this type of behaviour in paints, but it also applies to drilling mud, toothpaste, margarine, mustard and clay suspensions.

Bingham model. It is the most celebrated viscoplastic incompressible fluid model. A Bingham fluid exhibits a yield stress τ_y, as the Casson model, but its dynamical features look quite different. Indeed, if a shear stress less than the yield stress is applied to the fluid, it behaves like a (rigid) solid, whereas if a shear stress greater than the yield stress is applied, it starts to move like a Newtonian fluid. The constitutive assumption is then expressed as a relation where the strain rate is given as a function of the stress. Instead of the usual relations that hold for generalized (incompressible) Newtonian fluids,

$$\operatorname{tr}\mathbf{T} = -3\bar{p}, \qquad \mathbf{T}_0 = \tilde{\beta}_0(\theta, \mathcal{I}_2(\mathbf{D}_0))\mathbf{D}_0$$

in a Bingham fluid the latter is replaced by

$$\mathbf{D}_0 = \tilde{\delta}_0(\theta, \mathcal{I}_2(\mathbf{T}_0))\mathbf{T}_0 \tag{5.38}$$

where $\tilde{\delta}_0$ is expected to be a non-negative valued function. In particular, letting $\tau = \sqrt{\mathcal{I}_2(\mathbf{T}_0)}$ and τ_y its yielding value, we may choose

$$\delta_0 = \begin{cases} \dfrac{1}{2\mu_\infty}(1 - \tau_y/\tau) & \text{if } \tau > \tau_y, \\ 0 & \text{otherwise} \end{cases}$$

where the viscosity μ_∞ is a constitutive functions of θ and represents the limiting viscosity of the fluid when $\tau \gg \tau_y$. As is apparent, the constitutive relation cannot be reversed for all values of τ. Nevertheless, if $\tau > \tau_y$ from (5.38) we infer

$$\tau = \tau_y + 2\mu_\infty\sqrt{\mathcal{I}_2(\mathbf{D}_0)}$$

and then we may write

$$\beta_0(\theta, \mathcal{I}_2(\mathbf{D}_0)) = 2\mu_\infty + \tau_y/\sqrt{\mathcal{I}_2(\mathbf{D}_0)}$$

which looks like the Casson model for high strain rate values. The main difference occurs when $\tau < \tau_y$: according to Bingham the fluid moves as a rigid solid, whereas according to Casson the fluid merely exhibits a very high viscosity.

Geometry has a relevant role in the Bingham fluid dynamics. Indeed, in simple shear, the shear stress is constant throughout the flow field, but in more complex flows, such as flow in a straight pipe, the shear stress is not constant and may be regions where the fluid is flowing (since the yield criterion is reached) while in other regions the shear stress lies below the yielding limit and the fluid moves as a rigid body ($\mathbf{D} = \mathbf{0}$).

Generalized Bingham model. Generalizations of the Bingham model are obtained by choosing $\tilde{\delta}_0$ in (5.38) in the form

$$\delta_0 = \begin{cases} \dfrac{1}{\kappa_\infty \tau}\,(1-\tau_y/\tau)^{1/\sigma} & \text{if } \tau > \tau_y, \\ 0 & \text{otherwise} \end{cases}$$

where κ_∞ is a constitutive functions of θ. Although we cannot reverse this relation, we may proceed as before: if $\tau > \tau_y$ from (5.38) we infer

$$\tau = \tau_y + 2\mu_\infty [\mathcal{I}_2(\mathbf{D}_0)]^{\frac{\sigma}{2}}$$

where $\mu_\infty = \kappa_\infty^\sigma/2$, and then we may write

$$\beta_0(\theta, \mathcal{I}_2(\mathbf{D}_0)) = 2\mu_\infty [\mathcal{I}_2(\mathbf{D}_0)]^{\frac{\sigma-1}{2}} + \tau_y/\sqrt{\mathcal{I}_2(\mathbf{D}_0)}$$

which looks like the Herschel–Bulkley model for high strain rate values. For $\sigma < 1$ the fluid is shear-thinning, whereas for $\sigma > 1$ the fluid is shear-thickening. As $\sigma = 1$ we get the Bingham model.

Pseudo-plastic model. In a lot of real biological materials as ketchup and mayonnaise the viscoplastic behaviour is quite similar, but the existence of a true yield stress has no experimental evidence.[18] For this reason and for ease in computing it is preferable to adopt a modified Bingham model (also termed *pseudo-plastic*) with index σ,

$$\operatorname{tr}\mathbf{T} = -3\bar{p}, \qquad \mathbf{T}_0 = \Big(2\mu_0 + \frac{\tau_y\,[1-\exp(-\sigma\sqrt{\mathcal{I}_2(\mathbf{D}_0)})]}{\sqrt{\mathcal{I}_2(\mathbf{D}_0)}}\Big)\mathbf{D}_0,$$

where, the yield-stress parameter τ_y has a conventional meaning, as in Casson and Herschel–Bulkley models. Indeed, the modified Bingham model turns out to be a special case of the generalized Newtonian model for incompressible fluids. Taking into account that this modified stress-stretching relation leads to

$$\tau = 2\mu_0\sqrt{\mathcal{I}_2(\mathbf{D}_0)} + \tau_y\left[1-\exp(-\sigma\sqrt{\mathcal{I}_2(\mathbf{D}_0)})\right],$$

we may conclude that it recovers the Newtonian model when $\sigma = 0$ and approaches the Bingham model as $\sigma \to \infty$.

Viscoelastic fluids: differential and rate-type models

Viscoelastic fluids are a common form of non-Stokesian fluid. They can exhibit a response that resembles that of an elastic solid under some circumstances, or the

[18] See, e.g. [30].

response of a viscous liquid under other circumstances. Typically, fluids that exhibit this behaviour are macromolecular in nature, such as polymeric fluids (melts and solutions) used to make plastic articles, food systems such as dough used to make bread and pasta, and biological fluids such as synovial fluids found in joints. In the case of polymeric fluids, the nature of polymeric molecules has a direct influence on the processing behaviour and in some cases on the performance of the polymer in a given application.

In order to describe a so complex behaviour, the stress–strain constitutive relation of a viscoelastic fluid involves memory effects: the actual value of the stress accounts for not only the actual value of the strain rate but also its past values. In the literature three different approaches have been devised in this concern. First, the stress at some instant may depend not only on the strain rate but also on its time derivatives at the same instant. This differential approach is suitable in modelling fluids with *short memory*, for instance fluids of second (or higher) grade, and must involve objective time derivatives. Alternately, the hereditary character of the fluid may be implicitly established through a rate-type relation (either linear or nonlinear) between the actual values of the stress, the strain rate and their time derivatives. Relations of this type have a somewhat complex structure whose compatibility with thermodynamics and the principle of frame-indifference is not obvious. For this reason, rate-type models will be scrutinized in a separate section. Finally, assuming that the actual value of the stress depends on the whole past deformation history. This is the so-called *integral* (or *long-memory*) *approach* and traces back to Maxwell, Boltzmann and Volterra. Models of this type are mainly linear and may be applied to both solid and fluids. Their constitutive structure falls outside this chapter, so their analysis will be carried out within the second part of this book concerning non-simple materials.

5.9 Models of Turbulence

Turbulence denotes a fluid motion characterized by chaotic changes in pressure and flow velocity. It is viewed in contrast to a laminar flow, which occurs when a fluid flows in parallel layers, with no disruption between those layers. With this view it is standard to describe fluids where the velocity and the pressure are decomposed into ensemble means and fluctuating parts. Also, there is a view that turbulence is caused by excessive kinetic energy in parts of a fluid flow, which overcomes the damping effect of the fluid viscosity; that is why the so-called k-equation is applied in the modelling of turbulence, k being a turbulent kinetic energy.

5.9.1 Incompressible Fluids

For simplicity consider an incompressible viscous fluid; the velocity field $\mathbf{v}$ is subject to

$$\nabla \cdot \mathbf{v} = 0, \qquad \rho \partial_t \mathbf{v} + \rho (\mathbf{v} \cdot \nabla)\mathbf{v} = -\nabla p + 2\mu \nabla \cdot \mathbf{D}.$$

The fields $\mathbf{v}$, p are (assumed to be) random functions of space and time. Let

$$\mathbf{v} = \mathbf{V} + \mathbf{u}, \qquad p = P + \pi,$$

where $\mathbf{u}$ and π are the fluctuating parts. It is assumed that

$$\overline{\mathbf{u}} = \mathbf{0}, \quad \overline{\partial_t \mathbf{u}} = \mathbf{0}, \quad \overline{\nabla \mathbf{u}} = \mathbf{0}, \qquad \overline{\pi} = 0, \quad \overline{\partial_t \pi} = 0, \quad \overline{\nabla \pi} = \mathbf{0}.$$

the superposed bar denoting the time average (over a suitably large period of time). Accordingly $\overline{\nabla \cdot \mathbf{u}} = 0$ and

$$0 = \nabla \cdot \mathbf{v} = \nabla \cdot (\mathbf{V} + \mathbf{u}), \quad 0 = \nabla \cdot \mathbf{V} + \overline{\nabla \cdot \mathbf{u}}, \quad \Longrightarrow \quad \nabla \cdot \mathbf{V} = 0, \; \nabla \cdot \mathbf{u} = 0.$$

Moreover,

$$(\mathbf{v} \cdot \nabla)\mathbf{v} = \nabla \cdot (\mathbf{v} \otimes \mathbf{v}), \quad \overline{(\mathbf{v} \cdot \nabla)\mathbf{v}} = \nabla \cdot \overline{\mathbf{v} \otimes \mathbf{v}} = (\mathbf{V} \cdot \nabla)\mathbf{V} + \nabla \cdot \overline{\mathbf{u} \otimes \mathbf{u}}.$$

Substituting $\mathbf{v}$ and p in the equation of motion and taking the time average we obtain

$$\rho \partial_t \mathbf{V} + \rho (\mathbf{V} \cdot \nabla)\mathbf{V} = -\nabla P + \nabla \cdot (2\mu \mathfrak{D} + \rho \boldsymbol{\mathcal{T}})$$

where $\mathfrak{D} = \mathrm{sym} \nabla \mathbf{V}$ is the mean stretching tensor and

$$\boldsymbol{\mathcal{T}} = -\overline{\mathbf{u} \otimes \mathbf{u}}.$$

Hence we can say that $\nabla \cdot \boldsymbol{\mathcal{T}}$ is the sole contribution of the turbulent fluctuations to the mean motion, with velocity $\mathbf{V}$, and this distinguishes a turbulent flow from a laminar counterpart. Accordingly $\rho \boldsymbol{\mathcal{T}}$ is an additional stress due to the turbulent fluctuations; $\boldsymbol{\mathcal{T}}$ is usually referred to as *Reynolds stress.*

The decomposition of the instantaneous properties into mean and fluctuating parts has resulted in the occurrence of a new unknown, the Reynolds stress, without no additional equations. The system of equations is not closed. To close it we need additional equations; there are various approaches, leading to different turbulence models, that result in the closure of the system. In so doing often recourse is made to the *turbulence kinetic energy* defined by

$$k = \tfrac{1}{2} \overline{\mathbf{u} \cdot \mathbf{u}} = -\tfrac{1}{2} \operatorname{tr} \boldsymbol{\mathcal{T}}.$$

Boussinesq model

Boussinesq closure identifies the Reynolds stress as

$$\boldsymbol{\tau} = 2\nu_T \mathfrak{D} - \tfrac{2}{3}k\mathbf{1},$$

where ν_T is the *kinetic eddy viscosity*. It follows that

$$\operatorname{tr} \mathfrak{D} = 0,$$

which is consistent in that $\operatorname{tr} \mathfrak{D} = \nabla \cdot \mathbf{V} = 0$. To make this assumption operative a balance equation for k is needed.

Spalart–Allmaras model

In this model k is not involved and the Reynolds stress is defined

$$\boldsymbol{\tau} = 2\nu_T \mathfrak{D}$$

though now ν_T is defined in terms of quantities subject to appropriate transport equations.

The k − ω Model

The kinetic eddy viscosity ν_T is defined as the ratio

$$\nu_T = \frac{k}{\omega},$$

with k, ω satisfying the evolution (or transport) equations

$$\partial_t k + \mathbf{V} \cdot \nabla k = \boldsymbol{\tau} \cdot \mathbf{L} - \beta^* k\omega + \nabla \cdot [(\nu + \tfrac{1}{2}\nu_T)\nabla k],$$

$$\partial_t \omega + \mathbf{V} \cdot \nabla \omega = \alpha \frac{\omega}{k} \boldsymbol{\tau} \cdot \mathbf{L} - \beta \omega^2 + \nabla \cdot [(\nu + \tfrac{1}{2}\nu_T)\nabla \omega],$$

with properly defined parameters $\alpha, \beta, \beta^*, \nu, \nu_T$. It is worth observing that both equations can be viewed as diffusion equations, say

$$\dot{k} = -\nabla \cdot \mathbf{h}, \qquad \mathbf{h} = -\gamma \nabla k,$$

supplemented by appropriate terms in $\boldsymbol{\tau}$.

5.9.2 *Compressible Fluids*

We let $p = -\operatorname{tr} \mathbf{T}/3$, $\mathbf{T} = \mathbf{T}_0 - p\mathbf{1}$ and write the balance equations in the Eulerian form

$$\partial_t \rho + \nabla \cdot (\rho \mathbf{v}) = 0,$$

$$\partial_t (\rho \mathbf{v}) + \nabla \cdot (\rho \mathbf{v} \otimes \mathbf{v}) = -\nabla p + \nabla \mathbf{T}_0,$$

$$\partial_t [\rho(\varepsilon + \tfrac{1}{2}\mathbf{v}^2)] + \nabla \cdot [\rho \mathbf{v}(h + \tfrac{1}{2}\mathbf{v}^2)] = \nabla \cdot (\mathbf{T}_0 \mathbf{v}) - \nabla \cdot \mathbf{q} + \rho r,$$

where $h = \varepsilon + p/\rho$ is the enthalpy density (see (2.45)).

The space-time dependence of ρ motivates a time averaging that traces back to Favre [167]. For any density, per unit mass, φ we define

$$\tilde{\varphi} := \frac{\overline{\rho\varphi}}{\overline{\rho}}, \qquad \varphi_1 := \varphi - \tilde{\varphi}.$$

It follows that $\overline{\varphi - \overline{\varphi}} = 0$ whereas $\overline{\varphi}_1 = \overline{\varphi - \tilde{\varphi}} \neq 0$. We then review the balance equations in terms of the means $\overline{\varphi}$, $\tilde{\varphi}$. In this connection observe

$$\rho\varphi = \overline{\rho\varphi} + \rho\varphi - \overline{\rho\varphi} = \overline{\rho}\tilde{\varphi} + \rho\varphi - \overline{\rho\varphi},$$

whence

$$\overline{\rho\varphi} = \overline{\rho}\tilde{\varphi}, \qquad \rho\varphi\psi = \rho(\tilde{\varphi} + \varphi_1)(\tilde{\vartheta} + \vartheta_1), \qquad \overline{\rho\varphi_1} = \overline{\rho\varphi} - \overline{\rho\tilde{\varphi}} = \overline{\rho}\tilde{\varphi} - \overline{\rho}\tilde{\varphi} = 0.$$

Consequently

$$\rho\varphi\vartheta = \rho(\tilde{\varphi} + \varphi_1)(\tilde{\vartheta} + \vartheta_1), \qquad \overline{\rho\varphi\vartheta} = \overline{\rho}\tilde{\varphi}\tilde{\vartheta} + \overline{\rho\varphi_1\vartheta_1}.$$

The structure of the balance equations suggests that we let

$$\rho = \overline{\rho} + \varrho, \qquad p = \overline{p} + \pi, \qquad \mathbf{q} = \overline{\mathbf{q}} + \zeta, \qquad \mathbf{T}_0 = \overline{\mathbf{T}}_0 + \boldsymbol{\mathcal{T}},$$

$$\mathbf{v} = \tilde{\mathbf{v}} + \mathbf{v}_1, \qquad \varepsilon = \tilde{\varepsilon} + \varepsilon_1, \qquad h = \tilde{h} + h_1, \qquad r = \tilde{r} + r_1,$$

often referred to as Reynolds decomposition and Favre decomposition, respectively. Substitution in the balance equations and evaluation of the average result in

$$\partial_t \overline{\rho} + \nabla \cdot (\overline{\rho}\tilde{\mathbf{v}}) = 0, \tag{5.39}$$

$$\partial_t (\overline{\rho}\tilde{\mathbf{v}}) + \nabla \cdot (\overline{\rho}\tilde{\mathbf{v}}\tilde{\mathbf{v}}) = -\nabla \overline{p} + \nabla \cdot \overline{\mathbf{T}}_0 - \nabla \cdot (\overline{\rho \mathbf{v}_1 \otimes \mathbf{v}_1}), \tag{5.40}$$

$$\begin{aligned} &\partial_t [\overline{\rho}(\tilde{\varepsilon} + \tfrac{1}{2}\tilde{\mathbf{v}} \cdot \tilde{\mathbf{v}}) + \tfrac{1}{2}\overline{\rho \mathbf{v}_1 \cdot \mathbf{v}_1}] + \nabla \cdot [\overline{\rho}\tilde{\mathbf{v}}(\tilde{h} + \tfrac{1}{2}\tilde{\mathbf{v}} \cdot \tilde{\mathbf{v}}) + \tfrac{1}{2}\overline{\rho \mathbf{v}_1 \cdot \mathbf{v}_1}\tilde{\mathbf{v}}] \\ &= \nabla \cdot [\overline{\mathbf{T}}_0 - \overline{\rho \mathbf{v}_1 \otimes \mathbf{v}_1}\mathbf{v}_1 - \overline{\mathbf{q}} - \overline{\rho h_1 \mathbf{v}_1} + \overline{\mathbf{T}_0 \mathbf{v}_1} - \tfrac{1}{2}\overline{\rho(\mathbf{v}_1 \cdot \mathbf{v}_1)\mathbf{v}_1}] + \overline{\rho}\tilde{r}. \end{aligned}$$

Two kinetic energies occur, namely the averaged kinetic energy (per unit mass) and the turbulent kinetic energy

$$\tfrac{1}{2}\bar{\rho}\,\tilde{\mathbf{v}}\cdot\tilde{\mathbf{v}}, \qquad \tfrac{1}{2}\overline{\rho\,\mathbf{v}_1\cdot\mathbf{v}_1}.$$

Also by analogy with the case of incompressible fluids, the kinetic energy of turbulent character is defined to be

$$k = \tfrac{1}{2}\widetilde{\mathbf{v}_1\cdot\mathbf{v}_1}$$

so that

$$\tfrac{1}{2}\overline{\rho\,\mathbf{v}_1\cdot\mathbf{v}_1} = \bar{\rho}k.$$

If attention is restricted to (5.39) and (5.40) then only the quantity $\overline{\rho\,\mathbf{v}_1\cdot\mathbf{v}_1}$ occurs of turbulent character. Still recourse is made to Reynolds stress $\boldsymbol{\tau}$ by letting

$$\bar{\rho}\boldsymbol{\tau} = -\overline{\rho\,\mathbf{v}_1\cdot\mathbf{v}_1} = 2\mu_T(\tilde{\mathfrak{D}} - \tfrac{1}{3}\nabla\cdot\tilde{\mathbf{v}}\mathbf{1}) - \tfrac{2}{3}\bar{\rho}k\,\mathbf{1},$$

where $\tilde{\mathfrak{D}} = \mathrm{sym}\nabla\tilde{\mathbf{v}}$. As a consequence, the effective stress in (5.40) is $\overline{\mathbf{T}}_0 + \bar{\rho}\boldsymbol{\tau}$ only and $\boldsymbol{\tau}$ involves k of turbulent character. Turbulent models are then completed by adding an equation for k.

Part III
Non-simple Materials

Chapter 6
Rate-Type Models

Spatial interaction and memory effects are among the possible features of non-simple materials. As to memory effects in fluids, the inadequacy of the classical Navier-Stokes theory to describe rheological complex fluids such as geological materials, liquid foams, and polymeric fluids has led to the development of several theories of non-Newtonian fluids; Chap. 7 describes memory effects via functionals on appropriate histories. To simplify the models, a natural idea is to reduce the knowledge from that of a whole history to that of a set of time derivatives. In this sense many materials models are given in the literature to describe the non-Newtonian behaviour. Here, among others, the elastic-plastic and the Kelvin-Voigt solids and the Bingham, the Maxwell-Wiechert, and the Jeffreys fluids are shown to derive through rheological models.

A material of grade n denotes a model where the constitutive functions depend on time derivatives of appropriate variables up to order n. Among such models of fluids, second grade fluids are most often investigated, which means that the pertinent variables are the velocity (gradient) and the derivative of the velocity (gradient), the Newtonian fluid being a fluid of grade 1. Especially in connection with fluids, the more generic expression *fluids of differential type* is used to denote fluids of higher-order grade. The restrictions placed by the objectivity principle and the second law of thermodynamics are investigated both in the Eulerian and in the Lagrangian description. In particular, the Oldroyd-B fluid and the White-Metzner fluid are examined in detail. Temperature-rate dependent thermoelastic materials are modelled and the effects on discontinuity waves are investigated.

Likewise, spatial interaction might be described by functionals on the fields of appropriate physical variables. Again for operative convenience, the models are often considered by functions of space derivatives up to an order n of appropriate variables (often the displacement).

A. Morro and C. Giorgi, *Mathematical Modelling of Continuum Physics*,
Modeling and Simulation in Science, Engineering and Technology,
https://doi.org/10.1007/978-3-031-20814-0_6

6.1 Rheological Models

The modelling of material behaviour is often established by combining elastic, viscous, and plastic properties. Now, springs, dashpots, and friction dampers are familiar mechanical devices which exhibit elastic, viscous, and plastic response, respectively. It is then natural that constitutive equations describing the material behaviour are often produced by relating stress and strain as are force and displacement in appropriate combinations of springs, dashpots, and friction sliders.

The combinations of these elements are usually denoted as rheological models. They are especially convenient in the description of substances that have a complex microstructure (polymers, biological materials, soft matter). The adjective rheological is motivated by the fact that rheology is the study of the flow of matter, primarily in a liquid state, but also as solids under conditions in which they respond in non-elastic way.

Incidentally, the term rheology was coined by E.C. Bingham. The term was inspired by the aphorism of Simplicius (often attributed to Heraclitus), *pánta rheî*, "everything flows", and was first used to describe the flow of liquids and the deformation of solids.

There are three basic rheological (one-dimensional) models: the linear elastic spring, the linear viscous damper, and the friction slider (or St. Venant element). It is customary to denote[1] the stress by σ and the strain by ε. The linear *elastic spring* is governed by Hooke's law; stress is directly proportional to strain so that, borrowing by the proportionality between force and displacement, we write

$$\sigma^{(s)} = k\,\varepsilon^{(s)}, \tag{6.1}$$

$k > 0$ being Young's modulus, the analogue of the spring elastic constant. For the *dashpot*,

$$\sigma^{(d)} = \eta\,\partial_t\varepsilon^{(d)}, \tag{6.2}$$

$\eta > 0$ being the viscosity coefficient. For the *friction slider*

$$\partial_t\varepsilon^{(f)} \begin{cases} = 0 & \text{if } |\sigma^{(f)}| < \sigma_y, \\ > 0 & \text{if } \sigma^{(f)} = \sigma_y, \\ < 0 & \text{if } \sigma^{(f)} = -\sigma_y, \end{cases} \tag{6.3}$$

σ_y being the sliding limit (or yield stress). This device is also called *rigid-perfectly plastic model*. The superscripts (s), (d) and (f) denote quantities (stress or strain) pertaining to springs, dashpots and friction elements, respectively (Figs. 6.1 and 6.2).

[1] To adhere to the literature, in rheological models σ is the stress and ε is the strain, and η is the viscosity coefficient, not to be confused with the entropy production, the internal energy, and the entropy densities.

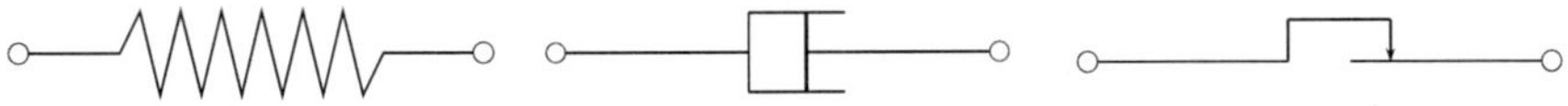

Fig. 6.1 A picture of spring, dashpot, and slider elements

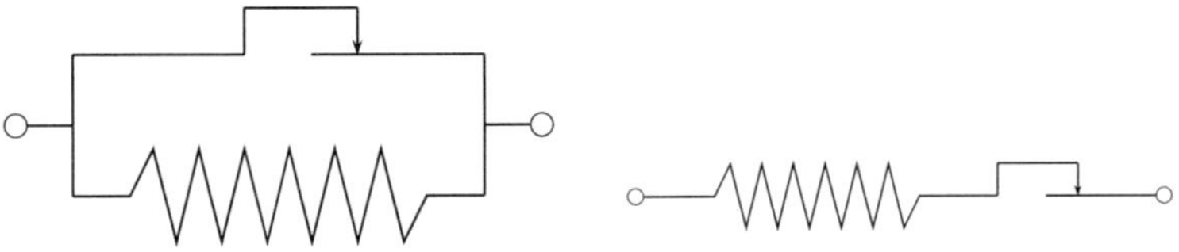

Fig. 6.2 A picture of the rigid-perfectly plastic and elastic-perfectly plastic units

6.1.1 *Rigid-Perfectly Plastic Model with Kinematic Hardening*

The arrangement of an elastic spring and a sliding frictional element in parallel leads to

$$\sigma = \sigma^{(s)} + \sigma^{(f)} = k\,\varepsilon^{(s)} + \sigma^{(f)}, \qquad \varepsilon = \varepsilon^{(s)} = \varepsilon^{(f)},$$

and then

$$\partial_t \varepsilon = \begin{cases} 0 & \text{if } |\sigma - k\varepsilon| < \sigma_y, \\ \frac{1}{k}\,\partial_t \sigma & \text{otherwise.} \end{cases} \tag{6.4}$$

6.1.2 *Elastic-Perfectly Plastic Solid*

The serial arrangement of an elastic spring and a sliding frictional element dates back to Reuss and gave rise to the additive decomposition of strain into elastic and plastic components. Namely, the stress is the same whereas the strain is shared by both components,

$$\varepsilon = \varepsilon^{(s)} + \varepsilon^{(f)} = \tfrac{1}{k}\,\sigma^{(s)} + \varepsilon^{(f)}, \qquad \sigma = \sigma^{(s)} = \sigma^{(f)},$$

and then

$$\partial_t \varepsilon \begin{cases} = \frac{1}{k}\,\partial_t \sigma & \text{if } |\sigma| < \sigma_y, \\ > 0 & \text{if } \sigma = \sigma_y, \\ < 0 & \text{if } \sigma = -\sigma_y. \end{cases} \tag{6.5}$$

By the standard entropy inequality, in isothermal conditions, we have the free-energy imbalance [216] $\rho_R \dot{\psi} \le \mathbf{S}_R \cdot \dot{\mathbf{E}}$; here we may state the imbalance in the form

$$\partial_t \Psi \le \sigma \partial_t \varepsilon, \tag{6.6}$$

where Ψ is the (free) energy. Taking Ψ as the energy of the linear spring,

$$\Psi = \tfrac{1}{2} k \left[\varepsilon^{(s)}\right]^2 = \begin{cases} \frac{1}{k}\sigma^2 & \text{if } |\sigma| < \sigma_y, \\ \frac{1}{k}\sigma_y^2 & \text{otherwise,} \end{cases}$$

it follows from (6.5) that the imbalance inequality holds, namely

$$\sigma \partial_t \varepsilon = \begin{cases} \frac{1}{k}\sigma \partial_t \sigma = \partial_t \Psi & \text{if } |\sigma| < \sigma_y, \\ \sigma_y \partial_t \varepsilon > 0 = \partial_t \Psi & \text{if } \sigma = \sigma_y, \\ -\sigma_y \partial_t \varepsilon > 0 = \partial_t \Psi & \text{if } \sigma = -\sigma_y. \end{cases}$$

6.1.3 Elastic-Plastic Model with Kinematical Hardening

The model consists of a spring and an elastic-perfectly plastic (EP) unit connected in parallel. Denote by $k^{(EP)}$, $\sigma^{(EP)}$, $\varepsilon^{(EP)}$ the quantities pertaining to the EP unit and by $k^{(s)}$, $\sigma^{(s)}$, $\varepsilon^{(s)}$ those pertaining to the spring in parallel. Since

$$\sigma = \sigma^{(EP)} + \sigma^{(s)} = \sigma^{(EP)} + k^{(s)}\varepsilon^{(s)}, \qquad \varepsilon = \varepsilon^{(EP)} = \varepsilon^{(s)},$$

we then obtain

$$\partial_t \sigma = \begin{cases} [k^{(s)} + k^{(EP)}]\, \partial_t \varepsilon & \text{if } |\sigma - k^{(s)}\varepsilon| < \sigma_y, \\ k^{(s)}\, \partial_t \varepsilon & \text{if } |\sigma - k^{(s)}\varepsilon| = \sigma_y. \end{cases} \tag{6.7}$$

It looks natural to take the energy for this model as the sum of the energies due to the two linear springs in the unit,

$$\Psi_{EP} = \frac{1}{2} k^{(s)} \varepsilon^2 + \frac{1}{2k^{(EP)}} \left(\sigma - k^{(s)}\varepsilon\right)^2.$$

In view of (6.7)

$$\partial_t \Psi_{EP} = k^{(s)} \varepsilon\, \partial_t \varepsilon + \frac{1}{k^{(EP)}} \left(\sigma - k^{(s)}\varepsilon\right) \begin{cases} k^{(EP)}\, \partial_t \varepsilon & \text{if } |\sigma - k^{(s)}\varepsilon| < \sigma_y, \\ 0 & \text{if } |\sigma - k^{(s)}\varepsilon| = \sigma_y. \end{cases}$$

and then the imbalance inequality (6.6) holds (Fig. 6.3).

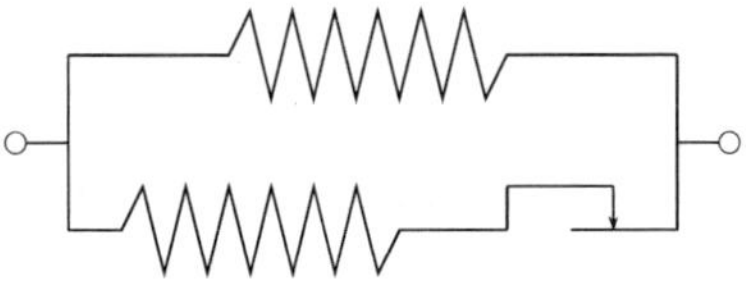

Fig. 6.3 A picture of the elastic-plastic unit with kinematical hardening

6.1.4 *Maxwell-Wiechert Fluid*

In the Maxwell, or Maxwell-Wiechert, (MW) model a spring and a dashpot are placed in series so that the strain ε is shared by both components whereas the stress σ is the same,

$$\varepsilon = \varepsilon^{(s)} + \varepsilon^{(d)}, \qquad \sigma = \sigma^{(s)} = \sigma^{(d)}.$$

Owing to (6.1)–(6.2), time differentiation of $\sigma^{(s)}$ and the summation $\partial_t \sigma^{(s)}/k + \sigma^{(d)}/\eta$ give

$$\frac{1}{k}\partial_t \sigma + \frac{1}{\eta}\sigma = \partial_t \varepsilon. \tag{6.8}$$

The response to a given stress loading follows by a direct integration to obtain

$$\varepsilon(t) - \varepsilon(t_0) = \frac{1}{k}[\sigma(t) - \sigma(t_0)] + \frac{1}{\eta}\int_{t_0}^{t} \sigma(s)ds.$$

Looking at a constant stress σ, the strain increases unboundedly; that is why the scheme is regarded as a fluid model. Moreover, it follows from (6.8) that $\sigma = \eta \partial_t \varepsilon$, which is the classical behaviour of a Newtonian fluid (Fig. 6.4).

Letting $\sigma(t_0) = \sigma_0$ and solving (6.8) in the unknown σ we find

$$\sigma(t) = \sigma_0 \exp\left[-(t - t_0)/\tau\right] + k\int_{t_0}^{t} \exp\left[-(t - s)/\tau\right]\partial_s \varepsilon(s)ds,$$

where $\tau = \eta/k$. At constant strain the stress decays with the time constant τ. We say that the model accounts for *relaxation* and hence that τ is the *relaxation time.* In particular, as $t \to \infty$ we have $\sigma \to 0$, a property that is characteristic of a fluid. Letting $t_0 \to -\infty$ we obtain

$$\sigma(t) = k\int_{-\infty}^{t} \exp\left[-(t - s)/\tau\right]\partial_s \varepsilon(s)ds.$$

By the change of variable $t - s = y$ we find

Fig. 6.4 The Maxwell-Wiechert fluid model

$$\sigma(t) = k \int_0^\infty \exp(-y/\tau)\, \partial_t \varepsilon(t-y) dy. \tag{6.9}$$

Thus $\sigma(t)$ depends linearly on the past values of the strain rate $\partial_t \varepsilon$ via a weighted integral with the memory kernel $k \exp(-y/\tau)$. Finally, an integration by parts yields

$$\sigma(t) = k\,\varepsilon(t) - \frac{k}{\tau} \int_0^\infty \exp(-y/\tau)\, \varepsilon(t-y) dy. \tag{6.10}$$

It is natural to take the energy as that of the spring,

$$\Psi_{MW} = \tfrac{1}{2} k\, [\varepsilon^{(s)}]^2 = \frac{1}{2k}\sigma^2.$$

Hence, the energy of the MW fluid is a function of σ. It follows from (6.8) that

$$\partial_t \Psi_{MW} = (1/k)\sigma \partial_t \sigma = \sigma \partial_t \varepsilon - (1/\eta)\sigma^2 \le \sigma \partial_t \varepsilon.$$

6.1.5 Bingham Model

The arrangement of a dashpot and a sliding frictional element in parallel gives rise to a rigid-perfectly viscoplastic unit usually denoted as *Bingham fluid model* or *Bingham plastic element*. A Bingham plastic is a viscoplastic material that behaves as a rigid body at low stresses but flows as a viscous fluid at high stress. Letting

$$\sigma = \sigma^{(d)} + \sigma^{(f)} = \eta\, \partial_t \varepsilon^{(d)} + \sigma^{(f)}, \qquad \varepsilon = \varepsilon^{(d)} = \varepsilon^{(f)},$$

we obtain

$$\partial_t \varepsilon = \begin{cases} 0 & \text{if } |\sigma| < \sigma_y, \\ \frac{1}{\eta}(\sigma - \sigma_y) & \text{if } \sigma > \sigma_y, \\ \frac{1}{\eta}(\sigma + \sigma_y) & \text{if } \sigma < -\sigma_y. \end{cases} \tag{6.11}$$

It is used as a mathematical model of mud flow along a pipe in drilling engineering. Another common example is toothpaste. If $\sigma_y = 0$ and the viscosity of the dashpot, η, has the Norton form

$$\eta = \eta_0 \left[\frac{\eta_0}{|\sigma|}\right]^{n-1}, \qquad n \in \mathbb{N},$$

η_0 being the kinematic viscosity, then the corresponding constitutive equation is named *Norton-Hoff model* (Fig. 6.5).

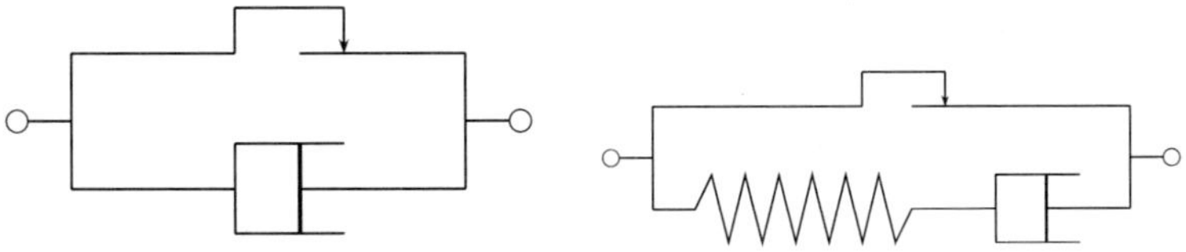

Fig. 6.5 A picture of the Bingham (left) and Bingham-Maxwell (right) units

6.1.6 *Bingham-Maxwell Model*

The model consists of a spring and a Bingham unit connected in series, but can also be achieved by arranging a Maxwell unit in parallel with a sliding frictional element (Fig. 6.5). Letting

$$\varepsilon = \varepsilon^{(s)} + \varepsilon^{(B)} = \tfrac{1}{k}\,\sigma^{(s)} + \varepsilon^{(B)}, \qquad \sigma = \sigma^{(s)} = \sigma^{(B)},$$

we obtain

$$\partial_t\varepsilon = \begin{cases} \tfrac{1}{k}\,\partial_t\sigma & \text{if } |\sigma| < \sigma_Y, \\ \tfrac{1}{k}\,\partial_t\sigma + \tfrac{1}{\eta}\,(\sigma - \sigma_Y) & \text{if } \sigma > \sigma_Y, \\ \tfrac{1}{k}\,\partial_t\sigma + \tfrac{1}{\eta}\,(\sigma + \sigma_Y) & \text{if } \sigma < -\sigma_Y, \end{cases} \tag{6.12}$$

When η takes the Norton form,

$$\eta = \eta_0\Big[\frac{\eta_0}{|\sigma| - \sigma_Y}\Big]^{n-1}, \qquad n \in \mathbb{N},$$

then the corresponding model is named *Bingham-Norton solid.* Rheological models of this kind are typically developed for metals and steel but are, to some extent, used to characterize viscoplastic effects in geomaterials.

A direct check shows that the imbalance inequality (6.6) holds with $\Psi_{BM} = \frac{1}{2k}\sigma^2$. Indeed, in view of (6.12),

$$\partial_t\Psi_{BM} = \frac{1}{k}\sigma\partial_t\sigma = \frac{1}{k}\sigma \begin{cases} k\,\partial_t\varepsilon & \text{if } |\sigma| < \sigma_Y, \\ k\,\partial_t\varepsilon - \tfrac{k}{\eta}\,(\sigma - \sigma_Y) & \text{if } \sigma > \sigma_Y, \\ k\,\partial_t\varepsilon - \tfrac{k}{\eta}\,(\sigma + \sigma_Y) & \text{if } \sigma < -\sigma_Y, \end{cases}$$

$$= \sigma\,\partial_t\varepsilon - \begin{cases} 0 & \text{if } |\sigma| < \sigma_Y, \\ \tfrac{1}{\eta}\,\sigma(\sigma - \sigma_Y) & \text{if } \sigma \geq \sigma_Y, \\ \tfrac{1}{\eta}\,\sigma(\sigma + \sigma_Y) & \text{if } \sigma \leq -\sigma_Y, \end{cases} \leq \sigma\,\partial_t\varepsilon,$$

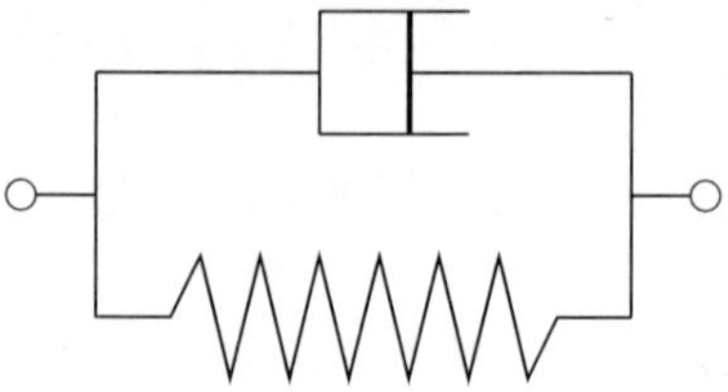

Fig. 6.6 The Kelvin-Voigt solid model

6.1.7 Kelvin-Voigt Solid

The model consists of a spring and a dashpot connected in parallel (Fig. 6.6). The analogue between σ, ε and the force-displacement pair indicates that the stress is shared by both components whereas the strain is the same, namely

$$\sigma = \sigma^{(s)} + \sigma^{(d)}, \qquad \varepsilon = \varepsilon^{(s)} = \varepsilon^{(d)}.$$

Hence the formal sum of (6.1) and (6.2) results in the constitutive equation

$$\sigma = k\,\varepsilon + \eta\,\partial_t \varepsilon. \tag{6.13}$$

The response to a given strain loading is then given directly. Instead, the response to stress loading is determined by solving the differential problem

$$\partial_t \varepsilon + \frac{k}{\eta}\,\varepsilon = \frac{1}{\eta}\,\sigma, \qquad \varepsilon(t_0) = \varepsilon_0,$$

in the unknown ε to obtain

$$\varepsilon(t) = \varepsilon_0 \exp\big[-(t-t_0)/\tau\big] + \frac{1}{\eta}\int_{t_0}^{t} \exp\big[-(t-s)/\tau\big]\,\sigma(s)ds, \tag{6.14}$$

where $\tau = \eta/k$ is referred to as *retardation time.* If the loading is a step stress, $\sigma(t) = \sigma_0 H(t - t_0)$, then

$$\varepsilon(t) = \varepsilon_0 \exp\big[-(t-t_0)/\tau\big] + \frac{\sigma_0}{k}\big\{1 - \exp\big[-(t-t_0)/\tau\big]\big\}.$$

Hence, in the Kelvin-Voigt (KV) model the creep compliance, that is the strain response per unit stress when $\varepsilon_0 = 0$, is given by

$$\varepsilon(t) = K(t-t_0)\sigma_0, \qquad K(t-t_0) = \frac{1}{k}\big\{1 - \exp\big[-(t-t_0)/\tau\big]\big\}.$$

The KV model is relative to a solid, as apparent by observing that, asymptotically, we recover the strain-stress relation of an elastic solid: as $t \to \infty$,

$$\varepsilon \to \sigma_0/k.$$

Letting $t_0 \to -\infty$ in (6.14), the change of variable $t - s = y$ yields

$$\varepsilon(t) = \frac{1}{\eta} \int_0^\infty \exp(-y/\tau)\, \sigma(t-y) dy,$$

and upon an integration by parts we obtain

$$\varepsilon(t) = \frac{1}{k}\sigma(t) - \frac{1}{k} \int_0^\infty \exp(-y/\tau)\, \partial_t \sigma(t-y) dy. \tag{6.15}$$

For the Kelvin-Voigt solid, it follows from (6.13) that the imbalance inequality (6.6) holds with

$$\Psi_{KV} = \tfrac{1}{2} k\, \varepsilon^2.$$

6.1.8 *Jeffreys' and Burgers' Fluids*

A more realistic model for a composite viscoelastic fluid is obtained by adding a second dashpot in parallel with a Maxwell-Wiechert (MW) unit. In the literature of viscoelastic fluids (including polymeric liquids) this model is known as the *Jeffreys fluid*[2] or the anti-Zener model.[3] Denote by $k^{(M)}$, $\eta^{(M)}$, $\sigma^{(M)}$, $\varepsilon^{(M)}$ the quantities pertaining to the MW unit and by $\eta^{(d)}$, $\sigma^{(d)}$, $\varepsilon^{(d)}$ those pertaining to the dashpot in parallel. The common value $\varepsilon = \varepsilon^{(M)} = \varepsilon^{(d)}$ of the strain is such that

$$\sigma^{(d)} = \eta^{(d)}\, \partial_t \varepsilon, \qquad \frac{1}{k^{(M)}} \partial_t \sigma^{(M)} + \frac{1}{\eta^{(M)}} \sigma^{(M)} = \partial_t \varepsilon.$$

Substitution of $\sigma^{(M)} = \sigma - \sigma^{(d)}$ and multiplication by $\eta^{(M)}$ yield

$$\sigma + \tau \partial_t \sigma = (\eta^{(M)} + \eta^{(d)}) \partial_t \varepsilon + \tau \eta^{(d)}\, \partial_t^2 \varepsilon, \tag{6.16}$$

where $\tau = \eta^{(M)}/k^{(M)}$ is the same relaxation time as MW unit. This equation involves three independent parameters, say $\eta^{(d)}$, $\eta^{(M)}$ and $k^{(M)}$ (or τ).

Let $\sigma(t_0) = \sigma_0$ and the strain on $[t_0, t]$ be given. By solving the differential equation (6.16) in the unknown σ we obtain

$$\sigma(t) = \sigma_0 \exp\big[-(t-t_0)/\tau\big] + \int_{t_0}^{t} \exp\big[-(t-s)/\tau\big]\big[\frac{1}{\tau}(\eta^{(M)} + \eta^{(d)}) \partial_s \varepsilon(s) + \eta^{(d)}\, \partial_s^2 \varepsilon(s)\big] ds.$$

---

[2] See e.g. [244] and the review paper by Bird and Wiest [47].

[3] See e.g. [79, 293].

An integration by parts results in

$$\sigma(t) = [\sigma_0 - \eta^{(d)}\,\partial_t\varepsilon_0]\exp\big[-(t-t_0)/\tau\big] + \eta^{(d)}\,\partial_t\varepsilon(t) + k^{(M)}\int_{t_0}^{t}\exp\big[-(t-s)/\tau\big]\partial_s\varepsilon(s)ds,$$

where $\partial_t\varepsilon_0 = \lim_{t\to t_0^+}\partial_t\varepsilon(t)$. As in the Maxwell-Wiechert fluid, also in this case at constant strain we have $\lim_{t\to\infty}\sigma(t) = 0$. Letting $t_0 \to -\infty$ we obtain

$$\sigma(t) = \eta^{(d)}\,\partial_t\varepsilon(t) + k^{(M)}\int_{-\infty}^{t}\exp\big[-(t-s)/\tau\big]\,\partial_s\varepsilon(s)ds.$$

The change of variable $t - s = y$ yields

$$\sigma(t) = \eta^{(d)}\,\partial_t\varepsilon(t) + k^{(M)}\int_0^{\infty}\exp(-y/\tau)\,\partial_t\varepsilon(t-y)dy. \tag{6.17}$$

Accordingly, $\sigma(t)$ is given by the present value of the strain rate $\partial_t\varepsilon(t)$ and by the values of its past history up to the present time t via a weighted integral with the memory kernel $k^{(M)}\exp(-y/\tau)$. Finally, an integration by parts shows that

$$\sigma(t) = \eta^{(d)}\,\partial_t\varepsilon(t) + k^{(M)}\,\varepsilon(t) - \frac{k^{(M)}}{\tau}\int_0^{\infty}\exp(-y/\tau)\,\varepsilon(t-y)dy. \tag{6.18}$$

Taking the energy for the Jeffreys fluid as that of the spring in the MW unit we obtain Ψ_J as a function of σ and $\partial_t\varepsilon$,

$$\Psi_J(\sigma, \partial_t\varepsilon) = \frac{1}{2k^{(M)}}\big[\sigma^{(M)}\big]^2 = \frac{1}{2k^{(M)}}(\sigma - \eta^{(d)}\partial_t\varepsilon)^2.$$

In view of (6.16) a direct check shows that the imbalance inequality holds,

$$\begin{aligned}\partial_t\Psi_J &= \frac{\tau}{\eta^{(M)}}(\sigma - \eta^{(d)}\partial_t\varepsilon)(\partial_t\sigma - \eta^{(d)}\partial_t^2\varepsilon) = \frac{1}{\eta^{(M)}}(\sigma - \eta^{(d)}\partial_t\varepsilon)(-\sigma + \eta^{(d)}\partial_t\varepsilon + \eta^{(M)}\partial_t\varepsilon)\\ &= -\frac{1}{\eta^{(M)}}(\sigma - \eta^{(d)}\partial_t\varepsilon)^2 - \eta^{(d)}\big[\partial_t\varepsilon\big]^2 + \sigma\partial_t\varepsilon \le \sigma\partial_t\varepsilon.\end{aligned}$$

Otherwise, we can define

$$\Psi_J^*(\sigma, \partial_t\varepsilon) = \frac{\eta^{(M)}}{2k^{(M)}(\eta^{(M)} + 2\eta^{(d)})}(\sigma - \eta^{(d)}\partial_t\varepsilon)^2,$$

and the imbalance inequality holds anyway. Indeed we have

$$\partial_t\Psi_J^* = -\frac{\sigma^2}{\eta^{(M)} + 2\eta^{(d)}} - \frac{\eta^{(d)}(\eta^{(M)} + \eta^{(d)})}{\eta^{(M)} + 2\eta^{(d)}}\big[\partial_t\varepsilon\big]^2 + \sigma\partial_t\varepsilon \le \sigma\partial_t\varepsilon.$$

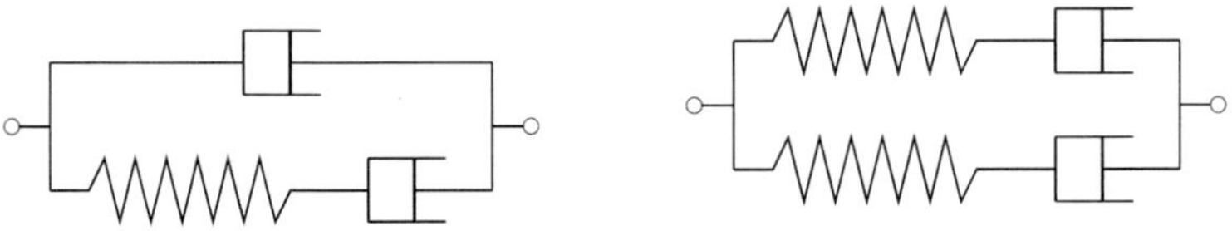

Fig. 6.7 The Jeffreys and Burgers models

Moreover, each function obtained by the convex combination of Ψ_J and Ψ_J^* satisfies the same inequality, in that

$$\partial_t\big[\lambda\Psi_J + (1-\lambda)\Psi_J^*\big] = \sigma\partial_t\varepsilon - \lambda\xi^2 - (1-\lambda)\zeta^2 \le \sigma\partial_t\varepsilon, \qquad \lambda \in [0,1],$$

where

$$\xi^2 = \frac{1}{\eta^{(M)}}(\sigma - \eta^{(d)}\partial_t\varepsilon)^2 + \eta^{(d)}\big[\partial_t\varepsilon\big]^2, \quad \zeta^2 = \frac{\sigma^2}{\eta^{(M)} + 2\eta^{(d)}} + \frac{\eta^{(d)}(\eta^{(M)} + \eta^{(d)})}{\eta^{(M)} + 2\eta^{(d)}}\big[\partial_t\varepsilon\big]^2.$$

We therefore conclude that there are an infinite number of energy functions for the Jeffreys fluid that verify the imbalance inequality (Fig. 6.7).

Burgers' fluid

A collection of two MW units in parallel is referred to as *Burgers' fluid model*. Since the stress is shared by both components whereas the strain is the same, then

$$\sigma = \sigma_1 + \sigma_2, \qquad \varepsilon = \varepsilon_1 = \varepsilon_2,$$

where, in force of (6.8),

$$\frac{\eta_i}{k_i}\partial_t\sigma_i + \sigma_i = \eta_i\partial_t\varepsilon_i. \qquad i = 1, 2. \tag{6.19}$$

Upon time differentiation and multiplication by η_j/k_j, $j \neq i$, the formal sum of (6.19) results in

$$\frac{\eta_1\eta_2}{k_1k_2}\partial_t^2\sigma + \frac{\eta_1}{k_1}\partial_t\sigma_2 + \frac{\eta_2}{k_2}\partial_t\sigma_1 = \frac{\eta_1\eta_2}{k_1k_2}(k_1 + k_2)\partial_t^2\varepsilon. \tag{6.20}$$

On the other hand the direct sum of (6.19) yields

$$\frac{\eta_1}{k_1}\partial_t\sigma_1 + \frac{\eta_2}{k_2}\partial_t\sigma_2 + \sigma = (\eta_1 + \eta_2)\partial_t\varepsilon.$$

and then

$$\frac{\eta_1}{k_1}\partial_t\sigma_2 + \frac{\eta_2}{k_2}\partial_t\sigma_1 = \Big(\frac{\eta_1}{k_1} + \frac{\eta_2}{k_2}\Big)\partial_t\sigma + \sigma - (\eta_1 + \eta_2)\partial_t\varepsilon.$$

In view of this identity, (6.20) gives the constitutive rate equation of Burgers' model

$$\frac{\eta_1\eta_2}{k_1k_2}\partial_t^2\sigma + \Big(\frac{\eta_1}{k_1}+\frac{\eta_2}{k_2}\Big)\partial_t\sigma + \sigma = (\eta_1+\eta_2)\partial_t\varepsilon + \frac{\eta_1\eta_2}{k_1k_2}(k_1+k_2)\partial_t^2\varepsilon. \qquad (6.21)$$

It might seem natural to take the energy for Burgers' model as that of the two springs. Yet, this quantity cannot be expressed as a function of ε, σ and their time derivatives. This in turn indicates that for involved materials the physical energy need not be the pertinent function of the imbalance inequality and hence the recourse to additional (internal) variables is justified.

An integral form of this constitutive equation can easily be obtained by adding constitutive relations of two MW elements in the form (6.9). We obtain

$$\sigma(t) = \int_0^\infty [k_1\exp(-y/\tau_1) + k_2\exp(-y/\tau_2)]\,\partial_t\varepsilon(t-y)dy, \qquad (6.22)$$

where $\tau_i = \eta_i/k_i$, $i = 1, 2$.

Various combinations of (KV, MW, SLS) units may result in models capable of reproducing the behaviour of real materials [220]. In particular, generalized Kelvin-Voigt and Maxwell-Wiechert models are some of the most well-known models to characterize the behavior of polymers [392].

6.1.9 *Standard Linear Solid*

A more involved model for solids is given by the so called *standard linear solid* (SLS) introduced by Zener [44]. It is obtained by adding a second spring in parallel with a MW unit. Denote by $k^{(M)}$, $\eta^{(M)}$, $\sigma^{(M)}$, $\varepsilon^{(M)}$ the quantities pertaining to the MW unit and by $k^{(s)}$, $\sigma^{(s)}$, $\varepsilon^{(s)}$ those pertaining to the spring in parallel. The common value $\varepsilon = \varepsilon^{(M)} = \varepsilon^{(s)}$ of the strain is such that

$$\sigma^{(s)} = k^{(s)}\varepsilon, \qquad \frac{1}{k^{(M)}}\partial_t\sigma^{(M)} + \frac{1}{\eta^{(M)}}\sigma^{(M)} = \partial_t\varepsilon.$$

Substitution of $\sigma^{(M)} = \sigma - \sigma^{(s)}$ and multiplication by $\eta^{(M)}$ yields

$$\sigma + \tau\partial_t\sigma = k^{(s)}\varepsilon + (k^{(s)} + k^{(M)})\tau\,\partial_t\varepsilon, \qquad (6.23)$$

where $\tau = \eta^{(M)}/k^{(M)}$ is the relaxation time. This equation involves three parameters, say $k^{(s)}$, $k^{(M)}$, and $\eta^{(M)}$ (or τ). That is why the SLS is also named *three parameter model* (Fig. 6.8).

The energy for the SLS is the sum of the energies of the two linear springs in the unit. Owing to (6.23) and letting

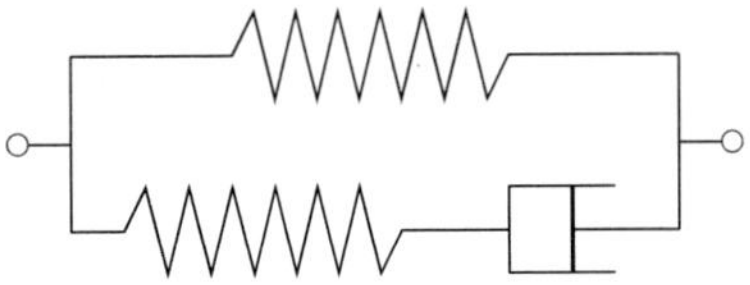

Fig. 6.8 The standard linear solid model

$$\Psi_{SLS}(\varepsilon,\sigma)=\frac{1}{2}k^{(s)}\,[\varepsilon^{(s)}]^2+\frac{1}{2k^{(M)}}\big[\sigma^{(M)}\big]^2=\frac{1}{2}k^{(s)}\,\varepsilon^2+\frac{1}{2k^{(M)}}(\sigma-k^{(s)}\varepsilon)^2,$$

we obtain

$$\begin{aligned}\partial_t\Psi_{SLS}&=k^{(s)}\,\varepsilon\,\partial_t\varepsilon+\frac{\tau}{\eta^{(M)}}(\sigma-k^{(s)}\varepsilon)(\partial_t\sigma-k^{(s)}\partial_t\varepsilon)\\&=k^{(s)}\,\varepsilon\,\partial_t\varepsilon+\frac{1}{\eta^{(M)}}(\sigma-k^{(s)}\varepsilon)(-\sigma+k^{(s)}\varepsilon+\eta^{(M)}\partial_t\varepsilon)\\&=-\frac{1}{\eta^{(M)}}(\sigma-k^{(s)}\varepsilon)^2+\sigma\partial_t\varepsilon\le\sigma\partial_t\varepsilon.\end{aligned}$$

so proving the imbalance inequality.

Assuming $\sigma(t_0)=\sigma_0$ and a given strain on $[t_0,t]$ we solve the differential equation (6.23) in the unknown σ to obtain

$$\sigma(t)=\sigma_0\exp(-(t-t_0)/\tau)+\int_{t_0}^{t}\exp(-(t-s)/\tau)\big[\frac{k^{(s)}}{\tau}\varepsilon(s)+(k^{(s)}+k^{(M)})\partial_s\varepsilon(s)\big]ds.$$

Let $\varepsilon_0=\varepsilon(t_0)$. Integrating by parts we find

$$\begin{aligned}\sigma(t)=[\sigma_0-(k^{(s)}+k^{(M)})\varepsilon_0]\exp(-(t-t_0)/\tau)+(k^{(s)}+k^{(M)})\varepsilon(t)\\-\frac{k^{(M)}}{\tau}\int_{t_0}^{t}\exp(-(t-s)/\tau)\varepsilon(s)ds.\end{aligned}\tag{6.24}$$

Likewise we can solve (6.23) in the unknown ε,

$$\varepsilon+\tau^*\partial_t\varepsilon=\frac{1}{k^{(s)}}\sigma+\frac{\tau}{k^{(s)}}\partial_t\sigma,$$

where $\tau^*=(1+k^{(M)}/k^{(s)})\tau$; τ^* is called retardation time. We obtain

$$\varepsilon(t)=\varepsilon_0\exp\big[-(t-t_0)/\tau^*\big]+\frac{1}{k^{(s)}\tau^*}\int_{t_0}^{t}\exp(-(t-s)/\tau^*)\big[\sigma(s)+\tau\partial_s\sigma(s)\big]ds.$$

An integration by parts and some rearrangements result in

$$\varepsilon(t) = \big(\varepsilon_0 - \frac{1}{k^{(s)} + k^{(M)}}\sigma_0\big) \exp\big[-(t-t_0)/\tau^*\big] + \frac{1}{k^{(s)} + k^{(M)}}\sigma(t) + \frac{k^{(M)}/k^{(s)}}{\tau^*(k^{(s)} + k^{(M)})} \int_{t_0}^{t} \exp(-(t-s)/\tau^*)\sigma(s)ds. \quad (6.25)$$

The SLS accounts for both relaxation and retardation. These features are shown by means of relaxation and creep tests, respectively. A relaxation test consists of monitoring the time-dependent stress $\sigma(t)$ resulting from the application of a steady strain. A creep test consists of measuring the time dependent strain $\varepsilon(t)$ from the application of a steady stress. It is worth looking at the results of relaxation and creep tests by applying (6.24) and (6.25), respectively.

Let $t_0 = 0$. At constant strain, $\varepsilon(t) = \varepsilon_0$ as $t > 0$, by (6.24) we find that the stress decays with time constant τ,

$$\sigma(t) = (\sigma_0 - k^{(s)}\varepsilon_0)\exp(-t/\tau) + k^{(s)}\varepsilon_0.$$

Asymptotically the model behaves like a spring: $\sigma \to k^{(s)}\varepsilon_0 =: \sigma_\infty$ as $t \to \infty$. Likewise, at constant stress, $\sigma(t) = \sigma_0$ as $t > 0$, the strain (6.25) decays with time constant $\tau^{(d)}$,

$$\varepsilon(t) = (\varepsilon_0 - \sigma_0/k^{(s)})\exp(-t/\tau^{(d)}) + \sigma_0/k^{(s)}.$$

Hence, as $t \to \infty$ we have $\varepsilon \to \sigma_0/k^{(s)} =: \varepsilon_\infty$. Creep and relaxation tests can then verify whether the model fits the data and accordingly provide the values of $k^{(s)}, \tau, \tau^{(d)}$.

Letting $t_0 \to -\infty$ in (6.24) and (6.25) we obtain

$$\sigma(t) = (k^{(s)} + k^{(M)})\varepsilon(t) - (k^{(M)}/\tau) \int_{-\infty}^{t} \exp(-(t-s)/\tau)\varepsilon(s)ds,$$

$$\varepsilon(t) = \frac{1}{k^{(s)} + k^{(M)}}\sigma(t) + \frac{k^{(M)}}{\tau^* k^{(s)}(k^{(s)} + k^{(M)})} \int_{-\infty}^{t} \exp(-(t-s)/\tau^*)\sigma(s)ds.$$

The change of variable $t - s = y$ yields

$$\sigma(t) = (k^{(s)} + k^{(M)})\varepsilon(t) - (k^{(M)}/\tau) \int_{0}^{\infty} \exp(-y/\tau)\,\varepsilon(t-y)dy,$$
$$\varepsilon(t) = \frac{1}{k^{(s)} + k^{(M)}}\sigma(t) + \frac{k^{(M)}}{\tau^* k^{(s)}(k^{(s)} + k^{(M)})} \int_{0}^{\infty} \exp(-y/\tau^*)\sigma(t-y)dy. \quad (6.26)$$

Finally, we integrate by parts to find

$$\sigma(t) = k^{(s)}\varepsilon(t) + k^{(M)} \int_{0}^{\infty} \exp(-y/\tau)\partial_t\varepsilon(t-y)dy, \quad (6.27)$$

$$\varepsilon(t) = \frac{1}{k^{(s)}}\sigma(t) - \frac{k^{(M)}}{k^{(s)}(k^{(s)} + k^{(M)})} \int_{0}^{\infty} \exp(-y/\tau^*)\partial_t\sigma(t-y)dy. \quad (6.28)$$

6.1.10 Generalized Models

A *generalized Kelvin-Voigt (GKV) model* is a collection of n KV units in series plus a single spring. In the j-th unit

$$\sigma_j = k_j \varepsilon_j + \eta_j \partial_t \varepsilon_j, \qquad j = 1, ..., n. \tag{6.29}$$

The possible single spring is governed by $\sigma_0 = k_0 \varepsilon_0$. To determine the constitutive equation, that is the relation between the common value of σ and $\varepsilon = \sum_{i=0}^{n} \varepsilon_i$, we first solve the differential equation (6.29) for ε_j, $j = 1, ..., n$, to obtain

$$\varepsilon_j(t) = \varepsilon_j(t_0) \exp[-(t - t_0)/\tau_j] + \frac{1}{\eta_j} \int_{t_0}^{t} \exp[-(t - s)/\tau_j] \sigma(s) ds.$$

where $\tau_j = \eta_j / k_j$. The limit as $t_0 \to -\infty$ results in

$$\varepsilon_j(t) = \frac{1}{\eta_j} \int_{-\infty}^{t} \exp[-(t - s)/\tau_j] \sigma(s) ds.$$

Hence, upon summation and the change of variable $t - s = y$, we find the constitutive equation

$$\varepsilon(t) = \frac{1}{k_0} \sigma(t) + \int_0^{\infty} \sum_{j=1}^{n} \frac{1}{\eta_j} \exp[-y/\tau_j] \sigma(t - y) dy. \tag{6.30}$$

Equation (6.30) is an example of constitutive relation where the strain ε, at time t, depends on the present value $\sigma(t)$ of the stress and on the history of σ up to time t. The function

$$\sum_{j=1}^{n} (1/\eta_j) \exp(-y/\tau_j)$$

is known in the literature as the Prony series and is referred to as the creep function of the GKV model. As is apparent, it is the sum of the creep functions of each KV unit.

A *generalized Maxwell-Wiechert (GMW) model* is a collection of n MW units in parallel plus a single spring. For any j-th MW unit,

$$\partial_t \varepsilon_j = \frac{1}{k_j} \partial_t \sigma_j + \frac{1}{\eta_j} \sigma_j, \qquad j = 1, ..., n, \tag{6.31}$$

whereas the single damper is governed by $\sigma_0 = \eta_0 \, \partial_t \varepsilon_0$. To find the relation between $\sigma = \sum_{i=0}^{n} \sigma_i$ and the common value of ε, we have to solve the differential equation (6.31) for σ_j, $j = 1, ..., n$. Letting $\tau_j = \eta_j / k_j$ and taking the limit as $t_0 \to -\infty$ we find that

$$\sigma_j(t) = k_j \int_{-\infty}^{t} \exp[-(t - s)/\tau_j] \partial_t \varepsilon(s) ds.$$

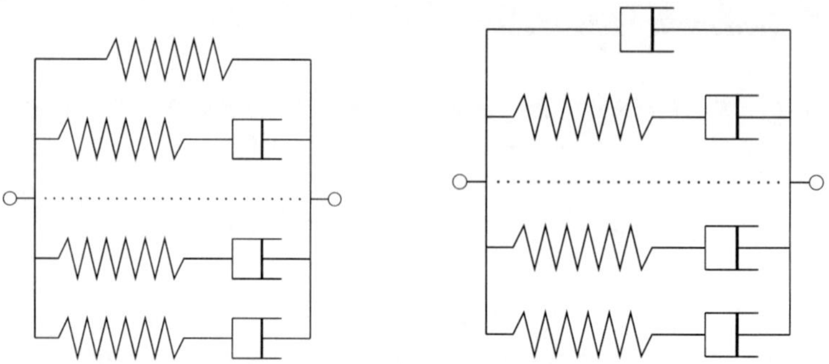

Fig. 6.9 The generalized Maxwell-Wiechert and Jeffreys models

If k_0 is the constant of the single damper then, upon summation and the change of variable $t - s = y$, it follows

$$\sigma(t) = k_0\, \varepsilon(t) + \int_0^\infty \sum_{j=1}^n k_j \exp[-y/\tau_j] \partial_t \varepsilon(t-y) dy,$$

whence, upon an integration by parts,

$$\sigma(t) = [k_0 + \sum_{j=1}^n k_j]\varepsilon(t) - \int_0^\infty \sum_{j=1}^n \frac{k_j}{\tau_j} \exp[-y/\tau_j]\varepsilon(t-y) dy. \tag{6.32}$$

The relaxation function of the GMW model

$$\sum_{j=1}^n (k_j/\tau_j) \exp[-y/\tau_j]$$

is the sum of the relaxation functions of each MW unit (Fig. 6.9).

We consider again a collection of n MW units connected in parallel with each other. If we add a single damper (rather than a single spring) in parallel we obtain the *generalized Jeffreys (or anti-Zener) model*. For any j-th MW unit,

$$\partial_t \varepsilon_j = \frac{1}{k_j} \partial_t \sigma_j + \frac{1}{\eta_j} \sigma_j, \qquad j = 1, ..., n, \tag{6.33}$$

whereas the single damper is governed by $\sigma_0 = \eta_0\, \partial_t \varepsilon_0$. To find the relation between $\sigma = \sum_{i=0}^n \sigma_i$ and the common value of $\varepsilon = \varepsilon_i$, $i = 0, 1, ..., n$, we have to solve the differential equation (6.33) for σ_j. Letting $\tau_j = \eta_j / k_j$ and taking the limit as $t_0 \to -\infty$ we find that

$$\sigma_j(t) = k_j \int_{-\infty}^t \exp[-(t-s)/\tau_j] \partial_t \varepsilon(s) ds, \qquad j = 1, ..., n.$$

Upon summation and the change of variable $t - s = y$, we obtain

$$\sigma(t) = \eta_0\, \partial_t \varepsilon(t) + \int_0^\infty \sum_{j=1}^n k_j \exp[-y/\tau_j] \partial_t \varepsilon(t-y) dy.$$

Finally, an integration by parts yields

$$\sigma(t) = \eta_0\, \partial_t \varepsilon(t) + \sum_{j=1}^n k_j \varepsilon(t) - \int_0^\infty \sum_{j=1}^n (k_j/\tau_j) \exp(-y/\tau_j) \varepsilon(t-y) dy.$$

We say that $\sum_{j=1}^n (k_j/\tau_j) \exp[-y/\tau_j]$ is the relaxation function of the generalized Jeffreys model. As with the GMW model, it is the sum of the relaxation functions of each unit connected in parallel.

The *generalized Zener model* is given by n SLS units connected in parallel. For any j-th SLS unit, since $\varepsilon_j = \varepsilon$ we have (see Sect. 6.1.9)

$$\sigma_j(t) = (k_j^{(s)} + k_j^{(M)}) \varepsilon(t) - (k_j^{(M)}/\tau_j) \int_0^\infty \exp(-y/\tau_j)\, \varepsilon(t-y) dy. \tag{6.34}$$

Upon summation we obtain

$$\sigma(t) = \sum_{j=1}^n \left[k_j^{(s)} + k_j^{(M)}\right] \varepsilon(t) - \int_0^\infty \sum_{j=1}^n (k_j^{(M)}/\tau_j) \exp(-y/\tau_j) \varepsilon(t-y) dy,$$

which matches (6.32) provided that $k_j = k_j^{(M)}$ and

$$k_0 = \sum_{j=1}^n k_j^{(s)}.$$

We then conclude that the generalized Zener model is equivalent to the GMW model.

6.1.11 Relaxation Modulus and Creep Compliance

It is customary to denote by *relaxation modulus* the function G on $[0, \infty)$ such that

$$\sigma(t) = \int_0^\infty G(s) \partial_t \varepsilon(t-s) ds.$$

Likewise, the *creep compliance* is the function J on $[0, \infty)$ such that

$$\varepsilon(t) = \int_0^\infty J(s) \partial_t \sigma(t-s) ds.$$

According to (6.9), (6.17) and (6.27) the relaxation moduli $G^{(M)}$, $G^{(J)}$ and $G^{(s)}$ for the Maxwell-Wiechert, Jeffreys and SLS models are

$$\begin{aligned} G^{(M)}(s) &= k\exp(-s/\tau), \\ G^{(J)}(s) &= \eta^{(d)}\delta(s) + k^{(M)}\exp(-s/\tau), \\ G^{(s)}(s) &= k^{(s)} + k^{(M)}\exp(-s/\tau). \end{aligned} \tag{6.35}$$

By (6.15) and (6.28) we find that the compliance moduli $J^{(K)}$ and $J^{(s)}$ for the Kelvin-Voigt and SLS models are

$$\begin{aligned} J^{(K)}(s) &= \frac{1}{k} - \frac{1}{k}\exp(-s/\tau), \\ J^{(s)}(s) &= \frac{1}{k^{(s)}} - \frac{k^{(M)}}{k^{(s)}(k^{(s)} + k^{(M)})}\exp(-s/\tau^*). \end{aligned} \tag{6.36}$$

Similar expressions can be recovered when generalized models are considered.

Rheological models constitute helpful suggestions in the modelling of material behaviours. In a sense they are rate-type models (see next section), in that the resulting constitutive equations involve derivatives of the pertinent variables σ, ε. Also, they may be quite involved, depending on the number of (KV, MW, SLS) units. Yet they are quite elementary in that they are linear and involve stress and strain in one-dimensional pictures. We now look for rate-type models in a more general setting.

6.2 Rate-Type Models of Fluids

Rate equations occur in various models of continuous bodies, mainly for materials with internal variables. Modelling the constitutive properties via internal variables may be appropriate to describe microstructural effects on the macromotion or memory features associated with motion and thermal behaviour of the body.

Internal variables trace back to Duhem [427], Sect. I.8, and Bridgman [66]. The first analysis within continuum thermodynamics is given in [99] while a review on the subject is given in [303]. For definiteness, suppose the state of a solid is characterized by deformation, temperature, and an internal scalar variable α so that the evolution is given by

$$\dot{\alpha} = f(\mathbf{F}, \theta, \nabla\theta, \alpha),$$

the stress $\mathbf{T}$, the free energy ψ, the entropy η, and the heat flux $\mathbf{q}$ too being functions of the variables $\mathbf{F}$, θ, $\nabla\theta$, α.

Assume the entropy flux is taken as $\mathbf{q}/\theta$ so that the Clausius-Duhem inequality reads

$$-\dot{\psi} - \eta\,\dot{\theta} + \frac{1}{\rho}\mathbf{T}\cdot\mathbf{D} - \frac{1}{\rho\theta}\mathbf{q}\cdot\nabla\theta \geq 0.$$

Upon computation of $\dot{\psi}$ and using the arbitrariness of $\dot{\overline{\nabla\theta}}$, $\dot{\theta}$, $\dot{\mathbf{F}}$ we find

$$\partial_{\nabla\theta}\psi = 0, \qquad \eta = -\partial_\theta\psi, \qquad \mathbf{T} = \rho\,\partial_{\mathbf{F}}\psi\,\mathbf{F}^T \in \mathrm{Sym},$$

$$-\partial_\alpha\psi\, f - \frac{1}{\rho\theta}\mathbf{q}\cdot\nabla\theta \geq 0.$$

If, in particular, f is independent of $\nabla\theta$ then we have

$$\partial_\alpha\psi\, f \leq 0,$$

that is a condition on $\partial_\alpha\psi$ and on the evolution function f. Things are more involved if α is a vector or a tensor function because $\dot{\alpha}$ is no longer an objective derivative.

6.2.1 *Rate Equation for Thermo-Viscous Fluids in the Eulerian Description*

Borrowing from the standard linear solid (6.23) and the Maxwell-Wiechert fluid (6.8) one might guess that in the three-dimensional setting Eqs. (6.23) and (6.8) become

$$\alpha\dot{\varepsilon} + \beta\varepsilon = \gamma\mathbf{T} + \delta\dot{\mathbf{T}}, \qquad \dot{\varepsilon} = c\mathbf{T} + d\dot{\mathbf{T}}.$$

However $\dot{\mathbf{T}}$ and $\dot{\varepsilon}$ are not objective. We might replace $\dot{\varepsilon}$ with $\mathbf{D}$, which is objective. Hence a conceptually-correct form of constitutive assumption might be to let the objective rate of $\mathbf{T}$ be a function of the chosen variables. Indeed, we regard the body as a fluid and hence it seems natural to consider the rate equation for the dissipative stress

$$\mathcal{T} = \mathbf{T} + p\mathbf{1}.$$

Likewise, borrowing from the wide literature on the Maxwell-Cattaneo equation for the heat flux[4]

$$\mathbf{q} + \tau\dot{\mathbf{q}} = -k\nabla\theta,$$

we might write the constitutive equation for $\mathbf{q}$ as a rate-type equation involving an objective derivative. It might seem natural to let $p = \hat{p}(\rho, \theta)$ be related to the free energy. We will see in a moment that the occurrence of rate equations results in more involved thermodynamic schemes, unless the fluid is incompressible.

[4] See, e.g., Sect. 7.3 and [402].

For definiteness and simplicity we take rate equations involving the corotational derivative. Hence we let

$$\overset{\circ}{\mathcal{T}} = \hat{\mathcal{T}}(\rho, \theta, \mathbf{D}, \nabla\theta, \mathcal{T}, \mathbf{q}), \qquad \overset{\circ}{\mathbf{q}} = \hat{\mathbf{q}}(\rho, \theta, \mathbf{D}, \nabla\theta, \mathcal{T}, \mathbf{q}).$$

We therefore let the free energy ψ and the entropy η depend on the variables ρ, θ, $\mathbf{D}$, $\nabla\theta$, $\mathcal{T}$, $\mathbf{q}$. The time derivatives of $\mathcal{T}$ and $\mathbf{q}$ take the form

$$\dot{\mathcal{T}} = \mathbf{W}\mathcal{T} - \mathcal{T}\mathbf{W} + \hat{\mathcal{T}}(\rho, \theta, \mathbf{D}, \nabla\theta, \mathcal{T}, \mathbf{q}),$$

$$\dot{\mathbf{q}} = \mathbf{W}\mathbf{q} + \hat{\mathbf{q}}(\rho, \theta, \mathbf{D}, \nabla\theta, \mathcal{T}, \mathbf{q}).$$

For convenience in calculations we consider separately the rate equations of $T = \operatorname{tr}\mathcal{T}$ and $\mathcal{T}_0 = \mathcal{T} - \frac{1}{3}T\,\mathbf{1}$; it follows

$$\dot{\mathcal{T}}_0 = \mathbf{W}\mathcal{T}_0 - \mathcal{T}_0\mathbf{W} + \hat{\mathcal{T}}_0(\rho, \theta, \mathbf{D}, \nabla\theta, \mathcal{T}, \mathbf{q}), \qquad \dot{T} = \hat{T}(\rho, \theta, \mathbf{D}, \nabla\theta, \mathcal{T}, \mathbf{q}).$$

Thermodynamic consistency

The constitutive functions ψ, η, $\hat{\mathcal{T}}_0$, $\hat{T}$, $\hat{\mathbf{q}}$, p depend on the variables

$$\rho, \theta, \mathbf{D}, \nabla\theta, \mathcal{T}_0, T, \mathbf{q}.$$

To examine the thermodynamic consistency of these constitutive assumptions we start with the entropy inequality

$$-(\dot{\psi} + \eta\dot{\theta}) + \frac{1}{\rho}\mathbf{T}\cdot\mathbf{D} - \frac{1}{\rho\theta}\mathbf{q}\cdot\nabla\theta \geq 0.$$

Upon computation of $\dot{\psi}$ and substitution we find the inequality

$$-(\partial_\theta\psi + \eta)\dot{\theta} + \rho\partial_\rho\psi\nabla\cdot\mathbf{v} - \partial_{\mathbf{D}}\psi\cdot\dot{\mathbf{D}} - \partial_{\nabla\theta}\psi\cdot\dot{\overline{\nabla\theta}} - \partial_{\mathcal{T}_0}\psi\cdot(\mathbf{W}\mathcal{T}_0 - \mathcal{T}_0\mathbf{W} + \hat{\mathcal{T}}_0)$$
$$-\partial_T\psi\,\hat{T} - \partial_{\mathbf{q}}\psi\cdot(\mathbf{W}\mathbf{q} + \hat{\mathbf{q}}) + \frac{1}{\rho}\mathcal{T}_0\cdot\mathbf{D}_0 + \frac{1}{3\rho}T\nabla\cdot\mathbf{v} - \frac{p}{\rho}\nabla\cdot\mathbf{v} - \frac{1}{\rho\theta}\mathbf{q}\cdot\nabla\theta \geq 0.$$

The arbitrariness of $\dot{\mathbf{D}}$, $\dot{\overline{\nabla\theta}}$, $\dot{\theta}$ implies that

$$\partial_{\mathbf{D}}\psi = \mathbf{0}, \qquad \partial_{\nabla\theta}\psi = \mathbf{0}, \qquad \eta = -\partial_\theta\psi.$$

Hence the entropy inequality simplifies to

$$\rho\partial_\rho\psi\nabla\cdot\mathbf{v} - \partial_{\mathcal{T}_0}\psi\cdot(\mathbf{W}\mathcal{T}_0 - \mathcal{T}_0\mathbf{W} + \hat{\mathcal{T}}_0) - \partial_{\mathcal{T}}\psi\,\hat{\mathcal{T}} - \partial_{\mathbf{q}}\psi\cdot(\mathbf{W}\mathbf{q} + \hat{\mathbf{q}})$$
$$+\frac{1}{\rho}\mathcal{T}_0\cdot\mathbf{D}_0 + \frac{1}{3\rho}\mathcal{T}\nabla\cdot\mathbf{v} - \frac{p}{\rho}\nabla\cdot\mathbf{v} - \frac{1}{\rho\theta}\mathbf{q}\cdot\nabla\theta \geq 0.$$

While this is required to hold for any pair of functions $\hat{\mathcal{T}}_0, \hat{\mathcal{T}}, \hat{\mathbf{q}}$ of $\rho, \theta, \mathbf{D}, \nabla\theta, \mathcal{T}, \mathbf{q}$, detailed restrictions follow by specifying the functions $\hat{\mathcal{T}}_0, \hat{\mathcal{T}}, \hat{\mathbf{q}}$. Here we define $\hat{\mathcal{T}}, \hat{\mathcal{T}}$, and $\hat{\mathbf{q}}$ by generalizing the classical Navier-Stokes and Fourier laws and to recover them in stationary conditions. We let

$$\hat{\mathcal{T}}_0 = -\frac{1}{\tau_T}[\mathcal{T}_0 - 2\mu\,\mathbf{D}_0], \qquad \hat{\mathcal{T}} = -\frac{1}{\tau_T}[\mathcal{T} - (2\mu + 3\lambda)\nabla\cdot\mathbf{v}], \qquad \hat{\mathbf{q}} = -\frac{1}{\tau_q}[\mathbf{q} + \kappa\,\nabla\theta],$$

where $\tau_T, \tau_q, \mu, \lambda, \kappa$ are functions of ρ and θ; moreover $\tau_T, \tau_q > 0$. Thus the evolution of $\mathcal{T}$ and $\mathbf{q}$ is governed by

$$\dot{\mathcal{T}}_0 = \mathbf{W}\mathcal{T}_0 - \mathcal{T}_0\mathbf{W} - \frac{1}{\tau_T}[\mathcal{T}_0 - 2\mu\mathbf{D}_0],\ \dot{\mathcal{T}} = -\frac{1}{\tau_T}[\mathcal{T} - (2\mu + 3\lambda)\nabla\cdot\mathbf{v}],\ \dot{\mathbf{q}} = \mathbf{W}\mathbf{q} - \frac{1}{\tau_q}[\mathbf{q} + \kappa\nabla\theta].$$

Upon substitution of the rates, the entropy inequality has the form

$$-\partial_{\mathcal{T}_0}\psi\cdot(\mathbf{W}\mathcal{T}_0 - \mathcal{T}_0\mathbf{W}) - \partial_{\mathbf{q}}\psi\cdot\mathbf{W}\mathbf{q} + (\frac{\kappa}{\tau_q}\partial_{\mathbf{q}}\psi - \frac{1}{\rho\theta}\mathbf{q})\cdot\nabla\theta + ... \geq 0,$$

where the dots stand for terms independent of $\mathbf{W}$ and $\nabla\theta$. Hence it follows that

$$\partial_{\mathbf{q}}\psi - \frac{\tau_q}{\rho\theta\kappa}\mathbf{q} = 0, \qquad \partial_{\mathcal{T}_0}\psi\mathcal{T}_0^T + (\partial_{\mathcal{T}_0}\psi)^T\mathcal{T}_0 + \partial_{\mathbf{q}}\psi\otimes\mathbf{q} \in \mathrm{Sym}.$$

Consequently

$$\psi = \tilde{\psi}(\rho, \theta, \mathcal{T}_0, \mathcal{T}) + \frac{\tau_q}{2\rho\theta\kappa}\mathbf{q}^2,$$

and therefore $\partial_{\mathbf{q}}\psi\otimes\mathbf{q} \in \mathrm{Sym}$. Further, since $\mathcal{T} \in \mathrm{Sym}$ then it follows

$$\partial_{\mathcal{T}_0}\psi\mathcal{T}_0 \in \mathrm{Sym}.$$

The reduced inequality can be written

$$[\rho\partial_\rho\psi - \frac{2\mu + 3\lambda}{\tau_T}\partial_{\mathcal{T}}\psi - \frac{p}{\rho} + \frac{1}{3\rho}\mathcal{T}]\nabla\cdot\mathbf{v} + (\frac{1}{\rho}\mathcal{T}_0 - \frac{2\mu}{\tau_T}\partial_{\mathcal{T}_0}\psi)\cdot\mathbf{D}_0$$
$$+\frac{1}{\tau_T}(\partial_{\mathcal{T}_0}\psi\cdot\mathcal{T}_0 + \partial_{\mathcal{T}}\psi\,\mathcal{T}) + \frac{1}{\tau_q}\partial_{\mathbf{q}}\psi\cdot\mathbf{q} \geq 0$$

and implies that

$$\rho\partial_\rho\psi - \frac{2\mu + 3\lambda}{\tau_T}\partial_{\mathcal{T}}\psi - \frac{p}{\rho} + \frac{1}{3\rho}\mathcal{T} = 0, \qquad \frac{1}{\rho}\boldsymbol{\mathcal{T}}_0 - \frac{2\mu}{\tau_T}\partial_{\boldsymbol{\mathcal{T}}_0}\psi = \mathbf{0},$$

$$\frac{1}{\tau_T}(\partial_{\boldsymbol{\mathcal{T}}_0}\psi \cdot \boldsymbol{\mathcal{T}}_0 + \partial_{\mathcal{T}}\psi\, \mathcal{T}) + \frac{1}{\tau_q}\partial_{\mathbf{q}}\psi \cdot \mathbf{q} \geq 0.$$

The first equation shows an indeterminacy between p and $\mathcal{T}/3$ in that $p - \mathcal{T}/3$ occurs in the thermodynamic restriction. It seems natural to let

$$p = \rho^2\partial_\rho\psi,$$

as it happens formally in thermoelastic fluids. Here, though, ψ depends also on $\boldsymbol{\mathcal{T}}_0, \mathcal{T}, \mathbf{q}$ and hence we cannot require that p be a function of ρ, θ. A direct integration of

$$\partial_{\mathcal{T}}\psi = \frac{\tau_T}{3\rho(2\mu + 3\lambda)}\mathcal{T}, \qquad \partial_{\boldsymbol{\mathcal{T}}_0}\psi = \frac{\tau_T}{2\,\rho\,\mu}\boldsymbol{\mathcal{T}}_0$$

yields

$$\psi = \Psi(\rho, \theta) + \frac{\tau_T}{4\,\rho\,\mu}\boldsymbol{\mathcal{T}}_0 \cdot \boldsymbol{\mathcal{T}}_0 + \frac{\tau_T}{6\,\rho(2\mu + 3\lambda)}\mathcal{T}^2 + \frac{\tau_q}{2\,\rho\,\theta\,\kappa}\mathbf{q}^2.$$

Thus the inequality is

$$\frac{1}{2\rho\mu}\boldsymbol{\mathcal{T}}_0 \cdot \boldsymbol{\mathcal{T}}_0 + \frac{1}{3\rho(2\mu + 3\lambda)}\mathcal{T}^2 + \frac{1}{\rho\theta\kappa}\mathbf{q}^2 \geq 0.$$

The independence of $\boldsymbol{\mathcal{T}}_0, \mathcal{T}, \mathbf{q}$ implies that

$$\mu > 0, \qquad 2\mu + 3\lambda > 0, \qquad \kappa > 0.$$

We have thus obtained the classical thermodynamic requirements on the shear viscosity μ, the bulk viscosity $\mu + 3\lambda/2$, and the heat conductivity κ.

6.2.2 Invariant Fields and Objective Rates

Let $\boldsymbol{\mathcal{T}} = \mathbf{T} + p\mathbf{1}$. The invariance of the free energy ψ under Euclidean transformations is made obvious by letting ψ depend on $\boldsymbol{\mathcal{T}}$ and $\mathbf{q}$ through invariants. Let $\mathbf{R}$ be the rotation matrix associated with the deformation gradient and given by the polar decomposition, $\mathbf{F} = \mathbf{RU}$. Under a Euclidean transformation with rotation matrix $\mathbf{Q}$

$$\mathbf{R}^T\boldsymbol{\mathcal{T}}\mathbf{R} \to (\mathbf{QR})^T\mathbf{Q}\boldsymbol{\mathcal{T}}\mathbf{Q}^T(\mathbf{QR}) = \mathbf{R}^T\boldsymbol{\mathcal{T}}\mathbf{R}, \qquad \mathbf{R}^T\mathbf{q} \to (\mathbf{QR})^T\mathbf{Qq} = \mathbf{R}^T\mathbf{q}.$$

Hence $\mathbf{R}^T\boldsymbol{\mathcal{T}}\mathbf{R}$ and $\mathbf{R}^T\mathbf{q}$ are invariant. Denote by

$$\mathfrak{T} := \mathbf{R}^T \boldsymbol{\mathcal{T}} \mathbf{R}, \qquad \mathcal{Q} := \mathbf{R}^T \mathbf{q}$$

the two invariants. To distinguish formally the behaviour in dilatations and deformations we decompose $\boldsymbol{\mathcal{T}}$ in the trace-free part $\boldsymbol{\mathcal{T}}_0$ and the trace $\mathcal{T}$,

$$\boldsymbol{\mathcal{T}} = \boldsymbol{\mathcal{T}}_0 + \tfrac{1}{3}\mathcal{T}\,\mathbf{1}.$$

Correspondingly the invariant tensor $\mathfrak{T}$ satisfies

$$\mathfrak{T} = \mathfrak{T}_0 + \tfrac{1}{3}\mathfrak{T}\,\mathbf{1}, \qquad \mathfrak{T} = \operatorname{tr}\mathfrak{T} = \mathcal{T}, \qquad \mathfrak{T}_0 = \mathbf{R}^T \boldsymbol{\mathcal{T}}_0 \mathbf{R}.$$

Let $\mathbf{Z} = \dot{\mathbf{R}}\mathbf{R}^T$ and hence $\dot{\mathbf{R}} = \mathbf{Z}\mathbf{R}$. Observe

$$\dot{\mathfrak{T}} = \mathbf{R}^T(\dot{\boldsymbol{\mathcal{T}}} - \mathbf{Z}\boldsymbol{\mathcal{T}} - \boldsymbol{\mathcal{T}}\mathbf{Z}^T)\mathbf{R}, \qquad \dot{\mathcal{Q}} = \mathbf{R}^T(\dot{\mathbf{q}} - \mathbf{Z}\mathbf{q})\mathbf{R},$$

whence

$$\dot{\mathfrak{T}} = \mathbf{R}^T \overset{\odot}{\boldsymbol{\mathcal{T}}} \mathbf{R}, \qquad \dot{\mathcal{Q}} = \mathbf{R}^T \overset{\odot}{\mathbf{q}}.$$

The time derivative of the invariants $\mathfrak{T}$, $\mathcal{Q}$ is the transform of the (objective) Green-Naghdi derivatives $\overset{\odot}{\boldsymbol{\mathcal{T}}}$, $\overset{\odot}{\mathbf{q}}$.

Let $(\rho, \theta, \mathfrak{T}, \mathcal{Q}, \mathbf{D}, \nabla\theta)$ be the set of independent variables. Hence the free energy ψ, the entropy η, the pressure p, and the derivatives $\overset{\odot}{\boldsymbol{\mathcal{T}}}$, $\overset{\odot}{\mathbf{q}}$ are assumed to be functions of $(\rho, \theta, \mathfrak{T}, \mathcal{Q}, \mathbf{D}, \nabla\theta)$. Evaluation of $\dot{\psi}$ and substitution in the entropy inequality result in

$$\rho(\partial_\rho\psi - p/\rho^2)\dot{\rho} + \rho(\partial_\theta\psi + \eta)\dot{\theta} + \rho\partial_{\mathfrak{T}}\psi \cdot \dot{\mathfrak{T}} + \rho\partial_{\mathcal{Q}}\psi \cdot \dot{\mathcal{Q}} - \rho\partial_{\mathbf{D}}\psi \cdot \dot{\mathbf{D}} + \partial_{\nabla\theta}\psi \cdot \dot{\overline{\nabla\theta}}$$
$$+\boldsymbol{\mathcal{T}} \cdot \mathbf{D} - \frac{1}{\theta}\mathbf{q} \cdot \nabla\theta \geq 0.$$

By the standard arguments it follows

$$\partial_{\mathbf{D}}\psi = \mathbf{0}, \qquad \partial_{\nabla\theta}\psi = \mathbf{0}, \qquad \eta = -\partial_\theta\psi.$$

Again we observe $\dot{\rho}$ is not independent of $\mathbf{D}$. Yet we can assume

$$p = \rho^2\partial_\rho\psi,$$

as we did in the previous section and as it happens formally in thermoelastic fluids. Here, though, we cannot require that p be a function of ρ, θ. As a consequence, the inequality reduces to

$$\rho\partial_{\mathfrak{T}}\psi \cdot \dot{\mathfrak{T}} + \rho\partial_{\mathcal{Q}}\psi \cdot \dot{\mathcal{Q}} + \boldsymbol{\mathcal{T}} \cdot \mathbf{D} - \frac{1}{\theta}\mathbf{q} \cdot \nabla\theta \geq 0.$$

If ψ depends separately on $\boldsymbol{\mathfrak{T}}_0$ and $\mathfrak{T}$ then the reduced inequality becomes

$$\tfrac{1}{3}\mathfrak{T}\,\nabla\cdot\mathbf{v}+\boldsymbol{\mathcal{T}}_0\cdot\mathbf{D}_0-\frac{1}{\theta}\mathbf{q}\cdot\nabla\theta-\rho\partial_{\mathfrak{T}}\psi\cdot\dot{\mathfrak{T}}+\rho\partial_{\boldsymbol{\mathfrak{T}}_0}\psi\cdot\dot{\boldsymbol{\mathfrak{T}}}_0-\rho\partial_{\boldsymbol{\mathcal{Q}}}\psi\cdot\dot{\boldsymbol{\mathcal{Q}}}\geq 0.$$

By $\boldsymbol{\mathfrak{T}}_0=\mathbf{R}^T\boldsymbol{\mathcal{T}}_0\mathbf{R}$ and $\boldsymbol{\mathcal{Q}}=\mathbf{R}^T\mathbf{q}$ we have

$$\partial_{\boldsymbol{\mathcal{T}}_0}\psi=\mathbf{R}\partial_{\boldsymbol{\mathfrak{T}}_0}\psi\mathbf{R}^T,\quad \dot{\boldsymbol{\mathfrak{T}}}_0=\mathbf{R}^T\,\overset{\odot}{\boldsymbol{\mathcal{T}}}_0\,\mathbf{R},\quad \partial_{\boldsymbol{\mathfrak{T}}_0}\psi\cdot\dot{\boldsymbol{\mathfrak{T}}}_0=\partial_{\boldsymbol{\mathcal{T}}_0}\psi\cdot\overset{\odot}{\boldsymbol{\mathcal{T}}}_0,$$

$$\partial_{\mathbf{q}}\psi=\mathbf{R}\partial_{\boldsymbol{\mathcal{Q}}}\psi,\quad \dot{\boldsymbol{\mathcal{Q}}}=\mathbf{R}^T\,\overset{\odot}{\mathbf{q}},\quad \partial_{\boldsymbol{\mathcal{Q}}}\psi\cdot\dot{\boldsymbol{\mathcal{Q}}}=\partial_{\mathbf{q}}\psi\cdot\overset{\odot}{\mathbf{q}}\,.$$

Hence we formally consider $\mathcal{T},\boldsymbol{\mathcal{T}},\mathbf{q}$ as independent variables in place of $\mathfrak{T},\boldsymbol{\mathfrak{T}},\boldsymbol{\mathcal{Q}}$ and take the reduced inequality in the form

$$\tfrac{1}{3}\mathcal{T}\,\nabla\cdot\mathbf{v}+\boldsymbol{\mathcal{T}}_0\cdot\mathbf{D}_0-\frac{1}{\theta}\mathbf{q}\cdot\nabla\theta-\rho\partial_{\mathcal{T}}\psi\cdot\dot{\mathcal{T}}-\rho\partial_{\boldsymbol{\mathcal{T}}_0}\psi\cdot\overset{\odot}{\boldsymbol{\mathcal{T}}}_0-\rho\partial_{\mathbf{q}}\psi\cdot\overset{\odot}{\mathbf{q}}\geq 0. \quad (6.37)$$

We now consider rate equations for $\boldsymbol{\mathcal{T}}_0$, $\mathcal{T}$, $\mathbf{q}$ and show two views to determine the consequences of (6.37).

Rate equations and thermodynamic restrictions

For definiteness we let $\boldsymbol{\mathcal{T}}_0$, $\mathcal{T}$, $\mathbf{q}$ be given by the rate equations

$$\tau_T\dot{\mathcal{T}}+\mathcal{T}=(3\lambda+2\mu)\nabla\cdot\mathbf{v},\qquad \tau_T\,\overset{\odot}{\boldsymbol{\mathcal{T}}}_0+\boldsymbol{\mathcal{T}}_0=2\mu\mathbf{D}_0,\qquad \tau_q\,\overset{\odot}{\mathbf{q}}+\mathbf{q}=-\kappa\nabla\theta, \quad (6.38)$$

so that the Navier-Stokes-Fourier scheme is recovered in stationary conditions; in terms of $\boldsymbol{\mathfrak{T}}_0$, $\mathfrak{T}$, $\boldsymbol{\mathcal{Q}}$ the rate equations are

$$\tau_T\dot{\mathfrak{T}}+\mathfrak{T}=(3\lambda+2\mu)\nabla\cdot\mathbf{v},\qquad \tau_T\dot{\boldsymbol{\mathfrak{T}}}_0+\boldsymbol{\mathfrak{T}}_0=2\mu\mathbf{R}\mathbf{D}_0\mathbf{R}^T,\qquad \tau_q\dot{\boldsymbol{\mathcal{Q}}}+\boldsymbol{\mathcal{Q}}=-\kappa\mathbf{R}\nabla\theta.$$

Substitute $\nabla\cdot\mathbf{v}$, $\mathbf{D}_0$, $\nabla\theta$ from (6.38) into (6.37) to obtain

$$\begin{aligned}(-\rho\partial_{\mathcal{T}}\psi+\frac{\tau_T}{3(2\mu+3\lambda)}\mathcal{T})\dot{\mathcal{T}}+(-\rho\partial_{\boldsymbol{\mathcal{T}}_0}\psi+\frac{\tau_T}{2\mu}\boldsymbol{\mathcal{T}}_0)\cdot\overset{\odot}{\boldsymbol{\mathcal{T}}}_0+(-\rho\partial_{\mathbf{q}}\psi+\frac{\tau_q}{\kappa\theta}\mathbf{q})\cdot\overset{\odot}{\mathbf{q}}\\ +\frac{1}{3(2\mu+3\lambda)}\mathcal{T}^2+\frac{1}{2\mu}\boldsymbol{\mathcal{T}}_0\cdot\boldsymbol{\mathcal{T}}_0+\frac{1}{\kappa\theta}\mathbf{q}\cdot\mathbf{q}\geq 0.\end{aligned}$$

Since $\nabla\cdot\mathbf{v}$, $\mathbf{D}_0$, $\nabla\theta$ are allowed to take arbitrary values then the same is true for $\dot{\mathcal{T}},\overset{\odot}{\boldsymbol{\mathcal{T}}}_0,\overset{\odot}{\mathbf{q}}$. Hence the inequality holds if and only if

$$-\rho\partial_{\mathcal{T}}\psi+\frac{\tau_T}{3(2\mu+3\lambda)}\mathcal{T}=0,\qquad -\rho\partial_{\boldsymbol{\mathcal{T}}_0}\psi+\frac{\tau_T}{2\mu}\boldsymbol{\mathcal{T}}_0=\mathbf{0},\qquad -\rho\partial_{\mathbf{q}}\psi+\frac{\tau_q}{\kappa\theta}\mathbf{q}=\mathbf{0},$$

$$\frac{1}{3(2\mu+3\lambda)}\mathcal{T}^2 + \frac{1}{2\mu}\boldsymbol{\mathcal{T}}_0 \cdot \boldsymbol{\mathcal{T}}_0 + \frac{1}{\kappa\theta}\mathbf{q}\cdot\mathbf{q} \geq 0.$$

The inequality then implies the classical restrictions

$$\mu > 0, \qquad 2\mu + 3\lambda > 0, \qquad \kappa > 0. \tag{6.39}$$

By direct integrations, the free energy takes the form

$$\psi = \Psi(\rho,\theta) + \frac{\tau_T}{6\rho(2\mu+3\lambda)}\mathcal{T}^2 + \frac{\tau_T}{4\rho\mu}\boldsymbol{\mathcal{T}}_0 \cdot \boldsymbol{\mathcal{T}}_0 + \frac{1}{2\rho\kappa\theta}\mathbf{q}\cdot\mathbf{q}. \tag{6.40}$$

Since τ_T and τ_q are assumed to be positive then the free energy has a minimum at equilibrium, that is when $\mathcal{T}, \boldsymbol{\mathcal{T}}_0$, and $\mathbf{q}$ are zero.

If, instead, we substitute $\dot{\mathcal{T}}, \overset{\odot}{\boldsymbol{\mathcal{T}}}_0, \overset{\odot}{\mathbf{q}}$ from (6.38) into (6.37) and use the arbitrariness of $\nabla\cdot\mathbf{v}$, $\mathbf{D}_0$, $\nabla\theta$ we eventually find

$$\frac{\rho(2\mu+3\lambda)}{\tau_T}\partial_{\mathcal{T}}\psi = \tfrac{1}{3}\mathcal{T}, \quad \frac{2\rho\mu}{\tau_T}\partial_{\boldsymbol{\mathcal{T}}_0}\psi = \boldsymbol{\mathcal{T}}_0, \quad \frac{\rho\kappa}{\tau_q}\partial_{\mathbf{q}}\psi = \frac{1}{\theta}\mathbf{q},$$

$$\frac{1}{\tau_T}\partial_{\mathcal{T}}\psi\,\mathcal{T} + \frac{1}{\tau_T}\partial_{\boldsymbol{\mathcal{T}}_0}\psi\cdot\boldsymbol{\mathcal{T}}_0 + \frac{1}{\tau_q}\partial_{\mathbf{q}}\psi\cdot\mathbf{q} \geq 0.$$

The usual integrations provide the free energy function and hence the restrictions on μ, $2\mu+3\lambda$, κ; the results (6.39) and (6.40) follow again.

6.2.3 *Rate Equations in the Material Description*

It is crucial for the next developments that
The material fields $\mathbf{q}_R$ and $\mathbf{T}_{RR}$ *are invariant under changes of frame.*
To check these invariance properties we observe that under the change of frame $\mathcal{F} \to \mathcal{F}^*$ we have

$$\mathbf{q}_R^* = J\,\mathbf{Q}\,\mathbf{q}(\mathbf{QF})^{-T} = J\,\mathbf{Q}\,\mathbf{q}\,\mathbf{F}^{-T}\mathbf{Q} = J\,\mathbf{Q}^T\mathbf{Q}\,\mathbf{q}\,\mathbf{F}^{-T} = \mathbf{q}_R$$

and

$$\mathbf{T}_{RR}^* = J(\mathbf{QF})^{-1}\mathbf{QTQ}^T(\mathbf{QF})^{-T} = J\mathbf{F}^{-1}\mathbf{Q}^T\mathbf{QTQ}^T\mathbf{QF}^{-T} = J\mathbf{F}^{-1}\mathbf{TF}^{-T} = \mathbf{T}_{RR}.$$

Thus the time derivatives $\dot{\mathbf{q}}_R$ and $\dot{\mathbf{T}}_{RR}$ too are invariant under a change of frame.

Also by analogy with the composite models of springs and dashpots, we let the elastic and viscous behaviours of the solid be described by a stress $\mathbf{Y}$, subject to a rate equation, and by the complementary part $\mathbf{T}_{RR} - \mathbf{Y}$; it is natural to view $\mathbf{Y}$ as the dissipative stress and $\mathbf{T}_{RR} - \mathbf{Y}$ as the elastic stress. We then assume that $\mathbf{Y}$ and $\mathbf{q}_R$ are governed by rate equations of the form

$$\dot{\mathbf{q}}_R = -\lambda(\mathbf{q}_R + \mathbf{K}\nabla_R \theta), \tag{6.41}$$

$$\dot{\mathbf{Y}} = -\Lambda(\mathbf{Y} - \mathbf{G}\dot{\mathbf{C}}). \tag{6.42}$$

We assume that the scalars λ, Λ, the second-order tensor $\mathbf{K}$ and the fourth-order tensor $\mathbf{G}$ are invariant so that the rate equations (6.41) and (6.42) are invariant. We assume that $\lambda \geq \lambda_0 > 0$, $\Lambda \geq \Lambda_0 > 0$ and that $\mathbf{K}$ and $\mathbf{G}$ are invertible. In stationary conditions

$$\mathbf{q}_R = -\mathbf{K}\,\nabla_R\,\theta, \qquad \mathbf{Y} = \mathbf{G}\,\dot{\mathbf{C}}.$$

Hence $\mathbf{K}$ and $\mathbf{G}$ can be viewed as the conductivity tensor and the viscosity tensor (loss modulus) in R.

We now examine the compatibility of (6.41) and (6.42) with the second law of thermodynamics via the Clausius-Duhem inequality (2.43). We let the extra-entropy flux be zero and show that the scheme proves to be compatible with the second law. Let ψ, η, $\mathbf{q}_R$, $\mathbf{T}_{RR}$ be given by continuous functions of the set of independent variables

$$\Gamma = (\theta, \mathbf{C}, \mathbf{q}_R, \mathbf{Y}, \nabla_R\theta, \dot{\mathbf{C}}).$$

Moreover let ψ be a C^2 function. The scalars λ, Λ and the tensors $\mathbf{K}$, $\mathbf{G}$ are allowed to depend on the temperature θ. Inequality (2.43) then reads

$$-\rho_R(\partial_\theta\psi + \eta)\dot{\theta} + (\tfrac{1}{2}\mathbf{T}_{RR} - \rho_R\partial_{\mathbf{C}}\psi - \rho_R\Lambda\mathbf{G}^T\partial_{\mathbf{Y}}\psi)\cdot\dot{\mathbf{C}} + (\rho_R\,\lambda\mathbf{K}^T\partial_{\mathbf{q}_R}\psi - \frac{1}{\theta}\mathbf{q}_R)\cdot\nabla_R\theta$$
$$+\rho_R\lambda\partial_{\mathbf{q}_R}\psi\cdot\mathbf{q}_R + \rho_R\Lambda\partial_{\mathbf{Y}}\psi\cdot\mathbf{Y} - \rho_R\partial_{\nabla_R\theta}\psi\cdot\nabla_R\dot{\theta} - \rho_R\partial_{\dot{\mathbf{C}}}\psi\cdot\ddot{\mathbf{C}} \geq 0.$$

The arbitrariness of the values $\nabla_R\dot{\theta}$, $\ddot{\mathbf{C}}$, $\dot{\theta}$ implies that

$$\partial_{\nabla_R\theta}\psi = 0, \qquad \partial_{\dot{\mathbf{C}}}\psi = 0, \qquad \eta = -\partial_\theta\psi.$$

Moreover we make the identification

$$\mathbf{Y} := \mathbf{T}_{RR} - 2\rho_R\partial_{\mathbf{C}}\psi.$$

Hence the inequality reduces to

$$(\tfrac{1}{2}\mathbf{Y} - \rho_R\Lambda\mathbf{G}^T\partial_{\mathbf{Y}}\psi)\cdot\dot{\mathbf{C}} + (\rho_R\,\lambda\mathbf{K}^T\partial_{\mathbf{q}_R}\psi - \frac{1}{\theta}\mathbf{q}_R)\cdot\nabla_R\theta + \rho_R\lambda\partial_{\mathbf{q}_R}\psi\cdot\mathbf{q}_R + \rho_R\Lambda\partial_{\mathbf{Y}}\psi\cdot\mathbf{Y} \geq 0.$$

The arbitrariness of $\dot{\mathbf{C}}$ and $\nabla_R\,\theta$ implies that

$$\rho_R\Lambda\mathbf{G}^T\partial_{\mathbf{Y}}\psi = \tfrac{1}{2}\mathbf{Y}, \qquad \rho_R\,\lambda\,\mathbf{K}^T\partial_{\mathbf{q}_R}\psi = \frac{1}{\theta}\mathbf{q}_R, \tag{6.43}$$

$$\rho_R \lambda \partial_{\mathbf{q}_R} \psi \cdot \mathbf{q}_R + \rho_R \Lambda \partial_{\mathbf{Y}} \psi \cdot \mathbf{Y} \geq 0. \tag{6.44}$$

Since $\mathbf{G}$ and $\mathbf{K}$ are invertible we can write

$$\partial_{\mathbf{Y}} \psi = \frac{1}{2\rho_R \Lambda} \mathbf{G}^{-T} \mathbf{Y}, \qquad \partial_{\mathbf{q}_R} \psi = \frac{1}{\rho_R \lambda \theta} \mathbf{K}^{-T} \mathbf{q}_R$$

and observe

$$\partial_{\mathbf{Y}} \partial_{\mathbf{Y}} \psi = \frac{1}{2\rho_R \Lambda} \mathbf{G}^{-T}, \qquad \partial_{\mathbf{q}_R} \partial_{\mathbf{q}_R} \psi = \frac{1}{\rho_R \lambda \theta} \mathbf{K}^{-T}.$$

The symmetry of $\partial_{\mathbf{Y}} \partial_{\mathbf{Y}} \psi$ and $\partial_{\mathbf{q}_R} \partial_{\mathbf{q}_R} \psi$ implies the symmetry of $\mathbf{G}$ and $\mathbf{K}$,

$$\mathbf{G} = \mathbf{G}^T, \qquad \mathbf{K} = \mathbf{K}^T.$$

Consequently, by integration of Eqs. (6.43) we have

$$\psi = \Psi(\theta, \mathbf{C}) + \frac{1}{4\rho_R \Lambda} \mathbf{Y} \cdot \mathbf{G}^{-1} \mathbf{Y} + \frac{1}{2\rho_R \lambda \theta} \mathbf{q}_R \cdot \mathbf{K}^{-1} \mathbf{q}_R.$$

Hence inequality (6.44) results in

$$\mathbf{Y} \cdot \mathbf{G}^{-1} \mathbf{Y} \geq 0, \qquad \mathbf{q}_R \cdot \mathbf{K}^{-1} \mathbf{q}_R \geq 0.$$

This means that the viscosity tensor $\mathbf{G}$ and the conductivity tensor $\mathbf{K}$ are required to be positive definite.

We now look for the rate equations, in the spatial description, associated with (6.41) and (6.42). First we observe that, by viewing ψ as a function of $\mathbf{F}$ through $\mathbf{C}$, the chain rule implies

$$\partial_{\mathbf{F}} \psi = 2\, \mathbf{F}\, \partial_{\mathbf{C}} \psi.$$

Thus

$$\rho\, \partial_{\mathbf{F}} \psi\, \mathbf{F}^T = 2\, \rho\, \mathbf{F}\, \partial_{\mathbf{C}} \psi\, \mathbf{F}^T$$

and hence

$$2\, \rho_R \partial_{\mathbf{C}} \psi = \rho\, J\, \mathbf{F}^{-1} \partial_{\mathbf{F}} \psi = J\, \mathbf{F}^{-1} (\rho\, \partial_{\mathbf{F}} \psi\, \mathbf{F}^T) \mathbf{F}^{-T}.$$

This in turn requires that

$$\mathbf{Y} := \mathbf{T}_{RR} - 2\, \rho_R\, \partial_{\mathbf{C}} \psi = J\, \mathbf{F}^{-1} (\mathbf{T} - \rho\, \partial_{\mathbf{F}} \psi\, \mathbf{F}^T) \mathbf{F}^{-T}.$$

Hence

$$\mathcal{T} := \mathbf{T} - \rho\, \partial_{\mathbf{F}} \psi\, \mathbf{F}^T$$

is the dissipative stress in the spatial description (dissipative Cauchy stress).

Substitution of

$$\mathbf{q}_R = J\,\mathbf{F}^{-1}\mathbf{q}, \qquad \mathbf{Y} = J\,\mathbf{F}^{-1}\mathcal{T}\mathbf{F}^{-T}$$

and use of

$$\dot{J} = J\,\nabla\cdot\mathbf{v}, \qquad \overline{\mathbf{F}^{-1}}^{\,\cdot} = -\mathbf{F}^{-1}\,\dot{\mathbf{F}}\,\mathbf{F}^{-1}$$

yield

$$\dot{\mathbf{q}}_R = J\,\mathbf{F}^{-1}\,(\dot{\mathbf{q}} - \mathbf{L}\mathbf{q} + \nabla\cdot\mathbf{v}\,\mathbf{q}),$$

$$\dot{\mathbf{Y}} = J\mathbf{F}^{-1}(\dot{\mathcal{T}} + \nabla\cdot\mathbf{v}\mathcal{T} - \mathbf{L}\mathcal{T} - \mathcal{T}\mathbf{L}^T)\mathbf{F}^{-T}.$$

By (6.41) and (6.42) we obtain

$$\dot{\mathbf{q}} - \mathbf{L}\mathbf{q} + \nabla\cdot\mathbf{v}\,\mathbf{q} + \lambda\mathbf{q} = -\frac{\lambda}{J}\mathbf{F}\,\mathbf{K}\,\nabla_R\theta,$$

$$\dot{\mathcal{T}} - \mathbf{L}\mathcal{T} - \mathcal{T}\mathbf{L}^T + \nabla\cdot\mathbf{v}\mathcal{T} + \Lambda\mathcal{T} = \frac{\Lambda}{J}\mathbf{F}\,\mathbf{G}\dot{\mathbf{C}}\,\mathbf{F}^T.$$

To express the right-hand sides by spatial gradients we use the identities

$$\mathbf{F}\,\nabla_R\theta = \mathbf{F}\,\mathbf{F}^T\,\nabla\theta, \qquad \overline{\mathbf{F}^T\mathbf{F}}^{\,\cdot} = \mathbf{F}^T\,(\mathbf{L} + \mathbf{L}^T)\mathbf{F}.$$

Hence we have

$$\dot{\mathbf{q}} - \mathbf{L}\mathbf{q} + \nabla\cdot\mathbf{v}\,\mathbf{q} + \lambda\mathbf{q} = -\frac{\lambda}{J}\mathbf{F}\,\mathbf{K}\,\mathbf{F}^T\,\nabla\theta, \tag{6.45}$$

$$\dot{\mathcal{T}} - \mathbf{L}\mathcal{T} - \mathcal{T}\mathbf{L}^T + \nabla\cdot\mathbf{v}\mathcal{T} + \Lambda\mathcal{T} = \frac{2\Lambda}{J}\mathbf{F}\,\mathsf{G}\,\mathbf{F}^T\mathbf{D}\mathbf{F}\,\mathbf{F}^T. \tag{6.46}$$

If we approximate $\mathbf{F}$ on the right as the unit tensor we find

$$\dot{\mathbf{q}} - \mathbf{L}\mathbf{q} + \nabla\cdot\mathbf{v}\,\mathbf{q} + \lambda\mathbf{q} = -\lambda\,\mathbf{K}\nabla\theta$$

$$\dot{\mathcal{T}} - \mathbf{L}\mathcal{T} - \mathcal{T}\mathbf{L}^T + \nabla\cdot\mathbf{v}\mathcal{T} + \Lambda\mathcal{T} = 2\Lambda\,\mathsf{G}\,\mathbf{D}.$$

The rate equations (6.45) and (6.46) are compatible with thermodynamics, the compatibility being proved within the material description. Thus the Truesdell rates are compatible with thermodynamics and this follows by the identities

$$\dot{\mathbf{q}}_R = J\mathbf{F}^{-1}\,\overset{\square}{\mathbf{q}}, \qquad \dot{\mathbf{Y}} = J\mathbf{F}^{-1}\,\overset{\square}{\mathcal{T}}\,\mathbf{F}^{-T}.$$

In undeformable bodies ($\mathbf{L} = \mathbf{0}$, $J = 1$) Eq. (6.45) reduces to the classical Maxwell-Cattaneo model for heat conduction. In deformable materials we see that the Truesdell derivatives for the rate equations are compatible with thermodynamics if appropriate terms involving $\nabla\theta$ and $\mathbf{D}$ are considered.

To determine $\mathcal{T}$ and $\mathbf{q}$ we go back to (6.42) and (6.41). Let λ, $\mathbf{K}$, Λ, $\mathbf{G}$ be known functions of time. Then by (6.42) we find

$$\mathbf{Y}(t) = \int_{-\infty}^{t} \Lambda(\tau)\mathbf{G}(\tau)\exp(-\int_{\tau}^{t}\Lambda(s)ds)\dot{\mathbf{C}}(\tau)d\tau,$$

where the assumption $\Lambda \geq \Lambda_0 > 0$ is used. Since $\mathbf{Y} = J\,\mathbf{F}^{-1}\mathcal{T}\,\mathbf{F}^{-T}$ it follows

$$\mathcal{T}(t) = \frac{1}{J(t)}\mathbf{F}(t)\int_{-\infty}^{t} \Lambda(\tau)\mathbf{G}(\tau)\exp(-\int_{\tau}^{t}\Lambda(s)ds)\dot{\mathbf{C}}(\tau)d\tau\,\mathbf{F}^{T}(t).$$

The analogue holds for $\mathbf{q}$ with obvious changes.

As for Eqs. (6.42) and (6.41) we observe that the right-hand sides are written by analogy with Fourier's and Navier-Stokes' laws. Yet, in the spatial description the rate equation, with the rate $\overset{\square}{\mathbf{q}}$, is

$$\overset{\square}{\mathbf{q}} = -\lambda(\mathbf{q} + \kappa\nabla\theta).$$

Now, $\mathbf{q}$ and $\nabla\theta$ have different transformation laws, from $\mathcal{R}$ to R,

$$\mathbf{q}_R = J\mathbf{F}^{-1}\mathbf{q}, \qquad \nabla_R\,\theta = \mathbf{F}^{T}\nabla\theta.$$

Hence

$$J\mathbf{F}^{-1}(\mathbf{q} + \kappa\nabla\theta) = \mathbf{q}_R + \kappa J\mathbf{C}^{-1}\nabla_R\,\theta$$

and the rate equation becomes

$$\dot{\mathbf{q}}_R = \lambda(\mathbf{q}_R + \mathbf{K}\,\nabla_R\,\theta), \qquad \mathbf{K} = \kappa J\mathbf{C}^{-1}.$$

Remarkably, this shows that an isotropic conductivity in the spatial description produces an anisotropic conductivity in the reference configuration, this being produced by the different properties of $\mathbf{q}$ and $\nabla\theta$. Upon integration we find that

$$\mathbf{q}_R(t) = \int_{-\infty}^{t} \kappa(\tau)J(\tau)\exp(-\int_{\tau}^{t}\lambda(s)ds)\mathbf{C}^{-1}(\tau)\nabla_R\theta(\tau)d\tau$$

whence

$$\mathbf{q}(t) = \frac{1}{J(t)}\mathbf{F}(t)\int_{-\infty}^{t} \kappa(\tau)J(\tau)\exp(-\int_{\tau}^{t}\lambda(s)ds)\mathbf{F}^{-1}(\tau)\nabla\theta(\tau)d\tau.$$

It is of interest that this solution is independent on the chosen reference configuration. For, let

$$\mathbf{F} = \tilde{\mathbf{F}}\mathbf{F}_0,$$

$\mathbf{F}_0$ being the deformation gradient of the intermediate configuration $\tilde{\mathrm{R}}$ relative to the reference configuration R. Let $\tilde{J} = \det \tilde{\mathbf{F}}$. Upon substitution it follows that

$$\mathbf{q}(t) = \frac{1}{\tilde{J}(t)}\tilde{\mathbf{F}}(t) \int_{-\infty}^{t} \kappa(\tau)\tilde{J}(\tau) \exp(-\int_{\tau}^{t} \lambda(s)ds)\tilde{\mathbf{F}}^{-1}(\tau)\nabla\theta(\tau).$$

Hence $\mathbf{q}$ is formally invariant under the change $\mathrm{R} \to \tilde{\mathrm{R}}$ of the reference configuration.

6.2.4 Nonlinear Rate-Type Models of Fluids

Based on the classical models of Navier-Stokes and Fourier we show how nonlinear rate-type models of fluids by letting the entropy production σ be an appropriate constitutive function. Hence we apply the standard Clausius-Duhem inequality where σ occurs in the form

$$-\rho(\dot{\psi} + \eta\dot{\theta}) + \mathbf{T}\cdot\mathbf{D} - \frac{1}{\theta}\mathbf{q}\cdot\nabla\theta = \theta\sigma \geq 0.$$

Remarkable rate-type equations are shown to follow by letting some time derivatives be mutually dependent.

Heat-conducting fluids

Heat-conducting fluids are modelled by letting the constitutive functions depend on the mass density ρ, the temperature θ, and possibly their derivatives. For formal convenience we use the Jacobian $J = \det \mathbf{F} = \rho_R/\rho$ as an independent variable in place of ρ. Hence we start with models where

$$\Xi = (J, \theta, \dot{J}, \dot{\theta}, \nabla\theta)$$

is the set of variables and let $\psi, \eta, \mathbf{T}, \mathbf{q}, \sigma$ be functions of Ξ. Indeed, to begin with, we let

$$\mathbf{T} = -p\mathbf{1}, \qquad \mathbf{q} = -\kappa\nabla\theta,$$

thus accounting for Pascal's principle about pressure and Fourier's law about $\mathbf{q}$. The Clausius-Duhem inequality can be written in the form

$$-\rho(\dot{\psi} + \eta\dot{\theta}) - p\,\mathrm{tr}\,\mathbf{D} = \theta\zeta, \qquad \zeta = \sigma - \sigma_F, \qquad \sigma_F := \frac{\kappa}{\theta^2}|\nabla\theta|^2,$$

ζ being the entropy production due to mechanical dissipation. Since $\zeta = \sigma$ when $\nabla\theta = \mathbf{0}$ then $\zeta \geq 0$.

Compute $\dot{\psi}$ and substitute in the Clausius-Duhem inequality to find

$$(\partial_J \psi + p/\rho_R)\dot{J} + (\partial_\theta \psi + \eta)\dot{\theta} + \partial_{\dot{J}}\psi \ddot{J} + \partial_{\dot{\theta}}\psi\ddot{\theta} + \partial_{\nabla\theta}\psi \cdot \dot{\overline{\nabla\theta}} = -J\theta\zeta/\rho_R.$$

The arbitrariness of $\ddot{\theta}$, $\ddot{J}$, and $\dot{\overline{\nabla\theta}}$ implies

$$\partial_{\dot{\theta}}\psi = 0, \qquad \partial_{\dot{J}}\psi = 0, \qquad \partial_{\nabla\theta}\psi = \mathbf{0}.$$

Consequently $\psi = \psi(J, \theta)$ and

$$(\partial_J \psi + p/\rho_R)\dot{J} + (\partial_\theta \psi + \eta)\dot{\theta} = -J\theta\zeta/\rho_R. \tag{6.47}$$

Heat-conducting thermoelastic fluids

We consider thermoelastic fluids characterized by $\zeta = 0$ while p, η, in addition to ψ are independent of $\nabla\theta$. Hence $\psi(J, \theta), \eta(J, \theta), p(J, \theta)$ are required to satisfy the equality

$$(\partial_J \psi + p/\rho_R)\dot{J} + (\partial_\theta \psi + \eta)\dot{\theta} = 0. \tag{6.48}$$

If $\dot{J}$ and $\dot{\theta}$ are mutually independent then it follows

$$\eta = -\partial_\theta \psi, \qquad p = -\rho_R \partial_J \psi;$$

the relation for p is usually referred to as the equation of state of the fluid. By way of example, let $\psi = \psi_0(\theta) - \alpha(\theta)\ln J$. Hence $\rho = \rho_R/J$ and

$$\eta = -\psi_0'(\theta) + \alpha'(\theta)\ln J, \qquad p = \alpha(\theta)\rho.$$

If, in particular, $\psi_0(\theta) = c_v\theta(1 - \ln\theta)$, $\alpha(\theta) = \beta\theta$ we find

$$\eta = c_v \ln\theta + \beta \ln J, \qquad pJ = \rho_R \beta\theta.$$

In terms of the specific volume $v = 1/\rho = J/\rho_R$ we have[5]

$$\eta = c_v \ln\theta + \beta \ln \rho_R v, \qquad pv = \beta\theta, \qquad \varepsilon = \psi + \theta\eta = c_v\theta.$$

If, instead, $\dot{\theta}$ and $\dot{J}$ are not mutually independent then η and p are no longer given by partial derivatives of ψ. Now, if

$$\partial_\theta(\partial_J \psi + p/\rho_R) = \partial_J(\partial_\theta\psi + \eta), \qquad \text{i.e.} \quad \partial_\theta p = \rho_R \partial_J \eta,$$

[5] Equation $pv = \beta\theta$ is the equation of state of an ideal gas, with $\beta = R/M$, R being the gas constant and M the molar mass.

it follows that Eq. (6.48) is integrable; there is then a function, Φ say, such that $\Phi(\theta, J)$ is constant thus resulting in a connection between θ and J. As an example we let

$$\eta = -\psi_0'(\theta) + \alpha'(\theta)\ln J, \qquad p = \rho_R\alpha(\theta)/J$$

and find

$$\Phi = \psi(\theta, J) - \psi_0(\theta) + \alpha(\theta)\ln J.$$

If $p + \rho_R\partial_J\psi \neq 0$ or $\eta + \partial_\theta\psi \neq 0$ we can write the relations

$$\dot{J} = -\frac{\partial_\theta\psi + \eta}{\partial_J\psi + p/\rho_R}\dot{\theta} \qquad \text{or} \qquad \dot{\theta} = -\frac{\partial_J\psi + p/\rho_R}{\partial_\theta\psi + \eta}\dot{J}.$$

Dissipative thermoelastic fluids

The dissipative character is related to a nonzero value of $\zeta > 0$. If, e.g., $\eta + \partial_\theta\psi \neq 0$ then we can write

$$\dot{\theta} = -\frac{\partial_J\psi + p/\rho_R}{\partial_\theta\psi + \eta}\dot{J} - \frac{J\theta\zeta}{\partial_\theta\psi + \eta}.$$

For appropriate dependences of ζ this relation allows the description of hysteretic phenomena among θ and J, the pressure p being a parameter.

Thermoviscous fluids

Stricter dissipative properties and more general models are described by letting an extra stress and an extra heat flux occur so that

$$\mathbf{T} = -p\,\mathbf{1} + \mathbf{S}, \qquad \mathbf{q} = -\kappa\nabla\theta + \mathbf{y}.$$

By means of the decomposition $\mathbf{S} = \mathbf{S}_0 + \frac{1}{3}(\mathrm{tr}\,\mathbf{S})\mathbf{1}$ the Clausius-Duhem inequality becomes

$$-\rho(\dot{\psi} + \eta\dot{\theta}) + (-p + \tfrac{1}{3}\mathrm{tr}\,\mathbf{S})\mathrm{tr}\,\mathbf{D} + \mathbf{S}_0\cdot\mathbf{D}_0 - \frac{1}{\theta}\mathbf{y}\cdot\nabla\theta = \theta\zeta, \tag{6.49}$$

where $\zeta = \sigma - \kappa|\nabla\theta|^2$.

Consider the referential quantities $\mathbf{T}_{RR} = J\mathbf{F}^{-1}\mathbf{T}\mathbf{F}^{-T}$, $\mathbf{q}_R = J\mathbf{F}^{-1}\mathbf{q}$, $\dot{\mathbf{E}} = \mathbf{F}^T\mathbf{D}\mathbf{F}$, $\nabla_R\theta = \mathbf{F}^T\nabla\theta$ and observe

$$\mathrm{tr}\,\mathbf{D} = \mathbf{C}^{-1}\cdot\dot{\mathbf{E}} = \mathrm{tr}\,(\mathbf{C}^{-1}\dot{\mathbf{E}}), \quad \mathbf{T}\cdot\mathbf{D} = J\mathbf{T}_{RR}\cdot\dot{\mathbf{E}}, \quad |\nabla\theta|^2 = \nabla_R\theta\cdot\mathbf{C}^{-1}\nabla_R\theta.$$

Hence we can express the stress tensor, the heat flux, and the Clausius-Duhem inequality in the form

$$\mathbf{T}_{RR} = -pJ\mathbf{C}^{-1} + \mathbf{S}_{RR}, \qquad \mathbf{q}_R = -\kappa J\mathbf{C}^{-1}\nabla_R\theta + \mathbf{y}_R, \tag{6.50}$$

$$-\rho_R(\dot{\psi} + \eta\dot{\theta}) - pJ\mathrm{tr}\,(\mathbf{C}^{-1}\dot{\mathbf{E}}) + \mathbf{S}_{RR}\cdot\dot{\mathbf{E}} - \frac{1}{\theta}\mathbf{y}_R\cdot\nabla_R\theta = J\theta\zeta,$$

where $\zeta = \sigma - \sigma_F$, $\sigma_F = \kappa\nabla_R\theta\cdot\mathbf{C}^{-1}\nabla_R\theta/\theta^2$.

Some models are now sketched by having recourse to appropriate sets of variables Ξ, free energies ψ, entropy productions σ.

The Navier-Stokes-Fourier fluid

Heat conduction is modelled by $\mathbf{q} = -\kappa\nabla\theta$ and then $\mathbf{y} = \mathbf{0}$. The fluid is compressible and hence we let

$$\Xi = (J, \theta, \dot{J}, \dot{\theta}, \mathbf{D}_0, \nabla\theta)$$

be the set of variables. Time differentiation of $\psi(\Xi)$ and substitution in (6.49) yield

$$J^{-1}(\rho_R\partial_J\psi + p - \tfrac{1}{3}\mathrm{tr}\,\mathbf{S})\dot{J} + \rho(\partial_\theta\psi + \eta)\dot{\theta} - \mathbf{S}_0\cdot\mathbf{D}_0 + \rho\partial_{\dot{J}}\psi\ddot{J} + \rho\partial_{\mathbf{D}_0}\psi\cdot\dot{\mathbf{D}}_0 + \rho\partial_{\nabla\theta}\psi\cdot\dot{\overline{\nabla\theta}} = -\theta\zeta,$$

where $\zeta = \zeta(\Xi)$. The arbitrariness of $\ddot{\theta}$, $\ddot{J}$, $\dot{\mathbf{D}}_0$, and $\dot{\overline{\nabla\theta}}$ implies that ψ is independent of $\dot{\theta}$, $\dot{J}$, $\mathbf{D}_0$, and $\nabla\theta$. Hence $\psi = \psi(J, \theta)$ and the Clausius-Duhem inequality simplifies to

$$J^{-1}(\rho_R\partial_J\psi + p - \tfrac{1}{3}\mathrm{tr}\,\mathbf{S})\dot{J} + \rho(\partial_\theta\psi + \eta)\dot{\theta} - \mathbf{S}_0\cdot\mathbf{D}_0 = -\theta\zeta. \tag{6.51}$$

The Navier-Stokes-Fourier fluid is characterized by

$$\zeta = \zeta_{NS} := \frac{1}{\theta}(\lambda + \tfrac{2}{3}\mu)(\mathrm{tr}\,\mathbf{D})^2 + \frac{2\mu}{\theta}|\mathbf{D}_0|^2, \qquad \sigma = \zeta_{NS} + \frac{\kappa}{\theta^2}|\nabla\theta|^2,$$

where μ, $3\lambda + 2\mu$, κ are non-negative valued functions of J, θ. If $\dot{\theta}$ and $\dot{J}$ are mutually independent and p is independent of $\dot{J}$ then it follows from (6.51) that

$$p = -\rho_R\partial_J\psi, \qquad \eta = -\partial_\theta\psi, \qquad \tfrac{1}{3}(\mathrm{tr}\,\mathbf{S})(\mathrm{tr}\,\mathbf{D}) + \mathbf{S}_0\cdot\mathbf{D}_0 = \theta\zeta_{NS}.$$

This in turn implies that

$$\mathrm{tr}\,\mathbf{S} = (3\lambda + 2\mu)\mathrm{tr}\,\mathbf{D}, \qquad \mathbf{S}_0 = 2\mu\mathbf{D}_0$$

and then

$$\mathbf{T} = (\rho_R\partial_J\psi + \lambda\,\mathrm{tr}\,\mathbf{D})\mathbf{1} + 2\mu\mathbf{D}.$$

Incompressibility implies $\rho = \rho_R$ and hence $J = 1$, $\mathrm{tr}\,\mathbf{D} = \nabla\cdot\mathbf{v} = 0$. Hence the set of variables reduces to

$$\Xi = (\theta, \dot{\theta}, \mathbf{D}_0, \nabla\theta)$$

and then $\psi = \psi(\theta)$, $\zeta_{NS} = (2\mu/\theta)|\mathbf{D}_0|^2$ and p is unknown while

$$\mathbf{T} = -p\,\mathbf{1} + 2\mu\mathbf{D}, \qquad \mu(\theta) \geq 0.$$

6.3 Rate-Type Models of Solids

In this section the memory properties of (viscoelastic) solids are modelled by rate-type equations. We argue within the referential description and let $\mathbf{T}_{RR}$ and $\mathbf{q}_R$ be given by (6.50). To establish constitutive rate equations we insert $\dot{\mathbf{S}}_{RR}$ and $\dot{\mathbf{y}}_R$ among the variables so that

$$\Xi_R = (\theta, \mathbf{E}, \mathbf{S}_{RR}, \mathbf{y}_R, \dot{\theta}, \dot{\mathbf{E}}, \dot{\mathbf{S}}_{RR}, \dot{\mathbf{y}}_R, \nabla_R\theta)$$

is the set of variables. We then let ψ, η, p, and σ be functions of Ξ_R. The Clausius-Duhem inequality takes the form

$$\rho_R(\partial_\theta\psi + \eta)\dot{\theta} + [\rho_R\partial_{\mathbf{E}}\psi + pJ(\mathbf{1}+2\mathbf{E})^{-1} - \mathbf{S}_{RR}]\cdot\dot{\mathbf{E}} + \rho_R\partial_{\mathbf{S}_{RR}}\psi\cdot\dot{\mathbf{S}}_{RR} + \rho_R\partial_{\mathbf{y}_R}\psi\cdot\dot{\mathbf{y}}_R$$
$$+\rho_R\partial_{\dot{\theta}}\psi\ddot{\theta} + \rho_R\partial_{\dot{\mathbf{E}}}\cdot\ddot{\mathbf{E}} + \rho_R\partial_{\dot{\mathbf{S}}_{RR}}\psi\cdot\ddot{\mathbf{S}}_{RR} + \rho_R\partial_{\dot{\mathbf{y}}_R}\psi\cdot\ddot{\mathbf{y}}_R\rho_R\partial_{\nabla_R\theta}\psi\cdot\nabla_R\dot{\theta} + \frac{1}{\theta}\mathbf{y}_R\cdot\nabla_R\theta = -J\theta\zeta,$$

where $\zeta = \sigma - (\kappa/\theta^2)\nabla_R\theta\cdot(\mathbf{1}+2\mathbf{E})^{-1}\nabla_R\theta$. The arbitrariness of $\nabla_R\dot{\theta}$, $\ddot{\theta}$, $\ddot{\mathbf{E}}$, $\ddot{\mathbf{S}}_{RR}$, $\ddot{\mathbf{y}}_R$, and $\dot{\theta}$ implies that ψ is independent of $\nabla_R\theta$, $\dot{\theta}$, $\dot{\mathbf{E}}$, $\dot{\mathbf{S}}_{RR}$, $\dot{\mathbf{y}}_R$ and η is related to ψ in the classical way so that

$$\psi = \psi(\theta, \mathbf{E}, \mathbf{S}_{RR}, \mathbf{y}_R), \qquad \eta = -\partial_\theta\psi.$$

Though cross-coupling mechanical-thermal effects are thermodynamically allowed, we restrict attention to the case when $\dot{\mathbf{E}}$ and $\dot{\mathbf{S}}_{RR}$ are independent of $\dot{\mathbf{y}}_R$ and $\nabla_R\theta$. Hence it follows

$$[\rho_R\partial_{\mathbf{E}}\psi + pJ(\mathbf{1}+2\mathbf{E})^{-1} - \mathbf{S}_{RR}]\cdot\dot{\mathbf{E}} + \rho_R\partial_{\mathbf{S}_{RR}}\psi\cdot\dot{\mathbf{S}}_{RR} = -J\theta\sigma_{\text{mech}}, \tag{6.52}$$

$$\rho_R\partial_{\mathbf{y}_R}\psi\cdot\dot{\mathbf{y}}_R + \frac{1}{\theta}\mathbf{y}_R\cdot\nabla_R\theta - \frac{\kappa J}{\theta}\nabla_R\theta\cdot\mathbf{C}^{-1}\nabla_R\theta = -J\theta\sigma_{\text{th}}, \tag{6.53}$$

with $\sigma = \sigma_{\text{mech}} + \sigma_{\text{th}}$, $\sigma_{\text{mech}} \geq 0$, $\sigma_{\text{th}} \geq 0$.

6.3.1 The Kelvin-Voigt-Fourier Solid

For definiteness we now establish a class of models where both mechanical dissipation and heat conduction occur. Look at (6.52) and let

$$\sigma_{\text{mech}} = \frac{\nu}{\theta J}|\dot{\mathbf{E}}|^n, \qquad \nu \geq 0, \quad n \in \mathbb{N},$$

where ν is possibly dependent on θ and $\mathbf{E}$. As the simplest model assume $\partial_{\mathbf{S}_{RR}}\psi = \mathbf{0}$ and hence, in light of the representation formula with $\mathbf{Z} = \mathbf{S}_{RR} - \rho_R \partial_{\mathbf{E}}\psi - pJ(\mathbf{1} + 2\mathbf{E})^{-1}$, $\mathbf{N} = \dot{\mathbf{E}}/|\dot{\mathbf{E}}|$ and $\mathbf{G} = \mathbf{0}$ it follows from (6.52) that

$$\mathbf{S}_{RR} = \rho_R \partial_{\mathbf{E}}\psi + pJ(\mathbf{1} + 2\mathbf{E})^{-1} + \nu|\dot{\mathbf{E}}^{n-2}|\,\dot{\mathbf{E}}.$$

Upon substitution of $\mathbf{S}_{RR}$ in (6.50) we find

$$\mathbf{T}_{RR} = \rho_R \partial_{\mathbf{E}}\psi(\theta, \mathbf{E}) + \nu|\dot{\mathbf{E}}|^{n-2}\dot{\mathbf{E}}.$$

If $n = 1$ and $\rho_R \partial_{\mathbf{E}}\psi(\theta, \mathbf{E}) = \mathsf{C}\mathbf{E}$ then we have the linear Kelvin-Voigt model

$$\mathbf{T}_{RR} = \mathsf{C}\mathbf{E} + \nu\dot{\mathbf{E}}.$$

If, rather, we let

$$\sigma_{\text{mech}} = \frac{1}{\theta J}\dot{\mathbf{E}} \cdot \mathsf{N}\dot{\mathbf{E}},$$

where N is a fourth-order tensor then we have

$$\mathbf{T}_{RR} = \mathsf{C}\mathbf{E} + \mathsf{N}\dot{\mathbf{E}}.$$

We now look for a non-linear model with entropy productions

$$\sigma_{\text{mech}} = \frac{\nu}{\theta J}|\dot{\mathbf{E}}|^n, \qquad \sigma_{\text{th}} = \sigma_F + \frac{\lambda}{\theta^2 J}|\nabla_R\theta|^m, \qquad \nu, \lambda \geq 0, \quad n, m \geq 2$$

where ν and λ are possibly dependent on the temperature θ. For definiteness we assume $\partial_{\mathbf{S}_{RR}}\psi, \partial_{\mathbf{y}_R}\psi = \mathbf{0}$. We then apply the representation formula with $\mathbf{Z} = \mathbf{S}_{RR} - \rho_R \partial_{\mathbf{E}}\psi - pJ(\mathbf{1} + 2\mathbf{E})^{-1}$, $\mathbf{N} = \dot{\mathbf{E}}/|\dot{\mathbf{E}}|$, and $\mathbf{G} = \mathbf{0}$ or $\mathbf{Z} = \mathbf{y}_R - \kappa J\mathbf{C}^{-1}\nabla_R\theta$, $\mathbf{N} = \nabla_R\theta/|\nabla_R\theta|$, and $\mathbf{w} = \mathbf{0}$.
It follows from (6.52) and (6.53) that

$$\mathbf{S}_{RR} = \rho_R \partial_{\mathbf{E}}\psi + pJ(\mathbf{1} + 2\mathbf{E})^{-1} + \nu|\dot{\mathbf{E}}^{n-2}|\,\dot{\mathbf{E}},$$

$$\mathbf{y}_R = (\kappa J\nabla_R\theta \cdot \mathbf{C}^{-1}\nabla_R\theta - J\theta^2\sigma_{\text{th}})\frac{\nabla_R\theta}{|\nabla_R\theta|^2} = -\lambda|\nabla_R\theta|^{m-2}\nabla_R\theta.$$

Upon substitution of $\mathbf{S}_{RR}$ and $\mathbf{y}_R$ we find $\mathbf{T}_{RR}$ and $\mathbf{q}_R$ in the form

$$\mathbf{T}_{RR} = \rho_R \partial_{\mathbf{E}}\psi + \nu|\dot{\mathbf{E}}|^{n-2}\dot{\mathbf{E}}, \qquad \mathbf{q}_R = -\kappa J\mathbf{C}^{-1}\nabla_R\theta - \lambda|\nabla_R\theta|^{m-2}\nabla_R\theta. \tag{6.54}$$

If $n = m = 2$ then we have the linear relations

$$\mathbf{T}_{RR} = \rho_R \partial_{\mathbf{E}} \psi + \nu \dot{\mathbf{E}}, \qquad \mathbf{q}_R = -\kappa J \mathbf{C}^{-1} \nabla_R \theta - \lambda \nabla_R \theta.$$

We may rightly refer to Eqs. (6.54) as the non-linear Kelvin-Voigt-Fourier solid.

6.3.2 Thermo-Viscoelastic Materials

To describe thermo-viscoelastic materials we let

$$\Xi_R = (\theta, \mathbf{E}, \mathbf{T}_{RR}, \mathbf{q}_R, \dot{\mathbf{E}}, \dot{\mathbf{T}}_{RR}, \dot{\mathbf{q}}_R, \nabla_R \theta)$$

be the set of variables; the single variables $\dot{\mathbf{E}}, \dot{\mathbf{T}}_{RR}, \dot{\mathbf{q}}_R, \nabla_R \theta$ need not be mutually independent.

We consider the Clausius-Duhem inequality. By the standard arguments we observe that the arbitrariness of $\dot{\theta}$, $\nabla_R \dot{\theta}$, $\ddot{\mathbf{E}}$, $\ddot{\mathbf{T}}_{RR}$, $\ddot{\mathbf{q}}_R$ implies

$$\psi = \psi(\theta, \mathbf{E}, \mathbf{T}_{RR}, \mathbf{q}_R), \qquad \eta = -\partial_\theta \psi,$$

and hence the Clausius-Duhem inequality reduces to

$$(\rho_R \partial_{\mathbf{E}} \psi - \mathbf{T}_{RR}) \cdot \dot{\mathbf{E}} + \rho_R \partial_{\mathbf{T}_{RR}} \psi \cdot \dot{\mathbf{T}}_{RR} + \rho_R \partial_{\mathbf{q}_R} \psi \cdot \dot{\mathbf{q}}_R + \frac{\mathbf{q}_R}{\theta} \cdot \nabla_R \theta = -\theta \sigma_R \le 0,$$

where $\sigma_R = J\sigma$.

Models of thermo-viscoelastic materials are characterized by non-negative (referential) entropy productions $\sigma^R_{\text{mech}} = J\sigma_{\text{mech}}$ and $\sigma^R_{\text{th}} = J\sigma_{\text{th}}$ independent of $\dot{\mathbf{q}}_R$, $\nabla_R \theta$, and $\dot{\mathbf{E}}$, $\dot{\mathbf{T}}_{RR}$. Let

$$\sigma^R_{\text{mech}} = \sigma_R(\theta, \mathbf{E}, \mathbf{T}_{RR}, \mathbf{q}_R, \dot{\mathbf{E}}, \dot{\mathbf{T}}_{RR}, \mathbf{0}, \mathbf{0}), \qquad \sigma^R_{\text{th}} = \sigma_R(\theta, \mathbf{E}, \mathbf{T}_{RR}, \mathbf{q}_R, \mathbf{0}, \mathbf{0}, \dot{\mathbf{q}}_R, \nabla_R \theta)$$

and assume $\dot{\mathbf{E}}$, $\dot{\mathbf{T}}_{RR}$ are independent of $\dot{\mathbf{q}}_R$, $\nabla_R \theta$. Hence the Clausius-Duhem inequality splits into the two inequalities

$$(\rho_R \partial_{\mathbf{E}} \psi - \mathbf{T}_{RR}) \cdot \dot{\mathbf{E}} + \rho_R \partial_{\mathbf{T}_{RR}} \psi \cdot \dot{\mathbf{T}}_{RR} = -\theta \sigma^R_{\text{mech}} \le 0, \tag{6.55}$$

$$\rho_R \partial_{\mathbf{q}_R} \psi \cdot \dot{\mathbf{q}}_R + \frac{\mathbf{q}_R}{\theta} \cdot \nabla_R \theta = -\theta \sigma^R_{\text{th}} \le 0, \tag{6.56}$$

Moreover assume $\partial_{\mathbf{T}_{RR}} \psi, \partial_{\mathbf{q}_R} \psi \ne \mathbf{0}$. Applying the representation formula (3.15) we obtain from (6.55) and (6.56) that

$$\dot{\mathbf{T}}_{RR} = \mathsf{C}_{RR} \dot{\mathbf{E}} - \mathbf{P}_{RR} + (\mathsf{I} - \mathbf{N} \otimes \mathbf{N}) \mathbf{G}_{RR}, \qquad \dot{\mathbf{q}}_R = -\mathbf{K}_R \nabla_R \theta - \mathbf{Q}_R + (\mathbf{1} - \mathbf{n} \otimes \mathbf{n}) \mathbf{w}_R, \tag{6.57}$$

where

$$\mathbf{N} = \frac{\partial_{\mathbf{T}_{RR}}\psi}{|\partial_{\mathbf{T}_{RR}}\psi|}, \quad \mathbf{C}_{RR} = \mathbf{N} \otimes \frac{\mathbf{T}_{RR} - \rho_R \partial_{\mathbf{E}}\psi}{\rho_R|\partial_{\mathbf{T}_{RR}}\psi|}, \quad \mathbf{P}_{RR} = \frac{\theta\sigma^R_{\text{mech}}\mathbf{N}}{\rho_R|\partial_{\mathbf{T}_{RR}}\psi|},$$

$$\mathbf{n} = \frac{\partial_{\mathbf{q}_R}\psi}{|\partial_{\mathbf{q}_R}\psi|}, \quad \mathbf{K}_R = \mathbf{n} \otimes \frac{\mathbf{q}_R}{\rho_R\theta|\partial_{\mathbf{q}_R}\psi|}, \quad \mathbf{Q}_R = \frac{\theta\sigma^R_{\text{th}}\mathbf{n}}{\rho_R|\partial_{\mathbf{q}_R}\psi|},$$

$\mathbf{G}_{RR}$ being an arbitrary tensor and $\mathbf{w}_R$ an arbitrary vector.

If, instead, $\partial_{\mathbf{T}_{RR}}\psi, \partial_{\mathbf{q}_R}\psi = \mathbf{0}$ then (6.55) and (6.56) reduce to

$$(\mathbf{T}_{RR} - \rho_R\partial_{\mathbf{E}}\psi) \cdot \dot{\mathbf{E}} = \theta\sigma^R_{\text{mech}}, \qquad \mathbf{q}_R \cdot \nabla_R\theta = -\theta^2\sigma^R_{\text{th}}. \tag{6.58}$$

As a possible consequence for $\mathbf{T}_{RR}$ let M be a symmetric non-singular fourth-order tensor with $\mathsf{M}\dot{\mathbf{E}} \neq \mathbf{0}$. Letting $\mathbf{N} = \mathsf{M}\dot{\mathbf{E}}/|\mathsf{M}\dot{\mathbf{E}}|$ and $\mathbf{Z} = \mathsf{M}^{-1}[\mathbf{T}_{RR} - \rho_R\partial_{\mathbf{E}}\psi]$, and $\mathbf{G} = \mathbf{0}$ by the representation formula we have

$$\mathbf{T}_{RR} = \rho_R\partial_{\mathbf{E}}\psi + \frac{\theta\sigma^R_{\text{mech}}}{|\mathsf{M}\dot{\mathbf{E}}|^2}\mathsf{M}^2\dot{\mathbf{E}}.$$

As for the second relation in (6.58) we observe that Fourier's law $\mathbf{q}_R = -\boldsymbol{\kappa}\nabla_R\theta$, $\boldsymbol{\kappa}$ being a positive-definite second-order tensor, is allowed with $\theta^2\sigma^R_{\text{th}} = \nabla_R\theta \cdot \boldsymbol{\kappa}\nabla_R\theta$. The same result follows by using the representation formula with $\mathbf{N} = \mathbf{A}\nabla_R\theta/|\mathbf{A}\nabla_R\theta|$, $\mathbf{Z} = \mathbf{A}^{-1}\mathbf{q}_R$, and $\mathbf{w} = \mathbf{0}$. We find

$$\mathbf{q}_R = -\theta^2\sigma^R_{\text{th}}\mathbf{A}^2\nabla_R\theta/|\mathbf{A}\nabla_R\theta|^2.$$

Fourier's law is recovered by taking $\boldsymbol{\kappa} = \mathbf{A}^{-1}\mathbf{q}_R$ and $\sigma^R_{\text{th}} = \nabla_R\theta \cdot \boldsymbol{\kappa}\nabla_R\theta/\theta^2$.

A class of thermo-viscoelastic models is established by the selection of quite a general pair of free energy ψ and entropy dissipation σ.

Let M be a non-singular, fully symmetric, fourth-order tensor, $\mathbf{M}$ a non-singular, symmetric, second-order tensor and $\mathcal{G}$ a smooth function from Sym to Sym. Consider the free energy ψ given by

$$\rho_R\psi = \rho_R\psi_0(\theta) + \int_{\mathbf{0}}^{\mathbf{E}}\mathcal{G}(\mathbf{u})d\mathbf{u} + \tfrac{1}{2}[\mathbf{T}_{RR} - \mathcal{G}(\mathbf{E})] \cdot \mathsf{M}(\theta)[\mathbf{T}_{RR} - \mathcal{G}(\mathbf{E})] + \tfrac{1}{2}\mathbf{q}_R \cdot \mathbf{M}(\theta)\mathbf{q}_R,$$

where $\mathbf{u} \in$ Sym. Observe that

$$\rho_R\partial_{\mathbf{T}_{RR}}\psi = \mathsf{M}[\mathbf{T}_{RR} - \mathcal{G}(\mathbf{E})], \quad \rho_R\partial_{\mathbf{E}}\psi = \mathcal{G}(\mathbf{E}) + [\mathcal{G}'(\mathbf{E})]^T\mathsf{M}[\mathbf{T}_{RR} - \mathcal{G}(\mathbf{E})], \quad \rho_R\partial_{\mathbf{q}_R}\psi = \mathbf{M}\mathbf{q}_R.$$

Moreover we let

$$\sigma^R_{\text{mech}} = \frac{1}{\theta\tau_T}[\mathbf{T}_{RR} - \mathcal{G}(\mathbf{E})] \cdot \mathsf{M}[\mathbf{T}_{RR} - \mathcal{G}(\mathbf{E})], \qquad \sigma^R_{\text{th}} = \frac{1}{\theta\tau_q}\mathbf{q}_R \cdot \mathbf{M}\mathbf{q}_R, \qquad \tau_T, \tau_q > 0.$$

Hence we take

$$\mathbf{N} = \frac{\mathsf{M}[\mathbf{T}_{RR} - \mathcal{G}(\mathbf{E})]}{|\mathsf{M}[\mathbf{T}_{RR} - \mathcal{G}(\mathbf{E})]|}, \qquad \mathbf{n} = \frac{\mathbf{M}\mathbf{q}_R}{|\mathbf{M}\mathbf{q}_R|}$$

and apply (6.57) to obtain

$$\dot{\mathbf{T}}_{RR} = [\mathsf{M}^{-1} + \mathcal{G}'(\mathbf{E})]\dot{\mathbf{E}} - \frac{1}{\tau_T}[\mathbf{T}_{RR} - \mathcal{G}(\mathbf{E})], \qquad \dot{\mathbf{q}}_R = -\frac{1}{\theta}\mathbf{M}^{-1}\nabla_R\theta - \frac{1}{\tau_q}\mathbf{q}_R, \tag{6.59}$$

where we have chosen

$$\mathbf{G}_{RR} = [\mathsf{M}^{-1} + \mathcal{G}'(\mathbf{E})]\dot{\mathbf{E}} - \frac{1}{\tau_T}[\mathbf{T}_{RR} - \mathcal{G}(\mathbf{E})], \qquad \mathbf{w}_R = -\frac{1}{\theta}\mathbf{M}^{-1}\nabla_R\theta - \frac{1}{\tau_q}\mathbf{q}_R.$$

A linear viscoelastic model
If the function $\mathcal{G}$ is linear, namely $\mathcal{G}(\mathbf{E}) = \mathsf{G}_\infty\mathbf{E}$, it follows that

$$\dot{\mathbf{T}}_{RR} + \frac{1}{\tau_T}(\mathbf{T}_{RR} - \mathsf{G}_\infty\mathbf{E}) = \mathsf{G}_0\dot{\mathbf{E}}, \qquad \dot{\mathbf{q}}_R + \frac{1}{\tau_q}\mathbf{q}_R = -\boldsymbol{\kappa}\nabla_R\theta, \tag{6.60}$$

where $\mathsf{G}_0 = \mathsf{G}_\infty + \mathsf{M}^{-1}$ and $\boldsymbol{\kappa} = \mathbf{M}^{-1}/\theta$; G_0 and G_∞ stand for the usual elastic and relaxation moduli while $\boldsymbol{\kappa}$ stands for the conductivity tensor. If $\mathsf{G}_\infty > \mathbf{0}$ the rate equations (6.60) represent the standard linear solid (or Zener) model with Maxwell-Cattaneo heat equation. If $\mathsf{G}_\infty = \mathbf{0}$ we have the Maxwell model for fluids

$$\dot{\mathbf{T}}_{RR} + \frac{1}{\tau_T}\mathbf{T}_{RR} = \mathsf{G}_0\dot{\mathbf{E}}, \qquad \dot{\mathbf{q}}_R + \frac{1}{\tau_q}\mathbf{q}_R = -\boldsymbol{\kappa}\nabla_R\theta.$$

The classical Maxwell model is recovered through the linear approximation, $\mathbf{T}_{RR} \simeq \mathbf{T}$, $\mathbf{E} \simeq \boldsymbol{\varepsilon}$.

6.3.3 *Thermo-Viscoplastic Models*

One-dimensional constitutive models for viscoplasticity, based on spring-dashpot-slider elements include the perfectly viscoplastic solid, the elastic viscoplastic solid, and the elasto-viscoplastic hardening solid. As with the rheological models, the elements may be connected in series (additivity of the strain) or in parallel (additivity of the stress). Many of these one-dimensional models can be generalized to three dimensions for the small strain regime. As a relevant example we now examine the following model.

The Bingham–Maxwell model

This model describes the behaviour of an elastic—perfectly viscoplastic solid. It is based on an elementary rheological unit where the sliding friction element and the dashpot are arranged in parallel and then connected in series to an elastic spring. The similarity with Maxwell's and Bingham's models makes it to be referred to as Bingham-Maxwell model.

Letting $\Xi_R = (\theta, \mathbf{E}, \mathbf{T}_{RR}, \mathbf{q}_R, \dot{\theta}, \dot{\mathbf{E}}, \dot{\mathbf{T}}_{RR}, \dot{\mathbf{q}}_R, \nabla_R \theta)$ be the set of variables we can argue in the standard way and assume that cross-coupling terms, between $\nabla_R \theta, \mathbf{q}_R$ and $\dot{\mathbf{E}}, \dot{\mathbf{T}}_{RR}$ are absent; we then get inequality (6.55). Moreover assume $\partial_{\mathbf{E}} \psi = \mathbf{0}$. Inequality (6.55) then reads

$$\mathbf{T}_{RR} \cdot \dot{\mathbf{E}} - \rho_R \partial_{\mathbf{T}_{RR}} \psi \cdot \dot{\mathbf{T}}_{RR} = \theta \sigma^R_{\text{mech}}.$$

Hence we apply the representation formula with $\mathbf{Z} = \dot{\mathbf{E}}, \mathbf{N} = \mathbf{T}_{RR}/|\mathbf{T}_{RR}|, \mathbf{G}_{RR} = \mathbf{0}$ to obtain

$$\dot{\mathbf{E}} = \Big(\dot{\mathbf{E}} \cdot \frac{\mathbf{T}_{RR}}{|\mathbf{T}_{RR}|}\Big) \frac{\mathbf{T}_{RR}}{|\mathbf{T}_{RR}|} = (\rho_R \partial_{\mathbf{T}_{RR}} \psi \cdot \dot{\mathbf{T}}_{RR} + \theta \sigma^R_{\text{mech}}) \frac{\mathbf{T}_{RR}}{|\mathbf{T}_{RR}|^2}$$

whence

$$\dot{\mathbf{E}} = \mathsf{H}_{RR} \dot{\mathbf{T}}_{RR} + \theta \sigma^R_{\text{mech}} \frac{\mathbf{T}_{RR}}{|\mathbf{T}_{RR}|^2}, \qquad \mathsf{H}_{RR} := \frac{\rho_R}{|\mathbf{T}_{RR}|^2} \mathbf{T}_{RR} \otimes \partial_{\mathbf{T}_{RR}} \psi. \tag{6.61}$$

The entropy production σ^R_{mech} is taken in the form

$$\sigma^R_{\text{mech}} = \begin{cases} 0 & \text{if } |\mathbf{T}_{RR}| \le S_y, \\ (1/\theta \lambda^n) |\mathbf{T}_{RR}|^n (|\mathbf{T}_{RR}| - S_y) & \text{otherwise,} \end{cases} \tag{6.62}$$

where λ, S_y are positive parameters. Equations (6.61)–(6.62), with $n > 1$, characterize the Bingham-Norton model [228]. If, instead, $n = 1$ and $S_y = 0$ the model reduces to

$$\dot{\mathbf{E}} = \mathsf{H}_{RR} \dot{\mathbf{T}}_{RR} + \frac{1}{\lambda} \mathbf{T}_{RR},$$

which may be viewed as a generalized Maxwell-Wiechert model. In any case, the free energy is given by a function $\psi(\theta, \mathbf{T}_{RR})$.

6.4 Higher-Order Rate Models

Non-Newtonian phenomena exhibited e.g. by asphalt and some biomaterials show properties that can be associated with different relaxation times or rather to higher-order differential equations.This view is often described by having recourse to the Burgers model [73] or the best known Oldroyd-B fluid model [348].

6.4.1 Burgers' Fluid

Let

$$\mathbf{T} = -p\mathbf{1} + \mathbf{S}.$$

In the Burgers model of an incompressible fluid the extra-stress $\mathbf{S}$ obeys the second-order rate equation

$$\mathbf{S} + \tau \overset{\triangledown}{\mathbf{S}} + \beta \overset{\triangledown\triangledown}{\mathbf{S}} = 2\mu\mathbf{D} + \nu \overset{\triangledown}{\mathbf{D}}. \tag{6.63}$$

Letting $\tau = \tau_1 + \tau_2$, $\beta = \tau_1\tau_2$, $\mu = \mu_1 + \mu_2$, $\nu = 2(\mu_1\tau_2 + \mu_2\tau_1)$, we can split (6.63) into the sum of two polymeric (viscoelastic) parts,

$$\mathbf{S} = \mathbf{S}_1 + \mathbf{S}_2, \qquad \mathbf{S}_1 + \tau_1 \overset{\triangledown}{\mathbf{S}}_1 = 2\mu_1\mathbf{D}, \qquad \mathbf{S}_2 + \tau_2 \overset{\triangledown}{\mathbf{S}}_2 = 2\mu_2\mathbf{D},$$

where μ_1, μ_2 and τ_1, τ_2 are thought as viscosity coefficients and relaxation times; they are expected to be positive.

The thermodynamic consistency of the Burgers model is investigated in the referential version. With $\mathbf{T} = -p\mathbf{1} + \mathbf{S}$ we associate $\mathbf{T}_{RR} = -Jp\mathbf{C}^{-1} + \mathbf{S}_{RR}$ where $\mathbf{S}_{RR}$ satisfies the differential equation

$$\mathbf{S}_{RR} + \tau\dot{\mathbf{S}}_{RR} + \beta\ddot{\mathbf{S}}_{RR} = 2\mu\dot{\mathbf{E}} + \nu\ddot{\mathbf{E}}. \tag{6.64}$$

Owing to incompressibility,

$$0 = \nabla \cdot \mathbf{v} = \mathbf{C}^{-1} \cdot \dot{\mathbf{E}}, \qquad \mathbf{S}_{RR} \cdot \dot{\mathbf{E}}.$$

Hence the Clausius-Duhem inequality can be written in the (referential) form

$$-\rho_R(\dot{\psi} + \eta)\dot{\theta} + \mathbf{S}_{RR} \cdot \dot{\mathbf{E}} = J\theta\sigma \geq 0.$$

We let the coefficients τ, μ, β, ν be parameterized by the temperature θ and the (constant) Jacobian J. Moreover we look at (6.64) as a rate equation of the form

$$\ddot{\mathbf{S}}_{RR} = \mathbf{f}(\theta, J, \mathbf{S}_{RR}, \dot{\mathbf{E}}, \dot{\mathbf{S}}_{RR}, \ddot{\mathbf{E}}).$$

Hence we investigate the thermodynamic consistency by letting $\theta, \mathbf{S}_{RR}, \dot{\mathbf{E}}, \dot{\mathbf{S}}_{RR}, \ddot{\mathbf{E}}$ be the independent variables. We then assume the free energy ψ has the form

$$\psi = \psi(\theta, \mathbf{S}_{RR}, \dot{\mathbf{E}}, \dot{\mathbf{S}}_{RR}, \ddot{\mathbf{E}}),$$

and the like for η and $\sigma_R = J\sigma$. Upon computation of $\dot{\psi}$ and substitution of $\dot{\psi}$ we find that

$$\partial_{\ddot{\mathbf{E}}}\psi = \mathbf{0}, \qquad \eta = -\partial_\theta\psi$$

and

$$-\rho_R\partial_{\mathbf{S}_{RR}}\psi\cdot\dot{\mathbf{S}}_{RR}-\rho_R\partial_{\dot{\mathbf{E}}}\psi\cdot\ddot{\mathbf{E}}-\rho_R\partial_{\dot{\mathbf{S}}_{RR}}\psi\cdot\mathbf{f}(\theta,J,\mathbf{S}_{RR},\dot{\mathbf{E}},\dot{\mathbf{S}}_{RR},\ddot{\mathbf{E}})+\mathbf{S}_{RR}\cdot\dot{\mathbf{E}}=\theta\sigma_R\geq 0. \tag{6.65}$$

Inequality (6.65) is the general thermodynamic restriction for the function **f**.

We now take **f** as given by the Burgers equation (6.64); substitution in (6.65) results in

$$\Big(\mathbf{S}_{RR}-\frac{2\mu\rho_R}{\beta}\partial_{\dot{\mathbf{S}}_{RR}}\psi\Big)\cdot\dot{\mathbf{E}}+\rho_R\Big(\frac{\tau}{\beta}\partial_{\dot{\mathbf{S}}_{RR}}\psi-\partial_{\mathbf{S}_{RR}}\psi\Big)\cdot\dot{\mathbf{S}}_{RR}-\rho_R\Big(\partial_{\dot{\mathbf{E}}}\psi+\frac{\nu}{\beta}\partial_{\dot{\mathbf{S}}_{RR}}\psi\Big)\cdot\ddot{\mathbf{E}}$$
$$+\rho_R\frac{1}{\beta}\partial_{\dot{\mathbf{S}}_{RR}}\psi\cdot\mathbf{S}_{RR}=\theta\sigma_R\geq 0.$$

The arbitrariness of $\ddot{\mathbf{E}}$ implies

$$\partial_{\dot{\mathbf{E}}}\psi=\frac{\nu}{\beta}\partial_{\dot{\mathbf{S}}_{RR}}\psi \tag{6.66}$$

and

$$\Big(\mathbf{S}_{RR}-\frac{2\mu\rho_R}{\beta}\partial_{\dot{\mathbf{S}}_{RR}}\psi\Big)\cdot\dot{\mathbf{E}}+\rho_R\Big(\frac{\tau}{\beta}\partial_{\dot{\mathbf{S}}_{RR}}\psi\psi-\partial_{\mathbf{S}_{RR}}\psi\Big)\cdot\dot{\mathbf{S}}_{RR}+\rho_R\frac{1}{\beta}\partial_{\dot{\mathbf{S}}_{RR}}\psi\cdot\mathbf{S}_{RR}=\theta\sigma_R\geq 0. \tag{6.67}$$

Further consequences follow once a detailed function ψ is selected. We take ψ in the quadratic form

$$\rho_R\psi=\rho_R\psi_0(\theta)+\frac{\alpha_1}{2}|\mathbf{S}_{RR}|^2+\frac{\alpha_2}{2}|\dot{\mathbf{S}}_{RR}|^2+\frac{\alpha_3}{2}|\dot{\mathbf{E}}|^2+\gamma_1\dot{\mathbf{S}}_{RR}\cdot\mathbf{S}_{RR}+\gamma_2\mathbf{S}_{RR}\cdot\dot{\mathbf{E}}+\gamma_3\dot{\mathbf{S}}_{RR}\cdot\dot{\mathbf{E}}. \tag{6.68}$$

Upon substitution of

$$\rho_R\partial_{\dot{\mathbf{E}}}\psi=\gamma_2\mathbf{S}_{RR}+\gamma_3\dot{\mathbf{S}}_{RR}+\alpha_3\dot{\mathbf{E}},\qquad \rho_R\partial_{\dot{\mathbf{S}}_{RR}}\psi=\gamma_1\mathbf{S}_{RR}+\alpha_2\dot{\mathbf{S}}_{RR}+\gamma_3\dot{\mathbf{E}},$$

into (6.66) it follows that

$$\gamma_2=-\frac{\nu}{\beta}\gamma_1,\qquad \gamma_3=-\frac{\nu}{\beta}\alpha_2,\qquad \alpha_3=-\frac{\nu}{\beta}\gamma_3.$$

Since also $\rho_R\partial_{\mathbf{S}_{RR}}\psi=\alpha_1\mathbf{S}_{RR}+\gamma_1\dot{\mathbf{S}}_{RR}+\gamma_2\dot{\mathbf{E}}$ then the reduced inequality can be written as

$$A_{11}|\mathbf{S}_{RR}|^2+A_{22}|\dot{\mathbf{S}}_{RR}|^2+A_{33}|\dot{\mathbf{E}}|^2+2A_{12}\dot{\mathbf{S}}_{RR}\cdot\mathbf{S}_{RR}+2A_{13}\mathbf{S}_{RR}\cdot\dot{\mathbf{E}}+2A_{23}\dot{\mathbf{S}}_{RR}\cdot\dot{\mathbf{E}}=\theta\sigma_R, \tag{6.69}$$

where

$$A_{11}=\frac{\gamma_1}{\beta},\qquad A_{22}=\frac{\tau\alpha_2}{\beta}-\gamma_1,\qquad A_{33}=-\frac{2\mu\gamma_3}{\beta},$$

$$2A_{12} = \frac{\tau\gamma_1}{\beta} + \frac{\alpha_2}{\beta} - \alpha_1, \qquad 2A_{13} = 1 + \frac{\gamma_3}{\beta} - \frac{2\mu\gamma_1}{\beta}, \qquad 2A_{23} = \frac{\tau\gamma_3}{\beta} - \gamma_2 - \frac{2\mu\alpha_2}{\beta}.$$

Thus σ_R is positive definite if the matrix A is positive definite. This is apparently the case if A is a diagonal matrix with positive diagonal entries namely when

$$\gamma_1 > 0, \quad \gamma_3 > 0, \quad \alpha_2 > \frac{\beta\gamma_1}{\tau}, \quad \alpha_1 = \frac{\tau\gamma_1}{\beta} + \frac{\alpha_2}{\beta}, \quad \gamma_3 = 2\mu\gamma_1 - \beta, \quad \gamma_2 = \frac{\tau\gamma_3}{\beta} - \frac{2\mu\alpha_2}{\beta}.$$

Necessary and sufficient conditions for the definiteness of σ_R follow by evaluating the principal minors of A, which are required to be positive. Here we state the result. **Proposition.** *The matrix A is positive definite, and hence the Burgers model (6.64) with free energy (6.68) is thermodynamically consistent, if and only if one of the following cases occur,*

$$\nu = 0,\ \mu > 0,\ \beta < 0; \quad \nu \neq 0,\ \mu = 0,\ \tau/\nu \geq 0; \quad \nu \neq 0,\ \mu > 0,\ \nu\tau \geq 2\mu\beta.$$

The three thermodynamically-consistent models are then represented as follows,

$$\mathbf{S}_{RR} + \tau\dot{\mathbf{S}}_{RR} + \beta\ddot{\mathbf{S}}_{RR} = 2\mu\dot{\mathbf{E}}, \qquad \beta < 0,\ \mu, > 0,$$

$$\mathbf{S}_{RR} + \tau\dot{\mathbf{S}}_{RR} + \beta\ddot{\mathbf{S}}_{RR} = \nu\ddot{\mathbf{E}}, \qquad \tau/\nu \geq 0,$$

$$\mathbf{S}_{RR} + \tau\dot{\mathbf{S}}_{RR} + \beta\ddot{\mathbf{S}}_{RR} = 2\mu\dot{\mathbf{E}} + \nu\ddot{\mathbf{E}}, \qquad \mu > 0,\ \nu\tau > 2\mu\beta.$$

More general, possibly non-linear, functions $\mathbf{f}(\theta, J, \mathbf{S}_{RR}, \dot{\mathbf{E}}, \dot{\mathbf{S}}_{RR}, \ddot{\mathbf{E}})$ compatible with thermodynamics follow from (6.65) and the representation formula (A.8). Letting $\mathbf{N} = \partial_{\dot{\mathbf{S}}_{RR}}\psi/|\partial_{\dot{\mathbf{S}}_{RR}}\psi|$ we find that

$$\ddot{\mathbf{S}}_{RR} = (\mathbf{S}_{RR}\cdot\dot{\mathbf{E}} - \rho_R\partial_{\mathbf{S}_{RR}}\psi\cdot\dot{\mathbf{S}}_{RR} - \rho_R\partial_{\dot{\mathbf{E}}}\psi\cdot\ddot{\mathbf{E}} - \theta\sigma_R)\frac{\partial_{\dot{\mathbf{S}}_{RR}}\psi}{\rho_R|\partial_{\dot{\mathbf{S}}_{RR}}\psi|^2} + (\mathbf{1} - \frac{\partial_{\dot{\mathbf{S}}_{RR}}\psi}{|\partial_{\dot{\mathbf{S}}_{RR}}\psi|}\otimes\frac{\partial_{\dot{\mathbf{S}}_{RR}}\psi}{|\partial_{\dot{\mathbf{S}}_{RR}}\psi|})\mathbf{G}, \tag{6.70}$$

$\mathbf{G}$ being any second-order tensor function of $\theta, J, \mathbf{S}_{RR}, \dot{\mathbf{E}}, \dot{\mathbf{S}}_{RR}, \ddot{\mathbf{E}}$. In the special case $\sigma_R = 0$ and $\mathbf{G} = \mathbf{0}$ we have

$$\ddot{\mathbf{S}}_{RR} = (\mathbf{S}_{RR}\cdot\dot{\mathbf{E}} - \rho_R\partial_{\mathbf{S}_{RR}}\psi\cdot\dot{\mathbf{S}}_{RR} - \rho_R\partial_{\dot{\mathbf{E}}}\psi\cdot\ddot{\mathbf{E}})\frac{\partial_{\dot{\mathbf{S}}_{RR}}\psi}{\rho_R|\partial_{\dot{\mathbf{S}}_{RR}}\psi|^2}.$$

As a check of the consistency of the whole approach we now ask whether (6.70) leads to (6.64) as a possible case, provided, of course, ψ and σ_R are thermodynamically admissible. To fix ideas we consider the first case where $\nu = 0$ and $\mu > 0$, $\beta < 0$; hence the expected equation is

$$\beta\ddot{\mathbf{S}}_{RR} = -\mathbf{S}_{RR} - \tau\dot{\mathbf{S}}_{RR} + 2\mu\dot{\mathbf{E}}. \tag{6.71}$$

If $\nu = 0$ then ψ and σ_R have the form

$$\rho_R\psi = \rho_R\psi_0(\theta) + \frac{\tau}{4\mu}|\mathbf{S}_{RR}|^2 + \frac{\beta}{2\mu}\dot{\mathbf{S}}_{RR}\cdot\mathbf{S}_{RR}, \quad \theta\sigma_R = \frac{1}{2\mu}|\mathbf{S}_{RR}|^2 - \frac{\beta}{2\mu}|\dot{\mathbf{S}}_{RR}|^2.$$

Thus

$$\mathbf{N} := \frac{\partial_{\dot{\mathbf{S}}_{RR}}\psi}{|\dot{\mathbf{S}}_{RR}\psi|} = \frac{\mathbf{S}_{RR}}{|\mathbf{S}_{RR}|} =: \hat{\mathbf{S}}, \qquad \frac{\partial_{\dot{\mathbf{S}}_{RR}}\psi}{\rho_R|\partial\dot{\mathbf{S}}_{RR}\psi|^2} = \frac{2\mu}{\beta}\frac{\hat{\mathbf{S}}}{|\mathbf{S}_{RR}|}.$$

Substitution in (6.70) yields

$$\beta\ddot{\mathbf{S}}_{RR} = (\hat{\mathbf{S}}\otimes\hat{\mathbf{S}})[2\mu\dot{\mathbf{E}} - \tau\dot{\mathbf{S}}_{RR}] - \mathbf{S}_{RR} + \beta(\mathbf{I} - \hat{\mathbf{S}}\otimes\hat{\mathbf{S}})\mathbf{G}.$$

Letting $\beta\mathbf{G} = 2\mu\dot{\mathbf{E}} - \tau\dot{\mathbf{S}}_{RR}$ we obtain Eq. (6.71).

While the representation formula, here (6.70), is likely to generate a class of non-linear equations, the (admissible) linear equations of rate-type models appear to be special cases.

6.4.2 *Oldroyd-B Fluid*

The Oldroyd-B model describes an incompressible fluid consisting of a solution of a (viscoelastic) polymer in a viscous solvent. Accordingly the stress $\mathbf{T}$ is (taken as) the sum of two terms,

$$\mathbf{T} = 2\mu_s\mathbf{D} + \mathbf{Y},$$

$2\mu_s\mathbf{D}$ being the viscous stress within the solvent and $\mathbf{Y}$ the viscoelastic stress of the polymer. The stress $\mathbf{Y}$ is assumed to be given by the rate equation

$$\mathbf{Y} + \tau_1 \overset{\triangledown}{\mathbf{Y}} = 2\mu_p\mathbf{D}. \tag{6.72}$$

If μ_s is constant then we obtain

$$\overset{\triangledown}{\mathbf{T}} = 2\mu_s \overset{\triangledown}{\mathbf{D}} + \overset{\triangledown}{\mathbf{Y}} = 2\mu_s \overset{\triangledown}{\mathbf{D}} + \frac{1}{\tau_1}(2\mu_p\mathbf{D} - \mathbf{T} + 2\mu_s\mathbf{D}).$$

Hence we have the *Oldroyd-B fluid model*[6]

$$\mathbf{T} + \tau_1 \overset{\triangledown}{\mathbf{T}} = 2\mu(\mathbf{D} + \tau_2 \overset{\triangledown}{\mathbf{D}}) \tag{6.73}$$

where $\mu = \mu_p + \mu_s$ and $\tau_2 = \mu_s\tau_1/\mu$.

As an aside, by assumption the Oldroyd model applies to incompressible fluids. For compressible fluids the corresponding model should involve the Truesdell rate

[6] The name Oldroyd-B is due to the chosen derivative; the alternative model Oldroyd-A involved the Cotter-Rivlin derivative, considered inappropriate.

$\overset{\square}{\mathbf{Y}}$ rather than the upper convected derivative $\overset{\triangledown}{\mathbf{Y}}$. Formally, the constitutive equation for $\boldsymbol{T}$ is obtained by replacing the time derivative, within the Maxwell model, with the upper convected derivative. That is why we can refer to equation (6.72) as *upper convected Maxwell* model. Moreover, (6.73) can be viewed as the direct extension of the Jeffreys model where the upper convected derivative replaces the time derivative. Hence the model (6.73) is also referred to as the *upper convected Jeffreys* model.

If we let the scalar μ_s be a function of, e.g., the temperature θ then

$$\overset{\triangledown}{\mathbf{T}}= 2\mu_s \overset{\triangledown}{\mathbf{D}} +2\partial_\theta \mu_s \dot{\theta}\, \mathbf{D}+ \overset{\triangledown}{\mathbf{Y}}$$

and (6.73) has to be replaced by

$$\mathbf{T} + \tau_1 \overset{\triangledown}{\mathbf{T}}= 2(\mu + \tau_1 \partial_\theta \mu_s \dot{\theta})\mathbf{D} + 2\mu\tau_2 \overset{\triangledown}{\mathbf{D}} .$$

Nonlinear models occur of polymer melts that may be viewed as generalizations of (6.72). Doi-Edwards theory for polymer melts results in the differential equation

$$\boldsymbol{T} + \tau \overset{\triangledown}{\mathbf{Y}} + \frac{2\tau}{3G}(\mathbf{D} \cdot \boldsymbol{T})\mathbf{Y} - G\,\mathbf{1} = 2\mu\mathbf{D}$$

for the stress $\mathbf{Y}$ determined by the stretching $\mathbf{D}$. The time τ, the viscosity μ and the pressure G are three parameters characterizing the model.

To investigate the thermodynamic consistency of (6.73) we start from its analog in the reference configuration,

$$\mathbf{T}_{RR} + \lambda \dot{\mathbf{T}}_{RR} = 2\mu\dot{\mathbf{E}} + \nu\ddot{\mathbf{E}}. \tag{6.74}$$

We assume that the free energy ψ and the entropy η are functions of the set of variables

$$\Xi_R = (\theta, \mathbf{E}, \mathbf{T}_{RR}, \dot{\theta}, \dot{\mathbf{E}}, \ddot{\mathbf{E}}).$$

The Clausius-Duhem inequality reads

$$\begin{aligned} &- \rho_R(\partial_\theta \psi + \eta)\dot{\theta} + (\mathbf{T}_{RR} - \rho_R \partial_{\mathbf{E}} \psi) \cdot \dot{\mathbf{E}} - \rho_R \partial_{\mathbf{T}_{RR}} \psi \cdot \dot{\mathbf{T}}_{RR} \\ &-\rho_R \partial_{\dot{\theta}} \psi \ddot{\theta} - \rho_R \partial_{\dot{\mathbf{E}}} \psi \cdot \ddot{\mathbf{E}} - \rho_R \partial_{\ddot{\mathbf{E}}} \psi \cdot \dddot{\mathbf{E}} = J\theta\sigma \geq 0. \end{aligned} \tag{6.75}$$

Assume η is independent of $\dot{\theta}$. Hence the arbitrariness of $\dddot{\mathbf{E}}$, $\ddot{\theta}$, and next of $\dot{\theta}$ implies

$$\partial_{\ddot{\mathbf{E}}} \psi = \mathbf{0}, \qquad \partial_{\dot{\theta}} \psi = 0, \qquad \eta = -\partial_\theta \psi.$$

For definiteness we also let $\partial_{\mathbf{E}} \psi = \mathbf{0}$. Hence (6.75) yields

$$\mathbf{T}_{RR} \cdot \dot{\mathbf{E}} - \rho_R \partial_{\mathbf{T}_{RR}} \psi \cdot \dot{\mathbf{T}}_{RR} - \rho_R \partial_{\dot{\mathbf{E}}} \psi \cdot \ddot{\mathbf{E}} = J\theta\sigma \geq 0. \tag{6.76}$$

For a generic model described by Ξ_R Eq. (6.76) is required to hold for pairs of functions $\psi(\theta, \mathbf{T}_{RR}, \dot{\mathbf{E}})$ and $\sigma(\Xi_R) \geq 0$. Yet here we investigate the thermodynamic consistency of (6.74). Hence, e.g., $\ddot{\mathbf{E}}$ is not arbitrary and we replace $\ddot{\mathbf{E}}$ from (6.74) to obtain

$$\nu \mathbf{T}_{RR} \cdot \dot{\mathbf{E}} - \rho \partial_{\dot{\mathbf{E}}} \psi \cdot \mathbf{T}_{RR} - \rho_R \lambda \partial_{\dot{\mathbf{E}}} \psi \cdot \dot{\mathbf{T}}_{RR} - \rho_R \nu \partial_{\mathbf{T}_{RR}} \psi \cdot \dot{\mathbf{T}}_{RR} + 2\mu \rho_R \partial_{\dot{\mathbf{E}}} \psi \cdot \dot{\mathbf{E}} = \nu J \theta \sigma.$$

We look for the function ψ in the form

$$\rho_R \psi = \rho_R \psi_0(\theta) + \gamma |\lambda \mathbf{T} - \nu \dot{\mathbf{E}}|^2,$$

where γ is a parameter, independent of $\mathbf{T}_{RR}$ and $\dot{\mathbf{E}}$, to be determined. Since

$$\rho_R \partial_{\mathbf{T}_{RR}} \psi = 2\gamma \lambda (\lambda \mathbf{T} - \nu \dot{\mathbf{E}}), \qquad \rho_R \partial_{\dot{\mathbf{E}}} \psi = -2\gamma \nu (\lambda \mathbf{T} - \nu \dot{\mathbf{E}}),$$

then we have

$$[1 - \gamma(2\nu + 4\mu\lambda)] \mathbf{T}_{RR} \cdot \dot{\mathbf{E}} + 2\gamma\lambda |\mathbf{T}_{RR}|^2 + 4\mu\gamma\nu |\dot{\mathbf{E}}|^2 = J\theta\sigma \geq 0.$$

The inequality holds if

$$\gamma = \frac{1}{2(\nu + 2\mu\lambda)}, \quad \gamma\lambda \geq 0, \qquad \mu\gamma\nu \geq 0.$$

Consequently

$$\psi = \psi_0(\theta) + \frac{1}{2(1 + 2\mu\lambda)} |\lambda \mathbf{T}_{RR} - \nu \dot{\mathbf{E}}|^2, \qquad J\theta\sigma = \frac{\lambda/\nu}{1 + 2\mu\lambda/\nu} |\mathbf{T}_{RR}|^2 + \frac{2\mu}{1 + 2\mu\lambda/\nu} |\dot{\mathbf{E}}|^2.$$

The thermodynamic consistency holds if[7] $\mu > 0$, $\lambda\nu > 0$.

6.4.3 *White-Metzner Fluid*

The White-Metzner fluid [182, 445] is similar to the Oldroyd-B model (6.73). It involves both the rate of the stress and the rate of the stretching. The coefficients μ, λ, ν are assumed to depend on $\mathbf{D}$ but λ and ν, viewed as relaxation times, are taken to be equal. Further the objective time derivative, of $\mathbf{T}$ and $\mathbf{D}$, is taken to be a combination of the Oldroyd and Cotter-Rivlin derivatives.

Here we follow the Eulerian description and consider the model equation in the form

$$\mathbf{T} + \lambda \overset{\circ}{\mathbf{T}} = 2\mu \mathbf{D} + \nu \overset{\circ}{\mathbf{D}}, \tag{6.77}$$

[7] As in other models, the (positive) sign of the relaxation times $1/\lambda$, $1/\nu$ is given by boundedness requirements of the solution; thermodynamics gives the (positive) sign of the product.

where the objective derivative is the corotational one. Consistent with the literature we let the fluid be incompressible. However the coefficients λ, ν are not assumed to be equal and we let them be given by functions $\mu(\theta, p, \mathbf{D}), \lambda(\theta, p), \nu(\theta, p)$, the pressure p being a parameter. To guarantee the boundedness of the solution $\mathbf{T}$ or $\mathbf{D}$ of (6.77) we assume $\lambda, \nu > 0$.

Let

$$\Xi = (\theta, \mathbf{T}, \mathbf{D}, \overset{\circ}{\mathbf{D}})$$

be the set of variables and consider ψ, η, and σ as functions of Ξ while $\overset{\circ}{\mathbf{T}}$ is given by (6.77). Upon substitution of $\dot{\psi}$ in the Clausius-Duhem inequality we have

$$-\rho(\partial_\theta \psi + \eta)\dot{\theta} - \rho \partial_{\mathbf{T}} \psi \cdot \dot{\mathbf{T}} - \rho \partial_{\mathbf{D}} \psi \cdot \dot{\mathbf{D}} - \rho \partial_{\overset{\circ}{\mathbf{D}}} \psi \cdot (\overset{\circ}{\mathbf{D}})^{\cdot} + \mathbf{T} \cdot \mathbf{D} = \theta\sigma \geq 0.$$

By (6.77) we can write $\dot{\mathbf{T}} = \overset{\circ}{\mathbf{T}} + \mathbf{WT} + \mathbf{TW}^T$ in the form

$$\dot{\mathbf{T}} = \frac{1}{\lambda}(2\mu\mathbf{D} + \nu\, \overset{\circ}{\mathbf{D}} - \mathbf{T}) + \mathbf{WT} + \mathbf{TW}^T. \tag{6.78}$$

By definition,

$$(\overset{\circ}{\mathbf{D}})^{\cdot} = \ddot{\mathbf{D}} - (\mathbf{WD})^{\cdot} - (\mathbf{DW}^T)^{\cdot}.$$

Hence $\ddot{\mathbf{D}}$ and $\dot{\theta}$ occur linearly in the Clausius-Duhem inequality; their arbitrariness implies

$$\partial_{\overset{\circ}{\mathbf{D}}} \psi = \mathbf{0}, \qquad \eta = -\partial_\theta \psi.$$

Substitution of $\dot{\mathbf{T}}$ from (6.78), with $\overset{\circ}{\mathbf{D}} = \dot{\mathbf{D}} - \mathbf{WD} - \mathbf{DW}^T$, makes the inequality in the form

$$-\rho\partial_{\mathbf{T}}\psi \cdot [\frac{1}{\lambda}(2\mu\mathbf{D} + \nu\dot{\mathbf{D}} - \nu\mathbf{WD} - \nu\mathbf{DW}^T - \mathbf{T}) + \mathbf{WT} + \mathbf{TW}^T] - \rho\partial_{\mathbf{D}}\psi \cdot \dot{\mathbf{D}} + \mathbf{T} \cdot \mathbf{D} = \theta\sigma \geq 0. \tag{6.79}$$

We now have to find a pair of functions ψ, σ satisfying inequality (6.79).

The occurrence of the terms $\mathbf{WD}$ and $\mathbf{WT}$ indicates that ψ might be a function of a combination of $\mathbf{D}$ and $\mathbf{T}$. Accordingly we let

$$\psi = \psi(\theta, p, \xi), \qquad \xi := |\lambda\mathbf{T} - \nu\mathbf{D}|^2.$$

Hence

$$\partial_{\mathbf{T}}\psi = 2\psi'(\lambda^2\mathbf{T} - \lambda\nu\mathbf{D}), \qquad \partial_{\mathbf{D}}\psi = 2\psi'(\nu^2\mathbf{D} - \lambda\nu\mathbf{T}),$$

where ψ' denotes the derivative with respect to ξ. Inequality (6.78) then becomes

$$-2\rho\psi'(\lambda\mathbf{T}-\nu\mathbf{D})\cdot(2\mu\mathbf{D}+\nu\dot{\mathbf{D}}-\mathbf{T})-2\rho\psi'(\nu^2\mathbf{D}-\lambda\nu\mathbf{T})\cdot\dot{\mathbf{D}}+\mathbf{T}\cdot\mathbf{D}$$
$$+2\rho\psi'(\lambda\mathbf{T}-\nu\mathbf{D})\cdot[\nu(\mathbf{W}\mathbf{D}+\mathbf{D}\mathbf{W}^T)-\lambda(\mathbf{W}\mathbf{T}+\mathbf{T}\mathbf{W}^T)]=\theta\sigma\geq 0$$

Observe that, by the symmetry of **D** and **T**,

$$\mathbf{D}\cdot(\mathbf{W}\mathbf{D})=\mathbf{D}\cdot(\mathbf{D}\mathbf{W}^T)=0,\quad \mathbf{D}\cdot(\mathbf{W}\mathbf{T})+\mathbf{T}\cdot(\mathbf{W}\mathbf{D})=(\mathbf{D}\mathbf{T}+\mathbf{T}\mathbf{D})\cdot\mathbf{W}=0,$$

and the like by interchanging **D** and **T**. Hence the occurrence of **W**, induced by the corotational derivative, results in a zero entropy production. Moreover we have

$$2\rho\psi'(\lambda\mathbf{T}-\nu\mathbf{D})\cdot\nu\dot{\mathbf{D}}+2\rho\psi'(\nu^2\mathbf{D}-\lambda\nu\mathbf{T})\cdot\dot{\mathbf{D}}=0.$$

The inequality then reduces to

$$\mathbf{T}\cdot\mathbf{D}(1-4\rho\mu\lambda\psi'-2\rho\nu\psi')+2\rho\lambda\psi'|\mathbf{T}|^2+4\rho\mu\nu\psi'|\mathbf{D}|^2=\theta\sigma\geq 0.$$

Hence it follows that

$$\psi'>0,\qquad \mu>0,\qquad |1-4\rho\mu\lambda\psi'-2\rho\nu\psi'|\leq 4\rho\sqrt{2\lambda\mu\nu}\,|\psi'|.$$

A sufficient condition is obtained by letting $1-4\rho\mu\lambda\psi'-2\rho\nu\psi'=0$ whence

$$\psi'=\frac{1}{2\rho(\nu+2\mu\lambda)},\qquad \psi=\psi_0(\theta,p)+\frac{1}{2\rho(\nu+2\mu\lambda)}|\lambda\mathbf{T}-\nu\mathbf{D}|^2.$$

Consequently the entropy production σ is given by

$$\theta\sigma=\frac{\lambda}{\nu+2\mu\lambda}|\mathbf{T}|^2+\frac{2\mu\nu}{\nu+2\mu\lambda}|\mathbf{D}|^2.$$

6.4.4 Wave Features in Higher-Order Rate Models

The Burgers model is usually applied in the Eulerian version so that

$$\mathbf{T}=-p\mathbf{1}+\mathbf{S},\qquad \mathbf{S}+\lambda\overset{\triangledown}{\mathbf{S}}+\beta\overset{\triangledown\triangledown}{\mathbf{S}}=2\mu\mathbf{D}+\nu\overset{\triangledown}{\mathbf{D}}.$$

In the Burgers model the extra-stress **S** describes an incompressible fluid and obeys the second-order rate equation

$$\mathbf{S}+\lambda\overset{\triangledown}{\mathbf{S}}+\beta\overset{\triangledown\triangledown}{\mathbf{S}}=2\mu\mathbf{D}+\nu\overset{\triangledown}{\mathbf{D}}.$$

Plane longitudinal waves in higher-order rate models are investigated with some approximations. The pressure is assumed to be uniform and hence no pressure contribution occurs in the equation of motion. Plane longitudinal waves denote that the spin **W** is zero. The equation of motion and the constitutive equation are considered in the linear approximation which means that $\overset{\nabla}{\mathbf{T}} \simeq \dot{\mathbf{T}} \simeq \partial_t \mathbf{T}$. In addition ρ is constant and $\dot{\mathbf{v}} \simeq \partial_t \mathbf{v}$, $\dot{\mathbf{T}} \simeq \partial_t \mathbf{T}$. Letting x be the coordinate in the direction of propagation we let the stress and the displacement depend on x and t in the form $\exp[i(kx - \omega t)]$. With these approximations, the equation of motion and the rate equation allow us to write

$$\rho \partial_t v = \partial_x T, \qquad T + \lambda \partial_t T + \beta \partial_t^2 T = \mu \partial_x v + \nu \partial_x \partial_t v.$$

Applying ∂_x to the second equation and replace $\partial_x T$ from the first one we have

$$\rho \partial_t v + \rho \lambda \partial_t^2 v + \rho \beta \partial_t^3 v = \mu \partial_x^2 v + \nu \partial_x^2 \partial_t v. \tag{6.80}$$

The dependence on $\exp[i(kx - \omega t)]$ of v in (6.80) implies that the wave number k and the angular frequency ω are subject to

$$i\rho\omega - \rho\lambda\omega^2 - i\rho\beta\omega^3 = \mu k^2 - i\nu k^2 \omega.$$

At high frequencies we have

$$\frac{k^2}{\omega^2} = \frac{\rho\beta}{\nu}.$$

Waves propagate with phase speed $\sqrt{\nu/\rho\beta}$, with $\nu\beta > 0$.

If $\beta = 0$ (Oldroyd model) then we have

$$\frac{k^2}{\omega^2} = -\rho \frac{i(\nu\lambda\omega - \mu/\omega) + \mu\lambda + \nu}{\mu^2 + \nu^2\omega^2}.$$

The real part is negative or positive according as $\lambda, \nu > 0$ or $\lambda, \nu < 0$. Let $k = k_r + ik_i$ and hence $k^2 = k_r^2 - k_i^2 + 2ik_r k_i$. It follows that $\Re k^2 > 0$, and hence $k_r^2 > k_i^2$, if $\lambda, \nu < 0$ and vice versa. Instead one expects that $\Im k^2 > 0$ so that $k_r k_i > 0$ (the wave amplitude decays while it propagates) and this is true only depending on the frequency range.

6.5 Wave Equation in Hereditary Fluids

Nonlinear acoustic wave propagation in viscous fluids with memory has been modelled with a third-order integro-differential equation. Here we outline the main steps leading to the model.

We follow the Navier-Stokes-Fourier model of the fluid and hence state the balance of mass, linear momentum, and energy in the form

$$\partial_t \varrho + \nabla \cdot (\rho \mathbf{v}) = 0,$$

$$\rho \partial_t \mathbf{v} + \rho(\mathbf{v} \cdot \nabla)\mathbf{v} = -\nabla \mathcal{P} + \mu \Delta \mathbf{v} + \zeta \nabla(\nabla \cdot \mathbf{v}),$$

$$\rho(\partial_t e + (\mathbf{v} \cdot \nabla) e = -\nabla \cdot \mathbf{q} + 2\mu \mathbf{D}_0 \cdot \mathbf{D}_0 + \zeta (\mathrm{tr}\, \mathbf{D})^2,$$

where $\zeta = \lambda + 2\mu/3)$. Let

$$p = p_0 + \mathscr{P}, \quad \rho = \rho_0 + \varrho, \quad \eta = \eta_0 + s, \quad \varepsilon = \varepsilon_0 + e, \quad \theta = \theta_0 + \vartheta,$$

where $p_0, \rho_0, \eta_0, \varepsilon_0, \theta_0$ are the values at equilibrium. We take $\mathcal{P}$ in the form

$$\mathcal{P} = A \frac{\varrho}{\rho_0} + \frac{B}{2} (\frac{\varrho}{\rho_0})^2 + \partial_\eta p\, s,$$

where $A = \rho_0 \partial_\rho p = \rho_0 c^2$, $B = \rho_0^2 \partial_\rho p$. Hence we can write

$$\mathcal{P} = c^2 \varrho + \frac{c^2}{\rho_0} \frac{B}{2A} \varrho^2 + \partial_\eta p\, s.$$

Finally we take the standard Fourier law in the form $\mathbf{q} = -\kappa \nabla \vartheta$.

6.6 Further Thermoelastic Models

There are thermodynamic schemes that lead to the heat conduction inequality

$$\mathbf{q} \cdot \nabla \theta \leq 0$$

as the reduced entropy inequality. This inequality in turn is apparently satisfied by Fourier's law $\mathbf{q} = -\kappa \nabla \theta$, $\kappa > 0$. If, further, the body is undeformable and the internal energy ε depends only on the temperature θ then the balance of energy results in the partial differential equation

$$\dot{\theta} = k \Delta \theta + g(\mathbf{x}, t)$$

where $k = \kappa/\rho c$, $c = d\varepsilon/d\theta$. An initial temperature field $\theta(\mathbf{x}, 0)$ with compact support is felt at any time $t > 0$ at every point outside the support; we say that the speed of propagation is infinite. This seems to be paradoxical from a physical point of view and is in contrast with low-temperature experimental observations.[8] To overcome this

[8] See, e.g., [86] and references therein.

drawback many schemes have been elaborated. The rate-type models of the previous section can be viewed within this line. Here we pursue this line and emphasize some features of the well-known Maxwell-Cattaneo equation and establish a generalized form. Next we review some uncommon approaches developed in the literature.

6.6.1 Maxwell-Cattaneo Equation; a Generalized Form

Physical heuristic arguments indicate [84, 306, 431] that in non-stationary processes Fourier's law of heat conduction in rigid solids might be replaced by the rate-type equation

$$\tau \dot{\mathbf{q}} + \mathbf{q} = -\kappa \nabla \theta, \qquad \tau > 0. \tag{6.81}$$

Equation (6.81) is named after Maxwell-Cattaneo (for short MC equation); it reduces to Fourier's law in stationary conditions ($\dot{\mathbf{q}} = \mathbf{0}$) or, formally, if the relaxation time τ vanishes. The thermodynamic consistency of (6.81) along with the balance of energy,

$$\rho \dot{\varepsilon} = -\nabla \cdot \mathbf{q} + \rho r,$$

is proved e.g. by letting θ, $\mathbf{q}$, $\nabla\theta$ be the set of independent variables (see Sect. 7.8.3). While κ and τ are often viewed as constants, a simple improvement is obtained by letting κ, τ be functions of θ. The fit of experimental data or physical aspects might indicate that a rate-type equation would be in order for an appropriate vector, say $\mathbf{w}$, in place of $\mathbf{q}$ [331]. Here we follow this idea by paralleling the model in [330].

We look for a generalized MC equation in the form

$$\dot{\mathbf{w}} + \mathbf{\Lambda} \mathbf{w} = -m \nabla \theta, \tag{6.82}$$

where $\mathbf{w}$ is the sought analogue of $\mathbf{q}$; the scalar m and the tensor $\mathbf{\Lambda}$ are allowed to depend on θ. Let Ψ, N be the free energy and the entropy per unit volume. We assume that $\Psi, N, \mathbf{q}$ are functions of $\theta, \mathbf{w}, \nabla\theta$ while $\mathbf{w}$ is subject to (6.82). The Clausius-Duhem inequality takes the form

$$(\partial_\theta \Psi + N)\dot{\theta} + \partial_{\mathbf{w}} \Psi \cdot \dot{\mathbf{w}} + \partial_{\nabla\theta} \Psi \cdot \nabla\dot{\theta} + \frac{1}{\theta}\mathbf{q} \cdot \nabla\theta \le 0.$$

Substitution of $\dot{\mathbf{w}}$ from (6.82) results in

$$(\partial_\theta \Psi + N)\dot{\theta} + (\frac{1}{\theta}\mathbf{q} - m\partial_{\mathbf{w}}\Psi) \cdot \nabla\theta - \partial_{\mathbf{w}}\Psi \cdot \mathbf{\Lambda}\mathbf{w} + \partial_{\nabla\theta}\Psi \cdot \nabla\dot{\theta} \le 0.$$

The arbitrariness of $\nabla\dot{\theta}$, $\dot{\theta}$ implies

$$\partial_{\nabla\theta}\Psi = \mathbf{0}, \qquad N = -\partial_\theta \Psi.$$

As for $\mathbf{q}(\theta, \mathbf{w}, \nabla\theta)$ we observe that a dependence of $\mathbf{q}$ on $\nabla\theta$ would result in a Fourier-like equation. Hence we consider $\mathbf{q}(\theta, \mathbf{w})$; the linearity of $\nabla\theta$ implies that

$$\frac{1}{\theta}\mathbf{q} = m\partial_{\mathbf{w}}\Psi, \qquad \partial_{\mathbf{w}}\Psi \cdot \mathbf{\Lambda w} \geq 0.$$

To make the model operative we have to determine the vector $\mathbf{w}$. We require the consistency between Fourier's law and the stationary version of (6.82),

$$\mathbf{q} = -\mathbf{K}\nabla\theta, \qquad \mathbf{w} = -m\mathbf{\Lambda}^{-1}\nabla\theta,$$

$\mathbf{K}$ being the conductivity tensor. Hence we assume

$$\mathbf{w} = m\mathbf{\Lambda}^{-1}\mathbf{K}^{-1}\mathbf{q}.$$

Consequently

$$\partial_{\mathbf{w}}\Psi = \frac{1}{\theta m^2}\mathbf{K\Lambda w}, \qquad \Psi = \Psi_0(\theta) + \frac{1}{2\theta m^2}\mathbf{w}\cdot\mathbf{K\Lambda w}$$

and hence

$$0 \leq \partial_{\mathbf{w}}\Psi \cdot \mathbf{\Lambda w} = \frac{1}{2\theta m^2}\mathbf{\Lambda w}\cdot\mathbf{K\Lambda w}.$$

Irrespective of the form of $\mathbf{\Lambda}$ it follows that $\mathbf{K}$ has to be positive definite, as expected.

We still need to determine $\mathbf{\Lambda}$ and m. By definition the energy $\mathcal{E}$ (per unit volume) is

$$\mathcal{E} = \Psi_0 - \theta\Psi_0' + \mathbf{w}\cdot\Big[\frac{\mathbf{K\Lambda}}{2\theta m^2} - \theta\Big(\frac{\mathbf{K\Lambda}}{2\theta m^2}\Big)'\Big]\mathbf{w},$$

the prime denoting the derivative with respect to θ. The energy then depends only on the temperature if

$$\frac{\mathbf{K\Lambda}}{2\theta m^2} - \theta\Big(\frac{\mathbf{K\Lambda}}{2\theta m^2}\Big)' = \mathbf{0},$$

which happens if

$$\frac{\mathbf{K\Lambda}}{2\theta m^2} = \theta\mathbf{\Gamma},$$

$\mathbf{\Gamma}$ being a constant tensor. Hence

$$\mathbf{\Lambda} = 2\theta^2 m^2\mathbf{K}^{-1}\mathbf{\Gamma}, \qquad \mathbf{w} = \frac{1}{2\theta^2 m}\mathbf{\Gamma}^{-1}\mathbf{q}.$$

The model is conclusive once the function m is selected.

6.6.2 Temperature-Rate Dependent Thermoelastic Materials

Borrowing from [288] and [52] we allow for the dependence on the temperature rate $\dot{\theta}$. Also to account for objectivity, it is convenient to follow the material formulation. Hence we let ψ_R, η_R, $\mathbf{T}_{RR}$, $\mathbf{q}_R$, $\mathbf{k}_R$ be functions of

$$\mathbf{C}, \theta, \nabla_R \theta, \dot{\theta}.$$

We assume that the free energy ψ_R and the extra-entropy $\mathbf{k}_R$ are differentiable while η_R, $\mathbf{T}_{RR}$, and $\mathbf{q}_R$ are continuous. To determine the thermodynamic restrictions we then consider the entropy inequality in the material formulation,

$$-(\dot{\psi}_R + \eta_R\dot{\theta}) + \tfrac{1}{2}\mathbf{T}_{RR} \cdot \dot{\mathbf{C}} - \frac{1}{\theta}\mathbf{q}_R \cdot \nabla_R \theta + \theta \nabla_R \cdot \mathbf{k}_R \geq 0.$$

Upon computation of $\dot{\psi}_R$ and $\nabla_R \cdot \mathbf{k}_R$ we obtain

$$-(\partial_\theta \psi_R + \eta_R)\dot{\theta} + (\tfrac{1}{2}\mathbf{T}_{RR} - \partial_{\mathbf{C}}\psi_R) \cdot \dot{\mathbf{C}} - \partial_{\nabla_R\theta}\psi_R \cdot \nabla_R \dot{\theta} - \partial_{\dot{\theta}}\psi_R\, \ddot{\theta} - \frac{1}{\theta}\mathbf{q}_R \cdot \nabla_R \theta$$
$$+\theta(\partial_{\mathbf{C}}\mathbf{k}_R \cdot \nabla_R \mathbf{C} + \partial_\theta \mathbf{k}_R \cdot \nabla_R \theta + \partial_{\nabla_R\theta}\mathbf{k}_R \cdot \nabla_R \nabla_R \theta + \partial_{\dot{\theta}}\mathbf{k}_R \cdot \nabla_R \dot{\theta}) \geq 0.$$

The arbitrariness of $\ddot{\theta}$, $\nabla_R \mathbf{C}$, $\nabla_R \nabla_R \theta$ and $\dot{\mathbf{C}}$, $\nabla_R \dot{\theta}$ implies that

$$\partial_{\dot{\theta}}\psi_R = 0, \qquad \partial_{\mathbf{C}}\mathbf{k}_R = \mathbf{0}, \qquad \partial_{\nabla_R\theta}\mathbf{k}_R \in \mathrm{Skw},$$

$$\mathbf{T}_{RR} = 2\,\partial_{\mathbf{C}}\psi_R, \qquad -\partial_{\nabla_R\theta}\psi_R + \theta\partial_{\dot{\theta}}\mathbf{k}_R = \mathbf{0}.$$

The entropy inequality reduces to

$$-(\partial_\theta \psi_R + \eta_R)\dot{\theta} - (\frac{1}{\theta}\mathbf{q}_R - \partial_\theta \mathbf{k}_R) \cdot \nabla_R \theta \geq 0.$$

As we see shortly, the qualitative novelty of this model is the dependence of the entropy on the temperature rate. Hence, for simplicity, we let $\mathbf{k}_R = \mathbf{0}$, which implies that

$$(\partial_\theta \psi_R + \eta_R)\dot{\theta} + \frac{1}{\theta}\mathbf{q}_R \cdot \nabla_R \theta \leq 0, \qquad \psi_R = \psi_R(\mathbf{C}, \theta).$$

Moreover we assume η is independent of $\nabla_R \theta$ and let

$$\eta_R = \eta_0(\mathbf{C}, \theta) + \lambda(\mathbf{C}, \theta)\dot{\theta}.$$

It follows that

$$\eta_0 = -\partial_\theta \psi_R, \qquad \lambda\dot{\theta}^2 + \frac{1}{\theta}\mathbf{q}_R \cdot \nabla_R \theta \leq 0.$$

Letting $\nabla_R \theta = \mathbf{0}$ we find the condition $\lambda \leq 0$. Fourier's law in the form

$$\mathbf{q}_R = -\mathbf{K}(\mathbf{C}, \theta)\nabla_R \theta,$$

$\mathbf{K}$ being positive definite completes a possible, thermodynamically consistent scheme. To sum up, in addition to Fourier's law we have

$$\psi_R = \psi_R(\mathbf{C}, \theta), \qquad \eta_R = -\partial_\theta \psi_R + \lambda(\mathbf{C}, \theta)\dot{\theta}, \qquad \mathbf{T}_{RR} = 2\,\partial_{\mathbf{C}} \psi_R.$$

Second-order waves

We now investigate the compatibility of this scheme with wave propagation at a finite speed. Let Σ be a time-dependent surface in the reference configuration R and denote by $\mathbf{n}_R$ the unit normal. The surface Σ is a second-order wave (or acceleration wave) if $\mathbf{x}$, $\dot{\mathbf{x}}$, $\mathbf{F}$, θ, $\dot{\theta}$, $\nabla_R \theta$ are continuous functions of $\mathbf{X}$ and t in a neighbourhood $\mathcal{N}$ of Σ, $\ddot{\mathbf{x}}$, $\dot{\mathbf{F}}$, $\ddot{\theta}$, $\nabla_R \dot{\theta}$, $\nabla_R \nabla_R \theta$ and all higher-order derivatives suffer jump discontinuities across Σ but are continuous functions of $\mathbf{X}$ and t in $\mathcal{N} \setminus \Sigma$.

As a consequence of the constitutive equations it follows that

$$[\![\psi_R]\!] = 0, \quad [\![\eta_R]\!] = 0, \quad [\![\mathbf{T}_{RR}]\!] = \mathbf{0}, \quad [\![\mathbf{q}_R]\!] = \mathbf{0}$$

across Σ. Define

$$\mathbf{a} := [\![\frac{d^2\mathbf{x}}{dn_R^2}]\!], \qquad \vartheta := [\![\frac{d^2\theta}{dn_R^2}]\!]$$

where $d/dn_R = \mathbf{n}_R \cdot \nabla_R$. Hugoniot's compatibility conditions imply that

$$[\![\nabla_R \mathbf{F}]\!] = \mathbf{a}\, \mathbf{n}_R \otimes \mathbf{n}_R, \qquad [\![\dot{\mathbf{F}}]\!] = -U_R\, \mathbf{a} \otimes \mathbf{n}_R, \qquad [\![\ddot{\mathbf{x}}]\!] = U_R^2\, \mathbf{a},$$

$$[\![\nabla_R \nabla_R \theta]\!] = \vartheta\, \mathbf{n}_R \otimes \mathbf{n}_R, \qquad [\![\nabla_R \dot{\theta}]\!] = -U_R\, \vartheta\, \mathbf{n}_R, \qquad [\![\ddot{\theta}]\!] = U_R^2\, \vartheta.$$

The discontinuities $\mathbf{a}$, ϑ are subject to the balance equations. We let ρ_R and $\mathbf{b}$ be continuous across Σ. Hence by the equation of motion $\rho_R \dot{\mathbf{v}} = \nabla_R \cdot \mathbf{T}_R + \rho_R \mathbf{b}$ it follows

$$\rho_R [\![\dot{\mathbf{v}}]\!] = [\![\nabla_R \cdot \mathbf{T}_R]\!].$$

Since $[\![\nabla_R \cdot \mathbf{T}_R]\!] = -[\![\dot{\mathbf{T}}_R]\!]\mathbf{n}_R$ then

$$-\, U_R^3 \rho_R \mathbf{a} = [\![\dot{\mathbf{T}}_R]\!]\mathbf{n}_R. \tag{6.83}$$

Moreover

$$\mathbf{T}_R = \mathbf{F}\mathbf{T}_{RR} = 2\mathbf{F}\partial_{\mathbf{C}} \psi_R(\mathbf{C}, \theta)$$

and then

$$\dot{\mathbf{T}}_R = \dot{\mathbf{F}}\mathbf{T}_{RR} + 2\mathbf{F}(\partial_{\mathbf{C}}^2 \psi_R \dot{\mathbf{C}} + \partial_\theta \partial_{\mathbf{C}} \psi_R \dot{\theta}).$$

Now,

$$[\![\dot{\mathbf{C}}]\!] = [\![\dot{\mathbf{F}}^T]\!]\mathbf{F} + \mathbf{F}^T[\![\dot{\mathbf{F}}]\!] = -U_R(\mathbf{n}_R \otimes \mathbf{F}^T\mathbf{a} + \mathbf{F}^T\mathbf{a} \otimes \mathbf{n}_R).$$

Let

$$\mathbf{A} = \partial^2_{\mathbf{C}}\psi_R;$$

by definition $\mathbf{A}$ enjoys the major and minor symmetries, $A_{HKMN} = A_{MNHK}$ and $A_{HKMN} = A_{KHMN} = A_{HKNM}$. Hence we find

$$[\![\dot{\mathbf{T}}_R]\!]\mathbf{n}_R = -U_R\mathbf{Q}\mathbf{a}$$

where

$$\mathbf{Q} = (\mathbf{n}_R \cdot \mathbf{T}_{RR}\mathbf{n}_R)\mathbf{1} + \mathbf{n}_R\mathbf{F}\mathbf{A}\mathbf{F}^T\mathbf{n}_R.$$

Consequently (6.83) implies

$$(\mathbf{Q} - \rho_R U_R^2\mathbf{1})\mathbf{a} = \mathbf{0}, \tag{6.84}$$

which is the analogue of the propagation condition for acceleration waves in elastic solids. This is so because $\partial_\theta\partial_{\mathbf{C}}\psi_R\dot{\theta}$ is continuous across Σ.

We now examine the balance of energy. Since

$$\varepsilon_R = \psi_R + \theta\eta_R = \psi_R - \theta\partial_\theta\psi_R + \theta\lambda\,\dot{\theta},$$

where $\lambda = \lambda(\mathbf{C}, \theta)$, we find

$$[\![\dot{\varepsilon}_R]\!] = \theta\lambda[\![\ddot{\theta}\,]\!].$$

Hence the balance of energy yields

$$\rho_R\theta\lambda[\![\ddot{\theta}\,]\!] = -[\![\nabla_R \cdot \mathbf{q}_R]\!] + \mathbf{T}_{RR} \cdot [\![\dot{\mathbf{C}}]\!].$$

Since

$$-[\![\nabla_R \cdot \mathbf{q}_R]\!] = \frac{1}{U_R}[\![\dot{\mathbf{q}}_R]\!] \cdot \mathbf{n}_R = -\frac{1}{U_R}\mathbf{n}_R \cdot \mathbf{K}[\![\nabla_R\dot{\theta}]\!] = \mathbf{n}_R \cdot \mathbf{K}\mathbf{n}_R\vartheta$$

then the balance of energy results in

$$(\rho_R\theta\lambda U_R^2 - \mathbf{n}_R \cdot \mathbf{K}\mathbf{n}_R)\vartheta = -2U_R(\mathbf{F}\mathbf{T}_{RR}\mathbf{n}_R) \cdot \mathbf{a}. \tag{6.85}$$

Two possibilities occur. If $\rho_R U_R^2$ is an eigenvalue of $\mathbf{Q}$, for an appropriate direction $\mathbf{n}_R$, then by (6.84) it follows that $\mathbf{a}$ propagates with that speed U_R. Correspondingly ϑ is determined by (6.85) in the form

$$\vartheta = -\frac{2U_R(\mathbf{F}\mathbf{T}_{RR}\mathbf{n}_R)}{\rho_R\theta\lambda U_R^2 - \mathbf{n}_R \cdot \mathbf{K}\mathbf{n}_R} \cdot \mathbf{a}.$$

Otherwise, if $\rho_R\theta\lambda U_R^2 - \mathbf{n}_R \cdot \mathbf{K}\mathbf{n}_R = 0$ then (6.85) implies $\mathbf{a} = \mathbf{0}$. However, $\lambda < 0$ and $\mathbf{K} > \mathbf{0}$ do not allow for this condition.

Thus the dependence of the entropy on $\dot{\theta}$, so that $\lambda \neq 0$, allows the propagation of thermal waves, $[\![\ddot{\theta}]\!] \neq 0$, with no contribution from heat conduction ($\mathbf{K} = \mathbf{0}$).

If, instead, $\lambda = 0$ then only waves occur with speed U_R such that $\rho_R U_R^2$ is an eigenvalue of $\mathbf{Q}$. In this case ϑ is given by

$$\vartheta = \frac{2U_R(\mathbf{F}\mathbf{T}_{RR}\mathbf{n}_R)}{\mathbf{n}_R \cdot \mathbf{K}\mathbf{n}_R} \cdot \mathbf{a}.$$

Remark 6.1 The model with $\lambda = 0$ might appear surprising in that it allows wave propagation with finite speed though without the seemingly crucial dependence of η_R on $\dot{\theta}$. Now, Eq. (6.85), with $\lambda = 0$, shows that ϑ is related to $\mathbf{a}$ and $\vartheta = 0$ if $\mathbf{a} = \mathbf{0}$. Moreover $\mathbf{a} = \mathbf{0}$ if the body is undeformable, $\dot{\mathbf{C}} = \mathbf{0}$. Hence the customary paradox of infinite speed of propagation is based also on the assumption that the body is undeformable.

6.6.3 Thermoelasticity Based on an Integral Variable

Two schemes about heat flows have been established by Green and Naghdi [210, 211], within a re-examination of the basic postulates of thermomechanics. Constitutive equations are given especially for the flow of heat, with particular reference to the propagation of thermal waves at finite speed. The schemes involve the so-called thermal displacement α and consider $\nabla\alpha$, instead of $\nabla\theta$, among the state variables. Both the absolute temperature and an empirical temperature are considered.[9] In essence, though, the content of thermal displacement is provided by restricting attention to a single temperature, the absolute one. That is why here we consider only the absolute temperature.

The thermal displacement α on $\mathrm{R} \times \mathbb{R}$ is defined by

$$\dot{\alpha} = \theta.$$

The thermal displacement gradient $\boldsymbol{\beta}$ is defined by

$$\boldsymbol{\beta} = \nabla_R\,\alpha.$$

It follows that

$$\dot{\boldsymbol{\beta}} = \nabla_R\,\dot{\alpha} = \nabla_R\,\theta = \mathbf{F}^T\nabla\theta.$$

Thermoelasticity without energy dissipation

[9] Similar approaches involving two temperatures are reviewed in [391].

As the constitutive assumption we assume that the free energy ψ, the entropy η, the Cauchy stress $\mathbf{T}$, and the heat flux $\mathbf{q}$ are functions of

$$\theta,\ \nabla_R \alpha,\ \mathbf{F};$$

the extra-entropy flux is assumed to be zero. The constitutive equations are required to satisfy the entropy inequality

$$-\rho(\dot{\psi}+\eta\dot{\theta})+\mathbf{T}\cdot\mathbf{D}-\frac{1}{\theta}\mathbf{q}\cdot\nabla\theta\geq 0.$$

Upon computation of $\dot{\psi}$ and substitution we find

$$-\rho(\partial_\theta\psi+\eta)\dot{\theta}-\rho\partial_{\nabla_R\alpha}\psi\cdot(\mathbf{F}^T\nabla\theta)-\rho(\partial_{\mathbf{F}}\psi\mathbf{F}^T)\cdot\mathbf{L}+\mathbf{T}\cdot\mathbf{D}-\frac{1}{\theta}\mathbf{q}\cdot\nabla\theta\geq 0.$$

The left-hand side is linear in $\dot{\theta}$, $\mathbf{W}$, $\mathbf{D}$, $\nabla\theta$. The arbitrariness of their values, independent of θ, $\nabla_R\alpha$, $\mathbf{F}$, implies that

$$\eta=-\partial_\theta\psi,\qquad \partial_{\mathbf{F}}\psi\mathbf{F}^T\in \mathrm{Sym},$$

$$\mathbf{T}=\rho\partial_{\mathbf{F}}\psi\mathbf{F}^T,\qquad \mathbf{q}=-\rho\mathbf{F}\partial_{\nabla_R\alpha}\psi.$$

By objectivity, ψ may depend on $\mathbf{F}$ via the invariant tensor $\mathbf{C}$. Hence we let

$$\psi=\tilde{\psi}(\theta,\nabla_R\alpha,\mathbf{C}).$$

Consequently it follows

$$\partial_{\mathbf{F}}\psi=\partial_{\mathbf{C}}\tilde{\psi}\partial_{\mathbf{F}}\mathbf{C}=2\mathbf{F}\partial_{\mathbf{C}}\tilde{\psi}$$

and then

$$\partial_{\mathbf{F}}\psi\mathbf{F}^T=2\mathbf{F}\partial_{\mathbf{C}}\tilde{\psi}\mathbf{F}^T.$$

The requirement $\partial_{\mathbf{F}}\psi\mathbf{F}^T\in \mathrm{Sym}$ is apparently satisfied and hence

$$\mathbf{T}=2\mathbf{F}\partial_{\mathbf{C}}\tilde{\psi}\mathbf{F}^T\in \mathrm{Sym},$$

as expected. The remaining conditions show that both η and $\mathbf{q}$ are derived by the free energy. Moreover, the entropy inequality results in an entropy equality and that is why we can say that this is a model of thermoelasticity without energy dissipation.

Of course, the vanishing of dissipation (entropy equality) implies that no inequality holds for the constitutive equations. This seems to be a weak point of the scheme. To fix ideas, let

$$\tilde{\psi}=\Psi(\theta,\mathbf{C})+\gamma(\theta,\mathbf{C})(\nabla_R\theta)^2;$$

no restriction follows from thermodynamics on the function $\gamma(\theta, \mathbf{C})$.

Second-order waves

Again we investigate the compatibility of the constitutive scheme with wave propagation at a finite speed though with a different characterization of second-order waves. Let σ be a time-dependent surface in the current configuration Ω and denote by $\mathbf{n}$ the unit normal. The surface σ is a second-order wave if $\mathbf{x}$, $\dot{\mathbf{x}}$, $\mathbf{F}$, α, $\dot{\alpha} = \theta$, $\nabla_R \alpha$ are continuous functions of $\mathbf{x}$ and t in a neighbourhood $\mathcal{N}$ of σ, $\ddot{\mathbf{x}}$, $\dot{\mathbf{F}}$, $\ddot{\alpha} = \dot{\theta}$, $\nabla_R \theta$, $\nabla_R \nabla_R \alpha$ and all higher-order derivatives suffer jump discontinuities across σ but are continuous functions of $\mathbf{x}$ and t in $\mathcal{N} \setminus \sigma$.

First we observe that if $[\![\phi]\!] = 0$ then

$$[\![\partial_t \phi]\!] = -u_n [\![\partial_n \phi]\!]$$

and, since $[\![\mathbf{v}]\!] = \mathbf{0}$,

$$[\![\dot{\phi}]\!] = -U [\![\partial_n \phi]\!],$$

where $\partial_n = \mathbf{n} \cdot \nabla$ is the normal derivative, u_n is the speed of displacement, and U is the local speed of propagation. Now, since $[\![\mathbf{F}]\!] = \mathbf{0}$ and $\dot{\mathbf{F}} = \nabla_R \mathbf{v} = \nabla \mathbf{v} \mathbf{F}$ then

$$[\![\dot{\mathbf{F}}]\!] = [\![\nabla \mathbf{v}]\!] \mathbf{F} = [\![\partial_n \mathbf{v}]\!] \otimes \mathbf{n} \mathbf{F}, \qquad [\![\dot{\mathbf{v}}]\!] = -U [\![\partial_n \mathbf{v}]\!],$$

$$[\![\nabla_R \theta]\!] = [\![\nabla \theta]\!] \mathbf{F} = [\![\partial_n \theta]\!] \mathbf{n} \mathbf{F}, \qquad [\![\dot{\theta}]\!] = -U [\![\partial_n \theta]\!].$$

Moreover

$$[\![\psi]\!] = 0, \qquad [\![\eta]\!] = 0, \qquad [\![\mathbf{T}]\!] = \mathbf{0}, \qquad [\![\mathbf{q}]\!] = \mathbf{0}.$$

We let

$$\mathbf{a} = [\![\partial_n \mathbf{v}]\!], \qquad \vartheta = [\![\partial_n \theta]\!]$$

be the representatives of the discontinuities at the wave.

Assume $[\![\mathbf{b}]\!] = \mathbf{0}$ and $[\![r]\!] = 0$. Since $[\![\rho]\!] = 0$ then the equation of motion results in

$$\rho [\![\dot{\mathbf{v}}]\!] = [\![\nabla \cdot \mathbf{T}]\!].$$

Now, since $[\![\mathbf{T}]\!] = \mathbf{0}$ then

$$[\![\nabla \cdot \mathbf{T}]\!] = [\![\partial_n \mathbf{T}]\!] \mathbf{n} = -\frac{1}{U} \{ \partial_{\mathbf{F}} \mathbf{T} [\![\dot{\mathbf{F}}]\!] + \partial_\theta \mathbf{T} [\![\dot{\theta}]\!] + \partial_{\nabla_R \alpha} \mathbf{T} [\![\nabla_R \theta]\!] \} \mathbf{n}.$$

Observe that

$$\partial_\theta \mathbf{T} [\![\dot{\theta}]\!] = -U \partial_\theta \mathbf{T} \vartheta.$$

Moreover, in suffix notation

$$\{(\partial_{\mathbf{F}}\mathbf{T}[\![\dot{\mathbf{F}}]\!])\mathbf{n}\}_i = n_j \partial_{F_{kP}} T_{ij} F^T_{Pl} n_l a_k =: Q_{ik} a_k = (\mathbf{Q}\mathbf{a})_i,$$

$$\{(\partial_{\nabla_R\alpha}\mathbf{T}[\![\nabla_R\theta]\!])\mathbf{n}\}_i = n_j \partial_{\alpha,P} T_{ij} [\![\theta_{,l}]\!] F_{lP} = n_j \partial_{\alpha,P} T_{ij} F^T_{Pl} n_l \vartheta =: c_i \vartheta.$$

Hence the equation of motion implies that

$$\rho U^2 \mathbf{a} = \mathbf{Q}\mathbf{a} + (\mathbf{c} - U\partial_\theta \mathbf{T})\vartheta. \tag{6.86}$$

We now investigate the consequences of the balance of energy. We have

$$\rho[\![\dot{\varepsilon}]\!] = \mathbf{T}\cdot[\![\mathbf{D}]\!] - [\![\nabla\cdot\mathbf{q}]\!].$$

Observe that

$$[\![\dot{\varepsilon}]\!] = \partial_{\mathbf{F}}\varepsilon\cdot[\![\dot{\mathbf{F}}]\!] + \partial_\theta\varepsilon[\![\dot{\theta}\,]\!] + \partial_{\nabla_R\alpha}\varepsilon\cdot[\![\nabla_R\theta]\!] = (\partial_{\mathbf{F}}\varepsilon\mathbf{F}^T\mathbf{n})\cdot\mathbf{a} + (\partial_{\nabla_R\alpha}\varepsilon\cdot\mathbf{n} - U\partial_\theta\varepsilon)\vartheta$$

and

$$\begin{aligned}[\![\nabla\cdot\mathbf{q}]\!] = [\![\partial_n\mathbf{q}]\!]\cdot\mathbf{n} = -\frac{1}{U}[\![\dot{\mathbf{q}}]\!]\cdot\mathbf{n} &= -\frac{1}{U}\{\partial_{\mathbf{F}}\mathbf{q}[\![\dot{\mathbf{F}}]\!] + \partial_\theta\mathbf{q}[\![\dot{\theta}]\!] + \partial_{\nabla_R\alpha}\mathbf{q}[\![\nabla_R\theta]\!]\}\\ &= \frac{1}{U}\{(\mathbf{n}\partial_{\mathbf{F}}\mathbf{q}\mathbf{F}^T\mathbf{n})\cdot\mathbf{a} + (\mathbf{n}\cdot\partial_{\nabla_R\alpha}\mathbf{q}\mathbf{F}^t\mathbf{n} - U\mathbf{n}\cdot\partial_\theta\mathbf{q})\vartheta\},\end{aligned}$$

$$[\![\mathbf{L}]\!] = [\![\nabla\mathbf{v}]\!] = \mathbf{a}\otimes\mathbf{n}, \qquad \mathbf{T}\cdot[\![\mathbf{D}]\!] = (\mathbf{T}\mathbf{n})\cdot\mathbf{a}.$$

Hence it follows

$$(\rho\partial_{\mathbf{F}}\varepsilon\mathbf{n} - \frac{1}{U}\mathbf{n}\partial_{\mathbf{F}}\mathbf{q}\mathbf{F}^T\mathbf{n} - \mathbf{T}\mathbf{n})\cdot\mathbf{a} + [\rho(\partial_{\nabla_R\alpha}\varepsilon\cdot\mathbf{n} - U\partial_\theta\varepsilon) - \frac{1}{U}\mathbf{n}\cdot\partial_{\nabla_R\alpha}\mathbf{q}\mathbf{F}^T\mathbf{n} - U\mathbf{n}\cdot\partial_\theta\mathbf{q}]\vartheta = 0. \tag{6.87}$$

Equations (6.86)–(6.87) constitute a linear homogeneous system in the unknowns $\mathbf{a}$, ϑ. A very simple scheme arises by letting

$$\psi = \psi_1(\theta) + \psi_2(\mathbf{F}) - \tfrac{1}{2}\kappa(\nabla_R\alpha)^2.$$

This free energy function implies that $\mathbf{c}$, $\partial_\theta\mathbf{T}$, $\partial_{\mathbf{F}}\varepsilon$, $\partial_{\nabla_R\alpha}\varepsilon$, and $\partial_\theta\mathbf{q}$ are zero. In addition let the body be free from stress and gradient of thermal displacement, $\mathbf{T} = \mathbf{0}$, $\nabla_R\alpha = \mathbf{0}$. The system simplifies to

$$(\rho U^2\mathbf{1} - \mathbf{Q})\mathbf{a} = \mathbf{0}, \qquad (\partial_\theta\varepsilon\, U^2 - \kappa\mathbf{n}\cdot\mathbf{F}\mathbf{F}^T\mathbf{n})\vartheta = 0.$$

The system splits into two uncoupled equations. The first one is satisfied by acceleration waves ($\vartheta = 0$) propagating with the speed U such that ρU^2 is an eigenvalue of $\mathbf{Q}$. The second one admits wave solutions if $\kappa > 0$, which is not implied by thermodynamics. In this case purely thermal waves ($\mathbf{a} = \mathbf{0}$) occur; they propagate with local speed U given by

$$U^2 = \frac{\kappa \mathbf{n} \cdot \mathbf{F}\mathbf{F}^T \mathbf{n}}{\partial_\theta \varepsilon}.$$

Thermoelastic solid, type III

Though within our approach and notation, we now review the scheme of heat conductor denoted as *type III* in [210]. Let

$$\theta, \nabla_R \alpha, \nabla\theta, \mathbf{F}$$

be the set of variables and let ψ, η, $\mathbf{T}$, $\mathbf{q}$ be given by constitutive functions. Differently from the thermoelastic solid without energy dissipation,[10] now the dependence on $\nabla\theta$ is considered. Upon substitution of $\dot{\psi}$ in the second-law inequality

$$-\rho(\dot{\psi} + \eta\dot{\theta}) + \mathbf{T} \cdot \mathbf{D} - \frac{1}{\theta}\mathbf{q} \cdot \nabla\theta \geq 0$$

we obtain

$$-\rho(\partial_\theta \psi + \eta)\dot{\theta} - \rho\partial_{\nabla_R \alpha}\psi \cdot \mathbf{F}^T \nabla\theta - \rho\partial_{\nabla\theta}\psi \cdot \overset{\cdot}{\overline{\nabla\theta}} - \rho(\partial_{\mathbf{F}}\psi\mathbf{F}^T) \cdot \mathbf{L} + \mathbf{T} \cdot \mathbf{D} - \frac{1}{\theta}\mathbf{q} \cdot \nabla\theta \geq 0.$$

The arbitrariness of $\overset{\cdot}{\overline{\nabla\theta}} = \nabla\dot{\theta} - \mathbf{L}\nabla\theta$ implies that

$$\partial_{\nabla\theta}\psi = \mathbf{0}.$$

Hence the arbitrariness of $\dot{\theta}$ and $\mathbf{L} = \mathbf{W} + \mathbf{D}$ requires that

$$\eta = -\partial_\theta \psi, \qquad \partial_{\mathbf{F}}\psi\mathbf{F}^T \in \mathrm{Sym}, \qquad \mathbf{T} = \rho\partial_{\mathbf{F}}\psi\mathbf{F}^T.$$

Thus it follows the reduced inequality

$$(\rho\mathbf{F}\partial_{\nabla_R \alpha}\psi + \frac{1}{\theta}\mathbf{q}) \cdot \nabla\theta \leq 0.$$

Since $\mathbf{q}$ depends on $\nabla\theta$, while ψ does not, then this inequality implies that

$$\mathbf{q} = -\theta\rho\,\mathbf{F}\partial_{\nabla_R \alpha}\psi + \tilde{\mathbf{q}},$$

where $\tilde{\mathbf{q}}$ is subject to the heat conduction inequality

$$\tilde{\mathbf{q}} \cdot \nabla\theta \leq 0.$$

[10] Denoted as type II in [210].

To fix ideas let

$$\psi = \Psi(\theta, \mathbf{F}) + \frac{1}{2}\gamma \nabla_R \alpha \cdot \nabla_R \alpha.$$

and

$$\tilde{\mathbf{q}} = -f\, \nabla\theta,$$

where γ and f are functions of θ, $\mathbf{F}$. Since $\nabla_R \alpha \cdot \nabla_R \alpha = \nabla\alpha \cdot \mathbf{B}\nabla\alpha$, where $\mathbf{B} = \mathbf{F}\mathbf{F}^T$, then

$$\mathbf{q} = -\gamma\theta\rho\mathbf{B}\nabla\alpha - f\nabla\theta,$$

and $\varepsilon = \psi - \theta\partial_\theta\psi$ is a function of θ, $\nabla_R\alpha$, $\mathbf{F}$. The energy equation makes

$$\rho\partial_\theta\varepsilon\dot{\theta} - f\Delta\theta$$

be equal to first-order terms in α and θ thus making the equation for θ parabolic. The equation becomes hyperbolic if $f = 0$ that is if the dependence on $\nabla\theta$ is ruled out, as it happens in the type II model.

Chapter 7
Materials with Memory

Real materials exhibit a continuing strain rate (creep) under persistent constant stress and decay of stress (relaxation) under constant strain. The analogue occurs for electric current or electric polarization and electric field. In many materials these effects are very pronounced while in others they may be negligible, the extent being possibly influenced by the temperature and being related to the time scale adopted. The theory of linear viscoelasticity encompasses those phenomena which can be described by a linear superposition of the effect of stress or strain. Viscoelasticity may be viewed as perhaps the leading model of material with long range, gradually fading memory. Electromagnetism in matter is likewise a subject described by fading memory.

After some preliminaries on the mathematical framework for fading memory, this chapter develops relevant models of materials and investigates the thermodynamic consistency and the wave propagation properties. First thermoelastic materials (with fading memory) are modelled through histories and summed histories. Next the wide topic of viscoelasticity is developed; the linear viscoelastic solid with emphasis on thermodynamic restrictions and free energies; viscoelastic solids with unbounded relaxation functions; viscoelastic fluids and in particular Boltzmann-like and Maxwell-like models. Attention is then addressed to electromagnetic materials: dielectrics with memory, magneto-viscoelastic materials. The connection between causality and the Kramers–Kronig relations is established. Finally, the modelling of memory properties via derivatives of fractional order is investigated; with particular reference to heat conductors, conceptual difficulties are shown for the thermodynamic consistency and wave properties.

7.1 Materials with Fading Memory

We consider functions on $\mathrm{R} \times \mathbb{R}$ and, to save writing, we usually omit writing R, as the space domain, and the pertinent point $\mathbf{X} \in \mathrm{R}$. A *process* is a function, say ϕ, on $\mathbb{R}$ with (scalar, vector, or tensor) values in a Hilbert space W: the symbol "·" denotes

A. Morro and C. Giorgi, *Mathematical Modelling of Continuum Physics*,
Modeling and Simulation in Science, Engineering and Technology,
https://doi.org/10.1007/978-3-031-20814-0_7

its inner product and $|\cdot|$ the corresponding norm. If ϕ is a process and $t \in \mathbb{R}$ then we can define

$$\phi^t(s) = \phi(t-s), \qquad s \in \mathbb{R}^+$$

and say that ϕ^t is the *history* of ϕ up to time t. The restriction of ϕ^t to $(0, \infty)$ is called the *past history* of ϕ up to time t and denoted by ϕ^t_r. For any history ϕ^t,

$$\phi^t(0) = \phi(t)$$

is called the present value of ϕ^t. Denote by $\phi(t_0)^\dagger$ the constant process with value $\phi(t_0)$,

$$\phi(t_0)^\dagger(t) = \phi(t_0), \qquad t \in \mathbb{R},$$

so that the constant history, up to time t_0, with value $\phi(t_0)$, is given by

$$\phi^{t_0 \dagger}(s) = \phi(t_0)^\dagger(t_0 - s) = \phi(t_0), \qquad s \in \mathbb{R}^+.$$

When $t > t_0$ the constant[1] continuation of ϕ^{t_0} up to t, also denoted by $\phi^t_{t_0}$, is the history defined by

$$\phi^t_{t_0}(s) = \begin{cases} \phi^{t_0}(0), & s \in [0, t-t_0), \\ \phi^{t_0}(s-t+t_0), & s \in [t-t_0, \infty). \end{cases}$$

Alternatively, if $\tau = t - t_0$ then the constant continuation up to $t_0 + \tau$ can be given as

$$\phi^{t_0+\tau}_{t_0}(s) = \begin{cases} \phi^{t_0}(0), & s \in [0, \tau), \\ \phi^{t_0}(s-\tau), & s \in [\tau, \infty). \end{cases}$$

Similarly, for a given $\tau > 0$ we define the τ–section $\phi^t_{(\tau)}$ of ϕ^t as

$$\phi^t_{(\tau)}(s) = \phi^t(s+\tau), \qquad s \in \mathbb{R}^+.$$

Since, by definition,

$$\phi^t_{(\tau)}(s) = \phi^{t-\tau}(s),$$

then $\phi^t_{(\tau)}$ is the history up to the earlier time $t - \tau$. Let $\mathcal{H}$ be the set of all admissible histories. It is required to contain all constant histories. Moreover, if $\phi^t \in \mathcal{H}$, then for any $\tau > 0$ both its constant continuation, $\phi^{t+\tau}_t$, and its τ–section, $\phi^t_{(\tau)}$, belong to $\mathcal{H}$.

General *materials with memory* are those in which a set of variables, say ψ, at a point $\mathbf{X} \in R$ and time t, are functionals of the histories of independent variables, say ϕ, up to time t. The response function ψ takes (scalar, vector, or tensor) values in a Hilbert space V. Obviously, the principle of causality is implied: the response value

[1] Not necessarily the statical continuation where $\phi^t_{t_0}(s) = 0$ as $s \in [0, t-t_0)$ [428].

at an instant t, $\psi(t)$, cannot depend on the values of the process ϕ subsequent to that instant. Hence we let

$$\psi(\mathbf{X}, t) = \Psi(\phi^t(\mathbf{X}, \cdot)), \tag{7.1}$$

where $\Psi : \mathcal{H} \to V$ is called response functional. For brevity we write

$$\psi(t) = \Psi(\phi^t),$$

it being understood that the response ψ and the history ϕ^t are considered at the same point $\mathbf{X} \in R$. For inhomogeneous bodies the constitutive functionals are allowed to depend explicitly on the position $\mathbf{X}$. For definiteness, in deformable non-isothermal bodies the response functionals provide the stress tensor $\mathbf{T}$, the heat flux $\mathbf{q}$, and the internal energy ε in terms of the histories of the deformation gradient $\mathbf{F}$, the temperature θ, and the temperature gradient $\nabla\theta$.

The study of these materials arises from the pioneering articles of Ludwig Boltzmann [55, 56] and Vito Volterra [436, 437]. However a statement like (7.1) raised some criticism in the scientific community due to the conceptual difficulty to accept the idea of a past history defined on an infinite time interval (when even the age of the universe is finite!). Really, the entire history of a quantity ϕ in a system can never be known. Then a constitutive relation of the memory type has an operative meaning only if additional assumptions are made. First, based on the model of viscoelasticity, the following property is usually assumed.

Relaxation property. The functional Ψ with values in a vector space V is said to have the relaxation property if

$$\lim_{t_0 \to -\infty} \Psi(\phi^t_{t_0}) = \Psi(\phi^{t_0 \dagger}).$$

In words, if ϕ is kept constant for a long time ($t \gg t_0$) then the initial history of ϕ becomes inessential as the response ψ is concerned.

Coleman and Mizel [100, 101] introduced the notion of *fading memory*. Precisely, they considered (7.1) with the further assumption that recent values of $\phi^t(s)$, with s around 0_+, rather than the remote values, $s \to \infty$, have the main effect on the present response value $\psi(t)$. In other words, the memory of the material is fading in time. Incidentally, this also gave the ultimate answer to the philosophical question of a memory of infinite duration. While the idea is quite intuitive, to make it operative, we need a precise mathematical statement. Indeed, the fading memory property can be established by endowing the space $\mathcal{H}$ of histories with a suitable norm. In essence, we first make a choice of the norm in the domain of Ψ such that two histories are close if they are close in the recent past. Then we assume the functional to be smooth in this topology, which amounts to letting $\psi(t)$ weakly dependent on remote values of ϕ.

Fading memory property. The material is said to have fading memory if the constitutive functionals have continuous Fréchet derivatives relative to an appropriate norm $\|\cdot\|_{\mathcal{H}}$ on a space $\mathcal{H}$ of histories ϕ^t. Specifically, the functional $\Psi : \mathcal{H} \to V$

is n-times Fréchet differentiable, at $\phi^t \in \mathcal{H}$, if there exist bounded linear operators $L_1, ..., L_n$, where $L_j : \underbrace{\mathcal{H} \otimes \mathcal{H} \otimes \cdots \otimes \mathcal{H}}_{j \text{ times}} \to V$, such that for any $\lambda^t \in \mathcal{H}$

$$\|\Psi(\phi^t + \lambda^t) - \Psi(\phi^t) - \sum_{j=1}^{n} L_j \underbrace{\lambda^t \otimes ... \otimes \lambda^t}_{j \text{ times}} \|_v = o(\|\lambda^t\|^n_{\mathcal{H}}).$$

This mathematical statement is made operative by the choice of the normed space $\mathcal{H}$.

7.1.1 Fading Memory Space of Histories

Let $k : \mathbb{R}^+ \to \mathbb{R}^{++}$, $k(0) = 1$, be a function which decays to zero for large s in such a way that,

$$k(s) = o(s^{-\alpha}) \quad \text{as} \quad s \to \infty$$

where $\alpha > 1$. A function satisfying these conditions is called an *influence function* of order α (or obliviator) and characterizes the rate at which the memory fades. For definiteness, to specify the smoothness of the functional Ψ, we introduce a weighted L^2-norm of ϕ^t, with weight k, defined as

$$\|\phi^t\|_k = \Big(\int_0^\infty |\phi^t(s)|^2 k(s) ds\Big)^{1/2} < \infty$$

where

$$|\phi^t(s)| = [\phi^t(s) \cdot \phi^t(s)]^{1/2},$$

the symbol "$\cdot$" denoting the inner product in the vector space W. If ϕ^t is a constant history, $\phi^t(s) = \phi_0$, $s \in \mathbb{R}^+$, then its norm is proportional to the norm of ϕ_0 in W,

$$\|\phi^t\|_k = |\phi_0| \Big(\int_0^\infty k(s) ds\Big)^{1/2} < \infty.$$

The collection of all histories with finite norm $\|\cdot\|_k$ forms a Banach space. Indeed, the natural definition of inner product between two histories ϕ^t, χ^t,

$$\langle \phi^t, \chi^t \rangle_k = \int_0^\infty \phi^t(s) \cdot \chi^t(s) k(s)\, ds$$

extends the Banach space to a Hilbert space, hereafter denoted by $S_k = L^2_k(\mathbb{R}^+, W)$. Any bounded, linear functional $L_0 : S_k \to \mathbb{R}$ may be given the form of an inner product so that for all histories $\lambda^t \in S_k$ there exists $g \in S_k$ satisfying

$$L_0[\lambda^t] = (g, \lambda^t)_k = \int_0^\infty g(s) \cdot \lambda^t(s)\, k(s)\, ds.$$

The choice $\mathcal{H} = S_k$ gives a definite mathematical meaning to the fading memory property. However, a stronger assumption on the influence function k is in order when we need to evaluate derivatives of the response functional. More precisely, an influence function k of order greater than $n+1$ is required in order that the response functional Ψ is defined and n-times Fréchet-differentiable in the memory space S_k.

Different choices of $\mathcal{H}$ are allowed. In particular, there can be materials where the present value takes a major role so that in fact we regard the response functional as dependent, separately, on the present value $\phi(t) \in W$ and the past (or reduced) history $\phi_r^t : \mathbb{R}^{++} \to W$. Hence, any history ϕ^t can be split into the pair $(\phi(t), \phi_r^t)$, so that the inner product for histories can be taken in the form

$$(\phi^t, \chi^t)_{\mathcal{H}} = \phi^t(0) \cdot \chi^t(0) + \langle \phi_r^t, \chi_r^t \rangle_k = \phi(t) \cdot \chi(t) + \int_0^\infty \phi(t-s) \cdot \chi(t-s)\, k(s)\, ds$$

and the associated norm turns out to be

$$\|\phi^t\|_{\mathcal{H}}^2 = |\phi^t(0)|^2 + \|\phi_r^t\|_k^2 = |\phi(t)|^2 + \int_0^\infty |\phi(t-s)|^2\, k(s)\, ds.$$

Hence the Hilbert space $\mathcal{H}$, called *fading memory space* of (total) histories, turns out to be the set of pairs $(\phi(t), \phi_r^t)$ belonging to $W \times S_k$. Further, $\mathcal{H}$ contains all constant histories: if $\phi^t(s) = \phi_0$, $s \in \mathbb{R}^+$, then

$$\|\phi^t\|_{\mathcal{H}}^2 = |\phi_0|^2 \big(1 + \int_0^\infty k(s) ds\big) < \infty.$$

A bounded, linear functional $L_1 : \mathcal{H} \to \mathbb{R}$ may be given the form of an inner product so that, for all histories λ^t such that $(\lambda^t(0), \lambda_r^t) \in \mathcal{H}$, there exists $(g_0, g) \in \mathcal{H}$ satisfying

$$L_1[\lambda^t(0), \lambda_r^t] = \big((g_0, g), (\lambda^t(0), \lambda_r^t)\big)_{\mathcal{H}},$$

namely

$$L_1[\lambda^t(0), \lambda_r^t] = g_0 \cdot \lambda^t(0) + \int_0^\infty g(s) \cdot \lambda^t(s)\, k(s)\, ds.$$

Of course

$$\|(g_0, g)\|_{\mathcal{H}}^2 = |g_0|^2 + \int_0^\infty |g(s)|^2 k(s)\, ds < \infty.$$

Letting

$$\mathcal{G}(s) = g_0 + \int_0^s g(\xi) k(\xi)\, d\xi \qquad s \in (0, +\infty),$$

we can write the linear functional in the form

$$L_1[\lambda^t] = \mathcal{G}(0) \cdot \lambda^t(0) + \int_0^\infty \mathcal{G}'(s) \cdot \lambda^t(s)\, ds = \mathcal{G}(0) \cdot \lambda(t) + \int_0^\infty \mathcal{G}'(s) \cdot \lambda(t-s)\, ds.$$

where $\mathcal{G}(0) = g_0$ and $\mathcal{G}' = gk$ is subject to the condition

$$\int_0^\infty |\mathcal{G}'(s)|^2 k^{-1}(s)\, ds < \infty.$$

Let $a \in W$, $\lambda : \mathbb{R}^{++} \to W$ and $\phi^t = (\phi(t), \phi^t_r) \in \mathcal{H}$. If the pair $(\phi(t) + a,\ \phi^t_r + \lambda) \in \mathcal{H}$, the functional Ψ on $\mathcal{H}$ is said to be continuously differentiable, at ϕ^t, if

$$\Psi(\phi(t) + a,\ \phi^t_r + \lambda) = \Psi(\phi(t),\ \phi^t_r) + D\Psi(\phi(t),\ \phi^t_r) \cdot a + d\Psi(\phi(t),\ \phi^t_r|\lambda) + o(|a| + \|\lambda\|)$$

where $d\Psi(\phi(t),\ \phi^t_r|\lambda)$ is the continuous Fréchet differential of Ψ, linear in λ, with respect to the second argument and $D\Psi$ can be viewed as the derivative with respect to the present value $\phi(t)$, so that we find it appropriate to let

$$D\Psi(\phi(t),\ \phi^t_r) := \partial_{\phi(t)}\Psi(\phi(t),\ \phi^t_r). \tag{7.2}$$

At a fixed point $\mathbf{X}$, the time derivative with respect to t of a history ϕ^t is denoted by $\dot{\phi}^t$ and is defined by

$$\dot{\phi}^t(s) := \frac{d}{dt}\phi(t-s) = -\frac{d}{ds}\phi^t(s).$$

This means that, at $s = 0$, we obtain $\dot{\phi}(t)$ as the left derivative $\dot{\phi}(t_-)$,

$$\dot{\phi}(t) = \lim_{s \to 0_+} \frac{\phi(t) - \phi(t-s)}{s} = \lim_{s \to 0_+} \frac{\phi^t(0) - \phi^t(s)}{s}. \tag{7.3}$$

To determine $\dot{\phi}(t_+)$ we need a continuation of ϕ^t that is a history $\phi^{t+\epsilon}$, $\epsilon > 0$. When the process ϕ is everywhere differentiable with respect to time, then for any $t, h \in \mathbb{R}$, $s \in \mathbb{R}^+$ we can write

$$\phi(t - s + h) = \phi(t-s) + \dot{\phi}(t-s)h + o(h).$$

Let the time dependence of the response be given by

$$\psi(t) = \Psi(\phi(t),\ \phi^t_r).$$

To compute the time derivative of ψ we observe

$$\begin{aligned}\psi(t+h)-\psi(t) &= \Psi(\phi(t)+\dot{\phi}(t)h+o(h),\ \phi_r^t+\dot{\phi}_r^t h+o(h))-\Psi(\phi(t),\ \phi_r^t)\\ &= h\,D\Psi(\phi(t),\ \phi_r^t)\dot{\phi}(t)+h\,d\Psi(\phi(t),\ \phi_r^t|\dot{\phi}_r^t)+o(h).\end{aligned}$$

In view of (7.2), the limit

$$\dot{\psi}(t)=\lim_{h\to 0}\frac{1}{h}\left[\Psi(\phi(t+h),\ \phi_r^{t+h})-\Psi(\phi(t),\ \phi_r^t)\right]$$

yields

$$\dot{\psi}(t)=D\Psi(\phi(t),\ \phi_r^t)\cdot\dot{\phi}(t)+d\Psi(\phi(t),\ \phi_r^t|\,\dot{\phi}_r^t).$$

Then, for practical purposes, we assume that the histories ϕ^t are time differentiable and the response functional Ψ is continuously differentiable in $\mathcal{H}$. This is the mathematical content of the fading memory assumption.

7.1.2 *Difference and Summed Histories*

A different approach[2] regards the response functional as dependent, separately, on the present value $\phi(t)$ and the *difference (or relative) history* ϕ_d^t, that is defined on $\mathbb{R}^+$ as the difference between the present value and the history values,

$$\phi_d^t(s):=\phi^t(0)-\phi^t(s)=\phi(t)-\phi(t-s).$$

In particular it follows that $\phi_d^t(0)=0$ and

$$\phi^t=\phi^{t\dagger}+\phi_d^t, \tag{7.4}$$

where $\phi^{t\dagger}$ is the constant history with value $\phi(t)$. As a consequence, any constant history has a null relative history. In this case we can take the inner product and the associated norm in the form

$$(\phi^t,\chi^t)_*=\phi^t(0)\cdot\chi^t(0)+\langle\phi_d^t,\chi_d^t\rangle_k=\phi(t)\cdot\chi(t)+\int_0^\infty\phi_d^t(s)\cdot\chi_d^t(s)\,k(s)\,ds$$

$$\|\phi^t\|_*^2=|\phi^t(0)|^2+\|\phi_d^t\|_k^2=|\phi(t)|^2+\int_0^\infty|\phi(t)-\phi(t-s)|^2\,k(s)\,ds. \tag{7.5}$$

The Hilbert space $\mathcal{H}_*$ endowed with this norm is the fading memory space of all pairs $(\phi(t),\phi_d^t)$ belonging to $W\times S_k^0$ where

$$S_k^0=\{f\in L_k^2(\mathbb{R}^+,W):f(0)=0\}.$$

---

[2] See, e.g. [102, 112].

The Like space $\mathcal{H}, \mathcal{H}_*$ contains all constant histories. Indeed, if $\phi^t(s) = \phi_0$, $s \in \mathbb{R}^+$, then $\|\phi^t\|_* = |\phi_0|$. Nevertheless, it is very different from $\mathcal{H}$. An advantage of this approach is that the history space $\mathfrak{H} = \{\phi^t : \mathbb{R}^+ \to W;\ (\phi(t), \phi_d^t) \in \mathcal{H}_*\}$ can be represented by virtue of (7.4) as a direct sum,

$$\mathfrak{H} = \mathfrak{W} \oplus S_k^0,$$

where $\mathfrak{W}$ is the (Hilbert) space of constant histories corresponding to W through an isometric isomorphism. Conversely, for any $(\chi_0, \chi) \in \mathcal{H}_*$ there exists $\phi^t \in \mathfrak{H}$ such that $\phi^t(s) = \chi_0 + \chi(s)$, $s \in \mathbb{R}^+$.

A bounded, linear functional $L_1^* : \mathcal{H}_* \to \mathbb{R}$ may be given the form of an inner product. For all histories $\lambda^t \in \mathfrak{H}$ such that $(\lambda^t(0), \lambda_d^t) \in \mathcal{H}_*$ there exists $(g_0^*, g^*) \in \mathcal{H}_*$ satisfying

$$L_1^*[(\lambda^t(0), \lambda_d^t)] = \big((g_0^*, g^*), (\lambda^t(0), \lambda_d^t)\big)_{\mathcal{H}^*},$$

namely

$$L_1^*[(\lambda^t(0), \lambda_d^t)] = g_0^* \cdot \lambda^t(0) + \int_0^\infty g^*(s) \cdot [\lambda(t) - \lambda(t-s)]\, k(s)\, ds.$$

Letting

$$\mathcal{G}(s) = g_0^* + \int_s^\infty g^*(\xi) k(\xi)\, d\xi \qquad s \in (0, +\infty),$$

we can write the linear functional in the form

$$L_1^*[\lambda^t] = \mathcal{G}_\infty \cdot \lambda(t) - \int_0^\infty \mathcal{G}'(s) \cdot [\lambda(t) - \lambda^t(s)]\, ds,$$

where $\mathcal{G}_\infty = \lim_{s\to\infty} \mathcal{G}(s) = g_0^*$ and $\mathcal{G}' = -g^* k$ is subject to the condition

$$\int_0^\infty |\mathcal{G}'(s)|^2 k^{-1}(s)\, ds < \infty.$$

At a fixed point $\mathbf{X}$, the time derivative with respect to t of a relative, or difference, history ϕ_d^t is defined by

$$\dot\phi_d^t(s) := \dot\phi(t) - \phi'(t-s)\partial_t(t-s) = \dot\phi(t) + \phi'(t-s)\partial_s(t-s) = \dot\phi(t) + \partial_s \phi_d^t(s),$$

where the prime denotes differentiation with respect to the whole argument. From the assumption

$$\psi(t) = \Psi(\phi_d^t)$$

it then follows that

$$\dot{\psi}(t)=d\Psi(\phi_d^t|\,\dot{\phi}^{t\dagger})+d\Psi\big(\phi_d^t\big|\,\partial_s\phi_d^t(s)\big),$$

where $\dot{\phi}^{t\dagger}$ is the constant history up to time t with value $\dot{\phi}(t)$. It is easy to check that there exists a functional $\mathcal{J}_{\dot{\phi}}$ such that

$$\dot{\psi}(t) = \mathcal{J}_{\dot{\phi}}(\phi_d^t)\dot{\phi}(t) + d\Psi\big(\phi_d^t\big|\partial_s\phi_d^t(s)\big).$$

There are models where the history is involved through functionals of summed histories.[3] Given the process ϕ on $\mathbb{R}$ and $t \in \mathbb{R}$, we denote by $\bar{\phi}^t : \mathbb{R}^+ \to W$ the function

$$\bar{\phi}^t(s) := \int_{t-s}^{t} \phi(\xi)d\xi = \int_0^s \phi^t(\sigma)d\sigma, \qquad s \in \mathbb{R}^+,$$

and say that $\bar{\phi}^t(s)$ is the *summed history* of ϕ^t on $[0, s]$. Of course, $\bar{\phi}^t(0) = 0$ and

$$\frac{d}{ds}\bar{\phi}^t(s) = \phi^t(s). \tag{7.6}$$

We observe that its time derivative with respect to t is given by the relative history,

$$\dot{\bar{\phi}}^t(s) = \frac{d}{dt}\int_{t-s}^{t} \phi(\xi)d\xi = \phi(t) - \phi(t-s) = \phi_d^t(s).$$

From the assumption

$$\psi(t) = \Psi(\bar{\phi}^t)$$

it then follows that

$$\dot{\psi}(t) = d\Psi(\bar{\phi}^t\,|\,\phi_d^t) = d\Psi(\bar{\phi}^t\,|\,\phi^{t\dagger}) - d\Psi(\bar{\phi}^t\,|\,\phi^t) = \mathcal{J}_\phi(\bar{\phi}^t)\phi(t) - d\Psi\big(\bar{\phi}^t\,|\,\partial_s\bar{\phi}^t(s)\big). \tag{7.7}$$

The fading memory space of all summed histories $\bar{\phi}^t \in S_k^0$ is an Hilbert space denoted by $\bar{\mathcal{H}}$. It contains all constant histories provided that k is an influence function of order $\alpha > 2$: if $\phi^t(s) = \phi_0$, $s \in \mathbb{R}^+$, then $\bar{\phi}^{t\dagger}(s) = \phi_0 s$ and

$$\|\bar{\phi}^{t\dagger}\|_k = |\phi_0|\big(\int_0^\infty k(s)s^2\,ds\big)^{1/2}.$$

A bounded, linear functional $\bar{L}_1 : \bar{\mathcal{H}} \to \mathbb{R}$ may take the form of an inner product. For all summed histories $\bar{\lambda}^t \in S_k^0$ there exists $g \in S_k^0$ satisfying

$$\bar{L}_1\bar{\lambda}^t = \int_0^\infty g(s)\cdot\bar{\lambda}^t(s)\,k(s)\,ds.$$

[3] See, e.g. [217].

Letting $\mathcal{G}' = gk$, we can write the linear functional in the form

$$\bar{L}_1\bar{\lambda}^t = \int_0^\infty \mathcal{G}'(s)\cdot\bar{\lambda}^t(s)\,ds = \int_0^\infty \mathcal{G}'(s)\cdot\int_0^s \lambda(t-\sigma)d\sigma\,ds,$$

where $\mathcal{G}'$ is subject to the condition $\mathcal{G}'(0) = 0$ and

$$\int_0^\infty |\mathcal{G}'(s)|^2 k^{-1}(s)\,ds < \infty.$$

Upon an integration by parts the linear functional $\bar{L}_1$ transforms into a linear functional on S_k, namely

$$\bar{L}_1\bar{\lambda}^t = L_0\lambda^t = -\int_0^\infty \check{\mathcal{G}}(s)\cdot\lambda(t-s)\,ds.$$

where $\check{\mathcal{G}}(s) = -\int_s^\infty g(\sigma)k(\sigma)\,d\sigma$.

7.1.3 Properties About Histories and Norms

Two properties about histories and norm are in order for later developments. We let ϕ^t be a history with scalar, vector, or tensor values as appropriate.

Proposition 7.1 *Given a time t, a history ϕ^t and an arbitrary value α, for every $\epsilon > 0$ there exist histories $\tilde{\phi}^t$ such that*

$$\tilde{\phi}(t) = \phi(t), \qquad \dot{\tilde{\phi}}(t) = \alpha,$$

$$\|\tilde{\phi}^t - \phi^t\|_k < \epsilon, \qquad \|\dot{\tilde{\phi}}^t - \dot{\phi}^t\|_k < \epsilon.$$

Proof To show this property (see [94]), for any fixed t we consider the process[4]

$$\tilde{\phi}(\tau) = \phi(\tau) - f(t-\tau)(\alpha - \dot{\phi}(t)), \qquad \tau \in (-\infty, t],$$

and the corresponding history

$$\tilde{\phi}^t(s) = \phi^t(s) - f(s)(\alpha - \dot{\phi}(t)), \qquad s \in [0, \infty).$$

Letting $f(0) = 0$ we find $\tilde{\phi}(t) = \phi(t)$. Now,

$$\dot{\tilde{\phi}}^t(s) = \dot{\phi}^t(s) + f'(s)(\alpha - \dot{\phi}(t)).$$

[4] Here $\phi(\tau) = \phi^t(t-\tau)$ and $\dot{\phi}(t)$ is given by (7.3).

Hence

$$\dot{\tilde{\phi}}(t) = f'(0)\,\alpha$$

and the requirement $\dot{\tilde{\phi}}(t) = \alpha$ is satisfied by letting $f'(0) = 1$. The pertinent norms are given by

$$\|\tilde{\phi}^t - \phi^t\|_k^2 = |\alpha - \dot{\phi}(t)|^2 \int_0^\infty f^2(s)k(s)ds,$$

$$\|\dot{\tilde{\phi}}^t(s) - \dot{\phi}^t(s)\|_k^2 = |\alpha - \dot{\phi}(t)|^2 \int_0^\infty f'^{\,2}(s)k(s)ds.$$

We can pick f such that the integrals in the right-hand sides are as small as we please. For definiteness, let f be given by

$$f(s) = b[1 - \exp(-s/b)],$$

so that $f(0) = 0,\ f'(0) = 1$. As a consequence,

$$\int_0^\infty f^2(s)k(s) \le b^2 \int_0^\infty k(s)ds,$$

and, by the Cauchy–Schwarz inequality,

$$\int_0^\infty f'^{\,2}(s)k(s)ds \le \Big[\int_0^\infty \exp(-4s/b)ds\Big]^{1/2} \Big[\int_0^\infty k^2(s)ds\Big]^{1/2} = \sqrt{b/4}\,\Big[\int_0^\infty k^2(s)ds\Big]^{1/2}.$$

Since $k(s) = o(s^{-1-\delta})$ then the integrals of k and k^2 are finite. Hence the choice of a suitably small b makes $\|\tilde{\phi}^t - \phi^t\|_k$ and $\|\dot{\tilde{\phi}}^t - \dot{\phi}^t\|_k$ smaller than the given ϵ. □

Proposition 7.2 *Given a time t and a history ϕ^t, for every $\epsilon > 0$ there exists a history $\tilde{\phi}^t$ such that*

$$\|\tilde{\phi}^t - \phi^t\|_k < \epsilon,$$

whereas $d\Psi(\tilde{\phi}^t|\dot{\tilde{\phi}}^t)$ takes arbitrary values.

Proof Let $\lambda \in \mathbb{R}$ and

$$\tilde{\phi}^t = \phi^t + \lambda\varphi^t.$$

A suitably small value of λ makes $\|\lambda\varphi^t\|_k$ as small as we please. Now,

$$d\Psi(\tilde{\phi}^t|\dot{\tilde{\phi}}^t) = d\Psi(\tilde{\phi}^t|\dot{\phi}^t) + \lambda d\Psi(\tilde{\phi}^t|\dot{\varphi}^t).$$

If $\varphi^t(s)$ depends on s via s/λ, namely $\varphi^t(s) = f(s/\lambda)$, then

$$\lambda d\Psi(\tilde{\phi}^t|\dot{\varphi}^t) = -d\Psi(\tilde{\phi}^t|f')$$

may take arbitrary values depending on f'. □

Likewise we can prove that given two histories θ^t, ϕ^t the functionals $d\Psi(\theta^t, \phi^t|\dot{\theta}^t)$ and $d\Psi(\theta^t, \phi^t|\dot{\phi}^t)$ can be regarded as arbitrary and independent of each other.

7.2 Thermoelastic Materials with Memory

A large domain of materials is accomplished by letting the independent variables be the histories $\mathbf{F}^t$, θ^t, $\nabla_R\theta^t$ of the deformation gradient $\mathbf{F}$, the temperature θ, and the material temperature gradient $\nabla_R\theta$. As is well investigated in the literature,[5] the dependence on the present value of $\nabla\theta$ is not compatible with wave propagation at a finite speed. It is then convenient to allow for both possibilities, namely for the set of variables

$$(\mathbf{F}(t), \theta(t), \nabla_R\,\theta(t), \mathbf{F}^t_r, \theta^t_r, \nabla_R\,\theta^t_r)$$

or

$$(\mathbf{F}(t), \theta(t), \mathbf{F}^t_r, \theta^t_r, \nabla_R\,\theta^t_r).$$

Not any arbitrary constitutive functional is admissible; the constitutive equations are subject to the principle of objectivity and the entropy inequality. We start to analyse the restrictions due to the principle of objectivity. For definiteness let the Cauchy stress $\mathbf{T}$ be objective and denote by $\hat{\mathbf{T}}$ the corresponding response functional. Let $\mathbf{x} = \chi(\mathbf{X}, t)$ be a motion and

$$\mathbf{T}(\mathbf{X}, t) = \hat{\mathbf{T}}(\mathbf{F}^t(\mathbf{X}))$$

be the corresponding tensor response functional. For a given change of frame, the corresponding motion is given by[6]

$$\mathbf{x}^* = \chi^*(\mathbf{X}, t^*) = \mathbf{c}(t) + \mathbf{Q}(t)\chi(\mathbf{X}, t), \qquad t^* = t - a,$$

where $\mathbf{c}$ is an arbitrary vector function, $\mathbf{Q}$ an arbitrary orthogonal tensor function, and a an arbitrary real number. Then, the associated tensor response functional is

$$\mathbf{T}^*(\mathbf{X}, t^*) = \mathbf{Q}(t)\mathbf{T}(\mathbf{X}, t)\mathbf{Q}^T(t).$$

For formal simplicity let $a = 0$. Then, $t^* = t$ and

$$\chi^{*t}(\mathbf{X}) = \mathbf{c}^t + \mathbf{Q}^t\chi^t(\mathbf{X}) \qquad \mathbf{F}^{*t}(\mathbf{X}) = \mathbf{Q}^t\mathbf{F}^t(\mathbf{X}).$$

Objectivity requires that

---

[5] See, e.g. [402].

[6] See (1.54).

$$\mathbf{Q}(t)\hat{\mathbf{T}}(\mathbf{F}^t(\mathbf{X}))\mathbf{Q}^T(t) = \hat{\mathbf{T}}(\mathbf{Q}^t\mathbf{F}^t(\mathbf{X})),$$

namely the history $\mathbf{Q}^t\mathbf{F}^t$, with values $\mathbf{Q}^t(s)\mathbf{F}^t(s) = \mathbf{Q}(t-s)\mathbf{F}(t-s)$, produces a tensor that is the transform of

$$\hat{\mathbf{T}}(\mathbf{F}^t(\mathbf{X})).$$

To save writing, we now omit specifying the obvious dependence on the selected point $\mathbf{X}$. By the polar decomposition theorem we can write $\mathbf{F}(t-s) = \mathbf{R}(t-s)\mathbf{U}(t-s)$, $\mathbf{R}$ being orthogonal. By the arbitrariness of $\mathbf{Q}$ we let $\mathbf{Q}(t-s) = \mathbf{R}^T(t-s)$. Hence we have

$$\begin{aligned}\hat{\mathbf{T}}(\mathbf{F}^t) &= \mathbf{Q}^T(t)\hat{\mathbf{T}}(\mathbf{Q}^t\mathbf{F}^t)\mathbf{Q}(t) = \mathbf{R}(t)\hat{\mathbf{T}}((\mathbf{R}^T)^t\mathbf{F}^t)\mathbf{R}^T(t)\\ &= \mathbf{F}(t)\mathbf{U}^{-1}(t)\hat{\mathbf{T}}((\mathbf{R}^T)^t\mathbf{R}^t\mathbf{U}^t)\mathbf{U}^{-1}(t)\mathbf{F}^T(t).\end{aligned}$$

Since $(\mathbf{R}^T)^t\mathbf{R}^t = \mathbf{1}$ and $\mathbf{C} = \mathbf{U}^2$, letting

$$\mathfrak{T}(\mathbf{C}^t) := \mathbf{U}^{-1}(t)\hat{\mathbf{T}}(\mathbf{U}^t)\mathbf{U}^{-1}(t),$$

we have

$$\hat{\mathbf{T}}(\mathbf{F}^t) = \mathbf{F}(t)\mathfrak{T}(\mathbf{C}^t)\mathbf{F}^T(t).$$

Apart from the occurrence of the present values $\mathbf{F}(t)$ and $\mathbf{F}^T(t)$ that operate on the components, objectivity requires that the functional dependence of $\mathbf{T}$ on the deformation history $\mathbf{F}^t$ is through $\mathbf{C}^t$.

We notice that materials with memory are often modelled within the material (Lagrangian) description. Indeed, the history of a given physical variable is naturally considered on the selected point $\mathbf{X}$; different points may have different histories. A further advantage is that the material gradient $\nabla_R\theta(t)$ is an invariant vector and so is its past history $\nabla_R\theta^t_r$ for any time t.

We now evaluate the restrictions on the constitutive functionals placed by the entropy inequality. In the material formulation we have

$$-(\dot{\psi}_R + \eta_R\dot{\theta}) + \tfrac{1}{2}\mathbf{T}_{RR}\cdot\dot{\mathbf{C}} - \frac{1}{\theta}\mathbf{q}_R\cdot\nabla_R\theta + \theta\nabla_R\cdot\mathbf{k}_R \geq 0.$$

We let $\psi_R, \eta_R, \mathbf{T}_{RR}, \mathbf{q}_R, \mathbf{k}_R$, at the position $\mathbf{X} \in \mathrm{R}$ and time t, be given by functionals $\Psi, N, \mathfrak{T}, \mathcal{Q}, \mathcal{K}$ on

$$\mathbf{\Lambda}(t) = (\mathbf{C}(t), \theta(t), \nabla_R\theta(t), \mathbf{C}^t_r, \theta^t_r, \nabla_R\theta^t_r),$$

or on

$$\mathbf{\Lambda}_0(t) = (\mathbf{C}(t), \theta(t), \mathbf{C}^t_r, \theta^t_r, \nabla_R\theta^t_r),$$

at $\mathbf{X}$. Let k on $\mathbb{R}^+$ be an influence function. The $\mathcal{H}$-norm of $\mathbf{\Lambda}$ is taken in the form

$$\|\mathbf{\Lambda}(t)\|_{\mathcal{H}} = [\mathbf{C}\cdot\mathbf{C} + \theta^2 + (\nabla_R\theta)^2]^{1/2}(t) + \Big[\int_0^\infty [\mathbf{C}^t\cdot\mathbf{C}^t + (\theta^t)^2 + \nabla_R\theta^t\cdot\nabla_R\theta^t](s)k(s)ds\Big]^{1/2}$$

or the analogous norm of $\mathbf{\Lambda}_0$ without the present value $(\nabla_R\theta)^2(t)$.

Let Ψ and $\mathcal{K}$ be continuously differentiable while N, $\mathfrak{T}$, $\mathcal{Q}$ are continuous. Hence

$$\dot{\psi}_R = \partial_{\mathbf{C}}\Psi(\mathbf{\Lambda})\cdot\dot{\mathbf{C}} + \partial_\theta\Psi(\mathbf{\Lambda})\dot{\theta} + \partial_{\nabla_R\theta}\Psi(\mathbf{\Lambda})\cdot\nabla_R\dot{\theta} + d\Psi(\mathbf{\Lambda}|\dot{\mathbf{C}}_r^t) + d\Psi(\mathbf{\Lambda}|\dot{\theta}_r^t) + d\Psi(\mathbf{\Lambda}|\nabla_R\dot{\theta}_r^t).$$

Upon substitution the entropy inequality takes the form

$$-(\partial_\theta\Psi(\mathbf{\Lambda}) + N(\mathbf{\Lambda}))\dot{\theta}(t) - \rho\partial_{\nabla_R\theta}\Psi(\mathbf{\Lambda})\cdot\nabla_R\dot{\theta}(t) + (\tfrac{1}{2}\mathfrak{T}(\mathbf{\Lambda}) - \partial_{\mathbf{C}}\Psi(\mathbf{\Lambda}))\cdot\dot{\mathbf{C}}(t)$$
$$-d\Psi(\mathbf{\Lambda}|\dot{\mathbf{C}}_r^t) - d\Psi(\mathbf{\Lambda}|\dot{\theta}_r^t) - d\Psi(\mathbf{\Lambda}|\nabla_R\dot{\theta}_r^t) - \frac{1}{\theta}\mathcal{Q}(\mathbf{\Lambda})\cdot\nabla_R\theta(t) + \theta\nabla_R\cdot\mathcal{K}(\mathbf{\Lambda}) \ge 0.$$

To determine the consequences of this inequality we need to establish the arbitrariness associated with the histories. By Proposition 7.1, we can say that $\dot{\mathbf{C}}$, $\dot{\theta}$, $\nabla_R\dot{\theta}$ at time t may take arbitrary values while the present values $\mathbf{C}$, θ, $\nabla_R\theta$, the past histories $\mathbf{C}_r^t$, θ_r^t, $\nabla_R\theta_r^t$ and their derivatives affect the functionals as small as desired. The arbitrariness of $\dot{\mathbf{C}}$, $\dot{\theta}$ and $\nabla_R\dot{\theta}$ requires that the inequality holds only if

$$\mathfrak{T}(\mathbf{\Lambda}) = 2\partial_{\mathbf{C}}\Psi(\mathbf{\Lambda}), \qquad N(\mathbf{\Lambda}) = -\partial_\theta\Psi(\mathbf{\Lambda}), \tag{7.8}$$

$$\partial_{\nabla_R\theta}\Psi(\mathbf{\Lambda}) = \mathbf{0}. \tag{7.9}$$

The mutual independence of the past histories $\mathbf{C}_r^t$, θ_r^t, $\nabla_R\theta_r^t$ (see Proposition 7.2) and the linearity and the arbitrariness of $\dot{\mathbf{C}}_r^t$, $\dot{\theta}_r^t$ imply that the remaining inequality holds only if

$$d\Psi(\mathbf{\Lambda}|\dot{\mathbf{C}}_r^t) \le 0, \qquad d\Psi(\mathbf{\Lambda}|\dot{\theta}_r^t) \le 0, \tag{7.10}$$

$$\rho d\Psi(\mathbf{\Lambda}|\nabla_R\dot{\theta}_r^t) + \frac{1}{\theta}\mathcal{Q}(\mathbf{\Lambda})\cdot\nabla_R\theta(t) - \theta\nabla_R\cdot\mathcal{K}(\mathbf{\Lambda}) \le 0. \tag{7.11}$$

Remark 7.1 If $\mathbf{\Lambda}_0$ is involved rather than $\mathbf{\Lambda}$, we simply find the relations (7.8), (7.10), and (7.11). Hence the difference is revealed only in the expression of the functional $\mathcal{Q}$, which depends on the actual value $\nabla_R\theta(t)$ or not.

In terms of referential quantities,

$$\sigma_R = \dot{\eta}_R - \frac{\rho_R r}{\theta} + \nabla_R\cdot(\mathbf{q}_R/\theta) + \nabla_R\cdot\mathbf{k}_R$$

is the (rate of) entropy production per unit volume in the reference configuration and the content of the second law is that $\sigma_R \ge 0$ for every process consistent with the balance equations. By the balance of energy we have

$$\sigma_R = -\frac{1}{\theta}(\dot{\psi}_R + \eta_R \dot{\theta}) + \frac{1}{2\theta}\mathbf{T}_{RR} \cdot \dot{\mathbf{C}} - \frac{1}{\theta^2}\mathbf{q}_R \cdot \nabla_R \theta + \nabla_R \cdot \mathbf{k}_R.$$

In view of (7.8) we find

$$\sigma_R = \gamma_R - \frac{1}{\rho\theta^2}\mathbf{q}_R \cdot \nabla_R \theta,$$

where

$$\gamma_R(t) = -\frac{1}{\theta}[d\Psi(\mathbf{\Lambda}|\dot{\mathbf{C}}_r^t) + d\Psi(\mathbf{\Lambda}|\dot{\theta}_r^t) + d\Psi(\mathbf{\Lambda}|\nabla_R \dot{\theta}_r^t)] + \nabla_R \cdot \mathbf{k}_R(t).$$

The entropy production σ_R consists of two terms; the former, γ_R, is called the internal dissipation and involves the dependence on the past histories $\mathbf{C}_r^t$, θ_r^t, $\nabla_R \theta_r^t$, while the latter, $-(1/\rho\theta^2)\mathbf{q}_R \cdot \nabla_R \theta$, represents the dissipation due to heat conduction. Since

$$\sigma_R \geq 0,$$

we can write

$$\frac{1}{\theta^2}\mathbf{q}_R \cdot \nabla_R \theta \leq \gamma_R.$$

In general, from this inequality we cannot conclude that

$$\mathbf{q}_R \cdot \nabla_R \theta \leq 0$$

whereas it is true that

$$\gamma_R = \sigma_R \geq 0 \quad \text{if} \quad \nabla_R \theta = \mathbf{0}.$$

Hereafter we let $\mathbf{k}_R = \mathbf{0}$ and this will prove to be consistent with the local nature of the constitutive equations. By (7.10) and (7.11) it follows that

$$\dot{\psi}_R \leq -\eta_R \dot{\theta} + \tfrac{1}{2}\mathbf{T}_{RR} \cdot \dot{\mathbf{C}} - \frac{1}{\theta}\mathbf{q}_R \cdot \nabla_R \theta.$$

Hence

$$\dot{\mathbf{C}} = \mathbf{0},\ \dot{\theta} = 0,\ \nabla_R \theta = \mathbf{0} \quad \Longrightarrow \quad \dot{\psi}_R \leq 0. \tag{7.12}$$

At a fixed point $\mathbf{X}$, if it happens that $\dot{\mathbf{C}}(t) = \mathbf{0}$, $\dot{\theta}(t) = 0$ and $\nabla_R \theta(t) = \mathbf{0}$ as $t \geq t_0$, then the free energy cannot increase, regardless of the histories $\mathbf{C}^{t_0}$, θ^{t_0}, $\nabla_R \theta^{t_0}$. Let ϕ be the triplet

$$\phi = (\mathbf{C}, \theta, \nabla_R \theta)$$

and, for any history ϕ^{t_0}, consider the constant continuation

$$\phi_{t_0}^{t_0+d}(s) = \begin{cases} \phi(t_0), & s \in [0, d), \\ \phi(t_0 - s + d), & s \in [d, \infty), \end{cases}$$

subject to $\nabla_R \theta(s) = \mathbf{0}$ as $s \in [0, d)$. Hence, for any $d > 0$, from (7.12) we have

$$\Psi(\phi_{t_0}^{t_0+d}) \leq \Psi(\phi^{t_0}).$$

Moreover,

$$\|\phi^{t_0\dagger} - \phi^{t_0+d}\|_k = \int_d^\infty |\phi(t_0) - \phi(t_0 - s + d)|^2 k(s)ds \leq |\phi_d^{t_0}|^2_{\max} \int_d^\infty k(s)ds,$$

where $|\phi_d^{t_0}|_{\max}$ is the maximum of $|\phi(t_0) - \phi(t_0 - \tau)|$ over $\tau \in [0, \infty)$. Since the integral of k on $[0, \infty)$ is finite, it follows that

$$\lim_{d\to\infty} \int_d^\infty k(s)ds = 0.$$

Hence ϕ^{t_0+d} approaches $\phi^{t_0\dagger}$ as $d \to \infty$. By the continuity of the functional Ψ we have

$$\Psi(\phi^{t_0\dagger}) = \lim_{d\to\infty} \Psi(\phi^{t_0+d}) \leq \Psi(\phi^{t_0}). \tag{7.13}$$

This result may be phrased as follows.

Proposition 7.3 *Let t be a given time. Among all histories $\phi^t = (\mathbf{C}^t, \theta^t, \nabla_R \theta^t)$ ending with the given values $\mathbf{C}(t) = \mathbf{C}$, $\theta(t) = \theta$ and $\nabla_R \theta(t) = \mathbf{0}$, the history corresponding to those constant values for all times, namely $\phi^\dagger = (\mathbf{C}^\dagger, \theta^\dagger, \mathbf{0}^\dagger)$, has the least free energy.*

Equations (7.8)–(7.11) give the restrictions placed by the second law on the constitutive functionals. The next sections examine them in connection with models of rigid heat conductors and linear viscoelastic solids. Throughout these sections we will continue to assume that the extra-entropy flux vanishes, $\mathbf{k}_R = \mathbf{0}$.

7.3 Rigid Heat Conductors

Due to the rigid character of the body we identify the position vector $\mathbf{x}$, of a point, with the reference position vector $\mathbf{X}$ and hence we let $\nabla_R = \nabla$. Then we put

$$\psi_R = \rho\psi, \quad \varepsilon_R = \rho\varepsilon, \quad \eta_R = \rho\eta, \quad \mathbf{q}_R = \mathbf{q}$$

7.3.1 Models Dependent on Thermal Histories

Consistent with the choice of $\mathbf{\Lambda}_0$, as the set of independent variables, and the thermodynamic restrictions (7.8)–(7.11), a simple model of (rigid) heat conductor with memory is established by restricting the independent variables to the present temperature and the past history of the temperature and the temperature gradient. Hence we let

$$\psi(t) = \Psi(\theta(t), \theta_r^t, \nabla\,\theta_r^t)$$

and, according to (7.8) and (7.11),

$$\eta(t) = N(\theta(t), \theta_r^t, \nabla\,\theta_r^t) = -\partial_\theta \Psi(\theta(t), \theta_r^t, \nabla\,\theta_r^t),$$

$$d\Psi(\theta(t), \theta_r^t, \nabla\,\theta_r^t \,|\, \dot{\theta}_r^t) + d\Psi(\theta(t), \nabla\,\theta_r^t | \nabla\,\dot{\theta}_r^t) + \frac{1}{\rho\theta}\mathcal{Q}(\theta(t), \nabla\,\theta_r^t) \cdot \nabla\theta(t) \le 0, \tag{7.14}$$

it being again understood that all quantities are evaluated at the same point $\mathbf{X}$.

For definiteness we might look at the model arising from the Maxwell–Cattaneo (MC) equation[7]

$$\dot{\mathbf{q}} + \frac{1}{\tau}\mathbf{q} = -\frac{\kappa}{\tau}\,\nabla\,\theta,$$

where τ and κ are assumed to be constants. By a direct integration we have

$$\mathbf{q}(t) = -\frac{\kappa}{\tau}\int_0^\infty \exp(-s/\tau)\nabla\,\theta(t-s)ds,$$

whence

$$\mathcal{Q}(\nabla\,\theta_r^t) = -\frac{\kappa}{\tau}\int_0^\infty \exp(-s/\tau)\nabla\,\theta(t-s)ds.$$

More generally, though still with a linear dependence on $\nabla\theta^t$, we allow for a possible anisotropy and let

$$\mathcal{Q}(\nabla\,\theta_r^t) = -\int_0^\infty \mathbf{K}(s)\nabla\,\theta(t-s)ds. \tag{7.15}$$

We assume

$$\mathbf{K}(s) = \mathbf{K}_0\,\beta(s),$$

where $\mathbf{K}_0$ is positive definite so that we can define $\mathbf{K}_0^{1/2}$. In addition, let $\beta(s) \ge 0$, $\beta(0) = 1$ and $\beta(s) \to 0$ as $s \to \infty$.

We now have to find the (or a) functional Ψ consistent with (7.14) and (7.15). For simplicity, we disregard the dependence on the temperature past history. First we look for Ψ in a double-integral form; we let

$$\Psi(\theta(t), \nabla\,\theta_r^t) = \psi_0(\theta(t)) + \tfrac{1}{2}\alpha(\theta(t))\big|\mathbf{K}_0^{1/2}\int_0^\infty \beta(s)\nabla\,\theta(t-s)ds\big|^2, \tag{7.16}$$

[7] See [319].

the function α being so far undetermined. Hence

$$d\Psi(\theta(t), \nabla\theta^t_r|\nabla_R\dot{\theta}^t_r) = \alpha(\theta(t))\mathbf{K}_0 \int_0^\infty \beta(s)\nabla\theta(t-s)ds \cdot \int_0^\infty \beta(\xi)\nabla\dot{\theta}(t-\xi)d\xi.$$

The observation that $\nabla\dot{\theta}(t-s) = -\partial_s\nabla\theta(t-s)$ and an integration by parts result in

$$d\Psi = \alpha(\theta(t)) \int_0^\infty \beta(s)\nabla\theta(t-s)ds \cdot \mathbf{K}_0\big[\nabla\theta(t) + \int_0^\infty \beta'(\xi)\nabla\theta(t-\xi)d\xi\big].$$

By substitution of $d\Psi$ and $\boldsymbol{Q}$ in (7.14) we have the inequality

$$(\alpha - \frac{1}{\rho\theta})\nabla\theta(t)\cdot\mathbf{K}_0 \int_0^\infty \beta(s)\nabla\theta^t(s)ds + \alpha\int_0^\infty \beta(s)\nabla\theta^t(s)ds \cdot \mathbf{K}_0 \int_0^\infty \beta'(\xi)\nabla\theta^t(\xi)d\xi \le 0.$$

Since, for any given $\nabla\theta(t)$, we may take the past history $\nabla\theta^t_r$ arbitrarily, it follows that

$$\alpha = \frac{1}{\rho\theta}, \qquad \int_0^\infty \beta(s)\nabla\theta(t-s)ds \cdot \mathbf{K}_0 \int_0^\infty \beta'(\xi)\nabla\theta(t-\xi)d\xi \le 0.$$

By the positive definiteness of $\mathbf{K}_0$, the assumptions on β and the arbitrariness of the past history $\nabla\theta^t_r$ it follows that the previous inequality is generally satisfied provided that

$$\beta'(s) \le -\lambda\beta(s), \quad s \in \mathbb{R}^+, \quad \lambda > 0.$$

A finite sum of decaying exponentials in time,

$$\beta(s) = \textstyle\sum_{h=1}^n c_h \exp(-\lambda_h s), \quad c_h > 0, \lambda_h > 0.$$

satisfies this condition by letting $\lambda = \min_h \lambda_h > 0$ and observing that

$$\beta'(s) = -\textstyle\sum_{h=1}^n c_h\lambda_h \exp(-\lambda_h s) \le -\lambda\sum_{h=1}^n c_h \exp(-\lambda_h s) = -\lambda\beta(s).$$

According to this choice,

$$\begin{aligned} \mathbf{q} &= -\mathbf{K}_0 \int_0^\infty \textstyle\sum_h c_h \exp(-\lambda_h s)\nabla\theta(t-s)ds, \\ \psi &= \psi_0(\theta) + \frac{1}{2\rho\theta}\big|\mathbf{K}_0^{1/2} \int_0^\infty \textstyle\sum_h c_h \exp(-\lambda_h s)\nabla\theta(t-s)ds\big|^2 \end{aligned} \tag{7.17}$$

are constitutive functionals consistent with the second law. This in turn implies that the entropy takes the form

$$\eta = -\psi_0'(\theta) + \frac{1}{2\rho\theta^2}\big|\mathbf{K}_0^{1/2} \int_0^\infty \textstyle\sum_h c_h \exp(-\lambda_h s)\nabla\theta(t-s)ds\big|^2. \tag{7.18}$$

Consistent with the general result (7.13), the free energy has a minimum at constant histories with zero past history of $\nabla \theta$.

Without changing the functional (7.15) for **q**, we might look for a different free energy functional in the single-integral form

$$\Psi(\theta(t), \nabla \theta_r^t) = \psi_0(\theta(t)) - \tfrac{1}{2}\int_0^\infty \beta^2(s)\nabla \theta(t-s)\cdot \hat{\mathbf{K}}\nabla \theta(t-s)ds,$$

where the tensor $\hat{\mathbf{K}}$ is to be determined. Since

$$d\Psi(\theta(t), \nabla \theta_r^t | \nabla \dot{\theta}_r^t) = -\int_0^\infty \nabla \theta(t-s)\cdot \beta^2(s)\mathbf{K}_* \nabla \dot{\theta}(t-s)ds$$

then the observation that $\nabla \dot{\theta}(t-s) = -\partial_s \nabla \theta(t-s)$ and an integration by parts yield

$$\begin{aligned} d\Psi &= \tfrac{1}{2}\int_0^\infty \beta^2(s)\partial_s\big[\nabla \theta^t(s)\cdot \hat{\mathbf{K}}\nabla \theta^t(s)\big]ds \\ &= -\tfrac{1}{2}\nabla \theta(t)\cdot \hat{\mathbf{K}}\nabla \theta(t) - \int_0^\infty \beta(s)\beta'(s)\nabla \theta^t(s)\cdot \hat{\mathbf{K}}\nabla \theta^t(s)ds. \end{aligned}$$

By substitution of $d\Psi$ and $\mathcal{Q}$ in (7.14) we have the inequality

$$-\tfrac{1}{2}\nabla\theta(t)\cdot \mathbf{K}_*\nabla\theta(t) - \int_0^\infty \beta(s)\beta'(s)\nabla \theta^t(s)\cdot \mathbf{K}_*\nabla \theta^t(s)ds \\ -\frac{1}{\rho\theta}\nabla \theta(t)\cdot \mathbf{K}_0\int_0^\infty \beta(s)\nabla \theta^t(s) \le 0.$$

For any given vector $\nabla \theta(t)$ we may take the past history $\nabla \theta_r^t$ arbitrarily, so that the last two integrals are as small as we please. The inequality then implies that $\hat{\mathbf{K}}$ be positive definite. Therefore, if now we take $\nabla \theta(t) = \mathbf{0}$ then we conclude that

$$\beta(s)\beta'(s) \ge 0.$$

This implies that β^2 is increasing, contrary to the assumption that $\beta(s) \to 0$ as $s \to \infty$. Hence only the double-integral representation of the free energy is admissible when models dependent on thermal histories are concerned.

Incidentally, as a generalization, we might consider a deformable heat conductor. By letting $(\mathbf{C}(t), \theta(t), \theta_r^t, \nabla_{\!R}\, \theta_r^t)$ be the set of independent variables then we find the relations (7.8) and the reduced inequality (7.14). We can parallel the previous procedure up to equations (7.17) and obtain the free energy in the form

$$\psi_R = \Psi(\mathbf{C}, \theta, \nabla_{\!R}\, \theta_r^t) = \Psi_0(\mathbf{C}, \theta) + \frac{1}{2\rho\theta}\Big|\mathbf{K}_0^{1/2}\int_0^\infty \textstyle\sum_h c_h \exp(-\lambda_h s)\nabla_{\!R}\, \theta(t-s)ds\Big|^2.$$

Consistent with the choice of $\mathbf{\Lambda}$ as independent variables of the rigid heat conductor, we might allow for the dependence on the present value of the temperature gradient, $\nabla\theta(t)$. As a consequence of (7.9), we would find that Ψ is independent of $\nabla\theta(t)$ and exploiting the entropy inequality we end up with the previous result (7.16) for Ψ. Instead of (7.15), however, we let

$$\mathbf{q}(t) = \mathcal{Q}(\nabla\,\theta(t), \nabla\,\theta_r^t) := -\tilde{\mathbf{K}}\nabla\,\theta(t) - \mathbf{K}_0 \int_0^\infty \beta(s)\nabla\,\theta(t-s)ds, \tag{7.19}$$

where $\beta(s) \geq 0$, $\beta(0) = 1$ and $\beta(s) \to 0$ as $s \to \infty$. Accordingly, by replacing $d\Psi$ and $\mathcal{Q}$ into (7.14) we obtain the inequality

$$\begin{aligned}\Big(\alpha - \frac{1}{\rho\theta}\Big)\nabla\,\theta(t)\cdot\mathbf{K}_0\int_0^\infty \beta(s)\nabla\,\theta^t(s)ds - \nabla\theta(t)\cdot\tilde{\mathbf{K}}\nabla\theta(t)\\ +\alpha\int_0^\infty \beta(s)\nabla\,\theta^t(s)ds\cdot\mathbf{K}_0\int_0^\infty \beta'(\xi)\nabla\,\theta^t(\xi)d\xi \leq 0.\end{aligned}$$

which is satisfied provided that $\mathbf{K}_0$ and $\tilde{\mathbf{K}}$ are positive definite and

$$\alpha = \frac{1}{\rho\theta}, \qquad \beta'(s) \leq -\lambda\beta(s), \quad s \in \mathbb{R}^+, \quad \lambda > 0.$$

We finally observe that a free energy functional in the single-integral form, as

$$\Psi(\theta(t), \nabla\,\theta_r^t) = \psi_0(\theta(t)) + \alpha(\theta(t))\int_0^\infty \beta^2(s)\nabla\,\theta(t-s)\cdot\tilde{\mathbf{K}}\nabla\,\theta(t-s)ds,$$

is compatible with (7.19) and the entropy inequality provided that α is properly chosen. Indeed, in this case (7.14) takes the form

$$\begin{aligned}\Big(\alpha - \frac{1}{\rho\theta}\Big)\nabla\theta(t)\cdot\tilde{\mathbf{K}}\nabla\theta(t) + 2\alpha\int_0^\infty \beta(s)\beta'(s)\nabla\,\theta^t(s)\cdot\tilde{\mathbf{K}}\nabla\,\theta^t(s)ds\\ -\frac{1}{\rho\theta}\nabla\,\theta(t)\cdot\mathbf{K}_0\int_0^\infty \beta(s)\nabla\,\theta^t(s) \leq 0.\end{aligned}$$

Since for any given vector $\nabla\,\theta(t)$ we may take the past history $\nabla\,\theta_r^t$ arbitrarily, the inequality then implies $\alpha = 1/\rho\theta$ and

$$2\int_0^\infty \beta(s)\beta'(s)\nabla\,\theta^t(s)\cdot\tilde{\mathbf{K}}\nabla\,\theta^t(s)ds - \nabla\,\theta(t)\cdot\mathbf{K}_0\int_0^\infty \beta(s)\nabla\,\theta^t(s) \leq 0.$$

Therefore, taking $\nabla\,\theta(t) = \mathbf{0}$ we conclude that $\beta(s)\beta'(s) \leq 0$, which is consistent with the assumption that $\beta(s) > 0$ and $\beta(s) \to 0$ as $s \to \infty$.

Temperature-rate waves

The next section shows a model of heat conductors in terms of summed histories [217]. Perhaps the main motivation for the introduction of summed histories was the observation that the classical equation of heat conduction,

$$\partial_t \theta = \alpha \Delta \theta, \qquad \alpha > 0,$$

is parabolic and the solution at any point in the body is felt immediately after at every other point (infinite speed of propagation). Really, also the model in terms of histories is adequate to find a finite speed of propagation [319]. Thermal waves are deferred to Sects. 7.8.3 and 7.6.2 where the models involved are developed in strictly analogous schemes of rate-type materials.

7.3.2 *Models Dependent on Summed Thermal Histories*

Following [217] we take as independent variables the present value of temperature $\theta(t)$, the summed history of temperature $\bar{\theta}^t$ and the summed history of the material temperature gradient $\nabla\bar{\theta}^t$,

$$\bar{\theta}^t(s) = \int_{t-s}^{t} \theta(\xi) d\xi, \qquad \nabla\bar{\theta}^t(s) = \int_{t-s}^{t} \nabla\theta(\xi) d\xi.$$

Let ψ, η, $\mathbf{q}$, at the position $\mathbf{x}$ and time t, be given by functionals Ψ, N, $\boldsymbol{\mathcal{Q}}$ on

$$\boldsymbol{\Lambda}_\theta(t) = (\theta(t), \bar{\theta}^t, \nabla\bar{\theta}^t),$$

at $\mathbf{x} \equiv \mathbf{X}$. In addition Ψ is assumed to be continuously differentiable while N and $\boldsymbol{\mathcal{Q}}$ are continuous. The functionals $\Psi, N, \boldsymbol{\mathcal{Q}}$ of $\psi, \eta, \mathbf{q}$ are subject to the entropy inequality

$$-(\dot{\psi} + \eta\dot{\theta}) - \frac{1}{\theta}\mathbf{q} \cdot \nabla\theta \geq 0,$$

where we let $\rho = 1$ for simplicity. Now,

$$\frac{d}{dt}\bar{\theta}^t(s) = \theta(t) - \theta^t(s) = \theta^t_d(s), \qquad \frac{d}{dt}\nabla\bar{\theta}^t(s) = \nabla\theta(t) - \nabla\theta^t(s) = \nabla\theta^t_d(s).$$

Hence, owing to (7.7),

$$\dot{\psi}(t) = \partial_\theta\Psi(\boldsymbol{\Lambda}_\theta)\dot{\theta}(t) + d\Psi(\boldsymbol{\Lambda}_\theta|\theta^{t\dagger}) - d\Psi(\boldsymbol{\Lambda}_\theta|\theta^t) + d\Psi(\boldsymbol{\Lambda}_\theta|\nabla\theta^{t\dagger}) - d\Psi(\boldsymbol{\Lambda}_\theta|\nabla\theta^t).$$

Let $\mathcal{J}_\theta$ and $\mathbf{J}_{\nabla\theta}$ be the functionals such that

$$d\Psi(\mathbf{\Lambda}_\theta|\theta^{t\dagger}) = \mathcal{J}_\theta(\mathbf{\Lambda}_\theta)\theta(t), \qquad d\Psi(\mathbf{\Lambda}_\theta|\nabla\theta^{t\dagger}) = \mathbf{J}_{\nabla\theta}(\mathbf{\Lambda}_\theta)\cdot\nabla\theta(t). \tag{7.20}$$

The entropy inequality can be written as

$$-(\partial_\theta\Psi+N)\dot{\theta}(t)-\mathcal{J}_\theta\,\theta(t)-\big(\mathbf{J}_{\nabla\theta}+\frac{1}{\theta}\mathbf{Q}\big)\cdot\nabla\theta(t)+d\Psi(\mathbf{\Lambda}_\theta|\theta^t)+d\Psi(\mathbf{\Lambda}_\theta|\nabla\theta^t)\geq 0.$$

For any values of $\theta(t)$ and $\nabla\theta(t)$ we can choose histories θ^t and $\nabla\theta^t$ such that $d\Psi(\mathbf{\Lambda}_\theta|\theta^t)$ and $d\Psi(\mathbf{\Lambda}_\theta|\nabla\theta^t)$ are as small as we please. In addition, the dependence on $\dot{\theta}(t)$ and $\nabla\theta(t)$ is linear and arbitrary. Hence the entropy inequality holds if and only if

$$N=-\partial_\theta\Psi, \qquad \mathbf{Q}=-\theta\mathbf{J}_{\nabla\theta}, \qquad \mathcal{J}_\theta(\mathbf{\Lambda}_\theta)\theta(t)-d\Psi(\mathbf{\Lambda}_\theta|\theta^t)-d\Psi(\mathbf{\Lambda}_\theta|\nabla\theta^t)\leq 0.$$

Hence, while the entropy is given by the classical partial derivative of the free energy, the entropy flux $\mathbf{Q}/\theta$ is given by the differential of the free energy relative to constant histories of the temperature gradient.

Quasilinear approximation

Let θ_* be some given absolute temperature and let $u\in U=(-\infty,1)$ denote the dimensionless variable

$$u=\frac{\theta-\theta_*}{\theta}$$

such that $u=0$ at $\theta=\theta_*$. By means of the *ballistic energy* (sometimes referred to as the *exergy potential* at constant pressure)

$$y=\varepsilon-\theta_*\eta=\psi+(\theta-\theta_*)\eta\ ,$$

the local form of the entropy inequality reads ($\rho=1$)

$$\dot{y}\leq-\nabla\cdot(u\mathbf{q})+ur\ . \tag{7.21}$$

Moreover, exploiting the dependence of y on θ through u

$$\partial_\theta y=(\theta-\theta_*)\partial_\theta\eta, \qquad \partial^2_\theta y=\partial_\theta\eta+(\theta-\theta_*)\partial^2_\theta\eta$$

Since $\partial_\theta\eta=c/\theta$, where $c>0$ is the specific heat, y has a local minimum with respect to temperature at $\theta=\theta_*$. By introducing the *rescaled free energy* [9]

$$\psi_*=y-u\,\varepsilon=\frac{\theta_*}{\theta}\psi$$

and observing that $\varepsilon=-\partial_u\psi_*$, inequality (7.21) can be written as

$$\dot{\psi}_* + \varepsilon\,\dot{u} \leq -\mathbf{q}\cdot\nabla u\ .$$

Hereafter we investigate this inequality for small values of u, that is, in a neighbourhood of the temperature θ_*. More precisely, we assume

$$\sup_{t\in\mathbb{R}} |u(t)| < \delta\quad ,\quad \sup_{t\in\mathbb{R}} \|\nabla u(t)\| < \delta\ ,$$

from which the following approximations are obtained

$$\theta = \theta_*(1+u) + O(\delta^2)\quad ,\quad \nabla\theta = \frac{\theta_*}{(1-u)^2}\nabla u = \theta_*\nabla u + O(\delta^2)\ . \tag{7.22}$$

Accordingly, we introduce the summed past histories $\bar{u}^t$ and $\nabla\bar{u}^t$

$$\bar{u}^t(\tau) = \int_0^\tau u(t-s)\,ds\ ,\qquad \nabla\bar{u}^t(\tau) = \int_0^\tau \nabla u(t-s)\,ds\ ,$$

belonging to the Hilbert spaces

$$\mathcal{H}_u := \{\bar{u}\in L^2_{k_1}(\mathbb{R}^+, U) : \bar{u}(0)=0\},\quad \mathcal{H}_{\nabla u} := \{\nabla\bar{u}\in L^2_{k_2}(\mathbb{R}^+,\mathbb{R}^3) : \nabla\bar{u}(0)=0\}$$

respectively. In this approximation scheme, the thermal variables $(\theta(t), \bar{\theta}^t, \nabla\bar{\theta}^t)$, can be replaced by $(u(t), \bar{u}^t, \nabla\bar{u}^t)$ whose norm in $\mathbb{R}\times\mathcal{H}_u\times\mathcal{H}_{\nabla u}$ is of order $O(\delta)$, namely

$$\sup_{t\in\mathbb{R}} |u(t)|^2 + \|\bar{u}^t\|^2_{\mathcal{H}_u} + \|\nabla\bar{u}^t\|^2_{\mathcal{H}_{\nabla u}} \leq C\delta^2\ . \tag{7.23}$$

Within this range, we are looking for thermodynamic restrictions. Letting

$$\mathbf{\Lambda}_u(t) = (u(t), \bar{u}^t, \nabla\bar{u}^t)$$

and assuming that ε, η, ψ_*, $\mathbf{q}$ are given by functionals $\hat{\varepsilon}$, $\hat{\eta}$, $\hat{\psi}_*$, $\hat{\mathbf{q}}$ on $\mathbf{\Lambda}_u$, we obtain

$$\dot{\psi}_*(t) = \partial_u\hat{\psi}_*(\mathbf{\Lambda}_u)\dot{u}(t) + \mathcal{J}_u(\mathbf{\Lambda}_u)\,u(t) + \mathbf{J}_{\nabla u}(\mathbf{\Lambda}_u)\cdot\nabla u(t) - d\hat{\psi}_*(\mathbf{\Lambda}_u|u^t) - d\hat{\psi}_*(\mathbf{\Lambda}_u|\nabla u^t),$$

where $\mathcal{J}_u$ and $\mathbf{J}_{\nabla u}$ have the analogue meaning of $\mathcal{J}_\theta$ and $\mathbf{J}_{\nabla\theta}$ in (7.20), respectively. As a consequence, the entropy inequality yields

$$(\partial_u\hat{\psi}_* + \varepsilon)\dot{u}(t) + (\mathbf{q}+\mathbf{J}_{\nabla u})\cdot\nabla u(t) + \mathcal{J}_u u(t) - d\hat{\psi}_*(\mathbf{\Lambda}_u|u^t) - d\hat{\psi}_*(\mathbf{\Lambda}_u|\nabla u^t) \leq 0\ ,$$

This implies

$$\hat{\varepsilon} = -\partial_u\hat{\psi}_*,\quad \hat{\mathbf{q}} = -\mathbf{J}_{\nabla u}, \tag{7.24}$$

$$\mathcal{J}_u(\mathbf{\Lambda}_u)u(t) - d\hat{\psi}_*(\mathbf{\Lambda}_u|u^t) - d\hat{\psi}_*(\mathbf{\Lambda}_u|\nabla u^t) \leq 0. \tag{7.25}$$

In order to obtain a *quasilinear* theory, according to the approximation assumptions (7.23) we consider constitutive relations for ε and $\mathbf{q}$ which depend at most linearly on $\mathbf{\Lambda}_u$,

$$\hat{\varepsilon}(\mathbf{\Lambda}_u) = c\,u(t) + \mathcal{F}[\bar{u}^t] + \mathcal{G}[\nabla\bar{u}^t], \quad \hat{\mathbf{q}}(\mathbf{\Lambda}_u) = \mathbf{r}\,u(t) + \boldsymbol{\mathcal{P}}[\bar{u}^t] + \boldsymbol{\mathcal{Q}}[\nabla\bar{u}^t]$$

where

$$\mathcal{F}[\bar{u}^t] = \int_0^\infty \alpha(s)\bar{u}^t(s)\,ds\,, \quad \boldsymbol{\mathcal{P}}[\bar{u}^t] = \int_0^\infty \boldsymbol{\kappa}(s)\bar{u}^t(s)\,ds\,,$$

$$\mathcal{G}[\nabla\bar{u}^t] = \int_0^\infty \boldsymbol{\beta}(s)\cdot\nabla\bar{u}^t(s)\,ds\,, \quad \boldsymbol{\mathcal{Q}}[\nabla\bar{u}^t] = \int_0^\infty \mathbf{K}(s)\nabla\bar{u}^t(s)\,ds\,,$$

are linear functionals on $\mathcal{H}_u$ and $\mathcal{H}_{\nabla u}$, respectively. Here $\mathbf{K}$ is tensor valued, whereas $\boldsymbol{\beta}$ and $\boldsymbol{\kappa}$ are vector valued. In particular, if the material is homogeneous and isotropic we can prove that the general form of linear constitutive relations is

$$\varepsilon(t) = \varepsilon_0 + c\,u(t) + \int_0^\infty \alpha(s)\bar{u}^t(s)\,ds, \qquad \mathbf{q}(t) = -\int_0^\infty K(s)\nabla\bar{u}^t(s)\,ds \tag{7.26}$$

Indeed, by virtue of the isotropy of the material, the constitutive relations are subject to

$$\mathcal{G} = 0\,, \quad \mathbf{r} = \mathbf{0}\,, \quad \boldsymbol{\mathcal{P}} = \mathbf{0}\,, \quad \mathbf{K} = -K\,\mathbf{1}\,.$$

In steady-state heat conduction, only constant histories are considered, for instance $u^t(s) = u^{t\dagger}$. Then

$$\varepsilon(t) = c_0\,u(t), \qquad \mathbf{q}(t) = -k\,\nabla u(t)$$

where $c_0 = c + \int_0^\infty \alpha(s)s\,ds > 0$ and $k = \int_0^\infty K(s)s\,ds > 0$ represent the stationary specific heat and thermal conductivity, respectively. In view of (7.24) and (7.26) the functional ψ_* can be represented as

$$\psi_*(\mathbf{\Lambda}_u) = -\tfrac{1}{2}c\,u^2(t) - u(t)\int_0^\infty \alpha(s)\bar{u}^t(s)\,ds$$
$$- \tfrac{1}{2}\int_0^\infty \alpha'(s)[\bar{u}^t(s)]^2\,ds + \tfrac{1}{2}\int_0^\infty K(s)\nabla\bar{u}^t(s)\cdot\nabla\bar{u}^t(s)\,ds.$$

We now prove that the reduced inequality (7.25) holds. First, in light of (7.6) we observe that ψ_* takes the form

$$\psi^*(\mathbf{\Lambda}_u) = -\tfrac{1}{2}c\,u^2(t) - \int_0^\infty \alpha(s)\bar{u}^t(s)[u(t) - u^t(s)]\,ds + \frac{1}{2}\int_0^\infty K(s)\nabla\bar{u}^t(s)\cdot\nabla\bar{u}^t(s)\,ds.$$

Hence, for any values of $u(t)$ and $\nabla u(t)$

$$\mathcal{J}_\theta(\mathbf{\Lambda}_u)\, u(t) := d\psi_*(\mathbf{\Lambda}_u | u^{t\dagger}) = -u^2(t) \int_0^\infty \alpha(s)\, ds + u(t) \int_0^\infty \alpha(s) u^t(s)\, ds,$$

$$\mathbf{J}_{\nabla u}(\mathbf{\Lambda}_u) \cdot \nabla u(t) := d\psi_*(\mathbf{\Lambda}_u | \nabla u^{t\dagger}) = \int_0^\infty K(s) \nabla \bar{u}^t(s)\, ds \cdot \nabla u(t).$$

Moreover we have

$$d\hat{\psi}_*(\mathbf{\Lambda}_u | u^t) = -u(t) \int_0^\infty \alpha(s) u^t(s)\, ds + \int_0^\infty \alpha(s) [u^t(s)]^2\, ds$$

and, by (7.6),

$$d\hat{\psi}_*(\mathbf{\Lambda}_u | \nabla u^t) = \int_0^\infty K(s) \nabla \bar{u}^t(s) \cdot \frac{d}{ds} \nabla \bar{u}^t(s)\, ds = -\tfrac{1}{2} \int_0^\infty K'(s) \nabla \bar{u}^t(s) \cdot \nabla \bar{u}^t(s)\, ds.$$

Consequently, the reduced inequality (7.25) can be written as

$$-\int_0^\infty \alpha(s) [u(t) - u^t(s)]^2\, ds + \tfrac{1}{2} \int_0^\infty K'(s) \nabla \bar{u}^t(s) \cdot \nabla \bar{u}^t(s)\, ds \le 0$$

and is satisfied provided that $\alpha(s) \ge 0$ and $K'(s) \le 0$, $s \in \mathbb{R}^+$.

Since $\bar{u}^t(0) = \nabla \bar{u}^t(0) = 0$, an integration by parts transforms (7.26) into linear functionals on the past histories, namely

$$\varepsilon(t) = c\, u(t) + \int_0^\infty m_\alpha(s) u^t(s)\, ds, \qquad \mathbf{q}(t) = -\int_0^\infty m_K(s) \nabla u^t(s)\, ds, \tag{7.27}$$

where

$$m_\alpha(s) = \int_s^\infty \alpha(\sigma)\, d\sigma, \qquad m_K(s) = \int_s^\infty K(\sigma)\, d\sigma.$$

7.3.3 Moore–Gibson–Thompson Temperature Equation

Applying (7.27) and assuming that $m_\alpha(0)$ and $m_K(0)$ are bounded, the linearized energy balance equation, $\partial_t \varepsilon + \nabla \cdot \mathbf{q} = 0$, gives

$$c\, \partial_t u(t) + \int_0^\infty m_\alpha(s) \partial_t u^t(s)\, ds = \int_0^\infty m_K(s) \Delta u^t(s)\, ds,$$

where $u \simeq (\theta - \theta_*)/\theta_*$ is the dimensionless relative temperature (see (7.22)). Upon some rearrangements this equation can be written as

$$c\, \partial_t^2 u(t) + m_\alpha(0) \partial_t u(t) + \int_0^\infty m'_\alpha(s) \partial_t u^t(s)\, ds = m_K(0) \Delta u(t) + \int_0^\infty m'_K(s) \Delta u^t(s)\, ds.$$

Since $m_\alpha(0), m_\kappa(0) > 0$, this equation shows a dissipative hyperbolic character and is associated with the constant wave speed $\sqrt{m_\kappa(0)/c}$.

In the special case where the heat flux and the internal energy kernels are in exponential forms,

$$m_\alpha(s) = m_\alpha(0)\exp(-s/\lambda_\alpha), \qquad m_\kappa(s) = m_\kappa(0)\exp(-s/\lambda_\kappa), \qquad \lambda_\alpha, \lambda_\kappa > 0,$$

from (7.27) it follows that

$$\lambda_\alpha \partial_t \varepsilon(t) + \varepsilon(t) = c\lambda_\alpha \partial_t u(t) + [\lambda_\alpha m_\alpha(0) + c]u, \qquad \lambda_\kappa \partial_t \mathbf{q}(t) + \mathbf{q}(t) = -\lambda_\kappa m_\kappa(0)\nabla u(t).$$

By eliminating ε and $\mathbf{q}$ from these equations and the linearized energy balance we find ([242], Eq. (5.7))

$$c\,\partial_t^3 u + \big[m_\alpha(0) + c(\frac{1}{\lambda_\kappa} + \frac{1}{\lambda_\alpha})\big]\partial_t^2 u + \frac{1}{\lambda_\kappa}\big[m_\alpha(0) + \frac{c}{\lambda_\alpha}\big]\partial_t u = m_\kappa(0)\Delta\partial_t u + \frac{1}{\lambda_\alpha} m_\kappa(0)\Delta u.$$

This equation is a third-order hyperbolic equation since discontinuities propagate with constant speed $\sqrt{m_\kappa(0)/c}$.

Recently, a similar result has been obtained in [370] by introducing a relaxation parameter in the Green–Naghdi type III model. The resulting constitutive equation of the heat flux vector is

$$\tau\partial_t^2 \mathbf{q}(t) + \partial_t \mathbf{q}(t) = k_* \nabla\theta(t) + k\nabla\partial_t\theta(t)$$

Comparing this relation with the linearized energy balance, with $\varepsilon = c\,\theta$, we have

$$\tau c\,\partial_t^3\theta + c\,\partial_t^2 u = k\Delta\partial_t\theta(t) + k_*\Delta\theta(t), \tag{7.28}$$

so that thermal waves propagate with finite speed $\sqrt{k/\tau c}$. This equation represents the linear version of a third-order wave equation describing the nonlinear propagation of sound in nonlinear acoustics, a model which is known in the literature as the Jordan–Moore–Gibson–Thompson equation [240, 247]. Hence, (7.28) is referred to as Moore–Gibson–Thompson temperature equation in [370] and its connection with linear viscoelasticity is outlined in [125].

7.4 Linear Viscoelasticity

Linear viscoelasticity[8] traces back to Boltzmann [55] who gave a linear theory based on the view that, at any point $\mathbf{x} \in \Omega \subset \mathscr{E}$ at a time t, the stress depends upon the strains which the point has experienced prior to time t. The influence of a previous

[8] See, e.g. [158, 276].

strain depends on the time elapsed since the strain occurred and is less for those strains occurring long ago, according to the fading memory principle described above. A superposition of the influence of previous strains holds, which means that the stress–strain relation is linear. This is the essence of the Boltzmann model of linear viscoelasticity.

Elastic solids and viscous liquids (or fluids) are two main types of simple materials. Among the mechanical aspects, the elastic solid has a definite shape and deforms under external forces, stores all energy that is obtained from external forces, restores the original shape once the force is removed. The viscous liquid, instead, has no definite shape and flows irreversibly under the action of external forces. Viscoelastic materials combine both liquid- and solid-like features. Accordingly, the theory of linear viscoelasticity can be profitably applied to both liquids and solids.

Quite often (see e.g. [50]) linear viscoelasticity is motivated through elementary models in terms of series of springs and dashpots as shown in Sect. 7.1. The Maxwell-Wiechert model, for instance, leads to (6.9), an integral relation where the dependence of the stress on the history of the strain rate can be viewed as the simplest model of a linear viscoelastic fluid. Furthermore, the standard linear solid—see (6.26)—accounts for both the dependence of the stress on the history of the strain and that of the strain on the history of the stress. In both cases the relaxation modulus turns out to be given by exponentials. For the sake of generality, here linear viscoelasticity is framed within the scheme of materials with fading memory.

7.4.1 Linear Viscoelastic Solids

Let $\mathbf{E}(\mathbf{X}, t)$ be the Green-St.Venant strain at $\mathbf{X} \in R$ at time t. We denote by $\mathbf{E}^t(\mathbf{X}, \cdot)$ the strain history experienced by the particle $\mathbf{X}$. For simplicity, hereafter we restrict our attention to homogeneous materials so that the dependence on $\mathbf{X}$ is understood and not written. Let $k \in L^1(\mathbb{R}^+)$ and consider the fading memory space

$$\mathcal{H} = \{\mathbf{E}^t : \mathbb{R}^+ \to \mathrm{Sym}, \int_0^\infty |\mathbf{E}^t(s)|^2 k(s)ds < \infty\}$$

endowed with the norm

$$\|\mathbf{E}^t\| = |\mathbf{E}(t)|^2 + \int_0^\infty |\mathbf{E}^t(s)|^2 k(s)ds.$$

Once we make the assumption that the constitutive equation for the stress $\mathbf{T}$ is given by a bounded, linear functional on $\mathcal{H}$ then the Riesz theorem gives the representation

$$\mathfrak{T}(\mathbf{E}^t) = \mathbf{G}_0\mathbf{E}(t) + \int_0^\infty \mathcal{G}(s)\mathbf{E}(t-s)ds.$$

The fourth-order tensor $\mathbf{G}_0$ is called an *instantaneous elastic modulus* and governs the response to instantaneous changes in strain, whereas the fourth-order tensor-valued function $\mathcal{G}$ on $\mathbb{R}^+$ is sometimes called *Boltzmann function* and is the memory kernel of the linear functional $\mathfrak{T}$. By the Cauchy–Schwarz inequality we can write

$$\Big|\int_0^\infty \mathcal{G}(s)\mathbf{E}(t-s)ds\Big| \le \Big(\int_0^\infty |\mathcal{G}(s)|^2 k^{-1}(s)ds\Big)^{1/2} \Big(\int_0^\infty k(s)|\mathbf{E}(t-s)|^2 ds\Big)^{1/2}.$$

The assumption that the response $\mathbf{T}$ be finite for any history $\mathbf{E}^t$ with a finite $\mathcal{H}$-norm requires that

$$\int_0^\infty |\mathcal{G}(s)|^2 k^{-1}(s)ds < \infty.$$

Moreover,

$$\Big|\int_0^\infty \mathcal{G}(s)\mathbf{E}(t-s)ds\Big| \le \sup_{s\in\mathbb{R}^+} |\mathbf{E}^t(s)| \int_0^\infty |\mathcal{G}(s)|ds.$$

Hence, to guarantee that all bounded histories belong to $\mathcal{H}$ we assume the

Property 1. *The function $\mathcal{G}$ belongs to $L^1(\mathbb{R}^+;\mathrm{Lin(Sym)})$.*

To adhere to the standard notation we let $\mathbf{G}:\mathbb{R}^+ \to \mathrm{Lin(Sym)}$ be such that

$$\mathbf{G}'(s) = \mathcal{G}(s), \qquad \mathbf{G}(0) = \mathbf{G}_0,$$

the prime denoting the derivative. Consistent with the notation for the Maxwell and linear solid models, the solution

$$\mathbf{G}(s) = \mathbf{G}_0 + \int_0^s \mathcal{G}(\xi)d\xi$$

is called *relaxation function* (or relaxation modulus). The limit of $\mathbf{G}(s)$, as $s \to \infty$, is supposed to exist and

$$\mathbf{G}_\infty = \lim_{s\to\infty} \mathbf{G}(s)$$

is called an *equilibrium elastic modulus* in that the constancy of the history $\mathbf{E}^t$, $\mathbf{E}^t(s) = \mathbf{E}$, $s \in \mathbb{R}^+$, gives

$$\mathbf{T}(t) = \mathbf{G}_\infty \mathbf{E}.$$

In a solid, a nonzero constant strain is supposed to induce a nonzero stress such that $\mathbf{T}\cdot\mathbf{E}$ is strictly positive. That is why in viscoelastic solids we let

$$\mathbf{G}_\infty > \mathbf{0},$$

whereas in viscoelastic fluids

$$\mathbf{G}_\infty = \mathbf{0}.$$

The generic constitutive equation of linear viscoelasticity can be written in the form

$$\mathbf{T}(t) = \mathfrak{T}(\mathbf{E}^t) := \mathbf{G}_0\mathbf{E}(t) + \int_0^\infty \mathbf{G}'(s)\mathbf{E}(t-s)ds. \tag{7.29}$$

Alternatively, we can write (7.29) in the form

$$\mathbf{T}(t) = \mathfrak{T}_*(\mathbf{E}(t), \mathbf{E}_d^t) := \mathbf{G}_\infty\mathbf{E}(t) - \int_0^\infty \mathbf{G}'(s)[\mathbf{E}(t) - \mathbf{E}(t-s)]ds. \tag{7.30}$$

where $\mathfrak{T}_*$ is a linear functional on $\mathcal{H}_*$ given by (7.5). If the strain function has compact support, or anyway $\lim_{s\to-\infty} \mathbf{E}(s) = \mathbf{0}$, then an obvious integration by parts yields

$$\mathbf{T}(t) = \int_0^\infty \mathbf{G}(s)\dot{\mathbf{E}}(t-s)ds, \tag{7.31}$$

the superposed dot denoting the time derivative with respect to t. We can also write

$$\mathbf{T}(t) = \mathbf{G}_\infty\mathbf{E}(t) - \int_0^\infty [\mathbf{G}(s) - \mathbf{G}_\infty]\dot{\mathbf{E}}(t-s)ds, \tag{7.32}$$

7.4.2 *Thermodynamic Restrictions for the Linear Viscoelastic Solid*

The entropy inequality

$$\rho\dot{\eta} + \nabla\cdot\frac{\mathbf{q}}{\theta} - \frac{\rho r}{\theta} + \nabla\cdot\mathbf{k} \geq 0$$

and the balance of energy allow us to write

$$\dot{\eta} + \frac{1}{\rho\theta}\mathbf{T}\cdot\mathbf{D} - \frac{1}{\theta}\dot{\varepsilon} - \frac{1}{\rho\theta^2}\mathbf{q}\cdot\nabla\theta + \frac{1}{\rho}\nabla\cdot\mathbf{k} \geq 0 \tag{7.33}$$

Let the material undergo a cycle as $t \in [0, d]$, which means that the argument of the constitutive functionals, say Λ, satisfies

$$\Lambda(0) = \Lambda(d).$$

Hence

$$\eta(\Lambda(d)) = \eta(\Lambda(0))$$

and the like for ε. As a consequence, integration of (7.33) gives

$$\int_0^d \Big\{\frac{1}{\rho\theta}\mathbf{T}\cdot\mathbf{D} - \frac{1}{\theta}\dot{\varepsilon} - \frac{1}{\rho\theta^2}\mathbf{q}\cdot\nabla\theta + \frac{1}{\rho}\nabla\cdot\mathbf{k}\Big\}dt \geq 0.$$

Let the material undergo an isothermal process, which means that θ is constant and uniform. Moreover let $\mathbf{k} = \mathbf{0}$, which has to be shown consistent with the whole inequality. Hence we conclude that, in isothermal processes, for every cycle

$$\int_0^d [\rho^{-1}\mathbf{T}\cdot\mathbf{D}](t)dt \geq 0.$$

Since $\dot{\mathbf{E}} = \frac{1}{2}\dot{\mathbf{C}} = \mathbf{F}^T\mathbf{D}\mathbf{F}$ we have

$$\frac{1}{\rho}\mathbf{T}\cdot\mathbf{D} = \frac{1}{\rho_R}\mathbf{T}_{RR}\cdot\dot{\mathbf{E}}.$$

Thus we can express the power in the form $\mathbf{T}\cdot\dot{\mathbf{E}}$ if $\mathbf{T}$ stands for the second Piola stress. Otherwise if $\mathbf{T}$ is the Cauchy stress then $\mathbf{T}\cdot\dot{\mathbf{E}}$ has to be viewed as a linear approximation (where $\mathbf{F} \simeq \mathbf{1}$) of the power $\mathbf{T}\cdot\mathbf{D}$.

Consistent with the scheme of linear viscoelasticity, here we express the stress power in the linearized form $\mathbf{T}\cdot\dot{\mathbf{E}}$. Moreover, the constitutive equation (7.29), or (7.31), does not involve the temperature. Accordingly, in linear viscoelasticity, the second law requires that

$$\int_0^d \mathbf{T}(t)\cdot\dot{\mathbf{E}}(t)dt \geq 0 \tag{7.34}$$

hold for any cycle.

The requirement (7.34) is a restriction on the relaxation functions compatible with thermodynamics. To find definite results[9] we consider oscillatory strain tensor evolutions of the form

$$\tilde{\mathbf{E}}(t) = \mathbf{E}_1\cos\omega t + \mathbf{E}_2\sin\omega t,$$

where $\omega > 0$ and $\mathbf{E}_1, \mathbf{E}_2 \in \mathrm{Sym}$. Hence

$$\tilde{\mathbf{E}}^t = \tilde{\mathbf{E}}^{t+d}, \qquad d = 2\pi/\omega.$$

Denote by the subscript $_s$ and $_c$ the half-range sine and cosine transform, respectively

$$\mathbf{G}'_s(\omega) = \int_0^\infty \mathbf{G}'(u)\sin\omega u\, du, \qquad \mathbf{G}'_c(\omega) = \int_0^\infty \mathbf{G}'(u)\cos\omega u\, du.$$

We can now derive a set of restrictions placed by the second-law inequality on the relaxation function $\mathbf{G}$.

---

[9] The sequel of this section re-visits Chap. 3 of [158].

Theorem 7.1 *Inequality (7.34) holds for any oscillatory function* $\tilde{\mathbf{E}}$, $\omega \in (0, \infty)$, *only if*

$$\mathsf{G}_0 = \mathsf{G}_0^T, \qquad \mathsf{G}_\infty = \mathsf{G}_\infty^T, \tag{7.35}$$

$$\mathsf{G}_s'(\omega) \leq 0, \qquad \omega \in (0, \infty), \tag{7.36}$$

$$|\mathbf{E}_1 \cdot (\mathsf{G}_c'(\omega) - \mathsf{G}_c'^T(\omega))\mathbf{E}_2|^2 \leq 2([\mathbf{E}_1 \cdot \mathsf{G}_s'(\omega)\mathbf{E}_1]^2 + [\mathbf{E}_2 \cdot \mathsf{G}_s'(\omega)\mathbf{E}_2]^2), \quad \mathbf{E}_1, \mathbf{E}_2 \in \mathrm{Sym}. \tag{7.37}$$

Proof The history $\tilde{\mathbf{E}}^t$ is periodic with period $d = 2\pi/\omega$. Upon substitution of $\tilde{\mathbf{E}}^t$ in (7.29), inequality (7.34) yields

$$\int_0^d \Big\{(-\omega\mathbf{E}_1 \sin\omega t + \omega\mathbf{E}_2 \cos\omega t) \cdot \big[\mathsf{G}_0(\mathbf{E}_1 \cos\omega t + \mathbf{E}_2 \sin\omega t)$$
$$+ \int_0^\infty \mathsf{G}'(s)[\mathbf{E}_1(\cos\omega t \cos\omega s + \sin\omega t \sin\omega s) + \mathbf{E}_2(\sin\omega t \cos\omega s - \cos\omega t \sin\omega s)\big]ds\Big\}dt \geq 0.$$

Integration with respect to t on $[0, d]$ results in

$$\mathbf{E}_1 \cdot [\mathsf{G}_0^T - \mathsf{G}_0]\mathbf{E}_2 - \mathbf{E}_1 \cdot \mathsf{G}_s'(\omega)\mathbf{E}_1 - \mathbf{E}_2 \cdot \mathsf{G}_s'(\omega)\mathbf{E}_2 - \mathbf{E}_1 \cdot [\mathsf{G}_c'(\omega) - \mathsf{G}_c'^T(\omega)]\mathbf{E}_2 \geq 0, \tag{7.38}$$

as $\omega > 0$. By Riemann-Lebesgue's lemma, the limit $\omega \to \infty$ makes the integrals vanish. Then the arbitrariness of $\mathbf{E}_1, \mathbf{E}_2 \in \mathrm{Sym}$ implies that $\mathsf{G}_0 = \mathsf{G}_0^T$ and we are left with

$$\mathbf{E}_1 \cdot \mathsf{G}_s'(\omega)\mathbf{E}_1 + \mathbf{E}_2 \cdot \mathsf{G}_s'(\omega)\mathbf{E}_2 + \mathbf{E}_1 \cdot [\mathsf{G}_c'(\omega) - \mathsf{G}_c'^T(\omega)]\mathbf{E}_2 \leq 0, \quad \omega > 0, \tag{7.39}$$

We now evaluate the limit as $\omega \to 0_+$. Since $\mathsf{G}' \in L^1(\mathbb{R}^+)$ then for every $\epsilon > 0$ there exists N such that

$$\Big|\int_N^\infty \mathsf{G}'(s) \sin\omega s \, ds\Big| \leq \int_N^\infty |\mathsf{G}'(s)| ds < \epsilon.$$

Now let $G = \int_0^\infty |\mathsf{G}'(s)| ds$ and observe

$$\Big|\int_0^N \mathsf{G}'(s) \sin\omega s \, ds\Big| \leq \int_0^N |\mathsf{G}'(s)||\sin\omega s| \, ds < \int_0^N |\mathsf{G}'(s)| \omega s \, ds.$$

Since $\omega \to 0_+$, we can choose $\omega < \epsilon/G\,N$ and hence

$$\Big|\int_0^N \mathsf{G}'(s) \sin\omega s \, ds\Big| < \epsilon.$$

As a consequence, $\forall\epsilon\ \exists\delta(=\epsilon/G\,N)$ such that $\omega<\delta$ implies

$$\Big|\int_0^\infty \mathbf{G}'(s)\sin\omega s\,ds\Big| \le \int_0^N |\mathbf{G}'(s)|\omega s ds + \int_N^\infty |\mathbf{G}'(s)|ds < 2\epsilon.$$

Likewise, since

$$\int_0^\infty \mathbf{G}'(s)\cos\omega s\,ds = \int_0^\infty \mathbf{G}'(s)ds + \int_0^\infty \mathbf{G}'(s)[\cos\omega s - 1]ds$$

and

$$|\cos\omega s - 1| = |(-\sin\omega\xi)\omega s| \le \omega s,\quad \xi\in(0,s),\qquad |\cos\omega s - 1|\le 2,\ s\in\mathbb{R},$$

then we find that, as $\omega<\epsilon/GN$,

$$\Big|\int_0^\infty \mathbf{G}'(s)[\cos\omega s - 1]ds\Big| \le \int_0^N |\mathbf{G}'(s)|\omega s\,ds + \int_N^\infty |\mathbf{G}'(s)||\cos\omega s - 1|ds < 3\epsilon.$$

Consequently

$$\lim_{\omega\to 0_+}\mathbf{G}'_s(\omega) = \mathbf{0},\qquad \lim_{\omega\to 0_+}\mathbf{G}'_c(\omega) = \int_0^\infty \mathbf{G}'(s)\,ds = \mathbf{G}_\infty - \mathbf{G}_0.$$

Accordingly, as $\omega\to 0_+$, inequality (7.39) implies

$$\mathbf{E}_1\cdot[\mathbf{G}_\infty - \mathbf{G}_\infty^T]\mathbf{E}_2 - \mathbf{E}_1\cdot[\mathbf{G}_0 - \mathbf{G}_0^T]\mathbf{E}_2 \le 0.$$

The symmetry of $\mathbf{G}_0$ reduces the inequality to

$$\mathbf{E}_1\cdot[\mathbf{G}_\infty - \mathbf{G}_\infty^T]\mathbf{E}_2 \le 0.$$

According to the arbitrariness of $\mathbf{E}_1,\mathbf{E}_2$ we first select $\mathbf{E}_2=\mathbf{E}_1$ and then $\mathbf{E}_2=-\mathbf{E}_1$ to find the symmetry of $\mathbf{G}_\infty$.

We now turn to (7.39). Let $\mathbf{E}_1=\mathbf{E}_2=\mathbf{E}$ and observe that, by the skew character of $\mathbf{G}'(s)-\mathbf{G}'^T(s)$, it follows $\mathbf{E}\cdot[\mathbf{G}'(s)-\mathbf{G}'^T(s)]\mathbf{E}=0$. Hence (7.39) reduces to

$$\mathbf{E}\cdot\mathbf{G}'_s(\omega)\mathbf{E}\le 0,\quad \omega>0, \tag{7.40}$$

that is (7.36). Finally, inequality (7.39) reads

$$\mathbf{E}_1\cdot\mathbf{G}'_s(\omega)\mathbf{E}_1 + \mathbf{E}_2\cdot\mathbf{G}'_s(\omega)\mathbf{E}_2 + \mathbf{E}_1\cdot[\mathbf{G}'_c(\omega) - \mathbf{G}'^T_c(\omega)]\mathbf{E}_2 \le 0,\quad \omega>0.$$

Since $\mathbf{G}'_s$ is negative semi-definite it follows that

$$\mathbf{G}_c'(\omega) - \mathbf{G}_c'^T(\omega)$$

is allowed to be nonzero provided that

$$\mathbf{E}_1 \cdot [\mathbf{G}_c'(\omega) - \mathbf{G}_c'^T(\omega)]\mathbf{E}_2 \le |\mathbf{E}_1 \cdot \mathbf{G}_s'(\omega)\mathbf{E}_1| + |\mathbf{E}_2 \cdot \mathbf{G}_s'(\omega)\mathbf{E}_2|.$$

Since $(a+b)^2 \le 2(a^2+b^2)$ then (7.37) holds. □

It is worth remarking that the Boltzmann function $\mathbf{G}'$ need not be symmetric. If we assume

$$\mathbf{G}'(u) = \mathbf{G}'^T(u), \qquad u \in \mathbb{R}^+,$$

then inequality (7.37) becomes redundant and the thermodynamic restrictions reduce to (7.35) and (7.36).

Inequality (7.36) in turn implies some further properties of the relaxation function. Upon extending the definition of $\mathbf{G}'$ to $\mathbb{R}^-$ as an odd function,

$$\mathbf{G}'(-u) = -\mathbf{G}'(u), \qquad u \in \mathbb{R}^+,$$

we let

$$\mathbf{G}_F'(\omega) = \int_{-\infty}^{\infty} \mathbf{G}'(u)\exp(-i\omega u)\,du, \qquad \omega \in \mathbb{R}.$$

We can then write the half-range sine transform of $\mathbf{G}'$ as

$$\mathbf{G}_s'(\omega) = \tfrac{1}{2}\int_{-\infty}^{\infty} \mathbf{G}'(u)\sin\omega u\,du = -\tfrac{1}{2}\Im\mathbf{G}_F'(\omega), \qquad \omega \in \mathbb{R}.$$

Hence

$$\mathbf{G}_s'(\omega) = -\mathbf{G}_s'(-\omega), \qquad \omega \in \mathbb{R}.$$

Thus we have the inversion formula

$$\mathbf{G}'(u) = \tfrac{1}{2\pi}\int_{-\infty}^{\infty} \mathbf{G}_F'(\omega)\exp(i\omega u)d\omega = \tfrac{1}{\pi}\int_{-\infty}^{\infty} \mathbf{G}_s'(\omega)\sin\omega u\,d\omega = \tfrac{2}{\pi}\int_{0}^{\infty} \mathbf{G}_s'(\omega)\sin\omega u\,d\omega.$$

Integration with respect to u gives

$$\mathbf{G}(u) - \mathbf{G}_0 = \tfrac{2}{\pi}\int_0^{\infty} \frac{1-\cos\omega u}{\omega}\mathbf{G}_s'(\omega)d\omega. \tag{7.41}$$

The negative definiteness of $\mathbf{G}_s'$ implies that

$$\mathbf{G}(u) - \mathbf{G}_0 < \mathbf{0}, \qquad u > 0. \tag{7.42}$$

The limit of (7.41) as $u \to \infty$ and that of the ratio

$$\frac{\mathbf{G}(u) - \mathbf{G}_0}{u}$$

as $u \to 0_+$ lead to

$$\mathbf{G}_0 - \mathbf{G}_\infty > \mathbf{0}, \qquad \mathbf{G}_0' := \lim_{u \to 0^+} \mathbf{G}'(u) \leq \mathbf{0}. \tag{7.43}$$

As a comment on (7.35), the symmetry of the instantaneous elastic modulus $\mathbf{G}_0$ was proved by Coleman [94], as a consequence of the second law in the form of the Clausius–Duhem inequality, whereas the symmetry of the equilibrium elastic modulus $\mathbf{G}_\infty$ was derived by Day [114] via the Clausius inequality. Inequality (7.36) was first derived by Graffi [202] in the case of isotropic materials by requiring that energy be dissipated in a period of a sinusoidal strain function.

Inequality (7.42), derived in [157], is in a sense related to a result by Day [114] who showed that, as a consequence of dissipativity, the relaxation function satisfies the condition

$$\mathbf{G}_0 - \mathbf{G}_\infty \geq \pm[\mathbf{G}(s) - \mathbf{G}_\infty]. \tag{7.44}$$

To establish the connection between the two inequalities, observe that the limit as $u \to \infty$ of (7.41) results in

$$\mathbf{G}_\infty - \mathbf{G}_0 = \tfrac{2}{\pi} \int_0^\infty \frac{1}{\omega} \mathbf{G}_s'(\omega) d\omega.$$

Hence it follows from (7.41) that

$$\mathbf{G}(u) - \mathbf{G}_\infty = -\tfrac{2}{\pi} \int_0^\infty \frac{\cos \omega u}{\omega} \mathbf{G}_s'(\omega) d\omega.$$

Now, for any strain $\mathbf{E}$ we have

$$\int_0^\infty \frac{1}{\omega} |\mathbf{E} \cdot \mathbf{G}_s'(\omega)\mathbf{E}| d\omega \geq \int_0^\infty \frac{|\cos \omega u|}{\omega} |\mathbf{E} \cdot \mathbf{G}_s'(\omega)\mathbf{E}| d\omega \geq \Big| \int_0^\infty \frac{\cos \omega u}{\omega} [\mathbf{E} \cdot \mathbf{G}_s'(\omega)\mathbf{E}] d\omega \Big|.$$

Consequently, using (7.40) we get

$$\tfrac{\pi}{2} \mathbf{E} \cdot [\mathbf{G}_0 - \mathbf{G}_\infty]\mathbf{E} = \mathbf{E} \cdot \int_0^\infty \tfrac{1}{\omega} [-\mathbf{G}_s'(\omega)] d\omega\, \mathbf{E} = \int_0^\infty \tfrac{1}{\omega} |\mathbf{E} \cdot \mathbf{G}_s'(\omega)\mathbf{E}|\, d\omega \geq \tfrac{\pi}{2} |\mathbf{E} \cdot [\mathbf{G}(u) - \mathbf{G}_\infty]\mathbf{E}|,$$

whence

$$\mathbf{E} \cdot [\mathbf{G}_0 - \mathbf{G}_\infty]\mathbf{E} \geq |\mathbf{E} \cdot [\mathbf{G}(u) - \mathbf{G}_\infty]\mathbf{E}|.$$

Inequality (7.44) is a short way of writing this result.

As we see in a moment, the restrictions (7.35) and (7.36) are also sufficient for the validity of the second law. In this connection, we prove the following preliminary

Lemma 7.1 *If* $\mathbf{E} : \mathbb{R} \to \mathrm{Sym}$ *is a periodic function with period* d *and* $\mathsf{K} \in \mathrm{Lin}(\mathrm{Sym})$ *is symmetric,* $\mathsf{K} = \mathsf{K}^T$, *then*

$$\int_0^d \dot{\mathbf{E}}(t) \cdot \mathsf{K}\mathbf{E}(t)dt = 0.$$

Proof Observe

$$\dot{\mathbf{E}}(t) \cdot \mathsf{K}\mathbf{E}(t) = [\mathbf{E}(t) \cdot \mathsf{K}\mathbf{E}(t)]\dot{} - \mathbf{E}(t) \cdot \mathsf{K}\dot{\mathbf{E}}(t) = [\mathbf{E}(t) \cdot \mathsf{K}\mathbf{E}(t)]\dot{} - \dot{\mathbf{E}}(t) \cdot \mathsf{K}^T\mathbf{E}(t).$$

Since $\mathsf{K}^T = \mathsf{K}$ then

$$2\dot{\mathbf{E}}(t) \cdot \mathsf{K}\mathbf{E}(t) = [\mathbf{E}(t) \cdot \mathsf{K}\mathbf{E}(t)]\dot{}.$$

Integration over $[0, d]$ gives

$$\int_0^d \dot{\mathbf{E}}(t) \cdot \mathsf{K}\mathbf{E}(t)dt = \tfrac{1}{2}\big[\mathbf{E}(t) \cdot \mathsf{K}\mathbf{E}(t)\big]_0^d = 0.$$ □

Theorem 7.2 *If* G *satisfies (7.35) and (7.36) for any* $\omega \in [0, \infty)$ *then the second-law inequality (7.34) holds for any periodic strain function.*

Proof Let $\mathbf{E} : \mathbb{R} \to \mathrm{Sym}$ be a periodic function with period d. Represent $\mathbf{E}$ through the Fourier series

$$\mathbf{E}(t) = \textstyle\sum_{k=0}^{\infty} \mathbf{A}_k \cos k\omega t + \mathbf{B}_k \sin k\omega t, \qquad \mathbf{A}_k, \mathbf{B}_k \in \mathrm{Sym},$$

where $\omega = 2\pi/d$. The history $\mathbf{E}^t$ is a cycle on $[0, d]$. Let

$$w := \int_0^d \mathfrak{T}(\mathbf{E}^t) \cdot \dot{\mathbf{E}}(t)dt,$$

where $\mathfrak{T}(\mathbf{E}^t)$ is the functional (7.29) of linear viscoelasticity. We have to show that $w \geq 0$ as a consequence of (7.35) and (7.36). Since $\mathbf{E}$ is periodic with period d then the symmetry of G_0 and the Lemma imply that

$$\int_0^d \dot{\mathbf{E}}(t) \cdot \mathsf{G}_0\mathbf{E}(t)dt = 0.$$

Hence

$$w = \int_0^d \dot{\mathbf{E}}(t) \cdot \Big(\int_0^\infty \mathsf{G}'(u)\mathbf{E}(t-u)du\Big)dt$$

Upon substitution of

$$\mathbf{E}^t(u) = \sum_{k=0}^{\infty} \mathbf{A}_k \cos k\omega(t-u) + \mathbf{B}_k \sin k\omega(t-u),$$

we have

$$w = \int_0^d \int_0^\infty \sum_{k=1}^{\infty}\sum_{h=0}^{\infty} k\omega[-\mathbf{A}_k \sin k\omega t + \mathbf{B}_k \cos k\omega t]$$
$$\cdot \mathbf{G}'(u)[\mathbf{A}_h \cos h\omega(t-u) + \mathbf{B}_h \sin h\omega(t-u)]du\,dt.$$

Term by term integration of the double series shows that the only nonzero terms are those with $h = k$. Integration in t times $k\omega$ provides $-\pi$. Hence we are left with

$$w = -\pi \sum_{k=1}^{\infty} \int_0^\infty \{\mathbf{A}_k \cdot \mathbf{G}'(u)\mathbf{A}_k + \mathbf{B}_k \cdot \mathbf{G}'(u)\mathbf{B}_k]\sin k\omega u + \mathbf{A}_k \cdot [\mathbf{G}'(u) - \mathbf{G}'^T(u)]\mathbf{B}_k \cos ku\}du$$

whence

$$w = -\pi \sum_{k=1}^{\infty} \{\mathbf{A}_k \cdot \mathbf{G}'_s(k\omega)\mathbf{A}_k + \mathbf{B}_k \cdot \mathbf{G}'_s(k\omega)\mathbf{B}_k + \mathbf{A}_k \cdot [\mathbf{G}'_c(k\omega) - \mathbf{G}'^T_c(k\omega)]\mathbf{B}_k\}.$$

By (7.37) each k-th term, in braces, of the series is non-positive and hence it follows that $w \geq 0$. □

The connection between (7.36) and the energy dissipation is emphasized in [276], Sect. 13, where the energy dissipated in one period of the strain function is determined. As a particular case of the previous evaluation of w, if $\mathbf{E}(t) = \mathbf{E}_0 \sin \omega t$, $d = 2\pi/\omega$, we have

$$w = -\pi \mathbf{E}_0 \cdot \mathbf{G}'_s(\omega)\mathbf{E}_0.$$

That is why $-\mathbf{G}'_s(\omega)$ is often called the *loss modulus.*

7.4.3 Free Energies and Minimal States

By the entropy inequality (7.33), in the approximation of linear isothermal viscoelasticity we can write

$$-\rho\dot{\psi} + \mathbf{T} \cdot \dot{\mathbf{E}} = \theta\sigma_R \geq 0 \tag{7.45}$$

ρ being regarded as constant. Since $\mathbf{T}$ is given by the functional (7.29) we take the free energy ψ as a functional of the present value $\mathbf{E}(t)$ and the past history $\mathbf{E}^t_r$. Hence

$$\psi(t) = \hat{\psi}(\mathbf{E}(t), \mathbf{E}^t_r).$$

Upon substitution in (7.45) we have

$$-\rho\partial_{\mathbf{E}}\hat{\psi}(\mathbf{E}(t), \mathbf{E}^t_r) \cdot \dot{\mathbf{E}}(t) - \rho\, d\hat{\psi}(\mathbf{E}(t), \mathbf{E}^t_r \,|\, \dot{\mathbf{E}}^t_r) + \mathfrak{T}(\mathbf{E}(t), \mathbf{E}^t_r) \cdot \dot{\mathbf{E}}(t) = \theta\sigma_R \geq 0.$$

For any history $\mathbf{E}^t$, we can take a strain history $\mathcal{E}^t$ such that $\dot{\mathcal{E}}(t)$ is arbitrary while

$$\mathcal{E}(t) = \mathbf{E}(t), \qquad \|\mathcal{E}^t - \mathbf{E}^t\|_{\mathcal{H}} < \epsilon, \qquad \|\dot{\mathcal{E}}^t - \dot{\mathbf{E}}^t\|_{\mathcal{H}} < \epsilon,$$

where ϵ is arbitrarily small. Accordingly the arbitrariness of the present value $\dot{\mathcal{E}}(t)$ implies that

$$\mathfrak{T} = \rho\, \partial_{\mathbf{E}} \hat{\psi}, \qquad d\hat{\psi}(\mathbf{E}(t), \mathbf{E}^t_r \,|\, \dot{\mathbf{E}}^t_r) = -\theta\sigma_R/\rho \leq 0.$$

In view of (7.29) we have

$$\rho\, \partial_{\mathbf{E}} \hat{\psi}(\mathbf{E}(t), \mathbf{E}^t_r) = \mathbf{G}_0 \mathbf{E}(t) + \int_0^\infty \mathbf{G}'(u)\mathbf{E}(t-u)du$$

whence

$$\rho\, \hat{\psi}(\mathbf{E}(t), \mathbf{E}^t_r) = \tfrac{1}{2}\mathbf{E}(t) \cdot \mathbf{G}_0 \mathbf{E}(t) + \mathbf{E}(t) \cdot \int_0^\infty \mathbf{G}'(u)\mathbf{E}(t-u)du + \rho\tilde{\psi}(\mathbf{E}^t_r). \quad (7.46)$$

In addition to $\rho\, \partial_{\mathbf{E}} \hat{\psi} = \mathfrak{T}$, the functional $\hat{\psi}$ has to satisfy $d\hat{\psi} \leq 0$ and to be such that, within the histories ending with given values of **E**, the functional corresponding to constant values of **E** for all times has the least value. This minimum property is consistent with the general statement for functionals of $\mathbf{F}^t$, θ^t. Yet we can see it very briefly. By (7.45) it follows that ψ is decreasing while **E** is constant. Let $\mathbf{E}^{t_0+d}_{t_0}$ be the constant continuation of $\mathbf{E}^{t_0}$ up to $t_0 + d$. Hence

$$\int_{t_0}^{t_0+d} \mathfrak{T}(\mathbf{E}^t_{t_0}) \cdot \dot{\mathbf{E}}(t)dt = 0, \qquad \psi(t_0 + d) \leq \psi(t_0).$$

If $d \to \infty$ then we have

$$\hat{\psi}(\mathbf{E}(t_0), \mathbf{E}^{t_0\dagger}) \leq \hat{\psi}(\mathbf{E}(t_0), \mathbf{E}^{t_0}_r).$$

Even if we assume a normalization condition at the null constant history,

$$\hat{\psi}(\mathbf{0}, \mathbf{0}^{t\dagger}) = 0, \quad (7.47)$$

the free energy functional $\hat{\psi}$ is non-unique[10]; this fact was first shown by Day [115]. For definiteness, after introducing the concept of a minimal state, we now determine some well-known expressions of this functional.

[10] See, e.g. [158], Sect. 3.5.

Minimal states

We consider the concept of a minimal state in the context of linear viscoelasticity. It might happen that two different initial past histories, $\mathbf{E}_1^{t_0}$ and $\mathbf{E}_2^{t_0}$, lead to the same strain and stress for $t \geq t_0$. From the viewpoint of the dynamics, such two different initial past histories are in a fact indistinguishable. This observation suggests that, rather than the past history $\mathbf{E}^t$, one should employ an alternative variable to describe the initial state of the system, satisfying the following natural minimum property: *two different initial states produce different evolutions for $t \geq 0$*. From the philosophical side, this means that the knowledge of $\mathbf{E}(t)$ for all $t \geq t_0$ determines in a unique way the initial state at t_0 of the problem, the only object that really influences the future dynamics.

Definition. *According to* [124], *two strain histories $\mathbf{E}_1^t$ and $\mathbf{E}_2^t$ are equivalent, or in the same minimal state, if and only if*

$$\mathbf{E}_1(t+s) = \mathbf{E}_2(t+s), \quad s \in \mathbb{R}^+ \tag{7.48}$$

provided that

$$\mathbf{J}^t(s, \mathbf{E}_1^t) = \mathbf{J}^t(s, \mathbf{E}_2^t), \quad s \in \mathbb{R}^+$$

where $\mathbf{J}^t$ is the linear functional

$$\mathbf{J}^t(s, \mathbf{E}^t) = \int_0^\infty \mathbf{G}'(s+u)\mathbf{E}^t(u)du. \tag{7.49}$$

A minimal state is therefore given by an equivalence class according to the equivalence relation so defined. The linearity of the functional $\mathbf{J}^t$ means that the requirement of the equivalence of $\mathbf{E}_1^t$ and $\mathbf{E}_2^t$ is the same as that $\mathbf{E}_1^t - \mathbf{E}_2^t$ be equivalent to the zero history. Thus, if the minimal state, including the zero history, is singleton (non-singleton) then all minimal states are singleton (non-singleton).

If we introduce the extra condition $\mathbf{E}_1(t) = \mathbf{E}_2(t)$ in the definition of a minimal state, then (7.48) follows from

$$\dot{\mathbf{E}}_1(t+s) = \dot{\mathbf{E}}_2(t+s), \quad s \in \mathbb{R}^+,$$

and, according to [129], we can introduce the minimal state functional as

$$\mathbf{I}^t(s, \mathbf{E}_d^t) = -\int_0^\infty \mathbf{G}'(s+u)\mathbf{E}_d^t(u)\,du = \int_0^\infty \mathbf{G}'(s+u)\big[\mathbf{E}^t(u) - \mathbf{E}(t)\big]\,du, \qquad s \geq 0.$$

For any given history $\mathbf{E}^t$, the function $\mathbf{I}^t(s, \cdot)$ fulfils the partial differential equation

$$\partial_t \mathbf{I}^t(s) = \partial_s \mathbf{I}^t(s) + \check{\mathbf{G}}(s)\dot{\mathbf{E}}(t), \qquad t, s \in \mathbb{R}^+.$$

Note that the state variables $\mathbf{I}^t$ and $\mathbf{J}^t$ are related by

$$\mathbf{I}^t(s, \mathbf{E}_d^t) = \mathbf{J}^t(s, \mathbf{E}^t) + \check{\mathbf{G}}(s)\mathbf{E}(t), \qquad u \geq 0,$$

where $\check{\mathbf{G}}(s) = \mathbf{G}(s) - \mathbf{G}_\infty$. Therefore, the stress–strain relation (7.30) takes the compact forms

$$\mathbf{T}(t) = \mathfrak{T}_*(\mathbf{E}(t), \mathbf{E}_d^t) = \mathbf{G}_\infty\mathbf{E}(t) + \mathbf{I}^t(0, \mathbf{E}_d^t) = \mathbf{G}_0\mathbf{E}(t) + \mathbf{J}^t(0, \mathbf{E}^t).$$

In order to define a free energy as a state functional, rather than $\mathbf{I}^t$, it seems more convenient to consider as a state the new variable [161]

$$\boldsymbol{\xi}^t(s, \mathbf{E}_d^t) = -\partial_s \mathbf{I}^t(s, \mathbf{E}_d^t) = \int_0^\infty \mathbf{G}''(s+u)\mathbf{E}_d^t(u)\, du, \qquad s \geq 0, \tag{7.50}$$

which, in turn, fulfils the linear transport equation

$$\partial_t \boldsymbol{\xi}^t(s) = \partial_s \boldsymbol{\xi}^t(s) - \mathbf{G}'(s)\dot{\mathbf{E}}(t), \qquad t, s \in \mathbb{R}^+. \tag{7.51}$$

We mention some further references on minimal states. In [161] the history and the state formulations are compared, showing some advantages of states. In [162, 199] the concept of a minimal state is applied to obtain a general representation of free energies in the frequency domain, including explicit expressions for the minimum and maximum free energies.

The Graffi-Volterra free energy

First we assume $\mathbf{G}' \in W^{1,1}(\mathbb{R}^+; \mathrm{Lin(Sym)})$ and

$$\mathbf{G}'(u) = \mathbf{G}'^T(u), \qquad \mathbf{G}'(u) \leq 0, \qquad \mathbf{G}''(u) \geq 0, \qquad u \in \mathbb{R}^+.$$

Letting

$$\rho\tilde{\psi}^G(\mathbf{E}_r^t) = -\tfrac{1}{2}\int_0^\infty \mathbf{E}(t-u) \cdot \mathbf{G}'(u)\mathbf{E}(t-u)du,$$

observing that

$$\mathbf{G}_\infty = \mathbf{G}_0 + \int_0^\infty \mathbf{G}'(u)du,$$

and using the symmetry of $\mathbf{G}'$, from (7.46) we obtain

$$\rho\,\hat{\psi}^G(\mathbf{E}(t),\, \mathbf{E}_d^t) = \tfrac{1}{2}\mathbf{E}(t) \cdot \mathbf{G}_\infty\mathbf{E}(t) - \tfrac{1}{2}\int_0^\infty \mathbf{E}_d^t(u) \cdot \mathbf{G}'(u)\mathbf{E}_d^t(u)\, du.$$

where $\mathbf{E}_d^t(u) = \mathbf{E}(t) - \mathbf{E}(t-u)$. Since $\mathbf{G}' \leq \mathbf{0}$, this functional is well defined in $\mathcal{H}_*^G \subset \mathcal{H}_*$, the Hilbert space endowed with the inner product and the associated norm in the form

$$(\mathbf{E}_1^t, \mathbf{E}_2^t)_{\mathcal{H}_*^G} := \mathbf{E}_1(t) \cdot \mathbf{E}_2(t) - \int_0^\infty \mathbf{E}_{1d}^t(s) \cdot \mathbf{G}'(s)\mathbf{E}_{2d}^t(s)\, ds,$$

$$\|\mathbf{E}^t\|_{\mathcal{H}_*^G}^2 := |\mathbf{E}(t)|^2 - \int_0^\infty \mathbf{E}_d^t(s) \cdot \mathbf{G}'(s)\mathbf{E}_d^t(s)\, ds.$$

The functional $\hat{\psi}^G$ represents a quadratic free energy. Indeed, it is apparent that $\hat{\psi}^G$ satisfies the minimum property at constant histories; for any history $\mathbf{E}^{t_0}$ we have

$$\rho\big[\hat{\psi}^G(\mathbf{E}^{t_0}) - \hat{\psi}^G(\mathbf{E}^{t_0\dagger})\big] = -\tfrac{1}{2}\int_0^\infty \mathbf{E}_d^{t_0}(u) \cdot \mathbf{G}'(u)\mathbf{E}_d^{t_0}(u)\, du \geq 0,$$

and the conclusion follows. To prove that $d\hat{\psi}^G(\mathbf{E}(t),\ \mathbf{E}_r^t \,|\, \dot{\mathbf{E}}_r^t) \leq 0$ we observe

$$\rho\, d\hat{\psi}^G(\mathbf{E}(t),\ \mathbf{E}_r^t \,|\, \dot{\mathbf{E}}_r^t) = \int_0^\infty \dot{\mathbf{E}}(t-u) \cdot \mathbf{G}'(u)[\mathbf{E}(t) - \mathbf{E}(t-u)]du.$$

Since $\dot{\mathbf{E}}(t-u) = -\partial_u \mathbf{E}(t-u)$, upon an integration by parts we obtain

$$\begin{aligned}\rho\, d\hat{\psi}^G &= \int_0^\infty \partial_u[\mathbf{E}(t) - \mathbf{E}(t-u)] \cdot \mathbf{G}'(u)[\mathbf{E}(t) - \mathbf{E}(t-u)]du \\ &= \tfrac{1}{2}[\mathbf{E}(t) - \mathbf{E}(-\infty)] \cdot \mathcal{G}_\infty[\mathbf{E}(t) - \mathbf{E}(-\infty)] - \tfrac{1}{2}\int_0^\infty \mathbf{E}_d^t(u) \cdot \mathbf{G}''(u)\mathbf{E}_d^t(u)\, du.\end{aligned}$$

The vanishing of $\mathcal{G}_\infty = \lim_{u\to\infty} \mathbf{G}'(u)$ and the positive semi-definiteness of $\mathbf{G}''$ imply that $d\hat{\psi}^G$ is a functional of the difference history $\mathbf{E}_d^t$ only, and $d\hat{\psi}^G \leq 0$. Moreover, if we assume the stronger condition[11]

$$\mathbf{G}''(s) + \lambda\mathbf{G}'(s) \geq 0, \quad s \in \mathbb{R}^+, \tag{7.52}$$

then it follows that

$$d\hat{\psi}^G \leq \tfrac{1}{2}\lambda\int_0^\infty \mathbf{E}_d^t(u) \cdot \mathbf{G}'(u)\mathbf{E}_d^t(u)\, du = -\lambda\big[\hat{\psi}^G(\mathbf{E}(t),\ \mathbf{E}_d^t) - \tfrac{1}{2}\mathbf{E}(t) \cdot \mathbf{G}_\infty\mathbf{E}(t)\big]. \tag{7.53}$$

The free energy functional $\hat{\psi}^G$ traces back to Volterra [436–438]. The thermodynamic admissibility was proved by Graffi [203–205]. This free energy is not in general a functional of the minimal state unless all minimal states are singleton [124]. It is however a positive definite functional on $\mathcal{H}_*^G$. By virtue of this property and of the inequality (7.53), energy functionals related to $\hat{\psi}^G$ have often been used in the literature to obtain interesting stability results in linear viscoelasticity [112, 198].

---

[11] This condition implies the exponential decay of the relaxation function [198].

The work function

Let $\mathbf{G}' \in W^{1,1}(\mathbb{R}^+; \mathrm{Lin}(\mathrm{Sym}))$ satisfy $\mathbf{G}'(u) = \mathbf{G}'^T(u)$, $u \in \mathbb{R}^+$, and the thermodynamic condition (7.36). Assuming that $u = |s_1 - s_2|$, $s_1, s_2 \in \mathbb{R}^+$, we have

$$\partial_{s_j} \mathbf{G}(|s_1 - s_2|) = (-1)^{j+1} \mathrm{sgn}\,(s_1 - s_2) \mathbf{G}'(|s_1 - s_2|). \tag{7.54}$$

Define

$$\mathbf{G}_{12}(|s_1 - s_2|) = \partial_{s_1} \partial_{s_2} \mathbf{G}(|s_1 - s_2|).$$

It follows that $\mathbf{G}_{12}$ has singular delta function behaviour at $s_1 = s_2$. Indeed,

$$\mathbf{G}_{12}(|s_1 - s_2|) = -2\delta(s_1 - s_2)\mathbf{G}'(|s_1 - s_2|) - \mathbf{G}''(|s_1 - s_2|) \tag{7.55}$$

and therefore $\mathbf{G}_{12}$ is not bounded. Nevertheless we can define

$$\rho\tilde{\psi}^W(\mathbf{E}_r^t) = \frac{1}{2} \int_0^\infty \int_0^\infty \mathbf{E}(t - s_1) \cdot \mathbf{G}_{12}(|s_1 - s_2|)\mathbf{E}(t - s_2)\, ds_1 d_2.$$

Lemma 7.2 *If* $\mathbf{E}^t \in L^2(\mathbb{R}^+; \mathrm{Sym})$ *the thermodynamic condition* (7.36) *implies* $\rho\tilde{\psi}^W \geq 0$.

Proof By (7.55) and the symmetry of $\mathbf{G}''$ we obtain

$$\begin{aligned}
\rho\tilde{\psi}^W(\mathbf{E}_r^t) &= -\int_0^\infty \int_0^\infty \mathbf{E}^t(s_1) \cdot \delta(s_1 - s_2)\mathbf{G}'(|s_1 - s_2|)\mathbf{E}^t(s_2)\, ds_1 ds_2 \\
&\quad - \frac{1}{2} \int_0^\infty \int_0^\infty \mathbf{E}^t(s_1) \cdot \mathbf{G}''(|s_1 - s_2|)\mathbf{E}^t(s_2)\, ds_1 ds_2 \\
&= -\int_0^\infty \mathbf{E}^t(s) \cdot \mathbf{G}'_0 \mathbf{E}^t(s)\, ds - \int_0^\infty \mathbf{E}^t(s_1) \cdot \int_0^{s_1} \mathbf{G}''(s_1 - s_2)\mathbf{E}^t(s_2)\, ds_1\, ds_2.
\end{aligned}$$

Upon extending the definition of $\mathbf{G}'$ and $\mathbf{E}^t$ to $\mathbb{R}^-$ by letting,

$$\mathbf{G}'(s) = \mathbf{E}^t(s) = \mathbf{0}, \qquad s < 0,$$

we can write

$$\rho\tilde{\psi}^W(\mathbf{E}_r^t) = -\int_{-\infty}^{\infty} \mathbf{E}^t(s) \cdot [\mathbf{G}'' + \mathbf{G}'_0\,\delta] *_F \mathbf{E}^t(s)\, ds,$$

where $*_F$ denotes the Fourier convolution,

$$[\mathbf{G}'' + \mathbf{G}'_0\,\delta] *_F \mathbf{E}^t(s) := \int_{-\infty}^{\infty} [\mathbf{G}''(s - \tau) + \mathbf{G}'_0\,\delta(s - \tau)]\mathbf{E}^t(\tau)\, d\tau.$$

Hence, since $\mathbf{G}'' + \mathbf{G}'_0\,\delta$ represents the distributional derivative of the extension of $\mathbf{G}'$ then by applying the Parseval–Plancherel theorem we have[12]

$$\rho\tilde{\psi}^W = -\frac{1}{2\pi}\int_{-\infty}^{\infty}[\mathbf{E}^t_F(\omega)]^*\cdot[\mathbf{G}''+\mathbf{G}'_0\,\delta]_F(\omega)\mathbf{E}^t_F(\omega)\,d\omega = -\frac{1}{2\pi}\int_{-\infty}^{\infty}[\mathbf{E}^t_F(\omega)]^*\cdot i\omega\mathbf{G}'_F(\omega)\mathbf{E}^t_F(\omega)\,d\omega,$$

Now, since $\mathbf{G}'$ and $\mathbf{E}^t$ vanish on $(-\infty, 0)$, their cosine transforms $\mathbf{G}'_c$, $\mathbf{E}^t_c$ are even functions while their sine transforms $\mathbf{G}'_s$, $\mathbf{E}^t_s$ are odd functions. Hence, (7.36) implies

$$\rho\tilde{\psi}^W = -\frac{1}{2\pi}\int_{-\infty}^{\infty}\omega\big[\mathbf{E}^t_c(\omega)\cdot\mathbf{G}'_s(\omega)\mathbf{E}^t_c(\omega)+\mathbf{E}^t_s(\omega)\cdot\mathbf{G}'_s(\omega)\mathbf{E}^t_s(\omega)\big]d\omega \geq 0. \qquad \square$$

We go back to the classical functional $\hat{\psi}$ in (7.46) and observe that since

$$\int_0^{\infty}\int_0^{\infty}\mathbf{G}_{12}(|s_1-s_2|)\,ds_1ds_2 = -\int_0^{\infty}\partial_{s_1}\mathbf{G}(|s_1-s_2|)\,ds_1 = \mathbf{G}_0-\mathbf{G}_\infty,$$

then it follows

$$\rho\,\hat{\psi}^W(\mathbf{E}(t),\,\mathbf{E}^t_d) = \frac{1}{2}\mathbf{E}(t)\cdot\mathbf{G}_\infty\mathbf{E}(t)+\frac{1}{2}\int_0^{\infty}\int_0^{\infty}\mathbf{E}^t_d(s_1)\cdot\mathbf{G}_{12}(|s_1-s_2|)\mathbf{E}^t_d(s_2)\,ds_1ds_2,$$

where the symmetry of $\mathbf{G}_{12}$ has been used. Apparently the functional $\hat{\psi}^W$ is quadratic in $\mathbf{E}$ and $\mathbf{E}^t_d$. To compute $d\hat{\psi}^W$ we observe

$$\begin{aligned}\rho\,d\hat{\psi}^W &= \int_0^{\infty}\int_0^{\infty}\dot{\mathbf{E}}(t-s_1)\cdot\mathbf{G}_{12}(|s_1-s_2|)[\mathbf{E}(t)-\mathbf{E}(t-s_2)]ds_1ds_2\\ &= \int_0^{\infty}\mathbf{G}'(s)\dot{\mathbf{E}}(t-s)\,ds\cdot\mathbf{E}(t)+\int_0^{\infty}\int_0^{\infty}\dot{\mathbf{E}}^t(s_1)\cdot\mathbf{G}_{12}(|s_1-s_2|)\mathbf{E}^t(s_2)\,ds_1ds_2.\end{aligned}$$

Upon changing the order in the double integral, an integration by parts yields

$$\begin{aligned}\rho\,d\hat{\psi}^W &= \int_0^{\infty}\mathbf{G}'(s)\dot{\mathbf{E}}(t-s)\,ds\cdot\mathbf{E}(t)-\int_0^{\infty}\dot{\mathbf{E}}^t(s_1)\cdot\mathbf{G}'(s_1)\mathbf{E}(t)ds_1\\ &\quad+\int_0^{\infty}\dot{\mathbf{E}}^t(s_1)\cdot\int_0^{\infty}\partial_{s_1}\mathbf{G}(|s_1-s_2|)\dot{\mathbf{E}}^t(s_2)\,ds_2\,ds_1.\end{aligned}$$

The first two integrals cancel each other while the third integral vanishes as a consequence of (7.54). Hence, $d\hat{\psi}^W = 0$ so that $\rho\dot{\psi}^W = \mathbf{T}\cdot\dot{\mathbf{E}}$. This is why $\psi^W(t) = \hat{\psi}^W(\mathbf{E}(t),\,\mathbf{E}^t_d)$ is named *work function*. The minimum property at constant histories can be proven as a consequence of the thermodynamic restriction $\mathbf{G}'_s \leq \mathbf{0}$. First we observe that

$$\rho\hat{\psi}^W(\mathbf{E}^{t_0\dagger}) = \frac{1}{2}\mathbf{E}(t_0)\cdot\mathbf{G}_\infty\mathbf{E}(t_0).$$

[12] As elsewhere, for any complex-valued quantity E, E^* denotes the complex conjugate.

Then applying the previous Lemma with $\mathbf{E}^t$ replaced by $\mathbf{E}_d^{t_0}$ we obtain

$$\rho[\hat{\psi}^W(\mathbf{E}^{t_0}) - \hat{\psi}^W(\mathbf{E}^{t_0\dagger})] = \frac{1}{2}\int_0^\infty\int_0^\infty \mathbf{E}_d^{t_0}(s_1)\cdot\mathbf{G}_{12}(|s_1-s_2|)\mathbf{E}_d^{t_0}(s_2)\,ds_1ds_2 \geq 0.$$

The work function ψ^W is not in general a functional of the minimal state [123]. It is however a positive definite functional on the Hilbert space $\mathcal{H}_*^W$ endowed with the inner product and the associated norm in the form

$$(\mathbf{E}_1^t, \mathbf{E}_2^t)_{\mathcal{H}_*^W} := \mathbf{E}_1(t)\cdot\mathbf{E}_2(t) + \int_0^\infty\int_0^\infty \mathbf{E}_{1d}^t(s_1)\cdot\mathbf{G}_{12}(|s_1-s_2|)\mathbf{E}_{2d}^t(s_2)\,ds_1ds_2,$$

$$\|\mathbf{E}^t\|_{\mathcal{H}_*^W}^2 := \mathbf{E}(t)\cdot\mathbf{E}(t) + \int_0^\infty\int_0^\infty \mathbf{E}_d^t((s_1)\cdot\mathbf{G}_{12}(|s_1-s_2|)\mathbf{E}_d^t(s_2)\,ds_1ds_2.$$

We finally note that ψ^W is sometimes referred to as *maximum free energy* [159, 162] since it is the maximum free energy for singleton materials. In particular, it can be shown that

$$\psi^G(t) \leq \psi^W(t).$$

Free energy as a functional of the minimal state

We recall that the Graffi-Volterra free energy holds under the assumptions that $\mathbf{G}' \in W^{1,1}(\mathbb{R}^+;\text{Lin(Sym)})$ and

$$\mathbf{G}'(u) = \mathbf{G}'^T(u), \qquad \mathbf{G}'(u) < 0, \qquad \mathbf{G}''(u) \geq 0, \qquad u \in \mathbb{R}^+.$$

With these conditions we now consider the functional

$$\rho\hat{\psi}^F(\mathbf{E}(t),\, \mathbf{E}_d^t) = \frac{1}{2}\mathbf{E}(t)\cdot\mathbf{G}_\infty\mathbf{E}(t) - \frac{1}{2}\int_0^\infty \boldsymbol{\xi}^t(s,\mathbf{E}_d^t)\cdot[\mathbf{G}']^{-1}(s)\boldsymbol{\xi}^t(s,\mathbf{E}_d^t)\,ds,$$

which involves the minimal state representation $\boldsymbol{\xi}^t(s,\mathbf{E}_d^t)$ as defined in (7.50). Since $\mathbf{G}' < \mathbf{0}$, this functional is well defined in $\mathcal{H}_*^F \subset \mathcal{H}_*$, the Hilbert space endowed with the inner product and the associated norm in the form

$$(\mathbf{E}_1^t, \mathbf{E}_2^t)_{\mathcal{H}_*^F} := \mathbf{E}_1(t)\cdot\mathbf{E}_2(t) - \int_0^\infty \boldsymbol{\xi}^t(s,\mathbf{E}_{1d}^t)\cdot[\mathbf{G}']^{-1}(s)\boldsymbol{\xi}^t(s,\mathbf{E}_{2d}^t)\,ds,$$

$$\|\mathbf{E}^t\|_{\mathcal{H}_*^F}^2 := |\mathbf{E}(t)|^2 - \int_0^\infty \boldsymbol{\xi}^t(s,\mathbf{E}_d^t)\cdot[\mathbf{G}']^{-1}(s)\boldsymbol{\xi}^t(\tau,\mathbf{E}_d^t)\,ds.$$

To prove that $\hat{\psi}^F$ is a (quadratic) free energy functional we first observe that

$$\boldsymbol{\xi}^t(s,\mathbf{E}_d^t) = -\mathbf{G}'(s)\mathbf{E}(t) - \int_0^\infty \mathbf{G}''(s+u)\mathbf{E}^t(u)\,du = -\mathbf{G}'(s)\mathbf{E}(t) - \partial_s \mathbf{J}^t(s,\mathbf{E}^t),$$

and then we can write

$$\rho\hat{\psi}^F = \tfrac{1}{2}\mathbf{E}(t)\cdot\mathbf{G}_0\mathbf{E}(t) + \mathbf{E}(t)\cdot\mathbf{J}^t(0,\mathbf{E}^t) - \tfrac{1}{2}\int_0^\infty \partial_s\mathbf{J}^t(s,\mathbf{E}^t)\cdot[\mathbf{G}']^{-1}(s)\partial_s\mathbf{J}^t(s,\mathbf{E}^t)\,ds.$$

Hence it follows

$$\rho\partial_{\mathbf{E}}\hat{\psi}^F(\mathbf{E}(t),\,\mathbf{E}_d^t) = \mathbf{G}_0\mathbf{E}(t) + \mathbf{J}^t(0,\mathbf{E}^t) = \mathbf{G}_\infty\mathbf{E}(t) + \mathbf{I}^t(0,\mathbf{E}_d^t) = \mathfrak{T}_*(\mathbf{E}(t),\mathbf{E}_d^t).$$

Since $\mathbf{G}' < 0$ then it is apparent that $\hat{\psi}^F$ enjoys the minimum property at constant histories. To evaluate $d\hat{\psi}^F$ we first observe

$$\rho\, d\hat{\psi}^F(\mathbf{E}(t),\,\mathbf{E}_r^t\,|\,\dot{\mathbf{E}}_r^t) := \rho\dot{\psi}^F(t) - \rho\partial_{\mathbf{E}}\hat{\psi}(\mathbf{E}(t),\,\mathbf{E}_r^t)\cdot\dot{\mathbf{E}}(t) = \rho\dot{\psi}^F(t) - \mathbf{T}(t)\cdot\dot{\mathbf{E}}(t).$$

Then (7.51) allows us to find

$$\begin{aligned}\rho\, d\hat{\psi}^F &= -\mathbf{I}^t(0)\cdot\dot{\mathbf{E}}(t) - \int_0^\infty \boldsymbol{\xi}^t(s)\cdot[\mathbf{G}']^{-1}(s)\partial_s\boldsymbol{\xi}^t(s)\,ds - \int_0^\infty \partial_s\mathbf{I}^t(s)\,ds\cdot\dot{\mathbf{E}}(t)\\ &= -\tfrac{1}{2}\int_0^\infty \boldsymbol{\xi}^t(s)\cdot\partial_s[\mathbf{G}']^{-1}(s)\boldsymbol{\xi}^t(s)\,ds = \int_0^\infty [\mathbf{G}']^{-1}\boldsymbol{\xi}^t(s)\cdot\mathbf{G}''(s)[\mathbf{G}']^{-1}\boldsymbol{\xi}^t(s)\,ds \ge 0.\end{aligned}$$

The property $d\hat{\psi}^F \ge 0$ completes the set of thermodynamic conditions for $\hat{\psi}^F$ to be a free energy functional. Incidentally, it can be shown that[13]

$$\psi^F(t) \le \psi^G(t).$$

7.4.4 Viscoelastic Solids with Unbounded Relaxation Functions

There are models in linear viscoelasticity where the kernel is unbounded; this is particularly the case of kernels in a power-law form [448]. Advances in mechanical testing hardware have revealed that many materials possessing a complex microstructure exhibit a characteristic power-law signature in their creep and relaxation. Examples of such materials include biological materials, gels, polymers, concrete, asphalt, ice (see [57] and refs therein). Constitutive properties are assessed by experimental data, possibly involving the stress due to time-harmonic strain. Next questions arise about the selection of the relaxation function on $[0,\infty)$ [90].

---

[13] See, e.g. [161].

Within this framework, linear viscoelastic solids are modelled by letting the kernel $\mathcal{G} = \mathbf{G}'$ belong to $L^1(\mathbb{R}^+; \mathrm{Lin(Sym)})$ in time. This condition implies the boundedness of the instantaneous elastic modulus

$$\mathbf{G}_0 := \lim_{s \to 0^+} \mathbf{G}(s) = \mathbf{G}_\infty + \int_0^\infty \mathcal{G}(s)\, ds$$

and agrees with the fading memory principle (at least in its weak form) as stated by Day [115]. In addition, it is usually understood that $\mathcal{G}_0 := \lim_{s\to 0^+} \mathcal{G}(0)$ is a bounded tensor, too.

In the literature various assumptions have been considered for the singularity of the relaxation function. For instance, some papers [128, 161, 187, 198] showed that a singular kernel can yield smoothing effects for the solution to the evolution problem, even if the gain in regularity cannot be derived without specifying the kind of singularity. Here, we consider a viscoelastic solid whose memory kernel $\mathcal{G} = \mathbf{G}'$ may exhibit an initial singularity [80, 197]. Here, we consider a viscoelastic solid whose memory kernel $\mathcal{G} = \mathbf{G}'$ may exhibit an initial singularity. If it is integrable then $\mathbf{G}_0$ is bounded, but $\mathcal{G}_0$ does not necessarily have to be. Otherwise, both are unbounded. In general, it is convenient to introduce

$$\check{\mathbf{G}}(s) = \mathbf{G}(s) - \mathbf{G}_\infty$$

and to express the stress–strain constitutive relation as in (7.32), namely

$$\mathbf{T}(t) := \mathfrak{T}(\mathbf{E}^t) = \mathbf{G}_\infty \mathbf{E}(t) + \int_0^\infty \check{\mathbf{G}}(s)\dot{\mathbf{E}}(t-s)ds, \tag{7.56}$$

together with the quite mild constitutive assumptions

$$\check{\mathbf{G}} \in L^1(\mathbb{R}^+; \mathrm{Lin(Sym)}), \qquad \mathbf{G}_\infty \text{ bounded.} \tag{7.57}$$

Hence we let both $\mathbf{G}_0$ and $\mathcal{G}_0$ be unbounded but require that the possible singularity of $\check{\mathbf{G}}$ at $s = 0$ be integrable. It is worth mentioning that viscoelastic solids are modelled by nonlinear functionals of $\dot{\mathbf{E}}^t$ in [361]; the linear term in [361] amounts to taking $\mathbf{G}_\infty = \mathbf{0}$ in (7.56).

For simplicity we restrict attention to the same set of histories $\mathcal{H}$ considered in Sect. 7.4.1. If the second law for isothermal processes is taken in the form (7.34), that is the inequality

$$\int_0^d \mathbf{T}(t) \cdot \dot{\mathbf{E}}(t)\, dt \geq 0$$

is assumed to hold for any non-constant cycle, then we obtain the following result [197].

Theorem 7.3 *The constitutive stress–strain relation* (7.56), *subject to conditions* (7.57), *satisfies the second law of thermodynamics only if*

$$\mathbf{G}_\infty^T = \mathbf{G}_\infty, \qquad \check{\mathbf{G}}_c(\omega) := \int_0^\infty \check{\mathbf{G}}(s)\cos\omega s\, ds > 0, \quad \forall \omega > 0. \tag{7.58}$$

Proof Let $\mathbf{E}_1, \mathbf{E}_2 \in \mathrm{Sym}$ and consider the periodic tensor function with period $d = 2\pi/\omega$

$$\tilde{\mathbf{E}}(t) = \mathbf{E}_1 \cos\omega t + \mathbf{E}_2 \sin\omega t, \qquad \omega > 0.$$

For any finite value of ω we have $\tilde{\mathbf{E}}^t \in \mathcal{H}$ and $\tilde{\mathbf{E}}^t = \tilde{\mathbf{E}}^{t+d}$, $d = 2\pi/\omega$. Substitution in (7.34) and integration with respect to t on $(0, d)$ yields

$$\mathbf{E}_2 \cdot (\mathbf{G}_\infty - \mathbf{G}_\infty^T)\mathbf{E}_1 + \omega\big[\mathbf{E}_1 \cdot \check{\mathbf{G}}_c(\omega)\mathbf{E}_1 + \mathbf{E}_2 \cdot \check{\mathbf{G}}_c(\omega)\mathbf{E}_2 + \mathbf{E}_2 \cdot \big(\check{\mathbf{G}}_s(\omega) - \check{\mathbf{G}}_s^T(\omega)\big)\mathbf{E}_1\big] > 0 \tag{7.59}$$

for any $\omega > 0$. The limit case $\omega \to 0$ and the arbitrariness of $\mathbf{E}_1$, $\mathbf{E}_2$ imply the symmetry of $\mathbf{G}_\infty$. Accordingly,

$$\omega\big[\mathbf{E}_1 \cdot \check{\mathbf{G}}_c(\omega)\mathbf{E}_1 + \mathbf{E}_2 \cdot \check{\mathbf{G}}_c(\omega)\mathbf{E}_2 + \mathbf{E}_2 \cdot \big(\check{\mathbf{G}}_s(\omega) - \check{\mathbf{G}}_s^T(\omega)\big)\mathbf{E}_1\big] > 0, \quad \omega > 0,$$

whence, letting $\mathbf{E}_1 = \mathbf{E}_2$ and exploiting the skew character of $\check{\mathbf{G}}_s(\omega) - \check{\mathbf{G}}_s^T(\omega)$, we conclude that $\check{\mathbf{G}}_c(\omega)$ must be positive definite for any nonzero, finite ω. □

The proof shows that the conditions (7.58) are necessary for the validity of the second law on cycles. Really, even in the particular case of time-harmonic histories, the second law leads to the inequality (7.59) which is stronger than (7.58) and involves also the half-range Fourier sine transform of $\check{\mathbf{G}}$. This difference disappears if we assume $\check{\mathbf{G}}(u) = \check{\mathbf{G}}^T(u)$, $u \in [0, \infty)$. In general, however, $\check{\mathbf{G}}$ need not be symmetric, and therefore (7.59) proves to be sufficient for the validity of the second law.

Applying the same procedure as in the previous section, a more general result can be established.

Theorem 7.4 *If* $\check{\mathbf{G}}$ *satisfies* (7.59), *and hence* (7.58), *then the second-law inequality* (7.34) *holds for every periodic history, with period* d.

Compatibility between the constitutive relation (7.56) with $\check{\mathbf{G}} \in L^1(\mathbb{R}^+)$, and that related to the standard relation (7.29) with $\mathbf{G}' \in L^1(\mathbb{R}^+)$ requires that the condition (7.59) be equivalent to (7.38), obviously by allowing $\check{\mathbf{G}} \in W^{1,1}(\mathbb{R}^+)$. This is so by virtue of the following properties.[14]

Lemma 7.3 *Let* $\check{\mathbf{G}} \in W^{1,1}(\mathbb{R}^+; \mathrm{Lin}(\mathrm{Sym}))$. *Then for all* $\omega \in \mathbb{R}$

$$\omega\check{\mathbf{G}}_c(\omega) = -\mathbf{G}_s'(\omega), \qquad \omega\check{\mathbf{G}}_s(\omega) = \mathbf{G}_0 - \mathbf{G}_\infty + \mathbf{G}_c'(\omega). \tag{7.60}$$

[14] See, for instance, [197].

In order to study existence, uniqueness and asymptotic behaviour of the solution to the dynamic problem, in [187] the Laplace transform method is applied when $\mathbf{G}$ has a singularity at $t = 0$ and satisfies (7.57) together with thermodynamic restrictions (7.58).

Free energy with an unbounded memory kernel

Assume that $\check{\mathbf{G}} \in L^1(\mathbb{R}^+)$ satisfies $\check{\mathbf{G}}^T(u) = \check{\mathbf{G}}(u)$, $u \in \mathbb{R}^+$, and the thermodynamic condition (7.58). Let $W(\mathbf{E}^t)$ be the work performed by the stress tensor (7.56) in connection with the history $\mathbf{E}^t$, namely

$$W(\mathbf{E}^t) := \int_{-\infty}^{t} \mathfrak{T}(\mathbf{E}^\tau) \cdot \dot{\mathbf{E}}(\tau)\, d\tau = \int_{-\infty}^{t} \big[\mathbf{G}_\infty \mathbf{E}(\tau) + \int_0^\infty \check{\mathbf{G}}(s)\dot{\mathbf{E}}(\tau - s) ds\big] \cdot \dot{\mathbf{E}}(\tau)\, d\tau.$$

Moreover, let $\lim_{\tau \to -\infty} \mathbf{E}(\tau) = \mathbf{0}$ and extend the definition of $\check{\mathbf{G}}$ to $\mathbb{R}^-$ by letting

$$\check{\mathbf{G}}(s) = \mathbf{0}, \qquad s < 0.$$

Thanks to the symmetry of $\mathbf{G}_\infty$ and $\check{\mathbf{G}}$ we can write [8]

$$\begin{aligned} W(\mathbf{E}^t) &= \int_{-\infty}^{t} \big[\mathbf{G}_\infty \mathbf{E}(\tau) + \int_{-\infty}^{\tau} \check{\mathbf{G}}(\tau - u)\dot{\mathbf{E}}(u) du\big] \cdot \dot{\mathbf{E}}(\tau)\, d\tau \\ &= \tfrac{1}{2}\, \mathbf{E}(t) \cdot \mathbf{G}_\infty \mathbf{E}(t) + \tfrac{1}{2} \int_{-\infty}^{t} \int_{-\infty}^{t} \dot{\mathbf{E}}(\tau) \cdot \check{\mathbf{G}}(|\tau - u|)\dot{\mathbf{E}}(u)\, du\, d\tau \\ &= \tfrac{1}{2}\, \mathbf{E}(t) \cdot \mathbf{G}_\infty \mathbf{E}(t) + \tfrac{1}{2} \int_0^\infty \int_0^\infty \dot{\mathbf{E}}(t - s_1) \cdot \check{\mathbf{G}}(|s_1 - s_2|)\dot{\mathbf{E}}(t - s_2)\, ds_1\, ds_2. \end{aligned} \tag{7.61}$$

This suggests that we consider the work function as an energy functional, namely

$$\rho\hat{\psi}^w(\mathbf{E}(t), \mathbf{E}^t_r) = \tfrac{1}{2}\, \mathbf{E}(t) \cdot \mathbf{G}_\infty \mathbf{E}(t) + \tfrac{1}{2} \int_0^\infty \int_0^\infty \dot{\mathbf{E}}^t(s_1) \cdot \check{\mathbf{G}}(|s_1 - s_2|)\dot{\mathbf{E}}^t(s_2)\, ds_1\, ds_2.$$

Consequently,

$$\rho\dot{\psi}^w(t) = \dot{W}(\mathbf{E}^t) = \mathfrak{T}(\mathbf{E}^t) \cdot \dot{\mathbf{E}}(t),$$

so proving that $d\hat{\psi}^w = 0$. Now, we continuously extend the history $\mathbf{E}^t$ to $\mathbb{R}^-$ by letting $\mathbf{E}^t(s) = \mathbf{E}^t(0) = \mathbf{E}(t)$, $s < 0$, so that $\dot{\mathbf{E}}(\tau) = 0$ for all $\tau > t$. Accordingly, (7.61) yields

$$\begin{aligned} \rho\hat{\psi}^w(\mathbf{E}(t), \mathbf{E}^t_r) = W(\mathbf{E}^t) &= \tfrac{1}{2}\mathbf{E}(t) \cdot \mathbf{G}_\infty \mathbf{E}(t) + \int_{-\infty}^{\infty} \dot{\mathbf{E}}(\tau) \cdot \int_{-\infty}^{\infty} \check{\mathbf{G}}(\tau - u)\dot{\mathbf{E}}(u)\, du\, d\tau \\ &= \tfrac{1}{2}\mathbf{E}(t) \cdot \mathbf{G}_\infty \mathbf{E}(t) + \int_{-\infty}^{\infty} \dot{\mathbf{E}}(\tau) \cdot \big[\check{\mathbf{G}} *_F \dot{\mathbf{E}}\big](u)\, du \end{aligned}$$

and applying the Parseval–Plancherel theorem it follows

$$\rho\hat{\psi}^w = \frac{1}{2}\mathbf{E}(t)\cdot\mathbf{G}_\infty\mathbf{E}(t) + \frac{1}{2\pi}\int_{-\infty}^{\infty}(i\omega\mathbf{E}_F)^*(\omega)\cdot\check{\mathbf{G}}_F(\omega)i\omega\mathbf{E}_F(\omega)\,d\omega.$$

Now, since $\check{\mathbf{G}}$ vanishes on $(-\infty, 0)$, its cosine transform $\check{\mathbf{G}}_c$ is an even function while the sine transform $\check{\mathbf{G}}_s$ is odd, and applying (7.58) we conclude that

$$\rho\hat{\psi}^w - \frac{1}{2}\mathbf{E}(t)\cdot\mathbf{G}_\infty\mathbf{E}(t) = \frac{1}{2\pi}\int_{-\infty}^{\infty}\omega^2\big[\mathbf{E}_c(\omega)\cdot\check{\mathbf{G}}_c(\omega)\mathbf{E}_c(\omega) + \mathbf{E}_s(\omega)\cdot\check{\mathbf{G}}_c(\omega)\mathbf{E}_s(\omega)\big]d\omega \geq 0,$$

so proving the minimum property at constant histories. On the other hand, extending the definition of $\mathbf{E}^t$ to $\mathbb{R}^-$ by letting $\mathbf{E}^t(s) = \mathbf{0}$, $s < 0$, and taking into account that $\dot{\mathbf{E}} + \mathbf{E}(t)\,\delta$ represents the distributional derivative of the extension of $\mathbf{E}$, we obtain

$$\begin{aligned}
W(\mathbf{E}^t) &= \frac{1}{2}\mathbf{E}(t)\cdot\mathbf{G}_\infty\mathbf{E}(t) + \int_{-\infty}^{\infty}\int_0^{\infty}\check{\mathbf{G}}(s)\dot{\mathbf{E}}(\tau-s)\,ds\cdot\big[\dot{\mathbf{E}}(\tau)+\delta(t-\tau)\mathbf{E}(t)\big]\,d\tau\\
&= \frac{1}{2}\mathbf{E}(t)\cdot\mathbf{G}_\infty\mathbf{E}(t) + \int_{-\infty}^{\infty}\int_{-\infty}^{\infty}\dot{\mathbf{E}}(\tau)\cdot\check{\mathbf{G}}(s)\dot{\mathbf{E}}(\tau-s)du\,d\tau + \int_0^{\infty}\check{\mathbf{G}}(s)\dot{\mathbf{E}}(t-s)ds\cdot\mathbf{E}(t)\\
&= \frac{1}{2}\mathbf{E}(t)\cdot\mathbf{G}_\infty\mathbf{E}(t) + \int_0^{\infty}\check{\mathbf{G}}(s)\dot{\mathbf{E}}(t-s)ds\cdot\mathbf{E}(t) + \int_{-\infty}^{\infty}\int_{-\infty}^{\infty}\dot{\mathbf{E}}(\tau)\cdot\check{\mathbf{G}}(\tau-u)\dot{\mathbf{E}}(u)du\,d\tau.
\end{aligned}$$

Since $\rho\hat{\psi}^w(\mathbf{E}(t), \mathbf{E}_r^t) = W(\mathbf{E}^t)$ we obtain

$$\rho\partial_{\mathbf{E}}\hat{\psi}^w(\mathbf{E}(t), \mathbf{E}_r^t) = \mathbf{G}_\infty\mathbf{E}(t) + \int_0^{\infty}\check{\mathbf{G}}(s)\dot{\mathbf{E}}(t-s)\,ds = \mathfrak{T}(\mathbf{E}^t).$$

This completes the proof that $\hat{\psi}^w$ represents a quadratic free energy. If $\check{\mathbf{G}}(s) = \check{\mathbf{G}}^T(s)$ then it is a positive definite functional on the Hilbert space $\mathcal{H}^w$ endowed with inner product and norm

$$(\mathbf{E}_1^t, \mathbf{E}_2^t)_{\mathcal{H}^w} := \mathbf{E}_1(t)\cdot\mathbf{E}_2(t) + \int_0^{\infty}\int_0^{\infty}\mathbf{E}_1^t(s_1)\cdot\check{\mathbf{G}}(|s_1-s_2|)\mathbf{E}_2^t(s_2)\,ds_1ds_2,$$

$$\|\mathbf{E}^t\|_{\mathcal{H}^w}^2 := \mathbf{E}(t)\cdot\mathbf{E}(t) + \int_0^{\infty}\int_0^{\infty}\mathbf{E}^t(s_1)\cdot\check{\mathbf{G}}(|s_1-s_2|)\mathbf{E}^t(s_2)\,ds_1ds_2.$$

Finally, assuming a regular symmetric kernel, $\check{\mathbf{G}} \in W^{2,1}(\mathbb{R}^+)$ and extending its definition to $\mathbb{R}^-$ by letting $\check{\mathbf{G}}(s) = \mathbf{0}$, $s < 0$, we can prove that ψ^w is equivalent to the work function ψ^W introduced in Sect. 7.4.3. Indeed, we first rewrite (7.61) as

$$\rho\hat{\psi}^w = \frac{1}{2}\mathbf{E}(t)\cdot\mathbf{G}_\infty\mathbf{E}(t) + \frac{1}{2}\int_{-\infty}^{t}\int_{-\infty}^{t}\dot{\mathbf{E}}(\tau)\cdot\check{\mathbf{G}}(|\tau-u|)\dot{\mathbf{E}}(u)\,du\,d\tau.$$

Repeated integrations by parts result in

$$\begin{aligned}
\rho\hat{\psi}^w &= \frac{1}{2}\mathbf{E}(t)\cdot\mathbf{G}_\infty\mathbf{E}(t) + \frac{1}{2}\int_{-\infty}^{t}\int_{-\infty}^{t}\dot{\mathbf{E}}(\tau)\cdot\check{\mathbf{G}}(|\tau-u|)\dot{\mathbf{E}}(u)\,du\,d\tau \\
&= \frac{1}{2}\mathbf{E}(t)\cdot\mathbf{G}_\infty\mathbf{E}(t) + \frac{1}{2}\int_{-\infty}^{t}\partial_\tau\mathbf{E}(\tau)\cdot\int_{-\infty}^{t}\check{\mathbf{G}}(|\tau-u|)\partial_u\mathbf{E}(u)\,du\,d\tau \\
&= \frac{1}{2}\mathbf{E}(t)\cdot\mathbf{G}_\infty\mathbf{E}(t) + \frac{1}{2}\mathbf{E}(t)\cdot\int_{-\infty}^{t}\check{\mathbf{G}}(t-u)\partial_u\mathbf{E}(u)\,du \\
&\quad - \frac{1}{2}\int_{-\infty}^{t}\mathbf{E}(\tau)\cdot\int_{-\infty}^{t}\left[\partial_\tau\check{\mathbf{G}}(|\tau-u|) + \check{\mathbf{G}}(0)\delta(\tau-u)\right]\partial_u\mathbf{E}(u)\,d\tau du \\
&= \frac{1}{2}\mathbf{E}(t)\cdot\mathbf{G}_\infty\mathbf{E}(t) + \mathbf{E}(t)\cdot\check{\mathbf{G}}(0)\mathbf{E}(t) + \mathbf{E}(t)\cdot\int_{-\infty}^{t}\mathbf{G}'(t-u)\mathbf{E}(u)\,du \\
&\quad - \int_{-\infty}^{t}\mathbf{E}(\tau)\cdot\int_{-\infty}^{t}\partial_\tau\check{\mathbf{G}}(|\tau-u|)\partial_u\mathbf{E}(u)\,d\tau du - \int_{-\infty}^{t}\mathbf{E}(\tau)\cdot\check{\mathbf{G}}(0)\partial_\tau\mathbf{E}(\tau)\,d\tau \\
&= \frac{1}{2}\mathbf{E}(t)\cdot\mathbf{G}_\infty\mathbf{E}(t) + \frac{1}{2}\mathbf{E}(t)\cdot\check{\mathbf{G}}(0)\mathbf{E}(t) + \mathbf{E}(t)\cdot\int_{0}^{\infty}\mathbf{G}'(s)\mathbf{E}^t(s)\,ds \\
&\quad - \mathbf{E}(t)\cdot\int_{-\infty}^{t}\partial_\tau\check{\mathbf{G}}(\tau-t)\mathbf{E}(\tau)\,d\tau + \int_{-\infty}^{t}\int_{-\infty}^{t}\mathbf{E}(\tau)\cdot\mathbf{G}_{12}(|\tau-u|)\mathbf{E}(u)\,d\tau du.
\end{aligned}$$

Since the extension of $\check{\mathbf{G}}$ vanishes on $\mathbb{R}^-$, it follows that

$$\int_{-\infty}^{t}\partial_\tau\check{\mathbf{G}}(\tau-t)\mathbf{E}(\tau)\,d\tau = \int_{0}^{\infty}\check{\mathbf{G}}'(-s)\mathbf{E}^t(s)\,ds = \mathbf{0},$$

and hence we can write $\rho\hat{\psi}^w$ in the form

$$\rho\hat{\psi}^w = \frac{1}{2}\mathbf{E}(t)\cdot\mathbf{G}_0\mathbf{E}(t) + \mathbf{E}(t)\cdot\int_{0}^{\infty}\mathbf{G}'(s)\mathbf{E}^t(s)\,ds + \frac{1}{2}\int_{0}^{\infty}\int_{0}^{\infty}\mathbf{E}^t(s_1)\cdot\mathbf{G}_{12}(|s_1-s_2|)\mathbf{E}^t(s_2)\,ds_1 ds_2.$$

7.4.5 Nonlinear Viscoelastic Models

Free energies with quadratic memory terms yield constitutive equations with linear memory terms. Here we show that free energies of nonlinear constitutive equations can be generated from known free energies associated with linear memory-dependent response functional $\mathfrak{T}$.

Let $\psi_0(t)$ be a free energy, at time t,

$$\psi_0(t) = \hat{\psi}_0(\mathbf{E}(t), \mathbf{E}_r^t).$$

The functional $\hat{\psi}_0$ is thermodynamically consistent in that is subject to the minimum property and the dissipation property,

$$\hat{\psi}_0(\mathbf{E}, \mathbf{E}_r^\dagger) \le \hat{\psi}_0(\mathbf{E}, \mathbf{E}_r^t), \quad d\hat{\psi}_0(\mathbf{E}, \mathbf{E}_r^t|\dot{\mathbf{E}}_r^t) \le 0, \tag{7.62}$$

and is associated with a stress tensor $\mathfrak{T}_0(\mathbf{E}(t), \mathbf{E}_r^t)$ in the form

$$\mathbf{T}_0(t) = \mathfrak{T}_0(\mathbf{E}(t), \mathbf{E}_r^t), \qquad \rho\partial_{\mathbf{E}}\hat{\psi}_0(\mathbf{E}(t), \mathbf{E}_r^t) = \mathfrak{T}_0(\mathbf{E}(t), \mathbf{E}_r^t). \tag{7.63}$$

Correspondingly the work function W_0 is defined by

$$W_0(t) = \int_{-\infty}^{t} \mathfrak{T}_0(\mathbf{E}(\tau), \mathbf{E}_r^\tau) \cdot \dot{\mathbf{E}}(\tau)d\tau.$$

The pair $\psi_0, \mathfrak{T}_0$ is subject to the entropy inequality so that

$$\rho\dot{\psi}_0(t) \le \mathfrak{T}_0(\mathbf{E}(t), \mathbf{E}_r^t) \cdot \dot{\mathbf{E}}(t) = \dot{W}_0.$$

The following theorem shows how a nonlinear functional can be generated by the pair $\{\psi_0, \mathfrak{T}_0\}$.

Theorem 7.5 *Let $\hat{\psi}$ be the functional*

$$\hat{\psi}(\mathbf{E}(t), \mathbf{E}_r^t) := f(\psi_0(t)) \tag{7.64}$$

where $f(x) : \mathbb{R}^+ \to \mathbb{R}^+$ is subject to

$$f(0) = 0, \qquad f'(x) \ge 0, \tag{7.65}$$

and ψ_0 satisfies (7.62) and (7.63). Then $\hat{\psi}$ is a free energy for the stress functional

$$\mathbf{T}(t) = \mathfrak{T}(\mathbf{E}(t), \mathbf{E}_r^t) := f'(\psi_0(t))\mathfrak{T}_0(\mathbf{E}(t), \mathbf{E}_r^t). \tag{7.66}$$

Proof Consider $\psi(t) = \hat{\psi}(\mathbf{E}(t), \mathbf{E}_r^t)$ given by (7.64). The definition (7.66) of $\mathfrak{T}$ and the relations (7.63) allow us to write

$$\rho\dot{\psi}(t) = f'(\psi_0(t))\, \rho\dot{\psi}_0(t) \le f'(\psi_0(t))\mathfrak{T}_0(\mathbf{E}(t), \mathbf{E}_r^t) \cdot \dot{\mathbf{E}}(t) = \mathfrak{T}(\mathbf{E}(t), \mathbf{E}_r^t) \cdot \dot{\mathbf{E}}(t).$$

Moreover, by (7.63), (7.65), and (7.66),

$$\rho\partial_{\mathbf{E}}\hat{\psi} = \rho f'(\psi_0)\, \partial_{\mathbf{E}}\hat{\psi}_0(\mathbf{E}(t), \mathbf{E}_r^t) = \mathfrak{T}(\mathbf{E}(t), \mathbf{E}_r^t).$$

Now, by (7.62) we have

$$d\hat{\psi}(\mathbf{E}, \mathbf{E}_r^t | \dot{\mathbf{E}}_r^t) = f'(\psi_0)\, d\hat{\psi}_0(\mathbf{E}, \mathbf{E}_r^t | \dot{\mathbf{E}}_r^t) \le 0.$$

As to the minimum property we observe that f is an increasing function. Hence $\hat{\psi}_0(\mathbf{E}, \mathbf{E}_r^\dagger) \le \hat{\psi}_0(\mathbf{E}, \mathbf{E}_r^t)$ implies

$$\hat{\psi}(\mathbf{E}, \mathbf{E}_r^\dagger) = f(\hat{\psi}_0(\mathbf{E}, \mathbf{E}_r^\dagger)) \le f(\hat{\psi}_0(\mathbf{E}, \mathbf{E}_r^t)) = \hat{\psi}(\mathbf{E}, \mathbf{E}_r^t). \qquad \square$$

This result allows us to establish free energies and nonlinear viscoelastic models as a superposition of linear viscoelastic solids [8].

Given a nonlinear smooth function $g : \mathbb{R}^+ \to \mathbb{R}^+$ such that

$$g(0) = 0, \quad g' \ge 0, \tag{7.67}$$

we can take f in the form

$$f(\psi_0) = \psi_0 + g(\psi_0).$$

In that case we obtain the nonlinear stress–strain relation in the form

$$\mathfrak{T}(\mathbf{E}(t), \mathbf{E}_r^t) = \mathfrak{T}_0(\mathbf{E}(t), \mathbf{E}_r^t) + g'(\psi_0(t))\mathfrak{T}_0(\mathbf{E}(t), \mathbf{E}_r^t), \qquad \psi(t) = \psi_0(t) + g(\psi_0(t)). \tag{7.68}$$

7.4.6 *Some Examples*

Let memory kernels $\mathbf{G}$ be the memory kernel of a Boltzmann-type model, satisfying the thermodynamic condition (7.36),

$$\mathbf{T}(t) = \mathfrak{T}_0(\mathbf{E}(t), \mathbf{E}_r^t) = \mathbf{G}_\infty \mathbf{E}(t) + \int_0^\infty \mathbf{G}'(s)\mathbf{E}_r^t(s)ds,$$

with $\mathbf{G}_\infty = \lim_{s\to\infty} \mathbf{G}(s)$, and let $\hat{\psi}_0$ be any related free energy functional. Thus, we have

$$\hat{\psi}_0(\mathbf{E}(t), \mathbf{E}_r^t) \ge 0, \quad \rho\partial_{\mathbf{E}}\hat{\psi}_0(\mathbf{E}(t), \mathbf{E}_r^t) = \mathfrak{T}_0(\mathbf{E}(t), \mathbf{E}_r^t) \quad \text{and} \quad \rho\dot{\psi}_0(t) \le \mathbf{T}_0(t) \cdot \dot{\mathbf{E}}(t).$$

With this pair of functionals $\hat{\psi}_0$, $\mathfrak{T}_0$ we can determine nonlinear functionals once the function f is selected. For definiteness we let $\hat{\psi}_0$ be Graffi's free energy and proceed as follows.

A nonlinear model based on Graffi's free energy

Let $\mathbf{T}_0$ be given by the linear functional (7.30),

$$\mathbf{T}_0(t) = \mathbf{G}_\infty \mathbf{E}(t) - \int_0^\infty \mathbf{G}'(s)\mathbf{E}_d^t(s)ds,$$

where

$$\mathbf{E}_d^t(u) = \mathbf{E}(t) - \mathbf{E}^t(u); \qquad \mathbf{G}'(u) = \mathbf{G}'^T(u), \quad \mathbf{G}'(u) \le 0, \quad \mathbf{G}''(u) \ge 0, \qquad u \in \mathbb{R}^+.$$

Correspondingly we choose $\hat{\psi}_0 = \hat{\psi}_G$, Graffi's free energy functional,

$$\rho \hat{\psi}_G(\mathbf{E}(t), \mathbf{E}_d^t) = \tfrac{1}{2}\mathbf{E}(t) \cdot \mathbf{G}_\infty \mathbf{E}(t) - \tfrac{1}{2} \int_0^\infty \mathbf{E}_d^t(u) \cdot \mathbf{G}'(u)\mathbf{E}_d^t(u)\, du.$$

If

$$f(\psi_0) = \psi_0 + \alpha\psi_0^2$$

then we have

$$\mathfrak{T}(\mathbf{E}^t, \mathbf{E}_r^t) = \mathfrak{T}_0(\mathbf{E}^t, \mathbf{E}_r^t) + 2\alpha\hat{\psi}_0(\mathbf{E}^t, \mathbf{E}_r^t)\mathfrak{T}_0(\mathbf{E}^t, \mathbf{E}_r^t)$$
$$= \mathbf{G}_\infty \mathbf{E}(t) - \int_0^\infty \mathbf{G}'(s)\mathbf{E}_d^t(s)ds + 2\alpha\hat{\psi}_0(\mathbf{E}(t), \mathbf{E}_r^t)[\mathbf{G}_\infty \mathbf{E}(t) - \int_0^\infty \mathbf{G}'(s)\mathbf{E}_d^t(s)ds]$$

and

$$\psi(\mathbf{E}(t), \mathbf{E}_r^t) = \tfrac{1}{2}[\mathbf{E}(t) \cdot \mathbf{G}_\infty \mathbf{E}(t) - \int_0^\infty \mathbf{E}_d^t(s) \cdot \mathbf{G}'(s)\mathbf{E}_d^t(s)ds]$$
$$+\alpha\tfrac{1}{4}[\mathbf{E}(t) \cdot \mathbf{G}_\infty \mathbf{E}(t) - \int_0^\infty \mathbf{E}_d^t(s) \cdot \mathbf{G}'(s)\mathbf{E}_d^t(s)ds]^2.$$

In terms of the product $[\mathbf{A} \otimes \mathbf{B}]\mathbf{C} = (\mathbf{B} \cdot \mathbf{C})\mathbf{A}$ for second-order tensors (see Sect. A.1.2) we can write the stress functional

$$\mathbf{T}(t) = \mathbf{G}_\infty \mathbf{E}(t) - \int_0^\infty \mathbf{G}'(s)\mathbf{E}_d^t(s)ds + \left[\mathbf{G}_\infty \mathbf{E}(t) \otimes \mathbf{G}_\infty \mathbf{E}(t)\right]\mathbf{E}(t)$$
$$- \left[\int_0^\infty \mathbf{G}'(s)\mathbf{E}_d^t(s)ds \otimes \mathbf{G}_\infty \mathbf{E}(t)\right] \mathbf{E}(t) - \int_0^\infty \left[\mathbf{G}_\infty \mathbf{E}(t) \otimes \mathbf{G}'(u)\mathbf{E}_d^t(u)\right]\mathbf{E}_d^t(u)\, du$$
$$+ \int_0^\infty \int_0^\infty \left[\mathbf{G}'(s)\mathbf{E}_d^t(s) \otimes \mathbf{G}'(u)\mathbf{E}_d^t(u)\right]\mathbf{E}_d^t(u)\, du ds.$$

Moreover, the corresponding rate of dissipation is

$$-\rho d\psi(t) = \tfrac{1}{2} \int_0^\infty \mathbf{E}_d^t(s) \cdot \mathbf{G}''(s)\mathbf{E}_r^t(s)\, ds\left[1 + \mathbf{E}(t) \cdot \mathbf{G}_\infty \mathbf{E}(t) - \int_0^\infty \mathbf{E}_d^t(s) \cdot \mathbf{G}'(s)\mathbf{E}_d^t(s)ds\right] \ge 0.$$

For isotropic viscoelastic materials the kernel G' and the relaxation modulus G_∞ take the form

$$\mathsf{G}'(s) = \lambda'(s)\mathbf{1} \odot \mathbf{1} + 2\mu'(s)\mathsf{I}, \qquad \mathsf{G}_\infty = \lambda_\infty \mathbf{1} \odot \mathbf{1} + 2\mu_\infty \mathsf{I},$$

where $\lambda', \mu' \in C^1 \cap L^1(\mathbb{R}^+; \mathbb{R})$ and

$$\lambda_\infty = \lim_{u\to\infty} \lambda(u), \qquad \lambda(u) = \lambda_0 + \int_0^u \lambda'(s)\, ds,$$

and the like for μ. Moreover, $\lambda', \mu' < 0$ and $\lambda'', \mu'' \geq 0$. Taking advantage of the decomposition,

$$\mathbf{E} = \tfrac{1}{3}(\mathrm{tr}\,\mathbf{E})\mathbf{1} + \mathbf{E}_0, \qquad \mathbf{T} = \tfrac{1}{3}(\mathrm{tr}\,\mathbf{T})\mathbf{1} + \mathbf{T}_0$$

where tr stands for the trace and the subscript $_0$ denotes the deviator of the tensor, we can write the nonlinear stress–strain relation in the form

$$\begin{aligned}\mathbf{T}_0(t) = {} & 2\mu_\infty \mathbf{E}_0(t)[1 + 4\mu_\infty |\mathbf{E}_0(t)|^2] + 2[1 + 4\mu_\infty |\mathbf{E}_0(t)|^2 \int_0^\infty \mu'(s)\mathbf{E}^t_{0r}(s)ds \\ & - 8\mu_\infty \mathbf{E}_0(t) \int_0^\infty \mu'(s)|\mathbf{E}^t_{0r}(s)|^2 ds - 8 \int_0^\infty \mu'(u)\mathbf{E}^t_{0r}(u)du \int_0^\infty \mu'(s)|\mathbf{E}^t_{0r}(s)|^2 ds,\end{aligned}$$

$$\begin{aligned}\mathrm{tr}\,\mathbf{T}(t) = {} & K_\infty \mathrm{tr}\,\mathbf{E}(t)[1+K_\infty|\mathrm{tr}\,\mathbf{E}(t)|^2]+[1+K_\infty|\mathrm{tr}\,\mathbf{E}(t)|^2 \int_0^\infty K'(s)\mathrm{tr}\,\mathbf{E}^t_r(s)ds \\ & - K_\infty \mathrm{tr}\,\mathbf{E}(t) \int_0^\infty K'(s)|\mathrm{tr}\,\mathbf{E}^t_r(s)|^2 ds - \int_0^\infty K'(u)\mathrm{tr}\,\mathbf{E}^t_r(u)du \int_0^\infty K'(s)|\mathrm{tr}\,\mathbf{E}^t_r(s)|^2 ds,\end{aligned}$$

where $K(s) = \lambda(s) + \frac{2}{3}\mu(s)$ and $K_\infty = \lambda_\infty + \frac{2}{3}\mu_\infty$ denote the bulk elastic kernel and the bulk relaxation modulus, respectively.

The related free energy takes the form

$$\begin{aligned}\psi(t) = {} & \tfrac{1}{2}K_\infty|\mathrm{tr}\,\mathbf{E}(t)|^2 + \mu_\infty|\mathbf{E}_0(t)|^2 - \tfrac{1}{2}\int_0^\infty K'(s)|\mathrm{tr}\,\mathbf{E}^t_r(s)|^2 ds - \int_0^\infty \mu'(s)|\mathbf{E}^t_{0r}(s)|^2 ds \\ & + \Big(\tfrac{1}{2}K_\infty|\mathrm{tr}\,\mathbf{E}(t)|^2 + \mu_\infty|\mathbf{E}_0(t)|^2 - \tfrac{1}{2}\int_0^\infty K'(s)|\mathrm{tr}\,\mathbf{E}^t_r(s)|^2 ds - \int_0^\infty \mu'(s)|\mathbf{E}^t_{0r}(s)|^2 ds\Big)^2.\end{aligned}$$

7.5 Viscous Fluids with Memory

There are fluids, e.g. polymer solutions, which exhibit two types of behaviour. They are viscous and flow like a Newtonian fluid. In addition, they can store or release energy, like an elastic solid. In modelling, this behaviour can be interpreted as the constitutive equations having some memory effect of the flow on the stress. The corresponding models are usually called viscoelastic fluids.

Models of fluids with memory require two essential features. As with any material with memory the pertinent quantities (position, velocity, ...) must be related to a fixed point $\mathbf{X} \in R$. In fluids, though, the Eulerian description is privileged as is the case for the velocity gradient in Navier–Stokes type models. To make these views operative we proceed as follows.

Let $\Omega \subset \mathscr{E}$ be the time-dependent region occupied by the fluid. At the time t the point $\mathbf{x} \in \Omega$ is occupied by the point $\mathbf{X} \in R$ such that $\mathbf{x} = \chi(\mathbf{X}, t)$. At any previous time $\tau \leq t$ the position of the point $\mathbf{X}$ is given by $\chi(\mathbf{X}, \tau)$. Hence the function

$$\chi(\mathbf{X}, \tau) = \chi(\chi^{-1}(\mathbf{x}, t), \tau)$$

denotes the position, at time $\tau \leq t$, of the point $(\mathbf{X})$ at $\mathbf{x} \in \Omega$ at time t. Given the times $\tau, t, \tau \leq t$, the function

$$\boldsymbol{\xi}(\mathbf{x}, t; \tau) := \chi(\chi^{-1}(\mathbf{x}, t), \tau), \qquad \mathbf{x} \in \Omega,$$

models an Eulerian description with the current region Ω as a reference configuration; in particular $\boldsymbol{\xi}(\mathbf{x}, t; t) = \mathbf{x}$. So $\partial_{\xi} = \nabla_{\xi}$ is the gradient in $\Omega(\tau)$ while ∇ and $\nabla_{\mathbf{X}}$ are the gradients in $\Omega(t)$ and R.

The relative displacement vector $\boldsymbol{\delta}$, with respect to the current configuration, is defined by

$$\boldsymbol{\delta}(\mathbf{x}, t; \tau) = \boldsymbol{\xi}(\mathbf{x}, t; \tau) - \mathbf{x} = \chi(\mathbf{X}, \tau) - \chi(\mathbf{X}, t);$$

it is $\boldsymbol{\delta}(\mathbf{x}, t; t) = \mathbf{0}$.

The velocity, at time $\tau \leq t$, of the point $\mathbf{X}$ is given by

$$\mathbf{V}(\mathbf{X}, \tau) := \partial_{\tau}\chi(\mathbf{X}, \tau) = \partial_{\tau}\boldsymbol{\xi}(\mathbf{x}, t; \tau).$$

Then the representation with respect to the current configuration is

$$\boldsymbol{v}(\mathbf{x}, t; \tau) := \partial_{\tau}\boldsymbol{\xi}(\mathbf{x}, t; \tau) = \partial_{\tau}\boldsymbol{\delta}(\mathbf{x}, t; \tau), \qquad \mathbf{x} \in \Omega, \tau \leq t. \tag{7.69}$$

Hence we have the Eulerian description of the velocity of the fluid in the form

$$\mathbf{v}(\boldsymbol{\xi}(\mathbf{x}, t; \tau) = \boldsymbol{v}(\mathbf{x}, t; \tau)$$

and we can define the velocity gradient in the form

$$\mathcal{D}(\mathbf{x}, t; \tau) := \operatorname{sym} \nabla \boldsymbol{v}(\boldsymbol{\xi}, t; \tau)$$

while

$$\partial_{\mathbf{x}}\boldsymbol{v} = \nabla_{\xi}\mathbf{v}\, \nabla\boldsymbol{\xi}$$

At $\tau = t$ we have

$$\upsilon(\mathbf{x}, t; t) = \mathbf{v}(\mathbf{x}, t), \qquad \mathcal{D}(\mathbf{x}, t; t) = \mathbf{D}(\mathbf{x}, t) = \mathrm{sym}\nabla\mathbf{v}(\mathbf{x}, t).$$

Alternatively we might compute $\mathcal{D}$ by means of the function $\mathbf{V}(\mathbf{X}, \tau)$,

$$\mathcal{D}(\mathbf{x}, t; \tau) = \mathrm{sym}\nabla_{\mathbf{X}}\mathbf{V}(\mathbf{X}, \tau).$$

Likewise the acceleration is determined by

$$\partial_\tau \upsilon(\mathbf{x}, t; \tau) = \partial_\tau \mathbf{v} + (\mathbf{v} \cdot \nabla_\xi)\mathbf{v},$$

where $\mathbf{v} = \mathbf{v}(\boldsymbol{\xi}(\mathbf{x}, t; \tau), \tau)$. At $\tau = t$ it follows

$$\partial_\tau \upsilon(\mathbf{x}, t; \tau)|_{\tau=t} = \partial_t \mathbf{v}(\mathbf{x}, t) + (\mathbf{v}(\mathbf{x}, t) \cdot \nabla)\mathbf{v}(\mathbf{x}, t).$$

7.5.1 *Incompressible Viscoelastic Fluids*

At any fixed point $\mathbf{x} \in \Omega$ and any time $t \geq 0$, the symmetric Cauchy stress tensor $\mathbf{T}$ of a linear incompressible viscoelastic fluid is given by $\mathbf{T} = -p\mathbf{1} + \mathcal{T}$ where $p = p(\mathbf{x}, t)$ is the pressure and

$$\mathcal{T}(\mathbf{x}, t) = 2 \int_0^\infty G_\mu(s)\, \mathcal{D}(\mathbf{x}, t - s) ds. \tag{7.70}$$

The memory kernel G_μ is called *relaxation shear modulus*. This constitutive equation implies that the stress response at $(\mathbf{x}, t)$ depends linearly on the values of the strain rate experienced in the instants preceding t at the point which is at $\mathbf{x}$ at time t. The incompressibility constraint is given by

$$\mathrm{tr}\,\mathcal{D}(\mathbf{x}, \tau) = \nabla \cdot \upsilon(\mathbf{x}, t; \tau) = 0, \qquad \mathbf{x} \in \Omega, \ \tau \leq t.$$

If $\mathcal{D}$ is nearly constant then $\mathcal{D}(\mathbf{x}, \tau) \approx \mathcal{D}(\mathbf{x}, t) = \mathbf{D}(\mathbf{x}, t)$, $\tau \leq t$, and we can use the approximation

$$\mathcal{T}(t) = 2\mu_0 \mathbf{D}(t), \qquad \mu_0 = \int_0^\infty G_\mu(s) ds.$$

where μ_0 represents the shear viscosity at constant strain rate.

Borrowing from linear viscoelasticity (cf. [18]), we consider

$$\mathfrak{E}(\mathbf{x}, \tau) = \mathrm{sym}\nabla\boldsymbol{\delta}(\mathbf{x}, t; \tau),$$

subject to the linearized incompressibility constraint

$$\text{tr}\,\mathfrak{E}(\mathbf{x}, t; \tau) = \nabla \cdot \boldsymbol{\delta}(\mathbf{x}, t; \tau) = 0, \qquad \mathbf{x} \in \Omega,\ \tau \leq t.$$

Then also $\mathfrak{E}(\mathbf{x}, t; t) = \mathbf{0}$ and in view of (7.69)

$$\partial_\tau \mathfrak{E}(\mathbf{x}, t; \tau) = \mathcal{D}(\mathbf{x}, t; \tau).$$

A formal integration by parts transforms (7.70) into

$$\mathcal{T}(\mathbf{x}, t) = 2 \int_0^\infty G'_\mu(s)\, \mathfrak{E}(\mathbf{x}, t; t - s) ds \tag{7.71}$$

The system of equations for the dynamics of an incompressible viscoelastic fluid is then obtained in the form

$$\begin{cases} \nabla \cdot \mathbf{v} = \nabla \cdot \boldsymbol{\delta} = 0, \\ \rho\big[\mathbf{v}_t + (\mathbf{v} \cdot \nabla)\, \mathbf{v}\big](t) + \int_0^\infty G'_\mu(s) \Delta \mathbf{u}^t(t - s)\, ds + \nabla p = \rho \mathbf{b}, \end{cases}$$

where ρ is the (constant) mass density and the dependence on $\mathbf{x}$ is understood and not written.

In a linear incompressible viscoelastic fluid of Jeffreys type the stress–strain rate constitutive relation consists of a Newtonian contribution and a viscoelastic contribution,[15]

$$\mathcal{T}(\mathbf{x}, t) = 2\mu \mathbf{D}(\mathbf{x}, t) + 2 \int_0^\infty G_\mu(s)\, \mathcal{D}^t(\mathbf{x}, t - s) ds, \tag{7.72}$$

which leads to the differential system

$$\begin{cases} \nabla \cdot \mathbf{v} = \nabla \cdot \mathbf{u} = 0, \\ \rho\big[\mathbf{v}_t + (\mathbf{v} \cdot \nabla)\, \mathbf{v}\big](t) - \mu \Delta \mathbf{v} + \int_0^\infty G'_\mu(s) \Delta \mathbf{u}^t(t - s)\, ds + \nabla p = \rho \mathbf{b}, \end{cases}$$

Approximation for small Weissenberg numbers

To express the equation of motion in terms of a single variable, we need to establish a relationship,[16] possibly approximate, between $\boldsymbol{\delta}$ and $\mathbf{v}$. We first observe that $\boldsymbol{\delta}$ satisfies the integral equation

$$\boldsymbol{\delta}(\mathbf{x}, t; \tau) = \int_\tau^t \mathbf{v}(\mathbf{x} - \mathbf{u}^t(\mathbf{x}, \xi), \xi) d\xi, \qquad \tau \leq t.$$

Indeed, for any fixed t and for all $\tau \leq t$ we have

[15] See [244].

[16] See, for instance, [186].

$$\int_{\tau}^{t} \mathbf{v}(\mathbf{x}+\boldsymbol{\delta}(\mathbf{x},t;\zeta)d\zeta = \int_{\tau}^{t} \upsilon(\mathbf{x},t;\zeta)d\zeta = \int_{\tau}^{t} \partial_\zeta \boldsymbol{\delta}(\mathbf{x},t;\zeta)d\zeta = -\boldsymbol{\delta}(\mathbf{x},t;\tau).$$

Assume the position is in a neighbourhood of the rest position $\mathbf{x}$. We can then make the first approximation

$$-\boldsymbol{\delta}(\mathbf{x},t;\tau) \approx \int_{\tau}^{t} \big[\mathbf{v}(\mathbf{x},\zeta) + (\boldsymbol{\delta}(\mathbf{x},t;\zeta)\cdot\nabla)\mathbf{v}(\mathbf{x},\zeta)\big]d\zeta, \qquad \tau \le t. \tag{7.73}$$

Upon a change of variables $s = t - \tau$ and $y = t - \zeta$ we have

$$-\boldsymbol{\delta}(\mathbf{x},t;t-s) \approx \int_{0}^{s} [\mathbf{v}(\mathbf{x},t-y) + (\boldsymbol{\delta}(\mathbf{x},t;t-y)\cdot\nabla)\mathbf{v}(\mathbf{x},t-y)]dy$$

Unfortunately, this integral relation is not explicit in $\mathbf{v} \to \boldsymbol{\delta}$. We then look for a further approximation. Let L and U be the characteristic length and velocity of the fluid flow, respectively. The ratio L/U is called *kinematic time* of the flow. Moreover, a *natural time* Λ can be defined in terms of the memory kernel, which represents in some sense the length of the memory. Following [18, p. 249] we assume Λ as defined in the form

$$\Lambda = \frac{\int_0^\infty s^2 G'_\mu(s)ds}{2\int_0^\infty sG'_\mu(s)ds}. \tag{7.74}$$

Hence, we consider the dimensionless space and time variables

$$\hat{\mathbf{x}} = \frac{\mathbf{x}}{L}, \qquad \hat{t} = \frac{Ut}{L}, \qquad \hat{\tau} = \frac{U\tau}{L}, \qquad \hat{s} = \frac{s}{\Lambda}$$

and introduce the *Weissenberg number* $\mathrm{We} = \Lambda U/L$. It follows

$$t - s = (\hat{t} - \mathrm{We}\,\hat{s})L/U.$$

Then, letting $\mathrm{We} = \epsilon$ and removing the hats, by recursion, the integral equation (7.73) yields the following power expansion of $\boldsymbol{\delta}(\mathbf{x},t;t-\epsilon s)$ with respect to ϵ

$$-\boldsymbol{\delta}(\mathbf{x},t;t-\epsilon s) \approx \epsilon\int_0^s \mathbf{v}(\mathbf{x},t-\epsilon y)dy - \epsilon^2\int_0^s\Big[\int_0^y \mathbf{v}(\mathbf{x},t-\epsilon\zeta)d\zeta\cdot\nabla\Big]\mathbf{v}(\mathbf{x},t-\epsilon y)dy + \epsilon^3\ldots$$

It is readily seen that the first-order approximation in the Weissenberg number gives

$$\boldsymbol{\delta}(\mathbf{x},t;t-\tau) \approx \int_0^\tau \mathbf{v}(x,t-\xi)d\xi, \qquad \tau \ge 0.$$

and then $\partial_\tau\boldsymbol{\delta}(\mathbf{x},t;t-\tau) \approx \mathbf{v}(x,t-\tau)$. This allows us to approximate $\boldsymbol{\mathcal{D}}$ with $\mathbf{D}$ at small Weissenberg numbers, that is when the natural time of the fluid, Λ, is small if compared to the kinematical time D/U of the flow.

7.5.2 Compressible Viscoelastic Fluids

Let $\mathbf{T} = -p\mathbf{1} + \mathcal{T} \in \mathrm{Sym}$ be the Cauchy stress. Compressible fluids are modelled through two scalar relaxation functions, μ and λ, named shear and bulk viscosity modulus; the stress $\mathcal{T}$ is determined in the form

$$\mathcal{T}(\mathbf{x}, t) = 2\int_0^\infty \mu(s)\,\mathcal{D}(\mathbf{x}, t; t-s)ds + \int_0^\infty \lambda(s)\,\mathrm{tr}\,\mathcal{D}(\mathbf{x}, t; t-s)\,ds. \tag{7.75}$$

For simplicity, we assume that $\mathcal{D}^t$ can be approximated by $\mathbf{D}^t$.

Thermodynamic consistency of a Boltzmann-like model

Consider a linear model of the viscous fluid in the form

$$\mathbf{T} = -p(\rho, \theta)\,\mathbf{1} + 2\int_0^\infty \mu(s)\mathbf{D}^t(s)ds + \int_0^\infty \lambda(s)(\mathrm{tr}\,\mathbf{D}^t)(s)ds;$$

the stress $\mathcal{T} = \mathbf{T} + p\mathbf{1}$ is a (linear) functional of the history of $\mathbf{D}$. We can then ask about the thermodynamic consistency of this model where the variables are the present values ρ, θ and the history $\mathbf{D}^t$. By means of the deviator $\mathbf{D}_0$ we can write

$$\mathbf{T} = -p(\rho, \theta)\,\mathbf{1} + 2\int_0^\infty \mu(s)\mathbf{D}_0^t(s)ds + \int_0^\infty \kappa(s)(\mathrm{tr}\,\mathbf{D}^t)(s)ds\,\mathbf{1}, \tag{7.76}$$

where $\kappa = \lambda + 2\mu/3$. In view of (7.76) we assume that ψ and η are given by functionals of ρ, θ, $\mathbf{D}_0^t$, $\mathrm{tr}\,\mathbf{D}^t$.

It follows from the entropy inequality

$$-\rho(\dot{\psi} + \eta\dot{\theta}) - \mathbf{T}\cdot\mathbf{D} \geq 0 \tag{7.77}$$

that as θ, ρ are constants and $\mathbf{D} = \mathbf{0}$ in the interval $t \in [t_0, \hat{t}]$ we have

$$\psi(\hat{t}) - \psi(t_0) \leq 0.$$

As $\hat{t} - t_0 \to \infty$ we find that

$$\psi(\rho, \theta, \mathbf{0}^\dagger) - \psi(\rho, \theta, \mathbf{D}^{t_0}) \leq 0;$$

the free energy has a minimum at the zero history of $\mathbf{D}$.

Time differentiation of $\psi(\rho, \theta, \mathbf{D}_0^t, \mathrm{tr}\,\mathbf{D}^t)$ and substitution in the entropy inequality result in

$$-\rho(\partial_\theta\psi + \eta)\dot{\theta} + (\rho^2\partial_\rho\psi - p)\nabla\cdot\mathbf{v} - \rho\, d\psi(\rho, \theta, \mathbf{D}_0^t|\dot{\mathbf{D}}_0^t, \mathrm{tr}\,\mathbf{D}^t) - \rho\, d\psi(\rho, \theta, \mathbf{D}_0^t, \mathrm{tr}\,\mathbf{D}^t|\mathrm{tr}\,\dot{\mathbf{D}}^t)$$
$$+\mathbf{D}_0 \cdot 2\int_0^\infty \mu(s)\,\mathbf{D}_0^t(s)ds + \mathrm{tr}\,\mathbf{D}\int_0^\infty \kappa(s)(\mathrm{tr}\,\mathbf{D}^t)(s)ds \geq 0.$$

Hence we have the classical relation $\eta = -\partial_\theta\psi$. The remaining inequality is a requirement on the functional ψ. Since $d\psi(\rho, \theta, \mathbf{D}_0^t|\dot{\mathbf{D}}_0^t, \mathrm{tr}\,\mathbf{D}^t)$ and $d\psi(\rho, \theta, \mathbf{D}_0^t, \mathrm{tr}\,\mathbf{D}^t|\mathrm{tr}\,\dot{\mathbf{D}}^t)$ are linear functionals on $\dot{\mathbf{D}}_0^t$ and $\mathrm{tr}\,\dot{\mathbf{D}}^t$ we can look for ψ in different ways. First, let

$$\psi = \Psi(\rho, \theta) + \beta\Big|\int_0^\infty \mu(s)\,\mathbf{D}_0^t(s)ds\Big|^2 + \gamma\Big(\int_0^\infty \kappa(s)\,\mathrm{tr}\,\mathbf{D}^t(s)ds\Big)^2, \tag{7.78}$$

β and γ being constants. Hence we obtain the inequality in the form

$$(\rho^2\partial_\rho\psi - p)\nabla\cdot\mathbf{v} - 2\beta\rho\int_0^\infty \mu(s)\mathbf{D}_0^t(s)ds \cdot \int_0^\infty \mu(s)\dot{\mathbf{D}}_0^t(s)ds$$
$$-2\gamma\rho\int_0^\infty \kappa(s)\mathrm{tr}\,\mathbf{D}^t(s)ds\int_0^\infty \kappa(s)\mathrm{tr}\,\dot{\mathbf{D}}^t(s)ds$$
$$+2\mathbf{D}_0 \cdot \int_0^\infty \mu(s)\mathbf{D}_0^t(s)ds + \mathrm{tr}\,\mathbf{D}\int_0^\infty \kappa(s)\mathrm{tr}\,\mathbf{D}^t(s)ds \geq 0.$$

Let $\mathrm{tr}\,\mathbf{D} = \nabla\cdot\mathbf{v} = 0$, at any time. The inequality simplifies to

$$-\beta\int_0^\infty \mu(s)\mathbf{D}_0^t(s)ds \cdot \int_0^\infty \mu(s)\dot{\mathbf{D}}_0^t(s)ds + \mathbf{D}_0 \cdot \int_0^\infty \mu(s)\mathbf{D}_0^t(s)ds \geq 0.$$

Upon an integration by parts we find

$$\int_0^\infty \mu(s)\dot{\mathbf{D}}_0^t(s)ds = -\int_0^\infty \mu(s)\partial_s\mathbf{D}_0^t(s)ds = \mu(0)\mathbf{D}_0(t) + \int_0^\infty \mu'(s)\mathbf{D}_0^t(s)ds.$$

Hence we have

$$-\beta\rho\int_0^\infty \mu(s)\mathbf{D}_0^t(s)ds \cdot \int_0^\infty \mu'(s)\mathbf{D}_0^t(s)ds + [1 - \beta\rho\mu(0)]\mathbf{D}_0(t) \cdot \int_0^\infty \mu(s)\mathbf{D}_0^t(s)ds \geq 0.$$

Now, $\mathbf{D}(t)$ and $\mathbf{D}^t$, as well as $\mathbf{D}_0(t)$ and $\hat{\mathbf{D}}_0^t$, are independent in that, for any value $\hat{\mathbf{D}}(t)$ and history $\mathbf{D}^t$ we can find a continuous function $\tilde{\mathbf{D}}^t$ such that

$$\tilde{\mathbf{D}}(t) = \hat{\mathbf{D}}(t), \qquad \|\tilde{\mathbf{D}}^t - \mathbf{D}^t\| < \epsilon,$$

ϵ being as small as we please. This implies that

$$\beta = \frac{1}{\rho\mu(0)} > 0, \qquad \mu(s)\mu'(s) \leq 0,\ s \in (0, \infty).$$

Now instead let $\mathbf{D}_0 = \mathbf{0}$. We have

$$(\rho^2 \partial_\rho \psi - p)\nabla \cdot \mathbf{v} - 2\gamma\rho \int_0^\infty \kappa(s) \mathrm{tr}\, \mathbf{D}^t(s) ds \int_0^\infty \kappa(s) \mathrm{tr}\, \dot{\mathbf{D}}^t(s) ds + \mathrm{tr}\, \mathbf{D}(t) \int_0^\infty \kappa(s) \mathrm{tr}\, \mathbf{D}^t(s) ds \geq 0.$$

Owing to the arbitrariness of $\nabla \cdot \mathbf{v} = \mathrm{tr}\, \mathbf{D}$ it follows

$$p = \rho^2 \partial_\rho \psi.$$

The inequality for $\mathrm{tr}\, \mathbf{D}$ is then exploited as the previous one for $\mathbf{D}_0$. As a result, it implies that

$$\gamma = \frac{1}{2\rho\kappa(0)}, \qquad \kappa(s)\kappa'(s) \leq 0, \; s \in (0, \infty).$$

The inequalities $\mu\mu' \leq 0$ and $\kappa\kappa' \leq 0$ are satisfied if $\mu'(s) = -\alpha_\mu \mu(s)$ and $\kappa'(s) = \alpha_\kappa \kappa(s)$, where $\alpha_\mu, \alpha_\kappa > 0$, and hence $\mu(s) = \mu(0) \exp(-\alpha_\mu s)$, $\kappa(s) = \kappa(0) \exp(-\alpha_\kappa s)$. The free energy

$$\psi = \Psi(\rho, \theta) + 2|\int_0^\infty \mu(0) \exp(-\alpha_\mu s) \mathbf{D}_0^t(s) ds|^2 + |\int_0^\infty \kappa(0) \exp(-\alpha_\kappa s) \mathrm{tr}\, \mathbf{D}^t(s) ds|^2$$

satisfies the minimum property, at $\mathbf{D}^t = \mathbf{0}$, and moreover complies with the reduced inequality

$$-\rho d\psi + \boldsymbol{T} \cdot \mathbf{D} \geq 0$$

for all histories $\mathbf{D}_0^t$, $\mathrm{tr}\, \mathbf{D}^t$.

An alternative scheme might be looked at with the free energy in the form

$$\psi(\rho, \theta, \mathbf{D}_0^t, \mathrm{tr}\, \mathbf{D}^t) = \Psi(\rho, \theta) + \beta \int_0^\infty \mu(s) |\mathbf{D}_0^t(s)|^2 ds + \gamma \int_0^\infty \kappa(s) |\mathrm{tr}\, \mathbf{D}^t(s)|^2 ds$$

which satisfies the minimum property provided $\beta, \gamma > 0$. However the corresponding entropy production

$$\begin{aligned}-\rho d\psi + \boldsymbol{T} \cdot \mathbf{D} = &-\beta\rho\mu(0)|\mathbf{D}_0(t)|^2 - \beta\rho \int_0^\infty \mu'(s) |\mathbf{D}_0^t(s)|^2 + \mathbf{D}_0(t) \cdot \int_0^\infty \mu(s) \mathbf{D}_0^t(s) ds \\ &-\gamma\rho\kappa(0)|\mathrm{tr}\, \mathbf{D}(t)|^2 - \gamma\rho \int_0^\infty \kappa'(s) |\mathrm{tr}\, \mathbf{D}^t(s)|^2 ds + \mathrm{tr}\, \mathbf{D}(t) \int_0^\infty \kappa(s) \mathrm{tr}\, \mathbf{D}^t(s) ds\end{aligned}$$

cannot have a positive sign for every history $\mathbf{D}^t$.

It is then of interest to investigate the requirements on the kernels μ, κ of the constitutive equation (7.76) without having recourse to the free energy functional.[17] Still let $\rho, \theta, \mathbf{D}^t$ the set of independent variables. Let C be a cycle occurring in the time interval $[0, d]$, so that $(\rho(0), \theta(0), \mathbf{D}^0) = (\rho(d), \theta(d), \mathbf{D}^d)$ and then $\psi(d) - \psi(0) =$

[17] In this way we determine necessary conditions.

0. If θ is kept constant then, upon integration on $t \in [0, d]$, the Clausius–Duhem inequality (7.77) leads to

$$\int_0^\infty \mathbf{T}(t) \cdot \mathbf{D}(t) dt \geq 0. \tag{7.79}$$

Let μ_c, κ_c denote the half-range cosine transform of $\mu, \kappa \in L^1(\mathbb{R}^+)$, e.g.

$$\mu_c(\omega) = \int_0^\infty \mu(s) \cos \omega s \, ds.$$

We then have the following

Theorem 7.6 *The constitutive stress–strain relation (7.76), at constant temperature, is consistent with (7.79) if and only if*

$$\mu_c(\omega) > 0, \qquad \kappa_c(\omega) > 0, \qquad \forall \omega > 0. \tag{7.80}$$

Observe that, in isothermal processes, p is a function of ρ and then

$$\int_0^d (p\mathbf{1} \cdot \mathbf{D})(t) dt = -\int_0^d [p(\rho(t))\dot{\rho}(t)/\rho(t)] dt.$$

Let $h(\rho)$ be a primitive of $p(\rho)/\rho$. Hence

$$\int_0^d [p(\rho(t))\dot{\rho}(t)/\rho(t)] dt = \int_0^d \dot{h}(\rho(t)) dt = h(\rho(d)) - h(\rho(0)) = 0.$$

Hence (7.79) amounts to

$$\int_0^\infty \mathcal{T}(t) \cdot \mathbf{D}(t) dt \geq 0, \qquad \mathcal{T}(t) = 2\int_0^\infty \mu(s)\mathbf{D}_0^t(s) ds + \int_0^\infty \kappa(s)(\mathrm{tr}\,\mathbf{D}^t)(s) ds \, \mathbf{1}.$$

To prove the theorem we select periodic histories in the form

$$\mathbf{D}(t) = \mathbf{D}_1 \cos \omega t + \mathbf{D}_2 \sin \omega t.$$

The proof proceeds as with viscoelastic solids with unbounded kernel, subject to $\mathsf{G}_\infty = \mathbf{0}$ and $\check{\mathsf{G}} = 2\mu\mathsf{I} + \lambda\mathbf{1} \otimes \mathbf{1}$, or directly as in [196] and [158] Sect. 3.7. □

In addition to thermodynamic consistency of the functional $\mathcal{T}$, with a free energy (7.78), we now look for a work-like function. Assume that $\mu_c, \kappa_c > 0$ and let

$$\psi(t) = \Psi(\rho(t), \theta(t)) + \psi^w(\mathbf{D}_r^t),$$

where $\psi^w(\mathbf{D}_r^t)$ is the work, per unit mass, performed by $\mathcal{T}(\mathbf{D}_r^t)$; we have

$$\rho\psi^w(\mathbf{D}_r^t) := \int_{-\infty}^{t} \mathbf{T}(\mathbf{D}_r^\tau)\cdot\mathbf{D}(\tau)d\tau$$
$$= \int_{-\infty}^{t} [2\mathbf{D}(\tau)\cdot\int_0^\infty \mu(s)\mathbf{D}_0(\tau-s)ds + \mathrm{tr}\,\mathbf{D}(\tau)\int_0^\infty \kappa(s)\mathrm{tr}\,\mathbf{D}(t-s)ds]d\tau$$

For technical convenience we extend the definition of μ, κ to $\mathbb{R}^-$ by letting

$$\mu(s)=0,\ \kappa(s)=0,\quad s<0.$$

Hence we can write

$$\rho\psi^w(\mathbf{D}_r^t) = \int_{-\infty}^{t} [2\mathbf{D}_0(\tau)\cdot\int_{-\infty}^{\tau} \mu(\tau-u)\mathbf{D}_0(u)du + \mathrm{tr}\,\mathbf{D}(\tau)\int_{-\infty}^{\tau} \kappa(\tau-u)\mathbf{D}_0(u)du]d\tau$$
$$= \int_{-\infty}^{t}\int_{-\infty}^{t} [\mathbf{D}_0(\tau)\cdot\mu(|\tau-u|)\mathbf{D}_0(u) + \tfrac{1}{2}\mathrm{tr}\,\mathbf{D}(\tau)\kappa(|\tau-u|)\mathrm{tr}\,\mathbf{D}(u)]du\,d\tau$$
$$= \int_0^\infty\int_0^\infty [\mathbf{D}_0^t(s_1)\cdot\mu(|s_1-s_2|)\mathbf{D}_0^t(s_2) + \tfrac{1}{2}\mathrm{tr}\,\mathbf{D}^t(s_1)\kappa(|s_1-s_2|)\mathrm{tr}\,\mathbf{D}^t(s_2)]ds_1\,ds_2.$$

By definition, $\rho\dot{\psi}^w = \boldsymbol{T}\cdot\mathbf{D}$. Thus the Clausius–Duhem inequality holds as an equality for $\psi = \Psi + \psi^w$ with

$$\eta = -\partial_\theta\Psi,\qquad p = \rho^2\partial_\rho\Psi,\qquad -\rho\dot{\psi}^w + \boldsymbol{T}\cdot\mathbf{D} = 0.$$

Since we know the functional $\psi^w(\mathbf{D}_r^t)$ we ask for the minimum property. Extend the history $\mathbf{D}^t$ to $\mathbb{R}^-$ by letting $\mathbf{D}(\tau) = \mathbf{0}$ when $\tau > t$. Hence we have

$$\rho\psi^w(\mathbf{D}_r^t) = 2\int_{-\infty}^{\infty} \mathbf{D}_0(\tau)\cdot\int_{-\infty}^{\infty} \mu(\tau-u)\mathbf{D}_0(u)du\,d\tau + \int_{-\infty}^{\infty} \mathrm{tr}\,\mathbf{D}(\tau)\cdot\int_{-\infty}^{\infty} \kappa(\tau-u)\mathrm{tr}\,\mathbf{D}(u)du\,d\tau$$
$$= \int_{-\infty}^{\infty} [2\mathbf{D}_0(\tau)\cdot(\mu*\mathbf{D}_0)(\tau) + \mathrm{tr}\,\mathbf{D}(\tau)(\kappa*\mathrm{tr}\,\mathbf{D})(\tau)]d\tau,$$

where $*$ denotes the Fourier convolution. Consequently we apply the Parseval–Plancherel theorem to find

$$\rho\psi^w(\mathbf{D}_r^t) = \frac{1}{\pi}\int_{-\infty}^{\infty} \{\mu_c(\omega)(|\mathbf{D}_{0c}(\omega)|^2 + |\mathbf{D}_{0s}(\omega)|^2) + 2\kappa_c(\omega)(|\mathrm{tr}\,\mathbf{D}_c(\omega)|^2 + \mathrm{tr}\,\mathbf{D}_s(\omega)|^2\}d\omega.$$

In view of (7.80) the minimum property at zero histories follows.

7.5.3 Maxwell-Like Viscoelastic Fluids

As outlined in this chapter, memory is naturally modelled by means of functionals. More elementary models of materials with memory are possible in terms of rate-type equations (see Chap. 6). Here we examine Maxwell's model as a rate-type scheme of viscoelastic fluids.

Maxwell's model is based on the differential equation

$$\tau\dot{\mathcal{T}} + \mathcal{T} = 2\mu\mathbf{D},$$

for the stress $\mathcal{T} = \mathbf{T} + p\mathbf{1}$; $\mathcal{T}$ and $\mathbf{D}$ are considered at the same time (and place). Upon integration we find that

$$\mathcal{T}(t) = 2(\mu/\tau) \int_0^\infty \exp(-s/\tau)\mathbf{D}(t-s)ds$$

and then $G(s) = (\mu/\tau)\exp(-s/\tau)$ is the relaxation modulus. Hence Maxwell's model proves to be a special case of Boltzmann's model. If $\mathbf{D}$ is nearly constant then

$$\mathcal{T}(t) \simeq 2(\mu/\tau)\Big[\int_0^\infty \exp(-s/\tau)ds\Big]\mathbf{D}(t) = 2\mu\mathbf{D}(t),$$

as with a Newtonian fluid.

Incidentally, the parameter τ is called the relaxation time and may be a measure of how long a fluid will remember.[18] Let t^* be a characteristic time scale (the ratio of a characteristic length L and a characteristic speed U). The quantity

$$\mathrm{We} = \frac{\tau}{t^*}$$

is called the Weissenberg number. The bigger is the Weissenberg number ($\tau \gg t^*$) the more the fluid behaves like an elastic solid ($\dot{\mathcal{T}} \simeq 2(\mu/\tau)\mathbf{D}$), the smaller is We the more the fluid behaves like a Newtonian fluid ($\mathcal{T} \simeq 2\mu\mathbf{D}$).

Both Boltzmann's model and Maxwell's model are generalized to obtain non-linear properties. Relative to Maxwell's model we first observe that the (total) time derivative is not objective. An objective time derivative is in order to replace $\dot{\mathcal{T}}$ and the Oldroyd derivative $\overset{\triangledown}{\mathcal{T}}$, the Cotter-Rivlin derivative $\overset{\vartriangle}{\mathcal{T}}$, and the corotational derivative $\overset{\circ}{\mathcal{T}}$ are of interest (for incompressible materials). Also linear superposition of any of them can be used; in the Johnson–Segalman model [236] the derivative

$$\overset{\diamond}{\mathcal{T}}_\zeta = (1 - \tfrac{1}{2}\zeta)\,\overset{\triangledown}{\mathcal{T}} + \tfrac{1}{2}\zeta\,\overset{\vartriangle}{\mathcal{T}}$$

is involved. Other non-linearities are inserted by adding objective nonlinear terms like $\kappa\mathcal{T}^2$ (Giesekus model) or $\kappa(\mathrm{tr}\,\mathcal{T})\mathcal{T}$ (Pan-Thien-Tanner model). A generic superposition (of $\overset{\triangledown}{\mathcal{T}}$ and $\overset{\vartriangle}{\mathcal{T}}$) can be written in the form

[18] This is consistent with the definition of natural time (7.74), in that $\Lambda = \tau$.

$$\overset{\diamond}{\mathcal{T}} = \overset{\circ}{\mathcal{T}} + \nu(\mathcal{T}\mathbf{D} + \mathbf{D}\mathcal{T}) = \dot{\mathcal{T}} - \mathbf{W}\mathcal{T} + \mathcal{T}\mathbf{W} + \nu(\mathcal{T}\mathbf{D} + \mathbf{D}\mathcal{T}).$$

Thermodynamic consistency of some generalized Maxwell models

Memory and non-linearity are modelled via the generalized Maxwell model for $\mathcal{T}$ involving a generic objective time derivative. For definiteness we consider the Truesdell time derivative,

$$\overset{\diamond}{\mathcal{T}} = \dot{\mathcal{T}} - \mathbf{L}\mathcal{T} - \mathcal{T}\mathbf{L}^T + (\nabla \cdot \mathbf{v})\mathcal{T}$$

and examine the rate equation

$$\mathbf{f}(\mathcal{T}) + \alpha \overset{\diamond}{\mathcal{T}} = 2\mu\mathbf{D} + \lambda(\mathrm{tr}\,\mathbf{D})\mathbf{1}. \tag{7.81}$$

where $\mathbf{f} : \mathrm{Sym} \to \mathrm{Sym}$ and $\alpha > 0$.

To investigate the thermodynamic consistency of (7.81) we restrict attention to non-conducting fluids and write the entropy inequality in the form

$$-\rho(\dot{\psi} + \eta\dot{\theta}) + \mathbf{T} \cdot \mathbf{D} \geq 0.$$

We let ρ, θ, $\mathcal{T}$, $\mathbf{D}$ be the independent variables. The free energy ψ is (continuously) differentiable while η is continuous. Time differentiation of ψ results in

$$-\rho(\partial_\theta\psi + \eta)\dot{\theta} + (p - \rho^2\partial_\rho\psi)\nabla \cdot \mathbf{v} - \rho\partial_{\mathcal{T}}\psi \cdot \dot{\mathcal{T}} - \rho\partial_{\mathbf{D}}\psi \cdot \dot{\mathbf{D}} + \mathcal{T} \cdot \mathbf{D} \geq 0.$$

Upon substitution of

$$\dot{\mathcal{T}} = \overset{\diamond}{\mathcal{T}} + \mathbf{L}\mathcal{T} + \mathcal{T}\mathbf{L}^T - (\nabla \cdot \mathbf{v})\mathcal{T} = \frac{1}{\alpha}[2\mu\mathbf{D} + \lambda(\nabla \cdot \mathbf{v})\mathbf{1} - \mathbf{f}(\mathcal{T})] + \mathbf{L}\mathcal{T} + \mathcal{T}\mathbf{L}^T - (\nabla \cdot \mathbf{v})\mathcal{T}$$

we obtain the inequality

$$\begin{aligned}-\rho(\partial_\theta\psi + \eta)\dot{\theta} + [\rho^2\partial_\rho\psi - p - \frac{\rho\lambda}{\alpha}\mathrm{tr}\,\partial_{\mathcal{T}}\psi + \rho\partial_{\mathcal{T}}\psi \cdot \mathcal{T}]\nabla \cdot \mathbf{v} + (\mathcal{T} - \frac{2\rho\mu}{\alpha}\partial_{\mathcal{T}}\psi) \cdot \mathbf{D} \\ + \frac{\rho}{\alpha}\partial_{\mathcal{T}}\psi \cdot \mathbf{f}(\mathcal{T}) - \rho\partial_{\mathcal{T}}\psi \cdot [(\mathbf{D} + \mathbf{W})\mathcal{T} + \mathcal{T}(\mathbf{D} - \mathbf{W})] - \rho\partial_{\mathbf{D}}\psi \cdot \dot{\mathbf{D}} \geq 0.\end{aligned}$$

The arbitrariness of $\dot{\mathbf{D}}$ and $\dot{\theta}$ implies that

$$\partial_{\mathbf{D}}\psi = \mathbf{0}, \qquad \eta = -\partial_\theta\psi.$$

Observe that

$$\rho\partial_{\mathcal{T}}\psi \cdot [\mathbf{W}\mathcal{T} - \mathcal{T}\mathbf{W}] = \rho[\partial_{\mathcal{T}}\psi\mathcal{T} - \mathcal{T}\partial_{\mathcal{T}}\psi] \cdot \mathbf{W}.$$

Since

$$\partial_{\mathcal{T}}\psi\mathcal{T} - \mathcal{T}\partial_{\mathcal{T}}\psi \in \mathrm{Skw}$$

then the arbitrariness of $\mathbf{W} \in \mathrm{Skw}$ implies that

$$\partial_{\mathcal{T}}\psi\mathcal{T} - \mathcal{T}\partial_{\mathcal{T}}\psi = \mathbf{0}. \tag{7.82}$$

Since ψ is independent of $\mathbf{D}$, letting $\mathbf{D} = \mathbf{0}$ we find

$$\partial_{\mathcal{T}}\psi \cdot \mathbf{f}(\mathcal{T}) \geq 0. \tag{7.83}$$

Now only linear terms in $\mathbf{D}$ are left. By using the decomposition $\mathbf{D} = \mathbf{D}_0 + \frac{1}{3}(\mathrm{tr}\,\mathbf{D})\mathbf{1}$ we can write

$$[\rho^2\partial_\rho\psi - p - \frac{\rho\lambda}{\alpha}\mathrm{tr}\,\partial_{\mathcal{T}}\psi + \rho\partial_{\mathcal{T}}\psi \cdot \mathcal{T} + \tfrac{1}{3}\mathrm{tr}\,(\mathcal{T} - \frac{2\rho\mu}{\alpha}\partial_{\mathcal{T}}\psi - 2\rho\partial_{\mathcal{T}}\psi\mathcal{T})]\nabla \cdot \mathbf{v}$$
$$+[\mathcal{T} - \frac{2\rho\mu}{\alpha}\partial_{\mathcal{T}}\psi - 2\rho\partial_{\mathcal{T}}\psi\mathcal{T}] \cdot \mathbf{D}_0 \geq 0.$$

The arbitrariness of $\nabla \cdot \mathbf{v}$ and $\mathbf{D}_0$ implies

$$\rho^2\partial_\rho\psi - p - \frac{\rho\lambda}{\alpha}\mathrm{tr}\,\partial_{\mathcal{T}}\psi + \rho\partial_{\mathcal{T}}\psi \cdot \mathcal{T} + \tfrac{1}{3}\mathrm{tr}\,(\mathcal{T} - \frac{2\rho\mu}{\alpha}\partial_{\mathcal{T}}\psi - 2\rho\partial_{\mathcal{T}}\psi\mathcal{T}) = 0, \tag{7.84}$$

$$\mathcal{T} - \frac{2\rho\mu}{\alpha}\partial_{\mathcal{T}}\psi - 2\rho\partial_{\mathcal{T}}\psi\mathcal{T} = \mathbf{0}. \tag{7.85}$$

Hence the admissible free energy functions are subject to (7.82)–(7.85). If p is a function of ρ and θ we may take

$$p = \rho^2\partial_\rho\psi.$$

Otherwise we might let (7.84) be the definition of p.

A simpler scheme arises by considering the corotational derivative $\overset{\circ}{\mathcal{T}}$ and replacing (7.81) by

$$\mathbf{f}(\mathcal{T}) + \alpha\,\overset{\circ}{\mathcal{T}} = 2\mu\mathbf{D} + \lambda(\mathrm{tr}\,\mathbf{D})\mathbf{1}. \tag{7.86}$$

By paralleling the previous derivation we find again that

$$\partial_{\mathbf{D}}\psi = \mathbf{0}, \qquad \eta = -\partial_\theta\psi, \qquad \partial_{\mathcal{T}}\psi\mathcal{T} - \mathcal{T}\partial_{\mathcal{T}}\psi = \mathbf{0}, \qquad \partial_{\mathcal{T}}\psi \cdot \mathbf{f}(\mathcal{T}) \geq 0. \tag{7.87}$$

Consequently it follows that

$$\rho^2\partial_\rho\psi - p - \frac{\rho\lambda}{\alpha}\mathrm{tr}\,\partial_{\mathcal{T}}\psi + \tfrac{1}{3}\mathrm{tr}\,(\mathcal{T} - \frac{2\rho\mu}{\alpha}\partial_{\mathcal{T}}\psi) = 0, \qquad (\mathcal{T} - \frac{2\rho\mu}{\alpha}\partial_{\mathcal{T}}\psi)_0 = \mathbf{0}.$$

Letting

$$p = \rho^2 \partial_\rho \psi - \frac{\rho\lambda}{\alpha} \mathrm{tr}\, \partial_{\mathcal{T}} \psi$$

we obtain

$$\mathcal{T} - \frac{2\rho\mu}{\alpha} \partial_{\mathcal{T}} \psi = \mathbf{0}.$$

The obvious integration results in

$$\psi(\rho, \theta, \mathcal{T}) = \Psi(\rho, \theta) + \frac{\alpha}{4\rho\mu} \mathcal{T} \cdot \mathcal{T}.$$

Moreover, the inequality in (7.87) holds if

$$\mathbf{f}(\mathcal{T}) = \beta(\rho, \theta, \mathcal{T})\mathcal{T}, \qquad \beta/\mu \geq 0.$$

An alternative scheme arises by considering the Lagrangian description of a generalized Maxwell model. Since $\mathbf{T} = -p\mathbf{1} + \mathcal{T}$ then we obtain the second Piola stress in the form

$$\mathbf{T}_{RR} = -Jp\mathbf{C}^{-1} + J\mathbf{F}^{-1}\mathcal{T}\mathbf{F}^{-T}.$$

For formal convenience let

$$\mathcal{T}_{RR} := J\mathbf{F}^{-1}\mathcal{T}\mathbf{F}^{-T}.$$

The power per unit volume then becomes

$$\mathbf{T} \cdot \mathbf{D} = \tfrac{1}{2}\mathbf{T}_{RR} \cdot \dot{\mathbf{C}} = -\tfrac{1}{2} Jp\, \mathbf{C}^{-1} \cdot \dot{\mathbf{C}} + \tfrac{1}{2}\mathcal{T}_{RR} \cdot \dot{\mathbf{C}}.$$

The stress $\mathcal{T}_{RR}$ is assumed to satisfy a rate equation

$$\tau\dot{\mathcal{T}}_{RR} + \mathbf{f}(\theta, \mathbf{C}, \mathcal{T}_{RR}) = \lambda(\theta)\dot{\mathbf{C}}. \tag{7.88}$$

If $\tau = 0$ then $\mathbf{f}(\theta, \mathbf{C}, \mathcal{T}_{RR}) = \lambda(\theta)\dot{\mathbf{C}}$ is a nonlinear description of a viscous material. If $\tau\dot{\mathcal{T}}_{RR}$ prevails then the equation can be viewed as representative of a thermoelastic material $\mathcal{T}_{RR} \simeq (\mathbf{C} - \mathbf{1})$.

We let θ, $\mathbf{C}$, $\mathcal{T}_{RR}$ be the independent variables, $\dot{\mathcal{T}}_{RR}$ being determined by the rate equation (7.88). Hence we consider the constitutive equation

$$\psi = \psi(\theta, \mathbf{C}, \mathcal{T}_{RR})$$

and the like for η and p. We then investigate the restrictions placed by the entropy inequality in the Lagrangian description,

$$-\rho_R(\dot{\psi} + \eta\dot{\theta}) + \tfrac{1}{2}\mathbf{T}_{RR} \cdot \dot{\mathbf{C}} \geq 0.$$

Upon evaluation of $\dot{\psi}$ and substitution of $\dot{\mathcal{T}}_{RR}$ from (7.88) we have

$$-\rho_R(\partial_\theta\psi + \eta)\dot{\theta} + (\tfrac{1}{2}\mathcal{T}_{RR} - \tfrac{1}{2}Jp\mathbf{C}^{-1} - \rho_R\partial_{\mathbf{C}}\psi - \frac{\rho_R\lambda}{\tau}\partial_{\mathcal{T}_{RR}}\psi)\cdot\dot{\mathbf{C}} + \rho_R\partial_{\mathcal{T}_{RR}}\psi\cdot\mathbf{f} \geq 0.$$

The arbitrariness of $\dot{\theta}$ and $\dot{\mathbf{C}}$ implies that

$$\eta = -\partial_\theta\psi, \qquad \tfrac{1}{2}\mathcal{T}_{RR} - \tfrac{1}{2}Jp\mathbf{C}^{-1} - \rho_R\partial_{\mathbf{C}}\psi - \frac{\rho_R\lambda}{\tau}\partial_{\mathcal{T}_{RR}}\psi = \mathbf{0}, \tag{7.89}$$

$$\rho_R\partial_{\mathcal{T}_{RR}}\psi\cdot\mathbf{f} \geq 0. \tag{7.90}$$

So far the pressure p is undetermined; given p, Eq. (7.89)$_2$ is a relation between ψ and $\mathcal{T}_{RR}$. For definiteness, let ψ depend on $\mathbf{C}$ via

$$(\det\mathbf{C})^{1/2} = \det\mathbf{F} = J.$$

Hence

$$\partial_{\mathbf{C}}\psi = \partial_J\psi\,\partial_{\mathbf{C}}J, \qquad \partial_{\mathbf{C}}J = \tfrac{1}{2}(\det\mathbf{C})^{-1/2}\partial_{\mathbf{C}}\det\mathbf{C} = \frac{1}{2J}(\det\mathbf{C})\mathbf{C}^{-1} = \tfrac{1}{2}J\mathbf{C}^{-1}.$$

Consequently we find

$$-\rho_R\partial_{\mathbf{C}}\psi = -\rho_R\partial_J\psi\,\tfrac{1}{2}J\mathbf{C}^{-1}.$$

In addition, since $J = \rho_R/\rho$ then

$$\partial_J\psi = -\partial_\rho\psi\,\rho_R/J^2.$$

Hence we obtain

$$\tfrac{1}{2}Jp\mathbf{C}^{-1} = \tfrac{1}{2}\rho^2 J\partial_\rho\psi\,\mathbf{C}^{-1}$$

whence it follows the classical relation

$$p = \rho^2\partial_\rho\psi.$$

The function ψ is subject to

$$\tfrac{1}{2}\mathcal{T}_{RR} - \frac{\rho_R\lambda}{\tau}\partial_{\mathcal{T}_{RR}}\psi = \mathbf{0} \tag{7.91}$$

and (7.90). In view of (7.91), inequality (7.90) holds if

$$\mathbf{f} = \mathbf{\Lambda}\mathcal{T}_{RR},$$

$\mathbf{\Lambda}/\lambda$ being a fourth-order positive definite tensor. In particular we may have

$$\boldsymbol{\Lambda} = \alpha \mathbf{1} \otimes \mathbf{1}.$$

Hence if $\dot{\boldsymbol{T}}_{RR} = \mathbf{0}$ then

$$\boldsymbol{T}_{RR} = \frac{\lambda}{\alpha}\dot{\mathbf{C}}, \qquad \boldsymbol{T}_{RR} \cdot \dot{\mathbf{C}} \geq 0,$$

which characterizes a dissipative model. Further, upon integration of (7.91) we find

$$\psi = \frac{\tau}{4\rho\lambda}\boldsymbol{T}_{RR} \cdot \boldsymbol{T}_{RR}.$$

Equation (7.88) is now considered in the linear case,

$$\tau\dot{\boldsymbol{T}}_{RR} + \alpha \boldsymbol{T}_{RR} = \lambda \dot{\mathbf{C}},$$

τ, α, λ being constants or known functions of time. Upon integration we find that

$$\boldsymbol{T}_{RR} = \int_{-\infty}^{t} \frac{\lambda}{\tau}(u)\dot{\mathbf{C}}(u)\exp[-\int_{u}^{t}\frac{\alpha}{\tau}(\xi)d\xi]du = \int_{0}^{\infty}\frac{\lambda}{\tau}(t-s)\dot{\mathbf{C}}(t-s)\exp[-\int_{t-s}^{t}\frac{\alpha}{\tau}(\xi)d\xi]ds.$$

Hence it follows that

$$\mathbf{T}(t) = -p(t)\mathbf{1} + \frac{1}{J(t)}\mathbf{F}(t)\int_{0}^{\infty}\frac{\lambda}{\tau}(t-s)\dot{\mathbf{C}}(t-s)\exp[-\int_{t-s}^{t}\frac{\alpha}{\tau}(\xi)d\xi]ds\,\mathbf{F}^{T}(t). \tag{7.92}$$

Since $\dot{\mathbf{C}} = 2\mathbf{F}^{T}\mathbf{D}\mathbf{F}$ then we find

$$\mathbf{T}(t) = -p(t)\mathbf{1} + \frac{1}{J(t)}\int_{0}^{\infty}\frac{\lambda}{\tau}(t-s)\mathbf{F}(t)\mathbf{F}^{T}(t-s)\mathbf{D}(t-s)\mathbf{F}(t-s)\mathbf{F}^{T}(t)\exp[-\int_{t-s}^{t}\frac{\alpha}{\tau}(\xi)d\xi]ds.$$

Upon an integration by parts the integral (7.92) can be written as

$$\mathbf{T}(t) = -p(t)\mathbf{1} + \frac{\lambda}{J\tau}(t)\mathbf{F}(t)[\mathbf{C}(t) - \mathbf{1}]\mathbf{F}^{T}(t)$$
$$-\frac{1}{J(t)}\mathbf{F}(t)\int_{0}^{\infty}[(\frac{\lambda}{\tau})^{\cdot} + \frac{\alpha}{\tau}](t-s)\exp[-\int_{t-s}^{t}\frac{\alpha}{\tau}(\xi)d\xi](\mathbf{C}-\mathbf{1})(t-s)ds\,\mathbf{F}^{T}(t).$$

Equation (7.92) determines the Cauchy stress $\mathbf{T}$ in terms of the history of $\dot{\mathbf{C}}$. Linear approximations follow by letting $J = 1$ and $\mathbf{F} = \mathbf{1}$. In that case

$$\mathbf{T}(t) = -p(t)\mathbf{1} + 2\int_{0}^{\infty}\frac{\lambda}{\tau}(t-s)\exp[-\int_{t-s}^{t}\frac{\alpha}{\tau}(\xi)d\xi]\mathbf{D}^{t}(s)ds.$$

We can then view

$$m(t,s) = 2\frac{\lambda}{\tau}(t-s)\exp[-\int_{t-s}^{t}\frac{\alpha}{\tau}(\xi)d\xi]$$

as the memory function and conclude that

$$\mathbf{T}(t) = -p(t)\mathbf{1} + \int_0^\infty m(t,s)\,\mathbf{D}^t(s)ds.$$

It is worth mentioning the BKZ constitutive equation [36],

$$\mathbf{T}(t) = -p(t)\mathbf{1} + \int_{-\infty}^{t} [\mathcal{F}^{-1}\partial_{\mathbf{E}}\mathcal{U}\,\mathcal{F}^{-T}](s)ds$$

where $\mathcal{F}$ is the relative deformation gradient,

$$\mathcal{F}_{ij}(s) = F_{iK}(s)F^{-1}_{Kj}(t),$$

and $\mathcal{U}$ is a function of the strain $\mathbf{E} = \frac{1}{2}(\mathbf{C} - \mathbf{1})$. No thermodynamic setting seems to have been established for the BKZ equation. The stress is assumed to be a linear functional of the strain history and the strain is measured with the current configuration as reference.

7.5.4 Acceleration Waves in Viscous Fluids with Memory

Here the properties of acceleration waves in viscous fluids are investigated. For definiteness and simplicity the fluid is assumed to be at a constant temperature and hence the pertinent balance equations are

$$\dot{\rho} = -\rho\nabla\cdot\mathbf{v}, \qquad \rho\dot{\mathbf{v}} = -\nabla p(\rho) + \nabla\cdot\mathcal{T}.$$

Two constitutive models for the (viscous) Cauchy stress $\mathcal{T}$ are considered in two cases according as the fluid is modelled by a rate-type equation or by a Boltzmann-like equation,

$$\mathcal{T} + \alpha\overset{\circ}{\mathcal{T}} = 2\mu\mathbf{D} + \lambda(\mathrm{tr}\,\mathbf{D})\mathbf{1}, \quad \mathcal{T} = 2\int_0^\infty \mu(s)\mathbf{D}(t-s)ds + \int_0^\infty \lambda(s)\mathrm{tr}\,\mathbf{D}(t-s)ds. \tag{7.93}$$

The pressure p and the scalars α, μ, ν are allowed to be parameterized by the temperature θ.

We now investigate the compatibility of this scheme with wave propagation at a finite speed. Let σ be a time-dependent surface, in the current configuration Ω, and denote by $\mathbf{n}$ the unit normal. The surface σ is a second-order wave (or acceleration wave) if

– $\mathbf{v}$, ρ are continuous functions of $\mathbf{x}$ and t in a neighbourhood $\mathcal{N}$ of σ,

– $\dot{\mathbf{v}}$, $\mathbf{L}$, $\dot{\rho}$, $\nabla\rho$ and all higher-order derivatives suffer jump discontinuities across σ but are continuous functions of $\mathbf{x}$ and t in $\mathcal{N}\setminus\sigma$.

Since $[\![\mathbf{v}]\!] = \mathbf{0}$ then Poisson's condition requires that

$$[\![p]\!] = 0, \qquad [\![\mathcal{T}]\!] = \mathbf{0}.$$

The continuity of p is consistent with the assumption $p = p(\rho)$, the continuity of $\mathcal{T}$ is also a consequence of Sect. 2.17.6 if the Boltzmann-like equation is chosen.

Let

$$\mathbf{a} := [\![\partial_n \mathbf{v}]\!], \qquad \varrho := [\![\partial_n \rho]\!]$$

where $\partial_n = \mathbf{n} \cdot \nabla$ and denote by $U = u_n - \mathbf{v} \cdot \mathbf{n}$ the local speed of propagation. The Hugoniot compatibility conditions and Maxwell's theorem imply that

$$[\![\dot{\rho}]\!] = -U[\![\partial_n \rho]\!], \qquad [\![\dot{\mathbf{v}}]\!] = -U[\![\partial_n \mathbf{v}]\!],$$

$$[\![\mathbf{D}]\!] = \tfrac{1}{2}(\mathbf{a} \otimes \mathbf{n} + \mathbf{n} \otimes \mathbf{a}), \qquad [\![\mathbf{W}]\!] = \tfrac{1}{2}(\mathbf{a} \otimes \mathbf{n} - \mathbf{n} \otimes \mathbf{a}), \qquad [\![\nabla \cdot \mathbf{v}]\!] = \mathbf{a} \cdot \mathbf{n}.$$

Maxwell-like models

The discontinuities $\mathbf{a}$, ϱ are subject to the balance equations. From the continuity equation $\dot{\rho} = -\rho \nabla \cdot \mathbf{v}$ it follows

$$U \varrho = \rho \mathbf{a} \cdot \mathbf{n}.$$

Now, by Maxwell's theorem

$$[\![\nabla p]\!] = p'[\![\nabla \rho]\!] = p' \varrho \mathbf{n}.$$

Moreover,

$$[\![\nabla \cdot \mathcal{T}]\!] = [\![\partial_n \mathcal{T}]\!]\mathbf{n} = -\frac{1}{U}[\![\dot{\mathcal{T}}]\!]\mathbf{n}.$$

Hence the equation of motion $\rho \dot{\mathbf{v}} = -\nabla p + \nabla \cdot \mathcal{T}$ implies that

$$\rho U^2 \mathbf{a} = \rho p'(\mathbf{a} \cdot \mathbf{n})\mathbf{n} + [\![\dot{\mathcal{T}}]\!]\mathbf{n}. \tag{7.94}$$

To compute $[\![\dot{\mathcal{T}}]\!]$ we look at the constitutive equation $(7.93)_1$ to obtain

$$[\![\dot{\mathcal{T}}]\!] = [\![\mathbf{W}]\!]\mathcal{T} - \mathcal{T}[\![\mathbf{W}]\!] + \frac{2\mu}{\alpha}[\![\mathbf{D}]\!] + \frac{\lambda}{\alpha}[\![\mathrm{tr}\, \mathbf{D}]\!]\mathbf{1}$$

whence

$$[\![\dot{\mathcal{T}}]\!] = \tfrac{1}{2}(\mathbf{a} \otimes \mathbf{n} - \mathbf{n} \otimes \mathbf{a})\mathcal{T} - \tfrac{1}{2}\mathcal{T}(\mathbf{a} \otimes \mathbf{n} - \mathbf{n} \otimes \mathbf{a}) + \frac{\mu}{\alpha}(\mathbf{a} \otimes \mathbf{n} + \mathbf{n} \otimes \mathbf{a}) + \frac{\lambda}{\alpha}(\mathbf{a} \cdot \mathbf{n})\mathbf{1}.$$

Upon substitution we find the propagation condition

$$\rho U^2 \mathbf{a} = \rho p'(\mathbf{a}\cdot\mathbf{n})\mathbf{n} + \tfrac{1}{2}(\mathbf{a}\otimes\mathbf{n} - \mathbf{n}\otimes\mathbf{a})\mathcal{T}\mathbf{n} - \tfrac{1}{2}\mathcal{T}(\mathbf{a} - (\mathbf{a}\cdot\mathbf{n})\mathbf{n}) + \frac{\mu}{\alpha}(\mathbf{a} + (\mathbf{a}\cdot\mathbf{n})\mathbf{n}) + \frac{\lambda}{\alpha}(\mathbf{a}\cdot\mathbf{n})\mathbf{n}.$$

Solutions for the vector $\mathbf{a}$ occur depending on $\mathcal{T}$ and hence on the state ahead of the wave. Two cases are considered. First let $\mathcal{T} = \beta\mathbf{1}$. Hence

$$(\mathbf{a}\otimes\mathbf{n} - \mathbf{n}\otimes\mathbf{a})\mathcal{T}\mathbf{n} - \mathcal{T}(\mathbf{a} - (\mathbf{a}\cdot\mathbf{n})\mathbf{n}) = \mathbf{0}$$

and the equation for $\mathbf{a}$ reduces to

$$\rho U^2 \mathbf{a} = \rho p'(\mathbf{a}\cdot\mathbf{n})\mathbf{n} + \frac{\mu}{\alpha}(\mathbf{a} + (\mathbf{a}\cdot\mathbf{n})\mathbf{n}) + \frac{\lambda}{\alpha}(\mathbf{a}\cdot\mathbf{n})\mathbf{n}.$$

Two solutions occur. Longitudinal waves, where $\mathbf{a} \propto \mathbf{n}$, happen with

$$U^2 = p' + \frac{2\mu + \lambda}{\rho\alpha}.$$

Transverse waves, where $\mathbf{a}\cdot\mathbf{n} = 0$, happen with

$$U^2 = \frac{2\mu}{\rho\alpha}.$$

The second case is $\mathcal{T} = T\mathbf{n}\otimes\mathbf{n}$, which represents a normal traction along $\mathbf{n}$. Hence

$$(\mathbf{a}\otimes\mathbf{n} - \mathbf{n}\otimes\mathbf{a})\mathcal{T}\mathbf{n} - \mathcal{T}(\mathbf{a} - (\mathbf{a}\cdot\mathbf{n})\mathbf{n}) = T\mathbf{a}_\perp,$$

where $\mathbf{a}_\perp = \mathbf{a} - (\mathbf{a}\cdot\mathbf{n})\mathbf{n}$. The propagation equation takes the form

$$\rho U^2 \mathbf{a} = \rho p'(\mathbf{a}\cdot\mathbf{n})\mathbf{n} + T\mathbf{a}_\perp + \frac{\mu}{\alpha}(\mathbf{a} + (\mathbf{a}\cdot\mathbf{n})\mathbf{n}) + \frac{\lambda}{\alpha}(\mathbf{a}\cdot\mathbf{n})\mathbf{n}.$$

Longitudinal waves still propagate with $U^2 = p' + (2\mu + \lambda)/\rho\alpha$ whereas transverse waves happen with

$$U^2 = p' + \frac{T}{\rho} + \frac{2\mu + \lambda}{\rho\alpha}. \tag{7.95}$$

Hence Eq. $(7.93)_1$ allows for the propagation of acceleration waves, the parameters μ, λ showing a role analogous to that of Lamé moduli in elasticity. The propagation mode is affected by the stress $\mathcal{T}$ ahead of the front σ and hence depends on the chosen derivative. As with (7.95) where T/ρ occurs, the speed U and the polarization $\mathbf{a}$ are affected by the type of objective time derivative as it follows with the generic derivative $\dot{\mathcal{T}} - \mathbf{W}\mathcal{T} + \mathcal{T}\mathbf{W} + \nu(\mathbf{D}\mathcal{T} + \mathcal{T}\mathbf{D})$.

Boltzmann-like models

Let $\mathcal{T}$ be given by the Boltzmann-like relation $(7.93)_2$. We observe that $[\![\mathcal{T}]\!] = \mathbf{0}$ while

$$[\![\dot{\mathcal{T}}]\!] = 2\mu(0)[\![\mathbf{D}]\!] + \lambda(0)[\![\mathrm{tr}\,\mathbf{D}]\!]\mathbf{1} = \mu(0)(\mathbf{a}\otimes\mathbf{n} + \mathbf{n}\otimes\mathbf{a}) + \lambda(0)(\mathbf{a}\cdot\mathbf{n})\mathbf{1}$$

and

$$[\![\dot{\mathcal{T}}]\!]\mathbf{n} = \mu(0)[\mathbf{a} + (\mathbf{a}\cdot\mathbf{n})\mathbf{n}] + \lambda(0)(\mathbf{a}\cdot\mathbf{n})\mathbf{n}.$$

Substitution in (7.94) yields

$$\rho U^2\mathbf{a} = \rho p'(\mathbf{a}\cdot\mathbf{n})\mathbf{n} + \mu(0)[\mathbf{a} + (\mathbf{a}\cdot\mathbf{n})\mathbf{n}] + \lambda(0)(\mathbf{a}\cdot\mathbf{n})\mathbf{n}.$$

Longitudinal waves, $\mathbf{a} \propto \mathbf{n}$, happen with

$$U^2 = p' + \frac{2\mu(0) + \lambda(0)}{\rho}$$

while transverse waves, $\mathbf{a}\cdot\mathbf{n} = 0$, happen with

$$U^2 = \frac{\mu(0)}{\rho}.$$

Measurements of wave propagation speeds allow us to determine the initial values $\mu(0)$, $\lambda(0)$ of the relaxation functions.

7.6 Electromagnetic Materials with Memory

We consider polarizable and magnetizable undeformable solids acted upon by electromagnetic fields. For simplicity the solid is taken to be macroscopically rigid so that we can select the reference frame at rest with the body. The constitutive properties are expressed by letting the polarization $\mathbf{P}$, the magnetization $\mathbf{M}$, and the density current $\mathbf{J}$ depend on the whole history of the electric field $\mathbf{E}$ and the magnetic field $\mathbf{H}$. Moreover, thermal properties are modelled by letting the constitutive functions depend on the histories θ^t and $\nabla\theta^t$. For the sake of generality we let

$$\Lambda = (\theta, \mathbf{E}, \mathbf{H}, \theta^t_r, \mathbf{E}^t_r, \mathbf{H}^t_r, \nabla\theta^t_r)$$

be the set of variables for the constitutive functionals of the free energy ψ, the entropy η, the polarization $\mathbf{P}$, the magnetization $\mathbf{M}$, the electric current $\mathbf{J}$, the heat flux $\mathbf{q}$, and the extra-entropy flux $\mathbf{k}$. The constitutive functionals are required to be consistent with the second-law inequality (2.85), that is

$$-\rho(\dot{\psi}+\eta\dot{\theta})+\mathbf{E}\cdot\dot{\mathbf{P}}+\mu_0\mathbf{H}\cdot\dot{\mathbf{M}}+\mathbf{E}\cdot\mathbf{J}-\frac{1}{\theta}\mathbf{q}\cdot\nabla\theta+\theta\nabla\cdot\mathbf{k}\geq 0.$$

For technical convenience we consider the free energy density[19] ϕ given by

$$\rho\phi=\rho\psi-\mathbf{E}\cdot\mathbf{P}-\mu_0\mathbf{H}\cdot\mathbf{M}.$$

Hence, since ρ is constant then we obtain the inequality in the form

$$-\rho(\dot{\phi}+\eta\dot{\theta})-\mathbf{P}\cdot\dot{\mathbf{E}}-\mu_0\mathbf{M}\cdot\dot{\mathbf{H}}+\mathbf{E}\cdot\mathbf{J}-\frac{1}{\theta}\mathbf{q}\cdot\nabla\theta+\theta\nabla\cdot\mathbf{k}\geq 0.$$

We now derive some thermodynamic restrictions placed on the constitutive functionals $\phi, \eta, \mathbf{P}, \mathbf{M}, \mathbf{J}, \mathbf{q}, \mathbf{k}$. We let ϕ be a differentiable functional of Λ; the entropy inequality can be written in the form

$$\begin{aligned}-\rho(\partial_\theta\phi+\eta)\dot{\theta}-(\rho\partial_{\mathbf{E}}\phi+\mathbf{P})\cdot\dot{\mathbf{E}}-(\rho\partial_{\mathbf{H}}\phi+\mu_0\mathbf{M})\cdot\dot{\mathbf{H}}+\mathbf{E}\cdot\mathbf{J}-\frac{1}{\theta}\mathbf{q}\cdot\nabla\theta+\theta\nabla\cdot\mathbf{k}\\ -\rho d\phi(\Lambda|\dot{\theta}^t_r)-\rho d\phi(\Lambda|\dot{\mathbf{E}}^t_r)-\rho d\phi(\Lambda|\dot{\mathbf{H}}^t_r)-\rho d\phi(\Lambda|\nabla\dot{\theta}^t_r)\geq 0.\end{aligned}$$

Formally, the extra-entropy flux $\mathbf{k}$ might depend on θ and θ^t_r so that

$$\nabla\cdot\mathbf{k}=\nabla\cdot\mathbf{k}(\theta,\theta^t_r);$$

other dependences on Λ would be ruled out by the entropy inequality in view of the arbitrariness of $\nabla\mathbf{E}$, $\nabla\mathbf{H}$, $\nabla\mathbf{E}^t_r$, $\nabla\mathbf{H}^t_r$. Yet the dependence of the vector $\mathbf{k}$ on the scalars θ and θ^t_r is ruled out unless anisotropic material vectors occur. Hence we let $\mathbf{k}=\mathbf{0}$.

As for the degree of arbitrariness of the electromagnetic fields we observe that Maxwell's equations are constraints on the admissible fields. Now

$$\nabla\times\mathbf{H}=\mathbf{J}+\epsilon_0\partial_t\mathbf{E}+\partial_t\mathbf{P},\qquad \nabla\times\mathbf{E}=-\partial_t\mathbf{B}$$

show that arbitrary values of $\partial_t\mathbf{E}=\dot{\mathbf{E}}$ and $\partial_t\mathbf{H}=\dot{\mathbf{H}}$ are allowed provided $\nabla\times\mathbf{H}$ and $\nabla\times\mathbf{E}$ satisfy the two equations. We can then assume that at a point $\mathbf{x}$ and at a time t the vectors $\nabla\times\mathbf{H}$ and $\nabla\times\mathbf{E}$ make Maxwell's equations to hold for arbitrarily selected $\dot{\mathbf{E}}$ and $\dot{\mathbf{B}}$. Furthermore, the set of variables Λ does not involve gradients of $\mathbf{E}$ and $\mathbf{H}$. As a consequence the selection of arbitrary (vector) values $\dot{\mathbf{E}}$, $\dot{\mathbf{H}}$ is consistent with Maxwell's equations and meanwhile does not affect the variables in Λ.

The arbitrariness of $\dot{\theta}$, $\dot{\mathbf{E}}$, $\dot{\mathbf{H}}$ implies that

$$\eta=-\partial_\theta\phi,\qquad \mathbf{P}=-\rho\partial_{\mathbf{E}}\phi,\qquad \mu_0\mathbf{M}=-\rho\partial_{\mathbf{H}}\phi.$$

[19] This function is considered also in [97, 98] where it is denoted by ζ and called free enthalpy density.

Hence it follows the dissipation inequality

$$\mathbf{E} \cdot \mathbf{J} - \frac{1}{\theta}\mathbf{q} \cdot \nabla\theta - \rho d\phi(\Lambda|\dot{\theta}_r^t) - \rho d\phi(\Lambda|\dot{\mathbf{E}}_r^t) - \rho d\phi(\Lambda|\dot{\mathbf{B}}_r^t) - \rho d\phi(\Lambda|\nabla\dot{\theta}_r^t) \geq 0.$$

For definiteness particular models are now investigated.

7.6.1 Dielectrics with Memory

A dielectric is a continuum that can be polarized by an applied electric field. When a dielectric is placed in an electric field, electric charges do not flow through the material as they do in an electrical conductor but only slightly shift from their average equilibrium positions causing dielectric polarization. We let $\mathbf{J} = \mathbf{0}$ as with electrical insulators.[20] It is assumed that $\mathbf{M} = \mathbf{0}, \mathbf{q} = \mathbf{0}$, and that the constitutive dependence on the field $\mathbf{H}$ and the temperature gradient $\nabla\theta$ is negligible. Hence $\Lambda = (\theta, \mathbf{E}, \theta_r^t, \mathbf{E}_r^t)$ is the set of independent variables and the constitutive properties are subject to

$$-\rho\dot{\phi} - \rho\eta\dot{\theta} - \mathbf{P} \cdot \dot{\mathbf{E}} \geq 0.$$

Consequently we find that

$$\eta = -\partial_\theta\phi, \qquad \mathbf{P} = -\rho\partial_{\mathbf{E}}\phi, \qquad d\phi(\Lambda|\dot{\theta}_r^t) + d\phi(\Lambda|\dot{\mathbf{E}}_r^t) \leq 0.$$

Moreover, given two histories $\theta^{t_0}, \mathbf{E}^{t_0}$ consider the constant continuations $\theta^t, \mathbf{E}^t$, as

$$\theta^t(s) = \begin{cases} \theta(t_0), & s \in [0, t - t_0), \\ \theta^{t_0}(s - t + t_0), & s \in [t - t_0, \infty), \end{cases}$$

and the like for $\mathbf{E}^t$. Hence

$$\dot{\phi}(\tau) \leq 0, \ \tau \in (t_0, t); \qquad \phi(t) \leq \phi(t_0).$$

As $t - t_0 \to \infty$ we have

$$\phi(\theta^\dagger(t_0), \mathbf{E}^\dagger(t_0)) \leq \phi(\theta^{t_0}, \mathbf{E}^{t_0}).$$

Among the histories ending with the values $\theta(t_0), \mathbf{E}(t_0)$ the constant histories make ϕ minimum.

In the simplest model, (linear) dielectrics are characterized by the constitutive equation

$$\mathbf{P} = \epsilon_0\chi\mathbf{E}$$

[20] Faraday introduced the term dielectric as synonymous with insulator.

where ϵ_0 is the permittivity of free space and χ is the dielectric susceptibility. Hence

$$\mathbf{D} = \epsilon_0 \epsilon_r \mathbf{E},$$

where $\epsilon_r = 1 + \chi$ is the relative permittivity. In anisotropic dielectrics the permittivity is a second-order tensor. To fix ideas we model dielectrics as linear materials with memory in the form

$$\mathbf{P}(t) = \boldsymbol{\Sigma}_0 \mathbf{E}(t) + \int_0^\infty \boldsymbol{\Sigma}'(s)\mathbf{E}(t-s)ds, \tag{7.96}$$

where $\boldsymbol{\Sigma}_0$ and $\boldsymbol{\Sigma}'$ are allowed to depend on the (present) temperature $\theta(t)$. Since

$$-\rho \partial_{\mathbf{E}} \partial_{\mathbf{E}} \phi = \boldsymbol{\Sigma}_0$$

then it follows that $\boldsymbol{\Sigma}_0 \in \mathrm{Sym}$. We assume that $\boldsymbol{\Sigma}' \in \mathrm{Sym}$ too and that

$$\lim_{s\to\infty} \boldsymbol{\Sigma}'(s) = \mathbf{0}.$$

We might observe that, by (7.96),

$$\rho\phi = \tfrac{1}{2}\mathbf{E} \cdot \boldsymbol{\Sigma}_0 \mathbf{E} + \mathbf{E} \cdot \int_0^\infty \mathbf{E}(t-s)ds + \rho\hat{\phi}(\mathbf{E}^t)$$

and determine $\hat{\phi}$ so that ϕ is consistent with thermodynamics. Instead, more conveniently, let

$$\boldsymbol{\Sigma}_\infty = \boldsymbol{\Sigma}_0 + \int_0^\infty \boldsymbol{\Sigma}'(s)ds \in \mathrm{Sym}$$

and write (7.96) in the form

$$\mathbf{P}(t) = \boldsymbol{\Sigma}_\infty \mathbf{E}(t) - \int_0^\infty \boldsymbol{\Sigma}'(s)[\mathbf{E}(t) - \mathbf{E}(t-s)]ds.$$

Then $\rho\phi$ can be taken in the form

$$\rho\phi = -\tfrac{1}{2}\mathbf{E} \cdot \boldsymbol{\Sigma}_\infty \mathbf{E} + \tfrac{1}{2}\int_0^\infty [\mathbf{E}(t) - \mathbf{E}(t-s)] \cdot \boldsymbol{\Sigma}'(s)[\mathbf{E}(t) - \mathbf{E}(t-s)]ds + \rho\tilde{\phi}$$

where $\rho\tilde{\phi}$ depends on θ. The functional ϕ is required to satisfy the minimum property at constant histories and

$$d\phi(\Lambda|\dot{\mathbf{E}}_r^t) \le 0. \tag{7.97}$$

The minimum property holds if and only if

$$\boldsymbol{\Sigma}' \geq \mathbf{0}.$$

Now notice that

$$\rho d\phi(\Lambda|\dot{\mathbf{E}}_r^t) = -\int_0^\infty \dot{\mathbf{E}}(t-s) \cdot \boldsymbol{\Sigma}'(s)[\mathbf{E}(t) - \mathbf{E}(t-s)]ds.$$

An integration by parts and the observation that

$$\big[[\mathbf{E}(t) - \mathbf{E}(t-s)] \cdot \boldsymbol{\Sigma}'(s)[\mathbf{E}(t) - \mathbf{E}(t-s)]\big]_0^\infty = 0$$

result in

$$\begin{aligned} \rho d\phi &= -\int_0^\infty \partial_s[\mathbf{E}(t) - \mathbf{E}(t-s)] \cdot \boldsymbol{\Sigma}'(s)[\mathbf{E}(t) - \mathbf{E}(t-s)]ds \\ &= \tfrac{1}{2}\int_0^\infty [\mathbf{E}(t) - \mathbf{E}(t-s)] \cdot \boldsymbol{\Sigma}''(s)[\mathbf{E}(t) - \mathbf{E}(t-s)]ds. \end{aligned}$$

In view of (7.97) it follows that

$$\boldsymbol{\Sigma}'' \leq \mathbf{0}.$$

Hence, by

$$\boldsymbol{\Sigma}'(s) = \boldsymbol{\Sigma}'(0) + \int_0^s \boldsymbol{\Sigma}''(\xi)d\xi$$

it follows that

$$\boldsymbol{\Sigma}'(0) = -\int_0^s \boldsymbol{\Sigma}''(\xi)d\xi > \mathbf{0}$$

and $\boldsymbol{\Sigma}'$ is a non-negative, decreasing tensor. Furthermore

$$\boldsymbol{\Sigma}(s) = \boldsymbol{\Sigma}(0) + \int_0^s \boldsymbol{\Sigma}'(\xi)d\xi$$

implies that

$$\boldsymbol{\Sigma}_\infty > \boldsymbol{\Sigma}_0.$$

It is worth remarking the analogy and a difference between viscoelasticity and dielectrics. A strictly formal analogy holds between stress–strain (7.29) and polarization-electric field (7.96) with the correspondence $\mathbf{G} \leftrightarrow \boldsymbol{\Sigma}$. Nevertheless we have

$$\mathbf{G}_\infty < \mathbf{G}_0, \qquad \boldsymbol{\Sigma}_\infty > \boldsymbol{\Sigma}_0.$$

These different inequalities follow from the terms

$$\mathbf{T} \cdot \dot{\mathbf{E}}, \qquad -\mathbf{P} \cdot \dot{\mathbf{E}},$$

and hence with different signs, in the entropy inequality.

7.6.2 Conductors with Memory

Both electric and thermal conduction are considered and the conductor is modelled by letting

$$\Lambda = (\theta, \mathbf{E}, \nabla\theta, \mathbf{E}^t_r, \nabla\theta^t_r)$$

be the independent variables so that[21] $\psi, \eta, \mathbf{J}, \mathbf{q}$ are given by functionals of Λ. The second-law inequality becomes

$$-\rho(\partial_\theta\psi + \eta)\dot{\theta} - \rho\partial_{\mathbf{E}}\psi \cdot \dot{\mathbf{E}} - \rho\partial_{\nabla\theta}\psi \cdot \nabla\dot{\theta} + \mathbf{E}\cdot\mathbf{J} - \frac{1}{\theta}\mathbf{q}\cdot\nabla\theta - \rho d\psi(\Lambda|\dot{\mathbf{E}}^t_r) - \rho d\psi(\Lambda|\nabla\dot{\theta}^t_r) \geq 0.$$

The arbitrariness of $\dot{\mathbf{E}}$, $\nabla\dot{\theta}$, and $\dot{\theta}$ implies

$$\partial_{\mathbf{E}}\psi = \mathbf{0}, \qquad \partial_{\nabla\theta}\psi = \mathbf{0}, \qquad \eta = -\partial_\theta\psi.$$

Hence it follows the dissipation inequality

$$\mathbf{E}\cdot\mathbf{J} - \frac{1}{\theta}\mathbf{q}\cdot\nabla\theta - \rho d\psi(\Lambda|\dot{\mathbf{E}}^t_r) - \rho d\psi(\Lambda|\nabla\dot{\theta}^t_r) \geq 0. \tag{7.98}$$

We remark that the occurrence of $\mathbf{E}$ and $\nabla\theta$, instead of time derivatives, does not allow ψ to enjoy the minimum property at constant histories.

We now look for possible functionals associated with $\mathbf{J}$, $\mathbf{q}$, and ψ. First observe that ψ cannot depend on the present values $\mathbf{E}$, $\nabla\theta$. Moreover functionals in the form of a quadratic integral (in $\mathbf{E}^t$) as

$$\rho\gamma(\mathbf{E}^t) = \tfrac{1}{2}\int_0^\infty \mathbf{E}^t(s)\cdot\mathbf{\Gamma}(s)\mathbf{E}^t(s)ds, \qquad \mathbf{\Gamma}(\infty) = \mathbf{0},$$

and the like for $\mathbf{q}$, would result in the inequality

$$\mathbf{E}\cdot(\mathbf{J} - \tfrac{1}{2}\mathbf{\Gamma}(0)\mathbf{E}) - \tfrac{1}{2}\int_0^\infty \mathbf{E}^t(s)\cdot\mathbf{\Gamma}'(s)\mathbf{E}^t(s)ds \geq 0$$

which holds if

$$\mathbf{J} = \tfrac{1}{2}\mathbf{\Gamma}(0)\mathbf{E}, \qquad \mathbf{\Gamma}'(s) \leq \mathbf{0},$$

possibly with $\mathbf{\Gamma}(0) \geq \mathbf{0}$. Though this is thermodynamically consistent, the constitutive equation for $\mathbf{J}$ does not model a material with memory.

[21] Since $\mathbf{P} = \mathbf{0}$ then $\phi = \psi$.

We then let $\boldsymbol{\Gamma}_0, \mathbf{K}_0 \in \mathrm{Sym}$ be positive definite and look for linear functionals of $\mathbf{E}^t$ and $\nabla\theta^t$,

$$\rho\psi = \rho\psi_0(\theta) + \tfrac{1}{2}\alpha(\theta)\big|\boldsymbol{\Gamma}_0^{1/2}\int_0^\infty \lambda(s)\mathbf{E}^t(s)ds\big|^2 + \tfrac{1}{2}\beta(\theta)\big|\mathbf{K}_0^{1/2}\int_0^\infty \nu(s)\nabla\theta^t(s)ds\big|^2, \tag{7.99}$$

where λ, ν are differentiable, $\lambda(\infty), \nu(\infty) = 0$ while α and β are positive-valued. Let $\rho d\psi$ stand for $\rho d\psi(\Lambda|\dot{\mathbf{E}}_r^t) + \rho d\psi(\Lambda|\nabla\dot{\theta}_r^t)$. Upon integrations by parts we find

$$\begin{aligned}\rho d\psi = \alpha \int_0^\infty \lambda(u)\dot{\mathbf{E}}^t(u)du \cdot \boldsymbol{\Gamma}_0 \int_0^\infty \lambda(s)\mathbf{E}^t(s)ds + \beta \int_0^\infty \nu(u)\nabla\dot{\theta}^t(u)du \cdot \mathbf{K}_0 \int_0^\infty \nu(s)\nabla\theta^t(s)ds \\ = \alpha\{\lambda(0)\mathbf{E} + \int_0^\infty \lambda'(u)\mathbf{E}^t(u)du\} \cdot \boldsymbol{\Gamma}_0 \int_0^\infty \lambda(u)\mathbf{E}^t(u)du \\ +\beta\{\nu(0)\nabla\theta + \int_0^\infty \nu'(u)\nabla\dot{\theta}^t(u)du\} \cdot \mathbf{K}_0 \int_0^\infty \nu(s)\nabla\theta^t(s)ds.\end{aligned}$$

Hence we can write the dissipation inequality (7.98) in the form

$$\begin{aligned}\mathbf{E} \cdot \{\mathbf{J} - \alpha\lambda(0)\boldsymbol{\Gamma}_0 \int_0^\infty \lambda(s)\mathbf{E}^t(s)ds\} - \nabla\theta \cdot \{\theta^{-1}\mathbf{q} + \beta\nu(0)\mathbf{K}_0 \int_0^\infty \nu(s)\nabla\theta^t(s)ds\} \\ -\alpha \int_0^\infty \lambda'(s)\mathbf{E}^t(s)ds \cdot \boldsymbol{\Gamma}_0 \int_0^\infty \lambda(s)\mathbf{E}^t(s)ds - \beta \int_0^\infty \nu'(s)\nabla\theta^t(s)ds \cdot \boldsymbol{\Gamma}_0 \int_0^\infty \nu(s)\nabla\theta^t(s)ds \ge 0.\end{aligned}$$

This inequality holds if and only if

$$\mathbf{J} = \alpha\lambda(0)\boldsymbol{\Gamma}_0 \int_0^\infty \lambda(s)\mathbf{E}^t(s)ds, \tag{7.100}$$

$$\mathbf{q} = -\theta\beta\nu(0)\mathbf{K}_0 \int_0^\infty \nu(s)\nabla\theta^t(s)ds, \tag{7.101}$$

$$\lambda'\lambda \le 0, \qquad \nu'\nu \le 0. \tag{7.102}$$

Inequalities (7.102) imply that

$$(\lambda^2)' \le 0, \qquad (\nu^2)' \le 0,$$

and hence that λ and ν do not change sign and that $|\lambda|$, $|\nu|$ are decreasing functions. This in turn implies that

$$\lambda(0)\lambda(s) > 0, \qquad \nu(0)\nu(s) > 0.$$

When $\mathbf{E}$ and $\nabla\theta$ are slowly variable in time we have

$$\mathbf{J} \simeq a(\theta)\boldsymbol{\Gamma}_0\mathbf{E}, \qquad \mathbf{q} \simeq -b(\theta)\mathbf{K}_0\nabla\theta,$$

where

$$a(\theta) = \alpha(\theta)\lambda(0)\int_0^\infty \lambda(s)ds > 0, \qquad b(\theta) = \theta\beta(\theta)\nu(0)\int_0^\infty \nu(s)ds > 0.$$

Nonlinear electric conductors

Borrowing from the previous section we now look for nonlinear models; for simplicity we restrict attention to electric conductors (and hence $\mathbf{q} = \mathbf{0}$). Let

$$\mathbf{w} = \boldsymbol{\Gamma}_0^{1/2}\int_0^\infty \lambda(s)\mathbf{E}^t(s)ds,$$

where

$$\mathrm{Sym} \ni \boldsymbol{\Gamma}_0 > \mathbf{0}, \qquad \lambda(s) \to 0 \text{ as } s \to \infty.$$

We look for the free energy in terms of $\mathbf{w}$ as an appropriate polynomial form, viz.

$$\rho\psi = \rho\psi_0(\theta) + \sum_{k=1}^n \alpha_k(\theta)\frac{1}{2k}(\mathbf{w}\cdot\mathbf{w})^k, \qquad \alpha_k \geq 0.$$

Hence $\psi = \psi(\theta, \mathbf{E}^t_r)$ and the entropy inequality is

$$-\rho(\partial_\theta\psi + \eta)\dot{\theta} + \mathbf{E}\cdot\mathbf{J} - \sum_{k=1}^n \alpha_k(\theta)|\mathbf{w}|^{2(k-1)}\mathbf{w}\cdot\dot{\mathbf{w}} \geq 0.$$

An integration by parts yields

$$\begin{aligned}\dot{\mathbf{w}} = \boldsymbol{\Gamma}_0^{1/2}\int_0^\infty \dot{\mathbf{E}}(t-s)ds &= \boldsymbol{\Gamma}_0^{1/2}\{-\big[\lambda(s)\mathbf{E}(t-s)\big]_0^\infty + \int_0^\infty \lambda'(s)\mathbf{E}(t-s)ds\} \\ &= \boldsymbol{\Gamma}_0^{1/2}\lambda(0)\mathbf{E}(t) + \boldsymbol{\Gamma}_0^{1/2}\int_0^\infty \lambda'(s)\mathbf{E}(t-s)ds.\end{aligned}$$

Hence we can write the entropy inequality in the form

$$\begin{aligned}&-\rho(\partial_\theta\psi + \eta)\dot{\theta} + \mathbf{E}\cdot(\mathbf{J} - \sum_{k=1}^n \alpha_k(\theta)\lambda(0)|\mathbf{w}|^{2(k-1)}\mathbf{w}\boldsymbol{\Gamma}_0^{1/2})\cdot\mathbf{E} \\ &-\sum_{k=1}^n \alpha_k(\theta)|\mathbf{w}|^{2(k-1)}\int_0^\infty \lambda(s)\mathbf{E}^t(s)ds\cdot\boldsymbol{\Gamma}_0\int_0^\infty \lambda'(u)\mathbf{E}^t(u)du \geq 0\end{aligned}$$

The arbitrariness of $\dot{\theta}$ and $\mathbf{E}$ implies

$$\eta = -\partial_\theta\psi, \qquad \mathbf{J} = \sum_{k=1}^n\alpha_k(\theta)\lambda(0)|\mathbf{w}|^{2(k-1)}\boldsymbol{\Gamma}_0\int_0^\infty \lambda(s)\mathbf{E}^t(s)ds,$$

$$\sum_{k=1}^{n} \alpha_k(\theta)|\mathbf{w}|^{2(k-1)} \int_0^\infty \lambda(s)\mathbf{E}^t(s)ds \cdot \mathbf{\Gamma}_0 \int_0^\infty \lambda'(u)\mathbf{E}^t(u)du \le 0.$$

The inequality holds if $\lambda\lambda' \le 0$. The density current $\mathbf{J}$ then shows a dependence on the history of $\mathbf{E}$ in the form

$$\mathbf{J} = \sum_{k=1}^{n} \alpha_k(\theta)\lambda(0)\Big|\int_0^\infty \lambda(s)\mathbf{E}^t(s)ds \cdot \mathbf{\Gamma}_0 \int_0^\infty \lambda(u)\mathbf{E}^t(u)du\Big|^{k-1}\mathbf{\Gamma}_0 \int_0^\infty \lambda(v)\mathbf{E}^t(v)dv. \tag{7.103}$$

If $n = 1$ then (7.103) simplifies to (7.100).

Waves in rigid conductors

Evolution equations in rigid conductors comprise Maxwell's equations, in the form

$$\nabla \cdot \mathbf{E} = q/\epsilon_0, \qquad \nabla \times \mathbf{H} = \mathbf{J} + \epsilon_0 \partial_t \mathbf{E},$$

$$\nabla \cdot \mathbf{H} = 0, \qquad \nabla \times \mathbf{E} + \mu_0 \partial_t \mathbf{H} = \mathbf{0},$$

and the balance of energy

$$\rho\dot{\varepsilon} = \mathbf{E} \cdot \mathbf{J} - \nabla \cdot \mathbf{q} + \rho r.$$

With the given functional (7.99) for the free energy ψ we find

$$\rho\varepsilon = \rho(\psi + \theta\eta) = \rho(\psi_0 - \theta\psi_0') + \tfrac{1}{2}(\alpha - \theta\alpha')\Big|\mathbf{\Gamma}_0^{1/2} \int_0^\infty \lambda(s)\mathbf{E}^t(s)ds\Big|^2$$
$$+\tfrac{1}{2}(\beta - \theta\beta')\Big|\mathbf{K}_0^{1/2} \int_0^\infty \nu(s)\nabla\theta^t(s)ds\Big|^2.$$

Hence it follows

$$\rho\dot{\varepsilon} = -\rho\theta\psi_0''\dot{\theta} - \tfrac{1}{2}\theta\alpha''\dot{\theta}\Big|\mathbf{\Gamma}_0^{1/2} \int_0^\infty \lambda(s)\mathbf{E}^t(s)ds\Big|^2 - \tfrac{1}{2}\theta\beta''\dot{\theta}\Big|\mathbf{K}_0^{1/2} \int_0^\infty \nu(s)\nabla\theta^t(s)ds\Big|^2$$
$$+(\alpha - \theta\alpha') \int_0^\infty \lambda(u)\dot{\mathbf{E}}^t(u)du \cdot \mathbf{\Gamma}_0 \int_0^\infty \lambda(s)\mathbf{E}^t(s)ds$$
$$+(\beta - \theta\beta') \int_0^\infty \nu(u)\nabla\dot{\theta}^t(u)du \cdot \mathbf{K}_0 \int_0^\infty \nu(s)\nabla\theta^t(s)ds.$$

We now look at solutions in the form of discontinuity waves characterized as follows. Let σ be a time-dependent surface in the region Ω and denote by $\mathbf{n}$ the unit normal. The surface σ is a discontinuity wave such that
$\mathbf{E}, \mathbf{H}, \theta$ are continuous functions of $\mathbf{x}$ and t in a neighbourhood $\mathcal{N}$ of σ,
$\dot{\mathbf{E}}, \dot{\mathbf{H}}, \dot{\theta}, \nabla\mathbf{E}, \nabla\mathbf{H}, \nabla\theta$ and all higher-order derivatives suffer jump discontinuities across σ but are continuous functions of $\mathbf{x}$ and t in $\mathcal{N} \setminus \sigma$.

Consistent with the model of conductors we let $q = 0$. Maxwell's equations and the geometrical conditions of compatibility imply

$$\mathbf{n} \cdot [\![\partial_n \mathbf{E}]\!] = 0, \qquad \mathbf{n} \times [\![\partial_n \mathbf{H}]\!] = \mu_0 [\![\mathbf{J}]\!] + \mu_0 \epsilon_0 [\![\dot{\mathbf{E}}]\!], \tag{7.104}$$

$$\mathbf{n} \cdot [\![\partial_n \mathbf{H}]\!] = 0, \qquad \mathbf{n} \times [\![\partial_n \mathbf{E}]\!] + \mu_0 [\![\dot{\mathbf{H}}]\!] = \mathbf{0}. \tag{7.105}$$

We assume that $[\![r]\!] = 0$ and hence the balance of energy implies

$$\rho [\![\dot{\varepsilon}]\!] = \mathbf{E} \cdot [\![\mathbf{J}]\!] - [\![\nabla \cdot \mathbf{q}]\!]. \tag{7.106}$$

By (7.100), the continuity of $\mathbf{E}$ and θ across σ implies the continuity of $\mathbf{J}$. Though $\nabla\theta$ suffers a jump discontinuity the integral in (7.101) is continuous and hence $\mathbf{q}$ is continuous too. Moreover, despite the jump of $\nabla\theta$ the integral $\int_0^\infty \nu(s) \nabla\theta^t(s) ds$ is continuous. Hence, upon an integration by parts we obtain

$$[\![\int_0^\infty \nu(s) \nabla\dot{\theta}^t(s) ds]\!] = \nu(0) [\![\nabla\theta]\!].$$

Now, by (7.101) it follows that

$$\dot{\mathbf{q}} = -(\beta + \theta\beta')\dot{\theta}\nu(0)\mathbf{K}_0 \int_0^\infty \nu(s)\nabla\theta^t(s) ds - \theta\beta\nu(0)\mathbf{K}_0 \int_0^\infty \nu(s)\nabla\dot{\theta}^t(s) ds.$$

Thus we obtain

$$[\![\dot{\mathbf{q}}]\!] = -\mathbf{h}[\![\dot{\theta}]\!] - \mathbf{K}[\![\nabla\theta]\!],$$

where

$$\mathbf{h} = (\beta + \theta\beta')\nu(0)\mathbf{K}_0 \int_0^\infty \nu(s)\nabla\theta^t(s) ds, \qquad \mathbf{K} = \theta\beta\nu^2(0)\mathbf{K}_0.$$

Likewise, from the expression of $\rho\dot{\varepsilon}$ it follows

$$\rho[\![\dot{\varepsilon}]\!] = \xi[\![\dot{\theta}]\!] + \boldsymbol{\gamma} \cdot [\![\nabla\theta]\!],$$

where

$$\xi := -\rho\theta\psi_0'' - \tfrac{1}{2}\theta\alpha'' \big|\boldsymbol{\Gamma}_0^{1/2} \int_0^\infty \lambda(s)\mathbf{E}^t(s) ds\big|^2 - \tfrac{1}{2}\theta\beta'' \big|\mathbf{K}_0^{1/2} \int_0^\infty \nu(s)\nabla\theta^t(s) ds\big|^2,$$

$$\boldsymbol{\gamma} = (\beta - \theta\beta')\nu(0)\mathbf{K}_0 \int_0^\infty \nu(s)\nabla\theta^t(s) ds.$$

We now apply the kinematical conditions of compatibility,

$$[\![\dot{f}]\!] = -U[\![\partial_n f]\!], \qquad [\![\dot{\mathbf{w}}]\!] \cdot \mathbf{n} = -U[\![\nabla \cdot \mathbf{w}]\!]$$

if f and $\mathbf{w}$ are continuous across σ, to obtain the equations for, e.g. $[\![\partial_n \mathbf{H}]\!]$ and $[\![\partial_n \theta]\!]$. As a consequence of (7.104) and (7.105) we have

$$\mathbf{n} \times (\mathbf{n} \times [\![\partial_n \mathbf{H}]\!]) = -[\![\partial_n \mathbf{H}]\!] = \epsilon_0 \mathbf{n} \times [\![\dot{\mathbf{E}}]\!] = -U\epsilon_0 \mathbf{n} \times [\![\partial_n \mathbf{E}]\!] = -U^2 \mu_0 \epsilon_0 [\![\partial_n \mathbf{H}]\!],$$

whence

$$(1 - U^2 \mu_0 \epsilon_0)[\![\partial_n \mathbf{H}]\!] = \mathbf{0}.$$

A nonzero discontinuity $[\![\partial_n \mathbf{H}]\!]$ can propagate at the speed $U = 1/\sqrt{\mu_0 \epsilon_0}$. Thus, the propagation condition for the electromagnetic discontinuities is not affected by the thermal conduction.

To investigate the existence of temperature-rate discontinuities, we consider Eq. (7.106) and apply the geometrical and kinematical conditions of compatibility to obtain

$$\xi [\![\dot{\theta}]\!] + \gamma \cdot [\![\nabla \theta]\!] = -[\![\nabla \cdot \mathbf{q}]\!] = U^{-1} [\![\dot{\mathbf{q}}]\!] \cdot \mathbf{n} = -U^{-1} (\mathbf{h} \cdot \mathbf{n} [\![\dot{\theta}]\!] + \mathbf{n} \cdot \mathbf{K} [\![\nabla \theta]\!])$$

whence

$$(\xi U^2 + 2\mathbf{p} \cdot \mathbf{n}\, U - \mathbf{n} \cdot \mathbf{K}\mathbf{n})[\![\partial_n \theta]\!] = 0,$$

where

$$\mathbf{p} = \tfrac{1}{2}(\mathbf{h} - \gamma) = \theta \beta' \nu(0) \mathbf{K}_0 \int_0^\infty \nu(s) \nabla \theta^t(s) ds.$$

Since $\mathbf{K}$ is positive definite, nonzero discontinuities $[\![\partial_n \theta]\!]$ can propagate at any of the two (positive) speeds

$$U = \frac{1}{\xi}(\sqrt{\mathbf{n} \cdot \mathbf{K}\mathbf{n} + (\mathbf{p} \cdot \mathbf{n})^2} \pm \mathbf{p} \cdot \mathbf{n}).$$

The past history $\nabla \theta_r^t$ affects the value of U via $\mathbf{p} \cdot \mathbf{n}$. The vector $\mathbf{p}$ is zero if the region ahead is kept at a uniform temperature up to the arrival of the wave or if β is constant. In these cases only one speed is allowed for the temperature-rate wave. The electric field affects the value of U via the value of ξ. If α and β are constants then the result for U coincides with that for the model in Sect. 7.3.1.

7.6.3 *Polarizable Conductors with Memory*

Simple physical pictures of polarization and magnetization may serve as an introduction to the modelling of electromagnetic materials with memory. Following the Lorentz model of polarization, while the electronic cloud of an atom can be regarded as centred at the atomic nucleus, as a consequence of an electric field $\mathbf{E}$ the mean position of the cloud is displaced at $\mathbf{d}$ relative to the nucleus. Let m be the mass of an electron. The restoring force is denoted by $-m\omega_0^2\mathbf{d}$ and $-m\gamma\dot{\mathbf{d}}$ is the frictional force. Hence the equation of motion becomes

$$m\ddot{\mathbf{d}} + m\gamma\dot{\mathbf{d}} + m\omega_0^2\mathbf{d} = -\mathfrak{q}\mathbf{E}$$

By Fourier transform,

$$f(t) \quad \to \quad f_F(\omega) = \int_{-\infty}^{\infty} f(t)\exp(-i\omega t)dt,$$

we obtain

$$m(-\omega^2 + i\omega\gamma + \omega_0^2)\mathbf{d}(\omega) = -\mathfrak{q}\mathbf{E}(\omega)$$

and hence

$$\mathbf{d}(\omega) = -\frac{\mathfrak{q}}{m}\frac{\mathbf{E}(\omega)}{\omega_0^2 - \omega^2 + i\omega\gamma},$$

where $\mathbf{d}(\omega)$ and $\mathbf{E}(\omega)$ stand for the Fourier transforms. Hence we get the dipole moment

$$\mathfrak{p}(\omega) = -\mathfrak{q}\,\mathbf{d}(\omega) = \frac{\mathfrak{q}^2}{m}\frac{\mathbf{E}(\omega)}{\omega_0^2 - \omega^2 + i\omega\gamma}.$$

Let $\alpha(\omega)$ be the polarizability of a single atom so that

$$\mathfrak{p}(\omega) = \alpha(\omega)\mathbf{E}(\omega), \qquad \alpha(\omega) = \frac{\mathfrak{q}^2}{m}\frac{1}{\omega_0^2 - \omega^2 + i\omega\gamma}.$$

Letting

$$\mathbf{P}(\omega) = n\mathfrak{p}(\omega),$$

n being the number density of atoms, we have

$$\mathbf{P}(\omega) = \frac{\mathfrak{q}^2}{m}\frac{1}{\omega_0^2 - \omega^2 + i\omega\gamma}\mathbf{E}(\omega).$$

Hence the relative permittivity is a function of the frequency ω.

Drude's model involves a non-interacting electron gas that flows in a metal with motionless positive ions. The equation of motion of electrons is

$$m\dot{\mathbf{v}} + m\gamma\mathbf{v} = -\mathfrak{q}\mathbf{E}.$$

Hence

$$\mathbf{v}(\omega) = -\frac{\mathfrak{q}}{m}\frac{\mathbf{E}(\omega)}{i\omega + \gamma}$$

and the conduction current $\mathbf{J}$ reads

$$\mathbf{J}(\omega) = -nq\mathbf{v} = \frac{nq^2}{m}\frac{1}{i\omega + \gamma}\mathbf{E}(\omega)$$

thus showing a frequency-dependent conductivity. In a non-polarizable material the displacement current $\mathbf{J}_d$ is given by

$$\mathbf{J}_d = \partial_t \mathbf{D} = \epsilon_0 \partial_t \mathbf{E}.$$

Hence the total current is expressed by

$$(\mathbf{J} + \mathbf{J}_d)(\omega) = [i\omega\epsilon_0 + \frac{nq^2}{m}\frac{1}{i\omega + \gamma}]\mathbf{E}(\omega).$$

The general feature of these models, as well as of more involved ones, is that the physical properties like permittivity and conductivity are frequency dependent. This feature holds in materials with memory too and then it is of interest to examine connections between materials with memory and frequency-dependent constitutive properties. As an example, let

$$\mathbf{J}(t) = \int_0^\infty \sigma(s)\mathbf{E}(t-s)ds.$$

By Fourier transform we have

$$\int_{-\infty}^{\infty} \mathbf{J}(t)\exp(-i\omega t)dt = \int_{-\infty}^{\infty} dt \exp(-i\omega t)\int_0^\infty \sigma(s)\mathbf{E}(t-s)ds = \sigma_F(\omega)\mathbf{E}_F(\omega),$$

where

$$\sigma_F(\omega) = \int_0^\infty \sigma(s)\exp(-i\omega s)ds.$$

The kernel σ on $\mathbb{R}^+$ is associated with a frequency-dependent transform $\sigma_F : \mathbb{R} \to \mathbb{C}$.

We consider a rigid body. Hence the mass density ρ is constant and the balance of energy simplifies to

$$\rho\dot{\varepsilon} = \mathbf{E}\cdot\mathbf{J} + \dot{\mathbf{P}}\cdot\mathbf{E} + \mu_0\dot{\mathbf{M}}\cdot\mathbf{H} - \nabla\cdot\mathbf{q} + \rho r.$$

From the standard entropy inequality $\rho\dot{\eta} \ge -\nabla\cdot(\mathbf{q}/\theta + \mathbf{k}) + \rho r/\theta$ it follows

$$-\rho(\dot{\varepsilon} + \theta\dot{\eta}) + \dot{\mathbf{P}}\cdot\mathbf{E} + \mu_0\dot{\mathbf{M}}\cdot\mathbf{H} + \mathbf{E}\cdot\mathbf{J} - \frac{1}{\theta}\mathbf{q}\cdot\nabla\theta + \theta\nabla\cdot\mathbf{k} \ge 0.$$

In terms of the free energy $\phi = \varepsilon - \theta\eta - (\mathbf{P}\cdot\mathbf{E} + \mu_0\mathbf{M}\cdot\mathbf{H})/\rho$ the entropy inequality can be written in the form

$$-\rho(\dot{\phi}+\eta\dot{\theta})-\mathbf{P}\cdot\dot{\mathbf{E}}-\mu_0\mathbf{M}\cdot\dot{\mathbf{H}}+\mathbf{E}\cdot\mathbf{J}-\frac{1}{\theta}\mathbf{q}\cdot\nabla\theta+\theta\nabla\cdot\mathbf{k}\geq 0. \qquad (7.107)$$

Denote by Ω the region occupied by the body. Let Γ be the 4-tuple

$$\Gamma=(\mathbf{E},\mathbf{H},\theta,\nabla\theta)$$

whose values are functions of $\mathbf{x}\in\Omega$ and $t\in\mathbb{R}$. Denote by $\mathcal{H}$ the vector space whose elements are the pairs of present values and past histories

$$(\Gamma(t),\Gamma_r^t),$$

for any $t\in\mathbb{R}$, the dependence of Γ on $\mathbf{x}\in\Omega$ being understood. For ease in writing we suppress the suffix r and let Γ^t denote the past history. The vector space $\mathcal{H}$ is endowed with the L_2 norm

$$\|(\Gamma(t),\Gamma^t)\|=\big(\Gamma^2(t)+\int_0^\infty[\Gamma^t(s)]^2h(s)ds\big)^{1/2},$$

where

$$\Gamma^2=\mathbf{E}^2+\mathbf{H}^2+\theta^2+(\nabla\theta)^2$$

and the like for $[\Gamma^t(s)]^2$ while h is positive and integrable on $(0,\infty)$.

As the constitutive assumption we let ϕ, η, $\mathbf{P}$, $\mathbf{M}$, $\mathbf{J}$, $\mathbf{q}$, $\mathbf{k}$, at $\mathbf{x}\in\Omega$ at time $t\in\mathbb{R}$, be given by functionals of

$$(\Gamma(t),\Gamma^t)=(\mathbf{E}(t),\mathbf{H}(t),\theta(t),\nabla\theta(t),\mathbf{E}^t,\mathbf{H}^t,\theta^t,\nabla\theta^t),$$

at the same point $\mathbf{x}$. Hence we say that, e.g. $\phi(t)$ is given by

$$\phi(t)=\hat{\phi}(\mathbf{E}(t),\mathbf{H}(t),\theta(t),\nabla\theta(t),\mathbf{E}^t,\mathbf{H}^t,\theta^t,\nabla\theta^t),$$

the dependence on $\mathbf{x}$ being understood. In words, ϕ, η, $\mathbf{P}$, $\mathbf{M}$, $\mathbf{J}$, $\mathbf{q}$, $\mathbf{k}$ depend on the present values and on the past histories of $\mathbf{E}$, $\mathbf{H}$, θ, $\nabla\theta$. We assume the functionals $\hat{\phi}$, $\hat{\eta}$, $\hat{\mathbf{P}}$, $\hat{\mathbf{M}}$, $\hat{\mathbf{J}}$, $\hat{\mathbf{q}}$, $\hat{\mathbf{k}}$, on $\mathcal{H}$, are continuous. Moreover we assume $\hat{\phi}$ is differentiable with respect to Γ and Fréchet differentiable with respect to Γ_r^t so that

$$\hat{\phi}(\Gamma(t)+\Lambda(t),\Gamma^t+\Lambda^t)=\hat{\phi}(\Gamma(t),\Gamma^t)+\partial_\Gamma\hat{\phi}\cdot\Lambda(t)+d\hat{\phi}(\Gamma(t),\Gamma^t|\Lambda^t)+o(\|(\Lambda(t),\Lambda^t)\|),$$

where $d\hat{\phi}(\cdot|\Lambda^t)$ is the Fréchet differential induced by the change of the past history. If $\Gamma(t)$ and Γ^t are differentiable, with respect to t, then

$$\dot{\phi}(t)=\partial_\Gamma\hat{\phi}\cdot\dot{\Gamma}(t)+d\hat{\phi}(\Gamma(t),\Gamma^t|\dot{\Gamma}^t)$$

or, more explicitly,

$$\dot{\phi}(t) = \partial_{\mathbf{E}}\hat{\phi} \cdot \dot{\mathbf{E}}(t) + \partial_{\mathbf{H}}\hat{\phi} \cdot \dot{\mathbf{H}}(t) + \partial_\theta\hat{\phi}\,\dot{\theta}(t) + \partial_{\nabla\theta}\hat{\phi} \cdot \nabla\dot{\theta}(t) + d\hat{\phi}(\Gamma(t), \Gamma^t|\dot{\mathbf{E}}^t, \dot{\mathbf{H}}^t, \dot{\theta}^t, \nabla\dot{\theta}^t).$$

To derive the thermodynamic restrictions we now examine the entropy inequality (7.107). Substitution of $\dot{\phi}$ results in

$$\begin{aligned}-\rho(\partial_\theta\hat{\phi} + \eta)\dot{\theta} - (\rho\partial_{\mathbf{E}}\hat{\phi} + \mathbf{P}) \cdot \dot{\mathbf{E}} - (\rho\partial_{\mathbf{H}}\hat{\phi} + \mu_0\mathbf{M}) \cdot \dot{\mathbf{H}} - \rho\partial_{\nabla\theta}\hat{\phi} \cdot \nabla\dot{\theta}\\ +\mathbf{E} \cdot \mathbf{J} - \frac{1}{\theta}\mathbf{q} \cdot \nabla\theta - \rho d\hat{\phi}(\Gamma(t), \Gamma^t|\dot{\mathbf{E}}^t, \dot{\mathbf{H}}^t, \dot{\theta}^t, \nabla\dot{\theta}^t) + \theta\nabla \cdot \mathbf{k} \geq 0,\end{aligned}$$

each term being evaluated at time t. The values $\dot{\theta}$, $\dot{\mathbf{E}}$, $\dot{\mathbf{H}}$, and $\nabla\dot{\theta}$ occur linearly. Now, at any $\mathbf{x} \in \Omega$ and $t \in \mathbb{R}$, these values can be chosen arbitrarily. This is so because, as is usually the case, $\dot{\theta}$ enters the balance of energy and the equation holds with a corresponding value of r. Likewise $\nabla\dot{\theta}$ can be chosen arbitrarily and the balance of energy holds with a corresponding value of ∇r. Moreover, $\dot{\mathbf{E}}$ and $\dot{\mathbf{H}}$ can be chosen arbitrarily since Maxwell's equations hold with appropriate values of $\nabla \times \mathbf{H}$ and $\nabla \times \mathbf{E}$, which do not enter the inequality. Hence the inequality has to hold with arbitrary values of $\dot{\theta}$, $\dot{\mathbf{E}}$, $\dot{\mathbf{H}}$, and $\nabla\dot{\theta}$. This happens if and only if

$$\partial_{\nabla\theta}\hat{\phi} = \mathbf{0},$$

$$\eta = -\partial_\theta\hat{\phi}, \qquad \mathbf{P} = -\rho\partial_{\mathbf{E}}\hat{\phi}, \qquad \mu_0\mathbf{M} = -\rho\partial_{\mathbf{H}}\hat{\phi}, \tag{7.108}$$

and

$$\mathbf{E} \cdot \mathbf{J} - \frac{1}{\theta}\mathbf{q} \cdot \nabla\theta - \rho d\hat{\phi}(\Gamma(t), \Gamma_r^t|\dot{\mathbf{E}}_r^t, \dot{\mathbf{H}}_r^t, \dot{\theta}_r^t, \nabla\dot{\theta}_r^t) + \theta\nabla \cdot \mathbf{k} \geq 0. \tag{7.109}$$

Further restrictions follow from the reduced inequality (7.109) if particular functionals are considered. Two remarkable models hold. First, let

$$\phi = \phi_0(\theta, \mathbf{E}, \mathbf{H}) + \frac{1}{2\rho\sigma(0)}\Big(\int_0^\infty \sigma(s)\mathbf{E}(t-s)ds\Big)^2 + \frac{1}{2\rho\kappa(0)\theta}\Big(\int_0^\infty \kappa(s)\nabla\theta(t-s)ds\Big)^2,$$

subject to $\sigma(\infty) = 0$, $\kappa(\infty) = 0$. Hence it follows

$$\eta = -\partial_\theta\phi_0 + \frac{1}{2\rho\kappa(0)\theta^2}\Big(\int_0^\infty \kappa(s)\nabla\theta(t-s)ds\Big)^2, \qquad \mathbf{P} = -\rho\partial_{\mathbf{E}}\phi_0, \qquad \mu_0\mathbf{M} = -\rho\partial_{\mathbf{H}}\phi_0$$

and

$$\begin{aligned}\mathbf{E} \cdot \mathbf{J} - \frac{1}{\theta}\mathbf{q} \cdot \nabla\theta + \theta\nabla \cdot \mathbf{k} - \frac{1}{\sigma(0)}\int_0^\infty \sigma(s)\mathbf{E}(t-s)ds \cdot \int_0^\infty \sigma(s)\dot{\mathbf{E}}(t-s)ds\\ -\frac{1}{\kappa(0)}\int_0^\infty \kappa(s)\nabla\theta(t-s)ds \cdot \int_0^\infty \kappa(s)\nabla\dot{\theta}(t-s)ds \geq 0.\end{aligned}$$

Notice that $\dot{\mathbf{E}}(t-s) = -\partial_s\mathbf{E}(t-s)$ and integrate by parts to obtain

$$\int_0^\infty \sigma(s)\dot{\mathbf{E}}(t-s)ds = \sigma(0)\mathbf{E}(t) - \int_0^\infty \sigma'(s)\mathbf{E}(t-s)ds$$

and the like for the integral with $\nabla\dot{\theta}$. The inequality can be written as

$$\mathbf{E}(t)\cdot\big(\mathbf{J}(t) - \int_0^\infty \sigma(s)\mathbf{E}(t-s)ds\big) - \nabla\theta(t)\cdot\big(\frac{\mathbf{q}}{\theta}(t) + \int_0^\infty \kappa(s)\nabla\theta(t-s)ds\big)$$
$$+\theta\nabla\cdot\mathbf{k} - \frac{1}{\sigma(0)}\int_0^\infty \sigma(s)\mathbf{E}(t-s)ds\cdot\int_0^\infty \sigma'(s)\mathbf{E}(t-s)ds$$
$$-\frac{1}{\kappa(0)\theta}\int_0^\infty \kappa(s)\nabla\theta(t-s)ds\cdot\int_0^\infty \kappa'(s)\nabla\theta(t-s)ds \geq 0.$$

This inequality holds if $\mathbf{k} = \mathbf{0}$ and

$$\sigma(s) = \sigma(0)\exp(-\alpha s),\ \alpha > 0,\ \sigma(0) > 0, \qquad \kappa(s) = \kappa(0)\exp(-\beta s),\ \beta > 0,\ \kappa(0) > 0$$

$$\mathbf{J}(t) = \int_0^\infty \sigma(s)\mathbf{E}(t-s)ds, \qquad \mathbf{q}(t) = -\int_0^\infty \kappa(s)\nabla\theta(t-s)ds.$$

Let $(0, \epsilon)$ be the support of σ and κ and

$$\tilde{\sigma} = \int_0^\epsilon \sigma(s)ds, \qquad \tilde{\kappa} = \int_0^\epsilon \kappa(s)ds.$$

The limit as $\epsilon \to 0$, that is the passage from fading memory to instantaneous response, yields

$$\mathbf{J}(t) \to \tilde{\sigma}\,\mathbf{E}(t), \qquad \mathbf{q}(t) \to -\tilde{\kappa}\,\nabla\theta(t),$$

thus recovering Ohm's law and Fourier's law.

Another model allows for memory effects also for the polarization $\mathbf{P}$ and the magnetization $\mathbf{M}$. The purpose is to obtain the constitutive relations

$$\mathbf{P}(t)=\mathbf{G}_0\mathbf{E}(t)+\int_0^\infty \mathbf{G}'(s)\mathbf{E}(t-s)ds, \qquad \mu_0\mathbf{M}(t) = \mathbf{A}_0\mathbf{H}(t) + \int_0^\infty \mathbf{A}'(s)\mathbf{H}(t-s)ds,$$

$\mathbf{G}_0, \mathbf{G}', \mathbf{A}_0, \mathbf{A}'$ being symmetric tensors. Let

$$\mathbf{G}_\infty = \mathbf{G}_0 + \int_0^\infty \mathbf{G}'(s)ds, \qquad \mathbf{A}_\infty = \mathbf{A}_0 + \int_0^\infty \mathbf{A}'(s)ds.$$

Borrowing from the previous scheme and that of linear viscoelasticity we consider

$$\rho\phi = \rho f(\theta) + \tfrac{1}{2}\{\mathbf{E}(t)\cdot\mathbf{G}_\infty\mathbf{E}(t) - \int_0^\infty[\mathbf{E}(t)-\mathbf{E}(t-s)]\cdot\mathbf{G}'(s)[\mathbf{E}(t)-\mathbf{E}(t-s)]ds\}$$
$$+\tfrac{1}{2}\{\mathbf{H}(t)\cdot\mathbf{A}_\infty\mathbf{H}(t) - \int_0^\infty[\mathbf{H}(t)-\mathbf{H}(t-s)]\cdot\mathbf{A}'(s)[\mathbf{H}(t)-\mathbf{H}(t-s)]ds\}$$
$$-\tfrac{1}{2}\boldsymbol{\Sigma}^{-1}(0)\Big(\int_0^\infty\boldsymbol{\Sigma}(s)\mathbf{E}(t-s)ds\Big)^2 - \frac{1}{2\theta}\mathbf{K}^{-1}(0)\Big(\int_0^\infty\mathbf{K}(s)\nabla\theta(t-s)ds\Big)^2.$$

By (7.108) and (7.109) it follows that

$$\mathbf{P}(t) = \mathbf{G}_\infty\mathbf{E}(t) - \int_0^\infty\mathbf{G}'(s)[\mathbf{E}(t)-\mathbf{E}(t-s)]ds = \mathbf{G}_0\mathbf{E}(t) + \int_0^\infty\mathbf{G}'(s)\mathbf{E}(t-s)ds,$$

$$\mu_0\mathbf{M}(t)=\mathbf{A}_\infty\mathbf{H}(t)-\int_0^\infty\mathbf{A}'(s)[\mathbf{H}(t)-\mathbf{H}(t-s)]ds = \mathbf{A}_0\mathbf{H}(t) + \int_0^\infty\mathbf{A}'(s)\mathbf{H}(t-s)ds,$$

and

$$-\int_0^\infty[\mathbf{E}(t)-\mathbf{E}(t-s)]\cdot\mathbf{G}'(s)\dot{\mathbf{E}}(t-s)ds - \int_0^\infty[\mathbf{H}(t)-\mathbf{H}(t-s)]\cdot\mathbf{A}'(s)\dot{\mathbf{H}}(t-s)ds$$
$$+\mathbf{E}\cdot\mathbf{J} - \frac{1}{\theta}\mathbf{q}\cdot\nabla\theta - \boldsymbol{\Sigma}^{-1}(0)\int_0^\infty\boldsymbol{\Sigma}(s)\mathbf{E}(t-s)ds\cdot\int_0^\infty\boldsymbol{\Sigma}(s)\dot{\mathbf{E}}(t-s)ds$$
$$-\frac{1}{\theta}\mathbf{K}^{-1}(0)\int_0^\infty\mathbf{K}(s)\nabla\theta(t-s)ds\cdot\int_0^\infty\mathbf{K}(s)\nabla\dot{\theta}(t-s)ds + \theta\nabla\cdot\mathbf{k} \ge 0.$$

Since

$$\dot{\mathbf{E}}(t-s) = \partial_s[\mathbf{E}(t)-\mathbf{E}(t-s)]$$

then an integration by parts yields

$$-\int_0^\infty[\mathbf{E}(t)-\mathbf{E}(t-s)]\cdot\mathbf{G}'(s)\dot{\mathbf{E}}(t-s)ds = \tfrac{1}{2}\int_0^\infty[\mathbf{E}(t)-\mathbf{E}(t-s)]\cdot\mathbf{G}''(s)[\mathbf{E}(t)-\mathbf{E}(t-s)]ds,$$

and likewise for the second integral. Further,

$$-\boldsymbol{\Sigma}^{-1}(0)\int_0^\infty\boldsymbol{\Sigma}(s)\mathbf{E}(t-s)ds\cdot\int_0^\infty\boldsymbol{\Sigma}(s)\dot{\mathbf{E}}(t-s)ds = -\mathbf{E}(t)\cdot\int_0^\infty\boldsymbol{\Sigma}(s)\mathbf{E}(t-s)ds$$
$$-\boldsymbol{\Sigma}^{-1}(0)\int_0^\infty\boldsymbol{\Sigma}(s)\mathbf{E}(t-s)ds\cdot\int_0^\infty\boldsymbol{\Sigma}'(s)\mathbf{E}(t-s)ds,$$

and likewise for the integrals on $\nabla\theta$. The inequality is then written as

$$\frac{1}{2}\int_0^\infty \mathbf{E}_d(t,s)\cdot \mathbf{G}''(s)\mathbf{E}_d(t,s)ds + \frac{1}{2}\int_0^\infty \nabla\theta_d(t,s)\cdot \mathbf{A}''(s)\nabla\theta_d(t,s)ds$$
$$-\boldsymbol{\Sigma}^{-1}(0)\int_0^\infty \boldsymbol{\Sigma}(s)\mathbf{E}(t-s)ds\cdot\int_0^\infty \boldsymbol{\Sigma}'(s)\mathbf{E}(t-s)ds$$
$$-\frac{1}{\theta}\mathbf{K}^{-1}(0)\int_0^\infty \mathbf{K}(s)\nabla\theta(t-s)ds\cdot\int_0^\infty \mathbf{K}'(s)\nabla\theta(t-s)ds$$
$$+\mathbf{E}\cdot(\mathbf{J}-\int_0^\infty \boldsymbol{\Sigma}(s)\mathbf{E}(t-s)ds)-\frac{1}{\theta}(\mathbf{q}+\int_0^\infty \mathbf{K}(s)\nabla\theta(t-s)ds)+\theta\nabla\cdot\mathbf{k}\geq 0,$$

where $\mathbf{E}_d(t,s) = \mathbf{E}(t) - \mathbf{E}(t-s)$, $\nabla\theta_d(t,s) = \nabla\theta(t) - \nabla\theta(t-s)$. The whole scheme holds if

$$\mathbf{G}''(s)\geq \mathbf{0}, \qquad \mathbf{A}''(s)\geq \mathbf{0},$$

$$\boldsymbol{\Sigma}'(s) = -\alpha\boldsymbol{\Sigma}(s), \quad \alpha>0, \quad \boldsymbol{\Sigma}(s)>0,$$

$$\mathbf{K}'(s) = -\beta\mathbf{K}(s), \quad \beta>0, \quad \mathbf{K}(s)>0$$

and

$$\mathbf{J} = \int_0^\infty \boldsymbol{\Sigma}(s)\mathbf{E}(t-s)ds, \qquad \mathbf{q} = -\int_0^\infty \mathbf{K}(s)\nabla\theta(t-s)ds.$$

The positive definiteness of the electrical conductivity tensor function $\boldsymbol{\Sigma}$ and of the thermal conductivity tensor function $\mathbf{K}$ generalize to memory functionals the well-known features of Ohm's law and Fourier's law. If further $\mathbf{G}$ and $\mathbf{A}$ have the form of exponentials,

$$\mathbf{G}(s) = \mathbf{G}_0\exp(-s/\tau_E), \qquad \mathbf{A}(s) = \mathbf{A}_0\exp(-s/\tau_B), \qquad \tau_E,\tau_B>0,$$

then $\mathbf{G}''(s)\geq\mathbf{0}$ and $\mathbf{A}''(s)\geq\mathbf{0}$ imply $\mathbf{G}(s)\geq\mathbf{0}$ and $\mathbf{A}(s)\geq\mathbf{0}$.

7.6.4 Magneto-Viscoelasticity

There are materials (magnetorheological elastomers) that change their mechanical behaviour in response to the application of a magnetic field. Depending on the preparation method, the resulting material may be isotropic or anisotropic.

The response of magnetization to an applied magnetic field is not instantaneous and hence the magnetoelastic properties are to be generalized by allowing for the viscoelastic nature. The time delay in response is essential in a proper modelling. Sometimes this delay effect is realized by superposing a viscous overstress on elastic deformation. A description is realized by allowing for an intermediate configuration; a viscous motion produces the intermediate configuration and next the elastic deformation leads to the current configuration. Both the deformation and the magnetic variables are the sum of an elastic and a viscous term, the viscous terms being determined by nonlinear rate-type equations [385]. Here we avoid the introduction of an intermediate configuration and let the viscoelastic scheme represent the elastic part via the instantaneous term and the dissipative (delayed) term via the memory integral.

Look at a magnetisable, deformable, and heat-conducting body. The balance of energy is written in the form

$$\rho\dot{\varepsilon} = \mu_0\rho\mathbf{H}\cdot\dot{\mathbf{m}} + \mathbf{T}\cdot\mathbf{L} - \nabla\cdot\mathbf{q} + \rho r$$

while the balance of angular momentum results in

$$\text{skw}\mathbf{T} = \text{skw}(\mu_0\mathbf{M}\otimes\mathbf{H}).$$

To allow for the interaction between the deformation and temperature fields with magnetization, in a scheme where both instantaneous response and delayed effects occur, we let the material properties be described by θ, $\mathbf{F}$, $\mathbf{H}$, $\nabla\theta$, both via the present values and the past histories. Indeed, we consider

$$\Gamma = (\theta, \mathbf{F}, \mathbf{H}, \nabla\theta, \mathbf{F}^t, \mathbf{H}^t, \nabla\theta^t)$$

be the set of independent variables. To guarantee the invariance we let ϕ depend on $\mathbf{F}, \mathbf{H}$ via the invariants $\mathfrak{E} = \frac{1}{2}(\mathbf{F}^T\mathbf{F} - \mathbf{1})$, $\mathfrak{H} = \mathbf{F}^T\mathbf{H}$; observe that $\mathfrak{H} = \mathbf{H}_R$ is the magnetic field in the reference configuration and is Euclidean invariant, $(\mathbf{F}^T\mathbf{H})^* = (\mathbf{QF})^T\mathbf{QH} = \mathbf{F}^T\mathbf{H}$. To make it apparent that the reference configuration is stress-free we use the strain $\mathfrak{E} = \frac{1}{2}(\mathbf{F}^T\mathbf{F} - \mathbf{1})$ as the deformation variable, in place of $\mathbf{F}$.

The second-law inequality and the balance of energy yield

$$-\rho(\dot{\phi} + \eta\dot{\theta}) + \mathbf{T}\cdot\mathbf{L} - \mu_0\mathbf{M}\cdot\dot{\mathbf{H}} - \frac{1}{\theta}\mathbf{q}\cdot\nabla\theta \geq 0,$$

where $\phi = \varepsilon - \theta\eta - \mu_0\mathbf{m}\cdot\mathbf{H}$; we restrict attention to models with zero extra-entropy flux. We let ϕ, η, $\mathbf{T}$, $\mathbf{q}$ depend on Γ.

The free energy ϕ is required to satisfy a minimum property. Let θ, $\mathbf{F}$, and $\mathbf{H}$ be defined on $\mathbb{R}$. Constant continuations of θ^{t_0}, $\mathbf{F}^{t_0}$, $\mathbf{H}^{t_0}$ are defined by

$$\tilde{\theta}^t(u) = \begin{cases} \theta(t_0) = \theta^{t_0}(0), & u \in [0, t - t_0), \\ \theta^{t_0}(u - t_0), & u \in [t - t_0, \infty), \end{cases} \qquad \tilde{\theta}(t) = \tilde{\theta}^t(0),$$

and the like for $\mathbf{F}^t$ and $\mathbf{H}^t$. In addition, we let $\nabla\theta = \mathbf{0}$ on $[t_0, t]$. Then it follows from the entropy inequality that

$$\dot{\phi}(\tau) \leq 0, \qquad \tau \in (t_0, t).$$

Hence we have

$$\phi(\tilde{\theta}(t), \tilde{\mathbf{F}}(t), \tilde{\mathbf{H}}(t), \mathbf{0}, \tilde{\mathbf{F}}^t, \tilde{\mathbf{H}}^t, \nabla\theta^t) \leq \phi(\theta(t_0), \mathbf{F}(t_0), \mathbf{H}(t_0), \mathbf{F}^{t_0}, \mathbf{H}^{t_0}, \mathbf{0}).$$

By the continuity of the functional, as $t - t_0 \to \infty$

$$\phi(\tilde{\theta}(t), \tilde{\mathbf{F}}(t), \tilde{\mathbf{H}}(t), \mathbf{0}, \tilde{\mathbf{F}}^t, \tilde{\mathbf{H}}^t, \nabla\theta^t) \to \phi(\theta(t_0), \mathbf{F}(t_0), \mathbf{H}(t_0), \mathbf{F}_0^\dagger, \mathbf{H}_0^\dagger, \mathbf{0}^\dagger),$$

where $\mathbf{F}_0^\dagger$ and $\mathbf{H}_0^\dagger$ are the constant histories $\mathbf{F}_0^\dagger(u) = \tilde{\mathbf{F}}(t_0) = \mathbf{F}_0$, $\mathbf{H}_0^\dagger(u) = \tilde{\mathbf{H}}(t_0) = \mathbf{H}_0$, up to time t, and $u \in [0, \infty)$. Hence

$$\phi(\theta(t_0), \mathbf{F}(t_0), \mathbf{H}(t_0), \mathbf{0}, \mathbf{F}_0^\dagger, \mathbf{H}_0^\dagger, \mathbf{0}^\dagger) \leq \phi(\theta(t_0), \mathbf{F}(t_0), \mathbf{H}(t_0), \mathbf{0}, \mathbf{F}^{t_0}, \mathbf{H}^{t_0}, \nabla\theta^{t_0}).$$

Hence, among all histories $\theta^t, \mathbf{F}^t, \mathbf{H}^t, \nabla\theta^t$ with given present values $\theta_0, \mathbf{F}_0, \mathbf{H}_0$ and $\nabla\theta(t_0) = \mathbf{0}$, none yields a smaller value of the free energy than that corresponding to the constant histories $\theta_0^\dagger, \mathbf{F}_0^\dagger, \mathbf{H}_0^\dagger$ and zero temperature gradient.

The free energy ϕ depends on Γ, through scalar invariants, and is supposed to be continuously differentiable. If ϕ depends on $\mathbf{F}$ and $\mathbf{H}$ via $\mathfrak{E}, \mathfrak{H}$ then the inequality becomes

$$-\rho(\partial_\theta\phi + \eta)\dot{\theta} - \rho\partial_{\mathfrak{E}}\phi \cdot \dot{\mathfrak{E}} - \rho\partial_{\mathfrak{H}}\phi \cdot \dot{\mathfrak{H}} - \rho\partial_{\nabla\theta}\phi \cdot \dot{\overline{\nabla\theta}} + \mathbf{T}\cdot\mathbf{L} - \mu_0\mathbf{M}\cdot\dot{\mathbf{H}} - \frac{1}{\theta}\mathbf{q}\cdot\nabla\theta$$
$$-\rho d\phi(\Gamma|\dot{\mathfrak{E}}^t) - \rho d\phi(\Gamma|\dot{\mathfrak{H}}^t) - \rho d\phi(\Gamma|\dot{\overline{\nabla\theta}}^t) \geq 0.$$

The arbitrariness of $\dot{\theta}, \dot{\overline{\nabla\theta}}$ implies

$$\partial_{\nabla\theta}\phi = \mathbf{0}, \qquad \eta = -\partial_\theta\phi.$$

Since $\dot{\mathfrak{E}} = \frac{1}{2}\dot{\mathbf{C}} = \mathbf{F}^T\mathfrak{D}\mathbf{F}$ and $\dot{\mathfrak{H}} = \mathbf{F}^T\mathbf{L}^T\mathbf{H} + \mathbf{F}^T\dot{\mathbf{H}}$ we can write the inequality in the form

$$(\mathbf{T} - \rho\mathbf{F}\partial_{\mathfrak{E}}\phi\mathbf{F}^T + \rho\mathbf{F}\partial_{\mathfrak{H}}\phi\otimes\mathbf{H})\cdot\mathfrak{D} + (\mathbf{T} + \rho\mathbf{F}\partial_{\mathfrak{H}}\phi\otimes\mathbf{H})\cdot\mathbf{W} - (\mu_0\mathbf{M} + \rho\mathbf{F}\partial_{\mathfrak{H}}\phi)\cdot\dot{\mathbf{H}}$$
$$-\frac{1}{\theta}\mathbf{q}\cdot\nabla\theta - \rho d\phi(\Gamma|\dot{\mathfrak{E}}^t) - \rho d\phi(\Gamma|\dot{\mathfrak{H}}^t) - \rho d\phi(\Gamma|\dot{\overline{\nabla\theta}}^t) \geq 0.$$

The arbitrariness of $\mathfrak{D}, \mathbf{W}, \dot{\mathbf{H}}$ implies

$$\mathbf{T} = \rho\mathbf{F}\partial_{\mathfrak{E}}\phi\mathbf{F}^T - \rho\mathbf{F}\partial_{\mathfrak{H}}\phi\otimes\mathbf{H}, \qquad \mu_0\mathbf{M} = -\rho\mathbf{F}\partial_{\mathfrak{H}}\phi.$$

Since $\mathbf{F}\partial_{\mathfrak{E}}\phi\mathbf{F}^T \in \mathrm{Sym}$ then $\mathrm{skw}\mathbf{T} = \mathrm{skw}(\mu_0\mathbf{M}\otimes\mathbf{H})$, as is required by the balance of angular momentum. The inequality then reduces to

$$-\frac{1}{\theta}\mathbf{q}\cdot\nabla\theta - \rho d\phi(\Gamma|\dot{\mathfrak{E}}^t) - \rho d\phi(\Gamma|\dot{\mathfrak{H}}^t) - \rho d\phi(\Gamma|\dot{\overline{\nabla\theta}}^t) \geq 0.$$

Restrictions placed by the reduced inequality are determined by considering some particular cases. First we suppose the temperature is uniform, $\nabla\theta^t = \mathbf{0}$. Nonlinear constitutive equations for **T** and **M** are established by looking for the partial derivatives

$$\partial_{\mathfrak{E}}\phi = \mathbf{G}_0(\theta, |\mathfrak{E}|)\mathfrak{E} + g(\theta)\int_0^\infty \mathbf{G}'(u)\mathfrak{E}(t-u)du, \qquad \mathbf{G}_0, \mathbf{G}' \in \mathrm{Lin},$$

$$\partial_{\mathfrak{H}}\phi = -\mathbf{A}_0(\theta, |\mathfrak{H}|)\mathfrak{H} - a(\theta)\int_0^\infty \mathbf{A}'(u)\mathfrak{H}(t-u)du, \qquad \mathbf{A}_0, \mathbf{A}' \in \mathrm{Lin}.$$

Now, given $\mathcal{G}(\theta, |\mathfrak{E}|) \in \mathrm{Lin}$, with fully symmetric values, we have

$$\partial_{\mathfrak{E}}\,\tfrac{1}{2}\mathfrak{E}\cdot\mathcal{G}\mathfrak{E} = \mathcal{G}\mathfrak{E} + \frac{1}{2|\mathfrak{E}|}(\mathfrak{E}\cdot\partial_{|\mathfrak{E}|}\mathcal{G}\mathfrak{E})\mathfrak{E}.$$

The analogue follows for $\mathcal{A}(\theta, |\mathfrak{H}|)$. Hence we obtain

$$\mathbf{G}_0 = \mathcal{G} + \frac{1}{2|\mathfrak{E}|}(\mathfrak{E}\cdot\partial_{|\mathfrak{E}|}\mathcal{G}\mathfrak{E})\mathbf{I}, \qquad \mathbf{A}_0 = \mathcal{A} + \frac{1}{2|\mathfrak{H}|}(\mathfrak{H}\cdot\partial_{|\mathfrak{H}|}\mathcal{A}\mathfrak{H})\mathbf{I},$$

$\mathbf{I}$ being the identity operator in Lin. The functional $\phi(\theta, \mathfrak{E}, \mathfrak{H}, \mathfrak{E}^t, \mathfrak{H}^t)$ then takes the form

$$\phi = \Phi(\theta, \mathfrak{E}^t, \mathfrak{H}^t) + \tfrac{1}{2}\mathfrak{E}\cdot\mathcal{G}\mathfrak{E} + g(\theta)\mathfrak{E}\cdot\int_0^\infty \mathbf{G}'(u)\mathfrak{E}(t-u)du - \tfrac{1}{2}\mathfrak{H}\cdot\mathcal{A}\mathfrak{H} - a(\theta)\mathfrak{H}\cdot\int_0^\infty \mathbf{A}'(u)\mathfrak{H}(t-u)du.$$

Among the possible free energies consider

$$\Phi(\theta, \mathfrak{E}^t, \mathbf{H}^t) = \phi_0(\theta) + \tfrac{1}{2}g(\theta)\int_0^\infty \mathfrak{E}^t(u)\cdot\mathbf{G}'(u)\mathfrak{E}^t(u)du - \tfrac{1}{2}a(\theta)\int_0^\infty \mathfrak{H}^t(u)\cdot\mathbf{A}'(u)\mathfrak{H}^t(u)du.$$

The whole functional $\phi(\theta, \mathfrak{E}, \mathfrak{H}, \mathfrak{E}^t, \mathfrak{H}^t)$ is required to satisfy

$$d\phi(\Gamma|\dot{\mathfrak{E}}^t) + d\phi(\Gamma|\dot{\mathfrak{H}}^t) \leq 0. \tag{7.110}$$

To this end observe that

$$\phi = \phi_0(\theta) + \tfrac{1}{2}\mathfrak{E}\cdot\mathcal{G}_\infty\mathfrak{E} - \tfrac{1}{2}g(\theta)\int_0^\infty[\mathfrak{E}(t) - \mathfrak{E}(t-u)]\cdot\mathbf{G}'(u)[\mathfrak{E}(t) - \mathfrak{E}(t-u)]du$$
$$-\tfrac{1}{2}\mathfrak{H}\cdot\mathcal{A}_\infty\mathfrak{H} + \tfrac{1}{2}a(\theta)\int_0^\infty[\mathfrak{H}(t) - \mathfrak{H}(t-u)]\cdot\mathbf{A}'(u)[\mathfrak{H}(t) - \mathfrak{H}(t-u)]du =: \phi_m. \tag{7.111}$$

It follows that

$$d\phi(\Gamma|\dot{\mathfrak{E}}^t) = -\tfrac{1}{2}g(\theta)\int_0^\infty [\mathfrak{E}(t) - \mathfrak{E}(t-u)] \cdot \mathbf{G}'(u)\dot{\mathfrak{E}}(t-u)]du$$
$$= \tfrac{1}{2}g(\theta)\{[\mathfrak{E}(t) - \mathfrak{E}(t-u)] \cdot \mathbf{G}'(u)[\mathfrak{E}(t) - \mathfrak{E}(t-u)]_0^\infty$$
$$- \int_0^\infty [\mathfrak{E}(t) - \mathfrak{E}(t-u)] \cdot \mathbf{G}''(u)[\mathfrak{E}(t) - \mathfrak{E}(t-u)]du\}.$$

The assumption $\mathbf{G}'(\infty) = \mathbf{0}$ implies

$$d\phi(\Gamma|\dot{\mathfrak{E}}^t) = -\tfrac{1}{2}g(\theta)\int_0^\infty [\mathfrak{E}(t) - \mathfrak{E}(t-u)] \cdot \mathbf{G}''(u)[\mathfrak{E}(t) - \mathfrak{E}(t-u)]du.$$

Likewise, letting $\mathbf{A}'(\infty) = \mathbf{0}$ we find

$$d\phi(\Gamma|\dot{\mathfrak{H}}^t) = \tfrac{1}{2}a(\theta)\int_0^\infty [\mathfrak{H}(t) - \mathfrak{H}(t-u)] \cdot \mathbf{G}''(u)[\mathfrak{H}(t) - \mathfrak{H}(t-u)]du.$$

Hence (7.110) holds if and only if

$$g(\theta)\mathbf{G}''(u) \geq \mathbf{0}, \qquad a(\theta)\mathbf{A}''(u) \leq \mathbf{0}$$

for all $u \in [0, \infty)$.

The minimum property of ϕ at constant histories $\mathfrak{E}^\dagger, \mathfrak{H}^\dagger$ holds if and only if

$$g(\theta)\mathbf{G}'(u) \leq \mathbf{0}, \qquad a(\theta)\mathbf{A}'(u) \geq \mathbf{0}.$$

Consequently the functional ϕ of (7.111) is thermodynamically consistent if $\mathbf{G}'$ and $\mathbf{G}''$, as well as $\mathbf{A}'$ and $\mathbf{A}''$, have opposite types of definiteness.

Now we let $\nabla\theta \neq \mathbf{0}$. Let $\mathbf{q}$ and ϕ depend on $\nabla\theta^t$ via the referential gradient $\nabla_R\theta^t = \nabla\theta^t\mathbf{F}^t$. For simplicity we look for models where the reduced inequality splits into two separate inequalities, which is anyway a sufficient condition for the validity of the second-law inequality. Hence we require that

$$\rho\theta\, d\phi(\Gamma|\nabla_R\dot{\theta}^t) + \frac{1}{\theta}\mathbf{q} \cdot \nabla\theta \geq 0.$$

Multiply by $J = \det \mathbf{F}$ and observe that, by means of the referential heat flux $\mathbf{Q} = J\mathbf{F}^{-T}\mathbf{q}$, we can write

$$\rho_R\theta d\phi(\Gamma|\nabla_R\dot{\theta}^t) + \mathbf{Q} \cdot \nabla_R\theta \leq 0. \tag{7.112}$$

For definiteness we look for the thermodynamic consistency of the constitutive equation

$$\mathbf{Q}(\theta(t), \nabla_R\,\theta^t) = \alpha(\theta)\mathbf{K}_0 \int_0^\infty \beta(u)\nabla_R\,\theta(t-u)du,$$

where $\mathbf{K}_0$ is supposed to be positive definite while α and β are so far undetermined; we let $\beta(\infty) = 0$. We let $\phi = \phi_m + \phi_c$ where ϕ_c is taken in the form

$$\phi_c = \alpha(\theta)\big|\mathbf{K}_0^{1/2} \int_0^\infty \beta(u)\nabla_R\theta(t-u)du\big|^2.$$

The minimum property of ϕ at $\nabla_R\theta \equiv \mathbf{0}$ implies that $\alpha > 0$. Inequality (7.112) can be written as

$$\rho_R\,\theta(\mathbf{K}_0 \int_0^\infty \beta(u)\nabla_R\,\theta(t-u)du)\cdot \int_0^\infty \beta(u)\nabla_R\,\dot\theta(t-u)du + (\mathbf{K}_0 \int_0^\infty \beta(u)\nabla_R\,\theta(t-u)du)\cdot\nabla_R\theta \le 0.$$

By integration by parts we have

$$\int_0^\infty \beta(u)\nabla_R\dot\theta(t-u)du = -\int_0^\infty \beta(u)\partial_u\nabla_R\theta(t-u)du$$

$$= -[\beta(u)\nabla_R\theta(t-u)]_0^\infty + \int_0^\infty \beta'(u)\nabla_R\theta(t-u)du = \beta(0)\nabla_R\theta(t) + \int_0^\infty \beta(u)\nabla_R\theta(t-u)du.$$

Hence the inequality can be written in the form

$$(\rho_R\theta\beta(0)+1)(\mathbf{K}_0 \int_0^\infty \beta(u)\nabla_R\theta(t-u)du)\cdot\nabla_R\theta(t)$$
$$+\rho_R\theta(\mathbf{K}_0 \int_0^\infty \beta(u)\nabla_R\theta(t-u)du)\cdot \int_0^\infty \beta'(u)\nabla_R\theta(t-u)du \le 0.$$

The arbitrariness of $\nabla_R\theta(t)$ implies

$$\beta(0) = -\frac{1}{\rho_R\theta}, \qquad \beta(u)\beta'(u) \le 0.$$

Since $\beta(0) < 0$ and $(\beta^2)' \le 0$ then $\beta(u) \le 0$ for all $u \in \mathbb{R}^+$. Consequently

$$\alpha(\theta)\mathbf{K}_0\beta(u) \le \mathbf{0}$$

as expected. Let

$$\hat\beta(u) = \frac{\beta(u)}{\beta(0)} \ge 0.$$

Hence

$$\mathcal{Q} = -\frac{\alpha(\theta)}{\rho_R\theta}\mathbf{K}_0 \int_0^\infty \hat\beta(u)\nabla_R\,\theta(t-u)du,$$

At the limit of short memory it follows

$$\mathcal{Q} = -\frac{\alpha(\theta)}{\rho_R \theta} \mathbf{K}_0 (\int_0^\infty \hat{\beta}(u) du) \nabla_R \theta.$$

In the spatial description we have

$$\mathbf{q} = -\frac{\alpha(\theta)}{J \rho_R \theta} \mathbf{F} \mathbf{K}_0 \int_0^\infty \beta(u) (\nabla \theta \mathbf{F})(t-u) du.$$

7.6.5 Rate Equations in the Eulerian Description

It is a crucial point of magnetoelasticity, as well as of magneto-viscoelasticity, that the stress tensor need not be symmetric. Hence the mechanical power $\mathbf{T} \cdot \mathbf{L}$ need not equal $\mathbf{T} \cdot \mathfrak{D}$ and moreover[22]

$$\mathbf{T} \cdot \mathbf{L} = J^{-1} (\mathbf{F} \mathbf{T}_{RR} \mathbf{F}^T) \cdot \mathbf{L} = J^{-1} \mathbf{T}_{RR} \cdot (\mathbf{F}^T \mathfrak{D} \mathbf{F}) + J^{-1} \mathbf{T}_{RR} \cdot (\mathbf{F}^T \mathbf{W} \mathbf{F}).$$

Since $\dot{\mathfrak{E}} = \mathbf{F}^T \mathfrak{D} \mathbf{F}$ then

$$\mathbf{T} \cdot \mathbf{L} = J^{-1} \mathbf{T}_{RR} \cdot \dot{\mathfrak{E}} + J^{-1} \mathbf{T}_{RR} \cdot (\mathbf{F}^T \mathbf{W} \mathbf{F}).$$

For simplicity we now ignore heat conduction. Hence in the Eulerian description we write the Clausius–Duhem inequality in the form

$$-\rho(\dot{\psi} + \eta \dot{\theta}) + \mathbf{T} \cdot \mathbf{L} - \mu_0 \mathbf{M} \cdot \dot{\mathbf{H}} = \theta \sigma \geq 0.$$

Since we look for rate equations, objectivity indicates that the independent variables are invariant so that their time derivatives be invariant too. Hence we assume that

$$\psi = \psi(\theta, \mathcal{T}, \mathfrak{E}, \mathfrak{H}),$$

where $\mathcal{T}$ is a stress-like, invariant, variable to be identified. Consequently the inequality can be written as

$$-\rho(\partial_\theta \psi + \eta)\dot{\theta} - \rho \partial_{\mathcal{T}} \psi \cdot \dot{\mathcal{T}} - \rho \partial_{\mathfrak{E}} \psi \cdot \dot{\mathfrak{E}} - \rho \partial_{\mathfrak{H}} \psi \cdot \dot{\mathfrak{H}} + \mathbf{T} \cdot \mathbf{L} - \mu_0 \mathbf{M} \cdot \dot{\mathbf{H}} \geq 0.$$

Since $\dot{\mathfrak{H}} = \mathbf{F}^T \mathbf{L}^T \mathbf{H} + \mathbf{F}^T \dot{\mathbf{H}}$ then we have

$$-\rho(\partial_\theta \psi + \eta)\dot{\theta} - \rho \partial_{\mathcal{T}} \psi \cdot \dot{\mathcal{T}} - \rho \partial_{\mathfrak{E}} \psi \cdot \dot{\mathfrak{E}} + (\mathbf{T} - \rho \mathbf{F} \partial_{\mathfrak{E}} \psi \mathbf{F}^T - \rho \mathbf{H} \otimes \mathbf{F} \partial_{\mathfrak{H}} \psi) \cdot (\mathbf{F}^{-T} \dot{\mathfrak{E}} \mathbf{F}^{-1})$$
$$+ (\mathbf{T} - \rho \mathbf{H} \otimes \mathbf{F} \partial_{\mathfrak{H}} \psi) \cdot \mathbf{W} - (\rho \mathbf{F} \partial_{\mathfrak{H}} \psi + \mu_0 \mathbf{M}) \cdot \dot{\mathbf{H}} \geq 0.$$

[22] In this section, $\mathfrak{D}$ is the stretching tensor, $\mathfrak{E}$ is the Green-St Venant tensor and $\mathfrak{H} = \mathbf{F}^T \mathbf{H}$.

The arbitrariness of $\dot{\theta}$, $\dot{\mathbf{H}}$, and $\mathbf{W}$ implies

$$\eta = -\partial_\theta \psi, \qquad \mu_0 \mathbf{M} = -\rho \mathbf{F} \partial_{\mathfrak{H}} \psi, \tag{7.113}$$

$$\mathrm{skw}(\mathbf{T} + \mu_0 \mathbf{H} \otimes \mathbf{M}) = \mathbf{0}. \tag{7.114}$$

Thus we replace $-\rho \mathbf{F} \partial_{\mathfrak{H}} \psi$ with $\mu_0 \mathbf{M}$ and write the remaining inequality in the form

$$-\rho \partial_{\mathcal{T}} \psi \cdot \dot{\mathcal{T}} + (-\rho \partial_{\mathfrak{E}} \psi + J^{-1} \mathbf{T}_{RR} + \mu_0 \mathbf{F}^{-1} \mathbf{H} \otimes \mathbf{F}^{-1} \mathbf{M}) \cdot \dot{\mathfrak{E}} \geq 0.$$

This result suggests that we define

$$\mathcal{T} := \mathbf{T}_{RR} + J \mu_0 \mathbf{F}^{-1} \mathbf{H} \otimes \mathbf{F}^{-1} \mathbf{M}$$

so that we have

$$-\rho_R \partial_{\mathcal{T}} \psi \cdot \dot{\mathcal{T}} + (\mathcal{T} - \rho \partial_{\mathfrak{E}} \psi) \cdot \dot{\mathfrak{E}} = J \theta \sigma \geq 0. \tag{7.115}$$

Observe that $\mathbf{F}^{-1}\mathbf{H}$ and $\mathbf{F}^{-1}\mathbf{M}$ are invariant vectors. This follows by a direct check,

$$(\mathbf{F}^{-1}\mathbf{H})^* = (\mathbf{QF})^{-1}\mathbf{QH} = \mathbf{F}^{-1}\mathbf{Q}^{-1}\mathbf{QH} = \mathbf{F}^{-1}\mathbf{H},$$

and the like for $\mathbf{F}^{-1}\mathbf{M}$. This in turn proves the invariance of $\mathcal{T}$. Moreover letting $\mathfrak{M} = J\mathbf{F}^{-1}\mathbf{M}$ we can write $(7.113)_2$ in the form

$$\mu_0 \mathfrak{M} = -\rho_R \partial_{\mathfrak{H}} \psi. \tag{7.116}$$

If $\dot{\mathcal{T}}$ and $\dot{\mathfrak{E}}$ are independent and $\sigma = 0$ then it follows from (7.115) that

$$\partial_{\mathcal{T}} \psi = \mathbf{0}, \qquad \mathcal{T} = \rho_R \partial_{\mathfrak{E}} \psi; \tag{7.117}$$

Equations (7.117) are said to characterize hyper-magnetoelastic materials. The results (7.116) and (7.117) allow us to write the incremental relations

$$\mu_0 \dot{\mathfrak{M}} = -\rho_R \partial_{\mathfrak{E}} \partial_{\mathfrak{H}} \psi \, \dot{\mathfrak{E}} - \rho_R \partial_{\mathfrak{H}} \partial_{\mathfrak{H}} \psi \, \dot{\mathfrak{H}}, \tag{7.118}$$

$$\dot{\mathcal{T}} = \rho_R \partial_{\mathfrak{E}} \partial_{\mathfrak{E}} \psi \, \dot{\mathfrak{E}} + \rho_R \partial_{\mathfrak{H}} \partial_{\mathfrak{E}} \psi \, \dot{\mathfrak{H}}. \tag{7.119}$$

If, again, $\sigma = 0$ but

$$\partial_{\mathcal{T}} \psi \neq \mathbf{0}, \qquad \partial_{\mathfrak{E}} \psi \neq \mathbf{0}$$

then we have

$$-\rho_R \partial_{\mathcal{T}} \psi \cdot \dot{\mathcal{T}} + (\mathcal{T} - \rho_R \partial_{\mathfrak{E}} \psi) \cdot \dot{\mathfrak{E}} = 0; \tag{7.120}$$

Equation (7.120) is said to characterize hypo-magnetoelastic materials. We can then apply the representation formula to $\dot{\mathcal{T}}$,

$$\dot{\mathcal{T}} = (\dot{\mathcal{T}} \cdot \mathbf{N})\mathbf{N} + (\mathbf{I} - \mathbf{N} \otimes \mathbf{N})\mathbf{G},$$

where $\mathbf{G}$ is any second-order tensor. Since $\partial_{\mathcal{T}}\psi \neq \mathbf{0}$ then we let $\mathbf{N} = \partial_{\mathcal{T}}\psi/|\partial_{\mathcal{T}}\psi|$. By (7.120) we have

$$\dot{\mathcal{T}} = \frac{(\mathcal{T} - \rho_R \partial_{\mathfrak{E}}\psi) \cdot \dot{\mathfrak{E}}}{\rho_R |\partial_{\mathcal{T}}\psi|^2} \partial_{\mathcal{T}}\psi + (\mathbf{I} - \frac{\partial_{\mathcal{T}}\psi}{|\partial_{\mathcal{T}}\psi|} \otimes \frac{\partial_{\mathcal{T}}\psi}{|\partial_{\mathcal{T}}\psi|})\mathbf{G}$$

or

$$\dot{\mathcal{T}} = \frac{1}{\rho_R |\partial_{\mathcal{T}}\psi|^2} [\partial_{\mathcal{T}}\psi \otimes (\mathcal{T} - \rho_R \partial_{\mathfrak{E}}\psi)]\dot{\mathfrak{E}} + (\mathbf{I} - \frac{\partial_{\mathcal{T}}\psi}{|\partial_{\mathcal{T}}\psi|} \otimes \frac{\partial_{\mathcal{T}}\psi}{|\partial_{\mathcal{T}}\psi|})\mathbf{G}$$

If, further, $\sigma \neq 0$ then the representation of $\dot{\mathcal{T}}$ generalizes to

$$\dot{\mathcal{T}} = \frac{(\mathcal{T} - \rho_R \partial_{\mathfrak{E}}\psi) \cdot \dot{\mathfrak{E}} - J\theta\sigma}{\rho_R |\partial_{\mathcal{T}}\psi|^2} \partial_{\mathcal{T}}\psi + (\mathbf{I} - \frac{\partial_{\mathcal{T}}\psi}{|\partial_{\mathcal{T}}\psi|} \otimes \frac{\partial_{\mathcal{T}}\psi}{|\partial_{\mathcal{T}}\psi|})\mathbf{G}. \tag{7.121}$$

Definite forms of (7.121) are now established by having in mind fluid or solid behaviours.

Fluids

We assume ψ and σ in the forms

$$\rho_R \psi = \rho_R \psi_0(\theta) + \tfrac{1}{2}\alpha|\mathcal{T}|^2 - \tfrac{1}{2}\xi|\mathfrak{H}|^2, \qquad \alpha, \xi > 0,$$

$$J\theta\sigma = \beta|\mathcal{T}|^2 + \gamma|\mathfrak{H}|^2, \qquad \beta, \gamma > 0.$$

The dependence of ψ on $\mathfrak{H}$ results in the magnetization relation

$$\mu_0 \mathfrak{M} = \xi \mathfrak{H}.$$

Since $\partial_{\mathfrak{E}}\psi = \mathbf{0}$ and $\rho_R \partial_{\mathcal{T}}\psi = \alpha\mathcal{T}$ then (7.121) can be written in the form

$$\dot{\mathcal{T}} = \frac{\mathcal{T} \otimes \mathcal{T}}{\alpha|\mathcal{T}|^2}\dot{\mathfrak{E}} - \frac{\beta}{\alpha}\mathcal{T} - \frac{\gamma|\mathfrak{H}|^2}{\alpha|\mathcal{T}|^2}\mathcal{T} + (\mathbf{I} - \frac{\mathcal{T}}{|\mathcal{T}|} \otimes \frac{\mathcal{T}}{|\mathcal{T}|})\mathbf{G}.$$

If, for definiteness, we let $\mathbf{G} = \dot{\mathfrak{E}}/\alpha$ then it follows

$$\dot{\mathcal{T}} = \frac{1}{\alpha}\dot{\mathfrak{E}} - \frac{\beta}{\alpha}\mathcal{T} - \frac{\gamma|\mathfrak{H}|^2}{\alpha|\mathcal{T}|^2}\mathcal{T}.$$

If, in particular, we let $\gamma = \gamma_0|\boldsymbol{T}|^2$ then we obtain

$$\dot{\boldsymbol{T}} = a\dot{\mathfrak{E}} - b\boldsymbol{T}, \qquad a = 1/\alpha > 0, \quad b = \beta/\alpha + \gamma_0|\mathfrak{H}|^2/\alpha > 0; \tag{7.122}$$

equation (7.122) is a generalization of the Maxwell (fluid) model, the parameter b being affected by the magnetic field $\mathfrak{H}$. In this connection we observe that, given $\boldsymbol{T}_0 = \boldsymbol{T}(0)$, if $|\mathfrak{H}|$ is assumed to be known (7.122) can be solved on $(0, t)$ to obtain

$$\boldsymbol{T}(t) = \boldsymbol{T}_0 \exp[-\int_0^t b(\lambda)d\lambda] + \int_0^t a(\tau)\exp[-\int_\tau^t b(\lambda)d\lambda]\dot{\mathfrak{E}}(\tau)d\tau.$$

If $b \geq b_0 > 0$ then $\boldsymbol{T}(-y)\exp[-\int_{-y}^0 b(\lambda)d\lambda$ vanishes as $y \to \infty$. Hence we let

$$\boldsymbol{T}_0 = \int_{-\infty}^0 a(\tau)\exp[-\int_\tau^0 b(\lambda)d\lambda]\dot{\tilde{\mathfrak{E}}}(\tau)d\tau,$$

where $\mathfrak{E}(\tau) = \tilde{\mathfrak{E}}(\tau)$, $\tau \leq 0$. Substituting the value of $\boldsymbol{T}_0$ we find

$$\boldsymbol{T}(t) = \int_{-\infty}^t a(\tau)\exp[-\int_\tau^t b(\lambda)d\lambda]\dot{\mathfrak{E}}(\tau)d\tau = \int_0^\infty a(t-s)\exp[-\int_0^s b(\xi)d\xi)]\dot{\mathfrak{E}}(t-s)ds. \tag{7.123}$$

As a comment we observe that the memory kernel of (7.123) has the form

$$G(\tau, t-\tau) = a(\tau)\exp[-\int_\tau^t b(\lambda)d\lambda];$$

it describes an aging effect in that the function G changes in time as a consequence of the factor $a(\tau)$. We get a standard memory kernel

$$G(t-\tau) = a_0 \exp[-\int_\tau^t b(\lambda)d\lambda]$$

if $a = a_0$ is constant.

Another model is obtained by letting σ be a viscous term in the form

$$J\theta\sigma = \nu|\dot{\mathfrak{E}}|^2 + \zeta|\dot{\mathfrak{H}}|^2, \qquad \nu, \zeta > 0.$$

It follows that

$$\dot{\boldsymbol{T}} = \frac{\boldsymbol{T}\otimes\boldsymbol{T}}{\alpha|\boldsymbol{T}|^2}\dot{\mathfrak{E}} - \frac{\nu|\dot{\mathfrak{E}}|^2 + \zeta|\dot{\mathfrak{H}}|^2}{\alpha|\boldsymbol{T}|^2}\boldsymbol{T} + (\mathbf{I} - \frac{\boldsymbol{T}}{|\boldsymbol{T}|}\otimes\frac{\boldsymbol{T}}{|\boldsymbol{T}|})\mathbf{G}. \tag{7.124}$$

If $\mathbf{G} = \dot{\mathfrak{E}}/\alpha$ then we have

$$\dot{\boldsymbol{T}} = \frac{1}{\alpha}\dot{\mathfrak{E}} - \frac{\nu|\dot{\mathfrak{E}}|^2 + \zeta|\dot{\mathfrak{H}}|^2}{\alpha|\boldsymbol{T}|^2}\boldsymbol{T}.$$

As to the influence of the magnetic field through $\dot{\mathfrak{H}}$ we observe that

$$\dot{\mathfrak{H}} = \mathbf{F}^T(\mathbf{L}^T\mathbf{H} + \dot{\mathbf{H}}) = \mathbf{F}^T(\overset{\circ}{\mathbf{H}} + \mathbf{D}\mathbf{H}),$$

$\overset{\circ}{\mathbf{H}}$ denoting the corotational derivative, $\overset{\circ}{\mathbf{H}} = \dot{\mathbf{H}} - \mathbf{W}\mathbf{H}$. Hence we have

$$|\dot{\mathfrak{H}}|^2 = (\overset{\circ}{\mathbf{H}} + \mathbf{D}\mathbf{H}) \cdot \mathbf{F}\mathbf{F}^T(\overset{\circ}{\mathbf{H}} + \mathbf{D}\mathbf{H})$$

and the representation (7.124) can be written in the form

$$\dot{\mathcal{T}} = \frac{\mathcal{T} \otimes \mathcal{T}}{\alpha|\mathcal{T}|^2}\dot{\mathfrak{E}} - \frac{\nu|\dot{\mathfrak{E}}|^2}{\alpha|\mathcal{T}|^2}\mathcal{T} - \frac{\zeta(\overset{\circ}{\mathbf{H}} + \mathbf{D}\mathbf{H}) \cdot \mathbf{F}\mathbf{F}^T(\overset{\circ}{\mathbf{H}} + \mathbf{D}\mathbf{H})}{\alpha|\mathcal{T}|^2}\mathcal{T} + (\mathsf{I} - \frac{\mathcal{T}}{|\mathcal{T}|} \otimes \frac{\mathcal{T}}{|\mathcal{T}|})\mathbf{G}.$$

7.6.6 *Solids*

Solids are characterized by a stress dependence such that, asymptotically, $\mathcal{T} = \mathbf{G}_\infty\mathfrak{E}$. Hence we let formally replace $\mathcal{T}$ of the fluid model with $\mathcal{T} - \mathbf{G}_\infty\mathfrak{E}$. Define

$$\rho_R\psi = \rho_R\psi_0(\theta) + \tfrac{1}{2}\mathfrak{E}\mathbf{G}_\infty\mathfrak{E} + \tfrac{1}{2}(\mathcal{T} - \mathbf{G}_\infty\mathfrak{E}) \cdot \mathbf{A}(\mathcal{T} - \mathbf{G}_\infty\mathfrak{E}) - \tfrac{1}{2}\mathfrak{H} \cdot \Xi\mathfrak{H}$$

and

$$J\theta\sigma = (\mathcal{T} - \mathbf{G}_\infty\mathfrak{E}) \cdot [(\beta + \mathfrak{H} \cdot \mathbf{\Gamma}\mathfrak{H})\mathbf{A}(\mathcal{T} - \mathbf{G}_\infty\mathfrak{E})], \qquad \beta > 0,$$

where $\Xi, \mathbf{\Gamma} \in \mathrm{Sym}^+$ and $\mathbf{A}$, $\mathbf{G}_\infty$ are fourth-order positive definite tensors. Observe that

$$\rho_R\partial_{\mathfrak{H}}\psi = -\Xi\mathfrak{H}, \qquad \rho_R\partial_{\mathcal{T}}\psi = \mathbf{A}(\mathcal{T} - \mathbf{G}_\infty\mathfrak{E}), \qquad \rho_R\partial_{\mathfrak{E}}\psi = \mathbf{G}_\infty\mathfrak{E} - \mathbf{G}_\infty^T\mathbf{A}(\mathcal{T} - \mathbf{G}_\infty\mathfrak{E}).$$

Moreover, by (7.116) it follows $\mathfrak{M} = J\mathbf{F}^{-1}\mathbf{M} = \mu_0^{-1}\Xi\mathfrak{H}$.

We now apply the representation (7.121) by letting

$$\mathbf{N} = \frac{\partial_{\mathcal{T}}\psi}{|\partial_{\mathcal{T}}\psi|} = \frac{\mathbf{A}(\mathcal{T} - \mathbf{G}_\infty\mathfrak{E})}{|\mathbf{A}(\mathcal{T} - \mathbf{G}_\infty\mathfrak{E})|}.$$

To within $(\mathsf{I} - \mathbf{N} \otimes \mathbf{N})\mathbf{G}$, the representation formula yields

$$\begin{aligned}\dot{\mathcal{T}} &\simeq \frac{[\mathcal{T} - \mathbf{G}_\infty\mathfrak{E} + \mathbf{G}_\infty^T\mathbf{A}(\mathcal{T} - \mathbf{G}_\infty\mathfrak{E})]\dot{\mathfrak{E}} - (\mathcal{T} - \mathbf{G}_\infty\mathfrak{E}) \cdot [\beta + \mathfrak{H}\mathbf{\Gamma}\mathfrak{H}]\mathbf{A}(\mathcal{T} - \mathbf{G}_\infty\mathfrak{E})}{|\mathbf{A}(\mathcal{T} - \mathbf{G}_\infty\mathfrak{E})|}\mathbf{N} \\ &= \frac{\mathbf{A}(\mathcal{T} - \mathbf{G}_\infty\mathfrak{E}) \cdot [\mathbf{A}^{-1} + \mathbf{G}_\infty\mathfrak{E})]\dot{\mathfrak{E}} - \mathbf{A}(\mathcal{T} - \mathbf{G}_\infty\mathfrak{E}) \cdot [\beta + \mathfrak{H}\mathbf{\Gamma}\mathfrak{H}](\mathcal{T} - \mathbf{G}_\infty\mathfrak{E})}{|\mathbf{A}(\mathcal{T} - \mathbf{G}_\infty\mathfrak{E})|}\mathbf{N}.\end{aligned}$$

Hence we have

$$\dot{\mathcal{T}} = (\mathbf{N} \otimes \mathbf{N})[\mathbf{A}^{-1} + \mathbf{G}_\infty]\dot{\mathfrak{E}} - (\mathbf{N} \otimes \mathbf{N})[\beta + \mathfrak{H} \cdot \mathbf{\Gamma}\mathfrak{H}](\mathcal{T} - \mathbf{G}_\infty\mathfrak{E}) + (\mathbf{I} - \mathbf{N} \otimes \mathbf{N})\mathbf{G}.$$

Choosing, e.g.,

$$\mathbf{G} = [\mathbf{A}^{-1} + \mathbf{G}_\infty]\dot{\mathfrak{E}} - [\beta + \mathfrak{H}\mathbf{\Gamma}\mathbf{H}](\mathcal{T} - \mathbf{G}_\infty\mathfrak{E})$$

we find

$$\dot{\mathcal{T}} = [\mathbf{A}^{-1} + \mathbf{G}_\infty]\dot{\mathfrak{E}} - [\beta + \mathfrak{H} \cdot \mathbf{\Gamma}\mathfrak{H}](\mathcal{T} - \mathbf{G}_\infty\mathfrak{E}),$$

whence

$$\dot{\mathcal{T}} - \mathbf{G}_\infty\dot{\mathfrak{E}} + [\beta + \mathfrak{H} \cdot \mathbf{\Gamma}\mathfrak{H}](\mathcal{T} - \mathbf{G}_\infty\mathfrak{E}) = \mathbf{A}^{-1}\dot{\mathfrak{E}}. \tag{7.125}$$

Equation (7.125) shows that $\mathcal{T} - \mathbf{G}_\infty\mathfrak{E}$ evolves with a relaxation time

$$\tau = \frac{1}{\beta + \mathfrak{H} \cdot \mathbf{\Gamma}\mathfrak{H}}.$$

Moreover, if $\dot{\mathfrak{E}} = \mathbf{0}$ then asymptotically we have

$$\mathcal{T} = \mathbf{G}_\infty\mathfrak{E}$$

as we expect for a solid model.

7.7 Causality and Kramers–Kronig Relations

Consider a generic response function of materials with memory,

$$R(t) = G_0 F(t) + \int_0^\infty G(u)F(t-u)du, \qquad G(u) \to 0 \quad \text{as} \quad u \to 0.$$

The restriction of the kernel G to $\mathbb{R}^+$, or rather its vanishing for $u < 0$, means that at time t only values of the cause F prior to that time enter in determining the response $R(t)$. If F is time harmonic,

$$F(t) = F_0 \exp(-i\omega t),$$

then

$$R(t) = \mathcal{G}(\omega)F(t), \qquad \mathcal{G}(\omega) := G_0 + \int_0^\infty G(u)\exp(i\omega u)du.$$

Let $\omega \in \mathbb{C}$. Letting $\omega = \omega_r + i\omega_i$ we have

$$\hat{G}(-\omega) = G_0 + \int_0^\infty G(u)\exp(-i\omega u)du = G_0 + \int_0^\infty G(u)\exp(-i\omega_r u + \omega_i u)du$$
$$= [G_0 + \int_0^\infty G(u)\exp(i\omega^* u)du]^*$$

whence

$$\hat{G}(-\omega) = \hat{G}^*(\omega^*).$$

Let $z = x + iy$. The function

$$\hat{G}(z) := \int_0^\infty G(u)\exp(izu)du = \int_0^\infty G(u)\exp(ixu)\exp(-yu)du$$

is bounded as $y \geq 0$. In addition, we assume $\hat{G}(z) \to 0$ as $|z| \to \infty$. Hence

$$\int_C \frac{\hat{G}(z)}{z - \omega} dz = 0$$

if $z = \omega$ is outside the contour C. We then choose the contour to trace the real axis, with a hump over the point $z = \omega \in \mathbb{R}$, and a semicircle in the upper half plane. By Cauchy's residue theorem we then find that

$$0 = \int_{-\infty}^\infty \frac{\hat{G}(x)}{x - \omega} dx - i\pi\hat{G}(\omega),$$

where

$$\int_{-\infty}^\infty = \lim_{\epsilon\to 0} \{ \int_{-\infty}^{-\epsilon} + \int_\epsilon^\infty \}$$

is also viewed as the principal value. Hence it follows

$$\hat{G}(\omega) = \frac{1}{i\pi} \int_{-\infty}^\infty \frac{\hat{G}(x)}{x - \omega} dx.$$

This in turn implies that the real and imaginary parts $\hat{G}_r$, $\hat{G}_i$ satisfy

$$\hat{G}_r(\omega) = \frac{1}{\pi} \int_{-\infty}^\infty \frac{\hat{G}_i(x)}{x - \omega} dx, \qquad \hat{G}_i(\omega) = -\frac{1}{\pi} \int_{-\infty}^\infty \frac{\hat{G}_r(x)}{x - \omega} dx. \tag{7.126}$$

For any function f on $\mathbb{R}$ the Hilbert transform $f_H(\omega)$ is defined by

$$f_H(\omega) = \frac{1}{\pi} \int_{-\infty}^\infty \frac{f(x)}{\omega - x} dx,$$

the integral being understood in the principal value sense. Thus, by (7.126) we can say that $\hat{G}_i$ is the Hilbert transform of $\hat{G}_r$ while $\hat{G}_r$ is the Hilbert transform of $-\hat{G}_i$.

By definition

$$\hat{G}(x) = \int_0^\infty G(u)\exp(ixu)du, \quad \hat{G}_r(x) = \int_0^\infty G(u)\cos(xu)du, \quad \hat{G}_i(x) = \int_0^\infty G(u)\sin(xu)du,$$

and hence

$$\hat{G}_r(x) = \hat{G}_r(-x), \quad \hat{G}_i(x) = -\hat{G}_i(-x), \qquad x \in \mathbb{R};$$

$\hat{G}_r(x)$ is even, $\hat{G}_i(x)$ is odd. Consider the relations in (7.126) and multiply the numerator and denominator of the integrand by $x + \omega$. In view of the even and odd properties we obtain

$$\hat{G}_r(\omega) = \tfrac{2}{\pi}\int_0^\infty \frac{x\hat{G}_i(x)}{x^2 - \omega^2}dx, \qquad \hat{G}_i(\omega) = -\tfrac{2}{\pi}\int_0^\infty \frac{\omega\hat{G}_r(x)}{x^2 - \omega^2}dx.$$

Since $\hat{G}_r = \mathcal{G}_r - G_0$ then we have

$$\mathcal{G}_r(\omega) - G_0 = \tfrac{2}{\pi}\int_0^\infty \frac{x\mathcal{G}_i(x)}{x^2 - \omega^2}dx, \qquad \mathcal{G}_i(\omega) = -\tfrac{2}{\pi}\int_0^\infty \frac{\omega[\mathcal{G}_r(x) - G_0]}{x^2 - \omega^2}dx. \quad (7.127)$$

Equations (7.127) are commonly referred to as Kramers–Kronig relations of the complex-valued response function $\mathcal{G}$.

7.8 Memory Models via Fractional Derivatives

In a letter dated September 30th, 1695, L'Hospital wrote to Leibniz asking him about a particular notation he had used for the nth derivative, D^n/Dx^n. L'Hospital posed the question to Leibniz, what would the result be if $n = 1/2$. Leibniz answered "An apparent paradox, from which one day useful consequences will be drawn". In these words fractional calculus is regarded to be born.

Among the mathematicians who first investigated fractional calculus, along with the mathematical properties, are Fourier, Euler and Laplace. Many definitions have been given for a non-integer-order derivative. Perhaps the most popular one is that of Riemann–Liouville. Most of the mathematical theory of fractional calculus was developed prior to the turn of the twentieth century. Yet it is in the past 100 years that the most intriguing laps in engineering and scientific application have been found.

7.8.1 Preliminaries on Fractional Derivatives

Understanding of definitions and use of fractional calculus become more immediate by a preliminary review on the gamma function, the beta function, and the Mittag-Leffler function.

For $\Re z > 0$ the gamma function $\Gamma(z)$ is defined by Euler's formula

$$\Gamma(z) = \int_0^\infty \exp(-t) t^{z-1} dt. \tag{7.128}$$

Differentiation with respect to z can be carried out and it follows that

$$\Gamma'(z) = \int_0^\infty \exp(-t) t^{z-1} \ln t dt.$$

Integration by parts of (7.128) yields

$$\int_0^\infty \exp(-t) t^{z-1} dt = \int_0^\infty \exp(-t) \frac{t^z}{z} dt$$

whence

$$\Gamma(z+1) = z\Gamma(z).$$

Since $\Gamma(1) = 1$ then

$$\Gamma(n+1) = n!$$

so that Γ is a generalized factorial function.

The beta function $B(p, q)$ is defined by

$$B(p, q) = \int_0^1 t^{p-1}(1-t)^{q-1} dt, \qquad \Re p > 0, \ \Re q > 0.$$

Replacing t by $1 - \tau$ we have

$$\int_0^1 t^{p-1}(1-t)^{q-1} dt = \int_0^1 (1-\tau)^{p-1}(\tau)^{q-1} d\tau$$

and hence

$$B(p, q) = B(q, p).$$

By the change of variable $t \to \tau : t = \tau/(1+\tau)$, $\tau \in [0, \infty)$, we find

$$B(p, q) = \int_0^\infty \frac{\tau^{p-1}}{(1+\tau)^{p+q}} d\tau.$$

It is possible to express the beta function in terms of gamma functions. To this end, letting $\Re p > 0$, $\Re q > 0$ we have

$$\Gamma(p)\Gamma(q) = \int_0^\infty \exp(-t)t^{p-1}dt \int_0^\infty \exp(-\tau)\tau^{q-1}d\tau = \int_0^\infty \exp(-t)t^{p-1}dt\big(t^q \int_0^\infty \exp(-tx)x^{q-1}dx\big)$$
$$= \int_0^\infty x^{q-1}dx \int_0^\infty \exp(-t(x+1))\, t^{p+q-1}dt = \int_0^\infty \frac{x^{q-1}}{(1+x)^{p+q}}dx \int_0^\infty \exp(-\tau)\tau^{p+q-1}d\tau.$$

Hence we can write

$$B(p,q) = \frac{\Gamma(p)\Gamma(q)}{\Gamma(p+q)}.$$

In terms of the gamma function, the Mittag-Leffler functions are defined by

$$E_\alpha(z) = \sum_{k=0}^\infty \frac{z^k}{\Gamma(\alpha k+1)}, \qquad \alpha \in \mathbb{C},\ \Re\alpha > 0,$$

$$E_{\alpha,\beta}(z) = \sum_{k=0}^\infty \frac{z^k}{\Gamma(\alpha k+\beta)}, \qquad \alpha, \beta \in \mathbb{C},\ \Re\alpha > 0, \Re\beta > 0.$$

7.8.2 Fractional Derivatives

We start from Cauchy's formula on iterated integrals,

$$\int_a^x \int_a^{x_1} \cdots \int_a^{x_{n-1}} f(x_n)dx_n...dx_2dx_1 = \frac{1}{(n-1)!}\int_a^x (x-y)^{n-1}f(y)dy,$$

where $a < x$ and $n \in \mathbb{N}$. The proof is established by induction. If $n = 1$ this formula holds trivially as

$$\int_a^x f(y)dy = \int_a^x f(y)dy.$$

Let

$$F(x_{n-1}) = \int_a^{x_{n-1}} f(x_n)dx_n.$$

Hence the validity for $n-1$ is expressed in the form

$$\int_a^x \int_a^{x_1} \cdots \int_a^{x_{n-1}} f(x_n)dx_n...dx_2dx_1 = \int_a^x \int_a^{x_1} \cdots \int_a^{x_{n-2}} F(x_{n-1})dx_{n-1}...dx_2dx_1$$
$$= \frac{1}{(n-2)!}\int_a^x (x-y)^{n-2}F(y)dy.$$

An integration by parts yields

$$\frac{1}{(n-2)!}\int_a^x (x-y)^{n-2}F(y)dy = \frac{1}{(n-1)!}\int_a^x (x-y)^{n-1}f(y)dy,$$

which completes the proof.

By Cauchy's formula, the right-hand side is the n-fold integral of f. Also, since $\Gamma(n) = (n-1)!$, we can define the (Riemann–Liouville) integral of order $\alpha > 0$ as

$$(I^{\alpha}f)(x) = \frac{1}{\Gamma(\alpha)}\int_a^x (x-y)^{\alpha-1}f(y)dy.$$

The *Riemann–Liouville* fractional derivative of order $\alpha \in (0,1)$ is defined for absolutely-integrable functions such that

$$(D^{\alpha}_{RL}f)(x) = \frac{d}{dx}(I^{1-\alpha}f)(x).$$

For $\alpha > 1$ the definition is extended as

$$(D^{\alpha}_{RL}f)(x) = \frac{d^n}{dx^n}(I^{n-\alpha}f)(x), \qquad n = [\alpha]+1,$$

n then being the smallest integer larger than α. Hence we have

$$(D^{\alpha}_{RL}f)(x) = \frac{d^n}{dx^n}\frac{1}{\Gamma(n-\alpha)}\int_a^x (x-y)^{n-\alpha-1}f(y)dy, \qquad \alpha \in [n-1,n).$$

For integer-order derivatives we find the customary derivative. Letting $\alpha = n-1$ we have

$$(D^{n-1}f)(x) = \frac{d^n}{dx^n}\frac{1}{\Gamma(1)}\int_a^x f(y)dy = f^{(n-1)}(x).$$

The *Caputo* fractional derivative is defined in the form

$$(D^{\alpha}_{C}f)(x) = (I^{n-\alpha}\frac{d^n f}{dx^n})(x) = \frac{1}{\Gamma(n-\alpha)}\int_a^x (x-y)^{n-\alpha-1}f^{(n)}(y)dy.$$

Direct differentiation and an integration by parts lead to

$$\frac{d}{dx}\int_a^x (x-y)^{n-\alpha-1}f(y)dy = (x-a)^{n-\alpha-1}f(a) + \int_a^x (x-y)^{n-\alpha-1}f'(y)dy.$$

Consequently

$$D^{\alpha}_{RL}f = D^{\alpha}_{C}f$$

for functions vanishing at the initial point, $f(a) = 0$.

The *Marchaud* fractional derivative of order α is defined as

$$(D^{\alpha}_{M} f)(x) = \frac{f(x)}{\Gamma(1-\alpha)(x-a)^{\alpha}} + \frac{\alpha}{\Gamma(1-\alpha)} \int_{a}^{x} \frac{f(x)-f(y)}{(x-y)^{\alpha+1}} dy.$$

For $a = -\infty$ the derivative reduces to

$$(D^{\alpha}_{M} f)(x) = \frac{\alpha}{\Gamma(1-\alpha)} \int_{-\infty}^{x} \frac{f(x)-f(y)}{(x-y)^{\alpha+1}} dy.$$

The change of variable, $y \to \tau = x - y$, results in

$$(D^{\alpha}_{M} f)(x) = \frac{\alpha}{\Gamma(1-\alpha)} \int_{0}^{\infty} \frac{f(x)-f(x-\tau)}{\tau^{\alpha+1}} d\tau.$$

The idea of Marchaud's derivative arises by the purpose of extending the Riemann–Liouville integral from $\alpha > 0$ to $\alpha < 0$. Now, if $-\alpha \in [-1, 0)$ then formally

$$(I^{-\alpha} f)(x) = \frac{1}{\Gamma(-\alpha)} \int_{a}^{x} (x-y)^{-\alpha-1} f(y) dy.$$

If we let $a \to -\infty$, by a change of variable we have

$$(I^{-\alpha} f)(x) = \frac{1}{\Gamma(-\alpha)} \int_{0}^{\infty} y^{-\alpha-1} f(x-y) dy.$$

However the integral diverges because of $1/y^{\alpha+1}$ as $y \to 0$. The idea is to subtract the divergent part and let

$$(D^{\alpha}_{M} f)(x) = \frac{\alpha}{\Gamma(1-\alpha)} \int_{0}^{\infty} \frac{f(x)-f(x-\tau)}{\tau^{\alpha+1}} d\tau.$$

Denote by $\lfloor \cdot \rfloor$ the floor function so that

$$\lfloor \alpha \rfloor = \max\{m \in \mathbb{Z} : m \leq \alpha\}.$$

Since $n = \lfloor \alpha \rfloor + 1$ it is convenient to write

$$(D^{\alpha}_{C} f)(x) = \frac{1}{\Gamma(\varpi)} \int_{a}^{x} (x-y)^{\lfloor \alpha \rfloor - \alpha} f^{(\lfloor \alpha \rfloor + 1)}(y) dy,$$

where

$$\varpi = 1 + \lfloor \alpha \rfloor - \alpha \in (0, 1].$$

It is worth remarking that $\lfloor \alpha \rfloor - \alpha \in (-1, 0]$, which is of interest in that the integral is bounded if $f^{(\lfloor \alpha \rfloor + 1)}$ is bounded as $y \to x$. Let $a \to -\infty$. By a change of variable

we have

$$(D_c^\alpha f)(x) = \frac{1}{\Gamma(\varpi)} \int_0^\infty y^{\lfloor\alpha\rfloor-\alpha} f^{(\lfloor\alpha\rfloor+1)}(x-y)dy.$$

We now check whether the fractional derivative and the ordinary derivative commute, $(DD_c^\alpha f)(x) = (D_c^\alpha Df)(x)$. By definition,

$$(DD_c^\alpha f)(x) = \frac{1}{\Gamma(\varpi)} \int_0^\infty y^{\lfloor\alpha\rfloor-\alpha} DD^{\lfloor\alpha\rfloor+1} f(x-y)dy,$$

$$(D_c^\alpha Df)(x) = \frac{1}{\Gamma(\varpi)} \int_0^\infty y^{\lfloor\alpha\rfloor-\alpha} D^{\lfloor\alpha\rfloor+1} Df(x-y)dy.$$

The commutation property then holds.

We now compute the Laplace transform of the fractional derivative. In this regard we restrict attention to functions vanishing on $(-\infty, 0)$ and then

$$(D_c^\alpha f)(x) = \frac{1}{\Gamma(\varpi)} \int_0^x y^{\lfloor\alpha\rfloor-\alpha} f^{(\lfloor\alpha\rfloor+1)}(x-y)dy = \frac{1}{\Gamma(\varpi)} \int_0^x (x-y)^{\lfloor\alpha\rfloor-\alpha} f^{(\lfloor\alpha\rfloor+1)}(y)dy.$$

For any $s \in \mathbb{R}^{++}$ we consider the Laplace transform of $D_c^\alpha f$,

$$\int_0^\infty (D_c^\alpha f)(x)\exp(-sx)dx = \frac{1}{\Gamma(\varpi)} \int_0^\infty \int_0^x (x-y)^{\lfloor\alpha\rfloor-\alpha} f^{(\lfloor\alpha\rfloor+1)}(y)\exp(-sx)dy\,dx.$$

Replace x by $x = y + (x - y)$ to obtain

$$\int_0^\infty (D_c^\alpha f)(x)\exp(-sx)dx = \frac{1}{\Gamma(\varpi)} \int_0^\infty \xi^{\lfloor\alpha\rfloor-\alpha} \exp(-s\xi)d\xi \int_0^\infty f^{(\lfloor\alpha\rfloor+1)}(y)\exp(-sy)dy.$$

The change of variable $z = s\,\xi$ allows us to write

$$\int_0^\infty \xi^{[\alpha]-\alpha} \exp(-s\xi)d\xi = s^{-\varpi} \int_0^\infty z^{\varpi-1} \exp(-z)dz = \Gamma(\varpi)s^{-\varpi}.$$

Hence we have

$$\int_0^\infty (D_c^\alpha f)(x)\exp(-sx)dx = s^{-\varpi} \int_0^\infty f^{\lfloor\alpha\rfloor+1}(y)\exp(-sy)dy.$$

The integral on the right-hand side is the Laplace transform of the $(\lfloor\alpha\rfloor + 1)$-th derivative of f. Denote by the subscript L the Laplace transform and recall the known property

$$(f_L^{(p)}(s) = s^p f_L(s) - \sum_{k=0}^{p-1} s^{p-1-k} f^{(k)}(0_+), \qquad p \in \mathbb{N}.$$

Letting $p = \lfloor\alpha\rfloor + 1$ we obtain

$$(D_C^\alpha f)_L(s) = s^\alpha f_L(s) - \sum_{k=0}^{\lfloor\alpha\rfloor} s^{\alpha-1-k} f^{(k)}(0_+).$$

7.8.3 Heat Conduction via Fractional Derivatives

The constitutive equation

$$\mathbf{q}(t) = -\int_0^\infty \kappa(s)\nabla\theta(t-s)ds \tag{7.129}$$

models linear heat conductors with memory. By a direct integration, it follows that the heat flux satisfying the MC equation (6.81) can be written

$$\mathbf{q}(t) = -\frac{k}{\tau}\int_0^\infty \exp(-s/\tau)\nabla\theta(t-s)ds,$$

which amounts to letting

$$\kappa(s) = \frac{k}{\tau}\exp(-s/\tau)$$

in (7.129).

In the literature fractional derivatives are widely applied to model memory effects in connection with equations similar to (6.81) mainly by identifying $\mathbf{q}$ with the diffusion flux and replacing $\nabla\theta$ with the density gradient [131, 226]. Applications are widely developed in viscoelasticity [292] often by referring to the rheological models. Here, for definiteness, we consider a model of heat conduction involving a fractional (time) derivative.

Assume the heat flux is characterized by the fractional differential equation [156]

$$\mathbf{q} + \tau_\alpha \partial_t^\alpha \mathbf{q} = -k\nabla\theta, \tag{7.130}$$

which becomes the MC equation when $\alpha = 1$. A vector $\mathbf{w}_\alpha$ arises naturally by the fractional derivative of $\mathbf{q}$. In this regard we let $\alpha \in [0, 1]$ so that the relation with the MC equation is immediate ($\alpha = 1$). By definition we take the α-th derivative in the form

$$\partial_t^\alpha \mathbf{q} = \frac{1}{\Gamma(\nu)}\int_0^\infty s^{-\alpha+\lfloor\alpha\rfloor}\,\partial_t^{\lfloor\alpha\rfloor+1}\mathbf{q}(t-s)ds, \qquad \nu = \lfloor\alpha\rfloor + 1 - \alpha. \tag{7.131}$$

Letting

$$\mathbf{w}_\alpha(t) := \frac{1}{\Gamma(\nu)}\int_0^\infty s^{-\alpha+\lfloor\alpha\rfloor}\,\partial_t^{\lfloor\alpha\rfloor}\mathbf{q}(t-s)ds \tag{7.132}$$

we have

$$\partial_t^\alpha \mathbf{q} = \partial_t \mathbf{w}_\alpha. \tag{7.133}$$

Hence it follows that (7.130) can be written in the form

$$\tau_\alpha \dot{\mathbf{w}}_\alpha + \mathbf{q} = -k\nabla\theta. \tag{7.134}$$

As $\alpha = 1$ Eq. (7.134) becomes the MC equation

$$\tau\dot{\mathbf{q}} + \mathbf{q} = -k\nabla\theta. \tag{7.135}$$

The parameters τ_α, τ, k are allowed to depend on the temperature θ.

We now investigate the compatibility of (7.134) and (7.135); while the MC equation (7.135) proves at once to be thermodynamically consistent the same is not true for (7.134).

Thermodynamic consistency

The entropy inequality is taken in the standard form

$$-\rho(\dot{\psi} + \eta\dot{\theta}) - \frac{1}{\theta}\mathbf{q} \cdot \nabla\theta \geq 0.$$

As to (7.135) let θ, $\mathbf{q}$, $\nabla\theta$ be the independent variables. Hence it follows

$$\rho(\partial_\theta\psi + \eta)\dot{\theta} + \rho\partial_{\mathbf{q}}\psi \cdot \dot{\mathbf{q}} + \rho\partial_{\nabla\theta}\psi \cdot \nabla\dot{\theta} + \frac{1}{\theta}\mathbf{q} \cdot \nabla\theta \leq 0$$

Substitution of $\dot{\mathbf{q}}$ from (7.135) results in

$$\rho(\partial_\theta\psi + \eta)\dot{\theta} + \rho\partial_{\nabla\theta}\psi \cdot \nabla\dot{\theta} - (\frac{k\rho}{\tau}\partial_{\mathbf{q}}\psi - \frac{1}{\theta}\mathbf{q}) \cdot \nabla\theta - \frac{\rho}{\tau}\partial_{\mathbf{q}}\psi \cdot \mathbf{q} \leq 0$$

The arbitrariness of $\nabla\dot{\theta}$, $\dot{\theta}$, $\nabla\theta$ implies

$$\partial_{\nabla\theta}\psi = \mathbf{0}, \qquad \eta = -\partial_\theta\psi, \qquad \partial_{\mathbf{q}}\psi = \frac{\tau}{k\rho\theta}\mathbf{q}$$

and hence

$$\partial_{\mathbf{q}}\psi \cdot \mathbf{q} \geq 0.$$

By the obvious integration we have

$$\psi = \Psi(\theta) + \frac{\tau}{2k\rho\theta}\mathbf{q}^2 \tag{7.136}$$

and the reduced inequality implies $k > 0$.

As to (7.134), let θ, $\mathbf{w}_\alpha$, $\nabla\theta$ be the independent variables. Hence

$$-\rho(\partial_\theta\psi + \eta)\dot{\theta} - \rho\partial_{\mathbf{w}_\alpha}\psi \cdot \dot{\mathbf{w}}_\alpha - \rho\partial_{\nabla\theta}\psi \cdot \nabla\dot{\theta} - \frac{1}{\theta}\mathbf{q} \cdot \nabla\theta \geq 0.$$

Substitution of $\dot{\mathbf{w}}_\alpha$ from (7.134) and use of the arbitrariness lead to

$$\partial_{\mathbf{w}_\alpha}\psi = \frac{\tau_\alpha}{\rho k\theta}\mathbf{q}, \qquad \frac{\rho}{\tau_\alpha}\partial_{\mathbf{w}_\alpha}\psi \cdot \mathbf{q} \geq 0.$$

By integration we find

$$\psi = \Psi(\theta) + \frac{\tau_\alpha}{2\rho k\theta}\mathbf{w}_\alpha \cdot \mathbf{q}$$

and $k > 0$. Yet $\mathbf{q}$ is not a variable and hence the dependence of ψ on $\mathbf{q}$ is contradictory. If, rather, we start with $\mathbf{q}$ as a variable then we might conclude $\partial_{\mathbf{q}}\psi = \mathbf{0}$ and this would lead again to the contradiction. The same problem arise if we replace $\mathbf{q}$ with $\mathbf{w}_\alpha$ in (7.134).

Compatibility is investigated in [156] by setting the model within equations with summed histories.

7.8.4 Wave Propagation Properties

Similarities and differences between the models (7.135) and (7.134) are now investigated.

We first examine the system of equations

$$\dot{\varepsilon} = -\nabla \cdot \mathbf{q} + r, \qquad \tau\dot{\mathbf{q}} + \mathbf{q} = -k\nabla\theta. \tag{7.137}$$

By (7.136) we have

$$\varepsilon = \Psi(\theta) - \theta\Psi'(\theta) + \lambda(\theta)\mathbf{q}^2,$$

where

$$\lambda = -\frac{\theta}{2\rho}\Big(\frac{\tau(\theta)}{\theta k(\theta)}\Big)'.$$

We look for the possibility of discontinuity wave propagation. Let θ, $\mathbf{q}$, r, and then $\varepsilon(\theta, \mathbf{q})$, $\tau(\theta)$, and $k(\theta)$, be continuous whereas $\dot{\theta}$, $\nabla\theta$, and $\dot{\varepsilon}$, $\dot{\mathbf{q}}$ suffer jump discontinuities across the wave front. The continuity of $\mathbf{q}$ and θ,

$$[\![\mathbf{q}]\!] = \mathbf{0}, \qquad [\![\theta]\!] = 0,$$

implies that time and space derivatives are related by

$$[\![\nabla \cdot \mathbf{q}]\!] = -\frac{1}{U}[\![\dot{\mathbf{q}}]\!] \cdot \mathbf{n}, \qquad [\![\nabla\theta]\!] = -\frac{1}{U}[\![\dot{\theta}]\!]\mathbf{n}.$$

Since $\partial_\theta \varepsilon$, $\partial_\mathbf{q}\varepsilon$, τ, k are continuous across the wave the system (7.137) yields

$$\partial_\theta\varepsilon[\![\dot{\theta}]\!] + \partial_\mathbf{q}\varepsilon \cdot [\![\dot{\mathbf{q}}]\!] = \frac{1}{U}[\![\dot{\mathbf{q}}]\!] \cdot \mathbf{n},$$

$$\tau[\![\dot{\mathbf{q}}]\!] + [\![\mathbf{q}]\!] = \frac{k}{U}[\![\dot{\theta}]\!]\mathbf{n}.$$

Hence we have

$$(\tau\partial_\theta\varepsilon\, U^2 + k\partial_\mathbf{q}\varepsilon \cdot \mathbf{n}\, U - k)[\![\dot{\theta}]\!] = 0.$$

Nontrivial waves, where $[\![\dot{\theta}]\!] \neq 0$, occur at the speed

$$U = \sqrt{k/\tau\partial_\theta\varepsilon}\Big\{\sqrt{1 + k(\partial_\mathbf{q}\varepsilon \cdot \mathbf{n})^2/4\tau\partial_\theta\varepsilon} - \tfrac{1}{2}\sqrt{k/\tau\partial_\theta\varepsilon}\;\partial_\mathbf{q}\varepsilon \cdot \mathbf{n}\Big\}.$$

Since $\partial_\mathbf{q}\varepsilon = 2\lambda\mathbf{q}$ then the speed U depends on the direction of propagation through $\mathbf{q} \cdot \mathbf{n}$. For waves propagating in a region at equilibrium, i.e. $\mathbf{q} = \mathbf{0}$, then

$$U = \sqrt{k/\tau\partial_\theta\varepsilon}.$$

We now investigate the system of equations within the scheme of fractional derivatives,

$$\dot{\varepsilon} = -\nabla \cdot \mathbf{q} + r, \qquad \tau_\alpha\partial_t^\alpha\mathbf{q} + \mathbf{q} = -k\nabla\theta. \tag{7.138}$$

Again we let θ, $\mathbf{q}$, and r be continuous whereas their derivatives suffer jump discontinuities at $\mathbf{x}$ at time t. Let $\alpha \in \mathbb{R}^+ \setminus \mathbb{N}$. To compute the discontinuity of $[\![\partial_t^\alpha\mathbf{q}]\!]$ we observe that, by (7.131),

$$\partial_t^\alpha\mathbf{q} = \frac{1}{\Gamma(\nu)}\int_0^\infty s^{-\beta}\,\partial_t^{\lfloor\alpha\rfloor+1}\mathbf{q}(t-s)ds, \quad \beta = \alpha - \lfloor\alpha\rfloor \in (0,1), \quad \nu = 1-\beta \in (0,1).$$

It is apparent that $\partial_t^\alpha\mathbf{q}$ is well-defined[23] for any bounded function $\partial_t^{\lfloor\alpha\rfloor+1}\mathbf{q}(t-s)$. We now need to compute the possible jump of $\partial_t^\alpha\mathbf{q}$ associated with the jump of $\partial_t^{\lfloor\alpha\rfloor+1}\mathbf{q}$.

For generality consider the integral

$$\int_0^\infty s^{-\beta}\,\mathbf{h}(t-s)ds$$

and let $\mathbf{h}$ suffer a jump discontinuity $[\![\mathbf{h}]\!]$ at time t_0 [156]. Hence, as $t > t_0$, we represent $\mathbf{h}$ in the form

[23] Moreover, we see that we cannot integrate by parts because $s^{-\beta}$ is unbounded as $s \to 0_+$.

$$\mathbf{h}(t-s) = \tilde{\mathbf{h}}(t-s) + [\![\mathbf{h}]\!] H(t-s-t_0),$$

where $\tilde{\mathbf{h}}$ is continuous in a neighbourhood of t_0 and H is the Heaviside step function. Consequently

$$\int_0^\infty s^{-\beta}\mathbf{h}(t-s)ds = \int_0^\infty s^{-\beta}\tilde{\mathbf{h}}(t-s)ds + [\![\mathbf{h}]\!] \int_0^{t-t_0} s^{-\beta}ds.$$

Since $\beta \in (0,1)$ then

$$\int_0^{t-t_0} s^{-\beta}ds = \frac{(t-t_0)^{1-\beta}}{1-\beta} \to 0 \quad \text{as} \quad t \to t_0^+.$$

Notice that, for any function $\mathbf{g}(\mathbf{x}, t)$, $[\![\mathbf{g}]\!](\mathbf{x}, t_0) = \mathbf{g}(\mathbf{x}, t_0^+) - \mathbf{g}(\mathbf{x}, t_0^-)$. Hence it follows that

$$[\![\int_0^\infty s^{-\beta}\mathbf{h}(t-s)ds]\!] = [\![\int_0^\infty s^{-\beta}\tilde{\mathbf{h}}(t-s)ds]\!] = \mathbf{0},$$

though $[\![\mathbf{h}]\!] \neq \mathbf{0}$. The crucial point for this result is that $\beta \in (0,1)$. This implies that $[\![\partial_t^\alpha \mathbf{h}]\!] = \mathbf{0}$ for any order $\alpha \in \mathbb{R}^+ \setminus \mathbb{N}$ though $\partial_t^{\lfloor\alpha\rfloor+1}\mathbf{h}$ has a jump discontinuity.

Back to Eqs. (7.138) we observe that

$$[\![\partial_t^\alpha \mathbf{q}]\!] = \mathbf{0}$$

though $[\![\partial_t^{\lfloor\alpha\rfloor+1}\mathbf{q}]\!] \neq \mathbf{0}$. Consequently it follows that

$$[\![\nabla\theta]\!] = \mathbf{0}.$$

The occurrence of the fractional derivative in (7.138) hinders the appearance of a discontinuity of $\nabla\theta$.

7.9 Viscoelastic Models of Fractional Order

A simple stress–strain relation of fractional order might be considered in the form[24]

$$\mathbf{T}_{RR}(t) = \mathsf{M}_v \partial_t^\alpha \mathbf{E}(t),$$

where $\mathsf{M}_v \in \text{Lin}(\text{Sym}, \text{Sym})$, $\mathsf{M}_v > \mathbf{0}$. For formal convenience let

$$\tilde{\mathsf{M}}(s) = \frac{1}{\Gamma(\lfloor\alpha\rfloor + 1 - \alpha)} \mathsf{M}_v s^{\lfloor\alpha\rfloor - \alpha}.$$

[24] For formal convenience, in this section ∂_t^α and ∂_t denote Lagrangian time derivatives.

Hence the constitutive equation might be written in the form

$$\mathbf{T}_{RR}(t) = \int_0^\infty \tilde{\mathbf{M}}(s)\partial_t^{\lfloor\alpha\rfloor+1}\mathbf{E}(t-s)ds.$$

Yet a stress–strain relation for solids should allow also for a instantaneous elastic response. In this regard we recall the Kelvin-Voigt model

$$\mathbf{T}_{RR}(t) = \mathbf{M}_\infty\mathbf{E}(t) + \mathbf{M}_v\dot{\mathbf{E}}(t), \qquad \mathbf{M}_\infty, \mathbf{M}_v > \mathbf{0}.$$

At any constant deformation process, $\mathbf{E}(t) = \mathbf{E}_0$, as $t \le t_0$, it follows $\mathbf{T}_{RR}(t) = \mathbf{M}_\infty\mathbf{E}_0$. This may suggest that we take the stress–strain relation in the form

$$\mathbf{T}_{RR}(t) = \mathbf{M}_\infty\mathbf{E}(t) + \mathbf{M}_v\partial_t^\alpha\mathbf{E}(t), \tag{7.139}$$

Hence we have

$$\mathbf{T}_{RR}(t) := \mathfrak{T}(\mathbf{E}(t), \mathbf{E}_r^t) = \mathbf{M}_\infty\mathbf{E}(t) + \int_0^\infty \tilde{\mathbf{M}}(s)\partial_t^{\lfloor\alpha\rfloor+1}\mathbf{E}^t(s)ds.$$

Letting $\alpha \in (n-1, n)$ and $\beta := \alpha - \lfloor\alpha\rfloor \in (0, 1)$ we consider $\mathbf{T}_{RR}$ in the form

$$\mathbf{T}_{RR}(t) = \mathbf{M}_\infty\mathbf{E}(t) + \int_0^\infty \tilde{\mathbf{M}}(s)\partial_t^n\mathbf{E}(t-s)ds, \qquad \tilde{\mathbf{M}}(s) = [\Gamma(1-\beta)]^{-1}\mathbf{M}_v s^{-\beta}. \tag{7.140}$$

Restrict attention to $n = 1$ and hence $\alpha \in (0, 1)$. The constitutive equation (7.140) looks like (7.56) but here $\tilde{\mathbf{M}} \in L^1_{loc}(\mathbb{R}^+)$ rather than $L^1(\mathbb{R}^+)$. By analogy with the procedure for (7.56) we obtain the following result.

Theorem 7.7 *As $n = 1$, the constitutive equation (7.140) is consistent with the second law of thermodynamics only if*

$$\mathbf{M}_v > \mathbf{0}, \quad \mathbf{M}_\infty = \mathbf{M}_\infty^T. \tag{7.141}$$

while[25] $\int_0^\infty s^{-\beta}\cos\omega s\, ds > 0,\ \forall\omega > 0.$

We now let

$$\mathbf{M}_v = \mathbf{M}_v^T, \quad \mathbf{M}_\infty = \mathbf{M}_\infty^T, \quad \lim_{\tau\to-\infty}\mathbf{E}(\tau) = \mathbf{0},$$

and look for the whole thermodynamic consistency of (7.140) as $n = 1$.

The work performed by the stress tensor in connection with the history $\mathbf{E}^t$ is

$$W(\mathbf{E}^t) = \int_{-\infty}^t \mathfrak{T}(\mathbf{E}^\tau)\cdot\partial_\tau\mathbf{E}(\tau)d\tau = \int_{-\infty}^t \Big[\mathbf{M}_\infty\mathbf{E}(\tau) + \int_0^\infty \tilde{\mathbf{M}}(s)\partial_\tau\mathbf{E}^\tau(s)ds\Big]\cdot\partial_\tau\mathbf{E}(\tau)d\tau.$$

[25] $\int_0^\infty s^{-\beta}\cos\omega s ds = \pi\omega^{\beta-1}/2\Gamma(\beta)\cos(\beta\pi/2).$

For technical convenience we extend $\tilde{\mathbf{M}}$ to $\mathbb{R}^-$ by letting

$$\tilde{\mathbf{M}}(s) = \mathbf{0} \quad \text{as} \quad s < 0.$$

A change of variables yields

$$\begin{aligned} W(\mathbf{E}^t) &= \int_{-\infty}^{t} [\mathbf{M}_\infty \mathbf{E}(\tau) + \int_{-\infty}^{\tau} \tilde{\mathbf{M}}(\tau - u)\partial_u \mathbf{E}(u) du] \cdot \partial_\tau \mathbf{E}(\tau) d\tau \\ &= \tfrac{1}{2}\mathbf{E}(t) \cdot \mathbf{M}_\infty \mathbf{E}(t) + \tfrac{1}{2} \int_{-\infty}^{t} \int_{-\infty}^{t} \partial_\tau \mathbf{E}(\tau) \cdot \tilde{\mathbf{M}}(|\tau - u|)\partial_u \mathbf{E}(u) du\, d\tau \qquad (7.142) \\ &= \tfrac{1}{2}\mathbf{E}(t) \cdot \mathbf{M}_\infty \mathbf{E}(t) + \tfrac{1}{2} \int_{-\infty}^{t} \int_{-\infty}^{t} \partial_t \mathbf{E}^t(s_1) \cdot \tilde{\mathbf{M}}(|s_1 - s_2|)\partial_t \mathbf{E}^t(s_2) ds_1\, ds_2. \end{aligned}$$

This suggests that we consider the work function as an energy functional, namely

$$\rho \hat{\psi}^w(\mathbf{E}(t), \mathbf{E}^t_r) = \tfrac{1}{2}\mathbf{E}(t) \cdot \mathbf{M}_\infty \mathbf{E}(t) + \tfrac{1}{2} \int_{-\infty}^{t} \partial_t \mathbf{E}^t(s_1) \cdot \tilde{\mathbf{M}}(|s_1 - s_2|)\partial_t \mathbf{E}^t(s_2) ds_1\, ds_2.$$

In order to prove that $\rho \partial_{\mathbf{E}} \hat{\psi}^w(\mathbf{E}(t), \mathbf{E}^t_r) = \mathfrak{T}(\mathbf{E}(t), \mathbf{E}^t_r)$ we extend the definition of to $\mathbb{R}^-$ by letting $\mathbf{E}^t(s) = \mathbf{0}$ as $s < 0$. Since $\dot{\mathbf{E}} + \mathbf{E}\delta(t)$ is the distributional derivative of the extension of $\mathbf{E}^t$ we obtain

$$\begin{aligned} \rho \hat{\psi}^w(t) = W(t) &= \tfrac{1}{2}\mathbf{E}(t) \cdot \mathbf{M}_\infty \mathbf{E}(t) + \int_{-\infty}^{\infty} \int_{0}^{\infty} \tilde{\mathbf{M}}(s)\partial_\tau \mathbf{E}^\tau(s) ds \cdot [\partial_\tau \mathbf{E}(\tau) + \delta(t - \tau)\mathbf{E}(t)] d\tau \\ &= \tfrac{1}{2}\mathbf{E}(t) \cdot \mathbf{M}_\infty \mathbf{E}(t) + \int_{-\infty}^{\infty} \int_{-\infty}^{\infty} \partial_\tau \mathbf{E}(\tau) \cdot \tilde{\mathbf{M}}(s)\partial_\tau \mathbf{E}^\tau(s) ds\, d\tau + \mathbf{E}(t) \cdot \int_{0}^{\infty} \tilde{\mathbf{M}}(s)\partial_t \mathbf{E}^t(s) ds \\ &= \tfrac{1}{2}\mathbf{E}(t) \cdot \mathbf{M}_\infty \mathbf{E}(t) + \mathbf{E}(t) \cdot \int_{0}^{\infty} \tilde{\mathbf{M}}(s)\partial_t \mathbf{E}^t(s) ds + \int_{-\infty}^{\infty} \int_{-\infty}^{\infty} \partial_\tau \mathbf{E}(\tau) \cdot \tilde{\mathbf{M}}(\tau - u)\partial_u \mathbf{E}(u) du\, d\tau. \end{aligned}$$

The last term is independent of the present value $\mathbf{E}(t)$ and the sought result

$$\rho \partial_{\mathbf{E}} \hat{\phi}^w = \mathbf{M}_\infty \mathbf{E}(t) + \int_{0}^{\infty} \tilde{\mathbf{M}}(s)\partial_t \mathbf{E}^t(s) ds$$

follows. Moreover we find

$$\begin{aligned} 0 &= \dot{W}(t) - \mathbf{T}_{RR}(t) \cdot \dot{\mathbf{E}}(t) = \rho \dot{\psi}^w(t) - \mathbf{T}_{RR}(t) \cdot \dot{\mathbf{E}}(t) \\ &= \rho \partial_{\mathbf{E}} \hat{\psi}^w(\mathbf{E}(t), \mathbf{E}^t_r) \cdot \dot{\mathbf{E}}(t) + \rho d\hat{\psi}^w(\mathbf{E}(t), \mathbf{E}^t_r | \dot{\mathbf{E}}^t_r) - \mathbf{T}_{RR}(t) \cdot \dot{\mathbf{E}}(t) = \rho d\hat{\psi}^w(\mathbf{E}(t), \mathbf{E}^t_r | \dot{\mathbf{E}}^t_r) \end{aligned}$$

and hence $d\hat{\psi}^w = 0$.

To check the minimum property at constant histories we extend by continuity the history $\mathbf{E}^t$ to $\mathbb{R}^-$ by letting $\mathbf{E}^t(s) = \mathbf{E}^t(0) = \mathbf{E}(t)$ as $s < 0$ so that $\dot{\mathbf{E}}(\tau) = \mathbf{0}$ as $\tau > t$. Accordingly by Eq. (7.142) we have

$$\rho\hat{\psi}^w(\mathbf{E}(t), \mathbf{E}_r^t) = W(t) = \tfrac{1}{2}\mathbf{E}(t) \cdot \mathbf{M}_\infty \mathbf{E}(t) + \int_{-\infty}^{\infty} \dot{\mathbf{E}}(\tau) du\, d\tau$$
$$= \tfrac{1}{2}\mathbf{E}(t) \cdot \mathbf{M}_\infty \mathbf{E}(t) + \int_{-\infty}^{\infty} \dot{\mathbf{E}}(\tau) \cdot [\tilde{\mathbf{M}} * \dot{\mathbf{E}}]\tau) d\tau,$$

where $*$ denotes the Fourier convolution. Accordingly by Parseval–Plancherel theorem we obtain

$$\rho\hat{\psi}^w = \tfrac{1}{2}\mathbf{E}(t) \cdot \mathbf{M}_\infty \mathbf{E}(t) + \tfrac{1}{2\pi} \int_{-\infty}^{\infty} (i\omega \mathbf{E}_F)^*(\omega) \tilde{\mathbf{M}}_F(\omega) i\omega \mathbf{E}_F(\omega) d\omega,$$

whence

$$\rho\hat{\psi}^w = \tfrac{1}{2}\mathbf{E}(t) \cdot \mathbf{M}_\infty \mathbf{E}(t) + \tfrac{1}{2\pi} \int_{-\infty}^{\infty} \omega^2 [\mathbf{E}_c(\omega) \tilde{\mathbf{M}}_c(\omega) \mathbf{E}_c(\omega) + \mathbf{E}_c(\omega) \tilde{\mathbf{M}}_c(\omega) \mathbf{E}_c(\omega)] d\omega.$$

Since $\tilde{\mathbf{M}}_c > 0$ and $\mathbf{E}_c, \mathbf{E}_s = \mathbf{0}$ for constant histories then the minimum property holds.

The thermodynamic consistency of (7.140) with $n > 1$ seems to be an open problem [333]. The present proof does not apply any longer. To look for a functional of the Graffi-Volterra type we might take $\tilde{\mathbf{M}}$ of finite extent, $\tilde{\mathbf{M}}(s) = \mathbf{0}$ as $s \geq d$. In this case, letting

$$\check{\mathbf{M}}(u) = \int_0^u \tilde{\mathbf{M}}(s) ds, \qquad \mathbf{M}_d = \int_0^d \tilde{\mathbf{M}}(s) ds,$$

we can integrate by parts to obtain

$$\int_0^\infty \tilde{\mathbf{M}}(s) \partial_t^n \mathbf{E}^t(s) ds = \int_0^d \tilde{\mathbf{M}}(s) \partial_t^n \mathbf{E}^t(s) ds = \int_0^d \check{\mathbf{M}}(s) \partial_t^{n+1} \mathbf{E}^t(s) ds,$$

thus avoiding the unboundedness of the kernel. Hence the condition $\mathbf{T}_{RR} = \rho \partial_{\mathbf{E}} \hat{\psi}$ $(\mathbf{E}(t), \mathbf{E}_r^t)$ holds with

$$\hat{\psi}(\mathbf{E}(t), \mathbf{E}_r^t) = \tfrac{1}{2}\mathbf{E}(t) \cdot \mathbf{M}_\infty \mathbf{E}(t) + \mathbf{E}(t) \cdot \int_0^d \check{\mathbf{M}}(s) \partial_t^{n+1} \mathbf{E}^t(s) ds + \rho \Psi(\theta, \mathbf{E}_r^t).$$

Though $\check{\mathbf{M}}$ is bounded, so far a functional Ψ with $n > 1$ seems to be unavailable.

Chapter 8
Aging and Higher-Order Grade Materials

This chapter investigates some aspects of non-simple materials. First, aging materials are examined. Aging is a general problem of the modelling of materials. Mathematically the problem is made formal by letting constitutive parameters depend on time. It is pointed out that the second law of thermodynamics might select physically admissible evolutions of the parameters. Next constitutive equations are considered which involve higher-order derivatives, in time or in space (materials of differential type). The dependence on time derivatives is an alternative to the functionals of histories when memory effects are of interest. Yet the time derivative is not objective and hence recourse is needed to objective time derivatives. The dependence on higher-order space derivatives is a way of modelling a non-simple behaviour where the response of the material, at a point, is influenced by appropriate fields in a neighbourhood of the point, and this is modelled by some space derivatives, often of the deformation gradient. In this framework materials of the Korteweg type are investigated. Materials of higher-order grade are often associated with an hyperstress tensor thus generalizing the form of the mechanical power. The thermodynamic consistency of a scheme with a balance of energy involving an hyperstress is investigated.

8.1 Aging

In general, aging of materials consists of any change in the constitutive properties. In essence there are two types of aging: a physical aging, which is due to thermodynamic processes, and a chemical aging, which is determined by chemical reactions. Although their origins differ, the macroscopic manifestations of physical and chemical aging may be modelled by a variation of the constitutive parameters.

In materials with memory, the response of a non-aging substance to an external stimulus changes with the passage of time in a way, described e.g. by the relaxation

A. Morro and C. Giorgi, *Mathematical Modelling of Continuum Physics*,
Modeling and Simulation in Science, Engineering and Technology,
https://doi.org/10.1007/978-3-031-20814-0_8

modulus or the creep compliance, that is independent of the time at which the experiment is started. Instead, the properties of aging materials change with the passage of time even in the absence of external agents. Hence two time scales are required to unambiguously describe the constitutive properties of an aging material. One time scale is needed to keep track of the time since the manufacturing of the material and the other time scale is needed to keep track of the time the material is under an external stimulus.

Aging is a time-dependent process. Aging degradation of mechanical structures reduce the strength of structures or their components. The aging degradation of the strength of a structure can be expressed by

$$R(t) = R_0 g(t),$$

where R_0 is the component capacity in the initial state (say at $t = 0$) and $g(t)$ is time dependent degradation function. The literature shows many examples of functions g. For instance aging degradation for steel is assumed to be given by

$$g(t) = 1 - \frac{1}{1 + a\exp(-bt)}, \tag{8.1}$$

where a and b are constant parameters [69]. Another example is related to the aging of polymers; a classical model of degradation is expressed by

$$g(t) = \exp[-(t/\tau)^\beta], \tag{8.2}$$

where β is a shape parameter and τ is the relaxation time [257].

It seems natural to view aging as a process which is related to the constitutive properties of the material. This suggests that we incorporate aging within the set of constitutive properties. Hence admissible aging processes are required to comply with the second law of thermodynamics. This view is embodied within a thermodynamic theory by comprising the time t among the independent variables. Hence if $\boldsymbol{\alpha}$ is the set of physical variables we let

$$\psi = \psi(\boldsymbol{\alpha}, t)$$

and the like for the constitutive equations of, e.g., η, $\mathbf{T}$, $\mathbf{q}$. The entropy inequality then becomes

$$-\rho(\partial_{\boldsymbol{\alpha}}\psi \cdot \dot{\boldsymbol{\alpha}} + \partial_t \psi) + \mathbf{T}\cdot\mathbf{D} - \frac{1}{\theta}\mathbf{q}\cdot\nabla\theta \geq 0$$

where $\partial_t \psi$ is the time derivative given by the explicit dependence on t that models the aging property. We now examine how aging effects may be modelled within the schemes of thermoelastic solids, rate-type materials, and thermo-viscoelastic materials.

8.2 Aging of Thermoelastic Solids

A thermoelastic solid is characterized by the constitutive equation

$$\mathbf{T}_{RR} = \mathsf{C}\mathbf{E} + \vartheta\mathbf{M}, \tag{8.3}$$

possibly with $\mathbf{T}$ in place of $\mathbf{T}_{RR}$ if the linear approximation is considered. For a non-aging material C and $\mathbf{M}$ are constant tensors. For an aging material it is natural to take C and $\mathbf{M}$ as dependent on time. This in turn suggests that the constitutive functions depend on

$$\mathbf{E}, \theta, \nabla\theta, t,$$

the dependence on t being thought as representing the changing properties of the material. Hence we let

$$\psi(\mathbf{E}, \theta, \nabla\theta, t)$$

be the constitutive function for the free energy ψ and the like for η, $\mathbf{T}$ and $\mathbf{q}$. The entropy inequality becomes

$$-\rho_R(\partial_\theta\psi + \eta)\dot{\theta} - \rho_R\partial_{\nabla_R\theta}\psi \cdot \overline{\nabla_R\theta}^{\boldsymbol{\cdot}} + (\mathbf{T}_{RR} - \rho_R\partial_{\mathbf{E}}\psi) \cdot \dot{\mathbf{E}} - \frac{1}{\theta}\mathbf{q}_R \cdot \nabla_R\theta - \rho_R\partial_t\psi \geq 0.$$

The arbitrariness of $\overline{\nabla_R\theta}^{\boldsymbol{\cdot}}, \dot{\theta}, \dot{\mathbf{E}}$ implies that

$$\partial_{\nabla_R\theta}\psi = \mathbf{0}, \qquad \eta = -\partial_\theta\psi, \qquad \mathbf{T}_{RR} = \rho_R\partial_{\mathbf{E}}\psi,$$

and

$$-\frac{1}{\theta}\mathbf{q}_R \cdot \nabla_R\theta - \rho_R\partial_t\psi \geq 0.$$

Since $\partial_{\nabla_R\theta}\psi = \mathbf{0}$ then $\partial_t\psi$ remains unaltered by letting $\nabla_R\theta = \mathbf{0}$, which implies that

$$\partial_t\psi \leq 0.$$

Restrict attention to the linear stress relation (8.3). To account for aging we let C and $\mathbf{M}$ depend on time. Since

$$\rho_R\psi = \Psi(\theta) + \tfrac{1}{2}\mathbf{E} \cdot \mathsf{C}\mathbf{E} + \vartheta\mathbf{M} \cdot \mathbf{E}$$

then we have

$$\rho_R\partial_t\psi = \tfrac{1}{2}\mathbf{E} \cdot \dot{\mathsf{C}}\mathbf{E} + \vartheta\dot{\mathbf{M}} \cdot \mathbf{E}.$$

For definiteness we look at isotropic solids so that

$$\mathbf{M} = m\mathbf{1}, \qquad \mathsf{C}\mathbf{E} = 2\mu\mathbf{E} + \lambda(\operatorname{tr}\mathbf{E})\mathbf{1} = 2\mu\mathbf{E}_0 + \kappa(\operatorname{tr}\mathbf{E})\mathbf{1},$$

μ, λ being the Lamé moduli and $\kappa = \lambda + 2\mu/3$, and

$$\mathbf{T}_{RR} = 2\mu \mathbf{E}_0 + \kappa(\mathrm{tr}\,\mathbf{E})\mathbf{1} + m\,\vartheta\,\mathbf{1}.$$

In stress-free conditions we have

$$\mathbf{E}_0 = \mathbf{0}, \qquad 0 = \kappa(\mathrm{tr}\,\mathbf{E}) + m\vartheta.$$

Since $\mathrm{tr}\,\mathbf{E} \simeq \nabla_R \cdot \mathbf{u}$ and $\nabla_R \cdot \mathbf{u}$ is the relative variation of the volume then $-m/\kappa$ is the coefficient of thermal expansion (in R). We assume that $m < 0$ so that, since $\kappa > 0$, the body expands when the temperature increases.

For isotropic solids we have

$$\rho_R \psi = \Psi(\theta) + \mu \mathbf{E}_0 \cdot \mathbf{E}_0 + \kappa(\mathrm{tr}\,\mathbf{E})^2 + m\vartheta \mathrm{tr}\,\mathbf{E}$$

and hence

$$\rho_R \partial_t \psi = \dot{\mu} \mathbf{E}_0 \cdot \mathbf{E}_0 + \dot{\kappa}(\mathrm{tr}\,\mathbf{E})^2 + \dot{m}\vartheta \mathrm{tr}\,\mathbf{E}.$$

In stress-free conditions $\vartheta = -\kappa\, \mathrm{tr}\,\mathbf{E}/m$. Consequently

$$\rho_R \partial_t \psi = \dot{\mu} \mathbf{E}_0 \cdot \mathbf{E}_0 + (\dot{\kappa} - \kappa \dot{m}/m)(\mathrm{tr}\,\mathbf{E})^2.$$

This implies that

$$\dot{\mu} \le 0, \qquad \dot{\kappa} - \frac{\kappa \dot{m}}{m} \le 0.$$

Let $\kappa > 0$. The second inequality can be given the form

$$\frac{\dot{\kappa}}{\kappa} + \frac{\dot{m}}{|m|} \le 0.$$

In a thermoelastic material, aging produces a decrease of μ. A joint decrease of μ, κ, and m is consistent with thermodynamics. Yet, since $m < 0$ then an increase of m is more realistic, $\dot{m} \ge 0$. This is consistent provided that

$$\frac{\dot{m}}{|m|} \le -\frac{\dot{\kappa}}{\kappa}, \qquad \dot{\kappa} \le 0.$$

The coefficient of thermal expansion $\alpha = -m/3\kappa$ satisfies

$$\dot{\alpha} = -\frac{m}{3\kappa}\Big(\frac{\dot{m}}{m} - \frac{\dot{\kappa}}{\kappa}\Big)$$

and hence

$$\dot{\alpha} \ge 0.$$

As a consequence of aging the ratio

$$\frac{\operatorname{tr}\mathbf{E}}{\vartheta}$$

increases; aging solids expand more and more per increment of temperature.

8.3 Aging of Rate-Type Materials

For definiteness we set up a model of a rate-type material by starting with a rheological model; next we establish a model of material within the continuum framework by the standard analogies stress-force and strain-elongation. We consider the rheological scalar model called *standard viscoelastic solid.* It consists of a Maxwell unit, where a spring and a dashpot are connected in series, placed in parallel with a single spring. Let ε_s, ε_d be the strain[1] of the spring and that of the dashpot of the Maxwell unit. The common strain ε of the elements in parallel is the sum of ε_s and ε_d,

$$\varepsilon = \varepsilon_s + \varepsilon_d.$$

Let σ_e be the stress on the isolated spring and σ_m the stress on the Maxwell unit. The total applied stress is the sum of them,

$$\sigma = \sigma_e + \sigma_m.$$

Let k, k_e be the elastic modulus of the spring in the Maxwell unit and γ the viscosity of the dashpot so that

$$\sigma_e = k_e\varepsilon, \qquad \sigma_m = k\varepsilon_s = \gamma\dot{\varepsilon}_d.$$

Hence we have

$$\varepsilon = \varepsilon_s + \varepsilon_d = \frac{1}{\alpha}\dot{\varepsilon}_d + \varepsilon_d, \qquad \alpha = \frac{k}{\gamma}. \tag{8.4}$$

Letting k and k_e be constants we obtain

$$\dot{\varepsilon} = \dot{\varepsilon}_s + \dot{\varepsilon}_d = \frac{1}{k}\dot{\sigma}_m + \frac{1}{\gamma}\sigma_m = \frac{1}{k}(\dot{\sigma} - k_e\dot{\varepsilon}) + \frac{1}{\gamma}(\sigma - k_e\varepsilon).$$

Let

$$g_0 = k_e + k, \qquad g_\infty = k_e.$$

We then find that the evolution of the standard viscoelastic solid is governed by

---

[1] Denoted $\varepsilon^{(s)}$ and $\varepsilon^{(d)}$ in Chap. 6.

$$\dot{\sigma} = g_0\dot{\varepsilon} - \alpha(\sigma - g_\infty \varepsilon). \tag{8.5}$$

A rate-type model of materials is considered by viewing the stress as the sum of an elastic part and a viscoelastic part; for formal simplicity we let the material be a non-conductor. In the referential scheme we let

$$\mathbf{T}_{RR} = \hat{\mathbf{T}}_{RR}(\theta, \mathbf{E}) + \boldsymbol{\mathcal{T}}_{RR}.$$

To account for aging we let the time t be one of the independent variables. Hence we let

$$\mathbf{T}_{RR} = \hat{\mathbf{T}}_{RR}(\theta, \mathbf{E}, t) + \boldsymbol{\mathcal{T}}_{RR}$$

and let $\theta, \mathbf{E}, \boldsymbol{\mathcal{T}}_{RR}, t$ be the set of independent variables. The entropy inequality reads

$$-\rho_R(\partial_\theta \psi + \eta)\dot{\theta} - \rho_R \partial_{\boldsymbol{\mathcal{T}}_{RR}} \psi \cdot \dot{\boldsymbol{\mathcal{T}}}_{RR} + (\hat{\mathbf{T}}_{RR} + \boldsymbol{\mathcal{T}}_{RR} - \rho_R \partial_{\mathbf{E}} \psi) \cdot \dot{\mathbf{E}} - \rho_R \partial_t \psi \ge 0.$$

To determine consequences of this inequality we need to know possible relations among $\dot{\boldsymbol{\mathcal{T}}}_{RR}$ and $\dot{\mathbf{E}}$.

We now let the stress $\boldsymbol{\mathcal{T}}_{RR}$ satisfy the analogue of the differential equation (8.5). Both a rate-type behaviour and possible anisotropies are modelled by assuming the relation

$$\tau \dot{\boldsymbol{\mathcal{T}}}_{RR} - \boldsymbol{\Gamma}_0 \dot{\mathbf{E}} + \boldsymbol{\mathcal{T}}_{RR} - \mathbf{G}_\infty \mathbf{E} = \mathbf{0},$$

where $\tau > 0$ may be viewed as a relaxation time. For formal convenience we let $\mathbf{G}_0 = \boldsymbol{\Gamma}_0/\tau$ so that we can write

$$\tau(\dot{\boldsymbol{\mathcal{T}}}_{RR} - \mathbf{G}_0 \dot{\mathbf{E}}) + \boldsymbol{\mathcal{T}}_{RR} - \mathbf{G}_\infty \mathbf{E} = \mathbf{0} \tag{8.6}$$

and assume $\mathbf{G}_0, \mathbf{G}_\infty$ are symmetric and positive-semidefinite fourth-order tensors on Sym. Both τ and $\mathbf{G}_0, \mathbf{G}_\infty$ are allowed to depend on the temperature θ.

As $\tau \to 0$ Eq. (8.6) simplifies to

$$\boldsymbol{\mathcal{T}}_{RR} = \mathbf{G}_\infty \mathbf{E},$$

whereby the material behaves elastically with $\mathbf{G}_\infty$ as elastic relaxation. If instead $\tau \to \infty$ then

$$\dot{\boldsymbol{\mathcal{T}}}_{RR} - \mathbf{G}_0 \dot{\mathbf{E}} + \frac{1}{\tau}(\boldsymbol{\mathcal{T}}_{RR} - \mathbf{G}_\infty \mathbf{E}) = \mathbf{0}$$

shows an instantaneous hypoelastic behaviour,

$$\dot{\boldsymbol{\mathcal{T}}}_{RR} = \mathbf{G}_0 \dot{\mathbf{E}},$$

$\mathbf{G}_0$ being the elastic modulus. Borrowing from the viscoelastic model we let

$$\mathsf{G}_0 > \mathsf{G}_\infty;$$

further $\mathsf{G}_\infty > \mathbf{0}$ for solids and $\mathsf{G}_\infty = \mathbf{0}$ for fluids.

Upon computation of $\dot{\psi}$ and substitution of $\dot{\mathcal{T}}_{RR}$ from (8.6) we obtain

$$-\rho_R(\partial_\theta\psi + \eta)\dot{\theta} + \frac{\rho_R}{\tau}\partial_{\mathcal{T}_{RR}}\psi \cdot (\mathcal{T}_{RR} - \mathsf{G}_\infty\mathbf{E}) + (\hat{\mathbf{T}}_{RR} + \mathcal{T}_{RR} - \rho_R\partial_{\mathbf{E}}\psi - \rho_R\mathsf{G}_0\partial_{\mathcal{T}_{RR}}\psi) \cdot \dot{\mathbf{E}} - \rho_R\partial_t\psi \geq 0.$$

The arbitrariness and the linearity of $\dot{\theta}$ and $\dot{\mathbf{E}}$ imply the classical relation $\eta = -\partial_\theta\psi$ and

$$\rho_R\partial_{\mathbf{E}}\psi = \hat{\mathbf{T}}_{RR} + \mathcal{T}_{RR} - \rho_R\mathsf{G}_0\partial_{\mathcal{T}_{RR}}\psi, \tag{8.7}$$

$$\frac{\rho_R}{\tau}\partial_{\mathcal{T}_{RR}}\psi \cdot (\mathcal{T}_{RR} - \mathsf{G}_\infty\mathbf{E}) - \rho_R\partial_t\psi \geq 0. \tag{8.8}$$

To determine the free-energy function we consider the inequality

$$\rho_R\partial_{\mathcal{T}_{RR}}\psi \cdot (\mathcal{T}_{RR} - \mathsf{G}_\infty\mathbf{E}) \geq 0, \tag{8.9}$$

which follows from (8.8) for non-aging materials or by letting the aging effect $\partial_t\psi$ as small as we please. Inequality (8.9) holds if

$$\rho_R\partial_{\mathcal{T}_{RR}}\psi = \mathsf{A}(\mathcal{T}_{RR} - \mathsf{G}_\infty\mathbf{E}), \tag{8.10}$$

A being a non-negative, symmetric fourth-order tensor, possibly dependent on θ, $\mathbf{E}$, $\mathbf{T}_{RR}$ and t. For definiteness, assume A is positive definite and depends only on θ and t. Moreover let ψ be a C^2 function. Now[2]

$$\rho_R\partial_{\mathbf{E}}\partial_{\mathcal{T}_{RR}}\psi = -\mathsf{A}\mathsf{G}_\infty.$$

Substitution of $\rho_R\partial_{\mathcal{T}_{RR}}\psi$ results in

$$\rho_R\partial_{\mathbf{E}}\psi = \hat{\mathbf{T}}_{RR} + \mathcal{T}_{RR} - \mathsf{G}_0\mathsf{A}(\mathcal{T}_{RR} - \mathsf{G}_\infty\mathbf{E}).$$

Hence we have

$$\rho_R\partial_{\mathcal{T}_{RR}}\partial_{\mathbf{E}}\psi = \mathsf{I} - \mathsf{G}_0\mathsf{A}.$$

The requirement $\partial_{\mathcal{T}_{RR}}\partial_{\mathbf{E}}\psi = (\partial_{\mathbf{E}}\partial_{\mathcal{T}_{RR}}\psi)^T$ then implies

$$\mathsf{I} - \mathsf{G}_0\mathsf{A} = -\mathsf{G}_\infty\mathsf{A}$$

whence

$$\mathsf{A} = (\mathsf{G}_0 - \mathsf{G}_\infty)^{-1}.$$

[2] For any pair of tensors $\mathbf{E}$, $\mathbf{T}$ we let $(\partial_{\mathbf{E}}\partial_{\mathbf{T}}\psi)_{hkij} = \partial_{E_{ij}}\partial_{T_{hk}}\psi$.

Thus A is positive definite if

$$\mathsf{G}_0 > \mathsf{G}_\infty.$$

The function ψ is subject to (8.10) and, since $\mathsf{G}_0 = \mathsf{G}_\infty + \mathsf{A}^{-1}$,

$$\begin{aligned}\rho_R \partial_{\mathbf{E}} \psi = \hat{\mathbf{T}}_{RR} + \mathcal{T}_{RR} - \mathsf{G}_0 \mathsf{A}(\mathcal{T}_{RR} - \mathsf{G}_\infty \mathbf{E}) &= \hat{\mathbf{T}}_{RR} + \mathcal{T}_{RR} - (\mathsf{G}_\infty + \mathsf{A}^{-1})\mathsf{A}(\mathcal{T}_{RR} - \mathsf{G}_\infty \mathbf{E}) \\ &= \hat{\mathbf{T}}_{RR} - \mathsf{G}_\infty \mathsf{A}(\mathcal{T}_{RR} - \mathsf{G}_\infty \mathbf{E}) + \mathsf{G}_\infty \mathbf{E}.\end{aligned}$$

Let $\psi_0(\theta, \mathbf{E})$ be subject to

$$\rho_R \partial_{\mathbf{E}} \psi_0 = \hat{\mathbf{T}}_{RR}.$$

Therefore upon integration of $\rho_R \partial_{\mathbf{E}} \psi$ and $\rho_R \partial_{\mathcal{T}_{RR}} \psi$ we find $\rho_R \psi$ in the form

$$\rho_R \psi = \rho_R \psi_0(\theta, \mathbf{E}) + \tfrac{1}{2}[\mathbf{E} \cdot \mathsf{G}_\infty \mathbf{E} + (\mathcal{T}_{RR} - \mathsf{G}_\infty \mathbf{E}) \cdot \mathsf{A}(\mathcal{T}_{RR} - \mathsf{G}_\infty \mathbf{E}). \tag{8.11}$$

Hence, to within the function ψ_0, the free energy $\rho_R \psi$ is a quadratic form and is positive definite provided G_∞ and A (and hence G_0) are positive definite.

8.3.1 Aging Properties

We now go back to the dissipation inequality (8.8). The aging properties are modelled by letting G_∞ and A depend on time. To save writing let

$$\mathbf{Y} = \mathcal{T}_{RR} - \mathsf{G}_\infty \mathbf{E}.$$

Hence we have

$$\rho_R \partial_t \psi = \tfrac{1}{2}\mathbf{E} \cdot \dot{\mathsf{G}}_\infty \mathbf{E} + \tfrac{1}{2}\mathbf{Y} \cdot \dot{\mathsf{A}} \mathbf{Y} - (\dot{\mathsf{G}}_\infty \mathsf{A} \mathbf{Y}) \cdot \mathbf{E}.$$

Inequality (8.8) then takes the form

$$\tau^{-1}\mathbf{Y} \cdot \mathsf{A}\mathbf{Y} - \tfrac{1}{2}\mathbf{E} \cdot \dot{\mathsf{G}}_\infty \mathbf{E} - \tfrac{1}{2}\mathbf{Y} \cdot \dot{\mathsf{A}}\mathbf{Y} + (\dot{\mathsf{G}}_\infty \mathsf{A}\mathbf{Y}) \cdot \mathbf{E} \geq 0. \tag{8.12}$$

To characterize the properties of G_∞ and A we establish the following result.

Theorem 8.1 *Let* $\mathsf{A} > \mathbf{0}, \mathsf{G}_\infty \geq \mathbf{0}$. *If*

$$\dot{\mathsf{G}}_\infty < \mathbf{0}, \qquad \dot{\mathsf{G}}_0 \geq -2\tau^{-1}(\mathsf{G}_0 - \mathsf{G}_\infty) \tag{8.13}$$

then the free energy (8.11) satisfies (8.8).

Proof Since $\mathsf{A} > \mathbf{0}$ and $\dot{\mathsf{G}}_\infty < \mathbf{0}$ then $-\dot{\mathsf{G}}_\infty \mathsf{A} > \mathbf{0}$. Let K be a fourth-order positive-definite tensor. Hence we can write

$$[\dot{\mathbf{G}}_\infty \mathbf{A}\mathbf{Y}] \cdot \mathbf{E} = [\mathbf{K}^{-1/2}\dot{\mathbf{G}}_\infty \mathbf{A}\mathbf{Y}] \cdot \mathbf{K}^{1/2}\mathbf{E}.$$

Let $\mathbf{K} = -\dot{\mathbf{G}}_\infty$. It follows

$$[\dot{\mathbf{G}}_\infty \mathbf{A}\mathbf{Y}] \cdot \mathbf{E} = [\mathbf{K}^{1/2}\mathbf{A}\mathbf{Y}] \cdot \mathbf{K}^{1/2}\mathbf{E}].$$

We now apply Cauchy's inequality

$$|\mathbf{M}\mathbf{E}_1 \cdot \mathbf{N}\mathbf{E}_2| \le \tfrac{1}{2}\mathbf{M}\mathbf{E}_1 \cdot \mathbf{M}\mathbf{E}_1 + \tfrac{1}{2}\mathbf{N}\mathbf{E}_2 \cdot \mathbf{N}\mathbf{E}_2, \quad \mathbf{E}_1, \mathbf{E}_2 \in \mathrm{Sym}, \quad \mathbf{M}, \mathbf{N} \in \mathrm{Lin(Sym)}$$

to obtain

$$\begin{aligned}[\mathbf{K}^{1/2}\mathbf{A}\mathbf{Y}] \cdot [\mathbf{K}^{1/2}\mathbf{E}] &\le \tfrac{1}{2}\mathbf{K}^{1/2}\mathbf{A}\mathbf{Y} \cdot \mathbf{K}^{1/2}\mathbf{A}\mathbf{Y} + \tfrac{1}{2}\mathbf{K}\mathbf{E} \cdot \mathbf{K}\mathbf{E} \\ &\le \tfrac{1}{2}[\mathbf{A}\mathbf{K}\mathbf{A}\mathbf{Y}] \cdot \mathbf{Y} + \tfrac{1}{2}[\mathbf{K}\mathbf{E}] \cdot \mathbf{E}.\end{aligned}$$

Consequently

$$\rho_R \partial_t \psi \le \tfrac{1}{2}\mathbf{Y} \cdot (\dot{\mathbf{A}} - \mathbf{A}\dot{\mathbf{G}}_\infty \mathbf{A})\mathbf{Y}$$

and

$$\tau^{-1}\mathbf{Y} \cdot \mathbf{A}\mathbf{Y} - \rho_R \partial_t \psi \ge \tau^{-1}\mathbf{Y} \cdot \mathbf{A}\mathbf{Y} - \tfrac{1}{2}\mathbf{Y} \cdot (\dot{\mathbf{A}} - \mathbf{A}\dot{\mathbf{G}}_\infty \mathbf{A})\mathbf{Y}.$$

Now, since $\mathbf{A}^{-1}\mathbf{A} = \mathbf{I}$ then

$$\dot{\overline{\mathbf{A}^{-1}}} = -\mathbf{A}^{-1}\dot{\mathbf{A}}\mathbf{A}^{-1}$$

and hence

$$\dot{\mathbf{A}} - \mathbf{A}\dot{\mathbf{G}}_\infty \mathbf{A} = \mathbf{A}[\mathbf{A}^{-1}\dot{\mathbf{A}}\mathbf{A}^{-1} - \dot{\mathbf{G}}_\infty]\mathbf{A} = -\mathbf{A}[\dot{\overline{\mathbf{A}^{-1}}} + \dot{\mathbf{G}}_\infty]\mathbf{A}$$

Since $\mathbf{A}^{-1} = \mathbf{G}_0 - \mathbf{G}_\infty$ then

$$\dot{\overline{\mathbf{A}^{-1}}} + \dot{\mathbf{G}}_\infty = \dot{\mathbf{G}}_0,$$

whence

$$\tau^{-1}\mathbf{Y} \cdot \mathbf{A}\mathbf{Y} - \rho_R \partial_t \psi \ge \mathbf{Y} \cdot [\tau^{-1}\mathbf{A} + \tfrac{1}{2}\mathbf{A}\dot{\mathbf{G}}_0 \mathbf{A}]\mathbf{Y}.$$

In view of (8.13) we find

$$\tau^{-1}\mathbf{Y} \cdot \mathbf{A}\mathbf{Y} - \rho_R \partial_t \psi \ge 0$$

and hence inequality (8.8) holds. □

In essence, this theorem tells us that $\mathbf{G}_0 > \mathbf{G}_\infty \ge \mathbf{0}$ allows for a positive-definite tensor $\mathbf{A}$ satisfying the dissipation inequality for non-aging materials. Furthermore it follows that the entropy inequality is guaranteed by non-increasing tensor functions $\mathbf{G}_0(t)$, $\mathbf{G}_\infty(t)$ while $\dot{\mathbf{G}}_0$ is bounded by (8.13). Indeed, integration of the second inequality in (8.13) on $[0, t]$ results in

$$\mathbf{G}_0(t) \geq \mathbf{G}_0(0)\exp(-2t/\tau) + \frac{2}{\tau}\int_0^t \mathbf{G}_\infty(s)\exp(-2(t-s)/\tau)ds;$$

$\mathbf{G}_0$ is allowed to decrease ($\dot{\mathbf{G}}_0 < \mathbf{0}$) bounded by the right-hand side. If $\mathbf{G}_\infty = \mathbf{0}$ then

$$\mathbf{G}_0(t) \geq \mathbf{G}_0(0)\exp(-2t/\tau). \tag{8.14}$$

This result for $\mathbf{G}_0$ may be contrasted with the models (8.1) and (8.2). By (8.1) it follows

$$\dot{g}(t) = -\frac{b\exp(-bt)}{[1 + a\exp(-bt)]^2} \geq -ba\exp(-bt),$$

consistent with (8.14). By (8.2) we find

$$\dot{g}(t) = -\frac{\beta}{\tau^\beta}t^{\beta-1}g(0)\exp[-(t/\tau)^\beta].$$

Consistency with (8.14) requires that $\beta \geq 1$.

The entropy inequality does not seem to place significant restrictions on the possible dependence of τ on time. Since $\mathbf{G}_0 = \mathbf{\Gamma}_0/\tau > \mathbf{0}$ if, e.g., $\mathbf{\Gamma}_0$ is constant then (8.13), i.e.

$$-\frac{1}{\tau^2}\mathbf{\Gamma}_0\dot{\tau} \geq -\frac{2}{\tau}(\mathbf{G}_0 - \mathbf{G}_\infty),$$

implies $\mathbf{G}_0\dot{\tau} \leq 2(\mathbf{G}_0 - \mathbf{G}_\infty)$. Any negative value of $\dot{\tau}$ would be admissible but also $\dot{\tau}\mathbf{I} \leq \mathbf{G}_0^{-1}(\mathbf{G}_0 - \mathbf{G}_\infty)$ would be allowed.

8.4 Aging of Thermo-Viscoelastic Materials

A model of thermo-viscoelastic material (with memory) is established by looking at the rheological Eq. (8.4),

$$\dot{\varepsilon}_d + \alpha\,\varepsilon_d = \alpha\,\varepsilon. \tag{8.15}$$

Viewing (8.15) as a differential equation for ε_d we can integrate on $[t_0, t]$ to obtain

$$\varepsilon_d(t) = \varepsilon_d(t_0)\exp[-\alpha(t - t_0)] + \int_{t_0}^t \alpha\exp[-\alpha(t-s)]\varepsilon(s)ds.$$

Let $\alpha > 0$. Hence the limit as $t_0 \to -\infty$ results in

$$\varepsilon_d(t) = \int_{-\infty}^t \alpha\exp[-\alpha(t-s)]\varepsilon(s)ds.$$

Since $\alpha = k/\gamma$ and

$$\sigma = k_e \varepsilon + k \varepsilon_s = (k_e + k)\varepsilon - k\varepsilon_d$$

we find the stress-strain relation

$$\sigma(t) = (k_e + k)\varepsilon(t) - \int_{-\infty}^{t} (k^2/\gamma) \exp[-\alpha(t-s)]\varepsilon(s)ds$$

which involves both the present value $\varepsilon(t)$ and the past history $\varepsilon(s)$, $s \in (-\infty, t)$. In terms of $g_0 = k_e + k$, $g_\infty = k_e$ we define

$$g(\xi) = g_\infty + (g_0 - g_\infty) \exp(-\alpha\xi).$$

Hence the stress σ can be written in the form

$$\sigma(t) = g_0 \varepsilon(t) - \int_{-\infty}^{t} \partial_s g(t-s)\varepsilon(s)ds.$$

This is a further argument supporting the model of viscoelastic solids via a Boltzmann-like representation of the stress-strain relation.

A tensorial representation of the stress-strain relation is defined as follows. Let

$$\mathscr{D} = \{(t, s) : t \in \mathbb{R}, s \in (-\infty, t]\}$$

or $\mathscr{D} = \{(t, s) : t \in \mathbb{R}, s \leq t\}$. Let $\mathsf{G} : \mathscr{D} \to \mathrm{Lin(Sym)}$ be the relaxation function such that the stress $\mathbf{T}$ is given by

$$\mathbf{T}(\mathbf{x}, t) = \mathsf{G}_0(t)\mathbf{E}(\mathbf{x}, t) - \int_{-\infty}^{t} \partial_s \mathsf{G}(t, s)\mathbf{E}(\mathbf{x}, s)ds, \qquad \mathsf{G}_0(t) := \mathsf{G}(t, t). \tag{8.16}$$

Upon a change of variable, $s \to u = t - s$, the t-dependent relaxation function

$$\tilde{\mathsf{G}} : \mathbb{R} \times \mathbb{R}^+ \to \mathrm{Lin(Sym)}, \qquad \tilde{\mathsf{G}}(t, u) = \mathsf{G}(t, t-u)$$

allows the stress-strain relation to be written in the form

$$\mathbf{T}(\mathbf{x}, t) = \mathsf{G}_0(t)\mathbf{E}(\mathbf{x}, t) + \int_0^{\infty} \partial_u \tilde{\mathsf{G}}(t, u)\mathbf{E}(\mathbf{x}, t-u)du,$$

where

$$\mathsf{G}_0(t) = \tilde{\mathsf{G}}(t, 0) = \mathsf{G}(t, t), \qquad \partial_u \tilde{\mathsf{G}}(t, u) = \partial_u \mathsf{G}(t, t-u) = -\partial_s \mathsf{G}(t, s).$$

Alternatively, letting

$$\mathsf{G}_\infty(t) = \lim_{u \to \infty} \tilde{\mathsf{G}}(t, u)$$

we can write

$$\mathbf{T}(\mathbf{x},t) = \mathbf{G}_\infty(t)\mathbf{E}(\mathbf{x},t) - \int_0^\infty \partial_u \tilde{\mathbf{G}}(t,u)[\mathbf{E}(\mathbf{x},t) - \mathbf{E}(\mathbf{x},t-u)]du.$$

For ease in writing let

$$\mathcal{G}(t,u) = -\partial_u \tilde{\mathbf{G}}(t,u), \qquad (t,u) \in \mathbb{R} \times \mathbb{R}^+;$$

the function $\mathcal{G}$ will be referred to as t-dependent *relaxation kernel.* Hence we have

$$\mathbf{T}(\mathbf{x},t) = \mathbf{G}_\infty(t)\mathbf{E}(\mathbf{x},t) + \int_0^\infty \mathcal{G}(t,u)[\mathbf{E}(\mathbf{x},t) - \mathbf{E}(\mathbf{x},t-u)]du. \tag{8.17}$$

We now restrict attention to the stress-strain relation (8.17). Consistent with the scalar case, in order to model viscoelastic solids the relaxation modulus $\mathbf{G}_\infty(t)$ is assumed to be positive definite for every $t \in \mathbb{R}$, namely

$$\mathbf{E} \cdot \mathbf{G}_\infty(t)\mathbf{E} > 0 \qquad \forall \mathbf{E} \in \mathrm{Sym}.$$

We remark that, for non-aging materials, $\mathbf{G}(t,s)$ and $\tilde{\mathbf{G}}(t,u)$ reduce to $\mathbf{G}(t-s)$ and $\mathbf{G}(u)$, respectively. In addition, $\partial_s \mathbf{G}(t,s)$ reduces to $\partial_s \mathbf{G}(t-s) = -\mathbf{G}'(t-s)$ while $\mathcal{G}(t,u) = -\partial_u \tilde{\mathbf{G}}(t,u)$ reduces to $-\mathbf{G}'(u)$. Hence, as expected, $\mathbf{G}_0(t)$ and $\mathbf{G}_\infty(t)$ reduce to the constant tensors $\mathbf{G}_0$ and $\mathbf{G}_\infty$.

The relaxation kernel is assumed to satisfy the following properties.

(M1) $(t,u) \mapsto \mathcal{G}(t,u) \in L^\infty(\mathscr{C})$ for every compact set $\mathscr{C} \subset \mathbb{R} \times \mathbb{R}^+$

(M2) For every fixed $t \in \mathbb{R}$ the map $u \mapsto \mathcal{G}(t,u)$ is positive semi-definite, absolutely continuous and summable on $\mathbb{R}^+$. Then, for every $t \in \mathbb{R}$,

$$\int_0^\infty \mathcal{G}(t,u)du = \mathbf{G}_0(t) - \mathbf{G}_\infty(t) \geq \mathbf{0}.$$

Moreover, the map is differentiable for all $t \in \mathbb{R}^+$ and

$$(t,u) \mapsto \partial_u \mathcal{G}(t,u) \in L^\infty(\mathscr{C})$$

for every compact set $\mathscr{C} \subset \mathbb{R} \times \mathbb{R}^+$.

(M3) For every fixed $u > 0$ the map $t \mapsto \mathcal{G}(t,u)$ is differentiable for all $t \in \mathbb{R}$. Moreover,

$$(t,u) \mapsto \partial_t \mathcal{G}(t,u) \in L^\infty(\mathscr{C})$$

for every compact set $\mathscr{C} \subset \mathbb{R} \times \mathbb{R}^+$.

In view of (M2) the t-dependent relaxation function $\tilde{\mathbf{G}}$ may be represented as

$$\tilde{\mathbf{G}}(t,u) = \mathbf{G}_0(t) - \int_0^u \mathcal{G}(t,\xi)d\xi.$$

If $\mathbf{E}(\mathbf{x}, \cdot)$ is a *fading strain history,* namely

$$\lim_{s\to-\infty} \mathbf{E}(\mathbf{x}, s) = \mathbf{0}, \qquad \mathbf{E}(\mathbf{x}, s) = \int_{-\infty}^{s} \partial_\xi \mathbf{E}(\mathbf{x}, \xi)d\xi,$$

an integration by parts of (8.16) results in

$$\mathbf{T}(\mathbf{x}, t) = \int_{-\infty}^{t} \mathsf{G}(t, s)\partial_s \mathbf{E}(\mathbf{x}, s)ds.$$

This representation allows $\mathsf{G}(t, s)$ to be unbounded at $s = t$. Consequently,

$$\mathbf{T}(\mathbf{x}, t) = \mathsf{G}_\infty(t)\mathbf{E}(\mathbf{x}, t) + \int_{-\infty}^{t} \check{\mathsf{G}}(t, s)\partial_s \mathbf{E}(\mathbf{x}, s)ds,$$

where

$$\check{\mathsf{G}}(t, s) = \mathsf{G}(t, s) - \mathsf{G}_\infty(t).$$

Let

$$\mathbf{E}_d^t(s) = \mathbf{E}(t) - \mathbf{E}(t - s) \qquad s \in [0, \infty)$$

be the relative strain history. Hence we can write $\mathbf{T}$ in the form

$$\mathbf{T}(\mathbf{x}, t) = \mathsf{G}_\infty(t)\mathbf{E}(\mathbf{x}, t) + \int_0^\infty \mathcal{G}(t, s)\mathbf{E}_d^t(s)ds = \mathsf{G}_0(t)\mathbf{E}(\mathbf{x}, t) - \int_0^\infty \mathcal{G}(t, s)\mathbf{E}(t - s)ds.$$

Any admissible free energy ψ is required to satisfy the entropy inequality

$$-\rho\dot{\psi} + \mathbf{T} \cdot \dot{\mathbf{E}} \geq 0$$

whence

$$(-\rho\partial_{\mathbf{E}}\psi + \mathbf{T}) \cdot \dot{\mathbf{E}} - \rho d\psi(\mathbf{E}(t), \mathbf{E}^t|\dot{\mathbf{E}}^t, t) - \rho\partial_t\psi \geq 0.$$

The arbitrariness of $\dot{\mathbf{E}}$ implies

$$\mathbf{T} = \rho\partial_{\mathbf{E}}\psi.$$

Hence we are left with the dissipation inequality

$$d\psi(\mathbf{E}(t), \mathbf{E}^t|\dot{\mathbf{E}}^t, t) + \partial_t\psi \leq 0. \tag{8.18}$$

The relation $\mathbf{T} = \rho\partial_{\mathbf{E}}\psi$ holds with the free energy in the Graffi-Volterra single-integral form,

$$\psi_G(\mathbf{E}(t), \mathbf{E}_d^t, t) = \tfrac{1}{2}\mathbf{E}(t) \cdot \mathsf{G}_\infty(t)\mathbf{E}(t) + \tfrac{1}{2}\int_0^\infty \mathbf{E}_d^t(s) \cdot \mathcal{G}(t, s)\mathbf{E}_d^t(s)ds.$$

Aging effects via the functions $\mathbf{G}_\infty(t)$ and $\mathcal{G}(t,s)$ are established by means of the following

Theorem 8.2 *The stress-strain relation (8.17) is consistent with the dissipation inequality (8.18) if*

$$\dot{\mathbf{G}}_\infty \le \mathbf{0} \quad \forall t \in \mathbb{R}, \qquad \partial_t \mathcal{G}(t,s) + \partial_s \mathcal{G}(t,s) \le \mathbf{0} \quad \forall (t,s) \in \mathbb{R} \times \mathbb{R}^+.$$

Proof Let $\psi = \psi_G$ and observe that

$$d\psi_G(\mathbf{E}(t), \mathbf{E}^t | \dot{\mathbf{E}}^t, t) = \int_0^\infty \partial_t \mathbf{E}^t(s) \cdot \mathcal{G}(t,s) \mathbf{E}^t_d(s) ds,$$

$$\partial_t \psi_G = \tfrac{1}{2} \mathbf{E}(t) \cdot \dot{\mathbf{G}}_\infty(t) \mathbf{E}(t) + \tfrac{1}{2} \int_0^\infty \mathbf{E}^t_d(s) \cdot \partial_t \mathcal{G}(t,s) \mathbf{E}^t_d(s) ds.$$

Upon an integration by parts, since $\mathcal{G}(t,s) \to \mathbf{0}$ as $s \to \infty$ and $\mathbf{E}^t_r(0) = \mathbf{0}$ then it follows

$$\int_0^\infty \partial_t \mathbf{E}^t(s) \cdot \mathcal{G}(t,s) \mathbf{E}^t_d(s) ds = -\int_0^\infty \partial_s \mathbf{E}^t_d(s) \cdot \mathcal{G}(t,s) \mathbf{E}^t_d(s) ds$$
$$= \tfrac{1}{2} \int_0^\infty \mathbf{E}^t_d(s) \cdot \partial_s \mathcal{G}(t,s) \mathbf{E}^t_d(s) ds.$$

Consequently

$$d\psi(\mathbf{E}(t), \mathbf{E}^t | \dot{\mathbf{E}}^t, t) + \partial_t \psi = \tfrac{1}{2} \mathbf{E}(t) \cdot \dot{\mathbf{G}}_\infty(t) \mathbf{E}(t) + \tfrac{1}{2} \int_0^\infty \mathbf{E}^t_d(s) \cdot [\partial_t \mathcal{G}(t,s) + \partial_s \mathcal{G}(t,s)] \mathbf{E}^t_d(s) ds$$

and inequality (8.18) follows. □

It is worth remarking that for non-aging materials $\mathcal{G}$ is independent of t in that $\mathcal{G}(t,s) = \mathcal{G}(s)$ whence

$$\partial_t \mathcal{G}(t,s) + \partial_s \mathcal{G}(t,s) = \partial_s \mathcal{G}(s) = -\partial_s^2 \tilde{\mathbf{G}}(s).$$

Hence the assumption $\partial_t \mathcal{G}(t,s) + \partial_s \mathcal{G}(t,s) \le 0$ reduces to the convexity of $\tilde{\mathbf{G}}$, $\partial_s^2 \tilde{\mathbf{G}}(s) \ge \mathbf{0}$.

A more restrictive assumption implies the validity of the dissipation inequality [105] as is shown in the following

Theorem 8.3 *The stress-strain relation (8.17) is consistent with the dissipation inequality (8.18) if*

$$\dot{\mathbf{G}}_\infty \le \mathbf{0} \quad \forall t \in \mathbb{R}, \qquad \partial_t \mathcal{G}(t,s) + \partial_s \mathcal{G}(t,s) \le -M(t) \mathcal{G}(t,s) \quad \forall (t,s) \in \mathbb{R} \times \mathbb{R}^+,$$

where $M(t) \ge 0$.

Proof Let $\psi = \psi_G$ and evaluate $d\psi_G, \partial_t\psi_G$ as in the previous theorem. Hence

$$d\psi(\mathbf{E}(t), \mathbf{E}^t, t|\dot{\mathbf{E}}^t) + \partial_t\psi = \tfrac{1}{2}\mathbf{E}(t)\cdot\dot{\mathbf{G}}_\infty(t)\mathbf{E}(t) + \tfrac{1}{2}\int_0^\infty \mathbf{E}_d^t(s)\cdot[\partial_t\mathcal{G}(t,s) + \partial_s\mathcal{G}(t,s)]\mathbf{E}_d^t(s)ds$$
$$\le \tfrac{1}{2}\mathbf{E}(t)\cdot\dot{\mathbf{G}}_\infty(t)\mathbf{E}(t) - \tfrac{1}{2}M(t)\int_0^\infty \mathbf{E}_d^t(s)\cdot\mathcal{G}(t,s)\mathbf{E}_d^t(s)ds.$$

The two assumptions imply the dissipation inequality (8.18). □

8.5 An Aging Model of Viscous Fluid with Memory

An aging model of viscous fluid with memory might be expressed by letting the memory function depend on time. Formally we might replace $m(s)$ of (7.76) with $\alpha^2(t)m(s)$, which means that the viscous property changes in time according to the function $\alpha^2(t)$. Hence let

$$\mathbf{T} = -p(\rho,\theta)\mathbf{1} + \alpha^2(t)\int_0^\infty m(s)\mathbf{D}(t-s)ds.$$

For formal convenience let $\alpha > 0$. Owing to the occurrence of $\alpha(t)$ we can view

$$\rho, \theta, \mathbf{D}^t, t$$

as the set of independent variables, the partial dependence on t being induced by $\alpha(t)$. Upon evaluation of $\dot{\psi}$ and substitution in the entropy inequality we find

$$-\rho(\partial_\theta\psi + \eta)\dot{\theta} + (\rho^2\partial_\rho\psi - p)\nabla\cdot\mathbf{v} - \rho\, d\psi(\rho,\theta,\mathbf{D}^t|\dot{\mathbf{D}}^t) - \rho\partial_t\psi$$
$$+\mathbf{D}\cdot\alpha^2(t)\int_0^\infty m(s)\,\mathbf{D}(t-s)ds \ge 0.$$

In addition to the classical relations $\eta = -\partial_\theta\psi$, $p = \rho^2\partial_\rho\psi$ we obtain the dissipation inequality

$$-\rho\, d\psi(\rho,\theta,\mathbf{D}^t|\dot{\mathbf{D}}^t) - \rho\partial_t\psi + \mathbf{D}\cdot\alpha^2(t)\int_0^\infty m(s)\,\mathbf{D}(t-s)ds \ge 0.$$

By taking the free energy function in the form

$$\psi = \Psi(\rho,\theta) + \beta\left|\alpha(t)\int_0^\infty m(s)\,\mathbf{D}(t-s)ds\right|^2$$

we have

$$\alpha(1-2\rho\beta m(0))\mathbf{D}\cdot\Big(\int_0^\infty m(s)\,\mathbf{D}(t-s)ds\Big)-2\beta\rho\dot{\alpha}\Big|\int_0^\infty m(s)\,\mathbf{D}(t-s)ds\Big|^2$$
$$-2\rho\beta\alpha\Big(\int_0^\infty m(s)\,\mathbf{D}(t-s)ds\Big)\cdot\Big(\int_0^\infty m'(s)\,\mathbf{D}(t-s)ds\Big)\geq 0.$$

The inequality holds if and only if $\beta = 1/2\rho m(0)$, $m(s) = m(0)\exp(-\gamma s)$, and, furthermore

$$\dot{\alpha} \leq 0.$$

This condition means that, by the aging effect the viscosity property described by the memory function $\alpha(t)m(s)$ decreases in time.

8.6 Damage

Continuum mechanics describes the damage as a degradation of the material. The degradation may result from different mechanisms typically related to the microstructure of the material and defects such as microcracks, growth of pores, rupture of fibres embedded in a matrix. Detailed descriptions may model the evolution of defects; here we look at macroscopic models of damage in terms of macroscopic variables.

The damage variable $\mathscr{D}$ is most often a scalar and $\mathscr{D} \in [0, 1]$; $\mathscr{D} = 1$ represents the undamaged state and $\mathscr{D} = 0$ corresponds to the fully degraded state. If $\mathscr{D}$ is expressed by an internal variable, say d, examples of dependences are

$$\mathscr{D} = 1 - d, \quad d \in [0, 1], \qquad \mathscr{D} = \exp(-\eta_d d), \quad d \in [0, \infty),$$

the internal variable d being positive and typically restricted by $\dot{d} \geq 0$.

Sometimes $\mathscr{D}$ is determined by a number of processes undergone by the material. For instance, if $N_i(S_i)$, $i = 1, 2, ..., N_D$, is the number of cyclic processes with amplitude S_i and $\sum_i N_i(S_i)$ leads to failure then a sequence $n_i(S_i)$ is said to produce a damage

$$\mathscr{D} = \sum_{i=1}^{N_D} \frac{n_i(S_i)}{N_i(S_i)}.$$

In a damaged linearly elastic body the effective Lamé constants are regarded to be $1 - \mathscr{D}$ times the values of the constants for an undamaged body,

$$\mathbf{T} = 2\mu(1-\mathscr{D})\varepsilon + \lambda(1-\mathscr{D})(\mathrm{tr}\,\varepsilon)\mathbf{1}.$$

In turn, the effective Young's modulus $\tilde{E}$ is the fraction $(1 - \mathscr{D})$ of the nominal value,

$$\tilde{E} = \frac{\mu(1-\mathscr{D})[2\mu(1-\mathscr{D})+3\lambda(1-\mathscr{D})}{\mu(1-\mathscr{D})[\mu(1-\mathscr{D})+\lambda(1-\mathscr{D})} = (1-\mathscr{D})\frac{\mu[2\mu+3\lambda]}{\mu+\lambda} = (1-\mathscr{D})E.$$

Consistently, Poisson's modulus $\nu = \lambda/2(\mu+\lambda)$ is left unchanged.

In the case of anisotropic damage, due to a distribution of cracks, the damage is modelled as a second-order tensor $\mathfrak{D}$ in the form [246]

$$\mathfrak{D} = \sum_i \omega_i \mathbf{n}_i \otimes \mathbf{n}_i,$$

where ω_i is the cracks density in the direction $\mathbf{n}_i$.

If $\theta, \varepsilon, \mathscr{D}$ are the independent variables then the Clausius-Duhem inequality[3]

$$-\rho(\partial_\theta\psi + \eta)\dot{\theta} + (\mathbf{T} - \rho\partial_\varepsilon\psi)\cdot\dot{\varepsilon} - \rho\partial_{\mathscr{D}}\psi\,\dot{\mathscr{D}} \geq 0$$

reduces to

$$\partial_{\mathscr{D}}\psi\,\dot{\mathscr{D}} \leq 0.$$

This suggested that $-\rho\partial_{\mathscr{D}}\psi\,\dot{\mathscr{D}}$ may be viewed as the dissipation of the material due to damage and hence that $F = -\rho\partial_{\mathscr{D}}\psi$ be the force due to the damage process; we let $F \geq 0$. Then it is introduced a damage yield function f in the form [296]

$$f(F, \mathscr{D}) = F - \kappa(\mathscr{D}),$$

where κ is a positive function representing the resistance of the material to the damage propagation. For the damage evolution it is assumed that

$$\dot{\mathscr{D}}\begin{cases} = 0 \text{ if } f \leq 0,\ \dot{f} < 0, \\ > 0 \text{ if } f = 0,\ \dot{f} = 0; \end{cases}$$

the damage $\mathscr{D}$ is constant if the material is below the yield limit and increases if the material is in the yield limit.

If $\rho\psi = \rho\psi_0(\theta) + \frac{1}{2}\varepsilon\cdot\mathsf{C}(\mathscr{D})\varepsilon$ then

$$F = -\varepsilon\cdot\partial_{\mathscr{D}}\mathsf{C}\varepsilon;$$

the assumption $F \geq 0$ means that

$$\partial_{\mathscr{D}}\mathsf{C} \leq \mathbf{0},$$

which is consistent with the decreasing property of the Lamé moduli.

[3] Here we take $\mathbf{T}\cdot\dot{\varepsilon}$ as the linear approximation of the power in elasticity.

8.7 Fluids of Higher-Order Grade

The inadequacy of the classical Navier-Stokes theory to describe the dynamics of rheologically complex fluids has led to the development of several theories of non-Newtonian fluids. Among the many models which have been used to model the non-Newtonian behaviour, the fluids of differential type have been developed at various levels of mathematical complexity. In general, fluid of differential type means that the constitutive equations involve appropriate derivatives of the stretching $\mathbf{D}$, in addition to $\mathbf{D}$ itself.

A fluid model of grade n denotes a model where time derivatives of the stretching $\mathbf{D}$ occurs up to the $(n-1)$-order. In this sense, a Newtonian fluid is a first-order fluid.

Among fluids of higher-order grade, the incompressible fluid of grade 2 has received an especially large amount of attention. Further, the fluid has been modelled via the Rivlin-Ericksen tensors. Other descriptions are equally admissible in connection with the objectivity principle.

As with the Rivlin-Ericksen tensors, we can define a sequence of kinematic tensors by letting

$$\mathcal{A}_{n+1} = \overset{\boxdot}{\mathcal{A}}_n, \qquad \mathcal{A}_1 = \mathbf{D},$$

$\boxdot$ denoting a generic objective time derivative. For definiteness we let $\overset{\boxdot}{\mathcal{A}}$ of any tensor $\mathcal{A}$ be defined by

$$\overset{\boxdot}{\mathcal{A}} = \dot{\mathcal{A}} - \mathbf{W}\mathcal{A} - \mathcal{A}\mathbf{W}^T + \nu_1\,\mathbf{D}\mathcal{A} + \nu_2\,\mathcal{A}\mathbf{D} + \xi\,\mathrm{tr}\,\mathbf{D}\,\mathcal{A},$$

ν_1, ν_2, ξ being scalar quantities; they are allowed to be given by functions of ρ and θ. For technical convenience we require that symmetry is preserved by the derivative in that

$$\mathcal{A} \in \mathrm{Sym} \quad \Longrightarrow \quad \overset{\boxdot}{\mathcal{A}} \in \mathrm{Sym}.$$

Since $\mathcal{A} \in \mathrm{Sym}$ implies $\dot{\mathcal{A}} \in \mathrm{Sym}$ and

$$\mathbf{W}\mathcal{A} + \mathcal{A}\mathbf{W}^T = \mathbf{W}\mathcal{A} + (\mathbf{W}\mathcal{A})^T \in \mathrm{Sym},$$

$$(\nu_1\,\mathbf{D}\mathcal{A} + \nu_2\,\mathcal{A}\mathbf{D})^T = \nu_1\,\mathcal{A}\mathbf{D} + \nu_2\,\mathbf{D}\mathcal{A}$$

it follows that symmetry is preserved for any $\mathcal{A}$ and $\mathbf{D}$ if and only if $\nu_1 = \nu_2$. Hence we define

$$\overset{\boxdot}{\mathcal{A}} := \dot{\mathcal{A}} - \mathbf{W}\mathcal{A} - \mathcal{A}\mathbf{W}^T + \nu(\mathbf{D}\mathcal{A} + \mathcal{A}\mathbf{D}) + \xi\,\mathrm{tr}\,\mathbf{D}\,\mathcal{A}.$$

8.7.1 Second Grade Fluid

For an easy comparison with the standard fluid of grade 2 in the literature we generalize the definition of fluid by letting the Cauchy stress $\mathbf{T}$ be given by

$$\mathbf{T} = -p\mathbf{1} + 2\mu\,\mathbf{D} + \lambda \mathrm{tr}\,\mathbf{D}\,\mathbf{1} + \beta_1\,\overset{\boxdot}{\mathbf{D}} + \beta_2\,\mathbf{D}\mathbf{D}. \tag{8.19}$$

The essential difference with the standard model of fluid of grade 2 is that the generic derivative $\boxdot$ is used for $\overset{\boxdot}{\mathbf{D}}$ while the Rivlin-Ericksen tensors involve the Cotter-Rivlin derivative. Moreover the fluid is allowed to be compressible and the compressibility effects occur explicitly in $\lambda \mathrm{tr}\,\mathbf{D}$ and $\xi \mathrm{tr}\,\mathbf{D}$ besides the implicit occurrence in $\mathbf{D}$.

Since Eq. (8.19) is an equation of differential type, and not of rate type, an obvious consistency suggests that we describe heat conduction via a Fourier-type law rather than a rate-type law.

To investigate the thermodynamic consistency of the stress function (8.19) we let

$$\rho,\ \theta,\ \mathbf{D},\ \overset{\boxdot}{\mathbf{D}},\ \nabla\theta$$

be the independent variables. The constitutive functions ψ, η, $\mathbf{q}$, and $\mathbf{T}$ are required to satisfy the second-law inequality

$$-\rho\dot{\psi} - \rho\eta\dot{\theta} + \mathbf{T}\cdot\mathbf{D} - \frac{1}{\theta}\mathbf{q}\cdot\nabla\theta \geq 0.$$

Substitution of $\dot{\psi}$ and (8.19) allow the inequality to be given the form

$$-\rho(\partial_\theta + \eta)\dot{\theta} + (\rho^2\partial_\rho\psi - p)\mathrm{tr}\,\mathbf{D} - \rho\partial_{\mathbf{D}}\psi\cdot\dot{\mathbf{D}} - \rho\partial_{\overset{\boxdot}{\mathbf{D}}}\psi\cdot(\overset{\boxdot}{\mathbf{D}})^{\cdot} - \rho\partial_{\nabla\theta}\psi\cdot(\nabla\theta)^{\cdot}$$
$$+2\mu\mathbf{D}\cdot\mathbf{D} + \lambda(\mathrm{tr}\,\mathbf{D})^2 + \beta_1\,\overset{\boxdot}{\mathbf{D}}\cdot\mathbf{D} + \beta_2(\mathbf{D}\mathbf{D})\cdot\mathbf{D} \geq 0.$$

Observe

$$(\overset{\boxdot}{\mathbf{D}})^{\cdot} = \ddot{\mathbf{D}} + [-\mathbf{W}\mathbf{D} - \mathbf{D}\mathbf{W}^T + 2\nu\mathbf{D}\mathbf{D} + \xi(\mathrm{tr}\,\mathbf{D})\mathbf{D}]^{\cdot},$$

and $(\nabla\theta)^{\cdot} = \nabla\dot{\theta} - \mathbf{L}^T\nabla\theta$. Hence any value of $(\overset{\boxdot}{\mathbf{D}})^{\cdot} \in \mathrm{Sym}$ is obtained by an appropriate value of $\ddot{\mathbf{D}}$ which, thanks to $\mathbf{b}$, can be taken to assume arbitrary values. Likewise any value of $(\nabla\theta)^{\cdot}$ is obtained by an appropriate value of $\nabla\dot{\theta}$. Of course $(\partial_\theta + \eta)\dot{\theta}$, $\partial_{\overset{\boxdot}{\mathbf{D}}}\psi\cdot(\overset{\boxdot}{\mathbf{D}})^{\cdot}$, and $\partial_{\nabla\theta}\psi\cdot(\nabla\theta)^{\cdot}$ are linear terms in $\dot{\theta}$, $(\overset{\boxdot}{\mathbf{D}})^{\cdot}$ and $(\nabla\theta)^{\cdot}$. The arbitrariness of $\dot{\theta}$, $(\overset{\boxdot}{\mathbf{D}})^{\cdot}$ and $(\nabla\theta)^{\cdot}$ implies that

$$\eta = -\partial_\theta\psi, \qquad \partial_{\overset{\boxdot}{\mathbf{D}}}\psi = 0, \qquad \partial_{\nabla\theta}\psi = 0.$$

Moreover, since only $\mathbf{q}$ can depend on $\nabla\theta$ it follows that

$$\mathbf{q} \cdot \nabla\theta \leq 0$$

and

$$(\rho^2 \partial_\rho \psi - p)\operatorname{tr}\mathbf{D} - \rho \partial_{\mathbf{D}} \psi \cdot \dot{\mathbf{D}} + 2\mu \mathbf{D} \cdot \mathbf{D} + \lambda(\operatorname{tr}\mathbf{D})^2 + \beta_1 \overset{\square}{\mathbf{D}} \cdot \mathbf{D} + \beta_2 (\mathbf{D}\mathbf{D}) \cdot \mathbf{D} \geq 0.$$

Observe

$$(\mathbf{W}\mathbf{D}) \cdot \mathbf{D} = (\mathbf{D}\mathbf{D}) \cdot \mathbf{W} = 0, \qquad (\mathbf{D}\mathbf{W}^T) \cdot \mathbf{D} = (\mathbf{D}\mathbf{D}) \cdot \mathbf{W} = 0$$

and hence

$$\overset{\square}{\mathbf{D}} \cdot \mathbf{D} = \dot{\mathbf{D}} \cdot \mathbf{D} + 2\nu (\mathbf{D}\mathbf{D}) \cdot \mathbf{D} + \xi (\operatorname{tr}\mathbf{D}) \mathbf{D} \cdot \mathbf{D}.$$

The reduced inequality takes the form

$$(\rho^2 \partial_\rho \psi - p + \xi \beta_1 \mathbf{D} \cdot \mathbf{D})\operatorname{tr}\mathbf{D} + (\beta_1 \mathbf{D} - \rho \partial_{\mathbf{D}} \psi) \cdot \dot{\mathbf{D}} + 2\mu \mathbf{D} \cdot \mathbf{D} + \lambda (\operatorname{tr}\mathbf{D})^2 + (2\nu\beta_1 + \beta_2)(\mathbf{D}\mathbf{D}) \cdot \mathbf{D} \geq 0.$$

The arbitrariness of $\dot{\mathbf{D}}$ implies

$$\beta_1 \mathbf{D} - \rho \partial_{\mathbf{D}} \psi = \mathbf{0}$$

and hence, upon integration,

$$\psi = \Psi(\rho, \theta) + \frac{\beta_1}{2\rho} \mathbf{D} \cdot \mathbf{D}. \tag{8.20}$$

To complete the set of thermodynamic restrictions we now compute the separate contributions of $\operatorname{tr}\mathbf{D}$ and the deviator $\mathbf{D}_0 = \mathbf{D} - \frac{1}{3}\operatorname{tr}\mathbf{D}\,\mathbf{1}$. Observe

$$\mathbf{D}\mathbf{D} = \mathbf{D}_0\mathbf{D}_0 + \tfrac{2}{3}\operatorname{tr}\mathbf{D}\,\mathbf{D}_0 + \tfrac{1}{9}(\operatorname{tr}\mathbf{D})^2 \mathbf{1},$$

$$(\mathbf{D}\mathbf{D}) \cdot \mathbf{D} = (\mathbf{D}_0\mathbf{D}_0) \cdot \mathbf{D}_0 + \operatorname{tr}\mathbf{D}\,\mathbf{D}_0 \cdot \mathbf{D}_0 + \tfrac{1}{9}(\operatorname{tr}\mathbf{D})^3$$

Hence the reduced inequality can be written

$$\begin{aligned}[\rho^2 \partial_\rho \psi - p + \xi\beta_1(\mathbf{D}_0 \cdot \mathbf{D}_0 + \tfrac{1}{3}(\operatorname{tr}\mathbf{D})^2)]\operatorname{tr}\mathbf{D} + 2\mu \mathbf{D}_0 \cdot \mathbf{D}_0 + (\tfrac{2}{3}\mu + \lambda)(\operatorname{tr}\mathbf{D})^2 \\ +(2\nu\beta_1 + \beta_2)[(\mathbf{D}_0\mathbf{D}_0) \cdot \mathbf{D}_0 + \operatorname{tr}\mathbf{D}\mathbf{D}_0 \cdot \mathbf{D}_0 + \tfrac{1}{9}(\operatorname{tr}\mathbf{D})^3] \geq 0.\end{aligned}$$

The occurrence of $(\mathbf{D}_0\mathbf{D}_0) \cdot \mathbf{D}_0$ implies the restriction

$$2\nu\beta_1 + \beta_2 = 0.$$

The remaining inequality holds if and only if

$$\mu \geq 0, \qquad \tfrac{2}{3}\mu + \lambda \geq 0,$$

$$\rho^2 \partial_\rho \psi - p + \xi \beta_1 (\mathbf{D}_0 \cdot \mathbf{D}_0 + \tfrac{1}{3}(\mathrm{tr}\,\mathbf{D})^2) = 0.$$

If p is assumed to be dependent only on ρ and θ then we set $\xi = 0$ and find the standard relation for fluids

$$p = \rho^2 \partial_\rho \psi.$$

Otherwise

$$p = \rho^2 \partial_\rho \psi + \xi \beta_1 (\mathbf{D}_0 \cdot \mathbf{D}_0 + \tfrac{1}{3}(\mathrm{tr}\,\mathbf{D})^2)$$

means that the dynamic pressure p depends also on the stretching $\mathbf{D}$ via the additive term $\nu \beta_1 \mathbf{D} \cdot \mathbf{D}$.

The restrictions on μ and λ are the classical ones of Newtonian fluids. Strictly speaking, thermodynamics does not place any restrictions on β_1, though the result (8.20) suggests that $\beta_1 > 0$ so that the free energy has a minimum at equilibrium (when $\mathbf{D} = \mathbf{0}$).

Much work has been developed in the literature on the incompressible fluid of grade 2 modelled by[4]

$$\mathbf{T} = -p\mathbf{1} + \mu \mathbf{A}_1 + \alpha_1 \mathbf{A}_2 + \alpha_2 \mathbf{A}_1 \mathbf{A}_1, \tag{8.21}$$

$\mathbf{A}_1, \mathbf{A}_2$ being the first two Rivlin-Ericksen tensors. The thermodynamic restrictions are $\mu \geq 0$, $\alpha_1 + \alpha_2 = 0$. The rest state proved to be unstable if $\alpha_1 < 0$ and meanwhile experiments show that $\alpha_1 \geq 0$, $\alpha_1 + \alpha_2 = 0$ are not satisfied.

Attempts to overcome this difficulty have been developed by looking at higher-order grade fluids [223, 297]. Here we show that, though within a model of grade 2, the parameters ν, ξ, occurring in the chosen derivative, allow us to avoid the joint restriction $\alpha_1 \geq 0$, $\alpha_1 + \alpha_2 = 0$.

First observe that since $\mathbf{A}_1 = 2\mathbf{D}$ then an immediate comparison between (8.19) and (8.21) indicates the identifications

$$\beta_1 = 2\alpha_1, \qquad \beta_2 = 4\alpha_2.$$

Consequently (8.20) can be written

$$\psi = \Psi(\rho, \theta) + \frac{\alpha_1}{\rho} \mathbf{D} \cdot \mathbf{D}$$

while

$$0 = 2\nu\beta_1 + \beta_2 = 4(\nu\alpha_1 + \alpha_2).$$

---

[4] See [141] and references therein.

Again we find it natural to set $\alpha_1 \geq 0$ so that ψ has a minimum at $\mathbf{D} = \mathbf{0}$. However,

$$\nu\alpha_1 + \alpha_2 = 0$$

implies that $\alpha_2 = -\nu\alpha_1$. In the Rivlin-Ericksen tensors $\nu = 1$. Here instead ν is allowed to take any value, positive or negative. Hence it follows that both α_1 and α_2 can be positive, not necessarily with $\alpha_2 = \alpha_1$.

8.7.2 *Third Grade Fluid*

In addition to fluids of second grade, third grade fluids have been investigated to account more properly for the effective shear viscosity (see [223] and refs therein). Here we consider the third grade fluid with two improvements, the fluid is compressible and the derivative need not be the Cotter-Rivlin one.

We consider the objective derivative $\overset{\square}{\boldsymbol{A}}$, of a tensor $\boldsymbol{A}$, defined by[5]

$$\overset{\square}{\boldsymbol{A}} = \dot{\boldsymbol{A}} - \mathbf{W}\boldsymbol{A} - \boldsymbol{A}\mathbf{W}^T + \nu(\mathbf{D}\boldsymbol{A} + \boldsymbol{A}\mathbf{D}).$$

The third grade fluid involves the stretching $\mathbf{D}$ and the derivatives

$$\overset{\square}{\mathbf{D}} = \dot{\mathbf{D}} - \mathbf{W}\mathbf{D} - \mathbf{D}\mathbf{W}^T + 2\nu\mathbf{D}\mathbf{D},$$

$$\overset{\square\square}{\mathbf{D}} = (\overset{\square}{\mathbf{D}})^{\cdot} - \mathbf{W}\overset{\square}{\mathbf{D}} - \overset{\square}{\mathbf{D}}\mathbf{W}^T + \nu(\mathbf{D}\overset{\square}{\mathbf{D}} + \overset{\square}{\mathbf{D}}\mathbf{D}).$$

We then let $\mathbf{T}$ be given by the constitutive relation

$$\mathbf{T} = -p\,\mathbf{1} + 2\mu\mathbf{D} + \lambda \mathrm{tr}\,\mathbf{D}\,\mathbf{1} + \alpha_1\overset{\square}{\mathbf{D}} + \alpha_2\mathbf{D}^2 + \beta_1\overset{\square\square}{\mathbf{D}} + \beta_2(\mathbf{D}\overset{\square}{\mathbf{D}} + \overset{\square}{\mathbf{D}}\mathbf{D}) + \beta_3(\mathrm{tr}\,\mathbf{D}^2)\mathbf{D}$$

which generalizes that in [171]. The viscosities μ, λ and the parameters $\alpha_1, \alpha_2, \beta_1, \beta_2, \beta_3$ are allowed to depend on the density ρ and the temperature θ.

By analogy with a misprint in [171], formula (1.2c) versus (4.1), we might have considered the last term as $\beta_3(\mathrm{tr}\,\mathbf{D}^2)\,\overset{\square}{\mathbf{D}}$. In such a case we would have found that, as a thermodynamic restriction, $\beta_3 = 0$.

The thermodynamic consistency is investigated by means of the entropy inequality and the possible restrictions are derived, by letting ψ, η, and $\mathbf{q}$ be functions of

$$\rho, \theta, \mathbf{D}, \overset{\square}{\mathbf{D}}, \overset{\square\square}{\mathbf{D}}, \nabla\theta.$$

---

[5] The Cotter-Rivlin derivative, which generates the Rivlin-Ericksen tensors, is the particular case $\nu = 1$.

Computation of $\dot{\psi}$ and substitution in the entropy inequality result in

$$-\rho(\partial_\theta\psi+\eta)\dot{\theta}+\rho^2\partial_\rho\psi\operatorname{tr}\mathbf{D}-\rho\partial_{\mathbf{D}}\psi\cdot\dot{\mathbf{D}}-\rho\partial_{\overset{\square}{\mathbf{D}}}\psi\cdot(\overset{\square}{\mathbf{D}})^{\cdot}-\rho\partial_{\overset{\square\square}{\mathbf{D}}}\psi\cdot(\overset{\square\square}{\mathbf{D}})^{\cdot}-\rho\partial_{\nabla\theta}\psi\cdot(\nabla\theta)^{\cdot}$$
$$+\mathbf{T}\cdot\mathbf{D}-\frac{1}{\theta}\mathbf{q}\cdot\nabla\theta\geq 0.$$

Observe that

$$\overset{\square\square}{\mathbf{D}}=(\overset{\square}{\mathbf{D}})^{\square}=(\overset{\square}{\mathbf{D}})^{\cdot}+\ldots=\ddot{\mathbf{D}}+\ldots$$

the dots denoting terms in $\mathbf{W}$, $\mathbf{D}$, $\dot{\mathbf{W}}$, $\dot{\mathbf{D}}$. Hence it follows

$$(\overset{\square\square}{\mathbf{D}})^{\cdot}=\dddot{\mathbf{D}}+\ldots$$

the dots denoting terms in $\mathbf{W}$, $\mathbf{D}$, $\dot{\mathbf{W}}$, $\dot{\mathbf{D}}$, $\ddot{\mathbf{W}}$, $\ddot{\mathbf{D}}$. The arbitrariness of $\dddot{\mathbf{D}}$, $(\nabla\theta)^{\cdot}$, $\dot{\theta}$ implies that

$$\partial_{\overset{\square\square}{\mathbf{D}}}\psi=\mathbf{0},\qquad \partial_{\nabla\theta}\psi=\mathbf{0},\qquad \eta=-\partial_\theta\psi.$$

Moreover, since only $\mathbf{q}$ can depend on $\nabla\theta$ then the function $\mathbf{q}$ is required to satisfy

$$\mathbf{q}\cdot\nabla\theta\leq 0.$$

We now exploit the reduced inequality

$$\rho^2\partial_\rho\psi\operatorname{tr}\mathbf{D}-\rho\partial_{\mathbf{D}}\psi\cdot\dot{\mathbf{D}}-\rho\partial_{\overset{\square}{\mathbf{D}}}\psi\cdot(\overset{\square}{\mathbf{D}})^{\cdot}+\mathbf{T}\cdot\mathbf{D}\geq 0.$$

Notice that

$$\mathbf{T}\cdot\mathbf{D}-\rho\partial_{\overset{\square}{\mathbf{D}}}\psi\cdot(\overset{\square}{\mathbf{D}})^{\cdot}=\beta_1\overset{\square\square}{\mathbf{D}}\cdot\mathbf{D}-\rho\partial_{\overset{\square}{\mathbf{D}}}\psi\cdot(\overset{\square}{\mathbf{D}})^{\cdot}+\ldots=(\beta_1\mathbf{D}-\rho\partial_{\overset{\square}{\mathbf{D}}}\psi)\cdot\ddot{\mathbf{D}}+\ldots$$

the dots containing terms independent of $\ddot{\mathbf{D}}$. The arbitrariness of $\ddot{\mathbf{D}}$ implies that

$$\beta_1\mathbf{D}-\rho\partial_{\overset{\square}{\mathbf{D}}}\psi=\mathbf{0}.$$

A direct integration yields

$$\psi=\hat{\psi}(\theta,\rho,\mathbf{D})+\frac{1}{\rho}\beta_1\overset{\square}{\mathbf{D}}\cdot\mathbf{D}.$$

The entropy inequality then reduces to

$$(\rho^2\partial_\rho\psi-p)\operatorname{tr}\mathbf{D}-\rho\partial_{\mathbf{D}}\hat{\psi}\cdot\dot{\mathbf{D}}-\beta_1\overset{\square}{\mathbf{D}}\cdot\dot{\mathbf{D}}+\beta_1\mathbf{D}\cdot[\overset{\square\square}{\mathbf{D}}-(\overset{\square}{\mathbf{D}})^{\cdot}]+\check{\mathbf{T}}\cdot\mathbf{D}\geq 0,$$

where $\check{\mathbf{T}} = \mathbf{T} + p\mathbf{1} - \beta_1 \overset{\boxdot\boxdot}{\mathbf{D}}$. We compute

$$\begin{aligned}\overset{\boxdot\boxdot}{\mathbf{D}} &= (\overset{\boxdot}{\mathbf{D}})^{\cdot} - \mathbf{W}\overset{\boxdot}{\mathbf{D}} - \overset{\boxdot}{\mathbf{D}}\mathbf{W}^T + \nu(\mathbf{D}\overset{\boxdot}{\mathbf{D}} + \overset{\boxdot}{\mathbf{D}}\mathbf{D}) \\ &= (\overset{\boxdot}{\mathbf{D}})^{\cdot} - \mathbf{W}(\dot{\mathbf{D}} - \mathbf{W}\mathbf{D} - \mathbf{D}\mathbf{W}^T + 2\nu\mathbf{D}\mathbf{D}) - (\dot{\mathbf{D}} - \mathbf{W}\mathbf{D} - \mathbf{D}\mathbf{W}^T + 2\nu\mathbf{D}\mathbf{D})\mathbf{W}^T \\ &\quad +\nu\mathbf{D}(\dot{\mathbf{D}} - \mathbf{W}\mathbf{D} - \mathbf{D}\mathbf{W}^T + 2\nu\mathbf{D}\mathbf{D}) + \nu(\dot{\mathbf{D}} - \mathbf{W}\mathbf{D} - \mathbf{D}\mathbf{W}^T + 2\nu\mathbf{D}\mathbf{D})\mathbf{D}.\end{aligned}$$

Hence

$$\begin{aligned}\mathbf{D} \cdot [\overset{\boxdot\boxdot}{\mathbf{D}} - (\overset{\boxdot}{\mathbf{D}})^{\cdot}] &= -\mathbf{D} \cdot [\mathbf{W}(\dot{\mathbf{D}} - \mathbf{W}\mathbf{D} - \mathbf{D}\mathbf{W}^T + 2\nu\mathbf{D}\mathbf{D})] \\ &\quad -\mathbf{D} \cdot [(\dot{\mathbf{D}} - \mathbf{W}\mathbf{D} - \mathbf{D}\mathbf{W}^T + 2\nu\mathbf{D}\mathbf{D})\mathbf{W}^T] \\ &\quad +\nu\mathbf{D} \cdot [\mathbf{D}(\dot{\mathbf{D}} - \mathbf{W}\mathbf{D} - \mathbf{D}\mathbf{W}^T + 2\nu\mathbf{D}\mathbf{D}) + (\dot{\mathbf{D}} - \mathbf{W}\mathbf{D} - \mathbf{D}\mathbf{W}^T + 2\nu\mathbf{D}\mathbf{D})\mathbf{D}].\end{aligned}$$

Some terms ($\mathbf{D} \cdot \mathbf{W}\mathbf{D}\mathbf{D}, \mathbf{D} \cdot \mathbf{D}\mathbf{D}\mathbf{W}^T, \mathbf{D} \cdot \mathbf{D}\mathbf{W}\mathbf{D}, \mathbf{D} \cdot \mathbf{D}\mathbf{D}\mathbf{W}^T, \mathbf{D} \cdot \mathbf{W}\mathbf{D}\mathbf{D}, \mathbf{D} \cdot \mathbf{D}\mathbf{D}\mathbf{W}^T$) are zero. It follows that

$$\mathbf{D} \cdot [\overset{\boxdot\boxdot}{\mathbf{D}} - (\overset{\boxdot}{\mathbf{D}})^{\cdot}] = -2(\mathbf{D}\dot{\mathbf{D}}) \cdot \mathbf{W} + 2(\mathbf{D}\mathbf{D}) \cdot (\mathbf{W}\mathbf{W}) + 2(\mathbf{D}\mathbf{W}) \cdot (\mathbf{W}\mathbf{D}) + 2\nu(\mathbf{D}\mathbf{D}) \cdot \dot{\mathbf{D}} + 4\nu^2\mathbf{D} \cdot (\mathbf{D}\mathbf{D}\mathbf{D}).$$

Likewise we find

$$\overset{\boxdot}{\mathbf{D}} \cdot \mathbf{D} = \mathbf{D} \cdot \dot{\mathbf{D}} + 2\nu(\mathbf{D}\mathbf{D}) \cdot \mathbf{D},$$

$$\overset{\boxdot}{\mathbf{D}} \cdot \dot{\mathbf{D}} = \dot{\mathbf{D}} \cdot \dot{\mathbf{D}} + 2(\mathbf{D}\dot{\mathbf{D}}) \cdot \mathbf{W} + 2\nu(\mathbf{D}\mathbf{D}) \cdot \dot{\mathbf{D}},$$

$$\begin{aligned}\check{\mathbf{T}} \cdot \mathbf{D} &= 2\mu\mathbf{D} \cdot \mathbf{D} + \lambda(\mathrm{tr}\,\mathbf{D})^2 + \alpha_1\mathbf{D} \cdot \dot{\mathbf{D}} + (2\alpha_1\nu + \alpha_2)\mathbf{D} \cdot (\mathbf{D}\mathbf{D}) \\ &\quad +\beta_2\mathbf{D} \cdot (\mathbf{D}\dot{\mathbf{D}} + \dot{\mathbf{D}}\mathbf{D}) + \beta_3(\mathbf{D} \cdot \mathbf{D})^2.\end{aligned}$$

Upon substitution in the entropy inequality we find

$$\begin{aligned}&(\rho^2\partial_\rho\psi - p)\mathrm{tr}\,\mathbf{D} - \rho\partial_{\mathbf{D}}\hat{\psi} \cdot \dot{\mathbf{D}} - 4\beta_1\mathbf{W} \cdot (\mathbf{D}\dot{\mathbf{D}}) + 2\mu\mathbf{D} \cdot \mathbf{D} + \lambda(\mathrm{tr}\,\mathbf{D})^2 + \beta_3(\mathbf{D} \cdot \mathbf{D})^2 \\ &\quad +\beta_1[2\mathbf{D} \cdot (\mathbf{W}\mathbf{W}\mathbf{D}) + 2\mathbf{D} \cdot (\mathbf{W}\mathbf{D}\mathbf{W}^T) + 4\nu^2\mathbf{D} \cdot (\mathbf{D}\mathbf{D}\mathbf{D}) - \dot{\mathbf{D}} \cdot \dot{\mathbf{D}}] \\ &\quad +\alpha_1\mathbf{D} \cdot \dot{\mathbf{D}} + (2\alpha_1\nu + \alpha_2)\mathbf{D} \cdot (\mathbf{D}\mathbf{D}) + 2\beta_2\dot{\mathbf{D}} \cdot (\mathbf{D}\mathbf{D}) + 4\nu\beta_2\mathbf{D}^2 \cdot \mathbf{D}^2 \geq 0\end{aligned}$$

The dependence on $\mathbf{W}$ is linear, in $4\beta_1\mathbf{W} \cdot (\mathbf{D}\dot{\mathbf{D}})$, and quadratic, in $\beta_1[2\mathbf{D} \cdot (\mathbf{W}\mathbf{W}\mathbf{D}) + 2\mathbf{D} \cdot (\mathbf{W}\mathbf{D}\mathbf{W}^T)]$. The arbitrariness of $\mathbf{W}$ and the linear term imply that

$$\beta_1 = 0.$$

The inequality reduces to

$$(\rho^2 \partial_\rho \psi - p)\text{tr}\,\mathbf{D} + (-\rho \partial_{\mathbf{D}} \hat{\psi} + \alpha_1 \mathbf{D} + 2\beta_2 \mathbf{D}\mathbf{D}) \cdot \dot{\mathbf{D}} + (2\alpha_1 \nu + \alpha_2)\mathbf{D} \cdot (\mathbf{D}\mathbf{D})$$
$$+2\mu \mathbf{D} \cdot \mathbf{D} + \lambda(\text{tr}\,\mathbf{D})^2 + \beta_3(\mathbf{D} \cdot \mathbf{D})^2 + 4\nu\beta_2 \mathbf{D}^2 \cdot \mathbf{D}^2 \geq 0.$$

The occurrence of $\dot{\mathbf{D}}$ implies that

$$-\rho \partial_{\mathbf{D}} \hat{\psi} + \alpha_1 \mathbf{D} + 2\beta_2 \mathbf{D}\mathbf{D} = \mathbf{0},$$

whence, by integration,

$$\psi = \hat{\psi}(\rho, \theta, \mathbf{D}) = \Psi(\rho, \theta) + \frac{\alpha_1}{2\rho} \mathbf{D} \cdot \mathbf{D} + \frac{2\beta_2}{3\rho} \mathbf{D} \cdot (\mathbf{D}\mathbf{D}). \tag{8.22}$$

The reduced inequality holds if and only if

$$p = \rho^2 \partial_\rho \psi, \qquad 2\alpha_1 \nu + \alpha_2 = 0,$$

$$\mu \geq 0, \qquad 2\mu + 3\lambda \geq 0, \qquad \beta_3 \geq 0.$$

The result (8.22) for the free energy may suggest further restrictions if we require that ψ has a minimum at $\mathbf{D} = \mathbf{0}$. Since $\mathbf{D} \cdot (\mathbf{D}\mathbf{D})$ is not definite then this requirement implies that we let $\beta_2 = 0$ while $\alpha_1 \geq 0$. Again, the arbitrary value of ν allows α_2 to be positive or negative. Hence the present model of compressible, third-grade fluid amounts to letting

$$\mathbf{T} = -p\mathbf{1} + 2\mu \mathbf{D} + \lambda(\text{tr}\,\mathbf{D})\mathbf{1} + \alpha_1 \overset{\square}{\mathbf{D}} + \alpha_2 \mathbf{D}^2 + \beta_3(\text{tr}\,\mathbf{D}^2)\mathbf{D}$$

subject to

$$2\alpha_1 \nu + \alpha_2 = 0, \qquad \alpha_1 \geq 0, \qquad \mu \geq 0, \qquad 2\mu + 3\lambda \geq 0, \qquad \beta_3 \geq 0.$$

8.8 Interaction Effects via Materials of Higher Order

Motivated by the purpose of describing spatial interaction effects of longer range, some constitutive models have been set up by means of various inequivalent approaches. A well-known model, due to Korteweg [258], describes a compressible fluid by letting the Cauchy stress $\mathbf{T}$ be given by

$$\mathbf{T} = (-p + \alpha \Delta \rho + \beta |\nabla \rho|^2)\mathbf{1} + \delta \nabla \rho \otimes \nabla \rho + \gamma \nabla^2 \rho,$$

where $\nabla^2 = \nabla\nabla$. The purpose of Korteweg's model was to describe capillarity effects.

In elastic materials a similar idea is applied by letting the stress depend not only on the strain (or the deformation gradient) but also on higher-order gradients of deformation. We then use the terminology of (elastic) material of grade n if the gradients of deformation are involved up to order n.

It is a drawback of higher-grade models that they are in general incompatible with the standard theory of thermodynamics. To overcome the incompatibility various approaches have been developed. In [142] a rate of supply of mechanical energy u, per unit area, is assumed to take place. The supply u is named *interstitial working* . By an analogue of Cauchy's theorem it follows that u depends linearly on the unit normal $\mathbf{n}$ via a vector field $\mathbf{u}$, $u = \mathbf{u} \cdot \mathbf{n}$, $\mathbf{u}$ being denoted as interstitial work flux, though conceptually we need to know that the heat supply is represented by the heat flux vector. Next, the balance of energy and the second law inequality are assumed in the form

$$\rho\dot{\varepsilon} = \mathbf{T} \cdot \mathbf{D} + \rho r - \nabla \cdot (\mathbf{q} - \mathbf{u}),$$

$$\rho\dot{\eta} \geq -\nabla \cdot \frac{\mathbf{q}}{\theta} + \frac{\rho r}{\theta}.$$

Physically $\mathbf{q}$ is the heat flux while $\mathbf{u}$ is a mechanical work flux. Formally, though, we might view $\mathbf{h} = \mathbf{q} - \mathbf{u}$ as the surface energy flux. If $\mathbf{q}/\theta$ is the entropy flux then

$$\frac{\mathbf{q}}{\theta} = \frac{\mathbf{h}}{\theta} + \frac{\mathbf{u}}{\theta}$$

might suggest that we interpret $\mathbf{u}/\theta$ as an extra-entropy flux, originated by longer range spatial interactions.

Borrowing from works of Toupin [421, 422] on elastic materials, involving a virtual-work principle, an approach has been developed where the stress power is not merely $\mathbf{T} \cdot \mathbf{D}$. Instead, the power comprises higher-order gradients of the velocity. For *second-gradient materials* the expression of the stress power is taken in the form

$$\mathbf{T} \cdot \mathbf{L} + \mathfrak{T} \cdot \nabla^2 \mathbf{v},$$

where $\mathbf{T}$ and $\mathfrak{T}$ (which is often named hyperstress) are basically seen as elements of the dual space of the first and second velocity gradient. Quite naturally it is usually stated that $\mathfrak{T}$ is symmetric with respect to the second and third index, the motivation being related to the occurrence of ∇^2. A skew-symmetric part would do no work for any motion of the body and then would appear to be useless. Instead, allowing for non-symmetric hyperstresses proves crucial to obtain thermodynamically consistent schemes [332].

As is shown in the next section a connection can be established between the modelling of longer range interactions via the interstitial working, as well as via the extra-entropy flux, and the expression of power via the hyperstress.

8.9 Modelling via the Interstitial Working

In addition to the standard equation of balance of mass and the equation of motion,

$$\dot{\rho} + \rho \nabla \cdot \mathbf{v} = 0, \qquad \rho \dot{\mathbf{v}} = \nabla \cdot \mathbf{T} + \rho \mathbf{b},$$

the balance of energy is taken in the form

$$\rho \dot{\varepsilon} = \mathbf{T} \cdot \mathbf{L} - \nabla \cdot \mathbf{q} + \nabla \cdot \mathbf{u} + \rho r$$

where $\mathbf{u}$ is the assumed interstitial work flux and $\mathbf{T}$ is allowed to be non-symmetric. Upon the standard assumption about the entropy inequality,

$$\rho \dot{\eta} \geq -\nabla \cdot \frac{\mathbf{q}}{\theta} + \frac{\rho r}{\theta},$$

it follows that

$$-\rho(\dot{\psi} + \eta \dot{\theta}) + \mathbf{T} \cdot \mathbf{L} + \nabla \cdot \mathbf{u} - \frac{1}{\theta} \mathbf{q} \cdot \nabla \theta \geq 0. \tag{8.23}$$

We let ψ, η, $\mathbf{T}$, $\mathbf{q}$, $\mathbf{u}$ be continuous functions of

$$\Gamma = (\mathbf{F}, \theta, \nabla_R \mathbf{F}, \nabla_R^2 \mathbf{F}, \nabla \theta, \dot{\mathbf{F}}).$$

Moreover we let ψ be differentiable.

To derive the restrictions placed by (8.23) it seems convenient to use the component form. To save writing we let

$$F_{iKH} = \partial_{X_K} \partial_{X_H} \chi_i, \quad F_{iKHR} = \partial_{X_K} \partial_{X_H} \partial_{X_R} \chi_i, \quad F_{iKHRS} = \partial_{X_K} \partial_{X_H} \partial_{X_R} \partial_{X_S} \chi_i$$

and the analogue for $\dot{F}_{iKH}$, $\dot{F}_{iKHR}$. Further let $\mathbf{g} = \nabla \theta$, $\mathbf{G} = \nabla_R \theta$. Since $\nabla = \mathbf{F}^{-T} \nabla_R$ we observe

$$g_i = F^{-1}_{Ki} G_K, \qquad \nabla \cdot \mathbf{u} = F^{-1}_{Ki} \partial_{X_K} u_i, \qquad \mathbf{q} \cdot \nabla \theta = F^{-1}_{Ki} q_i G_K$$

and hence, from $\mathbf{u} = \mathbf{u}(\mathbf{F}, \theta, \nabla_R \mathbf{F}, \nabla_R^2 \mathbf{F}, \nabla \theta, \dot{\mathbf{F}})$, we obtain

$$\begin{aligned} \partial_{X_K} u_i = {} & \partial_{F_{pQ}} u_i F_{pQK} + \partial_\theta u_i G_K + \partial_{F_{pQR}} u_i F_{pQRK} + \partial_{F_{pQRS}} u_i F_{pQRSK} \\ & + \partial_{g_p} u_i \partial_{X_K} (G_H F^{-1}_{Hp}) + \partial_{\dot{F}_{pQ}} u_i \dot{F}_{pQK}. \end{aligned}$$

Moreover we have

$$\partial_{X_K} (G_H F^{-1}_{Hp}) = (\partial_{X_K} \partial_{X_H} \theta) F^{-1}_{Hp} - G_H F^{-1}_{Hk} F_{kJK} F^{-1}_{Jp}.$$

Upon computation of $\dot{\psi}$ and $\nabla \cdot \mathbf{u}$ we can write inequality (8.23) in the form

$$(T_{ij} - \rho\partial_{F_{iK}}\psi F^T_{Kj})L_{ij} + (F^{-1}_{Ki}\partial_{\dot{F}_{jH}}u_i - \rho\partial_{F_{jHK}}\psi)\dot{F}_{jHK} - \rho\partial_{F_{iJHK}}\psi\dot{F}_{iJHK}$$
$$-\rho\partial_{g_i}\psi\,\dot{g}_i - \rho\partial_{\dot{F}_{iK}}\psi\ddot{F}_{iK} + [F^{-1}_{Hi}(\partial_\theta u_i - \frac{1}{\theta}q_i) - F^{-1}_{Ki}F^{-1}_{Hk}F^{-1}_{Jp}\partial_{g_p}u_i\,F_{kJK}]G_H$$
$$+F^{-1}_{Ki}\partial_{F_{PQ}}u_i\,F_{PQK} + F^{-1}_{Ki}\partial_{F_{PQR}}u_i\,F_{PQRK} + F^{-1}_{Ki}\partial_{F_{PQRS}}u_i\,F_{PQRSK}$$
$$-\rho(\partial_\theta\psi + \eta)\dot{\theta} + F^{-1}_{Ki}F^{-1}_{Hp}\partial_{g_p}u_i\,\partial_{X_K}\partial_{X_H}\theta \geq 0.$$

The arbitrariness of $\dot{\theta}$, $\nabla^2_R\dot{\mathbf{F}}$, $\dot{\overline{\nabla\theta}}$, $\ddot{\mathbf{F}}$, $\nabla^3_R\mathbf{F}$ implies

$$\eta = -\partial_\theta\psi,$$

$$\partial_{\nabla^2_R\mathbf{F}}\psi = \mathbf{0}, \quad \partial_{\nabla\theta}\psi = \mathbf{0}, \quad \partial_{\dot{\mathbf{F}}}\psi = \mathbf{0}, \quad \partial_{\nabla^2_R\mathbf{F}}\mathbf{u} = \mathbf{0}.$$

Moreover, the occurrence of $\nabla_R\dot{\mathbf{F}}$ implies

$$\mathbf{F}^{-1}\partial_{\dot{\mathbf{F}}}\mathbf{u} = \rho\partial_{\nabla_R\mathbf{F}}\psi. \tag{8.24}$$

Consequently, the free energy is required to be independent of $\nabla^2_R\mathbf{F}$, $\nabla\theta$, $\dot{\mathbf{F}}$ so that

$$\psi = \psi(\mathbf{F}, \theta, \nabla_R\mathbf{F}).$$

Relative to thermoelasticity, spatial interaction is allowed via the (admissible) dependence on $\nabla_R\mathbf{F}$. Meanwhile the entropy η is related to the free energy ψ as it happens in thermoelasticity.

The arbitrariness of $\nabla^2_R\theta \in \mathrm{Sym}$ implies that

$$F^{-1}_{Ki}F^{-1}_{Hp}\partial_{g_p}u_i \in \mathrm{Skw}.$$

The skew symmetry relative to K, H

$$F^{-1}_{Ki}F^{-1}_{Hp}\partial_{g_p}u_i = -F^{-1}_{Hi}F^{-1}_{Kp}\partial_{g_p}u_i$$

in turn implies that

$$\partial_{g_p}u_i = -\partial_{g_i}u_p, \qquad \partial_{\nabla\theta}\mathbf{u} \in \mathrm{Skw}.$$

Thus, since

$$F^{-1}_{Ki}F^{-1}_{Jp}F_{kJK} = F^{-1}_{Ji}F^{-1}_{Kp}F_{kJK}$$

then

$$F^{-1}_{Ki}F^{-1}_{Jp}\partial_{g_p}u_i\,F_{kJK} = 0, \quad \forall k = 1, 2, 3.$$

In view of (8.24), the vector function $\mathbf{u}$ satisfies

$$\partial_{\dot{\mathbf{F}}}\mathbf{u} = \rho\mathbf{F}\partial_{\nabla_R\mathbf{F}}\psi =: \mathbf{V}$$

where $\mathbf{V} = \mathbf{V}(\mathbf{F}, \theta, \nabla_R \mathbf{F})$ is a third-order tensor. The obvious integration leads to

$$\mathbf{u} = \mathbf{V}\dot{\mathbf{F}} + \hat{\mathbf{u}}(\mathbf{F}, \theta, \nabla_R \mathbf{F}, \nabla\theta),$$

where $(\mathbf{V}\dot{\mathbf{F}})_i = V_{ijH}\dot{F}_{jH}$. The reduced inequality reads

$$(T_{ij} - \rho\partial_{F_{iK}}\psi F^T_{Kj})L_{ij} + F^{-1}_{Hi}(\partial_\theta u_i - \frac{1}{\theta}q_i)G_H + F^{-1}_{Ki}\partial_{F_{PQ}}u_i\, F_{PQK} + F^{-1}_{Ki}\partial_{F_{PQR}}u_i\, F_{PQRK} \geq 0.$$

Simple non-trivial models arise by considering the following sufficient conditions for the validity of the second law. We let

$$\hat{\mathbf{u}} = \mathbf{0}, \qquad \mathbf{V} = \mathbf{V}(\mathbf{F}, \theta).$$

Hence the inequality simplifies to

$$(T_{ij} - \rho\partial_{F_{iK}}\psi F^T_{Kj})\dot{F}_{iH}F^{-1}_{Hj} + (\partial_\theta V_{ijK}\dot{F}_{jK} - \frac{1}{\theta}q_i)g_i + F^{-1}_{Ki}(\partial_{F_{PQ}}V_{ijH})F_{PQK}\dot{F}_{jH} \geq 0.$$

The occurrence of $\dot{F}_{jK}g_i$ allows for cross-coupling terms. Let

$$A_{iH} := (T_{ij} - \rho\partial_{F_{iK}}\psi F^T_{Kj})F^{-1}_{Hj} + F^{-1}_{Kj}(\partial_{F_{PQ}}V_{jiH})F_{PQK}.$$

The reduced inequality can be written in the form

$$A_{iH}\dot{F}_{iH} + \partial_\theta V_{ijK}\dot{F}_{jK}g_i - \frac{1}{\theta}q_i g_i \geq 0. \tag{8.25}$$

Inequality (8.25) is the thermodynamic restriction for the functions $\mathbf{T}, \mathbf{V}, \mathbf{q}$.

The stress $\mathbf{T}$ and the heat flux $\mathbf{q}$ are, so far, functions of the whole set of variables Γ. For definiteness, we now establish two solutions of inequality (8.25). First let $\mathbf{T}$ be independent of $\mathbf{g}$. Hence we may set $\mathbf{g} = \mathbf{0}$ and then the left hand side of the inequality reduces to

$$A_{iH}\dot{F}_{iH} \geq 0.$$

Since $\mathbf{T}$, and hence $\mathbf{A}$, may depend on $\dot{\mathbf{F}}$ then the inequality holds if

$$A_{iH} = B_{iHjK}\dot{F}_{jK}$$

where $\mathbf{B}$ is a fourth-order positive semi-definite tensor. Thus we have

$$T_{ij} = \rho\partial_{F_{iK}}\psi F^T_{Kj} - F^{-1}_{Kh}(\partial_{F_{PQ}}V_{hiH})F_{PQK}F_{jH} + F_{jH}B_{iHhK}\dot{F}_{hK}. \tag{8.26}$$

The reduced inequality then involves the heat flux $\mathbf{q}$,

$$(\partial_\theta V_{ijK}\dot{F}_{jK} - \frac{1}{\theta}q_i)g_i \geq 0$$

whence

$$q_i = \theta\partial_\theta V_{ijK}\dot{F}_{jK} - \kappa g_i,$$

κ being a non-negative function of Γ.

Otherwise, if $\mathbf{T}$ is allowed to depend (linearly) on $\mathbf{g} = \nabla\theta$ then a possible restriction is given by the pair of conditions

$$(A_{iH} + \partial_\theta V_{jiH}g_j)\dot{F}_{iH} \geq 0, \qquad q_i g_i \leq 0.$$

The first condition results in

$$T_{ij} = \rho\partial_{F_{iK}}\psi F^T_{Kj} - F^{-1}_{Ki}(\partial_{F_{pQ}}V_{jiH})F_{pQK}F_{jH} - F_{jH}\partial_\theta V_{kiH}g_k + F_{jH}B_{iHhK}\dot{F}_{hK}. \tag{8.27}$$

while the second one is the classical heat conduction inequality.

It is of interest that the occurrence of a non-zero tensor $\mathbf{V}$ is necessary to have the stress dependent on the second deformation gradient $\nabla_R\mathbf{F}$. If $\mathbf{V} = \mathbf{0}$ then, by (8.24), $\partial_{\nabla_R\mathbf{F}}\psi = \mathbf{0}$ and hence ψ depends only on $\mathbf{F}$ and θ. Accordingly, if $\mathbf{V} = \mathbf{0}$ then both (8.26) and (8.27) reduce to the stress of thermoelasticity.

The dependence of $\mathbf{T}$ and $\mathbf{u}$ on $\dot{\mathbf{F}}$ has to comply with objectivity. Consider $u_i = V_{ijH}\dot{F}_{jH}$ and, more conveniently the referential analogue

$$\mathbf{u}_R = J\mathbf{F}^{-1}\mathbf{u}, \qquad (\mathbf{u}_R)_K = V_{KjH}\dot{F}_{jH}, \qquad V_{KjH} = JF^{-1}_{Ki}V_{KjH}.$$

Under a change of frame we have

$$(V_{KjH}\dot{F}_{jH})^* = Q_{jp}V_{KpH}\overline{Q_{jh}F_{hH}}^{\cdot} = Q_{jp}V_{KpH}(\dot{Q}_{jh}F_{hH} + Q_{jh}\dot{F}_{hH}) = V_{KhH}\dot{F}_{hH} + (Q^T\dot{Q})_{ph}V_{KpH}F_{hH}.$$

The skew property of $Q^T\dot{Q}$ implies that $\mathbf{u}_R$ is invariant if and only if $\mathbf{V}$ satisfies

$$V_{KpH}F_{hH} = V_{KhH}F_{pH}.$$

8.9.1 *Modelling via the Extra-Entropy Flux*

It might seem more natural to keep the balance of energy unchanged without any supply of energy via the interstitial working and to use the statement of the second law by letting a possible extra-entropy flux occur. An objection against the use of the interstitial working is that this term occurs in the balance of energy but is not considered in connection with the entropy flux. We now review the modelling of second gradient materials by merely allowing for the occurrence of an extra-entropy flux in the statement of the second law.

The second law inequality is considered in the general form

$$\rho\dot{\eta} + \nabla\cdot(\frac{\mathbf{q}}{\theta} + \mathbf{k}) - \frac{\rho r}{\theta} \geq 0$$

while the balance of energy is given by

$$\rho\dot{\varepsilon} = \mathbf{T}\cdot\mathbf{L} - \nabla\cdot\mathbf{q} + \rho r.$$

Hence we obtain

$$-\rho(\dot{\psi} + \eta\dot{\theta}) + \mathbf{T}\cdot\mathbf{L} + \theta\nabla\cdot\mathbf{k} - \frac{1}{\theta}\mathbf{q}\cdot\nabla\theta \geq 0.$$

Relative to (8.23) the formal difference is the occurrence of $\theta\nabla\cdot\mathbf{k}$ instead of $\nabla\cdot\mathbf{u}$ in the Clausius-Duhem inequality. Hence the two schemes are not strictly equivalent unless θ is constant.

As before we let $\Gamma = (\mathbf{F}, \theta, \nabla_{\!R}\,\mathbf{F}, \nabla_{\!R}^2\,\mathbf{F}, \nabla\theta, \dot{\mathbf{F}})$ be the set of independent variables. Upon evaluation of $\dot{\psi}$ and $\nabla\cdot\mathbf{k}$ we can write the inequality in the form

$$\begin{aligned}
-\rho(\partial_\theta\psi + \eta)\dot{\theta} &+ (T_{ij} - \rho\partial_{F_{iK}}\psi F^T_{Kj})L_{ij} + (\theta F^{-1}_{Ki}\partial_{\dot{F}_{jH}}k_i - \rho\partial_{F_{jHK}}\psi)\dot{F}_{jHK} \\
&-\rho\partial_{F_{iJHK}}\psi\dot{F}_{iJHK} - \rho\partial_{g_i}\psi\,\dot{g}_i - \rho\partial_{\dot{F}_{iK}}\psi\ddot{F}_{iK} + \theta F^{-1}_{Ki}\partial_{F_{pQ}}k_i\,F_{pQK} \\
+\theta F^{-1}_{Ki}\partial_{F_{pQR}}k_i\,F_{pQRK} &+ \theta F^{-1}_{Ki}\partial_{F_{pQRS}}k_i\,F_{pQRSK} + \theta F^{-1}_{Ki}F^{-1}_{Hp}\partial_{g_p}k_i\,\partial_{X_K}\partial_{X_H}\theta \\
&+[F^{-1}_{Hi}(\theta\partial_\theta k_i - \frac{1}{\theta}q_i) - \theta F^{-1}_{Ki}F^{-1}_{Hh}F^{-1}_{Jp}\partial_{g_p}k_i\,F_{hJK}]\partial_{X_H}\theta \geq 0
\end{aligned}$$

By paralleling the previous procedure we obtain the following analogous results. By the linearity of $\dot{\theta}$, $\nabla_{\!R}^2\,\dot{\mathbf{F}}$, $\dot{\overline{\nabla\theta}}$, $\ddot{\mathbf{F}}$, $\nabla_{\!R}^3\,\mathbf{F}$ it follows

$$\eta = -\partial_\theta\psi,$$

$$\partial_{\nabla_{\!R}^2\mathbf{F}}\psi = \mathbf{0}, \quad \partial_{\nabla\theta}\psi = \mathbf{0}, \quad \partial_{\dot{\mathbf{F}}}\psi = \mathbf{0}, \quad \partial_{\nabla_{\!R}^2\mathbf{F}}\mathbf{k} = \mathbf{0}.$$

Now, the linearity of $\nabla_{\!R}\,\dot{\mathbf{F}}$ implies

$$\partial_{\dot{F}_{jH}}\mathbf{k}_i = \frac{\rho}{\theta}\partial_{F_{jHK}}\psi F_{iK}. \tag{8.28}$$

Integration of (8.28) yields

$$\mathbf{k} = \mathbf{K}(\mathbf{F}, \theta, \nabla_{\!R}\mathbf{F})\dot{\mathbf{F}} + \hat{\mathbf{k}}(\mathbf{F}, \theta, \nabla_{\!R}\mathbf{F}, \nabla\theta), \qquad K_{ijH} = \frac{\rho}{\theta}F_{iK}\partial_{F_{jHK}}\psi.$$

By the linearity of $\partial_{X_K}\partial_{X_H}\theta \in \mathrm{Sym}$ we have

$$F_{Ki}^{-1} F_{Hp}^{-1} \partial_{g_p} k_i \in \text{Skw},$$

whence

$$\partial_{\nabla\theta} \mathbf{k} \in \text{Skw}, \qquad F_{Ki}^{-1} F_{Hp}^{-1} \partial_{g_p} k_i F_{hHK} = 0.$$

The reduced inequality takes the form

$$(T_{ij} - \rho \partial_{F_{iK}} \psi F_{Kj}^T) L_{ij} + F_{Hi}^{-1} (\theta \partial_\theta k_i - \frac{1}{\theta} q_i) G_H + \theta F_{Ki}^{-1} \partial_{F_{PQ}} k_i \, F_{PQK} + \theta F_{Ki}^{-1} \partial_{F_{PQR}} k_i \, F_{PQRK} \geq 0.$$

We now examine the consequences of the assumption

$$\hat{\mathbf{k}} = \mathbf{0}, \qquad \mathbf{K} = \mathbf{K}(\mathbf{F}, \theta).$$

The inequality takes the form

$$A_{iH} \dot{F}_{iH} + \theta \, \partial_\theta K_{jiH} \, \dot{F}_{iH} \, g_j - \frac{1}{\theta} q_i g_i \geq 0,$$

where $A_{iH} = (T_{ij} - \rho \partial_{F_{iK}} \psi F_{Kj}^T) F_{Hj}^{-1} + \theta F_{Kh}^{-1} \partial_{F_{PQ}} K_{hiH} F_{PQK}$. The functions obtained within the scheme of the interstitial working occur here too with the formal replacement of $\partial_\theta \mathbf{V}$ with $\theta \partial_\theta \mathbf{K}$. The inequality holds if

$$A_{iH} = B_{iHjQ} \dot{F}_{jQ}, \qquad \theta^2 \partial_\theta K_{jiH} \, \dot{F}_{iH} - q_j = \kappa g_j$$

where B is a fourth-order non-negative tensor and κ is a non-negative scalar (or second-order tensor). Hence we have

$$T_{ij} = \rho \partial_{F_{iQ}} \psi F_{Qj}^T - \theta \, F_{Rh}^{-1} (\partial_{F_{PQ}} K_{hiH}) F_{PQR} F_{jH} + F_{jH} B_{iHPQ} \dot{F}_{PQ}$$

and

$$q_i = \theta^2 \partial_\theta K_{ijR} \dot{F}_{jR} - \kappa g_i.$$

Otherwise we can obtain

$$T_{ij} = \rho \partial_{F_{iQ}} \psi F_{Qj}^T - \theta \, F_{Rh}^{-1} (\partial_{F_{PQ}} K_{hiH}) F_{PQR} F_{jH} - \theta F_{jH} \partial_\theta K_{kiH} g_k + F_{jH} B_{iHPQ} \dot{F}_{PQ}$$

and Fourier's law for $\mathbf{q}$.

8.10 Materials of Korteweg Type

Paralleling the analysis of the previous section, we now consider (fluid) materials of Korteweg type. Hence our purpose is to see whether and how a stress function of the form

$$\mathbf{T} = \mathbf{T}(\rho, \theta, \nabla\rho, \nabla^2\rho, \nabla\theta, \mathbf{D})$$

is consistent with the principles of continuum thermodynamics; ∇^2 stands for $\nabla\nabla$. The dependence on $\nabla\theta$ and $\mathbf{D}$ means that heat conduction and viscosity are considered.

In view of the continuity equation $\dot{\rho} + \rho\nabla \cdot \mathbf{v} = 0$, the scalars $\dot{\rho}$, ρ, $\nabla \cdot \mathbf{v}$ are not linearly independent. For technical convenience we choose ρ and $\dot{\rho}$, but not $\nabla \cdot \mathbf{v}$, as independent quantities. Hence since $\mathbf{D} = \mathbf{D}_0 + \frac{1}{3}(\nabla \cdot \mathbf{v})\mathbf{1}$ a possible dependence on $\mathbf{D}$ is restricted to $\mathbf{D}_0$. We then let

$$\Gamma = (\rho, \theta, \nabla\rho, \nabla^2\rho, \nabla\theta, \dot{\rho}, \mathbf{D}_0)$$

be the set of independent variables and assume ψ, η, $\mathbf{T}$, $\mathbf{q}$, $\mathbf{k}$ are functions of Γ while ψ and $\mathbf{k}$ are differentiable. We let $\mathbf{T}$ be in the form

$$\mathbf{T} = -p(\rho, \theta)\mathbf{1} + \boldsymbol{T}.$$

Upon computation of $\dot{\psi}$ and $\nabla \cdot \mathbf{k}$ and substitution in the Clausius-Duhem inequality

$$-\rho(\dot{\psi} + \eta\dot{\theta}) + \mathbf{T} \cdot \mathbf{D} + \theta\nabla \cdot \mathbf{k} - \frac{1}{\theta}\mathbf{q} \cdot \nabla\theta \geq 0$$

we obtain

$$\begin{aligned} -\rho(\partial_\theta\psi + \eta)\dot{\theta} + (-\rho\partial_\rho\psi + p/\rho - \mathrm{tr}\,\boldsymbol{T}/3\rho)\dot{\rho} + \boldsymbol{T} \cdot \mathbf{L}_0 - \rho\partial_{\nabla\rho}\psi \cdot \dot{\overline{\nabla\rho}} - \rho\partial_{\nabla^2\rho}\psi \cdot \dot{\overline{\nabla^2\rho}} \\ -\rho\partial_{\nabla\theta}\psi \cdot \dot{\overline{\nabla\theta}} - \rho\partial_{\dot{\rho}}\psi\,\ddot{\rho} - \rho\partial_{\mathbf{D}_0}\psi \cdot \dot{\mathbf{D}}_0 + \theta\partial_\rho\mathbf{k} \cdot \nabla\rho + \theta\partial_{\nabla\rho}\mathbf{k} \cdot \nabla^2\rho \\ +\theta\partial_{\nabla^2\rho}\mathbf{k} \cdot \nabla^3\rho + \theta\partial_{\nabla\theta}\mathbf{k} \cdot \nabla^2\theta + \theta\partial_{\dot{\rho}}\mathbf{k} \cdot \nabla\dot{\rho} + \theta\partial_{\mathbf{D}_0}\mathbf{k} \cdot \nabla\mathbf{D}_0 + (\theta\partial_\theta\mathbf{k} - \frac{1}{\theta}\mathbf{q}) \cdot \nabla\theta \geq 0. \end{aligned}$$

We first observe that $\ddot{\rho}$, $\nabla^2\theta$, $\nabla^3\rho$ occur linearly. Their arbitrariness then implies that

$$\partial_{\dot{\rho}}\psi = 0, \quad \partial_{\nabla\theta}\mathbf{k} \in \mathrm{Skw}, \quad \partial_{\nabla^2\rho}\mathbf{k} \in \mathrm{Skw};$$

For simplicity we assume $\partial_{\nabla\theta}\mathbf{k} = \mathbf{0}$, $\partial_{\nabla^2\rho}\mathbf{k} = \mathbf{0}$. The inequality then reduces to

$$\begin{aligned} -\rho(\partial_\theta\psi + \eta)\dot{\theta} + (-\rho\partial_\rho\psi + p/\rho - \mathrm{tr}\,\boldsymbol{T}/3\rho)\dot{\rho} + \boldsymbol{T} \cdot \mathbf{L}_0 - \rho\partial_{\nabla\rho}\psi \cdot \dot{\overline{\nabla\rho}} \\ -\rho\partial_{\nabla^2\rho}\psi \cdot \dot{\overline{\nabla^2\rho}} - \rho\partial_{\nabla\theta}\psi \cdot \dot{\overline{\nabla\theta}} - \rho\partial_{\mathbf{D}_0}\psi \cdot \dot{\mathbf{D}}_0 + \theta\partial_\rho\mathbf{k} \cdot \nabla\rho \\ +\theta\partial_{\nabla\rho}\mathbf{k} \cdot \nabla^2\rho + \theta\partial_{\dot{\rho}}\mathbf{k} \cdot \nabla\dot{\rho} + \theta\partial_{\mathbf{D}_0}\mathbf{k} \cdot \nabla\mathbf{D}_0 + (\theta\partial_\theta\mathbf{k} - \frac{1}{\theta}\mathbf{q}) \cdot \nabla\theta \geq 0. \end{aligned}$$

To determine the extent of arbitrary quantities in the reduced inequality we need to exploit some identities. Since

$$\dot{\overline{\nabla\theta}} = \nabla\dot{\theta} - \mathbf{L}^T\nabla\theta$$

then the arbitrariness of $\nabla\dot{\theta}$ implies

$$\partial_{\nabla\theta}\psi = \mathbf{0}.$$

Moreover the arbitrariness of $\dot{\theta}$ implies the standard relation

$$\eta = -\partial_\theta \psi.$$

The identity $\dot{\overline{\nabla\rho}} = \nabla\dot{\rho} - \mathbf{L}^T\nabla\rho$ allows us to write

$$-\rho\partial_{\nabla_\rho}\psi \cdot \dot{\overline{\nabla\rho}} + \theta\partial_{\dot{\rho}}\mathbf{k}\cdot\nabla\dot{\rho} = (\theta\partial_{\dot{\rho}}\mathbf{k} - \rho\partial_{\nabla_\rho}\psi)\cdot\nabla\dot{\rho} + \rho\nabla\rho\otimes\partial_{\nabla_\rho}\psi\cdot\mathbf{L}.$$

The arbitrariness of $\nabla\dot{\rho}$ implies

$$\theta\partial_{\dot{\rho}}\mathbf{k} - \rho\partial_{\nabla_\rho}\psi = 0 \tag{8.29}$$

Observe that

$$\begin{aligned}\rho\nabla\rho\otimes\partial_{\nabla_\rho}\psi\cdot\mathbf{L} &= \rho\nabla\rho\otimes\partial_{\nabla_\rho}\psi\cdot\mathbf{L}_0 + \tfrac{1}{3}\rho\nabla\rho\cdot\partial_{\nabla_\rho}\psi\nabla\cdot\mathbf{v}\\ &= \rho\nabla\rho\otimes\partial_{\nabla_\rho}\psi\cdot\mathbf{L}_0 - \tfrac{1}{3}\nabla\rho\cdot\partial_{\nabla_\rho}\psi\dot{\rho}.\end{aligned}$$

Further we apply the identity[6]

$$\dot{\overline{\nabla^2\rho}} = \nabla^2\dot{\rho} - \nabla\rho\nabla^2\mathbf{v} - 2\,\mathrm{sym}\,\nabla^2\rho\,\mathbf{L}.$$

The arbitrariness of $\nabla^2\dot{\rho}$ and $\dot{\mathbf{D}}_0$ implies

$$\partial_{\nabla^2\rho}\psi = \mathbf{0}, \qquad \partial_{\nabla^2\mathbf{v}}\psi = \mathbf{0}, \qquad \partial_{\mathbf{L}_0}\psi = \mathbf{0}.$$

Consequently we obtain the reduced inequality

$$\begin{aligned}(-\rho\partial_\rho\psi + p/\rho - \mathrm{tr}\,\boldsymbol{T}/3\rho + \tfrac{1}{3}\nabla\rho\cdot\partial_{\nabla_\rho}\psi])\dot{\rho} + \boldsymbol{T}\cdot\mathbf{L}_0 - \rho\nabla\rho\otimes\partial_{\nabla_\rho}\psi\cdot\mathbf{L}_0\\ +\theta\partial_\rho\mathbf{k}\cdot\nabla\rho + \theta\partial_{\nabla_\rho}\mathbf{k}\cdot\nabla^2\rho + \theta\partial_{\mathbf{L}_0}\mathbf{k}\cdot\nabla\mathbf{L}_0 + (\theta\partial_\theta\mathbf{k} - \frac{1}{\theta}\mathbf{q})\cdot\nabla\theta \geq 0.\end{aligned} \tag{8.30}$$

As for $\rho\nabla\rho\otimes\partial_{\nabla_\rho}\psi\cdot\mathbf{L}_0$ in (8.30) observe that

$$\rho\nabla\rho\otimes\partial_{\nabla_\rho}\psi\cdot\mathbf{L}_0 = \rho\nabla\rho\otimes\partial_{\nabla_\rho}\psi\cdot\mathbf{D}_0 + \rho\nabla\rho\otimes\partial_{\nabla_\rho}\psi\cdot\mathbf{W}.$$

The arbitrariness of $\mathbf{W}$ implies

[6] In suffix notation $\dot{\overline{\partial_{x_i}\partial_{x_j}\rho}} = \partial_{x_i}\partial_{x_j}\dot{\rho} - (\partial_{x_k}\rho)\partial_{x_i}\partial_{x_j}v_k - (\partial_{x_i}\partial_{x_k}\rho)\partial_{x_j}v_k - (\partial_{x_j}\partial_{x_k}\rho)\partial_{x_i}v_k$.

$$\nabla\rho \otimes \partial_{\nabla\rho}\psi \in \mathrm{Sym}.$$

Hence it follows

$$\partial_{\nabla\rho}\psi = \alpha\nabla\rho.$$

Since ψ is a function of $\rho, \theta, \nabla\rho$ it follows that the dependence of ψ and α on $\nabla\rho$ is through $(\nabla\rho)^2$. As a further consequence, integration of (8.29) yields

$$\mathbf{k} = \frac{\rho}{\theta}\alpha(\rho, \theta, (\nabla\rho)^2)\nabla\rho\,\dot{\rho} + \tilde{\mathbf{k}}(\rho, \theta, \mathbf{D}_0). \tag{8.31}$$

Incidentally, the absence of $\mathbf{W}$ in the entropy inequality is consistent with objectivity.

In summary, compatibility with thermodynamics requires that $\psi = \psi(\rho, \theta, (\nabla\rho)^2)$, $\mathbf{k}$ is given by (8.31) and $\mathbf{T}, \mathbf{q}, \mathbf{k}, \psi$ are subject to

$$\begin{aligned}&(-\rho\partial_\rho\psi + p/\rho - \mathrm{tr}\,\mathcal{T}/3\rho + \tfrac{1}{3}\nabla\rho\cdot\partial_{\nabla\rho}\psi)\dot{\rho} + \mathcal{T} - \rho\nabla\rho\otimes\partial_{\nabla\rho}\psi)\cdot\mathbf{D}_0\\ &\quad+\theta\partial_\rho\mathbf{k}\cdot\nabla\rho + \theta\partial_{\nabla\rho}\mathbf{k}\cdot\nabla^2\rho + \theta\partial_{\mathbf{D}_0}\mathbf{k}\cdot\nabla\mathbf{D}_0 + (\theta\partial_\theta\mathbf{k} - \frac{1}{\theta}\mathbf{q})\cdot\nabla\theta \geq 0.\end{aligned}\tag{8.32}$$

For definiteness, we now establish a simple model based on the assumptions

$$\tilde{\mathbf{k}} = \mathbf{0}, \qquad \partial_{\nabla\theta}\mathcal{T} = \mathbf{0}, \qquad \partial_{\mathbf{D}_0}\mathrm{tr}\,\mathcal{T} = \mathbf{0},$$

and

$$\mathrm{tr}\,\mathcal{T} \to 0 \quad \text{as} \quad \nabla\cdot\mathbf{v} \to 0.$$

Since $\mathcal{T}$ and $\mathbf{k}$ (and ψ) are independent of $\nabla\theta$, then without any restriction on $\mathcal{T}$ and $\mathbf{k}$ we can examine (8.32) as $\nabla\theta = 0$ so that

$$(-\rho\partial_\rho\psi + p/\rho - \mathrm{tr}\,\mathcal{T}/3\rho + \tfrac{1}{3}\nabla\rho\cdot\partial_{\nabla\rho}\psi)\dot{\rho} + (\mathcal{T} - \rho\nabla\rho\otimes\partial_{\nabla\rho}\psi)\cdot\mathbf{D}_0 + \theta\partial_\rho\mathbf{k}\cdot\nabla\rho + \theta\partial_{\nabla\rho}\mathbf{k}\cdot\nabla^2\rho \geq 0.$$

Now observe

$$\partial_\rho\mathbf{k}\cdot\nabla\rho = \frac{1}{\theta}\beta(\nabla\rho)^2\dot{\rho}, \qquad \beta := \alpha + \rho\partial_\rho\alpha,$$

$$\partial_{\nabla\rho}\mathbf{k}\cdot\nabla^2\rho = \frac{\rho\dot{\rho}}{\theta}[\alpha\Delta\rho + \gamma\nabla\rho\otimes\nabla\rho\cdot\nabla^2\rho], \qquad \gamma := \partial_{(\nabla\rho)^2}\alpha.$$

The inequality then reads

$$\begin{aligned}&[-\rho\partial_\rho\psi + \frac{p}{\rho} - \frac{1}{3\rho}\mathrm{tr}\,\check{\mathbf{T}} + (\beta + \tfrac{1}{3}\alpha)(\nabla\rho)^2 + \rho\alpha\Delta\rho + \rho\gamma\nabla\rho\otimes\nabla\rho\cdot\nabla^2\rho]\dot{\rho}\\ &\qquad\qquad+(\check{\mathbf{T}} - \alpha\rho\nabla\rho\otimes\nabla\rho)\cdot\mathbf{D}_0 \geq 0\end{aligned}$$

The mutual independence of $\dot{\rho}$ and $\mathbf{D}_0$ implies that the two lines of the inequality are non-negative. Now, the assumption that $\mathrm{tr}\,\mathcal{T} \to 0$ as $\nabla\cdot\mathbf{v} \to 0$ (or $\dot{\rho} \to 0$) implies that the factor of $\dot\rho$, except tr $\mathcal{T}$, has to vanish. Hence we obtain the pressure p in the form

$$p = \bar p - \tfrac{1}{3}\alpha\,\rho(\nabla\rho)^2, \qquad \bar p := \rho^2\partial_\rho\psi - \beta\,\rho(\nabla\rho)^2 - \rho^2\alpha\Delta\rho - \rho^2\gamma(\nabla\rho\otimes\nabla\rho)\cdot\nabla^2\rho. \tag{8.33}$$

The inequalities

$$\mathrm{tr}\,\mathcal{T}\,\dot\rho \le 0, \qquad (\mathcal{T} - \alpha\rho\nabla\rho\otimes\nabla\rho)\cdot\mathbf{D}_0 \ge 0$$

imply

$$\tfrac{1}{3}\mathcal{T} = \xi\rho\dot\rho = \xi\nabla\cdot\mathbf{v}, \qquad \xi \ge 0,$$

$$(\mathcal{T} - \alpha\rho\nabla\rho\otimes\nabla\rho)_0 = 2\mu\mathbf{D}_0, \qquad \mu \ge 0,$$

where $(\;)_0$ denotes the deviator. Hence we have

$$\mathbf{T} = -p\mathbf{1} + \alpha\rho\nabla\rho\otimes\nabla\rho + \xi\,\mathrm{tr}\,\mathbf{D}\,\mathbf{1} + 2\mu\,\mathbf{D}_0. \tag{8.34}$$

The inequality involving the heat flux $\mathbf{q}$,

$$(\theta\partial_\theta\mathbf{k} - \frac{1}{\theta}\mathbf{q})\cdot\nabla\theta \ge 0$$

implies

$$\mathbf{q} = \theta^2\partial_\theta\mathbf{k} - \kappa\nabla\theta, \qquad \kappa \ge 0,$$

κ being a function of Γ. Since

$$\partial_\theta\mathbf{k} = \frac{\rho^2}{\theta^2}(\alpha - \theta\partial_\theta\alpha)\nabla\cdot\mathbf{v}\,\nabla\rho = -\frac{\rho}{\theta^2}(\alpha - \theta\partial_\theta\alpha)\dot\rho\,\nabla\rho$$

it follows

$$\mathbf{q} = (\alpha - \theta\partial_\theta\alpha)\rho^2\,\nabla\cdot\mathbf{v}\,\nabla\rho - \kappa\nabla\theta; \tag{8.35}$$

the scalar κ might be replaced with a non-negative second-order tensor.

These results deserve some comments. As for the pressure p, given by (8.33), the term $\rho^2\partial_\rho\psi$ is the classical (equilibrium) pressure of simple fluids while the other terms show the effects of the non-simple model described by gradients of the mass density. In addition, $\alpha\rho\nabla\rho\otimes\nabla\rho$ shows the anisotropic stress due to $\nabla\rho$. Of course $\xi\,\mathrm{tr}\,\mathbf{D}\,\mathbf{1} + 2\mu\,\mathbf{D}_0$ is the standard viscous Cauchy stress. Likewise, the heat flux is the sum of the classical Fourier term $-\kappa\nabla\theta$ and the term $(\alpha - \theta\partial_\theta\alpha)\rho^2\,\nabla\cdot\mathbf{v}\,\nabla\rho$ again induced by the gradient of the mass density.

It is of interest that a dual model, consistent with (8.32), holds with a contribution to $\mathbf{T}$ induced by $\nabla\theta$ while $\mathbf{q}$ is merely parallel to $-\nabla\theta$. To find the detailed expressions assume

$$\tilde{\mathbf{k}} = \mathbf{0}, \qquad \partial_{\dot\rho}\mathbf{q} = \mathbf{0}, \qquad \partial_{\mathbf{D}_0}\mathbf{q} = \mathbf{0}, \qquad \partial_{\mathbf{D}_0}\mathrm{tr}\,\mathcal{T} = \mathbf{0},$$

and, again, let $\mathrm{tr}\,\mathcal{T} \to 0$ as $\nabla \cdot \mathbf{v} = 0$. Hence inequality (8.32) splits into

$$[-\rho\partial_\rho\psi + \frac{p}{\rho} - \frac{1}{3\rho}\mathrm{tr}\,\mathcal{T} + (\beta + \tfrac{1}{3}\alpha)(\nabla\rho)^2 + \rho\alpha\Delta\rho + \rho\gamma\nabla\rho\otimes\nabla\rho\cdot\nabla^2\rho$$
$$-\frac{\rho}{\theta}(\alpha - \theta\partial_\theta\alpha)\nabla\theta\cdot\nabla\rho]\dot\rho + (\mathcal{T} - \alpha\rho\nabla\rho\otimes\nabla\rho)\cdot\mathbf{D}_0 \geq 0$$

$$\mathbf{q}\cdot\nabla\theta \leq 0.$$

Since $\dot\rho \to 0$ implies $\mathrm{tr}\,\mathcal{T} \to 0$ it follows that

$$p = \tilde{p} - \tfrac{1}{3}\alpha\,\rho(\nabla\rho)^2,$$

$$\tilde{p} := \rho^2\partial_\rho\psi - \beta\,\rho(\nabla\rho)^2 - \rho^2\alpha\Delta\rho - \rho^2\gamma(\nabla\rho\otimes\nabla\rho)\cdot\nabla^2\rho + \frac{\rho^2}{\theta}(\alpha - \theta\partial_\theta\alpha)\nabla\theta\cdot\nabla\rho.$$

Now since $\mathrm{tr}\,\mathcal{T}$ is independent of $\mathbf{D}_0$ we let $\mathbf{D}_0 = \mathbf{0}$ and the inequality implies

$$\tfrac{1}{3}\mathrm{tr}\,\mathcal{T} = \xi\nabla\cdot\mathbf{v}, \qquad \xi \geq 0,$$

while

$$0 \leq (\mathcal{T} - \alpha\rho\nabla\rho\otimes\nabla\rho)\cdot\mathbf{D}_0 = (\mathcal{T} - \alpha\rho\nabla\rho\otimes\nabla\rho)_0\cdot\mathbf{D}_0$$

results in

$$(\mathcal{T} - \alpha\rho\nabla\rho\otimes\nabla\rho)_0 = 2\mu\mathbf{D}_0, \qquad \mu \geq 0.$$

Hence

$$\mathbf{T} = -\tilde{p}\,\mathbf{1} + \alpha\rho\nabla\rho\otimes\nabla\rho + \xi\nabla\cdot\mathbf{v}\,\mathbf{1} + 2\mu\mathbf{D}_0.$$

Meanwhile $\mathbf{q}$ merely satisfies

$$\mathbf{q} = -\kappa\nabla\theta, \qquad \kappa = \kappa(\rho, \theta, \nabla\rho, \nabla^2\rho, \nabla\theta) \geq 0.$$

Hence, by means of the same term $\theta\partial_\theta\mathbf{k}\cdot\nabla\theta$, two inequivalent models are obtained. In the first case it results in an energy flux $(\alpha - \theta\partial_\theta\alpha)\rho^2\,\nabla\cdot\mathbf{v}\,\nabla\rho$, in the second case it gives rise to an additional pressure $(\rho^2/\theta)(\alpha - \theta\partial_\theta\alpha)\nabla\theta\cdot\nabla\rho$ while $\mathbf{q} = -\kappa\nabla\theta$.

8.11 Hyperstress and Materials of Higher-Order Grade

Equations of equilibrium and boundary conditions for materials of grade 2 are given by Toupin [421, 422] through a principle of virtual work. Borrowing from the work of E. and F. Cosserat [107] on material with couple-stresses, Toupin allows for hypertractions on the boundary of the pertinent region. The work, per unit area, of the hypertraction **H** is taken to be $H_i \delta F_{iK} N_K$, **N** being the unit normal. Hence it follows that the stress tensor can depend on $\nabla_R \mathbf{F}$. Hence the character of material with a higher-order grade is associated with the existence of an hypertraction.

Since higher-order grade materials are incompatible with the standard theory of thermodynamics, much research in the literature has been developed by modifying the expression of the power expended on the pertinent region $\mathcal{P}_t$. The mechanical power, per unit volume, is $\mathbf{T} \cdot \mathbf{L}$. This is so because the power of the traction, per unit area, is $\mathbf{v} \cdot \mathbf{Tn}$; the divergence theorem implies

$$\int_{\partial \mathcal{P}_t} \mathbf{v} \cdot \mathbf{Tn} da = \int_{\mathcal{P}_t} [\mathbf{v} \cdot (\nabla \cdot \mathbf{T}) + \mathbf{T} \cdot \mathbf{L}] dv.$$

The power $\mathbf{v} \cdot (\nabla \cdot \mathbf{T})$ cancels by means of the equation of motion while $\mathbf{T} \cdot \mathbf{L}$ is the power affecting the evolution of the internal energy. Also by analogy with [422], the mechanical power is generalized [180] by including the quantity $\mathbb{G} \cdot \nabla^2 \mathbf{v}$ so that the power is[7]

$$\mathbf{T} \cdot \mathbf{L} + \mathbb{G} \cdot \nabla^2 \mathbf{v}.$$

The third-order tensor $\mathbb{G}$ is named *hyperstress*. Obvious consistency requirements indicate that the introduction of the additive power $\mathbb{G} \cdot \nabla^2 \mathbf{v}$ is allowed only if the other balance equations are revised properly.

While $\mathbf{T} \cdot \mathbf{L}$, or $\mathbf{T} \cdot \mathbf{D}$, is well recognized to be the stress power, the theory of thermodynamics is modified (or generalized) by letting the power include an analogous term $\mathbb{G} \cdot \nabla^2 \mathbf{v}$, linear in the second gradient $\nabla^2 \mathbf{v}$, again with $\mathbb{G}$ a third-order hyperstress. Further generalizations occur with higher-order hyperstresses or with the temperature so that a power $a\dot{\theta}$ may occur.[8]

The power $\mathbf{T} \cdot \mathbf{L} + \mathbb{G} \cdot \nabla^2 \mathbf{v}$ might be framed within a scheme of balance equations as follows [180]. For any part $\mathcal{P}_t$ we can apply the divergence theorem to obtain

$$\int_{\mathcal{P}_t} \mathbf{T} \cdot \mathbf{L} dv = - \int_{\mathcal{P}_t} (\nabla \cdot \mathbf{T}) \cdot \mathbf{v} \, dv + \int_{\partial \mathcal{P}_t} \mathbf{v} \cdot \mathbf{Tn} \, dv.$$

Since

$$G_{ijk} \partial_{x_j} \partial_{x_k} v_i = \partial_{x_k} (G_{ijk} \partial_{x_j} v_i) - \partial_{x_j} (\partial_{x_k} G_{ijk} v_i) + (\partial_{x_j} \partial_{x_k} G_{ijk}) v_i$$

by applying the divergence theorem it follows

[7] In components, $\mathbb{G} \cdot \nabla^2 \mathbf{v} = G_{ijk} v_{i,jk}$.

[8] See, e.g., [83].

$$\int_{\mathcal{P}_t} \mathbf{T} \cdot \mathbf{L} + \mathbb{G} \cdot \nabla^2 \mathbf{v}\, dv = \int_{\mathcal{P}_t} [\nabla \cdot (\nabla \mathbb{G} - \mathbf{T})] \cdot \mathbf{v}\, dv + \int_{\partial \mathcal{P}_t} \mathbb{G}\mathbf{n} \cdot \mathbf{L} + [\mathbf{T}\mathbf{n} - (\nabla \cdot \mathbb{G})\mathbf{n}] \cdot \mathbf{v}\, da.$$

Denote by

$$\mathbf{P} = \mathbf{1} - \mathbf{n} \otimes \mathbf{n}$$

the projection onto the plane with normal $\mathbf{n}$ and

$$\partial_n = \mathbf{n} \cdot \nabla, \qquad \nabla_S = \mathbf{P}\nabla = \nabla - \mathbf{n}(\mathbf{n} \cdot \nabla)$$

the normal derivative and the surface gradient (relative to the surface S). Hence

$$\nabla \mathbf{v} = \nabla_S \mathbf{v} + \partial_n \mathbf{v} \otimes \mathbf{n}, \qquad L_{ij} = \partial_{x_j} v_i = n_j \partial_n v_i + (\nabla_S \mathbf{v})_{ij}.$$

Consequently

$$\mathbb{G}\mathbf{n} \cdot \mathbf{L} = \mathbb{G}\mathbf{n} \cdot \nabla_S \mathbf{v} + \partial_n \mathbf{v} \cdot [(\mathbb{G}\mathbf{n})\mathbf{n}]$$

and hence

$$\int_{\mathcal{P}_t} \mathbf{T} \cdot \mathbf{L} + \mathbb{G} \cdot \nabla^2 \mathbf{v}\, dv = \int_{\mathcal{P}_t} [\nabla \cdot (\nabla \mathbb{G} - \mathbf{T})] \cdot \mathbf{v}\, dv + \int_{\partial \mathcal{P}_t} \{\mathbb{G}\mathbf{n} \cdot \nabla_S \mathbf{v} + \partial_n \mathbf{v} \cdot [(\mathbb{G}\mathbf{n})\mathbf{n}] + [\mathbf{T}\mathbf{n} - (\nabla \cdot \mathbb{G})\mathbf{n}] \cdot \mathbf{v}\} da.$$

This splitting of the power justifies the balance of linear momentum,

$$\rho \dot{\mathbf{v}} = \nabla(\mathbf{T} - \nabla \cdot \mathbb{G}) + \rho \mathbf{b},$$

and the boundary conditions,

$$\mathbf{t}_S = \mathbf{T}\mathbf{n} + \mathbb{G}\mathbf{n} \cdot \nabla_S \mathbf{v} - (\nabla \cdot \mathbb{G})\mathbf{n}, \qquad \mathbf{m}_S = (\mathbb{G}\mathbf{n})\mathbf{n},$$

$\mathbf{t}_S$ and $\mathbf{m}_S$ being the traction and the *hypertraction.* However the thermodynamic restrictions are derived from the *free-energy imbalance*

$$\rho \dot{\psi} - \mathbf{T} \cdot \mathbf{D} - \mathbb{G} \cdot \nabla^2 \mathbf{v} \leq 0.$$

An example of constitutive equations is given for incompressible fluids in the form

$$T_{ij} = 2\mu D_{ij} - p\delta_{ij}, \qquad G_{ijk} = \eta_1 \partial_{x_j} \partial_{x_k} v_i + \eta_2 (\partial_{x_i} \partial_{x_j} v_k + \partial_{x_i} \partial_{x_k} v_j - \partial_{x_r} \partial_{x_r} v_i \delta_{jk}) - \pi_k \delta_{ij},$$

where p is a pressure while π is a hyperpressure vector.

8.11.1 Solids of Higher-Order Grade

Here we show that non-local models, in space and time, are possible by merely allowing for a non-zero extra-entropy production. Hence models of higher-order grade are allowed without the introduction of ad hoc terms like hyperstress or extra-energy supplies. For notational simplicity the scheme is restricted to first grade but models of higher-order grades can be established by the same approach.

We generalize the model of Sect. 8.9.1 by letting

$$\Gamma = (\mathbf{F}, \theta, \nabla_{R} \mathbf{F}, \nabla_{R}^{2} \mathbf{F}, \nabla_{R} \theta, \dot{\mathbf{F}}, \nabla_{R} \dot{\mathbf{F}}, \dot{\theta})$$

be the set of independent variables. Having chosen the material gradients, ∇_R, it is convenient to follow systematically the referential description. The entropy inequality takes the form

$$-\rho_R(\dot{\psi} + \eta\dot{\theta}) + \mathbf{T}_R \cdot \dot{\mathbf{F}} + \theta \nabla_R \cdot \mathbf{k}_R - \frac{1}{\theta}\mathbf{q}_R \cdot \nabla_R \theta \geq 0,$$

where $\mathbf{T}_R$ is the first Piola stress and $\mathbf{k}_R = J\mathbf{k}\mathbf{F}^{-T}$, $\mathbf{q}_R = J\mathbf{q}\mathbf{F}^{-T}$.

Upon computation of $\dot{\psi}$ and $\nabla_R \cdot \mathbf{k}_R$ we can write the inequality in the form

$$\begin{aligned}
&-\rho_R(\partial_\theta \psi + \eta)\dot{\theta} + (T^R_{iK} - \rho_R \partial_{F_{iK}} \psi)\dot{F}_{iK} - \rho_R \partial_{F_{iHK}} \psi \dot{F}_{iHK} - \rho_R \partial_{F_{iHKJ}} \psi \dot{F}_{iHKJ} \\
&\quad -\rho_R \partial_{\partial X_K} \psi \partial_{X_K} \dot{\theta} - \rho_R \partial_{\dot{F}_{iK}} \psi \ddot{F}_{iK} - \rho_R \partial_{\dot{F}_{iHK}} \psi \ddot{F}_{iHK} - \rho_R \partial_{\dot{\theta}} \psi \ddot{\theta} \\
&\quad +\theta \partial_{F_{iK}} k^R_H F_{iKH} + \theta \partial_\theta k^R_H \partial_{X_K} \theta + \theta \partial_{F_{iKJ}} k^R_H F_{iKJH} + \theta \partial_{F_{iKJI}} k^R_H F_{iKJIH} \\
&+\theta \partial_{\partial_{X_K} \theta} k^R_H \partial_{X_H} \partial_{X_K} \theta + \theta \partial_{\dot{F}_{iK}} k^R_H \dot{F}_{iKH} + \theta \partial_{\dot{F}_{iKJ}} k^R_H \dot{F}_{iKJH} + \theta \partial_{\dot{\theta}} k^R_H \partial_{X_H} \dot{\theta} - \frac{1}{\theta} q^R_K \partial_{X_K} \theta \geq 0
\end{aligned}$$

By the linearity (and arbitrariness) of $\ddot{\theta}$, $\ddot{\mathbf{F}}$, $\nabla_R \ddot{\mathbf{F}}$, $\nabla_R^3 \mathbf{F}$ it follows that

$$\partial_{\dot{\theta}} \psi = 0, \qquad \partial_{\dot{\mathbf{F}}} \psi = 0, \qquad \partial_{\nabla_R \dot{\mathbf{F}}} \psi = 0, \qquad \partial_{\nabla_R^2 \mathbf{F}} \mathbf{k}_R = 0.$$

Hence

$$\psi = \psi(\mathbf{F}, \theta, \nabla_R \mathbf{F}, \nabla_R^2 \mathbf{F}, \nabla_R \theta), \qquad \mathbf{k}_R = \mathbf{k}_R(\mathbf{F}, \theta, \nabla_R \mathbf{F}, \nabla_R \theta, \dot{\mathbf{F}}, \nabla_R \dot{\mathbf{F}}, \dot{\theta})$$

and

$$\begin{aligned}
&-\rho_R(\partial_\theta \psi + \eta)\dot{\theta} + (T^R_{iK} - \rho_R \partial_{F_{iK}} \psi)\dot{F}_{iK} - \rho_R \partial_{F_{iHK}} \psi \dot{F}_{iHK} - \rho_R \partial_{F_{iHKJ}} \psi \dot{F}_{iHKJ} \\
&\quad -\rho_R \partial_{\partial_{X_K} \theta} \psi \partial_{X_K} \theta + \theta \partial_{F_{iK}} k^R_H F_{iKH} + \theta \partial_\theta k^R_H \partial_{X_H} \theta + \theta \partial_{F_{iKJ}} k^R_H F_{iKJH} \\
&+\theta \partial_{\partial_{X_K} \theta} k^R_H \partial_{X_H} \partial_{X_K} \theta + \theta \partial_{\dot{F}_{iK}} k^R_H \dot{F}_{iKH} + \theta \partial_{\dot{F}_{iKJ}} k^R_H \dot{F}_{iKJH} + \theta \partial_{\dot{\theta}} k^R_H \partial_{X_H} \dot{\theta} - \frac{1}{\theta} q^R_K \partial_K \theta \geq 0
\end{aligned}$$

The linearity of $\nabla_R \dot{\theta}$ and $\nabla_R^2 \dot{\mathbf{F}}$ implies

$$-\rho_R \partial_{F_{iHKJ}} \psi + \theta \partial_{\dot{F}_{iKJ}} k^R_H = 0, \qquad -\rho_R \partial_{\partial_{X_H} \theta} \psi + \theta \partial_{\dot{\theta}} k^R_H = 0. \tag{8.36}$$

A direct integration of (8.36) gives

$$k_H^R = A_{iHKJ}\dot{F}_{iKJ} + c_H\dot{\theta} + \hat{k}_H, \qquad A_{iHKJ} := \frac{\rho_R}{\theta}\partial_{F_{iHKJ}}\psi, \quad c_H := \frac{\rho_R}{\theta}\partial_{\partial_{X_H}\theta}\psi,$$

where $\mathbf{A}$ and $\mathbf{c}$ depend on $(\mathbf{F}, \theta, \nabla_R\mathbf{F}, \nabla_R^2\mathbf{F}, \nabla_R\theta)$ while $\hat{\mathbf{k}}$ is a function of $(\mathbf{F}, \theta, \nabla_R\mathbf{F}, \nabla_R\theta, \dot{\mathbf{F}})$. Moreover the linearity, (arbitrariness) and symmetry of $\partial_{X_H}\partial_{X_K}\theta$ imply that

$$\partial_{\nabla_R\theta}\mathbf{k}_R \in \mathrm{Skw}.$$

It then follows the reduced inequality

$$-\rho_R(\partial_\theta\psi + \eta)\dot{\theta} + (T_{iK}^R - \rho_R\partial_{F_{iK}}\psi)\dot{F}_{iK} - \rho_R\partial_{F_{iHK}}\psi\dot{F}_{iHK} + \theta\partial_{F_{iK}}k_H^R F_{iKH}$$
$$+\theta\partial_\theta k_H^R\partial_{X_H}\theta + \theta\partial_{F_{iKJ}}k_H^R F_{iKJH} + \theta\partial_{\dot{F}_{iK}}k_H^R\dot{F}_{iKH} - \frac{1}{\theta}q_K^R\partial_{X_K}\theta \geq 0$$

Physically admissible properties are allowed provided this set of restrictions hold. Yet this scheme seems to be quite involved and we then look for particular constitutive functions allowing for the non-simple character provided by $\nabla_R\mathbf{F}$, $\nabla_R^2\mathbf{F}$, $\nabla_R\theta$. The first requirement in (8.36) implies that ψ depends at most linearly on $\nabla^2\mathbf{F}$. Consistent with the skew-symmetry of $\partial_{\nabla_R\theta}\mathbf{k}_R$ we let

$$\mathrm{Skw} \ni \boldsymbol{\Omega} = \partial_{\nabla_R\theta}\mathbf{k}_R.$$

Hence we simplify the constitutive properties by letting $\mathbf{A}$, $\mathbf{c}$, and $\boldsymbol{\Omega}$ be functions of $\mathbf{F}$, θ, by assuming that $\mathbf{T}_R$ and $\mathbf{q}_R$ be independent of $\nabla_R\dot{\mathbf{F}}$, $\dot{\theta}$, by letting

$$\mathbf{T}_R = \mathbf{T}_R(\mathbf{F}, \theta, \nabla_R\mathbf{F}, \nabla_R\theta, \dot{\mathbf{F}}), \qquad \mathbf{q}_R = \mathbf{q}_R(\mathbf{F}, \theta, \nabla_R\mathbf{F}, \nabla_R\theta, \dot{\mathbf{F}}),$$

and by taking the extra-entropy vector $\mathbf{k}_R$ in the special form

$$\mathbf{k}_H^R = A_{pHKJ}(\mathbf{F}, \theta)\dot{F}_{pKJ} + c_H(\mathbf{F}, \theta)\dot{\theta} + \Omega_{HK}(\mathbf{F}, \theta, \dot{\mathbf{F}})\partial_{X_K}\theta.$$

The reduced inequality then reads

$$[-\rho_R(\partial_\theta\psi + \eta) + \theta F_{iMH}\partial_{F_{iM}}c_H + \theta\partial_\theta c_H\partial_{X_H}\theta]\dot{\theta} + (T_{iK}^R - \rho_R\partial_{F_{iK}}\psi)\dot{F}_{iK}$$
$$+(-\rho_R\partial_{F_{pHK}}\psi + \theta F_{iMH}\partial_{F_{iM}}A_{pHKJ} + \theta\partial_\theta A_{pHKJ})\dot{F}_{pHK} + (\theta F_{iMH}\partial_{F_{iM}}\Omega_{HK} - \frac{1}{\theta}q_K^R)\partial_{X_K}\theta \geq 0.$$

The linearity of $\dot{\theta}$ and $\nabla_R\dot{\mathbf{F}}$ implies[9]

$$\eta = -\partial_\theta\psi + \frac{1}{\rho_R}(\theta F_{iMH}\partial_{F_{iM}}c_H + \theta\partial_\theta c_H\,\partial_{X_H}\theta), \tag{8.37}$$

$$-\rho_R\partial_{F_{pHK}}\psi + \theta F_{iMH}\partial_{F_{iM}}A_{pHKJ} + \theta\partial_\theta A_{pHKJ} = 0, \tag{8.38}$$

[9] Equation (8.37) is an uncommon case where entropy is not merely a derivative of a potential function.

and hence

$$(T^R_{iK} - \rho_R \partial_{F_{iK}} \psi) \dot{F}_{iK} + (\theta F_{iMH} \partial_{F_{iM}} \Omega_{HK} - \frac{1}{\theta} q^R_K) \partial_{X_K} \theta \geq 0.$$

We neglect cross-coupled terms $\nabla_R \theta \otimes \dot{\mathbf{F}}$. Hence the reduced inequality allows us to represent $\mathbf{T}_R$ and $\mathbf{q}_R$ in the form

$$T^R_{iK} = \rho_R \partial_{F_{iK}} \psi + \tilde{T}^R_{iK}, \qquad q^R_P = \theta \, F_{iMH} \partial_{F_{iM}} \Omega_{HP} - K_{PQ} \partial_{X_Q} \theta, \tag{8.39}$$

where $\mathbf{K}$ is a non-negative tensor while $\tilde{\mathbf{T}}_R$ is an additive Piola stress accounting for dissipation (viscosity). Indeed,

$$\tilde{\mathbf{T}}_R \cdot \dot{\mathbf{F}} \geq 0.$$

For definiteness we let

$$\tilde{\mathbf{T}}_R = [\alpha \mathbf{D} + \beta (\mathrm{tr}\, \mathbf{D}) \mathbf{1}] \mathbf{F}^{-T}.$$

Hence

$$\tilde{\mathbf{T}}_R \cdot \dot{\mathbf{F}} = [\alpha \mathbf{D} + \beta (\mathrm{tr}\, \mathbf{D}) \mathbf{1}] \mathbf{F}^{-T} \dot{\mathbf{F}} = \alpha |\mathbf{D}|^2 + \beta (\mathrm{tr}\, \mathbf{D})^2.$$

The inequality $\tilde{\mathbf{T}}^R \cdot \dot{\mathbf{F}} \geq 0$ holds provided only that $\alpha \geq 0$, $2\alpha + 3\beta \geq 0$.

As to the effects of the deformation gradient, since ψ is a function of $\mathbf{F}$, $\nabla_R \mathbf{F}$, $\nabla^2_R \mathbf{F}$ then by (8.39) it follows that $\mathbf{T}_R$ is a function of $\mathbf{F}$, $\nabla_R \mathbf{F}$, $\nabla^2_R \mathbf{F}$ and is allowed to depend linearly on $\dot{\mathbf{F}}$ via a non-negative tensor $\mathbf{M}$. Meanwhile Eq. (8.38) may be viewed as a constraint on the function ψ. Equation (8.37) implies that the entropy is allowed to depend on $\mathbf{F}$, θ, $\nabla_R \mathbf{F}$, $\nabla^2_R \mathbf{F}$, $\nabla_R \theta$.

If, instead, the occurrence of an hyperstress $\mathbb{G}$ is allowed so that the mechanical power is $\mathbf{T}_R \cdot \dot{\mathbf{F}} + \mathbb{G} \cdot \nabla_R \dot{\mathbf{F}}$ then the previous procedure can be repeated step by step to arrive at the reduced inequality

$$\begin{aligned} -\rho_R(\partial_\theta \psi + \eta)\dot{\theta} + (T^R_{iK} - \rho_R \partial_{F_{iK}} \psi) \dot{F}_{iK} + G_{iHK} \dot{F}_{iHK} - \rho_R \partial_{F_{iHK}} \psi \dot{F}_{iHK} + \theta \partial_{F_{iK}} k^R_H F_{iKH} \\ + \theta \partial_\theta k^R_H \partial_{X_H} \theta + \theta \partial_{F_{iKJ}} k^R_H F_{iKJH} + \theta \partial_{\dot{F}_{iK}} k^R_H \dot{F}_{iKH} - \frac{1}{\theta} q^R_K \partial_{X_K} \theta \geq 0. \end{aligned}$$

Then we find that (8.38) is replaced by

$$G_{pHK} = \rho_R \partial_{F_{pHK}} \psi - \theta F_{iMH} \partial_{F_{iM}} A_{pHKJ} - \theta \partial_\theta A_{pHKJ}. \tag{8.40}$$

8.12 A New Scheme Associated with the Hyperstress

We start with the power density

$$w := \mathbf{T}_R \cdot \dot{\mathbf{F}} + \mathbb{G} \cdot \nabla_R \dot{\mathbf{F}}$$

and assume w is objective.

First we show that w can be represented by appropriate surface and volume terms. Since

$$T^R_{hK}\dot{F}_{hK} + G_{hRS}\dot{F}_{hRS} = \partial_{X_K}(T^R_{hK}v_h) - \partial_{X_K}(T^R_{hK})v_h + \partial_{X_S}(G_{hRS}\dot{F}_{hR}) - (\partial_{X_S}G_{hRS})\dot{F}_{hR}$$

and

$$\partial_{X_K}(T^R_{hK}v_h) + \partial_{X_S}(G_{hRS}\dot{F}_{hR}) = \nabla_R \cdot (\mathbf{v}\mathbf{T}_R + \dot{\mathbf{F}}\mathbb{G})$$

then, for any region $\mathrm{P} \subset \mathrm{R}$, the power $\mathfrak{W} = \int_{\mathrm{P}} w\, dv_R$ can be written

$$\mathfrak{W} = \int_{\partial \mathrm{P}} [T^R_{hK} m_K v_h) + G_{hKQ} m_Q \dot{F}_{hK}] da_R - \int_{\mathrm{P}} [(\partial_{X_K} T^R_{hK}) v_h + (\partial_{X_Q} G_{hKQ})\dot{F}_{hK}] dv_R,$$

where $\mathbf{m} = \mathbf{n}_R$ is the unit outward normal in $\partial\mathrm{P}$, $\mathrm{P} \subset \mathrm{R}$. Now,

$$(\partial_{X_Q} G_{hKQ})\dot{F}_{hK} = \partial_{X_K}((\partial_{X_Q} G_{hKQ})v_h) - (\partial_{X_K}\partial_{X_Q} G_{hKQ})v_h.$$

Hence

$$\mathfrak{W} = \int_{\partial \mathrm{P}} [(T^R_{hK} - \partial_{X_Q} G_{hKQ}) m_K v_h) + G_{hKQ} m_Q \dot{F}_{hK}] da_R - \int_{\mathrm{P}} [\partial_{X_K}(T^R_{hK} - \partial_{X_Q} G_{hKQ}) v_h] dv_R.$$

Let

$$S_{hK} = T^R_{hK} - \partial_{X_Q} G_{hKQ}, \qquad f_Q = G_{hKQ}\dot{F}_{hK}.$$

The power $\mathfrak{W}$ then can be represented by

$$\mathfrak{W} = \int_{\partial \mathrm{P}} [S_{hK} m_K v_h + f_Q m_Q] da_R - \int_{\mathrm{P}} v_h \partial_{X_K} S_{hK} dv_R,$$

or, in absolute notation,

$$\mathfrak{W} = \int_{\partial \mathrm{P}} [\mathbf{v}\cdot\mathbf{S}\mathbf{m} + \mathbf{f}\cdot\mathbf{m}] da_R - \int_{\mathrm{P}} \mathbf{v}\cdot(\nabla_R \cdot \mathbf{S}) dv_R.$$

The identity

$$\nabla_R \cdot (\mathbf{v}\mathbf{S}) = \mathbf{S}\cdot\dot{\mathbf{F}} + \mathbf{v}\cdot(\nabla_R\cdot\mathbf{S})$$

and use of the divergence theorem allow $\mathfrak{W}$ to be written

$$\mathfrak{W} = \int_{\partial \mathrm{P}} \mathbf{f}\cdot\mathbf{m}\, da_R + \int_{\mathrm{P}} \mathbf{S}\cdot\dot{\mathbf{F}}\, dv_R.$$

According to this representation of $\mathfrak{W}$, it is natural to identify $\mathbf{S}$ with the effective Piola stress and to think of $\mathbf{f}$ as an energy flux vector which reminds the interstitial energy flux [142].

Associated with the effective Piola stress $\mathbf{S}$ is the Cauchy stress

$$\hat{\mathbf{T}} = \frac{1}{J}\mathbf{S}\mathbf{F}^T = \frac{1}{J}(\mathbf{T}_R - \nabla_R \cdot \mathbb{G})\mathbf{F}^T = \mathbf{T} - \frac{1}{J}(\nabla_R \cdot \mathbb{G})\mathbf{F}^T.$$

Based on the view that $\mathbf{S}$ and $\hat{\mathbf{T}}$ are the effective stress tensors, we now review the balance equations. First we assume the balance of linear momentum in the form

$$\rho_R \dot{\mathbf{v}} = \nabla_R \cdot \mathbf{S} + \rho_R \mathbf{b}.$$

The balance of angular momentum is satisfied if

$$\hat{\mathbf{T}} = \hat{\mathbf{T}}^T, \qquad \mathbf{S}\mathbf{F}^T = \mathbf{F}\mathbf{S}^T. \tag{8.41}$$

To see whether the condition (8.41) is plausible, we start from the required objectivity of the power w. Under the Euclidean transformation $\mathbf{x}^* = \mathbf{c} + \mathbf{Q}\mathbf{x}$, the transforms $\mathbf{F}^*$ and $\nabla_R \mathbf{F}^*$ take the forms

$$\mathbf{F}^* = \mathbf{Q}\mathbf{F}, \qquad \nabla_R \mathbf{F}^* = \mathbf{Q}\nabla_R \mathbf{F}$$

and hence

$$\dot{\overline{\mathbf{F}^*}} = \dot{\mathbf{Q}}\mathbf{F} + \mathbf{Q}\dot{\mathbf{F}}, \qquad \dot{\overline{\nabla_R \mathbf{F}^*}} = \dot{\mathbf{Q}}\nabla_R \mathbf{F} + \mathbf{Q}\nabla_R \dot{\mathbf{F}}.$$

Consequently the invariance $w^* = w$ results in

$$\mathbf{T}_R^* \cdot \dot{\mathbf{Q}}\mathbf{F} + \mathbb{G}^* \cdot \dot{\mathbf{Q}}\nabla_R \mathbf{F} + \mathbf{T}_R^* \cdot \mathbf{Q}\dot{\mathbf{F}} + \mathbb{G}^* \cdot \mathbf{Q}\nabla_R \dot{\mathbf{F}} = \mathbf{T}_R \cdot \dot{\mathbf{F}} + \mathbb{G} \cdot \nabla_R \dot{\mathbf{F}}.$$

The arbitrariness of $\dot{\mathbf{F}}$ and $\nabla_R \dot{\mathbf{F}}$ implies

$$\mathbf{T}_R = \mathbf{Q}^T \mathbf{T}_R^*, \qquad \mathbb{G} = \mathbf{Q}^T \mathbb{G}^*.$$

Substitution of $\mathbf{T}^* = \mathbf{Q}\mathbf{T}_R$ and $\mathbb{G}^* = \mathbf{Q}\mathbb{G}$ in the reduced equation

$$\mathbf{T}_R^* \cdot \dot{\mathbf{Q}}\mathbf{F} + \mathbb{G}^* \cdot \dot{\mathbf{Q}}\nabla_R \mathbf{F} = 0$$

leads to

$$\Omega_{lj}(T^R_{lK} F_{jK} + G_{lKP} F_{jKP}) = 0$$

where $\mathbf{\Omega} = \mathbf{Q}^T \dot{\mathbf{Q}} \in \mathrm{Skw}$. Hence it follows the symmetry condition

$$T^R_{lK} F_{jK} + G_{lKP} F_{jKP} = T^R_{jK} F_{lK} + G_{jKP} F_{lKP}. \tag{8.42}$$

We now go back to (8.41) or, in suffix notation,

$$T^R_{hK} F_{jK} - \partial_{X_P}(G_{hKP}) F_{jK} = T^R_{jK} F_{hK} - \partial_{X_P}(G_{jKP}) F_{hK}. \tag{8.43}$$

Proposition 8.1 *The objectivity of the stress power implies the symmetry of the effective Cauchy stress* $\hat{\mathbf{T}}$*; formally, the symmetry condition (8.42) implies (8.41).*

We follow the proof given in [332]. First we write the assumption (8.43) in the form

$$T^R_{hK}F_{jK} - \partial_{X_P}(G_{hKP}F_{jK}) + G_{hKP}F_{jKP} = T^R_{jK}F_{hK} - \partial_{X_P}(G_{jKP}F_{hK}) + G_{jKP}F_{hKP}.$$

In view of (8.42), the symmetry condition (8.43) holds if and only if

$$G_{hKP}F_{jK} = G_{jKP}F_{hK}.$$

Denote by M and N the symmetric and skew-symmetric parts of $\mathbb{G}$ relative to the capital indices,

$$G_{hKP} = M_{hKP} + N_{hKP}, \quad M_{hKP} = \tfrac{1}{2}(G_{hKP} + G_{hPK}), \quad N_{hKP} = \tfrac{1}{2}(G_{hKP} - G_{hPK}).$$

Hence we have

$$M_{hKP}F_{jK} + N_{hKP}F_{jK} = M_{jKP}F_{hK} + N_{jKP}F_{hK}$$

whence

$$M_{hKP}F_{jK}F_{iP} + N_{hKP}F_{jK}F_{iP} = M_{jKP}F_{hK}F_{iP} + N_{jKP}F_{hK}F_{iP}. \tag{8.44}$$

Let

$$D_{hji} := M_{hKP}F_{jK}F_{iP}, \qquad H_{hji} := N_{hKP}F_{jK}F_{iP}.$$

It follows that

$$D_{hji} = D_{hij}, \qquad H_{hji} = -H_{hij}$$

while, by (8.44),

$$D_{hji} + H_{hji} = D_{jhi} + H_{jhi}$$

or, equivalently,

$$D_{hji} - D_{jhi} = H_{jhi} - H_{hji}.$$

By interchanging j and i, and next h and j, we have

$$D_{hij} - D_{ihj} = H_{ihj} - H_{hij},$$

$$D_{jih} - D_{ijh} = H_{ijh} - H_{jih}.$$

By summing these three relations, in view of the symmetry of D and the skew-symmetry of H it follows

$$2D_{hij} - 2D_{ihj} = 2H_{jhi}$$

whence

$$M_{hKP} F_{iK} F_{jP} - M_{iKP} F_{hK} F_{jP} = N_{jKP} F_{hK} F_{iP}.$$

Upon multiplication by $F^{-1}_{Qh} F^{-1}_{Li}$ and summation of h and i yields

$$N_{jQL} = M_{hLP} F_{jP} F^{-1}_{Qh} - M_{iQP} F_{jP} F^{-1}_{Li}. \tag{8.45}$$

Hence, provided the skew part N of $\mathbb{G}$ is related to the symmetric part M by (8.45), the Cauchy stress $\hat{\mathbf{T}} = \mathbf{S}\mathbf{F}^T/J$ is symmetric. □

To sum up, the balance equations can be written in a consistent way in terms of the effective Piola stress $\mathbf{S}$ in the form

$$\rho_R \dot{\mathbf{v}} = \nabla_R \cdot \mathbf{S} + \rho_R \mathbf{b}, \qquad \mathbf{S}\mathbf{F}^T = \mathbf{F}\mathbf{S}^T,$$

$$\rho_R \dot{\varepsilon} = \mathbf{S} \cdot \dot{\mathbf{F}} + \nabla_R \cdot \mathbf{f} - \nabla_R \cdot \mathbf{q}_R + \rho_R r,$$

where $f_K = G_{hJK} \dot{F}_{hJ}$. Conceptually it emerges that the occurrence of a hyperstress $\mathbb{G}$, with stress power $\mathbb{G} \cdot \nabla_R \dot{\mathbf{F}}$, can be embodied in an effective Piola stress $\mathbf{S}$ provided an appropriate energy flux $\mathbf{f}$ is allowed to occur. We now investigate whether and how this scheme is compatible with thermodynamics.

Thermodynamic Restrictions

We follow the standard assumption on the second law of thermodynamics, expressed by the inequality

$$\rho_R \dot{\eta} \geq -\nabla_R \cdot \frac{\mathbf{q}_R}{\theta} - \nabla_R \cdot \mathbf{k}_R + \frac{\rho_R r}{\theta}.$$

Substitution of $\rho_R r - \nabla_R \cdot \mathbf{q}_R$ from the balance of energy results in the entropy inequality

$$-\rho_R(\dot{\psi} + \eta\dot{\theta}) + \mathbf{S} \cdot \dot{\mathbf{F}} + \nabla_R \cdot \mathbf{w} + \theta \nabla_R \cdot \mathbf{k}_R - \frac{1}{\theta} \mathbf{q}_R \cdot \nabla_R \theta \geq 0.$$

While the extra-entropy flux $\mathbf{k}_R$ is an unknown constitutive function, the expression of the energy flux $\mathbf{f}$ is known. Indeed, since $\mathbf{N} \cdot \nabla_R \dot{\mathbf{F}} = 0$ then

$$\nabla_R \cdot \mathbf{f} = \partial_{X_K}(G_{hJK} \dot{F}_{hJ}) = \partial_{X_K} G_{hJK} \dot{F}_{hJ} + M_{hJK} \dot{F}_{hJK} = (\nabla_R \mathbb{G}) \cdot \dot{\mathbf{F}} + \mathbf{M} \cdot \nabla_R \dot{\mathbf{F}}.$$

The entropy inequality can then be written

$$-\rho_R(\dot{\psi} + \eta\dot{\theta}) + (\mathbf{S} + \nabla_R \cdot \mathbb{G}) \cdot \dot{\mathbf{F}} + \mathbf{M} \cdot \nabla_R \dot{\mathbf{F}} + \theta \nabla_R \cdot \mathbf{k}_R - \frac{1}{\theta} \mathbf{q}_R \cdot \nabla_R \theta \geq 0.$$

Seemingly the skew part $\mathbf{N}$ of $\mathbb{G}$ does not enter the entropy inequality and this would be consistent with the vanishing of the power $\mathbf{N} \cdot \nabla_R \dot{\mathbf{F}}$. However $\nabla_R \cdot \mathbb{G}$ comprises $\nabla_R \cdot \mathbf{N}$ and this is consistent with the nonzero value of the power $(\nabla_R \cdot \mathbf{N}) \cdot \dot{\mathbf{F}}$.

We let ψ, η, $\mathbf{S}$, $\mathbb{G}$, $\mathbf{q}_R$ and $\mathbf{k}_R$ be given by functions of the set

$$(\mathbf{F}, \theta, \nabla_R \mathbf{F}, \nabla_R \theta, \dot{\mathbf{F}})$$

of independent variables. Now, to save writing, we observe that the role of $\nabla_R \cdot \mathbf{k}_R$ here proves inessential. We can prove that, as a consequence of the entropy inequality, $\partial_{\nabla_R \mathbf{F}} \mathbf{k}_R = \mathbf{0}$, $\partial_{\nabla_R \theta} \mathbf{k}_R = \mathbf{0}$, and $\partial_{\mathbf{F}} \mathbf{k}_R = \mathbf{0}$. Moreover, $\theta^2 \partial_\theta \mathbf{k}_R$ provides a physically inessential contribution to $\mathbf{q}_R$ as well as $\theta \partial_{\dot{\mathbf{F}}} \mathbf{k}_R$ is a possible minor correction to $\mathbf{M}$. Then no significant generality is lost by letting $\mathbf{k}_R = \mathbf{0}$.

Upon computation of $\dot{\psi}$ the entropy inequality takes the form

$$-\rho_R(\partial_\theta + \eta)\dot{\theta} + (-\rho_R \partial_{\mathbf{F}} \psi + \mathbf{S} + \nabla_R \cdot \mathbb{G}) \cdot \dot{\mathbf{F}} + (-\rho_R \partial_{\nabla_R \mathbf{F}} \psi + \mathbf{M}) \cdot \nabla_R \dot{\mathbf{F}}$$
$$-\rho_R \partial_{\nabla_R \theta} \psi \cdot \nabla_R \dot{\theta} - \rho_R \partial_{\dot{\mathbf{F}}} \psi \cdot \ddot{\mathbf{F}} - \frac{1}{\theta} \mathbf{q}_R \cdot \nabla_R \theta \geq 0.$$

The standard procedure shows that the entropy inequality holds if and only if

$$\partial_{\nabla_R \theta} \psi = 0, \qquad \partial_{\dot{\mathbf{F}}} \psi = 0, \qquad \eta = -\partial_\theta \psi, \qquad \mathbf{M} = \rho_R \partial_{\nabla_R \mathbf{F}} \psi \tag{8.46}$$

and

$$(-\rho_R \partial_{\mathbf{F}} \psi + \mathbf{S} + \nabla_R \cdot \mathbb{G}) \cdot \dot{\mathbf{F}} - \frac{1}{\theta} \mathbf{q}_R \cdot \nabla_R \theta \geq 0.$$

Since $\mathbf{N}$ is subject to (8.45) then also $\mathbf{N}$ and $\mathbb{G}$ are functions of $\mathbf{F}$, θ, $\nabla_R \mathbf{F}$ via $\partial_{\nabla_R \mathbf{F}} \psi$. Consequently only $\mathbf{S}$ and $\mathbf{q}_R$ are allowed to depend on $\nabla_R \theta$ and $\dot{\mathbf{F}}$. If cross-coupling terms are ignored then the reduced inequality implies that

$$\mathbf{S} = -\nabla_R \cdot \mathbb{G} + \rho_R \partial_{\mathbf{F}} \psi + \tilde{\mathbf{S}}, \qquad \mathbf{q}_R \cdot \nabla_R \theta \leq 0, \tag{8.47}$$

where $\tilde{\mathbf{S}}(\mathbf{F}, \theta, \nabla_R \mathbf{F}, \nabla_R \theta, \dot{\mathbf{F}})$ satisfies

$$\tilde{\mathbf{S}} \cdot \dot{\mathbf{F}} \geq 0$$

and hence models viscous effects. For definiteness we observe that

$$\tilde{\mathbf{S}} \cdot \dot{\mathbf{F}} = (\tilde{\mathbf{S}} \mathbf{F}^T) \cdot \mathbf{L} = (\tilde{\mathbf{S}} \mathbf{F}^T) \cdot \mathbf{D}.$$

Thus we can take the Newtonian model

$$\tilde{\mathbf{S}} \mathbf{F}^T = 2\mu \mathbf{D} + \lambda (\operatorname{tr} \mathbf{D}) \mathbf{1}, \qquad \mu \geq 0,\ 2\mu + 3\lambda \geq 0,$$

whence

$$\tilde{\mathbf{S}} \cdot \dot{\mathbf{F}} = 2\mu|\mathbf{D}|^2 + \lambda(\mathrm{tr}\,\mathbf{D})^2 \geq 0.$$

Quite similar results occur in the two procedures. Look at (8.46) and (8.47). Since $\mathbf{S} + \nabla_R \cdot \mathbb{G} = \mathbf{T}_R$ then it follows that

$$\mathbf{T}_R = \rho_R \partial_{\mathbf{F}} \psi + \tilde{\mathbf{S}}.$$

Moreover

$$\mathbb{G} = \rho_R \partial_{\nabla_R \mathbf{F}} \psi + \mathbf{N},$$

$\mathbf{N}$ being determined by $\mathbf{M}$ via (8.45).

In the previous subsection, $\mathbf{T}_R$ is given by (8.39) which coincides with (8.47). As for $\mathbb{G}$, the same term $\rho_R \partial_{\nabla_R \mathbf{F}} \psi$ arises in (8.40). The difference is given by the additional skew term $\mathbf{N}$ (8.45) which serves to get consistency of the whole scheme. Instead, in (8.40) the additional terms are related to the extra-entropy flux $\mathbf{k}_R$ and can be made to vanish if a simpler scheme is in order.

Chapter 9
Mixtures

Mixtures consist of different substances which keep their own properties. The individual molecules enjoy being near to each other so that, in the mathematical model, each point in the mixture may be occupied by all of the substances simultaneously. If the chemical structure of a substance changes when it enters the mixture then the mixture is said to be chemically reacting.

The easiest mixtures to deal with are gaseous mixtures in which gases readily mix. Fluid mixtures may be formed from fluids or from a fluid and gases that dissolve in the fluid or from a fluid and solids that dissolve in the fluid. In most cases, a fluid is preponderant and hence is called the solvent while the remaining constituents are called solutes. Both fluid mixtures and solid mixtures are examined.

The mixture is regarded as ideal in that the volumetric and energetic properties of the mixture are taken to be the linear combination of those of the constituents or species. The chapter provides a derivation of the balance equations by starting with global balances and allowing for interaction terms (growths and supplies). The second law is stated in terms of the constituents and is expressed by the non-negative value of the sum of the entropy productions. Details are given for the balance equations of the whole mixture as a single body; an analysis is given of the various forms of the second-law inequality (Gibbs equation) occurring in the literature. Diffusion is considered through various known models (Fick, Maxwell–Stefan, Cahn–Hilliard, Allen–Cahn). Moreover immiscible mixtures are investigated.

9.1 Kinematics

We consider n (fluid or solid) constituents occupying a time-dependent region $\Omega \subset \mathscr{E}$. The functions describing the evolution of the mixture have $\Omega \times \mathbb{R}$ as their common space-time domain. The subscripts $\alpha, \beta = 1, 2, ..., n$ label the quantities pertaining to

A. Morro and C. Giorgi, *Mathematical Modelling of Continuum Physics*,
Modeling and Simulation in Science, Engineering and Technology,
https://doi.org/10.1007/978-3-031-20814-0_9

the constituents. Hence ρ_α is the mass density, $\mathbf{v}_\alpha$ the velocity, ε_α the internal energy density, on $\Omega \times \mathbb{R}$, of the αth constituent. Fields pertaining to single constituents are named partial (or peculiar). Hence $\mathbf{v}_\alpha$ is a partial velocity.

The constituents are regarded as continuous bodies and are supposed to occupy the regions $\mathrm{R}_1, ..., \mathrm{R}_n$, in the three-dimensional space $\mathscr{E}$, in their reference configurations. The motion of the αth constituent is given by a function

$$\mathbf{x} = \chi_\alpha(\mathbf{X}_\alpha, t),$$

where $\mathbf{x} \in \Omega$ is the position vector at time t of the point at $\mathbf{X}_\alpha \in \mathrm{R}_\alpha$. For any region $\mathrm{P}_\alpha \subseteq \mathrm{R}_\alpha$ we denote by $\mathcal{P}^t_\alpha$ the region of space occupied by the αth constituent at time t,

$$\mathcal{P}^t_\alpha = \chi_\alpha(\mathrm{P}_\alpha, t) := \{\mathbf{x} \in \mathscr{E} : \mathbf{x} = \chi_\alpha(\mathbf{X}_\alpha, t), \ \mathbf{X}_\alpha \in \mathrm{P}_\alpha\}.$$

The whole region Ω occupied by the mixture is the union of $\chi_\alpha(\mathrm{R}_\alpha, t)$,

$$\Omega = \bigcup_\alpha \chi_\alpha(\mathrm{R}_\alpha, t).$$

Here and in sums it is tacitly understood that the label of the constituent, say α or β, runs over $1, 2, ..., n$.

The functions $\{\chi_\alpha\}$ are invertible, at each time t, so that $\mathbf{X}_\alpha = \chi_\alpha^{-1}(\mathbf{x}, t)$, $\mathbf{x} \in \mathcal{P}^t_\alpha$, and is twice continuously differentiable with respect to $\mathbf{X}_\alpha$ and t. Hence

$$\mathbf{v}_\alpha = \partial_t \chi_\alpha(\mathbf{X}_\alpha, t), \qquad \mathbf{a}_\alpha = \partial_t^2 \chi_\alpha(\mathbf{X}_\alpha, t)$$

are the velocity and acceleration of $\mathbf{X}_\alpha$ at time t. The symbol $\nabla_{\mathbf{X}_\alpha}$, or $\partial_{\mathbf{X}_\alpha}$, denotes the gradient with respect to $\mathbf{X}_\alpha$. The deformation gradient $\mathbf{F}_\alpha$ is then defined by[1]

$$\mathbf{F}_\alpha = \nabla_{\mathbf{X}_\alpha} \chi_\alpha, \qquad F^\alpha_{pK} = \partial_{X^\alpha_K} \chi^\alpha_p.$$

The velocity $\mathbf{v}_\alpha$ may be expressed as

$$\hat{\mathbf{v}}_\alpha(\mathbf{X}_\alpha, t) = \hat{\mathbf{v}}_\alpha(\chi_\alpha^{-1}(\mathbf{x}, t), t) =: \mathbf{v}_\alpha(\mathbf{x}, t).$$

The velocity gradient $\mathbf{L}_\alpha$ is defined by

$$\mathbf{L}_\alpha = \nabla \mathbf{v}_\alpha(\mathbf{x}, t), \qquad L^\alpha_{pq} = \partial_{x_q} v^\alpha_p(\mathbf{x}, t).$$

A backward prime affixed to a symbol with a subscript α denotes the αth peculiar time derivative, namely, the material derivative following the motion of the αth constituent. For instance, $\mathbf{a}_\alpha = \grave{\mathbf{v}}_\alpha$. Some properties are now established about peculiar time derivatives.

---

[1] To avoid too many subscripts, the subscript α labelling the constituent sometimes is denoted as a superscript.

Starting with the deformation gradient $\mathbf{F}_\alpha$, we have

$$\grave{\mathbf{F}}_\alpha = \partial_t \nabla_{\mathbf{X}_\alpha} \chi_\alpha(\mathbf{X}_\alpha, t) = \nabla_{\mathbf{X}_\alpha} \partial_t \chi_\alpha(\mathbf{X}_\alpha, t) = \nabla \mathbf{v}_\alpha(\mathbf{x}, t)\mathbf{F}_\alpha$$

whence

$$\grave{\mathbf{F}}_\alpha = \mathbf{L}_\alpha \mathbf{F}_\alpha. \tag{9.1}$$

By the invertibility of χ_α, $\mathbf{F}_\alpha$ is non-singular. Having in mind continuous motions we let

$$J_\alpha := \det \mathbf{F}_\alpha > 0.$$

The computation of the derivative yields

$$\grave{J}_\alpha = (\partial_{\mathbf{F}_\alpha} J_\alpha) \cdot \grave{\mathbf{F}}_\alpha = J_\alpha (\mathbf{F}_\alpha^{-1})^T \cdot \grave{\mathbf{F}}_\alpha.$$

Upon substitution of $\grave{\mathbf{F}}_\alpha$ from (9.1), since

$$(\mathbf{F}_\alpha^{-1})^T \cdot (\mathbf{L}_\alpha \mathbf{F}_\alpha) = \operatorname{tr} \mathbf{L}_\alpha = \nabla \cdot \mathbf{v}_\alpha,$$

we find

$$\grave{J}_\alpha = J_\alpha \nabla \cdot \mathbf{v}_\alpha. \tag{9.2}$$

Let $\mathbf{w}_\alpha$ and g_α be vector- and scalar-valued functions on $\mathrm{R}_\alpha \times \mathbb{R}$. Let

$$\breve{g}_\alpha(\mathbf{X}_\alpha, t) = \breve{g}_\alpha(\chi_\alpha^{-1}(\mathbf{x}, t), t) =: \tilde{g}_\alpha(\mathbf{x}, t), \tag{9.3}$$

and the like for $\mathbf{w}_\alpha$. Equation (9.3) establishes the connection between the Lagrangian description $\breve{g}_\alpha(\mathbf{X}_\alpha, t)$ and the Eulerian description $\tilde{g}_\alpha(\mathbf{x}, t)$ of a quantity g_α. Hence we have

$$\partial_{X_K^\alpha} \breve{g}_\alpha(\mathbf{X}_\alpha, t) = \partial_{x_p} \tilde{g}_\alpha(\mathbf{x}, t) F^\alpha_{pK}$$

and

$$\partial_t \breve{g}_\alpha(\mathbf{X}_\alpha, t) = \partial_t \tilde{g}_\alpha(\mathbf{x}, t) + \mathbf{v}_\alpha \cdot \nabla \tilde{g}_\alpha(\mathbf{x}, t).$$

To save writing, we usually omit the symbols $\breve{}$ and $\tilde{}$ over the function on $\mathrm{R}_\alpha \times \mathbb{R}$ and $\Omega \times \mathbb{R}$. Hence, in direct notation,

$$\nabla_{\mathbf{X}_\alpha} g_\alpha = \nabla g_\alpha \mathbf{F}_\alpha, \tag{9.4}$$

$$\grave{g}_\alpha = \partial_t g_\alpha + (\mathbf{v}_\alpha \cdot \nabla) g_\alpha, \tag{9.5}$$

Equation (9.5) provides the way of computing the peculiar derivative within the Eulerian description.

Upon substitution of (9.4) we obtain

$$\mathbf{w}_\alpha \cdot \nabla_{\mathbf{X}_\alpha} g_\alpha = w^\alpha_K \partial_{x_p} g_\alpha F^\alpha_{pK} = \partial_{x_p}(F^\alpha_{pK} w^\alpha_K g_\alpha) - g_\alpha \partial_{x_p}(F^\alpha_{pK} w^\alpha_K),$$

whence

$$\mathbf{w}_\alpha \cdot \nabla_{\mathbf{X}_\alpha} g_\alpha = \nabla \cdot (\mathbf{F}_\alpha \mathbf{w}_\alpha g_\alpha) - g_\alpha \nabla \cdot (\mathbf{F}_\alpha \mathbf{w}_\alpha). \tag{9.6}$$

For convenience in calculations we observe that

$$\overline{\nabla_{\mathbf{X}_\alpha} g_\alpha}^{\grave{}} = \partial_t \nabla_{\mathbf{X}_\alpha} \breve{g}_\alpha = \nabla_{\mathbf{X}_\alpha} \partial_t \breve{g}_\alpha = \nabla_{\mathbf{X}_\alpha} \grave{g}_\alpha$$

and hence we get the commutation property

$$\overline{\nabla_{\mathbf{X}_\alpha} g_\alpha}^{\grave{}} = \nabla_{\mathbf{X}_\alpha} \grave{g}_\alpha. \tag{9.7}$$

Let ρ_α be the mass density of the αth constituent, namely, the mass of constituent α per unit volume of the mixture. By the ideal character of the mixture we define the mass density ρ of the mixture as

$$\rho = \textstyle\sum_\alpha \rho_\alpha,$$

and the mass fraction ω_α as[2]

$$\omega_\alpha = \frac{\rho_\alpha}{\rho}.$$

Moreover we define the velocity of the mixture, or *barycentric velocity*,[3] $\mathbf{v}$ as

$$\mathbf{v} = \textstyle\sum_\alpha \omega_\alpha \mathbf{v}_\alpha.$$

Let M_α be the molecular weight (or mass) of the αth constituent. Hence $\mathcal{N}_\alpha = \rho_\alpha / M_\alpha$ is the αth number density of moles. The relation between the number densities $\{\mathcal{N}_\alpha\}$ and the mass fractions $\{\omega_\alpha\}$ is

$$\omega_\alpha = \frac{\mathcal{N}_\alpha M_\alpha}{\sum_\beta \mathcal{N}_\beta M_\beta}.$$

Define

$$x_\alpha = \frac{\mathcal{N}_\alpha}{\sum_\beta \mathcal{N}_\beta}$$

the mole fraction. In general $\omega_\alpha \neq x_\alpha$ and $\omega_\alpha \approx x_\alpha$ inasmuch as the molecular weights are nearly equal. Now define

---

[2] In continuum mechanics the mass fraction is often termed *concentration* and denoted by c_α (see, e.g. [64, 336, 427]). In chemistry c_α denotes the *molar concentration*, that is, the number of moles per unit volume. To avoid any misunderstandings, we avoid the use of c_α.

[3] Also named *mean velocity*.

$$M = \frac{\sum_\beta \mathcal{N}_\beta M_\beta}{\sum_\beta \mathcal{N}_\beta}$$

as the average molar weight of the mixture. Hence

$$\omega_\alpha = \frac{\mathcal{N}_\alpha M_\alpha}{\sum_\beta \mathcal{N}_\beta \, M} = x_\alpha \frac{M_\alpha}{M}.$$

While M_α is a constant, the average M is related to the composition of the mixture. The relation $\omega_\alpha = x_\alpha M_\alpha / M$ is then operative when the composition of the mixture is approximately constant (in time and space).

We denote by $\varphi_\alpha \in [0, 1]$ the volume fraction occupied by the constituent α. Unless there are voids we have

$$\sum_\alpha \varphi_\alpha = 1.$$

Sometimes the voids are considered as a constituent, other times they are not and then the sum over the constituents gives

$$\sum_\alpha \varphi_\alpha < 1.$$

Denote by $\mathbf{u}_\alpha$ the *diffusion velocity*,

$$\mathbf{u}_\alpha = \mathbf{v}_\alpha - \mathbf{v}.$$

Letting

$$\mathbf{h}_\alpha = \rho_\alpha \mathbf{u}_\alpha$$

we find that

$$\sum_\alpha \mathbf{h}_\alpha = \sum_\alpha \rho_\alpha (\mathbf{v}_\alpha - \mathbf{v}) = \mathbf{0}. \tag{9.8}$$

The true mass density[4] ρ^0_α of a constituent α is the mass of constituent α per unit volume of constituent α. Hence we have

$$\rho_\alpha = \rho^0_\alpha \varphi_\alpha.$$

By definition,

$$\omega_\alpha = \frac{\rho_\alpha}{\sum_\beta \rho_\beta} = \frac{\rho^0_\alpha \varphi_\alpha}{\sum_\beta \rho^0_\beta \varphi_\beta}.$$

Only if ρ^0_α is the same for all of the constituents, i.e. $\rho^0_\alpha = \rho^0$, we can say that

$$\omega_\alpha = \varphi_\alpha.$$

[4] Also denoted by $\rho_{\alpha R}$ in the literature.

If the constituents are incompressible then any true density ρ^0_α is a constant. In this case

$$\grave{\rho}_\alpha = \rho^0_\alpha \grave{\varphi}_\alpha.$$

9.2 Balance Equations

For any region $\mathrm{P}_\alpha \subseteq \mathrm{R}_\alpha$ we express balance laws, for the part of constituent α within P_α, in the form

$$\grave{\Big(\int_{\mathcal{P}^t_\alpha} g_\alpha dv\Big)} = \int_{\partial\mathcal{P}^t_\alpha} \mathbf{K}_\alpha \mathbf{n}\, da + \int_{\mathcal{P}^t_\alpha} s_\alpha\, dv + \int_{\mathcal{P}^t_\alpha} \gamma_\alpha\, dv. \tag{9.9}$$

This means that the content of g_α in $\mathcal{P}^t_\alpha$ changes in time because of boundary and volume terms, $\mathbf{K}_\alpha$ and s_α, as with single bodies, and also because of the interaction via γ_α of the αth constituent with the other ones. The fields $\{\gamma_\alpha\}$ are termed growths and are subject to

$$\textstyle\sum_\alpha \gamma_\alpha = 0.$$

To establish the local counterpart of the global balance equation (9.9) we observe

$$\int_{\mathcal{P}^t_\alpha} \tilde{g}_\alpha(\mathbf{x}, t)\, dv = \int_{\mathcal{P}_\alpha} \breve{g}_\alpha(\mathbf{X}_\alpha, t) \breve{J}_\alpha(\mathbf{X}_\alpha, t) dv_R$$

and hence, by (9.2),

$$\grave{\Big(\int_{\mathcal{P}^t_\alpha} g_\alpha dv\Big)} = \int_{\mathcal{P}^t_\alpha} (\grave{g}_\alpha + g_\alpha \nabla\cdot\mathbf{v}_\alpha)(\mathbf{x}, t) dv. \tag{9.10}$$

Moreover, by the divergence theorem,

$$\int_{\partial\mathcal{P}^t_\alpha} \mathbf{K}_\alpha \mathbf{n}\, da = \int_{\mathcal{P}^t_\alpha} \nabla\cdot\mathbf{K}_\alpha\, dv.$$

Consequently, by (9.9) the arbitrariness of $\mathcal{P}^t_\alpha$ and the continuity of the integrand in Ω, for any $t \in \mathbb{R}$, imply that

$$\grave{g}_\alpha + g_\alpha \nabla\cdot\mathbf{v}_\alpha - \nabla\cdot\mathbf{K}_\alpha - s_\alpha = \gamma_\alpha. \tag{9.11}$$

We can now state the balance equations. We postulate the balance equations for any constituent as those for a single body but allow for the pertinent interaction (relative to, e.g. mass, momentum, energy) with the other constituents of the mixture. Continuity and differentiability of the functions are assumed as needed.

Balance of mass.

Let ρ_α be the mass density so that

$$M(\mathcal{P}_\alpha^t) = \int_{\mathcal{P}_\alpha^t} \rho_\alpha \, dv$$

is the mass of constituent α in $\mathcal{P}_\alpha^t$. We denote by τ_α the mass growth (mass produced per unit volume and unit time) due to the pertinent chemical reactions and assume the balance of mass in the form

$$\grave{\Big(\int_{\mathcal{P}_\alpha^t} \rho_\alpha dv\Big)} = \int_{\mathcal{P}_\alpha^t} \tau_\alpha dv,$$

for any region $\mathcal{P}_\alpha^t \subseteq \Omega$, subject to

$$\textstyle\sum_\alpha \tau_\alpha = 0. \tag{9.12}$$

By applying (9.11) it follows that

$$\grave{\rho}_\alpha + \rho_\alpha \nabla \cdot \mathbf{v}_\alpha = \tau_\alpha. \tag{9.13}$$

For later convenience, we now consider functions of the form $g_\alpha = \rho_\alpha h_\alpha$. In view of (9.10) and (9.13) we find that

$$\grave{\Big(\int_{\mathcal{P}_\alpha^t} \rho_\alpha h_\alpha dv\Big)} = \int_{\mathcal{P}_\alpha^t} (\rho_\alpha \grave{h}_\alpha + \tau_\alpha h_\alpha) dv. \tag{9.14}$$

The result (9.14) is the content of the transport theorem for reacting constituents.

Balance of linear momentum.

The linear momentum $\mathbf{P}$ of the region $\mathcal{P}_\alpha^t$ is

$$\mathbf{P}(\mathcal{P}_\alpha^t) = \int_{\mathcal{P}_\alpha^t} \rho_\alpha \mathbf{v}_\alpha dv.$$

Letting $\mathbf{T}_\alpha$ be the partial, or peculiar, Cauchy stress tensor and $\mathbf{b}_\alpha$ the body force per unit mass we assume

$$\grave{\Big(\int_{\mathcal{P}_\alpha^t} \rho_\alpha \mathbf{v}_\alpha \, dv\Big)} - \int_{\partial\mathcal{P}_\alpha^t} \mathbf{T}_\alpha \mathbf{n} \, da - \int_{\mathcal{P}_\alpha^t} \rho_\alpha \mathbf{b}_\alpha \, dv = \int_{\mathcal{P}_\alpha^t} \mathbf{m}_\alpha dv, \tag{9.15}$$

where $\mathbf{m}_\alpha$ is the growth of linear momentum. The growths $\{\mathbf{m}_\alpha\}$ are subject to

$$\textstyle\sum_\alpha \mathbf{m}_\alpha = \mathbf{0}. \tag{9.16}$$

As with the general form (9.9), we find the local version of (9.15) as

$$\rho_\alpha \grave{\mathbf{v}}_\alpha + \grave{\rho}_\alpha \mathbf{v}_\alpha + \rho_\alpha \mathbf{v}_\alpha \nabla \cdot \mathbf{v}_\alpha - \nabla \cdot \mathbf{T}_\alpha - \rho_\alpha \mathbf{b}_\alpha = \mathbf{m}_\alpha.$$

In view of (9.13) we obtain

$$\rho_\alpha \dot{\mathbf{v}}_\alpha - \nabla \cdot \mathbf{T}_\alpha - \rho_\alpha \mathbf{b}_\alpha = \hat{\mathbf{p}}_\alpha, \tag{9.17}$$

where

$$\hat{\mathbf{p}}_\alpha = \mathbf{m}_\alpha - \tau_\alpha \mathbf{v}_\alpha. \tag{9.18}$$

Balance of angular momentum.

Let $\mathbf{r}$ be the position vector relative to a fixed (base) point. We define the angular momentum $\mathbf{L}(\mathcal{P}_\alpha^t)$ as

$$\mathbf{L}(\mathcal{P}_\alpha^t) = \int_{\mathcal{P}_\alpha^t} \mathbf{r} \times \rho_\alpha \mathbf{v}_\alpha \, dv.$$

For technical convenience, we represent the angular momentum growth as $\mathbf{r} \times \mathbf{m}_\alpha + \gamma_\alpha$. We then assume that the balance of angular momentum is expressed by

$$\left(\int_{\mathcal{P}_\alpha^t} \rho_\alpha \mathbf{r} \times \mathbf{v}_\alpha \, dv\right)^{\grave{}} - \int_{\partial\mathcal{P}_\alpha^t} \mathbf{r} \times \mathbf{T}_\alpha \mathbf{n} \, da - \int_{\mathcal{P}_\alpha^t} \rho_\alpha \mathbf{r} \times \mathbf{b}_\alpha \, dv = \int_{\mathcal{P}_\alpha^t} (\mathbf{r} \times \mathbf{m}_\alpha + \gamma_\alpha) dv. \tag{9.19}$$

The growths $\{\gamma_\alpha\}$ are subject to

$$\sum_\alpha \gamma_\alpha = \mathbf{0}$$

and hence, consistently,

$$\sum_\alpha (\mathbf{r} \times \mathbf{m}_\alpha + \gamma_\alpha) = \mathbf{0}.$$

Look at $\mathbf{r} \times \mathbf{T}_\alpha \mathbf{n}$ and observe

$$\mathbf{r} \times \mathbf{T}_\alpha \mathbf{n} = \varepsilon_{ijk} r_j T_{kh}^\alpha n_h \mathbf{e}_i.$$

Since

$$\partial_{x_h}(r_j T_{kh}^\alpha) = T_{kj}^\alpha + r_j (\nabla \cdot \mathbf{T}_\alpha)_k,$$

by means of the divergence theorem we find that

$$\int_{\partial\mathcal{P}_\alpha^t} (\mathbf{r} \times \mathbf{T}_\alpha \mathbf{n}) da = \int_{\mathcal{P}_\alpha^t} [(\mathbf{w}_\alpha + \mathbf{r} \times (\nabla \cdot \mathbf{T}_\alpha)] dv,$$

where

$$\mathbf{w}_\alpha = \varepsilon_{ijk} T_{kj}^\alpha \mathbf{e}_i.$$

Moreover, since $\dot{\mathbf{r}}$ is the derivative of $\mathbf{r}$ at $\mathbf{X}_\alpha$ fixed, $\dot{\mathbf{r}} = \mathbf{v}_\alpha$, then

$$\left(\int_{\mathcal{P}_\alpha^t} \rho_\alpha \mathbf{r} \times \mathbf{v}_\alpha \, dv\right)^{\grave{}} = \int_{\mathcal{P}_\alpha^t} \mathbf{r} \times (\rho_\alpha \dot{\mathbf{v}}_\alpha + \tau_\alpha \mathbf{v}_\alpha) dv.$$

Hence we find that

$$\int_{\mathcal{P}^t_\alpha} [\mathbf{r} \times (\rho_\alpha \dot{\mathbf{v}}_\alpha - \nabla \cdot \mathbf{T}_\alpha - \rho_\alpha \mathbf{b}_\alpha - \mathbf{m}_\alpha) - \mathbf{w}_\alpha - \gamma_\alpha] dv = \mathbf{0}.$$

By (9.17), the arbitrariness of $\mathcal{P}^t_\alpha$ and the continuity of the integrand imply that

$$\mathbf{w}_\alpha = -\gamma_\alpha.$$

Each stress tensor $\mathbf{T}_\alpha$ need not be symmetric. By the definition of $\mathbf{w}_\alpha$ we find that

$$T^\alpha_{pq} - T^\alpha_{qp} = \varepsilon_{pqi} \gamma^\alpha_i. \tag{9.20}$$

Let

$$\mathbf{T}_I = \sum_\alpha \mathbf{T}_\alpha.$$

Hence summation over α of (9.20) results in

$$\mathbf{T}_I - \mathbf{T}_I^T = \mathbf{0}. \tag{9.21}$$

Balance of energy.
Denote by ε_α the internal energy, per unit mass. Hence we define the (internal + kinetic) energy $E(\mathcal{P}^t_\alpha)$ as

$$E(\mathcal{P}^t_\alpha) = \int_{\mathcal{P}^t_\alpha} \rho_\alpha (\varepsilon_\alpha + \tfrac{1}{2} \mathbf{v}^2_\alpha) dv.$$

Moreover let $\mathbf{q}_\alpha$ be the heat flux vector and r_α the energy supply, per unit mass, in the αth constituent. We then assume that the balance of energy is expressed by

$$\Big(\int_{\mathcal{P}^t_\alpha} \rho_\alpha (\varepsilon_\alpha + \tfrac{1}{2} \mathbf{v}^2_\alpha) dv \Big)^{\grave{}} - \int_{\partial \mathcal{P}^t_\alpha} (\mathbf{v}_\alpha \cdot \mathbf{T}_\alpha - \mathbf{q}_\alpha) \cdot \mathbf{n}\, da - \int_{\mathcal{P}^t_\alpha} \rho_\alpha (\mathbf{b}_\alpha \cdot \mathbf{v}_\alpha + r_\alpha) dv = \int_{\mathcal{P}^t_\alpha} e_\alpha\, dv, \tag{9.22}$$

where the growths $\{e_\alpha\}$ are subject to

$$\sum_\alpha e_\alpha = 0.$$

In local form we obtain

$$\rho_\alpha \grave{\varepsilon}_\alpha + \tau_\alpha (\varepsilon_\alpha + \tfrac{1}{2} \mathbf{v}^2_\alpha) + \mathbf{v}_\alpha \cdot (\rho_\alpha \dot{\mathbf{v}}_\alpha - \nabla \cdot \mathbf{T}_\alpha - \rho_\alpha \mathbf{b}_\alpha) + \nabla \cdot \mathbf{q}_\alpha - \mathbf{T}_\alpha \cdot \mathbf{L}_\alpha + \rho_\alpha r_\alpha = e_\alpha.$$

Upon substitution from the equation of motion (9.17) we obtain the balance of internal energy in the form

$$\rho_\alpha \grave{\varepsilon}_\alpha + \tau_\alpha (\varepsilon_\alpha - \tfrac{1}{2} \mathbf{v}^2_\alpha) - \mathbf{T}_\alpha \cdot \mathbf{L}_\alpha + \nabla \cdot \mathbf{q}_\alpha - \rho_\alpha r_\alpha + \mathbf{m}_\alpha \cdot \mathbf{v}_\alpha = e_\alpha. \tag{9.23}$$

Equations (9.13), (9.17), and (9.23) are the (local) continuity equation, the equation of motion, and the energy equation with the peculiar derivative, namely, the time derivative at fixed $\mathbf{X}_\alpha$. To sum up, we can state them in the form

$$\begin{aligned}&\grave{\rho}_\alpha + \rho_\alpha \nabla \cdot \mathbf{v}_\alpha = \tau_\alpha\\&\rho_\alpha \grave{\mathbf{v}}_\alpha - \nabla \cdot \mathbf{T}_\alpha - \rho_\alpha \mathbf{b}_\alpha = \hat{\mathbf{p}}_\alpha,\\&\rho_\alpha \grave{\varepsilon}_\alpha - \mathbf{T}_\alpha \cdot \mathbf{L}_\alpha + \nabla \cdot \mathbf{q}_\alpha - \rho_\alpha r_\alpha = \hat{\varepsilon}_\alpha,\end{aligned} \tag{9.24}$$

where

$$\hat{\varepsilon}_\alpha = e_\alpha - \mathbf{m}_\alpha \cdot \mathbf{v}_\alpha - \tau_\alpha(\varepsilon_\alpha - \tfrac{1}{2}\mathbf{v}_\alpha^2). \tag{9.25}$$

Hence, $\tau_\alpha, \hat{\mathbf{p}}_\alpha, \hat{\varepsilon}_\alpha$ represent the interaction effects on mass, linear momentum, and energy, relative to the observer following the motion of the αth constituent. Consequently, $\tau_\alpha, \hat{\mathbf{p}}_\alpha, \hat{\varepsilon}_\alpha$ are objective quantities. Moreover, in light of the balance of angular momentum, it follows that $\mathbf{T}_I = \sum_\alpha \mathbf{T}_\alpha$ is symmetric.

Since Eq. (9.24) involves the peculiar derivative, we can view them as the (local) balance equations in the Lagrangian description. For completeness and later convenience, we now establish the corresponding equations in the Eulerian description.

9.2.1 Balance Equations in the Eulerian Description

The balance equations (9.13), (9.17), and (9.23) are now reviewed by applying Eq. (9.5) for the peculiar time derivative. Upon substitution of $\grave{\rho}_\alpha$ from (9.5) we have

$$\partial_t \rho_\alpha + \mathbf{v}_\alpha \cdot \nabla \rho_\alpha + \rho_\alpha \nabla \cdot \mathbf{v}_\alpha = \tau_\alpha.$$

Hence we can write

$$\partial_t \rho_\alpha + \nabla \cdot (\rho_\alpha \mathbf{v}_\alpha) = \tau_\alpha. \tag{9.26}$$

By (9.26), substitution of

$$\rho_\alpha = \rho\omega_\alpha, \qquad \mathbf{v}_\alpha = \mathbf{v} + \mathbf{u}_\alpha$$

results in

$$\rho(\partial_t \omega_\alpha + \mathbf{v} \cdot \nabla \omega_\alpha) + \omega_\alpha(\partial_t \rho + \nabla \cdot (\rho \mathbf{v})) + \nabla \cdot (\rho_\alpha \mathbf{u}_\alpha) = \tau_\alpha$$

whence

$$\rho\dot{\omega}_\alpha = -\nabla \cdot \mathbf{h}_\alpha + \tau_\alpha. \tag{9.27}$$

For any (scalar, vector, tensor) function h_α observe

$$\rho_\alpha \grave{h}_\alpha = \rho_\alpha \partial_t h_\alpha + \rho_\alpha \mathbf{v}_\alpha \cdot \nabla h_\alpha = \partial_t(\rho_\alpha h_\alpha) + \nabla \cdot (\rho_\alpha h_\alpha \mathbf{v}_\alpha) - h_\alpha[\partial_t \rho_\alpha + \nabla \cdot (\rho_\alpha \mathbf{v}_\alpha)].$$

In view of (9.26), we obtain

$$\rho_\alpha \grave{h}_\alpha = \partial_t(\rho_\alpha h_\alpha) + \nabla \cdot (\rho_\alpha h_\alpha \mathbf{v}_\alpha) - \tau_\alpha h_\alpha, \tag{9.28}$$

where $h_\alpha \mathbf{v}_\alpha$ has to be meant as a dyadic product $h_\alpha \otimes \mathbf{v}_\alpha$ if h_α is a vector or a tensor and

$$\nabla \cdot (\mathbf{u} \otimes \mathbf{v}) = \partial_{x_i}(u_j v_i)\mathbf{e}_j.$$

Equation (9.28) is an identity that yields a direct connection between Lagrangian and Eulerian descriptions. By identifying h_α of (9.28) with $\mathbf{v}_\alpha$ and ε_α we obtain the balance equations (9.24) in the form

$$\begin{aligned} &\partial_t \rho_\alpha + \nabla \cdot (\rho_\alpha \mathbf{v}_\alpha) = \tau_\alpha \\ &\partial_t(\rho_\alpha \mathbf{v}_\alpha) + \nabla \cdot (\rho_\alpha \mathbf{v}_\alpha \otimes \mathbf{v}_\alpha) - \nabla \cdot \mathbf{T}_\alpha - \rho_\alpha \mathbf{b}_\alpha = \mathbf{m}_\alpha, \\ &\partial_t(\rho_\alpha \varepsilon_\alpha) + \nabla \cdot (\rho_\alpha \varepsilon_\alpha \mathbf{v}_\alpha) - \mathbf{T}_\alpha \cdot \mathbf{L}_\alpha + \nabla \cdot \mathbf{q}_\alpha - \rho_\alpha r_\alpha = \lambda_\alpha, \end{aligned} \tag{9.29}$$

where $\mathbf{m}_\alpha = \hat{\mathbf{p}}_\alpha + \tau_\alpha \mathbf{v}_\alpha$, as in (9.18), and

$$\lambda_\alpha = e_\alpha - \mathbf{m}_\alpha \cdot \mathbf{v}_\alpha + \tfrac{1}{2}\tau_\alpha \mathbf{v}_\alpha^2.$$

Alternatively, we might replace $\grave{h}_\alpha$ from (9.28) in (9.14) to obtain

$$\rho_\alpha \grave{h}_\alpha + \tau_\alpha h_\alpha = \partial_t(\rho_\alpha h_\alpha) + \nabla \cdot (\rho_\alpha h_\alpha \mathbf{v}_\alpha)$$

and hence obtain the result of the transport theorem in the form

$$\grave{\left(\int_{\mathcal{P}_\alpha^t} \rho_\alpha h_\alpha dv \right)} = \int_{\mathcal{P}_\alpha^t} [\partial_t(\rho_\alpha h_\alpha) + \nabla \cdot (\rho_\alpha h_\alpha \mathbf{v}_\alpha)] dv. \tag{9.30}$$

By applying (9.30) to (9.15) and (9.22) we obtain at once the second equation of (9.29) and

$$\partial_t[\rho_\alpha(\varepsilon_\alpha + \tfrac{1}{2}\mathbf{v}_\alpha^2)] + \nabla \cdot [\rho_\alpha(\varepsilon_\alpha + \tfrac{1}{2}\mathbf{v}_\alpha^2)\mathbf{v}_\alpha - \mathbf{v}_\alpha \mathbf{T}_\alpha + \mathbf{q}_\alpha] - \rho_\alpha \mathbf{b}_\alpha \cdot \mathbf{v}_\alpha - \rho_\alpha r_\alpha = e_\alpha. \tag{9.31}$$

It is also of interest to compute $\grave{\overline{\nabla g_\alpha}}$ for any (scalar- or tensor-valued) function g_α on $\Omega \times \mathbb{R}$. Since

$$\grave{\mathbf{w}}_\alpha = (\partial_t + \mathbf{v}_\alpha \cdot \nabla)\mathbf{w}_\alpha$$

for any function $\mathbf{w}_\alpha$ on $\Omega \times \mathbb{R}$ then, for any C^2 function g_α we have

$$\grave{\overline{\nabla g_\alpha}} = (\partial_t + \mathbf{v}_\alpha \cdot \nabla)\nabla g_\alpha = \nabla \partial_t g_\alpha + (\mathbf{v}_\alpha \cdot \nabla)\nabla g_\alpha = \nabla \partial_t g_\alpha + \nabla(\mathbf{v}_\alpha \cdot \nabla g_\alpha) - \mathbf{L}_\alpha^T \nabla g_\alpha$$

whence

$$\grave{\overline{\nabla g_\alpha}} = \nabla \grave{g}_\alpha - \mathbf{L}_\alpha^T \nabla g_\alpha. \tag{9.32}$$

In connection with vectors we may need further identities. Look at a component, say w_i^α of a vector $\mathbf{w}_\alpha$ on $\Omega \times \mathbb{R}$. Then

$$\overset{\grave{}}{\overline{\partial_{x_k} w_i^\alpha}} = (\partial_t + v_h^\alpha \partial_{x_h})\partial_{x_k} w_i^\alpha = \partial_{x_k}\partial_t w_i^\alpha + \partial_{x_k}(v_h^\alpha \partial_{x_h} w_i^\alpha) - (\partial_{x_k} v_h^\alpha)\partial_{x_h} w_i^\alpha$$

whence

$$\overset{\grave{}}{\overline{\partial_{x_k} w_i^\alpha}} = \partial_{x_k} \overset{\grave{}}{\overline{w_i^\alpha}} - L_{hk}^\alpha \partial_{x_h} w_i^\alpha .$$

In compact form

$$\overset{\grave{}}{\overline{\nabla \otimes \mathbf{w}_\alpha}} = \nabla \otimes \grave{\mathbf{w}}_\alpha - \mathbf{L}_\alpha^T \nabla \otimes \mathbf{w}_\alpha .$$

Two particular significant cases arise. If we take the trace we get the identity for the divergence,

$$\overset{\grave{}}{\overline{\nabla \cdot \mathbf{w}_\alpha}} = \nabla \cdot \grave{\mathbf{w}}_\alpha - (\mathbf{L}_\alpha^T \nabla) \cdot \mathbf{w}_\alpha . \tag{9.33}$$

Instead the skew part yields

$$\overset{\grave{}}{\overline{\nabla \times \mathbf{w}_\alpha}} = \nabla \times \grave{\mathbf{w}}_\alpha - (\mathbf{L}_\alpha^T \nabla) \times \mathbf{w}_\alpha . \tag{9.34}$$

9.2.2 Mass Density in the Reference Configuration and Incompressibility

Denote by ϱ_α the mass density in the reference configuration R_α; it is also denoted as $\rho_{\alpha R}$ in the literature. Unlike what happens in a single body, in reacting mixtures the reference mass density ϱ_α depends on time. To clarify this dependence we start with the invariance of mass from $\mathrm{P}_\alpha \subseteq \mathrm{R}_\alpha$ to the image $\chi_\alpha(\mathrm{P}_\alpha, t)$,

$$\int_{\mathrm{P}_\alpha} \varrho_\alpha(\mathbf{X}_\alpha, t) dv_R = \int_{\chi_\alpha(\mathrm{P}_\alpha, t)} \rho_\alpha(\mathbf{x}, t) dv .$$

Under the change of (vector) variable $\mathbf{x} \in \chi_\alpha(\mathrm{P}_\alpha, t) \to \mathbf{X}_\alpha \in \mathrm{P}_\alpha$ the Jacobian is J_α and we have

$$\int_{\chi_\alpha(\mathrm{P}_\alpha, t)} \rho_\alpha(\mathbf{x}, t) dv = \int_{\mathrm{P}_\alpha} (\rho_\alpha J_\alpha)(\mathbf{X}_\alpha, t) dv_R .$$

Hence it follows

$$\int_{\mathrm{P}_\alpha} (\varrho_\alpha - \rho_\alpha J_\alpha)(\mathbf{X}_\alpha, t) dv_R = 0 .$$

The arbitrariness of P_α and the supposed continuity of the integrand implies that

$$\varrho_\alpha = \rho_\alpha J_\alpha \tag{9.35}$$

at any point $\mathbf{X}_\alpha \in \mathrm{P}_\alpha$ and time t. Upon time differentiation we obtain

$$\grave{\varrho}_\alpha = \tau_\alpha J_\alpha. \tag{9.36}$$

Equation (9.36) shows that, in non-reacting mixtures, where $\tau_\alpha = 0$, the reference density ϱ_α is constant. For reacting mixtures, instead, the reference density ϱ_α depends on time and (9.36) is the value of the time derivative.

For single bodies, incompressibility means that the mass density is constant at any point of the body, $\dot{\rho} = 0$, and hence incompressibility may be characterized by the constraint $\nabla \cdot \mathbf{v} = 0$. For a mixture, we may consider constituents that are regarded as intrinsically *incompressible*, namely, incompressible as single bodies. To make this operative, let ρ^0_α be the (constant) true density of constituent α, possibly dependent on $\mathbf{X}_\alpha$. Then letting φ_α be the volume fraction of constituent α we can write

$$\rho_\alpha = \rho^0_\alpha \varphi_\alpha$$

and, consequently,

$$\rho^0_\alpha \varphi_\alpha = \frac{\varrho_\alpha}{J_\alpha}.$$

Hence, it follows that $\varphi_\alpha J_\alpha$ is a function of the referential position $\mathbf{X}_\alpha$. For formal simplicity we may let $\varrho_\alpha = \rho^0_\alpha$. Hence, for incompressible constituents,

$$\varphi_\alpha J_\alpha = 1.$$

The continuity equation (9.13) implies

$$\grave{\varphi}_\alpha + \varphi_\alpha \nabla \cdot \mathbf{v}_\alpha = \frac{\tau_\alpha}{\rho^0_\alpha}, \qquad -\grave{J}_\alpha + J_\alpha \nabla \cdot \mathbf{v}_\alpha = \frac{\tau_\alpha J^2_\alpha}{\rho^0_\alpha}.$$

These equations are characteristic of incompressible constituents. The constraint $\nabla \cdot \mathbf{v}_\alpha = 0$ need not hold for incompressible constituents.

The assumption of saturation (no voids),

$$\textstyle\sum_\alpha \varphi_\alpha = 1,$$

implies that

$$\mathbf{0} = \textstyle\sum_\alpha \nabla \varphi_\alpha, \qquad 0 = \sum_\alpha \partial_t \varphi_\alpha = \sum_\alpha [-\nabla \cdot (\varphi_\alpha \mathbf{v}_\alpha) + \frac{\tau_\alpha}{\rho^0_\alpha}].$$

Upon substitution of $\mathbf{v}_\alpha = \mathbf{v} + \mathbf{u}_\alpha$ it follows that

$$\sum_\alpha (\nabla \varphi_\alpha \cdot \mathbf{u}_\alpha + \varphi_\alpha \nabla \cdot \mathbf{v}_\alpha - \frac{\tau_\alpha}{\rho^0_\alpha}) = 0. \tag{9.37}$$

9.3 Second Law of Thermodynamics

Let η_α be the entropy density (per unit mass) and θ_α the absolute temperature. We assume the balance of entropy in the form

$$\big(\int_{\mathcal{P}_\alpha^t} \rho_\alpha \eta_\alpha \, dv\big)\grave{} + \int_{\partial \mathcal{P}_\alpha^t} \mathbf{j}_\alpha \cdot \mathbf{n} \, da - \int_{\mathcal{P}_\alpha^t} \rho_\alpha s_\alpha \, dv = \int_{\mathcal{P}_\alpha^t} \sigma_\alpha \, dv, \tag{9.38}$$

where $\mathbf{j}_\alpha$ is the entropy flux and s_α the entropy supply. The entropy growth σ_α may account for two aspects, namely, dissipative effects within the αth constituent and interaction with the other constituents of the mixture. The local counterpart of (9.38) follows in the Lagrangian and Eulerian descriptions by using (9.14),

$$\rho_\alpha \grave{\eta}_\alpha + \tau_\alpha \eta_\alpha + \nabla \cdot \mathbf{j}_\alpha - \rho_\alpha s_\alpha = \sigma_\alpha, \tag{9.39}$$

and (9.30),

$$\partial_t(\rho_\alpha \eta_\alpha) + \nabla \cdot (\mathbf{j}_\alpha + \rho_\alpha \eta_\alpha \mathbf{v}_\alpha) - \rho_\alpha s_\alpha = \sigma_\alpha. \tag{9.40}$$

As with single bodies, the entropy supply s_α is identified with r_α/θ_α. The entropy growth σ_α can also be viewed as the entropy production, within the αth constituent.

As an assumption, we now state the second law of thermodynamics.[5] *The inequality*

$$\sum_\alpha \sigma_\alpha \geq 0 \tag{9.41}$$

must hold, at any point $\mathbf{x} \in \Omega$ *and for any time* $t \in \mathbb{R}$, *for any set of functions on* $\Omega \times \mathbb{R}$ *compatible with the balance equations.*

The entropy fluxes $\{\mathbf{j}_\alpha\}$ are unknown and have to be determined, along with the other constitutive functions, so that the second law is satisfied. For formal convenience, we let

$$\mathbf{j}_\alpha = \frac{\mathbf{q}_\alpha}{\theta_\alpha} + \mathbf{k}_\alpha,$$

the (unknown) vector $\mathbf{k}_\alpha$ representing the extra-entropy flux. As with the second law for single bodies, we may take the entropy productions $\{\sigma_\alpha\}$ as given by constitutive functions.

The analysis of compatibility of constitutive assumptions with the entropy inequality (9.41) is more direct if the peculiar derivative is involved. Hence, we look at (9.39) and write the entropy inequality (9.41) in the form

$$\sum_\alpha [\rho_\alpha \grave{\eta}_\alpha + \tau_\alpha \eta_\alpha + \nabla \cdot \frac{\mathbf{q}_\alpha}{\theta_\alpha} + \nabla \cdot \mathbf{k}_\alpha - \rho_\alpha \frac{r_\alpha}{\theta_\alpha}] \geq 0.$$

[5] Or axiom of dissipation, [427].

Substitution of $\nabla \cdot \mathbf{q}_\alpha - \rho_\alpha r_\alpha$ from (9.23) yields

$$\sum_\alpha \{\frac{1}{\theta_\alpha}[\rho_\alpha \theta_\alpha \dot{\eta}_\alpha + \tau_\alpha \theta_\alpha \eta_\alpha + e_\alpha - \mathbf{m}_\alpha \cdot \mathbf{v}_\alpha + \mathbf{T}_\alpha \cdot \mathbf{L}_\alpha - \rho_\alpha \grave{\varepsilon}_\alpha - \tau_\alpha(\varepsilon_\alpha - \tfrac{1}{2}\mathbf{v}_\alpha^2) - \frac{1}{\theta_\alpha}\mathbf{q}_\alpha \cdot \nabla\theta_\alpha + \theta_\alpha \nabla \cdot \mathbf{k}_\alpha]\} \geq 0.$$

Using the Helmholtz free energy

$$\psi_\alpha = \varepsilon_\alpha - \theta_\alpha \eta_\alpha$$

we can write the inequality in the form

$$\sum_\alpha \{\frac{1}{\theta_\alpha}[-\rho_\alpha(\grave{\psi}_\alpha + \eta_\alpha \grave{\theta}_\alpha) - \tau_\alpha(\psi_\alpha - \tfrac{1}{2}\mathbf{v}_\alpha^2) + e_\alpha - \mathbf{m}_\alpha \cdot \mathbf{v}_\alpha + \mathbf{T}_\alpha \cdot \mathbf{L}_\alpha - \frac{1}{\theta_\alpha}\mathbf{q}_\alpha \cdot \nabla\theta_\alpha + \theta_\alpha \nabla \cdot \mathbf{k}_\alpha]\} \geq 0. \tag{9.42}$$

By (9.25), inequality (9.42) can also be written in terms of the supplies $\hat{\varepsilon}_\alpha$ as follows:

$$\sum_\alpha \{\frac{1}{\theta_\alpha}[-\rho_\alpha(\grave{\psi}_\alpha + \eta_\alpha \grave{\theta}_\alpha) + \tau_\alpha \theta_\alpha \eta_\alpha + \hat{\varepsilon}_\alpha + \mathbf{T}_\alpha \cdot \mathbf{L}_\alpha - \frac{1}{\theta_\alpha}\mathbf{q}_\alpha \cdot \nabla\theta_\alpha + \theta_\alpha \nabla \cdot \mathbf{k}_\alpha]\} \geq 0. \tag{9.43}$$

Next we show how the balance equations of the constituents allow the derivation of the balance equations of the mixture as a single body. Here we look at the possibility of deriving constitutive properties of the mixture as a whole along with properties of the constituents. In this sense, it is of interest to look at (9.42), or (9.43), under the assumption that the mixture is subject to a single (common) temperature, $\theta_\alpha = \theta$, $\alpha = 1, 2, ..., n$. If this is viewed as a constraint we have

$$\grave{\theta}_\alpha = \partial_t \theta + \mathbf{v}_\alpha \cdot \nabla\theta = \dot{\theta} + (\mathbf{v}_\alpha - \mathbf{v}) \cdot \nabla\theta.$$

Consequently

$$-\sum_\alpha \frac{1}{\theta_\alpha}[\rho_\alpha \eta_\alpha \grave{\theta}_\alpha + \frac{1}{\theta_\alpha}\mathbf{q}_\alpha \cdot \nabla\theta_\alpha] \to -\frac{1}{\theta}\{\rho\eta\dot{\theta} + \sum_\alpha[\mathbf{q}_\alpha + \rho_\alpha \eta_\alpha(\mathbf{v}_\alpha - \mathbf{v})] \cdot \nabla\theta\}.$$

Likewise we use the continuity equation (9.27) and find that [61]

$$\sum_\alpha \rho_\alpha \grave{\psi}_\alpha = \sum_\alpha \rho_\alpha \dot{\psi}_\alpha + \rho_\alpha \mathbf{u}_\alpha \cdot \psi_\alpha] = \sum_\alpha [\rho\, \dot{\overline{\omega_\alpha \psi_\alpha}} - \rho\, \psi_\alpha \dot{\omega}_\alpha + \mathbf{h}_\alpha \cdot \nabla\psi_\alpha]$$
$$= \rho\dot{\psi}_I - \sum_\alpha [\psi_\alpha(\tau_\alpha - \nabla \cdot \mathbf{h}_\alpha) + \mathbf{h}_\alpha \cdot \nabla\psi_\alpha = \rho\dot{\psi}_I + \nabla \cdot \sum_\alpha \psi_\alpha \mathbf{h}_\alpha - \sum_\alpha \tau_\alpha \psi_\alpha.$$

Hence inequality (9.42) can be written in the form

$$-\rho(\dot{\psi}_I + \eta\dot{\theta}) - \frac{1}{\theta}\tilde{\mathbf{q}} \cdot \nabla\theta + \sum_\alpha [\mathbf{T}_\alpha \cdot \mathbf{L}_\alpha - \tfrac{1}{2}\tau_\alpha \mathbf{u}_\alpha^2 - \hat{\mathbf{p}}_\alpha \cdot \mathbf{u}_\alpha + \theta\nabla \cdot \mathbf{k}_\alpha - \nabla \cdot (\psi_\alpha \mathbf{h}_\alpha)] \geq 0, \tag{9.44}$$

where $\psi_I = \sum_\alpha \omega_\alpha \psi_\alpha$ and $\tilde{\mathbf{q}} = \sum_\alpha (\mathbf{q}_\alpha + \eta_\alpha \mathbf{h}_\alpha)$.

Yet, this inequality requires that we look for $\psi_I, \eta, \tilde{\mathbf{q}}$ as properties of the whole mixture while $\mathbf{T}_\alpha, \hat{\mathbf{p}}_\alpha, \mathbf{k}_\alpha$ are related to the constituents. Moreover, unless there is a mechanism equalling the temperatures, from a physical viewpoint the assumption that the constituents are subject to a common temperature merely means that the temperatures θ_α are almost equal thanks to the exchange of energy among the constituents. Hence, we find it reasonable to determine peculiar properties of the constituents and next to examine the remaining dissipation inequality within the assumption (or approximation) $\theta_\alpha = \theta$.

If $\theta_\alpha = \theta$ then we observe that $\sum_\alpha e_\alpha = 0$ and, in light of (9.42), we consider

$$\Phi := \sum_\alpha [\mathbf{m}_\alpha \cdot \mathbf{v}_\alpha - \tfrac{1}{2}\tau_\alpha \mathbf{v}_\alpha^2].$$

By (9.18) we have

$$\mathbf{m}_\alpha \cdot \mathbf{v}_\alpha - \tfrac{1}{2}\tau_\alpha \mathbf{v}_\alpha^2 = \hat{\mathbf{p}}_\alpha \cdot \mathbf{v}_\alpha + \tfrac{1}{2}\tau_\alpha \mathbf{v}_\alpha^2.$$

For any velocity $\mathbf{V}$ let

$$\check{\mathbf{v}}_\alpha = \mathbf{v}_\alpha - \mathbf{V}.$$

Hence, by (9.12) we have

$$\sum_\alpha \tau_\alpha \mathbf{v}_\alpha = \sum_\alpha \tau_\alpha \check{\mathbf{v}}_\alpha.$$

Consequently, upon replacing $\mathbf{v}_\alpha$ with $\mathbf{V} + \check{\mathbf{v}}_\alpha$ the constraint (9.16) results in

$$\mathbf{0} = \sum_\alpha \mathbf{m}_\alpha = \sum_\alpha (\hat{\mathbf{p}}_\alpha + \tau_\alpha \mathbf{v}_\alpha) = \sum_\alpha (\hat{\mathbf{p}}_\alpha + \tau_\alpha \check{\mathbf{v}}_\alpha). \tag{9.45}$$

By computing and using (9.12) and (9.45) we obtain

$$\Phi = \sum_\alpha (\hat{\mathbf{p}}_\alpha \cdot \mathbf{v}_\alpha + \tfrac{1}{2}\tau_\alpha \mathbf{v}_\alpha^2) = \sum_\alpha (\hat{\mathbf{p}}_\alpha \cdot \check{\mathbf{v}}_\alpha + \tfrac{1}{2}\tau_\alpha \check{\mathbf{v}}_\alpha^2).$$

Thus the scalar Φ is objective, namely, is invariant under a change of frame,

$$\mathbf{v}_\alpha \to \check{\mathbf{v}}_\alpha = \mathbf{v}_\alpha - \mathbf{V}.$$

In particular, we can take $\mathbf{V}$ as the barycentric velocity $\mathbf{v}$, and hence $\check{\mathbf{v}}_\alpha = \mathbf{u}_\alpha$.

9.3.1 *Principle of Phase Separation*

A classical rule for the formulation of constitutive equations is the principle of equipresence[6] which states that a variable present as an independent variable in one constitutive equation should be so present in all, unless the symmetry of the material, the principle of material objectivity, and the second law of thermodynamics require otherwise.

---

[6] See [429], pp. 703–704, and [95].

According to Passman and Nunziato [356], in mixtures (where) the individual constituents are clearly separated physically, it is plausible to think the material properties of the constituents as being separated from one another.[7] The principle of equipresence is then replaced by the principle of phase separation. By this principle, the material-specific-dependent variables of a given constituent (such as the stress $\mathbf{T}_\alpha$ and the internal energy ε_α) depend only on the independent variables of that constituent. The interaction terms (such as the interaction force $\mathbf{m}_\alpha$) depend on all the independent variables.

9.3.2 *Remarks on the Thermodynamic Restrictions*

Assume the constitutive functions have the form

$$\psi_\alpha = \psi_\alpha(\theta_\alpha, \nabla\theta_\alpha, \Upsilon_\alpha),$$

where Υ_α is a set of variables independent of $\theta_\alpha, \nabla\theta_\alpha, \grave{\theta}_\alpha, \overline{\nabla\theta_\alpha}^{\grave{}}$. Hence inequality (9.42) takes the form

$$\Sigma_\alpha\{\frac{1}{\theta_\alpha}[-\rho_\alpha(\partial_{\theta_\alpha}\psi_\alpha + \eta_\alpha)\grave{\theta}_\alpha - \rho_\alpha\partial_{\nabla\theta_\alpha}\psi_\alpha\, \overline{\nabla\theta_\alpha}^{\grave{}} + ...]\} \geq 0, \tag{9.46}$$

the dots indicating terms independent of $\grave{\theta}_\alpha$ and $\overline{\nabla\theta_\alpha}^{\grave{}}$. Since ε_α depends on θ_α and $\nabla\theta_\alpha$ then, by (9.23), we have

$$\rho_\alpha\partial_{\theta_\alpha}\varepsilon_\alpha\,\grave{\theta}_\alpha + \rho_\alpha\partial_{\nabla\theta_\alpha}\varepsilon_\alpha\,\overline{\nabla\theta_\alpha}^{\grave{}} + ... = \rho_\alpha r_\alpha.$$

At any point $\mathbf{x}$ and time t, $\grave{\theta}_\alpha$ can take any scalar value and meanwhile (9.23) holds by letting r_α take the appropriate value. Similarly, we can take the gradient of (9.23) and observe that the arbitrary value of $\overline{\nabla\theta_\alpha}^{\grave{}}$ is allowed by letting ∇r_α be appropriate. Consequently, inequality (9.46) has to hold for arbitrary values of $\grave{\theta}_\alpha$ and $\overline{\nabla\theta_\alpha}^{\grave{}}$. The linearity then implies that

$$\eta_\alpha = -\partial_{\theta_\alpha}\psi_\alpha, \qquad \partial_{\nabla\theta_\alpha}\psi_\alpha = 0.$$

9.4 Balance Equations for the Whole Mixture

It is natural to ask for the balance equations of a mixture as a single body and to look for the connection with the balance equations for the constituents. It is a

[7] See, also, [329].

guiding principle of the theory that the motion of the mixture is governed by the same equations as is a single body [427]. This principle in turn should indicate the appropriate definitions of the quantities pertaining to the whole mixture[8] in terms of the quantities related to the constituents. The starting point is the set of balance equations in Eulerian form (9.29) which enjoy the conceptual advantage of being referred to a single observer.

Summation of $(9.29)_1$ over α, the definitions

$$\rho = \sum_\alpha \rho_\alpha, \qquad \rho\mathbf{v} = \sum_\alpha \rho_\alpha \mathbf{v}_\alpha,$$

and the constraint (9.12) yields

$$\partial_t \rho + \nabla \cdot (\rho\mathbf{v}) = 0, \tag{9.47}$$

which is the continuity equation for single bodies with mass density ρ and velocity $\mathbf{v}$. The barycentric velocity $\mathbf{v}$ plays the role of the velocity of the mixture viewed as a single body.

Denote by a superposed dot the derivative following the motion defined by the velocity $\mathbf{v}$. For any (scalar or vector) function $\mathbf{w}(\mathbf{x}, t)$, we have

$$\dot{\mathbf{w}} = \partial_t \mathbf{w} + \mathbf{v} \cdot \nabla \mathbf{w}.$$

Since $\nabla \cdot (\rho\mathbf{v}) = \rho\nabla \cdot \mathbf{v} + (\mathbf{v} \cdot \nabla)\rho$ then (9.47) can be written in the form

$$\dot{\rho} + \rho\nabla \cdot \mathbf{v} = 0. \tag{9.48}$$

Following are some mathematical consequences of (9.47). First, in light of (9.13) we have

$$\dot{\rho}\omega_\alpha + \rho\dot{\omega}_\alpha = \overline{\rho\omega_\alpha}^{\,\cdot} = -\rho_\alpha \nabla \cdot (\mathbf{v} + \mathbf{u}_\alpha) + \tau_\alpha - (\mathbf{u}_\alpha \cdot \nabla)\rho_\alpha.$$

Consequently, in view of (9.48) it follows

$$\rho\dot{\omega}_\alpha = -\nabla \cdot (\rho_\alpha \mathbf{u}_\alpha) + \tau_\alpha. \tag{9.49}$$

Since

$$\partial_t(\rho\mathbf{w}) + \nabla \cdot (\rho\mathbf{w} \otimes \mathbf{v}) = \mathbf{w}[\partial_t \rho + \nabla \cdot (\rho\mathbf{v})] + \rho(\partial_t \mathbf{w} + \mathbf{v} \cdot \nabla \mathbf{w})$$

in view of (9.47) we find the identity

$$\partial_t(\rho\mathbf{w}) + \nabla \cdot (\rho\mathbf{w} \otimes \mathbf{v}) = \rho\dot{\mathbf{w}}. \tag{9.50}$$

---

[8] For example, internal energy, stress tensor, and heat flux.

We now examine Eq. $(9.29)_2$. Observe that replacing $\mathbf{v}_\alpha$ with $\mathbf{v}+\mathbf{u}_\alpha$ and accounting for the property (9.8) we have

$$\sum_\alpha \rho_\alpha \mathbf{v}_\alpha \otimes \mathbf{v}_\alpha = \rho \mathbf{v}\otimes\mathbf{v} + \sum_\alpha \rho_\alpha \mathbf{u}_\alpha \otimes \mathbf{u}_\alpha .$$

Hence summation of $(9.29)_2$ over α and using (9.16) we find

$$\partial_t(\rho\mathbf{v}) + \nabla\cdot(\rho\mathbf{v}\otimes\mathbf{v}) + \nabla\cdot\sum_\alpha \rho_\alpha \mathbf{u}_\alpha\otimes\mathbf{u}_\alpha - \nabla\cdot\mathbf{T}_{_I} - \rho\mathbf{b} = \mathbf{0}, \qquad (9.51)$$

where

$$\mathbf{T}_{_I} = \sum_\alpha \mathbf{T}_\alpha, \qquad \rho\mathbf{b} = \sum_\alpha \rho_\alpha \mathbf{b}_\alpha .$$

Letting

$$\mathbf{T} = \sum_\alpha (\mathbf{T}_\alpha - \rho_\alpha \mathbf{u}_\alpha \otimes \mathbf{u}_\alpha). \qquad (9.52)$$

and using (9.50) we can write Eq. (9.51) as

$$\rho\dot{\mathbf{v}} = \nabla\cdot\mathbf{T} + \rho\mathbf{b}. \qquad (9.53)$$

Equation (9.53) has the standard form of the equation of motion for a single body. The stress tensor $\mathbf{T}$ is given by (9.52) and is the result of the partial stresses $\{\mathbf{T}_\alpha\}$ and the additive contribution $-\sum_\alpha \rho_\alpha \mathbf{u}_\alpha\otimes\mathbf{u}_\alpha$ originated by the diffusion velocities $\{\mathbf{u}_\alpha\}$.

As a consequence of the balance of angular momentum Eq. (9.21) holds and states that $\mathbf{T}_{_I}$ is symmetric. Since

$$\mathbf{T} = \mathbf{T}_{_I} - \sum_\alpha \rho_\alpha \mathbf{u}_\alpha \otimes \mathbf{u}_\alpha$$

then the symmetry of $\mathbf{T}_{_I}$ implies that of $\mathbf{T}$.

To obtain the balance of energy we first consider (9.31) which involves the total energy density $\varepsilon_\alpha + \frac{1}{2}\mathbf{v}_\alpha^2$. Replace $\mathbf{v}_\alpha$ by $\mathbf{v}+\mathbf{u}_\alpha$ and then evaluate the sum of (9.31). In view of the constraint (9.8) and the definitions of $\mathbf{T}$ and $\mathbf{b}$ we can write the resulting equation as

$$\begin{aligned}&\partial_t(\tfrac{1}{2}\rho\mathbf{v}^2) + \nabla\cdot(\tfrac{1}{2}\rho\mathbf{v}^2\,\mathbf{v}) + \partial_t(\rho\varepsilon) + \nabla\cdot(\rho\varepsilon\mathbf{v}) - \nabla\cdot(\mathbf{v}\mathbf{T}) - \rho\mathbf{b}\cdot\mathbf{v}\\ &+\nabla\cdot\sum_\alpha[\mathbf{q}_\alpha - \mathbf{u}_\alpha\mathbf{T}_\alpha + \rho_\alpha(\varepsilon_\alpha + \tfrac{1}{2}\mathbf{u}_\alpha^2)\mathbf{u}_\alpha] - \sum_\alpha(\rho_\alpha r_\alpha + \rho_\alpha\mathbf{b}_\alpha\cdot\mathbf{u}_\alpha) = 0,\end{aligned}$$

where

$$\rho\varepsilon = \sum_\alpha \rho_\alpha(\varepsilon_\alpha + \tfrac{1}{2}\mathbf{u}_\alpha^2). \qquad (9.54)$$

Now, by using the identity (9.50) and the balance equation (9.53) we find that

$$\partial_t(\tfrac{1}{2}\rho\mathbf{v}^2) + \nabla\cdot(\tfrac{1}{2}\rho\mathbf{v}^2\,\mathbf{v}) - \nabla\cdot(\mathbf{v}\mathbf{T}) - \rho\mathbf{b}\cdot\mathbf{v} = \mathbf{v}(\rho\dot{\mathbf{v}} - \nabla\cdot\mathbf{T} - \rho\mathbf{b}) - \mathbf{T}\cdot\mathbf{D} = -\mathbf{T}\cdot\mathbf{D},$$

where the symmetry of $\mathbf{T}$ is used. Since

$$\partial_t(\rho\varepsilon) + \nabla\cdot(\rho\varepsilon\mathbf{v}) = \rho\dot{\varepsilon},$$

the remaining terms provide the standard balance of internal energy,

$$\rho\dot{\varepsilon} = \mathbf{T}\cdot\mathbf{D} - \nabla\cdot\mathbf{q} + \rho r, \tag{9.55}$$

provided only that we define $\mathbf{q}$ and r by the equations

$$\mathbf{q} = \textstyle\sum_\alpha[\mathbf{q}_\alpha - \mathbf{u}_\alpha\mathbf{T}_\alpha + \rho_\alpha(\varepsilon_\alpha + \frac{1}{2}\mathbf{u}_\alpha^2)\mathbf{u}_\alpha], \tag{9.56}$$

$$\rho r = \textstyle\sum_\alpha(\rho_\alpha r_\alpha + \rho_\alpha\mathbf{b}_\alpha\cdot\mathbf{u}_\alpha). \tag{9.57}$$

It is worth remarking that the quantities associated with the mixture are not merely the sum, or the average, of the corresponding quantities associated with the constituents. While

$$\rho = \textstyle\sum_\alpha\rho_\alpha, \qquad \mathbf{v} = \sum_\alpha\omega_\alpha\mathbf{v}_\alpha, \qquad \mathbf{b} = \sum_\alpha\omega_\alpha\mathbf{b}_\alpha,$$

analogous relations are not true for $\mathbf{T}$, ε, $\mathbf{q}$, and r as it follows from (9.52), (9.54), (9.56), and (9.57). Only if the diffusion velocities $\{\mathbf{u}_\alpha\}$ vanish, which is a trivial kinematic case, $\mathbf{T}$ and $\mathbf{q}$ are the sum while ε and r are the average.

If h_α and k_α are partial fields, the fields

$$h = \textstyle\sum_\alpha h_\alpha, \qquad k = \sum_\alpha\omega_\alpha k_\alpha$$

are named inner part of h and k. Only for the mass density and the velocity, the field associated with the whole mixture equals the inner part. Moreover, the inner part of ρ is merely the sum whereas that of $\mathbf{v}$ is the average.

It is worth deriving an approximated balance of energy occurring in the literature. Let the constituents be incompressible inviscid fluids so that $\mathbf{T}_\alpha = -p_\alpha\mathbf{1}$. The power $\mathbf{T}\cdot\mathbf{D}$ becomes

$$\mathbf{T}\cdot\mathbf{D} = -\textstyle\sum_\alpha(p_\alpha\mathbf{1} + \rho_\alpha\mathbf{u}_\alpha\otimes\mathbf{u}_\alpha)\cdot\mathbf{D} = -(\sum_\alpha p_\alpha)\nabla\cdot\mathbf{v} - (\sum_\alpha\rho_\alpha\mathbf{u}_\alpha\otimes\mathbf{u}_\alpha)\cdot\mathbf{D}.$$

If the last term is ignored and it is assumed that $\nabla\cdot\mathbf{v} = 0$ then $\mathbf{T}\cdot\mathbf{D} = 0$. In addition, if third-order terms $\frac{1}{2}\rho_\alpha\mathbf{u}_\alpha^2\,\mathbf{u}_\alpha$ in the expression of $\mathbf{q}$ are ignored then

$$\mathbf{q} = \mathbf{q}_I + \textstyle\sum_\alpha(p_\alpha\mathbf{u}_\alpha + \rho_\alpha\varepsilon_\alpha\mathbf{u}_\alpha).$$

In terms of the enthalpies[9] $\chi_\alpha = p_\alpha/\rho_\alpha + \varepsilon_\alpha$ we can write

---

[9] Often in the literature the specific enthalpy is denoted by h; here we do not use h_α to avoid ambiguities with the diffusion flux $\mathbf{h}_\alpha$.

$$\mathbf{q} = \mathbf{q}_I + \sum_\alpha \chi_\alpha \mathbf{h}_\alpha.$$

Moreover neglect the energy supply ρr. Within these approximations we obtain the balance of energy in the form

$$\rho \dot{\varepsilon} = -\nabla \cdot \mathbf{q}_I - \nabla \cdot \sum_\alpha \chi_\alpha \mathbf{h}_\alpha. \tag{9.58}$$

The entropy principle for the whole mixture

As a metaphysical principle, the balance equations of a mixture are the same as those for a single body [427]. The second law of thermodynamics (or entropy principle) can then be stated by assuming that there are an entropy density η (of the mixture), an entropy flux $\mathbf{j}$, and an entropy supply s such that the inequality

$$\rho \dot{\eta} + \nabla \cdot \mathbf{j} - \rho s \geq 0 \tag{9.59}$$

holds for the set of constitutive equations subject to the balance equations. To guarantee the consistency with (9.41), we consider (9.40) and sum over α to obtain

$$\partial_t \sum_\alpha \rho_\alpha \eta_\alpha + \nabla \cdot \sum_\alpha \rho_\alpha \eta_\alpha \mathbf{v}_\alpha + \nabla \cdot \sum_\alpha \mathbf{j}_\alpha - \sum_\alpha \rho_\alpha s_\alpha \geq 0.$$

This suggests that we define η by

$$\rho \eta = \sum_\alpha \rho_\alpha \eta_\alpha.$$

Upon replacing $\mathbf{v}_\alpha$ by $\mathbf{v} + \mathbf{u}_\alpha$ and using the identity (9.50) we find

$$\rho \dot{\eta} + \nabla \cdot \sum_\alpha (\mathbf{j}_\alpha + \rho_\alpha \eta_\alpha \mathbf{u}_\alpha) - \sum_\alpha \rho_\alpha s_\alpha \geq 0. \tag{9.60}$$

This means that the second law in the form (9.41) is consistent with the entropy principle (9.59) on condition that we make the identifications

$$\mathbf{j} = \sum_\alpha (\mathbf{j}_\alpha + \rho_\alpha \eta_\alpha \mathbf{u}_\alpha), \qquad \rho s = \sum_\alpha \rho_\alpha s_\alpha,$$

in addition to the definition of η as the average of the entropies $\{\eta_\alpha\}$.

The entropy flux $\mathbf{j}$ will be taken in the form $\mathbf{j} = \mathbf{q}/\theta + \mathbf{k}$ with $\mathbf{j}$ a constitutive function to be characterized by the second-law inequality. Instead s has to be defined and two definitions seem to be consistent but inequivalent. In the 70s, the literature (e.g. [61, 206, 334]) favoured the definition

$$s = s_r := \frac{1}{\theta} \sum_\alpha \omega_\alpha r_\alpha.$$

Alternatively we might assume

$$s = s_{rb} := \frac{1}{\theta}\sum_\alpha \omega_\alpha(r_\alpha + \mathbf{u}_\alpha \cdot \mathbf{b}_\alpha).$$

For instance, Müller [334] supports s_r on the basis of the kinetic theory of gases. Instead s_{rb} may be justified by both r_α and $\mathbf{u}_\alpha \cdot \mathbf{b}_\alpha$ being external powers.

We now show the consequences of the two definitions of s. Since we consider the whole mixture it is natural to assume that the constituents have a common temperature θ. Hence in both cases, upon multiplication of (9.60) by θ it follows

$$\rho\theta\dot{\eta} + \theta\nabla\cdot\mathbf{j} - \theta\sum_\alpha \rho_\alpha s_\alpha \geq 0. \tag{9.61}$$

Let $s = s_r$. Since the balance of internal energy (9.55) can be given the form

$$\rho\dot{\varepsilon} = \mathbf{T}\cdot\mathbf{D} - \nabla\cdot\mathbf{q} + \sum_\alpha \mathbf{h}_\alpha \cdot \mathbf{b}_\alpha + \sum_\alpha \rho_\alpha r_\alpha,$$

then the inequality can be written as

$$-\rho\dot{\varepsilon} + \rho\theta\dot{\eta} + \theta\nabla\cdot\mathbf{j} + \mathbf{T}\cdot\mathbf{D} - \nabla\cdot\mathbf{q} + \sum_\alpha \mathbf{h}_\alpha \cdot \mathbf{b}_\alpha \geq 0.$$

Letting, as is customary for single bodies, $\varepsilon = \psi + \theta\eta$ and $\mathbf{j} = \mathbf{q}/\theta + \mathbf{k}$ we find the second-law inequality in the form

$$-\rho\dot{\psi} - \rho\eta\dot{\theta} + \mathbf{T}\cdot\mathbf{D} - \frac{1}{\theta}\mathbf{q}\cdot\nabla\theta + \theta\nabla\cdot\mathbf{k} + \sum_\alpha \mathbf{h}_\alpha \cdot \mathbf{b}_\alpha \geq 0. \tag{9.62}$$

Let $s = s_{rb}$. The second-law inequality becomes

$$-\rho\dot{\psi} - \rho\eta\dot{\theta} + \mathbf{T}\cdot\mathbf{D} - \frac{1}{\theta}\mathbf{q}\cdot\nabla\theta + \theta\nabla\cdot\mathbf{k} \geq 0. \tag{9.63}$$

Inequalities (9.62) and (9.63) are the Clausius–Duhem inequalities for the whole mixture in terms of the Helmholtz free energy $\psi = \varepsilon - \theta\eta$. The difference between the two inequalities is that $\sum_\alpha \mathbf{h}_\alpha \cdot \mathbf{b}_\alpha$ occurs if we let $s = s_r$. This term vanishes, and the inequalities are equal, if $\mathbf{b}_\alpha$ is the same for all constituents, $\mathbf{b}_\alpha = \mathbf{b}$, in that

$$\sum_\alpha \mathbf{h}_\alpha \cdot \mathbf{b}_\alpha = (\sum_\alpha \mathbf{h}_\alpha)\cdot\mathbf{b} = 0.$$

This is so, e.g. if the body force is due to gravity, $\mathbf{b}_\alpha = \mathbf{g}$ being the gravity acceleration. Of course, no ambiguity arises if the second law is considered in the detailed form (9.42).

To conclude, the properties of the whole mixture, in terms of the partial quantities of the constituents, are given as follows:

$$\rho = \sum_\alpha \rho_\alpha, \quad \mathbf{v} = \sum_\alpha \omega_\alpha \mathbf{v}_\alpha, \quad \mathbf{b} = \sum_\alpha \omega_\alpha \mathbf{b}_\alpha, \quad \mathbf{T} = \sum_\alpha (\mathbf{T}_\alpha - \rho_\alpha \mathbf{u}_\alpha \otimes \mathbf{u}_\alpha),$$

$$\varepsilon = \sum_\alpha \omega_\alpha(\varepsilon_\alpha + \tfrac{1}{2}\mathbf{u}_\alpha^2), \quad \mathbf{q} = \sum_\alpha[\mathbf{q}_\alpha - \mathbf{u}_\alpha\mathbf{T}_\alpha + \rho_\alpha(\varepsilon_\alpha + \tfrac{1}{2}\mathbf{u}_\alpha^2)\mathbf{u}_\alpha], \quad r = \sum_\alpha \omega_\alpha(r_\alpha + \mathbf{b}_\alpha \cdot \mathbf{u}_\alpha)$$

$$\eta = \sum_\alpha \omega_\alpha \eta_\alpha, \quad s = \sum_\alpha \omega_\alpha s_\alpha, \quad \mathbf{j} = \sum_\alpha(\frac{\mathbf{q}_\alpha}{\theta_\alpha} + \mathbf{k}_\alpha + \omega_\alpha\eta_\alpha\mathbf{u}_\alpha), \quad \psi = \sum_\alpha \omega_\alpha(\psi_\alpha + \tfrac{1}{2}\mathbf{u}_\alpha^2).$$

Moreover it is

$$\sum_\alpha \omega_\alpha = 1, \quad \sum_\alpha \omega_\alpha \mathbf{u}_\alpha = \mathbf{0}, \quad \sum_\alpha \mathbf{m}_\alpha = \mathbf{0}, \quad \sum_\alpha e_\alpha = 0,$$

and

$$\mathbf{m}_\alpha = \hat{\mathbf{p}}_\alpha + \tau_\alpha \mathbf{v}_\alpha, \qquad e_\alpha = \hat{\varepsilon}_\alpha + \mathbf{m}_\alpha \cdot \mathbf{v}_\alpha + \tau_\alpha(\varepsilon_\alpha - \tfrac{1}{2}\mathbf{v}_\alpha^2).$$

The quantities

$$\mathbf{T}_I = \sum_\alpha \mathbf{T}_\alpha, \qquad \varepsilon_I = \sum_\alpha \omega_\alpha \varepsilon_\alpha, \qquad \mathbf{q}_I = \sum_\alpha \mathbf{q}_\alpha, \quad \psi_I = \sum_\alpha \omega_\alpha \psi_\alpha$$

are the *inner* parts of $\mathbf{T}$, ε, $\mathbf{q}$, ψ.

The second law of thermodynamics for mixtures is given different forms depending on whether we give evidence to the single constituents or we look at the mixture as a whole.

So far (9.41) and (9.59) should yield the same consequences on the constitutive models. In classical approaches though inequivalent constitutive models may arise merely because of the pertinent assumptions. Section 9.10 shows the derivation of some models with different thermodynamic assumptions.

9.4.1 Stoichiometry Requirements on the Mass Growths

In a chemical reaction, also the atomic substances making up the constituents are indestructible. This results in additional requirements on the mass growths $\{\tau_\alpha\}$. As an example, in a mixture of carbon, C; molecular oxygen, O_2; carbon dioxide, CO_2; and carbon monoxide, CO, the atoms of C and O are conserved.

Let M_α be the molecular weight of the αth constituent E_α. Hence τ_α/M_α is the number of moles of E_α produced or absorbed per unit time and unit volume.

Let T_i^α be the number of atoms A_i in the constituent E_α. Relative to the example, $A_1 = \mathrm{C}$, $A_2 = \mathrm{O}$, and

$$\{T_1^\alpha\} = (1, 0, 1, 1), \qquad \{T_2^\alpha\} = (0, 2, 2, 1).$$

The net molar rate of production of the ith atomic element results in

$$\sum_\alpha T_i^\alpha \frac{\tau_\alpha}{M_\alpha} = 0 \tag{9.64}$$

for every ith atomic substance. Equation (9.64) for any i is a set of restrictions on the growths $\{\tau_\alpha\}$.

Let W_i be the atomic weight of A_i. Hence

$$M_\alpha = \sum_i T_i^\alpha W_i.$$

Multiplying Eq. (9.64) by W_i and sum over i, we have

$$0 = \sum_i \sum_\alpha T_i^\alpha W_i \frac{\tau_\alpha}{M_\alpha} = \sum_\alpha (\sum_i T_i^\alpha W_i) \frac{\tau_\alpha}{M_\alpha} = \sum_\alpha M_\alpha \frac{\tau_\alpha}{M_\alpha} = \sum_\alpha \tau_\alpha;$$

we then find again that the growths $\{\tau_\alpha\}$ have a zero net value.

9.4.2 *Balance Equation for the Diffusion Flux*

We now follow a general scheme of diffusion flux, consistent with the notion that diffusion is relative motion of constituents. First let $\mathbf{v}_*$ be a reference velocity field. For generality, we let $\mathbf{v}_*$ stand for the velocity of any constituent or that of the mixture (barycentric velocity). We then consider the diffusion flux

$$\mathbf{h}_\alpha^* := \rho_\alpha (\mathbf{v}_\alpha - \mathbf{v}_*) \tag{9.65}$$

relative to $\mathbf{v}_*$. The interest in the flux $\mathbf{h}_\alpha^*$, relative to the velocity $\mathbf{v}_*$ of a constituent, is associated with the motion of the solute relative to the solvent or the motion of a constituent relative to markers in a crystal lattice[10] [390].

The evolution (or balance) equation of $\mathbf{h}_\alpha^*$ is inherited from the continuity equations and the equations of motion, (9.24) and (9.53), as appropriate. Next developments require a careful distinction between peculiar derivatives. Hence for functions g_α, g_* on $\Omega \times \mathbb{R}$ we let

$$\grave{g}_\alpha = (\partial_t + \mathbf{v}_\alpha \cdot \nabla) g_\alpha, \qquad \grave{g}_* = (\partial_t + \mathbf{v}_* \cdot \nabla) g_*.$$

Moreover, for any function on $\Omega \times \mathbb{R}$ we need the operator

$$D^* := \partial_t + \mathbf{v}_* \cdot \nabla,$$

namely, the time derivative relative to the observer that moves with velocity $\mathbf{v}_*$.

Let

$$\mathbf{u}_\alpha^* := \mathbf{v}_\alpha - \mathbf{v}_*.$$

It follows that

[10] In these cases, $\mathbf{v}_*$ is the velocity of the solvent or that of the markers.

$$D^* g_\alpha = (\partial_t + \mathbf{v}_* \cdot \nabla) g_\alpha = (\partial_t + \mathbf{v}_\alpha \cdot \nabla) g_\alpha - \mathbf{u}_\alpha^* \cdot \nabla g_\alpha.$$

Hence, for any function g_α on $\Omega \times \mathbb{R}$ we have the identity

$$D^* g_\alpha = \grave{g}_\alpha - \mathbf{u}_\alpha^* \cdot \nabla g_\alpha. \tag{9.66}$$

To evaluate $D^* \mathbf{h}_\alpha^*$ we first observe that, by means of (9.66) and (9.65),

$$D^* \mathbf{h}_\alpha^* = (\rho_\alpha \mathbf{v}_\alpha)\grave{} - \mathbf{u}_\alpha^* \cdot \nabla(\rho_\alpha \mathbf{v}_\alpha) - (\grave{\rho}_\alpha - \mathbf{u}_\alpha^* \cdot \nabla \rho_\alpha)\mathbf{v}_* - \rho_\alpha D^* \mathbf{v}_*.$$

Now, by (9.24) we have

$$(\rho_\alpha \mathbf{v}_\alpha)\grave{} = \nabla \cdot \mathbf{T}_\alpha + \rho_\alpha \mathbf{b}_\alpha + \hat{\mathbf{p}}_\alpha + \mathbf{v}_\alpha(-\rho_\alpha \nabla \cdot \mathbf{v}_\alpha + \tau_\alpha). \tag{9.67}$$

If $\mathbf{v}_* = \mathbf{v}_\beta$ then, by $(9.24)_2$,

$$\rho_\alpha D^* \mathbf{v}_* = \rho_\alpha \grave{\mathbf{v}}_\beta = \frac{\rho_\alpha}{\rho_\beta}(\nabla \cdot \mathbf{T}_\beta + \rho_\beta \mathbf{b}_\beta + \hat{\mathbf{p}}_\beta).$$

If, instead, $\mathbf{v}_* = \mathbf{v}$ then by (9.53) we have

$$\rho_\alpha D^* \mathbf{v}_* = \rho_\alpha \dot{\mathbf{v}} = \frac{\rho_\alpha}{\rho}(\nabla \cdot \mathbf{T} + \rho \mathbf{b}).$$

Let

$$\rho_* = \rho_\beta, \rho, \qquad \mathbf{T}_* = \mathbf{T}_\beta, \mathbf{T}, \qquad \mathbf{b}_* = \mathbf{b}_\beta, \mathbf{b}, \qquad \hat{\mathbf{p}}_* = \hat{\mathbf{p}}_\beta, \mathbf{0},$$

according as $\mathbf{v}_* = \mathbf{v}_\beta, \mathbf{v}$. Hence, in any case, we can write

$$\rho_\alpha D^* \mathbf{v}_* = \frac{\rho_\alpha}{\rho_*}(\nabla \cdot \mathbf{T}_* + \rho_* \mathbf{b}_* + \hat{\mathbf{p}}_*).$$

Upon the appropriate substitutions and some rearrangements, we obtain

$$\begin{aligned} D^* \mathbf{h}_\alpha^* = \nabla \cdot \mathbf{T}_\alpha - \frac{\rho_\alpha}{\rho_*} \nabla \cdot \mathbf{T}_* + \rho_\alpha(\mathbf{b}_\alpha - \mathbf{b}_*) + \hat{\mathbf{p}}_\alpha - \frac{\rho_\alpha}{\rho_*}\hat{\mathbf{p}}_* + \tau_\alpha \mathbf{u}_\alpha^* \\ - \mathbf{u}_\alpha^* \cdot \nabla(\rho_\alpha \mathbf{v}_\alpha) - \rho_\alpha \mathbf{u}_\alpha^* \nabla \cdot \mathbf{v}_\alpha + (\mathbf{u}_\alpha^* \cdot \nabla \rho_\alpha)\mathbf{v}_*. \end{aligned}$$

Now, replace $\mathbf{v}_\alpha$ with $\mathbf{v}_* + \mathbf{u}_\alpha^*$ and observe

$$-\mathbf{u}_\alpha^* \cdot \nabla(\rho_\alpha \mathbf{v}_\alpha) - \rho_\alpha \mathbf{u}_\alpha^* \nabla \cdot \mathbf{v}_\alpha = -\nabla \cdot (\rho_\alpha \mathbf{u}_\alpha^* \otimes \mathbf{u}_\alpha^*) - (\mathbf{u}_\alpha^* \cdot \nabla \rho_\alpha)\mathbf{v}_* - \rho_\alpha \mathbf{u}_\alpha^* \cdot \nabla \mathbf{v}_* - \rho_\alpha \mathbf{u}_\alpha^* \nabla \cdot \mathbf{v}_*.$$

Denote by $\mathbf{L}_*$ the gradient of the reference velocity field,

$$\mathbf{L}_* = \nabla \mathbf{v}_*.$$

Hence

$$-\rho_\alpha \mathbf{u}_\alpha^* \cdot \nabla \mathbf{v}_* - \rho_\alpha \mathbf{u}_\alpha^* \nabla \cdot \mathbf{v}_* = -[\mathbf{L}_* + (\nabla \cdot \mathbf{v}_*)\mathbf{1}]\mathbf{h}_\alpha^*.$$

Consequently,

$$-\mathbf{u}_\alpha^* \cdot \nabla(\rho_\alpha \mathbf{v}_\alpha) - \rho_\alpha \mathbf{u}_\alpha^* \nabla \cdot \mathbf{v}_\alpha + (\mathbf{u}_\alpha^* \cdot \nabla \rho_\alpha)\mathbf{v}_* = -\nabla \cdot (\rho_\alpha \mathbf{u}_\alpha^* \otimes \mathbf{u}_\alpha^*) - [\mathbf{L}_* + (\nabla \cdot \mathbf{v}_*)\mathbf{1}]\mathbf{h}_\alpha^*.$$

By (9.18) we have

$$\hat{\mathbf{p}}_\alpha - \frac{\rho_\alpha}{\rho_*}\hat{\mathbf{p}}_* + \tau_\alpha \mathbf{u}_\alpha^* = \mathbf{m}_\alpha - \frac{\rho_\alpha}{\rho_\beta}\mathbf{m}_\beta - (\tau_\alpha - \frac{\rho_\alpha}{\rho_\beta}\tau_\beta)\mathbf{v}_\beta, \qquad \text{if} \quad \mathbf{v}_* = \mathbf{v}_\beta$$

and

$$\hat{\mathbf{p}}_\alpha - \frac{\rho_\alpha}{\rho_*}\hat{\mathbf{p}}_* + \tau_\alpha \mathbf{u}_\alpha^* = \mathbf{m}_\alpha - \tau_\alpha \mathbf{v}, \qquad \text{if} \quad \mathbf{v}_* = \mathbf{v}.$$

Letting

$$\tau_* = \tau_\beta, 0, \qquad \mathbf{m}_* = \mathbf{m}_\beta, \mathbf{0},$$

according to $\mathbf{v}_* = \mathbf{v}_\beta, \mathbf{0}$, we can write

$$\hat{\mathbf{p}}_\alpha - \frac{\rho_\alpha}{\rho_*}\hat{\mathbf{p}}_* + \tau_\alpha \mathbf{u}_\alpha^* = \mathbf{m}_\alpha - \frac{\rho_\alpha}{\rho_*}\mathbf{m}_* - (\tau_\alpha - \frac{\rho_\alpha}{\rho_*}\tau_*)\mathbf{v}_*.$$

Hence we find that $\mathbf{h}_\alpha^*$ satisfies the balance equation

$$D^*\mathbf{h}_\alpha^* + (\mathbf{L}_* + \nabla \cdot \mathbf{v}_* \mathbf{1})\mathbf{h}_\alpha^* = \nabla \cdot (\mathbf{T}_\alpha - \rho_\alpha \mathbf{u}_\alpha^* \otimes \mathbf{u}_\alpha^*) - \frac{\rho_\alpha}{\rho_*}\nabla \cdot \mathbf{T}_* + \rho_\alpha(\mathbf{b}_\alpha - \mathbf{b}_*)$$
$$+\mathbf{m}_\alpha - \frac{\rho_\alpha}{\rho_*}\mathbf{m}_* - (\tau_\alpha - \frac{\rho_\alpha}{\rho_*}\tau_*)\mathbf{v}_*. \qquad (9.68)$$

The evolution (or balance) equation (9.68) with any reference velocity $\mathbf{v}_*$ was first given in [324]. The particular case of (9.68) with $\mathbf{v}_* = \mathbf{v}$,

$$\dot{\mathbf{h}}_\alpha + (\mathbf{L} + \nabla \cdot \mathbf{v}\mathbf{1})\mathbf{h}_\alpha = \nabla \cdot (\mathbf{T}_\alpha - \rho_\alpha \mathbf{u}_\alpha \otimes \mathbf{u}_\alpha) - \frac{\rho_\alpha}{\rho}\nabla \cdot \mathbf{T} + \rho_\alpha(\mathbf{b}_\alpha - \mathbf{b}) + \mathbf{m}_\alpha - \tau_\alpha \mathbf{v} \qquad (9.69)$$

is the balance equation of the diffusion flux $\mathbf{h}_\alpha$ relative to the barycentric reference. If we let $\mathbf{h}_\alpha = \rho\, \mathbf{p}_\alpha$ and observe

$$\dot{\mathbf{h}}_\alpha = \rho\, \dot{\mathbf{p}}_\alpha - \rho(\nabla \cdot \mathbf{v})\mathbf{p}_\alpha$$

then we find that (9.69) simplifies to the form

$$\rho\, \dot{\mathbf{p}}_\alpha + \rho\, \mathbf{L}\, \mathbf{p}_\alpha = \nabla \cdot (\mathbf{T}_\alpha - \rho_\alpha \mathbf{u}_\alpha \otimes \mathbf{u}_\alpha) - \frac{\rho_\alpha}{\rho}\nabla \cdot \mathbf{T} + \rho_\alpha(\mathbf{b}_\alpha - \mathbf{b}) + \mathbf{m}_\alpha - \tau_\alpha \mathbf{v}$$

that traces back to Müller [336].

Equation (9.68) for the diffusion flux $\mathbf{h}_\alpha^*$ is a first-order evolution equation with an apparent nonlinearity due to

$$\rho_\alpha \mathbf{u}_\alpha^* \otimes \mathbf{u}_\alpha^* = \frac{1}{\rho_\alpha} \mathbf{h}_\alpha^* \otimes \mathbf{h}_\alpha^*.$$

As $\alpha = 1, 2, ..., n$, Eqs. (9.68) constitute a coupled system of differential equations, the coupling being given by the velocity gradient $\mathbf{L}_*$, the stress $\mathbf{T}_*$, the mass growth τ_*, and the linear momentum growth $\mathbf{m}_*$. In addition, the mass densities $\{\rho_\alpha\}$ are unknowns too and the equations are operative once the constitutive equations for $\{\mathbf{T}_\alpha\}$, $\{\mathbf{m}_\alpha\}$ and $\{\tau_\alpha\}$ are given. It is worth remarking that only $n - 1$ equations of (9.69) are significant in that $\sum_\alpha \mathbf{h}_\alpha = \mathbf{0}$. This is consistent with the fact that summation of (9.69) over α produces a trivial identity.

In practical applications, the use of (9.68) and (9.69) is quite difficult in view of the occurrence of $\mathbf{L}_*$ (or $\mathbf{L}$) and the fact that they are a coupled system of equations. It is then natural to look for reasonable approximations. In this sense, we examine whether an appropriate approximation leads to Fick's law. For definiteness look at (9.69). First we assume that $\mathbf{b}_\alpha = \mathbf{b}$, as it happens for gravity. Moreover, we select the barycentric reference and hence take $\mathbf{v} = \mathbf{0}$ and also $\mathbf{L} = \mathbf{0}$. Further, we disregard the nonlinear term $\rho_\alpha \mathbf{u}_\alpha \otimes \mathbf{u}_\alpha$. The growth $\mathbf{m}_\alpha$ is assumed to be given by the velocity differences between constituents and proportional to the respective densities. We let

$$\mathbf{m}_\alpha = \xi \rho_\alpha \textstyle\sum_\beta \rho_\beta (\mathbf{v}_\beta - \mathbf{v}_\alpha),$$

ξ being a function of the temperature θ. Since

$$\textstyle\sum_\alpha \mathbf{m}_\alpha = \xi \sum_{\alpha,\beta} \rho_\alpha \rho_\beta (\mathbf{v}_\beta - \mathbf{v}_\alpha) = \mathbf{0}, \tag{9.70}$$

the constraint (9.16) is identically satisfied. Moreover, we can write $\mathbf{m}_\alpha$ in the form

$$\mathbf{m}_\alpha = \xi \rho_\alpha \textstyle\sum_\beta \rho_\beta (\mathbf{u}_\beta - \mathbf{u}_\alpha).$$

We now restrict attention to binary mixtures. Hence $\mathbf{h}_1 + \mathbf{h}_2 = \mathbf{0}$ and (9.70) implies

$$\mathbf{m}_1 = \xi \rho_1 \rho_2 (\mathbf{u}_2 - \mathbf{u}_1) = \xi (\rho_1 \mathbf{h}_2 - \rho_2 \mathbf{h}_1) = -\xi\, \rho\, \mathbf{h}_1$$

and

$$\mathbf{m}_2 = -\xi\, \rho\, \mathbf{h}_2.$$

Consequently, Eq. (9.69) simplifies to

$$\dot{\mathbf{h}}_\alpha + \xi\, \rho\, \mathbf{h}_\alpha = \nabla \cdot \mathbf{T}_\alpha - \omega_\alpha \nabla \cdot \mathbf{T}. \tag{9.71}$$

Further approximations are considered by restricting attention to stationary conditions, $\dot{\mathbf{h}}_\alpha = \mathbf{0}$, and regarding the constituents as inviscid fluids, so that

$$\mathbf{T}_\alpha = -p_\alpha \mathbf{1}, \qquad \mathbf{T} \simeq -p\mathbf{1}, \qquad p_\alpha = \omega_\alpha p.$$

Hence Eq. (9.71) reduces to

$$\mathbf{h}_\alpha = -\frac{p}{\xi\rho}\nabla\omega_\alpha, \tag{9.72}$$

which is just Fick's law for $\mathbf{h}_\alpha$ where $\kappa_\alpha = p/\xi\rho$. Yet, in ideal gases, see Sect. 9.6, $p_\alpha = x_\alpha p$, with x_α the αth mole fraction; it is $x_\alpha \approx \omega_\alpha$ if the molecular weights are nearly equal.

It is of interest to observe that in uniform conditions, namely, if $\mathbf{L}$, $\nabla\mathbf{T}_\alpha$, $\nabla\cdot\mathbf{T}$, and $\nabla\cdot(\rho_\alpha \mathbf{u}_\alpha \otimes \mathbf{u}_\alpha)$ vanish, the balance equation (9.69) reduces to

$$\dot{\mathbf{h}}_\alpha = -\xi\,\rho\,\mathbf{h}_\alpha.$$

This implies that $\mathbf{h}_\alpha$ decays in time as

$$\mathbf{h}_\alpha(t) = \mathbf{h}_\alpha(0)\exp(-\xi\,\rho\,t).$$

The interaction force $\mathbf{m}_\alpha$ results then in an exponential decay of the diffusion flux.

9.5 Constitutive Models for Fluid Mixtures

A wide set of models for fluid mixtures is obtained by letting the constitutive functions $\psi_\alpha, \eta_\alpha, \mathbf{T}_\alpha, \mathbf{q}_\alpha$ depend (continuously) on the variables

$$\Lambda_\alpha = (\theta_\alpha, \rho_\alpha, \nabla\theta_\alpha, \nabla\rho_\alpha, \mathbf{D}_\alpha).$$

The interaction terms $\{\tau_\alpha\}, \{e_\alpha\}, \{\mathbf{m}_\alpha\}, \{\mathbf{k}_\alpha\}$ are allowed to depend on the whole set of variables[11]

$$\{\theta_\alpha\}, \{\rho_\alpha\}, \{\nabla\theta_\alpha\}, \{\nabla\rho_\alpha\}, \{\mathbf{D}_\alpha\}, \{\mathbf{v}_\alpha - \mathbf{v}_\beta\}.$$

For mathematical convenience, the functions ψ_α and $\mathbf{k}_\alpha$ are assumed to be continuously differentiable.

We start from the second-law inequality (9.42), that is,

$$\Sigma_\alpha\{\frac{1}{\theta_\alpha}[-\rho_\alpha(\grave{\psi}_\alpha + \eta_\alpha\grave{\theta}_\alpha) - \tau_\alpha(\psi_\alpha - \tfrac{1}{2}\mathbf{v}_\alpha^2) + e_\alpha - \mathbf{m}_\alpha\cdot\mathbf{v}_\alpha + \mathbf{T}_\alpha\cdot\mathbf{L}_\alpha - \frac{1}{\theta_\alpha}\mathbf{q}_\alpha\cdot\nabla\theta_\alpha + \theta_\alpha\nabla\cdot\mathbf{k}_\alpha]\} \geq 0.$$

---

[11] In so doing we follow the principle of phase separation [356] rather than the principle of equipresence [427].

Upon evaluation of $\grave{\psi}_\alpha$ and substitution in (9.42), we find

$$\sum_\alpha \frac{1}{\theta_\alpha}[-\rho_\alpha(\partial_{\theta_\alpha}\psi_\alpha + \eta_\alpha)\grave{\theta}_\alpha - \rho_\alpha \partial_{\mathbf{D}_\alpha}\psi_\alpha \cdot \grave{\mathbf{D}}_\alpha - \rho_\alpha \partial_{\nabla\theta_\alpha}\psi_\alpha \cdot \grave{\overline{\nabla\theta_\alpha}} + ...] \geq 0,$$

the dots representing the remaining terms, which are independent of $\{\grave{\theta}_\alpha\}$, $\{\grave{\mathbf{D}}_\alpha\}$, and $\{\grave{\overline{\nabla\theta_\alpha}}\}$. The arbitrariness of the peculiar time derivatives $\{\grave{\theta}_\alpha\}$, $\{\grave{\mathbf{D}}_\alpha\}$, and $\{\grave{\overline{\nabla\theta_\alpha}}\}$ requires that

$$\psi_\alpha = \psi_\alpha(\theta_\alpha, \rho_\alpha, \nabla\rho_\alpha), \qquad \eta_\alpha = -\partial_{\theta_\alpha}\psi_\alpha,$$

for any $\alpha = 1, 2, ..., n$. Look now at the dependence on ρ_α and $\nabla\rho_\alpha$. Apply identity (9.32) to the function ρ_α and observe

$$-\frac{\rho_\alpha}{\theta_\alpha}\partial_{\rho_\alpha}\psi_\alpha\grave{\rho}_\alpha - \frac{\rho_\alpha}{\theta_\alpha}\partial_{\nabla\rho_\alpha}\psi_\alpha \cdot \grave{\overline{\nabla\rho_\alpha}} = -\frac{\rho_\alpha}{\theta_\alpha}\partial_{\rho_\alpha}\psi_\alpha\grave{\rho}_\alpha - \frac{\rho_\alpha}{\theta_\alpha}\partial_{\nabla\rho_\alpha}\psi_\alpha \cdot (\nabla\grave{\rho}_\alpha - \mathbf{L}_\alpha^T\nabla\rho_\alpha)$$
$$= -\nabla\cdot(\frac{\rho_\alpha}{\theta_\alpha}\partial_{\nabla\rho_\alpha}\psi_\alpha\grave{\rho}_\alpha) - \frac{\rho_\alpha}{\theta_\alpha}\delta_{\rho_\alpha}\psi_\alpha\,\grave{\rho}_\alpha + \frac{\rho_\alpha}{\theta_\alpha}(\nabla\rho_\alpha \otimes \partial_{\nabla\rho_\alpha}\psi_\alpha)\cdot\mathbf{L}_\alpha,$$

where

$$\delta_{\rho_\alpha}\psi_\alpha := \partial_{\rho_\alpha}\psi_\alpha - \frac{\theta_\alpha}{\rho_\alpha}\nabla\cdot(\frac{\rho_\alpha}{\theta_\alpha}\partial_{\nabla\rho_\alpha}\psi_\alpha)$$

is the (generalized) variational derivative of ψ_α with respect to ρ_α. Since

$$\grave{\rho}_\alpha = -\rho_\alpha\nabla\cdot\mathbf{v}_\alpha + \tau_\alpha = -\rho_\alpha\mathbf{1}\cdot\mathbf{L}_\alpha + \tau_\alpha,$$

upon some rearrangements we find the remaining inequality in the form

$$\sum_\alpha\{\frac{1}{\theta_\alpha}[(\mathbf{T}_\alpha + \rho_\alpha^2\delta_{\rho_\alpha}\psi_\alpha\mathbf{1} + \rho_\alpha\nabla\rho_\alpha\otimes\partial_{\nabla\rho_\alpha}\psi_\alpha)\cdot(\mathbf{D}_\alpha + \mathbf{W}_\alpha) - \tau_\alpha(\psi_\alpha + \rho_\alpha\delta_{\rho_\alpha}\psi_\alpha)$$
$$+e_\alpha - \mathbf{m}_\alpha\cdot\mathbf{v}_\alpha + \tfrac{1}{2}\tau_\alpha\mathbf{v}_\alpha^2 - \frac{1}{\theta_\alpha}\mathbf{q}_\alpha\cdot\nabla\theta_\alpha] + \nabla\cdot[\mathbf{k}_\alpha - \frac{\rho_\alpha}{\theta_\alpha}\partial_{\nabla\rho_\alpha}\psi_\alpha(-\rho_\alpha\nabla\cdot\mathbf{v}_\alpha + \tau_\alpha)]\} \geq 0.$$

The arbitrariness of $\mathbf{W}_\alpha$ implies that

$$\mathbf{T}_\alpha + \rho_\alpha\nabla\rho_\alpha\otimes\partial_{\nabla\rho_\alpha}\psi_\alpha \in \mathrm{Sym}.$$

Consequently,

$$\mathrm{skw}\mathbf{T}_\alpha = -\mathrm{skw}\rho_\alpha\nabla\rho_\alpha\otimes\partial_{\nabla\rho_\alpha}\psi_\alpha.$$

If ψ_α depends on $\nabla\rho_\alpha$ via $|\nabla\rho_\alpha|$ then

$$\rho_\alpha\nabla\rho_\alpha\otimes\partial_{\nabla\rho_\alpha}\psi_\alpha \in \mathrm{Sym}$$

and hence also

$$\mathbf{T}_\alpha \in \mathrm{Sym}.$$

The inequality so obtained indicates that, to within inessential divergence-free terms, we can take the extra-entropy flux $\mathbf{k}_\alpha$ in the form

$$\mathbf{k}_\alpha = \frac{\rho_\alpha}{\theta_\alpha}\partial_{\nabla\rho_\alpha}\psi_\alpha(-\rho_\alpha\nabla\cdot\mathbf{v}_\alpha + \tau_\alpha).$$

Let

$$p_\alpha := \rho_\alpha^2\delta_{\rho_\alpha}\psi_\alpha, \qquad \mu_\alpha := \psi_\alpha + \frac{p_\alpha}{\rho_\alpha}.$$

The function p_α may be viewed as the (thermodynamic) αth partial pressure in the mixture; if ψ_α is independent of $\nabla\rho_\alpha$ then $p_\alpha = \rho_\alpha^2\partial_{\rho_\alpha}\psi_\alpha$. Moreover, we regard μ_α as the αth *chemical potential*. The inequality then reduces to

$$\textstyle\sum_\alpha \frac{1}{\theta_\alpha}\{(\mathbf{T}_\alpha + p_\alpha\mathbf{1} + \rho_\alpha\nabla\rho_\alpha\otimes\partial_{\nabla\rho_\alpha}\psi_\alpha)\cdot\mathbf{D}_\alpha - \mu_\alpha\tau_\alpha + e_\alpha - \mathbf{m}_\alpha\cdot\mathbf{v}_\alpha + \frac{1}{2}\tau_\alpha\mathbf{v}_\alpha^2 - \frac{1}{\theta_\alpha}\mathbf{q}_\alpha\cdot\nabla\theta_\alpha\} \geq 0.$$

A sufficient condition is suggested by the fact that ψ_α, $\mathbf{T}_\alpha$, and $\mathbf{q}_\alpha$ are functions of variables pertaining to the αth constituent whereas τ_α, e_α, and $\mathbf{m}_\alpha$ depend on quantities pertaining to all of the constituents. Hence, we let

$$(\mathbf{T}_\alpha + p_\alpha\mathbf{1} + \rho_\alpha\nabla\rho_\alpha\otimes\partial_{\nabla\rho_\alpha}\psi_\alpha)\cdot\mathbf{D}_\alpha - \frac{1}{\theta_\alpha}\mathbf{q}_\alpha\cdot\nabla\theta_\alpha \geq 0, \tag{9.73}$$

$$\textstyle\sum_\alpha \frac{1}{\theta_\alpha}(-\mu_\alpha\tau_\alpha + e_\alpha - \mathbf{m}_\alpha\cdot\mathbf{v}_\alpha + \frac{1}{2}\tau_\alpha\mathbf{v}_\alpha^2) \geq 0. \tag{9.74}$$

Inequality (9.74) is a restriction on the functions μ_α, e_α, $\mathbf{m}_\alpha$, and τ_α. In addition to (9.74), the functions τ_α, $\mathbf{m}_\alpha$, e_α are subject to the constraints

$$\textstyle\sum_\alpha\tau_\alpha = 0, \qquad \sum_\alpha\mathbf{m}_\alpha = \mathbf{0}, \qquad \sum_\alpha e_\alpha = 0,$$

and the Galilean invariance of

$$\hat{\mathbf{p}}_\alpha = \mathbf{m}_\alpha - \tau_\alpha\mathbf{v}_\alpha, \qquad \hat{\varepsilon}_\alpha = e_\alpha - \hat{\mathbf{p}}_\alpha\cdot\mathbf{v}_\alpha - \tau_\alpha(\varepsilon_\alpha + \tfrac{1}{2}\mathbf{v}_\alpha^2).$$

Incidentally, Galilean invariant quantities means that, with respect to frames in relative motion with any velocity $\mathbf{V}$, they appear to be the same. By a change of frame, the velocities of the constituents undergo the transformation

$$\mathbf{v}_\alpha \;\to\; \mathbf{v}_\alpha + \mathbf{V}.$$

Incidentally, in connection with (9.74) we remark that

$$e_\alpha - \mathbf{m}_\alpha\cdot\mathbf{v}_\alpha + \tfrac{1}{2}\tau_\alpha\mathbf{v}_\alpha^2 = \hat{\varepsilon}_\alpha + \tau_\alpha\varepsilon_\alpha$$

and hence $e_\alpha - \mathbf{m}_\alpha \cdot \mathbf{v}_\alpha + \frac{1}{2}\tau_\alpha \mathbf{v}_\alpha^2$ is a Galilean invariant.

For definiteness, the analysis of restrictions is now developed in particular cases.

9.5.1 *Extent of Reaction and Law of Mass Action*

Assume the reacting mixture is at rest ($\mathbf{v}_\alpha = \mathbf{0}$) and exchanges of energy are negligible ($e_\alpha = 0$). Inequality (9.74) reduces to

$$\sum_\alpha \frac{\mu_\alpha}{\theta_\alpha} \tau_\alpha \le 0. \tag{9.75}$$

If the mixture consists of two constituents, 1 and 2, then $\tau_2 = -\tau_1$ and (9.75) reduces to

$$\left(\frac{\mu_1}{\theta_1} - \frac{\mu_2}{\theta_2}\right)\tau_1 \le 0.$$

The inequality holds if

$$\tau_1 = \lambda\left(\frac{\mu_2}{\theta_2} - \frac{\mu_1}{\theta_1}\right),$$

where λ is a positive function of $\theta_1, \theta_2, \rho_1, \rho_2$. Hence, the mass of constituent 1 increases if $\mu_2/\theta_2 > \mu_1/\theta_1$ and vice versa. Equilibrium occurs if $\mu_2/\theta_2 = \mu_1/\theta_1$. If the two phases are at the same temperature then equilibrium occurs when the two chemical potentials are equal,

$$\mu_1 = \mu_2. \tag{9.76}$$

We now derive some consequences of inequality (9.75) when several constituents occur. Look at the reaction

$$\sum_{\alpha=1}^{r} \nu_\alpha B_\alpha \leftrightarrows \sum_{\alpha=r+1}^{r+p} \nu_\alpha B_\alpha.$$

To fix ideas we let the reaction proceed from left to right so that $B_1, ..., B_r$ are the reactants and $B_{r+1}, ..., B_{r+p}$ are the products. Hence define the stoichiometric coefficients $\{\gamma_\alpha\}$ as

$$\sum_{\alpha=1}^{r+p} \gamma_\alpha B_\alpha = 0.$$

Of course we let $r + p = n$, the number of constituents.

Let M_α be the molar mass and $\mathcal{N}_\alpha$ the molar number density,[12] or number of moles per unit volume. Hence we have

$$\mathcal{N}_\alpha M_\alpha = \rho_\alpha.$$

[12] The molar number density, or molar concentration, is usually denoted by c_α in the chemical literature. We do not use the symbol c_α to avoid any confusion with the mass fraction ρ_α/ρ, often denoted by c_α within continuum mechanics.

Moreover, upon disregarding diffusion we can write the (approximate) relation

$$\dot{\rho}_\alpha = \tau_\alpha.$$

Denote by Ξ the extent of reaction or reaction coordinate. For definiteness we let $\Xi = 0$ denote the setting with zero moles of products. Let $\hat{\mathcal{N}}_\alpha$ denote the value of $\mathcal{N}_\alpha$ as $\Xi = 0$ and hence let $\hat{\mathcal{N}}_{r+1}, ..., \hat{\mathcal{N}}_{r+p} = 0$. Since diffusion is ignored then the molar number densities change because of the chemical reaction only and hence

$$\mathcal{N}_\alpha - \hat{\mathcal{N}}_\alpha = \gamma_\alpha \Xi.$$

Consequently we have

$$\tau_\alpha = \dot{\rho}_\alpha = \gamma_\alpha M_\alpha \dot{\Xi}.$$

Substitution of τ_α in (9.75) results in

$$\left(\sum_\alpha \gamma_\alpha \overline{\mu}_\alpha\right)\dot{\Xi} \leq 0,$$

where $\overline{\mu}_\alpha = M_\alpha \mu_\alpha / \theta_\alpha$ is the chemical potential[13] per mole. Hence it follows that

$$\dot{\Xi} = -\chi \sum_\alpha \gamma_\alpha \overline{\mu}_\alpha, \tag{9.77}$$

where χ is a non-negative valued function of $\{\theta_\alpha\}$, $\{\rho_\alpha\}$. The vanishing of $\sum_\alpha \gamma_\alpha \overline{\mu}_\alpha$ determines the equilibrium between products and reactants and may be viewed as the law of mass action ([337], Chap. 6).

As a remark we observe that, by the conservation of mass, $\sum_\alpha \tau_\alpha = 0$ and then we have

$$\sum_\alpha \gamma_\alpha M_\alpha = 0.$$

9.5.2 Non-reacting Mixtures with Several Temperatures

If $\tau_\alpha = 0$ then the growths $\mathbf{m}_\alpha = \hat{\mathbf{p}}_\alpha$ have to be both Galilean invariant and with vanishing sum. The vector function

$$\mathbf{m}_\alpha = \sum_\beta M_{\alpha\beta}(\mathbf{v}_\beta - \mathbf{v}_\alpha) \tag{9.78}$$

is apparently Galilean invariant. Moreover, provided $M_{\alpha\beta} = M_{\beta\alpha}$, it follows that

$$\sum_\alpha \mathbf{m}_\alpha = \sum_{\alpha,\beta} M_{\alpha\beta}(\mathbf{v}_\beta - \mathbf{v}_\alpha) = \mathbf{0}$$

[13] Rescaled by the temperature θ_α.

as a consequence of the symmetry of $M_{\alpha\beta}$ and the skew symmetry of $\mathbf{v}_\beta - \mathbf{v}_\alpha$, with respect to the indices α, β.

Different temperatures are likely to induce exchanges of energy. Hence we let

$$e_\alpha = \sum_\beta N_{\alpha\beta}(\theta_\beta - \theta_\alpha) + w_\alpha,$$

where $N_{\alpha\beta} = N_{\beta\alpha}$ and w_α is a scalar function to be determined. Now, $\hat{\varepsilon}_\alpha$ becomes

$$\hat{\varepsilon}_\alpha = \sum_\beta N_{\alpha\beta}(\theta_\beta - \theta_\alpha) + w_\alpha - \sum_\beta M_{\alpha\beta}(\mathbf{v}_\beta - \mathbf{v}_\alpha) \cdot \mathbf{v}_\alpha.$$

Hence $\hat{\varepsilon}_\alpha$ is invariant if so is

$$w_\alpha - \sum_\beta M_{\alpha\beta}(\mathbf{v}_\beta - \mathbf{v}_\alpha) \cdot \mathbf{v}_\alpha.$$

Upon replacing $\mathbf{v}_\alpha$ by $\mathbf{v} + \mathbf{u}_\alpha$ we conclude that w_α may be chosen in the form

$$w_\alpha = \sum_\beta M_{\alpha\beta}(\mathbf{v}_\beta - \mathbf{v}_\alpha) \cdot \mathbf{v}.$$

Hence we have

$$e_\alpha = \sum_\beta N_{\alpha\beta}(\theta_\beta - \theta_\alpha) + \sum_\beta M_{\alpha\beta}(\mathbf{v}_\beta - \mathbf{v}_\alpha) \cdot \mathbf{v}. \tag{9.79}$$

As expected, it follows that

$$\sum_\alpha e_\alpha = \sum_{\alpha,\beta} N_{\alpha\beta}(\theta_\beta - \theta_\alpha) + \left[\sum_{\alpha,\beta} M_{\alpha\beta}(\mathbf{v}_\beta - \mathbf{v}_\alpha)\right] \cdot \mathbf{v} = 0.$$

Finally, we have to examine the thermodynamic requirement

$$\sum_\alpha \left(\frac{1}{\theta_\alpha} e_\alpha - \frac{1}{\theta_\alpha} \mathbf{m}_\alpha \cdot \mathbf{v}_\alpha\right) \geq 0.$$

Upon some rearrangements, we can write

$$\sum_\alpha \left(\frac{1}{\theta_\alpha} e_\alpha - \frac{1}{\theta_\alpha} \mathbf{m}_\alpha \cdot \mathbf{v}_\alpha\right) = \sum_{\alpha<\beta} N_{\alpha\beta} \frac{(\theta_\beta - \theta_\alpha)^2}{\theta_\alpha \theta_\beta} + \sum_{\alpha<\beta} M_{\alpha\beta}(\mathbf{v}_\beta - \mathbf{v}_\alpha) \cdot \left(\frac{\mathbf{u}_\beta}{\theta_\beta} - \frac{\mathbf{u}_\alpha}{\theta_\alpha}\right).$$

The first sum, in the right-hand side, is obviously non-negative provided only that $N_{\alpha\beta} \geq 0$ for each pair α, β, $\beta \neq \alpha$. The second sum is shown at once to be non-negative in the case of binary mixtures. Since

$$\mathbf{v}_2 - \mathbf{v}_1 = \mathbf{u}_2 - \mathbf{u}_1, \qquad \rho_2 \mathbf{u}_2 = -\rho_1 \mathbf{u}_1$$

then

$$M_{12}(\mathbf{v}_2 - \mathbf{v}_1) \cdot \left(\frac{\mathbf{u}_2}{\theta_2} - \frac{\mathbf{u}_1}{\theta_1}\right) = M_{12}\left(\frac{\rho_1}{\rho_2} + 1\right)\left(\frac{\rho_1}{\rho_2 \theta_2} + \frac{1}{\theta_1}\right)\mathbf{u}_1^2 \geq 0$$

holds provided only that $M_{12} \geq 0$. Hence the model (9.78)–(9.79) is thermodynamically consistent for non-reacting binary mixtures.

As an aside, the positive values of $M_{\alpha\beta}$ mean that $M_{\alpha\beta}(\mathbf{v}_\beta - \mathbf{v}_\alpha)$ is a force, on the constituent α, in the direction of $\mathbf{v}_\beta - \mathbf{v}_\alpha$. The positive values of $N_{\alpha\beta}$ mean that $\theta_\beta > \theta_\alpha$ implies $N_{\alpha\beta}(\theta_\beta - \theta_\alpha) > 0$. Consequently $\theta_\beta > \theta_\alpha$ implies that thermal energy is transferred from the constituent at higher temperature (β) to the constituent at lower temperature (α).

9.5.3 Reacting Mixtures with a Single Temperature

The reacting mixture is considered in the limit case when all of the constituents have the same temperature θ. In this case the analysis of thermodynamic consistency for the chemical potentials $\{\mu_\alpha\}$ and the growths $\{\mathbf{m}_\alpha\}$ is greatly simplified.

Since $\sum_\alpha e_\alpha = 0$, inequality (9.74) becomes

$$\textstyle\sum_\alpha(-\mu_\alpha\tau_\alpha - \mathbf{m}_\alpha \cdot \mathbf{u}_\alpha + \frac{1}{2}\tau_\alpha \mathbf{v}_\alpha^2) \geq 0.$$

Replacing $\mathbf{m}_\alpha$ with $\hat{\mathbf{p}}_\alpha + \tau_\alpha \mathbf{v}_\alpha$ and $\mathbf{v}_\alpha$ with $\mathbf{v} + \mathbf{u}_\alpha$, we obtain

$$\textstyle\sum_\alpha(-\mu_\alpha\tau_\alpha - \mathbf{m}_\alpha \cdot \mathbf{u}_\alpha + \frac{1}{2}\tau_\alpha \mathbf{v}_\alpha^2) = \sum_\alpha[-(\mu_\alpha + \frac{1}{2}\mathbf{u}_\alpha^2)\tau_\alpha - \hat{\mathbf{p}}_\alpha \cdot \mathbf{u}_\alpha].$$

The inequality

$$\textstyle\sum_\alpha[(\mu_\alpha + \frac{1}{2}\mathbf{u}_\alpha^2)\tau_\alpha + \hat{\mathbf{p}}_\alpha \cdot \mathbf{u}_\alpha] \leq 0,$$

or similar ones [63], might suggest that thermodynamic models are considered with the separate inequalities

$$\textstyle\sum_\alpha(\mu_\alpha + \frac{1}{2}\mathbf{u}_\alpha^2)\tau_\alpha \leq 0, \qquad \sum_\alpha \hat{\mathbf{p}}_\alpha \cdot \mathbf{u}_\alpha \leq 0.$$

Yet the literature shows that only relations around equilibrium are considered and hence at zero diffusion velocities $\mathbf{u}_\alpha$.

We rather think that physically sound models are those compatible with the separate conditions

$$\textstyle\sum_\alpha \mu_\alpha\tau_\alpha \leq 0, \qquad \sum_\alpha(\hat{\mathbf{p}}_\alpha \cdot \mathbf{u}_\alpha + \frac{1}{2}\tau_\alpha \mathbf{u}_\alpha^2) = \sum_\alpha(\mathbf{m}_\alpha \cdot \mathbf{u}_\alpha - \frac{1}{2}\tau_\alpha \mathbf{v}_\alpha^2) \leq 0. \quad (9.80)$$

To establish a thermodynamically consistent model, we take $\mathbf{m}_\alpha$ in the form

$$\mathbf{m}_\alpha = \textstyle\sum_\beta M_{\alpha\beta}(\mathbf{v}_\beta - \mathbf{v}_\alpha) + \mathbf{w}_\alpha, \qquad \sum_\alpha \mathbf{w}_\alpha = \mathbf{0}.$$

Substitution of $\mathbf{m}_\alpha$ results in

$$-\textstyle\sum_{\alpha,\beta} M_{\alpha\beta}(\mathbf{v}_\beta - \mathbf{v}_\alpha) \cdot \mathbf{v}_\alpha - \sum_\alpha(\mathbf{w}_\alpha \cdot \mathbf{u}_\alpha - \frac{1}{2}\tau_\alpha \mathbf{v}_\alpha^2) \geq 0.$$

As before, provided $M_{\alpha\beta} = M_{\beta\alpha} \geq 0$ we have

$$-\sum_{\alpha,\beta} M_{\alpha\beta}(\mathbf{v}_\beta - \mathbf{v}_\alpha) \cdot \mathbf{v}_\alpha = \sum_{\alpha<\beta} M_{\alpha\beta}(\mathbf{v}_\beta - \mathbf{v}_\alpha)^2 \geq 0.$$

Hence inequality $(9.80)_2$ holds if

$$0 \leq \sum_\alpha (\tfrac{1}{2}\tau_\alpha \mathbf{v}_\alpha^2 - \mathbf{w}_\alpha \cdot \mathbf{u}_\alpha) = \sum_\alpha (\tfrac{1}{2}\tau_\alpha \mathbf{u}_\alpha^2 + \tau_\alpha \mathbf{v} \cdot \mathbf{u}_\alpha - \mathbf{w}_\alpha \cdot \mathbf{u}_\alpha).$$

For definiteness, we determine a solution for $\mathbf{w}_\alpha$ in the particular case

$$\sum_\alpha (\tfrac{1}{2}\tau_\alpha \mathbf{u}_\alpha^2 + \tau_\alpha \mathbf{v} \cdot \mathbf{u}_\alpha - \mathbf{w}_\alpha \cdot \mathbf{u}_\alpha) = 0.$$

Let

$$\mathbf{w}_\alpha = \tau_\alpha \mathbf{v} + \mathbf{c}_\alpha, \qquad \sum_\alpha \mathbf{c}_\alpha = 0.$$

Consequently, we have

$$\sum_\alpha \mathbf{c}_\alpha \cdot \mathbf{u}_\alpha = \tfrac{1}{2} \sum_\alpha \tau_\alpha \mathbf{u}_\alpha^2.$$

In binary mixtures, we find

$$\mathbf{c}_1 \cdot (\mathbf{u}_1 - \mathbf{u}_2) = \tfrac{1}{2}\tau_1(\mathbf{u}_1^2 - \mathbf{u}_2^2),$$

whence

$$\mathbf{c}_1 = \tfrac{1}{2}\tau_1(\mathbf{u}_1 + \mathbf{u}_2).$$

Likewise we find

$$\mathbf{c}_2 = \tfrac{1}{2}\tau_2(\mathbf{u}_1 + \mathbf{u}_2).$$

Hence we can write

$$\mathbf{c}_\alpha = \tfrac{1}{2}\tau_\alpha(\mathbf{u}_1 + \mathbf{u}_2), \qquad \alpha = 1, 2.$$

In conclusion, for binary mixtures

$$\mathbf{m}_\alpha = \sum_\beta M_{\alpha\beta}(\mathbf{v}_\beta - \mathbf{v}_\alpha) + \tau_\alpha \mathbf{v} + \tfrac{1}{2}\tau_\alpha(\mathbf{u}_1 + \mathbf{u}_2) \tag{9.81}$$

is a thermodynamically consistent model of the linear momentum growths $\mathbf{m}_1, \mathbf{m}_2$.

9.5.4 Gibbs Equations

We address attention to equations for the whole mixture and hence we let the constituents have a common temperature, θ. First some identities are derived which follow from the thermodynamic restrictions

$$\eta_\alpha = -\partial_\theta \psi_\alpha, \qquad p_\alpha = \rho_\alpha^2 \partial_{\rho_\alpha} \psi_\alpha.$$

Since $\psi_\alpha = \varepsilon_\alpha - \theta\eta_\alpha$ then we find

$$\eta_\alpha = -\partial_\theta(\varepsilon_\alpha - \theta\eta_\alpha) = -\partial_\theta\varepsilon_\alpha + \eta_\alpha + \theta\partial_\theta\eta_\alpha.$$

By $p_\alpha = \rho_\alpha^2 \partial_{\rho_\alpha}(\varepsilon_\alpha - \theta\eta_\alpha)$ it follows

$$\partial_{\rho_\alpha}\varepsilon_\alpha = \frac{p_\alpha}{\rho_\alpha^2} + \theta\partial_{\rho_\alpha}\eta_\alpha = \frac{p_\alpha}{\rho_\alpha^2} - \theta\partial_\theta\partial_{\rho_\alpha}\psi_\alpha = \frac{p_\alpha}{\rho_\alpha^2} - \theta\partial_\theta\frac{p_\alpha}{\rho_\alpha^2} = -\frac{1}{\rho_\alpha^2}\theta^2\partial_\theta(p_\alpha/\theta).$$

Consequently, since $\mu_\alpha = \psi_\alpha + p_\alpha/\rho_\alpha$ then

$$\partial_{\rho_\alpha}\mu_\alpha = \partial_{\rho_\alpha}\psi_\alpha + \frac{1}{\rho_\alpha}\partial_{\rho_\alpha}p_\alpha - \frac{1}{\rho_\alpha^2}p_\alpha = \frac{1}{\rho_\alpha}\partial_{\rho_\alpha}p_\alpha.$$

In summary, we have

$$\partial_\theta\varepsilon_\alpha = \theta\partial_\theta\eta_\alpha, \qquad \partial_{\rho_\alpha}\varepsilon_\alpha = -\frac{1}{\rho_\alpha^2}\theta^2\partial_\theta(p_\alpha/\theta), \qquad \partial_{\rho_\alpha}\mu_\alpha = \frac{1}{\rho_\alpha}\partial_{\rho_\alpha}p_\alpha. \tag{9.82}$$

We now establish properties of the whole mixture. Since $\dot{\varepsilon}_\alpha = \dot{\psi}_\alpha + \eta_\alpha\dot{\theta} + \theta\dot{\eta}_\alpha$ and then $\dot{\eta}_\alpha = (\dot{\varepsilon}_\alpha - \dot{\psi}_\alpha - \eta_\alpha\dot{\theta})/\theta$ it follows

$$\dot{\eta}_\alpha = \frac{1}{\theta}(\dot{\varepsilon}_\alpha - \frac{p_\alpha}{\rho_\alpha^2}\dot{\rho}_\alpha). \tag{9.83}$$

Observe

$$\mu_\alpha = \varepsilon_\alpha - \theta\eta_\alpha + \frac{p_\alpha}{\rho_\alpha}, \qquad \theta\eta_\alpha = \varepsilon_\alpha + \frac{p_\alpha}{\rho_\alpha} - \mu_\alpha.$$

Hence $\rho\eta = \sum_\alpha \rho_\alpha\eta_\alpha$ can be differentiated to obtain

$$\begin{aligned}\dot{\overline{\rho\eta}} &= \textstyle\sum_\alpha \rho_\alpha\dot{\eta}_\alpha + \dot{\rho}_\alpha\eta_\alpha = \frac{1}{\theta}\sum_\alpha[\rho_\alpha(\dot{\varepsilon}_\alpha - \frac{p_\alpha}{\rho_\alpha^2}\dot{\rho}_\alpha) + (\varepsilon_\alpha + \frac{p_\alpha}{\rho_\alpha} - \mu_\alpha)\dot{\rho}_\alpha] \\ &= \textstyle\frac{1}{\theta}\sum_\alpha[\rho_\alpha(\partial_{\rho_\alpha}\varepsilon_\alpha\dot{\rho}_\alpha + \partial_\theta\varepsilon_\alpha\dot{\theta}) + (\varepsilon_\alpha - \mu_\alpha)\dot{\rho}_\alpha] = \frac{1}{\theta}\sum_\alpha[\dot{\overline{\rho_\alpha\varepsilon_\alpha}} - \mu_\alpha\dot{\rho}_\alpha]\end{aligned}$$

whence

$$\dot{\overline{\rho\eta}} = \frac{1}{\theta}(\dot{\overline{\rho\varepsilon_I}} - \textstyle\sum_\alpha\mu_\alpha\dot{\rho}_\alpha). \tag{9.84}$$

Consequently, by use of $\rho_\alpha = \rho\omega_\alpha$ we obtain

$$\rho\dot{\eta} = \dot{\overline{\rho\eta}} - \eta\dot{\rho} = \frac{1}{\theta}[\rho\dot{\varepsilon}_I + \dot{\rho}\varepsilon_I - (\sum_\alpha \mu_\alpha\omega_\alpha)\dot{\rho} - \rho\sum_\alpha \mu_\alpha\dot{\omega}_\alpha] - \eta\dot{\rho}$$
$$= \frac{1}{\theta}[\rho\dot{\varepsilon}_I - \rho\sum_\alpha \mu_\alpha\dot{\omega}_\alpha + (\varepsilon_I - \eta\theta - \sum_\alpha \mu_\alpha\omega_\alpha)\dot{\rho}].$$

Let $p = \sum_\alpha p_\alpha$. Since

$$\sum_\alpha \mu_\alpha\omega_\alpha = \sum_\alpha(\psi_\alpha + \frac{p_\alpha}{\rho_\alpha})\omega_\alpha = \sum_\alpha[(\varepsilon_\alpha - \theta\eta_\alpha)\omega_\alpha + \frac{p_\alpha}{\rho}] = \varepsilon_I - \theta\eta + \frac{p}{\rho}$$

then

$$\sum_\alpha\dot{\mu}_\alpha\omega_\alpha + \mu_\alpha\dot{\omega}_\alpha = \dot{\varepsilon}_I - \dot{\theta}\eta - \theta\dot{\eta} + \frac{\dot{p}}{\rho} - \frac{1}{\rho^2}p\dot{\rho}.$$

Further,

$$\sum_\alpha\dot{\mu}_\alpha\omega_\alpha = \sum_\alpha\omega_\alpha[\dot{\psi}_\alpha + \frac{\dot{p}_\alpha}{\rho_\alpha} - \frac{p_\alpha}{\rho_\alpha^2}\dot{\rho}_\alpha] = -\eta\dot{\theta} + \frac{1}{\rho}\dot{p}.$$

Hence, it follows that

$$\dot{\eta} = \frac{1}{\theta}(\dot{\varepsilon}_I - \frac{p}{\rho^2}\dot{\rho} - \sum_\alpha \mu_\alpha\dot{\omega}_\alpha). \tag{9.85}$$

Equations (9.83)–(9.85) are referred to as Gibbs equations and express the derivative of the entropy; Eq. (9.83) is the Gibbs equation for the αth constituent; and Eqs. (9.84) and (9.85) are Gibbs equations for the mixture.

Incidentally, in binary mixtures $\dot{\omega}_1 + \dot{\omega}_2 = 0$ and hence (9.85) takes the form

$$\dot{\eta} = \frac{1}{\theta}[\dot{\varepsilon}_I - \frac{p}{\rho^2}\dot{\rho} - (\mu_2 - \mu_1)\dot{\omega}_2].$$

Further thermodynamic restrictions follow from (9.85) as a consequence of integrability conditions. For convenience in applications, it is worth letting θ, ρ, and $\{\omega_\alpha\}$ be the independent variables (state variables). Hence upon assuming $\varepsilon_I = \varepsilon_I(\theta, \rho, \{\omega_\alpha\})$ we can write (9.85) in the form

$$\dot{\eta} = \frac{1}{\theta}[\partial_\theta\varepsilon_I\dot{\theta} + (\partial_\rho\varepsilon_I - \frac{p}{\rho^2})\dot{\rho} + \sum_\alpha(\partial_{\omega_\alpha}\varepsilon_I - \mu_\alpha)\dot{\omega}_\alpha],$$

The integrability conditions

$$\partial_\rho\frac{1}{\theta}\partial_\theta\varepsilon_I = \partial_\theta\frac{1}{\theta}(\partial_\rho\varepsilon_I - \frac{p}{\rho^2}),\ \partial_{\omega_\alpha}\frac{1}{\theta}\partial_\theta\varepsilon_I = \partial_\theta\frac{1}{\theta}(\partial_{\omega_\alpha}\varepsilon_I - \mu_\alpha),\ \partial_{\omega_\alpha}\frac{1}{\theta}(\partial_\rho\varepsilon_I - \frac{p}{\rho^2}) = \partial_\rho\frac{1}{\theta}(\partial_{\omega_\alpha}\varepsilon_I - \mu_\alpha)$$

result in the relations

$$\partial_\rho\varepsilon_I = -\frac{\theta^2}{\rho^2}\partial_\theta\frac{p}{\theta}, \qquad \partial_{\omega_\alpha}\varepsilon_I = -\theta^2\partial_\theta\frac{\mu_\alpha}{\theta}, \qquad \partial_{\omega_\alpha}p = \rho^2\partial_\rho\mu_\alpha \tag{9.86}$$

for the constitutive functions ε_I, p, μ_α.

Let ϕ be the Gibbs free energy per unit mass,

$$\phi := \varepsilon_I + \frac{p}{\rho} - \theta\eta.$$

Hence

$$\dot{\phi} = \dot{\varepsilon}_I + \frac{1}{\rho}\dot{p} - \frac{p}{\rho^2}\dot{\rho} - \eta\dot{\theta} - \theta\dot{\eta}.$$

Substituting $\theta\dot{\eta}$ from the Gibbs relation (9.85) we find

$$\dot{\phi} = -\eta\dot{\theta} + \frac{1}{\rho}\dot{p} + \textstyle\sum_\alpha \mu_\alpha \dot{\omega}_\alpha, \tag{9.87}$$

which shows that

$$\mu_\alpha = \partial_{\omega_\alpha}\phi(\theta, p, \{\omega_\beta\}).$$

9.5.5 Second-Law Inequalities

There are various relations expressing the content of the second law for mixtures and this leads to different constitutive equations. For mixtures with several temperatures, the second-law inequality is expressed by (9.43),

$$\textstyle\sum_\alpha\{\frac{1}{\theta_\alpha}[-\rho_\alpha(\dot{\psi}_\alpha + \eta_\alpha\dot{\theta}_\alpha) - \tau_\alpha\eta_\alpha + \hat{\varepsilon}_\alpha + \mathbf{T}_\alpha \cdot \mathbf{L}_\alpha - \frac{1}{\theta_\alpha}\mathbf{q}_\alpha \cdot \nabla\theta_\alpha + \theta_\alpha\nabla\cdot\mathbf{k}_\alpha]\} \geq 0.$$

If instead the mixture is subject to a single temperature θ then the inequality can be written as

$$-\rho(\dot{\psi}_I + \eta\dot{\theta}) - \frac{1}{\theta}\tilde{\mathbf{q}}\cdot\nabla\theta + \textstyle\sum_\alpha[\mathbf{T}_\alpha\cdot\mathbf{L}_\alpha - \frac{1}{2}\tau_\alpha\mathbf{u}_\alpha^2 - \hat{\mathbf{p}}_\alpha\cdot\mathbf{u}_\alpha + \theta\nabla\cdot\mathbf{k}_\alpha - \nabla\cdot(\psi_\alpha\mathbf{h}_\alpha)] \geq 0.$$

Both these inequalities allow the analysis of the constitutive properties of the single constituents.

If the whole mixture is described as a continuous body then the second-law inequality is obtained in the classical form

$$-\rho\dot{\psi} - \rho\eta\dot{\theta} + \mathbf{T}\cdot\mathbf{D} - \frac{1}{\theta}\mathbf{q}\cdot\nabla\theta + \theta\nabla\cdot\mathbf{k} \geq 0,$$

if the entropy supply is taken as $s = \sum_\alpha \omega_\alpha(r_\alpha + \mathbf{u}_\alpha\cdot\mathbf{b}_\alpha)/\theta$. If, instead, the assumption $s = \sum_\alpha(\omega_\alpha r_\alpha)/\theta$ is made then the inequality becomes

$$-\rho\dot{\psi} - \rho\eta\dot{\theta} + \mathbf{T}\cdot\mathbf{D} - \frac{1}{\theta}\mathbf{q}\cdot\nabla\theta + \theta\nabla\cdot\mathbf{k} + \sum_\alpha \mathbf{h}_\alpha\cdot\mathbf{b}_\alpha \geq 0.$$

Sometimes the content of the second law is based on Gibbs equations, merely

$$\dot{\eta} = \frac{1}{\theta}(\dot{\varepsilon}_I - \frac{p}{\rho^2}\dot{\rho} - \sum_\alpha \mu_\alpha\dot{\omega}_\alpha).$$

The inequality arises by the determination of an appropriate flux $\mathbf{j}$ so that [245]

$$\rho\dot{\eta} + \nabla\cdot\mathbf{j} = \sigma \geq 0.$$

Now, $\varepsilon_I = \varepsilon - \sum_\alpha \omega_\alpha \mathbf{u}_\alpha^2$. To evaluate $\sum_\alpha(\frac{1}{2}\omega_\alpha\mathbf{u}_\alpha^2)^{\cdot}$ we observe

$$\tfrac{1}{2}\omega_\alpha\mathbf{u}_\alpha^2 = \frac{1}{2\rho^2\omega_\alpha}\mathbf{h}_\alpha^2.$$

Since $\dot{\rho} = -\rho\nabla\cdot\mathbf{v}$ and $\rho\dot{\omega}_\alpha = -\nabla\cdot\mathbf{h}_\alpha + \tau_\alpha$ then

$$\sum_\alpha(\tfrac{1}{2}\omega_\alpha\mathbf{u}_\alpha^2)^{\cdot} = \sum_\alpha[\omega_\alpha\mathbf{u}_\alpha^2\nabla\cdot\mathbf{v} - \frac{1}{2\rho}\mathbf{u}_\alpha^2(-\nabla\cdot\mathbf{h}_\alpha + \tau_\alpha) + \frac{1}{\rho\rho_\alpha}\mathbf{h}_\alpha\cdot\dot{\mathbf{h}}_\alpha].$$

Moreover, by (9.69) we obtain

$$-\sum_\alpha \frac{1}{\rho_\alpha}\mathbf{h}_\alpha\cdot\dot{\mathbf{h}}_\alpha = \sum_\alpha\{\rho_\alpha\mathbf{u}_\alpha\cdot(\mathbf{D} + \nabla\cdot\mathbf{v}\,\mathbf{1})\mathbf{u}_\alpha - \frac{1}{\rho_\alpha}\mathbf{h}_\alpha\cdot[\nabla\cdot(\mathbf{T}_\alpha - \rho_\alpha\mathbf{u}_\alpha\otimes\mathbf{u}_\alpha)]\}$$
$$-\sum_\alpha(\mathbf{h}_\alpha\cdot\mathbf{b}_\alpha + \hat{\mathbf{p}}_\alpha\cdot\mathbf{u}_\alpha + \tau_\alpha\mathbf{u}_\alpha^2).$$

Hence we find

$$-\rho\sum_\alpha(\tfrac{1}{2}\omega_\alpha\mathbf{u}_\alpha^2)^{\cdot} = -\sum_\alpha(\mathbf{h}_\alpha\cdot\mathbf{b}_\alpha + \hat{\mathbf{p}}_\alpha\cdot\mathbf{u}_\alpha + \tau_\alpha\mathbf{u}_\alpha^2)$$
$$+\sum_\alpha\{\tfrac{1}{2}\mathbf{u}_\alpha^2(-\nabla\cdot\mathbf{h}_\alpha + \tau_\alpha) + \rho_\alpha\mathbf{u}_\alpha\cdot\mathbf{D}\mathbf{u}_\alpha + \frac{1}{\rho_\alpha}\mathbf{h}_\alpha\cdot[\nabla\cdot(\mathbf{T}_\alpha - \rho_\alpha\mathbf{u}_\alpha\otimes\mathbf{u}_\alpha)]\}. \tag{9.88}$$

The thermodynamic consequences of the term $\sum_\alpha(\frac{1}{2}\omega_\alpha\mathbf{u}_\alpha^2)^{\cdot}$ are investigated in Sect. 9.10.1.

9.6 Mixtures of Ideal Gases

Consider a mixture of ideal gases. Denote by $\mathcal{N}_\alpha$ the number density (per unit volume) of moles of constituent α. The pressure–temperature relation is taken in the form of the equation of state of an ideal gas,

$$p_\alpha = \mathcal{N}_\alpha R\theta_\alpha,$$

where R is the universal gas constant. By Dalton's law, the pressure exerted by a mixture of gases is equal to the sum of the partial pressures of all of the components present in the mixture. By the theory of mixtures, $\mathbf{T} = \sum_\alpha(\mathbf{T}_\alpha - \rho_\alpha \mathbf{u}_\alpha \otimes \mathbf{u}_\alpha)$; if $\mathbf{T}_\alpha = -p_\alpha \mathbf{1}$ we obtain Dalton's law if the terms $\rho_\alpha \mathbf{u}_\alpha \otimes \mathbf{u}_\alpha$ are neglected.

For mixtures at a common temperature θ, we have

$$p_\alpha = \mathcal{N}_\alpha R\theta, \qquad p = \textstyle\sum_\alpha \mathcal{N}_\alpha R\theta, \qquad p_\alpha = x_\alpha p,$$

where $x_\alpha = \mathcal{N}_\alpha / \sum_\beta \mathcal{N}_\beta$ is the mole fraction. Letting M_α be the αth molecular weight, we have

$$\rho_\alpha = \mathcal{N}_\alpha M_\alpha, \qquad \dot{\rho}_\alpha = \dot{\mathcal{N}}_\alpha M_\alpha,$$

whence

$$\dot{\mathcal{N}}_\alpha = -\nabla \cdot (\mathbf{h}_\alpha / M_\alpha) - \mathcal{N}_\alpha \nabla \cdot \mathbf{v} + \frac{\tau_\alpha}{M_\alpha}, \qquad \dot{x}_\alpha = \frac{1}{\sum_\beta \mathcal{N}_\beta}\left[\dot{\mathcal{N}}_\alpha - x_\alpha \textstyle\sum_\beta \dot{\mathcal{N}}_\beta\right].$$

Let n_α be the number of moles. If the densities $\mathcal{N}_\alpha$ are uniform and V is the volume then $\mathcal{N}_\alpha = n_\alpha / V$ and

$$p_\alpha = \frac{n_\alpha}{V} R\theta, \qquad p = \frac{\sum_\alpha n_\alpha}{V} R\theta, \qquad p_\alpha = \frac{n_\alpha}{\sum_\beta n_\beta} p.$$

In terms of the mass density $\rho_\alpha = \mathcal{N}_\alpha M_\alpha$, we have

$$p_\alpha = \frac{R\theta_\alpha}{M_\alpha} \rho_\alpha.$$

At a common temperature

$$p = \textstyle\sum_\alpha p_\alpha = \sum_\alpha \dfrac{R\theta}{M_\alpha} \rho_\alpha, \qquad p_\alpha = \dfrac{\rho_\alpha / M_\alpha}{\sum_\beta \rho_\beta / M_\beta} p = x_\alpha p.$$

9.6.1 *Chemical Potentials*

We let $\psi_\alpha = \psi_\alpha(\theta_\alpha, \rho_\alpha)$ and hence $p_\alpha = \partial_{\rho_\alpha} \psi_\alpha$. The αth chemical potential is then given by

$$\mu_\alpha = \psi_\alpha + \frac{p_\alpha}{\rho_\alpha} = \psi_\alpha + \rho_\alpha \partial_{\rho_\alpha} \psi_\alpha.$$

The chemical potential μ_α is determined under the assumption that the constituents are ideal gases, namely,

$$p_\alpha = \frac{R\theta_\alpha}{M_\alpha}\rho_\alpha, \qquad p_\alpha = \rho_\alpha^2 \partial_{\rho_\alpha}\psi_\alpha.$$

Hence we have

$$\partial_{\rho_\alpha}\psi_\alpha = \frac{R\theta_\alpha}{M_\alpha}\frac{1}{\rho_\alpha}, \qquad \psi_\alpha(\theta_\alpha, \rho_\alpha) = \frac{R\theta_\alpha}{M_\alpha}\ln\frac{\rho_\alpha}{\rho_\alpha^0} + \hat{\psi}_\alpha(\theta_\alpha),$$

where $\hat{\psi}_\alpha$ is an undetermined function and ρ_α^0 is a reference mass density. As a consequence μ_α is given by

$$\mu_\alpha = \hat{\psi}_\alpha(\theta_\alpha) + \frac{R\theta_\alpha}{M_\alpha} + \frac{R\theta_\alpha}{M_\alpha}\ln\frac{\rho_\alpha}{\rho_\alpha^0}.$$

To determine $\hat{\psi}_\alpha$, we observe that the internal energy ε_α depends only on the temperature θ_α. This follows at once by accounting for the relation $\eta_\alpha = -\partial_{\theta_\alpha}\psi_\alpha$,

$$\varepsilon_\alpha = \psi_\alpha - \theta_\alpha\partial_{\theta_\alpha}\psi_\alpha = \hat{\psi}_\alpha(\theta_\alpha) - \theta_\alpha\hat{\psi}_\alpha'(\theta_\alpha).$$

Differentiating with respect to θ_α, we have

$$\hat{\psi}_\alpha''(\theta_\alpha) = -\frac{1}{\theta_\alpha}\varepsilon_\alpha'.$$

Let θ_0 be a reference temperature. Integration of $\hat{\psi}_\alpha''$ from θ_0 to θ_α results in

$$\hat{\psi}_\alpha'(\theta_\alpha) = -\int_{\theta_0}^{\theta_\alpha}\frac{1}{\xi}\varepsilon_\alpha'(\xi)d\xi + c_\alpha$$

and hence

$$\hat{\psi}_\alpha(\theta_\alpha) = -\int_{\theta_0}^{\theta_\alpha} d\xi \int_{\theta_0}^{\xi}\frac{1}{\zeta}\varepsilon_\alpha'(\zeta)d\zeta + c_\alpha(\theta_\alpha - \theta_0) + d_\alpha,$$

the symbols c_α, d_α denoting constants. Moreover, partial differentiation of $\psi_\alpha(\theta_\alpha, \rho_\alpha)$ with respect to θ_α determines the entropy,

$$\eta_\alpha(\theta_\alpha, \rho_\alpha) = -\frac{R}{M_\alpha}\ln\frac{\rho_\alpha}{\rho_\alpha^0} + \int_{\theta_0}^{\theta_\alpha}\frac{1}{\zeta}\varepsilon_\alpha'(\zeta)d\zeta - c_\alpha. \tag{9.89}$$

Back to the chemical potential, upon substitution of p_α and ψ_α, we find

$$\mu_\alpha = \frac{R\theta_\alpha}{M_\alpha}\left(1 + \ln\frac{\rho_\alpha}{\rho_\alpha^0}\right) - \int_{\theta_0}^{\theta_\alpha} d\xi \int_{\theta_0}^{\xi}\frac{1}{\zeta}\varepsilon_\alpha'(\zeta)d\zeta + c_\alpha(\theta_\alpha - \theta_0) + d_\alpha.$$

Let $p^0_\alpha, \rho^0_\alpha, \theta^0_\alpha$ be values of $p_\alpha, \rho_\alpha, \theta_\alpha$ in a reference state. For ideal gases, we have

$$p_\alpha = \frac{R\theta_\alpha}{M_\alpha}\rho_\alpha, \quad p^0_\alpha = \frac{R\theta^0_\alpha}{M_\alpha}\rho^0_\alpha, \quad \frac{\rho_\alpha}{\rho^0_\alpha} = \frac{p_\alpha}{p^0_\alpha}\frac{\theta^0_\alpha}{\theta_\alpha}.$$

Hence, we can write

$$\mu_\alpha = \mu^0_\alpha(\theta_\alpha) + \frac{R\theta_\alpha}{M_\alpha}\Big[\ln\frac{p_\alpha}{p^0_\alpha} - \ln\frac{\theta_\alpha}{\theta^0_\alpha}\Big], \tag{9.90}$$

where

$$\mu^0_\alpha(\theta_\alpha) = \frac{R\theta_\alpha}{M_\alpha} - \int_{\theta_0}^{\theta_\alpha} d\xi \int_{\theta_0}^{\xi} \frac{1}{\zeta}\varepsilon'_\alpha(\zeta)d\zeta + c_\alpha(\theta_\alpha - \theta_0) + d_\alpha.$$

If the temperature is constant then the relation (9.90) simplifies to

$$\mu_\alpha = \mu^0_\alpha(\theta_\alpha) + \frac{R\theta_\alpha}{M_\alpha}\ln\frac{p_\alpha}{p^0_\alpha},$$

which is commonly adopted for ideal gas mixtures [126].

Subject to the assumption of a common temperature $\theta_\alpha = \theta$, we now evaluate the effect of pressure on the extent of reaction. By (9.77), a reaction proceeds, $\dot{\Xi} > 0$, if

$$\textstyle\sum_\alpha \gamma_\alpha \overline{\mu}_\alpha < 0.$$

According to (9.90) this is the case if

$$\textstyle\sum_\alpha \gamma_\alpha M_\alpha \mu^0_\alpha(\theta) + R\theta \sum_\alpha \ln\frac{p_\alpha}{p_0} < 0.$$

If we take the view, common in chemistry, that p_α is the pressure p of the mixture times the mole fraction n_α/n,

$$p_\alpha = p\frac{n_\alpha}{n},$$

then the inequality becomes

$$\textstyle(\sum_\alpha \gamma_\alpha)R\theta \ln\frac{p}{p_0} + R\theta(\sum_\alpha \gamma_\alpha \ln\frac{n_\alpha}{n}) < -\sum_\alpha \gamma_\alpha M_\alpha \mu^0_\alpha(\theta).$$

At a fixed set of mole numbers $\{n_\alpha\}$ this inequality provides the values of p such that the reaction proceeds at a given temperature or the like for the values of θ at a given pressure. If $\sum_\alpha \gamma_\alpha > 0$ then we have

$$\ln\frac{p}{p_0} < -\frac{1}{(\sum_\alpha \gamma_\alpha)R\theta}\Big[R\theta \textstyle\sum_\alpha \gamma_\alpha \ln\frac{n_\alpha}{n} + \sum_\alpha M_\alpha \mu^0_\alpha(\theta)\Big],$$

which means that, at a given state $\{n_\alpha\}$ and temperature θ, the reaction is favoured by low pressures. If, instead, $\sum_\alpha \gamma_\alpha < 0$ then it follows

$$\ln \frac{p}{p_0} > \frac{1}{|\sum_\alpha \gamma_\alpha|\, R\theta}[R\theta \textstyle\sum_\alpha \gamma_\alpha \ln \frac{n_\alpha}{n} + \sum_\alpha M_\alpha \mu_\alpha^0(\theta)],$$

which indicates that the reaction is favoured by high pressures.

Temperature dependence of the chemical potential

There are processes, like in phase transitions, where the pressure is kept constant. In these cases, it is of greater interest to evaluate the effect of temperature on the chemical potential. Let the pressure be constant, $p_\alpha = p_\alpha^0$. Then by (9.89) η_α takes the form

$$\eta_\alpha = \frac{R}{M_\alpha} \ln \frac{\theta_\alpha}{\theta_0} + \int_{\theta_0}^{\theta_\alpha} \frac{1}{\zeta} \varepsilon_\alpha'(\zeta) d\zeta - c_\alpha,$$

while (9.90) reduces to

$$\mu_\alpha = \frac{R\theta_\alpha}{M_\alpha}(1 - \ln \frac{\theta_\alpha}{\theta_0}) - \int_{\theta_0}^{\theta_\alpha} d\xi \int_{\theta_0}^{\xi} \frac{1}{\zeta} \varepsilon_\alpha'(\zeta) d\zeta + c_\alpha(\theta_\alpha - \theta_0) + d_\alpha.$$

Hence, we have

$$\mu_\alpha' = -\frac{R}{M_\alpha} \ln \theta_\alpha \theta_0 - \int_{\theta_0}^{\theta_\alpha} \frac{1}{\zeta} \varepsilon_\alpha'(\zeta) d\zeta + c_\alpha = -\eta_\alpha.$$

We assume $\eta_\alpha > 0$ and then conclude that, at constant pressure, μ_α is a decreasing function of the temperature θ.

A strictly analogous conclusion follows by looking at the mixture as a single body with $\mathbf{T} = -p\mathbf{1}$ and $\mathbf{q} = \mathbf{0}, \mathbf{k} = \mathbf{0}$. The entropy inequality is then taken in the form

$$-\rho\dot{\psi} - \rho\eta\dot{\theta} - p\nabla \cdot \mathbf{v} = 0.$$

Since

$$\nabla \cdot \mathbf{v} = -\frac{\dot{\rho}}{\rho} = \frac{\dot{v}}{v}$$

then the entropy relation can be written in the form

$$-\dot{\psi} - \eta\dot{\theta} - p\dot{v} = 0,$$

whence

$$-\dot{\phi} - \eta\dot{\theta} + v\dot{p} = 0,$$

where $\phi = \psi + pv$ is the Gibbs free energy (per unit mass). At constant pressure

$$\dot{\phi} = -\eta\dot{\theta}, \qquad \phi' = -\eta.$$

The positive value of η implies that ϕ is a decreasing function of θ.

9.6.2 Mixing Entropy and Mixing Free Energy

Since $\eta_\alpha/\eta_\alpha^0 = (p_\alpha/p_\alpha^0)(\theta_\alpha^0/\theta_\alpha)$, by (9.89) we obtain the entropy η_α per unit mass in the form

$$\eta_\alpha = \frac{R}{M_\alpha}\ln\frac{\theta_\alpha}{\theta_0} + \int_{\theta_0}^{\theta_\alpha}\frac{1}{\zeta}\varepsilon'_\alpha(\zeta)d\zeta - c_\alpha - \frac{R}{M_\alpha}\ln\frac{p_\alpha}{p_\alpha^0}.$$

Assume the distribution is uniform, namely, the fields are independent of the position. Moreover, the mixture is subject to a common temperature, θ. The entropy per mole is then

$$M_\alpha\eta_\alpha = Rs_\alpha - R\ln\frac{p_\alpha}{p_0},$$

where

$$s_\alpha = \ln\frac{\theta_\alpha}{\theta_0} + \frac{M_\alpha}{R}\Big(\int_{\theta_0}^{\theta_\alpha}\frac{1}{\zeta}\varepsilon'_\alpha(\zeta)d\zeta - c_\alpha\Big).$$

The entropy of the constituent α is $n_\alpha M_\alpha\eta_\alpha$. The total entropy, S, is the sum,

$$S = R\sum_\alpha n_\alpha\Big(s_\alpha - \ln\frac{p_\alpha}{p_0}\Big).$$

Since the mixture is subject to a single temperature θ and occupies the volume V then

$$p_\alpha = \frac{n_\alpha}{V}R\theta = \frac{n_\alpha}{n}n\frac{R\theta}{V} = x_\alpha p,$$

where $n = \sum_\alpha n_\alpha$, $x_\alpha = n_\alpha/\sum_\beta n_\beta$, and p is the (total) pressure. Consequently, the partial pressure p_α,

$$p_\alpha = x_\alpha p = \frac{R\theta}{V/x_\alpha},$$

is the same as that obtained by confining the same gas, at temperature θ, in the volume V/x_α.

With this in mind we consider a conceptual experiment named after Gibbs. The volume V is partitioned in volumes $\{x_\alpha V\}$. The αth constituent, at the temperature θ, is placed within the volume $x_\alpha V$ and hence is subject to the pressure

$$p_\alpha = n_\alpha \frac{R\theta}{x_\alpha V} = n\frac{R\theta}{V} = p;$$

each gas is subject to the (same) pressure p, the total pressure of the previous derivation. Summing the corresponding entropies, we have

$$S_i = R\sum_\alpha n_\alpha (s_\alpha - \ln \frac{p}{p_0}).$$

Now let the gases diffuse uniformly in the volume V. At the end the system is the configuration of the previous case, with $p_\alpha = x_\alpha p$, and hence the entropy S_f takes the form

$$S_f = R\sum_\alpha n_\alpha (s_\alpha - \ln \frac{x_\alpha p}{p_0}).$$

Consequently

$$S_f - S_i = -R\sum_\alpha n_\alpha \ln x_\alpha.$$

The difference $S_f - S_i$ is the change of entropy due to diffusion and we call it mixing entropy. Since $x_\alpha \leq 1$ then $S_f - S_i$ is always positive.

Remark 9.1 The mixing entropy depends only on the fractions $\{x_\alpha\}$ and not on the properties of the gases. It might then be the case of the mixture of two portions of the same gas. A nonzero mixing entropy for the same gas is usually referred to as Gibbs' paradox.

Look now at the chemical potential (9.90). At a constant, common temperature θ we have

$$\mu_\alpha = \frac{R}{M_\alpha}\theta(1 + \ln \frac{x_\alpha p}{p_0}) - \gamma_\alpha,$$

where

$$\gamma_\alpha = \int_{\theta_0}^{\theta} d\xi \int_{\theta_0}^{\xi} \frac{1}{\zeta}\varepsilon'_\alpha(\zeta)d\zeta - c_\alpha(\theta - \theta_0) + d_\alpha.$$

For uniform mixtures, we compute the Gibbs free energy of the mixture as $G_i = \sum_\alpha n_\alpha M_\alpha \mu_\alpha$ to obtain

$$G_i = R\theta \sum_\alpha n_\alpha (1 + \ln \frac{x_\alpha p}{p_0} - \frac{M_\alpha}{R\theta}\gamma_\alpha).$$

By arguing as with the entropy we find that the Gibbs free energy after mixing, G_f differs from G_i for the occurrence of p instead of $x_\alpha p$. Hence, it follows

$$G_f - G_i = R\theta \sum_\alpha n_\alpha \ln x_\alpha. \tag{9.91}$$

Let H be the enthalpy, $H = G + \theta S$. It follows that

$$H_f - H_i = G_f - G_i + \theta(S_f - S_i) = 0.$$

For mixtures of ideal gases, the mixing enthalpy is zero. Intermolecular forces are mainly responsible of changes in the enthalpy of a mixture. Stronger attractive forces between the mixed molecules result in a lower enthalpy of the mixture.

9.7 Diffusion

Diffusion means relative motion between constituents of a mixture. The motion of a mixture, and hence the relative motion between constituents, may be determined by solving Eqs. (9.24) or (9.29) along with appropriate boundary and initial conditions. Yet, there are more direct ways of modelling, and then determining diffusion phenomena.

9.7.1 Dynamic Diffusion Equation

Models of diffusion are usually based on the balance equation (9.27) for the mass fraction ω_α and a constitutive equation for the mass flux $\mathbf{h}_\alpha$ (see Sect. 9.4.2), though never via the right equation (9.69) for $\mathbf{h}_\alpha$. More unusually, ω_α is taken to satisfy an evolution equation [215].

In all cases, the resulting differential equation for ω_α is strongly affected by constitutive assumptions. In this connection, we observe that the thermodynamic scheme for mixtures based on (9.42) does not involve the mass fluxes $\{\mathbf{h}_\alpha\}$. Consistently here we find that, according to the balance equations, the evolution of the mass densities $\{\rho_\alpha\}$ is free from the mass fluxes $\{\mathbf{h}_\alpha\}$.

We start with the balance of mass and linear momentum,

$$\partial_t \rho_\alpha = -\nabla \cdot (\rho_\alpha \mathbf{v}_\alpha) + \tau_\alpha,$$

$$\partial_t(\rho_\alpha \mathbf{v}_\alpha) + \nabla \cdot (\rho_\alpha \mathbf{v}_\alpha \otimes \mathbf{v}_\alpha) = \nabla \cdot \mathbf{T}_\alpha + \rho_\alpha \mathbf{b}_\alpha + \mathbf{m}_\alpha.$$

Partial time differentiation of the first equation, divergence of the second one, and substitution of $\nabla \cdot \partial_t(\rho_\alpha \mathbf{v}_\alpha)$ yield

$$\partial_t^2 \rho_\alpha = \nabla \cdot [\nabla \cdot (\rho_\alpha \mathbf{v}_\alpha \otimes \mathbf{v}_\alpha)] - \nabla \cdot (\nabla \cdot \mathbf{T}_\alpha) - \nabla \cdot (\rho_\alpha \mathbf{b}_\alpha) - \nabla \cdot \mathbf{m}_\alpha + \partial_t \tau_\alpha.$$

If the constituent is regarded as an inviscid fluid, $\mathbf{T}_\alpha = -p_\alpha \mathbf{1}$, it follows

$$\partial_t^2 \rho_\alpha = \nabla \cdot [\nabla \cdot (\rho_\alpha \mathbf{v}_\alpha \otimes \mathbf{v}_\alpha)] + \Delta p_\alpha - \nabla \cdot (\rho_\alpha \mathbf{b}_\alpha) - \nabla \cdot \mathbf{m}_\alpha + \partial_t \tau_\alpha.$$

We let $p_\alpha = p_\alpha(\rho_\alpha, \theta_\alpha)$ and hence

$$\nabla p_\alpha = \partial_{\rho_\alpha} p_\alpha \nabla \rho_\alpha + \partial_{\theta_\alpha} p_\alpha \nabla \theta_\alpha.$$

In the linear approximation, we neglect $\nabla \cdot [\nabla \cdot (\rho_\alpha \mathbf{v}_\alpha \otimes \mathbf{v}_\alpha)]$ and let

$$\Delta p_\alpha = \partial_{\rho_\alpha} p_\alpha \Delta \rho_\alpha + \partial_{\theta_\alpha} p_\alpha \Delta \theta_\alpha$$

to obtain

$$\partial_t^2 \rho_\alpha = \partial_{\rho_\alpha} p_\alpha \Delta \rho_\alpha + \partial_{\theta_\alpha} p_\alpha \Delta \theta_\alpha - \nabla \cdot (\rho_\alpha \mathbf{b}_\alpha) - \nabla \cdot \mathbf{m}_\alpha + \partial_t \tau_\alpha.$$

If, further, the temperature is assumed to be uniform, $\nabla \theta_\alpha = \mathbf{0}$, then

$$\partial_t^2 \rho_\alpha = \partial_{\rho_\alpha} p_\alpha \, \Delta \rho_\alpha - \nabla \cdot (\rho_\alpha \mathbf{b}_\alpha) - \nabla \cdot \mathbf{m}_\alpha + \partial_t \tau_\alpha. \tag{9.92}$$

Equation (9.92) is the αth equation of a system where $\nabla \cdot \mathbf{m}_\alpha$ and $\partial_t \tau_\alpha$ account for possible coupling terms. Since $\rho_\alpha = \rho \omega_\alpha$, if ρ is assumed to be constant then we have

$$\partial_t^2 \omega_\alpha = \partial_{\rho_\alpha} p_\alpha \, \Delta \omega_\alpha - \nabla \cdot (\omega_\alpha \mathbf{b}_\alpha) + (1/\rho)[\partial_t \tau_\alpha - \nabla \cdot \mathbf{m}_\alpha].$$

9.7.2 Fourier-Like and Rate-Type Diffusion Equations

The simplest and best known model of diffusion traces back to Fick [170] and is based on a strict analogy with the Fourier model of heat conduction. It is referred to the pertinent concentration (say mass fraction) ω_α and is expressed by the differential equation

$$\partial_t \omega_\alpha = \nabla \cdot (D_\alpha \nabla \omega_\alpha) + \zeta_\alpha, \tag{9.93}$$

where the parameter D_α is the diffusivity and ζ_α is related to the αth mass growth. Equation (9.93) and some generalizations are derived as follows.

Consider Eq. (9.27). If we take Fick's law (approximation) for $\mathbf{h}_\alpha$ in the form

$$\mathbf{h}_\alpha = -\kappa_\alpha \nabla \omega_\alpha, \tag{9.94}$$

κ_α being a constant, we obtain Eq. (9.93) with $D_\alpha = \kappa_\alpha/\rho$ and $\zeta_\alpha = \tau_\alpha/\rho$. If κ_α, and hence D_α, is not constant then (9.93) has to be generalized as

$$\dot{\omega}_\alpha = \frac{1}{\rho} \nabla \cdot (\kappa_\alpha \nabla \omega_\alpha) + \zeta_\alpha.$$

Other well-known models stem from the balance equation (9.27) along with an appropriate assumption on the diffusion flux $\mathbf{h}_\alpha$. Borrowing from the Ginzburg–Landau scheme [215] we let

$$\psi_\alpha(\omega_\alpha, \nabla\omega_\alpha) = f_\alpha(\omega_\alpha) + \tfrac{1}{2}\lambda_\alpha|\nabla\omega_\alpha|^2 \tag{9.95}$$

be the (Helmholtz) free energy of the αth constituent. Hence we take $\mathbf{h}_\alpha = -\kappa_\alpha\nabla\hat{\mu}_\alpha$, where $\hat{\mu}_\alpha$ is the variational derivative of ψ_α,

$$\hat{\mu}_\alpha = \delta_{\omega_\alpha}\psi_\alpha := \partial_{\omega_\alpha}\psi_\alpha - \nabla\cdot\partial_{\nabla\omega_\alpha}\psi_\alpha,$$

then, if λ_α is constant,

$$\hat{\mu}_\alpha = f'_\alpha - \lambda_\alpha\Delta\omega_\alpha.$$

Substitution of $\mathbf{h}_\alpha$ in (9.27) results in

$$\dot{\omega}_\alpha = \frac{1}{\rho}\nabla\cdot(\kappa_\alpha f''_\alpha\nabla\omega_\alpha) - \frac{\lambda_\alpha}{\rho}\Delta^2\omega_\alpha + \zeta_\alpha. \tag{9.96}$$

Equation (9.96) is usually referred to as the Cahn–Hilliard equation [75, 76]. It is a fourth-order partial differential equation. If the dependence of $\hat{\mu}_\alpha$ on $\nabla\omega_\alpha$ is ignored then the Cahn–Hilliard equation reduces formally to the classical second-order equation (9.93).

Instead of applying the balance equation (9.27) we might take the view that the evolution of ω_α is in fact a relaxation toward equilibrium governed by a parameter $\beta_\alpha > 0$ through

$$\beta_\alpha\dot{\omega}_\alpha = -\delta_{\omega_\alpha}\psi_\alpha. \tag{9.97}$$

If we assume again ψ_α in the form (9.95) then (9.97) reads

$$\beta_\alpha\dot{\omega}_\alpha = -\partial_{\omega_\alpha}\psi_\alpha + \nabla\cdot\partial_{\nabla\omega_\alpha}\psi_\alpha. \tag{9.98}$$

Equation (9.98) is a second-order partial differential equation even though the dependence of ψ_α on $\nabla\omega_\alpha$ is allowed; it is referred to as Ginzburg–Landau equation [87, 269] or also Allen–Cahn equation [6]. The lower order relative to the Cahn–Hilliard equation (9.96) is interpreted by saying that (9.98) describes the ordering of atoms within unit cells on a lattice whereas (9.96) describes the transport of atoms between unit cells.

Equations (9.93), (9.96), and (9.98) are well-renowned models in the literature of diffusion. Their consistency with thermodynamics is investigated in Sect. 2.7.

Both Fick's law (9.94) and the evolution equation (9.97) are approximations to (9.69) and (9.27). In a genuine thermodynamic theory of mixtures, Sect. 9.5, there is no need of approximations to the evolution equation for ω_α. This need occurs when we approximate the behaviour of a mixture with properties of the whole body, Sect. 9.10.

Diffusion and Boltzmann–Matano method

The homogeneous one-dimensional version of the diffusion equation (9.93) is sometimes investigated through the *Boltzmann–Matano method.* The mass fraction[14] ω is taken to satisfy the homogeneous equation

$$\partial_t \omega = \partial_x (D(\omega)\partial_x \omega).$$

It is a basic assumption (of the method) that ω depends on x and t via $\xi = x/2\sqrt{t}$. Hence

$$\partial_t \omega = -\frac{\xi}{2\sqrt{t}}\partial_\xi \omega, \qquad \partial_x \omega = \frac{1}{2\sqrt{t}}\partial_\xi \omega$$

and then

$$-\frac{\xi}{2\sqrt{t}}\partial_\xi \omega = \partial_x (D(\omega)\frac{1}{2\sqrt{t}}\partial_\xi \omega) = \frac{1}{2\sqrt{t}}(D(\omega)\partial_\xi \omega).$$

Thus ω satisfies the ordinary differential equation

$$-\xi \partial_\xi \omega = \partial_\xi (D(\omega)\partial_\xi \omega)$$

for the function $\omega(\xi)$. A trivial solution is given by $\partial_\xi \omega = 0$, that is, when $\omega(\xi)$ is constant. This is interpreted by saying that a concentration front propagates with $x \propto \sqrt{t}$ or $t \propto x^2$.

Now, if $\xi = x/2\sqrt{t}$ then $1/2\sqrt{t} = \xi/x$ and hence we cannot view t as independent of x any longer. Rather we have

$$\partial_x \frac{1}{2\sqrt{t}} = -\frac{\xi}{x^2} = -\frac{1}{x}\frac{1}{2\sqrt{t}}$$

and the differential equation can be written as

$$-\xi \partial_\xi \omega = \partial_\xi (D(\omega)\partial_\xi \omega) - \frac{1}{x} D(\omega)\partial_\xi \omega.$$

The trivial solution with $\partial_\xi \omega = 0$ still holds.

9.7.3 Maxwell–Stefan Diffusion Model

The literature shows various equations, under the name of Maxwell–Boltzmann model, which have been developed as approximations to the Boltzmann equation [307, 399]. For definiteness, we take the Maxwell–Stefan (MS) model as that arising from the assumption

---

[14] In this context, the mass fraction is most often denoted by c.

$$\mathbf{m}_\alpha = -\sum_\beta M_{\alpha\beta}(\frac{D_\alpha}{\rho_\alpha} - \frac{D_\beta}{\rho_\beta})\nabla \ln\theta + \sum_\beta M_{\alpha\beta}(\mathbf{v}_\beta - \mathbf{v}_\alpha), \tag{9.99}$$

where

$$M_{\alpha\beta} = \frac{x_\alpha x_\beta}{\mathcal{D}_{\alpha\beta}}.$$

The quantities $\mathcal{D}_{\alpha\beta}$ are usually referred to as MS diffusion coefficients while x_α is the αth mole fraction. It is required that $\mathcal{D}_{\alpha\beta} = \mathcal{D}_{\beta\alpha}$ so that $M_{\alpha\beta} = M_{\beta\alpha}$. By the symmetry of $M_{\alpha\beta}$ and the skew symmetry of $D_\alpha/\rho_\alpha - D_\beta/\rho_\beta$ and $\mathbf{v}_\beta - \mathbf{v}_\alpha$, relative to α and β, the model satisfies the required constraint on the growths $\{\mathbf{m}_\alpha\}$,

$$\sum_\alpha \mathbf{m}_\alpha = -\sum_{\alpha\beta} M_{\alpha\beta}(\frac{D_\alpha}{\rho_\alpha} - \frac{D_\beta}{\rho_\beta})\nabla \ln\theta + \sum_{\alpha\beta} M_{\alpha\beta}(\mathbf{v}_\beta - \mathbf{v}_\alpha) = \mathbf{0}.$$

The mass growths are zero and the constituents are subject to the same temperature. Consistency with thermodynamics is then guaranteed if the separate inequalities

$$\sum_\alpha (\mathbf{T}_\alpha + p_\alpha \mathbf{1} + \rho_\alpha \nabla\rho_\alpha \otimes \partial_{\nabla\rho_\alpha}\psi_\alpha) \cdot \mathbf{D}_\alpha \geq 0,$$

$$\sum_\alpha (\mathbf{m}_\alpha \cdot \mathbf{v}_\alpha + \frac{1}{\theta}\mathbf{q}_\alpha \cdot \nabla\theta) \leq 0 \tag{9.100}$$

hold. Only (9.100) is of interest for the MS model. Observe

$$\sum_{\alpha,\beta} M_{\alpha\beta}(\mathbf{v}_\beta - \mathbf{v}_\alpha) \cdot \mathbf{u}_\alpha = \sum_{\alpha,\beta} M_{\alpha\beta}(\mathbf{u}_\beta - \mathbf{u}_\alpha) \cdot \mathbf{u}_\alpha = -\tfrac{1}{2}\sum_{\alpha,\beta} M_{\alpha\beta}(\mathbf{u}_\beta - \mathbf{u}_\alpha)^2,$$

which is non-positive provided only that $M_{\alpha\beta} \geq 0$. Upon substitution of $\mathbf{m}_\alpha$ we then require that

$$-\sum_{\alpha,\beta} M_{\alpha\beta}(\frac{D_\alpha}{\rho_\alpha} - \frac{D_\beta}{\rho_\beta})\nabla \ln\theta \cdot \mathbf{u}_\alpha + \frac{1}{\theta}\sum_\alpha \mathbf{q}_\alpha \cdot \nabla\theta \leq 0.$$

This inequality holds if

$$\mathbf{q}_\alpha = -\kappa_\alpha \nabla\theta + [\sum_\beta M_{\alpha\beta}(\frac{D_\alpha}{\rho_\alpha} - \frac{D_\beta}{\rho_\beta})]\mathbf{u}_\alpha.$$

The first term is the classical Fourier-type constitutive function, subject to $\kappa_\alpha \geq 0$. The second one is uncommon in the literature and shows that the (partial) heat conduction $\mathbf{q}_\alpha$ is induced by the diffusion velocity $\mathbf{u}_\alpha$.

As a comment we observe that

$$\omega_\alpha\omega_\beta(\mathbf{v}_\beta - \mathbf{v}_\alpha) = \omega_\alpha\omega_\beta(\mathbf{u}_\beta - \mathbf{u}_\alpha) = \frac{\omega_\alpha\omega_\beta}{\rho}(\frac{\mathbf{h}_\beta}{\omega_\beta} - \frac{\mathbf{h}_\alpha}{\omega_\alpha}) = \frac{1}{\rho}(\omega_\alpha\mathbf{h}_\beta - \omega_\beta\mathbf{h}_\alpha).$$

That is why the term $M_{\alpha\beta}(\mathbf{v}_\beta - \mathbf{v}_\alpha)$ is often written in the form $\omega_\alpha \mathbf{h}_\beta - \omega_\beta \mathbf{h}_\alpha$.

9.7.4 Coupled Effects Among Diffusion and Heat Conduction

Fluid mixtures may show coupled effects such as the motion originated by a temperature gradient (Soret effect) or heat conduction affected by a gradient of mass density (Dufour effect). Here we investigate possible constitutive models allowing for coupled effects in mixtures and then generalizing the scheme of Sect. 9.5.2.

We consider a non-reacting mixture ($\tau_\alpha = 0$) with several temperatures and let

$$\rho_\alpha, \theta_\alpha, \mathbf{D}_\alpha, \nabla\rho_\alpha, \nabla\theta_\alpha$$

be the set of variables of the functions $\psi_\alpha, \eta_\alpha, \mathbf{T}_\alpha, \mathbf{q}_\alpha$. The interaction terms $\{\tau_\alpha\}, \{\mathbf{m}_\alpha\}, \{e_\alpha\}$ are allowed to depend on all of the variables $\{\rho_\beta\}, \{\theta_\beta\}, \{\mathbf{D}_\beta\}, \{\nabla\rho_\beta\}, \{\nabla\theta_\beta\}$. The stress $\mathbf{T}_\alpha$ is assumed to have the classical form $\mathbf{T}_\alpha = -p_\alpha(\rho_\alpha, \theta_\alpha)\mathbf{1} + \boldsymbol{\mathcal{T}}_\alpha$ where $\boldsymbol{\mathcal{T}}_\alpha \to \mathbf{0}$ as $\mathbf{D}_\alpha \to \mathbf{0}$. Moreover $\hat{\varepsilon}_\alpha$ is independent of $\{\mathbf{D}_\beta\}$. To save writing we let ψ_α depend only on ρ_α and θ_α. Moreover, we will see that the extra-entropy fluxes are zero. We let ψ_α be differentiable. Substituting $\grave{\psi}_\alpha$ into the entropy inequality (9.43) we find

$$\textstyle\sum_\alpha\{\frac{1}{\theta_\alpha}[-\rho_\alpha(\partial_{\theta_\alpha}\psi_\alpha + \eta_\alpha)\grave{\theta}_\alpha + (\rho_\alpha^2\partial_{\rho_\alpha}\psi_\alpha - p_\alpha)\nabla\cdot\mathbf{v}_\alpha + \tau_\alpha\theta_\alpha\eta_\alpha + \hat{\varepsilon}_\alpha + \boldsymbol{\mathcal{T}}_\alpha\cdot\mathbf{L}_\alpha - \frac{1}{\theta_\alpha}\mathbf{q}_\alpha\cdot\nabla\theta_\alpha]\} \geq 0.$$

The arbitrariness of $\grave{\theta}_\alpha, \nabla\cdot\mathbf{v}_\alpha, \mathbf{W}_\alpha = \mathrm{skw}\mathbf{L}_\alpha$ implies that

$$\eta_\alpha = -\partial_{\theta_\alpha}\psi_\alpha, \qquad p_\alpha = \rho_\alpha^2\partial_{\rho_\alpha}\psi_\alpha, \qquad \boldsymbol{\mathcal{T}}_\alpha \in \mathrm{Sym}.$$

The reduced inequality

$$\textstyle\sum_\alpha\{\frac{1}{\theta_\alpha}[\hat{\varepsilon}_\alpha + \boldsymbol{\mathcal{T}}\cdot\mathbf{D}_\alpha - \frac{1}{\theta_\alpha}\mathbf{q}_\alpha\cdot\nabla\theta_\alpha]\} \geq 0$$

is investigated in connection with particular constitutive functions. Let

$$\boldsymbol{\mathcal{T}}_\alpha = 2\mu_\alpha\mathbf{D}_\alpha + \lambda_\alpha \mathrm{tr}\,\mathbf{D}_\alpha\mathbf{1}, \qquad \mathbf{q}_\alpha = -\kappa_\alpha\nabla\theta_\alpha,$$

$$\mathbf{m}_\alpha = \hat{\mathbf{p}}_\alpha = \textstyle\sum_\beta[\xi_{\alpha\beta}(\mathbf{v}_\beta - \mathbf{v}_\alpha) + \gamma_{\alpha\beta}\nabla\theta_\beta],$$

$$\hat{\varepsilon}_\alpha = \textstyle\sum_\beta N_{\alpha\beta}(\theta_\beta - \theta_\alpha) - \sum_\beta M_{\alpha\beta}(\mathbf{v}_\beta - \mathbf{v}_\alpha)\cdot\mathbf{u}_\alpha.$$

A dependence of $\boldsymbol{\mathcal{T}}_\alpha$ on $\mathbf{W}_\alpha$ is not considered because of objectivity. We observe that the constraint

$$\mathbf{0} = \sum_\alpha \mathbf{m}_\alpha = \sum_{\alpha,\beta}[\xi_{\alpha\beta}(\mathbf{v}_\beta - \mathbf{v}_\alpha) + \gamma_{\alpha\beta}\nabla\theta_\beta]$$

holds for $\xi_{\alpha\beta} = \xi_{\beta\alpha}$ and

$$\sum_\alpha \gamma_{\alpha\beta} = 0.$$

This condition seems unrealistic in that different constituents would have opposite behaviours[15] (e.g. $\gamma_{1\beta} > 0$, $\gamma_{2\beta} < 0$). Replacing $\nabla\theta_\beta$ with $\nabla\theta_\beta - \nabla\theta_\alpha$, and hence $\gamma_{\alpha\beta} = \gamma_{\beta\alpha}$, would be of no interest in that of no effect in mixtures with a single temperature. Hence we let $\gamma_{\alpha\beta} = 0$. The constraint

$$0 = \sum_\alpha e_\alpha = \sum_\alpha \hat{\varepsilon}_\alpha + \mathbf{m}_\alpha \cdot \mathbf{v}_\alpha = \sum_{\alpha,\beta} N_{\alpha\beta}(\theta_\beta - \theta_\alpha) + \sum_{\alpha,\beta}(\mathbf{v}_\beta - \mathbf{v}_\alpha) \cdot (\xi_{\alpha\beta}\mathbf{v}_\alpha - M_{\alpha\beta}\mathbf{u}_\alpha)$$

implies that

$$N_{\alpha\beta} = N_{\beta\alpha}, \qquad \xi_{\alpha\beta} = M_{\alpha\beta}.$$

The reduced inequality then becomes

$$\sum_\alpha\{\frac{1}{\theta_\alpha}[\sum_\beta N_{\alpha\beta}(\theta_\beta - \theta_\alpha) - M_{\alpha\beta}(\mathbf{u}_\beta - \mathbf{u}_\alpha) \cdot \mathbf{u}_\alpha + 2\mu_\alpha \mathbf{D}_\alpha \cdot \mathbf{D}_\alpha + \lambda(\mathrm{tr}\,\mathbf{D}_\alpha)^2 + \kappa_\alpha|\nabla\theta_\alpha|^2]\} \geq 0$$

whence the classical inequalities $\mu_\alpha \geq 0$, $2\mu_\alpha + 3\lambda_\alpha \geq 0$, $\kappa_\alpha \geq 0$ follow. Moreover we observe

$$\sum_\alpha\{\frac{1}{\theta_\alpha}[\sum_\beta N_{\alpha\beta}(\theta_\beta - \theta_\alpha) - M_{\alpha\beta}(\mathbf{u}_\beta - \mathbf{u}_\alpha) \cdot \mathbf{u}_\alpha]\} = \sum_{\alpha<\beta}[N_{\alpha\beta}\frac{(\theta_\beta - \theta_\alpha)^2}{\theta_\alpha\theta_\beta} + M_{\alpha\beta}(\mathbf{u}_\beta - \mathbf{u}_\alpha) \cdot (\frac{\mathbf{u}_\beta}{\theta_\beta} - \frac{\mathbf{u}_\alpha}{\theta_\alpha})]\}.$$

The first term is apparently non-negative provided only that $N_{\alpha\beta} \geq 0$. The proof that also the second term is non-negative is immediate if $n = 2$ (see Sect. 9.5.2). It is worth remarking that when the temperature is uniform, $\theta_\beta = \theta$, $\forall\beta$, the first term vanishes and the second one is purely mechanical,

$$\sum_\alpha \frac{\hat{\varepsilon}_\alpha}{\theta} = \frac{1}{\theta}\sum_{\alpha<\beta} M_{\alpha\beta}(\mathbf{u}_\beta - \mathbf{u}_\alpha)^2.$$

This in turn shows that the coefficients $M_{\alpha\beta}$ are required to be non-negative.

Soret and Dufour effects

The Soret effect[16] denotes the feature that a temperature gradient provides mass density (or concentration) gradients. In the standard modelling of the Soret effect, the diffusion flux $\mathbf{h}_\alpha$ is given by a constitutive equation of the form [376]

$$\mathbf{h}_\alpha = -\rho\sum_\beta D_{\alpha\beta}\nabla\omega_\beta - \rho D_T \omega_\alpha(1 - \omega_\alpha)\nabla\theta.$$

[15] This crucial aspect is seen also in the Maxwell–Stefan term where the constraint holds thanks to the skew character of $D_\alpha/\rho_\alpha - D_\beta/\rho_\beta$.

[16] Named also Ludwig–Soret effect or thermal diffusion effect.

Here D_T is the thermodiffusion coefficient. This model equation indicates that the Soret effect is (mainly) related to the temperature gradient and not on the differences among the temperatures of the constituents.[17] Yet diffusion is governed by the equations of motion and the diffusion flux satisfies the differential equation (9.69). It might seem natural to model the Soret effect by letting

$$\mathbf{m}_\alpha \propto \nabla\theta_\alpha.$$

However, from a physical standpoint, the interaction force $\mathbf{m}_\alpha$ should involve differences (of velocities, of temperatures) rather than quantities pertaining to the constituent. This might be the case by letting

$$\mathbf{m}_\alpha = \textstyle\sum_\beta D_{\alpha\beta}(\nabla\theta_\beta - \nabla\theta_\alpha).$$

Yet the constraint $\sum_\alpha \mathbf{m}_\alpha = \mathbf{0}$ and the thermodynamic consistency rule out these proposals.

Since p_α as well as the parameters ν_α, λ_α are functions of θ_α, it follows that the motion of the αth constituent is affected by the temperature gradient $\nabla\theta_\alpha$. In view of the equation of motion (9.17), we have

$$\rho_\alpha \dot{\mathbf{v}}_\alpha = -\partial_{\theta_\alpha} p_\alpha \nabla\theta_\alpha - \partial_{\rho_\alpha} p_\alpha \nabla\rho_\alpha + \mathbf{m}_\alpha + \mathbf{f}_\alpha, \tag{9.101}$$

where $\mathbf{f}_\alpha = \nabla\cdot\boldsymbol{T}_\alpha + \rho_\alpha\mathbf{b}_\alpha - \tau_\alpha\mathbf{v}_\alpha$.

The Soret effect might be stated by saying that a mass flux is generated in regions where temperature gradients occur. We examine this feature in connection with the mass fluxes $\{\mathbf{h}_\alpha\}$ in *non-reacting binary mixtures*.

To establish both diffusion and heat-conduction models we go back to Sect. 9.5 and examine the entropy inequality. Consistency with thermodynamics indicates that the growths $\{\mathbf{m}_\alpha\}$ can be given the form

$$\mathbf{m}_\alpha = \sum_\beta M_{\alpha\beta}(\mathbf{v}_\beta - \mathbf{v}_\alpha), \qquad M_{\alpha\beta} = M_{\beta\alpha} \geq 0.$$

For definiteness, let $M_{\alpha\beta} = \xi\rho_\alpha\rho_\beta$ and hence

$$\mathbf{m}_\alpha = \xi\rho_\alpha \sum_\beta \rho_\beta(\mathbf{v}_\beta - \mathbf{v}_\alpha), \qquad \xi > 0.$$

The single terms $\alpha\beta$ are proportional to $\rho_\alpha\rho_\beta$; the force exchanged between α and β constituents is proportional to the mass densities that exchange linear momentum. This assumption is consistent identically with the constraint $\sum_\alpha \mathbf{m}_\alpha = \mathbf{0}$. It satisfies the thermodynamic requirement $\sum_\alpha \mathbf{m}_\alpha \cdot \mathbf{v}_\alpha \leq 0$ provided only that ξ is

[17] This view is consistent with the experimental results obtained firstly by the physician C.F.W. Ludwig [286].

non-negative; let $\xi > 0$. For binary mixtures,

$$\mathbf{m}_\alpha = \xi \rho_\alpha \rho_\beta (\mathbf{u}_\beta - \mathbf{u}_\alpha) = \xi \rho_\alpha \mathbf{h}_\beta - \xi \rho_\beta \mathbf{h}_\alpha = -\xi \rho \mathbf{h}_\alpha.$$

In stationary conditions, we neglect $\rho_\alpha \dot{\mathbf{v}}_\alpha$ and hence Eq. (9.101) can be written as

$$\mathbf{h}_\alpha = -\frac{\partial_{\theta_\alpha} p_\alpha}{\xi \rho} \nabla \theta_\alpha - \frac{\partial_{\rho_\alpha} p_\alpha}{\xi \rho} \nabla \rho_\alpha + \frac{1}{\xi \rho} \mathbf{f}_\alpha \tag{9.102}$$

thus ascribing to $\partial_{\theta_\alpha} p_\alpha / \xi \rho$ the meaning of thermodiffusion coefficient. In general, the nonzero diffusion flux $\mathbf{h}_\alpha$ comprises the contribution

$$-\frac{\partial_{\theta_\alpha} p_\alpha}{\xi \rho} \nabla \theta_\alpha$$

which accounts for the Soret effect.

By (9.56) the heat flux vector $\mathbf{q}$ can be given the form

$$\mathbf{q} = \textstyle\sum_\alpha [\mathbf{q}_\alpha - \mathbf{u}_\alpha \boldsymbol{T}_\alpha + (\varepsilon_\alpha + \frac{1}{2}\mathbf{u}_\alpha^2 + \rho_\alpha \partial_{\rho_\alpha} \psi_\alpha) \mathbf{h}_\alpha].$$

Let

$$\tilde{\chi}_\alpha = \varepsilon_\alpha + \tfrac{1}{2}\mathbf{u}_\alpha^2 + \rho_\alpha \partial_{\rho_\alpha} \psi_\alpha;$$

hence $\tilde{\chi}_\alpha$ is the standard enthalpy density plus the $\frac{1}{2}\mathbf{u}_\alpha^2$ term.

Now, $\mathbf{q}_\alpha = -\kappa_\alpha \nabla \theta_\alpha$, $\kappa_\alpha > 0$ implies $\mathbf{q}_\alpha \cdot \nabla \theta_\alpha \le 0$ and then is consistent with the entropy inequality. Upon substitution of $\mathbf{h}_\alpha$ we then find the model equation

$$\mathbf{q} = \textstyle\sum_\alpha [-(\kappa_\alpha + \frac{\tilde{\chi}_\alpha}{\xi \rho} \partial_{\theta_\alpha} p_\alpha) \nabla \theta_\alpha - \frac{\tilde{\chi}_\alpha}{\xi \rho} \partial_{\rho_\alpha} p_\alpha \nabla \rho_\alpha + \frac{\tilde{\chi}_\alpha}{\xi \rho} \mathbf{f}_\alpha - \mathbf{u}_\alpha \boldsymbol{T}_\alpha].$$

The Dufour effect, that is, the effect of concentration (or mass density) gradients on the heat flux, is then represented by

$$-\textstyle\sum_\alpha \frac{\tilde{\chi}_\alpha}{\xi \rho} \partial_{\rho_\alpha} p_\alpha \nabla \rho_\alpha.$$

9.7.5 Evolution Problems

A dynamic problem of a mixture is governed by the system (9.24) of partial differential equations in the unknowns

$$\rho_1, \ldots, \rho_n, \mathbf{v}_1, \ldots, \mathbf{v}_n, \theta_1, \ldots, \theta_n.$$

The stresses $\{\mathbf{T}_\alpha\}$, the heat fluxes $\{\mathbf{q}_\alpha\}$, the internal energies $\{\varepsilon_\alpha\}$, and the supplies $\{\tau_\alpha\}$, $\{\hat{\mathbf{p}}_\alpha\}$, $\{\hat{\varepsilon}_\alpha\}$ are supposed to be given by the thermodynamically consistent constitutive functions. In particular, if

$$\varepsilon_\alpha = \varepsilon_\alpha(\theta_\alpha, \rho_\alpha)$$

then the balance equations

$$\rho_\alpha \grave{\varepsilon}_\alpha = \mathbf{T}_\alpha \cdot \mathbf{L}_\alpha - \nabla \cdot \mathbf{q}_\alpha + \rho_\alpha r_\alpha - \mathbf{m}_\alpha \cdot \mathbf{v}_\alpha - \tau_\alpha(\varepsilon_\alpha - \tfrac{1}{2}\mathbf{v}_\alpha^2) + e_\alpha, \tag{9.103}$$

where $\alpha = 1, ..., n$, take the form

$$\rho_\alpha \partial_{\theta_\alpha} \varepsilon_\alpha \grave{\theta}_\alpha + \rho_\alpha \partial_{\rho_\alpha} \varepsilon_\alpha \grave{\rho}_\alpha = \mathbf{T}_\alpha \cdot \mathbf{L}_\alpha - \nabla \cdot \mathbf{q}_\alpha + \rho_\alpha r_\alpha - \mathbf{m}_\alpha \cdot \mathbf{v}_\alpha - \tau_\alpha(\varepsilon_\alpha - \tfrac{1}{2}\mathbf{v}_\alpha^2) + e_\alpha. \tag{9.104}$$

This makes apparent the structure of (9.24) as a system for $\{\rho_\alpha\}$, $\{\mathbf{v}_\alpha\}$, and $\{\theta_\alpha\}$. Moreover, once the fields $\{\rho_\alpha\}$, $\{\mathbf{v}_\alpha\}$, $\{\theta_\alpha\}$ are found, the diffusion fluxes $\{\mathbf{h}_\alpha\}$ are determined by solving (9.69).

Very often, however, it is physically sound, or at least convenient, to let the mixture be subject to a single temperature θ. Then the balance equations might seem redundant in view of the n equations of energy involving the single temperature. Now the single-temperature assumption is in fact a constraint, on the a priori distinct temperatures $\theta_1, ..., \theta_n$, which is realized by appropriate exchanges of energy $\{e_\alpha\}$ among the constituents. Hence, we can regard the balance equations (9.103) as a system of n equations in the n unknowns

$$\theta, e_1, ..., e_{n-1},$$

the growth e_n being given by the a priori constraint $\sum_\alpha e_\alpha = 0$. This view overcomes naturally the seeming closure problem connected to n balance equations in the single unknown θ. Indeed, if

$$\varepsilon_\alpha = \varepsilon_\alpha(\theta, \rho_\alpha)$$

then (9.103) reads

$$\rho_\alpha \partial_\theta \varepsilon_\alpha (\partial_t \theta + \mathbf{v}_\alpha \cdot \nabla\theta) + \rho_\alpha \partial_\theta \varepsilon_\alpha \grave{\rho}_\alpha = \mathbf{T}_\alpha \cdot \mathbf{L}_\alpha - \nabla \cdot \mathbf{q}_\alpha + \rho_\alpha r_\alpha - \mathbf{m}_\alpha \cdot \mathbf{v}_\alpha - \tau_\alpha(\varepsilon_\alpha - \tfrac{1}{2}\mathbf{v}_\alpha^2) + e_\alpha.$$

Summation over α and account of the constraint $\sum_\alpha e_\alpha = 0$ imply

$$\textstyle\sum_\alpha [\rho_\alpha \partial_\theta \varepsilon_\alpha (\partial_t \theta + \mathbf{v}_\alpha \cdot \nabla\theta) + \rho_\alpha \partial_\theta \varepsilon_\alpha \grave{\rho}_\alpha] = \sum_\alpha [\mathbf{T}_\alpha \cdot \mathbf{L}_\alpha - \nabla \cdot \mathbf{q}_\alpha + \rho_\alpha r_\alpha - \mathbf{m}_\alpha \cdot \mathbf{v}_\alpha - \tau_\alpha(\varepsilon_\alpha - \tfrac{1}{2}\mathbf{v}_\alpha^2)],$$

which can be viewed as the equation associated with the unknown temperature θ. Once θ is determined, equations

$$e_\alpha = \rho_\alpha \partial_\theta \varepsilon_\alpha (\partial_t \theta + \mathbf{v}_\alpha \cdot \nabla\theta) + \rho_\alpha \partial_\theta \varepsilon_\alpha \grave{\rho}_\alpha - \mathbf{T}_\alpha \cdot \mathbf{L}_\alpha + \nabla \cdot \mathbf{q}_\alpha - \rho_\alpha r_\alpha + \mathbf{m}_\alpha \cdot \mathbf{v}_\alpha + \tau_\alpha(\varepsilon_\alpha - \tfrac{1}{2}\mathbf{v}_\alpha^2)$$

determine the required energy growths that guarantee the common value of the temperature fields for the n constituents.

If the diffusion velocities are disregarded, in that

$$\partial_t\theta + \mathbf{v}_\alpha \cdot \nabla\theta \simeq \partial_t\theta + \mathbf{v} \cdot \nabla\theta = \dot{\theta},$$

then we have

$$\textstyle\sum_\alpha \rho_\alpha \partial_\theta \varepsilon_\alpha \dot{\theta} = \sum_\alpha [-\rho_\alpha \partial_\theta \varepsilon_\alpha \grave{\rho}_\alpha + \mathbf{T}_\alpha \cdot \mathbf{L}_\alpha - \nabla \cdot \mathbf{q}_\alpha + \rho_\alpha r_\alpha - \mathbf{m}_\alpha \cdot \mathbf{v}_\alpha - \tau_\alpha(\varepsilon_\alpha - \frac{1}{2}\mathbf{v}_\alpha^2)],$$

thus showing that

$$\textstyle\sum_\alpha \rho_\alpha \partial_\theta \varepsilon_\alpha = \partial_\theta \sum_\alpha \rho_\alpha \varepsilon_\alpha$$

is the equivalent (approximate) heat capacity of the mixture.

Sometimes a mean temperature is considered for mixtures with multiple temperatures. Going back to (9.104) and following the approximation

$$\grave{\theta}_\alpha \simeq \dot{\theta}_\alpha,$$

we can write

$$\textstyle\sum_\alpha \rho_\alpha \partial_\theta \varepsilon_\alpha \dot{\theta}_\alpha = \sum_\alpha [-\rho_\alpha \partial_\theta \varepsilon_\alpha \grave{\rho}_\alpha + \mathbf{T}_\alpha \cdot \mathbf{L}_\alpha - \nabla \cdot \mathbf{q}_\alpha + \rho_\alpha r_\alpha - \mathbf{m}_\alpha \cdot \mathbf{v}_\alpha - \tau_\alpha(\varepsilon_\alpha - \frac{1}{2}\mathbf{v}_\alpha^2)]. \tag{9.105}$$

Letting θ be defined by

$$\dot{\theta} = \frac{1}{\sum_\beta \rho_\beta \partial_\theta \varepsilon_\beta} \textstyle\sum_\alpha \rho_\alpha \partial_\theta \varepsilon_\alpha \dot{\theta}_\alpha$$

we can write (9.105) as

$$\textstyle\sum_\alpha \rho_\alpha \partial_\theta \varepsilon_\alpha \dot{\theta} = \sum_\alpha [-\rho_\alpha \partial_\theta \varepsilon_\alpha \grave{\rho}_\alpha + \mathbf{T}_\alpha \cdot \mathbf{L}_\alpha - \nabla \cdot \mathbf{q}_\alpha + \rho_\alpha r_\alpha - \mathbf{m}_\alpha \cdot \mathbf{v}_\alpha - \tau_\alpha(\varepsilon_\alpha - \frac{1}{2}\mathbf{v}_\alpha^2)]$$

and regard θ so defined as a convenient definition of mean temperature.

9.8 Solid Mixtures

The balance equations derived in Sect. 9.2 hold for any mixture irrespective of the nature of the constituents except for those with internal structure (e.g. micropolar continua). Yet a remark is in order about the mass density ϱ_α in the reference configuration R_α.

By (9.35) we conclude that in non-reacting mixtures

$$\rho_\alpha = \frac{\varrho_\alpha}{J_\alpha};$$

therefore, ρ_α is determined by J_α, and hence by $\mathbf{F}_\alpha$, because ϱ_α is a constant (at any $\mathbf{X}_\alpha$). Instead, in reacting mixtures ϱ_α is time dependent and hence ρ_α is dependent J_α.

Equations (9.35) and (9.36) motivate the insertion of the reference mass density ϱ_α (or of the mass density ρ_α) among the set of independent variables for the αth constituent. In single bodies, we have

$$\varrho = \rho J, \qquad \dot{\varrho} = 0,$$

so that ϱ is a constant (in time) possibly dependent on the point $\mathbf{X}$ in the reference configuration. Hence, at any point $\mathbf{X}$, the knowledge of $J = \det \mathbf{F}$ produces ρ to within the constant ϱ. We then regard ρ as determined by $\mathbf{F}$. For a constituent, instead, ρ_α is not determined by J_α in that, by (9.35), ϱ_α depends on t because of the chemical reaction. Consequently, J_α and ϱ_α, or J_α and ρ_α are independent of each other. For definiteness, we select ϱ_α and $\mathbf{F}_\alpha$ as independent variables.[18] To our mind this choice makes it more direct the passage from reacting to non-reacting mixtures.

9.8.1 Constitutive Assumptions and Thermodynamic Restrictions

Again we follow the principle of phase separation. For any constituent α, we consider as independent variables the temperature θ_α, the reference mass density ϱ_α, and the deformation gradient $\mathbf{F}_\alpha$, along with their gradients in the reference configuration. Hence we let the free energy ψ_α, the entropy η_α, the stress $\mathbf{T}_\alpha$, the heat flux $\mathbf{q}_\alpha$, and the extra-entropy flux $\mathbf{k}_\alpha$ be functions of the set of variables

$$\Gamma_\alpha := (\theta_\alpha, \varrho_\alpha, \mathbf{F}_\alpha, \nabla_{\mathbf{X}_\alpha}\theta_\alpha, \nabla_{\mathbf{X}_\alpha}\varrho_\alpha, \nabla_{\mathbf{X}_\alpha}\mathbf{F}_\alpha).$$

The interaction terms $\tau_\alpha, \hat{\mathbf{p}}_\alpha, \hat{\varepsilon}_\alpha$ are functions of $\Gamma_1, \Gamma_2, ..., \Gamma_n, \mathbf{u}_1, \mathbf{u}_2, ..., \mathbf{u}_n, \mathbf{v}_\alpha - \mathbf{v}_1, \mathbf{v}_\alpha - \mathbf{v}_2, ..., \mathbf{v}_\alpha - \mathbf{v}_n$. Moreover, the functions $\{\psi_\alpha\}$ are differentiable while the remaining functions are continuous.

By Sect. 9.3.2, the second law requires that ψ_α be independent of $\nabla_{\mathbf{X}_\alpha}\theta_\alpha$ and moreover

$$\eta_\alpha = -\partial_{\theta_\alpha}\psi_\alpha.$$

Hence inequality (9.43) becomes

$$\sum_\alpha \frac{1}{\theta_\alpha}[-\rho_\alpha \partial_{\varrho_\alpha}\psi_\alpha \grave{\varrho}_\alpha - \rho_\alpha \partial_{\mathbf{F}_\alpha}\psi_\alpha \cdot \grave{\mathbf{F}}_\alpha - \rho_\alpha \partial_{\nabla_{\mathbf{X}_\alpha}\varrho_\alpha}\psi_\alpha \cdot \nabla_{\mathbf{X}_\alpha}\grave{\varrho}_\alpha - \rho_\alpha \partial_{\nabla_{\mathbf{X}_\alpha}\mathbf{F}_\alpha}\psi_\alpha \cdot \nabla_{\mathbf{X}_\alpha}\grave{\mathbf{F}}_\alpha$$
$$+\tau_\alpha\theta_\alpha\eta_\alpha + \mathbf{T}_\alpha \cdot \mathbf{L}_\alpha - \frac{1}{\theta_\alpha}\mathbf{q}_\alpha \cdot \nabla\theta_\alpha + \hat{\varepsilon}_\alpha + \theta_\alpha \nabla \cdot \mathbf{k}_\alpha] \geq 0,$$

[18] This approach is developed in [327].

where use is made of the commutation property (9.7). In view of (9.1) and (9.36), we can write

$$\sum_\alpha \frac{1}{\theta_\alpha}[-\rho_\alpha \partial_{\varrho_\alpha}\psi_\alpha \, J_\alpha \tau_\alpha - \rho_\alpha \partial_{\mathbf{F}_\alpha}\psi_\alpha \cdot (\mathbf{L}_\alpha \mathbf{F}_\alpha) - \rho_\alpha \partial_{\nabla_{\mathbf{X}_\alpha}\varrho_\alpha}\psi_\alpha \cdot \nabla_{\mathbf{X}_\alpha}(J_\alpha \tau_\alpha) + \tau_\alpha \theta_\alpha \eta_\alpha$$
$$-\rho_\alpha \partial_{\nabla_{\mathbf{X}_\alpha}\mathbf{F}_\alpha}\psi_\alpha \cdot (\nabla_{\mathbf{X}_\alpha}(\mathbf{L}_\alpha \mathbf{F}_\alpha)) + \mathbf{T}_\alpha \cdot \mathbf{L}_\alpha - \frac{1}{\theta_\alpha}\mathbf{q}_\alpha \cdot \nabla\theta_\alpha + \hat{\varepsilon}_\alpha + \theta_\alpha \nabla \cdot \mathbf{k}_\alpha] \ge 0. \tag{9.106}$$

The field $\nabla_{\mathbf{X}_\alpha}(\mathbf{L}_\alpha \mathbf{F}_\alpha)$, at any point $\mathbf{x}$ and time t, can take arbitrary third-order tensor values and meanwhile the balance equations hold. This is so because

$$\varepsilon_\alpha = \psi_\alpha - \theta_\alpha \partial_{\theta_\alpha}\psi_\alpha = \varepsilon_\alpha(\theta_\alpha, \varrho_\alpha, \nabla_{\mathbf{X}_\alpha}\varrho_\alpha, \nabla_{\mathbf{X}_\alpha}\mathbf{F}_\alpha)$$

and $\nabla_{\mathbf{X}_\alpha}(\mathbf{L}_\alpha \mathbf{F}_\alpha)$ enters the balance of energy,

$$\rho_\alpha \partial_{\nabla_{\mathbf{X}_\alpha}\mathbf{F}_\alpha}\varepsilon_\alpha \cdot \nabla_{\mathbf{X}_\alpha}(\mathbf{L}_\alpha \mathbf{F}_\alpha) + \ldots = \mathbf{T}_\alpha \cdot \mathbf{L}_\alpha - \nabla \mathbf{q}_\alpha + \rho_\alpha r_\alpha + \hat{\varepsilon}_\alpha,$$

the dots indicating terms independent of $\nabla_{\mathbf{X}_\alpha}(\mathbf{L}_\alpha \mathbf{F}_\alpha)$. The conceptual arbitrariness of r_α allows $\nabla_{\mathbf{X}_\alpha}(\mathbf{L}_\alpha \mathbf{F}_\alpha)$ to be chosen arbitrarily while the balance of energy holds. Hence inequality (9.106) holds only if

$$\partial_{\nabla_{\mathbf{X}_\alpha}\mathbf{F}_\alpha}\psi_\alpha = 0$$

and hence

$$\psi_\alpha = \psi_\alpha(\theta_\alpha, \varrho_\alpha, \mathbf{F}_\alpha, \nabla_{\mathbf{X}_\alpha}\varrho_\alpha).$$

The linear dependence of (9.106) on $\mathbf{L}_\alpha$ allows us to conclude that $\mathbf{T}_\alpha$ is given by

$$\mathbf{T}_\alpha = \rho_\alpha \partial_{\mathbf{F}_\alpha}\psi_\alpha \mathbf{F}_\alpha^T. \tag{9.107}$$

Inequality (9.106) then simplifies to

$$\sum_\alpha \frac{1}{\theta_\alpha}[-\rho_\alpha \partial_{\varrho_\alpha}\psi_\alpha \, J_\alpha \tau_\alpha - \rho_\alpha \partial_{\nabla_{\mathbf{X}_\alpha}\varrho_\alpha}\psi_\alpha \cdot \nabla_{\mathbf{X}_\alpha}(J_\alpha \tau_\alpha)$$
$$+\tau_\alpha \theta_\alpha \eta_\alpha - \frac{1}{\theta_\alpha}\mathbf{q}_\alpha \cdot \nabla\theta_\alpha + \hat{\varepsilon}_\alpha + \theta_\alpha \nabla \cdot \mathbf{k}_\alpha] \ge 0. \tag{9.108}$$

We now establish sufficient conditions for (9.108) to hold. By (9.4), upon some rearrangements we have

$$\frac{1}{\theta_\alpha}\rho_\alpha \partial_{\nabla_{\mathbf{X}_\alpha}\varrho_\alpha}\psi_\alpha \cdot \nabla_{\mathbf{X}_\alpha}(J_\alpha \tau_\alpha) = \frac{1}{\theta_\alpha}\rho_\alpha \partial_{\nabla_{\mathbf{X}_\alpha}\varrho_\alpha}\psi_\alpha \mathbf{F}_\alpha^T \nabla(J_\alpha \tau_\alpha)$$
$$= \nabla \cdot (\frac{\varrho_\alpha \tau_\alpha}{\theta_\alpha}\mathbf{F}_\alpha \partial_{\nabla_{\mathbf{X}_\alpha}\varrho_\alpha}\psi_\alpha) - J_\alpha \tau_\alpha \nabla \cdot (\frac{\rho_\alpha}{\theta_\alpha}\mathbf{F}_\alpha \partial_{\nabla_{\mathbf{X}_\alpha}\varrho_\alpha}\psi_\alpha).$$

Consequently we take

$$\mathbf{k}_\alpha = \frac{\varrho_\alpha \tau_\alpha}{\theta_\alpha} \mathbf{F}_\alpha \partial_{\nabla_{\mathbf{X}_\alpha} \varrho_\alpha} \psi_\alpha.$$

Let

$$\mu_\alpha = \psi_\alpha + \varrho_\alpha \partial_{\varrho_\alpha} \psi_\alpha - \theta_\alpha J_\alpha \nabla \cdot (\frac{\rho_\alpha}{\theta_\alpha} \mathbf{F}_\alpha \partial_{\nabla_{\mathbf{X}_\alpha} \varrho_\alpha} \psi_\alpha) \tag{9.109}$$

and recall that $\hat{\varepsilon}_\alpha = e_\alpha - \mathbf{m}_\alpha \cdot \mathbf{v}_\alpha + \frac{1}{2}\tau_\alpha \mathbf{v}_\alpha^2 - \tau_\alpha \varepsilon_\alpha$. We can then write inequality (9.108) in the form

$$\sum_\alpha \frac{1}{\theta_\alpha}[-\mu_\alpha \tau_\alpha + e_\alpha - \mathbf{m}_\alpha \cdot \mathbf{v}_\alpha + \tfrac{1}{2}\tau_\alpha \mathbf{v}_\alpha^2 - \frac{1}{\theta_\alpha}\mathbf{q}_\alpha \cdot \nabla\theta_\alpha] \geq 0. \tag{9.110}$$

Though cross-coupling effects may occur, it is likely that no significant case is lost by letting separate inequalities hold in the form

$$\sum_\alpha \frac{\mu_\alpha}{\theta_\alpha}\tau_\alpha \leq 0, \qquad \sum_\alpha \frac{1}{\theta_\alpha}[e_\alpha - \mathbf{m}_\alpha \cdot \mathbf{v}_\alpha + \tfrac{1}{2}\tau_\alpha \mathbf{v}_\alpha^2] \geq 0, \qquad \sum_\alpha \mathbf{q}_\alpha \cdot \nabla \frac{1}{\theta_\alpha} \geq 0. \tag{9.111}$$

In the particular case where the constituents are subject to the same temperature θ these relations become

$$\sum_\alpha \mu_\alpha \tau_\alpha \leq 0, \qquad \sum_\alpha [\hat{\mathbf{p}}_\alpha \cdot \mathbf{u}_\alpha + \tfrac{1}{2}\tau_\alpha \mathbf{u}_\alpha^2] \leq 0, \qquad (\sum_\alpha \mathbf{q}_\alpha) \cdot \nabla\theta \leq 0. \tag{9.112}$$

Some comments are in order. First,

$$\varrho_\alpha \partial_{\varrho_\alpha} \psi_\alpha$$

is the analogue of p_α/ρ_α in fluids, with ϱ_α instead of ρ_α. Accordingly, μ_α as defined by (9.109) can be viewed as the chemical potential, as for fluids, with a correction due to the dependence of ψ_α on $\nabla_{\mathbf{X}_\alpha} \varrho_\alpha$. Consistently, $\sum_\alpha \tau_\alpha \mu_\alpha/\theta_\alpha$ governs the evolution of chemical reactions where the constituents have different temperatures.

The constitutive equation for the stress $\mathbf{T}_\alpha$ is subject to the principle of objectivity which asserts that the constitutive equations be independent of the observer.[19] We apply to any constituents the argument developed for a single body (Sect. 3.2).

Under a change of observer, such that

$$\chi_\alpha(\mathbf{X}_\alpha, t) \quad \longrightarrow \quad \mathbf{Q}\,\chi_\alpha(\mathbf{X}_\alpha, t)$$

the free energy ψ_α, as well as θ_α and η_α, is invariant while the deformation gradient is subject to

$$\mathbf{F}_\alpha \quad \longrightarrow \quad \mathbf{Q}\mathbf{F}_\alpha,$$

[19] See Sect. 3.2 and [428], Sects. 17–19.

for any $\alpha = 1, ..., n$. Hence objectivity requires that

$$\psi_\alpha(\theta_\alpha, \varrho_\alpha, \mathbf{F}_\alpha, \nabla_{\mathbf{X}_\alpha}\varrho_\alpha) = \psi_\alpha(\theta_\alpha, \varrho_\alpha, \mathbf{Q}\mathbf{F}_\alpha, \nabla_{\mathbf{X}_\alpha}\varrho_\alpha).$$

By the polar decomposition theorem, we can write

$$\mathbf{F}_\alpha = \mathbf{R}_\alpha \mathbf{U}_\alpha,$$

where $\mathbf{R}_\alpha$ is the rotation of $\mathbf{F}_\alpha$ and $\mathbf{U}_\alpha$ is the stretch. Hence choose $\mathbf{Q} = \mathbf{R}_\alpha^T$ and observe

$$\mathbf{R}_\alpha^T \mathbf{F}_\alpha = \mathbf{R}_\alpha^T \mathbf{R}_\alpha \mathbf{U}_\alpha$$

to obtain

$$\psi_\alpha(\theta_\alpha, \varrho_\alpha, \mathbf{F}_\alpha, \nabla_{\mathbf{X}_\alpha}\varrho_\alpha) = \psi_\alpha(\theta_\alpha, \varrho_\alpha, \mathbf{U}_\alpha, \nabla_{\mathbf{X}_\alpha}\varrho_\alpha).$$

Alternatively, we might take ψ_α as a function of the αth right Cauchy–Green strain tensor

$$\mathbf{C}_\alpha = \mathbf{U}_\alpha^2 = \mathbf{F}_\alpha^T \mathbf{F}_\alpha.$$

Since

$$\partial_{\mathbf{F}_\alpha}\psi_\alpha(\mathbf{C}_\alpha) = 2\mathbf{F}_\alpha \partial_{\mathbf{C}_\alpha}\psi_\alpha(\mathbf{C}_\alpha),$$

by (9.107) it follows that

$$\mathbf{T}_\alpha = 2\mathbf{F}_\alpha \partial_{\mathbf{C}_\alpha}\psi_\alpha \mathbf{F}_\alpha^T.$$

Hence, the stress in elastic constituents is symmetric.

Remark 9.2 By analogy with the observations in Sect. 5.5, we claim that the dependence of ψ_α on $\nabla_{\mathbf{X}_\alpha}\theta_\alpha$ and $\nabla_{\mathbf{X}_\alpha}\mathbf{F}_\alpha$ is admissible provided $\mathbf{k}_\alpha$, as well as the other constitutive functions, depend on $\grave{\theta}_\alpha$ and $\grave{\mathbf{F}}_\alpha$. In that case, we would find that

$$\mathbf{T}_\alpha = \rho_\alpha\big[\partial_{\mathbf{F}_\alpha}\psi_\alpha - \frac{\theta_\alpha}{\rho_\alpha}\nabla_{\mathbf{X}_\alpha}\cdot\big(\frac{\rho_\alpha}{\theta_\alpha}\partial_{\nabla_{\mathbf{X}_\alpha}\mathbf{F}_\alpha}\psi_\alpha\big)\big]\mathbf{F}_\alpha^T,$$

Now that we have established the constitutive equations in the Lagrangian description, it is worth looking at corresponding approximate equations in the Eulerian description. We have

$$\varrho_\alpha = \rho_\alpha J_\alpha$$

and

$$\mathbf{F}_\alpha \simeq \partial_{\mathbf{x}_\alpha}(\mathbf{x}_\alpha + \mathbf{u}_\alpha) = \mathbf{1} + \partial_{\mathbf{x}_\alpha}\mathbf{u}_\alpha, \qquad J_\alpha \simeq 1 + \nabla\cdot\mathbf{u}_\alpha, \qquad \varrho_\alpha \simeq \rho_\alpha(1 + \nabla\cdot\mathbf{u}_\alpha).$$

Hence, letting

$$\mathbf{H}_\alpha \simeq \partial_{\mathbf{x}_\alpha}\mathbf{u}_\alpha, \qquad \mathbf{E}_\alpha \simeq \tfrac{1}{2}(\mathbf{H}_\alpha + \mathbf{H}_\alpha^T),$$

we have

$$\mathbf{C}_\alpha = \mathbf{1} + \mathbf{H}_\alpha^T + \mathbf{H}_\alpha + \mathbf{H}_\alpha^T \mathbf{H}_\alpha \simeq \mathbf{1} + 2\,\mathbf{E}_\alpha,$$

the approximation being reasonable inasmuch as $|\partial_{\mathbf{x}_\alpha} \mathbf{u}_\alpha| \ll 1$. Since

$$\partial_{\mathbf{E}_\alpha} \psi_\alpha = 2\,\partial_{\mathbf{C}_\alpha} \psi_\alpha,$$

in the linear approximation we have

$$\psi_\alpha = \psi_\alpha(\theta_\alpha, \rho_\alpha, \mathbf{E}_\alpha)$$

and

$$\mathbf{T}_\alpha = \partial_{\mathbf{E}_\alpha} \psi_\alpha.$$

9.9 Immiscible Mixtures

An immiscible mixture is one in which the individual constituents of the mixture remain physically separate on some scale. This local separation of the constituents can be incorporated into the construction of a theory by means of appropriate constitutive assumptions. In [139], the immiscibility postulate is considered so that the free energy of a given constituent is assumed to depend only upon constitutive variables associated with that constituent. Here this view is adopted for the constitutive functions that are not related to interaction terms $(\tau_\alpha, \mathbf{m}_\alpha, e_\alpha)$ irrespective of the immiscibility property. To our mind, at the scale of interest, any region of the body contains separate parts of the constituents.

The (second) mathematical property of immiscible mixtures, modelling the local microstructure, is that the constitutive equations include a dependence on the volume fraction [62]. Now there is a decisive difference between the model of incompressible constituents and that of compressible ones.

Let the constituents be incompressible, which means that the true density ρ_α^0 of any constituent is constant. By

$$\varrho_\alpha = \rho_\alpha J_\alpha = \rho_\alpha^0 \varphi_\alpha J_\alpha$$

it follows that

$$\frac{\varrho_\alpha}{J_\alpha \varphi_\alpha} = \rho_\alpha^0$$

whence two variables among $\varrho_\alpha, J_\alpha, \varphi_\alpha$ are independent. This means that for solid constituents we may use

$$\mathbf{F}_\alpha, \varphi_\alpha$$

as independent variables. Moreover, because

$$\grave{J}_\alpha = J_\alpha \nabla \cdot \mathbf{v}_\alpha, \qquad \grave{\rho}_\alpha = -\rho_\alpha \nabla \cdot \mathbf{v}_\alpha + \tau_\alpha$$

then

$$\grave{\varrho}_\alpha = J_\alpha \tau_\alpha. \tag{9.113}$$

Equation (9.113) shows that ϱ_α is constant for non-reacting constituents and is a possible independent variable, in place of φ_α, for reacting constituents. The continuity equation for $\rho_\alpha = \varphi_\alpha \rho_\alpha^0$ results in

$$\grave{\varphi}_\alpha = -\varphi_\alpha \nabla \cdot \mathbf{v}_\alpha + \frac{1}{\rho_\alpha^0} \tau_\alpha. \tag{9.114}$$

Equation (9.114) is the evolution equation for φ_α and shows how φ_α is affected by the motion (via $\nabla \cdot \mathbf{v}_\alpha$) and the chemical reaction (via τ_α). Hence, for incompressible constituents, the evolution equation for φ_α is known and is equivalent to the continuity equation for the mass density ρ_α.

Things are no longer easy for compressible constituents. Though we can write again

$$\rho_\alpha = \varphi_\alpha \rho_\alpha^0,$$

the peculiar time derivative and use of the continuity equation yields

$$\grave{\varphi}_\alpha = -\varphi_\alpha \nabla \cdot \mathbf{v}_\alpha + \frac{1}{\rho_\alpha^0} \tau_\alpha - \varphi_\alpha \frac{\grave{\rho}_\alpha^0}{\rho_\alpha^0}. \tag{9.115}$$

Equation (9.115) is not operative in that we cannot determine $\grave{\rho}_\alpha^0$. We might take the approximation that $\varphi_\alpha \grave{\rho}_\alpha^0 / \rho_\alpha^0$ is negligibly small. In such a case (9.115) reduces to (9.114), the evolution equation for the volume fraction of incompressible constituents.

In [62], immiscible compressible constituents are modelled by letting $\grave{\varphi}_\alpha$ be given by a constitutive equation,

$$\grave{\varphi}_\alpha = \hat{\varphi}_\alpha(...),$$

the dots representing the independent variables. If the free energy is taken in the form

$$\psi_\alpha = \psi_\alpha(\varphi_\alpha, ...)$$

then, depending on the variables represented by the dots, the entropy inequality may lead to the reduced inequality

$$\textstyle\sum_\alpha \rho_\alpha \partial_{\varphi_\alpha} \psi_\alpha \, \hat{\varphi}_\alpha \le 0.$$

Hence $\hat{\varphi}_\alpha$ is required to be of the form

$$\hat{\varphi}_\alpha = -\sum_\beta \Lambda_{\alpha\beta} \rho_\beta \partial_{\varphi_\beta} \psi_\beta,$$

$\Lambda_{\alpha\beta}$ being a positive semi-definite tensor. This in turn indicates that the function $\hat{\varphi}_\alpha$ is related to the derivatives $\partial_{\varphi_\beta} \psi_\beta$. We might then obtain $\hat{\varphi}_\alpha$ by an appropriate assumption on $\partial_{\varphi_\beta} \psi_\beta$. However, the correct equation for φ_α is (9.115), which is of dynamic character and any constitutive equation may be viewed as an approximation for $\grave{\varphi}_\alpha$ as, e.g. (9.114).

9.10 Entropy Inequality and Models for the Whole Mixture

Though the balance equations of a mixture are required to be the same as those for a single body [427], practical applications to constitutive models may show significant differences. It is worth investigating how different constitutive restrictions follow from different ways of accounting for the underlying structure (mixture) of the whole body.

9.10.1 Gibbs Equations and Restrictions on the Diffusion Flux

Classical results follow from the assumption of the so-called Gibbs equation [61] in the form (9.85),

$$\theta \dot{\eta} = \dot{\varepsilon}_I - \frac{p}{\rho^2} \dot{\rho} - \sum_\alpha \mu_\alpha \dot{\omega}_\alpha,$$

where $p = \sum_\alpha p_\alpha$ is the pressure on the mixture and μ_α is the αth chemical potential. This equation amounts to assuming that η is a function of ε_I, ρ, and $\{\omega_\alpha\}$ with the obvious relation for the partial derivatives. As shown in Sect. 9.5.4, the Gibbs equations follow from the relations

$$\eta_\alpha = -\partial_\theta \psi_\alpha, \qquad p_\alpha = \rho_\alpha^2 \partial_{\rho_\alpha} \psi_\alpha,$$

which are obtained as a consequence of the second-law inequality.

Sometimes[20] the internal energy ε is considered instead of the inner part ε_I seemingly for technical convenience. By (9.54), the two energies are related by

$$\varepsilon = \varepsilon_I + \tfrac{1}{2} \sum_\alpha \omega_\alpha \mathbf{u}_\alpha^2,$$

the difference being due to a diffusion term. Hence,

[20] See, e.g. [118].

$$\rho\dot{\eta} = \frac{\rho}{\theta}\dot{\varepsilon} - \frac{\rho}{2\theta}\sum_\alpha (\omega_\alpha \mathbf{u}_\alpha^2)^{\cdot} - \frac{p}{\rho\theta}\dot{\rho} - \sum_\alpha \frac{\mu_\alpha}{\theta}\rho\dot{\omega}_\alpha. \tag{9.116}$$

We now show how entropy inequalities given in the literature follow from the Gibbs equation. As we see in a moment, a procedure is developed where the properties of the mixture as a whole ($\mathbf{T}$, $\mathbf{q}$, ε, $\mathbf{j}$) are connected to properties of the constituents (ω_α, $\mathbf{h}_\alpha$, μ_α). As to $\mathbf{h}_\alpha$ we observe that (9.69) might be regarded as a balance equation which expresses the derivative $\dot{\mathbf{h}}_\alpha$ in terms of $\mathbf{h}_\alpha$ and $\mathbf{T}_\alpha$, $\mathbf{T}$, $\mathbf{u}_\alpha$, $\mathbf{b}_\alpha$, $\mathbf{L}$, $\mathbf{m}_\alpha$, $\tau_\alpha\mathbf{v}$. Irrespective of specific constitutive assumptions we now examine the entropy inequality and look for sufficient conditions.

Entropy inequalities follow from (9.59), $\rho\dot{\eta} + \nabla\cdot\mathbf{j} - \rho s \geq 0$, by replacing $\rho\dot{\eta}$ from (9.116). Substitution of $\rho\dot{\varepsilon}$ and $\rho\dot{\omega}_\alpha$ from (9.55) and (9.27) allows (9.59) to be written in the form

$$\frac{1}{\theta}(\mathbf{T} + p\mathbf{1})\cdot\mathbf{D} - \frac{1}{2}\frac{\rho}{\theta}\sum_\alpha (\omega_\alpha \mathbf{u}_\alpha^2)^{\cdot} + \sum_\alpha \frac{\mu_\alpha}{\theta}\nabla\cdot\mathbf{h}_\alpha - \sum_\alpha \frac{\mu_\alpha}{\theta}\tau_\alpha - \frac{1}{\theta}\nabla\cdot\mathbf{q}$$
$$+\frac{\rho r_I}{\theta} + \frac{1}{\theta}\sum_\alpha \mathbf{b}_\alpha\cdot\mathbf{h}_\alpha + \nabla\cdot\mathbf{j} - \rho s \geq 0.$$

Upon some rearrangements, we obtain

$$\frac{1}{\theta}(\mathbf{T} + p\mathbf{1})\cdot\mathbf{D} - \frac{1}{2}\frac{\rho}{\theta}\sum_\alpha (\omega_\alpha \mathbf{u}_\alpha^2)^{\cdot} + \nabla\cdot[\mathbf{j} + \frac{1}{\theta}(\sum_\alpha \mu_\alpha \mathbf{h}_\alpha - \mathbf{q})] + \rho(\frac{r_I}{\theta} - s)$$
$$+\mathbf{q}\cdot\nabla\frac{1}{\theta} - \sum_\alpha \mathbf{h}_\alpha\cdot(\nabla\frac{\mu_\alpha}{\theta} - \frac{1}{\theta}\mathbf{b}_\alpha) - \frac{1}{\theta}\sum_\alpha \mu_\alpha\tau_\alpha \geq 0. \tag{9.117}$$

We first ignore $\sum_\alpha (\omega_\alpha \mathbf{u}_\alpha^2)^{\cdot}$. The entropy flux $\mathbf{j}$ is taken in the form

$$\mathbf{j} = \frac{1}{\theta}\mathbf{q} - \sum_\alpha \frac{\mu_\alpha}{\theta}\mathbf{h}_\alpha.$$

If we let $\rho s = (\rho r_I + \sum_\alpha \mathbf{b}_\alpha\cdot\mathbf{h}_\alpha)/\theta$ then the thermodynamic requirement is

$$(\mathbf{T} + p\mathbf{1})\cdot\mathbf{D} + \mathbf{q}\cdot\nabla\frac{1}{\theta} - \sum_\alpha \mathbf{h}_\alpha\cdot\nabla\frac{\mu_\alpha}{\theta} - \frac{1}{\theta}\sum_\alpha \mu_\alpha\tau_\alpha \geq 0. \tag{9.118}$$

Depending on the constitutive assumptions, inequality (9.118) allows for cross-coupling terms. Yet, sufficient conditions for the validity of (9.118) are given by the non-negative valuedness of each term. So $(\mathbf{T} + p\mathbf{1})\cdot\mathbf{D} \geq 0$ allows for viscosity effects whereas $\mathbf{q}\cdot\nabla(1/\theta) \geq 0$ models heat conduction. Moreover, $\sum_\alpha \mu_\alpha\tau_\alpha \geq 0$ governs the evolution of the constituents in chemical reactions. The remaining inequality

$$\mathbf{h}_\alpha\cdot\nabla\frac{\mu_\alpha}{\theta} \leq 0$$

shows the requirement on the diffusion flux $\mathbf{h}_\alpha$.

If instead $\rho s = \rho r_I/\theta$ then the requirement is

$$\mathbf{h}_\alpha \cdot (\nabla \frac{\mu_\alpha}{\theta} - \frac{1}{\theta}\mathbf{b}_\alpha) \le 0.$$

In both cases, the approach indicates for $\mathbf{h}_\alpha$ a Fick-like constitutive dependence

$$\mathbf{h}_\alpha = -\kappa_\alpha(\nabla \frac{\mu_\alpha}{\theta} - \frac{1}{\theta}\mathbf{b}_\alpha) \quad \text{or} \quad \mathbf{h}_\alpha = -\kappa_\alpha \nabla \frac{\mu_\alpha}{\theta}, \qquad \kappa_\alpha \ge 0. \tag{9.119}$$

More involved relations follow if the term $(\omega_\alpha \mathbf{u}_\alpha^2)^\cdot$ is taken into account. We observe that Eq. (9.119) commonly occurs in the literature though they are slightly contradictory in that $\mathbf{h}_\alpha$ is given by the differential equation (9.69).

9.10.2 Gibbs–Duhem Equation

Restrict attention to homogeneous mixtures. Let n_α be the number of moles of the αth constituent and let G, S, V, p be the Gibbs free energy, the entropy, the volume, and the pressure of the whole mixture. In classical thermodynamics, the chemical potential of the αth constituent is defined as the αth molal Gibbs function. To avoid formal contradictions denote by $\overline{\mu}_\alpha$ the chemical potential so defined. Let $G(\eta, p, \{n_\alpha\})$ be the Gibbs free energy function and define $\overline{\mu}_\alpha$ as

$$\overline{\mu}_\alpha = \partial_{n_\alpha} G.$$

This does not imply that (see [316], Chap. 11)

$$G = \textstyle\sum_\alpha n_\alpha \overline{\mu}_\alpha \tag{9.120}$$

but this is so if $\overline{\mu}_\alpha$ is independent of the mole numbers $\{n_\alpha\}$. Now,

$$\dot{G} = \partial_\theta G\, \dot{\theta} + \partial_p G\, \dot{p} + \textstyle\sum_\alpha \overline{\mu}_\alpha \dot{n}_\alpha$$

and, in view of (5.15),

$$\dot{G} = -S\, \dot{\theta} + V\, \dot{p} + \textstyle\sum_\alpha \overline{\mu}_\alpha \dot{n}_\alpha. \tag{9.121}$$

As a consequence of (9.120), we have

$$\dot{G} = \textstyle\sum_\alpha n_\alpha \dot{\overline{\mu}}_\alpha + \sum_\alpha \overline{\mu}_\alpha \dot{n}_\alpha.$$

Hence we find that

$$\textstyle\sum_\alpha n_\alpha \dot{\overline{\mu}}_\alpha = -S\, \dot{\theta} + V\, \dot{p}. \tag{9.122}$$

Equation (9.122) is usually referred to as *Gibbs–Duhem equation.* It determines the evolution of the chemical potentials $\{\overline{\mu}_\alpha\}$ in terms of the time dependence of θ and p.

Two remarks are in order about the Gibbs–Duhem equation. First, Eq. (9.122) is based on the assumption (9.120), which amounts to regarding $\overline{\mu}_\alpha$ as a function of θ and p, independent of the mole numbers. This assumption is motivated in [169] for dilute solutions, which means that the derivatives $\partial_{n_\alpha} G$ are evaluated at $n_\alpha = 0$. In general, we have to regard $\overline{\mu}_\alpha$ as dependent on the mole numbers and hence the function G in (9.120) is nonlinear with respect to the n_α's. Yet, (9.122) looks as a reasonable approximation in that we can write

$$V = \textstyle\sum_\alpha n_\alpha \overline{v}_\alpha, \qquad S = \textstyle\sum_\alpha n_\alpha \overline{\eta}_\alpha,$$

where the molal volume $\overline{v}_\alpha$ and the molal entropy $\overline{\eta}_\alpha$ are regarded as functions of θ and p. Hence all terms in (9.122) are linear in the mole numbers n_α. Second, divide (9.121) by the mass M of the mixture. Hence we have

$$\dot{\phi} = -\eta\dot{\theta} + \frac{1}{\rho}\dot{p} + \frac{1}{M}\textstyle\sum_\alpha \overline{\mu}_\alpha \dot{n}_\alpha.$$

Now,

$$\omega_\alpha = \frac{n_\alpha M_\alpha}{\sum_\beta n_\beta M_\beta},$$

where M_α is the αth mole mass. Upon differentiation,

$$\dot{\omega}_\alpha = \frac{\dot{n}_\alpha M_\alpha}{\sum_\beta n_\beta M_\beta} - \frac{n_\alpha M_\alpha}{(\sum_\beta n_\beta M_\beta)^2}\sum_\beta \dot{n}_\beta M_\beta.$$

This proves the inequivalence of (9.87) and (9.121).

Chapter 10
Micropolar Media

There are materials with internal structure (e.g. composites, polymers, liquid crystals, soil, and bone) for which a reasonable model should view the points no longer as purely geometric in character but as rigid-body particles or even deformable continua. Models accounting for rigid-body particles are within the theories of micropolar media; those allowing for the deformation of particles are regarded as micromorphic media.

The deformation of the micropolar continuum is described by ascribing, to each particle of the continuum, the position vector, and three orthonormal vectors (called directors) that describe the translations and the orientation changes of the particles. The balance laws of mass, linear momentum, angular momentum, and energy are derived by allowing for an orientational momentum, a body couple density, and a surface couple density. Next some constitutive models are developed, in particular some thermoviscous micropolar media. As a relevant topic, the micropolar model is applied to the description of liquid crystals, mainly the nematics. Mixtures of micropolar constituents are considered; the second law and the growth terms show interesting modelling subjects. As an application of the mixture theory, the model for nanofluids is established where a (solute) constituent has a natural microplar character.

10.1 Kinematics of Micropolar Media

For continua with significant microstructure contribution the classical Cauchy's theory of continuum mechanics is inadequate. At the end of the 19th century a generalization of the continuum model was set up by considering the rotational degrees of freedom of material particles. This material is named micropolar continuum; it is also named Cosserat continuum since the leading ideas are given in a monograph

A. Morro and C. Giorgi, *Mathematical Modelling of Continuum Physics*,
Modeling and Simulation in Science, Engineering and Technology,
https://doi.org/10.1007/978-3-031-20814-0_10

of the brothers Eugéne and François Cosserat. After the book by E. and F. Cosserat [107], linear and nonlinear theories have been developed; [149, 212, 222, 313, 421] and the book [151] show various steps of the theory and a general presentation of the subject.

A micropolar continuum is a continuous distribution of oriented material points (or particles). A material point is characterized by the position and the space orientation. As with any continuum, a point is identified by the position vector $\mathbf{X}$ in a chosen reference configuration R. The motion of a point is characterized by the position vector $\mathbf{x}$, relative to a fixed origin, and three vectors $\mathbf{k}_i$, $i = 1, 2, 3$, rigidly attached to the point, as functions of time,

$$\mathbf{x} = \chi(\mathbf{X}, t), \qquad \mathbf{k}_i = \hat{\mathbf{k}}_i(\mathbf{X}, t), \qquad \mathbf{X} \in \mathrm{R},\ t \in \mathbb{R}.$$

To avoid inessential difficulties we let the triad $\{\mathbf{k}_i\}$ satisfy

$$\mathbf{k}_i \cdot \mathbf{k}_j = \delta_{ij} \tag{10.1}$$

and hence the triad $\{\mathbf{k}_i\}$ is an orthonormal corotational basis. The function χ is assumed to be invertible at any time $t \in \mathbb{R}$. Moreover, χ and $\hat{\mathbf{k}}_i$ are assumed to be twice continuously differentiable. A superposed dot denotes time differentiation holding $\mathbf{X}$ fixed; hence as in classical kinematics we say that $\dot{\chi}$ is the velocity and $\ddot{\chi}$ is the acceleration.

Let

$$R_{iH}(\mathbf{X}, t) = \hat{\mathbf{k}}_i(\mathbf{X}, t) \cdot \hat{\mathbf{k}}_H(\mathbf{X}, 0)$$

be the (passive) rotation matrix. For formal simplicity, let $\mathbf{k}_i$ stand for $\hat{\mathbf{k}}_i(\mathbf{X}, t)$ and $\boldsymbol{\kappa}_H$ for $\hat{\mathbf{k}}_H(\mathbf{X}, 0)$. Since

$$(\mathbf{k}_i \cdot \boldsymbol{\kappa}_H)(\mathbf{k}_j \cdot \boldsymbol{\kappa}_H) = \mathbf{k}_i \cdot \mathbf{k}_j = \delta_{ij}$$

and

$$(\mathbf{k}_i \cdot \boldsymbol{\kappa}_H)(\mathbf{k}_i \cdot \boldsymbol{\kappa}_J) = \boldsymbol{\kappa}_H \cdot \boldsymbol{\kappa}_J = \delta_{HJ}$$

then the matrix R is orthogonal, $R\,R^T = R^T R = \mathbb{1}$. Hence R is invertible and $R^{-1} = R^T$. Consequently,

$$\dot{R}R^T + R\dot{R}^T = 0.$$

Letting

$$\mathcal{O}_{ij} := \dot{R}_{iH} R_{jH}$$

we have

$$\dot{R}_{iH} = \mathcal{O}_{ij} R_{jH}, \qquad \mathcal{O}_{ij} = \dot{R}_{iH} R_{jH} = -R_{iH}\dot{R}_{jH} = -\mathcal{O}_{ji}$$

and hence $\mathcal{O}$ is skew,

$$\mathcal{O} = -\mathcal{O}^T.$$

The matrix $\mathcal{O}$ is often referred to as gyration matrix.

The definition of R allows us to write

$$\mathbf{k}_i = R_{iH}\boldsymbol{\kappa}_H, \qquad \boldsymbol{\kappa}_H = R_{iH}\mathbf{k}_i.$$

We can view $\{\boldsymbol{\kappa}_H\} \to \{\mathbf{k}_i\}$ as a change of basis. The corresponding change of vector components $\{U_H\} \to \{u_i\}$ of a vector $\mathbf{u}$ is given by

$$u_i = \mathbf{u}\cdot\mathbf{k}_i = \mathbf{u}\cdot R_{iH}\boldsymbol{\kappa}_H = R_{iH}\mathbf{u}\cdot\boldsymbol{\kappa}_H, \qquad u_i = R_{iH}U_H.$$

Thus vector components and basis vectors obey the same transformation law.

Time differentiation of $\mathbf{k}_i$,

$$\dot{\mathbf{k}}_i = \dot{R}_{iH}\boldsymbol{\kappa}_H,$$

substitution for $\boldsymbol{\kappa}_H$, and the definition of $\mathcal{O}$ result in

$$\dot{\mathbf{k}}_i = \dot{R}_{iH}R_{hH}\mathbf{k}_h = \mathcal{O}_{ih}\mathbf{k}_h.$$

To each skew matrix we can associate a vector and vice versa. The axial vector, say[1] $\boldsymbol{\nu}$, associated with $\mathcal{O}$ is defined by

$$\nu_i = -\tfrac{1}{2}\epsilon_{ihk}\mathcal{O}_{hk},$$

and hence $\mathcal{O}_{hk} = -\epsilon_{hki}\nu_i$. In the literature, $\boldsymbol{\mathcal{O}}$ is called the *gyration tensor* and $\boldsymbol{\nu}$ is called the intrinsic angular velocity ([151], part I). Yet, in view of the results of Sect. 1.4, the angular velocity of the triad $\{\mathbf{k}_i\}$, and hence of the particle, is $\boldsymbol{\omega} = -\boldsymbol{\nu}$, namely

$$\omega_i = \tfrac{1}{2}\epsilon_{ihk}\mathcal{O}_{hk}, \qquad \mathcal{O}_{hk} = \epsilon_{hki}\omega_i.$$

By paralleling the procedure of Sect. 1.4, this definition of the angular velocity allows us to find the Poisson relation

$$\dot{\mathbf{k}}_p = \boldsymbol{\omega}\times\mathbf{k}_p, \qquad p = 1, 2, 3,$$

and hence the classical relation for any vector function $\mathbf{u}$,

$$\Big(\frac{d\mathbf{u}}{dt}\Big)_{\text{space}} = \Big(\frac{d\mathbf{u}}{dt}\Big)_{\text{body}} + \boldsymbol{\omega}\times\mathbf{u},$$

which reduces to the Poisson relation if $\mathbf{u}$ is any vector constant with respect to the body (particle). The derivative $(d\mathbf{u}/dt)_{\text{space}}$ is usually denoted by the superposed dot. If $\mathbf{u}$ is a vector along a direction rigidly attached to the point, and $|\mathbf{u}|$ is constant, then $(d\mathbf{u}/dt)_{\text{body}} = \mathbf{0}$ and $\mathbf{u}$ changes in time according to

[1] The symbol $\boldsymbol{\nu}$ for the axial vector in micropolar theories is customary in the literature.

$$\dot{\mathbf{u}} = \boldsymbol{\omega} \times \mathbf{u}.$$

The transformation $\{\boldsymbol{\kappa}_H\} \to \{\mathbf{k}_i\}$ can be described by the active rotation tensor $\boldsymbol{\mathcal{R}}$ such that

$$\boldsymbol{\mathcal{R}}\boldsymbol{\kappa}_H = \mathbf{k}_H, \qquad H = 1, 2, 3.$$

Then inner multiplying by $\boldsymbol{\kappa}_J$ we find

$$\boldsymbol{\kappa}_J \cdot \boldsymbol{\mathcal{R}}\boldsymbol{\kappa}_H = \mathbf{k}_H \cdot \boldsymbol{\kappa}_J, \qquad \mathcal{R}_{JH} = R_{HJ}, \qquad \mathcal{R} = R^T,$$

$\mathcal{R}_{JH}$ being the components in the fixed basis $\{\boldsymbol{\kappa}_H\}$. Under a change of frame (1.54) the vectors, $\mathbf{k}_H$ say, are subject to

$$\mathbf{k}_H = \boldsymbol{\mathcal{R}}\boldsymbol{\kappa}_H \;\to\; \mathbf{k}^*_H = \mathbf{Q}\mathbf{k}_H = \mathbf{Q}\boldsymbol{\mathcal{R}}\boldsymbol{\kappa}_H.$$

Hence $\mathbf{k}^*_H = \boldsymbol{\mathcal{R}}^*\boldsymbol{\kappa}_H$ implies that

$$\boldsymbol{\mathcal{R}}^* = \mathbf{Q}\boldsymbol{\mathcal{R}}.$$

To compute the time derivative of $\mathcal{R}$ we consider the identity

$$\dot{\mathcal{R}}_{JH} = \dot{\mathcal{R}}_{JL}\mathcal{R}_{KL}\mathcal{R}_{KH} = \mathcal{A}_{JK}\mathcal{R}_{KH},$$

whence

$$\dot{\mathcal{R}} = \mathcal{A}\mathcal{R}, \qquad \mathcal{A} = -\mathcal{A}^T. \tag{10.2}$$

Moreover, by (1.21),

$$\boldsymbol{\omega} = \mathbf{e}_L \tfrac{1}{2}\epsilon_{LHJ}\mathcal{A}_{JH}.$$

In connection with the objectivity principle, invariant strain measures are essential in the description of the constitutive functions. In this sense we consider the *wryness tensor* $\boldsymbol{\Gamma}$ defined by

$$\Gamma_{KL} = \tfrac{1}{2}\epsilon_{KMN}(\partial_{X_L}\mathcal{R}_{kM})\mathcal{R}_{kN},$$

in the fixed basis $\{\mathbf{e}_i\}$. Under a change of reference $\boldsymbol{\mathcal{R}} \to \mathbf{Q}\boldsymbol{\mathcal{R}}$. Hence

$$\Gamma^*_{KL} = \tfrac{1}{2}\epsilon_{KMN}\mathcal{Q}_{kh}(\partial_{X_L}\mathcal{R}_{hM}\mathcal{Q}_{kj}\mathcal{R}_{jN} = \tfrac{1}{2}\epsilon_{KMN}(\partial_{X_L}\mathcal{R}_{hM})\mathcal{Q}^T_{hk}\mathcal{Q}_{kj}\mathcal{R}_{jN} = \Gamma_{KL}.$$

The invariance of $\boldsymbol{\Gamma}$ implies the invariance, and hence the objectivity, of $\dot{\boldsymbol{\Gamma}}$.

For later use it is worth computing the time derivative of the wryness tensor. Since

$$\dot{\Gamma}_{KL} = \tfrac{1}{2}\epsilon_{KMN}[(\partial_{X_L}\dot{\mathcal{R}}_{kM})\mathcal{R}_{kN} + (\partial_{X_L}\mathcal{R}_{kM})\dot{\mathcal{R}}_{kN})]$$

and $\dot{\mathcal{R}} = \mathcal{A}\mathcal{R}$ we obtain

$$\dot{\Gamma}_{KL} = \frac{1}{2}\epsilon_{KMN}[(\partial_{X_L}\dot{\mathcal{R}}_{kM}\mathcal{R}_{kN} + (\partial_{X_L}\mathcal{R}_{kM})\dot{\mathcal{R}}_{kN}]$$
$$= \frac{1}{2}\epsilon_{KMN}[(\partial_{X_L}\mathcal{A}_{kj})\mathcal{R}_{jM}\mathcal{R}_{kN} + \mathcal{A}_{kj}(\partial_{X_L}\mathcal{R}_{kM})\mathcal{R}_{kN} + \mathcal{A}_{kj}(\partial_{X_L}\mathcal{R}_{kM})\mathcal{R}_{jN}].$$

Now,

$$\mathcal{A}_{kj}(\partial_{X_L}\mathcal{R}_{jM})\mathcal{R}_{kN} + \mathcal{A}_{kj}(\partial_{X_L}\mathcal{R}_{kM})\mathcal{R}_{jN} = \mathcal{A}_{kj}(\partial_{X_L}\mathcal{R}_{jM})\mathcal{R}_{kN} + \mathcal{A}_{jk}(\partial_{X_L}\mathcal{R}_{jM})\mathcal{R}_{kN}$$
$$= \mathcal{A}_{kj}[(\partial_{X_L}\mathcal{R}_{jM})\mathcal{R}_{kN} - (\partial_{X_L}\mathcal{R}_{jM})\mathcal{R}_{kN}] = 0.$$

As to the first term of $\dot{\Gamma}_{KL}$ we have

$$\epsilon_{KMN}\mathcal{R}_{jM}\mathcal{R}_{kN} = \epsilon_{ijk}\mathcal{R}_{iK}$$

and hence

$$\frac{1}{2}\epsilon_{KMN}(\partial_{X_L}\mathcal{A}_{kj})\mathcal{R}_{jM}\mathcal{R}_{kN} = \mathcal{R}_{iK}\partial_{X_L}\omega_i.$$

Consequently we find

$$\dot{\Gamma}_{KL} = \mathcal{R}_{iK}\partial_{X_L}\omega_i. \tag{10.3}$$

Consider the strain measure named *Cosserat deformation tensor* $\mathfrak{C}$ and defined by

$$\mathfrak{C}_{KL} = \mathcal{R}_{hK}F_{hL}.$$

Under a change of frame $\mathfrak{C}$ is invariant,

$$\mathfrak{C}^*_{KL} = \mathcal{R}^*_{hK}F^*_{hL} = Q_{hj}\mathcal{R}_{jK}Q_{hi}F_{iL} = \mathcal{R}_{jK}Q^T_{ih}Q_{hj}F_{iK} = \mathfrak{C}_{KL}.$$

The time derivative is then invariant too. Since $\dot{\mathbf{F}} = \mathbf{LF}$ and $\dot{\mathcal{R}} = \mathcal{A}\mathcal{R}$ we find

$$\dot{\mathfrak{C}}_{KL} = \mathcal{A}_{jh}\mathcal{R}_{hK}F_{jL}\mathcal{R}_{hK}L_{hj}F_{jL} = \mathcal{R}_{jK}(\mathcal{A}_{hj} + W_{jh} + D_{jh})F_{hL}.$$

Since

$$\mathcal{A}_{jh} = -\epsilon_{jhl}\omega_l, \qquad W_{jh} = -\epsilon_{jhl}w_l, \quad \mathbf{w} = \frac{1}{2}\nabla\times\mathbf{v}$$

we obtain

$$\dot{\mathfrak{C}}_{KL} = \mathcal{R}_{jK}D_{jh}F_{hL} + \epsilon_{hjl}\mathcal{R}_{jK}F_{hL}(w_l - \omega_l). \tag{10.4}$$

It is worth remarking that while the angular velocity $\boldsymbol{\omega}$ is non-objective both the difference $\mathbf{w} - \boldsymbol{\omega}$ (see 1.71) and the gradient $\nabla\boldsymbol{\omega}$ are objective. As to $\nabla\boldsymbol{\omega}$, since

$$\boldsymbol{\omega}^* = \mathbf{Q}\boldsymbol{\omega} + \mathbf{a}_\Omega$$

it follows that

$$(\nabla\boldsymbol{\omega})^* = \nabla^*\boldsymbol{\omega}^* = (\mathbf{Q}\nabla)(\mathbf{Q}\boldsymbol{\omega} + \mathbf{a}_\Omega) = \mathbf{Q}\nabla\boldsymbol{\omega}\mathbf{Q}^T.$$

Hence $\nabla\omega$ is an objective tensor. The referential gradient $\nabla_R\omega$ satisfies

$$(\nabla_R\omega)^* = \nabla_R\omega^* = \mathbf{Q}\nabla_R\omega$$

and hence is an objective vector.

10.1.1 Orientational Momentum

The rigid-body nature of the particles allows us to ascribe to each particle a micro-inertia tensor density $\mathbf{I}$, per unit mass, such that the orientational momentum, or intrinsic angular momentum, per unit mass $\boldsymbol{\sigma}$ is given by

$$\boldsymbol{\sigma} = \mathbf{I}\,\boldsymbol{\omega}.$$

In the corotational basis $\{\mathbf{k}_j\}$, $\mathbf{I}$ is represented by a constant symmetric matrix with entries I_{ij},

$$I_{ij} = \mathbf{k}_i \cdot \mathbf{I}\mathbf{k}_j, \qquad \mathbf{I} = \mathbf{k}_i I_{ij} \mathbf{k}_j.$$

Hence,

$$\boldsymbol{\sigma} = I_{ij}\omega_j \mathbf{k}_i.$$

If m is the mass of the particle then $m\mathbf{I}$ is the inertia tensor of the particle.

The angular momentum, per unit mass, is then defined as

$$\mathbf{r} \times \mathbf{v} + \boldsymbol{\sigma}$$

and the kinetic energy as

$$\tfrac{1}{2}(\mathbf{v}^2 + \boldsymbol{\omega} \cdot \boldsymbol{\sigma}).$$

For later convenience we evaluate the time derivative of $\boldsymbol{\sigma}$ and $\boldsymbol{\omega} \cdot \boldsymbol{\sigma}$. Relative to the corotational basis, since $\dot{\mathbf{k}}_i = \boldsymbol{\omega} \times \mathbf{k}_i$ it follows

$$\dot{\boldsymbol{\sigma}} = (I_{ij}\omega_j \mathbf{k}_i)^{\cdot} = I_{ij}\dot{\omega}_j \mathbf{k}_i + I_{ij}\omega_j \boldsymbol{\omega} \times \mathbf{k}_i.$$

Further,

$$\dot{\boldsymbol{\omega}} = \dot{\omega}_p \mathbf{k}_p + \omega_p \dot{\mathbf{k}}_p = \dot{\omega}_p \mathbf{k}_p + \omega_p \boldsymbol{\omega} \times \mathbf{k}_p = \dot{\omega}_p \mathbf{k}_p + \boldsymbol{\omega} \times \omega_p \mathbf{k}_p = \dot{\omega}_p \mathbf{k}_p$$

and

$$\mathbf{I}\,\dot{\boldsymbol{\omega}} = (\mathbf{k}_i I_{ij} \mathbf{k}_j)(\dot{\omega}_p \mathbf{k}_p) = \mathbf{k}_i I_{ij} \dot{\omega}_j.$$

Hence

$$\dot{\boldsymbol{\sigma}} = \mathbf{I}\,\dot{\boldsymbol{\omega}} + \boldsymbol{\omega} \times \mathbf{I}\,\boldsymbol{\omega}. \tag{10.5}$$

Thus it follows

$$(\boldsymbol{\omega} \cdot \boldsymbol{\sigma})^{\cdot} = \dot{\boldsymbol{\omega}} \cdot \boldsymbol{\sigma} + \boldsymbol{\omega} \cdot \dot{\boldsymbol{\sigma}} = \dot{\boldsymbol{\omega}} \cdot \mathbf{I}\boldsymbol{\omega} + \boldsymbol{\omega} \cdot \mathbf{I}\dot{\boldsymbol{\omega}}.$$

Owing to the symmetry of **I**, we have

$$\mathbf{u} \cdot \mathbf{I}\mathbf{v} = \mathbf{v} \cdot \mathbf{I}\mathbf{u}$$

for every pair of vectors **u**, **v**, and hence

$$(\boldsymbol{\omega} \cdot \boldsymbol{\sigma})^{\cdot} = 2\boldsymbol{\omega} \cdot \dot{\boldsymbol{\sigma}}. \tag{10.6}$$

It is worth pointing out that the micro-inertia matrix, with entries I_{ij}, relative to a generic (non-corotational) basis $\{\mathbf{e}_i\}$, satisfies a property usually referred to as *conservation of micro-inertia* [148],

$$\dot{I}_{jh} - \mathcal{O}_{jk} I_{hk} - \mathcal{O}_{hk} I_{jk} = 0. \tag{10.7}$$

To show this, we start with the law of transformation of the components of **I** under a change of orthonormal triad, $\{\boldsymbol{\kappa}_K\} \to \{\mathbf{e}_i\}$, $R_{iK} = \mathbf{e}_i \cdot \boldsymbol{\kappa}_K$. We have

$$\mathbf{e}_j \cdot \mathbf{I}\mathbf{e}_h = R_{jP} R_{hQ} \boldsymbol{\kappa}_P \cdot \mathbf{I}\boldsymbol{\kappa}_Q$$

and hence we can write

$$I_{jh} = R_{jP} R_{hQ} \hat{I}_{PQ}, \qquad \hat{I}_{PQ} := \boldsymbol{\kappa}_P \cdot \mathbf{I}\boldsymbol{\kappa}_Q,$$

as it happens for the components of any second-order tensor. Now, upon time differentiation we have

$$\dot{I}_{jh} = \dot{R}_{jP} R_{hQ} \hat{I}_{PQ} + R_{jP} \dot{R}_{hQ} \hat{I}_{PQ}.$$

Since $R_{iP} R_{hQ} \hat{I}_{PQ} = I_{ih}$, $R_{jP} R_{lQ} \hat{I}_{PQ} = I_{jl}$, use of the relations

$$\dot{R}_{jP} R_{iP} = \mathcal{O}_{ji}, \qquad \dot{R}_{hQ} R_{lQ} = \mathcal{O}_{hl},$$

allows us to write

$$\dot{I}_{jh} = \mathcal{O}_{ji} I_{hi} + \mathcal{O}_{hl} I_{jl}, \qquad h, j = 1, 2, 3.$$

Hence, in general, the rate Eq. (10.7) for the matrix I holds. If, however, the triad $\{\mathbf{e}_i\}$ is a corotational basis then the entries I_{ij} are constants and (10.7) reduces to

$$\mathcal{O}_{ji} I_{hi} + \mathcal{O}_{hl} I_{jl} = 0, \qquad h, j = 1, 2, 3.$$

10.2 Balance Laws

We let $\mathcal{P}_t$ be the spatial region that convects with the body so that $\mathcal{P}_t = \chi(\mathrm{P}, t)$ for some region P in the reference configuration. Reynold's transport theorem holds and is applied in the analysis of balance laws.

10.2.1 Balance Laws in the Spatial Description

The pertinent fields are described by functions of $\mathbf{x}, t$, with $\mathbf{x} \in \mathcal{P}_t \subseteq \Omega = \chi(\mathrm{R}, t)$.

Balance of Mass

The mass density ρ is required to obey the continuity equation

$$\partial_t \rho + \nabla \cdot (\rho \mathbf{v}) = 0.$$

Balance of Linear Momentum

Let $\mathbf{b}$ be the body force, per unit mass, and $\mathbf{t}$ the force per unit area. The balance of linear momentum is expressed by saying that

$$\frac{d}{dt} \int_{\mathcal{P}_t} \rho \mathbf{v}\, dv = \int_{\mathcal{P}_t} \rho \mathbf{b}\, dv + \int_{\partial \mathcal{P}_t} \mathbf{t}\, da.$$

Cauchy's theorem proves the existence of the stress tensor $\mathbf{T}$, such that $\mathbf{t}(\mathbf{x}, t, \mathbf{n}) = \mathbf{T}(\mathbf{x}, t)\mathbf{n}$. The arbitrariness of the region $\mathcal{P}_t$ implies the validity of the local form

$$\rho \dot{\mathbf{v}} = \nabla \cdot \mathbf{T} + \rho \mathbf{b}. \tag{10.8}$$

Balance of Angular Momentum

The internal structure of a micropolar body is made explicit by the orientational momentum (or intrinsic angular momentum) $\boldsymbol{\sigma}$, per unit mass. Let $\mathbf{r}$ be the position vector of a point relative to a fixed origin. So, in general, $\mathbf{r} = \mathbf{x} - \mathbf{d}$, $\mathbf{d}$ being a constant vector. We allow for contribution to couple terms by letting the occurrence of a body couple density $\mathbf{l}$, per unit mass, and a surface couple density $\mathbf{s}$, per unit area. Hence we state the balance of angular momentum in the form

$$\frac{d}{dt}\int_{\mathcal{P}_t} \rho(\mathbf{r}\times\mathbf{v}+\boldsymbol{\sigma})dv = \int_{\mathcal{P}_t} \rho(\mathbf{r}\times\mathbf{b}+\mathbf{l})dv + \int_{\partial\mathcal{P}_t} (\mathbf{r}\times\mathbf{t}+\mathbf{s})da.$$

Since $\mathbf{t} = \mathbf{Tn}$ then

$$\int_{\partial\mathcal{P}_t} \mathbf{r}\times\mathbf{t}\,da = \int_{\mathcal{P}_t} (\boldsymbol{\Upsilon} + \mathbf{r}\times\nabla\cdot\mathbf{T})dv$$

where $\boldsymbol{\Upsilon}$ is the vector given by $\Upsilon_i = \epsilon_{ijk}T_{kj}$. Hence we have

$$\int_{\mathcal{P}_t} [\rho(\mathbf{r}\times\dot{\mathbf{v}}+\dot{\boldsymbol{\sigma}}-\mathbf{r}\times\mathbf{b}-\mathbf{l}) - \boldsymbol{\Upsilon} - \mathbf{r}\times\nabla\cdot\mathbf{T}]dv = \int_{\partial\mathcal{P}_t} \mathbf{s}\,da.$$

In view of the equation of motion (10.8) it follows

$$\int_{\mathcal{P}_t} [\rho(\dot{\boldsymbol{\sigma}}-\mathbf{l}) - \boldsymbol{\Upsilon}]dv = \int_{\partial\mathcal{P}_t} \mathbf{s}\,da.$$

The assumed boundedness of $\rho(\dot{\boldsymbol{\sigma}}-\mathbf{l}) - \boldsymbol{\Upsilon}$ allows us to apply Cauchy's theorem and prove that there exists a tensor field $\mathbf{S}(\mathbf{x}, t)$ such that

$$\mathbf{s}(\mathbf{x}, t, \mathbf{n}) = \mathbf{S}(\mathbf{x}, t)\mathbf{n}.$$

The tensor $\mathbf{S}$ is named *couple stress tensor.* Substitution for $\mathbf{s}$ and use of the divergence theorem eventually results in

$$\rho\dot{\boldsymbol{\sigma}} = \boldsymbol{\Upsilon} + \rho\mathbf{l} + \nabla\cdot\mathbf{S}. \tag{10.9}$$

By analogy with rigid body dynamics we consider the inertia operator $\mathcal{I}$ of the particle so that $\mathcal{I}\boldsymbol{\omega}$ is the intrinsic angular momentum of the particle. If m is the mass of the particle then

$$\boldsymbol{\sigma} = \frac{1}{m}\mathcal{I}\boldsymbol{\omega}, \qquad \rho\boldsymbol{\sigma} = \mathcal{N}\mathcal{I}\boldsymbol{\omega},$$

$\mathcal{N} = \rho/m$ being the number density of particles. Let

$$\mathbf{I} = \frac{1}{m}\mathcal{I};$$

$\mathbf{I}$ is then the density of inertia tensor, per unit mass. Hence

$$\rho\mathbf{I} = \frac{\rho}{m}\mathcal{I} = \mathcal{N}\mathcal{I}, \qquad \boldsymbol{\sigma} = \mathbf{I}\,\boldsymbol{\omega}.$$

By (10.5) the balance Eq. (10.9) can be written in the form

$$\rho\mathbf{I}\,\dot{\boldsymbol{\omega}} + \boldsymbol{\omega}\times\rho\mathbf{I}\,\boldsymbol{\omega} = \boldsymbol{\Upsilon} + \rho\mathbf{l} + \nabla\cdot\mathbf{S}. \tag{10.10}$$

Balance of Energy

The micropolar character suggests that the kinetic energy consists of a contribution due to the intrinsic rotation, in addition to the kinetic energy associated with the macroscopic velocity field. Borrowing from the kinematics of rigid body motion we let

$$\rho\tfrac{1}{2}(\mathbf{v}^2 + \boldsymbol{\omega}\cdot\boldsymbol{\sigma})$$

be the kinetic energy and observe that $\boldsymbol{\omega}\cdot\boldsymbol{\sigma} = \boldsymbol{\omega}\cdot\mathbf{I}\,\boldsymbol{\omega}$. We assume that $\mathbf{l}$ and $\mathbf{s}$ exert a power $\rho\mathbf{l}\cdot\boldsymbol{\omega}$ per unit volume and $\mathbf{s}\cdot\boldsymbol{\omega}$ per unit area. In addition, an (internal) energy density ε, an energy supply r, per unit mass, and an energy flux h of non-mechanical character are allowed to occur. Hence we express the balance of energy in the form

$$\frac{d}{dt}\int_{\mathcal{P}_t}\rho[\varepsilon + \tfrac{1}{2}(\mathbf{v}^2 + \boldsymbol{\omega}\cdot\boldsymbol{\sigma})]dv = \int_{\mathcal{P}_t}\rho(\mathbf{b}\cdot\mathbf{v} + \mathbf{l}\cdot\boldsymbol{\omega} + r)dv + \int_{\partial\mathcal{P}_t}(\mathbf{t}\cdot\mathbf{v} + \mathbf{s}\cdot\boldsymbol{\omega} + h)da.$$

First we replace $\mathbf{t}$ with $\mathbf{Tn}$ and $\mathbf{s}$ with $\mathbf{Sn}$. Since

$$\nabla\cdot(\mathbf{vT}) = \mathbf{T}\cdot\mathbf{L} + \mathbf{v}\cdot(\nabla\cdot\mathbf{T}), \qquad \nabla\cdot(\boldsymbol{\omega}\mathbf{S}) = \mathbf{S}\cdot\nabla\boldsymbol{\omega} + \boldsymbol{\omega}\cdot(\nabla\cdot\mathbf{S}),$$

where $\mathbf{S}\cdot\nabla\boldsymbol{\omega} = S_{ij}\partial_{x_j}\omega_i$, the divergence theorem and use of (10.6), (10.8), and (10.9) imply that

$$\int_{\mathcal{P}_t}[\rho\dot{\varepsilon} - \rho r - \mathbf{T}\cdot\mathbf{L} + \boldsymbol{\omega}\cdot\boldsymbol{\Upsilon} - \mathbf{S}\cdot\nabla\boldsymbol{\omega}]dv = \int_{\partial\mathcal{P}_t} h\,da.$$

Again we apply Cauchy's theorem and find that the scalar field $h(\mathbf{x}, t, \mathbf{n})$ depends linearly on $\mathbf{n}$. Let

$$h(\mathbf{x}, t, \mathbf{n}) = -\mathbf{q}(\mathbf{x}, t)\cdot\mathbf{n}.$$

Substitution and use of the divergence theorem show that

$$\int_{\mathcal{P}_t}[\rho\dot{\varepsilon} - \rho r - \mathbf{T}\cdot\mathbf{L} + \boldsymbol{\omega}\cdot\boldsymbol{\Upsilon} - \mathbf{S}\cdot\nabla\boldsymbol{\omega} + \nabla\cdot\mathbf{q}]dv = 0.$$

Hence the arbitrariness of $\mathcal{P}_t$ allows us to conclude that the balance of energy implies the local balance equation[2]

$$\rho\dot{\varepsilon} = \mathbf{T}\cdot\mathbf{L} - \boldsymbol{\omega}\cdot\boldsymbol{\Upsilon} + \mathbf{S}\cdot\nabla\boldsymbol{\omega} - \nabla\cdot\mathbf{q} + \rho r.$$

Since

$$\Upsilon_i = \epsilon_{ikj}T^{\mathrm{skw}}_{jk}, \qquad T^{\mathrm{skw}}_{jk} = \tfrac{1}{2}\epsilon_{kjp}\Upsilon_p,$$

[2] There are many different equations in the literature. Equation (10.11) coincides, e.g., with (2.2.47) of [149] once we identify the present $\mathbf{T}$ and $\mathbf{q}$ with $\mathbf{T}^T$ and $-\mathbf{q}$ of [149].

then

$$\mathbf{T} \cdot \mathbf{W} - \boldsymbol{\Upsilon} \cdot \boldsymbol{\omega} = \boldsymbol{\Upsilon} \cdot (\mathbf{w} - \boldsymbol{\omega}), \qquad \mathbf{w} := \tfrac{1}{2}\nabla \times \mathbf{v};$$

$\mathbf{w}$ is the axial vector of $\mathbf{W}$ and is viewed as the angular velocity of the macroscopic motion. Consequently, $\boldsymbol{\Upsilon} \cdot (\mathbf{w} - \boldsymbol{\omega})$ is the power associated with the skew stress $\mathbf{T}^{\text{skw}}$ and $\mathbf{w} - \boldsymbol{\omega}$ is the relative angular velocity (among the fluid and the particle). Thus the balance of energy can be written in the form

$$\rho\dot{\varepsilon} = \mathbf{T} \cdot \mathbf{D} + \boldsymbol{\Upsilon} \cdot (\mathbf{w} - \boldsymbol{\omega}) + \mathbf{S} \cdot \nabla\boldsymbol{\omega} - \nabla \cdot \mathbf{q} + \rho r. \tag{10.11}$$

Remark 10.1 Though $\mathbf{w}$ and $\boldsymbol{\omega}$ are not objective, the difference $\mathbf{w} - \boldsymbol{\omega}$ and hence the power $\boldsymbol{\Upsilon} \cdot (\mathbf{w} - \boldsymbol{\omega})$ are objective. Conceptually non-objective powers might occur in that the balance equations hold in inertial frames. However, the objectivity of $\mathbf{w} - \boldsymbol{\omega}$ allows us to consider $\mathbf{w} - \boldsymbol{\omega}$ among the independent variables of the constitutive functions.

Second Law of Thermodynamics

Let η be the entropy density, per unit mass, and θ the absolute temperature, on $\Omega \times \mathbb{R}$. We follow the statement of the second law as in Sect. 2.6. Hence we have the inequality

$$\rho\dot{\eta} + \frac{1}{\theta}(\nabla \cdot \mathbf{q} - \rho r) - \frac{1}{\theta^2}\mathbf{q} \cdot \nabla\theta + \nabla \cdot \mathbf{k} \geq 0.$$

Upon substitution of $\nabla \cdot \mathbf{q} - \rho r$ from (10.11) we have

$$\rho\theta\dot{\eta} - \rho\dot{\varepsilon} + \mathbf{T} \cdot \mathbf{D} + \boldsymbol{\Upsilon} \cdot (\mathbf{w} - \boldsymbol{\omega}) + \mathbf{S} \cdot \nabla\boldsymbol{\omega} - \frac{1}{\theta}\mathbf{q} \cdot \nabla\theta + \theta\nabla \cdot \mathbf{k} \geq 0.$$

By means of the Helmholtz free energy $\psi = \varepsilon - \theta\eta$ we can write the inequality in the form

$$-\rho(\dot{\psi} + \eta\dot{\theta}) + \mathbf{T} \cdot \mathbf{D} + \boldsymbol{\Upsilon} \cdot (\mathbf{w} - \boldsymbol{\omega}) + \mathbf{S} \cdot \nabla\boldsymbol{\omega} - \frac{1}{\theta}\mathbf{q} \cdot \nabla\theta + \theta\nabla \cdot \mathbf{k} \geq 0. \tag{10.12}$$

The analysis of the restrictions placed by the inequality (10.12) is developed by regarding $\mathbf{b}, r$, and also $\mathbf{l}$ as quantities whose values make the balance of linear momentum, angular momentum, and energy hold for any $\dot{\mathbf{v}}, \dot{\boldsymbol{\omega}}, \dot{\varepsilon}$. Analogous claims hold for the space and time derivatives. Hence $\mathbf{b}, r$, and $\mathbf{l}$ are supposed to take the values that must be applied to support the pertinent process.

10.2.2 Balance Laws in the Material Description

The pertinent fields are described by functions of $\mathbf{X}, t$, with $\mathbf{X} \in \mathrm{P} \subseteq \mathrm{R}$. For any field $\mathbf{u}_R$ in R the divergence theorem reads

$$\int_{\partial \mathrm{P}} \mathbf{u}_R \cdot \mathbf{n}_R \, da_R = \int_{\mathrm{P}} \nabla_R \cdot \mathbf{u}_R \, dv_R.$$

Balance of Mass

The mass density ρ is required to obey the continuity equation, $\dot{\rho} + \rho \nabla \cdot \mathbf{v} = 0$, while the reference mass density ρ_R is constant and $\rho = \rho_R / J$.

Balance of Linear Momentum

The balance equation

$$\frac{d}{dt} \int_{\mathcal{P}_t} \rho \mathbf{v} \, dv = \int_{\mathcal{P}_t} \rho \mathbf{b} \, dv + \int_{\partial \mathcal{P}_t} \mathbf{T} \mathbf{n} \, da$$

in the reference configuration reads

$$\frac{d}{dt} \int_{\mathrm{P}} \rho_R \mathbf{v} \, dv_R = \int_{\mathrm{P}} \rho_R \mathbf{b} \, dv_R + \int_{\partial \mathrm{P}} \mathbf{T}_R \mathbf{n}_R \, da_R,$$

$\mathbf{T}_R$ being the first Piola stress,$\mathbf{T}_R = J \mathbf{T} \mathbf{F}^{-T}$. Hence, the arbitrariness of P implies that

$$\rho_R \dot{\mathbf{v}} = \rho_R \mathbf{b} + \nabla_R \cdot \mathbf{T}_R.$$

Balance of Angular Momentum

From

$$\frac{d}{dt} \int_{\mathcal{P}_t} \rho(\mathbf{r} \times \mathbf{v} + \boldsymbol{\sigma}) dv = \int_{\mathcal{P}_t} \rho(\mathbf{r} \times \mathbf{b} + \mathbf{l}) dv + \int_{\partial \mathcal{P}_t} (\mathbf{r} \times \mathbf{t} + \mathbf{s}) da,$$

since $\mathbf{t} = \mathbf{T}\mathbf{n}$ and $\mathbf{s} = \mathbf{S}\,\mathbf{n}$ then

$$\frac{d}{dt} \int_{\mathcal{P}_t} \rho(\mathbf{r} \times \mathbf{v} + \boldsymbol{\sigma}) dv = \int_{\mathcal{P}_t} \rho(\mathbf{r} \times \mathbf{b} + \mathbf{l}) dv + \int_{\partial \mathcal{P}_t} (\mathbf{r} \times \mathbf{T}\mathbf{n} + \mathbf{S}\mathbf{n}) da.$$

Now,

$$\int_{\partial \mathcal{P}_t} \mathbf{S}\mathbf{n} da = \int_{\partial \mathrm{P}} \mathbf{S}_R \mathbf{n}_R \, da_R, \qquad \mathbf{S}_R := J \mathbf{S} \mathbf{F}^{-T},$$

$\mathbf{S}_R$ being the *material couple stress tensor.* Hence, in the reference configuration,

$$\int_P [\rho_R(\mathbf{r} \times \dot{\mathbf{v}} + \dot{\boldsymbol{\sigma}} - \mathbf{r} \times \mathbf{b} - \mathbf{l})] dv_R = \int_{\partial P} (\mathbf{r} \times \mathbf{T}_R \mathbf{n}_R + \mathbf{S}_R \mathbf{n}_R) da_R.$$

Since

$$[\nabla_R \cdot (\mathbf{r} \times \mathbf{T}_R)]_i = \epsilon_{ihk} F_{hQ} T^R_{kQ} + (\mathbf{r} \times \nabla_R \cdot \mathbf{T}_R)_i$$

and

$$\epsilon_{ihk} F_{hQ} T^R_{kQ} = \epsilon_{ihk} J F_{hQ} T_{kl} F^{-T}_{lQ} = J \epsilon_{ihk} T_{kh} = J \Upsilon_i$$

then, in view of the equation of motion, $\rho_R \dot{\mathbf{v}} = \rho_R \mathbf{b} + \nabla_R \cdot \mathbf{T}_R$, it follows

$$\int_P [\rho_R(\dot{\boldsymbol{\sigma}} - \mathbf{l}) - J \boldsymbol{\Upsilon} - \nabla_R \cdot \mathbf{S}_R] dv_R = \mathbf{0}.$$

The arbitrariness of P implies the local form of the balance of orientational momentum,

$$\rho_R \dot{\boldsymbol{\sigma}} = \rho_R \mathbf{l} + J \boldsymbol{\Upsilon} + \nabla_R \cdot \mathbf{S}_R.$$

Balance of Energy

The global form of the balance of energy in the reference region P becomes

$$\frac{d}{dt} \int_P \rho_R [\varepsilon + \tfrac{1}{2}(\mathbf{v}^2 + \boldsymbol{\omega} \cdot \boldsymbol{\sigma})] dv_R = \int_P \rho_R (\mathbf{b} \cdot \mathbf{v} + \mathbf{l} \cdot \boldsymbol{\omega} + r) dv_R$$
$$+ \int_{\partial P} (\mathbf{v} \cdot \mathbf{T}_R \mathbf{n}_R + \boldsymbol{\omega} \cdot \mathbf{S}_R \mathbf{n}_R - \nabla_R \cdot \mathbf{q}_R) da_R.$$

The divergence theorem, the equation of motion and the balance of angular momentum allow us to find that the balance of energy results in

$$\int_P [\rho_R \dot{\varepsilon} - \mathbf{T}_R \cdot \dot{\mathbf{F}} - \mathbf{S}_R \cdot \nabla_R \boldsymbol{\omega} + J \boldsymbol{\Upsilon} \cdot \boldsymbol{\omega} - \nabla_R \cdot \mathbf{q}_R + \rho_R r] dv_R = 0.$$

Hence we get the local form

$$\rho_R \dot{\varepsilon} = \mathbf{T}_R \cdot \dot{\mathbf{F}} + \mathbf{S}_R \cdot \nabla_R \boldsymbol{\omega} - J \boldsymbol{\Upsilon} \cdot \boldsymbol{\omega} - \nabla_R \cdot \mathbf{q}_R + \rho_R r.$$

Observe that

$$\mathbf{T}_R \cdot \dot{\mathbf{F}} = \mathbf{T}_R \cdot (\mathbf{L}\mathbf{F}) = J \mathbf{T} \cdot \mathbf{D} + J \mathbf{T} \cdot \mathbf{W}, \qquad \mathbf{T} \cdot \mathbf{W} = \boldsymbol{\Upsilon} \cdot \mathbf{w}.$$

Consequently the balance of energy can be given the form

$$\rho_R \dot{\varepsilon} = J \mathbf{T} \cdot \mathbf{D} + \mathbf{S}_R \cdot \nabla_R \boldsymbol{\omega} + J \boldsymbol{\Upsilon} \cdot (\mathbf{w} - \boldsymbol{\omega}) - \nabla_R \cdot \mathbf{q}_R + \rho_R r.$$

Second Law of Thermodynamics

The entropy inequality

$$\frac{d}{dt}\int_{\mathcal{P}_t} \rho\eta dv \geq \int_{\mathcal{P}_t} \rho s dv - \int_{\partial\mathcal{P}_t} \mathbf{j}\cdot\mathbf{n} da$$

in the material description becomes

$$\frac{d}{dt}\int_{\mathrm{P}} \rho_R\eta dv_R \geq \int_{\mathrm{P}} \rho_R s dv_R - \int_{\partial\mathrm{P}} \mathbf{j}_R\cdot\mathbf{n}_R da_R$$

where

$$\mathbf{j}_R = J\,\mathbf{j}\,\mathbf{F}^{-T}$$

is the material entropy flux. We let $s = r/\theta$ and

$$\mathbf{j}_R = \frac{1}{\theta}\mathbf{q}_R + \mathbf{k}_R, \qquad \mathbf{k}_R = J\,\mathbf{k}\,\mathbf{F}^{-T},$$

$\mathbf{k}_R$ being the material extra-entropy flux. In the local form we have the entropy inequality

$$\rho_R\dot{\eta} \geq \frac{\rho_R r}{\theta} - \nabla_R\cdot\big(\frac{1}{\theta}\mathbf{q}_R + \mathbf{k}_R\big).$$

Hence

$$\rho_R\dot{\eta} + \frac{1}{\theta}(\nabla_R\cdot\mathbf{q}_R - \rho_R r) - \frac{1}{\theta^2}\mathbf{q}_R\cdot\nabla_R\theta + \nabla_R\cdot\mathbf{k}_R \geq 0.$$

Substitution of $\nabla_R\cdot\mathbf{q}_R - \rho_R r$ from the balance of energy results in

$$-\rho_R(\dot{\psi} + \eta\dot{\theta}) + \mathbf{T}_R\cdot\dot{\mathbf{F}} - J\boldsymbol{\Upsilon}\cdot\boldsymbol{\omega} + \mathbf{S}_R\cdot\nabla_R\boldsymbol{\omega} - \frac{1}{\theta}\mathbf{q}_R\cdot\nabla_R\theta + \theta\nabla_R\cdot\mathbf{k}_R \geq 0 \tag{10.13}$$

or

$$-\rho_R(\dot{\psi} + \eta\dot{\theta}) + J\mathbf{T}\cdot\mathbf{D} + \boldsymbol{\Upsilon}\cdot(\mathbf{w} - \boldsymbol{\omega}) + \mathbf{S}_R\cdot\nabla_R\boldsymbol{\omega} - \frac{1}{\theta}\mathbf{q}_R\cdot\nabla_R\theta + \theta\nabla_R\cdot\mathbf{k}_R \geq 0.$$

As with the spatial description, the analysis of the restrictions placed by the inequality (10.13) is developed by regarding $\mathbf{b}$, r, and also $\mathbf{l}$ as quantities whose values make the balance of linear momentum, angular momentum, and energy hold for any $\dot{\mathbf{v}}$, $\dot{\omega}$, $\dot{\varepsilon}$. Analogous claims hold for the space and time derivatives. Hence $\mathbf{b}$, r, and $\mathbf{l}$ are supposed to take the values that must be applied to support the pertinent process.

10.2.3 Thermoviscous Micropolar Media

Fluids

As a preliminary simple model of micropolar medium we consider a micropolar fluid, that is a fluid where the particles, or some particles within it, have the structure of rigid body. We then describe the constitutive properties of the fluid by letting the response functions ψ, η, $\mathbf{T}$, $\mathbf{q}$, and $\mathbf{k}$ depend on the set of variables

$$\rho, \theta, \mathbf{D}, \nabla\theta, \mathcal{R}, \mathbf{w} - \boldsymbol{\omega}, \nabla\boldsymbol{\omega}.$$

The possible dependence on $\mathcal{R}$, $\mathbf{w} - \boldsymbol{\omega}$, and $\nabla\boldsymbol{\omega}$ is peculiar to the microscopic structure; $\mathcal{R}$ may account for the orientation, $\mathbf{w} - \boldsymbol{\omega}$ for the rotational interaction between the particles and the macroscopic motion, $\nabla\boldsymbol{\omega}$ for the action of the couple stress tensor. Since ψ is invariant under changes of frame then ψ can depend on $\mathbf{D}$, $\nabla\theta$, $\mathcal{R}$, $\mathbf{w} - \boldsymbol{\omega}$, and $\nabla\boldsymbol{\omega}$ through their invariants.

Let ψ be differentiable. Moreover let $\mathbf{T} = -p\mathbf{1} + \boldsymbol{\mathcal{T}}$, where

$$\boldsymbol{\mathcal{T}} \to \mathbf{0} \quad \text{if} \quad \mathbf{L}, \boldsymbol{\omega}, \nabla\theta \to \mathbf{0}.$$

Hence the entropy inequality (10.12) can be written as

$$-\rho\{(\partial_\theta\psi + \eta)\dot{\theta} + \partial_{\nabla\theta}\psi \cdot \dot{\overline{\nabla\theta}} + \partial_{\mathbf{D}}\psi \cdot \dot{\mathbf{D}} + \partial_{\mathcal{R}_{ij}}\psi\dot{\mathcal{R}}_{ij} + \partial_{(\mathbf{w}-\boldsymbol{\omega})}\psi \cdot (\dot{\mathbf{w}} - \dot{\boldsymbol{\omega}}) + \partial_{\nabla\boldsymbol{\omega}}\psi \cdot \dot{\overline{\nabla\boldsymbol{\omega}}}\}$$
$$+(\rho^2\partial_\rho\psi - p)\nabla \cdot \mathbf{v} + \boldsymbol{\mathcal{T}} \cdot \mathbf{D} + \boldsymbol{\Upsilon} \cdot (\mathbf{w} - \boldsymbol{\omega}) + \mathbf{S} \cdot \nabla\boldsymbol{\omega} - \frac{1}{\theta}\mathbf{q} \cdot \nabla\theta + \theta\nabla \cdot \mathbf{k} \geq 0.$$

The arbitrariness of $\dot{\mathbf{D}}$, $\dot{\overline{\nabla\theta}}$, $\dot{\mathbf{w}} - \dot{\boldsymbol{\omega}}$, $\dot{\overline{\nabla\boldsymbol{\omega}}}$, and $\dot{\theta}$ implies

$$\partial_{\mathbf{D}}\psi = \mathbf{0}, \qquad \partial_{\nabla\theta}\psi = \mathbf{0}, \qquad \partial_{(\mathbf{w}-\boldsymbol{\omega})}\psi = \mathbf{0}, \qquad \partial_{\nabla\boldsymbol{\omega}}\psi = \mathbf{0}, \qquad \eta = -\partial_\theta\psi.$$

The model is consistent with the condition $\mathbf{k} = \mathbf{0}$. Since $\dot{\mathcal{R}} = \mathcal{A}\mathcal{R}$ and $\mathcal{A}_{jh} = -\epsilon_{jhl}\omega_l$ then it follows that

$$p = \rho^2\partial_\rho\psi.$$

The inequality then reduces to

$$-\rho\partial_{\mathcal{R}_{ij}}\psi\,(\mathcal{A}\mathcal{R})_{ij} + \boldsymbol{\mathcal{T}} \cdot \mathbf{D} + \boldsymbol{\Upsilon} \cdot (\mathbf{w} - \boldsymbol{\omega}) + \mathbf{S} \cdot \nabla\boldsymbol{\omega} - \frac{1}{\theta}\mathbf{q} \cdot \nabla\theta \geq 0.$$

Observe

$$-\rho\partial_{\mathcal{R}_{ij}}\psi\,(\mathcal{A}\mathcal{R})_{ij} = -\rho\partial_{\mathcal{R}_{ij}}\psi\,\mathcal{A}_{ip}\mathcal{R}_{pj} = \rho\partial_{\mathcal{R}_{ij}}\psi\,\mathcal{R}_{pj}\epsilon_{ipl}\omega_l = \boldsymbol{\gamma} \cdot \boldsymbol{\omega},$$

where

$$\gamma_l = \epsilon_{ipl}\,\rho\,\partial_{\mathcal{R}_{ij}}\psi\,\mathcal{R}^T_{jp}.$$

Hence the inequality can be written in the form

$$\boldsymbol{\gamma}\cdot\boldsymbol{\omega} + \boldsymbol{\Upsilon}\cdot(\mathbf{w}-\boldsymbol{\omega}) + \boldsymbol{\mathcal{T}}\cdot\mathbf{D} + \mathbf{S}\cdot\nabla\boldsymbol{\omega} - \frac{1}{\theta}\mathbf{q}\cdot\nabla\theta \geq 0. \tag{10.14}$$

Inequality (10.14) is exploited under the assumption that $\boldsymbol{\Upsilon}$ is linear in $\mathbf{w}-\boldsymbol{\omega}, \nabla\theta, \nabla\boldsymbol{\omega}$. Let

$$\boldsymbol{\Upsilon} = \Upsilon_1(\rho,\theta)(\mathbf{w}-\boldsymbol{\omega}) + \Upsilon_2(\rho,\theta)\nabla\theta + \Upsilon_3(\rho,\theta)\nabla\times\boldsymbol{\omega}.$$

As $\mathbf{D}$, $\nabla\boldsymbol{\omega}$, $\nabla\theta$ vanish, inequality (10.14) reduces to

$$\boldsymbol{\gamma}\cdot\boldsymbol{\omega} + \Upsilon_1(\mathbf{w}-\boldsymbol{\omega})^2 \geq 0.$$

This inequality holds only if

$$\boldsymbol{\gamma} = \mathbf{0}, \qquad \Upsilon_1 \geq 0.$$

For definiteness we now take $\boldsymbol{\mathcal{T}}$, $\mathbf{S}$, and $\mathbf{q}$ in linear forms. For any vector $\mathbf{u}$ we let $\Xi(\mathbf{u})$ be the skew tensor defined by $\Xi_{pq} = -\epsilon_{pq_i}u_i$. Hence we consider the constitutive equations

$$\begin{aligned}\boldsymbol{\mathcal{T}} &= 2\mu\mathbf{D} + \lambda(\operatorname{tr}\mathbf{D})\mathbf{1} + \varpi\nabla\boldsymbol{\omega} + \tfrac{1}{2}\Upsilon_1\,\Xi(\mathbf{w}-\boldsymbol{\omega}),\\ \mathbf{S} &= \alpha\,\Xi(\nabla\theta) + \beta\,\Xi(\mathbf{w}-\boldsymbol{\omega}) + \delta(\nabla\cdot\boldsymbol{\omega})\mathbf{1} + \nu\nabla\boldsymbol{\omega} + \sigma\mathbf{D},\\ \mathbf{q} &= -\kappa\nabla\theta + \xi(\mathbf{w}-\boldsymbol{\omega}) + \zeta\nabla\times\boldsymbol{\omega}.\end{aligned}$$

Inequality (10.14) then reads

$$\begin{aligned}&\Upsilon_1(\mathbf{w}-\boldsymbol{\omega})^2 + 2\mu\mathbf{D}\cdot\mathbf{D} + \lambda(\operatorname{tr}\mathbf{D})^2 + (\varpi+\sigma)\mathbf{D}\cdot\nabla\boldsymbol{\omega} + (\alpha-\zeta/\theta)\nabla\times\boldsymbol{\omega}\cdot\nabla\theta\\ &+\beta\nabla\times\boldsymbol{\omega}\cdot(\mathbf{w}-\boldsymbol{\omega}) + \delta(\nabla\cdot\boldsymbol{\omega})^2 + \nu|\nabla\boldsymbol{\omega}|^2 + \frac{1}{\theta}\kappa|\nabla\theta|^2 - \frac{1}{\theta}\xi(\mathbf{w}-\boldsymbol{\omega})\cdot\nabla\theta \geq 0.\end{aligned}$$

It follows that

$$\xi,\beta = 0, \quad \varpi = -\sigma, \quad \alpha = \zeta/\theta, \quad \mu \geq 0, 2\mu + 3\lambda \geq 0, \quad \nu \geq 0, \nu + 3\delta \geq 0, \quad \kappa \geq 0.$$

With these restrictions in mind, we can write the constitutive equations in the form

$$\begin{aligned}\boldsymbol{\mathcal{T}} &= 2\mu\mathbf{D} + \lambda(\operatorname{tr}\mathbf{D})\mathbf{1} - \sigma\nabla\boldsymbol{\omega} + \tfrac{1}{2}\Upsilon_1\,\Xi(\mathbf{w}-\boldsymbol{\omega}),\\ \mathbf{S} &= \alpha\,\Xi(\nabla\theta) + \delta(\nabla\cdot\boldsymbol{\omega})\mathbf{1} + \nu\nabla\boldsymbol{\omega} + \sigma\mathbf{D},\\ \mathbf{q} &= -\kappa\nabla\theta - \alpha\theta\nabla\times\boldsymbol{\omega}.\end{aligned}$$

The corresponding set of evolution equations follows from the balance equations. Observe

$$\partial_{x_j} \Xi_{ij}(\mathbf{w}-\omega) = -\epsilon_{ijk}\partial_{x_j}(w_k-\omega_k) = -[\nabla\times(\mathbf{w}-\omega)]_i, \quad \Xi(\mathbf{w}-\omega)\cdot\mathbf{D}=0,$$

$$\partial_{x_j}\Xi_{ij}(\nabla\theta) = -\epsilon_{ijk}\partial_{x_j}\partial_{x_k}\theta = 0,$$

$$\alpha\,\Xi(\nabla\theta)\cdot\nabla\omega - \nabla\cdot\zeta\nabla\times\omega = \alpha\nabla\theta\cdot\nabla\times\omega - \alpha\nabla\theta\cdot\nabla\times\omega = 0.$$

Hence we have

$$\dot\rho + \rho\nabla\cdot\mathbf{v} = 0,$$

$$\rho\dot{\mathbf{v}} = -\nabla p + \mu\Delta\mathbf{v} + (\mu+\lambda)\nabla\nabla\cdot\mathbf{v} - \sigma\Delta\omega - \tfrac{1}{2}\Upsilon_1\nabla\times(\mathbf{w}-\omega) + \rho\mathbf{b},$$

$$\rho\mathbf{I}\dot\omega + \rho\omega\times\mathbf{I}\omega = \tau_1(\mathbf{w}-\omega) + \delta\nabla(\nabla\cdot\omega) + \nu\Delta\omega + \tfrac{1}{2}\sigma(\Delta\mathbf{v}+\nabla(\nabla\cdot\mathbf{v})) + \rho\mathbf{l},$$

$$\rho\dot\varepsilon = -p\nabla\cdot\mathbf{v} + 2\mu\mathbf{D}\cdot\mathbf{D} + \lambda(\mathrm{tr}\,\mathbf{D})^2 + \delta(\nabla\cdot\omega)^2 + \nu|\nabla\omega|^2 + \kappa\Delta\theta + \rho r.$$

Solids

A micropolar solid is modelled so that elastic and thermo-viscous properties are allowed. We follow the material description and let the response functions of ψ, η, $\mathbf{T}_R$, $\mathbf{S}_R$, $\mathbf{q}_R$, depend on the set of variables $\mathbf{F}$, θ, $\mathbf{D}$, $\nabla_R\theta$, $\mathcal{R}$, $\nabla_R\mathcal{R}$, $\nabla_R\omega$. The free energy can depend only on invariants. Accordingly we consider the dependence on $\mathbf{F}$, $\mathcal{R}$, $\nabla_R\mathcal{R}$ via the Cosserat deformation tensor $\mathfrak{C}$ and the wryness tensor $\mathbf{\Gamma}$. The dependence on the Cosserat deformation tensor $\mathfrak{C}$ is devised to model the interaction between deformation and micromotion. The dependence on the wryness tensor $\mathbf{\Gamma}$ is considered so as to describe possible effects of the inhomogeneous rotation field. As we see in a moment, the dependence of ψ on $\mathbf{D}$, $\nabla_R\theta$, and $\nabla_R\omega$ is ruled out by the entropy inequality. Hence we let

$$\psi = \psi(\mathfrak{C}, \theta, \mathbf{D}, \nabla_R\theta, \mathbf{\Gamma}, \nabla_R\omega)$$

and suppose ψ is differentiable.

Upon computation of $\dot\psi$ and substitution in the entropy inequality (10.13) it follows

$$-\rho_R(\partial_\theta\psi+\eta)\dot\theta - \rho_R\partial_{\mathfrak{C}}\psi\cdot\dot{\mathfrak{C}} - \rho_R\partial_{\mathbf{D}}\psi\cdot\dot{\mathbf{D}} - \rho_R\partial_{\nabla_R\theta}\psi\cdot\nabla_R\dot\theta - \rho_R\partial_{\mathbf{\Gamma}}\psi\cdot\dot{\mathbf{\Gamma}} - \rho_R\partial_{\nabla_R\omega}\psi\cdot\nabla_R\dot\omega$$
$$+\mathbf{T}_R\cdot\dot{\mathbf{F}} - J\mathbf{\Upsilon}\cdot\omega + \mathbf{S}_R\cdot\nabla_R\omega - \frac{1}{\theta}\mathbf{q}_R\cdot\nabla_R\theta + \theta\nabla_R\cdot\mathbf{k}_R \ge 0.$$

In view of (10.4) and (10.3), the arbitrariness of $\dot{\mathbf{D}}$, $\nabla_R\dot\theta$, $\nabla_R\dot\omega$, $\dot\theta$ implies that

$$\partial_{\mathbf{D}}\psi = \mathbf{0}, \qquad \partial_{\nabla_R\theta}\psi = \mathbf{0}, \qquad \partial_{\nabla_R\omega}\psi = \mathbf{0}, \qquad \eta = -\partial_\theta\psi.$$

The inequality reduces to

$$-\rho_R \partial_{\mathfrak{C}} \psi \cdot \dot{\mathfrak{C}} - \rho_R \partial_{\boldsymbol{\Gamma}} \psi \cdot \dot{\boldsymbol{\Gamma}} + \mathbf{T}_R \cdot \dot{\mathbf{F}} - J \boldsymbol{\Upsilon} \cdot \boldsymbol{\omega} + \mathbf{S}_R \cdot \nabla_R \boldsymbol{\omega} - \frac{1}{\theta} \mathbf{q}_R \cdot \nabla_R \theta + \theta \nabla_R \cdot \mathbf{k}_R \geq 0.$$

We now investigate the mathematical structure of the powers in this inequality. By (10.4) we find

$$-\rho_R \partial_{\mathfrak{C}} \psi \cdot \dot{\mathfrak{C}} = -\rho_R \mathcal{R}_{jK} \partial_{\mathfrak{C}_{KL}} \psi F_{hL} D_{jh} - \rho_R \mathcal{R}_{jK} \partial_{\mathfrak{C}_{KL}} \psi F_{hL} \epsilon_{jhl} (w_l - \omega_l).$$

Let

$$\hat{\mathbf{T}} := \rho \mathbf{F} \partial_{\mathfrak{C}} \psi \mathcal{R}, \qquad \hat{\Upsilon}_l := \epsilon_{hjl} \hat{T}_{jh}.$$

Hence

$$-\rho_R \partial_{\mathfrak{C}} \psi \cdot \dot{\mathfrak{C}} = -J \hat{\mathbf{T}} \cdot \mathbf{D} - J \hat{\boldsymbol{\Upsilon}} \cdot (\mathbf{w} - \boldsymbol{\omega}).$$

By (10.3) we obtain

$$-\rho_R \partial_{\boldsymbol{\Gamma}} \psi \cdot \dot{\boldsymbol{\Gamma}} = -\rho_R \partial_{\Gamma_{KL}} \psi \mathcal{R}_{iK} \partial_{X_L} \omega_i = -\hat{\mathbf{S}}_R \cdot \nabla_R \boldsymbol{\omega}$$

where

$$\hat{\mathbf{S}}_R := \rho_R \mathcal{R} \partial_{\boldsymbol{\Gamma}} \psi.$$

Moreover

$$\mathbf{T}_R \cdot \dot{\mathbf{F}} = (\mathbf{T}_R \mathbf{F}^T) \cdot \mathbf{L} = J \mathbf{T} \cdot \mathbf{D} + J \boldsymbol{\Upsilon} \cdot \mathbf{w}.$$

We can then write the entropy inequality in the form

$$J(\mathbf{T} - \hat{\mathbf{T}}) \cdot \mathbf{D} + J(\boldsymbol{\Upsilon} - \hat{\boldsymbol{\Upsilon}}) \cdot (\mathbf{w} - \boldsymbol{\omega}) + (\mathbf{S}_R - \hat{\mathbf{S}}_R) \cdot \nabla_R \boldsymbol{\omega} - \frac{1}{\theta} \mathbf{q}_R \cdot \nabla_R \theta + \theta \nabla_R \cdot \mathbf{k}_R \geq 0.$$

Again, no significant generality is lost by letting $\mathbf{k}_R = \mathbf{0}$ in that we disregard the case $\mathbf{k}_R = \mathbf{k}_R(\theta)$. Apart from cross coupling terms we have

$$\boldsymbol{\Upsilon} - \hat{\boldsymbol{\Upsilon}} = \alpha(\mathbf{w} - \boldsymbol{\omega}), \qquad \mathbf{T} - \hat{\mathbf{T}} = 2\mu \mathbf{D} + \lambda (\mathrm{tr}\, \mathbf{D}) \mathbf{1} + \tfrac{1}{2} \alpha \boldsymbol{\Xi} (\mathbf{w} - \boldsymbol{\omega}),$$

$$\mathbf{S}_R - \hat{\mathbf{S}}_R = \xi \nabla_R \boldsymbol{\omega} + \zeta (\nabla_R \cdot \boldsymbol{\omega}) \mathbf{1},$$

$$\mathbf{q}_R = -\kappa \nabla_R \theta,$$

where

$$\alpha \geq 0, \quad \mu \geq 0, \quad 2\mu + 3\lambda \geq 0, \quad \xi \geq 0, \quad \xi + 3\zeta \geq 0, \quad \kappa \geq 0.$$

The stress $\mathbf{T}$ is the sum of a term $\hat{\mathbf{T}}$, induced by the dependence of ψ on $\mathfrak{C}$, and a dissipative term; $2\mu\mathbf{D} + \lambda(\mathrm{tr}\,\mathbf{D})\mathbf{1}$ is the classical viscous stress while $\frac{1}{2}\alpha\,\Xi\,(\mathbf{w} - \boldsymbol{\omega})$ is the interaction between the rotation of the single particle, $\boldsymbol{\omega}$, and the macroscopic motion, $\mathbf{w} = \frac{1}{2}\nabla \times \mathbf{v}$. Likewise the couple stress $\mathbf{S}_R$ is the sum of a term $\hat{\mathbf{S}}_R$, induced by $\boldsymbol{\Gamma}$, and a dissipative part related to $\nabla_R\boldsymbol{\omega}$.

If ψ is dependent on $\mathbf{F}$ also via $\mathbf{C} = \mathbf{F}^T\mathbf{F}$ then $\mathbf{T}$ has an additional term $2\mathbf{F}\,\partial_{\mathbf{C}}\psi\,\mathbf{F}^T$ which is a symmetric tensor.

10.3 Liquid Crystals

Liquid crystals are matter in a state that has properties between those of standard liquids and those of solids. The molecules (mesogens) are given a preferred direction, whose unit vector is called *director*. While a liquid crystal may flow like a liquid, in certain conditions the directors point along a common axis as in a solid crystal. Different types of liquid crystals can be distinguished by their different optical properties.

Liquid crystals (LCs) are divided in *thermotropic, lyotropic* and *metallotropic* phases. Thermotropic and lyotropic LCs consist of organic molecules. Thermotropic LCs exhibit a phase transition as temperature is changed whereas lyotropic LCs undergo phase transitions depending on the temperature but also on the concentration of the LC in a solvent. Metallotropic LCs consist of both organic and inorganic molecules.

The various liquid-crystal phases (mesophases) can be characterized by the type of ordering. One can distinguish positional order and orientational order. In the *nematic* phase the rod-shaped molecules have no positional order but they self-align to have long-range directional order with their long axes roughly parallel. Most nematics are uniaxial in that they have one axis that is longer and preferred, with the other two being equivalent. Some liquid crystals are biaxial nematics in that in addition to orienting their long axis they also orient along a secondary axis.

The *smectic* phase, which occurs at lower temperatures than the nematic, form well-defined layers that can slide over one another. The smectics are thus positionally ordered along one direction. In the smectic A phase the molecules are oriented along the layer normal, whereas in the smectic C phase they are tilted away from the layer normal.

The *chiral nematic*, or *cholesteric*, phase exhibits a twisting of the molecules perpendicular to the director. The chirality induces a finite azimuthal twist from one layer to the next thus producing a spiral twisting of the molecular axis along the layer normal. The chiral pitch is the distance over which the molecules undergo a full twist. The pitch changes when the temperature is altered or when other molecules are added to the liquid crystal host.

10.3.1 Modelling of Thermotropic Liquid Crystals

Thermotropic liquid crystals are modelled as micropolar media. The molecules (mesogens) are given a position $\mathbf{x}$ and an orientation; the orientation is characterized by an orthonormal principal triad $(\mathbf{k}_1, \mathbf{k}_2, \mathbf{k}_3)$ rigidly attached to the molecule. We let $\mathbf{k}_1$ be the director. We take the principal moments of inertia as

$$j_1 \leq j_2 \leq j_3 \qquad \text{or} \qquad j_1 \geq j_2 \geq j_3,$$

according as the liquid crystal consists of rod-like or disk-like molecules. If the spatial mean $\langle \mathbf{k}_1 \rangle$ of the director is nonzero we can define the unit vector $\mathbf{n}$ as

$$\mathbf{n} = \langle \mathbf{k}_1 \rangle / |\langle \mathbf{k}_1 \rangle|.$$

The scalar parameter $s = |\langle \mathbf{k}_1 \rangle|$ is called the *degree of orientation* and we have

$$\langle \mathbf{k}_1 \rangle = s\, \mathbf{n}.$$

Nematics can be easily aligned by an external magnetic or electric field. In smectics the molecules are arranged in (parallel) layers; let $\mathbf{N}$ be the normal to the layers. In smectic A LCs the molecular orientation $\mathbf{n}$ is perpendicular to the layers and hence $|\mathbf{n} \cdot \mathbf{N}| = 1$. In smectic C LCs the director $\mathbf{n}$ is tilted so that $|\mathbf{n} \cdot \mathbf{N}| < 1$.

In cholesterics the orientation $\mathbf{n}$ is constant within layers but $\mathbf{n} \cdot \mathbf{N}$ takes different values also relative to adjacent layers.

In isotropic LCs there are patches with transient positional as well as orientational order (as with smectics) and those with only transient orientational order (as with nematics). In these clusters $\mathbf{n}$, and possibly $\mathbf{N}$, are defined and vary as a function of space and time. The limit case of isotropic LCs is that where $s = 0$ while $\mathbf{n}$ and $\mathbf{N}$ are undefined.

The degree of orientation s can then be considered to describe the transition between isotropic phases and nematics.

Balance Equations

Owing to the micropolar character of LCs and the occurrence of electromagnetic fields we establish the balance equations on the basis of Sects. 2.16 and 10.2.

For generality, liquid crystals are allowed to be compressible. Hence the balance of mass is considered in the standard form

$$\dot{\rho} + \rho \nabla \cdot \mathbf{v} = 0.$$

The balance of linear momentum is taken in the form

$$\rho\dot{\mathbf{v}} = \nabla \cdot \mathbf{T} + \rho\mathbf{b}_{\text{mech}} + \rho\mathbf{b}_{\text{em}},$$

where, by (2.72),

$$\rho\mathbf{b}_{\text{em}} = (\mathbf{P} \cdot \nabla)\mathbf{E} + \mathbf{v} \times (\mathbf{P} \cdot \nabla)\mathbf{B} + \rho\dot{\boldsymbol{\pi}} \times \mathbf{B} + \mu_0(\mathbf{M} \cdot \nabla)\mathbf{H} \\ -\mathbf{v} \times (\mathbf{M} \cdot \nabla)\mathbf{D} - \mu_0\rho\dot{\mathbf{m}} \times \mathbf{D} + q\mathcal{E} + \mathcal{J} \times \mathbf{B},$$

while $\mathbf{b}_{\text{mech}}$ is the body force of mechanical character.

The balance of angular momentum (moment of momentum) is based on the view that a mechanical couple stress $\mathbf{S}$ may occur. The body couple may consist of a mechanical term $\mathbf{l}_{\text{mech}}$ and an electromagnetic one $\mathbf{l}_{\text{em}}$; by (2.73) we have

$$\rho\mathbf{l}_{\text{em}} = \mathbf{P} \times \mathcal{E} + \mu_0\mathbf{M} \times \mathbf{H}.$$

Hence the orientational momentum $\boldsymbol{\sigma}$ is subject to

$$\rho\dot{\boldsymbol{\sigma}} = \boldsymbol{\Upsilon} + \rho\mathbf{l}_{\text{mech}} + \nabla \cdot \mathbf{S} + \mathbf{P} \times \mathcal{E} + \mu_0\mathbf{M} \times \mathbf{H},$$

where $\boldsymbol{\Upsilon}$ is twice the axial vector of $\mathbf{T}$, $\Upsilon_i = \epsilon_{ijk}T_{kj}$.

Likewise we observe that the power $\rho\mathbf{l} \cdot \boldsymbol{\omega}$ splits into

$$\rho\mathbf{l} \cdot \boldsymbol{\omega} = (\rho\mathbf{l}_{\text{mech}} + \mathbf{P} \times \mathcal{E} + \mu_0\mathbf{M} \times \mathbf{H}) \cdot \boldsymbol{\omega}$$

while, by (2.75), an additional electromagnetic power

$$\mathcal{E} \cdot \mathcal{J} + \rho\mathcal{E} \cdot \dot{\boldsymbol{\pi}} + \rho\mu_0\mathcal{H} \cdot \dot{\mathbf{m}}$$

occurs. It follows that[3]

$$\rho\dot{\varepsilon} = \mathcal{E} \cdot \mathcal{J} + \rho\mathcal{E} \cdot \dot{\boldsymbol{\pi}} + \rho\mu_0\mathcal{H} \cdot \dot{\mathbf{m}} + \mathbf{T} \cdot \mathbf{D} + \boldsymbol{\Upsilon} \cdot (\mathbf{w} - \boldsymbol{\omega}) + \mathbf{S} \cdot \nabla\boldsymbol{\omega} - \nabla \cdot \mathbf{q} + \rho r. \tag{10.15}$$

Moreover the electromagnetic fields are required to satisfy Maxwell's equations while

$$\mathcal{E} = \mathbf{E} + \mathbf{v} \times \mathbf{B}, \qquad \mathcal{H} = \mathbf{H} - \mathbf{v} \times \mathbf{D}, \qquad \mathcal{J} = \mathbf{J} - \mathcal{N}q\mathbf{v}, \qquad \mathcal{N} = \rho/m.$$

LCs are then characterized by adding the constitutive equations. To this end we consider the electromagnetic fields, at the rest frame at the pertinent point $\mathbf{x}$ and time t, so that

$$\mathbf{E} = \mathcal{E}, \qquad \mathbf{H} = \mathcal{H}, \qquad \mathbf{J} = \mathcal{J}.$$

Hence, starting with the standard entropy inequality

[3] In suffix notation $\mathbf{T} \cdot \mathbf{L} + \mathbf{S} \cdot \nabla\boldsymbol{\omega} = T_{ij}\partial_{x_j}v_i + S_{ij}\partial_{x_j}\omega_i$. Moreover $\mathbf{D}$ is the stretching tensor, not the electric displacement.

$$\rho\dot{\eta} + \nabla \cdot \frac{\mathbf{q}}{\theta} - \frac{\rho r}{\theta} + \nabla \cdot \mathbf{k} \geq 0$$

and replacing $\nabla \cdot \mathbf{q} - \rho r$ from the balance of energy (10.15) we find

$$\rho\theta\dot{\eta} - \rho\dot{\varepsilon} + \mathbf{E} \cdot \mathbf{J} + \rho\mathbf{E} \cdot \dot{\pi} + \rho\mu_0\mathbf{H} \cdot \dot{\mathbf{m}} + \mathbf{T} \cdot \mathbf{D} + \boldsymbol{\Upsilon} \cdot (\mathbf{w} - \omega) + \mathbf{S} \cdot \nabla\omega - \frac{1}{\theta}\mathbf{q} \cdot \nabla\theta + \theta\nabla \cdot \mathbf{k} \geq 0.$$

For technical convenience we consider the function

$$\phi = \varepsilon - \theta\eta - \mathbf{E} \cdot \pi - \mu_0\mathbf{H} \cdot \mathbf{m}.$$

In terms of ϕ the second-law inequality can be written in the form

$$-\rho\dot{\phi} - \rho\eta\dot{\theta} - \mathbf{P} \cdot \dot{\mathbf{E}} - \mu_0\mathbf{M} \cdot \dot{\mathbf{H}} + \mathbf{E} \cdot \mathbf{J} + \mathbf{T} \cdot \mathbf{D} + \boldsymbol{\Upsilon} \cdot (\mathbf{w} - \omega) + \mathbf{S} \cdot \nabla\omega - \frac{1}{\theta}\mathbf{q} \cdot \nabla\theta + \theta\nabla \cdot \mathbf{k} \geq 0. \quad (10.16)$$

10.3.2 Director Field, Energy, and Objectivity

The energy ε of LCs might depend on the direction $\mathbf{n}$ of the director (of the mesogen) via the coupling with external fields. Also, the orientation $\mathbf{N}$ of possible layers might affect the energy via the coupling with the direction $\mathbf{n}$, through $\mathbf{n} \cdot \mathbf{N}$, and both $\nabla\mathbf{N}$ and $\nabla(\mathbf{n} \cdot \mathbf{N})$. The inner product $\mathbf{n} \cdot \mathbf{N}$ is an invariant, and hence objective, scalar. We examine the objectivity of the dependence on $\mathbf{n}$ and $\nabla\mathbf{n}$ (as well as on $\mathbf{N}$ and $\nabla\mathbf{N}$).

Concerning the dependence on $\mathbf{n}$, $\nabla\mathbf{n}$, as well as $\mathbf{N}$, $\nabla\mathbf{N}$, we have the following result.

A function $\varepsilon(\mathbf{n}, \nabla\mathbf{n})$ *is objective if and only if*

$$\epsilon_{jip}(\partial_{n_j}\varepsilon\, n_i + \partial_{n_{j,k}}\varepsilon\, n_{i,k} + \partial_{n_{k,j}}\varepsilon\, n_{k,i}) = 0, \qquad p = 1, 2, 3. \quad (10.17)$$

Proof Observe that a function $\varepsilon(\mathbf{n}, \nabla\mathbf{n})$ is objective if

$$\varepsilon(\mathbf{n}, \nabla\mathbf{n}) = \varepsilon(\mathbf{Q}\mathbf{n}, \mathbf{Q}\nabla\mathbf{Q}\mathbf{n}). \quad (10.18)$$

We define the vector function $\mathbf{n}(t)$ via the tensor function $\mathbf{Q}(t)$ associated with the change of frame,

$$\mathbf{n}(t) = \mathbf{Q}(t)\bar{\mathbf{n}}, \qquad (\nabla\mathbf{n})(t) = \mathbf{Q}(t)\nabla\mathbf{Q}(t)\bar{\mathbf{n}}.$$

Hence the objectivity requirement may also be stated in the form

$$\hat{\varepsilon}(t) := \varepsilon(\mathbf{n}(t), \nabla\mathbf{n}(t)) = \text{constant}.$$

Consequently $\varepsilon(\mathbf{n}, \nabla\mathbf{n})$ is objective if and only if $d\hat{\varepsilon}/dt = 0$ whence

$$\partial_{n_j}\varepsilon\, \dot{n}_j + \partial_{n_{j,k}}\varepsilon\, \overline{n_{j,k}}^{\cdot} = 0. \tag{10.19}$$

As usual we let $\boldsymbol{\Omega} = \dot{\mathbf{Q}}\mathbf{Q}^T$ and recall that $\boldsymbol{\Omega} \in \mathrm{Skw}$, $\mathbf{Q}\mathbf{Q}^T = \mathbf{Q}^T\mathbf{Q} = \mathbf{1}$. Now, upon some rearrangements we find

$$\dot{n}_j = \dot{Q}_{ji}\bar{n}_i = \dot{Q}_{ji}Q_{pi}Q_{ph}\bar{n}_h = \Omega_{jp}n_p,$$

$$\overline{n_{j,k}}^{\cdot} = \overline{Q_{kq}Q_{ji}}^{\cdot}\,\partial_{x_q}\bar{n}_i = \dot{Q}_{kq}Q_{lq}Q_{lp}\partial_{x_p}Q_{ji}\bar{n}_i + Q_{kq}\partial_{x_q}\dot{Q}_{ji}Q_{hi}Q_{hl}\bar{n}_l = \Omega_{kl}n_{j,l} + \Omega_{jh}n_{h,k}.$$

Hence we have

$$\partial_{n_j}\varepsilon\,\Omega_{jp}n_p + \partial_{n_{j,k}}\varepsilon(\Omega_{kl}n_{j,l} + \Omega_{jh}n_{h,k}) = 0.$$

Since $\boldsymbol{\Omega} \in \mathrm{Skw}$ then we can represent $\boldsymbol{\Omega}$ via a vector,[4] say $\mathbf{u}$,

$$\Omega_{jp} = \epsilon_{jpq}u_q.$$

Consequently we can write

$$\partial_{n_j}\varepsilon\,\epsilon_{jpq}u_qn_p + \partial_{n_{j,k}}\varepsilon(\epsilon_{klq}u_qn_{j,l} + \epsilon_{jhq}u_qn_{h,k}) = 0.$$

The arbitrariness of $\mathbf{u}$ implies

$$\epsilon_{jpq}\partial_{n_j}\varepsilon\, n_p + \epsilon_{klq}\partial_{n_{j,k}}\varepsilon\, n_{j,l} + \epsilon_{jhq}\partial_{n_{j,k}}\varepsilon\, n_{h,k} = 0, \quad q = 1, 2, 3.$$

Relabeling the summed indices we find (10.17).[5] □

The simplest energy function involves $\nabla\mathbf{n}$ in the form

$$\varepsilon = \propto |\nabla\mathbf{n}|^2.$$

We find the expression

$$(n_{i,k}\partial_{n_{j,k}} + n_{k,i}\partial_{n_{k,j}})(n_{p,q}n_{p,q}) = 2(n_{i,k}n_{j,k} + n_{k,i}n_{k,j})$$

which is symmetric relative to the indices i, j. Hence

$$\epsilon_{jip}(\partial_{n_{j,k}}\varepsilon\, n_{i,k} + \partial_{n_{k,j}}\varepsilon\, n_{k,i}) = 0, \qquad p = 1, 2, 3$$

and the requirement (10.17) is verified.

Among the possible functions $\varepsilon(\mathbf{n}, \nabla\mathbf{n})$ is the energy function

---

[4] Here $\mathbf{u}$ is the opposite of the axial vector of $\boldsymbol{\Omega}$.

[5] The requirement (10.17) traces back to Ericksen [145]; that proof shows only the necessity.

$$\varepsilon_{\text{OF}} = \kappa_1(\nabla\cdot\mathbf{n})^2 + \kappa_2(\mathbf{n}\cdot\nabla\times\mathbf{n})^2 + \kappa_3[(\mathbf{n}\cdot\nabla)\mathbf{n}]^2 + \kappa_4[(\nabla\mathbf{n})\cdot(\nabla\mathbf{n})^T - (\nabla\cdot\mathbf{n})^2] \tag{10.20}$$

usually referred to as Oseen-Frank energy density for a nematic [174, 351]. Upon a direct check it follows that the function (10.20) satisfies the objectivity requirement (10.17). Indeed, each term in (10.20) satisfies (10.17) and hence (10.20) is objective for any choice of the invariants $\kappa_1, \kappa_2, \kappa_3, \kappa_4$, possibly dependent on the temperature θ and the mass density ρ. Two functions depend only on $\nabla\mathbf{n}$ while two depend on $\mathbf{n}$ and $\nabla\mathbf{n}$. We compute the left hand side of (10.17) for $(\nabla\cdot\mathbf{n})^2$, $(\nabla\mathbf{n})\cdot(\nabla\mathbf{n})^T$, $(\mathbf{n}\cdot\nabla)\mathbf{n}$, $\mathbf{n}\cdot\nabla\times\mathbf{n})^2$ to obtain

$$\epsilon_{jip}[\partial_{n_j}(\nabla\cdot\mathbf{n})^2 n_i + \partial_{n_{j,k}}(\nabla\cdot\mathbf{n})^2 n_{i,k} + \partial_{n_{k,j}}(\nabla\cdot\mathbf{n})^2 n_{k,i}] = 2(\nabla\cdot\mathbf{n})\epsilon_{jip}(n_{i,j} + n_{j,i}) = 0,$$

$$\epsilon_{jip}\{\partial_{n_{j,k}}[(\nabla\mathbf{n})\cdot(\nabla\mathbf{n})^T]n_{i,k} + \partial_{n_{k,j}}[(\nabla\mathbf{n})\cdot(\nabla\mathbf{n})^T]n_{k,i}\} = 2\epsilon_{jip}(n_{k,j}n_{i,k} + n_{j,k}n_{k,i}) = 0,$$

$$\begin{aligned}\epsilon_{jip}\{\partial_{n_j}[(\mathbf{n}\cdot\nabla)\mathbf{n}]^2 n_i + \partial_{n_{j,k}}[(\mathbf{n}\cdot\nabla)\mathbf{n}]^2 n_{i,k} + \partial_{n_{k,j}}[(\mathbf{n}\cdot\nabla)\mathbf{n}]^2 n_{k,i}\}\\ = 2\epsilon_{jip}[n_q n_{k,q} n_i n_{k,j} + n_q n_{k,q} n_j n_{k,i} + n_q n_{j,q} n_k n_{i,k}] = 0,\end{aligned}$$

$$\begin{aligned}\epsilon_{jip}[\partial_{n_j}(\mathbf{n}\cdot\nabla\times\mathbf{n})^2 n_i + \partial_{n_{j,k}}(\mathbf{n}\cdot\nabla\times\mathbf{n})^2 n_{i,k} + \partial_{n_{k,j}}(\mathbf{n}\cdot\nabla\times\mathbf{n})^2 n_{k,i}]\\ = 2\,\mathbf{n}\cdot\nabla\times\mathbf{n}\,\epsilon_{jip}(n_i\epsilon_{jqr}n_{r,q} + n_r\epsilon_{rkj}n_{i,k} + n_r\epsilon_{rjk}n_{k,i}) = 0;\end{aligned}$$

in three cases the vanishing is apparent in that ϵ_{jip} multiplies a symmetric quantity in i and j, in the last case a direct calculation shows that the whole expression vanishes.

The director $\mathbf{n}$ is well-defined for nematics and cholesterics. A rotation matrix $\mathcal{R}$ is defined when a triad rigidly attached to the particles is available. Correspondingly, by (1.21) the angular velocity of the particle is given by $\omega_l = \frac{1}{2}\epsilon_{lhk}\dot{\mathcal{R}}_{ki}\mathcal{R}_{hi}$, relative to the fixed basis $\{\mathbf{e}_l\}$. As with any direction rigidly attached to the particle (mesogen), the unit vector $\mathbf{n}$ changes in time according to $\dot{\mathbf{n}} = \boldsymbol{\omega}\times\mathbf{n}$. In isotropic LCs $\mathbf{n}$ is undefined and $\mathcal{R}$ is undefined too.

10.4 Nematics

We now address attention to nematics where the director field $\mathbf{n}$ is defined, $|\mathbf{n}| = 1$. Moreover we let nematics be subject to an electric field $\mathbf{E}$. For simplicity we let the LC be non-magnetizable,[6] $\mathbf{M} = \mathbf{0}$. Hence we let the constitutive functions depend on the set of variables[7]

---

[6] In case of magnetizable bodies, the procedure for **M** and **B** is the dual of the present one for **P** and **E**.

[7] In principle, the dependence on $\mathbf{w} - \boldsymbol{\omega}$ is admissible in that $\mathbf{w} - \boldsymbol{\omega}$ is objective.

$$\rho, \theta, \mathbf{E}, \mathbf{D}, \nabla\theta, \mathbf{n}, \nabla\mathbf{n}, \mathbf{w} - \boldsymbol{\omega}, \nabla\boldsymbol{\omega},$$

where $\mathbf{D}$ is the stretching tensor.

The entropy inequality (10.16) becomes

$$-\rho(\partial_\theta\phi + \eta)\dot{\theta} - (\rho\partial_{\mathbf{E}}\phi + \mathbf{P}) \cdot \dot{\mathbf{E}} - \rho\partial_{\mathbf{D}}\phi \cdot \dot{\mathbf{D}} - \rho\partial_{\nabla\theta}\phi \cdot \overline{\nabla\theta}^{\cdot} - \rho\partial_{\nabla\mathbf{n}}\phi \cdot \overline{\nabla\mathbf{n}}^{\cdot} - \rho\partial_{\mathbf{w}-\boldsymbol{\omega}}\phi \cdot \overline{\mathbf{w} - \boldsymbol{\omega}}^{\cdot}$$
$$-\rho\partial_{\nabla\boldsymbol{\omega}}\phi \cdot \overline{\nabla\boldsymbol{\omega}}^{\cdot} - \rho\partial_{\mathbf{n}}\phi \cdot \dot{\mathbf{n}} + \boldsymbol{\Upsilon} \cdot (\mathbf{w} - \boldsymbol{\omega}) + (\mathbf{T} + \rho^2\partial_\rho\phi\mathbf{1}) \cdot \mathbf{D} + \mathbf{S} \cdot \nabla\boldsymbol{\omega} - \frac{1}{\theta}\mathbf{q} \cdot \nabla\theta + \theta\nabla \cdot \mathbf{k} \geq 0.$$

The arbitrariness of $\dot{\mathbf{D}}, \overline{\nabla\theta}^{\cdot}, \overline{\mathbf{w} - \boldsymbol{\omega}}^{\cdot}, \overline{\nabla\boldsymbol{\omega}}^{\cdot}, \dot{\theta}, \dot{\mathbf{E}}$ implies

$$\partial_{\mathbf{D}}\phi = \mathbf{0}, \quad \partial_{\nabla\theta}\phi = \mathbf{0}, \quad \partial_{\mathbf{w}-\boldsymbol{\omega}}\phi = \mathbf{0}, \quad \partial_{\nabla\boldsymbol{\omega}}\phi = \mathbf{0}, \quad \eta = -\partial_\theta\phi, \quad \mathbf{P} = -\rho\partial_{\mathbf{E}}\phi.$$

A thermodynamically consistent model is now derived where $\mathbf{k} = \mathbf{0}$. The entropy inequality simplifies to

$$-\rho\partial_{\nabla\mathbf{n}}\phi \cdot \overline{\nabla\mathbf{n}}^{\cdot} - \rho\partial_{\mathbf{n}}\phi \cdot \dot{\mathbf{n}} + (\mathbf{T} + \rho^2\partial_\rho\phi\mathbf{1}) \cdot \mathbf{D} + \boldsymbol{\Upsilon} \cdot (\mathbf{w} - \boldsymbol{\omega}) + \mathbf{S} \cdot \nabla\boldsymbol{\omega} - \frac{1}{\theta}\mathbf{q} \cdot \nabla\theta \geq 0.$$

Look at $\partial_{\nabla\mathbf{n}}\phi \cdot \overline{\nabla\mathbf{n}}^{\cdot} = \partial_{\nabla\mathbf{n}}\phi \cdot (\nabla\dot{\mathbf{n}} - \mathbf{L}^T\nabla\mathbf{n})$. In suffix notation, since $\dot{\mathbf{n}} = \boldsymbol{\omega} \times \mathbf{n}$ we have

$$\partial_{n_{i,j}}\phi\, \overline{n_{i,j}}^{\cdot} = \partial_{n_{i,j}}\phi(\partial_{x_j}\dot{n}_i - L_{kj}n_{i,k}) = \partial_{n_{i,j}}\phi[\epsilon_{ipq}(\omega_{p,j}n_q + \omega_p n_{q,j}) - L_{kj}n_{i,k}]$$

while

$$\partial_{\mathbf{n}}\phi \cdot \dot{\mathbf{n}} = \partial_{\mathbf{n}}\phi \cdot \boldsymbol{\omega} \times \mathbf{n} = \boldsymbol{\omega} \cdot \mathbf{n} \times \partial_{\mathbf{n}}\phi.$$

Hence we have

$$\rho\partial_{\mathbf{n}}\psi \cdot \dot{\mathbf{n}} + \rho\partial_{\nabla\mathbf{n}}\psi \cdot \overline{\nabla\mathbf{n}}^{\cdot} = \xi_k\,\omega_k + A_{ki}\omega_{k,i} - G_{ik}(D_{ik} + W_{ik}),$$

where

$$\xi_k = \rho\,\epsilon_{krs}[n_r\partial_{n_s}\phi + n_{r,j}\,\partial_{n_{s,j}}\phi], \quad A_{ki} = \rho\,\epsilon_{ksj}n_s\partial_{n_{j,i}}\phi, \quad G_{ik} = \rho\; n_{j,i}\partial_{n_{j,k}}\phi.$$

Moreover, let

$$g_i = \epsilon_{ipq}G_{qp}, \qquad G_{jk}^{\text{skw}} = \tfrac{1}{2}\epsilon_{kjp}g_p, \qquad \Xi_{jk}^{\text{skw}} = \tfrac{1}{2}\epsilon_{kjp}\xi_p;$$

observe $\mathbf{g} = 2\mathbf{a}_G$, $\mathbf{a}_G$ being the axial vector of $\mathbf{G}^{\text{skw}}$. We find that

$$G_{ik}^{\text{skw}} = \tfrac{1}{2}(\partial_{n_{j,k}}\phi\, n_{j,i} - \partial_{n_{j,i}}\phi\, n_{j,k}), \quad \Xi_{ik}^{\text{skw}} = \tfrac{1}{2}(n_k\partial_{n_i}\phi + n_{k,j}\partial_{n_{i,j}}\phi - n_i\partial_{n_k}\phi - n_{i,j}\partial_{n_{k,j}}\phi).$$

We can then write the Clausius-Duhem inequality in the form

$$- (\boldsymbol{\Upsilon} + \boldsymbol{\xi}) \cdot \boldsymbol{\omega} + (\mathbf{S} - \mathbf{A}) \cdot \nabla\boldsymbol{\omega} + (\mathbf{T}^{\text{sym}} + \mathbf{G}^{\text{sym}} + \rho^2 \partial_\rho \phi \mathbf{1}) \cdot \mathbf{D} + (\boldsymbol{\Upsilon} + \mathbf{g}) \cdot \mathbf{w} - \frac{1}{\theta} \mathbf{q} \cdot \nabla\theta \geq 0. \tag{10.21}$$

For definiteness, we let $\mathbf{S}$, sym$\mathbf{T}$, $\mathbf{q}$ be independent of $\mathbf{w} - \boldsymbol{\omega}$ and consider two cases about the vector $\boldsymbol{\Upsilon}$. First we let $\boldsymbol{\Upsilon}$ depend on $\mathbf{w} - \boldsymbol{\omega}$ so that $\boldsymbol{\Upsilon} \to \mathbf{0}$ as $\mathbf{w} - \boldsymbol{\omega} \to \mathbf{0}$. Hence inequality (10.21) holds only if $\mathbf{g} \cdot \mathbf{w} - \boldsymbol{\xi} \cdot \boldsymbol{\omega} \geq 0$; the arbitrariness of $\mathbf{w}$ and $\boldsymbol{\omega}$ requires that

$$\mathbf{g} = \mathbf{0}, \quad \boldsymbol{\xi} = \mathbf{0},$$

and hence $\mathbf{G}^{\text{skw}} = \mathbf{0}$. These conditions in turn imply

$$\partial_{n_{j,i}} \phi \, n_{j,k} - \partial_{n_{j,k}} \phi \, n_{j,i} = 0, \qquad n_i \partial_{n_k} \phi + n_{i,j} \partial_{n_{k,j}} \phi - n_k \partial_{n_i} \phi - n_{k,j} \partial_{n_{i,j}} \phi = 0 \tag{10.22}$$

$\forall i, k = 1, 2, 3$. The Clausius-Duhem inequality then reduces to

$$\boldsymbol{\Upsilon} \cdot (\mathbf{w} - \boldsymbol{\omega}) + (\mathbf{S} - \mathbf{A}) \cdot \nabla\boldsymbol{\omega} + (\mathbf{T}^{\text{sym}} + \mathbf{G}^{\text{sym}} + \rho^2 \partial_\rho \phi \mathbf{1}) \cdot \mathbf{D} - \frac{1}{\theta} \mathbf{q} \cdot \nabla\theta \geq 0. \tag{10.23}$$

Second, let $\boldsymbol{\Upsilon}$ be independent of $\mathbf{w} - \boldsymbol{\omega}$. Hence inequality (10.21) holds only if $(\boldsymbol{\Upsilon} + \mathbf{g}) \cdot \mathbf{w} - (\boldsymbol{\Upsilon} + \boldsymbol{\xi}) \cdot \boldsymbol{\omega} \geq 0$. The arbitrariness of $\mathbf{w}$ and $\boldsymbol{\omega}$ requires that

$$\mathbf{g} = -\boldsymbol{\Upsilon} = \boldsymbol{\xi}.$$

The vector condition $\mathbf{g} = \boldsymbol{\xi}$ results in

$$G_{ik}^{\text{skw}} = \tfrac{1}{2}[\partial_{n_{j,k}} \phi \, n_{j,i} - \partial_{n_{j,i}} \phi \, n_{j,k}] = \tfrac{1}{2}[n_k \partial_{n_i} \phi + n_{k,j} \partial_{n_{i,j}} \phi - n_i \partial_{n_k} \phi - n_{i,j} \partial_{n_{k,j}} \phi] = \Xi_{ik}^{\text{skw}} \tag{10.24}$$

$\forall i, k = 1, 2, 3$. The Clausius-Duhem inequality then implies

$$\boldsymbol{\Upsilon} = -\mathbf{g} = -2\mathbf{a}_{\mathbf{G}}$$

and reduces to

$$(\mathbf{S} - \mathbf{A}) \cdot \nabla\boldsymbol{\omega} + (\mathbf{T}^{\text{sym}} + \mathbf{G}^{\text{sym}} + \rho^2 \partial_\rho \phi \mathbf{1}) \cdot \mathbf{D} - \frac{1}{\theta} \mathbf{q} \cdot \nabla\theta \geq 0. \tag{10.25}$$

As to $\mathbf{T}^{\text{sym}}$ and $\mathbf{S}$ we let $\bar{\mathbf{T}}, \hat{\mathbf{T}} \in \text{Sym}$ and assume

$$\mathbf{T}^{\text{sym}} = \bar{\mathbf{T}}(\rho, \theta, \mathbf{n}, \nabla\mathbf{n}) + \hat{\mathbf{T}}, \qquad \mathbf{S} = \bar{\mathbf{S}}(\rho, \theta, \mathbf{n}, \nabla\mathbf{n}) + \hat{\mathbf{S}},$$

where $\hat{\mathbf{T}}$ and $\hat{\mathbf{S}}$ approach zero as $\mathbf{D}$, $\nabla\theta$, and $\nabla\boldsymbol{\omega}$ approach zero. As a consequence, (10.23) implies the relations

$$\bar{\mathbf{T}} = -\rho^2 \partial_\rho \phi \mathbf{1} - \mathbf{G}^{\text{sym}}, \qquad \bar{\mathbf{S}} = \mathbf{A},$$

and the dissipation inequality

$$\boldsymbol{\Upsilon} \cdot (\mathbf{w} - \boldsymbol{\omega}) + \hat{\mathbf{S}} \cdot \nabla \boldsymbol{\omega} + \hat{\mathbf{T}} \cdot \mathbf{D} - \frac{1}{\theta} \mathbf{q} \cdot \nabla \theta \geq 0. \tag{10.26}$$

Instead, if (10.24) holds then

$$\bar{\mathbf{T}} = -\rho^2 \partial_\rho \phi \mathbf{1} - \mathbf{G}^{\mathrm{sym}}, \qquad \bar{\mathbf{S}} = \mathbf{A},$$

and

$$\hat{\mathbf{S}} \cdot \nabla \boldsymbol{\omega} + \hat{\mathbf{T}} \cdot \mathbf{D} - \frac{1}{\theta} \mathbf{q} \cdot \nabla \theta \geq 0.$$

In both cases the constitutive equations

$$\hat{\mathbf{S}} = \alpha \nabla \boldsymbol{\omega} + \beta (\nabla \cdot \boldsymbol{\omega}) \mathbf{1} + \gamma \nabla \theta \otimes \nabla \theta,$$

$$\hat{\mathbf{T}} = 2\mu \mathbf{D} + \lambda (\nabla \cdot \mathbf{v}) \mathbf{1} + \nu \nabla \theta \otimes \nabla \theta,$$

$$\mathbf{q} = -\kappa \nabla \theta - \nu \nabla \theta \mathbf{D} - \gamma \nabla \theta \nabla \boldsymbol{\omega},$$

involving non-linear cross-coupling terms, are allowed if α, $\alpha + 3\beta$, μ, $2\mu + 3\lambda$, κ are non-negative. Moreover the dependence

$$\boldsymbol{\Upsilon} = \zeta (\mathbf{w} - \boldsymbol{\omega}), \qquad \zeta \geq 0,$$

is consistent with (10.26).

A joint dependence of $\hat{\mathbf{T}}$ on $\mathbf{n}$ and $\mathbf{D}$ is allowed in the form

$$\hat{\mathbf{T}} = 2\mu \mathbf{D} + \lambda (\nabla \cdot \mathbf{v}) \mathbf{1} + \alpha_1 (\mathbf{n} \otimes \mathbf{D}\mathbf{n} + \mathbf{D}\mathbf{n} \otimes \mathbf{n}) + \alpha_2 (\mathbf{n} \cdot \mathbf{D}\mathbf{n}) \mathbf{n} \otimes \mathbf{n}$$

where α_1, α_2 are functions of ρ, θ, $\mathbf{E}$ subject to $\alpha_1 \geq 0$, $\alpha_2 \geq -\alpha_1$. This is so because

$$\hat{\mathbf{T}} \cdot \mathbf{D} = 2\mu |\mathbf{D}_0|^2 + (\lambda + \tfrac{2}{3}\mu)(\nabla \cdot \mathbf{v})^2 + \alpha_1 |\mathbf{D}\mathbf{n}|^2 + \alpha_2 (\mathbf{n} \cdot \mathbf{D}\mathbf{n})^2.$$

We do not consider an additive stress of the form[8]

$$\overset{\circ}{\mathbf{n}} \otimes \mathbf{n} + \mathbf{n} \otimes \overset{\circ}{\mathbf{n}},$$

in that the additive contribution $2 \overset{\circ}{\mathbf{n}} \cdot \mathbf{D}\mathbf{n}$ to $\hat{\mathbf{T}} \cdot \mathbf{D}$ would be indefinite.

The thermodynamic requirements (10.22) follow from (10.24) when both $\mathbf{G}$ and $\boldsymbol{\Xi}$ vanish. Hence (10.24) is the general thermodynamic requirement on the dependence of ϕ on $\mathbf{n}$ and $\nabla \mathbf{n}$. For definiteness we now look at the function ϕ in the additive

---

[8] $\overset{\circ}{\mathbf{n}}$ denoting the corotational derivative of $\mathbf{n}$, i.e. $\overset{\circ}{\mathbf{n}} = \dot{\mathbf{n}} - \mathbf{W}\mathbf{n}$; such dependence is considered e.g. in [82].

form

$$\phi = \phi_0(\rho, \theta, \mathbf{E}) + \sigma f(\mathbf{n}, \nabla\mathbf{n}),$$

where σ may depend on $\rho, \theta, \mathbf{E}$. Hence (10.22) is a requirement on f. First let

$$f = |\nabla\mathbf{n}|^2.$$

Since

$$\partial_{n_{p,j}} |\nabla\mathbf{n}|^2 = 2\, n_{p,j}$$

then both requirements (10.22) hold and hence $\mathbf{g} = \boldsymbol{\xi} = \mathbf{0}$.

We now investigate the functions occurring in the Oseen-Frank energy density (10.20). Let $f = (\nabla \cdot \mathbf{n})^2$. Then both sides of (10.24) equal

$$\sigma \nabla \cdot \mathbf{n}(n_{k,i} - n_{i,k});$$

(10.24) holds and $G_{ik} = 2\rho\sigma\, \nabla \cdot \mathbf{n}\, n_{k,i}$.

Let $f = \nabla\mathbf{n} \cdot (\nabla\mathbf{n})^T$. Then both sides of (10.24) equal

$$\sigma(n_{j,i} n_{k,j} - n_{i,j} n_{j,k});$$

(10.24) holds and $G_{ik} = 2\rho\sigma n_{j,i} n_{k,j}$.

Let $f = |(\mathbf{n} \cdot \nabla)\mathbf{n}|^2$. We find that both sides of (10.24) equal

$$\sigma n_r n_{j,r} (n_k n_{j,i} - n_i n_{j,k});$$

again (10.24) holds while $G_{ik} = 2\rho\sigma n_k n_{j,i} (\mathbf{n} \cdot \nabla) n_j$.

Now let $f = (\mathbf{n} \cdot \nabla \times \mathbf{n})^2$. Since

$$\partial_{n_{j,k}} f\; n_{j,i} = 2(\mathbf{n} \cdot \nabla \times \mathbf{n}) n_{j,i} n_p \epsilon_{pkj},$$

$$n_k \partial_{n_i} f + n_{k,j} \partial_{n_{i,j}} f = 2(\mathbf{n} \cdot \nabla \times \mathbf{n})[n_k \epsilon_{ipq} n_{q,p} + n_{k,j} n_r \epsilon_{rji}$$

it follows that (10.24) does not hold.

For the thermodynamically consistent functions f so examined the stress $\mathbf{G} = \bar{\mathbf{T}} + p\mathbf{1}$, where $p = \rho^2 \partial_\rho \phi$, and the couple stress $\mathbf{A} = \bar{\mathbf{S}}$ take the following forms.

f	$\mathbf{G}$	$\mathbf{A}$
$\|\nabla\mathbf{n}\|^2$	$G_{ik} = 2\rho\sigma n_{j,i} n_{j,k}$	$A_{ik} = 2\rho\sigma\epsilon_{ksj} n_s n_{j,i}$
$(\nabla \cdot \mathbf{n})^2$	$G_{ik} = 2\rho\sigma n_{h,h} n_{k,i}$	$A_{ik} = 2\rho\sigma\epsilon_{isk} n_s$
$(\nabla\mathbf{n}) \cdot (\nabla\mathbf{n})^T$	$G_{ik} = 2\rho\sigma n_{j,i} n_{k,j}$	$A_{ik} = 2\rho\sigma\epsilon_{isj} n_s n_{k,j}$
$\|(\mathbf{n} \cdot \nabla)\mathbf{n}\|^2$	$G_{ik} = 2\rho\sigma n_q n_{j,q} n_k n_{j,i}$	$A_{ik} = 2\rho\sigma\epsilon_{isj} n_s n_p n_{j,p} n_k$

Consequently

$$f = |\nabla\mathbf{n}|^2 \implies \text{skw}\bar{\mathbf{T}} = \mathbf{0},$$

$$f = (\nabla \cdot \mathbf{n})^2 \implies (\mathrm{skw}\,\mathbf{T})_{ik} = \rho\sigma(\nabla \cdot \mathbf{n})(n_{k,i} - n_{i,k}),$$

$$f = (\nabla \mathbf{n}) \cdot (\nabla \mathbf{n})^T \implies (\mathrm{skw}\,\mathbf{T})_{ik} = \rho\sigma(n_{k,j}n_{j,i} - n_{i,j}n_{j,k}),$$

$$f = |(\mathbf{n} \cdot \nabla)\mathbf{n}|^2 \implies (\mathrm{skw}\,\bar{\mathbf{T}})_{ik} = \rho\sigma n_q n_{j,q}(n_k n_{j,i} - n_i n_{j,k}).$$

Incidentally, if $f = |\nabla \mathbf{n}|^2$ then $\boldsymbol{\Upsilon} = \mathbf{0}$ while if $f = (\nabla \cdot \mathbf{n})^2$ it follows that

$$\boldsymbol{\Upsilon} = 2\rho\sigma(\nabla \cdot \mathbf{n})\nabla \times \mathbf{n}.$$

The molecules may have a permanent electric dipole. Otherwise an electric field produces rearrangements of charges in the molecules such that an electric dipole results. In any case the polarization $\mathbf{P}$ depends on the orientation of the director $\mathbf{n}$ relative to the electric field. Moreover the polarization is invariant under the change $\mathbf{n} \to -\mathbf{n}$. Hence, in the approximation that $\mathbf{P}$ depends linearly on the electric field $\mathbf{E}$ we have

$$\mathbf{P} = \epsilon_0[\chi_\perp \mathbf{1} + (\chi_\parallel - \chi_\perp)\mathbf{n} \otimes \mathbf{n}]\mathbf{E}$$

whence

$$\mathbf{P} \cdot \mathbf{n} = \epsilon_0 \chi_\parallel \mathbf{E} \cdot \mathbf{n}, \qquad \mathbf{P} \times \mathbf{n} = \epsilon_0 \chi_\perp \mathbf{E} \times \mathbf{n}.$$

Accordingly, $\chi_\parallel, \chi_\perp$ are the electric susceptibilities in the direction of $\mathbf{n}$ and the orthogonal plane. In this case

$$\phi = \Phi(\rho, \theta) - \frac{1}{2\rho}\epsilon_0 \mathbf{E} \cdot [\chi_\perp \mathbf{1} + (\chi_\parallel - \chi_\perp)\mathbf{n} \otimes \mathbf{n}]\mathbf{E} + \sigma f(\mathbf{n}, \nabla \mathbf{n}).$$

Dynamic Equations

We now establish the evolution equations of nematics as a continuous distribution of molecules (mesogens). Since the nematic phase is allowed to be compressible the mass density ρ is required to satisfy the standard continuity equation,

$$\dot{\rho} + \rho \nabla \cdot \mathbf{v} = 0.$$

Let m be the molecular mass. Hence the mass density ρ and the number density $\mathcal{N}$ are related by

$$\rho = \mathcal{N} m.$$

For definiteness we let the net charge density and the magnetic moment of molecules be zero but let $\mathbf{p}$ be the nonzero electric dipole moment. Hence electromagnetic considerations suggest that

$$\rho \mathbf{b} = \mathcal{N} \mathbf{p} \cdot \nabla \mathbf{E}, \qquad \mathbf{l} = \mathcal{N} \mathbf{p} \times \mathbf{E}.$$

An additional term might be considered for $\mathbf{l}$ to model the tendency of molecules to align to the external field $\mathbf{E}$. Hence we assume

$$\mathbf{l} = \mathcal{N}\mathbf{p}\times\mathbf{E} - \zeta\mathcal{N}^2\mathbf{p}\times(\mathbf{p}\times\mathbf{E}), \qquad \zeta > 0.$$

For simplicity we let $\hat{\mathbf{S}} = \mathbf{0}$ and $\hat{\mathbf{T}} = 2\mu\mathbf{D} + \lambda(\nabla\cdot\mathbf{v})\mathbf{1}$ so that

$$\mathbf{T} = -p\mathbf{1} + 2\mu\mathbf{D} + \lambda(\nabla\cdot\mathbf{v})\mathbf{1} + \mathbf{G}, \qquad \mathbf{S} = \mathbf{A}.$$

Moreover $\boldsymbol{\Upsilon}$, which is twice the axial vector of $\mathbf{T}$, is given by

$$\Upsilon_i = \epsilon_{ijk}G_{kj}.$$

As shown in the previous table, the expressions of $\mathbf{G}$ and $\mathbf{A}$ depend on the selected free-energy dependence of ϕ on $\nabla\mathbf{n}$.

Let

$$\mathcal{I} = m\mathbf{I}$$

be the inertia tensor of a molecule and then $\mathbf{I}$ is the inertia tensor density, per unit mass. Moreover observe that

$$\mathbf{P} = \mathcal{N}\mathbf{p}$$

is just the polarization (per unit volume). The parameters μ, λ, γ are regarded as constants.

Since

$$T_{ik} = -p\delta_{ik} + 2\mu D_{ik} + \lambda(\nabla\cdot\mathbf{v})\delta_{ik} + G_{ik}$$

we can write the equation of motion (10.8) in the form

$$\rho\dot{v}_i = P_k E_{i,k} - p_{,i} + 2\mu v_{i,kk} + (\lambda + 2\mu)v_{k,ki} + G_{ik,k}.$$

The balance equation for the orientational momentum (10.10) can be given the form

$$\rho(\mathbf{I}\,\dot{\boldsymbol{\omega}} + \boldsymbol{\omega}\times\mathbf{I}\,\boldsymbol{\omega}) = \boldsymbol{\Upsilon} + \mathbf{P}\times\mathbf{E} - \zeta\mathbf{P}\times(\mathbf{P}\times\mathbf{E}) + \nabla\cdot\mathbf{S}. \tag{10.27}$$

The energy balance equation can be written as

$$\rho\dot{\varepsilon} = -p\nabla\cdot\mathbf{v} + 2\mu\mathbf{D}\cdot\mathbf{D} + \lambda(\nabla\cdot\mathbf{v})^2 + \boldsymbol{\Upsilon}\cdot(\mathbf{w} - \boldsymbol{\omega}) + \mathbf{S}\cdot\nabla\boldsymbol{\omega} - \nabla\cdot\mathbf{q} + \rho r,$$

where $\mathbf{w} = \frac{1}{2}\nabla\times\mathbf{v}$.

In general the vector $\boldsymbol{\Upsilon}$ can be given in the form

$$\Upsilon_i = \epsilon_{ijk}(G_{kj} + \hat{T}_{kj}),$$

where $\mathbf{G}$ is the stress tensor characterized by the dependence of the free energy ϕ on $\mathbf{n}$ and $\nabla\mathbf{n}$ while $\hat{\mathbf{T}}$ may involve appropriate dependences on $\nabla\omega$ and $\nabla\theta$.

10.4.1 Relation to Other Models Involving the Director Field

Among the first continuum models of LCs we mention those of Ericksen [146] and Leslie [277]. In these schemes the orientation of molecules is modelled via a director $\mathbf{n}$, $|\mathbf{n}| = 1$, and the form of molecules is described by a parameter, S, called *degree of orientation*. The main points of these approaches are summarized as follows.

The fluid is assumed to be incompressible and the equation of motion is taken in the standard form. Next a generalized scalar momentum P is considered subject to the balance equation

$$\dot{P} = \nabla \cdot \mathbf{t} + f_I + f_E,$$

where f_I and f_E represent internal and external body forces. To determine $\mathbf{n}$ it is assumed a vector equation

$$\pi = \nabla \cdot \mathcal{T} + \mathbf{g}_I + \mathbf{g}_E.$$

Next [146] the balance of energy is assumed in the form

$$\frac{d}{dt}\int_\Omega \rho(e + \tfrac{1}{2}\mathbf{v}^2)dv = \int_{\partial\Omega} [\mathbf{v}\cdot\mathbf{T} + \dot{\mathbf{n}}\cdot\boldsymbol{\Upsilon} + \dot{S}\mathcal{T} - \mathbf{q}]\cdot\mathbf{n}da + \int_\Omega [\mathbf{b}\cdot\mathbf{v} + f_E\dot{S} + \mathbf{f}_E\cdot\dot{\mathbf{n}} + \rho r]dv,$$

where e is said to include the rotational kinetic energy. Hence it follows

$$\rho\dot{e} = \mathbf{T}\cdot\mathbf{L} + \mathcal{T}\cdot\nabla\dot{\mathbf{n}} + \nabla\dot{S}\cdot\mathbf{t} + (\dot{P} - f_I)\dot{S} + (\dot{\pi} - \mathbf{g}_I)\cdot\dot{\mathbf{n}} - \nabla\cdot\mathbf{q} + \rho r.$$

Another approach (see, e.g., [277]) is based on balance laws for linear and angular momentum without appeal to generalized forces and moments associated with the director $\mathbf{N}$. Yet thermal and inertial terms are omitted, which simplifies the whole scheme. Here the essential points are outlined. The nematic is assumed to be incompressible and the balance of linear momentum takes the standard form, $\rho\dot{\mathbf{v}} = \nabla\cdot\mathbf{T} + \rho\mathbf{b}$. In our notation, the balance of angular momentum is assumed in the form

$$\frac{d}{dt}\int_\Omega \rho\mathbf{r}\times\mathbf{v}\,dv = \int_\Omega [\rho\mathbf{r}\times\mathbf{b} + \boldsymbol{\tau}]dv + \int_{\partial\Omega} (\mathbf{r}\times\mathbf{t} + \mathbf{s})da$$

whence it follows

$$\boldsymbol{\tau} + \boldsymbol{\Upsilon} + \nabla\cdot\mathbf{S} = \mathbf{0}. \tag{10.28}$$

Hence, neglecting the occurrence of the orientational momentum results in an equilibrium (or constraint) equation for $\boldsymbol{\tau}$, $\boldsymbol{\Upsilon}$, $\mathbf{S}$. Correspondingly, the balance of energy is assumed in the form

$$\frac{d}{dt}\int_{\Omega}\rho(e+\tfrac{1}{2}\mathbf{v}^2)dv = \int_{\partial\Omega}[\mathbf{t}\cdot\mathbf{v}+\mathbf{s}\cdot\boldsymbol{\omega}]da + \int_{\Omega}[\rho\mathbf{b}\cdot\mathbf{v}+\boldsymbol{\tau}\cdot\boldsymbol{\omega}-D]dv,$$

where D is said to be the rate of viscous dissipation. Thus it follows that

$$\rho\dot{e} = \mathbf{T}\cdot\mathbf{L}+\mathbf{S}\cdot\nabla\boldsymbol{\omega}-\boldsymbol{\Upsilon}\cdot\boldsymbol{\omega}-D.$$

Two comments are in order. The form of the mechanical power, $\mathbf{T}\cdot\mathbf{L}+\mathbf{S}\cdot\nabla\boldsymbol{\omega}-\boldsymbol{\Upsilon}\cdot\boldsymbol{\omega}$, is consistent with other approaches (and our's). Yet, the occurrence of (10.28) as an equilibrium condition, instead of the balance of orientational momentum, deprives the scheme of an equation for the time dependence of $\boldsymbol{\omega}$. Further, by arguing with the function $e(\mathbf{n}, \nabla\mathbf{n})$ it follows that dissipative terms in the balance of energy arise[9] subject to

$$\tilde{\mathbf{T}}\cdot\mathbf{L}+\tilde{\mathbf{S}}\cdot\nabla\boldsymbol{\omega}-\tilde{\boldsymbol{\Upsilon}}\cdot\boldsymbol{\omega} = D \ge 0.$$

So, the occurrence of D conceptually remedies the absence of the entropy inequality.

Borrowing from [146], the dynamic equations of nematic LCs have been considered [82] by assuming $|\mathbf{n}| = 1$, the incompressibility condition, $\nabla\cdot\mathbf{v} = 0$, the occurrence of an order parameter field s, and the evolution equations

$$\rho\dot{\mathbf{v}} = \nabla\cdot\mathbf{T}, \qquad \mathbf{n}\times\overset{\circ}{\mathbf{n}} = -\gamma\mathbf{n}\times\mathbf{D}\mathbf{n},$$

where

$$\mathbf{T} = -p\mathbf{1}+\alpha\mathbf{D}+\alpha_2(\mathbf{n}\otimes\mathbf{D}\mathbf{n}+\mathbf{D}\mathbf{n}\otimes\mathbf{n}+2\alpha_1(\mathbf{n}\cdot\mathbf{D}\mathbf{n})\mathbf{n}\otimes\mathbf{n}+\gamma_2(\overset{\circ}{\mathbf{n}}\otimes\mathbf{n}+\mathbf{n}\otimes\overset{\circ}{\mathbf{n}})+2\beta_1\dot{s}\mathbf{n}\otimes\mathbf{n},$$

the coefficients $\gamma, \alpha, \alpha_2, \alpha_1\gamma_2, \beta_1$ being functions of s. This scheme is consistent with thermodynamics if ϕ is a function of s while $\mathbf{T}$ is a function of s and $\mathbf{D}$; the entropy inequality simplifies to

$$-\rho\phi'\dot{s}+\mathbf{T}\cdot\mathbf{D} \ge 0,$$

that is

$$-\rho\phi'\dot{s}+\alpha|\mathbf{D}|^2+2\alpha_2|\mathbf{D}\mathbf{n}|^2+2\alpha_1(\mathbf{n}\cdot\mathbf{D}\mathbf{n})^2+2\gamma_2\overset{\circ}{\mathbf{n}}\cdot\mathbf{D}\mathbf{n}+2\beta_1\dot{s}\,\mathbf{n}\cdot\mathbf{D}\mathbf{n} \ge 0.$$

Since

$$\dot{\mathbf{n}}\cdot\mathbf{n} = 0, \qquad \overset{\circ}{\mathbf{n}}\cdot\mathbf{n} = \dot{\mathbf{n}}\cdot\mathbf{n}-\mathbf{n}\cdot\mathbf{W}\mathbf{n} = 0$$

then the evaluation of $\mathbf{n}\times(\mathbf{n}\times\overset{\circ}{\mathbf{n}})$ results in

$$\overset{\circ}{\mathbf{n}} = \gamma[\mathbf{D}\mathbf{n}-(\mathbf{n}\cdot\mathbf{D}\mathbf{n})\mathbf{n}].$$

---

[9] See, e.g., Eq. (3.12) of [277].

Upon substitution of $\overset{\circ}{\mathbf{n}}$ the inequality can be written as

$$(-\rho\phi' + 2\beta_1 \mathbf{n} \cdot \mathbf{Dn})\dot{s} + \alpha|\mathbf{D}|^2 + 2(\alpha_2 + \gamma_2\gamma)|\mathbf{Dn}|^2 + 2(\alpha_1 - \gamma_2\gamma)(\mathbf{n} \cdot \mathbf{Dn})^2 \geq 0.$$

The inequality holds if

$$\alpha_1, \alpha_2 > 0, \qquad |\gamma_2\gamma| < \alpha_1, \alpha_2$$

while s satisfies

$$\dot{s} = -\rho\phi' + 2\beta_1 \mathbf{n} \cdot \mathbf{Dn}.$$

This equation can be viewed as the evolution equation for the order parameter s.

The Landau-de Gennes model describes liquid crystals not via the director $\mathbf{n}$ but via an order parameter tensor $\mathbf{Q}$, whose entries are interpreted as the suitably normalized covariance matrix of the probability distribution in space of the unit vector $\mathbf{n}$ [48, 146] or the space average of a traceless dyadic product, $\sum_{i=1}^{N}(\mathbf{n}_i \otimes \mathbf{n}_i - \frac{1}{3}\mathbf{1})/N$ [13]. Otherwise, the order tensor $\mathbf{Q}$ is defined by analogy with the magnetic susceptibility ([117], p. 57).

10.5 Smectics and Cholesterics

The common feature of smectics and cholesterics is the layered structure; let $\mathbf{N}$ be the normal to the plane of the layer. In smectics the molecules have a common orientation within any layer; in smectic A and B LCs the molecules are parallel to $\mathbf{N}$, in smectic C LCs the molecules form a common angle with the normal within any layer and among the layers. In smectic A the molecular axes are not ordered within the planes, in smectics B the axes are packed in a hexagonal form. In cholesteric LCs the molecules form a common angle with the normal within any layer but the angle changes in passing from a layer to another. As a result, the mesogens show a helical structure.

To model smectics and cholesterics we keep denoting by $\mathbf{n}$ the director of the molecules. In smectics $\mathbf{N} \cdot \mathbf{n}$ is constant; $|\mathbf{N} \cdot \mathbf{n}| = 1$ in smectics A and B, $|\mathbf{N} \cdot \mathbf{n}| < 1$ in smectics C. Let α be the angle between $\mathbf{N}$ and $\mathbf{n}$, so that $\mathbf{N} \cdot \mathbf{n} = \cos\alpha$; $\alpha = 0$ in smectics A and B.

To set up any constitutive model we need a view on the energy. While the Oseen-Frank function $\varepsilon_{\mathrm{OF}}$ in (10.20) is a reference model of energy, smectic crystals require a more involved description. The Landau-de Gennes model is established by an assumed analogy between superconductors and smectics [116]. By this analogy, a complex-valued density modulation $\Psi(\mathbf{x})$ is introduced with the view that the absolute value is the amplitude of the smectic order and the phase describes the layer displacement. With this view the additional energy density is

$$\varepsilon_{dG} = |(i\nabla + q\mathbf{n})\Psi|^2 + \mu(-|\Psi|^2 + \tfrac{1}{2}|\Psi|^4),$$

where μ is a positive constant and, by [116],

$$\Psi(\mathbf{x}) = \rho_1 \exp(iq\mathbf{n}\cdot\mathbf{x}),$$

where q is the characteristic smectic layer number, $2\pi/q$ is the layer thickness. Various energy functions are referred to as Landau-de Gennes models. One involves the symmetric matrix $\mathbf{Q}$ in the form

$$\varepsilon(\mathbf{Q}) = f_e(\mathbf{Q}) + f_{LdG}(\mathbf{Q}),$$

where[10]

$$f_e(\nabla\mathbf{Q}) = \tfrac{1}{2}L_1|\nabla\mathbf{Q}|^2 + \tfrac{1}{2}L_2|\nabla\cdot\mathbf{Q}|^2 + \tfrac{1}{2}L_3\nabla\mathbf{Q}\cdot\nabla\mathbf{Q}^T,$$

$$f_{LdG}(\mathbf{Q}) = a\,\mathrm{tr}\,\mathbf{Q}^2 + \tfrac{2}{3}b\,\mathrm{tr}\,\mathbf{Q}^3 + \tfrac{1}{2}c\,(\mathrm{tr}\,\mathbf{Q}^2)^2.$$

Another one is based on the view that the energy density is the sum of three terms,

$$\varepsilon = \varepsilon_{\mathrm{layer}} + \varepsilon_{\mathrm{coupling}} + \varepsilon_{\mathrm{OF}},$$

where $\varepsilon_{\mathrm{layer}}$ is a function of $|\Psi|^2$ while $\varepsilon_{\mathrm{coupling}}$ is a function of $\nabla\Psi$.

Here we follow a scheme of continuum mechanics though in a unusual way. The smectic and cholesteric LCs are viewed as a sequence of thin parallel layers within an isotropic fluid. Each layer is viewed as a two-dimensional elastic continuum. The normal $\mathbf{N}$ to the layers is then constant, say $\mathbf{N} = \mathbf{e}_3$. Let x_1, x_2 describe the position within the layer. For technical convenience we adopt the Lagrangian description and denote by X_1, X_2 the pertinent coordinates for the layer, the third coordinate X_3 being a label for the layer. Hence the restriction of $\mathbf{F}$ and $\mathbf{C} = \mathbf{F}^T\mathbf{F}$ to the first minor 2×2 describes the deformation in a layer. Moreover let α be the angle between $\mathbf{N}$ and $\mathbf{n}$. The layered structure allows us to consider $\alpha(X_1, X_2)$ parameterized by X_3. The restriction e.g. to $\alpha(X_1, X_3)$ is expected to show the helical structure.

By the assumed elastic properties of the layers we let $\varepsilon_{\mathrm{layer}}$ be given in the form

$$\varepsilon_{\mathrm{layer}} = f_l(\nabla_R\mathbf{u}, \nabla_R\alpha),$$

where $\mathbf{u}$ is the displacement. While $\varepsilon_{\mathrm{coupling}}$ involves $|\nabla\Psi|^2$ in a Landau-de Gennes model, we simply let $\varepsilon_{\mathrm{coupling}}$ be a function of $\nabla\rho$. The Oseen-Frank energy is regarded as elastic energy. Here the elastic energy is considered via the dependence on $\mathbf{N}\times\nabla_R\mathbf{u}$. Consequently we let the free energy ψ and the other constitutive functions depend on the set of variables

$$\theta, \mathbf{E}, \nabla_R\alpha,$$

[10] In suffix notation, $\nabla\mathbf{Q}\cdot\nabla\mathbf{Q}^T = Q_{ik,j}Q_{ij,k}$.

where $\mathbf{E}$ is the Green-St Venant strain, $\mathbf{E} = \frac{1}{2}(\nabla_R \mathbf{u} + \nabla_R \mathbf{u}^T + \nabla_R \mathbf{u}^T \nabla_R \mathbf{u})$.

We assume the stress $\mathbf{T}$ is symmetric and hence $\hat{\mathbf{T}}_{RR} = J\mathbf{F}^{-1}\mathbf{T}\mathbf{F}^{-T}$ is symmetric too. The molecules in a layer may rotate, the angle α describing the rotation in the plane (X_1, X_3); let $\boldsymbol{\omega} = \dot{\alpha}\mathbf{e}_2$. Since $\mathbf{T} = \mathbf{T}^T$ then $\boldsymbol{\tau} = \mathbf{0}$. The material couple stress tensor $\mathbf{S}_R = J\mathbf{S}\mathbf{F}^{-T}$ allows us to write

$$J\mathbf{S} \cdot \nabla\boldsymbol{\omega} = \mathbf{S}_R \cdot \nabla_R \boldsymbol{\omega} = S^R_{iK} \partial_{X_K} \omega_i .$$

Thermodynamic restrictions are derived by means of the entropy inequality in the material form (10.13). For formal simplicity we neglect heat conduction effects and hence investigate the entropy inequality

$$-\rho_R(\dot{\psi} + \eta\dot{\theta}) + \mathbf{T}_{RR} \cdot \dot{\mathbf{E}} + \mathbf{S}_R \cdot \nabla_R \boldsymbol{\omega} + \theta\nabla_R \cdot \mathbf{k}_R \geq 0.$$

Upon substitution of $\dot{\psi}$ we have

$$-\rho_R(\partial_\theta\psi + \eta)\dot{\theta} - \rho_R(\partial_{\mathbf{E}}\psi + \mathbf{T}_{RR}) \cdot \dot{\mathbf{E}} - \rho_R\partial_{\nabla_R\alpha}\psi \cdot \nabla_R \dot{\alpha} + \mathbf{S}_R \cdot \nabla_R \boldsymbol{\omega} \geq 0.$$

Since $\boldsymbol{\omega} = \mathbf{e}_2\dot{\alpha}$ then

$$\mathbf{S}_R \cdot \nabla_R \boldsymbol{\omega} = \mathbf{e}_2\mathbf{S}_R \cdot \nabla_R \dot{\alpha}.$$

The arbitrariness of $\dot{\theta}$, $\dot{\mathbf{E}}$, $\nabla_R \dot{\alpha}$ implies that

$$\eta = -\partial_\theta\psi, \qquad \mathbf{T}_{RR} = \rho_R\partial_{\mathbf{E}}\psi, \qquad \mathbf{e}_2\mathbf{S}_R = \rho_R\partial_{\nabla_R\alpha}\psi.$$

10.6 Mixtures of Micropolar Constituents

There are cases where the occurrence of different constituents and the micropolar character indicate that a more appropriate model is that of a mixture of micropolar constituents rather than that of a single micropolar medium. For generality we now look at a mixture of micropolar continua. Later on we simplify the model by regarding only one constituent as strictly micropolar in character, which is the natural model for suspensions of particles in fluids.

We consider a mixture of n diffusing and chemically-reacting bodies $\mathcal{B}_\alpha$. The balances of mass and linear momentum are the same as those for non-polar constituents,

$$\partial_t\rho_\alpha + \nabla \cdot (\rho_\alpha\mathbf{v}_\alpha) = \tau_\alpha,$$

$$\partial_t(\rho_\alpha\mathbf{v}_\alpha) + \nabla \cdot (\rho_\alpha\mathbf{v}_\alpha \otimes \mathbf{v}_\alpha) = \nabla \cdot \mathbf{T}_\alpha + \rho_\alpha\mathbf{b}_\alpha + \mathbf{m}_\alpha,$$

in the Eulerian description, or

$$\grave{\rho}_\alpha + \rho_\alpha \nabla \cdot \mathbf{v}_\alpha = \tau_\alpha,$$

$$\rho_\alpha \grave{\mathbf{v}}_\alpha = \nabla \cdot \mathbf{T}_\alpha + \rho_\alpha \mathbf{b}_\alpha + \mathbf{m}_\alpha - \tau_\alpha \mathbf{v}_\alpha,$$

in the Lagrangian description. Of course, the growths $\{\tau_\alpha\}$, $\{\mathbf{m}_\alpha\}$ are subject to

$$\textstyle\sum_\alpha \tau_\alpha = 0, \qquad \sum_\alpha \mathbf{m}_\alpha = \mathbf{0}.$$

The balance of angular momentum is assumed in the form

$$\Big(\int_{\mathcal{P}^t_\alpha} \rho_\alpha (\mathbf{r} \times \mathbf{v}_\alpha + \boldsymbol{\sigma}_\alpha) dv\Big)^{\grave{}} = \int_{\partial\mathcal{P}^t_\alpha} \mathbf{r} \times (\mathbf{T}_\alpha \mathbf{n}) da + \int_{\mathcal{P}^t_\alpha} [\mathbf{r} \times (\rho_\alpha \mathbf{b}_\alpha + \mathbf{m}_\alpha) + \rho_\alpha \mathbf{l}_\alpha + \mathbf{c}_\alpha] dv,$$

where $\boldsymbol{\sigma}_\alpha$ is the orientational momentum, $\mathbf{l}_\alpha$ is the body couple density and $\mathbf{c}_\alpha$ is the growth of angular momentum, due to the action of the other constituents. We might also account for a surface couple, which then would be represented by a couple stress tensor applied to the normal $\mathbf{n}$, but this is neglected for formal simplicity. Since $\{\mathbf{c}_\alpha\}$ represents the set of exchanges we require that

$$\textstyle\sum_\alpha \mathbf{c}_\alpha = \mathbf{0}.$$

We recall that a balance equation of the form

$$\Big(\int_{\mathcal{P}^t_\alpha} \rho_\alpha \phi_\alpha dv\Big)^{\grave{}} = \int_{\mathcal{P}^t_\alpha} \xi_\alpha \, dv$$

implies that

$$\rho_\alpha \grave{\phi}_\alpha = \xi_\alpha - \tau_\alpha \phi_\alpha, \qquad \partial_t(\rho_\alpha \phi_\alpha) + \nabla \cdot (\rho_\alpha \phi_\alpha \mathbf{v}_\alpha) = \xi_\alpha.$$

Hence, since $\grave{\mathbf{r}}$ means the derivative at $\mathbf{X}_\alpha$ fixed, $\grave{\mathbf{r}} = \mathbf{v}_\alpha$ and the balance of angular momentum results in

$$\rho_\alpha (\mathbf{r} \times \grave{\mathbf{v}}_\alpha + \grave{\boldsymbol{\sigma}}_\alpha) = \mathbf{r} \times (\nabla \cdot \mathbf{T}_\alpha + \rho_\alpha \mathbf{b}_\alpha + \mathbf{m}_\alpha) + \rho_\alpha \mathbf{l}_\alpha + \boldsymbol{\Upsilon}_\alpha + \mathbf{c}_\alpha - \tau_\alpha (\mathbf{r} \times \mathbf{v}_\alpha + \boldsymbol{\sigma}_\alpha)$$

and

$$\partial_t [\rho_\alpha (\mathbf{r} \times \mathbf{v}_\alpha + \boldsymbol{\sigma}_\alpha)] + \nabla \cdot [\rho_\alpha (\mathbf{r} \times \mathbf{v}_\alpha + \boldsymbol{\sigma}_\alpha) \otimes \mathbf{v}_\alpha] = \mathbf{r} \times (\nabla \cdot \mathbf{T}_\alpha + \rho_\alpha \mathbf{b}_\alpha + \mathbf{m}_\alpha) + \rho_\alpha \mathbf{l}_\alpha + \boldsymbol{\Upsilon}_\alpha + \mathbf{c}_\alpha,$$

where

$$\boldsymbol{\Upsilon}_\alpha = \epsilon_{ijk} T^\alpha_{kj} \mathbf{e}_i.$$

Hence we find

$$\rho_\alpha \grave{\boldsymbol{\sigma}}_\alpha = \rho_\alpha \mathbf{l}_\alpha + \boldsymbol{\Upsilon}_\alpha + \mathbf{c}_\alpha - \tau_\alpha \boldsymbol{\sigma}_\alpha. \tag{10.29}$$

It is worth remarking that we might have postulated the balance of angular momentum by considering

$$\check{\mathbf{c}}_\alpha = \mathbf{c}_\alpha + \mathbf{r} \times \mathbf{m}_\alpha$$

as the growth of angular momentum. Since $\sum_\alpha \mathbf{m}_\alpha = \mathbf{0}$ then the requirements

$$\textstyle\sum_\alpha \mathbf{c}_\alpha = \mathbf{0}, \qquad \sum_\alpha \check{\mathbf{c}}_\alpha = \mathbf{0}$$

are equivalent. The use of $\check{\mathbf{c}}_\alpha$, instead of $\mathbf{c}_\alpha$, leads to the local balance of orientational momentum in the (equivalent) form

$$\rho_\alpha \dot{\boldsymbol{\sigma}}_\alpha = \rho_\alpha \mathbf{l}_\alpha + \boldsymbol{\Upsilon}_\alpha - \mathbf{r} \times \mathbf{m}_\alpha + \check{\mathbf{c}}_\alpha - \tau_\alpha \boldsymbol{\sigma}_\alpha.$$

It looks more suggestive to have the form (10.29) for the balance of orientational momentum.

As to the Eulerian description, we observe that

$$\begin{aligned}\nabla \cdot [\rho_\alpha (\mathbf{r} \times \mathbf{v}_\alpha) \otimes \mathbf{v}_\alpha]|_i &= \partial_{x_k} [\rho_\alpha (\mathbf{r} \times \mathbf{v}_\alpha)_i v_k^\alpha] = \partial_{x_k} (\rho_\alpha \epsilon_{ijp} r_j v_p^\alpha v_k^\alpha) \\ &= \epsilon_{ijp} r_j \partial_{x_k} (\rho_\alpha v_p^\alpha v_k^\alpha) + \rho_\alpha \epsilon_{ijp} v_p^\alpha v_k^\alpha \partial_{x_k} r_j \\ &= \mathbf{r} \times \nabla \cdot (\rho_\alpha \mathbf{v}_\alpha \otimes \mathbf{v}_\alpha)|_i + \epsilon_{ijp} (v_k^\alpha \partial_{x_k} r_j) \rho_\alpha v_p^\alpha\end{aligned}$$

and hence

$$\begin{aligned}\partial_t [\rho_\alpha (\mathbf{r} \times \mathbf{v}_\alpha)] + \nabla \cdot [\rho_\alpha (\mathbf{r} \times \mathbf{v}_\alpha) \otimes \mathbf{v}_\alpha] &= \mathbf{r} \times \partial_t (\rho_\alpha \mathbf{v}_\alpha) + \rho_\alpha \partial_t \mathbf{r} \times \mathbf{v}_\alpha \\ &\quad + \mathbf{r} \times \nabla (\rho_\alpha \mathbf{v}_\alpha \otimes \mathbf{v}_\alpha) + (\mathbf{v}_\alpha \cdot \nabla) \mathbf{r} \times \rho_\alpha \mathbf{v}_\alpha.\end{aligned}$$

Moreover, since $\partial_t \mathbf{r} + \mathbf{v}_\alpha \cdot \nabla \mathbf{r} = \mathbf{v}_\alpha$ then

$$\rho_\alpha (\partial_t \mathbf{r} + \mathbf{v}_\alpha \cdot \nabla \mathbf{r}) \times \mathbf{v}_\alpha = \rho_\alpha \mathbf{v}_\alpha \times \mathbf{v}_\alpha = \mathbf{0}.$$

As a consequence, we find that

$$\partial_t (\rho_\alpha \boldsymbol{\sigma}_\alpha) + \nabla \cdot (\rho_\alpha \boldsymbol{\sigma}_\alpha \otimes \mathbf{v}_\alpha) = \rho_\alpha \mathbf{l}_\alpha + \boldsymbol{\Upsilon}_\alpha + \mathbf{c}_\alpha. \tag{10.30}$$

Borrowing from the kinematics of rigid bodies we let $\mathbf{I}_\alpha$ be the symmetric inertia tensor (per unit mass) and assume that

$$\boldsymbol{\sigma}_\alpha = \mathbf{I}_\alpha \boldsymbol{\omega}_\alpha.$$

We also let

$$\rho \mathbf{I} = \textstyle\sum_\alpha \rho_\alpha \mathbf{I}_\alpha$$

define the mean inertia tensor $\mathbf{I}$. The vectors $\{\mathbf{k}_j^\alpha\}$ attached to the particles of the αth constituent evolve in time according to

$$\grave{\mathbf{k}}_j^\alpha = \boldsymbol{\omega}_\alpha \times \mathbf{k}_j^\alpha.$$

Hence we have

$$\grave{\boldsymbol{\sigma}}_\alpha = \mathbf{I}_\alpha \grave{\boldsymbol{\omega}}_\alpha + \boldsymbol{\omega}_\alpha \times \mathbf{I}_\alpha \boldsymbol{\omega}_\alpha.$$

As a consequence,

$$(\boldsymbol{\sigma}_\alpha \cdot \boldsymbol{\omega}_\alpha)\grave{} = 2\, \grave{\boldsymbol{\sigma}}_\alpha \cdot \boldsymbol{\omega}_\alpha.$$

The balance of energy is based on the view that the energy per unit mass, of any constituent α, comprises an internal energy ε_α, a kinetic energy $\mathbf{v}_\alpha^2/2$, and an orientational energy $\boldsymbol{\sigma}_\alpha \cdot \boldsymbol{\omega}_\alpha/2$. Hence we postulate the balance of energy in the form

$$\Big(\int_{\mathcal{P}_\alpha^t} \rho_\alpha(\varepsilon_\alpha + \tfrac{1}{2}\mathbf{v}_\alpha^2 + \tfrac{1}{2}\boldsymbol{\sigma}_\alpha \cdot \boldsymbol{\omega}_\alpha) dv\Big)\grave{} = \int_{\partial\mathcal{P}_\alpha^t} (\mathbf{v}_\alpha \mathbf{T}_\alpha - \mathbf{q}_\alpha) \cdot \mathbf{n}\, da$$
$$+ \int_{\mathcal{P}_\alpha^t} [\rho_\alpha \mathbf{v}_\alpha \cdot \mathbf{b}_\alpha + \rho_\alpha \mathbf{l}_\alpha \cdot \boldsymbol{\omega}_\alpha + \rho_\alpha r_\alpha + e_\alpha] dv,$$

where $\mathbf{q}_\alpha$ is the heat flux, r_α is the external energy supply and e_α is the growth of energy. As any set of growths, the set $\{e_\alpha\}$ is required to satisfy

$$\textstyle\sum_\alpha e_\alpha = 0.$$

In view of the equation of motion and of the balance of orientational momentum (10.29) we find the local form of the balance of energy in the form

$$\rho_\alpha \grave{\varepsilon}_\alpha = \mathbf{T}_\alpha \cdot \mathbf{L}_\alpha - \nabla \cdot \mathbf{q}_\alpha + \rho_\alpha r_\alpha - \boldsymbol{\Upsilon}_\alpha \cdot \boldsymbol{\omega}_\alpha + \hat{\varepsilon}_\alpha, \tag{10.31}$$

where

$$\hat{\varepsilon}_\alpha = e_\alpha - \tau_\alpha \varepsilon_\alpha - \mathbf{m}_\alpha \cdot \mathbf{v}_\alpha + \tfrac{1}{2}\tau_\alpha \mathbf{v}_\alpha^2 - \mathbf{c}_\alpha \cdot \boldsymbol{\omega}_\alpha + \tfrac{1}{2}\tau_\alpha \boldsymbol{\omega}_\alpha \cdot \mathbf{I}_\alpha \boldsymbol{\omega}_\alpha. \tag{10.32}$$

The energy supply $\hat{\varepsilon}_\alpha$ is the energy exchanged, per unit volume, by the αth constituent, with the other constituents, relative to the observer at the αth constituent. It is apparent how it consists of three contributions, related to internal energy, linear momentum and orientational momentum.

As with micropolar continua, we have

$$\mathbf{T}_\alpha \cdot \mathbf{L}_\alpha - \boldsymbol{\Upsilon}_\alpha \cdot \boldsymbol{\omega}_\alpha = \mathbf{T}_\alpha \cdot \mathbf{D}_\alpha + \mathbf{T}_\alpha \cdot \mathbf{W}_\alpha - \boldsymbol{\Upsilon}_\alpha \cdot \boldsymbol{\omega}_\alpha.$$

By means of the axial vector $\mathbf{w}_\alpha$ of $\mathbf{W}_\alpha$,

$$W_{ij}^\alpha = \epsilon_{kji} w_k^\alpha, \qquad \mathbf{w}_\alpha = \tfrac{1}{2}\nabla \times \mathbf{v}_\alpha,$$

we find that

$$\mathbf{T}_\alpha \cdot \mathbf{W}_\alpha - \boldsymbol{\Upsilon}_\alpha \cdot \boldsymbol{\omega}_\alpha = \boldsymbol{\Upsilon}_\alpha \cdot (\mathbf{w}_\alpha - \boldsymbol{\omega}_\alpha),$$

which may be viewed as the power of the skew part of the stress $\mathbf{T}_\alpha$ on the constituent α.

It is of interest that $\tau_\alpha \varepsilon_\alpha$ and $\tau_\alpha \frac{1}{2}(\mathbf{v}_\alpha^2 + \boldsymbol{\omega}_\alpha \cdot \mathbf{I}_\alpha \boldsymbol{\omega}_\alpha)$ occur in the energy supply $\hat{\varepsilon}_\alpha$ with different signs. If $\tau_\alpha > 0$ then $\tau_\alpha \varepsilon_\alpha$ is associated with a decrease of internal energy. Instead, $\tau_\alpha \frac{1}{2}(\mathbf{v}_\alpha^2 + \boldsymbol{\omega}_\alpha \cdot \mathbf{I}_\alpha \boldsymbol{\omega}_\alpha)$ produces an increase of internal energy in that mechanical kinetic energy results in internal energy.

By (10.31) we obtain the Eulerian description of the energy balance equation in the form

$$\begin{aligned}\partial_t(\rho_\alpha \varepsilon_\alpha) + \nabla \cdot (\rho_\alpha \varepsilon_\alpha \mathbf{v}_\alpha) = {} & \mathbf{T}_\alpha \cdot \mathbf{L}_\alpha - \nabla \cdot \mathbf{q}_\alpha + \rho_\alpha r_\alpha - \boldsymbol{\Upsilon}_\alpha \cdot \boldsymbol{\omega}_\alpha + e_\alpha \\ & - \mathbf{m}_\alpha \cdot \mathbf{v}_\alpha + \tfrac{1}{2}\tau_\alpha \mathbf{v}_\alpha^2 - \mathbf{c}_\alpha \cdot \boldsymbol{\omega}_\alpha + \tfrac{1}{2}\tau_\alpha \boldsymbol{\omega}_\alpha \cdot \mathbf{I}_\alpha \boldsymbol{\omega}_\alpha.\end{aligned}$$

10.6.1 Second Law Inequality

The statement of the second law parallels that for mixtures in Sect. 9.3. The balance of entropy in the Lagrangian and the Eulerian descriptions is given by

$$\rho_\alpha \grave{\eta}_\alpha + \tau_\alpha \eta_\alpha + \nabla \cdot \mathbf{j}_\alpha - \rho_\alpha s_\alpha = \sigma_\alpha,$$

and

$$\partial_t(\rho_\alpha \eta_\alpha) + \nabla \cdot (\mathbf{j}_\alpha + \rho_\alpha \eta_\alpha \mathbf{v}_\alpha) - \rho_\alpha s_\alpha = \sigma_\alpha,$$

where $\mathbf{j}_\alpha$ is the entropy flux and s_α is the entropy supply. The entropy growths $\{\sigma_\alpha\}$ enter the statement of the second law as follows.

The inequality

$$\textstyle\sum_\alpha \sigma_\alpha \geq 0 \tag{10.33}$$

must hold, at any point $\mathbf{x} \in \Omega$ *and for any time* $t \in \mathbb{R}$, *for any set of functions on* $\Omega \times \mathbb{R}$ *compatible with the balance equations.*

Again, the entropy supply s_α is identified with r_α/θ_α and the entropy flux $\mathbf{j}_\alpha$ is taken in the form

$$\mathbf{j}_\alpha = \frac{\mathbf{q}_\alpha}{\theta_\alpha} + \mathbf{k}_\alpha,$$

the (unknown) vector $\mathbf{k}_\alpha$ representing the extra-entropy flux.

The analysis of compatibility of constitutive assumptions with the entropy inequality (10.33) is more direct in the Lagrangian description. Hence we write the entropy inequality (10.33) in the form

$$\textstyle\sum_\alpha \frac{1}{\theta_\alpha}[\rho_\alpha \theta_\alpha \grave{\eta}_\alpha + \tau_\alpha \theta_\alpha \eta_\alpha + \nabla \cdot \mathbf{q}_\alpha - \rho_\alpha r_\alpha + \theta_\alpha \nabla \cdot \mathbf{k}_\alpha - \frac{1}{\theta_\alpha}\mathbf{q}_\alpha \cdot \nabla \theta_\alpha] \geq 0.$$

Substitution of $\nabla \cdot \mathbf{q}_\alpha - \rho_\alpha r_\alpha$ from (10.31) yields

$$\sum_\alpha \{\frac{1}{\theta_\alpha}[\rho_\alpha \theta_\alpha \grave{\eta}_\alpha + \tau_\alpha \theta_\alpha \eta_\alpha - \rho_\alpha \grave{\varepsilon}_\alpha + \mathbf{T}_\alpha \cdot \mathbf{L}_\alpha - \boldsymbol{\Upsilon}_\alpha \cdot \boldsymbol{\omega}_\alpha + \hat{\varepsilon}_\alpha - \frac{1}{\theta_\alpha}\mathbf{q}_\alpha \cdot \nabla \theta_\alpha + \theta_\alpha \nabla \cdot \mathbf{k}_\alpha]\} \geq 0.$$

In terms of the Helmholtz free energy

$$\psi_\alpha = \varepsilon_\alpha - \theta_\alpha \eta_\alpha,$$

the inequality can be written in the form

$$\sum_\alpha \{\frac{1}{\theta_\alpha}[-\rho_\alpha(\grave{\psi}_\alpha + \eta_\alpha \grave{\theta}_\alpha) + \mathbf{T}_\alpha \cdot \mathbf{L}_\alpha - \boldsymbol{\Upsilon}_\alpha \cdot \boldsymbol{\omega}_\alpha + \hat{\varepsilon}_\alpha + \tau_\alpha \theta_\alpha \eta_\alpha - \frac{1}{\theta_\alpha}\mathbf{q}_\alpha \cdot \nabla \theta_\alpha + \theta_\alpha \nabla \cdot \mathbf{k}_\alpha]\} \geq 0. \tag{10.34}$$

10.7 Nanofluids

Nanofluids are envisioned to describe suspensions of (solid) nanoparticles (nominally 1-100 nm in size) in conventional base fluids such as water, oils, or glycols. They are fundamentally characterized by the fact that Brownian agitation overcomes any settling motion due to gravity. The term nanofluid then indicates a mixture where the properties of both the nanoparticles and the base fluid contribute to the whole body. In this sense it is natural to view nanofluids as micropolar mixtures with two constituents: one constituent consists of the nanoparticles, the other one is the base fluid. Hence the nanoparticles are a micropolar constituent, the base fluid is a non-polar constituent. The non-polar character of the base fluid is made formal by letting the orientational momentum $\boldsymbol{\sigma}$ and the body couple (or torque) density $\mathbf{l}$ be zero. Moreover the constituents are supposed to be non-reacting, $\tau_\alpha = 0$.

Denote by the subscripts s and f the quantities pertaining to solid nanoparticles and base fluid. We let $\mathbf{b}_s = \mathbf{b}_f = \mathbf{g}$ be the gravity acceleration. Hence the balance equations are

$$\grave{\rho}_s + \rho_s \nabla \cdot \mathbf{v}_s = 0,$$

$$\grave{\rho}_f + \rho_f \nabla \cdot \mathbf{v}_f = 0,$$

$$\rho_s \grave{\mathbf{v}}_s = \nabla \cdot \mathbf{T}_s + \rho_s \mathbf{g} + \mathbf{m}_s,$$

$$\rho_f \grave{\mathbf{v}}_f = \nabla \cdot \mathbf{T}_f + \rho_f \mathbf{g} + \mathbf{m}_f,$$

$$\rho_s \grave{\boldsymbol{\sigma}}_s = \rho_s \mathbf{l}_s + \boldsymbol{\Upsilon}_s + \mathbf{c}_s,$$

$$\mathbf{0} = \boldsymbol{\Upsilon}_f + \mathbf{c}_f,$$

$$\rho_s \grave{\varepsilon}_s = \mathbf{T}_s \cdot \mathbf{L}_s - \nabla \cdot \mathbf{q}_s + \rho_s r_s - \boldsymbol{\Upsilon}_s \omega_s + e_s - \mathbf{m}_s \cdot \mathbf{v}_s - \mathbf{c}_s \cdot \omega_s,$$

$$\rho_f \grave{\varepsilon}_f = \mathbf{T}_f \cdot \mathbf{L}_f - \nabla \cdot \mathbf{q}_f + \rho_f r_f + e_f - \mathbf{m}_f \cdot \mathbf{v}_f,$$

subject to the constraints

$$\mathbf{m}_s + \mathbf{m}_f = \mathbf{0}, \qquad \mathbf{c}_s + \mathbf{c}_f = \mathbf{0}, \qquad e_s + e_f = 0.$$

The two equations of balance of orientational momentum imply that

$$\rho_s \grave{\boldsymbol{\sigma}}_s = \rho_s \mathbf{l}_s + \boldsymbol{\Upsilon}_s + \boldsymbol{\Upsilon}_f,$$

which means that the rotation of nanoparticles is governed by the (external) body couple density $\mathbf{l}_s$ and the skew part of the total stress, $\boldsymbol{\Upsilon}_s + \boldsymbol{\Upsilon}_f$. Due to the dispersion of the nanoparticles in the fluid, it is reasonable to assume that the nanoparticles are an inviscid, micropolar fluid while the base fluid is viscous.

Since ω_f is undefined it is consistent to let $\boldsymbol{\Upsilon}_f = \mathbf{0}$, which follows also as a thermodynamic requirement from (10.34) in view of the arbitrariness of $\mathbf{w}_f$. In turn it follows

$$\mathbf{c}_f = \mathbf{0}, \qquad \mathbf{c}_s = \mathbf{0}.$$

Though $\boldsymbol{\Upsilon}_s = \mathbf{0}$ would be allowed we assume $\boldsymbol{\Upsilon}_s = \zeta(\mathbf{w}_s - \omega_s)$. Hence we let

$$\mathbf{T}_s = -p_s \mathbf{1} + \mathbf{T}_s^{\mathrm{skw}}, \quad \boldsymbol{\Upsilon}_s = \zeta(\mathbf{w}_s - \omega_s), \qquad \mathbf{T}_f = -p_f \mathbf{1} + 2\mu_f \mathbf{D}_f + \lambda_f \mathrm{tr}\, \mathbf{D}_f \mathbf{1},$$

where $\mathbf{D}_f$ is the stretching of the fluid while p_s and p_f are the partial pressures. Hence we can write the balances of energy in the forms

$$\rho_s \grave{\varepsilon}_s = -p_s \nabla \cdot \mathbf{v}_s + \zeta(\mathbf{w}_s - \omega_s)^2 - \nabla \cdot \mathbf{q}_s + \rho_s r_s + e_s - \mathbf{m}_s \cdot \mathbf{v}_s,$$

$$\rho_f \grave{\varepsilon}_f = -p_f \nabla \cdot \mathbf{v}_f + \mathbf{T}_f \cdot \mathbf{D}_f - \nabla \cdot \mathbf{q}_f + \rho_f r_f - e_s - \mathbf{m}_f \cdot \mathbf{v}_f.$$

It is reasonable to regard $\mathbf{m}_s$ and $\mathbf{m}_f$ as interaction forces modelling the friction between the two constituents and then we let

$$\mathbf{m}_s = \nu(\mathbf{v}_f - \mathbf{v}_s) = -\mathbf{m}_f.$$

Hence it follows that

$$\rho_s \grave{\varepsilon}_s = -p_s \nabla \cdot \mathbf{v}_s + \zeta(\mathbf{w}_s - \omega_s)^2 - \nabla \cdot \mathbf{q}_s - \nu(\mathbf{v}_f - \mathbf{v}_s) \cdot \mathbf{v}_s + \rho_s r_s + e_s,$$

$$\rho_f \grave{\varepsilon}_f = -p_f \nabla \cdot \mathbf{v}_f + 2\mu_f \mathbf{D}_f \cdot \mathbf{D}_f + \lambda_f (\mathrm{tr}\, \mathbf{D}_f)^2 - \nabla \cdot \mathbf{q}_f + \nu(\mathbf{v}_f - \mathbf{v}_s) \cdot \mathbf{v}_f + \rho_f r_f - e_s.$$

This model is compatible with thermodynamics. Let

$$\psi_\alpha = \psi_\alpha(\theta_\alpha, \rho_\alpha)$$

and recall that, since $\tau_\alpha = 0$, then $\hat{\varepsilon}_\alpha$ simplifies to

$$\hat{\varepsilon}_\alpha = e_\alpha - \mathbf{m}_\alpha \cdot \mathbf{v}_\alpha.$$

Further,

$$-\mathbf{m}_s \cdot \mathbf{v}_s - \mathbf{m}_f \cdot \mathbf{v}_f = \nu(\mathbf{v}_f - \mathbf{v}_s)^2.$$

Hence inequality (10.34) yields

$$\eta_\alpha = -\partial_{\theta_\alpha}\psi_\alpha.$$

Let $\theta_s = \theta_f = \theta$. The entropy inequality reduces to

$$-\rho_s\partial_{\rho_s}\psi_s\grave{\rho}_s - \rho_f\partial_{\rho_f}\psi_f\grave{\rho}_f - p_s\nabla\cdot\mathbf{v}_s - p_f\nabla\cdot\mathbf{v}_f + 2\mu_f\mathbf{D}_f\cdot\mathbf{D}_f + \lambda_f(\mathrm{tr}\,\mathbf{D}_f)^2$$
$$+\zeta(\mathbf{w}_s - \boldsymbol{\omega}_s)^2 + \nu(\mathbf{v}_f - \mathbf{v}_s)^2 - \frac{1}{\theta}(\mathbf{q}_s + \mathbf{q}_f)\cdot\nabla\theta + \theta\nabla\cdot\mathbf{k} \geq 0.$$

Replace $\grave{\rho}_s$ and $\grave{\rho}_f$ with $-\rho_s\nabla\cdot\mathbf{v}_s$ and $-\rho_f\nabla\cdot\mathbf{v}_f$. The linearity of $\nabla\cdot\mathbf{v}_s$ and $\nabla\cdot\mathbf{v}_f$ implies that the standard relation has to hold for the partial pressures,

$$p_s = \rho_s^2\partial_{\rho_s}\psi_s, \qquad p_f = \rho_f^2\partial_{\rho_f}\psi_f.$$

The reduced inequality holds if $\mathbf{k} = \mathbf{0}$ and

$$\mu_f \geq 0, \qquad 2\mu_f + 3\lambda_f \geq 0, \qquad \nu \geq 0, \qquad \zeta \geq 0,$$

and

$$\mathbf{q}_s + \mathbf{q}_f = -\kappa\nabla\theta, \qquad \kappa \geq 0;$$

μ_f, λ_f may depend on $\theta, \rho_f, \nabla\theta, \mathbf{D}_f$ while ν, κ may depend on $\theta, \rho_s, \rho_f, \nabla\theta, \mathbf{D}_f$ and ζ may depend on θ, ρ_s.

Evolution problems consist in the determination of the fields $\rho_s, \rho_f, \mathbf{v}_s, \mathbf{v}_f, \boldsymbol{\omega}_s, \theta, e_s$.

10.7.1 *Brownian Diffusion and Thermophoresis*

Brownian diffusion results from continuous collisions between the nanoparticles and the molecules of the base fluid [72]. Correspondingly, the nanoparticle mass flux due to Brownian diffusion is modelled by letting a term of the form

$$\mathbf{h}_s = -\rho D \nabla \varphi,$$

where φ is the volume fraction of the nanoparticles. This term is merely the standard model of diffusion possibly replacing the volume fraction with the mass fraction.

In nanofluids it is customary to observe the thermophoretic effect, namely the appearance of a force on the particles when the temperature of the surrounding medium (gas) is not uniform. For particles whose diameters are smaller than the gas mean free path, the thermophoretic force depends on the temperature gradient in the surrounding gas molecules. Since a higher temperature means a larger kinetic energy of the molecules then the greater momentum received from the region at higher temperature results in a movement of the nanoparticles toward the cooler region. For particles whose diameters are larger than the gas mean free path the mechanism is more complicated because a temperature gradient is established within the particle. This in turn increases heat conduction and hence the temperature gradient in the gas decreases. In any case, the thermophoretic effect is modelled by letting the force, acting on the particles, be proportional to the temperature gradient in the gas.

On the whole, the diffusion flux $\mathbf{h}_s$ is taken to be of the form

$$\mathbf{h}_s = -\rho \omega_s \omega_f D' \nabla \theta - \rho D \nabla \omega_s,$$

where D' is the thermal diffusion coefficient or thermophoretic mobility and ω_s, ω_f are the mass fractions [72, 118].

Within the theory of mixtures the diffusion flux is required to be consistent with Eq. (9.68) and hence a model for the thermophoretic effect should be established via the interaction force $\mathbf{m}_s$. However, we cannot merely let

$$\mathbf{m}_s = \lambda \nabla \theta$$

because then we should let $\mathbf{m}_f = -\lambda \nabla \theta$ and account for the second law inequality. Instead, we can consider the standard model of the velocity differences and set

$$\mathbf{m}_s = M_{sf} \rho_s \rho_f (\mathbf{v}_f - \mathbf{v}_s).$$

As shown in Sect. 9.4.2, in binary mixtures we have the direct relation

$$\mathbf{m}_s = -M_{sf} \rho \mathbf{h}_s.$$

By (9.71) we have

$$\dot{\mathbf{h}}_s + M_{sf} \rho \mathbf{h}_s = \nabla \cdot \mathbf{T}_s - \omega_s \nabla \cdot \mathbf{T}.$$

In stationary conditions ($\dot{\mathbf{h}}_s = \mathbf{0}$) we have

$$M_{sf} \rho \mathbf{h}_s = -\nabla p_s - \omega_s \nabla \cdot \mathbf{T} \simeq -\nabla p_s,$$

the approximation being reasonable for small values of ω_s. Hence if $p_s = p_s(\rho_s, \theta)$ and $\rho = \rho_s + \rho_f$ is (assumed to be) constant then[11]

$$\nabla p_s = \partial_{\rho_s} p_s \, \rho \nabla \omega_s + \partial_\theta p_s \, \nabla \theta.$$

Accordingly

$$\mathbf{h}_s = -\frac{1}{M_{sf}\rho}(\rho \partial_{\rho_s} p_s \, \nabla \omega_s + \partial_\theta p_s \, \nabla \theta).$$

Upon the assumption of negligible values of $\nabla \omega_s$ the corresponding thermophoretic velocity $\mathbf{u}_s$ is then established as

$$\mathbf{u}_s = \frac{1}{\rho_s}\mathbf{h}_s = -\frac{1}{M_{sf}} \frac{\partial_\theta p_s}{\rho \rho_s} \nabla \theta.$$

These relations for the mass flux of nanoparticles and the thermophoretic velocity are found within the approximation of stationary fluxes in diluted nanofluids.

10.7.2 Model Approximations and Nanofluid Properties

In the current literature the properties of nanofluids are often determined by regarding the base fluid as incompressible while the particles have a finite extent. Hence, we let

$$\mathbf{T}_f = -p_f \mathbf{1} + 2\mu_f \mathbf{D}_f,$$

and say that, outside the particles, $\mathbf{v}_f$ is subject to

$$\nabla \cdot \mathbf{v}_f = 0$$

$$\rho_f \dot{\mathbf{v}}_f = -\nabla p_f + \mu_f \Delta \mathbf{v}_f.$$

The interaction force $\mathbf{m}_f$ is taken to be zero in the region occupied by the fluid. The interaction force and torque occur only at the surface of the particles. Hence the equation of motion for the particle is

$$\rho_s V_s \dot{\mathbf{v}}_s = \int_{\partial \mathcal{V}_s} \mathbf{T}_f \mathbf{n} \, da + \rho_s \, V_s \, \mathbf{g},$$

where $\mathcal{V}_s$ is the region occupied by the particle, with volume V_s, $\mathbf{g}$ is the gravity acceleration, $\rho_s \, V_s \, \mathbf{g}$ is the weight of the particle. The rotational motion, relative to the centre of mass $\mathbf{x}_0$, is governed by

[11] This approach parallels the argument given in connection with the Soret effect, see (9.102).

$$\rho_s V_s \dot{\boldsymbol{\sigma}}_s = \int_{\partial \mathcal{V}_s} (\mathbf{x} - \mathbf{x}_0) \times \mathbf{T}_f \mathbf{n}\, da + \rho_s V_s \mathbf{l}_s.$$

By the divergence and mean theorems we have

$$\int_{\partial \mathcal{V}_s} \mathbf{T}_f \mathbf{n}\, da = \int_{\mathcal{V}_s} \nabla \cdot \mathbf{T}_f\, dv = V_s \nabla \cdot \mathbf{T}_f(\tilde{\mathbf{x}}),$$

$\tilde{\mathbf{x}}$ being a point in $\mathcal{V}_s$. Dividing by V_s and letting the diameter of $\mathcal{V}_s$ approach zero we find the local equation

$$\rho_s \dot{\mathbf{v}}_s = \nabla \cdot \mathbf{T}_f + \rho_s \mathbf{g}.$$

Likewise,

$$\int_{\partial \mathcal{V}_s} (\mathbf{x} - \mathbf{x}_0) \times \mathbf{T}_f \mathbf{n}\, da = \int_{\mathcal{V}_s} [\boldsymbol{\Upsilon}_f + (\mathbf{x} - \mathbf{x}_0) \times \nabla \cdot \mathbf{T}_f] dv.$$

The components of $\boldsymbol{\Upsilon}_f + (\mathbf{x} - \mathbf{x}_0) \times \nabla \cdot \mathbf{T}_f$ are V_s times appropriate values in $\mathcal{V}_s$. Dividing by V_s and letting the diameter of $\mathcal{V}_s$ approach zero we find the expected relation

$$\rho_s \dot{\boldsymbol{\sigma}}_s = \boldsymbol{\Upsilon}_f + \rho_s \mathbf{l}_s.$$

For a finite domain $\mathcal{V}_s$ we might apply also forces and torques related to $\mathbf{v}_f - \mathbf{v}_s$.

We now go back to the mixture approach. If the fluid is incompressible then the properties of the nanofluid are consistently described by the volume fraction φ of the particles, and $1 - \varphi$ of the fluid. Letting ρ_f^0, ρ_s^0 be the constant intrinsic mass densities we have

$$\rho_s = \varphi \rho_s^0, \qquad \rho_f = (1 - \varphi) \rho_f^0.$$

However, while $\nabla \cdot \mathbf{v}_f = 0$ in the current literature in the fluid region, where $\varphi = 0$, in the mixture model we have to regard $\varphi \in [0, 1]$ then $\mathbf{v}_f$ and $\mathbf{v}_s$ need not be solenoidal. The continuity equations are given the form

$$\partial_t \varphi + \nabla \cdot (\varphi \mathbf{v}_s) = 0,$$

$$-\partial_t \varphi + \nabla \cdot ((1 - \varphi) \mathbf{v}_f) = 0,$$

whence

$$\nabla \cdot [\varphi \mathbf{v}_s + (1 - \varphi) \mathbf{v}_f] = 0.$$

The equations of motion read

$$\rho_s^0 \varphi \dot{\mathbf{v}}_s = -\nabla p_s + \varphi \rho_s^0 \mathbf{g} + \mathbf{m}_s,$$

$$\rho_f^0 \varphi \dot{\mathbf{v}}_f = -\nabla p_f + \mu_f (\Delta \mathbf{v}_f + \nabla \nabla \cdot \mathbf{v}_f) + (1 - \varphi) \rho_s^0 \mathbf{g} - \mathbf{m}_s,$$

where

$$p_s = p_s(\theta, \varphi\rho_s^0), \qquad p_f = p_f(\theta, (1-\varphi)\rho_f^0).$$

The balance of orientational momentum is written as

$$\mathbf{I}_s\dot{\boldsymbol{\omega}}_s + \boldsymbol{\omega}_s \times \mathbf{I}_s\boldsymbol{\omega}_s = \mathbf{l}_s.$$

Finally, look at the balance of energy under the assumption that the mixture has a single temperature θ. Regard ε_s and ε_f as functions of θ and the corresponding densities ρ_s and ρ_f. Since $p_s = \rho_s^2\partial_{\rho_s}\psi_s$ we have

$$\rho_s\partial_{\rho_s}\varepsilon_s\,\dot{\rho}_s = \rho_s\partial_{\rho_s}(\psi_s + \theta\eta_s)(-\rho_s\nabla\cdot\mathbf{v}_s) = -p_s\nabla\cdot\mathbf{v}_s - \rho_s^2\theta\partial_{\rho_s}\eta_s\nabla\cdot\mathbf{v}_s,$$

the analogue being true for ε_f and ρ_f. Hence summation of the energy equations and the constraint $e_s + e_f = 0$ yield

$$\begin{aligned}\partial_\theta(\rho_s\varepsilon_s + \rho_f\varepsilon_f)\dot{\theta} + \partial_\theta(\rho_s\varepsilon_s)\mathbf{u}_s\cdot\nabla\theta + \partial_\theta(\rho_f\varepsilon_f)\mathbf{u}_f\cdot\nabla\theta = \rho_s^2\theta\partial_{\rho_s}\eta_s\nabla\cdot\mathbf{v}_s + \rho_f^2\theta\partial_{\rho_f}\eta_f\nabla\cdot\mathbf{v}_f\\ -\nabla\cdot(\mathbf{q}_s + \mathbf{q}_f) + \rho_f r_f + \rho_s r_s + 2\mu_f\mathbf{D}_f\cdot\mathbf{D}_f + 2\lambda(\hat{\omega}_f - \omega_s)^2 + \nu(\mathbf{v}_f - \mathbf{v}_s)^2,\end{aligned}$$

where $\rho_s = \varphi\rho_s^0$ and $\rho_f = (1-\varphi)\rho_f^0$. One of the two balances of energy provides the energy growth e_s needed to keep the temperatures θ_s, θ_f at the common value θ. The equation for ε_s results in

$$\rho_s\partial_\theta\varepsilon_s\,(\partial_t + \mathbf{v}_s\cdot\nabla\theta) + \rho_s^2\partial_{\rho_s}\theta\eta_s\nabla\cdot\mathbf{v}_s = -\nabla\cdot\mathbf{q}_s + \rho_s r_s + e_s + \nu(\mathbf{v}_s - \mathbf{v}_f)\cdot\mathbf{v}_s + 2\lambda(\boldsymbol{\omega}_s - \hat{\boldsymbol{\omega}}_f)\cdot\boldsymbol{\omega}_s.$$

We have then a system of partial differential equations for the unknowns

$$\varphi, \mathbf{v}_s, \mathbf{v}_f, \boldsymbol{\omega}_s, \theta, e_s.$$

The density of the nanofluid ρ_{nf} is taken as

$$\rho_{nf} = \varphi\rho_s + (1-\varphi)\rho_f.$$

Now, apart from the terms

$$\partial_\theta(\rho_s\varepsilon_s)\mathbf{u}_s\cdot\nabla\theta, \quad \partial_\theta(\rho_f\varepsilon_f)\mathbf{u}_f\cdot\nabla\theta$$

arising from diffusion, the common value of the temperature, which is kept by an appropriate field e_s of energy growth, shows that the balance of energy holds as though the mixture had an equivalent internal energy

$$\rho_s\varepsilon_s + \rho_f\varepsilon_f.$$

We can then say that the heat capacity $\rho c = \rho\partial_\theta\varepsilon$ of the nanofluid is (approximately) given by[12]

$$\rho_{nf}c_{nf} = \varphi\rho_s^0 c_s + (1-\varphi)\rho_f^0 c_f.$$

Numerical values of the shear viscosity μ_{nf} of the nanofluid are taken to be given by the shear viscosity μ_f of the fluid by, e.g., the relation [67]

$$\nu_{nf} = \frac{\nu_f}{(1-\varphi)^{2.5}}$$

The thermal conductivity of the base fluid (e.g. water, oil) is very low. Usually the nanoparticles (e.g. Cu, Al_2O_3, TiO_2) have a very high conductivity, which results in an improved heat conduction even with low nanoparticle concentration. For definiteness, the thermal conductivity is approximated by the Maxwell-Garnett formula ([306], p. 364)

$$\kappa = \kappa_f \frac{\kappa_s + 2\kappa_f + 2\varphi(\kappa_s - \kappa_f)}{\kappa_s + 2\kappa_f - \varphi(\kappa_s - \kappa_f)},$$

valid for spherical particles. Heat conductivity may be further improved by letting the particles be anisotropic and controlling the orientation by means of the body couple density $\boldsymbol{\tau}_s$.

The literature shows a wide use of a very simple model where the nanofluid is viewed as an appropriate single fluid where micropolar effects are ignored ([72, 106]). Let $\mathbf{v}$ be the velocity of the nanofluid. The balance of mass and linear momentum are taken as in the case of a viscous fluid

$$\partial_t\rho + \nabla\cdot(\rho\mathbf{v}) = 0, \qquad \partial_t(\rho\mathbf{v}) + \nabla\cdot(\rho\mathbf{v}\otimes\mathbf{v}) = \nabla\cdot\mathbf{T} + \rho\mathbf{b},$$

where $\mathbf{b}$ is identified with the acceleration gravity while $\mathbf{T} = -p\mathbf{1} + 2\mu_{nf}\mathbf{D} + \frac{2}{3}\mu_{nf}\nabla\cdot\mathbf{v}\mathbf{1}$. The balance of energy is simplified to

$$c[\partial_t(\rho_s\theta) + \nabla\cdot(\rho_s\mathbf{v}\theta)] = \nabla\cdot(k\nabla\theta)$$

where c is the specific heat and k is the heat conductivity, of the nanofluid. The mixture character of the nanofluid is made formal by the diffusion equation in the form

$$\partial_t(\rho\omega) + \nabla\cdot(\rho\omega\mathbf{v}) = -\nabla\cdot\mathbf{J}_p,$$

where $\mathbf{J}_p = -\rho_{nf}(D_B\nabla\omega + D_T\nabla\theta)$ is the nanoparticles diffusion flux and D_B and D_T are the Brownian and thermophoretic diffusion coefficients. In this way a system is available for the unknowns ρ, $\mathbf{v}$, θ, and ω. Observe that

$$\partial_t(\rho\omega) + \nabla\cdot(\rho\omega\mathbf{v}) = \omega[\partial_t\rho + \nabla\cdot(\rho\mathbf{v})] + \rho[\partial_t\omega + \mathbf{v}\cdot\nabla\omega] = \rho\dot{\omega}.$$

---

[12] See, e.g., [254].

Hence the essential approximations of this system are

$$\varepsilon = c\theta, \quad \mathbf{q} = -k\nabla\theta, \quad \mathbf{J}_p = -\rho(D_B\nabla\omega + D_T\nabla\theta).$$

A further approximated model is sometimes used where the fluid is assumed to be incompressible and the convection steady. The balance of mass and linear momentum are written in the form

$$\nabla \cdot \mathbf{v} = 0, \qquad \rho\dot{\mathbf{v}} = -\nabla p + 2\mu\nabla \cdot \mathbf{D},$$

where p stands for the nanofluid pressure. The condition $\nabla \cdot \mathbf{v} = 0$ is questionable; even with incompressible constituents it is $\nabla \cdot \mathbf{v} \neq 0$. The balance of energy is based on the balance Eq. (9.58); only the nanoparticle flux is considered and hence the balance is taken in the form

$$\rho\dot{\varepsilon} = -\nabla \cdot (\mathbf{q}_l + \chi_s\mathbf{h}_s),$$

where $\chi_s = \varepsilon_s + p_s/\rho_s$ is the enthalpy of the particles. Moreover it is assumed that $\rho\dot{\varepsilon} = \rho c\dot{\theta}$ and

$$\mathbf{q}_l = -k\nabla\theta, \qquad \mathbf{q} = \mathbf{q}_l + \chi_s\mathbf{h}_s.$$

Upon substitution and the further approximation that $\nabla \cdot (\chi_s\mathbf{h}_s) \simeq \mathbf{h}_s \cdot \nabla\chi_s$ the balance of energy is written as [7, 72]

$$\rho c\dot{\theta} = \nabla \cdot (k\nabla\theta) - \mathbf{h}_s \cdot \nabla\chi_s.$$

Assuming χ_s is a function of θ completes the scheme.

Chapter 11
Porous Materials

A porous material (medium) is a substance that contains pores (voids), or spaces, between solid materials through which liquid or gas can flow. Examples of naturally occurring porous media include sand, soil, and some types of stone. Sponges, ceramics, and reticulated foam are also manufactured for porous materials. A porous material is characterized by its porosity, or the size of its pores. Materials with low porosity are less permeable and have smaller pores, making it more difficult for gas and/or liquid to flow through them.

It is natural to model porous materials as mixtures of a deformable solid and some fluids (gas, liquid). Also, the solid may be viewed as incompressible or compressible. With this in mind, the present approach to the modelling of porous materials is based directly on the theory of mixtures. Emphasis is given to the model consisting of a solid (skeletal material) and the remaining constituents as fluids. Next Darcy's law is shown to follow by thermodynamic considerations if stationary conditions hold and the Darcy–Forchheimer equation is reviewed. Within the theory of mixtures special models are established: materials with voids, materials with double porosity, and materials with a thermoelastic solid.

11.1 Porous Materials as Mixtures

To model porous materials we refer to a mixture of n constituents some of which are solids and the remaining ones are fluids. Hence, the kinematic scheme developed in Sect. 9.8 is adopted for the n constituents. The dependence on $(\mathbf{x}, t) \in \Omega \times \mathbb{R}$ is often understood and not written.

Let ρ_α represent the mass of the αth constituent per unit volume of the mixture. We refer to ρ_α as the bulk density. Meanwhile, we call the true density of the αth constituent, denoted by ρ_α^0, the mass of the αth constituent per unit volume of the αth

A. Morro and C. Giorgi, *Mathematical Modelling of Continuum Physics*,
Modeling and Simulation in Science, Engineering and Technology,
https://doi.org/10.1007/978-3-031-20814-0_11

constituent. Let φ_α be the volume fraction occupied by the αth constituent. Hence the two mass densities are related by

$$\rho_\alpha = \varphi_\alpha \, \rho_\alpha^0.$$

Since φ_α represents the volume of the αth constituent, per unit volume of the mixture, then

$$\sum_\alpha \varphi_\alpha \leq 1.$$

If equality holds we say that the mixture is saturated. Otherwise a void fraction occurs and the mixture is said to be unsaturated. If one of the constituents, say $\alpha = 1$, is a solid and the other constituents are fluids then

$$f = \sum_{\alpha=2}^n \varphi_\alpha$$

is the porosity.

Fluid mixtures are miscible if they form homogeneous solutions. Fluid mixtures are immiscible if the constituents rest separately also locally (like oil and water). Mathematically, we say that the mixture is miscible if the constitutive properties of the mixture are independent of the volume fractions. Instead we say that the mixture is immiscible if the properties are affected by the volume fractions φ_α.

The αth constituent is incompressible if the true density ρ_α^0 is a constant whereas ρ_α need not be a constant in that is affected by the motion of the constituent. The mixture is said to be incompressible whenever the true densities $\rho_1^0, \rho_2^0, ..., \rho_n^0$ are constants. Yet the mixture as a single body need not be incompressible even though $\rho_1^0, \rho_2^0, ..., \rho_n^0$ are constants.

It may be convenient to formulate the balance equations in terms of material coordinates of one of the constituents. For definiteness let the solid be the first constituent ($\alpha = 1$) and state the balance of mass relative to the observer following the motion of the solid. By the balance of mass

$$\partial_t \rho_\alpha + \nabla \cdot (\rho_\alpha \mathbf{v}_\alpha) = \tau_\alpha$$

we can write

$$\partial_t \rho_1 + \nabla \cdot (\rho_1 \mathbf{v}_1) = \tau_1$$

and, for $\beta = 2, 3, ..., n$,

$$\partial_t \rho_\beta + \nabla \cdot (\rho_\beta \mathbf{v}_1) + \nabla \cdot [\rho_\beta (\mathbf{v}_\beta - \mathbf{v}_1)] = \tau_\beta.$$

The operator

$$D_1 = \partial_t + \mathbf{v}_1 \cdot \nabla$$

is the peculiar time derivative relative to the first constituent. Hence we have

$$D_1\rho_\beta + \rho_\beta \nabla \cdot \mathbf{v}_1 + \nabla \cdot [\rho_\beta(\mathbf{v}_\beta - \mathbf{v}_1)] = \tau_\beta.$$

Multiplying by $J_1 = \det \mathbf{F}_1 > 0$ and observing that

$$D_1 J_1 = J_1 \nabla \cdot \mathbf{v}_1$$

we can write

$$D_1(\rho_\beta J_1) + J_1 \nabla \cdot [\rho_\beta(\mathbf{v}_\beta - \mathbf{v}_1)] = J_1 \tau_\beta.$$

For any vector field $\mathbf{h}(\mathbf{x}, t)$ the identity

$$J_1 \nabla \cdot \mathbf{h} = \nabla_{\mathbf{X}_1} \cdot (J_1 \mathbf{F}_1^{-1} \mathbf{h}) \tag{11.1}$$

holds where $\nabla_{\mathbf{X}_1}$ is the gradient relative to the reference coordinates $\mathbf{X}_1$ of the first constituent. Hence we have

$$D_1(J_1\rho_\beta) - \nabla_{\mathbf{X}_1} \cdot [J_1 \rho_\beta \mathbf{F}_1^{-1}(\mathbf{v}_\beta - \mathbf{v}_1)] = J_1 \tau_\beta. \tag{11.2}$$

Now, $J_1\rho_\beta$ is the mass of the βth constituent per unit volume in the reference configuration of the first constituent. The quantity $J_1\rho_\beta \mathbf{F}_1^{-1}(\mathbf{v}_\beta - \mathbf{v}_1)$ is the mass flux of the βth constituent into a material element of the first constituent.

The proof of (11.1) is based on a direct application of Nanson formula. For any region $\mathcal{P} \subset \Omega$ we can apply the divergence theorem to obtain

$$\int_{\partial \mathcal{P}} \mathbf{h} \cdot \mathbf{n}\, da = \int_{\mathcal{P}} \nabla \cdot \mathbf{h}\, dv = \int_{\mathrm{P}_1} J_1 \nabla \cdot \mathbf{h}\, dV_1,$$

where $\mathrm{P}_1 = \chi_1^{-1}(\mathcal{P})$. Moreover, by using Nanson formula and the divergence theorem we find

$$\int_{\partial \mathcal{P}} \mathbf{h} \cdot \mathbf{n}\, da = \int_{\partial \mathrm{P}_1} J_1 \mathbf{h} \cdot \mathbf{F}_1^{-T} \mathbf{n}_R\, dA_1 = \int_{\mathrm{P}_1} \nabla_{\mathbf{X}_1} \cdot J_1 \mathbf{F}_1^{-1} \mathbf{h}\, dV_1.$$

Consequently

$$\int_{\mathrm{P}_1} J_1 \nabla \cdot \mathbf{h}\, dV_1 = \int_{\mathrm{P}_1} \nabla_{\mathbf{X}_1} \cdot (J_1 \mathbf{F}_1^{-1} \mathbf{h})\, dV_1.$$

The arbitrariness of P_1 provides the result.

11.2 Constitutive Models of Porous Materials

For definiteness we let the first constituent (skeletal material) be a solid and the remaining constituents be fluids. The mixture is immiscible and hence we allow the constitutive properties to be affected by the volume fractions. Denote by the subscript

s the quantities pertaining to the solid (first constituent). Let $\varrho_s = \rho_s^0$ be the reference mass density of the solid,

$$\varrho_s = \rho_s J_s.$$

Hence we let

$$\psi_s = \psi_s(\theta_s, \varrho_s, \mathbf{F}_s, \varphi_s, \nabla\theta_s, \nabla\mathbf{F}_s), \qquad \psi_\beta = \psi_\beta(\theta_\beta, \rho_\beta, \varphi_\beta, \nabla\theta_\beta, \mathbf{D}_\beta),$$

and the like for $\mathbf{T}_s$, $\mathbf{T}_\beta$, $\mathbf{q}_s$, $\mathbf{q}_\beta$, where $\beta = 2, 3, ..., n$. The interaction terms τ_α, $\mathbf{m}_\alpha$, e_α are allowed to depend on all of the volume fractions and the velocity differences. Hence

$$\mathbf{m}_\alpha = \mathbf{m}_\alpha(\theta_s, \varrho_s, \mathbf{F}_s, \nabla\theta_s, \nabla\mathbf{F}_s, \theta_\beta, \rho_\beta, \varphi_\beta, \nabla\theta_\beta, \mathbf{D}_\beta, \mathbf{v}_\alpha - \mathbf{v}_\beta)$$

and the like for τ_α and e_α. Substitution in (9.42) results in

$$\sum\nolimits_\alpha\{-\frac{\rho_\alpha}{\theta_\alpha}(\partial_{\theta_\alpha}\psi_\alpha + \eta_\alpha)\grave{\theta}_\alpha - \frac{\rho_\alpha}{\theta_\alpha}\partial_{\nabla\theta_\alpha}\psi_\alpha \cdot \grave{\overline{\nabla\theta_\alpha}} - \frac{\rho_\beta}{\theta_\beta}\partial_{\mathbf{D}_\beta}\psi_\beta \cdot \grave{\mathbf{D}}_\alpha + ...\} \geq 0,$$

the dots representing quantities independent of $\grave{\theta}_\alpha$, $\grave{\overline{\nabla\theta_\alpha}}$, $\grave{\mathbf{D}}_\beta$. The linear dependence and arbitrariness of $\grave{\theta}_\alpha$, $\grave{\overline{\nabla\theta_\alpha}}$, $\grave{\mathbf{D}}_\beta$, at any point $\mathbf{x}$ and time t, imply that the entropy inequality holds only if the corresponding coefficients vanish, whence

$$\partial_{\mathbf{D}_\beta}\psi_\beta = \mathbf{0}, \qquad \partial_{\nabla\theta_\alpha}\psi_\alpha = \mathbf{0}, \qquad \eta_\alpha = -\partial_{\theta_\alpha}\psi_\alpha. \tag{11.3}$$

Since

$$\mathbf{L}_s = \grave{\mathbf{F}}_s\mathbf{F}_s^{-1},$$

the remaining inequality can be written as

$$-\frac{\rho_s}{\theta_s}(\partial_{\varrho_s}\psi_s\,\grave{\varrho}_s + \partial_{\mathbf{F}_s}\psi_s \cdot \grave{\mathbf{F}}_s + \partial_{\nabla\mathbf{F}_s}\psi_s \cdot \grave{\overline{\nabla\mathbf{F}_s}}) + \frac{1}{\theta_s}(\mathbf{T}_s\mathbf{F}_s^{-T}) \cdot \grave{\mathbf{F}}_s + ... \geq 0$$

the dots representing quantities independent of $\grave{\mathbf{F}}_s$ and $\grave{\overline{\nabla\mathbf{F}_s}}$. Since

$$\grave{\varrho}_s = J_s\tau_s$$

then $\partial_{\varrho_s}\psi_s\grave{\varrho}_s$ has to be examined along with the other terms involving τ_s. Moreover,

$$\grave{\overline{\nabla\mathbf{F}_s}} = \nabla\grave{\mathbf{F}}_s - \mathbf{L}_s^T\nabla\,\mathbf{F}_s.$$

The arbitrariness of $\nabla\grave{\mathbf{F}}_s$ implies that $\partial_{\nabla\mathbf{F}_s}\psi_s$ has to vanish. Moreover, the arbitrariness of $\grave{\mathbf{F}}_s$ implies that

$$\mathbf{T}_s = \rho_s\partial_{\mathbf{F}_s}\psi_s\,\mathbf{F}_s^T. \tag{11.4}$$

Granted (11.3) and (11.4), we now restrict the analysis of the remaining properties to the case where each constituent has the same temperature, say θ. Since

$$\psi_s = \psi_s(\theta, \varrho_s, \mathbf{F}_s, \varphi_s), \qquad \psi_\beta = \psi_\beta(\theta, \rho_\beta, \varphi_\beta),$$

in view of (11.3) and (11.4) we can write the entropy inequality (9.42) as

$$-\rho_s \partial_{\varphi_s} \psi_s \,\dot{\varphi}_s + \sum_{\beta=2}^n \mathbf{T}_\beta \cdot \mathbf{L}_\beta - \rho_\beta(\partial_{\rho_\beta}\psi_\beta \,\dot{\rho}_\beta + \partial_{\varphi_\beta}\psi_\beta \,\dot{\varphi}_\beta) - \rho_s \partial_{\varrho_s}\psi_s \, J_s \tau_s$$
$$+\sum_{\alpha=1}^n [-\tau_\alpha(\psi_\alpha - \tfrac{1}{2}\mathbf{v}_\alpha^2) + e_\alpha - \mathbf{m}_\alpha \cdot \mathbf{v}_\alpha - \frac{1}{\theta_\alpha}\mathbf{q}_\alpha \cdot \nabla\theta_\alpha + \theta_\alpha \nabla \cdot \mathbf{k}_\alpha] \geq 0.$$

Since $\sum_\alpha e_\alpha = 0$ then we let

$$\sum_\alpha [-\tau_\alpha \tfrac{1}{2}\mathbf{v}_\alpha^2 + \mathbf{m}_\alpha \cdot \mathbf{v}_\alpha] =: \Phi(\{\mathbf{v}_\alpha\})$$

as in the general case of mixtures.

The formal novelty of the modelling of porous media is the occurrence of the time derivative of the volume fractions φ_α. Very often the volume fractions are taken to obey a rate law as with materials with internal variables. The assumption then might be

$$\dot{\varphi}_\alpha = \hat{\varphi}_\alpha(\theta, \varrho_s, \mathbf{F}_s, \nabla\theta, \varphi_s, \nabla\mathbf{F}_s, \rho_\beta, \varphi_\beta, \mathbf{D}_\beta, \mathbf{v}_\alpha - \mathbf{v}_\beta).$$

Let

$$\mathbf{T}_\beta = -p_\beta(\theta, \rho_\beta, \varphi_\beta)\mathbf{1} + \tilde{\mathbf{T}}_\beta,$$

$\tilde{\mathbf{T}}_\beta$ approaching zero as $\mathbf{L}_\beta$ approaches zero. Substitution of $\dot{\rho}_\beta = -\rho_\beta \nabla \cdot \mathbf{v}_\beta + \tau_\beta$ yields

$$-\sum_\alpha \rho_\alpha \partial_{\varphi_\alpha}\psi_\alpha \,\hat{\varphi}_\alpha - \sum_\beta (p_\beta - \rho_\beta^2 \partial_{\rho_\beta}\psi_\beta)\nabla \cdot \mathbf{v}_\beta - \tau_s \psi_s - \varrho_s \partial_{\varrho_s}\psi_s \tau_s$$
$$-\sum_\beta \tau_\beta(\psi_\beta + \rho_\beta \partial_{\rho_\beta}\psi_\beta) - \Phi + \sum_\beta \tilde{\mathbf{T}}_\beta \cdot \mathbf{L}_\beta - \sum_\alpha (\mathbf{q}_\alpha \cdot \nabla\theta - \theta \nabla \cdot \mathbf{k}_\alpha) \geq 0$$

For definiteness, let $\nabla \cdot \mathbf{k}_\alpha$ and Φ be independent of the velocity gradients. Hence the inequality holds only if

$$p_\beta = \rho_\beta^2 \partial_{\rho_\beta}\psi_\beta, \qquad \beta = 2, 3, ..., n.$$

Let

$$\mu_\alpha = \begin{cases} \psi_s + \varrho_s \partial_{\varrho_s}\psi_s, & \alpha = 1, \\ \psi_\beta + p_\beta/\rho_\beta, & \alpha = \beta = 2, 3, ..., n. \end{cases}$$

The remaining inequality can be written as

$$-\sum_\alpha (\rho_\alpha \partial_{\varphi_\alpha}\psi_\alpha \,\hat{\varphi}_\alpha + \mu_\alpha \tau_\alpha) - \Phi + \sum_\beta \tilde{\mathbf{T}}_\beta \cdot \mathbf{L}_\beta - \sum_\alpha (\mathbf{q}_\alpha \cdot \nabla\theta - \theta \nabla \cdot \mathbf{k}_\alpha) \geq 0. \tag{11.5}$$

Simple models arise by letting each term in (11.5) be non-negative. Such is the case if $\mathbf{k}_\alpha = \mathbf{0}$, $\tilde{\mathbf{T}}_\beta$ and $\mathbf{q}_\alpha$ are given as in the Navier–Stokes–Fourier model, Φ is non-positive, and

$$\sum_\alpha \mu_\alpha \tau_\alpha \leq 0,$$

$$\sum_\alpha \rho_\alpha \partial_{\varphi_\alpha} \psi_\alpha \hat{\varphi}_\alpha \leq 0. \tag{11.6}$$

11.2.1 Darcy's Law

Darcy's law describes the flow of a fluid through a porous medium. The law was established by H. Darcy on the basis of experiments on the flow in stationary conditions. In homogeneous rectilinear flows, if $p(x_b) - p(x_a)$, $p(x_b) < p(x_a)$, is the pressure difference across a cylinder of section A and length L the mass Q flowing per unit time, from x_a to x_b, is

$$Q = \frac{kA}{\mu L}(p_b - p_a);$$

k is the permeability and μ is the viscosity of the fluid. In local form the law for the flux, or relative velocity,[1] $\mathbf{h}$ is assumed in the form

$$\mathbf{h} = -\frac{k}{\mu}\nabla p. \tag{11.7}$$

We now look for a derivation of Darcy's law within continuum mechanics.

The equation of motion for any (fluid) constituent reads

$$\rho_\beta \grave{\mathbf{v}}_\beta = \nabla \cdot \mathbf{T}_\beta + \rho_\beta \mathbf{b}_\beta + \mathbf{m}_\beta - \tau_\beta \mathbf{v}_\beta.$$

We specialize the model by letting $\tau_\beta = 0$ (non-reacting fluid) and letting $\mathbf{T}_\beta$ be a pressure tensor, $\mathbf{T}_\beta = -p_\beta \mathbf{1}$. In stationary conditions, we have

$$\mathbf{0} = -\nabla p_\beta + \rho_\beta \mathbf{b}_\beta + \mathbf{m}_\beta. \tag{11.8}$$

The standard model of the growths $\mathbf{m}_\beta$,

$$\mathbf{m}_\beta = \sum_\alpha \mathbf{M}_{\beta\alpha}(\mathbf{v}_\alpha - \mathbf{v}_\beta),$$

satisfies

$$\sum_\beta \mathbf{m}_\beta = \mathbf{0}, \qquad \Phi \leq 0$$

[1] In many contexts of application of Darcy's law the flux is denoted by $\mathbf{q}$.

provided each tensor $\mathbf{M}_{\beta\alpha}$ is symmetric and positive definite. This is so because $\mathbf{M}_{\beta\alpha}(\mathbf{v}_\alpha - \mathbf{v}_\beta)$ is skew in α, β and hence

$$\sum_\beta \mathbf{m}_\beta = \sum_{\alpha,\beta} \mathbf{M}_{\beta\alpha}(\mathbf{v}_\alpha - \mathbf{v}_\beta) = \mathbf{0}.$$

Moreover,

$$\sum_\beta \mathbf{m}_\beta \cdot \mathbf{u}_\beta = \sum_{\alpha,\beta} \mathbf{u}_\beta \cdot \mathbf{M}_{\alpha\beta}(\mathbf{u}_\alpha - \mathbf{u}_\beta) = \sum_{\alpha,\beta} \mathbf{u}_\beta \cdot \mathbf{M}_{\beta\alpha}\mathbf{u}_\alpha - \tfrac{1}{2}\sum_{\alpha,\beta} \mathbf{u}_\beta \cdot \mathbf{M}_{\beta\alpha}\mathbf{u}_\beta$$
$$-\tfrac{1}{2}\sum_{\alpha,\beta} \mathbf{u}_\alpha \cdot \mathbf{M}_{\beta\alpha}\mathbf{u}_\alpha = -\tfrac{1}{2}\sum_{\alpha,\beta} (\mathbf{u}_\alpha - \mathbf{u}_\beta) \cdot \mathbf{M}_{\beta\alpha}(\mathbf{u}_\alpha - \mathbf{u}_\beta).$$

Hence $\Phi \leq 0$ if and only if each tensor $\mathbf{M}_{\alpha\beta}$ is positive semi-definite.

The next step is the assumption that

$$\mathbf{M}_{\beta\alpha} \neq \mathbf{0} \quad \text{only if} \quad \alpha = s = 1.$$

Consequently

$$\mathbf{m}_\beta = \mathbf{M}_{\beta s}(\mathbf{v}_s - \mathbf{v}_\beta),$$

whence

$$\mathbf{v}_\beta - \mathbf{v}_s = -\mathbf{M}_{\beta s}^{-1}\mathbf{m}_\beta,$$

and, by (11.8),

$$\mathbf{v}_\beta - \mathbf{v}_s = -\mathbf{M}_{\beta s}^{-1}[-\nabla p_\beta + \rho_\beta \mathbf{b}_\beta].$$

It is reasonable to let $\mathbf{M}_{\beta s}$ be proportional to the volume fraction φ_β,

$$\mathbf{M}_{\beta s} = \varphi_\beta \mathbf{\Lambda}_\beta.$$

Hence, we can write

$$\mathbf{v}_\beta - \mathbf{v}_s = -\frac{\mathbf{\Lambda}_\beta}{\varphi_\beta}[\nabla p_\beta - \rho_\beta \mathbf{b}_\beta]. \tag{11.9}$$

The tensor $\mathbf{\Lambda}_\beta$ is called the *mobility* of phase β.

The local form of Darcy's law follows by neglecting the body force and letting $\mathbf{\Lambda}_\beta = \varphi_\beta(k/\mu)\mathbf{1}$ while $\mathbf{h} = \mathbf{v}_\beta - \mathbf{v}_s$.

11.2.2 Darcy-Like Equations

In modelling the unsaturated groundwater flow, it is common to apply Darcy's law along with some approximations. It is assumed that water and air flow within the soil but the flow equation for air is neglected and the air pressure is taken to be equal to the atmospheric pressure at the surface, p_A. The volume fraction of water, say φ_w, is also called the soil's moisture content (or water content). Let ρ_w^0 be the true[2] mass density of water so that

$$\rho_w = \varphi_w \rho_w^0$$

is the mass density. Look at the continuity equation

$$\partial_t(\varphi_w \rho_w^0) = -\nabla \cdot (\varphi_w \rho_w^0 \mathbf{v}_w).$$

The *pressure head* $\mathscr{P}$ in the water is defined by

$$\mathscr{P} = \frac{p_w - p_A}{\rho_w^0 g}.$$

Hence

$$\nabla p_w - \rho_w \mathbf{b}_w = g[\nabla(\rho_w^0 \mathscr{P}) + \varphi_w \rho_w^0 \nabla z].$$

Upon identifying $\mathbf{v}_w - \mathbf{v}_s$ with $\mathbf{v}_w$ and then substituting $\mathbf{v}_w$ from Darcy's law (11.9), we have

$$\partial_t(\varphi_w \rho_w^0) = \nabla \cdot \{\rho_w^0 g \mathbf{\Lambda}_w [\nabla(\rho_w^0 \mathscr{P}) + \varphi_w \rho_w^0 g \nabla z]\}.$$

Letting ρ_w^0 be taken as a constant, we have

$$\partial_t \varphi_w = \nabla \cdot [\mathbf{K}(\nabla \mathscr{P} + \varphi_w \nabla z)].$$

The tensor

$$\mathbf{K} = \rho_w^0 g \mathbf{\Lambda}_w$$

is called the hydraulic conductivity of the soil.

To get an equation in which $\mathscr{P}$ is the unique unknown it is assumed that φ_w is a function of $\mathscr{P}$ so that

$$C(\mathscr{P})\partial_t \mathscr{P} = \nabla \cdot [\mathbf{K}(\mathscr{P})(\nabla \mathscr{P} + \varphi_w(\mathscr{P})\nabla z)],$$

where $C(\mathscr{P}) = d\varphi_w / d\mathscr{P}$ is the *specific moisture capacity.* The one-dimensional version is known as Richards' equation [4, 378].

Equation (11.7) is improved to get a better fit in applications. In geological engineering, the flow is described as

$$\mathbf{h} = -\frac{k^{\text{eff}}}{\mu} \nabla p, \qquad k^{\text{eff}} = k(1 + b/p),$$

b being known as the Klinkenberg parameter; it shows that the flow decreases as the (mean) pressure p increases.

---

[2] Or intrinsic.

Nonlinear effects are modelled by the Darcy–Forchheimer equation. Qualitatively the idea is to allow for a nonlinear term in $\mathbf{q}$. In general the equation reads

$$\nabla p = -\frac{\mu}{k}\mathbf{h} + \rho\mathbf{b} - ck^{-1/2}|\mathbf{h}|\mathbf{h},$$

the body force $\mathbf{b}$ depending on the domain of application.

Two further equations are used in the literature. One is just

$$\tau\dot{\mathbf{h}} + \mathbf{h} = -(k/\mu)\nabla p,$$

with the same form of the Maxwell–Cattaneo equation for the heat flux. The other is Brinkman equation

$$-\beta\Delta\mathbf{h} + \mathbf{h} = -(k/\mu)\nabla p,$$

where β is a viscosity parameter. It is justified by viewing the relative velocity $\mathbf{h}$ as proportional to the divergence of the stress $-p\mathbf{1} + (\beta\mu/k)\nabla\mathbf{h}$.

11.3 Special Models of Porous Materials

Though the constraint (11.6) on the volume fractions looks quite general, operative models require that we specify the functions $\hat{\varphi}_\alpha$. For definiteness we restrict attention to incompressible constituents. As before, the first constituent is a solid.

In incompressible constituents the true density ρ^0_α is constant and this allows us to establish a simple evolution equation for the volume fraction φ_α. Since

$$\rho^0_\alpha \varphi_\alpha = \rho_\alpha = \frac{\varrho_\alpha}{J_\alpha},$$

by (9.13) we have

$$\dot{\varphi}_\alpha = -\varphi_\alpha \nabla\cdot\mathbf{v}_\alpha + \frac{1}{\rho^0_\alpha}\tau_\alpha. \tag{11.10}$$

Equation (11.10) is remarkable in that it shows how φ_α changes as a consequence of the chemical reactions, via τ_α, and the motion, via $\nabla\cdot\mathbf{v}_\alpha$. In incompressible constituents,

$$\frac{\varrho_\alpha}{J_\alpha\varphi_\alpha} = \rho^0_\alpha$$

is a constant and hence only two variables among ϱ_α, J_α, and φ_α are independent.

Let the remaining constituents, $\beta = 2, 3, ..., n$, be incompressible fluids. We allow the constitutive properties to be affected by the volume fractions and let[3]

[3] The model is developed in [326].

$$\psi_s = \psi_s(\theta_s, \mathbf{F}_s, \varphi_s, \nabla\theta_s, \nabla\mathbf{F}_s), \qquad \psi_\beta = \psi_\beta(\theta_\beta, \varphi_\beta, \nabla\theta_\beta, \mathbf{D}_\beta),$$

and the like for $\mathbf{T}_s, \mathbf{T}_\beta, \mathbf{q}_s, \mathbf{q}_\beta$, where $\beta = 2, 3, ..., n$. The interaction terms $\tau_\alpha, \mathbf{m}_\alpha, e_\alpha$ are allowed to depend on all of the volume fractions and the velocity differences. Hence

$$\mathbf{m}_\alpha = \mathbf{m}_\alpha(\theta_s, \mathbf{F}_s, \varphi_s, \nabla\theta_s, \nabla\mathbf{F}_s, \theta_\beta, \varphi_\beta, \nabla\theta_\beta, \mathbf{v}_\alpha - \mathbf{v}_\beta)$$

and the like for τ_α and e_α.

By arguing as in the previous section we find that, because of the entropy inequality (9.42), the free energy functions simplify to

$$\psi_s = \psi_s(\theta_s, \mathbf{F}_s, \varphi_s), \qquad \psi_\beta = \psi_\beta(\theta_\beta, \varphi_\beta)$$

while $\partial_{\theta_s}\psi_s = -\eta_s, \partial_{\theta_\beta}\psi_\beta = -\eta_\beta$. The entropy inequality then becomes

$$-\frac{\rho_s}{\theta_s}(\partial_{\mathbf{F}_s}\psi_s \cdot \grave{\mathbf{F}}_s + \partial_{\varphi_s}\psi_s\,\grave{\varphi}_s) - \sum_{\beta=2}^n \frac{\rho_\beta}{\theta_\beta}\partial_{\varphi_\beta}\psi_\beta\grave{\varphi}_\beta + \frac{1}{\theta_s}\mathbf{T}_s \cdot \mathbf{L}_s + \sum_{\beta=2}^n \frac{1}{\theta_\beta}\mathbf{T}_\beta \cdot \mathbf{L}_\beta$$
$$+\sum_{\alpha=1}^n \frac{1}{\theta_\alpha}(\tau_\alpha\theta_\alpha\eta_\alpha - \frac{1}{\theta_\alpha}\mathbf{q}_\alpha \cdot \nabla\theta_\alpha + \hat{\varepsilon}_\alpha + \theta_\alpha\nabla \cdot \mathbf{k}_\alpha) \geq 0.$$

In view of (11.10) and the relation $\grave{\mathbf{F}}_s = \mathbf{L}_s\mathbf{F}_s^T$, we obtain

$$\frac{1}{\theta_s}(\mathbf{T}_s - \rho_s\partial_{\mathbf{F}_s}\psi_s\mathbf{F}_s^T + \rho_s\varphi_s\partial_{\varphi_s}\psi_s\mathbf{1}) \cdot \mathbf{L}_s + \sum_{\beta=2}^n \frac{1}{\theta_\beta}(\mathbf{T}_\beta + \rho_\beta\varphi_\beta\partial_{\varphi_\beta}\psi_\beta\mathbf{1}) \cdot \mathbf{L}_\beta$$
$$+\sum_{\alpha=1}^n \frac{1}{\theta_\alpha}(\zeta_\alpha\tau_\alpha - \frac{1}{\theta_\alpha}\mathbf{q}_\alpha \cdot \nabla\theta_\alpha + \hat{\varepsilon}_\alpha + \theta_\alpha\nabla \cdot \mathbf{k}_\alpha) \geq 0,$$

where

$$\zeta_\alpha = \theta_\alpha\eta_\alpha - \varphi_\alpha\partial_{\varphi_\alpha}\psi_\alpha.$$

Now, $\mathbf{T}_s$ is independent of $\mathbf{L}_s$ whereas $\mathbf{T}_\beta$ is allowed to depend on $\nabla\theta_\beta$ and $\mathbf{D}_\beta$. For definiteness, we assume that $\mathbf{T}_\beta$ is expressed by

$$\mathbf{T}_\beta = \mathbf{T}_\beta^{\mathrm{nv}}(\theta_\beta, \varphi_\beta) + \mathbf{T}_\beta^{\mathrm{v}}(\theta_\beta, \varphi_\beta, \nabla\theta_\beta, \mathbf{D}_\beta),$$

where $\mathbf{T}_\beta^{\mathrm{v}}$ is subject to $\mathbf{T}_\beta^{\mathrm{v}} \to \mathbf{0}$ as $\mathbf{D}_\beta \to \mathbf{0}$. Moreover let

$$\hat{\mathbf{T}}_s = \mathbf{T}_s - \rho_s\partial_{\mathbf{F}_s}\psi_s\,\mathbf{F}_s^T + \rho_s\varphi_s\partial_{\varphi_s}\psi_s\mathbf{1}.$$

Consequently we find

$$\frac{1}{\theta_s}\hat{\mathbf{T}}_s \cdot \mathbf{L}_s + \sum_{\beta=2}^n \frac{1}{\theta_\beta}(\mathbf{T}_\beta^{\mathrm{nv}} + \mathbf{T}_\beta^{\mathrm{v}}) \cdot \mathbf{L}_\beta + \sum_{\alpha=1}^n \frac{1}{\theta_\alpha}(\zeta_\alpha\tau_\alpha - \frac{1}{\theta_\alpha}\mathbf{q}_\alpha \cdot \nabla\theta_\alpha + \hat{\varepsilon}_\alpha + \theta_\alpha\nabla \cdot \mathbf{k}_\alpha) \geq 0.$$

This inequality holds if

$$\hat{\mathbf{T}}_s = \mathbf{0}, \quad \mathbf{T}_\beta^{\mathrm{nv}} = -\rho_\beta \varphi_\beta \partial_{\varphi_\beta} \psi_\beta \mathbf{1},$$

and

$$\sum_{\beta=2}^{n} \frac{1}{\theta_\beta} \mathbf{T}_\beta^{\mathrm{v}} \cdot \mathbf{L}_\beta + \sum_{\alpha=1}^{n} \frac{1}{\theta_\alpha} (\zeta_\alpha \tau_\alpha - \frac{1}{\theta_\alpha} \mathbf{q}_\alpha \cdot \nabla \theta_\alpha + \hat{\varepsilon}_\alpha + \theta_\alpha \nabla \cdot \mathbf{k}_\alpha) \ge 0.$$

In particular, we can find the Navier–Stokes–Fourier model

$$\mathbf{T}_\beta^{\mathrm{v}} = 2\nu_\beta \mathbf{D}_\beta + \lambda_\beta \nabla \cdot \mathbf{v}_\beta \mathbf{1}, \quad \mathbf{q}_\alpha = -\kappa_\alpha \nabla \theta_\alpha,$$

where $\nu_\beta \ge 0$, $2\nu_\beta + 3\lambda_\beta \ge 0$, $\kappa_\alpha \ge 0$ and

$$\textstyle\sum_{\alpha=1}^{n} \frac{1}{\theta_\alpha} (\zeta_\alpha \tau_\alpha + \hat{\varepsilon}_\alpha + \theta_\alpha \nabla \cdot \mathbf{k}_\alpha) \ge 0.$$

To sum up, we can model the skeletal porous solid and the $n-1$ fluids by

$$\varepsilon_s = \psi_s(\theta_s, \mathbf{F}_s, \varphi_s) - \theta_s \partial_{\theta_s} \psi_s(\theta_s, \mathbf{F}_s, \varphi_s),$$

$$\varepsilon_\beta = \psi_\beta(\theta_\beta, \varphi_\beta) - \theta_\beta \partial_{\theta_\beta} \psi_\beta(\theta_\beta, \varphi_\beta),$$

$$\mathbf{T}_s = \rho_s \partial_{\mathbf{F}_s} \psi_s \mathbf{F}_s^T - \rho_s \varphi_s \partial_{\varphi_s} \psi_s \mathbf{1},$$

$$\mathbf{T}_\beta = -\rho_\beta \varphi_\beta \partial_{\varphi_\beta} \psi_\beta \mathbf{1} + 2\nu_\beta \mathbf{D}_\beta + \lambda_\beta \nabla \cdot \mathbf{v}_\beta \mathbf{1}, \qquad \mathbf{q}_\alpha = -\kappa_\alpha \nabla \theta_\alpha.$$

Both in solids and fluids we can view

$$p_\alpha := \rho_\alpha \varphi_\alpha \partial_{\varphi_\alpha} \psi_\alpha$$

as the (isotropic) pressure. However, in fluids this view is classical in that

$$\rho_\alpha \varphi_\alpha \partial_{\varphi_\alpha} \psi_\alpha = \rho_\alpha^2 \partial_{\rho_\alpha} \psi_\alpha.$$

In solids, instead, the view of $\rho_s \varphi_s \partial_{\varphi_s} \psi_s$ as the pressure is purely formal because the additive term $\rho_s \partial_{\mathbf{F}_s} \psi_s \mathbf{F}_s^T$ may comprise an isotropic tensor.

Look now at the case where each constituent has the same temperature, say θ. Hence, we have

$$\textstyle\sum_{\alpha=1}^{n} (\zeta_\alpha \tau_\alpha + \hat{\varepsilon}_\alpha + \theta_\alpha \nabla \cdot \mathbf{k}_\alpha) \ge 0.$$

Substitution of $\hat{\varepsilon}_\alpha$ from (9.25) results in

$$\sum_{\alpha=1}^{n}[(\zeta_\alpha - \varepsilon_\alpha)\tau_\alpha - \hat{\mathbf{p}}_\alpha \cdot \mathbf{u}_\alpha - \tfrac{1}{2}\tau_\alpha \mathbf{u}_\alpha^2 + \theta_\alpha \nabla \cdot \mathbf{k}_\alpha] \geq 0.$$

Since $\theta_\alpha = \theta$ we can write

$$\zeta_\alpha - \varepsilon_\alpha = -\psi_\alpha - \varphi_\alpha \partial_{\varphi_\alpha} \psi_\alpha,$$

whence

$$\zeta_\alpha - \varepsilon_\alpha = -(\psi_\alpha + p_\alpha/\rho_\alpha) =: -\mu_\alpha.$$

We may regard μ_α as the chemical potential. Since $\mathbf{k}_\alpha$ may be taken to be zero we obtain the standard form of the reduced inequality,

$$\sum_{\alpha=1}^{n}(\mu_\alpha \tau_\alpha + \hat{\mathbf{p}}_\alpha \cdot \mathbf{u}_\alpha + \tfrac{1}{2}\tau_\alpha \mathbf{u}_\alpha^2) \leq 0.$$

11.4 Materials with Voids

By analogy with the previous scheme of porous materials, we now look at porous solids in which the skeletal material is elastic and incompressible whereas the interstices are void of material. Owing to the voids, only the solid is a material constituent and hence we use a superposed dot to denote the time derivative following the motion of the solid.

The solid is regarded as incompressible so that

$$\rho = \varphi \rho^0,$$

the true mass density ρ^0 being constant.[4] We assume that the interaction terms between solid and voids are zero, namely, τ, $\mathbf{m}$, e are zero. Hence it follows that ρ and φ satisfy the same equation,

$$\dot{\rho} = -\rho \nabla \cdot \mathbf{v}, \qquad \dot{\varphi} = -\varphi \nabla \cdot \mathbf{v}.$$

The remaining balance equations are formally the classical ones for a single body, say

$$\rho \dot{\mathbf{v}} = -\nabla \cdot \mathbf{T} + \rho \mathbf{b},$$

$$\rho \dot{\varepsilon} = \mathbf{T} \cdot \mathbf{D} - \nabla \cdot \mathbf{q} + \rho r.$$

The effects of the voids on the solid are modelled by the constitutive equations. Let

$$\Gamma = (\theta, \mathbf{F}, \varphi, \nabla\theta, \nabla\varphi, \dot{\varphi})$$

[4] In this section, the subscript s is understood and not written.

be the set of variables so that $\psi, \eta, \varepsilon, \mathbf{T}, \mathbf{q}, \mathbf{k}$ are functions of Γ. The dependence on the time derivative is considered to allow for inertial properties in the evolution of φ. Let $\psi, \mathbf{q}, \mathbf{k}$ be differentiable and the remaining functions be continuous. The second-law inequality can be written as

$$-\rho(\partial_\theta\psi + \eta)\dot{\theta} - \rho\partial_{\mathbf{F}}\psi \cdot \dot{\mathbf{F}} - \rho\partial_\varphi\psi\,\dot{\varphi} - \rho\partial_{\nabla\theta}\psi \cdot \overline{\nabla\theta}^{\boldsymbol\cdot} - \rho\partial_{\nabla\varphi}\psi \cdot \overline{\nabla\varphi}^{\boldsymbol\cdot} - \rho\partial_{\dot{\varphi}}\psi\,\ddot{\varphi}$$
$$+\mathbf{T}\cdot\mathbf{D} - \frac{1}{\theta}\mathbf{q}\cdot\nabla\theta + \theta\nabla\cdot\mathbf{k} \geq 0.$$

By means of (9.32) we can replace $\overline{\nabla\theta}^{\boldsymbol\cdot}$ and $\overline{\nabla\varphi}^{\boldsymbol\cdot}$ to obtain

$$\begin{aligned} &-\rho(\partial_\theta\psi + \eta)\dot{\theta} - \rho\partial_{\mathbf{F}}\psi \cdot \dot{\mathbf{F}} - \rho\partial_\varphi\psi\,\dot{\varphi} - \rho\partial_{\nabla\theta}\psi \cdot \nabla\dot{\theta} - \rho\partial_{\nabla\varphi}\psi \cdot \nabla\dot{\varphi} - \rho\partial_{\dot{\varphi}}\psi\,\ddot{\varphi} \\ &\quad +\rho\partial_{\nabla\theta}\psi \cdot \mathbf{L}^T\nabla\theta + \rho\partial_{\nabla\varphi}\psi \cdot \mathbf{L}^T\nabla\varphi + \mathbf{T}\cdot\mathbf{D} - \frac{1}{\theta}\mathbf{q}\cdot\nabla\theta + \theta\nabla\cdot\mathbf{k} \geq 0. \end{aligned} \tag{11.11}$$

Time differentiation of $\dot{\varphi} = -\varphi\nabla\cdot\mathbf{v}$ and use of (9.33) yield

$$\ddot{\varphi} = -\dot{\varphi}\nabla\cdot\mathbf{v} - \varphi\nabla\cdot\dot{\mathbf{v}} + \varphi(\mathbf{L}^T\nabla)\cdot\mathbf{v}.$$

Now, at any point $\mathbf{x} \in \Omega$ and time t, $\nabla\cdot\dot{\mathbf{v}}$ can take any arbitrary value, independent of Γ and $\nabla\dot{\varphi}$, merely because $\nabla\cdot\mathbf{b}$ is regarded to be given by (the divergence of) the equation of motion $(9.24)_2$. Likewise $\dot{\theta}$ and $\nabla\dot{\theta}$ can take arbitrary values because $r(\mathbf{x}, t)$ and $\nabla r(\mathbf{x}, t)$ are regarded to be given by the energy balance equation $(9.24)_3$. Consequently, in view of the linear dependence on $\dot{\theta}$, $\nabla\dot{\theta}$, and $\ddot{\varphi}$, inequality (11.11) holds only if

$$\partial_\theta\psi + \eta = 0, \qquad \partial_{\nabla\theta}\psi = \mathbf{0}, \qquad \partial_{\dot{\varphi}}\psi = 0.$$

Divide the remaining inequality by θ. Upon some rearrangements we can write

$$\frac{1}{\theta}[-\rho\partial_{\mathbf{F}}\psi\,\mathbf{F}^T + \rho\varphi\partial_\varphi\psi\,\mathbf{1} - \varphi\theta\nabla\cdot(\frac{\rho}{\theta}\partial_{\nabla\varphi}\psi)\mathbf{1} + \rho\nabla\varphi\otimes\partial_{\nabla\varphi}\psi + \mathbf{T}]\cdot\mathbf{L}$$
$$-\frac{1}{\theta^2}\mathbf{q}\cdot\nabla\theta + \nabla\cdot(\mathbf{k} - \frac{\rho}{\theta}\partial_{\nabla\varphi}\psi\,\dot{\varphi}) \geq 0.$$

This suggests that we take

$$\mathbf{k} = \frac{\rho}{\theta}\partial_{\nabla\varphi}\psi\,\dot{\varphi}$$

and hence

$$[-\rho\partial_{\mathbf{F}}\psi\,\mathbf{F}^T + \rho\varphi\partial_\varphi\psi\,\mathbf{1} - \varphi\theta\nabla\cdot(\frac{\rho}{\theta}\partial_{\nabla\varphi}\psi)\mathbf{1} + \rho\nabla\varphi\otimes\partial_{\nabla\varphi}\psi + \mathbf{T}]\cdot\mathbf{L} - \frac{1}{\theta}\mathbf{q}\cdot\nabla\theta \geq 0.$$

The symmetry of $\mathbf{T}$ and the linearity of $\mathbf{L}$ imply that

$$-\rho\partial_{\mathbf{F}}\psi\,\mathbf{F}^T + \rho\nabla\varphi \otimes \partial_{\nabla\varphi}\psi \in \mathrm{Sym},$$

$$\mathbf{T} = \rho\partial_{\mathbf{F}}\psi\,\mathbf{F}^T - \rho\varphi\partial_\varphi\psi\,\mathbf{1} + \varphi\theta\nabla\cdot(\frac{\rho}{\theta}\partial_{\nabla\varphi}\psi)\mathbf{1} - \rho\nabla\varphi\otimes\partial_{\nabla\varphi}\psi \tag{11.12}$$

and

$$\mathbf{q}\cdot\nabla\theta \le 0,$$

whence

$$\mathbf{q} = -\kappa(\Gamma)\nabla\theta, \tag{11.13}$$

$\kappa(\Gamma)$ being a non-negative function. It is apparent that a dependence on $\mathbf{D}$ from the outset would allow us to describe also viscosity effects.

This simple model accounts for the dependence on the volume fraction φ of the solid, $1-\varphi$ being the porosity, and the volume fraction of the voids. Such dependence results in various effects. The stress tensor comprises $-\rho\varphi\partial_\varphi\psi\,\mathbf{1}$ as with elastic solids where the dependence on φ is allowed. The dependence on $\nabla\varphi$ produces additive terms,

$$\varphi\theta\nabla\cdot(\frac{\rho}{\theta}\partial_{\nabla\varphi}\psi)\mathbf{1} - \rho\nabla\varphi\otimes\partial_{\nabla\varphi}\psi,$$

which in principle may occur also in the model of single bodies. It is likely, though, that such dependence is more influential in materials with voids.

In a dynamic problem, the unknown functions are the volume fraction φ (or the mass density ρ), the velocity $\mathbf{v}$, and the temperature θ. They are determined by the differential equations

$$\dot\varphi = -\varphi\nabla\cdot\mathbf{v}, \qquad \rho\dot{\mathbf{v}} = -\nabla\cdot\mathbf{T} + \rho\mathbf{b}, \qquad \rho\dot\varepsilon = \mathbf{T}\cdot\mathbf{D} - \nabla\cdot\mathbf{q} + \rho r,$$

the constitutive equations (11.12), (11.13), and

$$\varepsilon = \varepsilon(\theta, \mathbf{F}, \varphi, \nabla\varphi),$$

along with appropriate initial-boundary-value data.

11.5 Porous Media with Double Porosity

Most frequently the flow in porous media is investigated by letting the body consist of a system of similar-sized pores connected by a single pore network as is done in the previous sections. Yet many geo-materials such as aggregated soils or fissured rocks show two or more dominant pore scales connected by multiple pore networks that display different orders of volume fractions. Here the porous medium is treated as a mixture of n constituents. The role of constituent refers to both the porous solid and the pore network. The modelling via the mixture theory makes it natural to describe the interacting processes. In particular, the mixture is reacting in that there can be a mass transfer between the pore networks.

We keep following the standard notation for mixtures and, for ease of reading, we review the balance equations. The local form of the balance of mass, linear momentum, and energy of the αth constituent can be written in the form (9.24), i.e.

$$\begin{aligned}
&\grave{\rho}_\alpha + \rho_\alpha \nabla \cdot \mathbf{v}_\alpha = \tau_\alpha, \\
&\rho_\alpha \grave{\mathbf{v}}_\alpha - \nabla \cdot \mathbf{T}_\alpha - \rho_\alpha \mathbf{b}_\alpha = \hat{\mathbf{p}}_\alpha, \\
&\rho_\alpha \grave{\varepsilon}_\alpha - \mathbf{T}_\alpha \cdot \mathbf{L}_\alpha + \nabla \cdot \mathbf{q}_\alpha - \rho_\alpha r_\alpha = \hat{\varepsilon}_\alpha.
\end{aligned}$$

In terms of the supplies τ_α, $\hat{\mathbf{p}}_\alpha$, $\hat{\varepsilon}_\alpha$, the overall balance for the mixture becomes

$$\textstyle\sum_\alpha \tau_\alpha = 0, \qquad \sum_\alpha \hat{\mathbf{p}}_\alpha + \tau_\alpha \mathbf{v}_\alpha = \mathbf{0}, \qquad \sum_\alpha \hat{\varepsilon}_\alpha + \hat{\mathbf{p}}_\alpha \cdot \mathbf{v}_\alpha + \tau_\alpha(\varepsilon_\alpha + \frac{1}{2}\mathbf{v}_\alpha^2) = 0.$$

As a simplifying assumption, we let $\mathbf{T}_\alpha$ be symmetric so that $\mathbf{T}_\alpha \cdot \mathbf{L}_\alpha = \mathbf{T}_\alpha \cdot \mathbf{D}_\alpha$. Moreover, we have in mind to investigate constitutive models where the extra-entropy fluxes $\{\mathbf{k}_\alpha\}$ vanish. Hence, the entropy inequality can be written as

$$\textstyle\sum_\alpha \{\frac{1}{\theta_\alpha}[-\rho_\alpha(\grave{\psi}_\alpha + \eta_\alpha \grave{\theta}_\alpha) - \tau_\alpha(\psi_\alpha - \frac{1}{2}\mathbf{v}_\alpha^2) + e_\alpha - \mathbf{m}_\alpha \cdot \mathbf{v}_\alpha + \mathbf{T}_\alpha \cdot \mathbf{D}_\alpha - \frac{1}{\theta_\alpha}\mathbf{q}_\alpha \cdot \nabla\theta_\alpha]\} \geq 0,$$

where $\{e_\alpha\}$ and $\{\mathbf{m}_\alpha\}$ are subject to

$$\textstyle\sum_\alpha e_\alpha = 0, \qquad \sum_\alpha \mathbf{m}_\alpha = \mathbf{0}.$$

For definiteness and simplicity, let all the constituents have the same temperature, $\theta_\alpha = \theta$. Hence, the entropy inequality can be given in the form

$$\textstyle\sum_\alpha [-\rho_\alpha(\grave{\psi}_\alpha + \eta_\alpha \grave{\theta}_\alpha) - \tau_\alpha \psi_\alpha - \hat{\mathbf{p}}_\alpha \cdot \mathbf{u}_\alpha - \frac{1}{2}\tau_\alpha \mathbf{u}_\alpha^2 + \mathbf{T}_\alpha \cdot \mathbf{D}_\alpha - \frac{1}{\theta}\mathbf{q}_\alpha \cdot \nabla\theta]\} \geq 0, \tag{11.14}$$

where, by (9.18), $\hat{\mathbf{p}}_\alpha = \mathbf{m}_\alpha - \tau_\alpha \mathbf{v}_\alpha$.

We now specify the properties of the porous material.

11.5.1 Porous Material with Undeformable Solid

There are two pore networks and the skeleton is a rigid material; since the solid is rigid the corresponding balance of linear momentum is trivially satisfied. The constituents 1 and 2 are the two pore networks. To fix ideas, the constituents 1 and 2 are the macro-pore and micro-pore networks. The porosities (or volume fractions) in the macro- and micro-pore networks are denoted by φ_1 and φ_2. The true (or seepage) velocity vector fields are denoted by $\mathbf{v}_1$ and $\mathbf{v}_2$ whereas

$$\tilde{\mathbf{v}}_\alpha = \varphi_\alpha \mathbf{v}_\alpha, \qquad \alpha = 1, 2,$$

are the discharge (or Darcy) velocities. If ρ^0 is the true mass density of the fluid then the mass densities in the macro- and micro-networks are

$$\rho_\alpha = \varphi_\alpha \rho^0, \qquad \alpha = 1, 2.$$

The fluid is incompressible and hence ρ^0 is constant. Since the skeleton is rigid the geometry of the pore networks is fixed and hence

$$\partial_t \varphi_\alpha = 0, \qquad \alpha = 1, 2.$$

Hence the continuity equation $\partial_t \rho_\alpha + \nabla \cdot (\rho_\alpha \mathbf{v}_\alpha) = \tau_\alpha$ can be written as

$$\rho^0 \nabla \cdot (\varphi_\alpha \mathbf{v}_\alpha) = \tau_\alpha.$$

Moreover,

$$\grave{\varphi}_\alpha = \mathbf{v}_\alpha \cdot \nabla \varphi_\alpha$$

and then each continuity equation results in (11.10),

$$\grave{\varphi}_\alpha = \frac{\tau_\alpha}{\rho^0} - \varphi_\alpha \nabla \cdot \mathbf{v}_\alpha.$$

It is worth pointing out that in a mixture of incompressible materials the velocity fields need not be divergence free. Irrespective of the value of τ_α, a nonzero value of $\nabla \cdot \mathbf{v}_\alpha$ is made possible by $\grave{\varphi}_\alpha \neq 0$.

It is customary to assume that the fluid can be exchanged between the two pore networks if there exists a sufficient difference of pressure between the networks. Formally, the mass transfer is often modelled by the relation

$$\tau_1(\mathbf{x}) = -\frac{\beta}{\mu} \rho^0 (p_1(\mathbf{x}) - p_2(\mathbf{x})), \tag{11.15}$$

where μ is the viscosity while β is a dimensionless characteristic of the porous medium. This relation traces back to Barenblatt et al. [29] and was derived by means of a dimensional analysis argument.

Thermodynamic restrictions on the constitutive equations are now derived, via the second-law inequality, by assuming that $\psi_\alpha, \eta_\alpha, \mathbf{T}_\alpha, \mathbf{q}_\alpha$ are functions of

$$\theta, \varphi_\alpha, \mathbf{D}_\alpha, \nabla\theta,$$

while $\hat{\mathbf{p}}_\alpha$ and τ_α may depend on quantities pertaining to both constituents. Moreover, we let

$$\mathbf{T}_\alpha = -p_\alpha(\theta, \varphi_\alpha)\mathbf{1} + \check{\mathbf{T}}_\alpha, \qquad \check{\mathbf{T}}_\alpha = O(\mathbf{D}_\alpha).$$

By a routine argument we can show that ψ_α is independent of $\mathbf{D}_\alpha$ and $\nabla\theta$. To save writing we then start with the constitutive assumption $\psi_\alpha = \psi_\alpha(\theta, \varphi_\alpha)$ and observe

$$\grave{\psi}_\alpha = \partial_\theta \psi_\alpha (\partial_t \theta + \mathbf{v}_\alpha \cdot \nabla\theta) + \partial_{\varphi_\alpha} \psi_\alpha \grave{\varphi}_\alpha.$$

Upon substitution in the second-law inequality (11.14), we have

$$\sum_\alpha [-\rho_\alpha(\partial_\theta \psi_\alpha + \eta_\alpha)\grave{\theta}_\alpha - \rho_\alpha \partial_{\varphi_\alpha} \psi_\alpha (\frac{\tau_\alpha}{\rho^0} - \varphi_\alpha \nabla \cdot \mathbf{v}_\alpha) - \tau_\alpha \psi_\alpha - p_\alpha \nabla \cdot \mathbf{v}_\alpha] + \zeta \geq 0,$$

where $\grave{\theta}_\alpha = \partial_t \theta + \mathbf{v}_\alpha \cdot \nabla\theta$ and

$$\zeta = \sum_\alpha [-\hat{\mathbf{p}}_\alpha \cdot \mathbf{u}_\alpha - \tfrac{1}{2}\tau_\alpha \mathbf{u}_\alpha^2 + \check{\mathbf{T}}_\alpha \cdot \mathbf{D}_\alpha - \frac{1}{\theta}\mathbf{q}_\alpha \cdot \nabla\theta].$$

The inequality holds if and only if

$$\eta_\alpha = -\partial_\theta \psi_\alpha, \qquad p_\alpha = \varphi_\alpha \rho_\alpha \partial_{\varphi_\alpha} \psi_\alpha,$$

and

$$\sum_\alpha \tau_\alpha (\psi_\alpha + \frac{p_\alpha}{\rho_\alpha}) - \zeta \leq 0.$$

If, further, $\tau_1 \, (= -\tau_2)$ is independent of $\{\mathbf{v}_\alpha\}$, $\{\mathbf{D}_\alpha\}$ and $\nabla\theta$ then the reduced inequality holds if and only if

$$\sum_\alpha \tau_\alpha (\psi_\alpha + \frac{p_\alpha}{\rho_\alpha}) \leq 0 \tag{11.16}$$

and $\zeta \geq 0$. Inequality (11.16) is the standard relation governing the exchanges of mass, $\psi_\alpha + p_\alpha/\rho_\alpha$ being the α-th chemical potential. If p_α/ρ_α prevails on ψ_α then we can take the approximate relation

$$\sum_\alpha \tau_\alpha \frac{p_\alpha}{\rho_\alpha} \leq 0,$$

whence

$$\tau_1 (\frac{p_1}{\rho_1} - \frac{p_2}{\rho_2}) \leq 0.$$

Consequently, τ_1 has the general form

$$\tau_1 = -\frac{\nu}{\rho^0}(\frac{p_1}{\varphi_1} - \frac{p_2}{\varphi_2}), \tag{11.17}$$

ν being any positive-valued function of the variables involved. The result (11.17) has a similarity and a difference to the Barenblatt-et-al relation (11.15). The similarity is due to the dependence of τ_1 (and τ_2) on the pressures in the fluid constituents. In this sense ν of (11.17) can be identified with $\beta\rho^0/\mu$ of (11.15). The difference is in the quantity $p_1/\varphi_1 - p_2/\varphi_2$ rather than $p_1 - p_2$. An insight in favour of (11.17) is that $p_\alpha \propto \rho_\alpha \propto \varphi_\alpha$. Hence it is the specific quantity p_α/φ_α that governs the exchange of mass.

As for the positive definiteness of ζ, the classical Navier–Stokes–Fourier model is allowed,

$$\mathbf{T}_\alpha = -p_\alpha \mathbf{1} + 2\mu \mathbf{D}_\alpha, \qquad \textstyle\sum_\alpha \mathbf{q}_\alpha = -\kappa \nabla\theta,$$

where μ and κ are positive. A non-negative value of $\sum_\alpha[-\hat{\mathbf{p}}_\alpha \cdot \mathbf{u}_\alpha - \frac{1}{2}\tau_\alpha \mathbf{u}_\alpha^2]$ for any pair of vectors $\mathbf{u}_1, \mathbf{u}_2$ follows by letting

$$\hat{\mathbf{p}}_\alpha = \mathbf{M} \textstyle\sum_\beta (\mathbf{u}_\beta - \mathbf{u}_\alpha) - \tau_\alpha \mathbf{u}_\alpha + \frac{1}{2}\tau_\alpha(\mathbf{u}_1 + \mathbf{u}_2),$$

$\mathbf{M}$ being any (symmetric) positive-definite tensor. It follows that $\sum_\alpha \hat{\mathbf{p}}_\alpha + \tau_\alpha \mathbf{v}_\alpha = \mathbf{0}$ and

$$\textstyle\sum_\alpha[-\hat{\mathbf{p}}_\alpha \cdot \mathbf{u}_\alpha - \frac{1}{2}\tau_\alpha \mathbf{u}_\alpha^2] = -(\mathbf{u}_1 - \mathbf{u}_2) \cdot \mathbf{M}(\mathbf{u}_1 - \mathbf{u}_2) \le 0.$$

11.5.2 Porous Material with a Thermoelastic Solid

The skeletal solid is now allowed to be deformable and, for definiteness, is supposed to be thermoelastic. The porous material is then modelled as a mixture with three constituents, 1 and 2 being the macro-pore and micro-pore fluids while the third constituent is the thermoelastic skeleton. The fluid is still incompressible and hence it is convenient to let $\rho_\alpha = \varphi_\alpha \rho^0$, $\alpha = 1, 2$. Moreover, Eq. (11.10) holds formally unchanged while now $\partial_t \varphi_\alpha \neq 0$. Here we let the subscript s denote the quantities pertaining to the solid (constituent 3).

The balance and constitutive equations are significantly affected by the constraints. To begin with, suppose the porous material is saturated. The saturation constraint

$$\varphi_s + \textstyle\sum_\alpha \varphi_\alpha = 1$$

implies that

$$\dot{\varphi}_s = \partial_t \varphi_s + (\mathbf{v}_s \cdot \nabla)\varphi_s = -\textstyle\sum_\alpha[\partial_t \varphi_\alpha + \mathbf{v}_s \cdot \nabla\varphi_\alpha] = -\sum_\alpha[\dot{\varphi}_\alpha + (\mathbf{v}_\alpha - \mathbf{v}_s) \cdot \nabla\varphi_\alpha].$$

In addition to be mathematically cumbersome, this constraint seems to be quite unrealistic from the physical point of view. Instead it seems of interest to let the true density of the solid be constant. Hence by $\rho_s = \varphi_s \rho_s^0$, where ρ_s^0 is constant, it follows that φ_s has to satisfy

$$\dot{\varphi}_s = -\varphi_s \nabla \cdot \mathbf{v}_s.$$

Thermodynamic restrictions

The constituents 1 and 2 are viscous fluids and hence their state variables are

$$\theta, \varphi_\alpha, \mathbf{D}_\alpha, \nabla\theta, \qquad \alpha = 1, 2.$$

The solid is thermoelastic and then the state variables of the solid are

$$\theta, \varphi_s, \mathbf{F}_s, \nabla\theta.$$

As in the previous section, we let the free energies ψ_α and ψ_s be independent on $\mathbf{D}_\alpha$ and $\nabla\theta$,

$$\psi_\alpha = \psi_\alpha(\theta, \varphi_\alpha), \quad \alpha = 1, 2; \qquad \psi_s = \psi_s(\theta, \varphi_s, \mathbf{F}_s).$$

Upon substitution in inequality (11.14) we obtain

$$-\rho_s(\partial_\theta\psi_s + \eta_s)\grave{\theta}_s + \textstyle\sum_\alpha[-\rho_\alpha(\partial_\theta\psi_\alpha + \eta_\alpha)\grave{\theta}_\alpha - \rho_\alpha\partial_{\varphi_\alpha}\psi_\alpha(\frac{\tau_\alpha}{\rho^0} - \varphi_\alpha\nabla\cdot\mathbf{v}_\alpha) - p_\alpha\nabla\cdot\mathbf{v}_\alpha]$$
$$+\rho_s\partial_{\varphi_s}\psi_s\,\varphi_s\nabla\cdot\mathbf{v}_s + (\mathbf{T}_s - \rho_s\partial_{\mathbf{F}_s}\psi_s\mathbf{F}_s^T)\cdot\mathbf{L}_s - \textstyle\sum_\alpha\tau_\alpha\psi_\alpha - \hat{\mathbf{p}}_s\cdot\mathbf{u}_s$$
$$+\textstyle\sum_\alpha[-\hat{\mathbf{p}}_\alpha\cdot\mathbf{u}_\alpha - \frac{1}{2}\tau_\alpha\mathbf{u}_\alpha^2 + \check{\mathbf{T}}_\alpha\cdot\mathbf{D}_\alpha] - \dfrac{1}{\theta}(\textstyle\sum_\alpha\mathbf{q}_\alpha + \mathbf{q}_s)\cdot\nabla\theta \geq 0.$$

By the standard argument about arbitrariness it follows that

$$\eta_s = -\partial_\theta\psi_s, \qquad \eta_\alpha = -\partial_\theta\psi_\alpha, \qquad p_\alpha = \rho_\alpha\varphi_\alpha\partial_{\varphi_\alpha}\psi_\alpha.$$

Consequently

$$\rho_\alpha\partial_{\varphi_\alpha}\psi_\alpha\frac{\tau_\alpha}{\rho^0} = \frac{p_\alpha}{\rho_\alpha}\tau_\alpha.$$

The inequality then simplifies to

$$\rho_s\partial_{\varphi_s}\psi_s\,\varphi_s\nabla\cdot\mathbf{v}_s + (\mathbf{T}_s - \rho_s\partial_{\mathbf{F}_s}\psi_s\mathbf{F}_s^T)\cdot\mathbf{L}_s - \textstyle\sum_\alpha\tau_\alpha(\psi_\alpha + \dfrac{p_\alpha}{\rho_\alpha}) - \hat{\mathbf{p}}_s\cdot\mathbf{u}_s$$
$$+\textstyle\sum_\alpha[-\hat{\mathbf{p}}_\alpha\cdot\mathbf{u}_\alpha - \frac{1}{2}\tau_\alpha\mathbf{u}_\alpha^2 + \check{\mathbf{T}}_\alpha\cdot\mathbf{D}_\alpha] - \dfrac{1}{\theta}(\textstyle\sum_\alpha\mathbf{q}_\alpha + \mathbf{q}_s)\cdot\nabla\theta \geq 0.$$

By the arbitrariness of $\mathbf{L}_s$ and disregarding cross-coupling terms ($\mathbf{D}_\alpha\nabla\theta$, $\mathbf{u}_\alpha\nabla\theta$ and so on) it follows

$$\mathbf{T}_s + \rho_s\varphi_s\partial_{\varphi_s}\psi_s\mathbf{1} - \rho_s\partial_{\mathbf{F}_s}\psi_s\,\mathbf{F}_s^T = \mathbf{0}, \tag{11.18}$$

$$\textstyle\sum_\alpha\tau_\alpha(\psi_\alpha + \dfrac{p_\alpha}{\rho_\alpha}) \leq 0, \qquad (\textstyle\sum_\alpha\mathbf{q}_\alpha + \mathbf{q}_s)\cdot\nabla\theta \leq 0, \qquad \check{\mathbf{T}}_\alpha\cdot\mathbf{D}_\alpha \geq 0,$$

$$\hat{\mathbf{p}}_s\cdot\mathbf{u}_s + \textstyle\sum_\alpha(\hat{\mathbf{p}}_\alpha\cdot\mathbf{u}_\alpha + \frac{1}{2}\tau_\alpha\mathbf{u}_\alpha^2) \leq 0. \tag{11.19}$$

Equation (11.18), namely,

$$\mathbf{T}_s = -\rho_s\varphi_s\partial_{\varphi_s}\psi_s\mathbf{1} + \rho_s\partial_{\mathbf{F}_s}\psi_s\,\mathbf{F}_s^T,$$

means that the stress tensor $\mathbf{T}_s$ of the solid is the sum of two terms, one is a pressure tensor, formally the pressure being $\rho_s\varphi_s\partial_{\varphi_s}\psi_s$, the other is the standard elastic stress. If ψ_s depends on $\mathbf{F}_s$ through $\mathbf{C}_s = \mathbf{F}_s^T\mathbf{F}_s$ then

$$\partial_{\mathbf{F}_s}\psi_s\mathbf{F}_s^T = 2\mathbf{F}_s\partial_{\mathbf{C}_s}\psi_s\mathbf{F}_s^T.$$

It is apparent that $\mathbf{F}_s\partial_{\mathbf{C}_s}\psi_s\mathbf{F}_s^T \in \mathrm{Sym}$. If, further, we observe that

$$\mathbf{F} = \mathbf{1} + \mathbf{H} = \mathbf{1} + \varepsilon\mathbf{F} = \mathbf{1} + \varepsilon(\mathbf{1} + \varepsilon\mathbf{F}),$$

then in the linear approximation with respect to ε we can write

$$\mathbf{C}_s - \mathbf{1} \simeq (\mathbf{1} + \varepsilon_s)(\mathbf{1} + \varepsilon_s) - \mathbf{1} \simeq 2\varepsilon_s.$$

Hence we have

$$2\mathbf{F}_s\partial_{\mathbf{C}_s}\mathbf{F}_s^T \simeq \partial_{\varepsilon_s}\psi_s.$$

Furthermore, if

$$\psi_s = \mu_s\varepsilon_s\cdot\varepsilon_s + \tfrac{1}{2}\lambda_s(\mathrm{tr}\,\varepsilon_s)^2,$$

then

$$\partial_{\varepsilon_s}\psi = 2\mu_s\varepsilon_s + \lambda_s(\mathrm{tr}\,\varepsilon_s)\mathbf{1}.$$

The result (11.16) holds here unchanged and this is consistent with the fact that now the solid is elastic but does not exchange mass with the fluid constituents. Hence Eq. (11.17) still follows.

Inequality (11.19) holds if the supplies of linear momentum $\{\hat{\mathbf{p}}_\beta\}$, $\beta = 1, 2, 3$, are given by

$$\hat{\mathbf{p}}_\beta = \textstyle\sum_\gamma \mathbf{M}_{\beta\gamma}(\mathbf{u}_\gamma - \mathbf{u}_\beta) - \tau_\beta\mathbf{u}_\beta + \tfrac{1}{2}\tau_\beta(\mathbf{u}_1 + \mathbf{u}_2), \tag{11.20}$$

where any tensor

$$\mathbf{M}_{\beta\gamma} = \mathbf{M}_{\gamma\beta}, \qquad \beta, \gamma = 1, 2, 3$$

is (symmetric) positive definite while $\tau_3 = 0$ and hence $\tau_1 + \tau_2 = 0$. To check the consistency of (11.20), we first observe that

$$\textstyle\sum_\beta \mathbf{m}_\beta = \sum_\beta \hat{\mathbf{p}}_\beta + \tau_\beta\mathbf{v}_\beta = \sum_{\beta,\gamma}\mathbf{M}_{\beta\gamma}(\mathbf{u}_\gamma - \mathbf{u}_\beta) + \sum_\beta \tau_\beta(\mathbf{v}_\beta - \mathbf{u}_\beta) + \tfrac{1}{2}\sum_\beta \tau_\beta(\mathbf{u}_1 + \mathbf{u}_2).$$

The sum $\sum_{\beta,\gamma}\mathbf{M}_{\beta\gamma}(\mathbf{u}_\gamma - \mathbf{u}_\beta)$ vanishes because $\mathbf{M}_{\beta\gamma}$ is symmetric, relative to β, γ, whereas $\mathbf{u}_\gamma - \mathbf{u}_\beta$ is skew. The other two sums are trivially zero. Hence $\sum_\beta \mathbf{m}_\beta = \mathbf{0}$ as is required. Now, to show that (11.19) holds, we compute

$$\sum_\beta[\hat{\mathbf{p}}_\beta \cdot \mathbf{u}_\beta + \tfrac{1}{2}\tau_\beta \mathbf{u}_\beta^2] = \sum_{\beta,\gamma} \mathbf{u}_\beta \cdot \mathbf{M}_{\beta\gamma}(\mathbf{u}_\gamma - \mathbf{u}_\beta) - \tfrac{1}{2}\sum_\alpha \tau_\alpha \mathbf{u}_\alpha^2 + \tfrac{1}{2}\sum_\alpha \tau_\alpha(\mathbf{u}_1 + \mathbf{u}_2)\cdot \mathbf{u}_\alpha.$$

Next we observe that

$$\begin{aligned}\sum_{\beta,\gamma} \mathbf{u}_\beta \cdot \mathbf{M}_{\beta\gamma}(\mathbf{u}_\gamma - \mathbf{u}_\beta) &= \tfrac{1}{2}[\sum_{\beta,\gamma} \mathbf{u}_\beta \cdot \mathbf{M}_{\beta\gamma}(\mathbf{u}_\gamma - \mathbf{u}_\beta) + \sum_{\beta,\gamma} \mathbf{u}_\gamma \cdot \mathbf{M}_{\beta\gamma}(\mathbf{u}_\beta - \mathbf{u}_\gamma)]\\ &= -\tfrac{1}{2}\sum_{\beta,\gamma}(\mathbf{u}_\gamma - \mathbf{u}_\beta)\cdot \mathbf{M}_{\beta\gamma}(\mathbf{u}_\gamma - \mathbf{u}_\beta) \le 0.\end{aligned}$$

The remaining terms provide a zero net result,

$$\tfrac{1}{2}\sum_\alpha \tau_\alpha(\mathbf{u}_1 + \mathbf{u}_2)\cdot \mathbf{u}_\alpha - \tfrac{1}{2}\sum_\alpha \tau_\alpha \mathbf{u}_\alpha^2 = \tfrac{1}{2}[\tau_1(\mathbf{u}_1 + \mathbf{u}_2)\cdot(\mathbf{u}_1 - \mathbf{u}_2) - \tau_1(\mathbf{u}_1^2 - \mathbf{u}_2^2)] = 0.$$

If, as is the case in some models exhibited in the literature, $\hat{\mathbf{p}}_s = \mathbf{0}$, the present scheme is still valid except for formal changes; (11.20) is nonzero for $\beta = 1, 2$ and hence $\sum_{\beta=1}^3$ becomes $\sum_{\beta=1}^2$.

Evolution equations

We now set up the evolution equations. Since we let the fluid and the solid be incompressible it follows that the convenient unknown functions are the volume fractions $\varphi_1, \varphi_2, \varphi_s$ and the velocities $\mathbf{v}_1, \mathbf{v}_2, \mathbf{v}_s$. Moreover, the temperature θ is a further unknown if the motion is not supposed to be isothermal.

The volume fractions are governed by the continuity equations

$$\dot{\varphi}_\alpha + \varphi_\alpha \nabla\cdot \mathbf{v}_\alpha = \frac{\tau_\alpha}{\rho^0}, \quad \alpha = 1, 2, \qquad \dot{\varphi}_s + \varphi_s \nabla\cdot \mathbf{v}_s = 0. \tag{11.21}$$

The equations of motion take the form

$$\rho^0\varphi_\alpha \dot{\mathbf{v}}_\alpha = \nabla\cdot \mathbf{T}_\alpha + \hat{\mathbf{p}}_\alpha + \rho^0\varphi_\alpha \mathbf{b}_\alpha, \quad \alpha = 1, 2, \qquad \rho_s^0\varphi_s \dot{\mathbf{v}}_s = \nabla\cdot \mathbf{T}_s + \hat{\mathbf{p}}_s + \rho_s^0 \mathbf{b}_s. \tag{11.22}$$

For a linear viscous fluid

$$\mathbf{T}_\alpha = -p_\alpha \mathbf{1} + 2\mu_\alpha \mathbf{D}_\alpha + \lambda_\alpha(\mathrm{tr}\,\mathbf{D}_\alpha)\mathbf{1},$$

and for a linear elastic solid

$$\mathbf{T}_s = -p_s \mathbf{1} + 2\bar{\mu}_\alpha \varepsilon_\alpha + \bar{\lambda}_\alpha(\mathrm{tr}\,\varepsilon_s)\mathbf{1}.$$

The literature shows quite different forms of evolution equations. If $\partial_t\varphi_\alpha = 0$, $\partial_t\varphi_s = 0$ then Eq. (11.21) becomes

$$\nabla\cdot(\varphi_\alpha \mathbf{v}_\alpha) = \frac{\tau_\alpha}{\rho^0}, \qquad \nabla\cdot(\varphi_s \mathbf{v}_s) = 0;$$

these equations coincide, e.g. with the continuity equations determined in [238]. Yet, more often, significantly different equations are considered.[5] First, $\tau_1/\rho^0 = -\tau_2/\rho^0$ is supposed to be a constant times the pressure difference

$$p_1 - p_2,$$

most likely because of the Barenblatt-et-al equation (11.15). Moreover, $\grave{\varphi}_\alpha$ (or perhaps $\dot{\varphi}_\alpha$) is taken to equal

$$-\gamma_\alpha \dot{p}_\alpha - \beta_\alpha \nabla \cdot \mathbf{v}_s,$$

where γ_α is said to be the compressibility of pores or fissures while β_α is a measure of the change of porosities. Finally, $\nabla \cdot \mathbf{v}_\alpha$ is evaluated via Darcy's law

$$\mathbf{v}_\alpha = -\frac{k_\alpha}{\mu_\alpha} \nabla p_\alpha$$

so that $\nabla \cdot \mathbf{v}_\alpha = -(k_\alpha/\mu_\alpha)\Delta p_\alpha$. Concerning the equations of motion, an effective stress concept is involved so that the pressures p_1, p_2 occur in $\mathbf{T}_s$ as the pressure tensor $(-\beta_1 p_1 - \beta_2 p_2)\mathbf{1}$ [109, 406]. Moreover, the forces $\hat{\mathbf{p}}_\alpha$ of the present approach are disregarded or are taken to be a scalar times $\varphi_\alpha^2 \mathbf{K}_\alpha^{-1} \mathbf{v}_\alpha$, $\mathbf{K}_\alpha$ being called the permeability [238]. This is questionable in relation to the consistency with the principle of objectivity and the constraint on $\sum_\alpha \hat{\mathbf{p}}_\alpha$.

[5] See, e.g. [172, 403, 406] and references therein.

Chapter 12
Electromagnetism of Continuous Media

This chapter is devoted to electromagnetism in deformable media. As with any model of material behaviour, the constitutive laws are required to be consistent with the second law of thermodynamics and to comply with the objectivity requirements.

Electromagnetism in the matter is obviously based on Maxwell's equations as the fundamental set of equations governing electromagnetic phenomena. Maxwell's equations are complemented by the balance equations where the electromagnetic fields enter the balance of linear momentum, angular momentum, and energy. Constitutive laws are needed to express the properties of the material and describe the interaction between the field and matter. Since the electromagnetic fields are dependent on the frame of reference, an appropriate approach to the modelling of electromagnetic media should involve the fields at rest with the body at the point under consideration. Hence attention is mainly restricted to the fields at the frame at rest, at the pertinent point.

The models developed in this chapter describe various material properties and applied electromagnetic fields. The general subject of electroelasticity allows us to investigate piezoelectric materials and deformable dielectrics. Magnetoelasticity allows the modelling of magnetic refrigeration. Micromagnetics is viewed as a particle-like scheme leading to rate-type equations for the magnetization. Ferrofluids are modelled as a mixture where the ferromagnetic particles are a micropolar constituent. A magnetoelastic material allows us to describe the effects of an applied magnetic field on the propagation of electric wave fields. Plasmas, i.e. ionized gases with free electrons and ions, are viewed as both binary mixtures and one-fluid thus leading to the magneto-hydrodynamic equations. Further chiral media are investigated through classical models of the literature to explain the phenomenon of optical activity. Ferrites are described as magnetic materials acted upon by a constant field; a superposed circularly polarized wave is found to propagate at different speeds depending on the direction of propagation.

A. Morro and C. Giorgi, *Mathematical Modelling of Continuum Physics*,
Modeling and Simulation in Science, Engineering and Technology,
https://doi.org/10.1007/978-3-031-20814-0_12

12.1 A Thermodynamic Setting for Electromagnetic Solids

A general scheme is set up for the interaction between electromagnetic fields and elastic deformations in polarizable materials. For generality, the scheme allows for electric conduction, heat conduction, and viscosity. Specific models are then established in the next sections[1] at the frame at rest, at the pertinent point.

The balance equations of linear momentum, angular momentum, and energy can be written in the form (Sect. 2.16.1)

$$\rho\dot{\mathbf{v}} = \nabla\cdot\mathbf{T} + \rho\mathbf{b} + (\mathbf{P}\cdot\nabla)\mathbf{E} + \mathbf{v}\times(\mathbf{P}\cdot\nabla)\mathbf{B} + \rho\dot{\pi}\times\mathbf{B}$$
$$+\mu_0(\mathbf{M}\cdot\nabla)\mathbf{H} - \mu_0\mathbf{v}\times(\mathbf{M}\cdot\nabla)\mathbf{D} - \mu_0\rho\dot{\mathbf{m}}\times\mathbf{D} + q\mathbf{E} + \mathbf{J}\times\mathbf{B},$$

$$\text{skw}\mathbf{T} = \text{skw}(\mathbf{P}\otimes\mathbf{E} + \mu_0\mathbf{M}\otimes\mathbf{H}),$$

$$\rho\dot{\varepsilon} = \mathbf{E}\cdot\mathbf{J} + \rho\mathbf{E}\cdot\dot{\pi} + \rho\mu_0\mathbf{H}\cdot\dot{\mathbf{m}} + \mathbf{T}\cdot\mathbf{L} - \nabla\cdot\mathbf{q} + \rho r, \tag{12.1}$$

where $\mathbf{b}$ is the mechanical body force, per unit mass. Owing to the choice of the co-moving observer, $\mathbf{E} = \mathcal{E}$, $\mathbf{J} = \mathcal{J}$.

Models of electromagnetic media are framed within thermodynamics by requiring that the constitutive equations be compatible with the second law expressed by the standard inequality

$$\rho\theta\dot{\eta} + \nabla\cdot\mathbf{q} - \rho r - \frac{1}{\theta}\mathbf{q}\cdot\nabla\theta + \theta\nabla\cdot\mathbf{k} \geq 0.$$

Upon substitution of $\nabla\cdot\mathbf{q} - \rho r$ we obtain

$$\rho\theta\dot{\eta} - \rho\dot{\varepsilon} + \mathbf{E}\cdot\mathbf{J} + \rho\dot{\pi}\cdot\mathbf{E} + \rho\mu_0\dot{\mathbf{m}}\cdot\mathbf{H} + \mathbf{T}\cdot\mathbf{L} + \theta\nabla\cdot\mathbf{k} \geq 0.$$

Using the Helmholtz free energy $\psi = \varepsilon - \theta\eta$ we can write the entropy inequality in the form

$$-\rho(\dot{\psi} + \eta\dot{\theta}) + \mathbf{E}\cdot\mathbf{J} + \rho\dot{\pi}\cdot\mathbf{E} + \rho\mu_0\dot{\mathbf{m}}\cdot\mathbf{H} + \mathbf{T}\cdot\mathbf{L} - \frac{1}{\theta}\mathbf{q}\cdot\nabla\theta + \theta\nabla\cdot\mathbf{k} \geq 0.$$

For technical convenience we consider the free energy ϕ defined by

$$\phi = \psi - \boldsymbol{\pi}\cdot\mathbf{E} - \mu_0\mathbf{m}\cdot\mathbf{H}.$$

Upon substitution of $\dot{\psi}$ we obtain the entropy inequality in the form

[1] To avoid obvious ambiguities, in this chapter $\mathbf{D}$ denotes the electric displacement (vector) and $\mathfrak{D}$ the stretching tensor.

$$- \rho(\dot{\phi} + \eta\dot{\theta}) + \mathbf{E}\cdot\mathbf{J} - \mathbf{P}\cdot\dot{\mathbf{E}} - \mu_0\mathbf{M}\cdot\dot{\mathbf{H}} + \mathbf{T}\cdot\mathbf{L} - \frac{1}{\theta}\mathbf{q}\cdot\nabla\theta + \theta\nabla\cdot\mathbf{k} \geq 0. \tag{12.2}$$

Quite a general model of electromagnetic medium might involve a set of constitutive equations where, e.g.

$$\phi, \eta, \mathbf{J}, \mathbf{P}, \mathbf{M}, \mathbf{T}, \mathbf{q}, \mathbf{k}$$

are functions of

$$\mathbf{F}, \theta, \mathbf{E}, \mathbf{H}, \nabla\theta$$

and their space and time derivatives. For instance, $\dot{\mathbf{F}}$ may occur to model viscous effects, $\nabla\theta$ to model heat conduction, $\nabla\mathbf{E}$ and $\dot{\mathbf{E}}$ to describe spatial interaction and memory. Also, $\mathbf{q}$ may occur among the independent variables if the constitutive equation for $\mathbf{q}$ is a rate-type equation. The free energy ϕ, whose time derivative occurs in (12.2), is required to be Euclidean invariant and hence can depend on the independent variables through their invariants.

In the analysis of the degrees of freedom allowed for the electromagnetic fields, Maxwell's equations play the role of constraints. In this connection it is worth remarking that, since $\mathbf{E}, \mathbf{B}, \mathbf{J}, \mathbf{P}, \mathbf{M}$ are relative to the co-moving observer, Maxwell's equations take the form

$$\nabla\cdot\mathbf{D} = q, \qquad \nabla\cdot\mathbf{B} = 0, \qquad \nabla\times\mathbf{H} = \mathbf{J} + \overset{\diamond}{\mathbf{D}}, \qquad \nabla\times\mathbf{E} = -\overset{\diamond}{\mathbf{B}}.$$

For later use we observe that Maxwell's equations for the co-moving observer can be given the form $\nabla\times\mathbf{H} = \mathbf{J} + \epsilon_0\overset{\diamond}{\mathbf{E}} + \overset{\diamond}{\mathbf{P}}$, $\nabla\times\mathbf{E} = -\mu_0(\overset{\diamond}{\mathbf{H}} + \overset{\diamond}{\mathbf{M}})$, in addition to $\nabla\cdot(\epsilon_0\mathbf{E} + \mathbf{P}) = q$, $\nabla\cdot(\mathbf{H} + \mathbf{M}) = 0$. Since $\overset{\diamond}{\mathbf{E}} = \dot{\mathbf{E}} - \mathbf{L}\mathbf{E} + (\nabla\cdot\mathbf{v})\mathbf{E}$, and the like for $\mathbf{H}$, appropriate values of $\nabla\times\mathbf{H}$ an $\nabla\times\mathbf{E}$ allow us to say that arbitrary prescribed values of $\dot{\mathbf{H}}$ and $\dot{\mathbf{E}}$, at any point $\mathbf{x}\in\mathcal{R}_t$ and any time t, are consistent with Maxwell's equations. Moreover, since $\epsilon_0\ddot{\mathbf{E}} = (\nabla\times\mathbf{H} - \mathbf{J} + \epsilon_0\mathbf{L}\mathbf{E} - \epsilon_0(\nabla\cdot\mathbf{v})\mathbf{E})^{\cdot}$, then arbitrarily prescribed values of $\dot{\mathbf{E}}$ and $\ddot{\mathbf{E}}$ are allowed, at any point $\mathbf{x}\in\mathcal{R}_t$ and any time t, by requiring that $\nabla\times\mathbf{H}$ and $(\nabla\times\mathbf{H})^{\cdot}$ take appropriate values.

Remarkably, for undeformable bodies, the recourse to the Poynting vector $\mathbf{E}\times\mathbf{H}$ as the electromagnetic energy flux allows us to state the balance of energy in the form

$$\rho\dot{\varepsilon} = -\nabla\cdot(\mathbf{q} + \mathbf{E}\times\mathbf{H}) + \rho r.$$

Since, by Maxwell's equations,

$$-\nabla\cdot(\mathbf{E}\times\mathbf{H}) = -(\nabla\times\mathbf{E})\cdot\mathbf{H} + \nabla\times\mathbf{H}\cdot\mathbf{E} = \dot{\mathbf{B}}\cdot\mathbf{H} + \mathbf{J}\cdot\mathbf{E} + \dot{\mathbf{D}}\cdot\mathbf{E},$$

then Eq. (12.1) is recovered provided we re-define the internal energy, $\rho\varepsilon \to \rho\varepsilon_{\text{new}} = \rho\varepsilon - \frac{1}{2}(\epsilon_0\mathbf{E}^2 + \mu_0\mathbf{H}^2)$.

12.2 Electroelasticity

The body is viewed as a deformable dielectric solid. We then assume the magnetization and the electric current are zero. Hence the electromagnetic fields **E**, **P**, **H**, **D**, **B** are

$$\mathbf{D} = \epsilon_0 \mathbf{E} + \mathbf{P}, \qquad \mathbf{B} = \mu_0 \mathbf{H}.$$

A wide range of material behaviours is established beyond the standard models of electroelasticity. For instance, flexoelectricity indicates that the dependence on the second gradient of deformation is allowed. By analogy with models involving the dependence on the polarization the dependence on the gradient of the electric field is also considered.

Heat conduction might be described via a scheme as in Fourier-type models where $\nabla\theta$ is an independent variable and $\mathbf{q}$ is given by a constitutive function. To get a model compatible with discontinuity wave propagation it is natural to describe heat conduction via a rate-type equation. Here, though, for formal simplicity we ignore heat conduction.

Hence the electroelastic solid might be modelled by letting

$$\mathbf{F}, \theta, \mathbf{E}, \nabla\mathbf{F}, \dot{\mathbf{F}}, \nabla\mathbf{E}, \dot{\mathbf{E}}$$

be the set of independent variables. The dependence on the time derivatives $\dot{\mathbf{F}}, \dot{\mathbf{E}}$ is motivated by requirements of thermodynamic consistency.

As to the free energy $\phi = \psi - \boldsymbol{\pi} \cdot \mathbf{E} - \mu_0 \mathbf{m} \cdot \mathbf{H}$, the constitutive equation is objective if the dependence is through invariants such as

$$\mathbf{C} = \mathbf{F}^T\mathbf{F}, \quad \nabla\mathbf{E} \cdot \nabla\mathbf{E}, \quad \mathfrak{E} = \mathbf{F}^T\mathbf{E}.$$

For definiteness we let

$$\phi = \phi(\mathbf{C}, \theta, \mathfrak{E}, \nabla_{\!R}\mathbf{C}, \dot{\mathbf{C}}, \nabla_{\!R}\mathfrak{E}, \dot{\mathfrak{E}}).$$

Observe that $\mathbf{F}^T\mathbf{E} = \mathbf{E}\mathbf{F}$ and hence $\mathfrak{E} = \mathbf{E}_R$, $\mathbf{E}_R$ being the electric field in the material description.

Since $\mathbf{J}$, $\mathbf{q}$, and $\mathbf{M}$ are assumed to vanish then the entropy inequality (12.2) simplifies to

$$-\rho(\dot{\phi} + \eta\dot{\theta}) - \mathbf{P} \cdot \dot{\mathbf{E}} + \mathbf{T} \cdot \mathbf{L} + \theta\nabla \cdot \mathbf{k} \geq 0.$$

Upon computation of $\dot{\phi}$ and substitution it follows

$$-\rho(\partial_\theta\phi + \eta)\dot{\theta} - \rho\partial_{\mathbf{C}}\phi \cdot \dot{\mathbf{C}} - \rho\partial_{\mathfrak{E}}\,\phi \cdot \dot{\mathfrak{E}} - \rho\partial_{\nabla_{\!R}\mathbf{C}}\phi \cdot \nabla_{\!R}\dot{\mathbf{C}} - \rho\partial_{\dot{\mathbf{C}}}\phi \cdot \ddot{\mathbf{C}}$$
$$-\rho\partial_{\nabla_{\!R}\mathfrak{E}}\phi \cdot \nabla_{\!R}\dot{\mathfrak{E}} - \rho\partial_{\dot{\mathfrak{E}}}\phi \cdot \ddot{\mathfrak{E}} - \mathbf{P} \cdot \dot{\mathbf{E}} + \mathbf{T} \cdot \mathbf{L} + \theta\nabla \cdot \mathbf{k} \geq 0.$$

The arbitrariness of $\ddot{\mathbf{C}}$, $\ddot{\mathfrak{E}}$, and $\dot{\theta}$ implies

$$\partial_{\dot{\mathbf{C}}}\phi = \mathbf{0}, \qquad \partial_{\dot{\mathfrak{E}}}\phi = \mathbf{0}, \qquad \eta = -\partial_\theta\phi.$$

Hence the inequality reduces to

$$-\rho\partial_{\mathbf{C}}\phi\cdot\dot{\mathbf{C}} - \rho\partial_{\mathfrak{E}}\phi\cdot\dot{\mathfrak{E}} - \rho\partial_{\nabla_R\mathbf{C}}\phi\cdot\nabla_R\dot{\mathbf{C}} - \rho\partial_{\nabla_R\mathfrak{E}}\phi\cdot\nabla_R\dot{\mathfrak{E}} - \mathbf{P}\cdot\dot{\mathbf{E}} + \mathbf{T}\cdot\mathbf{L} + \theta\nabla\cdot\mathbf{k} \geq 0.$$

Multiply by J/θ and observe

$$J\rho = \rho_R, \qquad J\nabla\cdot\mathbf{k} = \nabla_R\cdot\mathbf{k}_R, \qquad \mathbf{k}_R := J\mathbf{F}^{-1}\mathbf{k}.$$

We can then write

$$-\frac{\rho_R}{\theta}\partial_{\mathbf{C}}\phi\cdot\dot{\mathbf{C}} - \frac{\rho_R}{\theta}\partial_{\mathfrak{E}}\phi\cdot\dot{\mathfrak{E}} - \frac{\rho_R}{\theta}\partial_{\nabla_R\mathbf{C}}\phi\cdot\nabla_R\dot{\mathbf{C}} - \frac{\rho_R}{\theta}\partial_{\nabla_R\mathfrak{E}}\phi\cdot\nabla_R\dot{\mathfrak{E}} - \frac{J}{\theta}\mathbf{P}\cdot\dot{\mathbf{E}} + \frac{J}{\theta}\mathbf{T}\cdot\mathbf{L} + \nabla_R\cdot\mathbf{k}_R \geq 0,$$

whence

$$-\frac{\rho_R}{\theta}\delta_{\mathbf{C}}\phi\cdot\dot{\mathbf{C}} - \frac{\rho_R}{\theta}\delta_{\mathfrak{E}}\phi\cdot\dot{\mathfrak{E}} - \frac{J}{\theta}\mathbf{P}\cdot\dot{\mathbf{E}} + \frac{J}{\theta}\mathbf{T}\cdot\mathbf{L} + \nabla_R\cdot(\mathbf{k}_R - \frac{\rho_R}{\theta}\partial_{\nabla_R\mathbf{C}}\phi\dot{\mathbf{C}} - \frac{\rho_R}{\theta}\partial_{\nabla_R\mathfrak{E}}\phi\dot{\mathfrak{E}}) \geq 0,$$

where

$$\delta_{\mathbf{C}}\phi = \partial_{\mathbf{C}}\phi - \frac{\theta}{\rho_R}\nabla_R\cdot\left(\frac{\rho_R}{\theta}\partial_{\nabla_R\mathbf{C}}\phi\right), \qquad \delta_{\mathfrak{E}}\phi = \partial_{\mathfrak{E}}\phi - \frac{\theta}{\rho_R}\nabla_R\cdot\left(\frac{\rho_R}{\theta}\partial_{\nabla_R\mathfrak{E}}\phi\right)$$

can be viewed as the variational derivatives of ϕ with respect to $\mathbf{C}$ and $\mathfrak{E}$. This indicates that we let

$$\mathbf{k}_R = \frac{\rho_R}{\theta}(\partial_{\nabla_R\mathbf{C}}\phi\dot{\mathbf{C}} + \partial_{\nabla_R\mathfrak{E}}\phi\dot{\mathfrak{E}}).$$

The identities

$$\mathbf{\Lambda}\cdot\dot{\mathbf{C}} = \mathbf{\Lambda}\cdot(\dot{\mathbf{F}}^T\mathbf{F} + \mathbf{F}^T\dot{\mathbf{F}}) = 2(\mathbf{F}\mathbf{\Lambda}\mathbf{F}^T)\cdot\mathfrak{D},$$

$$\boldsymbol{\xi}\cdot\dot{\mathfrak{E}} = (\mathbf{F}\boldsymbol{\xi})\cdot\dot{\mathbf{E}} + (\mathbf{E}\otimes\boldsymbol{\xi})\cdot\dot{\mathbf{F}}, \qquad (\mathbf{E}\otimes\boldsymbol{\xi})\cdot\dot{\mathbf{F}} = [\mathbf{E}\otimes(\mathbf{F}\boldsymbol{\xi})]\cdot\mathbf{L},$$

hold for any symmetric tensor $\mathbf{\Lambda}$ and vector $\boldsymbol{\xi}$. Applying these relations to

$$\mathbf{\Lambda} = \delta_{\mathbf{C}}\phi, \qquad \boldsymbol{\xi} = \delta_{\mathfrak{E}}\phi$$

we can write the reduced inequality $-\rho\delta_{\mathbf{C}}\phi\cdot\dot{\mathbf{C}} - \rho\delta_{\mathfrak{E}}\phi\cdot\dot{\mathfrak{E}} - \mathbf{P}\cdot\dot{\mathbf{E}} + \mathbf{T}\cdot\mathbf{L} \geq 0$ in the form

$$[\mathbf{T} - 2\rho\mathbf{F}\,\delta_{\mathbf{C}}\phi\mathbf{F}^T - \rho\mathbf{E}\otimes(\mathbf{F}\delta_{\mathfrak{E}}\phi)]\cdot\mathfrak{D} + [\mathbf{T} - \rho\mathbf{E}\otimes(\mathbf{F}\delta_{\mathfrak{E}}\phi)]\cdot\mathbf{W} - (\mathbf{P} + \rho\mathbf{F}\delta_{\mathfrak{E}}\phi)\cdot\dot{\mathbf{E}} \geq 0.$$

The arbitrariness of $\mathbf{W}$ implies

$$\text{skw}[\mathbf{T} - \rho\mathbf{E}\otimes(\mathbf{F}\delta_{\mathfrak{E}}\phi)] = \mathbf{0}. \tag{12.3}$$

Let

$$\mathcal{T} = \mathbf{T} - 2\rho\mathbf{F}\,\delta_{\mathbf{C}}\phi\,\mathbf{F}^T - \rho\mathbf{E}\otimes(\mathbf{F}\delta_{\mathfrak{E}}\,\phi), \qquad \mathcal{P} = \mathbf{P} + \rho\mathbf{F}\delta_{\mathfrak{E}}\,\phi.$$

By (12.3) it follows that $\mathcal{T} \in \mathrm{Sym}$. The reduced inequality then reads

$$\mathcal{T}\cdot\mathfrak{D} - \mathcal{P}\cdot\dot{\mathbf{E}} \geq 0 \tag{12.4}$$

while

$$\mathbf{T} = 2\rho\mathbf{F}\,\delta_{\mathbf{C}}\phi\,\mathbf{F}^T + \rho\mathbf{E}\otimes(\mathbf{F}\delta_{\mathfrak{E}}\,\phi) + \mathcal{T}, \qquad \mathbf{P} = -\rho\mathbf{F}\delta_{\mathfrak{E}}\,\phi + \mathcal{P}.$$

Hence, $\mathcal{T}$ and $\mathcal{P}$ may depend on $\dot{\mathbf{C}}$ and $\dot{\mathfrak{E}}$ and are subject to (12.4). We can view $\mathcal{T}$ and $\mathcal{P}$ as the dissipative parts of the stress and the polarization. If, as is customary in electroelasticity, we neglect dissipative effects ($\mathcal{T} = \mathbf{0}, \mathcal{P} = \mathbf{0}$) then it follows

$$\mathbf{P} = -\rho\mathbf{F}\delta_{\mathfrak{E}}\,\phi, \tag{12.5}$$

$$\mathbf{T} = 2\rho\mathbf{F}\,\delta_{\mathbf{C}}\phi\,\mathbf{F}^T - \mathbf{E}\otimes\mathbf{P} \tag{12.6}$$

while (12.3) becomes

$$\mathrm{skw}[\mathbf{T} + \mathbf{E}\otimes\mathbf{P}] = \mathbf{0}. \tag{12.7}$$

The constraint (12.7) coincides with the balance equation of angular momentum. This shows that the dependence on $\mathbf{E}$ via the vector $\mathbf{F}^T\mathbf{E}$ is appropriate to satisfy the balance equation.

We wonder whether other dependencies might be consistent with the balance of angular momentum. For definiteness let ϕ depend on the electric field via the invariant scalars

$$\varkappa = \mathbf{E}^2, \qquad \sigma = |\nabla\mathbf{E}|^2, \qquad \varpi = |\dot{\mathbf{E}}|^2.$$

So we let

$$\phi = \tilde{\phi}(\mathbf{C}, \theta, \varkappa, \sigma, \varpi).$$

Computation of $\dot{\phi}$ and substitution in the entropy inequality yields

$$-\rho(\partial_\theta\tilde{\phi} + \eta)\dot{\theta} - 2\rho(\mathbf{F}\,\partial_{\mathbf{C}}\tilde{\phi}\,\mathbf{F}^T)\cdot\mathfrak{D} - 2\rho\partial_\varkappa\tilde{\phi}\mathbf{E}\cdot\dot{\mathbf{E}} - 2\rho\partial_\sigma\tilde{\phi}\nabla\mathbf{E}\cdot\dot{\overline{\nabla\mathbf{E}}} - 2\rho\partial_\varpi\tilde{\phi}\dot{\mathbf{E}}\cdot\ddot{\mathbf{E}}$$
$$-\mathbf{P}\cdot\dot{\mathbf{E}} + \mathbf{T}\cdot\mathbf{L} + \theta\nabla\cdot\mathbf{k} \geq 0.$$

The arbitrariness of $\ddot{\mathbf{E}}$ and $\dot{\theta}$ implies

$$\partial_\varpi\tilde{\phi} = 0, \qquad \eta = -\partial_\theta\tilde{\phi}.$$

Since $\dot{\overline{\nabla\mathbf{E}}} = \nabla\dot{\mathbf{E}} - (\mathbf{L}^T\nabla)\mathbf{E}$ then

$$\nabla\mathbf{E}\cdot\overline{\nabla\mathbf{E}}^{\cdot} = \partial_{x_i}E_j\partial_{x_i}\dot{E}_j - \partial_{x_i}E_jL_{ki}\partial_{x_k}E_j = \nabla\mathbf{E}\cdot\nabla\dot{\mathbf{E}} - (\nabla\mathbf{E}\otimes\nabla\mathbf{E})\cdot\mathfrak{D}.$$

Upon substitution we obtain the inequality in the form

$$[\mathbf{T} - 2\rho\,\mathbf{F}\,\partial_{\mathbf{C}}\tilde{\phi}\,\mathbf{F}^T + 2\rho\nabla\mathbf{E}\otimes\nabla\mathbf{E}]\cdot\mathfrak{D} - 2\rho\partial_{\varkappa}\tilde{\phi}\mathbf{E}\cdot\dot{\mathbf{E}} - 2\rho\partial_\sigma\tilde{\phi}\nabla\mathbf{E}\cdot\nabla\dot{\mathbf{E}} \\ -\mathbf{P}\cdot\dot{\mathbf{E}} + \mathbf{T}\cdot\mathbf{W} + \theta\nabla\cdot\mathbf{k} \geq 0$$

Divide by θ and observe

$$\frac{2\rho}{\theta}\partial_\sigma\tilde{\phi}\nabla\mathbf{E}\cdot\nabla\dot{\mathbf{E}} = \nabla\cdot(\frac{2\rho}{\theta}\partial_\sigma\tilde{\phi}\nabla\mathbf{E}\,\dot{\mathbf{E}}) - [\nabla\cdot(\frac{2\rho}{\theta}\partial_\sigma\tilde{\phi}\nabla\mathbf{E})]\cdot\dot{\mathbf{E}}.$$

Hence we have

$$\frac{1}{\theta}[\mathbf{T} - 2\rho\,\mathbf{F}\,\partial_{\mathbf{C}}\tilde{\phi}\,\mathbf{F}^T + 2\rho\nabla\mathbf{E}\otimes\nabla\mathbf{E}]\cdot\mathfrak{D} - \frac{1}{\theta}[\mathbf{P} + 2\rho\partial_{\varkappa}\tilde{\phi}\mathbf{E} - \theta\nabla\cdot(\frac{2\rho}{\theta}\partial_\sigma\tilde{\phi}\nabla\mathbf{E})]\cdot\dot{\mathbf{E}} \\ +\frac{1}{\theta}\mathbf{T}\cdot\mathbf{W} + \nabla\cdot(\mathbf{k} - \frac{2\rho}{\theta}\partial_\sigma\tilde{\phi}\nabla\mathbf{E}\,\dot{\mathbf{E}}) \geq 0.$$

We then let

$$\mathbf{k} = \frac{2\rho}{\theta}\partial_\sigma\tilde{\phi}\nabla\mathbf{E}\,\dot{\mathbf{E}}.$$

Moreover the arbitrariness of $\mathbf{W}$ implies

$$\mathrm{skw}\mathbf{T} = \mathbf{0}.$$

Hence letting

$$\tilde{\mathbf{T}} = \mathbf{T} - 2\rho\,\mathbf{F}\,\partial_{\mathbf{C}}\tilde{\phi}\,\mathbf{F}^T + 2\rho\nabla\mathbf{E}\otimes\nabla\mathbf{E}, \qquad \tilde{\mathbf{P}} = \mathbf{P} + 2\rho\partial_{\varkappa}\tilde{\phi}\mathbf{E} - \theta\nabla\cdot(\frac{2\rho}{\theta}\partial_\sigma\tilde{\phi}\nabla\mathbf{E}),$$

we conclude that

$$\mathbf{T} = 2\rho\,\mathbf{F}\,\partial_{\mathbf{C}}\tilde{\phi}\,\mathbf{F}^T - 2\rho\nabla\mathbf{E}\otimes\nabla\mathbf{E} + \tilde{\mathbf{T}}, \qquad \mathbf{P} = -2\rho\partial_{\varkappa}\tilde{\phi}\mathbf{E} + \theta\nabla\cdot(\frac{2\rho}{\theta}\partial_\sigma\tilde{\phi}\nabla\mathbf{E}) + \tilde{\mathbf{P}},$$

where $\tilde{\mathbf{T}} \in \mathrm{Sym}$ and $\tilde{\mathbf{P}}$ are subject to

$$\tilde{\mathbf{T}}\cdot\mathfrak{D} - \tilde{\mathbf{P}}\cdot\dot{\mathbf{E}} \geq 0.$$

This inequality is satisfied if $\tilde{\mathbf{T}}$ depends linearly on $\mathfrak{D}$, as with viscous materials, and $\tilde{\mathbf{P}}$ is parallel to $\dot{\mathbf{E}}$. Yet in electroelasticity the inequality is trivially satisfied by letting $\tilde{\mathbf{T}} = \mathbf{0}, \tilde{\mathbf{P}} = \mathbf{0}$.

The symmetry of $\mathbf{T}$, $\mathrm{skw}\mathbf{T} = \mathbf{0}$, looks inconsistent with the balance of angular momentum,

$$\mathrm{skw}(\mathbf{T} + \mathbf{E}\otimes\mathbf{P}) = \mathbf{0}.$$

In fact the dependence on $\varkappa$ produces a term parallel to $\mathbf{E}$ while the dependence on σ results in

$$\nabla(\frac{2\rho}{\theta}\partial_\sigma\tilde{\phi}) \cdot \nabla\mathbf{E} + (\frac{2\rho}{\theta}\partial_\sigma\tilde{\phi})\Delta\mathbf{E}.$$

Thus, in this scheme, consistency happens if ϕ depends on $\mathbf{E}$ only through $\varkappa = \mathbf{E}^2$, not through $\sigma = |\nabla\mathbf{E}|^2$. Consistency holds in general in terms of $\mathfrak{E}$ by substitution of $\dot{\mathbf{E}} = -\mathbf{L}^T\mathbf{E} + \mathbf{F}^{-T}\dot{\mathfrak{E}}$ and $\sigma = |\nabla_R\mathfrak{E}|^2$ (see Sect. 15.1).

12.3 Electroelastic Materials

Piezoelectric materials

Piezoelectricity means electricity resulting from mechanical stress. In a normal dielectric the polarization is zero in the absence of an electric field. In a piezoelectric solid, instead, a surface electric charge develops when the solid is subject to a mechanical stress, even in the absence of an electric field. This is called the direct piezoelectric effect. The effect is reversible and the inverse piezoelectric effect consists in a change of shape as a consequence of an applied electric field.

There are solids, called pyroelectric solids, where a change of temperature induces an electric polarization (change). Opposite effects on polarization are obtained by increasing or decreasing temperature. Pyroelectric crystals are a subset of piezoelectrics; all pyroelectric crystals are piezoelectrics but not all piezoelectrics show pyroelectricity.

The piezoelectric effect is understood as the linear electromechanical interaction between the mechanical and the electric behaviours. Hence, piezoelectricity is the combined effect of the linear electrical behaviour

$$\mathbf{D} = \epsilon\,\mathbf{E},$$

ϵ being the second-order permittivity tensor, and the linear stress–strain law

$$\mathbf{T} = \mathsf{C}\,\mathbf{S},$$

C being the elasticity fourth-order tensor and $\mathbf{S}$ being the infinitesimal strain tensor,[2]

$$\mathbf{S} = \mathrm{sym}\nabla\mathbf{u}.$$

Piezoelectricity is then described by letting

$$\mathbf{T} = \mathsf{C}\,\mathbf{S} + \mathbf{\Lambda}\,\mathbf{E},$$

[2] To avoid any ambiguities with the electric field $\mathbf{E}$ and the permittivity ϵ, here the infinitesimal strain is denoted by $\mathbf{S}$ and not by ε.

$$\mathbf{D} = \boldsymbol{\Gamma}\,\mathbf{S} + \epsilon\,\mathbf{E},$$

$\boldsymbol{\Lambda}, \boldsymbol{\Gamma}$ being third-order tensors.

First we ascertain the thermodynamic consistency of this description. As in the general scheme of electroelasticity, the material is taken to be non-conducting ($\mathbf{J} = \mathbf{0}$, $\mathbf{q} = \mathbf{0}$) and non-magnetizable ($\mathbf{M} = \mathbf{0}$). Moreover, for simplicity we first neglect effects described by the gradients $\nabla\mathbf{C}$ and $\nabla\mathbf{E}$. The dependence (of the free energy) on $\mathbf{F}, \theta, \mathbf{E}$ can be modelled via $\mathbf{C}, \theta, \mathbf{F}^T\mathbf{E}$ or $\mathbf{C}, \theta, \mathbf{E}^2$,

$$\phi = \hat{\phi}(\mathbf{C}, \theta, \mathfrak{E}), \qquad \phi = \tilde{\phi}(\mathbf{C}, \theta, \varkappa),$$

where $\mathfrak{E} = \mathbf{F}^T\mathbf{E}$ and $\varkappa = \mathbf{E}^2$. We then find that

$$\mathbf{P} = -\rho\mathbf{F}\partial_{\mathfrak{E}}\hat{\phi}, \qquad \mathbf{T} = 2\rho\mathbf{F}\partial_{\mathbf{C}}\hat{\phi}\mathbf{F}^T - \mathbf{E}\otimes\mathbf{P}, \tag{12.8}$$

or

$$\mathbf{P} = -2\rho\partial_{\varkappa}\tilde{\phi}\,\mathbf{E}, \qquad \mathbf{T} = 2\rho\,\mathbf{F}\partial_{\mathbf{C}}\tilde{\phi}\,\mathbf{F}^T. \tag{12.9}$$

Furthermore,

$$\eta = -\partial_\theta\hat{\phi} = -\partial_\theta\tilde{\phi}, \qquad \mathbf{k} = \mathbf{0}.$$

Equations (12.8) and (12.9) are nonlinear constitutive relations for $\mathbf{P}$ and $\mathbf{T}$ in electroelasticity. To model piezoelectricity we look at a linearized version and it is natural to choose the reference configuration as that around which we describe the (linearized) piezoelectric properties; as we see in a moment, the same scheme accounts also for pyroelectricity. Equations (12.9) are inadequate as the starting point for the model of piezoelectricity. The dependence on $\mathbf{E}^2$, rather than on $\mathbf{F}^T\mathbf{E}$ or $\mathbf{E}$, is reasonable if the material is isotropic, which is not the case of piezoelectricity.

We assume $\mathbf{P} \to \mathbf{0}$ as $\mathbf{E}$ and $\mathbf{C} - \mathbf{1}$ approach zero. Hence the stress $\mathbf{E}\otimes\mathbf{P}$ has to be neglected in a linear scheme of piezoelectricity. Since $\mathbf{C} = \mathbf{1}$ in the reference configuration, we regard $\mathbf{C} - \mathbf{1}$ as the convenient mechanical variable and observe

$$\mathbf{C} - \mathbf{1} = \nabla_R\mathbf{u} + (\nabla_R\mathbf{u})^T + o(|\nabla_R\mathbf{u}|) = \nabla\mathbf{u} + (\nabla\mathbf{u})^T + o(|\nabla_R\mathbf{u}|).$$

Hence, in the linear approximation

$$\mathbf{C} - \mathbf{1} \simeq 2\,\mathbf{S}, \qquad \mathfrak{E} = \mathbf{F}^T\,\mathbf{E} \simeq \mathbf{E}.$$

Let

$$\phi = g(\theta) + \tfrac{1}{8}(\mathbf{C}-\mathbf{1})\cdot\boldsymbol{\alpha}(\mathbf{C}-\mathbf{1}) + \tfrac{1}{2}(\mathbf{C}-\mathbf{1})\cdot\boldsymbol{\beta}\mathfrak{E} + \tfrac{1}{2}\Xi\cdot\boldsymbol{\gamma}\mathfrak{E} + \boldsymbol{\zeta}(\theta)\cdot\Xi =: \hat{\phi}(\mathbf{C}, \theta, \mathfrak{E})$$

where $\boldsymbol{\alpha}, \boldsymbol{\beta}, \boldsymbol{\gamma}$ are fourth-order, third-order, and second-order tensors. Moreover $\boldsymbol{\zeta}(\theta)$ is a temperature-dependent vector. Thus by (12.8) we have

$$\mathbf{T} = \rho\mathbf{F}[\tfrac{1}{2}\boldsymbol{\alpha}(\mathbf{C}-\mathbf{1}) + \boldsymbol{\beta}\mathfrak{E}], \qquad \mathbf{P} = -\rho\mathbf{F}[\tfrac{1}{2}\boldsymbol{\beta}^T(\mathbf{C}-\mathbf{1}) - \boldsymbol{\gamma}\mathfrak{E}] - \rho\mathbf{F}\boldsymbol{\zeta},$$

the term $\mathbf{E}\otimes\mathbf{P}$ in $\mathbf{T}$ being ignored because of linearization. Hence, in the linear approximation we have

$$\mathbf{T} = \rho_R\boldsymbol{\alpha}\mathbf{S} + \rho_R\boldsymbol{\beta}\mathbf{E}, \qquad \mathbf{P} = -\rho_R\boldsymbol{\beta}^T\mathbf{S} - \rho_R\boldsymbol{\gamma}\mathbf{E} - \rho_R\boldsymbol{\zeta}.$$

In the suffix notation,

$$T_{ij} = \rho_R\alpha_{ijhk}S_{hk} + \rho_R\beta_{ijk}E_k, \qquad P_k = -\rho_R\beta_{ijk}S_{ij} - \rho_R\gamma_{kh}E_h - \rho_R\zeta_k.$$

Since $\mathbf{D} = \epsilon_0\mathbf{E} + \mathbf{P}$ then we can write

$$D_k = -\rho_R\beta_{ijk}S_{ij} + (\epsilon_0\delta_{kh} - \rho_R\gamma_{kh})E_h - \rho_R\zeta_k.$$

We then identify $\rho_R\boldsymbol{\alpha}$ with the elasticity tensor C and $\epsilon_0\mathbf{1} - \rho_R\boldsymbol{\gamma}$ with the permittivity $\boldsymbol{\epsilon}$. Moreover we let

$$\boldsymbol{\Lambda} = \rho_R\boldsymbol{\beta}, \qquad \mathfrak{P} = -\rho_R\boldsymbol{\zeta},$$

so that we can write

$$\mathbf{T} = \mathsf{C}\,\mathbf{S} + \boldsymbol{\Lambda}\,\mathbf{E}, \qquad \mathbf{D} = -\boldsymbol{\Lambda}^T\mathbf{S} + \boldsymbol{\epsilon}\,\mathbf{E} + \mathfrak{P}$$

or

$$T_{ij} = C_{ijln}S_{ln} + \Lambda_{ijl}E_l, \qquad D_i = -\Lambda_{kmi}S_{km} + \epsilon_{ij}E_j + \mathfrak{P}_i.$$

By definition, and consistent with (12.8), it follows that $\boldsymbol{\alpha}$ enjoys major and minor symmetry properties,

$$\alpha_{ijhk} = \alpha_{jihk} = \alpha_{ijkh} = \alpha_{hkij},$$

the same then being true for C. The permittivity $\boldsymbol{\epsilon}$ is symmetric too.

The vector $\mathfrak{P}$ provides the polarization $\mathbf{P}$ when the applied electric field $\mathbf{E}$ and the stress $\mathbf{T}$ are zero. The variation of $\mathbf{P}$ on heating is reversed on cooling and this is consistent with

$$\dot{\mathbf{P}} = \partial_\theta\mathfrak{P}\,\dot{\theta}$$

when $\mathbf{T}$ and $\mathbf{E}$ are zero. Indeed, $\partial_\theta\mathfrak{P}$ can be viewed as the pyroelectric vector, the change of polarization per unit of variation of temperature.

The scheme so outlined shows that, in models of piezoelectric materials, the electric field should enter the free energy as a vector ($\mathbf{F}^T\mathbf{E}$) and not merely via scalars such as, e.g. $\varkappa = \mathbf{E}^2$. This is so because, in a quadratic free energy, no coupling would be possible between deformation and electric field. In a $\mathbf{E}^2$-dependent free energy we can write

$$\phi = g(\theta) + \tfrac{1}{8}(\mathbf{C}-\mathbf{1})\cdot\boldsymbol{\alpha}(\mathbf{C}-\mathbf{1}) + \lambda\mathbf{E}^2$$

and hence

$$\mathbf{T} = \mathbf{CS}, \qquad \mathbf{D} = \epsilon \mathbf{E},$$

without any piezoelectric and pyroelectric effect.

Mandel–Voigt notation

When dealing with symmetric tensors, Voigt notation allows us to reduce the order and correspondingly simplify the matrix form of calculations. We first observe that a symmetric tensor is characterized by 6 entries which can be viewed as the components of a 6-dimensional vector. Following the Mandel–Voigt notation, the entries of $S = S^T, T = T^T \in \mathbb{R}^{3\times3}$ are put in correspondence with $\tilde{S}, \tilde{T} \in \mathbb{R}^6$ in the form

$$\tilde{S} = (S_1, S_2, S_3, S_4, S_5, S_6) := (S_{11}, S_{22}, S_{33}, \sqrt{2}\, S_{23}, \sqrt{2}\, S_{13}, \sqrt{2}\, S_{12})$$

and the like for T. Moreover we let

$$[\tilde{\Lambda}_{\nu i}] = \begin{bmatrix} \Lambda_{111} & \Lambda_{112} & \Lambda_{113} \\ \Lambda_{221} & \Lambda_{222} & \Lambda_{223} \\ \Lambda_{331} & \Lambda_{332} & \Lambda_{333} \\ \sqrt{2}\,\Lambda_{231} & \sqrt{2}\,\Lambda_{232} & \sqrt{2}\,\Lambda_{233} \\ \sqrt{2}\,\Lambda_{131} & \sqrt{2}\,\Lambda_{132} & \sqrt{2}\,\Lambda_{133} \\ \sqrt{2}\,\Lambda_{121} & \sqrt{2}\,\Lambda_{122} & \sqrt{2}\,\Lambda_{123} \end{bmatrix},$$

$$[\tilde{C}_{\mu\nu}] = \begin{bmatrix} C_{1111} & C_{1122} & C_{1133} & \sqrt{2}\,C_{1123} & \sqrt{2}\,C_{1113} & \sqrt{2}\,C_{1112} \\ C_{2211} & C_{2222} & C_{2233} & \sqrt{2}\,C_{2223} & \sqrt{2}\,C_{2213} & \sqrt{2}\,C_{2212} \\ C_{3311} & C_{3322} & C_{3333} & \sqrt{2}\,C_{3323} & \sqrt{2}\,C_{3313} & \sqrt{2}\,C_{3312} \\ \sqrt{2}\,C_{2311} & \sqrt{2}\,C_{2322} & \sqrt{2}\,C_{2333} & 2\,C_{2323} & 2\,C_{2313} & 2\,C_{2312} \\ \sqrt{2}\,C_{1311} & \sqrt{2}\,C_{1322} & \sqrt{2}\,C_{1333} & 2\,C_{1323} & 2\,C_{1313} & 2\,C_{1312} \\ \sqrt{2}\,C_{1211} & \sqrt{2}\,C_{1222} & \sqrt{2}\,C_{1233} & 2\,C_{1223} & 2\,C_{1213} & 2\,C_{1212} \end{bmatrix}.$$

Hence we can write the linear equations in the form

$$\tilde{T}_\mu = \tilde{C}_{\mu\nu}\, \tilde{S}_\nu + \tilde{\Lambda}_{\mu l}\, E_l, \qquad D_i = -\tilde{\Lambda}^T_{i\nu}\, \tilde{S}_\nu + \epsilon_{ij}\, E_j.$$

In addition to the constitutive equations for **T** and **D** as (linear) functions of **S** and **E** we can change the independent variables based on suitable assumptions. In matrix notation we write

$$\tilde{T} = \tilde{C}\,\tilde{S} + \tilde{\Lambda}\, E, \qquad D = -\tilde{\Lambda}^T \tilde{S} + \epsilon\, E. \tag{12.10}$$

To obtain the corresponding equations with $\tilde{T}$ and E as independent variables we assume $\tilde{C}$ is invertible. Then $\tilde{C}^{-1}\tilde{T} = \tilde{S} + \tilde{C}^{-1}\tilde{\Lambda} E$ and hence

$$\tilde{S} = \tilde{C}^{-1}\tilde{T} - \tilde{C}^{-1}\tilde{\Lambda}E, \qquad D = -\tilde{\Lambda}^T\tilde{C}^{-1}\tilde{T} + (\epsilon + \tilde{\Lambda}^T\tilde{C}^{-1}\tilde{\Lambda})E. \tag{12.11}$$

Let $\mathcal{A} := \epsilon + \tilde{\Lambda}^T\tilde{C}^{-1}\tilde{\Lambda} \in \mathbb{R}^{3\times 3}$ be invertible. Consequently we can write

$$E = \mathcal{A}^{-1}D + \mathcal{A}^{-1}\tilde{\Lambda}^T\tilde{C}^{-1}\tilde{T}, \qquad \tilde{S} = \tilde{C}^{-1}(1\!\!1 - \tilde{\Lambda}\mathcal{A}^{-1}\tilde{\Lambda}^T\tilde{C}^{-1})\tilde{T} - \tilde{C}^{-1}\tilde{\Lambda}\mathcal{A}^{-1}D. \tag{12.12}$$

Finally, let ϵ be invertible. By (12.10) we find

$$\tilde{T} = (\tilde{C} + \tilde{\Lambda}\epsilon^{-1}\tilde{\Lambda}^T)\tilde{S} + \tilde{\Lambda}\epsilon^{-1}D, \qquad E = \epsilon^{-1}\tilde{\Lambda}^T\tilde{S} + \epsilon^{-1}D. \tag{12.13}$$

The full equivalence of Eqs. (12.10)–(12.13) holds provided $\tilde{C}$, ϵ, and $\epsilon + \tilde{\Lambda}^T\tilde{C}^{-1}\tilde{\Lambda}$ are invertible. Now, invertibility of the elasticity matrix $\tilde{C}$ and the permittivity matrix ϵ is a customary assumption. The invertibility of $\epsilon + \tilde{\Lambda}^T\tilde{C}^{-1}\tilde{\Lambda}$ is reasonable inasmuch as we can regard $\tilde{\Lambda}$ as a matrix with suitably small entries.

To account for pyroelectricity, relations (12.10) to (12.13) hold with the formal change $D_i \to D_i - \mathfrak{P}_i$.

According to the thermodynamic restrictions, in the linear approximation,

$$\partial_{\mathbf{E}}\hat{\phi} = -\frac{1}{\rho_R}\mathbf{P}, \qquad \partial_{\mathbf{S}}\hat{\phi} = \frac{1}{\rho_R}\mathbf{T}.$$

Hence it follows that

$$\partial_{\mathbf{S}}\mathbf{P} = -(\partial_{\mathbf{E}}\mathbf{T})^T.$$

Similar (Maxwell) relations follow from the linear equations (12.10)–(12.13). It is apparent that

$$\partial_E\tilde{T}(\tilde{S}, E) = -(\partial_{\tilde{S}}D)^T(\tilde{S}, E), \qquad \partial_E\tilde{S}(\tilde{T}, E) = (\partial_{\tilde{T}}D)^T(\tilde{T}, E),$$

$$\partial_{\tilde{T}}E(\tilde{T}, D) = -(\partial_D\tilde{S})^T(\tilde{T}, D), \qquad \partial_D\tilde{T}(\tilde{S}, D) = (\partial_{\tilde{S}}E)^T(\tilde{S}, D).$$

To change the independent variables from the original set, here $\tilde{S}$, E, to the other ones we could have performed appropriate Legendre transformations. Yet the direct procedure adopted here gives evidence to the assumption in passing from one set to another one.

Flexoelectricity and electrostriction

Flexoelectricity occurs in dielectrics and shows up as an electrical polarization induced by a strain gradient [127, 227, 394]. From the modelling viewpoint, flexoelectricity is closely related to piezoelectricity. Yet while piezoelectricity accounts for polarization due to a (uniform) strain, flexoelectricity models the occurring of polarization due to a non-uniform strain field.

Electrostriction occurs in all dielectrics; it describes the strain as a quadratic effect produced by the electric field (**E**) or the polarization (**P**). The electrostrictive effect is generally small when compared with that induced by piezoelectricity. However, there are single crystals/ceramics in which a high electrostrictive strain occurs because of the high dielectric response. These materials are referred to as relaxor ferroelectrics or relaxors. While the normal ferroelectrics[3] have a hysteresis loop that, at zero electric field, retains a large polarization, in the relaxor this zero field polarization is significantly smaller so that it is reasonable to regard relaxors as nonlinear paraelectrics with high values of permittivity [53, 383].

A description of flexoelectricity and electrostriction requires a scheme where both deformation and deformation gradient are among the independent variables. Hence we let $(\mathbf{C}, \nabla_R \mathbf{C}, \theta, \mathbf{E})$ be the set of independent variables. Moreover, as with piezoelectricity, we let the free energy ϕ depend on $\mathbf{E}$ via the Euclidean invariant $\mathfrak{E} = \mathbf{F}^T \mathbf{E}$. The thermodynamic analysis developed for the general scheme of electroelasticity applies as a particular case. If

$$\phi = \Phi(\mathbf{C}, \theta, \mathfrak{E}, \nabla_R \mathbf{C})$$

it follows that

$$\eta = -\partial_\theta \Phi, \qquad \mathbf{P} = -\rho \mathbf{F} \partial_{\mathfrak{E}} \Phi,$$

$$\mathbf{k} = \frac{\rho}{\theta} \partial_{\nabla_R \mathbf{C}} \Phi \, \dot{\mathbf{C}},$$

and

$$\mathbf{T} = 2\rho \mathbf{F} \, \delta_{\mathbf{C}} \Phi \, \mathbf{F}^T - \mathbf{E} \otimes \mathbf{P},$$

where

$$\delta_{\mathbf{C}} \Phi = \partial_{\mathbf{C}} \Phi - \frac{\theta}{\rho_R} \nabla_R \cdot (\frac{\rho_R}{\theta} \partial_{\nabla_R \mathbf{C}} \Phi).$$

It is worth remarking that to allow for a dependence, of $\mathbf{k}_R$, on $\dot{\mathbf{C}}$ and to obey the rule of equipresence, we should have started with $\mathbf{C}, \nabla_R \mathbf{C}, \theta, \mathfrak{E}$, and $\dot{\mathbf{C}}$ as the set of independent variables. We would have shown that Φ is in fact independent of $\dot{\mathbf{C}}$ and then we would have arrived to the same conclusions.

Also to adhere to the literature, we now consider **T** and **P** in the linear approximation. Since

$$\mathbf{F} \simeq \mathbf{1}, \qquad \mathbf{C} - \mathbf{1} \simeq 2\mathbf{S}, \qquad J \simeq 1, \qquad \rho \simeq \rho_R, \qquad \nabla_R \simeq \nabla,$$

we obtain

$$\mathbf{T} \simeq \rho \delta_{\mathbf{S}} \Phi - \mathbf{E} \otimes \mathbf{P},$$

while

[3] See Chap. 15.

$$\mathbf{P} \simeq -\rho\, \partial_{\mathbf{E}} \Phi.$$

For definiteness we let ρ, θ be uniform and assume

$$\rho\phi(\mathbf{S}, \theta, \mathbf{E}, \nabla\mathbf{S}) = \tfrac{1}{2}\mathbf{S}\cdot\mathsf{C}\mathbf{S} - \mathbf{S}\cdot\boldsymbol{\Lambda}\mathbf{E} - \nabla\mathbf{S}\cdot\boldsymbol{\Gamma}\mathbf{E} - \tfrac{1}{2}\mathbf{E}\cdot\gamma\mathbf{E} - \tfrac{1}{2}\nabla\mathbf{S}\cdot\Xi\nabla\mathbf{S},$$

or, in suffix notation,

$$\rho\phi = \tfrac{1}{2}S_{ij}C_{ijhk}S_{hk} - S_{ij}\Lambda_{ijh}E_h - (\partial_{x_h}S_{ij})\Gamma_{ijhk}E_k - \tfrac{1}{2}E_i\gamma_{ij}E_j - \tfrac{1}{2}(\partial_{x_i}S_{jk})\Xi_{ijkpqr}(\partial_{x_p}S_{qr}).$$

Further, assume that C, $\boldsymbol{\Lambda}$, γ, and $\boldsymbol{\Gamma}$ are tensors parameterized by ρ and θ and that C, γ, and Ξ are symmetric. Hence we obtain the constitutive equations

$$T_{ij} = C_{ijhk}S_{hk} - \Lambda_{ijh}E_h + \Gamma_{ijhk}\partial_{x_h}E_k + \Xi_{kijpqr}\partial_{x_k}\partial_{x_p}S_{qr} - E_iP_j,$$

$$P_i = S_{hk}\Lambda_{hki} + \Gamma_{kjhi}\partial_{x_h}S_{kj} + \gamma_{ij}E_j.$$

In compact notation

$$\mathbf{T} = \mathsf{C}\mathbf{S} - \boldsymbol{\Lambda}\mathbf{E} + \boldsymbol{\Gamma}\nabla\mathbf{E} - \Xi\nabla\nabla\mathbf{S} - \mathbf{E}\otimes\mathbf{P}, \tag{12.14}$$

$$\mathbf{P} = \gamma\mathbf{E} + \boldsymbol{\Lambda}^T\mathbf{S} + \boldsymbol{\Gamma}^T\nabla\mathbf{S}. \tag{12.15}$$

The dependence of $\mathbf{P}$ on $\nabla\mathbf{S}$ characterizes the flexoelectric effect [233, 343]. The dyadic product $\mathbf{E}\otimes\mathbf{P}$ is neglected in linear approximations. The model for piezoelectricity follows in the event $\boldsymbol{\Gamma} = \mathbf{0}$, $\Xi = \mathbf{0}$.

If $\boldsymbol{\Lambda}$ and $\boldsymbol{\Gamma}$ vanish then the interaction terms between $\mathbf{E}$ and $\mathbf{S}$ in the Gibbs free energy do not occur and $\rho\phi$ takes the more familiar form

$$\rho\phi = \tfrac{1}{2}\mathbf{S}\cdot\mathsf{C}\mathbf{S} - \tfrac{1}{2}\mathbf{E}\cdot\gamma\mathbf{E} - \tfrac{1}{2}\nabla\mathbf{S}\cdot\Xi\nabla\mathbf{S},$$

The literature[4] shows also the use of the version with $\mathbf{P}$ and $\mathbf{S}$ as independent variables. Since

$$\mathbf{E} = \gamma^{-1}[\mathbf{P} - \boldsymbol{\Lambda}^T\mathbf{S} - \boldsymbol{\Gamma}^T\mathbf{S}]$$

then substituting this expression of $\mathbf{E}$ in (12.14) we deduce the function of $\mathbf{T}$ in terms of $\mathbf{S}$, $\mathbf{P}$, $\nabla\mathbf{S}$, $\nabla\nabla\mathbf{S}$. Often the equation for $\mathbf{P}$ is given in terms of $\mathbf{T}$ and $\nabla\mathbf{S}$ (see, e.g. [343]). As the simplest stress–strain relation we take

$$\mathbf{S} = \mathsf{C}^{-1}\mathbf{T}.$$

Upon substitution we obtain the familiar equation of flexoelectricity,

[4] See, e.g. [408].

$$\mathbf{P} = \gamma\mathbf{E} + \boldsymbol{\Lambda}^T \mathsf{C}^{-1}\mathbf{T} + \boldsymbol{\Gamma}\nabla\mathbf{S},$$

where $\gamma\mathbf{E}$ is the classical dielectric term.

Quite naturally we wonder whether electrostriction is described by Eqs. (12.14)–(12.15). Since electrostriction denotes the occurrence of a strain as a (quadratic) effect of the electric field, we restrict attention to (12.14)–(12.15) in uniform conditions ($\nabla\mathbf{E} = \mathbf{0}$, $\nabla\mathbf{S} = \mathbf{0}$). If further the tensor $\boldsymbol{\Lambda}$ produces negligible effects then we can write

$$\mathbf{T} = \mathsf{C}\mathbf{S} - \mathbf{E} \otimes \mathbf{P}, \qquad \mathbf{P} = \gamma\mathbf{E}.$$

Since electrostriction is an effect of the electric field, not of the stress, we look at stress-free conditions, $\mathbf{T} = \mathbf{0}$, and find that

$$\mathbf{S} = \mathsf{C}^{-1}\mathbf{E} \otimes \gamma\mathbf{E}$$

or

$$S_{ij} = M_{ijpr} E_p E_r, \qquad M_{ijpr} := C^{-1}_{ijpq}\gamma_{qr}.$$

This result describes the electrostrictive effect and shows that the electrostrictive coefficients M_{ijpr} are given by the matrix product C^{-1} and γ. Hence we see that electrostriction is favoured by small values of the elastic coefficients and large values of the permittivity. This result in turn is consistent with the standard observation that high electrostrictive strains occur because of high dielectric permittivities.

12.4 Dielectrics with Polarization Gradient

Similar equations hold for dielectrics where the polarization gradient is involved to describe properties of a crystal lattice arising from the discrete structure.[5]

We assume $\mathbf{P}$ transforms as an objective vector, that is $\mathbf{P} \to \mathbf{QP}$ under a change of frame with rotation matrix $\mathbf{Q}$. Hence, to account for the dependence on the polarization, and the polarization gradient, we observe that both $\mathbf{F}^{-1}\mathbf{P}$ and $\mathbf{F}^T\mathbf{P}$ are objective. This is so because

$$\mathbf{F}^{-1}\mathbf{P} \;\to\; (\mathbf{QF})^{-1}\mathbf{QP} = \mathbf{F}^{-1}\mathbf{Q}^{-1}\mathbf{QP} = \mathbf{F}^{-1}\mathbf{P},$$

$$\mathbf{F}^T\mathbf{P} \;\to\; (\mathbf{QF})^T\mathbf{QP} = \mathbf{F}^T\mathbf{Q}^T\mathbf{QP} = \mathbf{F}^T\mathbf{P}.$$

For a non-magnetizable and non-conducting body the second-law inequality can be written in the form

---

[5] From the point of view of lattice theory of crystals, the polarization gradient is regarded to model, in the long-wave approximation, the shell-shell and core-shell interactions between atoms [17]; see also [311] and [404].

$$-\rho(\dot{\psi} + \eta\dot{\theta}) + \rho\dot{\boldsymbol{\pi}} \cdot \mathbf{E} + \mathbf{T} \cdot \mathbf{L} + \theta\nabla \cdot \mathbf{k} \geq 0.$$

where $\boldsymbol{\pi} = \mathbf{P}/\rho$. Multiply the inequality by J/θ and observe

$$J\nabla \cdot \mathbf{k} = \nabla_R \cdot \mathbf{k}_R, \qquad \mathbf{k}_R = J\mathbf{F}^{-1}\mathbf{k},$$

to obtain

$$-\frac{\rho_R}{\theta}(\dot{\psi} + \eta\dot{\theta}) + \frac{\rho_R}{\theta}\dot{\boldsymbol{\pi}} \cdot \mathbf{E} + \frac{J}{\theta}(\mathbf{T}\mathbf{F}^{-T}) \cdot \dot{\mathbf{F}} + \nabla_R \cdot \mathbf{k}_R \geq 0.$$

For definiteness look first at

$$\boldsymbol{\Pi} = \mathbf{F}^{-1}\boldsymbol{\pi} = \frac{1}{\rho}\mathbf{F}^{-1}\mathbf{P} = \frac{1}{\rho_R}J\mathbf{F}^{-1}\mathbf{P};$$

to within the scalar $1/\rho_R$, the vector $\boldsymbol{\Pi}$ equals the (invariant) polarization $\mathbf{P}_R$ in the reference configuration. Moreover we have

$$\dot{\boldsymbol{\Pi}} = \mathbf{F}^{-1}\dot{\boldsymbol{\pi}} - \mathbf{F}^{-1}\dot{\mathbf{F}}\mathbf{F}^{-1}\boldsymbol{\pi}.$$

Hence, for any vector function $\mathbf{w}$ we can write

$$\mathbf{w} \cdot \dot{\boldsymbol{\Pi}} = (\mathbf{F}^{-T}\mathbf{w}) \cdot \dot{\boldsymbol{\pi}} - (\mathbf{F}^{-T}\mathbf{w}) \otimes (\mathbf{F}^{-1}\boldsymbol{\pi}) \cdot \dot{\mathbf{F}}. \tag{12.16}$$

Let

$$(\mathbf{C}, \theta, \boldsymbol{\Pi}, \nabla_R\boldsymbol{\Pi})$$

be the set of independent variables. Computation of $\dot{\psi}$ and substitution in the entropy inequality results in

$$-\frac{\rho_R}{\theta}(\partial_\theta\psi + \eta)\dot{\theta} + \frac{\rho_R}{\theta}\dot{\boldsymbol{\pi}} \cdot \mathbf{E} - \frac{\rho_R}{\theta}\partial_{\boldsymbol{\Pi}}\psi \cdot \dot{\boldsymbol{\Pi}} - \frac{\rho_R}{\theta}\partial_{\mathbf{C}}\psi \cdot \dot{\mathbf{C}}$$
$$+ \frac{J}{\theta}(\mathbf{T}\mathbf{F}^{-T}) \cdot \dot{\mathbf{F}} - \frac{\rho_R}{\theta}\partial_{\nabla_R\boldsymbol{\Pi}}\psi \cdot \nabla_R\dot{\boldsymbol{\Pi}} + \nabla_R \cdot \mathbf{k}_R \geq 0.$$

The arbitrariness of $\dot{\theta}$ implies the classical relation

$$\eta = -\partial_\theta\psi.$$

Now observe

$$-\frac{\rho_R}{\theta}\partial_{\nabla_R\boldsymbol{\Pi}}\psi \cdot \nabla_R\dot{\boldsymbol{\Pi}} = -\nabla_R \cdot (\frac{\rho_R}{\theta}\partial_{\nabla_R\boldsymbol{\Pi}}\psi\, \dot{\boldsymbol{\Pi}}) + [\nabla_R \cdot (\frac{\rho_R}{\theta}\partial_{\nabla_R\boldsymbol{\Pi}}\psi)] \cdot \dot{\boldsymbol{\Pi}}.$$

Moreover, applying the identity (12.16) with $\mathbf{w} = \nabla_R \cdot (\rho_R\partial_{\nabla_R\boldsymbol{\Pi}}\psi/\theta)$ and $\mathbf{w} = \partial_{\boldsymbol{\Pi}}\psi$ we eventually can write the remaining entropy inequality in the form

$$\frac{\rho_R}{\theta}[\mathbf{E} - \mathbf{F}^{-T}\delta_{\mathbf{\Pi}}\psi] \cdot \dot{\pi} + \frac{\rho_R}{\theta}[\frac{1}{\rho}\mathbf{T}\mathbf{F}^{-T} - 2\mathbf{F}\partial_{\mathbf{C}}\psi + (\mathbf{F}^{-T}\delta_{\mathbf{\Pi}}\psi) \otimes (\mathbf{F}^{-1}\pi)] \cdot \dot{\mathbf{F}}$$
$$+\nabla_R \cdot (\mathbf{k}_R - \frac{\rho_R}{\theta}\partial_{\nabla_R\mathbf{\Pi}}\psi\,\dot{\mathbf{\Pi}}) \geq 0.$$

where the identities $\mathbf{T} \cdot \mathbf{L} = (\mathbf{T}\mathbf{F}^{-T}) \cdot \dot{\mathbf{F}}$ and $\partial_{\mathbf{C}}\psi \cdot \dot{\mathbf{C}} = 2(\mathbf{F}\partial_{\mathbf{C}}\psi) \cdot \dot{\mathbf{F}}$ are used and

$$\delta_{\mathbf{\Pi}}\psi = \partial_{\mathbf{\Pi}}\psi - \frac{\theta}{\rho_R}\nabla_R \cdot (\frac{\rho_R}{\theta}\partial_{\nabla_R\mathbf{\Pi}}\psi).$$

The inequality holds if

$$\mathbf{k}_R = \frac{\rho_R}{\theta}\partial_{\nabla_R\mathbf{\Pi}}\psi\,\dot{\mathbf{\Pi}}$$

$$\mathbf{E} = \mathbf{F}^{-T}\delta_{\mathbf{\Pi}}\psi, \qquad \frac{1}{\rho}\mathbf{T}\mathbf{F}^{-T} = 2\mathbf{F}\partial_{\mathbf{C}}\psi - (\mathbf{F}^{-T}\delta_{\mathbf{\Pi}}\psi) \otimes (\mathbf{F}^{-1}\pi). \tag{12.17}$$

Some comments on these results are in order. The dependence on the polarization gradient is compatible with thermodynamics if the extra-entropy flux is nonzero. Moreover, the free energy ψ is a potential for the electric field $\mathbf{E}$ through the joint dependence on $\mathbf{\Pi}$ and $\nabla_R\mathbf{\Pi}$. The tensor $\mathbf{T}\mathbf{F}^{-T}/\rho$ is just $1/\rho_R$ times the first Piola-Kirchhoff stress $\mathbf{T}_R$. Hence we can say that $\mathbf{T}_R$ and $\mathbf{T}$ follow from the potential ψ via the joint contributions of differentiations with respect to $\mathbf{C}$, $\mathbf{\Pi}$, $\nabla_R\mathbf{\Pi}$. Indeed, upon right multiplication by $\rho\mathbf{F}^T$ we have

$$\mathbf{T} = 2\rho\,\mathbf{F}\,\partial_{\mathbf{C}}\psi\mathbf{F}^T - \rho(\mathbf{F}^{-T}\delta_{\mathbf{\Pi}}\psi) \otimes \pi.$$

The observation that $\rho\pi = \mathbf{P}$ and $\mathbf{F}^{-T}\delta_{\mathbf{\Pi}}\psi = \mathbf{E}$ allows us to write $\mathbf{T}$ in the form

$$\mathbf{T} = 2\,\rho\,\mathbf{F}\partial_{\mathbf{C}}\psi\mathbf{F}^T - \mathbf{E} \otimes \mathbf{P}.$$

Hence we can view $\mathbf{T}$ as the sum of two tensors. One is the standard symmetric tensor $2\,\rho\,\mathbf{F}\partial_{\mathbf{C}}\psi\mathbf{F}^T$ of elasticity. The other one is the dyadic product $-\mathbf{E} \otimes \mathbf{P}$. Consequently

$$\mathrm{skw}\mathbf{T} = -\mathrm{skw}\mathbf{E} \otimes \mathbf{P} = \mathrm{skw}\mathbf{P} \otimes \mathbf{E}$$

as it must be in view of (2.73) and therefore confirming the same result (12.6) obtained within the model given by $\phi = \hat{\phi}(\mathbf{C}, \theta, \mathbf{F}^T\mathbf{E})$.

Things are quite similar if we replace $\mathbf{\Pi}$ by

$$\mathcal{P} = \mathbf{F}^T\pi = \frac{1}{\rho_R}J\mathbf{F}^T\mathbf{P}.$$

We parallel the previous procedure. Observe

$$\mathbf{w} \cdot \dot{\mathcal{P}} = (\mathbf{F}\mathbf{w}) \cdot \dot{\pi} + (\pi \otimes \mathbf{w}) \cdot \dot{\mathbf{F}}.$$

Take $\psi = \psi(\mathbf{C}, \theta, \mathcal{P}, \nabla_R \mathcal{P})$, compute $\dot{\psi}$ and substitute in the entropy inequality. It follows that the inequality holds if $\eta = -\partial_\theta \psi$ and

$$\mathbf{k}_R = \frac{\rho_R}{\theta} \partial_{\nabla_R \mathcal{P}} \psi \dot{\mathcal{P}},$$

$$\mathbf{E} = \mathbf{F} \delta_{\mathcal{P}} \psi, \qquad \mathbf{T} = 2\, \rho \mathbf{F} \partial_{\mathbf{C}} \psi \, \mathbf{F}^T + \mathbf{P} \otimes \mathbf{E}.$$

where $\delta_{\mathcal{P}} \psi = \partial_{\mathcal{P}} \psi - (\theta/\rho_R) \nabla_R \cdot ((\rho_R/\theta) \partial_{\nabla_R \mathcal{P}} \psi)$.

Often the literature describes the polarization effects by means of

$$(\mathbf{C}, \theta, \boldsymbol{\pi}, \nabla_R \boldsymbol{\pi})$$

as the independent variables. The essential difference is due to the direct dependence on $\boldsymbol{\pi}$ rather than on $\mathbf{F}^T \boldsymbol{\pi}$.

Upon computation of $\dot{\psi}$ and substitution in the entropy inequality we have

$$-\rho(\partial_\theta \psi + \eta)\dot{\theta} + \rho(\mathbf{E} - \partial_{\boldsymbol{\pi}} \psi) \cdot \dot{\boldsymbol{\pi}} - \rho \partial_{\nabla_R \boldsymbol{\pi}} \psi \cdot (\nabla_R \dot{\boldsymbol{\pi}}) + (\mathbf{T}\mathbf{F}^{-T} - 2\rho \mathbf{F} \partial_{\mathbf{C}} \psi) \cdot \dot{\mathbf{F}} + \theta \nabla \cdot \mathbf{k} \geq 0.$$

The arbitrariness of $\dot{\theta}$ and $\dot{\mathbf{F}}$ implies

$$\eta = -\partial_\theta \psi, \qquad \mathbf{T} = 2\rho \mathbf{F} \, \partial_{\mathbf{C}} \psi \, \mathbf{F}^T.$$

Now, for technical convenience multiply by J/θ. Hence, in view of the identity

$$-\frac{\rho_R}{\theta} \partial_{\nabla_R \boldsymbol{\pi}} \psi \cdot (\nabla_R \dot{\boldsymbol{\pi}}) = -\nabla_R \cdot (\frac{\rho_R}{\theta} \partial_{\nabla_R \boldsymbol{\pi}} \psi \, \dot{\boldsymbol{\pi}}) + [\nabla_R (\frac{\rho_R}{\theta} \partial_{\nabla_R \boldsymbol{\pi}} \psi)] \cdot \dot{\boldsymbol{\pi}}$$

and the definition $\delta_{\boldsymbol{\pi}} \psi = \partial_{\boldsymbol{\pi}} \psi - (\theta/\rho_R) \nabla_R \cdot ((\rho_R/\theta) \partial_{\nabla_R \boldsymbol{\pi}} \psi)$ we have

$$\frac{\rho_R}{\theta} [\mathbf{E} - \delta_{\boldsymbol{\pi}} \psi] \cdot \dot{\boldsymbol{\pi}} + \nabla_R \cdot [\mathbf{k}_R - \frac{\rho_R}{\theta} \partial_{\nabla_R \boldsymbol{\pi}} \psi \, \dot{\boldsymbol{\pi}}] \geq 0.$$

This inequality holds if

$$\mathbf{E} = \delta_{\boldsymbol{\pi}} \psi, \qquad \mathbf{k}_R = \frac{\rho_R}{\theta} \partial_{\nabla_R \boldsymbol{\pi}} \psi \, \dot{\boldsymbol{\pi}}.$$

Thus the model of dielectric is consistent with the dependence on the polarization gradient $\nabla_R \boldsymbol{\pi}$ if the thermodynamic restrictions

$$\eta = -\partial_\theta \psi, \qquad \mathbf{T} = 2\rho \mathbf{F} \, \partial_{\mathbf{C}} \psi \, \mathbf{F}^T, \qquad \mathbf{E} = \delta_{\boldsymbol{\pi}} \psi$$

hold.[6] In the linear approximation ($\rho \simeq \rho_R$, $\nabla_R \simeq \nabla$, $\mathbf{C} \simeq 2\mathbf{S}$) we have

[6] Here skw$\mathbf{T} = \mathbf{0}$. Since ψ is taken to depend directly on $\boldsymbol{\pi}$ and hence via the invariant $\boldsymbol{\pi}^2$, not e.g. via $\mathbf{F}^T \boldsymbol{\pi}$, then $\mathbf{E}$ is parallel to $\boldsymbol{\pi}$ and skw$\mathbf{E} \otimes \mathbf{P} = \mathbf{0}$, which is consistent with skw$\mathbf{T} = \mathbf{0}$.

$$\eta = -\partial_\theta \psi, \qquad \mathbf{T} = \rho_R\, \partial_{\mathbf{S}} \psi, \qquad \mathbf{E} = \delta_{\boldsymbol{\pi}} \psi.$$

The literature shows further approaches which are based on variational principles rather than on the second-law inequality. The variational principles trace back to a paper by Toupin [420] that has deeply influenced the subsequent research.[7] By means of the electric enthalpy $H = W - \mathbf{E} \cdot \mathbf{D}$, W being the energy density, the principle is assumed to assert that

$$-\delta \int_{V^*} H dv + \int_V (\mathbf{f} \cdot \delta \mathbf{u} + \mathbf{E} \cdot \delta \mathbf{P}) dv + \int_S \mathbf{t} \cdot \delta \mathbf{u}\, da$$

where V^* is a volume comprising V and $S = \partial V$ while $V^* \setminus V$ is an outer vacuum. Hence the variation of the electric enthalpy equals the work of the body force, the electric field, and the surface traction. In isothermal conditions, an appropriate enthalpy provides the equation of motion and[8]

$$\mathbf{E} = \partial_{\mathbf{P}} W - \nabla \cdot \partial_{\nabla \mathbf{P}} W,$$

where $W = W(\mathbf{S}, \mathbf{P}, \nabla \mathbf{P})$. Another approach establishes the balance equations by assigning a kinetic energy density $\frac{1}{2}\rho\, d\, \dot{\boldsymbol{\pi}}^2$ to the time-dependent polarization [304] thus providing a second-order differential equation for $\boldsymbol{\pi}$. A further approach [404] involves both a variational principle and micro-force fields. Following the work of Gurtin [215], a micro-force tensor $\boldsymbol{\xi}$, an internal micro-force $\boldsymbol{\tau}$ and an external micro-force $\boldsymbol{\gamma}$ are considered such that $\boldsymbol{\xi} \cdot (\mathbf{n} \otimes \dot{\mathbf{P}})$ is the power density expended across surfaces, while $\boldsymbol{\tau} \cdot \dot{\mathbf{P}}$ and $\boldsymbol{\gamma} \cdot \dot{\mathbf{P}}$ are the internal and external power densities (per unit volume).

12.5 Fluids in Electromagnetic Fields

We start from the balance of energy in the form (2.75) with the fields in the co-moving frame,

$$\rho \dot{\varepsilon} = \mathbf{E} \cdot \mathbf{J} + \rho (\mathbf{P}/\rho)^{\cdot} \cdot \mathbf{E} + \rho \mu_0 (\mathbf{M}/\rho)^{\cdot} \cdot \mathbf{H} + \mathbf{T} \cdot \mathbf{L} - \nabla \cdot \mathbf{q} + \rho r.$$

and consider the second-law inequality

$$\rho \theta \dot{\eta} + \nabla \cdot \mathbf{q} - \rho r - \frac{1}{\theta} \mathbf{q} \cdot \nabla \theta + \theta \nabla \cdot \mathbf{k} \geq 0.$$

Upon substitution of $\nabla \cdot \mathbf{q} - \rho r$ we obtain

---

[7] See, e.g. [290, 312].

[8] See Eq. (2.3) of [311].

$$\rho\theta\dot{\eta} - \rho\dot{\varepsilon} + \mathbf{E}\cdot\mathbf{J} + \rho(\mathbf{P}/\rho)^{\cdot}\cdot\mathbf{E} + \rho\mu_0(\mathbf{M}/\rho)^{\cdot}\cdot\mathbf{H} + \mathbf{T}\cdot\mathbf{L} + \theta\nabla\cdot\mathbf{k} \geq 0.$$

By means of the free energy $\psi = \varepsilon - \theta\eta$ we can write the second-law inequality in the form

$$-\rho(\dot{\psi} + \eta\dot{\theta}) + \mathbf{E}\cdot\mathbf{J} + \rho(\mathbf{P}/\rho)^{\cdot}\cdot\mathbf{E} + \rho\mu_0(\mathbf{M}/\rho)^{\cdot}\cdot\mathbf{H} + \mathbf{T}\cdot\mathbf{L} - \frac{1}{\theta}\mathbf{q}\cdot\nabla\theta + \theta\nabla\cdot\mathbf{k} \geq 0.$$

For technical convenience we consider the free energy (density) ϕ defined by

$$\phi = \psi - \frac{1}{\rho}\mathbf{P}\cdot\mathbf{E} - \frac{\mu_0}{\rho}\mathbf{M}\cdot\mathbf{H}.$$

Upon substitution of $\dot{\psi}$ we obtain the second-law inequality in the form

$$-\rho(\dot{\phi} + \eta\dot{\theta}) + \mathbf{E}\cdot\mathbf{J} - \mathbf{P}\cdot\dot{\mathbf{E}} - \mu_0\mathbf{M}\cdot\dot{\mathbf{H}} + \mathbf{T}\cdot\mathbf{L} - \frac{1}{\theta}\mathbf{q}\cdot\nabla\theta + \theta\nabla\cdot\mathbf{k} \geq 0. \tag{12.18}$$

For definiteness we consider heat-conducting, viscous fluids. We then let the constitutive equations have the form

$$\phi = \phi(\rho, \theta, \mathbf{E}, \mathbf{H}, \nabla\theta, \mathfrak{D}),$$

and the like for $\eta, \mathbf{P}, \mathbf{M}, \mathbf{J}, \mathbf{T}, \mathbf{q}$, where $\mathfrak{D}$ denotes the stretching tensor. For the present purpose there is no loss of generality in letting the extra-entropy flux $\mathbf{k}$ be zero. The function ϕ is differentiable while the other constitutive functions are continuous. Moreover we let

$$\mathbf{T} = -p(\rho, \theta)\mathbf{1} + \check{\mathbf{T}}(\rho, \theta, \mathbf{E}, \mathbf{H}, \nabla\theta, \mathfrak{D}), \qquad \check{\mathbf{T}} = O(\mathfrak{D}).$$

Upon substitution of $\dot{\phi}$ we have

$$\begin{aligned}-\rho[\partial_\rho\phi\,\dot{\rho} + (\partial_\theta\phi + \eta)\dot{\theta} - \rho[\partial_{\mathbf{E}}\phi\cdot\dot{\mathbf{E}} + \partial_{\mathbf{H}}\phi\cdot\dot{\mathbf{H}} + \partial_{\nabla\theta}\phi\cdot\dot{\overline{\nabla\theta}} + \partial_{\mathfrak{D}}\phi\cdot\dot{\mathfrak{D}}]\\ +\mathbf{E}\cdot\mathbf{J} - \mathbf{P}\cdot\dot{\mathbf{E}} - \mu_0\mathbf{M}\cdot\dot{\mathbf{H}} - p\nabla\cdot\mathbf{v} + \check{\mathbf{T}}\cdot\mathbf{L} - \frac{1}{\theta}\mathbf{q}\cdot\nabla\theta \geq 0.\end{aligned}$$

By the continuity equation

$$-p\nabla\cdot\mathbf{v} = \frac{p}{\rho}\dot{\rho}.$$

The arbitrariness of $\dot{\overline{\nabla\theta}}, \dot{\mathfrak{D}}, \dot{\theta}, \dot{\rho}$ implies that the inequality holds only if

$$\partial_{\nabla\theta}\phi = \mathbf{0}, \quad \partial_{\mathfrak{D}}\phi = \mathbf{0}, \quad \eta = -\partial_\theta\phi, \quad p = \rho^2\partial_\rho\phi.$$

At any point $\mathbf{x} \in \mathcal{R}_t$ and any time t, Maxwell's equations allow $\dot{\mathbf{E}}$ and $\dot{\mathbf{H}}$ take arbitrary vector values.[9] Hence the inequality holds only if

$$\mathbf{P} = -\rho\partial_{\mathbf{E}}\phi, \qquad \mu_0\mathbf{M} = -\rho\partial_{\mathbf{H}}\phi.$$

Incidentally, since $\rho\psi = \rho\phi + \mathbf{P}\cdot\mathbf{E} + \mu_0\mathbf{M}\cdot\mathbf{H}$, if $\mathbf{P}$ is independent of $\mathbf{H}$ and $\mathbf{M}$ is independent of $\mathbf{E}$ then it follows

$$\partial_{\mathbf{E}}\psi = \frac{1}{\rho}\mathbf{E}\,\partial_{\mathbf{E}}\mathbf{P}, \qquad \partial_{\mathbf{H}}\psi = \frac{\mu_0}{\rho}\mathbf{H}\,\partial_{\mathbf{H}}\mathbf{M};$$

$\partial_{E_i}\psi = (1/\rho)E_j\,\partial_{E_i}P_j$, $\partial_{H_i}\psi = (\mu_0/\rho)H_j\partial_{H_i}M_j$. As a consequence, by $\hat{\psi}(\rho,\theta,\mathbf{P},\mathbf{M})$ we have

$$\partial_{\mathbf{E}}\psi = \partial_{\mathbf{P}}\hat{\psi}\,\partial_{\mathbf{E}}\mathbf{P}, \qquad \partial_{\mathbf{H}}\psi = \partial_{\mathbf{M}}\hat{\psi}\,\partial_{\mathbf{H}}\mathbf{M}.$$

Hence, if $\det\partial_{\mathbf{E}}\mathbf{P} \neq 0$, $\det\partial_{\mathbf{H}}\mathbf{M} \neq 0$ it follows

$$\mathbf{E} = \rho\partial_{\mathbf{P}}\hat{\psi}, \qquad \mathbf{H} = \rho\partial_{\mathbf{M}}\hat{\psi},$$

in accordance with the classical Legendre transforms.

Now, $\mathbf{L} = \mathfrak{D} + \mathbf{W}$. Since $\mathbf{W}$ is arbitrary and occurs linearly (whereas $\mathfrak{D}$ does not) then the entropy inequality implies

$$\mathrm{skw}\check{\mathbf{T}} = 0, \qquad \mathbf{E}\cdot\mathbf{J} + \check{\mathbf{T}}\cdot\mathfrak{D} - \frac{1}{\theta}\mathbf{q}\cdot\nabla\theta \geq 0.$$

Since

$$\mathrm{skw}\mathbf{T} = \mathrm{skw}(\mathbf{P}\otimes\mathbf{E} + \mu_0\mathbf{M}\otimes\mathbf{H})$$

is required by the balance of angular momentum, the restriction $\mathbf{T} \in \mathrm{Sym}$ is allowed provided $\mathbf{P}\otimes\mathbf{E} + \mu_0\mathbf{M}\otimes\mathbf{H} \in \mathrm{Sym}$. If ϕ depends on $\mathbf{E}$ and $\mathbf{H}$ through the scalars $\mathbf{E}^2$, $\mathbf{H}^2$, and $\mathbf{E}\cdot\mathbf{H}$ then

$$\mathbf{P}\otimes\mathbf{E} + \mu_0\mathbf{M}\otimes\mathbf{H} = -\rho[2\partial_{\mathbf{E}^2}\phi\,\mathbf{E}\otimes\mathbf{E} + 2\partial_{\mathbf{H}^2}\phi\,\mathbf{H}\otimes\mathbf{H} + \partial_{\mathbf{E}\cdot\mathbf{H}}\phi\,(\mathbf{H}\otimes\mathbf{E} + \mathbf{E}\otimes\mathbf{H})] \in \mathrm{Sym},$$

and the balance of angular momentum holds.

If $\mathbf{J}$ and $\mathbf{q}$ are independent of $\mathfrak{D}$ then the entropy inequality reduces to

$$\check{\mathbf{T}}\cdot\mathfrak{D} \geq 0$$

and hence

---

[9] Maxwell's equations for the co-moving observer are $\nabla\times\mathbf{H} = \mathbf{J} + \epsilon_0\,\overset{\diamond}{\mathbf{E}} + \overset{\diamond}{\mathbf{P}} = \mathbf{J} + \epsilon_0(\dot{\mathbf{E}} - \mathbf{L}\mathbf{E} + (\nabla\cdot\mathbf{v})\mathbf{E}) + \overset{\diamond}{\mathbf{P}}$ and $\nabla\times\mathbf{E} = -\overset{\diamond}{\mathbf{B}} = -\mu_0(\dot{\mathbf{H}} - \mathbf{L}\mathbf{H} + (\nabla\cdot\mathbf{v})\mathbf{H}) - \mu_0\,\overset{\diamond}{\mathbf{M}}$. Hence appropriate values of $\nabla\times\mathbf{H}$ and $\nabla\times\mathbf{E}$, which do not occur in the independent variables, allow us to say that arbitrary values of $\dot{\mathbf{H}}$ and $\dot{\mathbf{E}}$ are consistent with Maxwell's equations.

$$\check{\mathbf{T}} = 2\mu\mathfrak{D} + \lambda(\nabla \cdot \mathbf{v})\mathbf{1}$$

where $\mu \geq 0$, $2\mu + 3\lambda \geq 0$. The viscosity coefficients μ, λ are allowed to depend on $|\mathfrak{D}|$ and $|\mathbf{E}|$, $|\mathbf{H}|$ thus providing models of electro-rheological fluids. If $\mathfrak{D} = \mathbf{0}$ it follows that the functions $\mathbf{J}$ and $\mathbf{q}$ are required to satisfy

$$\mathbf{E} \cdot \mathbf{J} - \frac{1}{\theta}\mathbf{q} \cdot \nabla\theta \geq 0. \tag{12.19}$$

Inequality (12.19) holds if

$$\mathbf{J} = \mathbf{A}_1\mathbf{E} + \mathbf{A}_2\nabla\theta, \qquad \mathbf{q} = \mathbf{A}_3\mathbf{E} + \mathbf{A}_4\nabla\theta,$$

where $\mathbf{A}_1$, ..., $\mathbf{A}_4$ are matrices possibly parameterized by ρ, θ, $\mathbf{E}$, $\mathbf{H}$, the 6×6 matrix

$$\mathbf{A} = \begin{bmatrix} \mathbf{A}_1 & \frac{1}{2}(\mathbf{A}_2 - \mathbf{A}_3^T/\theta) \\ \frac{1}{2}(\mathbf{A}_2 - \mathbf{A}_3^T/\theta) & -\mathbf{A}_4/\theta \end{bmatrix}$$

being positive definite. More general models are allowed by letting

$$\mathbf{J} = \mathbf{A}_1\mathbf{E} + \mathbf{A}_2\nabla\theta + a\mathbf{E} \times \mathbf{H} + b\nabla\theta \times \mathbf{H},$$

$$\mathbf{q} = \mathbf{A}_3\mathbf{E} + \mathbf{A}_4\nabla\theta + c\mathbf{E} \times \mathbf{H} + d\nabla\theta \times \mathbf{H},$$

where a, b, c, d are scalars subject to $b + c/\theta = 0$. For isotropic materials, the analogous relations with $\mathbf{J}$ and $\nabla\theta$ as independent variables are written in the form

$$\mathbf{E} = \mathbf{J}/\sigma + \alpha\nabla\theta + \mathcal{R}\mathbf{H} \times \mathbf{J} + \mathcal{N}\mathbf{H} \times \nabla\theta,$$

$$\mathbf{q} = \beta\mathbf{J} - \kappa\nabla\theta + \mathcal{S}\mathbf{H} \times \mathbf{J} + \mathcal{L}\mathbf{H} \times \nabla\theta.$$

The entropy inequality holds if[10]

$$\sigma > 0, \quad \kappa > 0, \quad (\theta\alpha - \beta)^2 \leq 4\kappa\theta/\sigma, \quad \mathcal{N} + \mathcal{S}/\theta = 0.$$

Each term models a well-known effect. The term $\mathcal{J}/\sigma$ models Ohm's law, $\mathbf{J}_e = \sigma\mathbf{E}$, σ being the electric conductivity. The term $\alpha\nabla\theta$ represents the thermoelectric effect which is the conversion of temperature differences to electric voltage; it is referred to as Thompson effect. The term $\beta\mathbf{J}$ is said to represent the Peltier effect while $\mathbf{q} = -\kappa\nabla\theta$ is Fourier's law of heat conduction. In addition, $\mathcal{R}\mathbf{B} \times \mathbf{J}$ is the electric field perpendicular to $\mathbf{J}$ and $\mathbf{H}$; it is called the Hall effect. Also $\mathcal{N}\mathbf{H} \times \nabla\theta$ is a further

[10] These inequalities follow from the observation that, for any two vectors $\mathbf{u}$, $\mathbf{v}$ and scalars a, b, c the requirement $a\mathbf{u}^2 + b\mathbf{u} \cdot \mathbf{v} + c\mathbf{v}^2 \geq 0$ holds if $a\mathbf{u}^2 - |b||\mathbf{u}||\mathbf{v}| + c\mathbf{v}^2 \geq 0$ and hence if $a \geq 0$, $c \geq 0$, $b^2 \leq 4ac$.

contribution to $\mathbf{E}$ and is called the Nernst effect. Moreover, $\mathcal{S}\mathbf{H}\times\mathbf{J}$ and $\mathcal{L}\mathbf{H}\times\nabla\theta$ are denoted as Ettingshausen and Leduc-Righi effects.

Let

$$F := \mathbf{E}\cdot\mathbf{J} - \frac{1}{\theta}\mathbf{q}\cdot\nabla\theta.$$

We can view F as given by a function $F(\mathbf{E}, \nabla\theta)$ parameterized by $\rho, \theta, \mathbf{H}$. By (12.19) it follows that

$$F(\mathbf{E}, \nabla\theta) \geq 0, \qquad F(\mathbf{0}, \mathbf{0}) = 0.$$

Hence F must have a minimum at $\mathbf{E} = \mathbf{0}$, $\nabla\theta = \mathbf{0}$. This implies that, if $\mathbf{J}$ and $\mathbf{q}$ are differentiable,

$$\mathbf{J}(\mathbf{0}, \mathbf{0}) = \partial_{\mathbf{E}} F(\mathbf{0}, \mathbf{0}) = \mathbf{0}, \qquad \mathbf{q}(\mathbf{0}, \mathbf{0}) = -\theta\partial_{\nabla\theta} F(\mathbf{0}, \mathbf{0}) = \mathbf{0}.$$

Both $\mathbf{J}$ and $\mathbf{q}$ vanish as $\mathbf{E} = \mathbf{0}$, $\nabla\theta = \mathbf{0}$. With this restriction we can represent F by Taylor's formula, relative to the variables $\mathbf{E}$, $\nabla\theta$, and find higher-order terms (corrections) to the above electromagnetic effects. These expansions, up to fourth-order, are given in [58, 59].

12.6 Magnetoelasticity

Magnetoelasticity models the interaction of magnetic fields with elastic solids. The scheme deals with the deformation and magnetization of elastic materials under external effects, namely applied loads and/or deformations and thermal and magnetic fields. Accordingly the free charge density q and the electric polarization $\mathbf{P}$ are neglected and hence $\mathbf{D} = \epsilon_0\mathbf{E}$, $\mathbf{B} = \mu_0(\mathbf{H} + \mathbf{M})$. A conduction current density $\mathbf{J}$ is allowed to occur. Again we regard the electromagnetic fields as those at the reference at rest with the pertinent point of the body.

The balance equations (2.72), (2.73), (2.75) simplify to

$$\rho\dot{\mathbf{v}} = \nabla\cdot\mathbf{T} + \rho\mathbf{b} + \mu_0(\mathbf{M}\cdot\nabla)\mathbf{H} - \epsilon_0\mu_0\rho\dot{\mathbf{m}}\times\mathbf{E} + \mu_0\mathbf{J}\times\mathbf{H},$$

$$\text{skw}\mathbf{T} = \text{skw}\mu_0(\mathbf{M}\otimes\mathbf{H}),$$

$$\rho\dot{\varepsilon} = \mathbf{E}\cdot\mathbf{J} + \mu_0\rho\mathbf{H}\cdot\dot{\mathbf{m}} + \mathbf{T}\cdot\mathbf{L} - \nabla\cdot\mathbf{q} + \rho r.$$

Hence the entropy inequality

$$\rho\dot{\eta} + \nabla\cdot\frac{\mathbf{q}}{\theta} - \frac{\rho r}{\theta} + \nabla\cdot\mathbf{k} \geq 0$$

and substitution of $\nabla\cdot\mathbf{q} - \rho r$ from the balance of energy result in

$$-\rho\dot{\varepsilon} + \rho\theta\dot{\eta} + \mathbf{E}\cdot\mathbf{J} + \mu_0\rho\mathbf{H}\cdot\dot{\mathbf{m}} + \mathbf{T}\cdot\mathbf{L} - \frac{1}{\theta}\mathbf{q}\cdot\nabla\theta + \theta\nabla\cdot\mathbf{k} \geq 0. \tag{12.20}$$

We find it natural to regard **H**, rather than **M** or **B**, as the variable of magnetic character within the set of independent variables. This suggests that we consider the free energy

$$\phi = \varepsilon - \theta\eta - \mu_0\mathbf{H}\cdot\mathbf{m}$$

and hence we can write the Clausius–Duhem inequality in the form

$$-\rho(\dot{\phi} + \eta\dot{\theta}) - \mu_0\mathbf{M}\cdot\dot{\mathbf{H}} + \mathbf{E}\cdot\mathbf{J} + \mathbf{T}\cdot\mathbf{L} - \frac{1}{\theta}\mathbf{q}\cdot\nabla\theta + \theta\nabla\cdot\mathbf{k} \geq 0.$$

We let

$$\Gamma = (\mathbf{F}, \theta, \nabla\theta, \mathbf{E}, \mathbf{H})$$

be the set of independent variables.[11] Accordingly we assume ϕ, η, **M**, **J**, **T**, **q**, **k** as functions of Γ. Moreover we let ϕ be continuously differentiable.

Objectivity requires that ϕ depend on invariants under Euclidean transformations. Now, we assume **H** is a vector in that

$$\mathbf{H} \to \mathbf{Q}\mathbf{H},$$

Q being the rotation tensor, $\mathbf{Q} \in SO(3)$. Since

$$\mathbf{F} \to \mathbf{Q}\mathbf{F}$$

then $\mathbf{F}^{-1}\mathbf{H}$, $J\mathbf{F}^{-1}\mathbf{H}$, $\mathbf{F}^T\mathbf{H}$, and $J\mathbf{F}^T\mathbf{H}$, are invariant vectors under a Euclidean transformation.[12] Of course also the scalars $\mathbf{H}\cdot\mathbf{H}$ and $\mathbf{H}\cdot\mathbf{F}\mathbf{F}^T\mathbf{H}$ are invariant,[13]

$$(\mathbf{Q}\mathbf{H})\cdot(\mathbf{Q}\mathbf{H}) = \mathbf{H}\cdot\mathbf{Q}^T\mathbf{Q}\mathbf{H} = \mathbf{H}\cdot\mathbf{H},$$

$$\mathbf{Q}\mathbf{H}\cdot\mathbf{Q}\mathbf{F}\mathbf{F}^T\mathbf{Q}^T\mathbf{Q}\mathbf{H} = \mathbf{H}\cdot\mathbf{Q}^T\mathbf{Q}\mathbf{F}\mathbf{F}^T\mathbf{H} = \mathbf{H}\cdot\mathbf{F}\mathbf{F}^T\mathbf{H}.$$

Instead, $\mathbf{H}\cdot\mathbf{C}\mathbf{H}$ is not invariant.

To fix ideas we let ϕ depend on **H** via the vector

$$\mathfrak{H} = \mathbf{F}^T\mathbf{H},$$

---

[11] As with electroelasticity we might allow for a dependence on $\mathfrak{D}$ so that dissipative mechanical effects are described.

[12] Incidentally, $J\mathbf{F}^{-1}\mathbf{B}$ is considered in [136] as an independent variable on the view that $J\mathbf{F}^{-1}\mathbf{B}$ is the magnetic induction in the reference configuration, and is regarded as a Lagrangian field.

[13] As well as $\mathbf{B}\cdot\mathbf{B}$ and $\mathbf{B}\cdot\mathbf{F}\mathbf{F}^T\mathbf{B}$; see, e.g. [152], p. 310.

and, as in electroelasticity, we let ϕ depend on $\mathbf{E}$ via $\mathfrak{E} = \mathbf{F}^T\mathbf{E}$. Hence we start with the free energy ϕ as given by[14]

$$\phi = \phi(\mathbf{C}, \theta, \nabla\theta, \mathfrak{E}, \mathfrak{H}).$$

Upon computation of $\dot{\phi}$ and substitution in the Clausius–Duhem inequality we have

$$-\rho(\partial_\theta\phi + \eta)\dot{\theta} - \rho\partial_{\mathbf{C}}\phi \cdot \dot{\mathbf{C}} - \rho\partial_{\nabla\theta}\phi \cdot \overline{\nabla\theta}^{\cdot} - \rho\partial_{\mathfrak{E}}\,\phi \cdot \dot{\mathfrak{E}} - \rho\partial_{\mathfrak{H}}\,\phi \cdot \dot{\mathfrak{H}}$$
$$-\mu_0\mathbf{M} \cdot \dot{\mathbf{H}} + \mathbf{E} \cdot \mathbf{J} + \mathbf{T} \cdot \mathbf{L} - \frac{1}{\theta}\mathbf{q} \cdot \nabla\theta + \theta\nabla \cdot \mathbf{k} \geq 0.$$

Observe

$$\dot{\mathfrak{H}} = \dot{\mathbf{F}}^T\mathbf{H} + \mathbf{F}^T\dot{\mathbf{H}} = \mathbf{F}^T\mathbf{L}^T\mathbf{H} + \mathbf{F}^T\dot{\mathbf{H}}$$

and the like for $\mathfrak{E}$. Moreover $\dot{\mathbf{C}} = 2\mathbf{F}^T\mathfrak{D}\mathbf{F}$. Substitution in the Clausius–Duhem inequality, the observation that no significant generality is lost by letting $\mathbf{k} = \mathbf{0}$, and some rearrangements yield

$$-\rho(\partial_\theta\phi + \eta)\dot{\theta} + [\mathbf{T} - 2\rho\mathbf{F}\partial_{\mathbf{C}}\phi\mathbf{F}^T - \rho(\mathbf{E} \otimes \mathbf{F}\partial_{\mathfrak{E}}\phi + \mathbf{H} \otimes \mathbf{F}\partial_{\mathfrak{H}}\,\phi)] \cdot \mathbf{L} - \rho\partial_{\nabla\theta}\phi \cdot \overline{\nabla\theta}^{\cdot}$$
$$-\rho(\mathbf{F}\partial_{\mathfrak{E}}\,\phi) \cdot \dot{\mathbf{E}} - (\rho\mathbf{F}\partial_{\mathfrak{H}}\,\phi + \mu_0\mathbf{M}) \cdot \dot{\mathbf{H}} + \mathbf{E} \cdot \mathbf{J} - \frac{1}{\theta}\mathbf{q} \cdot \nabla\theta \geq 0$$

The arbitrariness of $\dot{\theta}$, $\overline{\nabla\theta}^{\cdot}$ implies

$$\partial_{\nabla\theta}\phi = \mathbf{0}, \qquad \eta = -\partial_\theta\phi.$$

In view of Maxwell's equations, the values of $\dot{\mathbf{E}}$ and $\dot{\mathbf{H}}$ can be chosen arbitrarily; in addition, the values of $\dot{\mathbf{E}}$ and $\dot{\mathbf{H}}$ do not affect the remaining terms of the inequality, which depend on $\mathbf{F}$, θ, $\nabla\theta$, $\mathbf{E}$, $\mathbf{H}$. Hence the arbitrariness of $\dot{\mathbf{E}}$, $\dot{\mathbf{H}}$ implies

$$\partial_{\mathfrak{E}}\,\phi = \mathbf{0}, \qquad \mu_0\mathbf{M} = -\rho\mathbf{F}\partial_{\mathfrak{H}}\,\phi. \tag{12.21}$$

The inequality then reduces to

$$[\mathbf{T} - 2\rho\mathbf{F}\partial_{\mathbf{C}}\phi\mathbf{F}^T - \rho\mathbf{H} \otimes \mathbf{F}\partial_{\mathfrak{H}}\,\phi)] \cdot \mathfrak{D} + [\mathbf{T} - \rho\mathbf{H} \otimes \mathbf{F}\partial_{\mathfrak{H}}\,\phi)] \cdot \mathbf{W} + \mathbf{E} \cdot \mathbf{J} - \frac{1}{\theta}\mathbf{q} \cdot \nabla\theta \geq 0.$$

The arbitrariness of $\mathfrak{D}$ and $\mathbf{W}$ implies

$$\mathbf{T} = 2\rho\mathbf{F}\partial_{\mathbf{C}}\phi\mathbf{F}^T + \rho\mathbf{H} \otimes \mathbf{F}\partial_{\mathfrak{H}}\,\phi, \tag{12.22}$$

$$\mathbf{T} - \rho\mathbf{H} \otimes \mathbf{F}\partial_{\mathfrak{H}}\phi \in \mathrm{Sym}, \tag{12.23}$$

In view of (12.21), the requirement (12.23) can be written

---

[14] We observe that $\mathbf{F}^T\mathbf{H} = \mathbf{H}_R$, $\mathbf{F}^T\mathbf{E} = \mathbf{E}_R$.

$$\operatorname{skw} \mathbf{T} = \operatorname{skw} \mu_0 \mathbf{M} \otimes \mathbf{H},$$

in accordance with (2.73). We can then write the reduced inequality in the form

$$\mathbf{E} \cdot \mathbf{J} - \frac{1}{\theta} \mathbf{q} \cdot \nabla\theta \geq 0.$$

Possible models of $\mathbf{J}$ and $\mathbf{q}$ are given by[15]

$$\mathbf{J} = \mathbf{A}_1 \mathbf{E} + \mathbf{A}_2 \nabla\theta, \qquad \mathbf{q} = \mathbf{A}_3 \mathbf{E} + \mathbf{A}_4 \nabla\theta,$$

subject to the positive semi-definiteness of the matrix

$$\mathbf{A} = \begin{bmatrix} \mathbf{A}_1 & \frac{1}{2}(\mathbf{A}_2 - \mathbf{A}_3^T/\theta) \\ \frac{1}{2}(\mathbf{A}_2 - \mathbf{A}_3^T/\theta)^T & -\mathbf{A}_4/\theta \end{bmatrix},$$

$$\frac{1}{\theta} \det \mathbf{A}_1 \det \mathbf{A}_4 - \tfrac{1}{4}[\det(\mathbf{A}_2 - \mathbf{A}_3^T/\theta)]^2 \leq 0.$$

The simplest nontrivial model satisfying the inequality is given by

$$\mathbf{A}_1 = \sigma(J, \theta)\mathbf{1}, \qquad \mathbf{A}_4 = -\kappa(J, \theta)\mathbf{1},$$

while $\mathbf{A}_3 = 0$, $\mathbf{A}_4 = 0$, namely

$$\mathbf{J} = \sigma(J, \theta)\mathbf{E}, \qquad \mathbf{q} = -\kappa(J, \theta)\nabla\theta,$$

with $\sigma \geq 0$, $\kappa \geq 0$ while $J = \det \mathbf{F} = (\det \mathbf{C})^{1/2}$. Accordingly, σ and κ are required to be positive valued thus showing once again that the electric conductivity and the heat conductivity have to be positive.

As a comment on (12.22), we observe that $\mathbf{F}\partial_{\mathbf{C}}\phi\mathbf{F}^T \in \mathrm{Sym}$ and hence it follows that $\mathbf{T}$ is the sum of a mechanical part, $2\rho\mathbf{F}\partial_{\mathbf{C}}\phi\mathbf{F}^T \in \mathrm{Sym}$, and a magnetic part $-\mu_0\mathbf{H} \otimes \mathbf{M}$, which is the stress induced by the magnetic intensity $\mathbf{H}$ in the magnetizable body. The dependence of ϕ on $\mathfrak{H}$, e.g. $\mathfrak{H} = \mathbf{F}^T\mathbf{H}$ may be viewed as a modelling of the interaction between magnetic field and deformation.

Observe

$$\operatorname{skw} \mathbf{T} = \operatorname{skw} \mu_0 \mathbf{M} \otimes \mathbf{H}$$

arises from the balance of angular momentum. It is worth remarking that this condition holds unchanged if $\mathfrak{H}$ is taken to be one of the other invariant vectors. We have

$$\mathfrak{H} = \mathbf{F}^{-1}\mathbf{H}: \qquad \mathbf{T} - \rho\mathbf{H} \otimes (\mathbf{F}^{-T}\partial_{\mathfrak{H}}\phi) \in \mathrm{Sym}, \qquad \mu_0\mathbf{M} = -\rho\mathbf{F}^{-T}\partial_{\mathfrak{H}}\phi,$$

[15] Rate-type models of $\mathbf{q}$ and $\mathbf{J}$ require the insertion of $\mathbf{q}$ and $\mathbf{J}$ among the independent variables.

$$\mathfrak{H} = \mathbf{F}^T\mathbf{H}: \qquad \mathbf{T} - \rho\mathbf{H} \otimes (\mathbf{F}\partial_{\mathfrak{H}}\phi) \in \mathrm{Sym}, \qquad \mu_0\mathbf{M} = -\rho\mathbf{F}\partial_{\mathfrak{H}}\phi,$$

$$\mathfrak{H} = J\mathbf{F}^T\mathbf{H}: \qquad \mathbf{T} - \rho J\mathbf{H} \otimes (\mathbf{F}\partial_{\mathfrak{H}}\phi) \in \mathrm{Sym}, \qquad \mu_0\mathbf{M} = -\rho J\mathbf{F}\partial_{\mathfrak{H}}\phi.$$

In all of these cases the symmetry requirement $\mathbf{T} + \mu_0\mathbf{H} \otimes \mathbf{M} \in \mathrm{Sym}$ holds.

Things are consistent if ϕ depends on $\mathbf{H}$ through the scalar invariant $\alpha = \mathbf{H} \cdot \mathbf{H}$. We observe that the restriction to the pertinent terms in the Clausius–Duhem inequality result in

$$-\rho\partial_\alpha\phi\,\dot{\alpha} + [\mathbf{T} - 2\rho\mathbf{F}\partial_{\mathbf{C}}\phi\,\mathbf{F}^T] \cdot \mathbf{L} - \mu_0\mathbf{M} \cdot \dot{\mathbf{H}} \geq 0.$$

Since

$$\dot{\alpha} = 2\mathbf{H} \cdot \dot{\mathbf{H}}$$

then we can write

$$[\mathbf{T} - 2\rho\mathbf{F}\partial_{\mathbf{C}}\phi\,\mathbf{F}^T] \cdot \mathfrak{D} + \mathbf{T} \cdot \mathbf{W} - [\mu_0\mathbf{M} + 2\rho\partial_\alpha\phi\,\mathbf{H}] \cdot \dot{\mathbf{H}} \geq 0.$$

Hence it follows

$$\mathbf{T} \in \mathrm{Sym}, \qquad \mu_0\mathbf{M} = -2\rho\partial_\alpha\phi\,\mathbf{H}.$$

Since

$$\mu_0\mathbf{M} \otimes \mathbf{H} = -2\rho\partial_\alpha\phi\,\mathbf{H} \otimes \mathbf{H} \in \mathrm{Sym}$$

it follows that also the dependence on $\alpha = \mathbf{H} \cdot \mathbf{H}$ satisfies $\mathbf{T} + \mu_0\mathbf{H} \otimes \mathbf{M} \in \mathrm{Sym}$.

The literature addresses attention also to the scalar invariant $\varkappa = \mathbf{H} \cdot \mathbf{F}\mathbf{F}^T\mathbf{H}$. Now,

$$\mathbf{H} \cdot \mathbf{F}\mathbf{F}^T\mathbf{H} = (\mathbf{F}^T\mathbf{H}) \cdot (\mathbf{F}^T\mathbf{H})$$

and hence the previous analysis on the dependence on $\mathfrak{H} = \mathbf{F}^T\mathbf{H}$ applies. Yet, we may follow a different procedure and look at the pertinent terms in the Clausius–Duhem inequality,

$$-\rho\partial_\varkappa\phi\,\dot{\varkappa} + \mathbf{T} \cdot \mathbf{L} - \mu_0\mathbf{M} \cdot \dot{\mathbf{H}} \geq 0.$$

Let $\mathbf{u} = \mathbf{F}\mathbf{F}^T\mathbf{H}$. Since

$$\dot{\varkappa} = 2\mathbf{u} \cdot \dot{\mathbf{H}} + (\mathbf{H} \otimes \mathbf{u}) \cdot \mathbf{L} + (\mathbf{u} \otimes \mathbf{H}) \cdot \mathbf{L}^T$$

we can write the inequality in the form

$$[\mathbf{T} - 2\rho\mathbf{F}\partial_{\mathbf{C}}\phi\,\mathbf{F}^T - 2\rho\partial_\varkappa\phi\,\mathrm{sym}\mathbf{H} \otimes \mathbf{u}] \cdot \mathfrak{D} + [\mathbf{T} - 2\rho\partial_\varkappa\phi\,\mathrm{skw}\mathbf{H} \otimes \mathbf{u}] \cdot \mathbf{W} - [\mu_0\mathbf{M} + 2\rho\partial_\varkappa\phi\,\mathbf{u}] \cdot \dot{\mathbf{H}} \geq 0,$$

whence

$$\mathbf{T} - 2\rho\partial_\varkappa\phi\,\mathbf{H} \otimes \mathbf{u} \in \mathrm{Sym}, \qquad \mu_0\mathbf{M} = -2\rho\partial_\varkappa\phi\,\mathbf{u}.$$

Hence $\mathbf{T} + \mu_0 \mathbf{H} \otimes \mathbf{M} \in \mathrm{Sym}$.

Linear approximation

A simple scheme for applications follows if the fields involved justify a linear approximation. By linearizing, relative to $\mathbf{F} - \mathbf{1}$, we have

$$\mathbf{C} \simeq \mathbf{1} + 2\mathbf{S}, \qquad \mathbf{S} = \mathrm{sym}\nabla\mathbf{u}.$$

Also we let the initial values of $\mathbf{H}$ and $\mathbf{E}$ be zero and hence we linearize also with respect to $\mathbf{H}$ and $\mathbf{E}$. Moreover we let $\mathbf{M}$ vanish if both $\mathbf{H}$ and $\mathbf{S}$ approach zero. Hence $\mathbf{M} \otimes \mathbf{H}$ is a nonlinear term. The temperature θ is taken to be constant, $\theta = \theta_0$, in the reference configuration.

The constitutive relations (12.22) then simplify to

$$\mathbf{T} = \rho_0 \partial_{\mathbf{S}} \phi, \qquad \mu_0 \mathbf{M} = -\rho_0 \partial_{\mathbf{H}} \phi,$$

where $\rho_0 = \rho_R$. Consequently, in the linear approximation,

$$\partial_{\mathbf{S}} \mu_0 \mathbf{M} = -\partial_{\mathbf{H}} \mathbf{T},$$

or, in suffix notation,

$$\mu_0 \partial_{S_{ij}} M_k = -\partial_{H_k} T_{ij}.$$

Let $\vartheta = \theta - \theta_0$. The free energy ϕ is considered in the (quadratic) form

$$\rho_0\phi = \rho_0\phi_0 - \rho_0\eta_0\vartheta - \frac{\rho_0\beta}{2\theta_0}\vartheta^2 - \xi_{kl}\vartheta S_{kl} + \tfrac{1}{2} S_{kl}\Lambda_{klmn}S_{mn} + S_{kl}\gamma_{klm}H_m - \nu_k\vartheta H_k - \tfrac{1}{2} H_k\chi_{kl}H_l.$$

By the entropy relation and (12.21), (12.22) it follows

$$\eta = \eta_0 + \frac{\beta}{\theta_0}\vartheta + \frac{1}{\rho_0}\kappa_{kl}S_{kl} + \frac{\nu_k}{\rho_0}H_k,$$

$$T_{kl} = -\xi_{kl}\vartheta + \Lambda_{klmn}S_{mn} + \gamma_{klm}H_m,$$

$$\mu_0 M_k = -\gamma_{ijk}S_{ij} + \nu_k\vartheta + \chi_{kl}H_l,$$

while $\mathbf{q}$ and $\mathbf{J}$ are taken in the form

$$q_k = -\kappa_{kl}\partial_{x_l}\vartheta + \zeta_{kl}E_l, \qquad J_k = \sigma_{kl}E_l + \alpha_{kl}\partial_{x_l}\vartheta.$$

Let $\vartheta = 0$. Using the Mandel-Voigt notation we can write

$$\tilde{T} = \tilde{\Lambda}\tilde{S} + \tilde{\gamma}H, \qquad \mu_0 M = -\tilde{\gamma}^T\tilde{S} + \xi H,$$

where $\tilde{T}, \tilde{S} \in \mathbb{R}^6$, $H, M \in \mathbb{R}^3$, $\tilde{\Lambda} = \tilde{\Lambda}^T \in \mathbb{R}^{6\times 6}$, $\tilde{\gamma} \in \mathbb{R}^{6\times 3}$, $\chi = \chi^T \in R^{3\times 3}$. If, as is often the case, we regard the stress rather than the strain as an independent variable we observe

$$\tilde{S} = \tilde{\Lambda}^{-1}\tilde{T} - \tilde{\Lambda}^{-1}\tilde{\gamma}H, \qquad M = -\tilde{\gamma}^T\tilde{\Lambda}^{-1}\tilde{T} + \xi H.$$

Hence it follows that

$$\partial_H \tilde{S} = (\partial_{\tilde{T}} M)^T. \tag{12.24}$$

In isotropic materials the odd-order tensors vanish and the even-order tensors are expressed in terms of Kronecker deltas so that

$$\rho_0\phi = \rho_0\phi_0 - \rho_0\eta_0\vartheta - \frac{\rho_0\beta}{2\theta_0}\vartheta^2 - \xi\,\vartheta \mathrm{tr}\,\mathbf{S} + \mu\mathbf{S}\cdot\mathbf{S} + \tfrac{1}{2}\lambda(\mathrm{tr}\,\mathbf{S})^2 - \tfrac{1}{2}\chi\mathbf{H}^2.$$

$$\eta = \eta_0 + \frac{\beta}{\theta_0}\vartheta + \frac{\xi}{\rho_0}\mathrm{tr}\,\mathbf{S},$$

$$\mathbf{T} = (-\xi\vartheta + \lambda \mathrm{tr}\,\mathbf{S})\mathbf{1} + 2\mu\mathbf{S},$$

$$\mu_0\,\mathbf{M} = \chi\,\mathbf{H},$$

and

$$\mathbf{q} = -\kappa\,\nabla\vartheta + \zeta\,\mathbf{E}, \qquad \mathbf{J} = \sigma\,\mathbf{E} + \alpha\,\nabla\vartheta.$$

The requirement $\mathbf{E}\cdot\mathbf{J} - \mathbf{q}\cdot\nabla\theta/\theta \geq 0$ results in

$$\kappa \geq 0, \qquad \sigma \geq 0, \qquad 4\kappa\sigma \geq \theta(\alpha - \zeta/\theta)^2.$$

Magnetostriction is a property of ferromagnetic materials that causes them to change their shape when subjected to a magnetic field. The *Villari effect* consists in a change of magnetization as a consequence of an applied stress. Equation (12.24) means that the two effects are the inverse of each other and they are quantitatively equal.

There are approaches[16] where the sum of the stress parts is obtained upon an additive decomposition of the magnetic induction (or the electric field) in elastic and viscous internal variables while the deformation gradient is decomposed into the product of the corresponding parts, $\mathbf{F} = \mathbf{F}_e\mathbf{F}_v$.

[16] See, e.g. [385, 386].

12.7 Micromagnetics and Magnetism in Rigid Bodies

The magnetic behaviour of materials is modelled within micromagnetics by starting from the evolution of single spin under the action of a time-dependent external magnetic field. Following the classical picture of a charge, we start from the equation for (the expectation value of) the time-dependent spin $\mathbf{s}$

$$\dot{\mathbf{s}} = -\gamma \mathbf{s} \times \mathbf{B},$$

where

$$\gamma = \frac{|g|\mu_B}{\hbar},$$

$g < 0$ is the gyromagnetic ratio and μ_B is the Bohr magneton; if e is the charge and m is the mass then $\mu_B = e\hbar/2m$. The magnetic moment[17] $\mathfrak{m} = \gamma \mathbf{s}$ of an electron then satisfies

$$\dot{\mathfrak{m}} = -\gamma \mathfrak{m} \times \mathbf{B}.$$

Really this equation was introduced by Landau and Lifshitz [270] with $\mathbf{B}$ replaced by an effective field $\mathbf{H}_{\rm eff}$ which is the resultant of the external magnetic field $\mathbf{H}$, the demagnetizing field and also quantum mechanical effects due to the interaction between the magnetic moments (and γ comprising the permeability of free space μ_0). Therefore

$$\mathfrak{m} \cdot \dot{\mathfrak{m}} = 0$$

and hence $|\mathfrak{m}|$ would be constant in time. Moreover, if $\mathbf{B}$ is constant

$$\dot{\overline{\mathbf{B} \cdot \mathfrak{m}}} = \mathbf{B} \cdot \dot{\mathfrak{m}} = 0,$$

so that $\mathfrak{m}$ precesses about the $\mathbf{B}$ axis with constant frequency.

Consistent with Landau and Lifshitz [270], we replace $\mathbf{B}$ with $\mathbf{H}$, regarded as the effective field. Further, experiments show that beyond certain critical values of the applied magnetic field the magnetization saturates, becomes uniform and aligns parallel to the magnetic field. To incorporate this experimental fact from phenomenological grounds one can add a damping term. Landau and Lifshitz added a term in the evolution equation in the form

$$\dot{\mathfrak{m}} = -\gamma \mathfrak{m} \times \mathbf{H} - \alpha_{LL} \mathfrak{m} \times (\mathfrak{m} \times \mathbf{H}), \qquad \alpha_{LL} > 0. \tag{12.25}$$

Again $\mathfrak{m} \cdot \dot{\mathfrak{m}} = 0$ whereas, if $\mathbf{H}$ is constant,

$$|\mathbf{H}|\,|\mathfrak{m}|\dot{\overline{\cos\theta}} = \mathbf{H} \cdot \dot{\mathfrak{m}} = \alpha_{LL}[\mathfrak{m}^2\mathbf{H}^2 - (\mathfrak{m} \cdot \mathbf{H})^2] \geq 0,$$

[17] Often denoted by μ.

whence the angle of precession, θ, decreases in time. By arguing on a Rayleigh dissipation functional, Gilbert [191, 192] modified the evolution equation to a form that can be written

$$\dot{\mathfrak{m}} = -\gamma_G \mathfrak{m} \times \mathbf{H} + \alpha_G \mathfrak{m} \times \dot{\mathfrak{m}}, \tag{12.26}$$

$\gamma_G < 0$ and α_G being called the Gilbert gyromagnetic ratio and the Gilbert damping constant. Since, again, $\mathfrak{m} \cdot \dot{\mathfrak{m}} = 0$ (saturated dipoles) it follows

$$\mathfrak{m} \times \dot{\mathfrak{m}} = -\gamma_G \mathfrak{m} \times (\mathfrak{m} \times \mathbf{H}) - \alpha_G \mathfrak{m}^2 \dot{\mathfrak{m}}.$$

Hence a little algebra shows that (12.26) can be written

$$\dot{\mathfrak{m}} = -\frac{\gamma_G}{1 + \alpha_G^2 \mathfrak{m}^2} \mathfrak{m} \times \mathbf{H} - \frac{\alpha_G \gamma_G}{1 + \alpha_G^2 \mathfrak{m}^2} \mathfrak{m} \times (\mathfrak{m} \times \mathbf{H}). \tag{12.27}$$

This shows the formal equivalence of the Landau–Lifshitz equation and the Gilbert equation provided only we make the identifications

$$\gamma = \frac{\gamma_G}{1 + \alpha_G^2 \mathfrak{m}^2}, \qquad \alpha_{LL} = \frac{\alpha_G \gamma_G}{1 + \alpha_G^2 \mathfrak{m}^2},$$

and $\gamma_G < 0$. That is why Eq. (12.27) is referred to as Landau–Lifshitz–Gilbert equation (LLG for short). It is worth observing that $\alpha_G \to \infty$ implies $\alpha_{LL} \to \infty$. Hence, as $\alpha_G \to \infty$, in view of (12.27) we have $\dot{\mathfrak{m}} \to 0$ whereas $\alpha_{LL} \to \infty$ makes $\dot{\mathfrak{m}}$ undetermined.

While low temperatures, $\theta \ll \theta_C$, make the saturation assumption physically grounded, at relatively high temperatures, $\theta > \theta_C$, the saturation assumption is no longer valid. We then let

$$\mathfrak{m} = \mathrm{m}\, \mathbf{e},$$

$\mathbf{e}$ being the unit vector of $\mathfrak{m}$, and observe

$$\dot{\mathfrak{m}} = \dot{\mathrm{m}}\, \mathbf{e} + \mathrm{m}\, \dot{\mathbf{e}}$$

so that, geometrically, $\dot{\mathfrak{m}}$ consists of a vector parallel to $\mathfrak{m}$, $\dot{\mathrm{m}}\, \mathbf{e}$, and a vector perpendicular to $\mathfrak{m}$, $\mathrm{m}\, \dot{\mathbf{e}}$. Since $\dot{\mathfrak{m}}$ is no longer perpendicular to $\mathfrak{m}$ then it is natural to add a term parallel to $\mathfrak{m}$ in the evolution equation. This step traces back to Bloch [51, 185] and results in the Landau–Lifshitz–Bloch (LLB) equation

$$\dot{\mathfrak{m}} = -\gamma \mathfrak{m} \times \mathbf{H} - \gamma \alpha_\perp \mathfrak{m} \times (\mathfrak{m} \times \mathbf{H}) + \gamma \alpha_\parallel (\mathfrak{m} \cdot \mathbf{H}) \mathfrak{m},$$

where again $\mathbf{H}$ stands for the effective field. For definiteness the parameters $\alpha_\parallel$, $\alpha_\perp$ are taken to depend on temperature in the form

$$\alpha_{\|} = \frac{2}{3}\frac{\theta}{\theta_c}\lambda, \qquad \alpha_{\perp} = \begin{cases} \lambda(1 - \theta/3\theta_c) & \text{if } \theta < \theta_c, \\ \alpha_{\|}(\theta) & \text{if } \theta \geq \theta_c, \end{cases}$$

λ being a positive parameter.[18] Two remarks are in order. As $\theta \to 0$ LLB reduces to

$$\dot{\mathbf{m}} = -\gamma\boldsymbol{\mu} \times \mathbf{H} - \gamma\lambda\mathbf{m} \times (\mathbf{m} \times \mathbf{H}),$$

and hence the Landau–Lifshitz equation is recovered with $\alpha_{LL} = \gamma\lambda$. If, instead, $\theta > \theta_c$ then $\alpha_{\|} = \alpha_{\perp}$ and hence the evolution equation becomes

$$\dot{\mathbf{m}} = -\gamma\mathbf{m} \times \mathbf{H} + \gamma\alpha_{\perp}\mathbf{H}.$$

12.7.1 Rate Equations and Thermodynamic Restrictions

Borrowing from the evolution equations of micromagnetics we now set up a model for thermomagnetic solids. We assume the body is undeformable and hence regard the mass density, ρ, and the number density of dipoles, n, as constant. Let

$$\mathbf{M} = n\,\mathbf{m}.$$

The LLB equation indicates that we write the evolution equation

$$\dot{\mathbf{M}} = \alpha_1\mathbf{M} \times \mathbf{H} + \alpha_2\mathbf{M}^2\,\mathbf{H} + \alpha_3(\mathbf{M} \cdot \mathbf{H})\mathbf{M}, \tag{12.28}$$

where α_2 and α_3 arise as $\gamma\alpha_{\perp}/n$ and $\gamma(\alpha_{\|} - \alpha_{\perp})/n$. Since $\alpha_1, \alpha_2, \alpha_3$ are allowed to depend on temperature, Eq. (12.28) is a particular case of rate equations of the form

$$\dot{\mathbf{M}} = \mathcal{M}(\theta, \mathbf{H}, \mathbf{M}).$$

It is understood that the (Lagrangian) time derivative is relative to the observer at rest with the body, at the point under consideration.

In undeformable media, for the observer at rest the balance of energy (2.76) simplifies to

$$\rho\dot{\varepsilon} = \mathbf{E} \cdot \mathbf{J} + \mu_0\mathbf{H} \cdot \dot{\mathbf{M}} - \nabla \cdot \mathbf{q} + \rho r.$$

Hence the Clausius–Duhem inequality (2.85) can be written as

$$-\rho(\dot{\psi} + \eta\dot{\theta}) + \mathbf{E} \cdot \mathbf{J} + \mu_0\mathbf{H} \cdot \dot{\mathbf{M}} - \frac{1}{\theta}\mathbf{q} \cdot \nabla\theta + \theta\nabla \cdot \mathbf{k} \geq 0.$$

[18] λ is called thermal bath coupling constant.

The fields **E**, **B**, **H**, **M** are required to satisfy Maxwell's equations which, for rigid bodies, read

$$\nabla \times \mathbf{E} = -\dot{\mathbf{B}}, \qquad \nabla \cdot \mathbf{B} = 0, \qquad \nabla \times \mathbf{H} = \epsilon_0 \dot{\mathbf{E}} + \mathbf{J}, \qquad \nabla \cdot \mathbf{E} = q/\epsilon_0.$$

Moreover

$$\mathbf{D} = \epsilon_0 \mathbf{E}, \qquad \mathbf{B} = \mu_0(\mathbf{H} + \mathbf{M}).$$

These relations are to be viewed as constraints in the derivation of thermodynamic restrictions.

To allow for heat conduction and electric conduction we let $\nabla\theta$ and **E** be among the independent variables. Moreover we let a dependence on $\nabla\mathbf{M}$ occur so that, at a macroscopic level, we account for the exchange energy of the physical picture [192]. Hence to ensure the equipresence rule we take

$$\Gamma = (\theta, \nabla\theta, \mathbf{E}, \mathbf{H}, \mathbf{M}, \nabla\mathbf{M}, \nabla\nabla\mathbf{M})$$

as the set of independent variables.[19] So we let

$$\psi = \psi(\theta, \nabla\theta, \mathbf{E}, \mathbf{H}, \mathbf{M}, \nabla\mathbf{M}, \nabla\nabla\mathbf{M})$$

and the like for **J**, **q**, **k**, and

$$\dot{\mathbf{M}} = \mathcal{M}(\theta, \nabla\theta, \mathbf{E}, \mathbf{H}, \mathbf{M}, \nabla\mathbf{M}, \nabla\nabla\mathbf{M}).$$

Compute $\dot{\psi}$ and substitute in the Clausius–Duhem inequality to obtain

$$\begin{aligned}-\rho(\partial_\theta\psi + \eta)\dot{\theta} - \rho\partial_{\nabla\theta}\psi \cdot \nabla\dot{\theta} - \rho\partial_{\mathbf{E}}\psi \cdot \dot{\mathbf{E}} - \rho\partial_{\mathbf{H}}\psi \cdot \dot{\mathbf{H}} - \rho\partial_{\mathbf{M}}\psi \cdot \dot{\mathbf{M}} - \rho\partial_{\nabla\mathbf{M}}\psi \cdot (\nabla\dot{\mathbf{M}}) \\ -\rho\partial_{\nabla\nabla\mathbf{M}}\psi \cdot (\nabla\nabla\dot{\mathbf{M}}) + \mathbf{E}\cdot\mathbf{J} + \mathbf{B}\cdot\dot{\mathbf{M}} - \frac{1}{\theta}\mathbf{q}\cdot\nabla\theta + \theta\nabla\cdot\mathbf{k} \ge 0.\end{aligned}$$

The value of $\nabla\nabla\dot{\mathbf{M}}$ can be chosen arbitrarily without affecting the other terms within the inequality. The values of $\dot{\mathbf{E}}$ and $\dot{\mathbf{H}}$ too can be chosen arbitrarily. To show this, we observe that, given **J** by the constitutive equation, any value of $\dot{\mathbf{E}}$ is allowed subject to

$$\nabla \times \mathbf{H} = \epsilon_0 \dot{\mathbf{E}} + \mathbf{J}.$$

Hence, for any value of $\dot{\mathbf{E}}$ the constraint is satisfied by an appropriate value of $\nabla \times \mathbf{H}$ which does not enter the inequality. Likewise, any value of $\dot{\mathbf{H}}$ is allowed subject to

$$\nabla \times \mathbf{E} = -\mu_0(\dot{\mathbf{H}} + \dot{\mathbf{M}})$$

[19] Here we neglect from the start a dependence on $\dot{\mathbf{H}}$ which would allow the modelling of hysteretic materials. Hysteresis is developed in Sect. 15.3.

and $\nabla \times \mathbf{E}$ does not enter the inequality. Meanwhile, as is customary, we can give $\dot{\theta}$ and $\nabla\dot{\theta}$ arbitrary values without affecting the remaining terms of the inequality. The arbitrariness of $\nabla\nabla\dot{\mathbf{M}}$, $\nabla\dot{\theta}$, $\dot{\mathbf{E}}$, $\dot{\mathbf{H}}$, $\dot{\theta}$ implies

$$\partial_{\nabla\nabla\mathbf{M}}\psi = \mathbf{0}, \qquad \partial_{\nabla\theta}\psi = \mathbf{0}, \qquad \partial_{\mathbf{E}}\psi = \mathbf{0}, \qquad \partial_{\mathbf{H}}\psi = \mathbf{0}, \qquad \eta = -\partial_\theta\psi.$$

For formal convenience we write the remaining inequality in the form

$$-\frac{\rho}{\theta}\partial_{\mathbf{M}}\psi \cdot \dot{\mathbf{M}} - \frac{\rho}{\theta}\partial_{\nabla\mathbf{M}}\psi \cdot (\nabla\dot{\mathbf{M}}) + \frac{1}{\theta}\mathbf{E}\cdot\mathbf{J} + \frac{\mu_0}{\theta}\mathbf{H}\cdot\dot{\mathbf{M}} - \frac{1}{\theta^2}\mathbf{q}\cdot\nabla\theta + \nabla\cdot\mathbf{k} \geq 0.$$

In view of the identity

$$-\frac{1}{\theta}\partial_{\nabla\mathbf{M}}\psi \cdot (\nabla\dot{\mathbf{M}}) = -\nabla\cdot(\frac{1}{\theta}\partial_{\nabla\mathbf{M}}\psi\dot{\mathbf{M}}) + (\nabla\cdot\frac{1}{\theta}\partial_{\nabla\mathbf{M}}\psi)\cdot\dot{\mathbf{M}}$$

it follows

$$\frac{1}{\theta}(\mu_0\mathbf{H} - \rho\,\delta_{\mathbf{M}}\psi)\cdot\dot{\mathbf{M}} + \frac{1}{\theta}\mathbf{E}\cdot\mathbf{J} - \frac{1}{\theta^2}\mathbf{q}\cdot\nabla\theta + \nabla\cdot(\mathbf{k} - \frac{\rho}{\theta}\partial_{\nabla\mathbf{M}}\psi\dot{\mathbf{M}}) \geq 0,$$

where

$$\delta_{\mathbf{M}}\psi := \partial_{\mathbf{M}}\psi - \theta\,\nabla\cdot\partial_{\nabla\mathbf{M}}\frac{\psi}{\theta}.$$

This suggests that we let

$$\mathbf{k} = \frac{\rho}{\theta}\,\partial_{\nabla\mathbf{M}}\psi\,\dot{\mathbf{M}}$$

and hence

$$\mathbf{k} = \frac{\rho}{\theta}\,\partial_{\nabla\mathbf{M}}\psi\,\boldsymbol{\mathcal{M}}(\theta, \nabla\theta, \mathbf{E}, \mathbf{H}, \mathbf{M}, \nabla\mathbf{M}).$$

Consequently we are left with the reduced dissipation inequality

$$(\mu_0\mathbf{H} - \rho\,\delta_{\mathbf{M}}\psi)\cdot\dot{\mathbf{M}} + \mathbf{E}\cdot\mathbf{J} - \frac{1}{\theta}\mathbf{q}\cdot\nabla\theta \geq 0.$$

Irrespective of the chosen function $\boldsymbol{\mathcal{M}}$ for $\dot{\mathbf{M}}$, the reduced inequality holds if

$$(\mu_0\mathbf{H} - \rho\delta_{\mathbf{M}}\psi)\cdot\dot{\mathbf{M}} \geq 0, \tag{12.29}$$

$$\mathbf{E}\cdot\mathbf{J} - \frac{1}{\theta}\mathbf{q}\cdot\nabla\theta \geq 0.$$

We append some comments on these results. The operator $\delta_{\mathbf{M}}$ is the variational derivative, with respect to $\mathbf{M}$. If $\partial_{\nabla\mathbf{M}}\psi = 0$ then $\delta_{\mathbf{M}}\psi = \partial_{\mathbf{M}}\psi$. The occurrence of ψ/θ shows a feature that is related to the dependence on gradients, here $\nabla\mathbf{M}$. The ratio ψ/θ is called rescaled (or reduced) free energy. Rescaled potentials are found also in other frameworks such as in phase separation problems [9]. The result for $\mathbf{k}$ is consistent with the view that in phase-field theories the extra-entropy flux is linear in the time derivative of the phase variable [155]. Moreover, the flux is larger where the magnetization gradient is larger. The classical inequality $\mathbf{E} \cdot \mathbf{J} - \mathbf{q} \cdot \nabla\theta/\theta \geq 0$ is satisfied, e.g. by Ohm's law and Fourier's law,

$$\mathbf{J} = \sigma(\theta)\mathbf{E}, \quad \sigma > 0; \qquad \mathbf{q} = -\kappa(\theta)\nabla\theta, \quad \kappa > 0.$$

We now determine possible explicit functions $\mathcal{M}$ consistent with (12.29). Let

$$\mathbf{H}_{\text{eff}} := \mathbf{H} - \frac{\rho}{\mu_0}\delta_{\mathbf{M}}\,\psi.$$

Inequality (12.29) becomes

$$\mathbf{H}_{\text{eff}} \cdot \dot{\mathbf{M}} \geq 0. \tag{12.30}$$

To establish the compatibility with (12.30) of specific evolution equations it is convenient to examine first the constitutive equation

$$\dot{\mathbf{M}} = \mathbf{u} + \beta\mathbf{e} \times \dot{\mathbf{M}}, \tag{12.31}$$

where $\mathbf{u}$ is independent of $\dot{\mathbf{M}}$ and $\mathbf{e} = \mathbf{M}/|\mathbf{M}|$. Equation (12.31) generalizes the Gilbert equation (12.26). The following statement provides an equivalent explicit relation for $\dot{\mathbf{M}}$.
Equation (12.31) *is equivalent to*

$$\dot{\mathbf{M}} = \frac{1}{1+\beta^2}[\mathbf{u} + \beta^2(\mathbf{u}\cdot\mathbf{e})\mathbf{e} + \beta\mathbf{e}\times\mathbf{u}]. \tag{12.32}$$

To verify this, observe

$$\mathbf{e} \times \dot{\mathbf{M}} = \mathbf{e} \times \mathbf{u} + \beta[(\mathbf{e}\cdot\dot{\mathbf{M}})\mathbf{e} - \dot{\mathbf{M}}], \qquad \mathbf{e}\cdot\dot{\mathbf{M}} = \mathbf{e}\cdot\mathbf{u}.$$

Substitution in (12.31) and some rearrangements yield (12.32). □

For definiteness we now look at the evolution equation [40]

$$\dot{\mathbf{M}} = \alpha_1(\mathbf{H}_{\text{eff}}\cdot\mathbf{e})\mathbf{e} - \alpha_2\mathbf{e}\times(\mathbf{e}\times\mathbf{H}_{\text{eff}}) + \beta\mathbf{e}\times\dot{\mathbf{M}}, \tag{12.33}$$

which amounts to letting $\mathbf{u} = \alpha_1(\mathbf{H}_{\text{eff}}\cdot\mathbf{e})\mathbf{e} - \alpha_2\mathbf{e}\times(\mathbf{e}\times\mathbf{H}_{\text{eff}})$ in (12.31). We can then state the following property.
The evolution equation (12.33) *is compatible with* (12.30) *if and only if* $\alpha_1 \geq 0$, $\alpha_2 \geq 0$.
Observe that Eq. (12.33) is equivalent to

$$\dot{\mathbf{M}} = \alpha_1(\mathbf{H}_{\text{eff}} \cdot \mathbf{e})\mathbf{e} - \frac{\alpha_2}{1+\beta^2}[(\mathbf{H}_{\text{eff}} \cdot \mathbf{e})\mathbf{e} - \mathbf{H}_{\text{eff}}] + \frac{\alpha_2 \beta}{1+\beta^2}\mathbf{e} \times \mathbf{H}_{\text{eff}}. \tag{12.34}$$

Consequently,

$$\mathbf{H}_{\text{eff}} \cdot \dot{\mathbf{M}} = \alpha_1 |\mathbf{H}_{\text{eff}} \cdot \mathbf{e}|^2 + \frac{\alpha_2}{1+\beta^2}[|\mathbf{H}_{\text{eff}}|^2 - |\mathbf{H}_{\text{eff}} \cdot \mathbf{e}|^2].$$

Hence

$$\mathbf{H}_{\text{eff}} \cdot \dot{\mathbf{M}} \geq 0 \iff \alpha_1 \geq 0, \ \alpha_2 \geq 0$$

and the conclusion follows. □

It is worth remarking that the scalars $\alpha_1, \alpha_2, \beta$ are allowed to depend on $\theta, \mathbf{B}, \mathbf{M}$ provided only that the inequalities $\alpha_1 \geq 0, \ \alpha_2 \geq 0$ hold whereas no restriction is placed on β.

Equation (12.33) allows for nonzero values of $\mathbf{M} \cdot \dot{\mathbf{M}}$. Let $M = |\mathbf{M}|$ and hence

$$\mathbf{M} = M\mathbf{e}$$

so that

$$\dot{\mathbf{M}} = \dot{M}\mathbf{e} + M\dot{\mathbf{e}}.$$

In view of (12.33) we find

$$\dot{M} = \alpha_1 \mathbf{H}_{\text{eff}} \cdot \mathbf{e},$$

$$M\dot{\mathbf{e}} = -\frac{\alpha_2}{1+\beta^2}[(\mathbf{H}_{\text{eff}} \cdot \mathbf{e})\mathbf{e} - \mathbf{H}_{\text{eff}}] + \frac{\alpha_2 \beta}{1+\beta^2}\mathbf{e} \times \mathbf{H}_{\text{eff}}.$$

Since[20] $\alpha_1 > 0$ then M increases or decreases according as $\mathbf{H}_{\text{eff}} \cdot \mathbf{e}$ is positive or negative,

$$\dot{M} \lesseqgtr 0 \iff \mathbf{H}_{\text{eff}} \cdot \mathbf{e} \lesseqgtr 0.$$

Equilibrium magnetization is then characterized by

$$\mathbf{H} - \frac{\rho}{\mu_0} \delta_{\mathbf{M}} \frac{\psi}{\theta} = \mathbf{0}.$$

If ψ is independent of $\nabla \mathbf{M}$ then the equilibrium is given by

$$\mathbf{H} - \frac{\rho}{\mu_0} \partial_{\mathbf{M}} \psi = \mathbf{0}.$$

If we let

[20] We disregard the trivial case $\alpha_1 = 0$.

$$\psi = \Psi(\theta) + \frac{1}{2\rho}\chi(\theta)\mathbf{M}^2$$

then

$$\mathbf{H}_{\text{eff}} = \mathbf{H} - \frac{\chi}{\mu_0}\mathbf{M}$$

and we find

$$\dot{M} = \alpha_1(\mathbf{H} \cdot \mathbf{e} - \frac{\chi}{\mu_0}M).$$

Hence M increases or decreases according as $\mathbf{H} \cdot \mathbf{e}$ is larger or smaller than $\chi M/\mu_0$.

The evolution equation (12.34) can be viewed as strictly analogous to the LLB equation (12.28) with $\alpha_2\beta/(1+\beta^2)$ playing the role of the opposite of the gyromagnetic ratio. In addition, about (12.34), we have proved that α_1 and α_2 are non-negative as a consequence of the second law of thermodynamics.

We now go back to the possible expression of $\mathbf{H}_{\text{eff}}$ and hence of the free energy ψ. The physical picture suggests that there is an internal magnetic energy, quadratic in $\mathbf{M}$, and an energy related to exchange interactions, and hence proportional to $|\nabla\mathbf{M}|^2$. We then let

$$\psi = \Psi(\theta) - \frac{1}{2\rho}\mathbf{M} \cdot \boldsymbol{\mathcal{L}}(\theta)\mathbf{M} - \frac{1}{2\rho}a(\theta)|\nabla\mathbf{M}|^2.$$

Consequently

$$\mathbf{H}_{\text{eff}} = \frac{1}{\mu_0}[\mathbf{B} - \boldsymbol{\mathcal{L}}(\theta)\mathbf{M} - \theta\nabla \cdot (\frac{a(\theta)}{\theta}\nabla\mathbf{M})].$$

A detailed form of the tensor function $\boldsymbol{\mathcal{L}}(\theta)$ is given in Sect. 15.3 within the description of paramagnetic and ferromagnetic behaviours.

Micromagnetics and dissipation mechanisms

A different picture describes ferromagnets as two interacting continua, one with a mechanical structure and the other with a magnetic structure. It is a central point of the model that the dynamics is made formal by postulating the power expended in a typical process. The evolution equation for the magnetic moment $\mathbf{m}$ is assumed in the form [119, 365]

$$\dot{\mathbf{m}} = \gamma\mathbf{m} \times (\mathbf{f} + \mathbf{h} + \nabla \cdot \boldsymbol{\Lambda}).$$

whereby the interaction between the two continua is modelled by the force $\mathbf{h}$ and the couple stress $\boldsymbol{\Lambda}$. The body is assumed to be undeformable and the power expended is taken to be

$$\boldsymbol{\Lambda} \cdot \nabla\dot{\mathbf{m}} - \mathbf{h} \cdot \dot{\mathbf{m}},$$

where $\boldsymbol{\Lambda} \cdot \nabla\dot{\mathbf{m}} = \Lambda_{ij}\partial_{x_j}\dot{m}_i$. Hence the inequality

$$-\dot{\psi} - \mathbf{h} \cdot \dot{\mathbf{m}} + \mathbf{\Lambda} \cdot \nabla \dot{\mathbf{m}} \geq 0$$

is assumed to express the dissipation property. The free energy density ψ, per unit mass, is allowed to depend on $\mathbf{m}$ and $\nabla \mathbf{m}$ while $\mathbf{h}$ and $\mathbf{\Lambda}$ may depend also on $\dot{\mathbf{m}}$ and $\nabla \dot{\mathbf{m}}$. Upon computation of $\dot{\psi}$ and substitution we have

$$-(\partial_{\mathbf{m}} \psi + \mathbf{h}) \cdot \dot{\mathbf{m}} - (\partial_{\nabla \mathbf{m}} - \mathbf{\Lambda}) \cdot \nabla \dot{\mathbf{m}} \geq 0.$$

Let

$$\mathbf{h}^{\text{eq}} = -\partial_{\mathbf{m}} \psi, \qquad \mathbf{\Lambda}^{\text{eq}} = \partial_{\nabla \mathbf{m}} \psi$$

and

$$\mathbf{h}^{\text{vs}} = \mathbf{h} - \mathbf{h}^{\text{eq}}, \qquad \mathbf{\Lambda}^{\text{vs}} = \mathbf{\Lambda} - \mathbf{\Lambda}^{\text{eq}}.$$

So we can write the inequality in the form

$$-\mathbf{h}^{\text{vs}} \cdot \dot{\mathbf{m}} + \mathbf{\Lambda}^{\text{vs}} \cdot \nabla \dot{\mathbf{m}} \geq 0,$$

$\mathbf{h}^{\text{vs}}$ and $\mathbf{\Lambda}^{\text{vs}}$ playing the role of viscous terms. Let

$$\psi = \tfrac{1}{2}\beta(\mathbf{m} \cdot \mathbf{e})^2 + \tfrac{1}{2}\alpha|\nabla \mathbf{m}|^2, \qquad \alpha > 0,$$

where $\mathbf{e}$ is the unit vector of the applied field ($\mathbf{H}$). Hence

$$\mathbf{h}^{\text{eq}} = -\beta(\mathbf{m} \cdot \mathbf{e}), \qquad \mathbf{\Lambda}^{\text{eq}} = \alpha \nabla \mathbf{m}$$

while the viscous terms $\mathbf{h}^{\text{vs}}$, $\mathbf{\Lambda}^{\text{vs}}$ are chosen in the form

$$\mathbf{h}^{\text{vs}} = -\nu \dot{\mathbf{m}}, \quad \nu > 0; \qquad \mathbf{\Lambda}^{\text{vs}} = 0.$$

Upon substitution in the evolution equation for $\mathbf{m}$ we obtain the standard Gilbert equation

$$\gamma^{-1} \dot{\mathbf{m}} = \mathbf{m} \times (\mathbf{f} - \beta(\mathbf{m} \cdot \mathbf{e})\mathbf{e} + \alpha \Delta \mathbf{m}) - \nu\, \mathbf{m} \times \dot{\mathbf{m}}.$$

The vector $\mathbf{f} - \beta(\mathbf{m} \cdot \mathbf{e})\mathbf{e} + \alpha \Delta \mathbf{m}$ may then be viewed as the effective applied field acting on $\mathbf{m}$.

A simple variant of this scheme avoids the use of the couple stress. Start with

$$\gamma^{-1} \dot{\mathbf{m}} = \mathbf{m} \times (\mathbf{f} + \mathbf{h})$$

and state the dissipation inequality as

$$-\dot{\psi} - \mathbf{h} \cdot \dot{\mathbf{m}} + \theta \nabla \cdot \mathbf{k} \geq 0,$$

where $\mathbf{k}$ is the extra-entropy flux and θ is the constant temperature. As before, assume ψ is a function of $\mathbf{m}$ and $\nabla\mathbf{m}$ so that

$$-\partial_{\mathbf{m}}\psi \cdot \dot{\mathbf{m}} - \partial_{\nabla\mathbf{m}}\psi \cdot \nabla\dot{\mathbf{m}} - \mathbf{h}\cdot\dot{\mathbf{m}} + \theta\nabla\cdot\mathbf{k} \geq 0.$$

Since

$$\partial_{\nabla\mathbf{m}}\psi \cdot \nabla\dot{\mathbf{m}} = \nabla\cdot(\partial_{\nabla\mathbf{m}}\psi\,\dot{\mathbf{m}}) - (\nabla\cdot\partial_{\nabla\mathbf{m}}\psi)\cdot\dot{\mathbf{m}}$$

the inequality can be written

$$(-\partial_{\mathbf{m}}\psi + \nabla\cdot\partial_{\nabla\mathbf{m}}\psi - \mathbf{h})\cdot\dot{\mathbf{m}} + \nabla\cdot(\theta\mathbf{k} - \partial_{\nabla\mathbf{m}}\psi\,\dot{\mathbf{m}}) \geq 0.$$

The inequality holds if $\theta\mathbf{k} = \partial_{\nabla\mathbf{m}}\psi\,\dot{\mathbf{m}}$ and

$$\mathbf{h} = -\partial_{\mathbf{m}}\psi + \nabla\cdot\partial_{\nabla\mathbf{m}}\psi - \nu\dot{\mathbf{m}}, \qquad \nu > 0.$$

Assuming ψ as in the previous scheme we obtain (again) the Gilbert equation

$$\gamma^{-1}\dot{\mathbf{m}} = \mathbf{m}\times(\mathbf{f} - \beta(\mathbf{m}\cdot\mathbf{e})\mathbf{e} + \alpha\Delta\mathbf{m}) - \nu\,\mathbf{m}\times\dot{\mathbf{m}}.$$

12.7.2 Evolution Equations of Magnetization

Evolution equations are derived for the magnetization $\mathbf{M}$ by looking at $\mathbf{M}$ as the number density times the magnetic moment $\boldsymbol{\mu}$. The body under consideration is a rigid solid and $\mathbf{M}, \mathbf{H}$ are fields at the frame locally at rest with the body.

The simplest model originates from micromagnetics and is due to Landau and Lifshitz,

$$\dot{\mathbf{M}} = -\gamma\mathbf{M}\times\mathbf{H} \tag{12.35}$$

where $\mathbf{H}$ is the effective magnetic field in the frame (locally) at rest. It is apparent that, by (12.35), $\mathbf{M}\cdot\dot{\mathbf{M}} = 0$ and $\dot{\mathbf{M}}\cdot\mathbf{H} = 0$. Consequently two quantities are constant in time, $|\mathbf{M}|$ and, if $\mathbf{H}$ is constant, the angle between $\mathbf{M}$ and $\mathbf{H}$.

A dissipation term was then introduced by Landau and Lifshitz thus making the equation in the form

$$\dot{\mathbf{M}} = -\gamma\mathbf{M}\times\mathbf{H} - \alpha\mathbf{M}\times(\mathbf{M}\times\mathbf{H}). \tag{12.36}$$

Still $M = |\mathbf{M}|$ is constant. If, futher, $\mathbf{H}$ is constant then

$$(\mathbf{H}\cdot\mathbf{M})^{\cdot} = \mathbf{H}\cdot\dot{\mathbf{M}} = \alpha[M^2H^2 - (\mathbf{M}\cdot\mathbf{H})^2] > 0$$

as $\mathbf{M}\times\mathbf{H} \neq \mathbf{0}$; the angle between $\mathbf{M}$ and $\mathbf{H}$ decreases in time.

To account for dissipation Gilbert, considered a different term so that the evolution equation becomes

$$\dot{\mathbf{M}} = -\gamma_G \mathbf{M} \times \mathbf{H} + \beta \mathbf{M} \times \dot{\mathbf{M}}. \tag{12.37}$$

By (12.37) we have again $\mathbf{M} \cdot \dot{\mathbf{M}} = 0$. Hence the cross product of (12.37) with $\mathbf{M}$ and substitution of $\mathbf{M} \times \dot{\mathbf{M}}$, while $\mathbf{M} \cdot \dot{\mathbf{M}} = 0$, results in

$$\dot{\mathbf{M}} = -\frac{\gamma_G}{1+\beta^2 M^2} \mathbf{M} \times \mathbf{H} - \frac{\beta \gamma_G}{1+\beta^2 M^2} \mathbf{M} \times (\mathbf{M} \times \mathbf{H}).$$

Hence (12.37) is equivalent to (12.36) provided only that

$$\beta = \frac{\alpha}{\gamma}, \qquad \gamma_G = \gamma[1 + (\alpha/\gamma)^2 M^2].$$

Lately generalizations of (12.35) or (12.36) have been considered. We mention the equation

$$\dot{\mathbf{M}} = -\gamma \mathbf{M} \times \mathbf{H} + \mathbf{t}$$

where $\mathbf{t}$ denotes a torque with the role of forcing the system to approach to equilibrium. In this sense the Bloch equation specifies $\mathbf{t}$ so that

$$\dot{\mathbf{M}} = -\gamma \mathbf{M} \times \mathbf{H} - \frac{1}{\tau}[\mathbf{M} - \mathbf{M}_{\text{eq}}(\mathbf{H})], \tag{12.38}$$

where $[\mathbf{M} - \mathbf{M}_{\text{eq}}(\mathbf{H})]/\tau$ represents the forcing term toward equilibrium. The torque $\mathbf{t}$ has been specified in the form

$$\mathbf{t} = \lambda_1 \mathbf{H} - \lambda_2 \Delta \mathbf{H}.$$

A further equation has been considered in the form

$$\dot{\mathbf{M}} = -\gamma \mathbf{M} \times \mathbf{H} - \alpha_t \mathbf{M} \times (\mathbf{M} \times \mathbf{H}) + \alpha_l (\mathbf{M} \cdot \mathbf{H}) \mathbf{M}. \tag{12.39}$$

Both (12.38) and (12.39) are named after Bloch [40, 185]; α_t and α_l denote transverse and longitudinal contributions of $\mathbf{M}$ to $\dot{\mathbf{M}}$.

Further modifications of (12.35) have been considered as the equations modelling the evolution of Bloch walls between magnetic domains. Physically, the torque $\mathbf{t}$ is taken to represent the contribution of spin-polarized electric currents. Hence $\mathbf{t}$ is modelled in the form

$$\mathbf{t} = \mathbf{M} \times \mathbf{p} \tag{12.40}$$

$\mathbf{p}$ representing the polarization of the electric spin current. In, e.g. [412] the torque is expressed by

$$\mathbf{t} = -(\mathbf{u} \cdot \nabla) \mathbf{M} + \beta \mathbf{M} \times [(\mathbf{u} \cdot \nabla) \mathbf{M}], \tag{12.41}$$

$\mathbf{u}$ being a velocity related to the incident beam of spin polarized charges.

It is natural to ask about the thermodynamic consistency of the (evolution) equations (12.35)–(12.41). For simplicity consider a rigid and non-conducting solid, and let the Helmholtz free energy ψ, the entropy η, the entropy production σ, the time derivative $\dot{\mathbf{M}}$, and the extra-entropy flux be functions of the set of variables $\theta, \mathbf{H}, \mathbf{M}, \nabla\mathbf{M}$.

Time differentiation of $\psi(\theta, \mathbf{H}, \mathbf{M}, \nabla\mathbf{M})$ and substitution into the Clausius–Duhem inequality yields

$$-\rho(\partial_\theta\psi + \eta)\dot{\theta} - \rho\partial_{\mathbf{H}}\psi \cdot \dot{\mathbf{H}} + (\mu_0\mathbf{H} - \rho\partial_{\mathbf{M}}\psi) \cdot \dot{\mathbf{M}} - \rho\partial_{\nabla\mathbf{M}}\psi \cdot \nabla\dot{\mathbf{M}} + \theta\nabla \cdot \mathbf{k} = \theta\sigma \geq 0.$$

By the standard arguments we conclude that $\eta = -\partial_\theta\psi$. Differently from models of hysteresis where $\dot{\mathbf{H}}$ and $\dot{\mathbf{M}}$ are not independent, here we let $\dot{\mathbf{H}}$ and $\dot{\mathbf{M}}$ be independent. The arbitrariness of $\dot{\mathbf{H}}$ implies $\partial_{\mathbf{H}}\psi = \mathbf{0}$. Hence we obtain the inequality

$$(\mu_0\mathbf{H} - \rho\partial_{\mathbf{M}}\psi) \cdot \dot{\mathbf{M}} - \rho\partial_{\nabla\mathbf{M}}\psi \cdot \nabla\dot{\mathbf{M}} + \theta\nabla \cdot \mathbf{k} = \theta\sigma \geq 0. \tag{12.42}$$

Divide by θ and employ the identity

$$-\frac{\rho}{\theta}\partial_{\nabla\mathbf{M}}\psi \cdot \nabla\dot{\mathbf{M}} = -\nabla \cdot (\frac{\rho}{\theta}\partial_{\nabla\mathbf{M}}\psi\dot{\mathbf{M}}) + \dot{\mathbf{M}} \cdot \nabla \cdot \frac{\rho}{\theta}\partial_{\nabla\mathbf{M}}\psi$$

to obtain

$$\frac{1}{\theta}(\mu_0\mathbf{H} - \rho\delta_{\mathbf{M}}\psi) \cdot \dot{\mathbf{M}} + \nabla \cdot (\mathbf{k} - \frac{\rho}{\theta}\partial_{\nabla\mathbf{M}}\psi\dot{\mathbf{M}}) = \sigma$$

where $\delta_{\mathbf{M}}\psi = \partial_{\mathbf{M}}\psi - (\theta/\rho)\nabla \cdot [(\rho/\theta)\partial_{\nabla\mathbf{M}}\psi]$. Hence we let

$$\mathbf{k} = \frac{\rho}{\theta}\partial_{\nabla\mathbf{M}}\psi\dot{\mathbf{M}},$$

where $\dot{\mathbf{M}}$ is required to satisfy

$$(\mu_0\mathbf{H} - \rho\delta_{\mathbf{M}}\psi) \cdot \dot{\mathbf{M}} = \theta\sigma \geq 0. \tag{12.43}$$

Any relation

$$\dot{\mathbf{M}} = \mathbf{G}(\theta, \mathbf{H}, \mathbf{M})(\mu_0\mathbf{H} - \rho\delta_{\mathbf{M}}\psi), \qquad \mathbf{G}(\theta, \mathbf{H}, \mathbf{M}) \in \text{Psym},$$

obeys the reduced inequality (12.43).

If we address attention to Eqs. (12.35)–(12.41) we observe that they have the form

$$\dot{\mathbf{M}} = -\gamma\mathbf{M} \times \mathbf{H} + \mathbf{t}$$

with various forms of $\mathbf{t}$. This suggests that we take $\partial_{\nabla\mathbf{M}}\psi = \mathbf{0}$, define

$$\mathbf{w} := \mu_0\mathbf{H} - \rho\partial_{\mathbf{M}}\psi,$$

and assume $\partial_{\mathbf{M}}\psi = \alpha\mathbf{M} + \beta\mathbf{H}$. It follows

$$\mathbf{w} = (\mu_0 - \rho\beta)\mathbf{H} - \rho\alpha\mathbf{M}, \qquad \mathbf{w}\cdot(-\gamma\mathbf{M}\times\mathbf{H} + \boldsymbol{\tau}) = \mathbf{w}\cdot\mathbf{t} \geq 0.$$

Hence the reduced inequality (12.42) holds if

$$\dot{\mathbf{M}} = -\gamma\mathbf{M}\times\mathbf{H} + \xi\mathbf{M}\times\mathbf{w} + \nu\mathbf{H}\times\mathbf{w} + \mathbf{t}_d,$$

with arbitrary γ, ξ, ν while $\boldsymbol{\tau}_d$ is required to satisfy

$$\mathbf{w}\cdot\mathbf{t}_d = \theta\sigma \geq 0.$$

Since σ is required to be non-negative then we let $\mathbf{A}$ be a positive definite tensor and assume

$$\theta\sigma = \mathbf{t}_d\cdot\mathbf{A}\mathbf{t}_d$$

and regard $\boldsymbol{\tau}_d$ as a function of the chosen variables $\theta, \mathbf{H}, \mathbf{M}$. Then we have

$$0 = \mathbf{w}\cdot\mathbf{t}_d - \mathbf{t}_d\cdot\mathbf{A}\mathbf{t}_d = (\mathbf{w} - \mathbf{A}\mathbf{t}_d)\cdot\mathbf{t}_d.$$

This condition holds if

$$\mathbf{w} - \mathbf{A}\mathbf{t}_d = \mathbf{0}, \qquad \mathbf{t}_d = \mathbf{A}^{-1}\mathbf{w}.$$

Hence

$$\dot{\mathbf{M}} = -\gamma\mathbf{M}\times\mathbf{H} + \xi\mathbf{M}\times\mathbf{w} + \nu\mathbf{H}\times\mathbf{w} + \mathbf{A}^{-1}\mathbf{w} \tag{12.44}$$

is thermodynamically consistent. Equation (12.44) shows possible evolution equations of $\mathbf{M}$. The gyromagnetic term $\gamma\mathbf{M}\times\mathbf{H}$ is consistent provided $\partial_{\mathbf{M}}\psi$ embodies a term $\alpha\mathbf{M}$. The torques $\xi\mathbf{M}\times\mathbf{w}$, $\nu\mathbf{H}\times\mathbf{w}$ are allowed for arbitrary ξ and ν. The dissipative torque $\mathbf{t}_d$ is thermodynamically consistent if

$$\mathbf{t}_d = \mathbf{A}^{-1}(\mu_0\mathbf{H} - \rho\partial_{\mathbf{M}}\psi),$$

where $\mathbf{A}$ is positive definite. The torques $\gamma\mathbf{M}\times\mathbf{H}$, $\xi\mathbf{M}\times\mathbf{w}$, and $\nu\mathbf{H}\times\mathbf{w}$ are non-dissipative in that the corresponding entropy production is zero. Moreover, $\mathbf{M}\times\mathbf{w}$ and $\mathbf{H}\times\mathbf{w}$ are uncommon in the literature.

We might view $\mathbf{H} - \rho\partial_{\mathbf{M}}\psi/\mu_0 = \mathbf{w}/\mu_0$ as the effective field; both $\mathbf{M}\times\mathbf{w}$ and $\mathbf{H}\times\mathbf{w}$ have a decreasing influence inasmuch as $\mathbf{w}\to\mathbf{0}$. We might then think that $\dot{\mathbf{M}}$ approaches zero as far as $\mathbf{w}\to\mathbf{0}$ namely as $\mathbf{H}$ approaches the equilibrium value $\rho\partial_{\mathbf{M}}\psi/\mu_0$.

Since $\mathbf{w} = -\rho\alpha[\mathbf{M} - (\mu_0 - \rho\beta)\mathbf{H}/\rho\alpha]$ then Eq. (12.38) holds with

$$\mathbf{M}_{eq}(\mathbf{H}) = \frac{\mu_0 - \rho\beta}{\rho\alpha}\mathbf{H}.$$

As with (12.39), we find that

$$(\mu_0\mathbf{H} - \rho\alpha\mathbf{M}) \cdot \dot{\mathbf{M}} = (\mu_0 - \rho\beta)\alpha_t(\mathbf{M} \times \mathbf{H})^2 + (\mu_0 - \rho\beta)\alpha_t(\mathbf{M} \cdot \mathbf{H})^2 - \rho\alpha\alpha_l\mathbf{M}^2(\mathbf{M} \cdot \mathbf{H}).$$

Thermodynamic consistency holds if $\alpha_t \geq 0$ and $\alpha_l = 0$, which is the case of the Landau–Lifshitz equation (12.36).

The result for the dissipative torque $\mathbf{t}_d$ allows us to establish an interesting connection with the literature. Denote by $\mathbf{m}$ the unit vector of $\mathbf{M}$ and consider the tensor

$$\hat{\mathbf{A}} = \alpha_1\mathbf{m} \otimes \mathbf{m} + \alpha_2(\mathbf{1} - \mathbf{m} \otimes \mathbf{m}), \qquad \alpha_1, \alpha_2 > 0.$$

Hence, for any vector $\mathbf{u}$,

$$\hat{\mathbf{A}}\mathbf{u} = \alpha_1(\mathbf{m} \cdot \mathbf{u})\mathbf{m} + \alpha_2(\mathbf{u} - (\mathbf{m} \cdot \mathbf{u})\mathbf{m}) = \alpha_1(\mathbf{m} \cdot \mathbf{u})\mathbf{m} - \alpha_2\mathbf{m} \times (\mathbf{m} \times \mathbf{u}).$$

Incidentally

$$(\mathbf{m} \cdot \mathbf{u})\mathbf{m} = \mathbf{u}_{\parallel}, \qquad -\mathbf{m} \times (\mathbf{m} \times \mathbf{u}) = \mathbf{u}_{\perp}$$

are the parallel and perpendicular parts of $\mathbf{u}$ with respect to $\mathbf{m}$. Now let $\mathbf{A}^{-1} = \hat{\mathbf{A}}$ and then

$$\mathbf{t}_d = \mathbf{A}\mathbf{w} = \alpha_1(\mathbf{m} \cdot \mathbf{w})\mathbf{m} + \alpha_2(\mathbf{w} - (\mathbf{m} \cdot \mathbf{w})\mathbf{m}) = -\alpha_2\mathbf{w} + (\alpha_1 - \alpha_2)(\mathbf{m} \cdot \mathbf{w})\mathbf{m}.$$

Letting $\gamma, \nu = 0$ we can write Eq. (12.44) in the form[21]

$$\dot{\mathbf{M}} = \xi\mathbf{M} \times \mathbf{w} + \alpha_1(\mathbf{w} \cdot \mathbf{m})\mathbf{m} - \alpha_2\mathbf{m} \times (\mathbf{m} \times \mathbf{w}). \tag{12.45}$$

The coefficients α_1, α_2 have the dimensions of the inverse of time. Their reciprocals, say $\tau_{\parallel}, \tau_{\perp}$, are then named ferromagnetic longitudinal and transverse relaxation times. The larger are the values α_1, α_2 the faster is the damping of $\mathbf{M}_{\parallel}, \mathbf{M}_{\perp}$ to their equilibrium values.

Since $\dot{\mathbf{M}} = \dot{M}\mathbf{m} + M\dot{\mathbf{m}}$ and $\dot{\mathbf{m}} \cdot \mathbf{m} = 0$ then Eq. (12.45) splits into the pair of evolution equations

$$\begin{cases} \dot{\mathbf{m}} = -\gamma\mathbf{m} \times \mathbf{w} - (\alpha_2/M)\mathbf{m} \times (\mathbf{m} \times \mathbf{w}), \\ \dot{M} = \alpha_1\mathbf{w} \cdot \mathbf{m} \end{cases},$$

where $\mathbf{w} = \mu_0\mathbf{H} - \rho\partial_{\mathbf{M}}\psi$. The first equation governs the evolution of the unit vector and, except for $\mathbf{w}$ instead of $\mathbf{H}$, has the form of Eq. (12.36). The second equation tells us that the amplitude M may change if $\alpha_1 \neq 0$.

---

[21] An analogous result is obtained in [155], Eq. (4.7), by looking for thermodynamic restrictions on an equation like $\dot{\mathbf{M}} = \mathbf{t}_p$. Equation (12.45) is similar to (12.39), with $\mathbf{w}$ instead of $\mathbf{H}$.

12.8 Ferrofluids

Ferrofluids are colloidal liquids made of nanoscale ferromagnetic, or ferrimagnetic, particles, with typical dimensions of about 10 nm, suspended in a carrier fluid (usually an organic solvent or water). Large ferromagnetic particles can be ripped out of the homogeneous colloidal mixture, forming a separate clump of magnetic dust when exposed to strong magnetic fields. The magnetic attraction of nanoparticles is weak enough that the surfactant's Van der Waals force is sufficient to prevent magnetic clumping or agglomeration. Ferrofluids usually do not retain magnetization in the absence of an externally applied field.

Often the particles are made of magnetite, Fe_3O_4, coated with surfactant agents in order to prevent their aggregation. The particles have a symmetry axis of easy magnetization; denote by $\mathbf{e}$ the unit vector of the symmetry axis. In ionic ferrofluids the nanoparticles are electrically charged.

Quite naturally we may model ferrofluids by borrowing from the model of nanofluids. Yet two approaches can be developed. One describes ferrofluids as mixtures of two constituents: the fluid and the magnetic particles. The other approach models ferrofluids as micropolar media. Here we let the fluid be non-polar and the magnetic particles be solid. Moreover the ferrofluid is viewed as a mixture, the particles being viewed as a micropolar medium.

We follow the quasistatic approximation ($\partial_t \mathbf{B} = \mathbf{0}$, $\nabla \times \mathbf{H} = \mathbf{0}$) and neglect the polarization $\mathbf{P}$. We let q be the charge density and $\mathbf{J}$ the corresponding current density. We denote by the subscripts s and f the quantities pertaining to solid nanoparticles and base fluids. Let $\mathfrak{m}_s$ be the magnetic moment field of each particle. The occurrence of a magnetic field $\mathbf{H}$ results in the force $\mu_0(\mathfrak{m}_s \cdot \nabla)\mathbf{H}$, the body couple $\mu_0 \mathfrak{m}_s \times \mathbf{H}$, and the energy supply $\mu_0[\mathbf{H} \cdot \grave{\mathfrak{m}}_s + \mathbf{v}_s \cdot (\mathfrak{m}_s \cdot \nabla)\mathbf{H}]$, where $\grave{\mathfrak{m}}_s = \partial_t \mathfrak{m}_s + (\mathbf{v}_s \cdot \nabla)\mathfrak{m}_s$. Also, the electric charge shows up a force density $q\mathbf{E}$ and a power density $\mathbf{E} \cdot \mathbf{J}$; $\mathbf{E}, \mathbf{H}, \mathbf{J}$ are the fields at the reference of the co-moving observer. Moreover we let $\mathcal{N}_s$ be the number density of nanoparticles; $\mathcal{N}_s$ is related to ρ_s by $\mathcal{N}_s = \rho_s / m_s$, m_s being the mass of nanoparticles. We let

$$\mathbf{M} = \mathcal{N}_s \mathfrak{m}_s, \qquad \mathbf{m} = \frac{1}{\rho_s}\mathbf{M} = \frac{1}{m_s}\mathfrak{m}_s$$

be the magnetizations, per unit volume and unit mass, respectively. Hence the effects of magnetic moments, per unit volume, to body force, body couple, and energy supply are

$$\mu_0(\mathbf{M} \cdot \nabla)\mathbf{H}, \quad \mu_0 \mathbf{M} \times \mathbf{H}, \quad [\mu_0(\mathbf{M} \cdot \nabla)\mathbf{H}] \cdot \mathbf{v}_s + \mu_0 \rho_s \mathbf{H} \cdot \grave{\mathbf{m}},$$

where $\grave{\mathbf{m}} = \partial_t \mathbf{m} + (\mathbf{v}_s \cdot \nabla)\mathbf{m}$.

With this in mind the modelling of nanofluids is improved by letting the balance of mass, momentum, orientational momentum, and energy be given the form

$$\grave{\rho}_s + \rho_s \nabla \cdot \mathbf{v}_s = 0,$$

$$\grave{\rho}_f + \rho_s \nabla \cdot \mathbf{v}_f = 0,$$

$$\rho_s \grave{\mathbf{v}}_s = \nabla \cdot \mathbf{T}_s + \rho_s \mathbf{g} + \mu_0 (\mathbf{M} \cdot \nabla)\mathbf{H} + \mathbf{p}_s,$$

$$\rho_f \grave{\mathbf{v}}_f = \nabla \cdot \mathbf{T}_f + \rho_f \mathbf{g} + \mathbf{p}_f,$$

$$\rho_s \grave{\boldsymbol{\sigma}}_s = \mu_0 \mathbf{M} \times \mathbf{H} + \boldsymbol{\Upsilon}_s + \mathbf{c}_s,$$

$$\mathbf{0} = \boldsymbol{\Upsilon}_f + \mathbf{c}_f,$$

$$\rho_s \grave{\varepsilon}_s = \mathbf{T}_s \cdot \mathfrak{D}_s - \nabla \cdot \mathbf{q}_s + \rho_s r_s + \mathbf{J}_s \cdot \mathbf{E} + \mu_0 \rho_s \mathbf{H} \cdot \grave{\mathfrak{m}} + \boldsymbol{\Upsilon}_s \cdot (\mathbf{w}_s - \boldsymbol{\omega}_s) + e_s - \mathbf{p}_s \cdot \mathbf{v}_s - \mathbf{c}_s \cdot \boldsymbol{\omega}_s,$$

$$\rho_f \grave{\varepsilon}_f = \mathbf{T}_f \cdot \mathbf{L}_f - \nabla \cdot \mathbf{q}_f + \rho_f r_f + e_f - \mathbf{p}_f \cdot \mathbf{v}_f,$$

subject to the constraints

$$\mathbf{p}_s + \mathbf{p}_f = \mathbf{0}, \qquad \mathbf{c}_s + \mathbf{c}_f = \mathbf{0}, \qquad e_s + e_f = 0.$$

Adding the two equations of orientational momentum we find

$$\rho_s \grave{\boldsymbol{\sigma}}_s = \mu_0 \mathbf{M} \times \mathbf{H} + \boldsymbol{\Upsilon}_s + \boldsymbol{\Upsilon}_f,$$

which means that the rotation of nanoparticles is governed by the magnetic body couple density, $\mu_0 \mathbf{M} \times \mathbf{H}$, and the skew part of the total stress via $\boldsymbol{\Upsilon}_s + \boldsymbol{\Upsilon}_f$.

The system of balance equations is closed if the equation for the evolution of $\mathfrak{m}_s$ is given.

As with nanofluids, since $\boldsymbol{\omega}_f$ is undefined then it is consistent to let $\boldsymbol{\Upsilon}_f = \mathbf{0}$, which would follow also from the entropy inequality in view of the arbitrariness of $\mathbf{w}_f$. This implies that $\mathbf{c}_s = \mathbf{0}$, $\mathbf{c}_f = \mathbf{0}$. We then take $\boldsymbol{\Upsilon}_s$ in the form

$$\boldsymbol{\Upsilon}_s = \zeta(\mathbf{w}_s - \boldsymbol{\omega}_s),$$

where $\mathbf{w}_s = \frac{1}{2}\nabla \times \mathbf{v}_s$ is the axial vector of the spin $\mathbf{W}_s$. Hence a simple model is allowed with $\mathbf{T}_s, \mathbf{T}_f$ in the form

$$\mathbf{T}_s = -p_s \mathbf{1} + \mathbf{T}^{\text{skw}}, \; T^{\text{skw}}_{jk} = -\tfrac{1}{2}\epsilon_{jki}\Upsilon_i; \quad \mathbf{T}_f = -p_f \mathbf{1} + \mathbf{T}^{\text{vis}}_f(\mathfrak{D}_f), \; \mathbf{T}^{\text{vis}}_f = O(\mathfrak{D}_f), \tag{12.46}$$

where $p_s = p_s(\rho_s, \theta_s)$, $p_f = p_f(\rho_f, \theta_f)$.

The system of balance equations is closed if the equation for the evolution of $\boldsymbol{\mu}_s$, and hence of $\mathfrak{m}$, is given. A simple model arises by letting

$$\mathfrak{m}_s = \mathfrak{m}_s \mathbf{e}, \qquad \boldsymbol{\sigma}_s = \mathbf{I}\boldsymbol{\omega}_s,$$

$\mathbf{I}$ being the inertia tensor of particles, per unit mass, and $\mathbf{e}$ a unit vector attached to the particles, while $\mathfrak{m}_s$ is constant. Hence we have

$$\grave{\mathbf{e}}_s = \boldsymbol{\omega}_s \times \mathbf{e}_s, \qquad \grave{\mathfrak{m}}_s = \mathfrak{m}_s \grave{\mathbf{e}}_s,$$

$$\rho_s(\mathbf{I}\grave{\boldsymbol{\omega}}_s + \boldsymbol{\omega} \times \mathbf{I}\boldsymbol{\omega}_s) = \mu_0 \mathbf{M} \times \mathbf{H} + \zeta(\mathbf{w}_s - \boldsymbol{\omega}_s).$$

The thermodynamic consistency of this constitutive model is now examined by starting from the inequality[22]

$$\sum_\alpha \frac{1}{\theta_\alpha}[\rho_\alpha \theta_\alpha \grave{\eta}_\alpha + \nabla \cdot \mathbf{q}_\alpha - \rho_\alpha r_\alpha + \theta_\alpha \nabla \cdot \mathbf{k}_\alpha - \frac{1}{\theta_\alpha}\mathbf{q}_\alpha \cdot \nabla\theta_\alpha] \geq 0,$$

which is required to hold for any process compatible with the balance equations. Now, the energy balance equations can be written

$$\rho_\alpha \grave{\varepsilon}_\alpha = \mathbf{T}_\alpha \cdot \mathfrak{D}_\alpha - \nabla \cdot \mathbf{q}_\alpha + \rho_\alpha r_\alpha + [\mu_0 \rho_s \mathbf{H} \cdot \grave{\mathfrak{m}} + \mathbf{J} \cdot \mathbf{E} + \boldsymbol{\Upsilon}_s \cdot (\mathbf{w}_s - \boldsymbol{\omega}_s)]\delta_{s\alpha} + e_\alpha - \mathbf{p}_\alpha \cdot \mathbf{v}_\alpha - \mathbf{c}_\alpha \cdot \boldsymbol{\omega}_\alpha,$$

where $\delta_{sf} = 0$, $\delta_{ss} = 1$. Substituting $\nabla \cdot \mathbf{q}_\alpha - \rho_\alpha r_\alpha$ in the second-law inequality we obtain

$$\sum_\alpha \frac{1}{\theta_\alpha}[\rho_\alpha \theta_\alpha \grave{\eta}_\alpha - \rho_\alpha \grave{\varepsilon}_\alpha + \mathbf{T}_\alpha \cdot \mathfrak{D}_\alpha + e_\alpha - \mathbf{p}_\alpha \cdot \mathbf{v}_\alpha + \theta_\alpha \nabla \cdot \mathbf{k}_\alpha - \frac{1}{\theta_\alpha}\mathbf{q}_\alpha \cdot \nabla\theta_\alpha]$$
$$+ \frac{1}{\theta_s}[\mu_0 \rho_s \mathbf{H} \cdot \grave{\mathfrak{m}} + \mathbf{J} \cdot \mathbf{E} + \boldsymbol{\Upsilon}_s \cdot (\mathbf{w}_s - \boldsymbol{\omega}_s)] \geq 0.$$

Using the Helmholtz free energy $\psi_\alpha = \varepsilon_\alpha - \theta_\alpha \eta_\alpha$ we find

$$\sum_\alpha \frac{1}{\theta_\alpha}[-\rho_\alpha(\grave{\psi}_\alpha + \eta_\alpha \grave{\theta}_\alpha) + \mathbf{T}_\alpha \cdot \mathbf{L}_\alpha + e_\alpha - \mathbf{p}_\alpha \cdot \mathbf{v}_\alpha + \theta_\alpha \nabla \cdot \mathbf{k}_\alpha - \frac{1}{\theta_\alpha}\mathbf{q}_\alpha \cdot \nabla\theta_\alpha]$$
$$+ \frac{1}{\theta_s}[\mu_0 \rho_s \mathbf{H} \cdot \grave{\mathfrak{m}} + \mathbf{J} \cdot \mathbf{E} + \boldsymbol{\Upsilon}_s \cdot (\mathbf{w}_s - \boldsymbol{\omega}_s)] \geq 0.$$

The occurrence of the power $\mu_0 \rho_s \mathbf{H} \cdot \grave{\mathfrak{m}}$ suggests that we let

$$\phi_s = \psi_s - \mu_0 \mathbf{H} \cdot \mathbf{m}, \qquad \phi_f = \psi_f.$$

Hence we can write the inequality in the form

$$\sum_\alpha \frac{1}{\theta_\alpha}[-\rho_\alpha(\grave{\phi}_\alpha + \eta_\alpha \grave{\theta}_\alpha) + \mathbf{T}_\alpha \cdot \mathbf{L}_\alpha + e_\alpha - \mathbf{p}_\alpha \cdot \mathbf{v}_\alpha + \theta_\alpha \nabla \cdot \mathbf{k}_\alpha - \frac{1}{\theta_\alpha}\mathbf{q}_\alpha \cdot \nabla\theta_\alpha]$$
$$+ \frac{1}{\theta_s}[\mu_0 \mathbf{M} \cdot \grave{\mathbf{H}} + \mathbf{J} \cdot \mathbf{E} + \boldsymbol{\Upsilon}_s \cdot (\mathbf{w}_s - \boldsymbol{\omega}_s)] \geq 0,$$

---

[22] Here $\alpha = f, s$.

where $\grave{\mathbf{H}} = \partial_t \mathbf{H} + (\mathbf{v}_s \cdot \nabla)\mathbf{H}$.

Assume that $\phi_\alpha, \eta_\alpha, \mathbf{k}_\alpha, \mathbf{q}_\alpha$ depend on $\rho_\alpha, \theta_\alpha, \mathfrak{D}_\alpha, \nabla\theta_\alpha$, $\mathbf{E}$, and $\mathbf{H}$ while $\mathbf{T}_s, \mathbf{T}_f$ are given by (12.46) and $\mathbf{p}_s = -\mathbf{p}_f$ depends also on $\mathbf{v}_f - \mathbf{v}_s$. Upon computation of $\grave{\phi}_\alpha$ we find

$$-\sum_\alpha \frac{1}{\theta_\alpha}[\rho_\alpha(\partial_{\theta_\alpha}\phi_\alpha + \eta_\alpha)\grave{\theta}_\alpha + \rho_\alpha\,\partial_{\mathfrak{D}_\alpha}\phi_\alpha \cdot \grave{\mathfrak{D}}_\alpha + \rho_\alpha\,\partial_{\nabla\theta_\alpha}\phi_\alpha \cdot \overline{\nabla\theta_\alpha}^{\grave{}} + \rho_\alpha\partial_{\mathbf{E}}\phi_\alpha \cdot \grave{\mathbf{E}}] + ... \geq 0,$$

the dots denoting terms which are independent of $\grave{\mathfrak{D}}_\alpha$, $\overline{\nabla\theta_\alpha}^{\grave{}}$, $\grave{\mathbf{E}}$, and $\grave{\theta}_\alpha$, where $\grave{\mathbf{E}} = \partial_t\mathbf{E} + (\mathbf{v}_\alpha \cdot \nabla)\mathbf{E}$. Hence it follows

$$\partial_{\mathbf{D}_\alpha}\phi_\alpha = \mathbf{0}, \qquad \partial_{\nabla\theta_\alpha}\phi_\alpha = \mathbf{0}, \qquad \partial_{\mathbf{E}}\phi_\alpha = \mathbf{0}, \qquad \eta_\alpha = -\partial_{\theta_\alpha}\phi_\alpha.$$

The reduced inequality is investigated under the assumption that the temperatures θ_f, θ_s are equal. Since $\sum_\alpha e_\alpha = 0$ and $\sum_\alpha \mathbf{p}_\alpha = \mathbf{0}$ the inequality can be written as

$$\sum_\alpha[(\rho_\alpha^2\partial_{\rho_\alpha}\phi_\alpha - p_\alpha)\nabla\cdot\mathbf{v}_\alpha - \mathbf{p}_\alpha\cdot\mathbf{u}_\alpha + \theta\nabla\cdot\mathbf{k}_\alpha] - \frac{1}{\theta}\mathbf{q}\cdot\nabla\theta - (\rho_s\partial_{\mathbf{H}}\phi_s + \mu_0\mathbf{M})\cdot\grave{\mathbf{H}}$$
$$+\mathbf{J}\cdot\mathbf{E} + \mathbf{T}_f^{\mathrm{vis}}\cdot\mathfrak{D}_f + \boldsymbol{\Upsilon}_s\cdot(\mathbf{w}_s - \boldsymbol{\omega}_s) \geq 0,$$

where $\mathbf{q} = \sum_\alpha \mathbf{q}_\alpha$. The arbitrariness of $\grave{\mathbf{H}}$ implies

$$\mu_0\mathbf{M} = -\rho_s\partial_{\mathbf{H}}\phi_s.$$

Moreover, since $\mathbf{T}_f^{\mathrm{vis}} = O(\mathfrak{D}_f)$ letting $\mathbf{E} = \mathbf{0}$ and $\nabla\theta = \mathbf{0}$ we find

$$p_\alpha = \rho_\alpha^2\partial_{\rho_\alpha}\phi_\alpha.$$

The reduced inequality can then be written

$$\sum_\alpha[-\mathbf{p}_\alpha\cdot\mathbf{u}_\alpha + \theta\nabla\cdot\mathbf{k}_\alpha] - \frac{1}{\theta}\mathbf{q}\cdot\nabla\theta + \mathbf{J}\cdot\mathbf{E} + \mathbf{T}_f^{\mathrm{vis}}\cdot\mathfrak{D}_f + \boldsymbol{\Upsilon}_s\cdot(\mathbf{w}_s - \boldsymbol{\omega}_s) \geq 0.$$

Setting aside cross-coupling terms, we conclude that a thermodynamically consistent model is obtained by letting $\mathbf{k}_\alpha = \mathbf{0}$, $\alpha = f, s$, and

$$\mathbf{p}_s = \xi(\mathbf{v}_f - \mathbf{v}_s) = -\mathbf{p}_f, \quad \xi \geq 0, \qquad \mathbf{q} = -\kappa\nabla\theta, \quad \kappa \geq 0, \qquad \mathbf{J} = \sigma\mathbf{E}, \quad \sigma \geq 0,$$

$$\mathbf{T}_f = -\rho_f^2\partial_{\rho_f}\psi_f\mathbf{1} + 2\mu_f\mathfrak{D}_f + \lambda_f(\mathrm{tr}\,\mathfrak{D}_f)\mathbf{1}, \quad \mu_f \geq 0, 2\mu_f + 3\lambda_f \geq 0,$$

$$\boldsymbol{\Upsilon}_s = \zeta(\mathbf{w}_s - \boldsymbol{\omega}_s), \quad \zeta \geq 0,$$

$$\mathbf{T}_s = -\rho_s^2\partial_{\rho_s}\psi_s\,\mathbf{1} + \mathbf{T}_s^{\mathrm{skw}}, \qquad (\mathbf{T}_s^{\mathrm{skw}})_{ij} = \tfrac{1}{2}\epsilon_{jik}\zeta(\mathbf{w}_s - \boldsymbol{\omega}_s)_k.$$

Along with

$$\eta_\alpha = -\partial_{\theta_\alpha}\phi_\alpha, \qquad \mu_0\mathbf{M} = -\rho_s\partial_{\mathbf{H}}\phi_s,$$

these equations constitute a model for ferrofluids.

In a dynamic problem the magnetic field $\mathbf{H}$ and the electric field $\mathbf{E}$ are regarded as given fields and the unknown solution is the set of fields

$$\rho_s, \rho_f, \mathbf{v}_s, \mathbf{v}_f, \boldsymbol{\sigma}_s, \varepsilon_s, \varepsilon_f,$$

possibly with ε_s, ε_f replaced by the temperature θ as an unknown of the problem. Yet if $\mathbf{H}$ is given then we regard also $\mathbf{m}$ (or $\mathbf{M}$) as a field to be determined. Now, $\mu_0\mathbf{M} = -\rho_s\partial_{\mathbf{H}}\phi_s$ is a relation providing $\mathbf{M}$, at $(\mathbf{x}, t)$, in terms of $\mathbf{H}$ and other variables, at $(\mathbf{x}, t)$. More realistically, we might think that $\mathbf{M}$ and $\mathbf{H}$ are related by a rate-type equation. A model of this form follows by letting $\mathbf{m} = \mathbf{M}/\rho_s$ be a further variable and keeping ψ_s as the pertinent potential so that

$$-(\rho_s\partial_{\mathbf{m}}\psi_s - \mu_0\mathbf{H}) \cdot \dot{\mathbf{m}}$$

occurs in the second-law inequality. If the rate-type equation has the form

$$\dot{\mathbf{m}} = \mathcal{M}(\mathbf{H}, \mathbf{m})$$

then the thermodynamic requirement is

$$(\mu_0\mathbf{H} - \rho_s\partial_{\mathbf{m}}\psi_s) \cdot \mathcal{M}(\mathbf{H}, \mathbf{m}) \geq 0.$$

In a one-dimensional scheme we might assume

$$\dot{m} = -\frac{1}{\tau}(m - h(H)), \quad \tau > 0.$$

The thermodynamic requirement implies that

$$(\rho_s\partial_m\phi_s - \mu_0 H)(m - h(H)) \geq 0.$$

whence

$$\rho_s\partial_m\phi_s = \mu_0 H + \beta(m - h(H)), \quad \beta \geq 0.$$

Otherwise, if we let ϕ_s depend on both $\mathbf{H}$ and $\mathbf{m}$ then we find

$$\rho_s\partial_{\mathbf{m}}\phi_s \cdot \dot{\mathbf{m}} + (\rho_s\partial_{\mathbf{H}}\psi_s + \mu_0\mathbf{M}) \cdot \dot{\mathbf{H}} \leq 0.$$

This is a basic starting point for the modelling of hysteretic materials.

Relation to other approaches

It is worth comparing our scheme with other models of ferrofluids that appeared in the literature. The simplest model traces back to Neuringer and Rosensweig [342]. Ferrofluid is viewed as an incompressible magnetizable but nonconductive fluid, without any internal structures. A steady flow of a ferrofluid is then modelled by

$$\rho(\mathbf{v}\cdot\nabla)\mathbf{v} = -\nabla p + \mu\Delta\mathbf{v} + \mu_0(\mathbf{M}\cdot\nabla)\mathbf{H},$$

$$\nabla\cdot\mathbf{v} = 0, \qquad \nabla\times\mathbf{v} = \mathbf{0}, \qquad \nabla\cdot\mathbf{B} = 0.$$

In essence this model follows by letting $\mathbf{v}_s = \mathbf{v}_f$, adding the two equations of motion and neglecting $\rho_s\partial_t\mathbf{v}_s$, $(\rho_s + \rho_f)\mathbf{g}$.

The micropolar structure of ferrofluids is considered in [113] by letting $\boldsymbol{\sigma}_s = I\boldsymbol{\omega}_s$, where I is the scalar moment-of-inertia density. The occurrence of a couple stress tensor $\boldsymbol{\Lambda}$ is also allowed. Hence, in the present notation, the balance of orientational momentum becomes

$$\rho_s I\dot{\boldsymbol{\omega}}_s = \nabla\cdot\boldsymbol{\Lambda} + \mu_0\mathbf{M}\times\mathbf{H} + \boldsymbol{\Upsilon}_s + \boldsymbol{\Upsilon}_f.$$

Schliomis model [395] also accounts for the micropolar structure. However the effect of $\mathbf{H}$ on the angular momentum $\boldsymbol{\sigma}_s$ is viewed as though the torque $\mu_0\mathbf{M}\times\mathbf{H}$, in a time τ_s, would produce an angular momentum $\tau_s\mu_0\mathbf{M}\times\mathbf{H}$, in addition to $I\boldsymbol{\omega}_s$. Moreover the effect on $\mathbf{M}$ is taken as a constant M_0 times the unit vector of $\mathbf{H}$ plus a time τ_H times $\boldsymbol{\omega}_s\times\mathbf{M}$. Furthermore, the curl of $\frac{1}{2}\mu_0\mathbf{M}\times\mathbf{H}$ is taken to provide an equivalent body force. To sum up, by neglecting inertia the governing equations are taken in the form

$$-\nabla p + \mu_0(\mathbf{M}\cdot\nabla)\mathbf{H} + \mu\Delta\mathbf{v} + \frac{1}{2\tau_s}\nabla\times(\boldsymbol{\sigma}_s - I\boldsymbol{\omega}_s) = 0,$$

$$\boldsymbol{\sigma}_s = I\boldsymbol{\omega}_s + \tau_s\mu_0\mathbf{M}\times\mathbf{H}, \qquad \mathbf{M} = M_0\hat{\mathbf{H}} + \frac{\tau_H}{I}\boldsymbol{\sigma}_s\times\mathbf{M},$$

where $\hat{\mathbf{H}}$ is the unit vector of $\mathbf{H}$.

A model developed by Jenkins[23] [232] associates the micropolar character with the magnetization and is based on the laws of conservation obtained by Brown [70] for quasistatic processes in magnetic materials. The body is viewed as a single material with additional terms of magnetic character. Yet some original terms are considered. The balance of angular momentum involves an additional magnetic term $\mathbf{m}\times\dot{\mathbf{m}}$ and a magnetic stress tensor $\boldsymbol{\Lambda}$. The energy density involves a formal kinetic energy $\frac{1}{2}\beta\dot{\mathbf{m}}\cdot\dot{\mathbf{m}}$.

The balance of mass and linear momentum are given by the classical equations

[23] See also [166].

$$\dot{\rho} + \nabla \cdot (\rho \mathbf{v}) = 0,$$

$$\rho \dot{\mathbf{v}} = \nabla \cdot \mathbf{T} + \mu_0 (\mathbf{M} \cdot \nabla)\mathbf{H} + \rho \mathbf{b}.$$

The angular momentum, per unit volume, is assumed to be $\rho(\mathbf{r} \times \mathbf{v} + \beta \mathbf{m} \times \dot{\mathbf{m}})$, where β is a dimensional constant. Indeed, by analogy with micromagnetics (see Sect. 12.7), a balance law for the magnetization is assumed in the form

$$\frac{d}{dt} \int_{\mathcal{P}_t} \rho \beta \dot{\mathbf{m}}\, dv = \int_{\partial \mathcal{P}_t} \mathbf{\Lambda} \mathbf{n} da + \int_{\mathcal{P}_t} \rho(\mu_0 \mathbf{H} + \boldsymbol{\psi})\, dv$$

for any region $\mathcal{P}_t$ of the body. The tensor $\mathbf{\Lambda}$ is viewed as a magnetic stress tensor while $\boldsymbol{\psi}$ is called the intrinsic magnetic field. Hence the balance of angular momentum is assumed in the form

$$\frac{d}{dt} \int_{\mathcal{P}_t} \rho(\mathbf{r} \times \mathbf{v} + \beta \mathbf{m} \times \dot{\mathbf{m}}) dv = \int_{\partial \mathcal{P}_t} [\mathbf{r} \times (\mathbf{T}\mathbf{n}) + \mathbf{m} \times (\mathbf{\Lambda}\mathbf{n})] da$$
$$+ \int_{\mathcal{P}_t} \rho[\mu_0 \mathbf{m} \times \mathbf{H} + \mu_0 \mathbf{r} \times (\mathbf{m} \cdot \nabla)\mathbf{H} + \mathbf{r} \times \mathbf{b} + \mathbf{m} \times \boldsymbol{\psi}] dv.$$

The corresponding local form of the balance of angular momentum is

$$\rho(\mathbf{r} \times \dot{\mathbf{v}} + \beta \mathbf{m} \times \ddot{\mathbf{m}}) = \mathbf{\Upsilon} - (\mathbf{\Lambda} \nabla) \times \mathbf{m} + \mathbf{r} \times \nabla \cdot \mathbf{T} + \mathbf{m} \times \nabla \cdot \mathbf{\Lambda} + \mu_0 \mathbf{M} \times \mathbf{H} + \mathbf{r} \times \rho \mathbf{b} + \mu_0 \mathbf{r} \times (\mathbf{M} \cdot \nabla)\mathbf{H}$$

and hence, in view of the equation of motion,

$$\rho \beta \mathbf{m} \times \ddot{\mathbf{m}} = \mathbf{\Upsilon} - (\mathbf{\Lambda} \nabla) \times \mathbf{m} + \mathbf{m} \times \nabla \cdot \mathbf{\Lambda} + \mu_0 \mathbf{M} \times \mathbf{H} + \mathbf{M} \times \boldsymbol{\psi}.$$

The evolution equation

$$\rho \beta \ddot{\mathbf{m}} = \nabla \cdot \mathbf{\Lambda} + \rho \mu_0 \mathbf{H}$$

is allowed on condition that

$$\mathbf{\Upsilon} - (\mathbf{\Lambda} \nabla) \times \mathbf{m} + \mathbf{M} \times \boldsymbol{\psi} = \mathbf{0}.$$

Likewise the balance of energy is assumed to be expressed by

$$\frac{d}{dt} \int_{\mathcal{P}_t} \rho(\varepsilon + \tfrac{1}{2}\mathbf{v}^2 + \tfrac{1}{2}\beta \dot{\mathbf{m}}^2) dv = \int_{\partial \mathcal{P}_t} [\mathbf{v}\mathbf{T} + \dot{\mathbf{m}}\mathbf{\Lambda} - \mathbf{q}) \cdot \mathbf{n} da$$
$$+ \int_{\mathcal{P}_t} \rho[\mu_0 \mathbf{v} \cdot (\mathbf{m} \cdot \nabla)\mathbf{H} + \mathbf{v} \cdot \mathbf{b} + \mu_0 \dot{\mathbf{m}} \cdot \mathbf{H} + r] dv.$$

The local form is

$$\rho\dot{\varepsilon} + \rho\mathbf{v}\cdot\dot{\mathbf{v}} + \rho\beta\dot{\mathbf{m}}\cdot\ddot{\mathbf{m}} = \mathbf{T}\cdot\mathbf{L} + \mathbf{v}\cdot(\nabla\cdot\mathbf{T}) + (\mathbf{\Lambda}\nabla)\cdot\dot{\mathbf{m}} - \nabla\cdot\mathbf{q} + \dot{\mathbf{m}}\cdot(\nabla\cdot\mathbf{\Lambda})$$
$$+\mu_0\mathbf{v}\cdot(\mathbf{M}\cdot\nabla)\mathbf{H} + \rho\mathbf{v}\cdot\mathbf{b} + \rho\mu_0\dot{\mathbf{m}}\cdot\mathbf{H} + \rho r.$$

Upon substitution of $\rho\dot{\mathbf{v}}$ and $\rho\beta\ddot{\mathbf{m}}$ it follows that

$$\rho\dot{\varepsilon} = \mathbf{T}\cdot\mathbf{L} + (\mathbf{\Lambda}\nabla)\cdot\dot{\mathbf{m}} - \nabla\cdot\mathbf{q} + \rho\mu_0\dot{\mathbf{m}}\cdot\mathbf{H} + \rho r.$$

A simple model used for ferrofluids is based on the conditions

$$\nabla\cdot\mathbf{v} = 0, \quad \nabla\times\mathbf{H} = \mathbf{0}, \quad \nabla\cdot\mathbf{B} = 0, \quad \mathbf{M} = \chi\mathbf{H}$$

while

$$\mathbf{T} = -p\mathbf{1} + 2\mu\mathfrak{D} + \text{skw}\mathbf{T},$$

where $\mathfrak{D}$ is the stretching. Since $(\text{skw}\mathbf{T})_{jk} = \frac{1}{2}\epsilon_{kji}\Upsilon_i$ then

$$\partial_{x_k}(\text{skw}\mathbf{T})_{jk} = \tfrac{1}{2}\epsilon_{kji}\partial_{x_k}\Upsilon_i, \qquad \nabla\cdot(\text{skw}\mathbf{T}) = -\tfrac{1}{2}\nabla\times\mathbf{\Upsilon}.$$

For steady flows of ferrofluids the equation of motion becomes

$$-\nabla p + \mu\Delta\mathbf{v} + \mu_0(\mathbf{M}\cdot\nabla)\mathbf{H} - \nabla\times(\tfrac{1}{2}\mathbf{\Upsilon}) = \mathbf{0}$$

while $\mathbf{\Upsilon} = -\mu_0\mathbf{H}\times\mathbf{M}$, if $\ddot{\mathbf{m}}$ and $\mathbf{\Lambda}$ are neglected. This is consistent with the curl-term used in [298] where $\mathbf{H}$ is replaced with $\rho\alpha^2(\nabla\times\mathbf{v})\times\hat{\mathbf{M}}$, α being a suitable constant parameter and $\hat{\mathbf{M}}$ the unit vector of $\mathbf{M}$.

12.8.1 Magneto-Rheological Fluids

Magneto-rheological fluids respond to an applied magnetic field with a change in the rheological properties like, e.g. effective viscosity. Hence an electromagnetic control governs the behaviour of a mechanical system. While both electric and magnetic controls are of interest, in practice applications of the magnetic effects look more significant.

The interaction between the resulting magnetic dipoles causes the particles to form columnar structures parallel to the applied magnetic field. These chain-like structures hinder the motion of the fluid thus increasing the viscous properties. We then examine the action of the magnetic field on the properties of ferrofluids.

Assume the fluid flows between two infinite parallel fixed plates so that the flow is irrotational ($\nabla\times\mathbf{v} = \mathbf{0}$). So the balance of orientational momentum can be written as

$$\rho_s\dot{\boldsymbol{\sigma}}_s = \mu_0\mathbf{M}\times\mathbf{H} - \nu\boldsymbol{\omega}_s, \qquad \nu > 0.$$

Since $\grave{\boldsymbol{\sigma}}_s = \mathbf{I}\dot{\boldsymbol{\omega}}_s + \boldsymbol{\omega}_s \times \mathbf{I}\boldsymbol{\omega}_s$ then the inner product with $\boldsymbol{\omega}_s$ yields

$$\boldsymbol{\omega}_s \cdot \mathbf{I}\dot{\boldsymbol{\omega}}_s = \mu_0 \mathbf{M} \times \mathbf{H} \cdot \boldsymbol{\omega}_s - \nu \omega_s^2.$$

Now, $\boldsymbol{\omega}_s \cdot \mathbf{I}\dot{\boldsymbol{\omega}}_s = \frac{1}{2}(\boldsymbol{\omega}_s \cdot \mathbf{I}\boldsymbol{\omega}_s)\grave{}$ and then

$$\tfrac{1}{2}(\boldsymbol{\omega}_s \cdot \mathbf{I}\boldsymbol{\omega}_s)\grave{} = \mu_0 \mathbf{M} \times \mathbf{H} \cdot \boldsymbol{\omega}_s - \nu \omega_s^2.$$

This shows that the interaction of the fluid with the particles result in a decay of the kinetic energy $\frac{1}{2}(\boldsymbol{\omega}_s \cdot \mathbf{I}\boldsymbol{\omega}_s)$. Equilibrium of the particles, $\boldsymbol{\omega}_s = \mathbf{0}$, occurs when $\mathbf{M} \times \mathbf{H} = \mathbf{0}$ that is when the particle (axis) is aligned with the magnetic field $\mathbf{H}$.

12.9 Magneto-, Electro-, and Mechanical-Optical Effects

12.9.1 *Magneto-Optical Effects*

Magneto-optical effects are phenomena associated with electromagnetic waves propagating through a medium acted upon by a static magnetic field. In such materials, called gyrotropic or gyromagnetic, left- and right-rotating elliptical polarizations can propagate at different speeds. When light is transmitted through a layer of magneto-optic material, the plane of polarization can rotate and the speed of propagation depends on the polarization (Faraday effect). Discovered by Michael Faraday in 1845, the Faraday effect was the first experimental evidence that light and electromagnetism are related.

The Faraday effect is caused by left and right circularly polarized waves propagating at slightly different speeds, a property known as circular birefringence. Since a linear polarization can be decomposed into the superposition of two equal-amplitude circularly polarized components of opposite handedness and different phase, the effect of a relative phase shift, induced by the Faraday effect, is to rotate the orientation of the polarization.[24]

Mathematically, the Faraday effect is modelled by a linear relation between the electric field $\mathbf{E}$ and the time derivative $\dot{\mathbf{P}}$. Yet the occurrence of $\dot{\mathbf{P}}$ makes the constitutive equation non-objective. To overcome this difficulty we proceed in the following way.

Consider the electric field $\mathbf{E}_R$ and the polarization $\mathbf{P}_R$ in the reference configuration, $\mathbf{E}_R = \mathbf{E}\mathbf{F}$ and $\mathbf{P}_R = J\mathbf{F}^{-1}\mathbf{P}$. In addition to being referential fields, $\mathbf{E}_R$ and $\mathbf{P}_R$ are invariant under a change of frame. If $\mathbf{Q}$ is the time-dependent orthogonal tensor of the transformation we have

[24] As Faraday wrote in his daily notebook, "Still, I have at last succeeded in illuminating a magnetic curve or line of force, and in magnetizing a ray of light".

$$\mathbf{E}_R^* = (\mathbf{QE})(\mathbf{QF}) = \mathbf{E}(\mathbf{Q}^T\mathbf{Q})\mathbf{F} = \mathbf{EF} = \mathbf{E}_R,$$

$$\mathbf{P}_R^* = J(\mathbf{QF})^{-1}\mathbf{QP} = J\mathbf{F}^{-1}(\mathbf{Q}^T\mathbf{Q})\mathbf{P} = J\mathbf{F}^{-1}\mathbf{P} = \mathbf{P}_R.$$

Since $\mathbf{P}_R$ is invariant then $\dot{\mathbf{P}}_R$ is objective (really, is invariant) and hence $\dot{\mathbf{P}}_R$ can enter a constitutive equation. The same conclusion holds for[25] $\mathbf{F}^T\mathbf{P}$.

First we observe that, by (2.84), if $\mathbf{J}$, $\mathbf{q}$, $\mathbf{M}$, and $\mathbf{k}$ are zero then the entropy inequality can be written as

$$-\rho(\dot{\psi} + \eta\dot{\theta}) + \rho\mathbf{E}\cdot\dot{\boldsymbol{\pi}} + \mathbf{T}\cdot\mathbf{L} \geq 0,$$

where $\boldsymbol{\pi} = \mathbf{P}/\rho$. Now,

$$\boldsymbol{\pi} = \frac{1}{\rho J}\mathbf{F}\mathbf{P}_R = \frac{1}{\rho_R}\mathbf{F}\mathbf{P}_R, \qquad \dot{\boldsymbol{\pi}} = \frac{1}{\rho_R}(\dot{\mathbf{F}}\mathbf{P}_R + \mathbf{F}\dot{\mathbf{P}}_R).$$

Consequently,

$$\rho_R\mathbf{E}\cdot\dot{\boldsymbol{\pi}} = \mathbf{E}_R\cdot\dot{\mathbf{P}}_R + [\mathbf{E}\otimes(\mathbf{F}\mathbf{P}_R)]\cdot\mathbf{L}.$$

Hence, upon multiplying by J, the entropy inequality can be written in the form

$$-\rho_R(\dot{\psi} + \eta\dot{\theta}) + \mathbf{E}_R\cdot\dot{\mathbf{P}}_R + [J\mathbf{T} + \mathbf{E}\otimes(\mathbf{F}\mathbf{P}_R)]\cdot\mathbf{L} \geq 0.$$

We let $\psi, \eta, \mathbf{E}_R, \mathbf{T}$ depend on $\theta, \mathbf{P}_R, \dot{\mathbf{P}}_R, \mathbf{F}$. Indeed, since $\psi, \eta, \mathbf{E}_R$ are Euclidean invariants, they are assumed to depend on $\mathbf{F}$ via the Cauchy-Green tensor $\mathbf{C} = \mathbf{F}^T\mathbf{F}$. Since $\dot{\mathbf{C}} = 2\mathbf{F}^T\mathfrak{D}\mathbf{F}$, computing $\dot{\psi}$ and substituting we obtain the entropy inequality in the form

$$-\rho_R(\partial_\theta\psi + \eta)\dot{\theta} + (\mathbf{E}_R - \rho_R\partial_{\mathbf{P}_R}\psi)\cdot\dot{\mathbf{P}}_R - \rho_R\partial_{\dot{\mathbf{P}}_R}\psi\cdot\ddot{\mathbf{P}}_R + [J\mathbf{T} - 2\rho_R\mathbf{F}\partial_{\mathbf{C}}\psi\mathbf{F}^T + \mathbf{E}\otimes(\mathbf{F}\mathbf{P}_R)]\cdot\mathbf{L} \geq 0.$$

The arbitrariness of $\ddot{\mathbf{P}}_R$ and $\dot{\theta}$ implies

$$\partial_{\dot{\mathbf{P}}_R} = \mathbf{0}, \qquad \eta = -\partial_\theta\psi.$$

Moreover, the arbitrariness of $\mathbf{L}$ implies

$$\mathbf{T} = 2\rho\mathbf{F}\partial_{\mathbf{C}}\psi\mathbf{F}^T - \mathbf{E}\otimes(\mathbf{F}\mathbf{P}_R/J).$$

Since $\mathbf{F}\mathbf{P}_R/J = \mathbf{P}$ then

$$\mathbf{T} = 2\rho\mathbf{F}\partial_{\mathbf{C}}\psi\mathbf{F}^T - \mathbf{E}\otimes\mathbf{P}.$$

Hence $\mathbf{T}$ is the sum of a symmetric elastic stress $2\rho\mathbf{F}\partial_{\mathbf{C}}\psi\mathbf{F}^T$ and an electromagnetic stress $-\mathbf{E}\otimes\mathbf{P}$. Indeed,

---

[25] Under a change of reference with rotation tensor $\mathbf{Q}$, $\mathbf{F} \to \mathbf{QF}$ and $\mathbf{P} \to \mathbf{QP}$. As a consequence $\mathbf{F}^T\mathbf{P} \to (\mathbf{QF})^T\mathbf{QP} = \mathbf{F}^T\mathbf{P}$.

$$\text{skw}\mathbf{T} = -\text{skw}(\mathbf{E} \otimes \mathbf{P}) = \text{skw}(\mathbf{P} \otimes \mathbf{E}),$$

as expected.

The remaining inequality

$$(\mathbf{E}_R - \rho_R \partial_{\mathbf{P}_R} \psi) \cdot \dot{\mathbf{P}}_R \geq 0$$

holds if

$$\mathbf{E}_R = \rho_R \partial_{\mathbf{P}_R} \psi - \boldsymbol{\Gamma} \times \dot{\mathbf{P}}_R,$$

$\boldsymbol{\Gamma}$ being an invariant vector.

To derive a relation in terms of $\mathbf{E}$ and $\mathbf{P}$ we observe that

$$\partial_{\mathbf{P}} \psi = \partial_{\mathbf{P}_R} \psi J \mathbf{F}^{-1}.$$

So we have

$$\mathbf{E} = \rho \partial_{\mathbf{P}} \psi - (\boldsymbol{\Gamma} \times \dot{\mathbf{P}}_R) \mathbf{F}^{-1}.$$

The vector $\boldsymbol{\Gamma}$ is called the gyration vector. In isotropic materials $\boldsymbol{\Gamma}$ is determined by a given magnetic induction $\mathbf{B}_0$ so that

$$\boldsymbol{\Gamma} = h \mathbf{B}_0,$$

h being possibly a scalar function of $\mathbf{B}_0$.

We now apply the model to the analysis of linear wave propagation. We let ρ be constant and $\mathbf{F} = \mathbf{1}$. In addition, we let $\rho \partial_{\mathbf{P}} \psi = \alpha \mathbf{P}$. Hence we have

$$\mathbf{E} = \alpha \mathbf{P} - \boldsymbol{\Gamma} \times \dot{\mathbf{P}}.$$

The vector $\boldsymbol{\Gamma}$ so introduced is called the gyration vector. Our interest is in the case where the gyration vector is originated by a constant magnetic induction $\mathbf{B}_0$. For isotropic materials we let

$$\boldsymbol{\Gamma} = h \mathbf{B}_0,$$

h being possibly a scalar function of $\mathbf{B}_0$. Let $\mathbf{T}_{\text{el}}$ be the linear approximation of $2\rho \mathbf{F} \partial_{\mathbf{C}} \psi$ and hence assume

$$\mathbf{T}_{\text{el}} = \mu(\nabla \mathbf{u} + (\nabla \mathbf{u})^T) + \lambda(\nabla \cdot \mathbf{u})\mathbf{1}.$$

Since $\mathbf{T} = \mathbf{T}_{\text{el}} - \mathbf{E} \otimes \mathbf{P}$ then

$$\nabla \cdot \mathbf{T} = \nabla \cdot \mathbf{T}_{\text{el}} - \mathbf{E}(\nabla \cdot \mathbf{P}) - (\mathbf{P} \cdot \nabla)\mathbf{E}.$$

as is natural in formulations of constitutive equations, the value of $\mathbf{E}$ is meant as that at the frame at rest with the body. If we refer to the values at the laboratory frame then

$$\mathbf{E} \to \mathbf{E} + \mathbf{v} \times \mathbf{B}.$$

In view of the linear approximation, the terms $\mathbf{E}(\nabla \cdot \mathbf{P})$ and $(\mathbf{P} \cdot \nabla)\mathbf{E}$ are neglected in the equation of motion (2.72). Moreover, the force term $\rho\dot{\pi} \times \mathbf{B}$ is approximated by $\partial_t \mathbf{P} \times \mathbf{B}_0$. Hence the equation of motion, the constitutive equation for $\mathbf{P}$, i.e. $\mathbf{E} = \alpha\mathbf{P} - \boldsymbol{\Gamma} \times \dot{\mathbf{P}}$, and Maxwell's equations can be written in the form

$$\rho_0 \partial_t^2 \mathbf{u} = (\mu + \lambda)\nabla(\nabla \cdot \mathbf{u}) + \mu\Delta\mathbf{u} + \partial_t \mathbf{P} \times \mathbf{B}_0,$$

$$\mathbf{E} + \partial_t \mathbf{u} \times \mathbf{B}_0 - \alpha\mathbf{P} + h\mathbf{B}_0 \times \partial_t \mathbf{P} = \mathbf{0},$$

$$\nabla \times \mathbf{B}/\mu_0 - \epsilon_0 \partial_t \mathbf{E} - \partial_t \mathbf{P} = \mathbf{0}, \qquad \nabla \cdot (\epsilon_0 \mathbf{E} + \mathbf{P}) = 0,$$

$$\nabla \times \mathbf{E} + \partial_t \mathbf{B} = \mathbf{0}, \qquad \nabla \cdot \mathbf{B} = 0.$$

Magneto-optic effects are now described by having recourse to homogeneous plane-wave solutions of the system,

$$\mathbf{u}(\mathbf{x}, t) = \hat{\mathbf{u}} \exp[i(k\mathbf{n} \cdot \mathbf{x} - \omega t)], \qquad \mathbf{P} = \hat{\mathbf{P}} \exp[i(k\mathbf{n} \cdot \mathbf{x} - \omega t)]$$

and the like for $\mathbf{B} - \mathbf{B}_0$ and $\mathbf{E}$. Here $\hat{\mathbf{u}}, \hat{\mathbf{P}}$ are complex amplitude vectors, k is the wave number, and $\mathbf{n}$ is the unit vector of the propagation direction. Upon substitution of $\mathbf{u}, \mathbf{P}, \mathbf{B}, \mathbf{E}$ into the system of differential equations we find

$$\rho_0 \omega^2 \hat{\mathbf{u}} = (\mu + \lambda)k^2(\mathbf{n} \cdot \hat{\mathbf{u}})\mathbf{n} + \mu k^2 \hat{\mathbf{u}} + i\omega\hat{\mathbf{P}} \times \mathbf{B}_0, \tag{12.47}$$

$$\hat{\mathbf{E}} - i\omega\hat{\mathbf{u}} \times \mathbf{B}_0 - \alpha\hat{\mathbf{P}} - i\omega h\mathbf{B}_0 \times \hat{\mathbf{P}} = \mathbf{0}, \tag{12.48}$$

$$k\mathbf{n} \times \hat{\mathbf{B}}/\mu_0 + \omega\epsilon_0\hat{\mathbf{E}} + \omega\hat{\mathbf{P}} = \mathbf{0}, \qquad k\mathbf{n} \cdot (\epsilon_0\hat{\mathbf{E}} + \hat{\mathbf{P}}) = 0, \tag{12.49}$$

$$k\mathbf{n} \times \hat{\mathbf{E}} - \omega\hat{\mathbf{B}} = \mathbf{0}, \qquad k\mathbf{n} \cdot \hat{\mathbf{B}} = 0, \tag{12.50}$$

where $\hat{\mathbf{B}}$ is the amplitude vector of $\mathbf{B} - \mathbf{B}_0$.

Suggestive solutions emerge by restricting attention to waves propagating in the direction of $\mathbf{B}_0$. Let B_0 be the component of $\mathbf{B}_0$ along $\mathbf{n}$, namely $\mathbf{B}_0 = B_0\mathbf{n}$. Inner and vector multiplication of (12.47) by $\mathbf{n}$ gives the two equations

$$[\rho_0\omega^2 - (2\mu + \lambda)k^2]\mathbf{n} \cdot \hat{\mathbf{u}} = 0, \tag{12.51}$$

$$(\rho_0\omega^2 - \mu k^2)\mathbf{n} \times \hat{\mathbf{u}} - i\omega B_0\hat{\mathbf{P}} = \mathbf{0}. \tag{12.52}$$

Let $\hat{\mathbf{u}}$ be collinear with $\mathbf{n}$. Hence $\mathbf{n} \cdot \hat{\mathbf{u}} \neq 0$; by (12.51) the phase velocity $v_{\rm ph} = \omega/k$ is subject to

$$|v_{\rm ph}| = [(2\mu + \lambda)/\rho_0]^{1/2}.$$

By (12.52) it follows $\hat{\mathbf{P}} = \mathbf{0}$ and hence by (12.48) and (12.50) $\hat{\mathbf{E}} = \mathbf{0}$ and $\hat{\mathbf{B}} = \mathbf{0}$; Eq. (12.49) holds identically if $\hat{\mathbf{E}}$, $\hat{\mathbf{P}}$, and $\hat{\mathbf{B}}$ are zero. Hence the solution is the classical longitudinal sound wave ($\hat{\mathbf{u}} \parallel \mathbf{n}$); this wave is unaffected by the applied field $\mathbf{B}_0$.

We now let $\hat{\mathbf{u}}$ be orthogonal to $\mathbf{n}$. Substitution of $\hat{\mathbf{B}} = \mathbf{n} \times \hat{\mathbf{E}}k/\omega$ in (12.49) gives

$$\hat{\mathbf{P}} + (\epsilon_0 - k^2/\mu_0\omega^2)\hat{\mathbf{E}} = \mathbf{0}$$

whence

$$\hat{\mathbf{E}} = \gamma\hat{\mathbf{P}}, \qquad \gamma := \frac{\beta^2}{\epsilon_0(1-\beta^2)}, \qquad \beta = \frac{\omega/k}{c},$$

$c = 1/\sqrt{\mu_0\epsilon_0}$ being the light speed in free space. Substituting in (12.48) we find

$$(\alpha - \gamma)\hat{\mathbf{P}} + i\omega B_0[h\mathbf{n} \times \hat{\mathbf{P}} - \mathbf{n} \times \hat{\mathbf{u}}] = 0. \tag{12.53}$$

Inner multiply by $\mathbf{n} \times \hat{\mathbf{P}}$ and use the properties $\mathbf{n} \cdot \hat{\mathbf{P}} = 0$, $\hat{\mathbf{u}} \cdot \hat{\mathbf{P}} = 0$ to obtain

$$\omega h B_0 \hat{\mathbf{P}} \cdot \hat{\mathbf{P}} = 0.$$

Unless $\omega h B_0 = 0$ we have

$$\hat{\mathbf{P}} \cdot \hat{\mathbf{P}} = 0. \tag{12.54}$$

Since $\hat{\mathbf{P}}$ is complex-valued then we let $\hat{\mathbf{P}} = \hat{\mathbf{P}}_r + i\hat{\mathbf{P}}_i$ and observe that Eq. (12.54) implies

$$|\hat{\mathbf{P}}_r| = |\hat{\mathbf{P}}_i|, \qquad \hat{\mathbf{P}}_r \cdot \hat{\mathbf{P}}_i = 0,$$

$\hat{\mathbf{P}}_r$, $\hat{\mathbf{P}}_i$ denoting the real and imaginary parts of $\hat{\mathbf{P}}$. Let $\theta = k\mathbf{n} \cdot \mathbf{x} - \omega t$ and $\mathbf{e}_1$, $\mathbf{e}_2$ be the unit vectors of $\hat{\mathbf{P}}_r$, $\hat{\mathbf{P}}_i$, with $(\mathbf{e}_1, \mathbf{e}_2, \mathbf{n})$ an orthogonal triad. The function $\mathbf{P}(\mathbf{x}, t)$ can then be represented by

$$\mathbf{P}(\mathbf{x}, t) = (\hat{\mathbf{P}}_r + i\hat{\mathbf{P}}_i)\exp(i\theta) = P(\mathbf{e}_1 \cos\theta - \mathbf{e}_2 \sin\theta) + iP(\mathbf{e}_2 \cos\theta + \mathbf{e}_1 \sin\theta),$$

where $P = |\hat{\mathbf{P}}_r| = |\hat{\mathbf{P}}_i|$. Things are different depending on whether we look at the time behaviour of the wave at a fixed point or, at a fixed time, we look at the wave along the direction of propagation $\mathbf{n}$. As θ increases ($\mathbf{n} \cdot \mathbf{x}$ increases), if $(\mathbf{e}_1, \mathbf{e}_2, \mathbf{n})$ is a right-handed triad then the vector $\mathbf{P}$ (both $\hat{\mathbf{P}}_r$ and $\hat{\mathbf{P}}_i$) describes a circle in the clockwise sense and we say that $\mathbf{P}$ is a right circularly polarized wave. Likewise, as θ decreases[26] $\mathbf{P}$ is a left-circularly polarized wave. For definiteness, let θ increase.

[26] Namely, at fixed $\mathbf{x}$ as t increases or at fixed t in the reverse direction of propagation, $-\mathbf{n}$.

Conversely, if $\mathbf{P}$ is such that $(\mathbf{e}_1, \mathbf{e}_2, \mathbf{n})$ is a left-handed triad then $\mathbf{P}$ describes a left circularly polarized wave.

Observe

$$\mathbf{n} \times \hat{\mathbf{P}} = \pm i \hat{\mathbf{P}},$$

where the upper (lower) sign holds for left-handed (right-handed) triad $(\mathbf{e}_1, \mathbf{e}_2, \mathbf{n})$. Since

$$\mathbf{n} \times \hat{\mathbf{u}} = \frac{i\omega B_0}{\rho_0 \omega^2 - \mu k^2} \hat{\mathbf{P}},$$

and $\mathbf{n} \times \hat{\mathbf{P}} = \pm i \hat{\mathbf{P}}$ then (12.53) becomes

$$[\alpha - \gamma + \frac{B_0^2 v_{\text{ph}}^2}{\rho_0 (v_{\text{ph}}^2 - \mu/\rho_0)} \mp \omega B_0 h] \hat{\mathbf{P}} = \mathbf{0},$$

where $v_{\text{ph}} = \omega/k$. Waves occur if the propagation condition

$$\alpha - \gamma + \frac{B_0^2 v_{\text{ph}}^2}{\rho_0 (v_{\text{ph}}^2 - \mu/\rho_0)} \mp \omega B_0 h = 0 \tag{12.55}$$

holds. Equation (12.55) is the dispersion relation for transverse waves with a direction parallel to the applied magnetic field. The $\mp \omega B_0 h$ term means that left and right-circularly polarized waves propagate with different speeds; the upper sign refers to left-circularly polarized waves. This shows that the Faraday effect, that is the difference of the speeds depending on the polarization, is provided by the product $B_0 h$.

If $B_0 = 0$ then the propagation condition (12.55) simplifies to

$$\alpha - \gamma = 0, \qquad \beta^2 = \frac{1}{1 + \chi}.$$

Irrespective of the polarization, the quantity $B_0^2 v_{\text{ph}}^2 / \rho_0 (v_{\text{ph}}^2 - \mu/\rho_0)$ may account for the magnetoelastic dragging whereby the speed of a transverse sound wave traversing an elastic dielectric is lowered by the application of a magnetic field in the direction of propagation.

12.9.2 *Frequency Dependence, Closed Processes, Memory Functionals*

As shown in the previous section, elementary microscopic pictures of electromagnetic materials result in frequency-dependent conductivities and susceptibilities. More generally, we might consider constitutive relations in the form

$$\mathbf{P}_\omega(t) = \chi(\omega)\mathbf{E}_\omega(t), \qquad \mathbf{M}_\omega = \chi_m(\omega)\mathbf{H}_\omega(t), \tag{12.56}$$

where χ and χ_m are the complex-valued electric and magnetic tensor susceptibility.[27] Relations of the form (12.56) are applied to describe the response to electromagnetic wave propagation. Yet they are shown to follow from wave fields of the form

$$\mathbf{E}_\omega(\mathbf{x}, t) = \mathbf{E}_0 \exp[i(\mathbf{k}\cdot\mathbf{x} - \omega t)], \tag{12.57}$$

or the analogue for $\mathbf{H}$, where $\mathbf{E}_0$ is a complex vector. This function describes an electric field propagating in the direction of $\mathbf{k}$ with phase velocity $\omega/|\mathbf{k}|$. This is done with the convention that the physical fields are obtained by taking the real parts ([230], Chap. 7).

While it is well known that the complex exponential representation is convenient in calculations, here we show how definite conclusions can be obtained, without any convention, if a corresponding model is given in the time domain. For definiteness we look at $\mathbf{P}$ modelled by a memory functional of $\mathbf{E}$ in the form

$$\mathbf{P}(t) = \mathbf{G}_0\mathbf{E}(t) + \mathbf{G}_0' \int_0^\infty \chi(u)\mathbf{E}(t-u)du, \tag{12.58}$$

$\mathbf{G}_0, \mathbf{G}_0'$ being two second-order tensors while $\chi(u) \to 0$ as $u \to \infty$. Let

$$\mathbf{E}(t) = \mathbf{E}_1 \cos\alpha + \mathbf{E}_2 \sin\alpha, \qquad \alpha = \mathbf{k}\cdot\mathbf{x} - \omega t.$$

Substituting in (12.58) we find

$$\mathbf{P}(t) = \mathbf{G}_0(\mathbf{E}_1\cos\alpha + \mathbf{E}_2\sin\alpha) + \mathbf{G}_0'\mathbf{E}_1 \int_0^\infty \chi(u)\cos(\alpha - \omega u)du + \mathbf{G}_0'\mathbf{E}_2 \int_0^\infty \chi(u)\sin(\alpha - \omega u)du$$

Since

$$\int_0^\infty \chi(u)\cos(\alpha - \omega u)du = \chi_c(\omega)\cos\alpha + \chi_s(\omega)\sin\alpha,$$

$$\int_0^\infty \chi(u)\sin(\alpha - \omega u)du = \chi_c(\omega)\sin\alpha - \chi_s(\omega)\cos\alpha,$$

then

$$\mathbf{P}(t) = (\mathbf{G}_0 + \mathbf{G}_0'\chi_c)(\mathbf{E}_1\cos\alpha + \mathbf{E}_2\sin\alpha) - \mathbf{G}_0'\chi_s(\mathbf{E}_2\cos\alpha - \mathbf{E}_1\sin\alpha).$$

Moreover the power $\mathbf{P}\cdot\dot{\mathbf{E}}$ becomes

$$\mathbf{P}\cdot\dot{\mathbf{E}} = -\omega[(\mathbf{G}_0 + \mathbf{G}_0'\chi_c)(\mathbf{E}_1\cos\alpha + \mathbf{E}_2\sin\alpha) - \mathbf{G}_0'\chi_s(\mathbf{E}_2\cos\alpha - \mathbf{E}_1\sin\alpha)]\cdot[\mathbf{E}_2\cos\alpha - \mathbf{E}_1\sin\alpha].$$

[27] See, e.g. [274], Chap. 11.

We now examine the consequences of the second-law inequality for cyclic processes. We consider an undeformable dielectric at constant temperature; if $\mathbf{E}$ is cyclic on $[0, d]$ then, by (12.2), it follows

$$\int_0^d \mathbf{P}(t) \cdot \dot{\mathbf{E}}(t)\, dt \leq 0. \tag{12.59}$$

To proceed within the domain of complex-valued vectors we let

$$\mathbf{E}_0 = \mathbf{E}_1 + i\mathbf{E}_2,$$

$\mathbf{E}_1, \mathbf{E}_2$ being real-valued vectors, and observe that

$$\mathbf{E}_\omega = \mathbf{E}_1 \cos\alpha + \mathbf{E}_2 \sin\alpha + i(\mathbf{E}_2 \cos\alpha - \mathbf{E}_1 \sin\alpha).$$

Further, letting

$$\chi(\omega) = \mathbf{G}_0 + \mathbf{G}_0' \chi_c(\omega) + i\mathbf{G}_0' \chi_s(\omega) =: \chi_1 + i\chi_2$$

we have $\mathbf{P}_\omega(t) = \chi(\omega)\mathbf{E}_\omega(t)$ and

$$\mathbf{P}(t) = \Re \mathbf{P}_\omega(t).$$

Consequently

$$\mathbf{P}(t) \cdot \dot{\mathbf{E}}(t) = \Re \mathbf{P}_\omega(t) \cdot \Re \dot{\mathbf{E}}_\omega(t).$$

This allows us to examine the real, physical power $\mathbf{P} \cdot \dot{\mathbf{E}}$ within the domain of complex-valued vectors.

For any pair of complex vectors $\mathbf{u}, \mathbf{v}$ we have

$$\Re\mathbf{u} \cdot \Re\mathbf{v} = \tfrac{1}{2}(\mathbf{u} + \mathbf{u}^*) \cdot (\mathbf{v} + \mathbf{v}^*) = \tfrac{1}{4}(\mathbf{u} \cdot \mathbf{v} + \mathbf{u}^* \cdot \mathbf{v}^* + \mathbf{u}^* \cdot \mathbf{v} + \mathbf{u} \cdot \mathbf{v}^*).$$

Observe

$$\tfrac{1}{4}(\mathbf{u}^* \cdot \mathbf{v} + \mathbf{u} \cdot \mathbf{v}^*) = \tfrac{1}{2}\Re(\mathbf{u} \cdot \mathbf{v}^*).$$

If both $\mathbf{u}$ and $\mathbf{v}$ depend on time in the form (12.57) then

$$\int_0^{n2\pi/\omega} (\Re\mathbf{u} \cdot \Re\mathbf{v})(t)dt = \tfrac{1}{2}\Re \int_0^{n2\pi/\omega} (\mathbf{u} \cdot \mathbf{v}^*)(t)dt, \qquad n \in \mathbb{N}.$$

Letting $\mathbf{u} = \mathbf{P}_\omega$ and $\mathbf{v} = \dot{\mathbf{E}}_\omega$ we find

$$0 \geq \int_0^{n2\pi/\omega} (\mathbf{P} \cdot \dot{\mathbf{E}})(t)dt = \tfrac{1}{2} \int_0^{n2\pi/\omega} (\chi\mathbf{E}_\omega \cdot \dot{\mathbf{E}}_\omega^*)(t)dt = n\pi\{\mathbf{E}_2 \cdot (\chi_1 - \chi_1^T)\mathbf{E}_1 - (\mathbf{E}_1 \cdot \chi_2\mathbf{E}_1 + \mathbf{E}_2 \cdot \chi_2\mathbf{E}_2)\}.$$

The arbitrariness of $\mathbf{E}_1, \mathbf{E}_2$ implies

$$\chi_1 = \chi_1^T, \qquad \chi_2 \geq \mathbf{0}.$$

Absorption is absent, or the phenomenon is non-dissipative, if

$$\chi_2 \in \text{Skw}.$$

Hence the material allows for non-absorption if and only if

$$\chi^* = \chi^T. \tag{12.60}$$

Back to the constitutive equation we can write

$$\chi \mathbf{E}_\omega = \chi_1 \mathbf{E}_\omega + i\chi_2 \mathbf{E}_\omega;$$

the skew-symmetry of χ_2 implies that there is a vector $\mathbf{g}$, i.e. the axial vector of χ_2, such that

$$\chi_2 \mathbf{E}_\omega = \mathbf{g} \times \mathbf{E}_\omega.$$

So we can write

$$\mathbf{P}_\omega = \chi_1 \mathbf{E}_\omega + i\mathbf{g} \times \mathbf{E}_\omega.$$

The vector $\mathbf{g}$ is then referred to as the gyration vector and we observe that $i\mathbf{E}_\omega$ might be viewed as the time derivative of $\mathbf{E}_\omega/\omega$. A strictly analogous procedure would determine $\mathbf{E}_\omega$ in terms of $\mathbf{P}_\omega$.

As a comment, a specific model in the time domain, like (12.58), allows us to check that, e.g. $\mathbf{P} = \Re \mathbf{P}_\omega$ is the physical field. Otherwise we might take $\mathbf{P} = \Re \mathbf{P}_\omega$ by convention. Yet we see that $\chi_2 \in \text{Skw}$ holds if and only if $\mathbf{G}_0' \in \text{Skw}$. This however would contradict $\mathbf{G}_0' \chi_c(\omega) \in \text{Sym}$ unless $\chi_c(\omega) = 0$.

Relation to closed processes in the real domain

We now examine the restrictions placed by (12.59) on (12.58) without any recourse to complex-valued vectors. Let d be the duration of the cyclic process $\mathbf{E}(t)$. Having in mind Fourier's theorem we let

$$\mathbf{E}(t) = \mathbf{E}_k^c \cos k\omega t + \mathbf{E}_k^s \sin k\omega t, \qquad \omega = 2\pi/d, \quad k \in \mathbb{N} \cup \{0\},$$

where $\mathbf{E}_k^c$ and $\mathbf{E}_k^s$ can be represented in the base $\{\mathbf{e}_j\}$ as

$$\mathbf{E}_k^c = a_{kj}\mathbf{e}_j, \qquad \mathbf{E}_k^s = b_{kj}\mathbf{e}_j.$$

Substituting in (12.58) we have

$$\mathbf{P}(t) = (\mathbf{G}_0 + \mathbf{G}_0'\chi_c(k\omega))\mathbf{E}_k^c \cos k\omega t + (\mathbf{G}_0 + \mathbf{G}_0'\chi_s(k\omega)) \sin k\omega t.$$

Since

$$\dot{\mathbf{E}}(t) = k\omega(\mathbf{E}_k^s \cos k\omega t - \mathbf{E}_k^c \sin k\omega t)$$

then

$$\begin{aligned}\mathbf{P}\cdot\dot{\mathbf{E}} = k\omega\{&\cos^2 k\omega t[\mathbf{E}_k^s\cdot(\mathbf{G}_0+\mathbf{G}_0'\chi_c)\mathbf{E}_k^c - \mathbf{E}_k^s\cdot\mathbf{G}_0'\chi_s\mathbf{E}_k^s]\\ &- \sin^2 k\omega t[\mathbf{E}_k^c\cdot(\mathbf{G}_0+\mathbf{G}_0'\chi_c)\mathbf{E}_k^s + \mathbf{E}_k^c\cdot\mathbf{G}_0'\chi_s\mathbf{E}_k^c]\\ &+ \cos k\omega t\ \sin k\omega t[\mathbf{E}_k^s\cdot(\mathbf{G}_0+\mathbf{G}_0'\chi_c)\mathbf{E}_k^s + \mathbf{E}_k^s\cdot\mathbf{G}_0'\chi_s\mathbf{E}_k^c - \mathbf{E}_k^c\cdot(\mathbf{G}_0+\mathbf{G}_0'\chi_c)\mathbf{E}_k^c + \mathbf{E}_k^c\cdot\mathbf{G}_0'\chi_s\mathbf{E}_k^s]\}.\end{aligned}$$

For any k the function $\mathbf{E}(t)$ on $[0, 2\pi/|\omega|]$ is a closed process. Now,

$$\int_0^{2\pi/|\omega|} \cos^2 k\omega t\, dt = \frac{\pi}{k|\omega|},$$

and the like for $\sin^2 k\omega t$. Hence we find that

$$\int_0^{2\pi/|\omega|} \mathbf{P}(t)\cdot\dot{\mathbf{E}}(t)dt = k\pi\,\mathrm{sgn}\,\omega\{\mathbf{E}_k^s\cdot[\mathbf{G}_0+\mathbf{G}_0'\chi_c - (\mathbf{G}_0+\mathbf{G}_0'\chi_c)^T]\mathbf{E}_k^c - \mathbf{E}_k^s\cdot\mathbf{G}_0'\chi_s\mathbf{E}_k^s - \mathbf{E}_k^s\cdot\mathbf{G}_0'\chi_s\mathbf{E}_k^c\}.$$

In view of the arbitrariness of the vectors $\mathbf{E}_k^c$, $\mathbf{E}_k^s$, inequality (12.59) holds if and only if

$$\mathbf{G}_0 + \mathbf{G}_0'\chi_c \in \mathrm{Sym}, \qquad \omega\chi_s\,\mathrm{sym}\mathbf{G}_0' \geq \mathbf{0}. \tag{12.61}$$

The dielectric is non-dissipative if $\chi_s(k\omega) = 0$, which may be true only for particular frequencies ω.

The requirements (12.61) are necessary conditions; they follow by the second-law inequality in connection with closed processes. Hence we conclude that the constitutive equation (12.58) is physically admissible only if the conditions (12.61) hold. In general, for non-closed processes in undeformable dielectrics (12.58) is required to comply with

$$\rho(\dot{\phi} + \eta\dot{\theta}) + \mathbf{P}\cdot\dot{\mathbf{E}} \leq 0.$$

Hence, if $\phi, \eta, \mathbf{P}$ depend on $\theta, \mathbf{E}, \mathbf{E}^t$ we have

$$\rho(\partial_\theta\phi + \eta)\dot{\theta} + (\mathbf{P} + \rho\partial_{\mathbf{E}}\phi)\cdot\dot{\mathbf{E}} + d\phi(\theta, \mathbf{E}, \mathbf{E}^t|\dot{\mathbf{E}}^t) \leq 0.$$

It follows that

$$\eta = -\partial_\theta\phi, \qquad \mathbf{P} = -\rho\partial_{\mathbf{E}}\phi, \qquad d\phi(\theta, \mathbf{E}, \mathbf{E}^t|\dot{\mathbf{E}}^t) \leq 0.$$

Let $\mathbf{P}$ be given by (12.58) and, in view of (12.61), let

$$\mathbf{G}_0, \mathbf{G}_0' \in \mathrm{Sym}, \qquad \mathbf{G}'(u) = \mathbf{G}_0'\chi(u).$$

Hence the requirement $\mathbf{P} = -\rho\partial_{\mathbf{E}}\phi$ implies that

$$\rho\phi(t) = -\tfrac{1}{2}\mathbf{E}(t)\cdot\mathbf{G}_0\mathbf{E}(t) - \mathbf{E}(t)\cdot\int_0^\infty \mathbf{G}'(u)\mathbf{E}(t-u)du + \Phi(\theta(t), \mathbf{E}^t).$$

The free energy ϕ is also required to have a minimum property. Let $\tilde{\theta}, \tilde{\mathbf{E}}$ be temperature and electric field functions on $\mathbb{R}$. Constant continuations of $\tilde{\theta}^{t_0}, \tilde{\mathbf{E}}^{t_0}$ are defined by

$$\theta^t(u) = \begin{cases} \tilde{\theta}(t_0), u \in [0, t_0], \\ \tilde{\theta}^{t_0}(u - t_0), u \in [t_0, \infty), \end{cases} \qquad \mathbf{E}^t(u) = \begin{cases} \tilde{\mathbf{E}}(t_0), u \in [0, t_0], \\ \tilde{\mathbf{E}}^{t_0}(u - t_0), u \in [t_0, \infty). \end{cases}$$

It follows from $\dot{\phi} = d\phi(\theta, \mathbf{E}, \mathbf{E}^t|\dot{\mathbf{E}}^t) \le 0$ that

$$\phi(\theta(t), \mathbf{E}(t), \mathbf{E}^t) \le \phi(\tilde{\theta}(t_0), \tilde{\mathbf{E}}(t_0), \tilde{\mathbf{E}}^{t_0}).$$

By the continuity of the functional, as $t - t_0 \to \infty$

$$\phi(\theta(t), \mathbf{E}(t), \mathbf{E}^t) \to \phi(\theta(t_0), \mathbf{E}(t_0), \mathbf{E}_0^\dagger,$$

where $\mathbf{E}_0^\dagger$ is the constant history $\mathbf{E}_0^\dagger(u) = \tilde{\mathbf{E}}(t_0), u \in [0, \infty)$. Hence

$$\phi(\theta(t_0), \mathbf{E}(t_0), \mathbf{E}_0^\dagger) \le \phi(\tilde{\theta}(t_0), \tilde{\mathbf{E}}(t_0), \tilde{\mathbf{E}}^{t_0}).$$

Thus, among all histories $\mathbf{E}^t$ with given present value $\mathbf{E}_0$ none yields a smaller value of the free energy than that corresponding to the constant history $\mathbf{E}_0^\dagger$.

For definiteness we select the function Φ in the form

$$\Phi = \Phi_0(\theta(t)) + \tfrac{1}{2}\int_0^\infty \mathbf{E}(t-u)\cdot\mathbf{G}'(u)\mathbf{E}(t-u)du.$$

Consequently we find[28]

$$\rho\phi = \Phi_0(\theta) - \tfrac{1}{2}\mathbf{E}\cdot\mathbf{G}_0\mathbf{E} - \tfrac{1}{2}\mathbf{E}\cdot\Big(\int_0^\infty \mathbf{G}'(u)du\Big)\mathbf{E}$$
$$+\tfrac{1}{2}\mathbf{E}\cdot\Big(\int_0^\infty \mathbf{G}'(u)du\Big)\mathbf{E} - \mathbf{E}\cdot\int_0^\infty \mathbf{G}'(u)\mathbf{E}(t-u)du + \tfrac{1}{2}\int_0^\infty \mathbf{E}(t-u)\cdot\mathbf{G}'(u)\mathbf{E}(t-u)du$$
$$= \Phi_0(\theta) - \tfrac{1}{2}\mathbf{E}\cdot\mathbf{G}_\infty\mathbf{E} + \tfrac{1}{2}\int_0^\infty(\mathbf{E}(t) - \mathbf{E}(t-u))\cdot\mathbf{G}'(u)(\mathbf{E}(t) - \mathbf{E}(t-u))du,$$

where

$$\mathbf{G}_\infty = \mathbf{G}_0 + \int_0^\infty \mathbf{G}'(u)du.$$

[28] The dependence on the present values $\theta(t), \mathbf{E}(t)$ is denoted by $\theta, \mathbf{E}$.

Since ϕ has a minimum at constant histories then

$$\mathbf{G}'(u) \leq \mathbf{0}, \quad \forall u \in [0, \infty).$$

This in turn implies

$$\mathbf{G}_0 \geq \mathbf{G}_\infty.$$

Moreover, for constant histories

$$\mathbf{P} = \mathbf{G}_\infty \mathbf{E};$$

hence we conclude that

$$\mathbf{G}_\infty > \mathbf{0}.$$

The functional $\rho\phi$ is consistent with thermodynamics if the further requirement $d\phi(\theta, \mathbf{E}, \mathbf{E}^t|\dot{\mathbf{E}}^t) \leq 0$ holds. Now,

$$d\phi(\theta, \mathbf{E}, \mathbf{E}^t|\dot{\mathbf{E}}^t) = -\int_0^\infty \dot{\mathbf{E}}(t-u) \cdot \mathbf{G}'(u)(\mathbf{E}(t) - \mathbf{E}(t-u))du.$$

Observe $\dot{\mathbf{E}}(t-u) = -\partial_u \mathbf{E}(t-u)$; hence we obtain

$$\begin{aligned} d\phi(\theta, \mathbf{E}, \mathbf{E}^t|\dot{\mathbf{E}}^t) &= \int_0^\infty \partial_u(\mathbf{E}(t) - \mathbf{E}(t-u)) \cdot \mathbf{G}'(u)(\mathbf{E}(t) - \mathbf{E}(t-u))du \\ &= \tfrac{1}{2}[(\mathbf{E}(t) - \mathbf{E}(t-u)) \cdot \mathbf{G}'(u)(\mathbf{E}(t) - \mathbf{E}(t-u))]_0^\infty \\ &\quad -\tfrac{1}{2}\int_0^\infty [(\mathbf{E}(t) - \mathbf{E}(t-u)) \cdot \mathbf{G}''(u)(\mathbf{E}(t) - \mathbf{E}(t-u))]du. \end{aligned}$$

Since $\mathbf{G}'(u) \to \mathbf{0}$ as $u \to \infty$ then it follows

$$0 \geq d\phi(\theta, \mathbf{E}, \mathbf{E}^t|\dot{\mathbf{E}}^t) = -\tfrac{1}{2}\int_0^\infty [(\mathbf{E}(t) - \mathbf{E}(t-u)) \cdot \mathbf{G}''(u)(\mathbf{E}(t) - \mathbf{E}(t-u))]du.$$

This condition holds if and only if

$$\mathbf{G}'' = \mathbf{G}'_0 \chi' \geq \mathbf{0}.$$

Since $\mathbf{G}' \leq \mathbf{0}$ then we let $\mathbf{G}'_0 < \mathbf{0}$, $\chi(0) > 0$. Hence it follows $\chi' \leq 0$; the function $\chi(u)$ is a monotone decreasing function.

12.9.3 Electro-Optical Effects

Certain materials change their optical properties when subjected to an electric field. The electro-optical effect is the change of the refractive index n resulting from the application of a steady (dc) electric field. Physically this is caused by forces that distort the orientations or shapes of the molecules constituting the material. As a consequence the electric field controls the light. It happens that the material becomes birefringent, with different indices of refraction for light polarized parallel to or perpendicular to the applied field.

In modelling the electro-optical effects the customary assumptions are referred to as *Pockels effect* (n changes in proportion to the applied electric field) and *Kerr effect* (n changes in proportion to the square of the applied electric field). This indicates that we have to look for the electric polarization $\mathbf{P}$ as a nonlinear function of the electric field.

For ease of modelling we let the body be non-conducting ($\mathbf{J} = \mathbf{0}, \mathbf{q} = \mathbf{0}$) and non-magnetizable ($\mathbf{M} = \mathbf{0}$). Moreover it is consistent to let the extra-entropy flux be zero. Relative to an inertial frame, locally at rest, the second-law inequality (12.18) simplifies to

$$-\rho(\dot{\phi} + \eta\dot{\theta}) - \mathbf{P} \cdot \dot{\mathbf{E}} + \mathbf{T} \cdot \mathbf{L} \geq 0.$$

We model the material as thermo-electro-elastic and hence describe it by functions of the deformation gradient $\mathbf{F}$, the temperature θ, and the electric field $\mathbf{E}$. Borrowing from Sect. 12.2 we let ϕ be a Euclidean-invariant function of $\mathbf{C}$, θ, $\mathfrak{E}$. Hence we find that

$$\eta = -\partial_\theta \phi, \qquad \mathbf{P} = -\rho \mathbf{F} \partial_{\mathfrak{E}} \phi, \qquad \mathbf{T} = 2\rho \mathbf{F}\, \partial_{\mathbf{C}} \phi\, \mathbf{F}^T - \mathbf{E} \otimes \mathbf{P},$$

whence $\text{skw}[\mathbf{T} + \mathbf{E} \otimes \mathbf{P}] = \mathbf{0}$.

Let

$$\phi = \phi_0(\mathbf{C}, \theta) + \phi_e(\mathfrak{E}, \theta)$$

and, for definiteness, assume

$$\phi_e = -\frac{\epsilon_0}{\rho}[\tfrac{1}{2}\Xi^{(1)}(\theta) \cdot \mathfrak{E} \otimes \mathfrak{E} + \tfrac{1}{3}\Xi^{(2)}(\theta) \cdot \mathfrak{E} \otimes \mathfrak{E} \otimes \mathfrak{E} + \tfrac{1}{4}\Xi^{(3)}(\theta) \cdot \mathfrak{E} \otimes \mathfrak{E} \otimes \mathfrak{E} \otimes \mathfrak{E}],$$

where $\chi^{(i)}$, $i = 1, 2, 3$, are second-, third-, and fourth-order susceptibility tensors and, e.g. $\Xi^{(1)}(\theta) \cdot \mathfrak{E} \otimes \mathfrak{E} = \Xi^{(1)}_{HK} \mathfrak{E}_H \mathfrak{E}_K$. By definition, each $\Xi^{(i)}$ is fully symmetric. Hence it follows

$$\mathbf{P} = \epsilon_0 \mathbf{F}[\Xi^{(1)}(\theta)\mathfrak{E} + \Xi^{(2)}(\theta)\mathfrak{E} \otimes \mathfrak{E} + \Xi^{(3)}(\theta)\mathfrak{E} \otimes \mathfrak{E} \otimes \mathfrak{E}].$$

In terms of $\mathbf{E}$ we find

$$\mathbf{P} = \epsilon_0[\chi^{(1)}(\theta)\mathbf{E} + \chi^{(2)}(\theta)\mathbf{E} \otimes \mathbf{E} + \chi^{(3)}(\theta)\mathbf{E} \otimes \mathbf{E} \otimes \mathbf{E}],$$

where, e.g. $\chi^{(2)}_{ij} = F_{iH}F_{jK}\Xi^{(2)}_{HK}$. In many materials the third term $\chi^{(3)}\mathbf{E}\otimes\mathbf{E}\otimes\mathbf{E}$ is negligible in comparison with $\chi^{(2)}\mathbf{E}\otimes\mathbf{E}$. In such cases, occurring mainly in some crystals, we restrict attention to a Pockels medium characterized by

$$\mathbf{P} = \epsilon_0[\chi^{(1)} + \chi^{(2)}\mathbf{E}]\mathbf{E}.$$

If, instead, the material is a gas, a liquid, or a centro-symmetric crystal then we can restrict to a Kerr medium ($\chi^{(2)} = 0$) and hence

$$\mathbf{P} = \epsilon_0[\chi^{(1)} + \chi^{(3)}\mathbf{E}\otimes\mathbf{E}]\mathbf{E}.$$

To understand the birefringence properties associated with the Pockels and Kerr media, we first look at light in anisotropic non-magnetic media.

We assume

$$\mathbf{D} = \epsilon\mathbf{E}, \qquad \mathbf{B} = \mu_0\mathbf{H}$$

and let the dielectric permittivity tensor ϵ be real-valued and positive definite. Maxwell's equations[29]

$$\nabla\times\mathbf{E} = -\partial_t\mathbf{B}, \qquad \nabla\times\mathbf{B} = \mu_0\partial_t\mathbf{D}, \qquad \nabla\cdot\mathbf{D} = 0, \qquad \nabla\cdot\mathbf{B} = 0$$

indicate the absence of free charges and electric current. Let $\mathbf{E}$ be given by a plane wave of angular frequency ω and wave vector $\mathbf{k}$,

$$\mathbf{E}(\mathbf{x}, t) = \mathbf{E}_\omega \exp[i(\mathbf{k}\cdot\mathbf{x} - \omega t)].$$

Since

$$\mu_0\partial_t^2\mathbf{D} = -\nabla\times(\nabla\times\mathbf{E})$$

and $-\nabla\times(\nabla\times\mathbf{E}) = (\mathbf{k}\cdot\mathbf{E})\mathbf{k} - \mathbf{k}^2\mathbf{E}$ then it follows

$$\mathbf{k}^2\mathbf{E}_\omega - (\mathbf{k}\cdot\mathbf{E}_\omega)\mathbf{k} = \mu_0\omega^2(\epsilon\mathbf{E}_\omega). \tag{12.62}$$

In addition, $\nabla\cdot\mathbf{D} = 0$ implies

$$\mathbf{k}\cdot\mathbf{D} = 0.$$

whence $\mathbf{D}$ is perpendicular to the wave vector $\mathbf{k}$. Owing to the possibly anisotropic permittivity ϵ, the orthogonality condition need not hold for the electric field $\mathbf{E}$. If $\mathbf{k}\cdot\mathbf{E}_\omega \neq 0$ then plane waves occur if $\mathbf{k}$ satisfies the compatibility condition

$$\det[\mathbf{k}^2\mathbf{1} - \mathbf{k}\otimes\mathbf{k} - \mu_0\omega^2\epsilon] = 0. \tag{12.63}$$

[29] Here we let $\mathbf{F}$ be time independent and hence we may take the fields $\mathbf{E}$, $\mathbf{D}$, $\mathbf{H}$, $\mathbf{B}$ satisfy Maxwell's equations in the laboratory frame.

Equation (12.63) is called *Fresnel's equation* and determines the unknown function $\mathbf{k}(\omega)$.

If $\boldsymbol{\epsilon}$ is isotropic, $\boldsymbol{\epsilon} = \epsilon \mathbf{1}$, then $\mathbf{k} \cdot \mathbf{E}_\omega = 0$ and (12.62) reduces to

$$(-\mathbf{k}^2 + \mu_0 \epsilon \omega^2)\mathbf{E}_\omega = \mathbf{0}$$

thus showing that plane waves occur at any direction with phase speed

$$\frac{\omega^2}{\mathbf{k}^2} = \frac{1}{\mu_0 \epsilon},$$

possibly dependent on the frequency ω if ϵ depends on ω. In free space

$$\frac{\omega^2}{\mathbf{k}^2} = \frac{1}{\mu_0 \epsilon_0} = c^2.$$

Hence in matter

$$v^2 = \frac{\omega^2}{\mathbf{k}^2} = c^2 \frac{\epsilon_0}{\epsilon}.$$

Since $v^2/c^2 = 1/n^2$, n being the refractive index, we find

$$\epsilon = \epsilon_0 n^2.$$

For formal convenience represent $\boldsymbol{\epsilon}$ and $\mathbf{k}$ in the principal basis of $\boldsymbol{\epsilon}$. If $\epsilon_1, \epsilon_2, \epsilon_3$ are the eigenvalues of $\boldsymbol{\epsilon}$ then we can write

$$\boldsymbol{\epsilon} = \epsilon_0 \operatorname{diag}[n_1^2, n_2^2, n_3^2].$$

Hence $\det[-\mathbf{k}^2 \mathbf{1} + \mu_0 \omega^2 \boldsymbol{\epsilon} + \mathbf{k} \otimes \mathbf{k}] = 0$ becomes

$$\det \begin{bmatrix} -k_2^2 - k_3^2 + \omega^2 n_1^2/c^2 & k_1 k_2 & k_1 k_3 \\ k_1 k_2 & -k_1^2 - k_3^2 + \omega^2 n_2^2/c^2 & k_2 k_3 \\ k_1 k_3 & k_2 k_3 & -k_1^2 - k_2^2 + \omega^2 n_3^2/c^2 \end{bmatrix} = 0.$$

Upon computing the determinant and rearranging the terms we can write the equation in the form

$$\frac{\omega^4}{c^4} - \frac{\omega^2}{c^2}\left(\frac{k_1^2 + k_2^2}{n_3^2} + \frac{k_1^2 + k_3^2}{n_2^2} + \frac{k_2^2 + k_3^2}{n_1^2}\right) + \left(\frac{k_1^2}{n_2^2 n_3^2} + \frac{k_2^2}{n_1^2 n_3^2} + \frac{k_3^2}{n_1^2 n_2^2}\right)\mathbf{k}^2 = 0. \tag{12.64}$$

The simplest example of anisotropy is that of uniaxial materials.[30] For definiteness let $n_1 = n_2 \neq n_3$. In such a case we say that the third direction is the optical axis.

---

[30] Uniaxial materials comprise the largest family of birefringent specimens, including calcite, quartz, and ordered synthetic or biological structures.

As is customary in the literature we use the notation n_o, for n_1 and n_2, and n_e for n_3. The determinantal equation (12.64) can then be written in the factored form

$$\left(\frac{\mathbf{k}^2}{n_o^2} - \frac{\omega^2}{c^2}\right)\left(\frac{k_1^2 + k_2^2}{n_e^2} + \frac{k_3^2}{n_o^2} - \frac{\omega^2}{c^2}\right) = 0.$$

Both conditions

$$\frac{\mathbf{k}^2}{n_o^2} - \frac{\omega^2}{c^2} = 0, \qquad \frac{k_1^2 + k_2^2}{n_e^2} + \frac{k_3^2}{n_o^2} - \frac{\omega^2}{c^2} = 0$$

define surfaces in the space of wave vectors $\mathbf{k}$ that are allowed for a given ω. The first equation defines a sphere; this means that the refractive index is n_o regardless of the direction of $\mathbf{k}$ (ordinary rays). The second equation defines a spheroid symmetric about the optical axis and the wave speed depends on the direction of $\mathbf{k}$ (extraordinary rays). Along the optical axis the two rays propagate with the same speed,

$$\frac{\omega^2}{\mathbf{k}^2} = \frac{\omega^2}{k_3^2} = \frac{c^2}{n_o^2}.$$

If $\mathbf{k}$ is perpendicular to the optical axis ($k_3 = 0$) then the extraordinary ray propagates with speed

$$\frac{\omega^2}{\mathbf{k}^2} = \frac{\omega^2}{k_1^2 + k_2^2} = \frac{c^2}{n_e^2}.$$

Hence the extraordinary ray is the fast wave if $n_e < n_o$ (negative birefringence) and the slow wave if $n_e > n_o$.

To model Pockels and Kerr effects it is worth distinguishing the case when the light wave is superposed to an external electric field $\mathbf{E}_0$ (DC effect) from that when no external field is applied (AC effect). Let

$$\mathbf{E} = \mathbf{E}_0 + \tilde{\mathbf{E}} \exp[i(\mathbf{k} \cdot \mathbf{x} - \omega t)].$$

If $|\tilde{\mathbf{E}}| \ll |\mathbf{E}_0|$ we restrict attention to linear terms in $\tilde{\mathbf{E}}$ and write

$$\mathbf{P} \simeq \epsilon_0(\chi^{(1)} + 2\chi^{(2)}\mathbf{E}_0 + 3\chi^{(3)}\mathbf{E}_0 \otimes \mathbf{E}_0)\tilde{\mathbf{E}} \exp[i(\mathbf{k} \cdot \mathbf{x} - \omega t)],$$

which shows that

$$\chi_{\text{eq}} = \chi^{(1)} + 2\chi^{(2)}\mathbf{E}_0 + 3\chi^{(3)}\mathbf{E}_0 \otimes \mathbf{E}_0$$

is the equivalent susceptibility tensor and hence $\boldsymbol{\epsilon} = \epsilon_0(\mathbf{1} + \chi_{\text{eq}})$.

12.9.4 Mechanical-Optical Effects

The optical properties of a medium can be changed by mechanical agencies such as deformation and velocity gradients. A simple way of accounting for mechanical-optical effects is to allow the permittivity, and the susceptibility, depend on the strain and the spin. Moreover dependences on $\dot{\mathbf{E}}$ are in order; the thermodynamic analysis shows how this can be realized.

Again for ease of modelling we let the body be non-conducting ($\mathbf{J} = \mathbf{0}$, $\mathbf{q} = \mathbf{0}$) and non-magnetizable ($\mathbf{M} = \mathbf{0}$), and let $\mathbf{k} = \mathbf{0}$. Relative to an inertial frame, locally at rest, the second-law inequality (12.18) simplifies to

$$-\rho(\dot{\phi} + \eta\dot{\theta}) - \mathbf{P} \cdot \dot{\mathbf{E}} + \mathbf{T} \cdot \mathbf{L} \geq 0.$$

To account for mechanical-optical interactions we consider the reference electric field $\mathfrak{E} = \mathbf{F}^T \mathbf{E}$ and observe

$$\dot{\mathbf{E}} = \mathbf{F}^{-T} \dot{\mathfrak{E}} - \mathbf{L}^T \mathbf{E}.$$

Hence the inequality can be written in the form

$$-\rho(\dot{\phi} + \eta\dot{\theta}) - (\mathbf{F}^{-1}\mathbf{P}) \cdot \dot{\mathfrak{E}} + (\mathbf{E} \otimes \mathbf{P}) \cdot \mathbf{L} + \mathbf{T} \cdot \mathbf{L} \geq 0.$$

Having in mind some models occurring in the literature we let ϕ, η, $\mathbf{P}$, $\mathbf{T}$ depend on $\mathbf{C}$, θ, $\mathfrak{E}$, $\mathbf{E}^2$, $\dot{\mathfrak{E}}$. Since $\dot{\mathbf{C}} = 2\mathbf{F}^T\mathbf{D}\mathbf{F}$ and

$$\mathbf{E} \cdot \dot{\mathbf{E}} = \mathbf{E} \cdot [\mathbf{F}^{-T} \dot{\mathfrak{E}} - \mathbf{E} \cdot \mathbf{L}^T \mathbf{F}^{-T} \mathfrak{E}] = \mathbf{E} \cdot \mathbf{F}^{-T} \dot{\mathfrak{E}} - \mathbf{E} \cdot \mathbf{D}\mathbf{E}$$

we can write the entropy inequality in the form

$$\begin{aligned} &-\rho(\partial_\theta \phi + \eta)\dot{\theta} - (\rho\partial_{\mathfrak{E}}\phi + \mathbf{F}^{-1}\mathbf{P} + \rho\partial_{\mathbf{E}^2}\phi\, \mathbf{F}^{-1}\mathbf{E}) \cdot \dot{\mathfrak{E}} - \rho\partial_{\dot{\mathfrak{E}}}\phi \cdot \ddot{\mathfrak{E}} \\ &\quad +(\mathbf{T} - 2\rho\mathbf{F}\partial_{\mathbf{C}}\phi\mathbf{F}^T + \mathbf{E} \otimes \mathbf{P} + \rho\partial_{\mathbf{E}^2}\phi\, \mathbf{E} \otimes \mathbf{E}) \cdot \mathbf{L} \geq 0. \end{aligned}$$

The arbitrariness of $\dot{\theta}$, $\ddot{\mathfrak{E}}$, and $\mathbf{L}$ implies

$$\eta = -\partial_\theta \phi, \qquad \partial_{\dot{\mathfrak{E}}}\phi = \mathbf{0},$$

$$\mathbf{T} = 2\rho\mathbf{F}\partial_{\mathbf{C}}\phi\mathbf{F}^T - \mathbf{E} \otimes \mathbf{P} - \rho\partial_{\mathbf{E}^2}\phi\, \mathbf{E} \otimes \mathbf{E}, \tag{12.65}$$

$$(\rho\partial_{\mathfrak{E}}\phi + \mathbf{F}^{-1}\mathbf{P} + \rho\partial_{\mathbf{E}^2}\phi\, \mathbf{F}^{-1}\mathbf{E}) \cdot \dot{\mathfrak{E}} \leq 0. \tag{12.66}$$

By (12.65) it follows

$$\text{skw}\mathbf{T} = \text{skw}(\mathbf{P} \otimes \mathbf{E}),$$

that is the classical condition of balance of angular momentum. The reduced inequality (12.66) holds if

$$\mathbf{F}^{-1}\mathbf{P} = -\rho\partial_{\mathfrak{E}}\phi - \rho\partial_{\mathbf{E}^2}\phi\mathbf{F}^{-1}\mathbf{E} + \mathbf{\Lambda}(\mathbf{C}, \theta)\dot{\mathfrak{E}}.$$

If $\mathbf{\Lambda}$ is positive definite then this equation accounts for dissipation induced by $\dot{\mathfrak{E}}$. If $\mathbf{\Lambda} \in \mathrm{Skw}$ or $\mathbf{\Lambda} = \mathbf{0}$ then the model is regarded as non-dissipative. Anyway, since $\dot{\mathfrak{E}} = \mathbf{F}^T\mathbf{L}^T\mathbf{E} + \mathbf{F}^T\dot{\mathbf{E}}$ then we obtain

$$\mathbf{P} = -\rho\mathbf{F}\partial_{\mathfrak{E}}\phi - \rho\partial_{\mathbf{E}^2}\phi\mathbf{E} + \mathbf{F}\mathbf{\Lambda}\mathbf{F}^T\mathbf{L}^T\mathbf{E} + \mathbf{F}\mathbf{\Lambda}\mathbf{F}^T\dot{\mathbf{E}}.$$

If $\mathbf{\Lambda} \in \mathrm{Skw}$ then $\mathbf{F}\mathbf{\Lambda}\mathbf{F}^T \in \mathrm{Skw}$ too. In addition, we have a term in $\dot{\mathbf{E}}$ and one in $\mathbf{W}\mathbf{E}$, as we may find in the literature.

Let

$$\phi = \phi_0(\mathbf{C}, \theta) - \frac{\epsilon_0}{2\rho}[\mathfrak{E} \cdot \chi_0(\theta)\mathfrak{E} + a_1(\theta)\mathfrak{E} \cdot \mathfrak{E} + a_2(\theta)(\mathrm{tr}\, \mathbf{C})\mathbf{E}^2],$$

with $\mathrm{tr}\, \mathbf{C} = \mathbf{F} \cdot \mathbf{F}$. Hence we have

$$\mathbf{P} = \epsilon_0(\mathbf{F}\chi_0\mathbf{F}^T)\mathbf{E} + a_1(\mathbf{F}\mathbf{F}^T)\mathbf{E} + a_2(\mathrm{tr}\, \mathbf{C})\mathbf{E} + \mathbf{F}\mathbf{\Lambda}\mathbf{F}^T(\mathbf{D} - \mathbf{W})\mathbf{E} + \mathbf{F}\mathbf{\Lambda}\mathbf{F}^T\dot{\mathbf{E}}. \tag{12.67}$$

Hence the interaction of the electric field with deformation results in the susceptibility tensor

$$\chi = \mathbf{F}\chi_0\mathbf{F}^T + a_1\mathbf{F}\mathbf{F}^T + a_2(\mathrm{tr}\, \mathbf{C})\mathbf{1},$$

while $\mathbf{F}\mathbf{\Lambda}\mathbf{F}^T(\mathbf{D} - \mathbf{W})\mathbf{E} + \mathbf{F}\mathbf{\Lambda}\mathbf{F}^T\dot{\mathbf{E}}$ is a model for the interaction of the electric field with the velocity gradient field. Indeed, observe that

$$\mathbf{F}\mathbf{\Lambda}\mathbf{F}^T(\mathbf{D} - \mathbf{W})\mathbf{E} + \mathbf{F}\mathbf{\Lambda}\mathbf{F}^T\dot{\mathbf{E}} = \mathbf{F}\mathbf{\Lambda}\mathbf{F}^T\mathbf{D}\mathbf{E} + \mathbf{F}\mathbf{\Lambda}\mathbf{F}^T(\dot{\mathbf{E}} - \mathbf{W}\mathbf{E}).$$

Hence the dependence on the time derivative $\dot{\mathbf{E}}$ is through the corotational (objective) derivative $\dot{\mathbf{E}} - \mathbf{W}\mathbf{E}$ while the dependence on the velocity gradient is through $\mathbf{D}\mathbf{E}$. If $\mathbf{\Lambda} \in \mathrm{Skw}$ then, in terms of the axial vector $\boldsymbol{\lambda}$ of $\mathbf{\Lambda}$, we have

$$\mathbf{F}\mathbf{\Lambda}\mathbf{F}^T(\dot{\mathbf{E}} - \mathbf{W}\mathbf{E}) = \mathbf{F}[\boldsymbol{\lambda} \times \mathbf{F}^T(\dot{\mathbf{E}} - \mathbf{W}\mathbf{E})].$$

Equation (12.67) is a nonlinear version of the constitutive relation associated with the permittivity modelled in [274], Sect. 102. The linearized expression follows by observing

$$\mathbf{F} = \mathbf{1} + \nabla_{\mathbf{X}}\mathbf{u} = \mathbf{1} + \nabla\mathbf{u}(\mathbf{1} + \nabla\mathbf{u}\mathbf{F})$$

whence

$$\mathbf{F}\mathbf{F}^T = \mathbf{1} + 2\varepsilon + o(\nabla\mathbf{u}).$$

Consequently, to within linear terms we can write

$$\mathbf{P} = \epsilon_0[\chi_0 + (a_1 + 3a_2)\mathbf{1} + 2a_2(\mathrm{tr}\, \varepsilon)\mathbf{1}] + \mathbf{\Lambda}\mathbf{D}\mathbf{E} + \mathbf{\Lambda}(\dot{\mathbf{E}} - \mathbf{W}\mathbf{E}).$$

If $\mathbf{\Lambda} \in \text{Skw}$ then, in terms of the axial vector $\boldsymbol{\lambda}$, we have

$$\mathbf{\Lambda DE} + \mathbf{\Lambda}(\dot{\mathbf{E}} - \mathbf{WE}) = \boldsymbol{\lambda} \times \mathbf{DE} + \boldsymbol{\lambda} \times (\dot{\mathbf{E}} - \mathbf{WE}).$$

It is worth remarking that $\dot{\mathbf{E}} - \mathbf{WE}$ is the time derivative relative to a corotating observer.[31] Hence, relative to a corotating observer, if $\mathbf{\Lambda}$ and $\mathbf{W}$ are constants then time-harmonic waves $\mathbf{E} = \mathbf{E}_0 \exp(-i\omega t)$ are solutions and $\mathbf{\Lambda}(\dot{\mathbf{E}} - \mathbf{WE}) = -i\omega\boldsymbol{\lambda} \times \mathbf{E}$; moreover $\boldsymbol{\lambda}$ plays the role of gyration vector.

It is worth observing that sometimes [274] the permittivity tensor is assumed to be

$$\epsilon = \epsilon_0 + \nu\mathbf{D} + i\lambda\mathbf{W}.$$

A contribution $\nu\mathbf{D}$ to the permittivity occurs if $\mathbf{\Lambda} = \nu\mathbf{1}$. In such a case $\mathbf{\Lambda} \overset{\circ}{\mathbf{E}} = \nu\dot{\mathbf{E}} = -i\omega\nu\mathbf{E}$. A factored dependence on $\mathbf{W}$ seems to be in contradiction with thermodynamics. As with the dependence on $\mathbf{D}$, in suspensions and in colloidal solutions of anisotropically shaped particles, the stretching $\mathbf{D}$ is regarded to orienting the particles suspended in the fluid. Hence the suspension is optically anisotropic as a consequence of the velocity gradient and this behaviour is called the Maxwell effect. It is then appropriate to view the suspension as a structured continuum where the orientation of particles is affected by the velocity gradient.

12.9.5 Magnetocaloric and Electro-Caloric Effects

The magnetocaloric effect was discovered in 1881, when Warburg observed it in iron [441], and was explained independently by Debye [121] and Giauque [190]. They also suggested its practical use: the adiabatic demagnetisation used to reach temperatures lower than that of liquid helium, which had been the lowest achievable experimental temperature. One of the most notable examples of the magnetocaloric effect is in the chemical element gadolinium and some of its alloys. Gadolinium's temperature increases when it enters certain magnetic fields. When it leaves the magnetic field, the temperature drops. This effect is considerably stronger for the gadolinium alloy ($Gd_5Si_2Ge_2$) which has allowed scientists to approach to within one milliKelvin, one thousandth of a degree of absolute zero.

The magnetocaloric effect is then applied as an alternative technology for refrigeration. In essence, in a magnetic refrigeration process a decrease in the applied magnetic field allows the magnetic domains of a magnetocaloric material to become disoriented by the agitating action of the thermal energy present in the material. If the material is isolated (adiabatic process) the temperature drops as the domains absorb the thermal energy to perform their reorientation.

---

[31] Namely, an observer rotating with angular velocity $\mathbf{w}$ equal to the axial vector of $\mathbf{W}$, $w_i = -\frac{1}{2}\epsilon_{ijk} W_{jk}$.

Lately, a similar behaviour, called electro-caloric effect, was observed. In electro-caloric materials under adiabatic conditions, temperature increases when we apply an external electric field. When it leaves the electric field, the temperature drops. The electro-caloric effect is often considered to be the physical inverse of the pyroelectric effect. It should not be confused with the thermoelectric effect (specifically, the Peltier effect), in which a temperature difference occurs when a current is driven through an electric junction with two dissimilar conductors.

We now provide a thermodynamic setting for the magnetocaloric and the electro-caloric effects. For the sake of simplicity we look at a rigid body at rest in an inertial frame. The Clausius–Duhem inequality reads

$$-\rho(\dot{\psi} + \eta\dot{\theta}) + \mathbf{E} \cdot \mathbf{J} - \dot{\mathbf{P}} \cdot \mathbf{E} - \mu_0 \dot{\mathbf{M}} \cdot \mathbf{H} - \frac{1}{\theta}\mathbf{q} \cdot \nabla\theta \geq 0.$$

Assume $\psi = \tilde{\psi}(\mathbf{E}, \mathbf{H}, \theta)$ and, ignoring cross-coupling terms, $\mathbf{P} = \tilde{\mathbf{P}}(\mathbf{E}, \theta)$, $\mathbf{M} = \tilde{\mathbf{M}}(\mathbf{H}, \theta)$. Letting

$$\rho\phi = \rho\psi - \mathbf{P} \cdot \mathbf{E} - \mu_0 \mathbf{M} \cdot \mathbf{H}$$

we can write the Clausius–Duhem inequality in the form

$$\rho(\partial_\theta \phi + \eta)\dot{\theta} + (\rho\partial_{\mathbf{E}}\phi + \mathbf{P})\dot{\mathbf{E}} + (\rho\partial_{\mathbf{H}}\phi + \mu_0\mathbf{M})\dot{\mathbf{H}} - \mathbf{E} \cdot \mathbf{J} + \frac{1}{\theta}\mathbf{q} \cdot \nabla\theta \leq 0.$$

The values of $\dot{\mathbf{E}}$ and $\dot{\mathbf{H}}$ can be chosen arbitrarily thanks to the appropriate values of $\nabla \times \mathbf{H}$ and $\nabla \times \mathbf{E}$ which warrant the validity of Maxwell's equations and do not affect the constitutive equations. The arbitrariness of $\dot{\theta}$, $\dot{\mathbf{E}}$, $\dot{\mathbf{H}}$ implies that

$$\eta = -\partial_\theta \phi, \qquad \mathbf{P} = -\rho\partial_{\mathbf{E}}\phi, \qquad \mu_0\mathbf{M} = -\rho\partial_{\mathbf{H}}\phi \tag{12.68}$$

and

$$\mathbf{E} \cdot \mathbf{J} - \frac{1}{\theta}\mathbf{q} \cdot \nabla\theta \geq 0.$$

Look at the balance of energy,

$$\rho\dot{\varepsilon} = \mathbf{E} \cdot \mathbf{J} + \mathbf{E} \cdot \dot{\mathbf{P}} + \mu_0\mathbf{H} \cdot \dot{\mathbf{M}} - \nabla \cdot \mathbf{q} + \rho r. \tag{12.69}$$

Since

$$\rho\varepsilon = \rho\phi + \rho\theta\eta + \mathbf{E} \cdot \mathbf{P} + \mu_0\mathbf{H} \cdot \mathbf{M} \tag{12.70}$$

then, in view of (12.68), by (12.69) it follows

$$\rho\theta\dot{\eta} = \mathbf{E} \cdot \mathbf{J} - \nabla \cdot \mathbf{q} + \rho r.$$

Restrict attention to adiabatic processes, so that $\mathbf{q} = \mathbf{0}$, $r = 0$. Inasmuch as $\mathbf{E} \cdot \mathbf{J}$ is negligible the process can be regarded also as isentropic,

$$\dot{\eta} = 0.$$

Consequently

$$\partial_\theta \eta \dot{\theta} + \partial_{\mathbf{E}} \eta \cdot \dot{\mathbf{E}} + \partial_{\mathbf{H}} \eta \cdot \dot{\mathbf{H}} = 0$$

whence, by means of (12.68),

$$\dot{\theta} = -\frac{1}{\rho \partial_\theta \eta} (\partial_\theta \mathbf{P} \cdot \dot{\mathbf{E}} + \mu_0 \partial_\theta \mathbf{M} \cdot \dot{\mathbf{H}}). \tag{12.71}$$

It is worth observing that, by (12.70) and (12.68),

$$\rho \partial_\theta \varepsilon = \rho \theta \partial_\theta \eta + \mathbf{E} \cdot \partial_\theta \mathbf{P} + \mu_0 \mathbf{H} \cdot \partial_\theta \mathbf{M}$$

and hence $\rho \theta \partial_\theta \eta$ equals the specific heat, per unit volume, $\rho \partial_\theta \varepsilon$, apart from $\mathbf{E} \cdot \partial_\theta \mathbf{P}$ + $\mu_0 \mathbf{H} \cdot \partial_\theta \mathbf{M}$.

Assume **P** and **E**, as well as **M** and **H**, are aligned so that we may restrict attention to the components P, E, M, H in the common direction. A generalized Curie–Weiss law may be applied to paramagnets and simple ferroics (ferromagnets and ferroelectrics).[32] The electric and magnetic susceptibilities χ_e, χ_m are decreasing functions of the temperature θ,

$$P = \epsilon_0 \chi_e(\theta) E, \qquad M = \chi_m(\theta) H \qquad \chi_e' \le 0, \quad \chi_m' \le 0.$$

As θ is greater than the Curie temperature θ_c the Curie–Weiss law is assumed to hold,

$$\chi_e = \frac{C_e}{\theta - \theta_c}, \qquad \chi_m = \frac{C_m}{\theta - \theta_c}.$$

Consequently

$$\partial_\theta P \, \dot{E} = -\epsilon_0 \frac{C_e}{2(\theta - \theta_c)^2} (E^2)^{\cdot}, \qquad \partial_\theta M \, \dot{H} = -\frac{C_m}{2(\theta - \theta_c)^2} (H^2)^{\cdot}.$$

If the strengths of the externally applied fields E^2, H^2 decrease then

$$\partial_\theta P \, \dot{E} > 0, \qquad \partial_\theta M \, \dot{H} > 0.$$

Moreover since $\rho \theta \partial_\theta \eta = \rho \partial_\theta \varepsilon - E \cdot \partial_\theta P - \mu_0 H \cdot \partial_\theta M$, and $\rho \partial_\theta \varepsilon > 0$, for slowly varying fields $\rho \theta \partial_\theta \eta$ is positive yet smaller than the specific heat. Hence it follows from (12.71) that

---

[32] See [442], Eq. (6.62).

$$\dot{\theta} < 0\ (> 0) \quad \text{as} \quad (E^2)^{\cdot} < 0\ (> 0),\ (H^2)^{\cdot} < 0\ (> 0).$$

Magnetic refrigeration

To determine a connection between the changes of magnetic field and temperature we observe that by (12.68) it follows

$$\mu_0 \partial_\theta \mathbf{M} = \rho \partial_{\mathbf{H}} \eta.$$

For simplicity, disregard deformation and let the magnetic field **H** have a constant direction. Denote by H the component in the fixed direction and let η be given by a function $\eta(\theta, H)$. Then $\dot{\eta} = \partial_\theta \eta\, \dot{\theta} + \partial_H \eta\, \dot{H}$. In adiabatic conditions ($\mathbf{q} = \mathbf{0}$, $r = 0$) we have $\dot{\eta} = 0$ and hence

$$\partial_\theta \eta\, \dot{\theta} + \partial_H \eta\, \dot{H} = 0.$$

This shows the magnetocaloric effect,

$$\dot{\theta} = -\frac{1}{\partial_\theta \eta} \partial_H \eta\, \dot{H} = -\frac{\mu_0}{\rho \partial_\theta \eta} \partial_\theta M\, \dot{H},$$

and hence a variation of H induces a variation of temperature. In a process from H_{initial} to H_{final}, by a change of variable ($t \to H(t)$) we have

$$\theta_{\text{final}} - \theta_{\text{initial}} = -\int_{H_{\text{initial}}}^{H_{\text{final}}} \frac{1}{\rho \partial_\theta \eta} \partial_H \eta\, dH = -\int_{H_{\text{initial}}}^{H_{\text{final}}} \frac{\mu_0}{\rho \partial_\theta \eta} \partial_\theta M\, dH.$$

The integrand is a function of θ, H; often $1/\rho\, \partial_\theta \eta$ is denoted as θ / C, C being the specific heat.

Since $\partial_\theta \eta > 0$, two types of processes may occur.
If $\partial_\theta M = \rho \partial_H \eta / \mu_0 < 0$ then $\dot{H} > 0$ implies $\dot{\theta} > 0$; increasing the field H results in an increase of temperature (direct magnetocaloric effect).
If $\partial_\theta M = \rho \partial_H \eta / \mu_0 > 0$ then $\dot{H} > 0$ implies $\dot{\theta} < 0$; increasing the field H results in a decrease of temperature (inverse magnetocaloric effect).

For most magnetic materials $\partial_\theta M < 0$ and hence most often the direct effect occurs.

A process of refrigeration is associated with the direct effect. It is worth commenting on the physical picture of the process.
(1) Adiabatic magnetization. A magnetocaloric sample ($\partial_\theta M < 0$) is in an isolated environment. By applying the field H the magnetic dipoles tend to align, the magnetic entropy decreases. Since the whole entropy is constant ($\dot{\eta} = 0$) then the lattice entropy (and heat content) increases and the temperature increases.
(2) Isomagnetic transfer. While the field H is held constant, the added heat is removed.
(3) Adiabatic demagnetization. The external magnetic field is removed, the magnetic spin system becomes disordered and captures energy from the lattice, which decreases the temperature.

(4) Isomagnetic transfer. The magnetic field is held constant ($H = 0$) and the material is placed in thermal contact with the environment. Since the material is cooler then energy migrates into the magnetic material.

The entropy of a magnetic material can be expressed in the form

$$\eta(\theta, H) = \eta_m(\theta, H) + \eta_r(\theta) + \eta_e(\theta),$$

where η_m is the magnetic entropy, η_r is lattice entropy, and η_e is the electronic entropy (see, e.g. [416]). When a field H is applied under adiabatic conditions the total entropy remains unchanged. Since the magnetic entropy η_m decreases (ordered spin system) then lattice and electronic entropies increase to compensate. This in turn causes a temperature increase. When the external field is removed the magnetic spin system returns to the disordered arrangement by capturing energy from the lattice, which decreases the thermal entropy and the temperature decreases.

12.10 Plasmas

A plasma is a ionized gas with two electrically charged components, free electrons and ions. These particles generate electromagnetic fields through their elementary charges and currents. A macroscopic description of the plasma can be pursued in which emphasis is on its fluid nature. Depending on circumstances, this fluid description may be a one-fluid or a two-fluid approach.

The two-fluid description is more appropriate if plasma temperatures and energy densities are of interest; the electron and ion temperatures are often quite different due to the weak energy exchange rates between ions and electrons. Furthermore, the plasma might be weakly ionized and in this case the neutral particles (atoms) should be modelled as a further fluid.[33] In the two-fluid description the plasma is then modelled as a non-reacting binary fluid mixture; the ions and the electrons are treated as conducting fluids that are coupled through the interaction force and by Maxwell's equations. Instead, the one-fluid description is appropriate for dense ionized gases; electrons and ions move in such a way that no separation of charge occurs. Hence the motion can be described as a single conducting fluid with the usual variables of mass density, velocity, and pressure.

Two-fluid plasma theory

The subscripts e and i denote quantities pertaining to electrons and ions, respectively, while α stands for both e and i. Denote by n_e and n_i the number density of electrons and ions (per unit volume). Let m_e and $-e$ be the mass (at rest) and the electric charge of the electron. In addition, we assume that all ions are equal, with mass (at rest) m_i and electric charge ζe, where ζ is the number of electric charges per ion. Then the

[33] See [329] for a general model of plasmas as several chemically-reacting constituents.

mass and charge densities are given by

$$\rho_e = n_e m_e, \quad q_e = -n_e e, \qquad \rho_i = n_i m_i, \quad q_i = n_i \zeta e.$$

Each constituent of the plasma obeys the laws of conservation of mass and electric charge,

$$\partial_t \rho_\alpha + \nabla \cdot (\rho_\alpha \mathbf{v}_\alpha) = 0, \qquad \partial_t q_\alpha + \nabla \cdot (q_\alpha \mathbf{v}_\alpha) = 0, \tag{12.72}$$

both equations being accounted for by

$$\partial_t n_\alpha + \nabla \cdot (n_\alpha \mathbf{v}_\alpha) = 0. \tag{12.73}$$

Also, each constituent is subject to the balance of linear momentum,

$$\rho_\alpha \grave{\mathbf{v}}_\alpha = q_\alpha (\mathbf{E} + \mathbf{v}_\alpha \times \mathbf{B}) + \nabla \cdot \mathbf{T}_\alpha + \rho_\alpha \mathbf{b} + \mathbf{m}_\alpha, \tag{12.74}$$

or, equivalently,

$$\partial_t (\rho_\alpha \mathbf{v}_\alpha) + \nabla \cdot (\rho_\alpha \mathbf{v}_\alpha \otimes \mathbf{v}_\alpha) = q_\alpha (\mathbf{E} + \mathbf{v}_\alpha \times \mathbf{B}) + \nabla \cdot \mathbf{T}_\alpha + \rho_\alpha \mathbf{b} + \mathbf{m}_\alpha, \tag{12.75}$$

where $\mathbf{b}$, as is customary, is identified with the acceleration gravity. Here $\mathbf{m}_e$ and $\mathbf{m}_i$ are the interaction force densities, due to collisions, on electrons and ions. They are required to satisfy

$$\mathbf{m}_e + \mathbf{m}_i = \mathbf{0}. \tag{12.76}$$

Kinetic derivations of the balance equations indicate that [261]

$$\mathbf{m}_\alpha = \textstyle\sum_\beta \rho_\alpha (\mathbf{v}_\alpha - \mathbf{v}_\beta) \nu_{\alpha\beta}.$$

The requirement (12.76) would imply that $\rho_e \nu_{ei} = \rho_i \nu_{ie}$. The fields $\mathbf{E}$ and $\mathbf{B}$ are related to the charges and the motions by

$$\nabla \cdot \mathbf{E} = \textstyle\sum_\alpha q_\alpha, \qquad \nabla \times \mathbf{B} = \partial_t \mathbf{E} + \sum_\alpha q_\alpha \mathbf{v}_\alpha.$$

The balance of energy is written in the form[34]

$$\rho_\alpha (\varepsilon_\alpha + \tfrac{1}{2} \mathbf{v}_\alpha^2)\grave{} = \nabla \cdot (\mathbf{v}_\alpha \mathbf{T}_\alpha - \mathbf{q}_\alpha) + q_\alpha \mathbf{v}_\alpha \cdot \mathbf{E} + \rho_\alpha (r_\alpha + \mathbf{b} \cdot \mathbf{v}_\alpha) + l_\alpha,$$

where $\mathbf{q}_\alpha$ is the heat flux, r_α is the heat supply and l_α is the energy growth. The growths are required to satisfy

$$l_e + l_i = 0. \tag{12.77}$$

[34] Here the growths are denoted by l_α rather than e_α.

Equivalently we can write

$$\partial_t[\rho_\alpha(\varepsilon_\alpha + \tfrac{1}{2}\mathbf{v}_\alpha^2)] + \nabla\cdot[\rho_\alpha(\varepsilon_\alpha + \tfrac{1}{2}\mathbf{v}_\alpha^2)\mathbf{v}_\alpha] = \nabla\cdot(\mathbf{v}_\alpha\mathbf{T}_\alpha - \mathbf{q}_\alpha) + q_\alpha\mathbf{v}_\alpha\cdot\mathbf{E} + \rho_\alpha(r_\alpha + \mathbf{b}\cdot\mathbf{v}_\alpha) + l_\alpha. \tag{12.78}$$

By virtue of the equation of motion the balance of energy reduces to

$$\rho_\alpha\dot{\varepsilon}_\alpha = \mathbf{T}_\alpha\cdot\mathbf{L}_\alpha - \nabla\cdot\mathbf{q}_\alpha + \rho_\alpha r_\alpha + l_\alpha - \mathbf{m}_\alpha\cdot\mathbf{v}_\alpha. \tag{12.79}$$

12.10.1 One-Fluid Plasma Theory and Magnetohydrodynamics

A simpler model of plasma is obtained by using appropriate one-fluid variables. Consistent with the theory of mixtures we consider the mass density $\rho = \rho_e + \rho_i$ and the barycentric velocity $\mathbf{v} = (\rho_e\mathbf{v}_e + \rho_i\mathbf{v}_i)/\rho$. Moreover we consider the charge density and the current density

$$q = q_i + q_e = e(\zeta n_i - n_e), \qquad \mathbf{J} = q_e\mathbf{v}_e + q_i\mathbf{v}_i.$$

By adding Eqs. (12.72) over $\alpha = e, i$ we find

$$\partial_t\rho + \nabla\cdot(\rho\mathbf{v}) = 0, \qquad \partial_t q + \nabla\cdot\mathbf{J} = 0.$$

Moreover, adding (12.75) over α yields

$$\partial_t(\rho\mathbf{v}) + \nabla\cdot(\rho\mathbf{v}\otimes\mathbf{v}) = q\mathbf{E} + \mathbf{J}\times\mathbf{B} + \nabla\cdot\mathbf{T} + \rho\mathbf{b},$$

or,

$$\rho\dot{\mathbf{v}} = q\mathbf{E} + \mathbf{J}\times\mathbf{B} + \nabla\cdot\mathbf{T} + \rho\mathbf{b}, \tag{12.80}$$

where, as is general for mixtures,

$$\mathbf{T} = \textstyle\sum_\alpha(\mathbf{T}_\alpha - \rho_\alpha\mathbf{u}_\alpha\otimes\mathbf{u}_\alpha), \qquad \mathbf{u}_\alpha = \mathbf{v}_\alpha - \mathbf{v}.$$

The term $\mathbf{J}\times\mathbf{B}$ is referred to as *total Lorentz force.*

Since $\mathbf{J}$ involves the velocities of ions and electrons, the evolution of $\mathbf{J}$ is expected to be related to the equations of motion (12.75). For the sake of simplicity we look for the evolution equation under the assumption that the constituents are inviscid fluids so that

$$\mathbf{T}_\alpha = -p_\alpha\mathbf{1}.$$

We then regard $p = p_e + p_i$ as the pressure in the mixture (in the plasma) and write

$$\mathbf{T} = -p\mathbf{1} - \textstyle\sum_\alpha\rho_\alpha\mathbf{u}_\alpha\otimes\mathbf{u}_\alpha.$$

Multiply (12.74) by q_α/ρ_α to have

$$q_\alpha \dot{\mathbf{v}}_\alpha = \frac{q_\alpha^2}{\rho_\alpha}(\mathbf{E} + \mathbf{v}_\alpha \times \mathbf{B}) - \frac{q_\alpha}{\rho_\alpha}\nabla p_\alpha + q_\alpha \mathbf{b} + \frac{q_\alpha}{\rho_\alpha}\mathbf{m}_\alpha.$$

By virtue of the identity

$$q_\alpha \dot{\mathbf{v}}_\alpha = \partial_t(q_\alpha \mathbf{v}_\alpha) + \nabla \cdot (q_\alpha \mathbf{v}_\alpha \otimes \mathbf{v}_\alpha)$$

we compute the summation over α to obtain

$$\partial_t \mathbf{J} + \nabla \cdot \sum_\alpha q_\alpha \mathbf{v}_\alpha \otimes \mathbf{v}_\alpha = \sum_\alpha \frac{q_\alpha^2}{\rho_\alpha}(\mathbf{E} + \mathbf{v}_\alpha \times \mathbf{B}) - \sum_\alpha \frac{q_\alpha}{\rho_\alpha}\nabla p_\alpha + \sum_\alpha \frac{q_\alpha}{\rho_\alpha}\mathbf{m}_\alpha. \tag{12.81}$$

By replacing $\mathbf{v}_\alpha$ with $\mathbf{v} + \mathbf{u}_\alpha$ we find

$$\sum_\alpha q_\alpha \mathbf{v}_\alpha \otimes \mathbf{v}_\alpha = q\mathbf{v} \otimes \mathbf{v} + \mathbf{J} \otimes \mathbf{v} + \mathbf{v} \otimes \mathbf{J} + \sum_\alpha q_\alpha \mathbf{u}_\alpha \otimes \mathbf{u}_\alpha.$$

Hence, in view of the identity

$$\nabla \cdot (\mathbf{J} \otimes \mathbf{v} + \mathbf{v} \otimes \mathbf{J}) = \mathbf{J}\nabla \cdot \mathbf{v} + (\mathbf{v} \cdot \nabla)\mathbf{J} + \mathbf{v}\nabla \cdot \mathbf{J} + (\mathbf{J} \cdot \nabla)\mathbf{v},$$

we can write (12.81) in the form

$$\begin{aligned}\dot{\mathbf{J}} + \mathbf{L}\,\mathbf{J} + (\nabla \cdot \mathbf{v})\mathbf{J} + (\nabla \cdot \mathbf{J})\mathbf{v} + \nabla \cdot (q\mathbf{v} \otimes \mathbf{v}) &= \sum_\alpha \frac{q_\alpha^2}{\rho_\alpha}(\mathbf{E} + \mathbf{v}_\alpha \times \mathbf{B}) \\ &-\sum_\alpha \frac{q_\alpha}{\rho_\alpha}\nabla p_\alpha - \nabla \cdot \sum_\alpha q_\alpha \mathbf{u}_\alpha \otimes \mathbf{u}_\alpha + \sum_\alpha \frac{q_\alpha}{\rho_\alpha}\mathbf{m}_\alpha.\end{aligned} \tag{12.82}$$

Adding (12.78) over α and using the standard definitions of $\mathbf{T}$, $\mathbf{q}$, and ρr we obtain the balance of energy for the mixture (plasma) in the form

$$\rho \mathbf{v} \cdot \dot{\mathbf{v}} + \rho \dot{\varepsilon} = \nabla \cdot (\mathbf{T}\mathbf{v}) - \nabla \cdot \mathbf{q} + \rho r + \rho \mathbf{b} \cdot \mathbf{v} + \mathbf{J} \cdot \mathbf{E}.$$

In view of the equation of motion (12.80) it follows that

$$\rho \dot{\varepsilon} = \mathbf{T} \cdot \mathbf{D} - \nabla \cdot \mathbf{q} + \rho r + \mathbf{J} \cdot \mathbf{E} - \mathbf{v} \cdot (q\mathbf{E} + \mathbf{J} \times \mathbf{B}). \tag{12.83}$$

A suggestive form of (12.83) arises by using

$$\mathbf{h} = \mathbf{J} - q\mathbf{v}.$$

Substitution of $\mathbf{J}$ in (12.83) results in

$$\rho \dot{\varepsilon} = \mathbf{T} \cdot \mathbf{D} - \nabla \cdot \mathbf{q} + \rho r + \mathbf{h} \cdot (\mathbf{E} + \mathbf{v} \times \mathbf{B}). \tag{12.84}$$

Since $\mathbf{v}$ is the barycentric velocity,[35] then we can view $q\mathbf{v}$ as the convection current and $\mathbf{h}$ as the current relative to the barycentric framework. Quite consistently, the effective electric field $\mathbf{E} + \mathbf{v} \times \mathbf{B}$ is the field in the barycentric framework. It is of interest that (12.84) is consistent with the form obtained by means of the Poynting flux vector [98].

It is worth mentioning that Eqs. (12.80) and (12.84) look consistent with the corresponding equations in [88], once the current $\mathbf{h}$ is identified with $\mathbf{j}$ of [88]. Moreover, Eq. (9) of [88] amounts to letting $\rho\varepsilon = \frac{3}{2}p$ and observing that

$$\rho\dot{\varepsilon} = (\rho\varepsilon)^{\cdot} + \rho\varepsilon\nabla \cdot \mathbf{v}$$

so that (12.84) can be written as

$$(\tfrac{3}{2}p)^{\cdot} + \tfrac{5}{2}p\nabla \cdot \mathbf{v} = \mathbf{T}^d \cdot \mathbf{D} - \nabla \cdot \mathbf{q} + \rho r + \mathbf{h} \cdot (\mathbf{E} + \mathbf{v} \times \mathbf{B}), \qquad \mathbf{T}^d = \mathbf{T} + p\mathbf{1}.$$

If $\mathbf{T}^d$, $\mathbf{q}$, r, $\mathbf{E}$, and $\mathbf{B}$ vanish (adiabatic condition) then we have

$$0 = (\tfrac{3}{2}p)^{\cdot} + \tfrac{5}{2}p\nabla \cdot \mathbf{v} = \tfrac{3}{2}(\dot{p} - \tfrac{5}{3}p\dot{\rho}/\rho) = \tfrac{3}{2}\rho^{5/3}(p/\rho^{5/3})^{\cdot}.$$

This motivates why often the energy equation is taken in the form (adiabatic condition)

$$\frac{p}{\rho^{\gamma}} = \text{constant},$$

γ being the ratio of specific heats c_p/c_v; $\gamma = 5/3$ for monoatomic gases. Other times the equation is taken in the form

$$\mathbf{v} \cdot \nabla\eta = 0,$$

to state the constancy of entropy, in the adiabatic condition.

As is often reasonable, we now assume the plasma is *electrically neutral*, that is

$$q = q_e + q_i = 0, \qquad n_e = \zeta n_i.$$

Hence the equation of motion simplifies to

$$\partial_t(\rho\mathbf{v}) + \nabla \cdot (\rho\mathbf{v} \otimes \mathbf{v}) = \mathbf{J} \times \mathbf{B} + \nabla \cdot \mathbf{T} + \rho\mathbf{b},$$

where

$$\mathbf{J} = q_e(\mathbf{v}_e - \mathbf{v}_i) = n_e e(\mathbf{v}_i - \mathbf{v}_e).$$

Further, the neutrality assumption implies that $\nabla \cdot \mathbf{J} = 0$.

[35] Or, roughly, the velocity of ions.

Let

$$\gamma_e = \frac{e}{m_e}$$

be the charge-to-mass ratio of electrons. Hence it follows

$$\sum_\alpha \frac{q_\alpha^2}{\rho_\alpha} = |q_e|\gamma_e(1 + \zeta\frac{m_e}{m_i}), \qquad \sum_\alpha \frac{q_\alpha^2}{\rho_\alpha}\mathbf{v}_\alpha = |q_e|\gamma_e(\mathbf{v}_e + \zeta\frac{m_e}{m_i}\mathbf{v}_i),$$

$$-\sum_\alpha \frac{q_\alpha}{\rho_\alpha}\nabla p_\alpha = \gamma_e(\nabla p_e - \zeta\frac{m_e}{m_i}\nabla p_i), \qquad \sum_\alpha \frac{q_\alpha}{\rho_\alpha}\mathbf{m}_\alpha = -\gamma_e(1 + \zeta\frac{m_e}{m_i})\mathbf{m}_e.$$

Upon substitution in (12.82) we obtain the equation for $\mathbf{J}$ in the form

$$\dot{\mathbf{J}} + \mathbf{L}\,\mathbf{J} + (\nabla\cdot\mathbf{v})\mathbf{J} = |q_e|\gamma_e(1 + \zeta\frac{m_e}{m_i})\mathbf{E} + |q_e|\gamma_e(\mathbf{v}_e + \zeta\frac{m_e}{m_i}\mathbf{v}_i)\times\mathbf{B}$$
$$+\gamma_e(\nabla p_e - \zeta\frac{m_e}{m_i}\nabla p_i) - \nabla\cdot\sum_\alpha q_\alpha\mathbf{u}_\alpha\otimes\mathbf{u}_\alpha - \gamma_e(1 + \zeta\frac{m_e}{m_i})\mathbf{m}_e. \quad (12.85)$$

This equation may be viewed as the *generalized Ohm's law* for a plasma since it relates the current $\mathbf{J}$ to the electric field $\mathbf{E}$; the left-hand side has the same form of (9.69) for the mass flux $\mathbf{h}_\alpha$.

A simpler evolution equation for $\mathbf{J}$ follows as a consequence of further approximations. Since $m_i \geq 1836\, m_e$ then it is reasonable to neglect $\zeta m_e/m_i$ relative to unity. The pressures p_e, p_i are usually nearly equal and hence $p_e - \zeta(m_e/m_i)p_i \simeq p_e$. We then let $p = p_i + p_e \simeq 2p_e$. Equation (12.85) becomes

$$\dot{\mathbf{J}} + \mathbf{L}\,\mathbf{J} + (\nabla\cdot\mathbf{v})\mathbf{J} = |q_e|\gamma_e\mathbf{E} + |q_e|\gamma_e(\mathbf{v}_e + \zeta\frac{m_e}{m_i}\mathbf{v}_i)\times\mathbf{B} + \nabla\cdot\mathbf{S} - \gamma_e\mathbf{m}_e,$$

where

$$\mathbf{S} = \tfrac{1}{2}\gamma_e p\mathbf{1} - \sum_\alpha q_\alpha\mathbf{u}_\alpha\otimes\mathbf{u}_\alpha.$$

Since $\rho \simeq \rho_i$ then

$$\mathbf{v}_i + \frac{\zeta m_e}{m_i}\mathbf{v}_e \simeq \frac{1}{\rho}(\rho_i\mathbf{v}_i + n_i m_i\frac{\zeta m_e}{m_i}\mathbf{v}_e) = \frac{1}{\rho}(\rho_i\mathbf{v}_i + \rho_e\mathbf{v}_e) = \mathbf{v}.$$

Moreover,

$$\mathbf{v}_e + \frac{\zeta m_e}{m_i}\mathbf{v}_i = \mathbf{v}_i + \frac{\zeta m_e}{m_i}\mathbf{v}_e - (\mathbf{v}_i + \frac{\zeta m_e}{m_i}\mathbf{v}_e) + \mathbf{v}_e + \frac{\zeta m_e}{m_i}\mathbf{v}_i \simeq \mathbf{v} - (\mathbf{v}_i - \mathbf{v}_e)$$

and

$$|q_e|(\mathbf{v}_i - \mathbf{v}_e) = q_i\mathbf{v}_i + q_e\mathbf{v}_e = \mathbf{J}.$$

Hence we have

$$|q_e|\gamma_e(\mathbf{v}_e + \zeta\frac{m_e}{m_i}\mathbf{v}_i)\times\mathbf{B} = |q_e|\gamma_e\mathbf{v}\times\mathbf{B} - \gamma_e\mathbf{J}\times\mathbf{B}.$$

Further we let $\mathbf{m}_e$ be proportional to the velocity difference $\mathbf{v}_i - \mathbf{v}_e$,

$$\mathbf{m}_e = \beta(\mathbf{v}_i - \mathbf{v}_e) = \frac{\beta}{|q_e|}\mathbf{J}, \qquad \beta > 0.$$

Thus the generalized Ohm's law (12.85) simplifies to

$$\dot{\mathbf{J}} + \mathbf{L}\mathbf{J} + (\nabla\cdot\mathbf{v})\mathbf{J} + \frac{\gamma_e\beta}{|q_e|}\mathbf{J} = |q_e|\gamma_e(\mathbf{E} + \mathbf{v}\times\mathbf{B}) - \gamma_e\mathbf{J}\times\mathbf{B} + \nabla\cdot\mathbf{S}.$$

It is of interest to look at Ohm's law in the frame of ions, which amounts to neglecting terms in $\mathbf{v}_i$ and $\mathbf{v}$, relative to $\mathbf{v}_e$, and the derivatives $\mathbf{L}$ and $\nabla\cdot\mathbf{v}$. We also regard $\nabla\cdot(\sum_\alpha q_\alpha\mathbf{u}_\alpha\otimes\mathbf{u}_\alpha)$ negligible compared to $\gamma_e\nabla p_e$. Hence

$$\mathbf{J} = -|q_e|\mathbf{v}_e$$

and the equation for $\mathbf{J}$ reduces to

$$\dot{\mathbf{J}} + \frac{\beta}{|q_e|}\mathbf{J} = |q_e|(\mathbf{E} + \mathbf{v}\times\mathbf{B}) - \mathbf{J}\times\mathbf{B} + \nabla p_e.$$

In stationary conditions, $\dot{\mathbf{J}} = \mathbf{0}$, we have

$$\mathbf{J} = \sigma(\mathbf{E} + \mathbf{v}\times\mathbf{B}) - \frac{|q_e|}{\beta}\mathbf{J}\times\mathbf{B} + \frac{|q_e|}{\beta}\nabla p_e, \tag{12.86}$$

where $\sigma = |q_e|^2/\beta$ is the conductivity. In Drude's model of electric conduction, $\sigma = \gamma_e|q_e|\tau$, τ being the mean free time between collisions (of electrons).

It is worth remarking that Eq. (12.86) provides two classical physical effects, the *Seebeck effect* and the *Hall effect*. First, let p_e be a function of θ. Hence $\nabla p_e = p_e'\nabla\theta$ so that

$$\mathbf{J} = \sigma(\mathbf{E} + \mathbf{v}\times\mathbf{B}) - \frac{|q_e|}{\beta}\mathbf{J}\times\mathbf{B} + \sigma S\nabla\theta,$$

where S is the Seebeck coefficient (also called thermopower); here $S = p_e'/|q_e|$. The physical view is that the charged particles (here electrons) in the hotter region move faster than those in the colder region. Hence particles in the hotter region diffuse further and produce a higher density in the colder region. Hence $\mathbf{v}_e \propto -\nabla\theta$, $\mathbf{J} \propto \nabla\theta$. Because of the connection between temperature and electric current, the Seebeck effect, discovered by T.J. Seebeck in 1821, is called thermoelectric effect after H.C. Ørsted. The Hall effect, here expressed by $-(|q_e|/\beta)\mathbf{J}\times\mathbf{B}$, is the production of a voltage difference (the Hall voltage) across an electrical conductor, transverse to an electric current **J** in the conductor and to an applied magnetic field **B** (not parallel to **J**).

MHD equations

In magnetohydrodynamics (MHD for short) two further approximations are adopted. Both nonlinear terms $\sum_\alpha \rho_\alpha \mathbf{u}_\alpha \otimes \mathbf{u}_\alpha$ and $\sum_\alpha q_\alpha \mathbf{u}_\alpha \otimes \mathbf{u}_\alpha$ are neglected. The whole set of equations of MHD is then given by

$$\begin{gathered}\partial_t \rho + \nabla \cdot (\rho \mathbf{v}) = 0, \qquad \rho \dot{\mathbf{v}} = \mathbf{J} \times \mathbf{B} - \nabla p,\\ \dot{\mathbf{J}} + \mathbf{L}\,\mathbf{J} + (\nabla \cdot \mathbf{v})\mathbf{J} + (\gamma_e \beta / |q_e|)\mathbf{J} = |q_e| \gamma_e (\mathbf{E} + \mathbf{v} \times \mathbf{B}) - \gamma_e \mathbf{J} \times \mathbf{B} + \tfrac{1}{2}\gamma_e \nabla p,\\ \nabla \times \mathbf{E} = -\partial_t \mathbf{B}, \qquad \nabla \times \mathbf{B} = \mu_0 \mathbf{J} + \mu_0 \epsilon_0 \partial_t \mathbf{E}\end{gathered} \tag{12.87}$$

On the whole there are 13 equations in the 13 unknowns ρ, $\mathbf{v}$, $\mathbf{J}$, $\mathbf{E}$, and $\mathbf{B}$; it is understood that the pressure p is expressed in terms of the mass density ρ. Sometimes (see, e.g. [344]) the charge continuity equation $\partial_t q_e + \nabla \cdot \mathbf{J} = 0$ is added to the set of equations and then q is the 14-th unknown, though the electrical neutrality is assumed. If electrical neutrality is not assumed then the equation of motion comprises the force $q\,\mathbf{E}$ and Ohm's law is given by (12.82).

In the literature on MHD the term $\nabla \cdot (\mathbf{J} \otimes \mathbf{v} + \mathbf{v} \otimes \mathbf{J})$ is not considered and hence the left-hand-side of (12.87) is merely $\partial_t \mathbf{J} + \mathbf{J}/\tau$, where τ is an appropriate parameter, here $\tau = |q_e|/\gamma_e \beta$. In this sense the equation

$$\partial_t \mathbf{J} + \frac{1}{\tau}\mathbf{J} = |q_e| \gamma_e (\mathbf{E} + \mathbf{v} \times \mathbf{B}) - \gamma_e \mathbf{J} \times \mathbf{B} + \tfrac{1}{2}\gamma_e \nabla p$$

can be viewed as a rate-type equation for $\mathbf{J}$, similar to the Maxwell–Cattaneo equation, though with the partial time derivative. Perhaps that is why, on physical grounds, it is said that, for very low frequencies, one can ignore the $\partial_t \mathbf{J}$ term in the generalized Ohm's law, whereas for low temperatures the ∇p term can be ignored. In addition, it is assumed that the $\mathbf{J} \times \mathbf{B}$ term is negligible compared to the $\mathbf{v} \times \mathbf{B}$ term. Under these assumptions, Ohm's law becomes

$$\mathbf{J} = \sigma(\mathbf{E} + \mathbf{v} \times \mathbf{B}).$$

Hence the standard MHD is characterized by the set of equations

$$\begin{gathered}\partial_t \rho + \nabla \cdot (\rho \mathbf{v}) = 0, \qquad \rho \partial_t \mathbf{v} = \mathbf{J} \times \mathbf{B} - \nabla p,\\ \mathbf{J} = \sigma(\mathbf{E} + \mathbf{v} \times \mathbf{B}),\\ \nabla \times \mathbf{E} = -\partial_t \mathbf{B}, \qquad \nabla \times \mathbf{B} = \mu_0 \mathbf{J}.\end{gathered} \tag{12.88}$$

The set (12.88) comprises 13 equations in the 13 unknowns ρ, $\mathbf{v}$, $\mathbf{J}$, $\mathbf{E}$, and $\mathbf{B}$ while the pressure p is supposed to be a function of the density ρ. A comparison with kinetic derivations shows a satisfactory agreement. For instance, by (8.20) of [344], equality of the coefficients of $\mathbf{E}$ in the equation for $\partial_t \mathbf{J}$ requires that

$$|q_e|\gamma_e = \frac{\rho e^2}{m_e m_i},$$

which is true as far as $\rho \simeq n_i m_i = n_e m_i$.

A further approximation leads to *ideal MHD*. Finite values of $\mathbf{J}$ with infinite (very high) values of the conductivity σ are allowed if the effective electric field vanishes,

$$\mathbf{E} + \mathbf{v} \times \mathbf{B} = \mathbf{0}.$$

This in turn provides $\mathbf{E}$ in terms of $\mathbf{B}$,

$$\mathbf{E} = -\mathbf{v} \times \mathbf{B}.$$

The system of ideal MHD equations then reads

$$\begin{aligned} \partial_t \rho + \nabla \cdot (\rho \mathbf{v}) = 0, \qquad \rho \partial_t \mathbf{v} = \mathbf{J} \times \mathbf{B} - \nabla p, \\ \nabla \times (\mathbf{v} \times \mathbf{B}) = \partial_t \mathbf{B}, \qquad \nabla \times \mathbf{B} = \mu_0 \mathbf{J}. \end{aligned} \tag{12.89}$$

As to Eqs. (12.89) observe that the identity

$$(\nabla \times \mathbf{B}) \times \mathbf{B} = -\nabla \tfrac{1}{2} \mathbf{B}^2 + (\mathbf{B} \cdot \nabla)\mathbf{B}$$

allows the equation of motion to be written in the form

$$\rho \partial_t \mathbf{v} = -\nabla(p + \mathbf{B}^2/2\mu_0) + \frac{1}{\mu_0}(\mathbf{B} \cdot \nabla)\mathbf{B}. \tag{12.90}$$

If the field lines of $\mathbf{B}$ are straight and parallel, since $\nabla \cdot \mathbf{B} = 0$ then $(\mathbf{B} \cdot \nabla)\mathbf{B} = \mathbf{0}$. Thus the right-hand side reduces to $-\nabla(p + \mathbf{B}^2/2\mu_0)$. That is why $\mathbf{B}^2/2\mu_0$ is called the magnetic pressure (see [261], Sect. 3.8).

It is natural to expect that the standard Ohm's law is obtained from (12.82) via suitable approximations. As emphasized above, this is the case in stationary conditions, $\dot{\mathbf{J}} = \mathbf{0}$, with a negligible motion of the underlying body, $\mathbf{v} = \mathbf{0}$, negligible pressure gradients and negligible quadratic terms $q_\alpha \mathbf{u}_\alpha \otimes \mathbf{u}_\alpha$. We then let

$$\frac{q_e^2}{\rho_e}\mathbf{E} + \frac{q_e}{\rho_e}\mathbf{m}_e = 0.$$

If

$$\mathbf{m}_e = \lambda n_e n_i (\mathbf{v}_i - \mathbf{v}_e) \simeq -\lambda n_e n_i \mathbf{J}/q_e$$

then we have

$$\mathbf{J} = \sigma \mathbf{E}, \qquad \sigma = \frac{n_e e^2}{n_i \lambda}.$$

The term *generalized Ohm's law* is also used for the equation

$$\mathbf{E} = -\mathbf{v}\times\mathbf{B} + \frac{1}{\sigma}\mathbf{J} + \frac{1}{n_e e}(\mathbf{J}\times\mathbf{B} - \nabla p_e) + \frac{m_e}{n_e e^2}[\partial_t \mathbf{J} + \nabla\cdot(\mathbf{v}\otimes\mathbf{J} + \mathbf{J}\otimes\mathbf{v} - \frac{1}{n_e e}\mathbf{J}\otimes\mathbf{J})],$$

often referred to as Hall-MHD model [110, 397]. The comparison with (12.82) indicates the approximations

$$\sum_\alpha \frac{q_\alpha^2}{\rho_\alpha} \simeq \frac{q_e^2}{\rho_e} = \frac{n_e e^2}{m_e}, \qquad \sum_\alpha \frac{q_\alpha^2}{\rho_\alpha}\mathbf{v}_\alpha \simeq \frac{q_e^2}{\rho_e}\mathbf{v}_\alpha = \frac{n_e e^2}{m_e}\mathbf{v} + \frac{n_e e^2}{m_e}\mathbf{u}_e \simeq \frac{n_e e^2}{m_e}\mathbf{v} - \frac{e}{m_e}\mathbf{J}.$$

We know that $\mathbf{J}$ is defined to be $\sum_\alpha q_\alpha \mathbf{v}_\alpha \simeq q_e\mathbf{v}_e$. Hence it is quite ambiguous to use the same symbol, $\mathbf{J}$, to denote both $q_e\mathbf{v}_e$ and $q_e(\mathbf{v}_e - \mathbf{v}_i)$ unless we neglect the underlying (ion) motion. Rather, $\mathbf{J}/\sigma + (1/n_e e)\mathbf{J}\times\mathbf{B}$ could be replaced by

$$\frac{1}{\sigma}\mathbf{h} + \frac{1}{n_e e}\mathbf{h}\times\mathbf{B},$$

where $\mathbf{h} = q_e(\mathbf{v}_e - \mathbf{v}_i)$.

The balance of energy is sometimes written in the form [201]

$$\partial_t e + \nabla\cdot[(e + p + \mathbf{B}^2/2\mu_0)\mathbf{v}] = \mathbf{J}^2/\sigma,$$

where $e = \rho(\varepsilon + \frac{1}{2}\mathbf{v}^2)$. The flux $p\mathbf{v}$ as a contribution to the variation of the energy $\rho(\varepsilon + \frac{1}{2}\mathbf{v}^2)$ is standard, see (2.47). The occurrence of $(\mathbf{B}^2/2\mu_0)\mathbf{v}$ is less motivated. It emerges from (12.90) which merely motivates $\mathbf{B}^2/2\mu_0$ as an equivalent pressure to the effect of motion. The right-hand side term $\mathbf{J}^2/\sigma$ is just given by the power term $\mathbf{E}\cdot\mathbf{J}$ along with Ohm's law $\mathbf{E} = \mathbf{J}/\sigma$.

Thermodynamic consistency

Conceptually there are two scenarios for the statement of the second law according as we regard the plasma as a mixture or as a single (MHD) fluid. Within the mixture scenario we follow the assumption (9.41) and hence state that

$$\textstyle\sum_\alpha[\rho_\alpha\dot{\eta}_\alpha + \nabla\cdot\frac{\mathbf{q}_\alpha}{\theta_\alpha} + \nabla\cdot\mathbf{k}_\alpha - \rho_\alpha\frac{r_\alpha}{\theta_\alpha}] \geq 0$$

must hold for any set of constitutive functions compatible with the balance equations. Substitution of $\nabla\cdot\mathbf{q}_\alpha - \rho_\alpha r_\alpha$ from (12.79) results in

$$\textstyle\sum_\alpha\{\rho_\alpha\dot{\eta}_\alpha + \frac{1}{\theta_\alpha}(-\rho_\alpha + \mathbf{T}_\alpha\cdot\mathbf{L}_\alpha + l_\alpha - \mathbf{m}_\alpha\cdot\mathbf{v}_\alpha) - \frac{1}{\theta_\alpha^2}\mathbf{q}_\alpha\cdot\nabla\theta_\alpha + \nabla\cdot\mathbf{k}_\alpha\} \geq 0.$$

Using the free energy $\psi_\alpha = \varepsilon_\alpha - \theta_\alpha\eta_\alpha$ we can write the inequality in the form

$$\sum_\alpha \{\frac{1}{\theta_\alpha}[-\rho_\alpha(\dot{\psi}_\alpha + \eta_\alpha \dot{\theta}_\alpha) + \mathbf{T}_\alpha \cdot \mathbf{L}_\alpha + l_\alpha - \mathbf{m}_\alpha \cdot \mathbf{v}_\alpha] - \frac{1}{\theta_\alpha^2}\mathbf{q}_\alpha \cdot \nabla\theta_\alpha + \nabla \cdot \mathbf{k}_\alpha\} \geq 0.$$

A simple thermodynamically-consistent model is obtained by letting each constituent be a Navier–Stokes–Fourier fluid. Hence we let $\mathbf{k}_\alpha = \mathbf{0}$ and

$$\psi_\alpha = \psi_\alpha(\theta_\alpha, \rho_\alpha)$$

and set

$$p_\alpha = \rho_\alpha^2 \partial_{\rho_\alpha}\psi_\alpha, \qquad \eta_\alpha - \partial_{\theta_\alpha}\psi_\alpha,$$

$$\mathbf{T}_\alpha = -p_\alpha \mathbf{1} + 2\mu_\alpha \mathbf{D}_\alpha + \lambda_\alpha(\nabla \cdot \mathbf{v}_\alpha)\mathbf{1}, \qquad \mathbf{q}_\alpha = -\kappa_\alpha \nabla\theta_\alpha,$$

where $\mu_\alpha \geq 0,\ 2\mu_\alpha + 3\lambda_\alpha \geq 0$, and $\kappa_\alpha \geq 0$ while

$$\sum_\alpha \frac{1}{\theta_\alpha}(l_\alpha - \mathbf{m}_\alpha \cdot \mathbf{v}_\alpha) \geq 0.$$

As shown in Chap. 5, we can take $\mathbf{m}_\alpha$ and l_α in the form

$$\mathbf{m}_\alpha = \sum_\beta M_{\alpha\beta}(\mathbf{v}_\beta - \mathbf{v}_\alpha), \qquad l_\alpha = \sum_\beta N_{\alpha\beta}(\theta_\beta - \theta_\alpha) + \mathbf{v} \cdot \sum_\beta M_{\alpha\beta}(\mathbf{v}_\beta - \mathbf{v}_\alpha),$$

where M and N are symmetric matrices. Hence we have

$$\sum_\alpha \frac{1}{\theta_\alpha}(l_\alpha - \mathbf{m}_\alpha \cdot \mathbf{v}_\alpha) = \sum_{\alpha,\beta} \frac{N_{\alpha\beta}}{\theta_\alpha\theta_\beta}(\theta_\beta - \theta_\alpha)^2 + \sum_{\alpha,\beta} M_{\alpha\beta}(\mathbf{v}_\beta - \mathbf{v}_\alpha) \cdot \frac{\mathbf{u}_\alpha}{\theta_\alpha}.$$

The first term is non-negative provided $N_{\alpha\beta} \geq 0$, $\forall \alpha \neq \beta$. As with binary mixtures, it is $\rho_e \mathbf{u}_e + \rho_i \mathbf{u}_i = 0$ and hence we find that

$$\sum_{\alpha,\beta} M_{\alpha\beta}(\mathbf{v}_\beta - \mathbf{v}_\alpha) \cdot \frac{\mathbf{u}_\alpha}{\theta_\alpha} = M_{ei}(1 + \frac{\rho_e}{\rho_i})(\frac{1}{\theta_e} + \frac{\rho_e}{\rho_i \theta_i})\mathbf{u}_e^2.$$

Accordingly $M_{ei} = M_{ie} \geq 0$ guarantees the thermodynamic requirement.

The whole set of differential equations governing the evolution of the plasma, subject to given fields $\mathbf{E}$, $\mathbf{B}$, and $\mathbf{b}$, r_α is then expressed by

$$\dot{n}_\alpha + n_\alpha \nabla \cdot \mathbf{v}_\alpha = 0,$$

$$\rho_\alpha \dot{\mathbf{v}}_\alpha = q_\alpha(\mathbf{E} + \mathbf{v}_\alpha \times \mathbf{B}) - \nabla \cdot \mathbf{T}_\alpha + \rho_\alpha \mathbf{b} + \sum_\beta M_{\alpha\beta}(\mathbf{v}_\beta - \mathbf{v}_\alpha),$$

$$\rho_\alpha \dot{\varepsilon}_\alpha = \mathbf{T}_\alpha \cdot \mathbf{D}_\alpha - \nabla \cdot \mathbf{q}_\alpha + \rho_\alpha r_\alpha + \sum_\beta N_{\alpha\beta}(\theta_\beta - \theta_\alpha) - \mathbf{u}_\alpha \cdot \sum_\beta M_{\alpha\beta}(\mathbf{u}_\beta - \mathbf{u}_\alpha),$$

with $\alpha = e, i$. So we have a closed system of 10 equations in the 10 unknowns $n_e, n_i, \mathbf{v}_e, \mathbf{v}_i, \theta_e, \theta_i$. If $\mathbf{E}$ and $\mathbf{B}$ are unknown then we complete the system by inserting the Maxwell equations.

The MHD equations have different forms due to the various approximations made. The evolution of $\mathbf{J}$ is assumed to be given by an approximation of (12.85) where the right-hand side is replaced by a scalar times $\mathbf{E}'$, $\mathbf{E}' = \mathbf{E} + \mathbf{v} \times \mathbf{B}$. The evolution equation for $\mathbf{q}$, instead, is regarded as a constitutive equation. We then have to take an objective time rate of $\mathbf{q}$; for definiteness we adopt the Jaumann time rate. The whole system of equations is then taken in the form

$$\dot{\rho} + \rho \nabla \cdot \mathbf{v} = 0, \qquad \rho \dot{\mathbf{v}} = \mathbf{J} \times \mathbf{B} - \nabla p + \rho \mathbf{b},$$

$$\tau_J(\rho, \theta)(\dot{\mathbf{J}} + \mathbf{L}\mathbf{J} + (\nabla \cdot \mathbf{v})\mathbf{J}) + \mathbf{J} = \sigma(\rho, \theta)\mathbf{E}', \qquad \tau_q(\rho, \theta)(\dot{\mathbf{q}} - \mathbf{W}\mathbf{q}) + \mathbf{q} = -\kappa(\rho, \theta)\nabla\theta,$$

$$\rho \dot{\varepsilon} = \mathbf{T} \cdot \mathbf{D} - \nabla \cdot \mathbf{q} + \rho r + \mathbf{J} \cdot \mathbf{E}',$$

$$\nabla \times \mathbf{E} = -\partial_t \mathbf{B}, \qquad \nabla \times \mathbf{B} = \mu_0 \mathbf{J} + \mu_0 \epsilon_0 \partial_t \mathbf{E},$$

where $\tau_j, \tau_q > 0$. The fields $\mathbf{E}$ and $\mathbf{B}$ are subject only to Maxwell's equations; no constitutive properties are considered for them. The standard assumption on the entropy inequality and substitution of $-\nabla \cdot \mathbf{q} + \rho r$ from the balance of energy yield

$$-\rho(\dot{\psi} + \eta\dot{\theta}) - p\nabla \cdot \mathbf{v} + \mathbf{T}^d \cdot \mathbf{D} + \mathbf{J} \cdot \mathbf{E}' - \frac{1}{\theta}\mathbf{q} \cdot \nabla\theta \geq 0,$$

where $\mathbf{T}^d = \mathbf{T} + p(\rho, \theta)\mathbf{1}$. We let $\psi, \eta, \mathbf{T}$ be functions of

$$\rho, \theta, \mathbf{J}, \mathbf{q}, \mathbf{D}, \nabla\theta$$

while $\dot{\mathbf{J}}$ and $\dot{\mathbf{q}}$ are given by the corresponding rate equations. Moreover ψ is differentiable. Hence, upon evaluation of $\dot{\psi}$ and substitution of $\dot{\mathbf{J}}$ and $\dot{\mathbf{q}}$, the entropy inequality becomes

$$\begin{aligned}(-\rho\partial_\rho\psi + \frac{p}{\rho} - \tfrac{2}{3}\partial_{\mathbf{J}}\psi \cdot \mathbf{J})\dot{\rho} - \rho(\partial_\theta\psi + \eta)\dot{\theta} - \rho\partial_{\mathbf{D}}\psi \cdot \dot{\mathbf{D}} - \rho\partial_{\nabla\theta}\psi \cdot (\nabla\theta)^{\cdot} \\ +(\mathbf{T}^d + \rho\partial_{\mathbf{J}}\psi \otimes \mathbf{J} - \tfrac{1}{3}\rho\partial_{\mathbf{J}}\psi \cdot \mathbf{J}\,\mathbf{1}) \cdot \mathbf{D} - \frac{1}{\theta}\mathbf{q} \cdot \nabla\theta \\ -\rho\partial_{\mathbf{J}}\psi \cdot (-\mathbf{W}\mathbf{J} - \frac{1}{\tau_J}\mathbf{J} + \frac{\sigma}{\tau_J}\mathbf{E}') + \mathbf{J} \cdot \mathbf{E}' - \rho\partial_{\mathbf{q}}\psi \cdot (\mathbf{W}\mathbf{q} - \frac{1}{\tau_q}\mathbf{q} - \frac{\kappa}{\tau_q}\nabla\theta) \geq 0.\end{aligned}$$

We let $\mathbf{T}^d + \rho\partial_{\mathbf{J}}\psi \otimes \mathbf{J} - \frac{1}{3}\rho\partial_{\mathbf{J}}\psi \cdot \mathbf{J}\,\mathbf{1} = 0(|\mathbf{D}|)$. At any point $\mathbf{x}$ and time t the fields $\dot{\rho}, \dot{\theta}, \dot{\mathbf{D}}$, and $(\nabla\theta)^{\cdot}$ can take arbitrary values, independent of those of $\rho, \theta, \mathbf{D}, \mathbf{g}, \mathbf{J}, \mathbf{q}$. Hence the inequality holds only if

$$\partial_{\mathbf{D}}\psi = \mathbf{0}, \quad \partial_{\nabla\theta}\psi = \mathbf{0}, \quad p = \rho^2\partial_\rho\psi + \tfrac{2}{3}\rho\partial_{\mathbf{J}}\psi\cdot\mathbf{J}, \quad \eta = -\partial_\theta\psi.$$

The skew-symmetry of $\mathbf{W}$ implies that

$$\partial_{\mathbf{J}}\psi\otimes\mathbf{J} - \partial_{\mathbf{q}}\psi\otimes\mathbf{q} \in \mathrm{Sym}.$$

The remaining inequality holds only if

$$\frac{\sigma\rho}{\tau_J}\partial_{\mathbf{J}}\psi = \mathbf{J}, \qquad \frac{\kappa\rho}{\tau_q}\partial_{\mathbf{q}}\psi = \frac{1}{\theta}\mathbf{q},$$

$$\frac{\rho}{\tau_J}\partial_{\mathbf{J}}\psi\cdot\mathbf{J} + \frac{\rho}{\tau_q}\partial_{\mathbf{q}}\psi\cdot\mathbf{q} + (\mathbf{T}^d + \rho\partial_{\mathbf{J}}\psi\otimes\mathbf{J})\cdot\mathbf{D} \geq 0.$$

The obvious integrations yield

$$\psi = \Psi(\rho,\theta) + \frac{\tau_J}{2\rho\sigma}\mathbf{J}^2 + \frac{\tau_q}{2\rho\kappa\theta}\mathbf{q}^2.$$

Hence it follows that $\partial_{\mathbf{J}}\psi\otimes\mathbf{J},\ \partial_{\mathbf{q}}\psi\otimes\mathbf{q} \in \mathrm{Sym}$, separately, and that $\sigma, \kappa > 0$. The requirement $(\mathbf{T}^d + \rho\partial_{\mathbf{J}}\psi\otimes\mathbf{J} - \frac{1}{3}\rho\partial_{\mathbf{J}}\psi\cdot\mathbf{J}\,\mathbf{1})\cdot\mathbf{D} \geq 0$ holds if $\mathbf{T}^d$ comprises a viscous stress,

$$\mathbf{T}^d = -\rho(\partial_{\mathbf{J}}\psi\otimes\mathbf{J} - \tfrac{1}{3}\partial_{\mathbf{J}}\psi\cdot\mathbf{J}\,\mathbf{1}) + 2\mu(\rho,\theta)\mathbf{D} + \lambda(\rho,\theta)\nabla\cdot\mathbf{v}\,\mathbf{1}, \qquad \mu\geq 0,\ 2\mu+3\lambda\geq 0.$$

If, instead, as is often assumed in the thermodynamics of irreversible processes,

$$\tau\dot{\mathbf{J}} + \mathbf{J} = \sigma\mathbf{E}$$

then it follows that

$$p = \rho^2\partial_\rho\psi, \qquad \mathbf{T}^d = 2\mu(\rho,\theta)\mathbf{D} + \lambda(\rho,\theta)\nabla\cdot\mathbf{v}\,\mathbf{1}$$

while the remaining results hold unchanged, even though $\dot{\mathbf{J}}$ is replaced by the corotational derivative $\dot{\mathbf{J}} - \mathbf{W}\mathbf{J}$.

Linear plasma and MHD waves

In this section we follow linear approximations in the sense that we let the pertinent quantities, like n_e, $\mathbf{v}_e$, $\mathbf{E}$, be the sum of equilibrium values and disturbances, e.g.

$$n_e = n_0 + \tilde{n}(\mathbf{x}, t),$$

and examine the pertinent equations with only linear terms in the disturbances. We first examine oscillations and waves in an unmagnetized plasma, namely in a plasma where **E** and **B** are produced by the presence and the motion of electrons and ions.

We begin with the so-called *plasma oscillations* that is the motion of electrons under the action of the self-generated electric field. We let n_0 be the density at equilibrium, that is in the configuration $n_i = n_e = n_0$, $\mathbf{v}_e = \mathbf{0}$, $\mathbf{E} = \mathbf{0}$. The electric field **E** satisfies Poisson's equation

$$\nabla \cdot \mathbf{E} = \frac{q}{\epsilon_0}.$$

Hence the continuity equation, the equation of motion and Poisson's equation, in the linear approximation, result in

$$\partial_t \tilde{n} + n_0 \nabla \cdot \mathbf{v}_e = 0, \qquad \partial_t \mathbf{v}_e = -\gamma_e \mathbf{E}, \qquad \nabla \cdot \mathbf{E} = -\frac{\tilde{n} e}{\epsilon_0}.$$

The time derivative of the first equation and the divergence of the second one allows us to write

$$\partial_t^2 \tilde{n} = \frac{n_0 e^2}{\epsilon_0 m} \tilde{n}.$$

Hence the density $\tilde{n}$ oscillates with the (angular) frequency ω_p given by

$$\omega_p^2 = \frac{n_0 e^2}{\epsilon_0 m}.$$

The frequency $\omega_p > 0$ is called the *plasma frequency.*

We now look for electromagnetic waves by regarding **E** and **B** as produced within the plasma. The pertinent equations are Maxwell's equations,

$$\nabla \cdot \mathbf{B} = 0, \quad \nabla \cdot \mathbf{E} = -\frac{\tilde{n} e}{\epsilon_0}, \quad \nabla \times \mathbf{E} = -\partial_t \mathbf{B}, \quad \nabla \times \mathbf{B} = \mu_0 (\mathbf{J} + \epsilon_0 \partial_t \mathbf{E}),$$

and the approximations

$$\mathbf{J} = -n_0 e \mathbf{v}_e, \quad \partial_t \tilde{n} + n_0 \nabla \cdot \mathbf{v}_e = 0, \quad m_e \partial_t \mathbf{v}_e = -e(\mathbf{E} + \mathbf{v}_e \times \mathbf{B})$$

in the unknowns **E**, **B**, $\tilde{n}$, $\mathbf{v}_e$ (and **J**). The approximation of the equation of motion is, in a sense, consistent with (12.87). If we neglect $\mathbf{L}\mathbf{J} + (\nabla \cdot \mathbf{v})\mathbf{J} + (\gamma_e \beta / |q_e|)\mathbf{J}$ and $\gamma_e \nabla p_e$ in (12.87) then we are left with

$$\dot{\mathbf{J}} = |q_e| \gamma_e (\mathbf{E} + \mathbf{v}_e \times \mathbf{B}).$$

Upon substitution of $\mathbf{J} = -n_0 e \mathbf{v}_e$ we get the (linearized) equation of motion; $\mathbf{v}_e \times \mathbf{B}$ is in fact nonlinear and then has to be neglected too. By applying $\nabla \times$ to $m_e \partial_t \mathbf{v}_e$ we find that

$$m_e \partial_t \nabla \times \mathbf{v}_e = -e\nabla \times \mathbf{E} = e\partial_t \mathbf{B}.$$

To within an inessential function of the position $\mathbf{x}$ we can write

$$m_e \nabla \times \mathbf{v}_e = e\mathbf{B}.$$

Moreover by applying $\nabla\times$ to $\nabla \times \mathbf{B}$ and ∂_t to $\nabla \times \mathbf{E}$ we then obtain

$$\Delta \mathbf{B} = \mu_0 n_0 \frac{e^2}{m_e} \mathbf{B} + \mu_0 \epsilon_0 \partial_t^2 \mathbf{B}.$$

Time harmonic wave solutions $\exp[i(\mathbf{k}\cdot\mathbf{x} - \omega t)]$ occur subject to the dispersion relation

$$k^2 = \mu_0\epsilon_0(\omega^2 - \omega_p^2),$$

where $k = |\mathbf{k}|$. Since $\mu_0\epsilon_0 = 1/c^2$, c being the speed of light in free space, we can write

$$\omega^2 = k^2c^2 + \omega_p^2.$$

At high frequencies, $\omega \gg \omega_p$, it is $\omega \simeq ck$ so that waves propagate at the speed of light (in free space). Instead, the wave vector $\mathbf{k}$ becomes imaginary[36] when $|\omega| < \omega_p$; in this case the wave form is

$$\exp(-k_i \mathbf{n}\cdot\mathbf{x})\exp(-i\omega t)$$

where $k_i = |\mathbf{k}_i|$ and $\mathbf{n}$ is any unit vector. The wave does not propagate and the disturbance damps as $\exp(-k_i\mathbf{n}\cdot\mathbf{x})$. We say that the waves have a cut-off at $\omega = \omega_p$. The refractive index $n = c/(\omega/k)$ takes the form

$$n = \sqrt{1 - (\omega_p^2/\omega^2)}.$$

The relation

$$k^2 = \mu_0\omega^2\epsilon_0(1 - \omega_p^2/\omega^2)$$

may be regarded as a dispersion relation

$$k^2 = \mu_0\epsilon\omega^2$$

with the dielectric constant

$$\epsilon = \epsilon_0(1 - \omega_p^2/\omega^2).$$

---

[36] Waves with complex-valued wave vectors $\mathbf{k}$ are called inhomogeneous waves [85].

By allowing for a pressure gradient we find plasma oscillations in a form that is referred to as *Langmuir waves.* The whole set of (linearized) equations is

$$\nabla \cdot \mathbf{B} = 0, \quad \nabla \cdot \mathbf{E} = -\frac{\tilde{n}e}{\epsilon_0}, \quad \nabla \times \mathbf{E} = -\partial_t \mathbf{B}, \quad \nabla \times \mathbf{B} = \mu_0(\mathbf{J} + \epsilon_0 \partial_t \mathbf{E}),$$

and the approximations

$$\mathbf{J} = -n_0 e \mathbf{v}_e, \quad \partial_t \tilde{n} + n_0 \nabla \cdot \mathbf{v}_e = 0, \quad n_0 m_e \partial_t \mathbf{v}_e = -n_0 e \mathbf{E} - \nabla p_e.$$

Again the equation of motion, with the insertion of the pressure term $-\nabla p_e$ is consistent with (12.87) in the approximation

$$\partial_t \mathbf{J} = |q_e| \gamma_e \mathbf{E} + \gamma_e \nabla p_e$$

as it follows by replacing $\mathbf{J}$ and dividing by γ_e.

The pressure p_e is regarded as a function of $\rho_e = n_e m_e$, i.e. $p_e(\mathbf{x}, t) = \hat{p}_e(\rho_e(\mathbf{x}, t))$. Let $p_e(\mathbf{x}, t) = p_0 + \tilde{p}(\mathbf{x}, t)$ so that

$$n_0 m_e \partial_t \mathbf{v}_e = -n_0 e \mathbf{E} - \nabla \tilde{p}.$$

Compute the divergence and replace $\nabla \cdot \mathbf{v}_e = -\partial_t \tilde{n}/n_0$. Moreover let

$$\tilde{p} = c_s^2 \tilde{\rho}_e = c_s^2 m_e \tilde{n},$$

where $c_s^2 = dp_e/d\rho_e$ and c_s is the sound speed in the electron gas. Hence we find

$$-\partial_t^2 \tilde{n} = \omega_p^2 \tilde{n} - c_s^2 \Delta \tilde{n}.$$

Time harmonic wave solutions $\exp[i(\mathbf{k} \cdot \mathbf{x} - \omega t)]$ for $\tilde{n}$ occur with

$$\omega^2 = \omega_p^2 + k^2 c_s^2.$$

When the action of the electron pressure is taken into account, we say that "warm" electrons are considered. In the Bohm–Gross model, the dependence of p_e on n_e is taken in the form

$$\tilde{p} = 3\tilde{n} k_B \theta_e,$$

where k_B is Boltzmann's constant and θ_e is the uniform electron temperature. Hence it follows that $c_s^2 = 3k_B\theta_e/m_e$. Upon substitution we can write

$$\omega^2 = \omega_p^2 + (3k_B\theta_e/m_e)k^2,$$

usually referred to as the Bohm–Gross dispersion relation for Langmuir waves. Owing to the pressure term, longitudinal (Langmuir waves) occur as we see at once by the equation of motion,

$$i\,\omega\, n_0\, m_e\, \mathbf{v}_e = n_0\, e\, \mathbf{E} + i\, \tilde{p}\, \mathbf{k}.$$

More involved wave features occur in *magnetized plasmas* where the magnetic induction is the sum of a given constant field $\mathbf{B}_0$ and a space-time dependent $\boldsymbol{\mathcal{B}}$ which is produced in the plasma by the motion of electrons and ions. To fix ideas let $\mathbf{e}_3$ be the unit vector of $\mathbf{B}_0$. Since $0 = \nabla \cdot (\mathbf{B}_0 + \boldsymbol{\mathcal{B}}) = \nabla \cdot \boldsymbol{\mathcal{B}}$ then

$$\mathbf{k} \cdot \boldsymbol{\mathcal{B}} = 0, \qquad \boldsymbol{\mathcal{B}} = \mathcal{B}_1 \mathbf{e}_1 + \mathcal{B}_2 \mathbf{e}_2.$$

The continuity and the motion equations along with Faraday's and Ampere's equations are written in the form

$$\partial_t \tilde{n}_\alpha + n_{\alpha 0} \nabla \cdot \mathbf{v}_\alpha = 0, \qquad \partial_t \mathbf{v}_\alpha = \operatorname{sgn} q_\alpha\, \gamma_\alpha (\mathbf{E} + \mathbf{v}_\alpha \times \mathbf{B}_0),$$

$$\nabla \times \mathbf{E} = \partial_t \boldsymbol{\mathcal{B}}, \qquad \nabla \times \boldsymbol{\mathcal{B}} = \mu_0 \textstyle\sum_\alpha q_\alpha \mathbf{v}_\alpha + \mu_0 \epsilon_0 \partial_t \mathbf{E}.$$

Time differentiation of the equation of motion allows us to write

$$\partial_t^2 \mathbf{v}_\alpha = \operatorname{sgn} q_\alpha\, \gamma_\alpha (\partial_t \mathbf{E} + \partial_t \mathbf{v}_\alpha \times \mathbf{B}_0) = \operatorname{sgn}\, q_\alpha \gamma_\alpha \partial_t \mathbf{E} + \gamma_\alpha^2 \mathbf{E} \times \mathbf{B}_0 - \gamma_\alpha^2 B_0^2 \mathbf{v}_{\alpha\perp},$$

where $\mathbf{v}_{\alpha\perp} = \mathbf{v}_\alpha - (\mathbf{B}_0 \cdot \mathbf{v}_\alpha)\mathbf{B}_0 / B_0^2$ is the component of $\mathbf{v}_\alpha$ perpendicular to $\mathbf{B}_0$. Let

$$\omega_{c\alpha} = \gamma_\alpha B_0;$$

$\omega_{c\alpha}$ is called the *cyclotron frequency* (of the constituent α). Consequently, at a time-harmonic wave, $\mathbf{v}_\alpha \propto \exp(-i\omega t)$,

$$(\omega_{c\alpha}^2 - \omega^2)\mathbf{v}_{\alpha\perp} = -i\omega \operatorname{sgn} q_\alpha\, \gamma_\alpha \mathbf{E}_\perp + \gamma_\alpha^2 \mathbf{E} \times \mathbf{B}_0, \qquad -\omega^2 v_{\alpha 3} = -i\omega \operatorname{sgn} q_\alpha \gamma_\alpha E_3.$$

The components $v_{\alpha 1}$, $v_{\alpha 2}$, and $v_{\alpha 3}$ can then be given by

$$v_{e1} = -\frac{\gamma_e}{\omega} \frac{iE_1 + (\omega_{ce}/\omega)E_2}{1 - (\omega_{ce}/\omega)^2}, \qquad v_{e2} = \frac{\gamma_e}{\omega} \frac{(\omega_{ce}/\omega)E_1 - iE_2}{1 - (\omega_{ce}/\omega)^2}, \qquad v_{e3} = \frac{i\gamma_e}{\omega} E_3$$

and

$$v_{i1} = -\frac{\gamma_i}{\omega} \frac{-iE_1 + (\omega_{ci}/\omega)E_2}{1 - (\omega_{ci}/\omega)^2}, \qquad v_{i2} = \frac{\gamma_e}{\omega} \frac{(\omega_{ci}/\omega)E_1 + iE_2}{1 - (\omega_{ci}/\omega)^2}, \qquad v_{i3} = -\frac{i\gamma_i}{\omega} E_3$$

Now, with a view to Ampere's law $\nabla \times \boldsymbol{\mathcal{B}} = \mu_0(\mathbf{J} + \epsilon_0 \partial_t \mathbf{E})$, we observe that

$$\mathbf{J} + \epsilon_0 \partial_t \mathbf{E} = \sum_\alpha \rho_{\alpha 0} \gamma_\alpha \mathbf{v}_\alpha - i\omega\epsilon_0 \mathbf{E} = -i\omega\epsilon_0 \boldsymbol{\epsilon} \mathbf{E},$$

where

$$\epsilon = \begin{bmatrix} \epsilon_1 & i\epsilon_2 & 0 \\ -i\epsilon_2 & \epsilon_1 & 0 \\ 0 & 0 & \epsilon_3 \end{bmatrix}$$

and

$$\epsilon_1 = 1 + \frac{\omega_{pe}^2}{\omega_{ce}^2 - \omega^2} + \frac{\omega_{pi}^2}{\omega_{ci}^2 - \omega^2}, \quad \epsilon_2 = \frac{\omega_{ce}}{\omega}\frac{\omega_{pe}^2}{\omega_{ce}^2 - \omega^2} - \frac{\omega_{ci}}{\omega}\frac{\omega_{pi}^2}{\omega_{ci}^2 - \omega^2}, \quad \epsilon_3 = 1 - \frac{\omega_{pe}^2}{\omega^2} - \frac{\omega_{pi}^2}{\omega^2}.$$

The plasma can then be viewed as a dielectric with dielectric tensor $\epsilon_0 \epsilon$.

Since

$$\frac{\omega_{pi}}{\omega_{pe}} \simeq \sqrt{\frac{m_e}{m_i}}$$

in many applications, at high frequencies, the entries of ϵ can be simplified to

$$\epsilon_1 = 1 + \frac{\omega_{pe}^2}{\omega_{ce}^2 - \omega^2}, \qquad \epsilon_2 = \frac{\omega_{ce}}{\omega}\frac{\omega_{pe}^2}{\omega_{ce}^2 - \omega^2}, \qquad \epsilon_3 = 1 - \frac{\omega_{pe}^2}{\omega^2}.$$

We now look for wave solutions in an unbounded magnetized plasma. Faraday's and Ampere's equations are written in the form

$$\nabla \times \mathbf{E} = -\partial_t \mathcal{B}, \qquad \nabla \times \mathcal{B} = \mu_0 \epsilon_0 \epsilon \partial_t \mathbf{E},$$

where the dielectric tensor accounts for the motions of electrons and ions. Hence

$$\nabla \times (\nabla \times \mathbf{E}) = -\frac{1}{c^2} \epsilon \partial_t^2 \mathbf{E},$$

where $c^2 = 1/\mu_0 \epsilon_0$, c being the light speed in free space.

A time-harmonic plane wave is considered in the form

$$\mathbf{E}(\mathbf{x}, t) = \hat{\mathbf{E}} \exp[i(\mathbf{k} \cdot \mathbf{x} - \omega t)].$$

Upon substitution we see that $\mathbf{E}(\mathbf{x}, t)$ is a solution if

$$(\mathbf{k} \cdot \hat{\mathbf{E}})\mathbf{k} - \mathbf{k}^2 \hat{\mathbf{E}} = -\frac{\omega^2}{c^2} \epsilon \hat{\mathbf{E}}.$$

Maxwell's equations

$$\nabla \cdot \mathbf{E} = 0, \qquad \nabla \cdot \mathcal{B} = 0, \qquad \nabla \times \mathbf{E} = -\partial_t \mathcal{B}$$

imply that $\mathbf{E}$ and $\mathcal{B}$ are perpendicular to $\mathbf{k}$ and to each other,

$$\mathbf{k} \cdot \mathbf{E} = 0, \qquad \mathbf{k} \cdot \boldsymbol{\mathcal{B}} = 0, \qquad \mathbf{E} \cdot \boldsymbol{\mathcal{B}} = 0.$$

We let $\mathbf{k}$ be parallel to $\mathbf{B}_0 = B_0\mathbf{e}_3$ so that the waves propagate parallel to the magnetic field (along $\mathbf{B}_0$). Hence $E_3 = 0$, $\mathcal{B}_3 = 0$. The vector $\hat{\mathbf{E}}$ is then a solution to

$$\left[\frac{\omega^2}{c^2}\epsilon - \mathbf{k}^2\mathbf{1}\right]\hat{\mathbf{E}} = \mathbf{0},$$

with $\hat{E}_3 = 0$. Nontrivial values of $\hat{\mathbf{E}}$ are allowed by the condition

$$\det\left[\frac{\omega^2}{c^2}\epsilon - \mathbf{k}^2\mathbf{1}\right] = 0$$

restricted to the 2×2 submatrix. Though ϵ is complex valued it follows that $\mathbf{k}$ is real and $k^2 = \mathbf{k}^2$ takes the values

$$k_{\pm}^2 = \frac{\omega^2}{c^2}(\epsilon_1 \pm \epsilon_2).$$

Let

$$k_R = k_+ = \frac{\omega}{c}\Big(1 - \frac{\omega_{pe}^2}{\omega(\omega - \omega_{ce})}\Big)^{1/2}, \qquad k_L = k_- = \frac{\omega}{c}\Big(1 - \frac{\omega_{pe}^2}{\omega(\omega + \omega_{ce})}\Big)^{1/2}.$$

The corresponding eigenvectors $\hat{\mathbf{E}}_{R,L}$ are given by

$$\hat{\mathbf{E}} = \hat{E}_2(i\mathbf{e}_1 + \mathbf{e}_2), \qquad \mathbf{E}_R = \hat{E}_2\exp[i(\mathbf{k}_R \cdot \mathbf{x} - \omega t)](i\mathbf{e}_1 + \mathbf{e}_2),$$

$$\hat{\mathbf{E}} = \hat{E}_2(-i\mathbf{e}_1 + \mathbf{e}_2), \qquad \mathbf{E}_L = \hat{E}_2\exp[i(\mathbf{k}_L \cdot \mathbf{x} - \omega t)](-i\mathbf{e}_1 + \mathbf{e}_2),$$

Since $i = \exp(i\pi/2)$ we can write

$$\mathbf{E}_R = \hat{E}_2\{\exp[i(\mathbf{k}_R \cdot \mathbf{x} - (\omega t - \pi/2))]\mathbf{e}_1 + \exp[i(\mathbf{k}_R \cdot \mathbf{x} - \omega t)]\mathbf{e}_2)$$

and the like for $\mathbf{E}_L$ with $\omega t + \pi/2$. Take the real part. At a fixed place $\mathbf{x}$, as time increases we see that $\mathbf{E}_R$ is a vector, of constant amplitude, rotating in the clockwise direction. Hence $\mathbf{E}_R$ is a right circular polarized wave. Similarly, we find that $\mathbf{E}_L$ is a left circular polarized wave.

As a comment, the *circular polarization* of an electromagnetic wave is an electric field, with a constant magnitude, whose direction rotates with time at a steady rate in a plane perpendicular to the direction of the wave (in the plane $\mathbf{k} \cdot \mathbf{x}$ = constant). In electrodynamics the strength and direction of an electric field is defined by its electric field vector. In the case of a circularly polarized wave, the tip of the electric field vector, at a given point in space, describes a circle as time progresses. At any instant of time, the electric field vector of the wave describes a helix around the direction of propagation.

One well-known wave solution in a magnetized plasma is the *Alfvén wave.* In 1942 Alfvén recognized that a wave could propagate in an incompressible perfectly conducting fluid immersed in a strong magnetic field. In this model the magnetic field and the mass density provide the restoring force and the inertia. Here we show that this type of wave solution is allowed in a more general setting by allowing for the compressibility of the fluid and the restoring force of the pressure.

We apply the MHD equations (12.89), so that

$$\partial_t \rho + \nabla \cdot (\rho \mathbf{v}) = 0, \quad \rho \partial_t \mathbf{v} = \frac{1}{\mu_0}(\nabla \times \mathbf{B}) \times \mathbf{B} - \nabla p, \quad \nabla \times (\mathbf{v} \times \mathbf{B}) = \partial_t \mathbf{B}.$$

The pressure is a function of the density, possibly via the adiabatic law $p = K\rho^\gamma$, or rather a function $p(\rho, \eta)$, with the entropy η as a constant. We regard the plasma at equilibrium with mass density ρ_0 and magnetic induction $\mathbf{B}_0$. Then we imagine departures of $\mathbf{B}$, ρ, $\mathbf{v}$ from the equilibrium values in the form

$$\mathbf{B} = \mathbf{B}_0 + \boldsymbol{\mathcal{B}}(\mathbf{x}, t), \quad \rho = \rho_0 + \varrho(\mathbf{x}, t), \quad \mathbf{v} = \mathbf{v}(\mathbf{x}, t).$$

Upon substitution and linearization in the perturbations $\boldsymbol{\mathcal{B}}$, ϱ, $\mathbf{v}$ we have

$$\partial_t \varrho + \rho_0 \nabla \cdot \mathbf{v} = 0, \quad \rho_0 \partial_t \mathbf{v} + s^2 \nabla \varrho + \frac{1}{\mu_0} \mathbf{B}_0 \times (\nabla \times \boldsymbol{\mathcal{B}}) = \mathbf{0},$$
$$\partial_t \boldsymbol{\mathcal{B}} - \nabla \times (\mathbf{v} \times \mathbf{B_0},) = \mathbf{0},$$

where $s^2 = \partial_\rho p$. The gradient of the first equation and the (partial) time derivative of the second one allows us to write

$$\partial_t^2 \mathbf{v} - s^2 \nabla(\nabla \cdot \mathbf{v}) + \frac{1}{\rho_0 \mu_0} \mathbf{B}_0 \times \{\nabla \times [\nabla \times (\mathbf{v} \times \mathbf{B}_0)]\} = \mathbf{0}.$$

Look for $\mathbf{v}$ as a time-harmonic plane-wave solution,

$$\mathbf{v}(\mathbf{x}, t) = \hat{\mathbf{v}} \exp i(\mathbf{k} \cdot \mathbf{x} - \omega t),$$

with polarization $\hat{\mathbf{v}}$ and direction of propagation (wave vector) $\mathbf{k}$. Upon substitution we find that $\hat{\mathbf{v}}$ is required to satisfy

$$-\omega^2 \hat{\mathbf{v}} + s^2 (\mathbf{k} \cdot \hat{\mathbf{v}}) \mathbf{k} - \frac{1}{\rho_0 \mu_0} \mathbf{B}_0 \times \{\mathbf{k} \times [\mathbf{k} \times (\hat{\mathbf{v}} \times \mathbf{B}_0)]\} = \mathbf{0}.$$

Since

$$\mathbf{B}_0 \times \{\mathbf{k} \times [\mathbf{k} \times (\hat{\mathbf{v}} \times \mathbf{B}_0)]\} = -(\mathbf{k} \cdot \hat{\mathbf{v}}) \mathbf{B}_0^2 \mathbf{k} + \mathbf{B}_0 \cdot \mathbf{k}[(\mathbf{k} \cdot \hat{\mathbf{v}}) \mathbf{B}_0 - (\mathbf{B}_0 \cdot \mathbf{k}) \hat{\mathbf{v}} + (\mathbf{B}_0 \cdot \hat{\mathbf{v}}) \mathbf{k}],$$

letting

$$\mathbf{v}_A = \frac{\mathbf{B}_0}{\sqrt{\rho_0\mu_0}}$$

we can write

$$[-\omega^2 + (\mathbf{v}_A\cdot\mathbf{k})^2]\hat{\mathbf{v}} + [(s^2+v_A^2)(\mathbf{k}\cdot\hat{\mathbf{v}}) - (\mathbf{v}_A\cdot\mathbf{k})(\mathbf{v}_A\cdot\hat{\mathbf{v}})]\mathbf{k} - (\mathbf{v}_A\cdot\mathbf{k})(\mathbf{k}\cdot\hat{\mathbf{v}})\mathbf{v}_A = \mathbf{0}. \tag{12.91}$$

If $\mathbf{k}$ is *perpendicular* to $\mathbf{v}_A$ then (12.91) simplifies to

$$-\omega^2\hat{\mathbf{v}} + (s^2+v_A^2)(\mathbf{k}\cdot\hat{\mathbf{v}})\mathbf{k} = \mathbf{0}.$$

Hence the wave is longitudinal, $\hat{\mathbf{v}} \parallel \mathbf{k}$, and propagates with the phase speed $u = \omega/k$, $k = |\mathbf{k}|$, given by

$$u^2 = s^2 + v_A^2.$$

If instead $\mathbf{k}$ is *parallel* to $\mathbf{v}_A$ then

$$v_A^2(\mathbf{k}\cdot\hat{\mathbf{v}})\mathbf{k} - (\mathbf{v}_A\cdot\mathbf{k})(\mathbf{v}_A\cdot\hat{\mathbf{v}})\mathbf{k} - (\mathbf{v}_A\cdot\mathbf{k})(\mathbf{k}\cdot\hat{\mathbf{v}})\mathbf{v}_A = -\mathbf{k}^2(\mathbf{v}_A\cdot\hat{\mathbf{v}})\mathbf{v}_A$$

and (12.91) becomes

$$(k^2v_A^2 - \omega^2)\hat{\mathbf{v}} + (\frac{s^2}{v_A^2} - 1)k^2(\mathbf{v}_A\cdot\hat{\mathbf{v}})\mathbf{v}_A = \mathbf{0}.$$

Hence there are two solutions according as $\hat{\mathbf{v}}$ is parallel to $\mathbf{v}_A$ or is not. If $\hat{\mathbf{v}} \parallel \mathbf{v}_A$, and then $\hat{\mathbf{v}} \parallel \mathbf{k}$, it follows that

$$-\omega^2 + s^2k^2 = 0$$

and hence

$$u^2 = s^2.$$

This is a longitudinal wave, in fact a sound wave, unaffected by the magnetic induction $\mathbf{B}_0$. If, instead $\hat{\mathbf{v}}$ is not parallel to $\mathbf{v}_A$ then $k^2v_A^2 - \omega^2$ is required to vanish and this implies

$$u^2 = v_A^2.$$

Moreover it is required that $\mathbf{v}_A\cdot\hat{\mathbf{v}} = 0$ and this in turn implies that $\mathbf{k}\cdot\hat{\mathbf{v}} = 0$. As a consequence this is a transverse wave propagating with the Alfvén speed v_A. This is a true Alfvén wave and is a purely MHD phenomenon where only the magnetic field provides the restoring force.

12.11 Chiral Media and Optical Activity

Optical activity means that a linearly polarized plane of light will rotate as it passes through the medium, interacts with the state of an electromagnetic waves and couples selectively with either left or right circularly polarized component. Natural optical activity denotes that this behaviour occurs in materials without the application of appropriate external fields. Chiral materials are materials displaying natural optical activity. Our purpose here is to investigate some models of chiral materials.

The vector fields $\mathbf{E}$, $\mathbf{B}$, $\mathbf{D}$, $\mathbf{H}$ are assumed to be related by linear constitutive equations. In the simplest case

$$\mathbf{D} = \boldsymbol{\epsilon}\mathbf{E}, \qquad \mathbf{B} = \boldsymbol{\mu}\mathbf{H};$$

$\boldsymbol{\epsilon}$ is the permittivity and $\boldsymbol{\mu}$ is the permeability tensor. If $\boldsymbol{\epsilon} = \epsilon\mathbf{1}$ we say that the material is electrically isotropic and the like for $\boldsymbol{\mu}$. Crystals are described by symmetric tensors. If two eigenvalues are equal, e.g. $\mu_1 = \mu_2 \neq \mu_3$, then the crystal is said to be *uniaxial* and the principal axis with the unequal eigenvalue is called the optic axis. If the eigenvalues are unequal then the medium is said to be *biaxial*.

If cross-coupling terms occur between the electric and magnetic fields then we might have

$$\mathbf{D} = \boldsymbol{\epsilon}\mathbf{E} + \boldsymbol{\xi}\mathbf{H}, \qquad \mathbf{B} = \boldsymbol{\zeta}\mathbf{E} + \boldsymbol{\mu}\mathbf{H}.$$

The material is said to be biisotropic or bianisotropic according as both $\boldsymbol{\epsilon}$, $\boldsymbol{\mu}$ and $\boldsymbol{\xi}$, $\boldsymbol{\zeta}$ are isotropic or anisotropic.

Chiral media are modelled in a number of ways.[37] Isotropic chirality was modelled in the form

$$\mathbf{D} = \epsilon[\mathbf{E} + \beta\nabla\times\mathbf{E}], \qquad \mathbf{B} = \mu[\mathbf{H} + \beta\nabla\times\mathbf{H}]$$

by Drude, Born and Fedorov. Instead Condon modelled chirality in the form

$$\mathbf{D} = \epsilon\mathbf{E} - \beta\partial_t\mathbf{H}, \qquad \mathbf{B} = \mu\mathbf{H} + \beta\partial_t\mathbf{E}.$$

We might formally consider the analogue by letting $\mathbf{D}$, $\mathbf{H}$ be given by $\mathbf{E}$ and $\mathbf{B}$. Boys and Post constitutive equations are given by

$$\mathbf{D} = \epsilon\mathbf{E} + i\beta\mathbf{B}, \qquad \mathbf{H} = i\beta\mathbf{E} + \mathbf{B}/\mu.$$

Boys-Post equations look like the Condon model in the frequency domain if the time dependence is in the form $\propto \exp(-i\omega t)$.

Here we let $\mathbf{E}$ and $\mathbf{H}$ be independent variables and assume the body is undeformable. For definiteness we investigate the thermodynamic consistency of the constitutive equations

$$\mathbf{D} = \epsilon[\mathbf{E} + \beta\nabla\times\mathbf{E}], \qquad \mathbf{B} = \mu[\mathbf{H} + \gamma\nabla\times\mathbf{H}], \tag{12.92}$$

[37] See, e.g. [275], Chap. 3.

$$\mathbf{D} = \epsilon\mathbf{E} + \alpha\partial_t\mathbf{H}, \qquad \mathbf{B} = \mu\mathbf{H} + \lambda\partial_t\mathbf{E}, \tag{12.93}$$

the scalars ϵ, β, μ, γ, α, λ being possibly dependent on the temperature θ.

Since the body is undeformable then we let $\phi = \psi + \mathbf{E}\cdot\mathbf{P} + \mu_0\mathbf{H}\cdot\mathbf{M}$ and the entropy inequality (12.2) simplifies to

$$-\rho\dot{\phi} - \rho\eta\dot{\theta} - \mathbf{P}\cdot\dot{\mathbf{E}} - \mu_0\mathbf{M}\cdot\dot{\mathbf{H}} + \theta\nabla\cdot\mathbf{k} \geq 0 \tag{12.94}$$

while $\mathbf{E}$, $\mathbf{B}$, $\mathbf{D}$, $\mathbf{H}$ are subject to Maxwell's equations

$$\nabla\times\mathbf{E} = -\dot{\mathbf{B}}, \qquad \nabla\times\mathbf{H} = \dot{\mathbf{D}}.$$

To explore the consistency of equations (12.92) we let θ, $\mathbf{E}$, $\mathbf{H}$, $\nabla\times\mathbf{E}$, $\nabla\times\mathbf{H}$, $\dot{\mathbf{E}}$, $\dot{\mathbf{H}}$ be independent variables. Yet, if ϕ depends on $\dot{\mathbf{E}}$ and $\dot{\mathbf{H}}$ then $\dot{\phi}$ comprises $\partial_{\dot{\mathbf{E}}}\phi\cdot\ddot{\mathbf{E}} + \partial_{\dot{\mathbf{H}}}\phi\cdot\ddot{\mathbf{H}}$. Now, Maxwell's equations allow $\ddot{\mathbf{E}}$ and $\ddot{\mathbf{H}}$ to be given arbitrary values in that $\nabla\times\dot{\mathbf{E}}$ and $\nabla\times\dot{\mathbf{H}}$ do not enter the constitutive equations. The arbitrariness of $\ddot{\mathbf{E}}$ and $\ddot{\mathbf{H}}$ implies that $\partial_{\dot{\mathbf{E}}}\phi = \mathbf{0}$, $\partial_{\dot{\mathbf{H}}}\phi = \mathbf{0}$. Hence we start with a function $\phi(\theta, \mathbf{E}, \mathbf{H}, \nabla\times\mathbf{E}, \nabla\times\mathbf{H})$ for the free energy ϕ.

Computing $\dot{\phi}$ we have

$$-\rho(\partial_\theta\phi + \eta)\dot{\theta} - (\rho\partial_{\mathbf{E}}\phi + \mathbf{P})\cdot\dot{\mathbf{E}} - (\rho\partial_{\mathbf{H}}\phi + \mu_0\mathbf{M})\cdot\dot{\mathbf{H}} - \rho\partial_{\nabla\times\mathbf{E}}\phi\cdot\nabla\times\dot{\mathbf{E}} - \rho\partial_{\nabla\times\mathbf{H}}\phi\cdot\nabla\times\dot{\mathbf{H}} + \theta\nabla\cdot\mathbf{k} \geq 0.$$

The arbitrariness of $\dot{\theta}$ implies that

$$\eta = -\partial_\theta\phi.$$

Now, divide the inequality by θ,

$$-\frac{1}{\theta}(\rho\partial_{\mathbf{E}}\phi + \mathbf{P})\cdot\dot{\mathbf{E}} - \frac{1}{\theta}(\rho\partial_{\mathbf{H}}\phi + \mu_0\mathbf{M})\cdot\dot{\mathbf{H}} - \frac{\rho}{\theta}\partial_{\nabla\times\mathbf{E}}\phi\cdot\nabla\times\dot{\mathbf{E}} - \frac{\rho}{\theta}\partial_{\nabla\times\mathbf{H}}\phi\cdot\nabla\times\dot{\mathbf{H}} + \nabla\cdot\mathbf{k} \geq 0.$$

To determine the remaining degree of arbitrariness we notice that, for any two vector fields $\mathbf{w}$, $\mathbf{u}$, we have the identity

$$\mathbf{w}\cdot\nabla\times\mathbf{u} = \nabla\cdot(\mathbf{u}\times\mathbf{w}) + \mathbf{u}\cdot\nabla\times\mathbf{w}. \tag{12.95}$$

Hence

$$\frac{\rho}{\theta}\partial_{\nabla\times\mathbf{E}}\phi\cdot\nabla\times\dot{\mathbf{E}} = \nabla\cdot(\dot{\mathbf{E}}\times\frac{\rho}{\theta}\partial_{\nabla\times\mathbf{E}}\phi) + \dot{\mathbf{E}}\cdot\nabla\times\frac{\rho}{\theta}\partial_{\nabla\times\mathbf{E}}\phi$$

and the like for $\mathbf{H}$. Consequently we obtain the inequality

$$-\frac{1}{\theta}(\rho\partial_{\mathbf{E}}\phi + \mathbf{P} + \theta\nabla\times\frac{\rho}{\theta}\partial_{\nabla\times\mathbf{E}}\phi)\cdot\dot{\mathbf{E}} - \frac{1}{\theta}(\rho\partial_{\mathbf{H}}\phi + \mu_0\mathbf{M} + \theta\nabla\times\frac{\rho}{\theta}\partial_{\nabla\times\mathbf{H}}\phi)\cdot\dot{\mathbf{H}}$$
$$+\nabla\cdot(\mathbf{k} - \frac{\rho}{\theta}\dot{\mathbf{E}}\times\partial_{\nabla\times\mathbf{E}}\phi - \frac{\rho}{\theta}\dot{\mathbf{H}}\times\partial_{\nabla\times\mathbf{H}}\phi) \geq 0.$$

The inequality is linear in $\dot{\mathbf{E}}$ and $\dot{\mathbf{H}}$. It then follows that

$$\mathbf{P} = -\rho\partial_{\mathbf{E}}\phi - \theta\nabla \times \frac{\rho}{\theta}\partial_{\nabla\times\mathbf{E}}\phi, \qquad \mu_0\mathbf{M} = -\rho\partial_{\mathbf{H}}\phi - \theta\nabla \times \frac{\rho}{\theta}\partial_{\nabla\times\mathbf{H}}\phi,$$

and

$$\mathbf{k} = \frac{\rho}{\theta}\dot{\mathbf{E}} \times \partial_{\nabla\times\mathbf{E}}\phi + \frac{\rho}{\theta}\dot{\mathbf{H}} \times \partial_{\nabla\times\mathbf{H}}\phi.$$

We then look for the function ϕ which determines equations (12.92). Let

$$\phi = \Phi(\theta) + \tfrac{1}{2}c_1\mathbf{E}^2 + \tfrac{1}{2}c_2\mathbf{H}^2 + c_3\mathbf{E}\cdot\nabla\times\mathbf{E} + c_4\mathbf{H}\cdot\nabla\times\mathbf{H}$$

and assume θ is constant in space. Hence

$$\mathbf{P} = -\rho c_1\mathbf{E} - 2\rho c_3\nabla\times\mathbf{E}, \qquad \mu_0\mathbf{M} = -\rho c_2\mathbf{H} - 2\rho c_4\nabla\times\mathbf{H}.$$

Consequently, since $\mathbf{D} = \epsilon_0\mathbf{E} + \mathbf{P}$ and $\mathbf{B} = \mu_0(\mathbf{H} + \mathbf{M})$ then

$$\mathbf{D} = (\epsilon_0 - \rho c_1)\mathbf{E} - 2\rho c_3\nabla\times\mathbf{E}, \qquad \mathbf{B} = (\mu_0 - \rho c_2)\mathbf{H} - 2\rho c_4\nabla\times\mathbf{H}.$$

Upon the identification of the coefficients, Eqs. (12.92) follow. In this regard we observe that the condition $\gamma = \beta$, or

$$\frac{2\rho c_3}{\epsilon_0 - \rho c_1} = \frac{2\rho c_4}{\mu_0 - \rho c_2},$$

does not follow as a thermodynamic restriction.

The Condon model involves a dependence of $\mathbf{D}$ and $\mathbf{B}$ (and hence of $\mathbf{P}$ and $\mathbf{M}$) on the derivatives $\dot{\mathbf{E}}, \dot{\mathbf{H}}$. Let $\phi, \eta, \mathbf{P}, \mathbf{M}, \mathbf{k}$ be functions of $\theta, \mathbf{E}, \mathbf{H}, \dot{\mathbf{E}}, \dot{\mathbf{H}}$. The entropy inequality (12.94) becomes

$$-\rho(\partial_\theta\phi + \eta)\dot{\theta} - (\rho\partial_{\mathbf{E}}\phi + \mathbf{P})\cdot\dot{\mathbf{E}} - (\rho\partial_{\mathbf{H}}\phi + \mu_0\mathbf{M})\cdot\dot{\mathbf{H}} - \rho\partial_{\dot{\mathbf{E}}}\phi\cdot\ddot{\mathbf{E}} - \rho\partial_{\dot{\mathbf{H}}}\phi\cdot\ddot{\mathbf{H}} + \theta\nabla\cdot\mathbf{k} \geq 0.$$

By Maxwell's equations and the definition of $\mathbf{M}$ and $\mathbf{P}$ it follows that $\ddot{\mathbf{E}}$ and $\ddot{\mathbf{H}}$ can take arbitrary values independent of $\nabla\dot{\mathbf{E}}$ and $\nabla\dot{\mathbf{H}}$ (occurring in $\nabla\cdot\mathbf{k}$). Hence it follows

$$\partial_{\dot{\mathbf{E}}}\phi = \mathbf{0}, \qquad \partial_{\dot{\mathbf{H}}}\phi = \mathbf{0}.$$

Moreover it follows $\eta = -\partial_\theta\phi$. Thus $\nabla\cdot\mathbf{k} = 0$ and the entropy inequality reduces to

$$(\rho\partial_{\mathbf{E}}\phi + \mathbf{P})\cdot\dot{\mathbf{E}} + (\rho\partial_{\mathbf{H}}\phi + \mu_0\mathbf{M})\cdot\dot{\mathbf{H}} \leq 0.$$

The constitutive equations

$$\mathbf{P} = -\rho\partial_{\mathbf{E}}\phi - \lambda\dot{\mathbf{E}} - \varpi\dot{\mathbf{H}}, \qquad \mu_0\mathbf{M} = -\rho\partial_{\mathbf{H}}\phi - \xi\dot{\mathbf{H}} + \varpi\dot{\mathbf{E}},$$

where $\lambda, \xi \geq 0$, are then consistent with thermodynamics. The Condon model is obtained, in the particular case $\lambda, \xi = 0$, by letting $\partial_{\mathbf{E}}\phi \propto \mathbf{E}$ and $\partial_{\mathbf{H}}\phi \propto \mathbf{H}$.

12.11.1 Wave Solutions in Chiral Media

Consider the function

$$\mathbf{f}(\mathbf{x}, t) = (\xi \mathbf{i}_1 + i\zeta \mathbf{i}_2)\exp\{i[(\mathbf{k}_1 + i\mathbf{k}_2)\cdot \mathbf{x} - \omega t)]\},$$

where $\mathbf{k}_1$ and $\mathbf{k}_2$ are real-valued vectors. Then

$$\Re \mathbf{f} = \xi \exp(-\mathbf{k}_2 \cdot \mathbf{x})\cos(\mathbf{k}_1 \cdot \mathbf{x} - \omega t)\mathbf{i}_1 - \zeta \exp(-\mathbf{k}_2 \cdot \mathbf{x})\sin(\mathbf{k}_1 \cdot \mathbf{x} - \omega t)\mathbf{i}_2.$$

At any point $\mathbf{x}$, $\exp(-\mathbf{k}_2 \cdot \mathbf{x})$ is fixed. Then $\Re \mathbf{f}$ traces out an ellipse in the plane perpendicular to $\mathbf{k}_1$; the semi-major and semi-minor axes of the ellipse have lengths $a\exp(-\mathbf{k}_2 \cdot \mathbf{x})$ and $b\exp(-\mathbf{k}_2 \cdot \mathbf{x})$. As $\mathbf{k}_1 \cdot \mathbf{x}$ increases the ellipse is traced out in the clockwise direction if $\xi\zeta > 0$ and in the counterclockwise direction if $\xi\zeta < 0$.

We now investigate the optical activity properties associated with some models of chiral media.

Drude–Born–Fedorov model

To begin with we consider the model (12.92). Hence the fields $\mathbf{E}, \mathbf{B}, \mathbf{D}, \mathbf{H}$ are subject to Eqs. (12.92) and Maxwell's equations

$$\nabla \cdot \mathbf{D} = 0, \qquad \nabla \cdot \mathbf{B} = 0, \qquad \nabla \times \mathbf{E} = -\dot{\mathbf{B}}, \qquad \nabla \times \mathbf{H} = \dot{\mathbf{D}}.$$

Let $\mathbf{E}_0, \mathbf{H}_0$ be complex-valued vectors in the form

$$\mathbf{E}_0 = E_1\mathbf{e}_1 + iE_2\mathbf{e}_2, \qquad \mathbf{H}_0 = H_1\mathbf{e}_1 + iH_2\mathbf{e}_2,$$

where $E_1, E_2, H_1, H_2 \in \mathbb{R}$. We look for solutions of the form

$$\mathbf{E} = \mathbf{E}_0 \exp[i(k\mathbf{n}\cdot\mathbf{x} - \omega t)], \qquad \mathbf{H} = \mathbf{H}_0 \exp[i(k\mathbf{n}\cdot\mathbf{x} - \omega t)]. \tag{12.96}$$

These solutions represent plane waves propagating in the direction $\mathbf{n}$ with wave speed ω/k or slowness $s = k/\omega$. For definiteness we let $\mathbf{n} = \mathbf{e}_3$. We will find that the model allows for circularly polarized waves.

First we notice that, in view of (12.92), $\nabla \cdot \mathbf{D} = 0$ and $\nabla \cdot \mathbf{B} = 0$ imply $\nabla \cdot \mathbf{E} = 0, \nabla \cdot \mathbf{H} = 0$ and hence

$$\nabla \cdot \mathbf{E}_0 = 0, \quad \mathbf{n}\cdot\mathbf{E}_0 = 0, \qquad \nabla \cdot \mathbf{H}_0 = 0, \quad \mathbf{n}\cdot\mathbf{H}_0 = 0; \tag{12.97}$$

both **E** and **H** are orthogonal to the direction of propagation **n**. Now, $\nabla \times \mathbf{E} = -\dot{\mathbf{B}}$ and $\nabla \times \mathbf{H} = \dot{\mathbf{D}}$ result in

$$k\mathbf{n} \times \mathbf{E}_0 = \omega\mu[\mathbf{H}_0 + i\gamma k\mathbf{n} \times \mathbf{H}_0], \qquad k\mathbf{n} \times \mathbf{H}_0 = -\omega\epsilon[\mathbf{E}_0 + i\beta k\mathbf{n} \times \mathbf{E}_0]. \quad (12.98)$$

Since $\mathbf{n} \cdot \mathbf{E}_0 = 0$ and $\mathbf{n} \cdot \mathbf{H}_0 = 0$ then by applying the cross product $\mathbf{n}\times$ to the first equation in (12.98) we find

$$k\mathbf{E}_0 = -\omega\mu[\mathbf{n} \times \mathbf{H}_0 - i\gamma k\mathbf{H}_0].$$

Upon substitution of $\mathbf{E}_0$ in the second equation of (12.98) we obtain

$$a\mathbf{n} \times \mathbf{H}_0 + ib\mathbf{H}_0 = \mathbf{0},$$

where

$$a = k^2/\omega^2 - \mu\epsilon - \mu\epsilon\beta\gamma k^2, \qquad b = \mu\epsilon k(\gamma + \beta).$$

It follows that

$$aH_1 - bH_2 = 0, \qquad bH_1 - aH_2 = 0.$$

Nontrivial solutions occur if $a^2 - b^2 = 0$, whence

$$\frac{k^2}{\omega^2} = \mu\epsilon[1 + \beta\gamma k^2 \pm k(\beta + \gamma)].$$

If $s = s_+ = \{\mu\epsilon[1 + \beta\gamma k^2 + k(\beta + \gamma)]\}^{1/2}$ then $a = b$ and $H_1 = H_2$; the wave is left-circularly polarized. If $s = s_- = \{\mu\epsilon[1 + \beta\gamma k^2 - k(\beta + \gamma)]\}^{1/2}$ then $a = -b$ and $H_1 = -H_2$; the wave is right circularly polarized. Without the application of any appropriate external field, the Drude–Born–Fedorov model allows for the existence of two circularly polarized waves with different speeds of propagation.

Condon model

The fields **D**, **E**, **B**, **H** are subject to Maxwell's equations and the constitutive equations

$$\mathbf{D} = \epsilon\mathbf{E} - g\dot{\mathbf{H}}, \qquad \mathbf{B} = \mu\mathbf{H} + g\dot{\mathbf{E}}.$$

We look for solutions in the form (12.96). Hence Maxwell's equations $\nabla \times \mathbf{H} = \dot{\mathbf{D}}$ and $\nabla \times \mathbf{E} = \dot{\mathbf{B}}$ result in

$$k\mathbf{n} \times \mathbf{H}_0 = -\omega\epsilon\mathbf{E}_0 - ig\omega^2\mathbf{H}_0, \qquad k\mathbf{n} \times \mathbf{E}_0 = \mu\omega\mathbf{H}_0 - ig\omega^2\mathbf{E}_0.$$

Since Eqs. (12.97) still hold then by applying the cross product $\mathbf{n}\times$ to the first relation we have

$$-k\mathbf{H}_0 = -\omega\epsilon\mathbf{n} \times \mathbf{E}_0 - ig\omega^2\mathbf{n} \times \mathbf{H}_0.$$

Replacing $\mathbf{n} \times \mathbf{E}_0$ and next $\mathbf{E}_0$ we obtain

$$a\mathbf{H}_0 + ib\mathbf{n} \times \mathbf{H}_0 = \mathbf{0},$$

where

$$a = -k^2/\omega^2 + \mu\epsilon - g^2\omega^2, \qquad b = gk.$$

Hence it follows

$$aH_1 + bH_2 = 0, \qquad bH_1 + aH_2 = 0.$$

Nontrivial solutions occur if $a^2 - b^2 = 0$, whence

$$a = \pm b, \qquad -k^2/\omega^2 + \mu\epsilon - g^2\omega^2 = \pm gk.$$

If $a = b$ then $H_1 = H_2$ and the solution is a left (clockwise) circularly polarized wave; if $a = -b$ then $H_1 = -H_2$ and the solution is a right (counterclockwise) circularly polarized wave. Hence also the Condon model allows for the existence of two circularly polarized waves with different speeds of propagation.

12.12 Anisotropic Non-Dissipative Magnetic Materials (Ferrites)

Ferrites are examples of magnetic materials with sufficiently low losses in the required range of frequencies. They are then usually described by a complex-valued anisotropic permeability in the frequency domain. Here we show how the complex-valuedness arises from an appropriate model in the time domain.

Modelling of ferrites usually starts with the observation that the magnetic properties arise by the behaviour of spinning electrons, with magnetic dipole moment $\mathfrak{m}$ and angular momentum $\mathfrak{s}$, with

$$\mathfrak{m} = -\gamma\mathfrak{s}, \qquad \gamma = \frac{e}{m}$$

$-\gamma$ being called the gyromagnetic ratio; e and m are the (positive) charge and the mass of the electron. If a magnetic intensity $\mathbf{H}$ occurs (in matter) then a torque $\mu_0\mathfrak{m} \times \mathbf{H}$ is applied to the spinning electron. Hence $\mathfrak{s}$ and $\mathfrak{m}$ are subject to

$$\dot{\mathfrak{s}} = \mu_0\mathfrak{m} \times \mathbf{H}, \qquad \dot{\mathfrak{m}} = \gamma\mu_0\mathfrak{m} \times \mathbf{H}. \tag{12.99}$$

The solution $\mathfrak{m}(t)$ to Eq. (12.99) is a function $|\mathfrak{m}(t)|$ =constant. Further, if $\mathbf{H}$ is constant then also $\mathfrak{m} \cdot \mathbf{H}$ is constant; the angle, θ say, between $\mathfrak{m}$ and $\mathbf{H}$ is constant in that $(\mathfrak{m} \cdot \mathbf{H})^{\cdot} = \mathbf{H} \cdot \dot{\mathfrak{m}} = 0$. The vector $\mathfrak{m}$ rotates around $\mathbf{H}$ with a constant angle. Moreover, by a direct check we see that the transverse part of $\mathfrak{m}$ describes a circumference with constant angular frequency,

$$\omega_L = \gamma \mu_0 H,$$

known in the literature as *Larmor frequency.*

Letting N be the (constant) number of electrons per unit volume we have $\mathbf{M} = N\mathfrak{m}$. Hence

$$\dot{\mathbf{M}} = -\gamma \mu_0 \mathbf{M} \times \mathbf{H} = \gamma \mu_0 \mathbf{H} \times \mathbf{M}.$$

Let $\mathbf{H}$ be the sum of a constant field $\mathbf{H}_0$ and a time-dependent field $\mathfrak{H}$. Let $\mathbf{M} = \mathbf{M}_0 + \mathfrak{M}$, where $\mathbf{M}_0$ is the magnetization induced by $\mathbf{H}_0$. If $\mathbf{H} = \mathbf{H}_0$ then $\mathbf{M} = \mathbf{M}_0$ and

$$\mathbf{0} = \gamma \mu_0 \mathbf{H}_0 \times \mathbf{M}_0.$$

Hence $\mathbf{M}_0$ is parallel to $\mathbf{H}_0$; let $\mathbf{e}_3$ be the common unit vector of $\mathbf{H}_0$ and $\mathbf{M}_0$. If $\mathbf{H} = \mathbf{H}_0 + \mathfrak{H}$ then we have

$$\dot{\mathfrak{M}} = \gamma \mu_0 (\mathbf{H}_0 + \mathfrak{H}) \times (\mathbf{M}_0 + \mathfrak{M}) = \gamma \mu_0 [\mathbf{H}_0 \times \mathfrak{M} + \mathfrak{H} \times \mathbf{M}_0 + \mathfrak{H} \times \mathfrak{M}]. \quad (12.100)$$

If $\rho \partial_{\mathbf{M}} \psi = \alpha \mathbf{M}$, since $\mathbf{M}^2 = \mathfrak{M}^2 + 2M_0 \mathfrak{M} \cdot \mathbf{e}_3 + M_0^2$ then we have

$$\rho \psi = \tfrac{1}{2} \alpha (\mathfrak{M}^2 + 2M_0 \mathfrak{M} \cdot \mathbf{e}_3 + M_0^2) + \psi_0,$$

ψ_0 being independent of $\mathfrak{M}$. In light of (12.29) we have

$$(-\alpha \mathbf{M} + \mu_0 \mathbf{H}) \cdot \dot{\mathfrak{M}} \geq 0.$$

Substitution of $\dot{\mathfrak{M}}$ from (12.100) results in

$$\begin{aligned}[-\alpha(M_0 \mathbf{e}_3 + \mathfrak{M}) + \mu_0(H_0 \mathbf{e}_3 + \mathfrak{H})] \cdot \gamma \mu_0 (H_0 \mathbf{e}_3 \times \mathfrak{M} - M_0 \mathbf{e}_3 \times \mathfrak{M} + \mathfrak{H} \times \mathfrak{M}) \\ = \gamma \mu_0 (\alpha M_0 \mathbf{e}_3 \times \mathfrak{H} \cdot \mathfrak{M} + \mu_0 M_0 \mathbf{e}_3 \times \mathfrak{M} + \mu_0 H_0 \mathbf{e}_3 \cdot \mathfrak{H} \times \mathfrak{M})) \equiv 0\end{aligned}$$

for any value α. Wave solutions are determined by means of the linearized version of (12.100), i.e. by neglecting $\mathfrak{H} \times \mathfrak{M}$.

Here we review models compatible with the reduced inequality

$$(-\alpha \mathbf{M} + \mu_0 \mathbf{H}) \cdot \dot{\mathbf{M}} \geq 0.$$

Take the rate equation

$$\dot{\mathbf{M}} = \gamma \mu_0 \mathbf{H} \times \mathbf{M} - \lambda \mathbf{M} \times (\mathbf{M} \times \mathbf{H}) + \nu [(-\alpha \mathbf{M} + \mu_0 \mathbf{H}) \cdot \mathbf{M}] \mathbf{M},$$

where $\lambda, \nu \geq 0$, $\gamma \in \mathbb{R}$; equation $\dot{\mathbf{M}} = \gamma \mu_0 \mathbf{H} \times \mathbf{M}$ follows by letting $\lambda, \nu = 0$. We can split the effects of $\mathbf{M}$ by letting $\mathbf{M} = M\mathbf{m}$, $\mathbf{m}$ being the unit vector of $\mathbf{M}$. Hence

$$\dot{\mathbf{M}} = \dot{M}\mathbf{m} + M\dot{\mathbf{m}}, \qquad \mathbf{m} \cdot \dot{\mathbf{m}} = 0.$$

Accordingly we can write

$$\dot{M}\mathbf{m} + M\dot{\mathbf{m}} = \gamma\mu_0 M\mathbf{H} \times \mathbf{m} - \lambda M^2 \mathbf{m} \times (\mathbf{m} \times \mathbf{H}) + \nu M^2[(-\alpha M\mathbf{m} + \mu_0\mathbf{H}) \cdot \mathbf{m}]\mathbf{m}.$$

The orthogonality between $\mathbf{m}$ and $\dot{\mathbf{m}}$ allows us to split the effects of $\dot{\mathbf{M}}$ in the form

$$\dot{M} = -\alpha\nu M^3 + \mu_0\nu M^2\mathbf{H} \cdot \mathbf{m}, \qquad \dot{\mathbf{m}} = \gamma\mu_0 M\mathbf{H} \times \mathbf{m} + \lambda M^2 \mathbf{m} \times (\mathbf{H} \times \mathbf{m}).$$

Particular solutions are now considered.

• Constant solutions: $M(t) = M_0$, $\mathbf{m}(t) = \mathbf{m}_0$. They happen if

$$\mu_0\mathbf{H}_0 \cdot \mathbf{m}_0 = \alpha M_0, \qquad \mathbf{0} = \gamma\mu_0 M_0\mathbf{H}_0 \times \mathbf{m}_0 + \lambda M_0^2 \mathbf{m}_0 \times (\mathbf{H}_0 \times \mathbf{m}_0)$$

whence

$$\mu_0 H_0 = \alpha M_0, \qquad \mathbf{0} = \mathbf{H}_0 \times \mathbf{m}_0.$$

Constant solutions occur provided $\mathbf{m}_0$ is collinear to $\mathbf{H}_0$ and $M_0 = \mu_0 H_0/\alpha$.

• Solutions with constant magnitude, $M(t) = M_0$. They occur if

$$\mu_0\mathbf{H}_0 \cdot \mathbf{m}(t) = \alpha M_0, \qquad \dot{\mathbf{m}}(t) = \gamma\mu_0 M_0\mathbf{H}_0 \times \mathbf{m}(t) + \lambda M_0^2 \mathbf{m}(t) \times (\mathbf{H}_0 \times \mathbf{m}(t)).$$

Hence $\mathbf{m}$ rotates around $\mathbf{H}_0$ at a constant angle, say β, such that $\cos\beta = \alpha M_0/\mu_0 H_0$. Consequently these solutions occur if $M_0 \leq \mu_0 H_0/\alpha$.

• Solutions with constant direction, $\mathbf{m}(t) = \mathbf{m}_0$. They occur if

$$\dot{M}(t) = -\alpha\nu M^3(t) + \mu_0\nu M^2(t)\mathbf{H}_0 \cdot \mathbf{m}_0, \qquad \mathbf{0} = \gamma\mu_0 M(t)\mathbf{H}_0 \times \mathbf{m}_0 + \lambda M^2(t)\mathbf{m}_0 \times (\mathbf{H}_0 \times \mathbf{m}_0),$$

whence

$$\dot{M}(t) = -\alpha\nu M^3(t) + \mu_0\nu M^2(t) H_0, \qquad \mathbf{0} = \mathbf{H}_0 \times \mathbf{m}_0.$$

These solutions then occur provided $\mathbf{m}_0$ is parallel to $\mathbf{H}_0$.

As a remark observe that if we let $\mathbf{H} = H_0\mathbf{e}_3 + \mathfrak{H}$, $\mathbf{M} = M_0\mathbf{e}_3 + \mathfrak{M}$ the stationary solution $\mathbf{H} = H_0\mathbf{e}_3$, $\mathbf{M} = M_0\mathbf{e}_3$ holds only if $\alpha M_0 = \mu_0 H_0$ and then, consistently, α is required to equal $\mu_0 H_0/M_0$.

Wave solutions

Wave solutions in a ferrite medium are determined by modelling the body via linearized equations. Hence $\mathbf{D}$ and $\mathbf{E}$ are supposed to be related by $\mathbf{D} = \epsilon\mathbf{E}$ while $\mathfrak{M}$ is governed by the linearized version of (12.100), namely

$$\dot{\mathfrak{M}} = \gamma\mu_0(H_0\mathbf{e}_3 \times \mathfrak{M} - M_0\mathbf{e}_3 \times \mathfrak{H}). \tag{12.101}$$

In addition electromagnetic fields satisfy Maxwell's equations

$$\nabla \times \mathbf{H} = \epsilon \dot{\mathbf{E}}, \qquad \nabla \times \mathbf{E} = -\dot{\mathbf{B}}, \qquad \nabla \cdot \mathbf{B} = 0, \qquad \nabla \cdot \mathbf{E} = 0.$$

It follows from (12.101) that

$$0 = \dot{\mathfrak{M}} \cdot \mathbf{e}_3 = (\mathfrak{M} \cdot \mathbf{e}_3)^{\dot{}}.$$

We then let $\mathfrak{M} \cdot \mathbf{e}_3 = 0$.

Since $\nabla \cdot \mathbf{B} = 0$ then

$$0 = \mu_0 \nabla \cdot (\mathbf{H}_0 + \mathbf{M}_0) + \mu_0 \nabla \cdot (\mathfrak{H} + \mathfrak{M}) = \mu_0 \nabla \cdot (\mathfrak{H} + \mathfrak{M}).$$

Consistent with the requirement $\nabla \cdot (\mathfrak{H} + \mathfrak{M}) = 0$, the existence of wave solutions is now investigated in the complex-valued harmonic dependence

$$\mathfrak{H} = \mathfrak{H}_0 \exp[i(kz - \omega t)], \qquad \mathfrak{M} = \mathfrak{M}_0 \exp[i(kz - \omega t)], \tag{12.102}$$

where $\mathfrak{H} \cdot \mathbf{e}_3 = 0$, $\mathfrak{M} \cdot \mathbf{e}_3 = 0$, z being the Cartesian coordinate along $\mathbf{e}_3$. Hence $\mathfrak{H}_0 = \mathfrak{H}_1 \mathbf{e}_1 + \mathfrak{H}_2 \mathbf{e}_2$, $\mathfrak{H}_1, \mathfrak{H}_2 \in \mathbb{C}$, and the like for $\mathfrak{M}$. Hence $\nabla \cdot \mathfrak{H} = 0$, $\nabla \cdot \mathfrak{M} = 0$. In components, Eq. (12.101) becomes

$$-i\omega \mathfrak{M}_1 = \gamma \mu_0 (-H_0 \mathfrak{M}_2 + M_0 \mathfrak{H}_2), \qquad -i\omega \mathfrak{M}_2 = \gamma \mu_0 (H_0 \mathfrak{M}_1 - M_0 \mathfrak{H}_1).$$

Upon algebraic manipulations we find the magnetic susceptibility χ_m,

$$\mathfrak{M} = \chi_m \mathfrak{H}, \qquad \chi_m = \begin{bmatrix} \chi_1 & -i\chi_2 & 0 \\ i\chi_2 & \chi_1 & 0 \\ 0 & 0 & 0 \end{bmatrix}$$

where

$$\chi_1 = \frac{\omega_C \omega_M}{\omega_C^2 - \omega^2}, \qquad \chi_2 = \frac{\omega \omega_M}{\omega_C^2 - \omega^2}, \qquad \omega_C = \gamma \mu_0 H_0, \qquad \omega_M = \gamma \mu_0 M_0.$$

Hence it follows that

$$\mathbf{B} = \mu_0 \mathbf{H}_0 + \boldsymbol{\mu} \mathfrak{H}, \qquad \boldsymbol{\mu} = \mu_0 (\mathbf{1} + \chi_m),$$

that is the constitutive equation for the magnetic induction $\mathbf{B}$ subject to the assumptions (12.102) and (12.101).

To establish the existence of solutions in the form (12.102) we observe that $\mathbf{H}$ and $\mathbf{E}$ are now subject to

$$\nabla \times \mathbf{H} = -i\omega\epsilon \mathbf{E}, \qquad \nabla \times \mathbf{E} = i\omega \boldsymbol{\mu} \mathbf{H}.$$

Since $\nabla \cdot \mathbf{H} = \nabla \cdot (\mathbf{H}_0 + \mathfrak{H}) = 0$ then the curl of $\nabla \times \mathbf{H}$ leads to

$$\Delta \mathfrak{H} + \omega^2 \epsilon \mu \mathfrak{H} = \mathbf{0}$$

whence

$$(-k^2 + \omega^2 \epsilon \mu_1)\mathfrak{H}_1 - i\omega^2 \epsilon \mu_2 \mathfrak{H}_2 = 0, \qquad (-k^2 + \omega^2 \epsilon \mu_1)\mathfrak{H}_2 + i\omega^2 \epsilon \mu_2 \mathfrak{H}_1 = 0, \tag{12.103}$$

where $\mu_1 = \mu_0(1 + \chi_1)$, $\mu_2 = \mu_0 \chi_2$. Nontrivial solutions occur if

$$(k^2 - \omega^2 \epsilon \mu_1)^2 - \omega^4 \epsilon^2 \mu_2^2 = 0.$$

Hence there are two wavenumber solutions

$$k_\pm = \omega \sqrt{\epsilon(\mu_1 \pm \mu_2)}.$$

Correspondingly, Eq. (12.103) implies

$$\mathfrak{H}_2 = \pm i \mathfrak{H}_1;$$

accordingly k_+ and k_- are the wavenumber solutions of two positive and negative circularly polarized waves.

Incidentally, since

$$k_\pm = \omega \sqrt{\epsilon \mu_0} \sqrt{1 + \omega_M/(\omega_C \mp \omega)}$$

then the negatively polarized wave (k_-) propagate with speed ω/k_- for any value of ω. The positively polarized wave (k_+) propagate with speed ω/k_+ with a stopband. Since

$$1 + \frac{\omega_M}{\omega_C - \omega} = \frac{\omega_C + \omega_M - \omega}{\omega_C - \omega}$$

it follows that propagation is forbidden if $\omega \in [\omega_C, \omega_C + \omega_M]$.

A linearly polarized plane wave may be viewed as the superposition of two counter-rotating circularly polarized waves and vice versa. Formally this is expressed by the identity

$$w_1 \mathbf{e}_1 \exp[i(\mathbf{k} \cdot \mathbf{x} - \omega t)] = \tfrac{1}{2}(w_1 \mathbf{e}_1 + i w_1 \mathbf{e}_2) \exp[i(\mathbf{k} \cdot \mathbf{x} - \omega t)] + \tfrac{1}{2}(w_1 \mathbf{e}_1 - i w_1 \mathbf{e}_2) \exp[i(\mathbf{k} \cdot \mathbf{x} - \omega t)].$$

In a ferrite medium the two circularly polarized waves propagate at different speeds. If L is the length of the sample, say $z \in [0, L]$, then the outgoing wave is

$$\begin{aligned} &\tfrac{1}{2}(w_1 \mathbf{e}_1 + i w_1 \mathbf{e}_2) \exp[i(k_+ z - \omega t)] + \tfrac{1}{2}(w_1 \mathbf{e}_1 - i w_1 \mathbf{e}_2) \exp[i(k_- z - \omega t)] \\ &= \tfrac{1}{2}(w_1 \mathbf{e}_1 + i w_1 \mathbf{e}_2) \exp[i((k_+ - k_-) z - \omega t)] + w_1 \mathbf{e}_1 \exp[i(k_- z - \omega t)] \end{aligned}$$

at $z = L$. The result is a linearly polarized wave, in the direction of the incident one, plus a rotated wave, the angle being $(k_+ - k_-)L$.

Thermodynamic consistency

Let the body be undeformable and then look at the thermodynamic inequality

$$-\rho(\dot{\psi} + \eta\dot{\theta}) + \mathbf{E} \cdot \dot{\mathbf{P}} + \mu_0 \mathbf{H} \cdot \dot{\mathbf{M}} \geq 0.$$

We let ψ, η, $\mathbf{H}$ and $\mathbf{E}$ be functions of θ, $\mathbf{P}$, $\mathbf{M}$ while $\mathbf{M}$ is subject to the rate equation (12.100). By

$$-\rho(\partial_\theta\psi + \eta)\dot{\theta} + (\mathbf{E} - \rho\partial_{\mathbf{P}}\psi) \cdot \dot{\mathbf{P}} + (\mu_0\mathbf{H} - \rho\partial_{\mathbf{M}}\psi) \cdot \dot{\mathbf{M}} \geq 0.$$

The linearity and the arbitrariness of $\dot{\theta}$ and $\dot{\mathbf{P}}$ imply

$$\eta = -\partial_\theta\psi, \qquad \mathbf{E} = \rho\partial_{\mathbf{P}}\psi.$$

and the inequality reduces to

$$(\mu_0\mathbf{H} - \rho\partial_{\mathbf{M}}\psi) \cdot \dot{\mathbf{M}} \geq 0. \tag{12.104}$$

Assume

$$\rho\partial_{\mathbf{M}}\psi = \alpha\mathbf{M}.$$

Consistent with (12.100) we let $\mathbf{M} = M_0\mathbf{e}_3 + \mathfrak{M}$. Hence it follows

$$(\mu_0\mathbf{H} - \rho\partial_{\mathbf{M}}\psi) \cdot \dot{\mathbf{M}} = [\mu_0(H_0\mathbf{e}_3 + \mathfrak{H}) - \alpha(M_0\mathbf{e}_3 + \mathfrak{M})] \cdot \gamma\mu_0[H_0\mathbf{e}_3 \times \mathfrak{M} + \mathfrak{H} \times M_0\mathbf{e}_3 + \mathfrak{H} \times \mathfrak{M}]$$

It follows that

$$(\mu_0\mathbf{H} - \rho\partial_{\mathbf{M}}\psi) \cdot \dot{\mathbf{M}} = 0$$

holds identically for any values of α, H_0, M_0. Accordingly, the model (12.100) of ferrite can be viewed as non-dissipative. Since also α might be a function of H_0 then the model of ferrite is significantly nonlinear. Incidentally the free energy can be written in the form

$$\rho\psi = \Psi(\theta, \mathbf{P}) + \tfrac{1}{2}\alpha(M_0^2 + 2M_0\mathfrak{M} \cdot \mathbf{e}_3 + \mathfrak{M}^2).$$

12.12.1 Magnetization Model with Damping Effects

An improvement of the evolution law is due to Landau and Lifshitz [270] who established the equation

$$\dot{\mathbf{M}} = \gamma\mu_0\mathbf{H}_{\text{eff}} \times \mathbf{M} - \lambda\mathbf{M} \times (\mathbf{M} \times \mathbf{H}_{\text{eff}}), \tag{12.105}$$

where $\gamma, \lambda \geq 0$; γ is a constant while λ is allowed to depend on $|\mathbf{M}|$. The vector $\mathbf{H}_{\text{eff}}$ denotes the effective magnetic field and is defined as

$$\mathbf{H}_{\text{eff}} = \mathbf{H} - \partial_{\mathbf{M}}\varphi(\mathbf{M}) + (\mathbf{A} \cdot \nabla\nabla)\mathbf{M},$$

φ being a scalar function of $\mathbf{M}$. Physically, the field $(\mathbf{A} \cdot \nabla\nabla)\mathbf{M}$ is related to the exchange energy $\frac{1}{2}A_{ij}\partial_{x_i}\mathbf{M} \cdot \partial_{x_j}\mathbf{M}$ and $\varphi(\mathbf{M})$ is the anisotropy energy; e.g. $\varphi = \beta(\mathbf{M} \cdot \mathbf{e})^2$, $\mathbf{e}$ being called the direction of easy magnetization.

The term $\lambda\mathbf{M} \times (\mathbf{M} \times \mathbf{H}_{\text{eff}})$ produces a damping effect in the following sense. Observe that $|\mathbf{M}|$ is constant in time. This is so because

$$\mathbf{M} \cdot \dot{\mathbf{M}} = -\lambda\mathbf{M} \cdot [\mathbf{M} \times (\mathbf{M} \times \mathbf{H}_{\text{eff}})] = -\lambda\mathbf{M} \cdot [(\mathbf{M} \cdot \mathbf{H}_{\text{eff}})\mathbf{M} - \mathbf{M}^2\mathbf{H}_{\text{eff}}] = 0$$

and hence $\mathbf{M}^2$ is constant. Instead, the inner product with $\mathbf{H}_{\text{eff}}$ results in

$$\mathbf{H}_{\text{eff}} \cdot \dot{\mathbf{M}} = -\lambda\mathbf{H}_{\text{eff}} \cdot [(\mathbf{M} \cdot \mathbf{H}_{\text{eff}})\mathbf{M} - \mathbf{M}^2\mathbf{H}_{\text{eff}}] = \lambda\mathbf{M}^2[\mathbf{H}^2_{\text{eff}} - (\frac{\mathbf{M}}{|\mathbf{M}|} \cdot \mathbf{H}_{\text{eff}})^2] \geq 0.$$

Let $\mathbf{H}$ be constant. Then

$$0 \leq \mathbf{H}_{\text{eff}} \cdot \dot{\mathbf{M}} = |\mathbf{H}_{\text{eff}}|\,|\mathbf{M}|\,\dot{\overline{\cos\alpha}}$$

α being the angle between $\mathbf{H}$ and $\mathbf{M}$. Hence α decreases in time and this is the meaning of the damping. Having in mind Eq. (12.105) we now investigate possible evolution equations for the magnetization $\mathbf{M}$.

Consider an undeformable (non-conducting) solid. The entropy inequality can be written in the form

$$-\rho(\dot{\psi} + \eta\dot{\theta}) + \mathbf{E} \cdot \dot{\mathbf{P}} + \mu_0\mathbf{H} \cdot \dot{\mathbf{M}} + \theta\nabla \cdot \mathbf{k} \geq 0.$$

We let ψ, η, $\mathbf{k}$ and $\dot{\mathbf{M}}$ be given by constitutive functions of the set of variables

$$(\theta, \mathbf{P}, \mathbf{H}, \mathbf{M}, \nabla\mathbf{M}, \dot{\mathbf{M}}, \nabla\nabla\mathbf{M}).$$

The evolution equation for $\mathbf{M}$ is consistent with the objectivity requirement in that the solid is undeformable and the rate $\dot{\mathbf{M}}$ is relative to the observer at rest with the body. Equation (12.105) is expected to be among those consistent with thermodynamics.

Evaluation of $\dot{\psi}$ and substitution in the entropy inequality results in

$$-\rho(\partial_\theta\psi + \eta)\dot{\theta} - (\rho\partial_{\mathbf{P}}\psi - \mathbf{E}) \cdot \dot{\mathbf{P}} - \rho\partial_{\mathbf{H}}\psi \cdot \dot{\mathbf{H}} - (\rho\partial_{\mathbf{M}}\psi - \mu_0\mathbf{H}) \cdot \dot{\mathbf{M}} - \rho\partial_{\nabla\mathbf{M}}\psi \cdot \nabla\dot{\mathbf{M}}$$
$$-\rho\partial_{\dot{\mathbf{M}}}\psi \cdot \ddot{\mathbf{M}} - \rho\partial_{\nabla\nabla\mathbf{M}}\psi \cdot \nabla\nabla\dot{\mathbf{M}} + \theta\nabla \cdot \mathbf{k} \geq 0.$$

Maxwell's equations

$$\nabla \times \mathbf{E} = -\mu_0(\dot{\mathbf{H}} + \dot{\mathbf{M}}), \qquad \nabla \times \mathbf{H} = \epsilon_0 \dot{\mathbf{E}} + \dot{\mathbf{P}},$$

$$\nabla \cdot (\mathbf{H} + \mathbf{M}) = 0, \qquad \nabla \cdot (\epsilon_0 \mathbf{E} + \mathbf{P}) = 0$$

allows us to say that at any point $\mathbf{x}$ and time t the values $\dot{\mathbf{H}}$ and $\dot{\mathbf{P}}$ can be chosen arbitrarily; the consistency with Maxwell's equations is guaranteed by appropriate values of $\nabla \times \mathbf{E}$ and $\nabla \times \mathbf{H}$. The arbitrariness of $\dot{\theta}$, $\dot{\mathbf{H}}$, and $\dot{\mathbf{P}}$ implies

$$\eta = -\partial_\theta \psi, \qquad \partial_{\mathbf{H}} \psi = \mathbf{0}, \qquad \mathbf{E} = \rho \partial_{\mathbf{P}} \psi.$$

Since $\dot{\mathbf{M}}$ is up to now an undetermined constitutive function we merely assume that

$$\partial_{\dot{\mathbf{M}}} \psi = \mathbf{0}, \qquad \partial_{\nabla \nabla \mathbf{M}} \psi = \mathbf{0}.$$

Next we divide throughout by θ and observe

$$-\frac{\rho}{\theta} \partial_{\nabla \mathbf{M}} \psi \cdot \nabla \dot{\mathbf{M}} = -\nabla \cdot (\frac{\rho}{\theta} \partial_{\nabla \mathbf{M}} \psi \dot{\mathbf{M}}) + [\nabla \cdot (\frac{\rho}{\theta} \partial_{\nabla \mathbf{M}} \psi)] \dot{\mathbf{M}}.$$

Hence the inequality becomes

$$\frac{1}{\theta}[\mu_0 \mathbf{H} - \rho \partial_{\mathbf{M}} \psi + \theta \nabla \cdot (\frac{\rho}{\theta} \partial_{\nabla \mathbf{M}} \psi)] \cdot \dot{\mathbf{M}} + \nabla \cdot (\mathbf{k} - \frac{\rho}{\theta} \partial_{\nabla \mathbf{M}} \psi \dot{\mathbf{M}}) \geq 0.$$

We then let

$$\mathbf{k} = \frac{\rho}{\theta} \partial_{\nabla \mathbf{M}} \psi \dot{\mathbf{M}}.$$

Consequently the entropy inequality reduces to

$$[\mu_0 \mathbf{H} - \rho \partial_{\mathbf{M}} \psi + \theta \nabla \cdot (\frac{\rho}{\theta} \partial_{\nabla \mathbf{M}} \psi)] \cdot \dot{\mathbf{M}} \geq 0. \tag{12.106}$$

We look for dependences of $\dot{\mathbf{M}}$ on the independent variables satisfying inequality (12.106).

For any two vectors $\mathbf{v}$, $\mathbf{w}$,the inequality $\mathbf{v} \cdot \mathbf{w} \geq 0$ holds if

$$\mathbf{w} = \beta \mathbf{M} \times \mathbf{v} - \lambda \mathbf{M} \times (\mathbf{M} \times \mathbf{v}), \qquad \beta \in \mathbb{R}, \quad \lambda \in \mathbb{R}^+.$$

Indeed

$$\mathbf{v} \cdot \mathbf{w} = \lambda [\mathbf{M}^2 \mathbf{v}^2 - (\mathbf{M} \cdot \mathbf{v})^2]$$

and then $\lambda \geq 0$ makes $\mathbf{v} \cdot \mathbf{w} \geq 0$ to hold. Hence letting $\mathbf{w} = \dot{\mathbf{M}}$, $\mathbf{v} = \mu_0 \mathbf{H} - \rho \partial_{\mathbf{M}} \psi + \theta \nabla \cdot (\rho \partial_{\nabla \mathbf{M}} \psi / \theta)$ we conclude that inequality (12.106) holds and meanwhile the evolution equation (12.105) holds too with

$$\mathbf{H}_{\text{eff}} = \mathbf{H} - \frac{1}{\mu_0}[\rho\partial_{\mathbf{M}}\psi - \theta\nabla\cdot(\frac{\rho}{\theta}\partial_{\nabla\mathbf{M}}\psi)].$$

For definiteness, let

$$\psi = \psi_0(\theta) + \tfrac{1}{2}[c_1|\mathbf{M}|^2 + c_2(\mathbf{M}\cdot\mathbf{e})^2 + c_3\nabla\mathbf{M}\cdot\mathbf{A}\nabla\mathbf{M}]$$

where c_1, c_2 and $\mathbf{A} \in \text{Sym}^+$ may depend on θ while $\mathbf{e}$ is a possibly-privileged unit vector. For simplicity let ρ, θ be uniform. Upon substitution of ψ the field $\mathbf{H}_{\text{eff}}$ is found to have the form

$$\mathbf{H}_{\text{eff}} = \mathbf{H} - \frac{\rho}{\mu_0}[c_1\mathbf{M} + c_2|\mathbf{M}\cdot\mathbf{e}|\mathbf{e} + c_3(\mathbf{A}\cdot\nabla\nabla)\mathbf{M}].$$

Electromagnetic waves

Parallel fields $\mathbf{M}, \mathbf{H}$ produce a stationary solution, $\dot{\mathbf{M}} = \mathbf{0}$. Denote by $\mathbf{H} = H_0\mathbf{e}_3$, $\mathbf{M} = M_0\mathbf{e}_3$ the stationary solution and let

$$\mathbf{H} = H_0\mathbf{e}_3 + \mathfrak{H}, \qquad \mathbf{M} = M_0\mathbf{e}_3 + \mathfrak{M}.$$

The linear approximation to $\dot{\mathbf{M}} = \gamma\mu_0\mathbf{H}\times\mathbf{M} - \lambda\mathbf{M}\times(\mathbf{M}\times\mathbf{H})$ reads

$$\dot{\mathfrak{M}} = \gamma\mu_0(\mathbf{H}_0\times\mathfrak{M} + \mathfrak{H}\times\mathbf{M}_0) - \lambda(M_0H_0\mathfrak{M} - \mathbf{M}_0\cdot\mathfrak{M}\,\mathbf{H}_0 + \mathbf{M}_0\cdot\mathfrak{H}\,\mathbf{M}_0 - M_0^2\mathfrak{H}). \tag{12.107}$$

We compute $\dot{\mathfrak{M}}\cdot\mathbf{e}_3$ and find that

$$\dot{\mathfrak{M}}\cdot\mathbf{e}_3 = -\lambda(M_0H_0\mathfrak{M}\cdot\mathbf{e}_3 - M_0H_0\mathfrak{M}\cdot\mathbf{e}_3 + M_0^2\mathfrak{H}\cdot\mathbf{e}_3 - M_0^2\mathfrak{H}\cdot\mathbf{e}_3) = 0.$$

Hence we let $\mathfrak{M}\cdot\mathbf{e}_3 = 0$. Moreover, since $\nabla\cdot\mathbf{B} = 0$ then

$$\nabla\cdot(\mathfrak{H} + \mathfrak{M}) = 0.$$

So we let also $\mathfrak{H}\cdot\mathbf{e}_3 = 0$.

We now investigate the existence of time-harmonic solutions,

$$\mathfrak{H} = \mathfrak{H}_0\exp(-i\omega t), \qquad \mathfrak{M} = \mathfrak{M}_0\exp(-i\omega t),$$

where $\mathfrak{H}_0 = \mathfrak{H}_1\mathbf{e}_1 + \mathfrak{H}_2\mathbf{e}_2$ and the like for $\mathfrak{M}$. Upon substitution in (12.107) we find the system of equations

$$\begin{bmatrix} -i\omega + \lambda M_0H_0 & \gamma\mu_0H_0 \\ -\gamma\mu_0H_0 & -i\omega + \lambda M_0H_0 \end{bmatrix}\begin{bmatrix}\mathfrak{M}_1 \\ \mathfrak{M}_2\end{bmatrix} = \begin{bmatrix} \lambda M_0^2\mathfrak{H}_1 + \gamma\mu_0M_0\mathfrak{H}_2 \\ -\gamma\mu_0M_0\mathfrak{H}_1 + \lambda M_0^2\mathfrak{H}_2\end{bmatrix}.$$

We then compute $\mathfrak{M}$ in terms of $\mathfrak{H}$ to obtain

$$\begin{bmatrix} \mathfrak{M}_1 \\ \mathfrak{M}_2 \end{bmatrix} = \begin{bmatrix} \chi_1 & \chi_2 \\ -\chi_2 & \chi_1 \end{bmatrix} \begin{bmatrix} \mathfrak{H}_1 \\ \mathfrak{H}_2 \end{bmatrix},$$

where

$$\chi_1 = \frac{(-i\omega + \lambda M_0 H_0)\lambda M_0^2 + \gamma^2 \mu_0^2 H_0 M_0}{(\gamma \mu_0 H_0)^2 + (\lambda M_0 H_0)^2 - \omega^2 - 2i\omega\lambda M_0 H_0},$$

$$\chi_2 = \frac{(-i\omega + \lambda M_0 H_0)\gamma \mu_0 H_0 - \gamma \mu_0 H_0 \lambda M_0^2}{(\gamma \mu_0 H_0)^2 + (\lambda M_0 H_0)^2 - \omega^2 - 2i\omega\lambda M_0 H_0}.$$

and hence

$$\mathfrak{B} = \mu \mathfrak{H}, \qquad \mu = \mu_0 \begin{bmatrix} 1 + \chi_1 & \chi_2 \\ -\chi_2 & 1 + \chi_1 \end{bmatrix}.$$

The field $\mathfrak{H}$ satisfies the differential equation

$$\Delta \mathfrak{H} + \omega^2 \epsilon \mu \mathfrak{H} = \mathbf{0}.$$

Part IV
Hysteresis and Phase Transitions

Chapter 13
Plasticity

Plasticity describes the non-reversible deformation of a material in response to applied forces. The physical mechanisms that cause plastic, that is, non-reversible, deformation can vary widely. For many metals, tensile loading applied to a sample will cause it to behave in an elastic manner. However, once the load exceeds a threshold, the extension increases more rapidly than in the elastic region. Moreover, when the load is removed, some degree of extension will remain. Perfect plasticity is a property of materials to undergo irreversible deformation without any increase in stresses or loads. Models of plastic materials involve a combination of elastic and (perfect) plastic behaviours.

Consistent with the involved nature of plasticity this chapter exhibits well-known concepts (yield criteria) and classical models (gradient theories, Kröner decomposition). Next a decomposition-free approach to the modelling of plastic materials is developed through the essential role of the entropy production as a constitutive function. This allows a general scheme where the hyperelastic regime, the hypoelastic regime, and the hysteretic regime occur depending on the free energy and the entropy production. Definite examples are given of hysteretic models. Polymeric foams are also modelled as hysteretic materials.

13.1 Qualitative Aspects of the Stress–Strain Curve

Also to introduce a useful terminology, consider a tension test on a cylindrical specimen. Let L_0 and A_0 be the length and the cross-sectional area of the specimen; upon deformation they are changed to L and A. Let P denote the axial force and define

$$s = \frac{P}{A_0}, \qquad \sigma = \frac{P}{A};$$

A. Morro and C. Giorgi, *Mathematical Modelling of Continuum Physics*,
Modeling and Simulation in Science, Engineering and Technology,
https://doi.org/10.1007/978-3-031-20814-0_13

s and σ can be viewed as the Piola stress and the Cauchy stress. Let $\lambda = L/L_0$ and $e = \lambda - 1$. By increasing P (and hence s and σ) the $s - e$ curve is essentially linear as is in elastic bodies. Beyond the elastic limit, called *yield strength,* the specimen shows a monotone increasing dependence of σ and s on e, with a decisively lower slope; this phenomenon is called *strain hardening.* Moreover, beyond the yield strength a reduction of P results in an elastic unloading curve. Unloading to zero stress produces a permanent plastic strain e^p. At any point of this curve, upon increasing the stress (reloading) the stress–strain curve follows the elastic behaviour up to the point from which the unloading was initiated. By a further loading, the specimen follows the hardening curve. If, instead, P becomes a compression along the unloading curve then reverse yielding begins at the stress $\sigma_r (< 0)$. If σ_f is the value of stress from which the reversed deformation was initiated in the hardening curve, then it happens that

$$|\sigma_r| < \sigma_f;$$

the stress at which plastic flow recommences upon load reversal is smaller than that at which the reversed loading was initiated (Bauschinger effect). The closed interval $[\sigma_r, \sigma_f]$ is called the elastic range.

The rather complicate character of plasticity gives reasons of the various approaches set up to describe the corresponding phenomena. Many approaches are based on the decomposition of the displacement gradient $\mathbf{H} = \nabla_R \mathbf{u}$ as

$$\mathbf{H} = \mathbf{H}^e + \mathbf{H}^p,$$

where $\mathbf{H}^e$ is the elastic distortion and $\mathbf{H}^p$ is the plastic distortion. Hence, the elastic and plastic strains are considered in the form[1]

$$\mathcal{E}^e = \operatorname{sym} \mathbf{H}^e, \qquad \mathcal{E}^p = \operatorname{sym} \mathbf{H}^p.$$

In the simplest approach, it is assumed that $\operatorname{tr} \mathbf{H}^p = \operatorname{tr} \mathcal{E}^p = 0$, which holds for isochoric small deformations.

In the isothermal approximation, the Clausius–Duhem inequality, in the Lagrangian description, can be taken in the simple form

$$\rho_R \dot{\psi} - \mathbf{T}_R \cdot \dot{\mathbf{F}} \leq 0.$$

Since $\mathbf{T}_R = J\mathbf{T}\mathbf{F}^{-T}$, in the approximation of *small deformations* $\mathbf{T}_R \simeq J\mathbf{T} \in \mathrm{Sym}$. Hence[2]

$$\mathbf{T}_R \cdot \dot{\mathbf{F}} \simeq \mathbf{T}_R \cdot \dot{\mathcal{E}} = \mathbf{T}_R \cdot \dot{\mathcal{E}}^e + \mathbf{T}_R \cdot \dot{\mathcal{E}}^p.$$

Consequently, the Clausius–Duhem inequality can be written as

---

[1] To avoid ambiguities with the different strain tensors we observe that $\mathbf{E} = (\mathbf{C} - \mathbf{1})/2$ is the Green–St. Venant strain, $\mathcal{E} = \operatorname{sym}\mathbf{H}$ is the strain, and $\varepsilon = \operatorname{sym}\nabla\mathbf{u}$ is the infinitesimal strain.

[2] Henceforth the power $\mathbf{T}_R \cdot \dot{\mathcal{E}}$ is meant to be $J\mathbf{T} \cdot \dot{\varepsilon}$.

$$\rho_R \dot{\psi} - \mathbf{T}_R \cdot \dot{\mathcal{E}}^e - \mathbf{T}_R \cdot \dot{\mathcal{E}}^p \le 0.$$

Let $\mathbf{T}_0 := \operatorname{dev} \mathbf{T}_R$. The assumption[3] tr $\mathcal{E}^p = 0$ implies that

$$\mathbf{T}_R \cdot \mathcal{E}^p = \mathbf{T}_0 \cdot \dot{\mathcal{E}}^p.$$

If now it is assumed that energy dissipation is confined to the plastic behaviour then the Clausius–Duhem inequality splits into

$$\rho_R \dot{\psi} - \mathbf{T}_R \cdot \dot{\mathcal{E}}^e = 0, \qquad \mathbf{T}_0 \cdot \dot{\mathcal{E}}^p \ge 0. \tag{13.1}$$

A simple way to satisfy the inequality in (13.1) might be to take

$$\mathbf{T}_0 = \kappa \frac{\dot{\mathcal{E}}^p}{|\dot{\mathcal{E}}^p|}, \qquad \kappa \ge 0,$$

as $\dot{\mathcal{E}}^p \neq \mathbf{0}$. Yet the stress–strain curve indicates a more involved behaviour and this is realized by means of a scalar internal variable S, called the *hardening variable,* such that

$$\mathbf{T}_0 = \hat{\mathbf{T}}_0(\dot{\mathcal{E}}^p, S).$$

The scheme is completed by a constitutive equation for S; the evolution of S is assumed to be given by a rate equation

$$\dot{S} = h(\dot{\mathcal{E}}^p, S), \tag{13.2}$$

the function h providing the *hardening rate.*

13.1.1 Yield Criteria

Ductile materials can be stretched without breaking and drawn into thin wires; aluminum, copper, tin, mild steel, platinum, and lead are examples of ductile materials. *Yield points* are the points below which the body deforms elastically and beyond which the body deforms plastically with increase in stress applied. If the body is subject to a uniaxial tensile load it will start yielding, i.e. deforming plastically, when the stress reaches the yield stress, say σ_y. When the state of stress is triaxial a single stress does not characterize the yielding.

In general, we can assume that yield occurs at a point when some combination of the stress components reaches some critical value, say

---

[3] The plastic behaviour is usually ascribed to the microscopic structure and the occurrence of dislocations. The flow of dislocations is regarded to be isochoric and hence $\mathbf{H}^p$ is assumed to be deviatoric, tr $\mathbf{H}^p = 0$.

$$F(\mathbf{T}) = k,$$

$F(\mathbf{T})$ denoting a function of the components of $\mathbf{T}$ while k is a parameter to be determined experimentally. Alternatively we may express the dependence on $\mathbf{T}$ by means of the principal stresses T_1, T_2, T_3 and the principal directions $\mathbf{n}_1, \mathbf{n}_2, \mathbf{n}_3$. If the material is isotropic the function F is independent of the directions $\{\mathbf{n}_i\}$ and hence the yield criterion simplifies to

$$F(T_1, T_2, T_3) = k.$$

Since the principal invariants I_1, I_2, I_3 of stress are independent of the material orientation then we might write the condition in the form[4]

$$F(I_1, I_2, I_3) = k \qquad \text{or} \qquad F(I_1, J_2, J_3) = k,$$

where J_2, J_3 are the principal invariants of the deviator $\mathbf{T}_0$. Often it is assumed that the yield stress is independent of $I_1 = \text{tr}\,\mathbf{T}$ (hydrostatic stress) and hence the condition becomes

$$F(J_2, J_3) = k.$$

For definiteness observe that for ductile materials there are two main yield criteria, *von Mises* criterion and *Tresca* criterion. Let

$$T_m = \tfrac{1}{3}\text{tr}\,\mathbf{T} = \tfrac{1}{3}(T_1 + T_2 + T_3).$$

The scalar T_m is named the *hydrostatic stress;* it is a common assumption that T_m does not influence the yielding. Hence, the yield criterion is characterized by a condition on the deviator,

$$\mathbf{T}_0 = \mathbf{T} - T_m\mathbf{1}.$$

von Mises Criterion

The von Mises criterion is in fact the result of subsequent contributions of Lévy, von Mises, Prandtl, and Reuss. Indeed, it seems that the criterion was anticipated by Maxwell in 1856 [216]. We show that the criterion is in fact based on considerations about the strain energy.

Let $\{T_i\}$, $\{E_i\}$ be the principal stresses and strains. Based on linear elasticity we consider the energy, per unit volume, in the form

$$U_T = \tfrac{1}{2}T_iE_i.$$

[4] Though the function changes depending on the pertinent variables we keep using the same symbol F.

In view of the relations (4.10), we find

$$U_T = \frac{1}{2E}[T_1^2 + T_2^2 + T_3^2 - 2\nu(T_1T_2 + T_1T_3 + T_2T_3)].$$

Now consider what is called the hydrostatic energy, that is, the energy associated with the isotropic stress $T_m = \frac{1}{3}\sum_i T_i$. Replacing T_1, T_2, T_3 in U_T with T_m, we find

$$U_H = \frac{1-2\nu}{6E}[T_1^2 + T_2^2 + T_3^2 + 2(T_1T_2 + T_1T_3 + T_2T_3)].$$

Consequently, the distortion energy $U_D = U_T - U_H$ can be written in the form

$$U_D = \frac{1+\nu}{3E}[T_1^2 + T_2^2 + T_3^2 - T_1T_2 - T_1T_3 - T_2T_3] = \frac{1+\nu}{6E}[(T_1-T_2)^2 + (T_2-T_3)^2 + (T_2-T_3)^2].$$

In terms of U_D the (equivalent) von Mises stress is defined as

$$T_{\text{Mises}} := [T_1^2 + T_2^2 + T_3^2 - T_1T_2 - T_1T_3 - T_2T_3]^{1/2}$$

and the von Mises criterion is assumed in the form

$$T_{\text{Mises}} = Y,$$

Y being the yield stress. Equivalently, we have

$$(T_1 - T_2)^2 + (T_2 - T_3)^2 + (T_1 - T_3)^2 = 2Y^2. \tag{13.3}$$

To substantiate the von Mises criterion (13.3) we observe that an alternative statement is

$$|J_2| = k^2.$$

Since

$$|J_2| = \tfrac{1}{6}[(T_1 - T_2)^2 + (T_2 - T_3)^2 + (T_1 - T_3)^2]$$

then the criterion becomes

$$(T_1 - T_2)^2 + (T_2 - T_3)^2 + (T_1 - T_3)^2 = 6k^2.$$

In a uniaxial tension test, we have

$$T_{\max} = Y, \qquad 2T_{\max}^2 = 6k^2, \qquad k = Y/\sqrt{3}$$

and hence the von Mises criterion (13.3) follows.

Tresca Criterion

The Tresca criterion states that a material will yield if the maximum shear stress reaches a critical value. This is made formal by assuming

$$\max\{\tfrac{1}{2}|T_1 - T_2|, \tfrac{1}{2}|T_2 - T_3|, \tfrac{1}{2}|T_1 - T_3|\} = k.$$

In a simple tension test, we have $k = Y/2$. Hence we find the criterion

$$\max\{|T_1 - T_2|, |T_2 - T_3|, |T_1 - T_3|\} = Y. \tag{13.4}$$

It is of interest to contrast the two relations (13.3) and (13.4) in the plane stress condition $T_3 = 0$. Equation (13.4) simply becomes $\max|T_1 - T_2| = Y$. In the same plane geometry Eq. (13.3) becomes

$$(T_1 - T_2)^2 + T_2^2 + T_1^2 = 2Y^2.$$

Change the stress representation by taking a rotation of $\pi/4$ counterclockwise. We have

$$\hat{T}_1 = (1/\sqrt{2})(T_1 + T_2), \qquad \hat{T}_2 = (1/\sqrt{2})(T_1 - T_2).$$

Upon substitution we find

$$\frac{\hat{T}_2}{2/3} + \frac{\hat{T}_1}{2} = Y^2,$$

which is the equation of the ellipsis, centred at the origin, with major half-axis equal to $\sqrt{2}\,Y$, at $\theta = \pi/4$, and minor half-axis equal to $\sqrt{2/3}\,Y$.

13.2 Rate-Independent Scheme of Plastic Flow

Rate-independent models for plasticity are widely applied, at least at low temperatures.

Observe that a change in time scale is a transformation $t \to t^*$ of the form $t^* = \alpha t$, $\alpha > 0$; a function f is rate independent if $f(t^*) = f(t)$. Let $\dot{\boldsymbol{\mathcal{E}}}^p \neq \mathbf{0}$. The tensor

$$\mathbf{N}^p := \frac{\dot{\boldsymbol{\mathcal{E}}}^p}{|\dot{\boldsymbol{\mathcal{E}}}^p|}$$

is rate independent in that

$$\frac{d}{dt}\boldsymbol{\mathcal{E}}^p(\alpha t) = \alpha\dot{\boldsymbol{\mathcal{E}}}^p(t).$$

Apparently $\mathbf{N}^p$ is invariant. As for the function h in (13.2) we observe that

$$\alpha\dot{S} = h(\alpha\dot{\mathcal{E}}^p, S), \qquad \dot{S} = \alpha^{-1}h(\alpha\dot{\mathcal{E}}^p, S).$$

Hence the flow rule, for $\mathbf{T}_0$, and the hardening equation, for S, are rate independent if

$$\mathbf{T}_0 = \tilde{\mathbf{T}}_0(\mathbf{N}^p, S), \qquad \dot{S} = h(\mathbf{N}^p, S)\,|\dot{\mathcal{E}}^p|.$$

By the assumption $\mathbf{T}_0 \cdot \dot{\mathcal{E}}^p \geq 0$, dividing by $|\dot{\mathcal{E}}^p|$, we have

$$Y(\mathbf{N}^p, S) := \mathbf{N}^p \cdot \tilde{\mathbf{T}}_0(\mathbf{N}^p, S) \geq 0;$$

the scalar Y is called the *flow resistance.*

A further assumption is made on the function $\tilde{\mathbf{T}}_0$, namely,

$$\frac{\mathbf{T}_0}{|\mathbf{T}_0|} = \frac{\dot{\mathcal{E}}^p}{|\dot{\mathcal{E}}^p|} = \mathbf{N}^p,$$

referred to as the *codirectionality* of $\mathbf{T}_0$ and $\dot{\mathcal{E}}^p$. Consequently

$$|\mathbf{T}_0| = \mathbf{T}_0 \cdot \mathbf{N}^p = Y(\mathbf{N}^p, S)$$

and hence

$$\mathbf{T}_0 = Y(\mathbf{N}^p, S)\mathbf{N}^p.$$

A further restriction is taken on the functions $Y(\mathbf{N}^p, S)$ and $h(\mathbf{N}^p, S)$, namely, that they be independent of the flow direction $\mathbf{N}^p$. Hence, we have the *Mises flow equations*

$$\mathbf{T}_0 = Y(S)\mathbf{N}^p, \qquad \dot{S} = h(S)|\dot{\mathcal{E}}^p|.$$

These equations are operative once the initial condition for S is chosen. We let $Y(S) \neq S$ and state the initial-value problem in the form

$$\dot{S} = h(S)|\dot{\mathcal{E}}^p|, \qquad S(0) = 0. \tag{13.5}$$

Let e^p denote the *accumulated plastic strain* e^p defined by

$$\dot{e}^p = |\dot{\mathcal{E}}^p|, \qquad e^p(0) = 0,$$

the dependence on the position $\mathbf{X}$ being omitted. Hence

$$e^p(t) = \int_0^t |\dot{\mathcal{E}}^p|(\tau)d\tau \geq 0.$$

We now show that the hardening variable S can be related by a 1–1 correspondence with e^p. Consider the initial-value problem

$$\frac{dS}{de^p} = h(S), \qquad S(0) = 0.$$

To determine the function $S(t) = \hat{S}(e^p)$ we observe

$$\dot{S} = \frac{d\hat{S}}{de^p}\dot{e}^p = h(S)|\dot{\boldsymbol{\mathcal{E}}}^p|.$$

Hence the function $\hat{S}(e^p)$ is a solution of (13.5). This in turn allows us to let the flow resistance Y be a function of e^p,

$$Y = Y(e^p) = Y(\hat{S}(e^p)).$$

While $|\mathbf{T}_0| = Y$ if $\dot{\boldsymbol{\mathcal{E}}}^p \neq \mathbf{0}$, an additional assumption is considered to characterize no-flow states, that is, $|\mathbf{T}_0| \leq Y$ and $\dot{\boldsymbol{\mathcal{E}}}^p = \mathbf{0}$ for $|\mathbf{T}_0| < Y$. The model is then expressed in terms of the accumulated plastic strain e^p by means of the *Mises–Hill equations*

$$\mathbf{T}_0 = Y(e^p)\mathbf{N}^p \ \text{ for } \ \dot{\boldsymbol{\mathcal{E}}}^p \neq \mathbf{0}; \qquad \dot{e}^p = |\dot{\boldsymbol{\mathcal{E}}}^p|, \quad e^p(0) = 0,$$

$$|\mathbf{T}_0| \leq Y(e^p), \qquad \dot{\boldsymbol{\mathcal{E}}}^p = \mathbf{0} \ \text{ for } \ |\mathbf{T}_0| < Y(e^p).$$

13.2.1 Consistency of Mises–Hill Model with a Defect Energy

Instead of viewing the flow $\dot{\boldsymbol{\mathcal{E}}}^p$ as a purely dissipative term, namely, $\mathbf{T}_0 \cdot \dot{\boldsymbol{\mathcal{E}}}^p \geq 0$, we now generalize the scheme by letting the free energy ψ be the sum,

$$\psi = \psi^e + \psi^p,$$

of an elastic free energy ψ^e and a plastic free energy ψ^p, this being associated with the defect energy in the solid. For simplicity, it is assumed that

$$\rho_R \dot{\psi}^e = \mathbf{T} \cdot \dot{\boldsymbol{\mathcal{E}}}^e$$

and hence the Clausius–Duhem inequality reduces to

$$\rho_R \dot{\psi}^p - \mathbf{T}_0 \cdot \dot{\boldsymbol{\mathcal{E}}}^p \leq 0. \tag{13.6}$$

Let ψ^p be a function of e^p. It is assumed that

$$g(e^p) := \rho_R \frac{d\psi^p}{de^p}$$

is subject to

$$g(0) = 0, \qquad g(e^p) > 0 \quad \text{for} \quad e^p > 0.$$

Since $\dot{e}^p = |\dot{\mathcal{E}}^p|$ then

$$\rho_R \frac{d}{dt} \psi^p(e^p) = g(e^p)|\dot{\mathcal{E}}^p| \geq 0.$$

Moreover, since $|\dot{\mathcal{E}}^p| = \dot{\mathcal{E}}^p \cdot \mathbf{N}^p$ then inequality (13.6) becomes

$$(g\mathbf{N}^p - \mathbf{T}_0) \cdot \dot{\mathcal{E}}^p \leq 0.$$

Let

$$\mathbf{T}_0^{\text{dis}} := \mathbf{T}_0 - g\mathbf{N}^p.$$

The inequality becomes

$$\mathbf{T}_0^{\text{dis}} \cdot \dot{\mathcal{E}}^p \geq 0$$

and $\mathbf{T}_0^{\text{dis}}$ can be viewed as the dissipative part of the deviatoric stress $\mathbf{T}_0$; $\mathbf{T}_0^{\text{dis}}$ can then be taken in the form

$$\mathbf{T}_0^{\text{dis}} = Y_0 \mathbf{N}^p,$$

Y_0 being any positive-valued function of e^p. For the sake of simplicity we let Y_0 be constant. Consequently, we have

$$\mathbf{T}_0 = [Y_0 + g(e^p)]\mathbf{N}^p.$$

The Mises–Hill equation $\mathbf{T}_0 = Y(e^p)\mathbf{N}^p$ is recovered by simply letting $Y(e^p) = Y_0 + g(e^p)$.

13.3 A Temperature-Dependent Model

The Mises–Hill model may be generalized by letting the stress tensor and the free energy depend also on the temperature. In addition to provide a more precise model, the temperature dependence allows us to make a better distinction between dissipative and non-dissipative powers.[5]

Still within the approximation of small deformations, the stress power is written as $\mathbf{T}_R \cdot \dot{\mathcal{E}}$. We then omit the suffix $_R$ for $\mathbf{T}$ and write the Clausius–Duhem inequality in the form

$$-\rho_R(\dot{\psi} + \eta\dot{\theta}) + \mathbf{T} \cdot \dot{\mathcal{E}} - \frac{1}{\theta}\mathbf{q}_R \cdot \nabla_R \theta \geq 0.$$

[5] This section parallels Chap. 81 of [216].

We let $\mathcal{E} = \mathcal{E}^e + \mathcal{E}^p$ and use the accumulated plastic strain e^p to describe the plastic behaviour; since $\dot{e}^p = |\dot{\mathcal{E}}^p|$ then $\dot{\mathcal{E}}^p = \dot{e}^p \mathbf{N}^p$. Hence $\mathbf{T} \cdot \dot{\mathcal{E}} = \mathbf{T} \cdot \dot{\mathcal{E}}^e + \mathbf{T} \cdot \dot{\mathcal{E}}^p$ and

$$\mathbf{T} \cdot \dot{\mathcal{E}}^p = \tau \dot{e}^p, \qquad \tau := \mathbf{T}_0 \cdot \mathbf{N}^p,$$

so that τ, called the *resolved shear,* is the deviatoric stress resolved on the direction of plastic flow $\mathbf{N}^p$.

To determine the set of constitutive equations, we start with the assumption that

$$\psi = \psi(\mathcal{E}^e, e^p, \theta) = \psi^e(\mathcal{E}^e, \theta) + \psi^p(e^p, \theta)$$

and let $\mathbf{T}$, η, $\mathbf{q}_R$ depend on $\mathcal{E}^e$, e^p, θ, $\nabla_{\!R}\theta$. Instead, to allow for dissipativity, we let τ depend on e^p, θ, $\dot{e}^p$. The Clausius–Duhem inequality can be written as

$$-\rho_R(\partial_\theta \psi + \eta)\dot{\theta} - \rho_R(\partial_{\mathcal{E}^e}\psi^e - \mathbf{T}) \cdot \dot{\mathcal{E}}^e - \rho_R(\partial_{e^p}\psi^p - \tau)\dot{e}^p - \frac{1}{\theta}\mathbf{q}_R \cdot \nabla_{\!R}\theta \geq 0.$$

The arbitrariness of $\dot{\theta}$ and $\dot{\mathcal{E}}^e$ implies that

$$\eta = -\partial_\theta \psi, \qquad \mathbf{T} = \partial_{\mathcal{E}^e}\psi^e,$$

$$\rho_R(\partial_{e^p}\psi^p - \tau)\dot{e}^p + \frac{1}{\theta}\mathbf{q}_R \cdot \nabla_{\!R}\theta \leq 0.$$

If τ is independent of $\nabla_{\!R}\,\theta$ then it follows

$$(\tau - \partial_{e^p}\psi^p)\dot{e}^p \geq 0.$$

Let

$$\tau_{\text{dis}} := \tau - \partial_{e^p}\psi^p, \qquad \tau = \tau_{\text{dis}} + \partial_{e^p}\psi^p.$$

Hence

$$\tau_{\text{dis}}\dot{e}^p \geq 0.$$

Since $\dot{e}^p \geq 0$ then the inequality holds if and only if

$$\tau_{\text{dis}} \geq 0.$$

The positive valuedness of τ_{dis} is realized by a positive-valued function $Y_{\text{dis}}(e^p, \dot{e}^p, \theta)$ or $Y_{\text{dis}}(e^p, \theta)$.

Finally, letting $\dot{e}^p = 0$ we obtain the heat-conduction inequality

$$\mathbf{q}_R \cdot \nabla_{\!R}\theta \leq 0.$$

To complete the scheme we may require the codirectionality constraint,

$$\mathbf{T}_0 = |\mathbf{T}_0|\mathbf{N}^p.$$

Hence

$$\tau = |\mathbf{T}_0|, \qquad \mathbf{T}_0 = \tau\mathbf{N}^p = (Y_{\text{dis}} + \partial_{e^p}\psi^p)\mathbf{N}^p.$$

It is expected that the plastic stress power $\mathbf{T}_0 \cdot \dot{\boldsymbol{\mathcal{E}}}^p$ be non-negative. Now, the free energy ψ^p is taken to represent the defect energy and it is likely that the defect energy increases as the number of defects, described by e^p, increases. Hence, we let

$$\partial_{e^p}\psi^p \geq 0,$$

whence $\partial_{e^p}\psi^p\, \dot{e}^p \geq 0$. As a consequence

$$\mathbf{T}_0 \cdot \dot{\boldsymbol{\mathcal{E}}}^p = \tau\dot{e}^p = (Y_{\text{dis}} + \partial_{e^p}\psi^p)\dot{e}^p \geq 0.$$

It is worth checking the consequences of the model so established on the balance of energy. By

$$\rho_R\dot{\varepsilon} = \mathbf{T} \cdot \dot{\boldsymbol{\mathcal{E}}} - \nabla_R \cdot \mathbf{q}_R + \rho_R r$$

we have

$$\rho_R(\dot{\psi} + \dot{\theta}\eta + \theta\dot{\eta}) = \mathbf{T} \cdot \dot{\boldsymbol{\mathcal{E}}} - \nabla_R \cdot \mathbf{q}_R + \rho_R r.$$

Owing to the thermodynamic restrictions, we find

$$\rho_R\theta\dot{\eta} = \rho_R r - \nabla_R \cdot \mathbf{q}_R + (\tau - \rho_R\partial_{e^p}\psi^p)\dot{e}^p = \rho_R r - \nabla_R \cdot \mathbf{q}_R + Y_{\text{dis}}\dot{e}^p.$$

Hence

$$\rho_R\theta\dot{\eta} - (\rho_R r - \nabla_R \cdot \mathbf{q}_R) = Y_{\text{dis}}\dot{e}^p \geq 0;$$

the plastic flow produces a growth of entropy.

13.4 Gradient Theories of Plasticity

Gradient theories refer to models involving constitutive dependencies on the gradient of the plastic strain. This scheme is motivated by the need of justifying flow rules related to the size dependence.

Attempts at theories with size dependence were originated by the flow rule [2]

$$\mathbf{T}_0 = (Y(e^p) - \beta\Delta_R e^p)\mathbf{N}^p,$$

which implies that the resolved shear $\tau = \mathbf{T}_0 \cdot \mathbf{N}^p$ takes the form

$$\tau = Y(e^p) - \beta\Delta_R e^p,$$

Δ_R being the Laplacian relative to the reference coordinates. A thermodynamic scheme allowing for the dependence of the resolved shear on $\Delta_R e^p$ may be given by taking

$$\boldsymbol{\mathcal{E}}^e, e^p, \nabla_R e^p, \dot{e}^p$$

as the set of independent variables. Again we let

$$\mathbf{T}^e \cdot \dot{\boldsymbol{\mathcal{E}}}^e + \tau^p \cdot \dot{e}^p$$

be the power density, where $\tau^p = \mathbf{T}_0 \cdot \mathbf{N}^p$. In addition we let

$$\psi_R = \psi^e(\boldsymbol{\mathcal{E}}^e) + \psi^p(e^p, \nabla_R e^p)$$

be the reference free energy[6] and $\mathbf{T}^e = \mathbf{T}^e(\boldsymbol{\mathcal{E}}^e)$. Neglecting heat conduction, the second-law inequality is taken in the form

$$-\dot{\psi}_R + \mathbf{T} \cdot \dot{\boldsymbol{\mathcal{E}}}^e + \tau^p \dot{e}^p + \theta \nabla_R \cdot \mathbf{k}_R \geq 0,$$

where $\mathbf{k}_R$ is the reference entropy flux. Hence we have

$$(-\partial_{\boldsymbol{\mathcal{E}}^e} \psi^e + \mathbf{T}^e) \cdot \dot{\boldsymbol{\mathcal{E}}}^e + (-\partial_{e^p} \psi^p + \tau^p)\dot{e}^p - \partial_{\nabla_R e^p} \psi^p \cdot \nabla_R \dot{e}^p + \theta \nabla_R \cdot \mathbf{k}_R \geq 0.$$

Divide by θ and observe

$$\frac{1}{\theta} \partial_{\nabla_R e^p} \psi^p \cdot \nabla_R \dot{e}^p = \nabla_R \cdot (\frac{1}{\theta} \partial_{\nabla_R e^p} \psi^p \dot{e}^p) - [\nabla_R \cdot (\frac{1}{\theta} \partial_{\nabla_R e^p} \psi^p)]\dot{e}^p.$$

The inequality can then be written as

$$\frac{1}{\theta}(-\partial_{\boldsymbol{\mathcal{E}}^e} \psi^e + \mathbf{T}^e) \cdot \dot{\boldsymbol{\mathcal{E}}}^e + \frac{1}{\theta}[-\partial_{e^p} \psi^p + \theta \nabla_R \cdot (\frac{1}{\theta} \partial_{\nabla_R e^p} \psi^p) + \tau^p]\dot{e}^p + \nabla_R \cdot [\mathbf{k}_R - (\frac{1}{\theta} \partial_{\nabla_R e^p} \psi^p \dot{e}^p)] \geq 0.$$

Hence it follows that

$$\mathbf{T}^e = \partial_{\boldsymbol{\mathcal{E}}^e} \psi^e, \qquad \tau^p = \tau^p_{\text{en}} + \tau_{\text{dis}}, \qquad \mathbf{k}_R = \frac{1}{\theta} \partial_{\nabla_R e^p} \psi^p \dot{e}^p,$$

where

$$\tau^p_{\text{en}} = \partial_{e^p} \psi^p - \theta \nabla_R \cdot (\frac{1}{\theta} \partial_{\nabla_R e^p} \psi^p), \qquad \tau_{\text{dis}} \, \dot{e}^p \geq 0.$$

The flow rule is then given by

$$\tau = \tau_{\text{dis}}(e^p, \nabla_R e^p, \dot{e}^p) + \partial_{e^p} \psi^p - \theta \nabla_R \cdot (\frac{1}{\theta} \partial_{\nabla_R e^p} \psi^p).$$

---

[6] See [216], Sect. 89; ψ^e and ψ^p are said to represent the elastic strain energy and the defect energy, respectively.

Aifantis relation for τ is recovered by letting

$$\psi^p = \frac{1}{2}\beta|\nabla e^p|^2,$$

with $\beta > 0$ constant, and

$$\tau_{\text{dis}} = Y(e^p) > 0.$$

If the elastic stress is taken in the classical linear form

$$\mathbf{T}^e = 2\mu\mathbf{E}^e + \lambda(\text{tr}\,\mathbf{E}^e)\mathbf{1}$$

then the corresponding free energy is given by

$$\psi^e = \mu|\mathbf{E}^e|^2 + \lambda(\text{tr}\,\mathbf{E}^e)^2.$$

It is worth remarking that in [216], Sect. 89, the Aifantis flow rule is framed within a thermodynamic framework by introducing a *microscopic hyperstress* vector $\boldsymbol{\xi}^p$ so that the power comprises the additional term

$$\boldsymbol{\xi}^p \cdot \nabla\dot{e}^p.$$

The assumption $\boldsymbol{\xi}_{\text{dis}} = \mathbf{0}$ of [216], Sect. 89.4, makes $\boldsymbol{\xi}$ to coincide with $\partial_{\nabla_R e^p}\psi^p$ and hence, if $\nabla_R\theta = \mathbf{0}$, the result (89.44) of [216] coincides with the present expression of τ.

13.5 Model of Plasticity via the Kröner Decomposition

The additive decomposition $\mathbf{H} = \mathbf{H}^e + \mathbf{H}^p$ can be viewed as a linearization of the displacement gradient to apply in the case of small deformations. A modelling of large-deformation plasticity is based on the multiplicative decomposition

$$\mathbf{F} = \mathbf{F}^e\mathbf{F}^p; \tag{13.7}$$

$\mathbf{F}^e$ and $\mathbf{F}^p$ are said to represent the *elastic distortion* and the *plastic distortion.* The relation (13.7) is referred to as the *Kröner decomposition* [262]. The standard assumption $\det\mathbf{F} > 0$ is guaranteed by letting

$$\det\mathbf{F}^e > 0, \qquad \det\mathbf{F}^p > 0;$$

hence both $\mathbf{F}^e$ and $\mathbf{F}^p$ are invertible.

While $\mathbf{F} = \nabla_R\chi$ is the gradient of the motion χ, neither $\mathbf{F}^e$ nor $\mathbf{F}^p$ are the gradient of corresponding functions χ^e, χ^p. We can think that for vectors $\mathbf{Y}$ in a neighbourhood of $\mathbf{X}$, with $\mathbf{X}, \mathbf{Y} \in R$, the images $\mathbf{x}, \mathbf{y}$ satisfy

$$\mathbf{y} - \mathbf{x} = \mathbf{F}^e\mathbf{F}^p(\mathbf{Y} - \mathbf{X}) + o(|\mathbf{Y} - \mathbf{X}|).$$

Hence, upon neglecting $o(|\mathbf{Y} - \mathbf{X}|)$, $\mathbf{F}^p$ maps the vector $\mathbf{Y} - \mathbf{X}$ into $\hat{\mathbf{y}} - \hat{\mathbf{x}}$ and hence $\mathbf{F}^e$ maps $\hat{\mathbf{y}} - \hat{\mathbf{x}}$ into $\mathbf{y} - \mathbf{x}$. We can picture the transformation by viewing $\hat{\mathbf{y}}$ and $\hat{\mathbf{x}}$ as point in a structural space while $\mathbf{y}$ and $\mathbf{x}$ are points in $\chi(\mathrm{R}) \subset \mathscr{E}$.

Since[7]

$$\dot{\mathbf{F}} = \dot{\mathbf{F}}^e\mathbf{F}^p + \mathbf{F}^e\dot{\mathbf{F}}^p, \qquad \mathbf{F}^{-1} = \mathbf{F}^{p-1}\mathbf{F}^{e-1},$$

then $\mathbf{L} = \dot{\mathbf{F}}\mathbf{F}^{-1}$ is decomposed in the form

$$\mathbf{L} = (\dot{\mathbf{F}}^e\mathbf{F}^p + \mathbf{F}^e\dot{\mathbf{F}}^p)(\mathbf{F}^{p-1}\mathbf{F}^{e-1}) = \dot{\mathbf{F}}^e\mathbf{F}^{e-1} + \mathbf{F}^e(\dot{\mathbf{F}}^p\mathbf{F}^{p-1})\mathbf{F}^{e-1}.$$

So two velocity gradients can be defined,

$$\mathbf{L}^e = \dot{\mathbf{F}}^e\mathbf{F}^{e-1}, \qquad \mathbf{L}^p = \dot{\mathbf{F}}^p\mathbf{F}^{p-1},$$

and hence the decomposition of $\mathbf{L}$ is expressed as

$$\mathbf{L} = \mathbf{L}^e + \mathbf{F}^e\,\mathbf{L}^p\,\mathbf{F}^{e-1}.$$

We can then define the *elastic stretching* $\mathbf{D}^e$ and the *elastic spin* $\mathbf{W}^e$,

$$\mathbf{D}^e = \tfrac{1}{2}(\mathbf{L}^e + \mathbf{L}^{eT}), \qquad \mathbf{W}_e = \tfrac{1}{2}(\mathbf{L}^e - \mathbf{L}^{eT}).$$

Likewise, the *plastic stretching* $\mathbf{D}^p$ and the *plastic spin* $\mathbf{W}^p$ are defined in the form

$$\mathbf{D}^p = \tfrac{1}{2}(\mathbf{L}^p + \mathbf{L}^{pT}), \qquad \mathbf{W}^p = \tfrac{1}{2}(\mathbf{L}^p - \mathbf{L}^{pT}).$$

Hence

$$\mathbf{D} = \mathbf{D}^e + \mathrm{sym}(\mathbf{F}^e\mathbf{L}^p\mathbf{F}^{e-1}).$$

As with the approximation of small deformations, the plastic flow is assumed to be isochoric and hence $\mathbf{L}^p$ and $\mathbf{D}^p$ are assumed to be deviatoric, $\mathrm{tr}\,\mathbf{L}^p = \mathrm{tr}\,\mathbf{D}^p = 0$. We then let

$$\det\mathbf{F}^p = 1, \qquad J = \det\mathbf{F}^e, \qquad \dot{J} = J\,\mathrm{tr}\,\mathbf{D}^e.$$

The balance of linear and angular momentum is taken to be the standard ones,

$$\rho\dot{\mathbf{v}} = \rho\mathbf{b} + \nabla\cdot\mathbf{T}, \qquad \mathbf{T} = \mathbf{T}^T.$$

The balance of energy shows new questions induced by the multiplicative decomposition (13.7). Upon substitution, the power $\mathbf{T}\cdot\mathbf{D}$ becomes

[7] For a more compact notation, we let $\mathbf{F}^{e-1} := (\mathbf{F}^e)^{-1}$ and $\mathbf{F}^{e-T} := (\mathbf{F}^e)^{-T}$.

$$\mathbf{T} \cdot \mathbf{D} = \mathbf{T} \cdot \mathbf{D}^e + \mathbf{T} \cdot (\mathbf{F}^e \mathbf{L}^p \mathbf{F}^{e-1}). \tag{13.8}$$

Now, by

$$\mathbf{C}^e := \mathbf{F}^{eT} \mathbf{F}^e, \qquad \mathbf{E}^e := \tfrac{1}{2}(\mathbf{C}^e - \mathbf{1}), \qquad \dot{\mathbf{F}}^e := \mathbf{L}^e \mathbf{F}^e,$$

it follows

$$\dot{\mathbf{C}}^e = 2\, \mathbf{F}^{eT}\, \mathbf{D}^e\, \mathbf{F}^e, \qquad \dot{\mathbf{E}}^e = \mathbf{F}^{eT}\, \mathbf{D}^e\, \mathbf{F}^e.$$

Consequently,

$$\mathbf{T} \cdot \mathbf{D}^e = \mathbf{T} \cdot (\mathbf{F}^{e-T}\, \dot{\mathbf{E}}^e\, \mathbf{F}^{e-1}) = (\mathbf{F}^{e-1} \mathbf{T} \mathbf{F}^{e-T}) \cdot \dot{\mathbf{E}}^e.$$

The definition

$$\mathbf{T}^e := J \mathbf{F}^{e-1} \mathbf{T} \mathbf{F}^{e-T}$$

allows $\mathbf{T}^e$ to be viewed as the elastic counterpart of the standard second Piola stress; $\mathbf{T}^e$ is then called the *second Piola elastic stress*. Since $\mathbf{T} \in \mathrm{Sym}$ then also $\mathbf{T}^e \in \mathrm{Sym}$. Based on the power (13.8), we observe

$$\mathbf{T} \cdot (\mathbf{F}^e \mathbf{L}^p \mathbf{F}^{e-1}) = (\mathbf{F}^{eT} \mathbf{T} \mathbf{F}^{e-T}) \cdot \mathbf{L}^p.$$

Moreover

$$\mathrm{tr}\, (\mathbf{F}^{eT} \mathbf{T} \mathbf{F}^{e-T}) = \mathrm{tr}\, \mathbf{T},$$

and hence, since $\mathrm{tr}\, \mathbf{L}^p = 0$,

$$(\mathbf{F}^{eT} \mathbf{T} \mathbf{F}^{e-T}) \cdot \mathbf{L}^p = (\mathbf{F}^{eT} \mathbf{T}_0 \mathbf{F}^{e-T}) \cdot \mathbf{L}^p;$$

it is worth remarking that neither $\mathbf{F}^{eT} \mathbf{T} \mathbf{F}^{e-T}$ nor $\mathbf{F}^{eT} \mathbf{T}_0 \mathbf{F}^{e-T}$ are symmetric. Upon the requirement that

$$\mathbf{T} \cdot (\mathbf{F}^e \mathbf{L}^p \mathbf{F}^{e-1}) = J^{-1} \mathbf{T}^p \cdot \mathbf{L}^p, \qquad \mathrm{tr}\, \mathbf{L}^p = 0,$$

we find that the plastic stress $\mathbf{T}^p$ is given by

$$\mathbf{T}^p = J \mathbf{F}^{eT} \mathbf{T}_0 \mathbf{F}^{e-T}.$$

Now,

$$J \mathbf{F}^{eT} \mathbf{T}_0 \mathbf{F}^{e-T} = J (\mathbf{F}^{eT} \mathbf{T} \mathbf{F}^{e-T})_0 = J (\mathbf{F}^{eT} \mathbf{F}^e \mathbf{F}^{e-1} \mathbf{T} \mathbf{F}^{e-T})_0 = (\mathbf{C}^e \mathbf{T}^e)_0.$$

Letting

$$\mathbf{M}^e := \mathbf{C}^e \mathbf{T}^e = J \mathbf{F}^{eT} \mathbf{T} \mathbf{F}^{e-T}$$

we obtain

$$(\mathbf{M}^e)_0 \cdot \mathbf{L}^p = \mathbf{T}^p \cdot \mathbf{L}^p,$$

whence

$$\mathbf{T}^p = (\mathbf{M}^e)_0.$$

Consequently, the power (13.8) is expressed in the form

$$\mathbf{T} \cdot \mathbf{D} = J^{-1}\mathbf{T}^e \cdot \dot{\mathbf{E}}^e + J^{-1}(\mathbf{M}^e)_0 \cdot \mathbf{L}^p.$$

The stress tensor $\mathbf{M}^e$ is called the *Mandel stress* [295].

We neglect heat conduction and hence consider the entropy inequality

$$-\rho(\dot{\psi} + \eta\dot{\theta}) + J^{-1}\mathbf{T}^e \cdot \dot{\mathbf{E}}^e + J^{-1}\mathbf{T}^p \cdot \mathbf{L}^p \geq 0,$$

whence

$$-\rho_R(\dot{\psi} + \eta\dot{\theta}) + \mathbf{T}^e \cdot \dot{\mathbf{E}}^e + \mathbf{T}^p \cdot \mathbf{L}^p \geq 0.$$

The form of the power, $\mathbf{T}^e \cdot \dot{\mathbf{E}}^e + \mathbf{T}^p \cdot \mathbf{L}^p$, suggests that $\mathbf{T}^e$ might be derived from a potential whereas $\mathbf{T}^p$ might produce dissipative processes. A temperature-dependent model is obtained by letting

$$\psi = \hat{\psi}(\theta, \mathbf{E}^e), \qquad \mathbf{T}^e = \hat{\mathbf{T}}^e(\theta, \mathbf{E}^e), \qquad \mathbf{T}^p = \hat{\mathbf{T}}^p(\theta, \mathbf{L}^p, e^p).$$

The accumulated plastic strain e^p is now defined by

$$\dot{e}^p = |\mathbf{D}^e|,$$

to within an initial value. Evaluation of $\dot{\psi}$ and substitution into the entropy inequality result in

$$\rho_R(\partial_\theta\hat{\psi} + \eta)\dot{\theta} - (\rho_R\partial_{\mathbf{E}^e}\hat{\psi} - \mathbf{T}^e) \cdot \dot{\mathbf{E}}^e + \mathbf{T}^p \cdot \mathbf{L}^p \geq 0.$$

Since $\dot{\mathbf{E}}^e$ and $\mathbf{L}^p$ are independent of each other and both of them are independent of $\dot{\theta}$ then the inequality holds if and only if

$$\eta = -\partial_\theta\hat{\psi}, \qquad \mathbf{T}^e = \rho_R\partial_{\mathbf{E}^e}\hat{\psi},$$

and

$$\mathbf{T}^p \cdot \mathbf{L}^p \geq 0.$$

It is worth observing that $\mathbf{E}^e$ and e^p are objective. Furthermore, $\mathbf{L}^p$ is objective in that it is a map from the reference configuration to the structural space and then it is invariant under Euclidean transformations.

For definiteness, if ψ is considered in the form

$$\rho_R\psi = \mu|\mathbf{E}^e|^2 + \tfrac{1}{2}\lambda(\operatorname{tr}\mathbf{E}^e)^2,$$

which is adopted in the modelling of small elastic strains, it follows

$$\mathbf{T}^e = 2\mu\mathbf{E}^e + \lambda(\operatorname{tr}\mathbf{E}^e)\mathbf{1}.$$

Since $\mathbf{E}^e = \frac{1}{2}(\mathbf{C}^e - \mathbf{1})$ and $\mathbf{M}^e = \mathbf{C}^e\mathbf{T}^e$ then

$$\mathbf{M}^e = \mu(\mathbf{C}^{e2} - \mathbf{C}^e) + \tfrac{1}{2}(\operatorname{tr}\mathbf{C}^e - 3)\mathbf{C}^e.$$

In this approximation, it follows that $\mathbf{M}^e$ is symmetric, $\mathbf{M}^e = \mathbf{M}^{eT}$, and so is $\mathbf{T}^p$. Hence

$$\mathbf{T}^p \cdot \mathbf{L}^p = \mathbf{T}^p \cdot \mathbf{D}^p.$$

The plastic stress $\mathbf{T}^p$ (and the deviatoric Mandel stress $(\mathbf{M}^e)_0$) is taken to depend on $\mathbf{D}^p$ via $\mathbf{N}^p = \mathbf{D}^p/|\mathbf{D}^p|$ as $\mathbf{D}^p \neq \mathbf{0}$. The thermodynamic restriction $\mathbf{T}^p \cdot \mathbf{D}^p \geq 0$ then yields

$$Y(\mathbf{N}^p, e^p) := \mathbf{N}^p \cdot \mathbf{T}^p(\mathbf{N}^p, e^p) \geq 0.$$

If $Y(\mathbf{N}^p, e^p)$ is independent of $\mathbf{N}^p$ then

$$\mathbf{T}^p = Y(e^p)\mathbf{N}^p, \qquad \mathbf{D}^p \neq \mathbf{0}.$$

Along with

$$e^p(t) = \int_0^t |\mathbf{D}^p(\tau)|d\tau,$$

the evolution of $\mathbf{D}^p$ determines $\mathbf{T}^p = (\mathbf{M}^e)_0$.

13.6 A Decomposition-Free Thermodynamic Scheme

A wide class of viscoelastic and plastic materials is modelled by following an approach which is characterized by a thermodynamic potential and the entropy production; both the Helmholtz free energy ψ and its Legendre transform $\phi = \psi - \mathbf{E} \cdot \mathbf{T}_{RR}/\rho_R$ are applied as pertinent potentials. No decomposition, additive or multiplicative, is made in elastic and plastic distortions. No use is made of the linear approximation of $\mathbf{E}$.

We apply the second-law inequality in the standard form (Sect. 2.6),

$$\rho\dot{\eta} + \nabla \cdot \frac{\mathbf{q}}{\theta} - \frac{\rho r}{\theta} = \sigma \geq 0,$$

where the entropy production σ is taken to be a non-negative-valued constitutive function; for the present purposes, it is enough to identify $\mathbf{q}/\theta$ with the entropy flux. We then have

$$-\rho(\dot{\psi}+\eta\dot{\theta})+\mathbf{T}\cdot\mathbf{D}-\frac{1}{\theta}\mathbf{q}\cdot\nabla\theta=\theta\sigma,\quad -\rho_R(\dot{\psi}+\eta\dot{\theta})+\mathbf{T}_{RR}\cdot\dot{\mathbf{E}}-\frac{1}{\theta}\mathbf{q}_R\cdot\nabla_R\theta=J\theta\sigma,$$

according as we follow the Eulerian or the Lagrangian description.

Let

$$\Xi=(\theta,\mathbf{E},\mathbf{T}_{RR},\mathbf{q}_R,\nabla_R\theta,\dot{\mathbf{E}},\dot{\mathbf{T}}_{RR},\dot{\mathbf{q}}_R)$$

be the set of variables and take ψ, η, and σ be functions of Ξ. Computation of $\dot{\psi}$ and substitution in the Clausius–Duhem inequality allow us to find that ψ is independent of $\dot{\mathbf{E}}$, $\dot{\mathbf{T}}_{RR}$, $\dot{\mathbf{q}}_R$, $\nabla_R\theta$ while η is related to ψ as

$$\eta=-\partial_\theta\psi.$$

The Clausius–Duhem inequality then reduces to

$$(-\rho_R\partial_{\mathbf{E}}\psi+\mathbf{T}_{RR})\cdot\dot{\mathbf{E}}-\rho_R\partial_{\mathbf{T}_{RR}}\psi\cdot\dot{\mathbf{T}}_{RR}-\rho_R\partial_{\mathbf{q}_R}\psi\cdot\dot{\mathbf{q}}_R-\frac{1}{\theta}\mathbf{q}_R\cdot\nabla_R\theta=J\theta\sigma\geq 0.$$

We let $\dot{\mathbf{E}}$ and $\dot{\mathbf{T}}_{RR}$ be possibly related to each other but $\dot{\mathbf{q}}_R$, $\nabla_R\theta$ are independent of $\dot{\mathbf{E}}$, $\dot{\mathbf{T}}_{RR}$. Consequently it follows

$$(-\rho_R\partial_{\mathbf{E}}\psi+\mathbf{T}_{RR})\cdot\dot{\mathbf{E}}-\rho_R\partial_{\mathbf{T}_{RR}}\psi\cdot\dot{\mathbf{T}}_{RR}=J\theta\sigma_{ET}\geq 0, \tag{13.9}$$

$$-\rho_R\partial_{\mathbf{q}_R}\psi\cdot\dot{\mathbf{q}}_R-\frac{1}{\theta}\mathbf{q}_R\cdot\nabla_R\theta=J\theta\sigma_q\geq 0, \tag{13.10}$$

σ_{ET} and σ_q being entropy productions to be determined according to the constitutive model, $\sigma=\sigma_{ET}+\sigma_q$.

Inequality (13.10) models heat conduction. Let ψ depend on $\mathbf{q}_R$ via $\xi=|\mathbf{q}_R|^n$, $n\geq 2$. Hence

$$\partial_{\mathbf{q}_R}\psi=n\partial_\xi\psi|\mathbf{q}_R|^{n-2}\mathbf{q}_R$$

and inequality (13.10) becomes

$$(n\rho_R\partial_\xi\psi|\mathbf{q}_R|^{n-2}\,\dot{\mathbf{q}}_R+\frac{1}{\theta}\nabla_R\theta)\cdot\mathbf{q}_R=-J\theta\sigma_q\leq 0.$$

Letting

$$J\theta\sigma_q=\frac{1}{\kappa(\theta,J,\xi,|\nabla_R\theta|)}\mathbf{q}_R^2,\qquad \kappa>0,$$

we obtain

$$\tau\,\dot{\mathbf{q}}_R+\mathbf{q}_R=-\kappa\nabla_R\theta, \tag{13.11}$$

where $\tau = n\rho_R\, \theta\, \partial_\xi \psi |\mathbf{q}_R|^{n-2}$. Equation (13.11) is a generalization of the Maxwell–Cattaneo equation where τ plays the role of relaxation time. If $n = 2$ then

$$\tau = 2\kappa\rho_R\theta\partial_\xi\psi.$$

As to inequality (13.9) we observe that, in terms of the Legendre transform $\rho_R\phi = \rho_R\psi - \mathbf{T}_{RR} \cdot \mathbf{E}$, we have

$$-\rho_R\partial_{\mathbf{E}}\phi \cdot \dot{\mathbf{E}} - (\rho_R\partial_{\mathbf{T}_{RR}}\phi + \mathbf{E}) \cdot \dot{\mathbf{T}}_{RR} = J\theta\sigma_{ET} \geq 0. \tag{13.12}$$

It is worth observing that the free energy ϕ can be represented in the Eulerian description. Consider the energy $\mathbf{T}_{RR} \cdot \mathbf{E}$; by the definition of $\mathbf{T}_{RR}$ we have

$$\mathbf{T}_{RR} \cdot \mathbf{E} = J(\mathbf{F}^{-1}\mathbf{T}\mathbf{F}^{-T}) \cdot \mathbf{E} = J\mathbf{T} \cdot (\mathbf{F}^{-T}\mathbf{E}\mathbf{F}^{-1}).$$

Since $\mathbf{F}^{-T}\mathbf{E}\mathbf{F}^{-1} = \mathbf{E}_A$, the Eulerian–Almansi strain, then we conclude that

$$\mathbf{T}_{RR} \cdot \mathbf{E} = J\mathbf{T} \cdot \mathbf{E}_A, \qquad \rho_R\phi = \rho_R\psi - \mathbf{T}_{RR} \cdot \mathbf{E} \Longleftrightarrow \rho\phi = \rho\psi - \mathbf{T} \cdot \mathbf{E}_A.$$

Moreover,

$$\dot{\mathbf{E}} = \overline{\mathbf{F}^T\mathbf{E}_A\mathbf{F}}^{\boldsymbol{\cdot}} = \mathbf{F}^T(\mathbf{L}^T\mathbf{E}_A + \dot{\mathbf{E}}_A + \mathbf{E}_A\mathbf{L})\mathbf{F} = \mathbf{F}^T \overset{\triangle}{\mathbf{E}}_A \mathbf{F}.$$

Since $\overset{\triangle}{\mathbf{E}}_A = \mathbf{D}$ then we have the expected result

$$\mathbf{T}_{RR} \cdot \dot{\mathbf{E}} = J\mathbf{T} \cdot \mathbf{D}.$$

We now investigate a class of models consistent with the entropy inequality (13.9) and the objectivity principle.

13.6.1 Duhem-Like Solids

In the literature, Duhem-like models are characterized by differential equations of the form

$$\dot{x}(t) = f(x(t), u(t))g(\dot{u}(t)),$$

where x, u are vectors in $\mathbb{R}^n$. Models of this form, that traces back to Duhem, have been applied to represent friction and mechanical hysteresis (see, e.g. [291]).

With this picture in mind we now examine (13.9). The significant variables are θ, $\mathbf{E}$, $\mathbf{T}_{RR}$, $\dot{\mathbf{E}}$, $\dot{\mathbf{T}}_{RR}$ and then we observe that (13.9) can be specified by letting

$$J\theta\sigma_{ET} = \gamma_E(\theta, \mathbf{E}, \mathbf{T}_{RR}, \dot{\mathbf{E}}/|\dot{\mathbf{E}}|)|\dot{\mathbf{E}}|, \qquad \gamma_E \geq 0$$

or

$$J\theta\sigma_{ET} = \gamma_T(\theta, \mathbf{E}, \mathbf{T}_{RR}, \dot{\mathbf{T}}_{RR}/|\dot{\mathbf{T}}_{RR}|)|\dot{\mathbf{T}}_{RR}|, \qquad \gamma_T \geq 0.$$

Hence, inequality (13.9) can be written as an equality in the form

$$(\rho_R\partial_{\mathbf{E}}\psi - \mathbf{T}_{RR}) \cdot \dot{\mathbf{E}} + \rho_R\partial_{\mathbf{T}_{RR}}\psi \cdot \dot{\mathbf{T}}_{RR} = -\gamma_E|\dot{\mathbf{E}}| \leq 0 \tag{13.13}$$

or

$$(\rho_R\partial_{\mathbf{E}}\psi - \mathbf{T}_{RR}) \cdot \dot{\mathbf{E}} + \rho_R\partial_{\mathbf{T}_{RR}}\psi \cdot \dot{\mathbf{T}}_{RR} = -\gamma_T|\dot{\mathbf{T}}_{RR}| \leq 0. \tag{13.14}$$

Likewise, in terms of the free energy density ϕ, it follows from (13.12) that

$$\rho_R\partial_{\mathbf{E}}\phi \cdot \dot{\mathbf{E}} + (\rho_R\partial_{\mathbf{T}_{RR}}\phi + \mathbf{E}) \cdot \dot{\mathbf{T}}_{RR} = -\gamma_E|\dot{\mathbf{E}}| \leq 0, \tag{13.15}$$

$$\rho_R\partial_{\mathbf{E}}\phi \cdot \dot{\mathbf{E}} + (\rho_R\partial_{\mathbf{T}_{RR}}\phi + \mathbf{E}) \cdot \dot{\mathbf{T}}_{RR} = -\gamma_T|\dot{\mathbf{T}}_{RR}| \leq 0. \tag{13.16}$$

All these differential equations are invariant under the time transformation

$$t \to ct, \qquad c > 0,$$

since so are the coefficients γ_E and γ_T. Hence Eqs. (13.13)–(13.16) are rate independent.

Eulerian Scheme

It is natural to ask about the Eulerian analogue of Eq. (13.9) and the subsequent ones. If we start with $\hat{\psi}(\theta, \mathbf{E}_A, \mathbf{T})$ then the Clausius–Duhem inequality results in

$$-\rho\partial_{\mathbf{E}_A}\hat{\psi} \cdot \dot{\mathbf{E}}_A - \rho\partial_{\mathbf{T}}\hat{\psi} \cdot \dot{\mathbf{T}} + \mathbf{T} \cdot \mathbf{D} = \theta\sigma_{ET}. \tag{13.17}$$

Apart from the relation between $\dot{\mathbf{E}}_A$ and $\mathbf{D}$, which is considered shortly, we observe that Eq. (13.17) might be used to write a constitutive relation in a Duhem-like form involving $\dot{\mathbf{T}}$ and $\dot{\mathbf{E}}_A$. However, both $\dot{\mathbf{T}}$ and $\dot{\mathbf{E}}_A$ are not objective and hence (13.17), though formally correct, is inadequate. Instead Eq. (13.9) is objective in that $\mathbf{E}$ and $\mathbf{T}_{RR}$ are invariant. First recall that $\mathbf{E} = \mathbf{F}^T\mathbf{E}_A\mathbf{F}$ and then

$$\dot{\mathbf{E}} = \mathbf{F}^T \overset{\triangle}{\mathbf{E}}_A \mathbf{F}, \qquad \overset{\triangle}{\mathbf{E}}_A = \mathbf{D}, \qquad \mathbf{T}_{RR} \cdot \dot{\mathbf{E}} = \mathbf{T} \cdot \mathbf{D}.$$

Moreover, since $\mathbf{T}_{RR} = J\mathbf{F}^{-1}\mathbf{T}\mathbf{F}^{-T}$ then

$$\dot{\mathbf{T}}_{RR} = J\mathbf{F}^{-1} \overset{\square}{\mathbf{T}} \mathbf{F}^{-T}.$$

Now, if $\psi = \psi(\theta, \mathbf{E}(\mathbf{E}_A, \mathbf{F}), \mathbf{T}_{RR}(\mathbf{T}, \mathbf{F}))$ then we obtain

$$\partial_{\mathbf{T}}\psi = J\mathbf{F}^{-T}\partial_{\mathbf{T}_{RR}}\psi\mathbf{F}^{-1}, \qquad \partial_{\mathbf{E}_A}\psi = \mathbf{F}\partial_{\mathbf{E}}\psi\mathbf{F}^{T},$$

$$\rho\partial_{\mathbf{E}}\psi \cdot \dot{\mathbf{E}} = \rho\partial_{\mathbf{E}_A}\psi \cdot \mathbf{D}, \qquad \rho\partial_{\mathbf{T}_{RR}}\psi \cdot \dot{\mathbf{T}}_{RR} = \rho\partial_{\mathbf{T}}\psi \cdot \overset{\square}{\mathbf{T}}.$$

Hence substitution in (13.9) leads to

$$(-\rho\partial_{\mathbf{E}_A}\psi + \mathbf{T}) \cdot \mathbf{D} - \rho\partial_{\mathbf{T}}\psi \cdot \overset{\square}{\mathbf{T}} = \theta\sigma_{ET}, \tag{13.18}$$

where we take $\theta\sigma_{ET}$ in the form

$$\theta\sigma_{ET} = \tilde{\gamma}_E(\theta, \mathbf{E}_A, \mathbf{T}, \mathbf{D}/|\mathbf{D}|)|\mathbf{D}| \quad \text{or} \quad \theta\sigma_{ET} = \tilde{\gamma}_T(\theta, \mathbf{E}_A, \mathbf{T}, \overset{\square}{\mathbf{T}}/|\overset{\square}{\mathbf{T}}|)|\overset{\square}{\mathbf{T}}|.$$

So we have

$$(\rho\partial_{\mathbf{E}_A}\psi - \mathbf{T}) \cdot \mathbf{D} + \rho\partial_{\mathbf{T}}\psi \cdot \overset{\square}{\mathbf{T}} = -\tilde{\gamma}_E|\mathbf{D}|, \tag{13.19}$$

$$(\rho\partial_{\mathbf{E}_A}\psi - \mathbf{T}) \cdot \mathbf{D} + \rho\partial_{\mathbf{T}}\psi \cdot \overset{\square}{\mathbf{T}} = -\tilde{\gamma}_E|\overset{\square}{\mathbf{T}}|. \tag{13.20}$$

This shows that the constitutive properties of $\dot{\mathbf{T}}_{RR}$ in terms of $\dot{\mathbf{E}}$ have the strict analogue in $\overset{\square}{\mathbf{T}}$ in terms of $\mathbf{D}$.

The differences between (13.17) and (13.18) are obviously understood if we observe that ψ in (13.18) is a composite function, of $\mathbf{E}_A$ and $\mathbf{T}$, which depend also on $\mathbf{F}$. In (13.17), instead, $\hat{\psi}$ depends on $\mathbf{E}_A$ and $\mathbf{T}$ but is independent of $\mathbf{F}$.

13.6.2 Hyperelastic and Hypoelastic Solids

Look at Duhem-like solids without dissipation, namely, with $\sigma_{ET} = 0$. If $\sigma_{ET} = 0$ and $\partial_{\mathbf{T}_{RR}}\psi = \mathbf{0}$ then it follows from (13.13) that

$$(\rho_R\partial_{\mathbf{E}}\psi - \mathbf{T}_{RR}) \cdot \dot{\mathbf{E}} = 0.$$

The arbitrariness of $\dot{\mathbf{E}}$ implies

$$\mathbf{T}_{RR} = \rho_R\partial_{\mathbf{E}}\psi$$

and then

$$\mathbf{T} = \rho\mathbf{F}\partial_{\mathbf{E}}\psi\mathbf{F}^{T}.$$

This constitutive equation models *hyperelastic solids.*

We might have obtained (13.13) by assuming $\psi = \psi(\theta, \mathbf{E})$ from the start. If this is the case we have $\partial_{\mathbf{T}_{RR}}\psi = \mathbf{0}$ and (13.13) implies

$$\mathbf{T}_{RR} = \rho_R \partial_{\mathbf{E}} \psi, \qquad \dot{\mathbf{T}}_{RR} = \rho_R \partial^2_{\mathbf{E}} \psi(\mathbf{E})\dot{\mathbf{E}} + \rho_R \partial_\theta \partial_{\mathbf{E}} \psi\, \dot{\theta},$$

where $\partial^2_{\mathbf{E}}\psi$ is a fourth-order tensor with major and minor symmetries.

If instead $\gamma_E = 0$ but $\partial_{\mathbf{T}}\psi \neq \mathbf{0}$ (and $\partial_{\mathbf{T}_{RR}}\psi \neq \mathbf{0}$) then (13.9) can be written in the form

$$(\rho_R \partial_{\mathbf{E}}\psi - \mathbf{T}_{RR}) \cdot \dot{\mathbf{E}} + \rho_R \partial_{\mathbf{T}_{RR}}\psi \cdot \dot{\mathbf{T}}_{RR} = 0.$$

Consequently $\dot{\mathbf{T}}_{RR}$ can be given a linear representation in $\dot{\mathbf{E}}$.

To this end, we recall the representation formula (A.8) of tensors; for any tensor $\mathbf{Z}$ with a known value of the inner product $\mathbf{Z} \cdot \mathbf{N}$, and $\mathbf{N} \cdot \mathbf{N} = 1$, we have

$$\mathbf{Z} = (\mathbf{Z} \cdot \mathbf{N})\mathbf{N} + (\mathbf{I} - \mathbf{N} \otimes \mathbf{N})\mathbf{G}$$

for any tensor $\mathbf{G}$. Now, assume $|\partial_{\mathbf{T}_{RR}}\psi| \neq 0$ and let $\mathbf{N} = \partial_{\mathbf{T}_{RR}}\psi / |\partial_{\mathbf{T}_{RR}}\psi|$, $\mathbf{Z} = \dot{\mathbf{T}}_{RR}$, and $\mathbf{G} = \mathbf{G}_{RR}\dot{\mathbf{E}}$. Observe that, by (13.13) with $\gamma_E = 0$, we have

$$\dot{\mathbf{T}}_{RR} = \frac{1}{\rho_R}[(\mathbf{T}_{RR} - \rho_R \partial_{\mathbf{E}}\psi) \cdot \dot{\mathbf{E}}]\mathbf{N} + (\mathbf{I} - \mathbf{N} \otimes \mathbf{N})\mathbf{G}_{RR}\dot{\mathbf{E}},$$

where $\mathbf{G}_{RR}(\theta, \mathbf{E}, \mathbf{T}_{RR})$ is an arbitrary fourth-order tensor-valued function. Hence, letting

$$\mathbf{C}_{RR}(\theta, \mathbf{E}, \mathbf{T}_{RR}) = \mathbf{G}_{RR} + \frac{1}{\rho_R |\partial_{\mathbf{T}_{RR}}\psi|^2} \partial_{\mathbf{T}_{RR}}\psi \otimes (\mathbf{T}_{RR} - \rho_R \partial_{\mathbf{E}}\psi - \rho_R \mathbf{G}^T_{RR} \partial_{\mathbf{T}_{RR}}\psi), \tag{13.21}$$

we have

$$\dot{\mathbf{T}}_{RR} = \mathbf{C}_{RR}\dot{\mathbf{E}}. \tag{13.22}$$

This relation describes the constitutive properties of *hypoelastic* solids [345, 428] with a nonlinear tensor function $\mathbf{C}_{RR}$ of $\theta, \mathbf{E}, \mathbf{T}_{RR}$. By the arbitrariness of $\mathbf{G}_{RR}$ it follows that there are infinitely many hypoelastic tensors $\mathbf{C}_{RR}$ compatible with a given free energy ψ.

If, instead, a hypo-thermoelastic constitutive relation is given in advance in the form (13.22) then (13.13) can be written as

$$(\rho_R \partial_{\mathbf{E}}\psi - \mathbf{T}_{RR}) \cdot \dot{\mathbf{E}} + \rho_R \partial_{\mathbf{T}_{RR}}\psi \cdot \mathbf{C}_{RR}\dot{\mathbf{E}} = 0, \tag{13.23}$$

and the arbitrariness of $\dot{\mathbf{E}}$ implies

$$\mathbf{T}_{RR} - \rho_R \partial_{\mathbf{E}}\psi = \rho_R \mathbf{C}^T_{RR} \partial_{\mathbf{T}_{RR}}\psi. \tag{13.24}$$

In [310], the existence of a thermodynamic potential ψ satisfying (13.23) is investigated starting from (13.22) with a given tensor $\mathbf{C}_{RR}$.

In general, the fourth-order tensor $\mathbf{C}_{RR}$ enjoys the minor symmetries, not the major one. We now give some examples of the representation formula (13.22).

Example 1 Let

$$\rho_R \psi(\theta, \mathbf{T}_{RR}) = \rho_R \psi_0(\theta) + \frac{1}{2} \int_0^{|\mathbf{T}_{RR}|^2} \alpha(\theta, \xi)\, d\xi,$$

where α is a nonzero real-valued function. Then

$$\rho_R \partial_{\mathbf{T}_{RR}} \psi = \alpha(\theta, |\mathbf{T}_{RR}|^2)\mathbf{T}_{RR}, \quad \rho_R \partial_{\mathbf{E}} \psi = \mathbf{0}$$

so that choosing $\mathsf{G}_{RR} = \nu \mathsf{I}$ we have

$$\mathsf{C}_{RR}(\theta, \mathbf{T}_{RR}) = \nu \mathsf{I} + \Big(\frac{1}{\alpha} - \nu\Big)\frac{\mathbf{T}_{RR}}{|\mathbf{T}_{RR}|} \otimes \frac{\mathbf{T}_{RR}}{|\mathbf{T}_{RR}|},$$

which enjoys both minor and major symmetries. This tensor is positive definite provided that $1/\alpha \geq \nu > 0$. In particular, if we choose $\nu = 1/\alpha$ it follows that $\mathsf{C}_{RR} = (1/\alpha)\mathsf{I}$.

Example 2 Let

$$\rho_R \psi(\theta, \mathbf{E}, \mathbf{T}_{RR}) = \rho_R \psi_0(\theta) + \mathbf{T}_{RR} \cdot \mathbf{E} - \tfrac{1}{2}\alpha(\theta)|\mathbf{E}|^2 - \tfrac{1}{2} \int_0^{|\mathbf{E}|^2} \zeta \beta(\theta, \zeta)\, d\zeta,$$

where α, β are real-valued functions. Hence

$$\rho_R \partial_{\mathbf{T}_{RR}} \psi = \mathbf{E}, \quad \rho_R \partial_{\mathbf{E}} \psi = \mathbf{T}_{RR} - \alpha(\theta)\mathbf{E} - |\mathbf{E}|^2 \beta(\theta, |\mathbf{E}|^2)\mathbf{E}$$

so that choosing $\mathsf{G}_{RR} = \nu \mathsf{I}$ we have

$$\mathsf{C}_{RR}(\theta, \mathbf{E}) = \nu \mathsf{I} + \beta \mathbf{E} \otimes \mathbf{E} + (\alpha - \nu)\frac{\mathbf{E}}{|\mathbf{E}|} \otimes \frac{\mathbf{E}}{|\mathbf{E}|},$$

which enjoys both minor and major symmetries. This tensor is positive definite provided that $\nu \geq 0$, $\beta > 0$ and $\alpha \geq \nu$. In particular, if we choose $\nu = \alpha$ it follows (see [310], Sect. C)

$$\mathsf{C}_{RR}(\theta, \mathbf{E}) = \alpha \mathsf{I} + \beta \mathbf{E} \otimes \mathbf{E}.$$

The next example involves both $|\mathbf{E}|$ and $|\mathbf{T}_{RR}|$.

Example 3 Consider the free energy

$$\rho_R \psi(\theta, \mathbf{E}, \mathbf{T}_{RR}) = \rho_R \psi_0(\theta) + \alpha(\theta)\mathbf{T}_{RR} \cdot \mathbf{E} - \tfrac{1}{2}\beta(\theta)|\mathbf{E}|^2 + \tfrac{1}{2}\gamma(\theta)|\mathbf{T}_{RR}|^2,$$

where $\alpha(\theta) \neq 0$, $\beta(\theta) \neq 0$. Hence we have

$$\rho_R \partial_{\mathbf{T}_{RR}} \psi = \alpha(\theta)\mathbf{E} + \gamma(\theta)\mathbf{T}_{RR}, \quad \rho_R \partial_{\mathbf{E}} \psi = \alpha(\theta)\mathbf{T}_{RR} - \beta(\theta)\mathbf{E}.$$

Consequently, if $\gamma = (1-\alpha)\alpha/\beta$ and $\mathbf{G}_{RR} = \nu\mathbf{I}$ we have

$$\mathbf{C}_{RR}(\theta, \mathbf{E}, \mathbf{T}_{RR}) = \nu\mathbf{I} + \big(\frac{\beta}{\alpha} - \nu\big)\frac{\gamma\mathbf{T}_{RR} + \alpha\mathbf{E}}{|\gamma\mathbf{T}_{RR} + \alpha\mathbf{E}|} \otimes \frac{\gamma\mathbf{T}_{RR} + \alpha\mathbf{E}}{|\gamma\mathbf{T}_{RR} + \alpha\mathbf{E}|}$$

enjoys the major symmetry and is positive definite provided that $\beta/\alpha \geq \nu > 0$. If $\alpha = 1$ as in Example 2 then $\gamma = 0$.

A similar approach can be developed by starting from (13.12) with $\sigma_{ET} = 0$, namely,

$$\rho_R \partial_{\mathbf{E}}\phi \cdot \dot{\mathbf{E}} + \big(\rho_R \partial_{\mathbf{T}_{RR}}\phi + \mathbf{E}\big) \cdot \dot{\mathbf{T}}_{RR} = 0, \tag{13.25}$$

where $|\partial_{\mathbf{T}_{RR}}\phi| \neq 0$. Hence, by paralleling previous arguments and letting

$$\mathbf{K}_{RR}(\theta, \mathbf{E}, \mathbf{T}_{RR}) = \mathbf{J}_{RR} - \frac{1}{\rho_R |\partial_{\mathbf{E}}\phi|^2}\partial_{\mathbf{E}}\phi \otimes (\mathbf{E} + \rho_R \partial_{\mathbf{T}_{RR}}\phi + \mathbf{J}_{RR}^T\, \rho_R \partial_{\mathbf{E}}\phi), \tag{13.26}$$

where $\mathbf{J}_{RR}(\theta, \mathbf{E}, \mathbf{T}_{RR})$ is an arbitrary fourth-order tensor-valued function, we obtain

$$\dot{\mathbf{E}} = \mathbf{K}_{RR}\dot{\mathbf{T}}_{RR}. \tag{13.27}$$

Example 4 Let $\rho_R\phi(\theta, \mathbf{E}, \mathbf{T}_{RR}) = \rho_R\psi(\theta, \mathbf{T}_{RR}) - \mathbf{T}_{RR} \cdot \mathbf{E}$ and

$$\rho_R\psi(\theta, \mathbf{T}_{RR}) = \rho_R\psi_0(\theta) + \tfrac{1}{2}\alpha(\theta)|\mathbf{T}_{RR}|^2 + \tfrac{1}{2}\int_0^{|\mathbf{T}_{RR}|^2} u\beta(\theta, u)\, du$$

where α and β are real-valued functions. Hence, we obtain

$$\rho_R \partial_{\mathbf{T}_{RR}}\phi = \big[\alpha(\theta) + |\mathbf{T}_{RR}|^2\beta(\theta, |\mathbf{T}_{RR}|^2)\big]\mathbf{T}_{RR} - \mathbf{E}, \qquad \rho_R \partial_{\mathbf{E}}\phi = -\mathbf{T}_{RR}$$

so that letting $\mathbf{J}_{RR} = \mu\mathbf{I}$ we have

$$\mathbf{K}_{RR}(\theta, \mathbf{E}) = \mu\mathbf{I} + \beta(\theta, |\mathbf{T}_{RR}|^2)\mathbf{T}_{RR} \otimes \mathbf{T}_{RR} + (\alpha - \mu)\,\frac{\mathbf{T}_{RR}}{|\mathbf{T}_{RR}|} \otimes \frac{\mathbf{T}_{RR}}{|\mathbf{T}_{RR}|}.$$

The tensor $\mathbf{K}_{RR}$ enjoys the major symmetry and, choosing $\mu = \alpha$, reduces to

$$\mathbf{K}_{RR}(\theta, \mathbf{E}) = \alpha(\theta)\mathbf{I} + \beta(\theta, |\mathbf{T}_{RR}|^2)\mathbf{T}_{RR} \otimes \mathbf{T}_{RR}.$$

The resulting hypoelastic constitutive Eq. (13.27) is generated by the well-known Prandtl–Reuss work hardening plasticity theory provided that $\operatorname{tr}\mathbf{E} = \operatorname{tr}\mathbf{T}_{RR} = 0$ (see [310], Sect. C).

In the spatial description, using (13.19) with $\tilde{\gamma}_E = 0$, we obtain an expression equivalent to (13.22), namely,

$$\overset{\square}{\mathbf{T}} = \mathsf{C}\mathbf{D}, \qquad \mathsf{C}(\theta, \mathbf{E}_A, \mathbf{T}) = \mathsf{G} + \frac{1}{\rho|\partial_{\mathbf{T}}\psi|^2}\partial_{\mathbf{T}}\psi \otimes (\mathbf{T} - \rho\partial_{\mathbf{E}_A}\psi - \rho\mathsf{G}^T\partial_{\mathbf{T}}\psi). \tag{13.28}$$

After replacing ρ_R, $\mathbf{E}$, $\mathbf{T}_{RR}$, C_{RR} with ρ, $\mathbf{E}_A$, $\mathbf{T}$, C, respectively, we can translate all previous results into the spatial description.

13.6.3 Dissipative Duhem-Like Solids

The Duhem-like structure is considered by letting $\partial_{\mathbf{T}_{RR}}\psi = \mathbf{0}$, or $\partial_{\mathbf{E}}\phi = \mathbf{0}$, and allowing the model to be dissipative in that σ_{ET} is nonzero. Moreover, $\mathbf{T}_{RR}$, or $\mathbf{E}$, is still an independent variable.

If $\partial_{\mathbf{T}_{RR}}\psi = \mathbf{0}$, and $\gamma_E \neq 0$, by (13.13) we have

$$(\rho_R\partial_{\mathbf{E}}\psi - \mathbf{T}_{RR}) \cdot \dot{\mathbf{E}} = -\gamma_E|\dot{\mathbf{E}}| \leq 0.$$

To obtain a representation of $\mathbf{T}_{RR}$ we apply the representation formula (A.8) to $\mathbf{Z} = \rho_R\partial_{\mathbf{E}}\psi - \mathbf{T}_{RR}$ with $\mathbf{N} = \dot{\mathbf{E}}/|\dot{\mathbf{E}}|$ and, for simplicity, $\mathbf{G} = \mathbf{0}$, we find

$$\mathbf{T}_{RR} = \gamma_E\frac{\dot{\mathbf{E}}}{|\dot{\mathbf{E}}|} + \rho_R\partial_{\mathbf{E}}\psi(\theta, \mathbf{E}).$$

This equation is rate independent and the term $\gamma_E\dot{\mathbf{E}}/|\dot{\mathbf{E}}|$ may be viewed as a dissipative (viscous) effect. If, for instance, $\rho_R\psi$ is a quadratic function of $\mathbf{E}$, $\rho_R\psi = \frac{1}{2}\lambda(\theta)\mathbf{E}\cdot\mathbf{E}$, then

$$\mathbf{T} = \tilde{\gamma}_E\frac{\dot{\mathbf{E}}}{|\dot{\mathbf{E}}|} + \lambda(\theta)\mathbf{E},$$

which is similar to the model of Kelvin–Voigt viscoelastic solids. In the spatial description, if $\partial_{\mathbf{T}}\psi = \mathbf{0}$ we have

$$\mathbf{T} = \tilde{\gamma}_E\frac{\mathbf{D}}{|\mathbf{D}|} + \rho\partial_{\mathbf{E}_A}\psi(\theta, \mathbf{E}_A).$$

If instead we consider (13.16) and let $\partial_{\mathbf{E}}\phi = \mathbf{0}$ then we have

$$(\rho_R\partial_{\mathbf{T}_{RR}}\phi + \mathbf{E}) \cdot \dot{\mathbf{T}}_{RR} = -\gamma_T|\dot{\mathbf{T}}_{RR}|.$$

Applying the representation formula (A.8) with $\mathbf{N} = \dot{\mathbf{T}}_{RR}/|\dot{\mathbf{T}}_{RR}|$, $\mathbf{Z} = \rho_R\partial_{\mathbf{T}_{RR}}\phi + \mathbf{E}$, and $\mathbf{G} = \mathbf{0}$ we find the equation

$$\mathbf{E} = -\gamma_T\frac{\dot{\mathbf{T}}_{RR}}{|\dot{\mathbf{T}}_{RR}|} - \rho_R\partial_{\mathbf{T}_{RR}}\phi(\theta, \mathbf{T}_{RR}),$$

which is dissipative ($\gamma_T > 0$) and rate independent. If $\rho_R \phi$ has the quadratic form $\rho_R \phi = \frac{1}{2}\kappa(\theta)\mathbf{T}_{RR} \cdot \dot{\mathbf{T}}_{RR}$ then we find

$$\gamma_T \frac{\dot{\mathbf{T}}_{RR}}{|\dot{\mathbf{T}}_{RR}|} + \kappa(\theta)\mathbf{T}_{RR} = -\mathbf{E}. \tag{13.29}$$

Equation (13.29) is formally similar to the equation of a standard linear solid with a repulsive spring in parallel to the Maxwell unit. Letting $k^{(s)} = -k^{(M)}$ in (6.23) we find the analogue of (13.29).

13.6.4 Elastic–Plastic Models

Elastic–plastic models are characterized by an entropy production σ which depends linearly on either $|\dot{\mathbf{E}}|$ or $|\dot{\mathbf{T}}_{RR}|$. For definiteness, we consider (13.13), with $\partial_{\mathbf{T}_{RR}}\psi \neq \mathbf{0}$, or (13.16), with $\partial_{\mathbf{E}}\phi \neq \mathbf{0}$.

Assume $|\partial_{\mathbf{T}_{RR}}\psi| \neq 0$ and let

$$\mathbf{P}_{RR} = \frac{\partial_{\mathbf{T}_{RR}}\psi}{\rho_R |\partial_{\mathbf{T}_{RR}}\psi|^2}.$$

In view of the representation formula (A.8), we can write the constitutive Eq. (13.13) in the form

$$\dot{\mathbf{T}}_{RR} = \mathsf{C}_{RR}\dot{\mathbf{E}} - \gamma_E \mathbf{P}_{RR}|\dot{\mathbf{E}}|,$$

where C_{RR} is given in (13.21). In the spatial description, if $|\partial_{\mathbf{T}}\psi| \neq 0$ then (13.19) can be written as

$$\overset{\square}{\mathbf{T}} = \mathsf{C}\mathbf{D} - \tilde{\gamma}_E \mathbf{P}|\mathbf{D}|,$$

where C is given in (13.22) and

$$\mathbf{P} = \frac{\partial_{\mathbf{T}}\psi}{\rho |\partial_{\mathbf{T}}\psi|^2}.$$

A different class of elastic–plastic models follows from (13.16) by letting $\partial_{\mathbf{E}}\phi \neq \mathbf{0}$, which amounts to letting $\rho_R \partial_{\mathbf{E}}\psi - \mathbf{T}_{RR} \neq \mathbf{0}$, and

$$\mathbf{Q}_{RR}(\theta, \mathbf{E}, \mathbf{T}_{RR}) = \frac{\partial_{\mathbf{E}}\phi}{\rho_R |\partial_{\mathbf{E}}\phi|^2} = \frac{\rho_R \partial_{\mathbf{E}}\psi - \mathbf{T}_{RR}}{|\rho_R \partial_{\mathbf{E}}\psi - \mathbf{T}_{RR}|^2}.$$

In view of (A.8), we obtain

$$\dot{\mathbf{E}} = \mathsf{K}_{RR}\dot{\mathbf{T}}_{RR} - \gamma_T \mathbf{Q}_{RR}|\dot{\mathbf{T}}_{RR}|,$$

where K_{RR} is given in (13.26).

In a purely formal way, we might define

$$\dot{\mathbf{E}}^e = \mathbf{K}\dot{\mathbf{T}}_{RR}, \qquad \dot{\mathbf{E}}^p = -\gamma_T \mathbf{Q}_R |\dot{\mathbf{T}}_{RR}|$$

so that $\dot{\mathbf{E}} = \dot{\mathbf{E}}^e + \dot{\mathbf{E}}^p$. Moreover we might view $\dot{\mathbf{E}}^e$ and $\dot{\mathbf{E}}^p$ as the elastic and plastic strain rates. Yet, there is *no decomposition* $\mathbf{E} = \mathbf{E}^e + \mathbf{E}^p$ of the strain.

In the spatial description, we assume $\partial_{\mathbf{E}_A}\phi \neq \mathbf{0}$, and then $\mathbf{T} \neq \rho\partial_{\mathbf{E}_A}\psi$, and let

$$\mathbf{Q}(\theta, \mathbf{E}_A, \mathbf{T}) = \frac{\partial_{\mathbf{E}_A}\phi}{\rho|\partial_{\mathbf{E}_A}\phi|^2} = \frac{\partial_{\mathbf{E}_A}\psi - \mathbf{T}}{|\partial_{\mathbf{E}_A}\psi - \mathbf{T}|^2}.$$

Hence it follows from (13.20) that

$$\mathbf{D} = \mathsf{K}\overset{\square}{\mathbf{T}} - \tilde{\gamma}_T \mathbf{Q}|\overset{\square}{\mathbf{T}}|,$$

where K is obtained from K_{RR} by replacing $\rho_R, \mathbf{E}, \mathbf{T}_{RR}, \mathbf{J}_{RR}$ with $\rho, \mathbf{E}_A, \mathbf{T}, \mathbf{J}$. Again in a formal way, we can define

$$\mathbf{D}^e = \mathsf{K}\overset{\square}{\mathbf{T}}, \qquad \mathbf{D}^p = -\tilde{\gamma}_T \mathbf{Q}|\overset{\square}{\mathbf{T}}|$$

and say that $\mathbf{D} = \mathbf{D}^e + \mathbf{D}^p$; this however does not mean that the velocity field is decomposed in elastic and plastic parts.

13.6.5 One-Dimensional Models

Restrict attention to one-dimensional models associated with strain and stress in the direction $\mathbf{e}$ so that

$$\mathbf{E} = E\mathbf{e}\otimes\mathbf{e}, \qquad \mathbf{T}_{RR} = S\mathbf{e}\otimes\mathbf{e}.$$

The symbol S for the component of $\mathbf{T}_{RR}$ is consistent with the *engineering stress* considered in the literature as the ratio of the axial force over the reference area [216], p. 74. Let ψ be a function of E, S, parameterized by θ, and $\theta\sigma$ be a function of $E, S, \dot{E}, \dot{S}$. The generalized Clausius–Duhem inequality (13.9) is then written as

$$\partial_S\psi_R\,\dot{S} + (\partial_E\psi_R - S)\dot{E} = -J\theta\,\sigma(E, S, \dot{E}, \dot{S}), \tag{13.30}$$

where ψ_R is the referential free energy, $\psi_R = \rho_R\psi$. At constant temperature, $\dot{\psi}_R = \partial_E\psi_R\,\dot{E} + \partial_S\psi_R\,\dot{S}$ and hence integration of (13.30), as $t \in [t_1, t_2]$, along a closed curve in the $E - S$ plane results in

$$0 \geq \int_{t_1}^{t_2}[\partial_S\psi_R\,\dot{S} + (\partial_E\psi_R - S)\dot{E}]dt = -\int_{t_1}^{t_2} S\,\dot{E}\,dt = -\oint S\,dE,$$

$\oint$ denoting the integral along the closed curve. The positive value of $\oint S\,dE$ implies that the closed curve is run in the clockwise sense.

Consistent with (13.13) we let

$$J\theta\sigma = \gamma_E(E, S, \operatorname{sgn}\dot{E})|\dot{E}|, \qquad \gamma_E \geq 0.$$

Hence Eq. (13.30) is invariant under the time transformation $t \mapsto ct, c > 0$, and hence the associated model is rate independent because so is γ_E.

Look at time intervals where $\dot{E} \neq 0$. Since $\partial_S\psi_R \neq 0$, divide Eq. (13.30) by $\partial_S\psi_R\dot{E}$ to obtain

$$\frac{\dot{S}}{\dot{E}} = \frac{S - \partial_E\psi_R}{\partial_S\psi_R} - \frac{\gamma_E}{\partial_S\psi_R}\operatorname{sgn}\dot{E}. \tag{13.31}$$

Let

$$\chi_1 := \frac{S - \partial_E\psi_R}{\partial_S\psi_R}, \qquad \chi_2 := -\frac{\gamma_E}{\partial_S\psi_R}. \tag{13.32}$$

Hence the stress–strain slope can be written in the form

$$\frac{\dot{S}}{\dot{E}} = \chi_1 + \chi_2 \operatorname{sgn}\dot{E}, \tag{13.33}$$

where χ_1 is a function of E, S parameterized by θ, while χ_2 can also depend on $\operatorname{sgn}\dot{E}$.

Since γ_E depends on $\dot{E}$ only through sgn $\dot{E}$ then the right-hand side of (13.31) has the form

$$\frac{\dot{S}}{\dot{E}} = f(S, E).$$

Now if a stress–strain curve is piecewise described by the form $F(S, E) = 0$ then

$$\frac{\dot{S}}{\dot{E}} = -\frac{\partial_E F}{\partial_S F} = \frac{dS}{dE}.$$

Thus in (13.33) the ratio $\dot{S}/\dot{E}$ is the slope of the stress–strain curve and then is the *effective differential elastic modulus*.

If $\gamma_E = 0$, and hence $\chi_2 = 0$, then dS/dE is the slope of the hypoelastic curve; the assumption

$$\frac{dS}{dE} = \chi_1(\theta, E, S) > 0$$

means that the slope dS/dE is positive and the value at E depends also on S. Hence we regard the conditions

$$\gamma_E = 0, \qquad \partial_S\psi_R = 0, \qquad S = \partial_E\psi_R$$

as characterizing the elastic regime and regard χ_1 as the *differential elastic modulus.*

The slope of the stress–strain curve is assumed to be non-negative; since $\chi_1 > 0$ it is consistent to assume

$$\chi_1 + \chi_2 \operatorname{sgn} \dot{E} \geq 0, \qquad |\chi_2| \leq \chi_1. \tag{13.34}$$

The monotonicity condition (13.34) follows from physical arguments about the stress–strain curve but is not required by thermodynamics.

To establish models of elastic–plastic, rate-independent materials, we have to specify the functions χ_1 and χ_2. By definition, χ_1 is fully determined by the free energy ψ while χ_2 depends also on γ_E. Consequently once a free energy function ψ is selected, different models are determined by different functions γ_E. Indeed, since the hysteretic properties of the model depend on γ_E then we refer to γ_E as the *hysteretic function.*

The possible models are framed within the following scheme:

Hyperelastic regime	$\gamma_E = 0$	$\partial_S \psi_R = 0$	$S = \partial_E \psi_R$
Hypoelastic regime	$\gamma_E = 0$	$\partial_S \psi_R \neq 0$	$\dfrac{dS}{dE} = \dfrac{S - \partial_E \psi_R}{\partial_S \psi_R}$
Hysteretic regime	$\gamma_E \neq 0$	$\partial_S \psi_R \neq 0$	$\dfrac{dS}{dE} = \dfrac{S - \partial_E \psi_R}{\partial_S \psi_R} - \dfrac{\gamma_E}{\partial_S \psi_R} \operatorname{sgn} \dot{E}$

The next investigation deals with non-elastic bodies and hence is developed under the assumption that $\partial_S \psi_R \neq 0$.

A Formal Additive Decomposition of the Strain Rate

An additive decomposition of the strain rate, not the strain, follows by letting[8] $S - \partial_E \psi_R \neq 0$ and $J\theta\sigma = \gamma_T |\dot{S}|$; the function $\gamma_T \geq 0$ is non-negative and possibly dependent on E, S, $\operatorname{sgn} \dot{S}$ and parameterized by θ. Hence Eq. (13.30) becomes

$$\partial_S \psi_R \dot{S} + (\partial_E \psi_R - S)\dot{E} = -\gamma_T |\dot{S}|.$$

Dividing by $\partial_E \psi_R - S$ we have

$$\dot{E} = \xi_1 \dot{S} + \xi_2 |\dot{S}|, \qquad \xi_1 = \frac{1}{\chi_1} = \frac{\partial_S \psi_R}{S - \partial_E \psi_R}, \qquad \xi_2 = \frac{\gamma_T}{S - \partial_E \psi_R}.$$

This splitting mimics the standard relation $\dot{E} = \dot{E}^e + \dot{E}^p$ of the one-dimensional incremental theory of plasticity and suggests that we introduce an hardening variable S by means of the differential relation

[8] This amounts to assuming $\partial_E \phi_R \neq 0$, $\phi_R := \psi_R - ES$.

$$\dot{\mathsf{S}} = [\mathrm{sgn}\,(S - \partial_E \psi)]\dot{E}^p = [\mathrm{sgn}\,(S - \partial_E \psi)]\xi_2 |\dot{S}|,$$

with a zero initial value. This quantity turns out to be always non-negative and strictly related to the entropy production,

$$\dot{\mathsf{S}} = \frac{\gamma_T}{|S - \partial_E \psi_R|} |\dot{S}| = \frac{J\theta\sigma}{|S - \partial_E \psi_R|}. \tag{13.35}$$

In addition, whenever $\dot{S} \neq 0$, we can write

$$\frac{dE}{dS} = \xi_1 + \xi_2 \,\mathrm{sgn}\, \dot{S}. \tag{13.36}$$

Thus we require that

$$\xi_1 + \xi_2 \,\mathrm{sgn}\, \dot{S} \geq 0.$$

Some remarks are in order about energy dissipation. It is often asserted that the area of the hysteresis loop is related to the amount of energy dissipation. Now, in a cycle from t_1 to t_2

$$\oint S dE = \int_{t_1}^{t_2} J\theta\sigma dt.$$

The left-hand side is the area A of the loop. If we let $\sigma = \gamma_E |\dot{E}|$ then letting $J\theta\gamma_E$ roughly constant along the cycle we have

$$A \simeq J\theta\gamma_E \int_{t_1}^{t_2} |\dot{S}| dt = J\theta\gamma_E 2\Delta E,$$

ΔE being the variation of E. Hence

$$2J\theta\gamma_E \simeq \frac{A}{\Delta E}$$

is the approximate (vertical) thickness of the loop. Likewise, if $\sigma = \gamma_E |\dot{S}|$ then

$$2J\theta\gamma_T \simeq \frac{A}{\Delta S}$$

is the approximate (horizontal) thickness of the loop. Narrow (wide) loops are given by small (large) values of γ_E or γ_T.

13.6.6 The Helmholtz Free Energy

A model is characterized by χ_1, χ_2 and χ_1 follows from the free energy ψ_R. To select ψ_R we start from the generic assumption

$$\psi_R(E, S) = \mathcal{L}(S - \mathcal{G}(E)) + \mathcal{F}(S) + \mathcal{H}(E),$$

$\mathcal{L}, \mathcal{G}, \mathcal{F}$, and $\mathcal{H}$ being so far undetermined differentiable functions, possibly parameterized by the temperature θ. By (13.32) we find

$$\chi_1 = \frac{S + \mathcal{L}'(S - \mathcal{G}(E))\mathcal{G}'(E) - \mathcal{H}'(E)}{\mathcal{L}'(S - \mathcal{G}(E)) + \mathcal{F}'(S)}, \qquad \chi_2 = -\frac{\gamma_E}{\mathcal{L}'(S - \mathcal{G}(E)) + \mathcal{F}'(S)}.$$

For simplicity we let

$$\chi_1 = g(E) \geq 0,$$

g being characteristic of the model. Hence it follows

$$S - \mathcal{F}'(S)g(E) - \mathcal{L}'(S - \mathcal{G}(E))[g(E) - \mathcal{G}'(E)] = \mathcal{H}'(E).$$

A solution is given by

$$g(E) - \mathcal{G}'(E) = \alpha, \qquad \mathcal{F}'(S) = 0, \qquad \mathcal{H}'(E) = \mathcal{G}(E), \qquad \mathcal{L}'(S - \mathcal{G}(E)) = \frac{1}{\alpha}(S - \mathcal{G}(E)),$$

where α is a constant. Hence χ_2 is given by

$$\chi_2 = -\alpha \frac{\gamma_E}{S - \mathcal{G}(E)}.$$

Since $g \geq 0$ it is convenient to let $\alpha > 0$. The properties of the model are then summarized as follows.

- The free energy ψ_R is characterized by the function $\mathcal{G}(E)$ and parameterized by α (and the temperature θ) in the form

$$\psi_R(E, S) = \frac{1}{2\alpha}[S - \mathcal{G}(E)]^2 + \mathcal{H}(E), \qquad \mathcal{H}'(E) = \mathcal{G}(E); \tag{13.37}$$

- the functions χ_1 and χ_2 take the form

$$\chi_1 = \mathcal{G}'(E) + \alpha \geq 0, \qquad \chi_2 = -\alpha \frac{\gamma_E}{S - \mathcal{G}(E)}. \tag{13.38}$$

As a comment, the positive parameter α is the difference between the differential elastic modulus χ_1 and the slope of the elastic function $\mathcal{G}(E)$. The occurrence of both $\mathcal{G}$ and α allows a greater flexibility in the modelling of hysteretic materials.

13.6.7 Some Hysteretic Models

The hysteretic properties are closely related to γ_E. Here we show some examples induced by choices of γ_E.

Plastic Flow with Asymptotic Strength

Let $\mathcal{G}(E) = 0$ so that $\psi = S^2/2\alpha$ and $\chi_1 = \alpha$. Select the hysteretic function γ_E in the form

$$\gamma_E(S) = \frac{S^2}{S_u}, \qquad S_u > 0.$$

Hence (13.31) can be written as

$$\frac{dS}{dE} = \frac{\alpha}{S_u}(S_u - S \operatorname{sgn} \dot{E}).$$

Consequently

$$\frac{d}{dE}(S - S_u)^2 = 2\frac{\alpha}{S_u}(S - S_u)(S_u - S \operatorname{sgn} \dot{E}) \le 0$$

and then hysteresis loops are confined to the open strip $|S| < S_u$. No elastic region exists; at any point $\dot{E}$ affects the slope of the curve. Moreover $|S| < S_u$ implies that $dS/dE > 0$, which implies the monotonicity of the curve in any region. The model is characterized by the positive parameters α, S_u; since $\alpha = \chi_1$ then α is the slope of the elastic stress–strain path, while S_u is the ultimate tensile stress or the asymptotic strength as $|E| \to \infty$. To obtain a picture of the hysteresis loops, we consider the system of equations

$$\begin{cases} \dot{E} = \omega \mathfrak{E} \cos \omega t, \\ \dot{S} = \alpha(S_u \dot{E} - S|\dot{E}|)/S_u, \end{cases}$$

with initial values E_0, S_0. The rate independence of the model makes the loops to be independent of the angular frequency ω (Fig. 13.1).

Plastic Flow with Nonlinear Elastic Function

A more involved model is established by letting the elastic function $\mathcal{G}(E)$ be nonzero. Indeed we let $\mathcal{G}(E)$ be nonlinear,

$$\mathcal{G}(E) = \tanh(\kappa E), \qquad \kappa > 0,$$

and we let

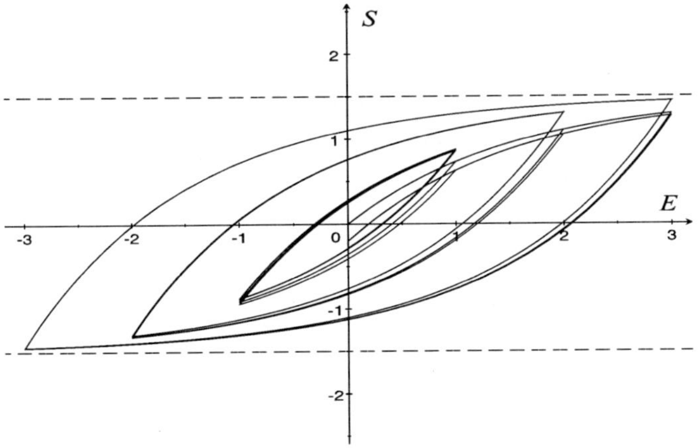

Fig. 13.1 Plastic model with asymptotic strength: hysteresis loops (solid) and asymptotic bounds $|S| = S_u$ (dashed) with $\alpha = 1$, $S_u = 1.5$, $\mathfrak{E} = 1, 2, 3$ and starting from (E_0, S_0) with $E_0 = 0$, $S_0 = 0, -0.1$

$$\gamma_E = \frac{1}{\lambda}[S - \tanh(\kappa E)]^2, \qquad \lambda > 0.$$

By (13.37) and (13.38) it follows that

$$\psi_R(E, S) = \frac{1}{2\alpha}[S - \tanh(\kappa E)]^2 + \frac{1}{\kappa}\ln[\cosh(\kappa E)],$$

$$\chi_1 = \kappa[1 - \tanh^2(\kappa E)] + \alpha, \qquad \chi_2 = -\alpha[S - \tanh(\kappa E)]\text{sgn}\,\dot{E}.$$

Accordingly, we have

$$\frac{dS}{dE} = \kappa[1 - \tanh^2(\kappa E)] + \frac{\alpha}{\lambda}\{\lambda - [S - \tanh(\kappa E)]\text{sgn}\,\dot{E}\}.$$

The elastic region reduces to the curve $S = \mathcal{G}(E)$. As $|E| \to \infty$ we have $|S| \to (\lambda + 1)_-$. The loops rely within the horizontal strip $|S| \le 1 + \lambda$. Examine the requirement

$$0 \le \chi_1 - |\chi_2| = \kappa[1 - \tanh^2(\kappa E)] + \alpha - \frac{\alpha}{\lambda}|S - \tanh(\kappa E)|.$$

As $E > 0$ the requirement is satisfied if

$$|S - \tanh(\kappa E)| \le \lambda + 1 - \tanh(\kappa E) \le \lambda + \frac{\lambda\kappa}{\alpha}[1 - \tanh(\kappa E)]$$

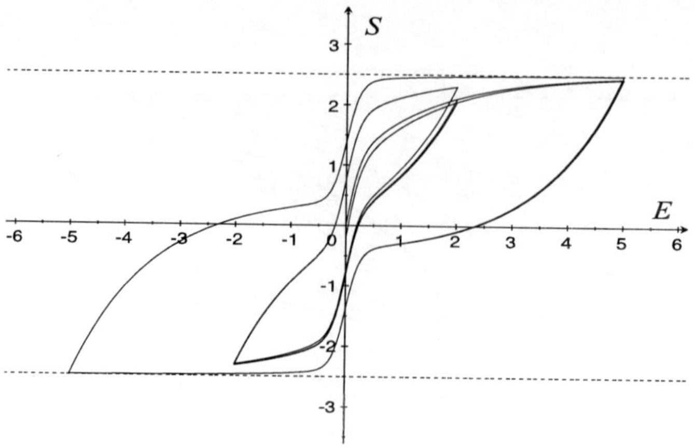

Fig. 13.2 Plastic model with nonlinear hardening. Hysteresis loops (solid) and yield-strength bounds $|S| = 1 + \lambda$ (dashed) are determined by letting $\alpha = 1$, $\kappa = 4$, $\lambda = 1.5$; the amplitude of the oscillations is $\mathfrak{E} = 2.5$

and hence if $\lambda\kappa \geq \alpha$. The same conclusion follows if $E < 0$. The Fig. 13.2 shows some loops determined with $\lambda\kappa/\alpha = 6$.

Elastic–Plastic Model

For simplicity, we consider a solid undergoing a linear behaviour in the elastic regime. Accordingly we let $\mathcal{G}(E) = 0$ and hence

$$\psi = \frac{S^2}{2\alpha}, \qquad \chi_1 = \alpha.$$

Let $S_u > S_y > 0$. The hysteretic function is chosen to be nonzero in the region $|S| \in [S_y, S_u]$ in the form

$$\gamma_E(S, \operatorname{sgn} \dot{E}) = \begin{cases} \frac{|S|-S_y}{S_u-S_y}|S| & \text{if} \quad |S| \geq S_y \quad \text{and} \quad S\dot{E} > 0, \\ 0 & \text{if} \quad |S| < S_y \quad \text{or} \quad |S| \geq S_y, \ S\dot{E} < 0. \end{cases}$$

Hence it follows

$$\frac{dS}{dE} = \begin{cases} \frac{\alpha}{S_u-S_y}(S_u - |S|) & \text{if} \quad |S| \geq S_y \quad \text{and} \quad S\dot{E} > 0, \\ \alpha & \text{if} \quad |S| < S_y \quad \text{or} \quad |S| \geq S_y, \ S\dot{E} < 0. \end{cases}$$

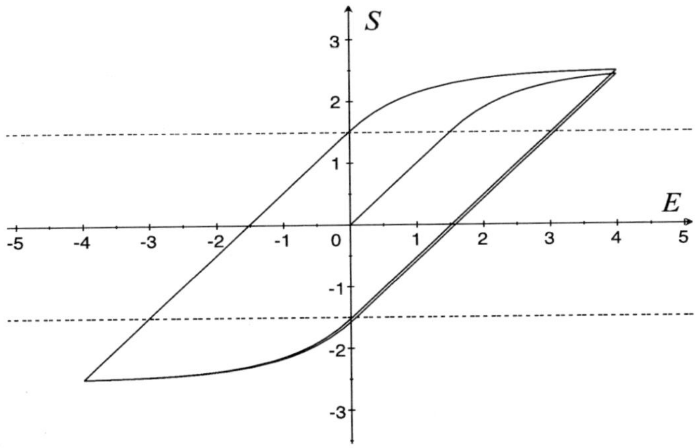

Fig. 13.3 Elastic–plastic model: hysteresis loops (solid) and yield-strength bounds $|S| = S_y$ (dashed) with $\alpha = 1$, $S_y = 1.5$, $S_u = 2.5$, $\mathfrak{E} = 4$ and starting from the origin $(E, S) = (0, 0)$

The loops are placed within the strip $S_y \le |S| \le S_u$. Within the open strip the body behaves elastically. Within the region $|S| \in [S_y, S_u]$ the material behaves elastically during unloading and plastically during loading (Fig. 13.3).

13.6.8 Models with Additive Decomposition of the Strain Rate

The model of E in terms of S as is given in (13.35) allows us to find connections with the one-dimensional incremental theory of plasticity [216, 349].

For simplicity, we first consider a linear strain hardening and a linear dependence in the elastic regime. Hence, we let $\mathcal{G}(E) = 0$ so that $\psi = S^2/2\alpha$. Since $\partial_E \psi = 0$ then by (13.35) we have $\xi_1 = 1/\alpha$ and $\xi_2 = \gamma_T/S$. The hysteretic function γ_T is assumed in the form

$$\gamma_T(S, \operatorname{sgn}\dot{S}) = \begin{cases} |S|/h \text{ if } & |S| \ge S_y \quad \text{and} \quad S\dot{S} > 0, \\ 0 & \text{otherwise,} \end{cases}$$

where $S_y, h > 0$ represent the yield strength and the hardening parameter, respectively. Consequently, the stress–strain evolution is given by

$$\frac{dE}{dS} = \begin{cases} 1/\alpha + 1/h \text{ if } & |S| \ge S_y \quad \text{and} \quad S\dot{S} > 0, \\ 1/\alpha & \text{otherwise.} \end{cases}$$

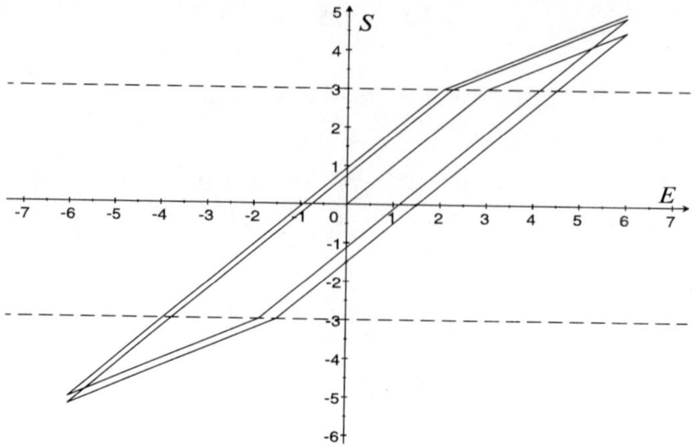

Fig. 13.4 Elastic–plastic model with linear strain hardening: hysteresis loops (solid) and yield-strength bounds $|S| = S_y$ (dashed) are determined by $\alpha = 1, h = 1, S_y = 3$

The elastic region is identified with the horizontal strip $\{(E, S) : |S| \le S_y\}$ (Fig. 13.4). The slopes of the stress–strain curves are α, in the elastic region and when $S\dot{S} < 0$, and $\alpha h/(\alpha + h) < \alpha$ when $|S| \ge S_y$, $S\dot{S} > 0$. The strain rate might be decomposed in the form

$$\dot{E}^e = \frac{1}{\alpha}\dot{S}, \qquad \dot{E}^p = \begin{cases} \dot{S}/h \text{ if } & |S| \ge S_y \text{ and } S\dot{S} > 0, \\ 0 & \text{otherwise.} \end{cases}$$

By (13.35) we define the hardening variable S on $[t_0, t]$, $\mathsf{S}(t_0) = 0$, via the rate equation

$$\dot{\mathsf{S}} = \frac{\gamma_T |\dot{S}|}{|S|} = \begin{cases} |\dot{S}|/h \text{ if } & |S| \ge S_y \text{ and } S\dot{S} > 0, \\ 0 & \text{otherwise.} \end{cases}$$

Hence $\dot{\mathsf{S}} \ge 0$ and

$$\mathsf{S}(t) = \int_{t_0}^{t} \dot{\mathsf{S}}(\tau) d\tau = \int_{t_0}^{t} \frac{\gamma_T(\tau)|\dot{S}(\tau)|}{|S(\tau)|} d\tau.$$

Consequently

$$\mathsf{S}(t) = \frac{1}{h} \int_{\mathcal{I}(t_0,t)} |\dot{S}(\tau)| d\tau,$$

where $\mathcal{I}(t_0, t) = \{\tau \in (t_0, t) : |S(\tau)| \ge S_y \text{ and } S(\tau)\dot{S}(\tau) > 0\}$.

A nonlinear strain hardening is now modelled by letting again $\mathcal{G}(E) = 0$ and hence $\psi = S^2/2\alpha$, $\xi_1 = 1/\alpha$. Letting $S_u > S_y > 0$ we choose the hysteretic function

$$\gamma_\tau(S, \operatorname{sgn}\dot{S}) = \begin{cases} \frac{|S|-S_y}{\alpha(S_u-|S|)}|S| & \text{if } |S| \geq S_y \text{ and } S\dot{S} > 0, \\ \frac{1}{\alpha} & \text{otherwise.} \end{cases}$$

Hence $\gamma_\tau > 0$ in the open strip $|S| \in (S_y, S_u)$ when $S\dot{S} > 0$. It follows that

$$\frac{dE}{dS} = \begin{cases} \frac{S_u-S_y}{\alpha(S_u-|S|)} & \text{if } |S| \geq S_y \text{ and } S\dot{S} > 0, \\ \frac{1}{\alpha} & \text{otherwise.} \end{cases}$$

The resulting loops have then the form of Fig. 13.3.

The additive decomposition of the strain rate is possible in the form

$$\dot{E}^e = \frac{1}{\alpha}\dot{S}, \qquad \dot{E}^p = \begin{cases} \frac{|S|-S_y}{\alpha(S_u-|S|)}\dot{S} & \text{if } |S| \geq S_y \text{ and } S\dot{S} > 0, \\ 0 & \text{otherwise.} \end{cases}$$

Moreover, the rate of the hardening variable S is given by

$$\dot{\mathsf{S}} = (\operatorname{sgn} S)\dot{E}^p = \begin{cases} \frac{|S|-S_y}{\alpha(S_u-|S|)}|\dot{S}| & \text{if } |S| \geq S_y \text{ and } S\dot{S} > 0, \\ 0 & \text{otherwise.} \end{cases}$$

Hence

$$\mathsf{S}(t) = \int_{\mathcal{I}(t_0,t)} \frac{|S(\tau)| - S_y}{\alpha(S_u - |S(\tau)|)}|\dot{S}(\tau)|d\tau,$$

where $\mathcal{I}(t_0, t) = \{\tau \in (t_0, t) : |S(\tau)| \geq S_y \text{ and } S(\tau)\dot{S}(\tau) > 0\}$.

13.7 Constitutive Models of Polymeric Foams

Polymeric foams can be viewed as porous media with a solid matrix while the pores are saturated with a gas. The importance of foams for industrial applications is connected with the energy-absorbing capacity; they undergo large deformation while maintaining a nearly constant stress value. Restrict attention to one-dimensional settings and denote by S and H the engineering stress and the engineering strain[9]; S is the axial force per unit undeformed area (Piola stress) while $H = \partial_X u \simeq (L - L_0)/L_0$, where L and L_0 are the current length and the reference length; by definition we have $H > -1$.

The stress–strain curve of polymeric foams generally shows three qualitative regions: an initial linearly elastic region, next a plasticity-like plateau region, and then a final densification region where the stress rises steeply.

The stress–strain relation has been investigated experimentally. We mention two models of $S = f(H)$ with five parameters, [20, 280]

[9] Denoted by σ and ε frequently in the literature.

$$f_1(H) = a\frac{\exp(\alpha H) - 1}{\exp(\beta H) + 1} + c[\exp(\gamma H) - 1],$$

$$f_2(H) = a\{1 - \exp[(-e/a)H(1 - H)^m]\} + b(\frac{H}{1 - H})^n,$$

where $a, c, \alpha, \beta, \gamma$ and a, b, e, m, n are the pertinent parameters.

Yet the dynamic loading shows a dependence of S also on the strain rate. To include strain-rate effects, the dependence has been improved in the form [341]

$$S = f(H)h(H, \dot{H}), \tag{13.39}$$

where f, h may depend on the temperature θ and the (initial) mass density ρ. The literature shows various forms of the functions f, h, possibly $f = f_1, f_2$, and, e.g.

$$h(H, \dot{H}) = (\dot{H}/\dot{H}_0)^{a+bH} \quad \text{and} \quad h(H, \dot{H}) = 1 + (a + bH)\ln(\dot{H}/\dot{H}_0),$$

where $\dot{H}_0$ is a reference strain rate (frequently $\dot{H}_0 = 10^{-3}\ \mathrm{s}^{-1}$).

The strain-rate dependence may indicate that also hysteretic effects occur. Hysteresis really happens as is checked experimentally [285]. Here we take the suggestion (13.39) for a constitutive equation $S(\theta, H, \dot{H})$ and look for thermodynamically consistent models of solid foams.

Following the literature we consider a one-dimensional setting and take the balance equation of energy in the form

$$\rho\dot{\varepsilon} = S\dot{H} - \partial_x q + \rho r.$$

The entropy inequality $\rho\dot{\eta} + \partial_x(q/\theta) = \sigma \geq 0$ leads to

$$-\rho(\dot{\psi} + \eta\dot{\theta}) + S\dot{H} - \frac{1}{\theta}q\partial_x\theta = \theta\sigma.$$

Let $\theta, H, S, \partial_x\theta, \dot{H}, \dot{S}$ be the set of independent variables and ψ, η, σ the constitutive functions; the joint dependence on $H, S, \dot{H}, \dot{S}$ is suggested by the need of modelling hysteretic effects. Upon evaluation of $\dot{\psi}$ and substitution, we find

$$-\rho(\partial_\theta\psi + \eta)\dot{\theta} - (\rho\partial_H\psi - S)\dot{H} - \rho\partial_S\psi\dot{S} - \rho\partial_{\partial_x\theta}\psi\overline{\partial_x\theta}^{\boldsymbol{\cdot}} - \rho\partial_{\dot{H}}\psi\ddot{H} - \rho\partial_{\dot{S}}\psi\ddot{S} - \frac{1}{\theta}q\partial_x\theta = \theta\sigma.$$

The arbitrariness of $\ddot{H}, \ddot{S}, \overline{\partial_x\theta}^{\boldsymbol{\cdot}}, \dot{\theta}$ implies that

$$\partial_{\dot{H}}\psi = 0, \quad \partial_{\dot{S}}\psi = 0, \quad \partial_{\partial_x\theta}\psi = 0, \quad \eta = -\partial_\theta\psi.$$

Since ψ is independent of $\partial_x\theta$ then letting $\partial_x\theta = 0$ we obtain

$$-(\rho\partial_H\psi - S)\dot{H} - \rho\partial_S\psi\dot{S} = \theta\sigma_{\text{mech}} \geq 0, \tag{13.40}$$

where $\sigma_{\text{mech}} = \sigma|_{\partial_x\theta=0}$. Since $\psi = \psi(\theta, H, S)$ then at any isothermal closed curve in $[t_1, t_2]$, $H(t_2) = H(t_1)$, $S(t_2) = S(t_1)$, we have

$$0 \leq \int_{t_1}^{t_2} \theta\sigma_{\text{mech}} dt = \int_{t_1}^{t_2} [-\rho\partial_S\psi\dot{S} + (S - \rho\partial_H\psi)\dot{H}]dt = \int_{t_1}^{t_2} S\,\dot{H}\,dt.$$

The non-negative value of $\int_{t_1}^{t_2} S\,\dot{H}\,dt$ along a closed curve implies that the curve is run in the clockwise sense.

We now apply Eq. (13.40) to derive a rate equation for the stress S. Let $\partial_S\psi \neq 0$ and assume

$$\theta\sigma_{\text{mech}} = \gamma(\theta, H, S, \dot{H})|\dot{H}|, \qquad \gamma \geq 0.$$

In view of (13.40), we can write

$$\dot{S} = \frac{(S - \rho\partial_H\psi)}{\rho\partial_H\psi}\dot{H} - \frac{\gamma}{\rho\partial_H\psi}|\dot{H}|, \tag{13.41}$$

and hence, except at the inversion points where $\dot{H} = 0$,

$$\frac{\dot{S}}{\dot{H}} = \frac{S - \rho\partial_H\psi}{\rho\partial_S\psi} - \frac{\gamma}{\rho\partial_S\psi}\operatorname{sgn}\dot{H}. \tag{13.42}$$

It is worth remarking that Eq. (13.41) is invariant under the time transformation

$$t \to ct, \qquad c > 0,$$

if γ is invariant and in particular if γ is independent of $\dot{H}$. Here instead we are interested in the dependence of γ on $\dot{H}$ and hence the present model describes a rate-dependent behaviour.

Since γ depends on $\dot{H}$ then (13.42) is a nonlinear rate equation for S. If, however, we restrict attention to numerical or experimental testing where $\dot{H}$ is constant then the right-hand side of (13.42) is a function of S, H and, as with (13.31),

$$\frac{\dot{S}}{\dot{H}} = \frac{dS}{dH}$$

is the slope of the stress–strain curve, parameterized by θ and $\dot{H}$. Alternatively, we observe that if $S = f(H, y)$ then, as y is constant, $\dot{S} = \partial_H f\ \dot{H}$. For the sake of generality we keep using the notation $\dot{S}/\dot{H}$ so that also a non-constant function $\dot{H}$ is considered.

In view of (13.42), we define

$$\chi_1 = \frac{S - \rho\partial_H\psi}{\rho\partial_S\psi}, \qquad \chi_2 = -\frac{\gamma}{\rho\partial_S\psi} \tag{13.43}$$

so that

$$\frac{\dot{S}}{\dot{H}} = \chi_1 + \chi_2 \operatorname{sgn} \dot{H}.$$

According to (13.41), the model is determined by means of the two functions $\psi(S, H)$, $\gamma(S, H, \dot{H})$, parameterized by θ. We start with the generic assumption

$$\psi(H, S) = \mathcal{L}(S - \mathcal{G}(H)) + \mathcal{F}(S) + \mathcal{H}(H).$$

Hence it follows

$$\chi_1 = \frac{S + \mathcal{L}'(S - \mathcal{G}(H))\mathcal{G}'(H) - \mathcal{H}'(H)}{\mathcal{L}'(S - \mathcal{G}(H)) + \mathcal{F}'(S)}, \qquad \chi_2 = -\frac{\gamma}{\mathcal{L}'(S - \mathcal{G}(H)) + \mathcal{F}'(S)}.$$

Let $\chi_1 = g(H)$. It follows from (13.43) that

$$S + \mathcal{L}'(S - \mathcal{G}(H))[\mathcal{G}'(H) - g(H)] - \mathcal{F}'(S)g(H) = \mathcal{H}'(H),$$

which holds if

$$\mathcal{L}'(S - \mathcal{G}(H)) = -\frac{1}{\alpha}[S - \mathcal{G}(H)], \qquad \mathcal{G}(H) = \alpha(H + 1), \qquad \mathcal{F}'(S) = \frac{1}{\alpha}S$$

and

$$\mathcal{H}'(H) = (H + 1)[\alpha - g(H)].$$

Hence the free energy ψ takes the form

$$\psi(H, S) = (H + 1)S - \int_0^H (1 + \xi)g(\xi)d\xi$$

and

$$\chi_1 = g(H), \qquad \chi_2 = -\frac{\gamma}{1 + H},$$

where $\gamma \geq 0$ and g has to be selected.

For definiteness and simplicity, in Fig. 13.5 we show an example of the loops obtained by solving the system of equations

$$\begin{cases} \dot{H} = \frac{1}{2}\omega\mathfrak{H} \sin \omega t, \\ \dot{S} = g(H)\dot{H} - \frac{\gamma}{1+H}|\dot{H}|, \end{cases}$$

with γ constant and $g(H) = f_1'(H)$; for definiteness we let

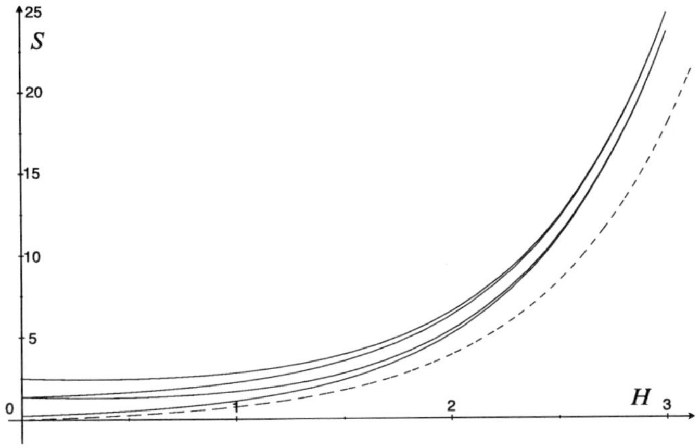

Fig. 13.5 Hysteresis loops (solid) and the graph of the function g (dashed) with $\mathfrak{H} = 3$, $\gamma = 0.2$, $H_0 = 0$, $S_0 = 2.5$

$$f_1(H) = 0.5\frac{\exp(3H) - 1}{\exp(H) + 1} + 0.2[\exp(1.5\,H) - 1].$$

The rate independence of the model allows the loops to be independent of the angular frequency ω.

Chapter 14
Superconductivity and Superfluidity

Superconductivity and superfluidity have a common origin due to the discovery of perfect conduction in mercury and superfluidity in helium. The uncommon properties are often said to be quantum-mechanical phenomena. However, the effects show on a large scale so that we can describe them as macroscopic phenomena.

In this chapter, superconductivity and superfluidity are developed in a continuum framework within the theory of mixtures. In superconductivity, the fluid is viewed as consisting of two constituents: normal electrons (satisfying Ohm's law) and superconducting electrons (they do not suffer any resistance). Indeed, within the London equations, the superconducting current satisfies a rate equation and hence follows from a free energy potential.

In superfluidity, the model involves two chemically reacting fluids, namely, the normal fluid and the superfluid (with zero entropy). The two fluids are viewed as a fluid mixture. A suitably approximated model which traces back to Landau allows the description of first and second sounds. Here the conjecture that second sound is generated by the normal heat flux is made formal by investigating the propagation of discontinuity waves governed by a Maxwell–Cattaneo objective equation.

14.1 Superconductors

14.1.1 Discoveries About Superconductivity

An outline is given of the main discoveries about superconductivity, starting from the discovery made by Kamerlingh Onnes in 1911; a detailed review and some

A. Morro and C. Giorgi, *Mathematical Modelling of Continuum Physics*,
Modeling and Simulation in Science, Engineering and Technology,
https://doi.org/10.1007/978-3-031-20814-0_14

mathematical models are given in [193] along with an exhaustive list of references on the subject.[1]

Drude's model of conductivity (1900) regarded metals as containing an electron gas responsible for the electrical conductivity. The known formula of conductivity $\sigma = e^2 n/m\nu$ is based on the view that the electrons are subject to the acceleration eE/m and that the mean velocity is $v = eE/m\nu$, where ν is the frequency of electron collisions with the lattice. Hence the current density $J = nev$, n being the number of electrons per unit volume, results in

$$J = \frac{ne^2}{\nu m} E$$

whence the result for $\sigma = J/E$. However the temperature dependences $n(\theta)$ and $\nu(\theta)$ remained quite unknown. This was a motivation for experiments in 1911 by Kamerlingh Onnes on the resistivity $1/\sigma$ of metals. Unexpected results appeared but these were unconvincing because the value was affected by the purity. Then Kamerlingh Onnes decided to study mercury and the resistance of the mercury sample dropped sharply at 4.15 K to an unmeasurably small value. It was natural that Kamerlingh Onnes would choose the name *superconductivity* for this phenomenon of perfect conductivity. Soon after the discovery by Kamerlingh Onnes many other metals were found to exhibit zero resistance when their temperatures were lowered below a characteristic value called the *critical temperature*,[2] θ_c.

In 1933 Meissner and Ochsenfeld studied the magnetic behaviour of superconductors in the presence of a magnetic field and found that, below their critical temperatures, the magnetic flux is expelled from the interior of the superconductor [309]. Moreover, at a fixed temperature $\theta \leq \theta_c$, these materials lost their superconducting property above a certain time-dependent critical magnetic field, $H_{0c}(\theta)$. Meissner and Ochsenfeld also discovered that, in the presence of a suitably weak magnetic field, by cooling the temperature below θ_c the field was expelled so that $\mathbf{H} = \mathbf{0}$ everywhere in the superconductor; this property is usually referred to as the *Meissner effect.* Modelling and explanations of superconductivity were developed by F. and H. London (1935) and by Bardeen, Cooper, and Schriffer (1957).

In 1962, Josephson [243] predicted a tunnelling current between two superconductors separated by a thin insulating barrier; the Josephson effect was next verified and many physical devices are based on this property.

In 1986, Bednorz and Müller [32] showed that superconductivity occurs in some materials at about 30 K. This was another success, in the history of superconductivity, that opened research and applications at (relatively) high temperatures.

Relative to the dependence of the critical magnetic induction on the temperature, two different behaviours are shown by superconductors. In *type I superconductors*

---

[1] In 1913 Kamerlingh Onnes was awarded the Nobel prize in physics for the study of matter at low temperatures and the liquefaction of helium. As remarked by Ginzburg [193], the superconductivity was first observed by Gills Holst who conducted measurements of memory resistance. Kamerlingh Onnes did not mention the name of Holst in the corresponding publication.

[2] Usually denoted by T_c.

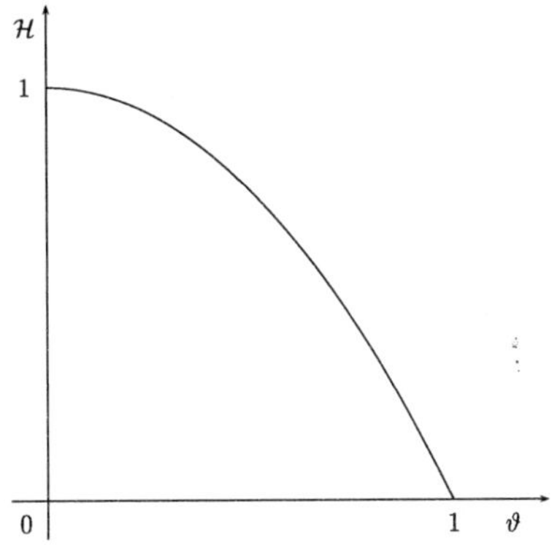

Fig. 14.1 The behaviour of the critical magnetic field with temperature was found empirically to be parabolic: $\mathcal{H}(\theta) = H_{0c}(\theta)/H_{0c}(0)$, $\vartheta = \theta/\theta_c$

the critical magnetic induction varies with temperature according to the approximate law

$$H_{0c}(\theta) = H_{0c}(0)[1 - (\theta/\theta_c)^2], \quad \theta \in [0, \theta_c].$$

As H_0 increases, with $0 \leq H_0 \leq H_{0c}(0)$, the temperature at which the material is no longer superconducting decreases. If $H_0 > H_{0c}(0)$ then the material never becomes superconducting at any temperature θ (Fig. 14.1).

By the 1950s it was clear that there is another class of superconductors then named *type II superconductors*. These materials are characterized by two critical fields, say H_{c1} and H_{c2} with $H_{c1} < H_{c2}$. As $\theta < \theta_c$, when the applied field is smaller than the lower critical field B_{c1} the material is superconducting and there is no flux penetration, as with type I superconductors. When the applied field is between H_{c1} and H_{c2} the material is in a mixed state, referred to as the vortex state, where the material has perfect conductivity but allows for partial flux penetration. If $H > H_{c2}$ then the material is a normal conductor.

14.1.2 The London Equations

In 1935, soon after the discovery of the Meissner effect, the brothers Heinz and Fritz London developed a phenomenological theory of superconductivity [281]. The London theory provides a remarkable scheme for the electrodynamics of superconductors. Among the consequences is the prediction of a nonzero penetration depth. Moreover, the London equations are likely to provide an idealized limit model of more involved equations of superconductivity.

The London equations are based on the so-called two-fluid model which traces back to Becker et al [31] and Gorter and Casimir [200]. The electrons of a superconducting material are divided in *normal* (as the electrons in a normal material scatter and suffer resistance to their motion) and *superconducting* (they cross the metal without suffering any resistance). Below the critical temperature θ_c the current consists of superconducting electrons and normal electrons. Above the critical temperature only normal electrons occur. Consistently, the current density $\mathbf{J}$ is supposed to be the sum of a normal and a superconducting part, that is,

$$\mathbf{J} = \mathbf{J}_n + \mathbf{J}_s,$$

the subscripts n and s denoting quantities pertaining to normal and superconducting electrons. While $\mathbf{J}_n$ is assumed to satisfy Ohm's law,

$$\mathbf{J}_n = \sigma \mathbf{E},$$

the superconducting part is assumed to satisfy an appropriate (London's) equation.

To obtain the whole set of London electrodynamic equations, we start with the view that the superconducting electrons are an incompressible, inviscid, charged fluid. The fluid is acted upon by an electric field $\mathbf{E}$ and a magnetic induction $\mathbf{B}$. The Lagrangian derivative $\dot{\mathbf{v}}_s$ can be written as

$$\dot{\mathbf{v}}_s = \partial_t \mathbf{v}_s + (\mathbf{v}_s \cdot \nabla)\mathbf{v}_s = \partial_t \mathbf{v}_s + \tfrac{1}{2}\nabla \mathbf{v}_s^2 - \mathbf{v}_s \times (\nabla \times \mathbf{v}_s).$$

Hence, letting $\gamma_e = e/m_e$ we can write the equation of motion in the form

$$\partial_t \mathbf{v}_s + \tfrac{1}{2}\nabla \mathbf{v}_s^2 - \mathbf{v}_s \times (\nabla \times \mathbf{v}_s) = -\frac{1}{\rho_s}\nabla p_s + \gamma_e(\mathbf{v}_s \times \mathbf{B}) + \gamma_e \mathbf{E},$$

whence

$$\partial_t \mathbf{v}_s + \tfrac{1}{2}\nabla \mathbf{v}_s^2 - \mathbf{v}_s \times \mathbf{w} = -\frac{1}{\rho_s}\nabla p_s + \gamma_e \mathbf{E}, \tag{14.1}$$

where

$$\mathbf{w} = \nabla \times \mathbf{v}_s + \gamma_e \mathbf{B}. \tag{14.2}$$

By applying the curl operator to (14.1) and using Faraday's law $\nabla \times \mathbf{E} = -\partial_t \mathbf{B}$ and (14.2) we obtain

$$\partial_t \mathbf{w} = \nabla \times (\mathbf{v}_s \times \mathbf{w}).$$

Of course $\nabla \cdot \mathbf{w} = 0$. Hence using the divergence and Stokes theorems we find

$$\int_{S_0} \mathbf{w} \cdot \mathbf{n}\, da = 0, \qquad \int_S (\mathbf{v}_s \times \mathbf{w}) \cdot \mathbf{n}\, da = \partial_t \int_{\partial S} \mathbf{w} \cdot \mathbf{t}\, ds$$

for any closed surface S_0 and open surface S. These relations are compatible with the assumption $\mathbf{w} = \mathbf{0}$, that is,

$$\nabla \times \mathbf{v}_s + \gamma_e \mathbf{B} = \mathbf{0}, \tag{14.3}$$

at any point and time. Consequently, since $\mathbf{J}_s = n_s e \mathbf{v}_s$ then it follows

$$\nabla \times (\Lambda \mathbf{J}_s) + \mathbf{B} = \mathbf{0}, \tag{14.4}$$

where

$$\Lambda = \frac{m}{e^2 n_s}.$$

Another equation follows by considering again the equation of motion and using Eq. (14.3) to obtain

$$\partial_t \mathbf{v}_s + \tfrac{1}{2}\nabla \mathbf{v}_s^2 = -\frac{1}{\rho_s}\nabla p_s + \frac{e}{m_e}\mathbf{E}.$$

Now two assumptions are made, namely, the nonlinear term $\nabla \mathbf{v}_s^2$ is neglected and the pressure p_s is taken to be zero. Hence, the equation of motion simplifies to

$$\partial_t \mathbf{v}_s = \frac{e}{m_e}\mathbf{E},$$

whence

$$\partial_t(\Lambda \mathbf{J}_s) = \mathbf{E}. \tag{14.5}$$

Equations (14.4) and (14.5) are the two *London equations* of superconductivity.

Some comments are in order. The original derivation of (14.5) was based on the equation of motion

$$m_e \dot{\mathbf{v}}_s = e\mathbf{E},$$

which upon substitution of $\mathbf{v}_s = \mathbf{J}_s / n_s e$ gives

$$(\Lambda \mathbf{J}_s)^{\cdot} = \mathbf{E},$$

the dot denoting an unspecified time derivative.

As a further comment, look at Faraday's law along with (14.5). We can write

$$\partial_t(\Lambda \mathbf{J}_s) = \mathbf{E}, \qquad \nabla \times \mathbf{E} = -\partial_t \mathbf{B}.$$

Applying the curl operator to the first equation and substituting $\nabla \times \mathbf{E}$, we have

$$\partial_t[\nabla \times (\Lambda \mathbf{J}_s) + \mathbf{B}] = \mathbf{0}. \tag{14.6}$$

In [281] not only the derivative is taken to be zero but also the function is assumed to vanish. If Λ is constant then (14.3) follows.

The whole set of differential equations governing the evolution of the superconducting body is given by Maxwell's equations,

$$\nabla \cdot \mathbf{D} = q, \qquad \nabla \times \mathbf{E} = -\partial_t \mathbf{B},$$

$$\nabla \cdot \mathbf{B} = 0, \qquad \nabla \times \mathbf{H} = \mathbf{J} + \partial_t \mathbf{D},$$

where $\mathbf{J} = \mathbf{J}_n + \mathbf{J}_s$ while

$$\mathbf{J}_n = \sigma \mathbf{E}, \qquad \nabla \times (\Lambda \mathbf{J}_s) + \mathbf{B} = \mathbf{0}, \qquad \partial_t (\Lambda \mathbf{J}_s) = \mathbf{E}.$$

The scalar Λ is now taken to be constant. The scheme is complete once the constitutive relations for **D** and **H** are made precise.

Penetration Depth and Meissner Effect

Assume for simplicity that time variations are slow enough that the displacement current $\partial_t \mathbf{D}$ is negligible. Hence we assume $\mathbf{B} = \mu \mathbf{H}$ and consider the approximate equation

$$\nabla \times \mathbf{B} = \mu \mathbf{J}.$$

Consistently, we assume that stationary conditions are allowed so that we take $\partial_t \mathbf{B} = \mathbf{0}$ whence

$$\nabla \times \mathbf{E} = \mathbf{0}, \qquad \nabla \times \mathbf{J}_n = \sigma \nabla \times \mathbf{E} = \mathbf{0}.$$

Hence, we can write

$$\nabla \times (\nabla \times \mathbf{B}) = \mu \nabla \times (\mathbf{J}_s + \mathbf{J}_n) = \mu \nabla \times \mathbf{J}_s = -\frac{\mu}{\Lambda} \mathbf{B}.$$

Thus, the magnetic induction satisfies the differential equation

$$\Delta \mathbf{B} = \alpha \mathbf{B}, \qquad \alpha := \frac{\mu}{\Lambda} = \frac{\mu n_s e^2}{m_e}. \tag{14.7}$$

This equation allows us to evaluate the magnetic induction in a superconductor. Address attention to the half-space $x \geq 0$ and assume that **B** depends on the single Cartesian coordinate x. The differential equation has the (bounded) solution

$$\mathbf{B}(x) = \mathbf{B}(0) \exp(-\sqrt{\alpha}\, x).$$

Thus the magnetic field is exponentially screened from the interior of the sample. Indeed,

$$\lambda_L := 1/\sqrt{\alpha} = \sqrt{m_e/\mu n_s e^2}$$

is called the London penetration depth. The same screen effect holds for the current $\mathbf{J}_s$. By $\nabla \times \mathbf{J}_s = -\mathbf{B}/\Lambda$ it follows that **B** and $\mathbf{J}_s$ are orthogonal to each other and

$$\mathbf{J}_s(x) = \mathbf{J}_s(0) \exp(-x/\lambda_L), \qquad |\mathbf{J}_s| = (\lambda_L/\Lambda)|\mathbf{B}|.$$

We now see how the London equations allow us to explain the Meissner effect and, meanwhile, that a superconductor is not merely a perfect conductor. Inside a perfect conductor the resistivity is zero and hence $\mathbf{E} = \mathbf{0}$. Consequently, by Faraday's law

$$\int_C \mathbf{E}\cdot\mathbf{t}ds = -\dot{\Phi}_B, \qquad \nabla\times\mathbf{E} = -\partial_t\mathbf{B},$$

it follows that the magnetic flux Φ_B through any loop in the material is constant or $\partial_t\mathbf{B} = \mathbf{0}$. Now, if a perfect conductor is cooled below its critical temperature, in the presence of an applied magnetic field, the field should be trapped in the interior of the conductor even after the field is removed ($\partial_t\mathbf{B} = \mathbf{0}$). If instead the field is applied after cooling below θ_c the field should be expelled from the (perfect) conductor. Yet, the experiments conducted by Meissner and Ochsenfeld showed that, at $\theta < \theta_c$, the magnetic field is expelled whether the field was applied before or after the material was cooled below the critical temperature. The difference between perfect conductors and superconductors occurs when the field is applied before the cooling below θ_c. Now, the reason why the field is expelled from the superconductor is due to the penetration depth that bounds $\mathbf{B}$ (and $\mathbf{J}_s$) to a layer of thickness λ_L which in turn is a consequence of (14.4). We can also describe the effect by saying that when a sample is placed in an external induction field $\mathbf{B}_0$ the sample acquires a magnetization $\mathbf{M}$. The magnetic induction $\mathbf{B}$ inside the sample is given by

$$\mathbf{B} = \mathbf{B}_0 + \mu_0\mathbf{M}.$$

In the superconducting state $\mathbf{B} = \mathbf{0}$ and then it follows that

$$\mathbf{M} = -\frac{1}{\mu_0}\mathbf{B}_0.$$

Whenever a material is in a superconducting state, the magnetization opposes the external magnetic induction and hence the corresponding susceptibility is negative as it happens for diamagnetic substances. Yet the vanishing of $\mathbf{B}$ within the superconductor may be viewed as the result of *induction surface currents* such that the produced magnetic induction $\mathbf{B}_s$, via $\nabla\times(\Lambda\mathbf{J}_s) + \mathbf{B}_s = \mathbf{0}$, makes the whole induction to vanish,

$$\mathbf{B} = \mathbf{B}_0 + \mathbf{B}_s = \mathbf{0}.$$

Thermodynamic Consistency of the London Equations

First, we observe that the London equations consist of the Maxwell equations, the splitting $\mathbf{J} = \mathbf{J}_n + \mathbf{J}_s$ of the current, and

$$\mathbf{J}_n = \sigma\mathbf{E}, \qquad \Lambda\partial_t\mathbf{J}_s = \mathbf{E}.$$

The curl of the last one gives (14.6). We now show that this set of equations is thermodynamically consistent. The London equations are then consistent with thermodynamics in that (14.4) follows from (14.6) as a particular case.

The superconductor is assumed to be macroscopically undeformable. Moreover, as with the London equations, we neglect nonlinear terms. The total time derivative $\dot{\mathbf{J}}$ coincides with the partial time derivative $\partial_t \mathbf{J}$. A state of the superconductor is the set of values

$$(\mathbf{J}_s, \mathbf{E}, \mathbf{H}, \theta, \nabla\theta).$$

The entropy (per unit mass) η, the Gibbs free energy[3] $\phi = \varepsilon - \theta\eta - \mathbf{P}\cdot\mathbf{E}/\rho - \mu_0 \mathbf{M}\cdot\mathbf{H}/\rho$, the normal current $\mathbf{J}_n$, the heat flux $\mathbf{q}$, and the time derivative $\partial_t \mathbf{J}_s$ are assumed to be functions of the state variables. Indeed, the constitutive functions for $\mathbf{J}_n$ and $\partial_t \mathbf{J}_s$ are just taken to be given by the London theory. The second-law inequality is taken in the form

$$-\rho(\dot{\phi} + \eta\dot{\theta}) + \mathbf{E}\cdot\mathbf{J} - \mathbf{P}\cdot\dot{\mathbf{E}} - \mu_0 \mathbf{M}\cdot\dot{\mathbf{H}} - \frac{1}{\theta}\mathbf{q}\cdot\nabla\theta \geq 0.$$

To save writing we let ϕ be independent of $\nabla\theta$. Upon evaluation of $\dot{\phi}$ and substitution we have

$$-\rho(\partial_\theta\phi + \eta)\dot{\theta} - (\rho\partial_{\mathbf{E}}\phi + \mathbf{P})\cdot\dot{\mathbf{E}} - (\rho\partial_{\mathbf{H}}\phi + \mu_0\mathbf{M})\cdot\dot{\mathbf{H}} - \rho\partial_{\mathbf{J}_s}\phi\cdot\dot{\mathbf{J}}_s + \mathbf{E}\cdot\mathbf{J} - \frac{1}{\theta}\mathbf{q}\cdot\nabla\theta \geq 0.$$

The arbitrariness of $\dot{\theta}$, $\dot{\mathbf{E}}$, $\dot{\mathbf{H}}$ implies that

$$\eta = -\partial_\theta\phi, \qquad \mathbf{P} = -\rho\partial_{\mathbf{E}}\phi, \qquad \mu_0\mathbf{M} = -\rho\partial_{\mathbf{H}}\phi.$$

Substitution of $\mathbf{J} = \mathbf{J}_n + \mathbf{J}_s$, $\mathbf{J}_n = \sigma\mathbf{E}$, and $\dot{\mathbf{J}}_s = \mathbf{E}/\Lambda$ yields the reduced inequality

$$(-\frac{\rho}{\Lambda}\partial_{\mathbf{J}_s}\phi + \mathbf{J}_s)\cdot\mathbf{E} + \sigma\mathbf{E}^2 - \frac{1}{\theta}\mathbf{q}\cdot\nabla\theta \geq 0.$$

The inequality holds if $\sigma \geq 0$ and

$$\mathbf{J}_s = \frac{\rho}{\Lambda}\partial_{\mathbf{J}_s}\phi, \qquad \mathbf{q} = -\kappa\nabla\theta, \quad \kappa \geq 0.$$

Hence, the London theory is thermodynamically consistent and, moreover, the Gibbs free energy takes the form

$$\phi = \frac{\Lambda}{2\rho}\mathbf{J}_s^2 + \Phi(\theta, \mathbf{E}, \mathbf{H})$$

while Fourier law for the heat flux is allowed.

We append some comments on the constitutive equation for the electric current $\mathbf{J}$. By

---

[3] In general we view the Gibbs free energy as a Legendre transform of the Helmholtz free energy $\psi = \varepsilon - \theta\eta$.

$$\dot{\mathbf{J}}_s = \frac{1}{\Lambda}\mathbf{E}, \qquad \mathbf{J}_n = \sigma\mathbf{E}, \qquad \mathbf{J} = \mathbf{J}_n + \mathbf{J}_s$$

it follows

$$\dot{\mathbf{J}} = \sigma\dot{\mathbf{E}} + \frac{1}{\Lambda}\mathbf{E}.$$

Applying the divergence operator we find that the charge density q satisfies the differential equation

$$\ddot{q} + \frac{\sigma}{\epsilon_0}\dot{q} + \frac{1}{\epsilon_0\Lambda}q = 0.$$

Any charge buildup in superconductors decays at the rate $\exp(-t/\tau)$, $\tau > 0$. F. London [283] argued that the decay is rapid in that τ is estimated to be about 10^{-12} sec. Yet the two values of τ are approximately

$$\tau_+ \simeq \epsilon_0/\sigma, \qquad \tau_- \simeq \sigma m_e/n_s^2 e.$$

Since $\sigma \to \infty$ as $\theta \to 0$ and $n_s \to 0$ or $\theta \to \theta_c$ it follows that τ_- becomes large as $\theta \to \theta_c$ and fluctuations should persist for long times. No such phenomena have ever been observed in superconductors as θ is close to zero or θ_c.

Taking the time derivative of Faraday's equation $\nabla \times \mathbf{B} = \mu\mathbf{J} + \mu\epsilon\partial_t\mathbf{E}$, we have

$$\nabla \times \partial_t\mathbf{B} = \mu\partial_t(\mathbf{J}_n + \mathbf{J}_s) + \mu\epsilon\partial_t^2\mathbf{E}.$$

Since $\partial_t\mathbf{B} = -\nabla \times \mathbf{E}$, $\partial_t\mathbf{J}_s = \mathbf{E}/\Lambda$, $\mathbf{J}_n = \sigma\mathbf{E}$, if the charge density is zero it follows

$$\Delta\mathbf{E} = \epsilon\partial_t^2\mathbf{E} + \mu\sigma\partial_t\mathbf{E} + (\mu/\Lambda)\mathbf{E}.$$

For slowly varying fields, we can neglect the time derivatives and consider the approximate equation

$$\Delta\mathbf{E} = \frac{1}{\lambda_L^2}\mathbf{E}, \qquad \frac{1}{\lambda_L^2} = \frac{\mu n_s e^2}{m_e}.$$

This equation implies that an electric field penetrates a distance λ_L, as a magnetic field does. However, by $\partial_t\mathbf{J}_s = \mathbf{E}/\Lambda$, the occurrence of a static electric field would imply an unbounded current $\mathbf{J}_s$, which is not observed experimentally.

This motivates alternative approaches to the modelling of superconductors.

14.1.3 An Alternative to London Equations

Maxwell's equations $\nabla \cdot \mathbf{B} = 0$ and $\nabla \times \mathbf{E} = -\partial_t\mathbf{B}$ allow $\mathbf{B}$ and $\mathbf{E}$ to be expressed by a vector potential $\mathbf{A}$ and a scalar potential φ in the form

$$\mathbf{B} = \nabla \times \mathbf{A}, \qquad \mathbf{E} = -\nabla\varphi - \partial_t \mathbf{A}.$$

The second London equation

$$\Lambda \nabla \times \Lambda \mathbf{J}_s + \mathbf{B} = \mathbf{0}$$

allows us to view $-\Lambda \mathbf{J}_s$ as a vector potential for $\mathbf{B}$. Now, since $\mathbf{B} = \nabla \times \mathbf{A}$, it follows that $\mathbf{A}$ and $-\Lambda \mathbf{J}_s$ are equal to within a gradient term,

$$\mathbf{J}_s = -\frac{1}{\Lambda}(\mathbf{A} + \nabla\Phi), \tag{14.8}$$

where, so far, Φ is any smooth scalar function of space and time. Upon partial time differentiation and substitution of $\mathbf{E}$ we have

$$\partial_t(\Lambda \mathbf{J}_s) = \mathbf{E} + \nabla(\varphi - \partial_t \Phi). \tag{14.9}$$

In stationary conditions, it follows from (14.9) that $\mathbf{E} = -\nabla\varphi$. This is consistent with the existence of a static electric field in a superconductor, given by the potential φ, which does not generate an electric current.

The observation that

$$\partial_t \mathbf{J}_s = \mathbf{0} \iff \mathbf{E} = -\nabla\varphi$$

is consistent with the existence of a static electric field in a superconductor, given by a potential φ, which does not generate an electric current. This is conceptually different from the London equation $\Lambda \partial_t \mathbf{J}_s = \mathbf{E}$ whereby an electric field produces an electric current. Accordingly, we now take (14.9) as one of the equations of superconductivity.

While $\Lambda \nabla \times \Lambda \mathbf{J}_s + \mathbf{B} = \mathbf{0}$ involves physical fields, the other equations involve the potentials Φ, φ and such fields are admissible provided the physical fields turn out to be invariant. We then look for conditions (gauges) such that the physical fields are left invariant by changes of Φ and φ.

As is customary, let the vector potential $\mathbf{A}$ and the scalar potential φ satisfy the Lorentz gauge

$$\nabla \cdot \mathbf{A} + \partial_t \varphi = 0.$$

Hence the divergence of (14.8) results in

$$\Lambda \nabla \cdot \mathbf{J}_s = \partial_t \varphi - \Delta\Phi.$$

If we let $\partial_t \Phi = \varphi$ it follows

$$\Lambda \nabla \cdot \mathbf{J}_s = \partial_t^2 \Phi - \Delta\Phi. \tag{14.10}$$

The condition of charge conservation $\nabla \cdot \mathbf{J}_s = 0$ is related to admissible potentials Φ in the form

$$\nabla \cdot \mathbf{J}_s = 0 \iff \partial_t^2 \Phi - \Delta\Phi = 0.$$

If the Coulomb gauge $\nabla \cdot \mathbf{A} = 0$ is assumed then (14.8) implies

$$\nabla \cdot \mathbf{J}_s = -\Delta\Phi/\Lambda$$

and hence the physical admissibility reads

$$\nabla \cdot \mathbf{J}_s = 0 \iff \Delta\Phi = 0.$$

Incidentally, with $\varphi = \partial_t \Phi$, time differentiation of (14.10) and the divergence of $\partial_t(\Lambda \mathbf{J}_s) = \mathbf{E}$ yield

$$\partial_t^2 \varphi - \Delta\varphi = \nabla \cdot \mathbf{E}.$$

Maxwell's equation

$$\nabla \cdot \mathbf{E} = \frac{q}{\epsilon_0}$$

then provides the well-known D'Alembert equation for the scalar potential φ,

$$\partial_t^2 \varphi - \Delta\varphi = \frac{q}{\epsilon_0}.$$

Let Ω be the region occupied by the superconductor. In view of (14.10), the boundary condition on $\partial\Omega$,

$$\mathbf{J}_s \cdot \mathbf{n} = 0,$$

results in

$$\int_\Omega (\partial_t^2 \Phi - \Delta\Phi) dv = 0.$$

The time derivative and the condition $\partial_t \Phi = \varphi$ lead to

$$\int_\Omega q \, dv = 0,$$

whence it follows that the superconductor is globally neutral.

14.1.4 A Nonlocal Model

In static situations, experiments show a magnetic penetration depth that is significantly larger than the prediction of London theory, in particular, in dirty samples with large scattering rates in the normal state. An explanation is provided on the basis of a nonlocal electromagnetic response of the superconductor. The underlying idea is that the quantum state of the electrons forming the superfluid cannot be arbitrarily localized. Pippard [363] introduced the coherence length ξ_0 as a measure of the minimum extent of electronic wavepackets and proposed that the local equation $\mathbf{J}_s = -\mathbf{A}/\Lambda$ be replaced by the nonlocal equation

$$\Lambda \mathbf{J}_s(\mathbf{x}) = -\frac{3}{4\pi\xi_0}\int_{\mathbb{R}^3} \exp(-r/\xi_0)\frac{\mathbf{r}\cdot\mathbf{A}(\mathbf{y})}{r^4}\mathbf{r}\, dv_y, \qquad \mathbf{r} = \mathbf{x} - \mathbf{y}, \quad r = |\mathbf{r}|.$$

The form of this equation was motivated by an earlier nonlocal generalization of Ohm's law. The main point is that the action of the field $\mathbf{A}$ is appreciable in a neighbourhood of radius ξ_0. Incidentally, if $\mathbf{A}$ does not change appreciably within the sphere $|\mathbf{y} - \mathbf{x}| \le \xi_0$ then

$$\Lambda \mathbf{J}_s(\mathbf{x}) \simeq -\frac{3}{4\pi\xi_0}\int_{\mathbb{R}^3} \exp(-r/\xi_0)\frac{\mathbf{r}\cdot\mathbf{A}(\mathbf{x})}{r^4}\mathbf{r}\, dv_y = -\mathbf{A}(\mathbf{x}).$$

In the presence of strong scattering, due to dirty samples, the electrons can be localized on the scale of the mean free path ℓ. Hence, in [363], the neighbourhood of appreciable influence is formally magnified by letting

$$\Lambda \mathbf{J}_s(\mathbf{x}) = -\frac{3}{4\pi\xi_0}\int_{\mathbb{R}^3} \exp(-r/\xi)\frac{\mathbf{r}\cdot\mathbf{A}(\mathbf{x})}{r^4}\mathbf{r}\, dv_y,$$

where

$$\frac{1}{\xi} = \frac{1}{\xi_0} + \frac{1}{\beta\ell}$$

and β is a constant of order unity. This expression is essentially the same as that obtained within the BCS [28] theory. In the dirty limit $\ell \ll \xi_0$, so that $\xi \simeq \beta\ell$, if $\mathbf{A}$ is appreciable only within $r \le \xi$ then it follows that

$$\Lambda \mathbf{J}_s(\mathbf{x}) = -\frac{\beta\ell}{\xi_0}\mathbf{A}(\mathbf{x}).$$

The penetration depth is then increased by a factor $\sqrt{\xi_0/\beta\ell}$.

14.1.5 Phase Transition Curve

The transition curve between the superconducting and the normal state shows that, as $\theta \le \theta_c$, the transition is affected by the magnetic field. For simplicity, let $\mathbf{E} = 0$ and hence consider the *magnetic Gibbs function* per unit mass

$$\phi(\theta, \mathbf{B}) = \varepsilon - \theta\eta - \mathbf{M}\cdot\mathbf{B}/\rho,$$

ρ being the constant mass density. Since

$$\mathbf{M} = -\rho \partial_{\mathbf{B}} \phi$$

then

$$\phi(\theta, \mathbf{B}) - \phi(\theta, 0) = \int_0^B \partial_{\mathbf{B}} \phi \cdot d\mathbf{B} = -\frac{1}{\rho} \int_0^B \mathbf{M} \cdot d\mathbf{B}.$$

We assume the dependence of **M** on **B** is linear,

$$\mathbf{M} = \frac{\chi}{\mu_0} \mathbf{B};$$

indeed conductors and superconductors exhibit diamagnetism so that $\chi < 0$, possibly $\chi = -1$. Let $\mathbf{B} = \mathbf{B}(\xi)$, $\xi \in [0, 1]$. Hence it follows

$$\phi(\theta, \mathbf{B}) - \phi(\theta, 0) = \frac{|\chi|}{\rho\mu_0} \int_0^1 \mathbf{B} \cdot \mathbf{B}' d\xi = \frac{|\chi|}{2\rho\mu_0} \mathbf{B}^2.$$

For the type I superconductor, as B varies from 0 to the critical value B_c, we have

$$\phi_s(\theta, B_c) - \phi_s(\theta, 0) = \frac{|\chi|}{2\rho\mu_0} B_c^2.$$

The transition curve is the graph of the function $B_c(\theta)$, $\theta \in (0, \theta_c]$. At the states $(\theta, B_c(\theta))$ the superconducting electrons are at equilibrium with normal electrons and hence we let

$$\phi_s(\theta, B_c(\theta)) = \phi_n(\theta, B_c(\theta)).$$

The normal phase is assumed to be nonmagnetic and then

$$\phi_n(\theta, B_c(\theta)) = \phi_n(\theta, 0).$$

Hence it follows that

$$\phi_n(\theta, 0) - \phi_s(\theta, 0) = \frac{|\chi|}{2\rho\mu_0} B_c^2.$$

Differentiation with respect to θ gives

$$\eta_n(\theta) - \eta_s(\theta) = -\frac{|\chi|}{\rho\mu_0} B_c \frac{dB_c}{d\theta}.$$

Now $\theta(\eta_n - \eta_s)$ is the latent heat L of transition from superconductor to normal conductor. Hence we have

$$L = -\frac{|\chi|}{\rho\mu_0} B_c \frac{dB_c}{d\theta}.$$

By the third law of thermodynamics, the entropy change associated with any condensed system undergoing a reversible isothermal process approaches zero as the temperature at which it is performed approaches 0 K. Now,

$$\eta_n(\theta) - \eta_s(\theta) \to 0 \quad \text{as} \quad \theta \to 0$$

requires that any transition curve satisfies

$$\frac{dB_c}{d\theta} \to 0 \quad \text{as} \quad \theta \to 0. \tag{14.11}$$

For the type II superconductor, the argument can be repeated except that the upper value of B_c is B_{c2}, that is, the value at the transition from the vortex state to the normal state.

14.1.6 Ginzburg–Landau Theory

The Ginzburg–Landau theory [194] deals with the intermediate phase between the normal state and the superconducting one in stationary conditions. The approach is variational in character in that it is based on an assumption about the free energy functional and the pertinent equations governing the system are the corresponding Euler–Lagrange equations.

Landau [266] had already suggested that the transition should be described by a complex-valued parameter Ψ, called the order parameter. The physical meaning of Ψ is specified by saying that $|\Psi|^2$ is the number density, n_s, of superconducting electrons. Hence $\Psi = 0$ means that the material is in the normal state, $\theta > \theta_c$, whereas $|\Psi| = 1$ corresponds to the state of a superconductor, $\theta = 0$.

Incidentally, at first Gorter and Casimir [200] elaborated a thermodynamic potential with a real-valued parameter. Later Ginzburg and Landau argued that the order parameter should be complex valued so as to make the theory gauge invariant.

The starting idea is that, at zero magnetic field, the (magnetic) free energy ϕ_0, as θ is around θ_c, is written in the form

$$\phi_0(\theta, \Psi) = -a(\theta)|\Psi|^2 + \tfrac{1}{2}b(\theta)|\Psi|^2.$$

Higher order terms in $|\Psi|^2$ are neglected, which means that the model holds for small values of $|\Psi|$. If a magnetic field occurs then the free energy comprises a magnetic energy density $\mathbf{B}^2/2\mu$. Letting Ω be the region occupied by the material we express the global free energy in the form

$$\Phi = \int_\Omega \phi(\theta, \Psi, \mathbf{B})dv = \int_\Omega [\phi_0(\theta, \Psi) + \frac{1}{2\mu}\mathbf{B}^2 + \frac{1}{2m_s}|i\hbar\nabla\Psi + e_s\mathbf{A}\Psi|^2]dv - \frac{1}{\mu}\int_{\partial\Omega} \mathbf{A} \times \mathbf{B}_{\text{ex}} \cdot \mathbf{n}\, da,$$

where m_s and $-e_s$ are the effective mass and charge of the superelectrons, $\mathbf{A}$ is the vector potential associated with $\mathbf{B}$, and $\hbar$ is Planck's constant. The vector $\mathbf{B}_{ex}$ denotes the external magnetic induction on the boundary $\partial\Omega$ and is subject to $\nabla \times \mathbf{B}_{ex} = \mathbf{0}$.

Assume that the free energy is stationary at equilibrium. The temperature θ is supposed to be fixed and consistently attention is restricted to quasi-static processes whereby $\mathbf{E} = \mathbf{0}$. Since the Ginzburg–Landau theory applies to stationary conditions we let

$$\nabla \times \mathbf{B} = \mu \mathbf{J}_s,$$

where $\mathbf{J}_s$ is considered instead of $\mathbf{J} = \mathbf{J}_n + \mathbf{J}_s$. Since $\mathbf{B} = \nabla \times \mathbf{A}$ we can write

$$\nabla \times (\nabla \times \mathbf{A}) = \mu \mathbf{J}_s.$$

The functional Φ involves Ψ and $\mathbf{A}$ as the unknown functions and hence we let $\mathbf{B}^2 = (\nabla \times \mathbf{A})^2$. We then derive the corresponding Euler–Lagrange equations. Observe that $|\Psi|^2 = \Psi^* \Psi$ is differentiable at $\Psi = 0$. It is standard to let

$$\partial_\Psi |\Psi|^2 = \Psi^*.$$

Moreover, observe

$$\nabla \cdot \partial_{\nabla \mathbf{A}} \tfrac{1}{2} (\nabla \times \mathbf{A})^2 = \nabla \times (\nabla \times \mathbf{A}).$$

Hence the Euler–Lagrange equations of ϕ,

$$\partial_\Psi \phi - \nabla \cdot \partial_{\nabla \Psi} \phi = 0, \qquad \partial_{\mathbf{A}} \phi - \nabla \cdot \partial_{\nabla \mathbf{A}} \phi = \mathbf{0},$$

can be written in the form

$$\frac{1}{2m_s} (i\hbar \nabla + e_s \mathbf{A})^2 \Psi - a\Psi + b|\Psi|^2 \Psi = 0, \tag{14.12}$$

$$\mathbf{J}_s = -i \frac{\hbar e_s}{2m_s} (\Psi^* \nabla \Psi - \Psi \nabla \Psi^*) - \frac{e_s^2}{m_s} |\Psi|^2 \mathbf{A}. \tag{14.13}$$

Equations (14.12) and (14.13) are called the Ginzburg–Landau equations (of superconductivity).

A connection with London equations now follows. Let $\Psi = |\Psi| \exp(i\alpha)$ so that Eq. (14.13) reads

$$\mathbf{J}_s = \frac{\hbar e_s}{m_s} |\Psi|^2 \nabla \alpha - \frac{e_s^2}{m_s} |\Psi|^2 \mathbf{A} = -\frac{1}{\Lambda} (-\frac{\hbar}{e_s} \nabla \alpha + \mathbf{A})$$

where $\Lambda = m_s / e_s^2 |\Psi|^2$. Applying the curl operator we obtain the London equation

$$\nabla \times (\Lambda \mathbf{J}_s) + \mathbf{B} = \mathbf{0}.$$

14.2 Superfluids

14.2.1 History of Superfluidity and Properties of Superfluids

Superfluidity is the characteristic property of a fluid with zero viscosity which therefore flows without loss of kinetic energy. When stirred, a superfluid forms cellular vortices that continue to rotate indefinitely. Superfluidity occurs in certain substances under special conditions. In particular, it occurs in two isotopes of helium when they are liquified by cooling to cryogenic temperatures.

Helium was first liquefied in 1908 by Kamerlingh Onnes who cooled it below the liquid/gas transition temperature of 4.2 K. Later, in 1927, Wolfke and Keesom realized that there is another phase transition at a lower temperature, 2.17 K. This phase transition had shown up via a discontinuity of the specific heat, whose curve as a function of temperature resembles the letter λ, whence the name of λ point for that transition. The two phases of liquid helium were termed helium I and helium II. The remarkable superfluid properties of helium II (strikingly low viscosity) were experimentally established by Kapitza [250] and independently by Allen and Misener [5] in 1938. Shortly afterwards Fritz London suggested that superfluidity could have some connection with Bose–Einstein condensation. London also realized that there might be a strong connection with superconductivity, which had been discovered many years before and which could be seen as superfluidity in the electron gas in a metal.

Next London [282] and Tisza [417, 418] suggested that liquid helium could be described by a two-fluid model, the condensed and non-condensed atoms being identified with the superfluid and normal components, respectively; natural helium consists essentially of isotope ^{4}He satisfying the mentioned properties.[4]

Some remarkable effects should be considered to set up a realistic model. The *thermomechanical effect,* discovered in 1938, is shown up as follows. Consider a flask filled with liquid helium. A differential container is placed inside the liquid. Throughout the liquid has a uniform concentration of normal fluid (and superfluid). Upon heating the liquid in the inner container, the concentration of normal component increases in that a fraction of superfluid becomes normal. By diffusion, superfluid from the colder region flows into the hotter (though $\theta < \theta_\lambda$) region while a flow of normal fluid in any direction, in the inner container, cannot occur because of (viscosity and) the channel resistance. The added superfluid in the hotter region partly becomes normal to get the equilibrium concentration. We then observe an increased mass of the liquid in the hotter region. The process stops when the pressure in the superfluid component is uniform across the channel.

The *mechanocaloric effect* was discovered in 1939. The apparatus consists of a round flask filled with a fine powder and liquid helium. The flask has an opening at the bottom. The superfluid component flows through the fine powder whereas the normal component, being viscous, cannot flow. As a result, the concentration of

---

[4] Helium has another isotope, ^{3}He. It becomes a superfluid at temperatures of about 2 mK.

the normal component increases above the powder. A thermometer inside the liquid shows an increase of the temperature. Indeed, the temperature increases up to the equilibrium value corresponding to the new concentration.

The *fountain effect* shows up in the following experience. A tube, with a fine capillary, is filled with a fine powder (superleak) and is immersed in liquid helium. By heating the powder, and hence the liquid in the tube, eventually the superfluid squirts out through the fine capillary at the top.

The *creeping effect* is also named Rollin effect after Bernard V. Rollin who discovered it in 1937. Consider a round flask containing liquid helium. If the flask is lowered into a bath of liquid helium then a (Rollin) film clings to the wall and gradually fills the flask. If, instead the flask is raised above the bath level then it empties out. In both cases, the fluid is able to flow against gravity, in the form of about 30 nm film, with a net flux from the higher level to the lower one. For the liquid film, the capillary forces (Van der Waals attraction) dominate the gravity and viscous forces. The rate of flow turns out to be independent of the height of the barrier and it increases as temperature decreases and hence the superfluid concentration increases. The rate is zero at $\theta = \theta_\lambda$ (absence of superfluid) and is constant below 1.5 K (superfluid concentration constant).

The effects described so far seem to indicate that liquid helium should be viewed as a binary mixture of two fluids, the normal fluid and the superfluid. Following this *two-fluid model*, denote by the subscripts n and s the quantities pertaining to the normal fluid and the superfluid. Hence ρ_n, $\mathbf{v}_n$, and ρ_s, $\mathbf{v}_s$ are the mass density, and the velocity of the two fluids. Also let $\rho = \rho_n + \rho_s$ be the density of helium. It is worth observing that, in stationary conditions, the two mass fractions ω_s, ω_n satisfy the conditions

$$\omega_s + \omega_n = 1, \qquad \omega_s \to 0 \text{ as } \theta \to \theta_\lambda^-, \qquad \omega_s \to 1 \text{ as } \theta \to 0^+.$$

Hence, as $\theta \in (0, \theta_\lambda)$, there is a free transition from the superfluid to the normal fluid and vice versa [337], Sect. 6.8. We then regard helium as a reacting mixture with a mass growth $\tau_s = -\tau_n$ given by a function of θ and ω_s, or θ and ω_n. The normal fluid is taken to be viscous, $\mathbf{T}_n = -p_n\mathbf{1} + 2\nu_n\mathbf{D}_n + \lambda_n\nabla\cdot\mathbf{v}_n\mathbf{1}$, whereas the superfluid is regarded as an ideal fluid, $\mathbf{T}_s = -p_s\mathbf{1}$, $\mathbf{q}_s = \mathbf{0}$. With this view, the balance equations can be written as

$$\partial_t\rho_n + \nabla\cdot(\rho_n\mathbf{v}_n) = -\tau_s,$$

$$\partial_t\rho_s + \nabla\cdot(\rho_s\mathbf{v}_s) = \tau_s,$$

$$\partial_t(\rho_n\mathbf{v}_n) + \nabla\cdot(\rho_n\mathbf{v}_n\otimes\mathbf{v}_n) - \nabla\cdot\mathbf{T}_n - \rho_n\mathbf{b}_n = -\mathbf{m}_s,$$

$$\partial_t(\rho_s\mathbf{v}_s) + \nabla\cdot(\rho_s\mathbf{v}_s\otimes\mathbf{v}_s) + \nabla p_s - \rho_s\mathbf{b}_s = \mathbf{m}_s,$$

$$\partial_t(\rho_n\varepsilon_n) + \nabla\cdot(\rho_n\varepsilon_n\mathbf{v}_n) - \mathbf{T}_n\cdot\mathbf{D}_n + \nabla\cdot\mathbf{q}_n - \rho_n r_n = -l_s,$$

$$\partial_t(\rho_s\varepsilon_s) + \nabla\cdot(\rho_s\varepsilon_s\mathbf{v}_s) + p_s\nabla\cdot\mathbf{v}_s - \rho_s r_s = l_s,$$

$\mathbf{m}_s$ and l_s being the growths of linear momentum and energy of the superfluid. The inviscid character of the superfluid component is modelled by assuming that it carries no heat or entropy ($\eta_s = 0$). Moreover $\mathbf{v}_s$ is curl free,[5]

$$\nabla \times \mathbf{v}_s = \mathbf{0}.$$

At any point of a boundary, with unit normal $\mathbf{n}$, the tangent component of $\mathbf{v}_s$ is unrestricted whereas $\mathbf{v}_s \cdot \mathbf{n} = 0$ and $\mathbf{v}_n = \mathbf{0}$.

Since we let the two fluids have a common temperature, $\theta_n = \theta_s = \theta$, the second-law inequality (9.42) can be written as

$$-\rho_n(\dot{\psi}_n + \eta_n\dot{\theta}_n) - \rho_s(\dot{\psi}_s + \eta_s\dot{\theta}_s) - \tau_s(\psi_s - \psi_n + \tfrac{1}{2}(\mathbf{v}_n^2 - \mathbf{v}_s^2))$$
$$-\mathbf{m}_s \cdot (\mathbf{v}_s - \mathbf{v}_n) + \mathbf{T}_n \cdot \mathbf{D}_n - p_s \nabla \cdot \mathbf{v}_s - \frac{1}{\theta_n}\mathbf{q}_n \cdot \nabla\theta_n \geq 0,$$

the extra-entropy fluxes $\mathbf{k}_n$, $\mathbf{k}_s$ being considered to be zero for the present purposes. For simplicity, we let

$$\psi_n = \psi_n(\theta_n, \rho_n), \qquad \psi_s = \psi_s(\theta_s, \rho_s), \qquad \mathbf{q}_n = -\kappa_n \nabla\theta_n.$$

Since

$$\dot{\rho}_n = -\rho_n \nabla \cdot \mathbf{v}_n - \tau_s, \qquad \dot{\rho}_s = -\rho_s \nabla \cdot \mathbf{v}_s + \tau_s$$

it follows that

$$-\rho_n(\partial_{\theta_n}\psi_n + \eta_n)\dot{\theta}_n - \rho_s(\partial_{\theta_s}\psi_s + \eta_s)\dot{\theta}_s + (\rho_n^2\partial_{\rho_n} - p_n)\nabla \cdot \mathbf{v}_n + (\rho_s^2\partial_{\rho_s}\psi_n - p_s)\nabla \cdot \mathbf{v}_s$$
$$-\tau_s(\psi_s + \rho_s\partial_{\rho_s}\psi_s - \psi_n - \rho_n\partial_{\rho_n}\psi_n + \tfrac{1}{2}(\mathbf{v}_n^2 - \mathbf{v}_s^2))$$
$$-\mathbf{m}_s \cdot (\mathbf{v}_s - \mathbf{v}_n) + 2\nu_n\mathbf{D}_n \cdot \mathbf{D}_n + \lambda_n(\nabla \cdot \mathbf{v}_n)^2 - \frac{1}{\theta_n}\kappa_n(\nabla\theta_n)^2 \geq 0.$$

As with the Navier–Stokes–Fourier's fluid we obtain

$$\eta_n = -\partial_{\theta_n}\psi_n, \quad \eta_s = -\partial_{\theta_s}\psi_s, \quad p_n = \rho_n^2\partial_{\rho_n}\psi_n, \quad p_s = \rho_s^2\partial_{\rho_s}\psi_s,$$

$$-\tau_s[\psi_s + \rho_s\partial_{\rho_s}\psi_s - \psi_n - \rho_n\partial_{\rho_n}\psi_n + \tfrac{1}{2}(\mathbf{v}_n^2 - \mathbf{v}_s^2)] - \mathbf{m}_s \cdot (\mathbf{v}_s - \mathbf{v}_n) \geq 0,$$

and $\nu_n \geq 0$, $2\nu_n + 3\lambda_n \geq 0$, $\kappa_n \geq 0$. Since $\rho_n\partial_{\rho_n}\psi_n = p_n/\rho_n$ and the analogue for p_s, let

$$\mu_n = \psi_n + \frac{p_n}{\rho_n}, \qquad \mu_s = \psi_s + \frac{p_s}{\rho_s}$$

be the chemical potentials. Since μ_n, μ_s are independent of the velocities then, borrowing from (9.81), we find that the two inequalities

[5] This is not exactly true, as we comment upon in Sect. 14.2.5.

$$(\mu_s - \mu_n)\tau_s \leq 0, \qquad \tfrac{1}{2}\tau_s(\mathbf{v}_n^2 - \mathbf{v}_s^2)) + \mathbf{m}_s \cdot (\mathbf{v}_s - \mathbf{v}_n) \leq 0$$

hold separately. The first inequality governs the time dependence of the transition. The second inequality holds with

$$\mathbf{m}_s = M(\mathbf{v}_n - \mathbf{v}_s) + \tau_s \mathbf{v} + \tfrac{1}{2}\tau_s(\mathbf{u}_n + \mathbf{u}_s),$$

where $M \geq 0$, whence

$$\tfrac{1}{2}\tau_s(\mathbf{v}_n^2 - \mathbf{v}_s^2)) + \mathbf{m}_s \cdot (\mathbf{v}_s - \mathbf{v}_n) = -M(\mathbf{v}_n - \mathbf{v}_s)^2.$$

As a comment to the equation for $\mathbf{m}_s$, we observe that the term $\tau_s \mathbf{v}$ vanishes in the barycentric reference. Moreover, since $\rho_n \mathbf{v}_n + \rho_s \mathbf{v}_s = \mathbf{0}$ it follows that $\mathbf{u}_s$ and $\mathbf{u}_n$ are antiparallel thus making $\mathbf{u}_n + \mathbf{u}_s$ rather small. It is then likely that $M(\mathbf{v}_n - \mathbf{v}_s)$ is the dominant term and this justifies the customary assumption $\mathbf{m}_s = M(\mathbf{v}_n - \mathbf{v}_s)$.

At equilibrium, the ratio ρ_s/ρ is a function of temperature, $\rho_s/\rho = f(\theta)$. It is then natural to let

$$\tau_s = \hat{\tau}_s(\rho_s/\rho - f(\theta)), \qquad y\,\hat{\tau}_s(y) \leq 0, \quad \hat{\tau}(0) = 0.$$

It is worth looking at other model equations which appeared in the literature. First, it is usually assumed that $\eta_s = 0$, which would imply $\psi_s = \psi_s(\rho_s)$. This assumption is motivated by quantum statistics which suggests that the superfluid is the prevailing phase around absolute zero and there the entropy vanishes according to Nernst's third law of thermodynamics.

Often[6] the equations of motion are written in the form

$$\rho_n \partial_t \mathbf{v}_n + \rho_n(\mathbf{v}_n \cdot \nabla)\mathbf{v}_n = -\omega_n \nabla p_n - \rho_s \eta \nabla\theta - \mathbf{m}_s + \nu \Delta \mathbf{v}_n, \tag{14.14}$$

$$\rho_s \partial_t \mathbf{v}_s + \rho_s(\mathbf{v}_s \cdot \nabla)\mathbf{v}_s = -\omega_s \nabla p_s + \rho_s \eta \nabla\theta + \mathbf{m}_s + \rho_s \mathbf{t}, \tag{14.15}$$

where the present notation is used and $\mathbf{t}$ is the tension. According to the literature, the terms $\omega\nabla p$ and $\pm\eta\nabla\theta$ arise from $-\nabla\mu/\rho$ as the force per unit mass. Indeed, by [10], Eq. (B1), Euler's equation of motion in a classical ideal fluid is $\partial_t \mathbf{v} + (\mathbf{v} \cdot \nabla)\mathbf{v} = -\nabla\mu$. Likewise, by [432], Sect. 2, the effective energy per atom in the condensate state is (the mass times) the chemical potential per unit mass and then[7]

$$\partial_t \mathbf{v}_s = -\nabla\mu.$$

Since $\partial_\theta \mu(\theta, p) = -\eta$, $\partial_p \mu(\theta, p) = 1/\rho$, then

$$\nabla\mu = -\eta\nabla\theta + \nabla p/\rho.$$

[6] See, e.g. [135] and [34, 221] for the so-called HVBK model.

[7] In Sect. 14.2.6, we point out how $-\nabla\mu$ can be viewed as the force acting on the superfluid on the basis of Landau's approach.

Hence Eqs. (14.14), (14.15) are claimed to follow. Yet, within the theory of mixtures we have

$$p_\alpha = \rho_\alpha^2 \partial_{\rho_\alpha} \psi_\alpha, \quad \eta_\alpha = -\partial_\theta \psi_\alpha, \quad \rho_\alpha \partial_{\rho_\alpha} \mu_\alpha = \partial_{\rho_\alpha} p_\alpha, \quad \rho_\alpha \partial_\theta \mu_\alpha = \partial_\theta p_\alpha - \rho_\alpha \eta_\alpha,$$

whence

$$\rho_\alpha \nabla \mu_\alpha = \nabla p_\alpha - \rho_\alpha \eta_\alpha \nabla \theta.$$

Now, (14.15) follows by letting $\nabla \mu_s = -\eta \nabla \theta + (1/\rho) \nabla p_s$, where ρ and η occur instead of ρ_s and η_s. In addition we observe that $\eta_s = 0$ is a common assumption in theories of liquid helium.

If Eqs. (14.14)–(14.15) are assumed then the second-law inequality reduces to

$$(-\mathbf{m}_s + \rho_s \eta \nabla \theta) \cdot (\mathbf{v}_s - \mathbf{v}_n) - \tfrac{1}{2} \tau_s (\mathbf{v}_n^2 - \mathbf{v}_s^2) - \frac{1}{\theta} \mathbf{q}_n \cdot \nabla \theta \geq 0$$

or

$$-\mathbf{m}_s \cdot (\mathbf{v}_s - \mathbf{v}_n) - \tfrac{1}{2} \tau_s (\mathbf{v}_n^2 - \mathbf{v}_s^2) - \frac{1}{\theta} (\mathbf{q}_n + \theta \rho_s \eta (\mathbf{v}_n - \mathbf{v}_s)) \cdot \nabla \theta \geq 0.$$

This inequality holds if [164]

$$-\mathbf{m}_s \cdot (\mathbf{v}_s - \mathbf{v}_n) - \tfrac{1}{2} \tau_s (\mathbf{v}_n^2 - \mathbf{v}_s^2) \geq 0, \quad \mathbf{q}_n = \theta \rho_s \eta (\mathbf{v}_n - \mathbf{v}_s) - \kappa_n \nabla \theta, \ \kappa_n \geq 0.$$

Though this constitutive scheme, and hence the heat flux

$$\theta \frac{\rho_s \rho_n}{\rho_s + \rho_n} \eta_n (\mathbf{v}_n - \mathbf{v}_s)$$

in the normal fluid, is formally consistent with the second-law inequality, to our mind, both the equation of motion in the form $\partial_t \mathbf{v}_s = -\nabla \mu$ and the view of $\pm \rho_s \eta \nabla \theta$ as growths of linear momentum are hardly convincing on physical grounds.

14.2.2 First and Second Sound

In helium below the lambda point, a wave propagates in addition to the customary sound wave and that is why it is called second sound. Second sound is observed via pulse propagation; the speed is close to zero near the lambda point, increasing to approximately 20 m/s around 1.8 K, about ten times slower than normal sound waves. At temperatures below 1 K, the speed of second sound in helium II increases as the temperature decreases [440]. Peshkov, in 1944, detected second sound waves and confirmed that its temperature variations are uncoupled from the pressure variations (first sound) [359].

The rather involved structure of the balance equations for helium II justifies some approximations; here we revisit an approach that traces back to Landau[8] [267, 268]. Following are the approximations: the normal fluid is inviscid, $\mathbf{T}_n = -p_n\mathbf{1}$; $\mathbf{m}_s = \mathbf{0}$ and $\mathbf{b}_s, \mathbf{b}_n = \mathbf{0}$; the two components are at equilibrium in that $\mu_s = \mu_n$; there is no thermal expansion so that $\partial_\theta\rho(\theta, p) = 0$, $\partial_p\eta(\theta, p) = 0$; $\rho = \bar{\rho} + \varrho$ and quadratic terms in $\varrho_n, \varrho_s, \mathbf{v}_n, \mathbf{v}_s$ are neglected.

Summation of the continuity equations and of the equations of motion leads to

$$\partial_t\rho + \nabla\cdot(\rho\mathbf{v}) = 0, \qquad \partial_t(\rho\mathbf{v}) + \nabla p = \mathbf{0},$$

the quantities $\nabla\cdot(\rho_n\mathbf{v}_n\otimes\mathbf{v}_n)$ and $\nabla\cdot(\rho_s\mathbf{v}_s\otimes\mathbf{v}_s)$ being neglected. Since $\rho = \rho(p)$ then

$$\partial_t\rho = \frac{d\rho}{dp}\partial_t p.$$

By linearization $d\rho/dp$ is evaluated at $p(\bar{\rho})$. By applying the time derivative and the divergence, respectively, and substituting $\partial_t\nabla\cdot(\rho\mathbf{v})$ we obtain

$$\partial_t^2 p - \frac{dp}{d\rho}\Delta p = 0 \tag{14.16}$$

whence it follows that perturbations of p and ρ propagate with speed $c = \sqrt{dp/d\rho}$ (first sound waves).

We now take the difference between the equations of motion with $\mathbf{m}_s = \mathbf{0}$, $\mathbf{b}_s = \mathbf{0}$, $\mathbf{b}_n = \mathbf{0}$. Upon linearization, $\partial_t(\rho\mathbf{v}) \simeq \bar{\rho}\partial_t\mathbf{v}$, letting

$$\mathbf{V}_s = \mathbf{v}_s - \mathbf{v}_n$$

we find

$$\partial_t\mathbf{V}_s + \frac{1}{\bar{\rho}_s}\nabla p_s - \frac{1}{\bar{\rho}_n}\nabla p_n = \mathbf{0}.$$

Since $\partial_{\rho_\alpha}p_\alpha = \rho_\alpha\partial_{\rho_\alpha}\mu_\alpha$ and $\partial_\theta\varepsilon_\alpha = \theta\partial_\theta\eta_\alpha$ then

$$\begin{aligned}\frac{1}{\rho_\alpha}\nabla p_\alpha = \frac{1}{\rho_\alpha}(\partial_{\rho_\alpha}p_\alpha\nabla\rho_\alpha + \partial_\theta p_\alpha\nabla\theta) &= \nabla\mu_\alpha - \partial_\theta\mu_\alpha\nabla\theta + \frac{1}{\rho_\alpha}\partial_\theta p_\alpha\nabla\theta\\ &= \nabla\mu_\alpha - \partial_\theta(\varepsilon_\alpha - \theta\eta_\alpha)\nabla\theta = \nabla\mu_\alpha + \eta_\alpha\nabla\theta\end{aligned}$$

whence, in view of the assumptions $\mu_s = \mu_n$ and $\eta_s = 0$,

$$\frac{1}{\rho_s}\nabla p_s - \frac{1}{\rho_n}\nabla p_n = \nabla(\mu_s - \mu_n) + (\eta_s - \eta_n)\nabla\theta = -\eta_n\nabla\theta = -\frac{\rho}{\rho_n}\eta\nabla\theta.$$

Consequently, upon linearization,

[8] See also [337], Sect. 6.8.1.2.

$$\bar{\rho}_n \partial_t \mathbf{V}_s = \bar{\rho}\bar{\eta}\nabla\theta. \tag{14.17}$$

The analysis of the balance of energy allows us to find a further equation for $\mathbf{V}_s$ thus leading to a differential equation for the temperature. By neglecting $\mathbf{u}_s^2$- and $\mathbf{u}_n^2$-terms we can express the balance of energy in the form

$$\partial_t(\rho\varepsilon) + \nabla\cdot(\rho\varepsilon\mathbf{v}) + \nabla\cdot\sum_\alpha \rho_\alpha(\varepsilon_\alpha + p_\alpha/\rho_\alpha)\mathbf{u}_\alpha + p\nabla\cdot\mathbf{v} = 0,$$

where $\rho\varepsilon \simeq \sum_\alpha \rho_\alpha\varepsilon_\alpha$ and $p \simeq \sum_\alpha p_\alpha$; hence $\varepsilon \simeq \varepsilon_I$. Since $\rho_s\mathbf{u}_s + \rho_n\mathbf{u}_n = 0$ then

$$\rho_s\mathbf{u}_s = \rho_s(\mathbf{v}_s - \mathbf{v}) = \rho_s(\mathbf{V}_s + \mathbf{u}_n) = \rho_s\mathbf{V}_s - (\rho_s/\rho_n)\rho_s\mathbf{u}_s$$

whence it follows the identity

$$\rho_s\mathbf{u}_s = \frac{\rho_s\rho_n}{\rho}\mathbf{V}_s.$$

Hence, upon the approximation $\partial_t(\rho\varepsilon) + \nabla\cdot(\rho\varepsilon\mathbf{v}) \simeq \bar{\rho}\partial_t\varepsilon$ and linearization the balance of energy yields

$$\bar{\rho}\,\partial_t\varepsilon + \frac{\bar{\rho}_s\bar{\rho}_n}{\bar{\rho}}(\bar{\varepsilon}_s + \frac{\bar{p}_s}{\bar{\rho}_s} - \bar{\varepsilon}_n - \frac{\bar{p}_n}{\bar{\rho}_n})\nabla\cdot\mathbf{V}_s + \bar{p}\,\nabla\cdot\mathbf{v} = 0. \tag{14.18}$$

Since $p\dot{\rho} = -\rho p\nabla\cdot\mathbf{n}$ then by (9.85) we have

$$\rho\dot{\varepsilon} + p\nabla\cdot\mathbf{v} = \rho\theta\dot{\eta} + \rho\textstyle\sum_\alpha\mu_\alpha\dot{\omega}_\alpha.$$

Moreover

$$\rho\,\dot{\omega}_s = -\nabla\cdot(\rho_s\mathbf{u}_s) = -\nabla\cdot\frac{\rho_s\rho_n}{\rho}\mathbf{V}_s.$$

Hence it follows

$$\rho\dot{\varepsilon} + p\nabla\cdot\mathbf{v} = \rho\theta\dot{\eta} + \rho(\mu_s - \mu_n)\dot{\omega}_s = \rho\theta\dot{\eta} - \nabla\cdot\frac{\rho_s\rho_n}{\rho}(\mu_s - \mu_n)\mathbf{V}_s$$

whence

$$\bar{\rho}\,\partial_t\varepsilon + \bar{p}\,\nabla\cdot\mathbf{v} = \bar{\rho}\bar{\theta}\partial_t\eta - \frac{\bar{\rho}_s\bar{\rho}_n}{\bar{\rho}}(\bar{\mu}_s - \bar{\mu}_n)\nabla\cdot\mathbf{V}_s.$$

Substitution of $\bar{\rho}\,\partial_t\varepsilon + \bar{p}\,\nabla\cdot\mathbf{v}$ in the balance of energy (14.18) and the assumption $\eta_s = 0$ show that

$$\partial_t\eta - \frac{\bar{\rho}_s}{\bar{\rho}}\bar{\eta}\,\nabla\cdot\mathbf{V}_s = 0.$$

Now,

$$\partial_t \eta = \partial_\theta \eta(\theta, p)\partial_t \theta + \partial_p \eta(\theta, p)\partial_t p, \qquad \partial_p \eta(\theta, p) = \frac{1}{\rho^2}\partial_\theta \rho(\theta, p).$$

The assumption that there is no thermal expansion, $\partial_\theta \rho(\theta, p) = 0$, and the observation that $\theta \partial_\theta \eta(\theta, p)$ is the specific heat at constant pressure, c_p, imply that

$$\partial_t \theta - \frac{\bar{\rho}_s \bar{\theta} \bar{\eta}}{\bar{\rho} c_p} \nabla \cdot \mathbf{V}_s = 0. \tag{14.19}$$

The divergence of (14.17) and the partial time derivative of (14.19) lead to

$$\partial_t^2 \theta - \frac{\rho_s \theta \eta^2}{\rho_n} \Delta \theta = 0, \tag{14.20}$$

where the superposed bar on $\rho_s, \rho_n, \theta, \eta$ are understood. This means that a temperature wave (second sound) propagates with speed $c = \sqrt{\rho_s \theta \eta^2 / \rho_n}$.

Remark 14.1 By (14.17) we might say that the temperature gradient $\nabla \theta$ acts as a driving force as with the HVBK model. Yet (14.17) involves the velocity difference $\mathbf{V}_s = \mathbf{v}_s - \mathbf{v}_n$ instead of $\mathbf{v}_s$, and $\rho \eta \nabla \theta$ is a force and not a growth. Moreover (14.17) arises because of the assumption $\mu_s = \mu_n$.

14.2.3 Discontinuity Waves

The second sound is a temperature wave. Equation (14.20) arises from (14.17) and (14.19) thus showing that the mechanism underlying second sound is the coupling between temperature and diffusion via $\nabla \theta$ and $\mathbf{V}_s$, the main physical role being played by the force $\rho \eta \nabla \theta$ in (14.17). Yet we can guess that the observed second sound is generated by the heat flux $\mathbf{q}_n$. Hence, we let $\mathbf{q}_n$ be governed by the (objective) equation

$$\dot{\mathbf{q}}_n - \mathbf{W}_n \mathbf{q}_n = -\frac{1}{\tau_n}(\mathbf{q}_n + \kappa_n \nabla \theta), \quad \tau_n > 0, \tag{14.21}$$

where $\mathbf{W}_n = \mathrm{skw}\mathbf{L}_n$. For simplicity we let $\mathbf{T}_n = -p_n \mathbf{1}$. The independent variables are $\theta_n, \rho_n, \theta_s, \rho_s, \mathbf{q}_n, \nabla \theta$. Assume that

$$\psi_n = \psi_n(\theta_n, \rho_n, \mathbf{q}_n), \qquad \psi_s = \psi_s(\theta_s, \rho_s).$$

Granted the restrictions $p_n = \rho_n^2 \partial_{\rho_n} \psi_n$, $p_s = \rho_s^2 \partial_{\rho_s} \psi_s$, $\eta_n = -\partial_\theta \psi_n$, $\eta_s = -\partial_\theta \psi_s$, we find that the entropy inequality simplifies to

$$-\rho_n \partial_{\mathbf{q}_n} \psi_n \cdot (\mathbf{W}_n \mathbf{q}_n - \frac{1}{\tau_n}\mathbf{q}_n) + \frac{\rho_n}{\tau_n} \partial_{\mathbf{q}_n} \psi_n \cdot \kappa_n \nabla \theta - \frac{1}{\theta}\mathbf{q}_n \cdot \nabla \theta \geq 0.$$

The arbitrariness of $\mathbf{W}_n \in \mathrm{Skw}$ and $\nabla\theta$ imply that

$$\partial_{\mathbf{q}_n}\psi_n \otimes \mathbf{q}_n \in \mathrm{Sym}, \quad \partial_{\mathbf{q}_n}\psi_n \cdot \mathbf{q}_n \geq 0, \quad \kappa_n \frac{\rho_n}{\tau_n}\partial_{\mathbf{q}_n}\psi_n = \frac{1}{\theta}\mathbf{q}_n.$$

A direct integration of the last equation (for $\partial_{\mathbf{q}_n}\psi_n$) leads to

$$\psi_n = \hat{\psi}_n(\theta, \rho_n) + \frac{\tau_n}{2\kappa_n\rho_n\theta}\mathbf{q}_n^2.$$

The symmetry of $\partial_{\mathbf{q}_n}\psi_n \otimes \mathbf{q}_n$ holds. Moreover, substitution of $\partial_{\mathbf{q}_n}\psi_n$ results in

$$\frac{\kappa_n}{\theta}\mathbf{q}_n^2 \geq 0$$

whence $\kappa_n \geq 0$. Hence the superfluid is governed by the free energy function

$$\psi_n(\theta, \rho_n, \mathbf{q}_n) = \hat{\psi}_n(\theta, \rho_n) + \frac{\tau_n}{2\kappa_n\rho_n\theta}\mathbf{q}_n^2, \qquad \kappa_n > 0.$$

We now look for discontinuity wave solutions. We let $\rho_n, \rho_s, \mathbf{v}_n, \mathbf{v}_s, \theta$ be continuous functions of $\mathbf{x}$ and t, jointly for all $\mathbf{x}$ in the space occupied by the body and $t \in \mathbb{R}$, whereas $\partial_t\rho_n, \partial_t\rho_s, \nabla\rho_n, \nabla\rho_s, \partial_t\mathbf{v}_n, \partial_t\mathbf{v}_s, \nabla\mathbf{v}_n, \nabla\mathbf{v}_s, \partial_t\theta, \nabla\theta$ and all higher order derivatives suffer jump discontinuities across a surface σ but are continuous, jointly in $\mathbf{x}$ and t, everywhere else. The external supplies $\mathbf{b}_n, \mathbf{b}_s, r_n$ and the growths $\tau_s, \mathbf{m}_s, l_s$ are assumed to be continuous across σ. Moreover, the two components are assumed to have the same temperature θ. This makes it redundant the energy equation for the superfluid. Since p_n and ε_n are functions of ρ_n and θ we have

$$\nabla p_n = \partial_{\rho_n}\nabla\rho_n + \partial_\theta p_n \nabla\theta$$

and the like for ε_n. Likewise,

$$\nabla p_s = \partial_{\rho_s}\nabla\rho_s + \partial_\theta p_s \nabla\theta,$$

while the heat flux $\mathbf{q}_n$ is governed by (14.21).

The state ahead of the wave is taken to be at rest ($\mathbf{v}_n, \mathbf{v}_s = \mathbf{0}$). As is customary, we let $[\![\cdot]\!]$ denote the jump across the wave. By applying the geometrical and kinematical compatibility conditions the balance equations result in[9]

$$-u_n[\![\partial_x\rho_n]\!] + \rho_n\mathbf{n}\cdot[\![\partial_x\mathbf{v}_n]\!] = 0,$$

$$-u_n[\![\partial_x\rho_s]\!] + \rho_s\mathbf{n}\cdot[\![\partial_x\mathbf{v}_s]\!] = 0,$$

[9] To avoid easy ambiguities among n for normal fluid and n for normal direction, we denote by ∂_x, instead of ∂_n, the normal derivative; u_n is the wave speed.

$$-\rho_n u_n [\![\partial_x \mathbf{v}_n]\!] + \mathbf{n}(\partial_{\rho_n} p_n [\![\partial_x \rho_n]\!] + \partial_\theta p_n [\![\partial_x \theta]\!]) = \mathbf{0},$$

$$-\rho_s u_n [\![\partial_x \mathbf{v}_s]\!] + \mathbf{n}(\partial_{\rho_s} p_s [\![\partial_x \rho_s]\!] + \partial_\theta p_s [\![\partial_x \theta]\!]) = \mathbf{0},$$

$$-\rho_n u_n (\partial_{\rho_n} \varepsilon_n [\![\partial_x \rho_n]\!] + \partial_\theta \varepsilon_n [\![\partial_x \theta]\!]) + (\rho_n \varepsilon_n + p_n)\mathbf{n} \cdot [\![\partial_x \mathbf{v}_n]\!] + \mathbf{n} \cdot [\![\partial_x \mathbf{q}_n]\!] = 0.$$

We now look at (14.21). Observe that

$$[\![\mathbf{W}_n]\!] = \tfrac{1}{2}([\![\partial_x \mathbf{v}_n]\!] \otimes \mathbf{n} - \mathbf{n} \otimes [\![\partial_x \mathbf{v}_n]\!])$$

and hence

$$[\![\mathbf{W}_n \mathbf{q}_n]\!] = \tfrac{1}{2}[\![\partial_x \mathbf{v}_n]\!]\, \mathbf{q}_n \cdot \mathbf{n} - \tfrac{1}{2}([\![\partial_x \mathbf{v}_n]\!] \cdot \mathbf{q}_n)\mathbf{n}.$$

Moreover, since $\mathbf{v}_n = \mathbf{0}$ then

$$[\![\dot{\mathbf{q}}_n]\!] = [\![\partial_t \mathbf{q}_n]\!] + [\![\mathbf{v}_n \cdot \nabla \mathbf{q}_n]\!] = -u_n [\![\partial_x \mathbf{q}_n]\!].$$

Consequently (14.21) leads to

$$-u_n [\![\partial_x \mathbf{q}_n]\!] - \frac{1}{2}\mathbf{n} \cdot \mathbf{q}_n [\![\partial_x \mathbf{v}]\!] + \tfrac{1}{2}([\![\partial_x \mathbf{v}]\!] \cdot \mathbf{q}_n)\mathbf{n} = -\frac{\kappa_n}{\tau_n}[\![\partial_x \theta]\!]\mathbf{n}. \tag{14.22}$$

This shows that the possible waves are affected by the selection of the time derivative, here the Jaumann-like derivative $\dot{\mathbf{q}}_n - \mathbf{W}_n \mathbf{q}_n$. Yet, the scalar character of the energy equation and the assumption that the normal fluid is inviscid, $\mathbf{T}_n = -p_n \mathbf{1}$, allow us to conclude that the possible waves are longitudinal, that is, $[\![\partial_n \mathbf{v}_n]\!]$ and $[\![\partial_n \mathbf{v}_s]\!]$ are parallel to $\mathbf{n}$. For simplicity, assume $\mathbf{q}_n = \mathbf{0}$ ahead of the wave. Thus (14.22) implies that a nonzero $[\![\partial_x \mathbf{q}_n]\!]$ is longitudinal, namely, $\parallel \mathbf{n}$. Hence we let

$$[\![\partial_x q_n]\!] = \frac{\kappa_n}{u_n \tau_n}[\![\partial_x \theta]\!].$$

Likewise it follows that

$$[\![\partial_x \mathbf{v}_n]\!] = [\![\partial_x v_n]\!]\mathbf{n}, \qquad [\![\partial_x \mathbf{v}_s]\!] = [\![\partial_x v_s]\!]\mathbf{n}$$

where $v_n = \mathbf{v}_n \cdot \mathbf{n}$, $v_s = \mathbf{v}_s \cdot \mathbf{n}$ and hence

$$\rho_n [\![\partial_x v_n]\!] = u_n [\![\partial_x \rho_n]\!], \qquad \rho_s [\![\partial_x v_s]\!] = u_n [\![\partial_x \rho_s]\!].$$

We then obtain the following system of equations:

$$\begin{bmatrix} -u_n^2 + \partial_{\rho_n} p_n & 0 & \partial_\theta p_n \\ 0 & -u_n^2 + \partial_{\rho_s} p_s & \partial_\theta p_s \\ (\varepsilon_n - \rho_n^2 \theta \partial_{\rho_n} \eta_n) u_n & 0 & -\rho_n u_n \partial_\theta \varepsilon_n + \kappa_n / \tau_n u_n \end{bmatrix} \begin{bmatrix} \partial_x \rho_n \\ \partial_x \rho_s \\ \partial_x \theta \end{bmatrix} = 0.$$

The possible speeds of propagation u_n are given by the propagation condition

$$(-u_n^2 + \partial_{\rho_s} p_s)[(-u_n^2 + \partial_{\rho_n} p_n)(-\rho_n u_n^2 \partial_\theta \varepsilon_n + \kappa_n/\tau_n) - \partial_\theta p_n(\varepsilon_n - \rho_n^2 \theta \partial_{\rho_n} \eta_n) u_n^2] = 0.$$

The immediate solution $u_n = \sqrt{\partial_{\rho_s} p_s}$ corresponds to a nonzero $[\![\partial_x \rho_s]\!]$ while $[\![\partial_x \rho_n]\!] = 0$, $[\![\partial_x \theta]\!] = 0$. The remaining equation

$$(-u_n^2 + \partial_{\rho_n} p_n)(-\rho_n u_n^2 \partial_\theta \varepsilon_n + \kappa_n/\tau_n) - \partial_\theta p_n(\varepsilon_n - \rho_n^2 \theta \partial_{\rho_n} \eta_n) u_n^2 = 0$$

yields the two speeds of other two modes. The modes are decoupled inasmuch as $\partial_\theta p_n$ and $\partial_\theta p_s$ are negligible. In this instance, the speeds

$$u_n = \sqrt{\partial_{\rho_n} p_n}, \qquad u_n = \sqrt{\kappa_n/\rho_n \tau_n \partial_\theta \varepsilon_n}$$

correspond to the nonzero discontinuities $[\![\partial_x \rho_n]\!]$ and $[\![\partial_x \theta]\!]$, respectively.

14.2.4 Mass Fractions and Chemical Potentials

The reduced entropy inequality for fluid mixtures holds if the chemical potentials, the mass growth, and the growth of linear momentum satisfy

$$(\mu_s - \mu_n)\tau_s \le 0, \qquad \tfrac{1}{2}(\mathbf{v}_n^2 - \mathbf{v}_s^2) + \mathbf{m}_s \cdot (\mathbf{v}_s - \mathbf{v}_n) \le 0.$$

For any growth τ_s, the function

$$\mathbf{m}_s = M(\mathbf{v}_n - \mathbf{v}_s) + \tfrac{1}{2}\tau_s(\mathbf{v}_s + \mathbf{v}_n)$$

makes the second inequality identically true. The inequality $(\mu_s - \mu_n)\tau_s \le 0$ holds if and only if $\operatorname{sgn}\tau_s = -\operatorname{sgn}(\mu_s - \mu_n)$. To determine an appropriate constitutive function for τ_s, we preliminarily observe that

$$\mu_s = \hat{\mu}_s(\rho_s, \theta), \qquad \mu_n = \hat{\mu}_n(\rho_n, \theta).$$

Let θ_λ be the threshold temperature; in helium $\theta_\lambda \simeq 2.1$ K. Restrict attention to the interval $0 < \theta \le \theta_\lambda$ where normal fluid and superfluid coexist. Experiments show that at rest, that is, when $\mathbf{v}_s = \mathbf{v}_n$, the ratio ρ_n/ρ is an increasing function, say $\alpha(\theta)$, subject to

$$\alpha(0) = 0, \qquad \alpha(\theta_\lambda) = 1.$$

We let

$$\mu_s(\rho_s, \theta) = \mu_0(\theta) + \tfrac{1}{2}\alpha(\theta)\rho_s, \qquad \mu_n(\rho_n, \theta) = \mu_0(\theta) + \tfrac{1}{2}(1 - \alpha(\theta))\rho_n.$$

Hence, we have

$$\mu_s - \mu_n = \tfrac{1}{2}[\alpha(\theta)\rho_s - (1-\alpha(\theta))\rho_n] = \tfrac{1}{2}\rho[\alpha(\theta) - \omega_n].$$

Thus

$$\operatorname{sgn}\tau_s = -\operatorname{sgn}(\mu_s - \mu_n) = \operatorname{sgn}(\omega_n - \alpha(\theta)).$$

Hence $\tau_s > 0$ and the mass density of the superfluid increases if $\omega_n > \alpha(\theta)$, and vice versa. If $\theta \to 0$ then $\mu_s \to \mu_0(0_+)$ and $\mu(0_+)$, since $\alpha \to 0$ and $\rho_n \to 0$. Likewise, $\mu_s = \mu_0(\theta_\lambda)$ and $\mu_n = \mu_0(\theta_\lambda)$ at $\theta = \theta_\lambda$. Moreover, the physical literature shows that α is well modelled by the function

$$\alpha(\theta) = (\theta/\theta_\lambda)^\delta, \qquad \delta \in [1, 1.5].$$

The thermodynamic restriction on τ_s and indications of experimental result suggest that we assume

$$\tau_s = -\gamma\omega_s(\mu_s - \mu_n), \qquad \gamma > 0,$$

so that, upon substitution of $\mu_s - \mu_n$, we can write

$$\tau_s = -\gamma\rho\omega_s[\alpha(\theta) - (1-\omega_s)]. \tag{14.23}$$

The result (14.23) applies if the two fluids are at rest, namely, they move with the same velocity. The occurrence of a nonzero relative velocity hinders the appearance of the superfluid [49]. To model this feature, appeal is usually made to Landau's critical velocity of the superfluid, $\mathbf{v}_\lambda$. Let θ^* be a temperature defined by

$$\theta^* = v^2\theta_\lambda, \qquad v^2 = |\mathbf{v}_s - \mathbf{v}_n|^2/|\mathbf{v}_\lambda|^2,$$

where $\mathbf{v}_\lambda$ is Landau's critical velocity possibly dependent on temperature. It seems more appropriate to think that $|\mathbf{v}_\lambda|^2$ is a temperature-dependent parameter; the parameter θ^* has the dimension of a temperature and depends on the velocities $\mathbf{v}_s$, $\mathbf{v}_n$ via the objective vector $\mathbf{v}_s - \mathbf{v}_n$. Now we formally replace θ in (14.23) by $(\theta + \theta^*)$. Hence it follows

$$\tau_s = -\gamma\rho\omega_s[\alpha(\theta + \theta^*) - (1-\omega_s)].$$

Since α is an increasing function then the derivative α' is positive and hence it follows

$$\frac{d\tau_s}{d\theta^*} = -\gamma\rho\omega_s\alpha' < 0.$$

This implies that the supply of superfluid τ_s decreases, and possibly becomes negative, if θ^* increases. If $|\mathbf{v}_s - \mathbf{v}_n| = |\mathbf{v}_\lambda|$ then $v^2 = 1$ and $\theta^* = \theta_\lambda$. Since $\alpha(\theta + \theta_\lambda) > 1$ then $\tau_s < 0$; at any temperature $\theta > 0$ the occurrence of $|\mathbf{v}_s - \mathbf{v}_n| = |\mathbf{v}_\lambda|$ forbids the (stable) existence of the superfluid.

In view of the equations of motion,

$$\partial_t(\rho_s \mathbf{v}_s) + \nabla \cdot (\rho_s \mathbf{v}_s \otimes \mathbf{v}_s) = -\nabla p_s + \rho_s \mathbf{g} + \mathbf{m}_s,$$

$$\partial_t(\rho_n \mathbf{v}_n) + \nabla \cdot (\rho_n \mathbf{v}_n \otimes \mathbf{v}_n) = -\nabla p_n + \rho_n \mathbf{g} - \mathbf{m}_s,$$

we try to describe the characteristic effects of superfluids. As a reasonable assumption, we let the pressures p_s, p_n be given by constitutive functions of the form

$$p_s = p_s(\rho_s, \theta), \qquad p_n = p_n(\rho_n, \theta).$$

At equilibrium, between superfluid and normal fluid, the fractions ω_s and ω_n are given by

$$\omega_n = (1 - \omega_s) = \alpha(\theta).$$

Now

$$\omega_s' = \frac{\rho\rho_s' - \rho_s\rho'}{\rho^2},$$

the prime denoting the derivative with respect to θ. We further assume that ρ is slightly variable, possibly constant, so that

$$\operatorname{sgn} \omega_s' = \operatorname{sgn} \rho_s'.$$

Hence, we have

$$\operatorname{sgn} \rho_s' = -\operatorname{sgn} \alpha'.$$

Along the equilibrium curve,

$$\grave{\eta}_s = (\partial_\theta \eta_s + \partial_{\rho_s} \eta_s \, \rho_s')\grave{\theta}.$$

By the standard thermodynamic restrictions

$$\eta_s = -\partial_\theta \psi_s, \qquad p_s = \rho_s^2 \partial_{\rho_s} \psi_s$$

we find

$$\partial_{\rho_s} \eta_s = -\frac{1}{\rho_s^2} \partial_\theta p_s, \qquad \partial_\theta \eta_s = \frac{1}{\theta} \partial_\theta \varepsilon_s = \frac{1}{\theta} c_v,$$

the specific heat c_v being assumed to be positive. Upon substitution we can write

$$\grave{\eta}_s = (\frac{c_v}{\theta} - \frac{1}{\rho_s^2} \partial_\theta p_s \, \rho_s')\grave{\theta}.$$

By the customary assumption that η_s be constant it follows that

$$\partial_\theta p_s = \frac{\rho_s^2 c_v}{\theta \rho_s'}.$$

The negative value of ρ_s' implies that

$$\partial_\theta p_s < 0. \tag{14.24}$$

The negative value of $\partial_\theta p_s$ is quite surprising in that we are accustomed to ideal gases where $\partial_\theta p = k\rho > 0$.

Fountain Effect

The motion of the superfluid occurs subject to the force

$$-\partial_{\rho_s} p_s \nabla \rho_s - \partial_\theta p_s \nabla \theta + \rho_s \mathbf{g} + \mathbf{m}_s.$$

Suppose $\nabla \theta$ is large enough that $-\partial_\theta p_s \nabla \theta$ is the dominant force. The fountain effect shows that the superfluid moves in the direction of the temperature gradient. This is fully consistent with, and explained by, the condition (14.24). First, if heat is added to the power in the tube then θ increases and p_s decreases; the pressure of the superfluid in the tube is lower than that in the bath. The superfluid then flows from the bath to the tube via the fine powder which impedes any flow of the normal fluid. The mass of the liquid in the tube increases and eventually the superfluid squirts out through the fine capillary at the top.

Thermomechanical Effect

A flask is filled with liquid helium and a heating coil is placed inside a differential container. When heat is added to the fluid the temperature inside the inner container increases (by $\Delta\theta$) and the concentration of the normal fluid increases too because of the phase transition ($\tau_s < 0$). Moreover, by (14.24), p_s decreases. Since p_s in the bath is larger than the superfluid flows through the narrow channel from the bath to the heat region. The normal fluid, instead, cannot flow through the channel in view of its viscosity. As a result, an amount of fluid is added to the heated region so that the gravitational pressure $\rho_s g h$ balances the decrease induced by temperature, $\partial_\theta p_s \Delta\theta$. The quantity $\rho_s g h$ is called thermomechanical pressure head. The height h is given by

$$h = \frac{|\partial_\theta p_s|}{\rho_s g} \Delta\theta.$$

Mechanocaloric Effect

A flask is filled with a fine powder and, above it, liquid helium. The flask has an opening at the bottom. The superfluid component flows through the fine powder and hence the concentration of normal fluid increases above the powder. The temperature of the fluid inside the flask increases, which is sensed by a thermometer placed in the fluid. The temperature increase is due to the fact that energy is conserved in that the superfluid has no thermal energy and hence no energy flows outside the flask. The conserved energy is then distributed over a smaller quantity of liquid helium thus resulting in a higher temperature.

Rollin (or Creeping) Effect

The liquid helium exhibits a property of clinging to the walls of the container. To show this effect, we consider a test tube filled with liquid helium. When the test tube is lowered into a liquid helium bath, a fluid film clings to the tube and gradually fills the tube. If, instead, the tube is raised above the bath level it empties out. The experiments show that in these films the capillary forces dominate the gravity and viscous forces. Moreover, if the rate of flow is independent of the height of flow, it increases with drop in temperature and vanishes at lambda point. To understand the phenomenon we first observe that, at the outset, when the container is just filled with helium, a force arises from the interaction between helium atoms and molecules of the container's wall (Van der Waals force); this draws an helium film upwards, parallel to the wall. Gravity is then overcome. The absence of viscosity favours the flow and hence we can think that the flow is made of the superfluid component. Although the superfluid component alone quits the container, ω_s cannot become zero in that the equilibrium at the existing temperature forces the transition from normal fluid to superfluid.

When the test tube is lowered into the bath then both flows (inward and outward) occur and the inward flow prevails because of pressure and temperature of the bath. Lower temperatures favour the flow due to the greater fraction of superfluid.

14.2.5 Rotation of Superfluids

Suppose that, at $\theta > \theta_\lambda$, a cylindrical container is made to rotate fast around the cylindrical axis with angular velocity[10] $\boldsymbol{\omega}$. The liquid helium, inside the container, rotates accordingly. If the liquid is cooled below θ_λ and then the container is stopped then the liquid continues to rotate with angular momentum

[10] To avoid ambiguities, the angular velocity is denoted by the vector $\boldsymbol{\omega}$; the scalars ω_s, ω_n still denote the mass fractions.

$$\mathbf{L} = \omega_s(\theta) I \boldsymbol{\omega},$$

where I is the moment of inertia around the cylindrical axis. This means that, upon cooling the temperature to $\theta < \theta_\lambda$, a fraction $\omega_s(\theta)$ of superfluid appears and the observed rotation can be explained as though the fraction ϖ_s of the superfluid rotates with the given angular velocity whereas the normal component is at rest; by viscosity, the container makes the normal fluid to be at rest.

We now let the container rotate slowly,[11] at $\theta > \theta_\lambda$, with angular velocity $\boldsymbol{\omega}$. Of course

$$\mathbf{L} = I\boldsymbol{\omega}.$$

We now cool the liquid, below θ_λ, while rotating the container. We observe that

$$\mathbf{L} = \omega_n I \boldsymbol{\omega}.$$

The picture in terms of the two fluids is as follows: the normal component keeps to rotate, with moment of inertia $\omega_n I$, whereas the superfluid stays at rest as though the superfluid could not assume arbitrarily small values of angular velocity. This feature is known as the *Hess–Fairbank effect* .

This effect is a direct motivation of the Onsager–Feynman quantization relation. The circulation of $\mathbf{v}_s$ along a superfluid vortex can assume a multiple of Planck's constant h divided by the mass of the vortex region,

$$\oint \mathbf{v}_s \cdot \mathbf{t} ds = n\frac{h}{m}.$$

For a uniform flow, we can write that, along a circle with radius R,

$$\mathbf{v}_s = \frac{n\hbar}{mR}\boldsymbol{\nu},$$

$\boldsymbol{\nu}$ being the unit normal to the vortex (plane) and $\hbar = h/2\pi$. This means that there is a critical value

$$\omega_c = \frac{\hbar}{mR^2}$$

of the angular velocity $|\boldsymbol{\omega}|$, below which there are no superfluid vortices. In practice, this is the case if $|\boldsymbol{\omega}| \ll \omega_c$; it is customary to take $|\boldsymbol{\omega}| < \omega_c/2$. Hence if $\omega < \omega_c/2$ the angular momentum is carried by the normal component, $\mathbf{L} = \omega_n(\theta) I \boldsymbol{\omega}$. If, instead, $|\boldsymbol{\omega}|$ is around $n\omega_c$ then

$$|\mathbf{L}| = I[\omega_n(\theta)|\boldsymbol{\omega}| + \omega_s(\theta) n\omega_c].$$

---

[11] As we see in a moment, slowly means $\omega < \omega_c/2$.

The angular velocity of the superfluid is kept constant, $n\omega_c$, for small variations of ω. The number n is called the *winding number.* For large values of n, $|\omega|$ and $n\omega_c$ are almost equal and hence it will appear as the whole liquid is rotating.

14.2.6 The Ginzburg–Landau Theory

Following Landau's view [268], liquid helium can be described as a ground state constituted by a (Bose–Einstein) condensate of helium atoms where excitations move as quasi-particles. As we see in a moment, a complex macroscopic wave function of the condensate allows us to introduce the mass density and velocity of the superfluid component and the energy flux associated with the elementary excitations.

Let

$$\psi(\mathbf{x}, t) = \varphi(\mathbf{x}, t) \exp(i\chi(\mathbf{x}, t))$$

be a scalar complex-valued function, φ, χ being real-valued functions. The starting point is the time-dependent Schrödinger equation,

$$i\hbar\partial_t\psi = -\frac{\hbar^2}{2m}\Delta\psi + V(\mathbf{x}, t)\psi,$$

where V is the potential and m is the mass of the particle. For slow variation of ψ at interatomic distances, the Gross–Pitaevskii equation [213, 364]

$$i\hbar\partial_t\psi = -\frac{\hbar^2}{2m}\Delta\psi + (\mu + V_0\varphi^2)\psi$$

is considered where μ is the chemical potential and V_0 is the potential of the interatomic forces. Now, $\psi\psi^* = \varphi^2$ is usually regarded as the number density n in the condensate state. Since $i\hbar\partial\varphi^2 = i\hbar\psi^*\partial_t\psi + i\hbar\psi\partial_t\psi^*$, computation of $\psi^*\partial_t\psi$, $\psi\partial_t\psi^*$, and substitution yields

$$\partial_t\varphi^2 = i\frac{\hbar}{2m}\nabla \cdot [\psi^*\nabla\psi - \psi\nabla\psi^*].$$

Hence

$$\mathbf{j} := -i\tfrac{1}{2}\hbar[\psi^*\nabla\psi - \psi\nabla\psi^*] = \hbar\varphi^2\nabla\chi$$

can be viewed as the particle flux vector so that the balance of mass is obtained in the form

$$m\partial_t\varphi^2 = -\nabla \cdot \mathbf{j}.$$

Since φ^2 is the number density then $m\varphi^2$ is the mass density. The form of $\mathbf{j}$ suggests that we define a velocity

$$\mathbf{u} := \frac{1}{m\varphi^2}\mathbf{j} = \frac{\hbar}{m}\nabla\chi.$$

Moreover, it seems natural to establish the equation of motion by taking the force as minus the gradient of the potential. Hence, neglecting the nonlinear term in φ we have [415]

$$m\partial_t\mathbf{u} = -\nabla(\mu + \tfrac{1}{2}m\mathbf{u}^2),$$

or, since $\nabla \times \mathbf{u} = \mathbf{0}$,

$$m\dot{\mathbf{u}} = -\nabla\mu,$$

where $\dot{\mathbf{u}} = \partial_t\mathbf{u} + (\mathbf{u}\cdot\nabla)\mathbf{u}$.

Standard models which appeared in the literature arise by identifying $\mathbf{u}$ with the velocity of the superfluid component. Moreover, this approach seems to be the motivation of the equation of motion where $-\nabla\mu$ is used as the force acting on the superfluid.

With some formal variants, this view is now developed to apply the Ginzburg–Landau theory of superconductivity [194] to superfluidity. We start formally with the complex-valued function

$$\psi = \varphi\exp(i\chi)$$

and make the identification

$$\omega_s = \varphi^2, \qquad \varphi^2 \le 1.$$

Next we let

$$\nabla \times \mathbf{v}_s = \mathbf{0}, \qquad \mathbf{v}_s = \frac{1}{m}\nabla\chi.$$

Further, observe

$$|\nabla\psi|^2 = \nabla\psi\cdot\nabla\psi^* = |\nabla\varphi|^2 + \varphi^2|\nabla\chi|^2.$$

Equilibrium values of ψ are determined as the stationary points of an appropriate functional. Look at the functional

$$\mathcal{L} = \tfrac{1}{2}\beta^2|\nabla\psi|^2 - \tfrac{1}{2}\alpha(\theta)|\psi|^2 + \tfrac{1}{4}|\psi|^4$$

whence

$$\mathcal{L} = \tfrac{1}{2}\beta^2|\nabla\varphi|^2 + \tfrac{1}{2}\varphi^2[\beta^2|\nabla\chi|^2 - \alpha(\theta)] + \tfrac{1}{4}\varphi^4.$$

The corresponding Euler–Lagrange equations

$$0 = \delta_\varphi\mathcal{L} = -\partial_\varphi\mathcal{L} - \nabla\cdot(\partial_{\nabla\varphi}\mathcal{L}),$$

$$\delta_\chi\mathcal{L} = -\partial_\chi\mathcal{L} - \nabla\cdot(\partial_{\nabla\chi}\mathcal{L})$$

are the Ginzburg–Landau equations

$$\varphi(\beta^2|\nabla\chi|^2 - \alpha(\theta) + \varphi^2) - \beta^2\Delta\varphi = 0,$$

$$\beta^2\nabla\cdot(\varphi^2\nabla\chi) = 0.$$

These equations are examined on the assumptions that φ is uniform, $\nabla\varphi = 0$, and the (velocity) vector

$$\mathbf{v} = \beta\nabla\chi$$

is constant. Hence $\nabla\cdot(\varphi^2\nabla\chi) = 0$ holds identically while the other equation reduces to

$$\varphi[\varphi^2 - \alpha(\theta) + v^2] = 0,$$

where $v = |\mathbf{v}|$. Look for solutions $\varphi(\theta, v)$. If $v^2 \geq \alpha(\theta)$ then only $\varphi(\theta, v) = 0$ is a real-valued solution; this means that the superfluid cannot occur. If, instead, $v^2 < \alpha(\theta)$ then three solutions hold,

$$\varphi_1 = 0, \qquad \varphi_{2,3} = \pm\sqrt{\alpha(\theta) - v^2}.$$

For definiteness, let

$$\alpha(\theta) = 1 - (\theta/\theta_\lambda)^{3/2}.$$

Hence $\varphi_{2,3}$ hold if $0 < \theta < \theta_\lambda$ and

$$\varphi_{2,3} = \pm\sqrt{1 - (\theta/\theta_\lambda)^{3/2} - v^2}, \qquad v^2 < 1 - (\theta/\theta_\lambda)^{3/2}.$$

Consequently, in addition to $\varphi = 0$, the number density

$$\varphi^2 = 1 - (\theta/\theta_\lambda)^{3/2} - v^2 > 0$$

is allowed if $v^2 < 1 - (\theta/\theta_\lambda)^{3/2}$.

Chapter 15
Ferroics

Ferroics is the general framework for ferroelectricity and ferromagnetism. Ferroelectrics are materials where the polarization exhibits nonlinear behaviour and hysteresis. They may have a strong permanent polarization but only if the temperature lies below a characteristic temperature, called the Curie temperature. In this chapter a nonlinear approach to ferroelectrics is investigated and then a general setting is established for the modelling of hysteretic effects. The leading idea is that the entropy production is, per se, a non-negative function here involving the time derivative of the electric field. Also with reference to the literature, the Ginzburg–Landau–Devonshire (GLD) theory is examined to describe the transition between linear paraelectric properties and hysteretic behaviours. In essence, the GLD theory characterizes the transition in terms of an order parameter; for the paraelectric–ferroelectric transition the order parameter is the (scalar) polarization. Next a nonlinear theory is developed for materials where the magnetization exhibits nonlinear behaviour and hysteresis. The leading idea is still to consider the entropy production as a non-negative function, here involving the time derivative of the magnetic field. As with ferroelectrics, the models account for the Curie temperature as a limit temperature for hysteretic effects.

15.1 Evolution Function and Hysteretic Effects in Ferroelectrics

One-dimensional models of ferroelectrics are investigated [103, 188, 433, 435] where the time derivative $\dot{P}$ of the polarization P is given by

$$\dot{P} = \hat{P}(P, E, \dot{E}).$$

A. Morro and C. Giorgi, *Mathematical Modelling of Continuum Physics*,
Modeling and Simulation in Science, Engineering and Technology,
https://doi.org/10.1007/978-3-031-20814-0_15

A special subclass is represented by the Duhem model[1]

$$\dot{P} = g(P, E)\dot{E} + f(P, E)|\dot{E}|. \tag{15.1}$$

Consistent with (15.1), here we follow the view that both the electric field and the electric polarization are among the independent variables. This view is related to the fact that an hysteresis loop is a closed curve surrounding a region in the $P - E$ plane or that $P(E)$ would not be a single-valued function. Moreover, for the sake of generality, the deformation of the material and the dependence on the temperature are allowed to occur.

We assume **M** and **J** are zero. The entropy inequality takes the form

$$-\rho(\dot{\psi} + \eta\dot{\theta}) + \rho\mathbf{E}\cdot\dot{\pi} + \mathbf{T}\cdot\mathbf{L} - \frac{1}{\theta}\mathbf{q}\cdot\nabla\theta + \theta\nabla\cdot\mathbf{k} \geq 0,$$

where $\pi = \mathbf{P}/\rho$ is the electric polarization density, per unit mass. In terms of the free energy $\phi = \psi - \mathbf{E}\cdot\pi = \varepsilon - \theta\eta - \mathbf{E}\cdot\pi$ the entropy inequality can be written as

$$-\rho(\dot{\phi} + \eta\dot{\theta}) - \rho\pi\cdot\dot{\mathbf{E}} + \mathbf{T}\cdot\mathbf{L} - \frac{1}{\theta}\mathbf{q}\cdot\nabla\theta + \theta\nabla\cdot\mathbf{k} \geq 0. \tag{15.2}$$

The ferroelectric hysteretic behaviour involves both the electric field **E** and the electric polarization density π. For generality, the temperature θ and the deformation gradient **F** are taken to affect the response of the material and, by analogy with the physical literature that incorporates the nearest neighbours in the GLD model [384], the dependence on the gradient of the polarization density, $\nabla\pi$, is also considered [404]. The dependence on $\dot{\pi}$ is allowed so that a consistent approach is obtained for the dependence on $\nabla\pi$ while a dependence on $\dot{\mathbf{E}}$ is in order to account for hysteretic effects and a dependence on the stretching tensor[2] $\mathfrak{D}$ describes dissipative effects. Hence we might take $\mathbf{F}, \mathbf{E}, \theta, \nabla\theta, \pi, \nabla\pi, \mathfrak{D}, \dot{\mathbf{E}}, \dot{\pi}$) as the set of independent variables for the constitutive functions ϕ, η, **T**, **q**, and **k**. However by the principle of objectivity the function ϕ has to be invariant under Euclidean transformations. Hence ϕ should depend only on Euclidean invariants. For definiteness, in addition to θ we take $\mathbf{C} = \mathbf{F}^T\mathbf{F}$, $\mathfrak{E} = \mathbf{F}^T\mathbf{E}$, $\mathfrak{P} = \mathbf{F}^{-1}\mathbf{P}$ and $\nabla\mathfrak{P}\cdot\nabla\mathfrak{P}$, $\nabla\theta\cdot\nabla\theta$, $\dot{\mathfrak{E}}$, $\dot{\mathfrak{P}}$ as appropriate invariant variables; ϕ and η should depend on scalars involving the set of variables[3]

$$\Gamma = (\theta, \mathbf{C}, \mathfrak{E}, \mathfrak{P}, \nabla\mathfrak{P}, \nabla\theta, \mathfrak{D}, \dot{\mathfrak{E}}, \dot{\mathfrak{P}}).$$

---

[1] Many models of this kind are developed also within the ferromagnetic context depending on the choice of the characteristic functions g and f [433].

[2] In this chapter $\mathfrak{D}$ is the stretching tensor, not to be confused with the electric displacement $\mathbf{D} = \epsilon_0\mathbf{E} + \mathbf{P}$.

[3] We might use as a variable the referential gradient $\nabla_R\mathfrak{P}$ rather than $\nabla\mathfrak{P}$, and the like for $\nabla_R\theta$ and $\nabla\theta$. In that case $(\nabla_R\mathfrak{P})^{\cdot} = \nabla_R\dot{\mathfrak{P}}$ and then a simpler expression of the stress tensor would arise.

Two remarks are in order. First, the invariance in $SO(3)$ holds for both $\mathbf{F}^T\mathbf{E}$ and $\mathbf{F}^{-1}\mathbf{E}$ and the like for $\mathbf{P}$; this common property is due algebraically to the fact that for rotation tensors $\mathbf{R}^{-1} = \mathbf{R}^T$. Secondly, $\mathbf{E}$ and $\mathbf{P}$ are the fields at the frame locally at rest which gives a clear physical meaning to the constitutive equations.

Computation of $\dot{\phi}$ and substitution in (15.2) result in

$$- \rho(\partial_\theta\phi + \eta)\dot{\theta} - \rho\partial_{\mathbf{C}}\phi \cdot \dot{\mathbf{C}} - \rho\partial_{\mathfrak{E}}\phi \cdot \dot{\mathfrak{E}} - \rho\partial_{\mathfrak{P}}\phi \cdot \dot{\mathfrak{P}} - \rho\partial_{\nabla\mathfrak{P}}\phi \cdot \dot{\overline{\nabla\mathfrak{P}}} - \rho\partial_{\nabla\theta}\phi \cdot \dot{\overline{\nabla\theta}}$$
$$-\rho\partial_{\mathfrak{D}}\phi \cdot \dot{\mathfrak{D}} + \mathbf{T}\cdot\mathbf{L} - \mathbf{P}\cdot\dot{\mathbf{E}} - \frac{1}{\theta}\mathbf{q}\cdot\nabla\theta + \theta\nabla\cdot\mathbf{k} \geq 0. \tag{15.3}$$

The arbitrariness of $\dot{\mathfrak{D}}$, $\dot{\overline{\nabla\theta}}$, $\dot{\theta}$ implies that

$$\partial_{\mathfrak{D}}\phi = \mathbf{0}, \qquad \partial_{\nabla\theta}\phi = \mathbf{0}, \qquad \eta = -\partial_\theta\phi.$$

Observe that $\dot{\mathbf{C}} = 2\mathbf{F}^T\mathfrak{D}\mathbf{F}$ and

$$\mathfrak{P}\cdot\dot{\mathfrak{E}} = (\mathbf{F}^{-1}\mathbf{P})\cdot(\dot{\mathbf{F}}^T\mathbf{E} + \mathbf{F}^T\dot{\mathbf{E}}) = (\mathbf{E}\otimes\mathbf{P})\cdot\mathbf{L} + \mathbf{P}\cdot\dot{\mathbf{E}}.$$

Hence substituting $\rho\pi\cdot\dot{\mathbf{E}}$ with $\mathfrak{P}\cdot\dot{\mathfrak{E}} - (\mathbf{E}\otimes\mathbf{P})\cdot\mathbf{L}$ and dividing by θ it follows from (15.3) that

$$\frac{1}{\theta}(\mathbf{T} - 2\rho\mathbf{F}\partial_{\mathbf{C}}\phi\mathbf{F}^T + \mathbf{E}\otimes\mathbf{P})\cdot\mathbf{L} - \frac{1}{\theta}(\rho\partial_{\mathfrak{E}}\phi + \mathfrak{P})\cdot\dot{\mathfrak{E}} - \frac{1}{\theta}\rho\partial_{\mathfrak{P}}\phi\cdot\dot{\mathfrak{P}}$$
$$-\frac{\rho}{\theta}\partial_{\nabla\mathfrak{P}}\phi\cdot(\nabla\dot{\mathfrak{P}} - \mathbf{L}^T\nabla\mathfrak{P}) - \frac{1}{\theta^2}\mathbf{q}\cdot\nabla\theta + \nabla\cdot\mathbf{k} \geq 0.$$

Consider the identities

$$\frac{\rho}{\theta}\partial_{\nabla\mathfrak{P}}\phi\cdot\nabla\dot{\mathfrak{P}} = \nabla\cdot\big[\frac{\rho}{\theta}(\partial_{\nabla\mathfrak{P}}\phi)\dot{\mathfrak{P}}\big] - \dot{\mathfrak{P}}[\nabla\cdot(\frac{\rho}{\theta}\partial_{\nabla\mathfrak{P}}\phi)],$$

$$\partial_{\nabla\mathfrak{P}}\phi\cdot(\mathbf{L}^T\nabla\mathfrak{P}) \equiv \partial_{x_h}\mathfrak{P}_K\partial_{\partial_{x_j}\mathfrak{P}_K}\phi L_{hj} = [\nabla\mathfrak{P}(\partial_{\nabla\mathfrak{P}}\phi)^T]\cdot\mathbf{L}$$

For formal convenience let

$$\boldsymbol{\mathcal{T}} := \mathbf{T} - 2\rho\mathbf{F}\partial_{\mathbf{C}}\phi\mathbf{F}^T + \rho\nabla\mathfrak{P}(\partial_{\nabla\mathfrak{P}}\phi)^T + \mathbf{E}\otimes\mathbf{P}, \qquad \delta_{\mathfrak{P}} = \partial_{\mathfrak{P}}\phi - \frac{\rho}{\theta}\nabla\cdot(\frac{\rho}{\theta}\partial_{\nabla\mathfrak{P}}\phi),$$

where $\delta_{\mathfrak{P}}$ is the standard variational derivative. Since ϕ is independent of $\mathfrak{D}$ it is natural to let $\boldsymbol{\mathcal{T}}$ depend on $\mathfrak{D}$ or to view $\boldsymbol{\mathcal{T}}$ as the dissipative stress. Upon obvious rearrangements we obtain

$$\frac{1}{\theta}\boldsymbol{\mathcal{T}}\cdot\mathbf{L} - \frac{1}{\theta}(\rho\partial_{\mathfrak{E}}\phi + \mathfrak{P})\cdot\dot{\mathfrak{E}} - \frac{1}{\theta}\rho\delta_{\mathfrak{P}}\phi\cdot\dot{\mathfrak{P}} - \frac{1}{\theta^2}\mathbf{q}\cdot\nabla\theta + \nabla\cdot(\mathbf{k} - \frac{\rho}{\theta}\partial_{\nabla\mathfrak{P}}\phi\dot{\mathfrak{P}}) \geq 0. \tag{15.4}$$

Hence we let

$$\mathbf{k} = \frac{\rho}{\theta} \partial_{\nabla \mathfrak{P}} \phi \dot{\mathfrak{P}}.$$

The arbitrariness of $\mathbf{W} = \text{skw}\mathbf{L}$ implies

$$\mathbf{T} + \rho \nabla \mathfrak{P} (\partial_{\nabla \mathfrak{P}} \phi)^T + \mathbf{E} \otimes \mathbf{P} \in \text{Sym}.$$

If ϕ is independent of $\nabla \mathfrak{P}$ then it follows that

$$\text{skw}\mathbf{T} = \text{skw}(\mathbf{P} \otimes \mathbf{E}),$$

that is the classical relation arising from the balance of angular momentum.

Back to inequality (15.4) we can write

$$\mathcal{T} \cdot \mathfrak{D} - (\rho \partial_{\mathfrak{E}} \phi + \mathfrak{P}) \cdot \dot{\mathfrak{E}} - \rho \delta_{\mathfrak{P}} \phi \cdot \dot{\mathfrak{P}} - \frac{1}{\theta} \mathbf{q} \cdot \nabla \theta \geq 0.$$

Letting $\mathfrak{D} = \mathbf{0}$ and $\nabla \theta = \mathbf{0}$ we have

$$(\rho \partial_{\mathfrak{E}} \phi + \mathfrak{P}) \cdot \dot{\mathfrak{E}} + \rho \delta_{\mathfrak{P}} \phi \cdot \dot{\mathfrak{P}} \leq 0. \tag{15.5}$$

The whole inequality holds if

$$\mathcal{T} \cdot \mathfrak{D} - \frac{1}{\theta} \mathbf{q} \cdot \nabla \theta \geq 0. \tag{15.6}$$

Inequality (15.6) holds in particular with the Navier–Stokes–Fourier scheme. This in turn means that

$$\mathbf{T} = 2\rho \mathbf{F} \partial_{\mathbf{C}} \phi \mathbf{F}^T - \rho \nabla \mathfrak{P} (\partial_{\nabla \mathfrak{P}} \phi)^T - \mathbf{E} \otimes \mathbf{P} + \mathcal{T};$$

the stress $\mathbf{T}$ is the sum of an electro-elastic part, $2\rho \mathbf{F} \partial_{\mathbf{C}} \phi \mathbf{F}^T - \rho \nabla \mathfrak{P} (\partial_{\nabla \mathfrak{P}} \phi)^T - \mathbf{E} \otimes \mathbf{P}$, and a dissipative part $\mathcal{T}$ subject to (15.6).

We now examine some consequences of (15.5). First, though the values $\mathfrak{E}$ and $\mathfrak{P}$ are taken as independent variables the derivatives $\dot{\mathfrak{E}}$ and $\dot{\mathfrak{P}}$ cannot be independent of each other. If $\dot{\mathfrak{E}}$ and $\dot{\mathfrak{P}}$ can be taken arbitrarily then (15.5) implies

$$\rho \partial_{\mathfrak{E}} \phi = -\mathfrak{P}, \qquad \delta_{\mathfrak{P}} \phi = \mathbf{0}.$$

Let $\partial_{\nabla \mathfrak{P}} \phi = \mathbf{0}$. Then $\delta_{\mathfrak{P}} \phi = \partial_{\mathfrak{P}} \phi$. Observe that $\mathfrak{P}/\rho = \mathbf{F}^{-1} \pi$. Further, ϕ depends on $\mathbf{E}$ and π only through $\mathfrak{E}$ and $\mathfrak{P}$, respectively. Hence we find

$$\partial_{\mathbf{E}} \phi = \mathbf{F} \partial_{\mathfrak{E}} \phi = -\pi, \qquad \partial_{\pi} \phi = \rho \mathbf{F}^{-T} \partial_{\mathfrak{P}} \phi = \mathbf{0}.$$

If ϕ is a C^2 function then it follows

$$\partial_\pi \partial_{\mathbf{E}} \phi = -\mathbf{1}, \qquad \partial_{\mathbf{E}} \partial_\pi \phi = \mathbf{0},$$

which is a contradiction. The contradiction is removed if π and $\mathbf{E}$ are not simultaneously independent variables when $\dot{\mathfrak{E}}$ and $\dot{\mathfrak{P}}$ are mutually independent.

By the definition of entropy production we can replace inequality (15.5) with

$$(\rho \partial_{\mathfrak{E}} \phi + \mathfrak{P}) \cdot \dot{\mathfrak{E}} + \rho \delta_{\mathfrak{P}} \phi \cdot \dot{\mathfrak{P}} = -\theta \sigma_{PE}, \tag{15.7}$$

where $\sigma_{PE} \geq 0$ is the entropy production determined by the evolution of $\mathfrak{P}$, $\mathfrak{E}$ when $\mathfrak{D} = \mathbf{0}$, $\nabla\theta = \mathbf{0}$. We can then view (15.7) as the evolution equation for $\mathfrak{E}$, $\mathfrak{P}$. Indeed, by the representation formula (A.7) for vectors with a given longitudinal part, and arbitrary transverse part, we can write

$$\dot{\mathfrak{P}} = (\dot{\mathfrak{P}} \cdot \mathbf{e})\mathbf{e} + (\mathbf{1} - \mathbf{e} \otimes \mathbf{e})\mathbf{w},$$

$\mathbf{w}$ being an arbitrary vector. Here we let $\mathbf{e} = \delta_{\mathfrak{P}} \phi / |\delta_{\mathfrak{P}} \phi|$ and use (15.7) to obtain

$$(\dot{\mathfrak{P}} \cdot \mathbf{e})\mathbf{e} = (\dot{\mathfrak{P}} \cdot \rho \delta_{\mathfrak{P}} \phi) \frac{\delta_{\mathfrak{P}} \phi}{\rho |\delta_{\mathfrak{P}} \phi|^2} = -[(\rho \partial_{\mathfrak{E}} \phi + \mathfrak{P}) \cdot \dot{\mathfrak{E}} + \theta \sigma_{PE}] \frac{\delta_{\mathfrak{P}} \phi}{\rho |\delta_{\mathfrak{P}} \phi|}.$$

Hence (A.7) yields

$$\dot{\mathfrak{P}} = -[(\rho \partial_{\mathfrak{E}} \phi + \mathfrak{P}) \cdot \dot{\mathfrak{E}} + \theta \sigma_{PE}] \frac{\delta_{\mathfrak{P}} \phi}{\rho |\delta_{\mathfrak{P}} \phi|^2} + (\mathbf{1} - \frac{\delta_{\mathfrak{P}} \phi}{|\delta_{\mathfrak{P}} \phi|} \otimes \frac{\delta_{\mathfrak{P}} \phi}{|\delta_{\mathfrak{P}} \phi|})\mathbf{w} \tag{15.8}$$

where $\mathbf{w}$ is any vector function of the set of variables Γ. This is the most general (non-linear) evolution equation for $\mathfrak{P}$ consistent with the second-law inequality. Instead the simplest model happens when $\sigma_{PE} = 0$ and $\mathbf{w} = \mathbf{0}$. In this case (15.8) reduces to

$$\dot{\mathfrak{P}} = -[(\rho \partial_{\mathfrak{E}} \phi + \mathfrak{P}) \cdot \dot{\mathfrak{E}}] \frac{\delta_{\mathfrak{P}} \phi}{\rho |\delta_{\mathfrak{P}} \phi|^2}.$$

An intermediate model follows by letting $\mathbf{w} = \mathbf{0}$ and $\mathfrak{P} = -\rho \partial_{\mathfrak{E}} \phi$. Thus Eq. (15.8) simplifies to

$$\dot{\mathfrak{P}} = -\frac{\theta \sigma_{PE}}{\rho |\delta_{\mathfrak{P}} \phi|^2} \delta_{\mathfrak{P}} \phi.$$

This result may be phrased by saying that $\delta_{\mathfrak{P}} \phi = \mathbf{0}$ expresses the equilibrium condition of the system and hence the deviation from equilibrium governs the evolution of the polarization π. This simpler approach leads to a Ginzburg–Landau evolution equation for the polarization which includes the Ginzburg–Landau–Devonshire model as a particular case and is close to the scheme employed in ferromagnetism for the magnetization dynamics [38, 155]. The resulting constitutive theory is enough to account for some features of hysteresis but is inadequate for a full description of hysteretic phenomena (major and minor loops, for instance) where the evolution of

$\mathfrak{P}$ on a time interval depends on the evolution of $\mathfrak{E}$ as well as on the initial values of $\mathfrak{E}$ and $\mathfrak{P}$.

A simple but fruitful model is obtained by letting $\partial_{\nabla\mathfrak{P}}\phi = \mathbf{0}$ and $\theta\sigma_{PE} = \hat{\xi}|\dot{\mathfrak{E}}|$ so that Eq. (15.7) takes the form

$$(\rho\partial_{\mathfrak{E}}\phi + \mathfrak{P}) \cdot \dot{\mathfrak{E}} + \rho\partial_{\mathfrak{P}}\phi \cdot \dot{\mathfrak{P}} = -\hat{\xi}|\dot{\mathfrak{E}}|. \tag{15.9}$$

This equation is now investigated in a one-dimensional setting.

15.1.1 One-Dimensional Models of Ferroelectric Hysteresis

With reference to (15.9) we now let $\mathbf{F}$ be constant and hence ρ is constant too. Since $\dot{\mathfrak{E}} = \mathbf{F}^T\dot{\mathbf{E}}$, $\dot{\mathfrak{P}} = \mathbf{F}^{-1}\dot{\mathbf{P}}$ and $\partial_{\mathfrak{E}}\phi = \mathbf{F}^{-1}\partial_{\mathbf{E}}\phi$, $\partial_{\mathfrak{P}}\phi = \mathbf{F}^T\partial_{\mathbf{P}}\phi$ then it follows that

$$(\partial_{\mathbf{E}}\Phi + \mathbf{P}) \cdot \dot{\mathbf{E}} + \partial_{\mathbf{P}}\Phi \cdot \dot{\mathbf{P}} = -\hat{\xi}|\mathbf{F}^T\dot{\mathbf{E}}|, \tag{15.10}$$

where $\Phi = \rho\phi$. Moreover assume $\mathbf{E}, \mathbf{P}$ are collinear and directed along a principal direction of $\mathbf{F}$. Thus we denote by E, P the components of $\mathbf{E}, \mathbf{P}$ in their common direction and Eq. (15.10) simplifies to

$$(P + \partial_E\Phi)\dot{E} + \partial_P\Phi\dot{P} = -\xi|\dot{E}| \le 0. \tag{15.11}$$

Equation (15.11) is invariant under the time transformation

$$t \to c\,t, \qquad c > 0,$$

and hence any associated model is rate-independent, if so is the function ξ.

It is a consequence of (15.11) that, along any closed curve C, in the E-P plane, as $t \in [0, T]$,

$$0 \ge \int_0^T [(P + \partial_E\Phi)\dot{E} + \partial_P\Phi\dot{P}]\,dt = \int_0^T [P\dot{E} + \dot{\Phi}]\,dt = \int_0^T P\dot{E}\,dt; \tag{15.12}$$

this implies that the closed curve C is run in the counterclockwise sense.

Except at stationary points where $\dot{E} = 0$, we can divide Eq. (15.11) by $\dot{E}$ to obtain

$$\frac{\dot{P}}{\dot{E}} = -\frac{P + \partial_E\Phi}{\partial_P\Phi} - \frac{\xi}{\partial_P\Phi}\,\text{sgn}\,\dot{E}. \tag{15.13}$$

Let θ be constant. If ξ depends on $\dot{E}$ at most through sgn $\dot{E}$ then the right-hand side of (15.13) is piecewise a function of P, E and $\dot{P}/\dot{E}$ is the derivative of P with respect to E, dP/dE, the slope of the $E - P$ curve. Let

$$\chi_1 = -\frac{P + \partial_E \Phi}{\partial_P \Phi}, \qquad \chi_2 = -\frac{\xi}{\partial_P \Phi}. \tag{15.14}$$

Hence Φ and χ_1 are functions of E and P while ξ is so far to be determined. The polarization P is then given by the differential equation

$$\frac{dP}{dE} = \chi_1 + \chi_2 \operatorname{sgn} \dot{E}. \tag{15.15}$$

In paraelectric materials a dielectric polarization occurs when an electric field is applied to the material and the material looses the polarization when the electric field is removed. If $\xi = 0$, and hence $\chi_2 = 0$, the model (15.15) describes a paraelectric behaviour where χ_1 represents the positive slope of the polarization curve that is (ϵ_o times) the *differential susceptibility*. We then require that

$$\chi_1 > 0, \qquad |\chi_2| \leq \chi_1. \tag{15.16}$$

Unlike the counterclockwise property (15.12), inequalities (15.16) are not a consequence of the second law; they are assumed as physically sound restrictions so that the polarization be a monotone function of the electric field.

We now look for explicit representations of χ_1 and χ_2. Since the hysteretic behaviour occurs below an appropriate temperature θ_c, called Curie temperature, we let

$$\xi = \gamma(E, P) h(\theta_c - \theta).$$

We refer to $\gamma(E, P)$ as the *hysteretic function.* We might identify h with the Heaviside function H. Rather, if hysteretic effects expressed by ξ have to become smaller and smaller as θ approaches θ_c from below, then it is more appropriate to identify h with the positive part of $1 - \theta/\theta_c$, namely

$$h(\theta_c - \theta) = \max\{0, 1 - \theta/\theta_c\}.$$

This identification is consistent with the feature that, as $\theta \to \theta_c$ from below, all loops shrink to lines as is the case in the paraelectric behaviour.

While χ_1 is fully determined by the potential Φ, χ_2 depends also on ξ and hence many different models can be established by using the same energy potential Φ. The function $\chi_2 = -\xi/\partial_P \Phi$ is related to the entropy production (l.h.s. of (15.11)) and hence describes the dissipative and hysteretic properties of the material.

To determine χ_1 and χ_2 and hence the model of ferroelectric materials, we begin with the assumption that there is a characteristic non-decreasing function $\mathcal{P}$ such that

$$P = \mathcal{P}(E) \quad \Rightarrow \quad \chi_2 = 0, \tag{15.17}$$

the dependence on the parameter (temperature) θ being understood and not written. We refer to $\mathcal{P}$ as the *anhysteretic function* and to the corresponding graph

$\{(E, \mathcal{P}(E)) : E \in \mathbb{R}\}$ as the *anhysteretic curve* [231]. Indeed, by (15.11) we might observe that

$$(P + \partial_E \Phi)\dot{E} + \partial_P \Phi \dot{P} = 0 \tag{15.18}$$

is the thermodynamic condition corresponding to $\xi = 0$. Hence

$$\frac{dP}{dE} = -\frac{P + \partial_E \Phi}{\partial_P \Phi} = \chi_1$$

represents the non-hysteretic behaviour and χ_1 can be viewed as the *anhysteretic differential susceptibility*. This in turn suggests that we let χ_1 be a non-negative function of E only, namely

$$\chi_1 = g(E) \geq 0. \tag{15.19}$$

Finally, the required counterclockwise property (15.12) of ferroelectric cycles suggests that

$$\chi_2(E, P) \begin{cases} > 0 & \text{if } P < \mathcal{P}(E), \\ = 0 & \text{if } P = \mathcal{P}(E), \\ < 0 & \text{if } P > \mathcal{P}(E). \end{cases} \tag{15.20}$$

The Hysteretic Gibbs Potential

We now select the Gibbs potential Φ in the general form

$$\Phi(E, P) = \mathcal{L}(P - \mathcal{G}(E)) + \mathcal{F}(E) - PE, \tag{15.21}$$

$\mathcal{L}$, $\mathcal{G}$, and $\mathcal{F}$ being so far undetermined differentiable functions, possibly dependent on θ. Thus $\Psi := \rho\psi = \mathcal{L} + \mathcal{F}$ is the Helmholtz potential. Upon substitution of $\partial_P \Phi$ and $\partial_E \Phi$ in (15.14) we obtain

$$\chi_1 = \frac{\mathcal{L}'(P - \mathcal{G}(E))\mathcal{G}'(E) - \mathcal{F}'}{\mathcal{L}'(P - \mathcal{G}(E)) - E},$$
$$\chi_2 = -\frac{\xi}{\mathcal{L}'(P - \mathcal{G}(E)) - E}.$$

By (15.19) it follows

$$\mathcal{L}'(P - \mathcal{G}(E))\mathcal{G}'(E) - \mathcal{F}' = g(E)[\mathcal{L}'(P - \mathcal{G}(E)) - E],$$

which implies

$$\mathcal{G}' = g, \qquad \mathcal{F}' = Eg(E).$$

Since χ_1 is assumed to be positive then $\mathcal{G}$ is a monotone increasing function. In order to determine $\mathcal{L}$, we observe that the simplest way to satisfy (15.20) is to let

$$\operatorname{sgn} \chi_2(E, P) = -\operatorname{sgn} [P - \mathcal{P}(E)].$$

Hence $\xi \geq 0$ implies that

$$-\operatorname{sgn} [P - \mathcal{P}(E)] = \operatorname{sgn} \chi_2 = -\operatorname{sgn} [\mathcal{L}' - E]$$

whence

$$\frac{P - \mathcal{P}(E)}{\mathcal{L}'(P - \mathcal{G}(E)) - E} \geq 0.$$

Thus we take

$$\frac{P - \mathcal{P}(E)}{\mathcal{L}'(P - \mathcal{G}(E)) - E} = \frac{1}{\alpha} > 0,$$

α being possibly parameterized by θ. It follows that

$$\mathcal{L}'(P - \mathcal{G}(E)) = \alpha[P - \mathcal{G}(E)], \qquad \mathcal{P}(E) = \mathcal{G}(E) + \frac{1}{\alpha} E.$$

Summarizing all these results we conclude that

1. to within an additive constant (possibly dependent on θ) the expression (15.21) of Φ only depends on g and α as follows

$$\Phi(E, P) = \tfrac{1}{2}\alpha(P - \mathcal{G}(E))^2 + \mathcal{F}(E) - PE,$$

where

$$\mathcal{G}(E) = \int_0^E g(y)\, dy, \qquad \mathcal{F}(E) = \int_0^E y\, g(y)\, dy,$$

2. the characteristic functions χ_1 and χ_2 are then given by

$$\chi_1 = g(E), \qquad \chi_2 = -\frac{\xi}{\alpha[P - \mathcal{P}(E)]},$$

where the anhysteretic function takes the form

$$\mathcal{P}(E) = \int_0^E [g(y) + 1/\alpha]\, dy$$

and $g(E) = \mathcal{P}'(E) - 1/\alpha$.

The Hysteretic Function γ

To determine an explicit representation of χ_2,

$$\chi_2 = \frac{\gamma(E, P)h(\theta_c - \theta)}{\alpha(\mathcal{P} - P)},$$

we observe that the only thermodynamic requirement on γ is the non-negative valuedness.

The anhysteretic susceptibility g, the positive parameter α, and the hysteretic function γ characterize the model of the material. With this in mind we now examine the hysteretic properties associated with a given Φ and four different functions γ.

I. $\gamma(E, P) = \gamma_0(E)|P - \mathcal{P}(E)|$
where $\gamma_0 = \beta[1 - \lambda \exp(-\sigma|E|)]$ and $\beta, \sigma > 0$, $\lambda \in (0, 1)$.

Let $\beta_h = \beta h(\theta - \theta_c)$. It follows from (15.15) that

$$\frac{dP}{dE} = g(E) - \frac{\beta_h}{\alpha}[1 - \lambda \exp(-\sigma|E|)]\text{sgn}\,[(P - \mathcal{P})\dot{E}]. \tag{15.22}$$

This model is characterized by the positive parameters α, β, λ, σ and the anhysteretic function $\mathcal{P}$. Since $h \geq 0$ then inequalities (15.16) reduce to

$$g(E) \geq \frac{\beta_h}{\alpha}(1 - \lambda) \geq 0.$$

To show the hysteretic features we now let E oscillate in time with angular frequency ω. Let $\mathcal{E}$ be the amplitude of the oscillations, $E(t) = \mathcal{E} \sin \omega t$. Hysteresis loops in the (E, P)-plane are obtained by determining $P(t)$ such that

$$\dot{P} = [\mathcal{P}'(E) - \frac{1}{\alpha}]\dot{E} - \frac{\beta_h}{\alpha}[1 - \lambda \exp(-\sigma|E|)]\text{sgn}\,[(P - \mathcal{P})]|\dot{E}|, \qquad \dot{E} = \omega\mathcal{E} \cos \omega t.$$

This is done by starting from different initial states (E_0, P_0). Since the model is rate-independent then ω may be chosen arbitrarily. To check the saturation condition, namely $dP/dE \to 0$ as $|E| \to \infty$, we observe that

$$(P - \mathcal{P})\dot{E} < 0 \quad \Rightarrow \quad \lim_{|E|\to\infty} \frac{dP}{dE} = \lim_{|E|\to\infty} g(E) + \frac{\beta_h}{\alpha} \geq \frac{\beta_h}{\alpha}.$$

Hence, even if $g \to 0$ as $|E| \to \infty$ the saturation property may occur (approximately) only if $\beta_h/\alpha \ll 1$, namely if α is constant and $\theta \approx \theta_c$.

II. $\gamma(E, P) = \gamma_0(P)|P - \mathcal{P}(E)|$,
where $\gamma_0 = \beta\,\text{sgn}\,[\sigma - |P|]$ and $\beta, \sigma > 0$.

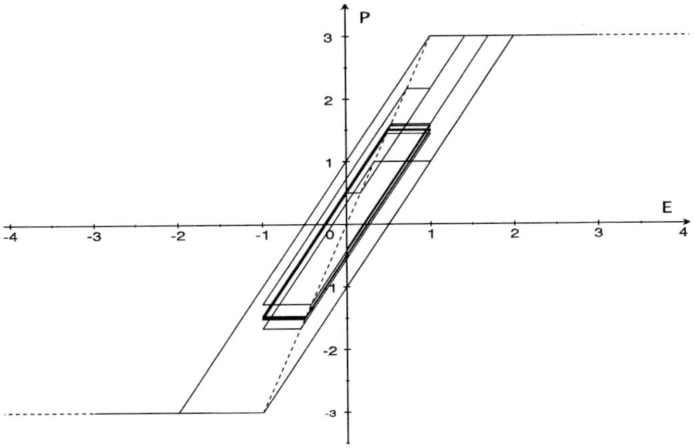

Fig. 15.1 Model II: hysteresis loops (solid) and anhysteretic curve (dashed) assuming $\mathcal{E} = 1, 3$ and starting from (E_0, P_0) with $E_0 = 0$, $P_0 = -1, -0.4, 0.5$

This model allows for the saturation of P at large values of $|E|$. With this purpose, restrict attention to the strip $|P| \le \sigma$ where $\gamma_0 > 0$. By (15.15) it follows that

$$\frac{dP}{dE} = g(E) - \frac{\beta_h}{\alpha} \operatorname{sgn}[(P - \mathcal{P})(\sigma - |P|)\dot{E}], \tag{15.23}$$

where $\beta_h = \beta h(\theta_c - \theta)$ and $g = \mathcal{P}' - 1/\alpha$. Again the hysteretic behaviour is modelled via $\mathcal{P}$ and the parameters α, β, σ.

Starting from different initial states (E_0, P_0) in the (E, P)-plane, hysteresis cycles are obtained by solving the equation

$$\dot{P} = [\mathcal{P}'(E) - \frac{1}{\alpha}]\dot{E} - \frac{\beta_h}{\alpha}\operatorname{sgn}[(P - \mathcal{P}(E))(\sigma - |P|)]|\dot{E}|, \qquad \dot{E} = \omega\mathcal{E}\cos\omega t.$$

In Fig. 15.1 cyclic processes with large amplitudes (major loops reaching saturation) and small amplitudes (minor loops) are depicted assuming $\mathcal{P}(E) = kE$ and $\alpha = \beta_h = 1/2$, $k = \sigma = 3$. As is apparent, here the anhysteretic curve is given by

$$P = \begin{cases} \mathcal{P}(E) & \text{if } |\mathcal{P}(E)| < \sigma, \\ \pm\sigma & \text{if } \mathcal{P}(E) = \pm\sigma, \end{cases}$$

and then the saturation property is achieved (see also [37]).

III. $\gamma(E, P) = \gamma_0[P - \mathcal{P}(E)]^2, \quad \gamma_0 > 0,$
where $g(E) = \mathcal{P}'(E) - 1/\alpha$.

Let $\gamma_h(\theta) = \gamma_0 h(\theta - \theta_c)$. The derivative dP/dE is then given by

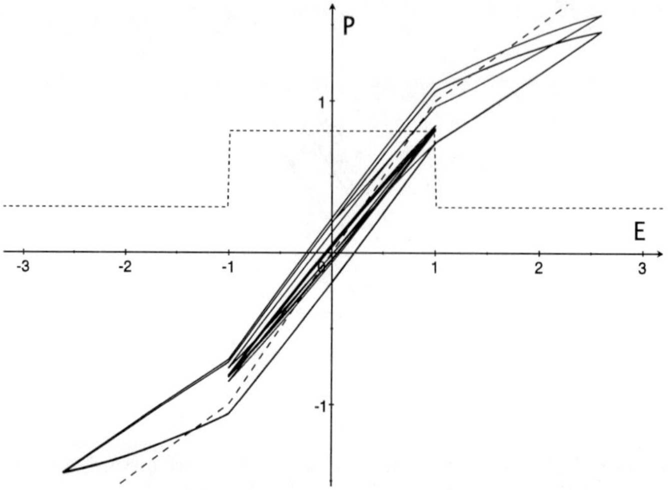

Fig. 15.2 Model III: hysteresis loops (solid), anhysteretic curve $\mathcal{P}$ (dashed) and graph of g (short dashed) with $\mathcal{E} = 1, 2.6$. The initial state E_0, P_0 has $E_0 = 0$ and $P_0 = -0.2, 0.2$

$$\frac{dP}{dE} = g(E) - \frac{\gamma_h}{\alpha}[P - \mathcal{P}(E)] \operatorname{sgn} \dot{E}. \tag{15.24}$$

This is a very simple model. Since $g = \mathcal{P}' - 1/\alpha$, it is characterized merely by the positive ratio γ_0/α and the anhysteretic function $\mathcal{P}$.

Hysteresis cycles are obtained by solving

$$\dot{P} = [\mathcal{P}'(E) - \frac{1}{\alpha}]\dot{E} - \frac{\gamma_h}{\alpha}[P - \mathcal{P}(E)]|\dot{E}|, \qquad \dot{E} = \omega \mathcal{E} \cos \omega t.$$

In Fig. 15.2 cyclic processes with large and small amplitudes are depicted assuming

$$\mathcal{P}(E) = \begin{cases} \frac{1}{2}(E - 1) & \text{if } x < -1, \\ E & \text{if } -1 \leq x \leq 1, \\ \frac{1}{2}(E + 1) & \text{if } x > 1, \end{cases}$$

and $\alpha = 5$, $\gamma_h = 5/2$.

According to Eq. (1.5) in [103], the condition

$$\int_E^\infty [f'(\zeta) - g(\zeta)] \exp(-\tau\zeta)\, d\zeta = \frac{1}{\alpha\tau} \exp(-\tau E) \geq 0, \qquad E \in [0, \infty)$$

is both necessary and sufficient for the validity of the assertion that the major loop has a counterclockwise orientation and is the boundary of the set of (attainable) states, i.e. the set of pairs (E, P) accessible, by appropriate variation of E, from the state

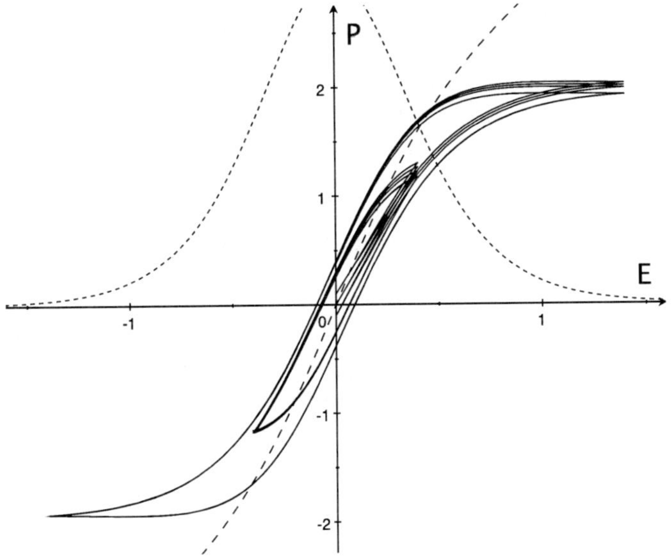

Fig. 15.3 Model IV: hysteresis loops (solid), anhysteretic curve $\mathcal{P}$ (dashed) and graph of g (short dashed) with $\mathcal{E} = 0.4, 1.4$. The initial state E_0, P_0 has $E_0 = 0$ and $P_0 = -0.1, 0, 0.1$

in which both E and P are zero. In addition, it follows from Eqs. (2.37) and (2.38) of [103] that the relation $f' \geq 0$ (hence, $g \geq 1/\alpha$) is necessary and sufficient for dP/dE to be non-negative.

Finally, we stress that the relation $f' - g \geq 0$ (cfr. [103], Eqs. (1.6) and (3.18)), which ensures that "the energy expended in a complete traversal of a primitive loop is never negative", is not enough to ensure thermodynamic compatibility. Indeed, previous arguments prove that a stronger condition is required in order to establish an energy potential Φ, namely $f' - g = \epsilon$ for some $\epsilon = 1/\alpha > 0$.

IV. $\gamma(E, P) = \gamma_0 g(E)[P - \mathcal{P}(E)]^2, \quad \gamma_0 > 0,$
where $g(E) = \mathcal{P}'(E) - 1/\alpha$.

Letting $\gamma_h(\theta) = \gamma_0 h(\theta_c - \theta)$, from (15.15) it follows that

$$\frac{dP}{dE} = g(E)[1 - \tau_h(P - \mathcal{P}(E)) \operatorname{sgn} \dot{E}], \quad \tau_h = \frac{\gamma_h}{\alpha} > 0.$$

The vanishing of g as $|E|$ approaches infinity is a way of modelling the saturation property of hysteresis. Hysteresis cycles are given by

$$\dot{P} = g(E)[\dot{E} - \tau_h(P - f(E))|\dot{E}|], \qquad \dot{E} = \omega\mathcal{E}\cos\omega t,$$

starting from different initial states (E_0, P_0). In Fig. 15.3 we use $\mathcal{P}(E) = 1.5(\tanh 2E + E)$ and $\alpha = 1.5$, $\tau_h = 1$. Then $g(E) = 3/\cosh^2 2E$.

In summary, Models I and II involve hysteretic functions γ which are proportional to $|P - \mathcal{P}(E)|$. In particular, Model I shows 'bilinear' hysteresis loops in connection with a linear anhysteretic function $\mathcal{P}$, whereas Model II accounts for the appearance of the saturation regime, too. In Model III and IV the hysteretic function γ is proportional to $[P - \mathcal{P}(E)]^2$. Model III shows hysteresis loops associated with a piecewise linear anhysteretic curve, whereas Model IV involves a nonlinear function $\mathcal{P}$ of the Langevin type. In addition, this model enjoys the saturation condition.

The essence of the present approach to the modelling of ferroelectric hysteresis is that the thermodynamic inequality makes the relation between $\dot{P}$ and $\dot{E}$ to consist of two parts. One is fully characterized by the free energy. The other one is related to the dissipative character of the hysteresis. Models compatible with thermodynamics can then be determined by appropriate selections of the free energy and of the dissipative part. While the anhysteretic function is fully determined by the free energy potential, the dissipative part of the model, characterized by χ_2, involves both the free energy influence, via the denominator $\partial_P \Phi$, and the non-negative purely-thermodynamic term ξ. For any free energy Φ then there are a family of dissipative parts characterized by the chosen function ξ.

Relation to Other Approaches

Among other schemes of ferroelectric materials, it is worth outlining an approach [404] based on the use of configurational forces [178, 179, 215]. The key assumption is that there is a micro-force system which consists of a vector stress $\boldsymbol{\xi}$ and scalar body forces π (internal) and γ (external). This system of forces is assumed to be subject to the micro-force balance

$$\int_{\partial \mathcal{P}_t} \boldsymbol{\xi} \cdot \mathbf{n}\, da + \int_{\mathcal{P}_t} (\pi + \gamma) dv = 0,$$

for each part $\mathcal{P}_t$ of the current configuration. Hence, for smooth functions $\boldsymbol{\xi}, \pi, \gamma$ it follows that

$$\nabla \cdot \boldsymbol{\xi} + \pi + \gamma = 0.$$

Moreover the power of these forces associated with the change of the order parameter, say ρ, is assumed to be [178, 179, 215]

$$\int_{\partial \mathcal{P}_t} \boldsymbol{\xi} \cdot \mathbf{n}\, \dot{\rho}\, da + \int_{\mathcal{P}_t} (\pi + \gamma) \dot{\rho}\, dv$$

and the second law is taken in the form of a dissipation inequality

$$\frac{d}{dt} \int_{\mathcal{P}_t} \psi_V\, dv \le \int_{\partial \mathcal{P}_t} \boldsymbol{\xi} \cdot \mathbf{n}\, \dot{\rho}\, da + \int_{\mathcal{P}_t} \gamma \dot{\rho}\, dv,$$

ψ_V being the free energy per unit volume.

Following a similar view, Su and Landis [404] consider a micro-force tensor $\boldsymbol{\xi}$ and micro-force vectors $\boldsymbol{\pi}$ (internal) and $\boldsymbol{\gamma}$ (external), subject to the equilibrium condition

$$\int_{\partial\mathcal{P}_t} \boldsymbol{\xi}\mathbf{n}\, da + \int_{\mathcal{P}_t} (\boldsymbol{\pi} + \boldsymbol{\gamma}) dv = 0, \qquad \nabla \cdot \boldsymbol{\xi} + \boldsymbol{\pi} + \boldsymbol{\gamma} = 0,$$

and assume that the corresponding power associated with the change of the polarization $\mathbf{P}$ is

$$\int_{\partial\mathcal{P}_t} (\boldsymbol{\xi}\mathbf{n}) \cdot \dot{\mathbf{P}}\, da + \int_{\mathcal{P}_t} (\boldsymbol{\pi} + \boldsymbol{\gamma}) \cdot \dot{\mathbf{P}}\, dv.$$

The second law is eventually taken to be expressed by the inequality

$$\dot{\psi}_V \leq \boldsymbol{\sigma} \cdot \dot{\boldsymbol{\varepsilon}} + \mathbf{E} \cdot \dot{\mathbf{D}} + \boldsymbol{\xi} \cdot \dot{\overline{\nabla \mathbf{P}}} - \boldsymbol{\pi} \cdot \dot{\mathbf{P}},$$

where $\boldsymbol{\sigma}$ is the Cauchy stress and, presumably, $\dot{\overline{\nabla \mathbf{P}}}$ is identified with $\nabla \dot{\mathbf{P}}$. Next upon the assumption that the independent variables are

$$\boldsymbol{\varepsilon}, \mathbf{D}, \mathbf{P}, \nabla \mathbf{P}, \dot{\mathbf{P}}$$

it is claimed that

$$\partial_{\dot{\mathbf{P}}} \psi_V = 0, \qquad \boldsymbol{\sigma} = \partial_{\boldsymbol{\varepsilon}} \psi_V, \qquad \mathbf{E} = \partial_{\mathbf{D}} \psi_V, \qquad \boldsymbol{\xi} = \partial_{\nabla \mathbf{P}} \psi_V$$

and

$$(\boldsymbol{\pi} + \partial_{\mathbf{P}} \psi_V) \cdot \dot{\mathbf{P}} \leq 0.$$

This inequality is satisfied by letting

$$\boldsymbol{\pi} = -\partial_{\mathbf{P}} \psi_V - \boldsymbol{\beta} \dot{\mathbf{P}},$$

where $\boldsymbol{\beta}$ is a positive definite tensor. Hence, in view of the equilibrium condition for micro-forces, it follows that

$$\nabla \cdot \partial_{\nabla \mathbf{P}} \psi_V - \partial_{\mathbf{P}} \psi_V + \boldsymbol{\gamma} = \boldsymbol{\beta} \dot{\mathbf{P}}.$$

15.2 Ginzburg–Landau–Devonshire Theory

Landau theory is a symmetry-based analysis of equilibrium near a phase transition [184]. Landau's symmetry-based theory was first applied to ferroelectricity by Ginzburg and next by Devonshire [86]. Here we just outline this Ginzburg–Landau–Devonshire (GLD) theory both in the isotropic case (that is appropriate for systems with spatially uniform polarization) and in connection with special anisotropic materials (transversely isotropic materials and materials whose lattice has cubic symmetry).

Look at $(12.17)_1$ and assume $\partial_{\nabla_R \Pi}\psi = 0$. Hence in stationary conditions

$$\mathbf{E} = \mathbf{F}^{-T}\,\partial_{\Pi}\psi.$$

Since

$$\mathbf{\Pi} = \frac{1}{\rho_R} J\mathbf{F}^{-1}\mathbf{P}, \qquad \partial_{\mathbf{P}}\psi = \partial_{\Pi}\psi\,\partial_{\mathbf{P}}\mathbf{\Pi} = \frac{1}{\rho}\mathbf{F}^{-T}\,\partial_{\Pi}\psi$$

then

$$\mathbf{E} = \rho\,\partial_{\mathbf{P}}\psi.$$

The effective electric field $\mathbf{E}^{\text{eff}}$ may be thought of as the sum of several contributions due to different phenomena. We assume that

$$\mathbf{E}^{\text{eff}} = \mathbf{E} + \mathbf{E}^{\text{int}} + \mathbf{E}^{\text{an}}, \tag{15.25}$$

where $\mathbf{E}$ is the external applied field, $\mathbf{E}^{\text{int}}$ is the electric field due to interaction of the polarization field with itself and $\mathbf{E}^{\text{an}}$ is due to anisotropic polarization (which strongly depends on the particle shape). By analogy with ferromagnetism we let

$$\mathbf{E}^{\text{int}} = -\mathsf{D}\,\pi\,, \qquad \pi = \mathbf{P}/\rho\,,$$

where D is a symmetric objective tensor. In absence of an external field, ferroelectric bodies tend to be polarized along precise directions which are called *easy directions*. This is mostly due to the depolarizing effect of spin-lattice coupling in crystals. Therefore, we take this phenomenon into account by adding the field

$$\mathbf{E}^{\text{an}} = -\mathsf{A}\,\pi\,,$$

where A is a symmetric, positive semi-definite and objective tensor depending on the shape of the material and on its electric anisotropy. Hence,

$$\mathbf{E}^{\text{eff}} = \mathbf{E} - \mathsf{L}\pi\,, \qquad \mathsf{L} = \mathsf{D} + \mathsf{A}; \tag{15.26}$$

the positive semi-definite tensor L may be referred to as *generalized depolarizing tensor.*

This representation of $\mathbf{E}^{\text{eff}}$ leads to the following explicit expression of the free energy density

$$\psi = \Psi_0(\mathbf{C},\theta) + \Psi_1(\mathbf{P},\theta),$$

where

$$\Psi_1(\mathbf{P},\theta) = \tfrac{1}{2}\mathbf{P}\cdot\mathsf{G}(\mathbf{P},\theta)\mathbf{P}, \tag{15.27}$$

and $\mathsf{G}(\mathbf{P}, \theta)$ is a objective symmetric tensor-valued function such that

$$\mathsf{G}(\mathbf{P}, \theta) + \tfrac{1}{2}\mathbf{P} \cdot \partial_{\mathbf{P}}\mathsf{G}(\mathbf{P}, \theta) = \mathsf{L}(\mathbf{P}, \theta).$$

The two main categories of ferroelectric materials are those undergoing a second-order transition, like triglycine sulphate, and those undergoing a first-order transition, like $BaTiO_3$ and other perovskites. Following are corresponding expressions of the free energy density.

15.2.1 Landau Free Energy

Second-order temperature-induced ferroelectric transitions are usually described by taking advantage of the fourth-order Landau–Ginzburg theory. If this is the case then D and A are assumed to be quadratic even functions of $\mathbf{P}$ and hence

$$\mathsf{D}(\mathbf{P}, \theta) = \big[\alpha\frac{\theta - \theta_c}{\theta_c} + \beta\mathbf{P}^2\big]\mathbf{1}, \tag{15.28}$$

where θ_c is the Curie temperature and α, β are positive constants.

Anisotropic properties of materials are characterized by their symmetry group. The largest symmetry group in three-dimensional space includes all orthogonal transformations, is denoted by $O(3)$ and is referred to as *isotropic*. In that case

$$\mathsf{A} = \mathbf{0}. \tag{15.29}$$

By virtue of (15.27) the free energy density reads

$$\psi = \Psi_0 + \frac{\alpha}{2}\frac{\theta - \theta_c}{\theta_c}\mathbf{P}^2 + \frac{\beta}{4}\mathbf{P}^4.$$

In stationary conditions we find that

$$\mathbf{E} = \rho\partial_{\mathbf{P}}\psi = \rho\left[\beta\,\mathbf{P}^2 - \alpha^*(\theta)\right]\mathbf{P}, \qquad \alpha^*(\theta) = \alpha\frac{\theta_c - \theta}{\theta_c}.$$

The polarization $\mathbf{P}$ is parallel or antiparallel to $\mathbf{E}$ depending on the sign of the quantity in square brackets. If $\theta > \theta_c$ then $\mathbf{P}$ is parallel to $\mathbf{E}$. Moreover, in the linear approximation

$$\mathbf{P} = \epsilon_0\chi\mathbf{E}, \qquad \chi = \frac{\theta_c}{\alpha\rho(\theta - \theta_c)},$$

the susceptibility χ being inversely proportional to $\theta - \theta_c$; in these conditions the material is said to be paraelectric. The inverse proportionality of χ on $\theta - \theta_c$ is said

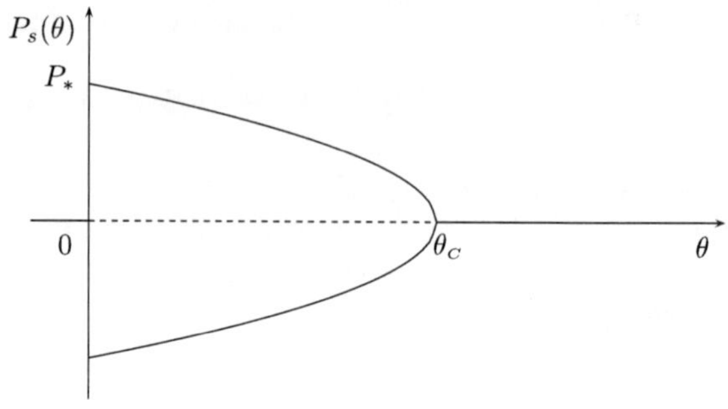

Fig. 15.4 Plot of the pitchfork bifurcation diagram with $\beta = \alpha = 1$

to represent the Curie–Weiss law. The critical temperature θ_c is the ferroelectric Curie temperature or the Curie point. For large values of $\mathbf{E}$, and hence of $\mathbf{P}$, so that $\alpha^* \ll \beta \mathbf{P}^2$, we have

$$\mathbf{P} \simeq (\beta \rho \mathbf{E}^2)^{-1/3}\mathbf{E}$$

Hence $|\mathbf{P}|$ increases slowly as $|\mathbf{E}|^{1/3}$. Yet, in the actual ferroelectrets $\mathbf{P}$ approaches a saturation value for large values of $|\mathbf{E}|$. This means that this simple scheme is inadequate to model the saturation.

If $\theta < \theta_c$ then the dependence of $\mathbf{P}$ on $\mathbf{E}$ is more involved. When the applied external field vanishes, $\mathbf{E} = \mathbf{0}$, the stationary condition reads $\left[\beta \mathbf{P}^2 - \alpha^*(\theta)\right]\mathbf{P} = \mathbf{0}$ so that we find $\mathbf{P} = \mathbf{0}$ and infinitely many pairs of nonzero spontaneous solutions,

$$\mathbf{P} = \pm P_s(\theta)\mathbf{e}, \qquad P_s(\theta) := \sqrt{\alpha^*(\theta)/\beta},$$

where $\mathbf{e}$ is a generic unit vector and $P_s(\theta)$ represents the *spontaneous polarization modulus* in the range $0 < \theta < \theta_C$. In addition, letting

$$P_* = P_s(0) = \sqrt{\alpha/\beta},$$

we conclude that $0 \leq P_s(\theta) < P_*$. The function P_s is decreasing on $(0, \theta_c)$. This suggests that we assume $\varphi = P/P_*$ as the *dimensionless order parameter* to describe second-order ferroelectric transitions so that the free energy takes the form

$$\psi = \Psi_0 + \frac{\beta P_*^4}{2}\left(\frac{\theta - \theta_C}{\theta_C}\varphi^2 + \frac{1}{2}\varphi^4\right).$$

In the (θ, P) plane the curve consisting of the two branches $P = \pm P_s(\theta)$ describes a super-critical pitchfork bifurcation (Fig. 15.4). This means that the spontaneous polarization varies continuously across the critical temperature.

We assume that, for a nonzero external field $\mathbf{E}$, $\mathbf{P}$ and $\mathbf{E}$ have a common direction; denote by P and E the pertinent components. Then we write

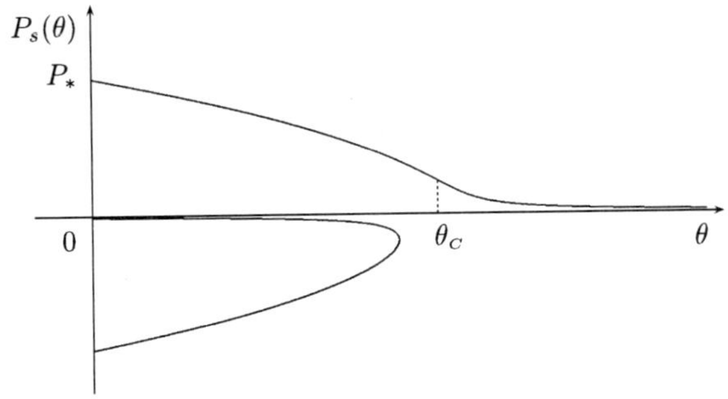

Fig. 15.5 The perturbed pitchfork bifurcation diagram with $\beta = \alpha = 1$ and $E = 0.05$

$$E = \rho\left[\beta P^2 - \alpha^*(\theta)\right] P, \qquad \alpha^*(\theta) = \alpha\frac{\theta_C - \theta}{\theta_C}.$$

For any given value of E, the corresponding curve in the (θ, P) plane consists of three branches and describes a perturbed super-critical pitchfork bifurcation (Fig. 15.5). The slope of the upper branch grows continuously from zero as the temperature drops below θ_c and this means that a second-order transition occurs.[4]

In particular, the differential susceptibility $\chi_d(P, \theta)$ varies according to the slope

$$\partial_E P = \frac{1}{\rho(3\beta P^2 - \alpha^*)}.$$

When $\theta > \theta_c$, $P_s(\theta)$ is negligible (Fig. 15.5) and we get the Curie–Weiss law

$$\chi_d(\theta) := \chi_d(P_s(\theta), \theta) \approx -\frac{1}{\rho\,\alpha^*(\theta)} = \frac{\theta_C}{\rho\,\alpha(\theta - \theta_c)} = \frac{C}{\theta - \theta_C}, \qquad C = \frac{\theta_C}{\rho\,\alpha}.$$

When $0 < \theta < \theta_c$, we have $P_s^2(\theta) = \alpha^*(\theta)/\beta$ so that

$$\chi_d(\theta) = \frac{1}{2\rho\,\alpha^*(\theta)} = \frac{C}{2(\theta_c - \theta)}.$$

The plot of $\chi_d(\theta)$ is given in Fig. 15.6.

In addition,

$$\chi_d \to \infty \quad \text{as} \quad P^2 \to \frac{\alpha(\theta_c - \theta)}{3\beta\theta_c} = \frac{1}{3}\,P_s^2(\theta).$$

[4] In a second-order phase transition the order parameter grows continuously from zero as the temperature drops below θ_C, [10], Chap. 2).

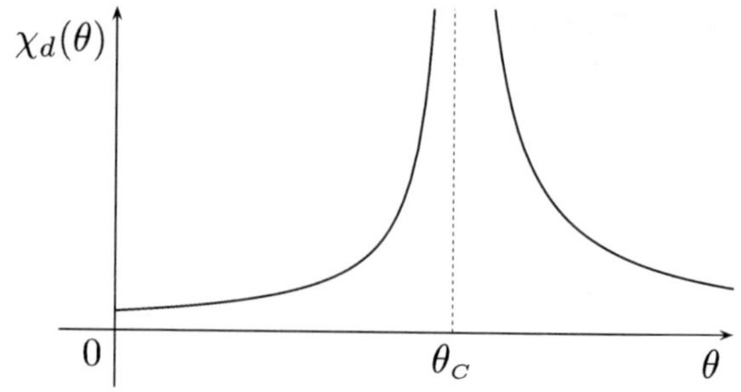

Fig. 15.6 The Landau differential susceptibility

Anisotropic Landau Free Energy

Though the isotropic Landau energy is widely applied and easy to compute, most ferroelectric materials are anisotropic. For instance, anisotropy in ferroelectrics is due to spin-lattice coupling in cubic crystals (as the perovskite system) and transversely isotropic symmetry in piezoelectric thin films. In the former case the group consisting of all proper rotations leaving invariant a cubic lattice is the rotational octahedral symmetry group. If, in addition, we allow reflections, we arrive at the so-called *octahedral symmetry group* which is denoted by O_h. Basically, when cubic anisotropy occurs then three privileged mutually orthogonal material directions exist. On the other case, a *transversely isotropic material* is characterized by symmetry with respect to one selected material direction referred to as principal direction. Properties of a transversely isotropic material remain unchanged by rotations about, and reflections from the planes orthogonal or parallel to, this direction. For the sake of definiteness, in the sequel we give the Landau free energy expressions related to octahedral and transversely isotropic symmetries.

Let $(\mathbf{i}_1, \mathbf{i}_2, \mathbf{i}_3)$ be a principal material direction basis and $\mathbf{S}_j = \mathbf{i}_j \otimes \mathbf{i}_j$, $j = 1, 2, 3$, be the related structural tensors so that $\mathbf{P} = \sum_{j=1}^{3} P_j \mathbf{i}_j$ and $\mathbf{S}_j \mathbf{P} = P_j \mathbf{i}_j$. After performing a suitable change of coordinates (in the frame locally at rest with the body), A can be expressed in diagonal form with respect to this basis. Then we assume

$$\mathsf{A} = \mathsf{A}_0 + \mathsf{A}_1 \mathsf{H}^2(\mathbf{P}),$$

where

$$\mathsf{A}_0 = \textstyle\sum_{j=1}^{3} \ell_j \mathbf{S}_j, \quad \mathsf{A}_1 = \sum_{j=1}^{3} \lambda_j \mathbf{S}_j, \quad \mathsf{H}(\mathbf{P}) = \sum_{j=1}^{3} P_j \mathbf{S}_j.$$

Coefficients ℓ_j and λ_j, $j = 1, 2, 3$, characterize the anisotropy of the body. Since $\mathbf{S}_j^n = \mathbf{S}_j$, $n \in \mathbb{N}$, and $\mathbf{S}_j \mathbf{S}_h = \mathbf{0}$ when $j \neq h$, it follows

$$\mathsf{H}^n(\mathbf{P}) = \textstyle\sum_{j=1}^{3} P_j^n \mathbf{S}_j$$

By (15.28) we have

Fig. 15.7 Plot of the fourth-order energy density $\Psi_1 = 1$ with transversely isotropic symmetry: $\ell, \lambda < 0$ (left) and $\ell, \lambda > 0$ (right)

$$\mathsf{L}(\mathbf{P}, \theta)\mathbf{P} = (\mathsf{D} + \mathsf{A})\mathbf{P} = \big(\alpha\frac{\theta - \theta_c}{\theta_c} + \beta P^2\big)\mathbf{P} + \textstyle\sum_{j=1}^3 P_j(\ell_j + \lambda_j P_j^2)\mathbf{i}_j.$$

Upon replacing this expression into $(15.27)_1$ we obtain

$$\Psi_1(\mathbf{P}, \theta) = \tfrac{1}{2}\big(\alpha\frac{\theta - \theta_c}{\theta_c} P^2 + \textstyle\sum_{j=1}^3 \ell_j P_j^2\big) + \tfrac{1}{4}\big(\beta P^4 + \textstyle\sum_{j=1}^3 \lambda_j P_j^4\big).$$

If the material is *transversely isotropic* (it has one easy direction, say $\mathbf{e}_3$) then $\ell_1 = \ell_2 = \ell$, $\ell_3 = 0$ and $\lambda_1 = \lambda_2 = \lambda$, $\lambda_3 = 0$. Hence,

$$\Psi_1 = \tfrac{1}{2}\big(\alpha\frac{\theta - \theta_c}{\theta_c} P^2 + \ell(P_1^2 + P_2^2)\big) + \tfrac{1}{4}\big(\beta P^4 + \lambda(P_1^4 + P_2^4)\big).$$

Provided that $\theta < \theta_c$, the energy density Ψ_1 takes nontrivial minima along either the easy direction, if $\ell, \lambda < 0$, or the easy plane, if $\ell, \lambda > 0$ (Fig. 15.7).

When the material lattice has *cubic symmetry*, then $\ell_1 = \ell_2 = \ell_3 = 0$ and $\lambda_1 = \lambda_2 = \lambda_3 = \lambda$. By introducing proper constants

$$\alpha_1 = -\frac{1}{2}\alpha^*(\theta) = \frac{\alpha}{2}\frac{\theta - \theta_c}{\theta_c}, \quad \alpha_{11} = \frac{\beta + \lambda}{4}, \quad \alpha_{12} = -\frac{\lambda}{2},$$

Equation (15.27) becomes the customary expression for the cubic system[5] O_3

$$\Psi_1 = \alpha_1 P^2 + \alpha_{11} P^4 + \alpha_{12}(P_1^2 P_2^2 + P_2^2 P_3^2 + P_1^2 P_3^2). \tag{15.30}$$

As for (15.30), we observe that α_{12} is negative, α_{11} is positive provided that $\lambda > -\beta$ and the sign of α_1 depends on θ. Then, below the Curie temperature, $\theta < \theta_c$, Ψ_1 takes nontrivial minima along the principal material directions provided that

[5] See, for instance, [78, 133] and in particular [77], Tables 1 and 2.

Fig. 15.8 Plot of the fourth-order energy density $\Psi_1 = 1$ with cubic symmetry: $\lambda > 0$ (left) and $\lambda < 0$ (right)

$\lambda > 0$, on the contrary it takes maxima along the same directions when $-\beta < \lambda < 0$ (see Fig. 15.8).

15.2.2 Landau–Devonshire Free Energy

According to the Landau-Ginzburg–Devonshire theory, to describe first-order phase-transitions in ferroelectric materials like $BaTiO_3$ and other perovskites a sixth-order approximation of the free energy density is required [133]. With this goal in mind, we assume that D and A are fourth-order even functions of $\mathbf{P}$, whose coefficients possibly depend on θ. In particular, we let

$$\mathsf{D}(\mathbf{P}, \theta) = \big[\alpha\frac{\theta - \theta_c}{\theta_c} + \beta\mathbf{P}^2 + \gamma\mathbf{P}^4\big]\mathbf{I}, \tag{15.31}$$

where α, γ are positive constants, whereas $\beta \in \mathbb{R}$.

If the material is isotropic then (15.29) holds. By virtue of (15.27) and (15.31) the free energy density reads

$$\Psi = \Psi_0 + \tfrac{1}{2}\alpha\frac{\theta - \theta_c}{\theta_c}\mathbf{P}^2 + \tfrac{1}{4}\beta\mathbf{P}^4 + \tfrac{1}{6}\gamma\mathbf{P}^6.$$

When the applied external field vanishes, $\mathbf{E} = \mathbf{0}$, from the stationary condition it follows

$$\mathbf{0} = \partial_{\mathbf{P}}\psi = \rho\big[\alpha\frac{\theta - \theta_c}{\theta_c} + \beta\mathbf{P}^2 + \gamma\mathbf{P}^4\big]\mathbf{P}.$$

Now, we separately discuss the cases $\beta \geq 0$ and $\beta < 0$.

• $\beta \geq 0$. It is easy to check that $\mathbf{P} = \mathbf{0}$ is the unique solution at equilibrium when $\theta \geq \theta_c$. Otherwise, letting $Y = \mathbf{P}^2$ we look for positive solutions of

$$\gamma Y^2 + \beta Y - \alpha^*(\theta) = 0, \qquad \alpha^*(\theta) = \alpha \frac{\theta_c - \theta}{\theta_c} > 0.$$

We obtain

$$Y_+(\theta) = \frac{-\beta + \sqrt{\beta^2 + 4\alpha^*(\theta)\gamma}}{2\gamma} > 0.$$

and then there exist infinitely many pairs of nonvanishing stationary solutions,

$$\mathbf{P} = \pm P_s(\theta)\mathbf{e}, \qquad P_s(\theta) := \sqrt{\frac{-\beta + \sqrt{\beta^2 + 4\alpha^*(\theta)\gamma}}{2\gamma}}$$

where $\mathbf{e}$ is a generic unit vector and $P_s(\theta)$ represents the *spontaneous polarization modulus* at $0 < \theta < \theta_c$. In addition, letting

$$P_* = P_s(0) = \sqrt{\frac{-\beta + \sqrt{\beta^2 + 4\alpha\gamma}}{2\gamma}},$$

we infer that $0 \leq P_s(\theta) < P_*$, P_s being a decreasing function on $(0, \theta_c]$. If this is the case, we conclude that at $\theta = \theta_c$ occurs a pitchfork bifurcation, as well as in the case of the fourth-order Landau free energy density. Namely, when $\beta \geq 0$ the properties of the model remain unchanged whether $\gamma > 0$ or $\gamma = 0$.

• $\beta < 0$. As in the previous case, $\mathbf{P} = \mathbf{0}$ is an equilibrium solution at any temperature. In addition, letting $Y = \mathbf{P}^2$ we look for positive solutions of

$$\gamma Y^2 - |\beta| Y - \alpha^*(\theta) = 0, \qquad \alpha^*(\theta) = \alpha \frac{\theta_c - \theta}{\theta_c}.$$

When $0 < \theta \leq \theta_c$ then $\alpha^*(\theta) \geq 0$ and we obtain only one positive value,

$$Y_+(\theta) = \frac{|\beta| + \sqrt{\beta^2 + 4\alpha^*(\theta)\gamma}}{2\gamma}.$$

Hence there exist infinitely many pairs of nonvanishing stationary solutions,

$$\mathbf{P} = \pm P_s(\theta)\mathbf{e}, \qquad P_s(\theta) := \sqrt{\frac{|\beta| + \sqrt{\beta^2 + 4\alpha^*(\theta)\gamma}}{2\gamma}}$$

where $\mathbf{e}$ is a generic unit vector and $P_s(\theta)$ represents the *spontaneous polarization modulus* at $\theta < \theta_c$. In particular we let

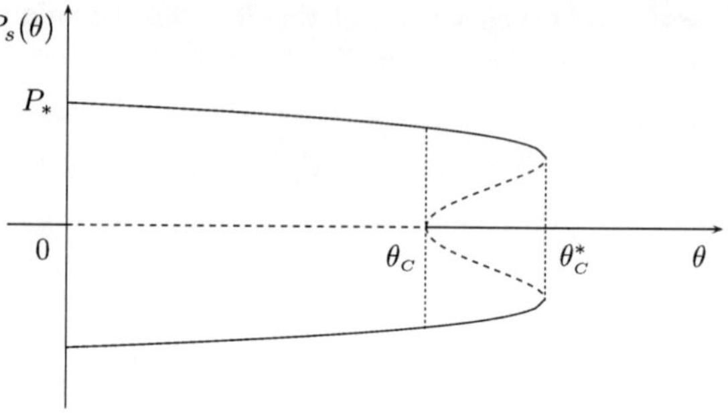

Fig. 15.9 Plot of the Devonshire bifurcation diagram with $\alpha = 1, \beta = -2, \gamma = 3$, and $\theta_C^*/\theta_C = 4/3$

$$P_* = P_s(0) = \sqrt{\frac{|\beta| + \sqrt{\beta^2 + 4\alpha\gamma}}{2\gamma}},$$

On the other hand, when $\theta > \theta_c$ then $\alpha^*(\theta) < 0$ and we get two positive values,

$$Y_+(\theta) = \frac{|\beta| + \sqrt{\beta^2 - 4|\alpha^*(\theta)|\gamma}}{2\gamma}, \qquad Y_-(\theta) = \frac{|\beta| - \sqrt{\beta^2 - 4|\alpha^*(\theta)|\gamma}}{2\gamma}.$$

After defining

$$\theta_c^* = \left(1 + \frac{\beta^2}{4\alpha\gamma}\right)\theta_c > \theta_c,$$

we can also write

$$Y_\pm(\theta) = \frac{|\beta|}{2\gamma} \pm \frac{\sqrt{\alpha(\theta_c^* - \theta)}}{\sqrt{\gamma\theta_c}}, \qquad P_s^\pm(\theta) := \sqrt{\frac{|\beta|}{2\gamma} \pm \frac{\sqrt{\alpha(\theta_c^* - \theta)}}{\sqrt{\gamma\theta_c}}}.$$

Hence, when $\theta_c \leq \theta \leq \theta_c^*$ there exist infinitely many quadruples of nonvanishing stationary solutions,

$$\mathbf{P} = \pm P_s^-(\theta)\mathbf{e}, \qquad \mathbf{P} = \pm P_s^+(\theta)\mathbf{e},$$

where $\mathbf{e}$ is a generic unit vector. We observe that the (stable) null solution suffers a jump bifurcation at $\theta = \theta_c$ as temperature decreases, whereas the stable branches collapse to 0 at $\theta = \theta_c^*$ as temperature increases. The corresponding diagram is given in Fig. 15.9.

We assume that, when the applied external field $\mathbf{E}$ does not vanish, $\mathbf{P}$ and $\mathbf{E}$ have a common direction and consider the pertinent components, P, E. Then we write

$$E = \rho\left[\gamma P^3 + \beta P^2 - \alpha^*(\theta)\right] P, \qquad \alpha^*(\theta) = \alpha\frac{\theta_c - \theta}{\theta_c}.$$

When $\beta < 0$ the corresponding curve in the (θ, P)-plane is given by Fig. 15.10.

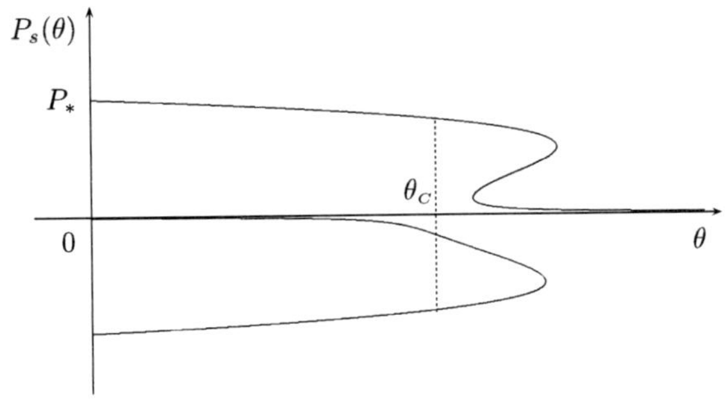

Fig. 15.10 The Devonshire perturbed bifurcation diagram with $\alpha = 1, \beta = -2, \gamma = 3$, and $E = 0.05$

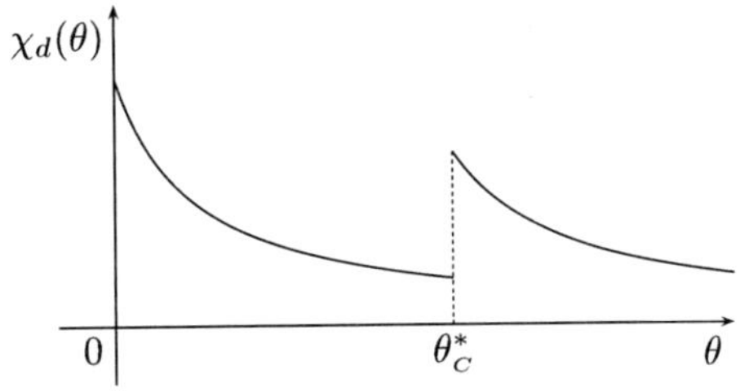

Fig. 15.11 The Devonshire differential susceptibility: $\theta_c^*/\theta_c = 4/3$

In particular, the differential susceptibility $\chi_d(P, \theta)$ varies according to the slope

$$\partial_{\mathcal{E}} P = \frac{1}{\rho[4\gamma P^4 + 3\beta P^2 - \alpha^*(\theta)]}.$$

When $\theta > \theta_c^*$, $P_s(\theta)$ is negligible (see Figs. 15.8 and 15.9) and we obtain

$$\chi_d(\theta) := \chi_d(P_s(\theta), \theta) \approx -\frac{1}{\rho\alpha^*(\theta)} = \frac{\theta_c}{\rho\alpha(\theta - \theta_c)} = \frac{C}{\theta - \theta_c}, \qquad C = \frac{\theta_c}{\rho\alpha}.$$

On the other hand, when $0 < \theta < \theta_c^*$, $P_s(\theta) \approx P_*$ and since $\beta < 0$ we have

$$\chi_d(\theta) \approx \frac{1}{\rho[4\gamma P_*^4 - 3|\beta| P_*^2 - \alpha^*(\theta)]} = \frac{C}{\theta + \theta_0},$$

where

$$\frac{\theta_0}{\theta_c} = 4\gamma P_*^4 - 3|\beta| P_*^2 - \alpha = \frac{\beta^2 + |\beta|\sqrt{\beta^2 + 4\alpha\gamma}}{2\gamma} + 3\alpha > 0.$$

The plot of $\chi_d(\theta)$ is given in Fig. 15.11. Because of the finite jump at the critical temperature, we conclude that a first-order phase transition occurs.

In addition,

$$\chi_d \to \infty \qquad \text{as} \qquad P^2 \to \frac{3|\beta|}{8\gamma} \pm \frac{\sqrt{\alpha(\theta_c^* - \theta)}}{2\sqrt{\gamma\theta_c}}.$$

The Anisotropic Case

Hereafter we construct $\mathbf{A}$ in order to account a material lattice with octahedral symmetries. We assume

$$\mathbf{A} = \mathbf{A}_0 + \mathbf{A}_1\mathbf{H}^2(\mathbf{P}) + \mathbf{A}_2 P^2\mathbf{H}^2(\mathbf{P}) + \mathbf{A}_3\mathbf{H}^4(\mathbf{P}),$$

where

$$\mathbf{A}_0 = \textstyle\sum_{j=1}^3 \ell_j \mathbf{S}_j, \quad \mathbf{A}_1 = \sum_{j=1}^3 \lambda_j \mathbf{S}_j, \quad \mathbf{A}_2 = \sum_{j=1}^3 \mu_j \mathbf{S}_j, \quad \mathbf{A}_3 = \sum_{j=1}^3 \nu_j \mathbf{S}_j.$$

The constants $\ell_j, \lambda_j, \mu_j, j = 1, 2, 3$, characterize the anisotropy of the body. Since $\mathbf{L} = \mathbf{D} + \mathbf{A}$, from (15.31)

$$\mathbf{L}(\mathbf{P}, \theta)\mathbf{P} = \big(\alpha\frac{\theta - \theta_C}{\theta_C} + \beta P^2 + \gamma P^4\big)\mathbf{P} + \textstyle\sum_{j=1}^3 P_j(\ell_j + \lambda_j P_j^2 + \mu_j P^2 P_j^2 + \nu_j P_j^4)\mathbf{i}_j.$$

Upon substitution of this expression into $(15.27)_1$ we have

$$\begin{aligned}\Psi_1(\mathbf{P}, \theta) &= \tfrac{1}{2}\big(\alpha\frac{\theta - \theta_c}{\theta_c}P^2 + \textstyle\sum_{j=1}^3 \ell_j P_j^2\big)\\ &\quad + \tfrac{1}{4}\big(\beta P^4 + \textstyle\sum_{j=1}^3 \lambda_j P_j^4\big) + \tfrac{1}{6}\big(\gamma P^6 + \textstyle\sum_{j=1}^3 \big[\mu_j P^2 P_j^4 + \nu_j P_j^6\big]\big).\end{aligned}$$

When the material lattice has cubic symmetry we have $\lambda_j = \lambda$, $\mu_j = \mu$, $\nu_j = \nu$ and $\ell_j = 0$, $j = 1, 2, 3$. In terms of the constants

$$\begin{aligned}&\alpha_1 = \tfrac{1}{2}\alpha\frac{\theta - \theta_c}{\theta_c}, \quad \alpha_{11} = \tfrac{1}{4}(\beta + \lambda), \quad \alpha_{12} = -\tfrac{1}{2}\lambda,\\ &\alpha_{111} = \tfrac{1}{6}(\gamma + \mu + \nu), \quad \alpha_{112} = -\tfrac{1}{6}(2\mu + 3\nu), \quad \alpha_{123} = -(\mu + \nu),\end{aligned}$$

we then obtain (see, e.g. [77], Table 1)

$$\begin{aligned}\Psi_1 &= \alpha_1 P^2 + \alpha_{11} P^4 + \alpha_{12}(P_1^2 P_2^2 + P_2^2 P_3^2 + P_1^2 P_3^2) + \alpha_{111} P^6\\ &\quad + \alpha_{112}[P_1^4(P_2^2 + P_3^2) + P_2^4(P_1^2 + P_3^2) + P_3^4(P_1^2 + P_2^2)] + \alpha_{123} P_1^2 P_2^2 P_3^2.\end{aligned}$$

We observe that α_{111} is positive provided that $\mu + \nu > -\gamma$, whereas the sign of α_{112} and α_{123} only depend on the values of μ and ν (see Fig. 15.12).

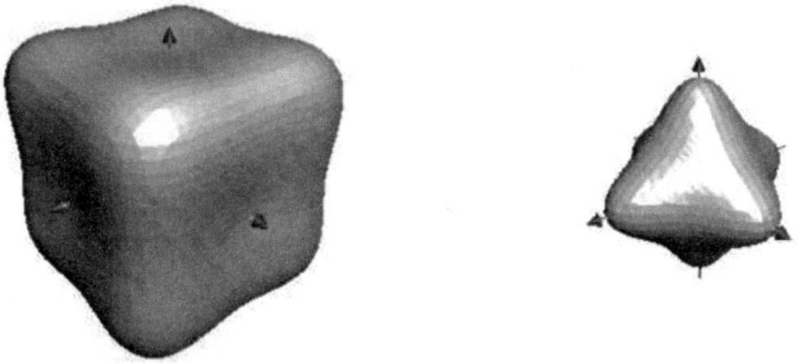

Fig. 15.12 Plot of the sixth-order energy density $\Psi_1 = 1$ with cubic symmetry: $\lambda > 0, \beta + \lambda < 0$ and $2\mu + 3\nu > 0$ (left) and $2\mu + 3\nu < 0$ (right)

15.3 Ferromagnetism

The characteristic property of *ferromagnetism* is that, below a certain temperature called the Curie temperature, θ_c, the material can possess a spontaneous magnetization in the absence of an applied magnetic field. Upon application of a weak magnetic field, the magnetization increases rapidly to a high value called the saturation magnetization, which is in general a function of temperature. Physically this behaviour is viewed as the consequence of the alignment of atomic magnetic moments in response to the applied magnetic field. That is why various approaches to ferromagnetism are based on a micromagnetic picture of the points of the body.

The combined effect of the atomic magnetic moments can give rise to a relatively large magnetization for a given applied field. Above the Curie temperature, a ferromagnetic substance behaves as paramagnetic in that its differential susceptibility χ is roughly constant and approaches the Curie–Weiss law (Sect. 15.3.3), namely χ is inversely proportional to $\theta - \theta_c$.

Some materials having atoms with unequal moments exhibit a special form of ferromagnetism below the Curie temperature called *ferrimagnetism.*

Upon the assumption that **M** and **H** have a common fixed alignment, the evolution equation was first set in the form[6]

$$\dot{M} = \alpha_0\mu_0 H - \alpha_1(\theta - \theta_c)M + \alpha_2 M^3 + \alpha_3\Delta M,$$

where $\alpha_0, \alpha_1, \alpha_2$ are positive parameters. This non-isothermal model accounts for the transition, at the Curie temperature θ_c, between the paramagnetic and the ferromagnetic behaviours. Yet the corresponding magnetization curve

[6] See, e.g. [274] p. 39, [260].

$$H = [a(\theta - \theta_c) + bM^2]M,$$

where $a = \alpha_1/\alpha_0\mu_0, b = \alpha_2/\alpha_0\mu_0$, shows that sgn $H = \pm$sgn M depending on the sign of $a(\theta - \theta_c) + bM^2$. If $\theta > \theta_c$ then **H** and **M** are parallel. Moreover, if M is small, so that $bM^2 \ll a(\theta - \theta_c)$, then the magnetic susceptibility χ_M is roughly constant,

$$M \simeq \chi_M H, \qquad \chi_M = \frac{1/a}{\theta - \theta_c},$$

consistent with the ferromagnetic Curie–Weiss law. If, instead, $\theta < \theta_c$ then $H = 0$ at $M = 0$ and

$$M^2 = \frac{a}{b}(\theta_c - \theta).$$

As $M^2 < (a/b)(\theta_c - \theta)$ we have sgn $M = -$sgn H. The value $[(a/b)(\theta_c - \theta)]^{1/2}$ is the spontaneous magnetization. As with ferroelectricity, this curve does not allow for saturation.

In 1907 Weiss developed a theory of ferromagnetic domains structure known as mean field theory. In essence he arranged the Langevin potential in order to describe the non-isothermal paramagnetic-ferromagnetic transition and to account for the saturation phenomena [256]. Weiss assumed that in ferromagnetic materials each dipole (elementary magnet) is subject not only to the applied field H but also to a field H_{mol}, called molecular field by Weiss, such that $H_{mol} = N_w M$, N_w being a material constant. Thus the effective magnetic field would be

$$H_e = H + N_w M.$$

The classical relation between M and H, obtained by statistical argument, is

$$M = M_{sat}\mathcal{L}(mH/k\theta),$$

where m is the absolute value of the magnetic moment, k is the Boltzmann constant and $\mathcal{L}$ is the Langevin function,

$$\mathcal{L}(x) = \coth x - \frac{1}{x}, \qquad x \in \mathbb{R}.$$

By the Weiss assumption the relation becomes

$$M = M_{sat}\mathcal{L}[m(H + N_w M)/k\theta].$$

The Langevin function is increasing and odd and moreover $\mathcal{L}(x) \to \pm 1$ as $x \to \pm\infty$. As a consequence the inverse $\mathcal{L}^{-1}$ maps $(-1, 1)$ into $\mathbb{R}$. By the Weiss assumption we can then write

$$\frac{H + \beta M}{\mu \theta} = \mathcal{L}^{-1}(M/M_{sat}),$$

where M_{sat} is the saturation value, so that $M \in (-M_{sat}, M_{sat})$, $\beta = N_w$, and $\mu = k/m$, k being the Boltzmann constant. Hence H is given by

$$H = \mu\theta\mathcal{L}^{-1}(M/M_{sat}) - \beta M. \tag{15.32}$$

For formal convenience we let $\xi = M/M_{sat}$ and then

$$H = \mathcal{H}(\xi), \qquad \mathcal{H}(\xi) := \mu\theta\mathcal{L}^{-1}(\xi) - \beta M_{sat}\xi, \qquad \xi \in (-1, 1).$$

If $\theta \to 0_+$ then

$$H = -\beta M,$$

which is the classical law of diamagnetism or paramagnetism according as $\beta > 0$ or $\beta < 0$.

Experimental evidence shows that there is a transition temperature θ_c for paramagnetic-ferromagnetic behaviour such that

$$\mathcal{H}'(0) > 0 \ \text{ as } \ \theta > \theta_c, \qquad \mathcal{H}'(0) \in (-\beta M_{sat}, 0) \ \text{ as } \ \theta < \theta_c.$$

Consistently, it is natural to let

$$\mu\theta_c(\mathcal{L}^{-1})'(0) - \beta M_{sat} = 0. \tag{15.33}$$

This requirement induces a relation among β, μ, and θ_c. To this end let

$$z = \frac{H + \beta M}{\mu\theta}$$

and observe

$$(\mathcal{L}^{-1})'(\xi) = \frac{1}{\mathcal{L}'(z(\xi))}.$$

Since $\mathcal{L}^{-1}(0) = 0$ then $z(0) = \mathcal{L}^{-1}(0) = 0$. Moreover $\mathcal{L}'(0) = 1/3$ and then

$$(\mathcal{L}^{-1})'(0) = \frac{1}{\mathcal{L}'(0)} = 3.$$

Hence Eq. (15.33) yields

$$\mu = \frac{\beta M_{sat}}{3\theta_c},$$

which relates the parameter μ to the transition temperature θ_c. Upon substitution in (15.32) we can write

$$\frac{1}{\beta M_{sat}} H = \frac{u+1}{3} \mathcal{L}^{-1}(\xi) - \xi, \qquad u := \frac{\theta - \theta_c}{\theta_c}, \qquad \xi = \frac{M}{M_{sat}}.$$

15.3.1 Thermodynamic Approach to Magnetic Hysteresis

The interaction between electromagnetic fields and thermal and electric conduction is developed in Sect. 12.6; here we restrict attention to non-conducting ($\mathbf{J} = \mathbf{0}, \mathbf{q} = \mathbf{0}$) and non-polarizable ($\mathbf{P} = \mathbf{0}$) solids. Ferromagnetic hysteretic effects are described by letting the pertinent properties depend on $\mathbf{M}, \mathbf{H}, \dot{\mathbf{H}}, \dot{\mathbf{M}}$. In a three-dimensional setting of deformable bodies we need to account for the time evolution in a objective way, i.e. via objective derivatives. We assume that $\mathbf{H}$ and $\mathbf{M}$ are vectors, in $SO(3)$, and hence say that

$$\mathfrak{H} = \mathbf{F}^T \mathbf{H}, \qquad \mathfrak{M} = \mathbf{F}^{-1} \mathbf{M}$$

are invariant vectors.[7] Denote by $\mathbf{m}$ the magnetic moment per unit mass and hence $\mathbf{M} = \rho \mathbf{m}$. Consider the balance of energy (2.75), whence

$$\rho \dot{\varepsilon} = \mu_0 \rho \mathbf{H} \cdot \dot{\mathbf{m}} + \mathbf{T} \cdot \mathbf{L} + \rho r,$$

and the entropy inequality

$$\rho \dot{\eta} - \frac{\rho r}{\theta} = \sigma \geq 0.$$

Letting

$$\phi = \varepsilon - \theta \eta - \mu_0 \mathbf{H} \cdot \mathbf{m}$$

we can write the entropy inequality in the form

$$-\rho(\dot{\phi} + \eta \dot{\theta}) - \mu_0 \mathbf{M} \cdot \dot{\mathbf{H}} + \mathbf{T} \cdot \mathbf{L} = \theta \sigma \geq 0. \tag{15.34}$$

Observe that

$$\mathfrak{M} \cdot \dot{\mathfrak{H}} = (\mathbf{F}^{-1}\mathbf{M}) \cdot (\mathbf{F}^T \mathbf{L}^T \mathbf{H} + \mathbf{F}^T \dot{\mathbf{H}}) = (\mathbf{H} \otimes \mathbf{M}) \cdot \mathbf{L} + \mathbf{M} \cdot \dot{\mathbf{H}}.$$

Substitution of $\mathbf{M} \cdot \dot{\mathbf{H}} = \mathfrak{M} \cdot \dot{\mathfrak{H}} - (\mathbf{H} \otimes \mathbf{M}) \cdot \mathbf{L}$ in (15.34) leads to

$$-\rho(\dot{\phi} + \eta \dot{\theta}) - \mu_0 \mathfrak{M} \cdot \dot{\mathfrak{H}} + (\mathbf{T} + \mu_0 \mathbf{H} \otimes \mathbf{M}) \cdot \mathbf{L} = \theta \sigma. \tag{15.35}$$

The constitutive properties are based on $\theta, \mathbf{F}, \mathbf{H}, \mathbf{M}, \dot{\mathbf{H}}, \dot{\mathbf{M}}$ as independent variables; $\mathbf{H}$ and $\mathbf{M}$ are fields relative to the frame locally at rest with the body. Since ϕ, η, σ are required to be Euclidean invariants then we assume that ϕ is a (differentiable) function of

[7] We recall that $\mathbf{F}^T \mathbf{H} = \mathbf{H}_R$ is the reference (material) intensity.

$$\Gamma = (\theta, \mathbf{C}, \mathfrak{H}, \mathfrak{M}, \mathfrak{D}, \dot{\mathfrak{H}}, \dot{\mathfrak{M}})$$

where the stretching tensor $\mathfrak{D}$ has to occur via invariant terms. We compute $\dot{\phi}$ and substitute in (15.35) to find

$$-\rho(\partial_\theta \phi + \eta)\dot{\theta} - \rho\partial_{\mathbf{C}}\phi \cdot \dot{\mathbf{C}} - \rho\partial_{\dot{\mathfrak{H}}}\phi \cdot \ddot{\mathfrak{H}} - \rho\partial_{\dot{\mathfrak{M}}}\phi \cdot \ddot{\mathfrak{M}} - \rho\partial_{\mathfrak{D}}\phi \cdot \dot{\mathfrak{D}}$$
$$-(\rho\partial_{\mathfrak{H}}\phi + \mu_0\mathfrak{M}) \cdot \dot{\mathfrak{H}} - \rho\partial_{\mathfrak{M}}\phi \cdot \dot{\mathfrak{M}} + (\mathbf{T} + \mu_0 \mathbf{H} \otimes \mathbf{M}) \cdot \mathbf{L} = \theta\sigma \geq 0.$$

The quantities $\ddot{\mathfrak{H}}, \ddot{\mathfrak{M}}, \dot{\mathfrak{D}}, \dot{\theta}$ occur linearly. Now $\ddot{\mathfrak{H}} = \ddot{\mathbf{F}}^T\mathbf{H} + 2\dot{\mathbf{F}}^T\dot{\mathbf{H}} + \mathbf{F}^T\ddot{\mathbf{H}}$ and the like for $\ddot{\mathfrak{M}}$. Maxwell's equations involve $\partial_t^2\mathbf{H}$ and $\partial_t^2\mathbf{M}$ in $\nabla \times \partial_t\mathbf{E} = -\mu_0(\partial_t^2\mathbf{H} + \partial_t^2\mathbf{M})$. Since $\partial_t\mathbf{E}$ can be taken arbitrarily then we may ascribe arbitrary values to $\partial_t^2\mathbf{H}$ and $\partial_t^2\mathbf{M}$; the corresponding value of $\nabla \times \partial_t\mathbf{E}$ satisfies Maxwell's equation and meanwhile does not affect the other terms in the inequality. Since the arbitrariness holds also for $\dot{\mathfrak{D}}$ and $\dot{\theta}$ we conclude that

$$\partial_{\dot{\mathfrak{H}}}\phi = \mathbf{0}, \qquad \partial_{\dot{\mathfrak{M}}}\phi = \mathbf{0}, \qquad \partial_{\mathfrak{D}}\phi = \mathbf{0}, \qquad \eta = -\partial_\theta\phi.$$

Since $\dot{\mathbf{C}} = 2\mathbf{F}^T\mathfrak{D}\mathbf{F}$ then the remaining inequality can be written in the form

$$[\mathbf{T} - 2\rho\mathbf{F}\partial_{\mathbf{C}}\phi\mathbf{F}^T + \mu_0\mathbf{H} \otimes \mathbf{M}] \cdot \mathbf{L} - (\rho\partial_{\mathfrak{H}}\phi + \mu_0\mathfrak{M}) \cdot \dot{\mathfrak{H}} - \rho\partial_{\mathfrak{M}}\phi \cdot \dot{\mathfrak{M}} = \theta\sigma \geq 0. \tag{15.36}$$

The arbitrariness of $\mathbf{W} = \text{skw}\mathbf{L}$ implies

$$\mathbf{T} + \mu_0\mathbf{H} \otimes \mathbf{M} \in \text{Sym}, \qquad \text{skw}\mathbf{T} = \text{skw}[\mu_0\mathbf{M} \otimes \mathbf{H}],$$

which coincides with the classical condition arising from the balance of angular momentum.

Since ϕ is independent of $\mathfrak{D}$ then we set $\mathfrak{D} = \mathbf{0}$ in (15.36) to obtain

$$0 \leq \theta\sigma_{MH} = -(\rho\partial_{\mathfrak{H}}\phi + \mu_0\mathfrak{M}) \cdot \dot{\mathfrak{H}} - \rho\partial_{\mathfrak{M}}\phi \cdot \dot{\mathfrak{M}}, \tag{15.37}$$

where σ_{MH} is the value of σ for any Γ with $\mathfrak{D} = \mathbf{0}$. Consequently

$$[\mathbf{T} - 2\rho\mathbf{F}\partial_{\mathbf{C}}\phi\mathbf{F}^T + \mu_0\mathbf{H} \otimes \mathbf{M}] \cdot \mathfrak{D} = \theta\sigma_T, \qquad \sigma_T := \sigma - \sigma_{MH}.$$

Let

$$\hat{\mathbf{T}}(\theta, \mathbf{F}, \mathbf{H}, \mathbf{M}) := 2\rho\mathbf{F}\partial_{\mathbf{C}}\phi\mathbf{F}^T - \mu_0\mathbf{H} \otimes \mathbf{M}, \qquad \mathcal{T} := \mathbf{T} - \hat{\mathbf{T}}$$

and then $\mathcal{T}(\theta, \mathbf{F}, \mathbf{H}, \mathbf{M}, \mathfrak{D})$ is subject to

$$\mathcal{T} \cdot \mathfrak{D} = \theta\sigma_T.$$

Mathematically we might assume $\mathcal{T} \cdot \mathfrak{D} \geq -\theta\sigma_{MH}$; in that case inequality (15.36) holds with negative values of $\mathcal{T} \cdot \mathfrak{D}$. More naturally, we assume

$$\mathcal{T} \cdot \mathfrak{D} = \theta\sigma_T \geq 0;$$

the classical viscous terms are examples of $\sigma_T > 0$.

We now investigate inequality (15.37). If $\mathfrak{M}$ is not an independent variable then $\partial_{\mathfrak{M}}\phi = \mathbf{0}$ and inequality (15.37) simplifies to

$$0 \leq \theta\sigma_{MH} = -(\rho\partial_{\mathfrak{H}}\phi + \mu_0\mathfrak{M}) \cdot \dot{\mathfrak{H}},$$

where σ_{MH} is a non-negative function of Γ. The magnetization $\mathfrak{M}$ is then given by a constitutive function. First let $\mathfrak{M}$ be independent of $\dot{\mathfrak{H}}$. The right-hand side is then linear in $\dot{\mathfrak{H}}$ and hence can take positive and negative values if $\rho_R\partial_{\mathfrak{H}}\phi + \mu_0\mathfrak{M} \neq \mathbf{0}$. Consequently the inequality holds if and only if

$$\sigma_{MH} = 0, \qquad \mu_0\mathfrak{M} = -\rho\partial_{\mathfrak{H}}\phi. \tag{15.38}$$

Since $\phi(\theta, \mathbf{C}, \mathfrak{H}) = \phi(\theta, \mathbf{C}, \mathbf{F}^T\mathbf{H})$ then

$$\partial_{\mathbf{H}}\phi = \mathbf{F}\partial_{\mathfrak{H}}\phi, \qquad \partial_{\mathfrak{H}}\phi = \mathbf{F}^{-1}\partial_{\mathbf{H}}\phi.$$

Replacing $\mathfrak{M} = \mathbf{F}^{-1}\mathbf{M}$ in (15.38) we obtain the expected (spatial) relation

$$\mu_0\mathbf{M} = -\rho\partial_{\mathbf{H}}\phi.$$

If instead $\mathfrak{M}$ is allowed to depend on $\dot{\mathfrak{H}}$ then letting $\mathcal{M} = \rho\partial_{\mathfrak{H}}\phi + \mu_0\mathfrak{M}$ we have

$$\mathcal{M} \cdot \dot{\mathfrak{H}} = \theta\sigma_{MH} \geq 0, \qquad \mu_0\mathfrak{M} = -\rho\partial_{\mathfrak{H}}\phi + \mathcal{M}.$$

We now let both $\mathfrak{M}$ and $\mathfrak{H}$ be independent variables with $\partial_{\mathfrak{M}}\phi \neq \mathbf{0}$, $\partial_{\mathfrak{H}}\phi \neq \mathbf{0}$. Equation (15.37) can be used to represent $\dot{\mathfrak{M}}$ in terms of the inner product $\rho\partial_{\mathfrak{M}}\phi \cdot \dot{\mathfrak{M}} = -\theta\sigma_{MH} - (\rho\partial_{\mathfrak{H}}\phi + \mu_0\mathfrak{M}) \cdot \dot{\mathfrak{H}}$. By means of (A.7), letting $\mathbf{e} = \partial_{\mathfrak{M}}\phi/|\partial_{\mathfrak{M}}\phi|$ we have

$$\dot{\mathfrak{M}} = -[\theta\sigma_{MH} + (\rho\partial_{\mathfrak{H}}\phi + \mu_0\mathfrak{M}) \cdot \dot{\mathfrak{H}}]\frac{\partial_{\mathfrak{M}}\phi}{\rho|\partial_{\mathfrak{M}}\phi|^2} + (\mathbf{1} - \frac{\partial_{\mathfrak{M}}\phi}{|\partial_{\mathfrak{M}}\phi|} \otimes \frac{\partial_{\mathfrak{M}}\phi}{|\partial_{\mathfrak{M}}\phi|})\mathbf{w},$$

$\mathbf{w}$ being any vector dependent on $\theta, \mathbf{F}, \mathbf{H}, \mathbf{M}, \mathfrak{D}, \dot{\mathfrak{H}}$.

If $\sigma_{MH} = 0$ then Eq. (15.37) reads

$$(\rho\partial_{\mathfrak{H}}\phi + \mu_0\mathfrak{M}) \cdot \dot{\mathfrak{H}} + \rho\partial_{\mathfrak{M}}\phi \cdot \dot{\mathfrak{M}} = 0.$$

The inner product $\rho\partial_{\mathfrak{M}}\phi \cdot \dot{\mathfrak{M}}$ is required to equal $-(\rho\partial_{\mathfrak{H}}\phi + \mu_0\mathfrak{M}) \cdot \dot{\mathfrak{H}}$. By the representation formula (A.7), letting $\mathbf{u} = \dot{\mathfrak{M}}$ and $\mathbf{e} = \partial_{\mathfrak{M}}\phi/|\partial_{\mathfrak{M}}\phi|$ and observing that

$$\dot{\mathfrak{M}} \cdot \mathbf{e} = -\frac{(\rho\partial_{\mathfrak{H}}\phi + \mu_0\mathfrak{M}) \cdot \dot{\mathfrak{H}}}{|\rho\partial_{\mathfrak{M}}\phi|}$$

we can represent $\dot{\mathfrak{M}}$ in the form

$$\dot{\mathfrak{M}} = -\frac{(\rho\partial_{\mathfrak{H}}\phi + \mu_0\mathfrak{M}) \cdot \dot{\mathfrak{H}}}{|\rho\partial_{\mathfrak{M}}\phi|^2}\partial_{\mathfrak{M}}\phi + (\mathbf{1} - \frac{1}{|\partial_{\mathfrak{M}}\phi|^2}\partial_{\mathfrak{M}}\phi \otimes \partial_{\mathfrak{M}}\phi)\mathbf{w},$$

where $\mathbf{w}$ is any vector function of $(\theta, \mathbf{F}, \mathbf{H}, \mathbf{M}, \mathfrak{D}, \dot{\mathfrak{H}})$. If $\mathbf{w} = \mathbf{G}\dot{\mathfrak{H}}$ then we can write

$$\dot{\mathfrak{M}} = -\frac{(\rho\partial_{\mathfrak{H}}\phi + \mu_0\mathfrak{M}) \cdot \dot{\mathfrak{H}}}{|\rho\partial_{\mathfrak{M}}\phi|^2}\partial_{\mathfrak{M}}\phi + (\mathbf{1} - \frac{1}{|\partial_{\mathfrak{M}}\phi|^2}\partial_{\mathfrak{M}}\phi \otimes \mathbf{G}^T\partial_{\mathfrak{M}}\phi)\dot{\mathfrak{H}}. \quad (15.39)$$

Subject to the assumption $\mathbf{G}$ is a function of θ, $\mathbf{F}$, $\mathbf{H}$, $\mathbf{M}$, Eq. (15.39) is quite a general rate-type relation for $\dot{\mathfrak{M}}$.

As a further scheme we may let $\mathfrak{M}, \dot{\mathfrak{M}}$ be among the independent variables and moreover let (15.37) hold with $\sigma_{MH} \neq 0$. For definiteness we assume

$$\theta\sigma_{MH} = \nu_H|\dot{\mathfrak{H}}|, \qquad \nu_H = \nu_H(\theta, \mathbf{C}, \mathfrak{H}, \mathfrak{M}) \geq 0.$$

We now show how this scheme is appropriate to the modelling of hysteresis.

15.3.2 One-Dimensional Models of Ferromagnetic Hysteresis

We now let $\mathbf{F}$ be constant and hence ρ is constant too. Since $\dot{\mathfrak{H}} = \mathbf{F}^T\dot{\mathbf{H}}$, $\dot{\mathfrak{M}} = \mathbf{F}^{-1}\dot{\mathbf{M}}$ and $\partial_{\mathfrak{H}}\phi = \mathbf{F}^{-1}\partial_{\mathbf{H}}\phi$, $\partial_{\mathfrak{M}}\phi = \mathbf{F}^T\partial_{\mathbf{M}}\phi$ then it follows from (15.37) that

$$(\rho\partial_{\mathbf{H}}\phi + \mu_0\mathbf{M}) \cdot \dot{\mathbf{H}} + \partial_{\mathbf{M}}\phi \cdot \dot{\mathbf{M}} = -\nu_H|\mathbf{F}^T\dot{\mathbf{H}}|. \quad (15.40)$$

Moreover assume $\mathbf{H}, \mathbf{M}$ are collinear and directed along a principal direction of $\mathbf{F}$. Thus we denote by H, M the components of $\mathbf{H}, \mathbf{M}$ in their common direction. Equation (15.37) then simplifies to

$$(\partial_H\phi + \mu M)\dot{H} + \partial_M\phi\dot{M} = -\nu|\dot{H}|, \quad (15.41)$$

where $\mu = \mu_0/\rho$ and $\nu = \nu_H|F|/\rho$ is a positive-valued function of θ, F, H, M. The temperature θ and the deformation gradient F are considered as known parameters. The approach developed here parallels the one in Sect. 15.1.1 for ferroelectric hysteresis.

Integration of (15.41) along any closed curve in the $H - M$ plane, as $t \in [t_1, t_2]$, results in

$$0 \geq \int_{t_1}^{t_2}[(\partial_H\phi + \mu M)\dot{H} + \partial_M\phi\dot{M}]dt = \int_{t_1}^{t_2}[\mu M\,\dot{H} + \dot{\phi}]dt = \mu\int_{t_1}^{t_2} M\,\dot{H}dt;$$

this implies that the closed curve is run in the counterclockwise sense.

Except at inversion points (where $\dot{H} = 0$), by (15.41) we have

$$\frac{\dot{M}}{\dot{H}} = -\frac{\partial_H \phi + \mu M}{\partial_M \phi} - \frac{\nu}{\partial_M \phi} \operatorname{sgn} \dot{H}. \tag{15.42}$$

If ν depends on $\dot{H}$ at most through sgn $\dot{H}$ then the right-hand side of (15.42) is piecewise a function of M, H and then $\dot{M}/\dot{H}$ equals the derivative of M with respect to H, dM/dH, the slope of the $H - M$ curve. Letting

$$\chi_1 = -\frac{\partial_H \phi + \mu M}{\partial_M \phi}, \qquad \chi_2 = -\frac{\nu}{\partial_M \phi} \tag{15.43}$$

we can write the differential equation

$$\frac{dM}{dH} = \chi_1 + \chi_2 \operatorname{sgn} \dot{H} \tag{15.44}$$

for the unknown M in terms of H. If $\chi_2 = 0$ then χ_1 represents the slope of the curve $M(H)$ of a paramagnetic substance; possibly the slope is not constant and depends on the values of M and H. We can then view χ_1 as the positive, differential, *magnetic susceptibility*. We then require that

$$\chi_1 > 0, \qquad |\chi_2| \leq \chi_1.$$

The second term $\chi_2 \operatorname{sgn} \dot{H}$ describes hysteretic effects in that the slope changes depending on the sign of $\dot{H}$. These inequalities look physically sound assumptions but are not required by thermodynamics.

To account for the Curie temperature θ_c, above which hysteresis does not occur, we let

$$h(\theta_c - \theta) = \max\{0, 1 - \theta/\theta_c\}$$

and assume

$$\nu = \gamma(H, M)\, h(\theta_c - \theta).$$

Since $\nu \geq 0$ then $\gamma(H, M)$ is required to be non-negative. As we see in a while, $\gamma(H, M)$ governs the hysteretic properties; that is why it is referred to as the hysteretic function.

By definition, χ_1 is fully determined by the Gibbs potential ϕ whereas χ_2 depends also on $\nu = \gamma h$. Hence different models are obtained by using the same function ϕ. The function χ_2 accounts for the entropy production and hence affects both the dissipative and the hysteretic properties of the material.

Equation (15.41) is linear in $\dot{H}, \dot{M}, |\dot{H}|$ and hence is invariant under the time transformation $t \to ct$, with $c > 0$. Any model, characterized by ϕ and γ, is then rate-independent.

We now establish some general features that serve as guidelines for the elaboration of ferromagnetic models. First we assume that a function

$$M = \mathcal{M}(H)$$

determines M when $\chi_2 = 0$. As such, the function $\mathcal{M}$ is free from hysteretic effects; it is then referred to as anhysteretic function and we say that $\{(H, \mathcal{M}(H)) : H \in \mathbb{R}\}$ is the anhysteretic curve. Consequently

$$\frac{d\mathcal{M}}{dH} = \chi_1$$

is the anhysteretic, differential, magnetic susceptibility; moreover we let χ_1 be a function of H. To obtain the counterclockwise property given by $\oint M\dot{H}dt \leq 0$ we may require that χ_2 satisfies

$$\chi_2(H, M) \begin{cases} > 0 \text{ if } M < \mathcal{M}(H), \\ = 0 \text{ if } M = \mathcal{M}(H), \\ < 0 \text{ if } M > \mathcal{M}(H). \end{cases} \tag{15.45}$$

To select the Gibbs potential ϕ we look for functions in the form

$$\phi(M; H) = \Lambda(M - \Sigma(H)) + \Upsilon(H) - \mu M H;$$

the functions $\Lambda, \Sigma, \Upsilon$ are so far undetermined, differentiable functions of H, M, parameterized by θ and F. We observe that since $\phi = \psi - \mu M H$ then $\Lambda + \Upsilon$ is just the Helmholtz free energy ψ. Upon substitution of $\partial_H \phi$ and $\partial_M \phi$ we obtain

$$\chi_1(H) = \frac{\Lambda'(M - \Sigma(H))\Sigma'(H) - \Upsilon'(H)}{\Lambda'(M - \Sigma(H)) - \mu H}, \qquad \chi_2 = -\frac{\gamma h}{\Lambda'(M - \Sigma(H)) - \mu H}.$$

The first condition of (15.43) results in

$$[\Lambda'(M - \Sigma(H)) - \mu H]\chi_1(H) = \Lambda'(M - \Sigma(H))\Sigma'(H) - \Upsilon'(H)$$

holds if

$$\Sigma' = \chi_1, \qquad \Upsilon' = \mu H \chi_1(H).$$

Since $\chi_1 > 0$ then Σ is a monotone increasing function of H. To determine Λ we observe that the requirement (15.45) is obeyed if

$$\chi_2(H, M)[M - \mathcal{M}(H)] \leq 0.$$

Since $\nu = \gamma h \geq 0$ it follows that

$$\frac{M - \mathcal{M}(H)}{\Lambda'(M - \Sigma(H)) - \mu H} \geq 0.$$

This condition in turn is satisfied if, for any positive parameter α, possibly dependent on F and θ,

$$\frac{M - \mathcal{M}(H)}{\Lambda'(M - \Sigma(H)) - \mu H} = \frac{1}{\alpha}$$

whence

$$\Lambda'(M - \Sigma(H)) = \alpha[M - \Sigma(H)], \qquad \mathcal{M}(H) = \Sigma(H) + \frac{1}{\alpha}\mu H.$$

Hence

$$\Lambda = \tfrac{1}{2}\alpha[M - \Sigma(H)]^2,$$

to within an additive constant possibly dependent on F and θ.

To sum up, thermodynamically-consistent models of one-dimensional ferromagnets are obtained by letting

$$\phi(M, H) = \tfrac{1}{2}\alpha[M - \Sigma(H)]^2 + \Upsilon(H) - \mu M H,$$

where

$$\Sigma(H) = \int_0^H \chi_1(y)dy, \qquad \Upsilon(H) = \mu \int_0^H y\,\chi_1(y)dy,$$

while

$$\chi_2 = \frac{\gamma(H, M)h(\theta_c - \theta)}{\alpha(\mathcal{M}(H) - M)}, \qquad \mathcal{M}(H) = \int_0^H [\chi_1(y) + \mu/\alpha]dy.$$

Hence models are characterized by the (anhysteretic) differential permittivity $\chi_1(H) > 0$ and the parameter $\alpha > 0$, and the hysteretic function $\gamma(H, M) \geq 0$.

Soft Iron Models

Let

$$\gamma(H, M) = \gamma_0[\mathcal{M}(H) - M]^2, \qquad \gamma_0 > 0.$$

Define

$$\gamma_h(\theta) = \gamma_0 h(\theta_c - \theta).$$

Hence we express $M' = dM/dH$ in the form

$$M' = \chi_1(H) + \tau[\mathcal{M}(H) - M]\,\mathrm{sgn}\,\dot{H}, \tag{15.46}$$

where

$$\chi_1 = \mathcal{M}' - \frac{\mu}{\alpha}, \qquad \tau_h = \frac{\gamma_h}{\alpha}.$$

This is a very simple model; it is characterized by the positive ratio γ_0/α and the anhysteretic function $\mathcal{M}$.

For definiteness we look for the modelling of the class of ferromagnetic soft materials, particularly those called isoperms, like, e.g. Fe-Ni-Al isoperms or Mn-Zn ferrites. Equation (15.46) is the analogue of a model for soft materials [104, 439]. To get a direct connection with the literature we let

$$f = \mathcal{M}, \qquad g = \chi_1,$$

and write the differential equation in the form

$$\frac{dM}{dH} = g(H) - \tau_h(M - f(H))\mathrm{sgn}\,\dot{H}, \qquad \tau_h > 0. \tag{15.47}$$

The differential equation (15.47) is consistent with thermodynamics; by (15.46) f and g are related by

$$g = f' - \frac{\mu}{\alpha}, \tag{15.48}$$

where α is an arbitrary positive constant, possibly dependent on θ.

As a particular case, we let g be piecewise constant and, correspondingly, f is piecewise linear, as devised in [103, 104]. For instance, assume $\mu = 1$, $\alpha > 2$, and

$$\mathcal{M}(H) = \begin{cases} \frac{1}{2}(H-1)+1 & \text{if } H < -1, \\ H & \text{if } -1 \le H \le 1, \\ \frac{1}{2}(H+1)-1 & \text{if } H > 1, \end{cases} \qquad g(H) = \begin{cases} 1 - 1/\alpha & \text{if } -1 \le H \le 1, \\ \frac{1}{2} - 1/\alpha & \text{if } |H| > 1, \end{cases}$$

In this case, hysteresis cycles are obtained by solving the system

$$\begin{cases} \dot{H} = \omega\mathcal{H}\cos\omega t \\ \dot{M} = \big[\mathcal{M}'(H) - 1/\alpha\big]\dot{H} + \tau_h\big[\mathcal{M}(H) - M\big]|\dot{H}|. \end{cases}$$

Figure 15.13 shows cyclic processes with large and small amplitudes, corresponding to $\alpha = 5$ and $\tau_h = 0.3$.

According to [103], the condition

$$\int_H^\infty [f'(\zeta) - g(\zeta)]\exp(-\tau\zeta)\,d\zeta = \frac{1}{\alpha\tau}\exp(-\tau H) \ge 0, \qquad H \in [0, \infty)$$

is both necessary and sufficient to warrant that the major loop has a counterclockwise orientation and is the boundary of the set of (attainable) states, i.e. the set of pairs

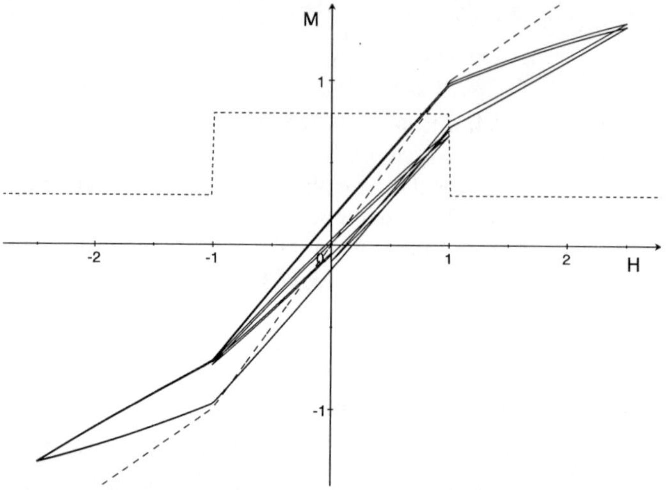

Fig. 15.13 Soft iron hysteresis loops (solid), anhysteretic curve $\mathcal{M}$ (dashed) and graph of g (short dashed) with $\mathcal{H} = 1, 2.5$. The initial states are $H_0 = 0$ and $M_0 = -0.1$

(H, M) accessible, by appropriate variation of H, from the state in which both H and M are zero.

Remark 15.1 As stated in [103], $f' \geq 0$, and hence $g \geq 1/\alpha$, is necessary and sufficient for dM/dH to be non-negative; the stronger condition $f' - g \geq 0$ (and hence, $1/\alpha \geq 0$) is needed to ensure that the energy expended in a complete traversal of a primitive loop is non-negative ([103] Eqs. (1.6) and (3.18)). This condition is not enough to ensure thermodynamic consistency. Moreover, the requirement

$$f' - g = 1/\alpha > \epsilon > 0,$$

in [103], implies that $f' - g$ cannot vanish, not even at the limit as $|H|$ goes to infinity and this prevents the model to exhibit the saturation property.

A model allowing for the saturation property is obtained as follows. Let

$$\gamma(H, M) = \gamma_0 g(H)[\mathcal{M}(H) - M]^2, \quad \gamma_0 > 0, \quad g(H) = \mathcal{M}'(H) - 1/\alpha.$$

Hence

$$\frac{dM}{dH} = g(H)[1 + \tau_h(\mathcal{M}(H) - M) \operatorname{sgn} \dot{H}].$$

The vanishing of χ as $|H|$ approaches infinity is a way of modelling the saturation property. Hysteresis cycles are obtained by solving the system

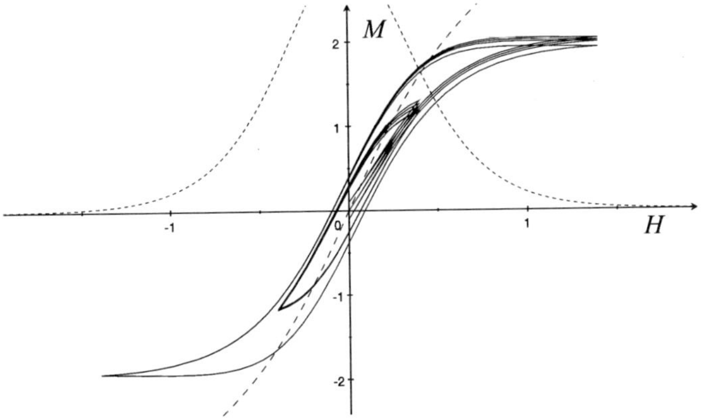

Fig. 15.14 Soft iron hysteresis loops with the saturation property (solid): anhysteretic curve $\mathcal{M}$ (dashed) and graph of g (short dashed) with $\mathcal{H} = 0.4, 1.4$. The initial state H_0, M_0 has $H_0 = 0$ and $M_0 = -0.1,\ 0,\ 0.1$

$$\begin{cases} \dot{H} = \omega \mathcal{H} \cos \omega t \\ \dot{M} = g(H)\{\dot{H} + \tau_h[\mathcal{M}(H) - M]|\dot{H}|\}, \end{cases}$$

and starting from different initial states (H_0, M_0). For instance, in Fig. 15.14 we use $\mathcal{M}(H) = 1.5(\tanh 2H + H)$ and $\alpha = 1.5$, $\tau_h = 1$. Then $g(H) = 3/(\cosh 2H)^2$.

15.3.3 Curie–Weiss Law

We give a simple physical motivation of the Curie–Weiss law in ferromagnetism.

In the physical literature the Weiss field is imagined as the magnetic field $\mathbf{H}_E$ whose effect would be to line up ionic and atomic magnetic moments so that a paramagnetic substance becomes the corresponding ferromagnetic substance. The orienting effect of the Weiss field is opposed by th thermal agitation. At the Curie temperature θ_c the interaction energy of the moment m, $\mu_0 m H_E$ is about the thermal energy $k\theta_c$. Let

$$H_E = \lambda M$$

where λ is said to be the Weiss field constant. The magnetization M is taken to be proportional to the magnetic intensity H and inversely proportional to the temperature θ. Thus we consider the whole magnetic intensity $H + \lambda M$ and say that

$$\frac{M}{H + \lambda M} = \frac{C}{\theta}, \qquad \chi = \frac{M}{H} = \frac{C}{\theta - \lambda C}.$$

Letting $\theta_c = \lambda C$ we find the Curie–Weiss law

$$\chi = \frac{C}{\theta - \theta_c}.$$

This indicates that the slope of the magnetization curve decreases inasmuch as the temperature θ increases. The singularity at $\theta = \theta_c$ is regarded as the indication that a non-zero magnetization occurs without any applied field H.

Chapter 16
Phase Transitions

A phase denotes a region of a body with physically uniform properties. Typical phases are the familiar states of matter, i.e. solid, liquid, and gas; in shape-memory alloys the phases are martensite and austenite. Phase transitions denote transformations of the substance in a phase to another phase. To describe a phase transition we need a physical quantity which governs the transformation; for instance, the temperature or the pressure. The description of phase transitions is realized essentially within two schemes: a sharp interface across which the constitutive properties suffer a sudden change or a diffuse space region where the pertinent fields vary continuously.

First the jump conditions are derived for the interface among two phases. Next details are determined for the liquid–vapour transition, thus finding the continuity of the Gibbs free energy at the interface and Clapeyron's equation. The brine channels formation is also developed. The transition in a finite layer is investigated through the mixture model. As a relevant application the solidification-melting of binary alloys is described. Finally, phase transitions in shape-memory alloy are modelled in terms of an order parameter related to the concentration of the mass fraction of martensite.

16.1 Jump Conditions at Interfaces

We first investigate phase transitions by considering different phases, e.g. solid and fluid, separated by a surface of zero thickness. Physical quantities such as density are assumed to be discontinuous across the interface. The conservation laws of mechanics and thermodynamics have their corresponding jump condition across the interface. In this respect we review the jump conditions developed in Sect. 2.17.2 and examine the physical consequences on the coexistence of different phases at opposite sides of the interface (Fig. 16.1).

A. Morro and C. Giorgi, *Mathematical Modelling of Continuum Physics*,
Modeling and Simulation in Science, Engineering and Technology,
https://doi.org/10.1007/978-3-031-20814-0_16

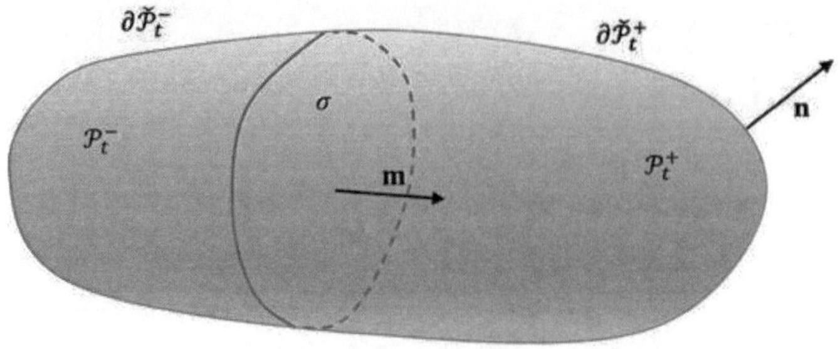

Fig. 16.1 The sharp interface σ separates two different phases in the regions $\mathcal{P}_t^+$ and $\mathcal{P}_t^-$

A time-dependent region $\mathcal{P}_t$ is convected by the body at hand. A surface (interface) σ is the common boundary of the two sub-regions $\mathcal{P}_t^-, \mathcal{P}_t^+$; the surface σ need not be convected. Denote by $\mathbf{m}$ the unit normal vector to σ from $\mathcal{P}_t^-$ to $\mathcal{P}_t^+$ and, for any field f, let $[\![f]\!]$ denote the jump $f_- - f_+$ of f across σ with f_-, f_+ the limit values at the pertinent point from $\mathcal{P}_t^-$ and $\mathcal{P}_t^+$. We observe that $\mathcal{P}_t = \mathcal{P}_t^- \cup \sigma \cup \mathcal{P}_t^+$ and $\partial\mathcal{P}_t = \partial\check{\mathcal{P}}_t^+ \cup \partial\check{\mathcal{P}}_t^-$ with $\partial\check{\mathcal{P}}_t^\pm = \partial\mathcal{P}_t^\pm \setminus \sigma$.

Consider a generic balance equation

$$\frac{d}{dt}\int_{\mathcal{P}_t} \rho\phi dv = \int_{\mathcal{P}_t} b dv + \int_{\partial\mathcal{P}_t} \mathbf{h}\mathbf{n} da,$$

where ρ is the mass density. Observe that, symbolically,

$$\int_{\mathcal{P}_t} dv = \int_{\mathcal{P}_t^-} dv + \int_{\mathcal{P}_t^+} dv, \qquad \int_{\partial\mathcal{P}_t} da = \int_{\partial\check{\mathcal{P}}_t^-} da + \int_{\partial\check{\mathcal{P}}_t^+} da.$$

By the generalized transport theorem we can write

$$\frac{d}{dt}\int_{\mathcal{P}_t^-} \rho\phi dv = \int_{\mathcal{P}_t^-} \rho\dot{\phi} dv + \int_{\sigma} (\rho\phi U)_- da, \qquad \frac{d}{dt}\int_{\mathcal{P}_t^+} \rho\phi dv = \int_{\mathcal{P}_t^+} \rho\dot{\phi} dv - \int_{\sigma} (\rho\phi U)_+ da,$$

where $U = (\boldsymbol{\nu} - \mathbf{v}) \cdot \mathbf{m}$ while $\boldsymbol{\nu}$ is the velocity of the point of σ and $\mathbf{v}$ is the limit of the velocity of the point of the body. Summing the two relations we arrive at

$$\int_{\mathcal{P}_t} \rho\dot{\phi} dv + \int_{\sigma} [\![\rho\phi U]\!] da = \int_{\partial\check{\mathcal{P}}_t^-} \mathbf{h}\mathbf{n} da + \int_{\partial\check{\mathcal{P}}_t^+} \mathbf{h}\mathbf{n} da.$$

Assume $\mathcal{P}_t$ is a pillbox region around σ and hence $\partial\check{\mathcal{P}}_t^-$ and $\partial\check{\mathcal{P}}_t^+$ approach σ with $\mathbf{n} = -\mathbf{m}$ at $\partial\check{\mathcal{P}}_t^-$ and $\mathbf{n} = \mathbf{m}$ at $\partial\check{\mathcal{P}}_t^+$. If $\rho\dot{\phi}$ is bounded then the limit as the thickness of the pillbox approaches zero yields

$$\int_{\sigma}\{[\![\rho\phi U]\!] + [\![\mathbf{h}]\!]\mathbf{m}\}da = 0.$$

The arbitrariness of σ, which is due to the arbitrariness of $\mathcal{P}_t$, and the continuity of the integrand over σ at both sides imply that

$$[\![\rho\phi U]\!] + [\![\mathbf{h}]\!]\mathbf{m} = 0$$

If, instead, the balance is expressed by an inequality,

$$\frac{d}{dt}\int_{\mathcal{P}_t} \rho\phi dv \geq \int_{\mathcal{P}_t} b dv + \int_{\partial\mathcal{P}_t} \mathbf{hn} da,$$

then it follows that

$$[\![\rho\phi U]\!] + [\![\mathbf{h}]\!]\mathbf{m} \geq 0.$$

The balances of mass, linear momentum, and energy are obtained by letting

$$\phi = 1, \mathbf{v}, \varepsilon + \tfrac{1}{2}\mathbf{v}^2; \qquad \mathbf{h} = 0, \mathbf{T}, \mathbf{Tv} - \mathbf{q},$$

where $\mathbf{T} \in \text{Sym}$. Hence we obtain the jump conditions

$$[\![\rho U]\!] = 0, \tag{16.1}$$

$$\rho U[\![\mathbf{v}]\!] + [\![\mathbf{T}]\!]\mathbf{m} = \mathbf{0}, \tag{16.2}$$

$$\rho U[\![\varepsilon + \tfrac{1}{2}\mathbf{v}^2]\!] + \mathbf{m}\cdot[\![\mathbf{Tv} - \mathbf{q}]\!] = 0. \tag{16.3}$$

The quantity ρU represents the relative (to the body) mass flow rate and Eq. (16.1) means that ρU is continuous across σ.

Equations (16.2) and (16.3) are expected to be invariant under a change of frame. Let $\mathbf{V}$ be the velocity of a new frame of reference and hence let $\mathbf{v} = \mathbf{u} + \mathbf{V}$. We have

$$\rho U[\![\tfrac{1}{2}\mathbf{v}^2]\!] + \mathbf{m}\cdot[\![\mathbf{Tv}]\!] = \rho U[\![\tfrac{1}{2}\mathbf{u}^2 + \mathbf{u}\cdot\mathbf{V}]\!] + \mathbf{m}\cdot[\![\mathbf{Tu}]\!] + \mathbf{m}\cdot[\![\mathbf{TV}]\!].$$

By (16.2) we have $\rho U[\![\mathbf{u}]\!] = -[\![\mathbf{T}]\!]\mathbf{m}$ and hence, since $\mathbf{T} \in \text{Sym}$,

$$\rho U[\![\mathbf{u}\cdot\mathbf{V}]\!] + \mathbf{m}\cdot[\![\mathbf{TV}]\!] = \mathbf{0}.$$

Consequently (16.3) yields

$$\rho U[\![\varepsilon + \tfrac{1}{2}\mathbf{u}^2]\!] + \mathbf{m}\cdot[\![\mathbf{Tu} - \mathbf{q}]\!] = 0,$$

and then it is invariant under a change of frame.

In general ρU is a parameter of the transition. If we know, or fix, U then we determine ρ; the invariance of ρU implies

$$\rho_- = \rho_+ \frac{U_+}{U_-}.$$

Now some consequences of (16.2)–(16.5) are derived on the assumption that the stress tensor is isotropic; we let $\mathbf{T} = -p\mathbf{1}$, thus possibly identifying p with the pressure.

Entropy inequality and entropy jump

The balance of entropy can be given the form

$$\frac{d}{dt}\int_{\mathcal{P}_t} \rho\eta dv = \int_{\mathcal{P}_t} \frac{\rho r}{\theta} dv - \int_{\partial\mathcal{P}_t} (\mathbf{q}/\theta + \mathbf{k})\cdot \mathbf{n}\, da + \int_{\mathcal{P}_t} \rho\gamma\, dv$$

where γ is the entropy production per unit mass. If $\mathbf{k} = \mathbf{0}$ and $\dot{\eta}, r, \gamma$ are bounded, and even more if $\gamma = 0$, then passing to the limit of zero thickness we have

$$\rho U[\![\eta]\!] - [\![\frac{\mathbf{q}}{\theta}]\!]\cdot \mathbf{m} = 0. \tag{16.4}$$

If, rather, the balance is

$$\frac{d}{dt}\int_{\mathcal{P}_t} \rho\eta dv \geq \int_{\mathcal{P}_t} \frac{\rho r}{\theta} dv + \int_{\partial\mathcal{P}_t} \frac{\mathbf{q}}{\theta}\cdot \mathbf{n}\, da$$

then across the interface

$$\rho U[\![\eta]\!] - [\![\frac{\mathbf{q}}{\theta}]\!]\cdot \mathbf{m} \geq 0. \tag{16.5}$$

16.2 Phase Transitions at Sharp Interfaces with $\mathbf{T} = -p\mathbf{1}$

Denote by $\mathrm{v} = \mathbf{v}\cdot\mathbf{m}$ and $q = \mathbf{q}\cdot\mathbf{m}$ the components of $\mathbf{v}$ and $\mathbf{q}$ along $\mathbf{m}$. Let $\mathbf{v} = \boldsymbol{\nu} + \mathbf{u}$. Hence we have

$$[\![\mathbf{v}]\!] = [\![\mathbf{u}]\!], \qquad [\![\tfrac{1}{2}\mathbf{v}^2]\!] = \boldsymbol{\nu}\cdot[\![\mathbf{u}]\!] + [\![\tfrac{1}{2}\mathbf{u}^2]\!].$$

Observe that $[\![\boldsymbol{\nu}]\!]\cdot\mathbf{m} = 0$ and hence, by the definition of $U = (\boldsymbol{\nu} - \mathbf{v})\cdot\mathbf{m}$, we obtain

$$[\![U]\!] = -[\![\mathrm{v}]\!].$$

Notice that

$$[\![p\mathbf{v}]\!] = [\![p\boldsymbol{\nu}]\!] - [\![\frac{p}{\rho}\rho(\boldsymbol{\nu} - \mathbf{v})]\!], \qquad [\![p\mathbf{v}]\!]\cdot\mathbf{m} = [\![p]\!]\boldsymbol{\nu}\cdot\mathbf{m} - [\![\frac{p}{\rho}]\!]\rho U$$

If $\mathbf{T} = -p\mathbf{1}$ then the jump conditions (16.2), (16.3) can be written in the form

$$\rho U[\![\mathbf{u}]\!] = [\![p]\!]\mathbf{m}, \qquad \rho U[\![\varepsilon + \tfrac{1}{2}\mathbf{u}^2]\!] - [\![\frac{p}{\rho}]\!]\rho U - [\![q]\!] = 0, \tag{16.6}$$

whence

$$\rho U[\![\varepsilon + \frac{p}{\rho} + \tfrac{1}{2}\mathbf{u}^2]\!] - [\![q]\!] = 0. \tag{16.7}$$

Hence to within the kinetic term $\frac{1}{2}\mathbf{u}^2$, relative to the observer at the interface, the balance of energy implies that the jump of q equals ρU times the jump of enthalpy.

If $U = 0$ then the interface is convected by the motion of the body. In this case $[\![p]\!] = 0$. The transverse parts of $\mathbf{v}$ and $\mathbf{q}$ are unrestricted while the longitudinal parts v and q are subject to

$$p[\![\mathrm{v}]\!] + [\![q]\!] = 0, \qquad [\![q/\theta]\!] \le 0.$$

Now let $\beta = \rho U \neq 0$. The jump $[\![\mathbf{v}]\!]$ is longitudinal and $\beta[\![\mathrm{v}]\!] = [\![p]\!]$. The unknowns $[\![\mathbf{v}]\!], [\![\theta]\!], [\![p]\!], [\![q]\!], [\![\eta]\!]$ occur nonlinearly. Let $[\![\theta]\!] = 0$. Equations (16.6) can be written in the form

$$\beta[\![\mathrm{v}]\!] = [\![p]\!], \qquad \beta[\![\varepsilon + \tfrac{1}{2}\mathrm{v}^2]\!] = [\![p\mathrm{v} + q]\!], \qquad \beta[\![\theta\eta]\!] - [\![q]\!] \ge 0.$$

If, further, $[\![p]\!] = 0$ then $[\![v]\!] = 0$, $[\![p\mathrm{v}]\!] = 0$ and

$$\beta[\![\varepsilon]\!] = [\![q]\!], \qquad \beta[\![\theta\eta]\!] - [\![q]\!] \ge 0.$$

Consequently $\beta[\![\varepsilon - \theta\eta]\!] \le 0$ whence

$$U[\![\psi]\!] \le 0. \tag{16.8}$$

Thus if $[\![\theta]\!] = 0, [\![p]\!] = 0$, then $U > 0$ is allowed if $\psi_- \le \psi_+$, namely if the Helmholtz free energy decreases across the interface.

Things are more involved if $[\![\theta]\!] \neq 0, [\![p]\!] \neq 0$. For any pair of fields α, $\mathbf{w}$, with α a scalar and $\mathbf{w}$ a scalar or a tensor of any order,

$$[\![\alpha\mathbf{w}]\!] = \alpha_+[\![\mathbf{w}]\!] + [\![\alpha]\!]\mathbf{w}_+ + [\![\alpha]\!][\![\mathbf{w}]\!], \qquad [\![\mathbf{w}]\!] = \frac{[\![\alpha\mathbf{w}]\!] - [\![\alpha]\!]\mathbf{w}_+}{\alpha_+ + [\![\alpha]\!]},$$

$$[\![\alpha\mathbf{w}]\!] = \mathbf{0} \quad \Longrightarrow \quad [\![\mathbf{w}]\!] = -\frac{[\![\alpha]\!]\mathbf{w}_+}{\alpha_+ + [\![\alpha]\!]}.$$

Hence $[\![\rho U]\!] = 0$ implies

$$[\![\rho]\!] = -\frac{\rho_+[\![U]\!]}{U_+ + [\![U]\!]} = \frac{\rho_+[\![\mathrm{v}]\!]}{U_+ - [\![\mathrm{v}]\!]}.$$

The specific volume v, such that $\rho v = 1$, satisfies

$$[\![v]\!] = -\frac{[\![\rho]\!]}{\rho_+(\rho_+ + [\![\rho]\!])}.$$

It then follows that

$$[\![\rho]\!] = \frac{\rho_+ [\![p]\!]}{\beta U_+ - [\![p]\!]}, \qquad [\![v]\!] = -\frac{[\![p]\!]}{\beta^2} \tag{16.9}$$

and

$$-\frac{1}{\beta^2} = [\![p]\!][\![v]\!] = -\frac{[\![p]\!][\![\rho]\!]}{\rho_+(\rho_+ + [\![\rho]\!]}.$$

By (16.9) we have

$$[\![p]\!][\![v]\!] \le 0, \qquad [\![p]\!][\![\rho]\!] \ge 0,$$

and

$$[\![pv]\!] = (p_+ + [\![p]\!])[\![v]\!] + v_+[\![p]\!] = (v_+ - \frac{p_+ + [\![p]\!]}{\beta^2})[\![p]\!].$$

Consider the balance of energy and observe

$$[\![p\mathrm{v}]\!] = (p_+ + [\![p]\!])\frac{[\![p]\!]}{\beta}, \qquad \beta[\![\tfrac{1}{2}\mathrm{v}^2]\!] = \beta\tfrac{1}{2}[\![\mathrm{v}]\!]^2 = \frac{[\![p]\!]^2}{2\beta},$$

where, for formal convenience, we have taken $\mathrm{v}_+ = 0$. Hence we have

$$\beta[\![\varepsilon]\!] = [\![p\mathrm{v}]\!] - \tfrac{1}{2}\beta[\![\mathrm{v}]\!]^2 + [\![q]\!] = (p_+ + \tfrac{1}{2}[\![p]\!])\frac{[\![p]\!]}{\beta} + [\![q]\!].$$

Consequently the possible values of β satisfy the equation

$$[\![\varepsilon]\!]\beta^2 - [\![q]\!]\beta - (p_+ + \tfrac{1}{2}[\![p]\!])[\![p]\!] = 0. \tag{16.10}$$

The values of ρU so determined are effective properties of the interface if

$$\rho U[\![\eta]\!] - [\![\frac{q}{\theta}]\!] \ge 0. \tag{16.11}$$

16.2.1 The Stefan Model

Among the models of phase transitions the Stefan model is well-known as the origin of an initial boundary value problem for a parabolic partial differential equation. The model is claimed to describe a solid–fluid phase transition [177, 434]; due to its simplicity it follows from (16.2)–(16.5) as a purely thermal problem. Let $[\![\mathbf{T}]\!] = \mathbf{0}$ and hence $[\![\mathbf{v}]\!] = \mathbf{0}$, $[\![\mathbf{v}^2]\!] = 0$.

If $[\![\mathbf{v}]\!] = \mathbf{0}$ then the requirement $[\![\rho U]\!] = 0$ results in

$$[\![\rho]\!] = 0,$$

which is reasonable in the ice-water transition.

The domain $\mathcal{P}_t$ is the interval $[0, d]$. Let $s(t) \in [0, d]$ be the position of the interface and then $U = \dot{s}$. Hence $\mathcal{P}_t^- = [0, s)$, $\mathcal{P}_t^+ = (s, d]$ are occupied by the fluid and solid phases. Let $x \in [0, d]$ and assume Fourier's law $q_\pm = -k_\pm \partial_x \theta$ in both phases. While (16.2) holds identically Eqs. (16.3)–(16.5) become

$$\rho U [\![\varepsilon]\!] + [\![k \partial_x \theta]\!] = 0, \qquad \rho U [\![\eta]\!] + [\![(k/\theta)\partial_x \theta]\!] \geq 0.$$

The requirement on $U[\![\eta]\!]$ is ignored. Since $[\![\mathbf{v}]\!] = \mathbf{0}$ we can view the two regions as rigid and hence $\rho[\![\varepsilon]\!]$ equals the latent heat λ. The temperature is assumed to be continuous across the interface. The whole problem for the function $\theta(x, t)$ then is represented by

$$c_- \partial_t \theta - k_- \partial_{xx} \theta = 0, \quad x \in (0, s); \qquad c_- \partial_t \theta - k_- \partial_{xx} \theta = 0, \quad x \in (s, d),$$

$$\lambda \dot{s} + [\![k \partial_x \theta]\!] = 0, \qquad \theta(s_-) = \theta(s_+), \qquad t \in (0, T),$$

along with the initial boundary conditions

$$k_- \partial_x \theta(0, t) = h_1(t), \; k_+ \partial_x \theta(d, t) = h_2(t), \; 0 < t < T; \quad \theta(x, 0) = \Theta(x), \; x \in (0, d),$$

and $s(0) = b \in [0, d]$

16.3 Liquid–Vapour Transition

The two phases are viewed as viscous and heat-conducting fluids. We then adopt the Navier–Stokes–Fourier model

$$\mathbf{T} = -p(\rho, \theta)\mathbf{1} + 2\mu \mathbf{D} + \lambda(\nabla \cdot \mathbf{v})\mathbf{1}, \qquad \mathbf{q} = -\kappa \nabla \theta.$$

Then Eqs. (16.2)–(16.5) become

$$\rho U [\![\mathbf{v}]\!] = [\![p]\!]\mathbf{m} - [\![2\mu \mathbf{D} + \lambda(\nabla \cdot \mathbf{v})\mathbf{1}]\!]\mathbf{m},$$

$$\rho U [\![\varepsilon + \tfrac{1}{2}\mathbf{v}^2]\!] = [\![p\mathbf{v}]\!] \cdot \mathbf{m} - [\![2\mu \mathbf{D}\mathbf{v} + \lambda(\nabla \cdot \mathbf{v})\mathbf{v}]\!] \cdot \mathbf{m} - [\![\kappa \partial_n \theta]\!],$$

$$\rho U[\![\eta]\!] + [\![\frac{\kappa}{\theta}\partial_n\theta]\!] \geq 0,$$

where $\partial_n = \mathbf{m} \cdot \nabla$.

For simplicity assume $\mathbf{D}_+ = D_+\mathbf{m} \otimes \mathbf{m}$ and hence $\nabla \cdot \mathbf{v}_+ = D_+$. It follows that a solution exists with $\mathbf{D}_- = D_-\mathbf{m} \otimes \mathbf{m}$ and v_- given by v_+ and

$$\rho U[\![\mathrm{v}]\!] = [\![p]\!] - [\![(2\mu + \lambda)D]\!].$$

If $[\![p]\!]$ is known or, possibly, $[\![p]\!] = 0$ then the following equation determines $[\![\varepsilon]\!]$ if $[\![\kappa\partial_n\theta]\!]$ is known. We find

$$\rho U[\![\varepsilon + \tfrac{1}{2}\mathrm{v}^2]\!] = [\![p\mathrm{v}]\!] - [\![(2\mu + \lambda)D\mathrm{v}]\!] - [\![\kappa\partial_n\theta]\!].$$

This shows that the system is determined if in any subregion we determine the fields $\rho, \mathbf{v}, \theta$ via the differential equations given by the balance equations

$$\rho\dot{\mathbf{v}} = -\nabla p + 2\mu\nabla \cdot \mathbf{D} + \lambda\nabla(\nabla \cdot \mathbf{v}) + \rho\mathbf{b},$$

$$\rho\dot{\varepsilon} = -p\nabla \cdot \mathbf{v} + 2\mu\mathbf{D} \cdot \mathbf{D} + \lambda(\nabla \cdot \mathbf{v})^2 + \rho r - \nabla \cdot \mathbf{q}.$$

Assume the fluids are inviscid or look for transitions where $\mathbf{D} = \mathbf{0}$ and $\partial_n\theta = 0$ in both phases. The jump conditions reduce to

$$\rho U(\mathrm{v}_- - \mathrm{v}_+) = p_- - p_+, \tag{16.12}$$

$$\rho U\{\varepsilon_- - \varepsilon_+ + \tfrac{1}{2}(\mathrm{v}_-^2 - \mathrm{v}_+^2)\} = p_-\mathrm{v}_- - p_+\mathrm{v}_+, \tag{16.13}$$

$$\rho U(\eta_- - \eta_+) \geq 0. \tag{16.14}$$

To fix ideas we let the region ahead (+) contain the liquid and the region behind (−) contain the vapour. Denote by the subscripts V and L the quantities pertaining to the vapour and the liquid phases. Hence $\eta_V > \eta_L$ implies $U > 0$ and this means that the interface moves toward the liquid phase or that the mass of the vapour phase increases.

To apply this approach to phase transitions we need detailed constitutive functions of the two phases.

Vapour

The vapour is modelled as an ideal gas (see Sect. 5.3). Hence we have[1]

$$p_V = K\theta\rho =: \rho^2\partial_\rho\psi_V(\theta, \rho), \qquad \psi_V(\theta, \rho) = K\theta \ln\frac{\rho}{\rho_0} + c_v\theta(1 - \ln\frac{\theta}{\theta_0}) + \varepsilon_0,$$

[1] The subscript V denotes quantities pertaining to the vapour.

$$\eta_v(\theta, \rho) := -\partial_\theta \psi_v = c_v \ln \frac{\theta}{\theta_0} - K \ln \frac{\rho}{\rho_0}, \qquad \varepsilon_v := \psi_v + \theta \eta_v = c_v \theta + \varepsilon_0,$$

where $K = R/M$. In terms of θ and p we have

$$\psi_v(\theta, p) = K\theta \ln \frac{p}{p_0} + c_v\theta - c_p\theta \ln \frac{\theta}{\theta_0} + \varepsilon_0, \quad \eta_v(\theta, p) = c_p \ln \frac{\theta}{\theta_0} - K \ln \frac{p}{p_0}. \tag{16.15}$$

The Gibbs free energy $\phi = \psi + p/\rho$ and the entropy $\eta = -\partial_\theta \phi$ have the form

$$\phi_v(\theta, p) = \psi_v(\theta, p) + \frac{p}{\rho} = K\theta \ln \frac{p}{p_0} + c_p\theta(1 - \ln \frac{\theta}{\theta_0}) + \varepsilon_0, \tag{16.16}$$

$$\eta_v(\theta, p) = c_p \ln \frac{\theta}{\theta_0} - K \ln \frac{p}{p_0}.$$

The internal energy ε and the enthalpy $h = \varepsilon + p/\rho$ have the form

$$\varepsilon_v = c_v\theta + \varepsilon_0, \qquad h_v = c_p\theta + \varepsilon_0.$$

16.3.1 Continuity of the Gibbs Free Energy

The latent heat (of vapourization) λ is the amount of energy that is needed to vapourize the unit mass of liquid at constant temperature. Consider the balance of energy for an inviscid fluid

$$\rho\dot{\varepsilon} = -p\nabla \cdot \mathbf{v} - \nabla \cdot \mathbf{q} + \rho r.$$

Since $\nabla \cdot \mathbf{v} = \dot{v}/v$ then we can write

$$\dot{\varepsilon} = -p\dot{v} - v\nabla \cdot \mathbf{q} + r.$$

Two models are now examined. First we let the transition occur in a layer of finite thickness and we consider the path from the + side to the − side as functions of time. Next we look at transitions across sharp interfaces.

Transitions across layers of finite thickness

Consider a process from a state ahead to the corresponding state behind, across the interface. If d is the duration of the process, integrating with respect to time in the interval $[0, d]$ we have

$$\int_0^d (\dot{\varepsilon} + p\dot{v})dt = \int_0^d (r - v\nabla \cdot \mathbf{q})dt = \lambda.$$

If, at constant temperature, P is the integral of $p(\theta, v)$, namely $\partial_v P(\theta, v) = p(\theta, v)$ then we have

$$\int_0^d (\dot{\varepsilon} + p\dot{v})dt = \varepsilon_V + P(\theta, v_V) - \varepsilon_L - P(\theta, v_L);$$

if p is constant then $P(\theta, v) = pv$. It is understood that ε_V, $P(\theta, v_V)$, ε_L, $P(\theta, v_L)$ are evaluated at the same point, but opposite sides of the interface, at constant temperature.

Let θ be constant within the layer. If the entropy production and the extra-entropy flux vanish then we can write

$$\theta\dot{\eta} = r - v\nabla \cdot \mathbf{q}. \tag{16.17}$$

Integration in time, over $[0, d]$, of (16.17) associated with the process placed at the interface, implies that the latent heat is also equal to $\theta(\eta_V - \eta_L)$. Consequently, at constant temperature, we can write the two representations

$$\lambda = \theta(\eta_V - \eta_L) = \varepsilon_V + P(\theta, v_V) - \varepsilon_L - P(\theta, v_L).$$

Consider now the Gibbs free energy in the form

$$\phi(\theta, v) = \varepsilon(\theta, v) - \theta\eta(\theta, v) + P(\theta, v).$$

We then have

$$\phi_V = \varepsilon_V - \theta\eta_V + P(\theta, v_V) = \varepsilon_L + P(\theta, v_L) + \lambda - (\theta\eta_L + \lambda) = \phi_L \tag{16.18}$$

and can state the following

Theorem 16.1 *Across a transition layer liquid–vapour, where the extra-entropy flux and the entropy production vanish while the temperature is constant, the Gibbs free energy ϕ satisfies the equality $\phi_V = \phi_L$.*

Transitions across sharp interfaces

Within an error $\rho U[\![\frac{1}{2}\mathbf{u}^2]\!]$, Eq. (16.7) can be written as

$$\rho U[\![\varepsilon + \frac{p}{\rho}]\!] - [\![q]\!] = 0, \tag{16.19}$$

where $q = \mathbf{q} \cdot \mathbf{m}$. If $\mathbf{k} = \mathbf{0}$ and $\rho\gamma$ vanishes, or is bounded, then (16.5) holds and then

$$\rho U[\![\eta]\!] = [\![q/\theta]\!].$$

If $[\![\theta]\!] = 0$ then

$$\rho U[\![\theta\eta]\!] = [\![q]\!].$$

Substitution of $[\![q]\!]$ in (16.19) yields

$$\rho U[\![\phi]\!] = 0, \qquad \phi = \varepsilon + \frac{p}{\rho} - \theta\eta. \tag{16.20}$$

Hence $U \neq 0$ implies $[\![\phi]\!] = 0$. This proves the following

Theorem 16.2 *At a non-stationary, sharp interface if the extra-entropy and the entropy production vanish then, to within an error* $[\![\frac{1}{2}(\mathbf{v} - \boldsymbol{\nu})^2]\!]$, *the continuity of the temperature,* $[\![\theta]\!] = 0$, *implies the continuity of the Gibbs free energy,* $[\![\phi]\!] = 0$.

We observe that the result $[\![\phi]\!] = 0$ is consistent with the equality of chemical potentials (9.76) in chemical reactions. Moreover, while $\eta_V(\theta, p)$ is expressed by (16.15) for any state of the vapour, the relation

$$\lambda(\theta, p) = \theta[\eta_V(\theta, p) - \eta_L(\theta, p)]$$

holds for the states (θ, p) at the interface. For definiteness, at $p = 1\,\text{atm}$, $\theta = 373.15\,\text{K}$ we have

$$\lambda = 40.67\ \text{kJ/mol}, \qquad \eta_V - \eta_L = 109\ \text{J/mol K}.$$

Clapeyron's equation

For a fluid the entropy η and the specific volume v are given by the Gibbs free energy ϕ via

$$\eta = -\partial_\theta\phi(\theta, p), \qquad v = \partial_p\phi(\theta, p).$$

These relations are assumed to hold for both vapour and liquid.

Let θ, p be constant (continuous) *across* the interface. The condition

$$\phi_V(\theta, p) = \phi_L(\theta, p) \tag{16.21}$$

results in the function $p = \hat{p}(\theta)$ characterizing the coexistence line. Differentiation of (16.21) with respect to θ yields

$$\partial_\theta\phi_V + \partial_p\phi_L\hat{p}' = \partial_\theta\phi_V + \partial_p\phi_L\hat{p}'.$$

Since $\partial_\theta\phi_V = -\eta_V$, $\partial_\theta\phi_L = -\eta_L$, $\partial_p\phi_V = v_V$, $\partial_p\phi_L = v_L$ then

$$\hat{p}' = -\frac{\partial_\theta\phi_V - \partial_\theta\phi_L}{\partial_p\phi_V - \partial_p\phi_L} = \frac{\eta_V - \eta_L}{v_V - v_L}.$$

Moreover, since $\eta_V - \eta_L = \lambda\theta$ then

$$\hat{p}' = \frac{\lambda}{\theta(v_V - v_L)}. \tag{16.22}$$

Equation (16.22), known as Clapeyron's equation, holds for any phase transition between continua where the stress is in fact a pressure tensor ($\mathbf{T} = -p\mathbf{1}$). Hence we can apply (16.22) to the transition ice-water provided we describe the stress as a pressure tensor. With the restriction to the pressure tensor, Eq. (16.22) applies to first-order phase transitions that take place at constant θ and p while the first-order derivatives of the Gibbs free energy, and hence η and v, suffer a jump discontinuity.

Two simple consequences are now derived from Clapeyron's equation (16.22). First we consider the water-vapour transition around the boiling point (p_B, θ_B). Since $v_V = 1677 \gg v_L = 1.043$ l/Kg, we may neglect v_L as compared to v_V. Moreover we may view the vapour as an ideal gas. Hence we have

$$p v_V = K\theta, \qquad p'(\theta) = \frac{\lambda}{\theta v_V(\theta, p)} = \frac{\lambda p}{K\theta^2},$$

where p is the pressure along the coexistence line. Assume λ is constant. Upon integration of the differential equation we have

$$\ln\frac{p}{p_B} = -\frac{\lambda}{K}\Big(\frac{1}{\theta} - \frac{1}{\theta_B}\Big), \qquad p = p_B \exp\Big[-\frac{\lambda}{K}\Big(\frac{1}{\theta} - \frac{1}{\theta_T}\Big)\Big].$$

Secondly, if stress in ice is modelled by $\mathbf{T} = -p\mathbf{1}$ then Clapeyron's equation applies also to the water-ice transition and we have

$$p'(\theta) = \frac{\lambda}{\theta(v_L - v_I)},$$

v_I being the ice specific volume. Since $v_L < v_I$ then $p'(\theta) < 0$. That is why, in the classical $\theta - p$ diagrams, the water-ice (coexistence) line has a negative large slope while the water-vapour line has a positive small slope. In addition, if the pressure on the ice is increased then the transition temperature decreases. This explains the experiment performed by placing a string on a block of ice and hanging two masses from the ends of the string. Sometime later, it is discovered that the string has passed through the ice without cutting it in half: the pressure caused by the string makes the ice melt just below the string, the string slides down, and the water freezes again just above the string.

Liquid

At the interface liquid and vapour coexist and, by (16.18), at any point of the interface we have $\phi_L = \phi_V$. We can then view $\phi_L - \phi_V = 0$ as the condition of equilibrium between liquid and vapour.

Denote the state of liquid and vapour by the pair (θ, p). The continuity of ϕ at the interface and the form (16.16) of ϕ_V suggests that we let

$$\phi_L(\theta, p) = \phi_V(\theta, p) - \kappa f(\theta, p) = K\theta \ln \frac{p}{p_0} + c_p\theta(1 - \ln \frac{\theta}{\theta_0} + \varepsilon_0 - \kappa f(\theta, p),$$

where $\kappa > 0$. Hence we obtain the entropy,

$$\eta_L(\theta, p) = -\partial_\theta \phi_L = \eta_V(\theta, p) + \kappa \partial_\theta f(\theta, p) = c_p \ln \frac{\theta}{\theta_0} - K \ln \frac{p}{p_0} + \kappa f(\theta, p),$$

and the enthalpy,

$$h_L(\theta, p) = \phi_L(\theta, p) + \theta\eta_L(\theta, p) = c_p\theta + \varepsilon_0 - \kappa[f(\theta, p) - \theta\partial_\theta f(\theta, p)]$$

Incidentally, since $\eta_V - \eta_L = \lambda/\theta$ then we find a hint for the determination of f, namely

$$\kappa\theta\partial_\theta f(\theta, p) = \frac{\lambda}{\theta};$$

the knowledge of $\lambda(\theta, p)$ at the interface determines $\partial_\theta f(\theta, p)$ at the interface.

Liquid–vapour coexistence line

For definiteness we let

$$f(\theta, p) = G(p) - F(\theta)$$

and require that $f = 0$ at the interface. Otherwise we might say that the continuity condition $\phi_L(\theta, p) = \phi_V(\theta, p)$ results in $0 = f(\theta, p) = G(p) - F(\theta)$. Anyway the coexistence line is supposed to take the form

$$G(p) = F(\theta), \tag{16.23}$$

where (θ, p) is any state at the interface. We now determine an approximate form of (16.23) to represent the water vapour coexistence line (or vapourization curve) from the triple point to the critical point.

Denote by T, B the triple point and the atmospheric boiling point. It is required that

$$G(p_T) = F(\theta_T), \qquad G(p_B) = F(\theta_B),$$

where $\theta_T = 273.16$ K, $\theta_B = 373.15$ K, $p_T = 611.657$ Pa, $p_B = 101.325 \cdot 10^3$ Pa. Since

$$\partial_p f(\theta, p) = G'(p), \qquad \partial_\theta f(\theta, p) = -F'(\theta),$$

then we have

$$\eta_V(\theta, p) - \eta_L(\theta, p) = \kappa F'(\theta), \qquad v_V - v_L = \kappa G'(p).$$

Let

$$G(p) = \ln \frac{p}{p_0}, \qquad F(\theta) = \alpha - \frac{\beta}{\theta - \gamma},$$

where p_0 is a reference pressure. The requirements

$$G(p_T) - F(\theta_T) = 0, \qquad G(p_B) - F(\theta_B) = 0$$

imply

$$\ln \frac{p_T}{p_0} - \alpha + \frac{\beta}{\theta_T - \gamma} = 0, \qquad \ln \frac{p_B}{p_0} - \alpha + \frac{\beta}{\theta_B - \gamma} = 0.$$

Hence it follows

$$\alpha = \ln \frac{p_T}{p_0} + \ln \frac{p_B}{p_T} \frac{\theta_B - \gamma}{\theta_B - \theta_T}, \qquad \beta = \ln \frac{p_B}{p_T} \frac{(\theta_B - \gamma)(\theta_T - \gamma)}{\theta_B - \theta_T}.$$

Consequently

$$\ln \frac{p}{p_T} = \ln \frac{p_B}{p_T} \frac{\theta_B - \gamma}{\theta_B - \theta_T} \frac{\theta - \theta_T}{\theta - \gamma}. \tag{16.24}$$

The parameter γ should be determined by a best fit procedure. An immediate simple estimate of γ is obtained by letting the curve $p = p(\theta)$ take the experimental value at the critical point. Letting θ_C and p_C be the values of θ and p at the critical point we find from (16.24) that

$$\frac{\theta_B - \gamma}{\theta_C - \gamma} = \frac{\theta_B - \theta_T}{\theta_C - \theta_T} \frac{\ln(p_C/p_T)}{\ln(p_B/p_B)} =: \delta < 1$$

whence

$$\gamma = \frac{\theta_B - \delta\theta_C}{1 - \delta}.$$

Since $\theta_C = 647.09\,\mathrm{K}$ and $p_C = 22.064 \cdot 10^6\,\mathrm{Pa}$ it follows that $\delta = 0.549$ and then

$$\gamma = 39.684\,\mathrm{K}.$$

Hence, letting $p_0 = p_T$ we find $\alpha = 17.042$, $\beta = 3978.855$. The coexistence line is then given the approximate formula

$$\ln \frac{p}{611.657} = 17.042 - \frac{3978.855}{\theta - 39.684},$$

with p in Pa and θ in K.

Evolution of the transition

The nonlinear system (16.12)–(16.14) gives insights into the effects of pressure and temperature on the transition. First we assume $U \neq 0$ whereby the relative mass flow-rate is nonzero. By (16.9), $\beta^2[\![v]\!] = -[\![p]\!]$. Since generally $[\![v]\!] \neq 0$ then $[\![p]\!] \neq 0$ and $[\![v]\!][\![p]\!] < 0$ or $[\![\rho]\!][\![p]\!] > 0$. In solid $\to$ liquid or liquid $\to$ vapour transitions most frequently $[\![v]\!] > 0$. Hence the transition is favoured by applying a smaller pressure in the final phase. In the ice $\to$ water transition $[\![\rho]\!] > 0$ and hence the transition is favoured with a greater pressure in the water phase.

Most often the transition occurs with a continuous pressure across the interface. We check how the jump conditions allow for transitions with $[\![p]\!] = 0$. Let $U = 0$, that is the interface is convected by the body. Hence

$$[\![\mathrm{v}]\!] = -[\![U]\!], \quad [\![p]\!] = \beta[\![\mathrm{v}]\!], \quad [\![p]\!] = -\beta^2[\![v]\!],$$

hold with $[\![\mathrm{v}]\!] = 0$ and $[\![p]\!] = 0$ while $[\![v]\!]$ and $[\![\rho]\!]$ are unrestricted. Since $[\![p]\!] = 0$ then the energy equation (16.10) holds identically. Finally, the entropy inequality is satisfied if

$$[\![\frac{q}{\theta}]\!] \leq 0.$$

16.3.2 Viscous Phases and Surface Tension

Based on the general jump conditions (16.1)–(16.5) we now evaluate the effects of viscosity of the phases and the curvature of the interface.

The two phases are viscous fluids described by

$$\mathbf{T}_{\pm} = -p_{\pm}\mathbf{1} + \mathcal{T}_{\pm},$$

where $\mathcal{T}$ is the viscous stress; we let

$$\mathcal{T} = 2\mu\mathbf{D} + \lambda(\mathrm{tr}\,\mathbf{D})\mathbf{1} + \hat{\mathcal{T}},$$

where $\hat{\mathcal{T}}$ stands for possibly nonlinear or higher-grade terms. The jump conditions (16.2), (16.3) become

$$\rho U[\![\mathbf{v}]\!] = [\![p]\!]\mathbf{m} - [\![\mathcal{T}]\!]\mathbf{m},$$

$$\rho U[\![\varepsilon + \tfrac{1}{2}\mathbf{v}^2]\!] = [\![p\mathrm{v}]\!] - \mathbf{m} \cdot [\![\mathcal{T}\mathbf{v}]\!] + \mathbf{m} \cdot [\![\mathbf{q}]\!].$$

This shows that the jump of $\mathbf{v}$ and ε are affected by the values of $\mathcal{T}$ and $\mathbf{q}$ at the sides of the interface and these in turn are induced by the solution v, $\mathbf{v}$, θ to the differential equations

$$\dot{v} = v\nabla \cdot \mathbf{v}, \quad \rho\dot{\mathbf{v}} = -\nabla p + \nabla \cdot \boldsymbol{T}, \quad \dot{\varepsilon} = -p\dot{v} + v\boldsymbol{T} \cdot \mathbf{D} + r - v\nabla \cdot \mathbf{q},$$

where $\boldsymbol{T}$, ε, $\mathbf{q}$ are meant to be replaced by the corresponding constitutive equations so that eventually we get a system of differential equations in the unknowns v, $\mathbf{v}$, θ.

We now wonder about the possible effect of $\boldsymbol{T}$ on the latent heat. Again we let ε, p, v be the values during the process (transition) from the liquid to the vapour phase or across the interface from the +side to the −side. If d is the duration of the process then

$$\int_0^d (\dot{\varepsilon} + p\dot{v})dt = \int_0^d v\boldsymbol{T} \cdot \mathbf{D}dt + \int_0^d (r - v\nabla \cdot \mathbf{q})dt.$$

Let P be the integral of p, $\partial_v P(\theta, v) = p(\theta, v)$. Then if, as we assume, the transition happens at constant temperature we have

$$\varepsilon_V + P(\theta, v_V) - \varepsilon_L + P(\theta, v_L) = \int_0^d v\boldsymbol{T} \cdot \mathbf{D}dt + \int_0^d (r - v\nabla \cdot \mathbf{q})dt.$$

Likewise by the entropy inequality we have

$$\dot{\eta} - (\frac{r}{\theta} - v\nabla \cdot \frac{\mathbf{q}}{\theta}) = v\boldsymbol{T} \cdot \mathbf{D} - \frac{v}{\theta^2}\mathbf{q} \cdot \nabla\theta$$

the right-hand side being the entropy production (per unit mass). Hence

$$\dot{\eta} = v\boldsymbol{T} \cdot \mathbf{D} + \frac{1}{\theta}(r - v\nabla \cdot \mathbf{q})$$

and then integrating we have

$$\theta(\eta_V - \eta_L) = \int_0^d v\boldsymbol{T} \cdot \mathbf{D}dt + \int_0^d (r - v\nabla \cdot \mathbf{q})dt.$$

It is then consistent to define the latent heat in the form

$$\lambda = \eta(\eta_V - \eta_L) = \varepsilon_V + P(\theta, v_V) - \varepsilon_L + P(\theta, v_L),$$

as with perfect fluids. Consequently we find again that

$$\phi_V = \phi_L$$

in the liquid–vapour transition.

In some models [15] the jump conditions are modified or improved by assuming that a surface tension σ at the interface produces an additional pressure $\sigma\kappa$, κ being the mean curvature of the interface. Formally we may use the present formulation by letting $\boldsymbol{T}\mathbf{m}$ include $\sigma\kappa\mathbf{m}$.

16.3.3 The Phase Rule

A phase containing n constituents is characterized by the mass fractions $\omega_1, ..., \omega_{n-1}$. To account for the mass conservation it is convenient to consider the dependences on the masses, $m_1, m_2, ...m_n$. If F phases may occur then we may have $(n-1)F$ different concentrations and 2 degrees of freedom, the temperature θ and the pressure p.

Let $\hat{\phi}_i(\omega_{i1}, \omega_{i2}, ..., \omega_{i(n-1)}) = \phi_i(m_{i1}, m_{i2}, ..., m_{in})$ be the Gibbs free energy of the ith phase. The Gibbs free energy of the whole system is the sum of the partial free energies, $\phi = \sum_h \phi_h$, and is assumed to have a minimum at equilibrium. If the kth constituent suffers the phase change $i \to j$ then

$$0 = \dot{\phi}_i + \dot{\phi}_j = \partial_{m_{ik}}\phi_i\, \dot{m}_{ik} + \partial_{m_{jk}}\phi_j\, \dot{m}_{jk}.$$

By the mass conservation, $\dot{m}_{ik} = -\dot{m}_{jk}$. Hence

$$(\partial_{m_{ik}}\phi_i - \partial_{m_{jk}}\phi_j)\dot{m}_{ik} = 0.$$

The arbitrariness of $\dot{m}_{ik}$ implies that

$$\partial_{m_{ik}}\phi_i = \partial_{m_{jk}}\phi_j, \qquad i \neq j = 1, 2, ..., F;\, k = 1, 2, ..., n.$$

Thus there are $n(F-1)$ equations with $2 + F(n-1)$ unknowns, θ, p, $\{\omega_{ik}\}$. The degree of variability, or the number of degrees of freedom, ν is given by

$$\nu = 2 + F(n-1) - n(F-1) = 2 + n - F.$$

The result $\nu = 2 + n - F$ is referred to as the (Gibbs) phase rule.

By way of example consider water in equilibrium with its vapour. We have $n = 1$ and $F = 2$; hence $\nu = 1$. We can choose arbitrarily the temperature but then the (equilibrium) pressure is uniquely determined.

At the triple point water is at equilibrium with ice and vapour. In this case $n = 1$ and $F = 3$. Thus $\nu = 0$. This means that there is no freedom of choice of the variables; the three phases can coexist only for a fixed value of the temperature and a fixed value of the pressure.

16.4 Solid–Fluid Transition

To model fluid–solid transitions we need to specify the constitutive equations for the stress. Fluids are reasonably viewed as thermoelastic materials where $\mathbf{T} = -p\mathbf{1}$ and $p = p(\theta, \rho)$. Solids are more involved. Ice models may be represented by

$$\mathbf{T} = -p\mathbf{1} + \mu(p)\mathbf{D} + \mathcal{T},$$

where $\mathcal{T}$ is a function of the Cotter–Rivlin derivative of $\mathbf{D}$ [14, 249]. Steel instead is modelled by a stress–strain relation with plastic effects [360]. If we denote generically with $\mathbf{T}_+$ the stress tensor in the solid phase while $\mathbf{T} = -p_-\mathbf{1}$ in the fluid phase then the jump conditions (16.2)–(16.3) become

$$\rho U[\![\mathbf{v}]\!] = -[\![\mathbf{T}]\!]\mathbf{m} = p_-\mathbf{m} + \mathbf{T}_+\mathbf{m},$$

$$\rho U[\![\varepsilon + \tfrac{1}{2}\mathbf{v}^2]\!] = -\mathbf{m}[\![\mathbf{T}\mathbf{v} - \mathbf{q}]\!] = p_-\mathbf{v}_- \cdot \mathbf{m} + \mathbf{m} \cdot \mathbf{T}_+\mathbf{v} + [\![\mathbf{q}]\!] \cdot \mathbf{m}.$$

If, instead, we let the fluid phase be viscous then $\mathbf{T} = -p\mathbf{1} + 2\mu\mathbf{D} + \lambda(\mathrm{tr}\,\mathbf{D})\mathbf{1}$ in the region behind $(-)$.

The solid might be modelled as linearly elastic and incompressible. We let $\mathcal{E} = \mathrm{sym}\mathbf{H} = \frac{1}{2}[\nabla_{\!R}\mathbf{u} + (\nabla_{\!R}\mathbf{u})^T]$ and assume (see Sect. 4.1.1)

$$\partial_{\mathcal{E}}\psi = 2\hat{\mu}(\theta)\mathcal{E}, \qquad \mathrm{tr}\,\mathcal{E} = \nabla_{\!R} \cdot \mathbf{u} = 0, \quad \mathrm{tr}\,\dot{\mathcal{E}} = \nabla_{\!R} \cdot \dot{\mathbf{u}} = 0.$$

Hence

$$\mathbf{T}_{RR} = -p\mathbf{1} + 2\mu\mathcal{E}, \qquad \mu = \rho_R\hat{\mu},$$

and, consistent with the linear approximation, we let

$$\mathbf{T} = -p\mathbf{1} + 2\mu\varepsilon, \qquad \mathrm{tr}\,\varepsilon = 0.$$

In any case, to determine ρU, $[\![\mathbf{v}]\!]$, $[\![\varepsilon]\!]$ we need knowing the possible jumps $[\![\mathbf{T}]\!]$ and $[\![\mathbf{q}]\!]$. Whenever the stress is qualitatively different in the two phases (e.g. pressure tensor in the fluid, elastic or viscous stress in the solid) the sharp interface looks inadequate. For instance we should have a stress jump $[\![\mathbf{T}]\!]$ that leads to $-p_-\mathbf{1}$ by starting from a non-isotropic $\mathbf{T}_+$. This motivates the recourse to placing a solid–fluid (or liquid–vapour) transition in a layer of finite thickness.[2]

Order parameter and phase field

The transition layer of finite thickness, instead of a sharp interface, is common within the phase field models. In the Landau theory of phase transitions [419] the free energy is taken as a function of the temperature and an order parameter, say e. The free energy ψ is then taken as a power expansion of e. At a sharp interface the function ψ depends on the values of e and θ at the interface. To account for non-local spatial effect and hence for a finite thickness of the interface the order parameter has to be a field, usually denoted as phase field in the literature.

There are cases where the phase field is advantageous for the modelling without taking a precise physical meaning. For instance the phase field is merely used to

[2] Sometimes named mushy region in the literature.

interpolate experimental data [444]. Further, a functional is assumed to depend on a phase variable, in addition to precise physical fields, and the corresponding differential equation is obtained by requiring the extremum property of the functional [9, 45].

In other cases the role of phase field is played by the concentration or mass fraction. In these cases though the field is subject to balance equations, in particular the consequences of the balance of mass. This view is pursued in the next section.

16.4.1 Transition in a Finite Layer—A Mixture Model

The solid–fluid transition is taken to occur in a layer of finite thickness. The two phases are regarded as the constituents of a mixture. Further, owing to the phase transition, the mixture is subject to mass exchanges between the constituents.

Denote by the indices F, S the quantities pertaining to the fluid and solid phases. Hence ρ_F, ρ_S are the mass densities, $\rho = \rho_F + \rho_S$ is the mass density of the mixture, and $\omega = \rho_F/\rho$ is the fluid mass fraction. If p is the pressure applied to the mixture we denote by p_F the partial pressure in the fluid and p_S that in the solid. The barycentric velocity $\mathbf{v}$ of the solid–fluid mixture is

$$\mathbf{v} = \omega\, \mathbf{v}_F + (1-\omega)\mathbf{v}_S.$$

Let $\mathbf{u}_F$, $\mathbf{u}_S$ be the diffusion velocities and $\mathbf{h}_F$, $\mathbf{h}_S$ the mass fluxes,

$$\mathbf{u}_F = \mathbf{v}_F - \mathbf{v}, \qquad \mathbf{u}_S = \mathbf{v}_S - \mathbf{v}, \qquad \mathbf{h}_F = \rho\omega\mathbf{u}_F, \qquad \mathbf{h}_S = \rho(1-\omega)\mathbf{u}_S;$$

by definition $\mathbf{h}_S + \mathbf{h}_F = \mathbf{0}$.

The phase transition results in an exchange of mass between the two phases as it happens in chemical reactions. This is expressed by the balances of mass,

$$\partial_t \rho_F + \nabla\cdot(\rho_F\mathbf{v}_F) = \tau, \qquad \partial_t \rho_S + \nabla\cdot(\rho_S\mathbf{v}_S) = -\tau,$$

where τ is the mass of fluid produced per unit volume and unit time. These balances can be expressed in terms of ρ and ω in the form

$$\rho\dot{\omega} = -\nabla\cdot\mathbf{h}_F + \tau, \qquad \dot{\rho} = -\rho\nabla\cdot\mathbf{v}. \tag{16.25}$$

The balances of linear momentum and energy are considered for the whole mixture in the standard form for continua,

$$\rho\dot{\mathbf{v}} = \nabla\cdot\mathbf{T} + \rho\mathbf{b}, \qquad \rho\dot{\varepsilon} = \mathbf{T}\cdot\mathbf{L} - \nabla\cdot\mathbf{q} + \rho r.$$

The entropy inequality is assumed by taking the entropy flux in the general form $\mathbf{j} = \mathbf{q}/\theta + \mathbf{k}$,

$$\rho\dot{\eta} \geq \frac{\rho r}{\theta} - \nabla\cdot(\frac{\mathbf{q}}{\theta} + \mathbf{k}).$$

Hence we have the Clausius–Duhem inequality in the form

$$-\rho(\dot{\psi} + \eta\dot{\theta}) + \mathbf{T}\cdot\mathbf{L} - \frac{1}{\theta}\mathbf{q}\cdot\nabla\theta + \theta\nabla\cdot\mathbf{k} \geq 0.$$

The constitutive equations are established for the functions $\psi, \eta, \mathbf{h}_F, \mathbf{T}, \mathbf{q}, \mathbf{k}$ in terms of the set of variables $\Gamma = (\theta, \rho, \omega, \nabla\theta, \nabla\omega, \mathbf{D}, \nabla\nabla\omega)$ while $\dot{\omega}$ and $\dot{\rho}$ are determined by (16.25). We assume $\psi(\Gamma)$ is differentiable. Computing $\dot{\psi}$ and substituting in the Clausius–Duhem inequality we have

$$-\rho(\partial_\theta\psi + \eta)\dot{\theta} - \rho\partial_\rho\psi\dot{\rho} - \rho\partial_\omega\psi\dot{\omega} - \rho\partial_{\nabla\theta}\psi\cdot\overset{\cdot}{\overline{\nabla\theta}} - \rho\partial_{\nabla\omega}\psi\cdot\overset{\cdot}{\overline{\nabla\omega}} - \rho\partial_{\mathbf{D}}\psi\cdot\dot{\mathbf{D}}$$
$$-\rho\partial_{\nabla\nabla\omega}\psi\cdot\overset{\cdot}{\overline{\nabla\nabla\omega}} + \mathbf{T}\cdot\mathbf{L} - \frac{1}{\theta}\mathbf{q}\cdot\nabla\theta + \theta\nabla\cdot\mathbf{k} \geq 0.$$

Recall that $\overset{\cdot}{\overline{\nabla\omega}} = \nabla\dot{\omega} - \mathbf{L}^T\nabla\omega$, (and the like for $\nabla\theta$), $\overset{\cdot}{\overline{\nabla\nabla\omega}} = \nabla\nabla\dot{\omega} - 2\mathrm{sym}\nabla\nabla\omega\mathbf{L}$, and $\dot{\rho} = -\rho\nabla\cdot\mathbf{v}$. Then we can write

$$-\rho(\partial_\theta\psi + \eta)\dot{\theta} - \rho\partial_\omega\psi\dot{\omega} - \rho\partial_{\nabla\omega}\psi\cdot\nabla\dot{\omega} - \rho\partial_{\nabla\nabla\omega}\psi\cdot(\nabla\nabla\dot{\omega} - 2\mathrm{sym}\nabla\nabla\omega\mathbf{L})$$
$$+[\mathbf{T} + \rho^2\partial_\rho\psi\mathbf{1} + \rho\nabla\theta\otimes\partial_{\nabla\theta}\psi + \rho\nabla\omega\otimes\partial_{\nabla\omega}\psi]\cdot(\mathbf{D} + \mathbf{W})$$
$$-\rho\partial_{\nabla\theta}\psi\cdot\nabla\dot{\theta} - \rho\partial_{\mathbf{D}}\psi\cdot\dot{\mathbf{D}} - \frac{1}{\theta}\mathbf{q}\cdot\nabla\theta + \theta\nabla\cdot\mathbf{k} \geq 0.$$

The arbitrariness of $\nabla\nabla\dot{\omega}$, $\nabla\dot{\theta}$, $\dot{\mathbf{D}}$, $\dot{\theta}$, and $\mathbf{W}$ implies

$$\partial_{\nabla\nabla\omega}\psi = \mathbf{0}, \qquad \partial_{\nabla\theta}\psi = \mathbf{0}, \qquad \partial_{\mathbf{D}}\psi = \mathbf{0}, \qquad \eta = -\partial_\theta\psi,$$

$$\mathbf{T} + \rho\nabla\omega\otimes\partial_{\nabla\omega}\psi \in \mathrm{Sym}.$$

Let $\mathcal{T} = \mathbf{T} + \rho^2\partial_\rho\psi\mathbf{1} + \rho\nabla\omega\otimes\partial_{\nabla\omega}\psi$. Upon dividing by θ we can write the remaining inequality in the form

$$-\frac{\rho}{\theta}\partial_\omega\psi\dot{\omega} - \frac{\rho}{\theta}\partial_{\nabla\omega}\psi\cdot\nabla\dot{\omega} + \frac{1}{\theta}\mathcal{T}\cdot\mathbf{D} - \frac{1}{\theta^2}\mathbf{q}\cdot\nabla\theta + \nabla\cdot\mathbf{k} \geq 0.$$

The identity

$$-\frac{\rho}{\theta}\partial_{\nabla\omega}\psi\cdot\nabla\dot{\omega} = -\nabla\cdot(\frac{\rho}{\theta}\partial_{\nabla\omega}\psi\dot{\omega}) + [\nabla\cdot(\frac{\rho}{\theta}\partial_{\nabla\omega}\psi)]\dot{\omega}$$

allows us to write

$$-\frac{\rho}{\theta}\delta_\omega\psi\dot{\omega} + \frac{1}{\theta}\mathcal{T}\cdot\mathbf{D} - \frac{1}{\theta^2}\mathbf{q}\cdot\nabla\theta + \nabla\cdot(\mathbf{k} - \frac{\rho}{\theta}\partial_{\nabla\omega}\psi\dot{\omega}) \geq 0,$$

where

$$\delta_\omega \psi = \partial_\omega \psi - \frac{\theta}{\rho} \nabla \cdot (\frac{\rho}{\theta} \partial_{\nabla\omega} \psi).$$

In light of $(16.25)_1$ we have

$$-\frac{\rho}{\theta} \delta_\omega \psi \, \dot{\omega} = -\frac{1}{\theta} \delta_\omega \psi \, \tau + \frac{1}{\theta} \delta_\omega \psi \, \nabla \cdot \mathbf{h}_F,$$

while

$$\frac{1}{\theta} \delta_\omega \psi \, \nabla \cdot \mathbf{h}_F = \nabla \cdot (\frac{1}{\theta} \delta_\omega \psi \, \mathbf{h}_F) - \mathbf{h}_F \cdot \nabla \frac{\delta_\omega \psi}{\theta}.$$

Hence the inequality can be written in the form

$$\frac{1}{\theta} \mathcal{T} \cdot \mathbf{D} - \frac{1}{\theta^2} \mathbf{q} \cdot \nabla \theta - \frac{1}{\theta} \delta_\omega \psi \, \tau - \mathbf{h}_F \cdot \nabla \frac{\delta_\omega \psi}{\theta} + \nabla \cdot (\mathbf{k} + \frac{1}{\theta} \delta_\omega \psi \; - \frac{\rho}{\theta} \partial_{\nabla\omega} \psi \, \dot{\omega}) \geq 0.$$

This inequality holds if

$$\mathbf{k} = -\frac{1}{\theta} \delta_\omega \psi \; + \frac{\rho}{\theta} \partial_{\nabla\omega} \psi \, \dot{\omega}$$

and

$$\mathcal{T} \cdot \mathbf{D} \geq 0, \qquad \mathbf{q} \cdot \nabla \theta \leq 0, \qquad \delta_\omega \psi \, \tau \leq 0, \qquad \mathbf{h}_F \cdot \nabla \frac{\delta_\omega \psi}{\theta} \leq 0.$$

Since ψ is a function of $(\theta, \rho, \omega, \nabla\omega)$ then ψ may depend on $\nabla\omega$ only through $|\nabla\omega|$. Consequently

$$\nabla\omega \otimes \partial_{\nabla\omega} \psi = \frac{\partial_{|\nabla\omega|} \psi}{|\nabla\omega|} \nabla\omega \otimes \nabla\omega \in \mathrm{Sym}$$

and hence $\mathbf{T} \in \mathrm{Sym}$. Moreover, the result for $\mathcal{T}$ implies that the stress $\mathbf{T}$ has the form

$$\mathbf{T} = -\rho^2 \partial_\rho \psi \mathbf{1} - \rho \frac{\partial_{|\nabla\omega|} \psi}{|\nabla\omega|} \nabla\omega \otimes \nabla\omega + \mathcal{T},$$

where $\mathcal{T}$ may be any (dissipative) stress function of $\mathbf{D}$ subject to $\mathcal{T} \cdot \mathbf{D} \geq 0$.

To compute $\delta_\omega \psi$ we observe that, by the theory of mixtures,

$$\rho\psi = \textstyle\sum_\alpha \rho_\alpha (\psi_\alpha + \frac{1}{2} \mathbf{u}_\alpha^2),$$

where α labels the peculiar (partial) terms; each ψ_α determines the partial pressure in the form $p_\alpha = \rho_\alpha^2 \partial_{\rho_\alpha} \psi_\alpha$. Here we neglect the diffusive terms $\mathbf{u}_\alpha^2$ and consider $\psi_F = \psi_F(\theta, \rho_F)$, $\psi_S = \psi_S(\theta, \rho_S)$ whence

$$\psi = \omega \psi_F(\theta, \rho\omega) + (1 - \omega) \psi_S(\theta, (1 - \omega)\rho).$$

It follows that

$$\partial_\omega \psi = \psi_F - \psi_S + \frac{p_F}{\rho_F} - \frac{p_S}{\rho_S} = \phi_F - \phi_S;$$

$\phi_F = \mu_F$ and $\phi_s = \mu_s$ are the Gibbs free energies or chemical potentials of the two phases. As with chemical reactions, $\tau\delta_\omega\psi \le 0$ means that equilibrium (i.e. $\tau = 0$) occurs when $\delta_\omega\psi = 0$ that is when

$$\phi_F = \phi_S.$$

Very often in the literature the governing equations are derived by starting from a (Ginzburg–Landau) functional for the entropy [45] or the free energy [54, 215] with a density function dependent on the parameter ω and a term $|\nabla\omega|^2$ to account for an interface energy. Here we can do the analogue, though consistent with the mixture model, by considering ψ in the form

$$\psi = \omega[\psi_F(\theta, \rho\omega) + \tfrac{1}{2}\gamma_F|\nabla\omega|^2] + (1-\omega)[\psi_s(\theta, \rho(1-\omega)) + \tfrac{1}{2}\gamma_s|\nabla\omega^2|].$$

If $\gamma_F = \gamma_s = \gamma$ then

$$\delta_\omega\psi = \phi_F - \phi_s - \frac{\theta}{\rho}\nabla\cdot(\frac{\rho}{\theta}\gamma\nabla\omega).$$

For formal simplicity let $\gamma_F = \gamma_s = \gamma$. Inequality $\tau\delta_\omega\psi \le 0$ leads to

$$\tau = -\kappa(\theta, \rho, \phi)\{\phi_F(\theta, \rho\omega) - \phi_s(\theta, \rho(1-\omega)) - \frac{\theta}{\rho}\nabla\cdot(\frac{\rho}{\theta}\gamma\nabla\omega)\}, \qquad \kappa \ge 0 \tag{16.26}$$

while $\mathbf{h}_F \cdot \nabla\delta_\omega\psi/\theta \le 0$ leads to

$$\mathbf{h}_F = -M\nabla\{\phi_F(\theta, \rho\omega) - \phi_s(\theta, \rho(1-\omega)) - \frac{\theta}{\rho}\nabla\cdot(\frac{\rho}{\theta}\gamma\nabla\omega)\}, \qquad M \ge 0. \tag{16.27}$$

We now let ρ, θ, γ be constants.

If $\tau \simeq \rho\dot\omega$ then (16.26) simplifies to

$$\dot\omega = -\frac{\kappa}{\rho}f(\omega) + \frac{\kappa\gamma}{\rho}\Delta\omega, \qquad f(\omega) := \phi_F(\theta, \rho\omega) - \phi_s(\theta, \rho(1-\omega)),$$

namely the Ginzburg–Landau equation. If instead we let $\rho\dot\omega \simeq -\nabla\cdot\mathbf{h}_F$ then (16.27) leads to

$$\dot\omega = \frac{1}{\rho\theta}\nabla\cdot M\nabla[f(\omega) - \gamma\Delta\omega],$$

namely the Cahn–Hilliard equation.

The $|\nabla\omega|^2$ term is considered as a negative contribution to the entropy density (e.g. $-\frac{1}{2}\epsilon_\omega^2|\nabla\omega|^2$ in [45]). We might then take γ as proportional to the temperature θ, $\gamma = \theta\gamma^*$, so that

$$\psi = f(\theta, \omega) + \tfrac{1}{2}\theta\gamma^*|\nabla\theta|^2, \qquad \eta = -\partial_\theta f - \tfrac{1}{2}\gamma^*|\nabla\omega|^2, \qquad \varepsilon = \psi + \theta\eta = f(\theta, \omega),$$

thus getting a scheme consistent with standard approaches [54, 215].

16.4.2 Evolution of the Mass Fraction

If the phase field is just the mass fraction then we have to account for the whole set of balance equations. The Ginzburg–Landau and Cahn–Hilliard equations may be of interest when the phase field is not merely the mass fraction or the volume fraction.

The evolution of the mass fraction ω is governed by the balance of energy. We start from the equation given by the theory of mixtures along with the approximations $\mathbf{T}_\alpha = -p_\alpha \mathbf{1}$, $p_\alpha = \omega_\alpha p$, and the neglect of the interaction terms $\hat{\varepsilon}_\alpha$. Hence we can write

$$\rho_\alpha \grave{\varepsilon}_\alpha = -p\omega_\alpha \nabla \cdot \mathbf{v}_\alpha - \nabla \cdot \mathbf{q}_\alpha + \rho_\alpha r_\alpha.$$

Since

$$\frac{\grave{v}_\alpha}{v_\alpha} = -\frac{\grave{\rho}_\alpha}{\rho_\alpha} = \nabla \cdot \mathbf{v}_\alpha - v_\alpha \tau_\alpha$$

then we have

$$\rho_\alpha \grave{\varepsilon}_\alpha = -p\omega_\alpha \Big(\frac{\grave{v}_\alpha}{v_\alpha} + v_\alpha \tau_\alpha\Big) - \nabla \cdot \mathbf{q}_\alpha + \rho_\alpha r_\alpha$$

whence

$$\grave{\varepsilon}_\alpha + p\omega_\alpha \grave{v}_\alpha = -p\omega_\alpha v_\alpha^2 \tau_\alpha - v_\alpha \nabla \cdot \mathbf{q}_\alpha + r_\alpha.$$

Observe that

$$\grave{\omega}_\alpha = \dot{\omega}_\alpha + \mathbf{u}_\alpha \cdot \nabla\omega_\alpha = -\frac{1}{\rho}\nabla \cdot (\rho_\alpha \mathbf{u}_\alpha) + \mathbf{u}_\alpha \cdot \nabla\omega_\alpha + \frac{1}{\rho}\tau_\alpha = -\frac{1}{\rho}\omega_\alpha \nabla \cdot (\rho \mathbf{u}_\alpha) + \frac{1}{\rho}\tau_\alpha.$$

Hence, at constant pressure,

$$p\omega_\alpha \grave{v}_\alpha = \grave{\overline{p\omega_\alpha v_\alpha}} - p\grave{\omega}_\alpha v_\alpha = \grave{\overline{p\omega_\alpha v_\alpha}} + \frac{p v_\alpha \omega_\alpha}{\rho}\nabla \cdot (\rho \mathbf{u}_\alpha) - \frac{p v_\alpha}{\rho}\tau_\alpha.$$

Notice that $h_\alpha = \varepsilon_\alpha + p\omega_\alpha v_\alpha$ is the enthalpy per unit mass. Hence upon substitution and neglect of the diffusive term $p v_\alpha \nabla \cdot (\rho \mathbf{u}_\alpha)$ we obtain

$$\rho_\alpha \grave{h}_\alpha \simeq -\nabla \cdot \mathbf{q}_\alpha + \rho_\alpha r_\alpha. \tag{16.28}$$

The sum of (16.28) over α gives the total heat supplied per unit volume. We now examine how (16.28) governs the time dependence of the mass fraction. To this purpose we still neglect the diffusive terms ($\grave{h}_\alpha \simeq \dot{h}_\alpha$) and sum over α to obtain[3]

$$\sum_\alpha \rho_\alpha \dot{h}_\alpha = -\nabla\cdot\mathbf{q} + \rho r, \qquad \mathbf{q} = \sum_\alpha \mathbf{q}_\alpha, \quad \rho r = \sum_\alpha \rho_\alpha r_\alpha.$$

Go back to fluid and solid constituents. Assume

$$h_F = h_F(\theta, \rho_F), \quad \rho_F = \rho\omega, \qquad h_s = h_s(\theta, \rho_S), \quad \rho_S = \rho(1-\omega).$$

If $\rho_F \simeq \rho_S$ then it is reasonable to assume ρ is constant and hence

$$\dot{\rho}_F = \rho\dot{\omega}, \qquad \dot{\rho}_S = -\rho\dot{\omega}.$$

Otherwise we have

$$\dot{\rho}_F = \omega\dot{\rho} - \nabla\cdot(\rho_F \mathbf{u}_F) + \tau$$

and we might exploit ρ_F instead of ω to describe the evolution of the transition.

Take ω to describe the transition and observe that (with $\dot{\rho}_F \simeq \rho\dot{\omega}$)

$$\rho_F \dot{h}_F + \rho_S \dot{h}_S = \rho(\rho_F \partial_{\rho_F} h_F - \rho_S \partial_{\rho_S} h_S)\dot{\omega}.$$

Hence we have

$$\dot{\omega} = \frac{1}{\rho(\rho_F \partial_{\rho_F} h_F - \rho_S \partial_{\rho_S} h_S)}(\rho r - \nabla\cdot\mathbf{q}). \tag{16.29}$$

16.5 Solidification-Melting of Binary Alloys

The transition solidification-melting is considered for alloys. For formal simplicity we restrict attention to binary (isomorphous) alloys and consider the continuum as a mixture of the two components A, B; as an example, we may have in mind the copper-nickel alloy [16]. The phase diagram shows three regions, call them Liquid, Liquid-Solid, Solid. The phase boundary which limits the liquid phase is called the *Liquidus* line; the line which limits the solid phase is called the *Solidus* line (Fig. 16.2).

Let θ_A, θ_B be the melting, or solidification, temperatures of A, B and take $\theta_A < \theta_B$. Consider the alloy, at a given composition, in the liquid phase, $\theta > \theta_L$. As the temperature θ decreases up to θ_L, on the Liquidus line, a first transition occurs where the liquid alloy ABl separates into A liquid and B solid, thanks to the corresponding absorption of energy. Once the transition is completed we have a mixture of A liquid and B solid. Upon decreasing the temperature the mixture reaches the state on the Solidus line, $\theta = \theta_S$. By subtracting energy, the transition occurs between A liquid - B solid and the solid alloy ABs.

---

[3] The literature shows various schemes where the evolution of the transition is given by the enthalpy, see e.g. [355].

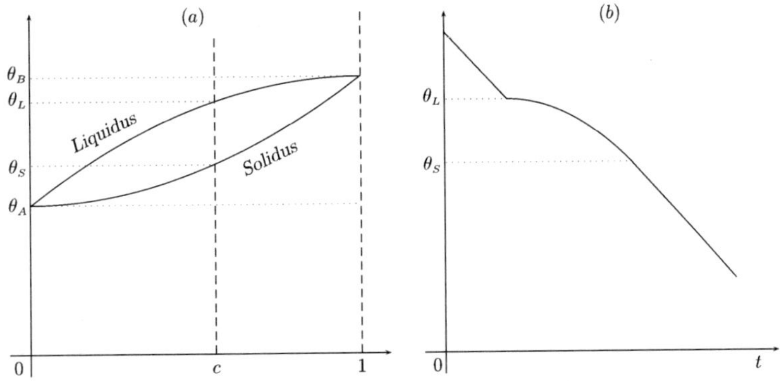

Fig. 16.2 **a** Liquidus-solidus diagram: the temperature θ is plotted versus c, the concentration of B. **b** Cooling curve: the temperature θ is plotted versus time t with constant heat removed

Both transitions can be modelled as a chemical reaction within a mixture of appropriate constituents. The transition at the Liquidus is viewed as a reaction between ABl, Al, Bs, with exchanges of atoms of B between ABl and Bs. Likewise, the transition at the Solidus is viewed as a reaction between Al, Bs, ABs with exchanges of atoms of A between Al and ABs. For definiteness here we describe the transition at the Liquidus and hence the chemical reaction between ABl, Al, Bs.

For simplicity we describe the mixture by means of the balances for a single body; the balance of energy and the entropy inequality are then taken in the standard form

$$\rho\dot{\varepsilon} = \mathbf{T} \cdot \mathbf{L} - \nabla \cdot \mathbf{q} + \rho r,$$

$$-\rho(\dot{\psi} + \eta\dot{\theta}) + \mathbf{T} \cdot \mathbf{L} - \frac{1}{\theta}\mathbf{q} \cdot \nabla\theta + \theta\nabla \cdot \mathbf{k} \geq 0.$$

For a more detailed description of the mixture we take advantage from the properties arising from the theory of mixtures. In the transition ABl $\Longleftrightarrow$ Al, Bs there is a mass exchange of B between ABl and Bs. Denote by the subscripts ABl, Bs, Al the quantities pertaining to ABl, Bs, Al and let

$$\rho = \rho_{ABl} + \rho_{BS} + \rho_{Al}, \qquad \rho\mathbf{v} = \rho_{ABl}\mathbf{v}_{ABl} + \rho_{BS}\mathbf{v}_{BS} + \rho_{Al}\mathbf{v}_{Al}.$$

We then consider the balances of mass, in the barycentric frame, in the form

$$\dot{\rho}_{ABl} = -\nabla \cdot (\rho_{ABl}\mathbf{u}_{ABl}) - \rho_{ABl}\nabla \cdot \mathbf{v} + \tau,$$

$$\dot{\rho}_{BS} = -\nabla \cdot (\rho_{BS}\mathbf{u}_{BS}) - \rho_{BS}\nabla \cdot \mathbf{v} - \tau,$$

$$\dot{\rho}_{Al} = -\nabla \cdot (\rho_{Al}\mathbf{u}_{Al}) - \rho_{Al}\nabla \cdot \mathbf{v},$$

where τ is the mass supply of B in ABl. Moreover the mass fractions $\omega_\alpha = \rho_\alpha/\rho$ are governed by

$$\rho\dot{\omega}_{ABl} = -\nabla\cdot(\rho_{ABl}\mathbf{u}_{ABl}) + \tau, \quad \rho\dot{\omega}_{Bs} = -\nabla\cdot(\rho_{Bs}\mathbf{u}_{Bs}) - \tau, \quad \rho\dot{\omega}_{Al} = -\nabla\cdot(\rho_{Al}\mathbf{u}_{Al}).$$

The constitutive properties of the mixture are described by the set of independent variables

$$\Gamma = (\theta, \{\rho_\alpha\}, \mathbf{D}, \nabla\theta, \{\nabla\rho_\alpha\}, \{\mathbf{u}_\alpha\}, \{\dot{\rho}_\alpha\},).$$

We let the free energy ψ, the entropy η, the stress $\mathbf{T}$, the heat flux $\mathbf{q}$, the extra-entropy flux $\mathbf{k}$ be continuous functions of Γ. Indeed, by the theory of mixtures, the free energy ψ is related to the peculiar free energies $\{\psi_\alpha\}$ of the constituents by

$$\rho\psi = \sum_\alpha \rho_\alpha(\psi_\alpha + \tfrac{1}{2}\mathbf{u}_\alpha^2).$$

Here we take the approximation that the quantities $\{\mathbf{u}_\alpha^2\}$ are negligible and hence we let

$$\rho\psi = \sum_\alpha \rho_\alpha\psi_\alpha(\theta, \rho_\alpha, \nabla\theta, \nabla\rho_\alpha, \mathbf{D}, \dot{\rho}_\alpha, \nabla\nabla\rho_\alpha), \qquad \alpha = \text{ABl, Bs, Al}.$$

Moreover ψ is taken to be continuously differentiable.

For a compact writing we first compute $\rho\dot{\psi}$ in the case where $\psi_\alpha = \psi_\alpha(\rho_\alpha)$. Then

$$\dot{\psi} = \sum_\alpha \overline{\omega_\alpha\psi_\alpha}^{\cdot} = \sum_\alpha \dot{\omega}_\alpha\psi_\alpha + \omega_\alpha\dot{\rho}_\alpha\partial_{\rho_\alpha}\psi_\alpha.$$

Since

$$\dot{\rho}_\alpha = -\nabla\cdot\mathbf{h}_\alpha - \rho_\alpha\nabla\cdot\mathbf{v} + \tau_\alpha$$

then

$$\rho\dot{\psi} = \sum_\alpha(\rho\dot{\omega}_\alpha\psi_\alpha + \rho_\alpha\partial_{\rho_\alpha}\psi_\alpha\dot{\rho}_\alpha) = \sum_\alpha\psi_\alpha(\tau_\alpha - \nabla\cdot\mathbf{h}_\alpha) + \sum_\alpha\rho_\alpha\partial_{\rho_\alpha}\psi_\alpha(\tau_\alpha - \nabla\cdot\mathbf{h}_\alpha - \rho_\alpha\nabla\cdot\mathbf{v}).$$

We let $\rho_\alpha^2\partial_{\rho_\alpha}\psi_\alpha = p_\alpha$ be the partial pressure and $p = \sum_\alpha p_\alpha$ the pressure in the mixture while $\mu_\alpha = \psi_\alpha + p_\alpha/\rho_\alpha$, $\mathbf{h}_\alpha = \rho_\alpha\mathbf{u}_\alpha$. It follows

$$\rho\dot{\psi} = \sum_\alpha[-\mu_\alpha\nabla\cdot\mathbf{h}_\alpha + \tau_\alpha\mu_\alpha - p\nabla\cdot\mathbf{v}].$$

With this in mind we now compute $\rho\dot{\psi}$ in more involved models. First we ignore spatial interaction effects via gradients of mass densities. Hence we take

$$\psi = \sum_\alpha \omega_\alpha\psi_\alpha(\theta, \rho_\alpha);$$

the dependence on $\mathbf{D}$ and $\nabla\theta$ would be ruled out by the Clausius–Duhem inequality; we omit the details. The Clausius–Duhem inequality can then be written in the form

$$-\rho(\partial_\theta\psi+\eta)\dot\theta+p\nabla\cdot\mathbf{v}-\sum_\alpha[\tau_\alpha\mu_\alpha-\mu_\alpha\nabla\cdot\mathbf{h}_\alpha)]+\mathbf{T}\cdot\mathbf{L}-\frac{1}{\theta}\mathbf{q}\cdot\nabla\theta+\theta\nabla\cdot\mathbf{k}\ge 0.$$

The arbitrariness of $\dot\theta$ implies that $\eta=-\partial_\theta\psi$. Moreover, since $\mathbf{L}=\mathbf{D}+\mathbf{W}$ then the arbitrariness of $\mathbf{W}$ implies $\mathbf{T}\in \mathrm{Sym}$. Now observe that $p\nabla\cdot\mathbf{v}+\mathbf{T}\cdot\mathbf{D}=(\mathbf{T}+p\mathbf{1})\cdot\mathbf{D}$. Divide the remaining inequality by θ,

$$-\frac{1}{\theta}\sum_\alpha\tau_\alpha\mu_\alpha+\frac{1}{\theta}\sum_\alpha\mu_\alpha\nabla\cdot\mathbf{h}_\alpha+\frac{1}{\theta}(\mathbf{T}+p\mathbf{1})\cdot\mathbf{D}-\frac{1}{\theta^2}\mathbf{q}\cdot\nabla\theta+\nabla\cdot\mathbf{k}\ge 0,$$

and use the identity

$$\frac{1}{\theta}\mu_\alpha\nabla\cdot\mathbf{h}_\alpha=\nabla\cdot\frac{1}{\theta}\mu_\alpha\mathbf{h}_\alpha-\mathbf{h}_\alpha\cdot\nabla\frac{\mu_\alpha}{\theta}$$

to obtain

$$-\frac{1}{\theta}\sum_\alpha\tau_\alpha\mu_\alpha+\sum_\alpha\mathbf{h}_\alpha\cdot\nabla\frac{\mu_\alpha}{\theta}+\frac{1}{\theta}(\mathbf{T}+p\mathbf{1})\cdot\mathbf{D}-\frac{1}{\theta^2}\mathbf{q}\cdot\nabla\theta+\nabla\cdot(\mathbf{k}+\frac{1}{\theta}\mu_\alpha\mathbf{h}_\alpha)\ge 0.$$

This inequality is satisfied if

$$\mathbf{T}=-p\mathbf{1}+\mathcal{T},\quad \mathcal{T}\cdot\mathbf{D}\ge 0,\qquad \sum_\alpha\mathbf{h}_\alpha\cdot\nabla\frac{\mu_\alpha}{\theta}\le 0,\qquad \sum_\alpha\tau_\alpha\mu_\alpha\le 0,$$

while $\mathbf{k}=-\sum_\alpha\mu_\alpha\mathbf{h}_\alpha/\theta$ and $\mathbf{q}\cdot\nabla\theta\le 0$. As expected $\sum_\alpha\tau_\alpha\mu_\alpha\le 0$ is consistent with (9.75). Here it follows that

$$\tau(\mu_{ABl}-\mu_{Bs})\le 0. \tag{16.30}$$

We may conclude that the transition occurs at the temperature θ_L such that

$$\mu_{ABl}(\theta_L)=\mu_{Bs}(\theta_L)$$

and does not occur at $\theta_B>\theta_L$. This in turn means that the alloy AB, with $\theta_A<\theta_B$, has a temperature of solidification which is lower than that of the metal B. This in turn shows that μ_{ABl} is a decreasing function of temperature. Moreover, inequality (16.30) implies that $\tau\ge 0$ if and only if $\mu_{ABl}-\mu_{Bs}\le 0$ and vice versa; matter flows toward the phase at a lower chemical potential. Inequality (16.30) indicates the direction of the transition; the value of τ is given by the balance of energy as with (16.29).

Strictly analogous conclusions hold if we consider the transition at the Solidus line between Al, Bs, and ABs.

Spatial interaction may be modelled by allowing for a dependence of ψ_α also on $\nabla\rho_\alpha, \dot\rho_\alpha, \nabla\nabla\rho_\alpha$. We find that $\partial_{\dot\rho_\alpha}\psi_\alpha=\mathbf{0}$, $\partial_{\nabla\nabla\rho_\alpha}\psi_\alpha=\mathbf{0}$ and

$$-\frac{1}{\theta}\textstyle\sum_\alpha \psi_\alpha(\tau_\alpha - \nabla\cdot\mathbf{h}_\alpha) - \sum_\alpha \frac{\rho_\alpha}{\theta}\partial_{\rho_\alpha}\psi_\alpha\dot{\rho}_\alpha - \sum_\alpha \frac{\rho_\alpha}{\theta}\partial_{\nabla\rho_\alpha}\psi_\alpha\cdot(\nabla\dot{\rho}_\alpha - \mathbf{L}^T\nabla\rho_\alpha)$$
$$+\frac{1}{\theta}\mathbf{T}\cdot\mathbf{L} - \frac{1}{\theta^2}\mathbf{q}\cdot\nabla\theta + \nabla\cdot\mathbf{k} \geq 0.$$

Applying the standard identity to $\sum_\alpha(\rho_\alpha/\theta)\partial_{\nabla\rho_\alpha}\psi_\alpha\cdot\nabla\dot{\rho}_\alpha$ we can write the inequality

$$-\frac{1}{\theta}\textstyle\sum_\alpha \psi_\alpha(\tau_\alpha - \nabla\cdot\mathbf{h}_\alpha) - \sum_\alpha \frac{\rho_\alpha}{\theta}\delta_{\rho_\alpha}\psi_\alpha\dot{\rho}_\alpha + \frac{1}{\theta}[\mathbf{T} + \sum_\alpha \rho_\alpha\nabla\rho_\alpha\otimes\partial_{\nabla\rho_\alpha}\psi_\alpha]\cdot\mathbf{L} - \frac{1}{\theta^2}\mathbf{q}\cdot\nabla\theta$$
$$+\nabla\cdot(\mathbf{k} - \textstyle\sum_\alpha \frac{\rho_\alpha}{\theta}\partial_{\nabla\rho_\alpha}\psi_\alpha\dot{\rho}_\alpha) \geq 0,$$

where $\delta_{\rho_\alpha}\psi_\alpha = \partial_{\rho_\alpha}\psi_\alpha - (\theta/\rho_\alpha)\nabla\cdot[(\rho_\alpha/\theta)\partial_{\nabla\rho_\alpha}\psi_\alpha]$. Now we replace $\dot{\rho}_\alpha$ and define

$$p_\alpha = \rho_\alpha^2\delta_{\rho_\alpha}\psi_\alpha, \qquad \mu_\alpha = \psi_\alpha + \rho_\alpha\delta_{\rho_\alpha}\psi_\alpha. \tag{16.31}$$

Letting $\mathcal{T} = \mathbf{T} + p\mathbf{1}$ we have

$$-\frac{1}{\theta}\textstyle\sum_\alpha \mu_\alpha(\tau_\alpha - \nabla\cdot\mathbf{h}_\alpha) + \frac{1}{\theta}[\mathcal{T} + \sum_\alpha \rho_\alpha(\nabla\rho_\alpha\otimes\partial_{\nabla\rho_\alpha}\psi_\alpha)]\cdot\mathbf{L} - \frac{1}{\theta^2}\mathbf{q}\cdot\nabla\theta + \nabla\cdot(\mathbf{k} - \sum_\alpha \frac{\rho_\alpha}{\theta}\psi_\alpha\dot{\rho}_\alpha) \geq 0$$

Applying the standard identity to $\sum_\alpha(\mu_\alpha/\theta)\nabla\cdot\mathbf{h}_\alpha$ we find that

$$-\frac{1}{\theta}\textstyle\sum_\alpha \mu_\alpha\tau_\alpha - \sum_\alpha \mathbf{h}_\alpha\cdot\nabla\frac{\mu_\alpha}{\theta} + \frac{1}{\theta}[\mathcal{T} + \sum_\alpha \rho_\alpha(\nabla\rho_\alpha\otimes\partial_{\nabla\rho_\alpha}\psi_\alpha)]\cdot\mathbf{L} - \frac{1}{\theta^2}\mathbf{q}\cdot\nabla\theta$$
$$+\nabla\cdot(\mathbf{k} - \textstyle\sum_\alpha \frac{\rho_\alpha}{\theta}\partial_{\nabla\rho_\alpha}\psi_\alpha\dot{\rho}_\alpha + \sum_\alpha \frac{\mu_\alpha}{\theta}\mathbf{h}_\alpha) \geq 0.$$

Again we find the condition $\sum_\alpha \mu_\alpha\tau_\alpha \leq 0$ though with the improved definition $(16.31)_2$ of the chemical potential μ_α. The arbitrariness of $\mathbf{W} = \text{skw}\mathbf{L}$ implies $\mathcal{T} + \sum_\alpha \rho_\alpha(\nabla\rho_\alpha\otimes\partial_{\nabla\rho_\alpha}\psi_\alpha) \in \text{Sym}$. If we assume $\mathcal{T} \in \text{Sym}$ then it follows $\partial_{\nabla\rho_\alpha}\psi_\alpha \propto \nabla\rho_\alpha$; this implies that ψ_α depends on $\nabla\rho_\alpha$ through $|\nabla\rho_\alpha|$.

16.6 Brine Channels Formation in Sea Ice

The presence of salt (solute) in water (solvent) results in the freezing point depression. This phenomenon is referred to as cryoscopic effect and depends only on the solute concentration.

During sea ice formation, salt is excluded from the solid phase. The dissolved salt makes the water of the ocean more dense. The salinity of the liquid phase thus increases and this lowers the freezing point. When the salinity is over 0.05 the solution is named brine. Nevertheless, at the end of the freezing process the salinity of the solution never reaches the unit value; the salt concentration has an upper limit which depends on temperature [263]. At room temperature ($\theta_r \simeq 291$ K) the solubility limit

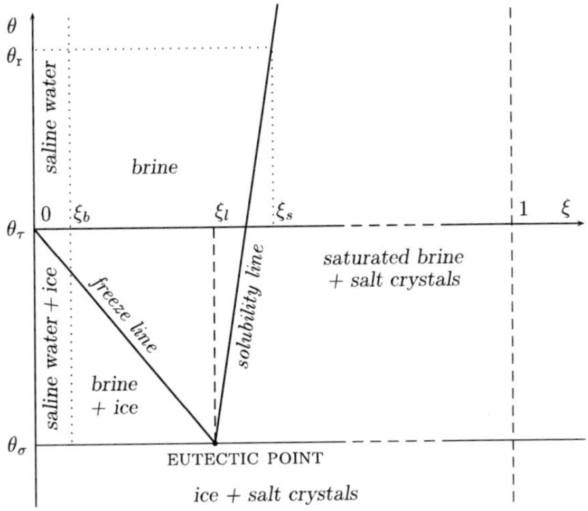

Fig. 16.3 Plot of the ice-saline water coexistence line (freeze line) and solubility line: $\xi_b = 0.05$ is the minimum salinity of the brine, $\xi_l = 0.233$ and $\xi_s = 0.27$ denote the solubility limits

of salt is about 0.27. This system then offers two phase transitions: ice—salt water and salt water—salt.

Let ρ_s, ρ_w be the mass densities of salt and water and ξ be the salt concentration in water,

$$\xi = \frac{\rho_s}{\rho_s + \rho_w}.$$

As $\xi > \xi_l$ equilibrium holds between saline water and saline water + salt through the solubility line (Fig. 16.3).

Ice—saline water transition

We let $\mathbf{T} = -p\mathbf{1}$ in both regions and hence consider the jump conditions in the form (16.6), namely

$$\rho U[\![\mathbf{v}]\!] = [\![p]\!]\mathbf{m}, \qquad \rho U[\![\varepsilon + \tfrac{1}{2}\mathbf{v}^2]\!] = [\![p\mathrm{v} + q]\!], \qquad \rho U[\![\eta]\!] - [\![q/\theta]\!] \geq 0.$$

We let $U > 0$ indicate that the ice region progresses. We may have the static solution $U = 0$ in which case $[\![p]\!] = 0$, $[\![\mathbf{v}]\!] = \mathbf{0}$ and the solution holds with $[\![q]\!] = 0$, $[\![\theta]\!] = 0$ while $\phi_i = \phi_{sw}$. If instead $[\![\theta]\!] = 0$ and $[\![p]\!] = 0$ then, by (16.8) it follows that > 0 if $\psi_i < \psi_{sw}$.

The coexistence line ice—saline water is the solution of the equality of free energies or chemical potentials, $\phi_i = \phi_{sw}$. It follows (experimentally) that the line is represented by

$$\theta(\xi) = \theta_\tau - \alpha_0 \xi, \qquad \alpha_0 = (\theta_\tau - \theta_\sigma)/\xi_l \simeq 91.3\,\mathrm{K}$$

The salinity lowers the freezing temperature up to $\omega_s = \xi_l \simeq 0.23$ and $\theta = \theta_\sigma \simeq$ 252 K.

Transition saline water – saline water and salt

The solubility line between saline water and saline water plus solid salt is found to be expressed by

$$\theta(\xi) = \theta_\sigma + \beta_0(\xi - \xi_l), \qquad \xi \in [\xi_l, \xi_s],$$

where $\beta_0 \simeq 975$ K.

Mixture model of the transition

If the two phases are not separated by a sharp interface then it is more appropriate to model saline water and ice as a mixture of three constituents occupying a time dependent region Ω [160]. Denote by the subscripts s, w, i the quantities pertaining to salt, water, and ice. Accordingly ρ_s, ρ_w, ρ_i are the mass densities of salt, water, and ice and

$$\rho = \rho_s + \rho_w + \rho_i$$

is the mass density of the mixture. Let ω_s, ω_i be the concentrations of salt and ice,

$$\omega_s = \frac{\rho_s}{\rho}, \qquad \omega_i = \frac{\rho_i}{\rho}.$$

Since $\xi = \rho_s/(\rho_s + \rho_w)$ then

$$\frac{1}{\omega_s} = \frac{\rho_i}{\rho_s} + \frac{\rho_s + \rho_w}{\rho_s} = \frac{\omega_i}{\omega_s} + \frac{1}{\xi}$$

whence

$$\omega_s = \xi(1 - \omega_i).$$

In a bounded (small) region it is reasonable to consider the constraint due to the mass of salt by letting

$$\frac{M_s}{M_s + M_\omega + M_i} \le \xi_s$$

the solubility limit in uniform conditions.

Let $\mathbf{v}_s, \mathbf{v}_w, \mathbf{v}_i$ be the velocity fields. The barycentric velocity $\mathbf{v}$ is written in the form

$$\mathbf{v} = \omega_s \mathbf{v}_s + \omega_i \mathbf{v}_i + (1 - \omega_s - \omega_i)\mathbf{v}_w.$$

Let $\mathbf{u}_s, \mathbf{u}_i$ the diffusion velocities and $\mathbf{h}_s, \mathbf{h}_i$ the mass fluxes,

$$\mathbf{u}_s = \mathbf{v}_s - \mathbf{v}, \quad \mathbf{u}_i = \mathbf{v}_i - \mathbf{v}, \quad \mathbf{h}_s = \rho\omega_s\mathbf{u}_s, \quad \mathbf{h}_i = \rho\omega_i\mathbf{u}_i.$$

In the transition ice—saline water the salt is conserved (in the phase of saline water). The salt then is subject to

$$\rho\dot{\omega}_s = -\nabla \cdot \mathbf{h}_s,$$

while

$$\int_\Omega \rho\omega_s dv = \text{constant}.$$

It follows that

$$0 = \frac{d}{dt}\int_\Omega \rho\omega_s dv = \int_\Omega \rho\dot{\omega}_s dv.$$

Hence we have

$$\int_{\partial\Omega} \mathbf{h}_s \cdot \mathbf{n} da = 0,$$

which is guaranteed by $\mathbf{h}_s \cdot \mathbf{n} = 0$ at $\partial\Omega$.

Ice and water suffer a phase transition which is modelled as a chemical reaction. First we consider the balance of mass of ice,

$$\partial_t \rho_i + \nabla \cdot (\rho\omega_i \mathbf{v}_i) = \tau,$$

τ being the mass of ice produced per unit time and unit volume. Otherwise we can write

$$\rho\dot{\omega}_i = -\nabla \cdot \mathbf{h}_i + \tau.$$

The continuity equation for water then can be written

$$\rho\dot{\omega}_i = \nabla \cdot \mathbf{h}_w + \tau.$$

The transition among saline water and saline water + salt is simply the formation of brine channels from the salt in excess of the solubility.

16.6.1 Transition Models via Minimization of Functionals

We now outline a variational approach developed in the literature. Let Ω be a spatial domain occupied by two constituents A, B constant mass densities and temperature. The equilibrium is assumed to be associated with the minimum of a functional

$$F_\Omega(\omega) = \int_\Omega F(\omega, \nabla\omega) dv,$$

F being a free energy [9, 68, 215] and ω is the mass fraction of a constituent, say A. For definiteness F_Ω is usually taken as the Landau–Ginzburg functional where

$$F(\omega, \nabla\omega) = f(\omega) + \tfrac{1}{2}\kappa|\nabla\omega|^2.$$

Observe that

$$\frac{dF_\Omega}{dt} = \int_\Omega [\partial_\omega f\, \dot\omega + \kappa\partial_{\nabla\omega} F \cdot \nabla\dot\omega] dv = \int_\Omega \delta_\omega F \dot\omega dv + \int_{\partial\Omega} \partial_{\nabla\omega} F \cdot \mathbf{n}\, \dot\omega da,$$

where $\delta_\omega = \partial_\omega - \nabla \cdot \partial_{\nabla\omega}$ is the variational derivative. The assumption $\mathbf{n} \cdot \nabla\omega = 0$ at $\partial\Omega$ makes the boundary term vanish if the Landau–Ginzburg functional is selected. Hence equilibrium is governed by the Euler–Lagrange equation $\delta_\omega F = 0$. If the mass of the single constituents is conserved then equilibrium is governed by

$$\partial_\omega f(\omega) - \nabla \cdot (\kappa\nabla\omega) = \lambda,$$

λ being the Lagrange multiplier connected with the mass conservation.

By the minimum property it follows that $dF_\Omega/dt \le 0$. Now,

$$0 \ge \frac{dF_\Omega}{dt} = \int_\Omega (\delta_\omega F)\dot\omega\, dv,$$

holds if

$$\dot\omega = -K(\omega, \theta)\delta_\omega F, \qquad K \ge 0, \tag{16.32}$$

in that

$$\frac{dF_\Omega}{dt} = -\int_\Omega K(\delta_\omega F)^2 dv.$$

If we let ω be governed by

$$\dot\omega = -\nabla \cdot \mathbf{h} \tag{16.33}$$

we assume

$$\mathbf{h} = -\hat{K}(\omega, \theta)\nabla\delta_\omega F, \qquad \hat{K} > 0,$$

and find

$$\frac{dF_\Omega}{dt} = -\int_\Omega \delta_\omega F\,(-\nabla \cdot \mathbf{h})dv = -\int_\Omega \delta_\omega \nabla \cdot (\tilde{K}\nabla\delta_\omega F)dv = -\int_\Omega \tilde{K}(\delta_\omega F)^2 dv$$

provided $\mathbf{n} \cdot \nabla\delta_\omega F = 0$ at $\partial\Omega$.

Two remarks are in order. First, it is claimed [9, 215] that $\delta_\omega F$ is the (relative) chemical potential. This can be justified, e.g. by letting $F = f$ as the Gibbs free energy in the form ([449], Chap. 17)

$$f = \sum_\alpha \omega_\alpha \mu_\alpha,$$

μ_α being the αth chemical potential. As $n = 2$, $\omega_1 = \omega$, $\omega_2 = -\omega$ and then

$$\partial_\omega f = \mu_1 - \mu_2$$

if μ_α is independent of ω_α. This independence looks an approximation. Secondly, the mass fractions are governed by the differential equations

$$\rho \dot{\omega}_\alpha = -\nabla \cdot (\rho_\alpha \mathbf{u}_\alpha) + \tau_\alpha$$

as is shown in Chap. 9. Hence (16.32) and (16.33) are used in place of the right Eqs. (9.27) and (9.69).

16.7 Phase-Field Model of Liquid-Solid Transitions

A simple model of phase transitions reduces significantly the set of fields describing the continuum by assuming that the body is undeformable. The two phases are incompressible solid and liquid; the mass density ρ is a parameter. The independent fields are then the temperature θ and a phase field φ; to fix ideas $\varphi = 0$ in the solid phase, $\varphi = 1$ in the liquid phase. The region occupied by the body, Ω, is time independent.

The balance of energy is of purely thermal character in the form

$$\rho \dot{\varepsilon} = \rho r - \nabla \cdot \mathbf{q}$$

and the second-law inequality is

$$\rho \dot{\eta} + \nabla \cdot (\frac{\mathbf{q}}{\theta} + \mathbf{k}) - \frac{\rho r}{\theta} = \sigma \geq 0.$$

Consequently we have

$$-\rho(\dot{\psi} + \eta \dot{\theta}) - \frac{1}{\theta}\mathbf{q} \cdot \nabla\theta + \theta \nabla \cdot \mathbf{k} = \theta\sigma \geq 0. \tag{16.34}$$

We look for a scheme where the phase field φ is governed by a rate-type equation. Let ψ, η, $\mathbf{q}$, $\mathbf{k}$, $\dot{\varphi}$ and σ be continuous functions of the independent variables θ, φ, $\nabla\theta$, $\nabla\varphi$, $\dot{\varphi}$, $\nabla\nabla\varphi$. Moreover let ψ be differentiable. Computing $\dot{\psi}$ and substituting in (16.34) we find

$$-\rho(\partial_\theta \psi + \eta)\dot{\theta} - \rho \partial_\varphi \psi \dot{\varphi} - \rho \partial_{\nabla\theta}\psi \cdot \nabla\dot{\theta} - \rho \partial_{\nabla\varphi}\psi \cdot \nabla\dot{\varphi} - \rho \partial_{\dot{\varphi}}\psi \ddot{\varphi} - \rho \partial_{\nabla\nabla\varphi}\psi \cdot \nabla\nabla\dot{\varphi} - \frac{1}{\theta}\mathbf{q} \cdot \nabla\theta + \theta\nabla \cdot \mathbf{k} \geq 0.$$

The arbitrariness of $\ddot{\varphi}$, $\nabla\nabla\dot{\varphi}$, $\nabla\dot{\theta}$, and $\dot{\theta}$ implies that

$$\partial_{\dot{\varphi}}\psi = 0, \qquad \partial_{\nabla\nabla\varphi}\psi = \mathbf{0}, \qquad \partial_{\nabla\theta}\psi = \mathbf{0}, \qquad \eta = -\partial_\theta \psi.$$

Consider the remaining inequality and divide by θ to have

$$-\frac{\rho}{\theta}\partial_\varphi\psi\dot{\varphi} - \frac{\rho}{\theta}\partial_{\nabla\varphi}\psi\cdot\nabla\dot{\varphi} - \frac{1}{\theta^2}\mathbf{q}\cdot\nabla\theta + \nabla\cdot\mathbf{k} \geq 0.$$

In view of the identity

$$-\frac{\rho}{\theta}\partial_{\nabla\varphi}\nabla\dot{\varphi} = -\nabla\cdot\big(\frac{\rho}{\theta}\partial_{\nabla\varphi}\psi\dot{\varphi}\big) + \big[\nabla\cdot\big(\frac{\rho}{\theta}\partial_{\nabla\varphi}\psi\big)\big]\dot{\varphi}$$

we can write

$$-\frac{\rho}{\theta}\delta_\varphi\psi\dot{\varphi} - \frac{1}{\theta^2}\mathbf{q}\cdot\nabla\theta + \nabla\cdot(\mathbf{k} - \frac{\rho}{\theta}\partial_{\nabla\varphi}\psi\,\dot{\varphi}) \geq 0,$$

where $\delta_\varphi\psi = \partial_\varphi\psi - (\theta/\rho)\nabla\cdot((\rho/\theta)\partial_{\nabla\varphi}\psi)$. Hence the extra-entropy flux $\mathbf{k}$ is given by

$$\mathbf{k} = \frac{\rho}{\theta}\partial_{\nabla\varphi}\psi\,\dot{\varphi}$$

while $\mathbf{q}$ satisfies the heat conduction inequality $\mathbf{q}\cdot\nabla\theta \leq 0$ and hence we take the Fourier-like law

$$\mathbf{q} = -\kappa(\theta,\varphi)\nabla\theta, \qquad \kappa(\theta,\varphi) \geq \mathbf{0}.$$

The sought function $\dot{\varphi}$ is subject to

$$\delta_\varphi\psi\dot{\varphi} \leq 0.$$

The simplest function $\dot{\varphi}$ is

$$\dot{\varphi} = -u(\theta,\varphi)\,\delta_\varphi\psi, \qquad u(\theta,\varphi) \geq \mathbf{0}.$$

Both κ and u are parameterized by ρ.

If

$$\psi = \psi_0(\theta,\varphi) + \tfrac{1}{2}h(\theta)|\nabla\varphi|^2$$

then

$$\delta_\varphi\psi = \partial_\varphi\psi_0 - \theta\nabla\cdot[(h(\theta)/\theta)\nabla\varphi]$$

and

$$\dot{\varphi} = -\Lambda[\partial_\varphi\psi_0 - h\Delta\varphi - \theta\nabla\varphi\cdot\nabla(h/\theta)].$$

Unless $h = c\theta$, the variational derivative $\delta_\varphi\psi$ contains a term $\nabla\varphi\cdot\nabla\theta$.

A physically motivated function ψ_0 is determined in terms of the internal energy ε. Since $\psi = \varepsilon - \theta\eta = \varepsilon + \theta\partial_\theta\psi$ then we have

$$\partial_\theta\psi - \frac{1}{\theta}\psi = -\frac{\varepsilon}{\theta}, \qquad \partial_\theta(\frac{1}{\theta}\psi) = -\frac{\varepsilon}{\theta^2}.$$

whence, for a chosen θ_0,

$$\psi(\theta, \varphi, \nabla\varphi) = -\theta \int_{\theta_0}^{\theta} \frac{\varepsilon(\zeta, \varphi, \nabla\varphi)}{\zeta^2} d\zeta + \theta\tilde{\psi}(\varphi, \nabla\varphi);$$

the occurrence of $\theta\hat{\psi}(\varphi, \nabla\varphi)$ leaves the internal energy ε invariant. Let θ_M be the melting temperature and

$$\varepsilon(\theta, \varphi, \nabla\varphi) = \hat{\varepsilon}(\theta, \varphi) + \tfrac{1}{2}g(\theta)|\nabla\varphi|^2.$$

Hence, letting $\theta_0 = \theta_M$ we can express ψ in the form

$$\psi(\theta, \varphi) = \theta\big[- \int_{\theta_M}^{\theta} \frac{\hat{\varepsilon}(\zeta, \varphi)}{\zeta^2} d\zeta + \hat{\psi}(\varphi, \nabla\varphi)\big] - \tfrac{1}{2}\theta G(\theta)|\nabla\varphi|^2,$$

where $\hat{\psi}(\varphi, \nabla\varphi)$ is so far an undetermined function while

$$G(\theta) = \int_{\theta_M}^{\theta} \frac{g(\zeta)}{\zeta^2} d\zeta$$

and hence $h(\theta) = -\theta G(\theta)$.

For a definite model we need the function $\varepsilon(\theta, \varphi)$. Let $\varepsilon_S, \varepsilon_L$ be the internal energy density of solid and liquid phases and then assume

$$\hat{\varepsilon}(\theta, \varphi) = \varepsilon_S(\theta) + f(\varphi)\lambda(\theta) = \varepsilon_L(\theta) + [f(\varphi) - 1]\lambda(\theta),$$

where $\lambda(\theta) = \varepsilon_L(\theta) - \varepsilon_S(\theta)$ and $\lambda(\theta_M)$ is the latent heat. The function $f(\varphi)$: $[0, 1] \to [0, 1]$ describes the transition and is subject to $f(0) = 0$, $f(1) = 1$; possibly $f(\varphi) = \varphi$. Hence $\psi_0(\theta, \varphi)$ can be written in the form

$$\psi_0(\theta, \varphi) = \theta\big[- \int_{\theta_M}^{\theta} \frac{\varepsilon_L(\zeta)}{\zeta^2} d\zeta - [f(\varphi) - 1]Q(\theta) + \tilde{\psi}(\varphi, \nabla\varphi)\big],$$

where

$$Q(\theta) = \int_{\theta_M}^{\theta} \frac{\lambda(\zeta)}{\zeta^2} d\zeta.$$

Let $\tilde{\psi}(\varphi, \nabla\varphi) = \frac{1}{2}\epsilon^2|\nabla\varphi|^2 + p(\varphi)$. The rate equation for φ then is given the form

$$\dot{\varphi} = u[\theta f'(\varphi)\mathcal{Q}(\theta) + \theta\epsilon^2\Delta\varphi - \theta p'(\varphi) + \nabla \cdot (G(\theta)\nabla\varphi)]. \quad (16.35)$$

Further details are necessary to determine space-time properties of temperature and phase field. Assume the internal energy in the liquid depends linearly on the temperature and hence let

$$\varepsilon_L(\theta) = \varepsilon_L(\theta_M) + c(\theta - \theta_M),$$

where c is the specific heat (at constant volume). The balance of energy, $\rho\dot{\varepsilon} = \nabla \cdot \kappa\nabla\theta$, results in

$$\rho\{[c + (f(\varphi) - 1)\lambda'(\theta)]\dot{\theta} + \lambda(\theta) f'(\varphi)\dot{\varphi}\} = -\nabla(\kappa\nabla\theta). \quad (16.36)$$

Equations (16.35) and (16.36) are the system of equations for θ and φ.

Geometrical aspects of the interfaces are modelled via the function $p(\varphi)$. At the melting temperature, G and $\mathcal{Q}$ vanish and the differential equation for φ shows that the stationary φ is given by

$$\epsilon^2\Delta\varphi - p'(\varphi) = 0.$$

16.8 Phase Transitions in SMA

A shape-memory alloy (SMA) is an alloy that can be deformed when cold and returns to the pre-deformed shape when heated. Nickel-titanium (Ni-Ti) is the most common alloy showing the SMA effect. This alloy can exist in two different phases, with three different crystal structures (phases): twinned martensite, detwinned martensite, and austenite. Martensite exists at lower temperatures, and austenite exists at higher temperatures.

Austenite, say $\mathcal{A}$, is a solid phase. Martensite can exist in two solid states: self-accommodated or twinned and oriented or detwinned, $\mathcal{M}^+, \mathcal{M}^-$. The thermomechanical phase transformation produces two effects, the shape memory effect and the pseudo-elastic effect. Pseudo-elasticity results in a hysteretic behaviour under a tensile test. As the temperature grows loops move away from the origin in the strain-stress plane. In the one-dimensional pseudo-elastic regime a mix of just two phases occurs, $\mathcal{A}$ and $\mathcal{M}^+$ in loading, $\mathcal{A}$ and $\mathcal{M}^-$ in unloading. Transitions are described by an order parameter φ such that $\varphi = 0$ in $\mathcal{A}$, $\varphi = 1$ in $\mathcal{M}^+$ and $\varphi = -1$ in $\mathcal{M}^-$. To simplify the description we use the parameter $\varphi^2 \in [0, 1]$ and hence we do not distinguish the content of martensite, $\mathcal{M}$, among $\mathcal{M}^+$ and $\mathcal{M}^-$.

The transition from martensite to austenite is induced by temperature and stress and no diffusion is involved. The SMA effect occurs because a temperature- or stress-induced transformation reverses deformation. Austenite transforms into martensite upon cooling; conversely, upon heating austenite transforms into austenite. Upon loading austenite transforms to martensite, upon unloading martensite reverts to austenite. If large stresses are applied a plastic behaviour happens. If the material is

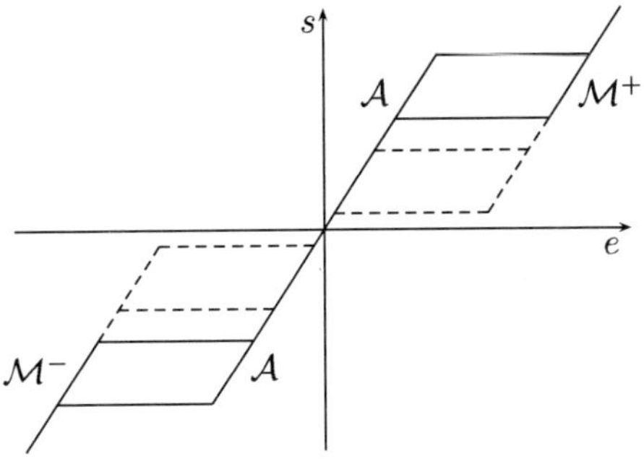

Fig. 16.4 SMA major hysteresis loops at $\theta_1 > \theta_{\mathcal{A}}$ (dashed) and $\theta_2 > \theta_1$ (solid)

unloaded before the plastic deformation, the hysteresis loop shows a reversible cycle (Fig. 16.4).

A SMA is viewed as a mixture of austenite and martensite phases. Let ρ_A and ρ_M be the mass densities of austenite and martensite and hence

$$\rho = \rho_A + \rho_M$$

the mass density of the alloy. We then let

$$\varphi^2 = \frac{\rho_M}{\rho}, \qquad \varphi^2 \in [0, 1],$$

be the mass fraction of martensite.

Restrict attention to homogeneous one-dimensional models by assuming that the reference mass density ρ_R keeps constant and the pertinent fields are uniaxial, namely $\mathbf{T}_{RR} = s\,\mathbf{e} \otimes \mathbf{e}$ and $\mathbf{E} = e\,\mathbf{e} \otimes \mathbf{e}$, $\mathbf{e}$ being a fixed unit vector; the scalar fields s and e are the engineering (Piola) stress and the engineering strain. The referential free enthalpy $\rho_R\phi = \rho_R\psi - s\,e$ is taken as a function of s, e parameterized by the temperature θ. For formal convenience we consider the referential densities

$$\varepsilon_R = \rho_R\varepsilon, \qquad \eta_R = \rho_R\eta, \qquad \psi_R = \rho_R\psi$$

and $q_R = \mathbf{q}_R \cdot \mathbf{e}$. The Clausius–Duhem inequality (2.43) can be written in the forms[4]

$$-(\dot\psi_R + \eta_R\dot\theta) + s\,\dot e - \frac{1}{\theta} q_R\, \partial_X\theta \ge 0, \qquad -(\dot\phi_R + \eta_R\dot\theta) - \dot s\, e - \frac{1}{\theta} q_R\, \partial_X\theta \ge 0,$$

depending on whether we base on the Helmholtz free energy ψ_R or the Gibbs free energy ϕ_R.

The transition from the martensitic to the austenitic phases may be induced by the temperature or the stress.

[4] The present approach is consistent with a zero extra-entropy flux.

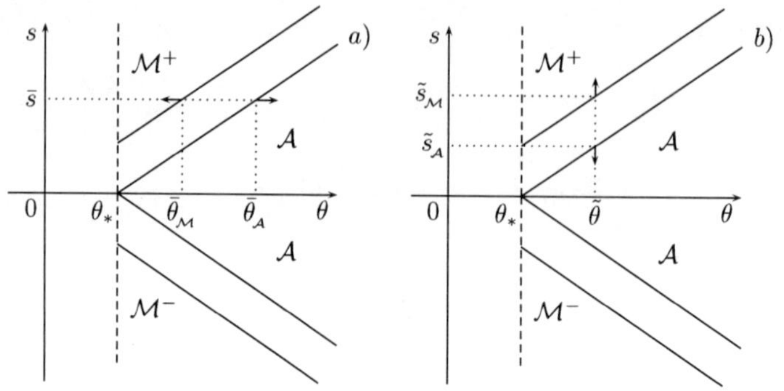

Fig. 16.5 Temperature-stress diagram: temperature induced (**a**) and stress induced (**b**)

To describe the temperature-induced transition at constant stress $s = \bar{s}$ we assume the alloy is in the martensitic phase up to $\theta = \bar{\theta}_{\mathcal{M}}$. Increasing the temperature produces an increase of the austenite fraction and the process is complete at $\theta > \bar{\theta}_{\mathcal{M}}$. Decreasing the temperature leaves the alloy unchanged up to $\bar{\theta}_{\mathcal{A}} < \bar{\theta}_{\mathcal{M}}$. Further decreasing of the temperature produces an increase of the martensite content at $\bar{\theta}_{\mathcal{A}}$ and the alloy goes back to being fully martensitic at lower temperatures (see Fig. 16.5a). A similar process occurs during stress-induced transitions at constant temperature $\theta = \tilde{\theta}$, as shown in Fig. 16.5b.

The occurrence of two distinct temperature thresholds, in the first case, and stress, in the second one, indicate that these processes have a hysteretic character. Yet a constitutive hysteretic relation between φ^2 and θ, but independent of e and s, may seem quite drastic. For definiteness we now neglect heat conduction and consider the Clausius–Duhem inequality in the form

$$-(\dot{\psi}_R + \eta_R \dot{\theta}) + s\,\dot{e} = -f \geq 0$$

where $f = J\theta\sigma$. In [371] ψ is a function of e, θ, and α, as the analogue of φ^2, while f is proportional to $|\dot{\alpha}|$ with coefficients dependent on $\operatorname{sgn}\dot{\alpha}$. By analogy with the hysteretic models of plasticity and ferroics we think that a model of SMA should involve a nonlinear relation between $\dot{s}$ and $\dot{e}$. We now let $\alpha = \varphi^2$ and assume

$$\Xi = (\theta, \alpha, s, e, \dot{\theta}, \dot{\alpha}, \dot{s}, \dot{e})$$

is the set of independent variables. By the standard arguments we find that ψ_R is in fact independent of $\dot{\theta}, \dot{\alpha}, \dot{s}, \dot{e}$. Hence the Clausius–Duhem inequality simplifies to

$$(\partial_\theta \psi_R + \eta_R)\dot{\theta} + \partial_\alpha \psi_R \dot{\alpha} + \partial_s \psi_R \dot{s} + (\partial_e \psi_R - s)\dot{e} = -f(\theta, \alpha, s, e, \dot{\theta}, \dot{\alpha}, \dot{s}, \dot{e}).$$

If, as is reasonable, we let $\theta = \theta_0$ be constant then we have

$$\partial_\alpha \psi_R \dot{\alpha} + \partial_s \psi_R \dot{s} + (\partial_e \psi_R - s)\dot{e} = -f_0,$$

where $f_0 = f(\theta_0, \alpha, s, 0, \dot{\alpha}, \dot{s})$. Hence s, e, α are required to satisfy a nonlinear first-order equation,

$$\dot{e} = \frac{\partial_\alpha \psi_R \dot{\alpha} + f_0(\dot{\alpha})}{s - \partial_e \psi_R} + \frac{\partial_s \psi_R \dot{s}}{s - \partial_e \psi_R}.$$

To simplify the relation among the unknowns, we assume that

$$e = e_{\rm el} + e_{\rm pl}$$

where $e_{\rm el}$ and $e_{\rm pl}$ are given by

$$\dot{e}_{\rm el} = \frac{\partial_s \psi_R \dot{s}}{s - \partial_e \psi_R}, \qquad \dot{e}_{\rm pl} = \frac{\partial_\alpha \psi_R \dot{\alpha} + f_0(\dot{\alpha})}{s - \partial_e \psi_R}.$$

Hence $e_{\rm el}$ plays the role of an elastic part of e while $e_{\rm pl}$ is the hysteretic part of e. Moreover

$$\frac{s - \partial_e \psi_R}{\partial_s \psi_R}$$

can be viewed as the elastic modulus [19].

The strain $e_{\rm pl}$ is related to the martensitic phases. In the deformation-induced transformation mechanisms (or tensile tests) the martensite $\mathcal{M}^+$ ($\mathcal{M}^-$) is produced if s is positive (negative). Hence we characterize $e_{\rm pl}$ and φ by assuming φ *takes the same sign of* s *and* $e_{\rm pl}$, *namely* $\varphi s \geq 0, \varphi e_{\rm pl} \geq 0$. To get detailed models we have merely to select the functions ψ_R and f_0.

A different approach follows by considering the Clausius–Duhem inequality in the form

$$\dot{\phi}_R + \eta_R \dot{\theta} + e\,\dot{s} = f \geq 0 \tag{16.37}$$

and assuming from the very beginning that a model of SMA should involve a decomposition of the deformation e into an elastic and a plastic part. For ease in writing, we now let $\alpha = \varphi^2$ and assume

$$e = e_{\rm el} + e_{\rm pl}, \qquad e_{\rm el} = g(s), \quad e_{\rm pl} = d(\theta)\gamma(\alpha)\,{\rm sgn}\, s \tag{16.38}$$

where $g : \mathbb{R} \to \mathbb{R}$ and $\gamma : [0, 1] \to [0, 1]$ are such that

$$g(0) = 0, \quad g'(s) > 0, \quad \gamma(0) = 0, \quad \gamma'(\alpha) \geq 0, \quad \gamma(1) = 1, \quad \gamma'(1) = 0, \quad \gamma''(1) < 0.$$

The set of independent variables is then

$$\Xi = (\theta, \alpha, s, \dot{\theta}, \dot{\alpha}, \dot{s}).$$

By the standard arguments we find that ϕ_R is in fact independent of $\dot{\theta}, \dot{\alpha}, \dot{s}$. Hence, using (16.38), the Clausius–Duhem inequality (16.37) simplifies to

$$(\partial_\theta \phi_R + \eta_R)\dot{\theta} + \partial_\alpha \phi_R \dot{\alpha} + [\partial_s \phi_R + g + d\gamma \operatorname{sgn} s]\dot{s} = f(\theta, \alpha, s, \dot{\theta}, \dot{\alpha}, \dot{s}).$$

If we let $\theta = \tilde{\theta}$ be constant then we have

$$\partial_\alpha \tilde{\phi}_R \dot{\alpha} + [\partial_s \tilde{\phi}_R + g + \tilde{d}\gamma \operatorname{sgn} s]\dot{s} = \tilde{f} \geq 0.$$

where $\tilde{\phi}_R(\alpha, s) = \phi_R(\tilde{\theta}, \alpha, s)$, $\tilde{d} = d(\tilde{\theta})$ and $\tilde{f}(\alpha, s, \dot{\alpha}, \dot{s}) = f(\tilde{\theta}, \alpha, s, 0, \dot{\alpha}, \dot{s})$. Hence, s and α are required to satisfy the following nonlinear first-order differential equation

$$\dot{\alpha} = -\frac{\partial_s \tilde{\phi}_R + g + \tilde{d}\gamma \operatorname{sgn} s}{\partial_\alpha \tilde{\phi}_R}\dot{s} + \frac{\tilde{f}(\alpha, s, \dot{\alpha}, \dot{s})}{\partial_\alpha \tilde{\phi}_R}. \tag{16.39}$$

which describes the hysteretic stress-induced phase transition at constant temperature after properly selecting the functions $\tilde{\phi}_R$ and $\tilde{f}$ (Figs. 16.6 and 16.7).

Otherwise, to describe temperature-induced transitions we let the stress $s = \bar{s}$ be constant; the Clausius–Duhem inequality (16.37) leads to

$$(\partial_\theta \bar{\phi}_R + \bar{\eta}_R)\dot{\theta} + \partial_\alpha \bar{\phi}_R \dot{\alpha} = \bar{f}(\theta, \alpha, s, \dot{\theta}, \dot{\alpha}, \dot{s}).$$

where $\bar{\phi}_R(\theta, \alpha) = \phi_R(\theta, \alpha, \bar{s}), \bar{\eta}_R(\theta, \alpha) = \eta_R(\theta, \alpha, \bar{s})$ and $\bar{f}(\theta, \alpha, \dot{\theta}, \dot{\alpha}) = f(\theta, \alpha, \bar{s}, \dot{\theta}, \dot{\alpha}, 0)$. Hence, θ and α are required to satisfy the following nonlinear first-order differential equation

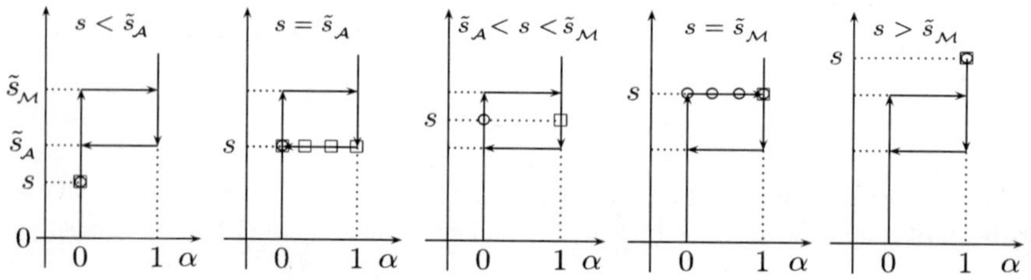

Fig. 16.6 Stress-induced transitions at a fixed temperature $\tilde{\theta} > \theta_*$: from left to right $\mathcal{A} \to \mathcal{M}^+$ (○), from right to left $\mathcal{M}^+ \to \mathcal{A}$ (□)

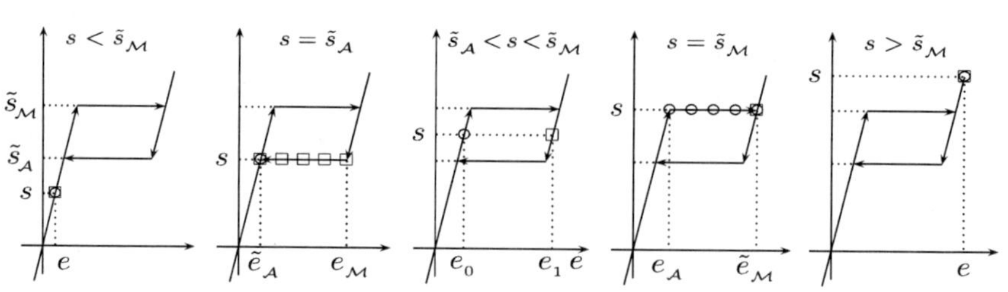

Fig. 16.7 Stress–strain loop in stress-induced transitions at a fixed temperature $\tilde{\theta} > \theta_*$: from left to right $\mathcal{A} \to \mathcal{M}^+$ (○), from right to left $\mathcal{M}^+ \to \mathcal{A}$ (□)

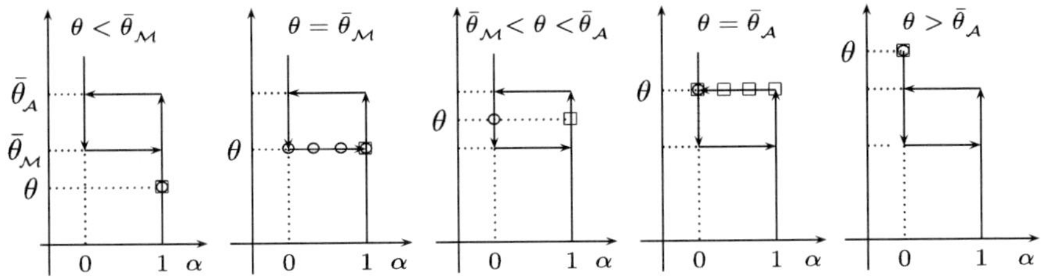

Fig. 16.8 Temperature-induced transitions at a fixed stress $\bar{s} > 0$: from right to left $\mathcal{A} \to \mathcal{M}^+$ (∘), from left to right $\mathcal{M}^+ \to \mathcal{A}$ (□)

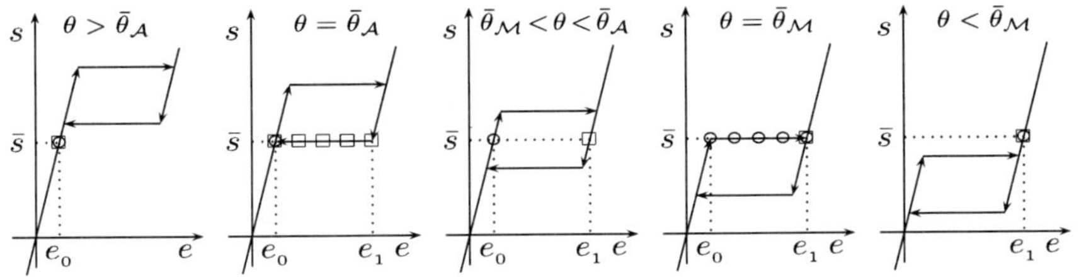

Fig. 16.9 Stress–strain loop in temperature-induced transitions at a fixed stress $\bar{s} > 0$: from left to right $\mathcal{A} \to \mathcal{M}^+$ (∘), from right to left $\mathcal{M}^+ \to \mathcal{A}$ (□)

$$\dot{\alpha} = -\frac{\partial_\theta \bar{\phi}_R + \bar{\eta}_R}{\partial_\alpha \bar{\phi}_R}\dot{\theta} + \frac{\bar{f}(\theta, \alpha, \dot{\theta}, \dot{\alpha})}{\partial_\alpha \bar{\phi}_R} \tag{16.40}$$

which describes the hysteretic temperature-induced phase transition at constant stress after properly selecting the functions $\bar{\phi}_R$, $\bar{\eta}$ and $\bar{f}$ (Figs. 16.8 and 16.9).

Appendix A
Notes on Vectors and Tensors

A.1 Vector and Tensor Algebra

The space under consideration is a three-dimensional Euclidean point space $\mathscr{E}$. Points are elements of $\mathscr{E}$ and are denoted by $\mathbf{x}$. Granted a choice of origin, we identify all points with the corresponding vectors from the origin.

A vector space over the reals $\mathbb{R}$ (or the complex numbers $\mathbb{C}$) is a set $\mathscr{V}$ together with two operations satisfying the properties

$$\mathbf{u}+\mathbf{v}=\mathbf{v}+\mathbf{u}, \quad \mathbf{u}+(\mathbf{v}+\mathbf{w})=(\mathbf{u}+\mathbf{v})+\mathbf{w}, \quad \exists \mathbf{0}:\mathbf{u}+\mathbf{0}=\mathbf{u}, \quad \exists -\mathbf{u}:\mathbf{u}+(-\mathbf{u})=\mathbf{0}$$

$$\alpha\beta\mathbf{u}=\alpha(\beta\mathbf{u}), \quad 1\mathbf{u}=\mathbf{u}, \quad (\alpha+\beta)\mathbf{u}=\alpha\mathbf{u}+\beta\mathbf{u}, \quad \alpha(\mathbf{u}+\mathbf{v})=\alpha\mathbf{u}+\alpha\mathbf{v},$$

for all $\mathbf{u}, \mathbf{v}, \mathbf{w} \in \mathscr{V}$ and $\alpha, \beta \in \mathbb{R}$. The length $|\mathbf{u}|$ of a vector $\mathbf{u}$ has the property of a norm,

$$|\mathbf{u}| \geq 0, \ |\mathbf{u}|=0 \Longleftrightarrow \mathbf{u}=\mathbf{0}; \quad |\mathbf{u}+\mathbf{v}| \leq |\mathbf{u}|+|\mathbf{v}|; \quad |\alpha\mathbf{u}|=|\alpha|\,|\mathbf{u}|,$$

$|\alpha|$ denoting the absolute value. The length $|\mathbf{u}|$ is also called the Euclidean norm or l_2-norm.

The inner product and cross product of vectors $\mathbf{u}, \mathbf{v}$ are denoted by

$$\mathbf{u}\cdot\mathbf{v}, \qquad \mathbf{u}\times\mathbf{v}.$$

By definition

$$\mathbf{u}\cdot\mathbf{v} := |\mathbf{u}|\,|\mathbf{v}|\cos\theta,$$

θ being the angle between $\mathbf{u}$ and $\mathbf{v}$; it follows that the inner product so defined satisfies

$$\mathbf{u}\cdot\mathbf{v}=\mathbf{v}\cdot\mathbf{u}, \qquad (\alpha\mathbf{u}+\beta\mathbf{v})\cdot\mathbf{w}=\alpha\mathbf{u}\cdot\mathbf{w}+\beta\mathbf{v}\cdot\mathbf{w}, \qquad \mathbf{u}\cdot\mathbf{u}\geq 0, \ =0 \Longleftrightarrow \mathbf{u}=\mathbf{0}.$$

A. Morro and C. Giorgi, *Mathematical Modelling of Continuum Physics*, Modeling and Simulation in Science, Engineering and Technology, https://doi.org/10.1007/978-3-031-20814-0

The magnitude (or length) $|\mathbf{u}|$ of $\mathbf{u}$ is related to the inner product by

$$|\mathbf{u}| = (\mathbf{u} \cdot \mathbf{u})^{1/2}.$$

Throughout the symbol $\mathbf{u}^2$ stands for $\mathbf{u} \cdot \mathbf{u} = |\mathbf{u}|^2$.

Let $\mathbf{n}$ be the unit vector, perpendicular to the plane containing $\mathbf{u}$ and $\mathbf{v}$, such that the angle θ between $\mathbf{u}$ and $\mathbf{v}$, in the counter-clockwise sense, is subject to $\theta \in (0, \pi)$; the direction of $\mathbf{n}$ is given also by the right-hand screw rule applied to $\mathbf{u}, \mathbf{v}, \mathbf{n}$. The cross product $\mathbf{u} \times \mathbf{v}$ is defined by

$$\mathbf{u} \times \mathbf{v} := |\mathbf{u}|\, |\mathbf{v}|\, \sin\theta\, \mathbf{n}$$

if $\theta \in (0, \pi)$ and $\mathbf{u} \times \mathbf{v} = \mathbf{0}$ if $\theta = 0, \pi$. A Cartesian coordinate frame for $\mathscr{E}$ consists of a chosen origin together with a positively oriented orthonormal basis $\{\mathbf{e}_1, \mathbf{e}_2, \mathbf{e}_3\}$. Throughout latin subscripts range over 1, 2, 3. Hence the orthonormal property is characterized by $\mathbf{e}_i \cdot \mathbf{e}_j = \delta_{ij}$, δ_{ij} denoting the Kronecker delta. The positively oriented property of the basis vectors implies that

$$\mathbf{e}_i \cdot (\mathbf{e}_j \times \mathbf{e}_k) = \epsilon_{ijk},$$

where ϵ_{ijk} is the alternating symbol,

$$\epsilon_{ijk} = \begin{cases} 1, & \text{if } i, j, k \text{ are an even permutation of } 1, 2, 3 \\ -1, & \text{if } i, j, k \text{ are an odd permutation of } 1, 2, 3 \\ 0, & \text{otherwise.} \end{cases}$$

It follows that the alternating symbol is invariant under cyclic permutations,

$$\epsilon_{ijk} = \epsilon_{jki} = \epsilon_{kij}.$$

Throughout the Einstein summation convention is understood. Since $\{\mathbf{e}_i\}$ is a basis then every vector $\mathbf{u}$ admits a unique representation,

$$\mathbf{u} = u_j\, \mathbf{e}_j,$$

the summation over $j = 1, 2, 3$ being understood. The scalars $\{u_j\}$ are called the Cartesian components of $\mathbf{u}$. By the orthonormal character of the basis it follows that

$$u_i = \mathbf{u} \cdot \mathbf{e}_i.$$

Likewise, letting $\mathbf{u} = \mathbf{e}_j \times \mathbf{e}_k$ we have

$$\mathbf{e}_j \times \mathbf{e}_k = [\mathbf{e}_i \cdot \mathbf{e}_j \times \mathbf{e}_k]\mathbf{e}_i = \epsilon_{ijk}\mathbf{e}_i.$$

In terms of the components, the inner and cross products are given by

$$\mathbf{u} \cdot \mathbf{v} = (u_i \mathbf{e}_i) \cdot (u_j \mathbf{e}_j) = u_i v_i$$

$$\mathbf{u} \times \mathbf{v} = (u_j \mathbf{e}_j) \times (v_k \mathbf{e}_k) = u_j v_k \mathbf{e}_j \times \mathbf{e}_k = \epsilon_{ijk} u_j v_k \mathbf{e}_i .$$

Accordingly,

$$(\mathbf{u} \times \mathbf{v})_i = \epsilon_{ijk} u_j v_k$$

and the mixed product $\mathbf{w} \cdot (\mathbf{u} \times \mathbf{v})$ is given by

$$\mathbf{w} \cdot (\mathbf{u} \times \mathbf{v}) = \epsilon_{ijk} w_i u_j v_k .$$

Since

$$\epsilon_{ijk} w_i u_j v_k = \epsilon_{jki} u_j v_k w_i = \epsilon_{kij} v_k w_i u_j$$

then

$$\mathbf{w} \cdot (\mathbf{u} \times \mathbf{v}) = \mathbf{u} \cdot (\mathbf{v} \times \mathbf{w}) = \mathbf{v} \cdot (\mathbf{w} \times \mathbf{u}),$$

a mixed product is invariant upon a cyclic permutation of the vectors or upon the interchange of the dot and cross products.

Some identities involving the cross product and the alternating symbol are often useful in vector calculus.[1] First

$$\mathbf{e}_i \times (\mathbf{e}_j \times \mathbf{e}_k) = \delta_{ik} \mathbf{e}_j - \delta_{ij} \mathbf{e}_k .$$

Hence

$$\mathbf{u} \times (\mathbf{v} \times \mathbf{w}) = (\mathbf{u} \cdot \mathbf{w})\mathbf{v} - (\mathbf{u} \cdot \mathbf{v})\mathbf{w}.$$

Now,

$$\mathbf{e}_i \cdot (\mathbf{e}_j \times \mathbf{e}_k)\, \mathbf{e}_i \cdot (\mathbf{e}_p \times \mathbf{e}_q) = (\mathbf{e}_j \times \mathbf{e}_k) \cdot (\mathbf{e}_p \times \mathbf{e}_q) = \mathbf{e}_p \cdot \mathbf{e}_q \times (\mathbf{e}_j \times \mathbf{e}_k) = \mathbf{e}_p \cdot (\delta_{qk} \mathbf{e}_j - \delta_{qj} \mathbf{e}_k).$$

Hence it follows the identity

$$\epsilon_{ijk} \epsilon_{ipq} = \delta_{jp} \delta_{kq} - \delta_{jq} \delta_{kp} . \tag{A.1}$$

Consequently,

$$\epsilon_{ijk} \epsilon_{ijq} = 2\delta_{kq} , \qquad \epsilon_{ijk} \epsilon_{ijk} = 6.$$

Since

$$\mathbf{e}_j \times \mathbf{e}_k = [\mathbf{e}_i \cdot (\mathbf{e}_j \times \mathbf{e}_k)]\mathbf{e}_i = \epsilon_{ijk} \mathbf{e}_i$$

---

[1] The detailed proofs are given in [325], pp. 6431–6440.

then

$$\epsilon_{pjk}\mathbf{e}_j \times \mathbf{e}_k = \epsilon_{pjk}\epsilon_{ijk}\mathbf{e}_i = 2\delta_{pi}\mathbf{e}_i = 2\mathbf{e}_p.$$

As a consequence it follows the identity

$$\mathbf{e}_i = \tfrac{1}{2}\epsilon_{ijk}\mathbf{e}_j \times \mathbf{e}_k.$$

A tensor, $\mathbf{S}$ say, is a linear mapping of vectors to vectors, $\mathbf{Su} = \mathbf{v}$. The linearity is expressed by

$$\mathbf{S}(\alpha\mathbf{u} + \beta\mathbf{v}) = \alpha\mathbf{Su} + \beta\mathbf{Sv},$$

for all vectors $\mathbf{u}$, $\mathbf{v}$ and scalars α, β. The sum of tensors, $\mathbf{S} + \mathbf{T}$, and the multiplication by a scalar, $\alpha\mathbf{S}$, are defined via the action on vectors, i.e.

$$(\mathbf{S} + \mathbf{T})\mathbf{u} = \mathbf{Su} + \mathbf{Tu}, \qquad (\alpha\mathbf{S})\mathbf{u} = \alpha\mathbf{Su},$$

for all vectors $\mathbf{u}$. The zero tensor $\mathbf{0}$ and the identity tensor $\mathbf{1}$ are defined by

$$\mathbf{0u} = \mathbf{0}, \qquad \mathbf{1u} = \mathbf{u}$$

for all vectors $\mathbf{u}$. The dyadic product $\mathbf{u} \otimes \mathbf{v}$ of two vectors $\mathbf{u}$ and $\mathbf{v}$ is a tensor defined by

$$(\mathbf{u} \otimes \mathbf{v})\mathbf{w} = (\mathbf{v} \cdot \mathbf{w})\mathbf{u}$$

for all vector $\mathbf{w}$.

Let $\mathbf{e}$ be a unit vector. Since

$$(\mathbf{e} \otimes \mathbf{e})\mathbf{u} = (\mathbf{u} \cdot \mathbf{e})\mathbf{e}$$

for any vector $\mathbf{u}$, then the tensor $\mathbf{e} \otimes \mathbf{e}$ maps each vector $\mathbf{u}$ to the projection $(\mathbf{u} \cdot \mathbf{e})\mathbf{e}$ of $\mathbf{u}$ in the direction of $\mathbf{e}$. Likewise,

$$(\mathbf{1} - \mathbf{e} \otimes \mathbf{e})\mathbf{u} = \mathbf{u} - (\mathbf{u} \cdot \mathbf{e})\mathbf{e}, \qquad \mathbf{e} \cdot (\mathbf{1} - \mathbf{e} \otimes \mathbf{e})\mathbf{u} = 0.$$

For any vector $\mathbf{u}$, the tensor $\mathbf{1} - \mathbf{e} \otimes \mathbf{e}$ maps $\mathbf{u}$ to the projection of $\mathbf{u}$ perpendicular to $\mathbf{e}$. Hence $\mathbf{e} \otimes \mathbf{e}$ and $\mathbf{1} - \mathbf{e} \otimes \mathbf{e}$ are tensors providing the projections onto $\mathbf{e}$ and onto the plane perpendicular to $\mathbf{e}$.

Incidentally, since

$$\mathbf{1u} = \mathbf{u} = (\mathbf{u} \cdot \mathbf{e}_i)\mathbf{e}_i = (\mathbf{e}_i \otimes \mathbf{e}_i)\mathbf{u}$$

then we have the representation of the identity,

$$\mathbf{1} = \mathbf{e}_i \otimes \mathbf{e}_i,$$

the summation over i being understood.

We now determine the components of a tensor. Given a tensor $\mathbf{S}$ and any vector $\mathbf{u}$ let $\mathbf{v} = \mathbf{S}\mathbf{u}$. Hence

$$v_k \mathbf{e}_k = \mathbf{S}(u_j \mathbf{e}_j) = u_j \mathbf{S}\mathbf{e}_j.$$

Inner multiplying by $\mathbf{e}_i$ we obtain

$$v_i = (\mathbf{e}_i \cdot \mathbf{S}\mathbf{e}_j) u_j.$$

Hence we have the component form of $\mathbf{v} = \mathbf{S}\mathbf{u}$ as

$$v_i = S_{ij} u_j$$

upon defining the components of $\mathbf{S}$ as

$$S_{ij} := \mathbf{e}_i \cdot \mathbf{S}\mathbf{e}_j.$$

Conversely, to obtain the tensor in terms of the components we observe

$$\mathbf{v} = v_i \mathbf{e}_i = (S_{ij} u_j)\mathbf{e}_i = (S_{ij}\mathbf{e}_j \cdot \mathbf{u})\mathbf{e}_i = (S_{ij}\mathbf{e}_i \otimes \mathbf{e}_j)\mathbf{u}$$

whence it follows $\mathbf{S}\mathbf{u} = (S_{ij}\mathbf{e}_i \otimes \mathbf{e}_j)\mathbf{u}$. By the arbitrariness of $\mathbf{u}$ we conclude that

$$\mathbf{S} = S_{ij}\mathbf{e}_i \otimes \mathbf{e}_j.$$

The matrix

$$S = [S_{ij}]$$

is the matrix associated with the tensor $\mathbf{S}$, relative to the selected basis $\{\mathbf{e}_i\}$.

The *transpose* of a tensor $\mathbf{S}$ is the unique tensor $\mathbf{S}^T$ such that

$$\mathbf{u} \cdot \mathbf{S}\mathbf{v} = \mathbf{v} \cdot \mathbf{S}\mathbf{u}$$

for all vectors $\mathbf{u}$ and $\mathbf{v}$. To find the connection between the matrices of $\mathbf{S}$ and $\mathbf{S}^T$ we observe

$$\mathbf{e}_i \cdot \mathbf{S}^T \mathbf{e}_j = \mathbf{e}_j \cdot \mathbf{S}\mathbf{e}_i$$

and hence

$$(\mathbf{S}^T)_{ij} = S_{ji}, \qquad \mathbf{S}^T = S_{ij}\mathbf{e}_j \otimes \mathbf{e}_i.$$

The matrix associated with $\mathbf{S}^T$ is the transpose of that associated with $\mathbf{S}$.

Applying the definition of transpose we find that

$$(\mathbf{S}+\mathbf{T})^T = \mathbf{S}^T + \mathbf{T}^T, \qquad (\mathbf{S}^T)^T = \mathbf{S}, \qquad \mathbf{1}^T = \mathbf{1}.$$

This is so because, by definition,

$$\mathbf{v}\cdot(\mathbf{S}+\mathbf{T})^T\mathbf{u}=\mathbf{u}\cdot(\mathbf{S}+\mathbf{T})\mathbf{v}=\mathbf{u}\cdot\mathbf{S}\mathbf{v}+\mathbf{u}\cdot\mathbf{T}\mathbf{v}=\mathbf{v}\cdot\mathbf{S}^T\mathbf{u}+\mathbf{v}\cdot\mathbf{T}^T\mathbf{u}=\mathbf{v}\cdot(\mathbf{S}^T+\mathbf{T}^T)\mathbf{u},$$

$$\mathbf{v}\cdot(\mathbf{S}^T)^T\mathbf{u}=\mathbf{u}\cdot\mathbf{S}^T\mathbf{v}=\mathbf{v}\cdot\mathbf{S}\mathbf{u},$$

$$\mathbf{v}\cdot\mathbf{1}^T\mathbf{u}=\mathbf{u}\cdot\mathbf{1}\mathbf{v}=\mathbf{v}\cdot\mathbf{u},$$

whence $(\mathbf{S}+\mathbf{T})^T\mathbf{u}=(\mathbf{S}^T+\mathbf{T}^T)\mathbf{u}$, $(\mathbf{S}^T)^T\mathbf{u}=\mathbf{S}\mathbf{u}$, $\mathbf{1}^T\mathbf{u}=\mathbf{u}$ and the conclusions follow.

A tensor $\mathbf{S}$ is *symmetric* if $\mathbf{S}=\mathbf{S}^T$ and is *skew* if $\mathbf{S}=-\mathbf{S}^T$. Let Sym be the set of symmetric tensors and Skw be the set of skew tensors. The identity

$$\mathbf{S}=\tfrac{1}{2}(\mathbf{S}+\mathbf{S}^T)+\tfrac{1}{2}(\mathbf{S}-\mathbf{S}^T)$$

and the observation that $\mathbf{S}+\mathbf{S}^T\in\mathrm{Sym}$, $\mathbf{S}-\mathbf{S}^T\in\mathrm{Skw}$ show that any tensor admits a (unique) decomposition into symmetric and skew parts,

$$\mathbf{S}=\mathrm{sym}\,\mathbf{S}+\mathrm{skw}\,\mathbf{S}.$$

The uniqueness of the decomposition follows by observing that if

$$\mathbf{S}=\mathbf{A}+\mathbf{B}=\hat{\mathbf{A}}+\hat{\mathbf{B}},\quad \mathbf{A},\hat{\mathbf{A}}\in\mathrm{Sym},\quad \mathbf{B},\hat{\mathbf{B}}\in\mathrm{Skw}$$

then

$$\mathbf{S}+\mathbf{S}^T=2\mathbf{A}=2\hat{\mathbf{A}},\qquad \mathbf{S}-\mathbf{S}^T=2\mathbf{B}=2\hat{\mathbf{B}},$$

whence $\mathbf{A}=\hat{\mathbf{A}}$, $\mathbf{B}=\hat{\mathbf{B}}$.

Given two tensors $\mathbf{S}$, $\mathbf{T}$ the *product* $\mathbf{ST}$ is defined by

$$(\mathbf{ST})\mathbf{u}:=\mathbf{S}(\mathbf{T}\mathbf{u}),$$

for all vectors $\mathbf{u}$. To determine the matrix associated with the product $\mathbf{ST}$ we observe

$$\mathbf{e}_i\cdot\mathbf{ST}\mathbf{e}_j=\mathbf{S}^T\mathbf{e}_i\cdot\mathbf{T}\mathbf{e}_j=S_{ik}\mathbf{e}_k\cdot T_{hj}\mathbf{e}_h=S_{ik}T_{hj}\mathbf{e}_k\cdot\mathbf{e}_h=S_{ik}T_{kj},$$

whence

$$(\mathbf{ST})_{ij}=S_{ik}T_{kj}.$$

Accordingly, the matrix associated with $\mathbf{ST}$ is obtained from S and T via the standard row-column rule for computing the product of matrices.

The symbol $\mathbf{S}^m$ denotes the m-times product,

$$\mathbf{S}^m=\underbrace{\mathbf{S}\mathbf{S}\cdots\mathbf{S}}_{m\text{ times}}.$$

In particular $\mathbf{S}^2 = \mathbf{S}\,\mathbf{S}$.

To determine the action of the transpose of the product we observe that

$$\mathbf{v} \cdot (\mathbf{ST})^T \mathbf{u} = \mathbf{u} \cdot (\mathbf{ST})\mathbf{v} = \mathbf{u} \cdot \mathbf{S}(\mathbf{Tv}) = (\mathbf{Tv}) \cdot \mathbf{S}^T \mathbf{u} = \mathbf{v} \cdot \mathbf{T}^T \mathbf{S}^T \mathbf{u},$$

and then

$$(\mathbf{ST})^T = \mathbf{T}^T \mathbf{S}^T.$$

In applications we may find tensors in the form $\mathbf{A}\,\mathbf{S}\,\mathbf{A}^T$. Since

$$\mathbf{v} \cdot \mathbf{ASA}^T \mathbf{u} = \mathbf{v} \cdot \mathbf{A}(\mathbf{SA}^T \mathbf{u}) = (\mathbf{SA}^T \mathbf{u}) \cdot \mathbf{A}^T \mathbf{v} = \mathbf{A}^T \mathbf{u}) \cdot \mathbf{S}^T \mathbf{A}^T \mathbf{v} = \mathbf{u} \cdot \mathbf{AS}^T \mathbf{A}^T \mathbf{v},$$

whence

$$(\mathbf{ASA}^T)^T = \mathbf{AS}^T \mathbf{A}^T.$$

As a consequence

$$\mathbf{S} \in \mathrm{Sym} \implies \mathbf{ASA}^T \in \mathrm{Sym}, \qquad \mathrm{sym}(\mathbf{ASA}^T) = \mathbf{A}(\mathrm{sym}\mathbf{S})\mathbf{A}^T.$$

The tensors may involve a dyadic product. Now,

$$\mathbf{S}(\mathbf{u} \otimes \mathbf{v})\mathbf{w} = \mathbf{S}[(\mathbf{u} \otimes \mathbf{v})\mathbf{w}] = \mathbf{S}(\mathbf{v} \cdot \mathbf{w})\mathbf{u} = (\mathbf{v} \cdot \mathbf{w})\mathbf{Su} = [(\mathbf{Su}) \otimes \mathbf{v}]\mathbf{w},$$

$$(\mathbf{u} \otimes \mathbf{v})\mathbf{Sw} = (\mathbf{u} \otimes \mathbf{v})(\mathbf{Sw}) = \mathbf{u}\,\mathbf{v} \cdot \mathbf{Sw} = \mathbf{u}\,\mathbf{S}^T \mathbf{v} \cdot \mathbf{w} = (\mathbf{u} \otimes \mathbf{S}^T \mathbf{v})\mathbf{w},$$

$$\mathbf{t} \cdot (\mathbf{u} \otimes \mathbf{v})^T \mathbf{w} = \mathbf{w} \cdot (\mathbf{u} \otimes \mathbf{v})\mathbf{t} = \mathbf{w} \cdot \mathbf{u}\;\mathbf{t} \cdot \mathbf{v} = \mathbf{t} \cdot (\mathbf{v} \otimes \mathbf{u})\mathbf{w}.$$

Accordingly,

$$\mathbf{S}(\mathbf{u} \otimes \mathbf{v}) = (\mathbf{Su}) \otimes \mathbf{v}, \qquad (\mathbf{u} \otimes \mathbf{v})\mathbf{S} = \mathbf{u} \otimes (\mathbf{S}^T \mathbf{v}), \qquad (\mathbf{u} \otimes \mathbf{v})^T = \mathbf{v} \otimes \mathbf{u}.$$

There is a 1-1 correspondence between skew tensors and *axial vectors*. If $\mathbf{\Lambda} \in \mathrm{Skw}$ then the axial vector $\mathbf{a}_\Lambda$ of $\mathbf{\Lambda}$ is defined by

$$\mathbf{\Lambda u} = \mathbf{a}_\Lambda \times \mathbf{u}$$

for all vectors $\mathbf{u}$. To determine the direct correspondence between $\mathbf{\Lambda}$ and $\mathbf{a}_\Lambda$ we observe that, by definition,

$$\Lambda_{pq} u_q = \epsilon_{piq} (\mathbf{a}_\Lambda)_i\, u_q$$

whence

$$\Lambda_{pq} = \epsilon_{piq} (\mathbf{a}_\Lambda)_i.$$

Conversely, multiplying by ϵ_{jpq} we have

$$\epsilon_{jpq}\,\Lambda_{pq} = \epsilon_{jpq}\,\epsilon_{piq}(\mathbf{a}_{\Lambda})_i = -2\,\delta_{ij}(\mathbf{a}_{\Lambda})_i = -2(\mathbf{a}_{\Lambda})_j,$$

whence

$$(\mathbf{a}_{\Lambda})_j = -\tfrac{1}{2}\,\epsilon_{jpq}\,\Lambda_{pq}.$$

The *trace* of a tensor $\mathbf{S}$ is a linear operation based on the definition

$$\text{tr}\,(\mathbf{e}_i \otimes \mathbf{e}_j) = \mathbf{e}_i \cdot \mathbf{e}_j.$$

Hence

$$\text{tr}\,\mathbf{S} = \text{tr}\,[S_{ij}\,\mathbf{e}_i \otimes \mathbf{e}_j] = S_{ij}\text{tr}\,(\mathbf{e}_i \otimes \mathbf{e}_j) = S_{ii}$$

and hence $\text{tr}\,\mathbf{S} = \text{tr}\,S$. Further,

$$\text{tr}\,\mathbf{1} = \text{tr}\,(\mathbf{e}_i \otimes \mathbf{e}_i) = 3.$$

If $\mathbf{S} \in \text{Skw}$ then $\text{tr}\,\mathbf{S} = 0$. A tensor $\mathbf{S}$ is said to be *traceless* if $\text{tr}\,\mathbf{S} = 0$.

For any tensor $\mathbf{S}$,

$$\mathbf{S}_0 := \mathbf{S} - \tfrac{1}{3}(\text{tr}\,\mathbf{S})\mathbf{1}$$

is traceless and $\mathbf{S}_0$ is said to be the *deviatoric part* (or deviator) of $\mathbf{S}$. The identity

$$\mathbf{S} = [\mathbf{S} - \tfrac{1}{3}(\text{tr}\,\mathbf{S})\mathbf{1}] + \tfrac{1}{3}(\text{tr}\,\mathbf{S})\mathbf{1}$$

means that each tensor admits the decomposition

$$\mathbf{S} = \mathbf{S}_0 + \tfrac{1}{3}(\text{tr}\,\mathbf{S})\mathbf{1}.$$

Given two tensors $\mathbf{S}, \mathbf{T}$ the inner product $\mathbf{S} \cdot \mathbf{T}$ is defined by[2]

$$\mathbf{S} \cdot \mathbf{T} = \text{tr}\,(\mathbf{S}^T\mathbf{T}).$$

It follows

$$\mathbf{S} \cdot \mathbf{T} = (\mathbf{S}^T\mathbf{T})_{jj} = S^T_{ji}T_{ij} = S_{ij}T_{ij}.$$

As a consequence,

$$\mathbf{S} \cdot \mathbf{T} = \mathbf{T} \cdot \mathbf{S}, \qquad \mathbf{1} \cdot \mathbf{S} = \text{tr}\,\mathbf{S}.$$

Some properties are useful about the inner product of tensors. First,

$$\mathbf{S} \in \text{Sym}, \quad \mathbf{W} \in \text{Skw} \quad \Longrightarrow \quad \mathbf{S} \cdot \mathbf{W} = 0;$$

in the geometrical language, $\mathbf{S}$ and $\mathbf{W}$ are orthogonal. This follows by observing that

---

[2] Often the notation $\mathbf{S} : \mathbf{T}$ is used instead of $\mathbf{S} \cdot \mathbf{T}$.

$$\mathbf{S} \cdot \mathbf{W} = \text{tr}\,(\mathbf{S}^T \mathbf{W}) = \text{tr}\,(\mathbf{SW}) = \text{tr}\,(\mathbf{WS}) = -\text{tr}\,(\mathbf{W}^T \mathbf{S}) = -\mathbf{W} \cdot \mathbf{S} = -\mathbf{S} \cdot \mathbf{W}.$$

Let

$$|\mathbf{S}| := \sqrt{\mathbf{S} \cdot \mathbf{S}} = \textstyle\sum_{i,j} S_{ij}^2.$$

Since

$$\mathbf{S} = \text{sym}\,\mathbf{S} + \text{skw}\,\mathbf{S}$$

and sym $\mathbf{S} \cdot$ skw $\mathbf{S} = 0$ then

$$|\mathbf{S}|^2 = |\text{sym}\,\mathbf{S}|^2 + |\text{skw}\,\mathbf{S}|^2.$$

Likewise, by the decomposition of $\mathbf{S}$ via the traceless part we have

$$|\mathbf{S}|^2 = |\mathbf{S}_0|^2 + \tfrac{1}{3}(\text{tr}\ \mathbf{S})^2.$$

A tensor $\mathbf{S}$ is invertible if there is a tensor $\mathbf{S}^{-1}$ such that

$$\mathbf{S}\mathbf{S}^{-1} = \mathbf{S}^{-1}\mathbf{S} = \mathbf{1},$$

$\mathbf{S}^{-1}$ being called the inverse of $\mathbf{S}$. If $\mathbf{S}$ and $\mathbf{T}$ are invertible then right multiplication of $(\mathbf{ST})^{-1}\mathbf{ST} = \mathbf{1}$ by $\mathbf{T}^{-1}$ and $\mathbf{S}^{-1}$ gives

$$(\mathbf{ST})^{-1} = \mathbf{T}^{-1}\mathbf{S}^{-1}.$$

Also by

$$\mathbf{S}^T(\mathbf{S}^{-1})^T = (\mathbf{S}^{-1}\mathbf{S})^T = \mathbf{1}$$

it follows that

$$\mathbf{S}^{-T} := (\mathbf{S}^{-1})^T = (\mathbf{S}^T)^{-1}.$$

The *determinant* of a tensor $\mathbf{S}$ is defined by [216]

$$\det \mathbf{S} = \frac{\mathbf{Su} \cdot (\mathbf{Sv} \times \mathbf{Sw})}{\mathbf{u} \cdot (\mathbf{v} \times \mathbf{w})}$$

for any basis $\mathbf{u}, \mathbf{v}, \mathbf{w}$. Owing to the geometrical meaning of the mixed product, $|\det \mathbf{S}|$ is the ratio of the volume of the parallelepiped defined by the vectors $\mathbf{Su}$, $\mathbf{Sv}$, and $\mathbf{Sw}$ to the volume of the parallelepiped defined by $\mathbf{u}$, $\mathbf{v}$, and $\mathbf{w}$. For simplicity we let $\mathbf{u}, \mathbf{v}, \mathbf{w}$ be the orthonormal triad and define

$$\det \mathbf{S} = \frac{\mathbf{Se}_i \cdot (\mathbf{Se}_j \times \mathbf{Se}_k)}{\mathbf{e}_i \cdot (\mathbf{e}_j \times \mathbf{e}_k)}$$

whence

$$\epsilon_{ijk} \det \mathbf{S} = \mathbf{S}\mathbf{e}_i \cdot (\mathbf{S}\mathbf{e}_j \times \mathbf{S}\mathbf{e}_k).$$

Since

$$\mathbf{S}\mathbf{e}_i = (S_{ph}\mathbf{e}_p \otimes \mathbf{e}_h)\mathbf{e}_i = S_{pi}\mathbf{e}_i$$

and the like for $\mathbf{S}\mathbf{e}_j$ and $\mathbf{S}\mathbf{e}_k$ we obtain

$$\epsilon_{ijk} \det \mathbf{S} = \epsilon_{pqr} S_{pi} S_{qj} S_{rk}.$$

Multiplying by ϵ_{ijk} and summing over i, j, k we find that

$$\det \mathbf{S} = \tfrac{1}{6}\epsilon_{ijk}\epsilon_{pqr} S_{pi} S_{qj} S_{rk}.$$

Otherwise, we let $i = 1, j = 2, k = 3$ to obtain

$$\det \mathbf{S} = \epsilon_{pqr} S_{p1} S_{q2} S_{r3} = \det S.$$

Accordingly, the determinant of the tensor equals the determinant of the corresponding matrix.

By using the equality of the determinants, $\det \mathbf{S} = \det S$, it follows that

$$\det \mathbf{S}^T = \det \mathbf{S}, \qquad \det(\mathbf{S}\,\mathbf{T}) = (\det \mathbf{S})(\det \mathbf{T}).$$

Hence by $\mathbf{S}^{-1}\mathbf{S} = \mathbf{1}$ it follows

$$\det \mathbf{S}^{-1} = (\det \mathbf{S})^{-1}.$$

Since $(\mathbf{S}^{-1})_{ij} = (S^{-1})_{ij}$ it follows that $\mathbf{S}$ is invertible if and only if $\det \mathbf{S} \neq 0$.

A tensor $\mathbf{Q}$ is *orthogonal* if

$$\mathbf{Q}\mathbf{u} \cdot \mathbf{Q}\mathbf{v} = \mathbf{u} \cdot \mathbf{v}$$

for all vectors $\mathbf{u}$, $\mathbf{v}$. In particular if $\mathbf{v} = \mathbf{u}$ it follows

$$(\mathbf{Q}\mathbf{u})^2 = \mathbf{u}^2$$

or $|\mathbf{Q}\mathbf{u}| = |\mathbf{u}|$; for any vector $\mathbf{u}$ the length of $\mathbf{Q}\mathbf{u}$ equals that of $\mathbf{u}$. Furthermore, if α is the angle between $\mathbf{u}$ and $\mathbf{v}$ while β is the angle between $\mathbf{Q}\mathbf{u}$ and $\mathbf{Q}\mathbf{v}$ then

$$\cos\beta = \frac{\mathbf{Q}\mathbf{u} \cdot \mathbf{Q}\mathbf{v}}{|\mathbf{Q}\mathbf{u}||\mathbf{Q}\mathbf{v}|} = \frac{\mathbf{u} \cdot \mathbf{v}}{|\mathbf{u}||\mathbf{v}|} = \cos\alpha;$$

the angle is left invariant upon the action of an orthogonal tensor.

Equivalently, orthogonal tensors $\mathbf{Q}$ are characterized by the condition

$$\mathbf{Q}^T\mathbf{Q} = \mathbf{Q}\mathbf{Q}^T = \mathbf{1}.$$

To prove the equivalence we first assume $\mathbf{Qu} \cdot \mathbf{Qv} = \mathbf{u} \cdot \mathbf{v}$ for all vectors $\mathbf{u}$, $\mathbf{v}$. Let $\mathbf{v} = \mathbf{u}$ so that

$$\mathbf{u} \cdot \mathbf{u} = \mathbf{Qu} \cdot \mathbf{Qu} = \mathbf{u} \cdot \mathbf{Q}^T\mathbf{Qu}$$

whence

$$\mathbf{u} \cdot (\mathbf{Q}^T\mathbf{Q} - \mathbf{1})\mathbf{u} = 0$$

for all $\mathbf{u}$. Let $\mathbf{T} = \mathbf{Q}^T\mathbf{Q} - \mathbf{1}$ and observe $\mathbf{T} \in \mathrm{Sym}$. We need showing that if $\mathbf{T} \in \mathrm{Sym}$ satisfies $\mathbf{u} \cdot \mathbf{Tu} = 0$ for any vector $\mathbf{u}$ then $\mathbf{T} = \mathbf{0}$. To this end we consider the identity

$$(\mathbf{u} + \mathbf{v}) \cdot \mathbf{T}(\mathbf{u} + \mathbf{v}) - \mathbf{u} \cdot \mathbf{Tu} - \mathbf{v} \cdot \mathbf{Tv} = 2\mathbf{u} \cdot \mathbf{Tv}$$

for any vectors $\mathbf{u}$, $\mathbf{v}$ and $\mathbf{T} \in \mathrm{Sym}$. The left-hand side is zero. Hence $\mathbf{u} \cdot \mathbf{Tv} = 0$ for any vectors $\mathbf{u}$, $\mathbf{v}$. This implies $\mathbf{T} = \mathbf{0}$. As a consequence

$$\mathbf{Q}^T\mathbf{Q} = \mathbf{1}.$$

This in turn implies that $\det \mathbf{Q} = \pm 1$, $\mathbf{Q}$ is invertible and $\mathbf{Q}^T = \mathbf{Q}^{-1}$. Conversely, if $\mathbf{Q}^T\mathbf{Q} = \mathbf{1}$ then

$$\mathbf{Qu} \cdot \mathbf{Qv} = \mathbf{u} \cdot \mathbf{Q}^T\mathbf{Qv} = \mathbf{u} \cdot \mathbf{v},$$

which completes the proof.

An orthogonal tensor $\mathbf{Q}$ is said to be a *rotation* if $\det \mathbf{Q} = 1$. If $\mathbf{Q}^T\mathbf{Q} = \mathbf{1}$ then $\mathbf{Q} \in \mathrm{Orth}$; if further $\det \mathbf{Q} = 1$ then $\mathbf{Q} \in \mathrm{Orth}^+$.

A scalar ω is an eigenvalue of a tensor $\mathbf{S}$ if there is a nonzero vector $\mathbf{v}$ such that

$$\mathbf{Sv} = \omega\mathbf{v}.$$

If $\mathbf{S}$ is symmetric then there is an orthonormal basis $\{\hat{\mathbf{e}}_i\}$ of eigenvectors of $\mathbf{S}$ such that

$$\mathbf{S} = \sum_{i=1}^{3} \omega_i \hat{\mathbf{e}}_i \otimes \hat{\mathbf{e}}_i,$$

ω_i being the eigenvalue of $\mathbf{S}$ associated with the eigenvector $\hat{\mathbf{e}}_i$. If $\mathbf{S}$ has two distinct eigenvalues, $\omega_1 \neq \omega_2 = \omega_3$ then

$$\omega_2\hat{\mathbf{e}}_2 \otimes \hat{\mathbf{e}}_2 + \omega_3\hat{\mathbf{e}}_3 \otimes \hat{\mathbf{e}}_3 = \omega_2(\hat{\mathbf{e}}_2 \otimes \hat{\mathbf{e}}_2 + \hat{\mathbf{e}}_3 \otimes \hat{\mathbf{e}}_3).$$

Since $\hat{\mathbf{e}}_2 \otimes \hat{\mathbf{e}}_2 + \hat{\mathbf{e}}_3 \otimes \hat{\mathbf{e}}_3 = \mathbf{1} - \hat{\mathbf{e}}_1 \otimes \hat{\mathbf{e}}_1$ then

$$\mathbf{S} = \omega_1\hat{\mathbf{e}}_1 \otimes \hat{\mathbf{e}}_1 + \omega_2(\mathbf{1} - \hat{\mathbf{e}}_1 \otimes \hat{\mathbf{e}}_1).$$

If $\omega_1 = \omega_2 = \omega_3 = \omega$ then

$$\mathbf{S} = \omega\mathbf{1};$$

in this case each vector is an eigenvector.

A tensor $\mathbf{S}$ is *positive-definite* if it is symmetric and

$$\mathbf{u} \cdot \mathbf{S}\mathbf{u} > 0$$

for all nonzero vectors $\mathbf{u}$. If $\mathbf{S}$ is positive-definite then letting $\hat{\mathbf{e}}_i$ be the i-th basis eigenvector we have

$$0 < \hat{\mathbf{e}}_i \cdot \mathbf{S}\hat{\mathbf{e}}_i = \omega_i$$

and hence each eigenvalue ω_i is positive. Conversely, if $\omega_1, \omega_2, \omega_3$ are positive then $\mathbf{u} = u_i \hat{\mathbf{e}}_i$ results in

$$\mathbf{u} \cdot \mathbf{S}\mathbf{u} = u_j \hat{\mathbf{e}}_j \cdot \mathbf{S} u_i \hat{\mathbf{e}}_i = \sum_{i,j} u_i u_j \omega_i \delta_{ij} = \sum_i \omega_i u_i^2.$$

Accordingly $\mathbf{S}$ is positive-definite if and only if the eigenvalues $\{\omega_i\}$ are positive.

Given two orthonormal bases $\{\mathbf{e}_j\}$, $\{\hat{\mathbf{e}}_i\}$ we can express $\hat{\mathbf{e}}_i$ relative to the basis $\{\mathbf{e}_j\}$ in the form

$$\hat{\mathbf{e}}_i = R_{ij} \mathbf{e}_j, \qquad R_{ij} = \hat{\mathbf{e}}_i \cdot \mathbf{e}_j.$$

Since

$$\delta_{ih} = \hat{\mathbf{e}}_i \cdot \hat{\mathbf{e}}_h = R_{ij} R_{hk} \mathbf{e}_j \cdot \mathbf{e}_k = R_{ij} R_{hj} = (RR^T)_{ih}$$

then $RR^T = \mathbb{1}$. Hence $R^T = R^{-1}$ and $R^T R = \mathbb{1}$. Accordingly R is orthogonal and, subject to $\det R = 1$, a rotation. Geometrically, R describes the (passive) rotation from $\mathbf{e}_j$ to $\hat{\mathbf{e}}_i$.

If R is the rotation matrix from $\{\mathbf{e}_j\}$ to $\{\hat{\mathbf{e}}_i\}$ so that

$$\hat{\mathbf{e}}_i = R_{ij} \mathbf{e}_j$$

then the matrices S, $\hat{S}$ of $\mathbf{S}$ relative to the bases $\{\mathbf{e}_j\}$, $\{\hat{\mathbf{e}}_i\}$ satisfy

$$S_{ij} = \mathbf{e}_i \cdot \mathbf{S}\mathbf{e}_j = R_{ip} \hat{\mathbf{e}}_p \cdot \mathbf{S} R_{jq} \hat{\mathbf{e}}_q = R_{ip} R_{jq} \hat{S}_{pq}.$$

Hence

$$S = R \hat{S} R^T.$$

If $\{\hat{\mathbf{e}}_i\}$ is the basis of eigenvectors then

$$\det \mathbf{S} = \det S = \det \hat{S} = \omega_1 \omega_2 \omega_3, \qquad \operatorname{tr} \mathbf{S} = \operatorname{tr} S = \operatorname{tr} \hat{S} = \omega_1 + \omega_2 + \omega_3.$$

As a consequence, if $\mathbf{S}$ is positive-definite then $\omega_i > 0$, $i = 1, 2, 3$, and hence

$$\det \mathbf{S} > 0, \qquad \operatorname{tr} \mathbf{S} > 0.$$

For any positive-definite tensor $\mathbf{S}$ there is a unique positive-definite tensor $\mathbf{U}$ such that

$$\mathbf{U}^2 = \mathbf{S};$$

the symbol $\sqrt{\mathbf{S}}$ is then used for $\mathbf{U}$ and $\sqrt{\mathbf{S}}$ is referred to as the square root of $\mathbf{S}$.

To prove the existence of $\sqrt{\mathbf{S}}$ we observe that if $\mathbf{S}$ is positive-definite then

$$\mathbf{S} = \textstyle\sum_i \omega_i \, \hat{\mathbf{e}}_i \otimes \hat{\mathbf{e}}_i, \qquad \omega_i > 0.$$

Let

$$\mathbf{U} = \textstyle\sum_i \sqrt{\omega_i} \, \hat{\mathbf{e}}_i \otimes \hat{\mathbf{e}}_i.$$

Now, for any vector $\mathbf{v}$

$$\begin{aligned}\mathbf{U}^2\mathbf{v} &= \textstyle\sum_{i,j} \sqrt{\omega_i}\,\sqrt{\omega_j}\, \hat{\mathbf{e}}_i \otimes \hat{\mathbf{e}}_i \; \hat{\mathbf{e}}_j \otimes \hat{\mathbf{e}}_j \, \mathbf{v} = \sum_{i,j} \sqrt{\omega_i}\,\sqrt{\omega_j}\, \hat{\mathbf{e}}_i \otimes \hat{\mathbf{e}}_i \; \hat{\mathbf{e}}_j \hat{v}_j \\ &= \textstyle\sum_{i,j} \sqrt{\omega_i}\,\sqrt{\omega_j}\, \hat{\mathbf{e}}_i \, \delta_{ij} \, \hat{v}_j = \sum_i \omega_i \, \hat{\mathbf{e}}_i \hat{v}_i = \sum_i \omega_i \, \hat{\mathbf{e}}_i \otimes \hat{\mathbf{e}}_i \, \mathbf{v}.\end{aligned}$$

Accordingly,

$$\textstyle\sum_{i,j} \sqrt{\omega_i}\,\sqrt{\omega_j}\, \hat{\mathbf{e}}_i \otimes \hat{\mathbf{e}}_i \; \hat{\mathbf{e}}_j \otimes \hat{\mathbf{e}}_j = \sum_i \omega_i \, \hat{\mathbf{e}}_i \otimes \hat{\mathbf{e}}_i$$

and

$$\mathbf{U}^2 = \mathbf{S}.$$

A.1.1 Eigenvalues and Eigenvectors

Let ω be an eigenvalue of $\mathbf{S}$ and $\mathbf{e}$ be the corresponding eigenvector. Then $(\mathbf{S} - \omega\mathbf{1})\mathbf{e} = 0$ and $\mathbf{S} - \omega\mathbf{1}$ is not invertible whence

$$\det(\mathbf{S} - \omega\mathbf{1}) = 0.$$

The corresponding matrix form,

$$\det(S - \omega \mathbb{1}) = 0$$

results in the *characteristic equation*

$$\omega^3 - a_1\omega^2 + a_2\omega - a_3 = 0, \tag{A.2}$$

where $a_1 = \operatorname{tr} S$, $a_3 = \det S$, while

$$a_2 = S_{22}S_{33} - S_{32}S_{23} + S_{11}S_{22} - S_{21}S_{12} + S_{11}S_{33} - S_{13}S_{31}$$

is the sum of the principal minors of S.

To get a simpler representation of a_2 we observe that

$$\text{tr } S^2 = S_{ij}S_{ji} = S_{11}^2 + S_{22}^2 + S_{33}^2 + 2S_{12}S_{21} + 2S_{13}S_{31} + 2S_{23}S_{32},$$

$$(\text{tr } S)^2 = S_{11}^2 + S_{22}^2 + S_{33}^2 + 2S_{11}S_{22} + 2S_{11}S_{33} + 2S_{22}S_{33}.$$

Hence

$$(\text{tr } S)^2 - \text{tr } S^2 = 2(S_{11}S_{22} - S_{12}S_{21}) + 2(S_{11}S_{33} - S_{13}S_{31}) + 2(S_{22}S_{33} - S_{23}S_{32}) = 2a_2.$$

Hence we can write

$$a_1 = \text{tr } S, \qquad a_2 = \tfrac{1}{2}[(\text{tr } S)^2 - \text{tr } S^2], \qquad a_3 = \det S.$$

Upon an orthogonal transformation, with matrix Q, the matrix S is transformed to $\hat{S}$ such that

$$\hat{S} = Q\,S\,Q^T, \qquad Q^T Q = Q Q^T = 1\!\!1.$$

It follows that

$$\text{tr } \hat{S} = Q_{ij}S_{jk}Q_{ik} = \delta_{jk}S_{jk} = \text{tr } S,$$

$$\text{tr } \hat{S}^2 = \text{tr } QSQ^T QSQ^T = \text{tr } QS^2Q^T = \text{tr } S^2,$$

$$\det \hat{S} = \det Q\,S\,Q^T = (\det Q)^2 \det S = \det S.$$

Accordingly, a_1, a_2, a_3 are invariant under the group of orthogonal transformations. That is why the symbols I_1, I_2, I_3 are currently used instead of a_1, a_2, a_3 and I_1, I_2, I_3 are called *principal invariants* of the matrix (and of the tensor).

Let $\mathbf{S}$ be a symmetric tensor and let $\mathbf{e}$ be an eigenvector of $\mathbf{S}$ with ω the corresponding eigenvalue. Hence $\mathbf{Se} = \omega\mathbf{e}$ and

$$\mathbf{S}^2\mathbf{e} = \mathbf{S}(\mathbf{Se}) = \omega\mathbf{Se} = \omega^2\mathbf{e}, \qquad \mathbf{S}^3\mathbf{e} = \omega^3\mathbf{e}.$$

Consequently

$$(\mathbf{S}^3 - I_1\mathbf{S}^2 + I_2\mathbf{S} - I_3\mathbf{1})\mathbf{e} = (\omega^3 - I_1\omega^2 + I_2\omega - I_3)\mathbf{e} = \mathbf{0}.$$

Since $\mathbf{S}$ is symmetric there is an orthonormal basis $\{\hat{\mathbf{e}}_i\}$ of eigenvectors of $\mathbf{S}$ and hence, for any vector $\mathbf{u} = \hat{u}_i\hat{\mathbf{e}}_i$,

$$(\mathbf{S}^3 - I_1\mathbf{S}^2 + I_2\mathbf{S} - I_3\mathbf{1})\mathbf{u} = \hat{u}_i(\mathbf{S}^3 - I_1\mathbf{S}^2 + I_2\mathbf{S} - I_3\mathbf{1})\hat{\mathbf{e}}_i = \textstyle\sum_i \hat{u}_i(\omega_i^3 - I_1\omega_i^2 + I_2\omega_i - I_3)\hat{\mathbf{e}}_i.$$

Since each eigenvalue ω_i satisfies the characteristic equation (A.2) then

$$(\mathbf{S}^3 - I_1\mathbf{S}^2 + I_2\mathbf{S} - I_3\mathbf{1})\mathbf{u} = \mathbf{0}$$

for any vector $\mathbf{u}$. This implies that

$$\mathbf{S}^3 - I_1\mathbf{S}^2 + I_2\mathbf{S} - I_3\mathbf{1} = \mathbf{0}. \tag{A.3}$$

By (A.3) we can state

Theorem A.1 (Cayley–Hamilton) *Every tensor satisfies its own characteristic equation.*

Equation (A.3) is naturally referred to as the *Cayley–Hamilton equation* for $\mathbf{S}$; it is the tensor analogue of the Cayley–Hamilton equation for matrices [183] which, though, need not require the symmetry of the matrix.

Incidentally, for symmetric tensors $\mathbf{S}$ the principal invariants are given by the eigenvalues in the form

$$I_1 = \omega_1 + \omega_2 + \omega_3, \qquad I_2 = \omega_1\omega_2 + \omega_2\omega_3 + \omega_1\omega_3, \qquad I_3 = \omega_1\omega_2\omega_3.$$

In connection with a tensor $\mathbf{S}$, define the quantities J_i, $i = 1, 2, 3$, as

$$J_i = \det S^i.$$

They are invariants in that

$$\det[\hat{S}^i] = \det[(QSQ^T)^i] = \det[QS^iQ^T] = \det S^i.$$

Indeed, they are named the *main invariants* of $\mathbf{S}$ [388].

Further invariants are defined as

$$\mathcal{I}_1 = \text{tr}\, S, \qquad \mathcal{I}_2 = \text{tr}\, S^2, \qquad \mathcal{I}_3 = \text{tr}\, S^3.$$

The invariance of tr S^i follows by observing that

$$\text{tr}\, \hat{S}^i = \text{tr}\,(QSQ^T)^i = \text{tr}\,(QS^iQ^T) = \text{tr}\,[(Q^TQ)S^i] = \text{tr}\, S^i.$$

The eigenvalues ω of an orthogonal tensor are subject to $|\omega| = 1$. If the orthogonal tensor is a rotation then one of the eigenvalues equals 1. To prove these assertions we consider $\mathbf{Q} \in \text{Orth}$ and let

$$\mathbf{Q}\mathbf{v} = \omega\mathbf{v}, \qquad \mathbf{v} \in \mathbb{C}^3.$$

Take the complex conjugate, denoted by $*$, and then the transpose to obtain

$$\mathbf{Q}\mathbf{v}^* = \omega^*\mathbf{v}^*, \qquad (\mathbf{v}^*)^T\mathbf{Q}^T = \omega^*(\mathbf{v}^*)^T.$$

Hence we find

$$(\mathbf{v}^*)^T\mathbf{Q}^T\mathbf{Q}\mathbf{v} = \omega\omega^*(\mathbf{v}^*)^T \cdot \mathbf{v}.$$

Since $\mathbf{Q} \in \text{Orth}$ then $\mathbf{Q}^T\mathbf{Q} = \mathbf{1}$ then it follows

$$(\mathbf{v}^*)^T \cdot \mathbf{v}(1 - \omega\omega^*) = 0.$$

As a consequence, $\mathbf{v} \neq \mathbf{0}$ implies

$$\omega\omega^* = 1,$$

and hence $|\omega| = 1$. The characteristic equation

$$\det(\mathbf{Q} - \omega\mathbf{1}) = -\omega^3 + I_1\omega^2 - I_2\omega + \det\mathbf{Q} = 0$$

is satisfied by at least one real value, say ω_1. The remaining roots ω_2, ω_3 are complex conjugate, $\omega_3 = \omega_2^*$. Now,

$$\det\mathbf{Q} = \omega_1\omega_2\omega_2^* = \omega_1$$

and hence $\omega_1 = \pm 1$. If $\det\mathbf{Q} = 1$ then $\omega_1 = 1$.

As a consequence, for any rotation $\mathbf{Q}$ there is a vector $\mathbf{v}$, associated with ω_1, such that

$$\mathbf{Q}\mathbf{v} = \mathbf{v};$$

the vector $\mathbf{v}$, which is invariant under $\mathbf{Q}$, defines the *rotation axis.*

Let $\mathbf{v}$ be a vector in $\mathbb{R}^3$ and $\mathbf{n}$ a unit vector along an axis of rotation. To determine the effect of a rotation on a vector we observe that any vector $\mathbf{v}$ can be decomposed as

$$\mathbf{v} = \mathbf{v}_\parallel + \mathbf{v}_\perp, \qquad \mathbf{v}_\parallel := (\mathbf{v}\cdot\mathbf{n})\mathbf{n}, \quad \mathbf{v}_\perp = -\mathbf{n}\times(\mathbf{n}\times\mathbf{v}),$$

$\mathbf{v}_\parallel$ and $\mathbf{v}_\perp$ being the parallel and the orthogonal parts of $\mathbf{v}$ relative to $\mathbf{n}$. Upon a rotation of $\mathbf{v}$ about $\mathbf{n}$ by an angle θ, the vector $\mathbf{v}_\parallel$ is left unchanged. Instead, if $\mathbf{i}_1, \mathbf{i}_2$ are unit vectors such that $\{\mathbf{i}_1, \mathbf{i}_2, \mathbf{n}\}$ is an orthonormal basis then

$$\hat{v}_1 = \cos\theta v_1 - \sin\theta v_2, \qquad \hat{v}_2 = -\sin\theta v_1 + \cos\theta v_2,$$

and hence

$$\hat{\mathbf{v}}_\perp = \cos\theta\mathbf{v}_\perp + \sin\theta\mathbf{n}\times\mathbf{v}_\perp.$$

As a consequence, upon the rotation, $\mathbf{v} \to \hat{\mathbf{v}}$ such that

$$\hat{\mathbf{v}} = \cos\theta\,\mathbf{v} + \sin\theta\,\mathbf{n}\times\mathbf{v} + (1 - \cos\theta)(\mathbf{n}\cdot\mathbf{v})\mathbf{n}. \qquad \text{(A.4)}$$

Equation (A.4) is usually referred to as *Rodrigues rotation formula*. The transformation $\mathbf{v} \to \hat{\mathbf{v}}$ given by (A.4) can be viewed as the effect of the rotation

$$\mathbf{Q} = \mathbf{n} \otimes \mathbf{n} + \cos\theta(\mathbf{1} - \mathbf{n} \otimes \mathbf{n}) + \sin\theta(\mathbf{i}_2 \otimes \mathbf{i}_1 - \mathbf{i}_1 \otimes \mathbf{i}_2).$$

Now,

$$I_1 = \operatorname{tr}\mathbf{Q} = 1 + 2\cos\theta, \quad I_2 = \tfrac{1}{2}[(\operatorname{tr}\mathbf{Q})^2 - \operatorname{tr}\mathbf{Q}^2] = 1 + 2\cos\theta, \quad I_3 = \det\mathbf{Q} = 1.$$

Since $I_1 = I_2 = \operatorname{tr}\mathbf{Q}$ then the Cayley–Hamilton theorem allows us to write

$$\mathbf{Q}^T = \mathbf{Q}^{-1} = \mathbf{Q}^2 - (\operatorname{tr}\mathbf{Q})\mathbf{Q} + (\operatorname{tr}\mathbf{Q})\mathbf{1}.$$

A.1.2 *Length and Induced Norm*

The cross product $\mathbf{u} \times \mathbf{v}$, the multiplication by a scalar $\alpha\mathbf{v}$, the product by a tensor $\mathbf{S}\mathbf{v}$, and the dyadic product $\mathbf{u} \otimes \mathbf{v}$ are linear operators of $\mathscr{V}$ into $\mathscr{V}$; an operator $\mathbf{A}$ is linear if

$$\mathbf{A}(\alpha\mathbf{u} + \beta\mathbf{v}) = \alpha\mathbf{A}\mathbf{u} + \beta\mathbf{A}\mathbf{v}$$

for all $\mathbf{u}, \mathbf{v} \in \mathscr{V}$ and all real α, β. A linear operator $\mathbf{A}$ is bounded if there is a constant M such that for all $\mathbf{v} \in \mathscr{V}$ we have $|\mathbf{A}\mathbf{v}| \le M|\mathbf{v}|$. The least such M is called the norm of $\mathbf{A}$, induced by the length of vectors. Thus

$$\|\mathbf{A}\| := \sup_{|\mathbf{v}|=1} |\mathbf{A}\mathbf{v}|.$$

Hence

$$|\mathbf{A}\mathbf{v}| \le \|\mathbf{A}\|\,|\mathbf{v}|.$$

An induced operator norm is sub-multiplicative in that $\|\mathbf{T}\mathbf{S}\| \le \|\mathbf{T}\|\,\|\mathbf{S}\|$. This follows at once by

$$|\mathbf{T}\mathbf{S}\mathbf{v}| \le \|\mathbf{T}\|\,\|\mathbf{S}\|\,|\mathbf{v}|, \qquad \|\mathbf{T}\mathbf{S}\| = \sup_{|\mathbf{v}|=1} |\mathbf{T}\mathbf{S}\mathbf{v}|.$$

The cross product $\mathbf{u} \times \mathbf{v}$ can be viewed as the operator $\mathbf{u}\times$ applied to $\mathbf{v}$. Since

$$|\mathbf{u} \times \mathbf{v}| \le |\mathbf{u}|\,|\mathbf{v}|,$$

and equality holds if $\mathbf{v}$ is perpendicular to $\mathbf{u}$, then $\|\mathbf{u}\times\| = |\mathbf{u}|$.

The norm of the dyadic product is found by observing that

$$|(\mathbf{u}\otimes\mathbf{w})\mathbf{v}| = |\mathbf{u}\,\mathbf{w}\cdot\mathbf{v}| = |\mathbf{u}|\,|\mathbf{w}\cdot\mathbf{v}| \le |\mathbf{u}|\,|\mathbf{w}|\,|\mathbf{v}|$$

while equality holds if $\mathbf{v} = \pm\mathbf{w}$. Hence it follows $|\mathbf{u}\otimes\mathbf{w}| = |\mathbf{u}|\,|\mathbf{w}|$.

To determine the norm of a tensor $\mathbf{S}$ we observe

$$\mathbf{S}\mathbf{v} = S_{ij}v_j\mathbf{e}_i, \qquad |\mathbf{S}\mathbf{v}| = [\textstyle\sum_i (S_{ij}v_j)^2]^{1/2}.$$

Now, $S_{ij}v_j$ can be viewed as the inner product (in $\mathbb{R}^3$) of the vectors (S_{i1}, S_{i2}, S_{i3}) and (v_1, v_2, v_3). The Cauchy–Schwarz inequality, as well as the inner product in $\mathscr{V}$, gives

$$|S_{ij}v_j| \le (\textstyle\sum_j S_{ij}^2)^{1/2}(\sum_h v_h^2)^{1/2}$$

and hence

$$|\mathbf{S}\mathbf{v}| = [\textstyle\sum_i (S_{ij}v_j)^2]^{1/2} \le [\sum_i(\sum_j S_{ij}^2)(\sum_h v_h^2)]^{1/2} = (\sum_{i,j} S_{ij}^2)^{1/2}(\sum_h v_h^2)^{1/2}.$$

Since $|\mathbf{v}| = (\sum_h v_h^2)^{1/2}$ then it follows

$$\|\mathbf{S}\| \le (\textstyle\sum_{i,j} S_{ij}^2)^{1/2}. \tag{A.5}$$

The same result emerges by observing that $\sum_{i,j} S_{ij}^2$ can be viewed as the inner product in $\mathbb{R}^9$ and applying the Cauchy–Schwarz inequality. Yet, if $\mathbf{S}$ is symmetric then we represent $\mathbf{v}$ in the basis of eigenvectors, $\mathbf{v} = \hat{v}_j\hat{\mathbf{e}}_j$, to obtain

$$|\mathbf{S}\mathbf{v}|^2 = (\mathbf{S}\mathbf{v})\cdot(\mathbf{S}\mathbf{v}) = \mathbf{S}\hat{v}_j\hat{\mathbf{e}}_j\cdot\mathbf{S}\hat{v}_h\hat{\mathbf{e}}_h = \textstyle\sum_{i,j}\omega_j\omega_h\hat{v}_j\hat{v}_h\hat{\mathbf{e}}_j\cdot\hat{\mathbf{e}}_h = \sum_j\omega_j^2\hat{v}_j^2 \le \omega_{\max}\sum_j\hat{v}_j^2,$$

where $\omega_{\max} = \max_j |\omega_j|$. Moreover equality holds if $\mathbf{v}$ is in the direction of the eigenvector associated with the largest eigenvalue. Hence, for symmetric tensors,

$$\|\mathbf{S}\| = \omega_{\max}; \tag{A.6}$$

the norm (A.6) is called the *spectral norm.* Accordingly, in general it is the condition (A.5) that holds true and hence we set

$$\|\mathbf{S}\| = (\textstyle\sum_{i,j} S_{ij}^2)^{1/2};$$

this norm is usually called the *Frobenius norm.*

There are several fourth-order tensors constructed from second-order ones. If $\mathbf{A}, \mathbf{B} \in \mathrm{Lin}$ the tensor products $\mathbf{A} \circledast \mathbf{B}$ and $\mathbf{A} \boxtimes \mathbf{B}$ are fourth-order tensors defined by

$$(\mathbf{A}\circledast\mathbf{B})\mathbf{C} := \mathbf{A}\,\mathbf{C}\,\mathbf{B},$$

$$(\mathbf{A} \boxtimes \mathbf{B})\mathbf{C} := \mathbf{A}\,\mathbf{C}\,\mathbf{B}^T,$$

for every $\mathbf{C} \in \text{Lin}$. In indicial notation,

$$(\mathbf{A} \circledast \mathbf{B})_{ijhk} = A_{ih}B_{kj}, \qquad (\mathbf{A} \boxtimes \mathbf{B})_{ijhk} = A_{ih}B_{jk}.$$

In the box product $\boxtimes$ the two tensors are ascribed the first and second pair of indices. Consistently we find

$$(\mathbf{A} \boxtimes \mathbf{B})(\mathbf{C} \boxtimes \mathbf{D}) = (\mathbf{AC}) \boxtimes (\mathbf{BD}),$$

$$(\mathbf{A} \boxtimes \mathbf{B})[\mathbf{u} \otimes \mathbf{v}] = (\mathbf{Au}) \otimes (\mathbf{Bv});$$

in indicial notation,

$$[(\mathbf{A} \boxtimes \mathbf{B})(\mathbf{C} \boxtimes \mathbf{D})]_{ijpq} = (\mathbf{A} \boxtimes \mathbf{B})_{ijhk}(\mathbf{C} \boxtimes \mathbf{D})_{hkpq} = A_{ih}B_{jk}C_{hk}D_{kq}$$
$$= (A_{ih}C_{hp})(B_{jk}D_{kq}) = (\mathbf{AC})_{ip}(\mathbf{BD})_{jq} = [(\mathbf{AC}) \boxtimes (\mathbf{BD})]_{ijpq}.$$

If $\mathbf{A} \in \text{Lin}$ then we let

$$(\mathbf{1} \boxtimes \mathbf{1})\mathbf{A} = \mathbf{1A1} = \mathbf{A}.$$

Hence

$$\mathsf{I} := \mathbf{1} \boxtimes \mathbf{1}$$

is the identity operator in Lin.

The transpose $(\mathbf{A} \boxtimes \mathbf{B})^T$ of $(\mathbf{A} \boxtimes \mathbf{B})$ is defined by

$$(\mathbf{A} \boxtimes \mathbf{B})^T = (\mathbf{A}^T \boxtimes \mathbf{B}^T).$$

Hence, for orthogonal tensors $\mathbf{Q}_1, \mathbf{Q}_2$ it follows

$$(\mathbf{Q}_1 \boxtimes \mathbf{Q}_2)^T(\mathbf{Q}_1 \boxtimes \mathbf{Q}_2) = (\mathbf{Q}_1^T \boxtimes \mathbf{Q}_2^T)(\mathbf{Q}_1 \boxtimes \mathbf{Q}_2) = (\mathbf{Q}_1^T\mathbf{Q}_1) \boxtimes (\mathbf{Q}_2^T \boxtimes \mathbf{Q}_2) = \mathbf{1} \boxtimes \mathbf{1} = \mathsf{I}.$$

For a fourth-order tensor C the eigenvalue problem involves finding a pair $(\lambda, \mathbf{A})$ of the eigenvalue λ and the (second-order) eigentensor $\mathbf{A}$ such that

$$\mathsf{C}\mathbf{A} = \lambda\mathbf{A}, \qquad C_{ijhk}A_{hk} = \lambda A_{ij}.$$

Let $\mathbf{n}$ be the rotation vector of $\mathbf{Q} \in \text{Orth}^+$, $\mathbf{Qn} = \mathbf{n}$, and let $(\lambda, \mathbf{m})$, $(\lambda^*, \mathbf{m}^*)$ denote the remaining eigenvalue-eigenvector pairs. We can then prove the following statement.

Let $\mathbf{Q} \in \text{Orth}^+$. *The eigenvalue-eigenvector pairs of* $\mathsf{Q} = \mathbf{Q} \boxtimes \mathbf{Q}$ *are* $(1, \mathbf{1})$, $(1, \mathbf{Q})$, $(1, \mathbf{Q}^2)$, $(\lambda, \mathbf{n} \otimes \mathbf{m})$, $(\lambda, \mathbf{m} \otimes \mathbf{n})$, $(\lambda^*, \mathbf{n} \otimes \mathbf{m}^*)$, $(\lambda^*, \mathbf{m}^* \otimes \mathbf{n})$, $(\lambda^2, \mathbf{m} \otimes \mathbf{m})$, $(\lambda^{*2}, \mathbf{m}^* \otimes \mathbf{m}^*)$.

The proof is given by a direct check. For instance,

$$\mathbf{Q1} = \mathbf{Q1Q}^T = \mathbf{QQ}^T = \mathbf{1}, \qquad \mathbf{QQ} = \mathbf{QQQ}^T = \mathbf{Q},$$

$$\mathbf{QQ}^2 = \mathbf{QQ}^2\mathbf{Q}^T = \mathbf{QQQQ}^T = \mathbf{QQ} = \mathbf{Q}^2,$$

$$\mathbf{Qn} \otimes \mathbf{m} = (\mathbf{Qn}) \otimes (\mathbf{Qm}) = \mathbf{n} \otimes \lambda\mathbf{m} = \lambda\mathbf{n} \otimes \mathbf{m},$$

$$\mathbf{Qm}^* \otimes \mathbf{m}^* = (\mathbf{Qm}^*) \otimes (\mathbf{Qm}^*) = \lambda^*\mathbf{m}^* \otimes \lambda^*\mathbf{m}^* = \lambda^{*2}\mathbf{m}^* \otimes \mathbf{m}^*.$$

The literature shows also a dyadic product between second-order tensors which is the strict analogue of dyadic product between vectors. If $\mathbf{A}, \mathbf{B}, \mathbf{C} \in \text{Lin}$ then $\mathbf{A} \otimes \mathbf{B}$ is defined by[3]

$$(\mathbf{A} \otimes \mathbf{B})\mathbf{C} = (\mathbf{B} \cdot \mathbf{C})\mathbf{A},$$

or

$$[(\mathbf{A} \otimes \mathbf{B})\mathbf{C}]_{ij} = (\mathbf{A} \otimes \mathbf{B})_{ijhk} C_{hk} = A_{ij} B_{hk} C_{hk}.$$

A.1.3 Representation Formulae for Vectors and Tensors

Let $\mathbf{e}$ be a unit vector. The tensors

$$\mathbf{e} \otimes \mathbf{e} \quad \text{and} \quad \mathbf{1} - \mathbf{e} \otimes \mathbf{e}$$

determine projections of any vector $\mathbf{u}$ onto $\mathbf{e}$, say $\mathbf{u}_\parallel$, and onto the plane perpendicular to $\mathbf{e}$, say $\mathbf{u}_\perp$. Hence

$$\mathbf{u} = \mathbf{u}_\parallel + \mathbf{u}_\perp = (\mathbf{e} \otimes \mathbf{e})\mathbf{u} + (\mathbf{1} - \mathbf{e} \otimes \mathbf{e})\mathbf{u}.$$

If $\mathbf{u}_\parallel = \mathbf{u} \cdot \mathbf{e}$ is fixed whereas $\mathbf{u}_\perp$ is undetermined then we can write

$$\mathbf{u} = (\mathbf{u} \cdot \mathbf{e})\mathbf{e} + (\mathbf{1} - \mathbf{e} \otimes \mathbf{e})\mathbf{w} \tag{A.7}$$

for any vector $\mathbf{w}$ in that

$$[(\mathbf{1} - \mathbf{e} \otimes \mathbf{e})\mathbf{w}] \cdot \mathbf{e} = \mathbf{w} \cdot \mathbf{e} - (\mathbf{e} \cdot \mathbf{e})\mathbf{e} \cdot \mathbf{w} = 0$$

and hence $(\mathbf{1} - \mathbf{e} \otimes \mathbf{e})\mathbf{w}$ is perpendicular to $\mathbf{e}$ for any vector $\mathbf{w}$.

The representation (A.7) has a strict analogue for tensors. Let $\mathbb{I}$ be the unit fourth-order tensor and $\mathbf{Y}$ a second-order tensor then

[3] This product is named fourth-order trace projection tensor [229]. The notation $\mathbf{A} \otimes \mathbf{B}$ is non-unique; the symbol $\mathbf{A} \odot \mathbf{B}$ is also used [234].

$$[(\mathbf{I} - \mathbf{N} \otimes \mathbf{N})\mathbf{Y}] \cdot \mathbf{N} = 0;$$

in words, the projection onto $\mathbf{N}$ of the orthogonal (or transverse) part of $\mathbf{Y}$ with respect to $\mathbf{N}$ is identically zero. Hence,

$$\mathbf{Y} = \mathbf{Y}_{\parallel} + \mathbf{Y}_{\perp}, \qquad \mathbf{Y}_{\parallel} := (\mathbf{Y} \cdot \mathbf{N})\mathbf{N}, \quad \mathbf{Y}_{\perp} := (\mathbf{I} - \mathbf{N} \otimes \mathbf{N})\mathbf{Y},$$

and $\mathbf{Y}_{\perp} \cdot \mathbf{N} = 0$. This representation is unique.

Now suppose $\mathbf{Z}$ is a second-order tensor such that $\mathbf{Z}_{\parallel}$ is known while $\mathbf{Z}_{\perp}$ is undetermined. Then we observe that

$$[(\mathbf{I} - \mathbf{N} \otimes \mathbf{N})\mathbf{G}] \cdot \mathbf{N} = 0,$$

with any second-order tensor $\mathbf{G}$ and then $(\mathbf{I} - \mathbf{N} \otimes \mathbf{N})\mathbf{G}$ is a possible representation of $\mathbf{Z}_{\perp}$. Hence to emphasize the unknown property of $\mathbf{Z}_{\perp}$ we represent $\mathbf{Z}$ in the form

$$\mathbf{Z} = (\mathbf{Z} \cdot \mathbf{N})\mathbf{N} + (\mathbf{I} - \mathbf{N} \otimes \mathbf{N})\mathbf{G} \tag{A.8}$$

for any (second-order) tensor $\mathbf{G}$.

As we can see in some contexts, thermodynamic requirements may result in the knowledge of the projections $\mathbf{u}_{\parallel}$ or $\mathbf{Z}_{\parallel}$ relative to appropriate vectors $\mathbf{e}$ or tensors $\mathbf{N}$. In those cases the representation formulae (A.7) or (A.8) apply.

A.2 Isotropic Tensors

By definition, isotropic tensors are characterized by the property that their components take the same values in all Cartesian coordinate systems. Let $\mathbf{Q} \neq \mathbf{1}$ be the rotation matrix between two coordinate systems. The components of a tensor of order n satisfy

$$A'_{i_1 \cdots i_n} = Q_{i_1 j_1} \cdots Q_{i_n j_n} A_{j_1 \cdots j_n}$$

The tensor is said to be isotropic if it is invariant,

$$A'_{i_1 \cdots i_n} = A_{i_1 \cdots i_n},$$

under any rotation $\mathbf{Q}$. Owing to invariance, all scalars are isotropic. A vector, say $\mathbf{w}$, is isotropic if

$$w'_i = Q_{ij} w_j = w_i.$$

This is true only if $\mathbf{Q} = \mathbf{1}$ and hence there are no isotropic vectors.

Among the admissible rotations are those associated with time-dependent differentiable matrices $\mathbf{Q}(t)$ such that $\mathbf{Q}(0) = \mathbf{1}$ and hence

$$\mathbf{Q}(t) = \mathbf{1} + \dot{\mathbf{Q}}(0)t + o(t).$$

The orthogonality of the function $\mathbf{Q}(t)$, namely $\mathbf{Q}(t)\mathbf{Q}^T(t) = \mathbf{1}$, for all $t \geq 0$, implies that

$$\dot{\mathbf{Q}}(t)\mathbf{Q}^T(t) = -\mathbf{Q}(t)\dot{\mathbf{Q}}^T(t) = -[\dot{\mathbf{Q}}(t)\mathbf{Q}^T(t)]^T$$

whence $\dot{\mathbf{Q}}(t)\mathbf{Q}^T(t) \in \text{Skw}$ for all $t \geq 0$. Since $\mathbf{Q}(0) = \mathbf{Q}^T(0) = \mathbf{1}$ then $\dot{\mathbf{Q}}(0) \in \text{Skw}$. Consequently there is a vector, say $\mathbf{w}$, such that

$$w_i = -\tfrac{1}{2}\epsilon_{ipq}\dot{Q}_{pq}, \qquad \dot{Q}_{pq} = \epsilon_{piq}w_i;$$

henceforth $\dot{\mathbf{Q}}$ stands for $\dot{\mathbf{Q}}(0)$.

Require that the second-order tensor $\mathbf{M}$ be isotropic,

$$M'_{ij} = Q_{ip}Q_{jq}M_{pq} = M_{ij}.$$

Substitution of $\mathbf{Q}$ results in

$$0 = (\dot{Q}_{ip}\delta_{jq} + \dot{Q}_{jq}\delta_{ip})M_{pq} = w_r(\epsilon_{irp}M_{pj} + \epsilon_{jrq}M_{iq}).$$

The arbitrariness of $\dot{\mathbf{Q}} \in \text{Skw}$, and hence that of $\mathbf{w}$, implies

$$\epsilon_{irp}M_{pj} + \epsilon_{jrq}M_{iq} = 0,$$

$\forall i, j, r = 1, 2, 3$. Multiply by ϵ_{irk} to obtain

$$0 = (\delta_{rr}\delta_{pk} - \delta_{rp}\delta_{rk})M_{pj} + (\delta_{ij}\delta_{kq} - \delta_{iq}\delta_{kj})M_{iq} = 2M_{kj} + M_{jk} - M_{qq}\delta_{jk}.$$

Hence

$$2M_{kj} + M_{jk} - M_{qq}\delta_{jk} = 0, \qquad 2M_{jk} + M_{kj} - M_{qq}\delta_{jk} = 0,$$

the second relation being obtained by interchanging j and k. Now, taking the difference it follows

$$M_{jk} - M_{kj} = 0.$$

Consequently we have

$$M_{jk} = \tfrac{1}{3}M_{qq}\delta_{jk}$$

and hence

$$M_{jk} = \lambda\,\delta_{jk}.$$

For a third-order tensor $\mathbf{N}$, isotropy means

$$N_{ijk} = N'_{ijk} = Q_{ip}Q_{jq}Q_{kr}N_{pqr}.$$

Consequently

$$N_{ijk} = (\delta_{ip} + \dot{Q}_{ip}t + o(t))(\delta_{jq} + \dot{Q}_{jq}t + o(t))(\delta_{kr} + \dot{Q}_{kr}t + o(t))N_{pqr}$$

whence it follows

$$\dot{Q}_{ip}N_{pjk} + \dot{Q}_{jq}N_{iqk} + \dot{Q}_{kr}N_{ijr} = 0.$$

Substitution of $\dot{Q}_{ip} = \epsilon_{iqp}w_q$ and the like for $\dot{Q}_{jq}$ and $\dot{Q}_{kr}$ results in

$$\epsilon_{isp}N_{pjk}w_s + \epsilon_{jsq}N_{iqk}w_s + \epsilon_{ksr}N_{ijr}w_s = 0$$

and, by the arbitrariness of $\mathbf{w}$,

$$\epsilon_{sip}N_{pjk} + \epsilon_{sjq}N_{iqk} + \epsilon_{skr}N_{ijr} = 0.$$

Upon multiplying by $\epsilon_{sih}, \epsilon_{sjh}, \epsilon_{skh}$ it follows

$$2N_{hjk} + N_{jhk} + N_{kjh} = N_{qqk}\delta_{hj} + N_{qjq}\delta_{hk},$$

$$2N_{ihk} + N_{hik} + N_{ikh} = N_{qqk}\delta_{hi} + N_{iqq}\delta_{hk},$$

$$2N_{ijh} + N_{hji} + N_{ihj} = N_{kjk}\delta_{hi} + N_{iqq}\delta_{hj}.$$

Multiply these equations by δ_{jk}, δ_{ik}, and δ_{ij}, respectively. We find that N_{hqq}, N_{qhq}, and N_{qqh} vanish for any $h = 1, 2, 3$. Hence, by relabelling the indices, it follows that

$$2N_{hik} + N_{ihk} + N_{kih} = 0, \quad 2N_{ihk} + N_{hik} + N_{ikh} = 0, \quad 2N_{ihk} + N_{khi} + N_{ikh} = 0.$$

The second and third equations imply

$$N_{hik} = N_{khi},$$

the matrix N is invariant upon a cyclic permutation. Hence the second equation results in

$$N_{ihk} = -N_{ikh}.$$

Accordingly, N
is invariant under a cyclic permutation,
changes sign when two indices are interchanged, and hence is zero if two indices are equal.
This means that N is a multiple of the Levi-Civita (or permutation) symbol,

$$N_{ihk} = \mu\, \epsilon_{ihk}.$$

For a fourth-order tensor $\mathbf{A}$, the isotropy condition is expressed by

$$A_{ijhk} = A'_{ijhk} = Q_{ip} Q_{jq} Q_{hr} Q_{ks} A_{pqrs}.$$

Substitution of $\mathbf{Q} = \mathbf{1} + \dot{\mathbf{Q}}t + o(t))$ results in

$$\dot{Q}_{ip} A_{pjhk} + \dot{Q}_{jp} A_{iphk} + \dot{Q}_{hp} A_{ijpk} + \dot{Q}_{kp} A_{ijhp} = 0,$$

whence

$$\epsilon_{mip} A_{pjhk} + \epsilon_{mjp} A_{iphk} + \epsilon_{mhp} A_{ijpk} + \epsilon_{mkp} A_{ijhp} = 0.$$

Multiply this equation by $\epsilon_{miq}, \epsilon_{mjq}, \epsilon_{mhq}, \epsilon_{mkq}$ to obtain

$$2A_{ijhk} + A_{jihk} + A_{hjik} + A_{kjhi} = A_{pphk}\delta_{ij} + A_{pjpk}\delta_{ih} + A_{pjhp}\delta_{ik},$$

$$2A_{ijhk} + A_{jihk} + A_{ihjk} + A_{ikhj} = A_{pphk}\delta_{ji} + A_{ippk}\delta_{jh} + A_{iphp}\delta_{jk},$$

$$2A_{ijhk} + A_{ijkh} + A_{hjik} + A_{ihjk} = A_{pjpk}\delta_{hi} + A_{ippk}\delta_{hj} + A_{ijpp}\delta_{hk},$$

$$2A_{ijhk} + A_{kjhi} + A_{ikhj} + A_{ijkh} = A_{pjhp}\delta_{ki} + A_{iphp}\delta_{kj} + A_{ijpp}\delta_{kh},$$

where q has been relabelled as i, j, h, k, respectively. Since $\mathbf{A}$ is an isotropic fourth-order tensor then any quantities with a pair of repeated indices, like A_{pphk}, are the components of a second-order isotropic tensor. Accordingly we let

$$A_{pphk} = \mu_{12}\delta_{hk}, \quad A_{pjpk} = \mu_{13}\delta_{jk}, \quad A_{pjhp} = \mu_{14}\delta_{jh},$$
$$A_{ippk} = \mu_{23}\delta_{ik}, \quad A_{iphp} = \mu_{24}\delta_{ih}, \quad A_{ijpp} = \mu_{34}\delta_{ij}.$$

Upon substitution, subtraction of the third and fourth equations from the summation of the first and second equation result in

$$A_{jihk} - A_{ijkh} = \mu_{12}\delta_{hk}\delta_{ij} - \mu_{34}\delta_{hk}\delta_{ij}.$$

Letting $i = j, h = k$ we find

$$\mu_{12} = \mu_{34} =: \alpha.$$

Consequently, by subtracting the fourth equation from the first one we obtain

$$A_{jihk} + A_{hjik} - A_{ikhj} - A_{ijkh} = (\mu_{13} - \mu_{24})\delta_{jk}\delta_{ih}.$$

Letting $j = k = i = h$ we find

$$\mu_{13} = \mu_{24} =: \beta.$$

We now subtract the second equation from the first one and use the condition $\mu_{13} = \mu_{24}$. It follows

$$A_{hjik} + A_{kjhi} - A_{ihjk} - A_{ikhj} = (\mu_{14} - \mu_{23})\delta_{jh}\delta_{ik}.$$

Letting $j = k = i = h$ we find

$$\mu_{14} = \mu_{23} =: \gamma.$$

Moreover we have

$$A_{jihk} = A_{ijkh}, \qquad A_{hjik} = A_{ikhj}. \tag{A.9}$$

In view of (A.9), the four conditions on A simplify to a single significant equation

$$2A_{ijhk} + A_{jihk} + A_{hjik} + A_{kjhi} = \lambda\delta_{hk}\delta_{ij} + \mu\delta_{jk}\delta_{ih} + \nu\delta_{jh}\delta_{ik}. \tag{A.10}$$

Now let $jhk \to hkj$ and $jhk \to kjh$ to obtain

$$2A_{ihkj} + A_{hikj} + A_{khij} + A_{jhki} = \lambda\delta_{kj}\delta_{ih} + \mu\delta_{hj}\delta_{ik} + \nu\delta_{hk}\delta_{ij}, \tag{A.11}$$

$$2A_{ikjh} + A_{kijh} + A_{jkih} + A_{hkji} = \lambda\delta_{jh}\delta_{ik} + \mu\delta_{kh}\delta_{ij} + \nu\delta_{kj}\delta_{ih}. \tag{A.12}$$

The sum of Eqs. (A.10)–(A.12) results in

$$\begin{aligned} 2(A_{ijhk} + A_{ihkj} + A_{ikjh}) + 3(A_{ijkh} + A_{ikhj} + A_{hikj}) \\ = (\lambda + \mu + \nu)(\delta_{hk}\delta_{ij} + \delta_{kj}\delta_{ih} + \delta_{jh}\delta_{ik}). \end{aligned} \tag{A.13}$$

The right-hand side is invariant under the interchange $h \leftrightarrow k$ (as well as under $i \leftrightarrow j$). The requirement that so be the left-hand side and use of (A.9), namely $A_{kihj} = A_{ikjh}$, $A_{ihjk} = A_{hikj}$, yield

$$\begin{aligned} &2(A_{ijhk} + A_{ihkj} + A_{ikjh}) + 3(A_{ijkh} + A_{ikhj} + A_{hikj}) \\ &= 2(A_{ijkh} + A_{ikhj} + A_{hikj}) + 3(A_{ijhk} + A_{ihkj} + A_{ikjh}). \end{aligned}$$

Hence letting

$$X = A_{ijhk} + A_{ihkj} + A_{ikjh}, \qquad Y = A_{ijkh} + A_{ikhj} + A_{hikj},$$

we have

$$2X + 3Y = 3X + 2Y$$

whence $X = Y$. Consequently (A.13) can be written in the form

$$5(A_{ijhk} + A_{ihkj} + A_{ikjh}) = (\alpha + \beta + \gamma)(\delta_{hk}\delta_{ij} + \delta_{kj}\delta_{ih} + \delta_{jh}\delta_{ik})$$

Upon substitution of $A_{ijhk} + A_{ihkj} + A_{ikjh}$ in (A.10) we find

$$A_{ijhk} = \lambda\delta_{ij}\delta_{hk} + \mu\delta_{ih}\delta_{jk} + \nu\delta_{ik}\delta_{jh} \tag{A.14}$$

where

$$\lambda = \tfrac{1}{10}(4\alpha - \beta - \gamma), \quad \mu = \tfrac{1}{10}(4\beta - \gamma - \alpha), \quad \nu = \tfrac{1}{10}(4\gamma - \alpha - \beta).$$

Equation (A.14) is the general representation of isotropic fourth-order tensors. Irrespective of the value of the parameters λ, μ, ν it follows that the major symmetry

$$A_{ijhk} = A_{hkij}$$

holds. The minor symmetries

$$A_{ijhk} = A_{jihk}, \qquad A_{ijhk} = A_{ijkh}$$

hold if and only if $\mu = \nu$ (and $\beta = \gamma$) whence

$$A_{ijhk} = \lambda\delta_{ij}\delta_{hk} + \mu(\delta_{ih}\delta_{jk} + \delta_{ik}\delta_{jh}). \tag{A.15}$$

A.3 Differentiation

A scalar, vector, or tensor function $\mathbf{w}(t)$, $t \in \mathcal{D} \subseteq \mathbb{R}$, has a limit as $t \to t_0$ if

$$\forall\epsilon > 0\ \exists\delta > 0 :\ 0 < |t - t_0| < \delta \implies |\mathbf{w}(t) - \mathbf{w}_0| < \epsilon,$$

the symbol $|\cdot|$ denoting the absolute value of scalars or the norm of vectors; $|\cdot|$ is replaced by $\|\cdot\|$ if $\mathbf{w}$ is a tensor. We then write

$$|\mathbf{w}(t) - \mathbf{w}_0| \to 0 \quad \text{as} \quad t \to t_0, \qquad \lim_{t\to t_0} \mathbf{w}(t) = \mathbf{w}_0.$$

If the definition holds with $\mathbf{w}_0$ replaced by $\mathbf{w}(t_0)$ we say that the function is continuous at t_0. If the scalar α, the vectors $\mathbf{u}$, $\mathbf{v}$, and the tensor $\mathbf{S}$ are continuous, at t, then so are $\alpha\mathbf{u}$, $\mathbf{u}\cdot\mathbf{v}$, $\mathbf{u}\times\mathbf{v}$, $\mathbf{u}\otimes\mathbf{v}$, $\mathbf{S}\mathbf{v}$. For instance, about $\mathbf{u}\cdot\mathbf{v}$,

$$\mathbf{u}(t+h)\cdot\mathbf{v}(t+h) - \mathbf{u}(t)\cdot\mathbf{v}(t) = [\mathbf{u}(t+h) - \mathbf{u}(t)]\cdot\mathbf{v}(t+h) + \mathbf{u}(t)\cdot[\mathbf{v}(t+h) - \mathbf{v}(t)],$$

$$\begin{aligned} |\mathbf{u}(t+h)\cdot\mathbf{v}(t+h) - \mathbf{u}(t)\cdot\mathbf{v}(t)| &\le |[\mathbf{u}(t+h) - \mathbf{u}(t)]\cdot\mathbf{v}(t+h)| + |\mathbf{u}(t)\cdot[\mathbf{v}(t+h) - \mathbf{v}(t)]| \\ &\le |\mathbf{u}(t+h) - \mathbf{u}(t)|\,|\mathbf{v}(t+h)| + |\mathbf{u}(t)|\,|\mathbf{v}(t+h) - \mathbf{v}(t)|. \end{aligned}$$

Since $|\mathbf{u}(t+h)-\mathbf{u}(t)| \to 0$, $|\mathbf{v}(t+h)-\mathbf{v}(t)| \to 0$ while $|\mathbf{v}(t+h)| \to |\mathbf{v}(t)|$ then the continuity of $\mathbf{u}\cdot\mathbf{v}$ follows. Likewise,

$$\mathbf{S}(t+h)\mathbf{v}(t+h)-\mathbf{S}(t)\mathbf{v}(t) = [\mathbf{S}(t+h)-\mathbf{S}(t)]\mathbf{v}(t+h)+\mathbf{S}(t)[\mathbf{v}(t+h)-\mathbf{v}(t)],$$

$$|\mathbf{S}(t+h)\mathbf{v}(t+h)-\mathbf{S}(t)\mathbf{v}(t)| \le \|\mathbf{S}(t+h)-\mathbf{S}(t)\|\,|\mathbf{v}(t+h)|+\|\mathbf{S}(t)\|\,|\mathbf{v}(t+h)-\mathbf{v}(t)|.$$

Since $\|\mathbf{S}(t+h)-\mathbf{S}(t)\| \to 0$ and $|\mathbf{v}(t+h)-\mathbf{v}(t)| \to 0$ as $h \to 0$ the continuity of $\mathbf{S}\mathbf{v}$ follows.

The derivative $\dot{\mathbf{w}}(t)$ is defined by the well-known limit of the difference quotient. If $\mathbf{w}$ is the function, the derivative is denoted by $\dot{\mathbf{w}}$ and

$$\dot{\mathbf{w}}(t) = \lim_{h\to 0}\frac{\mathbf{w}(t+h)-\mathbf{w}(t)}{h} \quad \text{or} \quad \lim_{h\to 0}\Big|\frac{\mathbf{w}(t+h)-\mathbf{w}(t)}{h}-\dot{\mathbf{w}}(t)\Big| = 0.$$

Hence the derivative of a scalar, vector, or tensor is a scalar, vector, or tensor. Also, if $\mathbf{w}$ is a point function, the difference $\mathbf{w}(t+h)-\mathbf{w}(t)$ is a vector and hence the derivative is a vector.

Let $\mathbf{u}$, $\mathbf{v}$ be differentiable vectors, α a differentiable scalar and $\mathbf{S}$, $\mathbf{T}$ differentiable tensors. Following are some differentiation rules,

$$\dot{\overline{\mathbf{u}\cdot\mathbf{v}}} = \dot{\mathbf{u}}\cdot\mathbf{v}+\mathbf{u}\cdot\dot{\mathbf{v}}, \qquad \dot{\overline{\mathbf{u}\times\mathbf{v}}} = \dot{\mathbf{u}}\times\mathbf{v}+\mathbf{u}\times\dot{\mathbf{v}}, \qquad \dot{\overline{\mathbf{u}\otimes\mathbf{v}}} = \dot{\mathbf{u}}\otimes\mathbf{v}+\mathbf{u}\otimes\dot{\mathbf{v}},$$

$$\dot{\overline{\alpha\,\mathbf{v}}} = \dot{\alpha}\,\mathbf{v}+\alpha\,\dot{\mathbf{v}}, \qquad \dot{\overline{\alpha\,\mathbf{S}}} = \dot{\alpha}\,\mathbf{S}+\alpha\,\dot{\mathbf{S}},$$

$$\dot{\overline{\mathbf{S}\,\mathbf{v}}} = \dot{\mathbf{S}}\,\mathbf{v}+\mathbf{S}\,\dot{\mathbf{v}}, \qquad \dot{\overline{\mathbf{T}\mathbf{S}}} = \dot{\mathbf{T}}\mathbf{S}+\mathbf{T}\dot{\mathbf{S}}.$$

For definiteness we consider the difference quotient of $\mathbf{S}\mathbf{v}$, at t, and find that, upon obvious rearrangements,

$$\Big|\frac{\mathbf{S}(t+h)\mathbf{v}(t+h)-\mathbf{S}(t)\mathbf{v}(t)}{h}-(\dot{\mathbf{S}}(t)\mathbf{v}(t)+\mathbf{S}(t)\dot{\mathbf{v}}(t))\Big| \le \Big\|\frac{\mathbf{S}(t+h)-\mathbf{S}(t)}{h}-\dot{\mathbf{S}}(t)\Big\|\,|\mathbf{v}(t)|$$
$$+\|\dot{\mathbf{S}}(t)\|\,|\mathbf{v}(t+h)-\mathbf{v}(t)|+\|\mathbf{S}(t)\|\,\Big|\frac{\mathbf{v}(t+h)-\mathbf{v}(t)}{h}-\dot{\mathbf{v}}(t)\Big|$$

Each term on the right-hand side approaches zero as $h \to 0$, which leads us to the sought result.

For constant vectors $\mathbf{u}$, $\mathbf{v}$ and constant tensors $\mathbf{S}$, $\mathbf{T}$ we have

$$\dot{\overline{\alpha\,\mathbf{u}+\beta\,\mathbf{v}}} = \dot{\alpha}\mathbf{u}+\dot{\beta}\mathbf{v}, \qquad \dot{\overline{\alpha\,\mathbf{S}+\beta\mathbf{T}}} = \dot{\alpha}\mathbf{S}+\dot{\beta}\mathbf{T}.$$

Accordingly, relative to the fixed orthonormal basis $\{\mathbf{e}_i\}$, the derivative of vectors and tensors is given by the derivatives of the components,

$$\dot{\mathbf{u}}(t) = \dot{u}_i \mathbf{e}_i, \qquad \dot{\mathbf{S}}(t) = \dot{S}_{ij} \mathbf{e}_i \otimes \mathbf{e}_j.$$

In applications we often need the derivative of a determinant. Since $\det \mathbf{S} = \det S$, we observe that for any matrix S, $\det S$ is a function of the entries S_{ij}. If the entries S_{ij} depend on t then

$$\overline{\det S}^{\boldsymbol{\cdot}} = \partial_{S_{ij}} \det S \, \dot{S}_{ij}.$$

Now, $\partial_{S_{ij}} \det S$ equals the cofactor of S_{ij} and then

$$\partial_{S_{ij}} \det S = \det S \, S^{-1}_{ji}. \tag{A.16}$$

As a consequence,

$$\overline{\det S}^{\boldsymbol{\cdot}} = \det S \, S^{-1}_{ji} \, \dot{S}_{ij} = \det S \operatorname{tr} (\dot{S} S^{-1}).$$

In tensor notation,

$$\overline{\det \mathbf{S}}^{\boldsymbol{\cdot}} = \det \mathbf{S} \operatorname{tr} (\dot{\mathbf{S}} \mathbf{S}^{-1}). \tag{A.17}$$

By the relation $\mathbf{S}^{-1}\mathbf{S} = \mathbf{1}$ for an invertible tensor $\mathbf{S}$, the derivative results in

$$\overline{\mathbf{S}^{-1}}^{\boldsymbol{\cdot}} \mathbf{S} + \mathbf{S}^{-1} \dot{\mathbf{S}} = \mathbf{0}$$

whence

$$\overline{\mathbf{S}^{-1}}^{\boldsymbol{\cdot}} = -\mathbf{S}^{-1} \dot{\mathbf{S}} \mathbf{S}^{-1}.$$

An orthogonal tensor $\mathbf{Q}$ satisfies $\mathbf{Q}^T\mathbf{Q} = \mathbf{Q}\mathbf{Q}^T = \mathbf{1}$. Upon differentiation it follows

$$\dot{\mathbf{Q}}\mathbf{Q}^T = -\mathbf{Q}\dot{\mathbf{Q}}^T.$$

Now, evaluating the transpose we have

$$(\dot{\mathbf{Q}}\mathbf{Q}^T)^T = \mathbf{Q}\dot{\mathbf{Q}}^T.$$

Hence it follows that $\dot{\mathbf{Q}}\mathbf{Q}^T = -(\dot{\mathbf{Q}}\mathbf{Q}^T)^T$, whence

$$\dot{\mathbf{Q}}\mathbf{Q}^T \in \mathrm{Skw}.$$

Functions of vectors and tensors

Let $\mathcal{U}, \mathcal{W}$ be finite-dimensional vector or tensor spaces and let $\mathbf{g} : \mathcal{U} \supseteq \mathcal{D} \to \mathcal{W}$; if $\mathbf{g}$ is a function of the position vector $\mathbf{x}$ then $\mathcal{U} = \mathcal{E}$. The function $\mathbf{g}$ is differentiable at $\mathbf{y} \in \mathcal{D}$ if the difference $\mathbf{g}(\mathbf{y}+\mathbf{h}) - \mathbf{g}(\mathbf{y})$ equals a linear function of $\mathbf{h}$ plus a $o(|\mathbf{h}|)$ term.[4] Precisely, $\mathbf{g}$ is differentiable at $\mathbf{y}$ if there exists a linear operator $\mathbf{G}(\mathbf{y}) : \mathcal{U} \to \mathcal{W}$ such that

$$\mathbf{g}(\mathbf{y}+\mathbf{u}) = \mathbf{g}(\mathbf{y}) + \mathbf{G}(\mathbf{y})\mathbf{h} + o(|\mathbf{h}|),$$

or, equivalently,

$$|\mathbf{g}(\mathbf{y}+\mathbf{h}) - \mathbf{g}(\mathbf{y}) - \mathbf{G}(\mathbf{y})\mathbf{h}| \to 0 \quad \text{as} \quad |\mathbf{h}| \to 0,$$

where $|\cdot|$ has to be replaced by the norm $\|\cdot\|$ if tensors are involved. The linear term $\mathbf{G}(\mathbf{y})\mathbf{h}$ is called strong (or Fréchet) differential of $\mathbf{g}$ at $\mathbf{y}$ and $\mathbf{G}(\mathbf{y})$ is called the strong (or Fréchet) derivative of $\mathbf{g}$ at $\mathbf{y}$. If it exists, the linear operator $\mathbf{G}$ is unique. For, if two operators $\mathbf{G}_1$ and $\mathbf{G}_2$ exist then $\mathbf{G}_1(\mathbf{y})\mathbf{h} + o(|\mathbf{h}|) = \mathbf{G}_2(\mathbf{y})\mathbf{h} + o(|\mathbf{h}|)$ implies

$$\mathbf{G}_1(\mathbf{y})\mathbf{h} - \mathbf{G}_2(\mathbf{y})\mathbf{h} = o(|\mathbf{h}|),$$

which holds only if $\mathbf{G}_1(\mathbf{y}) = \mathbf{G}_2(\mathbf{y})$. For scalar-valued function, $\mathcal{W} = \mathbb{R}$, the definition of differentiability becomes

$$\alpha(\mathbf{y}+\mathbf{h}) = \alpha(\mathbf{y}) + \mathbf{a}(\mathbf{y}) \cdot \mathbf{h} + o(|\mathbf{h}|).$$

The limit

$$d\mathbf{g}(\mathbf{y}, \mathbf{h}) = \lim_{t \to 0} \frac{\mathbf{g}(\mathbf{y}+t\mathbf{h}) - \mathbf{g}(\mathbf{y})}{t},$$

if it exists, is said to be the weak (or Gâteaux) differential of $\mathbf{g}$ at $\mathbf{y}$ in the direction of $\mathbf{h}$. If $d\mathbf{g}(\mathbf{y}, \mathbf{h})$ is linear in $\mathbf{h}$ then $D\mathbf{g}(\mathbf{y})$ such that

$$d\mathbf{g}(\mathbf{y}, \mathbf{h}) = D\mathbf{g}(\mathbf{y})\mathbf{h}$$

is called the weak (or Gâteaux) derivative of $\mathbf{g}$ at $\mathbf{y}$. Now, replace $\mathbf{u}$ by $t\mathbf{u}$ so that the strong differentiability gives

$$\mathbf{g}(\mathbf{y}+t\mathbf{h}) = \mathbf{g}(\mathbf{y}) + t\mathbf{G}(\mathbf{y})\mathbf{h} + o(t).$$

Hence it follows

$$D\mathbf{g}(\mathbf{y}) = \mathbf{G}(\mathbf{y});$$

strong differentiability implies weak differentiability and, moreover, the Fréchet derivative $\mathbf{G}(\mathbf{y})$ is equal to the Gâteaux derivative $D\mathbf{g}(\mathbf{y})$.

A Fréchet differentiable function is also continuous. Instead a Gâteaux differentiable function need not be Fréchet differentiable, neither need the function be continuous. Henceforth differentiability is meant as Fréchet differentiability.

[4] The little-oh $o(t)$ means a function such that $o(t)/t \to 0$ as $t \to 0$.

If $\mathscr{U} = \mathscr{V}$ is the space of position vectors then the derivatives **a** and **G** are said to be the gradients[5] of α and **g**,

$$\alpha(\mathbf{x}+\mathbf{h}) - \alpha(\mathbf{x}) = \nabla\alpha(\mathbf{x}) \cdot \mathbf{h} + o(|\mathbf{h}|),$$

$$\mathbf{g}(\mathbf{x}+\mathbf{h}) - \mathbf{g}(\mathbf{x}) = [\nabla\mathbf{g}(\mathbf{x})]\mathbf{h} + o(|\mathbf{h}|).$$

These definitions of gradients imply standard definitions in terms of partial derivatives. To show this, observe that

$$\partial_{x_i}\alpha(\mathbf{x}) = \lim_{h\to 0} \frac{\alpha(\mathbf{x}+h\mathbf{e}_i) - \alpha(\mathbf{x})}{h}$$

and the like for **g**. Since

$$\frac{\alpha(\mathbf{x}+\mathbf{h}) - \alpha(\mathbf{x}) - \nabla\alpha(\mathbf{x}) \cdot \mathbf{h}}{|\mathbf{h}|} \to 0$$

as $|\mathbf{h}| \to 0$ then letting $\mathbf{h} = h\mathbf{e}_i$, so that $|\mathbf{h}| = |h|$ and multiplying by $\operatorname{sgn} h$ we find

$$\frac{\alpha(\mathbf{x}+h\mathbf{e}_i) - \alpha(\mathbf{x}) - h\nabla\alpha(\mathbf{x}) \cdot \mathbf{e}_i}{h} \to 0$$

as $h \to 0$. Hence

$$\nabla\alpha(\mathbf{x}) \cdot \mathbf{e}_i = \lim_{h\to 0} \frac{\alpha(\mathbf{x}+h\mathbf{e}_i) - \alpha(\mathbf{x})}{h} = \partial_{x_i}\alpha(\mathbf{x}).$$

Likewise we obtain

$$[\nabla\mathbf{g}(\mathbf{x})]\mathbf{e}_i = \lim_{h\to 0} \frac{\mathbf{g}(\mathbf{x}+h\mathbf{e}_i) - \mathbf{g}(\mathbf{x})}{h} = \partial_{x_i}\mathbf{g}(\mathbf{x}).$$

Accordingly, the component forms of $\nabla\alpha$ and $\nabla\mathbf{g}$ provide the conventional definitions,

$$[\nabla\alpha]_i = \partial_{x_i}\alpha, \qquad [\nabla\mathbf{g}]_{ji} = \partial_{x_i} g_j.$$

Based on the result for the gradient, of scalar and vector functions, we define the divergence and curl of a vector **v** and the divergence of a tensor **T** as follows

$$\nabla \cdot \mathbf{v} = \partial_{x_i} v_i, \qquad (\nabla \times \mathbf{v})_i = \epsilon_{ijk}\partial_{x_j} v_k, \qquad (\nabla \cdot \mathbf{T})_i = \partial_{x_j} T_{ij}.$$

---

[5] Sometimes, e.g. in [216], the gradient is denoted by grad, grad α and grad **g**, while $\nabla\alpha$ and $\nabla\mathbf{g}$ denote the gradients with respect to the position **X** in the reference configuration. Here, instead, we use ∇ and ∇_R for the gradients with respect to the spatial position **x** and the reference position **X**.

If a vector function, say $\mathbf{v}$, is the gradient of a scalar function, $\mathbf{v} = \nabla\alpha$, then[6]

$$\nabla \cdot \mathbf{v} = \nabla \cdot \nabla\alpha =: \Delta\alpha, \qquad \Delta\alpha = \textstyle\sum_i \partial^2_{x_i}\alpha,$$

Δ being called the *Laplacian.* It often happens that we need to evaluate the curl of the curl. Now,

$$\begin{aligned}[\nabla \times (\nabla \times \mathbf{v})]_i &= \epsilon_{ijk}\partial_{x_j}(\nabla \times \mathbf{v})_k = \epsilon_{ijk}\epsilon_{kpq}\partial_{x_j}\partial_{x_p}v_q \\ &= (\delta_{ip}\delta_{jq} - \delta_{iq}\delta_{jp})\partial_{x_j}\partial_{x_p}v_q = \partial_{x_i}\partial_{x_j}v_j - \textstyle\sum_j \partial^2_{x_j}v_i\end{aligned}$$

and hence

$$\nabla \times (\nabla \times \mathbf{v}) = \nabla(\nabla \cdot \mathbf{v}) - \Delta\mathbf{v}.$$

If α is a scalar function of a tensor then α is said to be differentiable, at $\mathbf{S}$, if

$$\alpha(\mathbf{S} + \mathbf{H}) - \alpha(\mathbf{S}) - \mathbf{\Lambda}(\mathbf{S}) \cdot \mathbf{H} = o(\|\mathbf{H}\|).$$

Let $\mathbf{H} = H\,\mathbf{e}_i \otimes \mathbf{e}_j$. Then

$$\alpha(\mathbf{S} + H\,\mathbf{e}_i \otimes \mathbf{e}_j) - \alpha(\mathbf{S}) - H\,\mathbf{\Lambda}(\mathbf{S}) \cdot \mathbf{e}_i \otimes \mathbf{e}_j = o(H).$$

Now,

$$\mathbf{\Lambda} \cdot \mathbf{e}_i \otimes \mathbf{e}_j = \Lambda_{pq}\mathbf{e}_p \otimes \mathbf{e}_q \cdot \mathbf{e}_i \otimes \mathbf{e}_j.$$

Since

$$\mathbf{e}_p \otimes \mathbf{e}_q \cdot \mathbf{e}_i \otimes \mathbf{e}_j = \mathrm{tr}\,(\mathbf{e}_q \otimes \mathbf{e}_p)(\mathbf{e}_i \otimes \mathbf{e}_j) = \delta_{ip}\mathrm{tr}\,\mathbf{e}_q \otimes \mathbf{e}_j = \delta_{ip}\delta_{jq}.$$

As a consequence,

$$\mathbf{\Lambda} \cdot \mathbf{e}_i \otimes \mathbf{e}_j = \Lambda_{ij}$$

and, upon dividing by H,

$$\frac{\alpha(\mathbf{S} + H\,\mathbf{e}_i \otimes \mathbf{e}_j) - \alpha(\mathbf{S})}{H} - \mathbf{\Lambda}_{ij}(\mathbf{S}) \to 0$$

as $H \to 0$. Accordingly, we can write

$$\mathbf{\Lambda} = \partial_{\mathbf{S}}\,\alpha, \qquad [\partial_{\mathbf{S}}\,\alpha]_{ij} = \partial_{S_{ij}}\,\alpha.$$

Hence for differentiable functions α of vectors and tensors we have

$$\alpha(\mathbf{S} + \mathbf{H}) - \alpha(\mathbf{S}) = \partial_{\mathbf{S}}\alpha \cdot \mathbf{H} + o(\|\mathbf{H}\|), \qquad \alpha(\mathbf{v} + \mathbf{h}) - \alpha(\mathbf{v}) = \partial_{\mathbf{v}}\alpha \cdot \mathbf{h} + o(|\mathbf{h}|).$$

---

6 The notation ∂^n_x stands for the n-th order derivative with respect to x.

We now look for the derivative of composite functions. For definiteness, let $\alpha(\mathbf{v}(t))$ the composite function. To determine the derivative with respect to t we observe

$$\alpha(\mathbf{v}+\mathbf{h})-\alpha(\mathbf{v})=\partial_{\mathbf{v}}\alpha\cdot\mathbf{h}+o(|\mathbf{h}|),\qquad \mathbf{v}(t+\tau)-\mathbf{v}(t)=\dot{\mathbf{v}}\,\tau+o(\tau).$$

Letting $\mathbf{h}=\mathbf{v}(t+\tau)-\mathbf{v}(t)$ we have

$$\alpha(\mathbf{v}(t+\tau))-\mathbf{v}(t)=\partial_{\mathbf{v}}\alpha\cdot\dot{\mathbf{v}}\,\tau+o(\tau).$$

Dividing by τ and taking the limit as $\tau\to 0$ we obtain

$$\overline{\alpha(\mathbf{v}(t))}^{\,\cdot}=\partial_{\mathbf{v}}\alpha\cdot\dot{\mathbf{v}}.$$

The analogue holds for the dependence on scalars and tensors and for vector- and tensor-valued functions.

About functions of several variables, for definiteness look at $\alpha(\mathbf{u}(t),\mathbf{v}(t))$. Since $\alpha(\mathbf{u}+\mathbf{h},\mathbf{v}+\mathbf{k})-\alpha(\mathbf{u},\mathbf{v})=\alpha(\mathbf{u}+\mathbf{h},\mathbf{v}+\mathbf{k})-\alpha(\mathbf{u}+\mathbf{h},\mathbf{v})+\alpha(\mathbf{u}+\mathbf{h},\mathbf{v})-\alpha(\mathbf{u},\mathbf{v})$ then

$$\alpha(\mathbf{u}+\mathbf{h},\mathbf{v}+\mathbf{k})-\alpha(\mathbf{u},\mathbf{v})=\partial_{\mathbf{v}}\alpha(\mathbf{u}+\mathbf{h},\mathbf{v})\cdot\mathbf{k}+\partial_{\mathbf{u}}\alpha(\mathbf{u},\mathbf{v})\cdot\mathbf{h}+o(|\mathbf{k}|)+o(|\mathbf{h}|)$$

and

$$\mathbf{k}=\mathbf{v}(t+\tau)-\mathbf{v}(t)=\dot{\mathbf{v}}\,\tau+o(\tau),\qquad \mathbf{h}=\mathbf{u}(t+\tau)-\mathbf{u}(t)=\dot{\mathbf{u}}\,\tau+o(\tau).$$

If also $\partial_{\mathbf{v}}\alpha(\mathbf{u}+\mathbf{h},\mathbf{v})$ is continuous then we find that

$$\overline{\alpha(\mathbf{u},\mathbf{v})}^{\,\cdot}=\partial_{\mathbf{u}}\,\alpha\cdot\dot{\mathbf{u}}+\partial_{\mathbf{v}}\,\alpha\cdot\dot{\mathbf{v}}.$$

A.4 Integration

A line or curve C in space is a continuous mapping such that $\mathbb{R}\ni I\to\mathscr{E}$, I being an interval. The mapping may be given by parametric equations,

$$\mathbf{x}=\hat{\mathbf{x}}(\xi),\qquad \xi\in I,$$

ξ being a parameter. Let $I=[a,b]$. The curve is *simple* if, for any two values $\xi_1,\xi_2\in[a,b]$, it happens that $\hat{\mathbf{x}}(\xi_1)\neq\hat{\mathbf{x}}(\xi_2)$. The curve is *closed* if $\hat{\mathbf{x}}(a)=\hat{\mathbf{x}}(b)$. Assume $\hat{\mathbf{x}}(\xi)$ is differentiable and let $\hat{\mathbf{x}}'$ denote the derivative. The length of the curve C is given by

$$L=\int_a^b|\hat{\mathbf{x}}'|(\xi)d\xi.$$

The vector $\mathbf{x}'(\xi_0)$ is (parallel to the) tangent to the curve at $\hat{\mathbf{x}}(\xi_0)$. A curve is regular in $[a, b]$ if $|\hat{\mathbf{x}}'| > 0$ in $[a, b]$.

The length $s(\eta)$ of the arc, as $\xi \in [a, \eta]$, is then given by

$$s(\eta) = \int_a^\eta |\hat{\mathbf{x}}'|(\xi) d\xi.$$

It then follows that s is differentiable and

$$s'(\eta) = |\hat{\mathbf{x}}'|(\eta).$$

If f is a continuous function along C we define the line integral of f on C as

$$\int_a^b f(\hat{\mathbf{x}}(\xi)|\hat{\mathbf{x}}'(\xi)|d\xi. \tag{A.18}$$

Since $s(\xi)$ is invertible, by a change of variable we have

$$\int_a^b f(\hat{\mathbf{x}}(\xi)|\hat{\mathbf{x}}'(\xi)|d\xi = \int_{s_a}^{s_b} \tilde{f}(s)ds,$$

where $\tilde{f}(s) = f(\hat{\mathbf{x}}(\xi(s)))$, $s_a = s(a)$, $s_b = s(b)$.

The parametric representation $\hat{\mathbf{x}}(\xi)$ induces a positive sense on the curve. The symbol C is then meant as the curve in the selected positive sense. As a shorthand, the symbol

$$\int_C f\, dl$$

is used to mean the line integral (A.18).

It may happen that the function f is the inner product between a vector field $\mathbf{w}$ and the unit tangent vector $\mathbf{t} = \mathbf{x}'/|\mathbf{x}'|$, i.e. $f = \mathbf{w} \cdot \mathbf{t}$. It follows that

$$\int_a^b (\mathbf{w} \cdot \mathbf{t})(\xi)|\mathbf{x}'|(\xi)d\xi = \int_{s_a}^{s_b} \tilde{f}(s)ds.$$

In the literature, the symbol $\int_C \mathbf{w} \cdot d\mathbf{x}$ is also used to denote the integral of $\mathbf{w} \cdot \mathbf{t}$ along C. This quantity is defined by

$$\int_C \mathbf{w} \cdot d\mathbf{x} := \int_a^b \mathbf{w}(\hat{\mathbf{x}}(\xi)) \cdot \hat{\mathbf{x}}'(\xi)d\xi = \int_{s_a}^{s_b} (\mathbf{w} \cdot \mathbf{t})(s)ds.$$

Let D be an open connected set of the plane. A regular surface in $\mathscr{E}$ is the set of points $\mathbf{x}$ given by a mapping $\phi : D \to \mathscr{E}$, of class C^1, such that $\phi(u, v)$ is invertible for each $(u, v) \in D$ and the rank of the jacobian matrix

$$\frac{\partial(\phi_1, \phi_2, \phi_3)}{\partial(u, v)} = \begin{bmatrix} \partial_u\phi_1 & \partial_v\phi_1 \\ \partial_u\phi_2 & \partial_v\phi_2 \\ \partial_u\phi_3 & \partial_v\phi_3 \end{bmatrix}$$

equals 2. This means that the vectors $\partial_u\phi, \partial_v\phi$ are not parallel to each other and hence $\partial_u\phi \times \partial_v\phi \neq \mathbf{0}$.

Let S be a regular surface in $\mathscr{E}$ determined by a function $\phi : D \to \mathscr{E}$. If $f : S \to \mathbb{R}$ then the integral of f over the surface S is defined by the double integral

$$\int_S f\, da := \int_D f(\phi(u, v))|\partial_u\phi \times \partial_v\phi|(u, v) du\, dv.$$

Let $\mathbf{n}(u, v)$ be the unit normal to S (in the chosen direction). If $\mathbf{w} : S \to \mathscr{V}$ is a continuous vector field on S then the *flux* of $\mathbf{n}$ through s in the direction $\mathbf{n}$ is the integral

$$\int_S \mathbf{w} \cdot \mathbf{n}\, da.$$

Hence, by definition,

$$\int_S \mathbf{w} \cdot \mathbf{n}\, da = \int_D [\mathbf{w} \cdot \partial_u\phi \times \partial_v\phi](u, v)\, du\, dv.$$

Given a region $\Omega \subseteq \mathscr{E}$ we denote by $\int_\Omega f(\mathbf{x}) dv$ the volume (triple) integral of f in Ω. If f is a divergence we have the following result.

Theorem A.2 (Gauss, divergence) *Let Ω be a bounded region with boundary $\partial\Omega$. Given a vector field $\mathbf{w}$ on Ω if $\mathbf{n}$ is the unit outward normal to $\partial\Omega$ then*

$$\int_{\partial\Omega} \mathbf{w} \cdot \mathbf{n}\, da = \int_\Omega \nabla \cdot \mathbf{w}\, dv$$

Two interesting consequences follow via appropriate choices of $\mathbf{w}$. Let $\mathbf{w} = \mathbf{T}^T\mathbf{c}$, $\mathbf{c}$ being a constant vector field. Then

$$\mathbf{c} \cdot \int_{\partial\Omega} \mathbf{T}\mathbf{n}\, da = \int_{\partial\Omega} \mathbf{c} \cdot \mathbf{T}\mathbf{n}\, da = \int_{\partial\Omega} (\mathbf{T}^T\mathbf{c}) \cdot \mathbf{n}\, da = \int_\Omega \nabla \cdot (\mathbf{T}^T\mathbf{c}) dv = \mathbf{c} \cdot \int_\Omega \nabla \cdot \mathbf{T}\, dv,$$

where

$$(\nabla \cdot \mathbf{T})_i := \partial_{x_j} T_{ij}.$$

The arbitrariness of $\mathbf{c}$ allows us to conclude that

$$\int_{\partial\Omega} \mathbf{T}\mathbf{n}\, da = \int_\Omega \nabla \cdot \mathbf{T}\, dv.$$

As a further consequence, letting $\mathbf{T} = \alpha \mathbf{1}$ we find

$$\int_{\partial\Omega} \alpha \, \mathbf{n} \, da = \int_{\Omega} \nabla \alpha \, dv,$$

whose statement is often denoted as Green-Ostrogradskij theorem or gradient theorem.

Moreover, letting $\alpha = 1$ it follows

$$\int_{\partial\Omega} \mathbf{n} \, da = \mathbf{0},$$

the integral of the unit normal over a closed surface is zero.

By applying the divergence theorem we can derive further identities,

$$\int_{\partial\Omega} \mathbf{n} \cdot \nabla \times \mathbf{w} \, da = 0, \quad \int_{\partial\Omega} \mathbf{n} \times \mathbf{w} \, da = \int_{\Omega} \nabla \times \mathbf{w} \, dv, \quad \int_{\partial\Omega} \mathbf{w} \otimes \mathbf{n} \, da = \int_{\Omega} \nabla \mathbf{w} \, dv.$$

To show these identities we first observe that

$$\int_{\partial\Omega} \mathbf{n} \cdot \nabla \times \mathbf{w} \, da = \int_{\Omega} \nabla \cdot (\nabla \times \mathbf{w}) dv.$$

Since

$$\nabla \cdot (\nabla \times \mathbf{w}) = \epsilon_{ijk} \partial_{x_i} \partial_{x_j} w_k = 0$$

then the first identity follows.

Now, since $(\mathbf{n} \times \mathbf{w})_i = \epsilon_{ijk} n_j w_k$ then

$$\int_{\partial\Omega} (\mathbf{n} \times \mathbf{w})_i da = \int_{\Omega} \partial_{x_j} \epsilon_{ijk} w_k dv = \int_{\Omega} (\nabla \times \mathbf{w})_i dv$$

and the second identity follows.

Third, let $\mathbf{c}$ be a constant vector and consider $(\mathbf{n} \otimes \mathbf{w})\mathbf{c} = (\mathbf{c} \cdot \mathbf{w})\mathbf{n}$. Since $\nabla(\mathbf{c} \cdot \mathbf{w}) = \mathbf{c} \cdot (\nabla \mathbf{w})$ we obtain

$$\int_{\partial\Omega} (\mathbf{n} \otimes \mathbf{w})^T da \, \mathbf{c} = \mathbf{c} \int_{\Omega} \nabla \mathbf{w} \, dv = \int_{\Omega} (\nabla \mathbf{w})^T dv \, \mathbf{c},$$

whence the conclusion.

The boundary ∂S of a surface S is an oriented curve C; the positive sense of $C = \partial S$ is the one for which the region S remains on the left side as seen from the direction of the normal $\mathbf{n}$.

Theorem A.3 (Stokes) *Let S be a surface in $\mathscr{E}$ so that the boundary ∂S of S is a closed curve C oriented in the positive sense. Given a C^1 vector field on S then*

$$\int_{\partial S} \mathbf{w} \cdot d\mathbf{x} = \int_S \mathbf{n} \cdot \nabla \times \mathbf{w} \, da.$$

Here too appropriate choices of **w** lead to two further analogue theorems. Let $\mathbf{w} = \mathbf{Tc}$, **c** being a constant vector. By Stokes' theorem

$$\int_S \mathbf{n} \cdot \nabla \times (\mathbf{Tc}) da = \int_{\partial S} (\mathbf{Tc}) \cdot d\mathbf{x} = \mathbf{c} \cdot \int_a^b (\mathbf{Tx}')(\xi) d\xi.$$

Moreover,

$$\int_S \mathbf{n} \cdot \nabla \times (\mathbf{Tc}) da = c_p \int_S n_i \epsilon_{ijk} \partial_{x_j} T_{kp} da = \mathbf{c} \cdot \int_S (\nabla \times \mathbf{T}^T) \mathbf{n} \, da$$

where

$$(\nabla \times \mathbf{T}^T)_{pi} := (\nabla \times \mathbf{T})_{ip} := \epsilon_{ijk} \partial_{x_j} T_{kp}.$$

By the arbitrariness of **c**, with this notation we can write

$$\int_{\partial S} \mathbf{T} \, d\mathbf{x} = \int_S (\nabla \times \mathbf{T}^T) \mathbf{n} \, da.$$

If now $\mathbf{T} = \alpha \mathbf{1}$ we have

$$(\nabla \times \mathbf{T}^T)_{pi} = \epsilon_{ijk} \partial_{x_j} \alpha \delta_{kp} = \epsilon_{pij} \partial_{x_j} \alpha, \qquad c_p (\nabla \times \mathbf{T}^T)_{pi} n_i = \mathbf{c} \cdot (\mathbf{n} \times \nabla \alpha).$$

Hence it follows that

$$\int_{\partial S} \alpha \, d\mathbf{x} = \int_S (\mathbf{n} \times \nabla \alpha) \, da.$$

A.5 Harmonic Waves and Complex-Valued Functions

By appealing to Fourier analysis it is often claimed that no generality is lost by considering fields $\mathbf{f}(\mathbf{x}, t)$ which vary sinusoidally in time as, e.g.,

$$\mathbf{f}(\mathbf{x}, t) = \mathbf{f}_0(\mathbf{x}) \exp(-i\omega t)$$

or, in the form of plane wave,

$$\mathbf{f}(\mathbf{x}, t) = \mathbf{f}_0 \exp[i(\mathbf{k} \cdot \mathbf{x} - \omega t)].$$

It is then asserted that the real part, or rather the real and the imaginary parts, of such expressions is to be taken to obtain physical quantities (e.g. [230], Sect. 7.1; [1], Sect. 1.6.2). The dependence on $\exp[i(\mathbf{k} \cdot \mathbf{x} - \omega t)]$, possibly with $\mathbf{k} \in \mathbb{C}$, is standard and denotes a plane wave propagating in the direction $\Re\mathbf{k}$ with phase speed $\omega/|\Re\mathbf{k}|$.

The dependence on $\mathbf{x}$ and t via the exponential is remarkably convenient in calculations. Mathematically though we need to check that the solution to the pertinent equations is not affected by the use of complex-valued functions.

Let $\mathbf{f}(\mathbf{x}, t)$ be a complex-valued vector function of $\mathbf{x}$ and t. We can then write

$$\mathbf{f}(\mathbf{x}, t) = \mathbf{u}(\mathbf{x}, t) + i\mathbf{v}(\mathbf{x}, t),$$

where $\mathbf{u}$ and $\mathbf{v}$ are real-valued vector functions. Since

$$\partial_{\mathbf{x}}\mathbf{f}(\mathbf{x}, t) = \partial_{\mathbf{x}}\mathbf{u}(\mathbf{x}, t) + i\partial_{\mathbf{x}}\mathbf{v}(\mathbf{x}, t), \qquad \partial_t\mathbf{f}(\mathbf{x}, t) = \partial_t\mathbf{u}(\mathbf{x}, t) + i\partial_t\mathbf{v}(\mathbf{x}, t),$$

then it follows that

$$\Re\partial_{\mathbf{x}}\mathbf{f}(\mathbf{x}, t) = \partial_{\mathbf{x}}\Re\mathbf{f}(\mathbf{x}, t), \qquad \Re\partial_t\mathbf{f}(\mathbf{x}, t) = \partial_t\Re\mathbf{f}(\mathbf{x}, t).$$

Linear equations characterized by operators of the form

$$E(\partial_{\mathbf{x}}, \partial_t) := a_{ij}\,\partial_{x_j} + b\,\partial_t, \qquad a_{ij}, b \in \mathbb{R},$$

satisfy

$$\Re\{a_{ij}\,\partial_{x_j}\mathbf{f} + b\,\partial_t\,\mathbf{f}\} = a_{ij}\,\partial_{x_j}\Re\,\mathbf{f} + b\,\partial_t\Re\,\mathbf{f}$$

and the like for the imaginary part. Accordingly,

$$E(\partial_{\mathbf{x}}, \partial_t)\mathbf{f}(\mathbf{x}, t) = \mathbf{0} \implies E(\partial_{\mathbf{x}}, \partial_t)\Re\,\mathbf{f}(\mathbf{x}, t) = \mathbf{0}, \quad E(\partial_{\mathbf{x}}, \partial_t)\Im\,\mathbf{f}(\mathbf{x}, t) = \mathbf{0}.$$

This allows us to look for real-valued solutions $\Re\,\mathbf{f}(\mathbf{x}, t)$, or $\Im\,\mathbf{f}(\mathbf{x}, t)$, via the check that $E(\partial_{\mathbf{x}}, \partial_t)\mathbf{f}(\mathbf{x}, t) = \mathbf{0}$.

Consider the equation

$$\mathbf{P}(t) = \mathbf{G}_0\mathbf{E}(t) + \int_0^\infty \mathbf{G}(s)\mathbf{E}(t - s)ds + \mathbf{A}\partial_t\mathbf{E}(t), \tag{A.19}$$

where $\mathbf{G}_0, \mathbf{G}(s), \mathbf{A}$ are second-order real-valued tensors. If $\mathbf{E}$ is taken in the form $\mathbf{E}(t) = \mathbf{E}_0 \exp(-i\omega t)$, $\mathbf{E}_0$ being a complex-valued vector, then we expect that

$$\mathbf{P}(t) = \mathbf{P}_0 \exp(-i\omega t), \qquad \mathbf{P}_0 = \chi(\omega)\mathbf{E}_0,$$

$\chi(\omega)$ being a complex-valued tensor. Let

$$\mathbf{G}_c(\omega) := \int_0^\infty \mathbf{G}(u)\cos(\omega u)du, \qquad \mathbf{G}_s(\omega) := \int_0^\infty \mathbf{G}(u)\sin(\omega u)du.$$

Upon substitution of $\mathbf{E}(t) = \mathbf{E}_0 \exp(-i\omega t)$ in (A.19) it follows that

$$\chi(\omega) = \mathbf{G}_0 + \mathbf{G}_c(\omega) + i\mathbf{G}_s(\omega) + i\omega\mathbf{A}.$$

Instead, substitution of

$$\mathbf{E}(t) = \mathbf{E}_1 \cos(\omega t) + \mathbf{E}_2 \sin(\omega t) = \Re\mathbf{E}_0 \exp(-i\omega t), \qquad \mathbf{E}_0 = \mathbf{E}_1 + i\mathbf{E}_2,$$

in (A.19) results in

$$\mathbf{P}(t) = (\mathbf{G}_0 + \mathbf{G}_c(\omega))[\mathbf{E}_1 \cos(\omega t) + \mathbf{E}_2 \sin(\omega t)] + (\mathbf{G}_s(\omega) + \mathbf{A})[\mathbf{E}_1 \sin(\omega t) - \mathbf{E}_2 \cos(\omega t)].$$

The same result follows by computing

$$\Re[\chi(\omega)(\mathbf{E}_1 + i\mathbf{E}_2) \exp(-i\omega t)].$$

Accordingly, with linear operators of the form (A.19) the real value of $\mathbf{P}(t)$ coincides with the real part of $\chi(\omega)(\mathbf{E}_1 + i\mathbf{E}_2) \exp(-i\omega t)$.

This does not mean that $\Im\mathbf{P} = \mathbf{0}$. Indeed,

$$\Im\mathbf{P}(t) = (\mathbf{G}_0 + \mathbf{G}_c(\omega))(\mathbf{E}_2 \cos \omega t - \mathbf{E}_1 \sin \omega t) + (\mathbf{G}_s(\omega) + \mathbf{A})(\mathbf{E}_1 \cos(\omega t) + \mathbf{E}_2 \sin(\omega t)).$$

The recourse to complex-valued exponentials is customary in the investigation of plane wave propagation. An inhomogeneous plane wave is a function of the form

$$\mathbf{w}(\mathbf{x}, t) = \mathbf{w}_0 \exp[-\mathbf{k}_2 \cdot \mathbf{x} + i(\mathbf{k}_1 \cdot \mathbf{x} - \omega t)] \tag{A.20}$$

where $\mathbf{w}_0$ is a complex vector. Let $\mathbf{w}_0 = w_1\mathbf{e}_1 + iw_2\mathbf{e}_2$, $\mathbf{e}_1$ and $\mathbf{e}_2$ being perpendicular to $\mathbf{k}_1$. We have

$$\Re\mathbf{w} = \exp(-\mathbf{k}_2 \cdot \mathbf{x})[(w_1\mathbf{e}_1 \cos(\mathbf{k}_1 \cdot \mathbf{x} - \omega t) - w_2\mathbf{e}_2 \sin(\mathbf{k}_1 \cdot \mathbf{x} - \omega t)],$$
$$\Im\mathbf{w} = \exp(-\mathbf{k}_2 \cdot \mathbf{x})[(w_1\mathbf{e}_1 \sin(\mathbf{k}_1 \cdot \mathbf{x} - \omega t) + w_2\mathbf{e}_2 \cos(\mathbf{k}_1 \cdot \mathbf{x} - \omega t)].$$

At an inhomogeneous plane wave the surfaces of constant amplitude are the planes perpendicular to $\mathbf{k}_1$, the surfaces of constant phase are the planes perpendicular to $\mathbf{k}_2$. At a fixed point in space we have

$$\Re\mathbf{w} \propto w_1\mathbf{e}_1 \cos(\omega t - \xi) + w_2\mathbf{e}_2 \sin(\omega t - \xi), \qquad \xi = \mathbf{k}_1 \cdot \mathbf{x}.$$

If $w_1 w_2 > 0$ then, as time increases, the vector $\Re\mathbf{w}$ rotates in time by sweeping around an ellipse, at the angular frequency ω, in the counter-clockwise (or right-handed) sense. Indeed, $|w_1|$ and $|w_2|$ are the semi-major and semi-minor axes of the ellipse. If instead $w_1 w_2 < 0$ then $\Re\mathbf{w}$ sweeps around the same ellipse in the clockwise (or left-handed) sense. The same conclusions follow if we look at $\Im\mathbf{w}$. Apparently if $|w_1| = |w_2|$ then the ellipse becomes a circle. In that case the wave is said to be left (or right) circularly polarized[7] if $w_1 w_2 > 0$ (or $w_1 w_2 < 0$).

If instead $\mathbf{w}_0$ is real, say

$$\mathbf{w}_0 = w_1\mathbf{e}_1 + w_2\mathbf{e}_2,$$

then

$$\Re\mathbf{w} \propto (w_1\mathbf{e}_1 + w_2\mathbf{e}_2)\cos(\omega t - \xi), \qquad \Im\mathbf{w} \propto (w_1\mathbf{e}_1 + w_2\mathbf{e}_2)\sin(\omega t - \xi).$$

The vectors $\Re\mathbf{w}$ and $\Im\mathbf{w}$ oscillate along the line making the angle α such that $\tan\alpha = w_2/w_1$. Then we say that the wave is linearly polarized.

Consequently we can look for wave solutions in the complex form (A.20). Moreover, we conclude that the wave is linearly polarized if $\mathbf{w}_0$ is real and is elliptically polarized if $\mathbf{w}_0 = w_1\mathbf{e}_1 + i w_2\mathbf{e}_2$; in the counter-clockwise sense if $w_1 w_2 > 0$, in the clockwise sense if $w_1 w_2 < 0$.

It is worth examining the geometric aspects of solutions of the form

$$\mathbf{w} = a_1\mathbf{e}_1\cos\omega t + a_2\mathbf{e}_2\cos(\omega t + \delta),$$

δ being the phase difference between the two oscillations. By a direct check we find that

$$\Big(\frac{w_1}{a_1}\Big)^2 - 2\frac{w_1 w_2}{a_1 a_2}\cos\delta + \Big(\frac{w_2}{a_2}\Big)^2 = \sin^2\delta.$$

By means of the variables X, Y related to w_1, w_2 by

$$w_1 = X\cos\chi - Y\sin\chi, \qquad w_2 = X\sin\chi + Y\cos\chi$$

we obtain ([237], Sect. 6.4)

$$\big(X/a\big)^2 + \big(Y/b\big)^2 = 1,$$

where $a^2 = 2(a_1 a_2 \sin\delta)^2/(a_1^2 + a_2^2 - \alpha)$, $b^2 = 2(a_1 a_2\sin\delta)^2/(a_1^2 + a_2^2 + \alpha)$, $\alpha = (a_1^4 + a_2^4 + 2a_1^2 a_2^2\cos 2\delta)^{1/2}$. Except when $\delta = 0, \pi$ the composition of the two oscillations sweeps around an ellipse whose semi-major axis is rotated in the $(\mathbf{e}_1, \mathbf{e}_2)$ plane of the angle χ, $\tan 2\chi = (2a_1 a_2\cos\delta)/(a_1^2 - a_2^2)$. As $\delta = 0, \pi$ the vector $\mathbf{w}(t)$ describes a segment rotated of the angle $\chi = \pm\arctan a_2/a_1$.

If the argument of the two functions w_1, w_2 is no longer the same then the advantage of the complex exponential is lost.

---

[7] In the terminology of modern physics, left (right) circularly polarized waves are said to have positive (negative) helicity.

A.6 Fourier Transform

In dealing with materials with memory it is often convenient to make use of the Fourier transform. If $f \in L_1(\mathbb{R})$ the Fourier transform f_F is defined as

$$f_F(\omega) = \int_{-\infty}^{\infty} \exp(-i\omega t) f(t) dt.$$

By definition,

$$f_F(\omega) = \int_{-\infty}^{\infty} \cos(\omega t) f(t) dt - i \int_{-\infty}^{\infty} \sin(\omega t) f(t) dt.$$

The real and imaginary parts are usually called full-range cosine and sine transforms of f. In applications we more frequently deal with

$$f_c(\omega) := \int_{0}^{\infty} \cos(\omega t) f(t) dt, \qquad f_s(\omega) := \int_{0}^{\infty} \sin(\omega t) f(t) dt,$$

called half-range cosine and sine transforms of f.

The Fourier transform can be inverted in that, given f_F we determine f. Precisely, if $f \in L_1(\mathbb{R})$ is piecewise continuous and has finite one-sided derivatives at possible discontinuities then

$$f(t) = \frac{1}{2\pi} \int_{-\infty}^{\infty} f_F(\omega) \exp(i\omega t) d\omega.$$

The following result[8] is advantageous in passing from the time domain to the frequency domain.

Theorem A.4 (Parseval–Plancherel) *If* $f, g \in L_2(\mathbb{R})$ *then*

$$\int_{-\infty}^{\infty} |f(t)|^2(t) dt = \frac{1}{2\pi} \int_{-\infty}^{\infty} |f_F(\omega)|^2(\omega) d\omega,$$

$$\int_{-\infty}^{\infty} f(t) g(t) dt = \frac{1}{2\pi} \int_{-\infty}^{\infty} f_F(\omega) g_F(\omega) d\omega.$$

[8] See, e.g. [21].

References

1. J.D. Achenbach, *Wave Propagation in Elastic Solids* (North-Holland, Amsterdam, 1975)
2. E.C. Aifantis, The physics of plastic deformation. Int. J. Plast. **3**, 211–247 (1987)
3. H. Alfvén, Existence of electromagnetic-hydrodynamic waves. Nature **150**, 405–406 (1942)
4. M.B. Allen, Numerical modelling of multiphase porous media. Adv. Water Resour. **8**, 162–187 (1985)
5. J.F. Allen, A.D. Misener, Flow phenomena in liquid helium II. Nature **142**, 643–644 (1938)
6. S.M. Allen, J.W. Cahn, A macroscopic theory for antiphase boundary motion and its application to antiphase domain coarsening. Acta Metall. **27**, 1085–1095 (1979)
7. A.I. Alsabery, M.A. Ismael, A.J. Chamkha, I. Hashim, Mixed convection of Al_2 O_3-water nanofluid in a double lid-driven square cavity with a solid inner insert using Buongiorno's two-phase model. Int. J. Heat Mass Transf. **119**, 939–961 (2018)
8. G. Amendola, M. Fabrizio, J.M. Golden, *A linear memory model, in Thermodynamics of Materials with Memory* (Springer, Boston, MA, 2012)
9. H.W. Alt, I. Pawlow, A mathematical model of dynamics of non-isothermal phase separation. Physica D **59**, 389–416 (1992)
10. P.M. Anderson, Considerations on the flow of superfluid helium. Rev. Modern Phys. **38**, 298–310 (1966)
11. D.M. Anderson, G.B. McFadden, A.A. Wheeler, Diffuse-interface methods in fluid mechanics. Ann. Rev. Fluid Mech. **30**, 139–165 (1998)
12. E.C. Andrade, Viscosity of liquids. Nature **125**, 309–310 (1930)
13. D. Andrienko, Introduction to liquid crystals. J. Mol. Liq. **267**, 520–541 (2018)
14. M. Arcangioli, A. Farina, L. Fusi, G. Saccomandi, Constitutive equations with pressure-dependent rheological parameters for describing ice creep. J. Glaciol. **65**, 557–564 (2019)
15. R. Aris, *Vectors, Tensors, and the Basic Equations of Fluid Mechanics* (Prentice Hall, Englewood Cliffs, NJ, 1962)
16. M.F. Ashby, D.R.H. Jones, *Engineering Materials 2* (Elsevier, Amsterdam, 2006). ((ch. 3))
17. A. Askar, P. Lee, A. Cakmak, Lattice-dynamics approach to the theory of elastic dielectrics with polarization gradient. Phys. Rev. B **1**, 3525–3537 (1970)
18. G. Astarita, G. Marrucci, *Principles of Non-Newtonian Fluid Mechanics* (McGraw Hill, New York, 1974)
19. T. Atanackovic, I. Müller, A new form of the coherency energy in pseudoelasticity. Meccanica **30**, 467–474 (1995)

A. Morro and C. Giorgi, *Mathematical Modelling of Continuum Physics*, Modeling and Simulation in Science, Engineering and Technology,
https://doi.org/10.1007/978-3-031-20814-0

20. M. Avalle, G. Belingardi, R. Montanini, Characterization of polymeric structural foams under compressive impact loading by means of energy absorption diagram. Int. J. Impact Eng. **25**, 455–472 (2001)
21. L. Baggett, W. Fulks, *Fourier Analysis* (Anjou Press, Boulder, 1979)
22. S. Bair, The temperature and pressure dependence of viscosity and volume for two reference liquids. Lubr. Sci. **28**, 81–95 (2016)
23. F. Bampi, A. Morro, Gravity waves in water of variable depth. Nuovo Cimento **1C**, 377–388 (1978)
24. F. Bampi, A. Morro, Korteweg-de Vries equation and nonlinear waves. Lett. Nuovo Cimento **26**, 61–63 (1979)
25. F. Bampi, A. Morro, Water wave theories and variational principles. Nuovo Cimento **2C**, 352–366 (1979)
26. F. Bampi, A. Morro, Objectivity and objective time derivatives in continuum physics. Found. Phys. **10**, 905–920 (1980)
27. F. Bampi, A. Morro, The inverse problem of the calculus of variations applied to continuum physics. J. Math. Phys. **23**, 2312–2321 (1982)
28. J. Bardeen, L. Cooper, J.R. Schreiffer, Theory of superconductivity. Phys. Rev. **108**, 1175–1189 (1957)
29. G.I. Barenblatt, I.P. Zheltov, I.N. Kochina, Basic concepts in the theory of seepage of homogeneous liquids in fissured rocks. J. Appl. Math. Mech. **24**, 1286–1303 (1960)
30. H.A. Barnes, The yield stress - a review or '$\pi\alpha\nu\tau\alpha$ $\rho\epsilon\iota$- everything flows. J. Non-Newtonian Fluid Mech. **81**, 133–178 (1999)
31. R. Becker, G. Heller, F. Sauter, Uber die Stromverteilung in einer supraleitenden Kugel. Zeit. Phys. **85**, 772–787 (1933)
32. J.G. Bednorz, K.A. Müller, Possible high T_c superconductivity in the Ba-La-CuO system. Z. Phys. B **64**, 189–193 (1986)
33. N. Bellomo, M. Lo Schiavo, *Lecture Notes on Mathematical Theory of Generalized Boltzmann Models* (World Scientific, Singapore, 2000)
34. I.L. Bekarevich, I.M. Khalatnikov, Phenomenological derivation of the equation of vortex motion in He II. Sov. Phys. JETP **13**, 643–646 (1961)
35. E. Beltrami, Sull'interpretazione meccanica delle formole di Maxwell. Memorie Accad. Sci. Bologna **7**, 1–38 (1886)
36. B. Bernstein, E.A. Kearsley, L.J. Zapas, A study of stress relaxations with finite strain. Trans. Soc. Rheol. **7**, 391–410 (1963)
37. A. Berti, C. Giorgi, E. Vuk, Free energies in one-dimensional models of magnetic transitions with hysteresis. Nuovo Cimento B **125**, 371–394 (2010)
38. A. Berti, C. Giorgi, E. Vuk, Hysteresis and temperature-induced transitions in ferromagnetic materials. Appl. Math. Model. **39**, 820–837 (2015)
39. A. Berti, C. Giorgi, E. Vuk, Free energies and pseudo-elastic transitions for shape-memory alloys. Discrete Cont. Dyn. Syst. Sr. S **6**, 293–316 (2013)
40. A. Berti, C. Giorgi, Derivation of the Landau-Lifshitz-Bloch equation from continuum thermodynamics. Physica B **500**, 142–153 (2016)
41. V. Berti, M. Fabrizio, D. Grandi, Hysteresis and phase transitions for one-dimensional and three-dimensional models in shape memory alloys. J. Math. Phys. **51**, 062901 (2010)
42. A. Bertram, *Elasticity and Plasticity of Large Deformations: An Introduction* (Springer, Berlin, 2008)
43. Z. Bi, R.F. Sekerka, Phase-field model of solidification of a binary alloy. Physica A **261**, 95–106 (1998)
44. E.C. Bingham, An investigation of the laws of plastic flow. Bulle. Bur. Stand. **13**(2), 309–353 (1916)
45. M.A. Biot, *Mechanics of Incremental Deformation: Theory of Elasticity and Viscoelasticity of Initially Stressed Solids and Fluids, Including Thermodynamic Foundations and Applications to Finite Strain* (Wiley, New York, 1965)

46. R.B. Bird, J.M. Wiest, Constitutive equations for polymeric liquids. Annu. Rev. Fluid Mech. **27**, 169–193 (1995)
47. P. Biscari, G. Napoli, S. Turzi, Bulk and surface biaxiality in nematic liquid crystals. Phys. Rev. E **74**, 031708 (2006)
48. A.C. Biswas, R.K. Pathria, Critical velocity in superfluid helium. J. Low Temp. Phys. **28**, 151–156 (1977)
49. D.R. Bland, *The Theory of Linear Viscoelasticity* (Pergamon Press, Oxford, 1960)
50. F. Bloch, Nuclear induction. Phys. Rev. **70**, 460–474 (1946)
51. D.B. Bogy, P.M. Naghdi, On heat conduction and wave propagation in rigid solids. J. Math. Phys. **11**, 917–923 (1970)
52. A.A. Bokov, Z.-G. Ye, Recent progress in relaxor ferroelectrics with perovskite structure. J. Mater. Sci. **41**, 31–52 (2006)
53. P.C. Bollada, P.K. Jimack, A.M. Mullis, Multiphase field modelling of alloy solidification. Comp. Mat. Sci. **171**, 109085 (2020)
54. L. Boltzmann, Zur theorie der elastichen nachwirkung. Sitzgsber. Akad. Wiss. Wien **70**, 275–300 (1874)
55. L. Boltzmann, Zur Theorie der elastischen Nachwirkung. Wied. Ann. **5**, 430–432 (1878)
56. A. Bonfanti, J.L. Kaplan, G. Charras, A. Kabla, Fractional viscoelastic models for power-law materials. Soft Matter **16**, 6002–6020 (2020)
57. R. Borghesani, A. Morro, Thermodynamics and isotropy in thermal and electric conduction. Meccanica **4**, 63–69 (1974)
58. R. Borghesani, A. Morro, Thermodynamic restrictions on thermoelectric, thermomagnetic and galvanomagnetic coefficients. Meccanica **4**, 157–161 (1974)
59. J. Boussinesq, *Théorie Analitique de la Chaleur* (Gauthier-Villars, Paris, 1903)
60. R.M. Bowen, Theory of mixtures, in *Continuum Physics III*, ed. by A.C. Eringen (Academic Press, New York, 1976)
61. R.M. Bowen, Compressible porous media models by use of the theory of mixtures. Int. J. Eng. Sci. **20**, 697–735 (1982)
62. R.M. Bowen, P.J. Chen, On the effects of diffusion and chemical reactions on the growth and decay of acceleration waves. Arch. Rational Mech. Anal. **48**, 319–351 (1972)
63. R.M. Bowen, J.C. Wiese, Diffusion in mixtures of elastic materials. Int. J. Eng. Sci. **7**, 689–722 (1969)
64. A.F. Bower, *Applied Mechanics of Solids* (CRC Press, Boca Raton, 2009)
65. P.W. Bridgman, *The Nature of Thermodynamics* (Harvard University Press, Cambridge, MA, 1961)
66. H.C. Brinkman, The viscosity of concentrated suspensions and solutions. J. Chem. Phys. **20**, 571–581 (1952)
67. M. Brokate, J. Sprekels, *Hysteresis and Phase Transitions* (Springer, Berlin, 1996)
68. J.P. Broomfield, *Corrosion of Steel in Concrete; Understanding, Investigation, and Repair* (E. & F.N. Spon, London, 1997)
69. W.F. Brown Jr., *Magnetoelastic Interactions, Springer Tracts in Natural Philosophy*, vol. 9 (Springer, New York, 1966)
70. L. Brun, Ondes de choc finies dans les solides elastiques, in *Mechanical Waves in Solids*, ed. by J. Mandel, L. Brun (Springer, CISM-Udine, 1975)
71. J. Buongiorno, Convective transport in nanofluids. Trans. ASME **128**, 240–250 (2006)
72. J.M. Burgers, Mechanical considerations - Model systems - Phenomenological theories of relaxation and viscosity, in *First Report on Viscosity and Plasticity*, ed. by J.M. Burgers (Nordemann Publishing, New York, 1939), ch. 1
73. J.W. Cahn, J.E. Hilliard, Free energy of a nonuniform system I. Interfacial energy. J. Chem. Phys. **28**, 258–267 (1958)
74. J.C. Cahn, On spinodal decomposition. Acta Metall. **9**, 795–801 (1961)
75. J.C. Cahn, J.E. Hilliard, Spinodal decomposition: a reprise. Acta Metall. **19**, 151–161 (1971)
76. W. Cao, Constructing Landau-Ginzburg-Devonshire type models for ferroelectric systems based on symmetry. Ferroelectrics **375**, 28–39 (2008)

77. W. Cao, L.E. Cross, Theory of tetragonal twin structures in ferroelectric perovskites with a first-order phase transition. Phys. Rev. B **44**, 5–12 (1991)
78. M. Caputo, F. Mainardi, Linear models of dissipation in anelastic solids. La Rivista del Nuovo Cimento **1**, 161–198 (1971)
79. S. Carillo, C. Giorgi, Non-classical memory kernels in linear viscoelasticity, in *Viscoelastic and Viscoplastic Materials*. ed. by F. Mohamed (Croatia, El-Amin (INTECH, 2016)
80. D.E. Carlson, Linear Thermoelasticity, in *Encyclopedia of Physics*, ed. by C. Truesdell, vol. VIa/2 (Springer, Berlin, 1972)
81. M.C. Calderer, Mathematical problems in liquid crystal flows. Math. Meth. Appl. Sci. **17**, 171–188 (1994)
82. J.-M. Cardona, S. Forest, R. Sievert, Towards a theory of second grade thermoelasticity. Extracta Math. **14**, 127–140 (1999)
83. C. Cattaneo, Sulla conduzione del calore. Atti Sem. Mat. Fis. Univ. Modena **3**, 3–21 (1948)
84. G. Caviglia, A. Morro, *Inhomogenous Waves in Solids and Fluids* (World Scientific, Singapore, 1992)
85. G. Caviglia, A. Morro, B. Straughan, Thermoelasticity at cryogenic temperatures. Int. J. Non-Lin. Mech. **27**, 251–263 (1992)
86. S.-K. Chan, Steady-state kinetics of diffusionless first-order phase transformations. J. Chem. Phys. **67**, 5755–5762 (1977)
87. S. Chapman, T.G. Cowling, *The Mathematical Theory of Non-Uniform Gases* (Cambridge University Press, 1970), p. 363
88. P.J. Chen, Growth and decay of waves in solids, in *Encyclopedia of Physics*, vol. VIa/3, ed. by C. Truesdell (Springer, Berlin, 1973)
89. J. Ciambella, A. Paolone, S. Vidoli, Memory decay rates of viscoelastic solids: not too slow, but not too fast either. Rheol. Acta **50**, 661–674 (2011)
90. M. Ciarletta, B. Straughan, V. Tibullo, Explosive instabilities for a generalized second grade fluid. J. Math. Anal. Appl. **432**, 945–953 (2015)
91. H. Cohen, C.-C. Wang, On the response and symmetry of elastic materials with internal constraints. Arch. Rational Mech. Anal. **99**, 1–36 (1987)
92. B.D. Coleman, W. Noll, The thermodynamics of elastic materials with heat conduction and viscosity. Arch. Rational Mech. Anal. **13**, 167–178 (1963)
93. B.D. Coleman, Thermodynamics of materials with memory. Arch. Rational Mech. Anal. **17**, 1–46 (1964)
94. B.D. Coleman, V.J. Mizel, Existence of caloric equations of state in thermodynamics. J. Chem. Phys. **40**, 1116–1125 (1964)
95. B.D. Coleman, On thermodynamics, strain impulses, and viscoelasticity. Arch. Rational Mech. Anal. **17**, 230–254 (1964)
96. B.D. Coleman, E.H. Dill, On the thermodynamics of electromagnetic fields in materials with memory. Arch. Rational Mech. Anal. **41**, 132–162 (1971)
97. B.D. Coleman, E.H. Dill, Thermodynamics restrictions on the constitutive equations of electromagnetic theory. ZAMP **22**, 691–702 (1971)
98. B.D. Coleman, M.E. Gurtin, Thermodynamics with internal state variables. J. Chem. Phys. **47**, 597–613 (1967)
99. B.D. Coleman, V.J. Mizel, Norms and semi-groups in the theory of fading memory. Arch. Rational Mech. Anal. **23**, 87–123 (1967)
100. B.D. Coleman, V.J. Mizel, On the general theory of fading memory. Arch. Rational Mech. Anal. **29**, 18–31 (1968)
101. B.D. Coleman, W. Noll, Foundations of linear viscoelasticity. Rev. Mod. Phys. **33**, 239–249 (1961)
102. B.D. Coleman, M.L. Hodgdon, On a class of constitutive relations for ferromagnetic hysteresis. Arch. Rational Mech. Anal. **99**, 375–396 (1987)
103. B.D. Coleman, M.L. Hodgdon, A constitutive relation for rate-independent hysteresis in ferromagnetically soft materials. Int. J. Eng. Sci. **24**, 897–919 (1986)

104. M. Conti, V. Danese, C. Giorgi, V. Pata, A model of viscoelasticity with time-dependent memory kernels. Am. J. Math. **140**, 349–389 (2018)
105. M. Corcione, M. Cianfrini, A. Quintino, Two-phase mixture modeling of natural convection of nanofluids with temperature-dependent properties. Int. J. Therm. Sci. **71**, 182–195 (2013)
106. E. Cosserat, F. Cosserat, *Théorie des Corps Déformables* (Hermann, Paris, 1909)
107. B.A. Cotter, R.S. Rivlin, Tensors associated with time-dependent stress. Quart. Appl. Math. **13**, 177–182 (1955)
108. S.C. Cowin, Bone poroelasticity. J. Biomech. **32**, 217–238 (1999)
109. N.F. Cramer, *The Physics of Alfvén Waves* (Wiley, New York, 2001)
110. F.H. Crawford, *Heat, Thermodynamics, and Statistical Physics* (Rupert Hart-Davis, London, 1963)
111. C.M. Dafermos, Asymptotic stability in viscoelasticity. Arch. Rational Mech. Anal. **37**, 297–308 (1970)
112. J.S. Dahler, L.E. Scriven, Theory of structured continua I. General consideration of angular momentum and polarization. Proc. R. Soc. Lond. A **275**, 504–527 (1963)
113. W.A. Day, Restrictions on relaxation functions in linear viscoelasticity. Quart. J. Mech. Appl. Math. **24**, 487–497 (1971)
114. W.A. Day, *The Thermodynamics of Simple Materials with Fading Memory* (Springer, Berlin, 1972)
115. P.G. de Gennes, An analogy between superconductivity and smectics A. Solid State Commun. **10**, 753–756 (1972)
116. P.G. de Gennes, J. Prost, *The Physics of Liquid Crystals* (Clarendon Press, Oxford, 1998)
117. S.R. de Groot, P. Mazur, *Non-Equilibrium Thermodynamics* (Dover, New York, 1984)
118. A. De Simone, P. Podio Guidugli, On the continuum theory of deformable ferromagnetic solids. Arch. Rational Mech. Anal. **136**, 201–233 (1996)
119. E.A. de Souza Neto, D. Peric, D.R.J. Owen, *Computational Methods for Plasticity: Theory and Applications* (Wiley, New York, 2008); Appendix A
120. P. Debye, Einige Bemerkungen zur Magnetisierung bei tiefer Temperatur. Annalen der Physik **81**, 1154 (1926)
121. J.A. Dean, N.A. Lange, *Lange's Handbook of Chemistry*, 15th edn. (McGraw-Hill, 1999), p. 1436
122. G. Del Piero, L. Deseri, On the analytic expression of the free energy in linear viscoelasticity. J. Elast. **43**, 247–278 (1996)
123. G. Del Piero, L. Deseri, On the concepts of state and free energy in linear viscoelasticity. Arch. Rational Mech. Anal. **138**, 1–35 (1997)
124. F. Dell'Oro, V. Pata, On the Moore-Gibson-Thompson equation and its relation to linear viscoelasticity. Appl. Math. Optim. **76**, 641–655 (2017)
125. K.G. Denbigh, *The Principles of Chemical Equilibrium* (Cambridge University Press, Cambridge, 1981)
126. Q. Deng, M. Kammoun, A. Erturk, P. Sharma, Nanoscale flexoelectric energy harvesting. Int. J. Solids Struct. **51**, 3218–3225 (2014)
127. W. Desch, R. Grimmer, Singular relaxation moduli and smoothing in three-dimensional viscoelasticity. Trans. Am. Math. Soc. **314**, 381–404 (1989)
128. L. Deseri, M. Fabrizio, J.M. Golden, The concept of a minimal state in viscoelasticity: new free energies and applications to PDEs. Arch. Rational Mech. Anal. **181**, 43–96 (2006)
129. L. Deseri, G. Gentili, J.M. Golden, An explicit formula for the minimum free energy in linear viscoelasticity. J. Elast. **54**, 141–185 (1999)
130. L. Deseri, M. Zingales, A mechanical picture of fractional-order Darcy equation. Commun. Nonlinear Sci. Numer. Simulat. **20**, 940–949 (2015)
131. M. Destrade, G. Saccomandi, M. Vianello, Proper formulation of viscous dissipation for nonlinear solids. J. Acoust. Soc. Am. **133**, 1255–1259 (2013)
132. A.F. Devonshire, Theory of barium titanate. Phil. Mag. **40**, 1040–1063 (1949)
133. A.F. Devonshire, Theory of ferroelectrics. Adv. Phys. **3**, 85–130 (1954)

134. R.J. Donnelly, The two-fluid theory and second sound in liquid helium. Phys. Today **62**(10), 34–39 (2009)
135. A. Dorfmann, R.W. Ogden, Nonlinear magnetoelastic deformations. Q. J. Mech. Appl. Math. **57**, 599–622 (2004)
136. A. Dorfmann, R.W. Ogden, Nonlinear electroelasticity. Acta Mech. **174**, 167–183 (2005)
137. L. Dorfmann, R.W. Ogden, Nonlinear electroelasticity: material properties, continuum theory and applications. Proc. Roy. Soc. A **473**, 1–18 (2017)
138. D.S. Drumheller, A. Bedford, A thermodynamic theory for reacting immiscible mixtures. Arch. Rational Mech. Anal. **73**, 257–284 (1980)
139. J.E. Dunn, R.L. Fosdick, Thermodynamics, stability, and boundedness of fluids of complexity 2 and fluids of second grade. Arch. Rational Mech. Anal. **56**, 191–252 (1974)
140. J.E. Dunn, K.R. Rajagopal, Fluids of differential type: critical review and thermodynamic analysis. Int. J. Eng. Sci. **33**, 689–729 (1995)
141. J.E. Dunn, J. Serrin, On the thermodynamics of interstitial working. Arch. Rational Mech. Anal. **88**, 96–133 (1985)
142. A. Einstein, W.J. de Haas, Experimental proof of the existence of Ampère's molecular currents. Koninklijke Akademie van Wetenschappen te Amsterdam, Proceedings **18**, 696–711 (1915)
143. A. Engel, R. Friedrichs, On the electromagnetic force on a polarizable body. Am. J. Phys. **70**, 428–435 (2002)
144. J.L. Ericksen, Hydrostatic theory of liquid crystals. Arch. Rational Mech. Anal. **9**, 371–378 (1962)
145. J.L. Ericksen, Liquid crystals with a variable degree of orientation. Arch. Rational Mech. Anal. **113**, 97–120 (1991)
146. A.C. Eringen, *Mechanics of Continua* (Wiley, New York, 1967)
147. A.C. Eringen, A unified continuum theory of electrodynamics of liquid crystals. Int. J. Eng. Sci. **35**, 1137–1157 (1997)
148. A. Eringen, *Microcontinuum Field Theories: I (Foundations and Solids* (Springer, New York, 1999)
149. A.C. Eringen, Micropolar mixture theory of porous media. J. Appl. Phys. **94**, 4184–4190 (2003)
150. A.C. Eringen, C.B. Kafadar, Polar field theories, in *Continuum Physics IV*, ed. by A.C. Eringen (Academic, New York, 1976)
151. A.C. Eringen, G.A. Maugin, *Electrodynamics of Continua I (Foundations and Solid Media* (Springer, Berlin, 1990)
152. J.D. Eshelby, The determination of the elastic field of an ellipsoidal inclusion, and related problems. Phil. Trans. R. Soc. Lond. A **241**, 376–396 (1957)
153. J.D. Eshelby, The elastic energy-momentum tensor. J. Elast. **5**, 321–335 (1975)
154. M. Fabrizio, C. Giorgi, A. Morro, A thermodynamic approach to ferromagnetism and phase transitions. Int. J. Eng. Sci. **47**, 821–839 (2009)
155. M. Fabrizio, C. Giorgi, A. Morro, Modeling of heat conduction via fractional derivatives. Heat Mass Transf. **53**, 2785–2797 (2017)
156. M. Fabrizio, A. Morro, Viscoelastic relaxation functions compatible with thermodynamics. J. Elast. **19**, 63–75 (1988)
157. M. Fabrizio, A. Morro, *Mathematical Problems in Linear Viscoelasticity* (SIAM, Philadelphia, 1992)
158. M. Fabrizio, C. Giorgi, A. Morro, Internal dissipation, relaxation property, and free energy in materials with fading memory. J. Elast. **40**, 107–122 (1995)
159. M. Fabrizio, C. Giorgi, A. Morro, Solidification and separation in saline water. Discrete Contin. Dyn. Syst. Ser. S **9**, 139–155 (2016)
160. M. Fabrizio, C. Giorgi, V. Pata, A new approach to equations with memory. Arch. Rational Mech. Anal. **198**, 189–232 (2010)
161. M. Fabrizio, J.M. Golden, Maximum and minimum free energies for a linear viscoelastic material. Q. Appl. Math. **60**, 341–381 (2002)

162. M. Fabrizio, A. Morro, *Electromagnetism of Continuous Media* (Oxford Science Publications, Oxford, 2003)
163. M. Fabrizio, A. Morro, Thermodynamics and second sound in a two-fluid model of helium II; revisited. J. Non-Equilib. Thermodyn. **208**, 69–84 (2003)
164. R.M. Fano, L.J. Chu, R.B. Adler, *Electromagnetic Fields, Energy, and Forces* (Wiley, New york, 1960)
165. S.H. Faria, G.M. Kremer, A constitutive theory for ferrofluids. Cont. Mech. Thermodyn. **11**, 297–306 (1999)
166. A. Favre, *Statistical equations of turbulent gases, in Problems of Hydrodynamics and Continuum Mechanics* (SIAM, Philadelphia, 1969), pp.231–266
167. R.P. Feynman, R.B. Leighton, M. Sands, *The Feynman Lectures on Physics* (Addison, Reading Massachusetts, 1966)
168. E. Fermi, *Thermodynamics* (Dover, New York, 1956)
169. A. Fick, Über diffusion. Annalen der Physik **170**, 59–86 (1855)
170. R.L. Fosdick, K.R. Rajagopal, Thermodynamics and stability of fluids of third grade. Proc. R. Soc. Lond. A **339**, 351–377 (1980)
171. F. Franchi, B. Lazzari, R. Nibbi, B. Straughan, Uniqueness and decay in local thermal non-equilibrium double porosity thermoelasticity. Math. Meth. Appl. Sci. **41**, 6763–6771 (2018)
172. V. Francois-Lavet, F. Henrotte, L. Stainier, L. Noels, An energy-based variational model of ferromagnetic hysteresis for finite element computations. J. Comput. Appl. Math. **246**, 243–250 (2013)
173. F.C. Frank, On the theory of liquid crystals. Discuss. Faraday Soc. **25**, 19–28 (1958)
174. M. Frémond, Adhérence des solides. J. Méc. Théor. Appl. **6**, 383–407 (1987)
175. M. Frémond, B. Nedjar, Damage, gradient of damage and principle of virtual power. Int. J. Solids Struct. **33**, 1083–1103 (1996)
176. M. Frémond, *Phase Change in Mechanics* (Springer, Berlin, 2012); ch. 4
177. E. Fried, M.E. Gurtin, Continuum theory of thermally induced phase transitions based on an order parameter. Physica D **68**, 326–343 (1993)
178. E. Fried, M.E. Gurtin, Dynamic solid-solid phase transitions with phase characterized by an order parameter. Physica D **72**, 287–308 (1994)
179. E. Fried, M.E. Gurtin, Tractions, balances, and boundary conditions for nonsimple materials with application to liquid flow at small-length scales. Arch. Rational Mech. Anal. **182**, 513–554 (2006)
180. Y.C. Fung, *Biodynamics (Circulation* (Springer, New York, 1984)
181. R.E. Gaidos, R. Darby, Numerical simulation and change in type in the developing flow of a nonlinear elastic fluid. J. Non-Newton. Fluid Mech. **29**, 59–79 (1988)
182. F.R. Gantmacher, *The Theory of Matrices*, vol. 1 (Chelsea, New York, 1959)
183. F.R. Gantmacher, *The Theory of Matrices*, vol. 2 (Chelsea, New York, 1974)
184. D.A. Garanin, Fokker-Planck and Landau-Lifshitz-Bloch equations for classical ferromagnets. Phys. Rev. B **55**, 3050–3057 (1997)
185. S. Gatti, C. Giorgi, V. Pata, Navier-Stokes limit of Jeffreys type flows. Physica D **203**, 55–79 (2005)
186. G. Gentili, Regularity and stability for a viscoelastic material with a singular memory kernel. J. Elast. **37**, 139–156 (1995)
187. G. Gentili, C. Giorgi, A new model for rate-independent hysteresis in permanent magnets. Int. J. Eng. Sci. **39**, 1057–1090 (2001)
188. P. Germain, *Mécanique des Milieux Continus* (Masson, Paris, 1962)
189. W.F. Giauque, A thermodynamic treatment of certain magnetic effects. A proposed method of producing temperatures considerably below 1° absolute. J. Am. Chem. Soc. **49**, 1864–1870 (1927)
190. T.L. Gilbert, A Lagrangian formulation of the gyromagnetic equation of the magnetization field. Phys. Rev. B **100**, 1243 (1955) [Abstract only; full report, Armor Research Foundation Project No. A059, Supplementary Report, May 1, 1956] (unpublished)

191. T.L. Gilbert, A phenomenological theory of damping in ferromagnetic materials. IEEE Trans. Magn. **40**, 3443–3449 (2004)
192. V.L. Ginzburg, Superconductivity: the day before yesterday, yesterday, today, and tomorrow. J. Supercond. **13**, 665–677 (2000)
193. V.L. Ginzburg, L.D. Landau, On the theory of superconductivity, Zh. Eksp. Teor. Fiz. **20**, 1064–1082 (1950); English translation in Collected papers of L.D. Landau, ed. D. ter Haar (Pergamon, Oxford, 1965), pp. 546–568
194. C. Giorgi, Phase-field models for transition phenomena in materials with hysteresis. Discrete Cont. Dyn. Syst. Ser. S **8**, 693–722 (2015)
195. C. Giorgi, A. Morro, Extremum principles for viscoelastic fluids. Int. J. Eng. Sci. **29**, 807–817 (1991)
196. C. Giorgi, A. Morro, Viscoelastic solids with unbounded relaxation function. Contin. Mech. Thermodyn. **4**, 151–165 (1992)
197. C. Giorgi, J.E. Muñoz Rivera, V. Pata, Global attractors for a semilinear hyperbolic equation in viscoelasticity. J. Math. Anal. Appl. **260**, 83–99 (2001)
198. J.M. Golden, Free energies in the frequency domain: the scalar case. Q. Appl. Math. **58**, 127–150 (2000)
199. C.J. Gorter, H. Casimir, On supraconductivity I. Physica **1**, 306–320 (1934)
200. P.-A. Gourdain, The impact of the Hall term on tokamak plasmas, arXiv:1703.00987 physics.plasm.-ph
201. D. Graffi, Sui problemi dell'ereditarietà lineare. Nuovo Cimento A **5**, 53–71 (1928)
202. D. Graffi, Sull'espressione dell'energia libera nei materiali viscoelastici lineari. Ann. Mat. Pura Appl. **98**, 273–279 (1974)
203. D. Graffi, Sull'espressione analitica di alcune grandezze termodinamiche nei materiali con memoria. Rend. Sem. Mat. Univ. Padova **68**, 17–29 (1982)
204. D. Graffi, On the fading memory. Appl. Anal. **15**, 295–311 (1983)
205. A.E. Green, P.M. Naghdi, A dynamical theory of interacting continua. Int. J. Eng. Sci. **3**, 231–263 (1965)
206. A.E. Green, P.M. Naghdi, A general theory of an elastic-plastic continuum. Arch. Rational Mech. Anal. **18**, 251–281 (1965)
207. A.E. Green, P.M. Naghdi, A derivation of equations for wave propagation in water of variable depth. J. Fluid Mech. **78**, 237–246 (1976)
208. A.E. Green, P.M. Naghdi, On thermodynamics and the nature of the second law. Proc. R. Soc. Lond. A **357**, 253–270 (1977)
209. A.E. Green, P.M. Naghdi, A re-examination of the basic postulates of thermomechanics. Proc. R. Soc. Lond. A **432**, 171–194 (1991)
210. A.E. Green, P.M. Naghdi, Thermoelasticity without energy dissipation. J. Elast. **31**, 189–208 (1993)
211. G. Grioli, *Mathematical Theory of Elastic Equilibrium* (Springer, Berlin, 1962)
212. E.P. Gross, Hydrodynamics of a superfluid condensate. J. Math. Phys. **4**, 195–207 (1963)
213. A.E. Green, P.M. Naghdi, J.A. Trapp, Thermodynamics of a continuum with internal constraints. Int. J. Eng. Sci. **8**, 891–902 (1970)
214. M.E. Gurtin, Generalized Ginzburg-Landau and Cahn-Hilliard equations based on a microforce balance. Physica D **92**, 178–192 (1996)
215. M.E. Gurtin, E. Fried, L. Anand, *The Mechanics and Thermodynamics of Continua* (Cambridge University Press, 2010)
216. M.E. Gurtin, A.C. Pipkin, A general theory of heat conduction with finite wave speeds. Arch. Rational Mech. Anal. **31**, 113–126 (1968)
217. M.E. Gurtin, P. Podio Guidugli, The thermodynamics of constrained materials. Arch. Rational Mech. Anal. **51**, 192–208 (1973)
218. M.E. Gurtin, K. Spear, On the relationship between the logarithmic strain rate and the stretching tensor. Int. J. Solids Struct. **19**, 437–444 (1983)
219. D. Gutierrez-Lemini, *Engineering Viscoelasticity* (Springer, New York, 2014)

220. H.E. Hall, W.F. Vinen, The rotation of liquid helium II. The theory of mutual friction in uniformly rotating helium II. Proc. Roy. Soc. A **238**, 204–214 (1956)
221. S. Hassanpour, G.R. Heppler, Micropolar elasticity theory: a survey of linear isotropic equations, representative notations, and experimental investigations. Math. Mech. Solids **22**, 224–242 (2017)
222. T. Hayat, Y. Wang, K. Hutter, Flow of a fourth grade fluid. Math. Mod. Meth. Appl. Sci. **12**, 797–811 (2002)
223. S. Helgason, *Differential Geometry, Lie Groups, and Symmetric Spaces* (Academic Press, New york, 1978)
224. H. Hsiao, O.T. Bruhns, A.T.M. Meyers, Strain rates and material spins. J. Elast. **52**, 1–41 (1998)
225. R. Klages, G. Radons, I.M. Sokolov, *Anomalous Transport: Foundations and Applications* (Wiley, Weinheim, 2008)
226. V.L. Idenbom, E.B. Loginov, M.A. Osipov, The flexoelectric effect and the structure of crystals. Sov. Phys. Crystallogr. **26**, 656–658 (1981)
227. F. Irgens, *Continuum Mechanics* (Springer, New York, 2008)
228. M. Itskov, *Tensor Algebra and Tensor Analysis for Engineers with applications to Continuum Mechanics* (Springer, Berlin, 2009)
229. J.D. Jackson, *Classical Electrodynamics* (Wiley, New York, 1975)
230. B. Jayawardhana, R. Ouyang, V. Andrieu, Dissipativity of general Duhem hysteresis models, in *Proceedings of the 50th IEEE Conference on Decision and Control and European Control Conference (CDC-ECC)* (New York, 2011), pp. 3234–3239
231. J.T. Jenkins, A theory of magnetic fluids. Arch. Rational Mech. Anal. **46**, 42–60 (1972)
232. X. Jiang, W. Huang, S. Zhang, Flexoelectric nano-generator: materials, structures and devices. Nano Energy **2**, 1079–1092 (2013)
233. C.S. Jog, A concise proof of the representation theorem for fourth-order isotropic tensors. J. Elast. **85**, 119–124 (2006)
234. C.C. Johnson, *Field and Wave Electrodynamics* (McGraw-Hill, New York, 1965)
235. M. Johnson, D. Segalman, A model for viscoelastic fluid behavior which allows non-affine deformations. J. Non-Newton. Fluid Mech. **2**, 255–270 (1977)
236. D.S. Jones, *Acoustic and Electromagnetic Waves* (Clarendon, Oxford, 1986)
237. S.H.S. Joodat, K.B. Nakshatrala, R. Ballarini, Modeling flow in porous media with double porosity/permeability: a stabilized mixed formulation, error analysis, and numerical solutions. Comp. Methods Appl. Mech. Eng. **337**, 632–676 (2018)
238. P.M. Jordan, Growth and decay of acoustic acceleration waves in Darcy-type porous media. Proc. R. Soc. A **461**, 2749–2766 (2005)
239. P.M. Jordan, Second-sound phenomena in inviscid, thermally relaxing gases. Discrete Contin. Dyn. Syst. Ser. B **19**, 2189–2205 (2014)
240. P.M. Jordan, Second-sound propagation in rigid, nonlinear conductors. Mech. Res. Commun. **68**, 52–79 (2015)
241. D.D. Joseph, L. Preziosi, Heat waves. Rev. Mod. Phys. **61**, 41–73 (1989)
242. B.D. Josephson, Possible new effects in superconducting tunneling. Phys. Lett. **1**, 251–253 (1962)
243. D.D. Joseph, *Fluid Dynamics of Viscoelastic Liquids* (Springer, New York, 1990)
244. D. Jou, J. Casas-Vázquez, G. Lebon, *Extended Irreversible Thermodynamics* (Springer, Berlin, 2010)
245. M. Kachanov, Continuum model of medium with cracks. J. Eng. Mech. **106**, 1039–1051 (1980)
246. B. Kaltenbacher, I. Lasiecka, M.K. Pospieszalska, Well-posedness and exponential decay of the energy in the nonlinear Jordan-Moore-Gibson-Thompson equation arising in high intensity ultrasound. Math. Models Methods Appl. Sci. **22**, 1250035 (2012)
247. M. Kamlah, Ferroelectric and ferroelastic piezoceramics - modeling of electromechanical hysteresis phenomena. Cont. Mech. Thermodyn. **13**, 219–268 (2001)

248. K. Kannan, K.R. Rajagopal, A model for the flow of rock glaciers. Int. J. Non-Lin. Mech. **48**, 59–64 (2013)
249. P. Kapitza, Viscosity of liquid helium below the λ-point. Nature **141**, 74–76 (1938)
250. J.I. Kaplan, M.L. Glasser, Effective magnetic field in metals. Phys. Rev. **170**, 649–650 (1968)
251. A. Karma, Phase-field formulation for quantitative modeling of alloy solidification. Phys. Rev. Lett. **87**, 115701 (2001)
252. G. Karniadakis, A. Beskok, N. Aluru, *Microflows and Nanoflows (Fundamentals and Simulation* (Springer, Berlin, 2005)
253. K.H. Khanafer, K. Vafai, M. Lightstone, Buoyancy-driven heat transfer enhancement in a two-dimensional enclosure utilizing nanofluids. Int. J. Heat Mass Transf. **46**, 3639–3653 (2003)
254. A. Kholmetskii, O. Missevitch, T. Yarman, Electric/magnetic dipole in an electromagnetic field: force, torque and energy. Eur. Phys. J. Plus **129**, 215 (2014)
255. C. Kittel, *Introduction to Solid State Physics* (Wiley, New York, 1996)
256. R. Kohlrausch, Nachtrag fiber die Elastische Nachwirlmng beim Cocon- und Glasfaden, und die Hygroskopische Eigenschaft des Ersteren. Pogg. Ann. Phys. **72**(11), 393–425 (1847)
257. D.J. Korteweg, Sur la forme que prennent les équations du mouvement des fluides si l'on tient compte des forces capillaires causées par des variations de densité considérables mais continues et sur la théorie de la capillarité dans l'hypothèse d'une variation continue de la densité. Arch. Neerl. Sci. Exactes Ser. **II**(6), 1–24 (1901)
258. N.E. Kotchine, Sur la théorie des ondes de choc dans un fluide. Rend. Circ. Mat. Palermo **50**, 305–344 (1926)
259. A. Kovetz, *Electromagnetic Theory* (Oxford University Press, 2000)
260. N.A. Krall, A.W. Trivelpiece, *Principles of Plasma Physics* (McGraw-Hill, New York, 1973). ((ch. 3))
261. E. Kröner, Allgemeine kontinuumstheorie der versetzungen und eigenspannungen. Arch. Rational Mech. Anal. **4**, 273–334 (1960)
262. B. Kutschan, K. Morawetz, S. Gemming, Modeling and morphogenesis of brine channels in sea ice. Phys. Rev. E **81**, 036106 (2010)
263. R. Lakes, K.W. Wojciechowski, Negative compressibility, negative Poisson's ratio, and stability. Physica Status Solidi (b) **245**, 545–551 (2008)
264. L.D. Landau, Zur Theorie der Phasenumwandlungen. Phys. Z. Sowjet. **11**, 26–35 (1937)
265. L.D. Landau, Unbranched model of intermediate state. Phys. Z. Sowjet. **11**, 129–138 (1937)
266. L.D. Landau, The theory of superfluidity of helium II. J. Phys. USSR **5**, 71–73 (1941)
267. L.D. Landau, The theory of superfluidity of helium II. Phys. Rev. **60**, 356–358 (1941)
268. L.D. Landau, I.M. Khalatnikov, On the theory of superconductivity, in *Collected Papers of L.D. Landau*, ed. by D. Ter Haar (Pergamon, Oxford, 1965)
269. L.D. Landau, E.M. Lifshitz, On the theory of dispersion of magnetic permeability in ferromagnetic bodies. Phys. Z. Sowjetunion **8**, 135–169 (1935)
270. L.D. Landau, E.M. Lifshitz, *Statistical Physics* (Pergamon Press, Oxford, 1959)
271. L.D. Landau, E.M. Lifshitz, *Theory of Elasticity* (Pergamon, New York, 1986)
272. L.D. Landau, E.M. Lifshitz, *Fluid Mechanics* (Pergamon Press, Oxford, 1987)
273. L.D. Landau, E.M. Lifshitz, L.P. Pitaevskii, *Electrodynamics of Continuous Media* (Pergamon, Oxford, 1984)
274. A. Lakhtakia, *Beltrami Fields in Chiral Media* (World Scientific, Singapore, 1994)
275. M.J. Leitman, G.M. Fisher, The linear theory of viscoelasticity, in *Encyclopedia of Physics, vol. VIa/3*. ed. by C. Truesdell (Springer, Berlin, 1973), pp.1–123
276. F.M. Leslie, Continuum theory for nematic liquid crystals. Cont. Mech. Thermodyn. **4**, 167–175 (1992)
277. D.R. Lide (ed.), *CRC Handbook of Chemistry and Physics*, 86th edn. (CRC Press, Boca Raton, FL, 2005), pp. 8–116
278. I-S. Liu, Method of Lagrange multipliers for exploitation of the entropy principle. Arch. Rational Mech. Anal. **46**, 131–148 (1972)

279. Q. Liu, G. Subhash, A phenomenological constitutive model for foams under large deformation. Polym. Eng. Sci. **44**, 463–473 (2004)
280. F. London, H. London, Supraleitung und diamagnetismus. Physica **2**, 341–354 (1935)
281. F. London, On the Bose-Einstein condensation. Phys. Rev. **54**, 947–954 (1938)
282. F. London, *Superfluids* (Dover, New York, 1961)
283. K. Lonngren, A.C. Scott, *Solitons in Action* (Wiley, New york, 1978)
284. W.-Y. Lu, Compression and shear response of 3D printed foam pads, in *Mechanics of Additive and Advanced Manufacturing*, vol. 8, ed. by S. Kramer, J.L. Jordan, H. Jin, J. Carrol, A.M. Beese (Springer, New York 2019)
285. C. Ludwig, Diffusion Zwischen Ungleich Erwarmten Orten Gleich Zusammengestzter Losungen, Sitz. Ber Acad. Wiss. Wien Math-Naturw. Kl. **20**, 539 (1856)
286. A.T. McBride, B.D. Reddy, P. Steinmann, Dissipation-consistent modelling and classification of extended plasticity formulations. J. Mech. Phys. Solids **119**, 118–139 (2018)
287. M. Mc Carthy, Wave propagation in generalized thermoelasticity. Int. J. Eng. Sci. **10**, 593–602 (1972)
288. M. Mc Carthy, Singular surfaces and waves, in *Continuum Physics II*, ed. by A.C. Eringen (Wiley, New York, 1975)
289. R.M. Mc Meeking, C.M. Landis, S.M.A. Jimenez, A principle of virtual work for combined electrostatic and mechanical loading of materials. Int. J. Non-Lin. Mech. **42**, 831–838 (2007)
290. J.W. Macki, P. Nistri, P. Zecca, Mathematical models for hysteresis. SIAM Rev. **35**, 94–123 (1993)
291. F. Mainardi, *Fractional Calculus and Waves in Linear Viscoelasticity* (Imperial College Press, London, 2010)
292. F. Mainardi, G. Spada, Creep, relaxation and viscosity properties for basic fractional models in rheology. Eur. Phys. J. Spec. Top. **193**, 133–160 (2011)
293. J. Málek, K.R. Rajagopal, K. Tuma, Derivation of the variants of the Burgers model using a thermodynamic approach and appealing to the concept of evolving natural configurations. Fluids **3**(4), 1–18 (2018)
294. J. Mandel, Plasticité Classique et Viscoplasticité, CISM Courses and Lectures no. 97 (Springer, Berlin, 1972)
295. J.J. Marigo, Modeling of brittle and fatigue damage for elastic material by growth of microvoids. Eng. Fract. Mech. **21**, 861–874 (1985)
296. M. Massoudi, A. Vaidya, On some generalizations of the second grade fluid model. Nonlin. Anal. Real World Appl. **9**, 1169–1183 (2008)
297. G.A. Maugin, The principle of virtual power: application to coupled fields. Acta Mech. **35**, 1–70 (1980)
298. G.A. Maugin, *Continuum Mechanics of Electromagnetic Solids* (North-Holland, Amsterdam, 1988)
299. G.A. Maugin, Internal variables and dissipative structures. J. Non-Equilib. Thermodyn. **15**, 173–192 (1990)
300. G.A. Maugin, On the thermomechanics of continuous media with diffusion and/or weak nonlocality. Arch. Appl. Mech. **75**, 723–738 (2006)
301. G.A. Maugin, The principle of virtual power: from eliminating metaphysical forces to providing an efficient modelling tool. Continuum Mech. Thermodyn. **25**, 127–146 (2013)
302. G.A. Maugin, W. Muschik, Thermodynamics with internal variables. Part I. General concepts. J. Non-Equilib. Thermodyn. **19**, 217–249 (1994)
303. G.A. Maugin, J. Pouget, Electroacoustic equations for one-domain ferroelectric bodies. J. Acoust. Soc. Am. **68**, 575–587 (1980)
304. J.C. Maxwell, Constitution of bodies, Sci. Papers, College Arts Sci. Univ. Tokyo **2**, 616–624 (1877)
305. J.C. Maxwell, *A Treatise on Electricity and Magnetism (Oxford University Press, 1904), reprint by* (Dover, New York, 1954)
306. J.C. Maxwell, On the dynamical theory of gases, in *The Scientific Papers of J.C. Maxwell*, vol. 2, pp. 26–78 (1965)

307. H. Mehrer, *Diffusion in Solids* (Springer, Berlin, 2007)
308. W. Meissner, R. Ochsenfeld, Ein neuer effect bei eintritt der supraleitfähigkeit. Naturwissenshaften **21**, 787–788 (1933)
309. M. Mihâilescu-Suliciu, I. Suliciu, Energy for hypoelastic constitutive equations. Arch. Rational Mech. Anal. **70**, 168–179 (1979)
310. R.D. Mindlin, Polarization gradient in elastic dielectrics. Int. J. Solids Struct. **4**, 637–642 (1968)
311. R.D. Mindlin, Elasticity, piezoelectricity and crystal lattice dynamics. J. Elast. **2**, 217–282 (1972)
312. R.D. Mindlin, H.F. Tiersten, Effects of couple-stresses in linear elasticity. Arch. Rational Mech. Anal. **11**, 415–448 (1962)
313. C.W. Monroe, J. Newman, An introduction to the Onsager reciprocal relations. Chem. Eng. Educ. **41**, 233–238 (2007)
314. M. Mooney, A theory of large elastic deformation. J. Appl. Phys. **11**, 582–591 (1940)
315. M.J. Moran, H.N. Shapiro, *Fundamentals of Engineering Thermodynamics* (Wiley, Chichester, 2006)
316. J.J. Moreau, Sur les lois de frottement, de viscosité, et de plasticité. C.R. Acad. Sci. **271**, 608–611 (1970)
317. A. Morozov, S. Spagnolie, Introduction to complex fluids, in *Complex Fluids in Biological Systems. Experiment, Theory, and Computation*, ed. by S. Spagnolie (Springer, 2015)
318. A. Morro, Temperature waves in materials with memory. Meccanica **12**, 73–77 (1977)
319. A. Morro, Interaction of acoustic waves with shock waves in elastic solids. Zeit. Angew. Math. Phys. 822–827 (1978)
320. A. Morro, Interaction of waves with shocks in magneto-fluid-dynamics. Acta Mech. 197–213 (1980)
321. A. Morro, Jump relations and discontinuity waves in conductors with memory. Math. Comp. Model. **43**, 138–149 (2005)
322. A. Morro, A phase-field approach to non-isothermal transitions. Math. Comp. Model. **48**, 621–633 (2008)
323. A. Morro, Balance and constitutive equations for diffusion in mixtures of fluids. Meccanica **49**, 2109–2123 (2014)
324. A. Morro, Vector spaces and matrix algebra: vector and tensor algebra, in *Encyclopedia of Thermal Stresses*, vol. 11, ed. by R. Hetnarski (Springer, Berlin, 2014)
325. A. Morro, A mixture model of poroelastic materials. Eur. J. Mech. A/Solids **53**, 121–130 (2015)
326. A. Morro, Diffusion in mixtures of reacting thermoelastic solids. J. Elas. **123**, 59–84 (2016)
327. A. Morro, Modelling of viscoelastic materials and creep behaviour. Meccanica **52**, 3015–3021 (2017)
328. A. Morro, M. Romeo, Thermodynamic derivation of Saha's equation for a multitemperature plasma. J. Plasma Phys. **39**, 41–51 (1988)
329. A. Morro, T. Ruggeri, Second sound and internal energy in solids. Int. J. Non-Lin. Mech. **23**, 27–36 (1987)
330. A. Morro, T. Ruggeri, Non-equilibrium properties of solids obtained from second-sound measurements. J. Phys. C: Solid State Phys. **21**, 1743–1752 (1988)
331. A. Morro, M. Vianello, Interstitial energy flux and stress-power for second-gradient elasticity. Math. Mech. Solids **21**, 403–412 (2016)
332. A. Morro, Thermodynamic restrictions in linear viscoelasticity. Materials **15**, 2706 (2022)
333. I. Müller, A thermodynamic theory of mixtures of fluids. Arch. Rational Mech. Anal. **28**, 1–39 (1968)
334. I. Müller, The coldness, a universal function in thermoelastic bodies. Arch. Rational Mech. Anal. **41**, 319–332 (1971)
335. I. Müller, Thermodynamics of mixtures of fluids. J. Mécanique **14**, 267–303 (1975)
336. I. Müller, *Thermodynamics* (Pitman, London, 1985)
337. I. Müller, T. Ruggeri, *Extended Thermodynamics* (Springer, New York, 1993)

338. I. Müller, Thermodynamics of mixtures and phase field theory. Int. J. Solids Struct. **38**, 1105–1113 (2001)
339. I. Müller, S. Seelecke, Thermodynamic aspects of shape memory alloys. Math. Comp. Model. **34**, 1307–1355 (2001)
340. A. Nagy, W.L. Ko, U.S. Lindholm, Mechanical behavior of foamed materials under dynamic compression. J. Cell. Plast. **10**, 127–134 (1974)
341. J.L. Neuringer, R.E. Rosensweig, Ferrohydrodynamics. Phys. Fluids **7**, 1927–1937 (1964)
342. T.D. Nguyen, S. Mao, Y.-W. Yeh, P.K. Purohit, M.C. McAlpine, Nanoscale flexoelectricity. Adv. Mater. **25**, 946–974 (2013)
343. D.R. Nicholson, *Introduction to Plasma Theory* (Wiley, New York, 1983). ((ch. 8))
344. W. Noll, A new mathematical theory of simple materials. Arch. Rational Mech. Anal. **48**, 1–50 (2017)
345. A. Oberbeck, Uber die Wärmeleitung der Flüssigkeiten bei Berücksichtigung der Strömunggeninfolge von Temperaturdifferenzen. Ann. Phys. Chem. **7**, 271–292 (1879)
346. R.W. Ogden, *Non-Linear Elastic Deformations* (Dover, New York, 1998)
347. J. Oldroyd, On the formulation of rheological equations of state. Proc. Roy. Soc. Lond. Ser. A **200**, 523–541 (1950)
348. X. Oliver, C. Agelet de Saracibar, *Continuum Mechanics for Engineers (Theory and Problems* (UPC, Barcelona, 2017)
349. L. Onsager, Reciprocal relations in irreversible processes. Phys. Rev. **37**, 405–426 (1931)
350. W.C. Oseen, The theory of liquid crystals. Trans. Faraday Soc. **29**, 883–899 (1933)
351. M. Otténio, M. Destrade, R.W. Ogden, Incremental magnetoelasticdeformations, with application to surface instability. J. Elast. **90**, 19–42 (2008)
352. T.C. Papanastasiou, Flows of materials with yields. J. Rheology. **31**, 384–404 (1987)
353. Y.-S. Pao, K. Hutter, Electrodynamics for moving elastic solids and viscous fluids. Proc. IEEE **63**, 1011–1021 (1975)
354. J.R.G. Parkinson, D.F. Martin, A.J. Wells, R.F. Katz, Modelling binary alloy solidification with adaptive mesh refinement. J. Comput. Phys. **X5**, 100043 (2020)
355. S.L. Passman, J.W. Nunziato, A theory of multiphase mixtures, in *Rational Thermodynamics*, ed. by C. Truesdell (Springer, New York, 1984)
356. P. Penfield, H.A. Haus, *Electrodynamics of Moving Media* (MIT Press, Cambridge Mass, 1967)
357. O. Penrose, P.C. Fife, Thermodynamically consistent models of phase-field type for the kinetics of phase transitions. Physica D **43**, 44–62 (1990)
358. V.P. Peshkov, Second sound in helium II. Sov. Phys. JETP **11**, 580–584 (1960); J. Phys. USSR **8**, 381 (1944)
359. T.-H. Pham, J.J. Kim, S.-E. Kim, Estimating constitutive equation of structural steel using indentation. Int. J. Mech. Sci. **90**, 151–161 (2015)
360. A.C. Pipkin, Small finite deformations of viscoelastic solids. Rev. Mod. Phys. **38**, 1034–1041 (1964)
361. A.C. Pipkin, Constraints in linearly elastic materials. J. Elast. **6**, 179–193 (1986)
362. A.B. Pippard, L. Bragg, An experimental and theoretical study of relations between magnetic fields and current in superconductors. Proc. Roy. Soc. Lond. A **216**, 547–568 (1953)
363. L. Pitaevskii, Vortex lines in an imperfect Bose gas. J. Exp. Theor. Phys. **13**, 451–454 (1961)
364. P. Podio Guidugli, On dissipation mechanisms in micromagnetics. Eur. Phys. J. B **19**, 417–424 (2001)
365. P. Podio Guidugli, A primer in elasticity. J. Elast. **58**, 1–103 (2000)
366. P. Podio-Guidugli, *Continuum Thermodynamics* (Springer, Cham, 2019)
367. P. Podio Guidugli, M. Vianello, Constraint manifolds for isotropic solids. Arch. Rational Mech. Anal. **105**, 105–121 (1989)
368. E.M. Purcell, *Electricity and Magnetism* (McGraw-Hill, New York, 1965)
369. R. Quintanilla, Moore-Gibson-Thompson thermoelasticity. Math. Mech. Solids **24**, 4020–4031 (2019)

370. K.R. Rajagopal, A.R. Srinivasa, On the thermomechanics of shape memory wires. Z. angew. Math. Phys. **50**, 459–496 (1999)
371. K.R. Rajagopal, M. Ruzicka, Mathematical modelling of electrorheological materials. Cont. Mech. Thermodyn. **13**, 59–78 (2001)
372. K.R. Rajagopal, On implicit constitutive theories for fluids. J. Fluid Mech. **550**, 243–249 (2006)
373. K.R. Rajagopal, G. Saccomandi, L. Vergori, On the Oberbeck-Boussinesq approximation for fluids with pressure dependent viscosities. Nonlinear Anal. Real World Appl. **10**, 1139–1150 (2009)
374. K.R. Rajagopal, G. Saccomandi, L. Vergori, Flow of fluids with pressure- and shear-dependent viscosity down an inclined plane. J. Fluid Mech. **706**, 173–189 (2012)
375. M.A. Rahman, M.Z. Saghir, Thermodiffusion or Soret effect: historical review. Int. J. Heat Mass Transf. **73**, 693–705 (2014)
376. J.R. Reitz, F.J. Milford, R.W. Christy, *Foundations of Electromagnetic Theory* (Addison-Wesley, Reading Massachusetts, 1993)
377. L.A. Richards, Capillary conduction of liquids in porous media. Physics **1**, 318–333 (1931)
378. R.S. Rivlin, Large elastic deformations isotropic materials IV. Further developments of the general theory. Phil. Trans. R. Soc. A **241**, 379–397 (1948)
379. R.S. Rivlin, J.L. Ericksen, Stress-deformation relation for isotropic materials. J. Rational Mech. Anal. **4**, 323–425 (1955)
380. A. Romano, Proof that the formulations of the electrodynamics of moving bodies are equivalent. Arch. Rational Mech. Anal. **68**, 283–298 (1978)
381. D. Said Schicchi, A. Caggiano, M. Hunkel, E.A.B. Koenders, Thermodynamically consistent multiscale formulation of a thermo-mechanical problem with phase transformations. Cont. Mech. Thermodyn. **31**, 273–299 (2019)
382. G.A. Samara, The relaxational properties of compositionally disordered SBO_3 perovskites. J. Phys.: Condens. Matter **15**, R367–R411 (2003)
383. M. Sarjala, E.T. Seppälä, M.J. Alava, Dynamic hysteresis in ferroelectrics with quenched randomness. Physica B **40**, 418–421 (2008)
384. P. Saxena, M. Hossain, P. Steinmann, A theory of finite deformation magneto-viscoelasticity. Int. J. Solids Struct. **50**, 3886–3897 (2013)
385. P. Saxena, D.K. Vu, P. Steinmann, On rate-dependent dissipation effects in electro-elasticity. Int. J. Non-Lin. Mech. **62**, 1–11 (2014)
386. J.W. Schmelzer, E.D. Zanotto, V.M. Fokin, Pressure dependence of viscosity. J. Chem. Phys. **122**, 074511 (2005)
387. J. Schröder, P. Neff, *Poly-, Quasi- and Rank-One Convexity in Applied Mechanics* (Springer, Berlin, 2010)
388. J.F. Scott, Electrocaloric materials. Annu. Rev. Mater. Res. **41**, 229–240 (2011)
389. R.F. Sekerka, Similarity solutions for a binary diffusion couple with diffusivity and density dependent on composition. Prog. Mat. Sci. **49**, 511–536 (2004)
390. A. Sellitto, V.A. Cimmelli, D. Jou, *Mesoscopic Theories of Heat Transport in Nanosystems* (Springer, Berlin, 2016)
391. A. Serra-Aguila, J.M. Puigoriol-Forcada, G. Reyes et al., Viscoelastic models revisited: characteristics and interconversion formulas for generalized Kelvin-Voigt and Maxwell models. Acta Mech. Sin. **35**, 1191–1209 (2019)
392. J. Serrin, Mathematical principles of classical fluid mechanics, in *Encyclopedia of Physics, vol. VIII/1*. ed. by S. Flügge (Springer, Berlin, 1959), pp.125–263
393. N.D. Sharma, R. Maranganti, P. Sharma, On the possibility of piezoelectric nanocomposites without using piezoelectric materials. J. Mech. Phys. Solids **55**, 2328–2350 (2007)
394. M.I. Shliomis, Effective viscosity of magnetic suspensions. Soviet Phys. JETP **34**, 1291–1994 (1972)
395. W. Shockley, R.P. James, "Try Simplest Cases" Discovery of "Hidden Momentum" Forces on "Magnetic Currents". Phys. Rev. Lett. **18**, 876–879 (1967)
396. B.V. Somov, *Plasma Astrophysics* (Springer, Berlin, 2007). ((ch. 11))

397. A.J.C.B. de St. Venant, Établissement élémentaire des formules et équations générales de la theorie de l'élasticité des corpes solides. Appendix in: Résumé des Leçons données à l'École des Ponts et Chaussées sur l'Application de la Mécanique; De la Résistance des Corps Solides, par C.L.M.H. Navier (Paris, 1864)
398. J. Stefan, Über das Gleichgewicht und Bewegung, insbesondere die Diffusion von Gemishen. Sitzungsberichte der Kaiserlichen Akademie der Wissenschaften Wien **63**, 63–124 (1871)
399. J.J. Stoker, *Water Waves* (Wiley, New York, 1958)
400. J.A. Stratton, *Electromagnetic Theory* (McGraw-Hill, New York, 1962)
401. B. Straughan, *Heat Waves* (Springer, Berlin, 2011)
402. B. Straughan, Stability and uniqueness in double porosity elasticity. Int. J. Eng. Sci. **65**, 1–8 (2013)
403. Y. Su, C.M. Landis, Continuum thermodynamics of ferroelectric domain evolution: theory, finite element implementation, and application to domain wall pinning. J. Mech. Phys. Solids **55**, 280–305 (2007)
404. E.S. Suhubi, Thermoelastic Solids, in *Continuum Physics*, vol. II, ed. by A.C. Eringen (Academic, New York, 1975)
405. M. Svanadze, Plane waves and boundary value problems in the theory of elasticity for solids with double porosity. Acta Appl. Math. **122**, 461–471 (2012)
406. H. Szambolics, J.-Ch. Toussaint, A. Marty, I.M. Miron, L.D. Buda-Prejbeanu, Domain wall motion in ferromagnetic systems with perpendicular magnetization. J. Magn. Magn. Mater. **321**, 1912–1918 (2009)
407. A.K. Tagantsev, A.S. Yurkov, Flexoelectric effect in finite samples. J. Appl. Phys. **112**, 044103 (2012)
408. C.-T. Tai, On the presentation of Maxwell's theory. Proc. IEEE **60**, 936–945 (1972)
409. R. Taylor et al., Small particles, big impacts: a review of the diverse applications of nanofluids. J. Appl. Phys. **113**, 011301 (2013)
410. J.J. Thalakkottor, K. Mohseni, Unified slip boundary condition for fluid flows. Phys. Rev. E **94**, 023113 (2016)
411. A. Thiaville, Y. Nakatani, J. Miltat, Y. Suzuki, Micromagnetic understanding of current-driven domain wall motion in patterned nanowires. Europhys. Lett. **69**, 990–996 (2005)
412. N.L. Thomas, A.H. Windle, A theory of case II diffusion. Polymer **23**, 529–542 (1982)
413. H.F. Tiersten, On the nonlinear equations of thermo-electroelasticity. Int. J. Eng. Sci. **9**, 587–604 (1971)
414. D. Tilley, J. Tilley, *Superfluidity and Superconductivity* (IOP Publishing, Bristol, 1990)
415. A.M. Tishin, Y.I. Spichkin, *The Magnetocaloric Effect and Its Applications (Series in Condensed Matter Physics)* (Publishing Ltd., Institute of Physics, 2003)
416. L. Tisza, Sur la supraconductibilité thermique de l'hélium II liquide et la statistique de Bose-Einstein. C. R. Acad. Sci. **207**, 1035–1037 (1938)
417. L. Tisza, La viscosité de l'hélium liquide et la statistique de Bose-Einstein. C. R. Acad. Sci. **207**, 1186–1189 (1938)
418. J.C. Tolédano, P. Tolédano, *The Landau Theory of Phase Transitions* (World Scientific, Singapore, 1987)
419. R.A. Toupin, The elastic dielectric. J. Rational Mech. Anal. **5**, 849–915 (1956)
420. R.A. Toupin, Elastic materials with couple-stresses. Arch. Rational Mech. Anal. **11**, 385–414 (1962)
421. R.A. Toupin, Theory of elasticity with couple-stresses. Arch. Rational Mech. Anal. **17**, 85–112 (1964)
422. L.R.G. Treloar, *The Physics of Rubber Elasticity* (Clarendon Press, Oxford, 1958)
423. C. Truesdell, The simplest rate theory of pure elasticity. Comm. Pure Appl. Math. **8**, 123–132 (1955)
424. C. Truesdell, *Introduction à la Mécanique Rationelle des Milieux Continus* (Masson, Paris, 1974)
425. C. Truesdell, *A First Course in Rational Continuum Mechanics* (Academic, NewYork, 1977)
426. C. Truesdell, *Rational Thermodynamics* (Springer, New York, 1984). ((ch. 5))

427. C. Truesdell, W. Noll, The non-linear field theory of mechanics, in *Encyclopedia of Physics*, vol. III/3, ed. by S. Flügge (Springer, Berlin, 1965)
428. C. Truesdell, R.A. Toupin, The classical field theories, in *Encyclopedia of Physics*, vol. III/1, ed. by S. Flügge (Springer, Berlin, 1960)
429. L. Vaidman, Torque and force on a magnetic dipole. Am. J. Phys. **58**, 978–983 (1990)
430. M.P. Vernotte, Les paradoxes de la théorie continue de l'équation de la chaleur. Comptes Rendus de l'Académie des Sciences **246**, 3154–3155 (1958)
431. W.F. Vinen, J.J. Niemela, Quantum turbulence. J. Low Temp. Phys. **128**, 167–231 (2002)
432. A. Visintin, *Differential Models of Hysteresis* (Springer, Berlin, 1994)
433. A. Visintin, *Models of Phase Transition* (Birkäuser, Boston, 1996); ch. 4
434. A. Visintin, Mathematical models of hysteresis, in *The Science of Hysteresis*, ed. by G. Bertotti, I. Mayergoyz (Elsevier, 2006), pp. 1–123
435. V. Volterra, *Leçons sur les Fonctions des Lignes* (Gauthier-Villars, Paris, 1913)
436. V. Volterra, Sur la théorie mathématique des phénomènes héréditaires. J. Math. Pures Appl. **7**, 249–298 (1928)
437. V. Volterra, Energia nei fenomeni ereditari. Acta Pontificia Acad. Scient. **4**, 115–128 (1940)
438. J. Vörös, Modeling and identification of hysteresis using special forms of the Coleman-Hodgdon model. J. Electr. Eng. **60**, 100–105 (2009)
439. R.T. Wang, W.T. Wagner, R.J. Donnelly, Precision second-sound velocity measurements in helium II. J. Low Temp. Phys. **68**, 409–417 (1987)
440. E. Warburg, Magnetische Untersuchungen. Annalen der Physik **13**, 141–164 (1881)
441. J.G. Webster (ed.), *The Measurement, Instrumentation, and Sensors Handbook* (CRC Press (in cooperation with IEEE Press), Boca Raton, FL, 1999). 0849383471
442. J. Weingarten, Sulle superfici di discontinuità nella teoria dell'elasticità dei corpi solidi. Atti Accad. Naz. Lincei Cl. Sci. Fis. Mat. Natur. Rend. **10**, 57–60 (1901)
443. A. Wheeler, W. Boettinger, G. McFadden, Phase-field model for isothermal phase transitions in binary alloys. Phys. Rev. A **d5**, 7424–7440 (1992)
444. J.L. White, A.B. Metzner, Development of constitutive equations for polymeric melts and solutions. J. Appl. Polym. Sci. **7**, 1867–1889 (1963)
445. E. Wiechert, Gesetze der Elastichen Nachwirkung für constante Temperatur. Annalen der Physik **286**(335–348), 546–570 (1893)
446. L. Xu, S. Zhang, Magnetization dynamics at elevated temperatures. Physica E **45**, 72–76 (2012)
447. Y. Yu, P. Perdikaris, G.E. Karniadakis, Fractional modeling of viscoelasticity in 3D cerebral arteries and aneurysms. J. Comp. Phys. **323**, 219–242 (2016)
448. M.W. Zemansky, *Heat and Thermodynamics* (McGraw-Hill, New York, 1968)
449. C. Zener, *Elasticity and Anelasticity of Metals* (University of Chicago Press, Chicago, 1948)

Index

A. Morro and C. Giorgi, *Mathematical Modelling of Continuum Physics*, Modeling and Simulation in Science, Engineering and Technology, https://doi.org/10.1007/978-3-031-20814-0